PUB. 9. Volume II, 1981 Ed.

Page 6—Line 12; read:

$$d=\sqrt{\left(\frac{\tan\alpha}{0.0002419}\right)^2+\frac{H-h}{0.7349}}-\frac{\tan\alpha}{0.0002419},$$

(DMAHTC) 23/82

PUB 9 Volume II 1981 Ed

LAST NM 23/82

Page 397 – Table 41/Angle 0°25'
Column 120 substitute 2.72 for 3.72
(DMAHTC/MCNN)

45/87

AMERICAN PRACTICAL NAVIGATOR (BOWDITCH)

PUB 9 Volume II 1981 Ed LAST NM 45/87

Page 124–Line 4/R; nautical miles column, read:
60.026 vice 60.077
(DMAHTC/MCNN) 16/89

PUB 9 Volume II 1981 Ed LAST NM 16/89

Page 656–Line 8/read:
$10 \log_{10}(10^3) = 10 \times 3 = 30$ dB
(DMAHTC/MCNN) 23/89

Page 656–Line 10/read:
$10 \log_{10}(10^{-3}) = 10 \times (-3) = -30$ dB
(DMAHTC/MCNN) 23/89

Pub. No. 9

AMERICAN PRACTICAL NAVIGATOR

AN EPITOME OF NAVIGATION

ORIGINALLY BY

NATHANIEL BOWDITCH, LL. D.

Volume II

USEFUL TABLES

CALCULATIONS

GLOSSARY OF MARINE NAVIGATION

1981 Edition

PUBLISHED BY THE

DEFENSE MAPPING AGENCY HYDROGRAPHIC/TOPOGRAPHIC CENTER

For sale by authorized Sale Agents of the Defense Mapping Agency Office of Distribution Services

DMA STOCK NO. NVPUB9V2

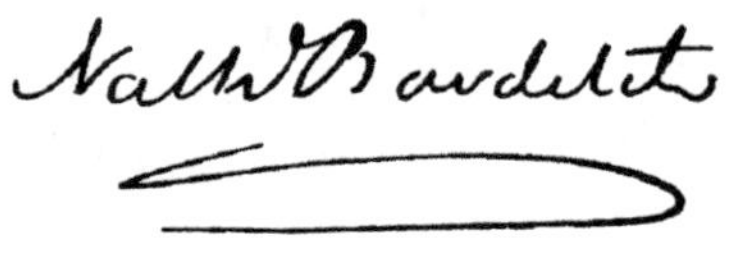

Last painting by Gilbert Stuart (1828). Considered by the family of Bowditch to be the best of various paintings made, although it was unfinished when the artist died.

NATHANIEL BOWDITCH

(1773–1838)

Nathaniel Bowditch was born on March 26, 1773, at Salem, Mass., fourth of the seven children of shipmaster Habakkuk Bowditch and his wife, Mary.

Since the migration of William Bowditch from England to the Colonies in the 17th century, the family had resided at Salem. Most of its sons, like those of other families in this New England seaport, had gone to sea, and many of them became shipmasters. Nathaniel Bowditch himself sailed as master on his last voyage, and two of his brothers met untimely deaths while pursuing careers at sea.

It is reported that Nathaniel Bowditch's father lost two ships at sea, and by late Revolutionary days he returned to the trade of cooper, which he had learned in his youth. This provided insufficient income to properly supply the needs of his growing family, and hunger and cold were often experienced. For many years the nearly destitute family received an annual grant of fifteen to twenty dollars from the Salem Marine Society. By the time Nathaniel had reached the age of ten, the family's poverty necessitated his leaving school and joining his father in the cooper's trade.

Nathaniel was unsuccessful as a cooper, and when he was about 12 years of age, he entered the first of two ship-chandlery firms by which he was employed. It was during the nearly ten years he was so employed that his great mind first attracted public attention. From the time he began school Bowditch had an all-consuming interest in learning, particularly mathematics. By his middle teens he was recognized in Salem as an authority on that subject. Salem being primarily a shipping town, most of the inhabitants sooner or later found their way to the ship chandler, and news of the brilliant young clerk spread until eventually it came to the attention of the learned men of his day. Impressed by his desire to educate himself, they supplied him with books that he might learn of the discoveries of other men. Since many of the best books were written by Europeans, Bowditch first taught himself their languages. French, Spanish, Latin, Greek, and German were among the two dozen or more languages and dialects he studied during his life. At the age of 16 he began the study of Newton's *Principia*, translating parts of it from the Latin. He even found an error in that classic, and though lacking the confidence to announce it at the time, he later published his findings and had them accepted.

During the Revolutionary War a privateer out of Beverly, a neighboring town to Salem, had taken as one of its prizes an English vessel which was carrying the philosophical library of a famed Irish scholar, Dr. Richard Kirwan. The books were brought to the Colonies and there bought by a group of educated Salem men who used them to found the Philosophical Library Company, reputed to have been the best library north of Philadelphia at the time. In 1791, when Bowditch was 18, two Harvard-educated ministers, Rev. John Prince and Rev. William Bentley, persuaded the Company to allow Bowditch the use of its library. Encouraged by these two men and a third—Nathan Read, an apothecary and also a Harvard man—Bowditch studied the works of the great men who had preceded him, especially the mathematicians and the astronomers. By the time he became of age, this knowledge, acquired before and after his long working hours and in his spare time, had made young Bowditch the outstanding mathematician in the Commonwealth, and perhaps in the country.

In the seafaring town of Salem, Bowditch was drawn to navigation early, learning the subject at the age of 13 from an old British sailor. A year later he began studying surveying, and in 1794 he assisted in a survey of the town. At 15 he devised an almanac reputed to have been of great accuracy. His other youthful accomplishments included the construction of a crude barometer and a sundial.

When Bowditch went to sea at the age of 21, it was as captain's writer and nominal second mate, the officer's berth being offered him because of his reputation as a scholar. Under Captain Henry Prince, the ship *Henry* sailed from Salem in the winter of 1795 on what was to be a year-long voyage to the Ile de Bourbon (now called Réunion) in the Indian Ocean.

Bowditch began his seagoing career when accurate time was not available to the average naval or merchant ship. A reliable marine chronometer had been invented some 60 years before, but the prohibitive cost, plus the long voyages without opportunity to check the error of the timepiece, made the large investment an impractical one. A system of determining longitude by "lunar distance," a method which did not require an accurate timepiece, was known, but this product of the minds of mathematicians and astronomers was so involved as to be beyond the capabilities of the uneducated seamen of that day. Consequently, ships navigated by a combination of dead reckoning and parallel sailing (a system of sailing north or south to the latitude of the destination and then east or west to the destination).

To Bowditch, the mathematical genius, computation of lunar distances was no mystery, of course, but he recognized the need for an easier method of working them in order to navigate ships more safely and efficiently. Through analysis and observation, he derived a new and simplified formula during his first trip.

John Hamilton Moore's *The New Practical Navigator* was the leading navigational text when Bowditch first went to sea, and had been for many years. Early in his first voyage, however, the captain's writer-second mate began turning up errors in Moore's book, and before long he found it necessary to recompute some of the tables he most often used in working his sights. Bowditch recorded the errors he found, and by the end of his second voyage, made in the higher capacity of supercargo, news of his findings in *The New Practical Navigator* had reached Edmund Blunt, a printer at Newburyport, Mass. At Blunt's request, Bowditch agreed to participate with other learned men in the preparation of an American edition of the thirteenth (1798) edition of Moore's work. The first American edition was published at Newburyport by Blunt in 1799. This edition corrected many of the errors that Moore had failed to correct. Although most of the errors were of little significance to practical navigation as they were errors in the fifth and sixth places of logarithm tables, some errors were significant. The most significant error was listing the year 1800 as a leap year in the table of the sun's declination. The consequence was that Moore gave the declination for March 1, 1800, as 7°11′. Since the actual value was 7°33′, the calculation of a meridian altitude would be in error by 22 minutes of latitude.

Bowditch's principal contribution to the first American edition was his chapter "The Method of finding the Longitude at Sea," which was his new method for computing the lunar distance. Following publication of the first American edition, Blunt obtained Bowditch's services in checking the American and English editions for further errors. Blunt then published a second American edition of Moore's thirteenth edition in 1800. When preparing a third American edition for the press, Blunt decided that Bowditch had revised Moore's work to such an extent that Bowditch should be named as author. The title was changed to *The New American Practical Navigator* and the book was published in 1802 as a first edition.

Bowditch made a total of five trips to sea, over a period of about nine years, his last as master and part owner of the three-masted *Putnam*. Homeward bound from a 13-month voyage to Sumatra and the Ile de France (now called Mauritius) the *Putman* approached Salem harbor on December 25, 1803, during a thick fog without having had a celestial observation since noon on the 24th. Relying upon his dead reckoning, Bowditch conned his wooden-hulled ship to the entrance of the rocky harbor, where he had the good fortune to get a momentary glimpse of Eastern Point, Cape Ann, enough to confirm his position. The *Putnam* proceeded in, past such hazards as "Bowditch's Ledge" (named after a great-grandfather who had wrecked his ship on the rock more than a century before) and anchored safely at 1900 that evening. Word of the daring feat, performed when other masters were hove-to outside the harbor, spread along the coast and added greatly to Bowditch's reputation. He was, indeed, the "practical navigator."

His standing as a mathematician and successful shipmaster earned him a lucrative (for those times) position ashore within a matter of weeks after his last voyage. He was installed as president of a Salem fire and marine insurance company, at the age of 30, and during the 20 years he held that position the company prospered. In 1823 he left Salem to take a similar position with a Boston insurance firm, serving that company with equal success until his death.

From the time he finished the "*Navigator*" until 1814, Bowditch's mathematical and scientific pursuits consisted of studies and papers on the orbits of comets, applications of Napier's rules, magnetic variation, eclipses, calculations on tides, and the charting of Salem harbor. In that year, however, he turned to what he considered the greatest work of his life, the translation into English of *Mécanique Céleste*, by Pierre Laplace. *Mècanique Cèleste* was a summary of all the then known facts about the workings of the heavens. Bowditch translated four of the five volumes before his death, and published them at his own expense. He gave many formula derivations which Laplace had not shown, and also included further discoveries following the time of publication. His work made this information available to American astronomers and enabled them to pursue their studies on the basis of that which was already known. Continuing his style of writing for the learner, Bowditch presented his English version of *Mècanique Cèleste* in such a manner that the student of mathematics could easily trace the steps involved in reaching the most complicated conclusions.

Shortly after the publication of *The New American Practical Navigator*, Harvard College honored its author with the presentation of the honorary degree of Master of Arts, and in 1816 the college made him an honorary Doctor of Laws. From the time the Harvard graduates of Salem first assisted him in his studies, Bowditch had a great interest in that college, and in 1810 he was elected one of its Overseers, a position he held until 1826, when he was elected to the Corporation. During 1826–27 he was the leader of a small group of men who saved the school from financial disaster by forcing necessary economies on the college's reluctant president. At one time Bowditch was offered a Professorship in Mathematics at Harvard but this, as well as similar offers from West Point and the University of Virginia, he declined. In all his life he was never known to have made a public speech or to have addressed any large group of people.

Many other honors came to Bowditch in recognition of his astronomical, mathematical, and marine accomplishments. He became a member of the American Academy of Arts and Sciences, the East India Marine Society, the Royal Academy of Edinburgh, the Royal Society of London, the Royal Irish Academy, the American Philosophical Society, the Connecticut Academy of Arts and Sciences, the Boston Marine Society, the Royal Astronomical Society, the Palermo Academy of Science, and the Royal Academy of Berlin.

Nathaniel Bowditch outlived all of his brothers and sisters by nearly 30 years. Death came to him on March 16, 1838, in his sixty-fifth year. The following eulogy by the Salem Marine Society indicates the regard in which this distinguished American was held by his contemporaries:

"In his death a public, a national, a human benefactor has departed. Not this community, nor our country only, but the whole world, has reason to do honor to his memory. When the voice of Eulogy shall be still, when the tear of Sorrow shall cease to flow, no monument will be needed to keep alive his memory among men; but as long as ships shall sail, the needle point to the north, and the stars go through their wonted courses in the heavens, the name of Dr. Bowditch will be revered as of one who helped his fellow-men in a time of need, who was and is a guide to them over the pathless ocean, and of one who forwarded the great interests of mankind."

The New American Practical Navigator was revised by Nathaniel Bowditch several times after 1802 for subsequent editions of the book. After his death, Jonathan Ingersoll Bowditch, a son who made several voyages, took up the work and his name appeared on the title page from the eleventh edition through the thirty-fifth, in 1867. In 1868 the newly organized U.S. Navy Hydrographic Office bought the copyright. Revisions have been made from time to time to keep the work in step with navigational improvements. The name has been altered to the *American Practical Navigator*, but the book is still commonly known as "Bowditch." A total of more than 850,000 copies has been printed in about 70 editions during the more than a century and a half since the book was first published in 1802. It has lived because it has combined the best thoughts of each generation of navigators, who have looked to it as their final authority.

PREFACE

This epitome of navigation has been maintained continuously since it was first published in 1802. The U.S. Navy maintained "Bowditch" from 1868 until 1972, when the Defense Mapping Agency Hydrographic Center was assigned the responsibility for its publication. In 1978 this responsibility was assigned to the Defense Mapping Agency Hydrographic/Topographic Center. This volume provides the tables, formulas, data, and instructions needed by the navigator to perform many of the computations associated with dead reckoning, piloting, and celestial navigation. This volume also provides a glossary which reflects the marine navigator's sphere of interests in such branches of science as astronomy, meteorology, oceanography, cartography, electronics, etc. An attempt has been made to define the terms in the language and from the viewpoint of the marine navigator. Only those meanings more or less directly related to navigation have been included.

In addition to the inclusion of a glossary, the differences between this edition and the 1975 edition may be briefly summarized as chapter rearrangements, deletion of those chapters the contents of which are considered to be covered adequately in the glossary, changes in certain tables due to changes in the reference ellipsoid (tables 5 and 6) or the refraction value used in the computations (tables 8, 9, and 22), the addition of a compact traverse table (tab. 38), two vertical angle tables (tables 39 and 41) for use in piloting, and a geographic range table (tab. 40).

The intent of the original author to provide a compendium of navigational material understandable to the mariner has been consistently followed. However, navigation is not presented as a mechanical process to be followed blindly. Rather, emphasis has been given to the fact that the aids provided by *science* can be used effectively to improve the *art* of navigation only if a well-informed person of mature judgment and experience is on hand to interpret information as it becomes available. Thus, the facts needed to perform the mechanics of navigation have been supplemented with additional material intended to help the navigator acquire perspective in meeting the various needs that arise.

Users should refer corrections, additions, and comments for improving this product to DIRECTOR, DEFENSE MAPPING AGENCY HYDROGRAPHIC/TOPOGRAPHIC CENTER, Washington, D.C. 20315, ATTN: Code PRH.

ACKNOWLEDGMENTS

1975 EDITION

Many institutions, organizations, groups, and individuals have assisted in the preparation of volume II, but all of the material has been edited by one individual to assure continuity and consistency. Particular acknowledgment is given the following: Dr. P. Kenneth Seidelmann, Assistant Director, Nautical Almanac Office, U.S. Naval Observatory, for his assistance in matters pertaining to the almanacs and related topics; Dr. William H. Guier for the improved table 9; Captain Wayne M. Waldo and Mr. Richard M. Plant, Maritime Institute of Technology and Graduate Studies, Linthicum Heights, Maryland, and Lieutenant Commander Guy P. Clark, U.S. Coast Guard, for their contributions to chapter XII; Rear Commander Allan E. Bayless, N, and Past Staff Commander O. B. Ellis, N, U.S. Power Squadrons, for their assistance in examining the text for errors; Commander Frank E. Bassett, U.S. Navy, Chairman, Lieutenant Commander Arthur J. Tuttle, U.S. Navy, and Lieutenant Commander Richard A. Smith, Royal Navy, of the Department of Navigation, Division of Naval Command and Management, United States Naval Academy, for their guidance in practical areas; the U.S. Power Squadrons for suggestions relating to the graph of article 1106 for height of tide determination, and table 35; and many individuals, especially the marine navigation instructors and experienced practicing navigators who have responded to our questionnaires and offered constructive suggestions or directed attention to errors in previous editions.

1981 EDITION

Those institutions, organizations, groups, and individuals who contributed to the 1975 edition of this volume have also contributed to this edition. For this edition particular acknowledgment is given to the following: Vice Commander Allan E. Bayless, N, when Director of Education, United States Power Squadrons, for his helpful guidance in matters pertaining to tables; Dr. Wayland C. Marlow for his technical assistance in matters pertaining to terrestrial refraction and improvements in vertical angle tables; for their assistance in the preparation of the glossary, Mr. Steacy D. Hicks, National Ocean Survey of the National Oceanic and Atmospheric Administration; Mr. Melvin E. Cruser, Naval Guided Missiles School, Dam Neck, Virginia Beach, Virginia; and Mr. Richard J. Sandifer, Associate Professor of Transportation, Maritime College, State University of New York, Fort Schuyler, Bronx, New York; and Captain John F. Underhill, Maritime Institute of Technology and Graduate Studies, Linthicum Heights, Maryland, for his technical assistance in matters pertaining to tables.

CONTENTS

PART ONE

USEFUL TABLES

PART ONE

USEFUL TABLES

PART ONE

USEFUL TABLES

EXPLANATION OF TABLES

Table 1. Conversion Angle.—The angles listed in this table are the differences between the great-circle and rhumb line (Mercator) directions between various points. The table is used in converting radio bearings to equivalent rhumb line bearings for plotting on a Mercator chart. The sign to be used is indicated at the bottom of the table.

The first one and one-half pages of the table, for differences of longitude of not more than 4°.5, and middle latitudes between 0° and 85°, is intended primarily for use in converting radio bearings observed near a coast. In high latitudes it may be needed for converting visual bearings of objects a considerable distance away.

In high latitudes it is not unusual to make use of bearings on objects a considerable distance from the vessel. Because of the rapid convergence of the meridians in these areas, such bearings are not correctly represented by straight lines on a Mercator chart. If this projection is used, the bearings should be corrected in the same manner that radio bearings are corrected, since both can be considered great circles. This part of the table is entered with (1) the middle latitude between the craft and radio station or object, and (2) the difference of longitude between these two points. Do not use this part of the table if the latitudes are of contrary name (one north, the other south), if the difference of latitude is more than 10°, or if the difference of longitude is more than 4°.5. Under any of these conditions, use the second part of the table. For this part, the entering arguments are (1) the latitude of the receiver, (2) the difference of longitude (DLo) between the two points involved, and (3) the latitude of the transmitter. For accurate results, use triple interpolation (art. 204).

Use of the table is explained in article 402.

The conversion angles on the first one and one-half pages of the table were computed by means of the formula:

$$\tan \text{conversion angle} = \sin \text{Lm} \tan \tfrac{1}{2} \text{DLo},$$

in which Lm is the middle latitude, and DLo is the difference of longitude. This formula is based upon the assumption that the plot of a great-circle track on a Mercator chart is symmetrical, the axis of symmetry being the perpendicular bisector of the rhumb line connecting the two points. No error of practical significance in ordinary navigation is introduced by this erroneous assumption over the range covered by the first one and one-half pages of the table.

The remainder of the table was computed by means of the formula: conversion angle=initial great-circle direction∼rhumb line direction. The initial great-circle directions were computed by means of the formula:

$$\text{hav C} = \csc \text{D} \sec \text{L}_1 [\text{hav co L}_2 - \text{hav (D} \sim \text{co L}_1)],$$

in which C is the initial great-circle direction, D is the great-circle distance in arc units, L_1 is the latitude of departure (or receiver), and L_2 is the latitude of the destination (or transmitter). The distance D was computed by means of the formula:

$$\text{hav D} = \text{hav DLo} \cos \text{L}_1 \cos \text{L}_2 + \text{hav (L}_1 \sim \text{L}_2),$$

in which the notation is the same as above.

The rhumb line directions were computed by means of the formula:

$$\tan \text{C} = \frac{\text{DLo}}{\text{m}},$$

in which C is the rhumb line direction, and $m = M_1 \sim M_2$, the meridional parts of the two latitudes.

Table 2. Conversion of Compass Points to Degrees.—In this table the compass is boxed to 128 quarter points and the equivalent angle is given in degrees, minutes, and seconds. The naming of the quarter points, as given here, is one of several systems that have been used.

Table 3. Traverse Table.—This table can be used in the solution of the usual problems encountered in any of the sailings except great-circle and composite. In providing the values of the difference of latitude and departure corresponding to distances up to 600 miles and for courses for every degree of the compass, table 3 is essentially a tabulation of the solutions of plane right triangles. Since the solutions are for integral values of the acute angle and the distance, interpolation for intermediate values may be required. Through appropriate interchanges of the headings of the columns, solutions for other than plane sailings can be effected. The interchanges of the headings of the different columns are summarized at the foot of each table opening.

The distance, difference of latitude, and departure columns are labeled Dist., D.Lat., and Dep., respectively.

For the solution of the plane right triangle, any number N in the distance column is the hypotenuse; the number opposite in the difference of latitude column is N times the cosine of the acute angle; and the other number opposite in the departure column is N times the sine of the acute angle. Or, the number in the column labeled D.Lat. is the value of the side adjacent and the number in the column labeled Dep. is the value of the side opposite the acute angle.

The use of this table and an alternative traverse table (tab. 38) in the solution of problems of the various sailings is given in chapter X.

Table 4. Table of Offsets.—This table gives the corrections to the straight line of position (LOP) as drawn on a chart or plotting sheet to provide a closer approximation to the arc of the circle of equal altitude, a small circle of radius equal to the zenith distance.

In adjusting the straight LOP to obtain a closer approximation of the arc of the circle of equal altitude, points on the LOP are offset at right angles to the LOP in the direction of the celestial body. The arguments for entering the table are the distance from the DR to the foot of the perpendicular and the altitude of the body.

The table was computed by means of the formulas:

$$R = 3438' \cot h$$

$$\sin \theta = D/R$$

$$X = R(1 - \cos \theta),$$

in which X is the offset, R is the radius of a circle of equal altitude for altitude h, and D is the distance from the intercept to the point on the LOP to be offset.

Table 5. Meridional Parts.—In this table the meridional parts used in the construction of Mercator charts and in Mercator sailing are tabulated to one decimal place for each minute of latitude from the equator to the poles. The use of the table is explained in article 1007.

The table was computed by means of the formula:

$$M = a \log_e 10 \log \tan \left(45° + \frac{L}{2}\right) - a\left(e^2 \sin L + \frac{e^4}{3} \sin^3 L + \frac{e^6}{5} \sin^5 L + \ldots\right),$$

in which

M is the number of meridional parts between the equator and the given latitude, a is the equatorial radius of the earth, expressed in minutes of arc of the equator, or

$$a = \frac{21{,}600}{2\pi} = 3437.74677078 \text{ (log} = 3.5362738827),$$

$\log_e$ is the natural (Naperian) logarithm, using the base e=2.71828182846,

$\log_e 10=2.30258509299$ (log=0.3622156886),

L is the latitude,

e is the eccentricity of the earth, or $\sqrt{2f-f^2}=0.0818188$ (log=8.087146894−10)

and

f is the flattening of the earth, or $f=\frac{1}{298.26}=0.00335278$ (log=7.474594943−10)

Using these values,

a $\log_e 10=7915.704468$ (log=3.8984895715)

$ae^2=23.01336392$ (log=1.361980094)

$\frac{ae^4}{3}=0.051352909$ (log=8.28943495−10) $\frac{ae^6}{5}=0.0002062636$ (log=6.685577407−10).

Hence, the formula becomes

$$M=7915.704468 \log \tan\left(45°+\frac{L}{2}\right)-23.0133633 \sin L - 0.051353 \sin^3 L - 0.000206 \sin^5 L \ldots$$

The constants used in this derivation and in the table are based upon the World Geodetic System (WGS) ellipsoid of 1972 (app. D).

Table 6. Length of a Degree of Latitude and Longitude.—This table gives the length of one degree of latitude and longitude at intervals of 1° from the equator to the poles. In the case of latitude, the values given are the lengths of the arcs extending half a degree on each side of the tabulated latitudes. Lengths are given in nautical miles, statute miles, feet, and meters.

The values were computed in meters, using the World Geodetic System ellipsoid of 1972 (app. D), and converted to other units by the factors given in appendix D. The following formulas were used:

$$M=111{,}132.92-559.82 \cos 2L+1.175 \cos 4L-0.0023 \cos 6L+\ldots$$

$$P=111{,}412.84 \cos L-93.5 \cos 3L+0.118 \cos 5L-\ldots$$

in which M is the length of 1° of the meridian (latitude), L is the latitude, and P is the length of 1° of the parallel (longitude).

Table 7. Distance of an Object by Two Bearings.—To determine the distance of an object as a vessel on a steady course passes it, observe the difference between the course and two bearings of the object, and note the time interval between bearings. Enter this table with the two differences. Multiply the distance run between bearings by the number in the first column to find the distance of the object at the time of the second bearing, and by the number in the second column to find the distance when abeam. Use of the table is explained in article 505.

The table was computed by solving plane oblique and right triangles (art. 142).

Table 8. Distance of the Horizon.—This table gives the distance, in nautical and statute miles, of the visible sea horizon for various heights of eye in feet and equivalent heights in meters. The actual distance varies somewhat as refraction changes. However, the error is generally less than that introduced by nonstandard atmospheric conditions. Also the formula used contains an approximation which introduces a small error at the greatest heights tabulated.

Use of the table is explained in article 1202.

The table was computed by means of the formula:

$$d=\sqrt{\frac{2\,r_o\,h_f}{6076.1\,\beta_o}},$$

in which d is the distance to the horizon in nautical miles; r_o is the mean radius of the earth, 3440.1 nautical miles; h_f is the height of eye in feet; and β_o is a parameter which characterizes the effect o terrestrial refraction.

This formula simplifies to:

$$d=1.17\sqrt{h_f}.$$

Table 9. Distance by Vertical Angle Measured Between Sea Horizon and Top of Object Beyond Sea Horizon.— This table provides means for determining the distance to an object of known height above sea level when the object lies *beyond* the sea horizon. The vertical angle between the top of the object and the visible (sea) horizon (the sextant altitude) is measured and corrected for index error and dip only. The table is entered with the difference in the height of the object and the height of eye of the observer and the corrected vertical angle; and the distance in nautical miles is taken directly from the table. An error may be introduced if refraction differs from the standard value used in the computation of the table. Use of the table is explained in article 503.

The table was computed by means of the formula:

$$d=\sqrt{\left(\frac{\tan \alpha}{0.0002419}\right)^2+\frac{H-h}{0.7349}}-\frac{\tan \alpha}{0.002419},$$

in which d is the distance in nautical miles, α is the corrected vertical angle, H is the height of the top of the object above sea level in feet, and h is the height of eye of the observer above sea level in feet. The constants 0.0002419 and 0.7349 are parameters which characterize the effect of terrestrial refraction.

Table 10. Direction and Speed of True Wind.—This table provides a means of converting apparent wind, observed aboard a moving craft, to true wind. To use the table, divide the apparent wind, in knots, by the vessel's speed, also in knots. This gives the apparent wind speed in units of ship's speed. Enter the table with this value and the difference between the heading and the apparent wind direction. The values taken from the table are (1) the difference between the heading and the true wind direction, and (2) the speed of the true wind in units of ship's speed. The true wind is on the same side as the apparent wind, and from a point farther aft. To convert wind speed in units of ship's speed to speed in knots, multiply by the vessel's speed in knots. The steadiness of the wind and the accuracy of its measurement are seldom sufficient to warrant interpolation in this table. If speed of the true wind and relative direction of the apparent wind are known, enter the column for direction of the apparent wind, and find the speed of the true wind, in units of ship's speed. The number to the left is the relative direction of the true wind. The number on the same line in the side columns is the speed of the apparent wind in units of ship's speed. Two solutions are possible if speed of the true wind is less than ship's speed.

The table was computed by solving the triangle involved in a graphical solution, using the formulas:

$$\tan \alpha=\frac{\sin B_A}{S_A-\cos B_A},$$

$$B_T=B_A+\alpha,$$

$$S_T=\frac{\sin B_A}{\sin \alpha},$$

in which α is an auxiliary angle, B_A is the difference between the heading and the apparent wind direction, S_A is the speed of the apparent wind in units of ship's speed, B_T is the difference between the heading and the true wind direction, and S_T is the speed of the true wind in units of ship's speed.

Table 11. Correction of Barometer Reading for Height Above Sea Level.—If simultaneous barometer readings at different heights are to be of maximum value in weather analysis, they should be converted to the corresponding readings at a standard height, usually sea level. To convert the observed barometer reading to this level, enter this table with the outside temperature and the height of the barometer above sea level. The height of a barometer is the height of its sensitive element; in the case of a mercurial barometer, this is the height of the free surface of mercury in the cistern. The correction taken from this table applies to the readings of any type barometer, and is always *added* to the observed readings, unless the barometer is below sea level.

The correction was computed by means of the formula:

$$C=29.92126\left[1-\frac{1}{\text{antilog}\left(\frac{0.0081350\,H}{T+0.00178308\,H}\right)}\right]$$

in which

C is the correction in inches of mercury,

H is the height of the barometer above sea level in feet, and,

T is the mean temperature, in degrees Rankine (degrees Fahrenheit plus 459°67), of the air between the barometer and sea level. At sea, the outside air temperature is sufficiently accurate for this purpose.

Table 12. Correction of Barometer Reading for Gravity.—The height of the mercury column of a mercurial barometer is affected by the force of gravity, which changes with latitude and is approximately equal along any parallel of latitude. The average gravitational force at latitude 45°32′40″ is used as the standard for calibration. This table provides a correction to convert the observed reading at any other latitude to the corresponding value at latitude 45°32′40″, so that it will have maximum value in weather analysis of an area. Enter the table with the latitude, take out the correction, and apply in accordance with the sign given. The correction does not apply to aneroid barometers.

The correction was computed by means of the formula:

$$C=B\,(-0.002637\cos 2L+0.000006\cos^2 2L-0.000050),$$

in which

C is the correction in inches,

B is the observed reading of the barometer (corrected for temperature and instrumental errors) in inches of mercury. This table was computed for a standard height of 30 inches, and

L is the latitude.

Table 13. Correction of Barometer Reading for Temperature.—Because of the difference in expansion of the mercury column of a mercurial barometer and that of the brass scale by which the height is measured, a correction should be applied to the reading when the temperature differs from the standard used for calibration of the instrument. To find the correction, enter this table with the temperature in degrees Fahrenheit, and the barometer reading. Apply the correction in accordance with the sign given. This correction does not apply to aneroid barometers.

The standard temperature used for calibration is 32° F for the mercury, and 62° F for the brass. The correction was computed by means of the formula:

$$C=-B\frac{m\,(T-32°)-l\,(T-62°)}{1+m\,(T-32°)},$$

in which

C is the correction in inches,

B is the observed reading of the barometer in inches of mercury,

m is the coefficient of cubical expansion of mercury = 0.0001010 cubic inches per degree F,

l is the coefficient of linear expansion of brass = 0.0000102 inches per degree F, and

T is the temperature of the attached thermometer in degrees F.

Substituting the values for m and l and simplifying:

$$C=-B\frac{T-28{.}^{\circ}630}{1.1123\,T+10978°}.$$

The minus sign before B indicates that the correction is negative if the temperature is more than 28°630.

Table 14. Conversion Table for Millibars, Inches of Mercury, and Millimeters of Mercury.—The reading of a barometer in inches or millimeters of mercury corresponding to a given reading in millibars can be found directly from this table.

The formula for the pressure in millibars is:

$$P=\frac{B_m Dg}{1000},$$

in which

P is the atmospheric pressure in millibars,

B_m is the height of the column of mercury in millimeters,

D is the density of mercury = 13.5951 grams per cubic centimeter, and

g is the standard value of gravity = 980.665 dynes. Substituting numerical values:

$$P=1.33322\,B_m,$$

and

$$B_m=\frac{P}{1.33322}=0.750064P.$$

Since one millimeter = 0.03937 inches,

$$B_i=\frac{0.03937\,P}{1.33322}=0.0295300\,P,$$

in which B_i is the height of the column of mercury, in inches.

Table 15. Conversion Table for Thermometer Scales.—Enter this table with temperature Fahrenheit, F; Celsius (centigrade), C; or Kelvin, K; and take out the corresponding readings on the other two temperature scales.

On the Fahrenheit scale, the freezing temperature of pure water at standard sea-level pressure is 32°, and the boiling point under the same conditions is considered 212°. The corresponding temperatures are 0° and 100°, respectively, on the Celsius scale and 273°15 and 373°15, respectively, on the Kelvin scale. The value of (−) 273°15 C for absolute zero, the starting point of the Kelvin scale, is the value recognized officially by the National Bureau of Standards of the United States.

The formulas for converting the reading of one scale to the corresponding values of the others, derived from the figures given above, are:

$$C=\frac{5}{9}(F-32^\circ)=K-273^\circ\!.15,$$

$$F=\frac{9}{5}C+32^\circ=\frac{9}{5}K-459^\circ\!.67,$$

$$K=\frac{5}{9}(F+459^\circ\!.67)=C+273^\circ\!.15,$$

in which all temperatures are in degrees.

Table 16. Relative Humidity.—To determine the relative humidity of the atmosphere, enter this table with the dry-bulb (air) temperature (F), and the *difference* between the dry-bulb and wet-bulb temperatures (F). The value taken from the table is the approximate percentage of relative humidity. If the dry-bulb and wet-bulb temperatures are the same, relative humidity is 100 percent.

The table was computed by means of the formula:

$$R=\frac{100e}{e_w},$$

in which

R is the approximate relative humidity in percent,

e is the ambient vapor pressure, and

e_w is the saturation vapor pressure over water at dry-bulb temperature.

Professor Ferrel's psychrometric formula was used for computation of e:

$$e=e'-\left[0.000367\,P\,(t-t')\left(1+\frac{t'-32^\circ}{1571}\right)\right],$$

in which

e is the ambient vapor pressure in millibars,

e' is the saturation vapor pressure in millibars at wet-bulb temperature with respect to water,

P is the atmospheric pressure (the millibar equivalent of 30 inches of mercury is used for this table),

t is the dry-bulb temperature in degrees Fahrenheit, and

t' is the wet-bulb temperature in degrees Fahrenheit.

The values of e_w were taken from the International Meteorological Organization Publication Number 79, 1951, table 2, pages 82–83.

Table 17. Dew Point.—To determine the dew point, enter this table with the dry-bulb (air) temperature (F), and the *difference* between the dry-bulb and wet-bulb temperatures (F). The value taken from the table is the dew point in degrees Fahrenheit. If the dry-bulb and wet-bulb temperatures are the same, the air is at or below the dew point.

The values given in this table were obtained (1) by determining the saturation vapor pressure e' for the given temperature T (in degrees Rankine) by means of the following formula:

$$\log_{10} e' = -7.90298\left(\frac{671.67}{T}-1\right)+5.02808\log_{10}\frac{671.67}{T}-1.3816\times10^{-7}\left(10^{11.344\left(1-\frac{T}{671.67}\right)}-1\right)$$
$$+8.1328\times10^{-3}\left(10^{-3.49149\left(\frac{671.67}{T}-1\right)}-1\right)+\log_{10}1013.246,$$

(2) by determining the ambient vapor pressure by means of Ferrel's formula (see explanation to table 16), (3) by substituting e for e' in the formula of (1) to obtain the temperature T of the wet bulb when saturation occurs (to the precision of table 17), and (4) by converting the wet-bulb temperature (T) to the dry-bulb temperature T′ by means of the equation:

$$T' = T + (t - t'),$$

where $(t-t')$ is the depression of the wet-bulb temperature. Tables evaluating e' in terms of T for use in steps (1) and (3) are given in International Meteorological Organization Publication Number 79, 1951, and the Smithsonian Meteorological Tables, Sixth Revised Edition, 1951.

Table 18. Speed Table for Measured Mile.—To find the speed of a vessel traversing a measured nautical mile in a given number of minutes and seconds of time, enter this table at the top or bottom with the number of minutes, and at either side with the number of seconds. The number taken from the table is speed in knots. Accurate results can be obtained by interpolating to the nearest 0.1 second.

This table was computed by means of the formula:

$$S = \frac{3600}{T},$$

in which S is speed in knots, and T is elapsed time in seconds.

Table 19. Speed, Time, and Distance.—To find the distance steamed at any given speed between 0.5 and 40 knots in any given number of minutes from 1 to 60, enter this table at the top with the speed, and at the left with the number of minutes. The number taken from the table is the distance in nautical miles. If hours are substituted for minutes, the tabulated distance should be multiplied by 60; if seconds are substituted for minutes, the tabulated distance should be divided by 60. Use of the table is explained in article 502.

The table was computed by means of the formula:

$$D = \frac{ST}{60},$$

in which D is distance in nautical miles, S is speed is knots, and T is elapsed time in minutes.

Table 20. Conversion Table for Nautical and Statute Miles.—This table gives the number of statute miles corresponding to any whole number of nautical miles from 1 to 100, and the number of nautical miles corresponding to any whole number of statute miles within the same range. The entering value can be multiplied by any power of 10, including negative powers, if the corresponding value of the other unit is multiplied by the same power. Thus, 2,700 nautical miles are equivalent to 3,107.1 statute miles, and 0.3 statute mile is equivalent to 0.2607 nautical mile. Hence, to find the number of statute miles equal to 2463.2 nautical miles:

Nautical miles	*Statute miles*
2400.0	2761.9
63.0	72.5
0.2	0.2
2463.2	2834.6

The table was computed by means of the conversion factors of appendix D:

1 nautical mile = 1.15077945 statute miles,
1 statute mile = 0.86897624 nautical mile.

Table 21. Conversion Table for Meters, Feet, and Fathoms.—The number of feet and fathoms corresponding to a given number of meters, and vice versa, can be taken directly from this table for any value of the entering argument from 1 to 120. The entering value can be multiplied by any power of 10, including negative powers, if the corresponding values of the other units are multiplied by the same power. Thus, 420 meters are equivalent to 1378.0 feet, and 11.2 fathoms are equivalent to 20.483 meters. Hence, to find the number of meters equal to 2163 feet:

Feet	*Meters*
2100	640
63	19
2163	659.

The table was computed by means of the relationships given in appendix D:

1 meter = 39.370079 inches,
1 foot = 12 inches,
1 fathom = 6 feet.

Approximately the same results would be obtained by using the direct conversion factors given in appendix D.

Table 22. Dip of the Sea Short of the Horizon.—If land, another vessel, or other obstruction is between the observer and the sea horizon, use the waterline of the obstruction as the horizontal reference for altitude measurements, and substitute dip from this table for the dip of the horizon (height of eye correction) given in the *Nautical Almanac* or other source. The values below the bold rules are for normal dip, the visible horizon being between the observer and the obstruction. Use of the table is explained in article 726.

The table was computed by means of the formula:

$$D_s = 60 \tan^{-1} \left(\frac{h_f}{6076.1 d_s} + \frac{\beta_o d_s}{2 r_o} \right)$$

in which D_s is the dip short of the sea horizon, in minutes of arc; h_f is the height of eye of the observer above sea level in feet; β is a parameter (0.8321) which characterizes the effect of terrestrial refraction; r_o is the mean radius of the earth, 3440.1 nautical miles; and d_s is the distance to the waterline of the obstruction in nautical miles. Use of the table is explained in article 726.

Table 23. Altitude Correction for Air Temperature.—This table provides a correction to be applied to the altitude of a celestial body when the air temperature varies from the 50° F used for determining mean refraction by means of the *Nautical Almanac*. For maximum accuracy, apply index correction and dip to sextant altitude first, obtaining apparent (rectified) altitude for use in entering this table. Enter the table with altitude and air temperature in degrees Fahrenheit. Apply the correction, in accordance with its tabulated sign, to altitude. Use of the table is explained principally in chapter VIII, and especially in articles 807 and 827.

The table was computed by means of formula:

$$\text{Correction} = R_m\left(1 - \frac{510}{460+T}\right),$$

in which R_m is mean refraction and T is temperature in degrees Fahrenheit.

Table 24. Altitude Correction for Atmospheric Pressure.—This table provides a correction to be applied to the altitude of a celestial body when the atmospheric pressure varies from the 29.83 inches (1010 millibars) used for determining mean refraction by means of the *Nautical Almanac*. For most accurate results, apply index correction and dip to sextant altitude first, obtaining apparent (rectified) altitude for use in entering this table. Enter the table with altitude and atmospheric pressure. Apply the correction to altitude, *adding* if the pressure is less than 29.83 inches and *subtracting* if it is more than 29.83 inches. Use of the table is explained principally in chapter VIII, and especially in articles 808 and 827.

The table was computed by means of the formula:

$$\text{Correction} = R_m\left(1 - \frac{P}{29.83}\right),$$

in which R_m is mean refraction and P is atmospheric pressure in inches of mercury.

Table 25. Meridian Angle and Altitude of a Body on the Prime Vertical Circle.—A celestial body having a declination of contrary name to the latitude does not cross the prime vertical above the celestial horizon, its nearest approach being at rising or setting.

If the declination and latitude are of the same name, and the declination is numerically greater, the body does not cross the prime vertical, but makes its nearest approach (in azimuth) when its meridian angle, east or west, and altitude are as shown in this table, these values being given in italics above the heavy line. At this time the body is stationary in azimuth.

If the declination and latitude are of the same name and numerically equal, the body passes through the zenith as it crosses both the celestial meridian and the prime vertical, as shown in the table.

If the declination and latitude are of the same name, and the declination is numerically less, the body crosses the prime vertical when its meridian angle, east or west, and altitude are as tabulated in vertical type below the heavy line.

The table is entered with declination of the celestial body and the latitude of the observer. Computed altitudes are given, no allowance having been made for refraction, dip, parallax, etc. The tabulated values apply to any celestial body, but values are not given for declination greater than 23° because the tabulated information is generally desired for the sun only. Use of the information given in this table is discussed in articles 710 and 721.

The table was computed by means of the following formulas, derived by Napier's rules (art. 142):

Nearest approach (in azimuth) to the prime vertical:

$$\csc h = \sin d \csc L,$$
$$\sec t = \tan d \cot L.$$

On the prime vertical:

$$\sin h = \sin d \csc L,$$
$$\cos t = \tan d \cot L.$$

In these formulas, h is the altitude, d is the declination, L is the latitude, t is the meridian angle.

Table 26. Latitude and Longitude Factors.—The latitude obtained by solution of an ex-meridian sight (art. 732) is inaccurate if the longitude used in determining the meridian angle is incorrect. Similarly, the longitude obtained by solution of a time sight (volume I) is inaccurate if the latitude used in the solution is incorrect, unless the celestial body is on the prime vertical. This table gives the errors resulting from unit errors in the assumed values used in the computations. There are two columns for each tabulated value of latitude. The first gives the latitude factor, f, which is the error in minutes of latitude for a one-minute error of longitude. The second gives the longitude factor, F, which is the error in minutes of longitude for a one-minute error of latitude. In each case, the total error is the factor multiplied by the number of minutes error in the assumed value. Although the factors were originally intended for use in correcting ex-meridian altitudes and time-sight longitudes, they have other uses which may suggest themselves.

The azimuth angle used for entering the table can be measured from *either* the north or south, through 90°; or it may be measured from the elevated pole, through 180°. If the celestial body is in the southeast (090°–180°) or northwest (270°–360°) quadrant, the f correction is applied to the northward if the correct longitude is east of that used in the solution, and to the southward if the correct longitude is west of that used; while the F correction is applied to the eastward if the correct latitude is north of that used in the solution, and to the westward if the correct latitude is south of that used. If the body is in the northeast (000°–090°) or southwest (180°–270°) quadrant, the correction is applied in the opposite direction. These rules apply in both north and south latitude.

The table was computed by means of the formulas:

$$f = \cos L \tan Z = \frac{1}{\sec L \cot Z} = \frac{1}{F},$$

$$F = \sec L \cot Z = \frac{1}{\cos L \tan Z} = \frac{1}{f},$$

in which f is the tabulated latitude factor, L is the latitude, Z is the azimuth angle, and F is the tabulated longitude factor.

Table 27. Amplitudes.—This table lists amplitudes of celestial bodies at rising and setting. Enter with the declination of the body and the latitude of the observer. The value taken from the table is the amplitude when the *center* of the body is on the *celestial* horizon. For the sun, this occurs when the lower limb is a little more than half a diameter above the visible horizon. For the moon it occurs when the upper limb is about on the horizon. Use the prefix E if the body is rising, and W if it is setting; use the suffix N or S to agree with the declination of the body. Table 28 can be used with reversed sign to correct the tabluations to the values for the visible horizon. Use of table 27 is explained in article 720.

The table was computed by means of the following formula, derived by Napier's rules (art. 142):

$$\sin A = \sec L \sin d,$$

in which A is the amplitude, L is the latitude of the observer, and d is the declination of the celestial body.

Table 28. Correction of Amplitude as Observed on the Visible Horizon.—This table contains a correction to be applied to the amplitude observed when the center of a celestial body is on the visible horizon, to obtain the corresponding amplitude when the center of the body is on the celestial horizon. For the sun, a planet, or a star, apply the correction in the direction *away* from the elevated pole, thus *increasing* the *azimuth angle*. For the moon apply *half* the correction *toward* the elevated pole. This correction can be applied in the opposite direction to a value taken from table 27, to find the corresponding amplitude when the center of a celestial body is on the visible horizon. The table was computed for a height of eye of 41 feet. For other heights normally encountered, the error is too small to be of practical significance in ordinary navigation. Use of the table is explained in article 720.

The values in the table were determined by computing the azimuth angle when the center of the celestial body is on the visible horizon, converting this to amplitude, and determining the difference between this value and the corresponding value from table 27. Computation of azimuth angle was made for an altitude of (−)0°42′.0, determined as follows:

Dip at 41 feet height of eye	(−) 6′.2
Refraction at (−)6′.2 alt.	(−)35′.3
Irradiation of horizon	(−) 0′.6
Parallax (value for sun)	(+) 0′.1
	(−)42′.0.

Azimuth angle was computed by means of the formula:

$$\cos Z = \frac{\sin d - \sin h \sin L}{\cos h \cos L},$$

in which Z is the azimuth angle, d is the declination of the celestial body, h is the altitude (−0°42′.0), and L is the latitude of the observer.

Table 29. Altitude Factor.—In one minute of time from meridian transit the altitude of a celestial body changes by the amount shown in this table if the altitude is between 6° and 86°, the latitude is not more than 60°, and the declination is not more than 63°. The values taken from this table are used to enter table 30 for solving reduction to the meridian (ex-meridian) problems.

For upper transit, use the left-hand pages if the declination and latitude are of the same name (both north or both south) and the right-hand pages if of contrary name. For lower transit, use the values below the heavy lines on the last three contrary-name pages. When a factor is taken from this part of the table, the correction from table 30 is *subtracted* from the observed altitude to obtain the corresponding meridian altitude. All other corrections are added.

The table was computed by means of the formula:

$$a = 1''.9635 \cos L \cos d \csc (L \sim d),$$

in which *a* is the change of altitude in one minute from meridian transit (the tabulated value), L is the latitude of the observer, and d is the declination of the celestial body.

This formula can be used to compute values outside the limits of the table, but is not accurate if the altitude is greater than 86°.

Table 30. Change of Altitude in Given Time from Meridian Transit.—Enter this table with the altitude factor from table 29 and the meridian angle, in either arc or time units, and take out the difference between the altitude at the given time and the altitude at meridian transit. Enter the table separately with whole numbers and tenths of a, interpolating for t if necessary, and add the two values to obtain the total difference. This total can be applied as a correction to observed altitude to obtain the corresponding meridian altitude, adding for upper transit and subtracting for lower transit.

The table was computed by means of the formula:

$$C=\frac{at^2}{60},$$

in which C is the tabulated difference to be used as a correction to observed altitude, in minutes of arc; a is the altitude factor from table 29, in seconds of arc; and t is the meridian angle, in minutes of time.

This formula should not be used for determining values beyond the limits of the table, unless reduced accuracy is acceptable.

Table 31. Natural Trigonometric Functions.—This table gives the values of natural sines, cosecants, tangents, cotangents, secants, and cosines of angles from 0° to 180°, at intervals of 1′. For angles between 0° and 45° use the column labels at the top and the minutes at the left; for angles between 45° and 90° use the column labels at the bottom and the minutes at the right; for angles between 90° and 135° use the column labels at the bottom and the minutes at the left; and for angles between 135° and 180° use the column labels at the top and the minutes at the right. These combinations are indicated by the arrows accompanying the figures representing the number of degrees. For angles between 180° and 360°, subtract 180° and proceed as indicated above to obtain the numerical values of the various functions.

Differences between consecutive entries are shown in the "Diff. 1′" column to the right of each column of values of a trigonometric function, as an aid to interpolation. These differences are one-half line out of step with the numbers to which they apply, as in a critical table. Each difference applies to the values half a line above and half a line below. To determine the correction to apply to the value for the smaller entering angle, multiply the difference by the number of tenths of a minute $\left(\text{or } \frac{\text{seconds}}{60}\right)$ of the entering angle. Note whether the function is increasing or decreasing, and add or subtract the correction as appropriate, so that the interpolated value lies between the two values between which interpolation is made.

The logarithms of values given in this table are given in table 33. The trigonometric functions are explained in article 139.

Table 32. Logarithms of Numbers.—The first page of this table gives the complete common logarithm (characteristic and mantissa) of numbers 1 through 250. The succeeding pages give the mantissa only of the common logarithm of any number. Values are given for four significant digits of entering values, the first three being in the left-hand column, and the fourth at the heading of one of the other columns. Thus, the mantissa of a three-digit number is given in the column headed 0, on the line with the given number; while the mantissa of a four-digit number is given in the column headed by the fourth digit, on the line with the first three digits. As an example, the mantissa of 328 is 51587, while that of 3.284 is 51640. For additional digits, interpolation should be used. The difference between each tabulated mantissa and the next larger tabulated mantissa is given in the "d" column to the right of the smaller mantissa. This difference can be used to enter the appropriate proportional parts ("Prop. parts") auxiliary table to interpolate for the fifth digit of the given number. If an accuracy of more than five significant digits is to be preserved in a computation, a table of logarithms to additional decimal places should be used. For a number of one or two digits, use the first page of the table or add zeros to make three digits. That is, the mantissa of 3, 30, and 300 is the same, 47712. Interpolation on the first page of the table is not recommended. The second part should be used for values not listed on the first page.

Additional information on the nature and use of logarithms is given in article 110.

Table 33. Logarithms of Trigonometric Functions.—This table gives the common logarithms (+10) of sines, cosecants, tangents, cotangents, secants, and cosines of angles from 0° to 180°, at intervals of 1′. For angles between 0° and 45° use the column labels at the top and the minutes at the left; for angles between 45° and 90° use the column labels at the bottom and the minutes at the right; for angles between 90° and 135° use the column labels at the bottom and the minutes at the left; and for angles between 135° and 180° use the column labels at the top and the minutes at the right. These combinations are indicated by the arrows accompanying the figures representing the number of degrees. For angles between 180° and 360°, subtract 180° and proceed as indicated above to obtain the numerical values of the various functions.

Differences between consecutive entries are shown in the "Diff. 1′" columns as in table 31, except that one difference column is used for both sines and cosecants, another for both tangents and cotangents, and a third for both secants and cosines. These differences, given as an aid to interpolation, are one-half line out of step with the numbers to which they apply, as in a critical table. Each difference applies to the values half a line above and half a line below. To determine the correction to apply to the value for the smaller entering angle, multiply the difference by the number of tenths of a minute $\left(\text{or } \frac{\text{seconds}}{60}\right)$ of the entering angle. Note whether the function is increasing or decreasing, and add or subtract the correction as appropriate, so that the interpolated value lies between the two values between which interpolation is made.

Natural trigonometric functions are given in table 31. The trigonometric functions, both natural and logarithmic, are explained in article 139.

Table 34. Haversines.—This table lists the common logarithms (+10) of haversines, and natural haversines, of angles from 0° to 360°, at intervals of 1′. For angles between 0° and 180° use the degrees as given at the tops of the columns and the minutes at the left; for angles between 180° and 360° use the degrees as given at the bottoms of the columns and the minutes at the right.

A haversine is half of a versed sine:

$$\text{hav } A = \tfrac{1}{2} \text{ ver } A = \tfrac{1}{2}(1 - \cos A) = \sin^2 \tfrac{1}{2} A.$$

It is further discussed in article 139.

Table 35. The Ageton Method.—The late Rear Admiral Arthur A. Ageton, USN, devised this method while serving in the grade of lieutenant. It was first published in 1931 as H.O. Pub. No. 211, *Dead Reckoning Altitude and Azimuth Table*.

The table is designed for the solution of the divided astronomical triangle with all formulas in terms of secants and cosecants. The table gives five-place log cosecants (labeled A) and log secants (labeled B), both multipled by 100,000 to eliminate the characteristic and decimal. These values are given in parallel columns for each 0.′5 of angle from 0° to 180°. The method can be used for solution from the dead reckoning or any other assumed position. The method is intended for use without interpolation. However, when the quantity K used in the solution or the meridian angle approaches 90°, the solution accuracy is significantly reduced. When the value of K is found to lie within the limits of 87°30′ and 92°30′, the maximum error in computed altitude without interpolation is about 30′. When interpolation is neglected outside these limits, the maximum error is about 6′. The interpolation of the log secant (B) of R, the auxiliary part used to divide the oblique spherical triangle into two right spherical triangles, from the log cosecant (A) of R *alone* reduces the maximum error in altitude to about 2′.0 when K is 90° and to about 0′.8 when K is 80°. When K is between 82° and 98°, it is good practice to either discard the sight or to interpolate the log secant (B) of R from the log cosecant (A) of R. Use of the table is explained in articles 718 and 1018.

Table 36. Time Zones, Zone Descriptions, and Suffixes.—The zone description and the single letter of the alphabet designating a time zone and sometimes used as a suffix to zone time for all time zones are given in this table.

Table 37. Natural and Numerical Chart Scales.—This table gives the numerical scale equivalents for various natural or fractional chart scales.

The scale of a chart is the ratio of a given distance on the chart to the actual distance which it represents on the earth. The scale may be expressed as a simple ratio or fraction, the natural scale. For example, 1:80,000 or $\frac{1}{80,000}$ means that one unit (such as an inch) on the chart represents 80,000 of the same unit on the surface of the earth. Or the scale may be expressed as a statement of that distance on the earth shown as one unit (usually an inch) on the chart, or vice versa. This is the numerical scale.

The table was computed using 72,913.39 inches per nautical mile and 63,360 inches per statute mile.

Table 38. Compact Traverse Table.—This table provides an alternative to the use of table 3 for the tabular solutions of the various sailings except great circle and composite. Two sets of column headings are given to indicate the corresponding values in different problems. Thus, if DLo is used as entering argument in the first column, p is taken from the second column, but if D is used as entering argument in the first column, l is taken from the second column and p is taken from the third column. When the top line of column headings is used, the individual table is selected by means of the *latitude*. When the second line is used, the table is selected by means of the *course*.

The entering argument can be multiplied by any power of 10, including negative powers (art. 102), if the corresponding values taken from the table are multiplied by the same power. Thus, using the table for course 205°, if D=6 miles, l=5′.438, and p=2.536 miles; but if D=600 miles, l=543′.8, and p=253.6 miles; or if D=0.6 mile, l=0′.5438, and p=0.2536 mile.

In this table, DLo is difference of longitude, p is departure, D is distance, l is difference of latitude, and m is meridional difference (difference of meridional parts at two latitudes). In the solution of any right triangle, D can be considered the hypotenuse, and l and p the other two sides. The angle is that opposite side p. Also, if m is one of the short sides of a plane right triangle, DLo is the other side if the given angle is that opposite DLo. If the two short sides are known, the angle opposite one of them can be determined from the tabulation on the right if the table is entered with the quotient found by dividing this side by the other short side.

The use of this table in the solution of problems of the various sailings is given in chapter X.

The top decimal in each individual table is a natural trigonometric function, the p, l column being cosines, the p column being sines, the DLo, D column being secants, and the DLo column being tangents. The decimals below the top line are multiples of the top value. The decimals in the center column to the right of the double line are natural tangents at intervals of 0°.1. For additional decimal places, use table 31.

Table 39. Distance by Vertical Angle Measured Between Waterline at Object and Sea Horizon Beyond Object.—This table provides means for determining the distance to an object lying within or short of the horizon when the height of eye of the observer is known. The vertical angle between the waterline at the object and the visible (sea) horizon beyond is measured and corrected for index error. The table is entered with the corrected vertical angle and the height of eye of the observer in feet; the distance in yards is taken directly from the table.

The table was computed from the formula:

$$\tan h_s = (A-B) \div (1+AB)$$

where

$$A = \frac{h}{d_s} + \frac{B_o d_s}{2r_o}$$

and

$$B = \sqrt{2\beta_o h/r_o}$$

in which β_o is a parameter (0.8279) which characterizes terrestrial refraction, r_o is the mean radius of the earth; 3440.1 nautical miles; h is the height of eye of the observer in feet; h_s is the observed vertical angle corrected for index error, and d_s is the distance to the waterline of the object in nautical miles. Use of the table is explained in article 503.

Table 40. Geographic Range.—This table gives the geographic range or the maximum distance at which the curvature of the earth permits a light to be seen from a particular height of eye without regard to the luminous intensity of the light. The geographic range depends upon the height of both the light and the eye of the observer. Use of the table is explained in article 1202.

The table was computed by means of the formula:

$$d = 1.17\sqrt{H} + 1.17\sqrt{h},$$

in which d is the geographic range in nautical miles, H is the height, in feet, of the light above sea level, and h is the height, in feet, of the eye of the observer above sea level.

Table 41. Distance by Vertical Angle Measured Between Waterline at Object and Top of Object.—This table provides means for determining the angle subtended by an object of known height lying at a particular distance within the observer's visible horizon or vice versa.

The table provides the solution of a plane right triangle having its right angle at the base of the observed object and its altitude coincident with the vertical dimension of the observed object. The solutions are based upon the following simplifying assumptions: (1) the eye of the observer is at sea level, (2) the sea surface between the observer and the object is flat, (3) atmospheric refraction is negligible, and (4) the waterline at the object is vertically below the peak of the object. The error due to the height of eye of the observer does not exceed 3 percent of the distance-off for sextant angles less than 20° and heights of eye less than one-third of the object height. The error due to the waterline not being below the peak of the object does not exceed 3 percent of the distance-off when the height of eye is less than one-third of the object height and the offset of the waterline from the base of the object is less than one-tenth of the distance-off. Errors due to earth's curvature and atmospheric refraction are negligible for cases of practical interest. Use of the table is explained in article 503.

TABLE 1

Conversion Angle

Mid lat.	Difference of longitude										Mid lat.
	0°	0°.5	1°	1°.5	2°	2°.5	3°	3°.5	4°	4°.5	
°	°	°	°	°	°	°	°	°	°	°	°
0	0.0	0.0	0.0	0.0	0.0	0.0	0.0	0.0	0.0	0.0	0
2	0.0	0.0	0.0	0.0	0.0	0.0	0.1	0.1	0.1	0.1	2
4	0.0	0.0	0.0	0.0	0.0	0.1	0.1	0.1	0.1	0.2	4
6	0.0	0.0	0.1	0.1	0.1	0.1	0.2	0.2	0.2	0.2	6
8	0.0	0.0	0.1	0.1	0.1	0.2	0.2	0.2	0.3	0.3	8
10	0.0	0.0	0.1	0.1	0.1	0.2	0.2	0.3	0.4	0.4	10
11	0.0	0.0	0.1	0.1	0.2	0.2	0.3	0.3	0.4	0.4	11
12	0.0	0.1	0.1	0.1	0.2	0.3	0.3	0.4	0.4	0.5	12
13	0.0	0.1	0.1	0.2	0.2	0.3	0.3	0.4	0.4	0.5	13
14	0.0	0.1	0.1	0.2	0.2	0.3	0.4	0.4	0.5	0.6	14
15	0.0	0.1	0.1	0.2	0.3	0.3	0.4	0.4	0.5	0.6	15
16	0.0	0.1	0.1	0.2	0.3	0.4	0.4	0.5	0.6	0.6	16
17	0.0	0.1	0.2	0.2	0.3	0.4	0.4	0.5	0.6	0.6	17
18	0.0	0.1	0.2	0.2	0.3	0.4	0.5	0.5	0.6	0.7	18
19	0.0	0.1	0.2	0.2	0.3	0.4	0.5	0.6	0.6	0.7	19
20	0.0	0.1	0.2	0.2	0.3	0.4	0.5	0.6	0.7	0.8	20
21	0.0	0.1	0.2	0.3	0.4	0.5	0.5	0.6	0.7	0.8	21
22	0.0	0.1	0.2	0.3	0.4	0.5	0.6	0.6	0.8	0.8	22
23	0.0	0.1	0.2	0.3	0.4	0.5	0.6	0.7	0.8	0.9	23
24	0.0	0.1	0.2	0.3	0.4	0.5	0.6	0.7	0.8	0.9	24
25	0.0	0.1	0.2	0.3	0.4	0.5	0.6	0.7	0.8	1.0	25
26	0.0	0.1	0.2	0.3	0.4	0.6	0.6	0.8	0.9	1.0	26
27	0.0	0.1	0.2	0.3	0.4	0.6	0.7	0.8	0.9	1.0	27
28	0.0	0.1	0.2	0.4	0.5	0.6	0.7	0.8	0.9	1.1	28
29	0.0	0.1	0.2	0.4	0.5	0.6	0.7	0.8	1.0	1.1	29
30	0.0	0.1	0.2	0.4	0.5	0.6	0.8	0.9	1.0	1.1	30
31	0.0	0.1	0.2	0.4	0.5	0.6	0.8	0.9	1.0	1.2	31
32	0.0	0.1	0.3	0.4	0.5	0.7	0.8	0.9	1.1	1.2	32
33	0.0	0.1	0.3	0.4	0.6	0.7	0.8	1.0	1.1	1.2	33
34	0.0	0.1	0.3	0.4	0.6	0.7	0.8	1.0	1.1	1.2	34
35	0.0	0.1	0.3	0.4	0.6	0.7	0.9	1.0	1.2	1.3	35
36	0.0	0.1	0.3	0.4	0.6	0.7	0.9	1.0	1.2	1.3	36
37	0.0	0.2	0.3	0.4	0.6	0.8	0.9	1.1	1.2	1.4	37
38	0.0	0.2	0.3	0.5	0.6	0.8	0.9	1.1	1.2	1.4	38
39	0.0	0.2	0.3	0.5	0.6	0.8	1.0	1.1	1.2	1.4	39
40	0.0	0.2	0.3	0.5	0.6	0.8	1.0	1.1	1.3	1.4	40
41	0.0	0.2	0.3	0.5	0.6	0.8	1.0	1.2	1.3	1.5	41
42	0.0	0.2	0.3	0.5	0.7	0.8	1.0	1.2	1.3	1.5	42
43	0.0	0.2	0.3	0.5	0.7	0.8	1.0	1.2	1.4	1.5	43
44	0.0	0.2	0.4	0.5	0.7	0.9	1.1	1.2	1.4	1.6	44
45	0.0	0.2	0.4	0.5	0.7	0.9	1.1	1.2	1.4	1.6	45
46	0.0	0.2	0.4	0.5	0.7	0.9	1.1	1.3	1.4	1.6	46
47	0.0	0.2	0.4	0.6	0.7	0.9	1.1	1.3	1.5	1.7	47
48	0.0	0.2	0.4	0.6	0.8	0.9	1.1	1.3	1.5	1.7	48
49	0.0	0.2	0.4	0.6	0.8	1.0	1.1	1.3	1.5	1.7	49
50	0.0	0.2	0.4	0.6	0.8	1.0	1.1	1.3	1.5	1.7	50
51	0.0	0.2	0.4	0.6	0.8	1.0	1.2	1.4	1.6	1.8	51
52	0.0	0.2	0.4	0.6	0.8	1.0	1.2	1.4	1.6	1.8	52
53	0.0	0.2	0.4	0.6	0.8	1.0	1.2	1.4	1.6	1.8	53
54	0.0	0.2	0.4	0.6	0.8	1.0	1.2	1.4	1.6	1.8	54
55	0.0	0.2	0.4	0.6	0.8	1.0	1.2	1.4	1.6	1.8	55
56	0.0	0.2	0.4	0.6	0.8	1.0	1.2	1.4	1.7	1.9	56
57	0.0	0.2	0.4	0.6	0.8	1.1	1.2	1.5	1.7	1.9	57
58	0.0	0.2	0.4	0.6	0.8	1.1	1.3	1.5	1.7	1.9	58
59	0.0	0.2	0.4	0.6	0.8	1.1	1.3	1.5	1.7	1.9	59
60	0.0	0.2	0.4	0.7	0.9	1.1	1.3	1.5	1.7	2.0	60

Receiver (latitude)	Transmitter (direction from receiver)	Correction Sign	Receiver (latitude)	Transmitter (direction from receiver)	Correction Sign
North	Eastward	+	South	Eastward	−
North	Westward	−	South	Westward	+

TABLE 1

Conversion Angle

Mid lat.	Difference of longitude 0°	0°.5	1°	1°.5	2°	2°.5	3°	3°.5	4°	4°.5	Mid lat.
°	°	°	°	°	°	°	°	°	°	°	°
61	0.0	0.2	0.4	0.7	0.9	1.1	1.3	1.5	1.7	2.0	61
62	0.0	0.2	0.4	0.7	0.9	1.1	1.3	1.5	1.8	2.0	62
63	0.0	0.2	0.4	0.7	0.9	1.1	1.3	1.6	1.8	2.0	63
64	0.0	0.2	0.4	0.7	0.9	1.1	1.3	1.6	1.8	2.0	64
65	0.0	0.2	0.5	0.7	0.9	1.1	1.4	1.6	1.8	2.0	65
66	0.0	0.2	0.5	0.7	0.9	1.1	1.4	1.6	1.8	2.1	66
67	0.0	0.2	0.5	0.7	0.9	1.2	1.4	1.6	1.8	2.1	67
68	0.0	0.2	0.5	0.7	0.9	1.2	1.4	1.6	1.9	2.1	68
69	0.0	0.2	0.5	0.7	0.9	1.2	1.4	1.6	1.9	2.1	69
70	0.0	0.2	0.5	0.7	0.9	1.2	1.4	1.6	1.9	2.1	70
71	0.0	0.2	0.5	0.7	0.9	1.2	1.4	1.7	1.9	2.1	71
72	0.0	0.2	0.5	0.7	1.0	1.2	1.4	1.7	1.9	2.1	72
73	0.0	0.2	0.5	0.7	1.0	1.2	1.4	1.7	1.9	2.2	73
74	0.0	0.2	0.5	0.7	1.0	1.2	1.4	1.7	1.9	2.2	74
75	0.0	0.2	0.5	0.7	1.0	1.2	1.4	1.7	1.9	2.2	75
76	0.0	0.2	0.5	0.7	1.0	1.2	1.5	1.7	1.9	2.2	76
77	0.0	0.2	0.5	0.7	1.0	1.2	1.5	1.7	1.9	2.2	77
78	0.0	0.2	0.5	0.7	1.0	1.2	1.5	1.7	2.0	2.2	78
79	0.0	0.2	0.5	0.7	1.0	1.2	1.5	1.7	2.0	2.2	79
80	0.0	0.2	0.5	0.7	1.0	1.2	1.5	1.7	2.0	2.2	80
81	0.0	0.2	0.5	0.7	1.0	1.2	1.5	1.7	2.0	2.2	81
82	0.0	0.2	0.5	0.7	1.0	1.2	1.5	1.7	2.0	2.2	82
83	0.0	0.2	0.5	0.7	1.0	1.2	1.5	1.7	2.0	2.2	83
84	0.0	0.2	0.5	0.7	1.0	1.2	1.5	1.7	2.0	2.2	84
85	0.0	0.2	0.5	0.7	1.0	1.2	1.5	1.7	2.0	2.2	85

DLo	0° Latitude of Receiver					5° Latitude of Receiver					DLo
	Latitude of Transmitter					Latitude of Transmitter					
	10°	5°	0°	5°	10°	5°	0°	5°	10°	15°	
°	°	°	°	°	°	°	°	°	°	°	°
0	0.0	0.0	0.0	0.0	0.0	0.0	0.0	0.0	0.0	0.0	0
5	0.3	0.3	0.0	0.3	0.3	0.1	0.0	0.3	0.5	0.5	5
10	0.5	0.3	0.0	0.3	0.5	0.0	0.1	0.4	0.7	0.9	10
15	0.6	0.3	0.0	0.3	0.6	0.0	0.3	0.7	1.0	1.3	15

DLo	10° Latitude of Receiver					15° Latitude of Receiver					DLo
	Latitude of Transmitter					Latitude of Transmitter					
	0°	5°	10°	15°	20°	5°	10°	15°	20°	25°	
°	°	°	°	°	°	°	°	°	°	°	°
0	0.0	0.0	0.0	0.0	0.0	0.0	0.0	0.0	0.0	0.0	0
5	0.1	0.2	0.5	0.7	0.7	0.4	0.4	0.7	0.9	0.9	5
10	0.4	0.6	0.9	1.2	1.3	0.8	1.0	1.3	1.6	1.7	10
15	0.7	1.0	1.3	1.6	1.9	1.4	1.6	2.0	2.3	2.5	15

Receiver (latitude)	Transmitter (direction from receiver)	Correction Sign	Receiver (latitude)	Transmitter (direction from receiver)	Correction Sign
North	Eastward	+	South	Eastward	–
North	Westward	–	South	Westward	+

TABLE 1

Conversion Angle

DLo	20° Latitude of Receiver					25° Latitude of Receiver					DLo
	Latitude of Transmitter					Latitude of Transmitter					
	10°	15°	20°	25°	30°	15°	20°	25°	30°	35°	
°	°	°	°	°	°	°	°	°	°	°	°
0	0.0	0.0	0.0	0.0	0.0	0.0	0.0	0.0	0.0	0.0	0
5	0.6	0.6	0.9	1.1	1.1	0.8	0.8	1.1	1.3	1.3	5
10	1.3	1.4	1.7	2.0	2.1	1.7	1.9	2.1	2.4	2.5	10
15	2.0	2.3	2.6	2.9	3.1	2.7	2.9	3.2	3.5	3.7	15

DLo	30° Latitude of Receiver					35° Latitude of Receiver					DLo
	Latitude of Transmitter					Latitude of Transmitter					
	20°	25°	30°	35°	40°	25°	30°	35°	40°	45°	
°	°	°	°	°	°	°	°	°	°	°	°
0	0.0	0.0	0.0	0.0	0.0	0.0	0.0	0.0	0.0	0.0	0
5	1.0	1.0	1.3	1.4	1.4	1.2	1.3	1.5	1.6	1.6	5
10	2.1	2.3	2.5	2.7	2.8	2.5	2.7	2.9	3.1	3.2	10
15	3.3	3.5	3.8	4.0	4.2	3.9	4.1	4.3	4.5	4.7	15

DLo	40° Latitude of Receiver					45° Latitude of Receiver					DLo
	Latitude of Transmitter					Latitude of Transmitter					
	30°	35°	40°	45°	50°	35°	40°	45°	50°	55°	
°	°	°	°	°	°	°	°	°	°	°	°
0	0.0	0.0	0.0	0.0	0.0	0.0	0.0	0.0	0.0	0.0	0
5	1.4	1.5	1.6	1.7	1.7	1.6	1.6	1.8	1.9	1.9	5
10	2.9	3.0	3.2	3.4	3.4	3.3	3.4	3.6	3.7	3.7	10
15	4.5	4.6	4.8	5.0	5.1	5.0	5.2	5.3	5.5	5.5	15
20	6.0	6.3	6.5	6.7	6.8	6.8	6.9	7.1	7.3	7.4	20
25	7.6	7.9	8.1	8.3	8.5	8.5	8.7	8.9	9.1	9.2	25
30	9.2	9.5	9.8	10.0	10.2	10.3	10.5	10.7	10.9	11.0	30

DLo	50° Latitude of Receiver					55° Latitude of Receiver					DLo
	Latitude of Transmitter					Latitude of Transmitter					
	40°	45°	50°	55°	60°	45°	50°	55°	60°	65°	
°	°	°	°	°	°	°	°	°	°	°	°
0	0.0	0.0	0.0	0.0	0.0	0.0	0.0	0.0	0.0	0.0	0
5	1.8	1.8	1.9	2.0	2.0	2.0	2.0	2.1	2.1	2.1	5
10	3.7	3.7	3.8	3.9	3.9	4.0	4.0	4.1	4.2	4.1	10
15	5.6	5.6	5.8	5.9	5.9	6.0	6.1	6.2	6.2	6.1	15
20	7.5	7.6	7.7	7.8	7.8	8.1	8.2	8.2	8.2	8.2	20
25	9.4	9.5	9.6	9.7	9.8	10.2	10.2	10.3	10.3	10.2	25
30	11.3	11.5	11.6	11.7	11.7	12.3	12.3	12.4	12.4	12.3	30

Receiver (latitude)	Transmitter (direction from receiver)	Correction Sign	Receiver (latitude)	Transmitter (direction from receiver)	Correction Sign
North	Eastward	+	South	Eastward	−
North	Westward	−	South	Westward	+

TABLE 1

Conversion Angle

DLo	60° Latitude of Receiver					65° Latitude of Receiver					DLo
	Latitude of Transmitter					Latitude of Transmitter					
	50°	55°	60°	65°	70°	55°	60°	65°	70°	75°	
°	°	°	°	°	°	°	°	°	°	°	°
0	0.0	0.0	0.0	0.0	0.0	0.0	0.0	0.0	0.0	0.0	0
5	2.2	2.1	2.2	2.2	2.1	2.3	2.3	2.3	2.2	2.2	5
10	4.3	4.3	4.3	4.3	4.2	4.6	4.6	4.5	4.5	4.3	10
15	6.5	6.5	6.5	6.5	6.3	6.9	6.9	6.8	6.7	6.5	15
20	8.7	8.7	8.7	8.6	8.5	9.3	9.2	9.1	8.9	8.6	20
25	10.9	10.9	10.9	10.8	10.6	11.6	11.5	11.4	11.1	10.8	25
30	13.1	13.1	13.1	12.9	12.7	14.0	13.8	13.7	13.4	12.9	30
35	15.4	15.4	15.3	15.1	14.8	16.3	16.2	16.0	15.6	15.1	35
40	17.7	17.6	17.5	17.3	17.0	18.7	18.5	18.3	17.9	17.2	40
45	19.9	19.9	19.7	19.5	19.1	21.1	20.9	20.6	20.1	19.4	45

DLo	70° Latitude of Receiver					75° Latitude of Receiver					DLo
	Latitude of Transmitter					Latitude of Transmitter					
	60°	65°	70°	75°	80°	65°	70°	75°	80°	85°	
°	°	°	°	°	°	°	°	°	°	°	°
0	0.0	0.0	0.0	0.0	0.0	0.0	0.0	0.0	0.0	0.0	0
5	2.4	2.4	2.4	2.3	2.1	2.6	2.5	2.4	2.3	2.0	5
10	4.9	4.8	4.7	4.5	4.3	5.1	5.0	4.8	4.6	4.0	10
15	7.3	7.2	7.1	6.8	6.4	7.7	7.5	7.3	6.8	6.1	15
20	9.8	9.6	9.4	9.1	8.6	10.3	10.0	9.7	9.1	8.1	20
25	12.2	12.0	11.8	11.4	10.7	12.9	12.6	12.1	11.4	10.1	25
30	14.7	14.5	14.1	13.6	12.8	15.5	15.1	14.5	13.7	12.1	30
35	17.2	16.9	16.5	15.9	15.0	18.1	17.6	16.9	15.9	14.1	35
40	19.7	19.4	18.9	18.2	17.1	20.7	20.1	19.4	18.2	16.2	40
45	22.2	21.8	21.3	20.5	19.3	23.3	22.7	21.8	20.5	18.2	45

DLo	80° Latitude of Receiver					80° Latitude of Receiver					DLo
	Latitude of Transmitter					Latitude of Transmitter					
	70°	75°	80°	85°	90°	70°	75°	80°	85°	90°	
°	°	°	°	°	°	°	°	°	°	°	°
0	0.0	0.0	0.0	0.0	0.0	49.6	47.5	44.6	39.4	0.0	90
5	2.7	2.6	2.6	2.2	0.0	52.4	50.2	47.1	41.6	0.0	95
10	5.4	5.2	4.9	4.4	0.0	55.2	52.9	49.6	43.7	0.0	100
15	8.1	7.8	7.4	6.6	0.0	58.1	55.7	52.1	45.9	0.0	105
20	10.9	10.5	9.9	8.8	0.0	61.0	58.4	54.6	48.1	0.0	110
25	13.6	13.1	12.3	11.0	0.0	63.8	61.1	57.1	50.2	0.0	115
30	16.3	15.7	14.8	13.2	0.0	66.7	63.8	59.6	52.4	0.0	120
35	19.0	18.3	17.3	15.4	0.0	69.6	66.6	62.1	54.5	0.0	125
40	21.8	20.9	19.7	17.5	0.0	72.6	69.3	64.7	56.7	0.0	130
45	24.5	23.6	22.2	19.7	0.0	75.5	72.1	67.2	58.8	0.0	135
50	27.3	26.2	24.7	21.9	0.0	78.5	74.9	69.7	60.9	0.0	140
55	30.0	28.9	27.2	24.1	0.0	81.4	77.7	72.2	63.0	0.0	145
60	32.8	31.5	29.6	26.3	0.0	84.4	80.5	74.8	65.1	0.0	150
65	35.6	34.2	32.1	28.5	0.0	87.4	83.2	77.3	67.2	0.0	155

Receiver (latitude)	Transmitter (direction from receiver)	Correction Sign	Receiver (latitude)	Transmitter (direction from receiver)	Correction Sign
North	Eastward	+	South	Eastward	−
North	Westward	−	South	Westward	+

TABLE 1

Conversion Angle

DLo	80° Latitude of Receiver					80° Latitude of Receiver					DLo
	Latitude of Transmitter					Latitude of Transmitter					
	70°	75°	80°	85°	90°	70°	75°	80°	85°	90°	
°	°	°	°	°	°	°	°	°	°	°	°
70	38.3	36.8	34.6	30.7	0.0	90.4	86.0	79.8	69.3	0.0	160
75	41.1	39.5	37.1	32.9	0.0	93.5	88.9	82.4	71.4	0.0	165
80	43.9	42.2	39.6	35.0	0.0	96.5	91.7	84.9	73.4	0.0	170
85	46.7	44.8	42.1	37.2	0.0	99.6	94.6	87.5	75.5	0.0	175
90	49.6	47.5	44.6	39.4	0.0	102.6	97.4	90.0	77.5	0.0	180

DLo	85° Latitude of Receiver				85° Latitude of Receiver				DLo
	Latitude of Transmitter				Latitude of Transmitter				
	75°	80°	85°	90°	75°	80°	85°	80°	
°	°	°	°	°	°	°	°	°	°
0	0.0	0.0	0.0	0.0	53.1	50.2	44.9	0.0	90
5	2.9	2.8	2.6	0.0	56.1	53.0	47.4	0.0	95
10	5.9	5.5	5.0	0.0	59.2	55.9	49.9	0.0	100
15	8.8	8.3	7.5	0.0	62.2	58.7	52.4	0.0	105
20	11.7	11.1	10.0	0.0	65.2	61.5	54.9	0.0	110
25	14.6	13.8	12.5	0.0	68.3	64.4	57.4	0.0	115
30	17.6	16.6	14.9	0.0	71.4	67.3	59.9	0.0	120
35	20.5	19.4	17.4	0.0	74.4	70.1	62.4	0.0	125
40	23.4	22.2	19.9	0.0	77.5	73.0	64.9	0.0	130
45	26.4	25.0	22.4	0.0	80.6	75.9	67.4	0.0	135
50	29.3	27.7	24.9	0.0	83.7	78.8	69.9	0.0	140
55	32.3	30.5	27.4	0.0	86.9	81.8	72.4	0.0	145
60	35.2	33.3	29.9	0.0	90.1	84.7	74.9	0.0	150
65	38.2	36.1	32.4	0.0	93.2	87.6	77.5	0.0	155
70	41.1	38.9	34.9	0.0	96.4	90.6	80.0	0.0	160
75	44.1	41.7	37.4	0.0	99.6	93.6	82.5	0.0	165
80	47.1	44.5	39.9	0.0	102.8	96.5	85.0	0.0	170
85	50.1	47.4	42.4	0.0	106.1	99.5	87.5	0.0	175
90	53.1	50.2	44.9	0.0	109.4	102.5	90.0	0.0	180

Receiver (latitude)	Transmitter (direction from receiver)	Correction Sign	Receiver (latitude)	Transmitter (direction from receiver)	Correction Sign
North	Eastward	+	South	Eastward	-
North	Westward	-	South	Westward	+

TABLE 2

Conversion of Compass Points to Degrees

	Points	Angular measure		Points	Angular measure
NORTH TO EAST		° ′ ″	**SOUTH TO WEST**		° ′ ″
North	0	0 00 00	South	16	180 00 00
N¼E	¼	2 48 45	S¼W	16¼	182 48 45
N½E	½	5 37 30	S½W	16½	185 37 30
N¾E	¾	8 26 15	S¾W	16¾	188 26 15
N by E	1	11 15 00	S by W	17	191 15 00
N by E¼E	1¼	14 03 45	S by W¼W	17¼	194 03 45
N by E½E	1½	16 52 30	S by W½W	17½	196 52 30
N by E¾E	1¾	19 41 15	S by W¾W	17¾	199 41 15
NNE	2	22 30 00	SSW	18	202 30 00
NNE¼E	2¼	25 18 45	SSW¼W	18¼	205 18 45
NNE½E	2½	28 07 30	SSW½W	18½	208 07 30
NNE¾E	2¾	30 56 15	SSW¾W	18¾	210 56 15
NE by N	3	33 45 00	SW by S	19	213 45 00
NE¾N	3¼	36 33 45	SW¾S	19¼	216 33 45
NE½N	3½	39 22 30	SW½S	19½	219 22 30
NE¼N	3¾	42 11 15	SW¼S	19¾	222 11 15
NE	4	45 00 00	SW	20	225 00 00
NE¼E	4¼	47 48 45	SW¼W	20¼	227 48 45
NE½E	4½	50 37 30	SW½W	20½	230 37 30
NE¾E	4¾	53 26 15	SW¾W	20¾	233 26 15
NE by E	5	56 15 00	SW by W	21	236 15 00
NE by E¼E	5¼	59 03 45	SW by W¼W	21¼	239 03 45
NE by E½E	5½	61 52 30	SW by W½W	21½	241 52 30
NE by E¾E	5¾	64 41 15	SW by W¾W	21¾	244 41 15
ENE	6	67 30 00	WSW	22	247 30 00
ENE¼E	6¼	70 18 45	WSW¼W	22¼	250 18 45
ENE½E	6½	73 07 30	WSW½W	22½	253 07 30
ENE¾E	6¾	75 56 15	WSW¾W	22¾	255 56 15
E by N	7	78 45 00	W by S	23	258 45 00
E¾N	7¼	81 33 45	W¾S	23¼	261 33 45
E½N	7½	84 22 30	W½S	23½	264 22 30
E¼N	7¾	87 11 15	W¼S	23¾	267 11 15
EAST TO SOUTH			**WEST TO NORTH**		
East	8	90 00 00	West	24	270 00 00
E¼S	8¼	92 48 45	W¼N	24¼	272 48 45
E½S	8½	95 37 30	W½N	24½	275 37 30
E¾S	8¾	98 26 15	W¾N	24¾	278 26 15
E by S	9	101 15 00	W by N	25	281 15 00
ESE¾E	9¼	104 03 45	WNW¾W	25¼	284 03 45
ESE½E	9½	106 52 30	WNW½W	25½	286 52 30
ESE¼E	9¾	109 41 15	WNW¼W	25¾	289 41 15
ESE	10	112 30 00	WNW	26	292 30 00
SE by E¾E	10¼	115 18 45	NW by W¾W	26¼	295 18 45
SE by E½E	10½	118 07 30	NW by W½W	26½	298 07 30
SE by E¼E	10¾	120 56 15	NW by W¼W	26¾	300 56 15
SE by E	11	123 45 00	NW by W	27	303 45 00
SE¾E	11¼	126 33 45	NW¾W	27¼	306 33 45
SE½E	11½	129 22 30	NW½W	27½	309 22 30
SE¼E	11¾	132 11 15	NW¼W	27¾	312 11 15
SE	12	135 00 00	NW	28	315 00 00
SE¼S	12¼	137 48 45	NW¼N	28¼	317 48 45
SE½S	12½	140 37 30	NW½N	28½	320 37 30
SE¾S	12¾	143 26 15	NW¾N	28¾	323 26 15
SE by S	13	146 15 00	NW by N	29	326 15 00
SSE¾E	13¼	149 03 45	NNW¾W	29¼	329 03 45
SSE½E	13½	151 52 30	NNW½W	29½	331 52 30
SSE¼E	13¾	154 41 15	NNW¼W	29¾	334 41 15
SSE	14	157 30 00	NNW	30	337 30 00
S by E¾E	14¼	160 18 45	N by W¾W	30¼	340 18 45
S by E½E	14½	163 07 30	N by W½W	30½	343 07 30
S by E¼E	14¾	165 56 15	N by W¼W	30¾	345 56 15
S by E	15	168 45 00	N by W	31	348 45 00
S¾E	15¼	171 33 45	N¾W	31¼	351 33 45
S½E	15½	174 22 30	N½W	31½	354 22 30
S¼E	15¾	177 11 15	N¼W	31¾	357 11 15
South	16	180 00 00	North	32	360 00 00

359° | 001°		TABLE 3		359° | 001°
181° | 179°	Traverse	1°	Table	181° | 179°

Dist.	D. Lat.	Dep.	Dist.	D. Lat.	Dep.	Dist.	D. Lat.	Dep.	Dist.	D. Lat.	Dep.	Dist.	D. Lat.	Dep.
1	1.0	0.0	61	61.0	1.1	121	121.0	2.1	181	181.0	3.2	241	241.0	4.2
2	2.0	0.0	62	62.0	1.1	22	122.0	2.1	82	182.0	3.2	42	242.0	4.2
3	3.0	0.1	63	63.0	1.1	23	123.0	2.1	83	183.0	3.2	43	243.0	4.2
4	4.0	0.1	64	64.0	1.1	24	124.0	2.2	84	184.0	3.2	44	244.0	4.3
5	5.0	0.1	65	65.0	1.1	25	125.0	2.2	85	185.0	3.2	45	245.0	4.3
6	6.0	0.1	66	66.0	1.2	26	126.0	2.2	86	186.0	3.2	46	246.0	4.3
7	7.0	0.1	67	67.0	1.2	27	127.0	2.2	87	187.0	3.3	47	247.0	4.3
8	8.0	0.1	68	68.0	1.2	28	128.0	2.2	88	188.0	3.3	48	248.0	4.3
9	9.0	0.2	69	69.0	1.2	29	129.0	2.3	89	189.0	3.3	49	249.0	4.3
10	10.0	0.2	70	70.0	1.2	30	130.0	2.3	90	190.0	3.3	50	250.0	4.4
11	11.0	0.2	71	71.0	1.2	131	131.0	2.3	191	191.0	3.3	251	251.0	4.4
12	12.0	0.2	72	72.0	1.3	32	132.0	2.3	92	192.0	3.4	52	252.0	4.4
13	13.0	0.2	73	73.0	1.3	33	133.0	2.3	93	193.0	3.4	53	253.0	4.4
14	14.0	0.2	74	74.0	1.3	34	134.0	2.3	94	194.0	3.4	54	254.0	4.4
15	15.0	0.3	75	75.0	1.3	35	135.0	2.4	95	195.0	3.4	55	255.0	4.5
16	16.0	0.3	76	76.0	1.3	36	136.0	2.4	96	196.0	3.4	56	256.0	4.5
17	17.0	0.3	77	77.0	1.3	37	137.0	2.4	97	197.0	3.4	57	257.0	4.5
18	18.0	0.3	78	78.0	1.4	38	138.0	2.4	98	198.0	3.5	58	258.0	4.5
19	19.0	0.3	79	79.0	1.4	39	139.0	2.4	99	199.0	3.5	59	259.0	4.5
20	20.0	0.3	80	80.0	1.4	40	140.0	2.4	200	200.0	3.5	60	260.0	4.5
21	21.0	0.4	81	81.0	1.4	141	141.0	2.5	201	201.0	3.5	261	261.0	4.6
22	22.0	0.4	82	82.0	1.4	42	142.0	2.5	02	202.0	3.5	62	262.0	4.6
23	23.0	0.4	83	83.0	1.4	43	143.0	2.5	03	203.0	3.5	63	263.0	4.6
24	24.0	0.4	84	84.0	1.5	44	144.0	2.5	04	204.0	3.6	64	264.0	4.6
25	25.0	0.4	85	85.0	1.5	45	145.0	2.5	05	205.0	3.6	65	265.0	4.6
26	26.0	0.5	86	86.0	1.5	46	146.0	2.5	06	206.0	3.6	66	266.0	4.6
27	27.0	0.5	87	87.0	1.5	47	147.0	2.6	07	207.0	3.6	67	267.0	4.7
28	28.0	0.5	88	88.0	1.5	48	148.0	2.6	08	208.0	3.6	68	268.0	4.7
29	29.0	0.5	89	89.0	1.6	49	149.0	2.6	09	209.0	3.6	69	269.0	4.7
30	30.0	0.5	90	90.0	1.6	50	150.0	2.6	10	210.0	3.7	70	270.0	4.7
31	31.0	0.5	91	91.0	1.6	151	151.0	2.6	211	211.0	3.7	271	271.0	4.7
32	32.0	0.6	92	92.0	1.6	52	152.0	2.7	12	212.0	3.7	72	272.0	4.7
33	33.0	0.6	93	93.0	1.6	53	153.0	2.7	13	213.0	3.7	73	273.0	4.8
34	34.0	0.6	94	94.0	1.6	54	154.0	2.7	14	214.0	3.7	74	274.0	4.8
35	35.0	0.6	95	95.0	1.7	55	155.0	2.7	15	215.0	3.8	75	275.0	4.8
36	36.0	0.6	96	96.0	1.7	56	156.0	2.7	16	216.0	3.8	76	276.0	4.8
37	37.0	0.6	97	97.0	1.7	57	157.0	2.7	17	217.0	3.8	77	277.0	4.8
38	38.0	0.7	98	98.0	1.7	58	158.0	2.8	18	218.0	3.8	78	278.0	4.9
39	39.0	0.7	99	99.0	1.7	59	159.0	2.8	19	219.0	3.8	79	279.0	4.9
40	40.0	0.7	100	100.0	1.7	60	160.0	2.8	20	220.0	3.8	80	280.0	4.9
41	41.0	0.7	101	101.0	1.8	161	161.0	2.8	221	221.0	3.9	281	281.0	4.9
42	42.0	0.7	02	102.0	1.8	62	162.0	2.8	22	222.0	3.9	82	282.0	4.9
43	43.0	0.8	03	103.0	1.8	63	163.0	2.8	23	223.0	3.9	83	283.0	4.9
44	44.0	0.8	04	104.0	1.8	64	164.0	2.9	24	224.0	3.9	84	284.0	5.0
45	45.0	0.8	05	105.0	1.8	65	165.0	2.9	25	225.0	3.9	85	285.0	5.0
46	46.0	0.8	06	106.0	1.8	66	166.0	2.9	26	226.0	3.9	86	286.0	5.0
47	47.0	0.8	07	107.0	1.9	67	167.0	2.9	27	227.0	4.0	87	287.0	5.0
48	48.0	0.8	08	108.0	1.9	68	168.0	2.9	28	228.0	4.0	88	288.0	5.0
49	49.0	0.9	09	109.0	1.9	69	169.0	2.9	29	229.0	4.0	89	289.0	5.0
50	50.0	0.9	10	110.0	1.9	70	170.0	3.0	30	230.0	4.0	90	290.0	5.1
51	51.0	0.9	111	111.0	1.9	171	171.0	3.0	231	231.0	4.0	291	291.0	5.1
52	52.0	0.9	12	112.0	2.0	72	172.0	3.0	32	232.0	4.0	92	292.0	5.1
53	53.0	0.9	13	113.0	2.0	73	173.0	3.0	33	233.0	4.1	93	293.0	5.1
54	54.0	0.9	14	114.0	2.0	74	174.0	3.0	34	234.0	4.1	94	294.0	5.1
55	55.0	1.0	15	115.0	2.0	75	175.0	3.1	35	235.0	4.1	95	295.0	5.1
56	56.0	1.0	16	116.0	2.0	76	176.0	3.1	36	236.0	4.1	96	296.0	5.2
57	57.0	1.0	17	117.0	2.0	77	177.0	3.1	37	237.0	4.1	97	297.0	5.2
58	58.0	1.0	18	118.0	2.1	78	178.0	3.1	38	238.0	4.2	98	298.0	5.2
59	59.0	1.0	19	119.0	2.1	79	179.0	3.1	39	239.0	4.2	99	299.0	5.2
60	60.0	1.0	20	120.0	2.1	80	180.0	3.1	40	240.0	4.2	300	300.0	5.2
Dist.	Dep.	D. Lat.	Dist.	Dep.	D. Lat.	Dist.	Dep.	D. Lat.	Dist.	Dep.	D. Lat.	Dist.	Dep.	D. Lat.

271° | 089°	89°
269° | 091°	

Dist.	D. Lat.	Dep.
N.	N×Cos.	N×Sin.
Hypotenuse	Side Adj.	Side Opp.

359° | 001°
181° | 179°

TABLE 3

Traverse **1°** Table

359° | 001°
181° | 179°

Dist.	D. Lat.	Dep.	Dist.	D. Lat.	Dep.	Dist.	D. Lat.	Dep.	Dist.	D. Lat.	Dep.	Dist.	D. Lat.	Dep.
301	301.0	5.3	361	360.9	6.3	421	420.9	7.3	481	480.9	8.4	541	540.9	9.4
02	302.0	5.3	62	361.9	6.3	22	421.9	7.4	82	481.9	8.4	42	541.9	9.5
03	303.0	5.3	63	362.9	6.3	23	422.9	7.4	83	482.9	8.4	43	542.9	9.5
04	304.0	5.3	64	363.9	6.4	24	423.9	7.4	84	483.9	8.4	44	543.9	9.5
05	305.0	5.3	65	364.9	6.4	25	424.9	7.4	85	484.9	8.5	45	544.9	9.5
06	306.0	5.3	66	365.9	6.4	26	425.9	7.4	86	485.9	8.5	46	545.9	9.5
07	307.0	5.4	67	366.9	6.4	27	426.9	7.5	87	486.9	8.5	47	546.9	9.5
08	308.0	5.4	68	367.9	6.4	28	427.9	7.5	88	487.9	8.5	48	547.9	9.6
09	309.0	5.4	69	368.9	6.4	29	428.9	7.5	89	488.9	8.5	49	548.9	9.6
10	310.0	5.4	70	369.9	6.5	30	429.9	7.5	90	489.9	8.6	50	549.9	9.6
311	311.0	5.4	371	370.9	6.5	431	430.9	7.5	491	490.9	8.6	551	550.9	9.6
12	312.0	5.4	72	371.9	6.5	32	431.9	7.5	92	491.9	8.6	52	551.9	9.6
13	313.0	5.5	73	372.9	6.5	33	432.9	7.6	93	492.9	8.6	53	552.9	9.7
14	314.0	5.5	74	373.9	6.5	34	433.9	7.6	94	493.9	8.6	54	553.9	9.7
15	315.0	5.5	75	374.9	6.5	35	434.9	7.6	95	494.9	8.6	55	554.9	9.7
16	316.0	5.5	76	375.9	6.6	36	435.9	7.6	96	495.9	8.7	56	555.9	9.7
17	317.0	5.5	77	376.9	6.6	37	436.9	7.6	97	496.9	8.7	57	556.9	9.7
18	318.0	5.5	78	377.9	6.6	38	437.9	7.6	98	497.9	8.7	58	557.9	9.7
19	319.0	5.6	79	378.9	6.6	39	438.9	7.7	99	498.9	8.7	59	558.9	9.8
20	320.0	5.6	80	379.9	6.6	40	439.9	7.7	500	499.9	8.7	60	559.9	9.8
321	321.0	5.6	381	380.9	6.6	441	440.9	7.7	501	500.9	8.7	561	560.9	9.8
22	322.0	5.6	82	381.9	6.7	42	441.9	7.7	02	501.9	8.8	62	561.9	9.8
23	323.0	5.6	83	382.9	6.7	43	442.9	7.7	03	502.9	8.8	63	562.9	9.8
24	324.0	5.7	84	383.9	6.7	44	443.9	7.7	04	503.9	8.8	64	563.9	9.8
25	325.0	5.7	85	384.9	6.7	45	444.9	7.8	05	504.9	8.8	65	564.9	9.9
26	326.0	5.7	86	385.9	6.7	46	445.9	7.8	06	505.9	8.8	66	565.9	9.9
27	327.0	5.7	87	386.9	6.8	47	446.9	7.8	07	506.9	8.9	67	566.9	9.9
28	328.0	5.7	88	387.9	6.8	48	447.9	7.8	08	507.9	8.9	68	567.9	9.9
29	328.9	5.7	89	388.9	6.8	49	448.9	7.8	09	508.9	8.9	69	568.9	9.9
30	329.9	5.8	90	389.9	6.8	50	449.9	7.9	10	509.9	8.9	70	569.9	9.9
331	330.9	5.8	391	390.9	6.8	451	450.9	7.9	511	510.9	8.9	571	570.9	10.0
32	331.9	5.8	92	391.9	6.8	52	451.9	7.9	12	511.9	9.0	72	571.9	10.0
33	332.9	5.8	93	392.9	6.9	53	452.9	7.9	13	512.9	9.0	73	572.9	10.0
34	333.9	5.8	94	393.9	6.9	54	453.9	7.9	14	513.9	9.0	74	573.9	10.0
35	334.9	5.8	95	394.9	6.9	55	454.9	7.9	15	514.9	9.0	75	574.9	10.0
36	335.9	5.9	96	395.9	6.9	56	455.9	8.0	16	515.9	9.0	76	575.9	10.1
37	336.9	5.9	97	396.9	6.9	57	456.9	8.0	17	516.9	9.0	77	576.9	10.1
38	337.9	5.9	98	397.9	6.9	58	457.9	8.0	18	517.9	9.0	78	577.9	10.1
39	338.9	5.9	99	398.9	7.0	59	458.9	8.0	19	518.9	9.1	79	578.9	10.1
40	339.9	5.9	400	399.9	7.0	60	459.9	8.0	20	519.9	9.1	80	579.9	10.1
341	340.9	6.0	401	400.9	7.0	461	460.9	8.0	521	520.9	9.1	581	580.9	10.1
42	341.9	6.0	02	401.9	7.0	62	461.9	8.1	22	521.9	9.1	82	581.9	10.2
43	342.9	6.0	03	402.9	7.0	63	462.9	8.1	23	522.9	9.1	83	582.9	10.2
44	343.9	6.0	04	403.9	7.1	64	463.9	8.1	24	523.9	9.1	84	583.9	10.2
45	344.9	6.0	05	404.9	7.1	65	464.9	8.1	25	524.9	9.2	85	584.9	10.2
46	345.9	6.0	06	405.9	7.1	66	465.9	8.1	26	525.9	9.2	86	585.9	10.2
47	346.9	6.1	07	406.9	7.1	67	466.9	8.2	27	526.9	9.2	87	586.9	10.2
48	347.9	6.1	08	407.9	7.1	68	467.9	8.2	28	527.9	9.2	88	587.9	10.3
49	348.9	6.1	09	408.9	7.1	69	468.9	8.2	29	528.9	9.2	89	588.9	10.3
50	349.9	6.1	10	409.9	7.2	70	469.9	8.2	30	529.9	9.2	90	589.9	10.3
351	350.9	6.1	411	410.9	7.2	471	470.9	8.2	531	530.9	9.3	591	590.9	10.3
52	351.9	6.1	12	411.9	7.2	72	471.9	8.2	32	531.9	9.3	92	591.9	10.3
53	352.9	6.2	13	412.9	7.2	73	472.9	8.3	33	532.9	9.3	93	592.9	10.3
54	353.9	6.2	14	413.9	7.2	74	473.9	8.3	34	533.9	9.3	94	593.9	10.4
55	354.9	6.2	15	414.9	7.2	75	474.9	8.3	35	534.9	9.3	95	594.9	10.4
56	355.9	6.2	16	415.9	7.3	76	475.9	8.3	36	535.9	9.4	96	595.9	10.4
57	356.9	6.2	17	416.9	7.3	77	476.9	8.3	37	536.9	9.4	97	596.9	10.4
58	357.9	6.2	18	417.9	7.3	78	477.9	8.3	38	537.9	9.4	98	597.9	10.4
59	358.9	6.3	19	418.9	7.3	79	478.9	8.4	39	538.9	9.4	99	598.9	10.5
60	359.9	6.3	20	419.9	7.3	80	479.9	8.4	40	539.9	9.4	600	599.9	10.5
Dist.	Dep.	D. Lat.	Dist.	Dep.	D. Lat.	Dist.	Dep.	D. Lat.	Dist.	Dep.	D. Lat.	Dist.	Dep.	D. Lat.

Dist.	D. Lat.	Dep.
D Lo	Dep.	
	m	D Lo

89°

271° | 089°
269° | 091°

358° | 002°
182° | 178°

TABLE 3

Traverse **2°** Table

358° | 002°
182° | 178°

Dist.	D. Lat.	Dep.	Dist.	D. Lat.	Dep.	Dist.	D. Lat.	Dep.	Dist.	D. Lat.	Dep.	Dist.	D. Lat.	Dep.
1	1.0	0.0	61	61.0	2.1	121	120.9	4.2	181	180.9	6.3	241	240.9	8.4
2	2.0	0.1	62	62.0	2.2	22	121.9	4.3	82	181.9	6.4	42	241.9	8.4
3	3.0	0.1	63	63.0	2.2	23	122.9	4.3	83	182.9	6.4	43	242.9	8.5
4	4.0	0.1	64	64.0	2.2	24	123.9	4.3	84	183.9	6.4	44	243.9	8.5
5	5.0	0.2	65	65.0	2.3	25	124.9	4.4	85	184.9	6.5	45	244.9	8.6
6	6.0	0.2	66	66.0	2.3	26	125.9	4.4	86	185.9	6.5	46	245.9	8.6
7	7.0	0.2	67	67.0	2.3	27	126.9	4.4	87	186.9	6.5	47	246.8	8.6
8	8.0	0.3	68	68.0	2.4	28	127.9	4.5	88	187.9	6.6	48	247.8	8.7
9	9.0	0.3	69	69.0	2.4	29	128.9	4.5	89	188.9	6.6	49	248.8	8.7
10	10.0	0.3	70	70.0	2.4	30	129.9	4.5	90	189.9	6.6	50	249.8	8.7
11	11.0	0.4	71	71.0	2.5	131	130.9	4.6	191	190.9	6.7	251	250.8	8.8
12	12.0	0.4	72	72.0	2.5	32	131.9	4.6	92	191.9	6.7	52	251.8	8.8
13	13.0	0.5	73	73.0	2.5	33	132.9	4.6	93	192.9	6.7	53	252.8	8.8
14	14.0	0.5	74	74.0	2.6	34	133.9	4.7	94	193.9	6.8	54	253.8	8.9
15	15.0	0.5	75	75.0	2.6	35	134.9	4.7	95	194.9	6.8	55	254.8	8.9
16	16.0	0.6	76	76.0	2.7	36	135.9	4.7	96	195.9	6.8	56	255.8	8.9
17	17.0	0.6	77	77.0	2.7	37	136.9	4.8	97	196.9	6.9	57	256.8	9.0
18	18.0	0.6	78	78.0	2.7	38	137.9	4.8	98	197.9	6.9	58	257.8	9.0
19	19.0	0.7	79	79.0	2.8	39	138.9	4.9	99	198.9	6.9	59	258.8	9.0
20	20.0	0.7	80	80.0	2.8	40	139.9	4.9	200	199.9	7.0	60	259.8	9.1
21	21.0	0.7	81	81.0	2.8	141	140.9	4.9	201	200.9	7.0	261	260.8	9.1
22	22.0	0.8	82	82.0	2.9	42	141.9	5.0	02	201.9	7.0	62	261.8	9.1
23	23.0	0.8	83	82.9	2.9	43	142.9	5.0	03	202.9	7.1	63	262.8	9.2
24	24.0	0.8	84	83.9	2.9	44	143.9	5.0	04	203.9	7.1	64	263.8	9.2
25	25.0	0.9	85	84.9	3.0	45	144.9	5.1	05	204.9	7.2	65	264.8	9.2
26	26.0	0.9	86	85.9	3.0	46	145.9	5.1	06	205.9	7.2	66	265.8	9.3
27	27.0	0.9	87	86.9	3.0	47	146.9	5.1	07	206.9	7.2	67	266.8	9.3
28	28.0	1.0	88	87.9	3.1	48	147.9	5.2	08	207.9	7.3	68	267.8	9.4
29	29.0	1.0	89	88.9	3.1	49	148.9	5.2	09	208.9	7.3	69	268.8	9.4
30	30.0	1.0	90	89.9	3.1	50	149.9	5.2	10	209.9	7.3	70	269.8	9.4
31	31.0	1.1	91	90.9	3.2	151	150.9	5.3	211	210.9	7.4	271	270.8	9.5
32	32.0	1.1	92	91.9	3.2	52	151.9	5.3	12	211.9	7.4	72	271.8	9.5
33	33.0	1.2	93	92.9	3.2	53	152.9	5.3	13	212.9	7.4	73	272.8	9.5
34	34.0	1.2	94	93.9	3.3	54	153.9	5.4	14	213.9	7.5	74	273.8	9.6
35	35.0	1.2	95	94.9	3.3	55	154.9	5.4	15	214.9	7.5	75	274.8	9.6
36	36.0	1.3	96	95.9	3.4	56	155.9	5.4	16	215.9	7.5	76	275.8	9.6
37	37.0	1.3	97	96.9	3.4	57	156.9	5.5	17	216.9	7.6	77	276.8	9.7
38	38.0	1.3	98	97.9	3.4	58	157.9	5.5	18	217.9	7.6	78	277.8	9.7
39	39.0	1.4	99	98.9	3.5	59	158.9	5.5	19	218.9	7.6	79	278.8	9.7
40	40.0	1.4	100	99.9	3.5	60	159.9	5.6	20	219.9	7.7	80	279.8	9.8
41	41.0	1.4	101	100.9	3.5	161	160.9	5.6	221	220.9	7.7	281	280.8	9.8
42	42.0	1.5	02	101.9	3.6	62	161.9	5.7	22	221.9	7.7	82	281.8	9.8
43	43.0	1.5	03	102.9	3.6	63	162.9	5.7	23	222.9	7.8	83	282.8	9.9
44	44.0	1.5	04	103.9	3.6	64	163.9	5.7	24	223.9	7.8	84	283.8	9.9
45	45.0	1.6	05	104.9	3.7	65	164.9	5.8	25	224.9	7.9	85	284.8	9.9
46	46.0	1.6	06	105.9	3.7	66	165.9	5.8	26	225.9	7.9	86	285.8	10.0
47	47.0	1.6	07	106.9	3.7	67	166.9	5.8	27	226.9	7.9	87	286.8	10.0
48	48.0	1.7	08	107.9	3.8	68	167.9	5.9	28	227.9	8.0	88	287.8	10.1
49	49.0	1.7	09	108.9	3.8	69	168.9	5.9	29	228.9	8.0	89	288.8	10.1
50	50.0	1.7	10	109.9	3.8	70	169.9	5.9	30	229.9	8.0	90	289.8	10.1
51	51.0	1.8	111	110.9	3.9	171	170.9	6.0	231	230.9	8.1	291	290.8	10.2
52	52.0	1.8	12	111.9	3.9	72	171.9	6.0	32	231.9	8.1	92	291.8	10.2
53	53.0	1.8	13	112.9	3.9	73	172.9	6.0	33	232.9	8.1	93	292.8	10.2
54	54.0	1.9	14	113.9	4.0	74	173.9	6.1	34	233.9	8.2	94	293.8	10.3
55	55.0	1.9	15	114.9	4.0	75	174.9	6.1	35	234.9	8.2	95	294.8	10.3
56	56.0	2.0	16	115.9	4.0	76	175.9	6.1	36	235.9	8.2	96	295.8	10.3
57	57.0	2.0	17	116.9	4.1	77	176.9	6.2	37	236.9	8.3	97	296.8	10.4
58	58.0	2.0	18	117.9	4.1	78	177.9	6.2	38	237.9	8.3	98	297.8	10.4
59	59.0	2.1	19	118.9	4.2	79	178.9	6.2	39	238.9	8.3	99	298.8	10.4
60	60.0	2.1	20	119.9	4.2	80	179.9	6.3	40	239.9	8.4	300	299.8	10.5
Dist.	Dep.	D. Lat.	Dist.	Dep.	D. Lat.	Dist.	Dep.	D. Lat.	Dist.	Dep.	D. Lat.	Dist.	Dep.	D. Lat.

272° | 088°
268° | 092°

88°

Dist.	D. Lat.	Dep.
N.	N×Cos.	N×Sin.
Hypotenuse	Side Adj.	Side Opp.

358°	002°
182°	178°

TABLE 3

Traverse **2°** Table

358°	002°
182°	178°

Dist.	D. Lat.	Dep.	Dist.	D. Lat.	Dep.	Dist.	D. Lat.	Dep.	Dist.	D. Lat.	Dep.	Dist.	D. Lat.	Dep.
301	300.8	10.5	361	360.8	12.6	421	420.7	14.7	481	480.7	16.8	541	540.7	18.9
02	301.8	10.5	62	361.8	12.6	22	421.7	14.7	82	481.7	16.8	42	541.7	18.9
03	302.8	10.6	63	362.8	12.7	23	422.7	14.7	83	482.7	16.8	43	542.7	19.0
04	303.8	10.6	64	363.8	12.7	24	423.7	14.8	84	483.7	16.9	44	543.7	19.0
05	304.8	10.6	65	364.8	12.7	25	424.7	14.8	85	484.7	16.9	45	544.7	19.0
06	305.8	10.7	66	365.8	12.8	26	425.7	14.9	86	485.7	16.9	46	545.7	19.1
07	306.8	10.7	67	366.8	12.8	27	426.7	14.9	87	486.7	17.0	47	546.7	19.1
08	307.8	10.7	68	367.8	12.8	28	427.7	14.9	88	487.7	17.0	48	547.7	19.1
09	308.8	10.8	69	368.8	12.9	29	428.7	15.0	89	488.7	17.0	49	548.7	19.2
10	309.8	10.8	70	369.8	12.9	30	429.7	15.0	90	489.7	17.1	50	549.7	19.2
311	310.8	10.8	371	370.8	12.9	431	430.7	15.0	491	490.7	17.1	551	550.7	19.2
12	311.8	10.9	72	371.8	13.0	32	431.7	15.1	92	491.7	17.1	52	551.7	19.3
13	312.8	10.9	73	372.8	13.0	33	432.7	15.1	93	492.7	17.2	53	552.7	19.3
14	313.8	10.9	74	373.8	13.0	34	433.7	15.1	94	493.7	17.2	54	553.7	19.3
15	314.8	11.0	75	374.8	13.1	35	434.7	15.2	95	494.7	17.2	55	554.7	19.4
16	315.8	11.0	76	375.8	13.1	36	435.7	15.2	96	495.7	17.3	56	555.7	19.4
17	316.8	11.0	77	376.8	13.1	37	436.7	15.2	97	496.7	17.3	57	556.7	19.4
18	317.8	11.1	78	377.8	13.2	38	437.7	15.3	98	497.7	17.3	58	557.7	19.5
19	318.8	11.1	79	378.8	13.2	39	438.7	15.3	99	498.7	17.4	59	558.7	19.5
20	319.8	11.2	80	379.8	13.2	40	439.7	15.3	500	499.7	17.4	60	559.7	19.5
321	320.8	11.2	381	380.8	13.3	441	440.7	15.4	501	500.7	17.5	561	560.7	19.6
22	321.8	11.2	82	381.8	13.3	42	441.7	15.4	02	501.7	17.5	62	561.7	19.6
23	322.8	11.3	83	382.8	13.3	43	442.7	15.4	03	502.7	17.5	63	562.7	19.6
24	323.8	11.3	84	383.8	13.4	44	443.7	15.5	04	503.7	17.6	64	563.7	19.7
25	324.8	11.3	85	384.8	13.4	45	444.7	15.5	05	504.7	17.6	65	564.7	19.7
26	325.8	11.4	86	385.8	13.5	46	445.7	15.6	06	505.7	17.6	66	565.7	19.8
27	326.8	11.4	87	386.8	13.5	47	446.7	15.6	07	506.7	17.7	67	566.7	19.8
28	327.8	11.4	88	387.8	13.5	48	447.7	15.6	08	507.7	17.7	68	567.7	19.8
29	328.8	11.5	89	388.8	13.6	49	448.7	15.7	09	508.7	17.7	69	568.7	19.9
30	329.8	11.5	90	389.8	13.6	50	449.7	15.7	10	509.7	17.8	70	569.7	19.9
331	330.8	11.5	391	390.8	13.6	451	450.7	15.7	511	510.7	17.9	571	570.7	19.9
32	331.8	11.6	92	391.8	13.7	52	451.7	15.8	12	511.7	17.9	72	571.7	20.0
33	332.8	11.6	93	392.8	13.7	53	452.7	15.8	13	512.7	17.9	73	572.7	20.0
34	333.8	11.6	94	393.8	13.7	54	453.7	15.8	14	513.7	17.9	74	573.7	20.0
35	334.8	11.7	95	394.8	13.8	55	454.7	15.9	15	514.7	18.0	75	574.6	20.1
36	335.8	11.7	96	395.8	13.8	56	455.7	15.9	16	515.7	18.0	76	575.6	20.1
37	336.8	11.7	97	396.8	13.8	57	456.7	15.9	17	516.7	18.0	77	576.6	20.1
38	337.8	11.8	98	397.8	13.9	58	457.7	16.0	18	517.7	18.1	78	577.6	20.2
39	338.8	11.8	99	398.8	13.9	59	458.7	16.0	19	518.7	18.1	79	578.6	20.2
40	339.8	11.9	400	399.8	13.9	60	459.7	16.0	20	519.7	18.1	80	579.6	20.2
341	340.8	11.9	401	400.8	14.0	461	460.7	16.1	521	520.7	18.2	581	580.6	20.3
42	341.8	11.9	02	401.8	14.0	62	461.7	16.1	22	521.7	18.2	82	581.6	20.3
43	342.8	12.0	03	402.8	14.0	63	462.7	16.1	23	522.7	18.3	83	582.6	20.3
44	343.8	12.0	04	403.8	14.1	64	463.7	16.2	24	523.7	18.3	84	583.6	20.4
45	344.8	12.0	05	404.8	14.1	65	464.7	16.2	25	524.7	18.3	85	584.6	20.4
46	345.8	12.1	06	405.8	14.2	66	465.7	16.2	26	525.7	18.4	86	585.6	20.5
47	346.8	12.1	07	406.8	14.2	67	466.7	16.3	27	526.7	18.4	87	586.6	20.5
48	347.8	12.1	08	407.8	14.2	68	467.7	16.3	28	527.7	18.4	88	587.6	20.5
49	348.8	12.2	09	408.8	14.3	69	468.7	16.4	29	528.7	18.5	89	588.6	20.6
50	349.8	12.2	10	409.8	14.3	70	469.7	16.4	30	529.7	18.5	90	589.6	20.6
351	350.8	12.2	411	410.7	14.3	471	470.7	16.4	531	530.7	18.5	591	590.6	20.6
52	351.8	12.3	12	411.7	14.4	72	471.7	16.5	32	531.7	18.6	92	591.6	20.7
53	352.8	12.3	13	412.7	14.4	73	472.7	16.5	33	532.7	18.6	93	592.6	20.7
54	353.8	12.3	14	413.7	14.4	74	473.7	16.5	34	533.7	18.6	94	593.6	20.7
55	354.8	12.4	15	414.7	14.5	75	474.7	16.6	35	534.7	18.7	95	594.6	20.8
56	355.8	12.4	16	415.7	14.5	76	475.7	16.6	36	535.7	18.7	96	595.6	20.8
57	356.8	12.4	17	416.7	14.5	77	476.7	16.6	37	536.7	18.7	97	596.6	20.8
58	357.8	12.5	18	417.7	14.6	78	477.7	16.7	38	537.7	18.8	98	597.6	20.9
59	358.8	12.5	19	418.7	14.6	79	478.7	16.7	39	538.7	18.8	99	598.6	20.9
60	359.8	12.5	20	419.7	14.6	80	479.7	16.7	40	539.7	18.8	600	599.6	20.9
Dist.	Dep.	D. Lat.	Dist.	Dep.	D. Lat.	Dist.	Dep.	D. Lat.	Dist.	Dep.	D. Lat.	Dist.	Dep.	D. Lat.

Dist.	D. Lat.	Dep.
D Lo	Dep.	
	m	D Lo

88°

272°	088°
268°	092°

TABLE 3

357° | 003°
183° | 177°

Traverse **3°** Table

357° | 003°
183° | 177°

Dist.	D. Lat.	Dep.	Dist.	D. Lat.	Dep.	Dist.	D. Lat.	Dep.	Dist.	D. Lat.	Dep.	Dist.	D. Lat.	Dep.
1	1.0	0.1	61	60.9	3.2	121	120.8	6.3	181	180.8	9.5	241	240.7	12.6
2	2.0	0.1	62	61.9	3.2	22	121.8	6.4	82	181.8	9.5	42	241.7	12.7
3	3.0	0.2	63	62.9	3.3	23	122.8	6.4	83	182.7	9.6	43	242.7	12.7
4	4.0	0.2	64	63.9	3.3	24	123.8	6.5	84	183.7	9.6	44	243.7	12.8
5	5.0	0.3	65	64.9	3.4	25	124.8	6.5	85	184.7	9.7	45	244.7	12.8
6	6.0	0.3	66	65.9	3.5	26	125.8	6.6	86	185.7	9.7	46	245.7	12.9
7	7.0	0.4	67	66.9	3.5	27	126.8	6.6	87	186.7	9.8	47	246.7	12.9
8	8.0	0.4	68	67.9	3.6	28	127.8	6.7	88	187.7	9.8	48	247.7	13.0
9	9.0	0.5	69	68.9	3.6	29	128.8	6.8	89	188.7	9.9	49	248.7	13.0
10	10.0	0.5	70	69.9	3.7	30	129.8	6.8	90	189.7	9.9	50	249.7	13.1
11	11.0	0.6	71	70.9	3.7	131	130.8	6.9	191	190.7	10.0	251	250.7	13.1
12	12.0	0.6	72	71.9	3.8	32	131.8	6.9	92	191.7	10.0	52	251.7	13.2
13	13.0	0.7	73	72.9	3.8	33	132.8	7.0	93	192.7	10.1	53	252.7	13.2
14	14.0	0.7	74	73.9	3.9	34	133.8	7.0	94	193.7	10.2	54	253.7	13.3
15	15.0	0.8	75	74.9	3.9	35	134.8	7.1	95	194.7	10.2	55	254.7	13.3
16	16.0	0.8	76	75.9	4.0	36	135.8	7.1	96	195.7	10.3	56	255.6	13.4
17	17.0	0.9	77	76.9	4.0	37	136.8	7.2	97	196.7	10.3	57	256.6	13.5
18	18.0	0.9	78	77.9	4.1	38	137.8	7.2	98	197.7	10.4	58	257.6	13.5
19	19.0	1.0	79	78.9	4.1	39	138.8	7.3	99	198.7	10.4	59	258.6	13.6
20	20.0	1.0	80	79.9	4.2	40	139.8	7.3	200	199.7	10.5	60	259.6	13.6
21	21.0	1.1	81	80.9	4.2	141	140.8	7.4	201	200.7	10.5	261	260.6	13.7
22	22.0	1.2	82	81.9	4.3	42	141.8	7.4	02	201.7	10.6	62	261.6	13.7
23	23.0	1.2	83	82.9	4.3	43	142.8	7.5	03	202.7	10.6	63	262.6	13.8
24	24.0	1.3	84	83.9	4.4	44	143.8	7.5	04	203.7	10.7	64	263.6	13.8
25	25.0	1.3	85	84.9	4.4	45	144.8	7.6	05	204.7	10.7	65	264.6	13.9
26	26.0	1.4	86	85.9	4.5	46	145.8	7.6	06	205.7	10.8	66	265.6	13.9
27	27.0	1.4	87	86.9	4.6	47	146.8	7.7	07	206.7	10.8	67	266.6	14.0
28	28.0	1.5	88	87.9	4.6	48	147.8	7.7	08	207.7	10.9	68	267.6	14.0
29	29.0	1.5	89	88.9	4.7	49	148.8	7.8	09	208.7	10.9	69	268.6	14.1
30	30.0	1.6	90	89.9	4.7	50	149.8	7.9	10	209.7	11.0	70	269.6	14.1
31	31.0	1.6	91	90.9	4.8	151	150.8	7.9	211	210.7	11.0	271	270.6	14.2
32	32.0	1.7	92	91.9	4.8	52	151.8	8.0	12	211.7	11.1	72	271.6	14.2
33	33.0	1.7	93	92.9	4.9	53	152.8	8.0	13	212.7	11.1	73	272.6	14.3
34	34.0	1.8	94	93.9	4.9	54	153.8	8.1	14	213.7	11.2	74	273.6	14.3
35	35.0	1.8	95	94.9	5.0	55	154.8	8.1	15	214.7	11.3	75	274.6	14.4
36	36.0	1.9	96	95.9	5.0	56	155.8	8.2	16	215.7	11.3	76	275.6	14.4
37	36.9	1.9	97	96.9	5.1	57	156.8	8.2	17	216.7	11.4	77	276.6	14.5
38	37.9	2.0	98	97.9	5.1	58	157.8	8.3	18	217.7	11.4	78	277.6	14.5
39	38.9	2.0	99	98.9	5.2	59	158.8	8.3	19	218.7	11.5	79	278.6	14.6
40	39.9	2.1	100	99.9	5.2	60	159.8	8.4	20	219.7	11.5	80	279.6	14.7
41	40.9	2.1	101	100.9	5.3	161	160.8	8.4	221	220.7	11.6	281	280.6	14.7
42	41.9	2.2	02	101.9	5.3	62	161.8	8.5	22	221.7	11.6	82	281.6	14.8
43	42.9	2.3	03	102.9	5.4	63	162.8	8.5	23	222.7	11.7	83	282.6	14.8
44	43.9	2.3	04	103.9	5.4	64	163.8	8.6	24	223.7	11.7	84	283.6	14.9
45	44.9	2.4	05	104.9	5.5	65	164.8	8.6	25	224.7	11.8	85	284.6	14.9
46	45.9	2.4	06	105.9	5.5	66	165.8	8.7	26	225.7	11.8	86	285.6	15.0
47	46.9	2.5	07	106.9	5.6	67	166.8	8.7	27	226.7	11.9	87	286.6	15.0
48	47.9	2.5	08	107.9	5.7	68	167.8	8.8	28	227.7	11.9	88	287.6	15.1
49	48.9	2.6	09	108.9	5.7	69	168.8	8.8	29	228.7	12.0	89	288.6	15.1
50	49.9	2.6	10	109.8	5.8	70	169.8	8.9	30	229.7	12.0	90	289.6	15.2
51	50.9	2.7	111	110.8	5.8	171	170.8	8.9	231	230.7	12.1	291	290.6	15.2
52	51.9	2.7	12	111.8	5.9	72	171.8	9.0	32	231.7	12.1	92	291.6	15.3
53	52.9	2.8	13	112.8	5.9	73	172.8	9.1	33	232.7	12.2	93	292.6	15.3
54	53.9	2.8	14	113.8	6.0	74	173.8	9.1	34	233.7	12.2	94	293.6	15.4
55	54.9	2.9	15	114.8	6.0	75	174.8	9.2	35	234.7	12.3	95	294.6	15.4
56	55.9	2.9	16	115.8	6.1	76	175.8	9.2	36	235.7	12.4	96	295.6	15.5
57	56.9	3.0	17	116.8	6.1	77	176.8	9.3	37	236.7	12.4	97	296.6	15.5
58	57.9	3.0	18	117.8	6.2	78	177.8	9.3	38	237.7	12.5	98	297.6	15.6
59	58.9	3.1	19	118.8	6.2	79	178.8	9.4	39	238.7	12.5	99	298.6	15.6
60	59.9	3.1	20	119.8	6.3	80	179.8	9.4	40	239.7	12.6	300	299.6	15.7
Dist.	Dep.	D. Lat.	Dist.	Dep.	D. Lat.	Dist.	Dep.	D. Lat.	Dist.	Dep.	D. Lat.	Dist.	Dep.	D. Lat.

273° | 087°
267° | 093°

87°

Dist.	D. Lat.	Dep.
N.	N×Cos.	N×Sin.
Hypotenuse	Side Adj.	Side Opp.

357° | 003° / 183° | 177°						TABLE 3 — Traverse 3° Table							357° | 003° / 183° | 177°	
Dist.	D. Lat.	Dep.	Dist.	D. Lat.	Dep.	Dist.	D. Lat.	Dep.	Dist.	D. Lat.	Dep.	Dist.	D. Lat.	Dep.
301	300.6	15.7	361	360.5	18.9	421	420.4	22.0	481	480.3	25.2	541	540.3	28.3
02	301.6	15.8	62	361.5	18.9	22	421.4	22.1	82	481.3	25.2	42	541.3	28.4
03	302.6	15.9	63	362.5	19.0	23	422.4	22.1	83	482.3	25.3	43	542.3	28.4
04	303.6	15.9	64	363.5	19.1	24	423.4	22.2	84	483.3	25.3	44	543.3	28.5
05	304.6	16.0	65	364.5	19.1	25	424.4	22.2	85	484.3	25.4	45	544.3	28.5
06	305.6	16.0	66	365.5	19.2	26	425.4	22.3	86	485.3	25.4	46	545.3	28.6
07	306.6	16.1	67	366.5	19.2	27	426.4	22.3	87	486.3	25.5	47	546.3	28.6
08	307.6	16.1	68	367.5	19.3	28	427.4	22.4	88	487.3	25.5	48	547.2	28.7
09	308.6	16.2	69	368.5	19.3	29	428.4	22.5	89	488.3	25.6	49	548.2	28.7
10	309.6	16.2	70	369.5	19.4	30	429.4	22.5	90	489.3	25.6	50	549.2	28.8
311	310.6	16.3	371	370.5	19.4	431	430.4	22.6	491	490.3	25.7	551	550.2	28.8
12	311.6	16.3	72	371.5	19.5	32	431.4	22.6	92	491.3	25.7	52	551.2	28.9
13	312.6	16.4	73	372.5	19.5	33	432.4	22.7	93	492.3	25.8	53	552.2	28.9
14	313.6	16.4	74	373.5	19.6	34	433.4	22.7	94	493.3	25.9	54	553.2	29.0
15	314.6	16.5	75	374.5	19.6	35	434.4	22.8	95	494.3	25.9	55	554.2	29.0
16	315.6	16.5	76	375.5	19.7	36	435.4	22.8	96	495.3	26.0	56	555.2	29.1
17	316.6	16.6	77	376.5	19.7	37	436.4	22.9	97	496.3	26.0	57	556.2	29.2
18	317.6	16.6	78	377.5	19.8	38	437.4	22.9	98	497.3	26.1	58	557.2	29.2
19	318.6	16.7	79	378.5	19.8	39	438.4	23.0	99	498.3	26.1	59	558.2	29.3
20	319.6	16.7	80	379.5	19.9	40	439.4	23.0	500	499.3	26.2	60	559.2	29.3
321	320.6	16.8	381	380.5	19.9	441	440.4	23.1	501	500.3	26.2	561	560.2	29.4
22	321.6	16.9	82	381.5	20.0	42	441.4	23.1	02	501.3	26.3	62	561.2	29.4
23	322.6	16.9	83	382.5	20.0	43	442.4	23.2	03	502.3	26.3	63	562.2	29.5
24	323.6	17.0	84	383.5	20.1	44	443.4	23.2	04	503.3	26.4	64	563.2	29.5
25	324.6	17.0	85	384.5	20.1	45	444.4	23.3	05	504.3	26.4	65	564.2	29.6
26	325.6	17.1	86	385.5	20.2	46	445.4	23.3	06	505.3	26.5	66	565.2	29.6
27	326.6	17.1	87	386.5	20.3	47	446.4	23.4	07	506.3	26.5	67	566.2	29.7
28	327.6	17.2	88	387.5	20.3	48	447.4	23.4	08	507.3	26.6	68	567.2	29.7
29	328.5	17.2	89	388.5	20.4	49	448.4	23.5	09	508.3	26.6	69	568.2	29.8
30	329.5	17.3	90	389.5	20.4	50	449.4	23.6	10	509.3	26.7	70	569.2	29.8
331	330.5	17.3	391	390.5	20.5	451	450.4	23.6	511	510.3	26.7	571	570.2	29.9
32	331.5	17.4	92	391.5	20.5	52	451.4	23.7	12	511.3	26.8	72	571.2	29.9
33	332.5	17.4	93	392.5	20.6	53	452.4	23.7	13	512.3	26.8	73	572.2	30.0
34	333.5	17.5	94	393.5	20.6	54	453.4	23.8	14	513.3	26.9	74	573.2	30.0
35	334.5	17.5	95	394.5	20.7	55	454.4	23.8	15	514.3	27.0	75	574.2	30.1
36	335.5	17.6	96	395.5	20.7	56	455.4	23.9	16	515.3	27.0	76	575.2	30.1
37	336.5	17.6	97	396.5	20.8	57	456.4	23.9	17	516.3	27.1	77	576.2	30.2
38	337.5	17.7	98	397.5	20.8	58	457.4	24.0	18	517.3	27.1	78	577.2	30.3
39	338.5	17.7	99	398.5	20.9	59	458.4	24.0	19	518.3	27.2	79	578.2	30.3
40	339.5	17.8	400	399.5	20.9	60	459.4	24.1	20	519.3	27.2	80	579.2	30.4
341	340.5	17.8	401	400.5	21.0	461	460.4	24.1	521	520.3	27.3	581	580.2	30.4
42	341.5	17.9	02	401.4	21.0	62	461.4	24.2	22	521.3	27.3	82	581.2	30.5
43	342.5	18.0	03	402.4	21.1	63	462.4	24.2	23	522.3	27.4	83	582.2	30.5
44	343.5	18.0	04	403.4	21.1	64	463.4	24.3	24	523.3	27.4	84	583.2	30.6
45	344.5	18.1	05	404.4	21.2	65	464.4	24.3	25	524.3	27.5	85	584.2	30.6
46	345.5	18.1	06	405.4	21.2	66	465.4	24.4	26	525.3	27.5	86	585.2	30.7
47	346.5	18.2	07	406.4	21.3	67	466.4	24.4	27	526.3	27.6	87	586.2	30.7
48	347.5	18.2	08	407.4	21.4	68	467.4	24.5	28	527.3	27.6	88	587.2	30.8
49	348.5	18.3	09	408.4	21.4	69	468.4	24.5	29	528.3	27.7	89	588.2	30.8
50	349.5	18.3	10	409.4	21.5	70	469.4	24.6	30	529.3	27.7	90	589.2	30.9
351	350.5	18.4	411	410.4	21.5	471	470.4	24.7	531	530.3	27.8	591	590.2	30.9
52	351.5	18.4	12	411.4	21.6	72	471.4	24.7	32	531.3	27.8	92	591.2	31.0
53	352.5	18.5	13	412.4	21.6	73	472.4	24.8	33	532.3	27.9	93	592.2	31.0
54	353.5	18.5	14	413.4	21.7	74	473.4	24.8	34	533.3	27.9	94	593.2	31.1
55	354.5	18.6	15	414.4	21.7	75	474.3	24.9	35	534.3	28.0	95	594.2	31.1
56	355.5	18.6	16	415.4	21.8	76	475.3	24.9	36	535.3	28.1	96	595.2	31.2
57	356.5	18.7	17	416.4	21.8	77	476.3	25.0	37	536.3	28.1	97	596.2	31.2
58	357.5	18.7	18	417.4	21.9	78	477.3	25.0	38	537.3	28.2	98	597.2	31.3
59	358.5	18.8	19	418.4	21.9	79	478.3	25.1	39	538.3	28.2	99	598.2	31.3
60	359.5	18.8	20	419.4	22.0	80	479.3	25.1	40	539.3	28.3	600	599.2	31.4
Dist.	Dep.	D. Lat.	Dist.	Dep.	D. Lat.	Dist.	Dep.	D. Lat.	Dist.	Dep.	D. Lat.	Dist.	Dep.	D. Lat.

Dist.	D. Lat.	Dep.
D Lo	Dep.	
	m	D Lo

87°

273° | 087°
267° | 093°

356° | 004° / 184° | 176°	TABLE 3 Traverse 4° Table	356° | 004° / 184° | 176°

Dist.	D. Lat.	Dep.	Dist.	D. Lat.	Dep.	Dist.	D. Lat.	Dep.	Dist.	D. Lat.	Dep.	Dist.	D. Lat.	Dep.
1	1.0	0.1	61	60.9	4.3	121	120.7	8.4	181	180.6	12.6	241	240.4	16.8
2	2.0	0.1	62	61.8	4.3	22	121.7	8.5	82	181.6	12.7	42	241.4	16.9
3	3.0	0.2	63	62.8	4.4	23	122.7	8.6	83	182.6	12.8	43	242.4	17.0
4	4.0	0.3	64	63.8	4.5	24	123.7	8.6	84	183.6	12.8	44	243.4	17.0
5	5.0	0.3	65	64.8	4.5	25	124.7	8.7	85	184.5	12.9	45	244.4	17.1
6	6.0	0.4	66	65.8	4.6	26	125.7	8.8	86	185.5	13.0	46	245.4	17.2
7	7.0	0.5	67	66.8	4.7	27	126.7	8.9	87	186.5	13.0	47	246.4	17.2
8	8.0	0.6	68	67.8	4.7	28	127.7	8.9	88	187.5	13.1	48	247.4	17.3
9	9.0	0.6	69	68.8	4.8	29	128.7	9.0	89	188.5	13.2	49	248.4	17.4
10	10.0	0.7	70	69.8	4.9	30	129.7	9.1	90	189.5	13.3	50	249.4	17.4
11	11.0	0.8	71	70.8	5.0	131	130.7	9.1	191	190.5	13.3	251	250.4	17.5
12	12.0	0.8	72	71.8	5.0	32	131.7	9.2	92	191.5	13.4	52	251.4	17.6
13	13.0	0.9	73	72.8	5.1	33	132.7	9.3	93	192.5	13.5	53	252.4	17.6
14	14.0	1.0	74	73.8	5.2	34	133.7	9.3	94	193.5	13.5	54	253.4	17.7
15	15.0	1.0	75	74.8	5.2	35	134.7	9.4	95	194.5	13.6	55	254.4	17.8
16	16.0	1.1	76	75.8	5.3	36	135.7	9.5	96	195.5	13.7	56	255.4	17.9
17	17.0	1.2	77	76.8	5.4	37	136.7	9.6	97	196.5	13.7	57	256.4	17.9
18	18.0	1.3	78	77.8	5.4	38	137.7	9.6	98	197.5	13.8	58	257.4	18.0
19	19.0	1.3	79	78.8	5.5	39	138.7	9.7	99	198.5	13.9	59	258.4	18.1
20	20.0	1.4	80	79.8	5.6	40	139.7	9.8	200	199.5	14.0	60	259.4	18.1
21	20.9	1.5	81	80.8	5.7	141	140.7	9.8	201	200.5	14.0	261	260.4	18.2
22	21.9	1.5	82	81.8	5.7	42	141.7	9.9	02	201.5	14.1	62	261.4	18.3
23	22.9	1.6	83	82.8	5.8	43	142.7	10.0	03	202.5	14.2	63	262.4	18.3
24	23.9	1.7	84	83.8	5.9	44	143.6	10.0	04	203.5	14.2	64	263.4	18.4
25	24.9	1.7	85	84.8	5.9	45	144.6	10.1	05	204.5	14.3	65	264.4	18.5
26	25.9	1.8	86	85.8	6.0	46	145.6	10.2	06	205.5	14.4	66	265.4	18.6
27	26.9	1.9	87	86.8	6.1	47	146.6	10.3	07	206.5	14.4	67	266.3	18.6
28	27.9	2.0	88	87.8	6.1	48	147.6	10.3	08	207.5	14.5	68	267.3	18.7
29	28.9	2.0	89	88.8	6.2	49	148.6	10.4	09	208.5	14.6	69	268.3	18.8
30	29.9	2.1	90	89.8	6.3	50	149.6	10.5	10	209.5	14.6	70	269.3	18.8
31	30.9	2.2	91	90.8	6.3	151	150.6	10.5	211	210.5	14.7	271	270.3	18.9
32	31.9	2.2	92	91.8	6.4	52	151.6	10.6	12	211.5	14.8	72	271.3	19.0
33	32.9	2.3	93	92.8	6.5	53	152.6	10.7	13	212.5	14.9	73	272.3	19.0
34	33.9	2.4	94	93.8	6.6	54	153.6	10.7	14	213.5	14.9	74	273.3	19.1
35	34.9	2.4	95	94.8	6.6	55	154.6	10.8	15	214.5	15.0	75	274.3	19.2
36	35.9	2.5	96	95.8	6.7	56	155.6	10.9	16	215.5	15.1	76	275.3	19.3
37	36.9	2.6	97	96.8	6.8	57	156.6	11.0	17	216.5	15.1	77	276.3	19.3
38	37.9	2.7	98	97.8	6.8	58	157.6	11.0	18	217.5	15.2	78	277.3	19.4
39	38.9	2.7	99	98.8	6.9	59	158.6	11.1	19	218.5	15.3	79	278.3	19.5
40	39.9	2.8	100	99.8	7.0	60	159.6	11.2	20	219.5	15.3	80	279.3	19.5
41	40.9	2.9	101	100.8	7.0	161	160.6	11.2	221	220.5	15.4	281	280.3	19.6
42	41.9	2.9	02	101.8	7.1	62	161.6	11.3	22	221.5	15.5	82	281.3	19.7
43	42.9	3.0	03	102.7	7.2	63	162.6	11.4	23	222.5	15.6	83	282.3	19.7
44	43.9	3.1	04	103.7	7.3	64	163.6	11.4	24	223.5	15.6	84	283.3	19.8
45	44.9	3.1	05	104.7	7.3	65	164.6	11.5	25	224.5	15.7	85	284.3	19.9
46	45.9	3.2	06	105.7	7.4	66	165.6	11.6	26	225.4	15.8	86	285.3	20.0
47	46.9	3.3	07	106.7	7.5	67	166.6	11.6	27	226.4	15.8	87	286.3	20.0
48	47.9	3.3	08	107.7	7.5	68	167.6	11.7	28	227.4	15.9	88	287.3	20.1
49	48.9	3.4	09	108.7	7.6	69	168.6	11.8	29	228.4	16.0	89	288.3	20.2
50	49.9	3.5	10	109.7	7.7	70	169.6	11.9	30	229.4	16.0	90	289.3	20.2
51	50.9	3.6	111	110.7	7.7	171	170.6	11.9	231	230.4	16.1	291	290.3	20.3
52	51.9	3.6	12	111.7	7.8	72	171.6	12.0	32	231.4	16.2	92	291.3	20.4
53	52.9	3.7	13	112.7	7.9	73	172.6	12.1	33	232.4	16.3	93	292.3	20.4
54	53.9	3.8	14	113.7	8.0	74	173.6	12.1	34	233.4	16.3	94	293.3	20.5
55	54.9	3.8	15	114.7	8.0	75	174.6	12.2	35	234.4	16.4	95	294.3	20.6
56	55.9	3.9	16	115.7	8.1	76	175.6	12.3	36	235.4	16.5	96	295.3	20.6
57	56.9	4.0	17	116.7	8.2	77	176.6	12.3	37	236.4	16.5	97	296.3	20.7
58	57.9	4.0	18	117.7	8.2	78	177.6	12.4	38	237.4	16.6	98	297.3	20.8
59	58.9	4.1	19	118.7	8.3	79	178.6	12.5	39	238.4	16.7	99	298.3	20.9
60	59.9	4.2	20	119.7	8.4	80	179.6	12.6	40	239.4	16.7	300	299.3	20.9
Dist.	Dep.	D. Lat.	Dist.	Dep.	D. Lat.	Dist.	Dep.	D. Lat.	Dist.	Dep.	D. Lat.	Dist.	Dep.	D. Lat.

274° | 086°
266° | 094°

86°

Dist.	D. Lat.	Dep.
N.	N×Cos.	N×Sin.
Hypotenuse	Side Adj.	Side Opp.

356° | 004°	TABLE 3	356° | 004°
184° | 176°	Traverse **4°** Table	184° | 176°

Dist.	D. Lat.	Dep.	Dist.	D. Lat.	Dep.	Dist.	D. Lat.	Dep.	Dist.	D. Lat.	Dep.	Dist.	D. Lat.	Dep.
301	300.3	21.0	361	360.1	25.2	421	420.0	29.4	481	479.8	33.6	541	539.7	37.7
02	301.3	21.1	62	361.1	25.2	22	421.0	29.4	82	480.8	33.6	42	540.7	37.8
03	302.3	21.1	63	362.1	25.3	23	422.0	29.5	83	481.8	33.7	43	541.7	37.9
04	303.3	21.2	64	363.1	25.4	24	423.0	29.6	84	482.8	33.8	44	542.7	37.9
05	304.3	21.3	65	364.1	25.5	25	424.0	29.6	85	483.8	33.8	45	543.7	38.0
06	305.3	21.3	66	365.1	25.5	26	425.0	29.7	86	484.8	33.9	46	544.7	38.1
07	306.3	21.4	67	366.1	25.6	27	426.0	29.8	87	485.8	34.0	47	545.7	38.2
08	307.2	21.5	68	367.1	25.7	28	427.0	29.9	88	486.8	34.0	48	546.7	38.2
09	308.2	21.6	69	368.1	25.7	29	428.0	29.9	89	487.8	34.1	49	547.7	38.3
10	309.2	21.6	70	369.1	25.8	30	429.0	30.0	90	488.8	34.2	50	548.7	38.4
311	310.2	21.7	371	370.1	25.9	431	430.0	30.1	491	489.8	34.3	551	549.7	38.4
12	311.2	21.8	72	371.1	25.9	32	430.9	30.1	92	490.8	34.3	52	550.7	38.5
13	312.2	21.8	73	372.1	26.0	33	431.9	30.2	93	491.8	34.4	53	551.7	38.6
14	313.2	21.9	74	373.1	26.1	34	432.9	30.3	94	492.8	34.5	54	552.7	38.6
15	314.2	22.0	75	374.1	26.2	35	433.9	30.3	95	493.8	34.5	55	553.6	38.7
16	315.2	22.1	76	375.1	26.2	36	434.9	30.4	96	494.8	34.6	56	554.6	38.8
17	316.2	22.1	77	376.1	26.3	37	435.9	30.5	97	495.8	34.7	57	555.6	38.9
18	317.2	22.2	78	377.1	26.4	38	436.9	30.6	98	496.8	34.7	58	556.6	38.9
19	318.2	22.3	79	378.1	26.4	39	437.9	30.6	99	497.8	34.8	59	557.6	39.0
20	319.2	22.3	80	379.1	26.5	40	438.9	30.7	500	498.8	34.9	60	558.6	39.1
321	320.2	22.4	381	380.1	26.6	441	439.9	30.8	501	499.8	34.9	561	559.6	39.1
22	321.2	22.5	82	381.1	26.6	42	440.9	30.8	02	500.8	35.0	62	560.6	39.2
23	322.2	22.5	83	382.1	26.7	43	441.9	30.9	03	501.8	35.1	63	561.6	39.3
24	323.2	22.6	84	383.1	26.8	44	442.9	31.0	04	502.8	35.2	64	562.6	39.3
25	324.2	22.7	85	384.1	26.9	45	443.9	31.0	05	503.8	35.2	65	563.6	39.4
26	325.2	22.7	86	385.1	26.9	46	444.9	31.1	06	504.8	35.3	66	564.6	39.5
27	326.2	22.8	87	386.1	27.0	47	445.9	31.2	07	505.8	35.4	67	565.6	39.6
28	327.2	22.9	88	387.1	27.1	48	446.9	31.3	08	506.8	35.4	68	566.6	39.6
29	328.2	22.9	89	388.1	27.1	49	447.9	31.3	09	507.8	35.5	69	567.6	39.7
30	329.2	23.0	90	389.0	27.2	50	448.9	31.4	10	508.8	35.6	70	568.6	39.8
331	330.2	23.1	391	390.0	27.3	451	449.9	31.5	511	509.8	35.6	571	569.6	39.8
32	331.2	23.2	92	391.0	27.3	52	450.9	31.5	12	510.8	35.7	72	570.6	39.9
33	332.2	23.2	93	392.0	27.4	53	451.9	31.6	13	511.8	35.8	73	571.6	40.0
34	333.2	23.3	94	393.0	27.5	54	452.9	31.7	14	512.7	35.9	74	572.6	40.0
35	334.2	23.4	95	394.0	27.6	55	453.9	31.7	15	513.7	35.9	75	573.6	40.1
36	335.2	23.4	96	395.0	27.6	56	454.9	31.8	16	514.7	36.0	76	574.6	40.2
37	336.2	23.5	97	396.0	27.7	57	455.9	31.9	17	515.7	36.1	77	575.6	40.2
38	337.2	23.6	98	397.0	27.8	58	456.9	31.9	18	516.7	36.1	78	576.6	40.3
39	338.2	23.6	99	398.0	27.8	59	457.9	32.0	19	517.7	36.2	79	577.6	40.4
40	339.2	23.7	400	399.0	27.9	60	458.9	32.1	20	518.7	36.3	80	578.6	40.5
341	340.2	23.8	401	400.0	28.0	461	459.9	32.2	521	519.7	36.3	581	579.6	40.5
42	341.2	23.9	02	401.0	28.0	62	460.9	32.2	22	520.7	36.4	82	580.6	40.6
43	342.2	23.9	03	402.0	28.1	63	461.9	32.3	23	521.7	36.5	83	581.6	40.7
44	343.2	24.0	04	403.0	28.2	64	462.9	32.4	24	522.7	36.6	84	582.6	40.7
45	344.2	24.1	05	404.0	28.3	65	463.9	32.4	25	523.7	36.6	85	583.6	40.8
46	345.2	24.1	06	405.0	28.3	66	464.9	32.5	26	524.7	36.7	86	584.6	40.9
47	346.2	24.2	07	406.0	28.4	67	465.9	32.6	27	525.7	36.8	87	585.6	40.9
48	347.2	24.3	08	407.0	28.5	68	466.9	32.6	28	526.7	36.8	88	586.6	41.0
49	348.1	24.3	09	408.0	28.5	69	467.9	32.7	29	527.7	36.9	89	587.6	41.1
50	349.1	24.4	10	409.0	28.6	70	468.9	32.8	30	528.7	37.0	90	588.6	41.2
351	350.1	24.5	411	410.0	28.7	471	469.9	32.9	531	529.7	37.0	591	589.6	41.2
52	351.1	24.6	12	411.0	28.7	72	470.9	32.9	32	530.7	37.1	92	590.6	41.3
53	352.1	24.6	13	412.0	28.8	73	471.8	33.0	33	531.7	37.2	93	591.6	41.4
54	353.1	24.7	14	413.0	28.9	74	472.8	33.1	34	532.7	37.2	94	592.6	41.4
55	354.1	24.8	15	414.0	28.9	75	473.8	33.1	35	533.7	37.3	95	593.6	41.5
56	355.1	24.8	16	415.0	29.0	76	474.8	33.2	36	534.7	37.4	96	594.5	41.6
57	356.1	24.9	17	416.0	29.1	77	475.8	33.3	37	535.7	37.5	97	595.5	41.6
58	357.1	25.0	18	417.0	29.2	78	476.8	33.3	38	536.7	37.5	98	596.5	41.7
59	358.1	25.0	19	418.0	29.2	79	477.8	33.4	39	537.7	37.6	99	597.5	41.8
60	359.1	25.1	20	419.0	29.3	80	478.8	33.5	40	538.7	37.7	600	598.5	41.9
Dist.	Dep.	D. Lat.	Dist.	Dep.	D. Lat.	Dist.	Dep.	D. Lat.	Dist.	Dep.	D. Lat.	Dist.	Dep.	D. Lat.

Dist.	D. Lat.	Dep.
D Lo	Dep.	
	m	D Lo

86°

274°	086°
266°	094°

355° | 005° / 185° | 175°	TABLE 3 — Traverse 5° Table	355° | 005° / 185° | 175°

Dist.	D. Lat.	Dep.	Dist.	D. Lat.	Dep.	Dist.	D. Lat.	Dep.	Dist.	D. Lat.	Dep.	Dist.	D. Lat.	Dep.
1	1.0	0.1	61	60.8	5.3	121	120.5	10.5	181	180.3	15.8	241	240.1	21.0
2	2.0	0.2	62	61.8	5.4	22	121.5	10.6	82	181.3	15.9	42	241.1	21.1
3	3.0	0.3	63	62.8	5.5	23	122.5	10.7	83	182.3	15.9	43	242.1	21.2
4	4.0	0.3	64	63.8	5.6	24	123.5	10.8	84	183.3	16.0	44	243.1	21.3
5	5.0	0.4	65	64.8	5.7	25	124.5	10.9	85	184.3	16.1	45	244.1	21.4
6	6.0	0.5	66	65.7	5.8	26	125.5	11.0	86	185.3	16.2	46	245.1	21.4
7	7.0	0.6	67	66.7	5.8	27	126.5	11.1	87	186.3	16.3	47	246.1	21.5
8	8.0	0.7	68	67.7	5.9	28	127.5	11.2	88	187.3	16.4	48	247.1	21.6
9	9.0	0.8	69	68.7	6.0	29	128.5	11.2	89	188.3	16.5	49	248.1	21.7
10	10.0	0.9	70	69.7	6.1	30	129.5	11.3	90	189.3	16.6	50	249.0	21.8
11	11.0	1.0	71	70.7	6.2	131	130.5	11.4	191	190.3	16.6	251	250.0	21.9
12	12.0	1.0	72	71.7	6.3	32	131.5	11.5	92	191.3	16.7	52	251.0	22.0
13	13.0	1.1	73	72.7	6.4	33	132.5	11.6	93	192.3	16.8	53	252.0	22.1
14	13.9	1.2	74	73.7	6.4	34	133.5	11.7	94	193.3	16.9	54	253.0	22.1
15	14.9	1.3	75	74.7	6.5	35	134.5	11.8	95	194.3	17.0	55	254.0	22.2
16	15.9	1.4	76	75.7	6.6	36	135.5	11.9	96	195.3	17.1	56	255.0	22.3
17	16.9	1.5	77	76.7	6.7	37	136.5	11.9	97	196.3	17.2	57	256.0	22.4
18	17.9	1.6	78	77.7	6.8	38	137.5	12.0	98	197.2	17.3	58	257.0	22.5
16	18.9	1.7	79	78.7	6.9	39	138.5	12.1	99	198.2	17.3	59	258.0	22.6
20	19.9	1.7	80	79.7	7.0	40	139.5	12.2	200	199.2	17.4	60	259.0	22.7
21	20.9	1.8	81	80.7	7.1	141	140.5	12.3	201	200.2	17.5	261	260.0	22.7
22	21.9	1.9	82	81.7	7.1	42	141.5	12.4	02	201.2	17.6	62	261.0	22.8
23	22.9	2.0	83	82.7	7.2	43	142.5	12.5	03	202.2	17.7	63	262.0	22.9
24	23.9	2.1	84	83.7	7.3	44	143.5	12.6	04	203.2	17.8	64	263.0	23.0
25	24.9	2.2	85	84.7	7.4	45	144.4	12.6	05	204.2	17.9	65	264.0	23.1
26	25.9	2.3	86	85.7	7.5	46	145.4	12.7	06	205.2	18.0	66	265.0	23.2
27	26.9	2.4	87	86.7	7.6	47	146.4	12.8	07	206.2	18.0	67	266.0	23.3
28	27.9	2.4	88	87.7	7.7	48	147.4	12.9	08	207.2	18.1	68	267.0	23.4
29	28.9	2.5	89	88.7	7.8	49	148.4	13.0	09	208.2	18.2	69	268.0	23.4
30	29.9	2.6	90	89.7	7.8	50	149.4	13.1	10	209.2	18.3	70	269.0	23.5
31	30.9	2.7	91	90.7	7.9	151	150.4	13.2	211	210.2	18.4	271	270.0	23.6
32	31.9	2.8	92	91.6	8.0	52	151.4	13.2	12	211.2	18.5	72	271.0	23.7
33	32.9	2.9	93	92.6	8.1	53	152.4	13.3	13	212.2	18.6	73	272.0	23.8
34	33.9	3.0	94	93.6	8.2	54	153.4	13.4	14	213.2	18.7	74	273.0	23.9
35	34.9	3.1	95	94.6	8.3	55	154.4	13.5	15	214.2	18.7	75	274.0	24.0
36	35.9	3.1	96	95.6	8.4	56	155.4	13.6	16	215.2	18.8	76	274.9	24.1
37	36.9	3.2	97	96.6	8.5	57	156.4	13.7	17	216.2	18.9	77	275.9	24.1
38	37.9	3.3	98	97.6	8.5	58	157.4	13.8	18	217.2	19.0	78	276.9	24.2
39	38.9	3.4	99	98.6	8.6	59	158.4	13.9	19	218.2	19.1	79	277.9	24.3
40	39.8	3.5	100	99.6	8.7	60	159.4	13.9	20	219.2	19.2	80	278.9	24.4
41	40.8	3.6	101	100.6	8.8	161	160.4	14.0	221	220.2	19.3	281	279.9	24.5
42	41.8	3.7	02	101.6	8.9	62	161.4	14.1	22	221.2	19.3	82	280.9	24.6
43	42.8	3.7	03	102.6	9.0	63	162.4	14.2	23	222.2	19.4	83	281.9	24.7
44	43.8	3.8	04	103.6	9.1	64	163.4	14.3	24	223.1	19.5	84	282.9	24.8
45	44.8	3.9	05	104.6	9.2	65	164.4	14.4	25	224.1	19.6	85	283.9	24.8
46	45.8	4.0	06	105.6	9.2	66	165.4	14.5	26	225.1	19.7	86	284.9	24.9
47	46.8	4.1	07	106.6	9.3	67	166.4	14.6	27	226.1	19.8	87	285.9	25.0
48	47.8	4.2	08	107.6	9.4	68	167.4	14.6	28	227.1	19.9	88	286.9	25.1
49	48.8	4.3	09	108.6	9.5	69	168.4	14.7	29	228.1	20.0	89	287.9	25.2
50	49.8	4.4	10	109.6	9.6	70	169.4	14.8	30	229.1	20.0	90	288.9	25.3
51	50.8	4.4	111	110.6	9.7	171	170.3	14.9	231	230.1	20.1	291	289.9	25.4
52	51.8	4.5	12	111.6	9.8	72	171.3	15.0	32	231.1	20.2	92	290.9	25.4
53	52.8	4.6	13	112.6	9.8	73	172.3	15.1	33	232.1	20.3	93	291.9	25.5
54	53.8	4.7	14	113.6	9.9	74	173.3	15.2	34	233.1	20.4	94	292.9	25.6
55	54.8	4.8	15	114.6	10.0	75	174.3	15.3	35	234.1	20.5	95	293.9	25.7
56	55.8	4.9	16	115.6	10.1	76	175.3	15.3	36	235.1	20.6	96	294.9	25.8
57	56.8	5.0	17	116.6	10.2	77	176.3	15.4	37	236.1	20.7	97	295.9	25.9
58	57.8	5.1	18	117.6	10.3	78	177.3	15.5	38	237.1	20.7	98	296.9	26.0
59	58.8	5.1	19	118.5	10.4	79	178.3	15.6	39	238.1	20.8	99	297.9	26.1
60	59.8	5.2	20	119.5	10.5	80	179.3	15.7	40	239.1	20.9	300	298.9	26.1
Dist.	Dep.	D. Lat.	Dist.	Dep.	D. Lat.	Dist.	Dep.	D. Lat.	Dist.	Dep.	D. Lat.	Dist.	Dep.	D. Lat.

275° | 085° / 265° | 095°	85°

Dist.	D. Lat.	Dep.
N.	N×Cos.	N×Sin.
Hypotenuse	Side Adj.	Side Opp.

355° | 005° / 185° | 175°	TABLE 3 Traverse 5° Table	355° | 005° / 185° | 175°

Dist.	D. Lat.	Dep.	Dist.	D. Lat.	Dep.	Dist.	D. Lat.	Dep.	Dist.	D. Lat.	Dep.	Dist.	D. Lat.	Dep.
301	299.9	26.2	361	359.6	31.5	421	419.4	36.7	481	479.2	41.9	541	538.9	47.2
02	300.9	26.3	62	360.6	31.6	22	420.4	36.8	82	480.2	42.0	42	539.9	47.2
03	301.8	26.4	63	361.6	31.6	23	421.4	36.9	83	481.2	42.1	43	540.9	47.3
04	302.8	26.5	64	362.6	31.7	24	422.4	37.0	84	482.2	42.2	44	541.9	47.4
05	303.8	26.6	65	363.6	31.8	25	423.4	37.1	85	483.2	42.3	45	542.9	47.5
06	304.8	26.7	66	364.6	31.9	26	424.4	37.1	86	484.2	42.4	46	543.9	47.6
07	305.8	26.8	67	365.6	32.0	27	425.4	37.2	87	485.1	42.4	47	544.9	47.7
08	306.8	26.8	68	366.6	32.1	28	426.4	37.3	88	486.1	42.5	48	545.9	47.8
09	307.8	26.9	69	367.6	32.2	29	427.4	37.4	89	487.1	42.6	49	546.9	47.8
10	308.8	27.0	70	368.6	32.2	30	428.4	37.5	90	488.1	42.7	50	547.9	47.9
311	309.8	27.1	371	369.6	32.3	431	429.4	37.6	491	489.1	42.8	551	548.9	48.0
12	310.8	27.2	72	370.6	32.4	32	430.4	37.7	92	490.1	42.9	52	549.9	48.1
13	311.8	27.3	73	371.6	32.5	33	431.4	37.7	93	491.1	43.0	53	550.9	48.2
14	312.8	27.4	74	372.6	32.6	34	432.3	37.8	94	492.1	43.1	54	551.9	48.3
15	313.8	27.5	75	373.6	32.7	35	433.3	37.9	95	493.1	43.1	55	552.9	48.4
16	314.8	27.5	76	374.6	32.8	36	434.3	38.0	96	494.1	43.2	56	553.9	48.5
17	315.8	27.6	77	375.6	32.9	37	435.3	38.1	97	495.1	43.3	57	554.9	48.5
18	316.8	27.7	78	376.6	33.0	38	436.3	38.2	98	496.1	43.4	58	555.9	48.6
19	317.8	27.8	79	377.6	33.0	39	437.3	38.3	99	497.1	43.5	59	556.9	48.7
20	318.8	27.9	80	378.6	33.1	40	438.3	38.3	500	498.1	43.6	60	557.9	48.8
321	319.8	28.0	381	379.6	33.2	441	439.3	38.4	501	499.1	43.7	561	558.9	48.9
22	320.8	28.1	82	380.5	33.3	42	440.3	38.5	02	500.1	43.8	62	559.9	49.0
23	321.8	28.2	83	381.5	33.4	43	441.3	38.6	03	501.1	43.8	63	560.9	49.1
24	322.8	28.2	84	382.5	33.5	44	442.3	38.7	04	502.1	43.9	64	561.9	49.2
25	323.8	28.3	85	383.5	33.6	45	443.3	38.8	05	503.1	44.0	65	562.9	49.3
26	324.8	28.4	86	384.5	33.6	46	444.3	38.9	06	504.1	44.1	66	563.8	49.3
27	325.8	28.5	87	385.5	33.7	47	445.3	39.0	07	505.1	44.2	67	564.8	49.4
28	326.8	28.6	88	386.5	33.8	48	446.3	39.0	08	506.1	44.3	68	565.8	49.5
29	327.7	28.7	89	387.5	33.9	49	447.3	39.1	09	507.1	44.4	69	566.8	49.6
30	328.7	28.8	90	388.5	34.0	50	448.3	39.2	10	508.1	44.4	70	567.8	49.7
331	329.7	28.8	391	389.5	34.1	451	449.3	39.3	511	509.1	44.5	571	568.8	49.8
32	330.7	28.9	92	390.5	34.2	52	450.3	39.4	12	510.1	44.6	72	569.8	49.9
33	331.7	29.0	93	391.5	34.3	53	451.3	39.5	13	511.0	44.7	73	570.8	49.9
34	332.7	29.1	94	392.5	34.3	54	452.3	39.6	14	512.0	44.8	74	571.8	50.0
35	333.7	29.2	95	393.5	34.4	55	453.3	39.7	15	513.0	44.9	75	572.8	50.1
36	334.7	29.3	96	394.5	34.5	56	454.3	39.7	16	514.0	45.0	76	573.8	50.2
37	335.7	29.4	97	395.5	34.6	57	455.3	39.8	17	515.0	45.1	77	574.8	50.3
38	336.7	29.5	98	396.5	34.7	58	456.3	39.9	18	516.0	45.1	78	575.8	50.4
39	337.7	29.6	99	397.5	34.8	59	457.3	40.0	19	517.0	45.2	79	576.8	50.5
40	338.7	29.6	400	398.5	34.9	60	458.2	40.1	20	518.0	45.3	80	577.8	50.6
341	339.7	29.7	401	399.5	34.9	461	459.2	40.2	521	519.0	45.4	581	578.8	50.6
42	340.7	29.8	02	400.5	35.0	62	460.2	40.3	22	520.0	45.5	82	579.8	50.7
43	341.7	29.9	03	401.5	35.1	63	461.2	40.4	23	521.0	45.6	83	580.8	50.8
44	342.7	30.0	04	402.5	35.2	64	462.2	40.4	24	522.0	45.7	84	581.8	50.9
45	343.7	30.1	05	403.5	35.3	65	463.2	40.5	25	523.0	45.8	85	582.8	51.0
46	344.7	30.2	06	404.5	35.4	66	464.2	40.6	26	524.0	45.8	86	583.8	51.1
47	345.7	30.2	07	405.5	35.5	67	465.2	40.7	27	525.0	45.9	87	584.8	51.2
48	346.7	30.3	08	406.4	35.6	68	466.2	40.8	28	526.0	46.0	88	585.8	51.2
49	347.7	30.4	09	407.4	35.6	69	467.2	40.9	29	527.0	46.1	89	586.8	51.3
50	348.7	30.5	10	408.4	35.7	70	468.2	41.0	30	528.0	46.2	90	587.8	51.4
351	349.7	30.6	411	409.4	35.8	471	469.2	41.1	531	529.0	46.3	591	588.8	51.5
52	350.7	30.7	12	410.4	35.9	72	470.2	41.1	32	530.0	46.4	92	589.7	51.6
53	351.7	30.8	13	411.4	36.0	73	471.2	41.2	33	531.0	46.5	93	590.7	51.7
54	352.7	30.9	14	412.4	36.1	74	472.2	41.3	34	532.0	46.5	94	591.7	51.8
55	353.6	30.9	15	413.4	36.2	75	473.2	41.4	35	533.0	46.6	95	592.7	51.9
56	354.6	31.0	16	414.4	36.3	76	474.2	41.5	36	534.0	46.7	96	593.7	51.9
57	355.6	31.1	17	415.4	36.3	77	475.2	41.6	37	535.0	46.8	97	594.7	52.0
58	356.6	31.2	18	416.4	36.4	78	476.2	41.7	38	536.0	46.9	98	595.7	52.1
59	357.6	31.3	19	417.4	36.5	79	477.2	41.7	39	536.9	47.0	99	596.7	52.2
60	358.6	31.4	20	418.4	36.6	80	478.2	41.8	40	537.9	47.1	600	597.7	52.3
Dist.	Dep.	D. Lat.	Dist.	Dep.	D. Lat.	Dist.	Dep.	D. Lat.	Dist.	Dep.	D. Lat.	Dist.	Dep.	D. Lat.

Dist.	D. Lat.	Dep.
D Lo	Dep.	
	m	D Lo

85°

275° | 085° / 265° | 095°

TABLE 3

354° | 006°
186° | 174°

Traverse **6°** Table

354° | 006°
186° | 174°

Dist.	D. Lat.	Dep.	Dist.	D. Lat.	Dep.	Dist.	D. Lat.	Dep.	Dist.	D. Lat.	Dep.	Dist.	D. Lat.	Dep.
1	1.0	0.1	61	60.7	6.4	121	120.3	12.6	181	180.0	18.9	241	239.7	25.2
2	2.0	0.2	62	61.7	6.5	22	121.3	12.8	82	181.0	19.0	42	240.7	25.3
3	3.0	0.3	63	62.7	6.6	23	122.3	12.9	83	182.0	19.1	43	241.7	25.4
4	4.0	0.4	64	63.6	6.7	24	123.3	13.0	84	183.0	19.2	44	242.7	25.5
5	5.0	0.5	65	64.6	6.8	25	124.3	13.1	85	184.0	19.3	45	243.7	25.6
6	6.0	0.6	66	65.6	6.9	26	125.3	13.2	86	185.0	19.4	46	244.7	25.7
7	7.0	0.7	67	66.6	7.0	27	126.3	13.3	87	186.0	19.5	47	245.6	25.8
8	8.0	0.8	68	67.6	7.1	28	127.3	13.4	88	187.0	19.7	48	246.6	25.9
9	9.0	0.9	69	68.6	7.2	29	128.3	13.5	89	188.0	19.8	49	247.6	26.0
10	9.9	1.0	70	69.6	7.3	30	129.3	13.6	90	189.0	19.9	50	248.6	26.1
11	10.9	1.1	71	70.6	7.4	131	130.3	13.7	191	190.0	20.0	251	249.6	26.2
12	11.9	1.3	72	71.6	7.5	32	131.3	13.8	92	190.9	20.1	52	250.6	26.3
13	12.9	1.4	73	72.6	7.6	33	132.3	13.9	93	191.9	20.2	53	251.6	26.4
14	13.9	1.5	74	73.6	7.7	34	133.3	14.0	94	192.9	20.3	54	252.6	26.6
15	14.9	1.6	75	74.6	7.8	35	134.4	14.1	95	193.9	20.4	55	253.6	26.7
16	15.9	1.7	76	75.6	7.9	36	135.3	14.2	96	194.9	20.5	56	254.6	26.8
17	16.9	1.8	77	76.6	8.0	37	136.2	14.3	97	195.9	20.6	57	255.6	26.9
18	17.9	1.9	78	77.6	8.2	38	137.2	14.4	98	196.9	20.7	58	256.6	27.0
19	18.9	2.0	79	78.6	8.3	39	138.2	14.5	99	197.9	20.8	59	257.6	27.1
20	19.9	2.1	80	79.6	8.4	40	139.2	14.6	200	198.9	20.9	60	258.6	27.2
21	20.9	2.2	81	80.6	8.5	141	140.2	14.7	201	199.9	21.0	261	259.6	27.3
22	21.9	2.3	82	81.6	8.6	42	141.2	14.8	02	200.9	21.1	62	260.6	27.4
23	22.9	2.4	83	82.5	8.7	43	142.2	14.9	03	201.9	21.2	63	261.6	27.5
24	23.9	2.5	84	83.5	8.8	44	143.2	15.1	04	202.9	21.3	64	262.6	27.6
25	24.9	2.6	85	84.5	8.9	45	144.2	15.2	05	203.9	21.4	65	263.5	27.7
26	25.9	2.7	86	85.5	9.0	46	145.2	15.3	06	204.9	21.5	66	264.5	27.8
27	26.9	2.8	87	86.5	9.1	47	146.2	15.4	07	205.9	21.6	67	265.5	27.9
28	27.8	2.9	88	87.5	9.2	48	147.2	15.5	08	206.9	21.7	68	266.5	28.0
29	28.8	3.0	89	88.5	9.3	49	148.2	15.6	09	207.9	21.8	69	267.5	28.1
30	29.8	3.1	90	89.5	9.4	50	149.2	15.7	10	208.8	22.0	70	268.5	28.2
31	30.8	3.2	91	90.5	9.5	151	150.2	15.8	211	209.8	22.1	271	269.5	28.3
32	31.8	3.3	92	91.5	9.6	52	151.2	15.9	12	210.8	22.2	72	270.5	28.4
33	32.8	3.4	93	92.5	9.7	53	152.2	16.0	13	211.8	22.3	73	271.5	28.5
34	33.8	3.6	94	93.5	9.8	54	153.2	16.1	14	212.8	22.4	74	272.5	28.6
35	34.8	3.7	95	94.5	9.9	55	154.2	16.2	15	213.8	22.5	75	273.5	28.7
36	35.8	3.8	96	95.5	10.0	56	155.1	16.3	16	214.8	22.6	76	274.5	28.8
37	36.8	3.9	97	96.5	10.1	57	156.1	16.4	17	215.8	22.7	77	275.5	29.0
38	37.8	4.0	98	97.5	10.2	58	157.1	16.5	18	216.8	22.8	78	276.5	29.1
39	38.8	4.1	99	98.5	10.3	59	158.1	16.6	19	217.8	22.9	79	277.5	29.2
40	39.8	4.2	100	99.5	10.5	60	159.1	16.7	20	218.8	23.0	80	278.5	29.3
41	40.8	4.3	101	100.4	10.6	161	160.1	16.8	221	219.8	23.1	281	279.5	29.4
42	41.8	4.4	02	101.4	10.7	62	161.1	16.9	22	220.8	23.2	82	280.5	29.5
43	42.8	4.5	03	102.4	10.8	63	162.1	17.0	23	221.8	23.3	83	281.4	29.6
44	43.8	4.6	04	103.4	10.9	64	163.1	17.1	24	222.8	23.4	84	282.4	29.7
45	44.8	4.7	05	104.4	11.0	65	164.1	17.2	25	223.8	23.5	85	283.4	29.8
46	45.7	4.8	06	105.4	11.1	66	165.1	17.4	26	224.8	23.6	86	284.4	29.9
47	46.7	4.9	07	106.4	11.2	67	166.1	17.5	27	225.8	23.7	87	285.4	30.0
48	47.7	5.0	08	107.4	11.3	68	167.1	17.6	28	226.8	23.8	88	286.4	30.1
49	48.7	5.1	09	108.4	11.4	69	168.1	17.7	29	227.7	23.9	89	287.4	30.2
50	49.7	5.2	10	109.4	11.5	70	169.1	17.8	30	228.7	24.0	90	288.4	30.3
51	50.7	5.3	111	110.4	11.6	171	170.1	17.9	231	229.7	24.1	291	289.4	30.4
52	51.7	5.4	12	111.4	11.7	72	171.1	18.0	32	230.7	24.3	92	290.4	30.5
53	52.7	5.5	13	112.4	11.8	73	172.1	18.1	33	231.7	24.4	93	291.4	30.6
54	53.7	5.6	14	113.4	11.9	74	173.0	18.2	34	232.7	24.5	94	292.4	30.7
55	54.7	5.7	15	114.4	12.0	75	174.0	18.3	35	233.7	24.6	95	293.4	30.8
56	55.7	5.9	16	115.4	12.1	76	175.0	18.4	36	234.7	24.7	96	294.4	30.9
57	56.7	6.0	17	116.4	12.2	77	176.0	18.5	37	235.7	24.8	97	295.4	31.0
58	57.7	6.1	18	117.4	12.3	78	177.0	18.6	38	236.7	24.9	98	296.4	31.1
59	58.7	6.2	19	118.3	12.4	79	178.0	18.7	39	237.7	25.0	99	297.4	31.3
60	59.7	6.3	20	119.3	12.5	80	179.0	18.8	40	238.7	25.1	300	298.4	31.4
Dist.	Dep.	D. Lat.	Dist.	Dep.	D. Lat.	Dist.	Dep.	D. Lat.	Dist.	Dep.	D. Lat.	Dist.	Dep.	D. Lat.

276° | 084°
264° | 096°

84°

Dist.	D. Lat.	Dep.
N.	N×Cos.	N×Sin.
Hypotenuse	Side Adj.	Side Opp.

354° | 006°
186° | 174°

TABLE 3

Traverse **6°** Table

354° | 006°
186° | 174°

Dist.	D. Lat.	Dep.	Dist.	D. Lat.	Dep.	Dist.	D. Lat.	Dep.	Dist.	D. Lat.	Dep.	Dist.	D. Lat.	Dep.
301	299.3	31.5	361	359.0	37.7	421	418.7	44.0	481	478.4	50.3	541	538.0	56.5
02	300.3	31.6	62	360.0	37.8	22	419.7	44.1	82	479.4	50.4	42	539.0	56.7
03	301.3	31.7	63	361.0	37.9	23	420.7	44.2	83	480.4	50.5	43	540.0	56.8
04	302.3	31.8	64	362.0	38.0	24	421.7	44.3	84	481.3	50.6	44	541.0	56.9
05	303.3	31.9	65	363.0	38.2	25	422.7	44.4	85	482.3	50.7	45	542.0	57.0
06	304.3	32.0	66	364.0	38.3	26	423.7	44.5	86	483.3	50.8	46	543.0	57.1
07	305.3	32.1	67	365.0	38.4	27	424.7	44.6	87	484.3	50.9	47	544.0	57.2
08	306.3	32.2	68	366.0	38.5	28	425.7	44.7	88	485.3	51.0	48	545.0	57.3
09	307.3	32.3	69	367.0	38.6	29	426.6	44.8	89	486.3	51.1	49	546.0	57.4
10	308.3	32.4	70	368.0	38.7	30	427.6	44.9	90	487.3	51.2	50	547.0	57.5
311	309.3	32.5	371	369.0	38.8	431	428.6	45.1	491	488.3	51.3	551	548.0	57.6
12	310.3	32.6	72	370.0	38.9	32	429.6	45.2	92	489.3	51.4	52	549.0	57.7
13	311.3	32.7	73	371.0	39.0	33	430.6	45.3	93	490.3	51.5	53	550.0	57.8
14	312.3	32.8	74	372.0	39.1	34	431.6	45.4	94	491.3	51.6	54	551.0	57.9
15	313.3	32.9	75	372.9	39.2	35	432.6	45.5	95	492.3	51.7	55	552.0	58.0
16	314.3	33.0	76	373.9	39.3	36	433.6	45.6	96	493.3	51.8	56	553.0	58.1
17	315.3	33.1	77	374.9	39.4	37	434.6	45.7	97	494.3	52.0	57	553.9	58.2
18	316.3	33.2	78	375.9	39.5	38	435.6	45.8	98	495.3	52.1	58	554.9	58.3
19	317.3	33.3	79	376.9	39.6	39	436.6	45.9	99	496.3	52.2	59	555.9	58.4
20	318.2	33.4	80	377.9	39.7	40	437.6	46.0	500	497.3	52.3	60	556.9	58.5
321	319.2	33.6	381	378.9	39.8	441	438.6	46.1	501	498.3	52.4	561	557.9	58.6
22	320.2	33.7	82	379.9	39.9	42	439.6	46.2	02	499.3	52.5	62	558.9	58.7
23	321.2	33.8	83	380.9	40.0	43	440.6	46.3	03	500.2	52.6	63	559.9	58.8
24	322.2	33.9	84	381.9	40.1	44	441.6	46.4	04	501.2	52.7	64	560.9	59.0
25	323.2	34.0	85	382.9	40.2	45	442.6	46.5	05	502.2	52.8	65	561.9	59.1
26	324.2	34.1	86	383.9	40.3	46	443.6	46.6	06	503.2	52.9	66	562.9	59.2
27	325.2	34.2	87	384.9	40.5	47	444.6	46.7	07	504.2	53.0	67	563.9	59.3
28	326.2	34.3	88	385.9	40.6	48	445.5	46.8	08	505.2	53.1	68	564.9	59.4
29	327.2	34.4	89	386.9	40.7	49	446.5	46.9	09	506.2	53.2	69	565.9	59.5
30	328.2	34.5	90	387.9	40.8	50	447.5	47.0	10	507.2	53.3	70	566.9	59.6
331	329.2	34.6	391	388.9	40.9	451	448.5	47.1	511	508.2	53.4	571	567.9	59.7
32	330.2	34.7	92	389.9	41.0	52	449.5	47.2	12	509.2	53.5	72	568.9	59.8
33	331.2	34.8	93	390.8	41.1	53	450.5	47.4	13	510.2	53.6	73	569.9	59.9
34	332.2	34.9	94	391.8	41.2	54	451.5	47.5	14	511.2	53.7	74	570.9	60.0
35	333.2	35.0	95	392.8	41.3	55	452.5	47.6	15	512.2	53.8	75	571.9	60.1
36	334.2	35.1	96	393.8	41.4	56	453.5	47.7	16	513.2	53.9	76	572.8	60.2
37	335.2	35.2	97	394.8	41.5	57	454.5	47.8	17	514.2	54.0	77	573.8	60.3
38	336.1	35.3	98	395.8	41.6	58	455.5	47.9	18	515.2	54.1	78	574.8	60.4
39	337.1	35.4	99	396.8	41.7	59	456.5	48.0	19	516.2	54.3	79	575.8	60.5
40	338.1	35.5	400	397.8	41.8	60	457.5	48.1	20	517.2	54.4	80	576.8	60.6
341	339.1	35.6	401	398.8	41.9	461	458.5	48.2	521	518.1	54.5	581	577.8	60.7
42	340.1	35.7	02	399.8	42.0	62	459.5	48.3	22	519.1	54.6	82	578.8	60.8
43	341.1	35.9	03	400.8	42.1	63	460.5	48.4	23	520.1	54.7	83	579.8	60.9
44	342.1	36.0	04	401.8	42.2	64	461.5	48.5	24	521.1	54.8	84	580.8	61.0
45	343.1	36.1	05	402.8	42.3	65	462.5	48.6	25	522.1	54.9	85	581.8	61.1
46	344.1	36.2	06	403.8	42.4	66	463.4	48.7	26	523.1	55.0	86	582.8	61.3
47	345.1	36.3	07	404.8	42.5	67	464.4	48.8	27	524.1	55.1	87	583.8	61.4
48	346.1	36.4	08	405.8	42.6	68	465.4	48.9	28	525.1	55.2	88	584.8	61.5
49	347.1	36.5	09	406.8	42.8	69	466.4	49.0	29	526.1	55.3	89	585.8	61.6
50	348.1	36.6	10	407.8	42.9	70	467.4	49.1	30	527.1	55.4	90	586.8	61.7
351	349.1	36.7	411	408.7	43.0	471	468.4	49.2	531	528.1	55.5	591	587.8	61.8
52	350.1	36.8	12	409.7	43.1	72	469.4	49.3	32	529.1	55.6	92	588.8	61.9
53	351.1	36.9	13	410.7	43.2	73	470.4	49.4	33	530.1	55.7	93	589.8	62.0
54	352.1	37.0	14	411.7	43.3	74	471.4	49.5	34	531.1	55.8	94	590.7	62.1
55	353.1	37.1	15	412.7	43.4	75	472.4	49.7	35	532.1	55.9	95	591.7	62.2
56	354.0	37.2	16	413.7	43.5	76	473.4	49.8	36	533.1	56.0	96	592.7	62.3
57	355.0	37.3	17	414.7	43.6	77	474.4	49.9	37	534.1	56.1	97	593.7	62.4
58	356.0	37.4	18	415.7	43.7	78	475.4	50.0	38	535.1	56.2	98	594.7	62.5
59	357.0	37.5	19	416.7	43.8	79	476.4	50.1	39	536.0	56.3	99	595.7	62.6
60	358.0	37.6	20	417.7	43.9	80	477.4	50.2	40	537.0	56.4	600	596.7	62.7
Dist.	Dep.	D. Lat.	Dist.	Dep.	D. Lat.	Dist.	Dep.	D. Lat.	Dist.	Dep.	D. Lat.	Dist.	Dep.	D. Lat.

Dist.	D. Lat.	Dep.
D Lo	Dep.	
	m	D Lo

84°

276° | 084°
264° | 096°

TABLE 3

353° | 007°
187° | 173°

Traverse **7°** Table

353° | 007°
187° | 173°

Dist.	D. Lat.	Dep.	Dist.	D. Lat.	Dep.	Dist.	D. Lat.	Dep.	Dist.	D. Lat.	Dep.	Dist.	D. Lat.	Dep.
1	1.0	0.1	61	60.5	7.4	121	120.1	14.7	181	179.7	22.1	241	239.2	29.4
2	2.0	0.2	62	61.5	7.6	22	121.1	14.9	82	180.6	22.2	42	240.2	29.5
3	3.0	0.4	63	62.5	7.7	23	122.1	15.0	83	181.6	22.3	43	241.2	29.6
4	4.0	0.5	64	63.5	7.8	24	123.1	15.1	84	182.6	22.4	44	242.2	29.7
5	5.0	0.6	65	64.5	7.9	25	124.1	15.2	85	183.6	22.5	45	243.2	29.9
6	6.0	0.7	66	65.5	8.0	26	125.1	15.4	86	184.6	22.7	46	244.2	30.0
7	6.9	0.9	67	66.5	8.2	27	126.1	15.5	87	185.6	22.8	47	245.2	30.1
8	7.9	1.0	68	67.5	8.3	28	127.0	15.6	88	186.6	22.9	48	246.2	30.2
9	8.9	1.1	69	68.5	8.4	29	128.0	15.7	89	187.6	23.0	49	247.1	30.3
10	9.9	1.2	70	69.5	8.5	30	129.0	15.8	90	188.6	23.2	50	248.1	30.5
11	10.9	1.3	71	70.5	8.7	131	130.0	16.0	191	189.6	23.3	251	249.1	30.6
12	11.9	1.5	72	71.5	8.8	32	131.0	13.1	92	190.6	23.4	52	250.1	30.7
13	12.9	1.6	73	72.5	8.9	33	132.0	16.2	93	191.6	23.5	53	251.1	30.8
14	13.9	1.7	74	73.4	9.0	34	133.0	16.3	94	192.6	23.6	54	252.1	31.0
15	14.9	1.8	75	74.4	9.1	35	134.0	16.5	95	193.5	23.8	55	253.1	31.1
16	15.9	1.9	76	75.4	9.3	36	135.0	16.6	96	194.5	23.9	56	254.1	31.2
17	16.9	2.1	77	76.4	9.4	37	136.0	16.7	97	195.5	24.0	57	255.1	31.3
18	17.9	2.2	78	77.4	9.5	38	137.0	16.8	98	196.5	24.1	58	256.1	31.4
19	18.9	2.3	79	78.4	9.6	39	138.0	16.9	99	197.5	24.3	59	257.1	31.6
20	19.9	2.4	80	79.4	9.7	40	139.0	17.1	200	198.5	24.4	60	258.1	31.7
21	20.8	2.6	81	80.4	9.9	141	139.9	17.2	201	199.5	24.5	261	259.1	31.8
22	21.8	2.7	82	81.4	10.0	42	140.9	17.3	02	200.5	24.6	62	260.0	31.9
23	22.8	2.8	83	82.4	10.1	43	141.9	17.4	03	201.5	24.7	63	261.0	32.1
24	23.8	2.9	84	83.4	10.2	44	142.9	17.5	04	202.5	24.9	64	262.0	32.2
25	24.8	3.0	85	84.4	10.4	45	143.9	17.7	05	203.5	25.0	65	263.0	32.3
26	25.8	3.2	86	85.4	10.5	46	144.9	17.8	06	204.5	25.1	66	264.0	32.4
27	26.8	3.3	87	86.4	10.6	47	145.9	17.9	07	205.5	25.2	67	265.0	32.5
28	27.8	3.4	88	87.3	10.7	48	146.9	18.0	08	206.4	25.3	68	266.0	32.7
29	28.8	3.5	89	88.3	10.8	49	147.9	18.2	09	207.4	25.5	69	267.0	32.8
30	29.8	3.7	90	89.3	11.0	50	148.9	18.3	10	208.4	25.6	70	268.0	32.9
31	30.8	3.8	91	90.3	11.1	151	149.9	18.4	211	209.4	25.7	271	269.0	33.0
32	31.8	3.9	92	91.3	11.2	52	150.9	18.5	12	210.4	25.8	72	270.0	33.1
33	32.8	4.0	93	92.3	11.3	53	151.9	18.6	13	211.4	26.0	73	271.0	33.3
34	33.7	4.1	94	93.3	11.5	54	152.9	18.8	14	212.4	26.1	74	272.0	33.4
35	34.7	4.3	95	94.3	11.6	55	153.8	18.9	15	213.4	26.2	75	273.0	33.5
36	35.7	4.4	96	95.3	11.7	56	154.8	19.0	16	214.4	26.3	76	273.9	33.6
37	36.7	4.5	97	96.3	11.8	57	155.8	19.1	17	215.4	26.4	77	274.9	33.8
38	37.7	4.6	98	97.3	11.9	58	156.8	19.3	18	216.4	26.6	78	275.9	33.9
39	38.7	4.8	99	98.3	12.1	59	157.8	19.4	19	217.4	26.7	79	276.9	34.0
40	39.7	4.9	100	99.3	12.2	60	158.8	19.5	20	218.4	26.8	80	277.9	34.1
41	40.7	5.0	101	100.2	12.3	161	159.8	19.6	221	219.4	26.9	281	278.9	34.2
42	41.7	5.1	02	101.2	12.4	62	160.8	19.7	22	220.3	27.1	82	279.9	34.4
43	42.7	5.2	03	102.2	12.6	63	161.8	19.9	23	221.3	27.2	83	280.9	34.5
44	43.7	5.4	04	103.2	12.7	64	162.8	20.0	24	222.3	27.3	84	281.9	34.6
45	44.7	5.5	05	104.2	12.8	65	163.8	20.1	25	223.3	27.4	85	282.9	34.7
46	45.7	5.6	06	105.2	12.9	66	164.8	20.2	26	224.3	27.5	86	283.9	34.9
47	46.6	5.7	07	106.2	13.0	67	165.8	20.4	27	225.3	27.7	87	284.9	35.0
48	47.6	5.8	08	107.2	13.2	68	166.7	20.5	28	226.3	27.8	88	285.9	35.1
49	48.6	6.0	09	108.2	13.3	69	167.7	20.6	29	227.3	27.9	89	286.8	35.2
50	49.6	6.1	10	109.2	13.4	70	168.7	20.7	30	228.3	28.0	90	287.8	35.3
51	50.6	6.2	111	110.2	13.5	171	169.7	20.8	231	229.3	28.2	291	288.8	35.5
52	51.6	6.3	12	111.2	13.6	72	170.7	21.0	32	230.3	28.3	92	289.8	35.6
53	52.6	6.5	13	112.2	13.8	73	171.7	21.1	33	231.3	28.4	93	290.8	35.7
54	53.6	6.6	14	113.2	13.9	74	172.7	21.2	34	232.3	28.5	94	291.8	35.8
55	54.6	6.7	15	114.1	14.0	75	173.7	21.3	35	233.2	28.6	95	292.8	36.0
56	55.6	6.8	16	115.1	14.1	76	174.7	21.4	36	234.2	28.8	96	293.8	36.1
57	56.6	6.9	17	116.1	14.3	77	175.7	21.6	37	235.2	28.9	97	294.8	36.2
58	57.6	7.1	18	117.1	14.4	78	176.7	21.7	38	236.2	29.0	98	295.8	36.3
59	58.6	7.2	19	118.1	14.5	79	177.7	21.8	39	237.2	29.1	99	296.8	36.4
60	59.6	7.3	20	119.1	14.6	80	178.7	21.9	40	238.2	29.2	300	297.8	36.6
Dist.	Dep.	D. Lat.	Dist.	Dep.	D. Lat.	Dist.	Dep.	D. Lat.	Dist.	Dep.	D. Lat.	Dist.	Dep.	D. Lat.

277° | 083°
263° | 097°

83°

Dist.	D. Lat.	Dep.
N.	N×Cos.	N×Sin.
Hypotenuse	Side Adj.	Side Opp.

353° | 007°
187° | 173°

TABLE 3

Traverse **7°** Table

353° | 007°
187° | 173°

Dist.	D. Lat.	Dep.	Dist.	D. Lat.	Dep.	Dist.	D. Lat.	Dep.	Dist.	D. Lat.	Dep.	Dist.	D. Lat.	Dep.
301	298.7	36.7	361	358.3	44.0	421	417.9	51.3	481	477.4	58.6	541	537.0	65.9
02	299.7	36.8	62	359.3	44.1	22	418.9	51.4	82	478.4	58.7	42	538.0	66.1
03	300.7	36.9	63	360.3	44.2	23	419.8	51.6	83	479.4	58.9	43	539.0	66.2
04	301.7	37.0	64	361.3	44.4	24	420.8	51.7	84	480.4	59.0	44	539.9	66.3
05	302.7	37.2	65	362.3	44.5	25	421.8	51.8	85	481.4	59.1	45	540.9	66.4
06	303.7	37.3	66	363.3	44.6	26	422.8	51.9	86	482.4	59.2	46	541.9	66.5
07	304.7	37.4	67	364.3	44.7	27	423.8	52.0	87	483.4	59.4	47	542.9	66.7
08	305.7	37.5	68	365.3	44.8	28	424.8	52.2	88	484.4	59.5	48	543.9	66.8
09	306.7	37.7	69	366.2	45.0	29	425.8	52.3	89	485.4	59.6	49	544.9	66.9
10	307.7	37.8	70	367.2	45.1	30	426.8	52.4	90	486.3	59.7	50	545.9	67.0
311	308.7	37.9	371	368.2	45.2	431	427.8	52.5	491	487.3	59.8	551	546.9	67.1
12	309.7	38.0	72	369.2	45.3	32	428.8	52.6	92	488.3	60.0	52	547.9	67.3
13	310.7	38.1	73	370.2	45.5	33	429.8	52.8	93	489.3	60.1	53	548.9	67.4
14	311.7	38.3	74	371.2	45.6	34	430.8	52.9	94	490.3	60.2	54	549.9	67.5
15	312.7	38.4	75	372.2	45.7	35	431.8	53.0	95	491.3	60.3	55	550.9	67.6
16	313.6	38.5	76	373.2	45.8	36	432.8	53.1	96	492.3	60.4	56	551.9	67.8
17	314.6	38.6	77	374.2	45.9	37	433.7	53.3	97	493.3	60.6	57	552.8	67.9
18	315.6	38.8	78	375.2	46.1	38	434.7	53.4	98	494.3	60.7	58	553.8	68.0
19	316.6	38.9	79	376.2	46.2	39	435.7	53.5	99	495.3	60.8	59	554.8	68.1
20	317.6	39.0	80	377.2	46.3	40	436.7	53.6	500	496.3	60.9	60	555.8	68.2
321	318.6	39.1	381	378.2	46.4	441	437.7	53.7	501	497.3	61.1	561	556.8	68.4
22	319.6	39.2	82	379.2	46.6	42	438.7	53.9	02	498.3	61.2	62	557.8	68.5
23	320.6	39.4	83	380.1	46.7	43	439.7	54.0	03	499.3	61.3	63	558.8	68.6
24	321.6	39.5	84	381.1	46.8	44	440.7	54.1	04	500.2	61.4	64	559.8	68.7
25	322.6	39.6	85	382.1	46.9	45	441.7	54.2	05	501.2	61.5	65	560.8	68.9
26	323.6	39.7	86	383.1	47.0	46	442.7	54.4	06	502.2	61.7	66	561.8	69.0
27	324.6	39.8	87	384.1	47.2	47	443.7	54.5	07	503.2	61.8	67	562.8	69.1
28	325.6	40.0	88	385.1	47.3	48	444.7	54.6	08	504.2	61.9	68	563.8	69.2
29	326.5	40.1	89	386.1	47.4	49	445.7	54.7	09	505.2	62.0	69	564.8	69.3
30	327.5	40.2	90	387.1	47.5	50	446.6	54.8	10	506.2	62.2	70	565.8	69.5
331	328.5	40.3	391	388.1	47.7	451	447.6	55.0	511	507.2	62.3	571	566.7	69.6
32	329.5	40.5	92	389.1	47.8	52	448.6	55.1	12	508.2	62.4	72	567.7	69.7
33	330.5	40.6	93	390.1	47.9	53	449.6	55.2	13	509.2	62.5	73	568.7	69.8
34	331.5	40.7	94	391.1	48.0	54	450.6	55.3	14	510.2	62.6	74	569.7	70.0
35	332.5	40.8	95	392.1	48.1	55	451.6	55.5	15	511.2	62.8	75	570.7	70.1
36	333.5	40.9	96	393.0	48.3	56	452.6	55.6	16	512.2	62.9	76	571.7	70.2
37	334.5	41.1	97	394.0	48.4	57	453.6	55.7	17	513.1	63.0	77	572.7	70.3
38	335.5	41.2	98	395.0	48.5	58	454.6	55.8	18	514.1	63.1	78	573.7	70.4
39	336.5	41.3	99	396.0	48.6	59	455.6	55.9	19	515.1	63.3	79	574.7	70.6
40	337.5	41.4	400	397.0	48.7	60	456.6	56.1	20	516.1	63.4	80	575.7	70.7
341	338.5	41.6	401	398.0	48.9	461	457.6	56.2	521	517.1	63.5	581	576.7	70.8
42	339.5	41.7	02	399.0	49.0	62	458.6	56.3	22	518.1	63.6	82	577.7	70.9
43	340.4	41.8	03	400.0	49.1	63	459.5	56.4	23	519.1	63.7	83	578.7	71.0
44	341.4	41.9	04	401.0	49.2	64	460.5	56.5	24	520.1	63.9	84	579.6	71.2
45	342.4	42.0	05	402.0	49.4	65	461.5	56.7	25	521.1	64.0	85	580.6	71.3
46	343.4	42.2	06	403.0	49.5	66	462.5	56.8	26	522.1	64.1	86	581.6	71.4
47	344.4	42.3	07	404.0	49.6	67	463.5	56.9	27	523.1	64.2	87	582.6	71.5
48	345.4	42.4	08	405.0	49.7	68	464.5	57.0	28	524.1	64.3	88	583.6	71.7
49	346.4	42.5	09	406.0	49.8	69	465.5	57.2	29	525.1	64.5	89	584.6	71.8
50	347.4	42.7	10	406.9	50.0	70	466.5	57.3	30	526.0	64.6	90	585.6	71.9
351	348.4	42.8	411	407.9	50.1	471	467.5	57.4	531	527.0	64.7	591	586.6	72.0
52	349.4	42.9	12	408.9	50.2	72	468.5	57.5	32	528.0	64.8	92	587.6	72.1
53	350.4	43.0	13	409.9	50.3	73	469.5	57.6	33	529.0	65.0	93	588.6	72.3
54	351.4	43.1	14	410.9	50.5	74	470.5	57.8	34	530.0	65.1	94	589.6	72.4
55	352.4	43.3	15	411.9	50.6	75	471.5	57.9	35	531.0	65.2	95	590.6	72.5
56	353.3	43.4	16	412.9	50.7	76	472.5	58.0	36	532.0	65.3	96	591.6	72.6
57	354.3	43.5	17	413.9	50.8	77	473.4	58.1	37	533.0	65.4	97	592.6	72.8
58	355.3	43.6	18	414.9	50.9	78	474.4	58.3	38	534.0	65.6	98	593.5	72.9
59	356.3	43.7	19	415.9	51.1	79	475.4	58.4	39	535.0	65.7	99	594.5	73.0
60	357.3	43.9	20	416.9	51.2	80	476.4	58.5	40	536.0	65.8	600	595.5	73.1
Dist.	Dep.	D. Lat.	Dist.	Dep.	D. Lat.	Dist.	Dep.	D. Lat.	Dist.	Dep.	D. Lat.	Dist.	Dep.	D. Lat.

Dist.	D. Lat.	Dep.
D Lo	Dep.	
	m	D Lo

83°

277° | 083°
263° | 097°

352° | 008° / 188° | 172°	TABLE 3 — Traverse 8° Table	352° | 008° / 188° | 172°

Dist.	D. Lat.	Dep.	Dist.	D. Lat.	Dep.	Dist.	D. Lat.	Dep.	Dist.	D. Lat.	Dep.	Dist.	D. Lat.	Dep.
1	1.0	0.1	61	60.4	8.5	121	119.8	16.8	181	179.2	25.2	241	238.7	33.5
2	2.0	0.3	62	61.4	8.6	22	120.8	17.0	82	180.2	25.3	42	239.6	33.7
3	3.0	0.4	63	62.4	8.8	23	121.8	17.1	83	181.2	25.5	43	240.6	33.8
4	4.0	0.6	64	63.4	8.9	24	122.8	17.3	84	182.2	25.6	44	241.6	34.0
5	5.0	0.7	65	64.4	9.0	25	123.8	17.4	85	183.2	25.7	45	242.6	34.1
6	5.9	0.8	66	65.4	9.2	26	124.8	17.5	86	184.2	25.9	46	243.6	34.2
7	6.9	1.0	67	66.3	9.3	27	125.8	17.7	87	185.2	26.0	47	244.6	34.4
8	7.9	1.1	68	67.3	9.5	28	126.8	17.8	88	186.2	26.2	48	245.6	34.5
9	8.9	1.3	69	68.3	9.6	29	127.7	18.0	89	187.2	26.3	49	246.6	34.7
10	9.9	1.4	70	69.3	9.7	30	128.7	18.1	90	188.2	26.4	50	247.6	34.8
11	10.9	1.5	71	70.3	9.9	131	129.7	18.2	191	189.1	26.6	251	248.6	34.9
12	11.9	1.7	72	71.3	10.0	32	130.7	18.4	92	190.1	26.7	52	249.5	35.1
13	12.9	1.8	73	72.3	10.2	33	131.7	18.5	93	191.1	26.9	53	250.5	35.2
14	13.9	1.9	74	73.3	10.3	34	132.7	18.6	94	192.1	27.0	54	251.5	35.3
15	14.9	2.1	75	74.3	10.4	35	133.7	18.8	95	193.1	27.1	55	252.5	35.5
16	15.8	2.2	76	75.3	10.6	36	134.7	18.9	96	194.1	27.3	56	253.5	35.6
17	16.8	2.4	77	76.3	10.7	37	135.7	19.1	97	195.1	27.4	57	254.5	35.8
18	17.8	2.5	78	77.2	10.9	38	136.7	19.2	98	196.1	27.6	58	255.5	35.9
19	18.8	2.6	79	78.2	11.0	39	137.7	19.3	99	197.1	27.7	59	256.5	36.0
20	19.8	2.8	80	79.2	11.1	40	138.6	19.5	200	198.1	27.8	60	257.5	36.2
21	20.8	2.9	81	80.2	11.3	141	139.6	19.6	201	199.0	28.0	261	258.5	36.3
22	21.8	3.1	82	81.2	11.4	42	140.6	19.8	02	200.0	28.1	62	259.5	36.5
23	22.8	3.2	83	82.2	11.6	43	141.6	19.9	03	201.0	28.3	63	260.4	36.6
24	23.8	3.3	84	83.2	11.7	44	142.6	20.0	04	202.0	28.4	64	261.4	36.7
25	24.8	3.5	85	84.2	11.8	45	143.6	20.2	05	203.0	28.5	65	262.4	36.9
26	25.7	3.6	86	85.2	12.0	46	144.6	20.3	06	204.0	28.7	66	263.4	37.0
27	26.7	3.8	87	86.2	12.1	47	145.6	20.5	07	205.0	28.8	67	264.4	37.2
28	27.7	3.9	88	87.1	12.2	48	146.6	20.6	08	206.0	28.9	68	265.4	37.3
29	28.7	4.0	89	88.1	12.4	49	147.5	20.7	09	207.0	29.1	69	266.4	37.4
30	29.7	4.2	90	89.1	12.5	50	148.5	20.9	10	208.0	29.2	70	267.4	37.6
31	30.7	4.3	91	90.1	12.7	151	149.5	21.0	211	208.9	29.4	271	268.4	37.7
32	31.7	4.5	92	91.1	12.8	52	150.5	21.2	12	209.9	29.5	72	269.4	37.9
33	32.7	4.6	93	92.1	12.9	53	151.5	21.3	13	210.9	29.6	73	270.3	38.0
34	33.7	4.7	94	93.1	13.1	54	152.5	21.4	14	211.9	29.8	74	271.3	38.1
35	34.7	4.9	95	94.1	13.2	55	153.5	21.6	15	212.9	29.9	75	272.3	38.3
36	35.6	5.0	96	95.1	13.4	56	154.5	21.7	16	213.9	30.1	76	273.3	38.4
37	36.6	5.1	97	96.1	13.5	57	155.5	21.9	17	214.9	30.2	77	274.3	38.6
38	37.6	5.3	98	97.0	13.6	58	156.5	22.0	18	215.9	30.3	78	275.3	38.7
39	38.6	5.4	99	98.0	13.8	59	157.5	22.1	19	216.9	30.5	79	276.3	38.8
40	39.6	5.6	100	99.0	13.9	60	158.4	22.3	20	217.9	30.6	80	277.3	39.0
41	40.6	5.7	101	100.0	14.1	161	159.4	22.4	221	218.8	30.8	281	278.3	39.1
42	41.6	5.8	02	101.0	14.2	62	160.4	22.5	22	219.8	30.9	82	279.3	39.2
43	42.6	6.0	03	102.0	14.3	63	161.4	22.7	23	220.8	31.0	83	280.2	39.4
44	43.6	6.1	04	103.0	14.5	64	162.4	22.8	24	221.8	31.2	84	281.2	39.5
45	44.6	6.3	05	104.0	14.6	65	163.4	23.0	25	222.8	31.3	85	282.2	39.7
46	45.6	6.4	06	105.0	14.8	66	164.4	23.1	26	223.8	31.5	86	283.2	39.8
47	46.5	6.5	07	106.0	14.9	67	165.4	23.2	27	224.8	31.6	87	284.2	39.9
48	47.5	6.7	08	106.9	15.0	68	166.4	23.4	28	225.8	31.7	88	285.2	40.1
49	48.5	6.8	09	107.9	15.2	69	167.4	23.5	29	226.8	31.9	89	286.2	40.2
50	49.5	7.0	10	108.9	15.3	70	168.3	23.7	30	227.8	32.0	90	287.2	40.4
51	50.5	7.1	111	109.9	15.4	171	169.3	23.8	231	228.8	32.1	291	288.2	40.5
52	51.5	7.2	12	110.9	15.6	72	170.3	23.9	32	229.7	32.3	92	289.2	40.6
53	52.5	7.4	13	111.9	15.7	73	171.3	24.1	33	230.7	32.4	93	290.1	40.8
54	53.5	7.5	14	112.9	15.9	74	172.3	24.2	34	231.7	32.6	94	291.1	40.9
55	54.5	7.7	15	113.9	16.0	75	173.3	24.4	35	232.7	32.7	95	292.1	41.1
56	55.5	7.8	16	114.9	16.1	76	174.3	24.5	36	233.7	32.8	96	293.1	41.2
57	56.4	7.9	17	115.9	16.3	77	175.3	24.6	37	234.7	33.0	97	294.1	41.3
58	57.4	8.1	18	116.9	16.4	78	176.3	24.8	38	235.7	33.1	98	295.1	41.5
59	58.4	8.2	19	117.8	16.6	79	177.3	24.9	39	236.7	33.3	99	296.1	41.6
60	59.4	8.4	20	118.8	16.7	80	178.2	25.1	40	237.7	33.4	300	297.1	41.8
Dist.	Dep.	D. Lat.	Dist.	Dep.	D. Lat.	Dist.	Dep.	D. Lat.	Dist.	Dep.	D. Lat.	Dist.	Dep.	D. Lat.

278° | 082°
262° | 098°

82°

Dist.	D. Lat.	Dep.
N.	N×Cos.	N×Sin.
Hypotenuse	Side Adj.	Side Opp.

352° | 008° / 188° | 172°	TABLE 3 Traverse 8° Table	352° | 008° / 188° | 172°

Dist.	D. Lat.	Dep.	Dist.	D. Lat.	Dep.	Dist.	D. Lat.	Dep.	Dist.	D. Lat.	Dep.	Dist.	D. Lat.	Dep.
301	298.1	41.9	361	357.5	50.2	421	416.9	58.6	481	476.3	66.9	541	535.7	75.3
02	299.1	42.0	62	358.5	50.4	22	417.9	58.7	82	477.3	67.1	42	536.7	75.4
03	300.1	42.2	63	359.5	50.5	23	418.9	58.9	83	478.3	67.2	43	537.7	75.6
04	301.0	42.3	64	360.5	50.7	24	419.9	59.0	84	479.3	67.4	44	538.7	75.7
05	302.0	42.4	65	361.4	50.8	25	420.9	59.1	85	480.3	67.5	45	539.7	75.8
06	303.0	42.6	66	362.4	50.9	26	421.9	59.3	86	481.3	67.6	46	540.7	76.0
07	304.0	42.7	67	363.4	51.1	27	422.8	59.4	87	482.3	67.8	47	541.7	76.1
08	305.0	42.9	68	364.4	51.2	28	423.8	59.6	88	483.3	67.9	48	542.7	76.3
09	306.0	43.0	69	365.4	51.4	29	424.8	59.7	89	484.2	68.1	49	543.7	76.4
10	307.0	43.1	70	366.4	51.5	30	425.8	59.8	90	485.2	68.2	50	544.6	76.5
311	308.0	43.3	371	367.4	51.6	431	426.8	60.0	491	486.2	68.3	551	545.6	76.7
12	309.0	43.4	72	368.4	51.8	32	427.8	60.1	92	487.2	68.5	52	546.6	76.8
13	310.0	43.6	73	369.4	51.9	33	428.8	60.3	93	488.2	68.6	53	547.6	77.0
14	310.9	43.7	74	370.4	52.1	34	429.8	60.4	94	489.2	68.8	54	548.6	77.1
15	311.9	43.8	75	371.4	52.2	35	430.8	60.5	95	490.2	68.9	55	549.6	77.2
16	312.9	44.0	76	372.3	52.3	36	431.8	60.7	96	491.2	69.0	56	550.6	77.4
17	313.9	44.1	77	373.3	52.5	37	432.7	60.8	97	492.2	69.2	57	551.6	77.5
18	314.9	44.3	78	374.3	52.6	38	433.7	61.0	98	493.2	69.3	58	552.6	77.7
19	315.9	44.4	79	375.3	52.7	39	434.7	61.1	99	494.1	69.6	59	553.6	77.8
20	316.9	44.5	80	376.3	52.9	40	435.7	61.2	500	495.1	69.6	60	554.6	77.9
321	317.9	44.7	381	377.3	53.0	441	436.7	61.4	501	496.1	69.7	561	555.5	78.1
22	318.9	44.8	82	378.3	53.2	42	437.7	61.5	02	497.1	69.9	62	556.5	78.2
23	319.9	45.0	83	379.3	53.3	43	438.7	61.7	03	498.1	70.0	63	557.5	78.4
24	320.8	45.1	84	380.3	53.4	44	439.7	61.8	04	499.1	70.2	64	558.5	78.5
25	321.8	45.2	85	381.3	53.6	45	440.7	61.9	05	500.1	70.3	65	559.5	78.6
26	322.8	45.4	86	382.2	53.7	46	441.7	62.1	06	501.1	70.4	66	560.5	78.8
27	323.8	45.5	87	383.2	53.9	47	442.6	62.2	07	502.1	70.6	67	561.5	78.9
28	324.8	45.6	88	384.2	54.0	48	443.6	62.3	08	503.1	70.7	68	562.5	79.1
29	325.8	45.8	89	385.2	54.1	49	444.6	62.5	09	504.0	70.8	69	563.5	79.2
30	326.8	45.9	90	386.2	54.3	50	445.6	62.6	10	505.0	71.0	70	564.5	79.3
331	327.8	46.1	391	387.2	54.4	451	446.6	62.8	511	506.0	71.1	571	565.4	79.5
32	328.8	46.2	92	388.2	54.6	52	447.6	62.9	12	507.0	71.3	72	566.4	79.6
33	329.8	46.3	93	389.1	54.7	53	448.6	63.0	13	508.0	71.4	73	567.4	79.7
34	330.7	46.5	94	390.1	54.8	54	449.6	63.2	14	509.0	71.5	74	568.4	79.9
35	331.7	46.6	95	391.1	55.0	55	450.6	63.3	15	510.0	71.7	75	569.4	80.0
36	332.7	46.8	96	392.1	55.1	56	451.6	63.5	16	511.0	71.8	76	570.4	80.2
37	333.7	46.9	97	393.1	55.3	57	452.6	63.6	17	512.0	72.0	77	571.4	80.3
38	334.7	47.0	98	394.1	55.4	58	453.5	63.7	18	513.0	72.1	78	572.4	80.4
39	335.7	47.2	99	395.1	55.5	59	454.5	63.9	19	513.9	72.2	79	573.4	80.6
40	336.7	47.3	400	396.1	55.7	60	455.5	64.0	20	514.9	72.4	80	574.4	80.7
341	337.7	47.5	401	397.1	55.8	461	456.5	64.2	521	515.9	72.5	581	575.3	80.9
42	338.7	47.6	02	398.1	55.9	62	457.5	64.3	22	516.9	72.6	82	576.3	81.0
43	339.7	47.7	03	399.1	56.1	63	458.5	64.4	23	517.9	72.8	83	577.3	81.1
44	340.7	47.9	04	400.1	56.2	64	459.5	64.6	24	518.9	72.9	84	578.3	81.3
45	341.6	48.0	05	401.1	56.4	65	460.5	64.7	25	519.9	73.1	85	579.3	81.4
46	342.6	48.2	06	402.0	56.5	66	461.5	64.9	26	520.9	73.2	86	580.3	81.6
47	343.6	48.3	07	403.0	56.6	67	462.5	65.0	27	521.9	73.3	87	581.3	81.7
48	344.6	48.4	08	404.0	56.8	68	463.4	65.1	28	522.9	73.5	88	582.3	81.8
49	345.6	48.6	09	405.0	56.9	69	464.4	65.3	29	523.9	73.6	89	583.3	82.0
50	346.6	48.7	10	406.0	57.1	70	465.4	65.4	30	524.8	73.8	90	584.3	82.1
351	347.6	48.8	411	407.0	57.2	471	466.4	65.6	531	525.8	73.9	591	585.2	82.3
52	348.6	49.0	12	408.0	57.3	72	467.4	65.7	32	526.8	74.0	92	586.2	82.4
53	349.6	49.1	13	409.0	57.5	73	468.4	65.8	33	527.8	74.2	93	587.2	82.5
54	350.6	49.3	14	410.0	57.6	74	469.4	66.0	34	528.8	74.3	94	588.2	82.7
55	351.5	49.4	15	411.0	57.8	75	470.4	66.1	35	529.8	74.5	95	589.2	82.8
56	352.5	49.5	16	411.9	57.9	76	471.4	66.2	36	530.8	74.6	96	590.2	82.9
57	353.5	49.7	17	412.9	58.0	77	472.4	66.4	37	531.8	74.7	97	591.2	83.1
58	354.5	49.8	18	413.9	58.2	78	473.3	66.5	38	532.8	74.9	98	592.2	83.2
59	355.5	50.0	19	414.9	58.3	79	474.3	66.7	39	533.8	75.0	99	593.2	83.4
60	356.5	50.1	20	415.9	58.5	80	475.3	66.8	40	534.7	75.2	600	594.2	83.5
Dist.	Dep.	D. Lat.	Dist.	Dep.	D. Lat.	Dist.	Dep.	D. Lat.	Dist.	Dep.	D. Lat.	Dist.	Dep.	D. Lat.

Dist.	D. Lat.	Dep.
D Lo	Dep.	
	m	D Lo

82°

278°	082°
262°	098°

351°	009°
189°	171°

TABLE 3

Traverse **9°** Table

351°	009°
189°	171°

Dist.	D. Lat.	Dep.	Dist.	D. Lat.	Dep.	Dist.	D. Lat.	Dep.	Dist.	D. Lat.	Dep.	Dist.	D. Lat.	Dep.
1	1.0	0.2	61	60.2	9.5	121	119.5	18.9	181	178.8	28.3	241	238.0	37.7
2	2.0	0.3	62	61.2	9.7	22	120.5	19.1	82	179.8	28.5	42	239.0	37.9
3	3.0	0.5	63	62.2	9.9	23	121.5	19.2	83	180.7	28.6	43	240.0	38.0
4	4.0	0.6	64	63.2	10.0	24	122.5	19.4	84	181.7	28.8	44	241.0	38.2
5	4.9	0.8	65	64.2	10.2	25	123.5	19.6	85	182.7	28.9	45	242.0	38.3
6	5.9	0.9	66	65.2	10.3	26	124.4	19.7	86	183.7	29.1	46	243.0	38.5
7	6.9	1.1	67	66.2	10.5	27	125.4	19.9	87	184.7	29.3	47	244.0	38.6
8	7.9	1.3	68	67.2	10.6	28	126.4	20.0	88	185.7	29.4	48	244.9	38.8
9	8.9	1.4	69	68.2	10.8	29	127.4	20.2	89	186.7	29.6	49	245.9	39.0
10	9.9	1.6	70	69.1	11.0	30	128.4	20.3	90	187.7	29.7	50	246.9	39.1
11	10.9	1.7	71	70.1	11.1	131	129.4	20.5	191	188.6	29.9	251	247.9	39.3
12	11.9	1.9	72	71.1	11.3	32	130.4	20.6	92	189.6	30.0	52	248.9	39.4
13	12.8	2.0	73	72.1	11.4	33	131.4	20.8	93	190.6	30.2	53	249.9	39.6
14	13.8	2.2	74	73.1	11.6	34	132.4	21.0	94	191.6	30.3	54	250.9	39.7
15	14.8	2.3	75	74.1	11.7	35	133.3	21.1	95	192.6	30.5	55	251.9	39.9
16	15.8	2.5	76	75.1	11.9	36	134.3	21.3	96	193.6	30.7	56	252.8	40.0
17	16.8	2.7	77	76.1	12.0	37	135.3	21.4	97	194.6	30.8	57	253.8	40.2
18	17.8	2.8	78	77.0	12.2	38	136.3	21.6	98	195.6	31.0	58	254.8	40.4
19	18.8	3.0	79	78.0	12.4	39	137.3	21.7	99	196.5	31.1	59	255.8	40.5
20	19.8	3.1	80	79.0	12.5	40	138.3	21.9	200	197.5	31.3	60	256.8	40.7
21	20.7	3.3	81	80.0	12.7	141	139.3	22.1	201	198.5	31.4	261	257.8	40.8
22	21.7	3.4	82	81.0	12.8	42	140.3	22.2	02	199.5	31.6	62	258.8	41.0
23	22.7	3.6	83	82.0	13.0	43	141.2	22.4	03	200.5	31.8	63	259.8	41.1
24	23.7	3.8	84	83.0	13.1	44	142.2	22.5	04	201.5	31.9	64	260.7	41.3
25	24.7	3.9	85	84.0	13.3	45	143.2	22.7	05	202.5	32.1	65	261.7	41.5
26	25.7	4.1	86	84.9	13.5	46	144.2	22.8	06	203.5	32.2	66	262.7	41.6
27	26.7	4.2	87	85.9	13.6	47	145.2	23.0	07	204.5	32.4	67	263.7	41.8
28	27.7	4.4	88	86.9	13.8	48	146.2	23.2	08	205.4	32.5	68	264.7	41.9
29	28.6	4.5	89	87.9	13.9	49	147.2	23.3	09	206.4	32.7	69	265.7	42.1
30	29.6	4.7	90	88.9	14.1	50	148.2	23.5	10	207.4	32.9	70	266.7	42.2
31	30.6	4.8	91	89.9	14.2	151	149.1	23.6	211	208.4	33.0	271	267.7	42.4
32	31.6	5.0	92	90.9	14.4	52	150.1	23.8	12	209.4	33.2	72	268.7	42.6
33	32.6	5.2	93	91.9	14.5	53	151.1	23.9	13	210.4	33.3	73	269.6	42.7
34	33.6	5.3	94	92.8	14.7	54	152.1	24.1	14	211.4	33.5	74	270.6	42.9
35	34.6	5.5	95	93.8	14.9	55	153.1	24.2	15	212.4	33.6	75	271.6	43.0
36	35.6	5.6	96	94.8	15.0	56	154.1	24.4	16	213.3	33.8	76	272.6	43.2
37	36.5	5.8	97	95.8	15.2	57	155.1	24.6	17	214.3	33.9	77	273.6	43.3
38	37.5	5.9	98	96.8	15.3	58	156.1	24.7	18	215.3	34.1	78	274.6	43.5
39	38.5	6.1	99	97.8	15.5	59	157.0	24.9	19	216.3	34.3	79	275.6	43.6
40	39.5	6.3	100	98.8	15.6	60	158.0	25.0	20	217.3	34.4	80	276.6	43.8
41	40.5	6.4	101	99.8	15.8	161	159.0	25.2	221	218.3	34.6	281	277.5	44.0
42	41.5	6.6	02	100.7	16.0	62	160.0	25.3	22	219.3	34.7	82	278.5	44.1
43	42.5	6.7	03	101.7	16.1	63	161.0	25.5	23	220.3	34.9	83	279.5	44.3
44	43.5	6.9	04	102.7	16.3	64	162.0	25.7	24	221.2	35.0	84	280.5	44.4
45	44.4	7.0	05	103.7	16.4	65	163.0	25.8	25	222.2	35.2	85	281.5	44.6
46	45.4	7.2	06	104.7	16.6	66	164.0	26.0	26	223.2	35.4	86	282.5	44.7
47	46.4	7.4	07	105.7	16.7	67	164.9	26.1	27	224.2	35.5	87	283.5	44.9
48	47.4	7.5	08	106.7	16.9	68	165.9	26.3	28	225.2	35.7	88	284.5	45.1
49	48.4	7.7	09	107.7	17.1	69	166.9	26.4	29	226.2	35.8	89	285.4	45.2
50	49.4	7.8	10	108.6	17.2	70	167.9	26.6	30	227.2	36.0	90	286.4	45.4
51	50.4	8.0	111	109.6	17.4	171	168.9	26.8	231	228.2	36.1	291	287.4	45.5
52	51.4	8.1	12	110.6	17.5	72	169.9	26.9	32	229.1	36.3	92	288.4	45.7
53	52.3	8.3	13	111.6	17.7	73	170.9	27.1	33	230.1	36.4	93	289.4	45.8
54	53.3	8.4	14	112.6	17.8	74	171.9	27.2	34	231.1	36.6	94	290.4	46.0
55	54.3	8.6	15	113.6	18.0	75	172.8	27.4	35	232.1	36.8	95	291.4	46.1
56	55.3	8.8	16	114.6	18.1	76	173.8	27.5	36	233.1	36.9	96	292.4	46.3
57	56.3	8.9	17	115.6	18.3	77	174.8	27.7	37	234.1	37.1	97	293.3	46.5
58	57.3	9.1	18	116.5	18.5	78	175.8	27.8	38	235.1	37.2	98	294.3	46.6
59	58.3	9.2	19	117.5	18.6	79	176.8	28.0	39	236.1	37.4	99	295.3	46.8
60	59.3	9.4	20	118.5	18.8	80	177.8	28.2	40	237.0	37.5	300	296.3	46.9
Dist.	Dep.	D. Lat.	Dist.	Dep.	D. Lat.	Dist.	Dep.	D. Lat.	Dist.	Dep.	D. Lat.	Dist.	Dep.	D. Lat.

279°	081°
261°	099°

81°

Dist.	D. Lat.	Dep.
N.	N×Cos.	N×Sin.
Hypotenuse	Side Adj.	Side Opp.

351°	009°
189°	171°

TABLE 3

Traverse **9°** Table

351°	009°
189°	171°

Dist.	D. Lat.	Dep.	Dist.	D. Lat.	Dep.	Dist.	D. Lat.	Dep.	Dist.	D. Lat.	Dep.	Dist.	D. Lat.	Dep.
301	297.3	47.1	361	356.6	56.5	421	415.8	65.9	481	475.1	75.2	541	534.3	84.6
02	298.3	47.2	62	357.5	56.6	22	416.8	66.0	82	476.1	75.4	42	535.3	84.8
03	299.3	47.4	63	358.5	56.8	23	417.8	66.2	83	477.1	75.6	43	536.3	84.9
04	300.3	47.6	64	359.5	56.9	24	418.8	66.3	84	478.0	75.7	44	537.3	85.1
05	301.2	47.7	65	360.5	57.1	25	419.8	66.5	85	479.0	75.9	45	538.3	85.3
06	302.2	47.9	66	361.5	57.3	26	420.8	66.6	86	480.0	76.0	46	539.3	85.4
07	303.2	48.0	67	362.5	57.4	27	421.7	66.8	87	481.0	76.2	47	540.3	85.6
08	304.2	48.2	68	363.5	57.6	28	422.7	67.0	88	482.0	76.3	48	541.3	85.7
09	305.2	48.3	69	364.5	57.7	29	423.7	67.1	89	483.0	76.5	49	542.2	85.9
10	306.2	48.5	70	365.4	57.9	30	424.7	67.3	90	484.0	76.7	50	543.2	86.0
311	307.2	48.7	371	366.4	58.0	431	425.7	67.4	491	485.0	76.8	551	544.2	86.2
12	308.2	48.8	72	367.4	58.2	32	426.7	67.6	92	485.9	77.0	52	545.2	86.4
13	309.1	49.0	73	368.4	58.4	33	427.7	67.7	93	486.9	77.1	53	546.2	86.5
14	310.1	49.1	74	369.4	58.5	34	428.7	67.9	94	487.9	77.3	54	547.2	86.7
15	311.1	49.3	75	370.4	58.7	35	429.6	68.0	95	488.9	77.4	55	548.2	86.8
16	312.1	49.4	76	371.4	58.8	36	430.6	68.2	96	489.9	77.6	56	549.2	87.0
17	313.1	49.6	77	372.4	59.0	37	431.6	68.4	97	490.9	77.7	57	550.1	87.1
18	314.1	49.7	78	373.3	59.1	38	432.6	68.5	98	491.9	77.9	58	551.1	87.3
19	315.1	49.9	79	374.3	59.3	39	433.6	68.7	99	492.9	78.1	59	552.1	87.4
20	316.1	50.1	80	375.3	59.4	40	434.6	68.8	500	493.8	78.2	60	553.1	87.6
321	317.0	50.2	381	376.3	59.6	441	435.6	69.0	501	494.8	78.4	561	554.1	87.8
22	318.0	50.4	82	377.3	59.8	42	436.6	69.1	02	495.8	78.5	62	555.1	87.9
23	319.0	50.5	83	378.3	59.9	43	437.5	69.3	03	496.8	78.7	63	556.1	88.1
24	320.0	50.7	84	379.3	60.1	44	438.5	69.5	04	497.8	78.8	64	557.1	88.2
25	321.0	50.8	85	380.3	60.2	45	439.5	69.6	05	498.8	79.0	65	558.0	88.4
26	322.0	51.0	86	381.2	60.4	46	440.5	69.8	06	499.8	79.2	66	559.0	88.5
27	323.0	51.2	87	382.2	60.5	47	441.5	69.9	07	500.8	79.3	67	560.0	88.7
28	324.0	51.3	88	383.2	60.7	48	442.5	70.1	08	501.7	79.5	68	561.0	88.9
29	324.9	51.5	89	384.2	60.9	49	443.5	70.2	09	502.7	79.6	69	562.0	89.0
30	325.9	51.6	90	385.2	61.0	50	444.5	70.4	10	503.7	79.8	70	563.0	89.2
331	326.9	51.8	391	386.2	61.2	451	445.4	70.6	511	504.7	79.9	571	564.0	89.3
32	327.9	51.9	92	387.2	61.3	52	446.4	70.7	12	505.7	80.1	72	565.0	89.5
33	328.9	52.1	93	388.2	61.5	53	447.4	70.9	13	506.7	80.3	73	565.9	89.6
34	329.9	52.2	94	389.1	61.6	54	448.4	71.0	14	507.7	80.4	74	566.9	89.8
35	330.9	52.4	95	390.1	61.8	55	449.4	71.2	15	508.7	80.6	75	567.9	89.9
36	331.9	52.6	96	391.1	61.9	56	450.4	71.3	16	509.6	80.7	76	568.9	90.1
37	332.9	52.7	97	392.1	62.1	57	451.4	71.5	17	510.6	80.9	77	569.9	90.3
38	333.8	52.9	98	393.1	62.3	58	452.4	71.6	18	511.6	81.0	78	570.9	90.4
39	334.8	53.0	99	394.1	62.4	59	453.3	71.8	19	512.6	81.2	79	571.9	90.6
40	335.8	53.2	400	395.1	62.6	60	454.3	72.0	20	513.6	81.3	80	572.9	90.7
341	336.8	53.3	401	396.1	62.7	461	455.3	72.1	521	514.6	81.5	581	573.8	90.9
42	337.8	53.5	02	397.1	62.9	62	456.3	72.3	22	515.6	81.7	82	574.8	91.0
43	338.8	53.7	03	398.0	63.0	63	457.3	72.4	23	516.6	81.8	83	575.8	91.2
44	339.8	53.8	04	399.0	63.2	64	458.3	72.6	24	517.5	82.0	84	576.8	91.4
45	340.8	54.0	05	400.0	63.4	65	459.3	72.7	25	518.5	82.1	85	577.8	91.5
46	341.7	54.1	06	401.0	63.5	66	460.3	72.9	26	519.5	82.3	86	578.8	91.7
47	342.7	54.3	07	402.0	63.7	67	461.3	73.1	27	520.5	82.4	87	579.8	91.8
48	343.7	54.4	08	403.0	63.8	68	462.2	73.2	28	521.5	82.6	88	580.8	92.0
49	344.7	54.6	09	404.0	64.0	69	463.2	73.4	29	522.5	82.8	89	581.7	92.1
50	345.7	54.8	10	405.0	64.1	70	464.2	73.5	30	523.5	82.9	90	582.7	92.3
351	346.7	54.9	411	405.9	64.3	471	465.2	73.7	531	524.5	83.1	591	583.7	92.5
52	347.7	55.1	12	406.9	64.5	72	466.2	73.8	32	525.5	83.2	92	584.7	92.6
53	348.7	55.2	13	407.9	64.6	73	467.2	74.0	33	526.4	83.4	93	585.7	92.8
54	349.6	55.4	14	408.9	64.8	74	468.2	74.1	34	527.4	83.5	94	586.7	92.9
55	350.6	55.5	15	409.9	64.9	75	469.2	74.3	35	528.4	83.7	95	587.7	93.1
56	351.6	55.7	16	410.9	65.1	76	470.1	74.5	36	529.4	83.8	96	588.7	93.2
57	352.6	55.8	17	411.9	65.2	77	471.1	74.6	37	530.4	84.0	97	589.6	93.4
58	353.6	56.0	18	412.9	65.4	78	472.1	74.8	38	531.4	84.2	98	590.6	93.5
59	354.6	56.2	19	413.8	65.5	79	473.1	74.9	39	532.4	84.3	99	591.6	93.7
60	355.6	56.3	20	414.8	65.7	80	474.1	75.1	40	533.4	84.5	600	592.6	93.9
Dist.	Dep.	D. Lat.	Dist.	Dep.	D. Lat.	Dist.	Dep.	D. Lat.	Dist.	Dep.	D. Lat.	Dist.	Dep.	D. Lat.

Dist.	D. Lat.	Dep.
D Lo	Dep.	
	m	D Lo

81°

279°	081°
261°	099°

350°	010°
190°	170°

TABLE 3

Traverse 10° Table

Dist.	D. Lat.	Dep.	Dist.	D. Lat.	Dep.	Dist.	D. Lat.	Dep.	Dist.	D. Lat.	Dep.	Dist.	D. Lat.	Dep.
1	1.0	0.2	61	60.1	10.6	121	119.2	21.0	181	178.3	31.4	241	237.3	41.8
2	2.0	0.3	62	61.1	10.8	22	120.1	21.2	82	179.2	31.6	42	238.3	42.0
3	3.0	0.5	63	62.0	10.9	23	121.1	21.4	83	180.2	31.8	43	239.3	42.2
4	3.9	0.7	64	63.0	11.1	24	122.1	21.5	84	181.2	32.0	44	240.3	42.4
5	4.9	0.9	65	64.0	11.3	25	123.1	21.7	85	182.2	32.1	45	241.3	42.5
6	5.9	1.0	66	65.0	11.5	26	124.1	21.9	86	183.2	32.3	46	242.3	42.7
7	6.9	1.2	67	66.0	11.6	27	125.1	22.1	87	184.2	32.5	47	243.2	42.9
8	7.9	1.4	68	67.0	11.8	28	126.1	22.2	88	185.1	32.6	48	244.2	43.1
9	8.9	1.6	69	68.0	12.0	29	127.0	22.4	89	186.1	32.8	49	245.2	43.2
10	9.8	1.7	70	68.9	12.2	30	128.0	22.6	90	187.1	33.0	50	246.2	43.4
11	10.8	1.9	71	69.9	12.3	131	129.0	22.7	191	188.1	33.2	251	247.2	43.6
12	11.8	2.1	72	70.9	12.5	32	130.0	22.9	92	189.1	33.3	52	248.2	43.8
13	12.8	2.3	73	71.9	12.7	33	131.0	23.1	93	190.1	33.5	53	249.2	43.9
14	13.8	2.4	74	72.9	12.8	34	132.0	23.3	94	191.1	33.7	54	250.1	44.1
15	14.8	2.6	75	73.9	13.0	35	132.9	23.4	95	192.0	33.9	55	251.1	44.3
16	15.8	2.8	76	74.8	13.2	36	133.9	23.6	96	193.0	34.0	56	252.1	44.5
17	16.7	3.0	77	75.8	13.4	37	134.9	23.8	97	194.0	34.2	57	253.1	44.6
18	17.7	3.1	78	76.8	13.5	38	135.9	24.0	98	195.0	34.4	58	254.1	44.8
19	18.7	3.3	79	77.8	13.7	39	136.9	24.1	99	196.0	34.6	59	255.1	45.0
20	19.7	3.5	80	78.8	13.9	40	137.9	24.3	200	197.0	34.7	60	256.1	45.1
21	20.7	3.6	81	79.8	14.1	141	138.9	24.5	201	197.9	34.9	261	257.0	45.3
22	21.7	3.8	82	80.8	14.2	42	139.8	24.7	02	198.9	35.1	62	258.0	45.5
23	22.7	4.0	83	81.7	14.4	43	140.8	24.8	03	199.9	35.3	63	259.0	45.7
24	23.6	4.2	84	82.7	14.6	44	141.8	25.0	04	200.9	35.4	64	260.0	45.8
25	24.6	4.3	85	83.7	14.8	45	142.8	25.2	05	201.9	35.6	65	261.0	46.0
26	25.6	4.5	86	84.7	14.9	46	143.8	25.4	06	202.9	35.8	66	262.0	46.2
27	26.6	4.7	87	85.7	15.1	47	144.8	25.5	07	203.9	35.9	67	262.9	46.4
28	27.6	4.9	88	86.7	15.3	48	145.8	25.7	08	204.8	36.1	68	263.9	46.5
29	28.6	5.0	89	87.6	15.5	49	146.7	25.9	09	205.8	36.3	69	264.9	46.7
30	29.5	5.2	90	88.6	15.6	50	147.7	26.0	10	206.8	36.5	70	265.9	46.9
31	30.5	5.4	91	89.6	15.8	151	148.7	26.2	211	207.8	36.6	271	266.9	47.1
32	31.5	5.6	92	90.6	16.0	52	149.7	26.4	12	208.8	36.8	72	267.9	47.2
33	32.5	5.7	93	91.6	16.1	53	150.7	26.6	13	209.8	37.0	73	268.9	47.4
34	33.5	5.9	94	92.6	16.3	54	151.7	26.7	14	210.7	37.2	74	269.8	47.6
35	34.5	6.1	95	93.6	16.5	55	152.6	26.9	15	211.7	37.3	75	270.8	47.8
36	35.5	6.3	96	94.5	16.7	56	153.6	27.1	16	212.7	37.5	76	271.8	47.9
37	36.4	6.4	97	95.5	16.8	57	154.6	27.3	17	213.7	37.7	77	272.8	48.1
38	37.4	6.6	98	96.5	17.0	58	155.6	27.4	18	214.7	37.9	78	273.8	48.3
39	38.4	6.8	99	97.5	17.2	59	156.6	27.6	19	215.7	38.0	79	274.8	48.4
40	39.4	6.9	100	98.5	17.4	60	157.6	27.8	20	216.7	38.2	80	275.7	48.6
41	40.4	7.1	101	99.5	17.5	161	158.6	28.0	221	217.6	38.4	281	276.7	48.8
42	41.4	7.3	02	100.5	17.7	62	159.5	28.1	22	218.6	38.5	82	277.7	49.0
43	42.3	7.5	03	101.4	17.9	63	160.5	28.3	23	219.6	38.7	83	278.7	49.1
44	43.3	7.6	04	102.4	18.1	64	161.5	28.5	24	220.6	38.9	84	279.7	49.3
45	44.3	7.8	05	103.4	18.2	65	162.5	28.7	25	221.6	39.1	85	280.7	49.5
46	45.3	8.0	06	104.4	18.4	66	163.5	28.8	26	222.6	39.2	86	281.7	49.7
47	46.3	8.2	07	105.4	18.6	67	164.5	29.0	27	223.6	39.4	87	282.6	49.8
48	47.3	8.3	08	106.4	18.8	68	165.4	29.2	28	224.5	39.6	88	283.6	50.0
49	48.3	8.5	09	107.3	18.9	69	166.4	29.3	29	225.5	39.8	89	284.6	50.2
50	49.2	8.7	10	108.3	19.1	70	167.4	29.5	30	226.5	39.9	90	285.6	50.4
51	50.2	8.9	111	109.3	19.3	171	168.4	29.7	231	227.5	40.1	291	286.6	50.5
52	51.2	9.0	12	110.3	19.4	72	169.4	29.9	32	228.5	40.3	92	287.6	50.7
53	52.2	9.2	13	111.3	19.6	73	170.4	30.0	33	229.5	40.5	93	288.5	50.9
54	53.2	9.4	14	112.3	19.8	74	171.4	30.2	34	230.4	40.6	94	289.5	51.1
55	54.2	9.6	15	113.3	20.0	75	172.3	30.4	35	231.4	40.8	95	290.5	51.2
56	55.1	9.7	16	114.2	20.1	76	173.3	30.6	36	232.4	41.0	96	291.5	51.4
57	56.1	9.9	17	115.2	20.3	77	174.3	30.7	37	233.4	41.2	97	292.5	51.6
58	57.1	10.1	18	116.2	20.5	78	175.3	30.9	38	234.4	41.3	98	293.5	51.7
59	58.1	10.2	19	117.2	20.7	79	176.3	31.1	39	235.4	41.5	99	294.5	51.9
60	59.1	10.4	20	118.2	20.8	80	177.3	31.3	40	236.4	41.7	300	295.4	52.1
Dist.	Dep.	D. Lat.	Dist.	Dep.	D. Lat.	Dist.	Dep.	D. Lat.	Dist.	Dep.	D. Lat.	Dist.	Dep.	D. Lat.

280°	080°
260°	100°

80°

Dist.	D. Lat.	Dep.
N.	N×Cos.	N×Sin.
Hypotenuse	Side Adj.	Side Opp.

350°	010°
190°	170°

TABLE 3

Traverse **10°** Table

Dist.	D. Lat.	Dep.	Dist.	D. Lat.	Dep.	Dist.	D. Lat.	Dep.	Dist.	D. Lat.	Dep.	Dist.	D. Lat.	Dep.
301	296.4	52.3	361	355.5	62.7	421	414.6	73.1	481	473.7	83.5	541	532.8	93.9
02	297.4	52.4	62	356.5	62.9	22	415.6	73.3	82	474.7	83.7	42	533.8	94.1
03	298.4	52.6	63	357.5	63.0	23	416.6	73.5	83	475.7	83.9	43	534.8	94.3
04	299.4	52.8	64	358.5	63.2	24	417.6	73.6	84	476.6	84.0	44	535.7	94.5
05	300.4	53.0	65	359.5	63.4	25	418.5	73.8	85	477.6	84.2	45	536.7	94.6
06	301.4	53.1	66	360.4	63.6	26	419.5	74.0	86	478.6	84.4	46	537.7	94.8
07	302.3	53.3	67	361.4	63.7	27	420.5	74.1	87	479.6	84.6	47	538.7	95.0
08	303.3	53.5	68	362.4	63.9	28	421.5	74.3	88	480.6	84.7	48	539.7	95.2
09	304.3	53.7	69	363.4	64.1	29	422.5	74.5	89	481.6	84.9	49	540.7	95.3
10	305.3	53.8	70	364.4	64.2	30	423.5	74.7	90	482.6	85.1	50	541.6	95.5
311	306.3	54.0	371	365.4	64.4	431	424.5	74.8	491	483.5	85.3	551	542.6	95.7
12	307.3	54.2	72	366.4	64.6	32	425.4	75.0	92	484.5	85.4	52	543.6	95.9
13	308.2	54.4	73	367.3	64.8	33	426.4	75.2	93	485.5	85.6	53	544.6	96.0
14	309.2	54.5	74	368.3	64.9	34	427.4	75.4	94	486.5	85.8	54	545.6	96.2
15	310.2	54.7	75	369.3	65.1	35	428.4	75.5	95	487.5	86.0	55	546.6	96.4
16	311.2	54.9	76	370.3	65.3	36	429.4	75.7	96	488.5	86.1	56	547.6	96.5
17	312.2	55.0	77	371.3	65.5	37	430.4	75.9	97	489.4	86.3	57	548.5	96.7
18	313.2	55.2	78	372.3	65.6	38	431.3	76.1	98	490.4	86.5	58	549.5	96.9
19	314.2	55.4	79	373.2	65.8	39	432.3	76.2	99	491.4	86.7	59	550.5	97.1
20	315.1	55.6	80	374.2	66.0	40	433.3	76.4	500	492.4	86.8	60	551.5	97.2
321	316.1	55.7	381	375.2	66.2	441	434.3	76.6	501	493.4	87.0	561	552.5	97.4
22	317.1	55.9	82	376.2	66.3	42	435.3	76.8	02	494.4	87.2	62	553.5	97.6
23	318.1	56.1	83	377.2	66.5	43	436.3	76.9	03	495.4	87.3	63	554.4	97.8
24	319.1	56.3	84	378.2	66.7	44	437.3	77.1	04	496.3	87.5	64	555.4	97.9
25	320.1	56.4	85	379.2	66.9	45	438.2	77.3	05	497.3	87.7	65	556.4	98.1
26	321.0	56.6	86	380.1	67.0	46	439.2	77.4	06	498.3	87.9	66	557.4	98.3
27	322.0	56.8	87	381.1	67.2	47	440.2	77.6	07	499.3	88.0	67	558.4	98.5
28	323.0	57.0	88	382.1	67.4	48	441.2	77.8	08	500.3	88.2	68	559.4	98.6
29	324.0	57.1	89	383.1	67.5	49	442.2	78.0	09	501.3	88.4	69	560.4	98.8
30	325.0	57.3	90	384.1	67.7	50	443.2	78.1	10	502.3	88.6	70	561.3	99.0
331	326.0	57.5	391	385.1	67.9	451	444.1	78.3	511	503.2	88.7	571	562.3	99.2
32	327.0	57.7	92	386.0	68.1	52	445.1	78.5	12	504.2	88.9	72	563.3	99.3
33	327.9	57.8	93	387.0	68.2	53	446.1	78.7	13	505.2	89.1	73	564.3	99.5
34	328.9	58.0	94	388.0	68.4	54	447.1	78.8	14	506.2	89.3	74	565.3	99.7
35	329.9	58.2	95	389.0	68.6	55	448.1	79.0	15	507.2	89.4	75	566.3	99.8
36	330.9	58.3	96	390.0	68.8	56	449.1	79.2	16	508.2	89.6	76	567.2	100.0
37	331.9	58.5	97	391.0	68.9	57	450.1	79.4	17	509.1	89.8	77	568.2	100.2
38	332.9	58.7	98	392.0	69.1	58	451.0	79.5	18	510.1	89.9	78	569.2	100.4
39	333.9	58.9	99	392.9	69.3	59	452.0	79.7	19	511.1	90.1	79	570.2	100.5
40	334.8	59.0	400	393.9	69.5	60	453.0	79.9	20	512.1	90.3	80	571.2	100.7
341	335.8	59.2	401	394.9	69.6	461	454.0	80.1	521	513.1	90.5	581	572.2	100.9
42	336.8	59.4	02	395.9	69.8	62	455.0	80.2	22	514.1	90.6	82	573.2	101.1
43	337.8	59.6	03	396.9	70.0	63	456.0	80.4	23	515.1	90.8	83	574.1	101.2
44	338.8	59.7	04	397.9	70.2	64	457.0	80.6	24	516.0	91.0	84	575.1	101.4
45	339.8	59.9	05	398.9	70.3	65	457.9	80.7	25	517.0	91.2	85	576.1	101.6
46	340.7	60.1	06	399.8	70.5	66	458.9	80.9	26	518.0	91.3	86	577.1	101.8
47	341.7	60.3	07	400.8	70.7	67	459.9	81.1	27	519.0	91.5	87	578.1	101.9
48	342.7	60.4	08	401.8	70.8	68	460.9	81.3	28	520.0	91.7	88	579.1	102.1
49	343.7	60.6	09	402.8	71.0	69	461.9	81.4	29	521.0	91.9	89	580.1	102.3
50	344.7	60.8	10	403.8	71.2	70	462.9	81.6	30	521.9	92.0	90	581.0	102.5
351	345.7	61.0	411	404.8	71.4	471	463.8	81.8	531	522.9	92.2	591	582.0	102.6
52	346.7	61.1	12	405.7	71.5	72	464.8	82.0	32	523.9	92.4	92	583.0	102.8
53	347.6	61.3	13	406.7	71.7	73	465.8	82.1	33	524.9	92.6	93	584.0	103.0
54	348.6	61.5	14	407.7	71.9	74	466.8	82.3	34	525.9	92.7	94	585.0	103.1
55	349.6	61.6	15	408.7	72.1	75	467.8	82.5	35	526.9	92.9	95	586.0	103.3
56	350.6	61.8	16	409.7	72.2	76	468.8	82.7	36	527.9	93.1	96	586.9	103.5
57	351.6	62.0	17	410.7	72.4	77	469.8	82.8	37	528.8	93.2	97	587.9	103.7
58	352.6	62.2	18	411.6	72.6	78	470.7	83.0	38	529.8	93.4	98	588.9	103.8
59	353.5	62.3	19	412.6	72.8	79	471.7	83.2	39	530.8	93.6	99	589.9	104.0
60	354.5	62.5	20	413.6	72.9	80	472.7	83.4	40	531.8	93.8	600	590.9	104.2
Dist.	Dep.	D. Lat.	Dist.	Dep.	D. Lat.	Dist.	Dep.	D. Lat.	Dist.	Dep.	D. Lat.	Dist.	Dep.	D. Lat.

Dist.	D. Lat.	Dep.
D Lo	Dep.	
	m	D Lo

80°

280°	080°
260°	100°

349°	011°
191°	169°

TABLE 3

Traverse 11° Table

Dist.	D. Lat.	Dep.	Dist.	D. Lat.	Dep.	Dist.	D. Lat.	Dep.	Dist.	D. Lat.	Dep.	Dist.	D. Lat.	Dep.
1	1.0	0.2	61	59.9	11.6	121	118.8	23.1	181	177.7	34.5	241	236.6	46.0
2	2.0	0.4	62	60.9	11.8	22	119.8	23.3	82	178.7	34.7	42	237.6	46.2
3	2.9	0.6	63	61.8	12.0	23	120.7	23.5	83	179.6	34.9	43	238.5	46.4
4	3.9	0.8	64	62.8	12.2	24	121.7	23.7	84	180.6	35.1	44	239.5	46.6
5	4.9	1.0	65	63.8	12.4	25	122.7	23.9	85	181.6	35.3	45	240.5	46.7
6	5.9	1.1	66	64.8	12.6	26	123.7	24.0	86	182.6	35.5	46	241.5	46.9
7	6.9	1.3	67	65.8	12.8	27	124.7	24.2	87	183.6	35.7	47	242.5	47.1
8	7.9	1.5	68	66.8	13.0	28	125.6	24.4	88	184.5	35.9	48	243.4	47.3
9	8.8	1.7	69	67.7	13.2	29	126.6	24.6	89	185.5	36.1	49	244.4	47.5
10	9.8	1.9	70	68.7	13.4	30	127.6	24.8	90	186.5	36.3	50	245.4	47.7
11	10.8	2.1	71	69.7	13.5	131	128.6	25.0	191	187.5	36.4	251	246.4	47.9
12	11.8	2.3	72	70.7	13.7	32	129.6	25.2	92	188.5	36.6	52	247.4	48.1
13	12.8	2.5	73	71.7	13.9	33	130.6	25.4	93	189.5	36.8	53	248.4	48.3
14	13.7	2.7	74	72.6	14.1	34	131.5	25.6	94	190.4	37.0	54	249.3	48.5
15	14.7	2.9	75	73.6	14.3	35	132.5	25.8	95	191.4	37.2	55	250.3	48.7
16	15.7	3.1	76	74.6	14.5	36	133.5	26.0	96	192.4	37.4	56	251.3	48.8
17	16.7	3.2	77	75.6	14.7	37	134.5	26.1	97	193.4	37.6	57	252.3	49.0
18	17.7	3.4	78	76.6	14.9	38	135.5	26.3	98	194.4	37.8	58	253.3	49.2
19	18.7	3.6	79	77.5	15.1	39	136.4	26.5	99	195.3	38.0	59	254.2	49.4
20	19.6	3.8	80	78.5	15.3	40	137.4	26.7	200	196.3	38.2	60	255.2	49.6
21	20.6	4.0	81	79.5	15.5	141	138.4	26.9	201	197.3	38.4	261	256.2	49.8
22	21.6	4.2	82	80.5	15.6	42	139.4	27.1	02	198.3	38.5	62	257.2	50.0
23	22.6	4.4	83	81.5	15.8	43	140.4	27.3	03	199.3	38.7	63	258.2	50.2
24	23.6	4.6	84	82.5	16.0	44	141.4	27.5	04	200.3	38.9	64	259.1	50.4
25	24.5	4.8	85	83.4	16.2	45	142.3	27.7	05	201.2	39.1	65	260.1	50.6
26	25.5	5.0	86	84.4	16.4	46	143.3	27.9	06	202.2	39.3	66	261.1	50.8
27	26.5	5.2	87	85.4	16.6	47	144.3	28.0	07	203.2	39.5	67	262.1	50.9
28	27.5	5.3	88	86.4	16.8	48	145.3	28.2	08	204.2	39.7	68	263.1	51.1
29	28.5	5.5	89	87.4	17.0	49	146.3	28.4	09	205.2	39.9	69	264.1	51.3
30	29.4	5.7	90	88.3	17.2	50	147.2	28.6	10	206.1	40.1	70	265.0	51.5
31	30.4	5.9	91	89.3	17.4	151	148.2	28.8	211	207.1	40.3	271	266.0	51.7
32	31.4	6.1	92	90.3	17.6	52	149.2	29.0	12	208.1	40.5	72	267.0	51.9
33	32.4	6.3	93	91.3	17.7	53	150.2	29.2	13	209.1	40.6	73	268.0	52.1
34	33.4	6.5	94	92.3	17.9	54	151.2	29.4	14	210.1	40.8	74	269.0	52.3
35	34.4	6.7	95	93.3	18.1	55	152.2	29.6	15	211.0	41.0	75	269.9	52.5
36	35.3	6.9	96	94.2	18.3	56	153.1	29.8	16	212.0	41.2	76	270.9	52.7
37	36.3	7.1	97	95.2	18.5	57	154.1	30.0	17	213.0	41.4	77	271.9	52.9
38	37.3	7.3	98	96.2	18.7	58	155.1	30.1	18	214.0	41.6	78	272.9	53.0
39	38.3	7.4	99	97.2	18.9	59	156.1	30.3	19	215.0	41.8	79	273.9	53.2
40	39.3	7.6	100	98.2	19.1	60	157.1	30.5	20	216.0	42.0	80	274.9	53.4
41	40.2	7.8	101	99.1	19.3	161	158.0	30.7	221	216.9	42.2	281	275.8	53.6
42	41.2	8.0	02	100.1	19.5	62	159.0	30.9	22	217.9	42.4	82	276.8	53.8
43	42.2	8.2	03	101.1	19.7	63	160.0	31.1	23	218.9	42.6	83	277.8	54.0
44	43.2	8.4	04	102.1	19.8	64	161.0	31.3	24	219.9	42.7	84	278.8	54.2
45	44.2	8.6	05	103.1	20.0	65	162.0	31.5	25	220.9	42.9	85	279.8	54.4
46	45.2	8.8	06	104.1	20.2	66	163.0	31.7	26	221.8	43.1	86	280.7	54.6
47	46.1	9.0	07	105.0	20.4	67	163.9	31.9	27	222.8	43.3	87	281.7	54.8
48	47.1	9.2	08	106.0	20.6	68	164.9	32.1	28	223.8	43.5	88	282.7	55.0
49	48.1	9.3	09	107.0	20.8	69	165.9	32.2	29	224.8	43.7	89	283.7	55.1
50	49.1	9.5	10	108.0	21.0	70	166.9	32.4	30	225.8	43.9	90	284.7	55.3
51	50.1	9.7	111	109.0	21.2	171	167.9	32.6	231	226.8	44.1	291	285.7	55.5
52	51.0	9.9	12	109.9	21.4	72	168.8	32.8	32	227.7	44.3	92	286.6	55.7
53	52.0	10.1	13	110.9	21.6	73	169.8	33.0	33	228.7	44.5	93	287.6	55.9
54	53.0	10.3	14	111.9	21.8	74	170.8	33.2	34	229.7	44.6	94	288.6	56.1
55	54.0	10.5	15	112.9	21.9	75	171.8	33.4	35	230.7	44.8	95	289.6	56.3
56	55.0	10.7	16	113.9	22.1	76	172.8	33.6	36	231.7	45.0	96	290.6	56.5
57	56.0	10.9	17	114.9	22.3	77	173.7	33.8	37	232.6	45.2	97	291.5	56.7
58	56.9	11.1	18	115.8	22.5	78	174.7	34.0	38	233.6	45.4	98	292.5	56.9
59	57.9	11.3	19	116.8	22.7	79	175.7	34.2	39	234.6	45.6	99	293.5	57.1
60	58.9	11.4	20	117.8	22.9	80	176.7	34.3	40	235.6	45.8	300	294.5	57.2
Dist.	Dep.	D. Lat.	Dist.	Dep.	D. Lat.	Dist.	Dep.	D. Lat.	Dist.	Dep.	D. Lat.	Dist.	Dep.	D. Lat.

281°	079°
259°	101°

79°

Dist.	D. Lat.	Dep.
N.	N×Cos.	N×Sin.
Hypotenuse	Side Adj.	Side Opp.

349°	011°
191°	169°

TABLE 3

Traverse 11° Table

349°	011°
191°	169°

Dist.	D. Lat.	Dep.	Dist.	D. Lat.	Dep.	Dist.	D. Lat.	Dep.	Dist.	D. Lat.	Dep.	Dist.	D. Lat.	Dep.
301	295.5	57.4	361	354.4	68.9	421	413.3	80.3	481	472.2	91.8	541	531.1	103.2
02	296.5	57.6	62	355.3	69.1	22	414.2	80.5	82	473.1	92.0	42	532.0	103.4
03	297.4	57.8	63	356.3	69.3	23	415.2	80.7	83	474.1	92.2	43	533.0	103.6
04	298.4	58.0	64	357.3	69.5	24	416.2	80.9	84	475.1	92.4	44	534.0	103.8
05	299.4	58.2	65	358.3	69.6	25	417.2	81.1	85	476.1	92.5	45	535.0	104.0
06	300.4	58.4	66	359.3	69.8	26	418.2	81.3	86	477.1	92.7	46	536.0	104.2
07	301.4	58.6	67	360.3	70.0	27	419.2	81.5	87	478.1	92.9	47	537.0	104.4
08	302.3	58.8	68	361.2	70.2	28	420.1	81.7	88	479.0	93.1	48	537.9	104.6
09	303.3	59.0	69	362.2	70.4	29	421.1	81.9	89	480.0	93.3	49	538.9	104.8
10	304.3	59.2	70	363.2	70.6	30	422.1	82.1	90	481.0	93.5	50	539.9	104.9
311	305.3	59.3	371	364.2	70.8	431	423.0	82.2	491	482.0	93.7	551	540.9	105.1
12	306.3	59.5	72	365.2	71.0	32	424.1	82.4	92	483.0	93.9	52	541.9	105.3
13	307.2	59.7	73	366.1	71.2	33	425.0	82.6	93	483.9	94.1	53	542.8	105.5
14	308.2	59.9	74	367.1	71.4	34	426.0	82.8	94	484.9	94.3	54	543.8	105.7
15	309.2	60.1	75	368.1	71.6	35	427.0	83.0	95	485.9	94.5	55	544.8	105.9
16	310.2	60.3	76	369.1	71.7	36	428.0	83.2	96	486.9	94.6	56	545.8	106.1
17	311.2	60.5	77	370.1	71.9	37	428.9	83.4	97	487.9	94.8	57	546.8	106.3
18	312.2	60.7	78	371.1	72.1	38	430.0	83.6	98	488.9	95.0	58	547.7	106.5
19	313.1	60.9	79	372.0	72.3	39	430.9	83.8	99	489.8	95.2	59	548.7	106.7
20	314.1	61.1	80	373.0	72.5	40	431.9	84.0	500	490.8	95.4	60	549.7	106.9
321	315.1	61.2	381	374.0	72.7	441	432.9	84.1	501	491.8	95.6	561	550.7	107.0
22	316.1	61.4	82	375.0	72.9	42	433.9	84.3	02	492.8	95.8	62	551.7	107.2
23	317.1	61.6	83	376.0	73.1	43	434.9	84.5	03	493.8	96.0	63	552.7	107.4
24	318.0	61.8	84	376.9	73.3	44	435.8	84.7	04	494.7	96.2	64	553.6	107.6
25	319.0	62.0	85	377.9	73.5	45	436.8	84.9	05	495.7	96.4	65	554.6	107.8
26	320.0	62.2	86	378.9	73.7	46	437.8	85.1	06	496.7	96.5	66	555.6	108.0
27	321.0	62.4	87	379.9	73.8	47	438.8	85.3	07	497.7	96.7	67	556.6	108.2
28	322.0	62.6	88	380.8	74.0	48	439.8	85.5	08	498.7	96.9	68	557.6	108.4
29	323.0	62.8	89	381.9	74.2	49	440.8	85.7	09	499.6	97.1	69	558.5	108.6
30	323.9	63.0	90	382.8	74.4	50	441.7	85.9	10	500.6	97.3	70	559.5	108.8
331	324.9	63.2	391	383.8	74.6	451	442.7	86.1	511	501.6	97.5	571	560.5	109.0
32	325.9	63.3	92	384.8	74.8	52	443.7	86.2	12	502.6	97.7	72	561.5	109.1
33	326.9	63.5	93	385.8	75.0	53	444.7	86.4	13	503.6	97.9	73	562.5	109.3
34	327.9	63.7	94	386.8	75.2	54	445.7	86.6	14	504.6	98.1	74	563.5	109.5
35	328.8	63.9	95	387.7	75.4	55	446.6	86.8	15	505.5	98.3	75	564.4	109.7
36	329.8	64.1	96	388.7	75.6	56	447.6	87.0	16	506.5	98.5	76	565.4	109.9
37	330.8	64.3	97	389.7	75.8	57	448.6	87.2	17	507.5	98.6	77	566.4	110.1
38	331.8	64.5	98	390.7	75.9	58	449.6	87.4	18	508.5	98.8	78	567.4	110.3
39	332.7	64.7	99	391.7	76.1	59	450.6	87.6	19	509.5	99.0	79	568.4	110.5
40	333.8	64.9	400	392.7	76.3	60	451.5	87.8	20	510.4	99.2	80	569.3	110.7
341	334.7	65.1	401	393.6	76.5	461	452.5	88.0	521	511.4	99.4	581	570.3	110.9
42	335.7	65.3	02	394.6	76.7	62	453.5	88.2	22	512.4	99.6	82	571.3	111.1
43	336.7	65.4	03	395.6	76.9	63	454.5	88.3	23	513.4	99.8	83	572.3	111.2
44	337.7	65.6	04	396.6	77.1	64	455.4	88.5	24	514.4	100.0	84	573.3	111.4
45	338.7	65.8	05	397.6	77.3	65	456.5	88.7	25	515.4	100.2	85	574.3	111.6
46	339.6	66.0	06	398.5	77.5	66	457.4	88.9	26	516.3	100.4	86	575.2	111.8
47	340.6	66.2	07	399.5	77.7	67	458.4	89.1	27	517.3	100.6	87	576.2	112.1
48	341.6	66.4	08	400.5	77.9	68	459.4	89.3	28	518.3	100.7	88	577.2	112.3
49	342.6	66.6	09	401.5	78.1	69	460.4	89.5	29	519.3	100.9	89	578.2	112.4
50	343.6	66.8	10	402.5	78.2	70	461.4	89.7	30	520.3	101.1	90	579.2	112.6
351	344.6	67.0	411	403.4	78.4	471	462.3	89.9	531	521.2	101.3	591	580.1	112.8
52	345.5	67.2	12	404.4	78.6	72	463.3	90.1	32	522.2	101.5	92	581.1	113.0
53	346.5	67.4	13	405.4	78.8	73	464.3	90.3	33	523.2	101.7	93	582.1	113.2
54	347.5	67.5	14	406.4	79.0	74	465.3	90.4	34	524.2	101.9	94	583.1	113.3
55	348.5	67.7	15	407.4	79.2	75	466.3	90.6	35	525.2	102.1	95	584.1	113.5
56	349.5	67.9	16	408.4	79.4	76	467.3	90.8	36	526.2	102.3	96	585.0	113.7
57	350.4	68.1	17	409.3	79.6	77	468.2	91.0	37	527.1	102.5	97	586.0	113.9
58	351.4	68.3	18	410.3	79.8	78	469.2	91.2	38	528.1	102.7	98	587.0	114.1
59	352.4	68.5	19	411.3	79.9	79	470.2	91.4	39	529.1	102.8	99	588.0	114.3
60	353.4	68.7	20	412.3	80.1	80	471.2	91.6	40	530.1	103.0	600	589.0	114.5
Dist.	Dep.	D. Lat.	Dist.	Dep.	D. Lat.	Dist.	Dep.	D. Lat.	Dist.	Dep.	D. Lat.	Dist.	Dep.	D. Lat.

Dist.	D. Lat.	Dep.
D Lo	Dep.	
	m	D Lo

79°

281°	079°
259°	101°

348° | 012° / 192° | 168°

TABLE 3

Traverse **12°** Table

348° | 012° / 192° | 168°

Dist.	D. Lat.	Dep.	Dist.	D. Lat.	Dep.	Dist.	D. Lat.	Dep.	Dist.	D. Lat.	Dep.	Dist.	D. Lat.	Dep.
1	1.0	0.2	61	59.7	12.7	121	118.4	25.2	181	177.0	37.6	241	235.7	50.1
2	2.0	0.4	62	60.6	12.9	22	119.3	25.4	82	178.0	37.8	42	236.7	50.3
3	2.9	0.6	63	61.6	13.1	23	120.3	25.6	83	179.0	38.0	43	237.7	50.5
4	3.9	0.8	64	62.6	13.3	24	121.3	25.8	84	180.0	38.3	44	238.7	50.7
5	4.9	1.0	65	63.6	13.5	25	122.3	26.0	85	181.0	38.5	45	239.6	50.9
6	5.9	1.2	66	64.6	13.7	23	123.2	26.2	86	181.9	38.7	46	240.6	51.1
7	6.8	1.5	67	65.5	13.9	27	124.2	26.4	87	182.9	38.9	47	241.6	51.4
8	7.8	1.7	68	66.5	14.1	28	125.2	26.6	88	183.9	39.1	48	242.6	51.6
9	8.8	1.9	69	67.5	14.3	29	126.2	26.8	89	184.9	39.3	49	243.6	51.8
10	9.8	2.1	70	68.5	14.6	30	127.2	27.0	90	185.8	39.5	50	244.5	52.0
11	10.8	2.3	71	69.4	14.8	131	128.1	27.2	191	186.8	39.7	251	245.5	52.2
12	11.7	2.5	72	70.4	15.0	32	129.1	27.4	92	187.8	39.9	52	246.5	52.4
13	12.7	2.7	73	71.4	15.2	33	130.1	27.7	93	188.8	40.1	53	247.5	52.6
14	13.7	2.9	74	72.4	15.4	34	131.1	27.9	94	189.8	40.3	54	248.4	52.8
15	14.7	3.1	75	73.4	15.6	35	132.0	28.1	95	190.7	40.5	55	249.4	53.0
16	15.7	3.3	76	74.3	15.8	36	133.0	28.3	96	191.7	40.8	56	250.4	53.2
17	16.6	3.5	77	75.3	16.0	37	134.0	28.5	97	192.7	41.0	57	251.4	53.4
18	17.6	3.7	78	76.3	16.2	38	135.0	28.7	98	193.7	41.2	58	252.4	53.6
19	18.6	4.0	79	77.3	16.4	39	136.0	28.9	99	194.7	41.4	59	253.3	53.8
20	19.6	4.2	80	78.3	16.6	40	136.9	29.1	200	195.6	41.6	60	254.3	54.1
21	20.5	4.4	81	79.2	16.8	141	137.9	29.3	201	196.6	41.8	261	255.3	54.3
22	21.5	4.6	82	80.2	17.0	42	138.9	29.5	02	197.6	42.0	62	256.3	54.5
23	22.5	4.8	83	81.2	17.3	43	139.9	29.7	03	198.6	42.2	63	257.3	54.7
24	23.5	5.0	84	82.2	17.5	44	140.9	29.9	04	199.5	42.4	64	258.2	54.9
25	24.5	5.2	85	83.1	17.7	45	141.8	30.1	05	200.5	42.6	65	259.2	55.1
26	25.4	5.4	86	84.1	17.9	46	142.8	30.4	06	201.5	42.8	66	260.2	55.3
27	26.4	5.6	87	85.1	18.1	47	143.8	30.6	07	202.5	43.0	67	261.2	55.5
28	27.4	5.8	88	86.1	18.3	48	144.8	30.8	08	203.5	43.2	68	262.1	55.7
29	28.4	6.0	89	87.1	18.5	49	145.7	31.0	09	204.4	43.5	69	263.1	55.9
30	29.3	6.2	90	88.0	18.7	50	146.7	31.2	10	205.4	43.7	70	264.1	56.1
31	30.3	6.4	91	89.0	18.9	151	147.7	31.4	211	206.4	43.9	271	265.1	56.3
32	31.3	6.7	92	90.0	19.1	52	148.7	31.6	12	207.4	44.1	72	266.1	56.6
33	32.3	6.9	93	91.0	19.3	53	149.7	31.8	13	208.3	44.3	73	267.0	56.8
34	33.3	7.1	94	91.9	19.5	54	150.6	32.0	14	209.3	44.5	74	268.0	57.0
35	34.2	7.3	95	92.9	19.8	55	151.6	32.2	15	210.3	44.7	75	269.0	57.2
36	35.2	7.5	96	93.9	20.0	56	152.6	32.4	16	211.3	44.9	76	270.0	57.4
37	36.2	7.7	97	94.9	20.2	57	153.6	32.6	17	212.3	45.1	77	270.9	57.6
38	37.2	7.9	98	95.9	20.4	58	154.5	32.9	18	213.2	45.3	78	271.9	57.8
39	38.1	8.1	99	96.8	20.6	59	155.5	33.1	19	214.2	45.5	79	272.9	58.0
40	39.1	8.3	100	97.8	20.8	60	156.5	33.3	20	215.2	45.7	80	273.9	58.2
41	40.1	8.5	101	98.8	21.0	161	157.5	33.5	221	216.2	45.9	281	274.9	58.4
42	41.1	8.7	02	99.8	21.2	62	158.5	33.7	22	217.1	46.2	82	275.8	58.6
43	42.1	8.9	03	100.7	21.4	63	159.4	33.9	23	218.1	46.4	83	276.8	58.8
44	43.0	9.1	04	101.7	21.6	64	160.4	34.1	24	219.1	46.6	84	277.8	59.0
45	44.0	9.4	05	102.7	21.8	65	161.4	34.3	25	220.1	46.8	85	278.8	59.3
46	45.0	9.6	06	103.7	22.0	66	162.4	34.5	26	221.1	47.0	86	279.8	59.5
47	46.0	9.8	07	104.7	22.2	67	163.4	34.7	27	222.0	47.2	87	280.7	59.7
48	47.0	10.0	08	105.7	22.5	68	164.3	34.9	28	223.0	47.4	88	281.7	59.9
49	47.9	10.2	09	106.6	22.7	69	165.3	35.1	29	224.0	47.6	89	282.7	60.1
50	48.9	10.4	10	107.6	22.9	70	166.3	35.3	30	225.0	47.8	90	283.7	60.3
51	49.9	10.6	111	108.6	23.1	171	167.3	35.6	231	226.0	48.0	291	284.6	60.5
52	50.9	10.8	12	109.6	23.3	72	168.2	35.8	32	226.9	48.2	92	285.6	60.7
53	51.8	11.0	13	110.5	23.5	73	169.2	36.0	33	227.9	48.4	93	286.6	60.9
54	52.8	11.2	14	111.5	23.7	74	170.2	36.2	34	228.9	48.7	94	287.6	61.1
55	53.8	11.4	15	112.5	23.9	75	171.2	36.4	35	229.9	48.9	95	288.6	61.3
56	54.8	11.6	16	113.5	24.1	76	172.2	36.6	36	230.8	49.1	96	289.5	61.5
57	55.8	11.9	17	114.4	24.3	77	173.1	36.8	37	231.8	49.3	97	290.5	61.7
58	56.7	12.1	18	115.4	24.5	78	174.1	37.0	38	232.8	49.5	98	291.5	62.0
59	57.7	12.3	19	116.4	24.7	79	175.1	37.2	39	233.8	49.7	99	292.5	62.2
60	58.7	12.5	20	117.4	24.9	80	176.1	37.4	40	234.8	49.9	300	293.4	62.4
Dist.	Dep.	D. Lat.	Dist.	Dep.	D. Lat.	Dist.	Dep.	D. Lat.	Dist.	Dep.	D. Lat.	Dist.	Dep.	D. Lat.

282° | 078° / 258° | 102°

78°

Dist.	D. Lat.	Dep.
N.	N×Cos.	N×Sin.
Hypotenuse	Side Adj.	Side Opp.

348° | 012° / 192° | 168°					TABLE 3 Traverse 12° Table							348° | 012° / 192° | 168°		
Dist.	D. Lat.	Dep.	Dist.	D. Lat.	Dep.	Dist.	D. Lat.	Dep.	Dist.	D. Lat.	Dep.	Dist.	D. Lat.	Dep.
301	294.4	62.6	361	353.1	75.0	421	411.8	87.5	481	470.5	100.0	541	529.2	112.5
02	295.4	62.8	62	354.1	75.2	22	412.8	87.7	82	471.5	100.2	42	530.2	112.7
03	296.4	63.0	63	355.1	75.4	23	413.8	87.9	83	472.4	100.4	43	531.1	112.9
04	297.4	63.2	64	356.0	75.7	24	414.7	88.1	84	473.4	100.6	44	532.1	113.1
05	298.3	63.4	65	357.0	75.9	25	415.7	88.3	85	474.4	100.8	45	533.1	113.3
06	299.3	63.6	66	358.0	76.1	26	416.7	88.6	86	475.4	101.0	46	534.1	113.5
07	300.3	63.8	67	359.0	76.3	27	417.7	88.8	87	476.4	101.3	47	535.0	113.7
08	301.3	64.0	68	360.0	76.5	28	418.6	89.0	88	477.3	101.5	48	536.0	113.9
09	302.2	64.2	69	360.9	76.7	29	419.6	89.2	89	478.3	101.7	49	537.0	114.1
10	303.2	64.5	70	361.9	76.9	30	420.6	89.4	90	479.3	101.9	50	538.0	114.4
311	304.2	64.7	371	362.9	77.1	431	421.6	89.6	491	480.3	102.1	551	539.0	114.6
12	305.2	64.9	72	363.9	77.3	32	422.6	89.8	92	481.2	102.3	52	539.9	114.8
13	306.2	65.1	73	364.8	77.6	33	423.5	90.0	93	482.2	102.5	53	540.9	115.0
14	307.1	65.3	74	365.8	77.8	34	424.5	90.2	94	483.2	102.7	54	541.9	115.2
15	308.1	65.5	75	366.8	78.0	35	425.5	90.4	95	484.2	102.9	55	542.9	115.4
16	309.1	65.7	76	367.8	78.2	36	426.5	90.6	96	485.2	103.1	56	543.9	115.6
17	310.1	65.9	77	368.8	78.4	37	427.5	90.9	97	486.1	103.3	57	544.8	115.8
18	311.1	66.1	78	369.7	78.6	38	428.4	91.1	98	487.1	103.5	58	545.8	116.0
19	312.0	66.3	79	370.7	78.8	39	429.4	91.3	99	488.1	103.7	59	546.8	116.2
20	313.0	66.5	80	371.7	79.0	40	430.4	91.5	500	489.1	104.0	60	547.8	116.4
321	314.0	66.7	381	372.7	79.2	441	431.4	91.7	501	490.1	104.2	561	548.7	116.6
22	315.0	66.9	82	373.7	79.4	42	432.3	91.9	02	491.0	104.4	62	549.7	116.8
23	315.9	67.2	83	374.6	79.6	43	433.3	92.1	03	492.0	104.6	63	550.7	117.1
24	316.9	67.4	84	375.6	79.8	44	434.3	92.3	04	493.0	104.8	64	551.7	117.3
25	317.9	67.6	85	376.6	80.0	45	435.3	92.5	05	494.0	105.0	65	552.7	117.5
26	318.9	67.8	86	377.6	80.3	46	436.3	92.7	06	494.9	105.2	66	553.6	117.7
27	319.9	68.0	87	378.5	80.5	47	437.2	92.9	07	495.9	105.4	67	554.6	117.9
28	320.8	68.2	88	379.5	80.7	48	438.2	93.1	08	496.9	105.6	68	555.6	118.1
29	321.8	68.4	89	380.5	80.9	49	439.2	93.4	09	497.9	105.8	69	556.6	118.3
30	322.8	68.6	90	381.5	81.1	50	440.2	93.6	10	498.9	106.0	70	557.5	118.5
331	323.8	68.8	391	382.5	81.3	451	441.1	93.8	511	499.8	106.2	571	558.5	118.7
32	324.7	69.0	92	383.4	81.5	52	442.1	94.0	12	500.8	106.5	72	559.5	118.9
33	325.7	69.2	93	384.4	81.7	53	443.1	94.2	13	501.8	106.7	73	560.5	119.1
34	326.7	69.4	94	385.4	81.9	54	444.1	94.4	14	502.8	106.9	74	561.5	119.3
35	327.7	69.7	95	386.4	82.1	55	445.1	94.6	15	503.7	107.1	75	562.4	119.5
36	328.7	69.9	96	387.3	82.3	56	446.0	94.8	16	504.7	107.3	76	563.4	119.8
37	329.6	70.1	97	388.3	82.5	57	447.0	95.0	17	505.7	107.5	77	564.4	120.0
38	330.6	70.3	98	389.3	82.7	58	448.0	95.2	18	506.7	107.7	78	565.4	120.2
39	331.6	70.5	99	390.3	83.0	59	449.0	95.4	19	507.7	107.9	79	566.3	120.4
40	332.6	70.7	400	391.3	83.2	60	449.9	95.6	20	508.7	108.1	80	567.3	120.6
341	333.5	70.9	401	392.2	83.4	461	450.9	95.8	521	509.6	108.3	581	568.3	120.8
42	334.5	71.1	02	393.2	83.6	62	451.9	96.1	22	510.6	108.5	82	569.3	121.0
43	335.5	71.3	03	394.2	83.8	63	452.9	96.3	23	511.6	108.7	83	570.3	121.2
44	336.5	71.5	04	395.2	84.0	64	453.9	96.5	24	512.5	108.9	84	571.2	121.4
45	337.5	71.7	05	396.2	84.2	65	454.8	96.7	25	513.5	109.2	85	572.2	121.6
46	338.4	71.9	06	397.1	84.4	66	455.8	96.9	26	514.5	109.4	86	573.2	121.8
47	339.4	72.1	07	398.1	84.6	67	456.8	97.1	27	515.5	109.6	87	574.2	122.0
48	340.4	72.4	08	399.1	84.8	68	457.8	97.3	28	516.5	109.8	88	575.2	122.3
49	341.4	72.6	09	400.1	85.0	69	458.8	97.5	29	517.4	110.0	89	576.1	122.5
50	342.4	72.8	10	401.0	85.2	70	459.7	97.7	30	518.4	110.2	90	577.1	122.7
351	343.3	73.0	411	402.0	85.5	471	460.7	97.9	531	519.4	110.4	591	578.1	122.9
52	344.3	73.2	12	403.0	85.7	72	461.7	98.1	32	520.4	110.6	92	579.1	123.1
53	345.3	73.4	13	404.0	85.9	73	462.7	98.3	33	521.4	110.8	93	580.0	123.3
54	346.3	73.6	14	405.0	86.1	74	463.6	98.6	34	522.3	111.0	94	581.0	123.5
55	347.2	73.8	15	405.9	86.3	75	464.6	98.8	35	523.3	111.2	95	582.0	123.7
56	348.2	74.0	16	406.9	86.5	76	465.6	99.0	36	524.3	111.4	96	583.0	123.9
57	349.2	74.2	17	407.9	86.7	77	466.6	99.2	37	525.3	111.6	97	584.0	124.1
58	350.2	74.4	18	408.9	86.9	78	467.6	99.4	38	526.2	111.9	98	584.9	124.3
59	351.2	74.6	19	409.8	87.1	79	468.5	99.6	39	527.2	112.1	99	585.9	124.5
60	352.1	74.8	20	410.8	87.3	80	469.5	99.8	40	528.2	112.3	600	586.9	124.7
Dist.	Dep.	D. Lat.	Dist.	Dep.	D. Lat.	Dist.	Dep.	D. Lat.	Dist.	Dep.	D. Lat.	Dist.	Dep.	D. Lat.

Dist.	D. Lat.	Dep.
D Lo	Dep.	
	m	D Lo

78°

282° | 078°
258° | 102°

347° | 013°
193° | 167°

TABLE 3

Traverse **13°** Table

347° | 013°
193° | 167°

Dist.	D. Lat.	Dep.	Dist.	D. Lat.	Dep.	Dist.	D. Lat.	Dep.	Dist.	D. Lat.	Dep.	Dist.	D. Lat.	Dep.
1	1.0	0.2	61	59.4	13.7	121	117.9	27.2	181	176.4	40.7	241	234.8	54.2
2	1.9	0.4	62	60.4	13.9	22	118.9	27.4	82	177.3	40.9	42	235.8	54.4
3	2.9	0.7	63	61.4	14.2	23	119.8	27.7	83	178.3	41.2	43	236.8	54.7
4	3.9	0.9	64	62.4	14.4	24	120.8	27.9	84	179.3	41.4	44	237.7	54.9
5	4.9	1.1	65	63.3	14.6	25	121.8	28.1	85	180.3	41.6	45	238.7	55.1
6	5.8	1.3	66	64.3	14.8	26	122.8	28.3	86	181.2	41.8	46	239.7	55.3
7	6.8	1.6	67	65.3	15.1	27	123.7	28.6	87	182.2	42.1	47	240.7	55.6
8	7.8	1.8	68	66.3	15.3	28	124.7	28.8	88	183.2	42.3	48	241.6	55.8
9	8.8	2.0	69	67.2	15.5	29	125.7	29.0	89	184.2	42.5	49	242.6	56.0
10	9.7	2.2	70	68.2	15.7	30	126.7	29.2	90	185.1	42.7	50	243.6	56.2
11	10.7	2.5	71	69.2	16.0	131	127.6	29.5	191	186.1	43.0	251	244.6	56.5
12	11.7	2.7	72	70.2	16.2	32	128.6	29.7	92	187.1	43.2	52	245.5	56.7
13	12.7	2.9	73	71.1	16.4	33	129.6	29.9	93	188.1	43.4	53	246.5	56.9
14	13.6	3.1	74	72.1	16.6	34	130.6	30.1	94	189.0	43.6	54	247.5	57.1
15	14.6	3.4	75	73.1	16.9	35	131.5	30.4	95	190.0	43.9	55	248.5	57.4
16	15.6	3.6	76	74.1	17.1	36	132.5	30.6	96	191.0	44.1	56	249.4	57.6
17	16.6	3.8	77	75.0	17.3	37	133.5	30.8	97	192.0	44.3	57	250.4	57.8
18	17.5	4.0	78	76.0	17.5	38	134.5	31.0	98	192.9	44.5	58	251.4	58.0
19	18.5	4.3	79	77.0	17.8	39	135.4	31.3	99	193.9	44.8	59	252.4	58.3
20	19.5	4.5	80	77.9	18.0	40	136.4	31.5	200	194.9	45.0	60	253.3	58.5
21	20.5	4.7	81	78.9	18.2	141	137.4	31.7	201	195.8	45.2	261	254.3	58.7
22	21.4	4.9	82	79.9	18.4	42	138.4	31.9	02	196.8	45.4	62	255.3	58.9
23	22.4	5.2	83	80.9	18.7	43	139.3	32.2	03	197.8	45.7	63	256.3	59.2
24	23.4	5.4	84	81.8	18.9	44	140.3	32.4	04	198.8	45.9	64	257.2	59.4
25	24.4	5.6	85	82.8	19.1	45	141.3	32.6	05	199.7	46.1	65	258.2	59.6
26	25.3	5.8	86	83.8	19.3	46	142.3	32.8	06	200.7	46.3	66	259.2	59.8
27	26.3	6.1	87	84.8	19.6	47	143.2	33.1	07	201.7	46.6	67	260.2	60.1
28	27.3	6.3	88	85.7	19.8	48	144.2	33.3	08	202.7	46.8	68	261.1	60.3
29	28.3	6.5	89	86.7	20.0	49	145.2	33.5	09	203.6	47.0	69	262.1	60.5
30	29.2	6.7	90	87.7	20.2	50	146.2	33.7	10	204.6	47.2	70	263.1	60.7
31	30.2	7.0	91	88.7	20.5	151	147.1	34.0	211	205.6	47.5	271	264.1	61.0
32	31.2	7.2	92	89.6	20.7	52	148.1	34.2	12	206.6	47.7	72	265.0	61.2
33	32.2	7.4	93	90.6	20.9	53	149.1	34.4	13	207.5	47.9	73	266.0	61.4
34	33.1	7.6	94	91.6	21.1	54	150.1	34.6	14	208.5	48.1	74	267.0	61.6
35	34.1	7.9	95	92.6	21.4	55	151.0	34.9	15	209.5	48.4	75	268.0	61.9
36	35.1	8.1	96	93.5	21.6	56	152.0	35.1	16	210.5	48.6	76	268.9	62.1
37	36.1	8.3	97	94.5	21.8	57	153.0	35.3	17	211.4	48.8	77	269.9	62.3
38	37.0	8.5	98	95.5	22.0	58	154.0	35.5	18	212.4	49.0	78	270.9	62.5
39	38.0	8.8	99	96.5	22.3	59	154.9	35.8	19	213.4	49.3	79	271.8	62.8
40	39.0	9.0	100	97.4	22.5	60	155.9	36.0	20	214.4	49.5	80	272.8	63.0
41	39.9	9.2	101	98.4	22.7	161	156.9	36.2	221	215.3	49.7	281	273.8	63.2
42	40.9	9.4	02	99.4	22.9	62	157.8	36.4	22	216.3	49.9	82	274.8	63.4
43	41.9	9.7	03	100.4	23.2	63	158.8	36.7	23	217.3	50.2	83	275.7	63.7
44	42.9	9.9	04	101.3	23.4	64	159.8	36.9	24	218.3	50.4	84	276.7	63.9
45	43.8	10.1	05	102.3	23.6	65	160.8	37.1	25	219.2	50.6	85	277.7	64.1
46	44.8	10.3	06	103.3	23.8	66	161.7	37.3	26	220.2	50.8	86	278.7	64.3
47	45.8	10.6	07	104.3	24.1	67	162.7	37.6	27	221.2	51.1	87	279.6	64.6
48	46.8	10.8	08	105.2	24.3	68	163.7	37.8	28	222.2	51.3	88	280.6	64.8
49	47.7	11.0	09	106.2	24.5	69	164.7	38.0	29	223.1	51.5	89	281.6	65.0
50	48.7	11.2	10	107.2	24.7	70	165.6	38.2	30	224.1	51.7	90	282.6	65.2
51	49.7	11.5	111	108.2	25.0	171	166.6	38.5	231	225.1	52.0	291	283.5	65.5
52	50.7	11.7	12	109.1	25.2	72	167.6	38.7	32	226.1	52.2	92	284.5	65.7
53	51.6	11.9	13	110.1	25.4	73	168.6	38.9	33	227.0	52.4	93	285.5	65.9
54	52.6	12.1	14	111.1	25.6	74	169.5	39.1	34	228.0	52.6	94	286.5	66.1
55	53.6	12.4	15	112.1	25.9	75	170.5	39.4	35	229.0	52.9	95	287.4	66.4
56	54.6	12.6	16	113.0	26.1	76	171.5	39.6	36	230.0	53.1	96	288.4	66.6
57	55.5	12.8	17	114.0	26.3	77	172.5	39.8	37	230.9	53.3	97	289.4	66.8
58	56.5	13.0	18	115.0	26.5	78	173.4	40.0	38	231.9	53.5	98	290.4	67.0
59	57.5	13.3	19	116.0	26.8	79	174.4	40.3	39	232.9	53.8	99	291.3	67.3
60	58.5	13.5	20	116.9	27.0	80	175.4	40.5	40	233.8	54.0	300	292.3	67.5
Dist.	Dep.	D. Lat.	Dist.	Dep.	D. Lat.	Dist.	Dep.	D. Lat.	Dist.	Dep.	D. Lat.	Dist.	Dep.	D. Lat.

283° | 077°
257° | 103°

77°

Dist.	D. Lat.	Dep.
N.	N×Cos.	N×Sin.
Hypotenuse	Side Adj.	Side Opp.

347°	013°
193°	167°

TABLE 3

Traverse **13°** Table

347°	013°
193°	167°

Dist.	D. Lat.	Dep.	Dist.	D. Lat.	Dep.	Dist.	D. Lat.	Dep.	Dist.	D. Lat.	Dep.	Dist.	D. Lat.	Dep.
301	293.3	67.7	361	351.7	81.2	421	410.2	94.7	481	468.7	108.2	541	527.1	121.7
02	294.3	67.9	62	352.7	81.4	22	411.2	94.9	82	469.6	108.4	42	528.1	121.9
03	295.2	68.2	63	353.7	81.7	23	412.2	95.2	83	470.6	108.7	43	529.1	122.1
04	296.2	68.4	64	354.7	81.9	24	413.1	95.4	84	471.6	108.9	44	530.1	122.4
05	297.2	68.6	65	355.6	82.1	25	414.1	95.6	85	472.6	109.1	45	531.0	122.6
06	298.2	68.8	66	356.6	82.3	26	415.1	95.8	86	473.5	109.3	46	532.0	122.8
07	299.1	69.1	67	357.6	82.6	27	416.1	96.1	87	474.5	109.6	47	533.0	123.0
08	300.1	69.3	68	358.6	82.8	28	417.0	96.3	88	475.5	109.8	48	534.0	123.3
09	301.1	69.5	69	359.5	83.0	29	418.0	96.5	89	476.5	110.0	49	534.9	123.5
10	302.1	69.7	70	360.5	83.2	30	419.0	96.7	90	477.4	110.2	50	535.9	123.7
311	303.0	70.0	371	361.5	83.5	431	420.0	97.0	491	478.4	110.5	551	536.9	123.9
12	304.0	70.2	72	362.5	83.7	32	420.9	97.2	92	479.4	110.7	52	537.9	124.2
13	305.0	70.4	73	363.4	83.9	33	421.9	97.4	93	480.4	110.9	53	538.8	124.4
14	306.0	70.6	74	364.4	84.1	34	422.9	97.6	94	481.3	111.1	54	539.8	124.6
15	306.9	70.9	75	365.4	84.4	35	423.9	97.9	95	482.3	111.4	55	540.8	124.8
16	307.9	71.1	76	366.4	84.6	36	424.8	98.1	96	483.3	111.6	56	541.7	125.1
17	308.9	71.3	77	367.3	84.8	37	425.8	98.3	97	484.3	111.8	57	542.7	125.3
18	309.8	71.5	78	368.3	85.0	38	426.8	98.5	98	485.2	112.0	58	543.7	125.5
19	310.8	71.8	79	369.3	85.3	39	427.7	98.8	99	486.2	112.3	59	544.7	125.7
20	311.8	72.0	80	370.3	85.5	40	428.7	99.0	500	487.2	112.5	60	545.6	126.0
321	312.8	72.2	381	371.2	85.7	441	429.7	99.2	501	488.2	112.7	561	546.6	126.2
22	313.7	72.4	82	372.2	85.9	42	430.7	99.4	02	489.1	112.9	62	547.6	126.4
23	314.7	72.7	83	373.2	86.2	43	431.6	99.7	03	490.1	113.2	63	548.6	126.6
24	315.7	72.9	84	374.2	86.4	44	432.6	99.9	04	491.1	113.4	64	549.5	126.9
25	316.7	73.1	85	375.1	86.6	45	433.6	100.1	05	492.1	113.6	65	550.5	127.1
26	317.6	73.3	86	376.1	86.8	46	434.6	100.3	06	493.0	113.8	66	551.5	127.3
27	318.6	73.6	87	377.1	87.1	47	435.5	100.6	07	494.0	114.1	67	552.5	127.5
28	319.6	73.8	88	378.1	87.3	48	436.5	100.8	08	495.0	114.3	68	553.4	127.8
29	320.6	74.0	89	379.0	87.5	49	437.5	101.0	09	496.0	114.5	69	554.4	128.0
30	321.5	74.2	90	380.0	87.7	50	438.5	101.2	10	496.9	114.7	70	555.4	128.2
331	322.5	74.5	391	381.0	88.0	451	439.4	101.5	511	497.9	115.0	571	556.4	128.4
32	323.5	74.7	92	382.0	88.2	52	440.4	101.7	12	498.9	115.2	72	557.3	128.7
33	324.5	74.9	93	382.9	88.4	53	441.4	101.9	13	499.9	115.4	73	558.3	128.9
34	325.4	75.1	94	383.9	88.6	54	442.4	102.1	14	500.8	115.6	74	559.3	129.1
35	326.4	75.4	95	384.9	88.9	55	443.3	102.4	15	501.8	115.8	75	560.3	129.3
36	327.4	75.6	96	385.9	89.1	56	444.3	102.6	16	502.8	116.1	76	561.2	129.6
37	328.4	75.8	97	386.8	89.3	57	445.3	102.8	17	503.7	116.3	77	562.2	129.8
38	329.3	76.0	98	387.8	89.5	58	446.3	103.0	18	504.7	116.5	78	563.2	130.0
39	330.3	76.3	99	388.8	89.8	59	447.2	103.3	19	505.7	116.7	79	564.2	130.2
40	331.3	76.5	400	389.7	90.0	60	448.2	103.5	20	506.7	117.0	80	565.1	130.5
341	332.3	76.7	401	390.7	90.2	461	449.2	103.7	521	507.6	117.2	581	566.1	130.7
42	333.2	76.9	02	391.7	90.4	62	450.2	103.9	22	508.6	117.5	82	567.1	130.9
43	334.2	77.2	03	392.7	90.7	63	451.1	104.2	23	509.6	117.6	83	568.1	131.1
44	335.2	77.4	04	393.6	90.9	64	452.1	104.4	24	510.6	117.9	84	569.0	131.4
45	336.2	77.6	05	394.6	91.1	65	453.1	104.6	25	511.5	118.1	85	570.0	131.6
46	337.1	77.8	06	395.6	91.3	66	454.1	104.8	26	512.5	118.3	86	571.0	131.8
47	338.1	78.1	07	396.6	91.6	67	455.0	105.1	27	513.5	118.5	87	572.0	132.0
48	339.1	78.3	08	397.5	91.8	68	456.0	105.3	28	514.5	118.8	88	572.9	132.3
49	340.1	78.5	09	398.5	92.0	69	457.0	105.5	29	515.4	119.0	89	573.9	132.5
50	341.0	78.7	10	399.5	92.2	70	458.0	105.7	30	516.4	119.2	90	574.9	132.7
351	342.0	79.0	411	400.5	92.5	471	458.9	106.0	531	517.4	119.4	591	575.9	132.9
52	343.0	79.2	12	401.4	92.7	72	459.9	106.2	32	518.4	119.7	92	576.8	133.2
53	344.0	79.4	13	402.4	92.9	73	460.9	106.4	33	519.3	119.9	93	577.8	133.4
54	344.9	79.6	14	403.4	93.1	74	461.9	106.6	34	520.3	120.1	94	578.8	133.6
55	345.9	79.9	15	404.4	93.4	75	462.8	106.9	35	521.3	120.3	95	579.8	133.8
56	346.9	80.1	16	405.3	93.6	76	463.8	107.1	36	522.3	120.6	96	580.7	134.1
57	347.9	80.3	17	406.3	93.8	77	464.8	107.3	37	523.2	120.8	97	581.7	134.3
58	348.8	80.5	18	407.3	94.0	78	465.7	107.5	38	524.2	121.0	98	582.7	134.5
59	349.8	80.8	19	408.3	94.3	79	466.7	107.8	39	525.2	121.2	99	583.6	134.7
60	350.8	81.0	20	409.2	94.5	80	467.7	108.0	40	526.2	121.5	600	584.6	135.0
Dist.	Dep.	D. Lat.	Dist.	Dep.	D. Lat.	Dist.	Dep.	D. Lat.	Dist.	Dep.	D. Lat.	Dist.	Dep.	D. Lat.

Dist.	D. Lat.	Dep.
D Lo	Dep.	
	m	D Lo

77°

283°	077°
257°	103°

346°	014°
194°	166°

TABLE 3

Traverse **14°** Table

346°	014°
194°	166°

Dist.	D. Lat.	Dep.	Dist.	D. Lat.	Dep.	Dist.	D. Lat.	Dep.	Dist.	D. Lat.	Dep.	Dist.	D. Lat.	Dep.
1	1.0	0.2	61	59.2	14.8	121	117.4	29.3	181	175.6	43.8	241	233.8	58.3
2	1.9	0.5	62	60.2	15.0	22	118.4	29.5	82	176.6	44.0	42	234.8	58.5
3	2.9	0.7	63	61.1	15.2	23	119.3	29.8	83	177.6	44.3	43	235.8	58.8
4	3.9	1.0	64	62.1	15.5	24	120.3	30.0	84	178.5	44.5	44	236.8	59.0
5	4.9	1.2	65	63.1	15.7	25	121.3	30.2	85	179.5	44.8	45	237.7	59.3
6	5.8	1.5	66	64.0	16.0	26	122.3	30.5	86	180.5	45.0	46	238.7	59.5
7	6.8	1.7	67	65.0	16.2	27	123.2	30.7	87	181.4	45.2	47	239.7	59.8
8	7.8	1.9	68	66.0	16.5	28	124.2	31.0	88	182.4	45.5	48	240.6	60.0
9	8.7	2.2	69	67.0	16.7	29	125.2	31.2	89	183.4	45.7	49	241.6	60.2
10	9.7	2.4	70	67.9	16.9	30	126.1	31.4	90	184.4	46.0	50	242.6	60.5
11	10.7	2.7	71	68.9	17.2	131	127.1	31.7	191	185.3	46.2	251	243.5	60.7
12	11.6	2.9	72	69.9	17.4	32	128.1	31.9	92	186.3	46.4	52	244.5	61.0
13	12.6	3.1	73	70.8	17.7	33	129.0	32.2	93	187.3	46.7	53	245.5	61.2
14	13.6	3.4	74	71.8	17.9	34	130.0	32.4	94	188.2	46.9	54	246.5	61.4
15	14.6	3.6	75	72.8	18.1	35	131.0	32.7	95	189.2	47.2	55	247.4	61.7
16	15.5	3.9	76	73.7	18.4	36	132.0	32.9	96	190.2	47.4	56	248.4	61.9
17	16.5	4.1	77	74.7	18.6	37	132.9	33.1	97	191.1	47.7	57	249.4	62.2
18	17.5	4.4	78	75.7	18.9	38	133.9	33.4	98	192.1	47.9	58	250.3	62.4
19	18.4	4.6	79	76.7	19.1	39	134.9	33.6	99	193.1	48.1	59	251.3	62.7
20	19.4	4.8	80	77.6	19.4	40	135.8	33.9	200	194.1	48.4	60	252.3	62.9
21	20.4	5.1	81	78.6	19.6	141	136.8	34.1	201	195.0	48.6	261	253.2	63.1
22	21.3	5.3	82	79.6	19.8	42	137.8	34.4	02	196.0	48.9	62	254.2	63.4
23	22.3	5.6	83	80.5	20.1	43	138.8	34.6	03	197.0	49.1	63	255.2	63.6
24	23.3	5.8	84	81.5	20.3	44	139.7	34.8	04	197.9	49.4	64	256.2	63.9
25	24.3	6.0	85	82.5	20.6	45	140.7	35.1	05	198.9	49.6	65	257.1	64.1
26	25.2	6.3	86	83.4	20.8	46	141.7	35.3	06	199.9	49.8	66	258.1	64.4
27	26.2	6.5	87	84.4	21.0	47	142.6	35.6	07	200.9	50.1	67	259.1	64.6
28	27.2	6.8	88	85.4	21.3	48	143.6	35.8	08	201.8	50.3	68	260.0	64.8
29	28.1	7.0	89	86.4	21.5	49	144.6	36.0	09	202.8	50.6	69	261.0	65.1
30	29.1	7.3	90	87.3	21.8	50	145.5	36.3	10	203.8	50.8	70	262.0	65.3
31	30.1	7.5	91	88.3	22.0	151	146.5	36.5	211	204.7	51.0	271	263.0	65.6
32	31.0	7.7	92	89.3	22.3	52	147.5	36.8	12	205.7	51.3	72	263.9	65.8
33	32.0	8.0	93	90.2	22.5	53	148.5	37.0	13	206.7	51.5	73	264.9	66.0
34	33.0	8.2	94	91.2	22.7	54	149.4	37.3	14	207.6	51.8	74	265.9	66.3
35	34.0	8.5	95	92.2	23.0	55	150.4	37.5	15	208.6	52.0	75	266.8	66.5
36	34.9	8.7	96	93.1	23.2	56	151.4	37.7	16	209.6	52.3	76	267.8	66.8
37	35.9	9.0	97	94.1	23.5	57	152.3	38.0	17	210.6	52.5	77	268.8	67.0
38	36.9	9.2	98	95.1	23.7	58	153.3	38.2	18	211.5	52.7	78	269.7	67.3
39	37.8	9.4	99	96.1	24.0	59	154.3	38.5	19	212.5	53.0	79	270.7	67.5
40	38.8	9.7	100	97.0	24.2	60	155.2	38.7	20	213.5	53.2	80	271.7	67.7
41	39.8	9.9	101	98.0	24.4	161	156.2	38.9	221	214.4	53.5	281	272.7	68.0
42	40.8	10.2	02	99.0	24.7	62	157.2	39.2	22	215.4	53.7	82	273.6	68.2
43	41.7	10.4	03	99.9	24.9	63	158.2	39.4	23	216.4	53.9	83	274.6	68.5
44	42.7	10.6	04	100.9	25.2	64	159.1	39.7	24	217.3	54.2	84	275.6	68.7
45	43.7	10.9	05	101.9	25.4	65	160.1	39.9	25	218.3	54.4	85	276.5	68.9
46	44.6	11.1	06	102.9	25.6	66	161.1	40.2	26	219.3	54.7	86	277.5	69.2
47	45.6	11.4	07	103.8	25.9	67	162.0	40.4	27	220.3	54.9	87	278.5	69.4
48	46.6	11.6	08	104.8	26.1	68	163.0	40.6	28	221.2	55.2	88	279.4	69.7
49	47.5	11.9	09	105.8	26.4	69	164.0	40.9	29	222.2	55.4	89	280.4	69.9
50	48.5	12.1	10	106.7	26.6	70	165.0	41.1	30	223.2	55.6	90	281.4	70.2
51	49.5	12.3	111	107.7	26.9	171	165.9	41.4	231	224.1	55.9	291	282.4	70.4
52	50.5	12.6	12	108.7	27.1	72	166.9	41.6	32	225.1	56.1	92	283.3	70.6
53	51.4	12.8	13	109.6	27.3	73	167.9	41.9	33	226.1	56.4	93	284.3	70.9
54	52.4	13.1	14	110.6	27.6	74	168.8	42.1	34	227.0	56.6	94	285.3	71.1
55	53.4	13.3	15	111.6	27.8	75	169.8	42.3	35	228.0	56.9	95	286.2	71.4
56	54.3	13.5	16	112.6	28.1	76	170.8	42.6	36	229.0	57.1	96	287.2	71.6
57	55.3	13.8	17	113.5	28.3	77	171.7	42.8	37	230.0	57.3	97	288.2	71.9
58	56.3	14.0	18	114.5	28.5	78	172.7	43.1	38	230.9	57.6	98	289.1	72.1
59	57.2	14.3	19	115.5	28.8	79	173.7	43.3	39	231.9	57.8	99	290.1	72.3
60	58.2	14.5	20	116.4	29.0	80	174.7	43.5	40	232.9	58.1	300	291.1	72.6
Dist.	Dep.	D. Lat.	Dist.	Dep.	D. Lat.	Dist.	Dep.	D. Lat.	Dist.	Dep.	D. Lat.	Dist.	Dep.	D. Lat.

284°	076°
256°	104°

76°

Dist.	D. Lat.	Dep.
N.	N×Cos.	N×Sin.
Hypotenuse	Side Adj.	Side Opp.

346°	014°
194°	166°

TABLE 3

Traverse **14°** Table

346°	014°
194°	166°

Dist.	D. Lat.	Dep.	Dist.	D. Lat.	Dep.	Dist.	D. Lat.	Dep.	Dist.	D. Lat.	Dep.	Dist.	D. Lat.	Dep.
301	292.1	72.8	361	350.3	87.3	421	408.5	101.8	481	466.7	116.4	541	524.9	130.9
02	293.0	73.1	62	351.2	87.6	22	409.5	102.1	82	467.7	116.6	42	525.9	131.1
03	294.0	73.3	63	352.2	87.8	23	410.4	102.3	83	468.7	116.8	43	526.9	131.4
04	295.0	73.5	64	353.2	88.1	24	411.4	102.6	84	469.6	117.1	44	527.8	131.6
05	295.9	73.8	65	354.2	88.3	25	412.4	102.8	85	470.6	117.3	45	528.8	131.8
06	296.9	74.0	66	355.1	88.5	26	413.3	103.1	86	471.6	117.6	46	529.8	132.1
07	297.9	74.3	67	356.1	88.8	27	414.3	103.3	87	472.5	117.8	47	530.8	132.3
08	298.9	74.5	68	357.1	89.0	28	415.3	103.5	88	473.5	118.1	48	531.7	132.6
09	299.8	74.8	69	358.0	89.3	29	416.3	103.8	89	474.5	118.3	49	532.7	132.8
10	300.8	75.0	70	359.0	89.5	30	417.2	104.0	90	475.4	118.5	50	533.7	133.1
311	301.8	75.2	371	360.0	89.8	431	418.2	104.3	491	476.4	118.8	551	534.6	133.3
12	302.7	75.5	72	361.0	90.0	32	419.2	104.5	92	477.4	119.0	52	535.6	133.5
13	303.7	75.7	73	361.9	90.2	33	420.1	104.8	93	478.4	119.3	53	536.6	133.8
14	304.6	76.0	74	362.9	90.5	34	421.1	105.0	94	479.3	119.5	54	537.5	134.0
15	305.6	76.2	75	363.9	90.7	35	422.0	105.2	95	480.3	119.8	55	538.5	134.3
16	306.6	76.4	76	364.8	91.0	36	423.0	105.5	96	481.3	120.0	56	539.5	134.5
17	307.6	76.7	77	365.8	91.2	37	424.0	105.7	97	482.2	120.2	57	540.5	134.8
18	308.6	76.9	78	366.8	91.4	38	425.0	106.0	98	483.2	120.5	58	541.4	135.0
19	309.5	77.2	79	367.7	91.7	39	426.0	106.2	99	484.2	120.7	59	542.4	135.2
20	310.5	77.4	80	368.7	91.9	40	426.9	106.4	500	485.1	121.0	60	543.4	135.5
321	311.5	77.7	381	369.7	92.2	441	427.9	106.7	501	486.1	121.2	561	544.3	135.7
22	312.4	77.9	82	370.7	92.4	42	428.9	106.9	02	487.1	121.4	62	545.3	136.0
23	313.4	78.1	83	371.6	92.7	43	429.8	107.2	03	488.1	121.7	63	546.3	136.2
24	314.4	78.4	84	372.6	92.9	44	430.8	107.4	04	489.0	121.9	64	547.2	136.4
25	315.3	78.6	85	373.6	93.1	45	431.8	107.7	05	490.0	122.2	65	548.2	136.7
26	316.3	78.9	86	374.5	98.4	46	432.8	107.9	06	491.0	122.4	66	549.2	136.9
27	317.3	79.1	87	375.5	93.6	47	433.7	108.1	07	491.9	122.7	67	550.2	137.2
28	318.3	79.4	88	376.4	93.9	48	434.7	108.4	08	492.9	122.9	68	551.1	137.4
29	319.2	79.6	89	377.4	94.1	49	435.7	108.6	09	493.9	123.1	69	552.1	137.7
30	320.2	79.8	90	378.4	94.3	50	436.6	108.9	10	494.9	123.4	70	553.1	137.9
331	321.2	80.1	391	379.4	94.6	451	437.6	109.1	511	495.8	123.6	571	554.0	138.1
32	322.1	80.3	92	380.4	94.8	52	438.6	109.3	12	496.8	123.9	72	555.0	138.4
33	323.1	80.6	93	381.3	95.1	53	439.5	109.6	13	497.8	124.1	73	556.0	138.6
34	324.1	80.8	94	382.3	95.3	54	440.5	109.8	14	498.7	124.3	74	556.9	138.9
35	325.0	81.0	95	383.3	95.6	55	441.5	110.1	15	499.7	124.6	75	557.9	139.1
36	326.0	81.3	96	384.2	95.8	56	442.5	110.3	16	500.7	124.8	76	558.9	139.3
37	327.0	81.5	97	385.2	96.0	57	443.4	110.6	17	501.6	125.1	77	559.9	139.6
38	328.0	81.8	98	386.2	96.3	58	444.4	110.8	18	502.6	125.3	78	560.8	139.8
39	328.9	82.0	99	387.1	96.5	59	445.4	111.0	19	503.6	125.6	79	561.8	140.1
40	329.9	82.3	400	388.1	96.8	60	446.3	111.3	20	504.6	125.8	80	562.8	140.3
341	330.8	82.5	401	389.1	97.0	461	447.3	111.5	521	505.5	126.0	581	563.7	140.6
42	331.8	82.7	02	390.1	97.3	62	448.3	111.8	22	506.5	126.3	82	564.7	140.8
43	332.8	83.0	03	391.0	97.5	63	449.2	112.0	23	507.5	126.5	83	565.7	141.0
44	333.8	83.2	04	392.0	97.7	64	450.2	112.3	24	508.4	126.8	84	566.7	141.3
45	334.8	83.5	05	393.0	98.0	65	451.2	112.5	25	509.4	127.0	85	567.6	141.5
46	335.7	83.7	06	393.9	98.2	66	452.2	112.7	26	510.4	127.3	86	568.6	141.8
47	336.7	83.9	07	394.9	98.5	67	453.1	113.0	27	511.3	127.5	87	569.6	142.0
48	337.7	84.2	08	395.9	98.7	68	454.1	113.2	28	512.3	127.7	88	570.5	142.3
49	338.6	84.4	09	396.9	98.9	69	455.1	113.5	29	513.3	128.0	89	571.5	142.5
50	339.6	84.7	10	397.8	99.2	70	456.0	113.7	30	514.3	128.2	90	572.5	142.7
351	340.6	84.9	411	398.8	99.4	471	457.0	113.9	531	515.2	128.5	591	573.4	143.0
52	341.5	85.2	12	399.8	99.7	72	458.0	114.2	32	516.2	128.7	92	574.4	143.2
53	342.5	85.4	13	400.7	99.9	73	458.9	114.4	33	517.2	128.9	93	575.4	143.5
54	343.5	85.6	14	401.7	100.2	74	459.9	114.7	34	518.1	129.2	94	576.4	143.7
55	344.5	85.9	15	402.7	100.4	75	460.9	114.9	35	519.1	129.4	95	577.3	143.9
56	345.4	86.1	16	403.6	100.6	76	461.9	115.2	36	520.1	129.7	96	578.3	144.2
57	346.4	86.4	17	404.6	100.9	77	462.8	115.4	37	521.0	129.9	97	579.3	144.4
58	347.4	86.6	18	405.6	101.1	78	463.8	115.6	38	522.0	130.2	98	580.2	144.7
59	348.3	86.8	19	406.6	101.4	79	464.8	115.9	39	523.0	130.4	99	581.2	144.9
60	349.3	87.1	20	407.5	101.6	80	465.7	116.1	40	524.0	130.6	600	582.2	145.2
Dist.	Dep.	D. Lat.	Dist.	Dep.	D. Lat.	Dist.	Dep.	D. Lat.	Dist.	Dep.	D. Lat.	Dist.	Dep.	D. Lat.

Dist.	D. Lat.	Dep.
D Lo	Dep.	
	m	D Lo

76°

284°	076°
256°	104°

345° | 015° / 195° | 165°	TABLE 3 — Traverse 15° Table	345° | 015° / 195° | 165°

Dist.	D. Lat.	Dep.	Dist.	D. Lat.	Dep.	Dist.	D. Lat.	Dep.	Dist.	D. Lat.	Dep.	Dist.	D. Lat.	Dep.
1	1.0	0.3	61	58.9	15.8	121	116.9	31.3	181	174.8	46.8	241	232.8	62.4
2	1.9	0.5	62	59.9	16.0	22	117.8	31.6	82	175.8	47.1	42	233.8	62.6
3	2.9	0.8	63	60.9	16.3	23	118.8	31.8	83	176.8	47.4	43	234.7	62.9
4	3.9	1.0	64	61.8	16.6	24	119.8	32.1	84	177.7	47.6	44	235.7	63.2
5	4.8	1.3	65	62.8	16.8	25	120.7	32.4	85	178.7	47.9	45	236.7	63.4
6	5.8	1.6	66	63.8	17.1	26	121.7	32.6	86	179.7	48.1	46	237.6	63.7
7	6.8	1.8	67	64.7	17.3	27	122.7	32.9	87	180.6	48.4	47	238.6	63.9
8	7.7	2.1	68	65.7	17.6	28	123.6	33.1	88	181.6	48.7	48	239.5	64.2
9	8.7	2.3	69	66.6	17.9	29	124.6	33.4	89	182.6	48.9	49	240.5	64.4
10	9.7	2.6	70	67.6	18.1	30	125.6	33.6	90	183.5	49.2	50	241.5	64.7
11	10.6	2.8	71	68.6	18.4	131	126.5	33.9	191	184.5	49.4	251	242.4	65.0
12	11.6	3.1	72	69.5	18.6	32	127.5	34.2	92	185.5	49.7	52	243.4	65.2
13	12.6	3.4	73	70.5	18.9	33	128.5	34.4	93	186.4	50.0	53	244.4	65.5
14	13.5	3.6	74	71.5	19.2	34	129.4	34.7	94	187.4	50.2	54	245.3	65.7
15	14.5	3.9	75	72.4	19.4	35	130.4	34.9	95	188.4	50.5	55	246.3	66.0
16	15.5	4.1	76	73.4	19.7	36	131.4	35.2	96	189.3	50.7	56	247.3	66.3
17	16.4	4.4	77	74.4	19.9	37	132.3	35.5	97	190.3	51.0	57	248.2	66.5
18	17.4	4.7	78	75.3	20.2	38	133.3	35.7	98	191.3	51.2	58	249.2	66.8
19	18.4	4.9	79	76.3	20.4	39	134.3	36.0	99	192.2	51.5	59	250.2	67.0
20	19.3	5.2	80	77.3	20.7	40	135.2	36.2	200	193.2	51.8	60	251.1	67.3
21	20.3	5.4	81	78.2	21.0	141	136.2	36.5	201	194.2	52.0	261	252.1	67.6
22	21.3	5.7	82	79.2	21.2	42	137.2	36.8	02	195.1	52.3	62	253.1	67.8
23	22.2	6.0	83	80.2	21.5	43	138.1	37.0	03	196.1	52.5	63	254.0	68.1
24	23.2	6.2	84	81.1	21.7	44	139.1	37.3	04	197.0	52.8	64	255.0	68.3
25	24.1	6.5	85	82.1	22.0	45	140.1	37.5	05	198.0	53.1	65	256.0	68.6
26	25.1	6.7	86	83.1	22.3	46	141.0	37.8	06	199.0	53.3	66	256.9	68.8
27	26.1	7.0	87	84.0	22.5	47	142.0	38.0	07	199.9	53.6	67	257.9	69.1
28	27.0	7.2	88	85.0	22.8	48	143.0	38.3	08	200.9	53.8	68	258.9	69.4
29	28.0	7.5	89	86.0	23.0	49	143.9	38.6	09	201.9	54.1	69	259.8	69.6
30	29.0	7.8	90	86.9	23.3	50	144.9	38.8	10	202.8	54.4	70	260.8	69.9
31	29.9	8.0	91	87.9	23.6	151	145.9	39.1	211	203.8	54.6	271	261.8	70.1
32	30.9	8.3	92	88.9	23.8	52	146.8	39.3	12	204.8	54.9	72	262.7	70.4
33	31.9	8.5	93	89.8	24.1	53	147.8	39.6	13	205.7	55.1	73	263.7	70.7
34	32.8	8.8	94	90.8	24.3	54	148.8	39.9	14	206.7	55.4	74	264.7	70.9
35	33.8	9.1	95	91.8	24.6	55	149.7	40.1	15	207.7	55.6	75	265.6	71.2
36	34.8	9.3	96	92.7	24.8	56	150.7	40.4	16	208.6	55.9	76	266.6	71.4
37	35.7	9.6	97	93.7	25.1	57	151.7	40.6	17	209.6	56.2	77	267.6	71.7
38	36.7	9.8	98	94.7	25.4	58	152.6	40.9	18	210.6	56.4	78	268.5	72.0
39	37.7	10.1	99	95.6	25.6	59	153.6	41.2	19	211.5	56.7	79	269.5	72.2
40	38.6	10.4	100	96.6	25.9	60	154.5	41.4	20	212.5	56.9	80	270.5	72.5
41	39.6	10.6	101	97.6	26.1	161	155.5	41.7	221	213.5	57.2	281	271.4	72.7
42	40.6	10.9	02	98.5	26.4	62	156.5	41.9	22	214.4	57.5	82	272.4	73.0
43	41.5	11.1	03	99.5	26.7	63	157.4	42.2	23	215.4	57.7	83	273.4	73.2
44	42.5	11.4	04	100.5	26.9	64	158.4	42.4	24	216.4	58.0	84	274.3	73.5
45	43.5	11.6	05	101.4	27.2	65	159.4	42.7	25	217.3	58.2	85	275.3	73.8
46	44.4	11.9	06	102.4	27.4	66	160.3	43.0	26	218.3	58.5	86	276.3	74.0
47	45.4	12.2	07	103.4	27.7	67	161.3	43.2	27	219.3	58.8	87	277.2	74.3
48	46.4	12.4	08	104.3	28.0	68	162.3	43.5	28	220.2	59.0	88	278.2	74.5
49	47.3	12.7	09	105.3	28.2	69	163.2	43.7	29	221.2	59.3	89	279.2	74.8
50	48.3	12.9	10	106.3	28.5	70	164.2	44.0	30	222.2	59.5	90	280.1	75.1
51	49.3	13.2	111	107.2	28.7	171	165.2	44.3	231	223.1	59.8	291	281.1	75.3
52	50.2	13.5	12	108.2	29.0	72	166.1	44.5	32	224.1	60.0	92	282.1	75.6
53	51.2	13.7	13	109.1	29.2	73	167.1	44.8	33	225.1	60.3	93	283.0	75.8
54	52.2	14.0	14	110.1	29.5	74	168.1	45.0	34	226.0	60.6	94	284.0	76.1
55	53.1	14.2	15	111.1	29.8	75	169.0	45.3	35	227.0	60.8	95	284.9	76.4
56	54.1	14.5	16	112.0	30.0	76	170.0	45.6	36	228.0	61.1	96	285.9	76.6
57	55.1	14.8	17	113.0	30.3	77	171.0	45.8	37	228.9	61.3	97	286.9	76.9
58	56.0	15.0	18	114.0	30.5	78	171.9	46.1	38	229.9	61.6	98	287.8	77.1
59	57.0	15.3	19	114.9	30.8	79	172.9	46.3	39	230.9	61.9	99	288.8	77.4
60	58.0	15.5	20	115.9	31.1	80	173.9	46.6	40	231.8	62.1	300	289.8	77.6
Dist.	Dep.	D. Lat.	Dist.	Dep.	D. Lat.	Dist.	Dep.	D. Lat.	Dist.	Dep.	D. Lat.	Dist.	Dep.	D. Lat.

285° | 075°
255° | 105°

75°

Dist.	D. Lat.	Dep.
N.	N×Cos.	N×Sin.
Hypotenuse	Side Adj.	Side Opp.

345°	015°
195°	165°

TABLE 3

Traverse **15°** Table

345°	015°
195°	165°

Dist.	D. Lat.	Dep.	Dist.	D. Lat.	Dep.	Dist.	D. Lat.	Dep.	Dist.	D. Lat.	Dep.	Dist.	D. Lat.	Dep.
301	290.7	77.9	361	348.7	93.4	421	406.7	109.0	481	464.6	124.5	541	522.6	140.0
02	291.7	78.2	62	349.7	93.7	22	407.6	109.2	82	465.6	124.8	42	523.5	140.3
03	292.7	78.4	63	350.6	94.0	23	408.6	109.5	83	466.5	125.0	43	524.5	140.5
04	293.6	78.7	64	351.6	94.2	24	409.6	109.7	84	467.5	125.3	44	525.5	140.8
05	294.6	78.9	65	352.6	94.5	25	410.5	110.0	85	468.5	125.5	45	526.4	141.1
06	295.6	79.2	66	353.5	94.7	26	411.5	110.3	86	469.4	125.8	46	527.4	141.3
07	296.5	79.5	67	354.5	95.0	27	412.4	110.5	87	470.4	126.0	47	528.4	141.6
08	297.5	79.7	68	355.5	95.2	28	413.4	110.8	88	471.4	126.3	48	529.3	141.8
09	298.5	80.0	69	356.4	95.5	29	414.4	111.0	89	472.3	126.6	49	530.3	142.1
10	299.4	80.2	70	357.4	95.8	30	415.3	111.3	90	473.3	126.8	50	531.3	142.4
311	300.4	80.5	371	358.4	96.0	431	416.3	111.6	491	474.3	127.1	551	532.2	142.6
12	301.4	80.8	72	359.3	96.3	32	417.3	111.8	92	475.2	127.3	52	533.2	142.9
13	302.3	81.0	73	360.3	96.5	33	418.2	112.1	93	476.2	127.6	53	534.2	143.1
14	303.3	81.3	74	361.3	96.8	34	419.2	112.3	94	477.2	127.9	54	535.1	143.4
15	304.3	81.5	75	362.2	97.1	35	420.2	112.6	95	478.1	128.1	55	536.1	143.6
16	305.2	81.8	76	363.2	97.3	36	421.1	112.8	96	479.1	128.4	56	537.1	143.9
17	306.2	82.0	77	364.2	97.6	37	422.1	113.1	97	480.1	128.6	57	538.0	144.2
18	307.2	82.3	78	365.1	97.8	38	423.1	113.4	98	481.0	128.9	58	539.0	144.4
19	308.1	82.6	79	366.1	98.1	39	424.0	113.6	99	482.0	129.2	59	540.0	144.7
20	309.1	82.8	80	367.1	98.4	40	425.0	113.9	500	483.0	129.4	60	540.9	144.9
321	310.1	83.1	381	368.0	98.6	441	426.0	114.1	501	483.9	129.7	561	541.9	145.2
22	311.0	83.3	82	369.0	98.9	42	426.9	114.4	02	484.9	129.9	62	542.9	145.5
23	312.0	83.6	83	369.9	99.1	43	427.9	114.7	03	485.9	130.2	63	543.8	145.7
24	313.0	83.9	84	370.9	99.4	44	428.9	114.9	04	486.8	130.4	64	544.8	146.0
25	313.9	84.1	85	371.9	99.6	45	429.8	115.2	05	487.8	130.7	65	545.7	146.2
26	314.9	84.4	86	372.8	99.9	46	430.8	115.4	06	488.8	131.0	66	546.7	146.5
27	315.9	84.6	87	373.8	100.2	47	431.8	115.7	07	489.7	131.2	67	547.7	146.8
28	316.8	84.9	88	374.8	100.4	48	432.7	116.0	08	490.7	131.5	68	548.6	147.0
29	317.8	85.2	89	375.7	100.7	49	433.7	116.2	09	491.7	131.7	69	549.6	147.3
30	318.8	85.4	90	376.7	100.9	50	434.7	116.5	10	492.6	132.0	70	550.6	147.5
331	319.7	85.7	391	377.7	101.2	451	435.6	116.7	511	493.6	132.3	571	551.5	147.8
32	320.7	85.9	92	378.6	101.5	52	436.6	117.0	12	494.6	132.5	72	552.5	148.0
33	321.7	86.2	93	379.6	101.7	53	437.6	117.2	13	495.5	132.8	73	553.5	148.3
34	322.6	86.4	94	380.6	102.0	54	438.5	117.5	14	496.5	133.0	74	554.4	148.6
35	323.6	86.7	95	381.5	102.2	55	439.5	117.8	15	497.5	133.3	75	555.4	148.8
36	324.6	87.0	96	382.5	102.5	56	440.5	118.0	16	498.4	133.6	76	556.4	149.1
37	325.5	87.2	97	383.5	102.8	57	441.4	118.3	17	499.4	133.8	77	557.3	149.3
38	326.5	87.5	98	384.4	103.0	58	442.4	118.5	18	500.3	134.1	78	558.3	149.6
39	327.4	87.7	99	385.4	103.3	59	443.4	118.8	19	501.3	134.3	79	559.3	149.8
40	328.4	88.0	400	386.4	103.5	60	444.3	119.1	20	502.3	134.6	80	560.2	150.1
341	329.4	88.3	401	387.3	103.8	461	445.3	119.3	521	503.2	134.8	581	561.2	150.4
42	330.3	88.5	02	388.3	104.0	62	446.3	119.6	22	504.2	135.1	82	562.2	150.6
43	331.3	88.8	03	389.3	104.3	63	447.2	119.8	23	505.2	135.4	83	563.1	150.9
44	332.3	89.0	04	390.2	104.6	64	448.2	120.1	24	506.1	135.6	84	564.1	151.2
45	333.2	89.3	05	391.2	104.8	65	449.2	120.4	25	507.1	135.9	85	565.1	151.4
46	334.2	89.6	06	392.2	105.1	66	450.1	120.6	26	508.1	136.1	86	566.0	151.6
47	335.2	89.8	07	393.1	105.3	67	451.1	120.9	27	509.0	136.4	87	567.0	151.9
48	336.1	90.1	08	394.1	105.6	68	452.1	121.1	28	510.0	136.7	88	568.0	152.2
49	337.1	90.3	09	395.1	105.9	69	453.0	121.4	29	511.0	136.9	89	568.9	152.4
50	338.1	90.6	10	396.0	106.1	70	454.0	121.6	30	511.9	137.2	90	569.9	152.7
351	339.0	90.8	411	397.0	106.4	471	455.0	121.9	531	512.9	137.4	591	570.9	153.0
52	341.0	91.1	12	398.0	106.6	72	455.9	122.2	32	513.9	137.7	92	571.8	153.2
53	341.0	91.4	13	398.9	106.9	73	456.9	122.4	33	514.8	138.0	93	572.8	153.5
54	341.9	91.6	14	399.9	107.2	74	457.8	122.7	34	515.8	138.2	94	573.8	153.7
55	342.9	91.9	15	400.9	107.4	75	458.8	122.9	35	516.8	138.5	95	574.7	154.0
56	343.9	92.1	16	401.8	107.7	76	459.8	123.2	36	517.7	138.7	96	575.7	154.3
57	344.8	92.4	17	402.8	107.9	77	460.7	123.5	37	518.7	139.0	97	576.7	154.5
58	345.8	92.7	18	403.8	108.2	78	461.7	123.7	38	519.7	139.2	98	577.6	154.8
59	346.8	92.9	19	404.7	108.4	79	462.7	124.0	39	520.6	139.5	99	578.6	155.0
60	347.7	93.2	20	405.7	108.7	80	463.6	124.2	40	521.6	139.8	600	579.6	155.3
Dist.	Dep.	D. Lat.	Dist.	Dep.	D. Lat.	Dist.	Dep.	D. Lat.	Dist.	Dep.	D. Lat.	Dist.	Dep.	D. Lat.

Dist.	D. Lat.	Dep.
D Lo	Dep.	
	m	D Lo

75°

285°	075°
255°	105°

344° | 016° / 196° | 164°	TABLE 3 — Traverse 16° Table	344° | 016° / 196° | 164°

Dist.	D. Lat.	Dep.	Dist.	D. Lat.	Dep.	Dist.	D. Lat.	Dep.	Dist.	D. Lat.	Dep.	Dist.	D. Lat.	Dep.
1	1.0	0.3	61	58.6	16.8	121	116.3	33.4	181	174.0	49.9	241	231.7	66.4
2	1.9	0.6	62	59.6	17.1	22	117.3	33.6	82	174.9	50.2	42	232.6	66.7
3	2.9	0.8	63	60.6	17.4	23	118.2	33.9	83	175.9	50.4	43	233.6	67.0
4	3.8	1.1	64	61.5	17.6	24	119.2	34.2	84	176.9	50.7	44	234.5	67.3
5	4.8	1.4	65	62.5	17.9	25	120.2	34.5	85	177.8	51.0	45	235.5	67.5
6	5.8	1.7	66	63.4	18.2	26	121.1	34.7	86	178.8	51.3	46	236.5	67.8
7	6.7	1.9	67	64.4	18.5	27	122.1	35.0	87	179.8	51.5	47	237.4	68.1
8	7.7	2.2	68	65.4	18.7	28	123.0	35.3	88	180.7	51.8	48	238.4	68.4
9	8.7	2.5	69	66.3	19.0	29	124.0	35.6	89	181.7	52.1	49	239.4	68.6
10	9.6	2.8	70	67.3	19.3	30	125.0	35.8	90	182.6	52.4	50	240.3	68.9
11	10.6	3.0	71	68.2	19.6	131	125.9	36.1	191	183.6	52.6	251	241.3	69.2
12	11.5	3.3	72	69.2	19.8	32	126.9	36.4	92	184.6	52.9	52	242.2	69.5
13	12.5	3.6	73	70.2	20.1	33	127.8	36.7	93	185.5	53.2	53	243.2	69.7
14	13.5	3.9	74	71.1	20.4	34	128.8	36.9	94	186.5	53.5	54	244.2	70.0
15	14.4	4.1	75	72.1	20.7	35	129.8	37.2	95	187.4	53.7	55	245.1	70.3
16	15.4	4.4	76	73.1	20.9	36	130.7	37.5	96	188.4	54.0	56	246.1	70.6
17	16.3	4.7	77	74.0	21.2	37	131.7	37.8	97	189.4	54.3	57	247.0	70.8
18	17.3	5.0	78	75.0	21.5	38	132.7	38.0	98	190.3	54.6	58	248.0	71.1
19	18.3	5.2	79	75.9	21.8	39	133.6	38.3	99	191.3	54.9	59	249.0	71.4
20	19.2	5.5	80	76.9	22.1	40	134.6	38.6	200	192.3	55.1	60	249.9	71.7
21	20.2	5.8	81	77.9	22.3	141	135.5	38.9	201	193.2	55.4	261	250.9	71.9
22	21.1	6.1	82	78.8	22.6	42	136.5	39.1	02	194.2	55.7	62	251.9	72.2
23	22.1	6.3	83	79.8	22.9	43	137.5	39.4	03	195.1	56.0	63	252.8	72.5
24	23.1	6.6	84	80.7	23.2	44	138.4	39.7	04	196.1	56.2	64	253.8	72.8
25	24.0	6.9	85	81.7	23.4	45	139.4	40.0	05	197.1	56.5	65	254.7	73.0
26	25.0	7.2	86	82.7	23.7	46	140.3	40.2	06	198.0	56.8	66	255.7	73.3
27	26.0	7.4	87	83.6	24.0	47	141.3	40.5	07	199.0	57.1	67	256.7	73.6
28	26.9	7.7	88	84.6	24.3	48	142.3	40.8	08	199.9	57.3	68	257.6	73.9
29	27.9	8.0	89	85.6	24.5	49	143.2	41.1	09	200.9	57.6	69	258.6	74.1
30	28.8	8.3	90	86.5	24.8	50	144.2	41.3	10	201.9	57.9	70	259.5	74.4
31	29.8	8.5	91	87.5	25.1	151	145.2	41.6	211	202.8	58.2	271	260.5	74.7
32	30.8	8.8	92	88.4	25.4	52	146.1	41.9	12	203.8	58.4	72	261.5	75.0
33	31.7	9.1	93	89.4	25.6	53	147.1	42.2	13	204.7	58.7	73	262.4	75.2
34	32.7	9.4	94	90.4	25.9	54	148.0	42.4	14	205.7	59.0	74	263.4	75.5
35	33.6	9.6	95	91.3	26.2	55	149.0	42.7	15	206.7	59.3	75	264.3	75.8
36	34.6	9.9	96	92.3	26.5	56	150.0	43.0	16	207.6	59.5	76	265.3	76.1
37	35.6	10.2	97	93.2	26.7	57	150.9	43.3	17	208.6	59.8	77	266.3	76.4
38	36.5	10.5	98	94.2	27.0	58	151.9	43.6	18	209.6	60.1	78	267.2	76.6
39	37.5	10.7	99	95.2	27.3	59	152.8	43.8	19	210.5	60.4	79	268.2	76.9
40	38.5	11.0	100	96.1	27.6	60	153.8	44.1	20	211.5	60.6	80	269.2	77.2
41	39.4	11.3	101	97.1	27.8	161	154.8	44.4	221	212.4	60.9	281	270.1	77.5
42	40.4	11.6	02	98.0	28.1	62	155.7	44.7	22	213.4	61.2	82	271.1	77.7
43	41.3	11.9	03	99.0	28.4	63	156.7	44.9	23	214.4	61.5	83	272.0	78.0
44	42.3	12.1	04	100.0	28.7	64	157.6	45.2	24	215.3	61.7	84	273.0	78.3
45	43.3	12.4	05	100.9	28.9	65	158.6	45.5	25	216.3	62.0	85	274.0	78.6
46	44.2	12.7	06	101.9	29.2	66	159.6	45.8	26	217.2	62.3	86	274.9	78.8
47	45.2	13.0	07	102.9	29.5	67	160.5	46.0	27	218.2	62.6	87	275.9	79.1
48	46.1	13.2	08	103.8	29.8	68	161.5	46.3	28	219.2	62.8	88	276.8	79.4
49	47.1	13.5	09	104.8	30.0	69	162.5	46.6	29	220.1	63.1	89	277.8	79.7
50	48.1	13.8	10	105.7	30.3	70	163.4	46.9	30	221.1	63.4	90	278.8	79.9
51	49.0	14.1	111	106.7	30.6	171	164.4	47.1	231	222.1	63.7	291	279.7	80.2
52	50.0	14.3	12	107.7	30.9	72	165.3	47.4	32	223.0	63.9	92	280.7	80.5
53	50.9	14.6	13	108.6	31.1	73	166.3	47.7	33	224.0	64.2	93	281.6	80.8
54	51.9	14.9	14	109.6	31.4	74	167.3	48.0	34	224.9	64.5	94	282.6	81.0
55	52.9	15.2	15	110.5	31.7	75	168.2	48.2	35	225.9	64.8	95	283.6	81.3
56	53.8	15.4	16	111.5	32.0	76	169.2	48.5	36	226.9	65.1	96	284.5	81.6
57	54.8	15.7	17	112.5	32.2	77	170.1	48.8	37	227.8	65.3	97	285.5	81.9
58	55.8	16.0	18	113.4	32.5	78	171.1	49.1	38	228.8	65.6	98	286.5	82.1
59	56.7	16.3	19	114.4	32.8	79	172.1	49.3	39	229.7	65.9	99	287.4	82.4
60	57.7	16.5	20	115.4	33.1	80	173.0	49.6	40	230.7	66.2	300	288.4	82.7
Dist.	Dep.	D. Lat.	Dist.	Dep.	D. Lat.	Dist.	Dep.	D. Lat.	Dist.	Dep.	D. Lat.	Dist.	Dep.	D. Lat.

286° | 074° / 254° | 106°

74°

Dist.	D. Lat.	Dep.
N.	N×Cos.	N×Sin.
Hypotenuse	Side Adj.	Side Opp.

344° | 016°	TABLE 3	344° | 016°
196° | 164°	Traverse **16°** Table	196° | 164°

Dist.	D. Lat.	Dep.	Dist.	D. Lat.	Dep.	Dist.	D. Lat.	Dep.	Dist.	D. Lat.	Dep.	Dist.	D. Lat.	Dep.
301	289.3	83.0	361	347.0	99.5	421	404.7	116.0	481	462.4	132.6	541	520.0	149.1
02	290.3	83.2	62	348.0	99.8	22	405.7	116.3	82	463.3	132.9	42	521.0	149.4
03	291.3	83.5	63	348.9	100.1	23	406.6	116.6	83	464.3	133.1	43	522.0	149.7
04	292.2	83.8	64	349.9	100.3	24	407.6	116.9	84	465.3	133.4	44	522.9	149.9
05	293.2	84.1	65	350.9	100.6	25	408.5	117.1	85	466.2	133.7	45	523.9	150.2
06	294.1	84.3	66	351.8	100.9	26	409.5	117.4	86	467.2	134.0	46	524.8	150.5
07	295.1	84.6	67	352.8	101.2	27	410.5	117.7	87	468.1	134.2	47	525.8	150.8
08	296.1	84.9	68	353.7	101.4	28	411.4	118.0	88	469.1	134.5	48	526.8	151.0
09	297.0	85.2	69	354.7	101.7	29	412.4	118.2	89	470.1	134.8	49	527.7	151.3
10	298.0	85.4	70	355.7	102.0	30	413.3	118.5	90	471.0	135.1	50	528.7	151.6
311	299.0	85.7	371	356.6	102.3	431	414.3	118.8	491	472.0	135.3	551	529.7	151.9
12	299.9	86.0	72	357.6	102.5	32	415.3	119.1	92	472.9	135.6	52	530.6	152.2
13	300.9	86.3	73	358.6	102.8	33	416.2	119.4	93	473.9	135.9	53	531.6	152.4
14	301.8	86.6	74	359.5	103.1	34	417.2	119.6	94	474.9	136.2	54	532.5	152.7
15	302.8	86.8	75	360.5	103.4	35	418.1	119.9	95	475.8	136.4	55	533.5	153.0
16	303.8	87.1	76	361.4	103.6	36	419.1	120.2	96	476.8	136.7	56	534.5	153.3
17	304.7	87.4	77	362.4	103.9	37	420.1	120.5	97	477.7	137.0	57	535.4	153.5
18	305.7	87.7	78	363.4	104.2	38	421.0	120.7	98	478.7	137.3	58	536.4	153.8
19	306.6	87.9	79	364.3	104.5	39	422.0	121.0	99	479.7	137.5	59	537.3	154.1
20	307.6	88.2	80	365.3	104.7	40	423.0	121.3	500	480.6	137.8	60	538.3	154.4
321	308.6	88.5	381	366.2	105.0	441	423.9	121.6	501	481.6	138.1	561	539.3	154.6
22	309.5	88.8	82	367.2	105.3	42	424.9	121.8	02	482.6	138.4	62	540.2	154.9
23	310.5	89.0	83	368.2	105.6	43	425.8	122.1	03	483.5	138.6	63	541.2	155.2
24	311.4	89.3	84	369.1	105.8	44	426.8	122.4	04	484.5	138.9	64	542.2	155.5
25	312.4	89.6	85	370.1	106.1	45	427.8	122.7	05	485.4	139.2	65	543.1	155.7
26	313.4	89.9	86	371.0	106.4	46	428.7	122.9	06	486.4	139.5	66	544.1	156.0
27	314.3	90.1	87	372.0	106.7	47	429.7	123.2	07	487.4	139.7	67	545.0	156.3
28	315.3	90.4	88	373.0	106.9	48	430.6	123.5	08	488.3	140.0	68	546.0	156.6
29	316.3	90.7	89	373.9	107.2	49	431.6	123.8	09	489.3	140.3	69	547.0	156.8
30	317.2	91.0	90	374.9	107.5	50	432.6	124.0	10	490.2	140.6	70	547.9	157.1
331	318.2	91.2	391	375.9	107.8	451	433.5	124.3	511	491.2	140.9	571	548.9	157.4
32	319.1	91.5	92	376.8	108.0	52	434.5	124.6	12	492.2	141.1	72	549.8	157.7
33	320.1	91.8	93	377.8	108.3	53	435.5	124.9	13	493.1	141.4	73	550.8	157.9
34	321.1	92.1	94	378.7	108.6	54	436.4	125.1	14	494.1	141.7	74	551.8	158.2
35	322.0	92.3	95	379.7	108.9	55	437.4	125.4	15	495.0	142.0	75	552.7	158.5
36	323.0	92.6	96	380.7	109.2	56	438.3	125.7	16	496.0	142.2	76	553.7	158.8
37	323.9	92.9	97	381.6	109.4	57	439.3	126.0	17	497.0	142.5	77	554.6	159.0
38	324.9	93.2	98	382.6	109.7	58	440.3	126.2	18	497.9	142.8	78	555.6	159.3
39	325.8	93.4	99	383.5	110.0	59	441.2	126.5	19	498.9	143.1	79	556.6	159.6
40	326.8	93.7	400	384.5	110.3	60	442.2	126.8	20	499.9	143.3	80	557.5	159.9
341	327.8	94.0	401	385.5	110.5	461	443.1	127.1	521	500.8	143.6	581	558.5	160.1
42	328.8	94.3	02	386.4	110.8	62	444.1	127.3	22	501.8	143.9	82	559.5	160.4
43	329.7	94.5	03	387.4	111.1	63	445.1	127.6	23	502.7	144.2	83	560.4	160.6
44	330.7	94.8	04	388.3	111.4	64	446.0	127.9	24	503.7	144.4	84	561.4	161.0
45	331.6	95.1	05	389.3	111.6	65	447.0	128.2	25	504.7	144.7	85	562.3	161.2
46	332.6	95.4	06	390.3	111.9	66	447.9	128.4	26	505.6	145.0	86	563.3	161.5
47	333.6	95.6	07	391.2	112.2	67	448.9	128.7	27	506.6	145.3	87	564.3	161.8
48	334.5	95.9	08	392.2	112.5	68	449.9	129.0	28	507.5	145.5	88	565.2	162.1
49	335.5	96.2	09	393.2	112.7	69	450.8	129.3	29	508.5	145.8	89	566.2	162.4
50	336.4	96.5	10	394.1	113.0	70	451.8	129.5	30	509.5	146.1	90	567.1	162.6
351	337.4	96.7	411	395.1	113.3	471	452.8	129.8	531	510.4	146.4	591	568.1	162.9
52	338.4	97.0	12	396.0	113.6	72	453.7	130.1	32	511.4	146.6	92	569.1	163.2
53	339.3	97.3	13	397.0	113.8	73	454.7	130.4	33	512.4	146.9	93	570.0	163.5
54	340.3	97.6	14	398.0	114.1	74	455.6	130.7	34	513.3	147.2	94	571.0	163.7
55	341.2	97.9	15	398.9	114.4	75	456.6	130.9	35	514.3	147.5	95	572.0	164.0
56	342.2	98.1	16	399.9	114.7	76	457.6	131.2	36	515.2	147.7	96	572.9	164.3
57	343.2	98.4	17	400.8	114.9	77	458.5	131.6	37	516.2	148.0	97	573.9	164.6
58	344.1	98.7	18	401.8	115.2	78	459.5	131.8	38	517.2	148.3	98	574.8	164.8
59	345.1	99.0	19	402.8	115.5	79	460.4	132.0	39	518.1	148.6	99	575.8	165.1
60	346.1	99.2	20	403.7	115.8	80	461.4	132.3	40	519.1	148.8	600	576.8	165.4
Dist.	Dep.	D. Lat.	Dist.	Dep.	D. Lat.	Dist.	Dep.	D. Lat.	Dist.	Dep.	D. Lat.	Dist.	Dep.	D. Lat.

Dist.	D. Lat.	Dep.
D Lo	Dep.	
	m	D Lo

74°

286° | 074°
254° | 106°

343° | 017° / 197° | 163°	TABLE 3	343° | 017° / 197° | 163°
	Traverse **17°** Table	

Dist.	D. Lat.	Dep.	Dist.	D. Lat.	Dep.	Dist.	D. Lat.	Dep.	Dist.	D. Lat.	Dep.	Dist.	D. Lat.	Dep.
1	1.0	0.3	61	58.3	17.8	121	115.7	35.4	181	173.1	52.9	241	230.5	70.5
2	1.9	0.6	62	59.3	18.1	22	116.7	35.7	82	174.0	53.2	42	231.4	70.8
3	2.9	0.9	63	60.2	18.4	23	117.6	36.0	83	175.0	53.5	43	232.4	71.0
4	3.8	1.2	64	61.2	18.7	24	118.6	36.3	84	176.0	53.8	44	233.3	71.3
5	4.8	1.5	65	62.2	19.0	25	119.5	36.5	85	176.9	54.1	45	234.3	71.6
6	5.7	1.8	66	63.1	19.3	26	120.5	36.8	86	177.9	54.4	46	235.3	71.9
7	6.7	2.0	67	64.1	19.6	27	121.5	37.1	87	178.8	54.7	47	236.2	72.2
8	7.7	2.3	68	65.0	19.9	28	122.4	37.4	88	179.8	55.0	48	237.2	72.5
9	8.6	2.6	69	66.0	20.2	29	123.4	37.7	89	180.7	55.3	49	238.1	72.8
10	9.6	2.9	70	66.9	20.5	30	124.3	38.0	90	181.7	55.6	50	239.1	73.1
11	10.5	3.2	71	67.9	20.8	131	125.3	38.3	191	182.7	55.8	251	240.0	73.4
12	11.5	3.5	72	68.9	21.1	32	126.2	38.6	92	183.6	56.1	52	241.0	73.7
13	12.4	3.8	73	69.8	21.3	33	127.2	38.9	93	184.6	56.4	53	241.9	74.0
14	13.4	4.1	74	70.8	21.6	34	128.1	39.2	94	185.5	56.7	54	242.9	74.3
15	14.3	4.4	75	71.7	21.9	35	129.1	39.5	95	186.5	57.0	55	243.9	74.6
16	15.3	4.7	76	72.7	22.2	36	130.1	39.8	96	187.4	57.3	56	244.8	74.8
17	16.3	5.0	77	73.6	22.5	37	131.0	40.1	97	188.4	57.6	57	245.8	75.1
18	17.2	5.3	78	74.6	22.8	38	132.0	40.3	98	189.3	57.9	58	246.7	75.4
19	18.2	5.6	79	75.5	23.1	39	132.9	40.6	99	190.3	58.2	59	247.7	75.7
20	19.1	5.8	80	76.5	23.4	40	133.9	40.9	200	191.3	58.5	60	248.6	76.0
21	20.1	6.1	81	77.5	23.7	141	134.8	41.2	201	192.2	58.8	261	249.6	76.3
22	21.0	6.4	82	78.4	24.0	42	135.8	41.5	02	193.2	59.1	62	250.6	76.6
23	22.0	6.7	83	79.4	24.3	43	136.8	41.8	03	194.1	59.4	63	251.5	76.9
24	23.0	7.0	84	80.3	24.6	44	137.7	42.1	04	195.1	59.6	64	252.5	77.2
25	23.9	7.3	85	81.3	24.9	45	138.7	42.4	05	196.0	59.9	65	253.4	77.5
26	24.9	7.6	86	82.2	25.1	46	139.6	42.7	06	197.0	60.2	66	254.4	77.8
27	25.8	7.9	87	83.2	25.4	47	140.6	43.0	07	198.0	60.5	67	255.3	78.1
28	26.8	8.2	88	84.2	25.7	48	141.5	43.3	08	198.9	60.8	68	256.3	78.4
29	27.7	8.5	89	85.1	26.0	49	142.5	43.6	09	199.9	61.1	69	257.2	78.6
30	28.7	8.8	90	86.1	26.3	50	143.4	43.9	10	200.8	61.4	70	258.2	78.9
31	29.6	9.1	91	87.0	26.6	151	144.4	44.1	211	201.8	61.7	271	259.2	79.2
32	30.6	9.4	92	88.0	26.9	52	145.4	44.4	12	202.7	62.0	72	260.1	79.5
33	31.6	9.6	93	88.9	27.2	53	146.3	44.7	13	203.7	62.3	73	261.1	79.8
34	32.5	9.9	94	89.9	27.5	54	147.3	45.0	14	204.6	62.6	74	262.0	80.1
35	33.5	10.2	95	90.8	27.8	55	148.2	45.3	15	205.6	62.9	75	263.0	80.4
36	34.4	10.5	96	91.8	28.1	56	149.2	45.6	16	206.6	63.2	76	263.9	80.7
37	35.4	10.8	97	92.8	28.4	57	150.1	45.9	17	207.5	63.4	77	264.9	81.0
38	36.3	11.1	98	93.7	28.7	58	151.1	46.2	18	208.5	63.7	78	265.9	81.3
39	37.3	11.4	99	94.7	28.9	59	152.1	46.5	19	209.4	64.0	79	266.8	81.6
40	38.3	11.7	100	95.6	29.2	60	153.0	46.8	20	210.4	64.3	80	267.8	81.9
41	39.2	12.0	101	96.6	29.5	161	154.0	47.1	221	211.3	64.6	281	268.7	82.2
42	40.2	12.3	02	97.5	29.8	62	154.9	47.4	22	212.3	64.9	82	269.7	82.4
43	41.1	12.6	03	98.5	30.1	63	155.9	47.7	23	213.3	65.2	83	270.6	82.7
44	42.1	12.9	04	99.5	30.4	64	156.8	47.9	24	214.2	65.5	84	271.6	83.0
45	43.0	13.2	05	100.4	30.7	65	157.8	48.2	25	215.2	65.8	85	272.5	83.3
46	44.0	13.4	06	101.4	31.0	66	158.7	48.5	26	216.1	66.1	86	273.5	83.6
47	44.9	13.7	07	102.3	31.3	67	159.7	48.8	27	217.1	66.4	87	274.5	83.9
48	45.9	14.0	08	103.3	31.6	68	160.7	49.1	28	218.0	66.7	88	275.4	84.2
49	46.9	14.3	09	104.2	31.9	69	161.6	49.4	29	219.0	67.0	89	276.4	84.5
50	47.8	14.6	10	105.2	32.2	70	162.6	49.7	30	220.0	67.2	90	277.3	84.8
51	48.8	14.9	111	106.1	32.5	171	163.5	50.0	231	220.9	67.5	291	278.3	85.1
52	49.7	15.2	12	107.1	32.7	72	164.5	50.3	32	221.9	67.8	92	279.2	85.4
53	50.7	15.5	13	108.1	33.0	73	165.4	50.6	33	222.8	68.1	93	280.2	85.7
54	51.6	15.8	14	109.0	33.3	74	166.4	50.9	34	223.8	68.4	94	281.2	86.0
55	52.6	16.1	15	110.0	33.6	75	167.4	51.2	35	224.7	68.7	95	282.1	86.2
56	53.6	16.4	16	110.9	33.9	76	168.3	51.5	36	225.7	69.0	96	283.1	86.5
57	54.5	16.7	17	111.9	34.2	77	169.3	51.7	37	226.6	69.3	97	284.0	86.8
58	55.5	17.0	18	112.8	34.5	78	170.2	52.0	38	227.6	69.6	98	285.0	87.1
59	56.4	17.2	19	113.8	34.8	79	171.2	52.3	39	228.6	69.9	99	285.9	87.4
60	57.4	17.5	20	114.8	35.1	80	172.1	52.6	40	229.5	70.2	300	286.9	87.7
Dist.	Dep.	D. Lat.	Dist.	Dep.	D. Lat.	Dist.	Dep.	D. Lat.	Dist.	Dep.	D. Lat.	Dist.	Dep.	D. Lat.

287° | 073°
253° | 107°

73°

Dist.	D. Lat.	Dep.
N.	N×Cos.	N×Sin.
Hypotenuse	Side Adj.	Side Opp.

343°	017°
197°	163°

TABLE 3

Traverse **17°** Table

343°	017°
197°	163°

Dist.	D. Lat.	Dep.	Dist.	D. Lat.	Dep.	Dist.	D. Lat.	Dep.	Dist.	D. Lat.	Dep.	Dist.	D. Lat.	Dep.
301	287. 8	88. 0	361	345. 2	105. 5	421	402. 6	123. 1	481	460. 0	140. 6	541	517. 4	158. 2
02	288. 8	88. 3	62	346. 2	105. 8	22	403. 6	123. 4	82	460. 9	140. 9	42	518. 3	158. 5
03	289. 8	88. 6	63	347. 1	106. 1	23	404. 5	123. 7	83	461. 9	141. 2	43	519. 3	158. 8
04	290. 7	88. 9	64	348. 1	106. 4	24	405. 5	124. 0	84	462. 9	141. 5	44	520. 2	159. 1
05	291. 7	89. 2	65	349. 1	106. 7	25	406. 4	124. 3	85	463. 8	141. 8	45	521. 2	159. 3
06	292. 6	89. 5	66	350. 0	107. 0	26	407. 4	124. 6	86	464. 8	142. 1	46	522. 1	159. 6
07	293. 6	89. 8	67	351. 0	107. 3	27	408. 3	124. 8	87	465. 7	142. 4	47	523. 1	159. 9
08	294. 5	90. 1	68	351. 9	107. 6	28	409. 3	125. 1	88	466. 7	142. 7	48	524. 1	160. 2
09	295. 5	90. 3	69	352. 9	107. 9	29	410. 3	125. 4	89	467. 6	143. 0	49	525. 0	160. 5
10	296. 5	90. 6	70	353. 8	108. 2	30	411. 2	125. 7	90	468. 6	143. 3	50	526. 0	160. 8
311	297. 4	90. 9	371	354. 8	108. 5	431	412. 2	126. 0	491	469. 5	143. 6	551	526. 9	161. 1
12	298. 4	91. 2	72	355. 7	108. 8	32	413. 1	126. 3	92	470. 5	143. 8	52	527. 9	161. 4
13	299. 3	91. 5	73	356. 7	109. 1	33	414. 1	126. 6	93	471. 5	144. 1	53	528. 8	161. 7
14	300. 3	91. 8	74	357. 7	109. 3	34	415. 0	126. 9	94	472. 4	144. 4	54	529. 8	162. 0
15	301. 2	92. 1	75	358. 6	109. 6	35	416. 0	127. 2	95	473. 4	144. 7	55	530. 7	162. 3
16	302. 2	92. 4	76	359. 6	109. 9	36	416. 9	127. 5	96	474. 3	145. 0	56	531. 7	162. 6
17	303. 1	92. 7	77	360. 5	110. 2	37	417. 9	127. 8	97	475. 3	145. 3	57	532. 7	162. 9
18	304. 1	93. 0	78	361. 5	110. 5	38	418. 9	128. 1	98	476. 2	145. 6	58	533. 6	163. 1
19	305. 1	93. 3	79	362. 4	110. 8	39	419. 8	128. 4	99	477. 2	145. 9	59	534. 6	163. 4
20	306. 0	93. 6	80	363. 4	111. 1	40	420. 8	128. 6	500	478. 2	146. 2	60	535. 5	163. 7
321	307. 0	93. 9	381	364. 4	111. 4	441	421. 7	128. 9	501	479. 1	146. 5	561	536. 5	164. 0
22	307. 9	94. 1	82	365. 3	111. 7	42	422. 7	129. 2	02	480. 1	146. 8	62	537. 4	164. 3
23	308. 9	94. 4	83	366. 3	112. 0	43	423. 6	129. 5	03	481. 0	147. 1	63	538. 4	164. 6
24	309. 8	94. 7	84	367. 2	112. 3	44	424. 6	129. 8	04	482. 0	147. 4	64	539. 4	164. 9
25	310. 8	95. 0	85	368. 2	112. 6	45	425. 6	130. 1	05	482. 9	147. 6	65	540. 3	165. 2
26	311. 8	95. 3	86	369. 1	112. 9	46	426. 5	130. 4	06	483. 9	147. 9	66	541. 3	165. 5
27	312. 7	95. 6	87	370. 1	113. 1	47	427. 5	130. 7	07	484. 8	148. 2	67	542. 2	165. 8
28	313. 6	95. 9	88	371. 0	113. 4	48	428. 4	131. 0	08	485. 8	148. 5	68	543. 2	166. 1
29	314. 6	96. 2	89	372. 0	113. 7	49	429. 4	131. 3	09	486. 8	148. 8	69	544. 1	166. 4
30	315. 5	96. 5	90	373. 0	114. 0	50	430. 3	131. 6	10	487. 7	149. 1	70	545. 1	166. 7
331	316. 5	96. 8	391	373. 9	114. 3	451	431. 3	131. 9	511	488. 7	149. 4	571	546. 1	166. 9
32	317. 5	97. 1	92	374. 9	114. 6	52	432. 2	132. 2	12	489. 6	149. 7	72	547. 0	167. 2
33	318. 4	97. 4	93	375. 8	114. 9	53	433. 2	132. 4	13	490. 6	150. 0	73	548. 0	167. 5
34	319. 4	97. 7	94	376. 8	115. 2	54	434. 2	132. 7	14	491. 5	150. 3	74	548. 9	167. 8
35	320. 4	97. 9	95	377. 7	115. 5	55	435. 1	133. 0	15	492. 5	150. 6	75	549. 9	168. 1
36	321. 3	98. 2	96	378. 7	115. 8	56	436. 1	133. 3	16	493. 5	150. 9	76	550. 8	168. 4
37	322. 3	98. 5	97	379. 7	116. 1	57	437. 0	133. 6	17	494. 4	151. 2	77	551. 8	168. 7
38	323. 2	98. 8	98	380. 6	116. 4	58	438. 0	133. 9	18	495. 4	151. 4	78	552. 7	169. 0
39	324. 2	99. 1	99	381. 6	116. 7	59	438. 9	134. 2	19	496. 3	151. 7	79	553. 7	169. 3
40	325. 1	99. 4	400	382. 5	116. 9	60	439. 9	134. 5	20	497. 3	152. 0	80	554. 7	169. 6
341	326. 1	99. 7	401	383. 5	117. 2	461	440. 9	134. 8	521	498. 2	152. 3	581	555. 6	169. 9
42	327. 1	100. 0	02	384. 4	117. 5	62	441. 8	135. 1	22	499. 2	152. 6	82	556. 6	170. 2
43	328. 0	100. 3	03	385. 4	117. 8	63	442. 8	135. 4	23	500. 1	152. 9	83	557. 5	170. 5
44	329. 0	100. 6	04	386. 3	118. 1	64	443. 7	135. 7	24	501. 1	153. 2	84	558. 5	170. 7
45	329. 9	100. 9	05	387. 3	118. 4	65	444. 7	136. 0	25	502. 1	153. 5	85	559. 4	171. 0
46	330. 8	101. 2	06	388. 3	118. 7	66	445. 6	136. 2	26	503. 0	153. 8	86	560. 4	171. 3
47	331. 8	101. 5	07	389. 2	119. 0	67	446. 6	136. 5	27	504. 0	154. 1	87	561. 4	171. 6
48	332. 8	101. 7	08	390. 2	119. 3	68	447. 6	136. 8	28	504. 9	154. 4	88	562. 3	171. 9
49	333. 8	102. 0	09	391. 1	119. 6	69	448. 5	137. 1	29	505. 9	154. 7	89	563. 3	172. 2
50	334. 7	102. 3	10	392. 1	119. 9	70	449. 5	137. 4	30	506. 8	155. 0	90	564. 2	172. 5
351	335. 7	102. 6	411	393. 0	120. 2	471	450. 4	137. 7	531	507. 8	155. 2	591	565. 2	172. 8
52	336. 6	102. 9	12	394. 0	120. 5	72	451. 4	138. 0	32	508. 8	155. 5	92	566. 1	173. 1
53	337. 6	103. 2	13	395. 0	120. 7	73	452. 3	138. 3	33	509. 7	155. 8	93	567. 1	173. 4
54	338. 5	103. 5	14	395. 9	121. 0	74	453. 3	138. 6	34	510. 7	156. 1	94	568. 0	173. 7
55	339. 5	103. 8	15	396. 9	121. 3	75	454. 2	138. 9	35	511. 6	156. 4	95	569. 0	174. 0
56	340. 4	104. 1	16	397. 8	121. 6	76	455. 2	139. 2	36	512. 6	156. 7	96	570. 0	174. 3
57	341. 4	104. 4	17	398. 8	121. 9	77	456. 2	139. 5	37	513. 5	157. 0	97	570. 9	174. 5
58	342. 4	104. 7	18	399. 7	122. 2	78	457. 1	139. 8	38	514. 5	157. 3	98	571. 9	174. 8
59	343. 3	105. 0	19	400. 7	122. 5	79	458. 1	140. 0	39	515. 4	157. 6	99	572. 8	175. 1
60	344. 3	105. 3	20	401. 6	122. 8	80	459. 0	140. 3	40	516. 4	157. 9	600	573. 8	175. 4
Dist.	Dep.	D. Lat.	Dist.	Dep.	D. Lat.	Dist.	Dep.	D. Lat.	Dist.	Dep.	D. Lat.	Dist.	Dep.	D. Lat.

Dist.	D. Lat.	Dep.
D Lo	Dep.	
	m	D Lo

73°

287°	073°
253°	107°

342° | 018°	TABLE 3	342° | 018°
198° | 162°	Traverse **18°** Table	198° | 162°

Dist.	D. Lat.	Dep.	Dist.	D. Lat.	Dep.	Dist.	D. Lat.	Dep.	Dist.	D. Lat.	Dep.	Dist.	D. Lat.	Dep.
1	1.0	0.3	61	58.0	18.9	121	115.1	37.4	181	172.1	55.9	241	229.2	74.5
2	1.9	0.6	62	59.0	19.2	22	116.0	37.7	82	173.1	56.2	42	230.2	74.8
3	2.9	0.9	63	59.9	19.5	23	117.0	38.0	83	174.0	56.6	43	231.1	75.1
4	3.8	1.2	64	60.9	19.8	24	117.9	38.3	84	175.0	56.9	44	232.1	75.4
5	4.8	1.5	65	61.8	20.1	25	118.9	38.6	85	175.9	57.2	45	233.0	75.7
6	5.7	1.9	66	62.8	20.4	26	119.8	38.9	86	176.9	57.5	46	234.0	76.0
7	6.7	2.2	67	63.7	20.7	27	120.8	39.2	87	177.8	57.8	47	234.9	76.3
8	7.6	2.5	68	64.7	21.0	28	121.7	39.6	88	178.8	58.1	48	235.9	76.6
9	8.6	2.8	69	65.6	21.3	29	122.7	39.9	89	179.7	58.4	49	236.8	76.9
10	9.5	3.1	70	66.6	21.6	30	123.6	40.2	90	180.7	58.7	50	237.8	77.3
11	10.5	3.4	71	67.5	21.9	131	124.6	40.5	191	181.7	59.0	251	238.7	77.6
12	11.4	3.7	72	68.5	22.2	32	125.5	40.8	92	182.6	59.3	52	239.7	77.9
13	12.4	4.0	73	69.4	22.6	33	126.5	41.1	93	183.6	59.6	53	240.6	78.2
14	13.3	4.3	74	70.4	22.9	34	127.4	41.4	94	184.5	59.9	54	241.6	78.5
15	14.3	4.6	75	71.3	23.2	35	128.4	41.7	95	185.5	60.3	55	242.5	78.8
16	15.2	4.9	76	72.3	23.5	36	129.3	42.0	96	186.4	60.6	56	243.5	79.1
17	16.2	5.3	77	73.2	23.8	37	130.3	42.3	97	187.4	60.9	57	244.4	79.4
18	17.1	5.6	78	74.2	24.1	38	131.2	42.6	98	188.3	61.2	58	245.4	79.7
19	18.1	5.9	79	75.1	24.4	39	132.2	43.0	99	189.3	61.5	59	246.3	80.0
20	19.0	6.2	80	76.1	24.7	40	133.1	43.3	200	190.2	61.8	60	247.3	80.3
21	20.0	6.5	81	77.0	25.0	141	134.1	43.6	201	191.2	62.1	261	248.2	80.7
22	20.9	6.8	82	78.0	25.3	42	135.1	43.9	02	192.1	62.4	62	249.2	81.0
23	21.9	7.1	83	78.9	25.6	43	136.0	44.2	03	193.1	62.7	63	250.1	81.3
24	22.8	7.4	84	79.9	26.0	44	137.0	44.5	04	194.0	63.0	64	251.1	81.6
25	23.8	7.7	85	80.8	26.3	45	137.9	44.8	05	195.0	63.3	65	252.0	81.9
26	24.7	8.0	86	81.8	26.6	46	138.9	45.1	06	195.9	63.7	66	253.0	82.2
27	25.7	8.3	87	82.7	26.9	47	139.8	45.4	07	196.9	64.0	67	253.9	82.5
28	26.6	8.7	88	83.7	27.2	48	140.8	45.7	08	197.8	64.3	68	254.9	82.8
29	27.6	9.0	89	84.6	27.5	49	141.7	46.0	09	198.8	64.6	69	255.8	83.1
30	28.5	9.3	90	85.6	27.8	50	142.7	46.4	10	199.7	64.9	70	256.8	83.4
31	29.5	9.6	91	86.5	28.1	151	143.6	46.7	211	200.7	65.2	271	257.7	83.7
32	30.4	9.9	92	87.5	28.4	52	144.6	47.0	12	201.6	65.5	72	258.7	84.1
33	31.4	10.2	93	88.4	28.7	53	145.5	47.3	13	202.6	65.8	73	259.6	84.4
34	32.3	10.5	94	89.4	29.0	54	146.5	47.6	14	203.5	66.1	74	260.6	84.7
35	33.3	10.8	95	90.4	29.4	55	147.4	47.9	15	204.5	66.4	75	261.5	85.0
36	34.2	11.1	96	91.3	29.7	56	148.4	48.2	16	205.4	66.7	76	262.5	85.3
37	35.2	11.4	97	92.3	30.0	57	149.3	48.5	17	206.4	67.1	77	263.4	85.6
38	36.1	11.7	98	93.2	30.3	58	150.3	48.8	18	207.3	67.4	78	264.4	85.9
39	37.1	12.1	99	94.2	30.6	59	151.2	49.1	19	208.3	67.7	79	265.3	86.2
40	38.0	12.4	100	95.1	30.9	60	152.2	49.4	20	209.2	68.0	80	266.3	86.5
41	39.0	12.7	101	96.1	31.2	161	153.1	49.8	221	210.2	68.3	281	267.2	86.8
42	39.9	13.0	02	97.0	31.5	62	154.1	50.1	22	211.1	68.6	82	268.2	87.1
43	40.9	13.3	03	98.0	31.8	63	155.0	50.4	23	212.1	68.9	83	269.1	87.5
44	41.8	13.6	04	98.9	32.1	64	156.0	50.7	24	213.0	69.2	84	270.1	87.8
45	42.8	13.9	05	99.9	32.4	65	156.9	51.0	25	214.0	69.5	85	271.1	88.1
46	43.7	14.2	06	100.8	32.8	66	157.9	51.3	26	214.9	69.8	86	272.0	88.4
47	44.7	14.5	07	101.8	33.1	67	158.8	51.6	27	215.9	70.1	87	273.0	88.7
48	45.7	14.8	08	102.7	33.4	68	159.8	51.9	28	216.8	70.5	88	273.9	89.0
49	46.6	15.1	09	103.7	33.7	69	160.7	52.2	29	217.8	70.8	89	274.9	89.3
50	47.6	15.5	10	104.6	34.0	70	161.7	52.5	30	218.7	71.1	90	275.8	89.6
51	48.5	15.8	111	105.6	34.3	171	162.6	52.8	231	219.7	71.4	291	276.8	89.9
52	49.5	16.1	12	106.5	34.6	72	163.6	53.2	32	220.6	71.7	92	277.7	90.2
53	50.4	16.4	13	107.5	34.9	73	164.5	53.5	33	221.6	72.0	93	278.7	90.5
54	51.4	16.7	14	108.4	35.2	74	165.5	53.8	34	222.5	72.3	94	279.6	90.9
55	52.3	17.0	15	109.4	35.5	75	166.4	54.1	35	223.5	72.6	95	280.6	91.2
56	53.3	17.3	16	110.3	35.8	76	167.4	54.4	36	224.4	72.9	96	281.5	91.5
57	54.2	17.6	17	111.3	36.2	77	168.3	54.7	37	225.4	73.2	97	282.5	91.8
58	55.2	17.9	18	112.2	36.5	78	169.3	55.0	38	226.4	73.5	98	283.4	92.1
59	56.1	18.2	19	113.2	36.8	79	170.2	55.3	39	227.3	73.9	99	284.4	92.4
60	57.1	18.5	20	114.1	37.1	80	171.2	55.6	40	228.3	74.2	300	285.3	92.7
Dist.	Dep.	D. Lat.	Dist.	Dep.	D. Lat.	Dist.	Dep.	D. Lat.	Dist.	Dep.	D. Lat.	Dist.	Dep.	D. Lat.

288° | 072°	**72°**
252° | 108°	

Dist.	D. Lat.	Dep.
N.	N×Cos.	N×Sin.
Hypotenuse	Side Adj.	Side Opp.

342°	018°
198°	162°

TABLE 3

Traverse **18°** Table

342°	018°
198°	162°

Dist.	D. Lat.	Dep.	Dist.	D. Lat.	Dep.	Dist.	D. Lat.	Dep.	Dist.	D. Lat.	Dep.	Dist.	D. Lat.	Dep.
301	286.3	93.0	361	343.3	111.6	421	400.4	130.1	481	457.5	148.6	541	514.5	167.2
02	287.2	93.3	62	344.3	111.9	22	401.3	130.4	82	458.4	148.9	42	515.5	167.5
03	288.2	93.6	63	345.2	112.2	23	402.3	130.7	83	459.4	149.3	43	516.4	167.8
04	289.1	93.9	64	346.2	112.5	24	403.2	131.0	84	460.3	149.6	44	517.4	168.1
05	290.1	94.3	65	347.1	112.8	25	404.2	131.3	85	461.3	149.9	45	518.3	168.4
06	291.0	94.6	66	348.1	113.1	26	405.2	131.6	86	462.2	150.2	46	519.3	168.7
07	292.0	94.9	67	349.0	113.4	27	406.1	132.0	87	463.2	150.5	47	520.2	169.0
08	292.9	95.2	68	350.0	113.7	28	407.1	132.3	88	464.1	150.8	48	521.2	169.3
09	293.9	95.5	69	350.9	114.0	29	408.0	132.6	89	465.1	151.1	49	522.1	169.7
10	294.8	95.8	70	351.9	114.3	30	409.0	132.9	90	466.0	151.4	50	523.1	170.0
311	295.8	96.1	371	352.8	114.6	431	409.9	133.2	491	467.0	151.7	551	524.0	170.3
12	296.7	96.4	72	353.8	115.0	32	410.9	133.5	92	467.9	152.0	52	525.0	170.6
13	297.7	96.7	73	354.7	115.3	33	411.8	133.8	93	468.9	152.3	53	525.9	170.9
14	298.6	97.0	74	355.7	115.6	34	412.8	134.1	94	469.8	152.7	54	526.9	171.2
15	299.6	97.3	75	356.6	115.9	35	413.7	134.4	95	470.8	153.0	55	527.8	171.5
16	300.5	97.6	76	357.6	116.2	36	414.7	134.7	96	471.7	153.3	56	528.8	171.8
17	301.5	98.0	77	358.5	116.5	37	415.6	135.0	97	472.7	153.6	57	529.7	172.1
18	302.4	98.3	78	359.5	116.8	38	416.6	135.3	98	473.6	153.9	58	530.7	172.4
19	303.4	98.6	79	360.5	117.1	39	417.5	135.7	99	474.6	154.2	59	531.6	172.7
20	304.3	98.9	80	361.4	117.4	40	418.5	136.0	500	475.5	154.5	60	532.6	173.0
321	305.3	99.2	381	362.4	117.7	441	419.4	136.3	501	476.5	154.8	561	533.5	173.4
22	306.2	99.5	82	363.3	118.0	42	420.4	136.6	02	477.4	155.1	62	534.5	173.7
23	307.2	99.8	83	364.3	118.4	43	421.3	136.9	03	478.4	155.4	63	535.4	174.0
24	308.2	100.1	84	365.2	118.7	44	422.3	137.2	04	479.3	155.7	64	536.4	174.3
25	309.1	100.4	85	366.2	119.0	45	423.2	137.5	05	480.3	156.1	65	537.3	174.6
26	310.0	100.7	86	367.1	119.3	46	424.2	137.8	06	481.2	156.4	66	538.3	174.9
27	311.0	101.0	87	368.1	119.6	47	425.1	138.1	07	482.2	156.7	67	539.2	175.2
28	311.9	101.4	88	369.0	119.9	48	426.1	138.4	08	483.1	157.0	68	540.2	175.5
29	312.9	101.7	89	370.0	120.2	49	427.0	138.7	09	484.1	157.3	69	541.2	175.8
30	313.8	102.0	90	370.9	120.5	50	428.0	139.1	10	485.0	157.6	70	542.1	176.1
331	314.8	102.3	391	371.9	120.8	451	428.9	139.4	511	486.0	157.9	571	543.1	176.4
32	315.8	102.6	92	372.8	121.1	52	429.9	139.7	12	486.9	158.2	72	544.0	176.8
33	316.7	102.9	93	373.8	121.4	53	430.8	140.0	13	487.9	158.5	73	545.0	177.1
34	317.7	103.2	94	374.7	121.8	54	431.8	140.3	14	488.8	158.8	74	545.9	177.4
35	318.6	103.5	95	375.7	122.1	55	432.7	140.6	15	489.8	159.1	75	546.9	177.7
36	319.6	103.8	96	376.6	122.4	56	433.7	140.9	16	490.7	159.5	76	547.8	178.0
37	320.5	104.1	97	377.6	122.7	57	434.6	141.2	17	491.7	159.8	77	548.8	178.3
38	321.5	104.6	98	378.5	123.0	58	435.6	141.5	18	492.6	160.1	78	549.7	178.6
39	322.4	104.8	99	379.5	123.3	59	436.5	141.8	19	493.6	160.4	79	550.7	178.9
40	323.4	105.1	400	380.4	123.6	60	437.5	142.1	20	494.5	160.7	80	551.6	179.2
341	324.3	105.4	401	381.4	123.9	461	438.4	142.5	521	495.5	161.0	581	552.6	179.5
42	325.3	105.7	02	382.3	124.2	62	439.4	142.8	22	496.5	161.3	82	553.5	179.8
43	326.2	106.0	03	383.3	124.5	63	440.3	143.1	23	497.4	161.6	83	554.5	180.2
44	327.2	106.3	04	384.2	124.8	64	441.3	143.4	24	498.4	161.9	84	555.4	180.5
45	328.1	106.6	05	385.2	125.2	65	442.2	143.7	25	499.3	162.2	85	556.4	180.8
46	329.1	106.9	06	386.1	125.5	66	443.2	144.0	26	500.3	162.5	86	557.3	181.1
47	330.0	107.2	07	387.1	125.8	67	444.1	144.3	27	501.2	162.9	87	558.3	181.4
48	331.0	107.5	08	388.0	126.1	68	445.1	144.6	28	502.2	163.2	88	559.2	181.7
49	331.9	107.8	09	389.0	126.4	69	446.0	144.9	29	503.1	163.5	89	560.2	182.0
50	332.9	108.2	10	389.9	126.7	70	447.0	145.2	30	504.1	163.8	90	561.1	182.3
351	333.8	108.5	411	390.9	127.0	471	447.9	145.5	531	505.0	164.1	591	562.1	182.6
52	334.8	108.8	12	391.8	127.3	72	448.9	145.9	32	506.0	164.4	92	563.0	182.9
53	335.7	109.1	13	392.8	127.6	73	449.8	146.2	33	506.9	164.7	93	564.0	183.2
54	336.7	109.4	14	393.7	127.9	74	450.8	146.5	34	507.9	165.0	94	564.9	183.6
55	337.6	109.7	15	394.7	128.2	75	451.8	146.8	35	508.8	165.3	95	565.9	183.9
56	338.6	110.0	16	395.6	128.6	76	452.7	147.1	36	509.8	165.6	96	566.8	184.2
57	339.5	110.3	17	396.6	128.9	77	453.7	147.4	37	510.7	165.9	97	567.8	184.5
58	340.5	110.6	18	397.5	129.2	78	454.6	147.7	38	511.7	166.3	98	568.7	184.8
59	341.4	110.9	19	398.5	129.5	79	455.6	148.0	39	512.6	166.6	99	569.7	185.1
60	342.4	111.2	20	399.4	129.8	80	456.5	148.3	40	513.6	166.9	600	570.6	185.4
Dist.	Dep.	D. Lat.	Dist.	Dep.	D. Lat.	Dist.	Dep.	D. Lat.	Dist.	Dep.	D. Lat.	Dist.	Dep.	D. Lat.

Dist.	D. Lat.	Dep.
D Lo	Dep.	
	m	D Lo

72°

288°	072°
252°	108°

341° | 019° / 199° | 161°					TABLE 3 Traverse 19° Table							341° | 019° / 199° | 161°		
Dist.	D. Lat.	Dep.	Dist.	D. Lat.	Dep.	Dist.	D. Lat.	Dep.	Dist.	D. Lat.	Dep.	Dist.	D. Lat.	Dep.
1	0.9	0.3	61	57.7	19.9	121	114.4	39.4	181	171.1	58.9	241	227.9	78.5
2	1.9	0.7	62	58.6	20.2	22	115.4	39.7	82	172.1	59.3	42	228.8	78.8
3	2.8	1.0	63	59.6	20.5	23	116.3	40.0	83	173.0	59.6	43	229.8	79.1
4	3.8	1.3	64	60.5	20.8	24	117.2	40.4	84	174.0	59.9	44	230.7	79.4
5	4.7	1.6	65	61.5	21.2	25	118.2	40.7	85	174.9	60.2	45	231.7	79.8
6	5.7	2.0	66	62.4	21.5	26	119.1	41.0	86	175.9	60.6	46	232.6	80.1
7	6.6	2.3	67	63.3	21.8	27	120.1	41.3	87	176.8	60.9	47	233.5	80.4
8	7.6	2.6	68	64.3	22.1	28	121.0	41.7	88	177.8	61.2	48	234.5	80.7
9	8.5	2.9	69	65.2	22.5	29	122.0	42.0	89	178.7	61.5	49	235.4	81.1
10	9.5	3.3	70	66.2	22.8	30	122.9	42.3	90	179.6	61.9	50	236.4	81.4
11	10.4	3.6	71	67.1	23.1	131	123.9	42.6	191	180.6	62.2	251	237.3	81.7
12	11.3	3.9	72	68.1	23.4	32	124.8	43.0	92	181.5	62.5	52	238.3	82.0
13	12.3	4.2	73	69.0	23.8	33	125.8	43.3	93	182.5	62.8	53	239.2	82.4
14	13.2	4.6	74	70.0	24.1	34	126.7	43.6	94	183.4	63.2	54	240.2	82.7
15	14.2	4.9	75	70.9	24.4	35	127.6	44.0	95	184.4	63.5	55	241.1	83.0
16	15.1	5.2	76	71.9	24.7	36	128.6	44.3	96	185.3	63.8	56	242.1	83.3
17	16.1	5.5	77	72.8	25.1	37	129.5	44.6	97	186.3	64.1	57	243.0	83.7
18	17.0	5.9	78	73.8	25.4	38	130.5	44.9	98	187.2	64.5	58	243.9	84.0
19	18.0	6.2	79	74.7	25.7	39	131.4	45.3	99	188.2	64.8	59	244.9	84.3
20	18.9	6.5	80	75.6	26.0	40	132.4	45.6	200	189.1	65.1	60	245.8	84.6
21	19.9	6.8	81	76.6	26.4	141	133.3	45.9	201	190.0	65.4	261	246.8	85.0
22	20.8	7.2	82	77.5	26.7	42	134.3	46.2	02	191.0	65.8	62	247.7	85.3
23	21.7	7.5	83	78.5	27.0	43	135.2	46.6	03	191.9	66.1	63	248.7	85.6
24	22.7	7.8	84	79.4	27.3	44	136.2	46.9	04	192.9	66.4	64	249.6	86.0
25	23.6	8.1	85	80.4	27.7	45	137.1	47.2	05	193.8	66.7	65	250.6	86.3
26	24.6	8.5	86	81.3	28.0	46	138.0	47.5	06	194.8	67.1	66	251.5	86.6
27	25.5	8.8	87	82.3	28.3	47	139.0	47.9	07	195.7	67.4	67	252.5	86.9
28	26.5	9.1	88	83.2	28.7	48	139.9	48.2	08	196.7	67.7	68	253.4	87.3
29	27.4	9.4	89	84.2	29.0	49	140.9	48.5	09	197.6	68.0	69	254.3	87.6
30	28.4	9.8	90	85.1	29.3	50	141.8	48.8	10	198.6	68.4	70	255.3	87.9
31	29.3	10.1	91	86.0	29.6	151	142.8	49.2	211	199.5	68.7	271	256.2	88.2
32	30.3	10.4	92	87.0	30.0	52	143.7	49.5	12	200.4	69.0	72	257.2	88.6
33	31.2	10.7	93	87.9	30.3	53	144.7	49.8	13	201.4	69.3	73	258.1	88.9
34	32.1	11.1	94	88.9	30.6	54	145.6	50.1	14	202.3	69.7	74	259.1	89.2
35	33.1	11.4	95	89.8	30.9	55	146.6	50.5	15	203.3	70.0	75	260.0	89.5
36	34.0	11.7	96	90.8	31.3	56	147.5	50.8	16	204.2	70.3	76	261.0	89.9
37	35.0	12.0	97	91.7	31.6	57	148.4	51.1	17	205.2	70.6	77	261.9	90.2
38	35.9	12.4	98	92.7	31.9	58	149.4	51.4	18	206.1	71.0	78	262.9	90.5
39	36.9	12.7	99	93.6	32.2	59	150.3	51.8	19	207.1	71.3	79	263.8	90.8
40	37.8	13.0	100	94.6	32.6	60	151.3	52.1	20	208.0	71.6	80	264.7	91.2
41	38.8	13.3	101	95.5	32.9	161	152.2	52.4	221	209.0	72.0	281	265.7	91.5
42	39.7	13.7	02	96.4	33.2	62	153.2	52.7	22	209.9	72.3	82	266.6	91.8
43	40.7	14.0	03	97.4	33.5	63	154.1	53.1	23	210.9	72.6	83	267.6	92.1
44	41.6	14.3	04	98.3	33.9	64	155.1	53.4	24	211.8	72.9	84	268.5	92.5
45	42.5	14.7	05	99.3	34.2	65	156.0	53.7	25	212.7	73.3	85	269.5	92.8
46	43.5	15.0	06	100.2	34.5	66	157.0	54.0	26	213.7	73.6	86	270.4	93.1
47	44.4	15.3	07	101.2	34.8	67	157.9	54.4	27	214.6	73.9	87	271.4	93.4
48	45.4	15.6	08	102.1	35.2	68	158.8	54.7	28	215.6	74.2	88	272.3	93.8
49	46.3	16.0	09	103.1	35.5	69	159.8	55.0	29	216.5	74.6	89	273.3	94.1
50	47.3	16.3	10	104.0	35.8	70	160.7	55.3	30	217.5	74.9	90	274.2	94.4
51	48.2	16.6	111	105.0	36.1	171	161.7	55.7	231	218.4	75.2	291	275.1	94.7
52	49.2	16.9	12	105.9	36.5	72	162.6	56.0	32	219.4	75.5	92	276.1	95.1
53	50.1	17.3	13	106.8	36.8	73	163.6	56.3	33	220.3	75.9	93	277.0	95.4
54	51.1	17.6	14	107.8	37.1	74	164.5	56.6	34	221.3	76.2	94	278.0	95.7
55	52.0	17.9	15	108.7	37.4	75	165.5	57.0	35	222.2	76.5	95	278.9	96.0
56	52.9	18.2	16	109.7	37.8	76	166.4	57.3	36	223.1	76.8	96	279.9	96.4
57	53.9	18.6	17	110.6	38.1	77	167.4	57.6	37	224.1	77.2	97	280.8	96.7
58	54.8	18.9	18	111.6	38.4	78	168.3	58.0	38	225.0	77.5	98	281.8	97.0
59	55.8	19.2	19	112.5	38.7	79	169.2	58.3	39	226.0	77.8	99	282.7	97.3
60	56.7	19.5	20	113.5	39.1	80	170.2	58.6	40	226.9	78.1	300	283.7	97.7
Dist.	Dep.	D. Lat.	Dist.	Dep.	D. Lat.	Dist.	Dep.	D. Lat.	Dist.	Dep.	D. Lat.	Dist.	Dep.	D. Lat.

289° | 071°
251° | 109°

71°

Dist.	D. Lat.	Dep.
N.	N×Cos.	N×Sin.
Hypotenuse	Side Adj.	Side Opp.

341°	019°
199°	161°

TABLE 3

Traverse **19°** Table

341°	019°
199°	161°

Dist.	D. Lat.	Dep.	Dist.	D. Lat.	Dep.	Dist.	D. Lat.	Dep.	Dist.	D. Lat.	Dep.	Dist.	D. Lat.	Dep.
301	284.6	98.0	361	341.3	117.5	421	398.1	137.1	481	454.8	156.6	541	511.5	176.1
02	285.5	98.3	62	342.3	117.9	22	399.0	137.4	82	455.7	156.9	42	512.5	176.5
03	286.5	98.6	63	343.2	118.2	23	400.0	137.7	83	456.7	157.2	43	513.4	176.8
04	287.4	99.0	64	344.2	118.5	24	400.9	138.0	84	457.6	157.6	44	514.4	177.1
05	288.4	99.3	65	345.1	118.8	25	401.8	138.4	85	458.6	157.9	45	515.3	177.4
06	289.3	99.6	66	346.1	119.2	26	402.8	138.7	86	459.5	158.2	46	516.3	177.8
07	290.3	99.9	67	347.0	119.5	27	403.7	139.0	87	460.5	158.6	47	517.2	178.1
08	291.2	100.3	68	348.0	119.8	28	404.7	139.3	88	461.4	158.9	48	518.1	178.4
09	292.2	100.6	69	348.9	120.1	29	405.6	139.7	89	462.4	159.2	49	519.1	178.7
10	293.1	100.9	70	349.8	120.5	30	406.6	140.0	90	463.3	159.5	50	520.0	179.1
311	294.1	101.3	371	350.8	120.8	431	407.5	140.3	491	464.2	159.9	551	521.0	179.4
12	295.0	101.6	72	351.7	121.1	32	408.5	140.6	92	465.2	160.2	52	521.9	179.7
13	295.9	101.9	73	352.7	121.4	33	409.4	141.0	93	466.1	160.5	53	522.9	180.0
14	296.9	102.2	74	353.6	121.8	34	410.4	141.3	94	467.1	160.8	54	523.8	180.4
15	297.8	102.6	75	354.6	122.1	35	411.3	141.6	95	468.0	161.2	55	524.8	180.7
16	298.8	102.9	76	355.5	122.4	36	412.2	141.9	96	469.0	161.5	56	525.7	181.0
17	299.7	103.2	77	356.5	122.7	37	413.2	142.3	97	469.9	161.8	57	526.7	181.3
18	300.7	103.5	78	357.4	123.1	38	414.1	142.6	98	470.9	162.1	58	527.6	181.7
19	301.6	103.8	79	358.4	123.4	39	415.1	142.9	99	471.8	162.5	59	528.5	182.0
20	302.6	104.2	80	359.3	123.7	40	416.0	143.3	500	472.8	162.8	60	529.5	182.3
321	303.5	104.5	381	360.2	124.0	441	417.0	143.6	501	473.7	163.1	561	530.4	182.6
22	304.5	104.8	82	361.2	124.4	42	417.9	143.9	02	474.7	163.4	62	531.4	183.0
23	305.4	105.2	83	362.1	124.7	43	418.9	144.2	03	475.6	163.8	63	532.3	183.3
24	306.3	105.5	84	363.1	125.0	44	419.8	144.6	04	476.5	164.1	64	533.3	183.6
25	307.3	105.8	85	364.0	125.3	45	420.8	144.9	05	477.5	164.4	65	534.2	183.9
26	308.2	106.1	86	365.0	125.7	46	421.7	145.2	06	478.4	164.7	66	535.2	184.3
27	309.2	106.5	87	365.9	126.0	47	422.6	145.5	07	479.4	165.1	67	536.1	184.6
28	310.1	106.8	88	366.9	126.3	48	423.6	145.9	08	480.3	165.4	68	537.1	184.9
29	311.1	107.1	89	367.8	126.6	49	424.5	146.2	09	481.3	165.7	69	538.0	185.2
30	312.0	107.4	90	368.8	127.0	50	425.5	146.5	10	482.2	166.0	70	538.9	185.6
331	313.0	107.8	391	369.7	127.3	451	426.4	146.8	511	483.2	166.4	571	539.9	185.9
32	313.9	108.1	92	370.6	127.6	52	427.4	147.2	12	484.1	166.7	72	540.8	186.2
33	314.9	108.4	93	371.6	127.9	53	428.3	147.5	13	485.1	167.0	73	541.8	186.6
34	315.8	108.7	94	372.5	128.3	54	429.3	147.8	14	486.0	167.3	74	542.7	186.9
35	316.7	109.1	95	373.5	128.6	55	430.2	148.1	15	486.9	167.7	75	543.7	187.2
36	317.7	109.4	96	374.4	128.9	56	431.2	148.5	16	487.9	168.0	76	544.6	187.5
37	318.6	109.7	97	375.4	129.3	57	432.1	148.8	17	488.8	168.3	77	545.6	187.9
38	319.6	110.0	98	376.3	129.6	58	433.0	149.1	18	489.7	168.6	78	546.5	188.2
39	320.5	110.4	99	377.3	129.9	59	434.0	149.4	19	490.7	169.0	79	547.5	188.5
40	321.5	110.7	400	378.2	130.2	60	434.9	149.8	20	491.6	169.3	80	548.4	188.8
341	322.4	111.0	401	379.2	130.6	461	435.9	150.1	521	492.6	169.6	581	549.3	189.2
42	323.4	111.3	02	380.1	130.9	62	436.8	150.4	22	493.6	169.9	82	550.3	189.5
43	324.3	111.7	03	381.0	131.2	63	437.8	150.7	23	494.5	170.3	83	551.2	189.8
44	325.3	112.0	04	382.0	131.5	64	438.7	151.1	24	495.5	170.6	84	552.2	190.1
45	326.2	112.3	05	382.9	131.9	65	439.7	151.4	25	496.4	170.9	85	553.1	190.5
46	327.1	112.6	06	383.9	132.2	66	440.6	151.7	26	497.3	171.2	86	554.1	190.8
47	328.1	113.0	07	384.8	132.5	67	441.6	152.0	27	498.3	171.6	87	555.0	191.1
48	329.0	113.3	08	385.8	132.8	68	442.5	152.4	28	499.2	171.9	88	556.0	191.4
49	330.0	113.6	09	386.7	133.2	69	443.4	152.7	29	500.2	172.2	89	556.9	191.8
50	330.9	113.9	10	387.7	133.5	70	444.4	153.0	30	501.1	172.6	90	557.9	192.0
351	331.9	114.3	411	388.6	133.8	471	445.3	153.3	531	502.1	172.9	591	558.8	192.4
52	332.8	114.6	12	389.6	134.1	72	446.3	153.7	32	503.0	173.2	92	559.7	192.7
53	333.8	114.9	13	390.5	134.5	73	447.2	154.0	33	504.0	173.5	93	560.7	193.1
54	334.7	115.3	14	391.4	134.8	74	448.2	154.3	34	504.9	173.9	94	561.6	193.4
55	335.7	115.6	15	392.4	135.1	75	449.1	154.6	35	505.9	174.2	95	562.6	193.7
56	336.6	115.9	16	393.3	135.4	76	450.1	155.0	36	506.8	174.5	96	563.5	194.0
57	337.6	116.2	17	394.3	135.8	77	451.0	155.3	37	507.7	174.8	97	564.5	194.4
58	338.5	116.6	18	395.2	136.1	78	452.0	155.6	38	508.7	175.2	98	565.4	194.7
59	339.4	116.9	19	396.2	136.4	79	452.9	155.9	39	509.6	175.5	99	566.4	195.0
60	340.4	117.2	20	397.1	136.7	80	453.8	156.3	40	510.6	175.8	600	567.3	195.3
Dist.	Dep.	D. Lat.	Dist.	Dep.	D. Lat.	Dist.	Dep.	D. Lat.	Dist.	Dep.	D. Lat.	Dist.	Dep.	D. Lat.

Dist.	D. Lat.	Dep.
D Lo	Dep.	
	m	D Lo

71°

289°	071°
251°	109°

340°	020°
200°	160°

TABLE 3

Traverse 20° Table

Dist.	D. Lat.	Dep.	Dist.	D. Lat.	Dep.	Dist.	D. Lat.	Dep.	Dist.	D. Lat.	Dep.	Dist.	D. Lat.	Dep.
1	0.9	0.3	61	57.3	20.9	121	113.7	41.4	181	170.1	61.9	241	226.5	82.4
2	1.9	0.7	62	58.3	21.2	22	114.6	41.7	82	171.0	62.2	42	227.4	82.8
3	2.8	1.0	63	59.2	21.5	23	115.6	42.1	83	172.0	62.6	43	228.3	83.1
4	3.8	1.4	64	60.1	21.9	24	116.5	42.4	84	172.9	62.9	44	229.3	83.5
5	4.7	1.7	65	61.1	22.2	25	117.5	42.8	85	173.8	63.3	45	230.2	83.8
6	5.6	2.1	66	62.0	22.6	26	118.4	43.1	86	174.8	63.6	46	231.2	84.1
7	6.6	2.4	67	63.0	22.9	27	119.3	43.4	87	175.7	64.0	47	232.1	84.5
8	7.5	2.7	68	63.9	23.3	28	120.3	43.8	88	176.7	64.3	48	233.0	84.8
9	8.5	3.1	69	64.8	23.6	29	121.2	44.1	89	177.6	64.6	49	234.0	85.2
10	9.4	3.4	70	65.8	23.9	30	122.2	44.5	90	178.5	65.0	50	234.9	85.5
11	10.3	3.8	71	66.7	24.3	131	123.1	44.8	191	179.5	65.3	251	235.9	85.8
12	11.3	4.1	72	67.7	24.6	32	124.0	45.1	92	180.4	65.7	52	236.8	86.2
13	12.2	4.4	73	68.6	25.0	33	125.0	45.5	93	181.4	66.0	53	237.7	86.5
14	13.2	4.8	74	69.5	25.3	34	125.9	45.8	94	182.3	66.4	54	238.7	86.9
15	14.1	5.1	75	70.5	25.7	35	126.9	46.2	95	183.2	66.7	55	239.6	87.2
16	15.0	5.5	76	71.4	26.0	36	127.8	46.5	96	184.2	67.0	56	240.6	87.6
17	16.0	5.8	77	72.4	26.3	37	128.7	46.9	97	185.1	67.4	57	241.5	87.9
18	16.9	6.2	78	73.3	26.7	38	129.7	47.2	98	186.1	67.7	58	242.4	88.2
19	17.9	6.5	79	74.2	27.0	39	130.6	47.5	99	187.0	68.1	59	243.4	88.6
20	18.8	6.8	80	75.2	27.4	40	131.6	47.9	200	187.9	68.4	60	244.3	88.9
21	19.7	7.2	81	76.1	27.7	141	132.5	48.2	201	188.9	68.7	261	245.3	89.3
22	20.7	7.5	82	77.1	28.0	42	133.4	48.6	02	189.8	69.1	62	246.2	89.6
23	21.6	7.9	83	78.0	28.4	43	134.4	48.9	03	190.8	69.4	63	247.1	90.0
24	22.6	8.2	84	78.9	28.7	44	135.3	49.3	04	191.7	69.8	64	248.1	90.3
25	23.5	8.6	85	79.9	29.1	45	136.3	49.6	05	192.6	70.1	65	249.0	90.6
26	24.4	8.9	86	80.8	29.4	46	137.2	49.9	06	193.6	70.5	66	250.0	91.0
27	25.4	9.2	87	81.8	29.8	47	138.1	50.3	07	194.5	70.8	67	250.9	91.3
28	26.3	9.6	88	82.7	30.1	48	139.1	50.6	08	195.5	71.1	68	251.8	91.7
29	27.3	9.9	89	83.6	30.4	49	140.0	51.0	09	196.4	71.5	69	252.8	92.0
30	28.2	10.3	90	84.6	30.8	50	140.9	51.3	10	197.3	71.8	70	253.7	92.3
31	29.1	10.6	91	85.5	31.1	151	141.9	51.6	211	198.3	72.2	271	254.7	92.7
32	30.1	10.9	92	86.5	31.5	52	142.8	52.0	12	199.2	72.5	72	255.6	93.0
33	31.0	11.3	93	87.4	31.8	53	143.8	52.3	13	200.2	72.9	73	256.5	93.4
34	31.9	11.6	94	88.3	32.1	54	144.7	52.7	14	201.1	73.2	74	257.5	93.7
35	32.9	12.0	95	89.3	32.5	55	145.7	53.0	15	202.0	73.5	75	258.4	94.1
36	33.8	12.3	96	90.2	32.8	56	146.6	53.4	16	203.0	73.9	76	259.4	94.4
37	34.8	12.7	97	91.2	33.2	57	147.5	53.7	17	203.9	74.2	77	260.3	94.7
38	35.7	13.0	98	92.1	33.5	58	148.5	54.0	18	204.9	74.6	78	261.2	95.1
39	36.6	13.3	99	93.0	33.9	59	149.4	54.4	19	205.8	74.9	79	262.2	95.4
40	37.6	13.7	100	94.0	34.2	60	150.4	54.7	20	206.7	75.2	80	263.1	95.8
41	38.5	14.0	101	94.9	34.5	161	151.3	55.1	221	207.7	75.6	281	264.1	96.1
42	39.5	14.4	02	95.8	34.9	62	152.2	55.4	22	208.6	75.9	82	265.0	96.4
43	40.4	14.7	03	96.8	35.2	63	153.2	55.7	23	209.6	76.3	83	265.9	96.8
44	41.3	15.0	04	97.7	35.6	64	154.1	56.1	24	210.5	76.6	84	266.9	97.1
45	42.3	15.4	05	98.7	35.9	65	155.0	56.4	25	211.4	77.0	85	267.8	97.5
46	43.2	15.7	06	99.6	36.3	66	156.0	56.8	26	212.4	77.3	86	268.8	97.8
47	44.2	16.1	07	100.5	36.6	67	156.9	57.1	27	213.3	77.6	87	269.7	98.2
48	45.1	16.4	08	101.5	36.9	68	157.9	57.5	28	214.2	78.0	88	270.6	98.5
49	46.0	16.8	09	102.4	37.3	69	158.8	57.8	29	215.2	78.3	89	271.6	98.8
50	47.0	17.1	10	103.4	37.6	70	159.7	58.1	30	216.1	78.7	90	272.5	99.2
51	47.9	17.4	111	104.3	38.0	171	160.7	58.5	231	217.1	79.0	291	273.5	99.5
52	48.9	17.8	12	105.2	38.3	72	161.6	58.8	32	218.0	79.3	92	274.4	99.9
53	49.8	18.1	13	106.2	38.6	73	162.6	59.2	33	218.9	79.7	93	275.3	100.2
54	50.7	18.5	14	107.1	39.0	74	163.5	59.5	34	219.9	80.0	94	276.3	100.6
55	51.7	18.8	15	108.1	39.3	75	164.4	59.9	35	220.8	80.4	95	277.2	100.9
56	52.6	19.2	16	109.0	39.7	76	165.4	60.2	36	221.8	80.7	96	278.1	101.2
57	53.6	19.5	17	109.9	40.0	77	166.3	60.5	37	222.7	81.1	97	279.1	101.6
58	54.5	19.8	18	110.9	40.4	78	167.3	60.9	38	223.6	81.4	98	280.0	101.9
59	55.4	20.2	19	111.8	40.7	79	168.2	61.2	39	224.6	81.7	99	281.0	102.3
60	56.4	20.5	20	112.8	41.0	80	169.1	61.6	40	225.5	82.1	300	281.9	102.6
Dist.	Dep.	D. Lat.	Dist.	Dep.	D. Lat.	Dist.	Dep.	D. Lat.	Dist.	Dep.	D. Lat.	Dist.	Dep.	D. Lat.

290°	070°
250°	110°

70°

Dist.	D. Lat.	Dep.
N.	N×Cos.	N×Sin.
Hypotenuse	Side Adj.	Side Opp.

340° | 020°
200° | 160°

TABLE 3

Traverse **20°** Table

340° | 020°
200° | 160°

Dist.	D. Lat.	Dep.	Dist.	D. Lat.	Dep.	Dist.	D. Lat.	Dep.	Dist.	D. Lat.	Dep.	Dist.	D. Lat.	Dep.
301	282.8	102.9	361	339.2	123.5	421	395.6	144.0	481	452.0	164.5	541	508.4	185.0
02	283.8	103.3	62	340.2	123.8	22	396.6	144.3	82	452.9	164.9	42	509.3	185.4
03	284.7	103.6	63	341.1	124.2	23	397.5	144.7	83	453.9	165.2	43	510.3	185.7
04	285.7	104.0	64	342.0	124.5	24	398.4	145.0	84	454.8	165.5	44	511.2	186.1
05	286.6	104.3	65	343.0	124.8	25	399.4	145.4	85	455.8	165.9	45	512.1	186.4
06	287.5	104.7	66	343.9	125.2	26	400.3	145.7	86	456.7	166.2	46	513.1	186.7
07	288.5	105.0	67	344.9	125.5	27	401.3	146.1	87	457.6	166.6	47	514.0	187.1
08	289.4	105.3	68	345.8	125.9	28	402.2	146.4	88	458.6	166.9	48	515.0	187.4
09	290.4	105.7	69	346.7	126.2	29	403.1	146.7	89	459.5	167.3	49	515.9	187.8
10	291.3	106.0	70	347.7	126.5	30	404.1	147.1	90	460.4	167.6	50	516.8	188.1
311	292.2	106.4	371	348.6	126.9	431	405.0	147.4	491	461.4	167.9	551	517.8	188.5
12	293.2	106.7	72	349.6	127.2	32	406.0	147.8	92	462.3	168.3	52	518.7	188.8
13	294.1	107.1	73	350.5	127.6	33	406.9	148.1	93	463.3	168.6	53	519.7	189.1
14	295.1	107.4	74	351.4	127.9	34	407.8	148.4	94	464.2	169.0	54	520.6	189.5
15	296.0	107.7	75	352.4	128.3	35	408.8	148.8	95	465.1	169.3	55	521.5	189.8
16	296.9	108.1	76	353.3	128.6	36	409.7	149.1	96	466.1	169.6	56	522.5	190.2
17	297.9	108.4	77	354.3	128.9	37	410.6	149.5	97	467.0	170.0	57	523.4	190.5
18	298.8	108.8	78	355.2	129.3	38	411.6	149.8	98	468.0	170.3	58	524.3	190.8
19	299.8	109.1	79	356.1	129.6	39	412.5	150.2	99	468.9	170.7	59	525.3	191.2
20	300.7	109.4	80	357.1	130.0	40	413.5	150.5	500	469.8	171.0	60	526.2	191.5
321	301.6	109.8	381	358.0	130.3	441	414.4	150.8	501	470.8	171.4	561	527.2	191.9
22	302.6	110.1	82	359.0	130.7	42	415.3	151.2	02	471.7	171.7	62	528.1	192.2
23	303.5	110.5	83	359.9	131.0	43	416.3	151.5	03	472.7	172.0	63	529.0	192.6
24	304.5	110.8	84	360.8	131.3	44	417.2	151.9	04	473.6	172.4	64	530.0	192.9
25	305.4	111.2	85	361.8	131.7	45	418.2	152.2	05	474.5	172.7	65	530.9	193.2
26	306.3	111.5	86	362.7	132.0	46	419.1	152.5	06	475.5	173.1	66	531.9	193.6
27	307.3	111.8	87	363.7	132.4	47	420.0	152.9	07	476.4	173.4	67	532.8	193.9
28	308.2	112.2	88	364.6	132.7	48	421.0	153.2	08	477.4	173.7	68	533.7	194.3
29	309.2	112.5	89	365.5	133.1	49	421.9	153.6	09	478.3	174.1	69	534.7	194.6
30	310.1	112.9	90	366.5	133.4	50	422.9	153.9	10	479.2	174.4	70	535.6	195.0
331	311.0	113.2	391	367.4	133.7	451	423.8	154.3	511	480.2	174.8	571	536.6	195.3
32	312.0	113.6	92	368.4	134.1	52	424.7	154.6	12	481.1	175.1	72	537.5	195.6
33	312.9	113.9	93	369.3	134.4	53	425.7	154.9	13	482.1	175.5	73	538.4	196.0
34	313.9	114.2	94	370.2	134.8	54	426.6	155.3	14	483.0	175.8	74	539.4	196.3
35	314.8	114.6	95	371.2	135.1	55	427.6	155.6	15	483.9	176.1	75	540.3	196.7
36	315.7	114.9	96	372.1	135.4	56	428.5	156.0	16	484.9	176.5	76	541.3	197.0
37	316.7	115.3	97	373.1	135.8	57	429.4	156.3	17	485.8	176.8	77	542.2	197.3
38	317.6	115.6	98	374.0	136.1	58	430.4	156.6	18	486.8	177.2	78	543.1	197.7
39	318.6	115.9	99	374.9	136.5	59	431.3	157.0	19	487.7	177.5	79	544.1	198.0
40	319.5	116.3	400	375.9	136.8	60	432.3	157.3	20	488.6	177.9	80	545.0	198.4
341	320.4	116.6	401	376.8	137.2	461	433.2	157.7	521	489.6	178.2	581	546.0	198.7
42	321.4	117.0	02	377.8	137.5	62	434.1	158.0	22	490.5	178.5	82	546.9	199.1
43	322.3	117.3	03	378.7	137.8	63	435.1	158.4	23	491.5	178.9	83	547.8	199.4
44	323.3	117.7	04	379.6	138.2	64	436.0	158.7	24	492.4	179.2	84	548.8	199.7
45	324.2	118.0	05	380.6	138.5	65	437.0	159.0	25	493.3	179.6	85	549.7	200.1
46	325.1	118.3	06	381.5	138.9	66	437.9	159.4	26	494.3	179.9	86	550.7	200.4
47	326.1	118.7	07	382.5	139.2	67	438.8	159.7	27	495.2	180.2	87	551.6	200.8
48	327.0	119.0	08	383.4	139.6	68	439.8	160.1	28	496.2	180.6	88	552.5	201.2
49	328.0	119.4	09	384.3	139.9	69	440.7	160.4	29	497.1	180.9	89	553.5	201.4
50	328.9	119.7	10	385.3	140.2	70	441.7	160.7	30	498.0	181.3	90	554.4	201.8
351	329.8	120.0	411	386.2	140.6	471	442.6	161.1	531	499.0	181.6	591	555.4	202.1
52	330.8	120.4	12	387.2	140.9	72	443.5	161.4	32	499.9	182.0	92	556.3	202.5
53	331.7	120.7	13	388.1	141.3	73	444.5	161.8	33	500.9	182.3	93	557.2	202.8
54	332.7	121.1	14	389.0	141.6	74	445.4	162.1	34	501.8	182.6	94	558.2	203.2
55	333.6	121.4	15	390.0	141.9	75	446.4	162.5	35	502.7	183.0	95	559.1	203.5
56	334.5	121.8	16	390.9	142.3	76	447.3	162.8	36	503.7	183.3	96	560.1	203.8
57	335.5	122.1	17	391.9	142.6	77	448.2	163.1	37	504.6	183.7	97	561.0	204.2
58	336.4	122.4	18	392.8	143.0	78	449.2	163.5	38	505.6	184.0	98	561.9	204.5
59	337.4	122.8	19	393.7	143.3	79	450.1	163.8	39	506.5	184.3	99	562.9	204.9
60	338.3	123.1	20	394.7	143.7	80	451.1	164.2	40	507.4	184.7	600	563.8	205.2
Dist.	Dep.	D. Lat.	Dist.	Dep.	D. Lat.	Dist.	Dep.	D. Lat.	Dist.	Dep.	D. Lat.	Dist.	Dep.	D. Lat.

Dist.	D. Lat.	Dep.
D Lo	Dep.	
	m	D Lo

70°

290° | 070°
250° | 110°

339° | 021° / 201° | 159°	TABLE 3 Traverse 21° Table	339° | 021° / 201° | 159°

Dist.	D. Lat.	Dep.	Dist.	D. Lat.	Dep.	Dist.	D. Lat.	Dep.	Dist.	D. Lat.	Dep.	Dist.	D. Lat.	Dep.
1	0.9	0.4	61	56.9	21.9	121	113.0	43.4	181	169.0	64.9	241	225.0	86.4
2	1.9	0.7	62	57.9	22.2	22	113.9	43.7	82	169.9	65.2	42	225.9	86.7
3	2.8	1.1	63	58.8	22.6	23	114.8	44.1	83	170.8	65.6	43	226.9	87.1
4	3.7	1.4	64	59.7	22.9	24	115.8	44.4	84	171.8	65.9	44	227.8	87.4
5	4.7	1.8	65	60.7	23.3	25	116.7	44.8	85	172.7	66.3	45	228.7	87.8
6	5.6	2.2	66	61.6	23.7	26	117.6	45.2	86	173.6	66.7	46	229.7	88.2
7	6.5	2.5	67	62.5	24.0	27	118.6	45.5	87	174.6	67.0	47	230.6	88.5
8	7.5	2.9	68	63.5	24.4	28	119.5	45.9	88	175.5	67.4	48	231.5	88.9
9	8.4	3.2	69	64.4	24.7	29	120.4	46.2	89	176.4	67.7	49	232.5	89.2
10	9.3	3.6	70	65.4	25.1	30	121.4	46.6	90	177.4	68.1	50	233.4	89.6
11	10.3	3.9	71	66.3	25.4	131	122.3	46.9	191	178.3	68.4	251	234.3	90.0
12	11.2	4.3	72	67.2	25.8	32	123.2	47.3	92	179.2	68.8	52	235.3	90.3
13	12.1	4.7	73	68.2	26.2	33	124.2	47.7	93	180.2	69.2	53	236.2	90.7
14	13.1	5.0	74	69.1	26.5	34	125.1	48.0	94	181.1	69.5	54	237.1	91.0
15	14.0	5.4	75	70.0	26.9	35	126.0	48.4	95	182.0	69.9	55	238.1	91.4
16	14.9	5.7	76	71.0	27.2	36	127.0	48.7	96	183.0	70.2	56	239.0	91.7
17	15.9	6.1	77	71.9	27.6	37	127.9	49.1	97	183.9	70.6	57	239.9	92.1
18	16.8	6.5	78	72.8	28.0	38	128.8	49.5	98	184.8	71.0	58	240.9	92.5
19	17.7	6.8	79	73.8	28.3	39	129.8	49.8	99	185.8	71.3	59	241.8	92.8
20	18.7	7.2	80	74.7	28.7	40	130.7	50.2	200	186.7	71.7	60	242.7	93.2
21	19.6	7.5	81	75.6	29.0	141	131.6	50.5	201	187.6	72.0	261	243.7	93.5
22	20.5	7.9	82	76.6	29.4	42	132.6	50.9	02	188.6	72.4	62	244.6	93.9
23	21.5	8.2	83	77.5	29.7	43	133.5	51.2	03	189.5	72.7	63	245.5	94.3
24	22.4	8.6	84	78.4	30.1	44	134.4	51.6	04	190.5	73.1	64	246.5	94.6
25	23.3	9.0	85	79.4	30.5	45	135.4	52.0	05	191.4	73.5	65	247.4	95.0
26	24.3	9.3	86	80.3	30.8	46	136.3	52.3	06	192.3	73.8	66	248.3	95.3
27	25.2	9.7	87	81.2	31.2	47	137.2	52.7	07	193.3	74.2	67	249.3	95.7
28	26.1	10.0	88	82.2	31.5	48	138.2	53.0	08	194.2	74.5	68	250.2	96.0
29	27.1	10.4	89	83.1	31.9	49	139.1	53.4	09	195.1	74.9	69	251.1	96.4
30	28.0	10.8	90	84.0	32.3	50	140.0	53.8	10	196.1	75.3	70	252.1	96.8
31	28.9	11.1	91	85.0	32.6	151	141.0	54.1	211	197.0	75.6	271	253.0	97.1
32	29.9	11.5	92	85.9	33.0	52	141.9	54.5	12	197.9	76.0	72	253.9	97.5
33	30.8	11.8	93	86.8	33.3	53	142.8	54.8	13	198.9	76.3	73	254.9	97.8
34	31.7	12.2	94	87.8	33.7	54	143.8	55.2	14	199.8	76.7	74	255.8	98.2
35	32.7	12.5	95	88.7	34.0	55	144.7	55.5	15	200.7	77.0	75	256.7	98.6
36	33.6	12.9	96	89.6	34.4	56	145.6	55.9	16	201.7	77.4	76	257.7	98.9
37	34.5	13.3	97	90.6	34.8	57	146.6	56.3	17	202.6	77.8	77	258.6	99.3
38	35.5	13.6	98	91.5	35.1	58	147.5	56.6	18	203.5	78.1	78	259.5	99.6
39	36.4	14.0	99	92.4	35.5	59	148.4	57.0	19	204.5	78.5	79	260.5	100.0
40	37.3	14.3	100	93.4	35.8	60	149.4	57.3	20	205.4	78.8	80	261.4	100.3
41	38.3	14.7	101	94.3	36.2	161	150.3	57.7	221	206.3	79.2	281	262.3	100.7
42	39.2	15.1	02	95.2	36.6	62	151.2	58.1	22	207.3	79.6	82	263.3	101.1
43	40.1	15.4	03	96.2	36.9	63	152.2	58.4	23	208.2	79.9	83	264.2	101.4
44	41.1	15.8	04	97.1	37.3	64	153.1	58.8	24	209.1	80.3	84	265.1	101.8
45	42.0	16.1	05	98.0	37.6	65	154.0	59.1	25	210.1	80.6	85	266.1	102.1
46	42.9	16.5	06	99.0	38.0	66	155.0	59.5	26	211.0	81.0	86	267.0	102.5
47	43.9	16.8	07	99.9	38.3	67	155.9	59.8	27	211.9	81.3	87	267.9	102.9
48	44.8	17.2	08	100.8	38.7	68	156.8	60.2	28	212.9	81.7	88	268.9	103.2
49	45.7	17.6	09	101.8	39.1	69	157.8	60.6	29	213.8	82.1	89	269.8	103.6
50	46.7	17.9	10	102.7	39.4	70	158.7	60.9	30	214.7	82.4	90	270.7	103.9
51	47.6	18.3	111	103.6	39.8	171	159.6	61.3	231	215.7	82.8	291	271.7	104.3
52	48.5	18.6	12	104.6	40.1	72	160.6	61.6	32	216.6	83.1	92	272.6	104.6
53	49.5	19.0	13	105.5	40.5	73	161.5	62.0	33	217.5	83.5	93	273.5	105.0
54	50.4	19.4	14	106.4	40.9	74	162.4	62.4	34	218.5	83.9	94	274.5	105.4
55	51.3	19.7	15	107.4	41.2	75	163.4	62.7	35	219.4	84.2	95	275.4	105.7
56	52.3	20.1	16	108.3	41.6	76	164.3	63.1	36	220.3	84.6	96	276.3	106.1
57	53.2	20.4	17	109.2	41.9	77	165.2	63.4	37	221.3	84.9	97	277.3	106.4
58	54.1	20.8	18	110.2	42.3	78	166.2	63.8	38	222.2	85.3	98	278.2	106.8
59	55.1	21.1	19	111.1	42.6	79	167.1	64.1	39	223.1	85.6	99	279.1	107.2
60	56.0	21.5	20	112.0	43.0	80	168.0	64.5	40	224.1	86.0	300	280.1	107.5
Dist.	Dep.	D. Lat.	Dist.	Dep.	D. Lat.	Dist.	Dep.	D. Lat.	Dist.	Dep.	D. Lat.	Dist.	Dep.	D. Lat.

291° | 069°
249° | 111°

69°

Dist.	D. Lat.	Dep.
N.	N×Cos.	N×Sin.
Hypotenuse	Side Adj.	Side Opp.

339°	021°
201°	159°

TABLE 3

Traverse 21° Table

339°	021°
201°	159°

Dist.	D. Lat.	Dep.	Dist.	D. Lat.	Dep.	Dist.	D. Lat.	Dep.	Dist.	D. Lat.	Dep.	Dist.	D. Lat.	Dep.
301	281.0	107.9	361	337.0	129.4	421	393.0	150.9	481	449.1	172.4	541	505.1	193.9
02	281.9	108.2	62	338.0	129.7	22	394.0	151.2	82	450.0	172.7	42	506.0	194.2
03	282.9	108.6	63	338.9	130.1	23	394.9	151.6	83	450.9	173.1	43	506.9	194.6
04	283.8	108.9	64	339.8	130.4	24	395.8	151.9	84	451.9	173.5	44	507.9	195.0
05	284.7	109.3	65	340.8	130.8	25	396.8	152.3	85	452.8	173.8	45	508.8	195.3
06	285.7	109.7	66	341.7	131.2	26	397.7	152.7	86	453.7	174.2	46	509.7	195.7
07	286.6	110.0	67	342.6	131.5	27	398.6	153.0	87	454.7	174.5	47	510.7	196.0
08	287.5	110.4	68	343.6	131.9	28	399.6	153.4	88	455.6	174.9	48	511.6	196.4
09	288.5	110.7	69	344.5	132.2	29	400.5	153.7	89	456.5	175.2	49	512.5	196.7
10	289.4	111.1	70	345.4	132.6	30	401.4	154.1	90	457.5	175.6	50	513.5	197.1
311	290.3	111.5	371	346.4	133.0	431	402.4	154.5	491	458.4	176.0	551	514.4	197.5
12	291.3	111.8	72	347.3	133.3	32	403.3	154.8	92	459.3	176.3	52	515.3	197.8
13	292.2	112.2	73	348.2	133.7	33	404.2	155.2	93	460.5	176.7	53	516.3	198.2
14	293.1	112.5	74	349.1	134.0	34	405.2	155.5	94	461.2	177.0	54	517.2	198.5
15	294.1	112.9	75	350.1	134.4	35	406.1	155.9	95	462.1	177.4	55	518.1	198.9
16	295.0	113.2	76	351.0	134.7	36	407.0	156.2	96	463.1	177.8	56	519.1	199.3
17	295.9	113.6	77	352.0	135.1	37	408.0	156.6	97	464.0	178.1	57	520.0	199.6
18	296.9	114.0	78	352.9	135.5	38	408.9	157.0	98	464.9	178.5	58	520.9	200.0
19	297.8	114.3	79	353.8	135.8	39	409.8	157.3	99	465.9	178.8	59	521.9	200.3
20	298.7	114.7	80	354.8	136.2	40	410.8	157.7	500	466.8	179.2	60	522.8	200.7
321	299.7	115.0	381	355.7	136.5	441	411.7	158.0	501	467.7	179.5	561	523.7	201.0
22	300.6	115.4	82	356.6	136.9	42	412.6	158.4	02	468.7	179.9	62	524.7	201.4
23	301.5	115.8	83	357.6	137.3	43	413.6	158.8	03	469.6	180.3	63	525.6	201.8
24	302.5	116.1	84	358.5	137.6	44	414.5	159.1	04	470.5	180.6	64	526.5	202.1
25	303.4	116.5	85	359.4	138.0	45	415.4	159.5	05	471.5	181.0	65	527.5	202.5
26	304.3	116.8	86	360.4	138.3	46	416.4	159.8	06	472.4	181.3	66	528.4	202.8
27	305.3	117.2	87	361.3	138.7	47	417.3	160.2	07	473.3	181.7	67	529.3	203.2
28	306.2	117.5	88	362.2	139.0	48	418.2	160.5	08	474.3	182.1	68	530.3	203.6
29	307.1	117.9	89	363.2	139.4	49	419.2	160.9	09	475.2	182.4	69	531.2	203.9
30	308.1	118.3	90	364.1	139.8	50	420.1	161.3	10	476.1	182.8	70	532.1	204.3
331	309.0	118.6	391	365.0	140.1	451	421.0	161.6	511	477.1	183.1	571	533.1	204.6
32	309.9	119.0	92	365.9	140.5	52	422.0	162.0	12	478.0	183.5	72	534.0	205.0
33	310.9	119.3	93	366.9	140.8	53	422.9	162.3	13	478.9	183.8	73	534.9	205.3
34	311.8	119.7	94	367.8	141.2	54	423.8	162.7	14	479.9	184.2	74	535.9	205.7
35	312.7	120.1	95	368.8	141.6	55	424.8	163.1	15	480.8	184.6	75	536.8	206.1
36	313.7	120.4	96	369.7	141.9	56	425.7	163.4	16	481.7	184.9	76	537.7	206.4
37	314.6	120.8	97	370.6	142.3	57	426.6	163.8	17	482.7	185.3	77	538.7	206.8
38	315.6	121.1	98	371.6	142.6	58	427.6	164.1	18	483.6	185.6	78	539.6	207.1
39	316.5	121.5	99	372.5	143.0	59	428.5	164.5	19	484.5	186.0	79	540.5	207.5
40	317.4	121.8	400	373.4	143.3	60	429.4	164.8	20	485.5	186.4	80	541.5	207.9
341	318.4	122.2	401	374.4	143.7	461	430.4	165.2	521	486.4	186.7	581	542.4	208.2
42	319.3	122.6	02	375.3	144.1	62	431.3	165.6	22	487.3	187.1	82	543.3	208.6
43	320.2	122.9	03	376.2	144.4	63	432.2	165.9	23	488.3	187.4	83	544.3	208.9
44	321.2	123.2	04	377.1	144.8	64	433.2	166.3	24	489.2	187.8	84	545.2	209.3
45	322.1	123.6	05	378.1	145.1	65	434.1	166.6	25	490.1	188.1	85	546.1	209.6
46	323.0	124.0	06	379.0	145.5	66	435.0	167.0	26	491.1	188.5	86	547.1	210.0
47	324.0	124.4	07	379.9	145.9	67	436.0	167.4	27	492.0	188.9	87	548.0	210.4
48	324.9	124.7	08	380.9	146.2	68	436.9	167.7	28	492.9	189.2	88	548.9	210.7
49	325.8	125.1	09	381.8	146.6	69	437.8	168.1	29	493.9	189.6	89	549.9	211.1
50	326.8	125.4	10	382.8	146.9	70	438.8	168.4	30	494.8	189.9	90	550.8	211.4
351	327.7	125.8	411	383.7	147.3	471	439.7	168.8	531	495.7	190.3	591	551.7	211.8
52	328.6	126.1	12	384.6	147.6	72	440.6	169.1	32	496.7	190.7	92	552.7	212.2
53	329.6	126.5	13	385.6	148.0	73	441.6	169.5	33	497.6	191.0	93	553.6	212.5
54	330.5	126.9	14	386.5	148.4	74	442.5	169.9	34	498.5	191.4	94	554.5	212.9
55	331.4	127.2	15	387.4	148.7	75	443.5	170.2	35	499.5	191.7	95	555.5	213.2
56	332.4	127.6	16	388.4	149.1	76	444.4	170.6	36	500.4	192.1	96	556.4	213.6
57	333.3	127.9	17	389.3	149.4	77	445.3	170.9	37	501.3	192.4	97	557.3	213.9
58	334.2	128.3	18	390.2	149.8	78	446.3	171.3	38	502.3	192.8	98	558.2	214.3
59	335.2	128.7	19	391.2	150.2	79	447.2	171.7	39	503.2	193.2	99	559.2	214.7
60	336.1	129.0	20	392.1	150.5	80	448.1	172.0	40	504.1	193.5	600	560.1	215.0
Dist.	Dep.	D. Lat.	Dist.	Dep.	D. Lat.	Dist.	Dep.	D. Lat.	Dist.	Dep.	D. Lat.	Dist.	Dep.	D. Lat.

Dist.	D. Lat.	Dep.
D Lo	Dep.	
	m	D Lo

69°

291°	069°
249°	111°

338° | 022° / 202° | 158°	TABLE 3 Traverse 22° Table	338° | 022° / 202° | 158°

Dist.	D. Lat.	Dep.	Dist.	D. Lat.	Dep.	Dist.	D. Lat.	Dep.	Dist.	D. Lat.	Dep.	Dist.	D. Lat.	Dep.
1	0.9	0.4	61	56.6	22.9	121	112.2	45.3	181	167.8	67.8	241	223.5	90.3
2	1.9	0.7	62	57.5	23.2	22	113.1	45.7	82	168.7	68.2	42	224.4	90.7
3	2.8	1.1	63	58.4	23.6	23	114.0	46.1	83	169.7	68.6	43	225.3	91.0
4	3.7	1.5	64	59.3	24.0	24	115.0	46.5	84	170.6	68.9	44	226.2	91.4
5	4.6	1.9	65	60.3	24.3	25	115.9	46.8	85	171.5	69.3	45	227.2	91.8
6	5.6	2.2	66	61.2	24.7	26	116.8	47.2	86	172.5	69.7	46	228.1	92.2
7	6.5	2.6	67	62.1	25.1	27	117.8	47.6	87	173.4	70.1	47	229.0	92.5
8	7.4	3.0	68	63.0	25.5	28	118.7	47.9	88	174.3	70.4	48	229.9	92.9
9	8.3	3.4	69	64.0	25.8	29	119.6	48.3	89	175.2	70.8	49	230.9	93.3
10	9.3	3.7	70	64.9	26.2	30	120.5	48.7	90	176.2	71.2	50	231.8	93.7
11	10.2	4.1	71	65.8	26.6	131	121.5	49.1	191	177.1	71.5	251	232.7	94.0
12	11.1	4.5	72	66.8	27.0	32	122.4	49.4	92	178.0	71.9	52	233.7	94.4
13	12.1	4.9	73	67.7	27.3	33	123.3	49.8	93	178.9	72.3	53	234.6	94.8
14	13.0	5.2	74	68.6	27.7	34	124.2	50.2	94	179.9	72.7	54	235.5	95.2
15	13.9	5.6	75	69.5	28.1	35	125.2	50.6	95	180.8	73.0	55	236.4	95.5
16	14.8	6.0	76	70.5	28.5	36	126.1	50.9	96	181.7	73.4	56	237.4	95.9
17	15.8	6.4	77	71.4	28.8	37	127.0	51.3	97	182.7	73.8	57	238.3	96.3
18	16.7	6.7	78	72.3	29.2	38	128.0	51.7	98	183.6	74.2	58	239.2	96.6
19	17.6	7.1	79	73.2	29.6	39	128.9	52.1	99	184.5	74.5	59	240.1	97.0
20	18.5	7.5	80	74.2	30.0	40	129.8	52.4	200	185.4	74.9	60	241.1	97.4
21	19.5	7.9	81	75.1	30.3	141	130.7	52.8	201	186.4	75.3	261	242.0	97.8
22	20.4	8.2	82	76.0	30.7	42	131.7	53.2	02	187.3	75.7	62	242.9	98.1
23	21.3	8.6	83	77.0	31.1	43	132.6	53.6	03	188.2	76.0	63	243.8	98.5
24	22.3	9.0	84	77.9	31.5	44	133.5	53.9	04	189.1	76.4	64	244.8	98.9
25	23.2	9.4	85	78.8	31.8	45	134.4	54.3	05	190.1	76.8	65	245.7	99.3
26	24.1	9.7	86	79.7	32.2	46	135.4	54.7	06	191.0	77.2	66	246.6	99.6
27	25.0	10.1	87	80.7	32.6	47	136.3	55.1	07	191.9	77.5	67	247.6	100.0
28	26.0	10.5	88	81.6	33.0	48	137.2	55.4	08	192.9	77.9	68	248.5	100.4
29	26.9	10.9	89	82.5	33.3	49	138.2	55.8	09	193.8	78.3	69	249.4	100.8
30	27.8	11.2	90	83.4	33.7	50	139.1	56.2	10	194.7	78.7	70	250.3	101.1
31	28.7	11.6	91	84.4	34.1	151	140.0	56.6	211	195.6	79.0	271	251.3	101.5
32	29.7	12.0	92	85.3	34.5	52	140.9	56.9	12	196.6	79.4	72	252.2	101.9
33	30.6	12.4	93	86.2	34.8	53	141.9	57.3	13	197.5	79.8	73	253.1	102.3
34	31.5	12.7	94	87.2	35.2	54	142.8	57.7	14	198.4	80.2	74	254.0	102.6
35	32.5	13.1	95	88.1	35.6	55	143.7	58.1	15	199.3	80.5	75	255.0	103.0
36	33.4	13.5	96	89.0	36.0	56	144.6	58.4	16	200.3	80.9	76	255.9	103.4
37	34.3	13.9	97	89.9	36.3	57	145.6	58.8	17	201.2	81.3	77	256.8	103.8
38	35.2	14.2	98	90.9	36.7	58	146.5	59.2	18	202.1	81.7	78	257.8	104.1
39	36.2	14.6	99	91.8	37.1	59	147.4	59.6	19	203.1	82.0	79	258.7	104.5
40	37.1	15.0	100	92.7	37.5	60	148.3	59.9	20	204.0	82.4	80	259.6	104.9
41	38.0	15.4	101	93.6	37.8	161	149.3	60.3	221	204.9	82.8	281	260.5	105.3
42	38.9	15.7	02	94.6	38.2	62	150.2	60.7	22	205.8	83.2	82	261.5	105.6
43	39.9	16.1	03	95.5	38.6	63	151.1	61.1	23	206.8	83.5	83	262.4	106.0
44	40.8	16.5	04	96.4	39.0	64	152.1	61.4	24	207.7	83.9	84	263.3	106.4
45	41.7	16.9	05	97.4	39.3	65	153.0	61.8	25	208.6	84.3	85	264.2	106.8
46	42.7	17.2	06	98.3	39.7	66	153.9	62.2	26	209.5	84.7	86	265.2	107.1
47	43.6	17.6	07	99.2	40.1	67	154.8	62.6	27	210.5	85.0	87	266.1	107.5
48	44.5	18.0	08	100.1	40.5	68	155.8	62.9	28	211.4	85.4	88	267.0	107.9
49	45.4	18.4	09	101.1	40.8	69	156.7	63.3	29	212.3	85.8	89	268.0	108.3
50	46.4	18.7	10	102.0	41.2	70	157.6	63.7	30	213.3	86.2	90	268.9	108.6
51	47.3	19.1	111	102.9	41.6	171	158.5	64.1	231	214.2	86.5	291	269.8	109.0
52	48.2	19.5	12	103.8	42.0	72	159.5	64.4	32	215.1	86.9	92	270.7	109.4
53	49.1	19.9	13	104.8	42.3	73	160.4	64.8	33	216.0	87.3	93	271.7	109.8
54	50.1	20.2	14	105.7	42.7	74	161.3	65.2	34	217.0	87.7	94	272.6	110.1
55	51.0	20.6	15	106.6	43.1	75	162.3	65.6	35	217.9	88.0	95	273.5	110.5
56	51.9	21.0	16	107.6	43.5	76	163.2	65.9	36	218.8	88.4	96	274.4	110.9
57	52.8	21.4	17	108.5	43.8	77	164.1	66.3	37	219.7	88.8	97	275.4	111.3
58	53.8	21.7	18	109.4	44.2	78	165.0	66.7	38	220.7	89.2	98	276.3	111.6
59	54.7	22.1	19	110.3	44.6	79	166.0	67.1	39	221.6	89.5	99	277.2	112.0
60	55.6	22.5	20	111.3	45.0	80	166.9	67.4	40	222.5	89.9	300	278.2	112.4
Dist.	Dep.	D. Lat.	Dist.	Dep.	D. Lat.	Dist.	Dep.	D. Lat.	Dist.	Dep.	D. Lat.	Dist.	Dep.	D. Lat.

292° | 068°
248° | 112°

68°

Dist.	D. Lat.	Dep.
N.	N×Cos.	N×Sin.
Hypotenuse	Side Adj.	Side Opp.

338°	022°
202°	158°

TABLE 3

Traverse **22°** Table

338°	022°
202°	158°

Dist.	D. Lat.	Dep.	Dist.	D. Lat.	Dep.	Dist.	D. Lat.	Dep.	Dist.	D. Lat.	Dep.	Dist.	D. Lat.	Dep.
301	279.8	112.8	361	334.7	135.2	421	390.3	157.7	481	446.0	180.2	541	501.6	202.7
02	280.0	113.1	62	335.6	135.6	22	391.3	158.1	82	446.9	180.6	42	502.5	203.0
03	280.9	113.5	63	336.6	136.0	23	392.2	158.5	83	447.8	180.9	43	503.5	203.4
04	281.9	113.9	64	337.5	136.4	24	393.1	158.8	84	448.8	181.3	44	504.4	203.8
05	282.8	114.3	65	338.4	136.7	25	394.1	159.2	85	449.7	181.7	45	505.3	204.2
06	283.7	114.6	66	339.3	137.1	26	395.0	159.6	86	450.6	182.1	46	506.2	204.5
07	284.6	115.0	67	340.3	137.5	27	395.9	160.0	87	451.5	182.4	47	507.2	204.9
08	285.6	115.4	68	341.2	137.9	28	396.8	160.3	88	452.5	182.8	48	508.1	205.3
09	286.5	115.8	69	342.1	138.2	29	397.8	160.7	89	453.4	183.2	49	509.0	205.7
10	287.4	116.1	70	343.1	138.6	30	398.7	161.1	90	454.3	183.6	50	510.0	206.0
311	288.4	116.5	371	344.0	139.0	431	399.6	161.5	491	455.2	184.0	551	510.9	206.4
12	289.3	116.9	72	344.9	139.4	32	400.5	161.8	92	456.2	184.3	52	511.8	206.8
13	290.2	117.3	73	345.8	139.7	33	401.5	162.2	93	457.1	184.7	53	512.7	207.2
14	291.1	117.6	74	346.8	140.1	34	402.4	162.6	94	458.0	185.1	54	513.7	207.5
15	292.1	118.0	75	347.7	140.5	35	403.3	163.0	95	459.0	185.4	55	514.6	207.9
16	293.0	118.4	76	348.6	140.9	36	404.3	163.3	96	459.9	185.8	56	515.5	208.3
17	293.9	118.8	77	349.5	141.2	37	405.2	163.7	97	460.8	186.2	57	516.4	208.7
18	294.8	119.1	78	350.5	141.6	38	406.1	164.1	98	461.7	186.6	58	517.4	209.0
19	295.8	119.5	79	351.4	142.0	39	407.0	164.5	99	462.7	186.9	59	518.3	209.4
20	296.7	119.9	80	352.3	142.4	40	408.0	164.8	500	463.6	187.3	60	519.2	209.8
321	297.6	120.2	381	353.3	142.7	441	408.9	165.2	501	464.5	187.7	561	520.2	210.2
22	298.6	120.6	82	354.2	143.1	42	409.8	165.6	02	465.4	188.1	62	521.1	210.5
23	299.5	121.0	83	355.1	143.5	43	410.7	166.0	03	466.4	188.4	63	522.0	210.9
24	300.4	121.4	84	356.0	143.8	44	411.7	166.3	04	467.3	188.8	64	522.9	211.3
25	301.3	121.7	85	357.0	144.2	45	412.6	166.7	05	468.2	189.2	65	523.9	211.7
26	302.3	122.1	86	357.9	144.6	46	413.5	167.1	06	469.2	189.6	66	524.8	212.0
27	303.2	122.5	87	358.8	145.0	47	414.5	167.4	07	470.1	189.9	67	525.7	212.4
28	304.1	122.9	88	359.7	145.3	48	415.4	167.8	08	471.0	190.3	68	526.6	212.8
29	305.0	123.2	89	360.7	145.7	49	416.3	168.2	09	471.9	190.7	69	527.6	213.2
30	306.0	123.6	90	361.6	146.1	50	417.2	168.6	10	472.9	191.0	70	528.5	213.5
331	306.9	124.0	391	362.5	146.5	451	418.2	168.9	511	473.8	191.4	571	529.4	213.9
32	307.8	124.4	92	363.5	146.8	52	419.1	169.3	12	474.7	191.8	72	530.3	214.3
33	308.8	124.7	93	364.4	147.2	53	420.0	169.7	13	475.6	192.2	73	531.3	214.6
34	309.7	125.1	94	365.3	147.6	54	420.9	170.1	14	476.6	192.5	74	532.2	215.0
35	310.6	125.5	95	366.2	148.0	55	421.9	170.4	15	477.5	192.9	75	533.1	215.4
36	311.5	125.9	96	367.2	148.3	56	422.8	170.8	16	478.4	193.3	76	534.1	215.8
37	312.5	126.2	97	368.1	148.7	57	423.7	171.2	17	479.4	193.7	77	535.0	216.1
38	313.4	126.6	98	369.0	149.1	58	424.7	171.6	18	480.3	194.0	78	535.9	216.5
39	314.3	127.0	99	369.9	149.5	59	425.6	171.9	19	481.2	194.4	79	536.8	216.9
40	315.2	127.4	400	370.9	149.8	60	426.5	172.3	20	482.1	194.8	80	537.8	217.3
341	316.2	127.7	401	371.8	150.2	461	427.4	172.7	521	483.1	195.2	581	538.7	217.6
42	317.1	128.1	02	372.7	150.6	62	428.4	173.1	22	484.0	195.5	82	539.6	218.0
43	318.0	128.5	03	373.7	151.0	63	429.3	173.4	23	484.9	195.9	83	540.5	218.4
44	319.0	128.9	04	374.6	151.3	64	430.2	173.8	24	485.8	196.3	84	541.5	218.8
45	319.9	129.2	05	375.5	151.7	65	431.1	174.2	25	486.8	196.7	85	542.4	219.1
46	320.8	129.6	06	376.4	152.1	66	432.1	174.6	26	487.7	197.0	86	543.3	219.5
47	321.7	130.0	07	377.4	152.5	67	433.0	174.9	27	488.6	197.4	87	544.3	219.9
48	322.7	130.4	08	378.3	152.8	68	433.9	175.3	28	489.6	197.8	88	545.2	220.3
49	323.6	130.7	09	379.2	153.2	69	434.8	175.7	29	490.5	198.2	89	546.1	220.6
50	324.5	131.1	10	380.1	153.6	70	435.8	176.1	30	491.4	198.5	90	547.0	221.0
351	325.4	131.5	411	381.1	154.0	471	436.7	176.4	531	492.3	198.9	591	548.0	221.4
52	326.4	131.9	12	382.0	154.3	72	437.6	176.8	32	493.3	199.3	92	548.9	221.8
53	327.3	132.2	13	382.9	154.7	73	438.6	177.2	33	494.2	199.7	93	549.8	222.1
54	328.2	132.6	14	383.9	155.1	74	439.5	177.6	34	495.1	200.0	94	550.7	222.5
55	329.2	133.0	15	384.8	155.5	75	440.4	177.9	35	496.0	200.4	95	551.7	222.9
56	330.1	133.4	16	385.7	155.8	76	441.3	178.3	36	497.0	200.8	96	552.6	223.3
57	331.0	133.7	17	386.6	156.2	77	442.3	178.7	37	497.9	201.2	97	553.5	223.6
58	332.0	134.1	18	387.6	156.6	78	443.2	179.1	38	498.8	201.5	98	554.5	224.0
59	332.9	134.5	19	388.5	157.0	79	444.1	179.4	39	499.8	201.9	99	555.4	224.4
60	333.8	134.9	20	389.4	157.3	80	445.0	179.8	40	500.7	202.3	600	556.3	224.8
Dist.	Dep.	D. Lat.	Dist.	Dep.	D. Lat.	Dist.	Dep.	D. Lat.	Dist.	Dep.	D. Lat.	Dist.	Dep.	D. Lat.

Dist.	D. Lat.	Dep.
D Lo	Dep.	
	m	D Lo

68°

292°	068°
248°	112°

337°	023°
203°	157°

TABLE 3

Traverse 23° Table

337°	023°
203°	157°

Dist.	D. Lat.	Dep.	Dist.	D. Lat.	Dep.	Dist.	D. Lat.	Dep.	Dist.	D. Lat.	Dep.	Dist.	D. Lat.	Dep.
1	0.9	0.4	61	56.2	23.8	121	111.4	47.3	181	166.6	70.7	241	221.8	94.2
2	1.8	0.8	62	57.1	24.2	22	112.3	47.7	82	167.5	71.1	42	222.8	94.6
3	2.8	1.2	63	58.0	24.6	23	113.2	48.1	83	168.5	71.5	43	223.7	94.9
4	3.7	1.6	64	58.9	25.0	24	114.1	48.5	84	169.4	71.9	44	224.6	95.3
5	4.6	2.0	65	59.8	25.4	25	115.1	48.8	85	170.3	72.3	45	225.5	95.7
6	5.5	2.3	66	60.8	25.8	26	116.0	49.2	86	171.2	72.7	46	226.4	96.1
7	6.4	2.7	67	61.7	26.2	27	116.9	49.6	87	172.1	73.1	47	227.4	96.5
8	7.4	3.1	68	62.6	26.6	28	117.8	50.0	88	173.1	73.5	48	228.3	96.9
9	8.3	3.5	69	63.5	27.0	29	118.7	50.4	89	174.0	73.8	49	229.2	97.3
10	9.2	3.9	70	64.4	27.4	30	119.7	50.8	90	174.9	74.2	50	230.1	97.7
11	10.1	4.3	71	65.4	27.7	131	120.6	51.2	191	175.8	74.6	251	231.0	98.1
12	11.0	4.7	72	66.3	28.1	32	121.5	51.6	92	176.7	75.0	52	232.0	98.5
13	12.0	5.1	73	67.2	28.5	33	122.4	52.0	93	177.7	75.4	53	232.9	98.9
14	12.9	5.5	74	68.1	28.9	34	123.3	52.4	94	178.6	75.8	54	233.8	99.2
15	13.8	5.9	75	69.0	29.3	35	124.3	52.7	95	179.5	76.2	55	234.7	99.6
16	14.7	6.3	76	70.0	29.7	36	125.2	53.1	96	180.4	76.6	56	235.6	100.0
17	15.6	6.6	77	70.9	30.1	37	126.1	53.5	97	181.3	77.0	57	236.6	100.4
18	16.6	7.0	78	71.8	30.5	38	127.0	53.9	98	182.3	77.4	58	237.5	100.8
19	17.5	7.4	79	72.7	30.9	39	128.0	54.3	99	183.2	77.8	59	238.4	101.2
20	18.4	7.8	80	73.6	31.3	40	128.9	54.7	200	184.1	78.1	60	239.3	101.6
21	19.3	8.2	81	74.6	31.6	141	129.8	55.1	201	185.0	78.5	261	240.3	102.0
22	20.3	8.6	82	75.5	32.0	42	130.7	55.5	02	185.9	78.9	62	241.2	102.4
23	21.2	9.0	83	76.4	32.4	43	131.6	55.9	03	186.9	79.3	63	242.1	102.8
24	22.1	9.4	84	77.3	32.8	44	132.6	56.3	04	187.8	79.7	64	243.0	103.2
25	23.0	9.8	85	78.2	33.2	45	133.5	56.7	05	188.7	80.1	65	243.9	103.5
26	23.9	10.2	86	79.2	33.6	46	134.4	57.0	06	189.6	80.5	66	244.9	103.9
27	24.9	10.5	87	80.1	34.0	47	135.3	57.4	07	190.5	80.9	67	245.8	104.3
28	25.8	10.9	88	81.0	34.4	48	136.2	57.8	08	191.5	81.3	68	246.7	104.7
29	26.7	11.3	89	81.9	34.8	49	137.2	58.2	09	192.4	81.7	69	247.6	105.1
30	27.6	11.7	90	82.8	35.2	50	138.1	58.6	10	193.3	82.1	70	248.5	105.5
31	28.5	12.1	91	83.8	35.6	151	139.0	59.0	211	194.2	82.4	271	249.5	105.9
32	29.5	12.5	92	84.7	35.9	52	139.9	59.4	12	195.1	82.8	72	250.4	106.3
33	30.4	12.9	93	85.6	36.3	53	140.8	59.8	13	196.1	83.2	73	251.3	106.7
34	31.3	13.3	94	86.5	36.7	54	141.8	60.2	14	197.0	83.6	74	252.2	107.1
35	32.2	13.7	95	87.4	37.1	55	142.7	60.6	15	197.9	84.0	75	253.1	107.5
36	33.1	14.1	96	88.4	37.5	56	143.6	61.0	16	198.8	84.4	76	254.1	107.8
37	34.1	14.5	97	89.3	37.9	57	144.5	61.3	17	199.7	84.8	77	255.0	108.2
38	35.0	14.8	98	90.2	38.3	58	145.4	61.7	18	200.7	85.2	78	255.9	108.6
39	35.9	15.2	99	91.1	38.7	59	146.4	62.1	19	201.6	85.6	79	256.8	109.0
40	36.8	15.6	100	92.1	39.1	60	147.3	62.5	20	202.5	86.0	80	257.7	109.4
41	37.7	16.0	101	93.0	39.5	161	148.2	62.9	221	203.4	86.4	281	258.7	109.8
42	38.7	16.4	02	93.9	39.9	62	149.1	63.3	22	204.4	86.7	82	259.6	110.2
43	39.6	16.8	03	94.8	40.2	63	150.0	63.7	23	205.3	87.1	83	260.5	110.6
44	40.5	17.2	04	95.7	40.6	64	151.0	64.1	24	206.2	87.5	84	261.4	111.0
45	41.4	17.6	05	96.7	41.0	65	151.9	64.5	25	207.1	87.9	85	262.3	111.4
46	42.3	18.0	06	97.6	41.4	66	152.8	64.9	26	208.0	88.3	86	263.3	111.7
47	43.3	18.4	07	98.5	41.8	67	153.7	65.3	27	209.0	88.7	87	264.2	112.1
48	44.2	18.8	08	99.4	42.2	68	154.6	65.6	28	209.9	89.1	88	265.1	112.5
49	45.1	19.1	09	100.3	42.6	69	155.6	66.0	29	210.8	89.5	89	266.0	112.9
50	46.0	19.5	10	101.3	43.0	70	156.5	66.4	30	211.7	89.9	90	266.9	113.3
51	46.9	19.9	111	102.2	43.4	171	157.4	66.8	231	212.6	90.3	291	267.9	113.7
52	47.9	20.3	12	103.1	43.8	72	158.3	67.2	32	213.6	90.6	92	268.8	114.1
53	48.8	20.7	13	104.0	44.2	73	159.2	67.6	33	214.5	91.0	93	269.7	114.5
54	49.7	21.1	14	104.9	44.5	74	160.2	68.0	34	215.4	91.4	94	270.6	114.9
55	50.6	21.5	15	105.9	44.9	75	161.1	68.4	35	216.3	91.8	95	271.5	115.3
56	51.5	21.9	16	106.8	45.3	76	162.0	68.8	36	217.2	92.2	96	272.5	115.7
57	52.5	22.3	17	107.7	45.7	77	162.9	69.2	37	218.2	92.6	97	273.4	116.0
58	53.4	22.7	18	108.6	46.1	78	163.8	69.6	38	219.1	93.0	98	274.3	116.4
59	54.3	23.1	19	109.5	46.5	79	164.8	69.9	39	220.0	93.4	99	275.2	116.8
60	55.2	23.4	20	110.5	46.9	80	165.7	70.3	40	220.9	93.8	300	276.2	117.2
Dist.	Dep.	D. Lat.	Dist.	Dep.	D. Lat.	Dist.	Dep.	D. Lat.	Dist.	Dep.	D. Lat.	Dist.	Dep.	D. Lat.

293°	067°
247°	113°

67°

Dist.	D. Lat.	Dep.
N.	N×Cos.	N×Sin.
Hypotenuse	Side Adj.	Side Opp.

337° | 023°		TABLE 3		337° | 023°
203° | 157°	Traverse	**23°**	Table	203° | 157°

Dist.	D. Lat.	Dep.	Dist.	D. Lat.	Dep.	Dist.	D. Lat.	Dep.	Dist.	D. Lat.	Dep.	Dist.	D. Lat.	Dep.
301	277.1	117.6	361	332.3	141.1	421	387.5	164.5	481	442.8	187.9	541	498.0	211.4
02	278.0	118.0	62	333.2	141.4	22	388.5	164.9	82	443.7	188.3	42	498.9	211.8
03	278.9	118.4	63	334.1	141.8	23	389.4	165.3	83	444.6	188.7	43	499.8	212.2
04	279.8	118.8	64	335.1	142.2	24	390.3	165.7	84	445.5	189.1	44	500.8	212.6
05	280.8	119.2	65	336.0	142.6	25	391.2	166.1	85	446.4	189.5	45	501.7	212.9
06	281.7	119.6	66	336.9	143.0	26	392.1	166.5	86	447.4	189.9	46	502.6	213.3
07	282.6	120.0	67	337.8	143.4	27	393.1	166.8	87	448.3	190.3	47	503.5	213.7
08	283.5	120.3	68	338.7	143.8	28	394.0	167.2	88	449.2	190.7	48	504.4	214.1
09	284.4	120.7	69	339.7	144.2	29	394.9	167.6	89	450.1	191.1	49	505.4	214.5
10	285.4	121.1	70	340.6	144.6	30	395.8	168.0	90	451.0	191.5	50	506.3	214.9
311	286.3	121.5	371	341.5	145.0	431	396.7	168.4	491	452.0	191.8	551	507.2	215.3
12	287.2	121.9	72	342.4	145.4	32	397.7	168.8	92	452.9	192.2	52	508.1	215.7
13	288.1	122.3	73	343.3	145.7	33	398.6	169.2	93	453.8	192.6	53	509.0	216.1
14	289.0	122.7	74	344.3	146.1	34	399.5	169.6	94	454.7	193.0	54	510.0	216.5
15	290.0	123.1	75	345.2	146.5	35	400.4	170.0	95	455.6	193.4	55	510.9	216.9
16	290.9	123.5	76	346.1	146.9	36	401.3	170.4	96	456.6	193.8	56	511.8	217.2
17	291.8	123.9	77	347.0	147.3	37	402.3	170.7	97	457.5	194.2	57	512.7	217.6
18	292.7	124.3	78	348.0	147.7	38	403.2	171.1	98	458.4	194.6	58	513.6	218.0
19	293.6	124.6	79	348.9	148.1	39	404.1	171.5	99	459.3	195.0	59	514.6	218.4
20	294.6	125.0	80	349.8	148.5	40	405.0	171.9	500	460.3	195.4	60	515.5	218.8
321	295.5	125.4	381	350.7	148.9	441	405.9	172.3	501	461.2	195.8	561	516.4	219.2
22	296.4	125.8	82	351.6	149.3	42	406.9	172.7	02	462.1	196.1	62	517.3	219.6
23	297.3	126.2	83	352.6	149.7	43	407.8	173.1	03	463.0	196.5	63	518.2	220.0
24	298.2	126.6	84	353.5	150.0	44	408.7	173.5	04	463.9	196.9	64	519.2	220.4
25	299.2	127.0	85	354.4	150.4	45	409.6	173.9	05	464.9	197.3	65	520.1	220.8
26	300.1	127.4	86	355.3	150.8	46	410.5	174.3	06	465.8	197.7	66	521.0	221.2
27	301.0	127.8	87	356.2	151.2	47	411.5	174.7	07	466.7	198.1	67	521.9	221.5
28	301.9	128.2	88	357.2	151.6	48	412.4	175.0	08	467.6	198.5	68	522.8	221.9
29	302.8	128.6	89	358.1	152.0	49	413.3	175.4	09	468.5	198.9	69	523.8	222.3
30	303.8	128.9	90	359.0	152.4	50	414.2	175.8	10	469.5	199.3	70	524.7	222.7
331	304.7	129.3	391	359.9	152.8	451	415.1	176.2	511	470.4	199.7	571	525.6	223.1
32	305.6	129.7	92	360.8	153.2	52	416.1	176.6	12	471.3	200.1	72	526.5	223.5
33	306.5	130.1	93	361.8	153.6	53	417.0	177.0	13	472.2	200.4	73	527.4	223.9
34	307.4	130.5	94	362.7	153.9	54	417.9	177.4	14	473.1	200.8	74	528.4	224.3
35	308.4	130.9	95	363.6	154.3	55	418.8	177.8	15	474.1	201.2	75	529.3	224.7
36	309.3	131.3	96	364.5	154.7	56	419.8	178.2	16	475.0	201.6	76	530.2	225.1
37	310.2	131.7	97	365.4	155.1	57	420.7	178.6	17	475.9	202.0	77	531.1	225.5
38	311.1	132.1	98	366.4	155.5	58	421.6	179.0	18	476.8	202.4	78	532.1	225.8
39	312.1	132.5	99	367.3	155.9	59	422.5	179.3	19	477.7	202.8	79	533.0	226.2
40	313.0	132.8	400	368.2	156.3	60	423.4	179.7	20	478.7	203.2	80	533.9	226.6
341	313.9	133.2	401	369.1	156.7	461	424.4	180.1	521	479.6	203.6	581	534.8	227.0
42	314.8	133.6	02	370.0	157.1	62	425.3	180.5	22	480.5	204.0	82	535.7	227.4
43	315.7	134.0	03	371.0	157.5	63	426.2	180.9	23	481.4	204.4	83	536.7	227.8
44	316.7	134.4	04	371.9	157.9	64	427.1	181.3	24	482.3	204.7	84	537.6	228.2
45	317.6	134.8	05	372.8	158.2	65	428.0	181.7	25	483.3	205.1	85	538.5	228.6
46	318.5	135.2	06	373.7	158.6	66	429.0	182.1	26	484.2	205.5	86	539.4	229.0
47	319.4	135.6	07	374.6	159.0	67	429.9	182.5	27	485.1	205.9	87	540.3	229.4
48	320.3	136.0	08	375.6	159.4	68	430.8	182.9	28	486.0	206.3	88	541.3	229.7
49	321.3	136.4	09	376.5	159.8	69	431.7	183.3	29	486.9	206.7	89	542.2	230.1
50	322.2	136.8	10	377.4	160.2	70	432.6	183.6	30	487.9	207.1	90	543.1	230.5
351	323.1	137.1	411	378.3	160.6	471	433.6	184.0	531	488.8	207.5	591	544.0	230.9
52	324.0	137.5	12	379.2	161.0	72	434.5	184.4	32	489.7	207.9	92	544.9	231.3
53	324.9	137.9	13	380.2	161.4	73	435.4	184.8	33	490.6	208.3	93	545.9	231.7
54	325.9	138.3	14	381.1	161.8	74	436.3	185.2	34	491.5	208.7	94	546.8	232.1
55	326.8	138.7	15	382.0	162.2	75	437.2	185.6	35	492.5	209.0	95	547.7	232.5
56	327.7	139.1	16	382.9	162.5	76	438.2	186.0	36	493.4	209.4	96	548.6	232.9
57	328.6	139.5	17	383.9	162.9	77	439.1	186.4	37	494.3	209.8	97	549.5	233.3
58	329.5	139.9	18	384.8	163.3	78	440.0	186.8	38	495.2	210.2	98	550.5	233.7
59	330.5	140.3	19	385.7	163.7	79	440.9	187.2	39	496.2	210.6	99	551.4	234.0
60	331.4	140.7	20	386.6	164.1	80	441.8	187.6	40	497.1	211.0	600	552.3	234.4
Dist.	Dep.	D. Lat.	Dist.	Dep.	D. Lat.	Dist.	Dep.	D. Lat.	Dist.	Dep.	D. Lat.	Dist.	Dep.	D. Lat.

Dist.	D. Lat.	Dep.
D Lo	Dep.	
	m	D Lo

67°

293°	067°
247°	113°

336° | 024°		TABLE 3		336° | 024°
204° | 156°	Traverse	24°	Table	204° | 156°

Dist.	D. Lat.	Dep.	Dist.	D. Lat.	Dep.	Dist.	D. Lat.	Dep.	Dist.	D. Lat.	Dep.	Dist.	D. Lat.	Dep.
1	0.9	0.4	61	55.7	24.8	121	110.5	49.2	181	165.4	73.6	241	220.2	98.0
2	1.8	0.8	62	56.6	25.2	22	111.5	49.6	82	166.3	74.0	42	221.1	98.4
3	2.7	1.2	63	57.6	25.6	23	112.4	50.0	83	167.2	74.4	43	222.0	98.8
4	3.7	1.6	64	58.5	26.0	24	113.3	50.4	84	168.1	74.8	44	222.9	99.2
5	4.6	2.0	65	59.4	26.4	25	114.2	50.8	85	169.0	75.2	45	223.8	99.7
6	5.5	2.4	66	60.3	26.8	26	115.1	51.2	86	169.9	75.7	46	224.7	100.1
7	6.4	2.8	67	61.2	27.3	27	116.0	51.7	87	170.8	76.1	47	225.6	100.5
8	7.3	3.3	68	62.1	27.7	28	116.9	52.1	88	171.7	76.5	48	226.6	100.9
9	8.2	3.7	69	63.0	28.1	29	117.8	52.5	89	172.7	76.9	49	227.5	101.3
10	9.1	4.1	70	63.9	28.5	30	118.8	52.9	90	173.6	77.3	50	228.4	101.7
11	10.0	4.5	71	64.9	28.9	131	119.7	53.3	191	174.5	77.7	251	229.3	102.1
12	11.0	4.9	72	65.8	29.3	32	120.6	53.7	92	175.4	78.1	52	230.2	102.5
13	11.9	5.3	73	66.7	29.7	33	121.5	54.1	93	176.3	78.5	53	231.1	102.9
14	12.8	5.7	74	67.6	30.1	34	122.4	54.5	94	177.2	78.9	54	232.0	103.3
15	13.7	6.1	75	68.5	30.5	35	123.3	54.9	95	178.1	79.3	55	233.0	103.7
16	14.6	6.5	76	69.4	30.9	36	124.2	55.3	96	179.1	79.7	56	233.9	104.1
17	15.5	6.9	77	70.3	31.3	37	125.2	55.7	97	180.0	80.1	57	234.8	104.5
18	16.4	7.3	78	71.3	31.7	38	126.1	56.1	98	180.9	80.5	58	235.7	104.9
19	17.4	7.7	79	72.2	32.1	39	127.0	56.5	99	181.8	80.9	59	236.6	105.3
20	18.3	8.1	80	73.1	32.5	40	127.9	56.9	200	182.7	81.3	60	237.5	105.8
21	19.2	8.5	81	74.0	32.9	141	128.8	57.3	201	183.6	81.8	261	238.4	106.2
22	20.1	8.9	82	74.9	33.4	42	129.7	57.8	02	184.5	82.2	62	239.3	106.6
23	21.0	9.4	83	75.8	33.8	43	130.6	58.2	03	185.4	82.6	63	240.3	107.0
24	21.9	9.8	84	76.7	34.2	44	131.6	58.6	04	186.4	83.0	64	241.2	107.4
25	22.8	10.2	85	77.7	34.6	45	132.5	59.0	05	187.3	83.4	65	242.1	107.8
26	23.8	10.6	86	78.6	35.0	46	133.4	59.4	06	188.2	83.8	66	243.0	108.2
27	24.7	11.0	87	79.5	35.4	47	134.3	59.8	07	189.1	84.2	67	243.9	108.6
28	25.6	11.4	88	80.4	35.8	48	135.2	60.2	08	190.0	84.6	68	244.8	109.0
29	26.5	11.8	89	81.3	36.2	49	136.1	60.6	09	190.9	85.0	69	245.7	109.4
30	27.4	12.2	90	82.2	36.6	50	137.0	61.0	10	191.8	85.4	70	246.7	109.8
31	28.3	12.6	91	83.1	37.0	151	137.9	61.4	211	192.8	85.8	271	247.6	110.2
32	29.2	13.0	92	84.0	37.4	52	138.9	61.8	12	193.7	86.2	72	248.5	110.6
33	30.1	13.4	93	85.0	37.8	53	139.8	62.2	13	194.6	86.6	73	249.4	111.0
34	31.1	13.8	94	85.9	38.2	54	140.7	62.6	14	195.5	87.0	74	250.3	111.4
35	32.0	14.2	95	86.8	38.6	55	141.6	63.0	15	196.4	87.4	75	251.2	111.9
36	32.9	14.6	96	87.7	39.0	56	142.5	63.5	16	197.3	87.9	76	252.1	112.3
37	33.8	15.0	97	88.6	39.5	57	143.4	63.9	17	198.2	88.3	77	253.1	112.7
38	34.7	15.5	98	89.5	39.9	58	144.3	64.3	18	199.2	88.7	78	254.0	113.1
39	35.6	15.9	99	90.4	40.3	59	145.3	64.7	19	200.1	89.1	79	254.9	113.5
40	36.5	16.3	100	91.4	40.7	60	146.2	65.1	20	201.0	89.5	80	255.8	113.9
41	37.5	16.7	101	92.3	41.1	161	147.1	65.5	221	201.9	89.9	281	256.7	114.3
42	38.4	17.1	02	93.2	41.5	62	148.0	65.9	22	202.8	90.3	82	257.6	114.7
43	39.3	17.5	03	94.1	41.9	63	148.9	66.3	23	203.7	90.7	83	258.5	115.1
44	40.2	17.9	04	95.0	42.3	64	149.8	66.7	24	204.6	91.1	84	259.4	115.5
45	41.1	18.3	05	95.9	42.7	65	150.7	67.1	25	205.5	91.5	85	260.4	115.9
46	42.0	18.7	06	96.8	43.1	66	151.6	67.5	26	206.5	91.9	86	261.3	116.3
47	42.9	19.1	07	97.7	43.5	67	152.6	67.9	27	207.4	92.3	87	262.2	116.7
48	43.9	19.5	08	98.7	43.9	68	153.5	68.3	28	208.3	92.7	88	263.1	117.1
49	44.8	19.9	09	99.6	44.3	69	154.4	68.7	29	209.2	93.1	89	264.0	117.5
50	45.7	20.3	10	100.5	44.7	70	155.3	69.1	30	210.1	93.5	90	264.9	118.0
51	46.6	20.7	111	101.4	45.1	171	156.2	69.6	231	211.0	94.0	291	265.8	118.4
52	47.5	21.2	12	102.3	45.6	72	157.1	70.0	32	211.9	94.4	92	266.8	118.8
53	48.4	21.6	13	103.2	46.0	73	158.0	70.4	33	212.9	94.8	93	267.7	119.2
54	49.3	22.0	14	104.1	46.4	74	159.0	70.8	34	213.8	95.2	94	268.6	119.6
55	50.2	22.4	15	105.1	46.8	75	159.9	71.2	35	214.7	95.6	95	269.5	120.0
56	51.2	22.8	16	106.0	47.2	76	160.8	71.6	36	215.6	96.0	96	270.4	120.4
57	52.1	23.2	17	106.9	47.6	77	161.7	72.0	37	216.5	96.4	97	271.3	120.8
58	53.0	23.6	18	107.8	48.0	78	162.6	72.4	38	217.4	96.8	98	272.2	121.2
59	53.9	24.0	19	108.7	48.4	79	163.5	72.8	39	218.3	97.2	99	273.2	121.6
60	54.8	24.4	20	109.6	48.8	80	164.4	73.2	40	219.3	97.6	300	274.1	122.0
Dist.	Dep.	D. Lat.	Dist.	Dep.	D. Lat.	Dist.	Dep.	D. Lat.	Dist.	Dep.	D. Lat.	Dist.	Dep.	D. Lat.

294° | 066°		
246° | 114°	66°	

Dist.	D. Lat.	Dep.
N.	N×Cos.	N×Sin.
Hypotenuse	Side Adj.	Side Opp.

336°	024°
204°	156°

TABLE 3

Traverse **24°** Table

336°	024°
204°	156°

Dist.	D. Lat.	Dep.	Dist.	D. Lat.	Dep.	Dist.	D. Lat.	Dep.	Dist.	D. Lat.	Dep.	Dist.	D. Lat.	Dep.
301	275.0	122.4	361	329.8	146.8	421	384.6	171.2	481	439.4	195.6	541	494.2	220.0
02	275.9	122.8	62	330.7	147.2	22	385.5	171.6	82	440.3	196.0	42	495.1	220.5
03	276.8	123.2	63	331.6	147.6	23	386.4	172.0	83	441.2	196.5	43	496.1	220.9
04	277.7	123.6	64	332.5	148.1	24	387.3	172.5	84	442.2	196.9	44	497.0	221.3
05	278.6	124.1	65	333.4	148.5	25	388.3	172.9	85	443.1	197.3	45	497.9	221.7
06	279.5	124.5	66	334.4	148.9	26	389.2	173.3	86	444.0	197.7	46	498.8	222.1
07	280.5	124.9	67	335.3	149.3	27	390.1	173.7	87	444.9	198.1	47	499.7	222.5
08	281.4	125.3	68	336.2	149.7	28	391.0	174.1	88	445.8	198.5	48	500.6	222.9
09	282.3	125.7	69	337.1	150.1	29	391.9	174.5	89	446.7	198.9	49	501.5	223.3
10	283.2	126.1	70	338.0	150.5	30	392.8	174.9	90	447.6	199.3	50	502.5	223.7
311	284.1	126.5	371	338.9	150.9	431	393.7	175.3	491	448.6	199.7	551	503.4	224.1
12	285.0	126.9	72	339.8	151.3	32	394.7	175.7	92	449.5	200.1	52	504.3	224.5
13	285.9	127.3	73	340.7	151.7	33	395.6	176.1	93	450.4	200.5	53	505.2	224.9
14	286.9	127.7	74	341.7	152.1	34	396.5	176.5	94	451.3	200.9	54	506.1	225.3
15	287.8	128.1	75	342.6	152.5	35	397.4	176.9	95	452.2	201.3	55	507.0	225.7
16	288.7	128.5	76	343.5	152.9	36	398.3	177.3	96	453.1	201.7	56	507.9	226.1
17	289.6	128.9	77	344.4	153.3	37	399.2	177.7	97	454.0	202.1	57	508.8	226.6
18	290.5	129.3	78	345.3	153.7	38	400.1	178.2	98	454.9	202.6	58	509.8	227.0
19	291.4	129.7	79	346.2	154.2	39	401.0	178.6	99	455.9	203.0	59	510.7	227.4
20	292.3	130.2	80	347.1	154.6	40	402.0	179.0	500	456.8	203.4	60	511.6	227.8
321	293.2	130.6	381	348.1	155.0	441	402.9	179.4	501	457.7	203.8	561	512.5	228.2
22	294.2	131.0	82	349.0	155.4	42	403.8	179.8	02	458.6	204.2	62	513.4	228.6
23	295.1	131.4	83	349.9	155.8	43	404.7	180.2	03	459.5	204.6	63	514.3	229.0
24	296.0	131.8	84	350.8	156.2	44	405.6	180.6	04	460.4	205.0	64	515.2	229.4
25	296.9	132.2	85	351.7	156.6	45	406.5	181.0	05	461.3	205.4	65	516.2	229.8
26	297.8	132.6	86	352.6	157.0	46	407.4	181.4	06	442.3	205.8	66	517.1	230.2
27	298.7	133.0	87	353.5	157.4	47	408.4	181.8	07	463.2	206.2	67	518.0	230.6
28	299.6	133.4	88	354.5	157.8	48	409.3	182.2	08	464.1	206.6	68	518.9	231.0
29	300.6	133.8	89	355.4	158.2	49	410.2	182.6	09	465.0	207.0	69	519.8	231.4
30	301.5	134.2	90	356.3	158.6	50	411.1	183.0	10	465.9	207.4	70	520.7	231.8
331	302.4	134.6	391	357.2	159.0	451	412.0	183.4	511	466.8	207.8	571	521.6	232.2
32	303.3	135.0	92	358.1	159.4	52	412.9	183.8	12	467.7	208.2	72	522.5	232.7
33	304.2	135.4	93	359.0	159.8	53	413.8	184.3	13	468.6	208.7	73	523.5	233.1
34	305.1	135.9	94	359.9	160.3	54	414.7	184.7	14	469.6	209.1	74	524.4	233.5
35	306.0	136.3	95	360.9	160.7	55	415.7	185.1	15	470.5	209.5	75	525.3	233.9
36	307.0	136.7	96	361.8	161.1	56	416.6	185.5	16	471.4	209.9	76	526.2	234.3
37	307.9	137.1	97	362.7	161.5	57	417.5	185.9	17	472.3	210.3	77	527.1	234.7
38	308.8	137.5	98	363.6	161.9	58	418.4	186.3	18	473.2	210.7	78	528.0	235.1
39	309.7	137.9	99	364.5	162.3	59	419.3	186.7	19	474.1	211.1	79	528.9	235.5
40	310.6	138.3	400	365.4	162.7	60	420.2	187.1	20	475.0	211.5	80	529.9	235.9
341	311.5	138.7	401	366.3	163.1	461	421.1	187.5	521	476.0	211.9	581	530.8	236.3
42	312.4	139.1	02	367.2	163.5	62	422.1	187.9	22	476.9	212.3	82	531.7	236.7
43	313.3	139.5	03	368.2	163.9	63	423.0	188.3	23	477.8	212.7	83	532.6	237.1
44	314.3	139.9	04	369.1	164.3	64	423.9	188.7	24	478.7	213.1	84	533.5	237.5
45	315.2	140.3	05	370.0	164.7	65	424.8	189.1	25	479.6	213.5	85	534.4	237.9
46	316.1	140.7	06	370.9	165.1	66	425.7	189.5	26	480.5	213.9	86	535.3	238.3
47	317.0	141.1	07	371.8	165.5	67	426.6	189.9	27	481.4	214.4	87	536.3	238.8
48	317.9	141.5	08	372.7	165.9	68	427.5	190.4	28	482.4	214.8	88	537.2	239.2
49	318.8	142.0	09	373.6	166.4	69	428.5	190.8	29	483.3	215.2	89	538.1	239.6
50	319.7	142.4	10	374.6	166.8	70	429.4	191.2	30	484.2	215.6	90	539.0	240.0
351	320.7	142.8	411	375.5	167.2	471	430.3	191.6	531	485.1	216.0	591	539.9	240.4
52	321.6	143.2	12	376.4	167.6	72	431.2	192.0	32	486.0	216.4	92	540.8	240.8
53	322.5	143.6	13	377.3	168.0	73	432.1	192.4	33	486.9	216.8	93	541.7	241.2
54	323.4	144.0	14	378.2	168.4	74	433.0	192.8	34	487.8	217.2	94	542.6	241.6
55	324.3	144.4	15	379.1	168.8	75	433.9	193.2	35	488.7	217.6	95	543.6	242.0
56	325.2	144.8	16	380.0	169.2	76	434.8	193.6	36	489.7	218.0	96	544.5	242.4
57	326.1	145.2	17	380.9	169.6	77	435.8	194.0	37	490.6	218.4	97	545.4	242.8
58	327.0	145.6	18	381.9	170.0	78	436.7	194.4	38	491.5	218.8	98	546.3	243.2
59	328.0	146.0	19	382.8	170.4	79	437.6	194.8	39	492.4	219.2	99	547.2	243.6
60	328.9	146.4	20	383.7	170.8	80	438.5	195.2	40	493.3	219.6	600	548.1	244.0
Dist.	Dep.	D. Lat.	Dist.	Dep.	D. Lat.	Dist.	Dep.	D. Lat.	Dist.	Dep.	D. Lat.	Dist.	Dep.	D. Lat.

Dist.	D. Lat.	Dep.
D Lo	Dep.	
	m	D Lo

66°

294°	066°
246°	114°

335° | 025° / 205° | 155°	TABLE 3 — Traverse 25° Table	335° | 025° / 205° | 155°

Dist.	D. Lat.	Dep.	Dist.	D. Lat.	Dep.	Dist.	D. Lat.	Dep.	Dist.	D. Lat.	Dep.	Dist.	D. Lat.	Dep.
1	0.9	0.4	61	55.3	25.8	121	109.7	51.1	181	164.0	76.5	241	218.4	101.9
2	1.8	0.8	62	56.2	26.2	22	110.6	51.6	82	164.9	76.9	42	219.3	102.3
3	2.7	1.3	63	57.1	26.6	23	111.5	52.0	83	165.9	77.3	43	220.2	102.7
4	3.6	1.7	64	58.0	27.0	24	112.4	52.4	84	166.8	77.8	44	221.1	103.1
5	4.5	2.1	65	58.9	27.5	25	113.3	52.8	85	167.7	78.2	45	222.0	103.5
6	5.4	2.5	66	59.8	27.9	26	114.2	53.2	86	168.6	78.6	46	223.0	104.0
7	6.3	3.0	67	60.7	28.3	27	115.1	53.7	87	169.5	79.0	47	223.9	104.4
8	7.3	3.4	68	61.6	28.7	28	116.0	54.1	88	170.4	79.5	48	224.8	104.8
9	8.2	3.8	69	62.5	29.2	29	116.9	54.5	89	171.3	79.9	49	225.7	105.2
10	9.1	4.2	70	63.4	29.6	30	117.8	54.9	90	172.2	80.3	50	226.6	105.7
11	10.0	4.6	71	64.3	30.0	131	118.7	55.4	191	173.1	80.7	251	227.5	106.1
12	10.9	5.1	72	65.3	30.4	32	119.6	55.8	92	174.0	81.1	52	228.4	106.5
13	11.8	5.5	73	66.2	30.9	33	120.5	56.2	93	174.9	81.6	53	229.3	106.9
14	12.7	5.9	74	67.1	31.3	34	121.4	56.6	94	175.8	82.0	54	230.2	107.3
15	13.6	6.3	75	68.0	31.7	35	122.4	57.1	95	176.7	82.4	55	231.1	107.8
16	14.5	6.8	76	68.9	32.1	36	123.3	57.5	96	177.6	82.8	56	232.0	108.2
17	15.4	7.2	77	69.8	32.5	37	124.2	57.9	97	178.5	83.3	57	232.9	108.6
18	16.3	7.6	78	70.7	33.0	38	125.1	58.3	98	179.4	83.7	58	233.8	109.0
19	17.2	8.0	79	71.6	33.4	39	126.0	58.7	99	180.4	84.1	59	234.7	109.5
20	18.1	8.5	80	72.5	33.8	40	126.9	59.2	200	181.3	84.5	60	235.6	109.9
21	19.0	8.9	81	73.4	34.2	141	127.8	59.6	201	182.2	84.9	261	236.5	110.3
22	19.9	9.3	82	74.3	34.7	42	128.7	60.0	02	183.1	85.4	62	237.5	110.7
23	20.8	9.7	83	75.2	35.1	43	129.6	60.4	03	184.0	85.8	63	238.4	111.1
24	21.8	10.1	84	76.1	35.5	44	130.5	60.9	04	184.9	86.2	64	239.3	111.6
25	22.7	10.6	85	77.0	35.9	45	131.4	61.3	05	185.8	86.6	65	240.2	112.0
26	23.6	11.0	86	77.9	36.3	46	132.3	61.7	06	186.7	87.1	66	241.1	112.4
27	24.5	11.4	87	78.8	36.8	47	133.2	62.1	07	187.6	87.5	67	242.0	112.8
28	25.4	11.8	88	79.8	37.2	48	134.1	62.5	08	188.5	87.9	68	242.9	113.3
29	26.3	12.3	89	80.7	37.6	49	135.0	63.0	09	189.4	88.3	69	243.8	113.7
30	27.2	12.7	90	81.6	38.0	50	135.9	63.4	10	190.3	88.7	70	244.7	114.1
31	28.1	13.1	91	82.5	38.5	151	136.9	63.8	211	191.2	89.2	271	245.6	114.5
32	29.0	13.5	92	83.4	38.9	52	137.8	64.2	12	192.1	89.6	72	246.5	115.0
33	29.9	13.9	93	84.3	39.3	53	138.7	64.7	13	193.0	90.0	73	247.4	115.4
34	30.8	14.4	94	85.2	39.7	54	139.6	65.1	14	193.9	90.4	74	248.3	115.8
35	31.7	14.8	95	86.1	40.1	55	140.5	65.5	15	194.9	90.9	75	249.2	116.2
36	32.6	15.2	96	87.0	40.6	56	141.4	65.9	16	195.8	91.3	76	250.1	116.6
37	33.5	15.6	97	87.9	41.0	57	142.3	66.4	17	196.7	91.7	77	251.0	117.1
38	34.4	16.1	98	88.8	41.4	58	143.2	66.8	18	197.6	92.1	78	252.0	117.5
39	35.3	16.5	99	89.7	41.8	59	144.1	67.2	19	198.5	92.6	79	252.9	117.9
40	36.3	16.9	100	90.6	42.3	60	145.0	67.6	20	199.4	93.0	80	253.8	118.3
41	37.2	17.3	101	91.5	42.7	161	145.9	68.0	221	200.3	93.4	281	254.7	118.8
42	38.1	17.7	02	92.4	43.1	62	146.8	68.5	22	201.2	93.8	82	255.6	119.2
43	39.0	18.2	03	93.3	43.5	63	147.7	68.9	23	202.1	94.2	83	256.5	119.6
44	39.9	18.6	04	94.3	44.0	64	148.6	69.3	24	203.0	94.7	84	257.4	120.0
45	40.8	19.0	05	95.2	44.4	65	149.5	69.7	25	203.9	95.1	85	258.3	120.4
46	41.7	19.4	06	96.1	44.8	66	150.4	70.2	26	204.8	95.5	86	259.2	120.9
47	42.6	19.9	07	97.0	45.2	67	151.4	70.6	27	205.7	95.9	87	260.1	121.3
48	43.5	20.3	08	97.9	45.6	68	152.3	71.0	28	206.6	96.4	88	261.0	121.7
49	44.4	20.7	09	98.8	46.1	69	153.2	71.4	29	207.5	96.8	89	261.9	122.1
50	45.3	21.1	10	99.7	46.5	70	154.1	71.8	30	208.5	97.2	90	262.8	122.6
51	46.2	21.6	111	100.6	46.9	171	155.0	72.3	231	209.4	97.6	291	263.7	123.0
52	47.1	22.0	12	101.5	47.3	72	155.9	72.7	32	210.3	98.0	92	264.6	123.4
53	48.0	22.4	13	102.4	47.8	73	156.8	73.1	33	211.2	98.5	93	265.5	123.8
54	48.9	22.8	14	103.3	48.2	74	157.7	73.5	34	212.1	98.9	94	266.5	124.2
55	49.8	23.2	15	104.2	48.6	75	158.6	74.0	35	213.0	99.3	95	267.4	124.7
56	50.8	23.7	16	105.1	49.0	76	159.5	74.4	36	213.9	99.7	96	268.3	125.1
57	51.7	24.1	17	106.0	49.4	77	160.4	74.8	37	214.8	100.2	97	269.2	125.5
58	52.6	24.5	18	106.9	49.9	78	161.3	75.2	38	215.7	100.6	98	270.1	125.9
59	53.5	24.9	19	107.9	50.3	79	162.2	75.6	39	216.6	101.0	99	271.0	126.4
60	54.4	25.4	20	108.8	50.7	80	163.1	76.1	40	217.5	101.4	300	271.9	126.8
Dist.	Dep.	D. Lat.	Dist.	Dep.	D. Lat.	Dist.	Dep.	D. Lat.	Dist.	Dep.	D. Lat.	Dist.	Dep.	D. Lat.

295° | 065° / 245° | 115°	65°	

Dist.	D. Lat.	Dep.
N.	N×Cos.	N×Sin.
Hypotenuse	Side Adj.	Side Opp.

335° | 025° / 205° | 155°						TABLE 3 Traverse **25°** Table						335° | 025° / 205° | 155°		
Dist.	D. Lat.	Dep.	Dist.	D. Lat.	Dep.	Dist.	D. Lat.	Dep.	Dist.	D. Lat.	Dep.	Dist.	D. Lat.	Dep.
301	272.8	127.2	361	327.2	152.6	421	381.6	177.9	481	435.9	203.3	541	490.3	228.6
02	273.7	127.6	62	328.0	153.0	22	382.5	178.3	82	436.8	203.7	42	491.2	229.1
03	274.6	128.1	63	329.0	153.4	23	383.4	178.8	83	437.7	204.1	43	492.1	229.5
04	275.5	128.5	64	329.9	153.8	24	384.3	179.2	84	438.7	204.5	44	493.0	229.9
05	276.4	128.9	65	330.8	154.3	25	385.2	179.6	85	439.6	204.9	45	493.9	230.3
06	277.3	129.3	66	331.7	154.7	26	386.1	180.0	86	440.5	205.4	46	494.8	230.7
07	278.2	129.7	67	332.6	155.1	27	387.0	180.5	87	441.4	205.8	47	495.8	231.2
08	279.1	130.2	68	333.5	155.5	28	387.9	180.9	88	442.3	206.2	48	496.7	231.6
09	280.0	130.6	69	334.4	155.9	29	388.8	181.3	89	443.2	206.6	49	497.6	232.0
10	281.0	131.0	70	335.3	156.4	30	389.7	181.7	90	444.1	207.1	50	498.5	232.4
311	281.9	131.4	371	336.2	156.8	431	390.6	182.1	491	445.0	207.5	551	499.4	232.9
12	282.8	131.9	72	337.1	157.2	32	391.5	182.6	92	445.9	207.9	52	500.3	233.3
13	283.7	132.3	73	338.1	157.6	33	392.4	183.0	93	446.8	208.4	53	501.2	233.7
14	284.6	132.7	74	339.0	158.1	34	393.3	183.4	94	447.7	208.8	54	502.1	234.1
15	285.5	133.1	75	339.9	158.5	35	394.2	183.8	95	448.6	209.2	55	503.0	234.6
16	286.4	133.5	76	340.8	158.9	36	395.2	184.3	96	449.5	209.6	56	503.9	235.0
17	287.3	134.0	77	341.7	159.3	37	396.1	184.7	97	450.4	210.0	57	504.8	235.4
18	288.2	134.4	78	342.5	159.7	38	397.0	185.1	98	451.3	210.4	58	505.7	235.8
19	289.1	134.8	79	343.5	160.2	39	397.9	185.5	99	452.2	210.9	59	506.6	236.2
20	290.0	135.2	80	344.4	160.6	40	398.8	186.0	500	453.2	211.3	60	507.5	236.7
321	290.9	135.7	381	345.3	161.0	441	399.6	186.3	501	454.1	211.7	561	508.4	237.1
22	291.8	136.1	82	346.2	161.4	42	400.6	186.8	02	455.0	212.2	62	509.3	237.5
23	292.7	136.5	83	347.1	161.9	43	401.5	187.2	03	455.9	212.6	63	510.3	237.9
24	293.6	136.9	84	348.0	162.3	44	402.4	187.6	04	456.8	213.0	64	511.2	238.4
25	294.6	137.4	85	348.9	162.7	45	403.3	188.1	05	457.7	213.4	65	512.1	238.8
26	295.5	137.8	86	349.8	163.1	46	404.2	188.5	06	458.6	213.8	66	513.0	239.2
27	296.4	138.2	87	350.7	163.6	47	405.1	188.9	07	459.5	214.3	67	513.9	239.6
28	297.3	138.6	88	351.6	164.0	48	406.0	189.3	08	460.4	214.7	68	514.8	240.0
29	298.2	139.0	89	352.6	164.4	49	406.9	189.8	09	461.3	215.1	69	515.7	240.5
30	299.1	139.5	90	353.5	164.8	50	407.8	190.2	10	462.2	215.5	70	516.6	240.9
331	300.0	139.9	391	354.4	165.2	451	408.7	190.6	511	463.1	216.0	571	517.5	241.3
32	300.9	140.3	92	355.3	165.7	52	409.7	191.0	12	464.0	216.4	72	518.4	241.7
33	301.8	140.7	93	356.2	166.1	53	410.6	191.4	13	464.9	216.8	73	519.3	242.2
34	302.7	141.2	94	357.1	166.5	54	411.5	191.7	14	465.8	217.2	74	520.2	242.6
35	303.6	141.6	95	358.0	166.9	55	412.4	192.3	15	466.7	217.6	75	521.1	243.0
36	304.5	142.0	96	358.9	167.4	56	413.3	192.7	16	467.7	218.1	76	522.0	243.4
37	305.4	142.4	97	359.8	167.8	57	414.1	193.1	17	468.6	218.5	77	522.9	243.9
38	306.3	142.8	98	360.7	168.2	58	415.1	193.6	18	469.5	218.9	78	523.8	244.3
39	307.2	143.3	99	361.6	168.6	59	416.0	194.0	19	470.4	219.3	79	524.8	244.7
40	308.1	143.7	400	362.5	169.0	60	416.9	194.4	20	471.8	219.8	80	525.7	245.1
341	309.1	144.1	401	363.4	169.5	461	417.8	194.8	521	472.2	220.2	581	526.6	245.5
42	310.0	144.5	02	364.3	169.9	62	418.7	195.2	22	473.1	220.6	82	527.5	246.0
43	310.9	145.0	03	365.2	170.3	63	419.6	195.7	23	474.0	221.0	83	528.4	246.4
44	311.8	145.4	04	366.1	170.7	64	420.5	196.1	24	474.9	221.5	84	529.3	246.8
45	312.7	145.8	05	367.1	171.2	65	421.4	196.5	25	475.8	221.9	85	530.2	247.2
46	313.6	146.2	06	368.0	171.6	66	422.3	196.9	26	476.7	222.3	86	531.1	247.7
47	314.5	146.6	07	368.9	172.0	67	423.2	197.4	27	477.6	222.7	87	532.0	248.1
48	315.4	147.1	08	369.8	172.4	68	424.2	197.8	28	478.5	223.1	88	532.9	248.5
49	316.3	147.5	09	370.7	172.9	69	425.1	198.2	29	479.4	223.6	89	533.8	248.9
50	317.2	147.9	10	371.6	173.3	70	426.0	198.6	30	480.3	224.0	90	534.7	249.3
351	318.1	148.3	411	372.5	173.7	471	426.9	199.1	531	481.2	224.4	591	535.6	249.8
52	319.0	148.8	12	373.4	174.1	72	427.8	199.5	32	482.1	224.8	92	536.5	250.2
53	319.9	149.2	13	374.3	174.5	73	428.7	199.9	33	483.1	225.3	92	537.4	250.6
54	320.8	149.6	14	375.2	175.0	74	429.6	200.3	34	482.0	225.7	94	538.3	251.0
55	321.7	150.0	15	376.1	175.4	75	430.5	200.7	35	484.9	226.1	95	539.3	251.5
56	322.6	150.5	16	377.0	175.8	76	431.4	201.2	36	485.8	226.5	96	540.2	251.9
57	323.6	150.9	17	377.9	176.2	77	432.3	201.6	37	486.7	226.9	97	541.1	252.3
58	324.5	151.3	18	378.8	176.7	78	433.2	202.0	38	487.6	227.4	98	542.0	252.7
59	325.4	151.7	19	379.7	177.1	79	434.1	202.4	39	488.5	227.8	99	542.9	253.1
60	326.3	152.1	20	380.6	177.5	80	435.0	202.9	40	489.4	228.2	600	543.8	253.6
Dist.	Dep.	D. Lat.	Dist.	Dep.	D. Lat.	Dist.	Dep.	D. Lat.	Dist.	Dep.	D. Lat.	Dist.	Dep.	D. Lat.

Dist.	D. Lat.	Dep.
D Lo	Dep.	
	m	D Lo

65°

295° | 065°

245° | 115°

334°	026°
206°	154°

TABLE 3

Traverse 26° Table

334°	026°
206°	154°

Dist.	D. Lat.	Dep.	Dist.	D. Lat.	Dep.	Dist.	D. Lat.	Dep.	Dist.	D. Lat.	Dep.	Dist.	D. Lat.	Dep.
1	0.9	0.4	61	54.8	26.7	121	108.8	53.0	181	162.7	79.3	241	216.6	105.6
2	1.8	0.9	62	55.7	27.2	22	109.7	53.5	82	163.6	79.8	42	217.5	106.1
3	2.7	1.3	63	56.6	27.6	23	110.6	53.9	83	164.5	80.2	43	218.4	106.5
4	3.6	1.8	64	57.5	28.1	24	111.5	54.4	84	165.4	80.7	44	219.3	107.0
5	4.5	2.2	65	58.4	28.5	25	112.3	54.8	85	166.3	81.1	45	220.2	107.4
6	5.4	2.6	66	59.3	28.9	26	113.2	55.2	86	167.2	81.5	46	221.1	107.8
7	6.3	3.1	67	60.2	29.4	27	114.1	55.7	87	168.1	82.0	47	222.0	108.3
8	7.2	3.5	68	61.1	29.8	28	115.0	56.1	88	169.0	82.4	48	222.9	108.7
9	8.1	3.9	69	62.0	30.2	29	115.9	56.5	89	169.9	82.9	49	223.8	109.2
10	9.0	4.4	70	62.9	30.7	30	116.8	57.0	90	170.8	83.3	50	224.7	109.6
11	9.9	4.8	71	63.8	31.1	131	117.7	57.4	191	171.7	83.7	251	225.6	110.0
12	10.8	5.3	72	64.7	31.6	32	118.6	57.9	92	172.6	84.2	52	226.5	110.5
13	11.7	5.7	73	65.6	32.0	33	119.5	58.3	93	173.5	84.6	53	227.4	110.9
14	12.6	6.1	74	66.5	32.4	34	120.4	58.7	94	174.4	85.0	54	228.3	111.3
15	13.5	6.6	75	67.4	32.9	35	121.3	59.2	95	175.3	85.5	55	229.2	111.8
16	14.4	7.0	76	68.3	33.3	36	122.2	59.6	96	176.2	85.9	56	230.1	112.2
17	15.3	7.5	77	69.2	33.8	37	123.1	60.1	97	177.1	86.4	57	231.0	112.7
18	16.2	7.9	78	70.1	34.2	38	124.0	60.5	98	178.0	86.8	58	231.9	113.1
19	17.1	8.3	79	71.0	34.6	39	124.9	60.9	99	178.9	87.2	59	232.8	113.5
20	18.0	8.8	80	71.9	35.1	40	125.8	61.4	200	179.8	87.7	60	233.7	114.0
21	18.9	9.2	81	72.8	35.5	141	126.7	61.8	201	180.7	88.1	261	234.6	114.4
22	19.8	9.6	82	73.7	35.9	42	127.6	62.2	02	181.6	88.6	62	235.5	114.9
23	20.7	10.1	83	74.6	36.4	43	128.5	62.7	03	182.5	89.0	63	236.4	115.3
24	21.6	10.5	84	75.5	36.8	44	129.4	63.1	04	183.4	89.4	64	237.3	115.7
25	22.5	11.0	85	76.4	37.3	45	130.3	63.6	05	184.3	89.9	65	238.2	116.2
26	23.4	11.4	86	77.3	37.7	46	131.2	64.0	06	185.2	90.3	66	239.1	116.6
27	24.3	11.8	87	78.2	38.1	47	132.1	64.4	07	186.1	90.7	67	240.0	117.0
28	25.2	12.3	88	79.1	38.6	48	133.0	64.9	08	186.9	91.2	68	240.9	117.5
29	26.1	12.7	89	80.0	39.0	49	133.9	65.3	09	187.8	91.6	69	241.8	117.9
30	27.0	13.2	90	80.9	39.5	50	134.8	65.8	10	188.7	92.1	70	242.7	118.4
31	27.9	13.6	91	81.8	39.9	151	135.7	66.2	211	189.6	92.5	271	243.6	118.8
32	28.8	14.0	92	82.7	40.3	52	136.6	66.6	12	190.5	92.9	72	244.5	119.2
33	29.7	14.5	93	83.6	40.8	53	137.5	67.1	13	191.4	93.4	73	245.4	119.7
34	30.6	14.9	94	84.5	41.2	54	138.4	67.5	14	192.3	93.8	74	246.3	120.1
35	31.5	15.3	95	85.4	41.6	55	139.3	67.9	15	193.2	94.2	75	247.2	120.6
36	32.4	15.8	96	86.3	42.1	56	140.2	68.4	16	194.1	94.7	76	248.1	121.0
37	33.3	16.2	97	87.2	42.5	57	141.1	68.8	17	195.0	95.1	77	249.0	121.4
38	34.2	16.7	98	88.1	43.0	58	142.0	69.3	18	195.9	95.6	78	249.9	121.9
39	35.1	17.1	99	89.0	43.4	59	142.9	69.7	19	196.8	96.0	79	250.8	122.3
40	36.0	17.5	100	89.9	43.8	60	143.8	70.1	20	197.7	96.4	80	251.7	122.7
41	36.9	18.0	101	90.8	44.3	161	144.7	70.6	221	198.6	96.9	281	252.6	123.2
42	37.7	18.4	02	91.7	44.7	62	145.6	71.0	22	199.5	97.3	82	253.5	123.6
43	38.6	18.8	03	92.6	45.2	63	146.5	71.5	23	200.4	97.8	83	254.4	124.1
44	39.5	19.3	04	93.5	45.6	64	147.4	71.9	24	201.3	98.2	84	255.3	124.5
45	40.4	19.7	05	94.4	46.0	65	148.3	72.3	25	202.2	98.6	85	256.2	124.9
46	41.3	20.2	06	95.3	46.5	66	149.2	72.8	26	203.1	99.1	86	257.1	125.4
47	42.2	20.6	07	96.2	46.9	67	150.1	73.2	27	204.0	99.5	87	258.0	125.8
48	43.1	21.0	08	97.1	47.3	68	151.0	73.6	28	204.9	99.9	88	258.9	126.3
49	44.0	21.5	09	98.0	47.8	69	151.9	74.1	29	205.8	100.4	89	259.8	126.7
50	44.9	21.9	10	98.9	48.2	70	152.8	74.5	30	206.7	100.8	90	260.7	127.1
51	45.8	22.4	111	99.8	48.7	171	153.7	75.0	231	207.6	101.3	291	261.5	127.6
52	46.7	22.8	12	100.7	49.1	72	154.6	75.4	32	208.5	101.7	92	262.4	128.0
53	47.6	23.2	13	101.6	49.5	73	155.5	75.8	33	209.4	102.1	93	263.3	128.4
54	48.5	23.7	14	102.5	50.0	74	156.4	76.3	34	210.3	102.6	94	264.2	128.9
55	49.4	24.1	15	103.4	50.4	75	157.3	76.7	35	211.2	103.0	95	265.1	129.3
56	50.3	24.5	16	104.3	50.9	76	158.2	77.2	36	212.1	103.5	96	266.0	129.8
57	51.2	25.0	17	105.2	51.3	77	159.1	77.6	37	213.0	103.9	97	266.9	130.2
58	52.1	25.4	18	106.1	51.7	78	160.0	78.0	38	213.9	104.3	98	267.8	130.6
59	53.0	25.9	19	107.0	52.2	79	160.9	78.5	39	214.8	104.8	99	268.7	131.1
60	53.9	26.3	20	107.9	52.6	80	161.8	78.9	40	215.7	105.2	300	269.6	131.5
Dist.	Dep.	D. Lat.	Dist.	Dep.	D. Lat.	Dist.	Dep.	D. Lat.	Dist.	Dep.	D. Lat.	Dist.	Dep.	D. Lat.

296°	064°
244°	116°

64°

Dist.	D. Lat.	Dep.
N.	N×Cos.	N×Sin.
Hypotenuse	Side Adj.	Side Opp.

334°	026°
206°	154°

TABLE 3

Traverse **26°** Table

334°	026°
206°	154°

Dist.	D. Lat.	Dep.	Dist.	D. Lat.	Dep.	Dist.	D. Lat.	Dep.	Dist.	D. Lat.	Dep.	Dist.	D. Lat.	Dep.
301	270.5	131.9	361	324.5	158.3	421	378.4	184.6	481	432.3	210.9	541	486.2	237.2
02	271.4	132.4	62	325.4	158.7	22	379.3	185.0	82	433.2	211.3	42	487.1	237.6
03	272.3	132.8	63	326.3	159.1	23	380.2	185.4	83	434.1	211.7	43	488.0	238.0
04	273.2	133.3	64	327.2	159.6	24	381.1	185.9	84	435.0	212.2	44	488.9	238.5
05	274.1	133.7	65	328.1	160.0	25	382.0	186.3	85	435.9	212.6	45	489.8	238.9
06	275.0	134.1	66	329.0	160.4	26	382.9	186.7	86	436.8	213.0	46	490.7	239.4
07	275.9	134.6	67	329.9	160.9	27	383.8	187.2	87	437.7	213.5	47	491.6	239.8
08	276.8	135.0	68	330.8	161.3	28	384.7	187.6	88	438.6	213.9	48	492.5	240.2
09	277.7	135.5	69	331.7	161.8	29	385.6	188.1	89	439.5	214.4	49	493.4	240.7
10	278.6	135.9	70	332.6	162.2	30	386.5	188.5	90	440.4	214.8	50	494.3	241.1
311	279.5	136.3	371	333.5	162.6	431	387.4	188.9	491	441.3	215.2	551	495.2	241.5
12	280.4	136.8	72	334.4	163.1	32	388.3	189.4	92	442.2	215.7	52	496.1	242.0
13	281.3	137.2	73	335.3	163.5	33	389.2	189.8	93	443.1	216.1	53	497.0	242.4
14	282.2	137.6	74	336.1	164.0	34	390.1	190.3	94	444.0	216.6	54	497.9	242.9
15	283.1	138.1	75	337.0	164.4	35	391.0	190.7	95	444.9	217.0	55	498.8	243.3
16	284.0	138.5	76	337.9	164.8	36	391.9	191.1	96	445.8	217.4	56	499.7	243.7
17	284.9	139.0	77	338.8	165.3	37	392.8	191.6	97	446.7	217.9	57	500.6	244.2
18	285.8	139.4	78	339.7	165.7	38	393.7	192.0	98	447.6	218.3	58	501.5	244.6
19	286.7	139.8	79	340.6	166.1	39	394.6	192.4	99	448.5	218.7	59	502.4	245.0
20	287.6	140.3	80	341.5	166.6	40	395.5	192.9	500	449.4	219.2	60	503.3	245.5
321	288.5	140.7	381	342.4	167.0	441	396.4	193.3	501	450.3	219.6	561	504.2	245.9
22	289.4	141.2	82	343.3	167.5	42	397.3	193.8	02	451.2	220.1	62	505.1	246.4
23	290.3	141.6	83	344.2	167.9	43	398.2	194.2	03	452.1	220.5	63	506.0	246.8
24	291.2	142.0	84	345.1	168.3	44	399.1	194.6	04	453.0	220.9	64	506.9	247.2
25	292.1	142.5	85	346.0	168.8	45	400.0	195.1	05	453.9	221.4	65	507.8	247.7
26	293.0	142.9	86	346.9	169.2	46	400.9	195.5	06	454.8	221.8	66	508.7	248.1
27	293.9	143.3	87	347.8	169.6	47	401.8	196.0	07	455.7	222.3	67	509.6	248.6
28	294.8	143.8	88	348.7	170.1	48	402.7	196.4	08	456.6	222.7	68	510.5	249.0
29	295.7	144.2	89	349.6	170.5	49	403.6	196.8	09	457.5	223.1	69	511.4	249.4
30	296.6	144.7	90	350.5	171.0	50	404.5	197.3	10	458.4	223.6	70	512.3	249.9
331	297.5	145.1	391	351.4	171.4	451	405.4	197.7	511	459.3	224.0	571	513.2	250.3
32	298.4	145.5	92	352.3	171.8	52	406.3	198.1	12	460.2	224.4	72	514.1	250.7
33	299.3	146.0	93	353.2	172.3	53	407.2	198.6	13	461.1	224.9	73	515.0	251.2
34	300.2	146.4	94	354.1	172.7	54	408.1	199.0	14	462.0	225.3	74	515.9	251.6
35	301.1	146.9	95	355.0	173.2	55	409.0	199.5	15	462.9	225.8	75	516.8	252.1
36	302.0	147.3	96	355.9	173.6	56	409.9	199.9	16	463.8	226.2	76	517.7	252.5
37	302.9	147.7	97	356.8	174.0	57	410.7	200.3	17	464.7	226.6	77	518.6	252.9
38	303.8	148.2	98	357.7	174.5	58	411.6	200.8	18	465.6	227.1	78	519.5	253.4
39	304.7	148.6	99	358.6	174.9	59	412.5	201.2	19	466.5	227.5	79	520.4	253.8
40	305.6	149.0	400	359.5	175.3	60	413.6	201.7	20	467.4	228.0	80	521.3	254.3
341	306.5	149.5	401	360.4	175.8	461	414.3	202.1	521	468.3	228.4	581	522.2	254.7
42	307.4	149.9	02	361.3	176.2	62	415.2	202.5	22	469.2	228.8	82	523.1	255.1
43	308.3	150.4	03	362.2	176.7	63	416.1	203.0	23	470.1	229.3	83	524.0	255.6
44	309.2	150.8	04	363.1	177.1	64	417.0	203.4	24	471.0	229.7	84	524.9	256.0
45	310.1	151.2	05	364.0	177.5	65	417.9	203.8	25	471.9	230.1	85	525.8	256.4
46	311.0	151.7	06	364.9	178.0	66	418.8	204.3	26	472.8	230.6	86	526.7	256.9
47	311.9	152.1	07	365.8	178.4	67	419.7	204.7	27	473.7	231.0	87	527.6	257.3
48	312.8	152.6	08	366.7	178.9	68	420.6	205.2	28	474.6	231.5	88	528.5	257.8
49	313.7	153.0	09	367.6	179.3	69	421.5	205.6	29	475.5	231.9	89	529.4	258.2
50	314.6	153.4	10	368.5	179.7	70	422.4	206.0	30	476.4	232.3	90	530.3	258.6
351	315.5	153.9	411	369.4	180.2	471	423.3	206.5	531	477.3	232.8	591	531.2	259.1
52	316.4	154.3	12	370.3	180.6	72	424.2	206.9	32	478.2	233.2	92	532.1	259.5
53	317.3	154.7	13	371.2	181.0	73	425.1	207.3	33	479.1	233.7	93	533.0	260.0
54	318.2	155.2	14	372.1	181.5	74	426.0	207.8	34	480.0	234.1	94	533.9	260.4
55	319.1	155.6	15	373.0	181.9	75	426.9	208.2	35	480.9	234.5	95	534.8	260.8
56	320.0	156.1	16	373.9	182.4	76	427.8	208.7	36	481.8	235.0	96	535.7	261.3
57	320.9	156.5	17	374.8	182.8	77	428.7	209.1	37	482.7	235.4	97	536.6	261.7
58	321.8	156.9	18	375.7	183.2	78	429.6	209.5	38	483.6	235.8	98	537.5	262.1
59	322.7	157.4	19	376.6	183.7	79	430.5	210.0	39	484.4	236.3	99	538.4	262.6
60	323.6	157.8	20	377.5	184.1	80	431.4	210.4	40	485.3	236.7	600	539.3	263.0
Dist.	Dep.	D. Lat.	Dist.	Dep.	D. Lat.	Dist.	Dep.	D. Lat.	Dist.	Dep.	D. Lat.	Dist.	Dep.	D. Lat.

Dist.	D. Lat.	Dep.
D Lo	Dep.	
	m	D Lo

64°

296°	064°
244°	116°

333° | 027°
207° | 153°

TABLE 3
Traverse **27°** Table

333° | 027°
207° | 153°

Dist.	D. Lat.	Dep.	Dist.	D. Lat.	Dep.	Dist.	D. Lat.	Dep.	Dist.	D. Lat.	Dep.	Dist.	D. Lat.	Dep.
1	0.9	0.5	61	54.4	27.7	121	107.8	54.9	181	161.3	82.2	241	214.7	109.4
2	1.8	0.9	62	55.2	28.1	22	108.7	55.4	82	162.2	82.6	42	215.6	109.9
3	2.7	1.4	63	56.1	28.6	23	109.6	55.8	83	163.1	83.1	43	216.5	110.3
4	3.6	1.8	64	57.0	29.1	24	110.5	56.3	84	163.9	83.5	44	217.4	110.8
5	4.5	2.3	65	57.9	29.5	25	111.4	56.7	85	164.8	84.0	45	218.3	111.2
6	5.3	2.7	66	58.8	30.0	26	112.3	57.2	86	165.7	84.4	46	219.2	111.7
7	6.2	3.2	67	59.7	30.4	27	113.2	57.7	87	166.6	84.9	47	220.1	112.1
8	7.1	3.6	68	60.6	30.9	28	114.0	58.1	88	167.5	85.4	48	221.0	112.6
9	8.0	4.1	69	61.5	31.3	29	114.9	58.6	89	168.4	85.8	49	221.9	113.0
10	8.9	4.5	70	62.4	31.8	30	115.8	59.0	90	169.3	86.3	50	222.8	113.5
11	9.8	5.0	71	63.3	32.2	131	116.7	59.5	191	170.2	86.7	251	223.6	114.0
12	10.7	5.4	72	64.2	32.7	32	117.6	59.9	92	171.1	87.2	52	224.5	114.4
13	11.6	5.9	73	65.0	33.1	33	118.5	60.4	93	172.0	87.6	53	225.4	114.9
14	12.5	6.4	74	65.9	33.6	34	119.4	60.8	94	172.9	88.1	54	226.3	115.3
15	13.4	6.8	75	66.8	34.0	35	120.3	61.3	95	173.7	88.5	55	227.2	115.8
16	14.3	7.3	76	67.7	34.5	36	121.2	61.7	96	174.6	89.0	56	228.1	116.2
17	15.1	7.7	77	68.6	35.0	37	122.1	62.2	97	175.5	89.4	57	229.0	116.7
18	16.0	8.2	78	69.5	35.4	38	123.0	62.7	98	176.4	89.9	58	229.9	117.1
19	16.9	8.6	79	70.4	35.9	39	123.8	63.1	99	177.3	90.3	59	230.8	117.6
20	17.8	9.1	80	71.3	36.3	40	124.7	63.6	200	178.2	90.8	60	231.7	118.0
21	18.7	9.5	81	72.2	36.8	141	125.6	64.0	201	179.1	91.3	261	232.6	118.5
22	19.6	10.0	82	73.1	37.2	42	126.5	64.5	02	180.0	91.7	62	233.4	118.9
23	20.5	10.4	83	74.0	37.7	43	127.4	64.9	03	180.9	92.2	63	234.3	119.4
24	21.4	10.9	84	74.8	38.1	44	128.3	65.4	04	181.8	92.6	64	235.2	119.9
25	22.3	11.3	85	75.7	38.6	45	129.2	65.8	05	182.7	93.1	65	236.1	120.3
26	23.2	11.8	86	76.6	39.0	46	130.1	66.3	06	183.5	93.5	66	237.0	120.8
27	24.1	12.3	87	77.5	39.5	47	131.0	66.7	07	184.4	94.0	67	237.9	121.2
28	24.9	12.7	88	78.4	40.0	48	131.9	67.2	08	185.3	94.4	68	238.8	121.7
29	25.8	13.2	89	79.3	40.4	49	132.8	67.6	09	186.2	94.9	69	239.7	122.1
30	26.7	13.6	90	80.2	40.9	50	133.7	68.1	10	187.1	95.3	70	240.6	122.6
31	27.6	14.1	91	81.1	41.3	151	134.5	68.6	211	188.0	95.8	271	241.5	123.0
32	28.5	14.5	92	82.0	41.8	52	135.4	69.0	12	188.9	96.2	72	242.4	123.5
33	29.4	15.0	93	82.9	42.2	53	136.3	69.5	13	189.8	96.7	73	243.2	123.9
34	30.3	15.4	94	83.8	42.7	54	137.2	69.9	14	190.7	97.2	74	244.1	124.4
35	31.2	15.9	95	84.6	43.1	55	138.1	70.4	15	191.6	97.6	75	245.0	124.8
36	32.1	16.3	96	85.5	43.6	56	139.0	70.8	16	192.5	98.1	76	245.9	125.3
37	33.0	16.8	97	86.4	44.0	57	139.9	71.3	17	193.3	98.5	77	246.8	125.8
38	33.9	17.3	98	87.3	44.5	58	140.8	71.7	18	194.2	99.0	78	247.7	126.2
39	34.7	17.7	99	88.2	44.9	59	141.7	72.2	19	195.1	99.4	79	248.6	126.7
40	35.6	18.2	100	89.1	45.4	60	142.6	72.6	20	196.0	99.9	80	249.5	127.1
41	36.5	18.6	101	90.0	45.9	161	143.5	73.1	221	196.9	100.3	281	250.4	127.6
42	37.4	19.1	02	90.9	46.3	62	144.3	73.5	22	197.8	100.8	82	251.3	128.0
43	38.3	19.5	03	91.8	46.8	63	145.2	74.0	23	198.7	101.2	83	252.2	128.5
44	39.2	20.0	04	92.7	47.2	64	146.1	74.5	24	199.6	101.7	84	253.0	128.9
45	40.1	20.4	05	93.6	47.7	65	147.0	74.9	25	200.5	102.1	85	253.9	129.4
46	41.0	20.9	06	94.4	48.1	66	147.9	75.4	26	201.4	102.6	86	254.8	129.8
47	41.9	21.3	07	95.3	48.6	67	148.8	75.8	27	202.3	103.1	87	255.7	130.3
48	42.8	21.8	08	96.2	49.0	68	149.7	76.3	28	203.1	103.5	88	256.6	130.7
49	43.7	22.2	09	97.1	49.5	69	150.6	76.7	29	204.0	104.0	89	257.5	131.2
50	44.6	22.7	10	98.0	49.9	70	151.5	77.2	30	204.9	104.4	90	258.4	131.7
51	45.4	23.2	111	98.9	50.4	171	152.4	77.6	231	205.8	104.9	291	259.3	132.1
52	46.3	23.6	12	99.8	50.8	72	153.3	78.1	32	206.7	105.3	92	260.2	132.6
53	47.2	24.1	13	100.7	51.3	73	154.1	78.5	33	207.6	105.8	93	261.1	133.0
54	48.1	24.5	14	101.6	51.8	74	155.0	79.0	34	208.5	106.2	94	262.0	133.5
55	49.0	25.0	15	102.5	52.2	75	155.9	79.4	35	209.4	106.7	95	262.8	133.9
56	49.9	25.4	16	103.4	52.7	76	156.8	79.9	36	210.3	107.1	96	263.7	134.4
57	50.8	25.9	17	104.2	53.1	77	157.7	80.4	37	211.2	107.6	97	264.6	134.8
58	51.7	26.3	18	105.1	53.6	78	158.6	80.8	38	212.1	108.0	98	265.5	135.3
59	52.6	26.8	19	106.0	54.0	79	159.5	81.3	39	213.0	108.5	99	266.4	135.7
60	53.5	27.2	20	106.9	54.5	80	160.4	81.7	40	213.8	109.0	300	267.3	136.2
Dist.	Dep.	D. Lat.	Dist.	Dep.	D. Lat.	Dist.	Dep.	D. Lat.	Dist.	Dep.	D. Lat.	Dist.	Dep.	D. Lat.

297° | 063°
243° | 117°

63°

Dist.	D. Lat.	Dep.
N.	N×Cos.	N×Sin.
Hypotenuse	Side Adj.	Side Opp.

333°	027°
207°	153°

TABLE 3

Traverse 27° Table

333°	027°
207°	153°

Dist.	D. Lat.	Dep.	Dist.	D. Lat.	Dep.	Dist.	D. Lat.	Dep.	Dist.	D. Lat.	Dep.	Dist.	D. Lat.	Dep.
301	268.2	136.7	361	321.7	163.9	421	375.1	191.1	481	428.6	218.4	541	482.0	245.6
02	269.1	137.1	62	322.5	164.3	22	376.0	191.6	82	429.5	218.8	42	482.9	246.1
03	270.0	137.6	63	323.4	164.8	23	376.9	192.0	83	430.4	219.3	43	483.8	246.5
04	270.9	138.0	64	324.3	165.3	24	377.8	192.5	84	431.2	219.7	44	484.7	247.0
05	271.8	138.5	65	325.2	165.7	25	378.7	192.9	85	432.1	220.2	45	485.6	247.4
06	272.6	138.9	66	326.1	166.2	26	379.6	193.4	86	433.0	220.6	46	486.5	247.9
07	273.5	139.4	67	327.0	166.6	27	380.5	193.9	87	433.9	221.1	47	487.4	248.3
08	274.4	139.8	68	327.9	167.1	28	381.4	194.3	88	434.8	221.5	48	488.3	248.8
09	275.3	140.3	69	328.8	167.5	29	382.2	194.8	89	435.7	222.0	49	489.2	249.2
10	276.2	140.7	70	329.7	168.0	30	383.1	195.2	90	436.6	222.5	50	490.1	249.7
311	277.1	141.2	371	330.6	168.4	431	384.0	195.7	491	437.5	222.9	551	490.9	250.1
12	278.0	141.6	72	331.5	168.9	32	384.9	196.1	92	438.4	223.4	52	491.8	250.6
13	278.9	142.1	73	332.3	169.3	33	385.8	196.6	93	439.3	223.8	53	492.7	251.1
14	279.8	142.6	74	333.2	169.8	34	386.7	197.0	94	440.2	224.3	54	493.6	251.5
15	280.7	143.0	75	334.1	170.2	35	387.6	197.5	95	441.0	224.7	55	494.5	252.0
16	281.6	143.5	76	335.0	170.7	36	388.5	197.9	96	441.9	225.2	56	495.4	252.4
17	282.4	143.9	77	335.9	171.2	37	389.4	198.4	97	442.8	225.6	57	496.3	252.9
18	283.3	144.4	78	336.8	171.6	38	390.3	198.8	98	443.7	226.1	58	497.2	253.3
19	284.2	144.8	79	337.7	172.1	39	391.2	199.3	99	444.6	226.5	59	498.1	253.8
20	285.1	145.3	80	338.6	172.5	40	392.0	199.8	500	445.5	227.0	60	499.0	254.2
321	286.0	145.7	381	339.5	173.0	441	392.9	200.2	501	446.4	227.4	561	499.9	254.7
22	286.9	146.2	82	340.4	173.4	42	393.8	200.7	02	447.3	227.9	62	500.7	255.1
23	287.8	146.6	83	341.3	173.9	43	394.7	201.1	03	448.2	228.4	63	501.6	255.6
24	288.7	147.1	84	342.1	174.3	44	395.6	201.6	04	449.0	228.8	64	502.5	256.1
25	289.6	147.5	85	343.0	174.8	45	396.5	202.0	05	450.0	229.3	65	503.4	256.5
26	290.5	148.0	86	343.9	175.2	46	397.4	202.5	06	450.8	229.7	66	504.3	257.0
27	291.4	148.5	87	344.8	175.7	47	398.3	202.9	07	451.7	230.2	67	505.2	257.4
28	292.3	148.9	88	345.7	176.1	48	399.2	203.4	08	452.6	230.6	68	506.1	257.9
29	293.1	149.4	89	346.6	176.6	49	400.1	203.8	09	453.5	231.1	69	507.0	258.3
30	294.0	149.8	90	347.5	177.1	50	401.0	204.3	10	454.4	231.5	70	507.9	258.8
331	294.9	150.3	391	348.4	177.5	451	401.8	204.7	511	455.3	232.0	571	508.8	259.2
32	295.8	150.7	92	349.3	178.0	52	402.7	205.2	12	456.2	232.4	72	509.6	259.7
33	296.7	151.2	93	350.2	178.4	53	403.6	205.7	13	457.1	232.9	73	510.5	260.1
34	297.6	151.6	94	351.1	178.9	54	404.5	206.1	14	458.0	233.4	74	511.4	260.6
35	298.5	152.1	95	351.9	179.3	55	405.4	206.6	15	458.9	233.8	75	512.3	261.0
36	299.4	152.5	96	352.8	179.8	56	406.3	207.0	16	459.8	234.3	76	513.2	261.5
37	300.3	153.0	97	353.7	180.2	57	407.2	207.5	17	460.7	234.7	77	514.1	262.0
38	301.2	153.4	98	354.6	180.7	58	408.1	207.9	18	461.5	235.2	78	515.0	262.4
39	302.1	153.9	99	355.5	181.1	59	409.0	208.4	19	462.4	235.6	79	515.9	262.9
40	302.9	154.4	400	356.4	181.6	60	409.9	208.8	20	463.3	236.1	80	516.8	263.4
341	303.8	154.8	401	357.3	182.1	461	410.8	209.3	521	464.2	236.5	581	517.7	263.8
42	304.7	155.3	02	358.2	182.5	62	411.6	209.7	22	465.1	237.0	82	518.6	264.2
43	305.6	155.7	03	359.1	183.0	63	412.5	210.2	23	466.0	237.4	83	519.5	264.7
44	306.5	156.2	04	360.0	183.4	64	413.4	210.7	24	466.9	237.9	84	520.3	265.1
45	307.4	156.6	05	360.9	183.9	65	414.3	211.1	25	467.8	238.3	85	521.2	265.6
46	308.3	157.1	06	361.8	184.3	66	415.2	211.6	26	468.7	238.8	86	522.1	266.0
47	309.2	157.5	07	362.6	184.8	67	416.1	212.0	27	469.6	239.3	87	523.0	266.5
48	310.1	158.0	08	363.5	185.2	68	417.0	212.5	28	470.5	239.7	88	523.9	266.9
49	311.0	158.4	09	364.4	185.7	69	417.9	212.9	29	471.3	240.2	89	524.8	267.4
50	311.9	158.9	10	365.3	186.1	70	418.8	213.4	30	472.2	240.6	90	525.7	267.9
351	312.7	159.4	411	366.2	186.6	471	419.7	213.8	531	473.1	241.1	591	526.6	268.3
52	313.6	159.8	12	367.1	187.0	72	420.6	214.3	32	474.0	241.5	92	527.5	268.8
53	314.5	160.3	13	368.0	187.5	73	421.4	214.7	33	474.9	242.0	93	528.4	269.2
54	315.4	160.7	14	368.9	188.0	74	422.3	215.2	34	475.8	242.4	94	529.3	269.7
55	316.3	161.2	15	369.8	188.4	75	423.2	215.6	35	476.7	242.9	95	530.1	270.1
56	317.2	161.6	16	370.7	188.9	76	424.1	216.1	36	477.6	243.3	96	531.0	270.6
57	318.1	162.1	17	371.5	189.3	77	425.0	216.6	37	478.5	243.8	97	531.9	271.0
58	319.0	162.5	18	372.4	189.8	78	425.9	217.0	38	479.4	244.2	98	532.8	271.5
59	319.9	163.0	19	373.3	190.2	79	426.8	217.5	39	480.3	244.7	99	533.7	271.9
60	320.8	163.4	20	374.2	190.7	80	427.7	217.9	40	481.1	245.2	600	534.6	272.4
Dist.	Dep.	D. Lat.	Dist.	Dep.	D. Lat.	Dist.	Dep.	D. Lat.	Dist.	Dep.	D. Lat.	Dist.	Dep.	D. Lat.

Dist.	D. Lat.	Dep.
D Lo	Dep.	
	m	D Lo

63°

297°	063°
243°	117°

332° | 028° / 208° | 152°	TABLE 3 Traverse 28° Table	332° | 028° / 208° | 152°

Dist.	D. Lat.	Dep.	Dist.	D. Lat.	Dep.	Dist.	D. Lat.	Dep.	Dist.	D. Lat.	Dep.	Dist.	D. Lat.	Dep.
1	0.9	0.5	61	53.9	28.6	121	106.8	56.8	181	159.8	85.0	241	212.8	113.1
2	1.8	0.9	62	54.7	29.1	22	107.7	57.3	82	160.7	85.4	42	213.7	113.6
3	2.6	1.4	63	55.6	29.6	23	108.6	57.7	83	161.6	85.9	43	214.6	114.1
4	3.5	1.9	64	56.5	30.0	24	109.5	58.2	84	162.5	86.4	44	215.4	114.6
5	4.4	2.3	65	57.4	30.5	25	110.4	58.7	85	163.3	86.9	45	216.3	115.0
6	5.3	2.8	66	58.3	31.0	26	111.3	59.2	86	164.2	87.3	46	217.2	115.5
7	6.2	3.3	67	59.2	31.5	27	112.1	59.6	87	165.1	87.8	47	218.1	116.0
8	7.1	3.8	68	60.0	31.9	28	113.0	60.1	88	166.0	88.3	48	219.0	116.4
9	7.9	4.2	69	60.9	32.4	29	113.9	60.6	89	166.9	88.7	49	219.9	116.9
10	8.8	4.7	70	61.8	32.9	30	114.8	61.0	90	167.8	89.2	50	220.7	117.4
11	9.7	5.2	71	62.7	33.3	131	115.7	61.5	191	168.6	89.7	251	221.6	117.8
12	10.6	5.6	72	63.6	33.8	32	116.5	62.0	92	169.5	90.1	52	222.5	118.3
13	11.5	6.1	73	64.5	34.3	33	117.4	62.4	93	170.4	90.6	53	223.4	118.8
14	12.4	6.6	74	65.3	34.7	34	118.3	62.9	94	171.3	91.1	54	224.3	119.2
15	13.2	7.0	75	66.2	35.2	35	119.2	63.4	95	172.2	91.5	55	225.2	119.7
16	14.1	7.5	76	67.1	35.7	36	120.1	63.8	96	173.1	92.0	56	226.0	120.2
17	15.0	8.0	77	68.0	36.1	37	121.0	64.3	97	173.9	92.5	57	226.9	120.7
18	15.9	8.5	78	68.9	36.6	38	121.8	64.8	98	174.8	93.0	58	227.8	121.1
19	16.8	8.9	79	69.8	37.1	39	122.7	65.3	99	175.7	93.4	59	228.7	121.6
20	17.7	9.4	80	70.6	37.6	40	123.6	65.7	200	176.6	93.9	60	229.6	122.1
21	18.5	9.9	81	71.5	38.0	141	124.5	66.2	201	177.5	94.4	261	230.4	122.5
22	19.4	10.3	82	72.4	38.5	42	125.4	66.7	02	178.4	94.8	62	231.3	123.0
23	20.3	10.8	83	73.3	39.0	43	126.3	67.1	03	179.2	95.3	63	232.2	123.5
24	21.2	11.3	84	74.2	39.4	44	127.1	67.6	04	180.1	95.8	64	233.1	123.9
25	22.1	11.7	85	75.1	39.9	45	128.0	68.1	05	181.0	96.2	65	234.0	124.4
26	23.0	12.2	86	75.9	40.4	46	128.9	68.5	06	181.9	96.7	66	234.9	124.9
27	23.8	12.7	87	76.8	40.8	47	129.8	69.0	07	182.8	97.2	67	235.7	125.3
28	24.7	13.1	88	77.7	41.3	48	130.7	69.5	08	183.7	97.7	68	236.6	125.8
29	25.6	13.6	89	78.6	41.8	49	131.6	70.0	09	184.5	98.1	69	237.5	126.3
30	26.5	14.1	90	79.5	42.3	50	132.4	70.4	10	185.4	98.6	70	238.4	126.8
31	27.4	14.6	91	80.3	42.7	151	133.3	70.9	211	186.3	99.1	271	239.3	127.2
32	28.3	15.0	92	81.2	43.2	52	134.2	71.4	12	187.2	99.5	72	240.2	127.7
33	29.1	15.5	93	82.1	43.7	53	135.1	71.8	13	188.1	100.0	73	241.0	128.2
34	30.0	16.0	94	83.0	44.1	54	136.0	72.3	14	189.0	100.5	74	241.9	128.6
35	30.9	16.4	95	83.9	44.6	55	136.9	72.8	15	189.8	100.9	75	242.8	129.1
36	31.8	16.9	96	84.8	45.1	56	137.7	73.2	16	190.7	101.4	76	243.7	129.6
37	32.7	17.4	97	85.6	45.5	57	138.6	73.7	17	191.6	101.9	77	244.6	130.0
38	33.6	17.8	98	86.5	46.0	58	139.5	74.2	18	192.5	102.3	78	245.5	130.5
39	34.4	18.3	99	87.4	46.5	59	140.4	74.6	19	193.4	102.8	79	246.3	131.0
40	35.3	18.8	100	88.3	46.9	60	141.3	75.1	20	194.2	103.3	80	247.2	131.5
41	36.2	19.2	101	89.2	47.4	161	142.2	75.6	221	195.1	103.8	281	248.1	131.9
42	37.1	19.7	02	90.1	47.9	62	143.0	76.1	22	196.0	104.2	82	249.0	132.4
43	38.0	20.2	03	90.9	48.4	63	143.9	76.5	23	196.9	104.7	83	249.9	132.9
44	38.8	20.7	04	91.8	48.8	64	144.8	77.0	24	197.8	105.2	84	250.8	133.3
45	39.7	21.1	05	92.7	49.3	65	145.7	77.5	25	198.7	105.6	85	251.6	133.8
46	40.6	21.6	06	93.6	49.8	66	146.6	77.9	26	199.5	106.1	86	252.5	134.3
47	41.5	22.1	07	94.5	50.2	67	147.5	78.4	27	200.4	106.6	87	253.4	134.7
48	42.4	22.5	08	95.4	50.7	68	148.3	78.9	28	201.3	107.0	88	254.3	135.2
49	43.3	23.0	09	96.2	51.2	69	149.2	79.3	29	202.2	107.5	89	255.2	135.7
50	44.1	23.5	10	97.1	51.6	70	150.1	79.8	30	203.1	108.0	90	256.1	136.1
51	45.0	23.9	111	98.0	52.1	171	151.0	80.3	231	204.0	108.4	291	256.9	136.6
52	45.9	24.4	12	98.9	52.6	72	151.9	80.7	32	204.8	108.9	92	257.8	137.1
53	46.8	24.9	13	99.8	53.1	73	152.7	81.2	33	205.7	109.4	93	258.7	137.6
54	47.7	25.4	14	100.7	53.5	74	153.6	81.7	34	206.6	109.9	94	259.6	138.0
55	48.6	25.8	15	101.5	54.0	75	154.5	82.2	35	207.5	110.3	95	260.5	138.5
56	49.4	26.3	16	102.4	54.5	76	155.4	82.6	36	208.4	110.8	96	261.4	139.0
57	50.3	26.8	17	103.3	54.9	77	156.3	83.1	37	209.3	111.3	97	262.2	139.4
58	51.2	27.2	18	104.2	55.4	78	157.2	83.6	38	210.1	111.7	98	263.1	139.9
59	52.1	27.7	19	105.1	55.9	79	158.0	84.0	39	211.0	112.2	99	264.0	140.4
60	53.0	28.2	20	106.0	56.3	80	158.9	84.5	40	211.9	112.7	300	264.9	140.8
Dist.	Dep.	D. Lat.	Dist.	Dep.	D. Lat.	Dist.	Dep.	D. Lat.	Dist.	Dep.	D. Lat.	Dist.	Dep.	D. Lat.

298° | 062°
242° | 118°

62°

Dist.	D. Lat.	Dep.
N.	N×Cos.	N×Sin.
Hypotenuse	Side Adj.	Side Opp.

332°	028°
208°	152°

TABLE 3

Traverse **28°** Table

332°	028°
208°	152°

Dist.	D. Lat.	Dep.	Dist.	D. Lat.	Dep.	Dist.	D. Lat.	Dep.	Dist.	D. Lat.	Dep.	Dist.	D. Lat.	Dep.
301	265.8	141.3	361	318.7	169.5	421	371.7	197.6	481	424.7	225.8	541	477.7	254.0
02	266.7	141.8	62	319.6	169.9	22	372.6	198.1	82	425.6	226.3	42	478.6	254.5
03	267.5	142.2	63	320.5	170.4	23	373.5	198.6	83	426.5	226.8	43	479.4	254.9
04	268.4	142.7	64	321.4	170.9	24	374.4	199.1	84	427.3	227.2	44	480.3	255.4
05	269.3	143.2	65	322.3	171.4	25	375.3	199.5	85	428.2	227.7	45	481.2	255.9
06	270.2	143.7	66	323.2	171.8	26	376.1	200.0	86	429.1	228.2	46	482.1	256.3
07	271.1	144.1	67	324.0	172.3	27	377.0	200.5	87	430.0	228.6	47	483.0	256.8
08	271.9	144.6	68	324.9	172.8	28	377.9	200.9	88	430.9	229.1	48	483.9	257.3
09	272.8	145.1	69	325.8	173.2	29	378.8	201.4	89	431.8	229.6	49	484.7	257.7
10	273.7	145.5	70	326.7	173.7	30	379.7	201.9	90	432.6	230.0	50	485.6	258.2
311	274.6	146.0	371	327.6	174.2	431	380.6	202.3	491	433.5	230.5	551	486.5	258.7
12	275.5	146.5	72	328.5	174.6	32	381.4	202.8	92	434.4	231.0	52	487.4	259.1
13	276.4	146.9	73	329.3	175.1	33	382.3	203.3	93	435.3	231.4	53	488.3	259.6
14	277.2	147.4	74	330.2	175.6	34	383.2	203.8	94	436.2	231.9	54	489.2	260.1
15	278.1	147.9	75	331.1	176.1	35	384.1	204.2	95	437.1	232.4	55	490.0	260.6
16	279.0	148.4	76	332.0	176.5	36	385.0	204.7	96	437.9	232.9	56	490.9	261.0
17	279.9	148.8	77	332.9	177.0	37	385.8	205.2	97	438.8	233.3	57	491.8	261.5
18	280.8	149.3	78	333.8	177.5	38	386.7	205.6	98	439.7	233.8	58	492.7	262.0
19	281.7	149.8	79	334.6	177.9	39	387.6	206.1	99	440.6	234.3	59	493.6	262.4
20	282.5	150.2	80	335.5	178.4	40	388.5	206.6	500	441.5	234.7	60	494.5	262.9
321	283.4	150.7	381	336.4	178.9	441	389.4	207.0	501	442.4	235.2	561	495.3	263.4
22	284.3	151.2	82	337.3	179.3	42	390.3	207.5	02	443.2	235.7	62	496.2	263.8
23	285.2	151.6	83	338.2	179.8	43	391.1	208.0	03	444.1	236.1	63	497.1	264.3
24	286.1	152.1	84	339.1	180.3	44	392.0	208.4	04	445.0	236.6	64	498.0	264.8
25	287.0	152.6	85	339.9	180.7	45	392.9	208.9	05	445.9	237.1	65	498.9	265.3
26	287.8	153.0	86	340.8	181.2	46	393.8	209.4	06	446.8	237.6	66	499.7	265.7
27	288.7	153.5	87	341.7	181.7	47	394.7	209.9	07	447.7	238.0	67	500.6	266.2
28	289.6	154.0	88	342.6	182.2	48	395.6	210.3	08	448.5	238.5	68	501.5	266.7
29	290.5	154.5	89	343.5	182.6	49	396.4	210.8	09	449.4	239.0	69	502.4	267.1
30	291.4	154.9	90	344.3	183.1	50	397.3	211.3	10	450.3	239.4	70	503.3	267.6
331	292.3	155.4	391	345.2	183.6	451	398.2	211.7	511	451.2	239.9	571	504.2	268.1
32	293.1	155.9	92	346.1	184.0	52	399.1	212.2	12	452.1	240.4	72	505.0	268.5
33	294.0	156.3	93	347.0	184.5	53	400.0	212.7	13	453.0	240.8	73	505.9	269.0
34	294.9	156.8	94	347.9	185.0	54	400.9	213.1	14	453.8	241.3	74	506.8	269.5
35	295.8	157.3	95	348.8	185.4	55	401.7	213.6	15	454.7	241.8	75	507.7	269.9
36	296.7	157.7	96	349.6	185.9	56	402.6	214.1	16	455.6	242.2	76	508.6	270.4
37	297.6	158.2	97	350.5	186.4	57	403.5	214.5	17	456.5	242.7	77	509.5	270.9
38	298.4	158.7	98	351.4	186.8	58	404.4	215.0	18	457.4	243.2	78	510.3	271.4
39	299.3	159.2	99	352.3	187.3	59	405.3	215.5	19	458.2	243.7	79	511.2	271.8
40	300.2	159.6	400	353.2	187.8	60	406.2	216.0	20	459.1	244.1	80	512.1	272.3
341	301.1	160.1	401	354.1	188.3	461	407.0	216.4	521	460.0	244.6	581	513.0	272.8
42	302.0	160.6	02	354.9	188.7	62	407.9	216.9	22	460.9	245.1	82	513.9	273.2
43	302.9	161.0	03	355.8	189.2	63	408.8	217.4	23	461.8	245.5	83	514.8	273.7
44	303.7	161.5	04	356.7	189.7	64	409.7	217.8	24	462.7	246.0	84	515.6	274.2
45	304.6	162.0	05	357.6	190.1	65	410.6	218.3	25	463.5	246.5	85	516.5	274.6
46	305.5	162.4	06	358.5	190.6	66	411.5	218.8	26	464.4	246.9	86	517.4	275.1
47	306.4	162.9	07	359.4	191.1	67	412.3	219.2	27	465.3	247.4	87	518.3	275.4
48	307.3	163.4	08	360.2	191.5	68	413.2	219.7	28	466.2	247.9	88	519.2	276.0
49	308.1	163.8	09	361.1	192.0	69	414.1	220.2	29	467.1	248.4	89	520.1	276.5
50	309.0	164.3	10	362.0	192.5	70	415.0	220.7	30	468.0	248.8	90	520.9	277.0
351	309.9	164.8	411	362.9	193.0	471	415.9	221.1	531	468.8	249.3	591	521.8	277.5
52	310.8	165.3	12	363.8	193.4	72	416.8	221.6	32	469.7	249.8	92	522.7	277.9
53	311.7	165.7	13	364.7	193.9	73	417.6	222.1	33	470.7	250.2	93	523.6	278.4
54	312.6	166.2	14	365.5	194.4	74	418.5	222.5	34	471.5	250.7	94	524.5	278.9
55	313.4	166.7	15	366.4	194.8	75	419.4	223.0	35	472.4	251.2	95	525.4	279.3
56	314.3	167.1	16	367.3	195.3	76	420.3	223.5	36	473.3	251.6	96	526.2	279.8
57	315.2	167.6	17	368.2	195.8	77	421.2	223.9	37	474.1	252.1	97	527.1	280.3
58	316.1	168.1	18	369.1	196.2	78	422.0	224.4	38	475.0	252.6	98	528.0	280.7
59	317.0	168.5	19	370.0	196.7	79	422.9	224.9	39	475.9	253.0	99	528.9	281.2
60	317.9	169.0	20	370.8	197.2	80	423.8	225.3	40	476.8	253.5	600	529.8	281.7
Dist.	Dep.	D. Lat.	Dist.	Dep.	D. Lat.	Dist.	Dep.	D. Lat.	Dist.	Dep.	D. Lat.	Dist.	Dep.	D. Lat.

Dist.	D. Lat.	Dep.
D Lo	Dep.	
	m	D Lo

62°

298°	062°
242°	118°

331° | 029° / 209° | 151°	TABLE 3 — Traverse 29° Table	331° | 029° / 209° | 151°

Dist.	D. Lat.	Dep.	Dist.	D. Lat.	Dep.	Dist.	D. Lat.	Dep.	Dist.	D. Lat.	Dep.	Dist.	D. Lat.	Dep.
1	0.9	0.5	61	53.4	29.6	121	105.8	58.7	181	158.3	87.8	241	210.8	116.8
2	1.7	1.0	62	54.2	30.1	22	106.7	59.1	82	159.2	88.2	42	211.7	117.3
3	2.6	1.5	63	55.1	30.5	23	107.6	59.6	83	160.1	88.7	43	212.5	117.8
4	3.5	1.9	64	56.0	31.0	24	108.5	60.1	84	160.9	89.2	44	213.4	118.3
5	4.4	2.4	65	56.9	31.5	25	109.3	60.6	85	161.8	89.7	45	214.3	118.8
6	5.2	2.9	66	57.7	32.0	26	110.2	61.1	86	162.7	90.2	46	215.2	119.3
7	6.1	3.4	67	58.6	32.5	27	111.1	61.6	87	163.6	90.7	47	216.0	119.7
8	7.0	3.9	68	59.5	33.0	28	112.0	62.1	88	164.4	91.1	48	216.9	120.2
9	7.9	4.4	69	60.3	33.5	29	112.8	62.5	89	165.3	91.6	49	217.8	120.7
10	8.7	4.8	70	61.2	33.9	30	113.7	63.0	90	166.2	92.1	50	218.7	121.2
11	9.6	5.3	71	62.1	34.4	131	114.6	63.5	191	167.1	92.6	251	219.5	121.7
12	10.5	5.8	72	63.0	34.9	32	115.4	64.0	92	167.9	93.1	52	220.4	122.2
13	11.4	6.3	73	63.8	35.4	33	116.3	64.5	93	168.8	93.6	53	221.3	122.7
14	12.2	6.8	74	64.7	35.9	34	117.2	65.0	94	169.7	94.1	54	222.2	123.1
15	13.1	7.3	75	65.6	36.4	35	118.1	65.4	95	170.6	94.5	55	223.0	123.6
16	14.0	7.8	76	66.5	36.8	36	118.9	65.9	96	171.4	95.0	56	223.9	124.1
17	14.9	8.2	77	67.3	37,3	37	119.8	66.4	97	172.3	95.5	57	224.8	124.6
18	15.7	8.7	78	68.2	37.8	38	120.7	66.9	98	173.2	96.0	58	225.7	125.1
19	16.6	9.2	79	69.1	38.3	39	121.6	67.4	99	174.0	96.5	59	226.5	125.6
20	17.5	9.7	80	70.0	38.8	40	122.4	67.9	200	174.9	97.0	60	227.4	126.1
21	18.4	10.2	81	70.8	39.3	141	123.3	68.4	201	175.8	97.4	261	228.3	126.5
22	19.2	10.7	82	71.7	39.8	42	124.2	68.8	02	176.7	97.9	62	229.2	127.0
23	20.1	11.2	83	72.6	40.2	43	125.1	69.3	03	177.5	98.4	63	230.0	127.5
24	21.0	11.6	84	73.5	40.7	44	125.9	69.8	04	178.4	98.9	64	230.9	128.0
25	21.9	12.1	85	74.3	41.2	45	126.8	70.3	05	179.3	99.4	65	231.8	128.5
26	22.7	12.6	86	75.2	41.7	46	127.7	70.8	06	180.2	99.9	66	232.6	129.0
27	23.6	13.1	87	76.1	42.2	47	128.6	71.3	07	181.0	100.4	67	233.5	129.4
28	24.5	13.6	88	77.0	42.7	48	129.4	71.8	08	181.9	100.8	68	234.4	129.9
29	25.4	14.1	89	77.8	43.1	49	130.3	72.2	09	182.8	101.3	69	235.3	130.4
30	26.2	14.5	90	78.7	43.6	50	131.2	72.7	10	183.7	101.8	70	236.1	130.9
31	27.1	15.0	91	79.6	44.1	151	132.1	73.2	211	184.5	102.3	271	237.0	131.4
32	28.0	15.5	92	80.5	44.6	52	132.9	73.7	12	185.4	102.8	72	237.9	131.9
33	28.9	16.0	93	81.3	45.1	53	133.8	74.2	13	186.3	103.3	73	238.8	132.4
34	29.7	16.5	94	82.2	45.6	54	134.7	74.7	14	187.2	103.7	74	239.6	132.8
35	30.6	17.0	95	83.1	46.1	55	135.6	75.1	15	188.0	104.2	75	240.5	133.3
36	31.5	17.5	96	84.0	46.5	56	136.4	75.6	16	188.9	104.7	76	241.4	133.8
37	32.4	17.9	97	84.8	47.0	57	137.3	76.1	17	189.8	105.2	77	242.3	134.3
38	33.2	18.4	98	85.7	47.5	58	138.2	76.6	18	190.7	105.7	78	243.1	134.8
39	34.1	18.9	99	86.6	48.0	59	139.1	77.1	19	191.5	106.2	79	244.0	135.3
40	35.0	19.4	100	87.5	48.5	60	139.9	77.6	20	192.4	106.7	80	244.9	135.7
41	35.9	19.9	101	88.3	49.0	161	140.8	78.1	221	193.3	107.1	281	245.8	136.2
42	36.7	20.4	02	89.2	49.5	62	141.7	78.5	22	194.2	107.6	82	246.6	136.7
43	37.6	20.8	03	90.1	49.9	63	142.6	79.0	23	195.0	108.1	83	247.5	137.2
44	38.5	21.3	04	91.0	50.4	64	143.4	79.5	24	195.9	108.6	84	248.4	137.7
45	39.4	21.8	05	91.8	50.9	65	144.3	80.0	25	196.8	109.1	85	249.3	138.2
46	40.2	22.3	06	92.7	51.4	66	145.2	80.5	26	197.7	109.6	86	250.1	138.7
47	41.1	22.8	07	93.6	51.9	67	146.1	81.0	27	198.5	110.1	87	251.0	139.1
48	42.0	23.3	08	94.5	52.4	68	146.9	81.4	28	199.4	110.5	88	251.9	139.6
49	42.9	23.8	09	95.3	52.8	69	147.8	81.9	29	200.3	111.0	89	252.8	140.1
50	43.7	24.2	10	96.2	53.3	70	148.7	82.4	30	201.2	111.5	90	253.6	140.6
51	44.6	24.7	111	97.1	53.8	171	149.6	82.9	231	202.0	112.0	291	254.5	141.1
52	45.5	25.2	12	98.0	54.3	72	150.4	83.4	32	202.9	112.5	92	255.4	141.6
53	46.4	25.7	13	98.8	54.8	73	151.3	83.9	33	203.8	113.0	93	256.3	142.0
54	47.2	26.2	14	99.7	55.3	74	152.2	84.4	34	204.7	113.4	94	257.1	142.5
55	48.1	26.7	15	100.6	55.8	75	153.1	84.8	35	205.5	113.9	95	258.0	143.0
56	49.0	27.1	16	101.5	56.2	76	153.9	85.3	36	206.4	114.4	96	258.9	143.5
57	49.9	27.6	17	102.3	56.7	77	154.8	85.8	37	207.3	114.9	97	259.8	144.0
58	50.7	28.1	18	103.2	57.2	78	155.7	86.3	38	208.2	115.4	98	260.6	144.5
59	51.6	28.6	19	104.1	57.7	79	156.6	86.8	39	209.0	115.9	99	261.5	145.0
60	52.5	29.1	20	105.0	58.2	80	157.4	87.3	40	209.9	116.4	300	262.4	145.4
Dist.	Dep.	D. Lat.	Dist.	Dep.	D. Lat.	Dist.	Dep.	D. Lat.	Dist.	Dep.	D. Lat.	Dist.	Dep.	D. Lat.

299° | 061°
241° | 119°

61°

Dist.	D. Lat.	Dep.
N.	N×Cos.	N×Sin.
Hypotenuse	Side Adj.	Side Opp.

331°	029°
209°	151°

TABLE 3

Traverse 29° Table

Dist.	D. Lat.	Dep.	Dist.	D. Lat.	Dep.	Dist.	D. Lat.	Dep.	Dist.	D. Lat.	Dep.	Dist.	D. Lat.	Dep.
301	263.3	145.9	361	315.7	175.0	421	368.2	204.1	481	420.7	233.2	541	473.2	262.3
02	264.1	146.4	62	316.6	175.5	22	369.1	204.6	82	421.6	233.7	42	474.0	262.8
03	265.0	146.9	63	317.5	176.0	23	370.0	205.1	83	422.4	234.2	43	474.9	263.3
04	265.9	147.4	64	318.4	176.5	24	370.8	205.6	84	423.3	234.6	44	475.8	263.7
05	266.8	147.9	65	319.2	177.0	25	371.7	206.0	85	424.2	235.1	45	476.7	264.2
06	267.6	148.4	66	320.1	177.4	26	372.6	206.5	86	425.1	235.6	46	477.5	264.7
07	268.5	148.8	67	321.0	177.9	27	373.5	207.0	87	425.9	236.1	47	478.4	265.2
08	269.4	149.3	68	321.9	178.4	28	374.3	207.5	88	426.8	236.6	48	479.3	265.7
09	270.3	149.8	69	322.7	178.9	29	375.2	208.0	89	427.7	237.1	49	480.2	266.2
10	271.1	150.3	70	323.6	179.4	30	376.1	208.5	90	428.6	237.6	50	481.0	266.6
311	272.0	150.8	371	324.5	179.9	431	377.0	209.0	491	429.4	238.0	551	481.9	267.1
12	272.9	151.3	72	325.4	180.3	32	377.8	209.4	92	430.3	238.5	52	482.8	267.6
13	273.8	151.7	73	326.2	180.8	33	378.7	209.9	93	431.2	239.0	53	483.7	268.1
14	274.6	152.2	74	327.1	181.3	34	379.6	210.4	94	432.1	239.5	54	484.5	268.6
15	275.5	152.7	75	328.0	181.8	35	380.5	210.9	95	432.9	240.0	55	485.4	269.1
16	276.4	153.2	76	328.9	182.3	36	381.3	211.4	96	433.8	240.5	56	486.3	269.6
17	277.3	153.7	77	329.7	182.8	37	382.2	211.9	97	434.7	241.0	57	487.2	270.0
18	278.1	154.2	78	330.6	183.3	38	383.1	212.3	98	435.6	241.4	58	488.0	270.5
19	279.0	154.7	79	331.5	183.7	39	384.0	212.8	99	436.4	241.9	59	488.9	271.0
20	279.9	155.1	80	332.4	184.2	40	384.8	213.3	500	437.3	242.4	60	489.8	271.5
321	280.8	155.6	381	333.2	184.7	441	385.7	213.8	501	438.2	242.9	561	490.7	272.0
22	281.6	156.1	82	334.1	185.2	42	386.6	214.3	02	439.1	243.4	62	491.5	272.5
23	282.5	156.6	83	335.0	185.7	43	387.5	214.8	03	439.9	243.9	63	492.4	272.9
24	283.4	157.1	84	335.9	186.2	44	388.3	215.3	04	440.8	244.3	64	493.3	273.4
25	284.3	157.6	85	336.7	186.7	45	389.2	215.7	05	441.7	244.8	65	494.2	273.9
26	285.1	158.0	86	337.6	187.1	46	390.0	216.2	06	442.6	245.3	66	495.0	274.4
27	286.0	158.5	87	338.5	187.6	47	391.0	216.7	07	443.4	245.8	67	495.9	274.9
28	286.9	159.0	88	339.4	188.1	48	391.8	217.2	08	444.3	246.3	68	496.8	275.4
29	287.7	159.5	89	340.2	188.6	49	392.7	217.7	09	445.2	246.8	69	497.7	275.9
30	288.6	160.0	90	341.1	189.1	50	393.6	218.2	10	446.1	247.3	70	498.5	276.3
331	289.5	160.5	391	342.0	189.6	451	394.5	218.6	511	446.9	247.7	571	499.4	276.8
32	290.4	161.0	92	342.9	190.0	52	395.3	219.1	12	447.8	248.2	72	500.3	277.3
33	291.2	161.4	93	343.7	190.5	53	396.2	219.6	13	448.7	248.7	73	501.2	277.8
34	292.1	161.9	94	344.6	191.0	54	397.1	220.1	14	449.6	249.2	74	502.0	278.3
35	293.0	162.4	95	345.5	191.5	55	398.0	220.6	15	450.4	249.7	75	502.9	278.8
36	293.9	162.9	96	346.3	192.0	56	398.8	221.1	16	451.3	250.2	76	503.8	279.3
37	294.7	163.4	97	347.2	192.5	57	399.7	221.6	17	452.2	250.6	77	504.7	279.7
38	295.6	163.9	98	348.1	193.0	58	400.6	222.0	18	453.1	251.1	78	505.5	280.2
39	296.5	164.4	99	349.0	193.4	59	401.5	222.5	19	453.9	251.6	79	506.4	280.7
40	297.4	164.8	400	349.8	193.9	60	402.3	223.0	20	454.8	252.1	80	507.3	281.2
341	298.2	165.3	401	350.7	194.4	461	403.2	223.5	521	455.7	252.6	581	508.2	281.7
42	299.1	165.8	02	351.6	194.9	62	404.0	224.0	22	456.6	253.1	82	509.0	282.2
43	300.0	166.3	03	352.5	195.4	63	404.9	224.5	23	457.4	253.6	83	509.9	282.6
44	300.9	166.8	04	353.3	195.9	64	405.8	225.0	24	458.3	254.0	84	510.7	283.1
45	301.7	167.3	05	354.2	196.3	65	406.7	225.4	25	459.2	254.5	85	511.7	283.6
46	302.6	167.7	06	355.1	196.8	66	407.5	225.9	26	460.0	255.0	86	512.5	284.1
47	303.5	168.2	07	356.0	197.3	67	408.4	226.4	27	460.9	255.5	87	513.4	284.6
48	304.4	168.7	08	356.8	197.8	68	409.3	226.9	28	461.8	256.0	88	514.3	285.1
49	305.2	169.2	09	357.7	198.3	69	410.2	227.4	29	462.7	256.5	89	515.2	285.6
50	306.1	169.7	10	358.6	198.8	70	411.0	227.9	30	463.5	256.9	90	516.0	286.0
351	307.0	170.2	411	359.5	199.3	471	411.9	228.3	531	464.4	257.4	591	516.9	286.5
52	307.9	170.7	12	360.3	199.7	72	412.8	228.8	32	465.3	257.9	92	517.8	287.0
53	308.7	171.1	13	361.2	200.2	73	413.7	229.3	33	466.2	258.4	93	518.6	287.5
54	309.6	171.6	14	362.1	200.7	74	414.5	229.8	34	467.0	258.9	94	519.5	288.0
55	310.5	172.1	15	363.0	201.2	75	415.4	230.3	35	467.9	259.4	95	520.4	288.5
56	311.4	172.6	16	363.8	201.7	76	416.3	230.8	36	468.8	259.9	96	521.3	288.9
57	312.2	173.1	17	364.7	202.2	77	417.2	231.3	37	469.6	260.3	97	522.1	289.4
58	313.1	173.6	18	365.6	202.7	78	418.0	231.7	38	470.5	260.8	98	523.0	289.9
59	314.0	174.0	19	366.5	203.1	79	418.9	232.2	39	471.4	261.3	99	523.9	290.4
60	314.9	174.5	20	367.3	203.6	80	419.8	232.7	40	472.3	261.8	600	524.8	290.9
Dist.	Dep.	D. Lat.	Dist.	Dep.	D. Lat.	Dist.	Dep.	D. Lat.	Dist.	Dep.	D. Lat.	Dist.	Dep.	D. Lat.

Dist.	D. Lat.	Dep.
D Lo	Dep.	
	m	D Lo

61°

299°	061°
241°	119°

330°	030°
210°	150°

TABLE 3

Traverse 30° Table

Dist.	D. Lat.	Dep.	Dist.	D. Lat.	Dep.	Dist.	D. Lat.	Dep.	Dist.	D. Lat.	Dep.	Dist.	D. Lat.	Dep.
1	0.9	0.5	61	52.8	30.5	121	104.8	60.5	181	156.8	90.5	241	208.7	120.5
2	1.7	1.0	62	53.7	31.0	22	105.7	61.0	82	157.6	91.0	42	209.6	121.0
3	2.6	1.5	63	54.6	31.5	23	106.5	61.5	83	158.5	91.5	43	210.4	121.5
4	3.5	2.0	64	55.4	32.0	24	107.4	62.0	84	159.3	92.0	44	211.3	122.0
5	4.3	2.5	65	56.3	32.5	25	108.3	62.5	85	160.2	92.5	45	212.2	122.5
6	5.2	3.0	66	57.2	33.0	26	109.1	63.0	86	161.1	93.0	46	213.0	123.0
7	6.1	3.5	67	58.0	33.5	27	110.0	63.5	87	161.9	93.5	47	213.9	123.5
8	6.9	4.0	68	58.9	34.0	28	110.9	64.0	88	162.8	94.0	48	214.8	124.0
9	7.8	4.5	69	59.8	34.5	29	111.7	64.5	89	163.7	94.5	49	215.6	124.5
10	8.7	5.0	70	60.6	35.0	30	112.6	65.0	90	164.5	95.0	50	216.5	125.0
11	9.5	5.5	71	61.5	35.5	131	113.4	65.5	191	165.4	95.5	251	217.4	125.5
12	10.4	6.0	72	62.4	36.0	32	114.3	66.0	92	166.3	96.0	52	218.2	126.0
13	11.3	6.5	73	63.2	36.5	33	115.2	66.5	93	167.1	96.5	53	219.1	126.5
14	12.1	7.0	74	64.1	37.0	34	116.0	67.0	94	168.0	97.0	54	220.0	127.0
15	13.0	7.5	75	65.0	37.5	35	116.9	67.5	95	168.9	97.5	55	220.8	127.5
16	13.9	8.0	76	65.8	38.0	36	117.8	68.0	96	169.7	98.0	56	221.7	128.0
17	14.7	8.5	77	66.7	38.5	37	118.6	68.5	97	170.6	98.5	57	222.6	128.5
18	15.6	9.0	78	67.5	39.0	38	119.5	69.0	98	171.5	99.0	58	223.4	129.0
19	16.5	9.5	79	68.4	39.5	39	120.4	69.5	99	172.3	99.5	59	224.3	129.5
20	17.3	10.0	80	69.3	40.0	40	121.2	70.0	200	173.2	100.0	60	225.2	130.0
21	18.2	10.5	81	70.1	40.5	141	122.1	70.5	201	174.1	100.5	261	226.0	130.5
22	19.1	11.0	82	71.0	41.0	42	123.0	71.0	02	174.9	101.0	62	226.9	131.0
23	19.9	11.5	83	71.9	41.5	43	123.8	71.5	03	175.8	101.5	63	227.8	131.5
24	20.8	12.0	84	72.7	42.0	44	124.7	72.0	04	176.7	102.0	64	228.6	132.0
25	21.7	12.5	85	73.6	42.5	45	125.6	72.5	05	177.5	102.5	65	229.5	132.5
26	22.5	13.0	86	74.5	43.0	46	126.4	73.0	06	178.4	103.0	66	230.4	133.0
27	23.4	13.5	87	75.3	43.5	47	127.3	73.5	07	179.3	103.5	67	231.2	133.5
28	24.2	14.0	88	76.2	44.0	48	128.2	74.0	08	180.1	104.0	68	232.1	134.0
29	25.1	14.5	89	77.1	44.5	49	129.0	74.5	09	181.0	104.5	69	233.0	134.5
30	26.0	15.0	90	77.9	45.0	50	129.9	75.0	10	181.9	105.0	70	233.8	135.0
31	26.8	15.5	91	78.8	45.5	151	130.8	75.5	211	182.7	105.5	271	234.7	135.5
32	27.7	16.0	92	79.7	46.0	52	131.6	76.0	12	183.6	106.0	72	235.6	136.0
33	28.6	16.5	93	80.5	46.5	53	132.5	76.5	13	184.5	106.5	73	236.4	136.5
34	29.4	17.0	94	81.4	47.0	54	133.4	77.0	14	185.3	107.0	74	237.3	137.0
35	30.3	17.5	95	82.3	47.5	55	134.2	77.5	15	186.2	107.5	75	238.2	137.5
36	31.2	18.0	96	83.1	48.0	56	135.1	78.0	16	187.1	108.0	76	239.0	138.0
37	32.0	18.5	97	84.0	48.5	57	136.0	78.5	17	187.9	108.5	77	239.9	138.5
38	32.9	19.0	98	84.9	49.0	58	136.8	79.0	18	188.8	109.0	78	240.8	139.0
39	33.8	19.5	99	85.7	49.5	59	137.7	79.5	19	189.7	109.5	79	241.6	139.5
40	34.6	20.0	100	86.6	50.0	60	138.6	80.0	20	190.5	110.0	80	242.5	140.0
41	35.5	20.5	101	87.5	50.5	161	139.4	80.5	221	191.4	110.5	281	243.4	140.5
42	36.4	21.0	02	88.3	51.0	62	140.3	81.0	22	192.3	111.0	82	244.2	141.0
43	37.2	21.5	03	89.2	51.5	63	141.2	81.5	23	193.1	111.5	83	245.1	141.5
44	38.1	22.0	04	90.1	52.0	64	142.0	82.0	24	194.0	112.0	84	246.0	142.0
45	39.0	22.5	05	90.9	52.5	65	142.9	82.5	25	194.9	112.5	85	246.8	142.5
46	39.8	23.0	06	91.8	53.0	66	143.8	83.0	26	195.7	113.0	86	247.7	143.0
47	40.7	23.5	07	92.7	53.5	67	144.6	83.5	27	196.6	113.5	87	248.5	143.5
48	41.6	24.0	08	93.5	54.0	68	145.5	84.0	28	197.5	114.0	88	249.4	144.0
49	42.4	24.5	09	94.4	54.5	69	146.4	84.5	29	198.3	114.5	89	250.3	144.5
50	43.3	25.0	10	95.3	55.0	70	147.2	85.0	30	199.2	115.0	90	251.1	145.0
51	44.2	25.5	111	96.1	55.5	171	148.1	85.5	231	200.1	115.5	291	252.0	145.5
52	45.0	26.0	12	97.0	56.0	72	149.0	86.0	32	200.9	116.0	92	252.9	146.0
53	45.9	26.5	13	97.9	56.5	73	149.8	86.5	33	201.8	116.5	93	253.7	146.5
54	46.8	27.0	14	98.7	57.0	74	150.7	87.0	34	202.6	117.0	94	254.6	147.0
55	47.6	27.5	15	99.6	57.5	75	151.6	87.5	35	203.5	117.5	95	255.5	147.5
56	48.5	28.0	16	100.5	58.0	76	152.4	88.0	36	204.4	118.0	96	256.3	148.0
57	49.4	28.5	17	101.3	58.5	77	153.3	88.5	37	205.2	118.5	97	257.2	148.5
58	50.2	29.0	18	102.2	59.0	78	154.2	89.0	38	206.1	119.0	98	258.1	149.0
59	51.1	29.5	19	103.1	59.5	79	155.0	89.5	39	207.0	119.5	99	258.9	149.5
60	52.0	30.0	20	103.9	60.0	80	155.9	90.0	40	207.8	120.0	300	259.8	150.0
Dist.	Dep.	D. Lat.	Dist.	Dep.	D. Lat.	Dist.	Dep.	D. Lat.	Dist.	Dep.	D. Lat.	Dist.	Dep.	D. Lat.

300°	060°
240°	120°

60°

Dist.	D. Lat.	Dep.
N.	N×Cos.	N×Sin.
Hypotenuse	Side Adj.	Side Opp.

330°	030°
210°	150°

TABLE 3

Traverse **30°** Table

330°	030°
210°	150°

Dist.	D. Lat.	Dep.	Dist.	D. Lat.	Dep.	Dist.	D. Lat.	Dep.	Dist.	D. Lat.	Dep.	Dist.	D. Lat.	Dep.
301	260.7	150.5	361	312.6	180.5	421	364.6	210.5	481	416.6	240.5	541	468.5	270.5
02	261.5	151.0	62	313.5	181.0	22	365.5	211.0	82	417.4	241.0	42	469.4	271.0
03	262.4	151.5	63	314.4	181.5	23	366.3	211.5	83	418.3	241.5	43	470.3	271.5
04	263.3	152.0	64	315.2	182.0	24	367.2	212.0	84	419.2	242.0	44	471.1	272.0
05	264.1	152.5	65	316.1	182.5	25	368.1	212.5	85	420.0	242.5	45	472.0	272.5
06	265.0	153.0	66	317.0	183.0	26	368.9	213.0	86	420.9	243.0	46	472.8	273.0
07	265.9	153.5	67	317.8	183.5	27	369.8	213.5	87	421.8	243.5	47	473.7	273.5
08	266.6	154.0	68	318.7	184.0	28	370.7	214.0	88	422.6	244.0	48	474.6	274.0
09	267.6	154.5	69	319.6	184.5	29	371.5	214.5	89	423.5	244.5	49	475.4	274.5
10	268.5	155.0	70	320.4	185.0	30	372.4	215.0	90	424.4	245.0	50	476.3	275.0
311	269.3	155.5	371	321.3	185.5	431	373.3	215.5	491	425.2	245.5	551	477.2	275.5
12	270.2	156.0	72	322.2	186.0	32	374.1	216.0	92	426.1	246.0	52	478.0	276.0
13	271.1	156.5	73	323.0	186.5	33	375.0	216.5	93	427.0	246.5	53	478.9	276.5
14	271.9	157.0	74	323.9	187.0	34	375.9	217.0	94	427.8	247.0	54	479.8	277.0
15	272.8	157.5	75	324.8	187.5	35	376.7	217.5	95	428.7	247.5	55	480.6	277.5
16	273.7	158.0	76	325.6	188.0	36	377.6	218.0	96	429.5	248.0	56	481.5	278.0
17	274.5	158.5	77	326.5	188.5	37	378.5	218.5	97	430.4	248.5	57	482.4	278.5
18	275.4	159.0	78	327.4	189.0	38	379.3	219.0	98	431.3	249.0	58	483.2	279.0
19	276.3	159.5	79	328.2	189.5	39	380.2	219.5	99	432.1	249.5	59	484.1	279.5
20	277.1	160.0	80	329.1	190.0	40	381.1	220.0	500	433.0	250.0	60	485.0	280.0
321	278.0	160.5	381	330.0	190.5	441	381.9	220.5	501	433.9	250.5	561	485.8	280.5
22	278.9	161.0	82	330.8	191.0	42	382.8	221.0	02	434.7	251.0	62	486.7	281.0
23	279.7	161.5	83	331.7	191.5	43	383.6	221.5	03	435.6	251.5	63	487.6	281.5
24	280.6	162.0	84	332.6	192.0	44	384.5	222.0	04	436.5	252.0	64	488.4	282.0
25	281.5	162.5	85	333.4	192.5	45	385.4	222.5	05	437.3	252.5	65	489.3	282.5
26	282.3	163.0	86	334.3	193.0	46	386.3	223.0	06	438.2	253.0	66	490.2	283.0
27	283.2	163.5	87	335.2	193.5	47	387.1	223.5	07	439.1	253.5	67	491.0	283.5
28	284.1	164.0	88	336.0	194.0	48	388.0	224.0	08	439.9	254.0	68	491.9	284.0
29	284.9	164.5	89	336.9	194.5	49	388.8	224.5	09	440.8	254.5	69	492.8	284.5
30	285.8	165.0	90	337.7	195.0	50	389.7	225.0	10	441.7	255.0	70	493.6	285.0
331	286.7	165.5	391	338.6	195.5	451	390.6	225.5	511	442.5	255.5	571	494.5	285.5
32	287.5	166.0	92	339.5	196.0	52	391.4	226.0	12	443.4	256.0	72	495.4	286.0
33	288.4	166.5	93	340.3	196.5	53	392.3	226.5	13	444.3	256.5	73	496.2	286.5
34	289.3	167.0	94	341.2	197.0	54	393.2	227.0	14	445.1	257.0	74	497.1	287.0
35	290.1	167.5	95	342.1	197.5	55	394.0	227.5	15	446.0	257.5	75	498.0	287.5
36	291.0	168.0	96	342.9	198.0	56	394.9	228.0	16	446.9	258.0	76	498.8	288.0
37	291.9	168.5	97	343.8	198.5	57	395.8	228.5	17	447.7	258.5	77	499.7	288.5
38	292.7	169.0	98	344.7	199.0	58	396.6	229.0	18	448.6	259.0	78	500.6	289.0
39	293.6	169.5	99	345.5	199.5	59	397.5	229.5	19	449.5	259.5	79	501.3	289.5
40	294.5	170.0	400	346.4	200.0	60	398.4	230.0	20	450.3	260.0	80	502.3	290.0
341	295.3	170.5	401	347.3	200.5	461	399.2	230.5	521	451.2	260.5	581	503.2	290.5
42	296.2	171.0	02	348.1	201.0	62	400.1	231.0	22	452.1	261.0	82	504.0	291.0
43	297.0	171.5	03	349.0	201.5	63	401.0	231.5	23	452.9	261.5	83	504.9	291.5
44	297.9	172.0	04	349.9	202.0	64	401.8	232.0	24	453.8	262.0	84	505.8	292.0
45	298.8	172.5	05	350.7	202.5	65	402.7	232.7	25	454.7	262.5	85	506.6	292.5
46	299.6	173.0	06	351.6	203.0	66	403.6	233.0	26	455.5	263.0	86	507.5	293.0
47	300.5	173.5	07	352.5	203.5	67	404.4	233.5	27	456.4	263.5	87	508.4	293.5
48	301.4	174.0	08	353.3	204.0	68	405.3	234.0	28	457.3	264.0	88	509.2	294.0
49	302.2	174.5	09	354.2	204.5	69	406.2	234.5	29	458.1	264.5	89	510.1	294.5
50	303.1	175.0	10	355.1	205.0	70	407.0	235.0	30	459.0	265.0	90	511.0	295.0
351	304.0	175.5	411	355.9	205.5	471	407.9	235.5	531	459.9	265.5	591	511.8	295.5
52	304.8	176.0	12	356.8	206.0	72	408.8	236.0	32	460.7	266.0	92	512.7	296.0
53	305.7	176.5	13	357.7	206.5	73	409.6	236.5	33	461.6	266.5	93	513.6	296.5
54	306.6	177.0	14	358.5	207.0	74	410.5	237.0	34	462.5	267.0	94	514.4	297.0
55	307.4	177.5	15	359.4	207.5	75	411.4	237.5	35	463.3	267.5	95	515.3	297.5
56	308.3	178.0	16	360.3	208.0	76	412.2	238.0	36	464.2	268.0	96	516.2	298.0
57	309.2	178.5	17	361.1	208.5	77	413.1	238.5	37	465.1	268.5	97	517.0	298.5
58	310.0	179.0	18	362.0	209.0	78	414.0	239.0	38	465.9	269.0	98	517.9	299.0
59	310.9	179.5	19	362.9	209.5	79	414.8	239.5	39	466.8	269.5	99	518.7	299.5
60	311.8	180.0	20	363.7	210.0	80	415.7	240.0	40	467.7	270.0	600	519.6	300.0
Dist.	Dep.	D. Lat.	Dist.	Dep.	D. Lat.	Dist.	Dep.	D. Lat.	Dist.	Dep.	D. Lat.	Dist.	Dep.	D. Lat.

Dist.	D. Lat.	Dep.
D Lo	Dep.	
	m	D Lo

60°

300°	060°
240°	120°

329° | 031°	TABLE 3	329° | 031°
211° | 149°	Traverse **31°** Table	211° | 149°

Dist.	D. Lat.	Dep.	Dist.	D. Lat.	Dep.	Dist.	D. Lat.	Dep.	Dist.	D. Lat.	Dep.	Dist.	D. Lat.	Dep.
1	0.9	0.5	61	52.3	31.4	121	103.7	62.3	181	155.1	93.2	241	206.6	124.1
2	1.7	1.0	62	53.1	31.9	22	104.6	62.8	82	156.0	93.7	42	207.4	124.6
3	2.6	1.5	63	54.0	32.4	23	105.4	63.3	83	156.9	94.3	43	208.3	125.2
4	3.4	2.1	64	54.9	33.0	24	106.3	63.9	84	157.7	94.8	44	209.1	125.7
5	4.3	2.6	65	55.7	33.5	25	107.1	64.4	85	158.6	95.3	45	210.0	126.2
6	5.1	3.1	66	56.6	34.0	26	108.0	64.9	86	159.4	95.8	46	210.9	126.7
7	6.0	3.6	67	57.4	34.5	27	108.9	65.4	87	160.3	96.3	47	211.7	127.2
8	6.9	4.1	68	58.3	35.0	28	109.7	65.9	88	161.1	96.8	48	212.6	127.7
9	7.7	4.6	69	59.1	35.5	29	110.6	66.4	89	162.0	97.3	49	213.4	128.2
10	8.6	5.2	70	60.0	36.1	30	111.4	67.0	90	162.9	97.9	50	214.3	128.8
11	9.4	5.7	71	60.9	36.6	131	112.3	67.5	191	163.7	98.4	251	215.1	129.3
12	10.3	6.2	72	61.7	37.1	32	113.1	68.0	92	164.6	98.9	52	216.0	129.8
13	11.1	6.7	73	62.6	37.6	33	114.0	68.5	93	165.4	99.4	53	216.9	130.3
14	12.0	7.2	74	63.4	38.1	34	114.9	69.0	94	166.3	99.9	54	217.7	130.8
15	12.9	7.7	75	64.3	38.6	35	115.7	69.5	95	167.1	100.4	55	218.6	131.3
16	13.7	8.2	76	65.1	39.1	36	116.6	70.0	96	168.0	100.9	56	219.4	131.8
17	14.6	8.8	77	66.0	39.7	37	117.4	70.6	97	168.9	101.5	57	220.3	132.4
18	15.4	9.3	78	66.9	40.2	38	118.3	71.1	98	169.7	102.0	58	221.1	132.9
19	16.3	9.8	79	67.7	40.7	39	119.1	71.6	99	170.6	102.5	59	222.0	133.4
20	17.1	10.3	80	68.6	41.2	40	120.0	72.1	200	171.4	103.0	60	222.9	133.9
21	18.0	10.8	81	69.4	41.7	141	120.9	72.6	201	172.3	103.5	261	223.7	134.4
22	18.9	11.3	82	70.3	42.2	42	121.7	73.1	02	173.1	104.0	62	224.6	134.9
23	19.7	11.8	83	71.1	42.7	43	122.6	73.7	03	174.0	104.6	63	225.4	135.5
24	20.6	12.4	84	72.0	43.3	44	123.4	74.2	04	174.9	105.1	64	226.3	136.0
25	21.4	12.9	85	72.9	43.8	45	124.3	74.7	05	175.7	105.6	65	227.1	136.5
26	22.3	13.4	86	73.7	44.3	46	125.1	75.2	06	176.6	106.1	66	228.0	137.0
27	23.1	13.9	87	74.6	44.8	47	126.0	75.7	07	177.4	106.6	67	228.9	137.5
28	24.0	14.4	88	75.4	45.3	48	126.9	76.2	08	178.3	107.1	68	229.7	138.0
29	24.9	14.9	89	76.3	45.8	49	127.7	76.7	09	179.1	107.6	69	230.6	138.5
30	25.7	15.5	90	77.1	46.4	50	128.6	77.3	10	180.0	108.2	70	231.4	139.1
31	26.6	16.0	91	78.0	46.9	151	129.4	77.8	211	180.9	108.7	271	232.3	139.6
32	27.4	16.5	92	78.9	47.4	52	130.3	78.3	12	181.7	109.2	72	233.1	140.1
33	28.3	17.0	93	79.7	47.9	53	131.1	78.8	13	182.6	109.7	73	234.0	140.6
34	29.1	17.5	94	80.6	48.4	54	132.0	79.3	14	183.4	110.2	74	234.9	141.1
35	30.0	18.0	95	81.4	48.9	55	132.9	79.8	15	184.3	110.7	75	235.7	141.6
36	30.9	18.5	96	82.3	49.4	56	133.7	80.3	16	185.1	111.2	76	236.6	142.2
37	31.7	19.1	97	83.1	50.0	57	134.6	80.9	17	186.0	111.8	77	237.4	142.7
38	32.6	19.6	98	84.0	50.5	58	135.4	81.4	18	186.9	112.3	78	238.3	143.2
39	33.4	20.1	99	84.9	51.0	59	136.3	81.9	19	187.7	112.8	79	239.1	143.7
40	34.3	20.6	100	85.7	51.5	60	137.1	82.4	20	188.6	113.3	80	240.0	144.2
41	35.1	21.1	101	86.6	52.0	161	138.0	82.9	221	189.4	113.8	281	240.9	144.7
42	36.0	21.6	02	87.4	52.5	62	138.9	83.4	22	190.3	114.3	82	241.7	145.2
43	36.9	22.1	03	88.3	53.0	63	139.7	84.0	23	191.1	114.9	83	242.6	145.8
44	37.7	22.7	04	89.1	53.6	64	140.6	84.5	24	192.0	115.4	84	243.4	146.3
45	38.6	23.2	05	90.0	54.1	65	141.4	85.0	25	192.9	115.9	85	244.3	146.8
46	39.4	23.7	06	90.9	54.6	66	142.3	85.5	26	193.7	116.4	86	245.1	147.3
47	40.3	24.2	07	91.7	55.1	67	143.1	86.0	27	194.6	116.9	87	246.0	147.8
48	41.1	24.7	08	92.6	55.6	68	144.0	86.5	28	195.4	117.4	88	246.9	148.3
49	42.0	25.2	09	93.4	56.1	69	144.9	87.0	29	196.3	117.9	89	247.7	148.8
50	42.9	25.8	10	94.3	56.7	70	145.7	87.6	30	197.1	118.5	90	248.6	149.4
51	43.7	26.3	111	95.1	57.2	171	146.6	88.1	231	198.0	119.0	291	249.4	149.9
52	44.6	26.8	12	96.0	57.7	72	147.4	88.6	32	198.9	119.5	92	250.3	150.4
53	45.4	27.3	13	96.9	58.2	73	148.3	89.1	33	199.7	120.0	93	251.2	150.9
54	46.3	27.8	14	97.7	58.7	74	149.1	89.6	34	200.6	120.5	94	252.0	151.4
55	47.1	28.3	15	98.6	59.2	75	150.0	90.1	35	201.4	121.0	95	252.9	151.9
56	48.0	28.8	16	99.4	59.7	76	150.9	90.6	36	202.3	121.5	96	253.7	152.5
57	48.9	29.4	17	100.3	60.3	77	151.7	91.2	37	203.1	122.1	97	254.6	153.0
58	49.7	29.9	18	101.1	60.8	78	152.6	91.7	38	204.0	122.6	98	255.4	153.5
59	50.6	30.4	19	102.0	61.3	79	153.4	92.2	39	204.9	123.1	99	256.3	154.0
60	51.4	30.9	20	102.9	61.8	80	154.3	92.7	40	205.7	123.6	300	257.1	154.5
Dist.	Dep.	D. Lat.	Dist.	Dep.	D. Lat.	Dist.	Dep.	D. Lat.	Dist.	Dep.	D. Lat.	Dist.	Dep.	D. Lat.

301° | 059°	**59°**
239° | 121°	

Dist.	D. Lat.	Dep.
N.	N×Cos.	N×Sin.
Hypotenuse	Side Adj.	Side Opp.

329° | 031°
211° | 149°

TABLE 3

Traverse **31°** Table

329° | 031°
211° | 149°

Dist.	D. Lat.	Dep.	Dist.	D. Lat.	Dep.	Dist.	D. Lat.	Dep.	Dist.	D. Lat.	Dep.	Dist.	D. Lat.	Dep.
301	258.0	155.0	361	309.4	185.9	421	360.9	216.8	481	412.3	247.7	541	463.7	278.6
02	258.9	155.5	62	310.3	186.4	22	361.7	217.3	82	413.2	248.2	42	464.6	279.2
03	259.7	156.1	63	311.2	187.0	23	362.6	217.9	83	414.0	248.8	43	465.4	279.7
04	260.6	156.6	64	312.0	187.5	24	363.4	218.4	84	414.9	249.3	44	466.3	280.2
05	261.4	157.1	65	312.9	188.0	25	364.3	218.9	85	415.7	249.8	45	467.2	280.7
06	262.3	157.6	66	313.7	188.5	26	365.2	219.4	86	416.6	250.3	46	468.0	281.2
07	263.2	158.1	67	314.6	189.0	27	366.0	219.9	87	417.4	250.8	47	468.9	281.7
08	264.0	158.6	68	315.4	189.5	28	366.9	220.4	88	418.3	251.3	48	469.7	282.2
09	264.9	159.1	69	316.3	190.0	29	367.7	221.0	89	419.2	251.9	49	470.6	282.8
10	265.7	159.7	70	317.2	190.6	30	368.6	221.5	90	420.0	252.4	50	471.4	283.3
311	266.6	160.2	371	318.0	191.1	431	369.4	222.0	491	420.9	252.9	551	472.3	283.8
12	267.4	160.7	72	318.9	191.6	32	370.3	222.5	92	421.7	253.4	52	473.2	284.3
13	268.3	161.2	73	319.7	192.1	33	371.2	223.0	93	422.6	253.9	53	474.0	284.8
14	269.2	161.7	74	320.6	192.6	34	372.0	223.5	94	423.4	254.4	54	474.9	285.3
15	270.0	162.2	75	321.4	193.1	35	372.9	224.0	95	424.3	254.9	55	475.7	285.8
16	270.9	162.8	76	322.3	193.7	36	373.7	224.6	96	425.2	255.5	56	476.6	286.4
17	271.7	163.3	77	323.2	194.2	37	374.6	225.1	97	426.0	256.0	57	477.4	286.9
18	272.6	163.8	78	324.0	194.7	38	375.4	225.6	98	426.9	256.5	58	478.3	287.4
19	273.4	164.3	79	324.9	195.2	39	376.3	226.1	99	427.7	257.0	59	479.2	287.9
20	274.3	164.8	80	325.7	195.7	40	377.2	226.6	500	428.6	257.5	60	480.0	288.4
321	275.2	165.3	381	326.6	196.2	441	378.0	227.1	501	429.4	258.0	561	480.9	288.9
22	276.0	165.8	82	327.4	196.7	42	378.9	227.6	02	430.3	258.5	62	481.7	289.5
23	276.9	166.4	83	328.3	197.3	43	379.7	228.2	03	431.2	259.1	63	482.6	290.0
24	277.7	166.9	84	329.2	197.8	44	380.6	228.7	04	432.0	259.6	64	483.4	290.5
25	278.6	167.4	85	330.0	198.3	45	381.4	229.2	05	432.9	260.1	65	484.3	291.0
26	279.4	167.9	86	330.9	198.8	46	382.3	229.7	06	433.7	260.6	66	485.2	291.5
27	280.3	168.4	87	331.7	199.3	47	383.2	230.2	07	434.6	261.1	67	486.0	292.0
28	281.2	168.9	88	332.6	199.8	48	384.0	230.7	08	435.4	261.6	68	486.9	292.5
29	282.0	169.4	89	333.4	200.3	49	384.9	231.3	09	436.3	262.2	69	487.7	293.1
30	282.9	170.0	90	334.3	200.9	50	385.7	231.8	10	437.2	262.7	70	488.6	293.6
331	283.7	170.5	391	335.2	201.4	451	386.6	232.3	511	438.0	263.2	571	489.4	294.1
32	284.6	171.0	92	336.0	201.9	52	387.4	232.8	12	438.9	263.7	72	490.3	294.6
33	285.4	171.5	93	336.9	202.4	53	388.3	233.3	13	439.7	264.2	73	491.2	295.1
34	286.3	172.0	94	337.7	202.9	54	389.2	233.8	14	440.6	264.7	74	492.0	295.6
35	287.2	172.5	95	338.6	203.4	55	390.0	234.3	15	441.4	265.2	75	492.9	296.1
36	288.0	173.1	96	339.4	204.0	56	390.9	234.9	16	442.3	265.8	76	493.7	296.7
37	288.9	173.6	97	340.3	204.5	57	391.7	235.4	17	443.2	266.3	77	494.6	297.2
38	289.7	174.1	98	341.2	205.0	58	392.6	235.9	18	444.0	266.8	78	495.4	297.7
39	290.6	174.6	99	342.0	205.5	59	393.4	236.4	19	444.9	267.3	79	496.3	298.2
40	291.4	175.1	400	342.9	206.0	60	394.3	236.9	20	445.7	267.8	80	497.2	298.7
341	292.3	175.6	401	343.7	206.5	461	395.2	237.4	521	446.6	268.3	581	498.0	299.2
42	293.2	176.1	02	344.6	207.0	62	396.0	237.9	22	447.4	268.8	82	498.9	299.8
43	294.0	176.7	03	345.4	207.6	63	396.9	238.5	23	448.3	269.4	83	499.7	300.3
44	294.9	177.2	04	346.3	208.1	64	397.7	239.0	24	449.2	269.9	84	500.6	300.8
45	295.7	177.7	05	347.2	208.6	65	398.6	239.5	25	450.0	270.4	85	501.4	301.3
46	296.6	178.2	06	348.0	209.1	66	399.4	240.0	26	450.9	270.9	86	502.3	301.8
47	297.4	178.7	07	348.9	209.6	67	400.3	240.5	27	451.7	271.4	87	503.2	302.3
48	298.3	179.2	08	349.7	210.1	68	401.2	241.0	28	452.6	271.9	88	504.0	302.8
49	299.2	179.7	09	350.6	210.7	69	402.0	241.6	29	453.4	272.5	89	504.9	303.4
50	300.0	180.3	10	351.4	211.2	70	402.9	242.1	30	454.3	273.0	90	505.7	303.9
351	300.9	180.8	411	352.3	211.7	471	403.7	242.6	531	455.2	273.5	591	506.6	304.4
52	301.7	181.3	12	353.2	212.2	72	404.6	243.1	32	456.0	274.0	92	507.4	304.9
53	302.6	181.8	13	354.0	212.7	73	405.4	243.6	33	456.9	274.5	93	508.3	305.4
54	303.4	182.3	14	354.9	213.2	74	406.3	244.1	34	457.7	275.0	94	509.2	305.9
55	304.3	182.8	15	355.7	213.7	75	407.2	244.6	35	458.6	275.5	95	510.0	306.4
56	305.2	183.4	16	356.6	214.3	76	408.0	245.2	36	459.4	276.1	96	510.9	307.0
57	306.0	183.9	17	357.4	214.8	77	408.9	245.7	37	460.3	276.6	97	511.7	307.5
58	306.9	184.4	18	358.3	215.3	78	409.7	246.2	38	461.2	277.1	98	512.6	308.0
59	307.7	184.9	19	359.2	215.8	79	410.6	246.7	39	462.0	277.6	99	513.4	308.5
60	308.6	185.4	20	360.0	216.3	80	411.4	247.2	40	462.9	278.1	600	514.3	309.0
Dist.	Dep.	D. Lat.	Dist.	Dep.	D. Lat.	Dist.	Dep.	D. Lat.	Dist.	Dep.	D. Lat.	Dist.	Dep.	D. Lat.

Dist.	D. Lat.	Dep.
D Lo	Dep.	
	m	D Lo

59°

301° | 059°
239° | 121°

328° | 032°		TABLE 3		328° | 032°
212° | 148°	Traverse	32°	Table	212° | 148°

Dist.	D. Lat.	Dep.	Dist.	D. Lat.	Dep.	Dist.	D. Lat.	Dep.	Dist.	D. Lat.	Dep.	Dist.	D. Lat.	Dep.
1	0.8	0.5	61	51.7	32.3	121	102.6	64.1	181	153.5	95.9	241	204.4	127.7
2	1.7	1.1	62	52.6	32.9	22	103.5	64.7	82	154.3	96.4	42	205.2	128.2
3	2.5	1.6	63	53.4	33.4	23	104.3	65.2	83	155.2	97.0	43	206.1	128.8
4	3.4	2.1	64	54.3	33.9	24	105.2	65.7	84	156.0	97.5	44	206.9	129.3
5	4.2	2.6	65	55.1	34.4	25	106.0	66.2	85	156.9	98.0	45	207.8	129.8
6	5.1	3.2	66	56.0	35.0	26	106.9	66.8	86	157.7	98.6	46	208.6	130.4
7	5.9	3.7	67	56.8	35.5	27	107.7	67.3	87	158.6	99.1	47	209.5	130.9
8	6.8	4.2	68	57.7	36.0	28	108.6	67.8	88	159.4	99.6	48	210.3	131.4
9	7.6	4.8	69	58.5	36.6	29	109.4	68.4	89	160.3	100.2	49	211.2	131.9
10	8.5	5.3	70	59.4	37.1	30	110.2	68.9	90	161.1	100.7	50	212.0	132.5
11	9.3	5.8	71	60.2	37.6	131	111.1	69.4	191	162.0	101.2	251	212.9	133.0
12	10.2	6.4	72	61.1	38.2	32	111.9	69.9	92	162.8	101.7	52	213.7	133.5
13	11.0	6.9	73	61.9	38.7	33	112.8	70.5	93	163.7	102.3	53	214.6	134.1
14	11.9	7.4	74	62.8	39.2	34	113.6	71.0	94	164.5	102.8	54	215.4	134.6
15	12.7	7.9	75	63.6	39.7	35	114.5	71.5	95	165.4	103.3	55	216.3	135.1
16	13.6	8.5	76	64.5	40.3	36	115.3	72.1	96	166.2	103.9	56	217.1	135.7
17	14.4	9.0	77	65.3	40.8	37	116.2	72.6	97	167.1	104.4	57	217.9	136.2
18	15.3	9.5	78	66.1	41.3	38	117.0	73.1	98	167.9	104.9	58	218.8	136.7
19	16.1	10.1	79	67.0	41.9	39	117.9	73.7	99	168.8	105.5	59	219.6	137.2
20	17.0	10.6	80	67.8	42.4	40	118.7	74.2	200	169.6	106.0	60	220.5	137.8
21	17.8	11.1	81	68.7	42.9	141	119.6	74.7	201	170.5	106.5	261	221.3	138.3
22	18.7	11.7	82	69.5	43.5	42	120.4	75.2	02	171.3	107.0	62	222.2	138.8
23	19.5	12.2	83	70.4	44.0	43	121.3	75.8	03	172.2	107.6	63	223.0	139.4
24	20.4	12.7	84	71.2	44.5	44	122.1	76.3	04	173.0	108.1	64	223.9	139.9
25	21.2	13.2	85	72.1	45.0	45	123.0	76.8	05	173.8	108.6	65	224.7	140.4
26	22.0	13.8	86	72.9	45.6	46	123.8	77.4	06	174.7	109.2	66	225.6	141.0
27	22.9	14.3	87	73.8	46.1	47	124.7	77.9	07	175.5	109.7	67	226.4	141.5
28	23.7	14.8	88	74.6	46.6	48	125.5	78.4	08	176.4	110.2	68	227.3	142.0
29	24.6	15.4	89	75.5	47.2	49	126.4	79.0	09	177.2	110.8	69	228.1	142.5
30	25.4	15.9	90	76.3	47.7	50	127.2	79.5	10	178.1	111.3	70	229.0	143.1
31	26.3	16.4	91	77.2	48.2	151	128.1	80.0	211	178.9	111.8	271	229.8	143.6
32	27.1	17.0	92	78.0	48.8	52	128.9	80.5	12	179.8	112.3	72	230.7	144.1
33	28.0	17.5	93	78.9	49.3	53	129.8	81.1	13	180.6	112.9	73	231.5	144.7
34	28.8	18.0	94	79.7	49.8	54	130.6	81.6	14	181.5	113.4	74	232.4	145.2
35	29.7	18.5	95	80.6	50.3	55	131.4	82.1	15	182.3	113.9	75	233.2	145.7
36	30.5	19.1	96	81.4	50.9	56	132.3	82.7	16	183.2	114.5	76	234.1	146.3
37	31.4	19.6	97	82.3	51.4	57	133.1	83.2	17	184.0	115.0	77	234.9	146.8
38	32.2	20.1	98	83.1	51.9	58	134.0	83.7	18	184.9	115.5	78	235.8	147.3
39	33.1	20.7	99	84.0	52.5	59	134.8	84.3	19	185.7	116.1	79	236.6	147.8
40	33.9	21.2	100	84.8	53.0	60	135.7	84.8	20	186.6	116.6	80	237.5	148.4
41	34.8	21.7	101	85.7	53.5	161	136.5	85.3	221	187.4	117.1	281	238.3	148.9
42	35.6	22.3	02	86.5	54.1	62	137.4	85.8	22	188.3	117.6	82	239.1	149.4
43	36.5	22.8	03	87.3	54.6	63	138.2	86.4	23	189.1	118.2	83	240.0	150.0
44	37.3	23.3	04	88.2	55.1	64	139.1	86.9	24	190.0	118.7	84	240.8	150.5
45	38.2	23.8	05	89.0	55.6	65	139.9	87.4	25	190.8	119.2	85	241.7	151.0
46	39.0	24.4	06	89.9	56.2	66	140.8	88.0	26	191.7	119.8	86	242.5	151.6
47	39.9	24.9	07	90.7	56.7	67	141.6	88.5	27	192.5	120.3	87	243.4	152.1
48	40.7	25.4	08	91.6	57.2	68	142.5	89.0	28	193.4	120.8	88	244.2	152.6
49	41.6	26.0	09	92.4	57.8	69	143.3	89.6	29	194.2	121.4	89	245.1	153.1
50	42.4	26.5	10	93.3	58.3	70	144.2	90.1	30	195.1	121.9	90	245.9	153.7
51	43.3	27.0	111	94.1	58.8	171	145.0	90.6	231	195.9	122.4	291	246.8	154.2
52	44.1	27.6	12	95.0	59.4	72	145.9	91.1	32	196.7	122.9	92	247.6	154.7
53	44.9	28.1	13	95.8	59.9	73	146.7	91.7	33	197.6	123.5	93	248.5	155.3
54	45.8	28.6	14	96.7	60.4	74	147.6	92.2	34	198.4	124.0	94	249.3	155.8
55	46.6	29.1	15	97.5	60.9	75	148.4	92.7	35	199.3	124.5	95	250.2	156.3
56	47.5	29.7	16	98.4	61.5	76	149.3	93.3	36	200.1	125.1	96	251.0	156.9
57	48.3	30.2	17	99.2	62.0	77	150.1	93.8	37	201.0	125.6	97	251.9	157.4
58	49.2	30.7	18	100.1	62.5	78	151.0	94.3	38	201.8	126.1	98	252.7	157.9
59	50.0	31.3	19	100.9	63.1	79	151.8	94.9	39	202.7	126.7	99	253.6	158.4
60	50.9	31.8	20	101.8	63.6	80	152.6	95.4	40	203.5	127.2	300	254.4	159.0
Dist.	Dep.	D. Lat.	Dist.	Dep.	D. Lat.	Dist.	Dep.	D. Lat.	Dist.	Dep.	D. Lat.	Dist.	Dep.	D. Lat.

302° | 058°	58°
238° | 122°	

Dist.	D. Lat.	Dep.
N.	N×Cos.	N×Sin.
Hypotenuse	Side Adj.	Side Opp.

328°	032°
212°	148°

TABLE 3

Traverse **32°** Table

328°	032°
212°	148°

Dist.	D. Lat.	Dep.	Dist.	D. Lat.	Dep.	Dist.	D. Lat.	Dep.	Dist.	D. Lat.	Dep.	Dist.	D. Lat.	Dep.
301	255.3	159.5	361	306.1	191.3	421	357.0	223.1	481	407.9	254.9	541	458.8	286.7
02	256.1	160.0	62	307.0	191.8	22	357.9	223.6	82	408.8	255.4	42	459.6	287.2
03	257.0	160.6	63	307.8	192.4	23	358.7	224.2	83	409.6	256.0	43	460.5	287.7
04	257.8	161.1	64	308.7	192.9	24	359.6	224.7	84	410.5	256.5	44	461.3	288.3
05	258.7	161.6	65	309.5	193.4	25	360.4	225.2	85	411.3	257.0	45	462.2	288.8
06	259.5	162.2	66	310.4	194.0	26	361.3	225.7	86	412.2	257.5	46	463.0	289.3
07	260.4	162.7	67	311.2	194.5	27	362.1	226.3	87	413.0	258.1	47	463.9	289.9
08	261.2	163.2	68	312.1	195.0	28	363.0	226.8	88	413.8	258.6	48	464.7	290.4
09	262.0	163.7	69	312.9	195.5	29	363.8	227.3	89	414.7	259.1	49	465.6	290.9
10	262.9	164.3	70	313.8	196.1	30	364.7	227.9	90	415.5	259.7	50	466.4	291.5
311	263.7	164.8	371	314.6	196.6	431	365.5	228.4	491	416.4	260.2	551	467.3	292.0
12	264.6	165.3	72	315.5	197.1	32	366.4	228.9	92	417.2	260.7	52	468.1	292.5
13	265.4	165.9	73	316.3	197.7	33	367.2	229.5	93	418.1	261.3	53	469.0	293.0
14	266.3	166.4	74	317.2	198.2	34	368.1	230.0	94	418.9	261.8	54	469.8	293.6
15	267.1	166.9	75	318.0	198.7	35	368.9	230.5	95	419.8	262.3	55	470.7	294.1
16	268.0	167.5	76	318.9	199.2	36	369.7	231.0	96	420.6	262.8	56	471.5	294.6
17	268.8	168.0	77	319.7	199.8	37	370.6	231.6	97	421.5	263.4	57	472.4	295.2
18	269.7	168.5	78	320.6	200.3	38	371.4	232.1	98	422.3	263.9	58	473.2	295.7
19	270.5	169.0	79	321.4	200.8	39	372.3	232.6	99	423.2	264.4	59	474.1	296.2
20	271.4	169.6	80	322.3	201.4	40	373.1	233.2	500	424.0	265.0	60	474.9	296.8
321	272.2	170.1	381	323.1	201.9	441	374.0	233.7	501	424.9	265.5	561	475.8	297.3
22	273.1	170.6	82	324.0	202.4	42	374.8	234.2	02	425.7	266.0	62	476.6	297.8
23	273.9	171.2	83	324.8	203.0	43	375.7	234.8	03	426.6	266.5	63	477.5	298.3
24	274.8	171.7	84	325.7	203.5	44	376.5	235.3	04	427.4	267.1	64	478.3	298.9
25	275.6	172.2	85	326.5	204.0	45	377.4	235.8	05	428.3	267.6	65	479.1	299.4
26	276.5	172.8	86	327.3	204.5	46	378.2	236.3	06	429.1	268.1	66	480.0	299.9
27	277.3	173.3	87	328.2	205.1	47	379.1	236.9	07	430.0	268.7	67	480.8	300.5
28	278.2	173.8	88	329.0	205.6	48	379.9	237.4	08	430.8	269.2	68	481.7	301.0
29	279.0	174.3	89	329.9	206.1	49	380.8	237.9	09	431.7	269.7	69	482.5	301.5
30	279.9	174.9	90	330.7	206.7	50	381.6	238.5	10	432.5	270.3	70	483.4	302.1
331	280.7	175.4	391	331.6	207.2	451	382.5	239.0	511	433.4	270.8	571	484.2	302.6
32	281.6	175.9	92	332.4	207.7	52	383.3	239.5	12	434.2	271.3	72	485.1	303.1
33	282.4	176.5	93	333.3	208.3	53	384.2	240.1	13	435.0	271.9	73	485.9	303.6
34	283.2	177.0	94	334.1	208.8	54	385.0	240.6	14	435.9	272.4	74	486.8	304.2
35	284.1	177.5	95	335.0	209.3	55	385.9	241.1	15	436.7	272.9	75	487.6	304.7
36	284.9	178.1	96	335.8	209.8	56	386.7	241.6	16	437.6	273.4	76	488.5	305.2
37	285.8	178.6	97	336.7	210.4	57	387.6	242.2	17	438.4	274.0	77	489.3	305.8
38	286.6	179.1	98	337.5	210.9	58	388.4	242.7	18	439.3	274.5	78	490.2	306.3
39	287.5	179.6	99	338.4	211.4	59	389.3	243.2	19	440.1	275.0	79	491.0	306.8
40	288.3	180.2	400	339.2	212.0	60	390.1	243.8	20	441.0	275.6	80	491.9	307.4
341	289.2	180.7	401	340.1	212.5	461	391.0	244.3	521	441.8	276.1	581	492.7	307.9
42	290.0	181.2	02	340.9	213.0	62	391.8	244.8	22	442.7	276.6	82	493.6	308.4
43	290.9	181.8	03	341.8	213.6	63	392.6	245.4	23	443.5	277.1	83	494.4	308.9
44	291.7	182.3	04	342.6	214.1	64	393.5	245.9	24	444.4	277.7	84	495.3	309.5
45	292.6	182.8	05	343.5	214.6	65	394.3	246.4	25	445.2	278.2	85	496.1	310.0
46	293.4	183.4	06	344.3	215.1	66	395.2	246.9	26	446.1	278.7	86	497.0	310.5
47	294.3	183.9	07	345.2	215.7	67	396.0	247.5	27	446.9	279.3	87	497.8	311.1
48	295.1	184.4	08	346.0	216.2	68	396.9	248.0	28	447.8	279.8	88	498.7	311.6
49	296.0	184.9	09	346.9	216.7	69	397.7	248.5	29	448.6	280.3	89	499.5	312.1
50	296.8	185.5	10	347.7	217.3	70	398.6	249.1	30	449.5	280.9	90	500.3	312.7
351	297.7	186.0	411	348.5	217.8	471	399.4	249.6	531	450.3	281.4	591	501.2	313.2
52	298.5	186.5	12	349.4	218.3	72	400.3	250.1	32	451.2	281.9	92	502.0	313.7
53	299.4	187.1	13	350.2	218.9	73	401.1	250.7	33	452.0	282.4	93	502.9	314.2
54	300.2	187.6	14	351.1	219.4	74	402.0	251.2	34	452.9	283.0	94	503.7	314.8
55	301.1	188.1	15	351.9	219.9	75	402.8	251.7	35	453.7	283.5	95	504.6	315.3
56	301.9	188.7	16	352.8	220.4	76	403.7	252.2	36	454.6	284.0	96	505.4	315.8
57	302.8	189.2	17	353.6	221.0	77	404.5	252.8	37	455.4	284.6	97	506.3	316.4
58	303.6	189.7	18	354.5	221.5	78	405.4	253.3	38	456.2	285.1	98	507.1	316.9
59	304.4	190.2	19	355.3	222.0	79	406.2	253.8	39	457.1	285.6	99	508.0	317.4
60	305.3	190.8	20	356.2	222.6	80	407.1	254.4	40	457.9	286.2	600	508.8	318.0
Dist.	Dep.	D. Lat.	Dist.	Dep.	D. Lat.	Dist.	Dep.	D. Lat.	Dist.	Dep.	D. Lat.	Dist.	Dep.	D. Lat.

Dist.	D. Lat.	Dep.
D Lo	Dep.	
	m	D Lo

58°

302°	058°
238°	122°

327° | 033°	TABLE 3	327° | 033°
213° | 147°	Traverse **33°** Table	213° | 147°

Dist.	D. Lat.	Dep.	Dist.	D. Lat.	Dep.	Dist.	D. Lat.	Dep.	Dist.	D. Lat.	Dep.	Dist.	D. Lat.	Dep.
1	0.8	0.5	61	51.2	33.2	121	101.5	65.9	181	151.8	98.6	241	202.1	131.3
2	1.7	1.1	62	52.0	33.8	22	102.3	66.4	82	152.6	99.1	42	203.0	131.8
3	2.5	1.6	63	52.8	34.3	23	103.2	67.0	83	153.5	99.7	43	203.8	132.3
4	3.4	2.2	64	53.7	34.9	24	104.0	67.5	84	154.3	100.2	44	204.6	132.9
5	4.2	2.7	65	54.5	35.4	25	104.8	68.1	85	155.2	100.8	45	205.5	133.4
6	5.0	3.3	66	55.4	35.9	26	105.7	68.6	86	156.0	101.3	46	206.3	134.0
7	5.9	3.8	67	56.2	36.5	27	106.5	69.2	87	156.8	101.8	47	207.2	134.5
8	6.7	4.4	68	57.0	37.0	28	107.3	69.7	88	157.7	102.4	48	208.0	135.1
9	7.5	4.9	69	57.9	37.6	29	108.2	70.3	89	158.5	102.9	49	208.8	135.6
10	8.4	5.4	70	58.7	38.1	30	109.0	70.8	90	159.3	103.5	50	209.7	136.2
11	9.2	6.0	71	59.5	38.7	131	109.9	71.3	191	160.2	104.0	251	210.5	136.7
12	10.1	6.5	72	60.4	39.2	32	110.7	71.9	92	161.0	104.6	52	211.3	137.2
13	10.9	7.1	73	61.2	39.8	33	111.5	72.4	93	161.9	105.1	53	212.2	137.8
14	11.7	7.6	74	62.1	40.3	34	112.4	73.0	94	162.7	105.7	54	213.0	138.3
15	12.6	8.2	75	62.9	40.8	35	113.2	73.5	95	163.5	106.2	55	213.9	138.9
16	13.4	8.7	76	63.7	41.4	36	114.1	74.1	96	164.4	106.7	56	214.7	139.4
17	14.3	9.3	77	64.6	41.9	37	114.9	74.6	97	165.2	107.3	57	215.5	140.0
18	15.1	9.8	78	65.4	42.5	38	115.7	75.2	98	166.1	107.8	58	216.4	140.5
19	15.9	10.3	79	66.3	43.0	39	116.6	75.7	99	166.9	108.4	59	217.2	141.1
20	16.8	10.9	80	67.1	43.6	40	117.4	76.2	200	167.7	108.9	60	218.1	141.6
21	17.6	11.4	81	67.9	44.1	141	118.3	76.8	201	168.6	109.5	261	218.9	142.2
22	18.5	12.0	82	68.8	44.7	42	119.1	77.3	02	169.4	110.0	62	219.7	142.7
23	19.3	12.5	83	69.6	45.2	43	119.9	77.9	03	170.3	110.6	63	220.6	143.2
24	20.1	13.1	84	70.4	45.7	44	120.8	78.4	04	171.1	111.1	64	221.4	143.8
25	21.0	13.6	85	71.3	46.3	45	121.6	79.0	05	171.9	111.7	65	222.2	144.3
26	21.8	14.2	86	72.1	46.8	46	122.4	79.5	06	172.8	112.2	66	223.1	144.9
27	22.6	14.7	87	73.0	47.4	47	123.3	80.1	07	173.6	112.7	67	223.9	145.4
28	23.5	15.2	88	73.8	47.9	48	124.1	80.6	08	174.4	113.3	68	224.8	146.0
29	24.3	15.8	89	74.6	48.5	49	125.0	81.2	09	175.3	113.8	69	225.6	146.5
30	25.2	16.3	90	75.5	49.0	50	125.8	81.7	10	176.1	114.4	70	226.4	147.1
31	26.0	16.9	91	76.3	49.6	151	126.6	82.2	211	177.0	114.9	271	227.3	147.6
32	26.8	17.4	92	77.2	50.1	52	127.5	82.8	12	177.8	115.5	72	228.1	148.1
33	27.7	18.0	93	78.0	50.7	53	128.3	83.3	13	178.6	116.0	73	229.0	148.7
34	28.5	18.5	94	78.8	51.2	54	129.2	83.9	14	179.5	116.6	74	229.8	149.2
35	29.4	19.1	95	79.7	51.7	55	130.0	84.4	15	180.3	117.1	75	230.6	149.8
36	30.2	19.6	96	80.5	52.3	56	130.8	85.0	16	181.2	117.6	76	231.5	150.3
37	31.0	20.2	97	81.4	52.8	57	131.7	85.5	17	182.0	118.2	77	232.3	150.9
38	31.9	20.7	98	82.2	53.4	58	132.5	86.1	18	182.8	118.7	78	233.2	151.4
39	32.7	21.2	99	83.0	53.9	59	133.3	86.6	19	183.7	119.3	79	234.0	152.0
40	33.5	21.8	100	83.9	54.5	60	134.2	87.1	20	184.5	119.8	80	234.8	152.5
41	34.4	22.3	101	84.7	55.0	161	135.0	87.7	221	185.3	120.4	281	235.7	153.0
42	35.2	22.9	02	85.5	55.6	62	135.9	88.2	22	186.2	120.9	82	236.5	153.6
43	36.1	23.4	03	86.4	56.1	63	136.7	88.8	23	187.0	121.5	83	237.3	154.1
44	36.9	24.0	04	87.2	56.6	64	137.5	89.3	24	187.9	122.0	84	238.2	154.7
45	37.7	24.5	05	88.1	57.2	65	138.4	89.9	25	188.7	122.5	85	239.0	155.2
46	38.6	25.1	06	88.9	57.7	66	139.2	90.4	26	189.5	123.1	86	239.9	155.8
47	39.4	25.6	07	89.7	58.3	67	140.1	91.0	27	190.4	123.6	87	240.7	156.3
48	40.3	26.1	08	90.6	58.8	68	140.9	91.5	28	191.2	124.2	88	241.5	156.9
49	41.1	26.7	09	91.4	59.4	69	141.7	92.0	29	192.1	124.7	89	242.4	157.4
50	41.9	27.2	10	92.3	59.9	70	142.6	92.6	30	192.9	125.3	90	243.2	157.9
51	42.8	27.8	111	93.1	60.5	171	143.4	93.1	231	193.7	125.8	291	244.1	158.5
52	43.6	28.3	12	93.9	61.0	72	144.3	93.7	32	194.6	126.4	92	244.9	159.0
53	44.4	28.9	13	94.8	61.5	73	145.1	94.2	33	195.4	126.9	93	245.7	159.6
54	45.3	29.4	14	95.6	62.1	74	145.9	94.8	34	196.2	127.4	94	246.6	160.1
55	46.1	30.0	15	96.4	62.6	75	146.8	95.3	35	197.1	128.0	95	247.4	160.7
56	47.0	30.5	16	97.3	63.2	76	147.6	95.9	36	197.9	128.5	96	248.2	161.2
57	47.8	31.0	17	98.1	63.7	77	148.4	96.4	37	198.8	129.1	97	249.1	161.8
58	48.6	31.6	18	99.0	64.3	78	149.3	96.9	38	199.6	129.6	98	249.9	162.3
59	49.5	32.1	19	99.8	64.8	79	150.1	97.5	39	200.4	130.2	99	250.8	162.8
60	50.3	32.7	20	100.6	65.4	80	151.0	98.0	40	201.3	130.7	300	251.6	163.4
Dist.	Dep.	D. Lat.	Dist.	Dep.	D. Lat.	Dist.	Dep.	D. Lat.	Dist.	Dep.	D. Lat.	Dist.	Dep.	D. Lat.

303° | 057°	**57°**
237° | 123°	

Dist.	D. Lat.	Dep.
N.	N×Cos.	N×Sin.
Hypotenuse	Side Adj.	Side Opp.

327°	033°
213°	147°

TABLE 3

Traverse 33° Table

327°	033°
213°	147°

Dist.	D. Lat.	Dep.	Dist.	D. Lat.	Dep.	Dist.	D. Lat.	Dep.	Dist.	D. Lat.	Dep.	Dist.	D. Lat.	Dep.
301	252.4	163.9	361	302.8	196.6	421	353.1	229.3	481	403.4	262.0	541	453.7	294.6
02	253.3	164.5	62	303.6	197.2	22	353.9	229.8	82	404.2	262.5	42	454.6	295.2
03	254.1	165.0	63	304.4	197.7	23	354.8	230.4	83	405.1	263.1	43	455.4	295.7
04	255.0	165.6	64	305.3	198.2	24	355.6	230.9	84	405.9	263.6	44	456.2	296.3
05	255.8	166.1	65	306.1	198.8	25	356.4	231.5	85	406.8	264.1	45	457.1	296.8
06	256.6	166.7	66	307.0	199.3	26	357.3	232.0	86	407.6	264.7	46	457.9	297.4
07	257.5	167.2	67	307.8	199.9	27	358.1	232.6	87	408.4	265.2	47	458.8	297.9
08	258.3	167.7	68	308.6	200.4	28	359.0	233.1	88	409.3	265.8	48	459.6	298.5
09	259.1	168.3	69	309.5	201.0	29	359.8	233.7	89	410.1	266.3	49	460.4	299.0
10	260.0	168.8	70	310.3	201.5	30	360.6	234.2	90	410.9	266.9	50	461.3	299.6
311	260.8	169.4	371	311.1	202.1	431	361.5	234.7	491	411.8	267.4	551	462.1	300.1
12	261.7	169.9	72	312.0	202.6	32	362.3	235.3	92	412.6	268.0	52	462.9	300.6
13	262.5	170.5	73	312.8	203.2	33	363.1	235.8	93	413.5	268.5	53	463.8	301.2
14	263.3	171.0	74	313.7	203.7	34	364.0	236.4	94	414.3	269.0	54	464.6	301.7
15	264.2	171.6	75	314.5	204.2	35	364.8	236.9	95	415.1	269.6	55	465.5	302.3
16	265.0	172.1	76	315.3	204.7	36	365.7	237.5	96	416.0	270.1	56	466.3	302.8
17	265.9	172.7	77	316.2	205.3	37	366.5	238.0	97	416.8	270.7	57	467.1	303.4
18	266.7	173.2	78	317.0	205.9	38	367.3	238.6	98	417.7	271.2	58	468.0	303.9
19	267.5	173.7	79	317.9	206.4	39	368.2	239.1	99	418.5	271.8	59	468.8	304.5
20	268.4	174.3	80	318.7	207.0	40	369.0	239.6	500	419.3	272.3	60	469.7	305.0
321	269.2	174.8	381	319.5	207.5	441	369.9	240.2	501	420.2	272.9	561	470.5	305.5
22	270.1	175.4	82	320.4	208.1	42	370.7	240.7	02	421.0	273.4	62	471.3	306.1
23	270.9	175.9	83	321.2	208.6	43	371.5	241.3	03	421.9	274.0	63	472.2	306.6
24	271.7	176.5	84	322.0	209.1	44	372.4	241.8	04	422.7	274.5	64	473.0	307.2
25	272.6	177.0	85	322.9	209.7	45	373.2	242.4	05	423.5	275.0	65	473.8	307.7
26	273.4	177.6	86	323.7	210.2	46	374.0	242.9	06	424.4	275.6	66	474.7	308.3
27	274.2	178.1	87	324.6	210.8	47	374.9	243.5	07	425.2	276.1	67	475.5	308.8
28	275.1	178.6	88	325.4	211.3	48	375.7	244.0	08	426.0	276.7	68	476.4	309.4
29	275.9	179.2	89	326.2	211.9	49	376.6	244.5	09	426.9	277.2	69	477.2	309.9
30	276.8	179.7	90	327.1	212.4	50	377.4	245.1	10	427.7	277.8	70	478.0	310.4
331	277.6	180.3	391	327.9	213.0	451	378.2	245.6	511	428.6	278.3	571	478.9	311.0
32	278.4	180.8	92	328.8	213.5	52	379.1	246.2	12	429.4	278.9	72	479.7	311.5
33	279.3	181.4	93	329.6	214.0	53	379.9	246.7	13	430.2	279.4	73	480.6	312.1
34	280.1	181.9	94	330.4	214.6	54	380.8	247.3	14	431.1	279.9	74	481.4	312.6
35	281.0	182.5	95	331.3	215.1	55	381.6	247.8	15	431.9	280.5	75	482.2	313.2
36	281.8	183.0	96	332.1	215.7	56	382.4	248.4	16	432.8	281.0	76	483.1	313.7
37	282.6	183.5	97	333.0	216.2	57	383.3	248.9	17	433.6	281.6	77	483.9	314.3
38	283.5	184.1	98	333.8	216.8	58	384.1	249.4	18	434.4	282.1	78	484.8	314.8
39	284.3	184.6	99	334.6	217.3	59	384.9	250.0	19	435.3	282.7	79	485.6	315.3
40	285.1	185.2	400	335.5	217.9	60	385.8	250.5	20	436.1	283.2	80	486.4	315.9
341	286.0	185.7	401	336.3	218.4	461	386.6	251.1	521	436.9	283.8	581	487.3	316.4
42	286.8	186.3	02	337.1	218.9	62	387.5	251.6	22	437.8	284.3	82	488.1	317.0
43	287.7	186.8	03	338.0	219.5	63	388.3	252.2	23	438.6	284.8	83	488.9	317.5
44	288.5	187.4	04	338.8	220.0	64	389.1	252.7	24	439.5	285.4	84	489.8	318.1
45	289.3	187.9	05	339.7	220.6	65	390.0	253.3	25	440.3	285.9	85	490.6	318.6
46	290.2	188.4	06	340.5	221.1	66	390.8	253.8	26	441.1	286.5	86	491.5	319.2
47	291.0	189.0	07	341.3	221.7	67	391.7	254.3	27	442.0	287.0	87	492.3	319.7
48	291.9	189.5	08	342.2	222.2	68	392.5	254.9	28	442.8	287.6	88	493.1	320.2
49	292.7	190.1	09	343.0	222.8	69	393.3	255.4	29	443.7	288.1	89	494.0	320.8
50	293.5	190.6	10	343.9	223.3	70	394.2	256.0	30	444.5	288.7	90	494.8	321.3
351	294.4	191.2	411	344.7	223.8	471	395.0	256.5	531	445.3	289.2	591	495.7	321.9
52	295.2	191.7	12	345.5	224.4	72	395.9	257.1	32	446.2	289.7	92	496.5	322.4
53	296.1	192.3	13	346.4	224.9	73	396.7	257.6	33	447.0	290.3	93	497.3	323.0
54	296.9	192.8	14	347.2	225.5	74	397.5	258.2	34	447.9	290.8	94	498.2	323.5
55	297.7	193.3	15	348.0	226.0	75	398.4	258.7	35	448.7	291.4	95	499.0	324.1
56	298.6	193.9	16	348.9	226.6	76	399.2	259.2	36	449.5	291.9	96	499.8	324.6
57	299.4	194.4	17	349.7	227.1	77	400.0	259.8	37	450.4	292.5	97	500.7	325.1
58	300.2	195.0	18	350.6	227.7	78	400.9	260.3	38	451.2	293.0	98	501.5	325.7
59	301.1	195.5	19	351.4	228.2	79	401.7	260.9	39	452.0	293.6	99	502.4	326.2
60	301.9	196.1	20	352.2	228.7	80	402.6	261.4	40	452.9	294.1	600	503.2	326.8
Dist.	Dep.	D. Lat.	Dist.	Dep.	D. Lat.	Dist.	Dep.	D. Lat.	Dist.	Dep.	D. Lat.	Dist.	Dep.	D. Lat.

Dist.	D. Lat.	Dep.
D Lo	Dep.	
	m	D Lo

57°

303°	057°
237°	123°

326°	034°
214°	146°

TABLE 3

Traverse 34° Table

326°	034°
214°	146°

Dist.	D. Lat.	Dep.	Dist.	D. Lat.	Dep.	Dist.	D. Lat.	Dep.	Dist.	D. Lat.	Dep.	Dist.	D. Lat.	Dep.
1	0.8	0.6	61	50.6	34.1	121	100.3	67.7	181	150.1	101.2	241	199.8	134.8
2	1.7	1.1	62	51.4	34.7	22	101.1	68.2	82	150.9	101.8	42	200.6	135.3
3	2.5	1.7	63	52.2	35.2	23	102.0	68.8	83	151.7	102.3	43	201.5	135.9
4	3.3	2.2	64	53.1	35.8	24	102.8	69.3	84	152.5	102.9	44	202.3	136.4
5	4.1	2.8	65	53.9	36.3	25	103.6	69.9	85	153.4	103.5	45	203.1	137.0
6	5.0	3.4	66	54.7	36.9	26	104.5	70.5	86	154.2	104.0	46	203.9	137.6
7	5.8	3.9	67	55.5	37.5	27	105.3	71.0	87	155.0	104.6	47	204.8	138.1
8	6.6	4.5	68	56.4	38.0	28	106.1	71.6	88	155.9	105.1	48	205.6	138.7
9	7.5	5.0	69	57.2	38.6	29	106.9	72.1	89	156.7	105.7	49	206.4	139.2
10	8.3	5.6	70	58.0	39.1	30	107.8	72.7	90	157.5	106.2	50	207.3	139.8
11	9.1	6.2	71	58.9	39.7	131	108.6	73.3	191	158.3	106.8	251	208.1	140.4
12	9.9	6.7	72	59.7	40.3	32	109.4	73.8	92	159.2	107.4	52	208.9	140.9
13	10.8	7.3	73	60.5	40.8	33	110.3	74.4	93	160.0	107.9	53	209.7	141.5
14	11.6	7.8	74	61.3	41.4	34	111.1	74.9	94	160.8	108.5	54	210.6	142.0
15	12.4	8.4	75	62.2	41.9	35	111.9	75.5	95	161.7	109.0	55	211.4	142.6
16	13.3	8.9	76	63.0	42.5	36	112.7	76.1	96	162.5	109.6	56	212.2	143.2
17	14.1	9.5	77	63.8	43.1	37	113.6	76.6	97	163.3	110.2	57	213.1	143.7
18	14.9	10.1	78	64.7	43.6	38	114.4	77.2	98	164.1	110.7	58	213.9	144.3
19	15.8	10.6	79	65.5	44.2	39	115.2	77.7	99	165.0	111.3	59	214.7	144.8
20	16.6	11.2	80	66.3	44.7	40	116.1	78.3	200	165.8	111.8	60	215.5	145.4
21	17.4	11.7	81	67.2	45.3	141	116.9	78.8	201	166.6	112.4	261	216.4	145.9
22	18.2	12.3	82	68.0	45.9	42	117.7	79.4	02	167.5	113.0	62	217.2	146.5
23	19.1	12.9	83	68.8	46.4	43	118.6	80.0	03	168.3	113.5	63	218.0	147.1
24	19.9	13.4	84	69.6	47.0	44	119.4	80.5	04	169.1	114.1	64	218.9	147.6
25	20.7	14.0	85	70.5	47.5	45	120.2	81.1	05	170.0	114.6	65	219.7	148.2
26	21.6	14.5	86	71.3	48.1	46	121.0	81.6	06	170.8	115.2	66	220.5	148.7
27	22.4	15.1	87	72.1	48.6	47	121.9	82.2	07	171.6	115.8	67	221.4	149.3
28	23.2	15.7	88	73.0	49.2	48	122.7	82.8	08	172.4	116.3	68	222.2	149.9
29	24.0	16.2	89	73.8	49.8	49	123.5	83.3	09	173.3	116.9	69	223.0	150.4
30	24.9	16.8	90	74.6	50.3	50	124.4	83.9	10	174.1	117.4	70	223.8	151.0
31	25.7	17.3	91	75.4	50.9	151	125.2	84.4	211	174.9	118.0	271	224.7	151.5
32	26.5	17.9	92	76.3	51.4	52	126.0	85.0	12	175.8	118.5	72	225.5	152.1
33	27.4	18.5	93	77.1	52.0	53	126.8	85.6	13	176.6	119.1	73	226.3	152.7
34	28.2	19.0	94	77.9	52.6	54	127.7	86.1	14	177.4	119.7	74	227.2	153.2
35	29.0	19.6	95	78.8	53.1	55	128.5	86.7	15	178.2	120.2	75	228.0	153.8
36	29.8	20.1	96	79.6	53.7	56	129.3	87.2	16	179.1	120.8	76	228.8	154.3
37	30.7	20.7	97	80.4	54.2	57	130.2	87.8	17	179.9	121.3	77	229.6	154.9
38	31.5	21.2	98	81.2	54.8	58	131.0	88.4	18	180.7	121.9	78	230.5	155.5
39	32.3	21.8	99	82.1	55.4	59	131.8	88.9	19	181.6	122.5	79	231.3	156.0
40	33.2	22.4	100	82.9	55.9	60	132.6	89.5	20	182.4	123.0	80	232.1	156.6
41	34.0	22.9	101	83.7	56.5	161	133.5	90.0	221	183.2	123.6	281	233.0	157.1
42	34.8	23.5	02	84.6	57.0	62	134.3	90.6	22	184.0	124.1	82	233.8	157.7
43	35.6	24.0	03	85.4	57.6	63	135.1	91.1	23	184.9	124.7	83	234.6	158.3
44	36.5	24.6	04	86.2	58.2	64	136.0	91.7	24	185.7	125.3	84	235.4	158.8
45	37.3	25.2	05	87.0	58.7	65	136.8	92.3	25	186.5	125.8	85	236.3	159.4
46	38.1	25.7	06	87.9	59.3	66	137.6	92.8	26	187.4	126.4	86	237.1	159.9
47	39.0	26.3	07	88.7	59.8	67	138.4	93.4	27	188.2	126.9	87	237.9	160.5
48	39.8	26.8	08	89.5	60.4	68	139.3	93.9	28	189.0	127.5	88	238.8	161.0
49	40.6	27.4	09	90.4	61.0	69	140.1	94.5	29	189.8	128.1	89	239.6	161.6
50	41.5	28.0	10	91.2	61.5	70	140.9	95.1	30	190.7	128.6	90	240.4	162.2
51	42.3	28.5	111	92.0	62.1	171	141.8	95.6	231	191.5	129.2	291	241.2	162.7
52	43.1	29.1	12	92.9	62.6	72	142.6	96.2	32	192.3	129.7	92	242.1	163.3
53	43.9	29.6	13	93.7	63.2	73	143.4	96.7	33	193.2	130.3	93	242.9	163.8
54	44.8	30.2	14	94.5	63.7	74	144.3	97.3	34	194.0	130.9	94	243.7	164.4
55	45.6	30.8	15	95.3	64.3	75	145.1	97.9	35	194.8	131.4	95	244.6	165.0
56	46.4	31.3	16	96.2	64.9	76	145.9	98.4	36	195.7	132.0	96	245.4	165.5
57	47.3	31.9	17	97.0	65.4	77	146.7	99.0	37	196.5	132.5	97	246.2	166.1
58	48.1	32.4	18	97.8	66.0	78	147.6	99.5	38	197.3	133.1	98	247.1	166.6
59	48.9	33.0	19	98.7	66.5	79	148.4	100.1	39	198.1	133.6	99	247.9	167.2
60	49.7	33.6	20	99.5	67.1	80	149.2	100.7	40	199.0	134.2	300	248.7	167.8
Dist.	Dep.	D. Lat.	Dist.	Dep.	D. Lat.	Dist.	Dep.	D. Lat.	Dist.	Dep.	D. Lat.	Dist.	Dep.	D. Lat.

304°	056°
236°	124°

56°

Dist.	D. Lat.	Dep.
N.	N×Cos.	N×Sin.
Hypotenuse	Side Adj.	Side Opp.

326°	034°
214°	146°

TABLE 3

Traverse **34°** Table

Dist.	D. Lat.	Dep.	Dist.	D. Lat.	Dep.	Dist.	D. Lat.	Dep.	Dist.	D. Lat.	Dep.	Dist.	D. Lat.	Dep.
301	249.5	168.3	361	299.3	201.9	421	349.0	235.4	481	398.8	269.0	541	448.5	302.5
02	250.4	168.9	62	300.1	202.4	22	349.9	236.0	82	399.6	269.5	42	449.3	303.1
03	251.2	169.4	63	300.9	203.0	23	350.7	236.5	83	400.4	270.1	43	450.2	303.6
04	252.0	170.0	64	301.8	203.5	24	351.5	237.1	84	401.3	270.6	44	451.0	304.2
05	252.9	170.6	65	302.6	204.1	25	352.3	237.7	85	402.1	271.2	45	451.8	304.8
06	253.7	171.1	66	303.4	204.7	26	353.2	238.2	86	402.9	271.8	46	452.7	305.3
07	254.5	171.7	67	304.3	205.2	27	354.0	238.8	87	403.7	272.3	47	453.5	305.9
08	255.3	172.2	68	305.1	205.8	28	354.8	239.3	88	404.6	272.9	48	454.3	306.4
09	256.2	172.8	69	305.9	206.3	29	355.7	239.9	89	405.4	273.4	49	455.1	307.0
10	257.0	173.3	70	306.7	206.9	30	356.5	240.5	90	406.2	274.0	50	456.0	307.6
311	257.8	173.9	371	307.6	207.5	431	357.3	241.0	491	407.1	274.6	551	456.8	308.1
12	258.7	174.5	72	308.4	208.0	32	358.1	241.6	92	407.9	275.1	52	457.6	308.7
13	259.5	175.0	73	309.2	208.6	33	359.0	242.1	93	408.7	275.7	53	458.5	309.2
14	260.3	175.6	74	310.1	209.1	34	359.8	242.7	94	409.5	276.2	54	459.3	309.8
15	261.1	176.1	75	310.9	209.7	35	360.6	243.2	95	410.4	276.8	55	460.1	310.4
16	262.0	176.7	76	311.7	210.3	36	361.5	243.8	96	411.2	277.4	56	460.9	310.9
17	262.8	177.3	77	312.5	210.8	37	362.3	244.4	97	412.0	277.9	57	461.8	311.5
18	263.6	177.8	78	313.4	211.4	38	363.1	244.9	98	412.9	278.5	58	462.6	312.0
19	264.5	178.4	79	314.2	211.9	39	364.0	245.5	99	413.7	279.0	59	463.4	312.6
20	265.3	178.9	80	315.0	212.5	40	364.8	246.0	500	414.5	279.6	60	464.3	313.1
321	266.1	179.5	381	315.9	213.1	441	365.6	246.6	501	415.3	280.2	561	465.1	313.7
22	267.0	180.1	82	316.7	213.6	42	366.4	247.2	02	416.2	280.7	62	465.9	314.3
23	267.8	180.6	83	317.5	214.2	43	367.3	247.7	03	417.0	281.3	63	466.7	314.8
24	268.6	181.2	84	318.4	214.7	44	368.1	248.3	04	417.8	281.8	64	467.6	315.5
25	269.4	181.7	85	319.2	215.3	45	368.9	248.8	05	418.7	282.4	65	468.4	315.9
26	270.3	182.3	86	320.0	215.8	46	369.8	249.4	06	419.5	283.0	66	469.2	316.5
27	271.1	182.9	87	320.8	216.4	47	370.6	250.0	07	420.3	283.5	67	470.1	317.1
28	271.9	183.4	88	321.7	217.0	48	371.4	250.5	08	421.2	284.1	68	470.9	317.6
29	272.8	184.0	89	322.5	217.5	49	372.2	251.1	09	422.0	284.6	69	471.7	318.2
30	273.6	184.5	90	323.3	218.1	50	373.1	251.6	10	422.8	285.2	70	472.6	318.7
331	274.4	185.1	391	324.2	218.6	451	373.9	252.2	511	423.6	285.9	571	473.4	319.3
32	275.2	185.7	92	325.0	219.2	52	374.7	252.8	12	424.5	286.3	72	474.2	319.9
33	276.1	186.2	93	325.8	219.8	53	375.6	253.3	13	425.3	286.9	73	475.0	320.4
34	276.9	186.8	94	326.6	220.3	54	376.4	253.9	14	426.1	287.4	74	475.9	321.0
35	277.7	187.3	95	327.5	220.9	55	377.2	254.4	15	427.0	288.0	75	476.7	321.5
36	278.6	187.9	96	328.3	221.4	56	378.0	255.0	16	427.8	288.5	76	477.5	322.1
37	279.4	188.4	97	329.1	222.0	57	378.9	255.6	17	428.6	289.1	77	478.4	322.7
38	280.2	189.0	98	330.0	222.6	58	379.7	256.1	18	429.4	289.7	78	479.2	323.2
39	281.0	189.6	99	330.8	223.1	59	380.5	256.7	19	430.3	290.2	79	480.0	323.8
40	281.9	190.1	400	331.6	223.7	60	381.4	257.2	20	431.1	290.8	80	480.8	324.3
341	282.7	190.7	401	332.4	224.2	461	382.2	257.8	521	431.9	291.3	581	481.7	324.9
42	283.5	191.2	02	333.3	224.8	62	383.0	258.3	22	432.8	291.9	82	482.5	325.4
43	284.4	191.8	03	334.1	225.4	63	383.8	258.9	23	433.6	292.5	83	483.3	326.0
44	285.2	192.4	04	334.9	225.9	64	384.7	259.5	24	434.4	293.0	84	484.2	326.6
45	286.0	192.9	05	335.8	226.5	65	385.5	260.0	25	435.2	293.6	85	485.0	327.1
46	286.8	193.5	06	336.6	227.0	66	386.3	260.6	26	436.1	294.1	86	485.8	327.7
47	287.7	194.0	07	337.4	227.6	67	387.2	261.1	27	436.9	294.7	87	486.6	328.2
48	288.5	194.6	08	338.2	228.2	68	388.0	261.7	28	437.7	295.3	88	487.5	328.8
49	289.3	195.2	09	339.1	228.7	69	388.8	262.3	29	438.6	295.8	89	488.3	329.4
50	290.2	195.7	10	339.9	229.3	70	389.6	262.8	30	439.4	296.4	90	489.1	329.9
351	291.0	196.3	411	340.7	229.8	471	390.5	263.4	531	440.2	296.9	591	490.0	330.5
52	291.8	196.8	12	341.6	230.4	72	391.3	263.9	32	441.0	297.5	92	490.8	331.0
53	292.7	197.4	13	342.4	230.9	73	392.1	264.5	33	441.9	298.0	93	491.6	331.6
54	293.5	198.0	14	343.2	231.5	74	393.0	265.1	34	442.7	298.6	94	492.4	332.2
55	294.3	198.5	15	344.1	232.1	75	393.8	265.6	35	443.5	299.2	95	493.3	332.7
56	295.1	199.1	16	344.9	232.6	76	394.6	266.2	36	444.4	299.7	96	494.1	333.3
57	296.0	199.6	17	345.7	233.2	77	395.5	266.7	37	445.3	300.3	97	494.9	333.8
58	296.8	200.2	18	346.5	233.7	78	396.3	267.3	38	446.0	300.8	98	495.8	334.4
59	297.6	200.8	19	347.4	234.3	79	397.1	267.9	39	446.9	301.4	99	496.6	335.0
60	298.5	201.3	20	348.2	234.9	80	397.9	268.4	40	447.7	302.0	600	497.4	335.5
Dist.	Dep.	D. Lat.	Dist.	Dep.	D. Lat.	Dist.	Dep.	D. Lat.	Dist.	Dep.	D. Lat.	Dist.	Dep.	D. Lat.

Dist.	D. Lat.	Dep.
D Lo	Dep.	
	m	D Lo

56°

304°	056°
236°	124°

325°	035°
215°	145°

TABLE 3

Traverse 35° Table

Dist.	D. Lat.	Dep.	Dist.	D. Lat.	Dep.	Dist.	D. Lat.	Dep.	Dist.	D. Lat.	Dep.	Dist.	D. Lat.	Dep.
1	0.8	0.6	61	50.0	35.0	121	99.1	69.4	181	148.3	103.8	241	197.4	138.2
2	1.6	1.1	62	50.8	35.6	22	99.9	70.0	82	149.1	104.4	42	198.2	138.8
3	2.5	1.7	63	51.6	36.1	23	100.8	70.5	83	149.9	105.0	43	199.1	139.4
4	3.3	2.3	64	52.4	36.7	24	101.6	71.1	84	150.7	105.5	44	199.9	140.0
5	4.1	2.9	65	53.2	37.3	25	102.4	71.7	85	151.5	106.1	45	200.7	140.5
6	4.9	3.4	66	54.1	37.9	26	103.2	72.3	86	152.4	106.7	46	201.5	141.1
7	5.7	4.0	67	54.9	38.4	27	104.0	72.8	87	153.2	107.3	47	202.3	141.7
8	6.6	4.6	68	55.7	39.0	28	104.9	73.4	88	154.0	107.8	48	203.1	142.2
9	7.4	5.2	69	56.5	39.6	29	105.7	74.0	89	154.8	108.4	49	204.0	142.8
10	8.2	5.7	70	57.3	40.2	30	106.5	74.6	90	155.6	109.0	50	204.8	143.4
11	9.0	6.3	71	58.2	40.7	131	107.3	75.1	191	156.5	109.6	251	205.6	144.0
12	9.8	6.9	72	59.0	41.3	32	108.1	75.7	92	157.3	110.1	52	206.4	144.5
13	10.6	7.5	73	59.8	41.9	33	108.9	76.3	93	158.1	110.7	53	207.2	145.1
14	11.5	8.0	74	60.6	42.4	34	109.8	76.9	94	158.9	111.3	54	208.1	145.7
15	12.3	8.6	75	61.4	43.0	35	110.6	77.4	95	159.7	111.8	55	208.9	146.3
16	13.1	9.2	76	62.3	43.6	36	111.4	78.0	96	160.6	112.4	56	209.7	146.8
17	13.9	9.8	77	63.1	44.2	37	112.2	78.6	97	161.4	113.0	57	210.5	147.4
18	14.7	10.3	78	63.9	44.7	38	113.0	79.2	98	162.2	113.6	58	211.3	148.0
19	15.6	10.9	79	64.7	45.3	39	113.9	79.7	99	163.0	114.1	59	212.2	148.6
20	16.4	11.5	80	65.5	45.9	40	114.7	80.3	200	163.8	114.7	60	213.0	149.1
21	17.2	12.0	81	66.4	46.5	141	115.5	80.9	201	164.6	115.3	261	213.8	149.7
22	18.0	12.6	82	67.2	47.0	42	116.3	81.4	02	165.5	115.9	62	214.6	150.3
23	18.8	13.2	83	68.0	47.6	43	117.1	82.0	03	166.3	116.4	63	215.4	150.9
24	19.7	13.8	84	68.8	48.2	44	118.0	82.6	04	167.1	117.0	64	216.3	151.4
25	20.5	14.3	85	69.6	48.8	45	118.8	83.2	05	167.9	117.6	65	217.1	152.0
26	21.3	14.9	86	70.4	49.3	46	119.6	83.7	06	168.7	118.2	66	217.9	152.6
27	22.1	15.5	87	71.3	49.9	47	120.4	84.3	07	169.6	118.7	67	218.7	153.1
28	22.9	16.1	88	72.1	50.5	48	121.2	84.9	08	170.4	119.3	68	219.5	153.7
29	23.8	16.6	89	72.9	51.0	49	122.1	85.5	09	171.2	119.9	69	220.4	154.3
30	24.6	17.2	90	73.7	51.6	50	122.9	86.0	10	172.0	120.5	70	221.2	154.9
31	25.4	17.8	91	74.5	52.2	151	123.7	86.6	211	172.8	121.0	271	222.0	155.4
32	26.2	18.4	92	75.4	52.8	52	124.5	87.2	12	173.7	121.6	72	222.8	156.0
33	27.0	18.9	93	76.2	53.3	53	125.3	87.8	13	174.5	122.2	73	223.6	156.6
34	27.9	19.5	94	77.0	53.9	54	126.1	88.3	14	175.3	122.7	74	224.4	157.2
35	28.7	20.1	95	77.8	54.5	55	127.0	88.9	15	176.1	123.3	75	225.3	157.7
36	29.5	20.6	96	78.6	55.1	56	127.8	89.5	16	176.9	123.9	76	226.1	158.3
37	30.3	21.2	97	79.5	55.6	57	128.6	90.1	17	177.8	124.5	77	226.9	158.9
38	31.1	21.8	98	80.3	56.2	58	129.4	90.6	18	178.6	125.0	78	227.7	159.5
39	31.9	22.4	99	81.1	56.8	59	130.2	91 2	19	179.4	125.6	79	228.5	160.0
40	32.8	22.9	100	81.9	57.4	60	131.1	91.8	20	180.2	126.2	80	229.4	160.6
41	33.6	23.5	101	82.7	57.9	161	131.9	92.3	221	181.0	126.8	281	230.2	161.2
42	34.4	24.1	02	83.6	58.5	62	132.7	92.9	22	181.9	127.3	82	231.0	161.7
43	35.2	24.7	03	84.4	59.1	63	133.5	93.5	23	182.7	127.9	83	231.8	162.3
44	36.0	25.2	04	85.2	59.7	64	134.3	94.1	24	183.5	128.5	84	232.6	162.9
45	36.9	25.8	05	86.0	60.2	65	135.2	94.6	25	184.3	129.1	85	233.5	163.5
46	37.7	26.4	06	86.8	60.8	66	136.0	95.2	26	185.1	129.6	86	234.3	164.0
47	38.5	27.0	07	87.6	61.4	67	136.8	95.8	27	185.9	130.2	87	235.1	164.6
48	39.3	27.5	08	88.5	61.9	68	137.6	96.4	28	186.8	130.8	88	235.9	165.2
49	40.1	28.1	09	89.3	62.5	69	138.4	96.9	29	187.6	131.3	89	236.7	165.8
50	41.0	28.7	10	90.1	63.1	70	139.3	97.5	30	188.4	131.9	90	237.6	166.3
51	41.8	29.3	111	90.9	63.7	171	140.1	98.1	231	189.2	132.5	291	238.4	166.9
52	42.6	29.8	12	91.7	64.2	72	140.9	98.7	32	190.0	133.1	92	239.2	167.5
53	43.4	30.4	13	92.6	64.8	73	141.7	99.2	33	190.9	133.6	93	240.0	168.1
54	44.2	31.0	14	93.4	65.4	74	142.5	99.8	34	191.7	134.2	94	240.8	168.6
55	45.1	31.5	15	94.2	66.0	75	143.4	100.4	35	192.5	134.8	95	241.6	169.2
56	45.9	32.1	16	95.0	66.5	76	144.2	100.9	36	193.3	135.4	96	242.5	169.8
57	46.7	32.7	17	95.8	67.1	77	145.0	101.5	37	194.1	135.9	97	243.3	170.4
58	47.5	33.3	18	96.7	67.7	78	145.8	102.1	38	195.0	136.5	98	244.1	170.9
59	48.3	33.8	19	97.5	68.3	79	146.6	102.7	39	195.8	137.1	99	244.9	171.5
60	49.1	34.4	20	98.3	68.8	80	147.4	103.2	40	196.6	137.7	300	245.7	172.1
Dist.	Dep.	D. Lat.	Dist.	Dep.	D. Lat.	Dist.	Dep.	D. Lat.	Dist.	Dep.	D. Lat.	Dist.	Dep.	D. Lat.

305°	055°
235°	125°

55°

Dist.	D. Lat.	Dep.
N.	N×Cos.	N×Sin.
Hypotenuse	Side Adj.	Side Opp.

325°	035°
215°	145°

TABLE 3

Traverse **35°** Table

325°	035°
215°	145°

Dist.	D. Lat.	Dep.	Dist.	D. Lat.	Dep.	Dist.	D. Lat.	Dep.	Dist.	D. Lat.	Dep.	Dist.	D. Lat.	Dep.
301	246.6	172.6	361	295.7	207.1	421	344.9	241.5	481	394.0	275.9	541	443.2	310.3
02	247.4	173.2	62	296.5	207.6	22	345.7	242.0	82	394.8	276.5	42	444.0	310.9
03	248.2	173.8	63	297.4	208.2	23	346.5	242.6	83	395.7	277.0	43	444.8	311.5
04	249.0	174.4	64	298.2	208.8	24	347.3	243.2	84	396.5	277.6	44	445.6	312.0
05	249.8	174.9	65	299.0	209.4	25	348.1	243.8	85	397.3	278.2	45	446.4	312.6
06	250.7	175.5	66	299.8	209.9	26	349.0	244.3	86	398.1	278.8	46	447.3	313.2
07	251.5	176.1	67	300.6	210.5	27	349.8	244.9	87	398.9	279.3	47	448.1	313.7
08	252.3	176.7	68	301.4	211.1	28	350.6	245.5	88	399.7	279.9	48	448.9	314.3
09	253.1	177.2	69	302.3	211.6	29	351.4	246.1	89	400.6	280.5	49	449.7	314.9
10	253.9	177.8	70	303.1	212.2	30	352.2	246.6	90	401.4	281.1	50	450.5	315.5
311	254.8	178.4	371	303.9	212.8	431	353.1	247.2	491	402.2	281.6	551	451.4	316.0
12	255.6	179.0	72	304.7	213.4	32	353.9	247.8	92	403.0	282.2	52	452.2	316.6
13	256.4	179.5	73	305.5	213.9	33	354.7	248.4	93	403.8	282.8	53	453.0	317.2
14	257.2	180.1	74	306.4	214.5	34	355.5	248.9	94	404.7	283.3	54	453.8	317.8
15	258.0	180.7	75	307.2	215.1	35	356.3	249.5	95	405.5	283.9	55	454.6	318.3
16	258.9	181.3	76	308.0	215.7	36	357.2	250.1	96	406.3	284.5	56	455.4	318.9
17	259.7	181.8	77	308.8	216.2	37	358.0	250.7	97	407.1	285.1	57	456.3	319.5
18	260.5	182.4	78	309.6	216.8	38	358.8	251.2	98	407.9	285.6	58	457.1	320.1
19	261.3	183.0	79	310.5	217.4	39	359.6	251.8	99	408.8	286.2	59	457.9	320.6
20	262.1	183.5	80	311.3	218.0	40	360.4	252.4	500	409.6	286.8	60	458.7	321.2
321	262.9	184.1	381	312.1	218.5	441	361.2	252.9	501	410.4	287.4	561	459.5	321.8
22	263.8	184.7	82	312.9	219.1	42	362.1	253.5	02	411.2	287.9	62	460.4	322.3
23	264.6	185.3	83	313.7	219.7	43	362.9	254.1	03	412.0	288.5	63	461.2	322.9
24	265.4	185.8	84	314.6	220.3	44	363.7	254.7	04	412.9	289.1	64	462.0	323.5
25	266.2	186.4	85	315.4	220.8	45	364.5	255.2	05	413.7	289.7	65	462.8	324.1
26	267.0	187.0	86	316.2	221.4	46	365.3	255.8	06	414.5	290.2	66	463.6	324.6
27	267.9	187.6	87	317.0	222.0	47	366.2	256.4	07	415.3	290.8	67	464.5	325.2
28	268.7	188.1	88	317.8	222.5	48	367.0	257.0	08	416.1	291.4	68	465.3	325.8
29	269.5	188.7	89	318.7	223.1	49	367.8	257.5	09	416.9	292.0	69	466.1	326.4
30	270.3	189.3	90	319.5	223.7	50	368.6	258.1	10	417.8	292.5	70	466.9	326.9
331	271.1	189.9	391	320.3	224.3	451	369.4	258.7	511	418.6	293.1	571	467.7	327.5
32	272.0	190.4	92	321.1	224.8	52	370.3	259.3	12	419.4	293.7	72	468.6	328.1
33	272.8	191.0	93	321.9	225.4	53	371.1	259.8	13	420.2	294.2	73	469.4	328.7
34	273.6	191.6	94	322.7	226.0	54	371.9	260.4	14	421.0	294.8	74	470.2	329.2
35	274.4	192.1	95	323.6	226.6	55	372.7	261.0	15	421.9	295.4	75	471.0	329.8
36	275.2	192.7	96	324.4	227.1	56	373.5	261.6	16	422.7	296.0	76	471.8	330.4
37	276.1	193.3	97	325.2	227.7	57	374.4	262.1	17	423.5	296.5	77	472.7	331.0
38	276.9	193.9	98	326.0	228.3	58	375.2	262.7	18	424.3	297.1	78	473.5	331.5
39	277.7	194.4	99	326.8	228.9	59	376.0	263.3	19	425.1	297.7	79	474.3	332.1
40	278.5	195.0	400	327.7	229.4	60	376.8	263.8	20	426.0	298.3	80	475.1	332.7
341	279.3	195.6	401	328.5	230.0	461	377.6	264.4	521	426.8	298.8	581	475.9	333.2
42	280.1	196.2	02	329.3	230.6	62	378.4	265.0	22	427.6	299.4	82	476.7	333.8
43	281.0	196.7	03	330.1	231.2	63	379.3	265.6	23	428.4	300.0	83	477.6	334.4
44	281.8	197.3	04	330.9	231.7	64	380.1	266.1	24	429.2	300.6	84	478.4	335.0
45	282.6	197.9	05	331.8	232.3	65	380.9	266.7	25	430.1	301.1	85	479.2	335.5
46	283.4	198.5	06	332.6	232.9	66	381.7	267.3	26	430.9	301.7	86	480.0	336.1
47	284.2	199.0	07	333.4	233.4	67	382.5	267.9	27	431.7	302.3	87	480.8	336.7
48	285.1	199.6	08	334.2	234.0	68	383.4	268.4	28	432.5	302.8	88	481.7	337.3
49	285.9	200.2	09	335.0	234.6	69	384.2	269.0	29	433.3	303.4	89	482.5	337.8
50	286.7	200.8	10	335.9	235.2	70	385.0	269.6	30	434.2	304.0	90	483.3	338.4
351	287.5	201.3	411	336.7	235.7	471	385.8	270.2	531	435.0	304.6	591	484.1	339.0
52	288.3	201.9	12	337.5	236.3	72	386.6	270.7	32	435.8	305.1	92	484.9	339.6
53	289.2	202.5	13	338.3	236.9	73	387.5	271.3	33	436.6	305.7	93	485.8	340.1
54	290.0	203.0	14	339.1	237.5	74	388.3	271.9	34	437.4	306.3	94	486.6	340.7
55	290.8	203.6	15	339.9	238.0	75	389.1	272.4	35	438.2	306.9	95	487.4	341.3
56	291.6	204.2	16	340.8	238.6	76	389.9	273.0	36	439.1	307.4	96	488.2	341.9
57	292.4	204.8	17	341.6	239.2	77	390.7	273.6	37	439.9	308.0	97	489.0	342.4
58	293.3	205.3	18	342.4	239.8	78	391.6	274.2	38	440.7	308.6	98	489.9	343.0
59	294.1	205.9	19	343.2	240.3	79	392.4	274.7	39	441.5	309.2	99	490.7	343.6
60	294.9	206.5	20	344.1	240.9	80	393.2	275.3	40	442.3	309.7	600	491.5	344.1
Dist.	Dep.	D. Lat.	Dist.	Dep.	D. Lat.	Dist.	Dep.	D. Lat.	Dist.	Dep.	D. Lat.	Dist.	Dep.	D. Lat.

Dist.	D. Lat.	Dep.
D Lo	Dep.	
	m	D Lo

55°

305°	055°
235°	125°

324° | 036°
216° | 144°

TABLE 3

Traverse 36° Table

324° | 036°
216° | 144°

Dist.	D. Lat.	Dep.	Dist.	D. Lat.	Dep.	Dist.	D. Lat.	Dep.	Dist.	D. Lat.	Dep.	Dist.	D. Lat.	Dep.
1	0.8	0.6	61	49.4	35.9	121	97.9	71.1	181	146.4	106.4	241	195.0	141.7
2	1.6	1.2	62	50.2	36.4	22	98.7	71.7	82	147.2	107.0	42	195.8	142.2
3	2.4	1.8	63	51.0	37.0	23	99.5	72.3	83	148.1	107.6	43	196.6	142.8
4	3.2	2.4	64	51.8	37.6	24	100.3	72.9	84	148.9	108.2	44	197.4	143.4
5	4.0	2.9	65	52.6	38.2	25	101.1	73.5	85	149.7	108.7	45	198.2	144.0
6	4.9	3.5	66	53.4	38.8	26	101.9	74.1	86	150.5	109.3	46	199.0	144.6
7	5.7	4.1	67	54.2	39.4	27	102.7	74.6	87	151.3	109.9	47	199.8	145.2
8	6.5	4.7	68	55.0	40.0	28	103.6	75.2	88	152.1	110.5	48	200.6	145.8
9	7.3	5.3	69	55.8	40.6	29	104.4	75.8	89	152.9	111.1	49	201.4	146.4
10	8.1	5.9	70	56.6	41.1	30	105.2	76.4	90	153.7	111.7	50	202.3	146.9
11	8.9	6.5	71	57.4	41.7	131	106.0	77.0	191	154.5	112.3	251	203.1	147.5
12	9.7	7.1	72	58.2	42.3	32	106.8	77.6	92	155.3	112.9	52	203.9	148.1
13	10.5	7.6	73	59.1	42.9	33	107.6	78.2	93	156.1	113.4	53	204.7	148.7
14	11.3	8.2	74	59.9	43.5	34	108.4	78.8	94	156.9	114.0	54	205.5	149.3
15	12.1	8.8	75	60.7	44.1	35	109.2	79.4	95	157.8	114.6	55	206.3	149.9
16	12.9	9.4	76	61.5	44.7	36	110.0	79.9	96	158.6	115.2	56	207.1	150.5
17	13.8	10.0	77	62.3	45.3	37	110.8	80.5	97	159.4	115.8	57	207.9	151.1
18	14.6	10.6	78	63.1	45.8	38	111.6	81.1	98	160.2	116.4	58	208.7	151.6
19	15.4	11.2	79	63.9	46.4	39	112.5	81.7	99	161.0	117.0	59	209.5	152.2
20	16.2	11.8	80	64.7	47.0	40	113.3	82.3	200	161.8	117.6	60	210.3	152.8
21	17.0	12.3	81	65.5	47.6	141	114.1	82.9	201	162.6	118.1	261	211.2	153.4
22	17.8	12.9	82	66.3	48.2	42	114.9	83.5	02	163.4	118.7	62	212.0	154.0
23	18.6	13.5	83	67.1	48.8	43	115.7	84.1	03	164.2	119.3	63	212.8	154.6
24	19.4	14.1	84	68.0	49.4	44	116.5	84.6	04	165.0	119.9	64	213.6	155.2
25	20.2	14.7	85	68.8	50.0	45	117.3	85.2	05	165.8	120.5	65	214.4	155.8
26	21.0	15.3	86	69.6	50.5	46	118.1	85.8	06	166.7	121.1	66	215.2	156.4
27	21.8	15.9	87	70.4	51.1	47	118.9	86.4	07	167.5	121.7	67	216.0	156.9
28	22.7	16.5	88	71.2	51.7	48	119.7	87.0	08	168.3	122.3	68	216.8	157.5
29	23.5	17.0	89	72.0	52.3	49	120.5	87.6	09	169.1	122.8	69	217.6	158.1
30	24.3	17.6	90	72.8	52.9	50	121.4	88.2	10	169.9	123.4	70	218.4	158.7
31	25.1	18.2	91	73.6	53.5	151	122.2	88.8	211	170.7	124.0	271	219.2	159.3
32	25.9	18.8	92	74.4	54.1	52	123.0	89.3	12	171.5	124.6	72	220.1	159.9
33	26.7	19.4	93	75.2	54.7	53	123.8	89.9	13	172.3	125.2	73	220.9	160.5
34	27.5	20.0	94	76.0	55.3	54	124.6	90.5	14	173.1	125.8	74	221.7	161.1
35	28.3	20.6	95	76.9	55.8	55	125.4	91.1	15	173.9	126.4	75	222.5	161.6
36	29.1	21.2	96	77.7	56.4	56	126.2	91.7	16	174.7	127.0	76	223.3	162.2
37	29.9	21.7	97	78.5	57.0	57	127.0	92.3	17	175.6	127.5	77	224.1	162.8
38	30.7	22.3	98	79.3	57.6	58	127.8	92.9	18	176.4	128.1	78	224.9	163.4
39	31.6	22.9	99	80.1	58.2	59	128.6	93.5	19	177.2	128.7	79	225.7	164.0
40	32.4	23.5	100	80.9	58.8	60	129.4	94.0	20	178.0	129.3	80	226.5	164.6
41	33.2	24.1	101	81.7	59.4	161	130.3	94.6	221	178.8	129.9	281	227.3	165.2
42	34.0	24.7	02	82.5	60.0	62	131.1	95.2	22	179.6	130.5	82	228.1	165.8
43	34.8	25.3	03	83.3	60.5	63	131.9	95.8	23	180.4	131.1	83	229.0	166.3
44	35.6	25.9	04	84.1	61.1	64	132.7	96.4	24	181.2	131.7	84	229.8	166.9
45	36.4	26.5	05	84.9	61.7	65	133.5	97.0	25	182.0	132.3	85	230.6	167.5
46	37.2	27.0	06	85.8	62.3	66	134.3	97.6	26	182.8	132.8	86	231.4	168.1
47	38.0	27.6	07	86.6	62.9	67	135.1	98.2	27	183.6	133.4	87	232.2	168.7
48	38.8	28.2	08	87.4	63.5	68	135.9	98.7	28	184.5	134.0	88	233.0	169.3
49	39.6	28.8	09	88.2	64.1	69	136.7	99.3	29	185.3	134.6	89	233.8	169.9
50	40.5	29.4	10	89.0	64.7	70	137.5	99.9	30	186.1	135.2	90	234.6	170.5
51	41.3	30.0	111	89.8	65.2	171	138.3	100.5	231	186.9	135.8	291	235.4	171.0
52	42.1	30.6	12	90.6	65.8	72	139.2	101.1	32	187.7	136.4	92	236.2	171.6
53	42.9	31.2	13	91.4	66.4	73	140.0	101.7	33	188.5	137.0	93	237.0	172.2
54	43.7	31.7	14	92.2	67.0	74	140.8	102.3	34	189.3	137.5	94	237.9	172.8
55	44.5	32.3	15	93.0	67.6	75	141.6	102.9	35	190.1	138.1	95	238.7	173.4
56	45.3	32.9	16	93.8	68.2	76	142.4	103.5	36	190.9	138.7	96	239.5	174.0
57	46.1	33.5	17	94.7	68.8	77	143.2	104.0	37	191.7	139.3	97	240.3	174.6
58	46.9	34.1	18	95.5	69.4	78	144.0	104.6	38	192.5	139.9	98	241.1	175.2
59	47.7	34.7	19	96.3	69.9	79	144.8	105.2	39	193.4	140.5	99	241.9	175.7
60	48.5	35.3	20	97.1	70.5	80	145.6	105.8	40	194.2	141.1	300	242.7	176.3
Dist.	Dep.	D. Lat.	Dist.	Dep.	D. Lat.	Dist.	Dep.	D. Lat.	Dist.	Dep.	D. Lat.	Dist.	Dep.	D. Lat.

306° | 054°
234° | 126°

54°

Dist.	D. Lat.	Dep.
N.	N×Cos.	N×Sin.
Hypotenuse	Side Adj.	Side Opp.

324° | 036°				TABLE 3								324° | 036°		
216° | 144°				Traverse **36°** Table								216° | 144°		
Dist.	D. Lat.	Dep.	Dist.	D. Lat.	Dep.	Dist.	D. Lat.	Dep.	Dist.	D. Lat.	Dep.	Dist.	D. Lat.	Dep.
301	243.5	176.9	361	292.1	212.2	421	340.6	247.5	481	389.1	282.7	541	437.7	318.0
02	244.3	177.5	62	292.9	212.8	22	341.4	248.0	82	389.9	283.3	42	438.5	318.6
03	245.1	178.1	63	293.7	213.4	23	342.2	248.6	83	390.8	283.9	43	439.3	319.2
04	245.9	178.7	64	294.5	214.0	24	343.0	249.2	84	391.6	284.5	44	440.2	319.8
05	246.8	179.3	65	295.3	214.5	25	343.8	249.8	85	392.4	285.1	45	440.9	320.3
06	247.6	179.9	66	296.1	215.1	26	344.6	250.4	86	393.2	285.7	46	441.7	320.9
07	248.4	180.5	67	296.9	215.7	27	345.5	251.0	87	394.0	286.3	47	442.5	321.5
08	249.2	181.0	68	297.7	216.3	28	346.3	251.6	88	394.8	286.8	48	443.3	322.1
09	250.0	181.6	69	298.5	216.9	29	347.1	252.2	89	395.6	287.4	49	444.2	322.7
10	250.8	182.2	70	299.3	217.5	30	347.9	252.7	90	396.4	288.0	50	445.0	323.3
311	251.6	182.8	371	300.1	218.1	431	348.7	253.3	491	397.2	288.6	551	445.8	323.9
12	252.4	183.4	72	301.0	218.7	32	349.5	253.9	92	398.0	289.2	52	446.6	324.5
13	253.2	184.0	73	301.8	219.2	33	350.3	254.5	93	398.8	289.8	53	447.4	325.0
14	254.0	184.6	74	302.6	219.8	34	351.1	255.1	94	399.7	290.4	54	448.2	325.6
15	254.8	185.2	75	303.4	220.4	35	351.9	255.7	95	400.5	291.0	55	449.0	326.2
16	255.6	185.7	76	304.2	221.0	36	352.7	256.3	96	401.3	291.5	56	449.8	326.8
17	256.5	186.3	77	305.0	221.6	37	353.5	256.9	97	402.1	292.1	57	450.6	327.4
18	257.3	186.9	78	305.8	222.2	38	354.3	257.4	98	402.9	292.7	58	451.4	328.0
19	258.1	187.5	79	306.6	222.8	39	355.2	258.0	99	403.7	293.3	59	452.2	328.6
20	258.9	188.1	80	307.4	223.4	40	356.0	258.6	500	404.5	293.9	60	453.0	329.2
321	259.7	188.7	381	308.2	223.9	441	356.8	259.2	501	405.3	294.5	561	453.9	329.7
22	260.5	189.3	82	309.0	224.5	42	357.6	259.8	02	406.1	295.1	62	454.7	330.3
23	261.3	189.9	83	309.9	225.1	43	358.4	260.4	03	406.9	295.7	63	455.5	330.9
24	262.1	190.4	84	310.7	225.7	44	359.2	261.0	04	407.7	296.2	64	456.3	331.5
25	262.9	191.0	85	311.5	226.3	45	360.0	261.6	05	408.6	296.8	65	457.1	332.1
26	263.7	191.6	86	312.3	226.9	46	360.8	262.2	06	409.4	297.4	66	457.9	332.7
27	264.5	192.2	87	313.1	227.5	47	361.6	262.7	07	410.2	298.0	67	458.7	333.3
28	265.4	192.8	88	313.9	228.1	48	362.4	263.3	08	411.0	298.6	68	459.5	333.9
29	266.2	193.4	89	314.7	228.6	49	363.2	263.9	09	411.8	299.2	69	460.3	334.4
30	267.0	194.0	90	315.5	229.2	50	364.1	264.5	10	412.6	299.8	70	461.1	335.0
331	267.8	194.6	391	316.3	229.8	451	364.9	265.1	511	413.4	300.4	571	461.9	335.6
32	268.6	195.1	92	317.1	230.4	52	365.7	265.7	12	414.2	300.9	72	462.8	336.2
33	269.4	195.7	93	317.9	231.0	53	366.5	266.3	13	415.0	301.5	73	463.6	336.8
34	270.2	196.3	94	318.8	231.6	54	367.3	266.9	14	415.8	302.1	74	464.4	337.4
35	271.0	196.9	95	319.6	232.2	55	368.1	267.6	15	416.6	302.7	75	465.2	338.0
36	271.8	197.5	96	320.4	232.8	56	368.9	268.0	16	417.5	303.3	76	466.0	338.6
37	272.6	198.1	97	321.2	233.4	57	369.7	268.6	17	418.3	303.9	77	466.8	339.2
38	273.4	198.7	98	322.0	233.9	58	370.5	269.2	18	419.1	304.5	78	467.6	339.7
39	274.3	199.3	99	322.8	234.5	59	371.3	269.8	19	419.9	305.1	79	468.4	340.3
40	275.1	199.8	400	323.6	235.1	60	372.1	270.4	20	420.7	305.6	80	469.2	340.9
341	275.9	200.4	401	324.4	235.7	461	373.0	271.0	521	421.5	306.2	581	470.0	341.5
42	276.7	201.0	02	325.2	236.3	62	373.8	271.6	22	422.3	306.8	82	470.8	342.1
43	277.5	201.6	03	326.0	236.9	63	374.6	272.1	23	423.1	307.4	83	471.7	342.7
44	278.3	202.2	04	326.9	237.5	64	375.4	272.7	24	423.9	308.0	84	472.5	343.3
45	279.1	202.8	05	327.7	238.1	65	376.2	273.3	25	424.7	308.6	85	473.3	343.9
46	279.9	203.4	06	328.5	238.7	66	377.0	273.9	26	425.5	309.2	86	474.1	344.4
47	280.7	204.0	07	329.3	239.2	67	377.8	274.5	27	426.4	309.8	87	474.9	345.0
48	281.5	204.5	08	330.1	239.8	68	378.6	275.1	28	427.2	310.4	88	475.7	345.6
49	282.3	205.1	09	330.9	240.4	69	379.4	275.7	29	428.0	310.9	89	476.5	346.2
50	283.2	205.7	10	331.7	241.0	70	380.2	276.3	30	428.8	311.5	90	477.3	346.8
351	284.0	206.3	411	332.5	241.6	471	381.1	276.8	531	429.6	312.1	591	478.1	347.4
52	284.8	206.9	12	333.3	242.2	72	381.9	277.4	32	430.4	312.7	92	478.9	348.0
53	285.6	207.5	13	334.1	242.8	73	382.7	278.0	33	431.2	313.3	93	479.7	348.6
54	286.4	208.1	14	334.9	243.3	74	383.5	278.6	34	432.0	313.9	94	480.6	349.1
55	287.2	208.7	15	335.7	243.9	75	384.3	279.2	35	432.8	314.5	95	481.4	349.7
56	288.0	209.3	16	336.6	244.5	76	385.1	279.8	36	433.6	315.1	96	482.2	350.3
57	288.8	209.8	17	337.4	245.1	77	385.9	280.4	37	434.4	315.6	97	483.0	350.9
58	289.6	210.4	18	338.2	245.7	78	386.7	281.0	38	435.3	316.2	98	483.8	351.5
59	290.4	211.0	19	339.0	246.3	79	387.5	281.5	39	436.1	316.8	99	484.6	352.1
60	291.2	211.6	20	339.8	246.9	80	388.3	282.1	40	436.9	317.4	600	485.4	352.7
Dist.	Dep.	D. Lat.	Dist.	Dep.	D. Lat.	Dist.	Dep.	D. Lat.	Dist.	Dep.	D. Lat.	Dist.	Dep.	D. Lat.

Dist.	D. Lat.	Dep.
D Lo	Dep.	
	m	D Lo

54°

306°	054°
234°	126°

323°	037°
217°	143°

TABLE 3

Traverse **37°** Table

323°	037°
217°	143°

Dist.	D. Lat.	Dep.	Dist.	D. Lat.	Dep.	Dist.	D. Lat.	Dep.	Dist.	D. Lat.	Dep.	Dist.	D. Lat.	Dep.
1	0.8	0.6	61	48.7	36.7	121	96.6	72.8	181	144.6	108.9	241	192.5	145.0
2	1.6	1.2	62	49.5	37.3	22	97.4	73.4	82	145.4	109.5	42	193.3	145.6
3	2.4	1.8	63	50.3	37.9	23	98.2	74.0	83	146.2	110.1	43	194.1	146.2
4	3.2	2.4	64	51.1	38.5	24	99.0	74.6	84	146.9	110.7	44	194.9	146.8
5	4.0	3.0	65	51.9	39.1	25	99.8	75.2	85	147.7	111.3	45	195.7	147.4
6	4.8	3.6	66	52.7	39.7	26	100.6	75.8	86	148.5	111.9	46	196.5	148.0
7	5.6	4.2	67	53.5	40.3	27	101.4	76.4	87	149.3	112.5	47	197.3	148.6
8	6.4	4.8	68	54.3	40.9	28	102.2	77.0	88	150.1	113.1	48	198.1	149.3
9	7.2	5.4	69	55.1	41.5	29	103.0	77.6	89	150.9	113.7	49	198.9	149.9
10	8.0	6.0	70	55.9	42.1	30	103.8	78.2	90	151.7	114.3	50	199.7	150.5
11	8.8	6.6	71	56.7	42.7	131	104.6	78.8	191	152.5	114.9	251	200.5	151.1
12	9.6	7.2	72	57.5	43.3	32	105.4	79.4	92	153.3	115.5	52	201.3	151.7
13	10.4	7.8	73	58.3	43.9	33	106.2	80.0	93	154.1	116.2	53	202.1	152.3
14	11.2	8.4	74	59.1	44.5	34	107.0	80.6	94	154.9	116.8	54	202.9	152.9
15	12.0	9.0	75	59.9	45.1	35	107.8	81.2	95	155.7	117.4	55	203.7	153.5
16	12.8	9.6	76	60.7	45.7	36	108.6	81.8	96	156.5	118.0	56	204.5	154.1
17	13.6	10.2	77	61.5	46.3	37	109.4	82.4	97	157.3	118.6	57	205.2	154.7
18	14.4	10.8	78	62.3	46.9	38	110.2	83.1	98	158.1	119.2	58	206.0	155.3
19	15.2	11.4	79	63.1	47.5	39	111.0	83.7	99	158.9	119.8	59	206.8	155.9
20	16.0	12.0	80	63.9	48.1	40	111.8	84.3	200	159.7	120.4	60	207.6	156.5
21	16.8	12.6	81	64.7	48.7	141	112.6	84.9	201	160.5	121.0	261	208.4	157.1
22	17.6	13.2	82	65.5	49.3	42	113.4	85.5	02	161.3	121.6	62	209.2	157.7
23	18.4	13.8	83	66.3	50.0	43	114.2	86.1	03	162.1	122.2	63	210.0	158.3
24	19.2	14.4	84	67.1	50.6	44	115.0	86.7	04	162.9	122.8	64	210.8	158.9
25	20.0	15.0	85	67.9	51.2	45	115.8	87.3	05	163.7	123.4	65	211.6	159.5
26	20.8	15.6	86	68.7	51.8	46	116.6	87.9	06	164.5	124.0	66	212.4	160.1
27	21.6	16.2	87	69.5	52.4	47	117.4	88.5	07	165.3	124.6	67	213.2	160.7
28	22.4	16.9	88	70.3	53.0	48	118.2	89.1	08	166.1	125.2	68	214.0	161.3
29	23.2	17.5	89	71.1	53.6	49	119.0	89.7	09	166.9	125.8	69	214.8	161.9
30	24.0	18.1	90	71.9	54.2	50	119.8	90.3	10	167.7	126.4	70	215.6	162.5
31	24.8	18.7	91	72.7	54.8	151	120.6	90.9	211	168.5	127.0	271	216.4	163.1
32	25.6	19.3	92	73.5	55.4	52	121.4	91.5	12	169.3	127.6	72	217.2	163.7
33	26.4	19.9	93	74.3	56.0	53	122.2	92.1	13	170.1	128.2	73	218.0	164.3
34	27.2	20.5	94	75.1	56.6	54	123.0	92.7	14	170.9	128.8	74	218.8	164.9
35	28.0	21.1	95	75.9	57.2	55	123.8	93.3	15	171.7	129.4	75	219.6	165.5
36	28.8	21.7	96	76.7	57.8	56	124.6	93.9	16	172.5	130.0	76	220.4	166.1
37	29.5	22.3	97	77.5	58.4	57	125.4	94.5	17	173.3	130.6	77	221.2	166.7
38	30.3	22.9	98	78.3	59.0	58	126.2	95.1	18	174.1	131.2	78	222.0	167.3
39	31.1	23.5	99	79.1	59.6	59	127.0	95.7	19	174.9	131.8	79	222.8	167.9
40	31.9	24.1	100	79.9	60.2	60	127.8	96.3	20	175.7	132.4	80	223.6	168.5
41	32.7	24.7	101	80.7	60.8	161	128.6	96.9	221	176.5	133.0	281	224.4	169.1
42	33.5	25.3	02	81.5	61.4	62	129.4	97.5	22	177.3	133.6	82	225.2	169.7
43	34.3	25.9	03	82.3	62.0	63	130.2	98.1	23	178.1	134.2	83	226.0	170.3
44	35.1	26.5	04	83.1	62.6	64	131.0	98.7	24	178.9	134.8	84	226.8	170.9
45	35.9	27.1	05	83.9	63.2	65	131.8	99.3	25	179.7	135.4	85	227.6	171.5
46	36.7	27.7	06	84.7	63.8	66	132.6	99.9	26	180.5	136.0	86	228.4	172.1
47	37.5	28.3	07	85.5	64.4	67	133.4	100.5	27	181.3	136.6	87	229.2	172.7
48	38.3	28.9	08	86.3	65.0	68	134.2	101.1	28	182.1	137.2	88	230.0	173.3
49	39.1	29.5	09	87.1	65.6	69	135.0	101.7	29	182.9	137.8	89	230.8	173.9
50	39.9	30.1	10	87.8	66.2	70	135.8	102.3	30	183.7	138.4	90	231.6	174.5
51	40.7	30.7	111	88.6	66.8	171	136.6	102.9	231	184.5	139.0	291	232.4	175.1
52	41.5	31.3	12	89.4	67.4	72	137.4	103.5	32	185.3	139.6	92	233.2	175.7
53	42.3	31.9	13	90.2	68.0	73	138.2	104.1	33	186.1	140.2	93	234.0	176.3
54	43.1	32.5	14	91.0	68.6	74	139.0	104.7	34	186.9	140.8	94	234.8	176.9
55	43.9	33.1	15	91.8	69.2	75	139.8	105.3	35	187.7	141.4	95	235.6	177.5
56	44.7	33.7	16	92.6	69.8	76	140.6	105.9	36	188.5	142.0	96	236.4	178.1
57	45.5	34.3	17	93.4	70.4	77	141.4	106.5	37	189.3	142.6	97	237.2	178.7
58	46.3	34.9	18	94.2	71.0	78	142.2	107.1	38	190.1	143.2	98	238.0	179.3
59	47.1	35.5	19	95.0	71.6	79	143.0	107.7	39	190.9	143.8	99	238.8	179.9
60	47.9	36.1	20	95.8	72.2	80	143.8	108.3	40	191.7	144.4	300	239.6	180.5
Dist.	Dep.	D. Lat.	Dist.	Dep.	D. Lat.	Dist.	Dep.	D. Lat.	Dist.	Dep.	D. Lat.	Dist.	Dep.	D. Lat.

307°	053°
233°	127°

53°

Dist.	D. Lat.	Dep.
N.	N×Cos.	N×Sin.
Hypotenuse	Side Adj.	Side Opp.

323° | 037° / 217° | 143°		TABLE 3 Traverse **37°** Table											323° | 037° / 217° | 143°	
Dist.	D. Lat.	Dep.	Dist.	D. Lat.	Dep.	Dist.	D. Lat.	Dep.	Dist.	D. Lat.	Dep.	Dist.	D. Lat.	Dep.
301	240.4	181.1	361	288.3	217.3	421	336.2	253.4	481	384.1	289.5	541	432.1	325.6
02	241.2	181.7	62	289.1	217.9	22	337.0	254.0	82	384.9	290.1	42	432.9	326.2
03	242.0	182.4	63	289.9	218.5	23	337.8	254.6	83	385.7	290.7	43	433.7	326.8
04	242.7	183.0	64	290.7	219.1	24	338.6	255.2	84	386.5	291.3	44	434.5	327.4
05	243.6	183.6	65	291.5	219.7	25	339.4	255.8	85	387.3	291.9	45	435.3	328.0
06	244.4	184.2	66	292.3	220.3	26	340.2	256.4	86	388.1	292.5	46	436.1	328.6
07	245.2	184.8	67	293.1	220.9	27	341.0	257.0	87	388.9	293.1	47	436.9	329.2
08	246.0	185.4	68	293.9	221.5	28	341.8	257.6	88	389.7	293.7	48	437.7	329.8
09	246.8	186.0	69	294.7	222.1	29	342.6	258.2	89	390.5	294.3	49	438.5	330.4
10	247.6	186.6	70	295.5	222.7	30	343.4	258.8	90	391.3	294.9	50	439.2	331.0
311	248.4	187.2	371	296.3	223.3	431	344.2	259.4	491	392.1	295.5	551	440.0	331.6
12	249.2	187.8	72	297.1	223.9	32	345.0	260.0	92	392.9	296.1	52	440.8	332.2
13	250.0	188.4	73	297.9	224.5	33	345.8	260.6	93	393.7	296.7	53	441.6	332.8
14	250.8	189.0	74	298.7	225.1	34	346.6	261.2	94	394.5	297.3	54	442.4	333.4
15	251.6	189.6	75	299.5	225.7	35	347.4	261.8	95	395.3	297.9	55	443.2	334.0
16	252.4	190.2	76	300.3	226.3	36	348.2	262.4	96	396.1	298.5	56	444.0	334.6
17	253.2	190.8	77	301.1	226.9	37	249.0	263.0	97	396.9	299.1	57	444.8	335.2
18	254.0	191.4	78	301.9	227.5	38	349.8	263.6	98	397.7	299.7	58	445.6	335.8
19	254.8	192.0	79	302.7	228.1	39	350.6	264.2	99	398.5	300.3	59	446.4	336.4
20	255.6	192.6	80	303.5	228.7	40	351.4	264.8	500	399.3	300.9	60	447.2	337.0
321	256.4	193.2	381	304.3	229.3	441	352.2	265.4	501	400.1	301.5	561	448.0	337.6
22	257.2	193.8	82	305.1	229.9	42	353.0	266.0	02	400.9	302.1	62	448.8	338.2
23	258.0	194.4	83	305.9	230.5	43	353.8	266.6	03	401.7	302.7	63	449.6	338.8
24	258.8	195.0	84	306.7	231.1	44	354.6	267.2	04	402.5	303.3	64	450.4	339.4
25	259.6	195.6	85	307.5	231.7	45	355.4	267.8	05	403.3	303.9	65	451.2	340.0
26	260.4	196.2	86	308.3	232.3	46	356.2	268.4	06	404.1	304.5	66	452.0	340.6
27	261.2	196.8	87	309.1	232.9	47	357.0	269.0	07	404.9	305.1	67	452.8	341.2
28	262.0	197.4	88	309.9	233.5	48	357.8	269.6	08	405.7	305.7	68	453.6	341.8
29	262.8	198.0	89	310.7	234.1	49	358.6	270.2	09	406.5	306.3	69	454.4	342.4
30	263.5	198.6	90	311.5	234.7	50	359.4	270.8	10	407.3	306.9	70	455.2	343.0
331	264.3	199.2	391	312.3	235.3	451	360.2	271.4	511	408.1	307.5	571	456.0	343.6
32	265.1	199.8	92	313.1	235.9	52	361.0	272.0	12	408.9	308.1	72	456.8	344.2
33	265.9	200.4	93	313.9	236.5	53	361.8	272.6	13	409.7	308.7	73	457.6	344.8
34	266.7	201.0	94	314.7	237.1	54	362.6	273.2	14	410.5	309.3	74	458.4	345.4
35	267.5	201.6	95	315.5	237.7	55	363.4	273.8	1E	411.3	309.9	75	459.2	346.0
36	268.3	202.2	96	316.3	238.3	56	364.2	274.4	16	412.1	310.5	76	460.0	346.6
37	269.1	202.8	97	317.1	238.9	57	365.0	275.0	17	412.9	311.1	77	460.8	347.2
38	269.9	203.4	98	317.9	239.5	58	365.8	275.6	18	413.7	311.7	78	461.6	347.8
39	270.7	204.0	99	318.7	240.1	59	366.6	276.2	19	414.5	312.3	79	462.4	348.5
40	271.5	204.6	400	319.5	240.7	60	367.4	276.8	20	415.3	312.1	80	463.2	349.1
341	272.3	205.2	401	320.3	241.3	461	368.2	277.4	521	416.1	313.5	581	464.0	349.7
42	273.1	205.8	02	321.1	241.9	62	369.0	278.0	22	416.9	314.1	82	464.8	350.3
43	273.9	206.4	03	321.9	242.5	63	369.8	278.6	23	417.7	314.7	83	465.6	350.9
44	274.7	207.0	04	322.6	243.1	64	370.6	279.2	24	418.5	315.4	84	466.4	351.5
45	275.5	207.6	05	323.4	243.7	65	371.4	279.8	25	419.3	316.0	85	467.2	352.1
46	276.3	208.2	06	324.2	244.3	66	372.2	280.4	26	420.1	316.6	86	468.0	352.7
47	277.1	208.8	07	325.0	244.9	67	373.0	281.0	27	420.9	317.2	87	468.8	353.3
48	277.9	209.4	08	325.8	245.5	68	373.8	281.6	28	421.7	317.8	88	469.6	353.9
49	278.7	210.0	09	326.6	246.1	69	374.6	282.3	29	422.5	318.4	89	470.4	354.5
50	279.5	210.6	10	327.4	246.7	70	375.4	282.9	30	423.3	319.0	90	471.2	355.1
351	280.3	211.2	411	328.2	247.3	471	376.2	283.5	531	424.1	319.6	591	472.0	355.7
52	281.1	211.8	12	329.0	247.9	72	377.0	284.1	32	424.9	320.2	92	472.8	356.3
53	281.9	212.4	13	329.8	248.5	73	377.8	284.7	33	425.7	320.8	93	473.6	356.9
54	282.7	213.0	14	330.6	249.2	74	378.6	285.3	34	426.5	321.4	94	474.4	357.5
55	283.5	213.6	15	331.4	249.8	75	379.4	285.9	35	427.3	322.0	95	475.2	358.1
56	284.3	214.2	16	332.2	250.4	76	380.2	286.5	36	428.1	322.6	96	476.0	358.7
57	285.1	214.8	17	333.0	251.0	77	380.9	287.1	37	428.9	323.2	97	476.8	359.3
58	285.9	215.4	18	333.8	251.6	78	381.7	287.7	38	429.7	323.8	98	477.6	359.9
59	286.7	216.1	19	334.6	252.2	79	382.5	288.3	39	430.5	324.4	99	478.4	360.5
60	287.5	216.7	20	335.4	252.8	80	383.3	288.9	40	431.3	325.0	600	479.2	361.1
Dist.	Dep.	D. Lat.	Dist.	Dep.	D. Lat.	Dist.	Dep.	D. Lat.	Dist.	Dep.	D. Lat.	Dist.	Dep.	D. Lat.

Dist.	D. Lat.	Dep.
D Lo	Dep.	
	m	D Lo

53°

307°	053°
233°	127°

322° | 038° / 218° | 142°		TABLE 3 Traverse **38°** Table										322° | 038° / 218° | 142°		
Dist.	D. Lat.	Dep.	Dist.	D. Lat.	Dep.	Dist.	D. Lat.	Dep.	Dist.	D. Lat.	Dep.	Dist.	D. Lat.	Dep.
1	0.8	0.6	61	48.1	37.6	121	95.3	74.5	181	142.6	111.4	241	189.9	148.4
2	1.6	1.2	62	48.9	38.2	22	96.1	75.1	82	143.4	112.1	42	190.7	149.0
3	2.4	1.8	63	49.6	38.8	23	96.9	75.7	83	144.2	112.7	43	191.5	149.6
4	3.2	2.5	64	50.4	39.4	24	97.7	76.3	84	145.0	113.3	44	192.3	150.2
5	3.9	3.1	65	51.2	40.0	25	98.5	77.0	85	145.8	113.9	45	193.1	150.8
6	4.7	3.7	66	52.0	40.6	26	99.3	77.6	86	146.6	114.5	46	193.9	151.5
7	5.5	4.3	67	52.8	41.2	27	100.1	78.2	87	147.4	115.1	47	194.6	152.1
8	6.3	4.9	68	53.6	41.9	28	100.9	78.8	88	148.1	115.7	48	195.4	152.7
9	7.1	5.5	69	54.4	42.5	29	101.7	79.4	89	148.9	116.4	49	196.2	153.3
10	7.9	6.2	70	55.2	43.1	30	102.4	80.0	90	149.7	117.0	50	197.0	153.9
11	8.7	6.8	71	55.9	43.7	131	103.2	80.7	191	150.5	117.6	251	197.8	154.5
12	9.5	7.4	72	56.7	44.3	32	104.0	81.3	92	151.3	118.2	52	198.6	155.1
13	10.2	8.0	73	57.5	44.9	33	104.8	81.9	93	152.1	118.8	53	199.4	155.8
14	11.0	8.6	74	58.3	45.6	34	105.6	82.5	94	152.9	119.4	54	200.2	156.4
15	11.8	9.2	75	59.1	46.2	35	106.4	83.1	95	153.7	120.1	55	200.9	157.0
16	12.6	9.9	76	59.9	46.8	36	107.2	83.7	96	154.5	120.7	56	201.7	157.6
17	13.4	10.5	77	60.7	47.4	37	108.0	84.3	97	155.2	121.3	57	202.5	158.2
18	14.2	11.1	78	61.5	48.0	38	108.7	85.0	98	156.0	121.9	58	203.3	158.8
19	15.0	11.7	79	62.3	48.6	39	109.5	85.6	99	156.8	122.5	59	204.1	159.5
20	15.8	12.3	80	63.0	49.3	40	110.3	86.2	200	157.6	123.1	60	204.9	160.1
21	16.5	12.9	81	63.8	49.9	141	111.1	86.8	201	158.4	123.7	261	205.7	160.7
22	17.3	13.5	82	64.6	50.5	42	111.9	87.4	02	159.2	124.4	62	206.5	161.3
23	18.1	14.2	83	65.4	51.1	43	112.7	88.0	03	160.0	125.0	63	207.2	161.9
24	18.9	14.8	84	66.2	51.7	44	113.5	88.7	04	160.8	125.6	64	208.0	162.5
25	19.7	15.4	85	67.0	52.3	45	114.3	89.3	05	161.5	126.2	65	208.8	163.2
26	20.5	16.0	86	67.8	52.9	46	115.0	89.9	06	162.3	126.8	66	209.6	163.8
27	21.3	16.6	87	68.6	53.6	47	115.8	90.5	07	163.1	127.4	67	210.4	164.4
28	22.1	17.2	88	69.3	54.2	48	116.6	91.1	08	163.9	128.1	68	211.2	165.0
29	22.9	17.9	89	70.1	54.8	49	117.4	91.7	09	164.7	128.7	69	212.0	165.6
30	23.6	18.5	90	70.9	55.4	50	118.2	92.3	10	165.5	129.3	70	212.8	166.2
31	24.4	19.1	91	71.7	56.0	151	119.0	93.0	211	166.3	129.9	271	213.6	166.8
32	25.2	19.7	92	72.5	56.6	52	119.8	93.6	12	167.1	130.5	72	214.3	167.5
33	26.0	20.3	93	73.3	57.3	53	120.6	94.2	13	167.8	131.1	73	215.1	168.1
34	26.8	20.9	94	74.1	57.9	54	121.4	94.8	14	168.6	131.8	74	215.9	168.7
35	27.6	21.5	95	74.9	58.5	55	122.1	95.4	15	169.4	132.4	75	216.7	169.3
36	28.4	22.2	96	75.6	59.1	56	122.9	96.0	16	170.2	133.0	76	217.5	169.9
37	29.2	22.8	97	76.4	59.7	57	123.7	96.7	17	171.0	133.6	77	218.3	170.5
38	29.9	23.4	98	77.2	60.3	58	124.5	97.3	18	171.8	134.2	78	219.1	171.2
39	30.7	24.0	99	78.0	61.0	59	125.3	97.9	19	172.6	134.8	79	219.9	171.8
40	31.5	24.6	100	78.8	61.6	60	126.1	98.5	20	173.4	135.4	80	220.6	172.4
41	32.3	25.2	101	79.6	62.2	161	126.9	99.1	221	174.2	136.1	281	221.4	173.0
42	33.1	25.9	02	80.4	62.8	62	127.7	99.7	22	174.9	136.7	82	222.2	173.6
43	33.9	26.5	03	81.2	63.4	63	128.4	100.4	23	175.7	137.3	83	223.0	174.2
44	34.7	27.1	04	82.0	64.0	64	129.2	101.0	24	176.5	137.9	84	223.8	174.8
45	35.5	27.7	05	82.7	64.6	65	130.0	101.6	25	177.3	138.5	85	224.6	175.5
46	36.2	28.3	06	83.5	65.3	66	130.8	102.2	26	178.1	139.1	86	225.4	176.1
47	37.0	28.9	07	84.3	65.9	67	131.6	102.8	27	178.9	139.8	87	226.2	176.7
48	37.8	29.6	08	85.1	66.5	68	132.4	103.4	28	179.7	140.4	88	226.9	177.3
49	38.6	30.2	09	85.9	67.1	69	133.2	104.0	29	180.5	141.0	89	227.7	177.9
50	39.4	30.8	10	86.7	67.7	70	134.0	104.7	30	181.2	141.6	90	228.5	178.5
51	40.2	31.4	111	87.5	68.3	171	134.7	105.3	231	182.0	142.2	291	229.3	179.2
52	41.0	32.0	12	88.3	69.0	72	135.5	105.9	32	182.8	142.8	92	230.1	179.8
53	41.8	32.6	13	89.0	69.6	73	136.3	106.5	33	183.6	143.4	93	230.9	180.4
54	42.6	33.2	14	89.8	70.2	74	137.1	107.1	34	184.4	144.1	94	231.7	181.0
55	43.3	33.9	15	90.6	70.8	75	137.9	107.7	35	185.2	144.7	95	232.5	181.6
56	44.1	34.5	16	91.4	71.4	76	138.7	108.4	36	186.0	145.3	96	233.3	182.2
57	44.9	35.1	17	92.2	72.0	77	139.5	109.0	37	186.8	145.9	97	234.0	182.9
58	45.7	35.7	18	93.0	72.6	78	140.3	109.6	38	187.5	146.5	98	234.8	183.5
59	46.5	36.3	19	93.8	73.3	79	141.1	110.2	39	188.3	147.1	99	235.6	184.1
60	47.3	36.9	20	94.6	73.9	80	141.8	110.8	40	189.1	147.8	300	236.4	184.7
Dist.	Dep.	D. Lat.	Dist.	Dep.	D. Lat.	Dist.	Dep.	D. Lat.	Dist.	Dep.	D. Lat.	Dist.	Dep.	D. Lat.
308° | 052° / 232° | 128°						**52°**								

Dist.	D. Lat.	Dep.
N.	N×Cos.	N×Sin.
Hypotenuse	Side Adj.	Side Opp.

322°	038°
218°	142°

TABLE 3

Traverse 38° Table

322°	038°
218°	142°

Dist.	D. Lat.	Dep.	Dist.	D. Lat.	Dep.	Dist.	D. Lat.	Dep.	Dist.	D. Lat.	Dep.	Dist.	D. Lat.	Dep.
301	237.2	185.3	361	284.5	222.3	421	331.8	259.2	481	379.0	296.1	541	426.3	333.1
02	238.0	185.9	62	285.3	222.9	22	332.5	259.8	82	379.8	296.7	42	427.1	333.7
03	238.8	186.6	63	286.0	223.5	23	333.3	260.4	83	380.6	297.4	43	427.9	334.3
04	239.6	187.2	64	286.8	224.1	24	334.1	261.0	84	381.4	298.0	44	428.7	334.9
05	240.3	187.8	65	287.6	224.7	25	334.9	261.7	85	382.2	298.6	45	429.5	335.5
06	241.1	188.4	66	288.4	225.3	26	335.7	262.3	86	383.0	299.2	46	430.3	336.2
07	241.9	189.0	67	289.2	225.9	27	336.5	262.9	87	383.8	299.8	47	431.0	336.8
08	242.7	189.6	68	290.0	226.6	28	337.3	263.5	88	384.5	300.4	48	431.8	337.4
09	243.5	190.2	69	290.8	227.2	29	338.1	264.1	89	385.3	301.1	49	432.6	338.0
10	244.3	190.9	70	291.6	227.8	30	338.8	264.7	90	386.1	301.7	50	433.4	338.6
311	245.1	191.5	371	292.4	228.4	431	339.6	265.4	491	386.9	302.3	551	434.2	339.2
12	245.9	192.1	72	293.1	229.0	32	340.4	266.0	92	387.7	302.9	52	435.0	339.8
13	246.6	192.7	73	293.9	229.6	33	341.2	266.6	93	388.5	303.5	53	435.8	340.5
14	247.4	193.3	74	294.7	230.3	34	342.0	267.2	94	389.3	304.1	54	436.6	341.1
15	248.2	193.9	75	295.5	230.9	35	342.8	267.8	95	390.1	304.8	55	437.3	341.7
16	249.0	194.5	76	296.3	231.5	36	343.6	268.4	96	390.9	305.4	56	438.1	342.3
17	249.8	195.2	77	297.1	232.1	37	344.4	269.0	97	391.6	306.0	57	438.9	342.9
18	250.6	195.8	78	297.9	232.7	38	345.1	269.7	98	392.4	306.6	58	439.7	343.5
19	251.4	196.4	79	298.7	233.3	39	345.9	270.3	99	393.2	307.2	59	440.5	344.2
20	252.2	197.0	80	299.4	234.0	40	346.7	270.9	500	394.0	307.8	60	441.3	344.8
321	253.0	197.6	381	300.2	234.6	441	347.5	271.5	501	394.8	308.4	561	442.1	345.4
22	253.7	198.2	82	301.0	235.2	42	348.3	272.1	02	395.6	309.1	62	442.9	346.0
23	254.5	198.9	83	301.8	235.8	43	349.1	272.7	03	396.4	309.7	63	443.7	346.6
24	255.3	199.5	84	302.6	236.4	44	349.9	273.4	04	397.2	310.3	64	444.4	347.2
25	256.1	200.1	85	303.4	237.0	45	350.7	274.0	05	397.9	310.9	65	445.2	347.8
26	256.9	200.7	86	304.2	237.6	46	351.5	274.6	06	398.7	311.5	66	446.0	348.5
27	257.7	201.3	87	305.0	238.3	47	352.2	275.2	07	399.5	312.1	67	446.8	349.1
28	258.5	201.9	88	305.7	238.9	48	353.0	275.8	08	400.3	312.8	68	447.6	349.7
29	259.3	202.6	89	306.5	239.5	49	353.8	276.4	09	401.1	313.4	69	448.4	350.3
30	260.0	203.2	90	307.3	240.1	50	354.6	277.0	10	401.9	314.0	70	449.2	350.9
331	260.8	203.8	391	308.1	240.7	451	355.4	277.7	511	402.7	314.6	571	450.0	351.5
32	261.6	204.4	92	308.9	241.3	52	356.2	278.3	12	403.5	315.2	72	450.7	352.2
33	262.4	205.0	93	309.7	242.0	53	357.0	278.9	13	404.2	315.8	73	451.5	352.8
34	263.2	205.6	94	310.5	242.6	54	357.8	279.5	14	405.0	316.5	74	452.3	353.4
35	264.0	206.2	95	311.3	243.2	55	358.5	280.1	15	405.8	317.1	75	453.1	354.0
36	264.8	206.9	96	312.1	243.8	56	359.3	280.7	16	406.6	317.7	76	453.9	354.6
37	265.6	207.5	97	312.8	244.4	57	360.1	281.4	17	407.4	318.3	77	454.7	355.2
38	266.3	208.1	98	313.6	245.0	58	360.9	282.0	18	408.2	318.9	78	455.5	355.7
39	267.1	208.7	99	314.4	245.6	59	361.7	282.6	19	409.0	319.5	79	456.3	356.5
40	267.9	209.3	400	315.2	246.3	60	362.5	283.2	20	409.8	320.1	80	457.0	357.1
341	268.7	209.9	401	316.0	246.9	461	363.3	283.8	521	410.6	320.8	581	457.8	357.7
42	269.5	210.6	02	316.8	247.5	62	364.1	284.4	22	411.3	321.4	82	458.6	358.3
43	270.3	211.2	03	317.6	248.1	63	364.8	285.1	23	412.1	322.0	83	459.4	358.9
44	271.1	211.8	04	318.4	248.7	64	365.6	285.7	24	412.9	322.6	84	460.2	359.5
45	271.9	212.4	05	319.1	249.3	65	366.4	286.3	25	413.7	323.2	85	461.0	360.2
46	272.7	213.0	06	319.9	250.0	66	367.2	286.9	26	414.5	323.8	86	461.8	360.8
47	273.4	213.6	07	320.7	250.6	67	368.0	287.5	27	415.3	324.5	87	462.6	361.4
48	274.2	214.3	08	321.5	251.2	68	368.8	288.1	28	416.1	325.1	88	463.4	362.0
49	275.0	214.9	09	322.3	251.8	69	369.6	288.7	29	416.9	325.7	89	464.1	362.6
50	275.8	215.5	10	323.1	252.4	70	370.4	289.4	30	417.6	326.3	90	464.9	363.2
351	276.6	216.1	411	323.9	253.0	471	371.2	290.0	531	418.4	326.9	591	465.7	363.7
52	277.4	216.7	12	324.7	253.7	72	371.9	290.6	32	419.2	327.5	92	466.5	364.5
53	278.2	217.3	13	325.5	254.3	73	372.7	291.2	33	420.0	328.1	93	467.3	365.1
54	279.0	217.9	14	326.2	254.9	74	373.5	291.8	34	420.8	328.8	94	468.1	365.7
55	279.7	218.6	15	327.0	255.5	75	374.3	292.4	35	421.6	329.4	95	468.9	366.3
56	280.5	219.2	16	327.8	256.1	76	375.1	293.1	36	422.4	330.0	96	469.7	366.9
57	281.3	219.8	17	328.6	256.7	77	375.9	293.7	37	423.2	330.6	97	470.4	367.5
58	282.1	220.4	18	329.4	257.3	78	376.7	294.3	38	424.0	331.2	98	471.2	368.2
59	282.9	221.0	19	330.2	258.0	79	377.5	294.9	39	424.7	331.8	99	472.0	368.8
60	283.7	221.6	20	331.0	258.6	80	378.2	295.5	40	425.5	332.5	600	472.8	369.4
Dist.	Dep.	D. Lat.	Dist.	Dep.	D. Lat.	Dist.	Dep.	D. Lat.	Dist.	Dep.	D. Lat.	Dist.	Dep.	D. Lat.

Dist.	D. Lat.	Dep.
D Lo	Dep.	
	m	D Lo

52°

308°	052°
232°	128°

321°	039°
219°	141°

TABLE 3

Traverse 39° Table

321°	039°
219°	141°

Dist.	D. Lat.	Dep.	Dist.	D. Lat.	Dep.	Dist.	D. Lat.	Dep.	Dist.	D. Lat.	Dep.	Dist.	D. Lat.	Dep.
1	0.8	0.6	61	47.4	38.4	121	94.0	76.1	181	140.7	113.9	241	187.3	151.7
2	1.6	1.3	62	48.2	39.0	22	94.8	76.8	82	141.4	114.5	42	188.1	152.3
3	2.3	1.9	63	49.0	39.6	23	95.6	77.4	83	142.2	115.2	43	188.8	152.9
4	3.1	2.5	64	49.7	40.3	24	96.4	78.0	84	143.0	115.8	44	189.6	153.6
5	3.9	3.1	65	50.5	40.9	25	97.1	78.7	85	143.8	116.4	45	190.4	154.2
6	4.7	3.8	66	51.3	41.5	26	97.9	79.3	86	144.5	117.1	46	191.2	154.8
7	5.4	4.4	67	52.1	42.2	27	98.7	79.9	87	145.3	117.7	47	192.0	155.4
8	6.2	5.0	68	52.8	42.8	28	99.5	80.6	88	146.1	118.3	48	192.7	156.1
9	7.0	5.7	69	53.6	43.4	29	100.3	81.2	89	146.9	118.9	49	193.5	156.7
10	7.8	6.3	70	54.4	44.1	30	101.0	81.8	90	147.7	119.6	50	194.3	157.3
11	8.5	6.9	71	55.2	44.7	131	101.8	82.4	191	148.4	120.2	251	195.1	158.0
12	9.3	7.6	72	56.0	45.3	32	102.6	83.1	92	149.2	120.8	52	195.8	158.6
13	10.1	8.2	73	56.7	45.9	33	103.4	83.7	93	150.0	121.5	53	196.6	159.2
14	10.9	8.8	74	57.5	46.6	34	104.1	84.3	94	150.8	122.1	54	197.4	159.8
15	11.7	9.4	75	58.3	47.2	35	104.9	85.0	95	151.5	122.7	55	198.2	160.5
16	12.4	10.1	76	59.1	47.8	36	105.7	85.6	96	152.3	123.3	56	198.9	161.1
17	13.2	10.7	77	59.8	48.5	37	106.5	86.2	97	153.1	124.0	57	199.7	161.7
18	14.0	11.3	78	60.6	49.1	38	107.2	86.8	98	153.9	124.6	58	200.5	162.4
19	14.8	12.0	79	61.4	49.7	39	108.0	87.5	99	154.7	125.2	59	201.3	163.0
20	15.5	12.6	80	62.2	50.3	40	108.8	88.1	200	155.4	125.9	60	202.1	163.6
21	16.3	13.2	81	62.9	51.0	141	109.6	88.7	201	156.2	126.5	261	202.8	164.3
22	17.1	13.8	82	63.7	51.6	42	110.4	89.4	02	157.0	127.1	62	203.6	164.9
23	17.9	14.5	83	64.5	52.2	43	111.1	90.0	03	157.8	127.8	63	204.4	165.5
24	18.7	15.1	84	65.3	52.9	44	111.9	90.6	04	158.5	128.4	64	205.2	166.1
25	19.4	15.7	85	66.1	53.5	45	112.7	91.3	05	159.3	129.0	65	205.9	166.8
26	20.2	16.4	86	66.8	54.1	46	113.5	91.9	06	160.1	129.6	66	206.7	167.4
27	21.0	17.0	87	67.6	54.8	47	114.2	92.5	07	160.9	130.3	67	207.5	168.0
28	21.8	17.6	88	68.4	55.4	48	115.0	93.1	08	161.6	130.9	68	208.3	168.7
29	22.5	18.3	89	69.2	56.0	49	115.8	93.8	09	162.4	131.5	69	209.1	169.3
30	23.3	18.9	90	69.9	56.6	50	116.6	94.4	10	163.2	132.2	70	209.8	169.9
31	24.1	19.5	91	70.7	57.3	151	117.3	95.0	211	164.0	132.8	271	210.6	170.5
32	24.9	20.1	92	71.5	57.9	52	118.1	95.7	12	164.8	133.4	72	211.4	171.2
33	25.6	20.8	93	72.3	58.5	53	118.9	96.3	13	165.5	134.0	73	212.2	171.8
34	26.4	21.4	94	73.1	59.2	54	119.7	96.9	14	166.3	134.7	74	212.9	172.4
35	27.2	22.0	95	73.8	59.8	55	120.5	97.5	15	167.1	135.3	75	213.7	173.1
36	28.0	22.7	96	74.6	60.4	56	121.2	98.2	16	167.9	135.9	76	214.5	173.7
37	28.8	23.3	97	75.4	61.0	57	122.0	98.8	17	168.6	136.6	77	215.3	174.3
38	29.5	23.9	98	76.2	61.7	58	122.8	99.4	18	169.4	137.2	78	216.0	175.0
39	30.3	24.5	99	76.9	62.3	59	123.6	100.1	19	170.2	137.8	79	216.8	175.6
40	31.1	25.2	100	77.7	62.9	60	124.3	100.7	20	171.0	138.5	80	217.6	176.2
41	31.9	25.8	101	78.5	63.6	161	125.1	101.3	221	171.7	139.1	281	218.4	176.8
42	32.6	26.4	02	79.3	64.2	62	125.9	101.9	22	172.5	139.7	82	219.2	177.5
43	33.4	27.1	03	80.0	64.8	63	126.7	102.6	23	173.3	140.3	83	219.9	178.1
44	34.2	27.7	04	80.8	65.4	64	127.5	103.2	24	174.1	141.0	84	220.7	178.7
45	35.0	28.3	05	81.6	66.1	65	128.2	103.8	25	174.9	141.6	85	221.5	179.4
46	35.7	28.9	06	82.4	66.7	66	129.0	104.5	26	175.6	142.2	86	222.3	180.0
47	36.5	29.6	07	83.2	67.3	67	129.8	105.1	27	176.4	142.9	87	223.0	180.6
48	37.3	30.2	08	83.9	68.0	68	130.6	105.7	28	177.2	143.5	88	223.8	181.2
49	38.1	30.8	09	84.7	68.6	69	131.3	106.4	29	178.0	144.1	89	224.6	181.9
50	38.9	31.5	10	85.5	69.2	70	132.1	107.0	30	178.7	144.7	90	225.4	182.5
51	39.6	32.1	111	86.3	69.9	171	132.9	107.6	231	179.5	145.4	291	226.1	183.1
52	40.4	32.7	12	87.0	70.5	72	133.7	108.2	32	180.3	146.0	92	226.9	183.8
53	41.2	33.4	13	87.8	71.1	73	134.4	108.9	33	181.1	146.6	93	227.7	184.4
54	42.0	34.0	14	88.6	71.7	74	135.2	109.5	34	181.9	147.3	94	228.5	185.0
55	42.7	34.6	15	89.4	72.4	75	136.0	110.1	35	182.6	147.9	95	229.3	185.6
56	43.5	35.2	16	90.1	73.0	76	136.8	110.8	36	183.4	148.5	96	230.0	186.3
57	44.3	35.9	17	90.9	73.6	77	137.6	111.4	37	184.2	149.1	97	230.8	186.9
58	45.1	36.5	18	91.7	74.3	78	138.3	112.0	38	185.0	149.8	98	231.6	187.5
59	45.9	37.1	19	92.5	74.9	79	139.1	112.6	39	185.7	150.4	99	232.4	188.2
60	46.6	37.8	20	93.3	75.5	80	139.9	113.3	40	186.5	151.0	300	233.1	188.8
Dist.	Dep.	D. Lat.	Dist.	Dep.	D. Lat.	Dist.	Dep.	D. Lat.	Dist.	Dep.	D. Lat.	Dist.	Dep.	D. Lat.

309°	051°
231°	129°

51°

Dist.	D. Lat.	Dep.
N.	N×Cos.	N×Sin.
Hypotenuse	Side Adj.	Side Opp.

321° | 039° / 219° | 141°						TABLE 3 Traverse 39° Table						321° | 039° / 219° | 141°		
Dist.	D. Lat.	Dep.	Dist.	D. Lat.	Dep.	Dist.	D. Lat.	Dep.	Dist.	D. Lat.	Dep.	Dist.	D. Lat.	Dep.
301	233.9	189.4	361	280.5	227.2	421	327.2	264.9	481	373.8	302.7	541	420.4	340.5
02	234.7	190.1	62	281.3	227.8	22	328.0	265.6	82	374.6	303.3	42	421.2	341.1
03	235.5	190.7	63	282.1	228.4	23	328.7	266.2	83	375.4	304.0	43	422.0	341.7
04	236.3	191.3	64	282.9	229.1	24	329.5	266.8	84	376.1	304.6	44	422.8	342.2
05	237.0	191.9	65	283.7	229.7	25	330.3	267.5	85	376.9	305.2	45	423.5	343.0
06	237.8	192.6	66	284.4	230.3	26	331.1	268.1	86	377.7	305.8	46	424.3	343.6
07	238.6	193.2	67	285.2	231.0	27	331.8	268.7	87	378.5	306.5	47	425.1	344.2
08	239.4	193.8	68	286.0	231.5	28	332.6	269.3	88	379.2	307.1	48	425.9	344.9
09	240.1	194.5	69	286.8	232.2	29	333.4	270.0	89	380.0	307.7	49	426.7	345.5
10	240.9	195.1	70	287.5	232.8	30	334.2	270.6	90	380.8	308.4	50	427.4	346.1
311	241.7	195.7	371	288.3	233.5	431	334.9	271.2	491	381.6	309.0	551	428.2	346.8
12	242.5	196.3	72	289.1	234.1	32	335.7	271.9	92	382.4	309.6	52	429.0	347.4
13	243.2	197.0	73	289.9	234.7	33	336.5	272.5	93	383.1	310.3	53	429.8	348.0
14	244.0	197.6	74	290.7	235.4	34	337.3	273.1	94	383.9	310.9	54	430.5	348.6
15	244.8	198.2	75	291.4	236.0	35	338.1	273.8	95	384.7	311.5	55	431.3	349.3
16	245.6	198.9	76	292.2	236.6	36	338.8	274.4	96	385.5	312.1	56	432.1	349.9
17	246.4	199.5	77	293.0	237.3	37	339.6	275.0	97	386.2	312.8	57	432.9	350.5
18	247.1	200.1	78	293.8	237.9	38	340.4	275.6	98	387.0	313.4	58	433.6	351.2
19	247.9	200.8	79	294.5	238.5	39	341.2	276.3	99	387.8	314.0	59	434.4	351.8
20	248.7	201.4	80	295.3	239.1	40	341.9	276.9	500	388.6	314.7	60	435.2	352.4
321	249.5	202.0	381	296.1	239.8	441	342.7	277.5	501	389.4	315.3	561	436.0	353.0
22	250.2	202.6	82	296.9	240.4	42	343.5	278.2	02	390.1	315.9	62	436.8	353.7
23	251.0	203.3	83	297.6	241.0	43	344.3	278.8	03	390.9	316.5	63	437.5	354.3
24	251.8	203.9	84	298.4	241.7	44	345.1	279.4	04	391.7	317.2	64	438.3	354.9
25	252.6	204.5	85	299.2	242.3	45	345.8	280.0	05	392.5	317.8	65	439.1	355.6
26	253.3	205.2	86	300.0	242.9	46	346.6	280.7	06	393.2	318.4	66	439.9	356.2
27	254.1	205.8	87	300.8	243.5	47	347.4	281.3	07	394.0	319.1	67	440.6	356.8
28	254.9	206.4	88	301.5	244.2	48	348.2	281.9	08	394.8	319.7	68	441.4	357.5
29	255.7	207.0	89	302.3	244.8	49	348.9	282.6	09	395.6	320.3	69	442.2	358.1
30	256.5	207.7	90	303.1	245.4	50	349.7	283.2	10	396.3	321.0	70	443.0	358.7
331	257.2	208.3	391	303.9	246.1	451	350.5	283.8	511	397.1	321.6	571	443.8	359.3
32	258.0	208.9	92	304.6	246.7	52	351.3	284.5	12	397.9	322.2	72	444.5	360.0
33	258.8	209.6	93	305.4	247.3	53	352.0	285.1	13	398.7	322.8	73	445.3	360.6
34	259.6	210.2	94	306.2	248.0	54	352.8	285.7	14	399.5	323.5	74	446.1	361.2
35	260.3	210.8	95	307.0	248.6	55	353.6	286.3	15	400.2	324.1	75	446.9	361.9
36	261.1	211.5	96	307.7	249.2	56	354.4	287.0	16	401.0	324.7	76	447.7	362.5
37	261.9	212.1	97	308.5	249.8	57	355.2	287.6	17	401.8	325.4	77	448.4	363.1
38	262.7	212.7	98	309.3	250.5	58	355.9	288.2	18	402.6	326.0	78	449.2	363.7
39	263.5	213.3	99	310.1	251.1	59	356.7	288.9	19	403.3	326.6	79	450.0	364.4
40	264.2	214.0	400	310.9	251.7	60	357.5	289.5	20	404.1	327.2	80	450.7	365.0
341	265.0	214.6	401	311.6	252.4	461	358.3	290.1	521	404.9	327.9	581	451.5	365.6
42	265.8	215.2	02	312.4	253.0	62	359.0	290.7	22	405.7	328.5	82	452.3	366.3
43	266.6	215.9	03	313.2	253.6	63	359.8	291.4	23	406.4	329.1	83	453.1	366.9
44	267.3	216.5	04	314.0	254.2	64	360.6	292.0	24	407.2	329.8	84	453.9	367.5
45	268.1	217.1	05	314.7	254.9	65	361.4	292.6	25	408.0	330.4	85	454.6	368.2
46	268.9	217.7	06	315.5	255.5	66	362.2	293.3	26	408.8	331.0	86	455.4	368.8
47	269.7	218.4	07	316.3	256.1	67	362.9	293.9	27	409.6	331.7	87	456.2	369.4
48	270.4	219.0	08	317.1	256.8	68	363.7	294.5	28	410.3	332.3	88	457.0	370.0
49	271.2	219.6	09	317.9	257.4	69	364.5	295.2	29	411.1	332.9	89	457.8	370.7
50	272.0	220.3	10	318.6	258.0	70	365.3	295.8	30	411.9	333.5	90	458.5	371.3
351	272.8	220.9	411	319.4	258.7	471	366.0	296.4	531	412.7	334.2	591	459.3	371.9
52	273.6	221.5	12	320.2	259.3	72	366.8	297.0	32	413.4	334.8	92	460.1	372.6
53	274.3	222.2	13	321.0	259.9	73	367.6	297.7	33	414.2	335.4	93	460.8	373.2
54	275.1	222.7	14	321.7	260.5	74	368.4	298.3	34	415.0	336.1	94	461.6	373.8
55	275.9	223.4	15	322.5	261.2	75	369.1	298.9	35	415.8	336.7	95	462.4	374.4
56	276.7	224.0	16	323.3	261.8	76	369.9	299.6	36	416.6	337.3	96	463.2	375.1
57	277.4	224.7	17	324.1	262.4	77	370.7	300.2	37	417.3	337.9	97	464.0	375.7
58	278.2	225.3	18	324.8	263.1	78	371.5	300.8	38	418.1	338.6	98	464.7	376.3
59	279.0	225.9	19	325.6	263.7	79	372.3	301.4	39	418.9	339.2	99	465.5	377.0
60	279.8	226.6	20	326.4	264.3	80	373.0	302.1	40	419.7	339.8	600	466.3	377.6
Dist.	Dep.	D. Lat.	Dist.	Dep.	D. Lat.	Dist.	Dep.	D. Lat.	Dist.	Dep.	D. Lat.	Dist.	Dep.	D. Lat.

Dist.	D. Lat.	Dep.
D Lo	Dep.	
	m	D Lo

51°

309°	051°
231°	129°

320° | 040°	TABLE 3	320° | 040°
220° | 140°	Traverse **40°** Table	220° | 140°

Dist.	D. Lat.	Dep.	Dist.	D. Lat.	Dep.	Dist.	D. Lat.	Dep.	Dist.	D. Lat.	Dep.	Dist.	D. Lat.	Dep.
1	0.8	0.6	61	46.7	39.2	121	92.7	77.8	181	138.7	116.3	241	184.6	154.9
2	1.5	1.3	62	47.5	39.9	22	93.5	78.4	82	139.4	117.0	42	185.4	155.6
3	2.3	1.9	63	48.3	40.5	23	94.2	79.1	83	140.2	117.6	43	186.1	156.2
4	3.1	2.6	64	49.0	41.1	24	95.0	79.7	84	141.0	118.3	44	186.9	156.8
5	3.8	3.2	65	49.8	41.8	25	95.8	80.3	85	141.7	118.9	45	187.7	157.5
6	4.6	3.9	66	50.6	42.4	26	96.5	81.0	86	142.5	119.6	46	188.4	158.1
7	5.4	4.5	67	51.3	43.1	27	97.3	81.6	87	143.3	120.2	47	189.2	158.8
8	6.1	5.1	68	52.1	43.7	28	98.1	82.3	88	144.0	120.8	48	190.0	159.4
9	6.9	5.8	69	52.9	44.4	29	98.8	82.9	89	144.8	121.5	49	190.7	160.1
10	7.7	6.4	70	53.6	45.0	30	99.6	83.6	90	145.5	122.1	50	191.5	160.7
11	8.4	7.1	71	54.4	45.6	131	100.4	84.2	191	146.3	122.8	251	192.3	161.3
12	9.2	7.7	72	55.2	46.3	32	101.1	84.8	92	147.1	123.4	52	193.0	162.0
13	10.0	8.4	73	55.9	46.9	33	101.9	85.5	93	147.8	124.1	53	193.8	162.6
14	10.7	9.0	74	56.7	47.6	34	102.6	86.1	94	148.6	124.7	54	194.6	163.3
15	11.5	9.6	75	57.5	48.2	35	103.4	86.8	95	149.4	125.3	55	195.3	163.9
16	12.3	10.3	76	58.2	48.9	36	104.2	87.4	96	150.1	126.0	56	196.1	164.6
17	13.0	10.9	77	59.0	49.5	37	104.9	88.1	97	150.9	126.6	57	196.9	165.2
18	13.8	11.6	78	59.8	50.1	38	105.7	88.7	98	151.7	127.3	58	197.6	165.8
19	14.6	12.2	79	60.5	50.8	39	106.5	89.3	99	152.4	127.9	59	198.4	166.5
20	15.3	12.9	80	61.3	51.4	40	107.2	90.0	200	153.2	128.6	60	199.2	167.1
21	16.1	13.5	81	62.0	52.1	141	108.0	90.6	201	154.0	129.2	261	199.9	167.8
22	16.9	14.1	82	62.8	52.7	42	108.8	91.3	02	154.7	129.8	62	200.7	168.4
23	17.6	14.8	83	63.6	53.4	43	109.5	91.9	03	155.5	130.5	63	201.5	169.1
24	18.4	15.4	84	64.3	54.0	44	110.3	92.6	04	156.3	131.1	64	202.2	169.7
25	19.2	16.1	85	65.1	54.6	45	111.1	93.2	05	157.0	131.8	65	203.0	170.3
26	19.9	16.7	86	65.9	55.3	46	111.8	93.8	06	157.8	132.4	66	203.8	171.0
27	20.7	17.4	87	66.6	55.9	47	112.6	94.5	07	158.6	133.1	67	204.5	171.6
28	21.4	18.0	88	67.4	56.6	48	113.4	95.1	08	159.3	133.7	68	205.3	172.3
29	22.2	18.6	89	68.2	57.2	49	114.1	95.8	09	160.1	134.3	69	206.1	172.9
30	23.0	19.3	90	68.9	57.9	50	114.9	96.4	10	160.9	135.0	70	206.8	173.6
31	23.7	19.9	91	69.7	58.5	151	115.7	97.1	211	161.6	135.6	271	207.6	174.2
32	24.5	20.6	92	70.5	59.1	52	116.4	97.7	12	162.4	136.3	72	208.4	174.8
33	25.3	21.2	93	71.2	59.8	53	117.2	98.3	13	163.2	136.9	73	209.1	175.5
34	26.0	21.9	94	72.0	60.4	54	118.0	99.0	14	163.9	137.6	74	209.9	176.1
35	26.8	22.5	95	72.8	61.1	55	118.7	99.6	15	164.7	138.2	75	210.7	176.8
36	27.6	23.1	96	73.5	61.7	56	119.5	100.3	16	165.5	138.8	76	211.4	177.4
37	28.3	23.8	97	74.3	62.4	57	120.3	100.9	17	166.2	139.5	77	212.2	178.1
38	29.1	24.4	98	75.1	63.0	58	121.0	101.6	18	167.0	140.1	78	213.0	178.7
39	29.9	25.1	99	75.8	63.6	59	121.8	102.2	19	167.8	140.8	79	213.7	179.3
40	30.6	25.7	100	76.6	64.3	60	122.6	102.8	20	168.5	141.4	80	214.5	180.0
41	31.4	26.4	101	77.4	64.9	161	123.3	103.5	221	169.3	142.1	281	215.3	180.6
42	32.2	27.0	02	78.1	65.6	62	124.1	104.1	22	170.1	142.7	82	216.0	181.3
43	32.9	27.6	03	78.9	66.2	63	124.9	104.8	23	170.8	143.3	83	216.8	181.9
44	33.7	28.3	04	79.7	66.8	64	125.6	105.4	24	171.6	144.0	84	217.6	182.6
45	34.5	28.9	05	80.4	67.5	65	126.4	106.1	25	172.4	144.6	85	218.3	183.2
46	35.2	29.6	06	81.2	68.1	66	127.2	106.7	26	173.1	145.3	86	219.1	183.8
47	36.0	30.2	07	82.0	68.8	67	127.9	107.3	27	173.9	145.9	87	219.9	184.5
48	36.8	30.9	08	82.7	69.4	68	128.7	108.0	28	174.7	146.6	88	220.6	185.1
49	37.5	31.5	09	83.5	70.1	69	129.5	108.6	29	175.4	147.2	89	221.4	185.8
50	38.3	32.1	10	84.3	70.7	70	130.2	109.3	30	176.2	147.8	90	222.2	186.4
51	39.1	32.8	111	85.0	71.3	171	131.0	109.9	231	177.0	148.5	291	222.9	187.1
52	39.8	33.4	12	85.8	72.0	72	131.8	110.6	32	177.7	149.1	92	223.7	187.7
53	40.6	34.1	13	86.6	72.6	73	132.5	111.2	33	178.5	149.8	93	224.5	188.3
54	41.4	34.7	14	87.3	73.3	74	133.3	111.8	34	179.3	150.4	94	225.2	189.0
55	42.1	35.4	15	88.1	73.9	75	134.1	112.5	35	180.0	151.1	95	226.0	189.6
56	42.9	36.0	16	88.9	74.6	76	134.8	113.1	36	180.8	151.7	96	226.7	190.3
57	43.7	36.6	17	89.6	75.2	77	135.6	113.8	37	181.6	152.3	97	227.5	190.9
58	44.4	37.3	18	90.4	75.8	78	136.4	114.4	38	182.3	153.0	98	228.3	191.6
59	45.2	37.9	19	91.2	76.5	79	137.1	115.1	39	183.1	153.6	99	229.0	192.2
60	46.0	38.6	20	91.9	77.1	80	137.9	115.7	40	183.9	154.3	300	229.8	192.8
Dist.	Dep.	D. Lat.	Dist.	Dep.	D. Lat.	Dist.	Dep.	D. Lat.	Dist.	Dep.	D. Lat.	Dist.	Dep.	D. Lat.

310° | 050°	**50°**
230° | 130°	

Dist.	D. Lat.	Dep.
N.	N×Cos.	N×Sin.
Hypotenuse	Side Adj.	Side Opp.

320°	040°
220°	140°

TABLE 3

Traverse **40°** Table

320°	040°
220°	140°

Dist.	D. Lat.	Dep.	Dist.	D. Lat.	Dep.	Dist.	D. Lat.	Dep.	Dist.	D. Lat.	Dep.	Dist.	D. Lat.	Dep.
301	230.6	193.5	361	276.5	232.1	421	322.5	270.6	481	368.5	309.2	541	414.4	347.7
02	231.3	194.1	62	277.3	232.7	22	323.3	271.3	82	369.2	309.8	42	415.2	348.4
03	232.1	194.8	63	278.1	233.3	23	324.0	271.9	83	370.0	310.5	43	416.0	349.0
04	232.9	195.4	64	278.8	234.0	24	324.8	272.5	84	370.8	311.1	44	416.7	349.7
05	233.6	196.1	65	279.6	234.6	25	325.6	273.2	85	371.5	311.8	45	417.5	350.3
06	234.4	196.7	66	280.4	235.3	26	326.3	273.8	86	372.3	312.4	46	418.3	351.0
07	235.2	197.3	67	281.1	235.9	27	327.1	274.5	87	373.1	313.0	47	419.0	351.6
08	235.9	198.0	68	281.9	236.5	28	327.9	275.1	88	373.8	313.7	48	419.8	352.2
09	236.7	198.6	69	282.7	237.2	29	328.6	275.8	89	374.6	314.3	49	420.6	352.9
10	237.5	199.3	70	283.4	237.8	30	329.4	276.4	90	375.4	315.0	50	421.3	353.5
311	238.2	199.9	371	284.2	238.5	431	330.2	277.0	491	376.1	315.6	551	422.1	354.2
12	239.0	200.5	72	285.0	239.1	32	330.9	277.7	92	376.9	316.3	52	422.9	354.8
13	239.8	201.2	73	285.7	239.8	33	331.7	278.3	93	377.7	316.9	53	423.6	355.5
14	240.5	201.8	74	286.5	240.4	34	332.5	279.0	94	378.4	317.5	54	424.4	356.1
15	241.3	202.5	75	287.3	241.0	35	333.2	279.6	95	379.2	318.2	55	425.2	356.7
16	242.1	203.1	76	288.0	241.7	36	334.0	280.3	96	380.0	318.8	56	425.9	357.4
17	242.8	203.8	77	288.8	242.3	37	334.8	280.9	97	380.7	319.5	57	426.7	358.0
18	243.6	204.4	78	289.6	243.0	38	335.5	281.5	98	381.5	320.1	58	427.5	358.7
19	244.4	205.0	79	290.3	243.6	39	336.3	282.2	99	382.3	320.8	59	428.2	359.3
20	245.1	205.7	80	291.1	244.3	40	337.1	282.8	500	383.0	321.4	60	429.0	360.0
321	245.9	206.3	381	291.9	244.9	441	337.8	283.5	501	383.8	322.0	561	429.8	360.6
22	246.7	207.0	82	292.6	245.5	42	338.6	284.1	02	384.6	322.7	62	430.5	361.2
23	247.4	207.6	83	293.4	246.2	43	339.4	284.8	03	385.3	323.3	63	431.3	361.9
24	248.2	208.3	84	294.2	246.8	44	340.1	285.4	04	386.1	324.0	64	432.0	362.5
25	249.0	208.9	85	294.9	247.5	45	340.9	286.0	05	386.9	324.6	65	432.8	363.2
26	249.7	209.5	86	295.7	248.1	46	341.7	286.7	06	387.6	325.3	66	433.6	363.8
27	250.5	210.2	87	296.5	248.8	47	342.4	287.3	07	388.4	325.9	67	434.3	364.5
28	251.3	210.8	88	297.2	249.4	48	343.2	288.0	08	389.2	326.5	68	435.1	365.1
29	252.0	211.5	89	298.0	250.0	49	344.0	288.6	09	389.9	327.2	69	435.9	365.7
30	252.8	212.1	90	298.8	250.7	50	344.7	289.3	10	390.7	327.8	70	436.6	366.4
331	253.6	212.8	391	299.5	251.3	451	345.5	289.9	511	391.4	328.5	571	437.4	367.0
32	254.3	213.4	92	300.3	252.0	52	346.3	290.5	12	392.2	329.1	72	438.2	367.7
33	255.1	214.0	93	301.1	252.6	53	347.0	291.2	13	393.0	329.8	73	438.9	368.3
34	255.9	214.7	94	301.8	253.3	54	347.8	291.8	14	393.7	330.4	74	439.7	369.0
35	256.6	215.3	95	302.6	253.9	55	348.6	292.5	15	394.5	331.0	75	440.5	369.6
36	257.4	216.0	96	303.4	254.5	56	349.3	293.1	16	395.3	331.7	76	441.2	370.2
37	258.2	216.6	97	304.1	255.2	57	350.1	293.8	17	396.0	332.3	77	442.0	370.9
38	258.9	217.3	98	304.9	255.8	58	350.8	294.4	18	396.8	333.0	78	442.8	371.5
39	259.7	217.9	99	305.7	256.5	59	351.6	295.0	19	397.6	333.6	79	443.5	372.2
40	260.5	218.5	400	306.4	257.1	60	352.4	295.7	20	398.3	334.2	80	444.3	372.8
341	261.2	219.2	401	307.2	257.8	461	353.1	296.3	521	399.1	334.9	581	445.1	373.5
42	262.0	219.8	02	307.9	258.4	62	353.9	297.0	22	399.9	335.5	82	445.8	374.1
43	262.8	220.5	03	308.7	259.0	63	354.7	297.6	23	400.6	336.2	83	446.6	374.7
44	263.5	221.1	04	309.5	259.7	64	355.4	298.3	24	401.4	336.8	84	447.4	375.4
45	264.3	221.8	05	310.2	260.3	65	356.2	298.9	25	402.2	337.5	85	448.1	376.0
46	265.1	222.4	06	311.0	261.0	66	357.0	299.5	26	402.9	338.1	86	448.9	376.7
47	265.8	223.0	07	311.8	261.6	67	357.7	300.2	27	403.7	338.7	87	449.7	377.3
48	266.6	223.7	08	312.5	262.3	68	358.5	300.8	28	404.5	339.4	88	450.4	378.0
49	267.3	224.3	09	313.3	262.9	69	359.3	301.5	29	405.2	340.0	89	451.2	378.6
50	268.1	225.0	10	314.1	263.5	70	360.0	302.1	30	406.0	340.7	90	452.0	379.2
351	268.9	225.6	411	314.8	264.2	471	360.8	302.8	531	406.8	341.3	591	452.7	379.9
52	269.6	226.3	12	315.6	264.8	72	361.6	303.4	32	407.5	342.0	92	453.5	380.5
53	270.4	226.9	13	316.4	265.5	73	362.3	304.0	33	408.3	342.6	93	454.3	381.2
54	271.2	227.5	14	317.1	266.1	74	363.1	304.7	34	409.1	343.2	94	455.0	381.8
55	271.9	228.2	15	317.9	266.8	75	363.9	305.3	35	409.8	343.9	95	455.8	382.5
56	272.7	228.8	16	318.7	267.4	76	364.6	306.0	36	410.6	344.5	96	456.6	383.1
57	273.5	229.5	17	319.4	268.0	77	365.4	306.6	37	411.4	345.2	97	457.3	383.7
58	274.2	230.1	18	320.2	268.7	78	366.2	307.3	38	412.1	345.8	98	458.1	384.4
59	275.0	230.8	19	321.0	269.3	79	366.9	307.9	39	412.9	346.5	99	458.9	385.0
60	275.8	231.4	20	321.7	270.0	80	367.7	308.5	40	413.7	347.1	600	459.6	385.7
Dist.	Dep.	D. Lat.	Dist.	Dep.	D. Lat.	Dist.	Dep.	D. Lat.	Dist.	Dep.	D. Lat.	Dist.	Dep.	D. Lat.

Dist.	D. Lat.	Dep.
D Lo	Dep.	
	m	D Lo

50°

310°	050°
230°	130°

319°	041°
221°	139°

TABLE 3

Traverse **41°** Table

319°	041°
221°	139°

Dist.	D. Lat.	Dep.	Dist.	D. Lat.	Dep.	Dist.	D. Lat.	Dep.	Dist.	D. Lat.	Dep.	Dist.	D. Lat.	Dep.
1	0.8	0.7	61	46.0	40.0	121	91.3	79.4	181	136.6	118.7	241	181.9	158.1
2	1.5	1.3	62	46.8	40.7	22	92.1	80.0	82	137.4	119.4	42	182.6	158.8
3	2.3	2.0	63	47.5	41.3	23	92.8	80.7	83	138.1	120.1	43	183.4	159.4
4	3.0	2.6	64	48.3	42.0	24	93.6	81.4	84	138.9	120.7	44	184.1	160.1
5	3.8	3.3	65	49.1	42.6	25	94.3	82.0	85	139.6	121.4	45	184.9	160.7
6	4.5	3.9	66	49.8	43.3	26	95.1	82.7	86	140.4	122.0	46	185.7	161.4
7	5.3	4.6	67	50.6	44.0	27	95.8	83.3	87	141.1	122.7	47	186.4	162.0
8	6.0	5.2	68	51.3	44.6	28	96.6	84.0	88	141.9	123.3	48	187.2	162.7
9	6.8	5.9	69	52.1	45.3	29	97.4	84.6	89	142.6	124.0	49	187.9	163.4
10	7.5	6.6	70	52.8	45.9	30	98.1	85.3	90	143.4	124.7	50	188.7	164.0
11	8.3	7.2	71	53.6	46.6	131	98.9	85.9	191	144.1	125.3	251	189.4	164.7
12	9.1	7.9	72	54.3	47.2	32	99.6	86.6	92	144.9	126.0	52	190.2	165.3
13	9.8	8.5	73	55.1	47.9	33	100.4	87.3	93	145.7	126.6	53	190.9	166.0
14	10.6	9.2	74	55.8	48.5	34	101.1	87.9	94	146.4	127.3	54	191.7	166.6
15	11.3	9.8	75	56.6	49.2	35	101.9	88.6	95	147.2	127.9	55	192.5	167.3
16	12.1	10.5	76	57.4	49.9	36	102.6	89.2	96	147.9	128.6	56	193.2	168.0
17	12.8	11.2	77	58.1	50.5	37	103.4	89.9	97	148.7	129.2	57	194.0	168.6
18	13.6	11.8	78	58.9	51.2	38	104.1	90.5	98	149.4	129.9	58	194.7	169.3
19	14.3	12.5	79	59.6	51.8	39	104.9	91.2	99	150.2	130.6	59	195.5	169.9
20	15.1	13.1	80	60.4	52.5	40	105.7	91.8	200	150.9	131.2	60	196.2	170.6
21	15.8	13.8	81	61.1	53.1	141	106.4	92.5	201	151.7	131.9	261	197.0	171.2
22	16.6	14.4	82	61.9	53.8	42	107.2	93.2	02	152.5	132.5	62	197.7	171.9
23	17.4	15.1	83	62.6	54.5	43	107.9	93.8	03	153.2	133.2	63	198.5	172.5
24	18.1	15.7	84	63.4	55.1	44	108.7	94.5	04	154.0	133.8	64	199.2	173.2
25	18.9	16.4	85	64.2	55.8	45	109.4	95.1	05	154.7	134.5	65	200.0	173.9
26	19.6	17.1	86	64.9	56.4	46	110.2	95.8	06	155.5	135.1	66	200.8	174.5
27	20.4	17.7	87	65.7	57.1	47	110.9	96.4	07	156.2	135.8	67	201.5	175.2
28	21.1	18.4	88	66.4	57.7	48	111.7	97.1	08	157.0	136.5	68	202.3	175.8
29	21.9	19.0	89	67.2	58.4	49	112.5	97.8	09	157.7	137.1	69	203.0	176.5
30	22.6	19.7	90	67.9	59.0	50	113.2	98.4	10	158.5	137.8	70	203.8	177.1
31	23.4	20.3	91	68.7	59.7	151	114.0	99.1	211	159.2	138.4	271	204.5	177.8
32	24.2	21.0	92	69.4	60.4	52	114.7	99.7	12	160.0	139.1	72	205.3	178.4
33	24.9	21.6	93	70.2	61.0	53	115.5	100.4	13	160.8	139.7	73	206.0	179.1
34	25.7	22.3	94	70.9	61.7	54	116.2	101.0	14	161.5	140.4	74	206.8	179.8
35	26.4	23.0	95	71.7	62.3	55	117.0	101.7	15	162.3	141.1	75	207.5	180.4
36	27.2	23.6	96	72.5	63.0	56	117.7	102.3	16	163.0	141.7	76	208.3	181.1
37	27.9	24.3	97	73.2	63.6	57	118.5	103.0	17	163.8	142.4	77	209.1	181.7
38	28.7	24.9	98	74.0	64.3	58	119.2	103.7	18	164.5	143.0	78	209.8	182.4
39	29.4	25.6	99	74.7	64.9	59	120.0	104.3	19	165.3	143.7	79	210.6	183.0
40	30.2	26.2	100	75.5	65.6	60	120.8	105.0	20	166.0	144.3	80	211.3	183.7
41	30.9	26.9	101	76.2	66.3	161	121.5	105.6	221	166.8	145.0	281	212.1	184.4
42	31.7	27.6	02	77.0	66.9	62	122.3	106.3	22	167.5	145.6	82	212.8	185.0
43	32.5	28.2	03	77.7	67.6	63	123.0	106.9	23	168.3	146.3	83	213.6	185.7
44	33.2	28.9	04	78.5	68.2	64	123.8	107.6	24	169.1	147.0	84	214.3	186.3
45	34.0	29.5	05	79.2	68.9	65	124.5	108.2	25	169.8	147.6	85	215.1	187.0
46	34.7	30.2	06	80.0	69.5	66	125.3	108.9	26	170.6	148.3	86	215.8	187.6
47	35.5	30.8	07	80.8	70.2	67	126.0	109.6	27	171.3	148.9	87	216.6	188.3
48	36.2	31.5	08	81.5	70.9	68	126.8	110.2	28	172.1	149.6	88	217.4	188.9
49	37.0	32.1	09	82.3	71.5	69	127.5	110.9	29	172.8	150.2	89	218.1	189.6
50	37.7	32.8	10	83.0	72.2	70	128.3	111.5	30	173.6	150.9	90	218.9	190.3
51	38.5	33.5	111	83.8	72.8	171	129.1	112.2	231	174.3	151.5	291	219.6	190.9
52	39.2	34.1	12	84.5	73.5	72	129.8	112.8	32	175.1	152.2	92	220.4	191.6
53	40.0	34.8	13	85.3	74.1	73	130.6	113.5	33	175.8	152.9	93	221.1	192.2
54	40.8	35.4	14	86.0	74.8	74	131.3	114.2	34	176.6	153.5	94	221.9	192.9
55	41.5	36.1	15	86.8	75.4	75	132.1	114.8	35	177.4	154.2	95	222.6	193.5
56	42.3	36.7	16	87.5	76.1	76	132.8	115.5	36	178.1	154.8	96	223.4	194.2
57	43.0	37.4	17	88.3	76.8	77	133.6	116.1	37	178.9	155.5	97	224.1	194.8
58	43.8	38.1	18	89.1	77.4	78	134.3	116.8	38	179.6	156.1	98	224.9	195.5
59	44.5	38.7	19	89.8	78.1	79	135.1	117.4	39	180.4	156.8	99	225.7	196.2
60	45.3	39.4	20	90.6	78.7	80	135.8	118.1	40	181.1	157.5	300	226.4	196.8
Dist.	Dep.	D. Lat.	Dist.	Dep.	D. Lat.	Dist.	Dep.	D. Lat.	Dist.	Dep.	D. Lat.	Dist.	Dep.	D. Lat.

311°	049°
229°	131°

49°

Dist.	D. Lat.	Dep.
N.	N×Cos.	N×Sin.
Hypotenuse	Side Adj.	Side Opp.

319°	041°
221°	139°

TABLE 3

Traverse 41° Table

319°	041°
221°	139°

Dist.	D. Lat.	Dep.	Dist.	D. Lat.	Dep.	Dist.	D. Lat.	Dep.	Dist.	D. Lat.	Dep.	Dist.	D. Lat.	Dep.
301	227.2	197.5	361	272.5	236.8	421	317.7	276.2	481	363.0	315.6	541	408.3	354.9
02	227.9	198.1	62	273.2	237.5	22	318.5	276.9	82	363.8	316.2	42	409.1	355.6
03	228.7	198.8	63	274.0	238.1	23	319.2	277.5	83	364.5	316.9	43	409.8	356.2
04	229.4	199.4	64	274.7	238.8	24	320.0	278.2	84	365.3	317.5	44	410.6	356.9
05	230.2	200.1	65	275.5	239.5	25	320.8	278.8	85	366.0	318.2	45	411.3	357.6
06	230.9	200.8	66	276.2	240.1	26	321.5	279.5	86	366.8	318.8	46	412.1	358.2
07	231.7	201.4	67	277.0	240.8	27	322.3	280.1	87	367.5	319.5	47	412.8	358.9
08	232.5	202.1	68	277.7	241.4	28	323.0	280.8	88	368.3	320.2	48	413.6	359.5
09	233.2	202.7	69	278.5	242.1	29	323.8	281.4	89	369.1	320.8	49	414.3	360.2
10	234.0	203.4	70	279.2	242.7	30	324.5	282.1	90	369.8	321.5	50	415.1	360.8
311	234.7	204.0	371	280.0	243.4	431	325.3	282.8	491	370.6	322.1	551	415.8	361.5
12	235.5	204.7	72	280.8	244.1	32	326.0	283.4	92	371.3	322.8	52	416.6	362.1
13	236.2	205.3	73	281.5	244.7	33	326.8	284.1	93	372.1	323.4	53	417.4	362.8
14	237.0	206.0	74	282.3	245.4	34	327.5	284.7	94	372.8	324.1	54	418.1	363.5
15	237.7	206.7	75	283.0	246.0	35	328.3	285.4	95	373.6	324.7	55	418.9	364.1
16	238.5	207.3	76	283.8	246.7	36	329.1	286.0	96	374.3	325.4	56	419.6	364.8
17	239.2	208.0	77	284.5	247.3	37	329.8	286.7	97	375.1	326.1	57	420.4	365.4
18	240.0	208.6	78	285.3	248.0	38	330.6	287.4	98	375.8	326.7	58	421.1	366.1
19	240.8	209.3	79	286.0	248.6	39	331.3	288.0	99	376.6	327.4	59	421.9	366.7
20	241.5	209.9	80	286.8	249.3	40	332.1	288.7	500	377.4	328.0	60	422.6	367.4
321	242.3	210.6	381	287.5	250.0	441	332.8	289.3	501	378.1	328.7	561	423.4	368.0
22	243.0	211.3	82	288.3	250.6	42	333.6	290.0	02	378.9	329.3	62	424.1	368.7
23	243.8	211.9	83	289.1	251.3	43	334.3	290.6	03	379.6	330.0	63	424.9	369.4
24	244.5	212.6	84	289.8	251.9	44	335.1	291.3	04	380.4	330.7	64	425.7	370.0
25	245.3	213.2	85	290.6	252.6	45	335.8	291.9	05	381.1	331.3	65	426.4	370.7
26	246.0	213.9	86	291.3	253.2	46	336.6	292.6	06	381.9	332.0	66	427.2	371.3
27	246.8	214.5	87	292.1	253.9	47	337.4	293.3	07	382.6	332.6	67	427.9	372.0
28	247.5	215.2	88	292.8	254.6	48	338.1	293.9	08	383.4	333.3	68	428.7	372.6
29	248.3	215.8	89	293.6	255.2	49	338.9	294.6	09	384.1	333.9	69	429.4	373.3
30	249.1	216.5	90	294.3	255.9	50	339.6	295.2	10	384.9	334.6	70	430.2	374.0
331	249.8	217.2	391	295.1	256.5	451	340.4	295.9	511	385.7	335.2	571	430.9	374.6
32	250.6	217.8	92	295.8	257.2	52	341.1	296.5	12	386.4	335.9	72	431.7	375.3
33	251.3	218.5	93	296.6	257.8	53	341.9	297.2	13	387.2	336.6	73	432.4	375.9
34	252.1	219.1	94	297.4	258.5	54	342.6	297.9	14	387.9	337.2	74	433.2	376.6
35	252.8	219.8	95	298.1	259.1	55	343.4	298.5	15	388.7	337.9	75	434.0	377.2
36	253.6	220.4	96	298.9	259.8	56	344.1	299.2	16	389.4	338.5	76	434.7	377.9
37	254.3	221.1	97	299.6	260.5	57	344.9	299.8	17	390.2	339.2	77	435.5	378.5
38	255.1	221.7	98	300.4	261.1	58	345.7	300.5	18	390.9	339.8	78	436.2	379.2
39	255.8	222.4	99	301.1	261.8	59	346.4	301.1	19	391.7	340.5	79	437.0	379.9
40	256.6	223.1	400	301.9	262.4	60	347.2	301.8	20	392.4	341.2	80	437.7	380.5
341	257.4	223.7	401	302.6	263.1	461	347.9	302.4	521	393.2	341.8	581	438.5	381.2
42	258.1	224.4	02	303.4	263.7	62	348.7	303.1	22	394.0	342.5	82	439.2	381.8
43	258.9	225.0	03	304.1	264.4	63	349.4	303.8	23	394.7	343.1	83	440.0	382.5
44	259.6	225.7	04	304.9	265.0	64	350.2	304.4	24	395.5	343.8	84	440.8	383.1
45	260.4	226.3	05	305.7	265.7	65	350.9	305.1	25	396.2	344.4	85	441.5	383.8
46	261.1	227.0	06	306.4	266.4	66	351.7	305.7	26	397.0	345.1	86	442.3	384.5
47	261.9	227.7	07	307.2	267.0	67	352.4	306.4	27	397.7	345.7	87	443.0	385.1
48	262.6	228.3	08	307.9	267.7	68	353.2	307.0	28	398.5	346.4	88	443.8	385.8
49	263.4	229.0	09	308.7	268.3	69	354.0	307.7	29	399.2	347.1	89	444.5	386.4
50	264.1	229.6	10	309.4	269.0	70	354.7	308.3	30	400.0	347.7	90	445.3	387.1
351	264.9	230.3	411	310.2	269.6	471	355.5	309.0	531	400.8	348.4	591	446.0	387.7
52	265.7	230.9	12	310.9	270.3	72	356.2	309.7	32	401.5	349.0	92	446.8	388.4
53	266.4	231.6	13	311.7	271.0	73	357.0	310.3	33	402.3	349.7	93	447.5	389.0
54	267.2	232.2	14	312.4	271.6	74	357.7	311.0	34	403.0	350.3	94	448.3	389.7
55	267.9	232.9	15	313.2	272.3	75	358.5	311.6	35	403.8	351.0	95	449.1	390.4
56	268.7	233.6	16	314.0	272.9	76	359.2	312.3	36	404.5	351.6	96	449.8	391.0
57	269.4	234.2	17	314.7	273.6	77	360.0	312.9	37	405.3	352.3	97	450.6	391.7
58	270.2	234.9	18	315.5	274.2	78	360.8	313.6	38	406.0	353.0	98	451.3	392.3
59	270.9	235.5	19	316.2	274.9	79	361.5	314.3	39	406.8	353.6	99	452.1	393.0
60	271.7	236.2	20	317.0	275.5	80	362.3	314.9	40	407.5	354.3	600	452.8	393.6
Dist.	Dep.	D. Lat.	Dist.	Dep.	D. Lat.	Dist.	Dep.	D. Lat.	Dist.	Dep.	D. Lat.	Dist.	Dep.	D. Lat.

Dist.	D. Lat.	Dep.
D Lo	Dep.	
	m	D Lo

49°

311°	049°
229°	131°

318°	042°
222°	138°

TABLE 3

Traverse **42°** Table

318°	042°
222°	138°

Dist.	D. Lat.	Dep.	Dist.	D. Lat.	Dep.	Dist.	D. Lat.	Dep.	Dist.	D. Lat.	Dep.	Dist.	D. Lat.	Dep.
1	0.7	0.7	61	45.3	40.8	121	89.9	81.0	181	134.5	121.1	241	179.1	161.3
2	1.5	1.3	62	46.1	41.5	22	90.7	81.6	82	135.3	121.8	42	179.8	161.9
3	2.2	2.0	63	46.8	42.2	23	91.4	82.3	83	136.0	122.5	43	180.6	162.6
4	3.0	2.7	64	47.6	42.8	24	92.1	83.0	84	136.7	123.1	44	181.3	163.3
5	3.7	3.3	65	48.3	43.5	25	92.9	83.6	85	137.5	123.8	45	182.1	163.9
6	4.5	4.0	66	49.0	44.2	26	93.6	84.3	86	138.2	124.5	46	182.8	164.6
7	5.2	4.7	67	49.8	44.8	27	94.4	85.0	87	139.0	125.1	47	183.6	165.3
8	5.9	5.4	68	50.5	45.5	28	95.1	85.6	88	139.7	125.8	48	184.3	165.9
9	6.7	6.0	69	51.3	46.2	29	95.9	86.3	89	140.5	126.5	49	185.0	166.6
10	7.4	6.7	70	52.0	46.8	30	96.6	87.0	90	141.2	127.1	50	185.8	167.3
11	8.2	7.4	71	52.8	47.5	131	97.4	87.7	191	141.9	127.8	251	186.5	168.0
12	8.9	8.0	72	53.5	48.2	32	98.1	88.3	92	142.7	128.5	52	187.3	168.6
13	9.7	8.7	73	54.2	48.8	33	98.8	89.0	93	143.4	129.1	53	188.0	169.3
14	10.4	9.4	74	55.0	49.5	34	99.6	89.7	94	144.2	129.8	54	188.8	170.0
15	11.1	10.0	75	55.7	50.2	35	100.3	90.3	95	144.9	130.5	55	189.5	170.6
16	11.9	10.7	76	56.5	50.9	36	101.1	91.0	96	145.7	131.1	56	190.2	171.3
17	12.6	11.4	77	57.2	51.5	37	101.8	91.7	97	146.4	131.8	57	191.0	172.0
18	13.4	12.0	78	58.0	52.2	38	102.6	92.3	98	147.1	132.5	58	191.7	172.6
19	14.1	12.7	79	58.7	52.9	39	103.3	93.0	99	147.9	133.2	59	192.5	173.3
20	14.9	13.4	80	59.5	53.5	40	104.0	93.7	200	148.6	133.8	60	193.2	174.0
21	15.6	14.1	81	60.2	54.2	141	104.8	94.3	201	149.4	134.5	261	194.0	174.6
22	16.3	14.7	82	60.9	54.9	42	105.5	95.0	02	150.1	135.2	62	194.7	175.3
23	17.1	15.4	83	61.7	55.5	43	106.3	95.7	03	150.9	135.8	63	195.4	176.0
24	17.8	16.1	84	62.4	56.2	44	107.0	96.4	04	151.6	136.5	64	196.2	176.7
25	18.6	16.7	85	63.2	56.9	45	107.8	97.0	05	152.3	137.2	65	196.9	177.3
26	19.3	17.4	86	63.9	57.5	46	108.5	97.7	06	153.1	137.8	66	197.7	178.0
27	20.1	18.1	87	64.7	58.2	47	109.2	98.4	07	153.8	138.5	67	198.4	178.7
28	20.8	18.7	88	65.4	58.9	48	110.0	99.0	08	154.6	139.2	68	199.2	179.3
29	21.6	19.4	89	66.1	59.6	49	110.7	99.7	09	155.3	139.8	69	199.9	180.0
30	22.3	20.1	90	66.9	60.2	50	111.5	100.4	10	156.1	140.5	70	200.6	180.7
31	23.0	20.7	91	67.6	60.9	151	112.2	101.0	211	156.8	141.2	271	201.4	181.3
32	23.8	21.4	92	68.4	61.6	52	113.0	101.7	12	157.5	141.9	72	202.1	182.0
33	24.5	22.1	93	69.1	62.2	53	113.7	102.4	13	158.3	142.5	73	202.9	182.7
34	25.3	22.8	94	69.9	62.9	54	114.4	103.0	14	159.0	143.2	74	203.6	183.3
35	26.0	23.4	95	70.6	63.6	55	115.2	103.7	15	159.8	143.9	75	204.4	184.0
36	26.8	24.1	96	71.3	64.2	56	115.9	104.4	16	160.5	144.5	76	205.1	184.7
37	27.5	24.8	97	72.1	64.9	57	116.7	105.1	17	161.3	145.2	77	205.9	185.3
38	28.2	25.4	98	72.8	65.6	58	117.4	105.7	18	162.0	145.9	78	206.6	186.0
39	29.0	26.1	99	73.6	66.2	59	118.2	106.4	19	162.7	146.5	79	207.3	186.7
40	29.7	26.8	100	74.3	66.9	60	118.9	107.1	20	163.5	147.2	80	208.1	187.4
41	30.5	27.4	101	75.1	67.6	161	119.6	107.7	221	164.2	147.9	281	208.8	188.0
42	31.2	28.1	02	75.8	68.3	62	120.4	108.4	22	165.0	148.5	82	209.6	188.7
43	32.0	28.8	03	76.5	68.9	63	121.1	109.1	23	165.7	149.2	83	210.3	189.4
44	32.7	29.4	04	77.3	69.6	64	121.9	109.7	24	166.5	149.9	84	211.1	190.0
45	33.4	30.1	05	78.0	70.3	65	122.6	110.4	25	167.2	150.6	85	211.8	190.7
46	34.2	30.8	06	78.8	70.9	66	123.4	111.1	26	168.0	151.2	86	212.5	191.4
47	34.9	31.4	07	79.5	71.6	67	124.1	111.7	27	168.7	151.9	87	213.3	192.0
48	35.7	32.1	08	80.3	72.3	68	124.8	112.4	28	169.4	152.6	88	214.0	192.7
49	36.4	32.8	09	81.0	72.9	69	125.6	113.1	29	170.2	153.2	89	214.8	193.4
50	37.2	33.5	10	81.7	73.6	70	126.3	113.8	30	170.9	153.9	90	215.5	194.0
51	37.9	34.1	111	82.5	74.3	171	127.1	114.4	231	171.7	154.6	291	216.3	194.7
52	38.6	34.8	12	83.2	74.9	72	127.8	115.1	32	172.4	155.2	92	217.0	195.4
53	39.4	35.5	13	84.0	75.6	73	128.6	115.8	33	173.2	155.9	93	217.7	196.1
54	40.1	36.1	14	84.7	76.3	74	129.3	116.4	34	173.9	156.6	94	218.5	196.7
55	40.9	36.8	15	85.5	77.0	75	130.1	117.1	35	174.6	157.2	95	219.2	197.4
56	41.6	37.5	16	86.2	77.6	76	130.8	117.8	36	175.4	157.9	96	220.0	198.1
57	42.4	38.1	17	86.9	78.3	77	131.5	118.4	37	176.1	158.6	97	220.7	198.7
58	43.1	38.8	18	87.7	79.0	78	132.3	119.1	38	176.9	159.3	98	221.5	199.4
59	43.8	39.5	19	88.4	79.6	79	133.0	119.8	39	177.6	159.9	99	222.2	200.1
60	44.6	40.1	20	89.2	80.3	80	133.8	120.4	40	178.4	160.6	300	222.9	200.7
Dist.	Dep.	D. Lat.	Dist.	Dep.	D. Lat.	Dist.	Dep.	D. Lat.	Dist.	Dep.	D. Lat.	Dist.	Dep.	D. Lat.

312°	048°
228°	132°

48°

Dist.	D. Lat.	Dep.
N.	N×Cos.	N×Sin.
Hypotenuse	Side Adj.	Side Opp.

318°	042°
222°	138°

TABLE 3

Traverse 42° Table

318°	042°
222°	138°

Dist.	D. Lat.	Dep.	Dist.	D. Lat.	Dep.	Dist.	D. Lat.	Dep.	Dist.	D. Lat.	Dep.	Dist.	D. Lat.	Dep.
301	223.7	201.4	361	268.3	241.6	421	312.9	281.7	481	357.5	321.9	541	402.0	362.0
02	224.4	202.1	62	269.0	242.2	22	313.6	282.4	82	358.2	322.5	42	402.8	362.7
03	225.2	202.7	63	269.8	242.9	23	314.4	283.0	83	358.9	323.2	43	403.5	363.3
04	225.9	203.4	64	270.5	243.6	24	315.1	283.7	84	359.7	323.9	44	404.3	364.0
05	226.7	204.1	65	271.2	244.2	25	315.8	284.4	85	360.4	324.5	45	405.0	364.7
06	227.4	204.8	66	272.0	244.9	26	316.6	285.1	86	361.2	325.2	46	405.8	365.3
07	228.1	205.4	67	272.7	245.6	27	317.3	285.7	87	361.9	325.9	47	406.5	366.0
08	228.9	206.1	68	273.5	246.2	28	318.1	286.4	88	362.7	326.5	48	407.2	366.7
09	229.6	206.8	69	274.2	246.9	29	318.8	287.1	89	363.4	327.2	49	408.0	367.4
10	230.4	207.4	70	275.0	247.6	30	319.6	287.7	90	364.1	327.9	50	408.7	368.0
311	231.1	208.1	371	275.7	248.2	431	320.3	288.4	491	364.9	328.5	551	409.5	368.7
12	231.9	208.8	72	276.4	248.9	32	321.0	289.1	92	365.6	329.2	52	410.2	369.4
13	232.6	209.4	73	277.2	249.6	33	321.8	289.7	93	366.4	329.9	53	411.0	370.0
14	233.3	210.1	74	277.9	250.3	34	322.5	290.4	94	367.1	330.6	54	411.7	370.7
15	234.1	210.8	75	278.7	250.9	35	323.3	291.1	95	367.9	331.2	55	412.4	371.4
16	234.8	211.4	76	279.4	251.6	36	324.0	291.7	96	368.6	331.9	56	413.2	372.0
17	235.6	212.1	77	280.2	252.3	37	324.8	292.4	97	369.3	332.6	57	413.9	372.7
18	236.3	212.8	78	280.9	252.9	38	325.5	293.1	98	370.1	333.2	58	414.7	373.4
19	237.1	213.5	79	281.7	253.6	39	326.2	293.7	99	370.8	333.9	59	415.4	374.0
20	237.8	214.1	80	282.4	254.3	40	327.0	294.4	500	371.6	334.6	60	416.2	374.7
321	238.5	214.8	381	283.1	254.9	441	327.7	295.1	501	372.3	335.2	561	416.9	375.4
22	239.3	215.5	82	283.9	255.6	42	328.5	295.8	02	373.1	335.9	62	417.6	376.1
23	240.0	216.1	83	284.6	256.3	43	329.2	296.4	03	373.8	336.6	63	418.4	376.7
24	240.8	216.8	84	285.4	256.9	44	330.0	297.1	04	374.5	337.2	64	419.1	377.4
25	241.5	217.5	85	286.1	257.6	45	330.7	297.8	05	375.3	337.9	65	419.9	378.1
26	242.3	218.1	86	286.9	258.3	46	331.4	298.4	06	376.0	338.6	66	420.6	378.7
27	243.0	218.8	87	287.6	259.0	47	332.2	299.1	07	376.8	339.2	67	421.4	379.4
28	243.8	219.5	88	288.3	259.6	48	332.9	299.8	08	377.5	339.9	68	422.1	380.1
29	244.5	220.1	89	289.1	260.3	49	333.7	300.4	09	378.3	340.6	69	422.8	380.7
30	245.2	220.8	90	289.8	261.0	50	334.4	301.1	10	379.0	341.3	70	423.6	381.4
331	246.0	221.5	391	290.6	261.6	451	335.2	301.8	511	379.7	341.9	571	424.3	382.1
32	246.7	222.2	92	291.3	262.3	52	335.9	302.4	12	380.5	342.6	72	425.1	382.7
33	247.5	222.8	93	292.1	263.0	53	336.6	303.1	13	381.2	343.3	73	425.8	383.4
34	248.2	223.5	94	292.8	263.6	54	337.4	303.8	14	382.0	343.9	74	426.6	384.1
35	249.0	224.2	95	293.5	264.3	55	338.1	304.5	15	382.7	344.6	75	427.3	384.8
36	249.7	224.8	96	294.3	265.0	56	338.9	305.1	16	383.5	345.3	76	428.1	385.4
37	250.4	225.5	97	295.0	265.6	57	339.6	305.8	17	384.2	345.9	77	428.8	386.1
38	251.2	226.2	98	295.8	266.3	58	340.4	306.5	18	384.9	346.6	78	429.5	386.8
39	251.9	226.8	99	296.5	267.0	59	341.1	307.1	19	385.7	347.3	79	430.3	387.4
40	252.7	227.5	400	297.3	267.7	60	341.8	307.8	20	386.4	347.9	80	431.0	388.1
341	253.4	228.2	401	298.0	268.3	461	342.6	308.5	521	387.2	348.6	581	431.8	388.8
42	254.2	228.8	02	298.7	269.0	62	343.3	309.1	22	387.9	349.3	82	432.5	389.4
43	254.9	229.5	03	299.5	269.7	63	344.1	309.8	23	388.7	350.0	83	433.3	390.1
44	255.6	230.2	04	300.2	270.3	64	344.8	310.5	24	389.4	350.6	84	434.0	390.8
45	256.4	230.9	05	301.0	271.0	65	345.6	311.1	25	390.2	351.3	85	434.7	391.4
46	257.1	231.5	06	301.7	271.7	66	346.3	311.8	26	390.9	352.0	86	435.5	392.1
47	257.9	232.2	07	302.5	272.3	67	347.0	312.5	27	391.6	352.6	87	436.2	392.8
48	258.6	232.9	08	303.2	273.0	68	347.8	313.2	28	392.4	353.3	88	437.0	393.4
49	259.4	233.5	09	303.9	273.7	69	348.5	313.8	29	393.1	354.0	89	437.7	394.1
50	260.1	234.2	10	304.7	274.3	70	349.3	314.5	30	393.9	354.6	90	438.5	394.8
351	260.8	234.9	411	305.4	275.0	471	350.0	315.2	531	394.6	355.3	591	439.2	395.5
52	261.6	235.5	12	306.2	275.7	72	350.8	315.8	32	395.4	356.0	92	439.9	396.1
53	262.3	236.2	13	306.9	276.4	73	351.5	316.5	33	396.1	356.6	93	440.7	396.8
54	263.1	236.9	14	307.7	277.0	74	352.3	317.2	34	396.8	357.3	94	441.4	397.5
55	263.8	237.5	15	308.4	277.7	75	353.0	317.8	35	397.6	358.0	95	442.2	398.1
56	264.6	238.2	16	309.1	278.4	76	353.7	318.5	36	398.3	358.7	96	442.9	398.8
57	265.3	238.9	17	309.9	279.0	77	354.5	319.2	37	399.1	359.3	97	443.7	399.5
58	266.0	239.5	18	310.6	279.7	78	355.2	319.8	38	399.8	360.0	98	444.4	400.1
59	266.8	240.2	19	311.4	280.4	79	356.0	320.5	39	400.6	360.7	99	445.1	400.8
60	267.5	240.9	20	312.1	281.0	80	356.7	321.2	40	401.3	361.3	600	445.9	401.5
Dist.	Dep.	D. Lat.	Dist.	Dep.	D. Lat.	Dist.	Dep.	D. Lat.	Dist.	Dep.	D. Lat.	Dist.	Dep.	D. Lat.

Dist.	D. Lat.	Dep.
D Lo	Dep.	
	m	D Lo

48°

312°	048°
228°	132°

317°	043°
223°	137°

TABLE 3

Traverse 43° Table

Dist.	D. Lat.	Dep.	Dist.	D. Lat.	Dep.	Dist.	D. Lat.	Dep.	Dist.	D. Lat.	Dep.	Dist.	D. Lat.	Dep.
1	0.7	0.7	61	44.6	41.6	121	88.5	82.5	181	132.4	123.4	241	176.3	164.4
2	1.5	1.4	62	45.3	42.3	22	89.2	83.2	82	133.1	124.1	42	177.0	165.0
3	2.2	2.0	63	46.1	43.0	23	90.0	83.9	83	133.8	124.8	43	177.7	165.7
4	2.9	2.7	64	46.8	43.6	24	90.7	84.6	84	134.6	125.5	44	178.5	166.4
5	3.7	3.4	65	47.5	44.3	25	91.4	85.2	85	135.3	126.2	45	179.2	167.1
6	4.4	4.1	66	48.3	45.0	26	92.2	85.9	86	136.0	126.9	46	179.9	167.8
7	5.1	4.8	67	49.0	45.7	27	92.9	86.6	87	136.8	127.5	47	180.6	168.5
8	5.9	5.5	68	49.7	46.4	28	93.6	87.3	88	137.5	128.2	48	181.4	169.1
9	6.6	6.1	69	50.5	47.1	29	94.3	88.0	89	138.2	128.9	49	182.1	169.8
10	7.3	6.8	70	51.2	47.7	30	95.1	88.7	90	139.0	129.6	50	182.8	170.5
11	8.0	7.5	71	51.9	48.4	131	95.8	89.3	191	139.7	130.3	251	183.6	171.2
12	8.8	8.2	72	52.7	49.1	32	96.5	90.0	92	140.4	130.9	52	184.3	171.9
13	9.5	8.9	73	53.4	49.8	33	97.3	90.7	93	141.2	131.6	53	185.0	172.5
14	10.2	9.5	74	54.1	50.5	34	98.0	91.4	94	141.9	132.3	54	185.8	173.2
15	11.0	10.2	75	54.9	51.1	35	98.7	92.1	95	142.6	133.0	55	186.5	173.9
16	11.7	10.9	76	55.6	51.8	36	99.5	92.8	96	143.3	133.7	56	187.2	174.6
17	12.4	11.6	77	56.3	52.5	37	100.2	93.4	97	144.1	134.4	57	188.0	175.3
18	13.2	12.3	78	57.0	53.2	38	100.9	94.1	98	144.8	135.0	58	188.7	176.0
19	13.9	13.0	79	57.8	53.9	39	101.7	94.8	99	145.5	135.7	59	189.4	176.6
20	14.6	13.6	80	58.5	54.6	40	102.4	95.5	200	146.3	136.4	60	190.2	177.3
21	15.4	14.3	81	59.2	55.2	141	103.1	96.2	201	147.0	137.1	261	190.9	178.0
22	16.1	15.0	82	60.0	55.9	42	103.9	96.8	02	147.7	137.8	62	191.6	178.7
23	16.8	15.7	83	60.7	56.6	43	104.6	97.5	03	148.5	138.4	63	192.3	179.4
24	17.6	16.4	84	61.4	57.3	44	105.3	98.2	04	149.2	139.1	64	193.1	180.0
25	18.3	17.0	85	62.2	58.0	45	106.0	98.9	05	149.9	139.8	65	193.8	180.7
26	19.0	17.7	86	62.9	58.7	46	106.8	99.6	06	150.7	140.5	66	194.5	181.4
27	19.7	18.4	87	63.6	59.3	47	107.5	100.3	07	151.4	141.2	67	195.3	182.1
28	20.5	19.1	88	64.4	60.0	48	108.2	100.9	08	152.1	141.9	68	196.0	182.8
29	21.2	19.8	89	65.1	60.7	49	109.0	101.6	09	152.9	142.5	69	196.7	183.5
30	21.9	20.5	90	65.8	61.4	50	109.7	102.3	10	153.6	143.2	70	197.5	184.1
31	22.7	21.1	91	66.6	62.1	151	110.4	103.0	211	154.3	143.9	271	198.2	184.8
32	23.4	21.8	92	67.3	62.7	52	111.2	103.7	12	155.0	144.6	72	198.9	185.5
33	24.1	22.5	93	68.0	63.4	53	111.9	104.3	13	155.8	145.3	73	199.7	186.2
34	24.9	23.2	94	68.7	64.1	54	112.6	105.0	14	156.5	145.9	74	200.4	186.9
35	25.6	23.9	95	69.5	64.8	55	113.4	105.7	15	157.2	146.6	75	201.1	187.5
36	26.3	24.6	96	70.2	65.5	56	114.1	106.4	16	158.0	147.3	76	201.9	188.2
37	27.1	25.2	97	70.9	66.2	57	114.8	107.1	17	158.7	148.0	77	202.6	188.9
38	27.8	25.9	98	71.7	66.8	58	115.6	107.8	18	159.4	148.7	78	203.3	189.6
39	28.5	26.6	99	72.4	67.5	59	116.3	108.4	19	160.2	149.4	79	204.0	190.3
40	29.3	27.3	100	73.1	68.2	60	117.0	109.1	20	160.9	150.0	80	204.8	191.0
41	30.0	28.0	101	73.9	68.9	161	117.7	109.8	221	161.6	150.7	281	205.5	191.6
42	30.7	28.6	02	74.6	69.6	62	118.5	110.5	22	162.4	151.4	82	206.2	192.3
43	31.4	29.3	03	75.3	70.2	63	119.2	111.2	23	163.1	152.1	83	207.0	193.0
44	32.2	30.0	04	76.1	70.9	64	119.9	111.8	24	163.8	152.8	84	207.7	193.7
45	32.9	30.7	05	76.8	71.6	65	120.7	112.5	25	164.6	153.4	85	208.4	194.4
46	33.6	31.4	06	77.5	72.3	66	121.4	113.2	26	165.3	154.1	86	209.2	195.1
47	34.4	32.1	07	78.3	73.0	67	122.1	113.9	27	166.0	154.8	87	209.9	195.7
48	35.1	32.7	08	79.0	73.7	68	122.9	114.6	28	166.7	155.5	88	210.6	196.4
49	35.8	33.4	09	79.7	74.3	69	123.6	115.3	29	167.5	156.2	89	211.4	197.1
50	36.6	34.1	10	80.4	75.0	70	124.3	115.9	30	168.2	156.9	90	212.1	197.8
51	37.3	34.8	111	81.2	75.7	171	125.1	116.6	231	168.9	157.5	291	212.8	198.5
52	38.0	35.5	12	81.9	76.4	72	125.8	117.3	32	169.7	158.2	92	213.6	199.1
53	38.8	36.1	13	82.6	77.1	73	126.5	118.0	33	170.4	158.9	93	214.3	199.8
54	39.5	36.8	14	83.4	77.7	74	127.3	118.7	34	171.1	159.6	94	215.0	200.5
55	40.2	37.5	15	84.1	78.4	75	128.0	119.3	35	171.9	160.3	95	215.7	201.2
56	41.0	38.2	16	84.8	79.1	76	128.7	120.0	36	172.6	161.0	96	216.5	201.9
57	41.7	38.9	17	85.6	79.8	77	129.4	120.7	37	173.3	161.6	97	217.2	202.6
58	42.4	39.6	18	86.3	80.5	78	130.2	121.4	38	174.1	162.3	98	217.9	203.2
59	43.1	40.2	19	87.0	81.2	79	130.9	122.1	39	174.8	163.0	99	218.7	203.9
60	43.9	40.9	20	87.8	81.8	80	131.6	122.8	40	175.5	163.7	300	219.4	204.6
Dist.	Dep.	D. Lat.	Dist.	Dep.	D. Lat.	Dist.	Dep.	D. Lat.	Dist.	Dep.	D. Lat.	Dist.	Dep.	D. Lat.

313°	047°
227°	133°

47°

Dist.	D. Lat.	Dep.
N.	N×Cos.	N×Sin.
Hypotenuse	Side Adj.	Side Opp.

317° | 043° / 223° | 137°						TABLE 3 Traverse 43° Table						317° | 043° / 223° | 137°		
Dist.	D. Lat.	Dep.	Dist.	D. Lat.	Dep.	Dist.	D. Lat.	Dep.	Dist.	D. Lat.	Dep.	Dist.	D. Lat.	Dep.
301	220.1	205.3	361	264.0	246.2	421	307.9	287.1	481	351.8	328.1	541	395.7	369.0
02	220.9	206.0	62	264.8	246.9	22	308.6	287.8	82	352.5	328.7	42	396.4	369.6
03	221.6	206.6	63	265.5	247.6	23	309.4	288.5	83	353.2	329.4	43	397.1	370.3
04	222.3	207.3	64	266.2	248.2	24	310.1	289.2	84	354.0	330.1	44	397.9	371.0
05	223.1	208.0	65	266.9	248.9	25	310.8	289.8	85	354.7	330.8	45	398.6	371.7
06	223.8	208.7	66	267.7	249.6	26	311.6	290.5	86	355.4	331.5	46	399.3	372.4
07	224.5	209.4	67	268.4	250.3	27	312.3	291.2	87	356.2	332.1	47	400.1	373.1
08	225.3	210.1	68	269.1	251.0	28	313.0	291.9	88	356.9	332.8	48	400.8	373.7
09	226.0	210.7	69	269.9	251.7	29	313.8	292.6	89	357.6	333.5	49	401.5	374.4
10	226.7	211.4	70	270.6	252.3	30	314.5	293.3	90	358.4	334.2	50	402.2	375.1
311	227.5	212.1	371	271.3	253.0	431	315.2	293.9	491	359.1	334.9	551	403.0	375.8
12	228.2	212.8	72	272.1	253.7	32	315.9	294.6	92	359.8	335.5	52	403.7	376.5
13	228.9	213.5	73	272.8	254.4	33	316.7	295.3	93	360.6	336.2	53	404.4	377.1
14	229.6	214.1	74	273.5	255.1	34	317.4	296.0	94	361.3	336.9	54	405.2	377.8
15	230.4	214.8	75	274.3	255.7	35	318.1	296.7	95	362.0	337.6	55	405.9	378.5
16	231.1	215.5	76	275.0	256.4	36	318.9	297.4	96	362.8	338.3	56	406.6	379.2
17	231.8	216.2	77	275.7	257.1	37	319.6	298.0	97	363.5	339.0	57	407.4	379.9
18	232.6	216.9	78	276.5	257.8	38	320.3	298.7	98	364.2	339.6	58	408.1	380.6
19	233.3	217.6	79	277.2	258.5	39	321.1	299.4	99	364.9	340.3	59	408.8	381.2
20	234.0	218.2	80	277.9	259.2	40	321.8	300.1	500	365.7	341.0	60	409.6	381.9
321	234.8	218.9	381	278.6	259.8	441	322.5	300.8	501	366.4	341.7	561	410.3	382.6
22	235.5	219.6	82	279.4	260.5	42	323.3	301.4	02	367.1	342.4	62	411.0	383.3
23	236.2	220.3	83	280.1	261.2	43	324.0	302.1	03	367.9	343.0	63	411.8	384.0
24	237.0	221.0	84	280.8	261.9	44	324.7	302.8	04	368.6	343.7	64	412.5	384.6
25	237.7	221.6	85	281.6	262.6	45	325.5	303.5	05	369.3	344.4	65	413.2	385.3
26	238.4	222.3	86	282.3	263.3	46	326.2	304.2	06	370.1	345.1	66	413.9	386.0
27	239.2	223.0	87	283.0	263.9	47	326.9	304.9	07	370.8	345.8	67	414.7	386.7
28	239.9	223.7	88	283.8	264.6	48	327.6	305.5	08	371.5	346.5	68	415.4	387.4
29	240.6	224.4	89	284.5	265.3	49	328.4	306.2	09	372.3	347.1	69	416.1	388.1
30	241.3	225.1	90	285.2	266.0	50	329.1	306.9	10	373.0	347.8	70	416.9	388.7
331	242.1	225.7	391	286.0	266.7	451	329.9	307.6	511	373.7	348.5	571	417.6	389.4
32	242.8	226.4	92	286.7	267.3	52	330.6	308.3	12	374.5	349.2	72	418.3	390.1
33	243.5	227.1	93	287.4	268.0	53	331.3	308.9	13	375.2	349.9	73	419.1	390.8
34	244.3	227.8	94	288.2	268.7	54	332.0	309.6	14	375.9	350.5	74	419.8	391.5
35	245.0	228.5	95	288.9	269.4	55	332.8	310.3	15	376.6	351.2	75	420.5	392.1
36	245.7	229.2	96	289.6	270.1	56	333.5	311.0	16	377.4	351.9	76	421.3	392.8
37	246.5	229.8	97	290.3	270.8	57	334.2	311.7	17	378.1	352.6	77	422.0	393.5
38	247.2	230.5	98	291.1	271.4	58	335.0	312.4	18	378.8	353.3	78	422.7	394.2
39	247.9	231.2	99	291.8	272.1	59	335.7	313.0	19	379.6	354.0	79	423.5	394.9
40	248.7	231.9	400	292.5	272.8	60	336.4	313.7	20	380.3	354.6	80	424.2	395.6
341	249.4	232.6	401	293.3	273.5	461	337.2	314.4	521	381.0	355.3	581	424.9	396.2
42	250.1	233.2	02	294.0	274.2	62	337.9	315.1	22	381.8	356.0	82	425.6	396.9
43	250.9	233.9	03	294.7	274.8	63	338.6	315.8	23	382.5	356.7	83	426.4	397.6
44	251.6	234.6	04	295.5	275.5	64	339.3	316.4	24	383.2	357.4	84	427.1	398.3
45	252.3	235.3	05	296.2	276.2	65	340.1	317.1	25	384.0	358.0	85	427.8	399.0
46	253.0	236.0	06	296.9	276.9	66	340.8	317.8	26	384.7	358.7	86	428.6	399.7
47	253.8	236.7	07	297.7	277.6	67	341.5	318.5	27	385.4	359.4	87	429.3	400.3
48	254.5	237.3	08	298.4	278.3	68	342.3	319.2	28	386.2	360.1	88	430.0	401.0
49	255.2	238.0	09	299.1	278.9	69	343.0	319.9	29	386.9	360.8	89	430.8	401.7
50	256.0	238.7	10	299.9	279.6	70	343.7	320.5	30	387.6	361.5	90	431.5	402.4
351	256.7	239.4	411	300.6	280.3	471	344.5	321.2	531	388.3	362.1	591	432.2	403.1
52	257.4	240.1	12	301.3	281.0	72	345.2	321.9	32	389.1	362.8	92	433.0	403.7
53	258.2	240.9	13	302.0	281.7	73	345.9	322.6	33	389.8	363.5	93	433.7	404.4
54	258.9	241.4	14	302.8	282.3	74	346.7	323.3	34	390.5	364.2	94	434.4	405.1
55	259.6	242.1	15	303.5	283.0	75	347.4	323.9	35	391.3	364.9	95	435.2	405.8
56	260.4	242.8	16	304.3	283.7	76	348.1	324.6	36	392.0	365.6	96	435.9	406.5
57	261.1	243.5	17	305.0	284.4	77	348.9	325.3	37	392.7	366.2	97	436.6	407.2
58	261.8	244.2	18	305.7	285.1	78	349.6	326.0	38	393.5	366.9	98	437.3	407.8
59	262.6	244.8	19	306.4	285.8	79	350.3	326.7	39	394.2	367.6	99	438.1	408.5
60	263.3	245.5	20	307.2	286.4	80	351.0	327.4	40	394.9	368.3	600	438.8	409.2
Dist.	Dep.	D. Lat.	Dist.	Dep.	D. Lat.	Dist.	Dep.	D. Lat.	Dist.	Dep.	D. Lat.	Dist.	Dep.	D. Lat.

Dist.	D. Lat.	Dep.
D Lo	Dep.	
	m	D Lo

47°

313° | 047°
227° | 133°

316° | 044° / 224° | 136°						TABLE 3 Traverse 44° Table							316° | 044° / 224° | 136°	
Dist.	D. Lat.	Dep.	Dist.	D. Lat.	Dep.	Dist.	D. Lat.	Dep.	Dist.	D. Lat.	Dep.	Dist.	D. Lat.	Dep.
1	0.7	0.7	61	43.9	42.4	121	87.0	84.1	181	130.2	125.7	241	173.4	167.4
2	1.4	1.4	62	44.6	43.1	22	87.8	84.7	82	130.9	126.4	42	174.1	168.1
3	2.2	2.1	63	45.3	43.8	23	88.5	85.4	83	131.6	127.1	43	174.8	168.8
4	2.9	2.8	64	46.0	44.5	24	89.2	86.1	84	132.4	127.8	44	175.5	169.5
5	3.6	3.5	65	46.8	45.2	25	89.9	86.8	85	133.1	128.5	45	176.2	170.2
6	4.3	4.2	66	47.5	45.8	26	90.6	87.5	86	133.8	129.2	46	177.0	170.9
7	5.0	4.9	67	48.2	46.5	27	91.4	88.2	87	134.5	129.9	47	177.7	171.6
8	5.8	5.6	68	48.9	47.2	28	92.1	88.9	88	135.2	130.6	48	178.4	172.3
9	6.5	6.3	69	49.6	47.9	29	92.8	89.6	89	136.0	131.3	49	179.1	173.0
10	7.2	6.9	70	50.4	48.6	30	93.5	90.3	90	136.7	132.0	50	179.8	173.7
11	7.9	7.6	71	51.1	49.3	131	94.2	91.0	191	137.4	132.7	251	180.6	174.4
12	8.6	8.3	72	51.8	50.0	32	95.0	91.7	92	138.1	133.4	52	181.3	175.1
13	9.4	9.0	73	52.5	50.7	33	95.7	92.4	93	138.8	134.1	53	182.0	175.7
14	10.1	9.7	74	53.2	51.4	34	96.4	93.1	94	139.6	134.8	54	182.7	176.4
15	10.8	10.4	75	54.0	52.1	35	97.1	93.8	95	140.3	135.5	55	183.4	177.1
16	11.5	11.1	76	54.7	52.8	36	97.8	94.5	96	141.0	136.2	56	184.2	177.8
17	12.2	11.8	77	55.4	53.5	37	98.5	95.2	97	141.7	136.8	57	184.9	178.5
18	12.9	12.5	78	56.1	54.2	38	99.3	95.9	98	142.4	137.5	58	185.6	179.2
19	13.7	13.2	79	56.8	54.9	39	100.0	96.6	99	143.1	138.2	59	186.3	179.9
20	14.4	13.9	80	57.5	55.6	40	100.7	97.3	200	143.9	138.9	60	187.0	180.6
21	15.1	14.6	81	58.3	56.3	141	101.4	97.9	201	144.6	139.6	261	187.7	181.3
22	15.8	15.3	82	59.0	57.0	42	102.1	98.6	02	145.3	140.3	62	188.5	182.0
23	16.5	16.0	83	59.7	57.7	43	102.9	99.3	03	146.0	141.0	63	189.2	182.7
24	17.3	16.7	84	60.4	58.4	44	103.6	100.0	04	146.7	141.7	64	189.9	183.4
25	18.0	17.4	85	61.1	59.0	45	104.3	100.7	05	147.5	142.4	65	190.6	184.1
26	18.7	18.1	86	61.9	59.7	46	105.0	101.4	06	148.2	143.1	66	191.3	184.8
27	19.4	18.8	87	62.6	60.4	47	105.7	102.1	07	148.9	143.8	67	192.1	185.5
28	20.1	19.5	88	63.3	61.1	48	106.5	102.8	08	149.6	144.5	68	192.8	186.2
29	20.9	20.1	89	64.0	61.8	49	107.2	103.5	09	150.3	145.2	69	193.5	186.9
30	21.6	20.8	90	64.7	62.5	50	107.9	104.2	10	151.1	145.9	70	194.2	187.6
31	22.3	21.5	91	65.5	63.2	151	108.6	104.9	211	151.8	146.6	271	194.9	188.3
32	23.0	22.2	92	66.2	63.9	52	109.3	105.6	12	152.5	147.3	72	195.7	188.9
33	23.7	22.9	93	66.9	64.6	53	110.1	106.3	13	153.2	148.0	73	196.4	189.6
34	24.5	23.6	94	67.6	65.3	54	110.8	107.0	14	153.9	148.7	74	197.1	190.3
35	25.2	24.3	95	68.3	66.0	55	111.5	107.7	15	154.7	149.4	75	197.8	191.0
36	25.9	25.0	96	69.1	66.7	56	112.2	108.4	16	155.4	150.0	76	198.5	191.7
37	26.6	25.7	97	69.8	67.4	57	112.9	109.1	17	156.1	150.7	77	199.3	192.4
38	27.3	26.4	98	70.5	68.1	58	113.7	109.8	18	156.8	151.4	78	200.0	193.1
39	28.1	27.1	99	71.2	68.8	59	114.4	110.5	19	157.5	152.1	79	200.7	193.8
40	28.8	27.8	100	71.9	69.5	60	115.1	111.1	20	158.3	152.8	80	201.4	194.5
41	29.5	28.5	101	72.7	70.2	161	115.8	111.8	221	159.0	153.5	281	202.1	195.2
42	30.2	29.2	02	73.4	70.9	62	116.5	112.5	22	159.7	154.2	82	202.9	195.9
43	30.9	29.9	03	74.1	71.5	63	117.3	113.2	23	160.4	154.9	83	203.6	196.6
44	31.7	30.6	04	74.8	72.2	64	118.0	113.9	24	161.1	155.6	84	204.3	197.3
45	32.4	31.3	05	75.5	72.9	65	118.7	114.6	25	161.9	156.3	85	205.0	198.0
46	33.1	32.0	06	76.3	73.6	66	119.4	115.3	26	162.6	157.0	86	205.7	198.7
47	33.8	32.6	07	77.0	74.3	67	120.1	116.0	27	163.3	157.7	87	206.5	199.4
48	34.5	33.3	08	77.7	75.0	68	120.8	116.7	28	164.0	158.4	88	207.2	200.1
49	35.2	34.0	09	78.4	75.7	69	121.6	117.4	29	164.7	159.1	89	207.9	200.8
50	36.0	34.7	10	79.1	76.4	70	122.3	118.1	30	165.4	159.8	90	208.6	201.5
51	36.7	35.4	111	79.8	77.1	171	123.0	118.8	231	166.2	160.5	291	209.3	202.1
52	37.4	36.1	12	80.6	77.8	72	123.7	119.5	32	166.9	161.2	92	210.0	202.8
53	38.1	36.8	13	81.3	78.5	73	124.4	120.2	33	167.6	161.9	93	210.8	203.5
54	38.8	37.5	14	82.0	79.2	74	125.2	120.9	34	168.3	162.6	94	211.5	204.2
55	39.6	38.2	15	82.7	79.9	75	125.9	121.6	35	169.0	163.2	95	212.2	204.9
56	40.3	38.9	16	83.4	80.6	76	126.6	122.3	36	169.8	163.9	96	212.9	205.6
57	41.0	39.6	17	84.2	81.3	77	127.3	123.0	37	170.5	164.6	97	213.6	206.3
58	41.7	40.3	18	84.9	82.0	78	128.0	123.6	38	171.2	165.3	98	214.4	207.0
59	42.4	41.0	19	85.6	82.7	79	128.8	124.3	39	171.9	166.0	99	215.1	207.7
60	43.2	41.7	20	86.3	83.4	80	129.5	125.0	40	172.6	166.7	300	215.8	208.4
Dist.	Dep.	D. Lat.	Dist.	Dep.	D. Lat.	Dist.	Dep.	D. Lat.	Dist.	Dep.	D. Lat.	Dist.	Dep.	D. Lat.

314° | 046°
226° | 134°

46°

Dist.	D. Lat.	Dep.
N.	N×Cos.	N×Sin.
Hypotenuse	Side Adj.	Side Opp.

316°	044°
224°	136°

TABLE 3

Traverse 44° Table

Dist.	D. Lat.	Dep.	Dist.	D. Lat.	Dep.	Dist.	D. Lat.	Dep.	Dist.	D. Lat.	Dep.	Dist.	D. Lat.	Dep.
301	216.5	209.1	361	259.7	250.8	421	302.8	292.5	481	346.0	334.1	541	389.2	375.8
02	217.2	209.8	62	260.4	251.5	22	303.6	293.1	82	346.7	334.8	42	389.9	376.5
03	218.0	210.5	63	261.1	252.2	23	304.3	293.8	83	347.4	335.5	43	390.6	377.2
04	218.7	211.2	64	261.8	252.9	24	305.0	294.5	84	348.2	336.2	44	391.3	377.9
05	219.4	211.9	65	262.6	253.6	25	305.7	295.2	85	348.9	336.9	45	392.0	378.6
06	220.1	212.6	66	263.3	254.2	26	306.4	295.9	86	349.6	337.6	46	392.8	379.3
07	220.8	213.3	67	264.0	254.9	27	307.2	296.6	87	350.3	338.3	47	393.5	380.0
08	221.6	214.0	68	264.7	255.6	28	307.9	297.3	88	351.0	339.0	48	394.2	380.7
09	222.3	214.6	69	265.4	256.3	29	308.6	298.0	89	351.7	339.7	49	394.9	381.4
10	223.0	215.3	70	266.2	257.0	30	309.3	298.7	90	352.5	340.4	50	395.6	382.1
311	223.7	216.0	371	266.9	257.7	431	310.0	299.4	491	353.2	341.1	551	396.4	382.8
12	224.4	216.7	72	267.6	258.4	32	310.8	300.1	92	353.9	341.8	52	397.1	383.5
13	225.2	217.4	73	268.3	259.1	33	311.5	300.8	93	354.6	342.5	53	397.8	384.1
14	225.9	218.1	74	269.0	259.8	34	312.2	301.5	94	355.4	343.2	54	398.5	384.8
15	226.6	218.8	75	269.8	260.5	35	312.9	302.2	95	356.1	343.9	55	399.2	385.5
16	227.3	219.5	76	270.5	261.2	36	313.6	302.9	96	356.8	344.6	56	400.0	386.2
17	228.0	220.2	77	271.2	261.9	37	314.4	303.6	97	357.5	345.2	57	400.7	386.9
18	228.8	220.9	78	271.9	262.6	38	315.1	304.3	98	358.2	345.9	58	401.4	387.6
19	229.5	221.6	79	272.6	263.3	39	315.8	305.0	99	359.0	346.6	59	402.1	388.3
20	230.2	222.3	80	273.3	264.0	40	316.6	305.6	500	359.7	347.3	60	402.8	389.0
321	230.9	223.0	381	274.1	264.7	441	317.2	306.3	501	360.4	348.0	561	403.5	389.7
22	231.6	223.7	82	274.8	265.4	42	317.9	307.0	02	361.1	348.7	62	404.3	390.4
23	232.3	224.4	83	275.5	266.1	43	318.7	307.7	03	361.8	349.4	63	405.0	391.1
24	233.1	225.1	84	276.2	266.7	44	319.4	308.4	04	362.5	350.1	64	405.7	391.8
25	233.8	225.8	85	276.9	267.4	45	320.1	309.1	05	363.3	350.8	65	406.4	392.5
26	234.5	226.5	86	277.7	268.1	46	320.8	309.8	06	364.0	351.5	66	407.1	393.2
27	235.2	227.2	87	278.4	268.8	47	321.5	310.5	07	364.7	352.2	67	407.9	393.9
28	235.9	227.8	88	279.1	269.5	48	322.3	311.2	08	365.4	352.9	68	408.6	394.6
29	236.7	228.5	89	279.8	270.2	49	323.0	311.9	09	366.1	353.6	69	409.3	395.3
30	237.4	229.2	90	280.5	270.9	50	323.7	312.6	10	366.9	354.3	70	410.0	396.0
331	238.1	229.9	391	281.3	271.6	451	324.4	313.3	511	367.6	355.0	571	410.7	396.6
32	238.8	230.6	92	282.0	272.3	52	325.1	314.0	12	368.3	355.7	72	411.5	397.3
33	239.5	231.3	93	282.7	273.0	53	325.9	314.7	13	369.0	356.4	73	412.2	398.0
34	240.3	232.0	94	283.4	273.7	54	326.6	315.4	14	369.7	357.1	74	412.9	398.7
35	241.0	232.7	95	284.1	274.4	55	327.3	316.1	15	370.5	357.7	75	413.6	399.4
36	241.7	233.4	96	284.9	275.1	56	328.0	316.8	16	371.2	358.4	76	414.3	400.1
37	242.4	234.1	97	285.6	275.8	57	328.7	317.5	17	371.9	359.1	77	415.1	400.8
38	243.1	234.8	98	286.3	276.5	58	329.5	318.2	18	372.6	359.8	78	415.8	401.5
39	243.9	235.5	99	287.0	277.2	59	330.2	318.8	19	373.3	360.5	79	416.5	402.2
40	244.6	236.2	400	287.7	277.9	60	330.9	319.5	20	374.1	361.2	80	417.2	402.9
341	245.3	236.9	401	288.5	278.6	461	331.6	320.2	521	374.8	361.9	581	417.9	403.6
42	246.0	237.6	02	289.2	279.3	62	332.3	320.9	22	375.5	362.6	82	418.7	404.3
43	246.7	238.3	03	289.9	279.9	63	333.1	321.6	23	376.2	363.3	83	419.4	405.0
44	247.5	239.0	04	290.6	280.6	64	333.8	322.3	24	376.9	364.0	84	420.1	405.7
45	248.2	239.7	05	291.3	281.3	65	334.5	323.0	25	377.7	364.7	85	420.8	406.4
46	248.9	240.4	06	292.1	282.0	66	335.2	323.7	26	378.4	365.4	86	421.5	407.1
47	249.6	241.0	07	292.8	282.7	67	335.9	324.4	27	379.1	366.1	87	422.3	407.8
48	250.3	241.7	08	293.5	283.4	68	336.7	325.1	28	379.8	366.8	88	423.0	408.5
49	251.0	242.4	09	294.2	284.1	69	337.4	325.8	29	380.5	367.5	89	423.7	409.2
50	251.8	243.1	10	294.9	284.8	70	338.1	326.5	30	381.3	368.2	90	424.4	409.8
351	252.5	243.8	411	295.6	285.5	471	338.8	327.2	531	382.0	368.9	591	425.1	410.5
52	253.2	244.5	12	296.4	286.2	72	339.5	327.9	32	382.7	369.6	92	425.8	411.2
53	253.9	245.2	13	297.1	286.9	73	340.2	328.6	33	383.4	370.3	93	426.6	411.9
54	254.6	245.9	14	297.8	287.6	74	341.0	329.3	34	384.1	370.9	94	427.3	412.6
55	255.4	246.6	15	298.5	288.3	75	341.7	330.0	35	384.8	371.6	95	428.0	413.3
56	256.1	247.3	16	299.2	289.0	76	342.4	330.7	36	385.6	372.3	96	428.7	414.0
57	256.8	248.0	17	300.0	289.7	77	343.1	331.4	37	386.3	373.0	97	429.4	414.7
58	257.5	248.7	18	300.7	290.4	78	343.8	332.0	38	387.0	373.7	98	430.2	415.4
59	258.2	249.4	19	301.4	291.1	79	344.6	332.7	39	387.7	374.4	99	430.9	416.1
60	259.0	250.1	20	302.1	291.8	80	345.3	333.4	40	388.4	375.1	600	431.6	416.8
Dist.	Dep.	D. Lat.	Dist.	Dep.	D. Lat.	Dist.	Dep.	D. Lat.	Dist.	Dep.	D. Lat.	Dist.	Dep.	D. Lat.

Dist.	D. Lat.	Dep.
D Lo	Dep.	
	m	D Lo

46°

314°	046°
226°	134°

315° | 045°
225° | 135°

TABLE 3

Traverse 45° Table

315° | 045°
225° | 135°

Dist.	D. Lat.	Dep.	Dist.	D. Lat.	Dep.	Dist.	D. Lat.	Dep.	Dist.	D. Lat.	Dep.	Dist.	D. Lat.	Dep.
1	0.7	0.7	61	43.1	43.1	121	85.6	85.6	181	128.0	128.0	241	170.4	170.4
2	1.4	1.4	62	43.8	43.8	22	86.3	86.3	82	128.7	128.7	42	171.1	171.1
3	2.1	2.1	63	44.5	44.5	23	87.0	87.0	83	129.4	129.4	43	171.8	171.8
4	2.8	2.8	64	45.3	45.3	24	87.7	87.7	84	130.1	130.1	44	172.5	172.5
5	3.5	3.5	65	46.0	46.0	25	88.4	88.4	85	130.8	130.8	45	173.2	173.2
6	4.2	4.2	66	46.7	46.7	26	89.1	89.1	86	131.5	131.5	46	173.9	173.9
7	4.9	4.9	67	47.4	47.4	27	89.8	89.8	87	132.2	132.2	47	174.7	174.7
8	5.7	5.7	68	48.1	48.1	28	90.5	90.5	88	132.9	132.9	48	175.4	175.4
9	6.4	6.4	69	48.8	48.8	29	91.2	91.2	89	133.6	133.6	49	176.1	176.1
10	7.1	7.1	70	49.5	49.5	30	91.9	91.9	90	134.4	134.4	50	176.8	176.8
11	7.8	7.8	71	50.2	50.2	131	92.6	92.6	191	135.1	135.1	251	177.5	177.5
12	8.5	8.5	72	50.9	50.9	32	93.3	93.3	92	135.8	135.8	52	178.2	178.2
13	9.2	9.2	73	51.6	51.6	33	94.0	94.0	93	136.5	136.5	53	178.9	178.9
14	9.9	9.9	74	52.3	52.3	34	94.8	94.8	94	137.2	137.2	54	179.6	179.6
15	10.6	10.6	75	53.0	53.0	35	95.5	95.5	95	137.9	137.9	55	180.3	180.3
16	11.3	11.3	76	53.7	53.7	36	96.2	96.2	96	138.6	138.6	56	181.0	181.0
17	12.0	12.0	77	54.4	54.4	37	96.9	96.9	97	139.3	139.3	57	181.7	181.7
18	12.7	12.7	78	55.2	55.2	38	97.6	97.6	98	140.0	140.0	58	182.4	182.4
19	13.4	13.4	79	55.9	55.9	39	98.3	98.3	99	140.7	140.7	59	183.1	183.1
20	14.1	14.1	80	56.6	56.6	40	99.0	99.0	200	141.4	141.4	60	183.8	183.8
21	14.8	14.8	81	57.3	57.3	141	99.7	99.7	201	142.1	142.1	261	184.6	184.6
22	15.6	15.6	82	58.0	58.0	42	100.4	100.4	02	142.8	142.8	62	185.3	185.3
23	16.3	16.3	83	58.7	58.7	43	101.1	101.1	03	143.5	143.5	63	186.0	186.0
24	17.0	17.0	84	59.4	59.4	44	101.8	101.8	04	144.2	144.2	64	186.7	186.7
25	17.7	17.7	85	60.1	60.1	45	102.5	102.5	05	145.0	145.0	65	187.4	187.4
26	18.4	18.4	86	60.8	60.8	46	103.2	103.2	06	145.7	145.7	66	188.1	188.1
27	19.1	19.1	87	61.5	61.5	47	103.9	103.9	07	146.4	146.4	67	188.8	188.8
28	19.8	19.8	88	62.2	62.2	48	104.7	104.7	08	147.1	147.1	68	189.5	189.5
29	20.5	20.5	89	62.9	62.9	49	105.4	105.4	09	147.8	147.8	69	190.2	190.2
30	21.2	21.2	90	63.6	63.6	50	106.1	106.1	10	148.5	148.5	70	190.9	190.9
31	21.9	21.9	91	64.3	64.3	151	106.8	106.8	211	149.2	149.2	271	191.6	191.6
32	22.6	22.6	92	65.1	65.1	52	107.5	107.5	12	149.9	149.9	72	192.3	192.3
33	23.3	23.3	93	65.8	65.8	53	108.2	108.2	13	150.6	150.6	73	193.0	193.0
34	24.0	24.0	94	66.5	66.5	54	108.9	108.9	14	151.3	151.3	74	193.7	193.7
35	24.7	24.7	95	67.2	67.2	55	109.6	109.6	15	152.0	152.0	75	194.5	194.5
36	25.5	25.5	96	67.9	67.9	56	110.3	110.3	16	152.7	152.7	76	195.2	195.2
37	26.2	26.2	97	68.6	68.6	57	111.0	111.0	17	153.4	153.4	77	195.9	195.9
38	26.9	26.9	98	69.3	69.3	58	111.7	111.7	18	154.1	154.1	78	196.6	196.6
39	27.6	27.6	99	70.0	70.0	59	112.4	112.4	19	154.9	154.9	79	197.3	197.3
40	28.3	28.3	100	70.7	70.7	60	113.1	113.1	20	155.6	155.6	80	198.0	198.0
41	29.0	29.0	101	71.4	71.4	161	113.8	113.8	221	156.3	156.3	281	198.7	198.7
42	29.7	29.7	02	72.1	72.1	62	114.6	114.6	22	157.0	157.0	82	199.4	199.4
43	30.4	30.4	03	72.8	72.8	63	115.3	115.3	23	157.7	157.7	83	200.1	200.1
44	31.1	31.1	04	73.5	73.5	64	116.0	116.0	24	158.4	158.4	84	200.8	200.8
45	31.8	31.8	05	74.2	74.2	65	116.7	116.7	25	159.1	159.1	85	201.5	201.5
46	32.5	32.5	06	75.0	75.0	66	117.4	117.4	26	159.8	159.8	86	202.2	202.2
47	33.2	33.2	07	75.7	75.7	67	118.1	118.1	27	160.5	160.5	87	202.9	202.9
48	33.9	33.9	08	76.4	76.4	68	118.8	118.8	28	161.2	161.2	88	203.6	203.6
49	34.6	34.6	09	77.1	77.1	69	119.5	119.5	29	161.9	161.9	89	204.4	204.4
50	35.4	35.4	10	77.8	77.8	70	120.2	120.2	30	162.6	162.6	90	205.1	205.1
51	36.1	36.1	111	78.5	78.5	171	120.9	120.9	231	163.3	163.3	291	205.8	205.8
52	36.8	36.8	12	79.2	79.2	72	121.6	121.6	32	164.0	164.0	92	206.5	206.5
53	37.5	37.5	13	79.9	79.9	73	122.3	122.3	33	164.8	164.8	93	207.2	207.2
54	38.2	38.2	14	80.6	80.6	74	123.0	123.0	34	165.5	165.5	94	207.9	207.9
55	38.9	38.9	15	81.3	81.3	75	123.7	123.7	35	166.2	166.2	95	208.6	208.6
56	39.6	39.6	16	82.0	82.0	76	124.5	124.5	36	166.9	166.9	96	209.3	209.3
57	40.3	40.3	17	82.7	82.7	77	125.2	125.2	37	167.6	167.6	97	210.0	210.0
58	41.0	41.0	18	83.4	83.4	78	125.9	125.9	38	168.3	168.3	98	210.7	210.7
59	41.7	41.7	19	84.1	84.1	79	126.6	126.6	39	169.0	169.0	99	211.4	211.4
60	42.4	42.4	20	84.9	84.9	80	127.3	127.3	40	169.7	169.7	300	212.1	212.1
Dist.	Dep.	D. Lat.	Dist.	Dep.	D. Lat.	Dist.	Dep.	D. Lat.	Dist.	Dep.	D. Lat.	Dist.	Dep.	D. Lat.

315° | 045°
225° | 135°

45°

Dist.	D. Lat.	Dep.
N.	N×Cos.	N×Sin.
Hypotenuse	Side Adj.	Side Opp.

315°	045°
225°	135°

TABLE 3

Traverse **45°** Table

315°	045°
225°	135°

Dist.	D. Lat.	Dep.	Dist.	D. Lat.	Dep.	Dist.	D. Lat.	Dep.	Dist.	D. Lat.	Dep.	Dist.	D. Lat.	Dep.
301	212.8	212.8	361	255.3	255.3	421	297.7	297.7	481	340.1	340.1	541	382.5	382.5
02	213.5	213.5	62	256.0	256.0	22	298.4	298.4	82	340.8	340.8	42	383.2	383.3
03	214.3	214.3	63	256.7	256.7	23	299.1	299.1	83	341.5	341.5	43	383.9	384.0
04	215.0	215.0	64	257.4	257.4	24	299.8	299.8	84	342.2	342.2	44	384.7	384.7
05	215.7	215.7	65	258.1	258.1	25	300.5	300.5	85	342.9	342.9	45	385.4	385.4
06	216.4	216.4	66	258.8	258.8	26	301.2	301.2	86	343.7	343.6	46	386.1	386.1
07	217.1	217.1	67	259.5	259.5	27	301.9	301.9	87	344.4	344.3	47	386.8	386.8
08	217.8	217.8	68	260.2	260.2	28	302.6	302.6	88	345.1	345.1	48	387.5	387.5
09	218.5	218.5	69	260.9	260.9	29	303.4	303.3	89	345.8	345.8	49	388.2	388.2
10	219.2	219.2	70	261.6	261.6	30	304.1	304.1	90	346.5	346.5	50	388.9	388.9
311	219.9	219.9	371	262.3	262.3	431	304.8	304.8	491	347.2	347.2	551	389.6	389.6
12	220.6	220.6	72	263.0	263.0	32	305.5	305.5	92	347.9	347.9	52	390.3	390.3
13	221.3	221.3	73	263.8	263.8	33	306.2	306.2	93	348.6	348.6	53	391.0	391.0
14	222.0	222.0	74	264.5	264.5	34	306.9	306.9	94	349.3	349.3	54	391.7	391.7
15	222.7	222.7	75	265.2	265.2	35	307.6	307.6	95	350.0	350.0	55	392.4	392.4
16	223.4	223.4	76	265.9	265.9	36	308.3	308.3	96	350.7	350.7	56	393.1	393.2
17	224.2	224.2	77	266.6	266.6	37	309.0	309.0	97	351.4	351.4	57	393.9	393.9
18	224.9	224.9	78	267.3	267.3	38	309.7	309.7	98	352.1	352.1	58	394.6	394.6
19	225.6	225.6	79	268.0	268.0	39	310.4	310.4	99	352.8	352.8	59	395.3	395.3
20	226.3	226.3	80	268.7	268.7	40	311.1	311.1	500	353.6	353.6	60	396.0	396.0
321	227.0	227.0	381	269.4	269.4	441	311.8	311.8	501	354.3	354.3	561	396.7	396.7
22	227.7	227.7	82	270.1	270.1	42	312.5	312.5	02	355.0	355.0	62	397.4	397.4
23	228.4	228.4	83	270.8	270.8	43	313.3	313.2	03	355.7	355.7	63	398.1	398.1
24	229.1	229.1	84	271.5	271.5	44	314.0	314.0	04	356.4	356.4	64	398.8	398.8
25	229.8	229.8	85	272.2	272.2	45	314.7	314.7	05	357.1	357.1	65	399.5	399.5
26	230.5	230.5	86	272.9	272.9	46	315.4	315.4	06	357.8	357.8	66	400.2	400.2
27	231.2	231.2	87	273.7	273.7	47	316.1	316.1	07	358.5	358.5	67	400.9	400.9
28	231.9	231.9	88	274.4	274.4	48	316.8	316.8	08	359.2	359.2	68	401.6	401.6
29	232.6	232.6	89	275.1	275.1	49	317.5	317.5	09	359.9	359.9	69	402.3	402.3
30	233.3	233.3	90	275.8	275.8	50	318.2	318.2	10	360.6	360.6	70	403.0	403.1
331	234.1	234.1	391	276.5	276.5	451	318.9	318.9	511	361.3	361.3	571	403.8	403.8
32	234.8	234.8	92	277.2	277.2	52	319.6	319.6	12	362.0	362.0	72	404.5	404.5
33	235.5	235.5	93	277.9	277.9	53	320.3	320.3	13	362.7	362.7	73	405.2	405.2
34	236.2	236.2	94	278.6	278.6	54	321.0	321.0	14	363.5	363.5	74	405.9	405.9
35	236.9	236.9	95	279.3	279.3	55	321.7	321.7	15	364.2	364.2	75	406.6	406.6
36	237.6	237.6	96	280.0	280.0	56	322.4	322.4	16	364.9	364.9	76	407.3	407.3
37	238.3	238.3	97	280.7	280.7	57	323.2	323.1	17	365.6	365.6	77	408.0	408.0
38	239.0	239.0	98	281.4	281.4	58	323.9	323.9	18	366.3	366.3	78	408.7	408.7
39	239.7	239.7	99	282.1	282.1	59	324.6	324.6	19	367.0	367.0	79	409.4	409.4
40	240.4	240.4	400	282.8	282.8	60	325.3	325.3	20	367.7	367.7	80	410.1	410.1
341	241.1	241.1	401	283.6	283.5	461	326.0	326.0	521	368.4	368.4	581	410.8	410.8
42	241.8	241.8	02	284.3	284.3	62	326.7	326.7	22	369.1	369.1	82	411.5	411.5
43	242.5	242.5	03	285.0	285.0	63	327.4	327.4	23	369.8	369.8	83	412.2	412.2
44	243.2	243.2	04	285.7	285.7	64	328.1	328.1	24	370.5	370.5	84	412.9	413.0
45	244.0	244.0	05	286.4	286.4	65	328.8	328.8	25	371.2	371.2	85	413.7	413.7
46	244.7	244.7	06	287.1	287.1	66	329.5	329.5	26	371.9	371.9	86	414.4	414.4
47	245.4	245.4	07	287.8	287.8	67	330.2	330.2	27	372.6	372.6	87	415.1	415.1
48	246.1	246.1	08	288.5	288.5	68	330.9	330.9	28	373.4	373.4	88	415.8	415.8
49	246.8	246.8	09	289.2	289.2	69	331.6	331.6	29	374.1	374.1	89	416.5	416.5
50	247.5	247.5	10	289.9	289.9	70	332.3	332.3	30	374.8	374.8	90	417.2	417.2
351	248.2	248.2	411	290.6	290.6	471	333.1	333.0	531	375.5	375.5	591	417.9	417.9
52	248.9	248.9	12	291.3	291.3	72	333.8	333.8	32	376.2	376.2	92	418.6	418.6
53	249.6	249.6	13	292.0	292.0	73	334.5	334.5	33	376.9	376.9	93	419.3	419.3
54	250.3	250.3	14	292.7	292.7	74	335.2	335.2	34	377.6	377.6	94	420.0	420.0
55	251.0	251.0	15	293.5	293.4	75	335.9	335.9	35	378.3	378.3	95	420.7	420.7
56	251.7	251.7	16	294.2	294.2	76	336.6	336.6	36	379.0	379.0	96	421.4	421.4
57	252.4	252.4	17	294.9	294.9	77	337.3	337.3	37	379.7	379.7	97	422.1	422.1
58	253.1	253.1	18	295.6	295.6	78	338.0	338.0	38	380.4	380.4	98	422.8	422.8
59	253.9	253.9	19	296.3	296.3	79	338.7	338.7	39	381.1	381.1	99	423.6	423.6
60	254.6	254.6	20	297.0	297.0	80	339.4	339.4	40	381.8	381.8	600	424.3	424.3
Dist.	Dep.	D. Lat.	Dist.	Dep.	D. Lat.	Dist.	Dep.	D. Lat.	Dist.	Dep.	D. Lat.	Dist.	Dep.	D. Lat.

Dist.	D. Lat.	Dep.
D Lo	Dep.	
	m	D Lo

45°

315°	045°
225°	135°

TABLE 4

Table of Offsets

ALT.	DISTANCE ALONG POSITION LINE FROM INTERCEPT										ALT.
	00′	05′	10′	15′	20′	25′	30′	35′	40′	45′	
	OFFSETS										
0°	0′.0	0′.0	0′.0	0′.0	0′.0	0′.0	0′.0	0′.0	0′.0	0′.0	0°
30	0.0	0.0	0.0	0.0	0.0	0.1	0.1	0.1	0.1	0.2	30
40	0.0	0.0	0.0	0.0	0.1	0.1	0.1	0.2	0.2	0.3	40
50	0.0	0.0	0.0	0.0	0.1	0.1	0.2	0.2	0.3	0.3	50
55	0.0	0.0	0.0	0.0	0.1	0.1	0.2	0.3	0.3	0.4	55
60	0.0	0.0	0.0	0.1	0.1	0.2	0.2	0.3	0.4	0.5	60
62	0.0	0.0	0.0	0.1	0.1	0.2	0.2	0.3	0.4	0.5	62
64	0.0	0.0	0.0	0.1	0.1	0.2	0.3	0.4	0.5	0.6	64
66	0.0	0.0	0.0	0.1	0.1	0.2	0.3	0.4	0.5	0.7	66
68	0.0	0.0	0.0	0.1	0.1	0.2	0.3	0.4	0.6	0.7	68
70	0.0	0.0	0.0	0.1	0.2	0.2	0.4	0.5	0.6	0.8	70
71	0.0	0.0	0.0	0.1	0.2	0.3	0.4	0.5	0.7	0.9	71
72	0.0	0.0	0.0	0.1	0.2	0.3	0.4	0.5	0.7	0.9	72
73	0.0	0.0	0.0	0.1	0.2	0.3	0.4	0.6	0.8	1.0	73
74	0.0	0.0	0.1	0.1	0.2	0.3	0.5	0.6	0.8	1.0	74
75	0.0	0.0	0.1	0.1	0.2	0.3	0.5	0.7	0.9	1.1	75
76	0.0	0.0	0.1	0.1	0.2	0.4	0.5	0.7	0.9	1.2	76
77	0.0	0.0	0.1	0.1	0.3	0.4	0.6	0.8	1.0	1.3	77
78	0.0	0.0	0.1	0.2	0.3	0.4	0.6	0.8	1.1	1.4	78
79	0.0	0.0	0.1	0.2	0.3	0.5	0.7	0.9	1.2	1.5	79
80.0	0.0	0.0	0.1	0.2	0.3	0.5	0.7	1.0	1.3	1.7	80.0
80.5	0.0	0.0	0.1	0.2	0.3	0.5	0.8	1.1	1.4	1.8	80.5
81.0	0.0	0.0	0.1	0.2	0.4	0.6	0.8	1.1	1.5	1.9	81.0
81.5	0.0	0.0	0.1	0.2	0.4	0.6	0.9	1.2	1.6	2.0	81.5
82.0	0.0	0.0	0.1	0.2	0.4	0.6	0.9	1.3	1.7	2.1	82.0
82.5	0.0	0.0	0.1	0.2	0.4	0.7	1.0	1.4	1.8	2.2	82.5
83.0	0.0	0.0	0.1	0.3	0.5	0.7	1.1	1.5	1.9	2.4	83.0
83.5	0.0	0.0	0.1	0.3	0.5	0.8	1.2	1.6	2.0	2.6	83.5
84.0	0.0	0.0	0.1	0.3	0.5	0.9	1.2	1.7	2.2	2.8	84.0
84.5	0.0	0.0	0.2	0.3	0.6	1.0	1.4	1.9	2.4	3.1	84.5
85.0	0.0	0.0	0.2	0.4	0.7	1.0	1.5	2.1	2.7	3.4	85.0
85.5	0.0	0.0	0.2	0.4	0.7	1.2	1.7	2.3	3.0	3.8	85.5
86.0	0.0	0.1	0.2	0.5	0.8	1.3	1.9	2.6	3.4	4.3	86.0
86.5	0.0	0.1	0.2	0.5	1.0	1.5	2.2	2.9	3.8	4.9	86.5
87.0	0.0	0.1	0.3	0.6	1.1	1.7	2.5	3.4	4.5	5.7	87.0
87.5	0.0	0.1	0.3	0.8	1.3	2.1	3.0	4.1	5.4	6.9	87.5
88.0	0.0	0.1	0.4	0.9	1.7	2.7	3.8	5.2	6.9	8.8	88.0
88.5	0.0	0.2	0.6	1.3	2.3	3.5	5.1	7.1	9.4	12.1	88.5
89.0	0.0	0.3	0.8	1.9	3.4	5.5	8.0	11.3	15.3	20.3	89.0

TABLE 5

Meridional Parts

Lat.	0°	1°	2°	3°	4°	5°	6°	7°	8°	9°	Lat.
′											′
0	0.0	59.6	119.2	178.9	238.6	298.4	358.3	418.2	478.4	538.6	0
1	1.0	60.6	20.2	79.9	39.6	299.4	59.3	19.2	79.4	39.6	1
2	2.0	61.6	21.2	80.9	40.6	300.4	60.3	20.2	80.4	40.6	2
3	3.0	62.6	22.2	81.9	41.6	01.4	61.3	21.2	81.4	41.7	3
4	4.0	63.6	23.2	82.9	42.6	02.4	62.2	22.2	82.4	42.7	4
5	5.0	64.6	124.2	183.9	243.6	303.4	363.2	423.2	483.4	543.7	5
6	6.0	65.6	25.2	84.8	44.6	04.4	64.2	24.2	84.4	44.7	6
7	7.0	66.6	26.2	85.8	45.6	05.4	65.2	25.3	85.4	45.7	7
8	7.9	67.5	27.2	86.8	46.6	06.4	66.2	26.3	86.4	46.7	8
9	8.9	68.5	28.2	87.8	47.6	07.4	67.2	27.3	87.4	47.7	9
10	9.9	69.5	129.2	188.8	248.5	308.3	368.2	428.3	488.4	548.7	10
11	10.9	70.5	30.2	89.8	49.5	09.3	69.2	29.3	89.4	49.7	11
12	11.9	71.5	31.1	90.8	50.5	10.3	70.2	30.3	90.4	50.7	12
13	12.9	72.5	32.1	91.8	51.5	11.3	71.2	31.3	91.4	51.7	13
14	13.9	73.5	33.1	92.8	52.5	12.3	72.2	32.3	92.4	52.7	14
15	14.9	74.5	134.1	193.8	253.5	313.3	373.2	433.3	493.4	553.7	15
16	15.9	75.5	35.1	94.8	54.5	14.3	74.2	34.3	94.4	54.7	16
17	16.9	76.5	36.1	95.8	55.5	15.3	75.2	35.3	95.4	55.7	17
18	17.9	77.5	37.1	96.8	56.5	16.3	76.2	36.3	96.4	56.7	18
19	18.9	78.5	38.1	97.8	57.5	17.3	77.2	37.3	97.4	57.8	19
20	19.9	79.5	139.1	198.8	258.5	318.3	378.2	438.3	498.4	558.8	20
21	20.9	80.5	40.1	199.8	59.5	19.3	79.2	39.3	499.4	59.8	21
22	21.9	81.5	41.1	200.8	60.5	20.3	80.2	40.3	500.4	60.8	22
23	22.8	82.5	42.1	01.8	61.5	21.3	81.2	41.3	01.4	61.8	23
24	23.8	83.4	43.1	02.8	62.5	22.3	82.2	42.3	02.5	62.8	24
25	24.8	84.4	144.1	203.8	263.5	323.3	383.2	443.3	503.5	563.8	25
26	25.8	85.4	45.1	04.7	64.5	24.3	84.2	44.3	04.5	64.8	26
27	26.8	86.4	46.1	05.7	65.5	25.3	85.2	45.3	05.5	65.8	27
28	27.8	87.4	47.1	06.7	66.5	26.3	86.2	46.3	06.5	66.8	28
29	28.8	88.4	48.0	07.7	67.5	27.3	87.2	47.3	07.5	67.8	29
30	29.8	89.4	149.0	208.7	268.5	328.3	388.2	448.3	508.5	568.8	30
31	30.8	90.4	50.0	09.7	69.5	29.3	89.2	49.3	09.5	69.8	31
32	31.8	91.4	51.0	10.7	70.5	30.3	90.2	50.3	10.5	70.8	32
33	32.8	92.4	52.0	11.7	71.5	31.3	91.2	51.3	11.5	71.9	33
34	33.8	93.4	53.0	12.7	72.5	32.3	92.2	52.3	12.5	72.9	34
35	34.8	94.4	154.0	213.7	273.5	333.3	393.2	453.3	513.5	573.9	35
36	35.8	95.4	55.0	14.7	74.5	34.3	94.2	54.3	14.5	74.9	36
37	36.8	96.4	56.0	15.7	75.4	35.3	95.2	55.3	15.5	75.9	37
38	37.7	97.4	57.0	16.7	76.4	36.3	96.2	56.3	16.5	76.9	38
39	38.7	98.4	58.0	17.7	77.4	37.3	97.2	57.3	17.5	77.9	39
40	39.7	99.3	159.0	218.7	278.4	338.3	398.2	458.3	518.5	578.9	40
41	40.7	100.3	60.0	19.7	79.4	39.3	399.2	59.3	19.5	79.9	41
42	41.7	01.3	61.0	20.7	80.4	40.3	400.2	60.3	20.5	80.9	42
43	42.7	02.3	62.0	21.7	81.4	41.3	01.2	61.3	21.5	81.9	43
44	43.7	03.3	63.0	22.7	82.4	42.3	02.2	62.3	22.5	82.9	44
45	44.7	104.3	164.0	223.7	283.4	343.3	403.2	463.3	523.6	583.9	45
46	45.7	05.3	65.0	24.7	84.4	44.3	04.2	64.3	24.6	85.0	46
47	46.7	06.3	65.9	25.6	85.4	45.3	05.2	65.3	25.6	86.0	47
48	47.7	07.3	66.9	26.6	86.4	46.3	06.2	66.3	26.6	87.0	48
49	48.7	08.3	67.9	27.6	87.4	47.3	07.2	67.3	27.6	88.0	49
50	49.7	109.3	168.9	228.6	288.4	348.3	408.2	468.3	528.6	589.0	50
51	50.7	10.3	69.9	29.6	89.4	49.3	09.2	69.3	29.6	90.0	51
52	51.7	11.3	70.9	30.6	90.4	50.3	10.2	70.3	30.6	91.0	52
53	52.6	12.3	71.9	31.6	91.4	51.3	11.2	71.3	31.6	92.0	53
54	53.6	13.3	72.9	32.6	92.4	52.3	12.2	72.3	32.6	93.0	54
55	54.6	114.3	173.9	233.6	293.4	353.3	413.2	473.3	533.6	594.0	55
56	55.6	15.2	74.9	34.6	94.4	54.3	14.2	74.4	34.6	95.0	56
57	56.6	16.2	75.9	35.6	95.4	55.3	15.2	75.4	35.6	96.0	57
58	57.6	17.2	76.9	36.6	96.4	56.3	16.2	76.4	36.6	97.1	58
59	58.6	18.2	77.9	37.6	97.4	57.3	17.2	77.4	37.6	98.1	59
60	59.6	119.2	178.9	238.6	298.4	358.3	418.2	478.4	538.6	599.1	60
Lat.	0°	1°	2°	3°	4°	5°	6°	7°	8°	9°	Lat.

TABLE 5

Meridional Parts

Lat.	10°	11°	12°	13°	14°	15°	16°	17°	18°	19°	Lat.
′											′
0	599. 1	659. 7	720. 5	781. 6	842. 9	904. 5	966. 4	1028. 6	1091. 1	1154. 0	0
1	600. 1	60. 7	21. 6	82. 6	43. 9	05. 5	67. 4	29. 6	92. 1	55. 0	1
2	01. 1	61. 7	22. 6	83. 6	45. 0	06. 6	68. 4	30. 7	93. 2	56. 1	2
3	02. 1	62. 7	23. 6	84. 7	46. 0	07. 6	69. 5	31. 7	94. 2	57. 1	3
4	03. 1	63. 7	24. 6	85. 7	47. 0	08. 6	70. 5	32. 7	95. 3	58. 2	4
5	604. 1	664. 8	725. 6	786. 7	848. 0	909. 6	971. 6	1033. 8	1096. 3	1159. 3	5
6	05. 1	65. 8	26. 6	87. 7	49. 1	10. 7	72. 6	34. 8	97. 4	60. 3	6
7	06. 1	66. 8	27. 6	88. 7	50. 1	11. 7	73. 6	35. 9	98. 4	61. 4	7
8	07. 1	67. 8	28. 7	89. 8	51. 1	12. 7	74. 7	36. 9	1099. 5	62. 4	8
9	08. 2	68. 8	29. 7	90. 8	52. 1	13. 8	75. 7	37. 9	1100. 5	63. 5	9
10	609. 2	669. 8	730. 7	791. 8	853. 2	914. 8	976. 7	1039. 0	1101. 6	1164. 5	10
11	10. 2	70. 8	31. 7	92. 8	54. 2	15. 8	77. 8	40. 0	02. 6	65. 6	11
12	11. 2	71. 9	32. 7	93. 8	55. 2	16. 9	78. 8	41. 1	03. 7	66. 6	12
13	12. 2	72. 9	33. 7	94. 9	56. 2	17. 9	79. 8	42. 1	04. 7	67. 7	13
14	13. 2	73. 9	34. 8	95. 9	57. 3	18. 9	80. 9	43. 1	05. 7	68. 7	14
15	614. 2	674. 9	735. 8	796. 9	858. 3	919. 9	981. 9	1044. 2	1106. 8	1169. 8	15
16	15. 2	75. 9	36. 8	97. 9	59. 3	21. 0	82. 9	45. 2	07. 8	70. 8	16
17	16. 2	76. 9	37. 8	798. 9	60. 3	22. 0	84. 0	46. 3	08. 9	71. 9	17
18	17. 2	77. 9	38. 8	800. 0	61. 4	23. 0	85. 0	47. 3	09. 9	72. 9	18
19	18. 3	78. 9	39. 8	01. 0	62. 4	24. 1	86. 0	48. 3	11. 0	74. 0	19
20	619. 3	680. 0	740. 9	802. 0	863. 4	925. 1	987. 1	1049. 4	1112. 0	1175. 0	20
21	20. 3	81. 0	41. 9	03. 0	64. 4	26. 1	88. 1	50. 4	13. 1	76. 1	21
22	21. 3	82. 0	42. 9	04. 1	65. 5	27. 2	89. 1	51. 5	14. 1	77. 1	22
23	22. 3	83. 0	43. 9	05. 1	66. 5	28. 2	90. 2	52. 5	15. 2	78. 2	23
24	23. 3	84. 0	44. 9	06. 1	67. 5	29. 2	91. 2	53. 5	16. 2	79. 3	24
25	624. 3	685. 0	746. 0	807. 1	868. 5	930. 2	992. 3	1054. 6	1117. 3	1180. 3	25
26	25. 3	86. 0	47. 0	08. 1	69. 6	31. 3	93. 3	55. 6	18. 3	81. 4	26
27	26. 3	87. 1	48. 0	09. 2	70. 6	32. 3	94. 3	56. 7	19. 4	82. 4	27
28	27. 3	88. 1	49. 0	10. 2	71. 6	33. 3	95. 4	57. 7	20. 4	83. 5	28
29	28. 4	89. 1	50. 0	11. 2	72. 6	34. 4	96. 4	58. 8	21. 5	84. 5	29
30	629. 4	690. 1	751. 0	812. 2	873. 7	935. 4	997. 4	1059. 8	1122. 5	1185. 6	30
31	30. 4	91. 1	52. 1	13. 2	74. 7	36. 4	98. 5	60. 8	23. 5	86. 6	31
32	31. 4	92. 1	53. 1	14. 3	75. 7	37. 5	999. 5	61. 9	24. 6	87. 7	32
33	32. 4	93. 1	54. 1	15. 3	76. 8	38. 5	1000. 5	62. 9	25. 6	88. 7	33
34	33. 4	94. 1	55. 1	16. 3	77. 8	39. 5	01. 6	64. 0	26. 7	89. 8	34
35	634. 4	695. 2	756. 1	817. 3	878. 8	940. 6	1002. 6	1065. 0	1127. 7	1190. 9	35
36	35. 4	96. 2	57. 1	18. 4	79. 8	41. 6	03. 7	66. 0	28. 8	91. 9	36
37	36. 4	97. 2	58. 2	19. 4	80. 9	42. 6	04. 7	67. 1	29. 8	93. 0	37
38	37. 4	98. 2	59. 2	20. 4	81. 9	43. 7	05. 7	68. 1	30. 9	94. 0	38
39	38. 5	99. 2	60. 2	21. 4	82. 9	44. 7	06. 8	69. 2	31. 9	95. 1	39
40	639. 5	700. 2	761. 2	822. 5	883. 9	945. 7	1007. 8	1070. 2	1133. 0	1196. 1	40
41	40. 5	01. 2	62. 2	23. 5	85. 0	46. 8	08. 8	71. 3	34. 0	97. 2	41
42	41. 5	02. 3	63. 3	24. 5	86. 0	47. 8	09. 9	72. 3	35. 1	98. 2	42
43	42. 5	03. 3	64. 3	25. 5	87. 0	48. 8	10. 9	73. 4	36. 1	1199. 3	43
44	43. 5	04. 3	65. 3	26. 5	88. 1	49. 9	12. 0	74. 4	37. 2	1200. 4	44
45	644. 5	705. 3	766. 3	827. 6	889. 1	950. 9	1013. 0	1075. 4	1138. 2	1201. 4	45
46	45. 5	06. 3	67. 3	28. 6	90. 1	51. 9	14. 0	76. 5	39. 3	02. 5	46
47	46. 5	07. 3	68. 4	29. 6	91. 1	52. 9	15. 1	77. 5	40. 3	03. 5	47
48	47. 6	08. 4	69. 4	30. 6	92. 2	54. 0	16. 1	78. 6	41. 4	04. 6	48
49	48. 6	09. 4	70. 4	31. 7	93. 2	55. 0	17. 1	79. 6	42. 4	05. 6	49
50	649. 6	710. 4	771. 4	832. 7	894. 2	956. 0	1018. 2	1080. 7	1143. 5	1206. 7	50
51	50. 6	11. 4	72. 4	33. 7	95. 2	57. 1	19. 2	81. 7	44. 5	07. 7	51
52	51. 6	12. 4	73. 4	34. 7	96. 3	58. 1	20. 3	82. 7	45. 6	08. 8	52
53	52. 6	13. 4	74. 5	35. 8	97. 3	59. 1	21. 3	83. 8	46. 6	09. 9	53
54	53. 6	14. 4	75. 5	36. 8	98. 3	60. 2	22. 3	84. 8	47. 7	10. 9	54
55	654. 6	715. 5	776. 5	837. 8	899. 4	961. 2	1023. 4	1085. 9	1148. 7	1212. 0	55
56	55. 7	16. 5	77. 5	38. 8	900. 4	62. 2	24. 4	86. 9	49. 8	13. 0	56
57	56. 7	17. 5	78. 5	39. 8	01. 4	63. 3	25. 5	88. 0	50. 8	14. 1	57
58	57. 7	18. 5	79. 6	40. 9	02. 4	64. 3	26. 5	89. 0	51. 9	15. 2	58
59	58. 7	19. 5	80. 6	41. 9	03. 5	65. 3	27. 5	90. 1	52. 9	16. 2	59
60	659. 7	720. 5	781. 6	842. 9	904. 5	966. 4	1028. 6	1091. 1	1154. 0	1217. 3	60
Lat.	10°	11°	12°	13°	14°	15°	16°	17°	18°	19°	Lat.

TABLE 5

Meridional Parts

Lat.	20°	21°	22°	23°	24°	25°	26°	27°	28°	29°	Lat.
′											′
0	1217.3	1280.9	1345.1	1409.6	1474.7	1540.3	1606.4	1673.1	1740.4	1808.3	0
1	18.3	82.0	46.1	10.7	75.8	41.4	07.5	74.2	41.5	09.4	1
2	19.4	83.1	47.2	11.8	76.9	42.5	08.6	75.3	42.6	10.5	2
3	20.4	84.1	48.3	12.9	78.0	43.6	09.7	76.4	43.7	11.7	3
4	21.5	85.2	49.4	14.0	79.0	44.7	10.8	77.5	44.9	12.8	4
5	1222.6	1286.3	1350.4	1415.0	1480.1	1545.8	1611.9	1678.6	1746.0	1814.0	5
6	23.6	87.3	51.5	16.1	81.2	46.9	13.2	79.8	47.1	15.1	6
7	24.7	88.4	52.6	17.2	82.3	48.0	14.1	80.9	48.2	16.2	7
8	25.7	89.5	53.6	18.3	83.4	49.0	15.2	82.0	49.4	17.4	8
9	26.8	90.5	54.7	19.4	84.5	50.1	16.3	83.1	50.5	18.5	9
10	1227.9	1291.6	1355.8	1420.4	1485.6	1551.2	1617.5	1684.2	1751.6	1819.7	10
11	28.9	92.7	56.9	21.5	86.7	52.3	18.6	85.4	52.8	20.8	11
12	30.0	93.7	57.9	22.6	87.8	53.4	19.7	86.5	53.9	21.9	12
13	31.0	94.8	59.0	23.7	88.9	54.5	20.8	87.6	55.0	23.1	13
14	32.1	95.9	60.1	24.8	89.9	55.6	21.9	88.7	56.1	24.2	14
15	1233.1	1296.9	1361.2	1425.9	1491.0	1556.7	1623.0	1689.8	1757.3	1825.4	15
16	34.2	98.0	62.2	26.9	92.1	57.8	24.1	90.9	58.4	26.5	16
17	35.3	1299.1	63.3	28.0	93.2	58.9	25.2	92.1	59.5	27.6	17
18	36.3	1300.1	64.4	29.1	94.3	60.0	26.3	93.2	60.7	28.8	18
19	37.4	01.2	65.5	30.2	95.4	61.1	27.4	94.3	61.8	29.9	19
20	1238.4	1302.3	1366.5	1431.3	1496.5	1562.2	1628.5	1695.4	1762.9	1831.1	20
21	39.5	03.3	67.6	32.3	97.6	63.3	29.7	96.5	64.1	32.2	21
22	40.6	04.4	68.7	33.4	98.7	64.4	30.8	97.7	65.2	33.3	22
23	41.6	05.5	69.8	34.5	1499.8	65.5	31.9	98.8	66.3	34.5	23
24	42.7	06.5	70.8	35.6	1500.9	66.6	33.0	1699.9	67.4	35.6	24
25	1243.7	1307.6	1371.9	1436.7	1502.0	1567.7	1634.1	1701.0	1768.6	1836.8	25
26	44.8	08.7	73.0	37.8	03.0	68.8	35.2	02.1	69.7	37.9	26
27	45.9	09.7	74.1	38.8	04.1	69.9	36.3	03.3	70.8	39.1	27
28	46.9	10.8	75.1	39.9	05.2	71.0	37.4	04.4	72.0	40.2	28
29	48.0	11.9	76.2	41.0	06.3	72.2	38.5	05.5	73.1	41.3	29
30	1249.1	1312.9	1377.3	1442.1	1507.4	1573.3	1639.6	1706.6	1774.2	1842.5	30
31	50.1	14.0	78.4	43.2	08.5	74.4	40.8	07.8	75.4	43.6	31
32	51.2	15.1	79.4	44.3	09.6	75.5	41.9	08.9	76.5	44.8	32
33	52.2	16.2	80.5	45.4	10.7	76.6	43.0	10.0	77.6	45.9	33
34	53.3	17.2	81.6	46.4	11.8	77.7	44.1	11.1	78.8	47.1	34
35	1254.4	1318.3	1382.7	1447.5	1512.9	1578.8	1645.2	1712.2	1779.9	1848.2	35
36	55.4	19.4	83.7	48.6	14.0	79.9	46.3	13.4	81.0	49.3	36
37	56.5	20.4	84.8	49.7	15.1	81.0	47.4	14.5	82.2	50.5	37
38	57.5	21.5	85.9	50.8	16.2	82.1	48.5	15.6	83.3	51.6	38
39	58.6	22.6	87.0	51.9	17.3	83.2	49.7	16.7	84.4	52.8	39
40	1259.7	1323.6	1388.1	1453.0	1518.3	1584.3	1650.8	1717.9	1785.6	1853.9	40
41	60.7	24.7	89.1	54.0	19.4	85.4	51.9	19.0	86.7	55.1	41
42	61.8	25.8	90.2	55.1	20.5	86.5	53.0	20.1	87.8	56.2	42
43	62.9	26.9	91.3	56.2	21.6	87.6	54.1	21.2	89.0	57.4	43
44	63.9	27.9	92.4	57.3	22.7	88.7	55.2	22.3	90.1	58.5	44
45	1265.0	1329.0	1393.4	1458.4	1523.8	1589.8	1656.3	1723.5	1791.2	1859.7	45
46	66.1	30.1	94.5	59.5	24.9	90.9	57.5	24.6	92.4	60.8	46
47	67.1	31.1	95.6	60.6	26.0	92.0	58.6	25.7	93.5	61.9	47
48	68.2	32.2	96.7	61.6	27.1	93.1	59.7	26.8	94.6	63.1	48
49	69.2	33.3	97.8	62.7	28.2	94.2	60.8	28.0	95.8	64.2	49
50	1270.3	1334.3	1398.8	1463.8	1529.3	1595.3	1661.9	1729.1	1796.9	1865.4	50
51	71.4	35.4	1399.9	64.9	30.4	96.4	63.0	30.2	98.0	66.5	51
52	72.4	36.5	1401.0	66.0	31.5	97.5	64.1	31.3	1799.2	67.7	52
53	73.5	37.6	02.1	67.1	32.6	98.6	65.3	32.5	1800.3	68.8	53
54	74.6	38.6	03.2	68.2	33.7	1599.7	66.4	33.6	01.5	70.0	54
55	1275.6	1339.7	1404.2	1469.3	1534.8	1600.8	1667.5	1734.7	1802.6	1871.1	55
56	76.7	40.8	05.3	70.3	35.9	02.0	68.6	35.8	03.7	72.3	56
57	77.8	41.8	06.4	71.4	37.0	03.1	69.7	37.0	04.9	73.4	57
58	78.8	42.9	07.5	72.5	38.1	04.2	70.8	38.1	06.0	74.6	58
59	79.9	44.0	08.6	73.6	39.2	05.3	71.9	39.2	07.1	75.7	59
60	1280.9	1345.1	1409.6	1474.7	1540.3	1606.4	1673.1	1740.4	1808.3	1876.9	60
Lat.	20°	21°	22°	23°	24°	25°	26°	27°	28°	29°	Lat.

TABLE 5

Meridional Parts

Lat.	30°	31°	32°	33°	34°	35°	36°	37°	38°	39°	Lat.
′											′
0	1876.9	1946.2	2016.2	2087.0	2158.6	2231.1	2304.5	2378.8	2454.1	2530.4	0
1	78.0	47.3	17.4	88.2	59.8	32.3	05.7	80.0	55.3	31.7	1
2	79.2	48.5	18.5	89.4	61.0	33.5	06.9	81.3	56.6	33.0	2
3	80.3	49.6	19.7	90.5	62.2	34.7	08.1	82.5	57.9	34.3	3
4	81.5	50.8	20.9	91.7	63.4	35.9	09.4	83.8	59.1	35.6	4
5	1882.6	1952.0	2022.1	2092.9	2164.6	2237.2	2310.6	2385.0	2460.4	2536.8	5
6	83.8	53.1	23.2	94.1	65.8	38.4	11.8	85.3	61.7	38.1	6
7	84.9	54.3	24.4	95.3	67.0	39.6	13.1	87.5	62.9	39.4	7
8	86.1	55.4	25.6	96.5	68.2	40.8	14.3	88.8	64.2	40.7	8
9	87.2	56.6	26.8	97.7	69.4	42.0	15.5	90.0	65.5	42.0	9
10	1888.4	1957.8	2027.9	2098.9	2170.6	2243.2	2316.8	2391.3	2466.7	2543.3	10
11	89.5	58.9	29.1	2100.1	71.8	44.5	18.0	92.5	68.0	44.5	11
12	90.7	60.1	30.3	01.2	73.0	45.7	19.2	93.8	69.3	45.8	12
13	91.8	61.3	31.5	02.4	74.2	46.9	20.5	95.0	70.5	47.1	13
14	93.0	62.4	32.6	03.6	75.4	48.1	21.7	96.3	71.8	48.4	14
15	1894.1	1963.6	2033.8	2104.8	2176.6	2249.3	2322.9	2397.5	2473.1	2549.7	15
16	95.3	64.8	35.0	06.0	77.8	50.6	24.2	2398.8	74.3	51.0	16
17	96.4	65.9	36.2	07.2	79.0	51.8	25.4	2400.0	75.6	52.3	17
18	97.6	67.1	37.3	08.4	80.3	53.0	26.6	01.3	76.9	53.6	18
19	98.7	68.2	38.5	09.6	81.5	54.2	27.9	02.5	78.1	54.8	19
20	1899.9	1969.4	2039.7	2110.8	2182.7	2255.4	2329.1	2403.8	2479.4	2556.1	20
21	1901.0	70.6	40.9	12.0	83.9	56.7	30.4	05.0	80.7	57.4	21
22	02.2	71.7	42.1	13.1	85.1	57.9	31.6	06.3	82.0	58.7	22
23	03.3	72.9	43.2	14.3	86.3	59.1	32.8	07.5	83.2	60.0	23
24	04.5	74.1	44.4	15.5	87.5	60.3	34.1	08.8	84.5	61.3	24
25	1905.6	1975.2	2045.6	2116.7	2188.7	2261.5	2335.3	2410.0	2485.8	2562.6	25
26	06.8	76.4	46.8	17.9	89.9	62.8	36.5	11.3	87.0	63.9	26
27	08.0	77.6	47.9	19.1	91.1	64.0	37.8	12.5	88.3	65.1	27
28	09.1	78.7	49.1	20.3	92.3	65.2	39.0	13.8	89.6	66.4	28
29	10.3	79.9	50.3	21.5	93.5	66.4	40.3	15.0	90.9	67.7	29
30	1911.4	1981.1	2051.5	2122.7	2194.7	2267.6	2341.5	2416.3	2492.1	2569.0	30
31	12.6	82.2	52.7	23.9	95.9	68.9	42.7	17.6	93.4	70.3	31
32	13.7	83.4	53.8	25.1	97.1	70.1	44.0	18.8	94.7	71.6	32
33	14.9	84.6	55.0	26.3	98.4	71.3	45.2	20.1	95.9	72.9	33
34	16.0	85.7	56.2	27.5	2199.6	72.5	46.4	21.3	97.2	74.2	34
35	1917.2	1986.9	2057.4	2128.7	2200.8	2273.8	2347.7	2422.6	2498.5	2575.5	35
36	18.4	88.1	58.6	29.9	02.0	75.0	48.9	23.8	2499.8	76.8	36
37	19.5	89.2	59.7	31.1	03.2	76.2	50.2	25.1	2501.0	78.1	37
38	20.7	90.4	60.9	32.2	04.4	77.4	51.4	26.3	02.3	79.4	38
39	21.8	91.6	62.1	33.4	05.6	78.7	52.6	27.6	03.6	80.6	39
40	1923.0	1992.8	2063.3	2134.6	2206.8	2279.9	2353.9	2428.9	2504.9	2581.9	40
41	24.1	93.9	64.5	35.8	08.0	81.1	55.1	30.1	06.1	83.2	41
42	25.3	95.1	65.7	37.0	09.2	82.3	56.4	31.4	07.4	84.5	42
43	26.4	96.3	66.8	38.2	10.5	83.6	57.6	32.6	08.7	85.8	43
44	27.6	97.4	68.0	39.4	11.7	84.8	58.9	33.9	10.0	87.1	44
45	1928.8	1998.6	2069.2	2140.6	2212.9	2286.0	2360.1	2435.2	2511.2	2588.4	45
46	29.9	1999.8	70.4	41.8	14.1	87.2	61.3	36.4	12.5	89.7	46
47	31.1	2000.9	71.6	43.0	15.3	88.5	62.6	37.7	13.8	91.0	47
48	32.2	02.1	72.8	44.2	16.5	89.7	63.8	38.9	15.1	92.3	48
49	33.4	03.3	73.9	45.4	17.7	90.9	65.1	40.2	16.4	93.6	49
50	1934.6	2004.5	2075.1	2146.6	2218.9	2292.2	2366.3	2441.5	2517.6	2594.9	50
51	35.7	05.6	76.3	47.8	20.1	93.4	67.6	42.7	18.9	96.2	51
52	36.9	06.8	77.5	49.0	21.4	94.6	68.8	44.0	20.2	97.5	52
53	38.0	08.0	78.7	50.2	22.6	95.8	70.0	45.2	21.5	2598.8	53
54	39.2	09.1	79.9	51.4	23.8	97.1	71.2	46.5	22.8	2600.1	54
55	1940.4	2010.3	2081.1	2152.6	2225.0	2298.3	2372.5	2447.8	2524.0	2601.4	55
56	41.5	11.5	82.2	53.8	26.2	2299.5	73.8	49.0	25.3	02.7	56
57	42.7	12.7	83.4	55.0	27.4	2300.8	75.0	50.3	26.6	04.0	57
58	43.8	13.8	84.6	56.2	28.6	02.0	76.3	51.6	27.9	05.3	58
59	45.0	15.0	85.8	57.4	29.9	03.2	77.5	52.8	29.2	06.6	59
60	1946.2	2016.2	2087.0	2158.6	2231.1	2304.5	2378.8	2454.1	2530.4	2607.9	60
Lat.	30°	31°	32°	33°	34°	35°	36°	37°	38°	39°	Lat.

TABLE 5

Meridional Parts

Lat.	40°	41°	42°	43°	44°	45°	46°	47°	48°	49°	Lat.
′											′
0	2607. 9	2686. 5	2766. 3	2847. 4	2929. 8	3013. 6	3099. 0	3185. 9	3274. 4	3364. 7	0
1	09. 2	87. 8	67. 6	48. 7	31. 2	15. 1	3100. 4	87. 3	75. 9	66. 2	1
2	10. 5	89. 1	69. 0	50. 1	32. 6	16. 5	01. 8	88. 8	77. 4	67. 7	2
3	11. 8	90. 4	70. 3	51. 5	34. 0	17. 9	03. 3	90. 2	78. 9	69. 3	3
4	13. 1	91. 8	71. 7	52. 8	35. 4	19. 3	04. 7	91. 7	80. 4	70. 8	4
5	2614. 4	2693. 1	2773. 0	2854. 2	2936. 7	3020. 7	3106. 2	3193. 2	3281. 9	3372. 3	5
6	15. 7	94. 4	74. 3	55. 6	38. 1	22. 1	07. 6	94. 6	83. 4	73. 8	6
7	17. 0	95. 7	75. 7	55. 9	39. 5	23. 5	09. 0	96. 1	84. 8	75. 3	7
8	18. 3	97. 1	77. 0	58. 3	40. 9	24. 9	10. 5	97. 6	86. 3	76. 9	8
9	19. 6	98. 4	78. 4	59. 7	42. 3	26. 4	11. 9	3199. 0	87. 8	78. 4	9
10	2620. 9	2699. 7	2779. 7	2861. 0	2943. 7	3027. 8	3113. 3	3200. 5	3289. 3	3379. 9	10
11	22. 2	2701. 0	81. 1	62. 4	45. 1	29. 2	14. 8	02. 0	90. 8	81. 4	11
12	23. 5	02. 3	82. 4	63. 8	46. 5	30. 6	16. 2	03. 4	92. 3	83. 0	12
13	24. 8	03. 7	83. 8	65. 1	47. 9	32. 0	17. 7	04. 9	93. 8	84. 5	13
14	26. 1	05. 0	85. 1	66. 5	49. 2	33. 4	19. 1	06. 4	95. 3	86. 0	14
15	2627. 4	2706. 3	2786. 4	2867. 9	2950. 6	3034. 8	3120. 5	3207. 8	3296. 8	3387. 5	15
16	28. 7	07. 6	87. 8	69. 2	52. 0	36. 3	22. 0	09. 3	98. 3	89. 1	16
17	30. 0	09. 0	89. 1	70. 6	53. 4	37. 7	23. 4	10. 8	3299. 8	90. 6	17
18	31. 3	10. 3	90. 5	72. 0	54. 8	39. 1	24. 9	12. 2	3301. 3	92. 1	18
19	32. 6	11. 6	91. 8	73. 3	56. 2	40. 5	26. 3	13. 7	02. 8	93. 7	19
20	2634. 0	2713. 0	2793. 2	2874. 7	2957. 6	3041. 9	3127. 8	3215. 2	3304. 3	3395. 2	20
21	35. 3	14. 3	94. 5	76. 1	59. 0	43. 3	29. 2	16. 7	05. 8	96. 7	21
22	36. 6	15. 6	95. 9	77. 4	60. 4	44. 8	30. 6	18. 1	07. 3	98. 3	22
23	37. 9	16. 9	97. 2	78. 8	61. 8	46. 2	32. 1	19. 6	08. 8	3399. 8	23
24	39. 2	18. 3	98. 6	80. 2	63. 2	47. 6	33. 5	21. 1	10. 3	3401. 3	24
25	2640. 5	2719. 6	2799. 9	2881. 6	2964. 6	3049. 0	3135. 0	3222. 5	3311. 8	3402. 8	25
26	41. 8	20. 9	2801. 3	82. 9	66. 0	50. 4	36. 4	24. 0	13. 3	04. 4	26
27	43. 1	22. 2	02. 6	84. 3	67. 4	51. 9	37. 9	25. 5	14. 8	05. 9	27
28	44. 4	23. 6	04. 0	85. 7	68. 8	53. 3	39. 3	27. 0	16. 3	07. 4	28
29	45. 7	24. 9	05. 3	87. 1	70. 2	54. 7	40. 8	28. 4	17. 8	09. 0	29
30	2647. 0	2726. 2	2806. 7	2888. 4	2971. 5	3056. 1	3142. 2	3229. 9	3319. 3	3410. 5	30
31	48. 3	27. 6	08. 0	89. 8	72. 9	57. 5	43. 7	31. 4	20. 8	12. 1	31
32	49. 7	28. 9	09. 4	91. 2	74. 3	59. 0	45. 1	32. 9	22. 3	13. 6	32
33	51. 0	30. 2	10. 7	92. 5	75. 7	60. 4	46. 6	34. 4	23. 8	15. 1	33
34	52. 3	31. 6	12. 1	93. 9	77. 1	61. 8	48. 0	35. 8	25. 3	16. 7	34
35	2653. 6	2732. 9	2813. 4	2895. 3	2978. 5	3063. 2	3149. 5	3237. 3	3326. 9	3418. 2	35
36	54. 9	34. 2	14. 8	96. 7	79. 9	64. 7	50. 9	38. 8	28. 4	19. 7	36
37	56. 2	35. 6	16. 1	98. 1	81. 3	66. 1	52. 4	40. 3	29. 9	21. 3	37
38	57. 5	36. 9	17. 5	2899. 4	82. 7	67. 5	53. 8	41. 7	31. 4	22. 8	38
39	58. 8	38. 2	18. 9	2900. 8	84. 1	68. 9	55. 3	43. 2	32. 9	24. 4	39
40	2660. 2	2739. 6	2820. 2	2902. 2	2985. 5	3070. 4	3156. 7	3244. 7	3334. 4	3425. 9	40
41	61. 5	40. 9	21. 6	03. 6	86. 9	71. 8	58. 2	46. 2	35. 9	27. 4	41
42	62. 8	42. 2	22. 9	04. 9	88. 3	73. 2	59. 6	47. 7	37. 4	29. 0	42
43	64. 1	43. 6	24. 3	06. 3	89. 7	74. 6	61. 1	49. 1	38. 9	30. 5	43
44	65. 4	44. 9	25. 6	07. 7	91. 1	76. 1	62. 5	50. 6	40. 4	32. 1	44
45	2666. 7	2746. 2	2827. 0	2909. 1	2992. 6	3077. 5	3164. 0	3252. 1	3342. 0	3433. 6	45
46	68. 0	47. 6	28. 3	10. 5	94. 0	78. 9	65. 4	53. 6	43. 5	35. 2	46
47	69. 4	48. 9	29. 7	11. 8	95. 4	80. 4	66. 9	55. 1	45. 0	36. 7	47
48	70. 7	50. 2	31. 1	13. 2	96. 8	81. 8	68. 4	56. 6	46. 5	38. 2	48
49	72. 0	51. 6	32. 4	14. 6	98. 2	83. 2	69. 8	58. 0	48. 0	39. 8	49
50	2673. 3	2752. 9	2833. 8	2916. 0	2999. 6	3084. 6	3171. 3	3259. 5	3349. 5	3441. 3	50
51	74. 6	54. 2	35. 1	17. 4	3001. 0	86. 1	72. 7	61. 0	51. 0	42. 9	51
52	75. 9	55. 6	36. 5	18. 7	02. 4	87. 5	74. 2	62. 5	52. 5	44. 4	52
53	77. 3	56. 9	37. 9	20. 1	03. 8	88. 9	75. 6	64. 0	54. 1	46. 0	53
54	78. 6	58. 3	39. 2	21. 5	05. 2	90. 4	77. 1	65. 5	55. 6	47. 5	54
55	2679. 9	2759. 6	2840. 6	2922. 9	3006. 6	3091. 8	3178. 6	3267. 0	3357. 1	3449. 1	55
56	81. 2	60. 9	41. 9	24. 3	08. 0	93. 2	80. 0	68. 4	58. 6	50. 6	56
57	82. 5	62. 3	43. 3	25. 7	09. 4	94. 7	81. 5	69. 9	60. 1	52. 2	57
58	83. 8	63. 6	44. 7	27. 0	10. 8	96. 1	82. 9	71. 4	61. 7	53. 7	58
59	85. 2	65. 0	46. 0	28. 4	12. 2	97. 5	84. 4	72. 9	63. 2	55. 3	59
60	2686. 5	2766. 3	2847. 4	2929. 8	3013. 6	3099. 0	3185. 9	3274. 4	3364. 7	3456. 8	60
Lat.	40°	41°	42°	43°	44°	45°	46°	47°	48°	49°	Lat.

TABLE 5

Meridional Parts

Lat.	50°	51°	52°	53°	54°	55°	56°	57°	58°	59°	Lat.
′											′
0	3456.8	3550.9	3647.0	3745.4	3846.0	3949.1	4054.8	4163.3	4274.8	4389.4	0
1	58.4	52.5	48.7	47.0	47.7	50.8	56.6	65.1	76.6	91.3	1
2	59.9	54.1	50.3	48.7	49.4	52.6	58.4	67.0	78.5	93.3	2
3	61.5	55.7	51.9	50.3	51.1	54.3	60.2	68.8	80.4	95.2	3
4	63.0	57.2	53.5	52.0	52.8	56.1	61.9	70.6	82.3	97.1	4
5	3464.6	3558.8	3655.1	3753.7	3854.5	3957.8	4063.7	4172.5	4284.2	4399.1	5
6	66.1	60.4	56.8	55.3	56.2	59.5	65.5	74.3	86.1	4401.0	6
7	67.7	62.0	58.4	57.0	57.9	61.3	67.3	76.1	88.0	03.0	7
8	69.2	63.6	60.0	58.6	59.6	63.0	69.1	78.0	89.8	04.9	8
9	70.8	65.2	61.6	60.3	61.3	64.8	70.9	79.8	91.7	06.9	9
10	3472.4	3566.8	3663.3	3762.0	3863.0	3966.5	4072.7	4181.7	4293.6	4408.8	10
11	73.9	68.4	64.9	63.6	64.7	68.3	74.5	83.5	95.5	10.8	11
12	75.5	70.0	66.5	65.3	66.4	70.0	76.3	85.3	97.4	12.7	12
13	77.0	71.5	68.1	67.0	68.1	71.8	78.1	87.2	4299.3	14.7	13
14	78.6	73.1	69.8	68.6	69.8	73.5	79.9	89.0	4301.2	16.6	14
15	3480.2	3574.7	3671.4	3770.3	3871.5	3975.3	4081.7	4190.9	4303.1	4418.6	15
16	81.7	76.3	73.0	72.0	73.2	77.0	83.4	92.7	05.0	20.5	16
17	83.3	77.9	74.7	73.6	74.9	78.8	85.2	94.6	06.9	22.5	17
18	84.8	79.5	76.3	75.3	76.7	80.5	87.0	96.4	08.8	24.4	18
19	86.4	81.1	77.9	77.0	78.4	82.3	88.8	98.3	10.7	26.4	19
20	3488.0	3582.7	3679.6	3778.6	3880.1	3984.0	4090.6	4200.1	4312.6	4428.3	20
21	89.5	84.3	81.2	80.3	81.8	85.8	92.4	02.0	14.5	30.3	21
22	91.1	85.9	82.8	82.0	83.5	87.5	94.2	03.8	16.4	32.3	22
23	92.6	87.5	84.5	83.7	85.2	89.3	96.0	05.7	18.3	34.2	23
24	94.2	89.1	86.1	85.3	86.9	91.0	97.8	07.5	20.2	36.2	24
25	3495.8	3590.7	3687.7	3787.0	3888.6	3992.8	4099.7	4209.4	4322.1	4438.1	25
26	97.3	92.3	89.4	88.7	90.4	94.6	4101.5	11.2	24.0	40.1	26
27	3498.9	93.9	91.0	90.3	92.1	96.3	03.3	13.1	25.9	42.1	27
28	3500.5	95.5	92.6	92.0	93.8	98.1	05.1	14.9	27.8	44.0	28
29	02.0	97.1	94.3	93.7	95.5	3999.8	06.9	16.8	29.7	46.0	29
30	3503.6	3598.7	3695.9	3795.4	3897.2	4001.6	4108.7	4218.6	4331.7	4448.0	30
31	05.2	3600.3	97.6	97.1	3898.9	03.4	10.5	20.5	33.6	49.9	31
32	06.7	01.9	3699.2	3798.7	3900.7	05.1	12.3	22.4	35.5	51.9	32
33	08.3	03.5	3700.8	3800.4	02.4	06.9	14.1	24.2	37.4	53.9	33
34	09.9	05.1	02.5	02.1	04.1	08.7	15.9	26.1	39.3	55.8	34
35	3511.5	3606.7	3704.1	3803.8	3905.8	4010.4	4117.7	4227.9	4341.2	4457.8	35
36	13.0	08.3	05.8	05.4	07.5	12.2	19.5	29.8	43.1	59.8	36
37	14.6	09.9	07.4	07.1	09.3	14.0	21.4	31.7	45.1	61.7	37
38	16.2	11.5	09.0	08.8	11.0	15.7	23.2	33.5	47.0	63.7	38
39	17.7	13.1	10.7	10.5	12.7	17.5	25.0	35.4	48.9	65.7	39
40	3519.3	3614.8	3712.3	3812.2	3914.4	4019.3	4126.8	4237.3	4350.8	4467.7	40
41	20.9	16.4	14.0	13.9	16.2	21.0	28.6	39.1	52.7	69.6	41
42	22.5	18.0	15.6	15.5	17.9	22.8	30.4	41.0	54.6	71.6	42
43	24.0	19.6	17.3	17.2	19.6	24.6	32.3	42.9	56.6	73.6	43
44	25.6	21.2	18.9	18.9	21.3	26.3	34.1	44.7	58.5	75.6	44
45	3527.2	3622.8	3720.6	3820.6	3923.1	4028.1	4135.9	4246.6	4360.4	4477.6	45
46	28.8	24.4	22.2	22.3	24.8	29.9	37.7	48.5	62.3	79.5	46
47	30.3	26.0	23.9	24.0	26.5	31.7	39.5	50.3	64.3	81.5	47
48	31.9	27.6	25.5	25.7	28.3	33.4	41.4	52.2	66.2	83.5	48
49	33.5	29.2	27.2	27.4	30.0	35.2	43.2	54.1	68.1	85.5	49
50	3535.1	3630.9	3728.8	3829.1	3931.7	4037.0	4145.0	4256.0	4370.0	4487.5	50
51	36.7	32.5	30.5	30.7	33.5	38.8	46.8	57.8	72.0	89.5	51
52	38.2	34.1	32.1	32.4	35.2	40.5	48.7	59.7	73.9	91.5	52
53	39.8	35.7	33.8	34.1	36.9	42.3	50.5	61.6	75.8	93.5	53
54	41.4	37.3	35.4	35.8	38.7	44.1	52.3	63.5	77.8	95.4	54
55	3543.0	3638.9	3737.1	3837.5	3940.4	4045.9	4154.1	4265.3	4379.7	4497.4	55
56	44.6	40.6	38.7	39.2	42.1	47.7	56.0	67.2	81.6	4499.4	56
57	46.1	42.2	40.4	40.9	43.9	49.4	57.8	69.1	83.6	4501.4	57
58	47.7	43.8	42.0	42.6	45.6	51.2	59.6	71.0	85.5	03.4	58
59	49.3	45.4	43.7	44.3	47.3	53.0	61.5	72.9	87.4	05.4	59
60	3550.9	3647.0	3745.4	3846.0	3949.1	4054.8	4163.3	4274.8	4389.4	4507.4	60
Lat.	50°	51°	52°	53°	54°	55°	56°	57°	58°	59°	Lat.

TABLE 5

Meridional Parts

Lat.	60°	61°	62°	63°	64°	65°	66°	67°	68°	69°	Lat.
′											′
0	4507. 4	4629. 1	4754. 6	4884. 4	5018. 7	5157. 9	5302. 4	5452. 8	5609. 4	5773. 0	0
1	09. 4	31. 1	56. 8	86. 6	21. 0	60. 3	04. 9	55. 3	12. 1	75. 8	1
2	11. 4	33. 2	58. 9	88. 8	23. 3	62. 6	07. 4	57. 9	14. 8	78. 6	2
3	13. 4	35. 2	61. 0	91. 0	25. 5	65. 0	09. 8	60. 5	17. 4	81. 4	3
4	15. 4	37. 3	63. 1	93. 2	27. 8	67. 4	12. 3	63. 0	20. 1	84. 2	4
5	4517. 4	4639. 4	4765. 3	4895. 4	5030. 1	5169. 7	5314. 7	5465. 6	5622. 8	5787. 0	5
6	19. 4	41. 4	67. 4	97. 6	32. 4	72. 1	17. 2	68. 2	25. 5	89. 8	6
7	21. 4	43. 5	69. 5	4899. 8	34. 7	74. 5	19. 7	70. 7	28. 2	92. 6	7
8	23. 4	45. 6	71. 7	4902. 0	37. 0	76. 9	22. 1	73. 3	30. 8	95. 4	8
9	25. 4	47. 6	73. 8	04. 2	39. 3	79. 2	24. 6	75. 9	33. 5	5798. 2	9
10	4527. 4	4649. 7	4776. 0	4906. 5	5041. 6	5181. 6	5327. 1	5478. 4	5636. 2	5801. 0	10
11	29. 4	51. 8	78. 1	08. 7	43. 8	84. 0	29. 6	81. 0	38. 9	03. 8	11
12	31. 4	53. 9	80. 2	10. 9	46. 1	86. 4	32. 0	83. 6	41. 6	06. 6	12
13	33. 4	55. 9	82. 4	13. 1	48. 4	88. 8	34. 5	86. 2	44. 3	09. 5	13
14	35. 5	58. 0	84. 5	15. 3	50. 7	99. 1	37. 0	88. 7	47. 0	12. 3	14
15	4537. 5	4660. 1	4786. 7	4917. 5	5053. 0	5193. 5	5339. 5	5491. 3	5649. 7	5815. 1	15
16	39. 5	62. 2	88. 8	19. 8	55. 3	95. 9	42. 0	93. 9	52. 4	17. 9	16
17	41. 5	64. 2	91. 0	22. 0	57. 6	5198. 3	44. 4	96. 5	55. 1	20. 7	17
18	43. 5	66. 3	93. 1	24. 2	59. 9	5200. 7	46. 9	5499. 1	57. 8	23. 6	18
19	45. 5	68. 4	95. 3	26. 4	62. 2	03. 1	49. 4	5501. 7	60. 5	26. 4	19
20	4547. 5	4670. 5	4797. 4	4928. 6	5064. 5	5205. 5	5351. 9	5504. 3	5663. 2	5829. 2	20
21	49. 6	72. 6	4799. 6	30. 9	66. 8	07. 9	54. 4	06. 9	65. 9	32. 1	21
22	51. 6	74. 6	4801. 7	33. 1	69. 2	10. 3	56. 9	09. 5	68. 6	34. 9	22
23	53. 6	76. 7	03. 9	35. 3	71. 5	12. 7	59. 4	12. 1	71. 3	37. 7	23
24	55. 6	78. 8	06. 0	37. 6	73. 8	15. 1	61. 9	14. 7	74. 0	40. 6	24
25	4557. 6	4680. 9	4808. 2	4939. 8	5076. 1	5217. 5	5364. 4	5517. 3	5676. 7	5843. 4	25
26	59. 7	83. 0	10. 3	42. 0	78. 4	19. 9	66. 9	19. 9	79. 4	46. 2	26
27	61. 7	85. 1	12. 5	44. 2	80. 7	22. 3	69. 4	22. 5	82. 2	49. 1	27
28	63. 7	87. 2	14. 6	46. 5	83. 0	24. 7	71. 9	25. 1	84. 9	51. 9	28
29	65. 7	89. 2	16. 8	48. 7	85. 3	27. 1	74. 4	27. 7	87. 6	54. 8	29
30	4567. 8	4691. 3	4819. 0	4951. 0	5087. 7	5229. 5	5376. 9	5530. 3	5690. 3	5857. 6	30
31	69. 8	93. 4	21. 1	53. 2	90. 0	31. 9	79. 4	32. 9	93. 1	60. 5	31
32	71. 8	95. 5	23. 3	55. 4	92. 3	34. 3	81. 9	35. 5	95. 8	63. 3	32
33	73. 8	97. 6	25. 5	57. 7	94. 6	36. 7	84. 4	38. 1	5698. 5	66. 2	33
34	75. 9	4699. 7	27. 6	59. 9	97. 0	39. 1	86. 9	40. 7	5701. 2	69. 1	34
35	4577. 9	4701. 8	4829. 8	4962. 2	5099. 3	5241. 6	5389. 4	5543. 6	5704. 0	5871. 9	35
36	79. 9	03. 9	32. 0	64. 4	5101. 6	44. 0	91. 9	46. 0	06. 7	74. 8	36
37	82. 0	06. 0	34. 1	66. 7	03. 9	46. 4	94. 4	48. 6	09. 5	77. 7	37
38	84. 0	08. 1	36. 3	68. 9	06. 3	48. 8	97. 0	51. 2	12. 2	80. 5	38
39	86. 1	10. 2	38. 5	71. 2	08. 6	51. 2	5399. 5	53. 9	14. 9	83. 4	39
40	4588. 1	4712. 3	4840. 7	4973. 4	5110. 9	5253. 7	5402. 0	5556. 5	5717. 7	5886. 3	40
41	90. 1	14. 4	42. 8	75. 7	13. 3	56. 1	04. 5	59. 1	20. 4	89. 2	41
42	92. 2	16. 5	45. 0	77. 9	15. 6	58. 5	07. 0	61. 7	23. 2	92. 0	42
43	94. 2	18. 6	47. 2	80. 2	17. 9	60. 9	09. 6	64. 4	25. 9	94. 9	43
44	96. 3	20. 7	49. 4	82. 4	20. 3	63. 4	12. 1	67. 0	28. 7	5897. 8	44
45	4598. 3	4722. 9	4851. 5	4984. 7	5122. 6	5265. 8	5414. 6	5569. 7	5731. 4	5900. 7	45
46	4600. 3	25. 0	53. 7	86. 9	25. 0	68. 2	17. 2	72. 3	34. 2	03. 6	46
47	02. 4	27. 1	55. 9	89. 2	27. 3	70. 7	19. 7	74. 9	37. 0	06. 5	47
48	04. 4	29. 2	58. 1	91. 5	29. 7	73. 1	22. 2	77. 6	39. 7	09. 4	48
49	06. 5	31. 3	60. 3	93. 7	32. 0	75. 5	24. 8	80. 2	42. 5	12. 3	49
50	4608. 5	4733. 4	4862. 5	4996. 0	5134. 4	5278. 0	5427. 3	5582. 9	4745. 3	5915. 2	50
51	10. 6	35. 5	64. 6	4998. 3	36. 7	80. 4	29. 8	85. 5	48. 0	18. 1	51
52	12. 6	37. 6	66. 8	5000. 5	39. 1	82. 9	32. 4	88. 2	50. 8	21. 0	52
53	14. 7	39. 8	69. 0	02. 8	41. 4	85. 3	34. 9	90. 8	53. 6	23. 9	53
54	16. 7	41. 9	71. 2	05. 1	43. 8	87. 7	37. 5	93. 5	56. 3	26. 8	54
55	4618. 8	4744. 0	4873. 4	5007. 3	5146. 1	5290. 2	5440. 0	5596. 1	5759. 1	5929. 7	55
56	20. 8	46. 1	75. 6	09. 6	48. 5	92. 6	42. 6	5598. 8	61. 9	32. 6	56
57	22. 9	48. 3	77. 8	11. 9	50. 8	95. 1	45. 1	5601. 4	64. 7	35. 5	57
58	24. 9	50. 4	80. 0	14. 1	53. 2	5297. 5	47. 7	04. 1	67. 5	38. 4	58
59	27. 0	52. 5	82. 2	16. 4	55. 5	5300. 0	50. 2	06. 8	70. 2	41. 3	59
60	4629. 1	4754. 6	4884. 4	5018. 7	5157. 9	5302. 4	5452. 8	5609. 4	5773. 0	5944. 2	60
Lat.	60°	61°	62°	63°	64°	65°	66°	67°	68°	69°	Lat.

TABLE 5

Meridional Parts

Lat.	70°	71°	72°	73°	74°	75°	76°	77°	78°	79°	Lat.
′											′
0	5944.2	6123.9	6312.9	6512.4	6723.6	6948.1	7187.7	7444.7	7722.0	8023.1	0
1	47.2	27.0	16.1	15.8	27.2	51.9	91.8	49.2	26.8	28.3	1
2	50.1	30.0	19.4	19.2	30.8	55.8	7196.0	53.6	31.6	33.6	2
3	53.0	33.1	22.6	22.6	34.5	59.7	7200.1	58.1	36.5	38.8	3
4	56.0	36.2	25.9	26.1	38.1	63.5	04.3	62.6	41.3	44.1	4
5	5958.9	6139.3	6329.1	6529.5	6741.8	6967.4	7208.4	7467.0	7746.1	8049.4	5
6	61.8	42.4	32.4	32.9	45.4	71.3	12.6	71.5	51.0	54.6	6
7	64.8	45.4	35.6	36.4	49.0	75.2	16.7	76.0	55.8	59.9	7
8	67.7	48.5	38.9	39.8	52.7	79.1	20.9	80.5	60.7	65.2	8
9	70.6	51.6	42.1	43.3	56.4	83.0	25.1	85.0	65.6	70.5	9
10	5973.6	6154.7	6345.4	6546.7	6760.0	6986.9	7229.3	7489.5	7770.4	8075.9	10
11	76.5	57.8	48.7	50.2	63.7	90.8	33.4	94.0	75.3	81.2	11
12	79.5	60.9	51.9	53.6	67.4	94.7	37.6	7498.5	80.2	86.5	12
13	82.4	64.0	55.2	57.1	71.0	6998.6	41.8	7503.0	85.1	91.9	13
14	85.4	67.1	58.5	60.5	74.7	7002.5	46.0	07.5	90.0	8097.2	14
15	5988.3	6170.2	6361.7	6564.0	6778.4	7006.5	7250.2	7512.0	7794.9	8102.6	15
16	91.3	73.3	65.0	67.5	82.1	10.4	54.4	16.6	7799.8	07.9	16
17	94.3	76.5	68.3	71.0	85.8	14.3	58.6	21.1	04.7	13.3	17
18	5997.2	79.6	71.6	74.4	89.4	18.3	62.9	25.6	09.6	18.7	18
19	6000.2	82.7	74.9	77.9	93.1	22.2	67.1	30.2	14.6	24.1	19
20	6003.2	6185.8	6378.2	6581.4	6796.8	7026.2	7271.3	7534.8	7819.5	8129.5	20
21	06.1	88.9	81.5	84.9	6800.5	30.1	75.5	39.3	24.5	34.9	21
22	09.1	92.1	84.8	88.4	04.3	34.1	79.8	43.9	29.4	40.3	22
23	12.1	95.2	88.1	91.9	08.0	38.0	84.0	48.5	34.4	45.7	23
24	15.0	6198.3	91.4	95.4	11.7	42.0	88.3	53.0	39.3	51.1	24
25	6018.0	6201.5	6394.7	6598.9	6815.4	7045.9	7292.5	7557.6	7844.3	8156.6	25
26	21.0	04.6	6398.0	6602.4	19.1	49.9	7296.8	62.2	49.3	62.0	26
27	24.0	07.7	6401.3	05.9	22.8	53.9	7301.1	66.8	54.3	67.5	27
28	27.0	10.9	04.6	09.4	26.6	57.9	05.3	71.4	59.3	72.9	28
29	30.0	14.0	07.9	12.9	30.3	61.9	09.6	76.0	64.3	78.4	29
30	6033.0	6217.2	6411.3	6616.4	6834.1	7065.9	7313.9	7580.6	7869.3	8183.9	30
31	36.0	20.3	14.6	19.9	37.8	69.8	18.2	85.3	74.3	89.4	31
32	39.0	23.5	17.9	23.5	41.5	73.8	22.4	89.9	79.3	8194.9	32
33	42.0	26.6	21.2	27.0	45.3	77.9	26.7	94.5	84.4	8200.4	33
34	45.0	29.8	24.6	30.5	49.0	81.9	31.0	7599.2	89.4	05.9	34
35	6048.0	6233.0	6427.9	6634.1	6852.8	7085.9	7335.4	7603.8	7894.5	8211.4	35
36	51.0	36.1	31.3	37.6	56.6	89.9	39.7	08.5	7899.5	17.0	36
37	54.0	39.3	34.6	41.1	60.3	93.9	44.0	13.1	7904.6	22.5	37
38	57.0	42.5	37.9	44.7	64.1	7097.9	48.3	17.8	09.7	28.1	38
39	60.0	45.6	41.3	48.2	67.9	7102.0	52.6	22.5	14.7	33.6	39
40	6063.0	6248.8	6444.6	6651.8	6871.7	7106.0	7357.0	7627.1	7919.8	8239.2	40
41	66.0	52.0	48.0	55.3	75.4	10.0	61.3	31.8	24.9	44.8	41
42	69.1	55.2	51.4	58.9	79.2	14.1	65.6	36.5	30.0	50.4	42
43	72.1	58.3	54.7	62.5	83.0	18.1	70.0	41.2	35.1	56.0	43
44	75.1	61.5	58.1	66.0	86.8	22.2	74.3	45.9	40.2	61.6	44
45	6078.1	6264.7	6461.5	6669.6	6890.6	7126.2	7378.7	7650.6	7945.3	8267.2	45
46	81.2	67.9	64.8	73.2	94.4	30.3	83.1	55.3	50.5	72.8	46
47	84.2	71.1	68.2	76.7	6898.2	34.4	87.4	60.0	55.6	78.4	47
48	87.3	74.3	71.6	80.3	02.0	38.5	91.8	64.8	60.7	84.1	48
49	90.3	77.5	75.0	83.9	05.8	42.5	7396.2	69.5	65.9	89.7	49
50	6093.3	6280.7	6478.3	6687.5	6909.7	7146.6	7400.6	7674.2	7971.1	8295.4	50
51	96.4	83.9	81.7	91.1	13.5	50.7	05.0	79.0	76.2	8301.0	51
52	6099.4	87.1	85.1	94.7	17.3	54.8	09.4	83.7	81.4	06.7	52
53	6102.5	90.3	88.5	6698.3	21.1	58.9	13.8	88.5	86.6	12.4	53
54	05.5	93.6	91.9	6701.9	25.0	63.0	18.2	93.3	91.8	18.1	54
55	6108.6	6296.8	6495.3	6705.5	6928.8	7167.1	7422.6	7698.0	7997.0	8323.8	55
56	11.6	6300.0	6498.7	09.1	32.6	71.2	27.0	7702.8	8002.2	29.5	56
57	14.7	03.2	02.1	12.7	36.5	75.3	31.4	07.6	07.4	35.3	57
58	17.8	06.4	05.5	16.3	40.3	79.4	35.9	12.4	12.6	41.0	58
59	20.8	09.7	09.0	20.0	44.2	83.6	40.3	17.2	17.8	46.7	59
60	6123.9	6312.9	6512.4	6723.6	6948.1	7187.7	7444.7	7722.0	8023.1	8352.5	60
Lat.	70°	71°	72°	73°	74°	75°	76°	77°	78°	79°	Lat.

TABLE 5

Meridional Parts

Lat.	80°	81°	82°	83°	84°	85°	86°	87°	88°	89°	Lat.
′											′
0	8352. 5	8716. 3	9122. 6	9582. 9	10113. 9	10741. 6	11509. 5	12499. 1	13893. 4	16276. 5	0
1	58. 2	22. 7	29. 8	91. 1	123. 5	753. 1	523. 9	518. 2	922. 2	334. 3	1
2	64. 0	29. 1	37. 0	9599. 4	133. 1	764. 7	538. 3	537. 5	951. 2	393. 0	2
3	69. 8	35. 5	44. 2	9607. 6	142. 8	776. 2	552. 8	556. 9	13980. 4	452. 8	3
4	75. 6	41. 9	51. 5	15. 9	152. 4	787. 8	567. 3	576. 4	14009. 9	513. 7	4
5	8381. 4	8748. 4	9158. 7	9624. 2	10162. 1	10799. 5	11581. 9	12596. 0	14039. 7	16575. 6	5
6	87. 2	54. 8	66. 0	32. 5	171. 8	811. 2	596. 6	615. 7	069. 7	638. 7	6
7	93. 0	61. 3	73. 3	40. 8	181. 6	822. 9	611. 3	635. 5	100. 0	703. 0	7
8	8398. 9	67. 8	80. 6	49. 2	191. 3	834. 7	626. 1	655. 4	130. 6	768. 5	8
9	8404. 7	74. 3	87. 9	57. 6	201. 1	846. 5	641. 0	675. 5	161. 4	835. 2	9
10	8410. 5	8780. 8	9195. 2	9666. 0	10211. 0	10858. 3	11655. 9	12695. 7	14192. 6	16903. 3	10
11	16. 4	87. 3	9202. 6	74. 4	220. 8	870. 2	670. 9	715. 9	224. 0	16972. 8	11
12	22. 3	8793. 8	09. 9	82. 8	230. 7	882. 1	686. 0	736. 4	255. 6	17043. 6	12
13	28. 1	8800. 4	17. 3	91. 3	240. 6	894. 1	701. 1	756. 9	287. 6	116. 0	13
14	34. 0	06. 9	24. 7	9699. 7	250. 5	906. 1	716. 3	777. 5	319. 9	189. 9	14
15	8439. 9	8813. 5	9232. 1	9708. 2	10260. 5	10918. 2	11731. 5	12798. 3	14352. 5	17265. 5	15
16	45. 8	20. 1	39. 5	16. 8	270. 5	930. 3	746. 9	819. 2	385. 4	342. 8	16
17	51. 8	26. 7	47. 0	25. 3	280. 5	942. 4	762. 3	840. 3	418. 6	421. 8	17
18	57. 7	33. 3	54. 4	33. 9	290. 6	954. 6	777. 7	861. 4	452. 2	502. 7	18
19	63. 6	39. 9	61. 9	42. 4	300. 7	966. 8	793. 2	882. 7	486. 0	585. 5	19
20	8469. 6	8846. 5	9269. 4	9751. 0	10310. 8	10979. 1	11808. 8	12904. 1	14520. 3	17670. 4	20
21	75. 5	53. 2	76. 9	59. 7	320. 9	10991. 4	824. 5	925. 7	554. 8	757. 5	21
22	81. 5	59. 8	84. 4	68. 3	331. 1	11003. 8	840. 3	947. 4	589. 7	846. 8	22
23	87. 5	66. 5	91. 9	77. 0	341. 3	016. 2	856. 1	969. 2	625. 0	17938. 4	23
24	93. 5	73. 2	9299. 5	85. 7	351. 5	028. 6	872. 0	12991. 2	660. 6	18032. 6	24
25	8499. 5	8879. 9	9307. 0	9794. 4	10361. 8	11041. 1	11887. 9	13013. 3	14696. 6	18129. 5	25
26	8505. 5	86. 6	14. 6	9803. 1	372. 1	053. 6	904. 0	035. 6	733. 0	229. 1	26
27	11. 5	8893. 3	22. 2	11. 9	382. 4	066. 2	920. 1	058. 0	769. 8	331. 8	27
28	17. 5	8900. 0	29. 8	20. 6	392. 7	078. 8	936. 3	080. 5	806. 9	437. 6	28
29	23. 6	06. 8	37. 5	29. 4	403. 1	091. 5	952. 5	103. 2	844. 5	546. 7	29
30	8529. 6	8913. 5	9345. 1	9838. 3	10413. 6	11104. 2	11968. 9	13126. 1	14882. 5	18659. 4	30
31	35. 7	20. 3	52. 8	47. 1	424. 0	117. 0	11985. 3	149. 1	920. 9	776. 0	31
32	41. 8	27. 1	60. 5	56. 0	434. 5	129. 8	12001. 8	172. 2	959. 8	18896. 6	32
33	47. 9	33. 9	68. 2	64. 9	445. 0	142. 7	018. 4	195. 6	14999. 1	19021. 6	33
34	54. 0	40. 7	75. 9	73. 8	455. 5	155. 6	035. 0	219. 0	15038. 8	151. 4	34
35	8560. 1	8947. 5	9383. 7	9882. 7	10466. 1	11168. 6	12051. 8	13242. 7	15079. 0	19286. 2	35
36	66. 2	54. 3	91. 4	9891. 7	476. 7	181. 6	068. 6	266. 5	119. 7	426. 5	36
37	72. 3	61. 2	9399. 2	9900. 6	487. 4	194. 6	085. 5	290. 4	160. 9	572. 9	37
38	78. 4	68. 1	9407. 0	09. 7	498. 0	207. 7	102. 5	314. 6	202. 6	725. 7	38
39	84. 6	74. 9	14. 8	18. 7	508. 7	220. 9	119. 5	338. 9	244. 7	19885. 6	39
40	8590. 7	8981. 8	9422. 6	9927. 7	10519. 5	11234. 1	12136. 7	13363. 3	15287. 5	20053. 3	40
41	8596. 9	88. 7	30. 4	36. 8	530. 3	247. 4	153. 9	388. 0	330. 7	229. 7	41
42	8603. 1	8995. 7	38. 3	45. 9	541. 1	260. 7	171. 3	412. 8	374. 5	415. 5	42
43	09. 3	9002. 6	46. 2	55. 0	551. 9	274. 0	188. 7	437. 8	418. 9	612. 0	43
44	15. 5	09. 5	54. 1	64. 2	562. 8	287. 5	206. 2	463. 0	463. 8	820. 4	44
45	8621. 7	9016. 5	9462. 0	9973. 4	10573. 7	11300. 9	12223. 8	13488. 4	15509. 3	21042. 3	45
46	27. 9	23. 5	69. 9	82. 6	584. 6	314. 4	241. 5	513. 9	555. 5	279. 5	46
47	34. 2	30. 5	77. 9	9991. 8	595. 6	328. 0	259. 2	539. 7	602. 3	534. 2	47
48	40. 4	37. 5	85. 8	10001. 0	606. 6	341. 6	277. 1	565. 7	649. 7	21809. 4	48
49	46. 7	44. 5	9493. 8	010. 3	617. 7	355. 3	295. 1	591. 8	697. 8	22108. 5	49
50	8652. 9	9051. 5	9501. 8	10019. 6	10628. 8	11369. 1	12313. 1	13618. 2	15746. 5	22436. 2	50
51	59. 2	58. 6	09. 9	028. 9	639. 9	382. 8	331. 3	644. 7	796. 0	22798. 4	51
52	65. 5	65. 6	17. 9	038. 3	651. 1	396. 7	349. 5	671. 5	846. 2	23203. 3	52
53	71. 8	72. 7	26. 0	047. 6	662. 3	410. 6	367. 9	698. 4	897. 1	662. 4	53
54	78. 1	79. 8	34. 0	057. 0	673. 5	424. 6	386. 3	725. 6	15948. 8	24192. 3	54
55	8684. 5	9086. 9	9542. 1	10066. 4	10684. 8	11438. 6	12404. 8	13753. 0	16001. 3	24819. 1	55
56	90. 8	9094. 0	50. 3	075. 9	696. 1	452. 6	423. 5	780. 6	054. 6	25586. 2	56
57	8697. 2	9101. 1	58. 4	085. 4	707. 4	466. 8	442. 2	808. 5	108. 8	26575. 1	57
58	8703. 5	08. 3	66. 6	094. 9	718. 8	481. 0	461. 1	836. 5	163. 8	27969. 0	58
59	09. 9	15. 4	74. 7	10104. 4	730. 2	495. 2	480. 0	864. 8	219. 7	30351. 9	59
60	8716. 3	9122. 6	9582. 9	10113. 9	10741. 6	11509. 5	12499. 1	13893. 4	16276. 5	-------	60
Lat.	80°	81°	82°	83°	84°	85°	86°	87°	88°	89°	Lat.

TABLE 6

Length of a Degree of Latitude and Longitude

Lat.	Degree of latitude				Degree of longitude				Lat.
	Nautical miles	Statute miles	Feet	Meters	Nautical miles	Statute miles	Feet	Meters	
°									°
0	59.705	68.708	362 776	110 574	60.108	69.171	365 221	111 319	0
1	.706	.708	778	575	60.099	69.160	365 166	111 303	1
2	.706	.709	781	576	60.071	69.129	365 000	111 252	2
3	.707	.710	786	577	60.077	69.077	364 724	111 168	3
4	.708	.711	794	580	59.962	69.003	364 338	110 050	4
5	59.710	68.713	362 804	110 583	59.880	68.909	363 841	110 899	5
6	.712	.715	816	586	59.781	68.794	363 234	110 714	6
7	.714	.718	831	591	59.663	68.659	362 517	110 495	7
8	.717	.721	847	596	59.527	68.502	361 690	110 243	8
9	.720	.725	866	601	59.373	68.325	360 754	109 958	9
10	59.723	68.728	362 886	110 608	59.201	68.127	359 709	109 639	10
11	.727	.733	909	615	59.011	67.908	358 555	109 288	11
12	.731	.738	934	622	58.803	67.669	357 292	108 903	12
13	.736	.743	961	630	58.577	67.409	355 921	108 485	13
14	.740	.748	990	639	58.334	67.129	354 442	108 034	14
15	59.746	68.754	363 021	110 649	58.073	66.829	352 856	107 550	15
16	.751	.760	053	659	57.794	66.508	351 163	107 034	16
17	.757	.767	088	669	57.498	66.167	349 363	106 486	17
18	.763	.774	125	680	57.184	65.806	347 457	105 905	18
19	.769	.781	163	692	56.853	65.425	345 446	105 292	19
20	59.776	68.788	363 203	110 704	56.505	65.025	343 330	104 647	20
21	.782	.796	245	717	56.140	64.604	341 110	103 970	21
22	.790	.805	288	730	55.757	64.164	338 786	103 262	22
23	.797	.813	333	744	55.358	63.705	336 360	102 523	23
24	.805	.822	380	758	54.942	63.226	333 831	101 752	24
25	59.813	68.831	363 428	110 773	54.509	62.727	331 201	100 950	25
26	.821	.840	478	788	54.059	62.210	328 470	100 118	26
27	.829	.850	529	804	53.593	61.674	325 639	99 255	27
28	.838	.860	581	819	53.111	61.119	322 709	98 362	28
29	.847	.870	634	836	52.613	60.546	319 681	97 439	29
30	59.856	68.881	363 689	110 852	52.098	59.954	316 556	96 486	30
31	.865	.891	745	869	51.568	59.344	313 334	95 504	31
32	.874	.902	802	887	51.022	58.715	310 017	94 493	32
33	.884	.913	860	904	50.461	58.069	306 605	93 453	33
34	.893	.924	919	922	49.884	57.405	303 100	92 385	34
35	59.903	68.935	363 978	110 941	49.292	56.724	299 502	91 288	35
36	.913	.947	364 039	959	48.684	56.025	295 813	90 164	36
37	.923	.958	100	978	48.062	55.309	292 033	89 012	37
38	.933	.970	162	996	47.426	54.577	288 164	87 832	38
39	.944	.982	224	111 015	46.774	53.827	284 207	86 626	39
40	59.954	68.994	364 287	111 035	46.109	53.061	280 163	85 394	40
41	.964	69.006	350	054	45.429	52.279	276 034	84 135	41
42	.975	.018	414	073	44.736	51.481	271 820	82 851	42
43	.985	.030	477	093	44.029	50.667	267 523	81 541	43
44	.996	.042	541	112	43.308	49.838	263 144	80 206	44
45	60.006	69.054	364 605	111 132	42.574	48.993	258 634	78 847	45

TABLE 6

Length of a Degree of Latitude and Longitude

Lat.	Degree of latitude				Degree of longitude				Lat.
	Nautical miles	Statute miles	Feet	Meters	Nautical miles	Statute miles	Feet	Meters	
°									°
45	60. 006	69. 054	364 605	111 132	42. 574	48. 993	258 684	78 847	45
46	. 017	. 066	670	151	41. 827	48. 133	254 145	77 463	46
47	. 027	. 078	734	171	41. 067	47. 259	249 527	76 056	47
48	. 038	. 090	798	190	40. 294	46. 370	244 834	74 625	48
49	. 048	. 103	861	210	39. 510	45. 467	240 065	73 172	49
50	60. 059	69. 115	364 925	111 229	38. 713	44. 550	235 222	71 696	50
51	. 069	. 126	988	248	37. 904	43. 619	230 307	70 198	51
52	. 080	. 138	365 050	267	37. 083	42. 675	225 321	68 678	52
53	. 090	. 150	112	286	36. 251	41. 717	220 266	67 137	53
54	. 100	. 162	174	305	35. 408	40. 747	215 144	65 576	54
55	60. 110	69. 173	365 235	111 323	34. 554	39. 764	209 954	63 994	55
56	. 120	. 185	295	342	33. 689	38. 769	204 701	62 393	56
57	. 129	. 196	354	360	32. 814	37. 762	199 384	60 772	57
58	. 139	. 207	412	378	31. 929	36. 743	194 005	59 133	58
59	. 149	. 218	469	395	31. 034	35. 713	188 567	57 475	59
60	60. 158	69. 228	365 526	111 412	30. 130	34. 672	183 071	55 800	60
61	. 167	. 239	581	429	29. 216	33. 621	177 518	54 107	61
62	. 176	. 249	635	446	28. 293	32. 559	171 910	52 398	62
63	. 184	. 259	688	462	27. 361	31. 487	166 249	50 673	63
64	. 193	. 269	739	477	26. 421	30. 405	160 537	48 932	64
65	60. 201	69. 278	365 789	111 493	25. 473	29. 314	154 775	47 176	65
66	. 209	. 287	838	507	24. 517	28. 213	148 966	45 405	66
67	. 217	. 296	885	522	23. 553	27. 104	143 110	43 620	67
68	. 224	. 305	931	536	22. 582	25. 987	137 210	41 822	68
69	. 232	. 313	975	549	21. 604	24. 861	131 267	40 010	69
70	60. 239	69. 321	366 017	111 562	20. 619	23. 728	125 284	38 187	70
71	. 245	. 329	058	574	19. 628	22. 587	119 262	36 351	71
72	. 252	. 336	096	586	18. 631	21. 440	113 203	34 504	72
73	. 258	. 343	133	597	17. 628	20. 286	107 109	32 647	73
74	. 264	. 350	169	608	16. 619	19. 125	100 981	30 779	74
75	60. 269	69. 356	366 202	111 618	15. 606	17. 959	94 823	28 902	75
76	. 274	. 362	233	628	14. 587	16. 787	88 635	27 016	76
77	. 279	. 368	262	637	13. 564	15. 610	82 419	25 121	77
78	. 284	. 373	290	645	12. 537	14. 428	76 178	23 219	78
79	. 288	. 378	315	653	11. 506	13. 241	69 913	21 310	79
80	60. 292	69. 382	366 338	111 660	10. 472	12. 051	63 627	19 393	80
81	. 295	. 386	359	666	9. 434	10. 856	57 321	17 471	81
82	. 298	. 390	378	672	8. 393	9. 658	50 997	15 544	82
83	. 301	. 393	395	677	7. 350	8. 458	44 657	13 611	83
84	. 303	. 396	409	682	6. 304	7. 254	38 303	11 675	84
85	60. 305	69. 398	366 422	111 685	5. 256	6. 049	31 937	9 735	85
86	. 307	. 400	432	688	4. 207	4. 841	25 562	7 791	86
87	. 308	. 401	440	691	3. 156	3. 632	19 178	5 846	87
88	. 309	. 403	445	693	2. 105	2. 422	12 789	3 898	88
89	. 310	. 403	449	694	1. 053	1. 211	6 395	1 949	89
90	60. 310	69. 403	366 450	111 694	0. 000	0. 000	0	0	90

TABLE 7

Distance of an Object by Two Bearings

Difference between the course and second bearing	Difference between the course and first bearing													
	20°		22°		24°		26°		28°		30°		32°	
°														
30	1.97	0.98												
32	1.64	0.87	2.16	1.14										
34	1.41	0.79	1.80	1.01	2.34	1.31								
36	1.24	0.73	1.55	0.91	1.96	1.15	2.52	1.48						
38	1.11	0.68	1.36	0.84	1.68	1.04	2.11	1.30	2.70	1.66				
40	1.00	0.64	1.21	0.78	1.48	0.95	1.81	1.16	2.26	1.45	2.88	1.85		
42	0.91	0.61	1.10	0.73	1.32	0.88	1.59	1.06	1.94	1.30	2.40	1.61	3.05	2.04
44	0.84	0.58	1.00	0.69	1.19	0.83	1.42	0.98	1.70	1.18	2.07	1.44	2.55	1.77
46	0.78	0.56	0.92	0.66	1.09	0.78	1.28	0.92	1.52	1.09	1.81	1.30	2.19	1.58
48	0.73	0.54	0.85	0.64	1.00	0.74	1.17	0.87	1.37	1.02	1.62	1.20	1.92	1.43
50	0.68	0.52	0.80	0.61	0.93	0.71	1.08	0.83	1.25	0.96	1.46	1.12	1.71	1.31
52	0.65	0.51	0.75	0.59	0.87	0.68	1.00	0.79	1.15	0.91	1.33	1.05	1.55	1.22
54	0.61	0.49	0.71	0.57	0.81	0.66	0.93	0.76	1.07	0.87	1.23	0.99	1.41	1.14
56	0.58	0.48	0.67	0.56	0.77	0.64	0.88	0.73	1.00	0.83	1.14	0.95	1.30	1.08
58	0.56	0.47	0.64	0.54	0.73	0.62	0.83	0.70	0.94	0.80	1.07	0.90	1.21	1.03
60	0.53	0.46	0.61	0.53	0.69	0.60	0.78	0.68	0.89	0.77	1.00	0.87	1.13	0.98
62	0.51	0.45	0.58	0.51	0.66	0.58	0.75	0.66	0.84	0.74	0.94	0.83	1.06	0.94
64	0.49	0.44	0.56	0.50	0.63	0.57	0.71	0.64	0.80	0.72	0.89	0.80	1.00	0.90
66	0.48	0.43	0.54	0.49	0.61	0.56	0.68	0.62	0.76	0.70	0.85	0.78	0.95	0.87
68	0.46	0.43	0.52	0.48	0.59	0.54	0.66	0.61	0.73	0.68	0.81	0.75	0.90	0.84
70	0.45	0.42	0.50	0.47	0.57	0.53	0.63	0.59	0.70	0.66	0.78	0.73	0.86	0.81
72	0.43	0.41	0.49	0.47	0.55	0.52	0.61	0.58	0.68	0.64	0.75	0.71	0.82	0.78
74	0.42	0.41	0.48	0.46	0.53	0.51	0.59	0.57	0.65	0.63	0.72	0.69	0.79	0.76
76	0.41	0.40	0.46	0.45	0.52	0.50	0.57	0.56	0.63	0.61	0.70	0.67	0.76	0.74
78	0.40	0.39	0.45	0.44	0.50	0.49	0.56	0.54	0.61	0.60	0.67	0.66	0.74	0.72
80	0.39	0.39	0.44	0.44	0.49	0.48	0.54	0.53	0.60	0.59	0.65	0.64	0.71	0.70
82	0.39	0.38	0.43	0.43	0.48	0.47	0.53	0.52	0.58	0.57	0.63	0.63	0.69	0.69
84	0.38	0.38	0.42	0.42	0.47	0.47	0.52	0.51	0.57	0.56	0.62	0.61	0.67	0.67
86	0.37	0.37	0.42	0.42	0.46	0.46	0.51	0.51	0.55	0.55	0.60	0.60	0.66	0.65
88	0.37	0.37	0.41	0.41	0.45	0.45	0.50	0.50	0.54	0.54	0.59	0.59	0.64	0.64
90	0.36	0.36	0.40	0.40	0.45	0.45	0.49	0.49	0.53	0.53	0.58	0.58	0.62	0.62
92	0.36	0.36	0.40	0.40	0.44	0.44	0.48	0.48	0.52	0.52	0.57	0.57	0.61	0.61
94	0.36	0.35	0.39	0.39	0.43	0.43	0.47	0.47	0.51	0.51	0.56	0.55	0.60	0.60
96	0.35	0.35	0.39	0.39	0.43	0.43	0.47	0.46	0.51	0.50	0.55	0.54	0.59	0.59
98	0.35	0.35	0.39	0.38	0.42	0.42	0.46	0.46	0.50	0.50	0.54	0.53	0.58	0.57
100	0.35	0.34	0.38	0.38	0.42	0.41	0.46	0.45	0.49	0.49	0.53	0.52	0.57	0.56
102	0.35	0.34	0.38	0.37	0.42	0.41	0.45	0.44	0.49	0.48	0.53	0.51	0.56	0.55
104	0.34	0.33	0.38	0.37	0.41	0.40	0.45	0.43	0.48	0.47	0.52	0.50	0.56	0.54
106	0.34	0.33	0.38	0.36	0.41	0.39	0.45	0.43	0.48	0.46	0.52	0.50	0.55	0.53
108	0.34	0.32	0.38	0.36	0.41	0.39	0.44	0.42	0.48	0.45	0.51	0.49	0.55	0.52
110	0.34	0.32	0.37	0.35	0.41	0.38	0.44	0.41	0.47	0.44	0.51	0.48	0.54	0.51
112	0.34	0.32	0.37	0.35	0.41	0.38	0.44	0.41	0.47	0.44	0.50	0.47	0.54	0.50
114	0.34	0.31	0.37	0.34	0.41	0.37	0.44	0.40	0.47	0.43	0.50	0.46	0.54	0.49
116	0.34	0.31	0.38	0.34	0.41	0.37	0.44	0.39	0.47	0.42	0.50	0.45	0.53	0.48
118	0.35	0.31	0.38	0.33	0.41	0.36	0.44	0.39	0.47	0.41	0.50	0.44	0.53	0.47
120	0.35	0.30	0.38	0.33	0.41	0.36	0.44	0.38	0.47	0.41	0.50	0.43	0.53	0.46
122	0.35	0.30	0.38	0.32	0.41	0.35	0.44	0.37	0.47	0.40	0.50	0.42	0.53	0.45
124	0.35	0.29	0.38	0.32	0.41	0.34	0.44	0.37	0.47	0.39	0.50	0.42	0.53	0.44
126	0.36	0.29	0.39	0.31	0.42	0.34	0.45	0.36	0.47	0.38	0.50	0.41	0.53	0.43
128	0.36	0.28	0.39	0.31	0.42	0.33	0.45	0.35	0.48	0.38	0.50	0.40	0.53	0.42
130	0.36	0.28	0.39	0.30	0.42	0.32	0.45	0.35	0.48	0.37	0.51	0.39	0.54	0.41
132	0.37	0.27	0.40	0.30	0.43	0.32	0.46	0.34	0.48	0.36	0.51	0.38	0.54	0.40
134	0.37	0.27	0.40	0.29	0.43	0.31	0.46	0.33	0.49	0.35	0.52	0.37	0.54	0.39
136	0.38	0.26	0.41	0.28	0.44	0.30	0.47	0.32	0.49	0.34	0.52	0.36	0.55	0.38
138	0.39	0.26	0.42	0.28	0.45	0.30	0.47	0.32	0.50	0.33	0.53	0.35	0.55	0.37
140	0.39	0.25	0.42	0.27	0.45	0.29	0.48	0.31	0.51	0.33	0.53	0.34	0.56	0.36
142	0.40	0.25	0.43	0.27	0.46	0.28	0.49	0.30	0.51	0.32	0.54	0.33	0.56	0.35
144	0.41	0.24	0.44	0.26	0.47	0.28	0.50	0.29	0.52	0.31	0.55	0.32	0.57	0.34
146	0.42	0.24	0.45	0.25	0.48	0.27	0.51	0.28	0.53	0.30	0.56	0.31	0.58	0.32
148	0.43	0.23	0.46	0.25	0.49	0.26	0.52	0.27	0.54	0.29	0.57	0.30	0.59	0.31
150	0.45	0.22	0.48	0.24	0.50	0.25	0.53	0.26	0.55	0.28	0.58	0.29	0.60	0.30
152	0.46	0.22	0.49	0.23	0.52	0.24	0.54	0.25	0.57	0.27	0.59	0.28	0.61	0.29
154	0.48	0.21	0.50	0.22	0.53	0.23	0.56	0.24	0.58	0.25	0.60	0.26	0.62	0.27
156	0.49	0.20	0.52	0.21	0.55	0.22	0.57	0.23	0.60	0.24	0.62	0.25	0.64	0.26
158	0.51	0.19	0.54	0.20	0.57	0.21	0.59	0.22	0.61	0.23	0.63	0.24	0.66	0.25
160	0.53	0.18	0.56	0.19	0.59	0.20	0.61	0.21	0.63	0.22	0.65	0.22	0.67	0.23

TABLE 7

Distance of an Object by Two Bearings

Difference between the course and second bearing	Difference between the course and first bearing													
	34°		36°		38°		40°		42°		44°		46°	
°														
44	3. 22	2. 24												
46	2. 69	1. 93	3. 39	2. 43										
48	2. 31	1. 72	2. 83	2. 10	3. 55	2. 63								
50	2. 03	1. 55	2. 43	1. 86	2. 96	2. 27	3. 70	2. 84						
52	1. 81	1. 43	2. 13	1. 68	2. 54	2. 01	3. 09	2. 44	3. 85	3. 04				
54	1. 63	1. 32	1. 90	1. 54	2. 23	1. 81	2. 66	2. 15	3. 22	2. 60	4. 00	3. 24		
56	1. 49	1. 24	1. 72	1. 42	1. 99	1. 65	2. 33	1. 93	2. 77	2. 29	3. 34	2. 77	4. 14	3. 43
58	1. 37	1. 17	1. 57	1. 33	1. 80	1. 53	2. 08	1. 76	2. 43	2. 06	2. 87	2. 44	3. 46	2. 93
60	1. 28	1. 10	1. 45	1. 25	1. 64	1. 42	1. 88	1. 63	2. 17	1. 88	2. 52	2. 18	2. 97	2. 57
62	1. 19	1. 05	1. 34	1. 18	1. 51	1. 34	1. 72	1. 52	1. 96	1. 73	2. 25	1. 98	2. 61	2. 30
64	1. 12	1. 01	1. 25	1. 13	1. 40	1. 26	1. 58	1. 42	1. 79	1. 61	2. 03	1. 83	2. 33	2. 09
66	1. 06	0. 96	1. 18	1. 07	1. 31	1. 20	1. 47	1. 34	1. 65	1. 51	1. 85	1. 69	2. 10	1. 92
68	1. 00	0. 93	1. 11	1. 03	1. 23	1. 14	1. 37	1. 27	1. 53	1. 42	1. 71	1. 58	1. 92	1. 78
70	0. 95	0. 89	1. 05	0. 99	1. 16	1. 09	1. 29	1. 21	1. 43	1. 34	1. 58	1. 49	1. 77	1. 66
72	0. 91	0. 86	1. 00	0. 95	1. 10	1. 05	1. 21	1. 15	1. 34	1. 27	1. 48	1. 41	1. 64	1. 56
74	0. 87	0. 84	0. 95	0. 92	1. 05	1. 01	1. 15	1. 10	1. 26	1. 21	1. 39	1. 34	1. 53	1. 47
76	0. 84	0. 81	0. 91	0. 89	1. 00	0. 97	1. 09	1. 06	1. 20	1. 16	1. 31	1. 27	1. 44	1. 40
78	0. 80	0. 79	0. 88	0. 86	0. 96	0. 94	1. 04	1. 02	1. 14	1. 11	1. 24	1. 22	1. 36	1. 33
80	0. 78	0. 77	0. 85	0. 83	0. 92	0. 91	1. 00	0. 98	1. 09	1. 07	1. 18	1. 16	1. 28	1. 27
82	0. 75	0. 75	0. 82	0. 81	0. 89	0. 88	0. 96	0. 95	1. 04	1. 03	1. 13	1. 12	1. 22	1. 21
84	0. 73	0. 73	0. 79	0. 79	0. 86	0. 85	0. 93	0. 92	1. 00	0. 99	1. 08	1. 07	1. 17	1. 16
86	0. 71	0. 71	0. 77	0. 77	0. 83	0. 83	0. 89	0. 89	0. 96	0. 96	1. 04	1. 04	1. 12	1. 12
88	0. 69	0. 69	0. 75	0. 75	0. 80	0. 80	0. 86	0. 86	0. 93	0. 93	1. 00	1. 00	1. 08	1. 07
90	0. 67	0. 67	0. 73	0. 73	0. 78	0. 78	0. 84	0. 84	0. 90	0. 90	0. 97	0. 97	1. 04	1. 04
92	0. 66	0. 66	0. 71	0. 71	0. 76	0. 76	0. 82	0. 82	0. 87	0. 87	0. 93	0. 93	1. 00	1. 00
94	0. 65	0. 64	0. 69	0. 69	0. 74	0. 74	0. 79	0. 79	0. 85	0. 85	0. 91	0. 90	0. 97	0. 97
96	0. 63	0. 63	0. 68	0. 67	0. 73	0. 72	0. 78	0. 77	0. 83	0. 82	0. 88	0. 88	0. 94	0. 93
98	0. 62	0. 62	0. 67	0. 66	0. 71	0. 70	0. 76	0. 75	0. 81	0. 80	0. 86	0. 85	0. 91	0. 90
100	0. 61	0. 60	0. 65	0. 64	0. 70	0. 69	0. 74	0. 73	0. 79	0. 78	0. 84	0. 83	0. 89	0. 88
102	0. 60	0. 59	0. 64	0. 63	0. 68	0. 67	0. 73	0. 71	0. 77	0. 76	0. 82	0. 80	0. 87	0. 85
104	0. 60	0. 58	0. 63	0. 61	0. 67	0. 65	0. 72	0. 69	0. 76	0. 74	0. 80	0. 78	0. 85	0. 82
106	0. 59	0. 57	0. 63	0. 60	0. 66	0. 64	0. 70	0. 68	0. 74	0. 72	0. 79	0. 76	0. 83	0. 80
108	0. 58	0. 55	0. 62	0. 59	0. 66	0. 62	0. 69	0. 66	0. 73	0. 70	0. 77	0. 74	0. 81	0. 77
110	0. 58	0. 54	0. 61	0. 57	0. 65	0. 61	0. 68	0. 64	0. 72	0. 68	0. 76	0. 71	0. 80	0. 75
112	0. 57	0. 53	0. 61	0. 56	0. 64	0. 59	0. 68	0. 63	0. 71	0. 66	0. 75	0. 69	0. 79	0. 73
114	0. 57	0. 52	0. 60	0. 55	0. 63	0. 58	0. 67	0. 61	0. 70	0. 64	0. 74	0. 68	0. 78	0. 71
116	0. 56	0. 51	0. 60	0. 54	0. 63	0. 57	0. 66	0. 60	0. 70	0. 63	0. 73	0. 66	0. 77	0. 69
118	0. 56	0. 50	0. 59	0. 52	0. 63	0. 55	0. 66	0. 58	0. 69	0. 61	0. 72	0. 64	0. 76	0. 67
120	0. 56	0. 49	0. 59	0. 51	0. 62	0. 54	0. 65	0. 57	0. 68	0. 59	0. 72	0. 62	0. 75	0. 65
122	0. 56	0. 47	0. 59	0. 50	0. 62	0. 53	0. 65	0. 55	0. 68	0. 58	0. 71	0. 60	0. 74	0. 63
124	0. 56	0. 46	0. 59	0. 49	0. 62	0. 51	0. 65	0. 54	0. 68	0. 56	0. 71	0. 58	0. 74	0. 61
126	0. 56	0. 45	0. 59	0. 48	0. 62	0. 50	0. 64	0. 52	0. 67	0. 54	0. 70	0. 57	0. 73	0. 59
128	0. 56	0. 44	0. 59	0. 46	0. 62	0. 49	0. 64	0. 51	0. 67	0. 53	0. 70	0. 55	0. 73	0. 57
130	0. 56	0. 43	0. 59	0. 45	0. 62	0. 47	0. 64	0. 49	0. 67	0. 51	0. 70	0. 53	0. 72	0. 55
132	0. 56	0. 42	0. 59	0. 44	0. 62	0. 46	0. 64	0. 48	0. 67	0. 50	0. 70	0. 52	0. 72	0. 54
134	0. 57	0. 41	0. 59	0. 43	0. 62	0. 45	0. 64	0. 46	0. 67	0. 48	0. 69	0. 50	0. 72	0. 52
136	0. 57	0. 40	0. 60	0. 41	0. 62	0. 43	0. 65	0. 45	0. 67	0. 47	0. 70	0. 48	0. 72	0. 50
138	0. 58	0. 39	0. 60	0. 40	0. 63	0. 42	0. 65	0. 43	0. 67	0. 45	0. 70	0. 47	0. 72	0. 48
140	0. 58	0. 37	0. 61	0. 39	0. 63	0. 40	0. 65	0. 42	0. 68	0. 43	0. 70	0. 45	0. 72	0. 46
142	0. 59	0. 36	0. 61	0. 38	0. 63	0. 39	0. 66	0. 41	0. 68	0. 42	0. 70	0. 43	0. 72	0. 45
144	0. 60	0. 35	0. 62	0. 36	0. 64	0. 38	0. 66	0. 39	0. 68	0. 40	0. 71	0. 41	0. 73	0. 43
146	0. 60	0. 34	0. 63	0. 35	0. 65	0. 36	0. 67	0. 37	0. 69	0. 39	0. 71	0. 40	0. 73	0. 41
148	0. 61	0. 32	0. 63	0. 34	0. 66	0. 35	0. 68	0. 36	0. 70	0. 37	0. 72	0. 38	0. 74	0. 39
150	0. 62	0. 31	0. 64	0. 32	0. 66	0. 33	0. 68	0. 34	0. 70	0. 35	0. 72	0. 36	0. 74	0. 37
152	0. 63	0. 30	0. 65	0. 31	0. 67	0. 32	0. 69	0. 33	0. 71	0. 33	0. 73	0. 34	0. 75	0. 35
154	0. 65	0. 28	0. 67	0. 29	0. 68	0. 30	0. 70	0. 31	0. 72	0. 32	0. 74	0. 32	0. 76	0. 33
156	0. 66	0. 27	0. 68	0. 28	0. 70	0. 28	0. 72	0. 29	0. 73	0. 30	0. 75	0. 30	0. 77	0. 31
158	0. 67	0. 25	0. 69	0. 26	0. 71	0. 27	0. 73	0. 27	0. 74	0. 28	0. 76	0. 28	0. 78	0. 29
160	0. 69	0. 24	0. 71	0. 24	0. 73	0. 25	0. 74	0. 25	0. 76	0. 26	0. 77	0. 26	0. 79	0. 27

TABLE 7

Distance of an Object by Two Bearings

Difference between the course and second bearing	Difference between the course and first bearing													
	48°		50°		52°		54°		56°		58°		60°	
°														
58	4.28	3.63												
60	3.57	3.10	4.41	3.82										
62	3.07	2.71	3.68	3.25	4.54	4.01								
64	2.70	2.42	3.17	2.85	3.79	3.41	4.66	4.19						
66	2.40	2.20	2.78	2.54	3.26	2.98	3.89	3.55	4.77	4.36				
68	2.17	2.01	2.48	2.30	2.86	2.65	3.34	3.10	3.99	3.71	4.88	4.53		
70	1.98	1.86	2.24	2.10	2.55	2.39	2.94	2.76	3.43	3.22	4.08	3.83	4.99	4.69
72	1.83	1.74	2.04	1.94	2.30	2.19	2.62	2.49	3.01	2.86	3.51	3.33	4.17	3.96
74	1.70	1.63	1.88	1.81	2.10	2.02	2.37	2.27	2.68	2.58	3.08	2.96	3.58	3.44
76	1.58	1.54	1.75	1.70	1.94	1.88	2.16	2.10	2.42	2.35	2.74	2.66	3.14	3.05
78	1.49	1.45	1.63	1.60	1.80	1.76	1.99	1.95	2.21	2.16	2.48	2.43	2.80	2.74
80	1.40	1.38	1.53	1.51	1.68	1.65	1.85	1.82	2.04	2.01	2.26	2.23	2.53	2.49
82	1.33	1.32	1.45	1.43	1.58	1.56	1.72	1.71	1.89	1.87	2.08	2.06	2.31	2.29
84	1.26	1.26	1.37	1.36	1.49	1.48	1.62	1.61	1.77	1.76	1.93	1.92	2.13	2.12
86	1.21	1.20	1.30	1.30	1.41	1.41	1.53	1.52	1.66	1.65	1.81	1.80	1.98	1.97
88	1.16	1.16	1.24	1.24	1.34	1.34	1.45	1.45	1.56	1.56	1.70	1.70	1.84	1.84
90	1.11	1.11	1.19	1.19	1.28	1.28	1.38	1.38	1.48	1.48	1.60	1.60	1.73	1.73
92	1.07	1.07	1.14	1.14	1.23	1.23	1.31	1.31	1.41	1.41	1.52	1.52	1.63	1.63
94	1.03	1.03	1.10	1.10	1.18	1.17	1.26	1.26	1.35	1.34	1.44	1.44	1.55	1.54
96	1.00	0.99	1.06	1.06	1.13	1.13	1.21	1.20	1.29	1.28	1.38	1.37	1.47	1.47
98	0.97	0.96	1.03	1.02	1.10	1.08	1.16	1.15	1.24	1.23	1.32	1.31	1.41	1.39
100	0.94	0.93	1.00	0.98	1.06	1.04	1.12	1.11	1.19	1.18	1.27	1.25	1.35	1.33
102	0.92	0.90	0.97	0.95	1.03	1.01	1.09	1.06	1.15	1.13	1.22	1.19	1.29	1.27
104	0.90	0.87	0.95	0.92	1.00	0.97	1.06	1.02	1.12	1.08	1.18	1.14	1.25	1.21
106	0.88	0.84	0.92	0.89	0.97	0.94	1.03	0.99	1.09	1.04	1.14	1.10	1.20	1.16
108	0.86	0.82	0.90	0.86	0.95	0.90	1.00	0.95	1.05	1.00	1.11	1.05	1.17	1.11
110	0.84	0.79	0.88	0.83	0.93	0.87	0.98	0.92	1.02	0.96	1.08	1.01	1.13	1.06
112	0.83	0.77	0.87	0.80	0.91	0.84	0.95	0.88	1.00	0.93	1.05	0.97	1.10	1.02
114	0.81	0.74	0.85	0.78	0.89	0.82	0.93	0.85	0.98	0.89	1.02	0.93	1.07	0.98
116	0.80	0.72	0.84	0.75	0.88	0.79	0.92	0.82	0.96	0.85	1.00	0.90	1.04	0.94
118	0.79	0.70	0.83	0.73	0.86	0.76	0.90	0.79	0.94	0.83	0.98	0.86	1.02	0.90
120	0.78	0.68	0.82	0.71	0.85	0.74	0.89	0.77	0.91	0.80	0.96	0.83	1.00	0.87
122	0.77	0.66	0.81	0.68	0.84	0.71	0.87	0.74	0.90	0.77	0.95	0.80	0.98	0.83
124	0.77	0.63	0.80	0.66	0.83	0.69	0.86	0.71	0.90	0.74	0.93	0.77	0.96	0.80
126	0.76	0.61	0.79	0.64	0.82	0.66	0.85	0.69	0.88	0.71	0.91	0.74	0.95	0.77
128	0.75	0.59	0.78	0.62	0.81	0.64	0.84	0.66	0.87	0.69	0.90	0.71	0.93	0.74
130	0.75	0.57	0.78	0.60	0.81	0.62	0.83	0.64	0.86	0.66	0.89	0.68	0.92	0.71
132	0.75	0.56	0.77	0.57	0.80	0.59	0.83	0.61	0.85	0.64	0.88	0.66	0.91	0.68
134	0.74	0.54	0.77	0.55	0.80	0.57	0.82	0.59	0.85	0.61	0.87	0.63	0.90	0.65
136	0.74	0.52	0.77	0.53	0.80	0.55	0.82	0.57	0.84	0.58	0.87	0.60	0.89	0.62
138	0.74	0.50	0.77	0.51	0.79	0.53	0.81	0.54	0.84	0.56	0.86	0.58	0.89	0.59
140	0.74	0.48	0.77	0.49	0.79	0.51	0.81	0.52	0.83	0.54	0.86	0.55	0.88	0.57
142	0.74	0.46	0.77	0.47	0.79	0.49	0.81	0.50	0.83	0.51	0.85	0.52	0.87	0.54
144	0.75	0.44	0.77	0.45	0.79	0.46	0.81	0.48	0.83	0.49	0.85	0.50	0.87	0.51
146	0.75	0.42	0.77	0.43	0.79	0.44	0.81	0.45	0.83	0.46	0.85	0.47	0.87	0.49
148	0.76	0.40	0.77	0.41	0.79	0.42	0.81	0.43	0.83	0.44	0.85	0.45	0.87	0.46
150	0.76	0.38	0.78	0.39	0.80	0.40	0.81	0.41	0.83	0.42	0.85	0.42	0.87	0.43
152	0.77	0.36	0.78	0.37	0.80	0.38	0.82	0.38	0.83	0.39	0.85	0.40	0.87	0.41
154	0.77	0.34	0.79	0.35	0.81	0.35	0.82	0.36	0.84	0.37	0.85	0.37	0.87	0.38
156	0.78	0.32	0.80	0.32	0.81	0.33	0.83	0.34	0.84	0.34	0.86	0.35	0.87	0.35
158	0.79	0.30	0.81	0.30	0.82	0.31	0.83	0.31	0.85	0.32	0.86	0.32	0.87	0.33
160	0.80	0.27	0.82	0.28	0.83	0.28	0.84	0.29	0.85	0.29	0.86	0.30	0.88	0.30

TABLE 7

Distance of an Object by Two Bearings

Difference between the course and second bearing	Difference between the course and first bearing															
	62°		64°		66°		68°		70°		72°		74°		76°	
°																
72	5.08	4.84														
74	4.25	4.08	5.18	4.98												
76	3.65	3.54	4.32	4.19	5.26	5.10										
78	3.20	3.13	3.72	3.63	4.39	4.30	5.34	5.22								
80	2.86	2.81	3.26	3.21	3.78	3.72	4.46	4.39	5.41	5.33						
82	2.58	2.56	2.91	2.88	3.31	3.28	3.83	3.80	4.52	4.48	5.48	5.42				
84	2.36	2.34	2.63	2.61	2.96	2.94	3.36	3.35	3.88	3.86	4.57	4.55	5.54	5.51		
86	2.17	2.17	2.40	2.39	2.67	2.66	3.00	2.99	3.41	3.40	3.93	3.92	4.62	4.61	5.59	5.57
88	2.01	2.01	2.21	2.21	2.44	2.44	2.71	2.71	3.04	3.04	3.45	3.45	3.97	3.97	4.67	4.66
90	1.88	1.88	2.05	2.05	2.25	2.25	2.48	2.48	2.75	2.75	3.08	3.08	3.49	3.49	4.01	4.01
92	1.77	1.76	1.91	1.91	2.08	2.08	2.28	2.28	2.51	2.51	2.78	2.78	3.11	3.11	3.52	3.52
94	1.67	1.66	1.80	1.79	1.95	1.94	2.12	2.11	2.31	2.30	2.54	2.53	2.81	2.80	3.14	3.13
96	1.58	1.57	1.70	1.69	1.83	1.82	1.97	1.96	2.14	2.13	2.34	2.33	2.57	2.55	2.84	2.82
98	1.50	1.49	1.61	1.59	1.72	1.71	1.85	1.84	2.00	1.98	2.17	2.15	2.36	2.34	2.59	2.56
100	1.43	1.41	1.53	1.51	1.63	1.61	1.75	1.72	1.88	1.85	2.03	2.00	2.19	2.16	2.39	2.35
102	1.37	1.34	1.46	1.43	1.55	1.52	1.66	1.62	1.77	1.73	1.90	1.86	2.05	2.00	2.21	2.16
104	1.32	1.28	1.40	1.36	1.48	1.44	1.58	1.53	1.68	1.63	1.79	1.74	1.92	1.87	2.07	2.01
106	1.27	1.22	1.34	1.29	1.42	1.37	1.51	1.45	1.60	1.54	1.70	1.63	1.81	1.74	1.94	1.87
108	1.23	1.17	1.29	1.23	1.37	1.30	1.44	1.37	1.53	1.45	1.62	1.54	1.72	1.63	1.83	1.74
110	1.19	1.12	1.25	1.17	1.32	1.24	1.39	1.30	1.46	1.37	1.54	1.45	1.64	1.54	1.74	1.63
112	1.15	1.07	1.21	1.12	1.27	1.18	1.33	1.24	1.40	1.30	1.48	1.37	1.56	1.45	1.65	1.53
114	1.12	1.02	1.17	1.07	1.23	1.12	1.29	1.18	1.35	1.24	1.42	1.30	1.50	1.37	1.58	1.44
116	1.09	0.98	1.14	1.03	1.19	1.07	1.25	1.12	1.31	1.17	1.37	1.23	1.44	1.29	1.51	1.36
118	1.07	0.94	1.11	0.98	1.16	1.02	1.21	1.07	1.26	1.12	1.32	1.17	1.38	1.22	1.45	1.28
120	1.04	0.90	1.08	0.94	1.13	0.98	1.18	1.02	1.23	1.06	1.28	1.11	1.34	1.16	1.40	1.21
122	1.02	0.86	1.06	0.90	1.10	0.93	1.15	0.97	1.19	1.01	1.24	1.05	1.29	1.10	1.35	1.14
124	1.00	0.83	1.04	0.86	1.08	0.89	1.12	0.93	1.16	0.96	1.21	1.00	1.25	1.04	1.31	1.08
126	0.98	0.79	1.02	0.82	1.05	0.85	1.09	0.88	1.13	0.92	1.18	0.95	1.22	0.99	1.27	1.02
128	0.97	0.76	1.00	0.79	1.03	0.82	1.07	0.84	1.11	0.87	1.15	0.90	1.19	0.94	1.23	0.97
130	0.95	0.73	0.98	0.75	1.02	0.78	1.05	0.80	1.09	0.83	1.12	0.86	1.16	0.89	1.20	0.92
132	0.94	0.70	0.97	0.72	1.00	0.74	1.03	0.77	1.06	0.79	1.10	0.82	1.13	0.84	1.17	0.87
134	0.93	0.67	0.96	0.69	0.99	0.71	1.01	0.73	1.04	0.75	1.08	0.77	1.11	0.80	1.14	0.82
136	0.92	0.64	0.95	0.66	0.97	0.68	1.00	0.69	1.03	0.71	1.06	0.74	1.09	0.76	1.12	0.78
138	0.91	0.61	0.94	0.63	0.96	0.64	0.99	0.66	1.01	0.68	1.04	0.70	1.07	0.72	1.10	0.74
140	0.90	0.58	0.93	0.60	0.95	0.61	0.97	0.63	1.00	0.64	1.03	0.66	1.05	0.68	1.08	0.70
142	0.90	0.55	0.92	0.57	0.94	0.58	0.96	0.59	0.99	0.61	1.01	0.62	1.04	0.64	1.06	0.65
144	0.89	0.52	0.91	0.54	0.93	0.55	0.96	0.56	0.98	0.57	1.00	0.59	1.02	0.60	1.05	0.62
146	0.89	0.50	0.91	0.51	0.93	0.52	0.95	0.53	0.97	0.54	0.99	0.55	1.01	0.57	1.03	0.58
148	0.89	0.47	0.90	0.48	0.92	0.49	0.94	0.50	0.96	0.51	0.98	0.52	1.00	0.53	1.02	0.54
150	0.88	0.44	0.90	0.45	0.92	0.46	0.94	0.47	0.95	0.48	0.97	0.49	0.99	0.50	1.01	0.50
152	0.88	0.41	0.90	0.42	0.92	0.43	0.93	0.44	0.95	0.45	0.97	0.45	0.98	0.46	1.00	0.47
154	0.88	0.39	0.90	0.39	0.91	0.40	0.93	0.41	0.94	0.41	0.96	0.42	0.98	0.43	0.99	0.43
156	0.89	0.36	0.90	0.37	0.91	0.37	0.93	0.38	0.94	0.38	0.96	0.39	0.97	0.39	0.99	0.40
158	0.89	0.33	0.90	0.34	0.91	0.34	0.93	0.35	0.94	0.35	0.95	0.36	0.97	0.36	0.98	0.37
160	0.89	0.30	0.90	0.31	0.91	0.31	0.93	0.32	0.94	0.32	0.95	0.33	0.96	0.33	0.98	0.33

TABLE 7

Distance of an Object by Two Bearings

Difference between the course and second bearing	Difference between the course and first bearing															
	78°		80°		82°		84°		86°		88°		90°		92°	
°																
88	5.63	5.63														
90	4.70	4.70	5.67	5.67												
92	4.04	4.04	4.74	4.73	5.70	5.70										
94	3.55	3.54	4.07	4.06	4.76	4.75	5.73	5.71								
96	3.17	3.15	3.57	3.55	4.09	4.07	4.78	4.76	5.74	5.71						
98	2.86	2.83	3.19	3.16	3.59	3.56	4.11	4.07	4.80	4.75	5.76	5.70				
100	2.61	2.57	2.88	2.84	3.20	3.16	3.61	3.55	4.12	4.06	4.81	4.73	5.76	5.67		
102	2.40	2.35	2.63	2.57	2.90	2.83	3.22	3.15	3.62	3.54	4.13	4.04	4.81	4.70	5.76	5.63
104	2.23	2.16	2.42	2.35	2.64	2.56	2.91	2.82	3.23	3.13	3.63	3.52	4.13	4.01	4.81	4.66
106	2.08	2.00	2.25	2.16	2.43	2.34	2.65	2.55	2.92	2.80	3.23	3.11	3.63	3.49	4.13	3.97
108	1.96	1.86	2.10	2.00	2.26	2.15	2.45	2.33	2.66	2.53	2.92	2.78	3.24	3.08	3.63	3.45
110	1.85	1.73	1.97	1.85	2.11	1.98	2.27	2.13	2.45	2.31	2.67	2.51	2.92	2.75	3.23	3.04
112	1.75	1.62	1.86	1.72	1.98	1.83	2.12	1.96	2.28	2.11	2.46	2.28	2.67	2.48	2.92	2.71
114	1.66	1.52	1.76	1.61	1.87	1.71	1.99	1.82	2.12	1.94	2.28	2.08	2.46	2.25	2.67	2.44
116	1.59	1.43	1.68	1.51	1.77	1.59	1.88	1.69	2.00	1.79	2.13	1.91	2.28	2.05	2.46	2.21
118	1.52	1.34	1.60	1.41	1.68	1.49	1.78	1.57	1.88	1.66	2.00	1.76	2.13	1.88	2.28	2.01
120	1.46	1.27	1.53	1.33	1.61	1.39	1.69	1.47	1.78	1.54	1.89	1.63	2.00	1.73	2.13	1.84
122	1.41	1.19	1.47	1.25	1.54	1.31	1.62	1.37	1.70	1.44	1.79	1.52	1.89	1.60	2.00	1.70
124	1.36	1.13	1.42	1.18	1.48	1.23	1.55	1.28	1.62	1.34	1.70	1.41	1.79	1.48	1.89	1.56
126	1.32	1.06	1.37	1.11	1.43	1.15	1.48	1.20	1.55	1.26	1.62	1.31	1.70	1.38	1.79	1.45
128	1.28	1.01	1.33	1.04	1.38	1.08	1.43	1.13	1.49	1.17	1.55	1.23	1.62	1.28	1.70	1.34
130	1.24	0.95	1.29	0.98	1.33	1.02	1.38	1.06	1.44	1.10	1.49	1.14	1.56	1.19	1.62	1.24
132	1.21	0.90	1.25	0.93	1.29	0.96	1.34	0.99	1.39	1.03	1.44	1.07	1.49	1.11	1.55	1.16
134	1.18	0.85	1.22	0.88	1.26	0.90	1.30	0.93	1.34	0.97	1.39	1.00	1.44	1.04	1.49	1.07
136	1.15	0.80	1.19	0.83	1.22	0.85	1.26	0.88	1.30	0.90	1.34	0.93	1.39	0.97	1.44	1.00
138	1.13	0.76	1.16	0.78	1.19	0.80	1.23	0.82	1.27	0.85	1.30	0.87	1.35	0.90	1.39	0.93
140	1.11	0.71	1.14	0.73	1.17	0.75	1.20	0.77	1.23	0.79	1.27	0.82	1.31	0.84	1.34	0.86
142	1.09	0.67	1.12	0.69	1.14	0.70	1.17	0.72	1.20	0.74	1.24	0.76	1.27	0.78	1.30	0.80
144	1.07	0.63	1.10	0.64	1.12	0.66	1.15	0.67	1.18	0.69	1.21	0.71	1.24	0.73	1.27	0.75
146	1.05	0.59	1.08	0.60	1.10	0.62	1.13	0.63	1.15	0.64	1.18	0.66	1.21	0.67	1.24	0.69
148	1.04	0.55	1.06	0.56	1.08	0.57	1.11	0.59	1.13	0.60	1.15	0.61	1.18	0.62	1.21	0.64
150	1.03	0.51	1.05	0.52	1.07	0.53	1.09	0.54	1.11	0.55	1.13	0.57	1.15	0.58	1.18	0.59
152	1.02	0.48	1.04	0.49	1.05	0.49	1.07	0.50	1.09	0.51	1.11	0.52	1.13	0.53	1.15	0.54
154	1.01	0.44	1.02	0.45	1.04	0.46	1.06	0.46	1.08	0.47	1.09	0.48	1.11	0.49	1.13	0.50
156	1.00	0.41	1.01	0.41	1.03	0.42	1.05	0.43	1.06	0.43	1.08	0.44	1.09	0.45	1.11	0.45
158	0.99	0.37	1.01	0.38	1.02	0.38	1.03	0.39	1.05	0.39	1.06	0.40	1.08	0.40	1.09	0.41
160	0.99	0.34	1.00	0.34	1.01	0.35	1.02	0.35	1.04	0.35	1.05	0.36	1.06	0.36	1.08	0.37

Difference between the course and second bearing	94°		96°		98°		100°		102°		104°		106°		108°	
°																
104	5.74	5.57														
106	4.80	4.61	5.73	5.51												
108	4.12	3.92	4.78	4.55	5.70	5.42										
110	3.62	3.40	4.11	3.86	4.76	4.48	5.67	5.33								
112	3.23	2.99	3.61	3.35	4.09	3.80	4.74	4.40	5.63	5.22						
114	2.92	2.66	3.22	2.94	3.59	3.28	4.07	3.72	4.70	4.30	5.59	5.10				
116	2.66	2.39	2.91	2.61	3.20	2.88	3.57	3.21	4.04	3.63	4.67	4.19	5.54	4.98		
118	2.45	2.17	2.65	2.34	2.90	2.56	3.19	2.81	3.55	3.13	4.01	3.54	4.62	4.08	5.48	4.84
120	2.28	1.97	2.45	2.12	2.64	2.29	2.88	2.49	3.17	2.74	3.52	3.05	3.97	3.44	4.57	3.96
122	2.12	1.80	2.27	1.92	2.43	2.06	2.63	2.23	2.86	2.43	3.14	2.66	3.49	2.96	3.93	3.33
124	2.00	1.65	2.12	1.76	2.26	1.87	2.42	2.01	2.61	2.16	2.84	2.35	3.11	2.58	3.45	2.86
126	1.88	1.52	1.99	1.61	2.11	1.71	2.25	1.82	2.40	1.95	2.59	2.10	2.81	2.27	3.08	2.49
128	1.78	1.41	1.88	1.48	1.98	1.56	2.10	1.65	2.23	1.76	2.39	1.88	2.57	2.02	2.78	2.19
130	1.70	1.30	1.78	1.36	1.87	1.43	1.97	1.51	2.08	1.60	2.21	1.70	2.36	1.81	2.54	1.94
132	1.62	1.20	1.69	1.26	1.77	1.32	1.86	1.38	1.96	1.45	2.07	1.54	2.19	1.63	2.34	1.74
134	1.55	1.12	1.62	1.16	1.68	1.21	1.76	1.27	1.85	1.33	1.94	1.40	2.05	1.47	2.17	1.56
136	1.49	1.04	1.55	1.07	1.61	1.12	1.68	1.16	1.75	1.22	1.83	1.27	1.92	1.34	2.03	1.41
138	1.44	0.96	1.49	0.99	1.54	1.03	1.60	1.07	1.66	1.11	1.74	1.16	1.81	1.21	1.90	1.27
140	1.39	0.89	1.43	0.92	1.48	0.95	1.53	0.98	1.59	1.02	1.65	1.06	1.72	1.10	1.79	1.15
142	1.34	0.83	1.38	0.85	1.43	0.88	1.47	0.91	1.52	0.94	1.58	0.97	1.64	1.01	1.70	1.05
144	1.30	0.77	1.34	0.79	1.38	0.81	1.42	0.83	1.46	0.86	1.51	0.89	1.56	0.92	1.62	0.95
146	1.27	0.71	1.30	0.73	1.33	0.75	1.37	0.77	1.41	0.79	1.45	0.81	1.50	0.84	1.54	0.86
148	1.23	0.65	1.26	0.67	1.29	0.69	1.33	0.70	1.36	0.72	1.40	0.74	1.44	0.76	1.48	0.78
150	1.20	0.60	1.23	0.61	1.26	0.63	1.29	0.64	1.32	0.66	1.35	0.67	1.38	0.69	1.42	0.71
152	1.18	0.55	1.20	0.56	1.22	0.57	1.25	0.59	1.28	0.60	1.31	0.61	1.34	0.63	1.37	0.64
154	1.15	0.50	1.17	0.51	1.19	0.52	1.22	0.53	1.24	0.54	1.27	0.56	1.29	0.57	1.32	0.58
156	1.13	0.46	1.15	0.47	1.17	0.47	1.19	0.48	1.21	0.49	1.23	0.50	1.25	0.51	1.28	0.52
158	1.11	0.42	1.13	0.42	1.14	0.43	1.16	0.44	1.18	0.44	1.20	0.45	1.22	0.46	1.24	0.47
160	1.09	0.37	1.11	0.38	1.12	0.38	1.14	0.39	1.15	0.39	1.17	0.40	1.19	0.41	1.21	0.41

TABLE 7

Distance of an Object by Two Bearings

Difference between the course and second bearing.	Difference between the course and first bearing.													
°	110°		112°		114°		116°		118°		120°		122°	
120	5. 41	4. 69												
122	4. 52	3. 83	5. 34	4. 53										
124	3. 88	3. 22	4. 46	3. 70	5. 26	4. 36								
126	3. 41	2. 76	3. 83	3. 10	4. 39	3. 55	5. 18	4. 19						
128	3. 04	2. 40	3. 36	2. 65	3. 78	2. 98	4. 32	3. 41	5. 08	4. 01				
130	2. 75	2. 10	3. 00	2. 30	3. 31	2. 54	3. 72	2. 85	4. 25	3. 25	4. 99	3. 82		
132	2. 51	1. 86	2. 71	2. 01	2. 96	2. 20	3. 26	2. 42	3. 65	2. 71	4. 17	3. 10	4. 88	3. 63
134	2. 31	1. 66	2. 48	1. 78	2. 67	1. 92	2. 91	2. 09	3. 20	2. 30	3. 58	2. 57	4. 08	2. 93
136	2. 14	1. 49	2. 28	1. 58	2. 44	1. 69	2. 63	1. 83	2. 86	1. 98	3. 14	2. 18	3. 51	2. 44
138	2. 00	1. 34	2. 12	1. 42	2. 25	1. 50	2. 40	1. 61	2. 58	1. 73	2. 80	1. 88	3. 08	2. 06
140	1. 88	1. 21	1. 97	1. 27	2. 08	1. 34	2. 21	1. 42	2. 36	1. 52	2. 53	1. 63	2. 74	1. 76
142	1. 77	1. 09	1. 85	1. 14	1. 95	1. 20	2. 05	1. 26	2. 17	1. 34	2. 31	1. 42	2. 48	1. 53
144	1. 68	0. 99	1. 75	1. 03	1. 83	1. 07	1. 91	1. 13	2. 01	1. 18	2. 13	1. 25	2. 26	1. 33
146	1. 60	0. 89	1. 66	0. 93	1. 72	0. 96	1. 80	1. 01	1. 88	1. 05	1. 98	1. 10	2. 08	1. 17
148	1. 53	0. 81	1. 58	0. 84	1. 63	0. 87	1. 70	0. 90	1. 77	0. 94	1. 84	0. 98	1. 93	1. 03
150	1. 46	0. 73	1. 51	0. 75	1. 55	0. 78	1. 61	0. 80	1. 67	0. 83	1. 73	0. 87	1. 81	0. 90
152	1. 40	0. 66	1. 44	0. 68	1. 48	0. 70	1. 53	0. 72	1. 58	0. 74	1. 63	0. 77	1. 70	0. 80
154	1. 35	0. 59	1. 39	0. 61	1. 42	0. 62	1. 46	0. 64	1. 50	0. 66	1. 55	0. 68	1. 60	0. 70
156	1. 31	0. 53	1. 33	0. 54	1. 37	0. 56	1. 40	0. 57	1. 43	0. 58	1. 47	0. 60	1. 52	0. 62
158	1. 26	0. 47	1. 29	0. 48	1. 32	0. 49	1. 34	0. 50	1. 37	0. 51	1. 41	0. 53	1. 44	0. 54
160	1. 23	0. 42	1. 25	0. 43	1. 27	0. 43	1. 29	0. 44	1. 32	0. 45	1. 35	0. 46	1. 38	0. 47
	124°		126°		128°		130°		132°		134°		136°	
134	4. 77	3. 43												
136	3. 99	2. 77	4. 66	3. 23										
138	3. 43	2. 29	3. 89	2. 60	4. 54	3. 04								
140	3. 01	1. 93	3. 34	2. 15	3. 79	2. 44	4. 41	2. 84						
142	2. 68	1. 65	2. 94	1. 81	3. 26	2. 01	3. 68	2. 27	4. 28	2. 63				
144	2. 42	1. 42	2. 62	1. 54	2. 86	1. 68	3. 17	1. 86	3. 57	2. 10	4. 14	2. 43		
146	2. 21	1. 24	2. 37	1. 32	2. 55	1. 43	2. 78	1. 55	3. 07	1. 72	3. 46	1. 93	4. 00	2. 24
148	2. 04	1. 08	2. 16	1. 14	2. 30	1. 22	2. 48	1. 31	2. 70	1. 43	2. 97	1. 58	3. 34	1. 77
150	1. 89	0. 95	1. 99	0. 99	2. 10	1. 05	2. 24	1. 12	2. 40	1. 20	2. 61	1. 30	2. 87	1. 44
152	1. 77	0. 83	1. 85	0. 87	1. 94	0. 91	2. 04	0. 96	2. 17	1. 02	2. 33	1. 09	2. 52	1. 18
154	1. 66	0. 73	1. 72	0. 76	1. 80	0. 79	1. 88	0. 83	1. 98	0. 87	2. 10	0. 92	2. 25	0. 99
156	1. 56	0. 64	1. 62	0. 66	1. 68	0. 68	1. 75	0. 71	1. 83	0. 74	1. 92	0. 78	2. 03	0. 83
158	1. 48	0. 56	1. 53	0. 57	1. 58	0. 59	1. 63	0. 61	1. 70	0. 64	1. 77	0. 66	1. 85	0. 69
160	1. 41	0. 48	1. 45	0. 49	1. 49	0. 51	1. 53	0. 52	1. 58	0. 54	1. 64	0. 56	1. 71	0. 58
	138°		140°		142°		144°		146°		148°		150°	
148	3. 85	2. 04												
150	3. 22	1. 61	3. 70	1. 85										
152	2. 77	1. 30	3. 09	1. 45	3. 55	1. 66								
154	2. 43	1. 06	2. 66	1. 16	2. 96	1. 30	3. 38	1. 48						
156	2. 17	0. 88	2. 33	0. 95	2. 54	1. 04	2. 83	1. 15	3. 22	1. 31				
158	1. 96	0. 73	2. 08	0. 78	2. 23	0. 84	2. 43	0. 91	2. 69	1. 01	3. 05	1. 14		
160	1. 79	0. 61	1. 88	0. 64	1. 99	0. 68	2. 13	0. 73	2. 31	0. 79	2. 55	0. 87	2. 88	0. 98

TABLE 8

Distance of the Horizon

Height Feet	Nautical miles	Statute miles	Height *meters*	Height Feet	Nautical miles	Statute miles	Height *meters*
1	1.2	1.3	.30	120	12.8	14.7	36.58
2	1.7	1.9	.61	125	13.1	15.1	38.10
3	2.0	2.3	.91	130	13.3	15.4	39.62
4	2.3	2.7	1.22	135	13.6	15.6	41.15
5	2.6	3.0	1.52	140	13.8	15.9	42.67
6	2.9	3.3	1.83	145	14.1	16.2	44.20
7	3.1	3.6	2.13	150	14.3	16.5	45.72
8	3.3	3.8	2.44	160	14.8	17.0	48.77
9	3.5	4.0	2.74	170	15.3	17.6	51.82
10	3.7	4.3	3.05	180	15.7	18.1	54.86
11	3.9	4.5	3.35	190	16.1	18.6	57.91
12	4.1	4.7	3.66	200	16.5	19.0	60.96
13	4.2	4.9	3.96	210	17.0	19.5	64.01
14	4.4	5.0	4.27	220	17.4	20.0	67.06
15	4.5	5.2	4.57	230	17.7	20.4	70.10
16	4.7	5.4	4.88	240	18.1	20.9	73.15
17	4.8	5.6	5.18	250	18.5	21.3	76.20
18	5.0	5.7	5.49	260	18.9	21.7	79.25
19	5.1	5.9	5.79	270	19.2	22.1	82.30
20	5.2	6.0	6.10	280	19.6	22.5	85.34
21	5.4	6.2	6.40	290	19.9	22.9	88.39
22	5.5	6.3	6.71	300	20.3	23.3	91.44
23	5.6	6.5	7.01	310	20.6	23.7	94.49
24	5.7	6.6	7.32	320	20.9	24.1	97.54
25	5.9	6.7	7.62	330	21.3	24.5	100.58
26	6.0	6.9	7.92	340	21.6	24.8	103.63
27	6.1	7.0	8.23	350	21.9	25.2	106.68
28	6.2	7.1	8.53	360	22.2	25.5	109.73
29	6.3	7.3	8.84	370	22.5	25.9	112.78
30	6.4	7.4	9.14	380	22.8	26.2	115.82
31	6.5	7.5	9.45	390	23.1	26.6	118.87
32	6.6	7.6	9.75	400	23.4	26.9	121.92
33	6.7	7.7	10.06	410	23.7	27.3	124.97
34	6.8	7.9	10.36	420	24.0	27.6	128.02
35	6.9	8.0	10.67	430	24.3	27.9	131.06
36	7.0	8.1	10.97	440	24.5	28.2	134.11
37	7.1	8.2	11.28	450	24.8	28.6	137.16
38	7.2	8.3	11.58	460	25.1	28.9	140.21
39	7.3	8.4	11.89	470	25.4	29.2	143.26
40	7.4	8.5	12.19	480	25.6	29.5	146.30
41	7.5	8.6	12.50	490	25.9	29.8	149.35
42	7.6	8.7	12.80	500	26.2	30.1	152.40
43	7.7	8.8	13.11	510	26.4	30.4	155.45
44	7.8	8.9	13.41	520	26.7	30.7	158.50
45	7.8	9.0	13.72	530	26.9	31.0	161.54
46	7.9	9.1	14.02	540	27.2	31.3	164.59
47	8.0	9.2	14.33	550	27.4	31.6	167.64
48	8.1	9.3	14.63	560	27.7	31.9	170.69
49	8.2	9.4	14.94	570	27.9	32.1	173.74
50	8.3	9.5	15.24	580	28.2	32.4	176.78
55	8.7	10.0	16.76	590	28.4	32.7	179.83
60	9.1	10.4	18.29	600	28.7	33.0	182.88
65	9.4	10.9	19.81	620	29.1	33.5	188.98
70	9.8	11.3	21.34	640	29.5	34.1	195.07
75	10.1	11.7	22.86	660	30.1	34.6	201.17
80	10.5	12.0	24.38	680	30.5	35.1	207.26
85	10.8	12.4	25.91	700	31.0	35.6	213.36
90	11.1	12.8	27.43	720	31.4	36.1	219.46
95	11.4	13.1	28.96	740	31.8	36.6	225.55
100	11.7	13.5	30.48	760	32.3	37.1	231.65
105	12.0	13.8	32.00	780	32.7	37.6	237.74
110	12.3	14.1	33.53	800	33.1	38.1	243.84
115	12.5	14.4	35.05	820	33.5	38.6	249.94

TABLE 9

Distance by Vertical Angle

Measured Between Sea Horizon and Top of Object Beyond Sea Horizon

Angle	Difference in feet between height of object and height of eye of observer										Angle
	25	30	35	40	45	50	60	70	80	90	
° ′	*Miles*	*Miles*	*Miles*	*Miles*	*Miles*	*Miles*	*Miles*	*Miles*	*Miles*	*Miles*	° ′
−0 04	12.4	12.8	13.2	13.6	14.0	14.4	15.0	15.7	16.3	16.9	−0 04
−0 03	10.5	10.9	11.4	11.8	12.2	12.6	13.3	14.0	14.6	15.2	−0 03
−0 02	8.7	9.2	9.7	10.2	10.6	11.0	11.8	12.5	13.1	13.7	−0 02
−0 01	7.2	7.7	8.2	8.7	9.1	9.5	10.3	11.0	11.7	12.3	−0 01
0 00	5.8	6.4	6.9	7.4	7.8	8.2	9.0	9.7	10.4	11.1	0 00
0 01	4.8	5.3	5.8	6.3	6.7	7.1	7.9	8.6	9.3	9.9	0 01
0 02	3.9	4.4	4.9	5.4	5.8	6.2	6.9	7.6	8.3	8.9	0 02
0 03	3.3	3.7	4.2	4.6	5.0	5.4	6.1	6.8	7.4	8.0	0 03
0 04	2.8	3.2	3.6	4.0	4.4	4.7	5.4	6.1	6.7	7.3	0 04
0 05	2.4	2.8	3.1	3.5	3.9	4.2	4.8	5.5	6.0	6.6	0 05
0 06	2.1	2.4	2.8	3.1	3.4	3.7	4.3	4.9	5.5	6.0	0 06
0 07	1.8	2.2	2.5	2.8	3.1	3.4	3.9	4.5	5.0	5.5	0 07
0 08	1.6	1.9	2.2	2.5	2.8	3.1	3.6	4.1	4.6	5.0	0 08
0 09	1.5	1.7	2.0	2.3	2.5	2.8	3.3	3.8	4.2	4.7	0 09
0 10	1.3	1.6	1.8	2.1	2.3	2.6	3.0	3.5	3.9	4.3	0 10
0 15	.9	1.1	1.3	1.5	1.6	1.8	2.1	2.5	2.8	3.1	0 15
0 20	.7	.8	1.0	1.1	1.2	1.4	1.6	1.9	2.2	2.4	0 20
0 25	.6	.7	.8	.9	1.0	1.1	1.3	1.5	1.8	2.0	0 25
0 30	.5	.6	.7	.7	.8	.9	1.1	1.3	1.5	1.7	0 30
0 35		.5	.6	.6	.7	.8	1.0	1.1	1.3	1.4	0 35
0 40			.5	.6	.6	.7	.8	1.0	1.1	1.3	0 40
0 45				.5	.6	.6	.7	.9	1.0	1.1	0 45
0 50				.5	.5	.6	.7	.8	.9	1.0	0 50
0 55					.5	.5	.6	.7	.8	.9	0 55
1 00						.5	.6	.7	.8	.8	1 00
1 10							.5	.6	.6	.7	1 10
1 20								.5	.6	.6	1 20
1 30									.5	.6	1 30
1 40									.5	.5	1 40
1 50										.5	1 50

TABLE 9

Distance by Vertical Angle

Measured Between Sea Horizon and Top of Object Beyond Sea Horizon

Angle	Difference in feet between height of object and height of eye of observer											Angle
	100	120	140	160	180	200	250	300	350	400	450	
° ′	*Miles*	*Miles*	*Miles*	*Miles*	*Miles*	*Miles*	*Miles*	*Miles*	*Miles*	*Miles*	*Miles*	° ′
0 00	11.7	12.8	13.8	14.8	15.7	16.5	18.4	20.2	21.8	23.3	24.7	0 00
0 0	10.5	11.6	12.7	13.6	14.5	15.3	17.3	19.0	20.7	22.2	23.6	0 01
0 02	9.5	10.6	11.6	12.5	13.4	14.3	16.2	17.9	19.6	21.0	22.5	0 02
0 03	8.6	9.7	10.7	11.6	12.5	13.3	15.2	16.9	18.5	20.0	21.4	0 03
0 04	7.8	8.8	9.8	10.7	11.6	12.4	14.3	16.0	17.5	19.0	20.4	0 04
0 05	7.1	8.1	9.0	9.9	10.8	11.5	13.4	15.1	16.6	18.1	19.5	0 05
0 06	6.5	7.5	8.4	9.2	10.0	10.8	12.6	14.2	15.8	17.2	18.6	0 06
0 07	6.0	6.9	7.7	8.6	9.4	10.1	11.9	13.5	15.0	16.4	17.7	0 07
0 08	5.5	6.4	7.2	8.0	8.8	9.5	11.2	12.8	14.2	15.6	16.9	0 08
0 09	5.1	5.9	6.7	7.5	8.2	8.9	10.6	12.1	13.5	14.9	16.2	0 09
0 10	4.7	5.5	6.3	7.0	7.7	8.4	10.0	11.5	12.9	14.2	15.5	0 10
0 11	4.4	5.2	5.9	6.6	7.3	7.9	9.5	10.9	12.3	13.6	14.8	0 11
0 12	4.1	4.8	5.5	6.2	6.9	7.5	9.0	10.4	11.7	13.0	14.2	0 12
0 13	3.9	4.6	5.2	5.9	6.5	7.1	8.5	9.9	11.2	12.5	13.6	0 13
0 14	3.6	4.3	4.9	5.6	6.2	6.7	8.1	9.5	10.7	11.9	13.1	0 14
0 15	3.4	4.1	4.7	5.3	5.8	6.4	7.8	9.0	10.3	11.5	12.6	0 15
0 20	2.7	3.2	3.7	4.2	4.6	5.1	6.3	7.4	8.4	9.5	10.5	0 20
0 25	2.2	2.6	3.0	3.4	3.8	4.2	5.2	6.2	7.1	8.0	8.9	0 25
0 30	1.8	2.2	2.6	2.9	3.2	3.6	4.4	5.3	6.1	6.9	7.7	0 30
0 35	1.6	1.9	2.2	2.5	2.8	3.1	3.9	4.6	5.3	6.0	6.7	0 35
0 40	1.4	1.7	1.9	2.2	2.5	2.8	3.4	4.1	4.7	5.4	6.0	0 40
0 45	1.2	1.5	1.7	2.0	2.2	2.5	3.1	3.6	4.2	4.8	5.4	0 45
0 50	1.1	1.3	1.6	1.8	2.0	2.2	2.8	3.3	3.8	4.4	4.9	0 50
0 55	1.0	1.2	1.4	1.6	1.8	2.0	2.5	3.0	3.5	4.0	4.5	0 55
1 00	.9	1.1	1.3	1.5	1.7	1.9	2.3	2.8	3.2	3.7	4.1	1 00
1 10	.8	1.0	1.1	1.3	1.4	1.6	2.0	2.4	2.8	3.2	3.6	1 10
1 20	.7	.8	1.0	1.1	1.3	1.4	1.8	2.1	2.4	2.8	3.1	1 20
1 30	.6	.8	.9	1.0	1.1	1.2	1.6	1.9	2.2	2.5	2.8	1 30
1 40	.6	.7	.8	.9	1.0	1.1	1.4	1.7	2.0	2.2	2.5	1 40
1 50	.5	.6	.7	.8	.9	1.0	1.3	1.5	1.8	2.0	2.3	1 50
2 00	.5	.6	.7	.8	.8	.9	1.2	1.4	1.6	1.9	2.1	2 00
2 30		.5	.5	.6	.7	.8	.9	1.1	1.3	1.5	1.7	2 30
3 00				.5	.6	.6	.8	.9	1.1	1.3	1.4	3 00
3 30					.5	.5	.7	.8	.9	1.1	1.2	3 30
4 00						.5	.6	.7	.8	.9	1.1	4 00
4 30							.5	.6	.7	.8	.9	4 30
5 00							.5	.6	.7	.8	.8	5 00
6 00								.5	.5	.6	.7	6 00
7 00									.5	.5	.6	7 00
8 00										.5	.5	8 00
10 00												10 00

TABLE 9

Distance by Vertical Angle

Measured Between Sea Horizon and Top of Object Beyond Sea Horizon

Angle	Difference in feet between height of object and height of eye of observer											Angle
	500	600	700	800	900	1000	1200	1400	1600	1800	2000	
° ′	*Miles*	*Miles*	*Miles*	*Miles*	*Miles*	*Miles*	*Miles*	*Miles*	*Miles*	*Miles*	*Miles*	° ′
0 05	20.8	23.2	25.4	27.5	29.5	31.4	34.8	38.0	41.0	43.8	46.5	0 05
0 06	19.8	22.3	24.5	26.6	28.5	30.4	33.8	37.0	40.0	42.8	45.4	0 06
0 07	19.0	21.4	23.6	25.6	27.6	29.4	32.9	36.0	39.0	41.8	44.4	0 07
0 08	18.2	20.5	22.7	24.7	26.7	28.5	31.9	35.1	38.0	40.8	43.4	0 08
0 09	17.4	19.7	21.9	23.9	25.8	27.6	31.0	34.1	37.0	39.8	42.5	0 09
0 10	16.7	19.0	21.1	23.1	25.0	26.8	30.1	33.2	36.2	38.9	41.5	0 10
0 11	16.0	18.3	20.4	22.3	24.2	26.0	29.3	32.4	35.3	38.0	40.6	0 11
0 12	15.4	17.6	19.6	21.6	23.4	25.2	28.5	31.5	34.4	37.1	39.7	0 12
0 13	14.8	16.9	19.0	20.9	22.7	24.4	27.7	30.7	33.6	36.3	38.8	0 13
0 14	14.2	16.3	18.3	20.2	22.0	23.7	26.9	30.0	32.8	35.4	38.0	0 14
0 15	13.7	15.8	17.7	19.6	21.3	23.0	26.2	29.2	32.0	34.6	37.2	0 15
0 17	12.7	14.7	16.6	18.4	20.1	21.7	24.8	27.8	30.5	33.1	35.6	0 17
0 20	11.4	13.3	15.1	16.8	18.4	20.0	23.0	25.8	28.4	31.0	33.4	0 20
0 25	9.7	11.4	13.0	14.6	16.1	17.5	20.3	22.9	25.4	27.8	30.1	0 25
0 30	8.4	9.9	11.4	12.8	14.2	15.5	18.1	20.5	22.9	25.2	27.4	0 30
0 35	7.4	8.8	10.1	11.4	12.6	13.9	16.3	18.5	20.7	22.9	24.9	0 35
0 40	6.6	7.8	9.0	10.2	11.4	12.5	14.7	16.9	18.9	20.9	22.9	0 40
0 45	6.0	7.1	8.2	9.3	10.3	11.4	13.4	15.4	17.3	19.2	21.1	0 45
0 50	5.4	6.4	7.5	8.5	9.4	10.4	12.3	14.2	16.0	17.7	19.5	0 50
0 55	5.0	5.9	6.8	7.8	8.7	9.6	11.4	13.1	14.8	16.5	18.1	0 55
1 00	4.6	5.5	6.3	7.2	8.0	8.9	10.5	12.2	13.8	15.3	16.9	1 00
1 10	3.9	4.7	5.5	6.2	7.0	7.7	9.2	10.6	12.1	13.5	14.9	1 10
1 20	3.5	4.2	4.8	5.5	6.2	6.8	8.1	9.4	10.7	12.0	13.2	1 20
1 30	3.1	3.7	4.3	4.9	5.5	6.1	7.3	8.5	9.6	10.8	11.9	1 30
1 40	2.8	3.3	3.9	4.4	5.0	5.5	6.6	7.7	8.7	9.8	10.8	1 40
1 50	2.5	3.0	3.6	4.1	4.5	5.0	6.0	7.0	8.0	9.0	9.9	1 50
2 00	2.3	2.8	3.3	3.7	4.2	4.6	5.5	6.5	7.4	8.2	9.1	2 00
2 30	1.9	2.2	2.6	3.0	3.4	3.7	4.5	5.2	5.9	6.7	7.4	2 30
3 00	1.6	1.9	2.2	2.5	2.8	3.1	3.7	4.4	5.0	5.6	6.2	3 00
3 30	1.3	1.6	1.9	2.1	2.4	2.7	3.2	3.7	4.3	4.8	5.3	3 30
4 00	1.2	1.4	1.6	1.9	2.1	2.3	2.8	3.3	3.7	4.2	4.7	4 00
5 00	.9	1.1	1.3	1.5	1.7	1.9	2.3	2.6	3.0	3.4	3.7	5 00
6 00	.8	.9	1.1	1.3	1.4	1.6	1.9	2.2	2.5	2.8	3.1	6 00
7 00	.7	.8	.9	1.1	1.2	1.3	1.6	1.9	2.1	2.4	2.7	7 00
8 00	.6	.7	.8	.9	1.1	1.2	1.4	1.6	1.9	2.1	2.3	8 00
10 00	.5	.6	.7	.7	.8	.9	1.1	1.3	1.5	1.7	1.9	10 00
12 00		.5	.5	.6	.7	.8	.9	1.1	1.2	1.4	1.5	12 00
15 00				.5	.6	.6	.7	.9	1.0	1.1	1.2	15 00
20 00						.5	.5	.6	.7	.8	.9	20 00
25 00								.5	.6	.6	.7	25 00
30 00									.5	.5	.6	30 00

TABLE 10
Direction and Speed of True Wind in Units of Ship's Speed

Apparent wind speed	Difference between the heading and apparent wind direction										Apparent wind speed
	0°		10°		20°		30°		40°		
	°		°		°		°		°		
0.0	180	1.00	180	1.00	180	1.00	180	1.00	180	1.00	0.0
0.1	180	0.90	179	0.90	178	0.91	177	0.91	176	0.93	0.1
0.2	180	0.80	178	0.80	175	0.81	173	0.83	171	0.86	0.2
0.3	180	0.70	176	0.71	172	0.73	169	0.76	166	0.79	0.3
0.4	180	0.60	173	0.61	168	0.64	163	0.68	160	0.74	0.4
0.5	180	0.50	170	0.51	162	0.56	156	0.62	152	0.70	0.5
0.6	180	0.40	166	0.42	155	0.48	148	0.57	144	0.66	0.6
0.7	180	0.30	159	0.33	145	0.42	138	0.53	136	0.65	0.7
0.8	180	0.20	147	0.25	132	0.37	127	0.50	127	0.64	0.8
0.9	180	0.10	126	0.19	117	0.34	116	0.50	118	0.66	0.9
1.0	calm	0.00	95	0.17	100	0.35	105	0.52	110	0.68	1.0
1.1	0	0.10	66	0.21	85	0.38	95	0.55	103	0.72	1.1
1.2	0	0.20	49	0.28	73	0.43	86	0.60	96	0.78	1.2
1.3	0	0.30	39	0.36	64	0.50	79	0.66	90	0.84	1.3
1.4	0	0.40	33	0.45	57	0.57	73	0.73	85	0.90	1.4
1.5	0	0.50	29	0.54	51	0.66	68	0.81	81	0.98	1.5
1.6	0	0.60	26	0.64	47	0.74	64	0.89	78	1.05	1.6
1.7	0	0.70	24	0.74	44	0.83	61	0.97	75	1.13	1.7
1.8	0	0.80	22	0.83	42	0.93	58	1.06	72	1.22	1.8
1.9	0	0.90	21	0.93	40	1.02	56	1.15	70	1.30	1.9
2.0	0	1.00	20	1.03	38	1.11	54	1.24	68	1.39	2.0
2.5	0	1.50	17	1.52	32	1.60	47	1.71	60	1.85	2.5
3.0	0	2.00	15	2.02	29	2.09	43	2.19	56	2.32	3.0
3.5	0	2.50	14	2.52	28	2.58	41	2.68	53	2.81	3.5
4.0	0	3.00	13	3.02	26	3.08	39	3.17	51	3.30	4.0
4.5	0	3.50	13	3.52	25	3.58	38	3.67	50	3.79	4.5
5.0	0	4.00	12	4.02	25	4.08	37	4.16	49	4.28	5.0
6.0	0	5.00	12	5.02	24	5.07	36	5.16	47	5.27	6.0
7.0	0	6.00	12	6.02	23	6.07	35	6.15	46	6.27	7.0
8.0	0	7.00	11	7.02	23	7.07	34	7.15	45	7.26	8.0
9.0	0	8.00	11	8.02	22	8.07	34	8.15	44	8.26	9.0
10.0	0	9.00	11	9.02	22	9.06	33	9.15	44	9.26	10.0

Apparent wind speed	50°		60°		70°		80°		90°		Apparent wind speed
	°		°		°		°		°		
0.0	180	1.00	180	1.00	180	1.00	180	1.00	180	1.00	0.0
0.1	175	0.94	175	0.95	174	0.97	174	0.99	174	1.00	0.1
0.2	170	0.88	169	0.92	169	0.95	168	0.99	169	1.02	0.2
0.3	164	0.84	163	0.89	163	0.94	163	0.99	163	1.04	0.3
0.4	158	0.80	157	0.87	156	0.94	157	1.01	158	1.08	0.4
0.5	151	0.78	150	0.87	150	0.95	152	1.04	153	1.12	0.5
0.6	143	0.77	143	0.87	145	0.97	147	1.07	149	1.17	0.6
0.7	136	0.77	137	0.89	139	1.01	142	1.12	145	1.22	0.7
0.8	128	0.78	131	0.92	134	1.05	138	1.17	141	1.28	0.8
0.9	121	0.81	125	0.95	129	1.09	134	1.22	138	1.35	0.9
1.0	115	0.85	120	1.00	125	1.15	130	1.29	135	1.41	1.0
1.1	109	0.89	115	1.05	121	1.21	127	1.35	132	1.49	1.1
1.2	104	0.95	111	1.11	118	1.27	124	1.42	130	1.56	1.2
1.3	99	1.01	107	1.18	114	1.34	121	1.50	128	1.64	1.3
1.4	95	1.08	104	1.25	112	1.42	119	1.57	126	1.72	1.4
1.5	92	1.15	101	1.32	109	1.49	117	1.65	124	1.80	1.5
1.6	89	1.23	98	1.40	107	1.57	115	1.73	122	1.89	1.6
1.7	86	1.31	96	1.48	105	1.65	113	1.82	120	1.97	1.7
1.8	84	1.39	94	1.56	103	1.73	111	1.90	119	2.06	1.8
1.9	81	1.47	92	1.65	101	1.82	110	1.99	118	2.15	1.9
2.0	79	1.56	90	1.73	100	1.91	108	2.07	117	2.24	2.0
2.5	72	2.01	83	2.18	94	2.35	103	2.53	112	2.69	2.5
3.0	68	2.48	79	2.65	89	2.82	99	2.99	108	3.16	3.0
3.5	65	2.96	76	3.12	87	3.29	96	3.47	106	3.64	3.5
4.0	63	3.44	74	3.61	84	3.78	94	3.95	104	4.12	4.0
4.5	61	3.93	72	4.09	83	4.26	93	4.44	103	4.61	4.5
5.0	60	4.42	71	4.58	81	4.75	92	4.93	101	5.10	5.0
6.0	58	5.41	69	5.57	79	5.74	90	5.91	99	6.08	6.0
7.0	57	6.40	68	6.56	78	6.72	88	6.90	98	7.07	7.0
8.0	56	7.40	67	7.55	77	7.72	87	7.89	97	8.06	8.0
9.0	55	8.39	66	8.54	76	8.71	86	8.88	96	9.06	9.0
10.0	55	9.39	65	9.54	76	9.70	86	9.88	96	10.01	10.0

TABLE 10

Direction and Speed of True Wind in Units of Ship's Speed

Apparent wind speed	Difference between the heading and apparent wind direction										Apparent wind speed
	90°		100°		110°		120°		130°		
	°		°		°		°		°		
0.0	180	1.00	180	1.00	180	1.00	180	1.00	180	1.00	0.0
0.1	174	1.00	174	1.02	175	1.04	175	1.05	176	1.07	0.1
0.2	169	1.02	169	1.05	170	1.08	171	1.11	172	1.14	0.2
0.3	163	1.04	164	1.09	166	1.14	167	1.18	169	1.21	0.3
0.4	158	1.08	160	1.14	162	1.20	164	1.25	166	1.29	0.4
0.5	153	1.12	156	1.19	158	1.26	161	1.32	164	1.38	0.5
0.6	149	1.17	152	1.25	155	1.33	158	1.40	162	1.46	0.6
0.7	145	1.22	148	1.32	152	1.40	156	1.48	160	1.55	0.7
0.8	141	1.28	145	1.38	149	1.48	154	1.56	158	1.63	0.8
0.9	138	1.35	143	1.46	147	1.56	152	1.65	156	1.72	0.9
1.0	135	1.41	140	1.53	145	1.64	150	1.73	155	1.81	1.0
1.1	132	1.49	138	1.61	143	1.72	148	1.82	154	1.90	1.1
1.2	130	1.56	136	1.69	141	1.81	147	1.91	153	2.00	1.2
1.3	128	1.64	134	1.77	140	1.89	146	2.00	152	2.09	1.3
1.4	126	1.72	132	1.86	138	1.98	145	2.09	151	2.18	1.4
1.5	124	1.80	130	1.94	137	2.07	143	2.18	150	2.28	1.5
1.6	122	1.89	129	2.03	136	2.16	142	2.27	149	2.37	1.6
1.7	120	1.97	128	2.12	135	2.25	141	2.36	148	2.46	1.7
1.8	119	2.06	127	2.21	134	2.34	141	2.46	147	2.56	1.8
1.9	118	2.15	125	2.30	133	2.43	140	2.55	147	2.66	1.9
2.0	117	2.24	124	2.39	132	2.52	139	2.65	146	2.75	2.0
2.5	112	2.69	120	2.85	128	2.99	136	3.12	144	3.23	2.5
3.0	108	3.16	117	3.32	126	3.47	134	3.61	142	3.72	3.0
3.5	106	3.64	115	3.80	124	3.96	132	4.09	140	4.21	3.5
4.0	104	4.12	113	4.29	122	4.44	131	4.58	139	4.71	4.0
4.5	103	4.61	112	4.78	121	4.93	130	5.07	138	5.20	4.5
5.0	101	5.10	111	5.27	120	5.42	129	5.57	138	5.69	5.0
6.0	99	6.08	109	6.25	118	6.41	128	6.56	137	6.69	6.0
7.0	98	7.07	108	7.24	117	7.40	127	7.55	136	7.68	7.0
8.0	97	8.06	107	8.23	116	8.39	126	8.54	135	8.68	8.0
9.0	96	9.06	106	9.23	116	9.39	125	9.54	135	9.67	9.0
10.0	96	10.01	106	10.22	115	10.39	125	10.54	134	10.67	10.0

Apparent wind speed	140°		150°		160°		170°		180°		Apparent wind speed
	°		°		°		°		°		
0.0	180	1.00	180	1.00	180	1.00	180	1.00	180	1.00	0.0
0.1	177	1.08	177	1.09	178	1.09	179	1.10	180	1.10	0.1
0.2	174	1.16	175	1.18	177	1.19	178	1.20	180	1.20	0.2
0.3	171	1.24	173	1.27	175	1.29	178	1.30	180	1.30	0.3
0.4	169	1.33	172	1.36	174	1.38	177	1.40	180	1.40	0.4
0.5	167	1.42	170	1.45	173	1.48	177	1.50	180	1.50	0.5
0.6	165	1.51	169	1.55	173	1.58	176	1.60	180	1.60	0.6
0.7	164	1.60	168	1.64	172	1.68	176	1.69	180	1.70	0.7
0.8	162	1.69	167	1.74	171	1.77	176	1.79	180	1.80	0.8
0.9	161	1.79	166	1.84	171	1.87	175	1.89	180	1.90	0.9
1.0	160	1.88	165	1.93	170	1.97	175	1.99	180	2.00	1.0
1.1	159	1.97	164	2.03	170	2.07	175	2.09	180	2.10	1.1
1.2	158	2.07	164	2.13	169	2.17	175	2.19	180	2.20	1.2
1.3	157	2.16	163	2.22	169	2.27	174	2.29	180	2.30	1.3
1.4	157	2.26	162	2.32	168	2.36	174	2.39	180	2.40	1.4
1.5	156	2.36	162	2.42	168	2.46	174	2.49	180	2.50	1.5
1.6	155	2.45	161	2.52	168	2.56	174	2.59	180	2.60	1.6
1.7	155	2.55	161	2.61	167	2.66	174	2.69	180	2.70	1.7
1.8	154	2.65	161	2.71	167	2.76	174	2.79	180	2.80	1.8
1.9	154	2.74	160	2.81	167	2.86	173	2.89	180	2.90	1.9
2.0	153	2.84	160	2.91	167	2.96	173	2.99	180	3.00	2.0
2.5	151	3.33	158	3.40	166	3.46	173	3.49	180	3.50	2.5
3.0	150	3.82	157	3.90	165	3.95	172	3.99	180	4.00	3.0
3.5	149	4.31	157	4.39	164	4.45	172	4.49	180	4.50	3.5
4.0	148	4.81	156	4.89	164	4.95	172	4.99	180	5.00	4.0
4.5	147	5.31	155	5.39	164	5.45	172	5.49	180	5.50	4.5
5.0	146	5.80	155	5.89	163	5.95	172	5.99	180	6.00	5.0
6.0	145	6.80	154	6.88	163	6.95	171	6.99	180	7.00	6.0
7.0	145	7.79	154	7.88	162	7.95	171	7.99	180	8.00	7.0
8.0	144	8.79	153	8.88	162	8.95	171	8.99	180	9.00	8.0
9.0	144	9.79	153	9.88	162	9.95	171	9.99	180	10.00	9.0
10.0	143	10.78	153	10.88	162	10.95	171	10.98	180	11.00	10.0

TABLE 11

Correction of Barometer Reading for Height Above Sea Level

All barometers. All values positive.

Height in feet	Outside temperature in degrees Fahrenheit													Height in feet
	−20°	−10°	0°	10°	20°	30°	40°	50°	60°	70°	80°	90°	100°	
	Inches	*Inches*	*Inches*	*Inches*	*Inches*	*Inches*	*Inches*	*Inches*	*Inches*	*Inches*	*Inches*	*Inches*	*Inches*	
5	0. 01	0. 01	0. 01	0. 01	0. 01	0. 01	0. 01	0. 01	0. 01	0. 01	0. 01	0. 01	0. 01	5
10	0. 01	0. 01	0. 01	0. 01	0. 01	0. 01	0. 01	0. 01	0. 01	0. 01	0. 01	0. 01	0. 01	10
15	0. 02	0. 02	0. 02	0. 02	0. 02	0. 02	0. 02	0. 02	0. 02	0. 02	0. 02	0. 02	0. 02	15
20	0. 03	0. 02	0. 02	0. 02	0. 02	0. 02	0. 02	0. 02	0. 02	0. 02	0. 02	0. 02	0. 02	20
25	0. 03	0. 03	0. 03	0. 03	0. 03	0. 03	0. 03	0. 03	0. 03	0. 03	0. 03	0. 03	0. 03	25
30	0. 04	0. 04	0. 04	0. 04	0. 04	0. 03	0. 03	0. 03	0. 03	0. 03	0. 03	0. 03	0. 03	30
35	0. 04	0. 04	0. 04	0. 04	0. 04	0. 04	0. 04	0. 04	0. 04	0. 04	0. 04	0. 04	0. 04	35
40	0. 05	0. 05	0. 05	0. 05	0. 05	0. 05	0. 04	0. 04	0. 04	0. 04	0. 04	0. 04	0. 04	40
45	0. 06	0. 06	0. 05	0. 05	0. 05	0. 05	0. 05	0. 05	0. 05	0. 05	0. 05	0. 05	0. 05	45
50	0. 06	0. 06	0. 06	0. 06	0. 06	0. 06	0. 06	0. 06	0. 05	0. 05	0. 05	0. 05	0. 05	50
55	0. 07	0. 07	0. 07	0. 07	0. 06	0. 06	0. 06	0. 06	0. 06	0. 06	0. 06	0. 06	0. 06	55
60	0. 08	0. 07	0. 07	0. 07	0. 07	0. 07	0. 07	0. 07	0. 06	0. 06	0. 06	0. 06	0. 06	60
65	0. 08	0. 08	0. 08	0. 08	0. 08	0. 07	0. 07	0. 07	0. 07	0. 07	0. 07	0. 07	0. 07	65
70	0. 09	0. 09	0. 09	0. 08	0. 08	0. 08	0. 08	0. 08	0. 08	0. 07	0. 07	0. 07	0. 07	70
75	0. 10	0. 09	0. 09	0. 09	0. 09	0. 09	0. 08	0. 08	0. 08	0. 08	0. 08	0. 08	0. 08	75
80	0. 10	0. 10	0. 10	0. 10	0. 09	0. 09	0. 09	0. 09	0. 09	0. 08	0. 08	0. 08	0. 08	80
85	0. 11	0. 11	0. 10	0. 10	0. 10	0. 10	0. 10	0. 09	0. 09	0. 09	0. 09	0. 09	0. 09	85
90	0. 11	0. 11	0. 11	0. 11	0. 11	0. 10	0. 10	0. 10	0. 10	0. 10	0. 09	0. 09	0. 09	90
95	0. 12	0. 12	0. 12	0. 11	0. 11	0. 11	0. 11	0. 10	0. 10	0. 10	0. 10	0. 10	0. 10	95
100	0. 13	0. 12	0. 12	0. 12	0. 12	0. 11	0. 11	0. 11	0. 11	0. 11	0. 10	0. 10	0. 10	100
105	0. 13	0. 13	0. 13	0. 13	0. 12	0. 12	0. 12	0. 12	0. 11	0. 11	0. 11	0. 11	0. 11	105
110	0. 14	0. 14	0. 13	0. 13	0. 13	0. 13	0. 12	0. 12	0. 12	0. 12	0. 11	0. 11	0. 11	110
115	0. 15	0. 14	0. 14	0. 14	0. 13	0. 13	0. 13	0. 13	0. 12	0. 12	0. 12	0. 12	0. 12	115
120	0. 15	0. 15	0. 15	0. 14	0. 14	0. 14	0. 13	0. 13	0. 13	0. 13	0. 12	0. 12	0. 12	120
125	0. 16	0. 16	0. 15	0. 15	0. 15	0. 14	0. 14	0. 14	0. 13	0. 13	0. 13	0. 13	0. 12	125

TABLE 12

Correction of Barometer Reading for Gravity

Mercurial barometers only.

Latitude	Correction	Latitude	Correction	Latitude	Correction	Latitude	Correction
°	*Inches*	°	*Inches*	°	*Inches*	°	*Inches*
0	−0. 08	25	−0. 05	50	+0. 01	75	+0. 07
5	−0. 08	30	−0. 04	55	+0. 03	80	+0. 07
10	−0. 08	35	−0. 03	60	+0. 04	85	+0. 08
15	−0. 07	40	−0. 02	65	+0. 05	90	+0. 08
20	−0. 06	45	0. 00	70	+0. 06		

TABLE 13

Correction of Barometer Reading for Temperature

Mercurial barometers only.

Temp. F	Height of barometer in inches								Temp. F
	27.5	28.0	28.5	29.0	29.5	30.0	30.5	31.0	
°	*Inches*	*Inches*	*Inches*	*Inches*	*Inches*	*Inches*	*Inches*	*Inches*	°
−20	+0.12	+0.12	+0.13	+0.13	+0.13	+0.13	+0.14	+0.14	−20
18	0.12	0.12	0.12	0.12	0.13	0.13	0.13	0.13	18
16	0.11	0.11	0.12	0.12	0.12	0.12	0.12	0.13	16
14	0.11	0.11	0.11	0.11	0.11	0.12	0.12	0.12	14
12	0.10	0.10	0.11	0.11	0.11	0.11	0.11	0.11	12
−10	+0.10	+0.10	+0.10	+0.10	+0.10	+0.11	+0.11	+0.11	−10
8	0.09	0.09	0.10	0.10	0.10	0.10	0.10	0.10	8
6	0.09	0.09	0.09	0.09	0.09	0.09	0.10	0.10	6
4	0.08	0.08	0.08	0.09	0.09	0.09	0.09	0.09	4
−2	0.08	0.08	0.08	0.08	0.08	0.08	0.09	0.09	−2
0	+0.07	+0.07	+0.07	+0.08	+0.08	+0.08	+0.08	+0.08	0
+2	0.07	0.07	0.07	0.07	0.07	0.07	0.07	0.08	+2
4	0.06	0.06	0.06	0.07	0.07	0.07	0.07	0.07	4
6	0.06	0.06	0.06	0.06	0.06	0.06	0.06	0.06	6
8	0.05	0.05	0.05	0.05	0.06	0.06	0.06	0.06	8
+10	+0.05	+0.05	+0.05	+0.05	+0.05	+0.05	+0.05	+0.05	+10
12	0.04	0.04	0.04	0.04	0.04	0.05	0.05	0.05	12
14	0.04	0.04	0.04	0.04	0.04	0.04	0.04	0.04	14
16	0.03	0.03	0.03	0.03	0.03	0.03	0.03	0.04	16
18	0.03	0.03	0.03	0.03	0.03	0.03	0.03	0.03	18
+20	+0.02	+0.02	+0.02	+0.02	+0.02	+0.02	+0.02	+0.02	+20
22	0.02	0.02	0.02	0.02	0.02	0.02	0.02	0.02	22
24	0.01	0.01	0.01	0.01	0.01	0.01	0.01	0.01	24
26	+0.01	+0.01	+0.01	+0.01	+0.01	+0.01	+0.01	+0.01	26
28	0.00	0.00	0.00	0.00	0.00	0.00	0.00	0.00	28
+30	0.00	0.00	0.00	0.00	0.00	0.00	0.00	0.00	+30
32	−0.01	−0.01	−0.01	−0.01	−0.01	−0.01	−0.01	−0.01	32
34	0.01	0.01	0.01	0.01	0.01	0.01	0.01	0.02	34
36	0.02	0.02	0.02	0.02	0.02	0.02	0.02	0.02	36
38	0.02	0.02	0.02	0.02	0.03	0.03	0.03	0.03	38
+40	−0.03	−0.03	−0.03	−0.03	−0.03	−0.03	−0.03	−0.03	+40
42	0.03	0.03	0.03	0.04	0.04	0.04	0.04	0.04	42
44	0.04	0.04	0.04	0.04	0.04	0.04	0.04	0.04	44
46	0.04	0.04	0.04	0.05	0.05	0.05	0.05	0.05	46
48	0.05	0.05	0.05	0.05	0.05	0.05	0.05	0.05	48
+50	−0.05	−0.05	−0.06	−0.06	−0.06	−0.06	−0.06	−0.06	+50
52	0.06	0.06	0.06	0.06	0.06	0.06	0.06	0.07	52
54	0.06	0.06	0.07	0.07	0.07	0.07	0.07	0.07	54
56	0.07	0.07	0.07	0.07	0.07	0.07	0.08	0.08	56
58	0.07	0.07	0.08	0.08	0.08	0.08	0.08	0.08	58
+60	−0.08	−0.08	−0.08	−0.08	−0.08	−0.09	−0.09	−0.09	+60
62	0.08	0.08	0.09	0.09	0.09	0.09	0.09	0.09	62
64	0.09	0.09	0.09	0.09	0.09	0.10	0.10	0.10	64
66	0.09	0.09	0.10	0.10	0.10	0.10	0.10	0.10	66
68	0.10	0.10	0.10	0.10	0.11	0.11	0.11	0.11	68
+70	−0.10	−0.10	−0.11	−0.11	−0.11	−0.11	−0.11	−0.12	+70
72	0.11	0.11	0.11	0.11	0.12	0.12	0.12	0.12	72
74	0.11	0.11	0.12	0.12	0.12	0.12	0.13	0.13	74
76	0.12	0.12	0.12	0.12	0.13	0.13	0.13	0.13	76
78	0.12	0.12	0.13	0.13	0.13	0.13	0.14	0.14	78
+80	−0.13	−0.13	−0.13	−0.13	−0.14	−0.14	−0.14	−0.14	+80
82	0.13	0.14	0.14	0.14	0.14	0.14	0.15	0.15	82
84	0.14	0.14	0.14	0.15	0.15	0.15	0.15	0.16	84
86	0.14	0.15	0.15	0.15	0.15	0.16	0.16	0.16	86
88	0.15	0.15	0.15	0.16	0.16	0.16	0.16	0.17	88
+90	−0.15	−0.16	−0.16	−0.16	−0.16	−0.17	−0.17	−0.17	+90
92	0.16	0.16	0.16	0.17	0.17	0.17	0.17	0.18	92
94	0.16	0.17	0.17	0.17	0.17	0.18	0.18	0.18	94
96	0.17	0.17	0.17	0.18	0.18	0.18	0.19	0.19	96
98	0.17	0.18	0.18	0.18	0.18	0.19	0.19	0.19	98
100	0.18	0.18	0.18	0.19	0.19	0.19	0.20	0.20	100

TABLE 14

Conversion Table for Millibars, Inches of Mercury, and Millimeters of Mercury

Millibars	Inches	Millimeters	Millibars	Inches	Millimeters	Millibars	Inches	Millimeters
900	26. 58	675. 1	960	28. 35	720. 1	1020	30. 12	765. 1
901	26. 61	675. 8	961	28. 38	720. 8	1021	30. 15	765. 8
902	26. 64	676. 6	962	28. 41	721. 6	1022	30. 18	766. 6
903	26. 67	677. 3	963	28. 44	722. 3	1023	30. 21	767. 3
904	26. 70	678. 1	964	28. 47	723. 1	1024	30. 24	768. 1
905	26. 72	678. 8	965	28. 50	723. 8	1025	30. 27	768. 8
906	26. 75	679. 6	966	28. 53	724. 6	1026	30. 30	769. 6
907	26. 78	680. 3	967	28. 56	725. 3	1027	30. 33	770. 3
908	26. 81	681. 1	968	28. 58	726. 1	1028	30. 36	771. 1
909	26. 84	681. 8	969	28. 61	726. 8	1029	30. 39	771. 8
910	26. 87	682. 6	970	28. 64	727. 6	1030	30. 42	772. 6
911	26. 90	683. 3	971	28. 67	728. 3	1031	30. 45	773. 3
912	26. 93	684. 1	972	28. 70	729. 1	1032	30. 47	774. 1
913	26. 96	684. 8	973	28. 73	729. 8	1033	30. 50	774. 8
914	26. 99	685. 6	974	28. 76	730. 6	1034	30. 53	775. 6
915	27. 02	686. 3	975	28. 79	731. 3	1035	30. 56	776. 3
916	27. 05	687. 1	976	28. 82	732. 1	1036	30. 59	777. 1
917	27. 08	687. 8	977	28. 85	732. 8	1037	30. 62	777. 8
918	27. 11	688. 6	978	28. 88	733. 6	1038	30. 65	778. 6
919	27. 14	689. 3	979	28. 91	734. 3	1039	30. 68	779. 3
920	27. 17	690. 1	980	28. 94	735. 1	1040	30. 71	780. 1
921	27. 20	690. 8	981	28. 97	735. 8	1041	30. 74	780. 8
922	27. 23	691. 6	982	29. 00	736. 6	1042	30. 77	781. 6
923	27. 26	692. 3	983	29. 03	737. 3	1043	30. 80	782. 3
924	27. 29	693. 1	984	29. 06	738. 1	1044	30. 83	783. 1
925	27. 32	693. 8	985	29. 09	738. 8	1045	30. 86	783. 8
926	27. 34	694. 6	986	29. 12	739. 6	1046	30. 89	784. 6
927	27. 37	695. 3	987	29. 15	740. 3	1047	30. 92	785. 3
928	27. 40	696. 1	988	29. 18	741. 1	1048	30. 95	786. 1
929	27. 43	696. 8	989	29. 21	741. 8	1049	30. 98	786. 8
930	27. 46	697. 6	990	29. 23	742. 6	1050	31. 01	787. 6
931	27. 49	698. 3	991	29. 26	743. 3	1051	31. 04	788. 3
932	27. 52	699. 1	992	29. 29	744. 1	1052	31. 07	789. 1
933	27. 55	699. 8	993	29. 32	744. 8	1053	31. 10	789. 8
934	27. 58	700. 6	994	29. 35	745. 6	1054	31. 12	790. 6
935	27. 61	701. 3	995	29. 38	746. 3	1055	31. 15	791. 3
936	27. 64	702. 1	996	29. 41	747. 1	1056	31. 18	792. 1
937	27. 67	702. 8	997	29. 44	747. 8	1057	31. 21	792. 8
938	27. 70	703. 6	998	29. 47	748. 6	1058	31. 24	793. 6
939	27. 73	704. 3	999	29. 50	749. 3	1059	31. 27	794. 3
940	27. 76	705. 1	1000	29. 53	750. 1	1060	31. 30	795. 1
941	27. 79	705. 8	1001	29. 56	750. 8	1061	31. 33	795. 8
942	27. 82	706. 6	1002	29. 59	751. 6	1062	31. 36	796. 6
943	27. 85	707. 3	1003	29. 62	752. 3	1063	31. 39	797. 3
944	27. 88	708. 1	1004	29. 65	753. 1	1064	31. 42	798. 1
945	27. 91	708. 8	1005	29. 68	753. 8	1065	31. 45	798. 8
946	27. 94	709. 6	1006	29. 71	754. 6	1066	31. 48	799. 6
947	27. 96	710. 3	1007	29. 74	755. 3	1067	31. 51	800. 3
948	27. 99	711. 1	1008	29. 77	756. 1	1068	31. 54	801. 1
949	28. 02	711. 8	1009	29. 80	756. 8	1069	31. 57	801. 8
950	28. 05	712. 6	1010	29. 83	757. 6	1070	31. 60	802. 6
951	28. 08	713. 3	1011	29. 85	758. 3	1071	31. 63	803. 3
952	28. 11	714. 1	1012	29. 88	759. 1	1072	31. 66	804. 1
953	28. 14	714. 8	1013	29. 91	759. 8	1073	31. 69	804. 8
954	28. 17	715. 6	1014	29. 94	760. 6	1074	31. 72	805. 6
955	28. 20	716. 3	1015	29. 97	761. 3	1075	31. 74	806. 3
956	28. 23	717. 1	1016	30. 00	762. 1	1076	31. 77	807. 1
957	28. 26	717. 8	1017	30. 03	762. 8	1077	31. 80	807. 8
958	28. 29	718. 6	1018	30. 06	763. 6	1078	31. 83	808. 6
959	28. 32	719. 3	1019	30. 09	764. 3	1079	31. 86	809. 3
960	28. 35	720. 1	1020	30. 12	765. 1	1080	31. 89	810. 1

TABLE 15

Conversion Tables for Thermometer Scales

F = Fahrenheit, C = Celsius (centigrade), K = Kelvin

F	C	K	F	C	K	C	F	K	K	F	C
°	°	°	°	°	°	°	°	°	°	°	°
−20	−28.9	244.3	+40	+4.4	277.6	−25	−13.0	248.2	250	−9.7	−23.2
19	28.3	244.8	41	5.0	278.2	24	11.2	249.2	251	7.9	22.2
18	27.8	245.4	42	5.6	278.7	23	9.4	250.2	252	6.1	21.2
17	27.2	245.9	43	6.1	279.3	22	7.6	251.2	253	4.3	20.2
16	26.7	246.5	44	6.7	279.8	21	5.8	252.2	254	2.5	19.2
−15	−26.1	247.0	+45	+7.2	280.4	−20	−4.0	253.2	255	−0.7	−18.2
14	25.6	247.6	46	7.8	280.9	19	2.2	254.2	256	+1.1	17.2
13	25.0	248.2	47	8.3	281.5	18	−0.4	255.2	257	2.9	16.2
12	24.4	248.7	48	8.9	282.0	17	+1.4	256.2	258	4.7	15.2
11	23.9	249.3	49	9.4	282.6	16	3.2	257.2	259	6.5	14.2
−10	−23.3	249.8	+50	+10.0	283.2	−15	+5.0	258.2	260	+8.3	−13.2
9	22.8	250.4	51	10.6	283.7	14	6.8	259.2	261	10.1	12.2
8	22.2	250.9	52	11.1	284.3	13	8.6	260.2	262	11.9	11.2
7	21.7	251.5	53	11.7	284.8	12	10.4	261.2	263	13.7	10.2
6	21.1	252.0	54	12.2	285.4	11	12.2	262.2	264	15.5	9.2
−5	−20.6	252.6	+55	+12.8	285.9	−10	+14.0	263.2	265	+17.3	−8.2
4	20.0	253.2	56	13.3	286.5	9	15.8	264.2	266	19.1	7.2
3	19.4	253.7	57	13.9	287.0	8	17.6	265.2	267	20.9	6.2
2	18.9	254.3	58	14.4	287.6	7	19.4	266.2	268	22.7	5.2
−1	18.3	254.8	59	15.0	288.2	6	21.2	267.2	269	24.5	4.2
0	−17.8	255.4	+60	+15.6	288.7	−5	+23.0	268.2	270	+26.3	−3.2
+1	17.2	255.9	61	16.1	289.3	4	24.8	269.2	271	28.1	2.2
2	16.7	256.5	62	16.7	289.8	3	26.6	270.2	272	29.9	1.2
3	16.1	257.0	63	17.2	290.4	2	28.4	271.2	273	31.7	−0.2
4	15.6	257.6	64	17.8	290.9	−1	30.2	272.2	274	33.5	+0.8
+5	−15.0	258.2	+65	+18.3	291.5	0	+32.0	273.2	275	+35.3	+1.8
6	14.4	258.7	66	18.9	292.0	+1	33.8	274.2	276	37.1	2.8
7	13.9	259.3	67	19.4	292.6	2	35.6	275.2	277	38.9	3.8
8	13.3	259.8	68	20.0	293.2	3	37.4	276.2	278	40.7	4.8
9	12.8	260.4	69	20.6	293.7	4	39.2	277.2	279	42.5	5.8
+10	−12.2	260.9	+70	+21.1	294.3	+5	+41.0	278.2	280	+44.3	+6.8
11	11.7	261.5	71	21.7	294.8	6	42.8	279.2	281	46.1	7.8
12	11.1	262.0	72	22.2	295.4	7	44.6	280.2	282	47.9	8.8
13	10.6	262.6	73	22.8	295.9	8	46.4	281.2	283	49.7	9.8
14	10.0	263.2	74	23.3	296.5	9	48.2	282.2	284	51.5	10.8
+15	−9.4	263.7	+75	+23.9	297.0	+10	+50.0	283.2	285	+53.3	+11.8
16	8.9	264.3	76	24.4	297.6	11	51.8	284.2	286	55.1	12.8
17	8.3	264.8	77	25.0	298.2	12	53.6	285.2	287	56.9	13.8
18	7.8	265.4	78	25.6	298.7	13	55.4	286.2	288	58.7	14.8
19	7.2	265.9	79	26.1	299.3	14	57.2	287.2	289	60.5	15.8
+20	−6.7	266.5	+80	+26.7	299.8	+15	+59.0	288.2	290	+62.3	+16.8
21	6.1	267.0	81	27.2	300.4	16	60.8	289.2	291	64.1	17.8
22	5.6	267.6	82	27.8	300.9	17	62.6	290.2	292	65.9	18.8
23	5.0	268.2	83	28.3	301.5	18	64.4	291.2	293	67.7	19.8
24	4.4	268.7	84	28.9	302.0	19	66.2	292.2	294	69.5	20.8
+25	−3.9	269.3	+85	+29.4	302.6	+20	+68.0	293.2	295	+71.3	+21.8
26	3.3	269.8	86	30.0	303.2	21	69.8	294.2	296	73.1	22.8
27	2.8	270.4	87	30.6	303.7	22	71.6	295.2	297	74.9	23.8
28	2.2	270.9	88	31.1	304.3	23	73.4	296.2	298	76.7	24.8
29	1.7	271.5	89	31.7	304.8	24	75.2	297.2	299	78.5	25.8
+30	−1.1	272.0	+90	+32.2	305.4	+25	+77.0	298.2	300	+80.3	+26.8
31	0.6	272.6	91	32.8	305.9	26	78.8	299.2	301	82.1	27.8
32	0.0	273.2	92	33.3	306.5	27	80.6	300.2	302	83.9	28.8
33	+0.6	273.7	93	33.9	307.0	28	82.4	301.2	303	85.7	29.8
34	1.1	274.3	94	34.4	307.6	29	84.2	302.2	304	87.5	30.8
+35	+1.7	274.8	+95	+35.0	308.2	+30	+86.0	303.2	305	+89.3	+31.8
36	2.2	275.4	96	35.6	308.7	31	87.8	304.2	306	91.1	32.8
37	2.8	275.9	97	36.1	309.3	32	89.6	305.2	307	92.9	33.8
38	3.3	276.5	98	36.7	309.8	33	91.4	306.2	308	94.7	34.8
39	3.9	277.0	99	37.2	310.4	34	93.2	307.2	309	96.5	35.8
+40	+4.4	277.6	+100	+37.8	310.9	+35	+95.0	308.2	310	+98.3	+36.8

TABLE 16

Relative Humidity

Dry-bulb temp. F	Difference between dry-bulb and wet-bulb temperatures														Dry-bulb temp. F
	1°	2°	3°	4°	5°	6°	7°	8°	9°	10°	11°	12°	13°	14°	
°	%	%	%	%	%	%	%	%	%	%	%	%	%	%	°
−20	7														−20
18	14														18
16	21														16
14	27														14
12	32														12
−10	37														−10
8	41	2													8
6	45	9													6
4	49	16													4
−2	52	22													−2
0	56	28													0
+2	59	33	7												+2
4	62	37	14												4
6	64	42	20												6
8	67	46	25	5											8
+10	69	50	30	11											+10
12	71	53	35	17											12
14	73	56	40	23	7										14
16	76	60	44	28	13										16
18	77	62	48	33	19	4									18
+20	79	65	51	37	24	10									+20
22	81	68	55	42	29	16	4								22
24	83	70	58	45	33	21	10								24
26	85	73	61	49	38	26	15	4							26
28	86	75	64	53	42	31	20	10							28
+30	88	77	66	56	45	35	25	15	6						+30
32	89	79	69	59	49	39	30	20	11	2					32
34	90	81	71	62	52	43	34	25	16	8					34
36	91	82	73	64	55	47	38	29	21	13	5				36
38	91	83	74	66	58	50	42	33	25	18	10	2			38
+40	92	84	76	68	60	52	45	37	30	22	15	7			+40
42	92	84	77	69	62	54	47	40	33	26	19	12	5		42
44	92	85	78	70	63	56	49	43	36	29	23	17	10	4	44
46	93	86	79	72	65	58	52	45	39	32	26	20	14	8	46
48	93	86	79	73	66	60	54	47	41	35	29	24	18	12	48
+50	93	87	80	74	68	61	55	49	44	38	32	27	21	16	+50
52	94	87	81	75	69	63	57	51	46	40	35	29	24	19	52
54	94	88	82	76	70	64	59	53	48	42	37	32	27	22	54
56	94	88	82	77	71	65	60	55	50	44	39	35	30	25	56
58	94	88	83	77	72	67	61	56	51	46	42	37	32	28	58
+60	94	89	83	78	73	68	63	58	53	48	43	39	34	30	+60
62	95	89	84	79	74	69	64	59	54	50	45	41	37	32	62
64	95	89	84	79	74	70	65	60	56	51	47	43	38	34	64
66	95	90	85	80	75	71	66	61	57	53	49	44	40	36	66
68	95	90	85	81	76	71	67	63	58	54	50	46	42	38	68
+70	95	90	86	81	77	72	68	64	59	55	51	48	44	40	+70
72	95	91	86	82	77	73	69	65	61	57	53	49	45	42	72
74	95	91	86	82	78	74	69	65	62	58	54	50	47	43	74
76	95	91	87	82	78	74	70	66	63	59	55	51	48	45	76
78	96	91	87	83	79	75	71	67	63	60	56	53	49	46	78
+80	96	91	87	83	79	75	72	68	64	61	57	54	50	47	+80
82	96	92	88	84	80	76	72	69	65	62	58	55	52	48	82
84	96	92	88	84	80	76	73	69	66	62	59	56	53	49	84
86	96	92	88	84	81	77	73	70	67	63	60	57	54	51	86
88	96	92	88	85	81	77	74	71	67	64	61	58	55	52	88
+90	96	92	89	85	81	78	74	71	68	65	61	58	55	52	+90
92	96	92	89	85	82	78	75	72	68	65	62	59	56	53	92
94	96	93	89	85	82	79	75	72	69	66	63	60	57	54	94
96	96	93	89	86	82	79	76	73	70	67	64	61	58	55	96
98	96	93	89	86	83	79	76	73	70	67	64	61	59	56	98
+100	96	93	90	86	83	80	77	74	71	68	65	62	59	57	+100

TABLE 16
Relative Humidity

Dry-bulb temp. F	15°	16°	17°	18°	19°	20°	21°	22°	23°	24°	25°	26°	27°	28°	Dry-bulb temp. F
	Difference between dry-bulb and wet-bulb temperatures														
°	%	%	%	%	%	%	%	%	%	%	%	%	%	%	°
+46	2														+46
48	7	1													48
+50	10	5													+50
52	14	9	4												52
54	17	12	7	3											54
56	20	16	11	7	2										56
58	23	19	14	10	6	2									58
+60	26	21	17	13	9	5	1								+60
62	28	24	20	16	12	8	4	1							62
64	30	26	22	19	15	11	8	4							64
66	32	29	25	21	17	14	10	7	4						66
68	34	31	27	23	20	16	13	10	7	3					68
+70	36	33	29	26	22	19	16	12	9	6	3				+70
72	38	34	31	28	24	21	18	15	12	9	6	3			72
74	40	36	33	30	26	23	20	17	14	11	8	6	3		74
76	41	38	35	31	28	25	22	19	16	14	11	8	5	3	76
78	43	39	36	33	30	27	24	21	18	16	13	10	8	5	78
+80	44	41	38	35	32	29	26	23	20	18	15	13	10	8	+80
82	45	42	39	36	33	30	28	25	22	20	17	15	12	10	82
84	46	43	40	38	35	32	29	27	24	21	19	17	14	12	84
86	48	45	42	39	36	33	31	28	26	23	21	18	16	14	86
88	49	46	43	40	37	35	32	30	27	25	22	20	18	16	88
+90	50	47	44	41	39	36	34	31	29	26	24	22	19	17	+90
92	51	48	45	42	40	37	35	32	30	28	25	23	21	19	92
94	51	49	46	44	41	39	36	34	31	29	27	25	23	20	94
96	52	50	47	45	42	40	37	35	33	30	28	26	24	22	96
98	53	51	48	45	43	41	38	36	34	32	29	27	25	23	98
+100	54	51	49	46	44	42	39	37	35	33	31	29	27	25	+100

Dry-bulb temp. F	29°	30°	31°	32°	33°	34°	35°	36°	37°	38°	39°	40°	41°	42°	Dry-bulb temp. F
	Difference between dry-bulb and wet-bulb temperatures														
°	%	%	%	%	%	%	%	%	%	%	%	%	%	%	°
+78	3														+78
+80	5	3													+80
82	7	5	3	1											82
84	10	7	5	3	1										84
86	11	9	7	5	3	1									86
88	13	11	9	7	5	3	1								88
+90	15	13	11	9	7	5	3	1							+90
92	17	15	13	11	9	7	5	3	1						92
94	18	16	14	12	11	9	7	5	3	2					94
96	20	18	16	14	12	10	9	7	5	4	2				96
98	21	19	17	16	14	12	10	9	7	5	4	2	1		98
+100	23	21	19	17	15	14	12	10	9	7	5	4	2	1	+100

TABLE 17

Dew Point

Dry-bulb temp. F	1°	2°	3°	4°	5°	6°	7°	8°	9°	10°	11°	12°	13°	14°	Dry-bulb temp. F
	Difference between dry-bulb and wet-bulb temperatures														
°	°	°	°	°	°	°	°	°	°	°	°	°	°	°	°
−20															−20
18	−52														18
16	45														16
14	39														14
12	34														12
−10	−29														−10
8	25	−75													8
6	22	50													6
4	18	39													4
−2	15	32													−2
0	−12	−26													0
+2	9	21	−49												+2
4	6	16	35												4
6	3	12	27												6
8	−1	9	20	−50											8
+10	+2	−5	−15	−34											+10
12	5	−2	10	24											12
14	7	+1	6	17											14
16	10	4	−2	11											16
18	12	7	+1	6											18
+20	+15	+10	+5	−2											+20
22	17	13	8	+2											22
24	20	16	11	6											24
26	22	18	14	10	+4	−4									26
28	24	21	17	13	8	+1	−8	−22							28
+30	+27	+24	+20	+16	+11	+6	−1	−12	−31						+30
32	29	26	23	19	15	10	+4	−4	16	−47					32
34	32	29	26	22	18	14	9	+2	−7	22					34
36	34	31	28	25	22	18	13	7	0	11	−30				36
38	36	33	31	28	25	21	17	12	+6	−2	14	−42			38
+40	+38	+35	+33	+30	+27	+24	+20	+16	+11	+4	−4	−18	−79		+40
42	40	38	35	33	30	27	23	19	15	10	+3	−7	23		42
44	42	40	37	35	32	29	26	23	19	14	9	+2	−9	−29	44
46	44	42	40	37	35	32	29	26	22	18	13	7	0	11	46
48	46	44	42	40	37	35	32	29	26	22	18	13	+6	−2	48
+50	+48	+46	+44	+42	+40	+37	+35	+32	+29	+25	+21	+17	+12	+5	+50
52	50	48	46	44	42	40	37	35	32	29	25	21	17	11	52
54	52	50	49	47	44	42	40	37	35	32	28	25	21	16	54
56	54	53	51	49	47	45	42	40	37	35	32	28	25	21	56
58	56	55	53	51	49	47	45	43	40	38	35	32	28	25	58
+60	+58	+57	+55	+53	+51	+49	+47	+45	+43	+40	+38	+35	+32	+28	+60
62	60	59	57	55	54	52	50	48	45	43	41	38	35	32	62
64	62	61	59	57	56	54	52	50	48	46	43	41	38	35	64
66	64	63	61	60	58	56	54	52	50	48	46	44	41	39	66
68	67	65	63	62	60	58	57	55	53	51	49	46	44	42	68
+70	+69	+67	+66	+64	+62	+61	+59	+57	+55	+53	+51	+49	+47	+45	+70
72	71	69	68	66	64	63	61	59	58	56	54	52	50	47	72
74	73	71	70	68	67	65	63	62	60	58	56	54	52	50	74
76	75	73	72	70	69	67	66	64	62	61	59	57	55	53	76
78	77	75	74	72	71	69	68	66	65	63	61	59	57	55	78
+80	+79	+77	+76	+74	+73	+72	+70	+68	+67	+65	+64	+62	+60	+58	+80
82	81	79	78	77	75	74	72	71	69	67	66	64	62	61	82
84	83	81	80	79	77	76	74	73	71	70	68	67	65	63	84
86	85	83	82	81	79	78	76	75	74	72	70	69	67	66	86
88	87	85	84	83	81	80	79	77	76	74	73	71	70	68	88
+90	+89	+87	+86	+85	+84	+82	+81	+79	+78	+76	+75	+73	+72	+70	+90
92	91	89	88	87	86	84	83	82	80	79	77	76	74	73	92
94	93	92	90	89	88	86	85	84	82	81	79	78	76	75	94
96	95	94	92	91	90	88	87	86	84	83	82	80	79	77	96
98	97	96	94	93	92	91	89	88	87	85	84	82	81	80	98
+100	+99	+98	+96	+95	+94	+93	+91	+90	+89	+87	+86	+85	+83	+82	+100

TABLE 17
Dew Point

Dry-bulb temp. F	Difference between dry-bulb and wet-bulb temperatures														Dry-bulb temp. F
	15°	16°	17°	18°	19°	20°	21°	22°	23°	24°	25°	26°	27°	28°	
°	°	°	°	°	°	°	°	°	°	°	°	°	°	°	°
+46	−36														+46
48	14	−45													48
+50	−3	−17	−78												+50
52	+4	−5	21												52
54	10	+3	−7	−25											54
56	16	10	+2	−8	−29										56
58	20	16	10	+2	−10	−34									58
+60	+25	+20	+15	+9	+1	−11	−39								+60
62	29	25	20	15	9	+1	−12	−45							62
64	32	29	25	20	15	9	0	−13	−52						64
66	36	33	29	25	21	15	+9	0	−14	−59					66
68	39	36	33	29	25	21	16	+9	0	−14	−68				68
+70	+42	+39	+36	+33	+30	+26	+21	+16	+9	0	−14	−76			+70
72	45	43	40	37	34	30	26	22	16	+10	+1	−14	−77		72
74	48	46	43	40	37	34	31	27	22	17	10	+1	−13	−70	74
76	51	48	46	44	41	38	35	31	27	23	17	11	+2	−12	76
78	53	51	49	47	44	41	38	35	32	28	23	18	11	+3	78
+80	+56	+54	+52	+50	+47	+45	+42	+39	+36	+32	+28	+24	+19	+12	+80
82	59	57	55	53	50	48	45	43	40	37	33	29	25	20	82
84	61	59	57	55	53	51	49	46	43	41	37	34	30	26	84
86	64	62	60	58	56	54	52	49	47	44	41	38	35	31	86
88	66	64	63	61	59	57	55	52	50	48	45	42	39	36	88
+90	+69	+67	+65	+63	+62	+60	+58	+55	+53	+51	+48	+46	+43	+40	+90
92	71	69	68	66	64	62	60	58	56	54	52	49	47	44	92
94	73	72	70	68	67	65	63	61	59	57	55	52	50	47	94
96	76	74	73	71	69	67	66	64	62	60	58	56	53	51	96
98	78	77	75	73	72	70	68	67	65	63	61	59	57	54	98
+100	+80	+79	+77	+76	+74	+73	+71	+69	+67	+66	+64	+62	+60	+57	+100

Dry-bulb temp. F	Difference between dry-bulb and wet-bulb temperatures														Dry-bulb temp. F
	29°	30°	31°	32°	33°	34°	35°	36°	37°	38°	39°	40°	41°	42°	
°	°	°	°	°	°	°	°	°	°	°	°	°	°	°	°
+76	−61														+76
78	−11	−53													78
+80	+4	−10	−45												+80
82	13	+5	−8	−39											82
84	20	14	+6	−6	−33										84
86	27	21	15	+7	−4	−28									86
88	32	27	22	16	+9	−2	−23								88
+90	+36	+33	+28	+24	+18	+10	0	−18							+90
92	41	37	34	30	25	19	+12	+2	−14						92
94	45	42	38	35	31	26	20	13	+4	−10					94
96	48	46	43	39	36	32	27	22	15	+6	−7	−43			96
98	52	49	47	44	40	37	33	28	23	17	+9	−4	−30		98
+100	+55	+53	+50	+47	+45	+41	+38	+34	+30	+25	+19	+11	0	−21	+100

TABLE 18
Speed Table for Measured Mile

Sec.	Minutes												Sec.
	1	2	3	4	5	6	7	8	9	10	11	12	
	Knots	*Knots*	*Knots*	*Knots*	*Knots*	*Knots*	*Knots*	*Knots*	*Knots*	*Knots*	*Knots*	*Knots*	
0	60.000	30.000	20.000	15.000	12.000	10.000	8.571	7.500	6.667	6.000	5.455	5.000	0
1	59.016	29.752	19.890	14.938	11.960	9.972	8.551	7.484	6.654	5.990	5.446	4.993	1
2	58.065	29.508	19.780	14.876	11.921	9.945	8.531	7.469	6.642	5.980	5.438	4.986	2
3	57.143	29.268	19.672	14.815	11.881	9.917	8.511	7.453	6.630	5.970	5.430	4.979	3
4	56.250	29.032	19.565	14.754	11.842	9.890	8.491	7.438	6.618	5.960	5.422	4.972	4
5	55.385	28.800	19.459	14.694	11.803	9.863	8.471	7.423	6.606	5.950	5.414	4.966	5
6	54.545	28.571	19.355	14.634	11.765	9.836	8.451	7.407	6.593	5.941	5.405	4.959	6
7	53.731	28.346	19.251	14.575	11.726	9.809	8.431	7.392	6.581	5.931	5.397	4.952	7
8	52.941	28.125	19.149	14.516	11.688	9.783	8.411	7.377	6.569	5.921	5.389	4.945	8
9	52.174	27.907	19.048	14.458	11.650	9.756	8.392	7.362	6.557	5.911	5.381	4.938	9
10	51.429	27.692	18.947	14.400	11.613	9.730	8.372	7.347	6.545	5.902	5.373	4.932	10
11	50.704	27.481	18.848	14.343	11.576	9.704	8.353	7.332	6.534	5.892	5.365	4.925	11
12	50.000	27.273	18.750	14.286	11.538	9.677	8.333	7.317	6.522	5.882	5.357	4.918	12
13	49.315	27.068	18.653	14.229	11.502	9.651	8.314	7.302	6.510	5.873	5.349	4.911	13
14	48.649	26.866	18.557	14.173	11.465	9.626	8.295	7.287	6.498	5.863	5.341	4.905	14
15	48.000	26.667	18.462	14.118	11.429	9.600	8.276	7.273	6.486	5.854	5.333	4.898	15
16	47.368	26.471	18.367	14.062	11.392	9.574	8.257	7.258	6.475	5.844	5.325	4.891	16
17	46.753	26.277	18.274	14.008	11.356	9.549	8.238	7.243	6.463	5.835	5.318	4.885	17
18	46.154	26.087	18.182	13.953	11.321	9.524	8.219	7.229	6.452	5.825	5.310	4.878	18
19	45.570	25.899	18.090	13.900	11.285	9.499	8.200	7.214	6.440	5.816	5.302	4.871	19
20	45.000	25.714	18.000	13.846	11.250	9.474	8.182	7.200	6.429	5.806	5.294	4.865	20
21	44.444	25.532	17.910	13.793	11.215	9.449	8.163	7.186	6.417	5.797	5.286	4.858	21
22	43.902	25.352	17.822	13.740	11.180	9.424	8.145	7.171	6.406	5.788	5.279	4.852	22
23	43.373	25.175	17.734	13.688	11.146	9.399	8.126	7.157	6.394	5.778	5.271	4.845	23
24	42.857	25.000	17.647	13.636	11.111	9.375	8.108	7.143	6.383	5.769	5.263	4.839	24
25	42.353	24.828	17.561	13.585	11.077	9.351	8.090	7.129	6.372	5.760	5.255	4.832	25
26	41.860	24.658	17.476	13.534	11.043	9.326	8.072	7.115	6.360	5.751	5.248	4.826	26
27	41.379	24.490	17.391	13.483	11.009	9.302	8.054	7.101	6.349	5.742	5.240	4.819	27
28	40.909	24.324	17.308	13.433	10.976	9.278	8.036	7.087	6.338	5.732	5.233	4.813	28
29	40.449	24.161	17.225	13.383	10.942	9.254	8.018	7.073	6.327	5.723	5.225	4.806	29
30	40.000	24.000	17.143	13.333	10.909	9.231	8.000	7.059	6.316	5.714	5.217	4.800	30
31	39.560	23.841	17.062	13.284	10.876	9.207	7.982	7.045	6.305	5.705	5.210	4.794	31
32	39.130	23.684	16.981	13.235	10.843	9.184	7.965	7.031	6.294	5.696	5.202	4.787	32
33	38.710	23.529	16.901	13.187	10.811	9.160	7.947	7.018	6.283	5.687	5.195	4.781	33
34	38.298	23.377	16.822	13.139	10.778	9.137	7.930	7.004	6.272	5.678	5.187	4.775	34
35	37.895	23.226	16.744	13.091	10.746	9.114	7.912	6.990	6.261	5.669	5.180	4.768	35
36	37.500	23.077	16.667	13.043	10.714	9.091	7.895	6.977	6.250	5.660	5.172	4.762	36
37	37.113	22.930	16.590	12.996	10.682	9.068	7.877	6.963	6.239	5.651	5.165	4.756	37
38	36.735	22.785	16.514	12.950	10.651	9.045	7.860	6.950	6.228	5.643	5.158	4.749	38
39	36.364	22.642	16.438	12.903	10.619	9.023	7.843	6.936	6.218	5.634	5.150	4.743	39
40	36.000	22.500	16.364	12.857	10.588	9.000	7.826	6.923	6.207	5.625	5.143	4.737	40
41	35.644	22.360	16.290	12.811	10.557	8.978	7.809	6.910	6.196	5.616	5.136	4.731	41
42	35.294	22.222	16.216	12.766	10.526	8.955	7.792	6.897	6.186	5.607	5.128	4.724	42
43	34.951	22.086	16.143	12.721	10.496	8.933	7.775	6.883	6.175	5.599	5.121	4.718	43
44	34.615	21.951	16.071	12.676	10.465	8.911	7.759	6.870	6.164	5.590	5.114	4.712	44
45	34.286	21.818	16.000	12.632	10.435	8.889	7.742	6.857	6.154	5.581	5.106	4.706	45
46	33.962	21.687	15.929	12.587	10.405	8.867	7.725	6.844	6.143	5.573	5.099	4.700	46
47	33.645	21.557	15.859	12.544	10.375	8.845	7.709	6.831	6.133	5.564	5.092	4.694	47
48	33.333	21.429	15.789	12.500	10.345	8.824	7.692	6.818	6.122	5.556	5.085	4.688	48
49	33.028	21.302	15.721	12.457	10.315	8.802	7.676	6.805	6.112	5.547	5.078	4.681	49
50	32.727	21.176	15.652	12.414	10.286	8.780	7.660	6.792	6.102	5.538	5.070	4.675	50
51	32.432	21.053	15.584	12.371	10.256	8.759	7.643	6.780	6.091	5.530	5.063	4.669	51
52	32.143	20.930	15.517	12.329	10.227	8.738	7.627	6.767	6.081	5.521	5.056	4.663	52
53	31.858	20.809	15.451	12.287	10.198	8.717	7.611	6.754	6.071	5.513	5.049	4.657	53
54	31.579	20.690	15.385	12.245	10.169	8.696	7.595	6.742	6.061	5.505	5.042	4.651	54
55	31.304	20.571	15.319	12.203	10.141	8.675	7.579	6.729	6.050	5.496	5.035	4.645	55
56	31.034	20.455	15.254	12.162	10.112	8.654	7.563	6.716	6.040	5.488	5.028	4.639	56
57	30.769	20.339	15.190	12.121	10.084	8.633	7.547	6.704	6.030	5.479	5.021	4.633	57
58	30.508	20.225	15.126	12.081	10.056	8.612	7.531	6.691	6.020	5.471	5.014	4.627	58
59	30.252	20.112	15.063	12.040	10.028	8.592	7.516	6.679	6.010	5.463	5.007	4.621	59
60	30.000	20.000	15.000	12.000	10.000	8.571	7.500	6.667	6.000	5.455	5.000	4.615	60
Sec.	1	2	3	4	5	6	7	8	9	10	11	12	Sec.

TABLE 19

Speed, Time, and Distance

Minutes	Speed in knots																Minutes
	0.5	1.0	1.5	2.0	2.5	3.0	3.5	4.0	4.5	5.0	5.5	6.0	6.5	7.0	7.5	8.0	
	Miles	*Miles*	*Miles*	*Miles*	*Miles*	*Miles*	*Miles*	*Miles*	*Miles*	*Miles*	*Miles*	*Miles*	*Miles*	*Miles*	*Miles*	*Miles*	
1	0.0	0.0	0.0	0.0	0.0	0.0	0.1	0.1	0.1	0.1	0.1	0.1	0.1	0.1	0.1	0.1	1
2	0.0	0.0	0.0	0.1	0.1	0.1	0.1	0.1	0.2	0.2	0.2	0.2	0.2	0.2	0.2	0.3	2
3	0.0	0.0	0.1	0.1	0.1	0.2	0.2	0.2	0.2	0.2	0.3	0.3	0.3	0.4	0.4	0.4	3
4	0.0	0.1	0.1	0.1	0.2	0.2	0.2	0.3	0.3	0.3	0.4	0.4	0.4	0.5	0.5	0.5	4
5	0.0	0.1	0.1	0.2	0.2	0.2	0.3	0.3	0.4	0.4	0.5	0.5	0.5	0.6	0.6	0.7	5
6	0.0	0.1	0.2	0.2	0.2	0.3	0.4	0.4	0.4	0.5	0.6	0.6	0.6	0.7	0.8	0.8	6
7	0.1	0.1	0.2	0.2	0.3	0.4	0.4	0.5	0.5	0.6	0.6	0.7	0.8	0.8	0.9	0.9	7
8	0.1	0.1	0.2	0.3	0.3	0.4	0.5	0.5	0.6	0.7	0.7	0.8	0.9	0.9	1.0	1.1	8
9	0.1	0.2	0.2	0.3	0.4	0.4	0.5	0.6	0.7	0.8	0.8	0.9	1.0	1.0	1.1	1.2	9
10	0.1	0.2	0.2	0.3	0.4	0.5	0.6	0.7	0.8	0.8	0.9	1.0	1.1	1.2	1.2	1.3	10
11	0.1	0.2	0.3	0.4	0.5	0.6	0.6	0.7	0.8	0.9	1.0	1.1	1.2	1.3	1.4	1.5	11
12	0.1	0.2	0.3	0.4	0.5	0.6	0.7	0.8	0.9	1.0	1.1	1.2	1.3	1.4	1.5	1.6	12
13	0.1	0.2	0.3	0.4	0.5	0.6	0.8	0.9	1.0	1.1	1.2	1.3	1.4	1.5	1.6	1.7	13
14	0.1	0.2	0.4	0.5	0.6	0.7	0.8	0.9	1.0	1.2	1.3	1.4	1.5	1.6	1.8	1.9	14
15	0.1	0.2	0.4	0.5	0.6	0.8	0.9	1.0	1.1	1.2	1.4	1.5	1.6	1.8	1.9	2.0	15
16	0.1	0.3	0.4	0.5	0.7	0.8	0.9	1.1	1.2	1.3	1.5	1.6	1.7	1.9	2.0	2.1	16
17	0.1	0.3	0.4	0.6	0.7	0.8	1.0	1.1	1.3	1.4	1.6	1.7	1.8	2.0	2.1	2.3	17
18	0.2	0.3	0.4	0.6	0.8	0.9	1.0	1.2	1.4	1.5	1.6	1.8	2.0	2.1	2.2	2.4	18
19	0.2	0.3	0.5	0.6	0.8	1.0	1.1	1.3	1.4	1.6	1.7	1.9	2.1	2.2	2.4	2.5	19
20	0.2	0.3	0.5	0.7	0.8	1.0	1.2	1.3	1.5	1.7	1.8	2.0	2.2	2.3	2.5	2.7	20
21	0.2	0.4	0.5	0.7	0.9	1.0	1.2	1.4	1.6	1.8	1.9	2.1	2.3	2.4	2.6	2.8	21
22	0.2	0.4	0.6	0.7	0.9	1.1	1.3	1.5	1.6	1.8	2.0	2.2	2.4	2.6	2.8	2.9	22
23	0.2	0.4	0.6	0.8	1.0	1.2	1.3	1.5	1.7	1.9	2.1	2.3	2.5	2.7	2.9	3.1	23
24	0.2	0.4	0.6	0.8	1.0	1.2	1.4	1.6	1.8	2.0	2.2	2.4	2.6	2.8	3.0	3.2	24
25	0.2	0.4	0.6	0.8	1.0	1.2	1.5	1.7	1.9	2.1	2.3	2.5	2.7	2.9	3.1	3.3	25
26	0.2	0.4	0.6	0.9	1.1	1.3	1.5	1.7	2.0	2.2	2.4	2.6	2.8	3.0	3.2	3.5	26
27	0.2	0.4	0.7	0.9	1.1	1.4	1.6	1.8	2.0	2.2	2.5	2.7	2.9	3.2	3.4	3.6	27
28	0.2	0.5	0.7	0.9	1.2	1.4	1.6	1.9	2.1	2.3	2.6	2.8	3.0	3.3	3.5	3.7	28
29	0.2	0.5	0.7	1.0	1.2	1.4	1.7	1.9	2.2	2.4	2.7	2.9	3.1	3.4	3.6	3.9	29
30	0.2	0.5	0.8	1.0	1.2	1.5	1.8	2.0	2.2	2.5	2.8	3.0	3.2	3.5	3.8	4.0	30
31	0.3	0.5	0.8	1.0	1.3	1.6	1.8	2.1	2.3	2.6	2.8	3.1	3.4	3.6	3.9	4.1	31
32	0.3	0.5	0.8	1.1	1.3	1.6	1.9	2.1	2.4	2.7	2.9	3.2	3.5	3.7	4.0	4.3	32
33	0.3	0.6	0.8	1.1	1.4	1.6	1.9	2.2	2.5	2.8	3.0	3.3	3.6	3.8	4.1	4.4	33
34	0.3	0.6	0.8	1.1	1.4	1.7	2.0	2.3	2.6	2.8	3.1	3.4	3.7	4.0	4.2	4.5	34
35	0.3	0.6	0.9	1.2	1.5	1.8	2.0	2.3	2.6	2.9	3.2	3.5	3.8	4.1	4.4	4.7	35
36	0.3	0.6	0.9	1.2	1.5	1.8	2.1	2.4	2.7	3.0	3.3	3.6	3.9	4.2	4.5	4.8	36
37	0.3	0.6	0.9	1.2	1.5	1.8	2.2	2.5	2.8	3.1	3.4	3.7	4.0	4.3	4.6	4.9	37
38	0.3	0.6	1.0	1.3	1.6	1.9	2.2	2.5	2.8	3.2	3.5	3.8	4.1	4.4	4.8	5.1	38
39	0.3	0.6	1.0	1.3	1.6	2.0	2.3	2.6	2.9	3.2	3.6	3.9	4.2	4.6	4.9	5.2	39
40	0.3	0.7	1.0	1.3	1.7	2.0	2.3	2.7	3.0	3.3	3.7	4.0	4.3	4.7	5.0	5.3	40
41	0.3	0.7	1.0	1.4	1.7	2.0	2.4	2.7	3.1	3.4	3.8	4.1	4.4	4.8	5.1	5.5	41
42	0.4	0.7	1.0	1.4	1.8	2.1	2.4	2.8	3.2	3.5	3.8	4.2	4.6	4.9	5.2	5.6	42
43	0.4	0.7	1.1	1.4	1.8	2.2	2.5	2.9	3.2	3.6	3.9	4.3	4.7	5.0	5.4	5.7	43
44	0.4	0.7	1.1	1.5	1.8	2.2	2.6	2.9	3.3	3.7	4.0	4.4	4.8	5.1	5.5	5.9	44
45	0.4	0.8	1.1	1.5	1.9	2.2	2.6	3.0	3.4	3.8	4.1	4.5	4.9	5.2	5.6	6.0	45
46	0.4	0.8	1.2	1.5	1.9	2.3	2.7	3.1	3.4	3.8	4.2	4.6	5.0	5.4	5.8	6.1	46
47	0.4	0.8	1.2	1.6	2.0	2.4	2.7	3.1	3.5	3.9	4.3	4.7	5.1	5.5	5.9	6.3	47
48	0.4	0.8	1.2	1.6	2.0	2.4	2.8	3.2	3.6	4.0	4.4	4.8	5.2	5.6	6.0	6.4	48
49	0.4	0.8	1.2	1.6	2.0	2.4	2.9	3.3	3.7	4.1	4.5	4.9	5.3	5.7	6.1	6.5	49
50	0.4	0.8	1.2	1.7	2.1	2.5	2.9	3.3	3.8	4.2	4.6	5.0	5.4	5.8	6.2	6.7	50
51	0.4	0.8	1.3	1.7	2.1	2.6	3.0	3.4	3.8	4.2	4.7	5.1	5.5	6.0	6.4	6.8	51
52	0.4	0.9	1.3	1.7	2.2	2.6	3.0	3.5	3.9	4.3	4.8	5.2	5.6	6.1	6.5	6.9	52
53	0.4	0.9	1.3	1.8	2.2	2.6	3.1	3.5	4.0	4.4	4.9	5.3	5.7	6.2	6.6	7.1	53
54	0.4	0.9	1.4	1.8	2.2	2.7	3.2	3.6	4.0	4.5	5.0	5.4	5.8	6.3	6.8	7.2	54
55	0.5	0.9	1.4	1.8	2.3	2.8	3.2	3.7	4.1	4.6	5.0	5.5	6.0	6.4	6.9	7.3	55
56	0.5	0.9	1.4	1.9	2.3	2.8	3.3	3.7	4.2	4.7	5.1	5.6	6.1	6.5	7.0	7.5	56
57	0.5	1.0	1.4	1.9	2.4	2.8	3.3	3.8	4.3	4.8	5.2	5.7	6.2	6.6	7.1	7.6	57
58	0.5	1.0	1.4	1.9	2.4	2.9	3.4	3.9	4.4	4.8	5.3	5.8	6.3	6.8	7.2	7.7	58
59	0.5	1.0	1.5	2.0	2.5	3.0	3.4	3.9	4.4	4.9	5.4	5.9	6.4	6.9	7.4	7.9	59
60	0.5	1.0	1.5	2.0	2.5	3.0	3.5	4.0	4.5	5.0	5.5	6.0	6.5	7.0	7.5	8.0	60

TABLE 19

Speed, Time, and Distance

Minutes	Speed in knots																Minutes
	8.5	9.0	9.5	10.0	10.5	11.0	11.5	12.0	12.5	13.0	13.5	14.0	14.5	15.0	15.5	16.0	
	Miles	*Miles*	*Miles*	*Miles*	*Miles*	*Miles*	*Miles*	*Miles*	*Miles*	*Miles*	*Miles*	*Miles*	*Miles*	*Miles*	*Miles*	*Miles*	
1	0.1	0.2	0.2	0.2	0.2	0.2	0.2	0.2	0.2	0.2	0.2	0.2	0.2	0.2	0.3	0.3	1
2	0.3	0.3	0.3	0.3	0.4	0.4	0.4	0.4	0.4	0.4	0.4	0.5	0.5	0.5	0.5	0.5	2
3	0.4	0.4	0.5	0.5	0.5	0.6	0.6	0.6	0.6	0.6	0.7	0.7	0.7	0.8	0.8	0.8	3
4	0.6	0.6	0.6	0.7	0.7	0.7	0.8	0.8	0.8	0.9	0.9	0.9	1.0	1.0	1.0	1.1	4
5	0.7	0.8	0.8	0.8	0.9	0.9	1.0	1.0	1.0	1.1	1.1	1.2	1.2	1.2	1.3	1.3	5
6	0.8	0.9	1.0	1.0	1.0	1.1	1.2	1.2	1.2	1.3	1.4	1.4	1.4	1.5	1.6	1.6	6
7	1.0	1.0	1.1	1.2	1.2	1.3	1.3	1.4	1.5	1.5	1.6	1.6	1.7	1.8	1.8	1.9	7
8	1.1	1.2	1.3	1.3	1.4	1.5	1.5	1.6	1.7	1.7	1.8	1.9	1.9	2.0	2.1	2.1	8
9	1.3	1.4	1.4	1.5	1.6	1.6	1.7	1.8	1.9	2.0	2.0	2.1	2.2	2.2	2.3	2.4	9
10	1.4	1.5	1.6	1.7	1.8	1.8	1.9	2.0	2.1	2.2	2.2	2.3	2.4	2.5	2.6	2.7	10
11	1.6	1.6	1.7	1.8	1.9	2.0	2.1	2.2	2.3	2.4	2.5	2.6	2.7	2.8	2.8	2.9	11
12	1.7	1.8	1.9	2.0	2.1	2.2	2.3	2.4	2.5	2.6	2.7	2.8	2.9	3.0	3.1	3.2	12
13	1.8	2.0	2.1	2.2	2.3	2.4	2.5	2.6	2.7	2.8	2.9	3.0	3.1	3.2	3.4	3.5	13
14	2.0	2.1	2.2	2.3	2.4	2.6	2.7	2.8	2.9	3.0	3.2	3.3	3.4	3.5	3.6	3.7	14
15	2.1	2.2	2.4	2.5	2.6	2.8	2.9	3.0	3.1	3.2	3.4	3.5	3.6	3.8	3.9	4.0	15
16	2.3	2.4	2.5	2.7	2.8	2.9	3.1	3.2	3.3	3.5	3.6	3.7	3.9	4.0	4.1	4.3	16
17	2.4	2.6	2.7	2.8	3.0	3.1	3.3	3.4	3.5	3.7	3.8	4.0	4.1	4.2	4.4	4.5	17
18	2.6	2.7	2.8	3.0	3.2	3.3	3.4	3.6	3.8	3.9	4.0	4.2	4.4	4.5	4.6	4.8	18
19	2.7	2.8	3.0	3.2	3.3	3.5	3.6	3.8	4.0	4.1	4.3	4.4	4.6	4.8	4.9	5.1	19
20	2.8	3.0	3.2	3.3	3.5	3.7	3.8	4.0	4.2	4.3	4.5	4.7	4.8	5.0	5.2	5.3	20
21	3.0	3.2	3.3	3.5	3.7	3.8	4.0	4.2	4.4	4.6	4.7	4.9	5.1	5.2	5.4	5.6	21
22	3.1	3.3	3.5	3.7	3.8	4.0	4.2	4.4	4.6	4.8	5.0	5.1	5.3	5.5	5.7	5.9	22
23	3.3	3.4	3.6	3.8	4.0	4.2	4.4	4.6	4.8	5.0	5.2	5.4	5.6	5.8	5.9	6.1	23
24	3.4	3.6	3.8	4.0	4.2	4.4	4.6	4.8	5.0	5.2	5.4	5.6	5.8	6.0	6.2	6.4	24
25	3.5	3.8	4.0	4.2	4.4	4.6	4.8	5.0	5.2	5.4	5.6	5.8	6.0	6.2	6.5	6.7	25
26	3.7	3.9	4.1	4.3	4.6	4.8	5.0	5.2	5.4	5.6	5.8	6.1	6.3	6.5	6.7	6.9	26
27	3.8	4.0	4.3	4.5	4.7	5.0	5.2	5.4	5.6	5.8	6.1	6.3	6.5	6.8	7.0	7.2	27
28	4.0	4.2	4.4	4.7	4.9	5.1	5.4	5.6	5.8	6.1	6.3	6.5	6.8	7.0	7.2	7.5	28
29	4.1	4.4	4.6	4.8	5.1	5.3	5.6	5.8	6.0	6.3	6.5	6.8	7.0	7.2	7.5	7.7	29
30	4.2	4.5	4.8	5.0	5.2	5.5	5.8	6.0	6.2	6.5	6.8	7.0	7.2	7.5	7.8	8.0	30
31	4.4	4.6	4.9	5.2	5.4	5.7	5.9	6.2	6.5	6.7	7.0	7.2	7.5	7.8	8.0	8.3	31
32	4.5	4.8	5.1	5.3	5.6	5.9	6.1	6.4	6.7	6.9	7.2	7.5	7.7	8.0	8.3	8.5	32
33	4.7	5.0	5.2	5.5	5.8	6.0	6.3	6.6	6.9	7.2	7.4	7.7	8.0	8.2	8.5	8.8	33
34	4.8	5.1	5.4	5.7	6.0	6.2	6.5	6.8	7.1	7.4	7.6	7.9	8.2	8.5	8.8	9.1	34
35	5.0	5.2	5.5	5.8	6.1	6.4	6.7	7.0	7.3	7.6	7.9	8.2	8.5	8.8	9.0	9.3	35
36	5.1	5.4	5.7	6.0	6.3	6.6	6.9	7.2	7.5	7.8	8.1	8.4	8.7	9.0	9.3	9.6	36
37	5.2	5.6	5.9	6.2	6.5	6.8	7.1	7.4	7.7	8.0	8.3	8.6	8.9	9.2	9.6	9.9	37
38	5.4	5.7	6.0	6.3	6.6	7.0	7.3	7.6	7.9	8.2	8.6	8.9	9.2	9.5	9.8	10.1	38
39	5.5	5.8	6.2	6.5	6.8	7.2	7.5	7.8	8.1	8.4	8.8	9.1	9.4	9.8	10.1	10.4	39
40	5.7	6.0	6.3	6.7	7.0	7.3	7.7	8.0	8.3	8.7	9.0	9.3	9.7	10.0	10.3	10.7	40
41	5.8	6.2	6.5	6.8	7.2	7.5	7.9	8.2	8.5	8.9	9.2	9.6	9.9	10.2	10.6	10.9	41
42	6.0	6.3	6.6	7.0	7.4	7.7	8.0	8.4	8.8	9.1	9.4	9.8	10.2	10.5	10.8	11.2	42
43	6.1	6.4	6.8	7.2	7.5	7.9	8.2	8.6	9.0	9.3	9.7	10.0	10.4	10.8	11.1	11.5	43
44	6.2	6.6	7.0	7.3	7.7	8.1	8.4	8.8	9.2	9.5	9.9	10.3	10.6	11.0	11.4	11.7	44
45	6.4	6.8	7.1	7.5	7.9	8.2	8.6	9.0	9.4	9.8	10.1	10.5	10.9	11.2	11.6	12.0	45
46	6.5	6.9	7.3	7.7	8.0	8.4	8.8	9.2	9.6	10.0	10.4	10.7	11.1	11.5	11.9	12.3	46
47	6.7	7.0	7.4	7.8	8.2	8.6	9.0	9.4	9.8	10.2	10.6	11.0	11.4	11.8	12.1	12.5	47
48	6.8	7.2	7.6	8.0	8.4	8.8	9.2	9.6	10.0	10.4	10.8	11.2	11.6	12.0	12.4	12.8	48
49	6.9	7.4	7.8	8.2	8.6	9.0	9.4	9.8	10.2	10.6	11.0	11.4	11.8	12.2	12.7	13.1	49
50	7.1	7.5	7.9	8.3	8.8	9.2	9.6	10.0	10.4	10.8	11.2	11.7	12.1	12.5	12.9	13.3	50
51	7.2	7.6	8.1	8.5	8.9	9.4	9.8	10.2	10.6	11.0	11.5	11.9	12.3	12.8	13.2	13.6	51
52	7.4	7.8	8.2	8.7	9.1	9.5	10.0	10.4	10.8	11.3	11.7	12.1	12.6	13.0	13.4	13.9	52
53	7.5	8.0	8.4	8.8	9.3	9.7	10.2	10.6	11.0	11.5	11.9	12.4	12.8	13.2	13.7	14.1	53
54	7.6	8.1	8.6	9.0	9.4	9.9	10.4	10.8	11.2	11.7	12.2	12.6	13.0	13.5	14.0	14.4	54
55	7.8	8.2	8.7	9.2	9.6	10.1	10.5	11.0	11.5	11.9	12.4	12.8	13.3	13.8	14.2	14.7	55
56	7.9	8.4	8.9	9.3	9.8	10.3	10.7	11.2	11.7	12.1	12.6	13.1	13.5	14.0	14.5	14.9	56
57	8.1	8.6	9.0	9.5	10.0	10.4	10.9	11.4	11.9	12.4	12.8	13.3	13.8	14.2	14.7	15.2	57
58	8.2	8.7	9.2	9.7	10.2	10.6	11.1	11.6	12.1	12.6	13.0	13.5	14.0	14.5	15.0	15.5	58
59	8.4	8.8	9.3	9.8	10.3	10.8	11.3	11.8	12.3	12.8	13.3	13.8	14.3	14.8	15.2	15.7	59
60	8.5	9.0	9.5	10.0	10.5	11.0	11.5	12.0	12.5	13.0	13.5	14.0	14.5	15.0	15.5	16.0	60

TABLE 19

Speed, Time, and Distance

Minutes	Speed in knots																Minutes
	16.5	17.0	17.5	18.0	18.5	19.0	19.5	20.0	20.5	21.0	21.5	22.0	22.5	23.0	23.5	24.0	
	Miles	*Miles*	*Miles*	*Miles*	*Miles*	*Miles*	*Miles*	*Miles*	*Miles*	*Miles*	*Miles*	*Miles*	*Miles*	*Miles*	*Miles*	*Miles*	
1	0.3	0.3	0.3	0.3	0.3	0.3	0.3	0.3	0.3	0.4	0.4	0.4	0.4	0.4	0.4	0.4	1
2	0.6	0.6	0.6	0.6	0.6	0.6	0.6	0.7	0.7	0.7	0.7	0.7	0.8	0.8	0.8	0.8	2
3	0.8	0.8	0.9	0.9	0.9	1.0	1.0	1.0	1.0	1.0	1.1	1.1	1.1	1.2	1.2	1.2	3
4	1.1	1.1	1.2	1.2	1.2	1.3	1.3	1.3	1.4	1.4	1.4	1.5	1.5	1.5	1.6	1.6	4
5	1.4	1.4	1.5	1.5	1.5	1.6	1.6	1.7	1.7	1.8	1.8	1.8	1.9	1.9	2.0	2.0	5
6	1.6	1.7	1.8	1.8	1.8	1.9	2.0	2.0	2.0	2.1	2.2	2.2	2.2	2.3	2.4	2.4	6
7	1.9	2.0	2.0	2.1	2.2	2.2	2.3	2.3	2.4	2.4	2.5	2.6	2.6	2.7	2.7	2.8	7
8	2.2	2.3	2.3	2.4	2.5	2.5	2.6	2.7	2.7	2.8	2.9	2.9	3.0	3.1	3.1	3.2	8
9	2.5	2.6	2.6	2.7	2.8	2.8	2.9	3.0	3.1	3.2	3.2	3.3	3.4	3.4	3.5	3.6	9
10	2.8	2.8	2.9	3.0	3.1	3.2	3.2	3.3	3.4	3.5	3.6	3.7	3.8	3.8	3.9	4.0	10
11	3.0	3.1	3.2	3.3	3.4	3.5	3.6	3.7	3.8	3.8	3.9	4.0	4.1	4.2	4.3	4.4	11
12	3.3	3.4	3.5	3.6	3.7	3.8	3.9	4.0	4.1	4.2	4.3	4.4	4.5	4.6	4.7	4.8	12
13	3.6	3.7	3.8	3.9	4.0	4.1	4.2	4.3	4.4	4.6	4.7	4.8	4.9	5.0	5.1	5.2	13
14	3.8	4.0	4.1	4.2	4.3	4.4	4.6	4.7	4.8	4.9	5.0	5.1	5.2	5.4	5.5	5.6	14
15	4.1	4.2	4.4	4.5	4.6	4.8	4.9	5.0	5.1	5.2	5.4	5.5	5.6	5.8	5.9	6.0	15
16	4.4	4.5	4.7	4.8	4.9	5.1	5.2	5.3	5.5	5.6	5.7	5.9	6.0	6.1	6.3	6.4	16
17	4.7	4.8	5.0	5.1	5.2	5.4	5.5	5.7	5.8	6.0	6.1	6.2	6.4	6.5	6.7	6.8	17
18	5.0	5.1	5.2	5.4	5.6	5.7	5.8	6.0	6.2	6.3	6.4	6.6	6.8	6.9	7.0	7.2	18
19	5.2	5.4	5.5	5.7	5.9	6.0	6.2	6.3	6.5	6.6	6.8	7.0	7.1	7.3	7.4	7.6	19
20	5.5	5.7	5.8	6.0	6.2	6.3	6.5	6.7	6.8	7.0	7.2	7.3	7.5	7.7	7.8	8.0	20
21	5.8	6.0	6.1	6.3	6.5	6.6	6.8	7.0	7.2	7.4	7.5	7.7	7.9	8.0	8.2	8.4	21
22	6.0	6.2	6.4	6.6	6.8	7.0	7.2	7.3	7.5	7.7	7.9	8.1	8.2	8.4	8.6	8.8	22
23	6.3	6.5	6.7	6.9	7.1	7.3	7.5	7.7	7.9	8.0	8.2	8.4	8.6	8.8	9.0	9.2	23
24	6.6	6.8	7.0	7.2	7.4	7.6	7.8	8.0	8.2	8.4	8.6	8.8	9.0	9.2	9.4	9.6	24
25	6.9	7.1	7.3	7.5	7.7	7.9	8.1	8.3	8.5	8.8	9.0	9.2	9.4	9.6	9.8	10.0	25
26	7.2	7.4	7.6	7.8	8.0	8.2	8.4	8.7	8.9	9.1	9.3	9.5	9.8	10.0	10.2	10.4	26
27	7.4	7.6	7.9	8.1	8.3	8.6	8.8	9.0	9.2	9.4	9.7	9.9	10.1	10.4	10.6	10.8	27
28	7.7	7.9	8.2	8.4	8.6	8.9	9.1	9.3	9.6	9.8	10.0	10.3	10.5	10.7	11.0	11.2	28
29	8.0	8.2	8.5	8.7	8.9	9.2	9.4	9.7	9.9	10.2	10.4	10.6	10.9	11.1	11.4	11.6	29
30	8.2	8.5	8.8	9.0	9.2	9.5	9.8	10.0	10.2	10.5	10.8	11.0	11.2	11.5	11.8	12.0	30
31	8.5	8.8	9.0	9.3	9.6	9.8	10.1	10.3	10.6	10.8	11.1	11.4	11.6	11.9	12.1	12.4	31
32	8.8	9.1	9.3	9.6	9.9	10.1	10.4	10.7	10.9	11.2	11.5	11.7	12.0	12.3	12.5	12.8	32
33	9.1	9.4	9.6	9.9	10.2	10.4	10.7	11.0	11.3	11.6	11.8	12.1	12.4	12.6	12.9	13.2	33
34	9.4	9.6	9.9	10.2	10.5	10.8	11.0	11.3	11.6	11.9	12.2	12.5	12.8	13.0	13.3	13.6	34
35	9.6	9.9	10.2	10.5	10.8	11.1	11.4	11.7	12.0	12.2	12.5	12.8	13.1	13.4	13.7	14.0	35
36	9.9	10.2	10.5	10.8	11.1	11.4	11.7	12.0	12.3	12.6	12.9	13.2	13.5	13.8	14.1	14.4	36
37	10.2	10.5	10.8	11.1	11.4	11.7	12.0	12.3	12.6	13.0	13.3	13.6	13.9	14.2	14.5	14.8	37
38	10.4	10.8	11.1	11.4	11.7	12.0	12.4	12.7	13.0	13.3	13.6	13.9	14.2	14.6	14.9	15.2	38
39	10.7	11.0	11.4	11.7	12.0	12.4	12.7	13.0	13.3	13.6	14.0	14.3	14.6	15.0	15.3	15.6	39
40	11.0	11.3	11.7	12.0	12.3	12.7	13.0	13.3	13.7	14.0	14.3	14.7	15.0	15.3	15.7	16.0	40
41	11.3	11.6	12.0	12.3	12.6	13.0	13.3	13.7	14.0	14.4	14.7	15.0	15.4	15.7	16.1	16.4	41
42	11.6	11.9	12.2	12.6	13.0	13.3	13.6	14.0	14.4	14.7	15.0	15.4	15.8	16.1	16.4	16.8	42
43	11.8	12.2	12.5	12.9	13.3	13.6	14.0	14.3	14.7	15.0	15.4	15.8	16.1	16.5	16.8	17.2	43
44	12.1	12.5	12.8	13.2	13.6	13.9	14.3	14.7	15.0	15.4	15.8	16.1	16.5	16.9	17.2	17.6	44
45	12.4	12.8	13.1	13.5	13.9	14.2	14.6	15.0	15.4	15.8	16.1	16.5	16.9	17.2	17.6	18.0	45
46	12.6	13.0	13.4	13.8	14.2	14.6	15.0	15.3	15.7	16.1	16.5	16.9	17.2	17.6	18.0	18.4	46
47	12.9	13.3	13.7	14.1	14.5	14.9	15.3	15.7	16.1	16.4	16.8	17.2	17.6	18.0	18.4	18.8	47
48	13.2	13.6	14.0	14.4	14.8	15.2	15.6	16.0	16.4	16.8	17.2	17.6	18.0	18.4	18.8	19.2	48
49	13.5	13.9	14.3	14.7	15.1	15.5	15.9	16.3	16.7	17.2	17.6	18.0	18.4	18.8	19.2	19.6	49
50	13.8	14.2	14.6	15.0	15.4	15.8	16.2	16.7	17.1	17.5	17.9	18.3	18.8	19.2	19.6	20.0	50
51	14.0	14.4	14.9	15.3	15.7	16.2	16.6	17.0	17.4	17.8	18.3	18.7	19.1	19.6	20.0	20.4	51
52	14.3	14.7	15.2	15.6	16.0	16.5	16.9	17.3	17.8	18.2	18.6	19.1	19.5	19.9	20.4	20.8	52
53	14.6	15.0	15.5	15.9	16.3	16.8	17.2	17.7	18.1	18.6	19.0	19.4	19.9	20.3	20.8	21.2	53
54	14.8	15.3	15.8	16.2	16.6	17.1	17.6	18.0	18.4	18.9	19.4	19.8	20.2	20.7	21.2	21.6	54
55	15.1	15.6	16.0	16.5	17.0	17.4	17.9	18.3	18.8	19.2	19.7	20.2	20.6	21.1	21.5	22.0	55
56	15.4	15.9	16.3	16.8	17.3	17.7	18.2	18.7	19.1	19.6	20.1	20.5	21.0	21.5	21.9	22.4	56
57	15.7	16.2	16.6	17.1	17.6	18.0	18.5	19.0	19.5	20.0	20.4	20.9	21.4	21.8	22.3	22.8	57
58	16.0	16.4	16.9	17.4	17.9	18.4	18.8	19.3	19.8	20.3	20.8	21.3	21.8	22.2	22.7	23.2	58
59	16.2	16.7	17.2	17.7	18.2	18.7	19.2	19.7	20.2	20.6	21.1	21.6	22.1	22.6	23.1	23.6	59
60	16.5	17.0	17.5	18.0	18.5	19.0	19.5	20.0	20.5	21.0	21.5	22.0	22.5	23.0	23.5	24.0	60

TABLE 19

Speed, Time, and Distance

Minutes	Speed in knots																Minutes
	24.5	25.0	25.5	26.0	26.5	27.0	27.5	28.0	28.5	29.0	29.5	30.0	30.5	31.0	31.5	32.0	
	Miles	*Miles*	*Miles*	*Miles*	*Miles*	*Miles*	*Miles*	*Miles*	*Miles*	*Miles*	*Miles*	*Miles*	*Miles*	*Miles*	*Miles*	*Miles*	
1	0.4	0.4	0.4	0.4	0.4	0.4	0.5	0.5	0.5	0.5	0.5	0.5	0.5	0.5	0.5	0.5	1
2	0.8	0.8	0.8	0.9	0.9	0.9	0.9	0.9	1.0	1.0	1.0	1.0	1.0	1.0	1.0	1.1	2
3	1.2	1.2	1.3	1.3	1.3	1.4	1.4	1.4	1.4	1.4	1.5	1.5	1.5	1.6	1.6	1.6	3
4	1.6	1.7	1.7	1.7	1.8	1.8	1.8	1.9	1.9	1.9	2.0	2.0	2.0	2.1	2.1	2.1	4
5	2.0	2.1	2.1	2.2	2.2	2.2	2.3	2.3	2.4	2.4	2.5	2.5	2.5	2.6	2.6	2.7	5
6	2.4	2.5	2.6	2.6	2.6	2.7	2.8	2.8	2.8	2.9	3.0	3.0	3.0	3.1	3.2	3.2	6
7	2.9	2.9	3.0	3.0	3.1	3.2	3.2	3.3	3.3	3.4	3.4	3.5	3.6	3.6	3.7	3.7	7
8	3.3	3.3	3.4	3.5	3.5	3.6	3.7	3.7	3.8	3.9	3.9	4.0	4.1	4.1	4.2	4.3	8
9	3.7	3.8	3.8	3.9	4.0	4.0	4.1	4.2	4.3	4.4	4.4	4.5	4.6	4.6	4.7	4.8	9
10	4.1	4.2	4.2	4.3	4.4	4.5	4.6	4.7	4.8	4.8	4.9	5.0	5.1	5.2	5.2	5.3	10
11	4.5	4.6	4.7	4.8	4.9	5.0	5.0	5.1	5.2	5.3	5.4	5.5	5.6	5.7	5.8	5.9	11
12	4.9	5.0	5.1	5.2	5.3	5.4	5.5	5.6	5.7	5.8	5.9	6.0	6.1	6.2	6.3	6.4	12
13	5.3	5.4	5.5	5.6	5.7	5.8	6.0	6.1	6.2	6.3	6.4	6.5	6.6	6.7	6.8	6.9	13
14	5.7	5.8	6.0	6.1	6.2	6.3	6.4	6.5	6.6	6.8	6.9	7.0	7.1	7.2	7.4	7.5	14
15	6.1	6.2	6.4	6.5	6.6	6.8	6.9	7.0	7.1	7.2	7.4	7.5	7.6	7.8	7.9	8.0	15
16	6.5	6.7	6.8	6.9	7.1	7.2	7.3	7.5	7.6	7.7	7.9	8.0	8.1	8.3	8.4	8.5	16
17	6.9	7.1	7.2	7.4	7.5	7.6	7.8	7.9	8.1	8.2	8.4	8.5	8.6	8.8	8.9	9.1	17
18	7.4	7.5	7.6	7.8	8.0	8.1	8.2	8.4	8.6	8.7	8.8	9.0	9.2	9.3	9.4	9.6	18
19	7.8	7.9	8.1	8.2	8.4	8.6	8.7	8.9	9.0	9.2	9.3	9.5	9.7	9.8	10.0	10.1	19
20	8.2	8.3	8.5	8.7	8.8	9.0	9.2	9.3	9.5	9.7	9.8	10.0	10.2	10.3	10.5	10.7	20
21	8.6	8.8	8.9	9.1	9.3	9.4	9.6	9.8	10.0	10.2	10.3	10.5	10.7	10.8	11.0	11.2	21
22	9.0	9.2	9.4	9.5	9.7	9.9	10.1	10.3	10.4	10.6	10.8	11.0	11.2	11.4	11.6	11.7	22
23	9.4	9.6	9.8	10.0	10.2	10.4	10.5	10.7	10.9	11.1	11.3	11.5	11.7	11.9	12.1	12.3	23
24	9.8	10.0	10.2	10.4	10.6	10.8	11.0	11.2	11.4	11.6	11.8	12.0	12.2	12.4	12.6	12.8	24
25	10.2	10.4	10.6	10.8	11.0	11.2	11.5	11.7	11.9	12.1	12.3	12.5	12.7	12.9	13.1	13.3	25
26	10.6	10.8	11.0	11.3	11.5	11.7	11.9	12.1	12.4	12.6	12.8	13.0	13.2	13.4	13.6	13.9	26
27	11.0	11.2	11.5	11.7	11.9	12.2	12.4	12.6	12.8	13.0	13.3	13.5	13.7	14.0	14.2	14.4	27
28	11.4	11.7	11.9	12.1	12.4	12.6	12.8	13.1	13.3	13.5	13.8	14.0	14.2	14.5	14.7	14.9	28
29	11.8	12.1	12.3	12.6	12.8	13.0	13.3	13.5	13.8	14.0	14.3	14.5	14.7	15.0	15.2	15.5	29
30	12.2	12.5	12.8	13.0	13.2	13.5	13.8	14.0	14.2	14.5	14.8	15.0	15.2	15.5	15.8	16.0	30
31	12.7	12.9	13.2	13.4	13.7	14.0	14.2	14.5	14.7	15.0	15.2	15.5	15.8	16.0	16.3	16.5	31
32	13.1	13.3	13.6	13.9	14.1	14.4	14.7	14.9	15.2	15.5	15.7	16.0	16.3	16.5	16.8	17.1	32
33	13.5	13.8	14.0	14.3	14.6	14.8	15.1	15.4	15.7	16.0	16.2	16.5	16.8	17.0	17.3	17.6	33
34	13.9	14.2	14.4	14.7	15.0	15.3	15.6	15.9	16.2	16.4	16.7	17.0	17.3	17.6	17.8	18.1	34
35	14.3	14.6	14.9	15.2	15.5	15.8	16.0	16.3	16.6	16.9	17.2	17.5	17.8	18.1	18.4	18.7	35
36	14.7	15.0	15.3	15.6	15.9	16.2	16.5	16.8	17.1	17.4	17.7	18.0	18.3	18.6	18.9	19.2	36
37	15.1	15.4	15.7	16.0	16.3	16.6	17.0	17.3	17.6	17.9	18.2	18.5	18.8	19.1	19.4	19.7	37
38	15.5	15.8	16.2	16.5	16.8	17.1	17.4	17.7	18.0	18.4	18.7	19.0	19.3	19.6	20.0	20.3	38
39	15.9	16.2	16.6	16.9	17.2	17.6	17.9	18.2	18.5	18.8	19.2	19.5	19.8	20.2	20.5	20.8	39
40	16.3	16.7	17.0	17.3	17.7	18.0	18.3	18.7	19.0	19.3	19.7	20.0	20.3	20.7	21.0	21.3	40
41	16.7	17.1	17.4	17.8	18.1	18.4	18.8	19.1	19.5	19.8	20.2	20.5	20.8	21.2	21.5	21.9	41
42	17.2	17.5	17.8	18.2	18.6	18.9	19.2	19.6	20.0	20.3	20.6	21.0	21.4	21.7	22.0	22.4	42
43	17.6	17.9	18.3	18.6	19.0	19.4	19.7	20.1	20.4	20.8	21.1	21.5	21.9	22.2	22.6	22.9	43
44	18.0	18.3	18.7	19.1	19.4	19.8	20.2	20.5	20.9	21.3	21.6	22.0	22.4	22.7	23.1	23.5	44
45	18.4	18.8	19.1	19.5	19.9	20.2	20.6	21.0	21.4	21.8	22.1	22.5	22.9	23.2	23.6	24.0	45
46	18.8	19.2	19.6	19.9	20.3	20.7	21.1	21.5	21.8	22.2	22.6	23.0	23.4	23.8	24.2	24.5	46
47	19.2	19.6	20.0	20.4	20.8	21.2	21.5	21.9	22.3	22.7	23.1	23.5	23.9	24.3	24.7	25.1	47
48	19.6	20.0	20.4	20.8	21.2	21.6	22.0	22.4	22.8	23.2	23.6	24.0	24.4	24.8	25.2	25.6	48
49	20.0	20.4	20.8	21.2	21.6	22.0	22.5	22.9	23.3	23.7	24.1	24.5	24.9	25.3	25.7	26.1	49
50	20.4	20.8	21.2	21.7	22.1	22.5	22.9	23.3	23.8	24.2	24.6	25.0	25.4	25.8	26.2	26.7	50
51	20.8	21.2	21.7	22.1	22.5	23.0	23.4	23.8	24.2	24.6	25.1	25.5	25.9	26.4	26.8	27.2	51
52	21.2	21.7	22.1	22.5	23.0	23.4	23.8	24.3	24.7	25.1	25.6	26.0	26.4	26.9	27.3	27.7	52
53	21.6	22.1	22.5	23.0	23.4	23.8	24.3	24.7	25.2	25.6	26.1	26.5	26.9	27.4	27.8	28.3	53
54	22.0	22.5	23.0	23.4	23.8	24.3	24.8	25.2	25.6	26.1	26.6	27.0	27.4	27.9	28.4	28.8	54
55	22.5	22.9	23.4	23.8	24.3	24.8	25.2	25.7	26.1	26.6	27.0	27.5	28.0	28.4	28.9	29.3	55
56	22.9	23.3	23.8	24.3	24.7	25.2	25.7	26.1	26.6	27.1	27.5	28.0	28.5	28.9	29.4	29.9	56
57	23.3	23.8	24.2	24.7	25.2	25.6	26.1	26.6	27.1	27.6	28.0	28.5	29.0	29.4	29.9	30.4	57
58	23.7	24.2	24.6	25.1	25.6	26.1	26.6	27.1	27.6	28.0	28.5	29.0	29.5	30.0	30.4	30.9	58
59	24.1	24.6	25.1	25.6	26.1	26.6	27.0	27.5	28.0	28.5	29.0	29.5	30.0	30.5	31.0	31.5	59
60	24.5	25.0	25.5	26.0	26.5	27.0	27.5	28.0	28.5	29.0	29.5	30.0	30.5	31.0	31.5	32.0	60

TABLE 19

Speed, Time, and Distance

Minutes	32.5	33.0	33.5	34.0	34.5	35.0	35.5	36.0	36.5	37.0	37.5	38.0	38.5	39.0	39.5	40.0	Minutes
	Speed in knots																
	Miles	*Miles*	*Miles*	*Miles*	*Miles*	*Miles*	*Miles*	*Miles*	*Miles*	*Miles*	*Miles*	*Miles*	*Miles*	*Miles*	*Miles*	*Miles*	
1	0.5	0.6	0.6	0.6	0.6	0.6	0.6	0.6	0.6	0.6	0.6	0.6	0.6	0.6	0.7	0.7	1
2	1.1	1.1	1.1	1.1	1.2	1.2	1.2	1.2	1.2	1.2	1.2	1.3	1.3	1.3	1.3	1.3	2
3	1.6	1.6	1.7	1.7	1.7	1.8	1.8	1.8	1.8	1.8	1.9	1.9	1.9	2.0	2.0	2.0	3
4	2.2	2.2	2.2	2.3	2.3	2.3	2.4	2.4	2.4	2.5	2.5	2.5	2.6	2.6	2.6	2.7	4
5	2.7	2.8	2.8	2.8	2.9	2.9	3.0	3.0	3.0	3.1	3.1	3.2	3.2	3.2	3.3	3.3	5
6	3.2	3.3	3.4	3.4	3.4	3.5	3.6	3.6	3.6	3.7	3.8	3.8	3.8	3.9	4.0	4.0	6
7	3.8	3.8	3.9	4.0	4.0	4.1	4.1	4.2	4.3	4.3	4.4	4.4	4.5	4.6	4.6	4.7	7
8	4.3	4.4	4.5	4.5	4.6	4.7	4.7	4.8	4.9	4.9	5.0	5.1	5.1	5.2	5.3	5.3	8
9	4.9	5.0	5.0	5.1	5.2	5.2	5.3	5.4	5.5	5.6	5.6	5.7	5.8	5.8	5.9	6.0	9
10	5.4	5.5	5.6	5.7	5.8	5.8	5.9	6.0	6.1	6.2	6.2	6.3	6.4	6.5	6.6	6.7	10
11	6.0	6.0	6.1	6.2	6.3	6.4	6.5	6.6	6.7	6.8	6.9	7.0	7.1	7.2	7.2	7.3	11
12	6.5	6.6	6.7	6.8	6.9	7.0	7.1	7.2	7.3	7.4	7.5	7.6	7.7	7.8	7.9	8.0	12
13	7.0	7.2	7.3	7.4	7.5	7.6	7.7	7.8	7.9	8.0	8.1	8.2	8.3	8.4	8.6	8.7	13
14	7.6	7.7	7.8	7.9	8.0	8.2	8.3	8.4	8.5	8.6	8.8	8.9	9.0	9.1	9.2	9.3	14
15	8.1	8.2	8.4	8.5	8.6	8.8	8.9	9.0	9.1	9.2	9.4	9.5	9.6	9.8	9.9	10.0	15
16	8.7	8.8	8.9	9.1	9.2	9.3	9.5	9.6	9.7	9.9	10.0	10.1	10.3	10.4	10.5	10.7	16
17	9.2	9.4	9.5	9.6	9.8	9.9	10.1	10.2	10.3	10.5	10.6	10.8	10.9	11.0	11.2	11.3	17
18	9.8	9.9	10.0	10.2	10.4	10.5	10.6	10.8	11.0	11.1	11.2	11.4	11.6	11.7	11.8	12.0	18
19	10.3	10.4	10.6	10.8	10.9	11.1	11.2	11.4	11.6	11.7	11.9	12.0	12.2	12.4	12.5	12.7	19
20	10.8	11.0	11.2	11.3	11.5	11.7	11.8	12.0	12.2	12.3	12.5	12.7	12.8	13.0	13.2	13.3	20
21	11.4	11.6	11.7	11.9	12.1	12.2	12.4	12.6	12.8	13.0	13.1	13.3	13.5	13.6	13.8	14.0	21
22	11.9	12.1	12.3	12.5	12.6	12.8	13.0	13.2	13.4	13.6	13.8	13.9	14.1	14.3	14.5	14.7	22
23	12.5	12.6	12.8	13.0	13.2	13.4	13.6	13.8	14.0	14.2	14.4	14.6	14.8	15.0	15.1	15.3	23
24	13.0	13.2	13.4	13.6	13.8	14.0	14.2	14.4	14.6	14.8	15.0	15.2	15.4	15.6	15.8	16.0	24
25	13.5	13.8	14.0	14.2	14.4	14.6	14.8	15.0	15.2	15.4	15.6	15.8	16.0	16.2	16.5	16.7	25
26	14.1	14.3	14.5	14.7	15.0	15.2	15.4	15.6	15.8	16.0	16.2	16.5	16.7	16.9	17.1	17.3	26
27	14.6	14.8	15.1	15.3	15.5	15.8	16.0	16.2	16.4	16.6	16.9	17.1	17.3	17.6	17.8	18.0	27
28	15.2	15.4	15.6	15.9	16.1	16.3	16.6	16.8	17.0	17.3	17.5	17.7	18.0	18.2	18.4	18.7	28
29	15.7	16.0	16.2	16.4	16.7	16.9	17.2	17.4	17.6	17.9	18.1	18.4	18.6	18.8	19.1	19.3	29
30	16.2	16.5	16.8	17.0	17.2	17.5	17.8	18.0	18.2	18.5	18.8	19.0	19.2	19.5	19.8	20.0	30
31	16.8	17.0	17.3	17.6	17.8	18.1	18.3	18.6	18.9	19.1	19.4	19.6	19.9	20.2	20.4	20.7	31
32	17.3	17.6	17.9	18.1	18.4	18.7	18.9	19.2	19.5	19.7	20.0	20.3	20.5	20.8	21.1	21.3	32
33	17.9	18.2	18.4	18.7	19.0	19.2	19.5	19.8	20.1	20.4	20.6	20.9	21.2	21.4	21.7	22.0	33
34	18.4	18.7	19.0	19.3	19.6	19.8	20.1	20.4	20.7	21.0	21.2	21.5	21.8	22.1	22.4	22.7	34
35	19.0	19.2	19.5	19.8	20.1	20.4	20.7	21.0	21.3	21.6	21.9	22.2	22.5	22.8	23.0	23.3	35
36	19.5	19.8	20.1	20.4	20.7	21.0	21.3	21.6	21.9	22.2	22.5	22.8	23.1	23.4	23.7	24.0	36
37	20.0	20.4	20.7	21.0	21.3	21.6	21.9	22.2	22.5	22.8	23.1	23.4	23.7	24.0	24.4	24.7	37
38	20.6	20.9	21.2	21.5	21.8	22.2	22.5	22.8	23.1	23.4	23.8	24.1	24.4	24.7	25.0	25.3	38
39	21.1	21.4	21.8	22.1	22.4	22.8	23.1	23.4	23.7	24.0	24.4	24.7	25.0	25.4	25.7	26.0	39
40	21.7	22.0	22.3	22.7	23.0	23.3	23.7	24.0	24.3	24.7	25.0	25.3	25.7	26.0	26.3	26.7	40
41	22.2	22.6	22.9	23.2	23.6	23.9	24.3	24.6	24.9	25.3	25.6	26.0	26.3	26.6	27.0	27.3	41
42	22.8	23.1	23.4	23.8	24.2	24.5	24.8	25.2	25.6	25.9	26.2	26.6	27.0	27.3	27.6	28.0	42
43	23.3	23.6	24.0	24.4	24.7	25.1	25.4	25.8	26.2	26.5	26.9	27.2	27.6	28.0	28.3	28.7	43
44	23.8	24.2	24.6	24.9	25.3	25.7	26.0	26.4	26.8	27.1	27.5	27.9	28.2	28.6	29.0	29.3	44
45	24.4	24.8	25.1	25.5	25.9	26.2	26.6	27.0	27.4	27.8	28.1	28.5	28.9	29.2	29.6	30.0	45
46	24.9	25.3	25.7	26.1	26.4	26.8	27.2	27.6	28.0	28.4	28.8	29.1	29.5	29.9	30.3	30.7	46
47	25.5	25.8	26.2	26.6	27.0	27.4	27.8	28.2	28.6	29.0	29.4	29.8	30.2	30.6	30.9	31.3	47
48	26.0	26.4	26.8	27.2	27.6	28.0	28.4	28.8	29.2	29.6	30.0	30.4	30.8	31.2	31.6	32.0	48
49	26.5	27.0	27.4	27.8	28.2	28.6	29.0	29.4	29.8	30.2	30.6	31.0	31.4	31.8	32.3	32.7	49
50	27.1	27.5	27.9	28.3	28.8	29.2	29.6	30.0	30.4	30.8	31.2	31.7	32.1	32.5	32.9	33.3	50
51	27.6	28.0	28.5	28.9	29.3	29.8	30.2	30.6	31.0	31.4	31.9	32.3	32.7	33.2	33.6	34.0	51
52	28.2	28.6	29.0	29.5	29.9	30.3	30.8	31.2	31.6	32.1	32.5	32.9	33.4	33.8	34.2	34.7	52
53	28.7	29.2	29.6	30.0	30.5	30.9	31.4	31.8	32.2	32.7	33.1	33.6	34.0	34.4	34.9	35.3	53
54	29.2	29.7	30.2	30.6	31.0	31.5	32.0	32.4	32.8	33.3	33.8	34.2	34.6	35.1	35.6	36.0	54
55	29.8	30.2	30.7	31.2	31.6	32.1	32.5	33.0	33.5	33.9	34.4	34.8	35.3	35.8	36.2	36.7	55
56	30.3	30.8	31.3	31.7	32.2	32.7	33.1	33.6	34.1	34.5	35.0	35.5	35.9	36.4	36.9	37.3	56
57	30.9	31.4	31.8	32.3	32.8	33.2	33.7	34.2	34.7	35.2	35.6	36.1	36.6	37.0	37.5	38.0	57
58	31.4	31.9	32.4	32.9	33.4	33.8	34.3	34.8	35.3	35.8	36.2	36.7	37.2	37.7	38.2	38.7	58
59	32.0	32.4	32.9	33.4	33.9	34.4	34.9	35.4	35.9	36.4	36.9	37.4	37.9	38.4	38.8	39.3	59
60	32.5	33.0	33.5	34.0	34.5	35.0	35.5	36.0	36.5	37.0	37.5	38.0	38.5	39.0	39.5	40.0	60

TABLE 20

Conversion Table for Nautical and Statute Miles

1 nautical mile=6,076.11548 . . . feet

1 statute mile=5,280 feet

Nautical miles to statute miles				Statute miles to nautical miles			
Nautical miles	Statute miles	Nautical miles	Statute miles	Statute miles	Nautical miles	Statute miles	Nautical miles
1	1. 151	51	58. 690	1	0. 869	51	44. 318
2	2. 302	52	59. 841	2	1. 738	52	45. 187
3	3. 452	53	60. 991	3	2. 607	53	46. 056
4	4. 603	54	62. 142	4	3. 476	54	46. 925
5	5. 754	55	63. 293	5	4. 345	55	47. 794
6	6. 905	56	64. 444	6	5. 214	56	48. 663
7	8. 055	57	65. 594	7	6. 083	57	49. 532
8	9. 206	58	66. 745	8	6. 952	58	50. 401
9	10. 357	59	67. 896	9	7. 821	59	51. 270
10	11. 508	60	69. 047	10	8. 690	60	52. 139
11	12. 659	61	70. 198	11	9. 559	61	53. 008
12	13. 809	62	71. 348	12	10. 428	62	53. 877
13	14. 960	63	72. 499	13	11. 297	63	54. 746
14	16. 111	64	73. 650	14	12. 166	64	55. 614
15	17. 262	65	74. 801	15	13. 035	65	56. 483
16	18. 412	66	75. 951	16	13. 904	66	57. 352
17	19. 563	67	77. 102	17	14. 773	67	58. 221
18	20. 714	68	78. 253	18	15. 642	68	59. 090
19	21. 865	69	79. 404	19	16. 511	69	59. 959
20	23. 016	70	80. 555	20	17. 380	70	60. 828
21	24. 166	71	81. 705	21	18. 249	71	61. 697
22	25. 317	72	82. 856	22	19. 117	72	62. 566
23	26. 468	73	84. 007	23	19. 986	73	63. 435
24	27. 619	74	85. 158	24	20. 855	74	64. 304
25	28. 769	75	86. 308	25	21. 724	75	65. 173
26	29. 920	76	87. 459	26	22. 593	76	66. 042
27	31. 071	77	88. 610	27	23. 462	77	66. 911
28	32. 222	78	89. 761	28	24. 331	78	67. 780
29	33. 373	79	90. 912	29	25. 200	79	68. 649
30	34. 523	80	92. 062	30	26. 069	80	69. 518
31	35. 674	81	93. 213	31	26. 938	81	70. 387
32	36. 825	82	94. 364	32	27. 807	82	71. 256
33	37. 976	83	95. 515	33	28. 676	83	72. 125
34	39. 127	84	96. 665	34	29. 545	84	72. 994
35	40. 277	85	97. 816	35	30. 414	85	73. 863
36	41. 428	86	98. 967	36	31. 283	86	74. 732
37	42. 579	87	100. 118	37	32. 152	87	75. 601
38	43. 730	88	101. 269	38	33. 021	88	76. 470
39	44. 880	89	102. 419	39	33. 890	89	77. 339
40	46. 031	90	103. 570	40	34. 759	90	78. 208
41	47. 182	91	104. 721	41	35. 628	91	79. 077
42	48. 333	92	105. 872	42	36. 497	92	79. 946
43	49. 484	93	107. 022	43	37. 366	93	80. 815
44	50. 634	94	108. 173	44	38. 235	94	81. 684
45	51. 785	95	109. 324	45	39. 104	95	82. 553
46	52. 936	96	110. 475	46	39. 973	96	83. 422
47	54. 087	97	111. 626	47	40. 842	97	84. 291
48	55. 237	98	112. 776	48	41. 711	98	85. 160
49	56. 388	99	113. 927	49	42. 580	99	86. 029
50	57. 539	100	115. 078	50	43. 449	100	86. 898

TABLE 21

Conversion Table for Meters, Feet, and Fathoms

Meters	Feet	Fathoms	Meters	Feet	Fathoms	Feet	Meters	Feet	Meters	Fathoms	Meters	Fathoms	Meters
1	3.28	0.55	61	200.13	33.36	1	0.30	61	18.59	1	1.83	61	111.56
2	6.56	1.09	62	203.41	33.90	2	0.61	62	18.90	2	3.66	62	113.39
3	9.84	1.64	63	206.69	34.45	3	0.91	63	19.20	3	5.49	63	115.21
4	13.12	2.19	64	209.97	35.00	4	1.22	64	19.51	4	7.32	64	117.04
5	16.40	2.73	65	213.25	35.54	5	1.52	65	19.81	5	9.14	65	118.87
6	19.69	3.28	66	216.54	36.09	6	1.83	66	20.12	6	10.97	66	120.70
7	22.97	3.83	67	219.82	36.64	7	2.13	67	20.42	7	12.80	67	122.53
8	26.25	4.37	68	223.10	37.18	8	2.44	68	20.73	8	14.63	68	124.36
9	29.53	4.92	69	226.38	37.73	9	2.74	69	21.03	9	16.46	69	126.19
10	32.81	5.47	70	229.66	38.28	10	3.05	70	21.34	10	18.29	70	128.02
11	36.09	6.01	71	232.94	38.82	11	3.35	71	21.64	11	20.12	71	129.84
12	39.37	6.56	72	236.22	39.37	12	3.66	72	21.95	12	21.95	72	131.67
13	42.65	7.11	73	239.50	39.92	13	3.96	73	22.25	13	23.77	73	133.50
14	45.93	7.66	74	242.78	40.46	14	4.27	74	22.56	14	25.60	74	135.33
15	49.21	8.20	75	246.06	41.01	15	4.57	75	22.86	15	27.43	75	137.16
16	52.49	8.75	76	249.34	41.56	16	4.88	76	23.16	16	29.26	76	138.99
17	55.77	9.30	77	252.62	42.10	17	5.18	77	23.47	17	31.09	77	140.82
18	59.06	9.84	78	255.91	42.65	18	5.49	78	23.77	18	32.92	78	142.65
19	62.34	10.39	79	259.19	43.20	19	5.79	79	24.08	19	34.75	79	144.48
20	65.62	10.94	80	262.47	43.74	20	6.10	80	24.38	20	36.58	80	146.30
21	68.90	11.48	81	265.75	44.29	21	6.40	81	24.69	21	38.40	81	148.13
22	72.18	12.03	82	269.03	44.84	22	6.71	82	24.99	22	40.23	82	149.96
23	75.46	12.58	83	272.31	45.38	23	7.01	83	25.30	23	42.06	83	151.79
24	78.74	13.12	84	275.59	45.93	24	7.32	84	25.60	24	43.89	84	153.62
25	82.02	13.67	85	278.87	46.48	25	7.62	85	25.91	25	45.72	85	155.45
26	85.30	14.22	86	282.15	47.03	26	7.92	86	26.21	26	47.55	86	157.28
27	88.58	14.76	87	285.43	47.57	27	8.23	87	26.52	27	49.38	87	159.11
28	91.86	15.31	88	288.71	48.12	28	8.53	88	26.82	28	51.21	88	160.93
29	95.14	15.86	89	291.99	48.67	29	8.84	89	27.13	29	53.04	89	162.76
30	98.43	16.40	90	295.28	49.21	30	9.14	90	27.43	30	54.86	90	164.59
31	101.71	16.95	91	298.56	49.76	31	9.45	91	27.74	31	56.69	91	166.42
32	104.99	17.50	92	301.84	50.31	32	9.75	92	28.04	32	58.52	92	168.25
33	108.27	18.04	93	305.12	50.85	33	10.06	93	28.35	33	60.35	93	170.08
34	111.55	18.59	94	308.40	51.40	34	10.36	94	28.65	34	62.18	94	171.91
35	114.83	19.14	95	311.68	51.95	35	10.67	95	28.96	35	64.01	95	173.74
36	118.11	19.69	96	314.96	52.49	36	10.97	96	29.26	36	65.84	96	175.56
37	121.39	20.23	97	318.24	53.04	37	11.28	97	29.57	37	67.67	97	177.39
38	124.67	20.78	98	321.52	53.59	38	11.58	98	29.87	38	69.49	98	179.22
39	127.95	21.33	99	324.80	54.13	39	11.89	99	30.18	39	71.32	99	181.05
40	131.23	21.87	100	328.08	54.68	40	12.19	100	30.48	40	73.15	100	182.88
41	134.51	22.42	101	331.36	55.23	41	12.50	101	30.78	41	74.98	101	184.71
42	137.80	22.97	102	334.65	55.77	42	12.80	102	31.09	42	76.81	102	186.54
43	141.08	23.51	103	337.93	56.32	43	13.11	103	31.39	43	78.64	103	188.37
44	144.36	24.06	104	341.21	56.87	44	13.41	104	31.70	44	80.47	104	190.20
45	147.64	24.61	105	344.49	57.41	45	13.72	105	32.00	45	82.30	105	192.02
46	150.92	25.15	106	347.77	57.96	46	14.02	106	32.31	46	84.12	106	193.85
47	154.20	25.70	107	351.05	58.51	47	14.33	107	32.61	47	85.95	107	195.68
48	157.48	26.25	108	354.33	59.06	48	14.63	108	32.92	48	87.78	108	197.51
49	160.76	26.79	109	357.61	59.60	49	14.94	109	33.22	49	89.61	109	199.34
50	164.04	27.34	110	360.89	60.15	50	15.24	110	33.53	50	91.44	110	201.17
51	167.32	27.89	111	364.17	60.70	51	15.54	111	33.83	51	93.27	111	203.00
52	170.60	28.43	112	367.45	61.24	52	15.85	112	34.14	52	95.10	112	204.83
53	173.88	28.98	113	370.73	61.79	53	16.15	113	34.44	53	96.93	113	206.65
54	177.17	29.53	114	374.02	62.34	54	16.46	114	34.75	54	98.76	114	208.48
55	180.45	30.07	115	377.30	62.88	55	16.76	115	35.05	55	100.58	115	210.31
56	183.73	30.62	116	380.58	63.43	56	17.07	116	35.36	56	102.41	116	212.14
57	187.01	31.17	117	383.86	63.98	57	17.37	117	35.66	57	104.24	117	213.97
58	190.29	31.71	118	387.14	64.52	58	17.68	118	35.97	58	106.07	118	215.80
59	193.57	32.26	119	390.42	65.07	59	17.98	119	36.27	59	107.90	119	217.63
60	196.85	32.81	120	393.70	65.62	60	18.29	120	36.58	60	109.73	120	219.46

TABLE 22

Dip of the Sea Short of the Horizon

Distance	5 (1.5)	10 (3.0)	15 (4.6)	20 (6.1)	25 (7.6)	30 (9.1)	35 (10.7)	40 (12.2)	45 (13.7)	50 (15.2)	Distance
	Height of eye above the sea, in feet and (meters)										
Miles	′	′	′	′	′	′	′	′	′	′	*Miles*
0. 2	14. 2	28. 4	42. 5	56. 7	70. 8	84. 9	99. 1	113. 2	127. 3	141. 5	0. 2
0. 3	9. 6	19. 0	28. 4	37. 8	47. 3	56. 7	66. 1	75. 6	85. 0	94. 4	0. 3
0. 4	7. 2	14. 3	21. 4	28. 5	35. 5	42. 6	49. 7	56. 7	63. 8	70. 9	0. 4
0. 5	5. 9	11. 5	17. 2	22. 8	28. 5	34. 2	39. 8	45. 5	51. 1	56. 8	0. 5
0. 6	5. 0	9. 7	14. 4	19. 1	23. 8	28. 5	33. 3	38. 0	42. 7	47. 4	0. 6
0. 7	4. 3	8. 4	12. 4	16. 5	20. 5	24. 5	28. 6	32. 6	36. 7	40. 7	0. 7
0. 8	3. 9	7. 4	10. 9	14. 5	18. 0	21. 5	25. 1	28. 6	32. 2	35. 7	0. 8
0. 9	3. 5	6. 7	9. 8	12. 9	16. 1	19. 2	22. 4	25. 5	28. 7	31. 8	0. 9
1. 0	3. 2	6. 1	8. 9	11. 7	14. 6	17. 4	20. 2	23. 0	25. 9	28. 7	1. 0
1. 1	3. 0	5. 6	8. 2	10. 7	13. 3	15. 9	18. 5	21. 0	23. 6	26. 2	1. 1
1. 2	2. 9	5. 2	7. 6	9. 9	12. 3	14. 6	17. 0	19. 4	21. 7	24. 1	1. 2
1. 3	2. 7	4. 9	7. 1	9. 2	11. 4	13. 6	15. 8	17. 9	20. 1	22. 3	1. 3
1. 4	2. 6	4. 6	6. 6	8. 7	10. 7	12. 7	14. 7	16. 7	18. 8	20. 8	1. 4
1. 5	2. 5	4. 4	6. 3	8. 2	10. 1	11. 9	13. 8	15. 7	17. 6	19. 5	1. 5
1. 6	2. 4	4. 2	6. 0	7. 7	9. 5	11. 3	13. 0	14. 8	16. 6	18. 3	1. 6
1. 7	2. 4	4. 0	5. 7	7. 4	9. 0	10. 7	12. 4	14. 0	15. 7	17. 3	1. 7
1. 8	2. 3	3. 9	5. 5	7. 0	8. 6	10. 2	11. 7	13. 3	14. 9	16. 5	1. 8
1. 9	2. 3	3. 8	5. 3	6. 7	8. 2	9. 7	11. 2	12. 7	14. 2	15. 7	1. 9
2. 0	2. 2	3. 7	5. 1	6. 5	7. 9	9. 3	10. 7	12. 1	13. 6	15. 0	2. 0
2. 1	2. 2	3. 6	4. 9	6. 3	7. 6	9. 0	10. 3	11. 7	13. 0	14. 3	2. 1
2. 2	2. 2	3. 5	4. 8	6. 1	7. 3	8. 6	9. 9	11. 2	12. 5	13. 8	2. 2
2. 3	2. 2	3. 4	4. 6	5. 9	7. 1	8. 3	9. 6	10. 8	12. 0	13. 3	2. 3
2. 4	2. 2	3. 4	4. 5	5. 7	6. 9	8. 1	9. 2	10. 4	11. 6	12. 8	2. 4
2. 5	2. 2	3. 3	4. 4	5. 6	6. 7	7. 8	9. 0	10. 1	11. 2	12. 4	2. 5
2. 6	2. 2	3. 3	4. 3	5. 4	6. 5	7. 6	8. 7	9. 8	10. 9	12. 0	2. 6
2. 7	2. 2	3. 2	4. 3	5. 3	6. 4	7. 4	8. 5	9. 5	10. 6	11. 6	2. 7
2. 8	2. 2	3. 2	4. 2	5. 2	6. 2	7. 2	8. 2	9. 2	10. 3	11. 3	2. 8
2. 9	2. 2	3. 2	4. 1	5. 1	6. 1	7. 1	8. 0	9. 0	10. 0	11. 0	2. 9
3. 0	2. 2	3. 1	4. 1	5. 0	6. 0	6. 9	7. 8	8. 8	9. 7	10. 7	3. 0
3. 1	2. 2	3. 1	4. 0	4. 9	5. 9	6. 8	7. 7	8. 6	9. 5	10. 4	3. 1
3. 2	2. 2	3. 1	4. 0	4. 9	5. 8	6. 6	7. 5	8. 4	9. 3	10. 2	3. 2
3. 3	2. 2	3. 1	3. 9	4. 8	5. 7	6. 5	7. 4	8. 2	9. 1	9. 9	3. 3
3. 4	2. 2	3. 1	3. 9	4. 7	5. 6	6. 4	7. 2	8. 1	8. 9	9. 7	3. 4
3. 5	2. 2	3. 1	3. 9	4. 7	5. 5	6. 3	7. 1	7. 9	8. 7	9. 5	3. 5
3. 6	2. 2	3. 1	3. 9	4. 6	5. 4	6. 2	7. 0	7. 8	8. 6	9. 4	3. 6
3. 7	2. 2	3. 1	3. 8	4. 6	5. 4	6. 1	6. 9	7. 7	8. 4	9. 2	3. 7
3. 8	2. 2	3. 1	3. 8	4. 6	5. 3	6. 0	6. 8	7. 5	8. 3	9. 0	3. 8
3. 9	2. 2	3. 1	3. 8	4. 5	5. 2	6. 0	6. 7	7. 4	8. 1	8. 9	3. 9
4. 0	2. 2	3. 1	3. 8	4. 5	5. 2	5. 9	6. 6	7. 3	8. 0	8. 7	4. 0
4. 1	2. 2	3. 1	3. 8	4. 5	5. 2	5. 8	6. 5	7. 2	7. 9	8. 6	4. 1
4. 2	2. 2	3. 1	3. 8	4. 4	5. 1	5. 8	6. 5	7. 1	7. 8	8. 5	4. 2
4. 3	2. 2	3. 1	3. 8	4. 4	5. 1	5. 7	6. 4	7. 1	7. 7	8. 4	4. 3
4. 4	2. 2	3. 1	3. 8	4. 4	5. 0	5. 7	6. 3	7. 0	7. 6	8. 3	4. 4
4. 5	2. 2	3. 1	3. 8	4. 4	5. 0	5. 6	6. 3	6. 9	7. 5	8. 2	4. 5
4. 6	2. 2	3. 1	3. 8	4. 4	5. 0	5. 6	6. 2	6. 8	7. 4	8. 1	4. 6
4. 7	2. 2	3. 1	3. 8	4. 4	5. 0	5. 6	6. 2	6. 8	7. 4	8. 0	4. 7
4. 8	2. 2	3. 1	3. 8	4. 4	4. 9	5. 5	6. 1	6. 7	7. 3	7. 9	4. 8
4. 9	2. 2	3. 1	3. 8	4. 3	4. 9	5. 5	6. 1	6. 7	7. 2	7. 8	4. 9
5. 0	2. 2	3. 1	3. 8	4. 3	4. 9	5. 5	6. 0	6. 6	7. 2	7. 7	5. 0
5. 5	2. 2	3. 1	3. 8	4. 3	4. 9	5. 4	5. 9	6. 4	6. 9	7. 4	5. 5
6. 0	2. 2	3. 1	3. 8	4. 3	4. 9	5. 3	5. 8	6. 3	6. 7	7. 2	6. 0
6. 5	2. 2	3. 1	3. 8	4. 3	4. 9	5. 3	5. 7	6. 2	6. 6	7. 1	6. 5
7. 0	2. 2	3. 1	3. 8	4. 3	4. 9	5. 3	5. 7	6. 1	6. 5	7. 0	7. 0
7. 5	2. 2	3. 1	3. 8	4. 3	4. 9	5. 3	5. 7	6. 1	6. 5	6. 9	7. 5
8. 0	2. 2	3. 1	3. 8	4. 3	4. 9	5. 3	5. 7	6. 1	6. 5	6. 9	8. 0
8. 5	2. 2	3. 1	3. 8	4. 3	4. 9	5. 3	5. 7	6. 1	6. 5	6. 9	8. 5
9. 0	2. 2	3. 1	3. 8	4. 3	4. 9	5. 3	5. 7	6. 1	6. 5	6. 9	9. 0
9. 5	2. 2	3. 1	3. 8	4. 3	4. 9	5. 3	5. 7	6. 1	6. 5	6. 9	9. 5
10. 0	2. 2	3. 1	3. 8	4. 3	4. 9	5. 3	5. 7	6. 1	6. 5	6. 9	10. 0

TABLE 22

Dip of the Sea Short of the Horizon

Distance	55 (16.8)	60 (18.3)	65 (19.8)	70 (21.3)	75 (22.9)	80 (24.4)	85 (25.9)	90 (27.4)	95 (29.0)	100 (30.5)	Distance
	Height of eye above the sea, in feet and (meters)										
Miles	′	′	′	′	′	′	′	′	′	′	*Miles*
0.2	155.6	169.7	183.3	197.9	212.0	226.1	240.2	254.2	268.3	282.3	0.2
0.3	103.8	113.3	122.7	132.1	141.6	151.0	160.4	169.9	179.3	188.7	0.3
0.4	77.9	85.0	92.1	99.2	106.2	113.3	120.3	127.4	134.5	141.5	0.4
0.5	62.4	68.1	73.8	79.4	85.1	90.7	96.4	102.0	107.7	113.3	0.5
0.6	52.1	56.8	61.5	66.3	71.0	75.7	80.4	85.1	89.8	94.5	0.6
0.7	44.7	48.8	52.8	56.9	60.9	64.9	69.0	73.0	77.1	81.1	0.7
0.8	39.2	42.8	46.3	49.8	53.4	56.9	60.4	64.0	67.5	71.1	0.8
0.9	34.9	38.1	41.2	44.4	47.5	50.7	53.8	56.9	60.1	63.2	0.9
1.0	31.5	34.4	37.2	40.0	42.8	45.7	48.5	51.3	54.2	57.0	1.0
1.1	28.7	31.3	33.9	36.5	39.0	41.6	44.2	46.7	49.3	51.9	1.1
1.2	26.4	28.8	31.1	33.5	35.9	38.2	40.6	42.9	45.3	47.6	1.2
1.3	24.5	26.7	28.8	31.0	33.2	35.4	37.5	39.7	41.9	44.1	1.3
1.4	22.8	24.8	26.8	28.9	30.9	32.9	34.9	37.0	39.0	41.0	1.4
1.5	21.4	23.3	25.1	27.0	28.9	30.8	32.7	34.6	36.5	38.3	1.5
1.6	20.1	21.9	23.6	25.4	27.2	29.0	30.7	32.5	34.3	36.0	1.6
1.7	19.0	20.7	22.3	24.0	25.7	27.3	29.0	30.7	32.3	34.0	1.7
1.8	18.0	19.6	21.2	22.8	24.3	25.9	27.5	29.0	30.6	32.2	1.8
1.9	17.2	18.7	20.1	21.6	23.1	24.6	26.1	27.6	29.1	30.6	1.9
2.0	16.4	17.8	19.2	20.6	22.0	23.5	24.9	26.3	27.7	29.1	2.0
2.1	15.7	17.0	18.4	19.7	21.1	22.4	23.8	25.1	26.5	27.8	2.1
2.2	15.1	16.3	17.6	18.9	20.2	21.5	22.8	24.1	25.3	26.6	2.2
2.3	14.5	15.7	16.9	18.2	19.4	20.6	21.9	23.1	24.3	25.6	2.3
2.4	14.0	15.1	16.3	17.5	18.7	19.9	21.0	22.2	23.4	24.6	2.4
2.5	13.5	14.6	15.7	16.9	18.0	19.1	20.3	21.4	22.5	23.7	2.5
2.6	13.0	14.1	15.2	16.3	17.4	18.5	19.6	20.7	21.8	22.8	2.6
2.7	12.6	13.7	14.7	15.8	16.8	17.9	18.9	20.0	21.0	22.1	2.7
2.8	12.3	13.3	14.3	15.3	16.3	17.3	18.3	19.3	20.4	21.4	2.8
2.9	11.9	12.9	13.9	14.9	15.8	16.8	17.8	18.8	19.7	20.7	2.9
3.0	11.6	12.6	13.5	14.4	15.4	16.3	17.3	18.2	19.2	20.1	3.0
3.1	11.3	12.2	13.2	14.1	15.0	15.9	16.8	17.7	18.6	19.5	3.1
3.2	11.1	11.9	12.8	13.7	14.6	15.5	16.4	17.2	18.1	19.0	3.2
3.3	10.8	11.7	12.5	13.4	14.2	15.1	15.9	16.8	17.7	18.5	3.3
3.4	10.6	11.4	12.2	13.1	13.9	14.7	15.6	16.4	17.2	18.1	3.4
3.5	10.3	11.2	12.0	12.8	13.6	14.4	15.2	16.0	16.8	17.6	3.5
3.6	10.1	10.9	11.7	12.5	13.3	14.1	14.9	15.6	16.4	17.2	3.6
3.7	9.9	10.7	11.5	12.2	13.0	13.8	14.5	15.3	16.1	16.8	3.7
3.8	9.8	10.5	11.3	12.0	12.7	13.5	14.2	15.0	15.7	16.5	3.8
3.9	9.6	10.3	11.1	11.8	12.5	13.2	14.0	14.7	15.4	16.1	3.9
4.0	9.4	10.1	10.9	11.6	12.3	13.0	13.7	14.4	15.1	15.8	4.0
4.1	9.3	10.0	10.7	11.4	12.1	12.7	13.4	14.1	14.8	15.5	4.1
4.2	9.2	9.8	10.5	11.2	11.8	12.5	13.2	13.9	14.5	15.2	4.2
4.3	9.0	9.7	10.3	11.0	11.7	12.3	13.0	13.6	14.3	14.9	4.3
4.4	8.9	9.5	10.2	10.8	11.5	12.1	12.8	13.4	14.0	14.7	4.4
4.5	8.8	9.4	10.0	10.7	11.3	11.9	12.6	13.2	13.8	14.4	4.5
4.6	8.7	9.3	9.9	10.5	11.1	11.8	12.4	13.0	13.6	14.2	4.6
4.7	8.6	9.2	9.8	10.4	11.0	11.6	12.2	12.8	13.4	14.0	4.7
4.8	8.5	9.1	9.7	10.2	10.8	11.4	12.0	12.6	13.2	13.8	4.8
4.9	8.4	9.0	9.5	10.1	10.7	11.3	11.9	12.4	13.0	13.6	4.9
5.0	8.3	8.9	9.4	10.0	10.6	11.1	11.7	12.3	12.8	13.4	5.0
5.5	7.9	8.5	9.0	9.5	10.0	10.5	11.0	11.5	12.1	12.6	5.5
6.0	7.7	8.2	8.6	9.1	9.6	10.0	10.5	11.0	11.5	11.9	6.0
6.5	7.5	7.9	8.4	8.8	9.2	9.7	10.1	10.5	11.0	11.4	6.5
7.0	7.4	7.8	8.2	8.6	9.0	9.4	9.8	10.2	10.6	11.0	7.0
7.5	7.3	7.6	8.0	8.4	8.8	9.2	9.5	9.9	10.3	10.7	7.5
8.0	7.2	7.6	7.9	8.3	8.6	9.0	9.3	9.7	10.0	10.4	8.0
8.5	7.2	7.5	7.9	8.2	8.5	8.9	9.2	9.5	9.9	10.2	8.5
9.0	7.2	7.5	7.8	8.1	8.5	8.8	9.1	9.4	9.7	10.0	9.0
9.5	7.2	7.5	7.8	8.1	8.4	8.7	9.0	9.3	9.6	9.9	9.5
10.0	7.2	7.5	7.8	8.1	8.4	8.7	9.0	9.2	9.5	9.8	10.0

TABLE 23

Altitude Correction for Air Temperature

Altitude	Temperature—degrees Fahrenheit								Altitude
	−40	−30	−20	−10	0	+10	+20	+30	
° ′	′	′	′	′	′	′	′	′	° ′
−0 10	−7.9	−6.8	−5.8	−4.9	−4.0	−3.1	−2.3	−1.5	−0 10
0 00	7.4	6.4	5.5	4.6	3.8	2.9	2.2	1.4	0 00
+0 10	6.9	6.0	5.2	4.3	3.5	2.8	2.0	1.3	+0 10
0 20	6.6	5.7	4.9	4.1	3.3	2.6	1.9	1.2	0 20
0 30	6.1	5.3	4.6	3.8	3.1	2.4	1.8	1.2	0 30
+0 45	−5.7	−4.9	−4.2	−3.5	−2.9	−2.2	−1.6	−1.1	+0 45
1 00	5.2	4.5	3.9	3.2	2.6	2.1	1.5	1.0	1 00
1 20	4.7	4.1	3.5	2.9	2.4	1.9	1.4	0.9	1 20
1 40	4.3	3.7	3.2	2.7	2.2	1.7	1.2	0.8	1 40
2 00	3.9	3.4	2.9	2.4	2.0	1.6	1.1	0.7	2 00
+2 30	−3.4	−3.0	−2.6	−2.1	−1.8	−1.4	−1.0	−0.7	+2 30
3 00	3.1	2.7	2.3	1.9	1.6	1.2	0.9	0.6	3 00
4	2.5	2.2	1.9	1.6	1.3	1.0	0.7	0.5	4
5	2.1	1.8	1.6	1.3	1.1	0.8	0.6	0.4	5
6	1.8	1.6	1.4	1.1	0.9	0.7	0.5	0.3	6
+7	−1.6	−1.4	−1.2	−1.0	−0.8	−0.6	−0.5	−0.3	+7
8	1.4	1.2	1.0	0.9	0.7	0.6	0.4	0.3	8
9	1.3	1.1	0.9	0.8	0.6	0.5	0.4	0.2	9
10	1.1	1.0	0.8	0.7	0.6	0.5	0.3	0.2	10
15	0.8	0.7	0.6	0.5	0.4	0.3	0.2	0.1	15
+20	−0.6	−0.5	−0.4	−0.3	−0.3	−0.2	−0.2	−0.1	+20
30	0.4	0.3	0.3	0.2	0.2	0.1	0.1	0.1	30
50	0.2	0.1	0.1	0.1	0.1	0.1	0.0	0.0	50
70	0.1	0.1	0.1	0.1	0.0	0.0	0.0	0.0	70
90	0.0	0.0	0.0	0.0	0.0	0.0	0.0	0.0	90

Altitude	Temperature—degrees Fahrenheit								Altitude
	+40	+50	+60	+70	+80	+90	+100	+110	
° ′	′	′	′	′	′	′	′	′	° ′
−0 10	−0.7	0.0	+0.7	+1.4	+2.0	+2.7	+3.3	+3.9	−0 10
0 00	0.7	0.0	0.7	1.3	1.9	2.5	3.1	3.6	0 00
+0 10	0.6	0.0	0.6	1.2	1.8	2.4	2.9	3.4	+0 10
0 20	0.6	0.0	0.6	1.2	1.7	2.2	2.7	3.2	0 20
0 30	0.6	0.0	0.6	1.1	1.6	2.1	2.6	3.0	0 30
+0 45	−0.5	0.0	+0.5	+1.0	+1.5	+1.9	+2.4	+2.8	+0 45
1 00	0.5	0.0	0.5	0.9	1.4	1.8	2.2	2.6	1 00
1 20	0.4	0.0	0.4	0.8	1.2	1.6	2.0	2.3	1 20
1 40	0.4	0.0	0.4	0.8	1.1	1.5	1.8	2.1	1 40
2 00	0.4	0.0	0.4	0.7	1.0	1.3	1.6	1.9	2 00
+2 30	−0.3	0.0	+0.3	+0.6	+0.9	+1.2	+1.4	+1.7	+2 30
3 00	0.3	0.0	0.3	0.5	0.8	1.0	1.3	1.5	3 00
4	0.2	0.0	0.2	0.4	0.7	0.9	1.1	1.2	4
5	0.2	0.0	0.2	0.4	0.6	0.7	0.9	1.0	5
6	0.2	0.0	0.2	0.3	0.5	0.6	0.8	0.9	6
+7	−0.1	0.0	+0.1	+0.3	+0.4	+0.5	+0.7	+0.8	+7
8	0.1	0.0	0.1	0.2	0.4	0.5	0.6	0.7	8
9	0.1	0.0	0.1	0.2	0.3	0.4	0.5	0.6	9
10	0.1	0.0	0.1	0.2	0.3	0.4	0.5	0.6	10
15	0.1	0.0	0.1	0.1	0.2	0.3	0.3	0.4	15
+20	−0.1	0.0	+0.1	+0.1	+0.1	+0.2	+0.2	+0.3	+20
30	0.0	0.0	0.0	0.1	0.1	0.1	0.2	0.2	30
50	0.0	0.0	0.0	0.0	0.0	0.1	0.1	0.1	50
70	0.0	0.0	0.0	0.0	0.0	0.0	0.0	0.0	70
90	0.0	0.0	0.0	0.0	0.0	0.0	0.0	0.0	90

TABLE 24
Altitude Correction for Atmospheric Pressure

Altitude	Pressure in inches or millibars—**Subtract** correction from sextant or rectified altitude								Altitude
	31. 2	31. 0	30. 8	30. 6	30. 4	30. 2	30. 0	29. 8	
	1056.56	1049.78	1043.01	1036.24	1029.46	1022.69	1015.92	1009.15	
° ′	′	′	′	′	′	′	′	′	° ′
−0 10	−1. 7	−1. 4	−1. 2	−1. 0	−0. 7	−0. 5	−0. 2	0. 0	−0 10
0 00	1. 6	1. 4	1. 1	0. 9	0. 7	0. 4	0. 2	0. 0	0 00
+0 10	1. 5	1. 3	1. 1	0. 8	0. 6	0. 4	0. 2	0. 0	+0 10
0 20	1. 4	1. 2	1. 0	0. 8	0. 6	0. 4	0. 2	0. 0	0 20
0 30	1. 3	1. 1	0. 9	0. 7	0. 6	0. 4	0. 2	0. 0	0 30
+0 45	−1. 2	−1. 0	−0. 9	−0. 7	−0. 5	−0. 3	−0. 2	0. 0	+0 45
1 00	1. 1	1. 0	0. 8	0. 6	0. 5	0. 3	0. 1	0. 0	1 00
1 20	1. 0	0. 9	0. 7	0. 6	0. 4	0. 3	0. 1	0. 0	1 20
1 40	0. 9	0. 8	0. 7	0. 5	0. 4	0. 3	0. 1	0. 0	1 40
2 00	0. 8	0. 7	0. 6	0. 5	0. 4	0. 2	0. 1	0. 0	2 00
+2 30	−0. 7	−0. 6	−0. 5	−0. 4	−0. 3	−0. 2	−0. 1	0. 0	+2 30
3 00	0. 7	0. 6	0. 5	0. 4	0. 3	0. 2	0. 1	0. 0	3 00
4	0. 5	0. 5	0. 4	0. 3	0. 2	0. 1	0. 1	0. 0	4
5	0. 5	0. 4	0. 3	0. 3	0. 2	0. 1	0. 0	0. 0	5
6	0. 4	0. 3	0. 3	0. 2	0. 2	0. 1	0. 0	0. 0	6
+7	−0. 3	−0. 3	−0. 2	−0. 2	−0. 1	−0. 1	0. 0	0. 0	+7
8	0. 3	0. 3	0. 2	0. 2	0. 1	0. 1	0. 0	0. 0	8
9	0. 3	0. 2	0. 2	0. 2	0. 1	0. 1	0. 0	0. 0	9
10	0. 2	0. 2	0. 2	0. 1	0. 1	0. 1	0. 0	0. 0	10
15	0. 2	0. 1	0. 1	0. 1	0. 1	0. 0	0. 0	0. 0	15
+20	−0. 1	−0. 1	−0. 1	−0. 1	−0. 1	0. 0	0. 0	0. 0	+20
30	0. 1	0. 1	0. 1	0. 0	0. 0	0. 0	0. 0	0. 0	30
50	0. 0	0. 0	0. 0	0. 0	0. 0	0. 0	0. 0	0. 0	50
70	0. 0	0. 0	0. 0	0. 0	0. 0	0. 0	0. 0	0. 0	70
+90	0. 0	0. 0	0. 0	0. 0	0. 0	0. 0	0. 0	0. 0	+90

Altitude	Pressure in inches or millibars—**Add** correction to sextant or rectified altitude								Altitude
	29. 6	29. 4	29. 2	29. 0	28. 8	28. 6	28. 4	28. 2	
	1002.37	995.60	988.83	982.05	975.28	968.51	961.74	954.96	
° ′	′	′	′	′	′	′	′	′	° ′
−0 10	+0. 3	+0. 5	+0. 8	+1. 0	+1. 3	+1. 5	+1. 8	+2. 0	−0 10
0 00	0. 3	0. 5	0. 7	1. 0	1. 2	1. 4	1. 6	1. 9	0 00
+0 10	0. 2	0. 5	0. 7	0. 9	1. 1	1. 3	1. 5	1. 8	+0 10
0 20	0. 2	0. 4	0. 6	0. 8	1. 1	1. 3	1. 5	1. 7	0 20
0 30	0. 2	0. 4	0. 6	0. 8	1. 0	1. 2	1. 4	1. 6	0 30
+0 45	+0. 2	+0. 4	+0. 6	+0. 7	+0. 9	+1. 1	+1. 3	+1. 4	+0 45
1 00	0. 2	0. 3	0. 5	0. 7	0. 8	1. 0	1. 2	1. 3	1 00
1 20	0. 2	0. 3	0. 5	0. 6	0. 8	0. 9	1. 1	1. 2	1 20
1 40	0. 2	0. 3	0. 4	0. 6	0. 7	0. 8	1. 0	1. 1	1 40
2 00	0. 1	0. 3	0. 4	0. 5	0. 6	0. 8	0. 9	1. 0	2 00
+2 30	+0. 1	+0. 2	+0. 3	+0. 4	+0. 6	+0. 7	+0. 8	+0. 9	+2 30
3 00	0. 1	0. 2	0. 3	0. 4	0. 5	0. 6	0. 7	0. 8	3 00
4	0. 1	0. 2	0. 2	0. 3	0. 4	0. 5	0. 6	0. 6	4
5	0. 1	0. 1	0. 2	0. 3	0. 3	0. 4	0. 5	0. 5	5
6	0. 1	0. 1	0. 2	0. 2	0. 3	0. 3	0. 4	0. 5	6
+7	+0. 1	+0. 1	+0. 2	+0. 2	+0. 3	+0. 3	+0. 4	+0. 4	+7
8	0. 0	0. 1	0. 1	0. 2	0. 2	0. 3	0. 3	0. 4	8
9	0. 0	0. 1	0. 1	0. 2	0. 2	0. 2	0. 3	0. 3	9
10	0. 0	0. 1	0. 1	0. 1	0. 2	0. 2	0. 3	0. 3	10
15	0. 0	0. 1	0. 1	0. 1	0. 1	0. 1	0. 2	0. 2	15
+20	0. 0	0. 0	+0. 1	+0. 1	+0. 1	+0. 1	+0. 1	+0. 1	+20
30	0. 0	0. 0	0. 0	0. 0	0. 1	0. 1	0. 1	0. 1	30
50	0. 0	0. 0	0. 0	0. 0	0. 0	0. 0	0. 0	0. 0	50
70	0. 0	0. 0	0. 0	0. 0	0. 0	0. 0	0. 0	0. 0	70
+90	0. 0	0. 0	0. 0	0. 0	0. 0	0. 0	0. 0	0. 0	+90

TABLE 25

Meridian Angle and Altitude of a Body on the Prime Vertical Circle

Latitude	Declination (same name as latitude)												Latitude
	0°		1°		2°		3°		4°		5°		
	t	Alt.	t	Alt.	t	Alt.	t	Alt.	t	Alt.	t	Alt.	
°	°	°	°	°	°	°	°	°	°	°	°	°	°
0	—	—	*90.0*	*0.0*	*90.0*	*0.0*	*90.0*	*0.0*	*90.0*	*0.0*	*90.0*	*0.0*	0
1	90.0	0.0	0.0	90.0	*60.0*	*30.0*	*70.5*	*19.5*	*75.5*	*14.5*	*78.5*	*11.6*	1
2	90.0	0.0	60.0	30.0	0.0	90.0	*48.2*	*41.8*	*60.0*	*30.0*	*66.5*	*23.6*	2
3	90.0	0.0	70.5	19.5	48.2	41.8	0.0	90.0	*41.5*	*48.6*	*53.2*	*36.9*	3
4	90.0	0.0	75.5	14.5	60.0	30.0	41.5	48.6	0.0	90.0	*36.9*	*53.2*	4
5	90.0	0.0	78.5	11.6	66.5	23.6	53.2	36.9	36.9	53.2	0.0	90.0	5
6	90.0	0.0	80.4	9.6	70.6	19.5	60.1	30.0	48.3	41.9	33.7	56.5	6
7	90.0	0.0	81.8	8.2	73.5	16.6	64.7	25.4	55.3	34.9	44.6	45.7	7
8	90.0	0.0	82.9	7.2	75.6	14.5	68.1	22.1	60.2	30.1	51.5	38.8	8
9	90.0	0.0	83.7	6.4	77.3	12.9	70.7	19.5	63.8	26.5	56.5	33.9	9
10	90.0	0.0	84.3	5.8	78.6	11.6	72.7	17.5	66.6	23.7	60.3	30.1	10
11	90.0	0.0	84.8	5.2	79.7	10.5	74.4	15.9	68.9	21.4	63.3	27.2	11
12	90.0	0.0	85.3	4.8	80.5	9.7	75.7	14.6	70.8	19.6	65.7	24.8	12
13	90.0	0.0	85.7	4.4	81.3	8.9	76.9	13.5	72.4	18.1	67.7	22.8	13
14	90.0	0.0	86.0	4.1	81.9	8.3	77.9	12.5	73.7	16.8	69.5	21.1	14
15	90.0	0.0	86.3	3.9	82.5	7.7	78.7	11.7	74.9	15.6	70.9	19.7	15
16	90.0	0.0	86.5	3.6	83.0	7.3	79.5	10.9	75.9	14.7	72.2	18.4	16
17	90.0	0.0	86.7	3.4	83.4	6.9	80.1	10.3	76.8	13.8	73.4	17.3	17
18	90.0	0.0	86.9	3.2	83.8	6.5	80.7	9.8	77.6	13.0	74.4	16.4	18
19	90.0	0.0	87.1	3.1	84.2	6.2	81.2	9.3	78.3	12.4	75.3	15.5	19
20	90.0	0.0	87.3	2.9	84.5	5.9	81.7	8.8	78.9	11.8	76.1	14.8	20
21	90.0	0.0	87.4	2.8	84.8	5.6	82.2	8.4	79.5	11.2	76.8	14.1	21
22	90.0	0.0	87.5	2.7	85.0	5.3	82.5	8.0	80.0	10.7	77.5	13.5	22
23	90.0	0.0	87.6	2.6	85.3	5.1	82.9	7.7	80.5	10.3	78.1	12.9	23
24	90.0	0.0	87.8	2.5	85.5	4.9	83.2	7.4	81.0	9.9	78.7	12.4	24
25	90.0	0.0	87.9	2.4	85.7	4.7	83.5	7.1	81.4	9.5	79.2	11.9	25
26	90.0	0.0	87.9	2.3	85.9	4.6	83.8	6.9	81.8	9.2	79.7	11.5	26
27	90.0	0.0	88.0	2.2	86.1	4.4	84.1	6.6	82.1	8.8	80.1	11.1	27
28	90.0	0.0	88.1	2.1	86.2	4.3	84.3	6.4	82.4	8.5	80.5	10.7	28
29	90.0	0.0	88.2	2.1	86.4	4.1	84.6	6.2	82.8	8.3	80.9	10.4	29
30	90.0	0.0	88.3	2.0	86.5	4.0	84.8	6.0	83.0	8.0	81.3	10.0	30
31	90.0	0.0	88.3	1.9	86.7	3.9	85.0	5.8	83.3	7.8	81.6	9.7	31
32	90.0	0.0	88.4	1.9	86.8	3.8	85.2	5.7	83.6	7.6	82.0	9.5	32
33	90.0	0.0	88.5	1.8	86.9	3.7	85.4	5.5	83.8	7.4	82.3	9.2	33
34	90.0	0.0	88.5	1.8	87.0	3.6	85.5	5.4	84.0	7.2	82.5	9.0	34
35	90.0	0.0	88.6	1.7	87.1	3.5	85.7	5.2	84.3	7.0	82.8	8.7	35
36	90.0	0.0	88.6	1.7	87.2	3.4	85.9	5.1	84.5	6.8	83.1	8.5	36
37	90.0	0.0	88.7	1.7	87.3	3.3	86.0	5.0	84.7	6.7	83.3	8.3	37
38	90.0	0.0	88.7	1.6	87.4	3.2	86.2	4.9	84.9	6.5	83.6	8.1	38
39	90.0	0.0	88.8	1.6	87.5	3.2	86.3	4.8	85.0	6.4	83.8	8.0	39
40	90.0	0.0	88.8	1.6	87.6	3.1	86.4	4.7	85.2	6.2	84.0	7.8	40
41	90.0	0.0	88.8	1.5	87.7	3.0	86.5	4.6	85.4	6.1	84.2	7.6	41
42	90.0	0.0	88.9	1.5	87.8	3.0	86.7	4.5	85.5	6.0	84.4	7.5	42
43	90.0	0.0	88.9	1.5	87.9	2.9	86.8	4.4	85.7	5.9	84.6	7.3	43
44	90.0	0.0	89.0	1.4	87.9	2.9	86.9	4.3	85.8	5.8	84.8	7.2	44
45	90.0	0.0	89.0	1.4	88.0	2.8	87.0	4.2	86.0	5.7	85.0	7.1	45
46	90.0	0.0	89.0	1.4	88.1	2.8	87.1	4.2	86.1	5.6	85.2	7.0	46
47	90.0	0.0	89.1	1.4	88.1	2.7	87.2	4.1	86.3	5.5	85.3	6.8	47
48	90.0	0.0	89.1	1.3	88.2	2.7	87.3	4.0	86.4	5.4	85.5	6.7	48
49	90.0	0.0	89.1	1.3	88.3	2.7	87.4	4.0	86.5	5.3	85.6	6.6	49
50	90.0	0.0	89.2	1.3	88.3	2.6	87.5	3.9	86.6	5.2	85.8	6.5	50
52	90.0	0.0	89.2	1.3	88.4	2.5	87.7	3.8	86.9	5.1	86.1	6.4	52
54	90.0	0.0	89.3	1.2	88.5	2.5	87.8	3.7	87.1	4.9	86.4	6.2	54
56	90.0	0.0	89.3	1.2	88.7	2.4	88.0	3.6	87.3	4.8	86.6	6.0	56
58	90.0	0.0	89.4	1.2	88.7	2.4	88.1	3.5	87.5	4.7	86.9	5.9	58
60	90.0	0.0	89.4	1.2	88.8	2.3	88.3	3.5	87.7	4.6	87.1	5.8	60
65	90.0	0.0	89.5	1.1	89.1	2.2	88.6	3.3	88.1	4.4	87.7	5.5	65
70	90.0	0.0	89.6	1.1	89.3	2.1	88.9	3.2	88.5	4.3	88.2	5.3	70
75	90.0	0.0	89.7	1.0	89.5	2.1	89.2	3.1	88.9	4.1	88.7	5.2	75
80	90.0	0.0	89.8	1.0	89.6	2.0	89.5	3.0	89.3	4.1	89.1	5.1	80
85	90.0	0.0	89.9	1.0	89.8	2.0	89.7	3.0	89.6	4.0	89.6	5.0	85

Numbers in *italics* indicate nearest approach to prime vertical

TABLE 25

Meridian Angle and Altitude of a Body on the Prime Vertical Circle

Latitude	Declination (same name as latitude) 6° t	6° Alt.	7° t	7° Alt.	8° t	8° Alt.	9° t	9° Alt.	10° t	10° Alt.	11° t	11° Alt.	Latitude
°	°	°	°	°	°	°	°	°	°	°	°	°	°
0	*90. 0*	*0. 0*	*90. 0*	*0. 0*	*90. 0*	*0. 0*	*90. 0*	*0. 0*	*90. 0*	*0. 0*	*90. 0*	*0. 0*	0
1	*80. 4*	*9. 6*	*81. 8*	*8. 2*	*82. 9*	*7. 2*	*83. 7*	*6. 4*	*84. 3*	*5. 8*	*84. 8*	*5. 2*	1
2	*70. 6*	*19. 5*	*73. 5*	*16. 6*	*75. 6*	*14. 5*	*77. 3*	*12. 9*	*78. 6*	*11. 6*	*79. 7*	*10. 5*	2
3	*60. 1*	*30. 0*	*64. 7*	*25. 4*	*68. 1*	*22. 1*	*70. 7*	*19. 5*	*72. 7*	*17. 5*	*74. 4*	*15. 9*	3
4	*48. 3*	*41. 9*	*55. 3*	*34. 9*	*60. 2*	*30. 1*	*63. 8*	*26. 5*	*66. 6*	*23. 7*	*68. 9*	*21. 4*	4
5	*33. 7*	*56. 5*	*44. 6*	*45. 7*	*51. 5*	*38. 8*	*56. 5*	*33. 9*	*60. 3*	*30. 1*	*63. 3*	*27. 2*	5
6	0. 0	90. 0	*31. 1*	*59. 1*	*41. 6*	*48. 7*	*48. 4*	*41. 9*	*53. 4*	*37. 0*	*57. 3*	*33. 2*	6
7	31. 1	59. 1	0. 0	90. 0	*29. 1*	*61. 1*	*39. 2*	*51. 2*	*45. 9*	*44. 6*	*50. 8*	*39. 7*	7
8	41. 6	48. 7	29. 1	61. 1	0. 0	90. 0	*27. 5*	*62. 8*	*37. 2*	*53. 3*	*43. 7*	*46. 8*	8
9	48. 4	41. 9	39. 2	51. 2	27. 5	62. 8	0. 0	90. 0	*26. 1*	*64. 3*	*35. 4*	*55. 1*	9
10	53. 4	37. 0	45. 9	44. 6	37. 2	53. 3	26. 1	64. 3	0. 0	90. 0	*24. 9*	*65. 5*	10
11	57. 3	33. 2	50. 8	39. 7	43. 7	46. 8	35. 4	55. 1	24. 9	65. 5	0. 0	90. 0	11
12	60. 4	30. 2	54. 7	35. 9	48. 6	42. 0	41. 8	48. 8	33. 9	56. 6	23. 9	66. 6	12
13	62. 9	27. 7	57. 9	32. 8	52. 5	38. 2	46. 7	44. 1	40. 2	50. 5	32. 7	58. 0	13
14	65. 1	25. 6	60. 5	30. 2	55. 7	35. 1	50. 6	40. 3	45. 0	45. 9	38. 8	52. 1	14
15	66. 9	23. 8	62. 7	28. 1	58. 4	32. 5	53. 8	37. 2	48. 8	42. 1	43. 5	47. 5	15
16	68. 5	22. 3	64. 6	26. 2	60. 7	30. 3	56. 5	34. 6	52. 1	39. 0	47. 3	43. 8	16
17	69. 9	20. 9	66. 3	24. 6	62. 6	28. 4	58. 8	32. 3	54. 8	36. 4	50. 5	40. 7	17
18	71. 1	19. 8	67. 8	23. 2	64. 4	26. 8	60. 8	30. 4	57. 1	34. 2	53. 3	38. 1	18
19	72. 2	18. 7	69. 1	22. 0	65. 9	25. 3	62. 6	28. 7	59. 2	32. 2	55. 6	35. 9	19
20	73. 2	17. 8	70. 3	20. 9	67. 3	24. 0	64. 2	27. 2	61. 0	30. 5	57. 7	33. 9	20
21	74. 1	17. 0	71. 3	19. 9	68. 5	22. 9	65. 6	25. 9	62. 7	29. 0	59. 6	32. 2	21
22	74. 9	16. 2	72. 3	19. 0	69. 6	21. 8	66. 9	24. 7	64. 1	27. 6	61. 2	30. 6	22
23	75. 7	15. 5	73. 2	18. 2	70. 7	20. 9	68. 1	23. 6	65. 5	26. 4	62. 7	29. 2	23
24	76. 3	14. 9	74. 0	17. 4	71. 6	20. 0	69. 2	22. 6	66. 7	25. 3	64. 1	28. 0	24
25	77. 0	14. 3	74. 7	16. 8	72. 5	19. 2	70. 1	21. 7	67. 8	24. 3	65. 4	26. 8	25
26	77. 6	13. 8	75. 4	16. 1	73. 3	18. 5	71. 1	20. 9	68. 8	23. 3	66. 5	25. 8	26
27	78. 1	13. 3	76. 1	15. 6	74. 0	17. 9	71. 9	20. 2	69. 8	22. 5	67. 6	24. 9	27
28	78. 6	12. 9	76. 6	15. 0	74. 7	17. 2	72. 7	19. 5	70. 6	21. 7	68. 6	24. 0	28
29	79. 1	12. 5	77. 2	14. 6	75. 3	16. 7	73. 4	18. 8	71. 5	21. 0	69. 5	23. 2	29
30	79. 5	12. 1	77. 7	14. 1	75. 9	16. 2	74. 1	18. 2	72. 2	20. 3	70. 3	22. 4	30
31	79. 9	11. 7	78. 2	13. 7	76. 5	15. 7	74. 7	17. 7	72. 9	19. 7	71. 1	21. 7	31
32	80. 3	11. 4	78. 7	13. 3	77. 0	15. 2	75. 3	17. 2	73. 6	19. 1	71. 9	21. 1	32
33	80. 7	11. 1	79. 1	12. 9	77. 5	14. 8	75. 9	16. 7	74. 2	18. 6	72. 6	20. 5	33
34	81. 0	10. 8	79. 5	12. 6	78. 0	14. 4	76. 4	16. 2	74. 8	18. 1	73. 3	20. 0	34
35	81. 4	10. 5	79. 9	12. 3	78. 4	14. 0	76. 9	15. 8	75. 4	17. 6	73. 9	19. 4	35
36	81. 7	10. 2	80. 3	12. 0	78. 8	13. 7	77. 4	15. 4	76. 0	17. 2	74. 5	18. 9	36
37	82. 0	10. 0	80. 6	11. 7	79. 3	13. 4	77. 9	15. 1	76. 5	16. 8	75. 1	18. 5	37
38	82. 3	9. 8	81. 0	11. 4	79. 6	13. 1	78. 3	14. 7	77. 0	16. 4	75. 6	18. 1	38
39	82. 5	9. 6	81. 3	11. 2	80. 0	12. 8	78. 7	14. 4	77. 4	16. 0	76. 1	17. 6	39
40	82. 8	9. 4	81. 6	10. 9	80. 4	12. 5	79. 1	14. 1	77. 9	15. 7	76. 6	17. 3	40
41	83. 1	9. 2	81. 9	10. 7	80. 7	12. 2	79. 5	13. 8	78. 3	15. 3	77. 1	16. 9	41
42	83. 3	9. 0	82. 2	10. 5	81. 0	12. 0	79. 9	13. 5	78. 7	15. 0	77. 5	16. 6	42
43	83. 5	8. 8	82. 4	10. 3	81. 3	11. 8	80. 2	13. 3	79. 1	14. 8	78. 0	16. 2	43
44	83. 8	8. 7	82. 7	10. 1	81. 6	11. 6	80. 6	13. 0	79. 5	14. 5	78. 4	15. 9	44
45	84. 0	8. 5	82. 9	9. 9	81. 9	11. 4	80. 9	12. 8	79. 8	14. 2	78. 8	15. 7	45
46	84. 2	8. 4	83. 2	9. 8	82. 2	11. 2	81. 2	12. 6	80. 2	14. 0	79. 2	15. 4	46
47	84. 4	8. 2	83. 4	9. 6	82. 5	11. 0	81. 5	12. 4	80. 5	13. 7	79. 6	15. 1	47
48	84. 6	8. 1	83. 7	9. 4	82. 7	10. 8	81. 8	12. 2	80. 9	13. 5	79. 9	14. 9	48
49	84. 8	8. 0	83. 9	9. 3	83. 0	10. 6	82. 1	12. 0	81. 2	13. 3	80. 3	14. 6	49
50	84. 9	7. 8	84. 1	9. 2	83. 2	10. 5	82. 4	11. 8	81. 5	13. 1	80. 6	14. 4	50
52	85. 3	7. 6	84. 5	8. 9	83. 7	10. 2	82. 9	11. 5	82. 1	12. 7	81. 3	14. 0	52
54	85. 6	7. 4	84. 9	8. 7	84. 1	9. 9	83. 4	11. 1	82. 6	12. 4	81. 9	13. 6	54
56	85. 9	7. 2	85. 2	8. 5	84. 6	9. 7	83. 9	10. 9	83. 2	12. 1	82. 5	13. 3	56
58	86. 2	7. 1	85. 6	8. 3	85. 0	9. 4	84. 3	10. 6	83. 7	11. 8	83. 0	13. 0	58
60	86. 5	6. 9	85. 9	8. 1	85. 3	9. 2	84. 8	10. 4	84. 2	11. 6	83. 6	12. 7	60
65	87. 2	6. 6	86. 7	7. 7	86. 2	8. 8	85. 8	9. 9	85. 3	11. 0	84. 8	12. 2	65
70	87. 8	6. 4	87. 4	7. 5	87. 1	8. 5	86. 7	9. 6	86. 3	10. 6	85. 9	11. 7	70
75	88. 4	6. 2	88. 1	7. 2	87. 8	8. 3	87. 6	9. 3	87. 3	10. 4	87. 0	11. 4	75
80	88. 9	6. 1	88. 8	7. 1	88. 6	8. 1	88. 4	9. 1	88. 2	10. 2	88. 0	11. 2	80
85	89. 5	6. 0	89. 4	7. 0	89. 3	8. 0	89. 2	9. 0	89. 1	10. 0	89. 0	11. 0	85

Numbers in *italics* indicate nearest approach to prime vertical

TABLE 25

Meridian Angle and Altitude of a Body on the Prime Vertical Circle

Latitude	Declination (same name as latitude) 12°		13°		14°		15°		16°		17°		Latitude
	t	Alt.	t	Alt.	t	Alt.	t	Alt.	t	Alt.	t	Alt.	
°	°	°	°	°	°	°	°	°	°	°	°	°	°
0	*90.0*	*0.0*	*90.0*	*0.0*	*90.0*	*0.0*	*90.0*	*0.0*	*90.0*	*0.0*	*90.0*	*0.0*	0
1	*85.3*	*4.8*	*85.7*	*4.4*	*86.0*	*4.1*	*86.3*	*3.9*	*86.5*	*3.6*	*86.7*	*3.4*	1
2	*80.5*	*9.7*	*81.3*	*8.9*	*81.9*	*8.3*	*82.5*	*7.7*	*83.0*	*7.3*	*83.4*	*6.9*	2
3	*75.7*	*14.6*	*76.9*	*13.5*	*77.9*	*12.5*	*78.7*	*11.7*	*79.5*	*10.9*	*80.1*	*10.3*	3
4	*70.8*	*19.6*	*72.4*	*18.1*	*73.7*	*16.8*	*74.9*	*15.6*	*75.9*	*14.7*	*76.8*	*13.8*	4
5	*65.7*	*24.8*	*67.7*	*22.8*	*69.5*	*21.1*	*70.9*	*19.7*	*72.2*	*18.4*	*73.4*	*17.3*	5
6	*60.4*	*30.2*	*62.9*	*27.7*	*65.1*	*25.6*	*66.9*	*23.8*	*68.5*	*22.3*	*69.9*	*20.9*	6
7	*54.7*	*35.9*	*57.9*	*32.8*	*60.5*	*30.2*	*62.7*	*28.1*	*64.6*	*26.2*	*66.3*	*24.6*	7
8	*48.6*	*42.0*	*52.5*	*38.2*	*55.7*	*35.1*	*58.4*	*32.5*	*60.7*	*30.3*	*62.6*	*28.4*	8
9	*41.8*	*48.8*	*46.7*	*44.1*	*50.6*	*40.3*	*53.8*	*37.2*	*56.5*	*34.6*	*58.8*	*32.3*	9
10	*33.9*	*56.6*	*40.2*	*50.5*	*45.0*	*45.9*	*48.8*	*42.1*	*52.1*	*39.0*	*54.8*	*36.4*	10
11	*23.9*	*66.6*	*32.7*	*58.0*	*38.8*	*52.1*	*43.5*	*47.5*	*47.3*	*43.8*	*50.5*	*40.7*	11
12	0.0	90.0	*23.0*	*67.6*	*31.5*	*59.3*	*37.5*	*53.4*	*42.2*	*49.0*	*46.0*	*45.3*	12
13	23.0	67.6	0.0	90.0	*22.2*	*68.4*	*30.5*	*60.4*	*36.4*	*54.7*	*41.0*	*50.3*	13
14	31.5	59.3	22.2	68.4	0.0	90.0	*21.5*	*69.2*	*29.6*	*61.4*	*35.4*	*55.8*	14
15	37.5	53.4	30.5	60.4	21.5	69.2	0.0	90.0	*20.9*	*69.9*	*28.8*	*62.3*	15
16	42.2	49.0	36.4	54.7	29.6	61.4	20.9	69.9	0.0	90.0	*20.3*	*70.5*	16
17	46.0	45.3	41.0	50.3	35.4	55.8	28.8	62.3	20.3	70.5	0.0	90.0	17
18	49.1	42.3	44.7	46.7	39.9	51.5	34.4	56.9	28.1	63.1	19.8	71.1	18
19	51.9	39.7	47.9	43.7	43.6	48.0	38.9	52.7	33.6	57.8	27.4	63.9	19
20	54.3	37.4	50.6	41.1	46.8	45.0	42.6	49.2	38.0	53.7	32.9	58.7	20
21	56.4	35.5	53.0	38.9	49.5	42.4	45.7	46.2	41.7	50.3	37.2	54.7	21
22	58.3	33.7	55.2	36.9	51.9	40.2	48.5	43.7	44.8	47.4	40.8	51.3	22
23	60.0	32.1	57.1	35.1	54.0	38.3	50.9	41.5	47.5	44.9	43.9	48.4	23
24	61.5	30.7	58.8	33.6	55.9	36.5	53.0	39.5	49.9	42.7	46.6	46.0	24
25	62.9	29.5	60.3	32.2	57.7	34.9	54.9	37.8	52.1	40.7	49.0	43.8	25
26	64.2	28.3	61.7	30.9	59.3	33.5	56.7	36.2	54.0	39.0	51.2	41.8	26
27	65.3	27.3	63.1	29.7	60.7	32.2	58.3	34.8	55.8	37.4	53.1	40.1	27
28	66.4	26.3	64.3	28.6	62.0	31.0	59.7	33.5	57.4	36.0	54.9	38.5	28
29	67.5	25.4	65.4	27.6	63.3	29.9	61.1	32.3	58.8	34.6	56.5	37.1	29
30	68.4	24.6	66.4	26.7	64.4	28.9	62.3	31.2	60.2	33.5	58.0	35.8	30
31	69.3	23.8	67.4	25.9	65.5	28.0	63.5	30.2	61.5	32.4	59.4	34.6	31
32	70.1	23.1	68.3	25.1	66.5	27.2	64.6	29.2	62.7	31.3	60.7	33.5	32
33	70.9	22.4	69.2	24.4	67.4	26.4	65.6	28.4	63.8	30.4	61.9	32.5	33
34	71.6	21.8	70.0	23.7	68.3	25.6	66.6	27.6	64.8	29.5	63.0	31.5	34
35	72.3	21.3	70.7	23.1	69.1	24.9	67.5	26.8	65.8	28.7	64.1	30.6	35
36	73.0	20.7	71.5	22.5	69.9	24.3	68.4	26.1	66.8	28.0	65.1	29.8	36
37	73.6	20.2	72.2	21.9	70.7	23.7	69.2	25.5	67.6	27.3	66.1	29.1	37
38	74.2	19.7	72.8	21.4	71.4	23.1	69.9	24.9	68.5	26.6	67.0	28.4	38
39	74.8	19.3	73.4	20.9	72.1	22.6	70.7	24.3	69.3	26.0	67.8	27.7	39
40	75.3	18.9	74.0	20.5	72.7	22.1	71.4	23.7	70.0	25.4	68.6	27.1	40
41	75.8	18.5	74.6	20.1	73.3	21.6	72.0	23.2	70.7	24.8	69.4	26.5	41
42	76.3	18.1	75.1	19.6	73.9	21.2	72.7	22.8	71.4	24.3	70.2	25.9	42
43	76.8	17.7	75.7	19.3	74.5	20.8	73.3	22.3	72.1	23.8	70.9	25.4	43
44	77.3	17.4	76.2	18.9	75.0	20.4	73.9	21.9	72.7	23.4	71.5	24.9	44
45	77.7	17.1	76.7	18.5	75.6	20.0	74.5	21.5	73.3	22.9	72.2	24.4	45
46	78.2	16.8	77.1	18.2	76.1	19.7	75.0	21.1	73.9	22.5	72.8	24.0	46
47	78.6	16.5	77.6	17.9	76.6	19.3	75.5	20.7	74.5	22.1	73.4	23.6	47
48	79.0	16.2	78.0	17.6	77.0	19.0	76.0	20.4	75.0	21.8	74.0	23.2	48
49	79.4	16.0	78.4	17.3	77.5	18.7	76.5	20.1	75.6	21.4	74.6	22.8	49
50	79.7	15.7	78.8	17.1	77.9	18.4	77.0	19.7	76.1	21.1	75.1	22.4	50
52	80.4	15.3	79.6	16.6	78.8	17.9	77.9	19.2	77.1	20.5	76.2	21.8	52
54	81.1	14.9	80.3	16.1	79.6	17.4	78.8	18.7	78.0	19.9	77.2	21.2	54
56	81.7	14.5	81.0	15.7	80.3	17.0	79.6	18.2	78.8	19.4	78.1	20.7	56
58	82.4	14.2	81.7	15.4	81.0	16.6	80.4	17.8	79.7	19.0	79.0	20.2	58
60	83.0	13.9	82.3	15.1	81.7	16.2	81.1	17.4	80.5	18.6	79.8	19.7	60
65	84.3	13.3	83.8	14.4	83.3	15.5	82.8	16.6	82.3	17.7	81.8	18.8	65
70	85.6	12.8	85.2	13.8	84.8	14.9	84.4	16.0	84.0	17.1	83.6	18.1	70
75	86.7	12.4	86.5	13.5	86.2	14.5	85.9	15.5	85.6	16.6	85.3	17.6	75
80	87.9	12.2	87.7	13.2	87.5	14.2	87.3	15.2	87.1	16.3	86.9	17.3	80
85	88.9	12.0	88.8	13.1	88.8	14.1	88.7	15.1	88.6	16.1	88.5	17.1	85

Numbers in *italics* indicate nearest approach to prime vertical

TABLE 25

Meridian Angle and Altitude of a Body on the Prime Vertical Circle

Latitude	Declination (same name as latitude)												Latitude
	18°		19°		20°		21°		22°		23°		
	t	Alt.	t	Alt.	t	Alt.	t	Alt.	t	Alt.	t	Alt.	
°	°	°	°	°	°	°	°	°	°	°	°	°	°
0	*90.0*	*0.0*	*90.0*	*0.0*	*90.0*	*0.0*	*90.0*	*0.0*	*90.0*	*0.0*	*90.0*	*0.0*	0
1	*86.9*	*3.2*	*87.1*	*3.1*	*87.3*	*2.9*	*87.4*	*2.8*	*87.5*	*2.7*	*87.6*	*2.6*	1
2	*83.8*	*6.5*	*84.2*	*6.2*	*84.5*	*5.9*	*84.8*	*5.6*	*85.0*	*5.3*	*85.3*	*5.1*	2
3	*80.7*	*9.8*	*81.2*	*9.3*	*81.7*	*8.8*	*82.2*	*8.4*	*82.5*	*8.0*	*82.9*	*7.7*	3
4	*77.6*	*13.0*	*78.3*	*12.4*	*78.9*	*11.8*	*79.5*	*11.2*	*80.0*	*10.7*	*80.5*	*10.3*	4
5	*74.4*	*16.4*	*75.3*	*15.5*	*76.1*	*14.8*	*76.8*	*14.1*	*77.5*	*13.5*	*78.1*	*12.9*	5
6	*71.1*	*19.8*	*72.2*	*18.7*	*73.2*	*17.8*	*74.1*	*17.0*	*74.9*	*16.2*	*75.7*	*15.5*	6
7	*67.8*	*23.2*	*69.1*	*22.0*	*70.3*	*20.9*	*71.3*	*19.9*	*72.3*	*19.0*	*73.2*	*18.2*	7
8	*64.4*	*26.8*	*65.9*	*25.3*	*67.3*	*24.0*	*68.5*	*22.9*	*69.6*	*21.8*	*70.7*	*20.9*	8
9	*60.8*	*30.4*	*62.6*	*28.7*	*64.2*	*27.2*	*65.6*	*25.9*	*66.9*	*24.7*	*68.1*	*23.6*	9
10	*57.1*	*34.2*	*59.2*	*32.2*	*61.0*	*30.5*	*62.7*	*29.0*	*64.1*	*27.6*	*65.5*	*26.4*	10
11	*53.3*	*38.1*	*55.6*	*35.9*	*57.7*	*33.9*	*59.6*	*32.2*	*61.2*	*30.6*	*62.7*	*29.2*	11
12	*49.1*	*42.3*	*51.9*	*39.7*	*54.3*	*37.4*	*56.4*	*35.5*	*58.3*	*33.7*	*60.0*	*32.1*	12
13	*44.7*	*46.7*	*47.9*	*43.7*	*50.6*	*41.1*	*53.0*	*38.9*	*55.2*	*36.9*	*57.0*	*35.2*	13
14	*39.9*	*51.5*	*43.6*	*48.0*	*46.8*	*45.0*	*49.5*	*42.5*	*51.9*	*40.2*	*54.0*	*38.3*	14
15	*34.4*	*56.9*	*38.9*	*52.7*	*42.6*	*49.2*	*45.7*	*46.2*	*48.5*	*43.7*	*50.9*	*41.5*	15
16	*28.1*	*63.1*	*33.6*	*57.8*	*38.0*	*53.7*	*41.7*	*50.3*	*44.8*	*47.4*	*47.5*	*44.9*	16
17	*19.8*	*71.1*	*27.4*	*63.9*	*32.9*	*58.7*	*37.2*	*54.7*	*40.8*	*51.3*	*43.9*	*48.4*	17
18	0.0	90.0	*19.3*	*71.7*	*26.8*	*64.6*	*32.2*	*59.6*	*36.5*	*55.6*	*40.1*	*52.3*	18
19	19.3	71.7	0.0	90.0	*18.9*	*72.2*	*26.2*	*65.3*	*31.5*	*60.4*	*35.8*	*56.4*	19
20	26.8	64.6	18.9	72.2	0.0	90.0	*18.5*	*72.6*	*25.7*	*65.9*	*31.0*	*61.1*	20
21	32.2	59.6	26.2	65.3	18.5	72.6	0.0	90.0	*18.2*	*73.1*	*25.3*	*66.5*	21
22	36.5	55.6	31.5	60.4	25.7	65.9	18.2	73.1	0.0	90.0	*17.9*	*73.5*	22
23	40.1	52.3	35.8	56.4	31.0	61.1	25.3	66.5	17.9	73.5	0.0	90.0	23
24	43.1	49.4	39.3	53.2	35.2	57.2	30.4	61.8	24.8	67.1	17.6	73.9	24
25	45.8	47.0	42.4	50.4	38.7	54.0	34.6	58.0	30.0	62.4	24.5	67.6	25
26	48.2	44.8	45.1	48.0	41.7	51.3	38.1	54.8	34.1	58.7	29.5	63.0	26
27	50.4	42.9	47.5	45.8	44.4	48.9	41.1	52.1	37.5	55.6	33.6	59.4	27
28	52.3	41.2	49.6	43.9	46.8	46.8	43.8	49.8	40.5	52.9	37.0	56.3	28
29	54.1	39.6	51.6	42.2	49.0	44.9	46.2	47.7	43.2	50.6	40.0	53.7	29
30	55.8	38.2	53.4	40.6	50.9	43.2	48.3	45.8	45.6	48.5	42.7	51.4	30
31	57.3	36.9	55.0	39.2	52.7	41.6	50.3	44.1	47.7	46.7	45.1	49.3	31
32	58.7	35.7	56.6	37.9	54.4	40.2	52.1	42.6	49.7	45.0	47.2	47.5	32
33	60.0	34.6	58.0	36.7	55.9	38.9	53.8	41.1	51.5	43.5	49.2	45.8	33
34	61.2	33.5	59.3	35.6	57.3	37.7	55.3	39.9	53.2	42.1	51.0	44.3	34
35	62.4	32.6	60.5	34.6	58.7	36.6	56.8	38.7	54.8	40.8	52.7	42.9	35
36	63.4	31.7	61.7	33.6	59.9	35.6	58.1	37.6	56.2	39.6	54.3	41.7	36
37	64.5	30.9	62.8	32.8	61.1	34.6	59.4	36.5	57.6	38.5	55.7	40.5	37
38	65.4	30.1	63.9	31.9	62.2	33.7	60.6	35.6	58.9	37.5	57.1	39.4	38
39	66.3	29.4	64.8	31.2	63.3	32.9	61.7	34.7	60.1	36.5	58.4	38.4	39
40	67.2	28.7	65.8	30.4	64.3	32.1	62.8	33.9	61.2	35.6	59.6	37.4	40
41	68.1	28.1	66.7	29.8	65.2	31.4	63.8	33.1	62.3	34.8	60.8	36.6	41
42	68.8	27.5	67.5	29.1	66.2	30.7	64.8	32.4	63.3	34.0	61.9	35.7	42
43	69.6	26.9	68.3	28.5	67.0	30.1	65.7	31.7	64.3	33.3	62.9	35.0	43
44	70.3	26.4	69.1	27.9	67.9	29.5	66.6	31.1	65.3	32.6	63.9	34.2	44
45	71.0	25.9	69.9	27.4	68.7	28.9	67.4	30.5	66.2	32.0	64.9	33.5	45
46	71.7	25.4	70.6	26.9	69.4	28.4	68.2	29.9	67.0	31.4	65.8	32.9	46
47	72.4	25.0	71.3	26.4	70.2	27.9	69.0	29.3	67.9	30.8	66.7	32.3	47
48	73.0	24.6	71.9	26.0	70.9	27.4	69.8	28.8	68.7	30.3	67.5	31.7	48
49	73.6	24.2	72.6	25.6	71.6	26.9	70.5	28.3	69.4	29.8	68.3	31.2	49
50	74.2	23.8	73.2	25.2	72.2	26.5	71.2	27.9	70.2	29.3	69.1	30.7	50
52	75.3	23.1	74.4	24.4	73.5	25.7	72.5	27.1	71.6	28.4	70.6	29.7	52
54	76.3	22.5	75.5	23.7	74.7	25.0	73.8	26.3	72.9	27.6	72.0	28.9	54
56	77.3	21.9	76.6	23.1	75.8	24.4	75.0	25.6	74.2	26.9	73.4	28.1	56
58	78.3	21.4	77.6	22.6	76.9	23.8	76.1	25.0	75.4	26.2	74.6	27.4	58
60	79.2	20.9	78.5	22.1	77.9	23.3	77.2	24.4	76.5	25.6	75.8	26.8	60
65	81.3	19.9	80.8	21.1	80.2	22.2	79.7	23.3	79.1	24.4	78.6	25.5	65
70	83.2	19.2	82.8	20.3	82.4	21.3	82.0	22.4	81.5	23.5	81.1	24.6	70
75	85.0	18.7	84.7	19.7	84.4	20.7	84.1	21.8	83.8	22.8	83.5	23.9	75
80	86.7	18.3	86.5	19.3	86.3	20.3	86.1	21.3	85.9	22.4	85.7	23.4	80
85	88.4	18.1	88.3	19.1	88.2	20.1	88.1	21.1	88.0	22.1	87.9	23.1	85

Numbers in *italics* indicate nearest approach to prime vertical

TABLE 26

Latitude and Longitude Factors
f, the change of latitude for a unit change in longitude
F, the change of longitude for a unit change in latitude

Azimuth angle	Latitude 0°		2°		4°		6°		8°		Azimuth angle
	f	F	f	F	f	F	f	F	f	F	
°											°
0	0. 00	—	0. 00	—	0. 00	—	0. 00	—	0. 00	—	180
1	0. 02	57. 29	0. 02	57. 32	0. 02	57. 43	0. 02	57. 61	0. 02	57. 85	179
2	0. 03	28. 64	0. 03	28. 65	0. 03	28. 71	0. 03	28. 79	0. 03	28. 92	178
3	0. 05	19. 08	0. 05	19. 09	0. 05	19. 13	0. 05	19. 19	0. 05	19. 27	177
4	0. 07	14. 30	0. 07	14. 31	0. 07	14. 34	0. 07	14. 38	0. 07	14. 44	176
5	0. 09	11. 43	0. 09	11. 44	0. 09	11. 46	0. 09	11. 49	0. 09	11. 54	175
6	0. 11	9. 51	0. 11	9. 52	0. 10	9. 54	0. 10	9. 57	0. 10	9. 61	174
7	0. 12	8. 14	0. 12	8. 15	0. 12	8. 16	0. 12	8. 19	0. 12	8. 22	173
8	0. 14	7. 12	0. 14	7. 12	0. 14	7. 13	0. 14	7. 15	0. 14	7. 18	172
9	0. 16	6. 31	0. 16	6. 32	0. 16	6. 33	0. 16	6. 35	0. 16	6. 38	171
10	0. 18	5. 67	0. 18	5. 68	0. 18	5. 69	0. 18	5. 70	0. 17	5. 73	170
12	0. 21	4. 70	0. 21	4. 71	0. 21	4. 72	0. 21	4. 73	0. 21	4. 75	168
14	0. 25	4. 01	0. 25	4. 01	0. 25	4. 02	0. 25	4. 03	0. 25	4. 05	166
16	0. 29	3. 49	0. 29	3. 49	0. 29	3. 50	0. 28	3. 51	0. 28	3. 52	164
18	0. 32	3. 08	0. 32	3. 08	0. 32	3. 08	0. 32	3. 10	0. 32	3. 11	162
20	0. 36	2. 75	0. 36	2. 75	0. 36	2. 75	0. 36	2. 76	0. 36	2. 77	160
22	0. 40	2. 48	0. 40	2. 48	0. 40	2. 48	0. 40	2. 49	0. 40	2. 50	158
24	0. 45	2. 25	0. 44	2. 25	0. 44	2. 25	0. 44	2. 26	0. 44	2. 27	156
26	0. 49	2. 05	0. 49	2. 05	0. 49	2. 05	0. 49	2. 06	0. 48	2. 07	154
28	0. 53	1. 88	0. 53	1. 88	0. 53	1. 88	0. 53	1. 89	0. 53	1. 90	152
30	0. 58	1. 73	0. 58	1. 73	0. 57	1. 74	0. 57	1. 74	0. 57	1. 75	150
32	0. 62	1. 60	0. 62	1. 60	0. 62	1. 60	0. 62	1. 61	0. 62	1. 62	148
34	0. 67	1. 48	0. 67	1. 48	0. 67	1. 49	0. 67	1. 49	0. 67	1. 50	146
36	0. 73	1. 38	0. 73	1. 38	0. 72	1. 38	0. 72	1. 38	0. 72	1. 39	144
38	0. 78	1. 28	0. 78	1. 28	0. 78	1. 28	0. 78	1. 29	0. 78	1. 29	142
40	0. 84	1. 19	0. 84	1. 19	0. 84	1. 19	0. 83	1. 20	0. 83	1. 20	140
42	0. 90	1. 11	0. 90	1. 11	0. 90	1. 11	0. 90	1. 12	0. 89	1. 12	138
44	0. 97	1. 04	0. 97	1. 04	0. 96	1. 04	0. 96	1. 04	0. 96	1. 05	136
46	1. 04	0. 97	1. 04	0. 97	1. 03	0. 97	1. 03	0. 97	1. 03	0. 98	134
48	1. 11	0. 90	1. 11	0. 90	1. 11	0. 90	1. 11	0. 90	1. 10	0. 91	132
50	1. 19	0. 84	1. 19	0. 84	1. 19	0. 84	1. 19	0. 84	1. 18	0. 85	130
52	1. 28	0. 78	1. 28	0. 78	1. 28	0. 78	1. 27	0. 79	1. 27	0. 79	128
54	1. 38	0. 73	1. 38	0. 73	1. 37	0. 73	1. 37	0. 73	1. 36	0. 73	126
56	1. 48	0. 67	1. 48	0. 67	1. 48	0. 68	1. 47	0. 68	1. 47	0. 68	124
58	1. 60	0. 62	1. 60	0. 63	1. 60	0. 63	1. 59	0. 63	1. 58	0. 63	122
60	1. 73	0. 58	1. 73	0. 58	1. 73	0. 58	1. 72	0. 58	1. 72	0. 58	120
62	1. 88	0. 53	1. 88	0. 53	1. 88	0. 53	1. 87	0. 53	1. 86	0. 54	118
64	2. 05	0. 49	2. 05	0. 49	2. 05	0. 49	2. 04	0. 49	2. 03	0. 49	116
66	2. 25	0. 45	2. 24	0. 45	2. 24	0. 45	2. 23	0. 45	2. 22	0. 45	114
68	2. 48	0. 40	2. 47	0. 40	2. 47	0. 40	2. 46	0. 40	2. 45	0. 41	112
70	2. 75	0. 36	2. 75	0. 36	2. 74	0. 36	2. 73	0. 37	2. 72	0. 37	110
72	3. 08	0. 32	3. 08	0. 33	3. 07	0. 33	3. 06	0. 33	3. 05	0. 33	108
74	3. 49	0. 29	3. 49	0. 29	3. 48	0. 29	3. 47	0. 29	3. 45	0. 29	106
76	4. 01	0. 25	4. 01	0. 25	4. 00	0. 25	3. 99	0. 25	3. 97	0. 25	104
78	4. 70	0. 21	4. 70	0. 21	4. 69	0. 21	4. 68	0. 21	4. 66	0. 21	102
80	5. 67	0. 18	5. 67	0. 18	5. 66	0. 18	5. 64	0. 18	5. 62	0. 18	100
81	6. 31	0. 16	6. 31	0. 16	6. 30	0. 16	6. 28	0. 16	6. 25	0. 16	99
82	7. 12	0. 14	7. 11	0. 14	7. 10	0. 14	7. 07	0. 14	7. 05	0. 14	98
83	8. 14	0. 12	8. 14	0. 12	8. 12	0. 12	8. 10	0. 12	8. 07	0. 12	97
84	9. 51	0. 11	9. 51	0. 11	9. 49	0. 11	9. 46	0. 11	9. 42	0. 11	96
85	11. 43	0. 09	11. 42	0. 09	11. 40	0. 09	11. 37	0. 09	11. 32	0. 09	95
86	14. 30	0. 07	14. 29	0. 07	14. 27	0. 07	14. 22	0. 07	14. 16	0. 07	94
87	19. 08	0. 05	19. 07	0. 05	19. 03	0. 05	18. 98	0. 05	18. 91	0. 05	93
88	28. 64	0. 03	28. 62	0. 03	28. 57	0. 03	28. 48	0. 03	28. 36	0. 03	92
89	57. 29	0. 02	57. 26	0. 02	57. 15	0. 02	56. 98	0. 02	56. 73	0. 02	91
90	—	0. 00	—	0. 00	—	0. 00	—	0. 00	—	0. 00	90
	0°		2°		4°		6°		8°		

Correction to latitude = f × error in longitude

Correction to longitude = F × error in latitude

TABLE 26

Latitude and Longitude Factors

f, the change of latitude for a unit change in longitude

F, the change of longitude for a unit change in latitude

Azimuth angle	Latitude 10° f	10° F	12° f	12° F	14° f	14° F	16° f	16° F	18° f	18° F	Azimuth angle
°											°
0	0. 00	—	0. 00	—	0. 00	—	0. 00	—	0. 00	—	180
1	0. 02	58. 17	0. 02	58. 57	0. 02	59. 04	0. 02	59. 60	0. 02	60. 24	179
2	0. 03	29. 08	0. 03	29. 28	0. 03	29. 51	0. 03	29. 79	0. 03	30. 11	178
3	0. 05	19. 38	0. 05	19. 51	0. 05	19. 67	0. 05	19. 85	0. 05	20. 06	177
4	0. 07	14. 52	0. 07	14. 62	0. 07	14. 74	0. 07	14. 88	0. 07	15. 04	176
5	0. 09	11. 61	0. 09	11. 69	0. 08	11. 78	0. 08	11. 89	0. 08	12. 02	175
6	0. 10	9. 66	0. 10	9. 73	0. 10	9. 81	0. 10	9. 90	0. 10	10. 00	174
7	0. 12	8. 27	0. 12	8. 33	0. 12	8. 39	0. 12	8. 47	0. 12	8. 56	173
8	0. 14	7. 22	0. 14	7. 27	0. 14	7. 33	0. 14	7. 40	0. 13	7. 48	172
9	0. 16	6. 41	0. 15	6. 45	0. 15	6. 51	0. 15	6. 57	0. 15	6. 64	171
10	0. 17	5. 76	0. 17	5. 80	0. 17	5. 85	0. 17	5. 90	0. 17	5. 96	170
12	0. 21	4. 78	0. 21	4. 81	0. 21	4. 85	0. 20	4. 89	0. 20	4. 95	168
14	0. 25	4. 07	0. 24	4. 10	0. 24	4. 13	0. 24	4. 17	0. 24	4. 22	166
16	0. 28	3. 54	0. 28	3. 56	0. 28	3. 59	0. 28	3. 63	0. 27	3. 67	164
18	0. 32	3. 13	0. 32	3. 15	0. 32	3. 17	0. 31	3. 20	0. 31	3. 24	162
20	0. 36	2. 79	0. 36	2. 81	0. 35	2. 83	0. 35	2. 86	0. 35	2. 89	160
22	0. 40	2. 51	0. 40	2. 53	0. 39	2. 55	0. 39	2. 57	0. 38	2. 60	158
24	0. 44	2. 28	0. 44	2. 30	0. 43	2. 32	0. 43	2. 34	0. 42	2. 36	156
26	0. 48	2. 08	0. 48	2. 10	0. 47	2. 11	0. 47	2. 13	0. 46	2. 16	154
28	0. 52	1. 91	0. 52	1. 92	0. 52	1. 94	0. 51	1. 96	0. 51	1. 98	152
30	0. 57	1. 76	0. 56	1. 77	0. 56	1. 78	0. 56	1. 80	0. 55	1. 82	150
32	0. 62	1. 63	0. 61	1. 64	0. 61	1. 65	0. 60	1. 66	0. 59	1. 68	148
34	0. 66	1. 50	0. 66	1. 52	0. 65	1. 53	0. 65	1. 54	0. 64	1. 56	146
36	0. 72	1. 40	0. 71	1. 41	0. 70	1. 42	0. 70	1. 43	0. 69	1. 45	144
38	0. 77	1. 30	0. 76	1. 31	0. 76	1. 32	0. 75	1. 33	0. 74	1. 35	142
40	0. 83	1. 21	0. 82	1. 22	0. 81	1. 23	0. 81	1. 24	0. 80	1. 25	140
42	0. 88	1. 13	0. 88	1. 14	0. 88	1. 14	0. 87	1. 15	0. 85	1. 17	138
44	0. 95	1. 05	0. 94	1. 06	0. 94	1. 07	0. 93	1. 08	0. 92	1. 09	136
46	1. 02	0. 98	1. 01	0. 99	1. 01	1. 00	1. 00	1. 01	0. 99	1. 02	134
48	1. 10	0. 91	1. 09	0. 92	1. 08	0. 93	1. 07	0. 94	1. 06	0. 95	132
50	1. 17	0. 85	1. 17	0. 86	1. 16	0. 87	1. 15	0. 87	1. 13	0. 88	130
52	1. 26	0. 79	1. 25	0. 80	1. 24	0. 80	1. 23	0. 81	1. 22	0. 82	128
54	1. 36	0. 74	1. 35	0. 74	1. 34	0. 75	1. 32	0. 76	1. 31	0. 76	126
56	1. 46	0. 68	1. 45	0. 69	1. 44	0. 69	1. 43	0. 70	1. 41	0. 71	124
58	1. 58	0. 63	1. 57	0. 64	1. 55	0. 64	1. 54	0. 65	1. 52	0. 66	122
60	1. 71	0. 59	1. 69	0. 59	1. 68	0. 60	1. 67	0. 60	1. 65	0. 61	120
62	1. 85	0. 54	1. 84	0. 54	1. 83	0. 55	1. 81	0. 55	1. 79	0. 56	118
64	2. 02	0. 50	2. 01	0. 50	1. 99	0. 50	1. 97	0. 51	1. 95	0. 51	116
66	2. 21	0. 45	2. 20	0. 46	2. 18	0. 46	2. 16	0. 46	2. 14	0. 47	114
68	2. 44	0. 41	2. 42	0. 41	2. 40	0. 42	2. 38	0. 42	2. 35	0. 42	112
70	2. 71	0. 37	2. 69	0. 37	2. 67	0. 37	2. 64	0. 38	2. 61	0. 38	110
72	3. 03	0. 33	3. 01	0. 33	2. 99	0. 33	2. 96	0. 34	2. 93	0. 34	108
74	3. 43	0. 29	3. 41	0. 29	3. 38	0. 30	3. 35	0. 30	3. 32	0. 30	106
76	3. 95	0. 25	3. 92	0. 25	3. 89	0. 26	3. 86	0. 26	3. 81	0. 26	104
78	4. 63	0. 22	4. 60	0. 22	4. 56	0. 22	4. 52	0. 22	4. 47	0. 22	102
80	5. 59	0. 18	5. 55	0. 18	5. 50	0. 18	5. 45	0. 18	5. 39	0. 18	100
81	6. 22	0. 16	6. 18	0. 16	6. 13	0. 16	6. 07	0. 16	6. 01	0. 17	99
82	7. 01	0. 14	6. 96	0. 14	6. 90	0. 14	6. 84	0. 15	6. 77	0. 15	98
83	8. 02	0. 12	7. 97	0. 13	7. 90	0. 13	7. 83	0. 13	7. 75	0. 13	97
84	9. 37	0. 11	9. 31	0. 11	9. 23	0. 11	9. 15	0. 11	9. 05	0. 11	96
85	11. 25	0. 09	11. 18	0. 09	11. 09	0. 09	10. 99	0. 09	10. 87	0. 09	95
86	14. 08	0. 07	13. 99	0. 07	13. 88	0. 07	13. 75	0. 07	13. 60	0. 07	94
87	18. 79	0. 05	18. 66	0. 05	18. 51	0. 05	18. 34	0. 05	18. 15	0. 05	93
88	28. 20	0. 03	28. 01	0. 04	27. 79	0. 04	27. 53	0. 04	27. 23	0. 04	92
89	56. 42	0. 02	56. 04	0. 02	55. 59	0. 02	55. 07	0. 02	54. 49	0. 02	91
90	—	0. 00	—	0. 00	—	0. 00	—	0. 00	—	0. 00	90
	10°		12°		14°		16°		18°		

Correction to latitude = f × error in longitude

Correction to longitude = F × error in latitude

TABLE 26

Latitude and Longitude Factors
f, the change of latitude for a unit change in longitude
F, the change of longitude for a unit change in latitude

Azimuth angle	Latitude 20° f	F	22° f	F	24° f	F	26° f	F	28° f	F	Azimuth angle
°											°
0	0.00	—	0.00	—	0.00	—	0.00	—	0.00	—	180
1	0.02	60.97	0.02	61.79	0.02	62.71	0.02	63.74	0.02	64.88	179
2	0.03	30.47	0.03	30.89	0.03	31.35	0.03	31.86	0.03	32.43	178
3	0.05	20.31	0.05	20.58	0.05	20.89	0.05	21.23	0.05	21.61	177
4	0.07	15.22	0.06	15.42	0.06	15.65	0.06	15.91	0.06	16.20	176
5	0.08	12.16	0.08	12.33	0.08	12.51	0.08	12.72	0.08	12.95	175
6	0.10	10.12	0.10	10.26	0.10	10.41	0.09	10.59	0.09	10.78	174
7	0.12	8.67	0.11	8.78	0.11	8.91	0.11	9.06	0.11	9.22	173
8	0.13	7.57	0.13	7.67	0.13	7.79	0.13	7.92	0.12	8.06	172
9	0.15	6.72	0.15	6.81	0.14	6.91	0.14	7.02	0.14	7.15	171
10	0.17	6.03	0.16	6.12	0.16	6.21	0.16	6.31	0.16	6.42	170
12	0.20	5.01	0.20	5.07	0.19	5.15	0.19	5.23	0.19	5.33	168
14	0.23	4.27	0.23	4.33	0.23	4.39	0.22	4.46	0.22	4.54	166
16	0.27	3.71	0.27	3.76	0.26	3.82	0.26	3.88	0.25	3.95	164
18	0.30	3.28	0.30	3.32	0.30	3.37	0.29	3.42	0.29	3.49	162
20	0.34	2.92	0.34	2.96	0.33	3.01	0.33	3.06	0.32	3.11	160
22	0.38	2.63	0.38	2.67	0.37	2.71	0.36	2.75	0.36	2.80	158
24	0.42	2.39	0.41	2.42	0.41	2.46	0.40	2.50	0.39	2.54	156
26	0.46	2.18	0.45	2.21	0.45	2.24	0.44	2.28	0.43	2.32	154
28	0.50	2.00	0.49	2.03	0.49	2.06	0.48	2.09	0.47	2.13	152
30	0.54	1.84	0.53	1.87	0.53	1.90	0.52	1.93	0.51	1.96	150
32	0.59	1.70	0.58	1.73	0.57	1.75	0.56	1.78	0.55	1.81	148
34	0.63	1.58	0.63	1.60	0.62	1.62	0.61	1.65	0.60	1.68	146
36	0.68	1.47	0.67	1.48	0.66	1.51	0.65	1.53	0.64	1.56	144
38	0.74	1.36	0.72	1.38	0.71	1.40	0.70	1.42	0.69	1.45	142
40	0.79	1.27	0.78	1.28	0.77	1.30	0.75	1.33	0.74	1.35	140
42	0.85	1.18	0.83	1.20	0.82	1.22	0.81	1.24	0.79	1.26	138
44	0.91	1.10	0.90	1.12	0.88	1.13	0.87	1.15	0.85	1.17	136
46	0.97	1.03	0.96	1.04	0.95	1.06	0.93	1.07	0.91	1.09	134
48	1.04	0.96	1.03	0.97	1.02	0.99	1.00	1.00	0.98	1.02	132
50	1.12	0.89	1.10	0.91	1.09	0.92	1.07	0.93	1.05	0.95	130
52	1.20	0.83	1.19	0.84	1.17	0.85	1.15	0.87	1.13	0.88	128
54	1.29	0.77	1.28	0.78	1.26	0.79	1.24	0.81	1.22	0.82	126
56	1.39	0.72	1.38	0.73	1.35	0.74	1.33	0.75	1.31	0.76	124
58	1.50	0.66	1.48	0.67	1.46	0.68	1.44	0.70	1.41	0.71	122
60	1.63	0.61	1.61	0.62	1.58	0.63	1.56	0.64	1.53	0.65	120
62	1.77	0.57	1.74	0.57	1.72	0.58	1.69	0.59	1.66	0.60	118
64	1.93	0.52	1.90	0.53	1.87	0.53	1.84	0.54	1.81	0.55	116
66	2.11	0.47	2.08	0.48	2.05	0.49	2.02	0.50	1.98	0.50	114
68	2.33	0.43	2.30	0.44	2.26	0.44	2.23	0.45	2.18	0.46	112
70	2.58	0.39	2.55	0.39	2.51	0.40	2.47	0.40	2.43	0.41	110
72	2.89	0.35	2.85	0.35	2.81	0.36	2.77	0.36	2.72	0.37	108
74	3.28	0.31	3.23	0.31	3.19	0.31	3.14	0.32	3.08	0.33	106
76	3.77	0.27	3.72	0.27	3.66	0.27	3.61	0.28	3.54	0.28	104
78	4.42	0.23	4.36	0.23	4.30	0.23	4.23	0.24	4.15	0.24	102
80	5.33	0.19	5.26	0.19	5.18	0.19	5.10	0.20	5.01	0.20	100
81	5.93	0.17	5.86	0.17	5.77	0.17	5.68	0.18	5.58	0.18	99
82	6.69	0.15	6.60	0.15	6.50	0.15	6.40	0.16	6.28	0.16	98
83	7.65	0.13	7.55	0.13	7.44	0.13	7.32	0.14	7.19	0.14	97
84	8.94	0.11	8.82	0.11	8.69	0.12	8.55	0.12	8.40	0.12	96
85	10.74	0.09	10.60	0.09	10.44	0.10	10.26	0.10	10.09	0.10	95
86	13.44	0.07	13.26	0.08	13.07	0.08	12.86	0.08	12.63	0.08	94
87	17.93	0.06	17.69	0.06	17.43	0.06	17.15	0.06	16.85	0.06	93
88	26.91	0.04	26.55	0.04	26.16	0.04	25.74	0.04	25.28	0.04	92
89	53.84	0.02	53.12	0.02	52.33	0.02	51.50	0.02	50.58	0.02	91
90	—	0.00	—	0.00	—	0.00	—	0.00	—	0.00	90
	20°		22°		24°		26°		28°		

Correction to latitude = f × error in longitude | Correction to longitude = F × error in latitude

TABLE 26

Latitude and Longitude Factors

f, the change of latitude for a unit change in longitude

F, the change of longitude for a unit change in latitude

Azimuth angle	Latitude 30° f	30° F	32° f	32° F	34° f	34° F	36° f	36° F	38° f	38° F	Azimuth angle
°											°
0	0. 00	—	0. 00	—	0. 00	—	0. 00	—	0. 00	—	180
1	0. 02	66. 15	0. 01	67. 56	0. 01	69. 10	0. 01	70. 81	0. 01	72. 70	179
2	0. 03	33. 07	0. 03	33. 77	0. 03	34. 54	0. 03	35. 40	0. 03	36. 34	178
3	0. 05	22. 03	0. 05	22. 50	0. 04	23. 02	0. 04	23. 59	0. 04	24. 21	177
4	0. 06	16. 51	0. 06	16. 86	0. 06	17. 25	0. 06	17. 68	0. 06	18. 15	176
5	0. 08	13. 20	0. 07	13. 48	0. 07	13. 79	0. 07	14. 13	0. 07	14. 50	175
6	0. 09	10. 99	0. 09	11. 22	0. 09	11. 48	0. 09	11. 76	0. 08	12. 07	174
7	0. 11	9. 40	0. 10	9. 60	0. 10	9. 82	0. 10	10. 07	0. 10	10. 34	173
8	0. 12	8. 22	0. 12	8. 39	0. 12	8. 58	0. 11	8. 79	0. 11	9. 03	172
9	0. 14	7. 29	0. 13	7. 45	0. 13	7. 62	0. 13	7. 80	0. 12	8. 01	171
10	0. 15	6. 55	0. 15	6. 69	0. 15	6. 84	0. 14	7. 01	0. 14	7. 20	170
12	0. 18	5. 43	0. 18	5. 55	0. 18	5. 67	0. 17	5. 82	0. 17	5. 97	168
14	0. 22	4. 63	0. 21	4. 73	0. 21	4. 84	0. 20	4. 96	0. 20	5. 09	166
16	0. 25	4. 03	0. 24	4. 11	0. 24	4. 21	0. 23	4. 31	0. 23	4. 43	164
18	0. 28	3. 55	0. 28	3. 63	0. 27	3. 71	0. 26	3. 80	0. 26	3. 91	162
20	0. 32	3. 17	0. 31	3. 24	0. 30	3. 31	0. 29	3. 40	0. 29	3. 49	160
22	0. 35	2. 86	0. 34	2. 92	0. 34	2. 99	0. 33	3. 06	0. 32	3. 14	158
24	0. 39	2. 59	0. 38	2. 65	0. 37	2. 71	0. 36	2. 78	0. 35	2. 85	156
26	0. 42	2. 37	0. 41	2. 42	0. 40	2. 47	0. 40	2. 53	0. 38	2. 60	154
28	0. 46	2. 17	0. 45	2. 22	0. 44	2. 27	0. 43	2. 32	0. 42	2. 39	152
30	0. 50	2. 00	0. 49	2. 04	0. 48	2. 09	0. 47	2. 14	0. 45	2. 20	150
32	0. 54	1. 85	0. 53	1. 89	0. 52	1. 93	0. 51	1. 98	0. 49	2. 03	148
34	0. 58	1. 71	0. 57	1. 75	0. 56	1. 79	0. 55	1. 83	0. 53	1. 88	146
36	0. 63	1. 59	0. 62	1. 62	0. 60	1. 66	0. 59	1. 70	0. 57	1. 75	144
38	0. 68	1. 48	0. 66	1. 51	0. 65	1. 54	0. 63	1. 58	0. 62	1. 62	142
40	0. 72	1. 38	0. 71	1. 41	0. 69	1. 44	0. 68	1. 47	0. 66	1. 51	140
42	0. 78	1. 28	0. 76	1. 31	0. 75	1. 34	0. 73	1. 37	0. 71	1. 41	138
44	0. 84	1. 20	0. 82	1. 22	0. 80	1. 25	0. 78	1. 28	0. 76	1. 31	136
46	0. 90	1. 11	0. 88	1. 14	0. 86	1. 16	0. 84	1. 19	0. 82	1. 23	134
48	0. 96	1. 04	0. 94	1. 06	0. 92	1. 09	0. 90	1. 11	0. 88	1. 14	132
50	1. 03	0. 97	1. 01	0. 99	0. 99	1. 01	0. 96	1. 04	0. 94	1. 06	130
52	1. 11	0. 90	1. 09	0. 92	1. 06	0. 94	1. 04	0. 97	1. 01	0. 99	128
54	1. 19	0. 84	1. 16	0. 86	1. 14	0. 88	1. 11	0. 90	1. 08	0. 92	126
56	1. 28	0. 78	1. 26	0. 79	1. 23	0. 81	1. 20	0. 83	1. 17	0. 86	124
58	1. 39	0. 72	1. 36	0. 74	1. 33	0. 75	1. 30	0. 77	1. 26	0. 79	122
60	1. 49	0. 67	1. 47	0. 68	1. 44	0. 70	1. 40	0. 71	1. 37	0. 73	120
62	1. 63	0. 61	1. 59	0. 63	1. 56	0. 64	1. 52	0. 66	1. 48	0. 67	118
64	1. 78	0. 56	1. 74	0. 57	1. 70	0. 59	1. 66	0. 60	1. 62	0. 62	116
66	1. 95	0. 51	1. 91	0. 52	1. 85	0. 54	1. 82	0. 55	1. 77	0. 56	114
68	2. 14	0. 47	2. 10	0. 48	2. 05	0. 49	2. 00	0. 50	1. 95	0. 51	112
70	2. 38	0. 42	2. 33	0. 43	2. 28	0. 44	2. 22	0. 45	2. 17	0. 46	110
72	2. 67	0. 38	2. 61	0. 38	2. 55	0. 39	2. 50	0. 40	2. 43	0. 41	108
74	3. 02	0. 33	2. 96	0. 34	2. 89	0. 35	2. 82	0. 35	2. 75	0. 36	106
76	3. 47	0. 29	3. 40	0. 29	3. 33	0. 30	3. 25	0. 31	3. 16	0. 32	104
78	4. 07	0. 24	3. 99	0. 25	3. 90	0. 26	3. 81	0. 26	3. 71	0. 27	102
80	4. 91	0. 20	4. 81	0. 21	4. 70	0. 21	4. 59	0. 22	4. 47	0. 22	100
81	5. 47	0. 18	5. 35	0. 19	5. 24	0. 19	5. 11	0. 20	4. 98	0. 20	99
82	6. 16	0. 16	6. 03	0. 17	5. 90	0. 17	5. 76	0. 17	5. 61	0. 18	98
83	7. 05	0. 14	6. 91	0. 14	6. 75	0. 15	6. 59	0. 15	6. 42	0. 16	97
84	8. 24	0. 12	8. 07	0. 12	7. 89	0. 13	7. 70	0. 13	7. 50	0. 13	96
85	9. 90	0. 10	9. 69	0. 10	9. 48	0. 11	9. 25	0. 11	9. 01	0. 11	95
86	12. 39	0. 08	12. 13	0. 08	11. 86	0. 08	11. 57	0. 09	11. 27	0. 09	94
87	16. 52	0. 06	16. 18	0. 06	15. 82	0. 06	15. 44	0. 06	15. 04	0. 07	93
88	24. 80	0. 04	24. 28	0. 04	23. 74	0. 04	23. 17	0. 04	22. 57	0. 04	92
89	49. 61	0. 02	48. 58	0. 02	47. 50	0. 02	46. 36	0. 02	45. 14	0. 02	91
90	—	0. 00	—	0. 00	—	0. 00	—	0. 00	—	0. 00	90
	30°		32°		34°		36°		38°		

Correction to latitude = f × error in longitude

Correction to longitude = F × error in latitude

TABLE 26

Latitude and Longitude Factors

f, the change of latitude for a unit change in longitude

F, the change of longitude for a unit change in latitude

Azimuth angle	Latitude 40° f	40° F	42° f	42° F	44° f	44° F	46° f	46° F	48° f	48° F	Azimuth angle
°											°
0	0.00	—	0.00	—	0.00	—	0.00	—	0.00	—	180
1	0.01	74.79	0.01	77.09	0.01	79.64	0.01	82.47	0.01	85.62	179
2	0.03	37.38	0.03	38.53	0.03	39.81	0.02	41.22	0.02	42.80	178
3	0.04	24.91	0.04	25.68	0.04	26.53	0.04	27.47	0.03	28.52	177
4	0.05	18.67	0.05	19.24	0.05	19.88	0.05	20.59	0.05	21.37	176
5	0.07	14.92	0.07	15.38	0.06	15.89	0.06	16.45	0.06	17.08	175
6	0.08	12.42	0.08	12.80	0.08	13.23	0.07	13.70	0.07	14.22	174
7	0.09	10.63	0.09	10.96	0.09	11.32	0.08	11.72	0.08	12.17	173
8	0.11	9.29	0.10	9.57	0.10	9.89	0.10	10.24	0.09	10.63	172
9	0.12	8.24	0.12	8.50	0.11	8.78	0.11	9.09	0.11	9.44	171
10	0.14	7.40	0.13	7.63	0.13	7.88	0.12	8.16	0.12	8.48	170
12	0.16	6.14	0.16	6.33	0.15	6.54	0.15	6.77	0.14	7.03	168
14	0.19	5.24	0.19	5.40	0.18	5.58	0.17	5.77	0.17	5.99	166
16	0.22	4.55	0.21	4.69	0.21	4.85	0.20	5.02	0.19	5.21	164
18	0.25	4.02	0.24	4.14	0.23	4.28	0.23	4.43	0.22	4.60	162
20	0.28	3.59	0.27	3.70	0.26	3.82	0.25	3.95	0.24	4.11	160
22	0.31	3.23	0.30	3.33	0.29	3.44	0.28	3.56	0.27	3.70	158
24	0.34	2.93	0.33	3.02	0.32	3.12	0.31	3.23	0.30	3.36	156
26	0.37	2.68	0.36	2.76	0.35	2.85	0.34	2.95	0.33	3.06	154
28	0.41	2.45	0.40	2.53	0.38	2.61	0.37	2.71	0.36	2.81	152
30	0.44	2.26	0.43	2.33	0.41	2.41	0.40	2.49	0.39	2.59	150
32	0.48	2.09	0.46	2.15	0.45	2.22	0.43	2.30	0.42	2.39	148
34	0.52	1.93	0.50	1.99	0.49	2.06	0.47	2.13	0.45	2.22	146
36	0.56	1.80	0.54	1.85	0.52	1.91	0.50	1.98	0.49	2.06	144
38	0.60	1.67	0.58	1.72	0.56	1.78	0.54	1.84	0.52	1.91	142
40	0.64	1.56	0.63	1.60	0.60	1.66	0.58	1.71	0.56	1.78	140
42	0.69	1.45	0.67	1.49	0.65	1.54	0.63	1.60	0.60	1.66	138
44	0.74	1.35	0.72	1.39	0.69	1.44	0.67	1.49	0.65	1.55	136
46	0.79	1.26	0.77	1.30	0.74	1.34	0.72	1.39	0.69	1.44	134
48	0.85	1.17	0.83	1.21	0.80	1.25	0.77	1.30	0.74	1.35	132
50	0.91	1.09	0.88	1.13	0.86	1.17	0.83	1.21	0.80	1.25	130
52	0.98	1.02	0.95	1.05	0.92	1.09	0.89	1.12	0.86	1.17	128
54	1.05	0.95	1.02	0.98	0.99	1.01	0.96	1.05	0.92	1.09	126
56	1.14	0.88	1.10	0.91	1.07	0.94	1.03	0.97	0.99	1.01	124
58	1.23	0.82	1.19	0.84	1.15	0.87	1.11	0.90	1.07	0.93	122
60	1.33	0.75	1.29	0.78	1.25	0.80	1.20	0.83	1.16	0.86	120
62	1.44	0.69	1.40	0.72	1.35	0.74	1.31	0.77	1.26	0.79	118
64	1.57	0.64	1.52	0.66	1.48	0.68	1.42	0.70	1.37	0.73	116
66	1.72	0.58	1.67	0.60	1.62	0.62	1.56	0.64	1.50	0.66	114
68	1.90	0.53	1.84	0.54	1.78	0.56	1.72	0.58	1.66	0.60	112
70	2.10	0.47	2.04	0.49	1.98	0.51	1.91	0.52	1.84	0.54	110
72	2.36	0.42	2.29	0.44	2.21	0.45	2.14	0.47	2.06	0.49	108
74	2.67	0.37	2.59	0.39	2.51	0.40	2.42	0.41	2.33	0.43	106
76	3.07	0.32	2.98	0.34	2.89	0.35	2.79	0.36	2.68	0.37	104
78	3.60	0.28	3.50	0.29	3.38	0.29	3.27	0.31	3.15	0.32	102
80	4.34	0.23	4.22	0.24	4.08	0.24	3.94	0.25	3.80	0.26	100
81	4.84	0.21	4.69	0.21	4.54	0.22	4.39	0.23	4.23	0.24	99
82	5.45	0.18	5.29	0.19	5.12	0.19	4.94	0.20	4.76	0.21	98
83	6.24	0.16	6.05	0.16	5.86	0.17	5.66	0.18	5.45	0.18	97
84	7.29	0.14	7.07	0.14	6.84	0.15	6.61	0.15	6.37	0.16	96
85	8.75	0.11	8.49	0.12	8.22	0.12	7.94	0.13	7.65	0.13	95
86	10.95	0.09	10.63	0.09	10.29	0.10	9.94	0.10	9.57	0.10	94
87	14.62	0.07	14.18	0.07	13.73	0.07	13.26	0.08	12.77	0.08	93
88	21.94	0.05	21.28	0.05	20.60	0.05	19.89	0.05	19.16	0.05	92
89	43.98	0.02	42.58	0.02	41.21	0.02	39.80	0.02	38.34	0.03	91
90	—	0.00	—	0.00	—	0.00	—	0.00	—	0.00	90
	40°		42°		44°		46°		48°		

Correction to latitude = f × error in longitude

Correction to longitude = F × error in latitude

TABLE 26

Latitude and Longitude Factors

f, the change of latitude for a unit change in longitude

F, the change of longitude for a unit change in latitude

Azimuth angle	Latitude 50° f	50° F	52° f	52° F	54° f	54° F	56° f	56° F	58° f	58° F	Azimuth angle
°											°
0	0.00	—	0.00	—	0.00	—	0.00	—	0.00	—	180
1	0.01	89.13	0.01	93.05	0.01	97.47	0.01	102.45	0.01	108.11	179
2	0.02	44.55	0.02	46.51	0.02	48.72	0.02	51.21	0.02	54.04	178
3	0.03	29.68	0.03	30.99	0.03	32.46	0.03	34.12	0.03	36.01	177
4	0.04	22.25	0.04	23.23	0.04	24.33	0.04	25.57	0.04	26.99	176
5	0.06	17.78	0.05	18.57	0.05	19.45	0.05	20.44	0.05	21.57	175
6	0.07	14.80	0.06	15.45	0.06	16.19	0.06	17.01	0.06	17.95	174
7	0.08	12.67	0.08	13.23	0.07	13.86	0.07	14.56	0.06	15.37	173
8	0.09	11.07	0.08	11.56	0.08	12.11	0.08	12.72	0.07	13.43	172
9	0.10	9.82	0.10	10.26	0.09	10.74	0.09	11.29	0.08	11.91	171
10	0.11	8.82	0.11	9.21	0.10	9.65	0.10	10.14	0.09	10.70	170
12	0.14	7.32	0.13	7.64	0.13	8.00	0.12	8.41	0.11	8.88	168
14	0.16	6.24	0.15	6.51	0.15	6.82	0.14	7.17	0.13	7.57	166
16	0.18	5.42	0.18	5.66	0.17	5.93	0.16	6.24	0.15	6.58	164
18	0.21	4.79	0.20	5.00	0.19	5.24	0.18	5.50	0.17	5.81	162
20	0.23	4.27	0.22	4.46	0.21	4.67	0.20	4.91	0.19	5.19	160
22	0.26	3.85	0.25	4.02	0.24	4.21	0.23	4.43	0.21	4.67	158
24	0.29	3.49	0.27	3.65	0.26	3.82	0.25	4.02	0.24	4.24	156
26	0.31	3.19	0.30	3.33	0.29	3.49	0.27	3.66	0.26	3.87	154
28	0.34	2.93	0.33	3.05	0.31	3.20	0.30	3.36	0.28	3.55	152
30	0.37	2.69	0.36	2.81	0.34	2.95	0.32	3.10	0.31	3.27	150
32	0.40	2.49	0.38	2.60	0.37	2.72	0.35	2.86	0.33	3.02	148
34	0.43	2.31	0.42	2.41	0.40	2.52	0.38	2.65	0.36	2.80	146
36	0.47	2.14	0.45	2.24	0.43	2.34	0.41	2.46	0.39	2.60	144
38	0.50	1.99	0.48	2.08	0.46	2.18	0.44	2.29	0.41	2.41	142
40	0.54	1.85	0.52	1.94	0.49	2.03	0.47	2.13	0.44	2.25	140
42	0.58	1.73	0.56	1.80	0.53	1.89	0.50	1.99	0.48	2.09	138
44	0.62	1.61	0.59	1.68	0.57	1.76	0.54	1.85	0.51	1.95	136
46	0.67	1.50	0.64	1.57	0.61	1.64	0.58	1.73	0.55	1.82	134
48	0.71	1.40	0.68	1.46	0.65	1.53	0.62	1.61	0.59	1.70	132
50	0.77	1.31	0.73	1.36	0.70	1.43	0.67	1.50	0.63	1.58	130
52	0.82	1.22	0.79	1.27	0.75	1.33	0.72	1.40	0.68	1.47	128
54	0.88	1.13	0.85	1.18	0.81	1.23	0.77	1.30	0.73	1.37	126
56	0.95	1.05	0.91	1.10	0.87	1.15	0.83	1.21	0.79	1.27	124
58	1.03	0.97	0.99	1.01	0.94	1.06	0.89	1.12	0.85	1.18	122
60	1.11	0.90	1.07	0.94	1.02	0.98	0.97	1.03	0.92	1.09	120
62	1.21	0.83	1.16	0.86	1.11	0.90	1.05	0.95	1.00	1.00	118
64	1.32	0.76	1.26	0.79	1.20	0.83	1.15	0.87	1.09	0.92	116
66	1.44	0.69	1.38	0.72	1.32	0.76	1.26	0.79	1.19	0.84	114
68	1.59	0.63	1.52	0.65	1.45	0.69	1.38	0.72	1.31	0.76	112
70	1.77	0.57	1.69	0.59	1.61	0.62	1.54	0.65	1.45	0.68	110
72	1.98	0.51	1.89	0.53	1.81	0.55	1.72	0.58	1.63	0.61	108
74	2.24	0.45	2.15	0.46	2.05	0.49	1.95	0.51	1.85	0.54	106
76	2.58	0.39	2.47	0.40	2.36	0.42	2.24	0.45	2.13	0.47	104
78	3.02	0.33	2.90	0.34	2.77	0.36	2.63	0.38	2.49	0.40	102
80	3.65	0.27	3.49	0.29	3.33	0.30	3.17	0.31	3.01	0.33	100
81	4.06	0.25	3.89	0.26	3.71	0.27	3.53	0.28	3.35	0.30	99
82	4.57	0.22	4.38	0.23	4.18	0.24	3.98	0.25	3.77	0.26	98
83	5.24	0.19	5.01	0.20	4.79	0.21	4.55	0.22	4.32	0.23	97
84	6.12	0.16	5.86	0.17	5.59	0.18	5.32	0.19	5.04	0.20	96
85	7.35	0.14	7.04	0.14	6.72	0.15	6.39	0.16	6.06	0.16	95
86	9.19	0.11	8.81	0.11	8.41	0.12	8.00	0.12	7.58	0.13	94
87	12.27	0.08	11.75	0.08	11.22	0.09	10.67	0.09	10.11	0.10	93
88	18.41	0.05	17.63	0.06	16.83	0.06	16.01	0.06	15.17	0.07	92
89	36.83	0.03	35.27	0.03	33.68	0.03	32.04	0.03	30.36	0.03	91
90	—	0.00	—	0.00	—	0.00	—	0.00	—	0.00	90
	50°		52°		54°		56°		58°		

Correction to latitude = f × error in longitude

Correction to longitude = F × error in latitude

TABLE 26

Latitude and Longitude Factors
f, the change of latitude for a unit change in longitude
F, the change of longitude for a unit change in latitude

Azimuth angle	Latitude										Azimuth angle
	60°		62°		64°		66°		68°		
	f	F	f	F	f	F	f	F	f	F	
°											°
0	0.00	—	0.00	—	0.00	—	0.00	—	0.00	—	180
1	0.01	114.58	0.01	122.03	0.01	130.69	0.01	140.85	0.01	152.93	179
2	0.02	57.27	0.02	61.00	0.02	65.32	0.01	70.40	0.01	76.44	178
3	0.03	38.16	0.02	40.64	0.02	43.53	0.02	46.91	0.02	50.94	177
4	0.03	28.60	0.03	30.46	0.03	32.62	0.03	35.16	0.03	38.18	176
5	0.04	22.86	0.04	24.35	0.04	26.07	0.04	28.10	0.03	30.51	175
6	0.05	19.03	0.05	20.27	0.05	21.70	0.04	23.39	0.04	25.40	174
7	0.06	16.29	0.06	17.35	0.05	18.58	0.05	20.02	0.05	21.74	173
8	0.07	14.23	0.07	15.16	0.06	16.23	0.06	17.49	0.05	18.99	172
9	0.08	12.63	0.07	13.45	0.07	14.40	0.06	15.52	0.06	16.85	171
10	0.09	11.34	0.08	12.08	0.08	12.94	0.07	13.94	0.07	15.14	170
12	0.11	9.41	0.10	10.02	0.09	10.73	0.09	11.57	0.08	12.56	168
14	0.12	8.02	0.12	8.54	0.11	9.15	0.10	9.86	0.09	10.71	166
16	0.14	6.97	0.13	7.43	0.13	7.96	0.12	8.57	0.11	9.31	164
18	0.16	6.15	0.15	6.56	0.14	7.02	0.13	7.57	0.12	8.22	162
20	0.18	5.49	0.17	5.85	0.16	6.27	0.15	6.75	0.14	7.33	160
22	0.20	4.95	0.19	5.27	0.18	5.65	0.16	6.09	0.15	6.61	158
24	0.22	4.49	0.21	4.78	0.20	5.12	0.18	5.52	0.17	6.00	156
26	0.24	4.10	0.23	4.37	0.21	4.68	0.20	5.04	0.18	5.47	154
28	0.27	3.76	0.25	4.01	0.23	4.29	0.22	4.62	0.20	5.02	152
30	0.29	3.46	0.27	3.69	0.25	3.95	0.23	4.26	0.22	4.62	150
32	0.31	3.20	0.29	3.41	0.27	3.65	0.25	3.93	0.23	4.27	148
34	0.34	2.96	0.32	3.16	0.30	3.38	0.27	3.65	0.25	3.96	146
36	0.36	2.75	0.34	2.93	0.32	3.14	0.30	3.38	0.27	3.67	144
38	0.39	2.56	0.37	2.73	0.34	2.92	0.32	3.15	0.29	3.42	142
40	0.42	2.38	0.39	2.54	0.37	2.72	0.34	2.93	0.31	3.18	140
42	0.45	2.22	0.42	2.37	0.39	2.53	0.37	2.73	0.34	2.96	138
44	0.48	2.07	0.45	2.21	0.42	2.36	0.39	2.55	0.36	2.76	136
46	0.52	1.93	0.49	2.06	0.45	2.20	0.42	2.37	0.39	2.58	134
48	0.56	1.80	0.52	1.92	0.49	2.05	0.45	2.21	0.42	2.40	132
50	0.60	1.68	0.56	1.79	0.52	1.91	0.48	2.06	0.45	2.24	130
52	0.64	1.56	0.60	1.66	0.56	1.78	0.52	1.92	0.48	2.09	128
54	0.69	1.45	0.65	1.55	0.60	1.66	0.56	1.79	0.52	1.94	126
56	0.74	1.35	0.70	1.44	0.65	1.54	0.60	1.66	0.56	1.80	124
58	0.80	1.25	0.75	1.33	0.70	1.43	0.65	1.54	0.60	1.67	122
60	0.87	1.15	0.81	1.23	0.76	1.32	0.70	1.42	0.65	1.54	120
62	0.94	1.06	0.88	1.13	0.82	1.21	0.76	1.31	0.70	1.42	118
64	1.03	0.97	0.96	1.04	0.90	1.11	0.83	1.20	0.77	1.30	116
66	1.12	0.89	1.05	0.95	0.98	1.02	0.91	1.09	0.84	1.19	114
68	1.24	0.81	1.16	0.86	1.09	0.92	1.01	0.99	0.93	1.08	112
70	1.37	0.73	1.29	0.78	1.20	0.83	1.12	0.89	1.03	0.97	110
72	1.54	0.65	1.44	0.69	1.35	0.74	1.25	0.80	1.15	0.87	108
74	1.74	0.57	1.64	0.61	1.53	0.65	1.42	0.70	1.31	0.77	106
76	2.01	0.50	1.88	0.53	1.76	0.57	1.63	0.61	1.50	0.67	104
78	2.35	0.42	2.21	0.45	2.06	0.48	1.91	0.52	1.76	0.57	102
80	2.84	0.35	2.66	0.38	2.49	0.40	2.31	0.43	2.12	0.47	100
81	3.16	0.32	2.96	0.34	2.77	0.36	2.57	0.39	2.37	0.42	99
82	3.56	0.28	3.34	0.30	3.12	0.32	2.89	0.35	2.67	0.38	98
83	4.07	0.25	3.82	0.26	3.57	0.28	3.31	0.30	3.05	0.33	97
84	4.76	0.21	4.47	0.22	4.17	0.24	3.87	0.26	3.56	0.28	96
85	5.72	0.17	5.37	0.19	5.01	0.20	4.65	0.22	4.28	0.23	95
86	7.15	0.14	6.71	0.15	6.27	0.16	5.82	0.17	5.36	0.19	94
87	9.54	0.10	8.96	0.11	8.36	0.12	7.76	0.13	7.15	0.14	93
88	14.32	0.07	13.44	0.07	12.55	0.08	11.65	0.09	10.73	0.09	92
89	28.65	0.03	26.90	0.04	25.11	0.04	23.30	0.04	21.46	0.05	91
90	—	0.00	—	0.00	—	0.00	—	0.00	—	0.00	90
	60°		62°		64°		66°		68°		

Correction to latitude = f × error in longitude | Correction to longitude = F × error in latitude

TABLE 27
Amplitudes

Latitude	Declination 0°.0	0°.5	1°.0	1°.5	2°.0	2°.5	3°.0	3°.5	4°.0	4°.5	5°.0	5°.5	6°.0	Latitude
°	°	°	°	°	°	°	°	°	°	°	°	°	°	°
0	0.0	0.5	1.0	1.5	2.0	2.5	3.0	3.5	4.0	4.5	5.0	5.5	6.0	0
10	0.0	0.5	1.0	1.5	2.0	2.5	3.0	3.6	4.1	4.6	5.1	5.6	6.1	10
15	0.0	0.5	1.0	1.6	2.1	2.6	3.1	3.6	4.1	4.7	5.2	5.7	6.2	15
20	0.0	0.5	1.1	1.6	2.1	2.7	3.2	3.7	4.3	4.8	5.3	5.9	6.4	20
25	0.0	0.6	1.1	1.7	2.2	2.8	3.3	3.9	4.4	5.0	5.5	6.1	6.6	25
30	0.0	0.6	1.2	1.7	2.3	2.9	3.5	4.0	4.6	5.2	5.8	6.4	6.9	30
32	0.0	0.6	1.2	1.8	2.4	2.9	3.5	4.1	4.7	5.3	5.9	6.5	7.1	32
34	0.0	0.6	1.2	1.8	2.4	3.0	3.6	4.2	4.8	5.4	6.0	6.6	7.2	34
36	0.0	0.6	1.2	1.9	2.5	3.1	3.7	4.3	4.9	5.6	6.2	6.8	7.4	36
38	0.0	0.6	1.3	1.9	2.5	3.2	3.8	4.4	5.1	5.7	6.4	7.0	7.6	38
40	0.0	0.7	1.3	2.0	2.6	3.3	3.9	4.6	5.2	5.9	6.5	7.2	7.8	40
42	0.0	0.7	1.3	2.0	2.7	3.4	4.0	4.7	5.4	6.1	6.7	7.4	8.1	42
44	0.0	0.7	1.4	2.1	2.8	3.5	4.2	4.9	5.6	6.3	7.0	7.7	8.4	44
46	0.0	0.7	1.4	2.2	2.9	3.6	4.3	5.0	5.8	6.5	7.2	7.9	8.7	46
48	0.0	0.7	1.5	2.2	3.0	3.7	4.5	5.2	6.0	6.7	7.5	8.2	9.0	48
50	0.0	0.8	1.6	2.3	3.1	3.9	4.7	5.4	6.2	7.0	7.8	8.6	9.4	50
51	0.0	0.8	1.6	2.4	3.2	4.0	4.8	5.6	6.4	7.2	8.0	8.8	9.6	51
52	0.0	0.8	1.6	2.4	3.2	4.1	4.9	5.7	6.5	7.3	8.1	9.0	9.8	52
53	0.0	0.8	1.7	2.5	3.3	4.2	5.0	5.8	6.7	7.5	8.3	9.2	10.0	53
54	0.0	0.9	1.7	2.6	3.4	4.3	5.1	6.0	6.8	7.7	8.5	9.4	10.2	54
55	0.0	0.9	1.7	2.6	3.5	4.4	5.2	6.1	7.0	7.9	8.7	9.6	10.5	55
56	0.0	0.9	1.8	2.7	3.6	4.5	5.4	6.3	7.2	8.1	9.0	9.9	10.8	56
57	0.0	0.9	1.8	2.8	3.7	4.6	5.5	6.4	7.4	8.3	9.2	10.1	11.1	57
58	0.0	0.9	1.9	2.8	3.8	4.7	5.7	6.6	7.6	8.5	9.5	10.4	11.4	58
59	0.0	1.0	1.9	2.9	3.9	4.9	5.8	6.8	7.8	8.8	9.7	10.7	11.7	59
60	0.0	1.0	2.0	3.0	4.0	5.0	6.0	7.0	8.0	9.0	10.0	11.1	12.1	60
61	0.0	1.0	2.1	3.1	4.1	5.2	6.2	7.2	8.3	9.3	10.3	11.4	12.5	61
62	0.0	1.1	2.1	3.2	4.3	5.3	6.4	7.5	8.5	9.6	10.7	11.8	12.9	62
63	0.0	1.1	2.2	3.3	4.4	5.5	6.6	7.7	8.8	10.0	11.1	12.2	13.3	63
64	0.0	1.1	2.3	3.4	4.6	5.7	6.9	8.0	9.2	10.3	11.5	12.6	13.8	64
65.0	0.0	1.2	2.4	3.6	4.7	5.9	7.1	8.3	9.5	10.7	11.9	13.1	14.3	65.0
65.5	0.0	1.2	2.4	3.6	4.8	6.0	7.3	8.5	9.7	10.9	12.1	13.4	14.6	65.5
66.0	0.0	1.2	2.5	3.7	4.9	6.2	7.4	8.6	9.9	11.1	12.4	13.6	14.9	66.0
66.5	0.0	1.3	2.5	3.8	5.0	6.3	7.5	8.8	10.1	11.3	12.6	13.9	15.2	66.5
67.0	0.0	1.3	2.6	3.8	5.1	6.4	7.7	9.0	10.3	11.6	12.9	14.2	15.5	67.0
67.5	0.0	1.3	2.6	3.9	5.2	6.5	7.9	9.2	10.5	11.8	13.2	14.5	15.9	67.5
68.0	0.0	1.3	2.7	4.0	5.3	6.7	8.0	9.4	10.7	12.1	13.5	14.8	16.2	68.0
68.5	0.0	1.4	2.7	4.1	5.5	6.8	8.2	9.6	11.0	12.4	13.8	15.2	16.6	68.5
69.0	0.0	1.4	2.8	4.2	5.6	7.0	8.4	9.8	11.2	12.6	14.1	15.5	17.0	69.0
69.5	0.0	1.4	2.9	4.3	5.7	7.2	8.6	10.0	11.5	12.9	14.4	15.9	17.4	69.5
70.0	0.0	1.5	2.9	4.4	5.9	7.3	8.8	10.3	11.8	13.3	14.8	16.3	17.8	70.0
70.5	0.0	1.5	3.0	4.5	6.0	7.5	9.0	10.5	12.1	13.6	15.1	16.7	18.2	70.5
71.0	0.0	1.5	3.1	4.6	6.2	7.7	9.3	10.8	12.4	13.9	15.5	17.1	18.7	71.0
71.5	0.0	1.6	3.2	4.7	6.3	7.9	9.5	11.1	12.7	14.3	15.9	17.6	19.2	71.5
72.0	0.0	1.6	3.2	4.9	6.5	8.1	9.8	11.4	13.0	14.7	16.4	18.1	19.8	72.0
72.5	0.0	1.7	3.3	5.0	6.7	8.3	10.0	11.7	13.4	15.1	16.8	18.6	20.3	72.5
73.0	0.0	1.7	3.4	5.1	6.9	8.6	10.3	12.1	13.8	15.6	17.3	19.1	20.9	73.0
73.5	0.0	1.8	3.5	5.3	7.1	8.8	10.6	12.4	14.2	16.0	17.9	19.7	21.6	73.5
74.0	0.0	1.8	3.6	5.4	7.3	9.1	10.9	12.8	14.7	16.5	18.4	20.3	22.3	74.0
74.5	0.0	1.9	3.7	5.6	7.5	9.4	11.3	13.2	15.1	17.1	19.0	21.0	23.0	74.5
75.0	0.0	1.9	3.9	5.8	7.7	9.7	11.7	13.6	15.6	17.6	19.7	21.7	23.8	75.0
75.5	0.0	2.0	4.0	6.0	8.0	10.0	12.1	14.1	16.2	18.3	20.4	22.5	24.7	75.5
76.0	0.0	2.1	4.1	6.2	8.3	10.4	12.5	14.6	16.8	18.9	21.1	23.3	25.6	76.0
76.5	0.0	2.1	4.3	6.4	8.6	10.8	13.0	15.2	17.4	19.6	21.9	24.2	26.6	76.5
77.0	0.0	2.2	4.4	6.7	8.9	11.2	13.5	15.7	18.1	20.4	22.8	25.2	27.7	77.0

TABLE 27

Amplitudes

Latitude	Declination 6.°0	6.°5	7.°0	7.°5	8.°0	8.°5	9.°0	9.°5	10.°0	10.°5	11.°0	11.°5	12.°0	Latitude
°	°	°	°	°	°	°	°	°	°	°	°	°	°	°
0	6.0	6.5	7.0	7.5	8.0	8.5	9.0	9.5	10.0	10.5	11.0	11.5	12.0	0
10	6.1	6.6	7.1	7.6	8.1	8.6	9.1	9.6	10.2	10.7	11.2	11.7	12.2	10
15	6.2	6.7	7.2	7.8	8.3	8.8	9.3	9.8	10.4	10.9	11.4	11.9	12.4	15
20	6.4	6.9	7.5	8.0	8.5	9.0	9.6	10.1	10.6	11.2	11.7	12.2	12.8	20
25	6.6	7.2	7.7	8.3	8.8	9.4	9.9	10.5	11.0	11.6	12.2	12.7	13.3	25
30	6.9	7.5	8.1	8.7	9.2	9.8	10.4	11.0	11.6	12.1	12.7	13.3	13.9	30
32	7.1	7.7	8.3	8.9	9.4	10.0	10.6	11.2	11.8	12.4	13.0	13.6	14.2	32
34	7.2	7.8	8.5	9.1	9.7	10.3	10.9	11.5	12.1	12.7	13.3	13.9	14.5	34
36	7.4	8.0	8.7	9.3	9.9	10.5	11.1	11.8	12.4	13.0	13.6	14.3	14.9	36
38	7.6	8.3	8.9	9.5	10.2	10.8	11.5	12.1	12.7	13.4	14.0	14.7	15.3	38
40	7.8	8.5	9.2	9.8	10.5	11.1	11.8	12.4	13.1	13.8	14.4	15.1	15.7	40
42	8.1	8.8	9.4	10.1	10.8	11.5	12.1	12.8	13.5	14.2	14.9	15.6	16.2	42
44	8.4	9.1	9.8	10.5	11.2	11.9	12.6	13.3	14.0	14.7	15.4	16.1	16.8	44
46	8.7	9.4	10.1	10.8	11.6	12.3	13.0	13.7	14.5	15.2	15.9	16.7	17.4	46
48	9.0	9.7	10.5	11.2	12.0	12.8	13.5	14.3	15.0	15.8	16.6	17.3	18.1	48
50	9.4	10.1	10.9	11.7	12.5	13.3	14.1	14.9	15.7	16.5	17.3	18.1	18.9	50
51	9.6	10.4	11.2	12.0	12.8	13.6	14.4	15.2	16.0	16.8	17.7	18.5	19.3	51
52	9.8	10.6	11.4	12.2	13.1	13.9	14.7	15.6	16.4	17.2	18.1	18.9	19.7	52
53	10.0	10.8	11.7	12.5	13.4	14.2	15.1	15.9	16.8	17.6	18.5	19.3	20.2	53
54	10.2	11.1	12.0	12.8	13.7	14.6	15.4	16.3	17.2	18.1	18.9	19.8	20.7	54
55	10.5	11.4	12.3	13.2	14.0	14.9	15.8	16.7	17.6	18.5	19.4	20.3	21.3	55
56	10.8	11.7	12.6	13.5	14.4	15.3	16.2	17.2	18.1	19.0	20.0	20.9	21.8	56
57	11.1	12.0	12.9	13.9	14.8	15.7	16.7	17.6	18.6	19.6	20.5	21.5	22.4	57
58	11.4	12.3	13.3	14.3	15.2	16.2	17.2	18.1	19.1	20.1	21.1	22.1	23.1	58
59	11.7	12.7	13.7	14.7	15.7	16.7	17.7	18.7	19.7	20.7	21.7	22.8	23.8	59
60	12.1	13.1	14.1	15.1	16.2	17.2	18.2	19.3	20.3	21.4	22.4	23.5	24.6	60
61	12.5	13.5	14.6	15.6	16.7	17.8	18.8	19.9	21.0	22.1	23.2	24.3	25.4	61
62	12.9	14.0	15.0	16.1	17.2	18.4	19.5	20.6	21.7	22.8	24.0	25.1	26.3	62
63	13.3	14.4	15.6	16.7	17.9	19.0	20.2	21.3	22.5	23.7	24.9	26.0	27.3	63
64	13.8	15.0	16.2	17.3	18.5	19.7	20.9	22.1	23.3	24.6	25.8	27.1	28.3	64
65.0	14.3	15.5	16.8	18.0	19.2	20.5	21.7	23.0	24.3	25.5	26.8	28.1	29.5	65.0
65.5	14.6	15.8	17.1	18.3	19.6	20.9	22.2	23.5	24.8	26.1	27.4	28.7	30.1	65.5
66.0	14.9	16.2	17.4	18.7	20.0	21.3	22.6	23.9	25.3	26.6	28.0	29.4	30.7	66.0
66.5	15.2	16.5	17.8	19.1	20.4	21.8	23.1	24.5	25.8	27.2	28.6	30.0	31.4	66.5
67.0	15.5	16.8	18.2	19.5	20.9	22.2	23.6	25.0	26.4	27.8	29.2	30.7	32.1	67.0
67.5	15.9	17.2	18.6	19.9	21.3	22.7	24.1	25.5	27.0	28.4	29.9	31.4	32.9	67.5
68.0	16.2	17.6	19.0	20.4	21.8	23.2	24.7	26.1	27.6	29.1	30.6	32.2	33.7	68.0
68.5	16.6	18.0	19.4	20.9	22.3	23.8	25.3	26.8	28.3	29.8	31.4	33.0	34.6	68.5
69.0	17.0	18.4	19.9	21.4	22.9	24.4	25.9	27.4	29.0	30.6	32.2	33.8	35.5	69.0
69.5	17.4	18.9	20.4	21.9	23.4	25.0	26.5	28.1	29.7	31.4	33.0	34.7	36.4	69.5
70.0	17.8	19.3	20.9	22.4	24.0	25.6	27.2	28.9	30.5	32.2	33.9	35.7	37.4	70.0
70.5	18.2	19.8	21.4	23.0	24.6	26.3	27.9	29.6	31.3	33.1	34.9	36.7	38.5	70.5
71.0	18.7	20.3	22.0	23.6	25.3	27.0	28.7	30.5	32.2	34.0	35.9	37.8	39.7	71.0
71.5	19.2	20.9	22.6	24.3	26.0	27.8	29.5	31.3	33.2	35.1	37.0	38.9	40.9	71.5
72.0	19.8	21.5	23.2	25.0	26.8	28.6	30.4	32.3	34.2	36.1	38.1	40.2	42.3	72.0
72.5	20.3	22.1	23.9	25.7	27.6	29.4	31.3	33.3	35.3	37.3	39.4	41.5	43.7	72.5
73.0	20.9	22.8	24.6	26.5	28.4	30.4	32.3	34.4	36.4	38.6	40.7	43.0	45.3	73.0
73.5	21.6	23.5	25.4	27.4	29.3	31.4	33.4	35.5	37.7	39.9	42.2	44.6	47.1	73.5
74.0	22.3	24.2	26.2	28.3	30.3	32.4	34.6	36.8	39.0	41.4	43.8	46.3	49.0	74.0
74.5	23.0	25.1	27.1	29.3	31.4	33.6	35.8	38.1	40.5	43.0	45.6	48.2	51.1	74.5
75.0	23.8	25.9	28.1	30.3	32.5	34.8	37.2	39.6	42.1	44.8	47.5	50.4	53.4	75.0
75.5	24.7	26.9	29.1	31.4	33.8	36.2	38.7	41.2	43.9	46.7	49.6	52.8	56.1	75.5
76.0	25.6	27.9	30.2	32.7	35.1	37.7	40.3	43.0	45.9	48.9	52.1	55.5	59.3	76.0
76.5	26.6	29.0	31.5	34.0	36.6	39.3	42.1	45.0	48.1	51.3	54.8	58.7	63.0	76.5
77.0	27.7	30.2	32.8	35.5	38.2	41.1	44.1	47.2	50.5	54.1	58.0	62.4	67.6	77.0

TABLE 27

Amplitudes

Latitude	Declination 12°0	12°5	13°0	13°5	14°0	14°5	15°0	15°5	16°0	16°5	17°0	17°5	18°0	Latitude
°	°	°	°	°	°	°	°	°	°	°	°	°	°	°
0	12.0	12.5	13.0	13.5	14.0	14.5	15.0	15.5	16.0	16.5	17.0	17.5	18.0	0
10	12.2	12.7	13.2	13.7	14.2	14.7	15.2	15.7	16.3	16.8	17.3	17.8	18.3	10
15	12.4	12.9	13.5	14.0	14.5	15.0	15.5	16.1	16.6	17.1	17.6	18.1	18.7	15
20	12.8	13.3	13.9	14.4	14.9	15.5	16.0	16.5	17.1	17.6	18.1	18.7	19.2	20
25	13.3	13.8	14.4	14.9	15.5	16.0	16.6	17.1	17.7	18.3	18.8	19.4	19.9	25
30	13.9	14.5	15.1	15.6	16.2	16.8	17.4	18.0	18.6	19.1	19.7	20.3	20.9	30
32	14.2	14.8	15.4	16.0	16.6	17.2	17.8	18.4	19.0	19.6	20.2	20.8	21.4	32
34	14.5	15.1	15.7	16.4	17.0	17.6	18.2	18.8	19.4	20.0	20.7	21.3	21.9	34
36	14.9	15.5	16.1	16.8	17.4	18.0	18.7	19.3	19.9	20.6	21.2	21.8	22.5	36
38	15.3	15.9	16.6	17.2	17.9	18.5	19.2	19.8	20.5	21.1	21.8	22.4	23.1	38
40	15.7	16.4	17.1	17.7	18.4	19.1	19.7	20.4	21.1	21.8	22.4	23.1	23.8	40
41	16.0	16.7	17.3	18.0	18.7	19.4	20.1	20.8	21.4	22.1	22.8	23.5	24.2	41
42	16.2	16.9	17.6	18.3	19.0	19.7	20.4	21.1	21.8	22.5	23.2	23.9	24.6	42
43	16.5	17.2	17.9	18.6	19.3	20.0	20.7	21.4	22.1	22.9	23.6	24.3	25.0	43
44	16.8	17.5	18.2	18.9	19.7	20.4	21.1	21.8	22.5	23.3	24.0	24.7	25.4	44
45	17.1	17.8	18.5	19.3	20.0	20.7	21.5	22.2	22.9	23.7	24.4	25.2	25.9	45
46	17.4	18.2	18.9	19.6	20.4	21.1	21.9	22.6	23.4	24.1	24.9	25.7	26.4	46
47	17.7	18.5	19.3	20.0	20.8	21.5	22.3	23.1	23.8	24.6	25.4	26.2	26.9	47
48	18.1	18.9	19.6	20.4	21.2	22.0	22.8	23.5	24.3	25.1	25.9	26.7	27.5	48
49	18.5	19.3	20.1	20.8	21.6	22.4	23.2	24.0	24.8	25.7	26.5	27.3	28.1	49
50	18.9	19.7	20.5	21.3	22.1	22.9	23.7	24.6	25.4	26.2	27.1	27.9	28.7	50
51	19.3	20.1	20.9	21.8	22.6	23.4	24.3	25.1	26.0	26.8	27.7	28.5	29.4	51
52	19.7	20.6	21.4	22.3	23.1	24.0	24.9	25.7	26.6	27.5	28.3	29.2	30.1	52
53	20.2	21.1	21.9	22.8	23.7	24.6	25.5	26.4	27.3	28.2	29.1	30.0	30.9	53
54	20.7	21.6	22.5	23.4	24.3	25.2	26.1	27.0	28.0	28.9	29.8	30.8	31.7	54
55	21.3	22.2	23.1	24.0	24.9	25.9	26.8	27.8	28.7	29.7	30.6	31.6	32.6	55
56	21.8	22.8	23.7	24.7	25.6	26.6	27.6	28.5	29.5	30.5	31.5	32.5	33.5	56
57	22.4	23.4	24.4	25.4	26.4	27.4	28.4	29.4	30.4	31.4	32.5	33.5	34.6	57
58	23.1	24.1	25.1	26.1	27.2	28.2	29.2	30.3	31.3	32.4	33.5	34.6	35.7	58
59	23.8	24.8	25.9	27.0	28.0	29.1	30.2	31.3	32.4	33.5	34.6	35.7	36.9	59
60	24.6	25.7	26.7	27.8	28.9	30.1	31.2	32.3	33.5	34.6	35.8	37.0	38.2	60
61	25.4	26.5	27.6	28.8	29.9	31.1	32.3	33.5	34.6	35.9	37.1	38.3	39.6	61
62	26.3	27.5	28.6	29.8	31.0	32.2	33.5	34.7	36.0	37.2	38.5	39.8	41.2	62
63	27.3	28.5	29.7	30.9	32.2	33.5	34.8	36.1	37.4	38.7	40.1	41.5	42.9	63
64	28.3	29.6	30.9	32.2	33.5	34.8	36.2	37.6	39.0	40.4	41.8	43.3	44.8	64
65.0	29.5	30.8	32.2	33.5	34.9	36.3	37.8	39.2	40.7	42.2	43.8	45.4	47.0	65.0
65.5	30.1	31.5	32.9	34.3	35.7	37.1	38.6	40.1	41.7	43.2	44.8	46.5	48.2	65.5
66.0	30.7	32.1	33.6	35.0	36.5	38.0	39.5	41.1	42.7	44.3	46.0	47.7	49.4	66.0
66.5	31.4	32.9	34.3	35.8	37.3	38.9	40.5	42.1	43.7	45.4	47.2	48.9	50.8	66.5
67.0	32.1	33.6	35.1	36.7	38.3	39.9	41.5	43.2	44.9	46.6	48.4	50.3	52.3	67.0
67.5	32.9	34.4	36.0	37.6	39.2	40.9	42.6	44.3	46.1	47.9	49.8	51.8	53.9	67.5
68.0	33.7	35.3	36.9	38.6	40.2	41.9	43.7	45.5	47.4	49.3	51.3	53.4	55.6	68.0
68.5	34.6	36.2	37.9	39.6	41.3	43.1	44.9	46.8	48.8	50.8	52.9	55.1	57.5	68.5
69.0	35.5	37.2	38.9	40.6	42.5	44.3	46.2	48.2	50.3	52.4	54.7	57.0	59.6	69.0
69.5	36.4	38.2	40.0	41.8	43.7	45.6	47.7	49.7	51.9	54.2	56.6	59.2	61.9	69.5
70.0	37.4	39.3	41.1	43.0	45.0	47.1	49.2	51.4	53.7	56.1	58.7	61.5	64.6	70.0
70.5	38.5	40.4	42.4	44.4	46.4	48.6	50.8	53.2	55.7	58.3	61.1	64.3	67.8	70.5
71.0	39.7	41.7	43.7	45.8	48.0	50.3	52.7	55.2	57.8	60.7	63.9	67.5	71.7	71.0
71.5	40.9	43.0	45.1	47.4	49.7	52.1	54.7	57.4	60.3	63.5	67.1	71.4	76.9	71.5
72.0	42.3	44.5	46.7	49.1	51.5	54.1	56.9	59.9	63.1	66.8	71.1	76.7	90.0	72.0
72.5	43.7	46.0	48.4	50.9	53.6	56.4	59.4	62.7	66.4	70.8	76.5	90.0		72.5
73.0	45.3	47.8	50.3	53.0	55.8	58.9	62.3	66.1	70.5	76.3	90.0			73.0
73.5	47.1	49.6	52.4	55.3	58.4	61.8	65.7	70.2	76.0	90.0				73.5
74.0	49.0	51.7	54.7	57.9	61.4	65.3	69.9	75.8	90.0					74.0
74.5	51.1	54.1	57.3	60.9	64.9	69.5	75.6	90.0						74.5

TABLE 27

Amplitudes

Latitude	Declination 18°0	18°5	19°0	19°5	20°0	20°5	21°0	21°5	22°0	22°5	23°0	23°5	24°0	Latitude
°	°	°	°	°	°	°	°	°	°	°	°	°	°	°
0	18.0	18.5	19.0	19.5	20.0	20.5	21.0	21.5	22.0	22.5	23.0	23.5	24.0	0
10	18.3	18.8	19.3	19.8	20.3	20.8	21.3	21.8	22.4	22.9	23.4	23.9	24.4	10
15	18.7	19.2	19.7	20.2	20.7	21.3	21.8	22.3	22.8	23.3	23.9	24.4	24.9	15
20	19.2	19.7	20.3	20.8	21.3	21.9	22.4	23.0	23.5	24.0	24.6	25.1	25.6	20
25	19.9	20.5	21.1	21.6	22.2	22.7	23.3	23.9	24.4	25.0	25.5	26.1	26.7	25
30	20.9	21.5	22.1	22.7	23.3	23.9	24.4	25.0	25.6	26.2	26.8	27.4	28.0	30
32	21.4	22.0	22.6	23.2	23.8	24.4	25.0	25.6	26.2	26.8	27.4	28.0	28.7	32
34	21.9	22.5	23.1	23.7	24.4	25.0	25.6	26.2	26.9	27.5	28.1	28.7	29.4	34
36	22.5	23.1	23.7	24.4	25.0	25.7	26.3	26.9	27.6	28.2	28.9	29.5	30.2	36
38	23.1	23.7	24.4	25.1	25.7	26.4	27.1	27.7	28.4	29.1	29.7	30.4	31.1	38
40	23.8	24.5	25.2	25.8	26.5	27.2	27.9	28.6	29.3	30.0	30.7	31.4	32.1	40
41	24.2	24.9	25.6	26.3	26.9	27.6	28.3	29.1	29.8	30.5	31.2	31.9	32.6	41
42	24.6	25.3	26.0	26.7	27.4	28.1	28.8	29.5	30.3	31.0	31.7	32.5	33.2	42
43	25.0	25.7	26.4	27.2	27.9	28.6	29.3	30.1	30.8	31.6	32.3	33.0	33.8	43
44	25.4	26.2	26.9	27.6	28.4	29.1	29.9	30.6	31.4	32.1	32.9	33.7	34.4	44
45	25.9	26.7	27.4	28.2	28.9	29.7	30.5	31.2	32.0	32.8	33.5	34.3	35.1	45
46	26.4	27.2	27.9	28.7	29.5	30.3	31.1	31.8	32.6	33.4	34.2	35.0	35.8	46
47	26.9	27.7	28.5	29.3	30.1	30.9	31.7	32.5	33.3	34.1	35.0	35.8	36.6	47
48	27.5	28.3	29.1	29.9	30.7	31.6	32.4	33.2	34.0	34.9	35.7	36.6	37.4	48
49	28.1	28.9	29.8	30.6	31.4	32.3	33.1	34.0	34.8	35.7	36.6	37.4	38.3	49
50	28.7	29.6	30.4	31.3	32.1	33.0	33.9	34.8	35.6	36.5	37.4	38.3	39.3	50
51	29.4	30.3	31.2	32.0	32.9	33.8	34.7	35.6	36.5	37.5	38.4	39.3	40.3	51
52	30.1	31.0	31.9	32.8	33.7	34.7	35.6	36.5	37.5	38.4	39.4	40.4	41.3	52
53	30.9	31.8	32.8	33.7	34.6	35.6	36.5	37.5	38.5	39.5	40.5	41.5	42.5	53
54	31.7	32.7	33.6	34.6	35.6	36.6	37.6	38.6	39.6	40.6	41.7	42.7	43.8	54
55	32.6	33.6	34.6	35.6	36.6	37.6	38.7	39.7	40.8	41.9	42.9	44.0	45.2	55
56	33.5	34.6	35.6	36.7	37.7	38.8	39.9	41.0	42.1	43.2	44.3	45.5	46.7	56
57	34.6	35.6	36.7	37.8	38.9	40.0	41.1	42.3	43.5	44.6	45.8	47.1	48.3	57
58	35.7	36.8	37.9	39.1	40.2	41.4	42.6	43.8	45.0	46.2	47.5	48.8	50.1	58
59	36.9	38.0	39.2	40.4	41.6	42.8	44.1	45.4	46.7	48.0	49.3	50.7	52.2	59
60.0	38.2	39.4	40.6	41.9	43.2	44.5	45.8	47.1	48.5	49.9	51.4	52.9	54.4	60.0
60.5	38.9	40.1	41.4	42.7	44.0	45.3	46.7	48.1	49.5	51.0	52.5	54.1	55.7	60.5
61.0	39.6	40.9	42.2	43.5	44.9	46.3	47.7	49.1	50.6	52.1	53.7	55.3	57.0	61.0
61.5	40.4	41.7	43.0	44.4	45.8	47.2	48.7	50.2	51.7	53.3	55.0	56.7	58.5	61.5
62.0	41.2	42.5	43.9	45.3	46.8	48.2	49.8	51.3	52.9	54.6	56.3	58.1	60.0	62.0
62.5	42.0	43.4	44.8	46.3	47.8	49.3	50.9	52.5	54.2	56.0	57.8	59.7	61.7	62.5
63.0	42.9	44.3	45.8	47.3	48.9	50.5	52.1	53.8	55.6	57.5	59.4	61.4	63.6	63.0
63.5	43.8	45.3	46.9	48.4	50.0	51.7	53.4	55.2	57.1	59.1	61.1	63.4	65.7	63.5
64.0	44.8	46.4	48.0	49.6	51.3	53.0	54.8	56.7	58.7	60.8	63.0	65.5	68.1	64.0
64.5	45.9	47.5	49.1	50.8	52.6	54.4	56.3	58.4	60.5	62.7	65.2	67.9	70.9	64.5
65.0	47.0	48.7	50.4	52.2	54.0	56.0	58.0	60.1	62.4	64.9	67.6	70.7	74.2	65.0
65.5	48.2	49.9	51.7	53.6	55.6	57.6	59.8	62.1	64.6	67.3	70.4	74.1	78.8	65.5
66.0	49.4	51.3	53.2	55.2	57.2	59.4	61.8	64.3	67.1	70.2	73.9	78.6	90.0	66.0
66.5	50.8	52.7	54.7	56.8	59.1	61.4	64.0	66.8	70.0	73.7	78.5	90.0		66.5
67.0	52.3	54.3	56.4	58.7	61.1	63.7	66.5	69.7	73.5	78.4	90.0			67.0
67.5	53.9	56.0	58.3	60.7	63.3	66.2	69.5	73.3	78.2	90.0				67.5
68.0	55.6	57.9	60.4	63.0	65.9	69.2	73.1	78.1	90.0					68.0
68.5	57.5	60.0	62.7	65.6	68.9	72.9	77.9	90.0						68.5
69.0	59.6	62.3	65.3	68.7	72.6	77.7	90.0							69.0
69.5	61.9	65.0	68.4	72.4	77.6	90.0								69.5
70.0	64.6	68.1	72.2	77.4	90.0									70.0
70.5	67.8	71.9	77.2	90.0										70.5
71.0	71.7	77.1	90.0											71.0
71.5	76.9	90.0												71.5
72.0	90.0													72.0

TABLE 28

Correction of Amplitude as Observed on the Visible Horizon

Latitude	Declination 0°	2°	4°	6°	8°	10°	12°	14°	16°	18°	20°	22°	24°	Latitude
°	°	°	°	°	°	°	°	°	°	°	°	°	°	°
0	0.0	0.0	0.0	0.0	0.0	0.0	0.0	0.0	0.0	0.0	0.0	0.0	0.0	0
10	0.1	0.1	0.1	0.1	0.1	0.1	0.1	0.1	0.1	0.1	0.1	0.1	0.1	10
15	0.2	0.2	0.2	0.2	0.2	0.2	0.2	0.2	0.2	0.2	0.2	0.2	0.2	15
20	0.3	0.3	0.3	0.3	0.3	0.3	0.3	0.3	0.3	0.3	0.3	0.3	0.3	20
25	0.3	0.3	0.3	0.3	0.3	0.4	0.3	0.3	0.3	0.3	0.3	0.3	0.3	25
30	0.4	0.4	0.4	0.4	0.5	0.4	0.4	0.4	0.4	0.4	0.4	0.5	0.5	30
32	0.4	0.4	0.4	0.4	0.5	0.4	0.4	0.4	0.4	0.4	0.5	0.5	0.5	32
34	0.5	0.5	0.5	0.5	0.5	0.5	0.5	0.5	0.5	0.5	0.5	0.5	0.5	34
36	0.5	0.5	0.5	0.5	0.5	0.5	0.5	0.5	0.6	0.5	0.6	0.6	0.6	36
38	0.6	0.6	0.6	0.6	0.6	0.6	0.6	0.6	0.6	0.6	0.6	0.6	0.6	38
40	0.6	0.6	0.6	0.6	0.6	0.6	0.6	0.6	0.6	0.6	0.7	0.7	0.7	40
42	0.6	0.6	0.6	0.6	0.7	0.7	0.7	0.7	0.7	0.7	0.7	0.7	0.7	42
44	0.7	0.7	0.7	0.6	0.6	0.7	0.7	0.7	0.8	0.8	0.8	0.8	0.9	44
46	0.7	0.7	0.7	0.7	0.7	0.8	0.8	0.8	0.8	0.8	0.8	0.9	0.9	46
48	0.8	0.8	0.8	0.8	0.8	0.8	0.8	0.8	0.9	0.9	1.0	1.0	1.0	48
50	0.8	0.8	0.8	0.8	0.9	0.9	0.9	0.9	0.9	1.0	1.0	1.1	1.0	50
51	0.8	0.8	0.8	0.8	0.9	0.9	0.9	0.9	0.9	1.0	1.1	1.1	1.1	51
52	0.9	0.9	0.9	0.9	0.9	0.9	1.0	1.0	1.0	1.1	1.1	1.1	1.3	52
53	0.9	0.9	0.9	0.9	0.9	0.9	1.0	1.0	1.0	1.1	1.2	1.2	1.3	53
54	1.0	1.0	1.0	1.0	1.0	1.0	1.1	1.1	1.1	1.2	1.2	1.3	1.3	54
55	1.0	1.0	1.0	1.0	1.1	1.1	1.0	1.2	1.2	1.2	1.3	1.3	1.4	55
56	1.0	1.0	1.0	1.0	1.1	1.1	1.2	1.2	1.2	1.3	1.3	1.4	1.5	56
57	1.1	1.1	1.1	1.1	1.1	1.1	1.2	1.2	1.3	1.3	1.4	1.5	1.7	57
58	1.1	1.1	1.1	1.1	1.2	1.2	1.2	1.2	1.4	1.4	1.5	1.6	1.8	58
59	1.2	1.2	1.2	1.2	1.2	1.2	1.3	1.3	1.3	1.4	1.6	1.7	1.9	59
60	1.2	1.2	1.2	1.2	1.2	1.3	1.3	1.4	1.4	1.5	1.7	1.9	2.2	60
61	1.3	1.3	1.3	1.3	1.3	1.3	1.4	1.5	1.6	1.7	1.8	2.0	2.4	61
62	1.3	1.3	1.3	1.3	1.4	1.4	1.5	1.6	1.6	1.7	1.9	2.3	2.6	62
63	1.3	1.4	1.4	1.4	1.4	1.5	1.5	1.6	1.7	1.9	2.1	2.5	3.3	63
64	1.4	1.4	1.4	1.5	1.5	1.6	1.7	1.7	1.8	2.1	2.3	2.9	4.3	64
65.0	1.5	1.5	1.5	1.6	1.6	1.6	1.7	1.9	2.0	2.2	2.7	3.5	7.2	65.0
65.5	1.5	1.5	1.5	1.6	1.6	1.7	1.8	1.9	2.1	2.3	2.8	3.9		65.5
66.0	1.6	1.6	1.6	1.6	1.7	1.7	1.9	2.0	2.1	2.5	3.1	4.4		66.0
66.5	1.6	1.6	1.6	1.7	1.7	1.8	1.9	2.1	2.3	2.6	3.3	5.4		66.5
67.0	1.7	1.7	1.7	1.7	1.7	1.8	2.0	2.1	2.3	2.8	3.6	7.5		67.0
67.5	1.7	1.7	1.7	1.7	1.8	1.9	2.0	2.2	2.5	2.9	4.1			67.5
68.0	1.7	1.8	1.8	1.8	1.9	2.0	2.1	2.3	2.6	3.2	4.7			68.0
68.5	1.8	1.8	1.8	1.8	2.0	2.0	2.2	2.4	2.8	3.5	5.7			68.5
69.0	1.8	1.8	1.9	1.9	1.9	2.1	2.2	2.5	2.9	3.8	7.9			69.0
69.5	1.9	1.9	1.9	1.9	2.1	2.2	2.4	2.6	3.2	4.3				69.5
70.0	1.9	1.9	1.9	2.0	2.1	2.3	2.5	2.8	3.4	5.0				70.0
70.5	2.0	2.0	2.0	2.2	2.2	2.4	2.6	3.0	3.6	6.0				70.5
71.0	2.0	2.0	2.1	2.2	2.3	2.5	2.7	3.1	4.1	8.3				71.0
71.5	2.1	2.1	2.2	2.3	2.4	2.5	2.9	3.3	4.6					71.5
72.0	2.2	2.2	2.3	2.3	2.4	2.6	3.0	3.6	5.3					72.0
72.5	2.2	2.2	2.3	2.4	2.5	2.7	3.2	3.9	6.4					72.5
73.0	2.3	2.3	2.4	2.5	2.7	2.9	3.4	4.4	8.9					73.0
73.5	2.4	2.4	2.5	2.6	2.8	3.0	3.6	4.9						73.5
74.0	2.4	2.4	2.5	2.7	2.9	3.3	3.8	5.6						74.0
74.5	2.5	2.6	2.7	2.8	3.0	3.4	4.2	6.8						74.5
75.0	2.6	2.7	2.8	2.9	3.2	3.7	4.7	9.3						75.0
75.5	2.7	2.8	2.8	3.0	3.3	3.9	5.3							75.5
76.0	2.8	2.8	2.9	3.2	3.5	4.2	5.6							76.0
76.5	2.9	3.0	3.1	3.3	3.7	4.5	7.3							76.5
77.0	3.0	3.1	3.2	3.5	4.0	5.1	10.2							77.0

For the **sun, a planet,** or a **star,** apply the correction to the observed amplitude in the direction **away** from the elevated pole. For the **moon** apply **half** the correction **toward** the elevated pole.

TABLE 29

Altitude Factor

a, the change of altitude in one minute from meridian transit; used for entering table 30

Latitude	Declination **same** name as latitude, **upper** transit: **add** correction to observed altitude												Latitude
	0°	1°	2°	3°	4°	5°	6°	7°	8°	9°	10°	11°	
°	"	"	"	"	"	"	"	"	"	"	"	"	°
0					28. 1	22. 4	18. 7	16. 0	14. 0	12. 4	11. 1	10. 1	0
1						28. 0	22. 4	18. 6	16. 0	13. 9	12. 4	11. 1	1
2							28. 0	22. 3	18. 6	15. 9	13. 9	12. 3	2
3								27. 9	22. 3	18. 5	15. 8	13. 8	3
4	28. 1								27. 8	22. 2	18. 5	15. 8	4
5	22. 4	28. 0								27. 7	22. 1	18. 4	5
6	18. 7	22. 4	28. 0								27. 6	22. 0	6
7	16. 0	18. 6	22. 3	27. 9								27. 4	7
8	14. 0	16. 0	18. 6	22. 3	27. 8								8
9	12. 4	13. 9	15. 9	18. 5	22. 2	27. 7							9
10	11. 1	12. 4	13. 9	15. 8	18. 5	22. 1	27. 6						10
11	10. 1	11. 1	12. 3	13. 8	15. 8	18. 4	22. 0	27. 4					11
12	9. 2	10. 1	11. 1	12. 3	13. 8	15. 7	18. 3	21. 9	27. 3				12
13	8. 5	9. 2	10. 0	11. 0	12. 2	13. 7	15. 6	18. 2	21. 7	27. 1			13
14	7. 9	8. 5	9. 2	10. 0	10. 9	12. 1	13. 6	15. 5	18. 0	21. 6	26. 9		14
15	7. 3	7. 8	8. 4	9. 1	9. 9	10. 9	12. 1	13. 5	15. 4	17. 9	21. 4	26. 7	15
16	6. 8	7. 3	7. 8	8. 4	9. 1	9. 8	10. 8	12. 0	13. 4	15. 3	17. 8	21. 3	16
17	6. 4	6. 8	7. 2	7. 8	8. 3	9. 0	9. 8	10. 7	11. 9	13. 3	15. 2	17. 6	17
18	6. 0	6. 4	6. 8	7. 2	7. 7	8. 3	8. 9	9. 7	10. 6	11. 8	13. 2	15. 0	18
19	5. 7	6. 0	6. 3	6. 7	7. 2	7. 6	8. 2	8. 9	9. 6	10. 6	11. 7	13. 1	19
20	5. 4	5. 7	6. 0	6. 3	6. 7	7. 1	7. 6	8. 1	8. 8	9. 5	10. 5	11. 6	20
21	5. 1	5. 4	5. 6	5. 9	6. 3	6. 6	7. 0	7. 5	8. 1	8. 7	9. 5	10. 4	21
22	4. 9	5. 1	5. 3	5. 6	5. 9	6. 2	6. 6	7. 0	7. 5	8. 0	8. 6	9. 4	22
23	4. 6	4. 8	5. 0	5. 3	5. 5	5. 8	6. 1	6. 5	6. 9	7. 4	7. 9	8. 5	23
24	4. 4	4. 6	4. 8	5. 0	5. 2	5. 5	5. 8	6. 1	6. 4	6. 8	7. 3	7. 8	24
25	4. 2	4. 4	4. 6	4. 7	5. 0	5. 2	5. 4	5. 7	6. 0	6. 4	6. 8	7. 2	25
26	4. 0	4. 2	4. 3	4. 5	4. 7	4. 9	5. 1	5. 4	5. 7	6. 0	6. 3	6. 7	26
27	3. 9	4. 0	4. 1	4. 3	4. 5	4. 7	4. 9	5. 1	5. 3	5. 6	5. 9	6. 2	27
28	3. 7	3. 8	4. 0	4. 1	4. 3	4. 4	4. 6	4. 8	5. 0	5. 3	5. 5	5. 8	28
29	3. 5	3. 7	3. 8	3. 9	4. 1	4. 2	4. 4	4. 6	4. 7	5. 0	5. 2	5. 5	29
30	3. 4	3. 5	3. 6	3. 7	3. 9	4. 0	4. 2	4. 3	4. 5	4. 7	4. 9	5. 1	30
31	3. 3	3. 4	3. 5	3. 6	3. 7	3. 8	4. 0	4. 1	4. 3	4. 4	4. 6	4. 8	31
32	3. 1	3. 2	3. 3	3. 4	3. 5	3. 7	3. 8	3. 9	4. 1	4. 2	4. 4	4. 6	32
33	3. 0	3. 1	3. 2	3. 3	3. 4	3. 5	3. 6	3. 7	3. 9	4. 0	4. 2	4. 3	33
34	2. 9	3. 0	3. 1	3. 2	3. 2	3. 3	3. 4	3. 6	3. 7	3. 8	3. 9	4. 1	34
35	2. 8	2. 9	3. 0	3. 0	3. 1	3. 2	3. 3	3. 4	3. 5	3. 6	3. 7	3. 9	35
36	2. 7	2. 8	2. 8	2. 9	3. 0	3. 1	3. 2	3. 3	3. 4	3. 5	3. 6	3. 7	36
37	2. 6	2. 7	2. 7	2. 8	2. 9	2. 9	3. 0	3. 1	3. 2	3. 3	3. 4	3. 5	37
38	2. 5	2. 6	2. 6	2. 7	2. 8	2. 8	2. 9	3. 0	3. 0	3. 2	3. 2	3. 3	38
39	2. 4	2. 5	2. 5	2. 6	2. 7	2. 7	2. 8	2. 9	2. 9	3. 0	3. 1	3. 2	39
40	2. 3	2. 4	2. 4	2. 5	2. 6	2. 6	2. 7	2. 7	2. 8	2. 9	3. 0	3. 0	40
41	2. 3	2. 3	2. 4	2. 4	2. 5	2. 5	2. 6	2. 6	2. 7	2. 8	2. 8	2. 9	41
42	2. 2	2. 2	2. 3	2. 3	2. 4	2. 4	2. 5	2. 5	2. 6	2. 6	2. 7	2. 8	42
43	2. 1	2. 1	2. 2	2. 2	2. 3	2. 3	2. 4	2. 4	2. 5	2. 5	2. 6	2. 7	43
44	2. 0	2. 1	2. 1	2. 1	2. 2	2. 2	2. 3	2. 3	2. 4	2. 4	2. 5	2. 5	44
45	2. 0	2. 0	2. 0	2. 1	2. 1	2. 2	2. 2	2. 2	2. 3	2. 3	2. 4	2. 4	45
46	1. 9	1. 9	2. 0	2. 0	2. 0	2. 1	2. 1	2. 2	2. 2	2. 2	2. 3	2. 3	46
47	1. 8	1. 9	1. 9	1. 9	2. 0	2. 0	2. 0	2. 1	2. 1	2. 1	2. 2	2. 2	47
48	1. 8	1. 8	1. 8	1. 9	1. 9	1. 9	2. 0	2. 0	2. 0	2. 1	2. 1	2. 1	48
49	1. 7	1. 7	1. 8	1. 8	1. 8	1. 8	1. 9	1. 9	1. 9	2. 0	2. 0	2. 1	49
50	1. 6	1. 7	1. 7	1. 7	1. 8	1. 8	1. 8	1. 8	1. 9	1. 9	1. 9	2. 0	50
51	1. 6	1. 6	1. 6	1. 7	1. 7	1. 7	1. 7	1. 8	1. 8	1. 8	1. 9	1. 9	51
52	1. 5	1. 6	1. 6	1. 6	1. 6	1. 6	1. 7	1. 7	1. 7	1. 8	1. 8	1. 8	52
53	1. 5	1. 5	1. 5	1. 5	1. 6	1. 6	1. 6	1. 6	1. 7	1. 7	1. 7	1. 7	53
54	1. 4	1. 4	1. 5	1. 5	1. 5	1. 5	1. 5	1. 6	1. 6	1. 6	1. 6	1. 7	54
55	1. 4	1. 4	1. 4	1. 4	1. 5	1. 5	1. 5	1. 5	1. 5	1. 6	1. 6	1. 6	55
56	1. 3	1. 3	1. 4	1. 4	1. 4	1. 4	1. 4	1. 4	1. 5	1. 5	1. 5	1. 5	56
57	1. 3	1. 3	1. 3	1. 3	1. 3	1. 4	1. 4	1. 4	1. 4	1. 4	1. 4	1. 5	57
58	1. 2	1. 2	1. 3	1. 3	1. 3	1. 3	1. 3	1. 3	1. 3	1. 4	1. 4	1. 4	58
59	1. 2	1. 2	1. 2	1. 2	1. 2	1. 3	1. 3	1. 3	1. 3	1. 3	1. 3	1. 3	59
60	1. 1	1. 1	1. 2	1. 2	1. 2	1. 2	1. 2	1. 2	1. 2	1. 2	1. 3	1. 3	60
Latitude	0°	1°	2°	3°	4°	5°	6°	7°	8°	9°	10°	11°	Latitude
	Declination **same** name as latitude, **upper** transit: **add** correction to observed altitude												

TABLE 29

Altitude Factor

a, the change of altitude in one minute from meridian transit; used for entering table 30

Latitude	Declination **contrary** name to latitude, **upper** transit: **add** correction to observed altitude												Latitude
	0°	1°	2°	3°	4°	5°	6°	7°	8°	9°	10°	11°	
°	″	″	″	″	″	″	″	″	″	″	″	″	°
0					28.1	22.4	18.7	16.0	14.0	12.4	11.1	10.1	0
1				28.1	22.4	18.7	16.0	14.0	12.4	11.2	10.1	9.3	1
2			28.1	22.4	18.7	16.0	14.0	12.5	11.2	10.2	9.3	8.6	2
3		28.1	22.4	18.7	16.0	14.0	12.5	11.2	10.2	9.3	8.6	8.0	3
4	28.1	22.4	18.7	16.0	14.0	12.5	11.2	10.2	9.3	8.6	8.0	7.4	4
5	22.4	18.7	16.0	14.0	12.5	11.2	10.2	9.3	8.6	8.0	7.4	7.0	5
6	18.7	16.0	14.0	12.5	11.2	10.2	9.3	8.6	8.0	7.5	7.0	6.6	6
7	16.0	14.0	12.4	11.2	10.2	9.3	8.6	8.0	7.5	7.0	6.6	6.2	7
8	14.0	12.4	11.2	10.2	9.3	8.6	8.0	7.5	7.0	6.6	6.2	5.9	8
9	12.4	11.2	10.2	9.3	8.6	8.0	7.5	7.0	6.6	6.2	5.9	5.6	9
10	11.1	10.1	9.3	8.6	8.0	7.4	7.0	6.6	6.2	5.9	5.6	5.3	10
11	10.1	9.3	8.6	8.0	7.4	7.0	6.6	6.2	5.9	5.6	5.3	5.1	11
12	9.2	8.5	7.9	7.4	7.0	6.5	6.2	5.9	5.6	5.3	5.0	4.8	12
13	8.5	7.9	7.4	6.9	6.5	6.2	5.8	5.6	5.3	5.0	4.8	4.6	13
14	7.9	7.4	6.9	6.5	6.2	5.8	5.5	5.3	5.0	4.8	4.6	4.4	14
15	7.3	6.9	6.5	6.1	5.8	5.5	5.3	5.0	4.8	4.6	4.4	4.2	15
16	6.8	6.5	6.1	5.8	5.5	5.2	5.0	4.8	4.6	4.4	4.2	4.1	16
17	6.4	6.1	5.8	5.5	5.2	5.0	4.8	4.6	4.4	4.2	4.1	3.9	17
18	6.0	5.7	5.5	5.2	5.0	4.8	4.6	4.4	4.2	4.1	3.9	3.8	18
19	5.7	5.4	5.2	4.9	4.7	4.5	4.4	4.2	4.0	3.9	3.8	3.6	19
20	5.4	5.1	4.9	4.7	4.5	4.3	4.2	4.0	3.9	3.8	3.6	3.5	20
21	5.1	4.9	4.7	4.5	4.3	4.2	4.0	3.9	3.7	3.6	3.5	3.4	21
22	4.9	4.7	4.5	4.3	4.1	4.0	3.9	3.7	3.6	3.5	3.4	3.3	22
23	4.6	4.4	4.3	4.1	4.0	3.8	3.7	3.6	3.5	3.4	3.3	3.2	23
24	4.4	4.2	4.1	3.9	3.8	3.7	3.6	3.5	3.4	3.3	3.2	3.1	24
25	4.2	4.1	3.9	3.8	3.7	3.5	3.4	3.3	3.2	3.1	3.1	3.0	25
26	4.0	3.9	3.8	3.6	3.5	3.4	3.3	3.2	3.1	3.0	3.0	2.9	26
27	3.9	3.7	3.6	3.5	3.4	3.3	3.2	3.1	3.0	2.9	2.9	2.8	27
28	3.7	3.6	3.5	3.4	3.3	3.2	3.1	3.0	2.9	2.8	2.8	2.7	28
29	3.5	3.4	3.3	3.2	3.1	3.1	3.0	2.9	2.8	2.8	2.7	2.6	29
30	3.4	3.3	3.2	3.1	3.0	3.0	2.9	2.8	2.7	2.7	2.6	2.5	30
31	3.3	3.2	3.1	3.0	2.9	2.9	2.8	2.7	2.6	2.6	2.5	2.5	31
32	3.2	3.1	3.0	2.9	2.8	2.8	2.7	2.6	2.6	2.5	2.5	2.4	32
33	3.0	2.9	2.9	2.8	2.7	2.7	2.6	2.5	2.5	2.4	2.4	2.3	33
34	2.9	2.8	2.8	2.7	2.6	2.6	2.5	2.5	2.4	2.4	2.3	2.3	34
35	2.8	2.7	2.7	2.6	2.5	2.5	2.4	2.4	2.3	2.3	2.2	2.2	35
36	2.7	2.6	2.6	2.5	2.5	2.4	2.4	2.3	2.3	2.2	2.2	2.1	36
37	2.6	2.5	2.5	2.4	2.4	2.3	2.3	2.2	2.2	2.2	2.1	2.1	37
38	2.5	2.5	2.4	2.4	2.3	2.3	2.2	2.2	2.1	2.1	2.1	2.0	38
39	2.4	2.4	2.3	2.3	2.2	2.2	2.1	2.1	2.1	2.0	2.0	2.0	39
40	2.3	2.3	2.2	2.2	2.2	2.1	2.1	2.0	2.0	2.0	1.9	1.9	40
41	2.3	2.2	2.2	2.1	2.1	2.1	2.0	2.0	1.9	1.9	1.9	1.8	41
42	2.2	2.1	2.1	2.1	2.0	2.0	2.0	1.9	1.9	1.9	1.8	1.8	42
43	2.1	2.1	2.0	2.0	2.0	1.9	1.9	1.9	1.8	1.8	1.8	1.7	43
44	2.0	2.0	2.0	1.9	1.9	1.9	1.8	1.8	1.8	1.7	1.7	1.7	44
45	2.0	1.9	1.9	1.9	1.8	1.8	1.8	1.7	1.7	1.7	1.7	1.6	45
46	1.9	1.9	1.8	1.8	1.8	1.7	1.7	1.7	1.7	1.6	1.6	1.6	46
47	1.8	1.8	1.8	1.7	1.7	1.7	1.7	1.6	1.6	1.6	1.6	1.6	47
48	1.8	1.7	1.7	1.7	1.7	1.6	1.6	1.6	1.6	1.6	1.5	1.5	48
49	1.7	1.7	1.7	1.6	1.6	1.6	1.6	1.5	1.5	1.5	1.5	1.5	49
50	1.6	1.6	1.6	1.6	1.6	1.5	1.5	1.5	1.5	1.5	1.4	1.4	50
51	1.6	1.6	1.6	1.5	1.5	1.5	1.5	1.5	1.4	1.4	1.4	1.4	51
52	1.5	1.5	1.5	1.5	1.5	1.4	1.4	1.4	1.4	1.4	1.4	1.3	52
53	1.5	1.5	1.4	1.4	1.4	1.4	1.4	1.4	1.3	1.3	1.3	1.3	53
54	1.4	1.4	1.4	1.4	1.4	1.3	1.3	1.3	1.3	1.3	1.3	1.3	54
55	1.4	1.4	1.3	1.3	1.3	1.3	1.3	1.3	1.3	1.2	1.2	1.2	55
56	1.3	1.3	1.3	1.3	1.3	1.3	1.2	1.2	1.2	1.2	1.2	1.2	56
57	1.3	1.3	1.3	1.2	1.2	1.2	1.2	1.2	1.2	1.2	1.1	1.1	57
58	1.2	1.2	1.2	1.2	1.2	1.2	1.2	1.1	1.1	1.1	1.1	1.1	58
59	1.2	1.2	1.2	1.2	1.1	1.1	1.1	1.1	1.1	1.1	1.1	1.1	59
60	1.1	1.1	1.1	1.1	1.1	1.1	1.1	1.1	1.0	1.0	1.0	1.0	60
Latitude	0°	1°	2°	3°	4°	5°	6°	7°	8°	9°	10°	11°	Latitude
	Declination **contrary** name to latitude, **upper** transit: **add** correction to observed altitude												

TABLE 29

Altitude Factor

a, the change of altitude in one minute from meridian transit; used for entering table 30

Latitude	Declination **same** name as latitude, **upper** transit: **add** correction to observed altitude													Latitude
	12°	13°	14°	15°	16°	17°	18°	19°	20°	21°	22°	23°	24°	
°	″	″	″	″	″	″	″	″	″	″	″	″	″	°
0	9.2	8.5	7.9	7.3	6.8	6.4	6.0	5.7	5.4	5.1	4.9	4.6	4.4	0
1	10.1	9.2	8.5	7.8	7.3	6.8	6.4	6.0	5.7	5.4	5.1	4.8	4.6	1
2	11.1	10.0	9.2	8.4	7.8	7.2	6.8	6.3	6.0	5.6	5.3	5.0	4.8	2
3	12.3	11.0	10.0	9.1	8.4	7.8	7.2	6.7	6.3	5.9	5.6	5.3	5.0	3
4	13.8	12.2	10.9	9.9	9.1	8.3	7.7	7.2	6.7	6.3	5.9	5.5	5.2	4
5	15.7	13.7	12.1	10.9	9.8	9.0	8.3	7.6	7.1	6.6	6.2	5.8	5.5	5
6	18.3	15.6	13.6	12.1	10.8	9.8	8.9	8.2	7.6	7.0	6.6	6.1	5.8	6
7	21.9	18.2	15.5	13.5	12.0	10.7	9.7	8.9	8.1	7.5	7.0	6.5	6.1	7
8	27.3	21.7	18.0	15.4	13.4	11.9	10.6	9.6	8.8	8.1	7.5	6.9	6.4	8
9		27.1	21.6	17.9	15.3	13.3	11.8	10.6	9.5	8.7	8.0	7.4	6.8	9
10			26.9	21.4	17.8	15.2	13.2	11.7	10.5	9.5	8.6	7.9	7.3	10
11				26.7	21.3	17.6	15.0	13.1	11.6	10.4	9.4	8.5	7.8	11
12					26.5	21.1	17.5	14.9	13.0	11.5	10.3	9.3	8.4	12
13						26.2	20.9	17.3	14.8	12.8	11.3	10.1	9.2	13
14							26.0	20.7	17.1	14.6	12.7	11.2	10.0	14
15								25.7	20.4	16.9	14.4	12.5	11.1	15
16	26.5								25.4	20.2	16.7	14.3	12.4	16
17	21.1	26.2								25.1	20.0	16.5	14.1	17
18	17.5	20.9	26.0								24.8	19.7	16.3	18
19	14.9	17.3	20.7	25.7								24.5	19.5	19
20	13.0	14.8	17.1	20.4	25.4								24.2	20
21	11.5	12.8	14.6	16.9	20.2	25.1								21
22	10.3	11.3	12.7	14.4	16.7	20.0	24.8							22
23	9.3	10.1	11.2	12.5	14.3	16.5	19.7	24.5						23
24	8.4	9.2	10.0	11.1	12.4	14.1	16.3	19.5	24.2					24
25	7.7	8.3	9.0	9.9	10.9	12.2	13.9	16.1	19.2	23.8				25
26	7.1	7.6	8.2	8.9	9.8	10.8	12.1	13.7	15.9	18.9	23.5			26
27	6.6	7.0	7.5	8.1	8.8	9.6	10.6	11.9	13.5	15.6	18.6	23.1		27
28	6.2	6.5	7.0	7.4	8.0	8.7	9.5	10.5	11.7	13.3	15.4	18.3	22.7	28
29	5.7	6.1	6.4	6.9	7.3	7.9	8.6	9.4	10.3	11.5	13.1	15.1	18.0	29
30	5.4	5.7	6.0	6.4	6.8	7.2	7.8	8.4	9.2	10.1	11.3	12.8	14.9	30
31	5.1	5.3	5.6	5.9	6.3	6.7	7.1	7.7	8.3	9.0	10.0	11.1	12.6	31
32	4.8	5.0	5.2	5.5	5.8	6.2	6.5	7.0	7.5	8.1	8.9	9.8	10.9	32
33	4.5	4.7	4.9	5.1	5.4	5.7	6.1	6.4	6.9	7.4	8.0	8.7	9.6	33
34	4.3	4.4	4.6	4.8	5.1	5.3	5.6	5.9	6.3	6.8	7.3	7.8	8.6	34
35	4.0	4.2	4.4	4.5	4.7	5.0	5.2	5.5	5.8	6.2	6.6	7.1	7.7	35
36	3.8	4.0	4.1	4.3	4.5	4.7	4.9	5.1	5.4	5.7	6.1	6.5	7.0	36
37	3.6	3.8	3.9	4.0	4.2	4.4	4.6	4.8	5.0	5.3	5.6	6.0	6.4	37
38	3.4	3.6	3.7	3.8	4.0	4.1	4.3	4.5	4.7	4.9	5.2	5.5	5.8	38
39	3.3	3.4	3.5	3.6	3.8	3.9	4.0	4.2	4.4	4.6	4.8	5.1	5.4	39
40	3.1	3.2	3.3	3.4	3.6	3.7	3.8	4.0	4.1	4.3	4.5	4.7	5.0	40
41	3.0	3.1	3.2	3.3	3.4	3.5	3.6	3.7	3.9	4.0	4.2	4.4	4.6	41
42	2.9	2.9	3.0	3.1	3.2	3.3	3.4	3.5	3.7	3.8	4.0	4.1	4.3	42
43	2.7	2.8	2.9	3.0	3.0	3.1	3.2	3.3	3.5	3.6	3.7	3.9	4.0	43
44	2.6	2.7	2.7	2.8	2.9	3.0	3.1	3.2	3.3	3.4	3.5	3.6	3.8	44
45	2.5	2.6	2.6	2.7	2.8	2.8	2.9	3.0	3.1	3.2	3.3	3.4	3.5	45
46	2.4	2.4	2.5	2.6	2.6	2.7	2.8	2.8	2.9	3.0	3.1	3.2	3.3	46
47	2.3	2.3	2.4	2.4	2.5	2.6	2.6	2.7	2.8	2.9	2.9	3.0	3.1	47
48	2.2	2.2	2.3	2.3	2.4	2.4	2.5	2.6	2.6	2.7	2.8	2.9	3.0	48
49	2.1	2.1	2.2	2.2	2.3	2.3	2.4	2.4	2.5	2.6	2.6	2.7	2.8	49
50	2.0	2.0	2.1	2.1	2.2	2.2	2.3	2.3	2.4	2.4	2.5	2.6	2.6	50
51	1.9	2.0	2.0	2.0	2.1	2.1	2.2	2.2	2.3	2.3	2.4	2.4	2.5	51
52	1.8	1.9	1.9	1.9	2.0	2.0	2.1	2.1	2.1	2.2	2.2	2.3	2.4	52
53	1.8	1.8	1.8	1.9	1.9	1.9	2.0	2.0	2.0	2.1	2.1	2.2	2.2	53
54	1.7	1.7	1.7	1.8	1.8	1.8	1.9	1.9	1.9	2.0	2.0	2.1	2.1	54
55	1.6	1.6	1.7	1.7	1.7	1.8	1.8	1.8	1.9	1.9	1.9	2.0	2.0	55
56	1.5	1.6	1.6	1.6	1.6	1.7	1.7	1.7	1.8	1.8	1.8	1.9	1.9	56
57	1.5	1.5	1.5	1.5	1.6	1.6	1.6	1.6	1.7	1.7	1.7	1.8	1.8	57
58	1.4	1.4	1.5	1.5	1.5	1.5	1.5	1.6	1.6	1.6	1.6	1.7	1.7	58
59	1.4	1.4	1.4	1.4	1.4	1.5	1.5	1.5	1.5	1.5	1.6	1.6	1.6	59
60	1.3	1.3	1.3	1.3	1.4	1.4	1.4	1.4	1.4	1.5	1.5	1.5	1.5	60
Latitude	12°	13°	14°	15°	16°	17°	18°	19°	20°	21°	22°	23°	24°	Latitude
	Declination **same** name as latitude, **upper** transit: **add** correction to observed altitude													

TABLE 29

Altitude Factor

a, the change of altitude in one minute from meridian transit; used for entering table 30

Latitude	Declination contrary name to latitude, upper transit: add correction to observed altitude													Latitude
	12°	13°	14°	15°	16°	17°	18°	19°	20°	21°	22°	23°	24°	
°	"	"	"	"	"	"	"	"	"	"	"	"	"	°
0	9.2	8.5	7.9	7.3	6.8	6.4	6.0	5.7	5.4	5.1	4.9	4.6	4.4	0
1	8.5	7.9	7.4	6.9	6.5	6.1	5.7	5.4	5.1	4.9	4.7	4.4	4.2	1
2	7.9	7.4	6.9	6.5	6.1	5.8	5.5	5.2	4.9	4.7	4.5	4.3	4.1	2
3	7.4	6.9	6.5	6.1	5.8	5.5	5.2	4.9	4.7	4.5	4.3	4.1	3.9	3
4	7.0	6.5	6.2	5.8	5.5	5.2	5.0	4.7	4.5	4.3	4.1	4.0	3.8	4
5	6.5	6.2	5.8	5.5	5.2	5.0	4.8	4.5	4.3	4.2	4.0	3.8	3.7	5
6	6.2	5.8	5.5	5.3	5.0	4.8	4.6	4.4	4.2	4.0	3.9	3.7	3.6	6
7	5.9	5.6	5.3	5.0	4.8	4.6	4.4	4.2	4.0	3.9	3.7	3.6	3.5	7
8	5.6	5.3	5.0	4.8	4.6	4.4	4.2	4.0	3.9	3.7	3.6	3.5	3.4	8
9	5.3	5.0	4.8	4.6	4.4	4.2	4.1	3.9	3.8	3.6	3.5	3.4	3.3	9
10	5.0	4.8	4.6	4.4	4.2	4.1	3.9	3.8	3.6	3.5	3.4	3.3	3.2	10
11	4.8	4.6	4.4	4.2	4.1	3.9	3.8	3.6	3.5	3.4	3.3	3.2	3.1	11
12	4.6	4.4	4.3	4.1	3.9	3.8	3.7	3.5	3.4	3.3	3.2	3.1	3.0	12
13	4.4	4.3	4.1	3.9	3.8	3.7	3.5	3.4	3.3	3.2	3.1	3.0	2.9	13
14	4.2	4.1	3.9	3.8	3.7	3.5	3.4	3.3	3.2	3.1	3.0	2.9	2.8	14
15	4.1	3.9	3.8	3.7	3.5	3.4	3.3	3.2	3.1	3.0	2.9	2.8	2.8	15
16	3.9	3.8	3.7	3.5	3.4	3.3	3.2	3.1	3.0	2.9	2.8	2.8	2.7	16
17	3.8	3.7	3.5	3.4	3.3	3.2	3.1	3.0	2.9	2.8	2.8	2.7	2.6	17
18	3.7	3.5	3.4	3.3	3.2	3.1	3.0	2.9	2.9	2.8	2.7	2.6	2.5	18
19	3.5	3.4	3.3	3.2	3.1	3.0	2.9	2.9	2.8	2.7	2.6	2.6	2.5	19
20	3.4	3.3	3.2	3.1	3.0	2.9	2.9	2.8	2.7	2.6	2.6	2.5	2.4	20
21	3.3	3.2	3.1	3.0	2.9	2.8	2.8	2.7	2.6	2.6	2.5	2.4	2.4	21
22	3.2	3.1	3.0	2.9	2.8	2.8	2.7	2.6	2.6	2.5	2.4	2.4	2.3	22
23	3.1	3.0	2.9	2.8	2.8	2.7	2.6	2.6	2.5	2.4	2.4	2.3	2.3	23
24	3.0	2.9	2.8	2.8	2.7	2.6	2.5	2.5	2.4	2.4	2.3	2.3	2.2	24
25	2.9	2.8	2.7	2.7	2.6	2.5	2.5	2.4	2.4	2.3	2.3	2.2	2.2	25
26	2.8	2.7	2.7	2.6	2.5	2.5	2.4	2.4	2.3	2.3	2.2	2.1	2.1	26
27	2.7	2.7	2.6	2.5	2.5	2.4	2.4	2.3	2.2	2.2	2.1	2.1	2.1	27
28	2.6	2.6	2.5	2.5	2.4	2.3	2.3	2.2	2.2	2.1	2.1	2.1	2.0	28
29	2.6	2.5	2.4	2.4	2.3	2.3	2.2	2.2	2.1	2.1	2.0	2.0	2.0	29
30	2.5	2.4	2.4	2.3	2.3	2.2	2.2	2.1	2.1	2.0	2.0	2.0	1.9	30
31	2.4	2.4	2.3	2.3	2.2	2.2	2.1	2.1	2.0	2.0	2.0	1.9	1.9	31
32	2.3	2.3	2.2	2.2	2.2	2.1	2.1	2.0	2.0	1.9	1.9	1.9	1.8	32
33	2.3	2.2	2.2	2.1	2.1	2.1	2.0	2.0	1.9	1.9	1.9	1.8	1.8	33
34	2.2	2.2	2.1	2.1	2.0	2.0	2.0	1.9	1.9	1.9	1.8	1.8	1.8	34
35	2.2	2.1	2.1	2.0	2.0	2.0	1.9	1.9	1.8	1.8	1.8	1.7	1.7	35
36	2.1	2.1	2.0	2.0	1.9	1.9	1.9	1.8	1.8	1.8	1.7	1.7	1.7	36
37	2.0	2.0	2.0	1.9	1.9	1.9	1.8	1.8	1.8	1.7	1.7	1.7	1.6	37
38	2.0	1.9	1.9	1.9	1.8	1.8	1.8	1.8	1.7	1.7	1.7	1.6	1.6	38
39	1.9	1.9	1.9	1.8	1.8	1.8	1.7	1.7	1.7	1.6	1.6	1.6	1.6	39
40	1.9	1.8	1.8	1.8	1.7	1.7	1.7	1.7	1.6	1.6	1.6	1.6	1.5	40
41	1.8	1.8	1.8	1.7	1.7	1.7	1.6	1.6	1.6	1.6	1.5	1.5	1.5	41
42	1.8	1.7	1.7	1.7	1.7	1.6	1.6	1.6	1.6	1.5	1.5	1.5	1.5	42
43	1.7	1.7	1.7	1.6	1.6	1.6	1.6	1.5	1.5	1.5	1.5	1.4	1.4	43
44	1.7	1.6	1.6	1.6	1.6	1.5	1.5	1.5	1.5	1.5	1.4	1.4	1.4	44
45	1.6	1.6	1.6	1.5	1.5	1.5	1.5	1.5	1.4	1.4	1.4	1.4	1.4	45
46	1.6	1.6	1.5	1.5	1.5	1.5	1.4	1.4	1.4	1.4	1.4	1.3	1.3	46
47	1.5	1.5	1.5	1.5	1.4	1.4	1.4	1.4	1.4	1.3	1.3	1.3	1.3	47
48	1.5	1.5	1.4	1.4	1.4	1.4	1.4	1.4	1.3	1.3	1.3	1.3	1.3	48
49	1.4	1.4	1.4	1.4	1.4	1.3	1.3	1.3	1.3	1.3	1.3	1.2	1.2	49
50	1.4	1.4	1.4	1.3	1.3	1.3	1.3	1.3	1.3	1.3	1.2	1.2	1.2	50
51	1.4	1.3	1.3	1.3	1.3	1.3	1.3	1.2	1.2	1.2	1.2	1.2	1.2	51
52	1.3	1.3	1.3	1.3	1.3	1.3	1.2	1.2	1.2	1.2	1.2	1.1	1.1	52
53	1.3	1.3	1.3	1.2	1.2	1.2	1.2	1.2	1.2	1.2	1.1	1.1	1.1	53
54	1.2	1.2	1.2	1.2	1.2	1.2	1.2	1.1	1.1	1.1	1.1	1.1	1.1	54
55	1.2	1.2	1.2	1.2	1.1	1.1	1.1	1.1	1.1	1.1	1.1	1.1	1.1	55
56	1.2	1.1	1.1	1.1	1.1	1.1	1.1	1.1	1.1	1.1	1.0	1.0	1.0	56
57	1.1	1.1	1.1	1.1	1.1	1.1	1.1	1.0	1.0	1.0	1.0	1.0	1.0	57
58	1.1	1.1	1.1	1.1	1.0	1.0	1.0	1.0	1.0	1.0	1.0	1.0	1.0	58
59	1.1	1.0	1.0	1.0	1.0	1.0	1.0	1.0	1.0	1.0	1.0	0.9	0.9	59
60	1.0	1.0	1.0	1.0	1.0	1.0	1.0	0.9	0.9	0.9	0.9	0.9	0.9	60
Latitude	12°	13°	14°	15°	16°	17°	18°	19°	20°	21°	22°	23°	24°	Latitude
	Declination contrary name to latitude, upper transit: add correction to observed altitude													

TABLE 29

Altitude Factor

a, the change of altitude in one minute from meridian transit; used for entering table 30

Declination **same** name as latitude, **upper** transit: **add** correction to observed altitude

Latitude	25°	26°	27°	28°	29°	30°	31°	32°	33°	34°	35°	36°	37°	Latitude
°	″	″	″	″	″	″	″	″	″	″	″	″	″	°
0	4.2	4.0	3.9	3.7	3.5	3.4	3.3	3.1	3.0	2.9	2.8	2.7	2.6	0
1	4.4	4.2	4.0	3.8	3.7	3.5	3.4	3.2	3.1	3.0	2.9	2.8	2.7	1
2	4.6	4.3	4.1	4.0	3.8	3.6	3.5	3.3	3.2	3.1	3.0	2.8	2.7	2
3	4.7	4.5	4.3	4.1	3.9	3.7	3.6	3.4	3.3	3.2	3.0	2.9	2.8	3
4	5.0	4.7	4.5	4.3	4.1	3.9	3.7	3.5	3.4	3.3	3.1	3.0	2.9	4
5	5.2	4.9	4.7	4.4	4.2	4.0	3.8	3.7	3.5	3.3	3.2	3.1	3.0	5
6	5.4	5.1	4.9	4.6	4.4	4.2	4.0	3.8	3.6	3.5	3.3	3.2	3.0	6
7	5.7	5.4	5.1	4.8	4.6	4.3	4.1	3.9	3.7	3.6	3.4	3.3	3.1	7
8	6.0	5.7	5.3	5.0	4.8	4.5	4.3	4.1	3.9	3.7	3.5	3.4	3.2	8
9	6.4	6.0	5.6	5.3	5.0	4.7	4.4	4.2	4.0	3.8	3.6	3.5	3.3	9
10	6.8	6.3	5.9	5.5	5.2	4.9	4.6	4.4	4.2	3.9	3.8	3.6	3.4	10
11	7.2	6.7	6.2	5.8	5.5	5.1	4.8	4.6	4.3	4.1	3.9	3.7	3.5	11
12	7.7	7.1	6.6	6.2	5.8	5.4	5.1	4.8	4.5	4.3	4.0	3.8	3.6	12
13	8.3	7.6	7.1	6.5	6.1	5.7	5.3	5.0	4.7	4.4	4.2	4.0	3.8	13
14	9.1	8.2	7.6	7.0	6.4	6.0	5.6	5.2	4.9	4.6	4.4	4.1	3.9	14
15	9.9	8.9	8.1	7.4	6.9	6.4	5.9	5.5	5.2	4.8	4.5	4.3	4.0	15
16	10.9	9.8	8.8	8.0	7.3	6.8	6.3	5.8	5.4	5.1	4.8	4.5	4.2	16
17	12.2	10.8	9.6	8.7	7.9	7.2	6.7	6.2	5.7	5.3	5.0	4.7	4.4	17
18	13.9	12.1	10.6	9.5	8.6	7.8	7.1	6.6	6.1	5.6	5.2	4.9	4.6	18
19	16.1	13.7	11.9	10.5	9.4	8.4	7.7	7.0	6.4	6.0	5.5	5.1	4.8	19
20	19.2	15.9	13.5	11.7	10.3	9.2	8.3	7.5	6.9	6.3	5.8	5.4	5.0	20
21	23.8	18.9	15.6	13.3	11.5	10.2	9.1	8.2	7.4	6.8	6.2	5.7	5.3	21
22		23.5	18.6	15.4	13.1	11.3	10.0	8.9	8.0	7.3	6.6	6.1	5.6	22
23			23.1	18.3	15.1	12.8	11.1	9.8	8.7	7.9	7.1	6.5	6.0	23
24				22.7	18.0	14.9	12.6	10.9	9.6	8.6	7.7	7.0	6.4	24
25					22.3	17.7	14.6	12.4	10.7	9.4	8.4	7.5	6.8	25
26						21.9	17.4	14.3	12.1	10.5	9.2	8.2	7.4	26
27							21.5	17.0	14.0	11.9	10.3	9.1	8.1	27
28								21.1	16.7	13.8	11.7	10.1	8.9	28
29	22.3								20.6	16.3	13.5	11.4	9.9	29
30	17.7	21.9								20.2	16.0	13.2	11.1	30
31	14.6	17.4	21.5								19.8	15.6	12.9	31
32	12.4	14.3	17.0	21.1								19.3	15.3	32
33	10.7	12.1	14.0	16.7	20.6								18.9	33
34	9.4	10.5	11.9	13.8	16.3	20.2								34
35	8.4	9.2	10.3	11.7	13.5	16.0	19.8							35
36	7.5	8.2	9.1	10.1	11.4	13.2	15.6	19.3						36
37	6.8	7.4	8.1	8.9	9.9	11.1	12.9	15.3	18.9					37
38	6.2	6.7	7.2	7.9	8.7	9.6	10.9	12.6	14.9	18.4				38
39	5.7	6.1	6.5	7.1	7.7	8.5	9.4	10.6	12.2	14.5	17.9			39
40	5.3	5.6	6.0	6.4	6.9	7.5	8.2	9.2	10.4	11.9	14.1	17.4		40
41	4.9	5.2	5.5	5.8	6.2	6.7	7.3	8.0	8.9	10.1	11.6	13.8	17.0	41
42	4.5	4.8	5.0	5.3	5.7	6.1	6.6	7.1	7.8	8.7	9.8	11.3	13.4	42
43	4.2	4.4	4.6	4.9	5.2	5.5	5.9	6.4	6.9	7.6	8.5	9.5	11.0	43
44	3.9	4.1	4.3	4.5	4.8	5.1	5.4	5.8	6.2	6.7	7.4	8.2	9.3	44
45	3.7	3.8	4.0	4.2	4.4	4.7	4.9	5.2	5.6	6.0	6.6	7.2	8.0	45
46	3.5	3.6	3.7	3.9	4.1	4.3	4.5	4.8	5.1	5.4	5.9	6.4	7.0	46
47	3.3	3.4	3.5	3.6	3.8	4.0	4.2	4.4	4.6	4.9	5.3	5.7	6.2	47
48	3.1	3.2	3.3	3.4	3.5	3.7	3.9	4.0	4.3	4.5	4.8	5.1	5.5	48
49	2.9	3.0	3.1	3.2	3.3	3.4	3.6	3.7	3.9	4.1	4.4	4.6	5.0	49
50	2.7	2.8	2.9	3.0	3.1	3.2	3.3	3.5	3.6	3.8	4.0	4.2	4.5	50
51	2.6	2.6	2.7	2.8	2.9	3.0	3.1	3.2	3.4	3.5	3.7	3.9	4.1	51
52	2.4	2.5	2.6	2.6	2.7	2.8	2.9	3.0	3.1	3.2	3.4	3.6	3.7	52
53	2.3	2.3	2.4	2.5	2.5	2.6	2.7	2.8	2.9	3.0	3.1	3.3	3.4	53
54	2.2	2.2	2.3	2.3	2.4	2.5	2.5	2.6	2.7	2.8	2.9	3.0	3.2	54
55	2.0	2.1	2.1	2.2	2.3	2.3	2.4	2.4	2.5	2.6	2.7	2.8	2.9	55
56	1.9	2.0	2.0	2.1	2.1	2.2	2.2	2.3	2.4	2.4	2.5	2.6	2.7	56
57	1.8	1.9	1.9	2.0	2.0	2.0	2.1	2.2	2.2	2.3	2.3	2.4	2.5	57
58	1.7	1.8	1.8	1.8	1.9	1.9	2.0	2.0	2.1	2.1	2.2	2.3	2.3	58
59	1.6	1.7	1.7	1.7	1.8	1.8	1.9	1.9	1.9	2.0	2.0	2.1	2.2	59
60	1.6	1.6	1.6	1.6	1.7	1.7	1.7	1.8	1.8	1.9	1.9	2.0	2.0	60
Latitude	25°	26°	27°	28°	29°	30°	31°	32°	33°	34°	35°	36°	37°	Latitude

Declination **same** name as latitude, **upper** transit: **add** correction to observed altitude

TABLE 29

Altitude Factor

a, the change of altitude in one minute from meridian transit; used for entering table 30

Lati-tude	Declination **contrary** name to latitude, **upper** transit: **add** correction to observed altitude													Lati-tude
	25°	26°	27°	28°	29°	30°	31°	32°	33°	34°	35°	36°	37°	
°	"	"	"	"	"	"	"	"	"	"	"	"	"	°
0	4.2	4.0	3.9	3.7	3.5	3.4	3.3	3.1	3.0	2.9	2.8	2.7	2.6	0
1	4.1	3.9	3.7	3.6	3.4	3.3	3.2	3.1	2.9	2.8	2.7	2.6	2.6	1
2	3.9	3.8	3.6	3.5	3.3	3.2	3.1	3.0	2.9	2.8	2.7	2.6	2.5	2
3	3.8	3.6	3.5	3.4	3.2	3.1	3.0	2.9	2.8	2.7	2.6	2.5	2.4	3
4	3.7	3.5	3.4	3.3	3.2	3.0	2.9	2.8	2.7	2.6	2.6	2.5	2.4	4
5	3.6	3.4	3.3	3.2	3.1	3.0	2.9	2.8	2.7	2.6	2.5	2.4	2.3	5
6	3.4	3.3	3.2	3.1	3.0	2.9	2.8	2.7	2.6	2.5	2.4	2.4	2.3	6
7	3.3	3.2	3.1	3.0	2.9	2.8	2.7	2.6	2.5	2.5	2.4	2.3	2.2	7
8	3.2	3.1	3.0	2.9	2.8	2.7	2.7	2.6	2.5	2.4	2.3	2.3	2.2	8
9	3.1	3.0	2.9	2.9	2.8	2.7	2.6	2.5	2.4	2.4	2.3	2.2	2.2	9
10	3.1	3.0	2.9	2.8	2.7	2.6	2.5	2.5	2.4	2.3	2.2	2.2	2.1	10
11	3.0	2.9	2.8	2.7	2.6	2.5	2.5	2.4	2.3	2.3	2.2	2.1	2.1	11
12	2.9	2.8	2.7	2.6	2.6	2.5	2.4	2.3	2.3	2.2	2.2	2.1	2.0	12
13	2.8	2.7	2.7	2.6	2.5	2.4	2.4	2.3	2.2	2.2	2.1	2.1	2.0	13
14	2.7	2.7	2.6	2.5	2.4	2.4	2.3	2.3	2.2	2.1	2.1	2.0	2.0	14
15	2.7	2.6	2.5	2.5	2.4	2.3	2.3	2.2	2.1	2.1	2.0	2.0	1.9	15
16	2.6	2.5	2.5	2.4	2.3	2.3	2.2	2.2	2.1	2.0	2.0	1.9	1.9	16
17	2.5	2.5	2.4	2.3	2.3	2.2	2.2	2.1	2.1	2.0	2.0	1.9	1.9	17
18	2.5	2.4	2.4	2.3	2.2	2.2	2.1	2.1	2.0	2.0	1.9	1.9	1.8	18
19	2.4	2.4	2.3	2.2	2.2	2.1	2.1	2.0	2.0	1.9	1.9	1.8	1.8	19
20	2.4	2.3	2.3	2.2	2.1	2.1	2.0	2.0	1.9	1.9	1.9	1.8	1.8	20
21	2.3	2.3	2.2	2.1	2.1	2.0	2.0	2.0	1.9	1.9	1.8	1.8	1.7	21
22	2.3	2.2	2.2	2.1	2.1	2.0	2.0	1.9	1.9	1.8	1.8	1.7	1.7	22
23	2.2	2.2	2.1	2.1	2.0	2.0	1.9	1.9	1.8	1.8	1.8	1.7	1.7	23
24	2.2	2.1	2.1	2.0	2.0	1.9	1.9	1.8	1.8	1.8	1.7	1.7	1.6	24
25	2.1	2.1	2.0	2.0	1.9	1.9	1.8	1.8	1.8	1.7	1.7	1.6	1.6	25
26	2.1	2.0	2.0	1.9	1.9	1.9	1.8	1.8	1.7	1.7	1.7	1.6	1.6	26
27	2.0	2.0	1.9	1.9	1.9	1.8	1.8	1.7	1.7	1.7	1.6	1.6	1.6	27
28	2.0	1.9	1.9	1.9	1.8	1.8	1.7	1.7	1.7	1.6	1.6	1.6	1.5	28
29	1.9	1.9	1.9	1.8	1.8	1.7	1.7	1.7	1.6	1.6	1.6	1.5	1.5	29
30	1.9	1.8	1.8	1.8	1.7	1.7	1.7	1.6	1.6	1.6	1.5	1.5	1.5	30
31	1.8	1.8	1.8	1.7	1.7	1.7	1.6	1.6	1.6	1.5	1.5	1.5	1.5	31
32	1.8	1.8	1.7	1.7	1.7	1.6	1.6	1.6	1.5	1.5	1.5	1.5	1.4	32
33	1.8	1.7	1.7	1.7	1.6	1.6	1.6	1.5	1.5	1.5	1.5	1.4	1.4	33
34	1.7	1.7	1.7	1.6	1.6	1.6	1.5	1.5	1.5	1.5	1.4	1.4	1.4	34
35	1.7	1.7	1.6	1.6	1.6	1.5	1.5	1.5	1.5	1.4	1.4	1.4	1.4	35
36	1.6	1.6	1.6	1.6	1.5	1.5	1.5	1.5	1.4	1.4	1.4	1.4	1.3	36
37	1.6	1.6	1.6	1.5	1.5	1.5	1.5	1.4	1.4	1.4	1.4	1.3	1.3	37
38	1.6	1.5	1.5	1.5	1.5	1.5	1.4	1.4	1.4	1.4	1.3	1.3	1.3	38
39	1.5	1.5	1.5	1.5	1.4	1.4	1.4	1.4	1.4	1.3	1.3	1.3	1.3	39
40	1.5	1.5	1.5	1.4	1.4	1.4	1.4	1.3	1.3	1.3	1.3	1.3	1.2	40
41	1.5	1.4	1.4	1.4	1.4	1.4	1.3	1.3	1.3	1.3	1.3	1.2	1.2	41
42	1.4	1.4	1.4	1.4	1.4	1.3	1.3	1.3	1.3	1.2	1.2	1.2	1.2	42
43	1.4	1.4	1.4	1.3	1.3	1.3	1.3	1.3	1.2	1.2	1.2	1.2	1.2	43
44	1.4	1.4	1.3	1.3	1.3	1.3	1.3	1.2	1.2	1.2	1.2	1.2	1.2	44
45	1.3	1.3	1.3	1.3	1.3	1.2	1.2	1.2	1.2	1.2	1.2	1.1	1.1	45
46	1.3	1.3	1.3	1.3	1.2	1.2	1.2	1.2	1.2	1.2	1.1	1.1	1.1	46
47	1.3	1.3	1.2	1.2	1.2	1.2	1.2	1.2	1.1	1.1	1.1	1.1	1.1	47
48	1.2	1.2	1.2	1.2	1.2	1.2	1.1	1.1	1.1	1.1	1.1	1.1		48
49	1.2	1.2	1.2	1.2	1.2	1.1	1.1	1.1	1.1	1.1	1.1			49
50	1.2	1.2	1.2	1.1	1.1	1.1	1.1	1.1	1.1	1.1				50
51	1.2	1.1	1.1	1.1	1.1	1.1	1.1	1.1	1.0					51
52	1.1	1.1	1.1	1.1	1.1	1.1	1.0	1.0						52
53	1.1	1.1	1.1	1.1	1.0	1.0	1.0							53
54	1.1	1.0	1.0	1.0	1.0	1.0								54
55	1.0	1.0	1.0	1.0	1.0									55
56	1.0	1.0	1.0	1.0										56
57	1.0	1.0	1.0											57
58	1.0	0.9												58
59	0.9												0.8	59
60												0.8	0.8	60
Lati-tude	25°	26°	27°	28°	29°	30°	31°	32°	33°	34°	35°	36°	37°	Lati-tude
	Declination **same** name as latitude, **lower** transit: **subtract** correction from observed altitude													

TABLE 29

Altitude Factor

a, the change of altitude in one minute from meridian transit; used for entering table 30

Declination **same** name as latitude, **upper** transit: **add** correction to observed altitude

Latitude	38°	39°	40°	41°	42°	43°	44°	45°	46°	47°	48°	49°	50°	Latitude
°	"	"	"	"	"	"	"	"	"	"	"	"	"	°
0	2.5	2.4	2.3	2.3	2.2	2.1	2.0	2.0	1.9	1.8	1.8	1.7	1.7	0
1	2.6	2.5	2.4	2.3	2.2	2.2	2.1	2.0	1.9	1.9	1.8	1.7	1.7	1
2	2.6	2.5	2.4	2.4	2.3	2.2	2.1	2.0	2.0	1.9	1.8	1.8	1.7	2
3	2.7	2.6	2.5	2.4	2.3	2.2	2.2	2.1	2.0	1.9	1.9	1.8	1.7	3
4	2.8	2.7	2.6	2.5	2.4	2.3	2.2	2.1	2.0	2.0	1.9	1.8	1.8	4
5	2.8	2.7	2.6	2.5	2.4	2.3	2.2	2.2	2.1	2.0	1.9	1.9	1.8	5
6	2.9	2.8	2.7	2.6	2.5	2.4	2.3	2.2	2.1	2.0	2.0	1.9	1.8	6
7	3.0	2.9	2.7	2.6	2.5	2.4	2.3	2.2	2.2	2.1	2.0	1.9	1.8	7
8	3.1	2.9	2.8	2.7	2.6	2.5	2.4	2.3	2.2	2.1	2.0	1.9	1.9	8
9	3.2	3.0	2.9	2.8	2.7	2.5	2.4	2.3	2.2	2.2	2.1	2.0	1.9	9
10	3.3	3.1	3.0	2.8	2.7	2.6	2.5	2.4	2.3	2.2	2.1	2.0	1.9	10
11	3.4	3.2	3.1	2.9	2.8	2.7	2.6	2.4	2.3	2.2	2.1	2.1	2.0	11
12	3.5	3.3	3.1	3.0	2.9	2.7	2.6	2.5	2.4	2.3	2.2	2.1	2.0	12
13	3.6	3.4	3.2	3.1	2.9	2.8	2.7	2.6	2.4	2.3	2.2	2.1	2.0	13
14	3.7	3.5	3.3	3.2	3.0	2.9	2.7	2.6	2.5	2.4	2.3	2.2	2.1	14
15	3.8	3.6	3.4	3.3	3.1	3.0	2.8	2.7	2.6	2.4	2.3	2.2	2.1	15
16	4.0	3.8	3.6	3.4	3.2	3.0	2.9	2.8	2.6	2.5	2.4	2.3	2.2	16
17	4.1	3.9	3.7	3.5	3.3	3.1	3.0	2.8	2.7	2.6	2.4	2.3	2.2	17
18	4.3	4.1	3.8	3.6	3.4	3.2	3.1	2.9	2.8	2.6	2.5	2.4	2.3	18
19	4.5	4.2	4.0	3.7	3.5	3.3	3.2	3.0	2.8	2.7	2.6	2.4	2.3	19
20	4.7	4.4	4.1	3.9	3.7	3.5	3.3	3.1	2.9	2.8	2.6	2.5	2.4	20
21	4.9	4.6	4.3	4.0	3.8	3.6	3.4	3.2	3.0	2.9	2.7	2.6	2.4	21
22	5.2	4.8	4.5	4.2	4.0	3.7	3.5	3.3	3.1	2.9	2.8	2.6	2.5	22
23	5.5	5.1	4.7	4.4	4.1	3.9	3.6	3.4	3.2	3.0	2.9	2.7	2.6	23
24	5.8	5.4	5.0	4.6	4.3	4.0	3.8	3.5	3.3	3.1	3.0	2.8	2.6	24
25	6.2	5.7	5.3	4.9	4.5	4.2	3.9	3.7	3.5	3.3	3.1	2.9	2.7	25
26	6.7	6.1	5.6	5.2	4.8	4.4	4.1	3.8	3.6	3.4	3.2	3.0	2.8	26
27	7.2	6.5	6.0	5.5	5.0	4.6	4.3	4.0	3.7	3.5	3.3	3.1	2.9	27
28	7.9	7.1	6.4	5.8	5.3	4.9	4.5	4.2	3.9	3.6	3.4	3.2	3.0	28
29	8.7	7.7	6.9	6.2	5.7	5.2	4.8	4.4	4.1	3.8	3.5	3.3	3.1	29
30	9.6	8.5	7.5	6.7	6.1	5.5	5.1	4.7	4.3	4.0	3.7	3.4	3.2	30
31	10.9	9.4	8.2	7.3	6.6	5.9	5.4	4.9	4.5	4.2	3.9	3.6	3.3	31
32	12.6	10.6	9.2	8.0	7.1	6.4	5.8	5.2	4.8	4.4	4.0	3.7	3.5	32
33	14.9	12.2	10.4	8.9	7.8	6.9	6.2	5.6	5.1	4.6	4.3	3.9	3.6	33
34	18.4	14.5	11.9	10.1	8.7	7.6	6.7	6.0	5.4	4.9	4.5	4.1	3.8	34
35		17.9	14.1	11.6	9.8	8.5	7.4	6.6	5.9	5.3	4.8	4.4	4.0	35
36			17.4	13.8	11.3	9.5	8.2	7.2	6.4	5.7	5.1	4.6	4.2	36
37				17.0	13.4	11.0	9.3	8.0	7.0	6.2	5.5	5.0	4.5	37
38					16.5	13.0	10.7	9.0	7.7	6.8	6.0	5.3	4.8	38
39						16.0	12.6	10.3	8.7	7.5	6.5	5.8	5.1	39
40							15.5	12.2	10.0	8.4	7.2	6.3	5.6	40
41								15.0	11.8	9.7	8.1	7.0	6.1	41
42	16.5								14.5	11.4	9.3	7.9	6.7	42
43	13.0	16.0								14.0	11.0	9.0	7.6	43
44	10.7	12.6	15.5								13.6	10.6	8.7	44
45	9.0	10.3	12.2	15.0								13.1	10.2	45
46	7.7	8.7	10.0	11.8	14.5								12.6	46
47	6.8	7.5	8.4	9.7	11.4	14.0								47
48	6.0	6.5	7.2	8.1	9.3	11.0	13.6							48
49	5.3	5.8	6.3	7.0	7.9	9.0	10.6	13.1						49
50	4.8	5.1	5.6	6.1	6.7	7.6	8.7	10.2	12.6					50
51	4.3	4.6	5.0	5.4	5.9	6.5	7.3	8.4	9.9	12.1				51
52	3.9	4.2	4.5	4.8	5.2	5.7	6.3	7.0	8.0	9.5	11.6			52
53	3.6	3.8	4.0	4.3	4.6	5.0	5.4	6.0	6.7	7.7	9.1	11.1		53
54	3.3	3.5	3.7	3.9	4.1	4.4	4.8	5.2	5.8	6.5	7.4	8.7	10.6	54
55	3.0	3.2	3.3	3.5	3.7	4.0	4.3	4.6	5.0	5.5	6.2	7.1	8.3	55
56	2.8	2.9	3.1	3.2	3.4	3.6	3.8	4.1	4.4	4.8	5.3	5.9	6.8	56
57	2.6	2.7	2.8	2.9	3.1	3.2	3.4	3.6	3.9	4.2	4.6	5.0	5.6	57
58	2.4	2.5	2.6	2.7	2.8	2.9	3.1	3.3	3.5	3.7	4.0	4.4	4.8	58
59	2.2	2.3	2.4	2.5	2.6	2.7	2.8	3.0	3.1	3.3	3.6	3.8	4.2	59
60	2.1	2.1	2.2	2.3	2.4	2.5	2.6	2.7	2.8	3.0	3.2	3.4	3.6	60
Latitude	38°	39°	40°	41°	42°	43°	44°	45°	46°	47°	48°	49°	50°	Latitude

Declination **same** name as latitude, **upper** transit: **add** correction to observed altitude

TABLE 29

Altitude Factor

a, the change of altitude in one minute from meridian transit; used for entering table 30

Lati-tude	Declination **contrary** name to latitude, **upper** transit: **add** correction to observed altitude													Lati-tude
	38°	39°	40°	41°	42°	43°	44°	45°	46°	47°	48°	49°	50°	
°	"	"	"	"	"	"	"	"	"	"	"	"	"	°
0	2.5	2.4	2.3	2.3	2.2	2.1	2.0	2.0	1.9	1.8	1.8	1.7	1.7	0
1	2.5	2.4	2.3	2.2	2.1	2.1	2.0	1.9	1.9	1.8	1.7	1.7	1.6	1
2	2.4	2.3	2.3	2.2	2.1	2.0	2.0	1.9	1.8	1.8	1.7	1.7	1.6	2
3	2.4	2.3	2.2	2.1	2.1	2.0	1.9	1.9	1.8	1.8	1.7	1.6	1.6	3
4	2.3	2.2	2.2	2.1	2.0	2.0	1.9	1.8	1.8	1.7	1.7	1.6	1.6	4
5	2.3	2.2	2.1	2.1	2.0	1.9	1.9	1.8	1.8	1.7	1.6	1.6	1.5	5
6	2.2	2.2	2.1	2.0	2.0	1.9	1.8	1.8	1.7	1.7	1.6	1.6	1.5	6
7	2.2	2.1	2.0	2.0	1.9	1.9	1.8	1.8	1.7	1.6	1.6	1.5	1.5	7
8	2.1	2.1	2.0	1.9	1.9	1.8	1.8	1.7	1.7	1.6	1.6	1.5	1.5	8
9	2.1	2.0	2.0	1.9	1.9	1.8	1.8	1.7	1.6	1.6	1.6	1.5	1.5	9
10	2.1	2.0	1.9	1.9	1.8	1.8	1.7	1.7	1.6	1.6	1.5	1.5	1.4	10
11	2.0	2.0	1.9	1.8	1.8	1.7	1.7	1.6	1.6	1.6	1.5	1.5	1.4	11
12	2.0	1.9	1.9	1.8	1.8	1.7	1.7	1.6	1.6	1.5	1.5	1.4	1.4	12
13	1.9	1.9	1.8	1.8	1.7	1.7	1.6	1.6	1.6	1.5	1.5	1.4	1.4	13
14	1.9	1.9	1.8	1.8	1.7	1.7	1.6	1.6	1.5	1.5	1.4	1.4	1.4	14
15	1.9	1.8	1.8	1.7	1.7	1.6	1.6	1.6	1.5	1.5	1.4	1.4	1.4	15
16	1.8	1.8	1.7	1.7	1.7	1.6	1.6	1.5	1.5	1.4	1.4	1.4	1.3	16
17	1.8	1.8	1.7	1.7	1.6	1.6	1.5	1.5	1.5	1.4	1.4	1.4	1.3	17
18	1.8	1.7	1.7	1.6	1.6	1.6	1.5	1.5	1.4	1.4	1.4	1.3	1.3	18
19	1.7	1.7	1.7	1.6	1.6	1.5	1.5	1.5	1.4	1.4	1.4	1.3	1.3	19
20	1.7	1.7	1.6	1.6	1.6	1.5	1.5	1.4	1.4	1.4	1.3	1.3	1.3	20
21	1.7	1.6	1.6	1.6	1.5	1.5	1.5	1.4	1.4	1.4	1.3	1.3	1.3	21
22	1.7	1.6	1.6	1.5	1.5	1.5	1.4	1.4	1.4	1.3	1.3	1.3	1.2	22
23	1.6	1.6	1.6	1.5	1.5	1.4	1.4	1.4	1.3	1.3	1.3	1.3	1.2	23
24	1.6	1.6	1.5	1.5	1.5	1.4	1.4	1.4	1.3	1.3	1.3	1.2	1.2	24
25	1.6	1.5	1.5	1.5	1.4	1.4	1.4	1.3	1.3	1.3	1.2	1.2	1.2	25
26	1.6	1.5	1.5	1.5	1.4	1.4	1.4	1.3	1.3	1.3	1.2	1.2	1.2	26
27	1.5	1.5	1.5	1.4	1.4	1.4	1.3	1.3	1.3	1.2	1.2	1.2	1.2	27
28	1.5	1.5	1.4	1.4	1.4	1.3	1.3	1.3	1.3	1.2	1.2	1.2	1.1	28
29	1.5	1.4	1.4	1.4	1.4	1.3	1.3	1.3	1.2	1.2	1.2	1.2	1.1	29
30	1.5	1.4	1.4	1.4	1.3	1.3	1.3	1.2	1.2	1.2	1.2	1.1	1.1	30
31	1.4	1.4	1.4	1.3	1.3	1.3	1.3	1.2	1.2	1.2	1.2	1.1	1.1	31
32	1.4	1.4	1.3	1.3	1.3	1.3	1.2	1.2	1.2	1.2	1.1	1.1	1.1	32
33	1.4	1.4	1.3	1.3	1.3	1.2	1.2	1.2	1.2	1.1	1.1	1.1	1.1	33
34	1.4	1.3	1.3	1.3	1.3	1.2	1.2	1.2	1.2	1.1	1.1	1.1	1.1	34
35	1.3	1.3	1.3	1.3	1.2	1.2	1.2	1.2	1.1	1.1	1.1	1.1		35
36	1.3	1.3	1.3	1.2	1.2	1.2	1.2	1.1	1.1	1.1	1.1			36
37	1.3	1.3	1.2	1.2	1.2	1.2	1.2	1.1	1.1	1.1				37
38	1.3	1.2	1.2	1.2	1.2	1.2	1.1	1.1	1.1					38
39	1.2	1.2	1.2	1.2	1.2	1.1	1.1	1.1						39
40	1.2	1.2	1.2	1.2	1.1	1.1	1.1							40
41	1.2	1.2	1.2	1.1	1.1	1.1								41
42	1.2	1.2	1.1	1.1	1.1									42
43	1.2	1.1	1.1	1.1										43
44	1.1	1.1	1.1											44
45	1.1	1.1												45
46	1.1												0.9	46
47												0.9	0.9	47
48											0.9	0.9	0.9	48
49										0.9	0.9	0.9	0.8	49
50									0.9	0.9	0.9	0.8	0.8	50
51								0.9	0.9	0.9	0.8	0.8	0.8	51
52							0.9	0.9	0.9	0.8	0.8	0.8	0.8	52
53						0.9	0.9	0.8	0.8	0.8	0.8	0.8	0.8	53
54					0.9	0.9	0.8	0.8	0.8	0.8	0.8	0.8	0.8	54
55				0.9	0.8	0.8	0.8	0.8	0.8	0.8	0.8	0.8	0.7	55
56			0.8	0.8	0.8	0.8	0.8	0.8	0.8	0.8	0.8	0.7	0.7	56
57		0.8	0.8	0.8	0.8	0.8	0.8	0.8	0.8	0.8	0.7	0.7	0.7	57
58	0.8	0.8	0.8	0.8	0.8	0.8	0.8	0.8	0.8	0.7	0.7	0.7	0.7	58
59	0.8	0.8	0.8	0.8	0.8	0.8	0.8	0.7	0.7	0.7	0.7	0.7	0.7	59
60	0.8	0.8	0.8	0.8	0.8	0.7	0.7	0.7	0.7	0.7	0.7	0.7	0.7	60
Lati-tude	38°	39°	40°	41°	42°	43°	44°	45°	46°	47°	48°	49°	50°	Lati-tude

Declination **same** name as latitude, **lower** transit: **subtract** correction from observed altitude

TABLE 29

Altitude Factor

a, the change of altitude in one minute from meridian transit; used for entering table 30

Declination **same** name as latitude, **upper** transit: **add** correction to observed altitude

Latitude	51°	52°	53°	54°	55°	56°	57°	58°	59°	60°	61°	62°	63°	Latitude
°	″	″	″	″	″	″	″	″	″	″	″	″	″	°
0	1.6	1.5	1.5	1.4	1.4	1.3	1.3	1.2	1.2	1.1	1.1	1.0	1.0	0
1	1.6	1.6	1.5	1.4	1.4	1.3	1.3	1.2	1.2	1.2	1.1	1.1	1.0	1
2	1.6	1.6	1.5	1.5	1.4	1.4	1.3	1.3	1.2	1.2	1.1	1.1	1.0	2
3	1.7	1.6	1.5	1.5	1.4	1.4	1.3	1.3	1.2	1.2	1.1	1.1	1.0	3
4	1.7	1.6	1.6	1.5	1.5	1.4	1.3	1.3	1.2	1.2	1.1	1.1	1.0	4
5	1.7	1.7	1.6	1.5	1.5	1.4	1.4	1.3	1.3	1.2	1.1	1.1	1.1	5
6	1.7	1.7	1.6	1.5	1.5	1.4	1.4	1.3	1.3	1.2	1.2	1.1	1.1	6
7	1.8	1.7	1.6	1.6	1.5	1.4	1.4	1.3	1.3	1.2	1.2	1.1	1.1	7
8	1.8	1.7	1.7	1.6	1.5	1.5	1.4	1.4	1.3	1.2	1.2	1.1	1.1	8
9	1.8	1.8	1.7	1.6	1.6	1.5	1.4	1.4	1.3	1.3	1.2	1.1	1.1	9
10	1.9	1.8	1.7	1.6	1.6	1.5	1.4	1.4	1.3	1.3	1.2	1.2	1.1	10
11	1.9	1.8	1.7	1.7	1.6	1.5	1.5	1.4	1.3	1.3	1.2	1.2	1.1	11
12	1.9	1.8	1.8	1.7	1.6	1.6	1.5	1.4	1.4	1.3	1.2	1.2	1.1	12
13	2.0	1.9	1.8	1.7	1.6	1.6	1.5	1.4	1.4	1.3	1.3	1.2	1.1	13
14	2.0	1.9	1.8	1.7	1.7	1.6	1.5	1.5	1.4	1.3	1.3	1.2	1.2	14
15	2.0	1.9	1.9	1.8	1.7	1.6	1.5	1.5	1.4	1.3	1.3	1.2	1.2	15
16	2.1	2.0	1.9	1.8	1.7	1.6	1.6	1.5	1.4	1.4	1.3	1.2	1.2	16
17	2.1	2.0	1.9	1.8	1.8	1.7	1.6	1.5	1.5	1.4	1.3	1.3	1.2	17
18	2.2	2.1	2.0	1.9	1.8	1.7	1.6	1.5	1.5	1.4	1.3	1.3	1.2	18
19	2.2	2.1	2.0	1.9	1.8	1.7	1.6	1.6	1.5	1.4	1.4	1.3	1.2	19
20	2.3	2.1	2.0	1.9	1.9	1.8	1.7	1.6	1.5	1.4	1.4	1.3	1.2	20
21	2.3	2.2	2.1	2.0	1.9	1.8	1.7	1.6	1.5	1.5	1.4	1.3	1.2	21
22	2.4	2.2	2.1	2.0	1.9	1.8	1.7	1.6	1.6	1.5	1.4	1.3	1.3	22
23	2.4	2.3	2.2	2.1	2.0	1.9	1.8	1.7	1.6	1.5	1.4	1.4	1.3	23
24	2.5	2.4	2.2	2.1	2.0	1.9	1.8	1.7	1.6	1.5	1.5	1.4	1.3	24
25	2.6	2.4	2.3	2.2	2.0	1.9	1.8	1.7	1.6	1.6	1.5	1.4	1.3	25
26	2.6	2.5	2.3	2.2	2.1	2.0	1.9	1.8	1.7	1.6	1.5	1.4	1.3	26
27	2.7	2.6	2.4	2.3	2.1	2.0	1.9	1.8	1.7	1.6	1.5	1.4	1.4	27
28	2.8	2.6	2.5	2.3	2.2	2.1	2.0	1.8	1.7	1.6	1.5	1.5	1.4	28
29	2.9	2.7	2.5	2.4	2.3	2.1	2.0	1.9	1.8	1.7	1.6	1.5	1.4	29
30	3.0	2.8	2.6	2.5	2.3	2.2	2.0	1.9	1.8	1.7	1.6	1.5	1.4	30
31	3.1	2.9	2.7	2.5	2.4	2.2	2.1	2.0	1.9	1.7	1.6	1.5	1.4	31
32	3.2	3.0	2.8	2.6	2.4	2.3	2.2	2.0	1.9	1.8	1.7	1.6	1.5	32
33	3.4	3.1	2.9	2.7	2.5	2.4	2.2	2.1	1.9	1.8	1.7	1.6	1.5	33
34	3.5	3.2	3.0	2.8	2.6	2.4	2.3	2.1	2.0	1.9	1.7	1.6	1.5	34
35	3.7	3.4	3.1	2.9	2.7	2.5	2.3	2.2	2.0	1.9	1.8	1.7	1.6	35
36	3.9	3.6	3.3	3.0	2.8	2.6	2.4	2.3	2.1	2.0	1.8	1.7	1.6	36
37	4.1	3.7	3.4	3.2	2.9	2.7	2.5	2.3	2.2	2.0	1.9	1.7	1.6	37
38	4.3	3.9	3.6	3.3	3.0	2.8	2.6	2.4	2.2	2.1	1.9	1.8	1.7	38
39	4.6	4.2	3.8	3.5	3.2	2.9	2.7	2.5	2.3	2.1	2.0	1.8	1.7	39
40	5.0	4.5	4.0	3.7	3.3	3.1	2.8	2.6	2.4	2.2	2.0	1.9	1.8	40
41	5.4	4.8	4.3	3.9	3.5	3.2	2.9	2.7	2.5	2.3	2.1	1.9	1.8	41
42	5.9	5.2	4.6	4.1	3.7	3.4	3.1	2.8	2.6	2.4	2.2	2.0	1.9	42
43	6.5	5.7	5.0	4.4	4.0	3.6	3.2	2.9	2.7	2.5	2.3	2.1	1.9	43
44	7.3	6.3	5.4	4.8	4.3	3.8	3.4	3.1	2.8	2.6	2.3	2.2	2.0	44
45	8.4	7.0	6.0	5.2	4.6	4.1	3.6	3.3	3.0	2.7	2.4	2.2	2.0	45
46	9.9	8.0	6.7	5.8	5.0	4.4	3.9	3.5	3.1	2.8	2.6	2.3	2.1	46
47	12.1	9.5	7.7	6.5	5.5	4.8	4.2	3.7	3.3	3.0	2.7	2.4	2.2	47
48		11.6	9.1	7.4	6.2	5.3	4.6	4.0	3.6	3.2	2.8	2.6	2.3	48
49			11.1	8.7	7.1	5.9	5.0	4.4	3.8	3.4	3.0	2.7	2.4	49
50				10.6	8.3	6.8	5.6	4.8	4.2	3.6	3.2	2.9	2.6	50
51					10.2	7.9	6.4	5.4	4.6	4.0	3.5	3.0	2.7	51
52						9.7	7.6	6.1	5.1	4.3	3.8	3.3	2.9	52
53							9.2	7.2	5.9	4.9	4.1	3.6	3.1	53
54								8.8	6.8	5.5	4.6	3.9	3.4	54
55	10.2								8.3	6.5	5.3	4.3	3.7	55
56	7.9	9.7								7.9	6.1	5.0	4.1	56
57	6.4	7.6	9.2								7.4	5.8	4.7	57
58	5.4	6.1	7.2	8.8								7.0	5.4	58
59	4.6	5.1	5.9	6.8	8.3								6.6	59
60	4.0	4.3	4.9	5.5	6.5	7.9								60
Latitude	51°	52°	53°	54°	55°	56°	57°	58°	59°	60°	61°	62°	63°	Latitude

Declination **same** name as latitude, **upper** transit: **add** correction to observed altitude

TABLE 29

Altitude Factor

a, the change of altitude in one minute from meridian transit; used for entering table 30

Latitude	Declination **contrary** name to latitude, **upper** transit: **add** correction to observed altitude													Latitude
	51°	52°	53°	54°	55°	56°	57°	58°	59°	60°	61°	62°	63°	
°	″	″	″	″	″	″	″	″	″	″	″	″	″	°
0	1.6	1.5	1.5	1.4	1.4	1.3	1.3	1.2	1.2	1.1	1.1	1.0	1.0	0
1	1.6	1.5	1.5	1.4	1.4	1.3	1.3	1.2	1.2	1.1	1.1	1.0	1.0	1
2	1.5	1.5	1.4	1.4	1.3	1.3	1.3	1.2	1.2	1.1	1.1	1.0	1.0	2
3	1.5	1.5	1.4	1.4	1.3	1.3	1.2	1.2	1.1	1.1	1.1	1.0	1.0	3
4	1.5	1.5	1.4	1.4	1.3	1.3	1.2	1.2	1.1	1.1	1.1	1.0	1.0	4
5	1.5	1.4	1.4	1.3	1.3	1.3	1.2	1.2	1.1	1.1	1.0	1.0	1.0	5
6	1.5	1.4	1.4	1.3	1.3	1.2	1.2	1.2	1.1	1.1	1.0	1.0	1.0	6
7	1.4	1.4	1.4	1.3	1.3	1.2	1.2	1.1	1.1	1.1	1.0	1.0	0.9	7
8	1.4	1.4	1.3	1.3	1.3	1.2	1.2	1.1	1.1	1.1	1.0	1.0	0.9	8
9	1.4	1.4	1.3	1.3	1.2	1.2	1.2	1.1	1.1	1.0	1.0	1.0	0.9	9
10	1.4	1.4	1.3	1.3	1.2	1.2	1.1	1.1	1.1	1.0	1.0	1.0	0.9	10
11	1.4	1.3	1.3	1.3	1.2	1.2	1.1	1.1	1.1	1.0	1.0	1.0	0.9	11
12	1.4	1.3	1.3	1.2	1.2	1.2	1.1	1.1	1.1	1.0	1.0	0.9	0.9	12
13	1.3	1.3	1.3	1.2	1.2	1.2	1.1	1.1	1.0	1.0	1.0	0.9	0.9	13
14	1.3	1.3	1.3	1.2	1.2	1.1	1.1	1.1	1.0	1.0	1.0	0.9	0.9	14
15	1.3	1.3	1.2	1.2	1.2	1.1	1.1	1.1	1.0	1.0	1.0	0.9	0.9	15
16	1.3	1.3	1.2	1.2	1.1	1.1	1.1	1.0	1.0	1.0	0.9	0.9	0.9	16
17	1.3	1.2	1.2	1.2	1.1	1.1	1.1	1.0	1.0	1.0	0.9	0.9	0.9	17
18	1.3	1.2	1.2	1.2	1.1	1.1	1.1	1.0	1.0	1.0	0.9	0.9	0.9	18
19	1.2	1.2	1.2	1.1	1.1	1.1	1.0	1.0	1.0	1.0	0.9	0.9	0.9	19
20	1.2	1.2	1.2	1.1	1.1	1.1	1.0	1.0	1.0	0.9	0.9	0.9	0.8	20
21	1.2	1.2	1.2	1.1	1.1	1.1	1.0	1.0	1.0	0.9	0.9	0.9	0.8	21
22	1.2	1.2	1.1	1.1	1.1	1.0	1.0	1.0	1.0	0.9	0.9	0.9		22
23	1.2	1.2	1.1	1.1	1.1	1.0	1.0	1.0	0.9	0.9	0.9			23
24	1.2	1.1	1.1	1.1	1.1	1.0	1.0	1.0	0.9	0.9				24
25	1.2	1.1	1.1	1.1	1.0	1.0	1.0	1.0	0.9					25
26	1.1	1.1	1.1	1.1	1.0	1.0	1.0	0.9						26
27	1.1	1.1	1.1	1.0	1.0	1.0	1.0							27
28	1.1	1.1	1.1	1.0	1.0	1.0								28
29	1.1	1.1	1.0	1.0	1.0									29
30	1.1	1.1	1.0	1.0										30
31	1.1	1.0	1.0											31
32	1.1	1.0												32
33	1.1												0.8	33
34												0.8	0.7	34
35											0.8	0.8	0.7	35
36										0.8	0.8	0.8	0.7	36
37									0.8	0.8	0.8	0.7	0.7	37
38								0.8	0.8	0.8	0.8	0.7	0.7	38
39							0.8	0.8	0.8	0.8	0.8	0.7	0.7	39
40						0.8	0.8	0.8	0.8	0.8	0.8	0.7	0.7	40
41					0.9	0.8	0.8	0.8	0.8	0.8	0.7	0.7	0.7	41
42				0.9	0.8	0.8	0.8	0.8	0.8	0.8	0.7	0.7	0.7	42
43			0.9	0.9	0.8	0.8	0.8	0.8	0.8	0.7	0.7	0.7	0.7	43
44		0.9	0.9	0.8	0.8	0.8	0.8	0.8	0.8	0.7	0.7	0.7	0.7	44
45	0.9	0.9	0.8	0.8	0.8	0.8	0.8	0.8	0.7	0.7	0.7	0.7	0.7	45
46	0.9	0.9	0.8	0.8	0.8	0.8	0.8	0.8	0.7	0.7	0.7	0.7	0.7	46
47	0.9	0.8	0.8	0.8	0.8	0.8	0.8	0.7	0.7	0.7	0.7	0.7	0.6	47
48	0.8	0.8	0.8	0.8	0.8	0.8	0.7	0.7	0.7	0.7	0.7	0.7	0.6	48
49	0.8	0.8	0.8	0.8	0.8	0.7	0.7	0.7	0.7	0.7	0.7	0.6	0.6	49
50	0.8	0.8	0.8	0.8	0.7	0.7	0.7	0.7	0.7	0.7	0.7	0.6	0.6	50
51	0.8	0.8	0.8	0.8	0.7	0.7	0.7	0.7	0.7	0.7	0.7	0.6	0.6	51
52	0.8	0.8	0.8	0.7	0.7	0.7	0.7	0.7	0.7	0.7	0.6	0.6	0.6	52
53	0.8	0.8	0.7	0.7	0.7	0.7	0.7	0.7	0.7	0.6	0.6	0.6	0.6	53
54	0.8	0.7	0.7	0.7	0.7	0.7	0.7	0.7	0.6	0.6	0.6	0.6	0.6	54
55	0.7	0.7	0.7	0.7	0.7	0.7	0.7	0.7	0.6	0.6	0.6	0.6	0.6	55
56	0.7	0.7	0.7	0.7	0.7	0.7	0.7	0.6	0.6	0.6	0.6	0.6	0.6	56
57	0.7	0.7	0.7	0.7	0.7	0.7	0.6	0.6	0.6	0.6	0.6	0.6	0.6	57
58	0.7	0.7	0.7	0.7	0.7	0.6	0.6	0.6	0.6	0.6	0.6	0.6	0.6	58
59	0.7	0.7	0.7	0.6	0.6	0.6	0.6	0.6	0.6	0.6	0.6	0.6	0.5	59
60	0.7	0.7	0.6	0.6	0.6	0.6	0.6	0.6	0.6	0.6	0.6	0.6	0.5	60
Latitude	51°	52°	53°	54°	55°	56°	57°	58°	59°	60°	61°	62°	63°	Latitude
	Declination **same** name as latitude, **lower** transit: **subtract** correction from observed altitude													

TABLE 30

Change of Altitude in Given Time from Meridian Transit

a (table 29)	t, meridian angle														a (table 29)
	5′	10′	15′	20′	25′	30′	35′	40′	45′	50′	55′	1°00′	1°05′	1°10′	
	0m 20s	0m 40s	1m 00s	1m 20s	1m 40s	2m 00s	2m 20s	2m 40s	3m 00s	3m 20s	3m 40s	4m 00s	4m 20s	4m 40s	
″	′	′	′	′	′	′	′	′	′	′	′	′	′	′	″
0. 1	0. 0	0. 0	0. 0	0. 0	0. 0	0. 0	0. 0	0. 0	0. 0	0. 0	0. 0	0. 0	0. 0	0. 0	0. 1
0. 2	0. 0	0. 0	0. 0	0. 0	0. 0	0. 0	0. 0	0. 0	0. 0	0. 0	0. 0	0. 1	0. 1	0. 1	0. 2
0. 3	0. 0	0. 0	0. 0	0. 0	0. 0	0. 0	0. 0	0. 0	0. 0	0. 1	0. 1	0. 1	0. 1	0. 1	0. 3
0. 4	0. 0	0. 0	0. 0	0. 0	0. 0	0. 0	0. 0	0. 0	0. 1	0. 1	0. 1	0. 1	0. 1	0. 1	0. 4
0. 5	0. 0	0. 0	0. 0	0. 0	0. 0	0. 0	0. 0	0. 1	0. 1	0. 1	0. 1	0. 1	0. 2	0. 2	0. 5
0. 6	0. 0	0. 0	0. 0	0. 0	0. 0	0. 0	0. 1	0. 1	0. 1	0. 1	0. 1	0. 2	0. 2	0. 2	0. 6
0. 7	0. 0	0. 0	0. 0	0. 0	0. 0	0. 0	0. 1	0. 1	0. 1	0. 1	0. 2	0. 2	0. 2	0. 3	0. 7
0. 8	0. 0	0. 0	0. 0	0. 0	0. 0	0. 1	0. 1	0. 1	0. 1	0. 1	0. 2	0. 2	0. 3	0. 3	0. 8
0. 9	0. 0	0. 0	0. 0	0. 0	0. 0	0. 1	0. 1	0. 1	0. 1	0. 2	0. 2	0. 2	0. 3	0. 3	0. 9
1. 0	0. 0	0. 0	0. 0	0. 0	0. 0	0. 1	0. 1	0. 1	0. 2	0. 2	0. 2	0. 3	0. 3	0. 4	1. 0
2. 0	0. 0	0. 0	0. 0	0. 1	0. 1	0. 1	0. 2	0. 2	0. 3	0. 4	0. 4	0. 5	0. 6	0. 7	2. 0
3. 0	0. 0	0. 0	0. 0	0. 1	0. 1	0. 2	0. 3	0. 4	0. 4	0. 6	0. 7	0. 8	0. 9	1. 1	3. 0
4. 0	0. 0	0. 0	0. 1	0. 1	0. 2	0. 3	0. 4	0. 5	0. 6	0. 7	0. 9	1. 1	1. 3	1. 5	4. 0
5. 0	0. 0	0. 0	0. 1	0. 1	0. 2	0. 3	0. 5	0. 6	0. 8	0. 9	1. 1	1. 3	1. 6	1. 8	5. 0
6. 0	0. 0	0. 0	0. 1	0. 2	0. 3	0. 4	0. 5	0. 7	0. 9	1. 1	1. 3	1. 6	1. 9	2. 2	6. 0
7. 0	0. 0	0. 1	0. 1	0. 2	0. 3	0. 5	0. 6	0. 8	1. 0	1. 3	1. 6	1. 9	2. 2	2. 5	7. 0
8. 0	0. 0	0. 1	0. 1	0. 2	0. 4	0. 5	0. 7	0. 9	1. 2	1. 5	1. 8	2. 1	2. 5	2. 9	8. 0
9. 0	0. 0	0. 1	0. 2	0. 3	0. 4	0. 6	0. 8	1. 1	1. 4	1. 7	2. 0	2. 4	2. 8	3. 3	9. 0
10. 0	0. 0	0. 1	0. 2	0. 3	0. 5	0. 7	0. 9	1. 2	1. 5	1. 9	2. 2	2. 7	3. 1	3. 6	10. 0
11. 0	0. 0	0. 1	0. 2	0. 3	0. 5	0. 7	1. 0	1. 3	1. 6	2. 0	2. 5	2. 9	3. 4	4. 0	11. 0
12. 0	0. 0	0. 1	0. 2	0. 4	0. 6	0. 8	1. 1	1. 4	1. 8	2. 2	2. 7	3. 2	3. 8	4. 4	12. 0
13. 0	0. 0	0. 1	0. 2	0. 4	0. 6	0. 9	1. 2	1. 5	2. 0	2. 4	2. 9	3. 5	4. 1	4. 7	13. 0
14. 0	0. 0	0. 1	0. 2	0. 4	0. 6	0. 9	1. 3	1. 7	2. 1	2. 6	3. 1	3. 7	4. 4	5. 1	14. 0
15. 0	0. 0	0. 1	0. 2	0. 4	0. 7	1. 0	1. 4	1. 8	2. 2	2. 8	3. 4	4. 0	4. 7	5. 4	15. 0
16. 0	0. 0	0. 1	0. 3	0. 5	0. 7	1. 1	1. 5	1. 9	2. 4	3. 0	3. 6	4. 3	5. 0	5. 8	16. 0
17. 0	0. 0	0. 1	0. 3	0. 5	0. 8	1. 1	1. 5	2. 0	2. 6	3. 1	3. 8	4. 5	5. 3	6. 2	17. 0
18. 0	0. 0	0. 1	0. 3	0. 5	0. 8	1. 2	1. 6	2. 1	2. 7	3. 3	4. 0	4. 8	5. 6	6. 5	18. 0
19. 0	0. 0	0. 1	0. 3	0. 6	0. 9	1. 3	1. 7	2. 3	2. 8	3. 5	4. 3	5. 1	5. 9	6. 9	19. 0
20. 0	0. 0	0. 1	0. 3	0. 6	0. 9	1. 3	1. 8	2. 4	3. 0	3. 7	4. 5	5. 3	6. 3	7. 3	20. 0
21. 0	0. 0	0. 2	0. 4	0. 6	1. 0	1. 4	1. 9	2. 5	3. 2	3. 9	4. 7	5. 6	6. 6	7. 6	21. 0
22. 0	0. 0	0. 2	0. 4	0. 7	1. 0	1. 5	2. 0	2. 6	3. 3	4. 1	4. 9	5. 9	6. 9	8. 0	22. 0
23. 0	0. 0	0. 2	0. 4	0. 7	1. 1	1. 5	2. 1	2. 7	3. 4	4. 3	5. 2	6. 1	7. 2	8. 3	23. 0
24. 0	0. 0	0. 2	0. 4	0. 7	1. 1	1. 6	2. 2	2. 8	3. 6	4. 4	5. 4	6. 4	7. 5	8. 7	24. 0
25. 0	0. 0	0. 2	0. 4	0. 7	1. 2	1. 7	2. 3	3. 0	3. 8	4. 6	5. 6	6. 7	7. 8	9. 1	25. 0
26. 0	0. 0	0. 2	0. 4	0. 8	1. 2	1. 7	2. 4	3. 1	3. 9	4. 8	5. 8	6. 9	8. 1	9. 4	26. 0
27. 0	0. 0	0. 2	0. 4	0. 8	1. 2	1. 8	2. 4	3. 2	4. 0	5. 0	6. 0	7. 2	8. 4	9. 8	27. 0
28. 0	0. 1	0. 2	0. 5	0. 8	1. 3	1. 9	2. 5	3. 3	4. 2	5. 2	6. 3	7. 5	8. 8	10. 2	28. 0

Caution.—If this table is entered with the meridian angle of the Moon in arc units, such units should correspond to the meridian angle in time units as given in the Increments and Corrections section of the *Nautical Almanac*.

TABLE 30

Change of Altitude in Given Time from Meridian Transit

a (table 29)	t, meridian angle														a (table 29)
	1°15′	1°20′	1°25′	1°30′	1°35′	1°40′	1°45′	1°50′	1°55′	2°00′	2°05′	2°10′	2°15′	2°20′	
	5m 00s	5m 20s	5m 40s	6m 00s	6m 20s	6m 40s	7m 00s	7m 20s	7m 40s	8m 00s	8m 20s	8m 40s	9m 00s	9m 20s	
″	′	′	′	′	′	′	′	′	′	′	′	′	′	′	″
0.1	0.0	0.0	0.1	0.1	0.1	0.1	0.1	0.1	0.1	0.1	0.1	0.1	0.1	0.1	0.1
0.2	0.1	0.1	0.1	0.1	0.1	0.1	0.2	0.2	0.2	0.2	0.2	0.3	0.3	0.3	0.2
0.3	0.1	0.1	0.2	0.2	0.2	0.2	0.2	0.3	0.3	0.3	0.3	0.4	0.4	0.4	0.3
0.4	0.2	0.2	0.2	0.2	0.3	0.3	0.3	0.4	0.4	0.4	0.5	0.5	0.5	0.6	0.4
0.5	0.2	0.2	0.3	0.3	0.3	0.4	0.4	0.5	0.5	0.5	0.6	0.6	0.7	0.7	0.5
0.6	0.2	0.3	0.3	0.4	0.4	0.4	0.5	0.5	0.6	0.6	0.7	0.8	0.8	0.9	0.6
0.7	0.3	0.3	0.4	0.4	0.5	0.5	0.6	0.6	0.7	0.7	0.8	0.9	0.9	1.0	0.7
0.8	0.3	0.4	0.4	0.5	0.5	0.6	0.7	0.7	0.8	0.9	0.9	1.0	1.1	1.2	0.8
0.9	0.4	0.4	0.5	0.5	0.6	0.7	0.7	0.8	0.9	1.0	1.0	1.1	1.2	1.3	0.9
1.0	0.4	0.5	0.5	0.6	0.7	0.7	0.8	0.9	1.0	1.1	1.2	1.3	1.4	1.5	1.0
2.0	0.8	0.9	1.1	1.2	1.3	1.5	1.6	1.8	2.0	2.1	2.3	2.5	2.7	2.9	2.0
3.0	1.2	1.4	1.6	1.8	2.0	2.2	2.4	2.7	2.9	3.2	3.5	3.8	4.0	4.4	3.0
4.0	1.7	1.9	2.1	2.4	2.7	3.0	3.3	3.6	3.9	4.3	4.6	5.0	5.4	5.8	4.0
5.0	2.1	2.4	2.7	3.0	3.3	3.7	4.1	4.5	4.9	5.3	5.8	6.3	6.8	7.3	5.0
6.0	2.5	2.8	3.2	3.6	4.0	4.4	4.9	5.4	5.9	6.4	6.9	7.5	8.1	8.7	6.0
7.0	2.9	3.3	3.7	4.2	4.7	5.2	5.7	6.3	6.9	7.5	8.1	8.8	9.4	10.2	7.0
8.0	3.3	3.8	4.3	4.8	5.3	5.9	6.5	7.2	7.8	8.5	9.3	10.0	10.8	11.6	8.0
9.0	3.8	4.3	4.8	5.4	6.0	6.7	7.4	8.1	8.8	9.6	10.4	11.3	12.2	13.1	9.0
10.0	4.2	4.7	5.4	6.0	6.7	7.4	8.2	9.0	9.8	10.7	11.6	12.5	13.5	14.5	10.0
11.0	4.6	5.2	5.9	6.6	7.4	8.1	9.0	9.9	10.8	11.7	12.7	13.8	14.8	16.0	11.0
12.0	5.0	5.7	6.4	7.2	8.0	8.9	9.8	10.8	11.8	12.8	13.9	15.0	16.2	17.4	12.0
13.0	5.4	6.2	7.0	7.8	8.7	9.6	10.6	11.7	12.7	13.9	15.0	16.3	17.6	18.9	13.0
14.0	5.8	6.6	7.5	8.4	9.4	10.4	11.4	12.5	13.7	14.9	16.2	17.5	18.9	20.3	14.0
15.0	6.2	7.1	8.0	9.0	10.0	11.1	12.2	13.4	14.7	16.0	17.4	18.8	20.2	21.8	15.0
16.0	6.7	7.6	8.6	9.6	10.7	11.9	13.1	14.3	15.7	17.1	18.5	20.0	21.6	23.2	16.0
17.0	7.1	8.1	9.1	10.2	11.4	12.6	13.9	15.2	16.7	18.1	19.7	21.3	23.0	24.7	17.0
18.0	7.5	8.5	9.6	10.8	12.0	13.3	14.7	16.1	17.6	19.2	20.8	22.5	24.3	26.1	18.0
19.0	7.9	9.0	10.2	11.4	12.7	14.1	15.5	17.0	18.6	20.3	22.0	23.8			19.0
20.0	8.3	9.5	10.7	12.0	13.4	14.8	16.3	17.9	19.6	21.3	23.1				20.0
21.0	8.8	10.0	11.2	12.6	14.0	15.6	17.2	18.8	20.6						21.0
22.0	9.2	10.4	11.8	13.2	14.7	16.3	18.0	19.7	21.6						22.0
23.0	9.6	10.9	12.3	13.8	15.4	17.0	18.8	20.6							23.0
24.0	10.0	11.4	12.8	14.4	16.0	17.8	19.6	21.5							24.0
25.0	10.4	11.9	13.4	15.0	16.7	18.5	20.4								25.0
26.0	10.8	12.3	13.9	15.6	17.4	19.3									26.0
27.0	11.2	12.8	14.4	16.2	18.0	20.0									27.0

Caution.—If this table is entered with the meridian angle of the Moon in arc units, such units should correspond to the meridian angle in time units as given in the Increments and Corrections section of the *Nautical Almanac.*

TABLE 30

Change of Altitude in Given Time from Meridian Transit

a (table 29)	t, meridian angle														a (table 29)
	2°25′	2°30′	2°35′	2°40′	2°45′	2°50′	2°55′	3°00′	3°05′	3°10′	3°15′	3°20′	3°25′	3°30′	
	9m 40s	10m 00s	10m 20s	10m 40s	11m 00s	11m 20s	11m 40s	12m 00s	12m 20s	12m 40s	13m 00s	13m 20s	13m 40s	14m 00s	
″	′	′	′	′	′	′	′	′	′	′	′	′	′	′	″
0. 1	0. 2	0. 2	0. 2	0. 2	0. 2	0. 2	0. 2	0. 2	0. 3	0. 3	0. 3	0. 3	0. 3	0. 3	0. 1
0. 2	0. 3	0. 3	0. 4	0. 4	0. 4	0. 4	0. 5	0. 5	0. 5	0. 5	0. 6	0. 6	0. 6	0. 7	0. 2
0. 3	0. 5	0. 5	0. 5	0. 6	0. 6	0. 6	0. 7	0. 7	0. 8	0. 8	0. 8	0. 9	0. 9	1. 0	0. 3
0. 4	0. 6	0. 7	0. 7	0. 8	0. 8	0. 9	0. 9	1. 0	1. 0	1. 1	1. 1	1. 2	1. 2	1. 3	0. 4
0. 5	0. 8	0. 8	0. 9	0. 9	1. 0	1. 1	1. 1	1. 2	1. 3	1. 3	1. 4	1. 5	1. 6	1. 6	0. 5
0. 6	0. 9	1. 0	1. 1	1. 1	1. 2	1. 3	1. 4	1. 4	1. 5	1. 6	1. 7	1. 8	1. 9	2. 0	0. 6
0. 7	1. 1	1. 2	1. 2	1. 3	1. 4	1. 5	1. 6	1. 7	1. 8	1. 9	2. 0	2. 1	2. 2	2. 3	0. 7
0. 8	1. 2	1. 3	1. 4	1. 5	1. 6	1. 7	1. 8	1. 9	2. 0	2. 1	2. 3	2. 4	2. 5	2. 6	0. 8
0. 9	1. 4	1. 5	1. 6	1. 7	1. 8	1. 9	2. 0	2. 2	2. 3	2. 4	2. 5	2. 7	2. 8	2. 9	0. 9
1. 0	1. 6	1. 7	1. 8	1. 9	2. 0	2. 1	2. 3	2. 4	2. 5	2. 7	2. 8	3. 0	3. 1	3. 3	1. 0
2. 0	3. 1	3. 3	3. 6	3. 8	4. 0	4. 3	4. 5	4. 8	5. 1	5. 3	5. 6	5. 9	6. 2	6. 5	2. 0
3. 0	4. 7	5. 0	5. 3	5. 7	6. 0	6. 4	6. 8	7. 2	7. 6	8. 0	8. 4	8. 9	9. 3	9. 8	3. 0
4. 0	6. 2	6. 7	7. 1	7. 6	8. 1	8. 6	9. 1	9. 6	10. 1	10. 7	11. 3	11. 9	12. 5	13. 1	4. 0
5. 0	7. 8	8. 3	8. 9	9. 5	10. 1	10. 7	11. 3	12. 0	12. 7	13. 4	14. 1	14. 8	15. 6	16. 3	5. 0
6. 0	9. 3	10. 0	10. 7	11. 4	12. 1	12. 8	13. 6	14. 4	15. 2	16. 0	16. 9	17. 8	18. 7	19. 6	6. 0
7. 0	10. 9	11. 7	12. 5	13. 3	14. 1	15. 0	15. 9	16. 8	17. 7	18. 7	19. 7	20. 7	21. 8	22. 9	7. 0
8. 0	12. 5	13. 3	14. 2	15. 2	16. 1	17. 1	18. 1	19. 2	20. 3	21. 4	22. 5	23. 7	24. 9	26. 1	8. 0
9. 0	14. 0	15. 0	16. 0	17. 1	18. 2	19. 3	20. 4	21. 6	22. 8	24. 1	25. 4	26. 7	28. 0	29. 4	9. 0
10. 0	15. 6	16. 7	17. 8	19. 0	20. 2	21. 4	22. 7	24. 0	25. 4	26. 7	28. 2	29. 6			10. 0
11. 0	17. 1	18. 3	19. 6	20. 9	22. 2	23. 5	25. 0	26. 4	27. 9	29. 4					11. 0
12. 0	18. 7	20. 0	21. 4	22. 8	24. 2	25. 7	27. 2	28. 8							12. 0
13. 0	20. 2	21. 7	23. 1	24. 7	26. 2	27. 8	29. 5								13. 0
14. 0	21. 8	23. 3	24. 9	26. 5	28. 2	30. 0									14. 0
15. 0	23. 4	25. 0	26. 7	28. 4	30. 2										15. 0
16. 0	24. 9	26. 7	28. 5	30. 3											16. 0
17. 0	26. 5	28. 3	30. 3												17. 0

a (table 29)	t, meridian angle														a (table 29)
	3°35′	3°40′	3°45′	3°50′	3°55′	4°00′	4°05′	4°10′	4°15′	4°20′	4°25′	4°30′	4°35′	4°40′	
	14m 20s	14m 40s	15m 00s	15m 20s	15m 40s	16m 00s	16m 20s	16m 40s	17m 00s	17m 20s	17m 40s	18m 00s	18m 20s	18m 40s	
″	′	′	′	′	′	′	′	′	′	′	′	′	′	′	″
0. 1	0. 3	0. 4	0. 4	0. 4	0. 4	0. 4	0. 4	0. 5	0. 5	0. 5	0. 5	0. 5	0. 6	0. 6	0. 1
0. 2	0. 7	0. 7	0. 8	0. 8	0. 8	0. 9	0. 9	0. 9	1. 0	1. 0	1. 0	1. 1	1. 1	1. 2	0. 2
0. 3	1. 0	1. 1	1. 1	1. 2	1. 2	1. 3	1. 3	1. 4	1. 4	1. 5	1. 6	1. 6	1. 7	1. 7	0. 3
0. 4	1. 4	1. 4	1. 5	1. 6	1. 6	1. 7	1. 8	1. 9	1. 9	2. 0	2. 1	2. 2	2. 2	2. 3	0. 4
0. 5	1. 7	1. 8	1. 9	2. 0	2. 0	2. 1	2. 2	2. 3	2. 4	2. 5	2. 6	2. 7	2. 8	2. 9	0. 5
0. 6	2. 1	2. 2	2. 2	2. 4	2. 5	2. 6	2. 7	2. 8	2. 9	3. 0	3. 1	3. 2	3. 4	3. 5	0. 6
0. 7	2. 4	2. 5	2. 6	2. 7	2. 9	3. 0	3. 1	3. 2	3. 4	3. 5	3. 6	3. 8	3. 9	4. 1	0. 7
0. 8	2. 7	2. 9	3. 0	3. 1	3. 3	3. 4	3. 6	3. 7	3. 9	4. 0	4. 2	4. 3	4. 5	4. 6	0. 8
0. 9	3. 1	3. 2	3. 4	3. 5	3. 7	3. 8	4. 0	4. 2	4. 3	4. 5	4. 7	4. 9	5. 0	5. 2	0. 9
1. 0	3. 4	3. 6	3. 8	3. 9	4. 1	4. 3	4. 4	4. 6	4. 8	5. 0	5. 2	5. 4	5. 6	5. 8	1. 0
2. 0	6. 8	7. 2	7. 5	7. 8	8. 2	8. 5	8. 9	9. 3	9. 6	10. 0	10. 4	10. 8	11. 2	11. 6	2. 0
3. 0	10. 3	10. 8	11. 2	11. 8	12. 3	12. 8	13. 3	13. 9	14. 4	15. 0	15. 6	16. 2	16. 8	17. 4	3. 0
4. 0	13. 7	14. 3	15. 0	15. 7	16. 4	17. 1	17. 8	18. 5	19. 3	20. 0	20. 8	21. 6	22. 4	23. 2	4. 0
5. 0	17. 1	17. 9	18. 8	19. 6	20. 5	21. 3	22. 2	23. 1	24. 1	25. 0	26. 0	27. 0	28. 0	29. 0	5. 0
6. 0	20. 5	21. 5	22. 5	23. 5	24. 5	25. 6	26. 7	27. 8							6. 0
7. 0	24. 0	25. 1	26. 2	27. 4											7. 0
8. 0	27. 4	28. 7	30. 0												8. 0

Caution.—If this table is entered with the meridian angle of the Moon in arc units, such units should correspond to the meridian angle in time units as given in the Increments and Corrections section of the *Nautical Almanac.*

TABLE 30

Change of Altitude in Given Time from Meridian Transit

a (table 29)	t, meridian angle														a (table 29)
	4°45′	4°50′	4°55′	5°00′	5°05′	5°10′	5°15′	5°20′	5°25′	5°30′	5°35′	5°40′	5°45′	5°50′	
	19m 00s	19m 20s	19m 40s	20m 00s	20m 20s	20m 40s	21m 00s	21m 20s	21m 40s	22m 00s	22m 20s	22m 40s	23m 00s	23m 20s	
″	′	′	′	′	′	′	′	′	′	′	′	′	′	′	″
0.1	0.6	0.6	0.6	0.7	0.7	0.7	0.7	0.8	0.8	0.8	0.8	0.9	0.9	0.9	0.1
0.2	1.2	1.2	1.3	1.3	1.4	1.4	1.5	1.5	1.6	1.6	1.7	1.7	1.8	1.8	0.2
0.3	1.8	1.9	1.9	2.0	2.1	2.1	2.2	2.3	2.3	2.4	2.5	2.6	2.6	2.7	0.3
0.4	2.4	2.5	2.6	2.7	2.8	2.8	2.9	3.0	3.1	3.2	3.3	3.4	3.5	3.6	0.4
0.5	3.0	3.1	3.2	3.3	3.4	3.6	3.7	3.8	3.9	4.0	4.2	4.3	4.4	4.5	0.5
0.6	3.6	3.7	3.9	4.0	4.1	4.3	4.4	4.6	4.7	4.8	5.0	5.1	5.3	5.4	0.6
0.7	4.2	4.4	4.5	4.7	4.8	5.0	5.1	5.3	5.5	5.6	5.8	6.0	6.2	6.4	0.7
0.8	4.8	5.0	5.2	5.3	5.5	5.7	5.9	6.1	6.3	6.5	6.7	6.9	7.1	7.3	0.8
0.9	5.4	5.6	5.8	6.0	6.2	6.4	6.6	6.8	7.0	7.3	7.5	7.7	7.9	8.2	0.9
1.0	6.0	6.2	6.4	6.7	6.9	7.1	7.4	7.6	7.8	8.1	8.3	8.6	8.8	9.1	1.0
2.0	12.0	12.5	12.9	13.3	13.8	14.2	14.7	15.2	15.6	16.1	16.6	17.1	17.6	18.1	2.0
3.0	18.0	18.7	19.3	20.0	20.7	21.4	22.0	22.8	23.5	24.2	24.9	25.7	26.4	27.2	3.0
4.0	24.1	24.9	25.8	26.7	27.6	28.5	29.4	30.3	31.3						4.0

a (table 29)	t, meridian angle														a (table 29)
	5°55′	6°00′	6°05′	6°10′	6°15′	6°20′	6°25′	6°30′	6°35′	6°40′	6°45′	6°50′	6°55′	7°00′	
	23m 40s	24m 00s	24m 20s	24m 40s	25m 00s	25m 20s	25m 40s	26m 00s	26m 20s	26m 40s	27m 00s	27m 20s	27m 40s	28m 00s	
″	′	′	′	′	′	′	′	′	′	′	′	′	′	′	″
0.1	0.9	1.0	1.0	1.0	1.0	1.1	1.1	1.1	1.2	1.2	1.2	1.2	1.3	1.3	0.1
0.2	1.9	1.9	2.0	2.0	2.1	2.1	2.2	2.3	2.3	2.4	2.4	2.5	2.6	2.6	0.2
0.3	2.8	2.9	3.0	3.0	3.1	3.2	3.3	3.4	3.5	3.6	3.6	3.7	3.8	3.9	0.3
0.4	3.7	3.8	3.9	4.1	4.2	4.3	4.4	4.5	4.6	4.7	4.9	5.0	5.1	5.2	0.4
0.5	4.7	4.8	4.9	5.1	5.2	5.3	5.5	5.6	5.8	5.9	6.1	6.2	6.4	6.5	0.5
0.6	5.6	5.8	5.9	6.1	6.2	6.4	6.6	6.8	6.9	7.1	7.3	7.5	7.7	7.8	0.6
0.7	6.5	6.7	6.9	7.1	7.3	7.5	7.7	7.9	8.1	8.3	8.5	8.7	8.9	9.1	0.7
0.8	7.5	7.7	7.9	8.1	8.3	8.6	8.8	9.0	9.2	9.5	9.7	10.0	10.2	10.5	0.8
0.9	8.4	8.6	8.9	9.1	9.4	9.6	9.9	10.1	10.4	10.7	10.9	11.2	11.5	11.8	0.9
1.0	9.3	9.6	9.9	10.1	10.4	10.7	11.0	11.3	11.6	11.9	12.2	12.5	12.8	13.1	1.0
2.0	18.7	19.2	19.7	20.3	20.8	21.4	22.0	22.5	23.1	23.7	24.3	24.9	25.5	26.1	2.0
3.0	28.0	28.8	29.6	30.4											3.0

Caution.—If this table is entered with the meridian angle of the Moon in arc units, such units should correspond to the meridian angle in time units as given in the Increments and Corrections section of the *Nautical Almanac.*

TABLE 31

Natural Trigonometric Functions

0° → ↓ ′	sin	Diff. 1′	csc	Diff. 1′	tan	Diff. 1′	cot	Diff. 1′	sec	Diff. 1′	cos	Diff. 1′	← 179° ↓ ′
0	0.00000	29	∞	—	0.00000	29	∞	—	1.00000	0	1.00000	0	60
1	.00029	29	3437.75	1718.88	.00029	29	3437.75	1718.88	.00000	0	.00000	0	59
2	.00058	29	1718.87	572.95	.00058	29	1718.87	572.95	.00000	0	.00000	0	58
3	.00087	29	1145.92	286.483	.00087	29	1145.92	286.484	.00000	0	.00000	0	57
4	.00116	29	859.437	171.887	.00116	29	859.436	171.887	.00000	0	.00000	0	56
5	0.00145	30	687.550	114.592	0.00145	30	687.549	114.592	1.00000	0	1.00000	0	55
6	.00175	29	572.958	81.851	.00175	29	572.957	81.851	.00000	0	.00000	0	54
7	.00204	29	491.107	61.388	.00204	29	491.106	61.388	.00000	0	.00000	0	53
8	.00233	29	429.719	47.747	.00233	29	429.718	47.747	.00000	0	.00000	0	52
9	.00262	29	381.972	38.197	.00262	29	381.971	38.197	.00000	0	.00000	0	51
10	0.00291	29	343.775	31.252	0.00291	29	343.774	31.253	1.00000	1	1.00000	1	50
11	.00320	29	312.523	26.044	.00320	29	312.521	26.043	.00001	0	0.99999	0	49
12	.00349	29	286.479	22.036	.00349	29	286.478	22.037	.00001	0	.99999	0	48
13	.00378	29	264.443	18.889	.00378	29	264.441	18.889	.00001	0	.99999	0	47
14	.00407	29	245.554	16.370	.00407	29	245.552	16.370	.00001	0	.99999	0	46
15	0.00436	29	229.184	14.324	0.00436	29	229.182	14.324	1.00001	0	0.99999	0	45
16	.00465	30	214.860	12.639	.00465	30	214.858	12.639	.00001	0	.99999	0	44
17	.00495	29	202.221	11.234	.00495	29	202.219	11.235	.00001	0	.99999	0	43
18	.00524	29	190.987	10.052	.00524	29	190.984	10.052	.00001	1	.99999	1	42
19	.00553	29	180.935	9.047	.00553	29	180.932	9.047	.00002	0	.99998	0	41
20	0.00582	29	171.888	8.185	0.00582	29	171.885	8.185	1.00002	0	0.99998	0	40
21	.00611	29	163.703	7.441	.00611	29	163.700	7.441	.00002	0	.99998	0	39
22	.00640	29	156.262	6.794	.00640	29	156.259	6.794	.00002	0	.99998	0	38
23	.00669	29	149.468	6.227	.00669	29	149.465	6.228	.00002	0	.99998	0	37
24	.00698	29	143.241	5.730	.00698	29	143.237	5.730	.00002	1	.99998	1	36
25	0.00727	29	137.511	5.289	0.00727	29	137.507	5.288	1.00003	0	0.99997	0	35
26	.00756	29	132.222	4.897	.00756	29	132.219	4.898	.00003	0	.99997	0	34
27	.00785	29	127.325	4.547	.00785	30	127.321	4.547	.00003	0	.99997	0	33
28	.00814	30	122.778	4.234	.00815	29	122.774	4.234	.00003	1	.99997	1	32
29	.00844	29	118.544	3.951	.00844	29	118.540	3.951	.00004	0	.99996	0	31
30	0.00873	29	114.593	3.696	0.00873	29	114.589	3.697	1.00004	0	0.99996	0	30
31	.00902	29	110.897	3.466	.00902	29	110.892	3.466	.00004	0	.99996	0	29
32	.00931	29	107.431	3.255	.00931	29	107.426	3.255	.00004	1	.99996	1	28
33	.00960	29	104.176	3.064	.00960	29	104.171	3.064	.00005	0	.99995	0	27
34	.00989	29	101.112	2.8890	.00989	29	101.107	2.8891	.00005	0	.99995	0	26
35	0.01018	29	98.2230	2.7283	0.01018	29	98.2179	2.7284	1.00005	0	0.99995	0	25
36	.01047	29	95.4947	2.5808	.01047	29	95.4895	2.5810	.00005	1	.99995	1	24
37	.01076	29	92.9139	2.4450	.01076	29	92.9085	2.4452	.00006	0	.99994	0	23
38	.01105	29	90.4689	2.3197	.01105	30	90.4633	2.3197	.00006	0	.99994	0	22
39	.01134	30	88.1492	2.2036	.01135	29	88.1436	2.2038	.00006	1	.99994	1	21
40	0.01164	29	85.9456	2.0961	0.01164	29	85.9398	2.0963	1.00007	0	0.99993	0	20
41	.01193	29	83.8495	1.9963	.01193	29	83.8435	1.9965	.00007	0	.99993	0	19
42	.01222	29	81.8532	1.9035	.01222	29	81.8470	1.9036	.00007	1	.99993	1	18
43	.01251	29	79.9497	1.8170	.01251	29	79.9434	1.8171	.00008	0	.99992	0	17
44	.01280	29	78.1327	1.7361	.01280	29	78.1263	1.7363	.00008	1	.99992	1	16
45	0.01309	29	76.3966	1.6607	0.01309	29	76.3900	1.6608	1.00009	0	0.99991	0	15
46	.01338	29	74.7359	1.5901	.01338	29	74.7292	1.5902	.00009	0	.99991	0	14
47	.01367	29	73.1458	1.5237	.01367	29	73.1390	1.5239	.00009	1	.99991	1	13
48	.01396	29	71.6221	1.4616	.01396	29	71.6151	1.4618	.00010	0	.99990	0	12
49	.01425	29	70.1605	1.4031	.01425	30	70.1533	1.4032	.00010	1	.99990	1	11
50	0.01454	29	68.7574	1.3481	0.01455	29	68.7501	1.3482	1.00011	0	0.99989	0	10
51	.01483	30	67.4093	1.2963	.01484	29	67.4019	1.2964	.00011	0	.99989	0	9
52	.01513	29	66.1130	1.2473	.01513	29	66.1055	1.2475	.00011	1	.99989	1	8
53	.01542	29	64.8657	1.2011	.01542	29	64.8580	1.2013	.00012	0	.99988	0	7
54	.01571	29	63.6646	1.1574	.01571	29	63.6567	1.1575	.00012	1	.99988	1	6
55	0.01600	29	62.5072	1.1161	0.01600	29	62.4992	1.1163	1.00013	0	0.99987	0	5
56	.01629	29	61.3911	1.0770	.01629	29	61.3829	1.0771	.00013	1	.99987	1	4
57	.01658	29	60.3141	1.0398	.01658	29	60.3058	1.0399	.00014	0	.99986	0	3
58	.01687	29	59.2743	1.0045	.01687	29	59.2659	1.0047	.00014	1	.99986	1	2
59	.01716	29	58.2698	0.9711	.01716	30	58.2612	0.9712	.00015	0	.99985	0	1
60	0.01745		57.2987		0.01746		57.2900		1.00015		0.99985		0
↑ 90° →	cos	Diff. 1′	sec	Diff. 1′	cot	Diff. 1′	tan	Diff. 1′	csc	Diff. 1′	sin	Diff. 1′	← 89° ↑

TABLE 31

Natural Trigonometric Functions

1°→ ↓ ′	sin	Diff. 1′	csc	Diff. 1′	tan	Diff. 1′	cot	Diff. 1′	sec	Diff. 1′	cos	Diff. 1′	←178° ↓ ′
0	0. 01745	29	57. 2987	9392	0. 01746	29	57. 2900	9394	1. 00015	1	0. 99985	1	60
1	. 01774	29	56. 3595	9090	. 01775	29	56. 3506	9091	. 00016	0	. 99984	0	59
2	. 01803	29	55. 4505	8800	. 01804	29	55. 4415	8802	. 00016	1	. 99984	1	58
3	. 01832	30	54. 5705	8526	. 01833	29	54. 5613	8527	. 00017	0	. 99983	0	57
4	. 01862	29	53. 7179	8263	. 01862	29	53. 7086	8265	. 00017	1	. 99983	1	56
5	0. 01891	29	52. 8916	8013	0. 01891	29	52. 8821	8014	1. 00018	0	0. 99982	0	55
6	. 01920	29	52. 0903	7774	. 01920	29	52. 0807	7775	. 00018	1	. 99982	1	54
7	. 01949	29	51. 3129	7545	. 01949	29	51. 3032	7547	. 00019	1	. 99981	1	53
8	. 01978	29	50. 5584	7326	. 01978	29	50. 5485	7328	. 00020	0	. 99980	0	52
9	. 02007	29	49. 8258	7117	. 02007	29	49. 8157	7118	. 00020	1	. 99980	1	51
10	0. 02036	29	49. 1141	6917	0. 02036	30	49. 1039	6918	1. 00021	0	0. 99979	0	50
11	. 02065	29	48. 4224	6724	. 02066	29	48. 4121	6726	. 00021	1	. 99979	1	49
12	. 02094	29	47. 7500	6540	. 02095	29	47. 7395	6542	. 00022	1	. 99978	1	48
13	. 02123	29	47. 0960	6364	. 02124	29	47. 0853	6364	. 00023	0	. 99977	0	47
14	. 02152	29	46. 4596	6193	. 02153	29	46. 4489	6195	. 00023	1	. 99977	1	46
15	0. 02181	30	45. 8403	6031	0. 02182	29	45. 8294	6033	1. 00024	0	0. 99976	0	45
16	. 02211	29	45. 2372	5874	. 02211	29	45. 2261	5875	. 00024	1	. 99976	1	44
17	. 02240	29	44. 6498	5723	. 02240	29	44. 6386	5725	. 00025	1	. 99975	1	43
18	. 02269	29	44. 0775	5579	. 02269	29	44. 0661	5580	. 00026	0	. 99974	0	42
19	. 02298	29	43. 5196	5439	. 02298	30	43. 5081	5440	. 00026	1	. 99974	1	41
20	0. 02327	29	42. 9757	5305	0. 02328	29	42. 9641	5306	1. 00027	1	0. 99973	1	40
21	. 02356	29	42. 4452	5175	. 02357	29	42. 4335	5177	. 00028	0	. 99972	0	39
22	. 02385	29	41. 9277	5050	. 02386	29	41. 9158	5052	. 00028	1	. 99972	1	38
23	. 02414	29	41. 4227	4931	. 02415	29	41. 4106	4932	. 00029	1	. 99971	1	37
24	. 02443	29	40. 9296	4814	. 02444	29	40. 9174	4816	. 00030	1	. 99970	1	36
25	0. 02472	29	40. 4482	4702	0. 02473	29	40. 4358	4703	1. 00031	0	0. 99969	0	35
26	. 02501	29	39. 9780	4595	. 02502	29	39. 9655	4596	. 00031	1	. 99969	1	34
27	. 02530	30	39. 5185	4489	. 02531	29	39. 5059	4491	. 00032	1	. 99968	1	33
28	. 02560	29	39. 0696	4389	. 02560	29	39. 0568	4391	. 00033	1	. 99967	1	32
29	. 02589	29	38. 6307	4291	. 02589	30	38. 6177	4292	. 00034	0	. 99966	0	31
30	0. 02618	29	38. 2016	4198	0. 02619	29	38. 1885	4199	1. 00034	1	0. 99966	1	30
31	. 02647	29	37. 7818	4105	. 02648	29	37. 7686	4107	. 00035	1	. 99965	1	29
32	. 02676	29	37. 3713	4018	. 02677	29	37. 3579	4019	. 00036	1	. 99964	1	28
33	. 02705	29	36. 9695	3932	. 02706	29	36. 9560	3933	. 00037	0	. 99963	0	27
34	. 02734	29	36. 5763	3849	. 02735	29	36. 5627	3851	. 00037	1	. 99963	1	26
35	0. 02763	29	36. 1914	3769	0. 02764	29	36. 1776	3770	1. 00038	1	0. 99962	1	25
36	. 02792	29	35. 8145	3691	. 02793	29	35. 8006	3693	. 00039	1	. 99961	1	24
37	. 02821	29	35. 4454	3616	. 02822	29	35. 4313	3618	. 00040	1	. 99960	1	23
38	. 02850	29	35. 0838	3543	. 02851	30	35. 0695	3544	. 00041	0	. 99959	0	22
39	. 02879	29	34. 7295	3472	. 02881	29	34. 7151	3473	. 00041	1	. 99959	1	21
40	0. 02908	30	34. 3823	3403	0. 02910	29	34. 3678	3405	1. 00042	1	0. 99958	1	20
41	. 02938	29	34. 0420	3337	. 02939	29	34. 0273	3338	. 00043	1	. 99957	1	19
42	. 02967	29	33. 7083	3271	. 02968	29	33. 6935	3273	. 00044	1	. 99956	1	18
43	. 02996	29	33. 3812	3209	. 02997	29	33. 3662	3210	. 00045	1	. 99955	1	17
44	. 03025	29	33. 0603	3148	. 03026	29	33. 0452	3149	. 00046	1	. 99954	1	16
45	0. 03054	29	32. 7455	3088	0. 03055	29	32. 7303	3090	1. 00047	1	0. 99953	1	15
46	. 03083	29	32. 4367	3030	. 03084	30	32. 4213	3032	. 00048	0	. 99952	0	14
47	. 03112	29	32. 1337	2975	. 03114	29	32. 1181	2976	. 00048	1	. 99952	1	13
48	. 03141	29	31. 8362	2920	. 03143	29	31. 8205	2921	. 00049	1	. 99951	1	12
49	. 03170	29	31. 5442	2866	. 03172	29	31. 5284	2868	. 00050	1	. 99950	1	11
50	0. 03199	29	31. 2576	2815	0. 03201	29	31. 2416	2817	1. 00051	1	0. 99949	1	10
51	. 03228	29	30. 9761	2765	. 03230	29	30. 9599	2766	. 00052	1	. 99948	1	9
52	. 03257	29	30. 6996	2716	. 03259	29	30. 6833	2717	. 00053	1	. 99947	1	8
53	. 03286	30	30. 4280	2668	. 03288	29	30. 4116	2670	. 00054	1	. 99946	1	7
54	. 03316	29	30. 1612	2622	. 03317	29	30. 1446	2623	. 00055	1	. 99945	1	6
55	0. 03345	29	29. 8990	2576	0. 03346	30	29. 8823	2578	1. 00056	1	0. 99944	1	5
56	. 03374	29	29. 6414	2533	. 03376	29	29. 6245	2534	. 00057	1	. 99943	1	4
57	. 03403	29	29. 3881	2489	. 03405	29	29. 3711	2491	. 00058	1	. 99942	1	3
58	. 03432	29	29. 1392	2448	. 03434	29	29. 1220	2449	. 00059	1	. 99941	1	2
59	. 03461	29	28. 8944	2407	. 03463	29	28. 8771	2408	. 00060	1	. 99940	1	1
60	0. 03490		28. 6537		0. 03492		28. 6363		1. 00061		0. 99939		0
↑ 91°→	cos	Diff. 1′	sec	Diff. 1′	cot	Diff. 1′	tan	Diff. 1′	csc	Diff. 1′	sin	Diff. 1′	←88° ↑

TABLE 31

Natural Trigonometric Functions

2°→ ↓	sin	Diff. 1′	csc	Diff. 1′	tan	Diff. 1′	cot	Diff. 1′	sec	Diff. 1′	cos	Diff. 1′	←177° ↓
′													′
0	0. 03490	29	28. 6537	2367	0. 03492	29	28. 6363	2369	1. 00061	1	0. 99939	1	60
1	. 03519	29	. 4170	2328	. 03521	29	. 3994	2330	. 00062	1	. 99938	1	59
2	. 03548	29	28. 1842	2291	. 03550	29	28. 1664	2292	. 00063	1	. 99937	1	58
3	. 03577	29	27. 9551	2253	. 03579	30	27. 9372	2255	. 00064	1	. 99936	1	57
4	. 03606	29	. 7298	2218	. 03609	29	. 7117	2218	. 00065	1	. 99935	1	56
5	0. 03635	29	27. 5080	2182	0. 03638	29	27. 4899	2184	1. 00066	1	0. 99934	1	55
6	. 03664	29	. 2898	2148	. 03667	29	. 2715	2149	. 00067	1	. 99933	1	54
7	. 03693	30	27. 0750	2114	. 03696	29	27. 0566	2116	. 00068	1	. 99932	1	53
8	. 03723	29	26. 8636	2081	. 03725	29	26. 8450	2083	. 00069	1	. 99931	1	52
9	. 03752	29	. 6555	2050	. 03754	29	. 6367	2051	. 00070	2	. 99930	1	51
10	0. 03781	29	26. 4505	2018	0. 03783	29	26. 4316	2020	1. 00072	1	0. 99929	2	50
11	. 03810	29	. 2487	1988	. 03812	30	. 2296	1989	. 00073	1	. 99927	1	49
12	. 03839	29	26. 0499	1957	. 03842	29	26. 0307	1959	. 00074	1	. 99926	1	48
13	. 03868	29	25. 8542	1929	. 03871	29	25. 8348	1930	. 00075	1	. 99925	1	47
14	. 03897	29	. 6613	1900	. 03900	29	. 6418	1901	. 00076	1	. 99924	1	46
15	0. 03926	29	25. 4713	1872	0. 03929	29	25. 4517	1873	1. 00077	1	0. 99923	1	45
16	. 03955	29	. 2841	1844	. 03958	29	. 2644	1846	. 00078	1	. 99922	1	44
17	. 03984	29	25. 0997	1818	. 03987	29	25. 0798	1820	. 00079	2	. 99921	2	43
18	. 04013	29	24. 9179	1792	. 04016	30	24. 8978	1793	. 00081	1	. 99919	1	42
19	. 04042	29	. 7387	1766	. 04046	29	. 7185	1767	. 00082	1	. 99918	1	41
20	0. 04071	29	24. 5621	1741	0. 04075	29	24. 5418	1743	1. 00083	1	0. 99917	1	40
21	. 04100	29	. 3880	1716	. 04104	29	. 3675	1718	. 00084	1	. 99916	1	39
22	. 04129	30	. 2164	1693	. 04133	29	. 1957	1694	. 00085	2	. 99915	2	38
23	. 04159	29	24. 0471	1669	. 04162	29	24. 0263	1670	. 00087	1	. 99913	1	37
24	. 04188	29	23. 8802	1646	. 04191	29	23. 8593	1648	. 00088	1	. 99912	1	36
25	0. 04217	29	23. 7156	1623	0. 04220	30	23. 6945	1624	1. 00089	1	0. 99911	1	35
26	. 04246	29	. 5533	1601	. 04250	29	. 5321	1603	. 00090	1	. 99910	1	34
27	. 04275	29	. 3932	1580	. 04279	29	. 3718	1581	. 00091	2	. 99909	2	33
28	. 04304	29	. 2352	1558	. 04308	29	. 2137	1560	. 00093	1	. 99907	1	32
29	. 04333	29	23. 0794	1538	. 04337	29	23. 0577	1539	. 00094	1	. 99906	1	31
30	0. 04362	29	22. 9256	1517	0. 04366	29	22. 9038	1519	1. 00095	2	0. 99905	1	30
31	. 04391	29	. 7739	1498	. 04395	29	. 7519	1499	. 00097	1	. 99904	2	29
32	. 04420	29	. 6241	1477	. 04424	30	. 6020	1479	. 00098	1	. 99902	1	28
33	. 04449	29	. 4764	1459	. 04454	29	. 4541	1460	. 00099	1	. 99901	1	27
34	. 04478	29	. 3305	1440	. 04483	29	. 3081	1441	. 00100	2	. 99900	2	26
35	0. 04507	29	22. 1865	1421	0. 04512	29	22. 1640	1423	1. 00102	1	0. 99898	1	25
36	. 04536	29	22. 0444	1403	. 04541	29	22. 0217	1404	. 00103	1	. 99897	1	24
37	. 04565	29	21. 9041	1385	. 04570	29	21. 8813	1387	. 00104	2	. 99896	2	23
38	. 04594	29	. 7656	1368	. 04599	29	. 7426	1370	. 00106	1	. 99894	1	22
39	. 04623	30	. 6288	1351	. 04628	30	. 6056	1352	. 00107	1	. 99893	1	21
40	0. 04653	29	21. 4937	1334	0. 04658	29	21. 4704	1335	1. 00108	2	0. 99892	2	20
41	. 04682	29	. 3603	1318	. 04687	29	. 3369	1320	. 00110	1	. 99890	1	19
42	. 04711	29	. 2285	1301	. 04716	29	. 2049	1302	. 00111	2	. 99889	1	18
43	. 04740	29	21. 0984	1286	. 04745	29	21. 0747	1287	. 00113	1	. 99888	2	17
44	. 04769	29	20. 9698	1270	. 04774	29	20. 9460	1272	. 00114	1	. 99886	1	16
45	0. 04798	29	20. 8428	1254	0. 04803	30	20. 8188	1256	1. 00115	2	0. 99885	2	15
46	. 04827	29	. 7174	1240	. 04833	29	. 6932	1241	. 00117	1	. 99883	1	14
47	. 04856	29	. 5934	1225	. 04862	29	. 5691	1226	. 00118	2	. 99882	1	13
48	. 04885	29	. 4709	1210	. 04891	29	. 4465	1212	. 00120	1	. 99881	2	12
49	. 04914	29	. 3499	1196	. 04920	29	. 3253	1197	. 00121	1	. 99879	1	11
50	0. 04943	29	20. 2303	1182	0. 04949	29	20. 2056	1184	1. 00122	2	0. 99878	2	10
51	. 04972	29	20. 1121	1169	. 04978	29	20. 0872	1170	. 00124	1	. 99876	1	9
52	. 05001	29	19. 9952	1154	. 05007	30	19. 9702	1156	. 00125	2	. 99875	2	8
53	. 05030	29	. 8798	1142	. 05037	29	. 8546	1143	. 00127	1	. 99873	1	7
54	. 05059	29	. 7656	1128	. 05066	29	. 7403	1130	. 00128	2	. 99872	2	6
55	0. 05088	29	19. 6528	1116	0. 05095	29	19. 6273	1117	1. 00130	1	0. 99870	1	5
56	. 05117	29	. 5412	1103	. 05124	29	. 5156	1105	. 00131	2	. 99869	2	4
57	. 05146	29	. 4309	1091	. 05153	29	. 4051	1092	. 00133	1	. 99867	1	3
58	. 05175	30	. 3218	1078	. 05182	30	. 2959	1080	. 00134	2	. 99866	2	2
59	. 05205	29	. 2140	1067	. 05212	29	. 1879	1068	. 00136	1	. 99864	1	1
60	0. 05234		19. 1073		0. 05241		19. 0811		1. 00137		0. 99863		0
↑ 92°→	cos	Diff. 1′	sec	Diff. 1′	cot	Diff. 1′	tan	Diff. 1′	csc	Diff. 1′	sin	Diff. 1′	←87° ↑

TABLE 31
Natural Trigonometric Functions

3° → ↓ ′	sin	Diff. 1′	csc	Diff. 1′	tan	Diff. 1′	cot	Diff. 1′	sec	Diff. 1′	cos	Diff. 1′	← 176° ↓ ′
0	0. 05234	29	19. 1073	1054	0. 05241	29	19. 0811	1056	1. 00137	2	0. 99863	2	60
1	. 05263	29	19. 0019	1044	. 05270	29	18. 9755	1044	. 00139	1	. 99861	1	59
2	. 05292	29	18. 8975	1031	. 05299	29	. 8711	1033	. 00140	2	. 99860	2	58
3	. 05321	29	. 7944	1021	. 05328	29	. 7678	1022	. 00142	1	. 99858	1	57
4	. 05350	29	. 6923	1009	. 05357	30	. 6656	1011	. 00143	2	. 99857	2	56
5	0. 05379	29	18. 5914	999	0. 05387	29	18. 5645	1000	1. 00145	2	0. 99855	1	55
6	. 05408	29	. 4915	988	. 05416	29	. 4645	990	. 00147	1	. 99854	2	54
7	. 05437	29	. 3927	977	. 05445	29	. 3655	978	. 00148	2	. 99852	1	53
8	. 05466	29	. 2950	967	. 05474	29	. 2677	969	. 00150	1	. 99851	2	52
9	. 05495	29	. 1983	957	. 05503	30	. 1708	958	. 00151	2	. 99849	2	51
10	0. 05524	29	18. 1026	947	0. 05533	29	18. 0750	948	1. 00153	2	0. 99847	1	50
11	. 05553	29	18. 0079	937	. 05562	29	17. 9802	939	. 00155	1	. 99846	2	49
12	. 05582	29	17. 9142	927	. 05591	29	. 8863	929	. 00156	2	. 99844	2	48
13	. 05611	29	. 8215	917	. 05620	29	. 7934	919	. 00158	1	. 99842	1	47
14	. 05640	29	. 7298	909	. 05649	29	. 7015	909	. 00159	2	. 99841	2	46
15	0. 05669	29	17. 6389	899	0. 05678	30	17. 6106	901	1. 00161	2	0. 99839	1	45
16	. 05698	29	. 5490	890	. 05708	29	. 5205	891	. 00163	1	. 99838	2	44
17	. 05727	29	. 4600	880	. 05737	29	. 4314	882	. 00164	2	. 99836	2	43
18	. 05756	29	. 3720	872	. 05766	29	. 3432	874	. 00166	2	. 99834	1	42
19	. 05785	29	. 2848	864	. 05795	29	. 2558	865	. 00168	1	. 99833	2	41
20	0. 05814	30	17. 1984	854	0. 05824	30	17. 1693	856	1. 00169	2	0. 99831	2	40
21	. 05844	29	. 1130	847	. 05854	29	17. 0837	847	. 00171	2	. 99829	2	39
22	. 05873	29	17. 0283	837	. 05883	29	16. 9990	840	. 00173	2	. 99827	1	38
23	. 05902	29	16. 9446	830	. 05912	29	. 9150	831	. 00175	1	. 99826	2	37
24	. 05931	29	. 8616	822	. 05941	29	. 8319	823	. 00176	2	. 99824	2	36
25	0. 05960	29	16. 7794	813	0. 05970	29	16. 7496	815	1. 00178	2	0. 99822	1	35
26	. 05989	29	. 6981	806	. 05999	30	. 6681	807	. 00180	2	. 99821	2	34
27	. 06018	29	. 6175	798	. 06029	29	. 5874	799	. 00182	1	. 99819	2	33
28	. 06047	29	. 5377	790	. 06058	29	. 5075	792	. 00183	2	. 99817	2	32
29	. 06076	29	. 4587	783	. 06087	29	. 4283	784	. 00185	2	. 99815	2	31
30	0. 06105	29	16. 3804	775	0. 06116	29	16. 3499	777	1. 00187	2	0. 99813	1	30
31	. 06134	29	. 3029	768	. 06145	30	. 2722	770	. 00189	1	. 99812	2	29
32	. 06163	29	. 2261	761	. 06175	29	. 1952	762	. 00190	2	. 99810	2	28
33	. 06192	29	. 1500	754	. 06204	29	. 1190	755	. 00192	2	. 99808	2	27
34	. 06221	29	16. 0746	747	. 06233	29	16. 0435	748	. 00194	2	. 99806	2	26
35	0. 06250	29	15. 9999	739	0. 06262	29	15. 9687	742	1. 00196	2	0. 99804	1	25
36	. 06279	29	. 9260	733	. 06291	30	. 8945	734	. 00198	2	. 99803	2	24
37	. 06308	29	. 8527	726	. 06321	29	. 8211	728	. 00200	1	. 99801	2	23
38	. 06337	29	. 7801	720	. 06350	29	. 7483	721	. 00201	2	. 99799	2	22
39	. 06366	29	. 7081	713	. 06379	29	. 6762	714	. 00203	2	. 99797	2	21
40	0. 06395	29	15. 6368	707	0. 06408	30	15. 6048	708	1. 00205	2	0. 99795	2	20
41	. 06424	29	. 5661	700	. 06438	29	. 5340	702	. 00207	2	. 99793	1	19
42	. 06453	29	. 4961	694	. 06467	29	. 4638	695	. 00209	2	. 99792	2	18
43	. 06482	29	. 4267	688	. 06496	29	. 3943	689	. 00211	2	. 99790	2	17
44	. 06511	29	. 3579	681	. 06525	29	. 3254	683	. 00213	2	. 99788	2	16
45	0. 06540	29	15. 2898	676	0. 06554	30	15. 2571	678	1. 00215	1	0. 99786	2	15
46	. 06569	29	. 2222	669	. 06584	29	. 1893	671	. 00216	2	. 99784	2	14
47	. 06598	29	. 1553	664	. 06613	29	. 1222	665	. 00218	2	. 99782	2	13
48	. 06627	29	. 0889	658	. 06642	29	15. 0557	659	. 00220	2	. 99780	2	12
49	. 06656	29	15. 0231	652	. 06671	29	14. 9898	654	. 00222	2	. 99778	2	11
50	0. 06685	29	14. 9579	647	0. 06700	30	14. 9244	648	1. 00224	2	0. 99776	2	10
51	. 06714	29	. 8932	641	. 06730	29	. 8596	642	. 00226	2	. 99774	2	9
52	. 06743	30	. 8291	635	. 06759	29	. 7954	637	. 00228	2	. 99772	2	8
53	. 06773	29	. 7656	630	. 06788	29	. 7317	632	. 00230	2	. 99770	2	7
54	. 06802	29	. 7026	625	. 06817	30	. 6685	626	. 00232	2	. 99768	2	6
55	0. 06831	29	14. 6401	619	0. 06847	29	14. 6059	621	1. 00234	2	0. 99766	2	5
56	. 06860	29	. 5782	614	. 06876	29	. 5438	615	. 00236	2	. 99764	2	4
57	. 06889	29	. 5168	609	. 06905	29	. 4823	611	. 00238	2	. 99762	2	3
58	. 06918	29	. 4559	604	. 06934	29	. 4212	605	. 00240	2	. 99760	2	2
59	. 06947	29	. 3955	599	. 06963	30	. 3607	600	. 00242	2	. 99758	2	1
60	0. 06976		14. 3356		0. 06993		14. 3007		1. 00244		0. 99756		0
↑ 93° →	cos	Diff. 1′	sec	Diff. 1′	cot	Diff. 1′	tan	Diff. 1′	csc	Diff. 1′	sin	Diff. 1′	← 86° ↑

TABLE 31

Natural Trigonometric Functions

4° → ↓ ′	sin	Diff. 1′	csc	Diff. 1′	tan	Diff. 1′	cot	Diff. 1′	sec	Diff. 1′	cos	Diff. 1′	← 175° ↓ ′
0	0. 06976	29	14. 3356	594	0. 06993	29	14. 3007	596	1. 00244	2	0. 99756	2	60
1	. 07005	29	. 2762	589	. 07022	29	. 2411	590	. 00246	2	. 99754	2	59
2	. 07034	29	. 2173	584	. 07051	29	. 1821	586	. 00248	2	. 99752	2	58
3	. 07063	29	. 1589	579	. 07080	30	. 1235	580	. 00250	2	. 99750	2	57
4	. 07092	29	. 1010	575	. 07110	29	. 0655	576	. 00252	2	. 99748	2	56
5	0. 07121	29	14. 0435	570	0. 07139	29	14. 0079	572	1. 00254	3	0. 99746	2	55
6	. 07150	29	13. 9865	565	. 07168	29	13. 9507	567	. 00257	2	. 99744	2	54
7	. 07179	29	. 9300	561	. 07197	30	. 8940	562	. 00259	2	. 99742	2	53
8	. 07208	29	. 8739	556	. 07227	29	. 8378	557	. 00261	2	. 99740	2	52
9	. 07237	29	. 8183	552	. 07256	29	. 7821	554	. 00263	2	. 99738	2	51
10	0. 07266	29	13. 7631	547	0. 07285	29	13. 7267	548	1. 00265	2	0. 99736	2	50
11	. 07295	29	. 7084	543	. 07314	30	. 6719	545	. 00267	2	. 99734	3	49
12	. 07324	29	. 6541	539	. 07344	29	. 6174	540	. 00269	2	. 99731	2	48
13	. 07353	29	. 6002	534	. 07373	29	. 5634	536	. 00271	3	. 99729	2	47
14	. 07382	29	. 5468	531	. 07402	29	. 5098	532	. 00274	2	. 99727	2	46
15	0. 07411	29	13. 4937	526	0. 07431	30	13. 4566	527	1. 00276	2	0. 99725	2	45
16	. 07440	29	. 4411	522	. 07461	29	. 4039	524	. 00278	2	. 99723	2	44
17	. 07469	29	. 3889	518	. 07490	29	. 3515	519	. 00280	2	. 99721	2	43
18	. 07498	29	. 3371	514	. 07519	29	. 2996	516	. 00282	2	. 99719	3	42
19	. 07527	29	. 2857	510	. 07548	30	. 2480	511	. 00284	3	. 99716	2	41
20	0. 07556	29	13. 2347	506	0. 07578	29	13. 1969	508	1. 00287	2	0. 99714	2	40
21	. 07585	29	. 1841	502	. 07607	29	. 1461	503	. 00289	2	. 99712	2	39
22	. 07614	29	. 1339	499	. 07636	29	. 0958	500	. 00291	2	. 99710	2	38
23	. 07643	29	. 0840	494	. 07665	30	13. 0458	496	. 00293	3	. 99708	3	37
24	. 07672	29	13. 0346	491	. 07695	29	12. 9962	493	. 00296	2	. 99705	2	36
25	0. 07701	29	12. 9855	487	0. 07724	29	12. 9469	488	1. 00298	2	0. 99703	2	35
26	. 07730	29	. 9368	484	. 07753	29	. 8981	485	. 00300	2	. 99701	2	34
27	. 07759	29	. 8884	480	. 07782	30	. 8496	482	. 00302	3	. 99699	3	33
28	. 07788	29	. 8404	476	. 07812	29	. 8014	478	. 00305	2	. 99696	2	32
29	. 07817	29	. 7928	473	. 07841	29	. 7536	474	. 00307	2	. 99694	2	31
30	0. 07846	29	12. 7455	469	0. 07870	29	12. 7062	471	1. 00309	3	0. 99692	3	30
31	. 07875	29	. 6986	466	. 07899	30	. 6591	467	. 00312	2	. 99689	2	29
32	. 07904	29	. 6520	463	. 07929	29	. 6124	464	. 00314	2	. 99687	2	28
33	. 07933	29	. 6057	459	. 07958	29	. 5660	461	. 00316	2	. 99685	2	27
34	. 07962	29	. 5598	456	. 07987	30	. 5199	457	. 00318	3	. 99683	3	26
35	0. 07991	29	12. 5142	452	0. 08017	29	12. 4742	454	1. 00321	2	0. 99680	2	25
36	. 08020	29	. 4690	449	. 08046	29	. 4288	450	. 00323	3	. 99678	2	24
37	. 08049	29	. 4241	446	. 08075	29	. 3838	448	. 00326	2	. 99676	3	23
38	. 08078	29	. 3795	443	. 08104	30	. 3390	444	. 00328	2	. 99673	2	22
39	. 08107	29	. 3352	439	. 08134	29	. 2946	441	. 00330	3	. 99671	3	21
40	0. 08136	29	12. 2913	437	0. 08163	29	12. 2505	438	1. 00333	2	0. 99668	2	20
41	. 08165	29	. 2476	433	. 08192	29	. 2067	435	. 00335	2	. 99666	2	19
42	. 08194	29	. 2043	431	. 08221	30	. 1632	431	. 00337	3	. 99664	3	18
43	. 08223	29	. 1612	427	. 08251	29	. 1201	429	. 00340	2	. 99661	2	17
44	. 08252	29	. 1185	424	. 08280	29	. 0772	426	. 00342	3	. 99659	2	16
45	0. 08281	29	12. 0761	421	0. 08309	30	12. 0346	423	1. 00345	2	0. 99657	3	15
46	. 08310	29	12. 0340	419	. 08339	29	11. 9923	419	. 00347	3	. 99654	2	14
47	. 08339	29	11. 9921	415	. 08368	29	. 9504	417	. 00350	2	. 99652	3	13
48	. 08368	29	. 9506	413	. 08397	30	. 9087	414	. 00352	2	. 99649	2	12
49	. 08397	29	. 9093	409	. 08427	29	. 8673	411	. 00354	3	. 99647	3	11
50	0. 08426	29	11. 8684	407	0. 08456	29	11. 8262	409	1. 00357	2	0. 99644	2	10
51	. 08455	29	. 8277	404	. 08485	29	. 7853	405	. 00359	3	. 99642	3	9
52	. 08484	29	. 7873	402	. 08514	30	. 7448	403	. 00362	2	. 99639	2	8
53	. 08513	29	. 7471	398	. 08544	29	. 7045	400	. 00364	3	. 99637	2	7
54	. 08542	29	. 7073	396	. 08573	29	. 6645	397	. 00367	2	. 99635	3	6
55	0. 08571	29	11. 6677	393	0. 08602	30	11. 6248	395	1. 00369	3	0. 99632	2	5
56	. 08600	29	. 6284	391	. 08632	29	. 5853	392	. 00372	2	. 99630	3	4
57	. 08629	29	. 5893	388	. 08661	29	. 5461	389	. 00374	3	. 99627	2	3
58	. 08658	29	. 5505	385	. 08690	30	. 5072	387	. 00377	2	. 99625	3	2
59	. 08687	29	. 5120	383	. 08720	29	. 4685	384	. 00379	3	. 99622	3	1
60	0. 08716		11. 4737		0. 08749		11. 4301		1. 00382		0. 99619		0
↑ 94° →	cos	Diff. 1′	sec	Diff. 1′	cot	Diff. 1′	tan	Diff. 1′	csc	Diff. 1′	sin	Diff. 1′	← 85° ↑

TABLE 31

Natural Trigonometric Functions

5°→ ↓ ′	sin	Diff. 1′	csc	Diff. 1′	tan	Diff. 1′	cot	Diff. 1′	sec	Diff. 1′	cos	Diff. 1′	←174° ↓ ′
0	0.08716	29	11.4737	380	0.08749	29	11.4301	382	1.00382	3	0.99619	2	60
1	.08745	29	.4357	378	.08778	29	.3919	379	.00385	2	.99617	3	59
2	.08774	29	.3979	375	.08807	30	.3540	377	.00387	3	.99614	2	58
3	.08803	28	.3604	373	.08837	29	.3163	374	.00390	2	.99612	3	57
4	.08831	29	.3231	370	.08866	29	.2789	372	.00392	3	.99609	2	56
5	0.08860	29	11.2861	368	0.08895	30	11.2417	369	1.00395	2	0.99607	3	55
6	.08889	29	.2493	365	.08925	29	.2048	367	.00397	3	.99604	2	54
7	.08918	29	.2128	363	.08954	29	.1681	365	.00400	3	.99602	3	53
8	.08947	29	.1765	361	.08983	30	.1316	362	.00403	2	.99599	3	52
9	.08976	29	.1404	359	.09013	29	.0954	360	.00405	3	.99596	2	51
10	0.09005	29	11.1045	356	0.09042	29	11.0594	357	1.00408	3	0.99594	3	50
11	.09034	29	.0689	353	.09071	30	11.0237	355	.00411	2	.99591	3	49
12	.09063	29	11.0336	352	.09101	29	10.9882	353	.00413	3	.99588	2	48
13	.09092	29	10.9984	349	.09130	29	.9529	351	.00416	3	.99586	3	47
14	.09121	29	.9635	347	.09159	30	.9178	349	.00419	2	.99583	3	46
15	0.09150	29	10.9288	345	0.09189	29	10.8829	346	1.00421	3	0.99580	2	45
16	.09179	29	.8943	343	.09218	29	.8483	344	.00424	3	.99578	3	44
17	.09208	29	.8600	340	.09247	30	.8139	342	.00427	2	.99575	3	43
18	.09237	29	.8260	339	.09277	29	.7797	340	.00429	3	.99572	2	42
19	.09266	29	.7921	336	.09306	29	.7457	338	.00432	3	.99570	3	41
20	0.09295	29	10.7585	334	0.09335	30	10.7119	336	1.00435	3	0.99567	3	40
21	.09324	29	.7251	332	.09365	29	.6783	333	.00438	2	.99564	2	39
22	.09353	29	.6919	330	.09394	29	.6450	332	.00440	3	.99562	3	38
23	.09382	29	.6589	328	.09423	30	.6118	329	.00443	3	.99559	3	37
24	.09411	29	.6261	326	.09453	29	.5789	327	.00446	3	.99556	3	36
25	0.09440	29	10.5935	324	0.09482	29	10.5462	326	1.00449	2	0.99553	2	35
26	.09469	29	.5611	322	.09511	30	.5136	323	.00451	3	.99551	3	34
27	.09498	29	.5289	320	.09541	29	.4813	322	.00454	3	.99548	3	33
28	.09527	29	.4969	319	.09570	30	.4491	319	.00457	3	.99545	3	32
29	.09556	29	.4650	316	.09600	29	.4172	318	.00460	3	.99542	2	31
30	0.09585	29	10.4334	314	0.09629	29	10.3854	316	1.00463	2	0.99540	3	30
31	.09614	28	.4020	312	.09658	30	.3538	314	.00465	3	.99537	3	29
32	.09642	29	.3708	311	.09688	29	.3224	311	.00468	3	.99534	3	28
33	.09671	29	.3397	308	.09717	29	.2913	311	.00471	3	.99531	3	27
34	.09700	29	.3089	307	.09746	30	.2602	308	.00474	3	.99528	2	26
35	0.09729	29	10.2782	305	0.09776	29	10.2294	306	1.00477	3	0.99526	3	25
36	.09758	29	.2477	303	.09805	29	.1988	305	.00480	2	.99523	3	24
37	.09787	29	.2174	301	.09834	30	.1683	302	.00482	3	.99520	3	23
38	.09816	29	.1873	300	.09864	29	.1381	301	.00485	3	.99517	3	22
39	.09845	29	.1573	298	.09893	30	.1080	300	.00488	3	.99514	3	21
40	0.09874	29	10.1275	296	0.09923	29	10.0780	297	1.00491	3	0.99511	3	20
41	.09903	29	.0979	294	.09952	29	.0483	296	.00494	3	.99508	2	19
42	.09932	29	.0685	293	.09981	30	10.0187	2939	.00497	3	.99506	3	18
43	.09961	29	.0392	291	.10011	29	9.98931	2924	.00500	3	.99503	3	17
44	.09990	29	.0101	2887	.10040	29	.96007	2906	.00503	3	.99500	3	16
45	0.10019	29	9.98123	2875	0.10069	30	9.93101	2890	1.00506	3	0.99497	3	15
46	.10048	29	.95248	2859	.10099	29	.90211	2873	.00509	3	.99494	3	14
47	.10077	29	.92389	2842	.10128	30	.87338	2856	.00512	3	.99491	3	13
48	.10106	29	.89547	2825	.10158	29	.84482	2841	.00515	3	.99488	3	12
49	.10135	29	.86722	2810	.10187	29	.81641	2824	.00518	3	.99485	3	11
50	0.10164	28	9.83912	2793	0.10216	30	9.78817	2808	1.00521	3	0.99482	3	10
51	.10192	29	.81119	2778	.10246	29	.76009	2792	.00524	3	.99479	3	9
52	.10221	29	.78341	2762	.10275	30	.73217	2776	.00527	3	.99476	3	8
53	.10250	29	.75579	2746	.10305	29	.70441	2761	.00530	3	.99473	3	7
54	.10279	29	.72833	2730	.10334	29	.67680	2745	.00533	3	.99470	3	6
55	0.10308	29	9.70103	2716	0.10363	30	9.64935	2730	1.00536	3	0.99467	3	5
56	.10337	29	.67387	2700	.10393	29	.62205	2715	.00539	3	.99464	3	4
57	.10366	29	.64687	2685	.10422	30	.59490	2699	.00542	3	.99461	3	3
58	.10395	29	.62002	2670	.10452	29	.56791	2685	.00545	3	.99458	3	2
59	.10424	29	.59332	2655	.10481	29	.54106	2670	.00548	3	.99455	3	1
60	0.10453		9.56677		0.10510		9.51436		1.00551		0.99452		0
↑ 95°→	cos	Diff. 1′	sec	Diff. 1′	cot	Diff. 1′	tan	Diff. 1′	csc	Diff. 1′	sin	Diff. 1′	←84° ↑

TABLE 31

Natural Trigonometric Functions

6° → ↓ ′	sin	Diff. 1′	csc	Diff. 1′	tan	Diff. 1′	cot	Diff. 1′	sec	Diff. 1′	cos	Diff. 1′	← 173° ↓ ′
0	0. 10453	29	9. 56677	2640	0. 10510	30	9. 51436	2655	1. 00551	3	0. 99452	3	60
1	. 10482	29	. 54037	2626	. 10540	29	. 48781	2640	. 00554	3	. 99449	3	59
2	. 10511	29	. 51411	2611	. 10569	30	. 46141	2626	. 00557	3	. 99446	3	58
3	. 10540	29	. 48800	2597	. 10599	29	. 43515	2611	. 00560	3	. 99443	3	57
4	. 10569	28	. 46203	2583	. 10628	29	. 40904	2597	. 00563	3	. 99440	3	56
5	0. 10597	29	9. 43620	2568	0. 10657	30	9. 38307	2583	1. 00566	3	0. 99437	3	55
6	. 10626	29	. 41052	2555	. 10687	29	. 35724	2569	. 00569	4	. 99434	3	54
7	. 10655	29	. 38497	2540	. 10716	30	. 33155	2556	. 00573	3	. 99431	3	53
8	. 10684	29	. 35957	2527	. 10746	29	. 30599	2541	. 00576	3	. 99428	4	52
9	. 10713	29	. 33430	2513	. 10775	30	. 28058	2528	. 00579	3	. 99424	3	51
10	0. 10742	29	9. 30917	2500	0. 10805	29	9. 25530	2514	1. 00582	3	0. 99421	3	50
11	. 10771	29	. 28417	2486	. 10834	29	. 23016	2500	. 00585	3	. 99418	3	49
12	. 10800	29	. 25931	2472	. 10863	30	. 20516	2488	. 00588	4	. 99415	3	48
13	. 10829	29	. 23459	2460	. 10893	29	. 18028	2474	. 00592	3	. 99412	3	47
14	. 10858	29	. 20999	2446	. 10922	30	. 15554	2461	. 00595	3	. 99409	3	46
15	0. 10887	29	9. 18553	2433	0. 10952	29	9. 13093	2447	1. 00598	3	0. 99406	4	45
16	. 10916	29	. 16120	2421	. 10981	30	. 10646	2435	. 00601	3	. 99402	3	44
17	. 10945	28	. 13699	2407	. 11011	29	. 08211	2422	. 00604	4	. 99399	3	43
18	. 10973	29	. 11292	2395	. 11040	30	. 05789	2410	. 00608	3	. 99396	3	42
19	. 11002	29	. 08897	2382	. 11070	29	. 03379	2396	. 00611	3	. 99393	3	41
20	0. 11031	29	9. 06515	2369	0. 11099	29	9. 00983	2385	1. 00614	3	0. 99390	4	40
21	. 11060	29	. 04146	2358	. 11128	30	8. 98598	2371	. 00617	4	. 99386	3	39
22	. 11089	29	9. 01788	2344	. 11158	29	. 96227	2360	. 00621	3	. 99383	3	38
23	. 11118	29	8. 99444	2333	. 11187	30	. 93867	2347	. 00624	3	. 99380	3	37
24	. 11147	29	. 97111	2320	. 11217	29	. 91520	2335	. 00627	3	. 99377	3	36
25	0. 11176	29	8. 94791	2309	0. 11246	30	8. 89185	2323	1. 00630	4	0. 99374	4	35
26	. 11205	29	. 92482	2296	. 11276	29	. 86862	2311	. 00634	3	. 99370	3	34
27	. 11234	29	. 90186	2285	. 11305	30	. 84551	2299	. 00637	3	. 99367	3	33
28	. 11263	28	. 87901	2273	. 11335	29	. 82252	2288	. 00640	4	. 99364	4	32
29	. 11291	29	. 85628	2261	. 11364	30	. 79964	2275	. 00644	3	. 99360	3	31
30	0. 11320	29	8. 83367	2249	0. 11394	29	8. 77689	2264	1. 00647	3	0. 99357	3	30
31	. 11349	29	. 81118	2238	. 11423	29	. 75425	2253	. 00650	4	. 99354	3	29
32	. 11378	29	. 78880	2227	. 11452	30	. 73172	2241	. 00654	3	. 99351	4	28
33	. 11407	29	. 76653	2215	. 11482	29	. 70931	2230	. 00657	3	. 99347	3	27
34	. 11436	29	. 74438	2204	. 11511	30	. 68701	2219	. 00660	4	. 99344	3	26
35	0. 11465	29	8. 72234	2193	0. 11541	29	8. 66482	2207	1. 00664	3	0. 99341	4	25
36	. 11494	29	. 70041	2182	. 11570	30	. 64275	2197	. 00667	4	. 99337	3	24
37	. 11523	29	. 67859	2171	. 11600	29	. 62078	2185	. 00671	3	. 99334	3	23
38	. 11552	28	. 65688	2160	. 11629	30	. 59893	2175	. 00674	3	. 99331	4	22
39	. 11580	29	. 63528	2149	. 11659	29	. 57718	2163	. 00677	4	. 99327	3	21
40	0. 11609	29	8. 61379	2138	0. 11688	30	8. 55555	2153	1. 00681	3	0. 99324	4	20
41	. 11638	29	. 59241	2128	. 11718	29	. 53402	2143	. 00684	4	. 99320	3	19
42	. 11667	29	. 57113	2117	. 11747	30	. 51259	2131	. 00688	3	. 99317	3	18
43	. 11696	29	. 54996	2107	. 11777	29	. 49128	2121	. 00691	4	. 99314	4	17
44	. 11725	29	. 52889	2096	. 11806	30	. 47007	2111	. 00695	3	. 99310	3	16
45	0. 11754	29	8. 50793	2086	0. 11836	29	8. 44896	2101	1. 00698	3	0. 99307	4	15
46	. 11783	29	. 48707	2075	. 11865	30	. 42795	2090	. 00701	4	. 99303	3	14
47	. 11812	28	. 46632	2066	. 11895	29	. 40705	2080	. 00705	3	. 99300	3	13
48	. 11840	29	. 44566	2055	. 11924	30	. 38625	2070	. 00708	4	. 99297	4	12
49	. 11869	29	. 42511	2045	. 11954	29	. 36555	2059	. 00712	3	. 99293	3	11
50	0. 11898	29	8. 40466	2035	0. 11983	30	8. 34496	2050	1. 00715	4	0. 99290	4	10
51	. 11927	29	. 38431	2026	. 12013	29	. 32446	2040	. 00719	3	. 99286	3	9
52	. 11956	29	. 36405	2015	. 12042	30	. 30406	2030	. 00722	4	. 99283	4	8
53	. 11985	29	. 34390	2006	. 12072	29	. 28376	2021	. 00726	4	. 99279	3	7
54	. 12014	29	. 32384	1996	. 12101	30	. 26355	2010	. 00730	3	. 99276	4	6
55	0. 12043	28	8. 30388	1986	0. 12131	29	8. 24345	2001	1. 00733	4	0. 99272	3	5
56	. 12071	29	. 28402	1977	. 12160	30	. 22344	1992	. 00737	3	. 99269	4	4
57	. 12100	29	. 26425	1968	. 12190	29	. 20352	1982	. 00740	4	. 99265	3	3
58	. 12129	29	. 24457	1957	. 12219	30	. 18370	1972	. 00744	3	. 99262	4	2
59	. 12158	29	. 22500	1949	. 12249	29	. 16398	1963	. 00747	4	. 99258	3	1
60	0. 12187		8. 20551		0. 12278		8. 14435		1. 00751		0. 99255		0
↑ 96° →	cos	Diff. 1′	sec	Diff. 1′	cot	Diff. 1′	tan	Diff. 1′	csc	Diff. 1′	sin	Diff. 1′	← 83° ↑

TABLE 31

Natural Trigonometric Functions

7° → ↓ ′	sin	Diff. 1′	csc	Diff. 1′	tan	Diff. 1′	cot	Diff. 1′	sec	Diff. 1′	cos	Diff. 1′	←172° ↓ ′
0	0.12187	29	8.20551	1939	0.12278	30	8.14435	1954	1.00751	4	0.99255	4	60
1	.12216	29	.18612	1931	.12308	30	.12481	1945	.00755	3	.99251	3	59
2	.12245	29	.16681	1921	.12338	29	.10536	1936	.00758	4	.99248	4	58
3	.12274	28	.14760	1911	.12367	30	.08600	1926	.00762	3	.99244	4	57
4	.12302	29	.12849	1903	.12397	29	.06674	1918	.00765	4	.99240	3	56
5	0.12331	29	8.10946	1894	0.12426	30	8.04756	1908	1.00769	4	0.99237	4	55
6	.12360	29	.09052	1885	.12456	29	.02848	1900	.00773	3	.99233	3	54
7	.12389	29	.07167	1876	.12485	30	8.00948	1890	.00776	4	.99230	4	53
8	.12418	29	.05291	1868	.12515	29	7.99058	1882	.00780	4	.99226	4	52
9	.12447	29	.03423	1858	.12544	30	.97176	1874	.00784	3	.99222	3	51
10	0.12476	28	8.01565	1851	0.12574	29	7.95302	1864	1.00787	4	0.99219	4	50
11	.12504	29	7.99714	1841	.12603	30	.93438	1856	.00791	4	.99215	4	49
12	.12533	29	.97873	1833	.12633	29	.91582	1848	.00795	4	.99211	3	48
13	.12562	29	.96040	1824	.12662	30	.89734	1839	.00799	3	.99208	4	47
14	.12591	29	.94216	1817	.12692	30	.87895	1831	.00802	4	.99204	4	46
15	0.12620	29	7.92399	1807	0.12722	29	7.86064	1822	1.00806	4	0.99200	3	45
16	.12649	29	.90592	1800	.12751	30	.84242	1814	.00810	3	.99197	4	44
17	.12678	28	.88792	1791	.12781	29	.82428	1806	.00813	4	.99193	4	43
18	.12706	29	.87001	1783	.12810	30	.80622	1797	.00817	4	.99189	3	42
19	.12735	29	.85218	1775	.12840	29	.78825	1790	.00821	4	.99186	4	41
20	0.12764	29	7.83443	1766	0.12869	30	7.77035	1781	1.00825	3	0.99182	4	40
21	.12793	29	.81677	1759	.12899	30	.75254	1774	.00828	4	.99178	3	39
22	.12822	29	.79918	1751	.12929	29	.73480	1765	.00832	4	.99175	4	38
23	.12851	29	.78167	1743	.12958	30	.71715	1758	.00836	4	.99171	4	37
24	.12880	28	.76424	1735	.12988	29	.69957	1749	.00840	4	.99167	4	36
25	0.12908	29	7.74689	1727	0.13017	30	7.68208	1742	1.00844	4	0.99163	3	35
26	.12937	29	.72962	1720	.13047	29	.66466	1734	.00848	3	.99160	4	34
27	.12966	29	.71242	1712	.13076	30	.64732	1727	.00851	4	.99156	4	33
28	.12995	29	.69530	1704	.13106	30	.63005	1718	.00855	4	.99152	4	32
29	.13024	29	.67826	1696	.13136	29	.61287	1712	.00859	4	.99148	4	31
30	0.13053	28	7.66130	1689	0.13165	30	7.59575	1703	1.00863	4	0.99144	3	30
31	.13081	29	.64441	1682	.13195	29	.57872	1696	.00867	4	.99141	4	29
32	.13110	29	.62759	1674	.13224	30	.56176	1689	.00871	4	.99137	4	28
33	.13139	29	.61085	1667	.13254	30	.54487	1681	.00875	3	.99133	4	27
34	.13168	29	.59418	1659	.13284	29	.52806	1674	.00878	4	.99129	4	26
35	0.13197	29	7.57759	1652	0.13313	30	7.51132	1667	1.00882	4	0.99125	3	25
36	.13226	28	.56107	1645	.13343	29	.49465	1659	.00886	4	.99122	4	24
37	.13254	29	.54462	1637	.13372	30	.47806	1652	.00890	4	.99118	4	23
38	.13283	29	.52825	1631	.13402	30	.46154	1645	.00894	4	.99114	4	22
39	.13312	29	.51194	1623	.13432	29	.44509	1638	.00898	4	.99110	4	21
40	0.13341	29	7.49571	1616	0.13461	30	7.42871	1631	1.00902	4	0.99106	4	20
41	.13370	29	.47955	1609	.13491	30	.41240	1624	.00906	4	.99102	4	19
42	.13399	28	.46346	1603	.13521	29	.39616	1617	.00910	4	.99098	4	18
43	.13427	29	.44743	1595	.13550	30	.37999	1610	.00914	4	.99094	3	17
44	.13456	29	.43148	1588	.13580	29	.36389	1603	.00918	4	.99091	4	16
45	0.13485	29	7.41560	1582	0.13609	30	7.34786	1596	1.00922	4	0.99087	4	15
46	.13514	29	.39978	1575	.13639	30	.33190	1590	.00926	4	.99083	4	14
47	.13543	29	.38403	1568	.13669	29	.31600	1582	.00930	4	.99079	4	13
48	.13572	28	.36835	1561	.13698	30	.30018	1576	.00934	4	.99075	4	12
49	.13600	29	.35274	1555	.13728	30	.28442	1569	.00938	4	.99071	4	11
50	0.13629	29	7.33719	1548	0.13758	29	7.26873	1563	1.00942	4	0.99067	4	10
51	.13658	29	.32171	1541	.13787	30	.25310	1556	.00946	4	.99063	4	9
52	.13687	29	.30630	1535	.13817	29	.23754	1550	.00950	4	.99059	4	8
53	.13716	28	.29095	1529	.13846	30	.22204	1543	.00954	4	.99055	4	7
54	.13744	29	.27566	1522	.13876	30	.20661	1536	.00958	4	.99051	4	6
55	0.13773	29	7.26044	1515	0.13906	29	7.19125	1531	1.00962	4	0.99047	4	5
56	.13802	29	.24529	1510	.13935	30	.17594	1523	.00966	4	.99043	4	4
57	.13831	29	.23019	1502	.13965	30	.16071	1518	.00970	5	.99039	4	3
58	.13860	29	.21517	1497	.13995	29	.14553	1511	.00975	4	.99035	4	2
59	.13889	28	.20020	1490	.14024	30	.13042	1505	.00979	4	.99031	4	1
60	0.13917		7.18530		0.14054		7.11537		1.00983		0.99027		0
↑ 97° →	cos	Diff. 1′	sec	Diff. 1′	cot	Diff. 1′	tan	Diff. 1′	csc	Diff. 1′	sin	Diff. 1′	←82° ↑

TABLE 31

Natural Trigonometric Functions

8° → ↓ ′	sin	Diff. 1′	csc	Diff. 1′	tan	Diff. 1′	cot	Diff. 1′	sec	Diff. 1′	cos	Diff. 1′	← 171° ↓ ′
0	0.13917	29	7.18530	1484	0.14054	30	7.11537	1499	1.00983	4	0.99027	4	60
1	.13946	29	.17046	1478	.14084	29	.10038	1492	.00987	4	.99023	4	59
2	.13975	29	.15568	1472	.14113	30	.08546	1487	.00991	4	.99019	4	58
3	.14004	29	.14096	1466	.14143	30	.07059	1480	.00995	4	.99015	4	57
4	.14033	28	.12630	1459	.14173	29	.05579	1474	.00999	5	.99011	5	56
5	0.14061	29	7.11171	1454	0.14202	30	7.04105	1468	1.01004	4	0.99006	4	55
6	.14090	29	.09717	1448	.14232	30	.02637	1463	.01008	4	.99002	4	54
7	.14119	29	.08269	1441	.14262	29	7.01174	1456	.01012	4	.98998	4	53
8	.14148	29	.06828	1436	.14291	30	6.99718	1450	.01016	4	.98994	4	52
9	.14177	28	.05392	1430	.14321	30	.98268	1445	.01020	4	.98990	4	51
10	0.14205	29	7.03962	1424	0.14351	30	6.96823	1438	1.01024	5	0.98986	4	50
11	.14234	29	.02538	1418	.14381	29	.95385	1433	.01029	4	.98982	4	49
12	.14263	29	7.01120	1412	.14410	30	.93952	1427	.01033	4	.98978	5	48
13	.14292	28	6.99708	1407	.14440	30	.92525	1421	.01037	4	.98973	4	47
14	.14320	29	.98301	1401	.14470	29	.91104	1416	.01041	5	.98969	4	46
15	0.14349	29	6.96900	1395	0.14499	30	6.89688	1410	1.01046	4	0.98965	4	45
16	.14378	29	.95505	1390	.14529	30	.88278	1404	.01050	4	.98961	4	44
17	.14407	29	.94115	1384	.14559	29	.86874	1399	.01054	5	.98957	4	43
18	.14436	28	.92731	1379	.14588	30	.85475	1393	.01059	4	.98953	5	42
19	.14464	29	.91352	1373	.14618	30	.84082	1388	.01063	4	.98948	4	41
20	0.14493	29	6.89979	1367	0.14648	30	6.82694	1382	1.01067	4	0.98944	4	40
21	.14522	29	.88612	1362	.14678	29	.81312	1376	.01071	5	.98940	4	39
22	.14551	29	.87250	1357	.14707	30	.79936	1372	.01076	4	.98936	5	38
23	.14580	28	.85893	1351	.14737	30	.78564	1365	.01080	4	.98931	4	37
24	.14608	29	.84542	1346	.14767	29	.77199	1361	.01084	5	.98927	4	36
25	0.14637	29	6.83196	1340	0.14796	30	6.75838	1355	1.01089	4	0.98923	4	35
26	.14666	29	.81856	1335	.14826	30	.74483	1350	.01093	4	.98919	5	34
27	.14695	28	.80521	1330	.14856	30	.73133	1344	.01097	5	.98914	4	33
28	.14723	29	.79191	1325	.14886	29	.71789	1339	.01102	4	.98910	4	32
29	.14752	29	.77866	1319	.14915	30	.70450	1334	.01106	5	.98906	4	31
30	0.14781	29	6.76547	1314	0.14945	30	6.69116	1329	1.01111	4	0.98902	5	30
31	.14810	28	.75233	1309	.14975	30	.67787	1324	.01115	4	.98897	4	29
32	.14838	29	.73924	1304	.15005	29	.66463	1319	.01119	5	.98893	4	28
33	.14867	29	.72620	1299	.15034	30	.65144	1313	.01124	4	.98889	5	27
34	.14896	29	.71321	1294	.15064	30	.63831	1308	.01128	5	.98884	4	26
35	0.14925	29	6.70027	1289	0.15094	30	6.62523	1304	1.01133	4	0.98880	4	25
36	.14954	28	.68738	1284	.15124	29	.61219	1298	.01137	5	.98876	5	24
37	.14982	29	.67454	1278	.15153	30	.59921	1294	.01142	4	.98871	4	23
38	.15011	29	.66176	1274	.15183	30	.58627	1288	.01146	5	.98867	4	22
39	.15040	29	.64902	1269	.15213	30	.57339	1284	.01151	4	.98863	5	21
40	0.15069	28	6.63633	1264	0.15243	29	6.56055	1278	1.01155	5	0.98858	4	20
41	.15097	29	.62369	1259	.15272	30	.54777	1274	.01160	4	.98854	5	19
42	.15126	29	.61110	1255	.15302	30	.53503	1269	.01164	5	.98849	4	18
43	.15155	29	.59855	1249	.15332	30	.52234	1264	.01169	4	.98845	4	17
44	.15184	28	.58606	1245	.15362	29	.50970	1260	.01173	5	.98841	5	16
45	0.15212	29	6.57361	1240	0.15391	30	6.49710	1254	1.01178	4	0.98836	4	15
46	.15241	29	.56121	1235	.15421	30	.48456	1250	.01182	5	.98832	5	14
47	.15270	29	.54886	1231	.15451	30	.47206	1245	.01187	4	.98827	4	13
48	.15299	28	.53655	1226	.15481	30	.45961	1241	.01191	5	.98823	5	12
49	.15327	29	.52429	1221	.15511	29	.44720	1236	.01196	4	.98818	4	11
50	0.15356	29	6.51208	1217	0.15540	30	6.43484	1231	1.01200	5	0.98814	5	10
51	.15385	29	.49991	1212	.15570	30	.42253	1227	.01205	4	.98809	4	9
52	.15414	28	.48779	1207	.15600	30	.41026	1222	.01209	5	.98805	5	8
53	.15442	29	.47572	1203	.15630	30	.39804	1217	.01214	5	.98800	4	7
54	.15471	29	.46369	1198	.15660	29	.38587	1213	.01219	4	.98796	5	6
55	0.15500	29	6.45171	1194	0.15689	30	6.37374	1209	1.01223	5	0.98791	4	5
56	.15529	28	.43977	1190	.15719	30	.36165	1204	.01228	5	.98787	5	4
57	.15557	29	.42787	1185	.15749	30	.34961	1200	.01233	4	.98782	4	3
58	.15586	29	.41602	1180	.15779	30	.33761	1195	.01237	5	.98778	5	2
59	.15615	28	.40422	1177	.15809	29	.32566	1191	.01242	5	.98773	4	1
60	0.15643		6.39245		0.15838		6.31375		1.01247		0.98769		0
↑ 98° →	cos	Diff. 1′	sec	Diff. 1′	cot	Diff. 1′	tan	Diff. 1′	csc	Diff. 1′	sin	Diff. 1′	← 81° ↑

TABLE 31

Natural Trigonometric Functions

9°→ ↓ ′	sin	Diff. 1′	csc	Diff. 1′	tan	Diff. 1′	cot	Diff. 1′	sec	Diff. 1′	cos	Diff. 1′	←170° ↓ ′
0	0.15643	29	6.39245	1172	0.15838	30	6.31375	1186	1.01247	4	0.98769	5	60
1	.15672	29	.38073	1167	.15868	30	.30189	1182	.01251	5	.98764	4	59
2	.15701	29	.36906	1163	.15898	30	.29007	1178	.01256	5	.98760	5	58
3	.15730	28	.35743	1159	.15928	30	.27829	1174	.01261	4	.98755	4	57
4	.15758	29	.34584	1155	.15958	30	.26655	1169	.01265	5	.98751	5	56
5	0.15787	29	6.33429	1150	0.15988	29	6.25486	1165	1.01270	5	0.98746	5	55
6	.15816	29	.32279	1146	.16017	30	.24321	1161	.01275	4	.98741	4	54
7	.15845	28	.31133	1142	.16047	30	.23160	1157	.01279	5	.98737	5	53
8	.15873	29	.29991	1138	.16077	30	.22003	1152	.01284	5	.98732	4	52
9	.15902	29	.28853	1134	.16107	30	.20851	1148	.01289	5	.98728	5	51
10	0.15931	28	6.27719	1129	0.16137	30	6.19703	1144	1.01294	4	0.98723	5	50
11	.15959	29	.26590	1126	.16167	29	.18559	1140	.01298	5	.98718	4	49
12	.15988	29	.25464	1121	.16196	30	.17419	1136	.01303	5	.98714	5	48
13	.16017	29	.24343	1117	.16226	30	.16283	1132	.01308	5	.98709	5	47
14	.16046	28	.23226	1113	.16256	30	.15151	1128	.01313	4	.98704	4	46
15	0.16074	29	6.22113	1109	0.16286	30	6.14023	1124	1.01317	5	0.98700	5	45
16	.16103	29	.21004	1106	.16316	30	.12899	1120	.01322	5	.98695	5	44
17	.16132	28	.19898	1101	.16346	30	.11779	1115	.01327	5	.98690	4	43
18	.16160	29	.18797	1097	.16376	29	.10664	1112	.01332	5	.98686	5	42
19	.16189	29	.17700	1093	.16405	30	.09552	1108	.01337	5	.98681	5	41
20	0.16218	28	6.16607	1090	0.16435	30	6.08444	1104	1.01342	4	0.98676	5	40
21	.16246	29	.15517	1085	.16465	30	.07340	1100	.01346	5	.98671	4	39
22	.16275	29	.14432	1082	.16495	30	.06240	1097	.01351	5	.98667	5	38
23	.16304	29	.13350	1077	.16525	30	.05143	1092	.01356	5	.98662	5	37
24	.16333	28	.12273	1074	.16555	30	.04051	1089	.01361	5	.98657	5	36
25	0.16361	29	6.11199	1070	0.16585	30	6.02962	1084	1.01366	5	0.98652	4	35
26	.16390	29	.10129	1067	.16615	30	.01878	1081	.01371	5	.98648	5	34
27	.16419	28	.09062	1062	.16645	29	6.00797	1077	.01376	5	.98643	5	33
28	.16447	29	.08000	1059	.16674	30	5.99720	1074	.01381	5	.98638	5	32
29	.16476	29	.06941	1055	.16704	30	.98646	1070	.01386	5	.98633	4	31
30	0.16505	28	6.05886	1052	0.16734	30	5.97576	1066	1.01391	4	0.98629	5	30
31	.16533	29	.04834	1047	.16764	30	.96510	1062	.01395	5	.98624	5	29
32	.16562	29	.03787	1044	.16794	30	.95448	1058	.01400	5	.98619	5	28
33	.16591	29	.02743	1041	.16824	30	.94390	1055	.01405	5	.98614	5	27
34	.16620	28	.01702	1036	.16854	30	.93335	1052	.01410	5	.98609	5	26
35	0.16648	29	6.00666	1033	0.16884	30	5.92283	1047	1.01415	5	0.98604	4	25
36	.16677	29	5.99633	1030	.16914	30	.91236	1045	.01420	5	.98600	5	24
37	.16706	28	.98603	1026	.16944	30	.90191	1040	.01425	5	.98595	5	23
38	.16734	29	.97577	1022	.16974	30	.89151	1037	.01430	5	.98590	5	22
39	.16763	29	.96555	1019	.17004	29	.88114	1034	.01435	5	.98585	5	21
40	0.16792	28	5.95536	1015	0.17033	30	5.87080	1029	1.01440	5	0.98580	5	20
41	.16820	29	.94521	1012	.17063	30	.86051	1027	.01445	5	.98575	5	19
42	.16849	29	.93509	1008	.17093	30	.85024	1023	.01450	5	.98570	5	18
43	.16878	28	.92501	1005	.17123	30	.84001	1019	.01455	5	.98565	4	17
44	.16906	29	.91496	1001	.17153	30	.82982	1016	.01460	6	.98561	5	16
45	0.16935	29	5.90495	998	0.17183	30	5.81966	1013	1.01466	5	0.98556	5	15
46	.16964	28	.89497	995	.17213	30	.80953	1009	.01471	5	.98551	5	14
47	.16992	29	.88502	991	.17243	30	.79944	1006	.01476	5	.98546	5	13
48	.17021	29	.87511	987	.17273	30	.78938	1002	.01481	5	.98541	5	12
49	.17050	28	.86524	985	.17303	30	.77936	999	.01486	5	.98536	5	11
50	0.17078	29	5.85539	981	0.17333	30	5.76937	996	1.01491	5	0.98531	5	10
51	.17107	29	.84558	977	.17363	30	.75941	992	.01496	5	.98526	5	9
52	.17136	28	.83581	975	.17393	30	.74949	989	.01501	5	.98521	5	8
53	.17164	29	.82606	971	.17423	30	.73960	986	.01506	6	.98516	5	7
54	.17193	29	.81635	968	.17453	30	.72974	982	.01512	5	.98511	5	6
55	0.17222	28	5.80667	964	0.17483	30	5.71992	979	1.01517	5	0.98506	5	5
56	.17250	29	.79703	961	.17513	30	.71013	976	.01522	5	.98501	5	4
57	.17279	29	.78742	959	.17543	30	.70037	973	.01527	5	.98496	5	3
58	.17308	28	.77783	954	.17573	30	.69064	970	.01532	5	.98491	5	2
59	.17336	29	.76829	952	.17603	30	.68094	966	.01537	6	.98486	5	1
60	0.17365		5.75877		0.17633		5.67128		1.01543		0.98481		0
↑ 99°→	cos	Diff. 1′	sec	Diff. 1′	cot	Diff. 1′	tan	Diff. 1′	csc	Diff. 1′	sin	Diff. 1′	←80° ↑

TABLE 31

Natural Trigonometric Functions

10°→ ↓ ′	sin	Diff. 1′	csc	Diff. 1′	tan	Diff. 1′	cot	Diff. 1′	sec	Diff. 1′	cos	Diff. 1′	←169° ↓ ′
0	0. 17365	28	5. 75877	948	0. 17633	30	5. 67128	963	1. 01543	5	0. 98481	5	60
1	. 17393	29	. 74929	946	. 17663	30	. 66165	960	. 01548	5	. 98476	5	59
2	. 17422	29	. 73983	942	. 17693	30	. 65205	957	. 01553	5	. 98471	5	58
3	. 17451	28	. 73041	939	. 17723	30	. 64248	953	. 01558	6	. 98466	5	57
4	. 17479	29	. 72102	936	. 17753	30	. 63295	951	. 01564	5	. 98461	6	56
5	0. 17508	29	5. 71166	932	0. 17783	30	5. 62344	947	1. 01569	5	0. 98455	5	55
6	. 17537	28	. 70234	930	. 17813	30	. 61397	945	. 01574	5	. 98450	5	54
7	. 17565	29	. 69304	927	. 17843	30	. 60452	941	. 01579	6	. 98445	5	53
8	. 17594	29	. 68377	923	. 17873	30	. 59511	938	. 01585	5	. 98440	5	52
9	. 17623	28	. 67454	921	. 17903	30	. 58573	935	. 01590	5	. 98435	5	51
10	0. 17651	29	5. 66533	917	0. 17933	30	5. 57638	932	1. 01595	6	0. 98430	5	50
11	. 17680	28	. 65616	915	. 17963	30	. 56706	929	. 01601	5	. 98425	5	49
12	. 17708	29	. 64701	911	. 17993	30	. 55777	926	. 01606	5	. 98420	6	48
13	. 17737	29	. 63790	909	. 18023	30	. 54851	924	. 01611	5	. 98414	5	47
14	. 17766	28	. 62881	905	. 18053	30	. 53927	920	. 01616	6	. 98409	5	46
15	0. 17794	29	5. 61976	903	0. 18083	30	5. 53007	917	1. 01622	5	0. 98404	5	45
16	. 17823	29	. 61073	899	. 18113	30	. 52090	914	. 01627	6	. 98399	5	44
17	. 17852	28	. 60174	897	. 18143	30	. 51176	912	. 01633	5	. 98394	5	43
18	. 17880	29	. 59277	894	. 18173	30	. 50264	908	. 01638	5	. 98389	6	42
19	. 17909	28	. 58383	890	. 18203	30	. 49356	905	. 01643	6	. 98383	5	41
20	0. 17937	29	5. 57493	888	0. 18233	30	5. 48451	903	1. 01649	5	0. 98378	5	40
21	. 17966	29	. 56605	885	. 18263	30	. 47548	900	. 01654	5	. 98373	5	39
22	. 17995	28	. 55720	883	. 18293	30	. 46648	897	. 01659	6	. 98368	6	38
23	. 18023	29	. 54837	879	. 18323	30	. 45751	894	. 01665	5	. 98362	5	37
24	. 18052	29	. 53958	877	. 18353	31	. 44857	891	. 01670	6	. 98357	5	36
25	0. 18081	28	5. 53081	873	0. 18384	30	5. 43966	888	1. 01676	5	0. 98352	5	35
26	. 18109	29	. 52208	871	. 18414	30	. 43078	886	. 01681	6	. 98347	6	34
27	. 18138	28	. 51337	869	. 18444	30	. 42192	883	. 01687	5	. 98341	5	33
28	. 18166	29	. 50468	865	. 18474	30	. 41309	880	. 01692	6	. 98336	5	32
29	. 18195	29	. 49603	863	. 18504	30	. 40429	877	. 01698	5	. 98331	6	31
30	0. 18224	28	5. 48740	859	0. 18534	30	5. 39552	875	1. 01703	6	0. 98325	5	30
31	. 18252	29	. 47881	858	. 18564	30	. 38677	872	. 01709	5	. 98320	5	29
32	. 18281	28	. 47023	854	. 18594	30	. 37805	869	. 01714	6	. 98315	5	28
33	. 18309	29	. 46169	852	. 18624	30	. 36936	866	. 01720	5	. 98310	6	27
34	. 18338	29	. 45317	849	. 18654	30	. 36070	864	. 01725	6	. 98304	5	26
35	0. 18367	28	5. 44468	846	0. 18684	30	5. 35206	861	1. 01731	5	0. 98299	5	25
36	. 18395	29	. 43622	844	. 18714	31	. 34345	858	. 01736	6	. 98294	6	24
37	. 18424	28	. 42778	841	. 18745	30	. 33487	856	. 01742	5	. 98288	5	23
38	. 18452	29	. 41937	838	. 18775	30	. 32631	853	. 01747	6	. 98283	6	22
39	. 18481	28	. 41099	836	. 18805	30	. 31778	850	. 01753	5	. 98277	5	21
40	0. 18509	29	5. 40263	833	0. 18835	30	5. 30928	848	1. 01758	6	0. 98272	5	20
41	. 18538	29	. 39430	830	. 18865	30	. 30080	845	. 01764	5	. 98267	6	19
42	. 18567	28	. 38600	828	. 18895	30	. 29235	842	. 01769	6	. 98261	5	18
43	. 18595	29	. 37772	825	. 18925	30	. 28393	840	. 01775	6	. 98256	6	17
44	. 18624	28	. 36947	823	. 18955	31	. 27553	838	. 01781	5	. 98250	5	16
45	0. 18652	29	5. 36124	820	0. 18986	30	5. 26715	835	1. 01786	6	0. 98245	5	15
46	. 18681	29	. 35304	818	. 19016	30	. 25880	832	. 01792	6	. 98240	6	14
47	. 18710	28	. 34486	815	. 19046	30	. 25048	830	. 01798	5	. 98234	5	13
48	. 18738	29	. 33671	812	. 19076	30	. 24218	827	. 01803	6	. 98229	6	12
49	. 18767	28	. 32859	810	. 19106	30	. 23391	825	. 01809	6	. 98223	5	11
50	0. 18795	29	5. 32049	808	0. 19136	30	5. 22566	822	1. 01815	5	0. 98218	6	10
51	. 18824	28	. 31241	805	. 19166	31	. 21744	819	. 01820	6	. 98212	5	9
52	. 18852	29	. 30436	802	. 19197	30	. 20925	818	. 01826	6	. 98207	6	8
53	. 18881	29	. 29634	801	. 19227	30	. 20107	814	. 01832	5	. 98201	5	7
54	. 18910	28	. 28833	797	. 19257	30	. 19293	813	. 01837	6	. 98196	6	6
55	0. 18938	29	5. 28036	795	0. 19287	30	5. 18480	809	1. 01843	6	0. 98190	5	5
56	. 18967	28	. 27241	793	. 19317	30	. 17671	808	. 01849	5	. 98185	6	4
57	. 18995	29	. 26448	790	. 19347	31	. 16863	805	. 01854	6	. 98179	5	3
58	. 19024	28	. 25658	788	. 19378	30	. 16058	802	. 01860	6	. 98174	6	2
59	. 19052	29	. 24870	786	. 19408	30	. 15256	801	. 01866	6	. 98168	5	1
60	0. 19081		5. 24084		0. 19438		5. 14455		1. 01872		0. 98163		0
↑ 100°→	cos	Diff. 1′	sec	Diff. 1′	cot	Diff. 1′	tan	Diff. 1′	csc	Diff. 1′	sin	Diff. 1′	↑ ←79°

TABLE 31

Natural Trigonometric Functions

11° → ↓ ′	sin	Diff. 1′	csc	Diff. 1′	tan	Diff. 1′	cot	Diff. 1′	sec	Diff. 1′	cos	Diff. 1′	← 168° ↓ ′
0	0.19081	28	5.24084	783	0.19438	30	5.14455	797	1.01872	5	0.98163	6	60
1	.19109	29	.23301	780	.19468	30	.13658	796	.01877	6	.98157	5	59
2	.19138	29	.22521	779	.19498	31	.12862	793	.01883	6	.98152	6	58
3	.19167	28	.21742	776	.19529	30	.12069	790	.01889	6	.98146	6	57
4	.19195	29	.20966	773	.19559	30	.11279	789	.01895	6	.98140	5	56
5	0.19224	28	5.20193	772	0.19589	30	5.10490	786	1.01901	5	0.98135	6	55
6	.19252	29	.19421	769	.19619	30	.09704	783	.01906	6	.98129	5	54
7	.19281	28	.18652	766	.19649	31	.08921	782	.01912	6	.98124	6	53
8	.19309	29	.17886	765	.19680	30	.08139	779	.01918	6	.98118	6	52
9	.19338	28	.17121	762	.19710	30	.07360	776	.01924	6	.98112	5	51
10	0.19366	29	5.16359	760	0.19740	30	5.06584	775	1.01930	6	0.98107	6	50
11	.19395	28	.15599	757	.19770	31	.05809	772	.01936	5	.98101	5	49
12	.19423	29	.14842	755	.19801	30	.05037	770	.01941	6	.98096	6	48
13	.19452	29	.14087	753	.19831	30	.04267	768	.01947	6	.98090	6	47
14	.19481	28	.13334	751	.19861	30	.03499	765	.01953	6	.98084	5	46
15	0.19509	29	5.12583	748	0.19891	30	5.02734	763	1.01959	6	0.98079	6	45
16	.19538	28	.11835	747	.19921	31	.01971	761	.01965	6	.98073	6	44
17	.19566	29	.11088	744	.19952	30	.01210	759	.01971	6	.98067	6	43
18	.19595	28	.10344	742	.19982	30	5.00451	756	.01977	6	.98061	5	42
19	.19623	29	.09602	739	.20012	30	4.99695	755	.01983	6	.98056	6	41
20	0.19652	28	5.08863	738	0.20042	31	4.98940	752	1.01989	6	0.98050	6	40
21	.19680	29	.08125	735	.20073	30	.98188	750	.01995	6	.98044	5	39
22	.19709	28	.07390	733	.20103	30	.97438	748	.02001	6	.98039	6	38
23	.19737	29	.06657	731	.20133	31	.96690	745	.02007	6	.98033	6	37
24	.19766	28	.05926	729	.20164	30	.95945	744	.02013	6	.98027	6	36
25	0.19794	29	5.05197	726	0.20194	30	4.95201	741	1.02019	6	0.98021	5	35
26	.19823	28	.04471	725	.20224	30	.94460	739	.02025	6	.98016	6	34
27	.19851	29	.03746	722	.20254	31	.93721	737	.02031	6	.98010	6	33
28	.19880	28	.03024	721	.20285	30	.92984	735	.02037	6	.98004	6	32
29	.19908	29	.02303	718	.20315	30	.92249	733	.02043	6	.97998	6	31
30	0.19937	28	5.01585	716	0.20345	31	4.91516	731	1.02049	6	0.97992	5	30
31	.19965	29	.00869	714	.20376	30	.90785	729	.02055	6	.97987	6	29
32	.19994	28	5.00155	712	.20406	30	.90056	726	.02061	6	.97981	6	28
33	.20022	29	4.99443	710	.20436	30	.89330	725	.02067	6	.97975	6	27
34	.20051	28	.98733	708	.20466	31	.88605	723	.02073	6	.97969	6	26
35	0.20079	29	4.98025	705	0.20497	30	4.87882	720	1.02079	6	0.97963	5	25
36	.20108	28	.97320	704	.20527	30	.87162	718	.02085	6	.97958	6	24
37	.20136	29	.96616	702	.20557	31	.86444	717	.02091	6	.97952	6	23
38	.20165	28	.95914	699	.20588	30	.85727	714	.02097	6	.97946	6	22
39	.20193	29	.95215	698	.20618	30	.85013	713	.02103	7	.97940	6	21
40	0.20222	28	4.94517	696	0.20648	31	4.84300	710	1.02110	6	0.97934	6	20
41	.20250	29	.93821	693	.20679	30	.83590	708	.02116	6	.97928	6	19
42	.20279	28	.93128	692	.20709	30	.82882	707	.02122	6	.97922	6	18
43	.20307	29	.92436	690	.20739	31	.82175	704	.02128	6	.97916	6	17
44	.20336	28	.91746	688	.20770	30	.81471	702	.02134	6	.97910	5	16
45	0.20364	29	4.91058	685	0.20800	30	4.80769	701	1.02140	6	0.97905	6	15
46	.20393	28	.90373	684	.20830	31	.80068	698	.02146	7	.97899	6	14
47	.20421	29	.89689	682	.20861	30	.79370	697	.02153	6	.97893	6	13
48	.20450	28	.89007	680	.20891	30	.78673	695	.02159	6	.97887	6	12
49	.20478	29	.88327	678	.20921	31	.77978	692	.02165	6	.97881	6	11
50	0.20507	28	4.87649	676	0.20952	30	4.77286	691	1.02171	7	0.97875	6	10
51	.20535	28	.86973	674	.20982	31	.76595	689	.02178	6	.97869	6	9
52	.20563	29	.86299	672	.21013	30	.75906	687	.02184	6	.97863	6	8
53	.20592	28	.85627	671	.21043	30	.75219	685	.02190	6	.97857	6	7
54	.20620	29	.84956	668	.21073	31	.74534	683	.02196	7	.97851	6	6
55	0.20649	28	4.84288	667	0.21104	30	4.73851	681	1.02203	6	0.97845	6	5
56	.20677	29	.83621	665	.21134	30	.73170	680	.02209	6	.97839	6	4
57	.20706	28	.82956	662	.21164	31	.72490	677	.02215	6	.97833	6	3
58	.20734	29	.82294	661	.21195	30	.71813	676	.02221	7	.97827	6	2
59	.20763	28	.81633	660	.21225	31	.71137	674	.02228	6	.97821	6	1
60	0.20791		4.80973		0.21256		4.70463		1.02234		0.97815		0
↑ 101° →	cos	Diff. 1′	sec	Diff. 1′	cot	Diff. 1′	tan	Diff. 1′	csc	Diff. 1′	sin	Diff. 1′	← 78° ↑

TABLE 31
Natural Trigonometric Functions

12°→ ↓	sin	Diff. 1′	csc	Diff. 1′	tan	Diff. 1′	cot	Diff. 1′	sec	Diff. 1′	cos	Diff. 1′	←167° ↓
′													′
0	0. 20791	29	4. 80973	657	0. 21256	30	4. 70463	672	1. 02234	6	0. 97815	6	60
1	. 20820	28	. 80316	655	. 21286	30	. 69791	670	. 02240	7	. 97809	6	59
2	. 20848	29	. 79661	654	. 21316	31	. 69121	669	. 02247	6	. 97803	6	58
3	. 20877	28	. 79007	652	. 21347	30	. 68452	666	. 02253	6	. 97797	6	57
4	. 20905	28	. 78355	650	. 21377	31	. 67786	665	. 02259	7	. 97791	7	56
5	0. 20933	29	4. 77705	648	0. 21408	30	4. 67121	663	1. 02266	6	0. 97784	6	55
6	. 20962	28	. 77057	646	. 21438	31	. 66458	661	. 02272	7	. 97778	6	54
7	. 20990	29	. 76411	645	. 21469	30	. 65797	659	. 02279	6	. 97772	6	53
8	. 21019	28	. 75766	643	. 21499	30	. 65138	658	. 02285	6	. 97766	6	52
9	. 21047	29	. 75123	641	. 21529	31	. 64480	655	. 02291	7	. 97760	6	51
10	0. 21076	28	4. 74482	639	0. 21560	30	4. 63825	654	1. 02298	6	0. 97754	6	50
11	. 21104	28	. 73843	638	. 21590	31	. 63171	653	. 02304	7	. 97748	6	49
12	. 21132	29	. 73205	636	. 21621	30	. 62518	650	. 02311	6	. 97742	7	48
13	. 21161	28	. 72569	634	. 21651	31	. 61868	649	. 02317	6	. 97735	6	47
14	. 21189	29	. 71935	632	. 21682	30	. 61219	647	. 02323	7	. 97729	6	46
15	0. 21218	28	4. 71303	630	0. 21712	31	4. 60572	645	1. 02330	6	0. 97723	6	45
16	. 21246	29	. 70673	629	. 21743	30	. 59927	644	. 02336	7	. 97717	6	44
17	. 21275	28	. 70044	627	. 21773	31	. 59283	642	. 02343	6	. 97711	6	43
18	. 21303	28	. 69417	626	. 21804	30	. 58641	640	. 02349	7	. 97705	7	42
19	. 21331	29	. 68791	624	. 21834	30	. 58001	638	. 02356	6	. 97698	6	41
20	0. 21360	28	4. 68167	622	0. 21864	31	4. 57363	637	1. 02362	7	0. 97692	6	40
21	. 21388	29	. 67545	620	. 21895	30	. 56726	635	. 02369	6	. 97686	6	39
22	. 21417	28	. 66925	618	. 21925	31	. 56091	633	. 02375	7	. 97680	7	38
23	. 21445	29	. 66307	617	. 21956	30	. 55458	632	. 02382	6	. 97673	6	37
24	. 21474	28	. 65690	616	. 21986	31	. 54826	630	. 02388	7	. 97667	6	36
25	0. 21502	28	4. 65074	613	0. 22017	30	4. 54196	628	1. 02395	7	0. 97661	6	35
26	. 21530	29	. 64461	612	. 22047	31	. 53568	627	. 02402	6	. 97655	7	34
27	. 21559	28	. 63849	611	. 22078	30	. 52941	625	. 02408	7	. 97648	6	33
28	. 21587	29	. 63238	608	. 22108	31	. 52316	623	. 02415	6	. 97642	6	32
29	. 21616	28	. 62630	607	. 22139	30	. 51693	622	. 02421	7	. 97636	6	31
30	0. 21644	28	4. 62023	606	0. 22169	31	4. 51071	620	1. 02428	7	0. 97630	7	30
31	. 21672	29	. 61417	604	. 22200	31	. 50451	619	. 02435	6	. 97623	6	29
32	. 21701	28	. 60813	602	. 22231	30	. 49832	617	. 02441	7	. 97617	6	28
33	. 21729	29	. 60211	600	. 22261	31	. 49215	615	. 02448	6	. 97611	7	27
34	. 21758	28	. 59611	599	. 22292	30	. 48600	614	. 02454	7	. 97604	6	26
35	0. 21786	28	4. 59012	598	0. 22322	31	4. 47986	612	1. 02461	7	0. 97598	6	25
36	. 21814	29	. 58414	595	. 22353	30	. 47374	610	. 02468	6	. 97592	7	24
37	. 21843	28	. 57819	595	. 22383	31	. 46764	609	. 02474	7	. 97585	6	23
38	. 21871	28	. 57224	592	. 22414	30	. 46155	607	. 02481	7	. 97579	6	22
39	. 21899	29	. 56632	591	. 22444	31	. 45548	606	. 02488	6	. 97573	7	21
40	0. 21928	28	4. 56041	590	0. 22475	30	4. 44942	604	1. 02494	7	0. 97566	6	20
41	. 21956	29	. 55451	588	. 22505	31	. 44338	603	. 02501	7	. 97560	7	19
42	. 21985	28	. 54863	586	. 22536	31	. 43735	601	. 02508	7	. 97553	6	18
43	. 22013	28	. 54277	585	. 22567	30	. 43134	600	. 02515	6	. 97547	6	17
44	. 22041	29	. 53692	583	. 22597	31	. 42534	598	. 02521	7	. 97541	7	16
45	0. 22070	28	4. 53109	582	0. 22628	30	4. 41936	596	1. 02528	7	0. 97534	6	15
46	. 22098	28	. 52527	580	. 22658	31	. 41340	595	. 02535	7	. 97528	7	14
47	. 22126	29	. 51947	579	. 22689	30	. 40745	593	. 02542	6	. 97521	6	13
48	. 22155	28	. 51368	577	. 22719	31	. 40152	592	. 02548	7	. 97515	7	12
49	. 22183	29	. 50791	575	. 22750	31	. 39560	591	. 02555	7	. 97508	6	11
50	0. 22212	28	4. 50216	574	0. 22781	30	4. 38969	588	1. 02562	7	0. 97502	6	10
51	. 22240	28	. 49642	573	. 22811	31	. 38381	588	. 02569	7	. 97496	7	9
52	. 22268	29	. 49069	571	. 22842	30	. 37793	586	. 02576	6	. 97489	6	8
53	. 22297	28	. 48498	570	. 22872	31	. 37207	584	. 02582	7	. 97483	7	7
54	. 22325	28	. 47928	568	. 22903	31	. 36623	583	. 02589	7	. 97476	6	6
55	0. 22353	29	4. 47360	567	0. 22934	30	4. 36040	581	1. 02596	7	0. 97470	7	5
56	. 22382	28	. 46793	565	. 22964	31	. 35459	580	. 02603	7	. 97463	6	4
57	. 22410	28	. 46228	564	. 22995	31	. 34879	579	. 02610	7	. 97457	7	3
58	. 22438	29	. 45664	562	. 23026	30	. 34300	577	. 02617	7	. 97450	6	2
59	. 22467	28	. 45102	561	. 23056	31	. 33723	575	. 02624	6	. 97444	7	1
60	0. 22495		4. 44541		0. 23087		4. 33148		1. 02630		0. 97437		0
↑ 102°→	cos	Diff. 1′	sec	Diff. 1′	cot	Diff. 1′	tan	Diff. 1′	csc	Diff. 1′	sin	Diff. 1′	←77° ↑

TABLE 31

Natural Trigonometric Functions

13°→ ↓ ′	sin	Diff. 1′	csc	Diff. 1′	tan	Diff. 1′	cot	Diff. 1′	sec	Diff. 1′	cos	Diff. 1′	←166° ↓ ′
0	0. 22495	28	4. 44541	559	0. 23087	30	4. 33148	575	1. 02630	7	0. 97437	7	60
1	. 22523	29	. 43982	558	. 23117	31	. 32573	572	. 02637	7	. 97430	6	59
2	. 22552	28	. 43424	557	. 23148	31	. 32001	571	. 02644	7	. 97424	7	58
3	. 22580	28	. 42867	555	. 23179	30	. 31430	570	. 02651	7	. 97417	6	57
4	. 22608	29	. 42312	553	. 23209	31	. 30860	569	. 02658	7	. 97411	7	56
5	0. 22637	28	4. 41759	553	0. 23240	31	4. 30291	567	1. 02665	7	0. 97404	6	55
6	. 22665	28	. 41206	550	. 23271	30	. 29724	565	. 02672	7	. 97398	7	54
7	. 22693	29	. 40656	550	. 23301	31	. 29159	564	. 02679	7	. 97391	7	53
8	. 22722	28	. 40106	548	. 23332	31	. 28595	563	. 02686	7	. 97384	6	52
9	. 22750	28	. 39558	546	. 23363	30	. 28032	561	. 02693	7	. 97378	7	51
10	0. 22778	29	4. 39012	546	0. 23393	31	4. 27471	560	1. 02700	7	0. 97371	6	50
11	. 22807	28	. 38466	543	. 23424	31	. 26911	559	. 02707	7	. 97365	7	49
12	. 22835	28	. 37923	543	. 23455	30	. 26352	557	. 02714	7	. 97358	7	48
13	. 22863	29	. 37380	541	. 23485	31	. 25795	556	. 02721	7	. 97351	6	47
14	. 22892	28	. 36839	540	. 23516	31	. 25239	554	. 02728	7	. 97345	7	46
15	0. 22920	28	4. 36299	538	0. 23547	31	4. 24685	553	1. 02735	7	0. 97338	7	45
16	. 22948	29	. 35761	537	. 23578	30	. 24132	552	. 02742	7	. 97331	6	44
17	. 22977	28	. 35224	535	. 23608	31	. 23580	550	. 02749	7	. 97325	7	43
18	. 23005	28	. 34689	535	. 23639	31	. 23030	549	. 02756	7	. 97318	7	42
19	. 23033	29	. 34154	532	. 23670	30	. 22481	548	. 02763	7	. 97311	7	41
20	0. 23062	28	4. 33622	532	0. 23700	31	4. 21933	546	1. 02770	7	0. 97304	6	40
21	. 23090	28	. 33090	530	. 23731	31	. 21387	545	. 02777	7	. 97298	7	39
22	. 23118	28	. 32560	529	. 23762	31	. 20842	544	. 02784	7	. 97291	7	38
23	. 23146	29	. 32031	528	. 23793	30	. 20298	542	. 02791	8	. 97284	6	37
24	. 23175	28	. 31503	526	. 23823	31	. 19756	541	. 02799	7	. 97278	7	36
25	0. 23203	28	4. 30977	525	0. 23854	31	4. 19215	540	1. 02806	7	0. 97271	7	35
26	. 23231	29	. 30452	523	. 23885	31	. 18675	538	. 02813	7	. 97264	7	34
27	. 23260	28	. 29929	523	. 23916	30	. 18137	537	. 02820	7	. 97257	6	33
28	. 23288	28	. 29406	521	. 23946	31	. 17600	536	. 02827	7	. 97251	7	32
29	. 23316	29	. 28885	519	. 23977	31	. 17064	534	. 02834	8	. 97244	7	31
30	0. 23345	28	4. 28366	519	0. 24008	31	4. 16530	533	1. 02842	7	0. 97237	7	30
31	. 23373	28	. 27847	517	. 24039	30	. 15997	532	. 02849	7	. 97230	7	29
32	. 23401	28	. 27330	516	. 24069	31	. 15465	531	. 02856	7	. 97223	6	28
33	. 23429	29	. 26814	514	. 24100	31	. 14934	529	. 02863	7	. 97217	7	27
34	. 23458	28	. 26300	513	. 24131	31	. 14405	528	. 02870	8	. 97210	7	26
35	0. 23486	28	4. 25787	512	0. 24162	31	4. 13877	527	1. 02878	7	0. 97203	7	25
36	. 23514	28	. 25275	511	. 24193	30	. 13350	525	. 02885	7	. 97196	7	24
37	. 23542	29	. 24764	509	. 24223	31	. 12825	524	. 02892	7	. 97189	7	23
38	. 23571	28	. 24255	509	. 24254	31	. 12301	523	. 02899	8	. 97182	6	22
39	. 23599	28	. 23746	507	. 24285	31	. 11778	522	. 02907	7	. 97176	7	21
40	0. 23627	29	4. 23239	505	0. 24316	31	4. 11256	520	1. 02914	7	0. 97169	7	20
41	. 23656	28	. 22734	505	. 24347	30	. 10736	520	. 02921	7	. 97162	7	19
42	. 23684	28	. 22229	503	. 24377	31	. 10216	517	. 02928	8	. 97155	7	18
43	. 23712	28	. 21726	502	. 24408	31	. 09699	517	. 02936	7	. 97148	7	17
44	. 23740	29	. 21224	501	. 24439	31	. 09182	516	. 02943	7	. 97141	7	16
45	0. 23769	28	4. 20723	499	0. 24470	31	4. 08666	514	1. 02950	8	0. 97134	7	15
46	. 23797	28	. 20224	499	. 24501	31	. 08152	513	. 02958	7	. 97127	7	14
47	. 23825	28	. 19725	497	. 24532	30	. 07639	512	. 02965	7	. 97120	7	13
48	. 23853	29	. 19228	495	. 24562	31	. 07127	511	. 02972	8	. 97113	7	12
49	. 23882	28	. 18733	495	. 24593	31	. 06616	509	. 02980	7	. 97106	6	11
50	0. 23910	28	4. 18238	494	0. 24624	31	4. 06107	508	1. 02987	7	0. 97100	7	10
51	. 23938	28	. 17744	492	. 24655	31	. 05599	507	. 02994	8	. 97093	7	9
52	. 23966	29	. 17252	491	. 24686	31	. 05092	506	. 03002	7	. 97086	7	8
53	. 23995	28	. 16761	490	. 24717	30	. 04586	505	. 03009	8	. 97079	7	7
54	. 24023	28	. 16271	489	. 24747	31	. 04081	503	. 03017	7	. 97072	7	6
55	0. 24051	28	4. 15782	487	0. 24778	31	4. 03578	502	1. 03024	8	0. 97065	7	5
56	. 24079	29	. 15295	486	. 24809	31	. 03076	502	. 03032	7	. 97058	7	4
57	. 24108	28	. 14809	486	. 24840	31	. 02574	500	. 03039	7	. 97051	7	3
58	. 24136	28	. 14323	484	. 24871	31	. 02074	498	. 03046	8	. 97044	7	2
59	. 24164	28	. 13839	482	. 24902	31	. 01576	498	. 03054	7	. 97037	7	1
60	0. 24192		4. 13357		0. 24933		4. 01078		1. 03061		0. 97030		0
↑ 103°→	cos	Diff. 1′	sec	Diff. 1′	cot	Diff. 1′	tan	Diff. 1′	csc	Diff. 1′	sin	Diff. 1′	←76° ↑

TABLE 31

Natural Trigonometric Functions

14°→ ↓ ′	sin	Diff. 1′	csc	Diff. 1′	tan	Diff. 1′	cot	Diff. 1′	sec	Diff. 1′	cos	Diff. 1′	←165° ↓ ′
0	0.24192	28	4.13357	482	0.24933	31	4.01078	496	1.03061	8	0.97030	7	60
1	.24220	29	.12875	481	.24964	31	.00582	496	.03069	7	.97023	8	59
2	.24249	28	.12394	479	.24995	31	4.00086	494	.03076	8	.97015	7	58
3	.24277	28	.11915	478	.25026	30	3.99592	493	.03084	7	.97008	7	57
4	.24305	28	.11437	477	.25056	31	.99099	492	.03091	8	.97001	7	56
5	0.24333	29	4.10960	476	0.25087	31	3.98607	490	1.03099	7	0.96994	7	55
6	.24362	28	.10484	475	.25118	31	.98117	490	.03106	8	.96987	7	54
7	.24390	28	.10009	474	.25149	31	.97627	488	.03114	7	.96980	7	53
8	.24418	28	.09535	472	.25180	31	.97139	488	.03121	8	.96973	7	52
9	.24446	28	.09063	472	.25211	31	.96651	486	.03129	8	.96966	7	51
10	0.24474	29	4.08591	470	0.25242	31	3.96165	485	1.03137	7	0.96959	7	50
11	.24503	28	.08121	469	.25273	31	.95680	484	.03144	8	.96952	7	49
12	.24531	28	.07652	468	.25304	31	.95196	483	.03152	7	.96945	8	48
13	.24559	28	.07184	467	.25335	31	.94713	481	.03159	8	.96937	7	47
14	.24587	28	.06717	466	.25366	31	.94232	481	.03167	8	.96930	7	46
15	0.24615	29	4.06251	465	0.25397	31	3.93751	480	1.03175	7	0.96923	7	45
16	.24644	28	.05786	464	.25428	31	.93271	478	.03182	8	.96916	7	44
17	.24672	28	.05322	462	.25459	31	.92793	477	.03190	7	.96909	7	43
18	.24700	28	.04860	462	.25490	31	.92316	477	.03197	8	.96902	8	42
19	.24728	28	.04398	460	.25521	31	.91839	475	.03205	8	.96894	7	41
20	0.24756	28	4.03938	459	0.25552	31	3.91364	474	1.03213	7	0.96887	7	40
21	.24784	29	.03479	459	.25583	31	.90890	473	.03220	8	.96880	7	39
22	.24813	28	.03020	457	.25614	31	.90417	472	.03228	8	.96873	7	38
23	.24841	28	.02563	456	.25645	31	.89945	471	.03236	8	.96866	8	37
24	.24869	28	.02107	455	.25676	31	.89474	470	.03244	7	.96858	7	36
25	0.24897	28	4.01652	454	0.25707	31	3.89004	468	1.03251	8	0.96851	7	35
26	.24925	29	.01198	453	.25738	31	.88536	468	.03259	8	.96844	7	34
27	.24954	28	.00745	452	.25769	31	.88068	467	.03267	8	.96837	8	33
28	.24982	28	4.00293	450	.25800	31	.87601	465	.03275	7	.96829	7	32
29	.25010	28	3.99843	450	.25831	31	.87136	465	.03282	8	.96822	7	31
30	0.25038	28	3.99393	449	0.25862	31	3.86671	463	1.03290	8	0.96815	8	30
31	.25066	28	.98944	447	.25893	31	.86208	463	.03298	8	.96807	7	29
32	.25094	28	.98497	447	.25924	31	.85745	461	.03306	7	.96800	7	28
33	.25122	29	.98050	446	.25955	31	.85284	460	.03313	8	.96793	7	27
34	.25151	28	.97604	444	.25986	31	.84824	460	.03321	8	.96786	8	26
35	0.25179	28	3.97160	444	0.26017	31	3.84364	458	1.03329	8	0.96778	7	25
36	.25207	28	.96716	442	.26048	31	.83906	457	.03337	8	.96771	7	24
37	.25235	28	.96274	442	.26079	31	.83449	457	.03345	8	.96764	8	23
38	.25263	28	.95832	440	.26110	31	.82992	455	.03353	7	.96756	7	22
39	.25291	29	.95392	440	.26141	31	.82537	454	.03360	8	.96749	7	21
40	0.25320	28	3.94952	438	0.26172	31	3.82083	453	1.03368	8	0.96742	8	20
41	.25348	28	.94514	438	.26203	32	.81630	453	.03376	8	.96734	7	19
42	.25376	28	.94076	436	.26235	31	.81177	451	.03384	8	.96727	8	18
43	.25404	28	.93640	436	.26266	31	.80726	450	.03392	8	.96719	7	17
44	.25432	28	.93204	434	.26297	31	.80276	449	.03400	8	.96712	7	16
45	0.25460	28	3.92770	433	0.26328	31	3.79827	449	1.03408	8	0.96705	8	15
46	.25488	28	.92337	433	.26359	31	.79378	447	.03416	8	.96697	7	14
47	.25516	29	.91904	431	.26390	31	.78931	446	.03424	8	.96690	8	13
48	.25545	28	.91473	431	.26421	31	.78485	445	.03432	7	.96682	7	12
49	.25573	28	.91042	429	.26452	31	.78040	445	.03439	8	.96675	8	11
50	0.25601	28	3.90613	429	0.26483	32	3.77595	443	1.03447	8	0.96667	7	10
51	.25629	28	.90184	428	.26515	31	.77152	443	.03455	8	.96660	7	9
52	.25657	28	.89756	426	.26546	31	.76709	441	.03463	8	.96653	8	8
53	.25685	28	.89330	426	.26577	31	.76268	440	.03471	8	.96645	7	7
54	.25713	28	.88904	425	.26608	31	.75828	440	.03479	8	.96638	8	6
55	0.25741	28	3.88479	423	0.26639	31	3.75388	438	1.03487	8	0.96630	7	5
56	.25769	29	.88056	423	.26670	31	.74950	438	.03495	8	.96623	8	4
57	.25798	28	.87633	422	.26701	32	.74512	437	.03503	8	.96615	7	3
58	.25826	28	.87211	421	.26733	31	.74075	435	.03511	9	.96608	8	2
59	.25854	28	.86790	420	.26764	31	.73640	435	.03520	8	.96600	7	1
60	0.25882		3.86370		0.26795		3.73205		1.03528		0.96593		0
↑ 104°→	cos	Diff. 1′	sec	Diff. 1′	cot	Diff. 1′	tan	Diff. 1′	csc	Diff. 1′	sin	Diff. 1′	↑ ←75°

TABLE 31
Natural Trigonometric Functions

15°→ ↓ ′	sin	Diff. 1′	csc	Diff. 1′	tan	Diff. 1′	cot	Diff. 1′	sec	Diff. 1′	cos	Diff. 1′	←164° ↓ ′
0	0.25882	28	3.86370	419	0.26795	31	3.73205	434	1.03528	8	0.96593	8	60
1	.25910	28	.85951	418	.26826	31	.72771	433	.03536	8	.96585	7	59
2	.25938	28	.85533	417	.26857	31	.72338	431	.03544	8	.96578	8	58
3	.25966	28	.85116	416	.26888	32	.71907	431	.03552	8	.96570	8	57
4	.25994	28	.84700	415	.26920	31	.71476	430	.03560	8	.96562	7	56
5	0.26022	28	3.84285	414	0.26951	31	3.71046	430	1.03568	8	0.96555	8	55
6	.26050	29	.83871	414	.26982	31	.70616	428	.03576	8	.96547	7	54
7	.26079	28	.83457	412	.27013	31	.70188	427	.03584	8	.96540	8	53
8	.26107	28	.83045	412	.27044	32	.69761	426	.03592	9	.96532	8	52
9	.26135	28	.82633	410	.27076	31	.69335	426	.03601	8	.96524	7	51
10	0.26163	28	3.82223	410	0.27107	31	3.68909	424	1.03609	8	0.96517	8	50
11	.26191	28	.81813	409	.27138	31	.68485	424	.03617	8	.96509	7	49
12	.26219	28	.81404	408	.27169	32	.68061	423	.03625	8	.96502	8	48
13	.26247	28	.80996	407	.27201	31	.67638	421	.03633	9	.96494	8	47
14	.26275	28	.80589	406	.27232	31	.67217	421	.03642	8	.96486	7	46
15	0.26303	28	3.80183	405	0.27263	31	3.66796	420	1.03650	8	0.96479	8	45
16	.26331	28	.79778	404	.27294	32	.66376	419	.03658	8	.96471	8	44
17	.26359	28	.79374	404	.27326	31	.65957	419	.03666	8	.96463	7	43
18	.26387	28	.78970	402	.27357	31	.65538	417	.03674	9	.96456	8	42
19	.26415	28	.78568	402	.27388	31	.65121	416	.03683	8	.96448	8	41
20	0.26443	28	3.78166	401	0.27419	32	3.64705	416	1.03691	8	0.96440	7	40
21	.26471	29	.77765	400	.27451	31	.64289	415	.03699	9	.96433	8	39
22	.26500	28	.77365	399	.27482	31	.63874	413	.03708	8	.96425	8	38
23	.26528	28	.76966	398	.27513	32	.63461	413	.03716	8	.96417	7	37
24	.26556	28	.76568	397	.27545	31	.63048	412	.03724	8	.96410	8	36
25	0.26584	28	3.76171	396	0.27576	31	3.62636	412	1.03732	9	0.96402	8	35
26	.26612	28	.75775	396	.27607	31	.62224	410	.03741	8	.96394	8	34
27	.26640	28	.75379	395	.27638	32	.61814	409	.03749	8	.96386	7	33
28	.26668	28	.74984	393	.27670	31	.61405	409	.03757	9	.96379	8	32
29	.26696	28	.74591	393	.27701	31	.60996	408	.03766	8	.96371	8	31
30	0.26724	28	3.74198	392	0.27732	32	3.60588	407	1.03774	9	0.96363	8	30
31	.26752	28	.73806	392	.27764	31	.60181	406	.03783	8	.96355	8	29
32	.26780	28	.73414	390	.27795	31	.59775	405	.03791	8	.96347	7	28
33	.26808	28	.73024	389	.27826	32	.59370	404	.03799	9	.96340	8	27
34	.26836	28	.72635	389	.27858	31	.58966	404	.03808	8	.96332	8	26
35	0.26864	28	3.72246	388	0.27889	32	3.58562	402	1.03816	9	0.96324	8	25
36	.26892	28	.71858	387	.27921	31	.58160	402	.03825	8	.96316	8	24
37	.26920	28	.71471	386	.27952	31	.57758	401	.03833	9	.96308	7	23
38	.26948	28	.71085	385	.27983	32	.57357	400	.03842	8	.96301	8	22
39	.26976	28	.70700	385	.28015	31	.56957	400	.03850	8	.96293	8	21
40	0.27004	28	3.70315	384	0.28046	31	3.56557	398	1.03858	9	0.96285	8	20
41	.27032	28	.69931	382	.28077	32	.56159	398	.03867	8	.96277	8	19
42	.27060	28	.69549	382	.28109	31	.55761	397	.03875	9	.96269	8	18
43	.27088	28	.69167	382	.28140	32	.55364	396	.03884	8	.96261	8	17
44	.27116	28	.68785	380	.28172	31	.54968	395	.03892	9	.96253	7	16
45	0.27144	28	3.68405	380	0.28203	31	3.54573	394	1.03901	8	0.96246	8	15
46	.27172	28	.68025	378	.28234	32	.54179	394	.03909	9	.96238	8	14
47	.27200	28	.67647	378	.28266	31	.53785	392	.03918	9	.96230	8	13
48	.27228	28	.67269	377	.28297	32	.53393	392	.03927	8	.96222	8	12
49	.27256	28	.66892	377	.28329	31	.53001	392	.03935	9	.96214	8	11
50	0.27284	28	3.66515	375	0.28360	31	3.52609	390	1.03944	8	0.96206	8	10
51	.27312	28	.66140	375	.28391	32	.52219	390	.03952	9	.96198	8	9
52	.27340	28	.65765	374	.28423	31	.51829	388	.03961	8	.96190	8	8
53	.27368	28	.65391	373	.28454	32	.51441	388	.03969	9	.96182	8	7
54	.27396	28	.65018	373	.28486	31	.51053	387	.03978	9	.96174	8	6
55	0.27424	28	3.64645	371	0.28517	32	3.50666	387	1.03987	8	0.96166	8	5
56	.27452	28	.64274	371	.28549	31	.50279	385	.03995	9	.96158	8	4
57	.27480	28	.63903	370	.28580	32	.49894	385	.04004	9	.96150	8	3
58	.27508	28	.63533	369	.28612	31	.49509	384	.04013	8	.96142	8	2
59	.27536	28	.63164	368	.28643	32	.49125	384	.04021	9	.96134	8	1
60	0.27564		3.62796		0.28675		3.48741		1.04030		0.96126		0
↑ 105°→	cos	Diff. 1′	sec	Diff. 1′	cot	Diff. 1′	tan	Diff. 1′	csc	Diff. 1′	sin	Diff. 1′	←74° ↑

TABLE 31

Natural Trigonometric Functions

16°→ ↓ ′	sin	Diff. 1′	csc	Diff. 1′	tan	Diff. 1′	cot	Diff. 1′	sec	Diff. 1′	cos	Diff. 1′	←163° ↓
0	0. 27564	28	3. 62796	368	0. 28675	31	3. 48741	382	1. 04030	9	0. 96126	8	60
1	. 27592	28	. 62428	367	. 28706	32	. 48359	382	. 04039	8	. 96118	8	59
2	. 27620	28	. 62061	366	. 28738	31	. 47977	381	. 04047	9	. 96110	8	58
3	. 27648	28	. 61695	365	. 28769	32	. 47596	380	. 04056	9	. 96102	8	57
4	. 27676	28	. 61330	365	. 28801	31	. 47216	379	. 04065	8	. 96094	8	56
5	0. 27704	27	3. 60965	364	0. 28832	32	3. 46837	379	1. 04073	9	0. 96086	8	55
6	. 27731	28	. 60601	363	. 28864	31	. 46458	378	. 04082	9	. 96078	8	54
7	. 27759	28	. 60238	362	. 28895	32	. 46080	377	. 04091	9	. 96070	8	53
8	. 27787	28	. 59876	362	. 28927	31	. 45703	376	. 04100	8	. 96062	8	52
9	. 27815	28	. 59514	360	. 28958	32	. 45327	376	. 04108	9	. 96054	8	51
10	0. 27843	28	3. 59154	360	0. 28990	31	3. 44951	375	1. 04117	9	0. 96046	9	50
11	. 27871	28	. 58794	360	. 29021	32	. 44576	374	. 04126	9	. 96037	8	49
12	. 27899	28	. 58434	358	. 29053	31	. 44202	373	. 04135	9	. 96029	8	48
13	. 27927	28	. 58076	358	. 29084	32	. 43829	373	. 04144	8	. 96021	8	47
14	. 27955	28	. 57718	357	. 29116	31	. 43456	372	. 04152	9	. 96013	8	46
15	0. 27983	28	3. 57361	356	0. 29147	32	3. 43084	371	1. 04161	9	0. 96005	8	45
16	. 28011	28	. 57005	356	. 29179	31	. 42713	370	. 04170	9	. 95997	8	44
17	. 28039	28	. 56649	355	. 29210	32	. 42343	370	. 04179	9	. 95989	8	43
18	. 28067	28	. 56294	354	. 29242	32	. 41973	369	. 04188	9	. 95981	9	42
19	. 28095	28	. 55940	353	. 29274	31	. 41604	368	. 04197	9	. 95972	8	41
20	0. 28123	27	3. 55587	353	0. 29305	32	3. 41236	367	1. 04206	8	0. 95964	8	40
21	. 28150	28	. 55234	351	. 29337	31	. 40869	367	. 04214	9	. 95956	8	39
22	. 28178	28	. 54883	352	. 29368	32	. 40502	366	. 04223	9	. 95948	8	38
23	. 28206	28	. 54531	350	. 29400	32	. 40136	365	. 04232	9	. 95940	9	37
24	. 28234	28	. 54181	350	. 29432	31	. 39771	365	. 04241	9	. 95931	8	36
25	0. 28262	28	3. 53831	349	0. 29463	32	3. 39406	364	1. 04250	9	0. 95923	8	35
26	. 28290	28	. 53482	348	. 29495	31	. 39042	363	. 04259	9	. 95915	8	34
27	. 28318	28	. 53134	347	. 29526	32	. 38679	362	. 04268	9	. 95907	9	33
28	. 28346	28	. 52787	347	. 29558	32	. 38317	362	. 04277	9	. 95898	8	32
29	. 28374	28	. 52440	346	. 29590	31	. 37955	361	. 04286	9	. 95890	8	31
30	0. 28402	27	3. 52094	346	0. 29621	32	3. 37594	360	1. 04295	9	0. 95882	8	30
31	. 28429	28	. 51748	344	. 29653	32	. 37234	359	. 04304	9	. 95874	9	29
32	. 28457	28	. 51404	344	. 29685	31	. 36875	359	. 04313	9	. 95865	8	28
33	. 28485	28	. 51060	344	. 29716	32	. 36516	358	. 04322	9	. 95857	8	27
34	. 28513	28	. 50716	342	. 29748	32	. 36158	358	. 04331	9	. 95849	8	26
35	0. 28541	28	3. 50374	342	0. 29780	31	3. 35800	357	1. 04340	9	0. 95841	9	25
36	. 28569	28	. 50032	341	. 29811	32	. 35443	356	. 04349	9	. 95832	8	24
37	. 28597	28	. 49691	341	. 29843	32	. 35087	355	. 04358	9	. 95824	8	23
38	. 28625	27	. 49350	340	. 29875	31	. 34732	355	. 04367	9	. 95816	9	22
39	. 28652	28	. 49010	339	. 29906	32	. 34377	354	. 04376	9	. 95807	8	21
40	0. 28680	28	3. 48671	338	0. 29938	32	3. 34023	353	1. 04385	9	0. 95799	8	20
41	. 28708	28	. 48333	338	. 29970	31	. 33670	353	. 04394	9	. 95791	9	19
42	. 28736	28	. 47995	337	. 30001	32	. 33317	352	. 04403	10	. 95782	8	18
43	. 28764	28	. 47658	337	. 30033	32	. 32965	351	. 04413	9	. 95774	8	17
44	. 28792	28	. 47321	335	. 30065	32	. 32614	350	. 04422	9	. 95766	9	16
45	0. 28820	27	3. 46986	335	0. 30097	31	3. 32264	350	1. 04431	9	0. 95757	8	15
46	. 28847	28	. 46651	335	. 30128	32	. 31914	349	. 04440	9	. 95749	9	14
47	. 28875	28	. 46316	333	. 30160	32	. 31565	349	. 04449	9	. 95740	8	13
48	. 28903	28	. 45983	333	. 30192	32	. 31216	348	. 04458	10	. 95732	8	12
49	. 28931	28	. 45650	333	. 30224	31	. 30868	347	. 04468	9	. 95724	9	11
50	0. 28959	28	3. 45317	331	0. 30255	32	3. 30521	347	1. 04477	9	0. 95715	8	10
51	. 28987	28	. 44986	331	. 30287	32	. 30174	345	. 04486	9	. 95707	9	9
52	. 29015	27	. 44655	331	. 30319	32	. 29829	346	. 04495	9	. 95698	8	8
53	. 29042	28	. 44324	329	. 30351	31	. 29483	344	. 04504	10	. 95690	9	7
54	. 29070	28	. 43995	329	. 30382	32	. 29139	344	. 04514	9	. 95681	8	6
55	0. 29098	28	3. 43666	329	0. 30414	32	3. 28795	343	1. 04523	9	0. 95673	9	5
56	. 29126	28	. 43337	327	. 30446	32	. 28452	343	. 04532	9	. 95664	8	4
57	. 29154	28	. 43010	327	. 30478	31	. 28109	342	. 04541	10	. 95656	9	3
58	. 29182	27	. 42683	327	. 30509	32	. 27767	341	. 04551	9	. 95647	8	2
59	. 29209	28	. 42356	326	. 30541	32	. 27426	341	. 04560	9	. 95639	9	1
60	0. 29237		3. 42030		0. 30573		3. 27085		1. 04569		0. 95630		0
↑ 106°→	cos	Diff. 1′	sec	Diff. 1′	cot	Diff. 1′	tan	Diff. 1′	csc	Diff. 1′	sin	Diff. 1′	←73° ↑

TABLE 31

Natural Trigonometric Functions

17° → ↓ ′	sin	Diff. 1′	csc	Diff. 1′	tan	Diff. 1′	cot	Diff. 1′	sec	Diff. 1′	cos	Diff. 1′	← 162° ↓ ′
0	0.29237	28	3.42030	325	0.30573	32	3.27085	340	1.04569	9	0.95630	8	60
1	.29265	28	.41705	324	.30605	32	.26745	339	.04578	10	.95622	9	59
2	.29293	28	.41381	324	.30637	32	.26406	339	.04588	9	.95613	8	58
3	.29321	27	.41057	323	.30669	31	.26067	338	.04597	9	.95605	9	57
4	.29348	28	.40734	323	.30700	32	.25729	337	.04606	10	.95596	8	56
5	0.29376	28	3.40411	322	0.30732	32	3.25392	337	1.04616	9	0.95588	9	55
6	.29404	28	.40089	321	.30764	32	.25055	336	.04625	10	.95579	8	54
7	.29432	28	.39768	320	.30796	32	.24719	336	.04635	9	.95571	9	53
8	.29460	27	.39448	320	.30828	32	.24383	334	.04644	9	.95562	8	52
9	.29487	28	.39128	320	.30860	31	.24049	335	.04653	10	.95554	9	51
10	0.29515	28	3.38808	319	0.30891	32	3.23714	333	1.04663	9	0.95545	9	50
11	.29543	28	.38489	318	.30923	32	.23381	333	.04672	10	.95536	8	49
12	.29571	28	.38171	317	.30955	32	.23048	333	.04682	9	.95528	9	48
13	.29599	27	.37854	317	.30987	32	.22715	331	.04691	9	.95519	8	47
14	.29626	28	.37537	316	.31019	32	.22384	331	.04700	10	.95511	9	46
15	0.29654	28	3.37221	316	0.31051	32	3.22053	331	1.04710	9	0.95502	9	45
16	.29682	28	.36905	315	.31083	32	.21722	330	.04719	10	.95493	8	44
17	.29710	27	.36590	314	.31115	32	.21392	329	.04729	9	.95485	9	43
18	.29737	28	.36276	314	.31147	31	.21063	329	.04738	10	.95476	9	42
19	.29765	28	.35962	313	.31178	32	.20734	328	.04748	9	.95467	8	41
20	0.29793	28	3.35649	313	0.31210	32	3.20406	327	1.04757	10	0.95459	9	40
21	.29821	28	.35336	311	.31242	32	.20079	327	.04767	9	.95450	9	39
22	.29849	27	.35025	312	.31274	32	.19752	326	.04776	10	.95441	8	38
23	.29876	28	.34713	310	.31306	32	.19426	326	.04786	9	.95433	9	37
24	.29904	28	.34403	311	.31338	32	.19100	325	.04795	10	.95424	9	36
25	0.29932	28	3.34092	309	0.31370	32	3.18775	324	1.04805	10	0.95415	8	35
26	.29960	27	.33783	309	.31402	32	.18451	324	.04815	9	.95407	9	34
27	.29987	28	.33474	308	.31434	32	.18127	323	.04824	10	.95398	9	33
28	.30015	28	.33166	308	.31466	32	.17804	323	.04834	9	.95389	9	32
29	.30043	28	.32858	307	.31498	32	.17481	322	.04843	10	.95380	8	31
30	0.30071	27	3.32551	307	0.31530	32	3.17159	321	1.04853	10	0.95372	9	30
31	.30098	28	.32244	305	.31562	32	.16838	321	.04863	9	.95363	9	29
32	.30126	28	.31939	306	.31594	32	.16517	320	.04872	10	.95354	9	28
33	.30154	28	.31633	305	.31626	32	.16197	320	.04882	9	.95345	8	27
34	.30182	27	.31328	304	.31658	32	.15877	319	.04891	10	.95337	9	26
35	0.30209	28	3.31024	303	0.31690	32	3.15558	318	1.04901	10	0.95328	9	25
36	.30237	28	.30721	303	.31722	32	.15240	318	.04911	9	.95319	9	24
37	.30265	27	.30418	303	.31754	32	.14922	317	.04920	10	.95310	9	23
38	.30292	28	.30115	301	.31786	32	.14605	317	.04930	10	.95301	8	22
39	.30320	28	.29814	302	.31818	32	.14288	316	.04940	10	.95293	9	21
40	0.30348	28	3.29512	300	0.31850	32	3.13972	316	1.04950	9	0.95284	9	20
41	.30376	27	.29212	300	.31882	32	.13656	315	.04959	10	.95275	9	19
42	.30403	28	.28912	300	.31914	32	.13341	314	.04969	10	.95266	9	18
43	.30431	28	.28612	299	.31946	32	.13027	314	.04979	10	.95257	9	17
44	.30459	27	.28313	298	.31978	32	.12713	313	.04989	9	.95248	8	16
45	0.30486	28	3.28015	298	0.32010	32	3.12400	313	1.04998	10	0.95240	9	15
46	.30514	28	.27717	297	.32042	32	.12087	312	.05008	10	.95231	9	14
47	.30542	28	.27420	297	.32074	32	.11775	311	.05018	10	.95222	9	13
48	.30570	27	.27123	296	.32106	33	.11464	311	.05028	10	.95213	9	12
49	.30597	28	.26827	296	.32139	32	.11153	311	.05038	9	.95204	9	11
50	0.30625	28	3.26531	294	0.32171	32	3.10842	310	1.05047	10	0.95195	9	10
51	.30653	27	.26237	295	.32203	32	.10532	309	.05057	10	.95186	9	9
52	.30680	28	.25942	294	.32235	32	.10223	309	.05067	10	.95177	9	8
53	.30708	28	.25648	293	.32267	32	.09914	308	.05077	10	.95168	9	7
54	.30736	27	.25355	293	.32299	32	.09606	308	.05087	10	.95159	9	6
55	0.30763	28	3.25062	292	0.32331	32	3.09298	307	1.05097	10	0.95150	8	5
56	.30791	28	.24770	292	.32363	33	.08991	306	.05107	9	.95142	9	4
57	.30819	27	.24478	291	.32396	32	.08685	306	.05116	10	.95133	9	3
58	.30846	28	.24187	290	.32428	32	.08379	306	.05126	10	.95124	9	2
59	.30874	28	.23897	290	.32460	32	.08073	305	.05136	10	.95115	9	1
60	0.30902		3.23607		0.32492		3.07768		1.05146		0.95106		0
↑ 107° →	cos	Diff. 1′	sec	Diff. 1′	cot	Diff. 1′	tan	Diff. 1′	csc	Diff. 1′	sin	Diff. 1′	← 72° ↑

TABLE 31

Natural Trigonometric Functions

18°→ ↓	sin	Diff. 1′	csc	Diff. 1′	tan	Diff. 1′	cot	Diff. 1′	sec	Diff. 1′	cos	Diff. 1′	←161° ↓
0	0.30902	27	3.23607	290	0.32492	32	3.07768	304	1.05146	10	0.95106	9	60
1	.30929	28	.23317	289	.32524	32	.07464	304	.05156	10	.95097	9	59
2	.30957	28	.23028	288	.32556	32	.07160	303	.05166	10	.95088	9	58
3	.30985	27	.22740	288	.32588	33	.06857	303	.05176	10	.95079	9	57
4	.31012	28	.22452	287	.32621	32	.06554	302	.05186	10	.95070	9	56
5	0.31040	28	3.22165	287	0.32653	32	3.06252	302	1.05196	10	0.95061	9	55
6	.31068	27	.21878	286	.32685	32	.05950	301	.05206	10	.95052	9	54
7	.31095	28	.21592	286	.32717	32	.05649	300	.05216	10	.95043	10	53
8	.31123	28	.21306	285	.32749	33	.05349	300	.05226	10	.95033	9	52
9	.31151	27	.21021	284	.32782	32	.05049	300	.05236	10	.95024	9	51
10	0.31178	28	3.20737	284	0.32814	32	3.04749	299	1.05246	10	0.95015	9	50
11	.31206	27	.20453	284	.32846	32	.04450	298	.05256	10	.95006	9	49
12	.31233	28	.20169	283	.32878	33	.04152	298	.05266	10	.94997	9	48
13	.31261	28	.19886	282	.32911	32	.03854	298	.05276	10	.94988	9	47
14	.31289	27	.19604	282	.32943	32	.03556	296	.05286	11	.94979	9	46
15	0.31316	28	3.19322	282	0.32975	32	3.03260	297	1.05297	10	0.94970	9	45
16	.31344	28	.19040	281	.33007	33	.02963	296	.05307	10	.94961	9	44
17	.31372	27	.18759	280	.33040	32	.02667	295	.05317	10	.94952	9	43
18	.31399	28	.18479	280	.33072	32	.02372	295	.05327	10	.94943	10	42
19	.31427	27	.18199	279	.33104	32	.02077	294	.05337	10	.94933	9	41
20	0.31454	28	3.17920	279	0.33136	33	3.01783	294	1.05347	10	0.94924	9	40
21	.31482	28	.17641	278	.33169	32	.01489	293	.05357	10	.94915	9	39
22	.31510	27	.17363	278	.33201	32	.01196	293	.05367	11	.94906	9	38
23	.31537	28	.17085	277	.33233	33	.00903	292	.05378	10	.94897	9	37
24	.31565	28	.16808	277	.33266	32	.00611	292	.05388	10	.94888	10	36
25	0.31593	27	3.16531	276	0.33298	32	3.00319	291	1.05398	10	0.94878	9	35
26	.31620	28	.16255	276	.33330	33	3.00028	290	.05408	10	.94869	9	34
27	.31648	27	.15979	275	.33363	32	2.99738	291	.05418	11	.94860	9	33
28	.31675	28	.15704	275	.33395	32	.99447	289	.05429	10	.94851	9	32
29	.31703	27	.15429	274	.33427	33	.99158	290	.05439	10	.94842	10	31
30	0.31730	28	3.15155	274	0.33460	32	2.98868	288	1.05449	10	0.94832	9	30
31	.31758	28	.14881	273	.33492	32	.98580	288	.05459	11	.94823	9	29
32	.31786	27	.14608	273	.33524	33	.98292	288	.05470	10	.94814	9	28
33	.31813	28	.14335	272	.33557	32	.98004	287	.05480	10	.94805	10	27
34	.31841	27	.14063	272	.33589	32	.97717	287	.05490	11	.94795	9	26
35	0.31868	28	3.13791	271	0.33621	33	2.97430	286	1.05501	10	0.94786	9	25
36	.31896	27	.13520	271	.33654	32	.97144	286	.05511	10	.94777	9	24
37	.31923	28	.13249	270	.33686	32	.96858	285	.05521	11	.94768	10	23
38	.31951	28	.12979	270	.33718	33	.96573	285	.05532	10	.94758	9	22
39	.31979	27	.12709	269	.33751	32	.96288	284	.05542	10	.94749	9	21
40	0.32006	28	3.12440	269	0.33783	33	2.96004	283	1.05552	11	0.94740	10	20
41	.32034	27	.12171	268	.33816	32	.95721	284	.05563	10	.94730	9	19
42	.32061	28	.11903	268	.33848	33	.95437	282	.05573	11	.94721	9	18
43	.32089	27	.11635	268	.33881	32	.95155	283	.05584	10	.94712	10	17
44	.32116	28	.11367	266	.33913	32	.94872	281	.05594	10	.94702	9	16
45	0.32144	27	3.11101	267	0.33945	33	2.94591	282	1.05604	11	0.94693	9	15
46	.32171	28	.10834	266	.33978	32	.94309	281	.05615	10	.94684	10	14
47	.32199	28	.10568	265	.34010	33	.94028	280	.05625	11	.94674	9	13
48	.32227	27	.10303	265	.34043	32	.93748	280	.05636	10	.94665	9	12
49	.32254	28	.10038	264	.34075	33	.93468	279	.05646	11	.94656	10	11
50	0.32282	27	3.09774	264	0.34108	32	2.93189	279	1.05657	10	0.94646	9	10
51	.32309	28	.09510	264	.34140	33	.92910	278	.05667	11	.94637	10	9
52	.32337	27	.09246	263	.34173	32	.92632	278	.05678	10	.94627	9	8
53	.32364	28	.08983	262	.34205	33	.92354	278	.05688	11	.94618	9	7
54	.32392	27	.08721	262	.34238	32	.92076	277	.05699	10	.94609	10	6
55	0.32419	28	3.08459	262	0.34270	33	2.91799	276	1.05709	11	0.94599	9	5
56	.32447	27	.08197	261	.34303	32	.91523	277	.05720	10	.94590	10	4
57	.32474	28	.07936	261	.34335	33	.91246	275	.05730	11	.94580	9	3
58	.32502	27	.07675	260	.34368	32	.90971	275	.05741	10	.94571	10	2
59	.32529	28	.07415	260	.34400	33	.90696	275	.05751	11	.94561	9	1
60	0.32557		3.07155		0.34433		2.90421		1.05762		0.94552		0
↑ 108°→	cos	Diff. 1′	sec	Diff. 1′	cot	Diff. 1′	tan	Diff. 1′	csc	Diff. 1′	sin	Diff. 1′	←71° ↑

TABLE 31
Natural Trigonometric Functions

19°→ ↓ ′	sin	Diff. 1′	csc	Diff. 1′	tan	Diff. 1′	cot	Diff. 1′	sec	Diff. 1′	cos	Diff. 1′	←160° ↓ ′
0	0. 32557	27	3. 07155	259	0. 34433	32	2. 90421	274	1. 05762	11	0. 94552	10	60
1	. 32584	28	. 06896	259	. 34465	33	. 90147	274	. 05773	10	. 94542	9	59
2	. 32612	27	. 06637	258	. 34498	32	. 89873	273	. 05783	11	. 94533	10	58
3	. 32639	28	. 06379	258	. 34530	33	. 89600	273	. 05794	11	. 94523	9	57
4	. 32667	27	. 06121	257	. 34563	33	. 89327	272	. 05805	10	. 94514	10	56
5	0. 32694	28	3. 05864	257	0. 34596	32	2. 89055	272	1. 05815	11	0. 94504	9	55
6	. 32722	27	. 05607	257	. 34628	33	. 88783	272	. 05826	10	. 94495	10	54
7	. 32749	28	. 05350	256	. 34661	32	. 88511	271	. 05836	11	. 94485	9	53
8	. 32777	27	. 05094	255	. 34693	33	. 88240	270	. 05847	11	. 94476	10	52
9	. 32804	28	. 04839	255	. 34726	32	. 87970	270	. 05858	11	. 94466	9	51
10	0. 32832	27	3. 04584	255	0. 34758	33	2. 87700	270	1. 05869	10	0. 94457	10	50
11	. 32859	28	. 04329	254	. 34791	33	. 87430	269	. 05879	11	. 94447	9	49
12	. 32887	27	. 04075	254	. 34824	32	. 87161	269	. 05890	11	. 94438	10	48
13	. 32914	28	. 03821	253	. 34856	33	. 86892	268	. 05901	10	. 94428	10	47
14	. 32942	27	. 03568	253	. 34889	33	. 86624	268	. 05911	11	. 94418	9	46
15	0. 32969	28	3. 03315	253	0. 34922	32	2. 86356	267	1. 05922	11	0. 94409	10	45
16	. 32997	27	. 03062	252	. 34954	33	. 86089	267	. 05933	11	. 94399	9	44
17	. 33024	27	. 02810	251	. 34987	33	. 85822	267	. 05944	11	. 94390	10	43
18	. 33051	28	. 02559	251	. 35020	32	. 85555	266	. 05955	10	. 94380	10	42
19	. 33079	27	. 02308	251	. 35052	33	. 85289	266	. 05965	11	. 94370	9	41
20	0. 33106	28	3. 02057	250	0. 35085	33	2. 85023	265	1. 05976	11	0. 94361	10	40
21	. 33134	27	. 01807	250	. 35118	32	. 84758	264	. 05987	11	. 94351	9	39
22	. 33161	28	. 01557	249	. 35150	33	. 84494	265	. 05998	11	. 94342	10	38
23	. 33189	27	. 01308	249	. 35183	33	. 84229	264	. 06009	11	. 94332	10	37
24	. 33216	28	. 01059	249	. 35216	32	. 83965	263	. 06020	10	. 94322	9	36
25	0. 33244	27	3. 00810	248	0. 35248	33	2. 83702	263	1. 06030	11	0. 94313	10	35
26	. 33271	27	. 00562	247	. 35281	33	. 83439	263	. 06041	11	. 94303	10	34
27	. 33298	28	. 00315	248	. 35314	32	. 83176	262	. 06052	11	. 94293	9	33
28	. 33326	27	3. 00067	246	. 35346	33	. 82914	261	. 06063	11	. 94284	10	32
29	. 33353	28	2. 99821	247	. 35379	33	. 82653	262	. 06074	11	. 94274	10	31
30	0. 33381	27	2. 99574	245	0. 35412	33	2. 82391	261	1. 06085	11	0. 94264	10	30
31	. 33408	28	. 99329	246	. 35445	32	. 82130	260	. 06096	11	. 94254	9	29
32	. 33436	27	. 99083	245	. 35477	33	. 81870	260	. 06107	11	. 94245	10	28
33	. 33463	27	. 98838	244	. 35510	33	. 81610	260	. 06118	11	. 94235	10	27
34	. 33490	28	. 98594	245	. 35543	33	. 81350	259	. 06129	11	. 94225	10	26
35	0. 33518	27	2. 98349	243	0. 35576	32	2. 81091	258	1. 06140	11	0. 94215	9	25
36	. 33545	28	. 98106	244	. 35608	33	. 80833	259	. 06151	11	. 94206	10	24
37	. 33573	27	. 97862	243	. 35641	33	. 80574	258	. 06162	11	. 94196	10	23
38	. 33600	27	. 97619	242	. 35674	33	. 80316	257	. 06173	11	. 94186	10	22
39	. 33627	28	. 97377	242	. 35707	33	. 80059	257	. 06184	11	. 94176	9	21
40	0. 33655	27	2. 97135	242	0. 35740	32	2. 79802	257	1. 06195	11	0. 94167	10	20
41	. 33682	28	. 96893	241	. 35772	33	. 79545	256	. 06206	11	. 94157	10	19
42	. 33710	27	. 96652	241	. 35805	33	. 79289	256	. 06217	11	. 94147	10	18
43	. 33737	27	. 96411	240	. 35838	33	. 79033	255	. 06228	11	. 94137	10	17
44	. 33764	28	. 96171	240	. 35871	33	. 78778	255	. 06239	11	. 94127	9	16
45	0. 33792	27	2. 95931	240	0. 35904	33	2. 78523	254	1. 06250	11	0. 94118	10	15
46	. 33819	27	. 95691	239	. 35937	32	. 78269	255	. 06261	11	. 94108	10	14
47	. 33846	28	. 95452	239	. 35969	33	. 78014	253	. 06272	11	. 94098	10	13
48	. 33874	27	. 95213	238	. 36002	33	. 77761	254	. 06283	12	. 94088	10	12
49	. 33901	28	. 94975	238	. 36035	33	. 77507	253	. 06295	11	. 94078	10	11
50	0. 33929	27	2. 94737	237	0. 36068	33	2. 77254	252	1. 06306	11	0. 94068	10	10
51	. 33956	27	. 94500	237	. 36101	33	. 77002	252	. 06317	11	. 94058	9	9
52	. 33983	28	. 94263	237	. 36134	33	. 76750	252	. 06328	11	. 94049	10	8
53	. 34011	27	. 94026	236	. 36167	32	. 76498	251	. 06339	11	. 94039	10	7
54	. 34038	27	. 93790	236	. 36199	33	. 76247	251	. 06350	12	. 94029	10	6
55	0. 34065	28	2. 93554	236	0. 36232	33	2. 75996	250	1. 06362	11	0. 94019	10	5
56	. 34093	27	. 93318	235	. 36265	33	. 75746	250	. 06373	11	. 94009	10	4
57	. 34120	27	. 93083	234	. 36298	33	. 75496	250	. 06384	11	. 93999	10	3
58	. 34147	28	. 92849	235	. 36331	33	. 75246	249	. 06395	12	. 93989	10	2
59	. 34175	27	. 92614	234	. 36364	33	. 74997	249	. 06407	11	. 93979	10	1
60	0. 34202		2. 92380		0. 36397		2. 74748		1. 06418		0. 93969		0
↑ 109°→	cos	Diff. 1′	sec	Diff. 1′	cot	Diff. 1′	tan	Diff. 1′	csc	Diff. 1′	sin	Diff. 1′	←70° ↑

TABLE 31
Natural Trigonometric Functions

20°→ ′ ↓	sin	Diff. 1′	csc	Diff. 1′	tan	Diff. 1′	cot	Diff. 1′	sec	Diff. 1′	cos	Diff. 1′	←159° ′ ↓
0	0.34202	27	2.92380	233	0.36397	33	2.74748	249	1.06418	11	0.93969	10	60
1	.34229	28	.92147	233	.36430	33	.74499	248	.06429	11	.93959	10	59
2	.34257	27	.91914	233	.36463	33	.74251	247	.06440	12	.93949	10	58
3	.34284	27	.91681	232	.36496	33	.74004	248	.06452	11	.93939	10	57
4	.34311	28	.91449	232	.36529	33	.73756	247	.06463	11	.93929	10	56
5	0.34339	27	2.91217	231	0.36562	33	2.73509	246	1.06474	12	0.93919	10	55
6	.34366	27	.90986	232	.36595	33	.73263	246	.06486	11	.93909	10	54
7	.34393	28	.90754	230	.36628	33	.73017	246	.06497	11	.93899	10	53
8	.34421	27	.90524	231	.36661	33	.72771	245	.06508	12	.93889	10	52
9	.34448	27	.90293	230	.36694	33	.72526	245	.06520	11	.93879	10	51
10	0.34475	28	2.90063	229	0.36727	33	2.72281	245	1.06531	11	0.93869	10	50
11	.34503	27	.89834	229	.36760	33	.72036	244	.06542	12	.93859	10	49
12	.34530	27	.89605	229	.36793	33	.71792	244	.06554	11	.93849	10	48
13	.34557	27	.89376	228	.36826	33	.71548	243	.06565	12	.93839	10	47
14	.34584	28	.89148	228	.36859	33	.71305	243	.06577	11	.93829	10	46
15	0.34612	27	2.88920	228	0.36892	33	2.71062	243	1.06588	12	0.93819	10	45
16	.34639	27	.88692	227	.36925	33	.70819	242	.06600	11	.93809	10	44
17	.34666	28	.88465	227	.36958	33	.70577	242	.06611	11	.93799	10	43
18	.34694	27	.88238	227	.36991	33	.70335	241	.06622	12	.93789	10	42
19	.34721	27	.88011	226	.37024	33	.70094	241	.06634	11	.93779	10	41
20	0.34748	27	2.87785	225	0.37057	33	2.69853	241	1.06645	12	0.93769	10	40
21	.34775	28	.87560	226	.37090	33	.69612	241	.06657	11	.93759	11	39
22	.34803	27	.87334	225	.37123	34	.69371	240	.06668	12	.93748	10	38
23	.34830	27	.87109	224	.37157	33	.69131	239	.06680	11	.93738	10	37
24	.34857	27	.86885	224	.37190	33	.68892	239	.06691	12	.93728	10	36
25	0.34884	28	2.86661	224	0.37223	33	2.68653	239	1.06703	12	0.93718	10	35
26	.34912	27	.86437	224	.37256	33	.68414	239	.06715	11	.93708	10	34
27	.34939	27	.86213	223	.37289	33	.68175	238	.06726	12	.93698	10	33
28	.34966	27	.85990	223	.37322	33	.67937	237	.06738	11	.93688	11	32
29	.34993	28	.85767	222	.37355	33	.67700	238	.06749	12	.93677	10	31
30	0.35021	27	2.85545	222	0.37388	34	2.67462	237	1.06761	12	0.93667	10	30
31	.35048	27	.85323	221	.37422	33	.67225	236	.06773	11	.93657	10	29
32	.35075	27	.85102	222	.37455	33	.66989	237	.06784	12	.93647	10	28
33	.35102	28	.84880	221	.37488	33	.66752	236	.06796	11	.93637	11	27
34	.35130	27	.84659	220	.37521	33	.66516	235	.06807	12	.93626	10	26
35	0.35157	27	2.84439	220	0.37554	34	2.66281	235	1.06819	12	0.93616	10	25
36	.35184	27	.84219	220	.37588	33	.66046	235	.06831	11	.93606	10	24
37	.35211	28	.83999	219	.37621	33	.65811	235	.06842	12	.93596	11	23
38	.35239	27	.83780	219	.37654	33	.65576	234	.06854	12	.93585	10	22
39	.35266	27	.83561	219	.37687	33	.65342	233	.06866	12	.93575	10	21
40	0.35293	27	2.83342	218	0.37720	34	2.65109	234	1.06878	11	0.93565	10	20
41	.35320	27	.83124	218	.37754	33	.64875	233	.06889	12	.93555	11	19
42	.35347	28	.82906	218	.37787	33	.64642	232	.06901	12	.93544	10	18
43	.35375	27	.82688	217	.37820	33	.64410	233	.06913	12	.93534	10	17
44	.35402	27	.82471	217	.37853	34	.64177	232	.06925	11	.93524	10	16
45	0.35429	27	2.82254	217	0.37887	33	2.63945	231	1.06936	12	0.93514	11	15
46	.35456	28	.82037	216	.37920	33	.63714	231	.06948	12	.93503	10	14
47	.35484	27	.81821	216	.37953	33	.63483	231	.06960	12	.93493	10	13
48	.35511	27	.81605	215	.37986	34	.63252	231	.06972	12	.93483	11	12
49	.35538	27	.81390	215	.38020	33	.63021	230	.06984	11	.93472	10	11
50	0.35565	27	2.81175	215	0.38053	33	2.62791	230	1.06995	12	0.93462	10	10
51	.35592	27	.80960	214	.38086	34	.62561	229	.07007	12	.93452	11	9
52	.35619	28	.80746	215	.38120	33	.62332	229	.07019	12	.93441	10	8
53	.35647	27	.80531	213	.38153	33	.62103	229	.07031	12	.93431	11	7
54	.35674	27	.80318	214	.38186	34	.61874	228	.07043	12	.93420	10	6
55	0.35701	27	2.80104	213	0.38220	33	2.61646	228	1.07055	12	0.93410	10	5
56	.35728	27	.79891	212	.38253	33	.61418	228	.07067	12	.93400	11	4
57	.35755	27	.79679	213	.38286	34	.61190	227	.07079	12	.93389	10	3
58	.35782	28	.79466	212	.38320	33	.60963	227	.07091	12	.93379	11	2
59	.35810	27	.79254	211	.38353	33	.60736	227	.07103	11	.93368	10	1
60	0.35837		2.79043		0.38386		2.60509		1.07114		0.93358		0
↑ 110°→	cos	Diff. 1′	sec	Diff. 1′	cot	Diff. 1′	tan	Diff. 1′	csc	Diff. 1′	sin	Diff. 1′	←69° ↑

TABLE 31

Natural Trigonometric Functions

21° ↓ ′	sin	Diff. 1′	csc	Diff. 1′	tan	Diff. 1′	cot	Diff. 1′	sec	Diff. 1′	cos	Diff. 1′	←158° ↓ ′
0	0.35837	27	2.79043	211	0.38386	34	2.60509	226	1.07114	12	0.93358	10	60
1	.35864	27	.78832	211	.38420	33	.60283	226	.07126	12	.93348	11	59
2	.35891	27	.78621	211	.38453	34	.60057	226	.07138	12	.93337	10	58
3	.35918	27	.78410	210	.38487	33	.59831	225	.07150	12	.93327	11	57
4	.35945	28	.78200	210	.38520	33	.59606	225	.07162	12	.93316	10	56
5	0.35973	27	2.77990	210	0.38553	34	2.59381	225	1.07174	12	0.93306	11	55
6	.36000	27	.77780	209	.38587	33	.59156	224	.07186	13	.93295	10	54
7	.36027	27	.77571	209	.38620	34	.58932	224	.07199	12	.93285	11	53
8	.36054	27	.77362	208	.38654	33	.58708	224	.07211	12	.93274	10	52
9	.36081	27	.77154	209	.38687	34	.58484	223	.07223	12	.93264	11	51
10	0.36108	27	2.76945	208	0.38721	33	2.58261	223	1.07235	12	0.93253	10	50
11	.36135	27	.76737	207	.38754	33	.58038	223	.07247	12	.93243	11	49
12	.36162	28	.76530	207	.38787	34	.57815	222	.07259	12	.93232	10	48
13	.36190	27	.76323	207	.38821	33	.57593	222	.07271	12	.93222	11	47
14	.36217	27	.76116	207	.38854	34	.57371	221	.07283	12	.93211	10	46
15	0.36244	27	2.75909	206	0.38888	33	2.57150	222	1.07295	12	0.93201	11	45
16	.36271	27	.75703	206	.38921	34	.56928	221	.07307	13	.93190	10	44
17	.36298	27	.75497	205	.38955	33	.56707	220	.07320	12	.93180	11	43
18	.36325	27	.75292	206	.38988	34	.56487	221	.07332	12	.93169	10	42
19	.36352	27	.75086	205	.39022	33	.56266	220	.07344	12	.93159	11	41
20	0.36379	27	2.74881	204	0.39055	34	2.56046	219	1.07356	12	0.93148	11	40
21	.36406	28	.74677	204	.39089	33	.55827	219	.07368	12	.93137	10	39
22	.36434	27	.74473	204	.39122	34	.55608	219	.07380	13	.93127	11	38
23	.36461	27	.74269	204	.39156	34	.55389	219	.07393	12	.93116	10	37
24	.36488	27	.74065	203	.39190	33	.55170	218	.07405	12	.93106	11	36
25	0.36515	27	2.73862	203	0.39223	34	2.54952	218	1.07417	12	0.93095	11	35
26	.36542	27	.73659	203	.39257	33	.54734	218	.07429	13	.93084	10	34
27	.36569	27	.73456	202	.39290	34	.54516	217	.07442	12	.93074	11	33
28	.36596	27	.73254	202	.39324	33	.54299	217	.07454	12	.93063	11	32
29	.36623	27	.73052	202	.39357	34	.54082	217	.07466	13	.93052	10	31
30	0.36650	27	2.72850	201	0.39391	34	2.53865	217	1.07479	12	0.93042	11	30
31	.36677	27	.72649	201	.39425	33	.53648	216	.07491	12	.93031	11	29
32	.36704	27	.72448	201	.39458	34	.53432	215	.07503	13	.93020	10	28
33	.36731	27	.72247	200	.39492	34	.53217	216	.07516	12	.93010	11	27
34	.36758	27	.72047	200	.39526	33	.53001	215	.07528	12	.92999	11	26
35	0.36785	27	2.71847	200	0.39559	34	2.52786	215	1.07540	13	0.92988	10	25
36	.36812	27	.71647	199	.39593	33	.52571	214	.07553	12	.92978	11	24
37	.36839	28	.71448	199	.39626	34	.52357	215	.07565	13	.92967	11	23
38	.36867	27	.71249	199	.39660	34	.52142	213	.07578	12	.92956	11	22
39	.36894	27	.71050	199	.39694	33	.51929	214	.07590	12	.92945	10	21
40	0.36921	27	2.70851	198	0.39727	34	2.51715	213	1.07602	13	0.92935	11	20
41	.36948	27	.70653	198	.39761	34	.51502	213	.07615	12	.92924	11	19
42	.36975	27	.70455	197	.39795	34	.51289	213	.07627	13	.92913	11	18
43	.37002	27	.70258	197	.39829	33	.51076	212	.07640	12	.92902	10	17
44	.37029	27	.70061	197	.39862	34	.50864	212	.07652	13	.92892	11	16
45	0.37056	27	2.69864	197	0.39896	34	2.50652	212	1.07665	12	0.92881	11	15
46	.37083	27	.69667	196	.39930	33	.50440	211	.07677	13	.92870	11	14
47	.37110	27	.69471	196	.39963	34	.50229	211	.07690	12	.92859	10	13
48	.37137	27	.69275	196	.39997	34	.50018	211	.07702	13	.92849	11	12
49	.37164	27	.69079	195	.40031	34	.49807	210	.07715	12	.92838	11	11
50	0.37191	27	2.68884	195	0.40065	33	2.49597	211	1.07727	13	0.92827	11	10
51	.37218	27	.68689	195	.40098	34	.49386	209	.07740	12	.92816	11	9
52	.37245	27	.68494	195	.40132	34	.49177	210	.07752	13	.92805	11	8
53	.37272	27	.68299	194	.40166	34	.48967	209	.07765	13	.92794	10	7
54	.37299	27	.68105	194	.40200	34	.48758	209	.07778	12	.92784	11	6
55	0.37326	27	2.67911	193	0.40234	33	2.48549	209	1.07790	13	0.92773	11	5
56	.37353	27	.67718	193	.40267	34	.48340	208	.07803	13	.92762	11	4
57	.37380	27	.67525	193	.40301	34	.48132	208	.07816	12	.92751	11	3
58	.37407	27	.67332	193	.40335	34	.47924	208	.07828	13	.92740	11	2
59	.37434	27	.67139	192	.40369	34	.47716	207	.07841	12	.92729	11	1
60	0.37461		2.66947		0.40403		2.47509		1.07853		0.92718		0
↑ 111°→	cos	Diff. 1′	sec	Diff. 1′	cot	Diff. 1′	tan	Diff. 1′	csc	Diff. 1′	sin	Diff. 1′	←68° ↑

TABLE 31

Natural Trigonometric Functions

22° → ↓ ′	sin	Diff. 1′	csc	Diff. 1′	tan	Diff. 1′	cot	Diff. 1′	sec	Diff. 1′	cos	Diff. 1′	← 157° ↓ ′
0	0. 37461	27	2. 66947	192	0. 40403	33	2. 47509	207	1. 07853	13	0. 92718	11	60
1	. 37488	27	. 66755	192	. 40436	34	. 47302	207	. 07866	13	. 92707	10	59
2	. 37515	27	. 66563	192	. 40470	34	. 47095	207	. 07879	13	. 92697	11	58
3	. 37542	27	. 66371	191	. 40504	34	. 46888	206	. 07892	12	. 92686	11	57
4	. 37569	26	. 66180	191	. 40538	34	. 46682	206	. 07904	13	. 92675	11	56
5	0. 37595	27	2. 65989	190	0. 40572	34	2. 46476	206	1. 07917	13	0. 92664	11	55
6	. 37622	27	. 65799	190	. 40606	34	. 46270	205	. 07930	13	. 92653	11	54
7	. 37649	27	. 65609	190	. 40640	34	. 46065	205	. 07943	12	. 92642	11	53
8	. 37676	27	. 65419	190	. 40674	33	. 45860	205	. 07955	13	. 92631	11	52
9	. 37703	27	. 65229	189	. 40707	34	. 45655	204	. 07968	13	. 92620	11	51
10	0. 37730	27	2. 65040	189	0. 40741	34	2. 45451	205	1. 07981	13	0. 92609	11	50
11	. 37757	27	. 64851	189	. 40775	34	. 45246	203	. 07994	12	. 92598	11	49
12	. 37784	27	. 64662	189	. 40809	34	. 45043	204	. 08006	13	. 92587	11	48
13	. 37811	27	. 64473	188	. 40843	34	. 44839	203	. 08019	13	. 92576	11	47
14	. 37838	27	. 64285	188	. 40877	34	. 44636	203	. 08032	13	. 92565	11	46
15	0. 37865	27	2. 64097	188	0. 40911	34	2. 44433	203	1. 08045	13	0. 92554	11	45
16	. 37892	27	. 63909	187	. 40945	34	. 44230	203	. 08058	13	. 92543	11	44
17	. 37919	27	. 63722	187	. 40979	34	. 44027	202	. 08071	13	. 92532	11	43
18	. 37946	27	. 63535	187	. 41013	34	. 43825	202	. 08084	13	. 92521	11	42
19	. 37973	26	. 63348	186	. 41047	34	. 43623	201	. 08097	12	. 92510	11	41
20	0. 37999	27	2. 63162	186	0. 41081	34	2. 43422	202	1. 08109	13	0. 92499	11	40
21	. 38026	27	. 62976	186	. 41115	34	. 43220	201	. 08122	13	. 92488	11	39
22	. 38053	27	. 62790	186	. 41149	34	. 43019	200	. 08135	13	. 92477	11	38
23	. 38080	27	. 62604	185	. 41183	34	. 42819	201	. 08148	13	. 92466	11	37
24	. 38107	27	. 62419	185	. 41217	34	. 42618	200	. 08161	13	. 92455	11	36
25	0. 38134	27	2. 62234	185	0. 41251	34	2. 42418	200	1. 08174	13	0. 92444	12	35
26	. 38161	27	. 62049	185	. 41285	34	. 42218	199	. 08187	13	. 92432	11	34
27	. 38188	27	. 61864	184	. 41319	34	. 42019	200	. 08200	13	. 92421	11	33
28	. 38215	26	. 61680	184	. 41353	34	. 41819	199	. 08213	13	. 92410	11	32
29	. 38241	27	. 61496	183	. 41387	34	. 41620	199	. 08226	13	. 92399	11	31
30	0. 38268	27	2. 61313	184	0. 41421	34	2. 41421	198	1. 08239	13	0. 92388	11	30
31	. 38295	27	. 61129	183	. 41455	35	. 41223	198	. 08252	13	. 92377	11	29
32	. 38322	27	. 60946	183	. 41490	34	. 41025	198	. 08265	13	. 92366	11	28
33	. 38349	27	. 60763	182	. 41524	34	. 40827	198	. 08278	13	. 92355	12	27
34	. 38376	27	. 60581	182	. 41558	34	. 40629	197	. 08291	14	. 92343	11	26
35	0. 38403	27	2. 60399	182	0. 41592	34	2. 40432	197	1. 08305	13	0. 92332	11	25
36	. 38430	26	. 60217	182	. 41626	34	. 40235	197	. 08318	13	. 92321	11	24
37	. 38456	27	. 60035	182	. 41660	34	. 40038	197	. 08331	13	. 92310	11	23
38	. 38483	27	. 59853	181	. 41694	34	. 39841	196	. 08344	13	. 92299	12	22
39	. 38510	27	. 59672	181	. 41728	35	. 39645	196	. 08357	13	. 92287	11	21
40	0. 38537	27	2. 59491	180	0. 41763	34	2. 39449	196	1. 08370	13	0. 92276	11	20
41	. 38564	27	. 59311	181	. 41797	34	. 39253	195	. 08383	14	. 92265	11	19
42	. 38591	26	. 59130	180	. 41831	34	. 39058	195	. 08397	13	. 92254	11	18
43	. 38617	27	. 58950	179	. 41865	34	. 38863	195	. 08410	13	. 92243	12	17
44	. 38644	27	. 58771	180	. 41899	34	. 38668	195	. 08423	13	. 92231	11	16
45	0. 38671	27	2. 58591	179	0. 41933	35	2. 38473	194	1. 08436	13	0. 92220	11	15
46	. 38698	27	. 58412	179	. 41968	34	. 38279	195	. 08449	14	. 92209	11	14
47	. 38725	27	. 58233	179	. 42002	34	. 38084	193	. 08463	13	. 92198	12	13
48	. 38752	26	. 58054	178	. 42036	34	. 37891	194	. 08476	13	. 92186	11	12
49	. 38778	27	. 57876	178	. 42070	35	. 37697	193	. 08489	14	. 92175	11	11
50	0. 38805	27	2. 57698	178	0. 42105	34	2. 37504	193	1. 08503	13	0. 92164	12	10
51	. 38832	27	. 57520	178	. 42139	34	. 37311	193	. 08516	13	. 92152	11	9
52	. 38859	27	. 57342	177	. 42173	34	. 37118	193	. 08529	13	. 92141	11	8
53	. 38886	26	. 57165	177	. 42207	35	. 36925	192	. 08542	14	. 92130	11	7
54	. 38912	27	. 56988	177	. 42242	34	. 36733	192	. 08556	13	. 92119	12	6
55	0. 38939	27	2. 56811	177	0. 42276	34	2. 36541	192	1. 08569	13	0. 92107	11	5
56	. 38966	27	. 56634	176	. 42310	35	. 36349	191	. 08582	14	. 92096	11	4
57	. 38993	27	. 56458	176	. 42345	34	. 36158	191	. 08596	13	. 92085	12	3
58	. 39020	26	. 56282	176	. 42379	34	. 35967	191	. 08609	14	. 92073	11	2
59	. 39046	27	. 56106	176	. 42413	34	. 35776	191	. 08623	13	. 92062	12	1
60	0. 39073		2. 55930		0. 42447		2. 35585		1. 08636		0. 92050		0
↑ 112° →	cos	Diff. 1′	sec	Diff. 1′	cot	Diff. 1′	tan	Diff. 1′	csc	Diff. 1′	sin	Diff. 1′	← 67° ↑

TABLE 31
Natural Trigonometric Functions

23°→ ↓ ′	sin	Diff. 1′	csc	Diff. 1′	tan	Diff. 1′	cot	Diff. 1′	sec	Diff. 1′	cos	Diff. 1′	←156° ↓ ′
0	0. 39073	27	2. 55930	175	0. 42447	35	2. 35585	190	1. 08636	13	0. 92050	11	60
1	. 39100	27	. 55755	175	. 42482	34	. 35395	190	. 08649	14	. 92039	11	59
2	. 39127	26	. 55580	175	. 42516	35	. 35205	190	. 08663	13	. 92028	12	58
3	. 39153	27	. 55405	174	. 42551	34	. 35015	190	. 08676	14	. 92016	11	57
4	. 39180	27	. 55231	174	. 42585	34	. 34825	189	. 08690	13	. 92005	11	56
5	0. 39207	27	2. 55057	174	0. 42619	35	2. 34636	189	1. 08703	14	0. 91994	12	55
6	. 39234	26	. 54883	174	. 42654	34	. 34447	189	. 08717	13	. 91982	11	54
7	. 39260	27	. 54709	173	. 42689	34	. 34258	189	. 08730	14	. 91971	12	53
8	. 39287	27	. 54536	173	. 42722	35	. 34069	188	. 08744	13	. 91959	11	52
9	. 39314	27	. 54363	173	. 42757	34	. 33881	188	. 08757	14	. 91948	12	51
10	0. 39341	26	2. 54190	173	0. 42791	35	2. 33693	188	1. 08771	13	0. 91936	11	50
11	. 39367	27	. 54017	172	. 42826	34	. 33505	188	. 08784	14	. 91925	11	49
12	. 39394	27	. 53845	173	. 42860	34	. 33317	187	. 08798	13	. 91914	12	48
13	. 39421	27	. 53672	172	. 42894	35	. 33130	187	. 08811	14	. 91902	11	47
14	. 39448	26	. 53500	171	. 42929	34	. 32943	187	. 08825	14	. 91891	12	46
15	0. 39474	27	2. 53329	172	0. 42963	35	2. 32756	186	1. 08839	13	0. 91879	11	45
16	. 39501	27	. 53157	171	. 42998	34	. 32570	187	. 08852	14	. 91868	12	44
17	. 39528	27	. 52986	171	. 43032	35	. 32383	186	. 08866	14	. 91856	11	43
18	. 39555	26	. 52815	170	. 43067	34	. 32197	185	. 08880	13	. 91845	12	42
19	. 39581	27	. 52645	171	. 43101	35	. 32012	186	. 08893	14	. 91833	11	41
20	0. 39608	27	2. 52474	170	0. 43136	34	2. 31826	185	1. 08907	13	0. 91822	12	40
21	. 39635	26	. 52304	170	. 43170	35	. 31641	185	. 08920	14	. 91810	11	39
22	. 39661	27	. 52134	169	. 43205	34	. 31456	185	. 08934	14	. 91799	12	38
23	. 39688	27	. 51965	170	. 43239	35	. 31271	185	. 08948	14	. 91787	12	37
24	. 39715	26	. 51795	169	. 43274	34	. 31086	184	. 08962	13	. 91775	11	36
25	0. 39741	27	2. 51626	169	0. 43308	35	2. 30902	184	1. 08975	14	0. 91764	12	35
26	. 39768	27	. 51457	168	. 43343	35	. 30718	184	. 08989	14	. 91752	11	34
27	. 39795	27	. 51289	169	. 43378	34	. 30534	183	. 09003	14	. 91741	12	33
28	. 39822	26	. 51120	168	. 43412	35	. 30351	184	. 09017	13	. 91729	11	32
29	. 39848	27	. 50952	168	. 43447	34	. 30167	183	. 09030	14	. 91718	12	31
30	0. 39875	27	2. 50784	167	0. 43481	35	2. 29984	183	1. 09044	14	0. 91706	12	30
31	. 39902	26	. 50617	168	. 43516	34	. 29801	182	. 09058	14	. 91694	11	29
32	. 39928	27	. 50449	167	. 43550	35	. 29619	182	. 09072	14	. 91683	12	28
33	. 39955	27	. 50282	167	. 43585	35	. 29437	183	. 09086	13	. 91671	11	27
34	. 39982	26	. 50115	167	. 43620	34	. 29254	181	. 09099	14	. 91660	12	26
35	0. 40008	27	2. 49948	166	0. 43654	35	2. 29073	182	1. 09113	14	0. 91648	12	25
36	. 40035	27	. 49782	166	. 43689	35	. 28891	181	. 09127	14	. 91636	11	24
37	. 40062	26	. 49616	166	. 43724	34	. 28710	182	. 09141	14	. 91625	12	23
38	. 40088	27	. 49450	166	. 43758	35	. 28528	180	. 09155	14	. 91613	12	22
39	. 40115	26	. 49284	165	. 43793	35	. 28348	181	. 09169	14	. 91601	11	21
40	0. 40141	27	2. 49119	165	0. 43828	34	2. 28167	180	1. 09183	14	0. 91590	12	20
41	. 40168	27	. 48954	165	. 43862	35	. 27987	181	. 09197	14	. 91578	12	19
42	. 40195	26	. 48789	165	. 43897	35	. 27806	180	. 09211	13	. 91566	11	18
43	. 40221	27	. 48624	165	. 43932	34	. 27626	179	. 09224	14	. 91555	12	17
44	. 40248	27	. 48459	164	. 43966	35	. 27447	180	. 09238	14	. 91543	12	16
45	0. 40275	26	2. 48295	164	0. 44001	35	2. 27267	179	1. 09252	14	0. 91531	12	15
46	. 40301	27	. 48131	164	. 44036	35	. 27088	179	. 09266	14	. 91519	11	14
47	. 40328	27	. 47967	163	. 44071	34	. 26909	179	. 09280	14	. 91508	12	13
48	. 40355	26	. 47804	164	. 44105	35	. 26730	178	. 09294	14	. 91496	12	12
49	. 40381	27	. 47640	163	. 44140	35	. 26552	178	. 09308	15	. 91484	12	11
50	0. 40408	26	2. 47477	163	0. 44175	35	2. 26374	178	1. 09323	14	0. 91472	11	10
51	. 40434	27	. 47314	162	. 44210	34	. 26196	178	. 09337	14	. 91461	12	9
52	. 40461	27	. 47152	163	. 44244	35	. 26018	178	. 09351	14	. 91449	12	8
53	. 40488	26	. 46989	162	. 44279	35	. 25840	177	. 09365	14	. 91437	12	7
54	. 40514	27	. 46827	162	. 44314	35	. 25663	177	. 09379	14	. 91425	11	6
55	0. 40541	26	2. 46665	161	0. 44349	35	2. 25486	177	1. 09393	14	0. 91414	12	5
56	. 40567	27	. 46504	162	. 44384	34	. 25309	177	. 09407	14	. 91402	12	4
57	. 40594	27	. 46342	161	. 44418	35	. 25132	176	. 09421	14	. 91390	12	3
58	. 40621	26	. 46181	161	. 44453	35	. 24956	176	. 09435	14	. 91378	12	2
59	. 40647	27	. 46020	161	. 44488	35	. 24780	176	. 09449	15	. 91366	11	1
60	0. 40674		2. 45859		0. 44523		2. 24604		1. 09464		0. 91355		0
↑ 113°→	cos	Diff. 1′	sec	Diff. 1′	cot	Diff. 1′	tan	Diff. 1′	csc	Diff. 1′	sin	Diff. 1′	←66° ↑

TABLE 31

Natural Trigonometric Functions

24°→ ↓	sin	Diff. 1′	csc	Diff. 1′	tan	Diff. 1′	cot	Diff. 1′	sec	Diff. 1′	cos	Diff. 1′	←155° ↓
′													′
0	0.40674	26	2.45859	160	0.44523	35	2.24604	176	1.09464	14	0.91355	12	60
1	.40700	27	.45699	160	.44558	35	.24428	176	.09478	14	.91343	12	59
2	.40727	26	.45539	161	.44593	34	.24252	175	.09492	14	.91331	12	58
3	.40753	27	.45378	159	.44627	35	.24077	175	.09506	14	.91319	12	57
4	.40780	26	.45219	160	.44662	35	.23902	175	.09520	15	.91307	12	56
5	0.40806	27	2.45059	159	0.44697	35	2.23727	174	1.09535	14	0.91295	12	55
6	.40833	27	.44900	159	.44732	35	.23553	175	.09549	14	.91283	11	54
7	.40860	26	.44741	159	.44767	35	.23378	174	.09563	14	.91272	12	53
8	.40886	27	.44582	159	.44802	35	.23204	174	.09577	15	.91260	12	52
9	.40913	26	.44423	159	.44837	35	.23030	173	.09592	14	.91248	12	51
10	0.40939	27	2.44264	158	0.44872	35	2.22857	174	1.09606	14	0.91236	12	50
11	.40966	26	.44106	158	.44907	35	.22683	173	.09620	15	.91224	12	49
12	.40992	27	.43948	158	.44942	35	.22510	173	.09635	14	.91212	12	48
13	.41019	26	.43790	157	.44977	35	.22337	173	.09649	14	.91200	12	47
14	.41045	27	.43633	157	.45012	35	.22164	172	.09663	15	.91188	12	46
15	0.41072	26	2.43476	158	0.45047	35	2.21992	173	1.09678	14	0.91176	12	45
16	.41098	27	.43318	156	.45082	35	.21819	172	.09692	15	.91164	12	44
17	.41125	26	.43162	157	.45117	35	.21647	172	.09707	14	.91152	12	43
18	.41151	27	.43005	157	.45152	35	.21475	171	.09721	14	.91140	12	42
19	.41178	26	.42848	156	.45187	35	.21304	172	.09735	15	.91128	12	41
20	0.41204	27	2.42692	156	0.45222	35	2.21132	171	1.09750	14	0.91116	12	40
21	.41231	26	.42536	156	.45257	35	.20961	171	.09764	15	.91104	12	39
22	.41257	27	.42380	155	.45292	35	.20790	171	.09779	14	.91092	12	38
23	.41284	26	.42225	155	.45327	35	.20619	170	.09793	15	.91080	12	37
24	.41310	27	.42070	156	.45362	35	.20449	171	.09808	14	.91068	12	36
25	0.41337	26	2.41914	154	0.45397	35	2.20278	170	1.09822	15	0.91056	12	35
26	.41363	27	.41760	155	.45432	35	.20108	170	.09837	14	.91044	12	34
27	.41390	26	.41605	155	.45467	35	.19938	169	.09851	15	.91032	12	33
28	.41416	27	.41450	154	.45502	36	.19769	170	.09866	14	.91020	12	32
29	.41443	26	.41296	154	.45538	35	.19599	169	.09880	15	.91008	12	31
30	0.41469	27	2.41142	154	0.45573	35	2.19430	169	1.09895	14	0.90996	12	30
31	.41496	26	.40988	153	.45608	35	.19261	169	.09909	15	.90984	12	29
32	.41522	27	.40835	154	.45643	35	.19092	169	.09924	15	.90972	12	28
33	.41549	26	.40681	153	.45678	35	.18923	168	.09939	14	.90960	12	27
34	.41575	27	.40528	153	.45713	35	.18755	168	.09953	15	.90948	12	26
35	0.41602	26	2.40375	153	0.45748	36	2.18587	168	1.09968	14	0.90936	12	25
36	.41628	27	.40222	152	.45784	35	.18419	168	.09982	15	.90924	13	24
37	.41655	26	.40070	152	.45819	35	.18251	167	.09997	15	.90911	12	23
38	.41681	26	.39918	152	.45854	35	.18084	168	.10012	14	.90899	12	22
39	.41707	27	.39766	152	.45889	35	.17916	167	.10026	15	.90887	12	21
40	0.41734	26	2.39614	152	0.45924	36	2.17749	167	1.10041	15	0.90875	12	20
41	.41760	27	.39462	151	.45960	35	.17582	166	.10056	15	.90863	12	19
42	.41787	26	.39311	152	.45995	35	.17416	167	.10071	14	.90851	12	18
43	.41813	27	.39159	151	.46030	35	.17249	166	.10085	15	.90839	13	17
44	.41840	26	.39008	151	.46065	36	.17083	166	.10100	15	.90826	12	16
45	0.41866	26	2.38857	150	0.46101	35	2.16917	166	1.10115	15	0.90814	12	15
46	.41892	27	.38707	151	.46136	35	.16751	166	.10130	14	.90802	12	14
47	.41919	26	.38556	150	.46171	35	.16585	165	.10144	15	.90790	12	13
48	.41945	27	.38406	150	.46206	36	.16420	165	.10159	15	.90778	12	12
49	.41972	26	.38256	150	.46242	35	.16255	165	.10174	15	.90766	13	11
50	0.41998	26	2.38106	149	0.46277	35	2.16090	165	1.10189	15	0.90753	12	10
51	.42024	27	.37957	149	.46312	36	.15925	165	.10204	14	.90741	12	9
52	.42051	26	.37808	150	.46348	35	.15760	164	.10218	15	.90729	12	8
53	.42077	27	.37658	149	.46383	35	.15596	164	.10233	15	.90717	13	7
54	.42104	26	.37509	148	.46418	36	.15432	164	.10248	15	.90704	12	6
55	0.42130	26	2.37361	149	0.46454	35	2.15268	164	1.10263	15	0.90692	12	5
56	.42156	27	.37212	148	.46489	36	.15104	164	.10278	15	.90680	12	4
57	.42183	26	.37064	148	.46525	35	.14940	163	.10293	15	.90668	13	3
58	.42209	26	.36916	148	.46560	35	.14777	163	.10308	15	.90655	12	2
59	.42235	27	.36768	148	.46595	36	.14614	163	.10323	15	.90643	12	1
60	0.42262		2.36620		0.46631		2.14451		1.10338		0.90631		0
↑ 114°→	cos	Diff. 1′	sec	Diff. 1′	cot	Diff. 1′	tan	Diff. 1′	csc	Diff. 1′	sin	Diff. 1′	←65° ↑

TABLE 31

Natural Trigonometric Functions

25° → ↓ ′	sin	Diff. 1′	csc	Diff. 1′	tan	Diff. 1′	cot	Diff. 1′	sec	Diff. 1′	cos	Diff. 1′	← 154° ↓ ′
0	0. 42262	26	2. 36620	147	0. 46631	35	2. 14451	163	1. 10338	15	0. 90631	13	60
1	. 42288	27	. 36473	148	. 46666	36	. 14288	163	. 10353	15	. 90618	12	59
2	. 42315	26	. 36325	147	. 46702	35	. 14125	162	. 10368	15	. 90606	12	58
3	. 42341	26	. 36178	147	. 46737	35	. 13963	162	. 10383	15	. 90594	12	57
4	. 42367	27	. 36031	146	. 46772	36	. 13801	162	. 10398	15	. 90582	13	56
5	0. 42394	26	2. 35885	147	0. 46808	35	2. 13639	162	1. 10413	15	0. 90569	12	55
6	. 42420	26	. 35738	146	. 46843	36	. 13477	161	. 10428	15	. 90557	12	54
7	. 42446	27	. 35592	146	. 46879	35	. 13316	162	. 10443	15	. 90545	13	53
8	. 42473	26	. 35446	146	. 46914	36	. 13154	161	. 10458	15	. 90532	12	52
9	. 42499	26	. 35300	146	. 46950	35	. 12993	161	. 10473	15	. 90520	13	51
10	0. 42525	27	2. 35154	145	0. 46985	36	2. 12832	161	1. 10488	15	0. 90507	12	50
11	. 42552	26	. 35009	146	. 47021	35	. 12671	160	. 10503	15	. 90495	12	49
12	. 42578	26	. 34863	145	. 47056	36	. 12511	161	. 10518	15	. 90483	13	48
13	. 42604	27	. 34718	145	. 47092	36	. 12350	160	. 10533	16	. 90470	12	47
14	. 42631	26	. 34573	144	. 47128	35	. 12190	160	. 10549	15	. 90458	12	46
15	0. 42657	26	2. 34429	145	0. 47163	36	2. 12030	159	1. 10564	15	0. 90446	13	45
16	. 42683	26	. 34284	144	. 47199	35	. 11871	160	. 10579	15	. 90433	12	44
17	. 42709	27	. 34140	144	. 47234	36	. 11711	159	. 10594	15	. 90421	13	43
18	. 42736	26	. 33996	144	. 47270	35	. 11552	160	. 10609	16	. 90408	12	42
19	. 42762	26	. 33852	144	. 47305	36	. 11392	159	. 10625	15	. 90396	13	41
20	0. 42788	27	2. 33708	143	0. 47341	36	2. 11233	158	1. 10640	15	0. 90383	12	40
21	. 42815	26	. 33565	143	. 47377	35	. 11075	159	. 10655	15	. 90371	13	39
22	. 42841	26	. 33422	144	. 47412	36	. 10916	158	. 10670	16	. 90358	12	38
23	. 42867	27	. 33278	143	. 47448	35	. 10758	158	. 10686	15	. 90346	12	37
24	. 42894	26	. 33135	142	. 47483	36	. 10600	158	. 10701	15	. 90334	13	36
25	0. 42920	26	2. 32993	143	0. 47519	36	2. 10442	158	1. 10716	15	0. 90321	12	35
26	. 42946	26	. 32850	142	. 47555	35	. 10284	158	. 10731	16	. 90309	13	34
27	. 42972	27	. 32708	142	. 47590	36	. 10126	157	. 10747	15	. 90296	12	33
28	. 42999	26	. 32566	142	. 47626	36	. 09969	158	. 10762	15	. 90284	13	32
29	. 43025	26	. 32424	142	. 47662	36	. 09811	157	. 10777	16	. 90271	12	31
30	0. 43051	26	2. 32282	142	0. 47698	35	2. 09654	156	1. 10793	15	0. 90259	13	30
31	. 43077	27	. 32140	141	. 47733	36	. 09498	157	. 10808	16	. 90246	13	29
32	. 43104	26	. 31999	141	. 47769	36	. 09341	157	. 10824	15	. 90233	12	28
33	. 43130	26	. 31858	141	. 47805	35	. 09184	156	. 10839	15	. 90221	13	27
34	. 43156	26	. 31717	141	. 47840	36	. 09028	156	. 10854	16	. 90208	12	26
35	0. 43182	27	2. 31576	140	0. 47876	36	2. 08872	156	1. 10870	15	0. 90196	13	25
36	. 43209	26	. 31436	141	. 47912	36	. 08716	156	. 10885	16	. 90183	12	24
37	. 43235	26	. 31295	140	. 47948	36	. 08560	155	. 10901	15	. 90171	13	23
38	. 43261	26	. 31155	140	. 47984	35	. 08405	155	. 10916	16	. 90158	12	22
39	. 43287	26	. 31015	140	. 48019	36	. 08250	156	. 10932	15	. 90146	13	21
40	0. 43313	27	2. 30875	140	0. 48055	36	2. 08094	155	1. 10947	16	0. 90133	13	20
41	. 43340	26	. 30735	139	. 48091	36	. 07939	154	. 10963	15	. 90120	12	19
42	. 43366	26	. 30596	139	. 48127	36	. 07785	155	. 10978	16	. 90108	13	18
43	. 43392	26	. 30457	139	. 48163	35	. 07630	154	. 10994	15	. 90095	13	17
44	. 43418	27	. 30318	139	. 48198	36	. 07476	155	. 11009	16	. 90082	12	16
45	0. 43445	26	2. 30179	139	0. 48234	36	2. 07321	154	1. 11025	16	0. 90070	13	15
46	. 43471	26	. 30040	139	. 48270	36	. 07167	153	. 11041	15	. 90057	12	14
47	. 43497	26	. 29901	138	. 48306	36	. 07014	154	. 11056	16	. 90045	13	13
48	. 43523	26	. 29763	138	. 48342	36	. 06860	154	. 11072	15	. 90032	13	12
49	. 43549	26	. 29625	138	. 48378	36	. 06706	153	. 11087	16	. 90019	12	11
50	0. 43575	27	2. 29487	138	0. 48414	36	2. 06553	153	1. 11103	16	0. 90007	13	10
51	. 43602	26	. 29349	138	. 48450	36	. 06400	153	. 11119	15	. 89994	13	9
52	. 43628	26	. 29211	137	. 48486	35	. 06247	153	. 11134	16	. 89981	13	8
53	. 43654	26	. 29074	137	. 48521	36	. 06094	152	. 11150	16	. 89968	12	7
54	. 43680	26	. 28937	137	. 48557	36	. 05942	152	. 11166	15	. 89956	13	6
55	0. 43706	27	2. 28800	137	0. 48593	36	2. 05790	153	1. 11181	16	0. 89943	13	5
56	. 43733	26	. 28663	137	. 48629	36	. 05637	152	. 11197	16	. 89930	12	4
57	. 43759	26	. 28526	136	. 48665	36	. 05485	152	. 11213	16	. 89918	13	3
58	. 43785	26	. 28390	137	. 48701	36	. 05333	151	. 11229	15	. 89905	13	2
59	. 43811	26	. 28253	136	. 48737	36	. 05182	152	. 11244	16	. 89892	13	1
60	0. 43837		2. 28117		0. 48773		2. 05030		1. 11260		0. 89879		0
↑ 115° →	cos	Diff. 1′	sec	Diff. 1′	cot	Diff. 1′	tan	Diff. 1′	csc	Diff. 1′	sin	Diff. 1′	← 64° ↑

TABLE 31

Natural Trigonometric Functions

26°→ ↓ ′	sin	Diff. 1′	csc	Diff. 1′	tan	Diff. 1′	cot	Diff. 1′	sec	Diff. 1′	cos	Diff. 1′	←153° ↓ ′
0	0. 43837	26	2. 28117	136	0. 48773	36	2. 05030	151	1. 11260	16	0. 89879	12	60
1	. 43863	26	. 27981	136	. 48809	36	. 04879	151	. 11276	16	. 89867	13	59
2	. 43889	27	. 27845	135	. 48845	36	. 04728	151	. 11292	16	. 89854	13	58
3	. 43916	26	. 27710	136	. 48881	36	. 04577	151	. 11308	15	. 89841	13	57
4	. 43942	26	. 27574	135	. 48917	36	. 04426	150	. 11323	16	. 89828	12	56
5	0. 43968	26	2. 27439	135	0. 48953	36	2. 04276	151	1. 11339	16	0. 89816	13	55
6	. 43994	26	. 27304	135	. 48989	37	. 04125	150	. 11355	16	. 89803	13	54
7	. 44020	26	. 27169	134	. 49026	36	. 03975	150	. 11371	16	. 89790	13	53
8	. 44046	26	. 27035	135	. 49062	36	. 03825	150	. 11387	16	. 89777	13	52
9	. 44072	26	. 26900	134	. 49098	36	. 03675	149	. 11403	16	. 89764	12	51
10	0. 44098	26	2. 26766	134	0. 49134	36	2. 03526	150	1. 11419	16	0. 89752	13	50
11	. 44124	27	. 26632	134	. 49170	36	. 03376	149	. 11435	16	. 89739	13	49
12	. 44151	26	. 26498	134	. 49206	36	. 03227	149	. 11451	16	. 89726	13	48
13	. 44177	26	. 26364	134	. 49242	36	. 03078	149	. 11467	16	. 89713	13	47
14	. 44203	26	. 26230	133	. 49278	37	. 02929	149	. 11483	16	. 89700	13	46
15	0. 44229	26	2. 26097	134	0. 49315	36	2. 02780	149	1. 11499	16	0. 89687	13	45
16	. 44255	26	. 25963	133	. 49351	36	. 02631	148	. 11515	16	. 89674	12	44
17	. 44281	26	. 25830	133	. 49387	36	. 02483	148	. 11531	16	. 89662	13	43
18	. 44307	26	. 25697	132	. 49423	36	. 02335	148	. 11547	16	. 89649	13	42
19	. 44333	26	. 25565	133	. 49459	36	. 02187	148	. 11563	16	. 89636	13	41
20	0. 44359	26	2. 25432	132	0. 49495	37	2. 02039	148	1. 11579	16	0. 89623	13	40
21	. 44385	26	. 25300	133	. 49532	36	. 01891	148	. 11595	16	. 89610	13	39
22	. 44411	26	. 25167	132	. 49568	36	. 01743	147	. 11611	16	. 89597	13	38
23	. 44437	27	. 25035	132	. 49604	36	. 01596	147	. 11627	16	. 89584	13	37
24	. 44464	26	. 24903	131	. 49640	37	. 01449	147	. 11643	16	. 89571	13	36
25	0. 44490	26	2. 24772	132	0. 49677	36	2. 01302	147	1. 11659	16	0. 89558	13	35
26	. 44516	26	. 24640	131	. 49713	36	. 01155	147	. 11675	16	. 89545	13	34
27	. 44542	26	. 24509	131	. 49749	37	. 01008	146	. 11691	17	. 89532	13	33
28	. 44568	26	. 24378	131	. 49786	36	. 00862	147	. 11708	16	. 89519	13	32
29	. 44594	26	. 24247	131	. 49822	36	. 00715	146	. 11724	16	. 89506	13	31
30	0. 44620	26	2. 24116	131	0. 49858	36	2. 00569	146	1. 11740	16	0. 89493	13	30
31	. 44646	26	. 23985	130	. 49894	37	. 00423	146	. 11756	16	. 89480	13	29
32	. 44672	26	. 23855	131	. 49931	36	. 00277	146	. 11772	17	. 89467	13	28
33	. 44698	26	. 23724	130	. 49967	37	2. 00131	145	. 11789	16	. 89454	13	27
34	. 44724	26	. 23594	130	. 50004	36	1. 99986	145	. 11805	16	. 89441	13	26
35	0. 44750	26	2. 23464	130	0. 50040	36	1. 99841	146	1. 11821	17	0. 89428	13	25
36	. 44776	26	. 23334	129	. 50076	37	. 99695	145	. 11838	16	. 89415	13	24
37	. 44802	26	. 23205	130	. 50113	36	. 99550	144	. 11854	16	. 89402	13	23
38	. 44828	26	. 23075	129	. 50149	36	. 99406	145	. 11870	16	. 89389	13	22
39	. 44854	26	. 22946	129	. 50185	37	. 99261	145	. 11886	17	. 89376	13	21
40	0. 44880	26	2. 22817	129	0. 50222	36	1. 99116	144	1. 11903	16	0. 89363	13	20
41	. 44906	26	. 22688	129	. 50258	37	. 98972	144	. 11919	17	. 89350	13	19
42	. 44932	26	. 22559	129	. 50295	36	. 98828	144	. 11936	16	. 89337	13	18
43	. 44958	26	. 22430	128	. 50331	37	. 98684	144	. 11952	16	. 89324	13	17
44	. 44984	26	. 22302	128	. 50368	36	. 98540	144	. 11968	17	. 89311	13	16
45	0. 45010	26	2. 22174	129	0. 50404	37	1. 98396	143	1. 11985	16	0. 89298	13	15
46	. 45036	26	. 22045	127	. 50441	36	. 98253	143	. 12001	17	. 89285	13	14
47	. 45062	26	. 21918	128	. 50477	37	. 98110	144	. 12018	16	. 89272	13	13
48	. 45088	26	. 21790	128	. 50514	36	. 97966	143	. 12034	17	. 89259	14	12
49	. 45114	26	. 21662	127	. 50550	37	. 97823	142	. 12051	16	. 89245	13	11
50	0. 45140	26	2. 21535	128	0. 50587	36	1. 97681	143	1. 12067	16	0. 89232	13	10
51	. 45166	26	. 21407	127	. 50623	37	. 97538	143	. 12083	17	. 89219	13	9
52	. 45192	26	. 21280	127	. 50660	36	. 97395	142	. 12100	17	. 89206	13	8
53	. 45218	25	. 21153	127	. 50696	37	. 97253	142	. 12117	16	. 89193	13	7
54	. 45243	26	. 21026	126	. 50733	36	. 97111	142	. 12133	17	. 89180	13	6
55	0. 45269	26	2. 20900	127	0. 50769	37	1. 96969	142	1. 12150	16	0. 89167	14	5
56	. 45295	26	. 20773	126	. 50806	37	. 96827	142	. 12166	17	. 89153	13	4
57	. 45321	26	. 20647	126	. 50843	36	. 96685	141	. 12183	16	. 89140	13	3
58	. 45347	26	. 20521	126	. 50879	37	. 96544	142	. 12199	17	. 89127	13	2
59	. 45373	26	. 20395	126	. 50916	37	. 96402	141	. 12216	17	. 89114	13	1
60	0. 45399		2. 20269		0. 50953		1. 96261		1. 12233		0. 89101		0
↑ 116°→	cos	Diff. 1′	sec	Diff. 1′	cot	Diff. 1′	tan	Diff. 1′	csc	Diff. 1′	sin	Diff. 1′	←63° ↑

TABLE 31
Natural Trigonometric Functions

27°→ ↓ ′	sin	Diff. 1′	csc	Diff. 1′	tan	Diff. 1′	cot	Diff. 1′	sec	Diff. 1′	cos	Diff. 1′	←152° ↓ ′
0	0. 45399	26	2. 20269	126	0. 50953	36	1. 96261	141	1. 12233	16	0. 89101	14	60
1	. 45425	26	. 20143	125	. 50989	37	. 96120	141	. 12249	17	. 89087	13	59
2	. 45451	26	. 20018	126	. 51026	37	. 95979	141	. 12266	17	. 89074	13	58
3	. 45477	26	. 19892	125	. 51063	36	. 95838	140	. 12283	16	. 89061	13	57
4	. 45503	26	. 19767	125	. 51099	37	. 95698	141	. 12299	17	. 89048	13	56
5	0. 45529	25	2. 19642	125	0. 51136	37	1. 95557	140	1. 12316	17	0. 89035	14	55
6	. 45554	26	. 19517	124	. 51173	36	. 95417	140	. 12333	16	. 89021	13	54
7	. 45580	26	. 19393	125	. 51209	37	. 95277	140	. 12349	17	. 89008	13	53
8	. 45606	26	. 19268	124	. 51246	37	. 95137	140	. 12366	17	. 88995	14	52
9	. 45632	26	. 19144	125	. 51283	36	. 94997	139	. 12383	17	. 88981	13	51
10	0. 45658	26	2. 19019	124	0. 51319	37	1. 94858	140	1. 12400	16	0. 88968	13	50
11	. 45684	26	. 18895	123	. 51356	37	. 94718	139	. 12416	17	. 88955	13	49
12	. 45710	26	. 18772	124	. 51393	37	. 94579	139	. 12433	17	. 88942	14	48
13	. 45736	26	. 18648	124	. 51430	37	. 94440	139	. 12450	17	. 88928	13	47
14	. 45762	25	. 18524	123	. 51467	36	. 94301	139	. 12467	17	. 88915	13	46
15	0. 45787	26	2. 18401	124	0. 51503	37	1. 94162	139	1. 12484	17	0. 88902	14	45
16	. 45813	26	. 18277	123	. 51540	37	. 94023	138	. 12501	17	. 88888	13	44
17	. 45839	26	. 18154	123	. 51577	37	. 93885	139	. 12518	16	. 88875	13	43
18	. 45865	26	. 18031	122	. 51614	37	. 93746	138	. 12534	17	. 88862	14	42
19	. 45891	26	. 17909	123	. 51651	37	. 93608	138	. 12551	17	. 88848	13	41
20	0. 45917	25	2. 17786	123	0. 51688	36	1. 93470	138	1. 12568	17	0. 88835	13	40
21	. 45942	26	. 17663	122	. 51724	37	. 93332	137	. 12585	17	. 88822	14	39
22	. 45968	26	. 17541	122	. 51761	37	. 93195	138	. 12602	17	. 88808	13	38
23	. 45994	26	. 17419	122	. 51798	37	. 93057	137	. 12619	17	. 88795	13	37
24	. 46020	26	. 17297	122	. 51835	37	. 92920	138	. 12636	17	. 88782	14	36
25	0. 46046	26	2. 17175	122	0. 51872	37	1. 92782	137	1. 12653	17	0. 88768	13	35
26	. 46072	25	. 17053	121	. 51909	37	. 92645	137	. 12670	17	. 88755	14	34
27	. 46097	26	. 16932	122	. 51946	37	. 92508	137	. 12687	17	. 88741	13	33
28	. 46123	26	. 16810	121	. 51983	37	. 92371	136	. 12704	17	. 88728	13	32
29	. 46149	26	. 16689	121	. 52020	37	. 92235	137	. 12721	17	. 88715	14	31
30	0. 46175	26	2. 16568	121	0. 52057	37	1. 92098	136	1. 12738	17	0. 88701	13	30
31	. 46201	25	. 16447	121	. 52094	37	. 91962	136	. 12755	17	. 88688	14	29
32	. 46226	26	. 16326	120	. 52131	37	. 91826	136	. 12772	17	. 88674	13	28
33	. 46252	26	. 16206	121	. 52168	37	. 91690	136	. 12789	18	. 88661	14	27
34	. 46278	26	. 16085	120	. 52205	37	. 91554	136	. 12807	17	. 88647	13	26
35	0. 46304	26	2. 15965	120	0. 52242	37	1. 91418	136	1. 12824	17	0. 88634	14	25
36	. 46330	25	. 15845	120	. 52279	37	. 91282	135	. 12841	17	. 88620	13	24
37	. 46355	26	. 15725	120	. 52316	37	. 91147	135	. 12858	17	. 88607	14	23
38	. 46381	26	. 15605	120	. 52353	37	. 91012	136	. 12875	17	. 88593	13	22
39	. 46407	26	. 15485	119	. 52390	37	. 90876	135	. 12892	18	. 88580	14	21
40	0. 46433	25	2. 15366	120	0. 52427	37	1. 90741	134	1. 12910	17	0. 88566	13	20
41	. 46458	26	. 15246	119	. 52464	37	. 90607	135	. 12927	17	. 88553	14	19
42	. 46484	26	. 15127	119	. 52501	37	. 90472	135	. 12944	17	. 88539	13	18
43	. 46510	26	. 15008	119	. 52538	37	. 90337	134	. 12961	18	. 88526	14	17
44	. 46536	25	. 14889	119	. 52575	38	. 90203	134	. 12979	17	. 88512	13	16
45	0. 46561	26	2. 14770	119	0. 52613	37	1. 90069	134	1. 12996	17	0. 88499	14	15
46	. 46587	26	. 14651	118	. 52650	37	. 89935	134	. 13013	18	. 88485	13	14
47	. 46613	26	. 14533	119	. 52687	37	. 89801	134	. 13031	17	. 88472	14	13
48	. 46639	25	. 14414	118	. 52724	37	. 89667	134	. 13048	17	. 88458	13	12
49	. 46664	26	. 14296	118	. 52761	37	. 89533	133	. 13065	18	. 88445	14	11
50	0. 46690	26	2. 14178	118	0. 52798	38	1. 89400	134	1. 13083	17	0. 88431	14	10
51	. 46716	26	. 14060	118	. 52836	37	. 89266	133	. 13100	17	. 88417	13	9
52	. 46742	25	. 13942	117	. 52873	37	. 89133	133	. 13117	18	. 88404	14	8
53	. 46767	26	. 13825	118	. 52910	37	. 89000	133	. 13135	17	. 88390	13	7
54	. 46793	26	. 13707	117	. 52947	38	. 88867	133	. 13152	18	. 88377	14	6
55	0. 46819	25	2. 13590	117	0. 52985	37	1. 88734	132	1. 13170	17	0. 88363	14	5
56	. 46844	26	. 13473	117	. 53022	37	. 88602	133	. 13187	18	. 88349	13	4
57	. 46870	26	. 13356	117	. 53059	37	. 88469	132	. 13205	17	. 88336	14	3
58	. 46896	25	. 13239	117	. 53096	38	. 88337	132	. 13222	17	. 88322	14	2
59	. 46921	26	. 13122	117	. 53134	37	. 88205	132	. 13239	18	. 88308	13	1
60	0. 46947		2. 13005		0. 53171		1. 88073		1. 13257		0. 88295		0
↑ 117°→	cos	Diff. 1′	sec	Diff. 1′	cot	Diff. 1′	tan	Diff. 1′	csc	Diff. 1′	sin	Diff. 1′	←62° ↑

TABLE 31

Natural Trigonometric Functions

28°→ ′ ↓	sin	Diff. 1′	csc	Diff. 1′	tan	Diff. 1′	cot	Diff. 1′	sec	Diff. 1′	cos	Diff. 1′	←151° ′ ↓
0	0.46947	26	2.13005	116	0.53171	37	1.88073	132	1.13257	18	0.88295	14	60
1	.46973	26	.12889	116	.53208	38	.87941	132	.13275	17	.88281	14	59
2	.46999	25	.12773	116	.53246	37	.87809	132	.13292	18	.88267	13	58
3	.47024	26	.12657	117	.53283	37	.87677	131	.13310	17	.88254	14	57
4	.47050	26	.12540	115	.53320	38	.87546	131	.13327	18	.88240	14	56
5	0.47076	25	2.12425	116	0.53358	37	1.87415	132	1.13345	17	0.88226	13	55
6	.47101	26	.12309	116	.53395	37	.87283	131	.13362	18	.88213	14	54
7	.47127	26	.12193	115	.53432	38	.87152	131	.13380	18	.88199	14	53
8	.47153	25	.12078	115	.53470	37	.87021	130	.13398	17	.88185	13	52
9	.47178	26	.11963	116	.53507	38	.86891	131	.13415	18	.88172	14	51
10	0.47204	25	2.11847	115	0.53545	37	1.86760	130	1.13433	18	0.88158	14	50
11	.47229	26	.11732	115	.53582	38	.86630	131	.13451	17	.88144	14	49
12	.47255	26	.11617	114	.53620	37	.86499	130	.13468	18	.88130	13	48
13	.47281	25	.11503	115	.53657	37	.86369	130	.13486	18	.88117	14	47
14	.47306	26	.11388	114	.53694	38	.86239	130	.13504	17	.88103	14	46
15	0.47332	26	2.11274	115	0.53732	37	1.86109	130	1.13521	18	0.88089	14	45
16	.47358	25	.11159	114	.53769	38	.85979	129	.13539	18	.88075	13	44
17	.47383	26	.11045	114	.53807	37	.85850	130	.13557	18	.88062	14	43
18	.47409	25	.10931	114	.53844	38	.85720	129	.13575	18	.88048	14	42
19	.47434	26	.10817	113	.53882	38	.85591	129	.13593	17	.88034	14	41
20	0.47460	26	2.10704	114	0.53920	37	1.85462	129	1.13610	18	0.88020	14	40
21	.47486	25	.10590	113	.53957	38	.85333	129	.13628	18	.88006	13	39
22	.47511	26	.10477	114	.53995	37	.85204	129	.13646	18	.87993	14	38
23	.47537	25	.10363	113	.54032	38	.85075	129	.13664	18	.87979	14	37
24	.47562	26	.10250	113	.54070	37	.84946	128	.13682	18	.87965	14	36
25	0.47588	26	2.10137	113	0.54107	38	1.84818	129	1.13700	18	0.87951	14	35
26	.47614	25	.10024	113	.54145	38	.84689	128	.13718	17	.87937	14	34
27	.47639	26	.09911	112	.54183	37	.84561	128	.13735	18	.87923	14	33
28	.47665	25	.09799	113	.54220	38	.84433	128	.13753	18	.87909	13	32
29	.47690	26	.09686	112	.54258	38	.84305	128	.13771	18	.87896	14	31
30	0.47716	25	2.09574	112	0.54296	37	1.84177	128	1.13789	18	0.87882	14	30
31	.47741	26	.09462	112	.54333	38	.84049	127	.13807	18	.87868	14	29
32	.47767	26	.09350	112	.54371	38	.83922	128	.13825	18	.87854	14	28
33	.47793	25	.09238	112	.54409	37	.83794	127	.13843	18	.87840	14	27
34	.47818	26	.09126	112	.54446	38	.83667	127	.13861	18	.87826	14	26
35	0.47844	25	2.09014	111	0.54484	38	1.83540	127	1.13879	18	0.87812	14	25
36	.47869	26	.08903	112	.54522	38	.83413	127	.13897	19	.87798	14	24
37	.47895	25	.08791	111	.54560	37	.83286	127	.13916	18	.87784	14	23
38	.47920	26	.08680	111	.54597	38	.83159	126	.13934	18	.87770	14	22
39	.47946	25	.08569	111	.54635	38	.83033	127	.13952	18	.87756	13	21
40	0.47971	26	2.08458	111	0.54673	38	1.82906	126	1.13970	18	0.87743	14	20
41	.47997	25	.08347	111	.54711	37	.82780	126	.13988	18	.87729	14	19
42	.48022	26	.08236	110	.54748	38	.82654	126	.14006	18	.87715	14	18
43	.48048	25	.08126	111	.54786	38	.82528	126	.14024	18	.87701	14	17
44	.48073	26	.08015	110	.54824	38	.82402	126	.14042	19	.87687	14	16
45	0.48099	25	2.07905	110	0.54862	38	1.82276	126	1.14061	18	0.87673	14	15
46	.48124	26	.07795	110	.54900	38	.82150	125	.14079	18	.87659	14	14
47	.48150	25	.07685	110	.54938	37	.82025	126	.14097	18	.87645	14	13
48	.48175	26	.07575	110	.54975	38	.81899	125	.14115	19	.87631	14	12
49	.48201	25	.07465	109	.55013	38	.81774	125	.14134	18	.87617	14	11
50	0.48226	26	2.07356	110	0.55051	38	1.81649	125	1.14152	18	0.87603	14	10
51	.48252	25	.07246	109	.55089	38	.81524	125	.14170	18	.87589	14	9
52	.48277	26	.07137	110	.55127	38	.81399	125	.14188	19	.87575	14	8
53	.48303	25	.07027	109	.55165	38	.81274	124	.14207	18	.87561	15	7
54	.48328	26	.06918	109	.55203	38	.81150	125	.14225	18	.87546	14	6
55	0.48354	25	2.06809	108	0.55241	38	1.81025	124	1.14243	19	0.87532	14	5
56	.48379	26	.06701	109	.55279	38	.80901	124	.14262	18	.87518	14	4
57	.48405	25	.06592	109	.55317	38	.80777	124	.14280	19	.87504	14	3
58	.48430	26	.06483	108	.55355	38	.80653	124	.14299	18	.87490	14	2
59	.48456	25	.06375	108	.55393	38	.80529	124	.14317	18	.87476	14	1
60	0.48481		2.06267		0.55431		1.80405		1.14335		0.87462		0
↑ 118°→	cos	Diff. 1′	sec	Diff. 1′	cot	Diff. 1′	tan	Diff. 1′	csc	Diff. 1′	sin	Diff. 1′	←61° ↑

TABLE 31
Natural Trigonometric Functions

29°→ ↓ ′	sin	Diff. 1′	csc	Diff. 1′	tan	Diff. 1′	cot	Diff. 1′	sec	Diff. 1′	cos	Diff. 1′	←150° ↓ ′
0	0. 48481	25	2. 06267	109	0. 55431	38	1. 80405	124	1. 14335	19	0. 87462	14	60
1	. 48506	26	. 06158	108	. 55469	38	. 80281	123	. 14354	18	. 87448	14	59
2	. 48532	25	. 06050	108	. 55507	38	. 80158	124	. 14372	19	. 87434	14	58
3	. 48557	26	. 05942	107	. 55545	38	. 80034	123	. 14391	18	. 87420	14	57
4	. 48583	25	. 05835	108	. 55583	38	. 79911	123	. 14409	19	. 87406	15	56
5	0. 48608	26	2. 05727	108	0. 55621	38	1. 79788	123	1. 14428	18	0. 87391	14	55
6	. 48634	25	. 05619	107	. 55659	38	. 79665	123	. 14446	19	. 87377	14	54
7	. 48659	25	. 05512	107	. 55697	39	. 79542	123	. 14465	18	. 87363	14	53
8	. 48684	26	. 05405	107	. 55736	38	. 79419	123	. 14483	19	. 87349	14	52
9	. 48710	25	. 05298	107	. 55774	38	. 79296	122	. 14502	19	. 87335	14	51
10	0. 48735	26	2. 05191	107	0. 55812	38	1. 79174	123	1. 14521	18	0. 87321	15	50
11	. 48761	25	. 05084	107	. 55850	38	. 79051	122	. 14539	19	. 87306	14	49
12	. 48786	25	. 04977	107	. 55888	38	. 78929	122	. 14558	18	. 87292	14	48
13	. 48811	26	. 04870	106	. 55926	38	. 78807	122	. 14576	19	. 87278	14	47
14	. 48837	25	. 04764	107	. 55964	39	. 78685	122	. 14595	19	. 87264	14	46
15	0. 48862	26	2. 04657	106	0. 56003	38	1. 78563	122	1. 14614	18	0. 87250	15	45
16	. 48888	25	. 04551	106	. 56041	38	. 78441	122	. 14632	19	. 87235	14	44
17	. 48913	25	. 04445	106	. 56079	38	. 78319	121	. 14651	19	. 87221	14	43
18	. 48938	26	. 04339	106	. 56117	39	. 78198	121	. 14670	19	. 87207	14	42
19	. 48964	25	. 04233	105	. 56156	38	. 78077	122	. 14689	18	. 87193	15	41
20	0. 48989	25	2. 04128	106	0. 56194	38	1. 77955	121	1. 14707	19	0. 87178	14	40
21	. 49014	26	. 04022	106	. 56232	38	. 77834	121	. 14726	19	. 87164	14	39
22	. 49040	25	. 03916	105	. 56270	39	. 77713	121	. 14745	19	. 87150	14	38
23	. 49065	25	. 03811	105	. 56309	38	. 77592	121	. 14764	18	. 87136	15	37
24	. 49090	26	. 03706	105	. 56347	38	. 77471	120	. 14782	19	. 87121	14	36
25	0. 49116	25	2. 03601	105	0. 56385	39	1. 77351	121	1. 14801	19	0. 87107	14	35
26	. 49141	25	. 03496	105	. 56424	38	. 77230	120	. 14820	19	. 87093	14	34
27	. 49166	26	. 03391	105	. 56462	39	. 77110	120	. 14839	19	. 87079	15	33
28	. 49192	25	. 03286	104	. 56501	38	. 76990	121	. 14858	19	. 87064	14	32
29	. 49217	25	. 03182	105	. 56539	38	. 76869	120	. 14877	19	. 87050	14	31
30	0. 49242	26	2. 03077	104	0. 56577	39	1. 76749	120	1. 14896	18	0. 87036	15	30
31	. 49268	25	. 02973	104	. 56616	38	. 76629	119	. 14914	19	. 87021	14	29
32	. 49293	25	. 02869	104	. 56654	39	. 76510	120	. 14933	19	. 87007	14	28
33	. 49318	26	. 02765	104	. 56693	38	. 76390	119	. 14952	19	. 86993	15	27
34	. 49344	25	. 02661	104	. 56731	38	. 76271	120	. 14971	19	. 86978	14	26
35	0. 49369	25	2. 02557	104	0. 56769	39	1. 76151	119	1. 14990	19	0. 86964	15	25
36	. 49394	25	. 02453	104	. 56808	38	. 76032	119	. 15009	19	. 86949	14	24
37	. 49419	26	. 02349	103	. 56846	39	. 75913	119	. 15028	19	. 86935	14	23
38	. 49445	25	. 02246	103	. 56885	38	. 75794	119	. 15047	19	. 86921	15	22
39	. 49470	25	. 02143	104	. 56923	39	. 75675	119	. 15066	19	. 86906	14	21
40	0. 49495	26	2. 02039	103	0. 56962	38	1. 75556	119	1. 15085	20	0. 86892	14	20
41	. 49521	25	. 01936	103	. 57000	39	. 75437	118	. 15105	19	. 86878	15	19
42	. 49546	25	. 01833	103	. 57039	39	. 75319	119	. 15124	19	. 86863	14	18
43	. 49571	25	. 01730	102	. 57078	38	. 75200	118	. 15143	19	. 86849	15	17
44	. 49596	26	. 01628	103	. 57116	39	. 75082	118	. 15162	19	. 86834	14	16
45	0. 49622	25	2. 01525	103	0. 57155	38	1. 74964	118	1. 15181	19	0. 86820	15	15
46	. 49647	25	. 01422	102	. 57193	39	. 74846	118	. 15200	19	. 86805	14	14
47	. 49672	25	. 01320	102	. 57232	39	. 74728	118	. 15219	20	. 86791	14	13
48	. 49697	26	. 01218	102	. 57271	38	. 74610	118	. 15239	19	. 86777	15	12
49	. 49723	25	. 01116	102	. 57309	39	. 74492	117	. 15258	19	. 86762	14	11
50	0. 49748	25	2. 01014	102	0. 57348	38	1. 74375	118	1. 15277	19	0. 86748	15	10
51	. 49773	25	. 00912	102	. 57386	39	. 74257	117	. 15296	19	. 86733	14	9
52	. 49798	26	. 00810	102	. 57425	39	. 74140	118	. 15315	20	. 86719	15	8
53	. 49824	25	. 00708	101	. 57464	39	. 74022	117	. 15335	19	. 86704	14	7
54	. 49849	25	. 00607	102	. 57503	38	. 73905	117	. 15354	19	. 86690	15	6
55	0. 49874	25	2. 00505	101	0. 57541	39	1. 73788	117	1. 15373	20	0. 86675	14	5
56	. 49899	25	. 00404	101	. 57580	39	. 73671	116	. 15393	19	. 86661	15	4
57	. 49924	26	. 00303	101	. 57619	38	. 73555	117	. 15412	19	. 86646	14	3
58	. 49950	25	. 00202	101	. 57657	39	. 73438	117	. 15431	20	. 86632	15	2
59	. 49975	25	. 00101	101	. 57696	39	. 73321	116	. 15451	19	. 86617	14	1
60	0. 50000		2. 00000		0. 57735		1. 73205		1. 15470		0. 86603		0
↑ 119°→	cos	Diff. 1′	sec	Diff. 1′	cot	Diff. 1′	tan	Diff. 1′	csc	Diff. 1′	sin	Diff. 1′	←60° ↑

TABLE 31

Natural Trigonometric Functions

30° → ↓ ′	sin	Diff. 1′	csc	Diff. 1′	tan	Diff. 1′	cot	Diff. 1′	sec	Diff. 1′	cos	Diff. 1′	← 149° ↓ ′
0	0.50000	25	2.00000	101	0.57735	39	1.73205	116	1.15470	19	0.86603	15	60
1	.50025	25	1.99899	100	.57774	39	.73089	116	.15489	20	.86588	15	59
2	.50050	26	.99799	101	.57813	38	.72973	116	.15509	19	.86573	14	58
3	.50076	25	.99698	100	.57851	39	.72857	116	.15528	20	.86559	15	57
4	.50101	25	.99598	100	.57890	39	.72741	116	.15548	19	.86544	14	56
5	0.50126	25	1.99498	100	0.57929	39	1.72625	116	1.15567	20	0.86530	15	55
6	.50151	25	.99398	100	.57968	39	.72509	116	.15587	19	.86515	14	54
7	.50176	25	.99298	100	.58007	39	.72393	115	.15606	20	.86501	15	53
8	.50201	26	.99198	100	.58046	39	.72278	115	.15626	19	.86486	15	52
9	.50227	25	.99098	100	.58085	39	.72163	116	.15645	20	.86471	14	51
10	0.50252	25	1.98998	99	0.58124	38	1.72047	115	1.15665	19	0.86457	15	50
11	.50277	25	.98899	100	.58162	39	.71932	115	.15684	20	.86442	15	49
12	.50302	25	.98799	99	.58201	39	.71817	115	.15704	20	.86427	14	48
13	.50327	25	.98700	99	.58240	39	.71702	114	.15724	19	.86413	15	47
14	.50352	25	.98601	99	.58279	39	.71588	115	.15743	20	.86398	14	46
15	0.50377	26	1.98502	99	0.58318	39	1.71473	115	1.15763	19	0.86384	15	45
16	.50403	25	.98403	99	.58357	39	.71358	114	.15782	20	.86369	15	44
17	.50428	25	.98304	99	.58396	39	.71244	115	.15802	20	.86354	14	43
18	.50453	25	.98205	98	.58435	39	.71129	114	.15822	19	.86340	15	42
19	.50478	25	.98107	99	.58474	39	.71015	114	.15841	20	.86325	15	41
20	0.50503	25	1.98008	98	0.58513	39	1.70901	114	1.15861	20	0.86310	15	40
21	.50528	25	.97910	99	.58552	39	.70787	114	.15881	20	.86295	14	39
22	.50553	25	.97811	98	.58591	40	.70673	113	.15901	19	.86281	15	38
23	.50578	25	.97713	98	.58631	39	.70560	114	.15920	20	.86266	15	37
24	.50603	25	.97615	98	.58670	39	.70446	114	.15940	20	.86251	14	36
25	0.50628	26	1.97517	97	0.58709	39	1.70332	113	1.15960	20	0.86237	15	35
26	.50654	25	.97420	98	.58748	39	.70219	113	.15980	20	.86222	15	34
27	.50679	25	.97322	98	.58787	39	.70106	114	.16000	19	.86207	15	33
28	.50704	25	.97224	97	.58826	39	.69992	113	.16019	20	.86192	14	32
29	.50729	25	.97127	98	.58865	40	.69879	113	.16039	20	.86178	15	31
30	0.50754	25	1.97029	97	0.58905	39	1.69766	113	1.16059	20	0.86163	15	30
31	.50779	25	.96932	97	.58944	39	.69653	112	.16079	20	.86148	15	29
32	.50804	25	.96835	97	.58983	39	.69541	113	.16099	20	.86133	14	28
33	.50829	25	.96738	97	.59022	39	.69428	112	.16119	20	.86119	15	27
34	.50854	25	.96641	97	.59061	40	.69316	113	.16139	20	.86104	15	26
35	0.50879	25	1.96544	96	0.59101	39	1.69203	112	1.16159	20	0.86089	15	25
36	.50904	25	.96448	97	.59140	39	.69091	112	.16179	20	.86074	15	24
37	.50929	25	.96351	96	.59179	39	.68979	113	.16199	20	.86059	14	23
38	.50954	25	.96255	97	.59218	40	.68866	112	.16219	20	.86045	15	22
39	.50979	25	.96158	96	.59258	39	.68754	111	.16239	20	.86030	15	21
40	0.51004	25	1.96062	96	0.59297	39	1.68643	112	1.16259	20	0.86015	15	20
41	.51029	25	.95966	96	.59336	40	.68531	112	.16279	20	.86000	15	19
42	.51054	25	.95870	96	.59376	39	.68419	111	.16299	20	.85985	15	18
43	.51079	25	.95774	96	.59415	39	.68308	112	.16319	20	.85970	14	17
44	.51104	25	.95678	95	.59454	40	.68196	111	.16339	20	.85956	15	16
45	0.51129	25	1.95583	96	0.59494	39	1.68085	111	1.16359	21	0.85941	15	15
46	.51154	25	.95487	95	.59533	40	.67974	111	.16380	20	.85926	15	14
47	.51179	25	.95392	96	.59573	39	.67863	111	.16400	20	.85911	15	13
48	.51204	25	.95296	95	.59612	39	.67752	111	.16420	20	.85896	15	12
49	.51229	25	.95201	95	.59651	40	.67641	111	.16440	20	.85881	15	11
50	0.51254	25	1.95106	95	0.59691	39	1.67530	111	1.16460	21	0.85866	15	10
51	.51279	25	.95011	95	.59730	40	.67419	110	.16481	20	.85851	15	9
52	.51304	25	.94916	95	.59770	39	.67309	111	.16501	20	.85836	15	8
53	.51329	25	.94821	95	.59809	40	.67198	110	.16521	20	.85821	15	7
54	.51354	25	.94726	94	.59849	39	.67088	110	.16541	21	.85806	14	6
55	0.51379	25	1.94632	95	0.59888	40	1.66978	111	1.16562	20	0.85792	15	5
56	.51404	25	.94537	94	.59928	39	.66867	110	.16582	20	.85777	15	4
57	.51429	25	.94443	94	.59967	40	.66757	110	.16602	21	.85762	15	3
58	.51454	25	.94349	95	.60007	39	.66647	109	.16623	20	.85747	15	2
59	.51479	25	.94254	94	.60046	40	.66538	110	.16643	20	.85732	15	1
60	0.51504		1.94160		0.60086		1.66428		1.16663		0.85717		0
↑ 120° →	cos	Diff. 1′	sec	Diff. 1′	cot	Diff. 1′	tan	Diff. 1′	csc	Diff. 1′	sin	Diff. 1′	← 59° ↑

TABLE 31

Natural Trigonometric Functions

31°→ ↓ ′	sin	Diff. 1′	csc	Diff. 1′	tan	Diff. 1′	cot	Diff. 1′	sec	Diff. 1′	cos	Diff. 1′	←148° ↓ ′
0	0. 51504	25	1. 94160	94	0. 60086	40	1. 66428	110	1. 16663	21	0. 85717	15	60
1	. 51529	25	. 94066	93	. 60126	39	. 66318	109	. 16684	20	. 85702	15	59
2	. 51554	25	. 93973	94	. 60165	40	. 66209	110	. 16704	21	. 85687	15	58
3	. 51579	25	. 93879	94	. 60205	40	. 66099	109	. 16725	20	. 85672	15	57
4	. 51604	24	. 93785	93	. 60245	39	. 65990	109	. 16745	21	. 85657	15	56
5	0. 51628	25	1. 93692	94	0. 60284	40	1. 65881	109	1. 16766	20	0. 85642	15	55
6	. 51653	25	. 93598	93	. 60324	40	. 65772	109	. 16786	20	. 85627	15	54
7	. 51678	25	. 93505	93	. 60364	39	. 65663	109	. 16806	21	. 85612	15	53
8	. 51703	25	. 93412	93	. 60403	40	. 65554	109	. 16827	21	. 85597	15	52
9	. 51728	25	. 93319	93	. 60443	40	. 65445	108	. 16848	20	. 85582	15	51
10	0. 51753	25	1. 93226	93	0. 60483	39	1. 65337	109	1. 16868	21	0. 85567	16	50
11	. 51778	25	. 93133	93	. 60522	40	. 65228	108	. 16889	20	. 85551	15	49
12	. 51803	25	. 93040	93	. 60562	40	. 65120	109	. 16909	21	. 85536	15	48
13	. 51828	24	. 92947	92	. 60602	40	. 65011	108	. 16930	20	. 85521	15	47
14	. 51852	25	. 92855	93	. 60642	39	. 64903	108	. 16950	21	. 85506	15	46
15	0. 51877	25	1. 92762	92	0. 60681	40	1. 64795	108	1. 16971	21	0. 85491	15	45
16	. 51902	25	. 92670	92	. 60721	40	. 64687	108	. 16992	20	. 85476	15	44
17	. 51927	25	. 92578	92	. 60761	40	. 64579	108	. 17012	21	. 85461	15	43
18	. 51952	25	. 92486	92	. 60801	40	. 64471	108	. 17033	21	. 85446	15	42
19	. 51977	25	. 92394	92	. 60841	40	. 64363	107	. 17054	21	. 85431	15	41
20	0. 52002	24	1. 92302	92	0. 60881	40	1. 64256	108	1. 17075	20	0. 85416	15	40
21	. 52026	25	. 92210	92	. 60921	39	. 64148	107	. 17095	21	. 85401	16	39
22	. 52051	25	. 92118	91	. 60960	40	. 64041	107	. 17116	21	. 85385	15	38
23	. 52076	25	. 92027	92	. 61000	40	. 63934	108	. 17137	21	. 85370	15	37
24	. 52101	25	. 91935	91	. 61040	40	. 63826	107	. 17158	20	. 85355	15	36
25	0. 52126	25	1. 91844	92	0. 61080	40	1. 63719	107	1. 17178	21	0. 85340	15	35
26	. 52151	24	. 91752	91	. 61120	40	. 63612	107	. 17199	21	. 85325	15	34
27	. 52175	25	. 91661	91	. 61160	40	. 63505	107	. 17220	21	. 85310	16	33
28	. 52200	25	. 91570	91	. 61200	40	. 63398	106	. 17241	21	. 85294	15	32
29	. 52225	25	. 91479	91	. 61240	40	. 63292	107	. 17262	21	. 85279	15	31
30	0. 52250	25	1. 91388	91	0. 61280	40	1. 63185	106	1. 17283	21	0. 85264	15	30
31	. 52275	24	. 91297	90	. 61320	40	. 63079	107	. 17304	21	. 85249	15	29
32	. 52299	25	. 91207	91	. 61360	40	. 62972	106	. 17325	21	. 85234	16	28
33	. 52324	25	. 91116	90	. 61400	40	. 62866	106	. 17346	21	. 85218	15	27
34	. 52349	25	. 91026	91	. 61440	40	. 62760	106	. 17367	21	. 85203	15	26
35	0. 52374	25	1. 90935	90	0. 61480	40	1. 62654	106	1. 17388	21	0. 85188	15	25
36	. 52399	24	. 90845	90	. 61520	41	. 62548	106	. 17409	21	. 85173	16	24
37	. 52423	25	. 90755	90	. 61561	40	. 62442	106	. 17430	21	. 85157	15	23
38	. 52448	25	. 90665	90	. 61601	40	. 62336	106	. 17451	21	. 85142	15	22
39	. 52473	25	. 90575	90	. 61641	40	. 62230	105	. 17472	21	. 85127	15	21
40	0. 52498	24	1. 90485	90	0. 61681	40	1. 62125	106	1. 17493	21	0. 85112	16	20
41	. 52522	25	. 90395	90	. 61721	40	. 62019	105	. 17514	21	. 85096	15	19
42	. 52547	25	. 90305	89	. 61761	40	. 61914	105	. 17535	21	. 85081	15	18
43	. 52572	25	. 90216	90	. 61801	41	. 61809	106	. 17556	21	. 85066	15	17
44	. 52597	24	. 90126	89	. 61842	40	. 61703	105	. 17577	21	. 85051	16	16
45	0. 52621	25	1. 90037	89	0. 61882	40	1. 61598	105	1. 17598	22	0. 85035	15	15
46	. 52646	25	. 89948	90	. 61922	40	. 61493	105	. 17620	21	. 85020	15	14
47	. 52671	25	. 89858	89	. 61962	41	. 61388	105	. 17641	21	. 85005	16	13
48	. 52696	24	. 89769	89	. 62003	40	. 61283	104	. 17662	21	. 84989	15	12
49	. 52720	25	. 89680	89	. 62043	40	. 61179	105	. 17683	21	. 84974	15	11
50	0. 52745	25	1. 89591	88	0. 62083	41	1. 61074	104	1. 17704	22	0. 84959	16	10
51	. 52770	24	. 89503	89	. 62124	40	. 60970	105	. 17726	21	. 84943	15	9
52	. 52794	25	. 89414	89	. 62164	40	. 60865	104	. 17747	21	. 84928	15	8
53	. 52819	25	. 89325	88	. 62204	41	. 60761	104	. 17768	22	. 84913	16	7
54	. 52844	25	. 89237	89	. 62245	40	. 60657	104	. 17790	21	. 84897	15	6
55	0. 52869	24	1. 89148	88	0. 62285	40	1. 60553	104	1. 17811	21	0. 84882	16	5
56	. 52893	25	. 89060	88	. 62325	41	. 60449	104	. 17832	22	. 84866	15	4
57	. 52918	25	. 88972	88	. 62366	40	. 60345	104	. 17854	21	. 84851	15	3
58	. 52943	24	. 88884	88	. 62406	40	. 60241	104	. 17875	21	. 84836	16	2
59	. 52967	25	. 88796	88	. 62446	41	. 60137	104	. 17896	22	. 84820	15	1
60	0. 52992		1. 88708		0. 62487		1. 60033		1. 17918		0. 84805		0
↑ 121°→	cos	Diff. 1′	sec	Diff. 1′	cot	Diff. 1′	tan	Diff. 1′	csc	Diff. 1′	sin	Diff. 1′	↑ ←58°

TABLE 31

Natural Trigonometric Functions

32°→ ↓ ′	sin	Diff. 1′	csc	Diff. 1′	tan	Diff. 1′	cot	Diff. 1′	sec	Diff. 1′	cos	Diff. 1′	←147° ↓ ′
0	0. 52992	25	1. 88708	88	0. 62487	40	1. 60033	103	1. 17918	21	0. 84805	16	60
1	. 53017	24	. 88620	88	. 62527	41	. 59930	104	. 17939	22	. 84789	15	59
2	. 53041	25	. 88532	87	. 62568	40	. 59826	103	. 17961	21	. 84774	15	58
3	. 53066	25	. 88445	88	. 62608	41	. 59723	103	. 17982	22	. 84759	16	57
4	. 53091	24	. 88357	87	. 62649	40	. 59620	103	. 18004	21	. 84743	15	56
5	0. 53115	25	1. 88270	87	0. 62689	41	1. 59517	103	1. 18025	22	0. 84728	16	55
6	. 53140	24	. 88183	88	. 62730	40	. 59414	103	. 18047	21	. 84712	15	54
7	. 53164	25	. 88095	87	. 62770	41	. 59311	103	. 18068	22	. 84697	16	53
8	. 53189	25	. 88008	87	. 62811	41	. 59208	103	. 18090	21	. 84681	15	52
9	. 53214	24	. 87921	87	. 62852	40	. 59105	103	. 18111	22	. 84666	16	51
10	0. 53238	25	1. 87834	86	0. 62892	41	1. 59002	102	1. 18133	22	0. 84650	15	50
11	. 53263	25	. 87748	87	. 62933	40	. 58900	103	. 18155	21	. 84635	16	49
12	. 53288	24	. 87661	87	. 62973	41	. 58797	102	. 18176	22	. 84619	15	48
13	. 53312	25	. 87574	86	. 63014	41	. 58695	102	. 18198	22	. 84604	16	47
14	. 53337	24	. 87488	87	. 63055	40	. 58593	103	. 18220	21	. 84588	15	46
15	0. 53361	25	1. 87401	86	0. 63095	41	1. 58490	102	1. 18241	22	0. 84573	16	45
16	. 53386	25	. 87315	86	. 63136	41	. 58388	102	. 18263	22	. 84557	15	44
17	. 53411	24	. 87229	87	. 63177	40	. 58286	102	. 18285	22	. 84542	16	43
18	. 53435	25	. 87142	86	. 63217	41	. 58184	101	. 18307	21	. 84526	15	42
19	. 53460	24	. 87056	86	. 63258	41	. 58083	102	. 18328	22	. 84511	16	41
20	0. 53484	25	1. 86970	85	0. 63299	41	1. 57981	102	1. 18350	22	0. 84495	15	40
21	. 53509	25	. 86885	86	. 63340	40	. 57879	101	. 18372	22	. 84480	16	39
22	. 53534	24	. 86799	86	. 63380	41	. 57778	102	. 18394	22	. 84464	16	38
23	. 53558	25	. 86713	86	. 63421	41	. 57676	101	. 18416	21	. 84448	15	37
24	. 53583	24	. 86627	85	. 63462	41	. 57575	101	. 18437	22	. 84433	16	36
25	0. 53607	25	1. 86542	85	0. 63503	41	1. 57474	102	1. 18459	22	0. 84417	15	35
26	. 53632	24	. 86457	86	. 63544	40	. 57372	101	. 18481	22	. 84402	16	34
27	. 53656	25	. 86371	85	. 63584	41	. 57271	101	. 18503	22	. 84386	16	33
28	. 53681	24	. 86286	85	. 63625	41	. 57170	101	. 18525	22	. 84370	15	32
29	. 53705	25	. 86201	85	. 63666	41	. 57069	100	. 18547	22	. 84355	16	31
30	0. 53730	24	1. 86116	85	0. 63707	41	1. 56969	101	1. 18569	22	0. 84339	15	30
31	. 53754	25	. 86031	85	. 63748	41	. 56868	101	. 18591	22	. 84324	16	29
32	. 53779	25	. 85946	85	. 63789	41	. 56767	100	. 18613	22	. 84308	16	28
33	. 53804	24	. 85861	84	. 63830	41	. 56667	101	. 18635	22	. 84292	15	27
34	. 53828	25	. 85777	85	. 63871	41	. 56566	100	. 18657	22	. 84277	16	26
35	0. 53853	24	1. 85692	84	0. 63912	41	1. 56466	100	1. 18679	22	0. 84261	16	25
36	. 53877	25	. 85608	85	. 63953	41	. 56366	101	. 18701	22	. 84245	15	24
37	. 53902	24	. 85523	84	. 63994	41	. 56265	100	. 18723	22	. 84230	16	23
38	. 53926	25	. 85439	84	. 64035	41	. 56165	100	. 18745	22	. 84214	16	22
39	. 53951	24	. 85355	84	. 64076	41	. 56065	99	. 18767	23	. 84198	16	21
40	0. 53975	25	1. 85271	84	0. 64117	41	1. 55966	100	1. 18790	22	0. 84182	15	20
41	. 54000	24	. 85187	84	. 64158	41	. 55866	100	. 18812	22	. 84167	16	19
42	. 54024	25	. 85103	84	. 64199	41	. 55766	100	. 18834	22	. 84151	16	18
43	. 54049	24	. 85019	84	. 64240	41	. 55666	99	. 18856	22	. 84135	15	17
44	. 54073	24	. 84935	83	. 64281	41	. 55567	100	. 18878	23	. 84120	16	16
45	0. 54097	25	1. 84852	84	0. 64322	41	1. 55467	99	1. 18901	22	0. 84104	16	15
46	. 54122	24	. 84768	83	. 64363	41	. 55368	99	. 18923	22	. 84088	16	14
47	. 54146	25	. 84685	84	. 64404	42	. 55269	99	. 18945	22	. 84072	15	13
48	. 54171	24	. 84601	83	. 64446	41	. 55170	99	. 18967	23	. 84057	16	12
49	. 54195	25	. 84518	83	. 64487	41	. 55071	99	. 18990	22	. 84041	16	11
50	0. 54220	24	1. 84435	83	0. 64528	41	1. 54972	99	1. 19012	22	0. 84025	16	10
51	. 54244	25	. 84352	83	. 64569	41	. 54873	99	. 19034	23	. 84009	15	9
52	. 54269	24	. 84269	83	. 64610	42	. 54774	99	. 19057	22	. 83994	16	8
53	. 54293	24	. 84186	83	. 64652	41	. 54675	99	. 19079	23	. 83978	16	7
54	. 54317	25	. 84103	83	. 64693	41	. 54576	98	. 19102	22	. 83962	16	6
55	0. 54342	24	1. 84020	82	0. 64734	41	1. 54478	99	1. 19124	22	0. 83946	16	5
56	. 54366	25	. 83938	83	. 64775	42	. 54379	98	. 19146	23	. 83930	15	4
57	. 54391	24	. 83855	82	. 64817	41	. 54281	98	. 19169	22	. 83915	16	3
58	. 54415	25	. 83773	83	. 64858	41	. 54183	98	. 19191	23	. 83899	16	2
59	. 54440	24	. 83690	82	. 64899	42	. 54085	99	. 19214	22	. 83883	16	1
60	0. 54464		1. 83608		0. 64941		1. 53986		1. 19236		0. 83867		0
↑ 122°→	cos	Diff. 1′	sec	Diff. 1′	cot	Diff. 1′	tan	Diff. 1′	csc	Diff. 1′	sin	Diff. 1′	←57° ↑

TABLE 31

Natural Trigonometric Functions

33°→ ↓ ′	sin	Diff. 1′	csc	Diff. 1′	tan	Diff. 1′	cot	Diff. 1′	sec	Diff. 1′	cos	Diff. 1′	←146° ↓ ′
0	0. 54464	24	1. 83608	82	0. 64941	41	1. 53986	98	1. 19236	23	0. 83867	16	60
1	. 54488	25	. 83526	82	. 64982	42	. 53888	97	. 19259	22	. 83851	16	59
2	. 54513	24	. 83444	82	. 65024	41	. 53791	98	. 19281	23	. 83835	16	58
3	. 54537	24	. 83362	82	. 65065	41	. 53693	98	. 19304	23	. 83819	15	57
4	. 54561	25	. 83280	82	. 65106	42	. 53595	98	. 19327	22	. 83804	16	56
5	0. 54586	24	1. 83198	82	0. 65148	41	1. 53497	97	1. 19349	23	0. 83788	16	55
6	. 54610	25	. 83116	82	. 65189	42	. 53400	98	. 19372	22	. 83772	16	54
7	. 54635	24	. 83034	81	. 65231	41	. 53302	97	. 19394	23	. 83756	16	53
8	. 54659	24	. 82953	82	. 65272	42	. 53205	98	. 19417	23	. 83740	16	52
9	. 54683	25	. 82871	81	. 65314	41	. 53107	97	. 19440	23	. 83724	16	51
10	0. 54708	24	1. 82790	81	0. 65355	42	1. 53010	97	1. 19463	22	0. 83708	16	50
11	. 54732	24	. 82709	82	. 65397	41	. 52913	97	. 19485	23	. 83692	16	49
12	. 54756	25	. 82627	81	. 65438	42	. 52816	97	. 19508	23	. 83676	16	48
13	. 54781	24	. 82546	81	. 65480	41	. 52719	97	. 19531	22	. 83660	15	47
14	. 54805	24	. 82465	81	. 65521	42	. 52622	97	. 19553	23	. 83645	16	46
15	0. 54829	25	1. 82384	81	0. 65563	41	1. 52525	96	1. 19576	23	0. 83629	16	45
16	. 54854	24	. 82303	81	. 65604	42	. 52429	97	. 19599	23	. 83613	16	44
17	. 54878	24	. 82222	80	. 65646	42	. 52332	97	. 19622	23	. 83597	16	43
18	. 54902	25	. 82142	81	. 65688	41	. 52235	96	. 19645	23	. 83581	16	42
19	. 54927	24	. 82061	80	. 65729	42	. 52139	96	. 19668	23	. 83565	16	41
20	0. 54951	24	1. 81981	81	0. 65771	42	1. 52043	97	1. 19691	22	0. 83549	16	40
21	. 54975	24	. 81900	80	. 65813	41	. 51946	96	. 19713	23	. 83533	16	39
22	. 54999	25	. 81820	80	. 65854	42	. 51850	96	. 19736	23	. 83517	16	38
23	. 55024	24	. 81740	81	. 65896	42	. 51754	96	. 19759	23	. 83501	16	37
24	. 55048	24	. 81659	80	. 65938	42	. 51658	96	. 19782	23	. 83485	16	36
25	0. 55072	25	1. 81579	80	0. 65980	41	1. 51562	96	1. 19805	23	0. 83469	16	35
26	. 55097	24	. 81499	80	. 66021	42	. 51466	96	. 19828	23	. 83453	16	34
27	. 55121	24	. 81419	79	. 66063	42	. 51370	95	. 19851	23	. 83437	16	33
28	. 55145	24	. 81340	80	. 66105	42	. 51275	96	. 19874	23	. 83421	16	32
29	. 55169	25	. 81260	80	. 66147	42	. 51179	95	. 19897	23	. 83405	16	31
30	0. 55194	24	1. 81180	79	0. 66189	41	1. 51084	96	1. 19920	24	0. 83389	16	30
31	. 55218	24	. 81101	80	. 66230	42	. 50988	95	. 19944	23	. 83373	17	29
32	. 55242	24	. 81021	79	. 66272	42	. 50893	96	. 19967	23	. 83356	16	28
33	. 55266	25	. 80942	80	. 66314	42	. 50797	95	. 19990	23	. 83340	16	27
34	. 55291	24	. 80862	79	. 66356	42	. 50702	95	. 20013	23	. 83324	16	26
35	0. 55315	24	1. 80783	79	0. 66398	42	1. 50607	95	1. 20036	23	0. 83308	16	25
36	. 55339	24	. 80704	79	. 66440	42	. 50512	95	. 20059	24	. 83292	16	24
37	. 55363	25	. 80625	79	. 66482	42	. 50417	95	. 20083	23	. 83276	16	23
38	. 55388	24	. 80546	79	. 66524	42	. 50322	94	. 20106	23	. 83260	16	22
39	. 55412	24	. 80467	79	. 66566	42	. 50228	95	. 20129	23	. 83244	16	21
40	0. 55436	24	1. 80388	79	0. 66608	42	1. 50133	95	1. 20152	24	0. 83228	16	20
41	. 55460	24	. 80309	78	. 66650	42	. 50038	94	. 20176	23	. 83212	17	19
42	. 55484	25	. 80231	79	. 66692	42	. 49944	95	. 20199	23	. 83195	16	18
43	. 55509	24	. 80152	78	. 66734	42	. 49849	94	. 20222	24	. 83179	16	17
44	. 55533	24	. 80074	79	. 66776	42	. 49755	94	. 20246	23	. 83163	16	16
45	0. 55557	24	1. 79995	78	0. 66818	42	1. 49661	95	1. 20269	23	0. 83147	16	15
46	. 55581	24	. 79917	78	. 66860	42	. 49566	94	. 20292	24	. 83131	16	14
47	. 55605	25	. 79839	78	. 66902	42	. 49472	94	. 20316	23	. 83115	17	13
48	. 55630	24	. 79761	79	. 66944	42	. 49378	94	. 20339	24	. 83098	16	12
49	. 55654	24	. 79682	78	. 66986	42	. 49284	94	. 20363	23	. 83082	16	11
50	0. 55678	24	1. 79604	77	0. 67028	43	1. 49190	93	1. 20386	24	0. 83066	16	10
51	. 55702	24	. 79527	78	. 67071	42	. 49097	94	. 20410	23	. 83050	16	9
52	. 55726	24	. 79449	78	. 67113	42	. 49003	94	. 20433	24	. 83034	17	8
53	. 55750	25	. 79371	78	. 67155	42	. 48909	93	. 20457	23	. 83017	16	7
54	. 55775	24	. 79293	77	. 67197	42	. 48816	94	. 20480	24	. 83001	16	6
55	0. 55799	24	1. 79216	78	0. 67239	43	1. 48722	93	1. 20504	23	0. 82985	16	5
56	. 55823	24	. 79138	77	. 67282	42	. 48629	93	. 20527	24	. 82969	16	4
57	. 55847	24	. 79061	77	. 67324	42	. 48536	94	. 20551	24	. 82953	17	3
58	. 55871	24	. 78984	78	. 67366	43	. 48442	93	. 20575	23	. 82936	16	2
59	. 55895	24	. 78906	77	. 67409	42	. 48349	93	. 20598	24	. 82920	16	1
60	0. 55919		1. 78829		0. 67451		1. 48256		1. 20622		0. 82904		0
↑ 123°→	cos	Diff. 1′	sec	Diff. 1′	cot	Diff. 1′	tan	Diff. 1′	csc	Diff. 1′	sin	Diff. 1′	←56° ↑

TABLE 31

Natural Trigonometric Functions

34°→ ′ ↓	sin	Diff. 1′	csc	Diff. 1′	tan	Diff. 1′	cot	Diff. 1′	sec	Diff. 1′	cos	Diff. 1′	←145° ′ ↓
0	0. 55919	24	1. 78829	77	0. 67451	42	1. 48256	93	1. 20622	23	0. 82904	17	60
1	. 55943	25	. 78752	77	. 67493	43	. 48163	93	. 20645	24	. 82887	16	59
2	. 55968	24	. 78675	77	. 67536	42	. 48070	93	. 20669	24	. 82871	16	58
3	. 55992	24	. 78598	77	. 67578	42	. 47977	92	. 20693	24	. 82855	16	57
4	. 56016	24	. 78521	76	. 67620	43	. 47885	93	. 20717	23	. 82839	17	56
5	0. 56040	24	1. 78445	77	0. 67663	42	1. 47792	93	1. 20740	24	0. 82822	16	55
6	. 56064	24	. 78368	77	. 67705	43	. 47699	92	. 20764	24	. 82806	16	54
7	. 56088	24	. 78291	76	. 67748	42	. 47607	93	. 20788	24	. 82790	17	53
8	. 56112	24	. 78215	77	. 67790	42	. 47514	92	. 20812	24	. 82773	16	52
9	. 56136	24	. 78138	76	. 67832	43	. 47422	92	. 20836	23	. 82757	16	51
10	0. 56160	24	1. 78062	76	0. 67875	42	1. 47330	92	1. 20859	24	0. 82741	17	50
11	. 56184	24	. 77986	76	. 67917	43	. 47238	92	. 20883	24	. 82724	16	49
12	. 56208	24	. 77910	77	. 67960	42	. 47146	93	. 20907	24	. 82708	16	48
13	. 56232	24	. 77833	76	. 68002	43	. 47053	91	. 20931	24	. 82692	17	47
14	. 56256	24	. 77757	76	. 68045	43	. 46962	92	. 20955	24	. 82675	16	46
15	0. 56280	25	1. 77681	75	0. 68088	42	1. 46870	92	1. 20979	24	0. 82659	16	45
16	. 56305	24	. 77606	76	. 68130	43	. 46778	92	. 21003	24	. 82643	17	44
17	. 56329	24	. 77530	76	. 68173	42	. 46686	91	. 21027	24	. 82626	16	43
18	. 56353	24	. 77454	76	. 68215	43	. 46595	92	. 21051	24	. 82610	17	42
19	. 56377	24	. 77378	75	. 68258	43	. 46503	92	. 21075	24	. 82593	16	41
20	0. 56401	24	1. 77303	76	0. 68301	42	1. 46411	91	1. 21099	24	0. 82577	16	40
21	. 56425	24	. 77227	75	. 68343	43	. 46320	91	. 21123	24	. 82561	17	39
22	. 56449	24	. 77152	75	. 68386	43	. 46229	92	. 21147	24	. 82544	16	38
23	. 56473	24	. 77077	76	. 68429	42	. 46137	91	. 21171	24	. 82528	17	37
24	. 56497	24	. 77001	75	. 68471	43	. 46046	91	. 21195	25	. 82511	16	36
25	0. 56521	24	1. 76926	75	0. 68514	43	1. 45955	91	1. 21220	24	0. 82495	17	35
26	. 56545	24	. 76851	75	. 68557	43	. 45864	91	. 21244	24	. 82478	16	34
27	. 56569	24	. 76776	75	. 68600	42	. 45773	91	. 21268	24	. 82462	16	33
28	. 56593	24	. 76701	75	. 68642	43	. 45682	90	. 21292	24	. 82446	17	32
29	. 56617	24	. 76626	74	. 68685	43	. 45592	91	. 21316	25	. 82429	16	31
30	0. 56641	24	1. 76552	75	0. 68728	43	1. 45501	91	1. 21341	24	0. 82413	17	30
31	. 56665	24	. 76477	75	. 68771	43	. 45410	90	. 21365	24	. 82396	16	29
32	. 56689	24	. 76402	74	. 68814	43	. 45320	91	. 21389	25	. 82380	17	28
33	. 56713	23	. 76328	75	. 68857	43	. 45229	90	. 21414	24	. 82363	16	27
34	. 56736	24	. 76253	74	. 68900	42	. 45139	90	. 21438	24	. 82347	17	26
35	0. 56760	24	1. 76179	74	0. 68942	43	1. 45049	91	1. 21462	25	0. 82330	16	25
36	. 56784	24	. 76105	74	. 68985	43	. 44958	90	. 21487	24	. 82314	17	24
37	. 56808	24	. 76031	75	. 69028	43	. 44868	90	. 21511	24	. 82297	16	23
38	. 56832	24	. 75956	74	. 69071	43	. 44778	90	. 21535	25	. 82281	17	22
39	. 56856	24	. 75882	74	. 69114	43	. 44688	90	. 21560	24	. 82264	16	21
40	0. 56880	24	1. 75808	74	0. 69157	43	1. 44598	90	1. 21584	25	0. 82248	17	20
41	. 56904	24	. 75734	73	. 69200	43	. 44508	90	. 21609	24	. 82231	17	19
42	. 56928	24	. 75661	74	. 69243	43	. 44418	89	. 21633	25	. 82214	16	18
43	. 56952	24	. 75587	74	. 69286	43	. 44329	90	. 21658	24	. 82198	17	17
44	. 56976	24	. 75513	73	. 69329	43	. 44239	90	. 21682	25	. 82181	16	16
45	0. 57000	24	1. 75440	74	0. 69372	44	1. 44149	89	1. 21707	24	0. 82165	17	15
46	. 57024	23	. 75366	73	. 69416	43	. 44060	90	. 21731	25	. 82148	16	14
47	. 57047	24	. 75293	74	. 69459	43	. 43970	89	. 21756	25	. 82132	17	13
48	. 57071	24	. 75219	73	. 69502	43	. 43881	89	. 21781	24	. 82115	17	12
49	. 57095	24	. 75146	73	. 69545	43	. 43792	89	. 21805	25	. 82098	16	11
50	0. 57119	24	1. 75073	73	0. 69588	43	1. 43703	89	1. 21830	25	0. 82082	17	10
51	. 57143	24	. 75000	73	. 69631	44	. 43614	89	. 21855	24	. 82065	17	9
52	. 57167	24	. 74927	73	. 69675	43	. 43525	89	. 21879	25	. 82048	16	8
53	. 57191	24	. 74854	73	. 69718	43	. 43436	89	. 21904	25	. 82032	17	7
54	. 57215	23	. 74781	73	. 69761	43	. 43347	89	. 21929	24	. 82015	16	6
55	0. 57238	24	1. 74708	73	0. 69804	43	1. 43258	89	1. 21953	25	0. 81999	17	5
56	. 57262	24	. 74635	73	. 69847	44	. 43169	89	. 21978	25	. 81982	17	4
57	. 57286	24	. 74562	72	. 69891	43	. 43080	88	. 22003	25	. 81965	16	3
58	. 57310	24	. 74490	73	. 69934	43	. 42992	89	. 22028	25	. 81949	17	2
59	. 57334	24	. 74417	72	. 69977	44	. 42903	88	. 22053	24	. 81932	17	1
60	0. 57358		1. 74345		0. 70021		1. 42815		1. 22077		0. 81915		0
↑ 124°→	cos	Diff. 1′	sec	Diff. 1′	cot	Diff. 1′	tan	Diff. 1′	csc	Diff. 1′	sin	Diff. 1′	←55° ↑

TABLE 31

Natural Trigonometric Functions

35° → ↓ ′	sin	Diff. 1′	csc	Diff. 1′	tan	Diff. 1′	cot	Diff. 1′	sec	Diff. 1′	cos	Diff. 1′	← 144° ↓ ′
0	0. 57358	23	1. 74345	73	0. 70021	43	1. 42815	89	1. 22077	25	0. 81915	16	60
1	. 57381	24	. 74272	72	. 70064	43	. 42726	88	. 22102	25	. 81899	17	59
2	. 57405	24	. 74200	72	. 70107	44	. 42638	88	. 22127	25	. 81882	17	58
3	. 57429	24	. 74128	72	. 70151	43	. 42550	88	. 22152	25	. 81865	17	57
4	. 57453	24	. 74056	73	. 70194	44	. 42462	88	. 22177	25	. 81848	16	56
5	0. 57477	24	1. 73983	72	0. 70238	43	1. 42374	88	1. 22202	25	0. 81832	17	55
6	. 57501	23	. 73911	71	. 70281	44	. 42286	88	. 22227	25	. 81815	17	54
7	. 57524	24	. 73840	72	. 70325	43	. 42198	88	. 22252	25	. 81798	16	53
8	. 57548	24	. 73768	72	. 70368	44	. 42110	88	. 22277	25	. 81782	17	52
9	. 57572	24	. 73696	72	. 70412	43	. 42022	88	. 22302	25	. 81765	17	51
10	0. 57596	23	1. 73624	72	0. 70455	44	1. 41934	87	1. 22327	25	0. 81748	17	50
11	. 57619	24	. 73552	71	. 70499	43	. 41847	88	. 22352	25	. 81731	17	49
12	. 57643	24	. 73481	72	. 70542	44	. 41759	87	. 22377	25	. 81714	16	48
13	. 57667	24	. 73409	71	. 70586	43	. 41672	88	. 22402	26	. 81698	17	47
14	. 57691	24	. 73338	71	. 70629	44	. 41584	87	. 22428	25	. 81681	17	46
15	0. 57715	23	1. 73267	72	0. 70673	44	1. 41497	88	1. 22453	25	0. 81664	17	45
16	. 57738	24	. 73195	71	. 70717	43	. 41409	87	. 22478	25	. 81647	16	44
17	. 57762	24	. 73124	71	. 70760	44	. 41322	87	. 22503	25	. 81631	17	43
18	. 57786	24	. 73053	71	. 70804	44	. 41235	87	. 22528	26	. 81614	17	42
19	. 57810	23	. 72982	71	. 70848	43	. 41148	87	. 22554	25	. 81597	17	41
20	0. 57833	24	1. 72911	71	0. 70891	44	1. 41061	87	1. 22579	25	0. 81580	17	40
21	. 57857	24	. 72840	71	. 70935	44	. 40974	87	. 22604	25	. 81563	17	39
22	. 57881	23	. 72769	71	. 70979	44	. 40887	87	. 22629	26	. 81546	16	38
23	. 57904	24	. 72698	70	. 71023	43	. 40800	86	. 22655	25	. 81530	17	37
24	. 57928	24	. 72628	71	. 71066	44	. 40714	87	. 22680	26	. 81513	17	36
25	0. 57952	24	1. 72557	70	0. 71110	44	1. 40627	87	1. 22706	25	0. 81496	17	35
26	. 57976	23	. 72487	71	. 71154	44	. 40540	86	. 22731	25	. 81479	17	34
27	. 57999	24	. 72416	70	. 71198	44	. 40454	87	. 22756	26	. 81462	17	33
28	. 58023	24	. 72346	71	. 71242	43	. 40367	86	. 22782	25	. 81445	17	32
29	. 58047	23	. 72275	70	. 71285	44	. 40281	86	. 22807	26	. 81428	16	31
30	0. 58070	24	1. 72205	70	0. 71329	44	1. 40195	86	1. 22833	25	0. 81412	17	30
31	. 58094	24	. 72135	70	. 71373	44	. 40109	87	. 22858	26	. 81395	17	29
32	. 58118	23	. 72065	70	. 71417	44	. 40022	86	. 22884	25	. 81378	17	28
33	. 58141	24	. 71995	70	. 71461	44	. 39936	86	. 22909	26	. 81361	17	27
34	. 58165	24	. 71925	70	. 71505	44	. 39850	86	. 22935	25	. 81344	17	26
35	0. 58189	23	1. 71855	70	0. 71549	44	1. 39764	85	1. 22960	26	0. 81327	17	25
36	. 58212	24	. 71785	70	. 71593	44	. 39679	86	. 22986	26	. 81310	17	24
37	. 58236	24	. 71715	69	. 71637	44	. 39593	86	. 23012	25	. 81293	17	23
38	. 58260	23	. 71646	70	. 71681	44	. 39507	86	. 23037	26	. 81276	17	22
39	. 58283	24	. 71576	70	. 71725	44	. 39421	85	. 23063	26	. 81259	17	21
40	0. 58307	23	1. 71506	69	0. 71769	44	1. 39336	86	1. 23089	25	0. 81242	17	20
41	. 58330	24	. 71437	69	. 71813	44	. 39250	85	. 23114	26	. 81225	17	19
42	. 58354	24	. 71368	70	. 71857	44	. 39165	86	. 23140	26	. 81208	17	18
43	. 58378	23	. 71298	69	. 71901	45	. 39079	85	. 23166	26	. 81191	17	17
44	. 58401	24	. 71229	69	. 71946	44	. 38994	85	. 23192	25	. 81174	17	16
45	0. 58425	24	1. 71160	69	0. 71990	44	1. 38909	85	1. 23217	26	0. 81157	17	15
46	. 58449	23	. 71091	69	. 72034	44	. 38824	86	. 23243	26	. 81140	17	14
47	. 58472	24	. 71022	69	. 72078	44	. 38738	85	. 23269	26	. 81123	17	13
48	. 58496	23	. 70953	69	. 72122	45	. 38653	85	. 23295	26	. 81106	17	12
49	. 58519	24	. 70884	69	. 72167	44	. 38568	84	. 23321	26	. 81089	17	11
50	0. 58543	24	1. 70815	69	0. 72211	44	1. 38484	85	1. 23347	26	0. 81072	17	10
51	. 58567	23	. 70746	69	. 72255	44	. 38399	85	. 23373	25	. 81055	17	9
52	. 58590	24	. 70677	68	. 72299	45	. 38314	85	. 23398	26	. 81038	17	8
53	. 58614	23	. 70609	69	. 72344	44	. 38229	84	. 23424	26	. 81021	17	7
54	. 58637	24	. 70540	68	. 72388	44	. 38145	85	. 23450	26	. 81004	17	6
55	0. 58661	23	1. 70472	69	0. 72432	45	1. 38060	84	1. 23476	26	0. 80987	17	5
56	. 58684	24	. 70403	68	. 72477	44	. 37976	85	. 23502	27	. 80970	17	4
57	. 58708	23	. 70335	68	. 72521	44	. 37891	84	. 23529	26	. 80953	17	3
58	. 58731	24	. 70267	69	. 72565	45	. 37807	85	. 23555	26	. 80936	17	2
59	. 58755	24	. 70198	68	. 72610	44	. 37722	84	. 23581	26	. 80919	17	1
60	0. 58779		1. 70130		0. 72654		1. 37638		1. 23607		0. 80902		0
↑ 125° →	cos	Diff. 1′	sec	Diff. 1′	cot	Diff. 1′	tan	Diff. 1′	csc	Diff. 1′	sin	Diff. 1′	← 54° ↑

TABLE 31

Natural Trigonometric Functions

36°→ ↓ ′	sin	Diff. 1′	csc	Diff. 1′	tan	Diff. 1′	cot	Diff. 1′	sec	Diff. 1′	cos	Diff. 1′	←143° ↓ ′
0	0.58779	23	1.70130	68	0.72654	45	1.37638	84	1.23607	26	0.80902	17	60
1	.58802	24	.70062	68	.72699	44	.37554	84	.23633	26	.80885	18	59
2	.58826	23	.69994	68	.72743	45	.37470	84	.23659	26	.80867	17	58
3	.58849	24	.69926	68	.72788	44	.37386	84	.23685	26	.80850	17	57
4	.58873	23	.69858	68	.72832	45	.37302	84	.23711	27	.80833	17	56
5	0.58896	24	1.69790	67	0.72877	44	1.37218	84	1.23738	26	0.80816	17	55
6	.58920	23	.69723	68	.72921	45	.37134	84	.23764	26	.80799	17	54
7	.58943	24	.69655	68	.72966	44	.37050	83	.23790	26	.80782	17	53
8	.58967	23	.69587	67	.73010	45	.36967	84	.23816	27	.80765	17	52
9	.58990	24	.69520	68	.73055	45	.36883	83	.23843	26	.80748	18	51
10	0.59014	23	1.69452	67	0.73100	44	1.36800	84	1.23869	26	0.80730	17	50
11	.59037	24	.69385	67	.73144	45	.36716	83	.23895	27	.80713	17	49
12	.59061	23	.69318	68	.73189	45	.36633	84	.23922	26	.80696	17	48
13	.59084	24	.69250	67	.73234	44	.36549	83	.23948	27	.80679	17	47
14	.59108	23	.69183	67	.73278	45	.36466	83	.23975	26	.80662	18	46
15	0.59131	23	1.69116	67	0.73323	45	1.36383	83	1.24001	27	0.80644	17	45
16	.59154	24	.69049	67	.73368	45	.36300	83	.24028	26	.80627	17	44
17	.59178	23	.68982	67	.73413	44	.36217	83	.24054	27	.80610	17	43
18	.59201	24	.68915	67	.73457	45	.36134	83	.24081	26	.80593	17	42
19	.59225	23	.68848	66	.73502	45	.36051	83	.24107	27	.80576	18	41
20	0.59248	24	1.68782	67	0.73547	45	1.35968	83	1.24134	26	0.80558	17	40
21	.59272	23	.68715	67	.73592	45	.35885	83	.24160	27	.80541	17	39
22	.59295	23	.68648	66	.73637	44	.35802	83	.24187	26	.80524	17	38
23	.59318	24	.68582	67	.73681	45	.35719	82	.24213	27	.80507	18	37
24	.59342	23	.68515	66	.73726	45	.35637	83	.24240	27	.80489	17	36
25	0.59365	24	1.68449	67	0.73771	45	1.35554	82	1.24267	26	0.80472	17	35
26	.59389	23	.68382	66	.73816	45	.35472	83	.24293	27	.80455	17	34
27	.59412	24	.68316	66	.73861	45	.35389	82	.24320	27	.80438	18	33
28	.59436	23	.68250	67	.73906	45	.35307	83	.24347	26	.80420	17	32
29	.59459	23	.68183	66	.73951	45	.35224	82	.24373	27	.80403	17	31
30	0.59482	24	1.68117	66	0.73996	45	1.35142	82	1.24400	27	0.80386	18	30
31	.59506	23	.68051	66	.74041	45	.35060	82	.24427	27	.80368	17	29
32	.59529	23	.67985	66	.74086	45	.34978	82	.24454	27	.80351	17	28
33	.59552	24	.67919	66	.74131	45	.34896	82	.24481	27	.80334	18	27
34	.59576	23	.67853	65	.74176	45	.34814	82	.24508	26	.80316	17	26
35	0.59599	23	1.67788	66	0.74221	46	1.34732	82	1.24534	27	0.80299	17	25
36	.59622	24	.67722	66	.74267	45	.34650	82	.24561	27	.80282	18	24
37	.59646	23	.67656	65	.74312	45	.34568	81	.24588	27	.80264	17	23
38	.59669	24	.67591	66	.74357	45	.34487	82	.24615	27	.80247	17	22
39	.59693	23	.67525	65	.74402	45	.34405	82	.24642	27	.80230	18	21
40	0.59716	23	1.67460	66	0.74447	45	1.34323	81	1.24669	27	0.80212	17	20
41	.59739	24	.67394	65	.74492	46	.34242	82	.24696	27	.80195	17	19
42	.59763	23	.67329	65	.74538	45	.34160	81	.24723	27	.80178	18	18
43	.59786	23	.67264	66	.74583	45	.34079	81	.24750	27	.80160	17	17
44	.59809	23	.67198	65	.74628	46	.33998	82	.24777	27	.80143	18	16
45	0.59832	24	1.67133	65	0.74674	45	1.33916	81	1.24804	28	0.80125	17	15
46	.59856	23	.67068	65	.74719	45	.33835	81	.24832	27	.80108	17	14
47	.59879	23	.67003	65	.74764	46	.33754	81	.24859	27	.80091	18	13
48	.59902	24	.66938	65	.74810	45	.33673	81	.24886	27	.80073	17	12
49	.59926	23	.66873	64	.74855	45	.33592	81	.24913	27	.80056	18	11
50	0.59949	23	1.66809	65	0.74900	46	1.33511	81	1.24940	27	0.80038	17	10
51	.59972	23	.66744	65	.74946	45	.33430	81	.24967	28	.80021	18	9
52	.59995	24	.66679	64	.74991	46	.33349	81	.24995	27	.80003	17	8
53	.60019	23	.66615	65	.75037	45	.33268	81	.25022	27	.79986	18	7
54	.60042	23	.66550	64	.75082	46	.33187	80	.25049	28	.79968	17	6
55	0.60065	24	1.66486	65	0.75128	45	1.33107	81	1.25077	27	0.79951	17	5
56	.60089	23	.66421	64	.75173	46	.33026	80	.25104	27	.79934	18	4
57	.60112	23	.66357	65	.75219	45	.32946	81	.25131	28	.79916	17	3
58	.60135	23	.66292	64	.75264	46	.32865	80	.25159	27	.79899	18	2
59	.60158	24	.66228	64	.75310	45	.32785	81	.25186	28	.79881	17	1
60	0.60182		1.66164		0.75355		1.32704		1.25214		0.79864		0
↑ 126°→	cos	Diff. 1′	sec	Diff. 1′	cot	Diff. 1′	tan	Diff. 1′	csc	Diff. 1′	sin	Diff. 1′	←53° ↑

TABLE 31

Natural Trigonometric Functions

37° → ↓ ′	sin	Diff. 1′	csc	Diff. 1′	tan	Diff. 1′	cot	Diff. 1′	sec	Diff. 1′	cos	Diff. 1′	←142° ↓ ′
0	0. 60182	23	1. 66164	64	0. 75355	46	1. 32704	80	1. 25214	27	0. 79864	18	60
1	. 60205	23	. 66100	64	. 75401	46	. 32624	80	. 25241	28	. 79846	17	59
2	. 60228	23	. 66036	64	. 75447	45	. 32544	80	. 25269	27	. 79829	18	58
3	. 60251	23	. 65972	64	. 75492	46	. 32464	80	. 25296	28	. 79811	18	57
4	. 60274	24	. 65908	64	. 75538	46	. 32384	80	. 25324	27	. 79793	17	56
5	0. 60298	23	1. 65844	64	0. 75584	45	1. 32304	80	1. 25351	28	0. 79776	18	55
6	. 60321	23	. 65780	63	. 75629	46	. 32224	80	. 25379	27	. 79758	17	54
7	. 60344	23	. 65717	64	. 75675	46	. 32144	80	. 25406	28	. 79741	18	53
8	. 60367	23	. 65653	64	. 75721	46	. 32064	80	. 25434	28	. 79723	17	52
9	. 60390	24	. 65589	63	. 75767	45	. 31984	80	. 25462	27	. 79706	18	51
10	0. 60414	23	1. 65526	64	0. 75812	46	1. 31904	79	1. 25489	28	0. 79688	17	50
11	. 60437	23	. 65462	63	. 75858	46	. 31825	80	. 25517	28	. 79671	18	49
12	. 60460	23	. 65399	64	. 75904	46	. 31745	79	. 25545	27	. 79653	18	48
13	. 60483	23	. 65335	63	. 75950	46	. 31666	80	. 25572	28	. 79635	17	47
14	. 60506	23	. 65272	63	. 75996	46	. 31586	79	. 25600	28	. 79618	18	46
15	0. 60529	24	1. 65209	63	0. 76042	46	1. 31507	80	1. 25628	28	0. 79600	17	45
16	. 60553	23	. 65146	63	. 76088	46	. 31427	79	. 25656	27	. 79583	18	44
17	. 60576	23	. 65083	63	. 76134	46	. 31348	79	. 25683	28	. 79565	18	43
18	. 60599	23	. 65020	63	. 76180	46	. 31269	79	. 25711	28	. 79547	17	42
19	. 60622	23	. 64957	63	. 76226	46	. 31190	80	. 25739	28	. 79530	18	41
20	0. 60645	23	1. 64894	63	0. 76272	46	1. 31110	79	1. 25767	28	0. 79512	18	40
21	. 60668	23	. 64831	63	. 76318	46	. 31031	79	. 25795	28	. 79494	17	39
22	. 60691	23	. 64768	63	. 76364	46	. 30952	79	. 25823	28	. 79477	18	38
23	. 60714	24	. 64705	62	. 76410	46	. 30873	78	. 25851	28	. 79459	18	37
24	. 60738	23	. 64643	63	. 76456	46	. 30795	79	. 25879	28	. 79441	17	36
25	0. 60761	23	1. 64580	62	0. 76502	46	1. 30716	79	1. 25907	28	0. 79424	18	35
26	. 60784	23	. 64518	63	. 76548	46	. 30637	79	. 25935	28	. 79406	18	34
27	. 60807	23	. 64455	62	. 76594	46	. 30558	78	. 25963	28	. 79388	17	33
28	. 60830	23	. 64393	63	. 76640	46	. 30480	79	. 25991	28	. 79371	18	32
29	. 60853	23	. 64330	62	. 76686	47	. 30401	78	. 26019	28	. 79353	18	31
30	0. 60876	23	1. 64268	62	0. 76733	46	1. 30323	79	1. 26047	28	0. 79335	17	30
31	. 60899	23	. 64206	62	. 76779	46	. 30244	78	. 26075	29	. 79318	18	29
32	. 60922	23	. 64144	63	. 76825	46	. 30166	79	. 26104	28	. 79300	18	28
33	. 60945	23	. 64081	62	. 76871	47	. 30087	78	. 26132	28	. 79282	18	27
34	. 60968	23	. 64019	62	. 76918	46	. 30009	78	. 26160	28	. 79264	17	26
35	0. 60991	24	1. 63957	62	0. 76964	46	1. 29931	78	1. 26188	28	0. 79247	18	25
36	. 61015	23	. 63895	61	. 77010	47	. 29853	78	. 26216	29	. 79229	18	24
37	. 61038	23	. 63834	62	. 77057	46	. 29775	79	. 26245	28	. 79211	18	23
38	. 61061	23	. 63772	62	. 77103	46	. 29696	78	. 26273	28	. 79193	17	22
39	. 61084	23	. 63710	62	. 77149	47	. 29618	77	. 26301	29	. 79176	18	21
40	0. 61107	23	1. 63648	61	0. 77196	46	1. 29541	78	1. 26330	28	0. 79158	18	20
41	. 61130	23	. 63587	62	. 77242	47	. 29463	78	. 26358	29	. 79140	18	19
42	. 61153	23	. 63525	61	. 77289	46	. 29385	78	. 26387	28	. 79122	17	18
43	. 61176	23	. 63464	62	. 77335	47	. 29307	78	. 26415	28	. 79105	18	17
44	. 61199	23	. 63402	61	. 77382	46	. 29229	77	. 26443	29	. 79087	18	16
45	0. 61222	23	1. 63341	62	0. 77428	47	1. 29152	78	1. 26472	28	0. 79069	18	15
46	. 61245	23	. 63279	61	. 77475	46	. 29074	77	. 26500	29	. 79051	18	14
47	. 61268	23	. 63218	61	. 77521	47	. 28997	78	. 26529	28	. 79033	17	13
48	. 61291	23	. 63157	61	. 77568	47	. 28919	77	. 26557	29	. 79016	18	12
49	. 61314	23	. 63096	61	. 77615	46	. 28842	78	. 26586	29	. 78998	18	11
50	0. 61337	23	1. 63035	61	0. 77661	47	1. 28764	77	1. 26615	28	0. 78980	18	10
51	. 61360	23	. 62974	61	. 77708	46	. 28687	77	. 26643	29	. 78962	18	9
52	. 61383	23	. 62913	61	. 77754	47	. 28610	77	. 26672	29	. 78944	18	8
53	. 61406	23	. 62852	61	. 77801	47	. 28533	77	. 26701	28	. 78926	18	7
54	. 61429	22	. 62791	61	. 77848	47	. 28456	77	. 26729	29	. 78908	17	6
55	0. 61451	23	1. 62730	61	0. 77895	46	1. 28379	77	1. 26758	29	0. 78891	18	5
56	. 61474	23	. 62669	60	. 77941	47	. 28302	77	. 26787	28	. 78873	18	4
57	. 61497	23	. 62609	61	. 77988	47	. 28225	77	. 26815	29	. 78855	18	3
58	. 61520	23	. 62548	61	. 78035	47	. 28148	77	. 26844	29	. 78837	18	2
59	. 61543	23	. 62487	60	. 78082	47	. 28071	77	. 26873	29	. 78819	18	1
60	0. 61566		1. 62427		0. 78129		1. 27994		1. 26902		0. 78801		0
↑ 127° →	cos	Diff. 1′	sec	Diff. 1′	cot	Diff. 1′	tan	Diff. 1′	csc	Diff. 1′	sin	Diff. 1′	← 52° ↑

TABLE 31
Natural Trigonometric Functions

38°→ ↓ ′	sin	Diff. 1′	csc	Diff. 1′	tan	Diff. 1′	cot	Diff. 1′	sec	Diff. 1′	cos	Diff. 1′	←141° ↓ ′
0	0. 61566	23	1. 62427	61	0. 78129	46	1. 27994	77	1. 26902	29	0. 78801	18	60
1	. 61589	23	. 62366	60	. 78175	47	. 27917	76	. 26931	29	. 78783	18	59
2	. 61612	23	. 62306	60	. 78222	47	. 27841	77	. 26960	28	. 78765	18	58
3	. 61635	23	. 62246	61	. 78269	47	. 27764	76	. 26988	29	. 78747	18	57
4	. 61658	23	. 62185	60	. 78316	47	. 27688	77	. 27017	29	. 78729	18	56
5	0. 61681	23	1. 62125	60	0. 78363	47	1. 27611	76	1. 27046	29	0. 78711	17	55
6	. 61704	22	. 62065	60	. 78410	47	. 27535	77	. 27075	29	. 78694	18	54
7	. 61726	23	. 62005	60	. 78457	47	. 27458	76	. 27104	29	. 78676	18	53
8	. 61749	23	. 61945	60	. 78504	47	. 27382	76	. 27133	29	. 78658	18	52
9	. 61772	23	. 61885	60	. 78551	47	. 27306	76	. 27162	29	. 78640	18	51
10	0. 61795	23	1. 61825	60	0. 78598	47	1. 27230	77	1. 27191	30	0. 78622	18	50
11	. 61818	23	. 61765	60	. 78645	47	. 27153	76	. 27221	29	. 78604	18	49
12	. 61841	23	. 61705	59	. 78692	47	. 27077	76	. 27250	29	. 78586	18	48
13	. 61864	23	. 61646	60	. 78739	47	. 27001	76	. 27279	29	. 78568	18	47
14	. 61887	22	. 61586	60	. 78786	48	. 26925	76	. 27308	29	. 78550	18	46
15	0. 61909	23	1. 61526	59	0. 78834	47	1. 26849	75	1. 27337	29	0. 78532	18	45
16	. 61932	23	. 61467	60	. 78881	47	. 26774	76	. 27366	30	. 78514	18	44
17	. 61955	23	. 61407	59	. 78928	47	. 26698	76	. 27396	29	. 78496	18	43
18	. 61978	23	. 61348	60	. 78975	47	. 26622	76	. 27425	29	. 78478	18	42
19	. 62001	23	. 61288	59	. 79022	48	. 26546	75	. 27454	29	. 78460	18	41
20	0. 62024	22	1. 61229	59	0. 79070	47	1. 26471	76	1. 27483	30	0. 78442	18	40
21	. 62046	23	. 61170	59	. 79117	47	. 26395	76	. 27513	29	. 78424	19	39
22	. 62069	23	. 61111	60	. 79164	48	. 26319	75	. 27542	30	. 78405	18	38
23	. 62092	23	. 61051	59	. 79212	47	. 26244	75	. 27572	29	. 78387	18	37
24	. 62115	23	. 60992	59	. 79259	47	. 26169	76	. 27601	29	. 78369	18	36
25	0. 62138	22	1. 60933	59	0. 79306	48	1. 26093	75	1. 27630	30	0. 78351	18	35
26	. 62160	23	. 60874	59	. 79354	47	. 26018	75	. 27660	29	. 78333	18	34
27	. 62183	23	. 60815	59	. 79401	48	. 25943	76	. 27689	30	. 78315	18	33
28	. 62206	23	. 60756	58	. 79449	47	. 25867	75	. 27719	29	. 78297	18	32
29	. 62229	22	. 60698	59	. 79496	48	. 25792	75	. 27748	30	. 78279	18	31
30	0. 62251	23	1. 60639	59	0. 79544	47	1. 25717	75	1. 27778	29	0. 78261	18	30
31	. 62274	23	. 60580	59	. 79591	48	. 25642	75	. 27807	30	. 78243	18	29
32	. 62297	23	. 60521	58	. 79639	47	. 25567	75	. 27837	30	. 78225	19	28
33	. 62320	22	. 60463	59	. 79686	48	. 25492	75	. 27867	29	. 78206	18	27
34	. 62342	23	. 60404	58	. 79734	47	. 25417	74	. 27896	30	. 78188	18	26
35	0. 62365	23	1. 60346	59	0. 79781	48	1. 25343	75	1. 27926	30	0. 78170	18	25
36	. 62388	23	. 60287	58	. 79829	48	. 25268	75	. 27956	29	. 78152	18	24
37	. 62411	22	. 60229	58	. 79877	47	. 25193	75	. 27985	30	. 78134	18	23
38	. 62433	23	. 60171	59	. 79924	48	. 25118	74	. 28015	30	. 78116	18	22
39	. 62456	23	. 60112	58	. 79972	48	. 25044	75	. 28045	30	. 78098	19	21
40	0. 62479	23	1. 60054	58	0. 80020	47	1. 24969	74	1. 28075	30	0. 78079	18	20
41	. 62502	22	. 59996	58	. 80067	48	. 24895	75	. 28105	29	. 78061	18	19
42	. 62524	23	. 59938	58	. 80115	48	. 24820	74	. 28134	30	. 78043	18	18
43	. 62547	23	. 59880	58	. 80163	48	. 24746	74	. 28164	30	. 78025	18	17
44	. 62570	22	. 59822	58	. 80211	47	. 24672	75	. 28194	30	. 78007	19	16
45	0. 62592	23	1. 59764	58	0. 80258	48	1. 24597	74	1. 28224	30	0. 77988	18	15
46	. 62615	23	. 59706	58	. 80306	48	. 24523	74	. 28254	30	. 77970	18	14
47	. 62638	22	. 59648	58	. 80354	48	. 24449	74	. 28284	30	. 77952	18	13
48	. 62660	23	. 59590	57	. 80402	48	. 24375	74	. 28314	30	. 77934	18	12
49	. 62683	23	. 59533	58	. 80450	48	. 24301	74	. 28344	30	. 77916	19	11
50	0. 62706	22	1. 59475	57	0. 80498	48	1. 24227	74	1. 28374	30	0. 77897	18	10
51	. 62728	23	. 59418	58	. 80546	48	. 24153	74	. 28404	30	. 77879	18	9
52	. 62751	23	. 59360	58	. 80594	48	. 24079	74	. 28434	30	. 77861	18	8
53	. 62774	22	. 59302	57	. 80642	48	. 24005	74	. 28464	31	. 77843	19	7
54	. 62796	23	. 59245	57	. 80690	48	. 23931	73	. 28495	30	. 77824	18	6
55	0. 62819	23	1. 59188	58	0. 80738	48	1. 23858	74	1. 28525	30	0. 77806	18	5
56	. 62842	22	. 59130	57	. 80786	48	. 23784	74	. 28555	30	. 77788	19	4
57	. 62864	23	. 59073	57	. 80834	48	. 23710	73	. 28585	30	. 77769	18	3
58	. 62887	22	. 59016	57	. 80882	48	. 23637	74	. 28615	31	. 77751	18	2
59	. 62909	23	. 58959	57	. 80930	48	. 23563	73	. 28646	30	. 77733	18	1
60	0. 62932		1. 58902		0. 80978		1. 23490		1. 28676		0. 77715		0
↑ 128°→	cos	Diff. 1′	sec	Diff. 1′	cot	Diff. 1′	tan	Diff. 1′	csc	Diff. 1′	sin	Diff. 1′	←51° ↑

TABLE 31

Natural Trigonometric Functions

39°→ ↓ ′	sin	Diff. 1′	csc	Diff. 1′	tan	Diff. 1′	cot	Diff. 1′	sec	Diff. 1′	cos	Diff. 1′	←140° ↓ ′
0	0.62932	23	1.58902	57	0.80978	49	1.23490	74	1.28676	30	0.77715	19	60
1	.62955	22	.58845	57	.81027	48	.23416	73	.28706	31	.77696	18	59
2	.62977	23	.58788	57	.81075	48	.23343	73	.28737	30	.77678	18	58
3	.63000	22	.58731	57	.81123	48	.23270	74	.28767	30	.77660	19	57
4	.63022	23	.58674	57	.81171	49	.23196	73	.28797	31	.77641	18	56
5	0.63045	23	1.58617	57	0.81220	48	1.23123	73	1.28828	30	0.77623	18	55
6	.63068	22	.58560	57	.81268	48	.23050	73	.28858	31	.77605	19	54
7	.63090	23	.58503	56	.81316	48	.22977	73	.28889	30	.77586	18	53
8	.63113	22	.58447	57	.81364	49	.22904	73	.28919	31	.77568	18	52
9	.63135	23	.58390	57	.81413	48	.22831	73	.28950	30	.77550	19	51
10	0.63158	22	1.58333	56	0.81461	49	1.22758	73	1.28980	31	0.77531	18	50
11	.63180	23	.58277	56	.81510	48	.22685	73	.29011	31	.77513	19	49
12	.63203	22	.58221	57	.81558	48	.22612	73	.29042	30	.77494	18	48
13	.63225	23	.58164	56	.81606	49	.22539	72	.29072	31	.77476	18	47
14	.63248	23	.58108	57	.81655	48	.22467	73	.29103	30	.77458	19	46
15	0.63271	22	1.58051	56	0.81703	49	1.22394	73	1.29133	31	0.77439	18	45
16	.63293	23	.57995	56	.81752	48	.22321	72	.29164	31	.77421	19	44
17	.63316	22	.57939	56	.81800	49	.22249	73	.29195	31	.77402	18	43
18	.63338	23	.57883	56	.81849	49	.22176	72	.29226	30	.77384	18	42
19	.63361	22	.57827	56	.81898	48	.22104	73	.29256	31	.77366	19	41
20	0.63383	23	1.57771	56	0.81946	49	1.22031	72	1.29287	31	0.77347	18	40
21	.63406	22	.57715	56	.81995	49	.21959	73	.29318	31	.77329	19	39
22	.63428	23	.57659	56	.82044	48	.21886	72	.29349	31	.77310	18	38
23	.63451	22	.57603	56	.82092	49	.21814	72	.29380	31	.77292	19	37
24	.63473	23	.57547	56	.82141	49	.21742	72	.29411	31	.77273	18	36
25	0.63496	22	1.57491	55	0.82190	48	1.21670	72	1.29442	31	0.77255	19	35
26	.63518	22	.57436	56	.82238	49	.21598	72	.29473	31	.77236	18	34
27	.63540	23	.57380	56	.82287	49	.21526	72	.29504	31	.77218	19	33
28	.63563	22	.57324	55	.82336	49	.21454	72	.29535	31	.77199	18	32
29	.63585	23	.57269	56	.82385	49	.21382	72	.29566	31	.77181	19	31
30	0.63608	22	1.57213	55	0.82434	49	1.21310	72	1.29597	31	0.77162	18	30
31	.63630	23	.57158	55	.82483	48	.21238	72	.29628	31	.77144	19	29
32	.63653	22	.57103	56	.82531	49	.21166	72	.29659	31	.77125	18	28
33	.63675	23	.57047	55	.82580	49	.21094	71	.29690	31	.77107	19	27
34	.63698	22	.56992	55	.82629	49	.21023	72	.29721	31	.77088	18	26
35	0.63720	22	1.56937	56	0.82678	49	1.20951	72	1.29752	32	0.77070	19	25
36	.63742	23	.56881	55	.82727	49	.20879	71	.29784	31	.77051	18	24
37	.63765	22	.56826	55	.82776	49	.20808	72	.29815	31	.77033	19	23
38	.63787	23	.56771	55	.82825	49	.20736	71	.29846	31	.77014	18	22
39	.63810	22	.56716	55	.82874	49	.20665	72	.29877	32	.76996	19	21
40	0.63832	22	1.56661	55	0.82923	49	1.20593	71	1.29909	31	0.76977	18	20
41	.63854	23	.56606	55	.82972	50	.20522	71	.29940	31	.76959	19	19
42	.63877	22	.56551	54	.83022	49	.20451	72	.29971	32	.76940	19	18
43	.63899	23	.56497	55	.83071	49	.20379	71	.30003	31	.76921	18	17
44	.63922	22	.56442	55	.83120	49	.20308	71	.30034	32	.76903	19	16
45	0.63944	22	1.56387	55	0.83169	49	1.20237	71	1.30066	31	0.76884	18	15
46	.63966	23	.56332	54	.83218	50	.20166	71	.30097	32	.76866	19	14
47	.63989	22	.56278	55	.83268	49	.20095	71	.30129	31	.76847	19	13
48	.64011	22	.56223	54	.83317	49	.20024	71	.30160	32	.76828	18	12
49	.64033	23	.56169	55	.83366	49	.19953	71	.30192	31	.76810	19	11
50	0.64056	22	1.56114	54	0.83415	50	1.19882	71	1.30223	32	0.76791	19	10
51	.64078	22	.56060	55	.83465	49	.19811	71	.30255	32	.76772	18	9
52	.64100	23	.56005	54	.83514	50	.19740	71	.30287	31	.76754	19	8
53	.64123	22	.55951	54	.83564	49	.19669	70	.30318	32	.76735	18	7
54	.64145	22	.55897	54	.83613	49	.19599	71	.30350	32	.76717	19	6
55	0.64167	23	1.55843	54	0.83662	50	1.19528	71	1.30382	31	0.76698	19	5
56	.64190	22	.55789	55	.83712	49	.19457	70	.30413	32	.76679	18	4
57	.64212	22	.55734	54	.83761	50	.19387	71	.30445	32	.76661	19	3
58	.64234	22	.55680	54	.83811	49	.19316	70	.30477	32	.76642	19	2
59	.64256	23	.55626	54	.83860	50	.19246	71	.30509	32	.76623	19	1
60	0.64279		1.55572		0.83910		1.19175		1.30541		0.76604		0
↑ 129°→	cos	Diff. 1′	sec	Diff. 1′	cot	Diff. 1′	tan	Diff. 1′	csc	Diff. 1′	sin	Diff. 1′	←50° ↑

TABLE 31

Natural Trigonometric Functions

40° → ↓ ′	sin	Diff. 1′	csc	Diff. 1′	tan	Diff. 1′	cot	Diff. 1′	sec	Diff. 1′	cos	Diff. 1′	← 139° ↓ ′
0	0. 64279	22	1. 55572	54	0. 83910	50	1. 19175	70	1. 30541	32	0. 76604	18	60
1	. 64301	22	. 55518	53	. 83960	49	. 19105	70	. 30573	32	. 76586	19	59
2	. 64323	23	. 55465	54	. 84009	50	. 19035	71	. 30605	31	. 76567	19	58
3	. 64346	22	. 55411	54	. 84059	49	. 18964	70	. 30636	32	. 76548	18	57
4	. 64368	22	. 55357	54	. 84108	50	. 18894	70	. 30668	32	. 76530	19	56
5	0. 64390	22	1. 55303	53	0. 84158	50	1. 18824	70	1. 30700	32	0. 76511	19	55
6	. 64412	23	. 55250	54	. 84208	50	. 18754	70	. 30732	32	. 76492	19	54
7	. 64435	22	. 55196	53	. 84258	49	. 18684	70	. 30764	32	. 76473	18	53
8	. 64457	22	. 55143	54	. 84307	50	. 18614	70	. 30796	33	. 76455	19	52
9	. 64479	22	. 55089	53	. 84357	50	. 18544	70	. 30829	32	. 76436	19	51
10	0. 64501	23	1. 55036	54	0. 84407	50	1. 18474	70	1. 30861	32	0. 76417	19	50
11	. 64524	22	. 54982	53	. 84457	50	. 18404	70	. 30893	32	. 76398	18	49
12	. 64546	22	. 54929	53	. 84507	49	. 18334	70	. 30925	32	. 76380	19	48
13	. 64568	22	. 54876	54	. 84556	50	. 18264	70	. 30957	32	. 76361	19	47
14	. 64590	22	. 54822	53	. 84606	50	. 18194	69	. 30989	33	. 76342	19	46
15	0. 64612	23	1. 54769	53	0. 84656	50	1. 18125	70	1. 31022	32	0. 76323	19	45
16	. 64635	22	. 54716	53	. 84706	50	. 18055	69	. 31054	32	. 76304	18	44
17	. 64657	22	. 54663	53	. 84756	50	. 17986	70	. 31086	33	. 76286	19	43
18	. 64679	22	. 54610	53	. 84806	50	. 17916	70	. 31119	32	. 76267	19	42
19	. 64701	22	. 54557	53	. 84856	50	. 17846	69	. 31151	32	. 76248	19	41
20	0. 64723	23	1. 54504	53	0. 84906	50	1. 17777	69	1. 31183	33	0. 76229	19	40
21	. 64746	22	. 54451	53	. 84956	50	. 17708	70	. 31216	32	. 76210	18	39
22	. 64768	22	. 54398	53	. 85006	51	. 17638	69	. 31248	33	. 76192	19	38
23	. 64790	22	. 54345	53	. 85057	50	. 17569	69	. 31281	32	. 76173	19	37
24	. 64812	22	. 54292	52	. 85107	50	. 17500	70	. 31313	33	. 76154	19	36
25	0. 64834	22	1. 54240	53	0. 85157	50	1. 17430	69	1. 31346	32	0. 76135	19	35
26	. 64856	22	. 54187	53	. 85207	50	. 17361	69	. 31378	33	. 76116	19	34
27	. 64878	23	. 54134	52	. 85257	51	. 17292	69	. 31411	32	. 76097	19	33
28	. 64901	22	. 54082	53	. 85308	50	. 17223	69	. 31443	33	. 76078	19	32
29	. 64923	22	. 54029	52	. 85358	50	. 17154	69	. 31476	33	. 76059	18	31
30	0. 64945	22	1. 53977	53	0. 85408	50	1. 17085	69	1. 31509	32	0. 76041	19	30
31	. 64967	22	. 53924	52	. 85458	51	. 17016	69	. 31541	33	. 76022	19	29
32	. 64989	22	. 53872	52	. 85509	50	. 16947	69	. 31574	33	. 76003	19	28
33	. 65011	22	. 53820	52	. 85559	50	. 16878	69	. 31607	33	. 75984	19	27
34	. 65033	22	. 53768	53	. 85609	51	. 16809	68	. 31640	32	. 75965	19	26
35	0. 65055	22	1. 53715	52	0. 85660	50	1. 16741	69	1. 31672	33	0. 75946	19	25
36	. 65077	23	. 53663	52	. 85710	51	. 16672	69	. 31705	33	. 75927	19	24
37	. 65100	22	. 53611	52	. 85761	50	. 16603	68	. 31738	33	. 75908	19	23
38	. 65122	22	. 53559	52	. 85811	51	. 16535	69	. 31771	33	. 75889	19	22
39	. 65144	22	. 53507	52	. 85862	50	. 16466	68	. 31804	33	. 75870	19	21
40	0. 65166	22	1. 53455	52	0. 85912	51	1. 16398	69	1. 31837	33	0. 75851	19	20
41	. 65188	22	. 53403	52	. 85963	51	. 16329	68	. 31870	33	. 75832	19	19
42	. 65210	22	. 53351	52	. 86014	50	. 16261	69	. 31903	33	. 75813	19	18
43	. 65232	22	. 53299	52	. 86064	51	. 16192	68	. 31936	33	. 75794	19	17
44	. 65254	22	. 53247	51	. 86115	51	. 16124	68	. 31969	33	. 75775	19	16
45	0. 65276	22	1. 53196	52	0. 86166	50	1. 16056	69	1. 32002	33	0. 75756	18	15
46	. 65298	22	. 53144	52	. 86216	51	. 15987	68	. 32035	33	. 75738	19	14
47	. 65320	22	. 53092	51	. 86267	51	. 15919	68	. 32068	33	. 75719	19	13
48	. 65342	22	. 53041	52	. 86318	50	. 15851	68	. 32101	33	. 75700	20	12
49	. 65364	22	. 52989	51	. 86368	51	. 15783	68	. 32134	34	. 75680	19	11
50	0. 65386	22	1. 52938	52	0. 86419	51	1. 15715	68	1. 32168	33	0. 75661	19	10
51	. 65408	22	. 52886	51	. 86470	51	. 15647	68	. 32201	33	. 75642	19	9
52	. 65430	22	. 52835	51	. 86521	51	. 15579	68	. 32234	33	. 75623	19	8
53	. 65452	22	. 52784	52	. 86572	51	. 15511	68	. 32267	34	. 75604	19	7
54	. 65474	22	. 52732	51	. 86623	51	. 15443	68	. 32301	33	. 75585	19	6
55	0. 65496	22	1. 52681	51	0. 86674	51	1. 15375	67	1. 32334	34	0. 75566	19	5
56	. 65518	22	. 52630	51	. 86725	51	. 15308	68	. 32368	33	. 75547	19	4
57	. 65540	22	. 52579	52	. 86776	51	. 15240	68	. 32401	33	. 75528	19	3
58	. 65562	22	. 52527	51	. 86827	51	. 15172	68	. 32434	34	. 75509	19	2
59	. 65584	22	. 52476	51	. 86878	51	. 15104	67	. 32468	33	. 75490	19	1
60	0. 65606		1. 52425		0. 86929		1. 15037		1. 32501		0. 75471		0
↑ 130° →	cos	Diff. 1′	sec	Diff. 1′	cot	Diff. 1′	tan	Diff. 1′	csc	Diff. 1′	sin	Diff. 1′	← 49° ↑

TABLE 31
Natural Trigonometric Functions

41° → ↓ ′	sin	Diff. 1′	csc	Diff. 1′	tan	Diff. 1′	cot	Diff. 1′	sec	Diff. 1′	cos	Diff. 1′	← 138° ↓ ′
0	0. 65606	22	1. 52425	51	0. 86929	51	1. 15037	68	1. 32501	34	0. 75471	19	60
1	. 65628	22	. 52374	51	. 86980	51	. 14969	67	. 32535	33	. 75452	19	59
2	. 65650	22	. 52323	50	. 87031	51	. 14902	68	. 32568	34	. 75433	19	58
3	. 65672	22	. 52273	51	. 87082	51	. 14834	67	. 32602	34	. 75414	19	57
4	. 65694	22	. 52222	51	. 87133	51	. 14767	68	. 32636	33	. 75395	20	56
5	0. 65716	22	1. 52171	51	0. 87184	52	1. 14699	67	1. 32669	34	0. 75375	19	55
6	. 65738	21	. 52120	51	. 87236	51	. 14632	67	. 32703	34	. 75356	19	54
7	. 65759	22	. 52069	50	. 87287	51	. 14565	67	. 32737	33	. 75337	19	53
8	. 65781	22	. 52019	51	. 87338	51	. 14498	68	. 32770	34	. 75318	19	52
9	. 65803	22	. 51968	50	. 87389	52	. 14430	67	. 32804	34	. 75299	19	51
10	0. 65825	22	1. 51918	51	0. 87441	51	1. 14363	67	1. 32838	34	0. 75280	19	50
11	. 65847	22	. 51867	50	. 87492	51	. 14296	67	. 32872	33	. 75261	20	49
12	. 65869	22	. 51817	51	. 87543	52	. 14229	67	. 32905	34	. 75241	19	48
13	. 65891	22	. 51766	50	. 87595	51	. 14162	67	. 32939	34	. 75222	19	47
14	. 65913	22	. 51716	51	. 87646	52	. 14095	67	. 32973	34	. 75203	19	46
15	0. 65935	21	1. 51665	50	0. 87698	51	1. 14028	67	1. 33007	34	0. 75184	19	45
16	. 65956	22	. 51615	50	. 87749	52	. 13961	67	. 33041	34	. 75165	19	44
17	. 65978	22	. 51565	50	. 87801	51	. 13894	66	. 33075	34	. 75146	20	43
18	. 66000	22	. 51515	50	. 87852	52	. 13828	67	. 33109	34	. 75126	19	42
19	. 66022	22	. 51465	50	. 87904	51	. 13761	67	. 33143	34	. 75107	19	41
20	0. 66044	22	1. 51415	51	0. 87955	52	1. 13694	67	1. 33177	34	0. 75088	19	40
21	. 66066	22	. 51364	50	. 88007	52	. 13627	66	. 33211	34	. 75069	19	39
22	. 66088	21	. 51314	49	. 88059	51	. 13561	67	. 33245	34	. 75050	20	38
23	. 66109	22	. 51265	50	. 88110	52	. 13494	66	. 33279	35	. 75030	19	37
24	. 66131	22	. 51215	50	. 88162	52	. 13428	67	. 33314	34	. 75011	19	36
25	0. 66153	22	1. 51165	50	0. 88214	51	1. 13361	66	1. 33348	34	0. 74992	19	35
26	. 66175	22	. 51115	50	. 88265	52	. 13295	67	. 33382	34	. 74973	20	34
27	. 66197	21	. 51065	50	. 88317	52	. 13228	66	. 33416	35	. 74953	19	33
28	. 66218	22	. 51015	49	. 88369	52	. 13162	66	. 33451	34	. 74934	19	32
29	. 66240	22	. 50966	50	. 88421	52	. 13096	67	. 33485	34	. 74915	19	31
30	0. 66262	22	1. 50916	50	0. 88473	51	1. 13029	66	1. 33519	35	0. 74896	20	30
31	. 66284	22	. 50866	49	. 88524	52	. 12963	66	. 33554	34	. 74876	19	29
32	. 66306	21	. 50817	50	. 88576	52	. 12897	66	. 33588	34	. 74857	19	28
33	. 66327	22	. 50767	49	. 88628	52	. 12831	66	. 33622	35	. 74838	20	27
34	. 66349	22	. 50718	49	. 88680	52	. 12765	66	. 33657	34	. 74818	19	26
35	0. 66371	22	1. 50669	50	0. 88732	52	1. 12699	66	1. 33691	35	0. 74799	19	25
36	. 66393	21	. 50619	49	. 88784	52	. 12633	66	. 33726	34	. 74780	20	24
37	. 66414	22	. 50570	49	. 88836	52	. 12567	66	. 33760	35	. 74760	19	23
38	. 66436	22	. 50521	50	. 88888	52	. 12501	66	. 33795	35	. 74741	19	22
39	. 66458	22	. 50471	49	. 88940	52	. 12435	66	. 33830	34	. 74722	19	21
40	0. 66480	21	1. 50422	49	0. 88992	53	1. 12369	66	1. 33864	35	0. 74703	20	20
41	. 66501	22	. 50373	49	. 89045	52	. 12303	65	. 33899	35	. 74683	19	19
42	. 66523	22	. 50324	49	. 89097	52	. 12238	66	. 33934	34	. 74664	20	18
43	. 66545	21	. 50275	49	. 89149	52	. 12172	66	. 33968	35	. 74644	19	17
44	. 66566	22	. 50226	49	. 89201	52	. 12106	65	. 34003	35	. 74625	19	16
45	0. 66588	22	1. 50177	49	0. 89253	53	1. 12041	66	1. 34038	35	0. 74606	20	15
46	. 66610	22	. 50128	49	. 89306	52	. 11975	66	. 34073	35	. 74586	19	14
47	. 66632	21	. 50079	49	. 89358	52	. 11909	65	. 34108	34	. 74567	19	13
48	. 66653	22	. 50030	49	. 89410	53	. 11844	66	. 34142	35	. 74548	20	12
49	. 66675	22	. 49981	48	. 89463	52	. 11778	65	. 34177	35	. 74528	19	11
50	0. 66697	21	1. 49933	49	0. 89515	52	1. 11713	65	1. 34212	35	0. 74509	20	10
51	. 66718	22	. 49884	49	. 89567	53	. 11648	66	. 34247	35	. 74489	19	9
52	. 66740	22	. 49835	48	. 89620	52	. 11582	65	. 34282	35	. 74470	19	8
53	. 66762	21	. 49787	49	. 89672	53	. 11517	65	. 34317	35	. 74451	20	7
54	. 66783	22	. 49738	48	. 89725	52	. 11452	65	. 34352	35	. 74431	19	6
55	0. 66805	22	1. 49690	49	0. 89777	53	1. 11387	66	1. 34387	36	0. 74412	20	5
56	. 66827	21	. 49641	48	. 89830	53	. 11321	65	. 34423	35	. 74392	19	4
57	. 66848	22	. 49593	49	. 89883	52	. 11256	65	. 34458	35	. 74373	20	3
58	. 66870	21	. 49544	48	. 89935	53	. 11191	65	. 34493	35	. 74353	19	2
59	. 66891	22	. 49496	48	. 89988	52	. 11126	65	. 34528	35	. 74334	20	1
60	0. 66913		1. 49448		0. 90040		1. 11061		1. 34563		0. 74314		0
↑ 131° →	cos	Diff. 1′	sec	Diff. 1′	cot	Diff. 1′	tan	Diff. 1′	csc	Diff. 1′	sin	Diff. 1′	← 48° ↑

TABLE 31

Natural Trigonometric Functions

42°→ ↓ ′	sin	Diff. 1′	csc	Diff. 1′	tan	Diff. 1′	cot	Diff. 1′	sec	Diff. 1′	cos	Diff. 1′	←137° ↓ ′
0	0. 66913	22	1. 49448	49	0. 90040	53	1. 11061	65	1. 34563	36	0. 74314	19	60
1	. 66935	21	. 49399	48	. 90093	53	. 10996	65	. 34599	35	. 74295	19	59
2	. 66956	22	. 49351	48	. 90146	53	. 10931	64	. 34634	35	. 74276	20	58
3	. 66978	21	. 49303	48	. 90199	52	. 10867	65	. 34669	35	. 74256	19	57
4	. 66999	22	. 49255	48	. 90251	53	. 10802	65	. 34704	36	. 74237	20	56
5	0. 67021	22	1. 49207	48	0. 90304	53	1. 10737	65	1. 34740	35	0. 74217	19	55
6	. 67043	21	. 49159	48	. 90357	53	. 10672	65	. 34775	36	. 74198	20	54
7	. 67064	22	. 49111	48	. 90410	53	. 10607	64	. 34811	35	. 74178	19	53
8	. 67086	21	. 49063	48	. 90463	53	. 10543	65	. 34846	36	. 74159	20	52
9	. 67107	22	. 49015	48	. 90516	53	. 10478	64	. 34882	35	. 74139	19	51
10	0. 67129	22	1. 48967	48	0. 90569	52	1. 10414	65	1. 34917	36	0. 74120	20	50
11	. 67151	21	. 48919	48	. 90621	53	. 10349	64	. 34953	35	. 74100	20	49
12	. 67172	22	. 48871	47	. 90674	53	. 10285	65	. 34988	36	. 74080	19	48
13	. 67194	21	. 48824	48	. 90727	54	. 10220	64	. 35024	36	. 74061	20	47
14	. 67215	22	. 48776	48	. 90781	53	. 10156	65	. 35060	35	. 74041	19	46
15	0. 67237	21	1. 48728	47	0. 90834	53	1. 10091	64	1. 35095	36	0. 74022	20	45
16	. 67258	22	. 48681	48	. 90887	53	. 10027	64	. 35131	36	. 74002	19	44
17	. 67280	21	. 48633	47	. 90940	53	. 09963	64	. 35167	36	. 73983	20	43
18	. 67301	22	. 48586	48	. 90993	53	. 09899	65	. 35203	35	. 73963	19	42
19	. 67323	21	. 48538	47	. 91046	53	. 09834	64	. 35238	36	. 73944	20	41
20	0. 67344	22	1. 48491	48	0. 91099	54	1. 09770	64	1. 35274	36	0. 73924	20	40
21	. 67366	21	. 48443	47	. 91153	53	. 09706	64	. 35310	36	. 73904	19	39
22	. 67387	22	. 48396	47	. 91206	53	. 09642	64	. 35346	36	. 73885	20	38
23	. 67409	21	. 48349	48	. 91259	54	. 09578	64	. 35382	36	. 73865	19	37
24	. 67430	22	. 48301	47	. 91313	53	. 09514	64	. 35418	36	. 73846	20	36
25	0. 67452	21	1. 48254	47	0. 91366	53	1. 09450	64	1. 35454	36	0. 73826	20	35
26	. 67473	22	. 48207	47	. 91419	54	. 09386	64	. 35490	36	. 73806	19	34
27	. 67495	21	. 48160	47	. 91473	53	. 09322	64	. 35526	36	. 73787	20	33
28	. 67516	22	. 48113	47	. 91526	54	. 09258	63	. 35562	36	. 73767	20	32
29	. 67538	21	. 48066	47	. 91580	53	. 09195	64	. 35598	36	. 73747	19	31
30	0. 67559	21	1. 48019	47	0. 91633	54	1. 09131	64	1. 35634	36	0. 73728	20	30
31	. 67580	22	. 47972	47	. 91687	53	. 09067	64	. 35670	37	. 73708	20	29
32	. 67602	21	. 47925	47	. 91740	54	. 09003	63	. 35707	36	. 73688	19	28
33	. 67623	22	. 47878	47	. 91794	53	. 08940	64	. 35743	36	. 73669	20	27
34	. 67645	21	. 47831	47	. 91847	54	. 08876	63	. 35779	36	. 73649	20	26
35	0. 67666	22	1. 47784	46	0. 91901	54	1. 08813	64	1. 35815	37	0. 73629	19	25
36	. 67688	21	. 47738	47	. 91955	53	. 08749	63	. 35852	36	. 73610	20	24
37	. 67709	21	. 47691	47	. 92008	54	. 08686	64	. 35888	36	. 73590	20	23
38	. 67730	22	. 47644	46	. 92062	54	. 08622	63	. 35924	37	. 73570	19	22
39	. 67752	21	. 47598	47	. 92116	54	. 08559	63	. 35961	36	. 73551	20	21
40	0. 67773	22	1. 47551	47	0. 92170	54	1. 08496	64	1. 35997	37	0. 73531	20	20
41	. 67795	21	. 47504	46	. 92224	53	. 08432	63	. 36034	36	. 73511	20	19
42	. 67816	21	. 47458	47	. 92277	54	. 08369	63	. 36070	37	. 73491	19	18
43	. 67837	22	. 47411	46	. 92331	54	. 08306	63	. 36107	36	. 73472	20	17
44	. 67859	21	. 47365	46	. 92385	54	. 08243	64	. 36143	37	. 73452	20	16
45	0. 67880	21	1. 47319	47	0. 92439	54	1. 08179	63	1. 36180	37	0. 73432	19	15
46	. 67901	22	. 47272	46	. 92493	54	. 08116	63	. 36217	36	. 73413	20	14
47	. 67923	21	. 47226	46	. 92547	54	. 08053	63	. 36253	37	. 73393	20	13
48	. 67944	21	. 47180	46	. 92601	54	. 07990	63	. 36290	37	. 73373	20	12
49	. 67965	22	. 47134	47	. 92655	54	. 07927	63	. 36327	36	. 73353	20	11
50	0. 67987	21	1. 47087	46	0. 92709	54	1. 07864	63	1. 36363	37	0. 73333	19	10
51	. 68008	21	. 47041	46	. 92763	54	. 07801	63	. 36400	37	. 73314	20	9
52	. 68029	22	. 46995	46	. 92817	55	. 07738	62	. 36437	37	. 73294	20	8
53	. 68051	21	. 46949	46	. 92872	54	. 07676	63	. 36474	37	. 73274	20	7
54	. 68072	21	. 46903	46	. 92926	54	. 07613	63	. 36511	37	. 73254	20	6
55	0. 68093	22	1. 46857	46	0. 92980	54	1. 07550	63	1. 36548	37	0. 73234	19	5
56	. 68115	21	. 46811	46	. 93034	54	. 07487	62	. 36585	37	. 73215	20	4
57	. 68136	21	. 46765	46	. 93088	55	. 07425	63	. 36622	37	. 73195	20	3
58	. 68157	22	. 46719	45	. 93143	54	. 07362	63	. 36659	37	. 73175	20	2
59	. 68179	21	. 46674	46	. 93197	55	. 07299	62	. 36696	37	. 73155	20	1
60	0. 68200		1. 46628		0. 93252		1. 07237		1. 36733		0. 73135		0
↑ 132°→	cos	Diff. 1′	sec	Diff. 1′	cot	Diff. 1′	tan	Diff. 1′	csc	Diff. 1′	sin	Diff. 1′	←47° ↑

TABLE 31

Natural Trigonometric Functions

43°→ ↓ ′	sin	Diff. 1′	csc	Diff. 1′	tan	Diff. 1′	cot	Diff. 1′	sec	Diff. 1′	cos	Diff. 1′	←136° ↓ ′
0	0. 68200	21	1. 46628	46	0. 93252	54	1. 07237	63	1. 36733	37	0. 73135	19	60
1	. 68221	21	. 46582	45	. 93306	54	. 07174	62	. 36770	37	. 73116	20	59
2	. 68242	22	. 46537	46	. 93360	55	. 07112	63	. 36807	37	. 73096	20	58
3	. 68264	21	. 46491	46	. 93415	54	. 07049	62	. 36844	37	. 73076	20	57
4	. 68285	21	. 46445	45	. 93469	55	. 06987	62	. 36881	38	. 73056	20	56
5	0. 68306	21	1. 46400	46	0. 93524	54	1. 06925	63	1. 36919	37	0. 73036	20	55
6	. 68327	22	. 46354	45	. 93578	55	. 06862	62	. 36956	37	. 73016	20	54
7	. 68349	21	. 46309	46	. 93633	55	. 06800	62	. 36993	37	. 72996	20	53
8	. 68370	21	. 46263	45	. 93688	54	. 06738	62	. 37030	38	. 72976	19	52
9	. 68391	21	. 46218	45	. 93742	55	. 06676	63	. 37068	37	. 72957	20	51
10	0. 68412	22	1. 46173	46	0. 93797	55	1. 06613	62	1. 37105	38	0. 72937	20	50
11	. 68434	21	. 46127	45	. 93852	54	. 06551	62	. 37143	37	. 72917	20	49
12	. 68455	21	. 46082	45	. 93906	55	. 06489	62	. 37180	38	. 72897	20	48
13	. 68476	21	. 46037	45	. 93961	55	. 06427	62	. 37218	37	. 72877	20	47
14	. 68497	21	. 45992	46	. 94016	55	. 06365	62	. 37255	38	. 72857	20	46
15	0. 68518	21	1. 45946	45	0. 94071	54	1. 06303	62	1. 37293	37	0. 72837	20	45
16	. 68539	22	. 45901	45	. 94125	55	. 06241	62	. 37330	38	. 72817	20	44
17	. 68561	21	. 45856	45	. 94180	55	. 06179	62	. 37368	38	. 72797	20	43
18	. 68582	21	. 45811	45	. 94235	55	. 06117	61	. 37406	37	. 72777	20	42
19	. 68603	21	. 45766	45	. 94290	55	. 06056	62	. 37443	38	. 72757	20	41
20	0. 68624	21	1. 45721	45	0. 94345	55	1. 05994	62	1. 37481	38	0. 72737	20	40
21	. 68645	21	. 45676	45	. 94400	55	. 05932	62	. 37519	37	. 72717	20	39
22	. 68666	22	. 45631	44	. 94455	55	. 05870	61	. 37556	38	. 72697	20	38
23	. 68688	21	. 45587	45	. 94510	55	. 05809	62	. 37594	38	. 72677	20	37
24	. 68709	21	. 45542	45	. 94565	55	. 05747	62	. 37632	38	. 72657	20	36
25	0. 68730	21	1. 45497	45	0. 94620	56	1. 05685	61	1. 37670	38	0. 72637	20	35
26	. 68751	21	. 45452	44	. 94676	55	. 05624	62	. 37708	38	. 72617	20	34
27	. 68772	21	. 45408	45	. 94731	55	. 05562	61	. 37746	38	. 72597	20	33
28	. 68793	21	. 45363	44	. 94786	55	. 05501	62	. 37784	38	. 72577	20	32
29	. 68814	21	. 45319	45	. 94841	55	. 05439	61	. 37822	38	. 72557	20	31
30	0. 68835	22	1. 45274	45	0. 94896	56	1. 05378	61	1. 37860	38	0. 72537	20	30
31	. 68857	21	. 45229	44	. 94952	55	. 05317	62	. 37898	38	. 72517	20	29
32	. 68878	21	. 45185	44	. 95007	55	. 05255	61	. 37936	38	. 72497	20	28
33	. 68899	21	. 45141	45	. 95062	56	. 05194	61	. 37974	38	. 72477	20	27
34	. 68920	21	. 45096	44	. 95118	55	. 05133	61	. 38012	39	. 72457	20	26
35	0. 68941	21	1. 45052	45	0. 95173	56	1. 05072	62	1. 38051	38	0. 72437	20	25
36	. 68962	21	. 45007	44	. 95229	55	. 05010	61	. 38089	38	. 72417	20	24
37	. 68983	21	. 44963	44	. 95284	56	. 04949	61	. 38127	38	. 72397	20	23
38	. 69004	21	. 44919	44	. 95340	55	. 04888	61	. 38165	39	. 72377	20	22
39	. 69025	21	. 44875	44	. 95395	56	. 04827	61	. 38204	38	. 72357	20	21
40	0. 69046	21	1. 44831	44	0. 95451	55	1. 04766	61	1. 38242	38	0. 72337	20	20
41	. 69067	21	. 44787	45	. 95506	56	. 04705	61	. 38280	39	. 72317	20	19
42	. 69088	21	. 44742	44	. 95562	56	. 04644	61	. 38319	38	. 72297	20	18
43	. 69109	21	. 44698	44	. 95618	55	. 04583	61	. 38357	39	. 72277	20	17
44	. 69130	21	. 44654	44	. 95673	56	. 04522	61	. 38396	38	. 72257	21	16
45	0. 69151	21	1. 44610	43	0. 95729	56	1. 04461	60	1. 38434	39	0. 72236	20	15
46	. 69172	21	. 44567	44	. 95785	56	. 04401	61	. 38473	39	. 72216	20	14
47	. 69193	21	. 44523	44	. 95841	56	. 04340	61	. 38512	38	. 72196	20	13
48	. 69214	21	. 44479	44	. 95897	55	. 04279	61	. 38550	39	. 72176	20	12
49	. 69235	21	. 44435	44	. 95952	56	. 04218	60	. 38589	39	. 72156	20	11
50	0. 69256	21	1. 44391	44	0. 96008	56	1. 04158	61	1. 38628	38	0. 72136	20	10
51	. 69277	21	. 44347	43	. 96064	56	. 04097	61	. 38666	39	. 72116	21	9
52	. 69298	21	. 44304	44	. 96120	56	. 04036	60	. 38705	39	. 72095	20	8
53	. 69319	21	. 44260	43	. 96176	56	. 03976	61	. 38744	39	. 72075	20	7
54	. 69340	21	. 44217	44	. 96232	56	. 03915	60	. 38783	39	. 72055	20	6
55	0. 69361	21	1. 44173	44	0. 96288	56	1. 03855	61	1. 38822	38	0. 72035	20	5
56	. 69382	21	. 44129	43	. 96344	56	. 03794	60	. 38860	39	. 72015	20	4
57	. 69403	21	. 44086	44	. 96400	57	. 03734	60	. 38899	39	. 71995	21	3
58	. 69424	21	. 44042	43	. 96457	56	. 03674	61	. 38938	39	. 71974	20	2
59	. 69445	21	. 43999	43	. 96513	56	. 03613	60	. 38977	39	. 71954	20	1
60	0. 69466		1. 43956		0. 96569		1. 03553		1. 39016		0. 71934		0
↑ 133°→	cos	Diff. 1′	sec	Diff. 1′	cot	Diff. 1′	tan	Diff. 1′	csc	Diff. 1′	sin	Diff. 1′	←46° ↑

TABLE 31

Natural Trigonometric Functions

44°→ ′ ↓	sin	Diff. 1′	csc	Diff. 1′	tan	Diff. 1′	cot	Diff. 1′	sec	Diff. 1′	cos	Diff. 1′	←135° ′ ↓
0	0.69466	21	1.43956	44	0.96569	56	1.03553	60	1.39016	39	0.71934	20	60
1	.69487	21	.43912	43	.96625	56	.03493	60	.39055	40	.71914	20	59
2	.69508	21	.43869	43	.96681	57	.03433	61	.39095	39	.71894	21	58
3	.69529	20	.43826	43	.96738	56	.03372	60	.39134	39	.71873	20	57
4	.69549	21	.43783	44	.96794	56	.03312	60	.39173	39	.71853	20	56
5	0.69570	21	1.43739	43	0.96850	57	1.03252	60	1.39212	39	0.71833	20	55
6	.69591	21	.43696	43	.96907	56	.03192	60	.39251	40	.71813	21	54
7	.69612	21	.43653	43	.96963	57	.03132	60	.39291	39	.71792	20	53
8	.69633	21	.43610	43	.97020	56	.03072	60	.39330	39	.71772	20	52
9	.69654	21	.43567	43	.97076	57	.03012	60	.39369	40	.71752	20	51
10	0.69675	21	1.43524	43	0.97133	56	1.02952	60	1.39409	39	0.71732	21	50
11	.69696	21	.43481	43	.97189	57	.02892	60	.39448	39	.71711	20	49
12	.69717	20	.43438	43	.97246	56	.02832	60	.39487	40	.71691	20	48
13	.69737	21	.43395	43	.97302	57	.02772	59	.39527	39	.71671	21	47
14	.69758	21	.43352	42	.97359	57	.02713	60	.39566	40	.71650	20	46
15	0.69779	21	1.43310	43	0.97416	56	1.02653	60	1.39606	40	0.71630	20	45
16	.69800	21	.43267	43	.97472	57	.02593	60	.39646	39	.71610	20	44
17	.69821	21	.43224	43	.97529	57	.02533	59	.39685	40	.71590	21	43
18	.69842	20	.43181	42	.97586	57	.02474	60	.39725	39	.71569	20	42
19	.69862	21	.43139	43	.97643	57	.02414	59	.39764	40	.71549	20	41
20	0.69883	21	1.43096	43	0.97700	56	1.02355	60	1.39804	40	0.71529	21	40
21	.69904	21	.43053	42	.97756	57	.02295	59	.39844	40	.71508	20	39
22	.69925	21	.43011	43	.97813	57	.02236	60	.39884	40	.71488	20	38
23	.69946	20	.42968	42	.97870	57	.02176	59	.39924	39	.71468	21	37
24	.69966	21	.42926	43	.97927	57	.02117	60	.39963	40	.71447	20	36
25	0.69987	21	1.42883	42	0.97984	57	1.02057	59	1.40003	40	0.71427	20	35
26	.70008	21	.42841	42	.98041	57	.01998	59	.40043	40	.71407	21	34
27	.70029	20	.42799	43	.98098	57	.01939	60	.40083	40	.71386	20	33
28	.70049	21	.42756	42	.98155	58	.01879	59	.40123	40	.71366	21	32
29	.70070	21	.42714	42	.98213	57	.01820	59	.40163	40	.71345	20	31
30	0.70091	21	1.42672	42	0.98270	57	1.01761	59	1.40203	40	0.71325	20	30
31	.70112	20	.42630	43	.98327	57	.01702	60	.40243	40	.71305	21	29
32	.70132	21	.42587	42	.98384	57	.01642	59	.40283	41	.71284	20	28
33	.70153	21	.42545	42	.98441	58	.01583	59	.40324	40	.71264	21	27
34	.70174	21	.42503	42	.98499	57	.01524	59	.40364	40	.71243	20	26
35	0.70195	20	1.42461	42	0.98556	57	1.01465	59	1.40404	40	0.71223	20	25
36	.70215	21	.42419	42	.98613	58	.01406	59	.40444	41	.71203	21	24
37	.70236	21	.42377	42	.98671	57	.01347	59	.40485	40	.71182	20	23
38	.70257	20	.42335	42	.98728	58	.01288	59	.40525	40	.71162	21	22
39	.70277	21	.42293	42	.98786	57	.01229	59	.40565	41	.71141	20	21
40	0.70298	21	1.42251	42	0.98843	58	1.01170	58	1.40606	40	0.71121	21	20
41	.70319	20	.42209	41	.98901	57	.01112	59	.40646	41	.71100	20	19
42	.70339	21	.42168	42	.98958	58	.01053	59	.40687	40	.71080	21	18
43	.70360	21	.42126	42	.99016	57	.00994	59	.40727	41	.71059	20	17
44	.70381	20	.42084	42	.99073	58	.00935	59	.40768	40	.71039	20	16
45	0.70401	21	1.42042	41	0.99131	58	1.00876	58	1.40808	41	0.71019	21	15
46	.70422	21	.42001	42	.99189	58	.00818	59	.40849	41	.70998	20	14
47	.70443	20	.41959	41	.99247	57	.00759	58	.40890	40	.70978	21	13
48	.70463	21	.41918	42	.99304	58	.00701	59	.40930	41	.70957	20	12
49	.70484	21	.41876	41	.99362	58	.00642	59	.40971	41	.70937	21	11
50	0.70505	20	1.41835	42	0.99420	58	1.00583	58	1.41012	41	0.70916	20	10
51	.70525	21	.41793	41	.99478	58	.00525	58	.41053	40	.70896	21	9
52	.70546	21	.41752	42	.99536	58	.00467	59	.41093	41	.70875	20	8
53	.70567	20	.41710	41	.99594	58	.00408	58	.41134	41	.70855	21	7
54	.70587	21	.41669	42	.99652	58	.00350	59	.41175	41	.70834	21	6
55	0.70608	20	1.41627	41	0.99710	58	1.00291	58	1.41216	41	0.70813	20	5
56	.70628	21	.41586	41	.99768	58	.00233	58	.41257	41	.70793	21	4
57	.70649	21	.41545	41	.99826	58	.00175	59	.41298	41	.70772	20	3
58	.70670	20	.41504	41	.99884	58	.00116	58	.41339	41	.70752	21	2
59	.70690	21	.41463	42	0.99942	58	.00058	58	.41380	41	.70731	20	1
60	0.70711		1.41421		1.00000		1.00000		1.41421		0.70711		0
↑ 134°→	cos	Diff. 1′	sec	Diff. 1′	cot	Diff. 1′	tan	Diff. 1′	csc	Diff. 1′	sin	Diff. 1′←45°	↑

TABLE 32

Logarithms of Numbers

1–250

No.	Log	No.	Log	No.	Log	No.	Log	No.	Log
1	0. 00000	51	1. 70757	101	2. 00432	151	2. 17898	201	2. 30320
2	0. 30103	52	1. 71600	102	2. 00860	152	2. 18184	202	2. 30535
3	0. 47712	53	1. 72428	103	2. 01284	153	2. 18469	203	2. 30750
4	0. 60206	54	1. 73239	104	2. 01703	154	2. 18752	204	2. 30963
5	0. 69897	55	1. 74036	105	2. 02119	155	2. 19033	205	2. 31175
6	0. 77815	56	1. 74819	106	2. 02531	156	2. 19312	206	2. 31387
7	0. 84510	57	1. 75587	107	2. 02938	157	2. 19590	207	2. 31597
8	0. 90309	58	1. 76343	108	2. 03342	158	2. 19866	208	2. 31806
9	0. 95424	59	1. 77085	109	2. 03743	159	2. 20140	209	2. 32015
10	1. 00000	60	1. 77815	110	2. 04139	160	2. 20412	210	2. 32222
11	1. 04139	61	1. 78533	111	2. 04532	161	2. 20683	211	2. 32428
12	1. 07918	62	1. 79239	112	2. 04922	162	2. 20952	212	2. 32634
13	1. 11394	63	1. 79934	113	2. 05308	163	2. 21219	213	2. 32838
14	1. 14613	64	1. 80618	114	2. 05690	164	2 21484	214	2. 33041
15	1. 17609	65	1. 81291	115	2. 06070	165	2. 21748	215	2. 33244
16	1. 20412	66	1. 81954	116	2. 06446	166	2. 22011	216	2. 33445
17	1. 23045	67	1. 82607	117	2. 06819	167	2. 22272	217	2. 33646
18	1. 25527	68	1. 83251	118	2. 07188	168	2. 22531	218	2. 33846
19	1. 27875	69	1. 83885	119	2. 07555	169	2. 22789	219	2. 34044
20	1. 30103	70	1. 84510	120	2. 07918	170	2. 23045	220	2. 34242
21	1. 32222	71	1. 85126	121	2. 08279	171	2. 23300	221	2. 34439
22	1. 34242	72	1. 85733	122	2. 08636	172	2. 23553	222	2. 34635
23	1. 36173	73	1. 86332	123	2. 08991	173	2. 23805	223	2. 34830
24	1. 38021	74	1. 86923	124	2. 09342	174	2. 24055	224	2. 35025
25	1. 39794	75	1. 87506	125	2. 09691	175	2. 24304	225	2. 35218
26	1. 41497	76	1. 88081	126	2. 10037	176	2. 24551	226	2. 35411
27	1. 43136	77	1. 88649	127	2. 10380	177	2. 24797	227	2. 35603
28	1. 44716	78	1. 89209	128	2. 10721	178	2. 25042	228	2. 35793
29	1. 46240	79	1. 89763	129	2. 11059	179	2. 25285	229	2. 35984
30	1. 47712	80	1. 90309	130	2. 11394	180	2. 25527	230	2. 36173
31	1. 49136	81	1. 90849	131	2. 11727	181	2. 25768	231	2. 36361
32	1. 50515	82	1. 91381	132	2. 12057	182	2. 26007	232	2. 36549
33	1. 51851	83	1. 91908	133	2. 12385	183	2. 26245	233	2. 36736
34	1. 53148	84	1. 92428	134	2. 12710	184	2. 26482	234	2. 36922
35	1. 54407	85	1. 92942	135	2. 13033	185	2. 26717	235	2. 37107
36	1. 55630	86	1. 93450	136	2. 13354	186	2. 26951	236	2. 37291
37	1. 56820	87	1. 93952	137	2. 13672	187	2. 27184	237	2. 37475
38	1. 57978	88	1. 94448	138	2. 13988	188	2. 27416	238	2. 37658
39	1. 59106	89	1. 94939	139	2. 14301	189	2. 27646	239	2. 37840
40	1. 60206	90	1. 95424	140	2. 14613	190	2. 27875	240	2. 38021
41	1. 61278	91	1. 95904	141	2. 14922	191	2. 28103	241	2. 38202
42	1. 62325	92	1. 96379	142	2. 15229	192	2. 28330	242	2. 38382
43	1. 63347	93	1. 96848	143	2. 15534	193	2. 28556	243	2. 38561
44	1. 64345	94	1. 97313	144	2. 15836	194	2. 28780	244	2. 38739
45	1. 65321	95	1. 97772	145	2. 16137	195	2. 29003	245	2. 38917
46	1. 66276	96	1. 98227	146	2. 16435	196	2. 29226	246	2. 39094
47	1. 67210	97	1. 98677	147	2. 16732	197	2. 29447	247	2. 39270
48	1. 68124	98	1. 99123	148	2. 17026	198	2. 29667	248	2. 39445
49	1. 69020	99	1. 99564	149	2. 17319	199	2. 29885	249	2. 39620
50	1. 69897	100	2. 00000	150	2. 17609	200	2. 30103	250	2. 39794

TABLE 32

Logarithms of Numbers

1000–1500

No.	0	d	1	d	2	d	3	d	4	d	5	d	6	d	7	d	8	d	9	d
100	00000	43	00043	44	00087	43	00130	43	00173	44	00217	43	00260	43	00303	43	00346	43	00389	43
101	00432	43	00475	43	00518	43	00561	43	00604	43	00647	42	00689	43	00732	43	00775	42	00817	43
102	00860	43	00903	42	00945	43	00988	42	01030	42	01072	43	01115	42	01157	42	01199	43	01242	42
103	01284	42	01326	42	01368	42	01410	42	01452	42	01494	42	01536	42	01578	42	01620	42	01662	41
104	01703	42	01745	42	01787	41	01828	42	01870	42	01912	41	01953	42	01995	41	02036	42	02078	41
105	02119	41	02160	42	02202	41	02243	41	02284	41	02325	41	02366	41	02407	42	02449	41	02490	41
106	02531	41	02572	40	02612	41	02653	41	02694	41	02735	41	02776	40	02816	41	02857	41	02898	40
107	02938	41	02979	40	03019	41	03060	40	03100	41	03141	40	03181	41	03222	40	03262	40	03302	40
108	03342	41	03383	40	03423	40	03463	40	03503	40	03543	40	03583	40	03623	40	03663	40	03703	40
109	03743	39	03782	40	03822	40	03862	40	03902	39	03941	40	03981	40	04021	39	04060	40	04100	39
110	04139	40	04179	39	04218	40	04258	39	04297	39	04336	40	04376	39	04415	39	04454	39	04493	39
111	04532	39	04571	39	04610	40	04650	39	04689	38	04727	39	04766	39	04805	39	04844	39	04883	39
112	04922	39	04961	38	04999	39	05038	39	05077	38	05115	39	05154	38	05192	39	05231	38	05269	39
113	05308	38	05346	39	05385	38	05423	38	05461	39	05500	38	05538	38	05576	38	05614	38	05652	38
114	05690	39	05729	38	05767	38	05805	38	05843	38	05881	37	05918	38	05956	38	05994	38	06032	38
115	06070	38	06108	37	06145	38	06183	38	06221	37	06258	38	06296	37	06333	38	06371	37	06408	38
116	06446	37	06483	38	06521	37	06558	37	06595	38	06633	37	06670	37	06707	37	06744	37	06781	38
117	06819	37	06856	37	06893	37	06930	37	06967	37	07004	37	07041	37	07078	37	07115	36	07151	37
118	07188	37	07225	37	07262	36	07298	37	07335	37	07372	36	07408	37	07445	37	07482	36	07518	37
119	07555	36	07591	37	07628	36	07664	36	07700	37	07737	36	07773	36	07809	37	07846	36	07882	36
120	07918	36	07954	36	07990	37	08027	36	08063	36	08099	36	08135	36	08171	36	08207	36	08243	36
121	08279	35	08314	36	08350	36	08386	36	08422	36	08458	35	08493	36	08529	36	08565	35	08600	36
122	08636	36	08672	35	08707	36	08743	35	08778	36	08814	35	08849	35	08884	36	08920	35	08955	36
123	08991	35	09026	35	09061	35	09096	36	09132	35	09167	35	09202	35	09237	35	09272	35	09307	35
124	09342	35	09377	35	09412	35	09447	35	09482	35	09517	35	09552	35	09587	34	09621	35	09656	35
125	09691	35	09726	34	09760	35	09795	35	09830	34	09864	35	09899	35	09934	34	09968	35	10003	34
126	10037	35	10072	34	10106	34	10140	35	10175	34	10209	34	10243	35	10278	34	10312	34	10346	34
127	10380	35	10415	34	10449	34	10483	34	10517	34	10551	34	10585	34	10619	34	10653	34	10687	34
128	10721	34	10755	34	10789	34	10823	34	10857	33	10890	34	10924	34	10958	34	10992	33	11025	34
129	11059	34	11093	33	11126	34	11160	33	11193	34	11227	34	11261	33	11294	33	11327	34	11361	33
130	11394	34	11428	33	11461	33	11494	34	11528	33	11561	33	11594	34	11628	33	11661	33	11694	33
131	11727	33	11760	33	11793	33	11826	34	11860	33	11893	33	11926	33	11959	33	11992	32	12024	33
132	12057	33	12090	33	12123	33	12156	33	12189	33	12222	32	12254	33	12287	33	12320	32	12352	33
133	12385	33	12418	32	12450	33	12483	33	12516	32	12548	33	12581	32	12613	33	12646	32	12678	32
134	12710	33	12743	32	12775	33	12808	32	12840	32	12872	33	12905	32	12937	32	12969	32	13001	32
135	13033	33	13066	32	13098	32	13130	32	13162	32	13194	32	13226	32	13258	32	13290	32	13322	32
136	13354	32	13386	32	13418	32	13450	31	13481	32	13513	32	13545	32	13577	32	13609	31	13640	32
137	13672	32	13704	31	13735	32	13767	32	13799	31	13830	32	13862	31	13893	32	13925	31	13956	32
138	13988	31	14019	32	14051	31	14082	32	14114	31	14145	31	14176	32	14208	31	14239	31	14270	31
139	14301	32	14333	31	14364	31	14395	31	14426	31	14457	32	14489	31	14520	31	14551	31	14582	31
140	14613	31	14644	31	14675	31	14706	31	14737	31	14768	31	14799	30	14829	31	14860	31	14891	31
141	14922	31	14953	30	14983	31	15014	31	15045	31	15076	30	15106	31	15137	31	15168	30	15198	31
142	15229	30	15259	31	15290	30	15320	31	15351	30	15381	31	15412	30	15442	31	15473	30	15503	31
143	15534	30	15564	30	15594	31	15625	30	15655	30	15685	30	15715	31	15746	30	15776	30	15806	30
144	15836	30	15866	31	15897	30	15927	30	15957	30	15987	30	16017	30	16047	30	16077	30	16107	30
145	16137	30	16167	30	16197	30	16227	29	16256	30	16286	30	16316	30	16346	30	16376	30	16406	29
146	16435	30	16465	30	16495	29	16524	30	16554	30	16584	29	16613	30	16643	30	16673	29	16702	30
147	16732	29	16761	30	16791	29	16820	30	16850	29	16879	30	16909	29	16938	29	16967	30	16997	29
148	17026	30	17056	29	17085	29	17114	29	17143	30	17173	29	17202	29	17231	29	17260	29	17289	30
149	17319	29	17348	29	17377	29	17406	29	17435	29	17464	29	17493	29	17522	29	17551	29	17580	29
150	17609	29	17638	29	17667	29	17696	29	17725	29	17754	28	17782	29	17811	29	17840	29	17869	29
No.	0	d	1	d	2	d	3	d	4	d	5	d	6	d	7	d	8	d	9	d

Prop. parts

	44	43	42	41	40	39	38	37	36	35	34	33
1	4	4	4	4	4	4	4	4	4	4	3	3
2	9	9	8	8	8	8	8	7	7	7	7	7
3	13	13	13	12	12	12	11	11	11	10	10	10
4	18	17	17	16	16	16	15	15	14	14	14	13
5	22	22	21	20	20	20	19	18	18	18	17	16
6	26	26	25	25	24	23	23	22	22	21	20	20
7	31	30	29	29	28	27	27	26	25	24	24	23
8	35	34	34	33	32	31	30	30	29	28	27	26
9	40	39	38	37	36	35	34	33	32	32	31	30

TABLE 32

Logarithms of Numbers

1500–2000

No.	0	d	1	d	2	d	3	d	4	d	5	d	6	d	7	d	8	d	9	d
150	17609	29	17638	29	17667	29	17696	29	17725	29	17754	28	17782	29	17811	29	17840	29	17869	29
151	17898	28	17926	29	17955	29	17984	29	18013	28	18041	29	18070	29	18099	28	18127	29	18156	28
152	18184	29	18213	28	18241	29	18270	28	18298	29	18327	28	18355	29	18384	28	18412	29	18441	28
153	18469	29	18498	28	18526	28	18554	29	18583	28	18611	28	18639	28	18667	29	18696	28	18724	28
154	18752	28	18780	28	18808	29	18837	28	18865	28	18893	28	18921	28	18949	28	18977	28	19005	28
155	19033	28	19061	28	19089	28	19117	28	19145	28	19173	28	19201	28	19229	28	19257	28	19285	27
156	19312	28	19340	28	19368	28	19396	28	19424	27	19451	28	19479	28	19507	28	19535	27	19562	28
157	19590	28	19618	27	19645	28	19673	27	19700	28	19728	28	19756	27	19783	28	19811	27	19838	28
158	19866	27	19893	28	19921	27	19948	28	19976	27	20003	27	20030	28	20058	27	20085	27	20112	28
159	20140	27	20167	27	20194	28	20222	27	20249	27	20276	27	20303	27	20330	28	20358	27	20385	27
160	20412	27	20439	27	20466	27	20493	27	20520	28	20548	27	20575	27	20602	27	20629	27	20656	27
161	20683	27	20710	27	20737	26	20763	27	20790	27	20817	27	20844	27	20871	27	20898	27	20925	27
162	20952	26	20978	27	21005	27	21032	27	21059	26	21085	27	21112	27	21139	26	21165	27	21192	27
163	21219	26	21245	27	21272	27	21299	26	21325	27	21352	26	21378	27	21405	26	21431	27	21458	26
164	21484	27	21511	26	21537	27	21564	26	21590	27	21617	26	21643	26	21669	27	21696	26	21722	26
165	21748	27	21775	26	21801	26	21827	27	21854	26	21880	26	21906	26	21932	26	21958	27	21985	26
166	22011	26	22037	26	22063	26	22089	26	22115	26	22141	26	22167	27	22194	26	22220	26	22246	26
167	22272	26	22298	26	22324	26	22350	26	22376	25	22401	26	22427	26	22453	26	22479	26	22505	26
168	22531	26	22557	26	22583	25	22608	26	22634	26	22660	26	22686	26	22712	25	22737	26	22763	26
169	22789	25	22814	26	22840	26	22866	25	22891	26	22917	26	22943	25	22968	26	22994	25	23019	26
170	23045	25	23070	26	23096	25	23121	26	23147	25	23172	26	23198	25	23223	26	23249	25	23274	26
171	23300	25	23325	25	23350	26	23376	25	23401	25	23426	26	23452	25	23477	25	23502	26	23528	25
172	23553	25	23578	25	23603	26	23629	25	23654	25	23679	25	23704	25	23729	25	23754	25	23779	26
173	23805	25	23830	25	23855	25	23880	25	23905	25	23930	25	23955	25	23980	25	24005	25	24030	25
174	24055	25	24080	25	24105	25	24130	25	24155	25	24180	24	24204	25	24229	25	24254	25	24279	25
175	24304	25	24329	24	24353	25	24378	25	24403	25	24428	24	24452	25	24477	25	24502	25	24527	24
176	24551	25	24576	25	24601	24	24625	25	24650	24	24674	25	24699	25	24724	24	24748	25	24773	24
177	24797	25	24822	24	24846	25	24871	24	24895	25	24920	24	24944	25	24969	24	24993	25	25018	24
178	25042	24	25066	25	25091	24	25115	24	25139	25	25164	24	25188	24	25212	25	25237	24	25261	24
179	25285	25	25310	24	25334	24	25358	24	25382	24	25406	25	25431	24	25455	24	25479	24	25503	24
180	25527	24	25551	24	25575	25	25600	24	25624	24	25648	24	25672	24	25696	24	25720	24	25744	24
181	25768	24	25792	24	25816	24	25840	24	25864	24	25888	24	25912	23	25935	24	25959	24	25983	24
182	26007	24	26031	24	26055	24	26079	23	26102	24	26126	24	26150	24	26174	24	26198	23	26221	24
183	26245	24	26269	24	26293	23	26316	24	26340	24	26364	23	26387	24	26411	24	26435	23	26458	24
184	26482	23	26505	24	26529	24	26553	23	26576	24	26600	23	26623	24	26647	23	26670	24	26694	23
185	26717	24	26741	23	26764	24	26788	23	26811	23	26834	24	26858	23	26881	24	26905	23	26928	23
186	26951	24	26975	23	26998	23	27021	24	27045	23	27068	23	27091	23	27114	24	27138	23	27161	23
187	27184	23	27207	24	27231	23	27254	23	27277	23	27300	23	27323	23	27346	24	27370	23	27393	23
188	27416	23	27439	23	27462	23	27485	23	27508	23	27531	23	27554	23	27577	23	27600	23	27623	23
189	27646	23	27669	23	27692	23	27715	23	27738	23	27761	23	27784	23	27807	23	27830	22	27852	23
190	27875	23	27898	23	27921	23	27944	23	27967	22	27989	23	28012	23	28035	23	28058	23	28081	22
191	28103	23	28126	23	28149	22	28171	23	28194	23	28217	23	28240	22	28262	23	28285	22	28307	23
192	28330	23	28353	22	28375	23	28398	23	28421	22	28443	23	28466	22	28488	23	28511	22	28533	23
193	28556	22	28578	23	28601	22	28623	23	28646	22	28668	23	28691	22	28713	22	28735	23	28758	22
194	28780	23	28803	22	28825	22	28847	23	28870	22	28892	22	28914	23	28937	22	28959	22	28981	22
195	29003	23	29026	22	29048	22	29070	22	29092	23	29115	22	29137	22	29159	22	29181	22	29203	23
196	29226	22	29248	22	29270	22	29292	22	29314	22	29336	22	29358	22	29380	23	29403	22	29425	22
197	29447	22	29469	22	29491	22	29513	22	29535	22	29557	22	29579	22	29601	22	29623	22	29645	22
198	29667	21	29688	22	29710	22	29732	22	29754	22	29776	22	29798	22	29820	22	29842	21	29863	22
199	29885	22	29907	22	29929	22	29951	22	29973	21	29994	22	30016	22	30038	22	30060	21	30081	22
200	30103	22	30125	21	30146	22	30168	22	30190	21	30211	22	30233	22	30255	21	30276	22	30298	22
No.	0	d	1	d	2	d	3	d	4	d	5	d	6	d	7	d	8	d	9	d

Prop. parts

	32	31	30	29	28	27	26	25	24	23	22	21
1	3	3	3	3	3	3	3	2	2	2	2	2
2	6	6	6	6	6	5	5	5	5	5	4	4
3	10	9	9	9	8	8	8	8	7	7	7	6
4	13	12	12	12	11	11	10	10	10	9	9	8
5	16	16	15	14	14	14	13	12	12	12	11	10
6	19	19	18	17	17	16	16	15	14	14	13	13
7	22	22	21	20	20	19	18	18	17	16	15	15
8	26	25	24	23	22	22	21	20	19	18	18	17
9	29	28	27	26	25	24	23	22	22	21	20	19

TABLE 32

Logarithms of Numbers

2000–2500

No.	0	d	1	d	2	d	3	d	4	d	5	d	6	d	7	d	8	d	9	d
200	30103	22	30125	21	30146	22	30168	22	30190	21	30211	22	30233	22	30255	21	30276	22	30298	22
201	30320	21	30341	22	30363	21	30384	22	30406	22	30428	21	30449	22	30471	21	30492	22	30514	21
202	30535	22	30557	21	30578	22	30600	21	30621	22	30643	21	30664	21	30685	22	30707	21	30728	22
203	30750	21	30771	21	30792	22	30814	21	30835	21	30856	22	30878	21	30899	21	30920	22	30942	21
204	30963	21	30984	22	31006	21	31027	21	31048	21	31069	22	31091	21	31112	21	31133	21	31154	21
205	31175	22	31197	21	31218	21	31239	21	31260	21	31281	21	31302	21	31323	22	31345	21	31366	21
206	31387	21	31408	21	31429	21	31450	21	31471	21	31492	21	31513	21	31534	21	31555	21	31576	21
207	31597	21	31618	21	31639	21	31660	21	31681	21	31702	21	31723	21	31744	21	31765	20	31785	21
208	31806	21	31827	21	31848	21	31869	21	31890	21	31911	20	31931	21	31952	21	31973	21	31994	21
209	32015	20	32035	21	32056	21	32077	21	32098	20	32118	21	32139	21	32160	21	32181	20	32201	21
210	32222	21	32243	20	32263	21	32284	21	32305	20	32325	21	32346	20	32366	21	32387	21	32408	20
211	32428	21	32449	20	32469	21	32490	20	32510	21	32531	21	32552	20	32572	21	32593	20	32613	21
212	32634	20	32654	21	32675	20	32695	20	32715	21	32736	20	32756	21	32777	20	32797	21	32818	20
213	32838	20	32858	21	32879	20	32899	20	32919	21	32940	20	32960	20	32980	21	33001	20	33021	20
214	33041	21	33062	20	33082	20	33102	20	33122	21	33143	20	33163	20	33183	20	33203	21	33224	20
215	33244	20	33264	20	33284	20	33304	21	33325	20	33345	20	33365	20	33385	20	33405	20	33425	20
216	33445	20	33465	21	33486	20	33506	20	33526	20	33546	20	33566	20	33586	20	33606	20	33626	20
217	33646	20	33666	20	33686	20	33706	20	33726	20	33746	20	33766	20	33786	20	33806	20	33826	20
218	33846	20	33866	19	33885	20	33905	20	33925	20	33945	20	33965	20	33985	20	34005	20	34025	19
219	34044	20	34064	20	34084	20	34104	20	34124	19	34143	20	34163	20	34183	20	34203	20	34223	19
220	34242	20	34262	20	34282	19	34301	20	34321	20	34341	20	34361	19	34380	20	34400	20	34420	19
221	34439	20	34459	20	34479	19	34498	20	34518	19	34537	20	34557	20	34577	19	34596	20	34616	19
222	34635	20	34655	19	34674	20	34694	19	34713	20	34733	20	34753	19	34772	20	34792	19	34811	19
223	34830	20	34850	19	34869	20	34889	19	34908	20	34928	19	34947	20	34967	19	34986	19	35005	20
224	35025	19	35044	20	35064	19	35083	19	35102	20	35122	19	35141	19	35160	20	35180	19	35199	19
225	35218	20	35238	19	35257	19	35276	19	35295	20	35315	19	35334	19	35353	19	35372	20	35392	19
226	35411	19	35430	19	35449	19	35468	20	35488	19	35507	19	35526	19	35545	19	35564	19	35583	20
227	35603	19	35622	19	35641	19	35660	19	35679	19	35698	19	35717	19	35736	19	35755	19	35774	19
228	35793	20	35813	19	35832	19	35851	19	35870	19	35889	19	35908	19	35927	19	35946	19	35965	19
229	35984	19	36003	18	36021	19	36040	19	36059	19	36078	19	36097	19	36116	19	36135	19	36154	19
230	36173	19	36192	19	36211	18	36229	19	36248	19	36267	19	36286	19	36305	19	36324	18	36342	19
231	36361	19	36380	19	36399	19	36418	18	36436	19	36455	19	36474	19	36493	18	36511	19	36530	19
232	36549	19	36568	18	36586	19	36605	19	36624	18	36642	19	36661	19	36680	18	36698	19	36717	19
233	36736	18	36754	19	36773	18	36791	19	36810	19	36829	18	36847	19	36866	18	36884	19	36903	19
234	36922	18	36940	19	36959	18	36977	19	36996	18	37014	19	37033	18	37051	19	37070	18	37088	19
235	37107	18	37125	19	37144	18	37162	19	37181	18	37199	19	37218	18	37236	18	37254	19	37273	18
236	37291	19	37310	18	37328	18	37346	19	37365	18	37383	18	37401	19	37420	18	37438	19	37457	18
237	37475	18	37493	18	37511	19	37530	18	37548	18	37566	19	37585	18	37603	18	37621	18	37639	19
238	37658	18	37676	18	37694	18	37712	19	37731	18	37749	18	37767	18	37785	18	37803	19	37822	18
239	37840	18	37858	18	37876	18	37894	18	37912	19	37931	18	37949	18	37967	18	37985	18	38003	18
240	38021	18	38039	18	38057	18	38075	18	38093	19	38112	18	38130	18	38148	18	38166	18	38184	18
241	38202	18	38220	18	38238	18	38256	18	38274	18	38292	18	38310	18	38328	18	38346	18	38364	18
242	38382	17	38399	18	38417	18	38435	18	38453	18	38471	18	38489	18	38507	18	38525	18	38543	18
243	38561	17	38578	18	38596	18	38614	18	38632	18	38650	18	38668	18	38686	17	38703	18	38721	18
244	38739	18	38757	18	38775	17	38792	18	38810	18	38828	18	38846	17	38863	18	38881	18	38899	18
245	38917	17	38934	18	38952	18	38970	17	38987	18	39005	18	39023	18	39041	17	39058	18	39076	18
246	39094	17	39111	18	39129	17	39146	18	39164	18	39182	17	39199	18	39217	18	39235	17	39252	18
247	39270	17	39287	18	39305	17	39322	18	39340	18	39358	17	39375	18	39393	17	39410	18	39428	17
248	39445	18	39463	17	39480	18	39498	17	39515	18	39533	17	39550	18	39568	17	39585	17	39602	18
249	39620	17	39637	18	39655	17	39672	18	39690	17	39707	17	39724	18	39742	17	39759	18	39777	17
250	39794	17	39811	18	39829	17	39846	17	39863	18	39881	17	39898	17	39915	18	39933	17	39950	17
No.	0	d	1	d	2	d	3	d	4	d	5	d	6	d	7	d	8	d	9	d

Prop. parts

	22
1	2
2	4
3	7
4	9
5	11
6	13
7	15
8	18
9	20

	21
1	2
2	4
3	6
4	8
5	10
6	13
7	15
8	17
9	19

	20
1	2
2	4
3	6
4	8
5	10
6	12
7	14
8	16
9	18

	19
1	2
2	4
3	6
4	8
5	10
6	11
7	13
8	15
9	17

	18
1	2
2	4
3	5
4	7
5	9
6	11
7	13
8	14
9	16

	17
1	2
2	3
3	5
4	7
5	8
6	10
7	12
8	14
9	15

TABLE 32

Logarithms of Numbers

2500–3000

No.	0	d	1	d	2	d	3	d	4	d	5	d	6	d	7	d	8	d	9	d
250	39794	17	39811	18	39829	17	39846	17	39863	18	39881	17	39898	17	39915	18	39933	17	39950	17
251	39967	18	39985	17	40002	17	40019	18	40037	17	40054	17	40071	17	40088	18	40106	17	40123	17
252	40140	17	40157	18	40175	17	40192	17	40209	17	40226	17	40243	18	40261	17	40278	17	40295	17
253	40312	17	40329	17	40346	18	40364	17	40381	17	40398	17	40415	17	40432	17	40449	17	40466	17
254	40483	17	40500	18	40518	17	40535	17	40552	17	40569	17	40586	17	40603	17	40620	17	40637	17
255	40654	17	40671	17	40688	17	40705	17	40722	17	40739	17	40756	17	40773	17	40790	17	40807	17
256	40824	17	40841	17	40858	17	40875	17	40892	17	40909	17	40926	17	40943	17	40960	16	40976	17
257	40993	17	41010	17	41027	17	41044	17	41061	17	41078	17	41095	16	41111	17	41128	17	41145	17
258	41162	17	41179	17	41196	16	41212	17	41229	17	41246	17	41263	17	41280	16	41296	17	41313	17
259	41330	17	41347	16	41363	17	41380	17	41397	17	41414	16	41430	17	41447	17	41464	17	41481	16
260	41497	17	41514	17	41531	16	41547	17	41564	17	41581	16	41597	17	41614	17	41631	16	41647	17
261	41664	17	41681	16	41697	17	41714	17	41731	16	41747	17	41764	16	41780	17	41797	17	41814	16
262	41830	17	41847	16	41863	17	41880	16	41896	17	41913	16	41929	17	41946	17	41963	16	41979	17
263	41996	16	42012	17	42029	16	42045	17	42062	16	42078	17	42095	16	42111	16	42127	17	42144	16
264	42160	17	42177	16	42193	17	42210	16	42226	17	42243	16	42259	16	42275	17	42292	16	42308	17
265	42325	16	42341	16	42357	17	42374	16	42390	16	42406	17	42423	16	42439	16	42455	17	42472	16
266	42488	16	42504	17	42521	16	42537	16	42553	17	42570	16	42586	16	42602	17	42619	16	42635	16
267	42651	16	42667	17	42684	16	42700	16	42716	16	42732	17	42749	16	42765	16	42781	16	42797	16
268	42813	17	42830	16	42846	16	42862	16	42878	16	42894	17	42911	16	42927	16	42943	16	42959	16
269	42975	16	42991	17	43008	16	43024	16	43040	16	43056	16	43072	16	43088	16	43104	16	43120	16
270	43136	6	43152	17	43169	16	43185	16	43201	16	43217	16	43233	16	43249	16	43265	16	43281	16
271	43297	16	43313	16	43329	16	43345	16	43361	16	43377	16	43393	16	43409	16	43425	16	43441	16
272	43457	16	43473	16	43489	16	43505	16	43521	16	43537	16	43553	16	43569	15	43584	16	43600	16
273	43616	16	43632	16	43648	16	43664	16	43680	16	43696	16	43712	15	43727	16	43743	16	43759	16
274	43775	16	43791	16	43807	16	43823	15	43838	16	43854	16	43870	16	43886	16	43902	15	43917	16
275	43933	16	43949	16	43965	16	43981	15	43996	16	44012	16	44028	16	44044	15	44059	16	44075	16
276	44091	16	44107	15	44122	16	44138	16	44154	16	44170	15	44185	16	44201	16	44217	15	44232	16
277	44248	16	44264	15	44279	16	44295	16	44311	15	44326	16	44342	16	44358	15	44373	16	44389	15
278	44404	16	44420	16	44436	15	44451	16	44467	16	44483	15	44498	16	44514	15	44529	16	44545	15
279	44560	16	44576	16	44592	15	44607	16	44623	15	44638	16	44654	15	44669	16	44685	15	44700	16
280	44716	15	44731	16	44747	15	44762	16	44778	15	44793	16	44809	15	44824	16	44840	15	44855	16
281	44871	15	44886	16	44902	15	44917	15	44932	16	44948	15	44963	16	44979	15	44994	16	45010	15
282	45025	15	45040	16	45056	15	45071	15	45086	16	45102	15	45117	16	45133	15	45148	15	45163	16
283	45179	15	45194	15	45209	16	45225	15	45240	15	45255	16	45271	15	45286	15	45301	16	45317	15
284	45332	15	45347	15	45362	16	45378	15	45393	15	45408	15	45423	16	45439	15	45454	15	45469	15
285	45484	16	45500	15	45515	15	45530	15	45545	16	45561	15	45576	15	45591	15	45606	15	45621	16
286	45637	15	45652	15	45667	15	45682	15	45697	15	45712	16	45728	15	45743	15	45758	15	45773	15
287	45788	15	45803	15	45818	16	45834	15	45849	15	45864	15	45879	15	45894	15	45909	15	45924	15
288	45939	15	45954	15	45969	15	45984	16	46000	15	46015	15	46030	15	46045	15	46060	15	46075	15
289	46090	15	46105	15	46120	15	46135	15	46150	15	46165	15	46180	15	46195	15	46210	15	46225	15
290	46240	15	46255	15	46270	15	46285	15	46300	15	46315	15	46330	15	46345	14	46359	15	46374	15
291	46389	15	46404	15	46419	15	46434	15	46449	15	46464	15	46479	15	46494	15	46509	14	46523	15
292	46538	15	46553	15	46568	15	46583	15	46598	15	46613	14	46627	15	46642	15	46657	15	46672	15
293	46687	15	46702	14	46716	15	46731	15	46746	15	46761	15	46776	14	46790	15	46805	15	46820	15
294	46835	15	46850	14	46864	15	46879	15	46894	15	46909	14	46923	15	46938	15	46953	14	46967	15
295	46982	15	46997	15	47012	14	47026	15	47041	15	47056	14	47070	15	47085	15	47100	14	47114	15
296	47129	15	47144	15	47159	14	47173	15	47188	14	47202	15	47217	15	47232	14	47246	15	47261	15
297	47276	14	47290	15	47305	14	47319	15	47334	15	47349	14	47363	15	47378	14	47392	15	47407	15
298	47422	14	47436	15	47451	14	47465	15	47480	14	47494	15	47509	15	47524	14	47538	15	47553	11
299	47567	15	47582	14	47596	15	47611	14	47625	15	47640	14	47654	15	47669	14	47683	15	47698	14
300	47712	15	47727	14	47741	15	47756	14	47770	14	47784	15	47799	14	47813	15	47828	14	47842	15
No.	0	d	1	d	2	d	3	d	4	d	5	d	6	d	7	d	8	d	9	d

Prop. parts

	18
1	2
2	4
3	5
4	7
5	9
6	11
7	13
8	14
9	16

	17
1	2
2	3
3	5
4	7
5	8
6	10
7	12
8	14
9	15

	16
1	2
2	3
3	5
4	6
5	8
6	10
7	11
8	13
9	14

	15
1	2
2	3
3	4
4	6
5	8
6	9
7	10
8	12
9	14

	14
1	1
2	3
3	4
4	6
5	7
6	8
7	10
8	11
9	13

TABLE 32

Logarithms of Numbers

3000–3500

No.	0	d	1	d	2	d	3	d	4	d	5	d	6	d	7	d	8	d	9	d
300	47712	15	47727	14	47741	15	47756	14	47770	14	47784	15	47799	14	47813	15	47828	14	47842	15
301	47857	14	47871	14	47885	15	47900	14	47914	15	47929	14	47943	15	47958	14	47972	14	47986	15
302	48001	14	48015	14	48029	15	48044	14	48058	15	48073	14	48087	14	48101	15	48116	14	48130	14
303	48144	15	48159	14	48173	14	48187	15	48202	14	48216	14	48230	14	48244	15	48259	14	48273	14
304	48287	15	48302	14	48316	14	48330	14	48344	15	48359	14	48373	14	48387	14	48401	15	48416	14
305	48430	14	48444	14	48458	15	48473	14	48487	14	48501	14	48515	15	48530	14	48544	14	48558	14
306	48572	14	48586	15	48601	14	48615	14	48629	14	48643	14	48657	14	48671	15	48686	14	48700	14
307	48714	14	48728	14	48742	14	48756	14	48770	15	48785	14	48799	14	48813	14	48827	14	48841	14
308	48855	14	48869	14	48883	14	48897	14	48911	15	48926	14	48940	14	48954	14	48968	14	48982	14
309	48996	14	49010	14	49024	14	49038	14	49052	14	49066	14	49080	14	49094	14	49108	14	49122	14
310	49136	14	49150	14	49164	14	49178	14	49192	14	49206	14	49220	14	49234	14	49248	14	49262	14
311	49276	14	49290	14	49304	14	49318	14	49332	14	49346	14	49360	14	49374	14	49388	14	49402	13
312	49415	14	49429	14	49443	14	49457	14	49471	14	49485	14	49499	14	49513	14	49527	14	49541	13
313	49554	14	49568	14	49582	14	49596	14	49610	14	49624	14	49638	13	49651	14	49665	14	49679	14
314	49693	14	49707	14	49721	13	49734	14	49748	14	49762	14	49776	14	49790	13	49803	14	49817	14
315	49831	14	49845	14	49859	13	49872	14	49886	14	49900	14	49914	13	49927	14	49941	14	49955	14
316	49969	13	49982	14	49996	14	50010	14	50024	13	50037	14	50051	14	50065	14	50079	13	50092	14
317	50106	14	50120	13	50133	14	50147	14	50161	13	50174	14	50188	14	50202	13	50215	14	50229	14
318	50243	13	50256	14	50270	14	50284	13	50297	14	50311	14	50325	13	50338	14	50352	13	50365	14
319	50379	14	50393	13	50406	14	50420	13	50433	14	50447	14	50461	13	50474	14	50488	13	50501	14
320	50515	14	50529	13	50542	14	50556	13	50569	14	50583	13	50596	14	50610	13	50623	14	50637	14
321	50651	13	50664	14	50678	13	50691	14	50705	13	50718	14	50732	13	50745	14	50759	13	50772	14
322	50786	13	50799	14	50813	13	50826	14	50840	13	50853	13	50866	14	50880	13	50893	14	50907	13
323	50920	14	50934	13	50947	14	50961	13	50974	13	50987	14	51001	13	51014	14	51028	13	51041	14
324	51055	13	51068	13	51081	14	51095	13	51108	13	51121	14	51135	13	51148	14	51162	13	51175	13
325	51188	14	51202	13	51215	13	51228	14	51242	13	51255	13	51268	14	51282	13	51295	13	51308	14
326	51322	13	51335	13	51348	14	51362	13	51375	13	51388	14	51402	13	51415	13	51428	13	51441	14
327	51455	13	51468	13	51481	14	51495	13	51508	13	51521	13	51534	14	51548	13	51561	13	51574	13
328	51587	14	51601	13	51614	13	51627	13	51640	14	51654	13	51667	13	51680	13	51693	13	51706	14
329	51720	13	51733	13	51746	13	51759	13	51772	14	51786	13	51799	13	51812	13	51825	13	51838	13
330	51851	14	51865	13	51878	13	51891	13	51904	13	51917	13	51930	13	51943	14	51957	13	51970	13
331	51983	13	51996	13	52009	13	52022	13	52035	13	52048	13	52061	14	52075	13	52088	13	52101	13
332	52114	13	52127	13	52140	13	52153	13	52166	13	52179	13	52192	13	52205	13	52218	13	52231	13
333	52244	13	52257	13	52270	14	52284	13	52297	13	52310	13	52323	13	52336	13	52349	13	52362	13
334	52375	13	52388	13	52401	13	52414	13	52427	13	52440	13	52453	13	52466	13	52479	13	52492	12
335	52504	13	52517	13	52530	13	52543	13	52556	13	52569	13	52582	13	52595	13	52608	13	52621	13
336	52634	13	52647	13	52660	13	52673	13	52686	13	52699	12	52711	13	52724	13	52737	13	52750	13
337	52763	13	52776	13	52789	13	52802	13	52815	12	52827	13	52840	13	52853	13	52866	13	52879	13
338	52892	13	52905	12	52917	13	52930	13	52943	13	52956	13	52969	13	52982	12	52994	13	53007	13
339	53020	13	53033	13	53046	12	53058	13	53071	13	53084	13	53097	13	53110	12	53122	13	53135	13
340	53148	13	53161	12	53173	13	53186	13	53199	13	53212	12	53224	13	53237	13	53250	13	53263	12
341	53275	13	53288	13	53301	13	53314	12	53326	13	53339	13	53352	12	53364	13	53377	13	53390	13
342	53403	12	53415	13	53428	13	53441	12	53453	13	53466	13	53479	12	53491	13	53504	13	53517	12
343	53529	13	53542	13	53555	12	53567	13	53580	13	53593	12	53605	13	53618	13	53631	12	53643	13
344	53656	12	53668	13	53681	13	53694	12	53706	13	53719	13	53732	12	53744	13	53757	12	53769	13
345	53782	12	53794	13	53807	13	53820	12	53832	13	53845	12	53857	13	53870	12	53882	13	53895	13
346	53908	12	53920	13	53933	12	53945	13	53958	12	53970	13	53983	12	53995	13	54008	12	54020	13
347	54033	12	54045	13	54058	12	54070	13	54083	12	54095	13	54108	12	54120	13	54133	12	54145	13
348	54158	12	54170	13	54183	12	54195	13	54208	12	54220	13	54233	12	54245	13	54258	12	54270	13
349	54283	12	54295	12	54307	13	54320	12	54332	13	54345	12	54357	13	54370	12	54382	12	54394	13
350	54407	12	54419	13	54432	12	54444	12	54456	13	54469	12	54481	13	54494	12	54506	12	54518	13
No.	0	d	1	d	2	d	3	d	4	d	5	d	6	d	7	d	8	d	9	d

Prop. parts

	15
1	2
2	3
3	4
4	6
5	8
6	9
7	10
8	12
9	14

	14
1	1
2	3
3	4
4	6
5	7
6	8
7	10
8	11
9	13

	13
1	1
2	3
3	4
4	5
5	6
6	8
7	9
8	10
9	12

	12
1	1
2	2
3	4
4	5
5	6
6	7
7	8
8	10
9	11

TABLE 32

Logarithms of Numbers

3500–4000

No.	0	d	1	d	2	d	3	d	4	d	5	d	6	d	7	d	8	d	9	d
350	54407	12	54419	13	54432	12	54444	12	54456	13	54469	12	54481	13	54494	12	54506	12	54518	13
351	54531	12	54543	12	54555	13	54568	12	54580	13	54593	12	54605	12	54617	13	54630	12	54642	12
352	54654	13	54667	12	54679	12	54691	13	54704	12	54716	12	54728	13	54741	12	54753	12	54765	12
353	54777	13	54790	12	54802	12	54814	13	54827	12	54839	12	54851	13	54864	12	54876	12	54888	12
354	54900	13	54913	12	54925	12	54937	12	54949	13	54962	12	54974	12	54986	12	54998	13	55011	12
355	55023	12	55035	12	55047	13	55060	12	55072	12	55084	12	55096	12	55108	13	55121	12	55133	12
356	55145	12	55157	12	55169	13	55182	12	55194	12	55206	12	55218	12	55230	12	55242	13	55255	12
357	55267	12	55279	12	55291	12	55303	12	55315	13	55328	12	55340	12	55352	12	55364	12	55376	12
358	55388	12	55400	13	55413	12	55425	12	55437	12	55449	12	55461	12	55473	12	55485	12	55497	12
359	55509	13	55522	12	55534	12	55546	12	55558	12	55570	12	55582	12	55594	12	55606	12	55618	12
360	55630	12	55642	12	55654	12	55666	12	55678	13	55691	12	55703	12	55715	12	55727	12	55739	12
361	55751	12	55763	12	55775	12	55787	12	55799	12	55811	12	55823	12	55835	12	55847	12	55859	12
362	55871	12	55883	12	55895	12	55907	12	55919	12	55931	12	55943	12	55955	12	55967	12	55979	12
363	55991	12	56003	12	56015	12	56027	11	56038	12	56050	12	56062	12	56074	12	56086	12	56098	12
364	56110	12	56122	12	56134	12	56146	12	56158	12	56170	12	56182	12	56194	11	56205	12	56217	12
365	56229	12	56241	12	56253	12	56265	12	56277	12	56289	12	56301	11	56312	12	56324	12	56336	12
366	56348	12	56360	12	56372	12	56384	12	56396	11	56407	12	56419	12	56431	12	56443	12	56455	12
367	56467	11	56478	12	56490	12	56502	12	56514	12	56526	12	56538	11	56549	12	56561	12	56573	12
368	56585	12	56597	11	56608	12	56620	12	56632	12	56644	12	56656	11	56667	12	56679	12	56691	12
369	56703	11	56714	12	56726	12	56738	12	56750	11	56761	12	56773	12	56785	12	56797	11	56808	12
370	56820	12	56832	12	56844	11	56855	12	56867	12	56879	12	56891	11	56902	12	56914	12	56926	11
371	56937	12	56949	12	56961	11	56972	12	56984	12	56996	12	57008	11	57019	12	57031	12	57043	11
372	57054	12	57066	12	57078	11	57089	12	57101	12	57113	11	57124	12	57136	12	57148	11	57159	12
373	57171	12	57183	11	57194	12	57206	11	57217	12	57229	12	57241	11	57252	12	57264	12	57276	11
374	57287	12	57299	11	57310	12	57322	12	57334	11	57345	12	57357	11	57368	12	57380	12	57392	11
375	57403	12	57415	11	57426	12	57438	11	57449	12	57461	12	57473	11	57484	12	57496	11	57507	12
376	57519	11	57530	12	57542	11	57553	12	57565	11	57576	12	57588	12	57600	11	57611	12	57623	11
377	57634	12	57646	11	57657	12	57669	11	57680	12	57692	11	57703	12	57715	11	57726	12	57738	11
378	57749	12	57761	11	57772	12	57784	11	57795	12	57807	11	57818	12	57830	11	57841	11	57852	12
379	57864	11	57875	12	57887	11	57898	12	57910	11	57921	12	57933	11	57944	11	57955	12	57967	11
380	57978	12	57990	11	58001	12	58013	11	58024	11	58035	12	58047	11	58058	12	58070	11	58081	11
381	58092	12	58104	11	58115	12	58127	11	58138	11	58149	12	58161	11	58172	12	58184	11	58195	11
382	58206	12	58218	11	58229	11	58240	12	58252	11	58263	11	58274	12	58286	11	58297	12	58309	11
383	58320	11	58331	12	58343	11	58354	11	58365	12	58377	11	58388	11	58399	11	58410	12	58422	11
384	58433	11	58444	12	58456	11	58467	11	58478	12	58490	11	58501	11	58512	12	58524	11	58535	11
385	58546	11	58557	12	58569	11	58580	11	58591	11	58602	12	58614	11	58625	11	58636	11	58647	12
386	58659	11	58670	11	58681	11	58692	12	58704	11	58715	11	58726	11	58737	12	58749	11	58760	11
387	58771	11	58782	12	58794	11	58805	11	58816	11	58827	11	58838	12	58850	11	58861	11	58872	11
388	58883	11	58894	12	58906	11	58917	11	58928	11	58939	11	58950	11	58961	12	58973	11	58984	11
389	58995	11	59006	11	59017	11	59028	12	59040	11	59051	11	59062	11	59073	11	59084	11	59095	11
390	59106	12	59118	11	59129	11	59140	11	59151	11	59162	11	59173	11	59184	11	59195	12	59207	11
391	59218	11	59229	11	59240	11	59251	11	59262	11	59273	11	59284	11	59295	11	59306	12	59318	11
392	59329	11	59340	11	59351	11	59362	11	59373	11	59384	11	59395	11	59406	11	59417	11	59428	11
393	59439	11	59450	11	59461	11	59472	11	59483	11	59494	12	59506	11	59517	11	59528	11	59539	11
394	59550	11	59561	11	59572	11	59583	11	59594	11	59605	11	59616	11	59627	11	59638	11	59649	11
395	59660	11	59671	11	59682	11	59693	11	59704	11	59715	11	59726	11	59737	11	59748	11	59759	11
396	59770	10	59780	11	59791	11	59802	11	59813	11	59824	11	59835	11	59846	11	59857	11	59868	11
397	59879	11	59890	11	59901	11	59912	11	59923	11	59934	11	59945	11	59956	10	59966	11	59977	11
398	59988	11	59999	11	60010	11	60021	11	60032	11	60043	11	60054	11	60065	11	60076	10	60086	11
399	60097	11	60108	11	60119	11	60130	11	60141	11	60152	11	60163	10	60173	11	60184	11	60195	11
400	60206	11	60217	11	60228	11	60239	10	60249	11	60260	11	60271	11	60282	11	60293	11	60304	10
No.	0	d	1	d	2	d	3	d	4	d	5	d	6	d	7	d	8	d	9	d

Prop. parts

	13
1	1
2	3
3	4
4	5
5	6
6	8
7	9
8	10
9	12

	12
1	1
2	2
3	4
4	5
5	6
6	7
7	8
8	10
9	11

	11
1	1
2	2
3	3
4	4
5	6
6	7
7	8
8	9
9	10

	10
1	1
2	2
3	3
4	4
5	5
6	6
7	7
8	8
9	9

TABLE 32

Logarithms of Numbers

4000–4500

No.	0	d	1	d	2	d	3	d	4	d	5	d	6	d	7	d	8	d	9	d
400	60206	11	60217	11	60228	11	60239	10	60249	11	60260	11	60271	11	60282	11	60293	11	60304	10
401	60314	11	60325	11	60336	11	60347	11	60358	11	60369	10	60379	11	60390	11	60401	11	60412	11
402	60423	10	60433	11	60444	11	60455	11	60466	11	60477	10	60487	11	60498	11	60509	11	60520	11
403	60531	10	60541	11	60552	11	60563	11	60574	10	60584	11	60595	11	60606	11	60617	10	60627	11
404	60638	11	60649	11	60660	10	60670	11	60681	11	60692	11	60703	10	60713	11	60724	11	60735	11
405	60746	10	60756	11	60767	11	60778	10	60788	11	60799	11	60810	11	60821	10	60831	11	60842	11
406	60853	10	60863	11	60874	11	60885	10	60895	11	60906	11	60917	10	60927	11	60938	11	60949	10
407	60959	11	60970	11	60981	10	60991	11	61002	11	61013	10	61023	11	61034	11	61045	10	61055	11
408	61066	11	61077	10	61087	11	61098	11	61109	10	61119	11	61130	10	61140	11	61151	11	61162	10
409	61172	11	61183	11	61194	10	61204	11	61215	10	61225	11	61236	11	61247	10	61257	11	61268	10
410	61278	11	61289	11	61300	10	61310	11	61321	10	61331	11	61342	10	61352	11	61363	11	61374	10
411	61384	11	61395	10	61405	11	61416	10	61426	11	61437	11	61448	10	61458	11	61469	10	61479	11
412	61490	10	61500	11	61511	10	61521	11	61532	10	61542	11	61553	10	61563	11	61574	10	61584	11
413	61595	11	61606	10	61616	11	61627	10	61637	11	61648	10	61658	11	61669	10	61679	11	61690	10
414	61700	11	61711	10	61721	10	61731	11	61742	10	61752	11	61763	10	61773	11	61784	10	61794	11
415	61805	10	61815	11	61826	10	61836	11	61847	10	61857	11	61868	10	61878	10	61888	11	61899	10
416	61909	11	61920	10	61930	11	61941	10	61951	11	61962	10	61972	10	61982	11	61993	10	62003	11
417	62014	10	62024	10	62034	11	62045	10	62055	11	62066	10	62076	10	62086	11	62097	10	62107	11
418	62118	10	62128	10	62138	11	62149	10	62159	11	62170	10	62180	10	62190	11	62201	10	62211	10
419	62221	11	62232	10	62242	10	62252	11	62263	10	62273	11	62284	10	62294	10	62304	11	62315	10
420	62325	10	62335	11	62346	10	62356	10	62366	11	62377	10	62387	10	62397	11	62408	10	62418	10
421	62428	11	62439	10	62449	10	62459	10	62469	11	62480	10	62490	10	62500	11	62511	10	62521	10
422	62531	11	62542	10	62552	10	62562	10	62572	11	62583	10	62593	10	62603	10	62613	11	62624	10
423	62634	10	62644	11	62655	10	62665	10	62675	10	62685	11	62696	10	62706	10	62716	10	62726	11
424	62737	10	62747	10	62757	10	62767	11	62778	10	62788	10	62798	10	62808	10	62818	11	62829	10
425	62839	10	62849	10	62859	11	62870	10	62880	10	62890	10	62900	10	62910	11	62921	10	62931	10
426	62941	10	62951	10	62961	11	62972	10	62982	10	62992	10	63002	10	63012	10	63022	11	63033	10
427	63043	10	63053	10	63063	10	63073	10	63083	11	63094	10	63104	10	63114	10	63124	10	63134	10
428	63144	11	63155	10	63165	10	63175	10	63185	10	63195	10	63205	10	63215	10	63225	11	63236	10
429	63246	10	63256	10	63266	10	63276	10	63286	10	63296	10	63306	11	63317	10	63327	10	63337	10
430	63347	10	63357	10	63367	10	63377	10	63387	10	63397	10	63407	10	63417	11	63428	10	63438	10
431	63448	10	63458	10	63468	10	63478	10	63488	10	63498	10	63508	10	63518	10	63528	10	63538	10
432	63548	10	63558	10	63568	11	63579	10	63589	10	63599	10	63609	10	63619	10	63629	10	63639	10
433	63649	10	63659	10	63669	10	63679	10	63689	10	63699	10	63709	10	63719	10	63729	10	63739	10
434	63749	10	63759	10	63769	10	63779	10	63789	10	63799	10	63809	10	63819	10	63829	10	63839	10
435	63849	10	63859	10	63869	10	63879	10	63889	10	63899	10	63909	10	63919	10	63929	10	63939	10
436	63949	10	63959	10	63969	10	63979	9	63988	10	63998	10	64008	10	64018	10	64028	10	64038	10
437	64048	10	64058	10	64068	10	64078	10	64088	10	64098	10	64108	10	64118	10	64128	9	64137	10
438	64147	10	64157	10	64167	10	64177	10	64187	10	64197	10	64207	10	64217	10	64227	10	64237	9
439	64246	10	64256	10	64266	10	64276	10	64286	10	64296	10	64306	10	64316	10	64326	9	64335	10
440	64345	10	64355	10	64365	10	64375	10	64385	10	64395	9	64404	10	64414	10	64424	10	64434	10
441	64444	10	64454	10	64464	9	64473	10	64483	10	64493	10	64503	10	64513	10	64523	9	64532	10
442	64542	10	64552	10	64562	10	64572	10	64582	9	64591	10	64601	10	64611	10	64621	10	64631	9
443	64640	10	64650	10	64660	10	64670	10	64680	9	64689	10	64699	10	64709	10	64719	10	64729	9
444	64738	10	64748	10	64758	10	64768	9	64777	10	64787	10	64797	10	64807	9	64816	10	64826	10
445	64836	10	64846	10	64856	9	64865	10	64875	10	64885	10	64895	9	64904	10	64914	10	64924	9
446	64933	10	64943	10	64953	10	64963	9	64972	10	64982	10	64992	10	65002	9	65011	10	65021	10
447	65031	9	65040	10	65050	10	65060	10	65070	9	65079	10	65089	10	65099	9	65108	10	65118	10
448	65128	9	65137	10	65147	10	65157	10	65167	9	65176	10	65186	10	65196	9	65205	10	65215	10
449	65225	9	65234	10	65244	10	65254	9	65263	10	65273	10	65283	9	65292	10	65302	10	65312	9
450	65321	10	65331	10	65341	9	65350	10	65360	9	65369	10	65379	10	65389	9	65398	10	65408	10
No.	0	d	1	d	2	d	3	d	4	d	5	d	6	d	7	d	8	d	9	d

Prop. parts

	11
1	1
2	2
3	3
4	4
5	6
6	7
7	8
8	9
9	10

	10
1	1
2	2
3	3
4	4
5	5
6	6
7	7
8	8
9	9

	9
1	1
2	2
3	3
4	4
5	4
6	5
7	6
8	7
9	8

TABLE 32

Logarithms of Numbers

4500–5000

No.	0	d	1	d	2	d	3	d	4	d	5	d	6	d	7	d	8	d	9	d
450	65321	10	65331	10	65341	9	65350	10	65360	9	65369	10	65379	10	65389	9	65398	10	65408	10
451	65418	9	65427	10	65437	10	65447	9	65456	10	65466	9	65475	10	65485	10	65495	9	65504	10
452	65514	9	65523	10	65533	10	65543	9	65552	10	65562	9	65571	10	65581	10	65591	9	65600	10
453	65610	9	65619	10	65629	10	65639	9	65648	10	65658	9	65667	10	65677	9	65686	10	65696	10
454	65706	9	65715	10	65725	9	65734	10	65744	9	65753	10	65763	9	65772	10	65782	10	65792	9
455	65801	10	65811	9	65820	10	65830	9	65839	10	65849	9	65858	10	65868	9	65877	10	65887	9
456	65896	10	65906	10	65916	9	65925	10	65935	9	65944	10	65954	9	65963	10	65973	9	65982	10
457	65992	9	66001	10	66011	9	66020	10	66030	9	66039	10	66049	9	66058	10	66068	9	66077	10
458	66087	9	66096	10	66106	9	66115	9	66124	10	66134	9	66143	10	66153	9	66162	10	66172	9
459	66181	10	66191	9	66200	10	66210	9	66219	10	66229	9	66238	9	66247	10	66257	9	66266	10
460	66276	9	66285	10	66295	9	66304	10	66314	9	66323	9	66332	10	66342	9	66351	10	66361	9
461	66370	10	66380	9	66389	9	66398	10	66408	9	66417	10	66427	9	66436	9	66445	10	66455	9
462	66464	10	66474	9	66483	9	66492	10	66502	9	66511	10	66521	9	66530	9	66539	10	66549	9
463	66558	9	66567	10	66577	9	66586	10	66596	9	66605	9	66614	10	66624	9	66633	9	66642	10
464	66652	9	66661	10	66671	9	66680	9	66689	10	66699	9	66708	9	66717	10	66727	9	66736	9
465	66745	10	66755	9	66764	9	66773	10	66783	9	66792	9	66801	10	66811	9	66820	9	66829	10
466	66839	9	66848	9	66857	10	66867	9	66876	9	66885	9	66894	10	66904	9	66913	9	66922	10
467	66932	9	66941	9	66950	10	66960	9	66969	9	66978	9	66987	10	66997	9	67006	9	67015	10
468	67025	9	67034	9	67043	9	67052	10	67062	9	67071	9	67080	9	67089	10	67099	9	67108	9
469	67117	10	67127	9	67136	9	67145	9	67154	10	67164	9	67173	9	67182	9	67191	10	67201	9
470	67210	9	67219	9	67228	9	67237	10	67247	9	67256	9	67265	9	67274	10	67284	9	67293	9
471	67302	9	67311	10	67321	9	67330	9	67339	9	67348	9	67357	10	67367	9	67376	9	67385	9
472	67394	9	67403	10	67413	9	67422	9	67431	9	67440	9	67449	10	67459	9	67468	9	67477	9
473	67486	9	67495	9	67504	10	67514	9	67523	9	67532	9	67541	9	67550	10	67560	9	67569	9
474	67578	9	67587	9	67596	9	67605	9	67614	10	67624	9	67633	9	67642	9	67651	9	67660	9
475	67669	10	67679	9	67688	9	67697	9	67706	9	67715	9	67724	9	67733	9	67742	10	67752	9
476	67761	9	67770	9	67779	9	67788	9	67797	9	67806	9	67815	10	67825	9	67834	9	67843	9
477	67852	9	67861	9	67870	9	67879	9	67888	9	67897	9	67906	10	67916	9	67925	9	67934	9
478	67943	9	67952	9	67961	9	67970	9	67979	9	67988	9	67997	9	68006	9	68015	9	68024	10
479	68034	9	68043	9	68052	9	68061	9	68070	9	68079	9	68088	9	68097	9	68106	9	68115	9
480	68124	9	68133	9	68142	9	68151	9	68160	9	68169	9	68178	9	68187	9	68196	9	68205	10
481	68215	9	68224	9	68233	9	68242	9	68251	9	68260	9	68269	9	68278	9	68287	9	68296	9
482	68305	9	68314	9	68323	9	68332	9	68341	9	68350	9	68359	9	68368	9	68377	9	68386	9
483	68395	9	68404	9	68413	9	68422	9	68431	9	68440	9	68449	9	68458	9	68467	9	68476	9
484	68485	9	68494	8	68502	9	68511	9	68520	9	68529	9	68538	9	68547	9	68556	9	68565	9
485	68574	9	68583	9	68592	9	68601	9	68610	9	68619	9	68628	9	68637	9	68646	9	68655	9
486	68664	9	68673	8	68681	9	68690	9	68699	9	68708	9	68717	9	68726	9	68735	9	68744	9
487	68753	9	68762	9	68771	9	68780	9	68789	8	68797	9	68806	9	68815	9	68824	9	68833	9
488	68842	9	68851	9	68860	9	68869	9	68878	8	68886	9	68895	9	68904	9	68913	9	68922	9
489	68931	9	68940	9	68949	9	68958	8	68966	9	68975	9	68984	9	68993	9	69002	9	69011	9
490	69020	8	69028	9	69037	9	69046	9	69055	9	69064	9	69073	9	69082	8	69090	9	69099	9
491	69108	9	69117	9	69126	9	69135	9	69144	8	69152	9	69161	9	69170	9	69179	9	69188	9
492	69197	8	69205	9	69214	9	69223	9	69232	9	69241	8	69249	9	69258	9	69267	9	69276	9
493	69285	9	69294	8	69302	9	69311	9	69320	9	69329	9	69338	8	69346	9	69355	9	69364	9
494	69373	8	69381	9	69390	9	69399	9	69408	9	69417	8	69425	9	69434	9	69443	9	69452	9
495	69461	8	69469	9	69478	9	69487	9	69496	8	69504	9	69513	9	69522	9	69531	8	69539	9
496	69548	9	69557	9	69566	8	69574	9	69583	9	69592	9	69601	8	69609	9	69618	9	69627	9
497	69636	8	69644	9	69653	9	69662	9	69671	8	69679	9	69688	9	69697	8	69705	9	69714	9
498	69723	9	69732	8	69740	9	69749	9	69758	9	69767	8	69775	9	69784	9	69793	8	69801	9
499	69810	9	69819	8	69827	9	69836	9	69845	9	69854	8	69862	9	69871	9	69880	8	69888	9
500	69897	9	69906	8	69914	9	69923	9	69932	8	69940	9	69949	9	69958	8	69966	9	69975	9
No.	0	d	1	d	2	d	3	d	4	d	5	d	6	d	7	d	8	d	9	d

Prop. parts

	10	9	8
1	1	1	1
2	2	2	2
3	3	3	2
4	4	4	3
5	5	4	4
6	6	5	5
7	7	6	6
8	8	7	6
9	9	8	7

TABLE 32

Logarithms of Numbers

5000–5500

No.	0	d	1	d	2	d	3	d	4	d	5	d	6	d	7	d	8	d	9	d
500	69897	9	69906	8	69914	9	69923	9	69932	8	69940	9	69949	9	69958	8	69966	9	69975	9
501	69984	8	69992	9	70001	9	70010	8	70018	9	70027	9	70036	8	70044	9	70053	9	70062	8
502	70070	9	70079	9	70088	8	70096	9	70105	9	70114	8	70122	9	70131	9	70140	8	70148	9
503	70157	8	70165	9	70174	9	70183	8	70191	9	70200	9	70209	8	70217	9	70226	8	70234	9
504	70243	9	70252	8	70260	9	70269	9	70278	8	70286	9	70295	8	70303	9	70312	9	70321	8
505	70329	9	70338	8	70346	9	70355	9	70364	8	70372	9	70381	8	70389	9	70398	8	70406	9
506	70415	9	70424	8	70432	9	70441	8	70449	9	70458	9	70467	8	70475	9	70484	8	70492	9
507	70501	8	70509	9	70518	8	70526	9	70535	9	70544	8	70552	9	70561	8	70569	9	70578	8
508	70586	9	70595	8	70603	9	70612	9	70621	8	70629	9	70638	8	70646	9	70655	9	70663	9
509	70672	8	70680	9	70689	8	70697	9	70706	8	70714	9	70723	8	70731	9	70740	9	70749	8
510	70757	9	70766	8	70774	9	70783	8	70791	9	70800	8	70808	9	70817	8	70825	9	70834	8
511	70842	9	70851	8	70859	9	70868	8	70876	9	70885	8	70893	9	70902	8	70910	9	70919	8
512	70927	8	70935	9	70944	8	70952	9	70961	8	70969	9	70978	8	70986	9	70995	8	71003	9
513	71012	8	71020	9	71029	8	71037	9	71046	8	71054	9	71063	8	71071	8	71079	9	71088	8
514	71096	9	71105	8	71113	9	71122	8	71130	9	71139	8	71147	8	71155	9	71164	8	71172	9
515	71181	8	71189	9	71198	8	71206	8	71214	9	71223	8	71231	9	71240	8	71248	9	71257	8
516	71265	8	71273	9	71282	8	71290	9	71299	8	71307	8	71315	9	71324	8	71332	9	71341	8
517	71349	8	71357	9	71366	8	71374	9	71383	8	71391	8	71399	9	71408	8	71416	9	71425	8
518	71433	8	71441	9	71450	8	71458	8	71466	9	71475	8	71483	9	71492	8	71500	8	71508	9
519	71517	8	71525	8	71533	9	71542	8	71550	9	71559	8	71567	8	71575	9	71584	8	71592	8
520	71600	9	71609	8	71617	8	71625	9	71634	8	71642	8	71650	9	71659	8	71667	8	71675	9
521	71684	8	71692	8	71700	9	71709	8	71717	8	71725	9	71734	8	71742	8	71750	9	71759	8
522	71767	8	71775	9	71784	8	71792	8	71800	9	71809	8	71817	8	71825	9	71834	8	71842	8
523	71850	8	71858	9	71867	8	71875	8	71883	9	71892	8	71900	8	71908	9	71917	8	71925	8
524	71933	8	71941	9	71950	8	71958	8	71966	9	71975	8	71983	8	71991	8	71999	9	72008	8
525	72016	8	72024	8	72032	9	72041	8	72049	8	72057	9	72066	8	72074	8	72082	8	72090	9
526	72099	8	72107	8	72115	8	72123	9	72132	8	72140	8	72148	8	72156	9	72165	8	72173	8
527	72181	8	72189	9	72198	8	72206	8	72214	8	72222	8	72230	9	72239	8	72247	8	72255	8
528	72263	9	72272	8	72280	8	72288	8	72296	8	72304	9	72313	8	72321	8	72329	8	72337	9
529	72346	8	72354	8	72362	8	72370	8	72378	9	72387	8	72395	8	72403	8	72411	8	72419	9
530	72428	8	72436	8	72444	8	72452	8	72460	9	72469	8	72477	8	72485	8	72493	8	72501	8
531	72509	9	72518	8	72526	8	72534	8	72542	8	72550	8	72558	9	72567	8	72575	8	72583	8
532	72591	8	72599	8	72607	9	72616	8	72624	8	72632	8	72640	8	72648	8	72656	9	72665	8
533	72673	8	72681	8	72689	8	72697	8	72705	8	72713	9	72722	8	72730	8	72738	8	72746	8
534	72754	8	72762	8	72770	9	72779	8	72787	8	72795	8	72803	8	72811	8	72819	8	72827	8
535	72835	8	72843	9	72852	8	72860	8	72868	8	72876	8	72884	8	72892	8	72900	8	72908	8
536	72916	9	72925	8	72933	8	72941	8	72949	8	72957	8	72965	8	72973	8	72981	8	72989	8
537	72997	9	73006	8	73014	8	73022	8	73030	8	73038	8	73046	8	73054	8	73062	8	73070	8
538	73078	8	73086	8	73094	8	73102	9	73111	8	73119	8	73127	8	73135	8	73143	8	73151	8
539	73159	8	73167	8	73175	8	73183	8	73191	8	73199	8	73207	8	73215	8	73223	8	73231	8
540	73239	8	73247	8	73255	8	73263	9	73272	8	73280	8	73288	8	73296	8	73304	8	73312	8
541	73320	8	73328	8	73336	8	73344	8	73352	8	73360	8	73368	8	73376	8	73384	8	73392	8
542	73400	8	73408	8	73416	8	73424	8	73432	8	73440	8	73448	8	73456	8	73464	8	73472	8
543	73480	8	73488	8	73496	8	73504	8	73512	8	73520	8	73528	8	73536	8	73544	8	73552	8
544	73560	8	73568	8	73576	8	73584	8	73592	8	73600	8	73608	8	73616	8	73624	8	73632	8
545	73640	8	73648	8	73656	8	73664	8	73672	7	73679	8	73687	8	73695	8	73703	8	73711	8
546	73719	8	73727	8	73735	8	73743	8	73751	8	73759	8	73767	8	73775	8	73783	8	73791	8
547	73799	8	73807	8	73815	8	73823	7	73830	8	73838	8	73846	8	73854	8	73862	8	73870	8
548	73878	8	73886	8	73894	8	73902	8	73910	8	73918	8	73926	7	73933	8	73941	8	73949	8
549	73957	8	73965	8	73973	8	73981	8	73989	8	73997	8	74005	8	74013	7	74020	8	74028	8
550	74036	8	74044	8	74052	8	74060	8	74068	8	74076	8	74084	8	74092	7	74099	8	74107	8
No.	0	d	1	d	2	d	3	d	4	d	5	d	6	d	7	d	8	d	9	d

Prop. parts

	9
1	1
2	2
3	3
4	4
5	4
6	5
7	6
8	7
9	8

	8
1	1
2	2
3	2
4	3
5	4
6	5
7	6
8	6
9	7

	7
1	1
2	1
3	2
4	3
5	4
6	4
7	5
8	6
9	6

TABLE 32

Logarithms of Numbers

5500–6000

No.	0	d	1	d	2	d	3	d	4	d	5	d	6	d	7	d	8	d	9	d
550	74036	8	74044	8	74052	8	74060	8	74068	8	74076	8	74084	8	74092	7	74099	8	74107	8
551	74115	8	74123	8	74131	8	74139	8	74147	8	74155	7	74162	8	74170	8	74178	8	74186	8
552	74194	8	74202	8	74210	8	74218	7	74225	8	74233	8	74241	8	74249	8	74257	8	74265	8
553	74273	7	74280	8	74288	8	74296	8	74304	8	74312	8	74320	7	74327	8	74335	8	74343	8
554	74351	8	74359	8	74367	7	74374	8	74382	8	74390	8	74398	8	74406	8	74414	7	74421	8
555	74429	8	74437	8	74445	8	74453	8	74461	7	74468	8	74476	8	74484	8	74492	8	74500	7
556	74507	8	74515	8	74523	8	74531	8	74539	8	74547	7	74554	8	74562	8	74570	8	74578	8
557	74586	7	74593	8	74601	8	74609	8	74617	7	74624	8	74632	8	74640	8	74648	8	74656	7
558	74663	8	74671	8	74679	8	74687	8	74695	7	74702	8	74710	8	74718	8	74726	7	74733	8
559	74741	8	74749	8	74757	7	74764	8	74772	8	74780	8	74788	8	74796	7	74803	8	74811	8
560	74819	8	74827	7	74834	8	74842	8	74850	8	74858	7	74865	8	74873	8	74881	8	74889	7
561	74896	8	74904	8	74912	8	74920	7	74927	8	74935	8	74943	7	74950	8	74958	8	74966	8
562	74974	7	74981	8	74989	8	74997	8	75005	7	75012	8	75020	8	75028	7	75035	8	75043	8
563	75051	8	75059	7	75066	8	75074	8	75082	7	75089	8	75097	8	75105	8	75113	7	75120	8
564	75128	8	75136	7	75143	8	75151	8	75159	7	75166	8	75174	8	75182	7	75189	8	75197	8
565	75205	8	75213	7	75220	8	75228	8	75236	7	75243	8	75251	8	75259	7	75266	8	75274	8
566	75282	7	75289	8	75297	8	75305	7	75312	8	75320	8	75328	7	75335	8	75343	8	75351	7
567	75358	8	75366	8	75374	7	75381	8	75389	8	75397	7	75404	8	75412	8	75420	7	75427	8
568	75435	7	75442	8	75450	8	75458	7	75465	8	75473	8	75481	7	75488	8	75496	8	75504	7
569	75511	8	75519	7	75526	8	75534	8	75542	7	75549	8	75557	8	75565	7	75572	8	75580	7
570	75587	8	75595	8	75603	7	75610	8	75618	8	75626	7	75633	8	75641	7	75648	8	75656	8
571	75664	7	75671	8	75679	7	75686	8	75694	8	75702	7	75709	8	75717	7	75724	8	75732	8
572	75740	7	75747	8	75755	7	75762	8	75770	8	75778	7	75785	8	75793	7	75800	8	75808	7
573	75815	8	75823	8	75831	7	75838	8	75846	7	75853	8	75861	7	75868	8	75876	8	75884	7
574	75891	8	75899	7	75906	8	75914	7	75921	8	75929	8	75937	7	75944	8	75952	7	75959	8
575	75967	7	75974	8	75982	7	75989	8	75997	8	76005	7	76012	8	76020	7	76027	8	76035	7
576	76042	8	76050	7	76057	8	76065	7	76072	8	76080	7	76087	8	76095	8	76103	7	76110	8
577	76118	7	76125	8	76133	7	76140	8	76148	7	76155	8	76163	7	76170	8	76178	7	76185	8
578	76193	7	76200	8	76208	7	76215	8	76223	7	76230	8	76238	7	76245	8	76253	7	76260	8
579	76268	7	76275	8	76283	7	76290	8	76298	7	76305	8	76313	7	76320	8	76328	7	76335	8
580	76343	7	76350	8	76358	7	76365	8	76373	7	76380	8	76388	7	76395	8	76403	7	76410	8
581	76418	7	76425	8	76433	7	76440	8	76448	7	76455	7	76462	8	76470	7	76477	8	76485	7
582	76492	8	76500	7	76507	8	76515	7	76522	8	76530	7	76537	8	76545	7	76552	7	76559	8
583	76567	7	76574	8	76582	7	76589	8	76597	7	76604	8	76612	7	76619	7	76626	8	76634	7
584	76641	8	76649	7	76656	8	76664	7	76671	7	76678	8	76686	7	76693	8	76701	7	76708	8
585	76716	7	76723	7	76730	8	76738	7	76745	8	76753	7	76760	8	76768	7	76775	7	76782	8
586	76790	7	76797	8	76805	7	76812	7	76819	8	76827	7	76834	8	76842	7	76849	7	76856	8
587	76864	7	76871	8	76879	7	76886	7	76893	8	76901	7	76908	8	76916	7	76923	7	76930	8
588	76938	7	76945	8	76953	7	76960	7	76967	8	76975	7	76982	7	76989	8	76997	7	77004	8
589	77012	7	77019	7	77026	8	77034	7	77041	7	77048	8	77056	7	77063	7	77070	8	77078	7
590	77085	8	77093	7	77100	7	77107	8	77115	7	77122	7	77129	8	77137	7	77144	7	77151	8
591	77159	7	77166	7	77173	8	77181	7	77188	7	77195	8	77203	7	77210	7	77217	8	77225	7
592	77232	8	77240	7	77247	7	77254	8	77262	7	77269	7	77276	7	77283	8	77291	7	77298	7
593	77305	8	77313	7	77320	7	77327	8	77335	7	77342	7	77349	8	77357	7	77364	7	77371	8
594	77379	7	77386	7	77393	8	77401	7	77408	7	77415	7	77422	8	77430	7	77437	7	77444	8
595	77452	7	77459	7	77466	8	77474	7	77481	7	77488	7	77495	8	77503	7	77510	7	77517	8
596	77525	7	77532	7	77539	7	77546	8	77554	7	77561	7	77568	8	77576	7	77583	7	77590	7
597	77597	8	77605	7	77612	7	77619	8	77627	7	77634	7	77641	7	77648	8	77656	7	77663	7
598	77670	7	77677	8	77685	7	77692	7	77699	7	77706	8	77714	7	77721	7	77728	7	77735	8
599	77743	7	77750	7	77757	7	77764	8	77772	7	77779	7	77786	7	77793	8	77801	7	77808	7
600	77815	7	77822	8	77830	7	77837	7	77844	7	77851	8	77859	7	77866	7	77873	7	77880	7
No.	0	d	1	d	2	d	3	d	4	d	5	d	6	d	7	d	8	d	9	d

Prop. parts

	8
1	1
2	2
3	2
4	3
5	4
6	5
7	6
8	6
9	7

	7
1	1
2	1
3	2
4	3
5	4
6	4
7	5
8	6
9	6

TABLE 32

Logarithms of Numbers

6000–6500

No.	0	d	1	d	2	d	3	d	4	d	5	d	6	d	7	d	8	d	9	d
600	77815	7	77822	8	77830	7	77837	7	77844	7	77851	8	77859	7	77866	7	77873	7	77880	7
601	77887	8	77895	7	77902	7	77909	7	77916	8	77924	7	77931	7	77938	7	77945	7	77952	8
602	77960	7	77967	7	77974	7	77981	7	77988	8	77996	7	78003	7	78010	7	78017	8	78025	7
603	78032	7	78039	7	78046	7	78053	8	78061	7	78068	7	78075	7	78082	7	78089	8	78097	7
604	78104	7	78111	7	78118	7	78125	7	78132	8	78140	7	78147	7	78154	7	78161	7	78168	8
605	78176	7	78183	7	78190	7	78197	7	78204	7	78211	8	78219	7	78226	7	78233	7	78240	7
606	78247	7	78254	8	78262	7	78269	7	78276	7	78283	7	78290	7	78297	8	78305	7	78312	7
607	78319	7	78326	7	78333	7	78340	7	78347	8	78355	7	78362	7	78369	7	78376	7	78383	7
608	78390	8	78398	7	78405	7	78412	7	78419	7	78426	7	78433	7	78440	7	78447	8	78455	7
609	78462	7	78469	7	78476	7	78483	7	78490	7	78497	7	78504	8	78512	7	78519	7	78526	7
610	78533	7	78540	7	78547	7	78554	7	78561	8	78569	7	78576	7	78583	7	78590	7	78597	7
611	78604	7	78611	7	78618	7	78625	8	78633	7	78640	7	78647	7	78654	7	78661	7	78668	7
612	78675	7	78682	7	78689	7	78696	8	78704	7	78711	7	78718	7	78725	7	78732	7	78739	7
613	78746	7	78753	7	78760	7	78767	7	78774	7	78781	8	78789	7	78796	7	78803	7	78810	7
614	78817	7	78824	7	78831	7	78838	7	78845	7	78852	7	78859	7	78866	7	78873	7	78880	8
615	78888	7	78895	7	78902	7	78909	7	78916	7	78923	7	78930	7	78937	7	78944	7	78951	7
616	78958	7	78965	7	78972	7	78979	7	78986	7	78993	7	79000	7	79007	7	79014	7	79021	8
617	79029	7	79036	7	79043	7	79050	7	79057	7	79064	7	79071	7	79078	7	79085	7	79092	7
618	79099	7	79106	7	79113	7	79120	7	79127	7	79134	7	79141	7	79148	7	79155	7	79162	7
619	79169	7	79176	7	79183	7	79190	7	79197	7	79204	7	79211	7	79218	7	79225	7	79232	7
620	79239	7	79246	7	79253	7	79260	7	79267	7	79274	7	79281	7	79288	7	79295	7	79302	7
621	79309	7	79316	7	79323	7	79330	7	79337	7	79344	7	79351	7	79358	7	79365	7	79372	7
622	79379	7	79386	7	79393	7	79400	7	79407	7	79414	7	79421	7	79428	7	79435	7	79442	7
623	79449	7	79456	7	79463	7	79470	7	79477	7	79484	7	79491	7	79498	7	79505	6	79511	7
624	79518	7	79525	7	79532	7	79539	7	79546	7	79553	7	79560	7	79567	7	79574	7	79581	7
625	79588	7	79595	7	79602	7	79609	7	79616	7	79623	7	79630	7	79637	7	79644	6	79650	7
626	79657	7	79664	7	79671	7	79678	7	79685	7	79692	7	79699	7	79706	7	79713	7	79720	7
627	79727	7	79734	7	79741	7	79748	6	79754	7	79761	7	79768	7	79775	7	79782	7	79789	7
628	79796	7	79803	7	79810	7	79817	7	79824	7	79831	6	79837	7	79844	7	79851	7	79858	7
629	79865	7	79872	7	79879	7	79886	7	79893	7	79900	6	79906	7	79913	7	79920	7	79927	7
630	79934	7	79941	7	79948	7	79955	7	79962	7	79969	6	79975	7	79982	7	79989	7	79996	7
631	80003	7	80010	7	80017	7	80024	6	80030	7	80037	7	80044	7	80051	7	80058	7	80065	7
632	80072	7	80079	6	80085	7	80092	7	80099	7	80106	7	80113	7	80120	7	80127	7	80134	6
633	80140	7	80147	7	80154	7	80161	7	80168	7	80175	7	80182	6	80188	7	80195	7	80202	7
634	80209	7	80216	7	80223	6	80229	7	80236	7	80243	7	80250	7	80257	7	80264	7	80271	6
635	80277	7	80284	7	80291	7	80298	7	80305	7	80312	6	80318	7	80325	7	80332	7	80339	7
636	80346	7	80353	6	80359	7	80366	7	80373	7	80380	7	80387	6	80393	7	80400	7	80407	7
637	80414	7	80421	7	80428	6	80434	7	80441	7	80448	7	80455	7	80462	6	80468	7	80475	7
638	80482	7	80489	7	80496	6	80502	7	80509	7	80516	7	80523	7	80530	6	80536	7	80543	7
639	80550	7	80557	7	80564	6	80570	7	80577	7	80584	7	80591	7	80598	6	80604	7	80611	7
640	80618	7	80625	7	80632	6	80638	7	80645	7	80652	7	80659	6	80665	7	80672	7	80679	7
641	80686	7	80693	6	80699	7	80706	7	80713	7	80720	6	80726	7	80733	7	80740	7	80747	7
642	80754	6	80760	7	80767	7	80774	7	80781	6	80787	7	80794	7	80801	7	80808	6	80814	7
643	80821	7	80828	7	80835	6	80841	7	80848	7	80855	7	80862	6	80868	7	80875	7	80882	7
644	80889	6	80895	7	80902	7	80909	7	80916	6	80922	7	80929	7	80936	7	80943	6	80949	7
645	80956	7	80963	6	80969	7	80976	7	80983	7	80990	6	80996	7	81003	7	81010	7	81017	6
646	81023	7	81030	7	81037	6	81043	7	81050	7	81057	7	81064	6	81070	7	81077	7	81084	6
647	81090	7	81097	7	81104	7	81111	6	81117	7	81124	7	81131	6	81137	7	81144	7	81151	7
648	81158	6	81164	7	81171	7	81178	6	81184	7	81191	7	81198	6	81204	7	81211	7	81218	6
649	81224	7	81231	7	81238	7	81245	6	81251	7	81258	7	81265	6	81271	7	81278	7	81285	6
650	81291	7	81298	7	81305	6	81311	7	81318	7	81325	6	81331	7	81338	7	81345	6	81351	7
No.	0	d	1	d	2	d	3	d	4	d	5	d	6	d	7	d	8	d	9	d

Prop. parts

	8
1	1
2	2
3	2
4	3
5	4
6	5
7	6
8	6
9	7

	7
1	1
2	1
3	2
4	3
5	4
6	4
7	5
8	6
9	6

	6
1	1
2	1
3	2
4	2
5	3
6	4
7	4
8	5
9	5

TABLE 32

Logarithms of Numbers

6500–7000

No.	0	d	1	d	2	d	3	d	4	d	5	d	6	d	7	d	8	d	9	d
650	81291	7	81298	7	81305	6	81311	7	81318	7	81325	6	81331	7	81338	7	81345	6	81351	7
651	81358	7	81365	6	81371	7	81378	7	81385	6	81391	7	81398	7	81405	6	81411	7	81418	7
652	81425	6	81431	7	81438	7	81445	6	81451	7	81458	7	81465	6	81471	7	81478	7	81485	6
653	81491	7	81498	7	81505	6	81511	7	81518	7	81525	6	81531	7	81538	6	81544	7	81551	7
654	81558	6	81564	7	81571	7	81578	6	81584	7	81591	7	81598	6	81604	7	81611	6	81617	7
655	81624	7	81631	6	81637	7	81644	7	81651	6	81657	7	81664	7	81671	6	81677	7	81684	6
656	81690	7	81697	7	81704	6	81710	7	81717	6	81723	7	81730	7	81737	6	81743	7	81750	7
657	81757	6	81763	7	81770	6	81776	7	81783	7	81790	6	81796	7	81803	6	81809	7	81816	7
658	81823	6	81829	7	81836	6	81842	7	81849	7	81856	6	81862	7	81869	6	81875	7	81882	7
659	81889	6	81895	7	81902	6	81908	7	81915	6	81921	7	81928	7	81935	6	81941	7	81948	6
660	81954	7	81961	7	81968	6	81974	7	81981	6	81987	7	81994	6	82000	7	82007	7	82014	6
661	82020	7	82027	6	82033	7	82040	6	82046	7	82053	7	82060	6	82066	7	82073	6	82079	7
662	82086	6	82092	7	82099	6	82105	7	82112	7	82119	6	82125	7	82132	6	82138	7	82145	6
663	82151	7	82158	6	82164	7	82171	7	82178	6	82184	7	82191	6	82197	7	82204	6	82210	7
664	82217	6	82223	7	82230	6	82236	7	82243	6	82249	7	82256	7	82263	6	82269	7	82276	6
665	82282	7	82289	6	82295	7	82302	6	82308	7	82315	6	82321	7	82328	6	82334	7	82341	6
666	82347	7	82354	6	82360	7	82367	6	82373	7	82380	7	82387	6	82393	7	82400	6	82406	7
667	82413	6	82419	7	82426	6	82432	7	82439	6	82445	7	82452	6	82458	7	82465	6	82471	7
668	82478	6	82484	7	82491	6	82497	7	82504	6	82510	7	82517	6	82523	7	82530	6	82536	7
669	82543	6	82549	7	82556	6	82562	7	82569	6	82575	7	82582	6	82588	7	82595	6	82601	6
670	82607	7	82614	6	82620	7	82627	6	82633	7	82640	6	82646	7	82653	6	82659	7	82666	6
671	82672	7	82679	6	82685	7	82692	6	82698	7	82705	6	82711	7	82718	6	82724	6	82730	7
672	82737	6	82743	7	82750	6	82756	7	82763	6	82769	7	82776	6	82782	7	82789	6	82795	7
673	82802	6	82808	6	82814	7	82821	6	82827	7	82834	6	82840	7	82847	6	82853	7	82860	6
674	82866	6	82872	7	82879	6	82885	7	82892	6	82898	7	82905	6	82911	7	82918	6	82924	6
675	82930	7	82937	6	82943	7	82950	6	82956	7	82963	6	82969	6	82975	7	82982	6	82988	7
676	82995	6	83001	7	83008	6	83014	6	83020	7	83027	6	83033	7	83040	6	83046	6	83052	7
677	83059	6	83065	7	83072	6	83078	7	83085	6	83091	6	83097	7	83104	6	83110	7	83117	6
678	83123	6	83129	7	83136	6	83142	7	83149	6	83155	6	83161	7	83168	6	83174	7	83181	6
679	83187	6	83193	7	83200	6	83206	7	83213	6	83219	6	83225	7	83232	6	83238	7	83245	6
680	83251	6	83257	7	83264	6	83270	6	83276	7	83283	6	83289	7	83296	6	83302	6	83308	7
681	83315	6	83321	6	83327	7	83334	6	83340	7	83347	6	83353	6	83359	7	83366	6	83372	6
682	83378	7	83385	6	83391	7	83398	6	83404	6	83410	7	83417	6	83423	6	83429	7	83436	6
683	83442	6	83448	7	83455	6	83461	6	83467	7	83474	6	83480	7	83487	6	83493	6	83499	7
684	83506	6	83512	6	83518	7	83525	6	83531	6	83537	7	83544	6	83550	6	83556	7	83563	6
685	83569	6	83575	7	83582	6	83588	6	83594	7	83601	6	83607	6	83613	7	83620	6	83626	6
686	83632	7	83639	6	83645	6	83651	7	83658	6	83664	6	83670	7	83677	6	83683	6	83689	7
687	83696	6	83702	6	83708	7	83715	6	83721	6	83727	7	83734	6	83740	6	83746	7	83753	6
688	83759	6	83765	6	83771	7	83778	6	83784	6	83790	7	83797	6	83803	6	83809	7	83816	6
689	83822	6	83828	7	83835	6	83841	6	83847	6	83853	7	83860	6	83866	6	83872	7	83879	6
690	83885	6	83891	6	83897	7	83904	6	83910	6	83916	7	83923	6	83929	6	83935	7	83942	6
691	83948	6	83954	6	83960	7	83967	6	83973	6	83979	6	83985	7	83992	6	83998	6	84004	7
692	84011	6	84017	6	84023	6	84029	7	84036	6	84042	6	84048	7	84055	6	84061	6	84067	6
693	84073	7	84080	6	84086	6	84092	6	84098	7	84105	6	84111	6	84117	6	84123	7	84130	6
694	84136	6	84142	6	84148	7	84155	6	84161	6	84167	6	84173	7	84180	6	84186	6	84192	6
695	84198	7	84205	6	84211	6	84217	6	84223	7	84230	6	84236	6	84242	6	84248	7	84255	6
696	84261	6	84267	6	84273	7	84280	6	84286	6	84292	6	84298	7	84305	6	84311	6	84317	6
697	84323	7	84330	6	84336	6	84342	6	84348	6	84354	7	84361	6	84367	6	84373	6	84379	7
698	84386	6	84392	6	84398	6	84404	6	84410	7	84417	6	84423	6	84429	6	84435	7	84442	6
699	84448	6	84454	6	84460	6	84466	7	84473	6	84479	6	84485	6	84491	6	84497	7	84504	6
700	84510	6	84516	6	84522	6	84528	7	84535	6	84541	6	84547	6	84553	6	84559	7	84566	6
No.	0	d	1	d	2	d	3	d	4	d	5	d	6	d	7	d	8	d	9	d

Prop. parts

	7
1	1
2	1
3	2
4	3
5	4
6	4
7	5
8	6
9	6

	6
1	1
2	1
3	2
4	2
5	3
6	4
7	4
8	5
9	5

TABLE 32

Logarithms of Numbers

7000–7500

No.	0	d	1	d	2	d	3	d	4	d	5	d	6	d	7	d	8	d	9	d
700	84510	6	84516	6	84522	6	84528	7	84535	6	84541	6	84547	6	84553	6	84559	7	84566	6
701	84572	6	84578	6	84584	6	84590	7	84597	6	84603	6	84609	6	84615	6	84621	7	84628	6
702	84634	6	84640	6	84646	6	84652	6	84658	7	84665	6	84671	6	84677	6	84683	6	84689	7
703	84696	6	84702	6	84708	6	84714	6	84720	6	84726	7	84733	6	84739	6	84745	6	84751	6
704	84757	6	84763	7	84770	6	84776	6	84782	6	84788	6	84794	6	84800	7	84807	6	84813	6
705	84819	6	84825	6	84831	6	84837	7	84844	6	84850	6	84856	6	84862	6	84868	6	84874	6
706	84880	7	84887	6	84893	6	84899	6	84905	6	84911	6	84917	7	84924	6	84930	6	84936	6
707	84942	6	84948	6	84954	6	84960	7	84967	6	84973	6	84979	6	84985	6	84991	6	84997	6
708	85003	6	85009	7	85016	6	85022	6	85028	6	85034	6	85040	6	85046	6	85052	6	85058	7
709	85065	6	85071	6	85077	6	85083	6	85089	6	85095	6	85101	6	85107	7	85114	6	85120	6
710	85126	6	85132	6	85138	6	85144	6	85150	6	85156	7	85163	6	85169	6	85175	6	85181	6
711	85187	6	85193	6	85199	6	85205	6	85211	6	85217	7	85224	6	85230	6	85236	6	85242	6
712	85248	6	85254	6	85260	6	85266	6	85272	6	85278	7	85285	6	85291	6	85297	6	85303	6
713	85309	6	85315	6	85321	6	85327	6	85333	6	85339	6	85345	7	85352	6	85358	6	85364	6
714	85370	6	85376	6	85382	6	85388	6	85394	6	85400	6	85406	6	85412	6	85418	7	85425	6
715	85431	6	85437	6	85443	6	85449	6	85455	6	85461	6	85467	6	85473	6	85479	6	85485	6
716	85491	6	85497	6	85503	6	85509	7	85516	6	85522	6	85528	6	85534	6	85540	6	85546	6
717	85552	6	85558	6	85564	6	85570	6	85576	6	85582	6	85588	6	85594	6	85600	6	85606	6
718	85612	6	85618	7	85625	6	85631	6	85637	6	85643	6	85649	6	85655	6	85661	6	85667	6
719	85673	6	85679	6	85685	6	85691	6	85697	6	85703	6	85709	6	85715	6	85721	6	85727	6
720	85733	6	85739	6	85745	6	85751	6	85757	6	85763	6	85769	6	85775	6	85781	7	85788	6
721	85794	6	85800	6	85806	6	85812	6	85818	6	85824	6	85830	6	85836	6	85842	6	85848	6
722	85854	6	85860	6	85866	6	85872	6	85878	6	85884	6	85890	6	85896	6	85902	6	85908	6
723	85914	6	85920	6	85926	6	85932	6	85938	6	85944	6	85950	6	85956	6	85962	6	85968	6
724	85974	6	85980	6	85986	6	85992	6	85998	6	86004	6	86010	6	86016	6	86022	6	86028	6
725	86034	6	86040	6	86046	6	86052	6	86058	6	86064	6	86070	6	86076	6	86082	6	86088	6
726	86094	6	86100	6	86106	6	86112	6	86118	6	86124	6	86130	6	86136	5	86141	6	86147	6
727	86153	6	86159	6	86165	6	86171	6	86177	6	86183	6	86189	6	86195	6	86201	6	86207	6
728	86213	6	86219	6	86225	6	86231	6	86237	6	86243	6	86249	6	86255	6	86261	6	86267	6
729	86273	6	86279	6	86285	6	86291	6	86297	6	86303	5	86308	6	86314	6	86320	6	86326	6
730	86332	6	86338	6	86344	6	86350	6	86356	6	86362	6	86368	6	86374	6	86380	6	86386	6
731	86392	6	86398	6	86404	6	86410	5	86415	6	86421	6	86427	6	86433	6	86439	6	86445	6
732	86451	6	86457	6	86463	6	86469	6	86475	6	86481	6	86487	6	86493	6	86499	5	86504	6
733	86510	6	86516	6	86522	6	86528	6	86534	6	86540	6	86546	6	86552	6	86558	6	86564	6
734	86570	6	86576	5	86581	6	86587	6	86593	6	86599	6	86605	6	86611	6	86617	6	86623	6
735	86629	6	86635	6	86641	5	86646	6	86652	6	86658	6	86664	6	86670	6	86676	6	86682	6
736	86688	6	86694	6	86700	5	86705	6	86711	6	86717	6	86723	6	86729	6	86735	6	86741	6
737	86747	6	86753	6	86759	5	86764	6	86770	6	86776	6	86782	6	86788	6	86794	6	86800	6
738	86806	6	86812	5	86817	6	86823	6	86829	6	86835	6	86841	6	86847	6	86853	6	86859	5
739	86864	6	86870	6	86876	6	86882	6	86888	6	86894	6	86900	6	86906	5	86911	6	86917	6
740	86923	6	86929	6	86935	6	86941	6	86947	6	86953	5	86958	6	86964	6	86970	6	86976	6
741	86982	6	86988	6	86994	5	86999	6	87005	6	87011	6	87017	6	87023	6	87029	6	87035	5
742	87040	6	87046	6	87052	6	87058	6	87064	6	87070	5	87075	6	87081	6	87087	6	87093	6
743	87099	6	87105	6	87111	5	87116	6	87122	6	87128	6	87134	6	87140	6	87146	5	87151	6
744	87157	6	87163	6	87169	6	87175	6	87181	5	87186	6	87192	6	87198	6	87204	6	87210	6
745	87216	5	87221	6	87227	6	87233	6	87239	6	87245	6	87251	5	87256	6	87262	6	87268	6
746	87274	6	87280	6	87286	5	87291	6	87297	6	87303	6	87309	6	87315	5	87320	6	87326	6
747	87332	6	87338	6	87344	5	87349	6	87355	6	87361	6	87367	6	87373	6	87379	5	87384	6
748	87390	6	87396	6	87402	6	87408	5	87413	6	87419	6	87425	6	87431	6	87437	5	87442	6
749	87448	6	87454	6	87460	6	87466	5	87471	6	87477	6	87483	6	87489	6	87495	5	87500	6
750	87506	6	87512	6	87518	5	87523	6	87529	6	87535	6	87541	6	87547	5	87552	6	87558	6
No.	0	d	1	d	2	d	3	d	4	d	5	d	6	d	7	d	8	d	9	d

Prop. parts

	7
1	1
2	1
3	2
4	3
5	4
6	4
7	5
8	6
9	6

	6
1	1
2	1
3	2
4	2
5	3
6	4
7	4
8	5
9	5

	5
1	0
2	1
3	2
4	2
5	2
6	3
7	4
8	4
9	4

TABLE 32

Logarithms of Numbers

7500–8000

No.	0	d	1	d	2	d	3	d	4	d	5	d	6	d	7	d	8	d	9	d
750	87506	6	87512	6	87518	5	87523	6	87529	6	87535	6	87541	6	87547	5	87552	6	87558	6
751	87564	6	87570	6	87576	5	87581	6	87587	6	87593	6	87599	5	87604	6	87610	6	87616	6
752	87622	6	87628	5	87633	6	87639	6	87645	6	87651	5	87656	6	87662	6	87668	6	87674	5
753	87679	6	87685	6	87691	6	87697	6	87703	5	87708	6	87714	6	87720	6	87726	5	87731	6
754	87737	6	87743	6	87749	5	87754	6	87760	6	87766	6	87772	5	87777	6	87783	6	87789	6
755	87795	5	87800	6	87806	6	87812	6	87818	5	87823	6	87829	6	87835	6	87841	5	87846	6
756	87852	6	87858	6	87864	5	87869	6	87875	6	87881	6	87887	5	87892	6	87898	6	87904	6
757	87910	5	87915	6	87921	6	87927	6	87933	5	87938	6	87944	6	87950	5	87955	6	87961	6
758	87967	6	87973	5	87978	6	87984	6	87990	6	87996	5	88001	6	88007	6	88013	5	88018	6
759	88024	6	88030	6	88036	5	88041	6	88047	6	88053	5	88058	6	88064	6	88070	6	88076	5
760	88081	6	88087	6	88093	5	88098	6	88104	6	88110	6	88116	5	88121	6	88127	6	88133	5
761	88138	6	88144	6	88150	6	88156	5	88161	6	88167	6	88173	5	88178	6	88184	6	88190	5
762	88195	6	88201	6	88207	6	88213	5	88218	6	88224	6	88230	5	88235	6	88241	6	88247	5
763	88252	6	88258	6	88264	6	88270	5	88275	6	88281	6	88287	5	88292	6	88298	6	88304	5
764	88309	6	88315	6	88321	5	88326	6	88332	6	88338	5	88343	6	88349	6	88355	5	88360	6
765	88366	6	88372	5	88377	6	88383	6	88389	6	88395	5	88400	6	88406	6	88412	5	88417	6
766	88423	6	88429	5	88434	6	88440	6	88446	5	88451	6	88457	6	88463	5	88468	6	88474	6
767	88480	5	88485	6	88491	6	88497	5	88502	6	88508	5	88513	6	88519	6	88525	5	88530	6
768	88536	6	88542	5	88547	6	88553	6	88559	5	88564	6	88570	6	88576	5	88581	6	88587	6
769	88593	5	88598	6	88604	6	88610	5	88615	6	88621	6	88627	5	88632	6	88638	5	88643	6
770	88649	6	88655	5	88660	6	88666	6	88672	5	88677	6	88683	6	88689	5	88694	6	88700	5
771	88705	6	88711	6	88717	5	88722	6	88728	6	88734	5	88739	6	88745	5	88750	6	88756	6
772	88762	5	88767	6	88773	6	88779	5	88784	6	88790	5	88795	6	88801	6	88807	5	88812	6
773	88818	6	88824	5	88829	6	88835	5	88840	6	88846	6	88852	5	88857	6	88863	5	88868	6
774	88874	6	88880	5	88885	6	88891	6	88897	5	88902	6	88908	5	88913	6	88919	6	88925	5
775	88930	6	88936	5	88941	6	88947	6	88953	5	88958	6	88964	5	88969	6	88975	6	88981	5
776	88986	6	88992	5	88997	6	89003	6	89009	5	89014	6	89020	5	89025	6	89031	6	89037	5
777	89042	6	89048	5	89053	6	89059	5	89064	6	89070	6	89076	5	89081	6	89087	5	89092	6
778	89098	6	89104	5	89109	6	89115	5	89120	6	89126	5	89131	6	89137	6	89143	5	89148	6
779	89154	5	89159	6	89165	5	89170	6	89176	6	89182	5	89187	6	89193	5	89198	6	89204	5
780	89209	6	89215	6	89221	5	89226	6	89232	5	89237	6	89243	5	89248	6	89254	6	89260	5
781	89265	6	89271	5	89276	6	89282	5	89287	6	89293	5	89298	6	89304	6	89310	5	89315	6
782	89321	5	89326	6	89332	5	89337	6	89343	5	89348	6	89354	6	89360	5	89365	6	89371	5
783	89376	6	89382	5	89387	6	89393	5	89398	6	89404	5	89409	6	89415	6	89421	5	89426	6
784	89432	5	89437	6	89443	5	89448	6	89454	5	89459	6	89465	5	89470	6	89476	5	89481	6
785	89487	5	89492	6	89498	6	89504	5	89509	6	89515	5	89520	6	89526	5	89531	6	89537	5
786	89542	6	89548	5	89553	6	89559	5	89564	6	89570	5	89575	6	89581	5	89586	6	89592	5
787	89597	6	89603	6	89609	5	89614	6	89620	5	89625	6	89631	5	89636	6	89642	5	89647	6
788	89653	5	89658	6	89664	5	89669	6	89675	5	89680	6	89686	5	89691	6	89697	5	89702	6
789	89708	5	89713	6	89719	5	89724	6	89730	5	89735	6	89741	5	89746	6	89752	5	89757	6
790	89763	5	89768	6	89774	5	89779	6	89785	5	89790	6	89796	5	89801	6	89807	5	89812	6
791	89818	5	89823	6	89829	5	89834	6	89840	5	89845	6	89851	5	89856	6	89862	5	89867	6
792	89873	5	89878	5	89883	6	89889	5	89894	6	89900	5	89905	6	89911	5	89916	6	89922	5
793	89927	6	89933	5	89938	6	89944	5	89949	6	89955	5	89960	6	89966	5	89971	6	89977	5
794	89982	6	89988	5	89993	5	89998	6	90004	5	90009	6	90015	5	90020	6	90026	5	90031	6
795	90037	5	90042	6	90048	5	90053	6	90059	5	90064	5	90069	6	90075	5	90080	6	90086	5
796	90091	6	90097	5	90102	6	90108	5	90113	6	90119	5	90124	5	90129	6	90135	5	90140	6
797	90146	5	90151	6	90157	5	90162	6	90168	5	90173	6	90179	5	90184	5	90189	6	90195	5
798	90200	6	90206	5	90211	6	90217	5	90222	5	90227	6	90233	5	90238	6	90244	5	90249	6
799	90255	5	90260	6	90266	5	90271	5	90276	6	90282	5	90287	6	90293	5	90298	6	90304	5
800	90309	5	90314	6	90320	5	90325	6	90331	5	90336	6	90342	5	90347	5	90352	6	90358	5
No.	0	d	1	d	2	d	3	d	4	d	5	d	6	d	7	d	8	d	9	d

Prop. parts

	6
1	1
2	1
3	2
4	2
5	3
6	4
7	4
8	5
9	5

	5
1	0
2	1
3	2
4	2
5	2
6	3
7	4
8	4
9	4

TABLE 32

Logarithms of Numbers

8000–8500

No.	0	d	1	d	2	d	3	d	4	d	5	d	6	d	7	d	8	d	9	d
800	90309	5	90314	6	90320	5	90325	6	90331	5	90336	6	90342	5	90347	5	90352	6	90358	5
801	90363	6	90369	5	90374	6	90380	5	90385	5	90390	6	90396	5	90401	6	90407	5	90412	5
802	90417	6	90423	5	90428	6	90434	5	90439	6	90445	5	90450	5	90455	6	90461	5	90466	6
803	90472	5	90477	5	90482	6	90488	5	90493	6	90499	5	90504	5	90509	6	90515	5	90520	6
804	90526	5	90531	5	90536	6	90542	5	90547	6	90553	5	90558	5	90563	6	90569	5	90574	6
805	90580	5	90585	5	90590	6	90596	5	90601	6	90607	5	90612	5	90617	6	90623	5	90628	6
806	90634	5	90639	5	90644	6	90650	5	90655	5	90660	6	90666	5	90671	6	90677	5	90682	5
807	90687	6	90693	5	90698	5	90703	6	90709	5	90714	6	90720	5	90725	5	90730	6	90736	5
808	90741	6	90747	5	90752	5	90757	6	90763	5	90768	5	90773	6	90779	5	90784	5	90789	6
809	90795	5	90800	6	90806	5	90811	5	90816	6	90822	5	90827	5	90832	6	90838	5	90843	6
810	90849	5	90854	5	90859	6	90865	5	90870	5	90875	6	90881	5	90886	5	90891	6	90897	5
811	90902	5	90907	6	90913	5	90918	6	90924	5	90929	5	90934	6	90940	5	90945	5	90950	6
812	90956	5	90961	5	90966	6	90972	5	90977	5	90982	6	90988	5	90993	5	90998	6	91004	5
813	91009	5	91014	6	91020	5	91025	5	91030	6	91036	5	91041	5	91046	6	91052	5	91057	5
814	91062	6	91068	5	91073	5	91078	6	91084	5	91089	5	91094	6	91100	5	91105	5	91110	6
815	91116	5	91121	5	91126	6	91132	5	91137	5	91142	6	91148	5	91153	5	91158	6	91164	5
816	91169	5	91174	6	91180	5	91185	5	91190	6	91196	5	91201	5	91206	6	91212	5	91217	5
817	91222	6	91228	5	91233	5	91238	5	91243	6	91249	5	91254	5	91259	6	91265	5	91270	5
818	91275	6	91281	5	91286	5	91291	6	91297	5	91302	5	91307	5	91312	6	91318	5	91323	5
819	91328	6	91334	5	91339	5	91344	6	91350	5	91355	5	91360	5	91365	6	91371	5	91376	5
820	91381	6	91387	5	91392	5	91397	6	91403	5	91408	5	91413	5	91418	6	91424	5	91429	5
821	91434	6	91440	5	91445	5	91450	5	91455	6	91461	5	91466	5	91471	6	91477	5	91482	5
822	91487	5	91492	6	91498	5	91503	5	91508	6	91514	5	91519	5	91524	5	91529	6	91535	5
823	91540	5	91545	6	91551	5	91556	5	91561	5	91566	6	91572	5	91577	5	91582	5	91587	6
824	91593	5	91598	5	91603	6	91609	5	91614	5	91619	5	91624	6	91630	5	91635	5	91640	5
825	91645	6	91651	5	91656	5	91661	5	91666	6	91672	5	91677	5	91682	5	91687	6	91693	5
826	91698	5	91703	6	91709	5	91714	5	91719	5	91724	6	91730	5	91735	5	91740	5	91745	6
827	91751	5	91756	5	91761	5	91766	6	91772	5	91777	5	91782	5	91787	6	91793	5	91798	5
828	91803	5	91808	6	91814	5	91819	5	91824	5	91829	5	91834	6	91840	5	91845	5	91850	5
829	91855	6	91861	5	91866	5	91871	5	91876	6	91882	5	91887	5	91892	5	91897	6	91903	5
830	91908	5	91913	5	91918	6	91924	5	91929	5	91934	5	91939	5	91944	6	91950	5	91955	5
831	91960	5	91965	6	91971	5	91976	5	91981	5	91986	5	91991	6	91997	5	92002	5	92007	5
832	92012	6	92018	5	92023	5	92028	5	92033	5	92038	6	92044	5	92049	5	92054	5	92059	6
833	92065	5	92070	5	92075	5	92080	5	92085	6	92091	5	92096	5	92101	5	92106	5	92111	6
834	92117	5	92122	5	92127	5	92132	5	92137	6	92143	5	92148	5	92153	5	92158	5	92163	6
835	92169	5	92174	5	92179	5	92184	5	92189	6	92195	5	92200	5	92205	5	92210	5	92215	6
836	92221	5	92226	5	92231	5	92236	5	92241	6	92247	5	92252	5	92257	5	92262	5	92267	6
837	92273	5	92278	5	92283	5	92288	5	92293	5	92298	6	92304	5	92309	5	92314	5	92319	5
838	92324	6	92330	5	92335	5	92340	5	92345	5	92350	5	92355	6	92361	5	92366	5	92371	5
839	92376	5	92381	6	92387	5	92392	5	92397	5	92402	5	92407	5	92412	6	92418	5	92423	5
840	92428	5	92433	5	92438	5	92443	6	92449	5	92454	5	92459	5	92464	5	92469	5	92474	6
841	92480	5	92485	5	92490	5	92495	5	92500	5	92505	6	92511	5	92516	5	92521	5	92526	5
842	92531	5	92536	6	92542	5	92547	5	92552	5	92557	5	92562	5	92567	5	92572	6	92578	5
843	92583	5	92588	5	92593	5	92598	5	92603	6	92609	5	92614	5	92619	5	92624	5	92629	5
844	92634	5	92639	6	92645	5	92650	5	92655	5	92660	5	92665	5	92670	5	92675	6	92681	5
845	92686	5	92691	5	92696	5	92701	5	92706	5	92711	5	92716	6	92722	5	92727	5	92732	5
846	92737	5	92742	5	92747	5	92752	6	92758	5	92763	5	92768	5	92773	5	92778	5	92783	5
847	92788	5	92793	6	92799	5	92804	5	92809	5	92814	5	92819	5	92824	5	92829	5	92834	6
848	92840	5	92845	5	92850	5	92855	5	92860	5	92865	5	92870	5	92875	6	92881	5	92886	5
849	92891	5	92896	5	92901	5	92906	5	92911	5	92916	5	92921	6	92927	5	92932	5	92937	5
850	92942	5	92947	5	92952	5	92957	5	92962	5	92967	6	92973	5	92978	5	92983	5	92988	5
No.	0	d	1	d	2	d	3	d	4	d	5	d	6	d	7	d	8	d	9	d

Prop. parts

	6
1	1
2	1
3	2
4	2
5	3
6	4
7	4
8	5
9	5

	5
1	0
2	1
3	2
4	2
5	2
6	3
7	4
8	4
9	4

TABLE 32

Logarithms of Numbers

8500–9000

No.	0	d	1	d	2	d	3	d	4	d	5	d	6	d	7	d	8	d	9	d
850	92942	5	92947	5	92952	5	92957	5	92962	5	92967	6	92973	5	92978	5	92983	5	92988	5
851	92993	5	92998	5	93003	5	93008	5	93013	5	93018	6	93024	5	93029	5	93034	5	93039	5
852	93044	5	93049	5	93054	5	93059	5	93064	5	93069	6	93075	5	93080	5	93085	5	93090	5
853	93095	5	93100	5	93105	5	93110	5	93115	5	93120	5	93125	6	93131	5	93136	5	93141	5
854	93146	5	93151	5	93156	5	93161	5	93166	5	93171	5	93176	5	93181	5	93186	6	93192	5
855	93197	5	93202	5	93207	5	93212	5	93217	5	93222	5	93227	5	93232	5	93237	5	93242	5
856	93247	5	93252	6	93258	5	93263	5	93268	5	93273	5	93278	5	93283	5	93288	5	93293	5
857	93298	5	93303	5	93308	5	93313	5	93318	5	93323	5	93328	6	93334	5	93339	5	93344	5
858	93349	5	93354	5	93359	5	93364	5	93369	5	93374	5	93379	5	93384	5	93389	5	93394	5
859	93399	5	93404	5	93409	5	93414	6	93420	5	93425	5	93430	5	93435	5	93440	5	93445	5
860	93450	5	93455	5	93460	5	93465	5	93470	5	93475	5	93480	5	93485	5	93490	5	93495	5
861	93500	5	93505	5	93510	5	93515	5	93520	6	93526	5	93531	5	93536	5	93541	5	93546	5
862	93551	5	93556	5	93561	5	93566	5	93571	5	93576	5	93581	5	93586	5	93591	5	93596	5
863	93601	5	93606	5	93611	5	93616	5	93621	5	93626	5	93631	5	93636	5	93641	5	93646	5
864	93651	5	93656	5	93661	5	93666	5	93671	5	93676	6	93682	5	93687	5	93692	5	93697	5
865	93702	5	93707	5	93712	5	93717	5	93722	5	93727	5	93732	5	93737	5	93742	5	93747	5
866	93752	5	93757	5	93762	5	93767	5	93772	5	93777	5	93782	5	93787	5	93792	5	93797	5
867	93802	5	93807	5	93812	5	93817	5	93822	5	93827	5	93832	5	93837	5	93842	5	93847	5
868	93852	5	93857	5	93862	5	93867	5	93872	5	93877	5	93882	5	93887	5	93892	5	93897	5
869	93902	5	93907	5	93912	5	93917	5	93922	5	93927	5	93932	5	93937	5	93942	5	93947	5
870	93952	5	93957	5	93962	5	93967	5	93972	5	93977	5	93982	5	93987	5	93992	5	93997	5
871	94002	5	94007	5	94012	5	94017	5	94022	5	94027	5	94032	5	94037	5	94042	5	94047	5
872	94052	5	94057	5	94062	5	94067	5	94072	5	94077	5	94082	4	94086	5	94091	5	94096	5
873	94101	5	94106	5	94111	5	94116	5	94121	5	94126	5	94131	5	94136	5	94141	5	94146	5
874	94151	5	94156	5	94161	5	94166	5	94171	5	94176	5	94181	5	94186	5	94191	5	94196	5
875	94201	5	94206	5	94211	5	94216	5	94221	5	94226	5	94231	5	94236	4	94240	5	94245	5
876	94250	5	94255	5	94260	5	94265	5	94270	5	94275	5	94280	5	94285	5	94290	5	94295	5
877	94300	5	94305	5	94310	5	94315	5	94320	5	94325	5	94330	5	94335	5	94340	5	94345	4
878	94349	5	94354	5	94359	5	94364	5	94369	5	94374	5	94379	5	94384	5	94389	5	94394	5
879	94399	5	94404	5	94409	5	94414	5	94419	5	94424	5	94429	4	94433	5	94438	5	94443	5
880	94448	5	94453	5	94458	5	94463	5	94468	5	94473	5	94478	5	94483	5	94488	5	94493	5
881	94498	5	94503	4	94507	5	94512	5	94517	5	94522	5	94527	5	94532	5	94537	5	94542	5
882	94547	5	94552	5	94557	5	94562	5	94567	4	94571	5	94576	5	94581	5	94586	5	94591	5
883	94596	5	94601	5	94606	5	94611	5	94616	5	94621	5	94626	4	94630	5	94635	5	94640	5
884	94645	5	94650	5	94655	5	94660	5	94665	5	94670	5	94675	5	94680	5	94685	4	94689	5
885	94694	5	94699	5	94704	5	94709	5	94714	5	94719	5	94724	5	94729	5	94734	4	94738	5
886	94743	5	94748	5	94753	5	94758	5	94763	5	94768	5	94773	5	94778	5	94783	4	94787	5
887	94792	5	94797	5	94802	5	94807	5	94812	5	94817	5	94822	5	94827	5	94832	4	94836	5
888	94841	5	94846	5	94851	5	94856	5	94861	5	94866	5	94871	5	94876	4	94880	5	94885	5
889	94890	5	94895	5	94900	5	94905	5	94910	5	94915	4	94919	5	94924	5	94929	5	94934	5
890	94939	5	94944	5	94949	5	94954	5	94959	4	94963	5	94968	5	94973	5	94978	5	94983	5
891	94988	5	94993	5	94998	4	95002	5	95007	5	95012	5	95017	5	95022	5	95027	5	95032	4
892	95036	5	95041	5	95046	5	95051	5	95056	5	95061	5	95066	5	95071	4	95075	5	95080	5
893	95085	5	95090	5	95095	5	95100	4	95105	4	95109	5	95114	5	95119	5	95124	5	95129	5
894	95134	5	95139	4	95143	5	95148	5	95153	5	95158	5	95163	5	95168	5	95173	4	95177	5
895	95182	5	95187	5	95192	5	95197	5	95202	5	95207	4	95211	5	95216	5	95221	5	95226	5
896	95231	5	95236	4	95240	5	95245	5	95250	5	95255	5	95260	5	95265	5	95270	4	95274	5
897	95279	5	95284	5	95289	5	95294	5	95299	4	95303	5	95308	5	95313	5	95318	5	95323	5
898	95328	4	95332	5	95337	5	95342	5	95347	5	95352	5	95357	4	95361	5	95366	5	95371	5
899	95376	5	95381	5	95386	4	95390	5	95395	5	95400	5	95405	5	95410	5	95415	4	95419	5
900	95424	5	95429	5	95434	5	95439	5	95444	4	95448	5	95453	5	95458	5	95463	5	95468	4
No.	0	d	1	d	2	d	3	d	4	d	5	d	6	d	7	d	8	d	9	d

Prop. parts

	6
1	1
2	1
3	2
4	2
5	3
6	4
7	4
8	5
9	5

	5
1	0
2	1
3	2
4	2
5	2
6	3
7	4
8	4
9	4

	4
1	0
2	1
3	1
4	2
5	2
6	2
7	3
8	3
9	4

TABLE 32

Logarithms of Numbers

9000–9500

No.	0	d	1	d	2	d	3	d	4	d	5	d	6	d	7	d	8	d	9	d
900	95424	5	95429	5	95434	5	95439	5	95444	4	95448	5	95453	5	95458	5	95463	5	95468	4
901	95472	5	95477	5	95482	5	95487	5	95492	5	95497	4	95501	5	95506	5	95511	5	95516	5
902	95521	4	95525	5	95530	5	95535	5	95540	5	95545	5	95550	4	95554	5	95559	5	95564	5
903	95569	5	95574	4	95578	5	95583	5	95588	5	95593	5	95598	4	95602	5	95607	5	95612	5
904	95617	5	95622	4	95626	5	95631	5	95636	5	95641	5	95646	4	95650	5	95655	5	95660	5
905	95665	5	95670	4	95674	5	95679	5	95684	5	95689	5	95694	4	95698	5	95703	5	95708	5
906	95713	5	95718	4	95722	5	95727	5	95732	5	95737	5	95742	4	95746	5	95751	5	95756	5
907	95761	5	95766	4	95770	5	95775	5	95780	5	95785	4	95789	5	95794	5	95799	5	95804	5
908	95809	4	95813	5	95818	5	95823	5	95828	4	95832	5	95837	5	95842	5	95847	5	95852	4
909	95856	5	95861	5	95866	5	95871	4	95875	5	95880	5	95885	5	95890	5	95895	4	95899	5
910	95904	5	95909	5	95914	4	95918	5	95923	5	95928	5	95933	5	95938	4	95942	5	95947	5
911	95952	5	95957	4	95961	5	95966	5	95971	5	95976	4	95980	5	95985	5	95990	5	95995	4
912	95999	5	96004	5	96009	5	96014	5	96019	4	96023	5	96028	5	96033	5	96038	4	96042	5
913	96047	5	96052	5	96057	4	96061	5	96066	5	96071	5	96076	4	96080	5	96085	5	96090	5
914	96095	4	96099	5	96104	5	96109	5	96114	4	96118	5	96123	5	96128	5	96133	4	96137	5
915	96142	5	96147	5	96152	4	96156	5	96161	5	96166	5	96171	4	96175	5	96180	5	96185	5
916	96190	4	96194	5	96199	5	96204	5	96209	4	96213	5	96218	5	96223	4	96227	5	96232	5
917	96237	5	96242	4	96246	5	96251	5	96256	5	96261	4	96265	5	96270	5	96275	5	96280	4
918	96284	5	96289	5	96294	4	96298	5	96303	5	96308	5	96313	4	96317	5	96322	5	96327	5
919	96332	4	96336	5	96341	5	96346	4	96350	5	96355	5	96360	5	96365	4	96369	5	96374	5
920	96379	5	96384	4	96388	5	96393	5	96398	4	96402	5	96407	5	96412	5	96417	4	96421	5
921	96426	5	96431	4	96435	5	96440	5	96445	5	96450	4	96454	5	96459	5	96464	4	96468	5
922	96473	5	96478	5	96483	4	96487	5	96492	5	96497	4	96501	5	96506	5	96511	4	96515	5
923	96520	5	96525	5	96530	4	96534	5	96539	5	96544	4	96548	5	96553	5	96558	4	96562	5
924	96567	5	96572	5	96577	4	96581	5	96586	5	96591	4	96595	5	96600	5	96605	4	96609	5
925	96614	5	96619	5	96624	4	96628	5	96633	5	96638	4	96642	5	96647	5	96652	4	96656	5
926	96661	5	96666	4	96670	5	96675	5	96680	5	96685	4	96689	5	96694	5	96699	4	96703	5
927	96708	5	96713	4	96717	5	96722	5	96727	4	96731	5	96736	5	96741	4	96745	5	96750	5
928	96755	4	96759	5	96764	5	96769	5	96774	4	96778	5	96783	5	96788	4	96792	5	96797	5
929	96802	4	96806	5	96811	5	96816	4	96820	5	96825	5	96830	4	96834	5	96839	5	96844	4
930	96848	5	96853	5	96858	4	96862	5	96867	5	96872	4	96876	5	96881	5	96886	4	96890	5
931	96895	5	96900	4	96904	5	96909	5	96914	4	96918	5	96923	5	96928	4	96932	5	96937	5
932	96942	4	96946	5	96951	5	96956	4	96960	5	96965	5	96970	4	96974	5	96979	5	96984	4
933	96988	5	96993	4	96997	5	97002	5	97007	4	97011	5	97016	5	97021	4	97025	5	97030	5
934	97035	4	97039	5	97044	5	97049	4	97053	5	97058	5	97063	4	97067	5	97072	5	97077	4
935	97081	5	97086	4	97090	5	97095	5	97100	4	97104	5	97109	5	97114	4	97118	5	97123	5
936	97128	4	97132	5	97137	5	97142	4	97146	5	97151	4	97155	5	97160	5	97165	4	97169	5
937	97174	5	97179	4	97183	5	97188	4	97192	5	97197	5	97202	4	97206	5	97211	5	97216	4
938	97220	5	97225	5	97230	4	97234	5	97239	4	97243	5	97248	5	97253	4	97257	5	97262	5
939	97267	4	97271	5	97276	4	97280	5	97285	5	97290	4	97294	5	97299	5	97304	4	97308	5
940	97313	4	97317	5	97322	5	97327	4	97331	5	97336	4	97340	5	97345	5	97350	4	97354	5
941	97359	5	97364	4	97368	5	97373	4	97377	5	97382	5	97387	4	97391	5	97396	4	97400	5
942	97405	5	97410	4	97414	5	97419	5	97424	4	97428	5	97433	4	97437	5	97442	5	97447	4
943	97451	5	97456	4	97460	5	97465	5	97470	4	97474	5	97479	4	97483	5	97488	5	97493	4
944	97497	5	97502	4	97506	5	97511	5	97516	4	97520	5	97525	4	97529	5	97534	5	97539	4
945	97543	5	97548	4	97552	5	97557	5	97562	4	97566	5	97571	4	97575	5	97580	5	97585	4
946	97589	5	97594	4	97598	5	97603	4	97607	5	97612	5	97617	4	97621	5	97626	4	97630	5
947	97635	5	97640	4	97644	5	97649	4	97653	5	97658	5	97663	4	97667	5	97672	4	97676	5
948	97681	4	97685	5	97690	5	97695	4	97699	5	97704	4	97708	5	97713	4	97717	5	97722	5
949	97727	4	97731	5	97736	4	97740	5	97745	4	97749	5	97754	5	97759	4	97763	5	97768	4
950	97772	5	97777	5	97782	4	97786	5	97791	4	97795	5	97800	4	97804	5	97809	4	97813	5
No.	0	d	1	d	2	d	3	d	4	d	5	d	6	d	7	d	8	d	9	d

Prop. parts

	5
1	0
2	1
3	2
4	2
5	2
6	3
7	4
8	4
9	4

	4
1	0
2	1
3	1
4	2
5	2
6	2
7	3
8	3
9	4

TABLE 32

Logarithms of Numbers

9500–10000

No.	0	d	1	d	2	d	3	d	4	d	5	d	6	d	7	d	8	d	9	d
950	97772	5	97777	5	97782	4	97786	5	97791	4	97795	5	97800	4	97804	5	97809	4	97813	5
951	97818	5	97823	4	97827	5	97832	4	97836	5	97841	4	97845	5	97850	5	97855	4	97859	5
952	97864	4	97868	5	97873	4	97877	5	97882	4	97886	5	97891	5	97896	4	97900	5	97905	4
953	97909	5	97914	4	97918	5	97923	5	97928	4	97932	5	97937	4	97941	5	97946	4	97950	5
954	97955	4	97959	5	97964	4	97968	5	97973	5	97978	4	97982	5	97987	4	97991	5	97996	4
955	98000	5	98005	4	98009	5	98014	5	98019	4	98023	5	98028	4	98032	5	98037	4	98041	5
956	98046	4	98050	5	98055	4	98059	5	98064	4	98068	5	98073	5	98078	4	98082	5	98087	4
957	98091	5	98096	4	98100	5	98105	4	98109	5	98114	4	98118	5	98123	4	98127	5	98132	5
958	98137	4	98141	5	98146	4	98150	5	98155	4	98159	5	98164	4	98168	5	98173	4	98177	5
959	98182	4	98186	5	98191	4	98195	5	98200	4	98204	5	98209	5	98214	4	98218	5	98223	4
960	98227	5	98232	4	98236	5	98241	4	98245	5	98250	4	98254	5	98259	4	98263	5	98268	4
961	98272	5	98277	4	98281	5	98286	4	98290	5	98295	4	98299	5	98304	4	98308	5	98313	5
962	98318	4	98322	5	98327	4	98331	5	98336	4	98340	5	98345	4	98349	5	98354	4	98358	5
963	98363	4	98367	5	98372	4	98376	5	98381	4	98385	5	98390	4	98394	5	98399	4	98403	5
964	98408	4	98412	5	98417	4	98421	5	98426	4	98430	5	98435	4	98439	5	98444	4	98448	5
965	98453	4	98457	5	98462	4	98466	5	98471	4	98475	5	98480	4	98484	5	98489	4	98493	5
966	98498	4	98502	5	98507	4	98511	5	98516	4	98520	5	98525	4	98529	5	98534	4	98538	5
967	98543	4	98547	5	98552	4	98556	5	98561	4	98565	5	98570	4	98574	5	98579	4	98583	5
968	98588	4	98592	5	98597	4	98601	4	98605	5	98610	4	98614	5	98619	4	98623	5	98628	4
969	98632	5	98637	4	98641	5	98646	4	98650	5	98655	4	98659	5	98664	4	98668	5	98673	4
970	98677	5	98682	4	98686	5	98691	4	98695	5	98700	4	98704	5	98709	4	98713	4	98717	5
971	98722	4	98726	5	98731	4	98735	5	98740	4	98744	5	98749	4	98753	5	98758	4	98762	5
972	98767	4	98771	5	98776	4	98780	4	98784	5	98789	4	98793	5	98798	4	98802	5	98807	4
973	98811	5	98816	4	98820	5	98825	4	98829	5	98834	4	98838	5	98843	4	98847	4	98851	5
974	98856	4	98860	5	98865	4	98869	5	98874	4	98878	5	98883	4	98887	5	98892	4	98896	4
975	98900	5	98905	4	98909	5	98914	4	98918	5	98923	4	98927	5	98932	4	98936	5	98941	4
976	98945	4	98949	5	98954	4	98958	5	98963	4	98967	5	98972	4	98976	5	98981	4	98985	4
977	98989	5	98994	4	98998	5	99003	4	99007	5	99012	4	99016	5	99021	4	99025	4	99029	5
978	99034	4	99038	5	99043	4	99047	5	99052	4	99056	5	99061	4	99065	4	99069	5	99074	4
979	99078	5	99083	4	99087	5	99092	4	99096	4	99100	5	99105	4	99109	5	99114	4	99118	5
980	99123	4	99127	4	99131	5	99136	4	99140	5	99145	4	99149	5	99154	4	99158	4	99162	5
981	99167	4	99171	5	99176	4	99180	5	99185	4	99189	4	99193	5	99198	4	99202	5	99207	4
982	99211	5	99216	4	99220	4	99224	5	99229	4	99233	5	99238	4	99242	5	99247	4	99251	4
983	99255	5	99260	4	99264	5	99269	4	99273	4	99277	5	99282	4	99286	5	99291	4	99295	5
984	99300	4	99304	4	99308	5	99313	4	99317	5	99322	4	99326	4	99330	5	99335	4	99339	5
985	99344	4	99348	4	99352	5	99357	4	99361	5	99366	4	99370	4	99374	5	99379	4	99383	5
986	99388	4	99392	4	99396	5	99401	4	99405	5	99410	4	99414	5	99419	4	99423	4	99427	5
987	99432	4	99436	5	99441	4	99445	4	99449	5	99454	4	99458	5	99463	4	99467	4	99471	5
988	99476	4	99480	4	99484	5	99489	4	99493	5	99498	4	99502	4	99506	5	99511	4	99515	5
989	99520	4	99524	4	99528	5	99533	4	99537	5	99542	4	99546	4	99550	5	99555	4	99559	5
990	99564	4	99568	4	99572	5	99577	4	99581	4	99585	5	99590	4	99594	5	99599	4	99603	4
991	99607	5	99612	4	99616	5	99621	4	99625	4	99629	5	99634	4	99638	4	99642	5	99647	4
992	99651	5	99656	4	99660	4	99664	5	99669	4	99673	4	99677	5	99682	4	99686	5	99691	4
993	99695	4	99699	5	99704	4	99708	4	99712	5	99717	4	99721	5	99726	4	99730	4	99734	5
994	99739	4	99743	4	99747	5	99752	4	99756	4	99760	5	99765	4	99769	5	99774	4	99778	4
995	99782	5	99787	4	99791	4	99795	5	99800	4	99804	4	99808	5	99813	4	99817	5	99822	4
996	99826	4	99830	5	99835	4	99839	4	99843	5	99848	4	99852	4	99856	5	99861	4	99865	5
997	99870	4	99874	4	99878	5	99883	4	99887	4	99891	5	99896	4	99900	4	99904	5	99909	4
998	99913	4	99917	5	99922	4	99926	4	99930	5	99935	4	99939	5	99944	4	99948	4	99952	5
999	99957	4	99961	4	99965	5	99970	4	99974	4	99978	5	99983	4	99987	4	99991	5	99996	4
1000	00000	4	00004	5	00009	4	00013	4	00017	5	00022	4	00026	4	00030	5	00035	4	00039	4
No.	0	d	1	d	2	d	3	d	4	d	5	d	6	d	7	d	8	d	9	d

Prop. parts

	5	4
1	0	0
2	1	1
3	2	1
4	2	2
5	2	2
6	3	2
7	4	3
8	4	3
9	4	4

TABLE 33

Logarithms of Trigonometric Functions

0° → ↓ ′	sin	Diff. 1′	csc	tan	Diff. 1′	cot	sec	Diff. 1′	cos	← 179° ↓ ′
0	∞	—	∞	∞	—	∞	10. 00000	0	10. 00000	60
1	6. 46373	30103	13. 53627	6. 46373	30103	13. 53627	. 00000	0	. 00000	59
2	. 76476	17609	. 23524	. 76476	17609	. 23524	. 00000	0	. 00000	58
3	6. 94085	12494	13. 05915	6. 94085	12494	13. 05915	. 00000	0	. 00000	57
4	7. 06579	9691	12. 93421	7. 06579	9691	12. 93421	. 00000	0	. 00000	56
5	7. 16270	7918	12. 83730	7. 16270	7918	12. 83730	10. 00000	0	10. 00000	55
6	. 24188	6694	. 75812	. 24188	6694	. 75812	. 00000	0	. 00000	54
7	. 30882	5800	. 69118	. 30882	5800	. 69118	. 00000	0	. 00000	53
8	. 36682	5115	. 63318	. 36682	5115	. 63318	. 00000	0	. 00000	52
9	. 41797	4576	. 58203	. 41797	4576	. 58203	. 00000	0	. 00000	51
10	7. 46373	4139	12. 53627	7. 46373	4139	12. 53627	10. 00000	0	10. 00000	50
11	. 50512	3779	. 49488	. 50512	3779	. 49488	. 00000	0	. 00000	49
12	. 54291	3476	. 45709	. 54291	3476	. 45709	. 00000	0	. 00000	48
13	. 57767	3218	. 42233	. 57767	3219	. 42233	. 00000	0	. 00000	47
14	. 60985	2997	. 39015	. 60986	2996	. 39014	. 00000	0	. 00000	46
15	7. 63982	2802	12. 36018	7. 63982	2803	12. 36018	10. 00000	0	10. 00000	45
16	. 66784	2633	. 33216	. 66785	2633	. 33215	. 00000	1	10. 00000	44
17	. 69417	2483	. 30583	. 69418	2482	. 30582	. 00001	0	9. 99999	43
18	. 71900	2348	. 28100	. 71900	2348	. 28100	. 00001	0	. 99999	42
19	. 74248	2227	. 25752	. 74248	2228	. 25752	. 00001	0	. 99999	41
20	7. 76475	2119	12. 23525	7. 76476	2119	12. 23524	10. 00001	0	9. 99999	40
21	. 78594	2021	. 21406	. 78595	2020	. 21405	. 00001	0	. 99999	39
22	. 80615	1930	. 19385	. 80615	1931	. 19385	. 00001	0	. 99999	38
23	. 82545	1848	. 17455	. 82546	1848	. 17454	. 00001	0	. 99999	37
24	. 84393	1773	. 15607	. 84394	1773	. 15606	. 00001	0	. 99999	36
25	7. 86166	1704	12. 13834	7. 86167	1704	12. 13833	10. 00001	0	9. 99999	35
26	. 87870	1639	. 12130	. 87871	1639	. 12129	. 00001	0	. 99999	34
27	. 89509	1579	. 10491	. 89510	1579	. 10490	. 00001	0	. 99999	33
28	. 91088	1524	. 08912	. 91089	1524	. 08911	. 00001	1	. 99999	32
29	. 92612	1472	. 07388	. 92613	1473	: 07387	. 00002	0	. 99998	31
30	7. 94084	1424	12. 05916	7. 94086	1424	12. 05914	10. 00002	0	9. 99998	30
31	. 95508	1379	. 04492	. 95510	1379	. 04490	. 00002	0	. 99998	29
32	. 96887	1336	. 03113	. 96889	1336	. 03111	. 00002	0	. 99998	28
33	. 98223	1297	. 01777	. 98225	1297	. 01775	. 00002	0	. 99998	27
34	7. 99520	1259	12. 00480	7. 99522	1259	12. 00478	. 00002	0	. 99998	26
35	8. 00779	1223	11. 99221	8. 00781	1223	11. 99219	10. 00002	0	9. 99998	25
36	. 02002	1190	. 97998	. 02004	1190	. 97996	. 00002	1	. 99998	24
37	. 03192	1158	. 96808	. 03194	1159	. 96806	. 00003	0	. 99997	23
38	. 04350	1128	. 95650	. 04353	1128	. 95647	. 00003	0	. 99997	22
39	. 05478	1100	. 94522	. 05481	1100	. 94519	. 00003	0	. 99997	21
40	8. 06578	1072	11. 93422	8. 06581	1072	11. 93419	10. 00003	0	9. 99997	20
41	. 07650	1046	. 92350	. 07653	1047	. 92347	. 00003	0	. 99997	19
42	. 08696	1022	. 91304	. 08700	1022	. 91300	. 00003	0	. 99997	18
43	. 09718	999	. 90282	. 09722	998	. 90278	. 00003	1	. 99997	17
44	. 10717	976	. 89283	. 10720	976	. 89280	. 00004	0	. 99996	16
45	8. 11693	954	11. 88307	8. 11696	955	11. 88304	10. 00004	0	9. 99996	15
46	. 12647	934	. 87353	. 12651	934	. 87349	. 00004	0	. 99996	14
47	. 13581	914	. 86419	. 13585	915	. 86415	. 00004	0	. 99996	13
48	. 14495	896	. 85505	. 14500	895	. 85500	. 00004	0	. 99996	12
49	. 15391	877	. 84609	. 15395	878	. 84605	. 00004	1	. 99996	11
50	8. 16268	860	11. 83732	8. 16273	860	11. 83727	10. 00005	0	9. 99995	10
51	. 17128	843	. 82872	. 17133	843	. 82867	. 00005	0	. 99995	9
52	. 17971	827	. 82029	. 17976	828	. 82024	. 00005	0	. 99995	8
53	. 18798	812	. 81202	. 18804	812	. 81196	. 00005	0	. 99995	7
54	. 19610	797	. 80390	. 19616	797	. 80384	. 00005	1	. 99995	6
55	8. 20407	782	11. 79593	8. 20413	782	11. 79587	10. 00006	0	9. 99994	5
56	. 21189	769	. 78811	. 21195	769	. 78805	. 00006	0	. 99994	4
57	. 21958	755	. 78042	. 21964	756	. 78036	. 00006	0	. 99994	3
58	. 22713	743	. 77287	. 22720	742	. 77280	. 00006	0	. 99994	2
59	. 23456	730	. 76544	. 23462	730	. 76538	. 00006	1	. 99994	1
60	8. 24186		11. 75814	8. 24192		11. 75808	10. 00007		9. 99993	0
↑ 90° →	cos	Diff. 1′	sec	cot	Diff. 1′	tan	csc	Diff. 1′	sin	← 89° ↑

TABLE 33

Logarithms of Trigonometric Functions

1° → ↓	sin	Diff. 1′	csc	tan	Diff. 1′	cot	sec	Diff. 1′	cos	← 178° ↓
′										′
0	8.24186	717	11.75814	8.24192	718	11.75808	10.00007	0	9.99993	60
1	.24903	706	.75097	.24910	706	.75090	.00007	0	.99993	59
2	.25609	695	.74391	.25616	696	.74384	.00007	0	.99993	58
3	.26304	684	.73696	.26312	684	.73688	.00007	1	.99993	57
4	.26988	673	.73012	.26996	673	.73004	.00008	0	.99992	56
5	8.27661	663	11.72339	8.27669	663	11.72331	10.00008	0	9.99992	55
6	.28324	653	.71676	.28332	654	.71668	.00008	0	.99992	54
7	.28977	644	.71023	.28986	643	.71014	.00008	0	.99992	53
8	.29621	634	.70379	.29629	634	.70371	.00008	1	.99992	52
9	.30255	624	.69745	.30263	625	.69737	.00009	0	.99991	51
10	8.30879	616	11.69121	8.30888	617	11.69112	10.00009	0	9.99991	50
11	.31495	608	.68505	.31505	607	.68495	.00009	1	.99991	49
12	.32103	599	.67897	.32112	599	.67888	.00010	0	.99990	48
13	.32702	590	.67298	.32711	591	.67289	.00010	0	.99990	47
14	.33292	583	.66708	.33302	584	.66698	.00010	0	.99990	46
15	8.33875	575	11.66125	8.33886	575	11.66114	10.00010	1	9.99990	45
16	.34450	568	.65550	.34461	568	.65539	.00011	0	.99989	44
17	.35018	560	.64982	.35029	561	.64971	.00011	0	.99989	43
18	.35578	553	.64422	.35590	553	.64410	.00011	0	.99989	42
19	.36131	547	.63869	.36143	546	.63857	.00011	1	.99989	41
20	8.36678	539	11.63322	8.36689	540	11.63311	10.00012	0	9.99988	40
21	.37217	533	.62783	.37229	533	.62771	.00012	0	.99988	39
22	.37750	526	.62250	.37762	527	.62238	.00012	1	.99988	38
23	.38276	520	.61724	.38289	520	.61711	.00013	0	.99987	37
24	.38796	514	.61204	.38809	514	.61191	.00013	0	.99987	36
25	8.39310	508	11.60690	8.39323	509	11.60677	10.00013	1	9.99987	35
26	.39818	502	.60182	.39832	502	.60168	.00014	0	.99986	34
27	.40320	496	.59680	.40334	496	.59666	.00014	0	.99986	33
28	.40816	491	.59184	.40830	491	.59170	.00014	1	.99986	32
29	.41307	485	.58693	.41321	486	.58679	.00015	0	.99985	31
30	8.41792	480	11.58208	8.41807	480	11.58193	10.00015	0	9.99985	30
31	.42272	474	.57728	.42287	475	.57713	.00015	1	.99985	29
32	.42746	470	.57254	.42762	470	.57238	.00016	0	.99984	28
33	.43216	464	.56784	.43232	464	.56768	.00016	0	.99984	27
34	.43680	459	.56320	.43696	460	.56304	.00016	1	.99984	26
35	8.44139	455	11.55861	8.44156	455	11.55844	10.00017	0	9.99983	25
36	.44594	450	.55406	.44611	450	.55389	.00017	0	.99983	24
37	.45044	445	.54956	.45061	446	.54939	.00017	1	.99983	23
38	.45489	441	.54511	.45507	441	.54493	.00018	0	.99982	22
39	.45930	436	.54070	.45948	437	.54052	.00018	0	.99982	21
40	8.46366	433	11.53634	8.46385	432	11.53615	10.00018	1	9.99982	20
41	.46799	427	.53201	.46817	428	.53183	.00019	0	.99981	19
42	.47226	424	.52774	.47245	424	.52755	.00019	0	.99981	18
43	.47650	419	.52350	.47669	420	.52331	.00019	1	.99981	17
44	.48069	416	.51931	.48089	416	.51911	.00020	0	.99980	16
45	8.48485	411	11.51515	8.48505	412	11.51495	10.00020	1	9.99980	15
46	.48896	408	.51104	.48917	408	.51083	.00021	0	.99979	14
47	.49304	404	.50696	.49325	404	.50675	.00021	0	.99979	13
48	.49708	400	.50292	.49729	401	.50271	.00021	1	.99979	12
49	.50108	396	.49892	.50130	397	.49870	.00022	0	.99978	11
50	8.50504	393	11.49496	8.50527	393	11.49473	10.00022	1	9.99978	10
51	.50897	390	.49103	.50920	390	.49080	.00023	0	.99977	9
52	.51287	386	.48713	.51310	386	.48690	.00023	0	.99977	8
53	.51673	382	.48327	.51696	383	.48304	.00023	1	.99977	7
54	.52055	379	.47945	.52079	380	.47921	.00024	0	.99976	6
55	8.52434	376	11.47566	8.52459	376	11.47541	10.00024	1	9.99976	5
56	.52810	373	.47190	.52835	373	.47165	.00025	0	.99975	4
57	.53183	369	.46817	.53208	370	.46792	.00025	1	.99975	3
58	.53552	367	.46448	.53578	367	.46422	.00026	0	.99974	2
59	.53919	363	.46081	.53945	363	.46055	.00026	0	.99974	1
60	8.54282		11.45718	8.54308		11.45692	10.00026		9.99974	0
↑ 91° →	cos	Diff. 1′	sec	cot	Diff. 1′	tan	csc	Diff. 1′	sin	↑ ← 88°

TABLE 33

Logarithms of Trigonometric Functions

2° → ↓ ′	sin	Diff. 1′	csc	tan	Diff. 1′	cot	sec	Diff. 1′	cos	← 177° ↓ ′
0	8.54282	360	11.45718	8.54308	361	11.45692	10.00026	1	9.99974	60
1	.54642	357	.45358	.54669	358	.45331	.00027	0	.99973	59
2	.54999	355	.45001	.55027	355	.44973	.00027	1	.99973	58
3	.55354	351	.44646	.55382	352	.44618	.00028	0	.99972	57
4	.55705	349	.44295	.55734	349	.44266	.00028	1	.99972	56
5	8.56054	346	11.43946	8.56083	346	11.43917	10.00029	0	9.99971	55
6	.56400	343	.43600	.56429	344	.43571	.00029	1	.99971	54
7	.56743	341	.43257	.56773	341	.43227	.00030	0	.99970	53
8	.57084	337	.42916	.57114	338	.42886	.00030	1	.99970	52
9	.57421	336	.42579	.57452	336	.42548	.00031	0	.99969	51
10	8.57757	332	11.42243	8.57788	333	11.42212	10.00031	1	9.99969	50
11	.58089	330	.41911	.58121	330	.41879	.00032	0	.99968	49
12	.58419	328	.41581	.58451	328	.41549	.00032	1	.99968	48
13	.58747	325	.41253	.58779	326	.41221	.00033	0	.99967	47
14	.59072	323	.40928	.59105	323	.40895	.00033	0	.99967	46
15	8.59395	320	11.40605	8.59428	321	11.40572	10.00033	1	9.99967	45
16	.59715	318	.40285	.59749	319	.40251	.00034	0	.99966	44
17	.60033	316	.39967	.60068	316	.39932	.00034	1	.99966	43
18	.60349	313	.39651	.60384	314	.39616	.00035	1	.99965	42
19	.60662	311	.39338	.60698	311	.39302	.00036	0	.99964	41
20	8.60973	309	11.39027	8.61009	310	11.38991	10.00036	1	9.99964	40
21	.61282	307	.38718	.61319	307	.38681	.00037	0	.99963	39
22	.61589	305	.38411	.61626	305	.38374	.00037	1	.99963	38
23	.61894	302	.38106	.61931	303	.38069	.00038	0	.99962	37
24	.62196	301	.37804	.62234	301	.37766	.00038	1	.99962	36
25	8.62497	298	11.37503	8.62535	299	11.37465	10.00039	0	9.99961	35
26	.62795	296	.37205	.62834	297	.37166	.00039	1	.99961	34
27	.63091	294	.36909	.63131	295	.36869	.00040	0	.99960	33
28	.63385	293	.36615	.63426	292	.36574	.00040	1	.99960	32
29	.63678	290	.36322	.63718	291	.36282	.00041	0	.99959	31
30	8.63968	288	11.36032	8.64009	289	11.35991	10.00041	1	9.99959	30
31	.64256	287	.35744	.64298	287	.35702	.00042	0	.99958	29
32	.64543	284	.35457	.64585	285	.35415	.00042	1	.99958	28
33	.64827	283	.35173	.64870	284	.35130	.00043	1	.99957	27
34	.65110	281	.34890	.65154	281	.34846	.00044	0	.99956	26
35	8.65391	279	11.34609	8.65435	280	11.34565	10.00044	1	9.99956	25
36	.65670	277	.34330	.65715	278	.34285	.00045	0	.99955	24
37	.65947	276	.34053	.65993	276	.34007	.00045	1	.99955	23
38	.66223	274	.33777	.66269	274	.33731	.00046	0	.99954	22
39	.66497	272	.33503	.66543	273	.33457	.00046	1	.99954	21
40	8.66769	270	11.33231	8.66816	271	11.33184	10.00047	1	9.99953	20
41	.67039	269	.32961	.67087	269	.32913	.00048	0	.99952	19
42	.67308	267	.32692	.67356	268	.32644	.00048	1	.99952	18
43	.67575	266	.32425	.67624	266	.32376	.00049	0	.99951	17
44	.67841	263	.32159	.67890	264	.32110	.00049	1	.99951	16
45	8.68104	263	11.31896	8.68154	263	11.31846	10.00050	1	9.99950	15
46	.68367	260	.31633	.68417	261	.31583	.00051	0	.99949	14
47	.68627	259	.31373	.68678	260	.31322	.00051	1	.99949	13
48	.68886	258	.31114	.68938	258	.31062	.00052	0	.99948	12
49	.69144	256	.30856	.69196	257	.30804	.00052	1	.99948	11
50	8.69400	254	11.30600	8.69453	255	11.30547	10.00053	1	9.99947	10
51	.69654	253	.30346	.69708	254	.30292	.00054	0	.99946	9
52	.69907	252	.30093	.69962	252	.30038	.00054	1	.99946	8
53	.70159	250	.29841	.70214	251	.29786	.00055	1	.99945	7
54	.70409	249	.29591	.70465	249	.29535	.00056	0	.99944	6
55	8.70658	247	11.29342	8.70714	248	11.29286	10.00056	1	9.99944	5
56	.70905	246	.29095	.70962	246	.29038	.00057	1	.99943	4
57	.71151	244	.28849	.71208	245	.28792	.00058	0	.99942	3
58	.71395	243	.28605	.71453	244	.28547	.00058	1	.99942	2
59	.71638	242	.28362	.71697	243	.28303	.00059	1	.99941	1
60	8.71880		11.28120	8.71940		11.28060	10.00060		9.99940	0
↑ 92° →	cos	Diff. 1′	sec	cot	Diff. 1′	tan	csc	Diff. 1′	sin	← 87° ↑

TABLE 33

Logarithms of Trigonometric Functions

3°→ ↓ ′	sin	Diff. 1′	csc	tan	Diff. 1′	cot	sec	Diff. 1′	cos	←176° ↓ ′
0	8. 71880	240	11. 28120	8. 71940	241	11. 28060	10. 00060	0	9. 99940	60
1	. 72120	239	. 27880	. 72181	239	. 27819	. 00060	1	. 99940	59
2	. 72359	238	. 27641	. 72420	239	. 27580	. 00061	1	. 99939	58
3	. 72597	237	. 27403	. 72659	237	. 27341	. 00062	0	. 99938	57
4	. 72834	235	. 27166	. 72896	236	. 27104	. 00062	1	. 99938	56
5	8. 73069	234	11. 26931	8. 73132	234	11. 26868	10. 00063	1	9. 99937	55
6	. 73303	232	. 26697	. 73366	234	. 26634	. 00064	0	. 99936	54
7	. 73535	232	. 26465	. 73600	232	. 26400	. 00064	1	. 99936	53
8	. 73767	230	. 26233	. 73832	231	. 26168	. 00065	1	. 99935	52
9	. 73997	229	. 26003	. 74063	229	. 25937	. 00066	0	. 99934	51
10	8. 74226	228	11. 25774	8. 74292	229	11. 25708	10. 00066	1	9. 99934	50
11	. 74454	226	. 25546	. 74521	227	. 25479	. 00067	1	. 99933	49
12	. 74680	226	. 25320	. 74748	226	. 25252	. 00068	0	. 99932	48
13	. 74906	224	. 25094	. 74974	225	. 25026	. 00068	1	. 99932	47
14	. 75130	223	. 24870	. 75199	224	. 24801	. 00069	1	. 99931	46
15	8. 75353	222	11. 24647	8. 75423	222	11. 24577	10. 00070	1	9. 99930	45
16	. 75575	220	. 24425	. 75645	222	. 24355	. 00071	0	. 99929	44
17	. 75795	220	. 24205	. 75867	220	. 24133	. 00071	1	. 99929	43
18	. 76015	219	. 23985	. 76087	219	. 23913	. 00072	1	. 99928	42
19	. 76234	217	. 23766	. 76306	219	. 23694	. 00073	1	. 99927	41
20	8. 76451	216	11. 23549	8. 76525	217	11. 23475	10. 00074	0	9. 99926	40
21	. 76667	216	. 23333	. 76742	216	. 23258	. 00074	1	. 99926	39
22	. 76883	214	. 23117	. 76958	215	. 23042	. 00075	1	. 99925	38
23	. 77097	213	. 22903	. 77173	214	. 22827	. 00076	1	. 99924	37
24	. 77310	212	. 22690	. 77387	213	. 22613	. 00077	0	. 99923	36
25	8. 77522	211	11. 22478	8. 77600	211	11. 22400	10. 00077	1	9. 99923	35
26	. 77733	210	. 22267	. 77811	211	. 22189	. 00078	1	. 99922	34
27	. 77943	209	. 22057	. 78022	210	. 21978	. 00079	1	. 99921	33
28	. 78152	208	. 21848	. 78232	209	. 21768	. 00080	0	. 99920	32
29	. 78360	208	. 21640	. 78441	208	. 21559	. 00080	1	. 99920	31
30	8. 78568	206	11. 21432	8. 78649	206	11. 21351	10. 00081	1	9. 99919	30
31	. 78774	205	. 21226	. 78855	206	. 21145	. 00082	1	. 99918	29
32	. 78979	204	. 21021	. 79061	205	. 20939	. 00083	0	. 99917	28
33	. 79183	203	. 20817	. 79266	204	. 20734	. 00083	1	. 99917	27
34	. 79386	202	. 20614	. 79470	203	. 20530	. 00084	1	. 99916	26
35	8. 79588	201	11. 20412	8. 79673	202	11. 20327	10. 00085	1	9. 99915	25
36	. 79789	201	. 20211	. 79875	201	. 20125	. 00086	1	. 99914	24
37	. 79990	199	. 20010	. 80076	201	. 19924	. 00087	0	. 99913	23
38	. 80189	199	. 19811	. 80277	199	. 19723	. 00087	1	. 99913	22
39	. 80388	197	. 19612	. 80476	198	. 19524	. 00088	1	. 99912	21
40	8. 80585	197	11. 19415	8. 80674	198	11. 19326	10. 00089	1	9. 99911	20
41	. 80782	196	. 19218	. 80872	196	. 19128	. 00090	1	. 99910	19
42	. 80978	195	. 19022	. 81068	196	. 18932	. 00091	0	. 99909	18
43	. 81173	194	. 18827	. 81264	195	. 18736	. 00091	1	. 99909	17
44	. 81367	193	. 18633	. 81459	194	. 18541	. 00092	1	. 99908	16
45	8. 81560	192	11. 18440	8. 81653	193	11. 18347	10. 00093	1	9. 99907	15
46	. 81752	192	. 18248	. 81846	192	. 18154	. 00094	1	. 99906	14
47	. 81944	190	. 18056	. 82038	192	. 17962	. 00095	1	. 99905	13
48	. 82134	190	. 17866	. 82230	190	. 17770	. 00096	0	. 99904	12
49	. 82324	189	. 17676	. 82420	190	. 17580	. 00096	1	. 99904	11
50	8. 82513	188	11. 17487	8. 82610	189	11. 17390	10. 00097	1	9. 99903	10
51	. 82701	187	. 17299	. 82799	188	. 17201	. 00098	1	. 99902	9
52	. 82888	187	. 17112	. 82987	188	. 17013	. 00099	1	. 99901	8
53	. 83075	186	. 16925	. 83175	186	. 16825	. 00100	1	. 99900	7
54	. 83261	185	. 16739	. 83361	186	. 16639	. 00101	1	. 99899	6
55	8. 83446	184	11. 16554	8. 83547	185	11. 16453	10. 00102	0	9. 99898	5
56	. 83630	183	. 16370	. 83732	184	. 16268	. 00102	1	. 99898	4
57	. 83813	183	. 16187	. 83916	184	. 16084	. 00103	1	. 99897	3
58	. 83996	181	. 16004	. 84100	182	. 15900	. 00104	1	. 99896	2
59	. 84177	181	. 15823	. 84282	182	. 15718	. 00105	1	. 99895	1
60	8. 84358		11. 15642	8. 84464		11. 15536	10. 00106		9. 99894	0
↑ 93°→	cos	Diff. 1′	sec	cot	Diff. 1′	tan	csc	Diff. 1′	sin	↑ ←86°

TABLE 33

Logarithms of Trigonometric Functions

4° → ′	sin	Diff. 1′	csc	tan	Diff. 1′	cot	sec	Diff. 1′	cos	← 175° ′
0	8.84358	181	11.15642	8.84464	182	11.15536	10.00106	1	9.99894	60
1	.84539	179	.15461	.84646	180	.15354	.00107	1	.99893	59
2	.84718	179	.15282	.84826	180	.15174	.00108	1	.99892	58
3	.84897	178	.15103	.85006	179	.14994	.00109	0	.99891	57
4	.85075	177	.14925	.85185	178	.14815	.00109	1	.99891	56
5	8.85252	177	11.14748	8.85363	177	11.14637	10.00110	1	9.99890	55
6	.85429	176	.14571	.85540	177	.14460	.00111	1	.99889	54
7	.85605	175	.14395	.85717	176	.14283	.00112	1	.99888	53
8	.85780	175	.14220	.85893	176	.14107	.00113	1	.99887	52
9	.85955	173	.14045	.86069	174	.13931	.00114	1	.99886	51
10	8.86128	173	11.13872	8.86243	174	11.13757	10.00115	1	9.99885	50
11	.86301	173	.13699	.86417	174	.13583	.00116	1	.99884	49
12	.86474	171	.13526	.86591	172	.13409	.00117	1	.99883	48
13	.86645	171	.13355	.86763	172	.13237	.00118	1	.99882	47
14	.86816	171	.13184	.86935	171	.13065	.00119	1	.99881	46
15	8.86987	169	11.13013	8.87106	171	11.12894	10.00120	1	9.99880	45
16	.87156	169	.12844	.87277	170	.12723	.00121	0	.99879	44
17	.87325	169	.12675	.87447	169	.12553	.00121	1	.99879	43
18	.87494	167	.12506	.87616	169	.12384	.00122	1	.99878	42
19	.87661	168	.12339	.87785	168	.12215	.00123	1	.99877	41
20	8.87829	166	11.12171	8.87953	167	11.12047	10.00124	1	9.99876	40
21	.87995	166	.12005	.88120	167	.11880	.00125	1	.99875	39
22	.88161	165	.11839	.88287	166	.11713	.00126	1	.99874	38
23	.88326	164	.11674	.88453	165	.11547	.00127	1	.99873	37
24	.88490	164	.11510	.88618	165	.11382	.00128	1	.99872	36
25	8.88654	163	11.11346	8.88783	165	11.11217	10.00129	1	9.99871	35
26	.88817	163	.11183	.88948	163	.11052	.00130	1	.99870	34
27	.88980	162	.11020	.89111	163	.10889	.00131	1	.99869	33
28	.89142	162	.10858	.89274	163	.10726	.00132	1	.99868	32
29	.89304	160	.10696	.89437	161	.10563	.00133	1	.99867	31
30	8.89464	161	11.10536	8.89598	162	11.10402	10.00134	1	9.99866	30
31	.89625	159	.10375	.89760	160	.10240	.00135	1	.99865	29
32	.89784	159	.10216	.89920	160	.10080	.00136	1	.99864	28
33	.89943	159	.10057	.90080	160	.09920	.00137	1	.99863	27
34	.90102	158	.09898	.90240	159	.09760	.00138	1	.99862	26
35	8.90260	157	11.09740	8.90399	158	11.09601	10.00139	1	9.99861	25
36	.90417	157	.09583	.90557	158	.09443	.00140	1	.99860	24
37	.90574	156	.09426	.90715	157	.09285	.00141	1	.99859	23
38	.90730	155	.09270	.90872	157	.09128	.00142	1	.99858	22
39	.90885	155	.09115	.91029	156	.08971	.00143	1	.99857	21
40	8.91040	155	11.08960	8.91185	155	11.08815	10.00144	1	9.99856	20
41	.91195	154	.08805	.91340	155	.08660	.00145	1	.99855	19
42	.91349	153	.08651	.91495	155	.08505	.00146	1	.99854	18
43	.91502	153	.08498	.91650	153	.08350	.00147	1	.99853	17
44	.91655	152	.08345	.91803	154	.08197	.00148	1	.99852	16
45	8.91807	152	11.08193	8.91957	153	11.08043	10.00149	1	9.99851	15
46	.91959	151	.08041	.92110	152	.07890	.00150	2	.99850	14
47	.92110	151	.07890	.92262	152	.07738	.00152	1	.99848	13
48	.92261	150	.07739	.92414	151	.07586	.00153	1	.99847	12
49	.92411	150	.07589	.92565	151	.07435	.00154	1	.99846	11
50	8.92561	149	11.07439	8.92716	150	11.07284	10.00155	1	9.99845	10
51	.92710	149	.07290	.92866	150	.07134	.00156	1	.99844	9
52	.92859	148	.07141	.93016	149	.06984	.00157	1	.99843	8
53	.93007	147	.06993	.93165	148	.06835	.00158	1	.99842	7
54	.93154	147	.06846	.93313	149	.06687	.00159	1	.99841	6
55	8.93301	147	11.06699	8.93462	147	11.06538	10.00160	1	9.99840	5
56	.93448	146	.06552	.93609	147	.06391	.00161	1	.99839	4
57	.93594	146	.06406	.93756	147	.06244	.00162	1	.99838	3
58	.93740	145	.06260	.93903	146	.06097	.00163	1	.99837	2
59	.93885	145	.06115	.94049	146	.05951	.00164	2	.99836	1
60	8.94030		11.05970	8.94195		11.05805	10.00166		9.99834	0
94° →	cos	Diff. 1′	sec	cot	Diff. 1′	tan	csc	Diff. 1′	sin	← 85°

TABLE 33

Logarithms of Trigonometric Functions

5° → '	sin	Diff. 1'	csc	tan	Diff. 1'	cot	sec	Diff. 1'	cos	← 174° '
0	8. 94030	144	11. 05970	8. 94195	145	11. 05805	10. 00166	1	9. 99834	60
1	. 94174	143	. 05826	. 94340	145	. 05660	. 00167	1	. 99833	59
2	. 94317	144	. 05683	. 94485	145	. 05515	. 00168	1	. 99832	58
3	. 94461	142	. 05539	. 94630	143	. 05370	. 00169	1	. 99831	57
4	. 94603	143	. 05397	. 94773	144	. 05227	. 00170	1	. 99830	56
5	8. 94746	141	11. 05254	8. 94917	143	11. 05083	10. 00171	1	9. 99829	55
6	. 94887	142	. 05113	. 95060	142	. 04940	. 00172	1	. 99828	54
7	. 95029	141	. 04971	. 95202	142	. 04798	. 00173	2	. 99827	53
8	. 95170	140	. 04830	. 95344	142	. 04656	. 00175	1	. 99825	52
9	. 95310	140	. 04690	. 95486	141	. 04514	. 00176	1	. 99824	51
10	8. 95450	139	11. 04550	8. 95627	140	11. 04373	10. 00177	1	9. 99823	50
11	. 95589	139	. 04411	. 95767	141	. 04233	. 00178	1	. 99822	49
12	. 95728	139	. 04272	. 95908	139	. 04092	. 00179	1	. 99821	48
13	. 95867	138	. 04133	. 96047	140	. 03953	. 00180	1	. 99820	47
14	. 96005	138	. 03995	. 96187	138	. 03813	. 00181	2	. 99819	46
15	8. 96143	137	11. 03857	8. 96325	139	11. 03675	10. 00183	1	9. 99817	45
16	. 96280	137	. 03720	. 96464	138	. 03536	. 00184	1	. 99816	44
17	. 96417	136	. 03583	. 96602	137	. 03398	. 00185	1	. 99815	43
18	. 96553	136	. 03447	. 96739	138	. 03261	. 00186	1	. 99814	42
19	. 96689	136	. 03311	. 96877	136	. 03123	. 00187	1	. 99813	41
20	8. 96825	135	11. 03175	8. 97013	137	11. 02987	10. 00188	2	9. 99812	40
21	. 96960	135	. 03040	. 97150	135	. 02850	. 00190	1	. 99810	39
22	. 97095	134	. 02905	. 97285	136	. 02715	. 00191	1	. 99809	38
23	. 97229	134	. 02771	. 97421	135	. 02579	. 00192	1	. 99808	37
24	. 97363	133	. 02637	. 97556	135	. 02444	. 00193	1	. 99807	36
25	8. 97496	133	11. 02504	8. 97691	134	11. 02309	10. 00194	2	9. 99806	35
26	. 97629	133	. 02371	. 97825	134	. 02175	. 00196	1	. 99804	34
27	. 97762	132	. 02238	. 97959	133	. 02041	. 00197	1	. 99803	33
28	. 97894	132	. 02106	. 98092	133	. 01908	. 00198	1	. 99802	32
29	. 98026	131	. 01974	. 98225	133	. 01775	. 00199	1	. 99801	31
30	8. 98157	131	11. 01843	8. 98358	132	11. 01642	10. 00200	2	9. 99800	30
31	. 98288	131	. 01712	. 98490	132	. 01510	. 00202	1	. 99798	29
32	. 98419	130	. 01581	. 98622	131	. 01378	. 00203	1	. 99797	28
33	. 98549	130	. 01451	. 98753	131	. 01247	. 00204	1	. 99796	27
34	. 98679	129	. 01321	. 98884	131	. 01116	. 00205	2	. 99795	26
35	8. 98808	129	11. 01192	8. 99015	130	11. 00985	10. 00207	1	9. 99793	25
36	. 98937	129	. 01063	. 99145	130	. 00855	. 00208	1	. 99792	24
37	. 99066	128	. 00934	. 99275	130	. 00725	. 00209	1	. 99791	23
38	. 99194	128	. 00806	. 99405	129	. 00595	. 00210	2	. 99790	22
39	. 99322	128	. 00678	. 99534	128	. 00466	. 00212	1	. 99788	21
40	8. 99450	127	11. 00550	8. 99662	129	11. 00338	10. 00213	1	9. 99787	20
41	. 99577	127	. 00423	. 99791	128	. 00209	. 00214	1	. 99786	19
42	. 99704	126	. 00296	8. 99919	127	11. 00081	. 00215	2	. 99785	18
43	. 99830	126	. 00170	9. 00046	128	10. 99954	. 00217	1	. 99783	17
44	8. 99956	126	11. 00044	. 00174	127	. 99826	. 00218	1	. 99782	16
45	9. 00082	125	10. 99918	9. 00301	126	10. 99699	10. 00219	1	9. 99781	15
46	. 00207	125	. 99793	. 00427	126	. 99573	. 00220	2	. 99780	14
47	. 00332	124	. 99668	. 00553	126	. 99447	. 00222	1	. 99778	13
48	. 00456	125	. 99544	. 00679	126	. 99321	. 00223	1	. 99777	12
49	. 00581	123	. 99419	. 00805	125	. 99195	. 00224	1	. 99776	11
50	9. 00704	124	10. 99296	9. 00930	125	10. 99070	10. 00225	2	9. 99775	10
51	. 00828	123	. 99172	. 01055	124	. 98945	. 00227	1	. 99773	9
52	. 00951	123	. 99049	. 01179	124	. 98821	. 00228	1	. 99772	8
53	. 01074	122	. 98926	. 01303	124	. 98697	. 00229	2	. 99771	7
54	. 01196	122	. 98804	. 01427	123	. 98573	. 00231	1	. 99769	6
55	9. 01318	122	10. 98682	9. 01550	123	10. 98450	10. 00232	1	9. 99768	5
56	. 01440	121	. 98560	. 01673	123	. 98327	. 00233	2	. 99767	4
57	. 01561	121	. 98439	. 01796	122	. 98204	. 00235	1	. 99765	3
58	. 01682	121	. 98318	. 01918	122	. 98082	. 00236	1	. 99764	2
59	. 01803	120	. 98197	. 02040	122	. 97960	. 00237	2	. 99763	1
60	9. 01923		10. 98077	9. 02162		10. 97838	10. 00239		9. 99761	0
↑ 95° →	cos	Diff. 1'	sec	cot	Diff. 1'	tan	csc	Diff. 1'	sin	← 84° ↑

TABLE 33
Logarithms of Trigonometric Functions

6°→ ↓	sin	Diff. 1′	csc	tan	Diff. 1′	cot	sec	Diff. 1′	cos	←173° ↓
′										′
0	9. 01923	120	10. 98077	9. 02162	121	10. 97838	10. 00239	1	9. 99761	60
1	. 02043	120	. 97957	. 02283	121	. 97717	. 00240	1	. 99760	59
2	. 02163	120	. 97837	. 02404	121	. 97596	. 00241	2	. 99759	58
3	. 02283	119	. 97717	. 02525	120	. 97475	. 00243	1	. 99757	57
4	. 02402	118	. 97598	. 02645	121	. 97355	. 00244	1	. 99756	56
5	9. 02520	119	10. 97480	9. 02766	119	10. 97234	10. 00245	2	9. 99755	55
6	. 02639	118	. 97361	. 02885	120	. 97115	. 00247	1	. 99753	54
7	. 02757	117	. 97243	. 03005	119	. 96995	. 00248	1	. 99752	53
8	. 02874	118	. 97126	. 03124	118	. 96876	. 00249	2	. 99751	52
9	. 02992	117	. 97008	. 03242	119	. 96758	. 00251	1	. 99749	51
10	9. 03109	117	10. 96891	9. 03361	118	10. 96639	10. 00252	1	9. 99748	50
11	. 03226	116	. 96774	. 03479	118	. 96521	. 00253	2	. 99747	49
12	. 03342	116	. 96658	. 03597	117	. 96403	. 00255	1	. 99745	48
13	. 03458	116	. 96542	. 03714	118	. 96286	. 00256	2	. 99744	47
14	. 03574	116	. 96426	. 03832	116	. 96168	. 00258	1	. 99742	46
15	9. 03690	115	10. 96310	9. 03948	117	10. 96052	10. 00259	1	9. 99741	45
16	. 03805	115	. 96195	. 04065	116	. 95935	. 00260	2	. 99740	44
17	. 03920	114	. 96080	. 04181	116	. 95819	. 00262	1	. 99738	43
18	. 04034	115	. 95966	. 04297	116	. 95703	. 00263	1	. 99737	42
19	. 04149	113	. 95851	. 04413	115	. 95587	. 00264	2	. 99736	41
20	9. 04262	114	10. 95738	9. 04528	115	10. 95472	10. 00266	1	9. 99734	40
21	. 04376	114	. 95624	. 04643	115	. 95357	. 00267	2	. 99733	39
22	. 04490	113	. 95510	. 04758	115	. 95242	. 00269	1	. 99731	38
23	. 04603	112	. 95397	. 04873	114	. 95127	. 00270	2	. 99730	37
24	. 04715	113	. 95285	. 04987	114	. 95013	. 00272	1	. 99728	36
25	9. 04828	112	10. 95172	9. 05101	113	10. 94899	10. 00273	1	9. 99727	35
26	. 04940	112	. 95060	. 05214	114	. 94786	. 00274	2	. 99726	34
27	. 05052	112	. 94948	. 05328	113	. 94672	. 00276	1	. 99724	33
28	. 05164	111	. 94836	. 05441	112	. 94559	. 00277	2	. 99723	32
29	. 05275	111	. 94725	. 05553	113	. 94447	. 00279	1	. 99721	31
30	9. 05386	111	10. 94614	9. 05666	112	10. 94334	10. 00280	2	9. 99720	30
31	. 05497	110	. 94503	. 05778	112	. 94222	. 00282	1	. 99718	29
32	. 05607	110	. 94393	. 05890	112	. 94110	. 00283	1	. 99717	28
33	. 05717	110	. 94283	. 06002	111	. 93998	. 00284	2	. 99716	27
34	. 05827	110	. 94173	. 06113	111	. 93887	. 00286	1	. 99714	26
35	9. 05937	109	10. 94063	9. 06224	111	10. 93776	10. 00287	2	9. 99713	25
36	. 06046	109	. 93954	. 06335	110	. 93665	. 00289	1	. 99711	24
37	. 06155	109	. 93845	. 06445	111	. 93555	. 00290	2	. 99710	23
38	. 06264	108	. 93736	. 06556	110	. 93444	. 00292	1	. 99708	22
39	. 06372	109	. 93628	. 06666	109	. 93334	. 00293	2	. 99707	21
40	9. 06481	108	10. 93519	9. 06775	110	10. 93225	10. 00295	1	9. 99705	20
41	. 06589	107	. 93411	. 06885	109	. 93115	. 00296	2	. 99704	19
42	. 06696	108	. 93304	. 06994	109	. 93006	. 00298	1	. 99702	18
43	. 06804	107	. 93196	. 07103	108	. 92897	. 00299	2	. 99701	17
44	. 06911	107	. 93089	. 07211	109	. 92789	. 00301	1	. 99699	16
45	9. 07018	106	10. 92982	9. 07320	108	10. 92680	10. 00302	2	9. 99698	15
46	. 07124	107	. 92876	. 07428	108	. 92572	. 00304	1	. 99696	14
47	. 07231	106	. 92769	. 07536	107	. 92464	. 00305	2	. 99695	13
48	. 07337	105	. 92663	. 07643	108	. 92357	. 00307	1	. 99693	12
49	. 07442	106	. 92558	. 07751	107	. 92249	. 00308	2	. 99692	11
50	9. 07548	105	10. 92452	9. 07858	106	10. 92142	10. 00310	1	9. 99690	10
51	. 07653	105	. 92347	. 07964	107	. 92036	. 00311	2	. 99689	9
52	. 07758	105	. 92242	. 08071	106	. 91929	. 00313	1	. 99687	8
53	. 07863	105	. 92137	. 08177	106	. 91823	. 00314	2	. 99686	7
54	. 07968	104	. 92032	. 08283	106	. 91717	. 00316	1	. 99684	6
55	9. 08072	104	10. 91928	9. 08389	106	10. 91611	10. 00317	2	9. 99683	5
56	. 08176	104	. 91824	. 08495	105	. 91505	. 00319	1	. 99681	4
57	. 08280	103	. 91720	. 08600	105	. 91400	. 00320	2	. 99680	3
58	. 08383	103	. 91617	. 08705	105	. 91295	. 00322	1	. 99678	2
59	. 08486	103	. 91514	. 08810	104	. 91190	. 00323	2	. 99677	1
60	9. 08589		10. 91411	9. 08914		10. 91086	10. 00325		9. 99675	0
↑ 96°→	cos	Diff. 1′	sec	cot	Diff. 1′	tan	csc	Diff. 1′	sin	↑ ←83°

TABLE 33
Logarithms of Trigonometric Functions

7°→ ↓ ′	sin	Diff. 1′	csc	tan	Diff. 1′	cot	sec	Diff. 1′	cos	←172° ↓ ′
0	9. 08589	103	10. 91411	9. 08914	105	10. 91086	10. 00325	1	9. 99675	60
1	. 08692	103	. 91308	. 09019	104	. 90981	. 00326	2	. 99674	59
2	. 08795	102	. 91205	. 09123	104	. 90877	. 00328	2	. 99672	58
3	. 08897	102	. 91103	. 09227	103	. 90773	. 00330	1	. 99670	57
4	. 08999	102	. 91001	. 09330	104	. 90670	. 00331	2	. 99669	56
5	9. 09101	101	10. 90899	9. 09434	103	10. 90566	10. 00333	1	9. 99667	55
6	. 09202	102	. 90798	. 09537	103	. 90463	. 00334	2	. 99666	54
7	. 09304	101	. 90696	. 09640	102	. 90360	. 00336	1	. 99664	53
8	. 09405	101	. 90595	. 09742	103	. 90258	. 00337	2	. 99663	52
9	. 09506	100	. 90494	. 09845	102	. 90155	. 00339	2	. 99661	51
10	9. 09606	101	10. 90394	9. 09947	102	10. 90053	10. 00341	1	9. 99659	50
11	. 09707	100	. 90293	. 10049	101	. 89951	. 00342	2	. 99658	49
12	. 09807	100	. 90193	. 10150	102	. 89850	. 00344	1	. 99656	48
13	. 09907	99	. 90093	. 10252	101	. 89748	. 00345	2	. 99655	47
14	. 10006	100	. 89994	. 10353	101	. 89647	. 00347	2	. 99653	46
15	9. 10106	99	10. 89894	9. 10454	101	10. 89546	10. 00349	1	9. 99651	45
16	. 10205	99	. 89795	. 10555	101	. 89445	. 00350	2	. 99650	44
17	. 10304	98	. 89696	. 10656	100	. 89344	. 00352	1	. 99648	43
18	. 10402	99	. 89598	. 10756	100	. 89244	. 00353	2	. 99647	42
19	. 10501	98	. 89499	. 10856	100	. 89144	. 00355	2	. 99645	41
20	9. 10599	98	10. 89401	9. 10956	100	10. 89044	10. 00357	1	9. 99643	40
21	. 10697	98	. 89303	. 11056	99	. 88944	. 00358	2	. 99642	39
22	. 10795	98	. 89205	. 11155	99	. 88845	. 00360	2	. 99640	38
23	. 10893	97	. 89107	. 11254	99	. 88746	. 00362	1	. 99638	37
24	. 10990	97	. 89010	. 11353	99	. 88647	. 00363	2	. 99637	36
25	9. 11087	97	10. 88913	9. 11452	99	10. 88548	10. 00365	2	9. 99635	35
26	. 11184	97	. 88816	. 11551	98	. 88449	. 00367	1	. 99633	34
27	. 11281	96	. 88719	. 11649	98	. 88351	. 00368	2	. 99632	33
28	. 11377	97	. 88623	. 11747	98	. 88253	. 00370	1	. 99630	32
29	. 11474	96	. 88526	. 11845	98	. 88155	. 00371	2	. 99629	31
30	9. 11570	96	10. 88430	9. 11943	97	10. 88057	10. 00373	2	9. 99627	30
31	. 11666	95	. 88334	. 12040	98	. 87960	. 00375	1	. 99625	29
32	. 11761	96	. 88239	. 12138	97	. 87862	. 00376	2	. 99624	28
33	. 11857	95	. 88143	. 12235	97	. 87765	. 00378	2	. 99622	27
34	. 11952	95	. 88048	. 12332	96	. 87668	. 00380	2	. 99620	26
35	9. 12047	95	10. 87953	9. 12428	97	10. 87572	10. 00382	1	9. 99618	25
36	. 12142	94	. 87858	. 12525	96	. 87475	. 00383	2	. 99617	24
37	. 12236	95	. 87764	. 12621	96	. 87379	. 00385	2	. 99615	23
38	. 12331	94	. 87669	. 12717	96	. 87283	. 00387	1	. 99613	22
39	. 12425	94	. 87575	. 12813	96	. 87187	. 00388	2	. 99612	21
40	9. 12519	93	10. 87481	9. 12909	95	10. 87091	10. 00390	2	9. 99610	20
41	. 12612	94	. 87388	. 13004	95	. 86996	. 00392	1	. 99608	19
42	. 12706	93	. 87294	. 13099	95	. 86901	. 00393	2	. 99607	18
43	. 12799	93	. 87201	. 13194	95	. 86806	. 00395	2	. 99605	17
44	. 12892	93	. 87108	. 13289	95	. 86711	. 00397	2	. 99603	16
45	9. 12985	93	10. 87015	9. 13384	94	10. 86616	10. 00399	1	9. 99601	15
46	. 13078	93	. 86922	. 13478	95	. 86522	. 00400	2	. 99600	14
47	. 13171	92	. 86829	. 13573	94	. 86427	. 00402	2	. 99598	13
48	. 13263	92	. 86737	. 13667	94	. 86333	. 00404	1	. 99596	12
49	. 13355	92	. 86645	. 13761	93	. 86239	. 00405	2	. 99595	11
50	9. 13447	92	10. 86553	9. 13854	94	10. 86146	10. 00407	2	9. 99593	10
51	. 13539	91	. 86461	. 13948	93	. 86052	. 00409	2	. 99591	9
52	. 13630	92	. 86370	. 14041	93	. 85959	. 00411	1	. 99589	8
53	. 13722	91	. 86278	. 14134	93	. 85866	. 00412	2	. 99588	7
54	. 13813	91	. 86187	. 14227	93	. 85773	. 00414	2	. 99586	6
55	9. 13904	90	10. 86096	9. 14320	92	10. 85680	10. 00416	2	9. 99584	5
56	. 13994	91	. 86006	. 14412	92	. 85588	. 00418	1	. 99582	4
57	. 14085	90	. 85915	. 14504	93	. 85496	. 00419	2	. 99581	3
58	. 14175	91	. 85825	. 14597	91	. 85403	. 00421	2	. 99579	2
59	. 14266	90	. 85734	. 14688	92	. 85312	. 00423	2	. 99577	1
60	9. 14356		10. 85644	9. 14780		10. 85220	10. 00425		9. 99575	0
↑ 97°→	cos	Diff. 1′	sec	cot	Diff. 1′	tan	csc	Diff. 1′	sin	↑ ←82°

TABLE 33

Logarithms of Trigonometric Functions

8°→ ↓ ′	sin	Diff. 1′	csc	tan	Diff. 1′	cot	sec	Diff. 1′	cos	←171° ↓ ′
0	9. 14356	89	10. 85644	9. 14780	92	10. 85220	10. 00425	1	9. 99575	60
1	. 14445	90	. 85555	. 14872	91	. 85128	. 00426	2	. 99574	59
2	. 14535	89	. 85465	. 14963	91	. 85037	. 00428	2	. 99572	58
3	. 14624	90	. 85376	. 15054	91	. 84946	. 00430	2	. 99570	57
4	. 14714	89	. 85286	. 15145	91	. 84855	. 00432	2	. 99568	56
5	9. 14803	88	10. 85197	9. 15236	91	10. 84764	10. 00434	1	9. 99566	55
6	. 14891	89	. 85109	. 15327	90	. 84673	. 00435	2	. 99565	54
7	. 14980	89	. 85020	. 15417	91	. 84583	. 00437	2	. 99563	53
8	. 15069	88	. 84931	. 15508	90	. 84492	. 00439	2	. 99561	52
9	. 15157	88	. 84843	. 15598	90	. 84402	. 00441	2	. 99559	51
10	9. 15245	88	10. 84755	9. 15688	89	10. 84312	10. 00443	1	9. 99557	50
11	. 15333	88	. 84667	. 15777	90	. 84223	. 00444	2	. 99556	49
12	. 15421	87	. 84579	. 15867	89	. 84133	. 00446	2	. 99554	48
13	. 15508	88	. 84492	. 15956	90	. 84044	. 00448	2	. 99552	47
14	. 15596	87	. 84404	. 16046	89	. 83954	. 00450	2	. 99550	46
15	9. 15683	87	10. 84317	9. 16135	89	10. 83865	10. 00452	2	9. 99548	45
16	. 15770	87	. 84230	. 16224	88	. 83776	. 00454	1	. 99546	44
17	. 15857	87	. 84143	. 16312	89	. 83688	. 00455	2	. 99545	43
18	. 15944	86	. 84056	. 16401	88	. 83599	. 00457	2	. 99543	42
19	. 16030	86	. 83970	. 16489	88	. 83511	. 00459	2	. 99541	41
20	9. 16116	87	10. 83884	9. 16577	88	10. 83423	10. 00461	2	9. 99539	40
21	. 16203	86	. 83797	. 16665	88	. 83335	. 00463	2	. 99537	39
22	. 16289	85	. 83711	. 16753	88	. 83247	. 00465	2	. 99535	38
23	. 16374	86	. 83626	. 16841	87	. 83159	. 00467	1	. 99533	37
24	. 16460	85	. 83540	. 16928	88	. 83072	. 00468	2	. 99532	36
25	9. 16545	86	10. 83455	9. 17016	87	10. 82984	10. 00470	2	9. 99530	35
26	. 16631	85	. 83369	. 17103	87	. 82897	. 00472	2	. 99528	34
27	. 16716	85	. 83284	. 17190	87	. 82810	. 00474	2	. 99526	33
28	. 16801	85	. 83199	. 17277	86	. 82723	. 00476	2	. 99524	32
29	. 16886	84	. 83114	. 17363	87	. 82637	. 00478	2	. 99522	31
30	9. 16970	85	10. 83030	9. 17450	86	10. 82550	10. 00480	2	9. 99520	30
31	. 17055	84	. 82945	. 17536	86	. 82464	. 00482	1	. 99518	29
32	. 17139	84	. 82861	. 17622	86	. 82378	. 00483	2	. 99517	28
33	. 17223	84	. 82777	. 17708	86	. 82292	. 00485	2	. 99515	27
34	. 17307	84	. 82693	. 17794	86	. 82206	. 00487	2	. 99513	26
35	9. 17391	83	10. 82609	9. 17880	85	10. 82120	10. 00489	2	9. 99511	25
36	. 17474	84	. 82526	. 17965	86	. 82035	. 00491	2	. 99509	24
37	. 17558	83	. 82442	. 18051	85	. 81949	. 00493	2	. 99507	23
38	. 17641	83	. 82359	. 18136	85	. 81864	. 00495	2	. 99505	22
39	. 17724	83	. 82276	. 18221	85	. 81779	. 00497	2	. 99503	21
40	9. 17807	83	10. 82193	9. 18306	85	10. 81694	10. 00499	2	9. 99501	20
41	. 17890	83	. 82110	. 18391	84	. 81609	. 00501	2	. 99499	19
42	. 17973	82	. 82027	. 18475	85	. 81525	. 00503	2	. 99497	18
43	. 18055	82	. 81945	. 18560	84	. 81440	. 00505	1	. 99495	17
44	. 18137	83	. 81863	. 18644	84	. 81356	. 00506	2	. 99494	16
45	9. 18220	82	10. 81780	9. 18728	84	10. 81272	10. 00508	2	9. 99492	15
46	. 18302	81	. 81698	. 18812	84	. 81188	. 00510	2	. 99490	14
47	. 18383	82	. 81617	. 18896	83	. 81104	. 00512	2	. 99488	13
48	. 18465	82	. 81535	. 18979	84	. 81021	. 00514	2	. 99486	12
49	. 18547	81	. 81453	. 19063	83	. 80937	. 00516	2	. 99484	11
50	9. 18628	81	10. 81372	9. 19146	83	10. 80854	10. 00518	2	9. 99482	10
51	. 18709	81	. 81291	. 19229	83	. 80771	. 00520	2	. 99480	9
52	. 18790	81	. 81210	. 19312	83	. 80688	. 00522	2	. 99478	8
53	. 18871	81	. 81129	. 19395	83	. 80605	. 00524	2	. 99476	7
54	. 18952	81	. 81048	. 19478	83	. 80522	. 00526	2	. 99474	6
55	9. 19033	80	10. 80967	9. 19561	82	10. 80439	10. 00528	2	9. 99472	5
56	. 19113	80	. 80887	. 19643	82	. 80357	. 00530	2	. 99470	4
57	. 19193	80	. 80807	. 19725	82	. 80275	. 00532	2	. 99468	3
58	. 19273	80	. 80727	. 19807	82	. 80193	. 00534	2	. 99466	2
59	. 19353	80	. 80647	. 19889	82	. 80111	. 00536	2	. 99464	1
60	9. 19433		10. 80567	9. 19971		10. 80029	10. 00538		9. 99462	0
↑ 98°→	cos	Diff. 1′	sec	cot	Diff. 1′	tan	csc	Diff. 1′	sin	←81° ↑

TABLE 33

Logarithms of Trigonometric Functions

9° → ↓ ′	sin	Diff. 1′	csc	tan	Diff. 1′	cot	sec	Diff. 1′	cos	← 170° ↓ ′
0	9.19433	80	10.80567	9.19971	82	10.80029	10.00538	2	9.99462	60
1	.19513	79	.80487	.20053	81	.79947	.00540	2	.99460	59
2	.19592	80	.80408	.20134	82	.79866	.00542	2	.99458	58
3	.19672	79	.80328	.20216	81	.79784	.00544	2	.99456	57
4	.19751	79	.80249	.20297	81	.79703	.00546	2	.99454	56
5	9.19830	79	10.80170	9.20378	81	10.79622	10.00548	2	9.99452	55
6	.19909	79	.80091	.20459	81	.79541	.00550	2	.99450	54
7	.19988	79	.80012	.20540	81	.79460	.00552	2	.99448	53
8	.20067	78	.79933	.20621	80	.79379	.00554	2	.99446	52
9	.20145	78	.79855	.20701	81	.79299	.00556	2	.99444	51
10	9.20223	79	10.79777	9.20782	80	10.79218	10.00558	2	9.99442	50
11	.20302	78	.79698	.20862	80	.79138	.00560	2	.99440	49
12	.20380	78	.79620	.20942	80	.79058	.00562	2	.99438	48
13	.20458	77	.79542	.21022	80	.78978	.00564	2	.99436	47
14	.20535	78	.79465	.21102	80	.78898	.00566	2	.99434	46
15	9.20613	78	10.79387	9.21182	79	10.78818	10.00568	3	9.99432	45
16	.20691	77	.79309	.21261	80	.78739	.00571	2	.99429	44
17	.20768	77	.79232	.21341	79	.78659	.00573	2	.99427	43
18	.20845	77	.79155	.21420	79	.78580	.00575	2	.99425	42
19	.20922	77	.79078	.21499	79	.78501	.00577	2	.99423	41
20	9.20999	77	10.79001	9.21578	79	10.78422	10.00579	2	9.99421	40
21	.21076	77	.78924	.21657	79	.78343	.00581	2	.99419	39
22	.21153	76	.78847	.21736	78	.78264	.00583	2	.99417	38
23	.21229	77	.78771	.21814	79	.78186	.00585	2	.99415	37
24	.21306	76	.78694	.21893	78	.78107	.00587	2	.99413	36
25	9.21382	76	10.78618	9.21971	78	10.78029	10.00589	2	9.99411	35
26	.21458	76	.78542	.22049	78	.77951	.00591	2	.99409	34
27	.21534	76	.78466	.22127	78	.77873	.00593	3	.99407	33
28	.21610	75	.78390	.22205	78	.77795	.00596	2	.99404	32
29	.21685	76	.78315	.22283	78	.77717	.00598	2	.99402	31
30	9.21761	75	10.78239	9.22361	77	10.77639	10.00600	2	9.99400	30
31	.21836	76	.78164	.22438	78	.77562	.00602	2	.99398	29
32	.21912	75	.78088	.22516	77	.77484	.00604	2	.99396	28
33	.21987	75	.78013	.22593	77	.77407	.00606	2	.99394	27
34	.22062	75	.77938	.22670	77	.77330	.00608	2	.99392	26
35	9.22137	74	10.77863	9.22747	77	10.77253	10.00610	2	9.99390	25
36	.22211	75	.77789	.22824	77	.77176	.00612	3	.99388	24
37	.22286	75	.77714	.22901	76	.77099	.00615	2	.99385	23
38	.22361	74	.77639	.22977	77	.77023	.00617	2	.99383	22
39	.22435	74	.77565	.23054	76	.76946	.00619	2	.99381	21
40	9.22509	74	10.77491	9.23130	76	10.76870	10.00621	2	9.99379	20
41	.22583	74	.77417	.23206	77	.76794	.00623	2	.99377	19
42	.22657	74	.77343	.23283	76	.76717	.00625	3	.99375	18
43	.22731	74	.77269	.23359	76	.76641	.00628	2	.99372	17
44	.22805	73	.77195	.23435	75	.76565	.00630	2	.99370	16
45	9.22878	74	10.77122	9.23510	76	10.76490	10.00632	2	9.99368	15
46	.22952	73	.77048	.23586	75	.76414	.00634	2	.99366	14
47	.23025	73	.76975	.23661	76	.76339	.00636	2	.99364	13
48	.23098	73	.76902	.23737	75	.76263	.00638	3	.99362	12
49	.23171	73	.76829	.23812	75	.76188	.00641	2	.99359	11
50	9.23244	73	10.76756	9.23887	75	10.76113	10.00643	2	9.99357	10
51	.23317	73	.76683	.23962	75	.76038	.00645	2	.99355	9
52	.23390	72	.76610	.24037	75	.75963	.00647	2	.99353	8
53	.23462	73	.76538	.24112	74	.75888	.00649	3	.99351	7
54	.23535	72	.76465	.24186	75	.75814	.00652	2	.99348	6
55	9.23607	72	10.76393	9.24261	74	10.75739	10.00654	2	9.99346	5
56	.23679	73	.76321	.24335	75	.75665	.00656	2	.99344	4
57	.23752	71	.76248	.24410	74	.75590	.00658	2	.99342	3
58	.23823	72	.76177	.24484	74	.75516	.00660	3	.99340	2
59	.23895	72	.76105	.24558	74	.75442	.00663	2	.99337	1
60	9.23967		10.76033	9.24632		10.75368	10.00665		9.99335	0
↑ 99° →	cos	Diff. 1′	sec	cot	Diff. 1′	tan	csc	Diff. 1′	sin	← 80° ↑

TABLE 33

Logarithms of Trigonometric Functions

10° → ↓ ′	sin	Diff. 1′	csc	tan	Diff. 1′	cot	sec	Diff. 1′	cos	← 169° ↓ ′
0	9. 23967	72	10. 76033	9. 24632	74	10. 75368	10. 00665	2	9. 99335	60
1	. 24039	71	. 75961	. 24706	73	. 75294	. 00667	2	. 99333	59
2	. 24110	71	. 75890	. 24779	74	. 75221	. 00669	3	. 99331	58
3	. 24181	72	. 75819	. 24853	73	. 75147	. 00672	2	. 99328	57
4	. 24253	71	. 75747	. 24926	74	. 75074	. 00674	2	. 99326	56
5	9. 24324	71	10. 75676	9. 25000	73	10. 75000	10. 00676	2	9. 99324	55
6	. 24395	71	. 75605	. 25073	73	. 74927	. 00678	3	. 99322	54
7	. 24466	70	. 75534	. 25146	73	. 74854	. 00681	2	. 99319	53
8	. 24536	71	. 75464	. 25219	73	. 74781	. 00683	2	. 99317	52
9	. 24607	70	. 75393	. 25292	73	. 74708	. 00685	2	. 99315	51
10	9. 24677	71	10. 75323	9. 25365	72	10. 74635	10. 00687	3	9. 99313	50
11	. 24748	70	. 75252	. 25437	73	. 74563	. 00690	2	. 99310	49
12	. 24818	70	. 75182	. 25510	72	. 74490	. 00692	2	. 99308	48
13	. 24888	70	. 75112	. 25582	73	. 74418	. 00694	2	. 99306	47
14	. 24958	70	. 75042	. 25655	72	. 74345	. 00696	3	. 99304	46
15	9. 25028	70	10. 74972	9. 25727	72	10. 74273	10. 00699	2	9. 99301	45
16	. 25098	70	. 74902	. 25799	72	. 74201	. 00701	2	. 99299	44
17	. 25168	69	. 74832	. 25871	72	. 74129	. 00703	3	. 99297	43
18	. 25237	70	. 74763	. 25943	72	. 74057	. 00706	2	. 99294	42
19	. 25307	69	. 74693	. 26015	71	. 73985	. 00708	2	. 99292	41
20	9. 25376	69	10. 74624	9. 26086	72	10. 73914	10. 00710	2	9. 99290	40
21	. 25445	69	. 74555	. 26158	71	. 73842	. 00712	3	. 99288	39
22	. 25514	69	. 74486	. 26229	72	. 73771	. 00715	2	. 99285	38
23	. 25583	69	. 74417	. 26301	71	. 73699	. 00717	2	. 99283	37
24	. 25652	69	. 74348	. 26372	71	. 73628	. 00719	3	. 99281	36
25	9. 25721	69	10. 74279	9. 26443	71	10. 73557	10. 00722	2	9. 99278	35
26	. 25790	68	. 74210	. 26514	71	. 73486	. 00724	2	. 99276	34
27	. 25858	69	. 74142	. 26585	70	. 73415	. 00726	3	. 99274	33
28	. 25927	68	. 74073	. 26655	71	. 73345	. 00729	2	. 99271	32
29	. 25995	68	. 74005	. 26726	71	. 73274	. 00731	2	. 99269	31
30	9. 26063	68	10. 73937	9. 26797	70	10. 73203	10. 00733	3	9. 99267	30
31	. 26131	68	. 73869	. 26867	70	. 73133	. 00736	2	. 99264	29
32	. 26199	68	. 73801	. 26937	71	. 73063	. 00738	2	. 99262	28
33	. 26267	68	. 73733	. 27008	70	. 72992	. 00740	3	. 99260	27
34	. 26335	68	. 73665	. 27078	70	. 72922	. 00743	2	. 99257	26
35	9. 26403	67	10. 73597	9. 27148	70	10. 72852	10. 00745	3	9. 99255	25
36	. 26470	68	. 73530	. 27218	70	. 72782	. 00748	2	. 99252	24
37	. 26538	67	. 73462	. 27288	69	. 72712	. 00750	2	. 99250	23
38	. 26605	67	. 73395	. 27357	70	. 72643	. 00752	3	. 99248	22
39	. 26672	67	. 73328	. 27427	69	. 72573	. 00755	2	. 99245	21
40	9. 26739	67	10. 73261	9. 27496	70	10. 72504	10. 00757	2	9. 99243	20
41	. 26806	67	. 73194	. 27566	69	. 72434	. 00759	3	. 99241	19
42	. 26873	67	. 73127	. 27635	69	. 72365	. 00762	2	. 99238	18
43	. 26940	67	. 73060	. 27704	69	. 72296	. 00764	3	. 99236	17
44	. 27007	66	. 72993	. 27773	69	. 72227	. 00767	2	. 99233	16
45	9. 27073	67	10. 72927	9. 27842	69	10. 72158	10. 00769	2	9. 99231	15
46	. 27140	66	. 72860	. 27911	69	. 72089	. 00771	3	. 99229	14
47	. 27206	67	. 72794	. 27980	69	. 72020	. 00774	2	. 99226	13
48	. 27273	66	. 72727	. 28049	68	. 71951	. 00776	3	. 99224	12
49	. 27339	66	. 72661	. 28117	69	. 71883	. 00779	2	. 99221	11
50	9. 27405	66	10. 72595	9. 28186	68	10. 71814	10. 00781	2	9. 99219	10
51	. 27471	66	. 72529	. 28254	69	. 71746	. 00783	3	. 99217	9
52	. 27537	65	. 72463	. 28323	68	. 71677	. 00786	2	. 99214	8
53	. 27602	66	. 72398	. 28391	68	. 71609	. 00788	3	. 99212	7
54	. 27668	66	. 72332	. 28459	68	. 71541	. 00791	2	. 99209	6
55	9. 27734	65	10. 72266	9. 28527	68	10. 71473	10. 00793	3	9. 99207	5
56	. 27799	65	. 72201	. 28595	67	. 71405	. 00796	2	. 99204	4
57	. 27864	66	. 72136	. 28662	68	. 71338	. 00798	2	. 99202	3
58	. 27930	65	. 72070	. 28730	68	. 71270	. 00800	3	. 99200	2
59	. 27995	65	. 72005	. 28798	67	. 71202	. 00803	2	. 99197	1
60	9. 28060		10. 71940	9. 28865		10. 71135	10. 00805		9. 99195	0
↑ 100° →	cos	Diff. 1′	sec	cot	Diff. 1′	tan	csc	Diff. 1′	sin	← 79° ↑

TABLE 33
Logarithms of Trigonometric Functions

11° → ↓ ′	sin	Diff. 1′	csc	tan	Diff. 1′	cot	sec	Diff. 1′	cos	← 168° ↓ ′
0	9.28060	65	10.71940	9.28865	68	10.71135	10.00805	3	9.99195	60
1	.28125	65	.71875	.28933	67	.71067	.00808	2	.99192	59
2	.28190	64	.71810	.29000	67	.71000	.00810	3	.99190	58
3	.28254	65	.71746	.29067	67	.70933	.00813	2	.99187	57
4	.28319	65	.71681	.29134	67	.70866	.00815	3	.99185	56
5	9.28384	64	10.71616	9.29201	67	10.70799	10.00818	2	9.99182	55
6	.28448	64	.71552	.29268	67	.70732	.00820	3	.99180	54
7	.28512	65	.71488	.29335	67	.70665	.00823	2	.99177	53
8	.28577	64	.71423	.29402	66	.70598	.00825	3	.99175	52
9	.28641	64	.71359	.29468	67	.70532	.00828	2	.99172	51
10	9.28705	64	10.71295	9.29535	66	10.70465	10.00830	3	9.99170	50
11	.28769	64	.71231	.29601	67	.70399	.00833	2	.99167	49
12	.28833	63	.71167	.29668	66	.70332	.00835	3	.99165	48
13	.28896	64	.71104	.29734	66	.70266	.00838	2	.99162	47
14	.28960	64	.71040	.29800	66	.70200	.00840	3	.99160	46
15	9.29024	63	10.70976	9.29866	66	10.70134	10.00843	2	9.99157	45
16	.29087	63	.70913	.29932	66	.70068	.00845	3	.99155	44
17	.29150	64	.70850	.29998	66	.70002	.00848	2	.99152	43
18	.29214	63	.70786	.30064	66	.69936	.00850	3	.99150	42
19	.29277	63	.70723	.30130	65	.69870	.00853	2	.99147	41
20	9.29340	63	10.70660	9.30195	66	10.69805	10.00855	3	9.99145	40
21	.29403	63	.70597	.30261	65	.69739	.00858	2	.99142	39
22	.29466	63	.70534	.30326	65	.69674	.00860	3	.99140	38
23	.29529	62	.70471	.30391	66	.69609	.00863	2	.99137	37
24	.29591	63	.70409	.30457	65	.69543	.00865	3	.99135	36
25	9.29654	62	10.70346	9.30522	65	10.69478	10.00868	2	9.99132	35
26	.29716	63	.70284	.30587	65	.69413	.00870	3	.99130	34
27	.29779	62	.70221	.30652	65	.69348	.00873	3	.99127	33
28	.29841	62	.70159	.30717	65	.69283	.00876	2	.99124	32
29	.29903	63	.70097	.30782	64	.69218	.00878	3	.99122	31
30	9.29966	62	10.70034	9.30846	65	10.69154	10.00881	2	9.99119	30
31	.30028	62	.69972	.30911	64	.69089	.00883	3	.99117	29
32	.30090	61	.69910	.30975	65	.69025	.00886	2	.99114	28
33	.30151	62	.69849	.31040	64	.68960	.00888	3	.99112	27
34	.30213	62	.69787	.31104	64	.68896	.00891	3	.99109	26
35	9.30275	61	10.69725	9.31168	65	10.68832	10.00894	2	9.99106	25
36	.30336	62	.69664	.31233	64	.68767	.00896	3	.99104	24
37	.30398	61	.69602	.31297	64	.68703	.00899	2	.99101	23
38	.30459	62	.69541	.31361	64	.68639	.00901	3	.99099	22
39	.30521	61	.69479	.31425	64	.68575	.00904	3	.99096	21
40	9.30582	61	10.69418	9.31489	63	10.68511	10.00907	2	9.99093	20
41	.30643	61	.69357	.31552	64	.68448	.00909	3	.99091	19
42	.30704	61	.69296	.31616	63	.68384	.00912	2	.99088	18
43	.30765	61	.69235	.31679	64	.68321	.00914	3	.99086	17
44	.30826	61	.69174	.31743	63	.68257	.00917	3	.99083	16
45	9.30887	60	10.69113	9.31806	64	10.68194	10.00920	2	9.99080	15
46	.30947	61	.69053	.31870	63	.68130	.00922	3	.99078	14
47	.31008	60	.68992	.31933	63	.68067	.00925	3	.99075	13
48	.31068	61	.68932	.31996	63	.68004	.00928	2	.99072	12
49	.31129	60	.68871	.32059	63	.67941	.00930	3	.99070	11
50	9.31189	61	10.68811	9.32122	63	10.67878	10.00933	3	9.99067	10
51	.31250	60	.68750	.32185	63	.67815	.00936	2	.99064	9
52	.31310	60	.68690	.32248	63	.67752	.00938	3	.99062	8
53	.31370	60	.68630	.32311	62	.67689	.00941	3	.99059	7
54	.31430	60	.68570	.32373	63	.67627	.00944	2	.99056	6
55	9.31490	59	10.68510	9.32436	62	10.67564	10.00946	3	9.99054	5
56	.31549	60	.68451	.32498	63	.67502	.00949	3	.99051	4
57	.31609	60	.68391	.32561	62	.67439	.00952	2	.99048	3
58	.31669	59	.68331	.32623	62	.67377	.00954	3	.99046	2
59	.31728	60	.68272	.32685	62	.67315	.00957	3	.99043	1
60	9.31788		10.68212	9.32747		10.67253	10.00960		9.99040	0
↑ 101° →	cos	Diff. 1′	sec	cot	Diff. 1′	tan	csc	Diff. 1′	sin	← 78° ↑

TABLE 33

Logarithms of Trigonometric Functions

12° → ↓ ′	sin	Diff. 1′	csc	tan	Diff. 1′	cot	sec	Diff. 1′	cos	← 167° ↓ ′
0	9.31788	59	10.68212	9.32747	63	10.67253	10.00960	2	9.99040	60
1	.31847	60	.68153	.32810	62	.67190	.00962	3	.99038	59
2	.31907	59	.68093	.32872	61	.67128	.00965	3	.99035	58
3	.31966	59	.68034	.32933	62	.67067	.00968	2	.99032	57
4	.32025	59	.67975	.32995	62	.67005	.00970	3	.99030	56
5	9.32084	59	10.67916	9.33057	62	10.66943	10.00973	3	9.99027	55
6	.32143	59	.67857	.33119	61	.66881	.00976	2	.99024	54
7	.32202	59	.67798	.33180	62	.66820	.00978	3	.99022	53
8	.32261	58	.67739	.33242	61	.66758	.00981	3	.99019	52
9	.32319	59	.67681	.33303	62	.66697	.00984	3	.99016	51
10	9.32378	59	10.67622	9.33365	61	10.66635	10.00987	2	9.99013	50
11	.32437	58	.67563	.33426	61	.66574	.00989	3	.99011	49
12	.32495	58	.67505	.33487	61	.66513	.00992	3	.99008	48
13	.32553	59	.67447	.33548	61	.66452	.00995	3	.99005	47
14	.32612	58	.67388	.33609	61	.66391	.00998	2	.99002	46
15	9.32670	58	10.67330	9.33670	61	10.66330	10.01000	3	9.99000	45
16	.32728	58	.67272	.33731	61	.66269	.01003	3	.98997	44
17	.32786	58	.67214	.33792	61	.66208	.01006	3	.98994	43
18	.32844	58	.67156	.33853	60	.66147	.01009	2	.98991	42
19	.32902	58	.67098	.33913	61	.66087	.01011	3	.98989	41
20	9.32960	58	10.67040	9.33974	60	10.66026	10.01014	3	9.98986	40
21	.33018	57	.66982	.34034	61	.65966	.01017	3	.98983	39
22	.33075	58	.66925	.34095	60	.65905	.01020	2	.98980	38
23	.33133	57	.66867	.34155	60	.65845	.01022	3	.98978	37
24	.33190	58	.66810	.34215	61	.65785	.01025	3	.98975	36
25	9.33248	57	10.66752	9.34276	60	10.65724	10.01028	3	9.98972	35
26	.33305	57	.66695	.34336	60	.65664	.01031	2	.98969	34
27	.33362	58	.66638	.34396	60	.65604	.01033	3	.98967	33
28	.33420	57	.66580	.34456	60	.65544	.01036	3	.98964	32
29	.33477	57	.66523	.34516	60	.65484	.01039	3	.98961	31
30	9.33534	57	10.66466	9.34576	59	10.65424	10.01042	3	9.98958	30
31	.33591	56	.66409	.34635	60	.65365	.01045	2	.98955	29
32	.33647	57	.66353	.34695	60	.65305	.01047	3	.98953	28
33	.33704	57	.66296	.34755	59	.65245	.01050	3	.98950	27
34	.33761	57	.66239	.34814	60	.65186	.01053	3	.98947	26
35	9.33818	56	10.66182	9.34874	59	10.65126	10.01056	3	9.98944	25
36	.33874	57	.66126	.34933	59	.65067	.01059	3	.98941	24
37	.33931	56	.66069	.34992	59	.65008	.01062	2	.98938	23
38	.33987	56	.66013	.35051	60	.64949	.01064	3	.98936	22
39	.34043	57	.65957	.35111	59	.64889	.01067	3	.98933	21
40	9.34100	56	10.65900	9.35170	59	10.64830	10.01070	3	9.98930	20
41	.34156	56	.65844	.35229	59	.64771	.01073	3	.98927	19
42	.34212	56	.65788	.35288	59	.64712	.01076	3	.98924	18
43	.34268	56	.65732	.35347	58	.64653	.01079	2	.98921	17
44	.34324	56	.65676	.35405	59	.64595	.01081	3	.98919	16
45	9.34380	56	10.65620	9.35464	59	10.64536	10.01084	3	9.98916	15
46	.34436	55	.65564	.35523	58	.64477	.01087	3	.98913	14
47	.34491	56	.65509	.35581	59	.64419	.01090	3	.98910	13
48	.34547	55	.65453	.35640	58	.64360	.01093	3	.98907	12
49	.34602	56	.65398	.35698	59	.64302	.01096	3	.98904	11
50	9.34658	55	10.65342	9.35757	58	10.64243	10.01099	3	9.98901	10
51	.34713	56	.65287	.35815	58	.64185	.01102	2	.98898	9
52	.34769	55	.65231	.35873	58	.64127	.01104	3	.98896	8
53	.34824	55	.65176	.35931	58	.64069	.01107	3	.98893	7
54	.34879	55	.65121	.35989	58	.64011	.01110	3	.98890	6
55	9.34934	55	10.65066	9.36047	58	10.63953	10.01113	3	9.98887	5
56	.34989	55	.65011	.36105	58	.63895	.01116	3	.98884	4
57	.35044	55	.64956	.36163	58	.63837	.01119	3	.98881	3
58	.35099	55	.64901	.36221	58	.63779	.01122	3	.98878	2
59	.35154	55	.64846	.36279	57	.63721	.01125	3	.98875	1
60	9.35209		10.64791	9.36336		10.63664	10.01128		9.98872	0
↑ 102° →	cos	Diff. 1′	sec	cot	Diff. 1′	tan	csc	Diff. 1′	sin	← 77° ↑

TABLE 33

Logarithms of Trigonometric Functions

13°→ ↓ ′	sin	Diff. 1′	csc	tan	Diff. 1′	cot	sec	Diff. 1′	cos	←166° ↓ ′
0	9.35209	54	10.64791	9.36336	58	10.63664	10.01128	3	9.98872	60
1	.35263	55	.64737	.36394	58	.63606	.01131	2	.98869	59
2	.35318	55	.64682	.36452	57	.63548	.01133	3	.98867	58
3	.35373	54	.64627	.36509	57	.63491	.01136	3	.98864	57
4	.35427	54	.64573	.36566	58	.63434	.01139	3	.98861	56
5	9.35481	55	10.64519	9.36624	57	10.63376	10.01142	3	9.98858	55
6	.35536	54	.64464	.36681	57	.63319	.01145	3	.98855	54
7	.35590	54	.64410	.36738	57	.63262	.01148	3	.98852	53
8	.35644	54	.64356	.36795	57	.63205	.01151	3	.98849	52
9	.35698	54	.64302	.36852	57	.63148	.01154	3	.98846	51
10	9.35752	54	10.64248	9.36909	57	10.63091	10.01157	3	9.98843	50
11	.35806	54	.64194	.36966	57	.63034	.01160	3	.98840	49
12	.35860	54	.64140	.37023	57	.62977	.01163	3	.98837	48
13	.35914	54	.64086	.37080	57	.62920	.01166	3	.98834	47
14	.35968	54	.64032	.37137	56	.62863	.01169	3	.98831	46
15	9.36022	53	10.63978	9.37193	57	10.62807	10.01172	3	9.98828	45
16	.36075	54	.63925	.37250	56	.62750	.01175	3	.98825	44
17	.36129	53	.63871	.37306	57	.62694	.01178	3	.98822	43
18	.36182	54	.63818	.37363	56	.62637	.01181	3	.98819	42
19	.36236	53	.63764	.37419	57	.62581	.01184	3	.98816	41
20	9.36289	53	10.63711	9.37476	56	10.62524	10.01187	3	9.98813	40
21	.36342	53	.63658	.37532	56	.62468	.01190	3	.98810	39
22	.36395	54	.63605	.37588	56	.62412	.01193	3	.98807	38
23	.36449	53	.63551	.37644	56	.62356	.01196	3	.98804	37
24	.36502	53	.63498	.37700	56	.62300	.01199	3	.98801	36
25	9.36555	53	10.63445	9.37756	56	10.62244	10.01202	3	9.98798	35
26	.36608	52	.63392	.37812	56	.62188	.01205	3	.98795	34
27	.36660	53	.63340	.37868	56	.62132	.01208	3	.98792	33
28	.36713	53	.63287	.37924	56	.62076	.01211	3	.98789	32
29	.36766	53	.63234	.37980	55	.62020	.01214	3	.98786	31
30	9.36819	52	10.63181	9.38035	56	10.61965	10.01217	3	9.98783	30
31	.36871	53	.63129	.38091	56	.61909	.01220	3	.98780	29
32	.36924	52	.63076	.38147	55	.61853	.01223	3	.98777	28
33	.36976	52	.63024	.38202	55	.61798	.01226	3	.98774	27
34	.37028	53	.62972	.38257	56	.61743	.01229	3	.98771	26
35	9.37081	52	10.62919	9.38313	55	10.61687	10.01232	3	9.98768	25
36	.37133	52	.62867	.38368	55	.61632	.01235	3	.98765	24
37	.37185	52	.62815	.38423	56	.61577	.01238	3	.98762	23
38	.37237	52	.62763	.38479	55	.61521	.01241	3	.98759	22
39	.37289	52	.62711	.38534	55	.61466	.01244	3	.98756	21
40	9.37341	52	10.62659	9.38589	55	10.61411	10.01247	3	9.98753	20
41	.37393	52	.62607	.38644	55	.61356	.01250	4	.98750	19
42	.37445	52	.62555	.38699	55	.61301	.01254	3	.98746	18
43	.37497	52	.62503	.38754	54	.61246	.01257	3	.98743	17
44	.37549	51	.62451	.38808	55	.61192	.01260	3	.98740	16
45	9.37600	52	10.62400	9.38863	55	10.61137	10.01263	3	9.98737	15
46	.37652	51	.62348	.38918	54	.61082	.01266	3	.98734	14
47	.37703	52	.62297	.38972	55	.61028	.01269	3	.98731	13
48	.37755	51	.62245	.39027	55	.60973	.01272	3	.98728	12
49	.37806	52	.62194	.39082	54	.60918	.01275	3	.98725	11
50	9.37858	51	10.62142	9.39136	54	10.60864	10.01278	3	9.98722	10
51	.37909	51	.62091	.39190	55	.60810	.01281	4	.98719	9
52	.37960	51	.62040	.39245	54	.60755	.01285	3	.98715	8
53	.38011	51	.61989	.39299	54	.60701	.01288	3	.98712	7
54	.38062	51	.61938	.39353	54	.60647	.01291	3	.98709	6
55	9.38113	51	10.61887	9.39407	54	10.60593	10.01294	3	9.98706	5
56	.38164	51	.61836	.39461	54	.60539	.01297	3	.98703	4
57	.38215	51	.61785	.39515	54	.60485	.01300	3	.98700	3
58	.38266	51	.61734	.39569	54	.60431	.01303	3	.98697	2
59	.38317	51	.61683	.39623	54	.60377	.01306	4	.98694	1
60	9.38368		10.61632	9.39677		10.60323	10.01310		9.98690	0
↑ 103°→	cos	Diff. 1′	sec	cot	Diff. 1′	tan	csc	Diff. 1′	sin	↑ ←76°

TABLE 33

Logarithms of Trigonometric Functions

14°→ ′	sin	Diff. 1′	csc	tan	Diff. 1′	cot	sec	Diff. 1′	cos	←165° ′
0	9. 38368	50	10. 61632	9. 39677	54	10. 60323	10. 01310	3	9. 98690	60
1	. 38418	51	. 61582	. 39731	54	. 60269	. 01313	3	. 98687	59
2	. 38469	50	. 61531	. 39785	53	. 60215	. 01316	3	. 98684	58
3	. 38519	51	. 61481	. 39838	54	. 60162	. 01319	3	. 98681	57
4	. 38570	50	. 61430	. 39892	53	. 60108	. 01322	3	. 98678	56
5	9. 38620	50	10. 61380	9. 39945	54	10. 60055	10. 01325	4	9. 98675	55
6	. 38670	51	. 61330	. 39999	53	. 60001	. 01329	3	. 98671	54
7	. 38721	50	. 61279	. 40052	54	. 59948	. 01332	3	. 98668	53
8	. 38771	50	. 61229	. 40106	53	. 59894	. 01335	3	. 98665	52
9	. 38821	50	. 61179	. 40159	53	. 59841	. 01338	3	. 98662	51
10	9. 38871	50	10. 61129	9. 40212	54	10. 59788	10. 01341	3	9. 98659	50
11	. 38921	50	. 61079	. 40266	53	. 59734	. 01344	4	. 98656	49
12	. 38971	50	. 61029	. 40319	53	. 59681	. 01348	3	. 98652	48
13	. 39021	50	. 60979	. 40372	53	. 59628	. 01351	3	. 98649	47
14	. 39071	50	. 60929	. 40425	53	. 59575	. 01354	3	. 98646	46
15	9. 39121	49	10. 60879	9. 40478	53	10. 59522	10. 01357	3	9. 98643	45
16	. 39170	50	. 60830	. 40531	53	. 59469	. 01360	4	. 98640	44
17	. 39220	50	. 60780	. 40584	52	. 59416	. 01364	3	. 98636	43
18	. 39270	49	. 60730	. 40636	53	. 59364	. 01367	3	. 98633	42
19	. 39319	50	. 60681	. 40689	53	. 59311	. 01370	3	. 98630	41
20	9. 39369	49	10. 60631	9. 40742	53	10. 59258	10. 01373	4	9. 98627	40
21	. 39418	49	. 60582	. 40795	52	. 59205	. 01377	3	. 98623	39
22	. 39467	50	. 60533	. 40847	53	. 59153	. 01380	3	. 98620	38
23	. 39517	49	. 60483	. 40900	52	. 59100	. 01383	3	. 98617	37
24	. 39566	49	. 60434	. 40952	53	. 59048	. 01386	4	. 98614	36
25	9. 39615	49	10. 60385	9. 41005	52	10. 58995	10. 01390	3	9. 98610	35
26	. 39664	49	. 60336	. 41057	52	. 58943	. 01393	3	. 98607	34
27	. 39713	49	. 60287	. 41109	52	. 58891	. 01396	3	. 98604	33
28	. 39762	49	. 60238	. 41161	53	. 58839	. 01399	4	. 98601	32
29	. 39811	49	. 60189	. 41214	52	. 58786	. 01403	3	. 98597	31
30	9. 39860	49	10. 60140	9. 41266	52	10. 58734	10. 01406	3	9. 98594	30
31	. 39909	49	. 60091	. 41318	52	. 58682	. 01409	3	. 98591	29
32	. 39958	48	. 60042	. 41370	52	. 58630	. 01412	4	. 98588	28
33	. 40006	49	. 59994	. 41422	52	. 58578	. 01416	3	. 98584	27
34	. 40055	48	. 59945	. 41474	52	. 58526	. 01419	3	. 98581	26
35	9. 40103	49	10. 59897	9. 41526	52	10. 58474	10. 01422	4	9. 98578	25
36	. 40152	48	. 59848	. 41578	51	. 58422	. 01426	3	. 98574	24
37	. 40200	49	. 59800	. 41629	52	. 58371	. 01429	3	. 98571	23
38	. 40249	48	. 59751	. 41681	52	. 58319	. 01432	3	. 98568	22
39	. 40297	49	. 59703	. 41733	51	. 58267	. 01435	4	. 98565	21
40	9. 40346	48	10. 59654	9. 41784	52	10. 58216	10. 01439	3	9. 98561	20
41	. 40394	48	. 59606	. 41836	51	. 58164	. 01442	3	. 98558	19
42	. 40442	48	. 59558	. 41887	52	. 58113	. 01445	4	. 98555	18
43	. 40490	48	. 59510	. 41939	51	. 58061	. 01449	3	. 98551	17
44	. 40538	48	. 59462	. 41990	51	. 58010	. 01452	3	. 98548	16
45	9. 40586	48	10. 59414	9. 42041	52	10. 57959	10. 01455	4	9. 98545	15
46	. 40634	48	. 59366	. 42093	51	. 57907	. 01459	3	. 98541	14
47	. 40682	48	. 59318	. 42144	51	. 57856	. 01462	3	. 98538	13
48	. 40730	48	. 59270	. 42195	51	. 57805	. 01465	4	. 98535	12
49	. 40778	47	. 59222	. 42246	51	. 57754	. 01469	3	. 98531	11
50	9. 40825	48	10. 59175	9. 42297	51	10. 57703	10. 01472	3	9. 98528	10
51	. 40873	48	. 59127	. 42348	51	. 57652	. 01475	4	. 98525	9
52	. 40921	47	. 59079	. 42399	51	. 57601	. 01479	3	. 98521	8
53	. 40968	48	. 59032	. 42450	51	. 57550	. 01482	3	. 98518	7
54	. 41016	47	. 58984	. 42501	51	. 57499	. 01485	4	. 98515	6
55	9. 41063	48	10. 58937	9. 42552	51	10. 57448	10. 01489	3	9. 98511	5
56	. 41111	47	. 58889	. 42603	50	. 57397	. 01492	3	. 98508	4
57	. 41158	47	. 58842	. 42653	51	. 57347	. 01495	4	. 98505	3
58	. 41205	47	. 58795	. 42704	51	. 57296	. 01499	3	. 98501	2
59	. 41252	48	. 58748	. 42755	50	. 57245	. 01502	4	. 98498	1
60	9. 41300		10. 58700	9. 42805		10. 57195	10. 01506		9. 98494	0
104°→	cos	Diff. 1′	sec	cot	Diff. 1′	tan	csc	Diff. 1′	sin	←75°

TABLE 33

Logarithms of Trigonometric Functions

15°→ ↓ ′	sin	Diff. 1′	csc	tan	Diff. 1′	cot	sec	Diff. 1′	cos	←164° ↓ ′
0	9. 41300	47	10. 58700	9. 42805	51	10. 57195	10. 01506	3	9. 98494	60
1	. 41347	47	. 58653	. 42856	50	. 57144	. 01509	3	. 98491	59
2	. 41394	47	. 58606	. 42906	51	. 57094	. 01512	4	. 98488	58
3	. 41441	47	. 58559	. 42957	50	. 57043	. 01516	3	. 98484	57
4	. 41488	47	. 58512	. 43007	50	. 56993	. 01519	4	. 98481	56
5	9. 41535	47	10. 58465	9. 43057	51	10. 56943	10. 01523	3	9. 98477	55
6	. 41582	46	. 58418	. 43108	50	. 56892	. 01526	3	. 98474	54
7	. 41628	47	. 58372	. 43158	50	. 56842	. 01529	4	. 98471	53
8	. 41675	47	. 58325	. 43208	50	. 56792	. 01533	3	. 98467	52
9	. 41722	46	. 58278	. 43258	50	. 56742	. 01536	4	. 98464	51
10	9. 41768	47	10. 58232	9. 43308	50	10. 56692	10. 01540	3	9. 98460	50
11	. 41815	46	. 58185	. 43358	50	. 56642	. 01543	4	. 98457	49
12	. 41861	47	. 58139	. 43408	50	. 56592	. 01547	3	. 98453	48
13	. 41908	46	. 58092	. 43458	50	. 56542	. 01550	3	. 98450	47
14	. 41954	47	. 58046	. 43508	50	. 56492	. 01553	4	. 98447	46
15	9. 42001	46	10. 57999	9. 43558	49	10. 56442	10. 01557	3	9. 98443	45
16	. 42047	46	. 57953	. 43607	50	. 56393	. 01560	4	. 98440	44
17	. 42093	47	. 57907	. 43657	50	. 56343	. 01564	3	. 98436	43
18	. 42140	46	. 57860	. 43707	49	. 56293	. 01567	4	. 98433	42
19	. 42186	46	. 57814	. 43756	50	. 56244	. 01571	3	. 98429	41
20	9. 42232	46	10. 57768	9. 43806	49	10. 56194	10. 01574	4	9. 98426	40
21	. 42278	46	. 57722	. 43855	50	. 56145	. 01578	3	. 98422	39
22	. 42324	46	. 57676	. 43905	49	. 56095	. 01581	4	. 98419	38
23	. 42370	46	. 57630	. 43954	50	. 56046	. 01585	3	. 98415	37
24	. 42416	45	. 57584	. 44004	49	. 55996	. 01588	3	. 98412	36
25	9. 42461	46	10. 57539	9. 44053	49	10. 55947	10. 01591	4	9. 98409	35
26	. 42507	46	. 57493	. 44102	49	. 55898	. 01595	3	. 98405	34
27	. 42553	46	. 57447	. 44151	50	. 55849	. 01598	4	. 98402	33
28	. 42599	45	. 57401	. 44201	49	. 55799	. 01602	3	. 98398	32
29	. 42644	46	. 57356	. 44250	49	. 55750	. 01605	4	. 98395	31
30	9. 42690	45	10. 57310	9. 44299	49	10. 55701	10. 01609	3	9. 98391	30
31	. 42735	46	. 57265	. 44348	49	. 55652	. 01612	4	. 98388	29
32	. 42781	45	. 57219	. 44397	49	. 55603	. 01616	3	. 98384	28
33	. 42826	46	. 57174	. 44446	49	. 55554	. 01619	4	. 98381	27
34	. 42872	45	. 57128	. 44495	49	. 55505	. 01623	4	. 98377	26
35	9. 42917	45	10. 57083	9. 44544	48	10. 55456	10. 01627	3	9. 98373	25
36	. 42962	46	. 57038	. 44592	49	. 55408	. 01630	4	. 98370	24
37	. 43008	45	. 56992	. 44641	49	. 55359	. 01634	3	. 98366	23
38	. 43053	45	. 56947	. 44690	48	. 55310	. 01637	4	. 98363	22
39	. 43098	45	. 56902	. 44738	49	. 55262	. 01641	3	. 98359	21
40	9. 43143	45	10. 56857	9. 44787	49	10. 55213	10. 01644	4	9. 98356	20
41	. 43188	45	. 56812	. 44836	48	. 55164	. 01648	3	. 98352	19
42	. 43233	45	. 56767	. 44884	49	. 55116	. 01651	4	. 98349	18
43	. 43278	45	. 56722	. 44933	48	. 55067	. 01655	3	. 98345	17
44	. 43323	44	. 56677	. 44981	48	. 55019	. 01658	4	. 98342	16
45	9. 43367	45	10. 56633	9. 45029	49	10. 54971	10. 01662	4	9. 98338	15
46	. 43412	45	. 56588	. 45078	48	. 54922	. 01666	3	. 98334	14
47	. 43457	45	. 56543	. 45126	48	. 54874	. 01669	4	. 98331	13
48	. 43502	44	. 56498	. 45174	48	. 54826	. 01673	3	. 98327	12
49	. 43546	45	. 56454	. 45222	49	. 54778	. 01676	4	. 98324	11
50	9. 43591	44	10. 56409	9. 45271	48	10. 54729	10. 01680	3	9. 98320	10
51	. 43635	45	. 56365	. 45319	48	. 54681	. 01683	4	. 98317	9
52	. 43680	44	. 56320	. 45367	48	. 54633	. 01687	4	. 98313	8
53	. 43724	45	. 56276	. 45415	48	. 54585	. 01691	3	. 98309	7
54	. 43769	44	. 56231	. 45463	48	. 54537	. 01694	4	. 98306	6
55	9. 43813	44	10. 56187	9. 45511	48	10. 54489	10. 01698	3	9. 98302	5
56	. 43857	44	. 56143	. 45559	47	. 54441	. 01701	4	. 98299	4
57	. 43901	45	. 56099	. 45606	48	. 54394	. 01705	4	. 98295	3
58	. 43946	44	. 56054	. 45654	48	. 54346	. 01709	3	. 98291	2
59	. 43990	44	. 56010	. 45702	48	. 54298	. 01712	4	. 98288	1
60	9. 44034		10. 55966	9. 45750		10. 54250	10. 01716		9. 98284	0
↑ 105°→	cos	Diff. 1′	sec	cot	Diff. 1′	tan	csc	Diff. 1′	sin	↑ ←74°

TABLE 33
Logarithms of Trigonometric Functions

16° → ′	sin	Diff. 1′	csc	tan	Diff. 1′	cot	sec	Diff. 1′	cos	← 163° ′
0	9. 44034	44	10. 55966	9. 45750	47	10. 54250	10. 01716	3	9. 98284	60
1	. 44078	44	. 55922	. 45797	48	. 54203	. 01719	4	. 98281	59
2	. 44122	44	. 55878	. 45845	47	. 54155	. 01723	4	. 98277	58
3	. 44166	44	. 55834	. 45892	48	. 54108	. 01727	3	. 98273	57
4	. 44210	43	. 55790	. 45940	47	. 54060	. 01730	4	. 98270	56
5	9. 44253	44	10. 55747	9. 45987	48	10. 54013	10. 01734	4	9. 98266	55
6	. 44297	44	. 55703	. 46035	47	. 53965	. 01738	3	. 98262	54
7	. 44341	44	. 55659	. 46082	48	. 53918	. 01741	4	. 98259	53
8	. 44385	43	. 55615	. 46130	47	. 53870	. 01745	4	. 98255	52
9	. 44428	44	. 55572	. 46177	47	. 53823	. 01749	3	. 98251	51
10	9. 44472	44	10. 55528	9. 46224	47	10. 53776	10. 01752	4	9. 98248	50
11	. 44516	43	. 55484	. 46271	48	. 53729	. 01756	4	. 98244	49
12	. 44559	43	. 55441	. 46319	47	. 53681	. 01760	3	. 98240	48
13	. 44602	44	. 55398	. 46366	47	. 53634	. 01763	4	. 98237	47
14	. 44646	43	. 55354	. 46413	47	. 53587	. 01767	4	. 98233	46
15	9. 44689	44	10. 55311	9. 46460	47	10. 53540	10. 01771	3	9. 98229	45
16	. 44733	43	. 55267	. 46507	47	. 53493	. 01774	4	. 98226	44
17	. 44776	43	. 55224	. 46554	47	. 53446	. 01778	4	. 98222	43
18	. 44819	43	. 55181	. 46601	47	. 53399	. 01782	3	. 98218	42
19	. 44862	43	. 55138	. 46648	46	. 53352	. 01785	4	. 98215	41
20	9. 44905	43	10. 55095	9. 46694	47	10. 53306	10. 01789	4	9. 98211	40
21	. 44948	44	. 55052	. 46741	47	. 53259	. 01793	3	. 98207	39
22	. 44992	43	. 55008	. 46788	47	. 53212	. 01796	4	. 98204	38
23	. 45035	42	. 54965	. 46835	46	. 53165	. 01800	4	. 98200	37
24	. 45077	43	. 54923	. 46881	47	. 53119	. 01804	4	. 98196	36
25	9. 45120	43	10. 54880	9. 46928	47	10. 53072	10. 01808	3	9. 98192	35
26	. 45163	43	. 54837	. 46975	46	. 53025	. 01811	4	. 98189	34
27	. 45206	43	. 54794	. 47021	47	. 52979	. 01815	4	. 98185	33
28	. 45249	43	. 54751	. 47068	46	. 52932	. 01819	4	. 98181	32
29	. 45292	42	. 54708	. 47114	46	. 52886	. 01823	3	. 98177	31
30	9. 45334	43	10. 54666	9. 47160	47	10. 52840	10. 01826	4	9. 98174	30
31	. 45377	42	. 54623	. 47207	46	. 52793	. 01830	4	. 98170	29
32	. 45419	43	. 54581	. 47253	46	. 52747	. 01834	4	. 98166	28
33	. 45462	42	. 54538	. 47299	47	. 52701	. 01838	3	. 98162	27
34	. 45504	43	. 54496	. 47346	46	. 52654	. 01841	4	. 98159	26
35	9. 45547	42	10. 54453	9. 47392	46	10. 52608	10. 01845	4	9. 98155	25
36	. 45589	43	. 54411	. 47438	46	. 52562	. 01849	4	. 98151	24
37	. 45632	42	. 54368	. 47484	46	. 52516	. 01853	3	. 98147	23
38	. 45674	42	. 54326	. 47530	46	. 52470	. 01856	4	. 98144	22
39	. 45716	42	. 54284	. 47576	46	. 52424	. 01860	4	. 98140	21
40	9. 45758	43	10. 54242	9. 47622	46	10. 52378	10. 01864	4	9. 98136	20
41	. 45801	42	. 54199	. 47668	46	. 52332	. 01868	3	. 98132	19
42	. 45843	42	. 54157	. 47714	46	. 52286	. 01871	4	. 98129	18
43	. 45885	42	. 54115	. 47760	46	. 52240	. 01875	4	. 98125	17
44	. 45927	42	. 54073	. 47806	46	. 52194	. 01879	4	. 98121	16
45	9. 45969	42	10. 54031	9. 47852	45	10. 52148	10. 01883	4	9. 98117	15
46	. 46011	42	. 53989	. 47897	46	. 52103	. 01887	3	. 98113	14
47	. 46053	42	. 53947	. 47943	46	. 52057	. 01890	4	. 98110	13
48	. 46095	41	. 53905	. 47989	46	. 52011	. 01894	4	. 98106	12
49	. 46136	42	. 53864	. 48035	45	. 51965	. 01898	4	. 98102	11
50	9. 46178	42	10. 53822	9. 48080	46	10. 51920	10. 01902	4	9. 98098	10
51	. 46220	42	. 53780	. 48126	45	. 51874	. 01906	4	. 98094	9
52	. 46262	41	. 53738	. 48171	46	. 51829	. 01910	3	. 98090	8
53	. 46303	42	. 53697	. 48217	45	. 51783	. 01913	4	. 98087	7
54	. 46345	41	. 53655	. 48262	45	. 51738	. 01917	4	. 98083	6
55	9. 46386	42	10. 53614	9. 48307	46	10. 51693	10. 01921	4	9. 98079	5
56	. 46428	41	. 53572	. 48353	45	. 51647	. 01925	4	. 98075	4
57	. 46469	42	. 53531	. 48398	45	. 51602	. 01929	4	. 98071	3
58	. 46511	41	. 53489	. 48443	46	. 51557	. 01933	4	. 98067	2
59	. 46552	42	. 53448	. 48489	45	. 51511	. 01937	3	. 98063	1
60	9. 46594		10. 53406	9. 48534		10. 51466	10. 01940		9. 98060	0
106° →	cos	Diff. 1′	sec	cot	Diff. 1′	tan	csc	Diff. 1′	sin	← 73°

TABLE 33

Logarithms of Trigonometric Functions

17° → ′	sin	Diff. 1′	csc	tan	Diff. 1′	cot	sec	Diff. 1′	cos	← 162° ′
0	9.46594	41	10.53406	9.48534	45	10.51466	10.01940	4	9.98060	60
1	.46635	41	.53365	.48579	45	.51421	.01944	4	.98056	59
2	.46676	41	.53324	.48624	45	.51376	.01948	4	.98052	58
3	.46717	41	.53283	.48669	45	.51331	.01952	4	.98048	57
4	.46758	42	.53242	.48714	45	.51286	.01956	4	.98044	56
5	9.46800	41	10.53200	9.48759	45	10.51241	10.01960	4	9.98040	55
6	.46841	41	.53159	.48804	45	.51196	.01964	4	.98036	54
7	.46882	41	.53118	.48849	45	.51151	.01968	3	.98032	53
8	.46923	41	.53077	.48894	45	.51106	.01971	4	.98029	52
9	.46964	41	.53036	.48939	45	.51061	.01975	4	.98025	51
10	9.47005	40	10.52995	9.48984	45	10.51016	10.01979	4	9.98021	50
11	.47045	41	.52955	.49029	44	.50971	.01983	4	.98017	49
12	.47086	41	.52914	.49073	45	.50927	.01987	4	.98013	48
13	.47127	41	.52873	.49118	45	.50882	.01991	4	.98009	47
14	.47168	41	.52832	.49163	44	.50837	.01995	4	.98005	46
15	9.47209	40	10.52791	9.49207	45	10.50793	10.01999	4	9.98001	45
16	.47249	41	.52751	.49252	44	.50748	.02003	4	.97997	44
17	.47290	40	.52710	.49296	45	.50704	.02007	4	.97993	43
18	.47330	41	.52670	.49341	44	.50659	.02011	3	.97989	42
19	.47371	40	.52629	.49385	45	.50615	.02014	4	.97986	41
20	9.47411	41	10.52589	9.49430	44	10.50570	10.02018	4	9.97982	40
21	.47452	40	.52548	.49474	45	.50526	.02022	4	.97978	39
22	.47492	41	.52508	.49519	44	.50481	.02026	4	.97974	38
23	.47533	40	.52467	.49563	44	.50437	.02030	4	.97970	37
24	.47573	40	.52427	.49607	45	.50393	.02034	4	.97966	36
25	9.47613	41	10.52387	9.49652	44	10.50348	10.02038	4	9.97962	35
26	.47654	40	.52346	.49696	44	.50304	.02042	4	.97958	34
27	.47694	40	.52306	.49740	44	.50260	.02046	4	.97954	33
28	.47734	40	.52266	.49784	44	.50216	.02050	4	.97950	32
29	.47774	40	.52226	.49828	44	.50172	.02054	4	.97946	31
30	9.47814	40	10.52186	9.49872	44	10.50128	10.02058	4	9.97942	30
31	.47854	40	.52146	.49916	44	.50084	.02062	4	.97938	29
32	.47894	40	.52106	.49960	44	.50040	.02066	4	.97934	28
33	.47934	40	.52066	.50004	44	.49996	.02070	4	.97930	27
34	.47974	40	.52026	.50048	44	.49952	.02074	4	.97926	26
35	9.48014	40	10.51986	9.50092	44	10.49908	10.02078	4	9.97922	25
36	.48054	40	.51946	.50136	44	.49864	.02082	4	.97918	24
37	.48094	39	.51906	.50180	43	.49820	.02086	4	.97914	23
38	.48133	40	.51867	.50223	44	.49777	.02090	4	.97910	22
39	.48173	40	.51827	.50267	44	.49733	.02094	4	.97906	21
40	9.48213	39	10.51787	9.50311	44	10.49689	10.02098	4	9.97902	20
41	.48252	40	.51748	.50355	43	.49645	.02102	4	.97898	19
42	.48292	40	.51708	.50398	44	.49602	.02106	4	.97894	18
43	.48332	39	.51668	.50442	43	.49558	.02110	4	.97890	17
44	.48371	40	.51629	.50485	44	.49515	.02114	4	.97886	16
45	9.48411	39	10.51589	9.50529	43	10.49471	10.02118	4	9.97882	15
46	.48450	40	.51550	.50572	44	.49428	.02122	4	.97878	14
47	.48490	39	.51510	.50616	43	.49384	.02126	4	.97874	13
48	.48529	39	.51471	.50659	44	.49341	.02130	4	.97870	12
49	.48568	39	.51432	.50703	43	.49297	.02134	5	.97866	11
50	9.48607	40	10.51393	9.50746	43	10.49254	10.02139	4	9.97861	10
51	.48647	39	.51353	.50789	44	.49211	.02143	4	.97857	9
52	.48686	39	.51314	.50833	43	.49167	.02147	4	.97853	8
53	.48725	39	.51275	.50876	43	.49124	.02151	4	.97849	7
54	.48764	39	.51236	.50919	43	.49081	.02155	4	.97845	6
55	9.48803	39	10.51197	9.50962	43	10.49038	10.02159	4	9.97841	5
56	.48842	39	.51158	.51005	43	.48995	.02163	4	.97837	4
57	.48881	39	.51119	.51048	44	.48952	.02167	4	.97833	3
58	.48920	39	.51080	.51092	43	.48908	.02171	4	.97829	2
59	.48959	39	.51041	.51135	43	.48865	.02175	4	.97825	1
60	9.48998		10.51002	9.51178		10.48822	10.02179		9.97821	0
↑ 107° →	cos	Diff. 1′	sec	cot	Diff. 1′	tan	csc	Diff. 1′	sin	← 72° ↑

TABLE 33
Logarithms of Trigonometric Functions

18° → ′	sin	Diff. 1′	csc	tan	Diff. 1′	cot	sec	Diff. 1′	cos	← 161° ′
0	9.48998	39	10.51002	9.51178	43	10.48822	10.02179	4	9.97821	60
1	.49037	39	.50963	.51221	43	.48779	.02183	5	.97817	59
2	.49076	39	.50924	.51264	42	.48736	.02188	4	.97812	58
3	.49115	38	.50885	.51306	43	.48694	.02192	4	.97808	57
4	.49153	39	.50847	.51349	43	.48651	.02196	4	.97804	56
5	9.49192	39	10.50808	9.51392	43	10.48608	10.02200	4	9.97800	55
6	.49231	38	.50769	.51435	43	.48565	.02204	4	.97796	54
7	.49269	39	.50731	.51478	42	.48522	.02208	4	.97792	53
8	.49308	39	.50692	.51520	43	.48480	.02212	4	.97788	52
9	.49347	38	.50653	.51563	43	.48437	.02216	5	.97784	51
10	9.49385	39	10.50615	9.51606	42	10.48394	10.02221	4	9.97779	50
11	.49424	38	.50576	.51648	43	.48352	.02225	4	.97775	49
12	.49462	38	.50538	.51691	43	.48309	.02229	4	.97771	48
13	.49500	39	.50500	.51734	42	.48266	.02233	4	.97767	47
14	.49539	38	.50461	.51776	43	.48224	.02237	4	.97763	46
15	9.49577	38	10.50423	9.51819	42	10.48181	10.02241	5	9.97759	45
16	.49615	39	.50385	.51861	42	.48139	.02246	4	.97754	44
17	.49654	38	.50346	.51903	43	.48097	.02250	4	.97750	43
18	.49692	38	.50308	.51946	42	.48054	.02254	4	.97746	42
19	.49730	38	.50270	.51988	43	.48012	.02258	4	.97742	41
20	9.49768	38	10.50232	9.52031	42	10.47969	10.02262	4	9.97738	40
21	.49806	38	.50194	.52073	42	.47927	.02266	5	.97734	39
22	.49844	38	.50156	.52115	42	.47885	.02271	4	.97729	38
23	.49882	38	.50118	.52157	43	.47843	.02275	4	.97725	37
24	.49920	38	.50080	.52200	42	.47800	.02279	4	.97721	36
25	9.49958	38	10.50042	9.52242	42	10.47758	10.02283	4	9.97717	35
26	.49996	38	.50004	.52284	42	.47716	.02287	5	.97713	34
27	.50034	38	.49966	.52326	42	.47674	.02292	4	.97708	33
28	.50072	38	.49928	.52368	42	.47632	.02296	4	.97704	32
29	.50110	38	.49890	.52410	42	.47590	.02300	4	.97700	31
30	9.50148	37	10.49852	9.52452	42	10.47548	10.02304	5	9.97696	30
31	.50185	38	.49815	.52494	42	.47506	.02309	4	.97691	29
32	.50223	38	.49777	.52536	42	.47464	.02313	4	.97687	28
33	.50261	37	.49739	.52578	42	.47422	.02317	4	.97683	27
34	.50298	38	.49702	.52620	41	.47380	.02321	5	.97679	26
35	9.50336	38	10.49664	9.52661	42	10.47339	10.02326	4	9.97674	25
36	.50374	37	.49626	.52703	42	.47297	.02330	4	.97670	24
37	.50411	38	.49589	.52745	42	.47255	.02334	4	.97666	23
38	.50449	37	.49551	.52787	42	.47213	.02338	5	.97662	22
39	.50486	37	.49514	.52829	41	.47171	.02343	4	.97657	21
40	9.50523	38	10.49477	9.52870	42	10.47130	10.02347	4	9.97653	20
41	.50561	37	.49439	.52912	41	.47088	.02351	4	.97649	19
42	.50598	37	.49402	.52953	42	.47047	.02355	5	.97645	18
43	.50635	38	.49365	.52995	42	.47005	.02360	4	.97640	17
44	.50673	37	.49327	.53037	41	.46963	.02364	4	.97636	16
45	9.50710	37	10.49290	9.53078	42	10.46922	10.02368	4	9.97632	15
46	.50747	37	.49253	.53120	41	.46880	.02372	5	.97628	14
47	.50784	37	.49216	.53161	41	.46839	.02377	4	.97623	13
48	.50821	37	.49179	.53202	42	.46798	.02381	4	.97619	12
49	.50858	38	.49142	.53244	41	.46756	.02385	5	.97615	11
50	9.50896	37	10.49104	9.53285	42	10.46715	10.02390	4	9.97610	10
51	.50933	37	.49067	.53327	41	.46673	.02394	4	.97606	9
52	.50970	37	.49030	.53368	41	.46632	.02398	5	.97602	8
53	.51007	36	.48993	.53409	41	.46591	.02403	4	.97597	7
54	.51043	37	.48957	.53450	42	.46550	.02407	4	.97593	6
55	9.51080	37	10.48920	9.53492	41	10.46508	10.02411	5	9.97589	5
56	.51117	37	.48883	.53533	41	.46467	.02416	4	.97584	4
57	.51154	37	.48846	.53574	41	.46426	.02420	4	.97580	3
58	.51191	36	.48809	.53615	41	.46385	.02424	5	.97576	2
59	.51227	37	.48773	.53656	41	.46344	.02429	4	.97571	1
60	9.51264		10.48736	9.53697		10.46303	10.02433		9.97567	0
108° →	cos	Diff. 1′	sec	cot	Diff. 1′	tan	csc	Diff. 1′	sin	← 71°

TABLE 33
Logarithms of Trigonometric Functions

19° → ↓	sin	Diff. 1′	csc	tan	Diff. 1′	cot	sec	Diff. 1′	cos	← 160° ↓
′										′
0	9. 51264	37	10. 48736	9. 53697	41	10. 46303	10. 02433	4	9. 97567	60
1	. 51301	37	. 48699	. 53738	41	. 46262	. 02437	5	. 97563	59
2	. 51338	36	. 48662	. 53779	41	. 46221	. 02442	4	. 97558	58
3	. 51374	37	. 48626	. 53820	41	. 46180	. 02446	4	. 97554	57
4	. 51411	36	. 48589	. 53861	41	. 46139	. 02450	5	. 97550	56
5	9. 51447	37	10. 48553	9. 53902	41	10. 46098	10. 02455	4	9. 97545	55
6	. 51484	36	. 48516	. 53943	41	. 46057	. 02459	5	. 97541	54
7	. 51520	37	. 48480	. 53984	41	. 46016	. 02464	4	. 97536	53
8	. 51557	36	. 48443	. 54025	40	. 45975	. 02468	4	. 97532	52
9	. 51593	36	. 48407	. 54065	41	. 45935	. 02472	5	. 97528	51
10	9. 51629	37	10. 48371	9. 54106	41	10. 45894	10. 02477	4	9. 97523	50
11	. 51666	36	. 48334	. 54147	40	. 45853	. 02481	4	. 97519	49
12	. 51702	36	. 48298	. 54187	41	. 45813	. 02485	5	. 97515	48
13	. 51738	36	. 48262	. 54228	41	. 45772	. 02490	4	. 97510	47
14	. 51774	37	. 48226	. 54269	40	. 45731	. 02494	5	. 97506	46
15	9. 51811	36	10. 48189	9. 54309	41	10. 45691	10. 02499	4	9. 97501	45
16	. 51847	36	. 48153	. 54350	40	. 45650	. 02503	5	. 97497	44
17	. 51883	36	. 48117	. 54390	41	. 45610	. 02508	4	. 97492	43
18	. 51919	36	. 48081	. 54431	40	. 45569	. 02512	4	. 97488	42
19	. 51955	36	. 48045	. 54471	41	. 45529	. 02516	5	. 97484	41
20	9. 51991	36	10. 48009	9. 54512	40	10. 45488	10. 02521	4	9. 97479	40
21	. 52027	36	. 47973	. 54552	41	. 45448	. 02525	5	. 97475	39
22	. 52063	36	. 47937	. 54593	40	. 45407	. 02530	4	. 97470	38
23	. 52099	36	. 47901	. 54633	40	. 45367	. 02534	5	. 97466	37
24	. 52135	36	. 47865	. 54673	41	. 45327	. 02539	4	. 97461	36
25	9. 52171	36	10. 47829	9. 54714	40	10. 45286	10. 02543	4	9. 97457	35
26	. 52207	35	. 47793	. 54754	40	. 45246	. 02547	5	. 97453	34
27	. 52242	36	. 47758	. 54794	41	. 45206	. 02552	4	. 97448	33
28	. 52278	36	. 47722	. 54835	40	. 45165	. 02556	5	. 97444	32
29	. 52314	36	. 47686	. 54875	40	. 45125	. 02561	4	. 97439	31
30	9. 52350	35	10. 47650	9. 54915	40	10. 45085	10. 02565	5	9. 97435	30
31	. 52385	36	. 47615	. 54955	40	. 45045	. 02570	4	. 97430	29
32	. 52421	35	. 47579	. 54995	40	. 45005	. 02574	5	. 97426	28
33	. 52456	36	. 47544	. 55035	40	. 44965	. 02579	4	. 97421	27
34	. 52492	35	. 47508	. 55075	40	. 44925	. 02583	5	. 97417	26
35	9. 52527	36	10. 47473	9. 55115	40	10. 44885	10. 02588	4	9. 97412	25
36	. 52563	35	. 47437	. 55155	40	. 44845	. 02592	5	. 97408	24
37	. 52598	36	. 47402	. 55195	40	. 44805	. 02597	4	. 97403	23
38	. 52634	35	. 47366	. 55235	40	. 44765	. 02601	5	. 97399	22
39	. 52669	36	. 47331	. 55275	40	. 44725	. 02606	4	. 97394	21
40	9. 52705	35	10. 47295	9. 55315	40	10. 44685	10. 02610	5	9. 97390	20
41	. 52740	35	. 47260	. 55355	40	. 44645	. 02615	4	. 97385	19
42	. 52775	36	. 47225	. 55395	39	. 44605	. 02619	5	. 97381	18
43	. 52811	35	. 47189	. 55434	40	. 44566	. 02624	4	. 97376	17
44	. 52846	35	. 47154	. 55474	40	. 44526	. 02628	5	. 97372	16
45	9. 52881	35	10. 47119	9. 55514	40	10. 44486	10. 02633	4	9. 97367	15
46	. 52916	35	. 47084	. 55554	39	. 44446	. 02637	5	. 97363	14
47	. 52951	35	. 47049	. 55593	40	. 44407	. 02642	5	. 97358	13
48	. 52986	35	. 47014	. 55633	40	. 44367	. 02647	4	. 97353	12
49	. 53021	35	. 46979	. 55673	39	. 44327	. 02651	5	. 97349	11
50	9. 53056	36	10. 46944	9. 55712	40	10. 44288	10. 02656	4	9. 97344	10
51	. 53092	34	. 46908	. 55752	39	. 44248	. 02660	5	. 97340	9
52	. 53126	35	. 46874	. 55791	40	. 44209	. 02665	4	. 97335	8
53	. 53161	35	. 46839	. 55831	39	. 44169	. 02669	5	. 97331	7
54	. 53196	35	. 46804	. 55870	40	. 44130	. 02674	4	. 97326	6
55	9. 53231	35	10. 46769	9. 55910	39	10. 44090	10. 02678	5	9. 97322	5
56	. 53266	35	. 46734	. 55949	40	. 44051	. 02683	5	. 97317	4
57	. 53301	35	. 46699	. 55989	39	. 44011	. 02688	4	. 97312	3
58	. 53336	34	. 46664	. 56028	39	. 43972	. 02692	5	. 97308	2
59	. 53370	35	. 46630	. 56067	40	. 43933	. 02697	4	. 97303	1
60	9. 53405		10. 46595	9. 56107		10. 43893	10. 02701		9. 97299	0
↑ 109° →	cos	Diff. 1′	sec	cot	Diff. 1′	tan	csc	Diff. 1′	sin	← 70° ↑

TABLE 33

Logarithms of Trigonometric Functions

20° → ↓ ′	sin	Diff. 1′	csc	tan	Diff. 1′	cot	sec	Diff. 1′	cos	← 159° ↓ ′
0	9. 53405	35	10. 46595	9. 56107	39	10. 43893	10. 02701	5	9. 97299	60
1	. 53440	35	. 46560	. 56146	39	. 43854	. 02706	5	. 97294	59
2	. 53475	34	. 46525	. 56185	39	. 43815	. 02711	4	. 97289	58
3	. 53509	35	. 46491	. 56224	40	. 43776	. 02715	5	. 97285	57
4	. 53544	34	. 46456	. 56264	39	. 43736	. 02720	4	. 97280	56
5	9. 53578	35	10. 46422	9. 56303	39	10. 43697	10. 02724	5	9. 97276	55
6	. 53613	34	. 46387	. 56342	39	. 43658	. 02729	5	. 97271	54
7	. 53647	35	. 46353	. 56381	39	. 43619	. 02734	4	. 97266	53
8	. 53682	34	. 46318	. 56420	39	. 43580	. 02738	5	. 97262	52
9	. 53716	35	. 46284	. 56459	39	. 43541	. 02743	5	. 97257	51
10	9. 53751	34	10. 46249	9. 56498	39	10. 43502	10. 02748	4	9. 97252	50
11	. 53785	34	. 46215	. 56537	39	. 43463	. 02752	5	. 97248	49
12	. 53819	35	. 46181	. 56576	39	. 43424	. 02757	5	. 97243	48
13	. 53854	34	. 46146	. 56615	39	. 43385	. 02762	4	. 97238	47
14	. 53888	34	. 46112	. 56654	39	. 43346	. 02766	5	. 97234	46
15	9. 53922	35	10. 46078	9. 56693	39	10. 43307	10. 02771	5	9. 97229	45
16	. 53957	34	. 46043	. 56732	39	. 43268	. 02776	4	. 97224	44
17	. 53991	34	. 46009	. 56771	39	. 43229	. 02780	5	. 97220	43
18	. 54025	34	. 45975	. 56810	39	. 43190	. 02785	5	. 97215	42
19	. 54059	34	. 45941	. 56849	38	. 43151	. 02790	4	. 97210	41
20	9. 54093	34	10. 45907	9. 56887	39	10. 43113	10. 02794	5	9. 97206	40
21	. 54127	34	. 45873	. 56926	39	. 43074	. 02799	5	. 97201	39
22	. 54161	34	. 45839	. 56965	39	. 43035	. 02804	4	. 97196	38
23	. 54195	34	. 45805	. 57004	38	. 42996	. 02808	5	. 97192	37
24	. 54229	34	. 45771	. 57042	39	. 42958	. 02813	5	. 97187	36
25	9. 54263	34	10. 45737	9. 57081	39	10. 42919	10. 02818	4	9. 97182	35
26	. 54297	34	. 45703	. 57120	38	. 42880	. 02822	5	. 97178	34
27	. 54331	34	. 45669	. 57158	39	. 42842	. 02827	5	. 97173	33
28	. 54365	34	. 45635	. 57197	38	. 42803	. 02832	5	. 97168	32
29	. 54399	34	. 45601	. 57235	39	. 42765	. 02837	4	. 97163	31
30	9. 54433	33	10. 45567	9. 57274	38	10. 42726	10. 02841	5	9. 97159	30
31	. 54466	34	. 45534	. 57312	39	. 42688	. 02846	5	. 97154	29
32	. 54500	34	. 45500	. 57351	38	. 42649	. 02851	4	. 97149	28
33	. 54534	33	. 45466	. 57389	39	. 42611	. 02855	5	. 97145	27
34	. 54567	34	. 45433	. 57428	38	. 42572	. 02860	5	. 97140	26
35	9. 54601	34	10. 45399	9. 57466	38	10. 42534	10. 02865	5	9. 97135	25
36	. 54635	33	. 45365	. 57504	39	. 42496	. 02870	4	. 97130	24
37	. 54668	34	. 45332	. 57543	38	. 42457	. 02874	5	. 97126	23
38	. 54702	33	. 45298	. 57581	38	. 42419	. 02879	5	. 97121	22
39	. 54735	34	. 45265	. 57619	39	. 42381	. 02884	5	. 97116	21
40	9. 54769	33	10. 45231	9. 57658	38	10. 42342	10. 02889	4	9. 97111	20
41	. 54802	34	. 45198	. 57696	38	. 42304	. 02893	5	. 97107	19
42	. 54836	33	. 45164	. 57734	38	. 42266	. 02898	5	. 97102	18
43	. 54869	34	. 45131	. 57772	38	. 42228	. 02903	5	. 97097	17
44	. 54903	33	. 45097	. 57810	39	. 42190	. 02908	5	. 97092	16
45	9. 54936	33	10. 45064	9. 57849	38	10. 42151	10. 02913	4	9. 97087	15
46	. 54969	34	. 45031	. 57887	38	. 42113	. 02917	5	. 97083	14
47	. 55003	33	. 44997	. 57925	38	. 42075	. 02922	5	. 97078	13
48	. 55036	33	. 44964	. 57963	38	. 42037	. 02927	5	. 97073	12
49	. 55069	33	. 44931	. 58001	38	. 41999	. 02932	5	. 97068	11
50	9. 55102	34	10. 44898	9. 58039	38	10. 41961	10. 02937	4	9. 97063	10
51	. 55136	33	. 44864	. 58077	38	. 41923	. 02941	5	. 97059	9
52	. 55169	33	. 44831	. 58115	38	. 41885	. 02946	5	. 97054	8
53	. 55202	33	. 44798	. 58153	38	. 41847	. 02951	5	. 97049	7
54	. 55235	33	. 44765	. 58191	38	. 41809	. 02956	5	. 97044	6
55	9. 55268	33	10. 44732	9. 58229	38	10. 41771	10. 02961	4	9. 97039	5
56	. 55301	33	. 44699	. 58267	37	. 41733	. 02965	5	. 97035	4
57	. 55334	33	. 44666	. 58304	38	. 41696	. 02970	5	. 97030	3
58	. 55367	33	. 44633	. 58342	38	. 41658	. 02975	5	. 97025	2
59	. 55400	33	. 44600	. 58380	38	. 41620	. 02980	5	. 97020	1
60	9. 55433		10. 44567	9. 58418		10. 41582	10. 02985		9. 97015	0
↑ 110° →	cos	Diff. 1′	sec	cot	Diff. 1′	tan	csc	Diff. 1′	sin	← 69° ↑

TABLE 33

Logarithms of Trigonometric Functions

21° → ↓ ′	sin	Diff. 1′	csc	tan	Diff. 1′	cot	sec	Diff. 1′	cos	← 158° ↓ ′
0	9.55433	33	10.44567	9.58418	37	10.41582	10.02985	5	9.97015	60
1	.55466	33	.44534	.58455	38	.41545	.02990	5	.97010	59
2	.55499	33	.44501	.58493	38	.41507	.02995	4	.97005	58
3	.55532	32	.44468	.58531	38	.41469	.02999	5	.97001	57
4	.55564	33	.44436	.58569	37	.41431	.03004	5	.96996	56
5	9.55597	33	10.44403	9.58606	38	10.41394	10.03009	5	9.96991	55
6	.55630	33	.44370	.58644	37	.41356	.03014	5	.96986	54
7	.55663	32	.44337	.58681	38	.41319	.03019	5	.96981	53
8	.55695	33	.44305	.58719	38	.41281	.03024	5	.96976	52
9	.55728	33	.44272	.58757	37	.41243	.03029	5	.96971	51
10	9.55761	32	10.44239	9.58794	38	10.41206	10.03034	4	9.96966	50
11	.55793	33	.44207	.58832	37	.41168	.03038	5	.96962	49
12	.55826	32	.44174	.58869	38	.41131	.03043	5	.96957	48
13	.55858	33	.44142	.58907	37	.41093	.03048	5	.96952	47
14	.55891	32	.44109	.58944	37	.41056	.03053	5	.96947	46
15	9.55923	33	10.44077	9.58981	38	10.41019	10.03058	5	9.96942	45
16	.55956	32	.44044	.59019	37	.40981	.03063	5	.96937	44
17	.55988	33	.44012	.59056	38	.40944	.03068	5	.96932	43
18	.56021	32	.43979	.59094	37	.40906	.03073	5	.96927	42
19	.56053	32	.43947	.59131	37	.40869	.03078	5	.96922	41
20	9.56085	33	10.43915	9.59168	37	10.40832	10.03083	5	9.96917	40
21	.56118	32	.43882	.59205	38	.40795	.03088	5	.96912	39
22	.56150	32	.43850	.59243	37	.40757	.03093	4	.96907	38
23	.56182	33	.43818	.59280	37	.40720	.03097	5	.96903	37
24	.56215	32	.43785	.59317	37	.40683	.03102	5	.96898	36
25	9.56247	32	10.43753	9.59354	37	10.40646	10.03107	5	9.96893	35
26	.56279	32	.43721	.59391	38	.40609	.03112	5	.96888	34
27	.56311	32	.43689	.59429	37	.40571	.03117	5	.96883	33
28	.56343	32	.43657	.59466	37	.40534	.03122	5	.96878	32
29	.56375	33	.43625	.59503	37	.40497	.03127	5	.96873	31
30	9.56408	32	10.43592	9.59540	37	10.40460	10.03132	5	9.96868	30
31	.56440	32	.43560	.59577	37	.40423	.03137	5	.96863	29
32	.56472	32	.43528	.59614	37	.40386	.03142	5	.96858	28
33	.56504	32	.43496	.59651	37	.40349	.03147	5	.96853	27
34	.56536	32	.43464	.59688	37	.40312	.03152	5	.96848	26
35	9.56568	31	10.43432	9.59725	37	10.40275	10.03157	5	9.96843	25
36	.56599	32	.43401	.59762	37	.40238	.03162	5	.96838	24
37	.56631	32	.43369	.59799	36	.40201	.03167	5	.96833	23
38	.56663	32	.43337	.59835	37	.40165	.03172	5	.96828	22
39	.56695	32	.43305	.59872	37	.40128	.03177	5	.96823	21
40	9.56727	32	10.43273	9.59909	37	10.40091	10.03182	5	9.96818	20
41	.56759	31	.43241	.59946	37	.40054	.03187	5	.96813	19
42	.56790	32	.43210	.59983	36	.40017	.03192	5	.96808	18
43	.56822	32	.43178	.60019	37	.39981	.03197	5	.96803	17
44	.56854	32	.43146	.60056	37	.39944	.03202	5	.96798	16
45	9.56886	31	10.43114	9.60093	37	10.39907	10.03207	5	9.96793	15
46	.56917	32	.43083	.60130	36	.39870	.03212	5	.96788	14
47	.56949	31	.43051	.60166	37	.39834	.03217	5	.96783	13
48	.56980	32	.43020	.60203	37	.39797	.03222	6	.96778	12
49	.57012	32	.42988	.60240	36	.39760	.03228	5	.96772	11
50	9.57044	31	10.42956	9.60276	37	10.39724	10.03233	5	9.96767	10
51	.57075	32	.42925	.60313	36	.39687	.03238	5	.96762	9
52	.57107	31	.42893	.60349	37	.39651	.03243	5	.96757	8
53	.57138	31	.42862	.60386	36	.39614	.03248	5	.96752	7
54	.57169	32	.42831	.60422	37	.39578	.03253	5	.96747	6
55	9.57201	31	10.42799	9.60459	36	10.39541	10.03258	5	9.96742	5
56	.57232	32	.42768	.60495	37	.39505	.03263	5	.96737	4
57	.57264	31	.42736	.60532	36	.39468	.03268	5	.96732	3
58	.57295	31	.42705	.60568	37	.39432	.03273	5	.96727	2
59	.57326	32	.42674	.60605	36	.39395	.03278	5	.96722	1
60	9.57358		10.42642	9.60641		10.39359	10.03283		9.96717	0
↑ 111° →	cos	Diff. 1′	sec	cot	Diff. 1′	tan	csc	Diff. 1′	sin	← 68° ↑

TABLE 33

Logarithms of Trigonometric Functions

22°→ ↓ ′	sin	Diff. 1′	csc	tan	Diff. 1′	cot	sec	Diff. 1′	cos	←157° ↓ ′
0	9. 57358	31	10. 42642	9. 60641	36	10. 39359	10. 03283	6	9. 96717	60
1	. 57389	31	. 42611	. 60677	37	. 39323	. 03289	5	. 96711	59
2	. 57420	31	. 42580	. 60714	36	. 39286	. 03294	5	. 96706	58
3	. 57451	31	. 42549	. 60750	36	. 39250	. 03299	5	. 96701	57
4	. 57482	32	. 42518	. 60786	37	. 39214	. 03304	5	. 96696	56
5	9. 57514	31	10. 42486	9. 60823	36	10. 39177	10. 03309	5	9. 96691	55
6	. 57545	31	. 42455	. 60859	36	. 39141	. 03314	5	. 96686	54
7	. 57576	31	. 42424	. 60895	36	. 39105	. 03319	5	. 96681	53
8	. 57607	31	. 42393	. 60931	36	. 39069	. 03324	6	. 96676	52
9	. 57638	31	. 42362	. 60967	37	. 39033	. 03330	5	. 96670	51
10	9. 57669	31	10. 42331	9. 61004	36	10. 38996	10. 03335	5	9. 96665	50
11	. 57700	31	. 42300	. 61040	36	. 38960	. 03340	5	. 96660	49
12	. 57731	31	. 42269	. 61076	36	. 38924	. 03345	5	. 96655	48
13	. 57762	31	. 42238	. 61112	36	. 38888	. 03350	5	. 96650	47
14	. 57793	31	. 42207	. 61148	36	. 38852	. 03355	5	. 96645	46
15	9. 57824	31	10. 42176	9. 61184	36	10. 38816	10. 03360	6	9. 96640	45
16	. 57855	30	. 42145	. 61220	36	. 38780	. 03366	5	. 96634	44
17	. 57885	31	. 42115	. 61256	36	. 38744	. 03371	5	. 96629	43
18	. 57916	31	. 42084	. 61292	36	. 38708	. 03376	5	. 96624	42
19	. 57947	31	. 42053	. 61328	36	. 38672	. 03381	5	. 96619	41
20	9. 57978	30	10. 42022	9. 61364	36	10. 38636	10. 03386	6	9. 96614	40
21	. 58008	31	. 41992	. 61400	36	. 38600	. 03392	5	. 96608	39
22	. 58039	31	. 41961	. 61436	36	. 38564	. 03397	5	. 96603	38
23	. 58070	31	. 41930	. 61472	36	. 38528	. 03402	5	. 96598	37
24	. 58101	30	. 41899	. 61508	36	. 38492	. 03407	5	. 96593	36
25	9. 58131	31	10. 41869	9. 61544	35	10. 38456	10. 03412	6	9. 96588	35
26	. 58162	30	. 41838	. 61579	36	. 38421	. 03418	5	. 96582	34
27	. 58192	31	. 41808	. 61615	36	. 38385	. 03423	5	. 96577	33
28	. 58223	30	. 41777	. 61651	36	. 38349	. 03428	5	. 96572	32
29	. 58253	31	. 41747	. 61687	35	. 38313	. 03433	5	. 96567	31
30	9. 58284	30	10. 41716	9. 61722	36	10. 38278	10. 03438	6	9. 96562	30
31	. 58314	31	. 41686	. 61758	36	. 38242	. 03444	5	. 96556	29
32	. 58345	30	. 41655	. 61794	36	. 38206	. 03449	5	. 96551	28
33	. 58375	31	. 41625	. 61830	35	. 38170	. 03454	5	. 96546	27
34	. 58406	30	. 41594	. 61865	36	. 38135	. 03459	6	. 96541	26
35	9. 58436	31	10. 41564	9. 61901	35	10. 38099	10. 03465	5	9. 96535	25
36	. 58467	30	. 41533	. 61936	36	. 38064	. 03470	5	. 96530	24
37	. 58497	30	. 41503	. 61972	36	. 38028	. 03475	5	. 96525	23
38	. 58527	30	. 41473	. 62008	35	. 37992	. 03480	6	. 96520	22
39	. 58557	31	. 41443	. 62043	36	. 37957	. 03486	5	. 96514	21
40	9. 58588	30	10. 41412	9. 62079	35	10. 37921	10. 03491	5	9. 96509	20
41	. 58618	30	. 41382	. 62114	36	. 37886	. 03496	6	. 96504	19
42	. 58648	30	. 41352	. 62150	35	. 37850	. 03502	5	. 96498	18
43	. 58678	31	. 41322	. 62185	36	. 37815	. 03507	5	. 96493	17
44	. 58709	30	. 41291	. 62221	35	. 37779	. 03512	5	. 96488	16
45	9. 58739	30	10. 41261	9. 62256	36	10. 37744	10. 03517	6	9. 96483	15
46	. 58769	30	. 41231	. 62292	35	. 37708	. 03523	5	. 96477	14
47	. 58799	30	. 41201	. 62327	35	. 37673	. 03528	5	. 96472	13
48	. 58829	30	. 41171	. 62362	36	. 37638	. 03533	6	. 96467	12
49	. 58859	30	. 41141	. 62398	35	. 37602	. 03539	5	. 96461	11
50	9. 58889	30	10. 41111	9. 62433	35	10. 37567	10. 03544	5	9. 96456	10
51	. 58919	30	. 41081	. 62468	36	. 37532	. 03549	6	. 96451	9
52	. 58949	30	. 41051	. 62504	35	. 37496	. 03555	5	. 96445	8
53	. 58979	30	. 41021	. 62539	35	. 37461	. 03560	5	. 96440	7
54	. 59009	30	. 40991	. 62574	35	. 37426	. 03565	6	. 96435	6
55	9. 59039	30	10. 40961	9. 62609	36	10. 37391	10. 03571	5	9. 96429	5
56	. 59069	29	. 40931	. 62645	35	. 37355	. 03576	5	. 96424	4
57	. 59098	30	. 40902	. 62680	35	. 37320	. 03581	6	. 96419	3
58	. 59128	30	. 40872	. 62715	35	. 37285	. 03587	5	. 96413	2
59	. 59158	30	. 40842	. 62750	35	. 37250	. 03592	5	. 96408	1
60	9. 59188		10. 40812	9. 62785		10. 37215	10. 03597		9. 96403	0
↑ 112°→	cos	Diff. 1′	sec	cot	Diff. 1′	tan	csc	Diff. 1′	sin	↑ ←67°

TABLE 33

Logarithms of Trigonometric Functions

23° → ↓ ′	sin	Diff. 1′	csc	tan	Diff. 1′	cot	sec	Diff. 1′	cos	← 156° ↓ ′
0	9.59188	30	10.40812	9.62785	35	10.37215	10.03597	6	9.96403	60
1	.59218	29	.40782	.62820	35	.37180	.03603	5	.96397	59
2	.59247	30	.40753	.62855	35	.37145	.03608	5	.96392	58
3	.59277	30	.40723	.62890	36	.37110	.03613	6	.96387	57
4	.59307	29	.40693	.62926	35	.37074	.03619	5	.96381	56
5	9.59336	30	10.40664	9.62961	35	10.37039	10.03624	6	9.96376	55
6	.59366	30	.40634	.62996	35	.37004	.03630	5	.96370	54
7	.59396	29	.40604	.63031	35	.36969	.03635	5	.96365	53
8	.59425	30	.40575	.63066	35	.36934	.03640	6	.96360	52
9	.59455	29	.40545	.63101	34	.36899	.03646	5	.96354	51
10	9.59484	30	10.40516	9.63135	35	10.36865	10.03651	6	9.96349	50
11	.59514	29	.40486	.63170	35	.36830	.03657	5	.96343	49
12	.59543	30	.40457	.63205	35	.36795	.03662	5	.96338	48
13	.59573	29	.40427	.63240	35	.36760	.03667	6	.96333	47
14	.59602	30	.40398	.63275	35	.36725	.03673	5	.96327	46
15	9.59632	29	10.40368	9.63310	35	10.36690	10.03678	6	9.96322	45
16	.59661	29	.40339	.63345	34	.36655	.03684	5	.96316	44
17	.59690	30	.40310	.63379	35	.36621	.03689	6	.96311	43
18	.59720	29	.40280	.63414	35	.36586	.03695	5	.96305	42
19	.59749	29	.40251	.63449	35	.36551	.03700	6	.96300	41
20	9.59778	30	10.40222	9.63484	35	10.36516	10.03706	5	9.96294	40
21	.59808	29	.40192	.63519	34	.36481	.03711	5	.96289	39
22	.59837	29	.40163	.63553	35	.36447	.03716	6	.96284	38
23	.59866	29	.40134	.63588	35	.36412	.03722	5	.96278	37
24	.59895	29	.40105	.63623	34	.36377	.03727	6	.96273	36
25	9.59924	30	10.40076	9.63657	35	10.36343	10.03733	5	9.96267	35
26	.59954	29	.40046	.63692	34	.36308	.03738	6	.96262	34
27	.59983	29	.40017	.63726	35	.36274	.03744	5	.96256	33
28	.60012	29	.39988	.63761	35	.36239	.03749	6	.96251	32
29	.60041	29	.39959	.63796	34	.36204	.03755	5	.96245	31
30	9.60070	29	10.39930	9.63830	35	10.36170	10.03760	6	9.96240	30
31	.60099	29	.39901	.63865	34	.36135	.03766	5	.96234	29
32	.60128	29	.39872	.63899	35	.36101	.03771	6	.96229	28
33	.60157	29	.39843	.63934	34	.36066	.03777	5	.96223	27
34	.60186	29	.39814	.63968	35	.36032	.03782	6	.96218	26
35	9.60215	29	10.39785	9.64003	34	10.35997	10.03788	5	9.96212	25
36	.60244	29	.39756	.64037	35	.35963	.03793	6	.96207	24
37	.60273	29	.39727	.64072	34	.35928	.03799	5	.96201	23
38	.60302	29	.39698	.64106	34	.35894	.03804	6	.96196	22
39	.60331	28	.39669	.64140	35	.35860	.03810	5	.96190	21
40	9.60359	29	10.39641	9.64175	34	10.35825	10.03815	6	9.96185	20
41	.60388	29	.39612	.64209	34	.35791	.03821	5	.96179	19
42	.60417	29	.39583	.64243	35	.35757	.03826	6	.96174	18
43	.60446	28	.39554	.64278	34	.35722	.03832	6	.96168	17
44	.60474	29	.39526	.64312	34	.35688	.03838	5	.96162	16
45	9.60503	29	10.39497	9.64346	35	10.35654	10.03843	6	9.96157	15
46	.60532	29	.39468	.64381	34	.35619	.03849	5	.96151	14
47	.60561	28	.39439	.64415	34	.35585	.03854	6	.96146	13
48	.60589	29	.39411	.64449	34	.35551	.03860	5	.96140	12
49	.60618	28	.39382	.64483	34	.35517	.03865	6	.96135	11
50	9.60646	29	10.39354	9.64517	35	10.35483	10.03871	6	9.96129	10
51	.60675	29	.39325	.64552	34	.35448	.03877	5	.96123	9
52	.60704	28	.39296	.64586	34	.35414	.03882	6	.96118	8
53	.60732	29	.39268	.64620	34	.35380	.03888	5	.96112	7
54	.60761	28	.39239	.64654	34	.35346	.03893	6	.96107	6
55	9.60789	29	10.39211	9.64688	34	10.35312	10.03899	6	9.96101	5
56	.60818	28	.39182	.64722	34	.35278	.03905	5	.96095	4
57	.60846	29	.39154	.64756	34	.35244	.03910	6	.96090	3
58	.60875	28	.39125	.64790	34	.35210	.03916	5	.96084	2
59	.60903	28	.39097	.64824	34	.35176	.03921	6	.96079	1
60	9.60931		10.39069	9.64858		10.35142	10.03927		9.96073	0
↑ 113° →	cos	Diff. 1′	sec	cot	Diff. 1′	tan	csc	Diff. 1′	sin	← 66° ↑

TABLE 33

Logarithms of Trigonometric Functions

24°→ ↓ ′	sin	Diff. 1′	csc	tan	Diff. 1′	cot	sec	Diff. 1′	cos	←155° ↓ ′
0	9.60931	29	10.39069	9.64858	34	10.35142	10.03927	6	9.96073	60
1	.60960	28	.39040	.64892	34	.35108	.03933	5	.96067	59
2	.60988	28	.39012	.64926	34	.35074	.03938	6	.96062	58
3	.61016	29	.38984	.64960	34	.35040	.03944	6	.96056	57
4	.61045	28	.38955	.64994	34	.35006	.03950	5	.96050	56
5	9.61073	28	10.38927	9.65028	34	10.34972	10.03955	6	9.96045	55
6	.61101	28	.38899	.65062	34	.34938	.03961	5	.96039	54
7	.61129	29	.38871	.65096	34	.34904	.03966	6	.96034	53
8	.61158	28	.38842	.65130	34	.34870	.03972	6	.96028	52
9	.61186	28	.38814	.65164	33	.34836	.03978	5	.96022	51
10	9.61214	28	10.38786	9.65197	34	10.34803	10.03983	6	9.96017	50
11	.61242	28	.38758	.65231	34	.34769	.03989	6	.96011	49
12	.61270	28	.38730	.65265	34	.34735	.03995	5	.96005	48
13	.61298	28	.38702	.65299	34	.34701	.04000	6	.96000	47
14	.61326	28	.38674	.65333	33	.34667	.04006	6	.95994	46
15	9.61354	28	10.38646	9.65366	34	10.34634	10.04012	6	9.95988	45
16	.61382	29	.38618	.65400	34	.34600	.04018	5	.95982	44
17	.61411	27	.38589	.65434	33	.34566	.04023	6	.95977	43
18	.61438	28	.38562	.65467	34	.34533	.04029	6	.95971	42
19	.61466	28	.38534	.65501	34	.34499	.04035	5	.95965	41
20	9.61494	28	10.38506	9.65535	33	10.34465	10.04040	6	9.95960	40
21	.61522	28	.38478	.65568	34	.34432	.04046	6	.95954	39
22	.61550	28	.38450	.65602	34	.34398	.04052	6	.95948	38
23	.61578	28	.38422	.65636	33	.34364	.04058	5	.95942	37
24	.61606	28	.38394	.65669	34	.34331	.04063	6	.95937	36
25	9.61634	28	10.38366	9.65703	33	10.34297	10.04069	6	9.95931	35
26	.61662	27	.38338	.65736	34	.34264	.04075	5	.95925	34
27	.61689	28	.38311	.65770	33	.34230	.04080	6	.95920	33
28	.61717	28	.38283	.65803	34	.34197	.04086	6	.95914	32
29	.61745	28	.38255	.65837	33	.34163	.04092	6	.95908	31
30	9.61773	27	10.38227	9.65870	34	10.34130	10.04098	5	9.95902	30
31	.61800	28	.38200	.65904	33	.34096	.04103	6	.95897	29
32	.61828	28	.38172	.65937	34	.34063	.04109	6	.95891	28
33	.61856	27	.38144	.65971	33	.34029	.04115	6	.95885	27
34	.61883	28	.38117	.66004	34	.33996	.04121	6	.95879	26
35	9.61911	28	10.38089	9.66038	33	10.33962	10.04127	5	9.95873	25
36	.61939	27	.38061	.66071	33	.33929	.04132	6	.95868	24
37	.61966	28	.38034	.66104	34	.33896	.04138	6	.95862	23
38	.61994	27	.38006	.66138	33	.33862	.04144	6	.95856	22
39	.62021	28	.37979	.66171	33	.33829	.04150	6	.95850	21
40	9.62049	27	10.37951	9.66204	34	10.33796	10.04156	5	9.95844	20
41	.62076	28	.37924	.66238	33	.33762	.04161	6	.95839	19
42	.62104	27	.37896	.66271	33	.33729	.04167	6	.95833	18
43	.62131	28	.37869	.66304	33	.33696	.04173	6	.95827	17
44	.62159	27	.37841	.66337	34	.33663	.04179	6	.95821	16
45	9.62186	28	10.37814	9.66371	33	10.33629	10.04185	5	9.95815	15
46	.62214	27	.37786	.66404	33	.33596	.04190	6	.95810	14
47	.62241	27	.37759	.66437	33	.33563	.04196	6	.95804	13
48	.62268	28	.37732	.66470	33	.33530	.04202	6	.95798	12
49	.62296	27	.37704	.66503	34	.33497	.04208	6	.95792	11
50	9.62323	27	10.37677	9.66537	33	10.33463	10.04214	6	9.95786	10
51	.62350	27	.37650	.66570	33	.33430	.04220	5	.95780	9
52	.62377	28	.37623	.66603	33	.33397	.04225	6	.95775	8
53	.62405	27	.37595	.66636	33	.33364	.04231	6	.95769	7
54	.62432	27	.37568	.66669	33	.33331	.04237	6	.95763	6
55	9.62459	27	10.37541	9.66702	33	10.33298	10.04243	6	9.95757	5
56	.62486	27	.37514	.66735	33	.33265	.04249	6	.95751	4
57	.62513	28	.37487	.66768	33	.33232	.04255	6	.95745	3
58	.62541	27	.37459	.66801	33	.33199	.04261	6	.95739	2
59	.62568	27	.37432	.66834	33	.33166	.04267	5	.95733	1
60	9.62595		10.37405	9.66867		10.33133	10.04272		9.95728	0
↑ 114°→	cos	Diff. 1′	sec	cot	Diff. 1′	tan	csc	Diff. 1′	sin	↑ ←65°

TABLE 33
Logarithms of Trigonometric Functions

25° → ↓ ′	sin	Diff. 1′	csc	tan	Diff. 1′	cot	sec	Diff. 1′	cos	← 154° ↓ ′
0	9. 62595	27	10. 37405	9. 66867	33	10. 33133	10. 04272	6	9. 95728	60
1	. 62622	27	. 37378	. 66900	33	. 33100	. 04278	6	. 95722	59
2	. 62649	27	. 37351	. 66933	33	. 33067	. 04284	6	. 95716	58
3	. 62676	27	. 37324	. 66966	33	. 33034	. 04290	6	. 95710	57
4	. 62703	27	. 37297	. 66999	33	. 33001	. 04296	6	. 95704	56
5	9. 62730	27	10. 37270	9. 67032	33	10. 32968	10. 04302	6	9. 95698	55
6	. 62757	27	. 37243	. 67065	33	. 32935	. 04308	6	. 95692	54
7	. 62784	27	. 37216	. 67098	33	. 32902	. 04314	6	. 95686	53
8	. 62811	27	. 37189	. 67131	32	. 32869	. 04320	6	. 95680	52
9	. 62838	27	. 37162	. 67163	33	. 32837	. 04326	6	. 95674	51
10	9. 62865	27	10. 37135	9. 67196	33	10. 32804	10. 04332	5	9. 95668	50
11	. 62892	26	. 37108	. 67229	33	. 32771	. 04337	6	. 95663	49
12	. 62918	27	. 37082	. 67262	33	. 32738	. 04343	6	. 95657	48
13	. 62945	27	. 37055	. 67295	32	. 32705	. 04349	6	. 95651	47
14	. 62972	27	. 37028	. 67327	33	. 32673	. 04355	6	. 95645	46
15	9. 62999	27	10. 37001	9. 67360	33	10. 32640	10. 04361	6	9. 95639	45
16	. 63026	26	. 36974	. 67393	33	. 32607	. 04367	6	. 95633	44
17	. 63052	27	. 36948	. 67426	32	. 32574	. 04373	6	. 95627	43
18	. 63079	27	. 36921	. 67458	33	. 32542	. 04379	6	. 95621	42
19	. 63106	27	. 36894	. 67491	33	. 32509	. 04385	6	. 95615	41
20	9. 63133	26	10. 36867	9. 67524	32	10. 32476	10. 04391	6	9. 95609	40
21	. 63159	27	. 36841	. 67556	33	. 32444	. 04397	6	. 95603	39
22	. 63186	27	. 36814	. 67589	33	. 32411	. 04403	6	. 95597	38
23	. 63213	26	. 36787	. 67622	32	. 32378	. 04409	6	. 95591	37
24	. 63239	27	. 36761	. 67654	33	. 32346	. 04415	6	. 95585	36
25	9. 63266	26	10. 36734	9. 67687	32	10. 32313	10. 04421	6	9. 95579	35
26	. 63292	27	. 36708	. 67719	33	. 32281	. 04427	6	. 95573	34
27	. 63319	26	. 36681	. 67752	33	. 32248	. 04433	6	. 95567	33
28	. 63345	27	. 36655	. 67785	32	. 32215	. 04439	6	. 95561	32
29	. 63372	26	. 36628	. 67817	33	. 32183	. 04445	6	. 95555	31
30	9. 63398	27	10. 36602	9. 67850	32	10. 32150	10. 04451	6	9. 95549	30
31	. 63425	26	. 36575	. 67882	33	. 32118	. 04457	6	. 95543	29
32	. 63451	27	. 36549	. 67915	32	. 32085	. 04463	6	. 95537	28
33	. 63478	26	. 36522	. 67947	33	. 32053	. 04469	6	. 95531	27
34	. 63504	27	. 36496	. 67980	32	. 32020	. 04475	6	. 95525	26
35	9. 63531	26	10. 36469	9. 68012	32	10. 31988	10. 04481	6	9. 95519	25
36	. 63557	26	. 36443	. 68044	33	. 31956	. 04487	6	. 95513	24
37	. 63583	27	. 36417	. 68077	32	. 31923	. 04493	7	. 95507	23
38	. 63610	26	. 36390	. 68109	33	. 31891	. 04500	6	. 95500	22
39	. 63636	26	. 36364	. 68142	32	. 31858	. 04506	6	. 95494	21
40	9. 63662	27	10. 36338	9. 68174	32	10. 31826	10. 04512	6	9. 95488	20
41	. 63689	26	. 36311	. 68206	33	. 31794	. 04518	6	. 95482	19
42	. 63715	26	. 36285	. 68239	32	. 31761	. 04524	6	. 95476	18
43	. 63741	26	. 36259	. 68271	32	. 31729	. 04530	6	. 95470	17
44	. 63767	27	. 36233	. 68303	33	. 31697	. 04536	6	. 95464	16
45	9. 63794	26	10. 36206	9. 68336	32	10. 31664	10. 04542	6	9. 95458	15
46	. 63820	26	. 36180	. 68368	32	. 31632	. 04548	6	. 95452	14
47	. 63846	26	. 36154	. 68400	32	. 31600	. 04554	6	. 95446	13
48	. 63872	26	. 36128	. 68432	33	. 31568	. 04560	6	. 95440	12
49	. 63898	26	. 36102	. 68465	32	. 31535	. 04566	7	. 95434	11
50	9. 63924	26	10. 36076	9. 68497	32	10. 31503	10. 04573	6	9. 95427	10
51	. 63950	26	. 36050	. 68529	32	. 31471	. 04579	6	. 95421	9
52	. 63976	26	. 36024	. 68561	32	. 31439	. 04585	6	. 95415	8
53	. 64002	26	. 35998	. 68593	33	. 31407	. 04591	6	. 95409	7
54	. 64028	26	. 35972	. 68626	32	. 31374	. 04597	6	. 95403	6
55	9. 64054	26	10. 35946	9. 68658	32	10. 31342	10. 04603	6	9. 95397	5
56	. 64080	26	. 35920	. 68690	32	. 31310	. 04609	7	. 95391	4
57	. 64106	26	. 35894	. 68722	32	. 31278	. 04616	6	. 95384	3
58	. 64132	26	. 35868	. 68754	32	. 31246	. 04622	6	. 95378	2
59	. 64158	26	. 35842	. 68786	32	. 31214	. 04628	6	. 95372	1
60	9. 64184		10. 35816	9. 68818		10. 31182	10. 04634		9. 95366	0
↑ 115° →	cos	Diff. 1′	sec	cot	Diff. 1′	tan	csc	Diff. 1′	sin	← 64° ↑

TABLE 33
Logarithms of Trigonometric Functions

26°→ ↓	sin	Diff. 1′	csc	tan	Diff. 1′	cot	sec	Diff. 1′	cos	←153° ↓
′										′
0	9.64184	26	10.35816	9.68818	32	10.31182	10.04634	6	9.95366	60
1	.64210	26	.35790	.68850	32	.31150	.04640	6	.95360	59
2	.64236	26	.35764	.68882	32	.31118	.04646	6	.95354	58
3	.64262	26	.35738	.68914	32	.31086	.04652	7	.95348	57
4	.64288	25	.35712	.68946	32	.31054	.04659	6	.95341	56
5	9.64313	26	10.35687	9.68978	32	10.31022	10.04665	6	9.95335	55
6	.64339	26	.35661	.69010	32	.30990	.04671	6	.95329	54
7	.64365	26	.35635	.69042	32	.30958	.04677	6	.95323	53
8	.64391	26	.35609	.69074	32	.30926	.04683	7	.95317	52
9	.64417	25	.35583	.69106	32	.30894	.04690	6	.95310	51
10	9.64442	26	10.35558	9.69138	32	10.30862	10.04696	6	9.95304	50
11	.64468	26	.35532	.69170	32	.30830	.04702	6	.95298	49
12	.64494	25	.35506	.69202	32	.30798	.04708	6	.95292	48
13	.64519	26	.35481	.69234	32	.30766	.04714	7	.95286	47
14	.64545	26	.35455	.69266	32	.30734	.04721	6	.95279	46
15	9.64571	25	10.35429	9.69298	31	10.30702	10.04727	6	9.95273	45
16	.64596	26	.35404	.69329	32	.30671	.04733	6	.95267	44
17	.64622	25	.35378	.69361	32	.30639	.04739	7	.95261	43
18	.64647	26	.35353	.69393	32	.30607	.04746	6	.95254	42
19	.64673	25	.35327	.69425	32	.30575	.04752	6	.95248	41
20	9.64698	26	10.35302	9.69457	31	10.30543	10.04758	6	9.95242	40
21	.64724	25	.35276	.69488	32	.30512	.04764	7	.95236	39
22	.64749	26	.35251	.69520	32	.30480	.04771	6	.95229	38
23	.64775	25	.35225	.69552	32	.30448	.04777	6	.95223	37
24	.64800	26	.35200	.69584	31	.30416	.04783	6	.95217	36
25	9.64826	25	10.35174	9.69615	32	10.30385	10.04789	7	9.95211	35
26	.64851	26	.35149	.69647	32	.30353	.04796	6	.95204	34
27	.64877	25	.35123	.69679	31	.30321	.04802	6	.95198	33
28	.64902	25	.35098	.69710	32	.30290	.04808	7	.95192	32
29	.64927	26	.35073	.69742	32	.30258	.04815	6	.95185	31
30	9.64953	25	10.35047	9.69774	31	10.30226	10.04821	6	9.95179	30
31	.64978	25	.35022	.69805	32	.30195	.04827	6	.95173	29
32	.65003	26	.34997	.69837	31	.30163	.04833	7	.95167	28
33	.65029	25	.34971	.69868	32	.30132	.04840	6	.95160	27
34	.65054	25	.34946	.69900	32	.30100	.04846	6	.95154	26
35	9.65079	25	10.34921	9.69932	31	10.30068	10.04852	7	9.95148	25
36	.65104	26	.34896	.69963	32	.30037	.04859	6	.95141	24
37	.65130	25	.34870	.69995	31	.30005	.04865	6	.95135	23
38	.65155	25	.34845	.70026	32	.29974	.04871	7	.95129	22
39	.65180	25	.34820	.70058	31	.29942	.04878	6	.95122	21
40	9.65205	25	10.34795	9.70089	32	10.29911	10.04884	6	9.95116	20
41	.65230	25	.34770	.70121	31	.29879	.04890	7	.95110	19
42	.65255	26	.34745	.70152	32	.29848	.04897	6	.95103	18
43	.65281	25	.34719	.70184	31	.29816	.04903	7	.95097	17
44	.65306	25	.34694	.70215	32	.29785	.04910	6	.95090	16
45	9.65331	25	10.34669	9.70247	31	10.29753	10.04916	6	9.95084	15
46	.65356	25	.34644	.70278	31	.29722	.04922	7	.95078	14
47	.65381	25	.34619	.70309	32	.29691	.04929	6	.95071	13
48	.65406	25	.34594	.70341	31	.29659	.04935	6	.95065	12
49	.65431	25	.34569	.70372	32	.29628	.04941	7	.95059	11
50	9.65456	25	10.34544	9.70404	31	10.29596	10.04948	6	9.95052	10
51	.65481	25	.34519	.70435	31	.29565	.04954	7	.95046	9
52	.65506	25	.34494	.70466	32	.29534	.04961	6	.95039	8
53	.65531	25	.34469	.70498	31	.29502	.04967	6	.95033	7
54	.65556	24	.34444	.70529	31	.29471	.04973	7	.95027	6
55	9.65580	25	10.34420	9.70560	32	10.29440	10.04980	6	9.95020	5
56	.65605	25	.34395	.70592	31	.29408	.04986	7	.95014	4
57	.65630	25	.34370	.70623	31	.29377	.04993	6	.95007	3
58	.65655	25	.34345	.70654	31	.29346	.04999	6	.95001	2
59	.65680	25	.34320	.70685	32	.29315	.05005	7	.94995	1
60	9.65705		10.34295	9.70717		10.29283	10.05012		9.94988	0
↑ 116°→	cos	Diff. 1′	sec	cot	Diff. 1′	tan	csc	Diff. 1′	sin	←63° ↑

TABLE 33

Logarithms of Trigonometric Functions

27° → ↓ ′	sin	Diff. 1′	csc	tan	Diff. 1′	cot	sec	Diff. 1′	cos	← 152° ↓ ′
0	9.65705	24	10.34295	9.70717	31	10.29283	10.05012	6	9.94988	60
1	.65729	25	.34271	.70748	31	.29252	.05018	7	.94982	59
2	.65754	25	.34246	.70779	31	.29221	.05025	6	.94975	58
3	.65779	25	.34221	.70810	31	.29190	.05031	7	.94969	57
4	.65804	24	.34196	.70841	32	.29159	.05038	6	.94962	56
5	9.65828	25	10.34172	9.70873	31	10.29127	10.05044	7	9.94956	55
6	.65853	25	.34147	.70904	31	.29096	.05051	6	.94949	54
7	.65878	24	.34122	.70935	31	.29065	.05057	7	.94943	53
8	.65902	25	.34098	.70966	31	.29034	.05064	6	.94936	52
9	.65927	25	.34073	.70997	31	.29003	.05070	7	.94930	51
10	9.65952	24	10.34048	9.71028	31	10.28972	10.05077	6	9.94923	50
11	.65976	25	.34024	.71059	31	.28941	.05083	6	.94917	49
12	.66001	24	.33999	.71090	31	.28910	.05089	7	.94911	48
13	.66025	25	.33975	.71121	32	.28879	.05096	6	.94904	47
14	.66050	25	.33950	.71153	31	.28847	.05102	7	.94898	46
15	9.66075	24	10.33925	9.71184	31	10.28816	10.05109	6	9.94891	45
16	.66099	25	.33901	.71215	31	.28785	.05115	7	.94885	44
17	.66124	24	.33876	.71246	31	.28754	.05122	7	.94878	43
18	.66148	25	.33852	.71277	31	.28723	.05129	6	.94871	42
19	.66173	24	.33827	.71308	31	.28692	.05135	7	.94865	41
20	9.66197	24	10.33803	9.71339	31	10.28661	10.05142	6	9.94858	40
21	.66221	25	.33779	.71370	31	.28630	.05148	7	.94852	39
22	.66246	24	.33754	.71401	30	.28599	.05155	6	.94845	38
23	.66270	25	.33730	.71431	31	.28569	.05161	7	.94839	37
24	.66295	24	.33705	.71462	31	.28538	.05168	6	.94832	36
25	9.66319	24	10.33681	9.71493	31	10.28507	10.05174	7	9.94826	35
26	.66343	25	.33657	.71524	31	.28476	.05181	6	.94819	34
27	.66368	24	.33632	.71555	31	.28445	.05187	7	.94813	33
28	.66392	24	.33608	.71586	31	.28414	.05194	7	.94806	32
29	.66416	25	.33584	.71617	31	.28383	.05201	6	.94799	31
30	9.66441	24	10.33559	9.71648	31	10.28352	10.05207	7	9.94793	30
31	.66465	24	.33535	.71679	30	.28321	.05214	6	.94786	29
32	.66489	24	.33511	.71709	31	.28291	.05220	7	.94780	28
33	.66513	24	.33487	.71740	31	.28260	.05227	6	.94773	27
34	.66537	25	.33463	.71771	31	.28229	.05233	7	.94767	26
35	9.66562	24	10.33438	9.71802	31	10.28198	10.05240	7	9.94760	25
36	.66586	24	.33414	.71833	30	.28167	.05247	6	.94753	24
37	.66610	24	.33390	.71863	31	.28137	.05253	7	.94747	23
38	.66634	24	.33366	.71894	31	.28106	.05260	6	.94740	22
39	.66658	24	.33342	.71925	30	.28075	.05266	7	.94734	21
40	9.66682	24	10.33318	9.71955	31	10.28045	10.05273	7	9.94727	20
41	.66706	25	.33294	.71986	31	.28014	.05280	6	.94720	19
42	.66731	24	.33269	.72017	31	.27983	.05286	7	.94714	18
43	.66755	24	.33245	.72048	30	.27952	.05293	7	.94707	17
44	.66779	24	.33221	.72078	31	.27922	.05300	6	.94700	16
45	9.66803	24	10.33197	9.72109	31	10.27891	10.05306	7	9.94694	15
46	.66827	24	.33173	.72140	30	.27860	.05313	7	.94687	14
47	.66851	24	.33149	.72170	31	.27830	.05320	6	.94680	13
48	.66875	24	.33125	.72201	30	.27799	.05326	7	.94674	12
49	.66899	23	.33101	.72231	31	.27769	.05333	7	.94667	11
50	9.66922	24	10.33078	9.72262	31	10.27738	10.05340	6	9.94660	10
51	.66946	24	.33054	.72293	30	.27707	.05346	7	.94654	9
52	.66970	24	.33030	.72323	31	.27677	.05353	7	.94647	8
53	.66994	24	.33006	.72354	30	.27646	.05360	6	.94640	7
54	.67018	24	.32982	.72384	31	.27616	.05366	7	.94634	6
55	9.67042	24	10.32958	9.72415	30	10.27585	10.05373	7	9.94627	5
56	.67066	24	.32934	.72445	31	.27555	.05380	6	.94620	4
57	.67090	23	.32910	.72476	30	.27524	.05386	7	.94614	3
58	.67113	24	.32887	.72506	31	.27494	.05393	7	.94607	2
59	.67137	24	.32863	.72537	30	.27463	.05400	7	.94600	1
60	9.67161		10.32839	9.72567		10.27433	10.05407		9.94593	0
↑ 117° →	cos	Diff. 1′	sec	cot	Diff. 1′	tan	csc	Diff. 1′	sin	← 62° ↑

TABLE 33

Logarithms of Trigonometric Functions

28°→ ↓	sin	Diff. 1′	csc	tan	Diff. 1′	cot	sec	Diff. 1′	cos	←151° ↓
′										′
0	9. 67161	24	10. 32839	9. 72567	31	10. 27433	10. 05407	6	9. 94593	60
1	. 67185	23	. 32815	. 72598	30	. 27402	. 05413	7	. 94587	59
2	. 67208	24	. 32792	. 72628	31	. 27372	. 05420	7	. 94580	58
3	. 67232	24	. 32768	. 72659	30	. 27341	. 05427	6	. 94573	57
4	. 67256	24	. 32744	. 72689	31	. 27311	. 05433	7	. 94567	56
5	9. 67280	23	10. 32720	9. 72720	30	10. 27280	10. 05440	7	9. 94560	55
6	. 67303	24	. 32697	. 72750	30	. 27250	. 05447	7	. 94553	54
7	. 67327	23	. 32673	. 72780	31	. 27220	. 05454	6	. 94546	53
8	. 67350	24	. 32650	. 72811	30	. 27189	. 05460	7	. 94540	52
9	. 67374	24	. 32626	. 72841	31	. 27159	. 05467	7	. 94533	51
10	9. 67398	23	10. 32602	9. 72872	30	10. 27128	10. 05474	7	9. 94526	50
11	. 67421	24	. 32579	. 72902	30	. 27098	. 05481	6	. 94519	49
12	. 67445	23	. 32555	. 72932	31	. 27068	. 05487	7	. 94513	48
13	. 67468	24	. 32532	. 72963	30	. 27037	. 05494	7	. 94506	47
14	. 67492	23	. 32508	. 72993	30	. 27007	. 05501	7	. 94499	46
15	9. 67515	24	10. 32485	9. 73023	31	10. 26977	10. 05508	7	9. 94492	45
16	. 67539	23	. 32461	. 73054	30	. 26946	. 05515	6	. 94485	44
17	. 67562	24	. 32438	. 73084	30	. 26916	. 05521	7	. 94479	43
18	. 67586	23	. 32414	. 73114	30	. 26886	. 05528	7	. 94472	42
19	. 67609	24	. 32391	. 73144	31	. 26856	. 05535	7	. 94465	41
20	9. 67633	23	10. 32367	9. 73175	30	10. 26825	10. 05542	7	9. 94458	40
21	. 67656	24	. 32344	. 73205	30	. 26795	. 05549	6	. 94451	39
22	. 67680	23	. 32320	. 73235	30	. 26765	. 05555	7	. 94445	38
23	. 67703	23	. 32297	. 73265	30	. 26735	. 05562	7	. 94438	37
24	. 67726	24	. 32274	. 73295	31	. 26705	. 05569	7	. 94431	36
25	9. 67750	23	10. 32250	9. 73326	30	10. 26674	10. 05576	7	9. 94424	35
26	. 67773	23	. 32227	. 73356	30	. 26644	. 05583	7	. 94417	34
27	. 67796	24	. 32204	. 73386	30	. 26614	. 05590	6	. 94410	33
28	. 67820	23	. 32180	. 73416	30	. 26584	. 05596	7	. 94404	32
29	. 67843	23	. 32157	. 73446	30	. 26554	. 05603	7	. 94397	31
30	9. 67866	24	10. 32134	9. 73476	31	10. 26524	10. 05610	7	9. 94390	30
31	. 67890	23	. 32110	. 73507	30	. 26493	. 05617	7	. 94383	29
32	. 67913	23	. 32087	. 73537	30	. 26463	. 05624	7	. 94376	28
33	. 67936	23	. 32064	. 73567	30	. 26433	. 05631	7	. 94369	27
34	. 67959	23	. 32041	. 73597	30	. 26403	. 05638	7	. 94362	26
35	9. 67982	24	10. 32018	9. 73627	30	10. 26373	10. 05645	6	9. 94355	25
36	. 68006	23	. 31994	. 73657	30	. 26343	. 05651	7	. 94349	24
37	. 68029	23	. 31971	. 73687	30	. 26313	. 05658	7	. 94342	23
38	. 68052	23	. 31948	. 73717	30	. 26283	. 05665	7	. 94335	22
39	. 68075	23	. 31925	. 73747	30	. 26253	. 05672	7	. 94328	21
40	9. 68098	23	10. 31902	9. 73777	30	10. 26223	10. 05679	7	9. 94321	20
41	. 68121	23	. 31879	. 73807	30	. 26193	. 05686	7	. 94314	19
42	. 68144	23	. 31856	. 73837	30	. 26163	. 05693	7	. 94307	18
43	. 68167	23	. 31833	. 73867	30	. 26133	. 05700	7	. 94300	17
44	. 68190	23	. 31810	. 73897	30	. 26103	. 05707	7	. 94293	16
45	9. 68213	24	10. 31787	9. 73927	30	10. 26073	10. 05714	7	9. 94286	15
46	. 68237	23	. 31763	. 73957	30	. 26043	. 05721	6	. 94279	14
47	. 68260	23	. 31740	. 73987	30	. 26013	. 05727	7	. 94273	13
48	. 68283	22	. 31717	. 74017	30	. 25983	. 05734	7	. 94266	12
49	. 68305	23	. 31695	. 74047	30	. 25953	. 05741	7	. 94259	11
50	9. 68328	23	10. 31672	9. 74077	30	10. 25923	10. 05748	7	9. 94252	10
51	. 68351	23	. 31649	. 74107	30	. 25893	. 05755	7	. 94245	9
52	. 68374	23	. 31626	. 74137	29	. 25863	. 05762	7	. 94238	8
53	. 68397	23	. 31603	. 74166	30	. 25834	. 05769	7	. 94231	7
54	. 68420	23	. 31580	. 74196	30	. 25804	. 05776	7	. 94224	6
55	9. 68443	23	10. 31557	9. 74226	30	10. 25774	10. 05783	7	9. 94217	5
56	. 68466	23	. 31534	. 74256	30	. 25744	. 05790	7	. 94210	4
57	. 68489	23	. 31511	. 74286	30	. 25714	. 05797	7	. 94203	3
58	. 68512	22	. 31488	. 74316	29	. 25684	. 05804	7	. 94196	2
59	. 68534	23	. 31466	. 74345	30	. 25655	. 05811	7	. 94189	1
60	9. 68557		10. 31443	9. 74375		10. 25625	10. 05818		9. 94182	0
↑ 118°→	cos	Diff. 1′	sec	cot	Diff. 1′	tan	csc	Diff. 1′	sin	←61° ↑

TABLE 33

Logarithms of Trigonometric Functions

29°→ ↓ ′	sin	Diff. 1′	csc	tan	Diff. 1′	cot	sec	Diff. 1′	cos	←150° ↓ ′
0	9.68557	23	10.31443	9.74375	30	10.25625	10.05818	7	9.94182	60
1	.68580	23	.31420	.74405	30	.25595	.05825	7	.94175	59
2	.68603	22	.31397	.74435	30	.25565	.05832	7	.94168	58
3	.68625	23	.31375	.74465	29	.25535	.05839	7	.94161	57
4	.68648	23	.31352	.74494	30	.25506	.05846	7	.94154	56
5	9.68671	23	10.31329	9.74524	30	10.25476	10.05853	7	9.94147	55
6	.68694	22	.31306	.74554	29	.25446	.05860	7	.94140	54
7	.68716	23	.31284	.74583	30	.25417	.05867	7	.94133	53
8	.68739	23	.31261	.74613	30	.25387	.05874	7	.94126	52
9	.68762	22	.31238	.74643	30	.25357	.05881	7	.94119	51
10	9.68784	23	10.31216	9.74673	29	10.25327	10.05888	7	9.94112	50
11	.68807	22	.31193	.74702	30	.25298	.05895	7	.94105	49
12	.68829	23	.31171	.74732	30	.25268	.05902	8	.94098	48
13	.68852	23	.31148	.74762	29	.25238	.05910	7	.94090	47
14	.68875	22	.31125	.74791	30	.25209	.05917	7	.94083	46
15	9.68897	23	10.31103	9.74821	30	10.25179	10.05924	7	9.94076	45
16	.68920	22	.31080	.74851	29	.25149	.05931	7	.94069	44
17	.68942	23	.31058	.74880	30	.25120	.05938	7	.94062	43
18	.68965	22	.31035	.74910	29	.25090	.05945	7	.94055	42
19	.68987	23	.31013	.74939	30	.25061	.05952	7	.94048	41
20	9.69010	22	10.30990	9.74969	29	10.25031	10.05959	7	9.94041	40
21	.69032	23	.30968	.74998	30	.25002	.05966	7	.94034	39
22	.69055	22	.30945	.75028	30	.24972	.05973	7	.94027	38
23	.69077	23	.30923	.75058	29	.24942	.05980	8	.94020	37
24	.69100	22	.30900	.75087	30	.24913	.05988	7	.94012	36
25	9.69122	22	10.30878	9.75117	29	10.24883	10.05995	7	9.94005	35
26	.69144	23	.30856	.75146	30	.24854	.06002	7	.93998	34
27	.69167	22	.30833	.75176	29	.24824	.06009	7	.93991	33
28	.69189	23	.30811	.75205	30	.24795	.06016	7	.93984	32
29	.69212	22	.30788	.75235	29	.24765	.06023	7	.93977	31
30	9.69234	22	10.30766	9.75264	30	10.24736	10.06030	7	9.93970	30
31	.69256	23	.30744	.75294	29	.24706	.06037	8	.93963	29
32	.69279	22	.30721	.75323	30	.24677	.06045	7	.93955	28
33	.69301	22	.30699	.75353	29	.24647	.06052	7	.93948	27
34	.69323	22	.30677	.75382	29	.24618	.06059	7	.93941	26
35	9.69345	23	10.30655	9.75411	30	10.24589	10.06066	7	9.93934	25
36	.69368	22	.30632	.75441	29	.24559	.06073	7	.93927	24
37	.69390	22	.30610	.75470	30	.24530	.06080	8	.93920	23
38	.69412	22	.30588	.75500	29	.24500	.06088	7	.93912	22
39	.69434	22	.30566	.75529	29	.24471	.06095	7	.93905	21
40	9.69456	23	10.30544	9.75558	30	10.24442	10.06102	7	9.93898	20
41	.69479	22	.30521	.75588	29	.24412	.06109	7	.93891	19
42	.69501	22	.30499	.75617	30	.24383	.06116	8	.93884	18
43	.69523	22	.30477	.75647	29	.24353	.06124	7	.93876	17
44	.69545	22	.30455	.75676	29	.24324	.06131	7	.93869	16
45	9.69567	22	10.30433	9.75705	30	10.24295	10.06138	7	9.93862	15
46	.69589	22	.30411	.75735	29	.24265	.06145	8	.93855	14
47	.69611	22	.30389	.75764	29	.24236	.06153	7	.93847	13
48	.69633	22	.30367	.75793	29	.24207	.06160	7	.93840	12
49	.69655	22	.30345	.75822	30	.24178	.06167	7	.93833	11
50	9.69677	22	10.30323	9.75852	29	10.24148	10.06174	7	9.93826	10
51	.69699	22	.30301	.75881	29	.24119	.06181	8	.93819	9
52	.69721	22	.30279	.75910	29	.24090	.06189	7	.93811	8
53	.69743	22	.30257	.75939	30	.24061	.06196	7	.93804	7
54	.69765	22	.30235	.75969	29	.24031	.06203	8	.93797	6
55	9.69787	22	10.30213	9.75998	29	10.24002	10.06211	7	9.93789	5
56	.69809	22	.30191	.76027	29	.23973	.06218	7	.93782	4
57	.69831	22	.30169	.76056	30	.23944	.06225	7	.93775	3
58	.69853	22	.30147	.76086	29	.23914	.06232	8	.93768	2
59	.69875	22	.30125	.76115	29	.23885	.06240	7	.93760	1
60	9.69897		10.30103	9.76144		10.23856	10.06247		9.93753	0
↑ 119°→	cos	Diff. 1′	sec	cot	Diff. 1′	tan	csc	Diff. 1′	sin	←60° ↑

TABLE 33

Logarithms of Trigonometric Functions

30° → ↓ ′	sin	Diff. 1′	csc	tan	Diff. 1′	cot	sec	Diff. 1′	cos	← 149° ↓ ′
0	9.69897	22	10.30103	9.76144	29	10.23856	10.06247	7	9.93753	60
1	.69919	22	.30081	.76173	29	.23827	.06254	8	.93746	59
2	.69941	22	.30059	.76202	29	.23798	.06262	7	.93738	58
3	.69963	21	.30037	.76231	30	.23769	.06269	7	.93731	57
4	.69984	22	.30016	.76261	29	.23739	.06276	7	.93724	56
5	9.70006	22	10.29994	9.76290	29	10.23710	10.06283	8	9.93717	55
6	.70028	22	.29972	.76319	29	.23681	.06291	7	.93709	54
7	.70050	22	.29950	.76348	29	.23652	.06298	7	.93702	53
8	.70072	21	.29928	.76377	29	.23623	.06305	8	.93695	52
9	.70093	22	.29907	.76406	29	.23594	.06313	7	.93687	51
10	9.70115	22	10.29885	9.76435	29	10.23565	10.06320	7	9.93680	50
11	.70137	22	.29863	.76464	29	.23536	.06327	8	.93673	49
12	.70159	21	.29841	.76493	29	.23507	.06335	7	.93665	48
13	.70180	22	.29820	.76522	29	.23478	.06342	8	.93658	47
14	.70202	22	.29798	.76551	29	.23449	.06350	7	.93650	46
15	9.70224	21	10.29776	9.76580	29	10.23420	10.06357	7	9.93643	45
16	.70245	22	.29755	.76609	30	.23391	.06364	8	.93636	44
17	.70267	21	.29733	.76639	29	.23361	.06372	7	.93628	43
18	.70288	22	.29712	.76668	29	.23332	.06379	7	.93621	42
19	.70310	22	.29690	.76697	28	.23303	.06386	8	.93614	41
20	9.70332	21	10.29668	9.76725	29	10.23275	10.06394	7	9.93606	40
21	.70353	22	.29647	.76754	29	.23246	.06401	8	.93599	39
22	.70375	21	.29625	.76783	29	.23217	.06409	7	.93591	38
23	.70396	22	.29604	.76812	29	.23188	.06416	7	.93584	37
24	.70418	21	.29582	.76841	29	.23159	.06423	8	.93577	36
25	9.70439	22	10.29561	9.76870	29	10.23130	10.06431	7	9.93569	35
26	.70461	21	.29539	.76899	29	.23101	.06438	8	.93562	34
27	.70482	22	.29518	.76928	29	.23072	.06446	7	.93554	33
28	.70504	21	.29496	.76957	29	.23043	.06453	8	.93547	32
29	.70525	22	.29475	.76986	29	.23014	.06461	7	.93539	31
30	9.70547	21	10.29453	9.77015	29	10.22985	10.06468	7	9.93532	30
31	.70568	22	.29432	.77044	29	.22956	.06475	8	.93525	29
32	.70590	21	.29410	.77073	28	.22927	.06483	7	.93517	28
33	.70611	22	.29389	.77101	29	.22899	.06490	8	.93510	27
34	.70633	21	.29367	.77130	29	.22870	.06498	7	.93502	26
35	9.70654	21	10.29346	9.77159	29	10.22841	10.06505	8	9.93495	25
36	.70675	22	.29325	.77188	29	.22812	.06513	7	.93487	24
37	.70697	21	.29303	.77217	29	.22783	.06520	8	.93480	23
38	.70718	21	.29282	.77246	28	.22754	.06528	7	.93472	22
39	.70739	22	.29261	.77274	29	.22726	.06535	8	.93465	21
40	9.70761	21	10.29239	9.77303	29	10.22697	10.06543	7	9.93457	20
41	.70782	21	.29218	.77332	29	.22668	.06550	8	.93450	19
42	.70803	21	.29197	.77361	29	.22639	.06558	7	.93442	18
43	.70824	22	.29176	.77390	28	.22610	.06565	8	.93435	17
44	.70846	21	.29154	.77418	29	.22582	.06573	7	.93427	16
45	9.70867	21	10.29133	9.77447	29	10.22553	10.06580	8	9.93420	15
46	.70888	21	.29112	.77476	29	.22524	.06588	7	.93412	14
47	.70909	22	.29091	.77505	28	.22495	.06595	8	.93405	13
48	.70931	21	.29069	.77533	29	.22467	.06603	7	.93397	12
49	.70952	21	.29048	.77562	29	.22438	.06610	8	.93390	11
50	9.70973	21	10.29027	9.77591	28	10.22409	10.06618	7	9.93382	10
51	.70994	21	.29006	.77619	29	.22381	.06625	8	.93375	9
52	.71015	21	.28985	.77648	29	.22352	.06633	7	.93367	8
53	.71036	22	.28964	.77677	29	.22323	.06640	8	.93360	7
54	.71058	21	.28942	.77706	28	.22294	.06648	8	.93352	6
55	9.71079	21	10.28921	9.77734	29	10.22266	10.06656	7	9.93344	5
56	.71100	21	.28900	.77763	28	.22237	.06663	8	.93337	4
57	.71121	21	.28879	.77791	29	.22209	.06671	7	.93329	3
58	.71142	21	.28858	.77820	29	.22180	.06678	8	.93322	2
59	.71163	21	.28837	.77849	28	.22151	.06686	7	.93314	1
60	9.71184		10.28816	9.77877		10.22123	10.06693		9.93307	0
↑ 120° →	cos	Diff. 1′	sec	cot	Diff. 1′	tan	csc	Diff. 1′	sin	← 59° ↑

TABLE 33
Logarithms of Trigonometric Functions

31°→ ↓ ′	sin	Diff. 1′	csc	tan	Diff. 1′	cot	sec	Diff. 1′	cos	←148° ↓ ′
0	9.71184	21	10.28816	9.77877	29	10.22123	10.06693	8	9.93307	60
1	.71205	21	.28795	.77906	29	.22094	.06701	8	.93299	59
2	.71226	21	.28774	.77935	28	.22065	.06709	7	.93291	58
3	.71247	21	.28753	.77963	29	.22037	.06716	8	.93284	57
4	.71268	21	.28732	.77992	28	.22008	.06724	7	.93276	56
5	9.71289	21	10.28711	9.78020	29	10.21980	10.06731	8	9.93269	55
6	.71310	21	.28690	.78049	28	.21951	.06739	8	.93261	54
7	.71331	21	.28669	.78077	29	.21923	.06747	7	.93253	53
8	.71352	21	.28648	.78106	29	.21894	.06754	8	.93246	52
9	.71373	20	.28627	.78135	28	.21865	.06762	8	.93238	51
10	9.71393	21	10.28607	9.78163	29	10.21837	10.06770	7	9.93230	50
11	.71414	21	.28586	.78192	28	.21808	.06777	8	.93223	49
12	.71435	21	.28565	.78220	29	.21780	.06785	8	.93215	48
13	.71456	21	.28544	.78249	28	.21751	.06793	7	.93207	47
14	.71477	21	.28523	.78277	29	.21723	.06800	8	.93200	46
15	9.71498	21	10.28502	9.78306	28	10.21694	10.06808	8	9.93192	45
16	.71519	20	.28481	.78334	29	.21666	.06816	7	.93184	44
17	.71539	21	.28461	.78363	28	.21637	.06823	8	.93177	43
18	.71560	21	.28440	.78391	28	.21609	.06831	8	.93169	42
19	.71581	21	.28419	.78419	29	.21581	.06839	7	.93161	41
20	9.71602	20	10.28398	9.78448	28	10.21552	10.06846	8	9.93154	40
21	.71622	21	.28378	.78476	29	.21524	.06854	8	.93146	39
22	.71643	21	.28357	.78505	28	.21495	.06862	7	.93138	38
23	.71664	21	.28336	.78533	29	.21467	.06869	8	.93131	37
24	.71685	20	.28315	.78562	28	.21438	.06877	8	.93123	36
25	9.71705	21	10.28295	9.78590	28	10.21410	10.06885	7	9.93115	35
26	.71726	21	.28274	.78618	29	.21382	.06892	8	.93108	34
27	.71747	20	.28253	.78647	28	.21353	.06900	8	.93100	33
28	.71767	21	.28233	.78675	29	.21325	.06908	8	.93092	32
29	.71788	21	.28212	.78704	28	.21296	.06916	7	.93084	31
30	9.71809	20	10.28191	9.78732	28	10.21268	10.06923	8	9.93077	30
31	.71829	21	.28171	.78760	29	.21240	.06931	8	.93069	29
32	.71850	20	.28150	.78789	28	.21211	.06939	8	.93061	28
33	.71870	21	.28130	.78817	28	.21183	.06947	7	.93053	27
34	.71891	20	.28109	.78845	29	.21155	.06954	8	.93046	26
35	9.71911	21	10.28089	9.78874	28	10.21126	10.06962	8	9.93038	25
36	.71932	20	.28068	.78902	28	.21098	.06970	8	.93030	24
37	.71952	21	.28048	.78930	29	.21070	.06978	8	.93022	23
38	.71973	21	.28027	.78959	28	.21041	.06986	7	.93014	22
39	.71994	20	.28006	.78987	28	.21013	.06993	8	.93007	21
40	9.72014	20	10.27986	9.79015	28	10.20985	10.07001	8	9.92999	20
41	.72034	21	.27966	.79043	29	.20957	.07009	8	.92991	19
42	.72055	20	.27945	.79072	28	.20928	.07017	7	.92983	18
43	.72075	21	.27925	.79100	28	.20900	.07024	8	.92976	17
44	.72096	20	.27904	.79128	28	.20872	.07032	8	.92968	16
45	9.72116	21	10.27884	9.79156	29	10.20844	10.07040	8	9.92960	15
46	.72137	20	.27863	.79185	28	.20815	.07048	8	.92952	14
47	.72157	20	.27843	.79213	28	.20787	.07056	8	.92944	13
48	.72177	21	.27823	.79241	28	.20759	.07064	7	.92936	12
49	.72198	20	.27802	.79269	28	.20731	.07071	8	.92929	11
50	9.72218	20	10.27782	9.79297	29	10.20703	10.07079	8	9.92921	10
51	.72238	21	.27762	.79326	28	.20674	.07087	8	.92913	9
52	.72259	20	.27741	.79354	28	.20646	.07095	8	.92905	8
53	.72279	20	.27721	.79382	28	.20618	.07103	8	.92897	7
54	.72299	21	.27701	.79410	28	.20590	.07111	8	.92889	6
55	9.72320	20	10.27680	9.79438	28	10.20562	10.07119	7	9.92881	5
56	.72340	20	.27660	.79466	29	.20534	.07126	8	.92874	4
57	.72360	21	.27640	.79495	28	.20505	.07134	8	.92866	3
58	.72381	20	.27619	.79523	28	.20477	.07142	8	.92858	2
59	.72401	20	.27599	.79551	28	.20449	.07150	8	.92850	1
60	9.72421		10.27579	9.79579		10.20421	10.07158		9.92842	0
↑ 121°→	cos	Diff. 1′	sec	cot	Diff. 1′	tan	csc	Diff. 1′	sin	↑ ←58°

TABLE 33

Logarithms of Trigonometric Functions

32° → ′	sin	Diff. 1′	csc	tan	Diff. 1′	cot	sec	Diff. 1′	cos	← 147° ′
0	9.72421	20	10.27579	9.79579	28	10.20421	10.07158	8	9.92842	60
1	.72441	20	.27559	.79607	28	.20393	.07166	8	.92834	59
2	.72461	21	.27539	.79635	28	.20365	.07174	8	.92826	58
3	.72482	20	.27518	.79663	28	.20337	.07182	8	.92818	57
4	.72502	20	.27498	.79691	28	.20309	.07190	7	.92810	56
5	9.72522	20	10.27478	9.79719	28	10.20281	10.07197	8	9.92803	55
6	.72542	20	.27458	.79747	29	.20253	.07205	8	.92795	54
7	.72562	20	.27438	.79776	28	.20224	.07213	8	.92787	53
8	.72582	20	.27418	.79804	28	.20196	.07221	8	.92779	52
9	.72602	20	.27398	.79832	28	.20168	.07229	8	.92771	51
10	9.72622	21	10.27378	9.79860	28	10.20140	10.07237	8	9.92763	50
11	.72643	20	.27357	.79888	28	.20112	.07245	8	.92755	49
12	.72663	20	.27337	.79916	28	.20084	.07253	8	.92747	48
13	.72683	20	.27317	.79944	28	.20056	.07261	8	.92739	47
14	.72703	20	.27297	.79972	28	.20028	.07269	8	.92731	46
15	9.72723	20	10.27277	9.80000	28	10.20000	10.07277	8	9.92723	45
16	.72743	20	.27257	.80028	28	.19972	.07285	8	.92715	44
17	.72763	20	.27237	.80056	28	.19944	.07293	8	.92707	43
18	.72783	20	.27217	.80084	28	.19916	.07301	8	.92699	42
19	.72803	20	.27197	.80112	28	.19888	.07309	8	.92691	41
20	9.72823	20	10.27177	9.80140	28	10.19860	10.07317	8	9.92683	40
21	.72843	20	.27157	.80168	27	.19832	.07325	8	.92675	39
22	.72863	20	.27137	.80195	28	.19805	.07333	8	.92667	38
23	.72883	19	.27117	.80223	28	.19777	.07341	8	.92659	37
24	.72902	20	.27098	.80251	28	.19749	.07349	8	.92651	36
25	9.72922	20	10.27078	9.80279	28	10.19721	10.07357	8	9.92643	35
26	.72942	20	.27058	.80307	28	.19693	.07365	8	.92635	34
27	.72962	20	.27038	.80335	28	.19665	.07373	8	.92627	33
28	.72982	20	.27018	.80363	28	.19637	.07381	8	.92619	32
29	.73002	20	.26998	.80391	28	.19609	.07389	8	.92611	31
30	9.73022	19	10.26978	9.80419	28	10.19581	10.07397	8	9.92603	30
31	.73041	20	.26959	.80447	27	.19553	.07405	8	.92595	29
32	.73061	20	.26939	.80474	28	.19526	.07413	8	.92587	28
33	.73081	20	.26919	.80502	28	.19498	.07421	8	.92579	27
34	.73101	20	.26899	.80530	28	.19470	.07429	8	.92571	26
35	9.73121	19	10.26879	9.80558	28	10.19442	10.07437	8	9.92563	25
36	.73140	20	.26860	.80586	28	.19414	.07445	9	.92555	24
37	.73160	20	.26840	.80614	28	.19386	.07454	8	.92546	23
38	.73180	20	.26820	.80642	27	.19358	.07462	8	.92538	22
39	.73200	19	.26800	.80669	28	.19331	.07470	8	.92530	21
40	9.73219	20	10.26781	9.80697	28	10.19303	10.07478	8	9.92522	20
41	.73239	20	.26761	.80725	28	.19275	.07486	8	.92514	19
42	.73259	19	.26741	.80753	28	.19247	.07494	8	.92506	18
43	.73278	20	.26722	.80781	27	.19219	.07502	8	.92498	17
44	.73298	20	.26702	.80808	28	.19192	.07510	8	.92490	16
45	9.73318	19	10.26682	9.80836	28	10.19164	10.07518	9	9.92482	15
46	.73337	20	.26663	.80864	28	.19136	.07527	8	.92473	14
47	.73357	20	.26643	.80892	27	.19108	.07535	8	.92465	13
48	.73377	19	.26623	.80919	28	.19081	.07543	8	.92457	12
49	.73396	20	.26604	.80947	28	.19053	.07551	8	.92449	11
50	9.73416	19	10.26584	9.80975	28	10.19025	10.07559	8	9.92441	10
51	.73435	20	.26565	.81003	27	.18997	.07567	8	.92433	9
52	.73455	19	.26545	.81030	28	.18970	.07575	9	.92425	8
53	.73474	20	.26526	.81058	28	.18942	.07584	8	.92416	7
54	.73494	19	.26506	.81086	27	.18914	.07592	8	.92408	6
55	9.73513	20	10.26487	9.81113	28	10.18887	10.07600	8	9.92400	5
56	.73533	19	.26467	.81141	28	.18859	.07608	8	.92392	4
57	.73552	20	.26448	.81169	27	.18831	.07616	8	.92384	3
58	.73572	19	.26428	.81196	28	.18804	.07624	9	.92376	2
59	.73591	20	.26409	.81224	28	.18776	.07633	8	.92367	1
60	9.73611		10.26389	9.81252		10.18748	10.07641		9.92359	0
122° →	cos	Diff. 1′	sec	cot	Diff. 1′	tan	csc	Diff. 1′	sin	← 57°

TABLE 33
Logarithms of Trigonometric Functions

33°→ ↓ ′	sin	Diff. 1′	csc	tan	Diff. 1′	cot	sec	Diff. 1′	cos	←146° ↓ ′
0	9. 73611	19	10. 26389	9. 81252	27	10. 18748	10. 07641	8	9. 92359	60
1	. 73630	20	. 26370	. 81279	28	. 18721	. 07649	8	. 92351	59
2	. 73650	19	. 26350	. 81307	28	. 18693	. 07657	8	. 92343	58
3	. 73669	20	. 26331	. 81335	27	. 18665	. 07665	9	. 92335	57
4	. 73689	19	. 26311	. 81362	28	. 18638	. 07674	8	. 92326	56
5	9. 73708	19	10. 26292	9. 81390	28	10. 18610	10. 07682	8	9. 92318	55
6	. 73727	20	. 26273	. 81418	27	. 18582	. 07690	8	. 92310	54
7	. 73747	19	. 26253	. 81445	28	. 18555	. 07698	9	. 92302	53
8	. 73766	19	. 26234	. 81473	27	. 18527	. 07707	8	. 92293	52
9	. 73785	20	. 26215	. 81500	28	. 18500	. 07715	8	. 92285	51
10	9. 73805	19	10. 26195	9. 81528	28	10. 18472	10. 07723	8	9. 92277	50
11	. 73824	19	. 26176	. 81556	27	. 18444	. 07731	9	. 92269	49
12	. 73843	20	. 26157	. 81583	28	. 18417	. 07740	8	. 92260	48
13	. 73863	19	. 26137	. 81611	27	. 18389	. 07748	8	. 92252	47
14	. 73882	19	. 26118	. 81638	28	. 18362	. 07756	9	. 92244	46
15	9. 73901	20	10. 26099	9. 81666	27	10. 18334	10. 07765	8	9. 92235	45
16	. 73921	19	. 26079	. 81693	28	. 18307	. 07773	8	. 92227	44
17	. 73940	19	. 26060	. 81721	27	. 18279	. 07781	8	. 92219	43
18	. 73959	19	. 26041	. 81748	28	. 18252	. 07789	9	. 92211	42
19	. 73978	19	. 26022	. 81776	27	. 18224	. 07798	8	. 92202	41
20	9. 73997	20	10. 26003	9. 81803	28	10. 18197	10. 07806	8	9. 92194	40
21	. 74017	19	. 25983	. 81831	27	. 18169	. 07814	9	. 92186	39
22	. 74036	19	. 25964	. 81858	28	. 18142	. 07823	8	. 92177	38
23	. 74055	19	. 25945	. 81886	27	. 18114	. 07831	8	. 92169	37
24	. 74074	19	. 25926	. 81913	28	. 18087	. 07839	9	. 92161	36
25	9. 74093	20	10. 25907	9. 81941	27	10. 18059	10. 07848	8	9. 92152	35
26	. 74113	19	. 25887	. 81968	28	. 18032	. 07856	8	. 92144	34
27	. 74132	19	. 25868	. 81996	27	. 18004	. 07864	9	. 92136	33
28	. 74151	19	. 25849	. 82023	28	. 17977	. 07873	8	. 92127	32
29	. 74170	19	. 25830	. 82051	27	. 17949	. 07881	8	. 92119	31
30	9. 74189	19	10. 25811	9. 82078	28	10. 17922	10. 07889	9	9. 92111	30
31	. 74208	19	. 25792	. 82106	27	. 17894	. 07898	8	. 92102	29
32	. 74227	19	. 25773	. 82133	28	. 17867	. 07906	8	. 92094	28
33	. 74246	19	. 25754	. 82161	27	. 17839	. 07914	9	. 92086	27
34	. 74265	19	. 25735	. 82188	27	. 17812	. 07923	8	. 92077	26
35	9. 74284	19	10. 25716	9. 82215	28	10. 17785	10. 07931	9	9. 92069	25
36	. 74303	19	. 25697	. 82243	27	. 17757	. 07940	8	. 92060	24
37	. 74322	19	. 25678	. 82270	28	. 17730	. 07948	8	. 92052	23
38	. 74341	19	. 25659	. 82298	27	. 17702	. 07956	9	. 92044	22
39	. 74360	19	. 25640	. 82325	27	. 17675	. 07965	8	. 92035	21
40	9. 74379	19	10. 25621	9. 82352	28	10. 17648	10. 07973	9	9. 92027	20
41	. 74398	19	. 25602	. 82380	27	. 17620	. 07982	8	. 92018	19
42	. 74417	19	. 25583	. 82407	28	. 17593	. 07990	8	. 92010	18
43	. 74436	19	. 25564	. 82435	27	. 17565	. 07998	9	. 92002	17
44	. 74455	19	. 25545	. 82462	27	. 17538	. 08007	8	. 91993	16
45	9. 74474	19	10. 25526	9. 82489	28	10. 17511	10. 08015	9	9. 91985	15
46	. 74493	19	. 25507	. 82517	27	. 17483	. 08024	8	. 91976	14
47	. 74512	19	. 25488	. 82544	27	. 17456	. 08032	9	. 91968	13
48	. 74531	18	. 25469	. 82571	28	. 17429	. 08041	8	. 91959	12
49	. 74549	19	. 25451	. 82599	27	. 17401	. 08049	9	. 91951	11
50	9. 74568	19	10. 25432	9. 82626	27	10. 17374	10. 08058	8	9. 91942	10
51	. 74587	19	. 25413	. 82653	28	. 17347	. 08066	9	. 91934	9
52	. 74606	19	. 25394	. 82681	27	. 17319	. 08075	8	. 91925	8
53	. 74625	19	. 25375	. 82708	27	. 17292	. 08083	9	. 91917	7
54	. 74644	18	. 25356	. 82735	27	. 17265	. 08092	8	. 91908	6
55	9. 74662	19	10. 25338	9. 82762	28	10. 17238	10. 08100	9	9. 91900	5
56	. 74681	19	. 25319	. 82790	27	. 17210	. 08109	8	. 91891	4
57	. 74700	19	. 25300	. 82817	27	. 17183	. 08117	9	. 91883	3
58	. 74719	18	. 25281	. 82844	27	. 17156	. 08126	8	. 91874	2
59	. 74737	19	. 25263	. 82871	28	. 17129	. 08134	9	. 91866	1
60	9. 74756		10. 25244	9. 82899		10. 17101	10. 08143		9. 91857	0
↑ 123°→	cos	Diff. 1′	sec	cot	Diff. 1′	tan	csc	Diff. 1′	sin	↑ ←56°

TABLE 33

Logarithms of Trigonometric Functions

34°→ ↓ ′	sin	Diff. 1′	csc	tan	Diff. 1′	cot	sec	Diff. 1′	cos	←145° ↓ ′
0	9. 74756	19	10. 25244	9. 82899	27	10. 17101	10. 08143	8	9. 91857	60
1	. 74775	19	. 25225	. 82926	27	. 17074	. 08151	9	. 91849	59
2	. 74794	18	. 25206	. 82953	27	. 17047	. 08160	8	. 91840	58
3	. 74812	19	. 25188	. 82980	28	. 17020	. 08168	9	. 91832	57
4	. 74831	19	. 25169	. 83008	27	. 16992	. 08177	8	. 91823	56
5	9. 74850	18	10. 25150	9. 83035	27	10. 16965	10. 08185	9	9. 91815	55
6	. 74868	19	. 25132	. 83062	27	. 16938	. 08194	8	. 91806	54
7	. 74887	19	. 25113	. 83089	28	. 16911	. 08202	9	. 91798	53
8	. 74906	18	. 25094	. 83117	27	. 16883	. 08211	8	. 91789	52
9	. 74924	19	. 25076	. 83144	27	. 16856	. 08219	9	. 91781	51
10	9. 74943	18	10. 25057	9. 83171	27	10. 16829	10. 08228	9	9. 91772	50
11	. 74961	19	. 25039	. 83198	27	. 16802	. 08237	8	. 91763	49
12	. 74980	19	. 25020	. 83225	27	. 16775	. 08245	9	. 91755	48
13	. 74999	18	. 25001	. 83252	28	. 16748	. 08254	8	. 91746	47
14	. 75017	19	. 24983	. 83280	27	. 16720	. 08262	9	. 91738	46
15	9. 75036	18	10. 24964	9. 83307	27	10. 16693	10. 08271	9	9. 91729	45
16	. 75054	19	. 24946	. 83334	27	. 16666	. 08280	8	. 91720	44
17	. 75073	18	. 24927	. 83361	27	. 16639	. 08288	9	. 91712	43
18	. 75091	19	. 24909	. 83388	27	. 16612	. 08297	8	. 91703	42
19	. 75110	18	. 24890	. 83415	27	. 16585	. 08305	9	. 91695	41
20	9. 75128	19	10. 24872	9. 83442	28	10. 16558	10. 08314	9	9. 91686	40
21	. 75147	18	. 24853	. 83470	27	. 16530	. 08323	8	. 91677	39
22	. 75165	19	. 24835	. 83497	27	. 16503	. 08331	9	. 91669	38
23	. 75184	18	. 24816	. 83524	27	. 16476	. 08340	9	. 91660	37
24	. 75202	19	. 24798	. 83551	27	. 16449	. 08349	8	. 91651	36
25	9. 75221	18	10. 24779	9. 83578	27	10. 16422	10. 08357	9	9. 91643	35
26	. 75239	19	. 24761	. 83605	27	. 16395	. 08366	9	. 91634	34
27	. 75258	18	. 24742	. 83632	27	. 16368	. 08375	8	. 91625	33
28	. 75276	18	. 24724	. 83659	27	. 16341	. 08383	9	. 91617	32
29	. 75294	19	. 24706	. 83686	27	. 16314	. 08392	9	. 91608	31
30	9. 75313	18	10. 24687	9. 83713	27	10. 16287	10. 08401	8	9. 91599	30
31	. 75331	19	. 24669	. 83740	28	. 16260	. 08409	9	. 91591	29
32	. 75350	18	. 24650	. 83768	27	. 16232	. 08418	9	. 91582	28
33	. 75368	18	. 24632	. 83795	27	. 16205	. 08427	8	. 91573	27
34	. 75386	19	. 24614	. 83822	27	. 16178	. 08435	9	. 91565	26
35	9. 75405	18	10. 24595	9. 83849	27	10. 16151	10. 08444	9	9. 91556	25
36	. 75423	18	. 24577	. 83876	27	. 16124	. 08453	9	. 91547	24
37	. 75441	18	. 24559	. 83903	27	. 16097	. 08462	8	. 91538	23
38	. 75459	19	. 24541	. 83930	27	. 16070	. 08470	9	. 91530	22
39	. 75478	18	. 24522	. 83957	27	. 16043	. 08479	9	. 91521	21
40	9. 75496	18	10. 24504	9. 83984	27	10. 16016	10. 08488	8	9. 91512	20
41	. 75514	19	. 24486	. 84011	27	. 15989	. 08496	9	. 91504	19
42	. 75533	18	. 24467	. 84038	27	. 15962	. 08505	9	. 91495	18
43	. 75551	18	. 24449	. 84065	27	. 15935	. 08514	9	. 91486	17
44	. 75569	18	. 24431	. 84092	27	. 15908	. 08523	8	. 91477	16
45	9. 75587	18	10. 24413	9. 84119	27	10. 15881	10. 08531	9	9. 91469	15
46	. 75605	19	. 24395	. 84146	27	. 15854	. 08540	9	. 91460	14
47	. 75624	18	. 24376	. 84173	27	. 15827	. 08549	9	. 91451	13
48	. 75642	18	. 24358	. 84200	27	. 15800	. 08558	9	. 91442	12
49	. 75660	18	. 24340	. 84227	27	. 15773	. 08567	8	. 91433	11
50	9. 75678	18	10. 24322	9. 84254	26	10. 15746	10. 08575	9	9. 91425	10
51	. 75696	18	. 24304	. 84280	27	. 15720	. 08584	9	. 91416	9
52	. 75714	19	. 24286	. 84307	27	. 15693	. 08593	9	. 91407	8
53	. 75733	18	. 24267	. 84334	27	. 15666	. 08602	9	. 91398	7
54	. 75751	18	. 24249	. 84361	27	. 15639	. 08611	8	. 91389	6
55	9. 75769	18	10. 24231	9. 84388	27	10. 15612	10. 08619	9	9. 91381	5
56	. 75787	18	. 24213	. 84415	27	. 15585	. 08628	9	. 91372	4
57	. 75805	18	. 24195	. 84442	27	. 15558	. 08637	9	. 91363	3
58	. 75823	18	. 24177	. 84469	27	. 15531	. 08646	9	. 91354	2
59	. 75841	18	. 24159	. 84496	27	. 15504	. 08655	9	. 91345	1
60	9. 75859		10. 24141	9. 84523		10. 15477	10. 08664		9. 91336	0
↑ 124°→	cos	Diff. 1′	sec	cot	Diff. 1′	tan	csc	Diff. 1′	sin	↑ ←55°

TABLE 33
Logarithms of Trigonometric Functions

35° → ↓ ′	sin	Diff. 1′	csc	tan	Diff. 1′	cot	sec	Diff. 1′	cos	← 144° ↓ ′
0	9.75859	18	10.24141	9.84523	27	10.15477	10.08664	8	9.91336	60
1	.75877	18	.24123	.84550	26	.15450	.08672	9	.91328	59
2	.75895	18	.24105	.84576	27	.15424	.08681	9	.91319	58
3	.75913	18	.24087	.84603	27	.15397	.08690	9	.91310	57
4	.75931	18	.24069	.84630	27	.15370	.08699	9	.91301	56
5	9.75949	18	10.24051	9.84657	27	10.15343	10.08708	9	9.91292	55
6	.75967	18	.24033	.84684	27	.15316	.08717	9	.91283	54
7	.75985	18	.24015	.84711	27	.15289	.08726	8	.91274	53
8	.76003	18	.23997	.84738	26	.15262	.08734	9	.91266	52
9	.76021	18	.23979	.84764	27	.15236	.08743	9	.91257	51
10	9.76039	18	10.23961	9.84791	27	10.15209	10.08752	9	9.91248	50
11	.76057	18	.23943	.84818	27	.15182	.08761	9	.91239	49
12	.76075	18	.23925	.84845	27	.15155	.08770	9	.91230	48
13	.76093	18	.23907	.84872	27	.15128	.08779	9	.91221	47
14	.76111	18	.23889	.84899	26	.15101	.08788	9	.91212	46
15	9.76129	17	10.23871	9.84925	27	10.15075	10.08797	9	9.91203	45
16	.76146	18	.23854	.84952	27	.15048	.08806	9	.91194	44
17	.76164	18	.23836	.84979	27	.15021	.08815	9	.91185	43
18	.76182	18	.23818	.85006	27	.14994	.08824	9	.91176	42
19	.76200	18	.23800	.85033	26	.14967	.08833	9	.91167	41
20	9.76218	18	10.23782	9.85059	27	10.14941	10.08842	9	9.91158	40
21	.76236	17	.23764	.85086	27	.14914	.08851	8	.91149	39
22	.76253	18	.23747	.85113	27	.14887	.08859	9	.91141	38
23	.76271	18	.23729	.85140	26	.14860	.08868	9	.91132	37
24	.76289	18	.23711	.85166	27	.14834	.08877	9	.91123	36
25	9.76307	17	10.23693	9.85193	27	10.14807	10.08886	9	9.91114	35
26	.76324	18	.23676	.85220	27	.14780	.08895	9	.91105	34
27	.76342	18	.23658	.85247	26	.14753	.08904	9	.91096	33
28	.76360	18	.23640	.85273	27	.14727	.08913	9	.91087	32
29	.76378	17	.23622	.85300	27	.14700	.08922	9	.91078	31
30	9.76395	18	10.23605	9.85327	27	10.14673	10.08931	9	9.91069	30
31	.76413	18	.23587	.85354	26	.14646	.08940	9	.91060	29
32	.76431	17	.23569	.85380	27	.14620	.08949	9	.91051	28
33	.76448	18	.23552	.85407	27	.14593	.08958	9	.91042	27
34	.76466	18	.23534	.85434	26	.14566	.08967	10	.91033	26
35	9.76484	17	10.23516	9.85460	27	10.14540	10.08977	9	9.91023	25
36	.76501	18	.23499	.85487	27	.14513	.08986	9	.91014	24
37	.76519	18	.23481	.85514	26	.14486	.08995	9	.91005	23
38	.76537	17	.23463	.85540	27	.14460	.09004	9	.90996	22
39	.76554	18	.23446	.85567	27	.14433	.09013	9	.90987	21
40	9.76572	18	10.23428	9.85594	26	10.14406	10.09022	9	9.90978	20
41	.76590	17	.23410	.85620	27	.14380	.09031	9	.90969	19
42	.76607	18	.23393	.85647	27	.14353	.09040	9	.90960	18
43	.76625	17	.23375	.85674	26	.14326	.09049	9	.90951	17
44	.76642	18	.23358	.85700	27	.14300	.09058	9	.90942	16
45	9.76660	17	10.23340	9.85727	27	10.14273	10.09067	9	9.90933	15
46	.76677	18	.23323	.85754	26	.14246	.09076	9	.90924	14
47	.76695	17	.23305	.85780	27	.14220	.09085	9	.90915	13
48	.76712	18	.23288	.85807	27	.14193	.09094	10	.90906	12
49	.76730	17	.23270	.85834	26	.14166	.09104	9	.90896	11
50	9.76747	18	10.23253	9.85860	27	10.14140	10.09113	9	9.90887	10
51	.76765	17	.23235	.85887	26	.14113	.09122	9	.90878	9
52	.76782	18	.23218	.85913	27	.14087	.09131	9	.90869	8
53	.76800	17	.23200	.85940	27	.14060	.09140	9	.90860	7
54	.76817	18	.23183	.85967	26	.14033	.09149	9	.90851	6
55	9.76835	17	10.23165	9.85993	27	10.14007	10.09158	10	9.90842	5
56	.76852	18	.23148	.86020	26	.13980	.09168	9	.90832	4
57	.76870	17	.23130	.86046	27	.13954	.09177	9	.90823	3
58	.76887	17	.23113	.86073	27	.13927	.09186	9	.90814	2
59	.76904	18	.23096	.86100	26	.13900	.09195	9	.90805	1
60	9.76922		10.23078	9.86126		10.13874	10.09204		9.90796	0
↑ 125° →	cos	Diff. 1′	sec	cot	Diff. 1′	tan	csc	Diff. 1′	sin	← 54° ↑

TABLE 33
Logarithms of Trigonometric Functions

36°→ ↓ ′	sin	Diff. 1′	csc	tan	Diff. 1′	cot	sec	Diff. 1′	cos	←143° ↓ ′
0	9.76922	17	10.23078	9.86126	27	10.13874	10.09204	9	9.90796	60
1	.76939	18	.23061	.86153	26	.13847	.09213	10	.90787	59
2	.76957	17	.23043	.86179	27	.13821	.09223	9	.90777	58
3	.76974	17	.23026	.86206	26	.13794	.09232	9	.90768	57
4	.76991	18	.23009	.86232	27	.13768	.09241	9	.90759	56
5	9.77009	17	10.22991	9.86259	26	10.13741	10.09250	9	9.90750	55
6	.77026	17	.22974	.86285	27	.13715	.09259	10	.90741	54
7	.77043	18	.22957	.86312	26	.13688	.09269	9	.90731	53
8	.77061	17	.22939	.86338	27	.13662	.09278	9	.90722	52
9	.77078	17	.22922	.86365	27	.13635	.09287	9	.90713	51
10	9.77095	17	10.22905	9.86392	26	10.13608	10.09296	10	9.90704	50
11	.77112	18	.22888	.86418	27	.13582	.09306	9	.90694	49
12	.77130	17	.22870	.86445	26	.13555	.09315	9	.90685	48
13	.77147	17	.22853	.86471	27	.13529	.09324	9	.90676	47
14	.77164	17	.22836	.86498	26	.13502	.09333	10	.90667	46
15	9.77181	18	10.22819	9.86524	27	10.13476	10.09343	9	9.90657	45
16	.77199	17	.22801	.86551	26	.13449	.09352	9	.90648	44
17	.77216	17	.22784	.86577	26	.13423	.09361	9	.90639	43
18	.77233	17	.22767	.86603	27	.13397	.09370	10	.90630	42
19	.77250	18	.22750	.86630	26	.13370	.09380	9	.90620	41
20	9.77268	17	10.22732	9.86656	27	10.13344	10.09389	9	9.90611	40
21	.77285	17	.22715	.86683	26	.13317	.09398	10	.90602	39
22	.77302	17	.22698	.86709	27	.13291	.09408	9	.90592	38
23	.77319	17	.22681	.86736	26	.13264	.09417	9	.90583	37
24	.77336	17	.22664	.86762	27	.13238	.09426	9	.90574	36
25	9.77353	17	10.22647	9.86789	26	10.13211	10.09435	10	9.90565	35
26	.77370	17	.22630	.86815	27	.13185	.09445	9	.90555	34
27	.77387	18	.22613	.86842	26	.13158	.09454	9	.90546	33
28	.77405	17	.22595	.86868	26	.13132	.09463	10	.90537	32
29	.77422	17	.22578	.86894	27	.13106	.09473	9	.90527	31
30	9.77439	17	10.22561	9.86921	26	10.13079	10.09482	9	9.90518	30
31	.77456	17	.22544	.86947	27	.13053	.09491	10	.90509	29
32	.77473	17	.22527	.86974	26	.13026	.09501	9	.90499	28
33	.77490	17	.22510	.87000	27	.13000	.09510	10	.90490	27
34	.77507	17	.22493	.87027	26	.12973	.09520	9	.90480	26
35	9.77524	17	10.22476	9.87053	26	10.12947	10.09529	9	9.90471	25
36	.77541	17	.22459	.87079	27	.12921	.09538	10	.90462	24
37	.77558	17	.22442	.87106	26	.12894	.09548	9	.90452	23
38	.77575	17	.22425	.87132	26	.12868	.09557	9	.90443	22
39	.77592	17	.22408	.87158	27	.12842	.09566	10	.90434	21
40	9.77609	17	10.22391	9.87185	26	10.12815	10.09576	9	9.90424	20
41	.77626	17	.22374	.87211	27	.12789	.09585	10	.90415	19
42	.77643	17	.22357	.87238	26	.12762	.09595	9	.90405	18
43	.77660	17	.22340	.87264	26	.12736	.09604	10	.90396	17
44	.77677	17	.22323	.87290	27	.12710	.09614	9	.90386	16
45	9.77694	17	10.22306	9.87317	26	10.12683	10.09623	9	9.90377	15
46	.77711	17	.22289	.87343	26	.12657	.09632	10	.90368	14
47	.77728	16	.22272	.87369	27	.12631	.09642	9	.90358	13
48	.77744	17	.22256	.87396	26	.12604	.09651	10	.90349	12
49	.77761	17	.22239	.87422	26	.12578	.09661	9	.90339	11
50	9.77778	17	10.22222	9.87448	27	10.12552	10.09670	10	9.90330	10
51	.77795	17	.22205	.87475	26	.12525	.09680	9	.90320	9
52	.77812	17	.22188	.87501	26	.12499	.09689	10	.90311	8
53	.77829	17	.22171	.87527	27	.12473	.09699	9	.90301	7
54	.77846	16	.22154	.87554	26	.12446	.09708	10	.90292	6
55	9.77862	17	10.22138	9.87580	26	10.12420	10.09718	9	9.90282	5
56	.77879	17	.22121	.87606	27	.12394	.09727	10	.90273	4
57	.77896	17	.22104	.87633	26	.12367	.09737	9	.90263	3
58	.77913	17	.22087	.87659	26	.12341	.09746	10	.90254	2
59	.77930	16	.22070	.87685	26	.12315	.09756	9	.90244	1
60	9.77946		10.22054	9.87711		10.12289	10.09765		9.90235	0
↑ 126°→	cos	Diff. 1′	sec	cot	Diff. 1′	tan	csc	Diff. 1′	sin	↑ ←53°

TABLE 33
Logarithms of Trigonometric Functions

37° → ′ ↓	sin	Diff. 1′	csc	tan	Diff. 1′	cot	sec	Diff. 1′	cos	← 142° ′ ↓
0	9.77946	17	10.22054	9.87711	27	10.12289	10.09765	10	9.90235	60
1	.77963	17	.22037	.87738	26	.12262	.09775	9	.90225	59
2	.77980	17	.22020	.87764	26	.12236	.09784	10	.90216	58
3	.77997	16	.22003	.87790	27	.12210	.09794	9	.90206	57
4	.78013	17	.21987	.87817	26	.12183	.09803	10	.90197	56
5	9.78030	17	10.21970	9.87843	26	10.12157	10.09813	9	9.90187	55
6	.78047	16	.21953	.87869	26	.12131	.09822	10	.90178	54
7	.78063	17	.21937	.87895	27	.12105	.09832	9	.90168	53
8	.78080	17	.21920	.87922	26	.12078	.09841	10	.90159	52
9	.78097	16	.21903	.87948	26	.12052	.09851	10	.90149	51
10	9.78113	17	10.21887	9.87974	26	10.12026	10.09861	9	9.90139	50
11	.78130	17	.21870	.88000	27	.12000	.09870	10	.90130	49
12	.78147	16	.21853	.88027	26	.11973	.09880	9	.90120	48
13	.78163	17	.21837	.88053	26	.11947	.09889	10	.90111	47
14	.78180	17	.21820	.88079	26	.11921	.09899	10	.90101	46
15	9.78197	16	10.21803	9.88105	26	10.11895	10.09909	9	9.90091	45
16	.78213	17	.21787	.88131	27	.11869	.09918	10	.90082	44
17	.78230	16	.21770	.88158	26	.11842	.09928	9	.90072	43
18	.78246	17	.21754	.88184	26	.11816	.09937	10	.90063	42
19	.78263	17	.21737	.88210	26	.11790	.09947	10	.90053	41
20	9.78280	16	10.21720	9.88236	26	10.11764	10.09957	9	9.90043	40
21	.78296	17	.21704	.88262	27	.11738	.09966	10	.90034	39
22	.78313	16	.21687	.88289	26	.11711	.09976	10	.90024	38
23	.78329	17	.21671	.88315	26	.11685	.09986	9	.90014	37
24	.78346	16	.21654	.88341	26	.11659	.09995	10	.90005	36
25	9.78362	17	10.21638	9.88367	26	10.11633	10.10005	10	9.89995	35
26	.78379	16	.21621	.88393	27	.11607	.10015	9	.89985	34
27	.78395	17	.21605	.88420	26	.11580	.10024	10	.89976	33
28	.78412	16	.21588	.88446	26	.11554	.10034	10	.89966	32
29	.78428	17	.21572	.88472	26	.11528	.10044	9	.89956	31
30	9.78445	16	10.21555	9.88498	26	10.11502	10.10053	10	9.89947	30
31	.78461	17	.21539	.88524	26	.11476	.10063	10	.89937	29
32	.78478	16	.21522	.88550	27	.11450	.10073	9	.89927	28
33	.78494	16	.21506	.88577	26	.11423	.10082	10	.89918	27
34	.78510	17	.21490	.88603	26	.11397	.10092	10	.89908	26
35	9.78527	16	10.21473	9.88629	26	10.11371	10.10102	10	9.89898	25
36	.78543	17	.21457	.88655	26	.11345	.10112	9	.89888	24
37	.78560	16	.21440	.88681	26	.11319	.10121	10	.89879	23
38	.78576	16	.21424	.88707	26	.11293	.10131	10	.89869	22
39	.78592	17	.21408	.88733	26	.11267	.10141	10	.89859	21
40	9.78609	16	10.21391	9.88759	27	10.11241	10.10151	9	9.89849	20
41	.78625	17	.21375	.88786	26	.11214	.10160	10	.89840	19
42	.78642	16	.21358	.88812	26	.11188	.10170	10	.89830	18
43	.78658	16	.21342	.88838	26	.11162	.10180	10	.89820	17
44	.78674	17	.21326	.88864	26	.11136	.10190	9	.89810	16
45	9.78691	16	10.21309	9.88890	26	10.11110	10.10199	10	9.89801	15
46	.78707	16	.21293	.88916	26	.11084	.10209	10	.89791	14
47	.78723	16	.21277	.88942	26	.11058	.10219	10	.89781	13
48	.78739	17	.21261	.88968	26	.11032	.10229	10	.89771	12
49	.78756	16	.21244	.88994	26	.11006	.10239	9	.89761	11
50	9.78772	16	10.21228	9.89020	26	10.10980	10.10248	10	9.89752	10
51	.78788	17	.21212	.89046	27	.10954	.10258	10	.89742	9
52	.78805	16	.21195	.89073	26	.10927	.10268	10	.89732	8
53	.78821	16	.21179	.89099	26	.10901	.10278	10	.89722	7
54	.78837	16	.21163	.89125	26	.10875	.10288	10	.89712	6
55	9.78853	16	10.21147	9.89151	26	10.10849	10.10298	9	9.89702	5
56	.78869	17	.21131	.89177	26	.10823	.10307	10	.89693	4
57	.78886	16	.21114	.89203	26	.10797	.10317	10	.89683	3
58	.78902	16	.21098	.89229	26	.10771	.10327	10	.89673	2
59	.78918	16	.21082	.89255	26	.10745	.10337	10	.89663	1
60	9.78934		10.21066	9.89281		10.10719	10.10347		9.89653	0
127° → ↑	cos	Diff. 1′	sec	cot	Diff. 1′	tan	csc	Diff. 1′	sin	← 52° ↑

TABLE 33

Logarithms of Trigonometric Functions

38° → ↓ '	sin	Diff. 1'	csc	tan	Diff. 1'	cot	sec	Diff. 1'	cos	← 141° ↓ '
0	9.78934	16	10.21066	9.89281	26	10.10719	10.10347	10	9.89653	60
1	.78950	17	.21050	.89307	26	.10693	.10357	10	.89643	59
2	.78967	16	.21033	.89333	26	.10667	.10367	9	.89633	58
3	.78983	16	.21017	.89359	26	.10641	.10376	10	.89624	57
4	.78999	16	.21001	.89385	26	.10615	.10386	10	.89614	56
5	9.79015	16	10.20985	9.89411	26	10.10589	10.10396	10	9.89604	55
6	.79031	16	.20969	.89437	26	.10563	.10406	10	.89594	54
7	.79047	16	.20953	.89463	26	.10537	.10416	10	.89584	53
8	.79063	16	.20937	.89489	26	.10511	.10426	10	.89574	52
9	.79079	16	.20921	.89515	26	.10485	.10436	10	.89564	51
10	9.79095	16	10.20905	9.89541	26	10.10459	10.10446	10	9.89554	50
11	.79111	17	.20889	.89567	26	.10433	.10456	10	.89544	49
12	.79128	16	.20872	.89593	26	.10407	.10466	10	.89534	48
13	.79144	16	.20856	.89619	26	.10381	.10476	10	.89524	47
14	.79160	16	.20840	.89645	26	.10355	.10486	10	.89514	46
15	9.79176	16	10.20824	9.89671	26	10.10329	10.10496	9	9.89504	45
16	.79192	16	.20808	.89697	26	.10303	.10505	10	.89495	44
17	.79208	16	.20792	.89723	26	.10277	.10515	10	.89485	43
18	.79224	16	.20776	.89749	26	.10251	.10525	10	.89475	42
19	.79240	16	.20760	.89775	26	.10225	.10535	10	.89465	41
20	9.79256	16	10.20744	9.89801	26	10.10199	10.10545	10	9.89455	40
21	.79272	16	.20728	.89827	26	.10173	.10555	10	.89445	39
22	.79288	16	.20712	.89853	26	.10147	.10565	10	.89435	38
23	.79304	15	.20696	.89879	26	.10121	.10575	10	.89425	37
24	.79319	16	.20681	.89905	26	.10095	.10585	10	.89415	36
25	9.79335	16	10.20665	9.89931	26	10.10069	10.10595	10	9.89405	35
26	.79351	16	.20649	.89957	26	.10043	.10605	10	.89395	34
27	.79367	16	.20633	.89983	26	.10017	.10615	10	.89385	33
28	.79383	16	.20617	.90009	26	.09991	.10625	11	.89375	32
29	.79399	16	.20601	.90035	26	.09965	.10636	10	.89364	31
30	9.79415	16	10.20585	9.90061	25	10.09939	10.10646	10	9.89354	30
31	.79431	16	.20569	.90086	26	.09914	.10656	10	.89344	29
32	.79447	16	.20553	.90112	26	.09888	.10666	10	.89334	28
33	.79463	15	.20537	.90138	26	.09862	.10676	10	.89324	27
34	.79478	16	.20522	.90164	26	.09836	.10686	10	.89314	26
35	9.79494	16	10.20506	9.90190	26	10.09810	10.10696	10	9.89304	25
36	.79510	16	.20490	.90216	26	.09784	.10706	10	.89294	24
37	.79526	16	.20474	.90242	26	.09758	.10716	10	.89284	23
38	.79542	16	.20458	.90268	26	.09732	.10726	10	.89274	22
39	.79558	15	.20442	.90294	26	.09706	.10736	10	.89264	21
40	9.79573	16	10.20427	9.90320	26	10.09680	10.10746	10	9.89254	20
41	.79589	16	.20411	.90346	25	.09654	.10756	11	.89244	19
42	.79605	16	.20395	.90371	26	.09629	.10767	10	.89233	18
43	.79621	15	.20379	.90397	26	.09603	.10777	10	.89223	17
44	.79636	16	.20364	.90423	26	.09577	.10787	10	.89213	16
45	9.79652	16	10.20348	9.90449	26	10.09551	10.10797	10	9.89203	15
46	.79668	16	.20332	.90475	26	.09525	.10807	10	.89193	14
47	.79684	15	.20316	.90501	26	.09499	.10817	10	.89183	13
48	.79699	16	.20301	.90527	26	.09473	.10827	11	.89173	12
49	.79715	16	.20285	.90553	25	.09447	.10838	10	.89162	11
50	9.79731	15	10.20269	9.90578	26	10.09422	10.10848	10	9.89152	10
51	.79746	16	.20254	.90604	26	.09396	.10858	10	.89142	9
52	.79762	16	.20238	.90630	26	.09370	.10868	10	.89132	8
53	.79778	15	.20222	.90656	26	.09344	.10878	10	.89122	7
54	.79793	16	.20207	.90682	26	.09318	.10888	11	.89112	6
55	9.79809	16	10.20191	9.90708	26	10.09292	10.10899	10	9.89101	5
56	.79825	15	.20175	.90734	25	.09266	.10909	10	.89091	4
57	.79840	16	.20160	.90759	26	.09241	.10919	10	.89081	3
58	.79856	16	.20144	.90785	26	.09215	.10929	11	.89071	2
59	.79872	15	.20128	.90811	26	.09189	.10940	10	.89060	1
60	9.79887		10.20113	9.90837		10.09163	10.10950		9.89050	0
↑ 128° →	cos	Diff. 1'	sec	cot	Diff. 1'	tan	csc	Diff. 1'	sin	← 51° ↑

TABLE 33
Logarithms of Trigonometric Functions

39° → ′	sin	Diff. 1′	csc	tan	Diff. 1′	cot	sec	Diff. 1′	cos	← 140° ′
0	9.79887	16	10.20113	9.90837	26	10.09163	10.10950	10	9.89050	60
1	.79903	15	.20097	.90863	26	.09137	.10960	10	.89040	59
2	.79918	16	.20082	.90889	25	.09111	.10970	10	.89030	58
3	.79934	16	.20066	.90914	26	.09086	.10980	11	.89020	57
4	.79950	15	.20050	.90940	26	.09060	.10991	10	.89009	56
5	9.79965	16	10.20035	9.90966	26	10.09034	10.11001	10	9.88999	55
6	.79981	15	.20019	.90992	26	.09008	.11011	11	.88989	54
7	.79996	16	.20004	.91018	25	.08982	.11022	10	.88978	53
8	.80012	15	.19988	.91043	26	.08957	.11032	10	.88968	52
9	.80027	16	.19973	.91069	26	.08931	.11042	10	.88958	51
10	9.80043	15	10.19957	9.91095	26	10.08905	10.11052	11	9.88948	50
11	.80058	16	.19942	.91121	26	.08879	.11063	10	.88937	49
12	.80074	15	.19926	.91147	25	.08853	.11073	10	.88927	48
13	.80089	16	.19911	.91172	26	.08828	.11083	11	.88917	47
14	.80105	15	.19895	.91198	26	.08802	.11094	10	.88906	46
15	9.80120	16	10.19880	9.91224	26	10.08776	10.11104	10	9.88896	45
16	.80136	15	.19864	.91250	26	.08750	.11114	11	.88886	44
17	.80151	15	.19849	.91276	25	.08724	.11125	10	.88875	43
18	.80166	16	.19834	.91301	26	.08699	.11135	10	.88865	42
19	.80182	15	.19818	.91327	26	.08673	.11145	11	.88855	41
20	9.80197	16	10.19803	9.91353	26	10.08647	10.11156	10	9.88844	40
21	.80213	15	.19787	.91379	25	.08621	.11166	10	.88834	39
22	.80228	16	.19772	.91404	26	.08596	.11176	11	.88824	38
23	.80244	15	.19756	.91430	26	.08570	.11187	10	.88813	37
24	.80259	15	.19741	.91456	26	.08544	.11197	10	.88803	36
25	9.80274	16	10.19726	9.91482	25	10.08518	10.11207	11	9.88793	35
26	.80290	15	.19710	.91507	26	.08493	.11218	10	.88782	34
27	.80305	15	.19695	.91533	26	.08467	.11228	11	.88772	33
28	.80320	16	.19680	.91559	26	.08441	.11239	10	.88761	32
29	.80336	15	.19664	.91585	25	.08415	.11249	10	.88751	31
30	9.80351	15	10.19649	9.91610	26	10.08390	10.11259	11	9.88741	30
31	.80366	16	.19634	.91636	26	.08364	.11270	10	.88730	29
32	.80382	15	.19618	.91662	26	.08338	.11280	11	.88720	28
33	.80397	15	.19603	.91688	25	.08312	.11291	10	.88709	27
34	.80412	16	.19588	.91713	26	.08287	.11301	11	.88699	26
35	9.80428	15	10.19572	9.91739	26	10.08261	10.11312	10	9.88688	25
36	.80443	15	.19557	.91765	26	.08235	.11322	10	.88678	24
37	.80458	15	.19542	.91791	25	.08209	.11332	11	.88668	23
38	.80473	16	.19527	.91816	26	.08184	.11343	10	.88657	22
39	.80489	15	.19511	.91842	26	.08158	.11353	11	.88647	21
40	9.80504	15	10.19496	9.91868	25	10.08132	10.11364	10	9.88636	20
41	.80519	15	.19481	.91893	26	.08107	.11374	11	.88626	19
42	.80534	16	.19466	.91919	26	.08081	.11385	10	.88615	18
43	.80550	15	.19450	.91945	26	.08055	.11395	11	.88605	17
44	.80565	15	.19435	.91971	25	.08029	.11406	10	.88594	16
45	9.80580	15	10.19420	9.91996	26	10.08004	10.11416	11	9.88584	15
46	.80595	15	.19405	.92022	26	.07978	.11427	10	.88573	14
47	.80610	15	.19390	.92048	25	.07952	.11437	11	.88563	13
48	.80625	16	.19375	.92073	26	.07927	.11448	10	.88552	12
49	.80641	15	.19359	.92099	26	.07901	.11458	11	.88542	11
50	9.80656	15	10.19344	9.92125	25	10.07875	10.11469	10	9.88531	10
51	.80671	15	.19329	.92150	26	.07850	.11479	11	.88521	9
52	.80686	15	.19314	.92176	26	.07824	.11490	11	.88510	8
53	.80701	15	.19299	.92202	25	.07798	.11501	10	.88499	7
54	.80716	15	.19284	.92227	26	.07773	.11511	11	.88489	6
55	9.80731	15	10.19269	9.92253	26	10.07747	10.11522	10	9.88478	5
56	.80746	16	.19254	.92279	25	.07721	.11532	11	.88468	4
57	.80762	15	.19238	.92304	26	.07696	.11543	10	.88457	3
58	.80777	15	.19223	.92330	26	.07670	.11553	11	.88447	2
59	.80792	15	.19208	.92356	25	.07644	.11564	11	.88436	1
60	9.80807		10.19193	9.92381		10.07619	10.11575		9.88425	0
↑ 129° →	cos	Diff. 1′	sec	cot	Diff. 1′	tan	csc	Diff. 1′	sin	← 50° ↑

TABLE 33

Logarithms of Trigonometric Functions

40° → ′ ↓	sin	Diff. 1′	csc	tan	Diff. 1′	cot	sec	Diff. 1′	cos	← 139° ′ ↓
0	9. 80807	15	10. 19193	9. 92381	26	10. 07619	10. 11575	10	9. 88425	60
1	. 80822	15	. 19178	. 92407	26	. 07593	. 11585	11	. 88415	59
2	. 80837	15	. 19163	. 92433	25	. 07567	. 11596	10	. 88404	58
3	. 80852	15	. 19148	. 92458	26	. 07542	. 11606	11	. 88394	57
4	. 80867	15	. 19133	. 92484	26	. 07516	. 11617	11	. 88383	56
5	9. 80882	15	10. 19118	9. 92510	25	10. 07490	10. 11628	10	9. 88372	55
6	. 80897	15	. 19103	. 92535	26	. 07465	. 11638	11	. 88362	54
7	. 80912	15	. 19088	. 92561	26	. 07439	. 11649	11	. 88351	53
8	. 80927	15	. 19073	. 92587	25	. 07413	. 11660	10	. 88340	52
9	. 80942	15	. 19058	. 92612	26	. 07388	. 11670	11	. 88330	51
10	9. 80957	15	10. 19043	9. 92638	25	10. 07362	10. 11681	11	9. 88319	50
11	. 80972	15	. 19028	. 92663	26	. 07337	. 11692	10	. 88308	49
12	. 80987	15	. 19013	. 92689	26	. 07311	. 11702	11	. 88298	48
13	. 81002	15	. 18998	. 92715	25	. 07285	. 11713	11	. 88287	47
14	. 81017	15	. 18983	. 92740	26	. 07260	. 11724	10	. 88276	46
15	9. 81032	15	10. 18968	9. 92766	26	10. 07234	10. 11734	11	9. 88266	45
16	. 81047	14	. 18953	. 92792	25	. 07208	. 11745	11	. 88255	44
17	. 81061	15	. 18939	. 92817	26	. 07183	. 11756	10	. 88244	43
18	. 81076	15	. 18924	. 92843	25	. 07157	. 11766	11	. 88234	42
19	. 81091	15	. 18909	. 92868	26	. 07132	. 11777	11	. 88223	41
20	9. 81106	15	10. 18894	9. 92894	26	10. 07106	10. 11788	11	9. 88212	40
21	. 81121	15	. 18879	. 92920	25	. 07080	. 11799	10	. 88201	39
22	. 81136	15	. 18864	. 92945	26	. 07055	. 11809	11	. 88191	38
23	. 81151	15	. 18849	. 92971	25	. 07029	. 11820	11	. 88180	37
24	. 81166	14	. 18834	. 92996	26	. 07004	. 11831	11	. 88169	36
25	9. 81180	15	10. 18820	9. 93022	26	10. 06978	10. 11842	10	9. 88158	35
26	. 81195	15	. 18805	. 93048	25	. 06952	. 11852	11	. 88148	34
27	. 81210	15	. 18790	. 93073	26	. 06927	. 11863	11	. 88137	33
28	. 81225	15	. 18775	. 93099	25	. 06901	. 11874	11	. 88126	32
29	. 81240	14	. 18760	. 93124	26	. 06876	. 11885	10	. 88115	31
30	9. 81254	15	10. 18746	9. 93150	25	10. 06850	10. 11895	11	9. 88105	30
31	. 81269	15	. 18731	. 93175	26	. 06825	. 11906	11	. 88094	29
32	. 81284	15	. 18716	. 93201	26	. 06799	. 11917	11	. 88083	28
33	. 81299	15	. 18701	. 93227	25	. 06773	. 11928	11	. 88072	27
34	. 81314	14	. 18686	. 93252	26	. 06748	. 11939	10	. 88061	26
35	9. 81328	15	10. 18672	9. 93278	25	10. 06722	10. 11949	11	9. 88051	25
36	. 81343	15	. 18657	. 93303	26	. 06697	. 11960	11	. 88040	24
37	. 81358	14	. 18642	. 93329	25	. 06671	. 11971	11	. 88029	23
38	. 81372	15	. 18628	. 93354	26	. 06646	. 11982	11	. 88018	22
39	. 81387	15	. 18613	. 93380	26	. 06620	. 11993	11	. 88007	21
40	9. 81402	15	10. 18598	9. 93406	25	10. 06594	10. 12004	11	9. 87996	20
41	. 81417	14	. 18583	. 93431	26	. 06569	. 12015	10	. 87985	19
42	. 81431	15	. 18569	. 93457	25	. 06543	. 12025	11	. 87975	18
43	. 81446	15	. 18554	. 93482	26	. 06518	. 12036	11	. 87964	17
44	. 81461	14	. 18539	. 93508	25	. 06492	. 12047	11	. 87953	16
45	9. 81475	15	10. 18525	9. 93533	26	10. 06467	10. 12058	11	9. 87942	15
46	. 81490	15	. 18510	. 93559	25	. 06441	. 12069	11	. 87931	14
47	. 81505	14	. 18495	. 93584	26	. 06416	. 12080	11	. 87920	13
48	. 81519	15	. 18481	. 93610	26	. 06390	. 12091	11	. 87909	12
49	. 81534	15	. 18466	. 93636	25	. 06364	. 12102	11	. 87898	11
50	9. 81549	14	10. 18451	9. 93661	26	10. 06339	10. 12113	10	9. 87887	10
51	. 81563	15	. 18437	. 93687	25	. 06313	. 12123	11	. 87877	9
52	. 81578	14	. 18422	. 93712	26	. 06288	. 12134	11	. 87866	8
53	. 81592	15	. 18408	. 93738	25	. 06262	. 12145	11	. 87855	7
54	. 81607	15	. 18393	. 93763	26	. 06237	. 12156	11	. 87844	6
55	9. 81622	14	10. 18378	9. 93789	25	10. 06211	10. 12167	11	9. 87833	5
56	. 81636	15	. 18364	. 93814	26	. 06186	. 12178	11	. 87822	4
57	. 81651	14	. 18349	. 93840	25	. 06160	. 12189	11	. 87811	3
58	. 81665	15	. 18335	. 93865	26	. 06135	. 12200	11	. 87800	2
59	. 81680	14	. 18320	. 93891	25	. 06109	. 12211	11	. 87789	1
60	9. 81694		10. 18306	9. 93916		10. 06084	10. 12222		9. 87778	0
↑ 130° →	cos	Diff. 1′	sec	cot	Diff. 1′	tan	csc	Diff. 1′	sin	← 49° ↑

TABLE 33
Logarithms of Trigonometric Functions

41° → ↓ ′	sin	Diff. 1′	csc	tan	Diff. 1′	cot	sec	Diff. 1′	cos	← 138° ↓ ′
0	9.81694	15	10.18306	9.93916	26	10.06084	10.12222	11	9.87778	60
1	.81709	14	.18291	.93942	25	.06058	.12233	11	.87767	59
2	.81723	15	.18277	.93967	26	.06033	.12244	11	.87756	58
3	.81738	14	.18262	.93993	25	.06007	.12255	11	.87745	57
4	.81752	15	.18248	.94018	26	.05982	.12266	11	.87734	56
5	9.81767	14	10.18233	9.94044	25	10.05956	10.12277	11	9.87723	55
6	.81781	15	.18219	.94069	26	.05931	.12288	11	.87712	54
7	.81796	14	.18204	.94095	25	.05905	.12299	11	.87701	53
8	.81810	15	.18190	.94120	26	.05880	.12310	11	.87690	52
9	.81825	14	.18175	.94146	25	.05854	.12321	11	.87679	51
10	9.81839	15	10.18161	9.94171	26	10.05829	10.12332	11	9.87668	50
11	.81854	14	.18146	.94197	25	.05803	.12343	11	.87657	49
12	.81868	14	.18132	.94222	26	.05778	.12354	11	.87646	48
13	.81882	15	.18118	.94248	25	.05752	.12365	11	.87635	47
14	.81897	14	.18103	.94273	26	.05727	.12376	11	.87624	46
15	9.81911	15	10.18089	9.94299	25	10.05701	10.12387	12	9.87613	45
16	.81926	14	.18074	.94324	26	.05676	.12399	11	.87601	44
17	.81940	15	.18060	.94350	25	.05650	.12410	11	.87590	43
18	.81955	14	.18045	.94375	26	.05625	.12421	11	.87579	42
19	.81969	14	.18031	.94401	25	.05599	.12432	11	.87568	41
20	9.81983	15	10.18017	9.94426	26	10.05574	10.12443	11	9.87557	40
21	.81998	14	.18002	.94452	25	.05548	.12454	11	.87546	39
22	.82012	14	.17988	.94477	26	.05523	.12465	11	.87535	38
23	.82026	15	.17974	.94503	25	.05497	.12476	11	.87524	37
24	.82041	14	.17959	.94528	26	.05472	.12487	12	.87513	36
25	9.82055	14	10.17945	9.94554	25	10.05446	10.12499	11	9.87501	35
26	.82069	15	.17931	.94579	25	.05421	.12510	11	.87490	34
27	.82084	14	.17916	.94604	26	.05396	.12521	11	.87479	33
28	.82098	14	.17902	.94630	25	.05370	.12532	11	.87468	32
29	.82112	14	.17888	.94655	26	.05345	.12543	11	.87457	31
30	9.82126	15	10.17874	9.94681	25	10.05319	10.12554	12	9.87446	30
31	.82141	14	.17859	.94706	26	.05294	.12566	11	.87434	29
32	.82155	14	.17845	.94732	25	.05268	.12577	11	.87423	28
33	.82169	15	.17831	.94757	26	.05243	.12588	11	.87412	27
34	.82184	14	.17816	.94783	25	.05217	.12599	11	.87401	26
35	9.82198	14	10.17802	9.94808	26	10.05192	10.12610	12	9.87390	25
36	.82212	14	.17788	.94834	25	.05166	.12622	11	.87378	24
37	.82226	14	.17774	.94859	25	.05141	.12633	11	.87367	23
38	.82240	15	.17760	.94884	26	.05116	.12644	11	.87356	22
39	.82255	14	.17745	.94910	25	.05090	.12655	11	.87345	21
40	9.82269	14	10.17731	9.94935	26	10.05065	10.12666	12	9.87334	20
41	.82283	14	.17717	.94961	25	.05039	.12678	11	.87322	19
42	.82297	14	.17703	.94986	26	.05014	.12689	11	.87311	18
43	.82311	15	.17689	.95012	25	.04988	.12700	12	.87300	17
44	.82326	14	.17674	.95037	25	.04963	.12712	11	.87288	16
45	9.82340	14	10.17660	9.95062	26	10.04938	10.12723	11	9.87277	15
46	.82354	14	.17646	.95088	25	.04912	.12734	11	.87266	14
47	.82368	14	.17632	.95113	26	.04887	.12745	12	.87255	13
48	.82382	14	.17618	.95139	25	.04861	.12757	11	.87243	12
49	.82396	14	.17604	.95164	26	.04836	.12768	11	.87232	11
50	9.82410	14	10.17590	9.95190	25	10.04810	10.12779	12	9.87221	10
51	.82424	15	.17576	.95215	25	.04785	.12791	11	.87209	9
52	.82439	14	.17561	.95240	26	.04760	.12802	11	.87198	8
53	.82453	14	.17547	.95266	25	.04734	.12813	12	.87187	7
54	.82467	14	.17533	.95291	26	.04709	.12825	11	.87175	6
55	9.82481	14	10.17519	9.95317	25	10.04683	10.12836	11	9.87164	5
56	.82495	14	.17505	.95342	26	.04658	.12847	12	.87153	4
57	.82509	14	.17491	.95368	25	.04632	.12859	11	.87141	3
58	.82523	14	.17477	.95393	25	.04607	.12870	11	.87130	2
59	.82537	14	.17463	.95418	26	.04582	.12881	12	.87119	1
60	9.82551		10.17449	9.95444		10.04556	10.12893		9.87107	0
↑ 131° →	cos	Diff. 1′	sec	cot	Diff. 1′	tan	csc	Diff. 1′	sin	← 48° ↑

TABLE 33

Logarithms of Trigonometric Functions

42°→ ↓ ′	sin	Diff. 1′	csc	tan	Diff. 1′	cot	sec	Diff. 1′	cos	←137° ↓ ′
0	9.82551	14	10.17449	9.95444	25	10.04556	10.12893	11	9.87107	60
1	.82565	14	.17435	.95469	26	.04531	.12904	11	.87096	59
2	.82579	14	.17421	.95495	25	.04505	.12915	12	.87085	58
3	.82593	14	.17407	.95520	25	.04480	.12927	11	.87073	57
4	.82607	14	.17393	.95545	26	.04455	.12938	12	.87062	56
5	9.82621	14	10.17379	9.95571	25	10.04429	10.12950	11	9.87050	55
6	.82635	14	.17365	.95596	26	.04404	.12961	11	.87039	54
7	.82649	14	.17351	.95622	25	.04378	.12972	12	.87028	53
8	.82663	14	.17337	.95647	25	.04353	.12984	11	.87016	52
9	.82677	14	.17323	.95672	26	.04328	.12995	12	.87005	51
10	9.82691	14	10.17309	9.95698	25	10.04302	10.13007	11	9.86993	50
11	.82705	14	.17295	.95723	25	.04277	.13018	12	.86982	49
12	.82719	14	.17281	.95748	26	.04252	.13030	11	.86970	48
13	.82733	14	.17267	.95774	25	.04226	.13041	12	.86959	47
14	.82747	14	.17253	.95799	26	.04201	.13053	11	.86947	46
15	9.82761	14	10.17239	9.95825	25	10.04175	10.13064	12	9.86936	45
16	.82775	13	.17225	.95850	25	.04150	.13076	11	.86924	44
17	.82788	14	.17212	.95875	26	.04125	.13087	11	.86913	43
18	.82802	14	.17198	.95901	25	.04099	.13098	12	.86902	42
19	.82816	14	.17184	.95926	26	.04074	.13110	11	.86890	41
20	9.82830	14	10.17170	9.95952	25	10.04048	10.13121	12	9.86879	40
21	.82844	14	.17156	.95977	25	.04023	.13133	12	.86867	39
22	.82858	14	.17142	.96002	26	.03998	.13145	11	.86855	38
23	.82872	13	.17128	.96028	25	.03972	.13156	12	.86844	37
24	.82885	14	.17115	.96053	25	.03947	.13168	11	.86832	36
25	9.82899	14	10.17101	9.96078	26	10.03922	10.13179	12	9.86821	35
26	.82913	14	.17087	.96104	25	.03896	.13191	11	.86809	34
27	.82927	14	.17073	.96129	26	.03871	.13202	12	.86798	33
28	.82941	14	.17059	.96155	25	.03845	.13214	11	.86786	32
29	.82955	13	.17045	.96180	25	.03820	.13225	12	.86775	31
30	9.82968	14	10.17032	9.96205	26	10.03795	10.13237	11	9.86763	30
31	.82982	14	.17018	.96231	25	.03769	.13248	12	.86752	29
32	.82996	14	.17004	.96256	25	.03744	.13260	12	.86740	28
33	.83010	13	.16990	.96281	26	.03719	.13272	11	.86728	27
34	.83023	14	.16977	.96307	25	.03693	.13283	12	.86717	26
35	9.83037	14	10.16963	9.96332	25	10.03668	10.13295	11	9.86705	25
36	.83051	14	.16949	.96357	26	.03643	.13306	12	.86694	24
37	.83065	13	.16935	.96383	25	.03617	.13318	12	.86682	23
38	.83078	14	.16922	.96408	25	.03592	.13330	11	.86670	22
39	.83092	14	.16908	.96433	26	.03567	.13341	12	.86659	21
40	9.83106	14	10.16894	9.96459	25	10.03541	10.13353	12	9.86647	20
41	.83120	13	.16880	.96484	26	.03516	.13365	11	.86635	19
42	.83133	14	.16867	.96510	25	.03490	.13376	12	.86624	18
43	.83147	14	.16853	.96535	25	.03465	.13388	12	.86612	17
44	.83161	13	.16839	.96560	26	.03440	.13400	11	.86600	16
45	9.83174	14	10.16826	9.96586	25	10.03414	10.13411	12	9.86589	15
46	.83188	14	.16812	.96611	25	.03389	.13423	12	.86577	14
47	.83202	13	.16798	.96636	26	.03364	.13435	11	.86565	13
48	.83215	14	.16785	.96662	25	.03338	.13446	12	.86554	12
49	.83229	13	.16771	.96687	25	.03313	.13458	12	.86542	11
50	9.83242	14	10.16758	9.96712	26	10.03288	10.13470	12	9.86530	10
51	.83256	14	.16744	.96738	25	.03262	.13482	11	.86518	9
52	.83270	13	.16730	.96763	25	.03237	.13493	12	.86507	8
53	.83283	14	.16717	.96788	26	.03212	.13505	12	.86495	7
54	.83297	13	.16703	.96814	25	.03186	.13517	11	.86483	6
55	9.83310	14	10.16690	9.96839	25	10.03161	10.13528	12	9.86472	5
56	.83324	14	.16676	.96864	26	.03136	.13540	12	.86460	4
57	.83338	13	.16662	.96890	25	.03110	.13552	12	.86448	3
58	.83351	14	.16649	.96915	25	.03085	.13564	11	.86436	2
59	.83365	13	.16635	.96940	26	.03060	.13575	12	.86425	1
60	9.83378		10.16622	9.96966		10.03034	10.13587		9.86413	0
↑ 132°→	cos	Diff. 1′	sec	cot	Diff. 1′	tan	csc	Diff. 1′	sin	↑ ←47°

TABLE 33
Logarithms of Trigonometric Functions

43° → ↓ ′	sin	Diff. 1′	csc	tan	Diff. 1′	cot	sec	Diff. 1′	cos	← 136° ↓ ′
0	9. 83378	14	10. 16622	9. 96966	25	10. 03034	10. 13587	12	9. 86413	60
1	. 83392	13	. 16608	. 96991	25	. 03009	. 13599	12	. 86401	59
2	. 83405	14	. 16595	. 97016	26	. 02984	. 13611	12	. 86389	58
3	. 83419	13	. 16581	. 97042	25	. 02958	. 13623	11	. 86377	57
4	. 83432	14	. 16568	. 97067	25	. 02933	. 13634	12	. 86366	56
5	9. 83446	13	10. 16554	9. 97092	26	10. 02908	10. 13646	12	9. 86354	55
6	. 83459	14	. 16541	. 97118	25	. 02882	. 13658	12	. 86342	54
7	. 83473	13	. 16527	. 97143	25	. 02857	. 13670	12	. 86330	53
8	. 83486	14	. 16514	. 97168	25	. 02832	. 13682	12	. 86318	52
9	. 83500	13	. 16500	. 97193	26	. 02807	. 13694	11	. 86306	51
10	9. 83513	14	10. 16487	9. 97219	25	10. 02781	10. 13705	12	9. 86295	50
11	. 83527	13	. 16473	. 97244	25	. 02756	. 13717	12	. 86283	49
12	. 83540	14	. 16460	. 97269	26	. 02731	. 13729	12	. 86271	48
13	. 83554	13	. 16446	. 97295	25	. 02705	. 13741	12	. 86259	47
14	. 83567	14	. 16433	. 97320	25	. 02680	. 13753	12	. 86247	46
15	9. 83581	13	10. 16419	9. 97345	26	10. 02655	10. 13765	12	9. 86235	45
16	. 83594	14	. 16406	. 97371	25	. 02629	. 13777	12	. 86223	44
17	. 83608	13	. 16392	. 97396	25	. 02604	. 13789	11	. 86211	43
18	. 83621	13	. 16379	. 97421	26	. 02579	. 13800	12	. 86200	42
19	. 83634	14	. 16366	. 97447	25	. 02553	. 13812	12	. 86188	41
20	9. 83648	13	10. 16352	9. 97472	25	10. 02528	10. 13824	12	9. 86176	40
21	. 83661	13	. 16339	. 97497	26	. 02503	. 13836	12	. 86164	39
22	. 83674	14	. 16326	. 97523	25	. 02477	. 13848	12	. 86152	38
23	. 83688	13	. 16312	. 97548	25	. 02452	. 13860	12	. 86140	37
24	. 83701	14	. 16299	. 97573	25	. 02427	. 13872	12	. 86128	36
25	9. 83715	13	10. 16285	9. 97598	26	10. 02402	10. 13884	12	9. 86116	35
26	. 83728	13	. 16272	. 97624	25	. 02376	. 13896	12	. 86104	34
27	. 83741	14	. 16259	. 97649	25	. 02351	. 13908	12	. 86092	33
28	. 83755	13	. 16245	. 97674	26	. 02326	. 13920	12	. 86080	32
29	. 83768	13	. 16232	. 97700	25	. 02300	. 13932	12	. 86068	31
30	9. 83781	14	10. 16219	9. 97725	25	10. 02275	10. 13944	12	9. 86056	30
31	. 83795	13	. 16205	. 97750	26	. 02250	. 13956	12	. 86044	29
32	. 83808	13	. 16192	. 97776	25	. 02224	. 13968	12	. 86032	28
33	. 83821	13	. 16179	. 97801	25	. 02199	. 13980	12	. 86020	27
34	. 83834	14	. 16166	. 97826	25	. 02174	. 13992	12	. 86008	26
35	9. 83848	13	10. 16152	9. 97851	26	10. 02149	10. 14004	12	9. 85996	25
36	. 83861	13	. 16139	. 97877	25	. 02123	. 14016	12	. 85984	24
37	. 83874	13	. 16126	. 97902	25	. 02098	. 14028	12	. 85972	23
38	. 83887	14	. 16113	. 97927	26	. 02073	. 14040	12	. 85960	22
39	. 83901	13	. 16099	. 97953	25	. 02047	. 14052	12	. 85948	21
40	9. 83914	13	10. 16086	9. 97978	25	10. 02022	10. 14064	12	9. 85936	20
41	. 83927	13	. 16073	. 98003	26	. 01997	. 14076	12	. 85924	19
42	. 83940	14	. 16060	. 98029	25	. 01971	. 14088	12	. 85912	18
43	. 83954	13	. 16046	. 98054	25	. 01946	. 14100	12	. 85900	17
44	. 83967	13	. 16033	. 98079	25	. 01921	. 14112	12	. 85888	16
45	9. 83980	13	10. 16020	9. 98104	26	10. 01896	10. 14124	12	9. 85876	15
46	. 83993	13	. 16007	. 98130	25	. 01870	. 14136	13	. 85864	14
47	. 84006	14	. 15994	. 98155	25	. 01845	. 14149	12	. 85851	13
48	. 84020	13	. 15980	. 98180	26	. 01820	. 14161	12	. 85839	12
49	. 84033	13	. 15967	. 98206	25	. 01794	. 14173	12	. 85827	11
50	9. 84046	13	10. 15954	9. 98231	25	10. 01769	10. 14185	12	9. 85815	10
51	. 84059	13	. 15941	. 98256	25	. 01744	. 14197	12	. 85803	9
52	. 84072	13	. 15928	. 98281	26	. 01719	. 14209	12	. 85791	8
53	. 84085	13	. 15915	. 98307	25	. 01693	. 14221	13	. 85779	7
54	. 84098	14	. 15902	. 98332	25	. 01668	. 14234	12	. 85766	6
55	9. 84112	13	10. 15888	9. 98357	26	10. 01643	10. 14246	12	9. 85754	5
56	. 84125	13	. 15875	. 98383	25	. 01617	. 14258	12	. 85742	4
57	. 84138	13	. 15862	. 98408	25	. 01592	. 14270	12	. 85730	3
58	. 84151	13	. 15849	. 98433	25	. 01567	. 14282	12	. 85718	2
59	. 84164	13	. 15836	. 98458	26	. 01542	. 14294	13	. 85706	1
60	9. 84177		10. 15823	9. 98484		10. 01516	10. 14307		9. 85693	0
↑ 133° →	cos	Diff. 1′	sec	cot	Diff. 1′	tan	csc	Diff. 1′	sin	← 46° ↑

TABLE 33
Logarithms of Trigonometric Functions

44° → ′	sin	Diff. 1′	csc	tan	Diff. 1′	cot	sec	Diff. 1′	cos	← 135° ′
0	9. 84177	13	10. 15823	9. 98484	25	10. 01516	10. 14307	12	9. 85693	60
1	. 84190	13	. 15810	. 98509	25	. 01491	. 14319	12	. 85681	59
2	. 84203	13	. 15797	. 98534	26	. 01466	. 14331	12	. 85669	58
3	. 84216	13	. 15784	. 98560	25	. 01440	. 14343	12	. 85657	57
4	. 84229	13	. 15771	. 98585	25	. 01415	. 14355	13	. 85645	56
5	9. 84242	13	10. 15758	9. 98610	25	10. 01390	10. 14368	12	9. 85632	55
6	. 84255	14	. 15745	. 98635	26	. 01365	. 14380	12	. 85620	54
7	. 84269	13	. 15731	. 98661	25	. 01339	. 14392	12	. 85608	53
8	. 84282	13	. 15718	. 98686	25	. 01314	. 14404	13	. 85596	52
9	. 84295	13	. 15705	. 98711	26	. 01289	. 14417	12	. 85583	51
10	9. 84308	13	10. 15692	9. 98737	25	10. 01263	10. 14429	12	9. 85571	50
11	. 84321	13	. 15679	. 98762	25	. 01238	. 14441	12	. 85559	49
12	. 84334	13	. 15666	. 98787	25	. 01213	. 14453	13	. 85547	48
13	. 84347	13	. 15653	. 98812	26	. 01188	. 14466	12	. 85534	47
14	. 84360	13	. 15640	. 98838	25	. 01162	. 14478	12	. 85522	46
15	9. 84373	12	10. 15627	9. 98863	25	10. 01137	10. 14490	13	9. 85510	45
16	. 84385	13	. 15615	. 98888	25	. 01112	. 14503	12	. 85497	44
17	. 84398	13	. 15602	. 98913	26	. 01087	. 14515	12	. 85485	43
18	. 84411	13	. 15589	. 98939	25	. 01061	. 14527	13	. 85473	42
19	. 84424	13	. 15576	. 98964	25	. 01036	. 14540	12	. 85460	41
20	9. 84437	13	10. 15563	9. 98989	26	10. 01011	10. 14552	12	9. 85448	40
21	. 84450	13	. 15550	. 99015	25	. 00985	. 14564	13	. 85436	39
22	. 84463	13	. 15537	. 99040	25	. 00960	. 14577	12	. 85423	38
23	. 84476	13	. 15524	. 99065	25	. 00935	. 14589	12	. 85411	37
24	. 84489	13	. 15511	. 99090	26	. 00910	. 14601	13	. 85399	36
25	9. 84502	13	10. 15498	9. 99116	25	10. 00884	10. 14614	12	9. 85386	35
26	. 84515	13	. 15485	. 99141	25	. 00859	. 14626	13	. 85374	34
27	. 84528	12	. 15472	. 99166	25	. 00834	. 14639	12	. 85361	33
28	. 84540	13	. 15460	. 99191	26	. 00809	. 14651	12	. 85349	32
29	. 84553	13	. 15447	. 99217	25	. 00783	. 14663	13	. 85337	31
30	9. 84566	13	10. 15434	9. 99242	25	10. 00758	10. 14676	12	9. 85324	30
31	. 84579	13	. 15421	. 99267	26	. 00733	. 14688	13	. 85312	29
32	. 84592	13	. 15408	. 99293	25	. 00707	. 14701	12	. 85299	28
33	. 84605	13	. 15395	. 99318	25	. 00682	. 14713	13	. 85287	27
34	. 84618	12	. 15382	. 99343	25	. 00657	. 14726	12	. 85274	26
35	9. 84630	13	10. 15370	9. 99368	26	10. 00632	10. 14738	12	9. 85262	25
36	. 84643	13	. 15357	. 99394	25	. 00606	. 14750	13	. 85250	24
37	. 84656	13	. 15344	. 99419	25	. 00581	. 14763	12	. 85237	23
38	. 84669	13	. 15331	. 99444	25	. 00556	. 14775	13	. 85225	22
39	. 84682	12	. 15318	. 99469	26	. 00531	. 14788	12	. 85212	21
40	9. 84694	13	10. 15306	9. 99495	25	10. 00505	10. 14800	13	9. 85200	20
41	. 84707	13	. 15293	. 99520	25	. 00480	. 14813	12	. 85187	19
42	. 84720	13	. 15280	. 99545	25	. 00455	. 14825	13	. 85175	18
43	. 84733	12	. 15267	. 99570	26	. 00430	. 14838	12	. 85162	17
44	. 84745	13	. 15255	. 99596	25	. 00404	. 14850	13	. 85150	16
45	9. 84758	13	10. 15242	9. 99621	25	10. 00379	10. 14863	12	9. 85137	15
46	. 84771	13	. 15229	. 99646	26	. 00354	. 14875	13	. 85125	14
47	. 84784	12	. 15216	. 99672	25	. 00328	. 14888	12	. 85112	13
48	. 84796	13	. 15204	. 99697	25	. 00303	. 14900	13	. 85100	12
49	. 84809	13	. 15191	. 99722	25	. 00278	. 14913	13	. 85087	11
50	9. 84822	13	10. 15178	9. 99747	26	10. 00253	10. 14926	12	9. 85074	10
51	. 84835	12	. 15165	. 99773	25	. 00227	. 14938	13	. 85062	9
52	. 84847	13	. 15153	. 99798	25	. 00202	. 14951	12	. 85049	8
53	. 84860	13	. 15140	. 99823	25	. 00177	. 14963	13	. 85037	7
54	. 84873	12	. 15127	. 99848	26	. 00152	. 14976	12	. 85024	6
55	9. 84885	13	10. 15115	9. 99874	25	10. 00126	10. 14988	13	9. 85012	5
56	. 84898	13	. 15102	. 99899	25	. 00101	. 15001	13	. 84999	4
57	. 84911	12	. 15089	. 99924	25	. 00076	. 15014	12	. 84986	3
58	. 84923	13	. 15077	. 99949	26	. 00051	. 15026	13	. 84974	2
59	. 84936	13	. 15064	9. 99975	25	. 00025	. 15039	12	. 84961	1
60	9. 84949		10. 15051	10. 00000		10. 00000	10. 15051		9. 84949	0
134° →	cos	Diff. 1′	sec	cot	Diff. 1′	tan	csc	Diff. 1′	sin	← 45°

TABLE 34

Haversines

′	0°		1°		2°		3°		4°		′
	Log Hav	Nat. Hav	Log Hav	Nat. Hav	Log Hav	Nat. Hav	Log Hav	Nat. Hav	Log Hav	Nat. Hav	
0	Inf. Neg.	0. 00000	5. 88168	0. 00008	6. 48371	0. 00030	6. 83584	0. 00069	7. 08564	0. 00122	60
1	2. 32539	. 00000	. 89604	. 00008	. 49092	. 00031	. 84065	. 00069	. 08925	. 00123	59
2	2. 92745	. 00000	. 91016	. 00008	. 49807	. 00031	. 84543	. 00070	. 09284	. 00124	58
3	3. 27963	. 00000	. 92406	. 00008	. 50516	. 00032	. 85019	. 00071	. 09642	. 00125	57
4	. 52951	. 00000	. 93774	. 00009	. 51219	. 00033	. 85492	. 00072	. 09999	. 00126	56
5	3. 72333	0. 00000	5. 95121	0. 00009	6. 51916	0. 00033	6. 85963	0. 00072	7. 10354	0. 00127	55
6	3. 88169	. 00000	. 96447	. 00009	. 52608	. 00034	. 86431	. 00073	. 10708	. 00128	54
7	4. 01559	. 00000	. 97753	. 00009	. 53295	. 00034	. 86897	. 00074	. 11060	. 00129	53
8	. 13157	. 00000	5. 99040	. 00010	. 53976	. 00035	. 87360	. 00075	. 11411	. 00130	52
9	. 23388	. 00000	6. 00308	. 00010	. 54652	. 00035	. 87821	. 00076	. 11760	. 00131	51
10	4. 32539	0. 00000	6. 01557	0. 00010	6. 55323	0. 00036	6. 88279	0. 00076	7. 12108	0. 00132	50
11	. 40818	. 00000	. 02789	. 00011	. 55988	. 00036	. 88735	. 00077	. 12455	. 00133	49
12	. 48375	. 00000	. 04004	. 00011	. 56649	. 00037	. 89188	. 00078	. 12800	. 00134	48
13	. 55328	. 00000	. 05202	. 00011	. 57304	. 00037	. 89639	. 00079	. 13144	. 00135	47
14	. 61765	. 00000	. 06384	. 00012	. 57955	. 00038	. 90088	. 00080	. 13486	. 00136	46
15	4. 67757	0. 00000	6. 07550	0. 00012	6. 58600	0. 00039	6. 90535	0. 00080	7. 13827	0. 00137	45
16	. 73363	. 00001	. 08700	. 00012	. 59241	. 00039	. 90979	. 00081	. 14167	. 00139	44
17	. 78629	. 00001	. 09836	. 00013	. 59878	. 00040	. 91421	. 00082	. 14506	. 00140	43
18	. 83594	. 00001	. 10956	. 00013	. 60509	. 00040	. 91860	. 00083	. 14843	. 00141	42
19	. 88290	. 00001	. 12063	. 00013	. 61136	. 00041	. 92298	. 00084	. 15179	. 00142	41
20	4. 92745	0. 00001	6. 13155	0. 00014	6. 61759	0. 00041	6. 92733	0. 00085	7. 15513	0. 00143	40
21	4. 96983	. 00001	. 14234	. 00014	. 62377	. 00042	. 93166	. 00085	. 15846	. 00144	39
22	5. 01024	. 00001	. 15300	. 00014	. 62991	. 00043	. 93597	. 00086	. 16178	. 00145	38
23	. 04885	. 00001	. 16353	. 00015	. 63600	. 00043	. 94026	. 00087	. 16509	. 00146	37
24	. 08581	. 00001	. 17393	. 00015	. 64205	. 00044	. 94453	. 00088	. 16839	. 00147	36
25	5. 12127	0. 00001	6. 18421	0. 00015	6. 64806	0. 00044	6. 94877	0. 00089	7. 17167	0. 00148	35
26	. 15534	. 00001	. 19437	. 00016	. 65403	. 00045	. 95300	. 00090	. 17494	. 00150	34
27	. 18812	. 00002	. 20441	. 00016	. 65996	. 00046	. 95720	. 00091	. 17820	. 00151	33
28	. 21971	. 00002	. 21433	. 00016	. 66585	. 00046	. 96139	. 00091	. 18144	. 00152	32
29	. 25019	. 00002	. 22415	. 00017	. 67170	. 00047	. 96555	. 00092	. 18468	. 00153	31
30	5. 27963	0. 00002	6. 23385	0. 00017	6. 67751	0. 00048	6. 96970	0. 00093	7. 18790	0. 00154	30
31	. 30811	. 00002	. 24345	. 00018	. 68328	. 00048	. 97382	. 00094	. 19111	. 00155	29
32	. 33569	. 00002	. 25294	. 00018	. 68901	. 00049	. 97793	. 00095	. 19430	. 00156	28
33	. 36242	. 00002	. 26233	. 00018	. 69470	. 00050	. 98201	. 00096	. 19749	. 00158	27
34	. 38835	. 00002	. 27162	. 00019	. 70036	. 00050	. 98608	. 00097	. 20066	. 00159	26
35	5. 41352	0. 00003	6. 28081	0. 00019	6. 70598	0. 00051	6. 99013	0. 00098	7. 20383	0. 00160	25
36	. 43799	. 00003	. 28991	. 00019	. 71157	. 00051	. 99416	. 00099	. 20698	. 00161	24
37	. 46179	. 00003	. 29891	. 00020	. 71712	. 00052	6. 99817	. 00100	. 21012	. 00162	23
38	. 48496	. 00003	. 30781	. 00020	. 72263	. 00053	7. 00216	. 00100	. 21325	. 00163	22
39	. 50752	. 00003	. 31663	. 00021	. 72811	. 00053	. 00613	. 00101	. 21636	. 00165	21
40	5. 52951	0. 00003	6. 32536	0. 00021	6. 73355	0. 00054	7. 01009	0. 00102	7. 21947	0. 00166	20
41	. 55095	. 00004	. 33400	. 00022	. 73896	. 00055	. 01403	. 00103	. 22256	. 00167	19
42	. 57189	. 00004	. 34256	. 00022	. 74434	. 00056	. 01795	. 00104	. 22565	. 00168	18
43	. 59232	. 00004	. 35103	. 00022	. 74969	. 00056	. 02185	. 00105	. 22872	. 00169	17
44	. 61229	. 00004	. 35943	. 00023	. 75500	. 00057	. 02573	. 00106	. 23178	. 00171	16
45	5. 63181	0. 00004	6. 36774	0. 00023	6. 76028	0. 00058	7. 02960	0. 00107	7. 23483	0. 00172	15
46	. 65090	. 00004	. 37597	. 00024	. 76552	. 00058	. 03345	. 00108	. 23787	. 00173	14
47	. 66958	. 00005	. 38412	. 00024	. 77074	. 00059	. 03729	. 00109	. 24090	. 00174	13
48	. 68787	. 00005	. 39220	. 00025	. 77592	. 00060	. 04110	. 00110	. 24392	. 00175	12
49	. 70578	. 00005	. 40021	. 00025	. 78108	. 00060	. 04490	. 00111	. 24693	. 00177	11
50	5. 72332	0. 00005	6. 40814	0. 00026	6. 78620	0. 00061	7. 04869	0. 00112	7. 24993	0. 00178	10
51	. 74052	. 00006	. 41600	. 00026	. 79129	. 00062	. 05245	. 00113	. 25292	. 00179	9
52	. 75739	. 00006	. 42379	. 00027	. 79636	. 00063	. 05620	. 00114	. 25590	. 00180	8
53	. 77394	. 00006	. 43151	. 00027	. 80139	. 00063	. 05994	. 00115	. 25886	. 00181	7
54	. 79017	. 00006	. 43916	. 00027	. 80640	. 00064	. 06366	. 00116	. 26182	. 00183	6
55	5. 80611	0. 00006	6. 44675	0. 00028	6. 81137	0. 00065	7. 06736	0. 00117	7. 26477	0. 00184	5
56	. 82176	. 00007	. 45427	. 00028	. 81632	. 00066	. 07105	. 00118	. 26771	. 00185	4
57	. 83713	. 00007	. 46172	. 00029	. 82124	. 00066	. 07472	. 00119	. 27064	. 00186	3
58	. 85224	. 00007	. 46911	. 00029	. 82614	. 00067	. 07837	. 00120	. 27355	. 00188	2
59	. 86709	. 00007	. 47644	. 00030	. 83100	. 00068	. 08201	. 00121	. 27646	. 00189	1
60	5. 88168	0. 00008	6. 48371	0. 00030	6. 83584	0. 00069	7. 08564	0. 00122	7. 27936	0. 00190	0
	359°		358°		357°		356°		355°		

TABLE 34

Haversines

′	5°		6°		7°		8°		9°		′
	Log Hav	Nat. Hav	Log Hav	Nat. Hav	Log Hav	Nat. Hav	Log Hav	Nat. Hav	Log Hav	Nat. Hav	
0	7. 27936	0. 00190	7. 43760	0. 00274	7. 57135	0. 00373	7. 68717	0. 00487	7. 78929	0. 00616	60
1	. 28225	. 00192	. 44001	. 00275	. 57341	. 00374	. 68897	. 00489	. 79089	. 00618	59
2	. 28513	. 00193	. 44241	. 00277	. 57547	. 00376	. 69077	. 00491	. 79249	. 00620	58
3	. 28800	. 00194	. 44480	. 00278	. 57752	. 00378	. 69257	. 00493	. 79409	. 00622	57
4	. 29086	. 00195	. 44719	. 00280	. 57957	. 00380	. 69437	. 00495	. 79568	. 00625	56
5	7. 29371	0. 00197	7. 44957	0. 00282	7. 58162	0. 00382	7. 69616	0. 00497	7. 79728	0. 00627	55
6	. 29655	. 00198	. 45194	. 00283	. 58366	. 00383	. 69794	. 00499	. 79886	. 00629	54
7	. 29938	. 00199	. 45431	. 00285	. 58569	. 00385	. 69972	. 00501	. 80045	. 00632	53
8	. 30220	. 00201	. 45667	. 00286	. 58772	. 00387	. 70150	. 00503	. 80203	. 00634	52
9	. 30502	. 00202	. 45903	. 00288	. 58974	. 00389	. 70328	. 00505	. 80361	. 00636	51
10	7. 30782	0. 00203	7. 46138	0. 00289	7. 59176	0. 00391	7. 70505	0. 00507	7. 80519	0. 00639	50
11	. 31062	. 00204	. 46372	. 00291	. 59378	. 00392	. 70682	. 00509	. 80677	. 00641	49
12	. 31340	. 00206	. 46605	. 00292	. 59579	. 00394	. 70858	. 00511	. 80834	. 00643	48
13	. 31618	. 00207	. 46838	. 00294	. 59779	. 00396	. 71034	. 00513	. 80991	. 00646	47
14	. 31895	. 00208	. 47071	. 00296	. 59979	. 00398	. 71210	. 00515	. 81147	. 00648	46
15	7. 32171	0. 00210	7. 47302	0. 00297	7. 60179	0. 00400	7. 71385	0. 00517	7. 81303	0. 00650	45
16	. 32446	. 00211	. 47533	. 00299	. 60378	. 00402	. 71560	. 00520	. 81459	. 00653	44
17	. 32720	. 00212	. 47764	. 00300	. 60577	. 00403	. 71735	. 00522	. 81615	. 00655	43
18	. 32994	. 00214	. 47994	. 00302	. 60775	. 00405	. 71909	. 00524	. 81771	. 00657	42
19	. 33266	. 00215	. 48223	. 00304	. 60973	. 00407	. 72083	. 00526	. 81926	. 00660	41
20	7. 33538	0. 00216	7. 48452	0. 00305	7. 61170	0. 00409	7. 72257	0. 00528	7. 82081	0. 00662	40
21	. 33809	. 00218	. 48680	. 00307	. 61367	. 00411	. 72430	. 00530	. 82235	. 00664	39
22	. 34079	. 00219	. 48907	. 00308	. 61564	. 00413	. 72603	. 00532	. 82390	. 00667	38
23	. 34348	. 00221	. 49134	. 00310	. 61760	. 00415	. 72775	. 00534	. 82544	. 00669	37
24	. 34616	. 00222	. 49360	. 00312	. 61955	. 00416	. 72948	. 00536	. 82698	. 00671	36
25	7. 34884	0. 00223	7. 49586	0. 00313	7. 62151	0. 00418	7. 73119	0. 00539	7. 82851	0. 00674	35
26	. 35150	. 00225	. 49811	. 00315	. 62345	. 00420	. 73291	. 00541	. 83004	. 00676	34
27	. 35416	. 00226	. 50036	. 00316	. 62540	. 00422	. 73462	. 00543	. 83157	. 00679	33
28	. 35681	. 00227	. 50259	. 00318	. 62733	. 00424	. 73633	. 00545	. 83310	. 00681	32
29	. 35945	. 00229	. 50483	. 00320	. 62927	. 00426	. 73803	. 00547	. 83463	. 00683	31
30	7. 36209	0. 00230	7. 50706	0. 00321	7. 63120	0. 00428	7. 73974	0. 00549	7. 83615	0. 00686	30
31	. 36471	. 00232	. 50928	. 00323	. 63312	. 00430	. 74143	. 00551	. 83767	. 00688	29
32	. 36733	. 00233	. 51149	. 00325	. 63504	. 00432	. 74313	. 00554	. 83918	. 00691	28
33	. 36994	. 00234	. 51370	. 00326	. 63696	. 00433	. 74482	. 00556	. 84070	. 00693	27
34	. 37254	. 00236	. 51591	. 00328	. 63887	. 00435	. 74651	. 00558	. 84221	. 00695	26
35	7. 37514	0. 00237	7. 51811	0. 00330	7. 64078	0. 00437	7. 74819	0. 00560	7. 84372	0. 00698	25
36	. 37773	. 00239	. 52030	. 00331	. 64269	. 00439	. 74988	. 00562	. 84522	. 00700	24
37	. 38030	. 00240	. 52249	. 00333	. 64458	. 00441	. 75155	. 00564	. 84672	. 00703	23
38	. 38288	. 00241	. 52467	. 00335	. 64648	. 00443	. 75323	. 00567	. 84822	. 00705	22
39	. 38544	. 00243	. 52685	. 00336	. 64837	. 00445	. 75490	. 00569	. 84972	. 00707	21
40	7. 38800	0. 00244	7. 52902	0. 00338	7. 65026	0. 00447	7. 75657	0. 00571	7. 85122	0. 00710	20
41	. 39054	. 00246	. 53119	. 00340	. 65214	. 00449	. 75824	. 00573	. 85271	. 00712	19
42	. 39309	. 00247	. 53335	. 00341	. 65402	. 00451	. 75990	. 00575	. 85420	. 00715	18
43	. 39562	. 00249	. 53550	. 00343	. 65590	. 00453	. 76156	. 00578	. 85569	. 00717	17
44	. 39815	. 00250	. 53766	. 00345	. 65777	. 00455	. 76321	. 00580	. 85717	. 00720	16
45	7. 40067	0. 00252	7. 53980	0. 00347	7. 65964	0. 00457	7. 76487	0. 00582	7. 85866	0. 00722	15
46	. 40318	. 00253	. 54194	. 00348	. 66150	. 00459	. 76652	. 00584	. 86014	. 00725	14
47	. 40568	. 00255	. 54407	. 00350	. 66336	. 00461	. 76816	. 00586	. 86161	. 00727	13
48	. 40818	. 00256	. 54620	. 00352	. 66521	. 00463	. 76981	. 00589	. 86309	. 00730	12
49	. 41067	. 00257	. 54833	. 00353	. 66706	. 00465	. 77145	. 00591	. 86456	. 00732	11
50	7. 41315	0. 00259	7. 55045	0. 00355	7. 66891	0. 00467	7. 77308	0. 00593	7. 86603	0. 00735	10
51	. 41563	. 00260	. 55256	. 00357	. 67075	. 00469	. 77472	. 00595	. 86750	. 00737	9
52	. 41810	. 00262	. 55467	. 00359	. 67259	. 00471	. 77635	. 00598	. 86896	. 00740	8
53	. 42056	. 00263	. 55677	. 00360	. 67443	. 00473	. 77798	. 00600	. 87042	. 00742	7
54	. 42301	. 00265	. 55887	. 00362	. 67626	. 00475	. 77960	. 00602	. 87188	. 00745	6
55	7. 42546	0. 00266	7. 56096	0. 00364	7. 67809	0. 00477	7. 78122	0. 00604	7. 87334	0. 00747	5
56	. 42790	. 00268	. 56305	. 00366	. 67991	. 00479	. 78284	. 00607	. 87480	. 00750	4
57	. 43034	. 00269	. 56513	. 00367	. 68173	. 00481	. 78446	. 00609	. 87625	. 00752	3
58	. 43277	. 00271	. 56721	. 00369	. 68355	. 00483	. 78607	. 00611	. 87770	. 00755	2
59	. 43519	. 00272	. 56928	. 00371	. 68536	. 00485	. 78768	. 00613	. 87915	. 00757	1
60	7. 43760	0. 00274	7. 57135	0. 00373	7. 68717	0. 00487	7. 78929	0. 00616	7. 88059	0. 00760	0
	354°		353°		352°		351°		350°		

TABLE 34

Haversines

	10°		11°		12°		13°		14°		
′	Log Hav	Nat. Hav	Log Hav	Nat. Hav	Log Hav	Nat. Hav	Log Hav	Nat. Hav	Log Hav	Nat. Hav	′
0	7.88059	0.00760	7.96315	0.00919	8.03847	0.01093	8.10772	0.01281	8.17179	0.01485	60
1	.88203	.00762	.96446	.00921	.03967	.01096	.10883	.01285	.17282	.01489	59
2	.88348	.00765	.96577	.00924	.04087	.01099	.10993	.01288	.17384	.01492	58
3	.88491	.00767	.96707	.00927	.04207	.01102	.11104	.01291	.17487	.01496	57
4	.88635	.00770	.96838	.00930	.04326	.01105	.11214	.01295	.17590	.01499	56
5	7.88778	0.00772	7.96968	0.00933	8.04446	0.01108	8.11324	0.01298	8.17692	0.01503	55
6	.88921	.00775	.97098	.00935	.04565	.01111	.11435	.01301	.17794	.01506	54
7	.89064	.00777	.97228	.00938	.04684	.01114	.11544	.01305	.17896	.01510	53
8	.89207	.00780	.97358	.00941	.04803	.01117	.11654	.01308	.17998	.01513	52
9	.89349	.00783	.97487	.00944	.04922	.01120	.11764	.01311	.18100	.01517	51
10	7.89491	0.00785	7.97617	0.00947	8.05041	0.01123	8.11873	0.01314	8.18202	0.01521	50
11	.89633	.00788	.97746	.00949	.05159	.01126	.11983	.01318	.18303	.01524	49
12	.89775	.00790	.97875	.00952	.05277	.01129	.12092	.01321	.18405	.01528	48
13	.89916	.00793	.98003	.00955	.05395	.01132	.12201	.01324	.18506	.01531	47
14	.90057	.00795	.98132	.00958	.05513	.01135	.12310	.01328	.18607	.01535	46
15	7.90198	0.00798	7.98260	0.00961	8.05631	0.01138	8.12419	0.01331	8.18709	0.01538	45
16	.90339	.00801	.98389	.00964	.05749	.01142	.12528	.01334	.18810	.01542	44
17	.90480	.00803	.98517	.00966	.05866	.01145	.12636	.01338	.18910	.01546	43
18	.90620	.00806	.98644	.00969	.05984	.01148	.12745	.01341	.19011	.01549	42
19	.90760	.00808	.98772	.00972	.06101	.01151	.12853	.01344	.19112	.01553	41
20	7.90900	0.00811	7.98899	0.00975	8.06218	0.01154	8.12961	0.01348	8.19212	0.01556	40
21	.91039	.00814	.99027	.00978	.06335	.01157	.13069	.01351	.19313	.01560	39
22	.91179	.00816	.99154	.00981	.06451	.01160	.13177	.01354	.19413	.01564	38
23	.91318	.00819	.99281	.00984	.06568	.01163	.13285	.01358	.19513	.01567	37
24	.91457	.00821	.99407	.00986	.06684	.01166	.13392	.01361	.19613	.01571	36
25	7.91596	0.00824	7.99534	0.00989	8.06800	0.01170	8.13500	0.01365	8.19713	0.01574	35
26	.91734	.00827	.99660	.00992	.06917	.01173	.13607	.01368	.19813	.01578	34
27	.91872	.00829	.99786	.00995	.07032	.01176	.13714	.01371	.19913	.01582	33
28	.92010	.00832	7.99912	.00998	.07148	.01179	.13822	.01375	.20012	.01585	32
29	.92148	.00835	8.00038	.01001	.07264	.01182	.13928	.01378	.20112	.01589	31
30	7.92286	0.00837	8.00163	0.01004	8.07379	0.01185	8.14035	0.01382	8.20211	0.01593	30
31	.92423	.00840	.00289	.01007	.07494	.01188	.14142	.01385	.20310	.01596	29
32	.92560	.00843	.00414	.01010	.07610	.01192	.14248	.01388	.20410	.01600	28
33	.92697	.00845	.00539	.01012	.07725	.01195	.14355	.01392	.20509	.01604	27
34	.92834	.00848	.00664	.01015	.07839	.01198	.14461	.01395	.20608	.01607	26
35	7.92970	0.00851	8.00788	0.01018	8.07954	0.01201	8.14567	0.01399	8.20706	0.01611	25
36	.93107	.00853	.00913	.01021	.08069	.01204	.14673	.01402	.20805	.01615	24
37	.93243	.00856	.01037	.01024	.08183	.01207	.14779	.01405	.20904	.01618	23
38	.93379	.00859	.01161	.01027	.08297	.01211	.14885	.01409	.21002	.01622	22
39	.93514	.00861	.01285	.01030	.08411	.01214	.14991	.01412	.21100	.01626	21
40	7.93650	0.00864	8.01409	0.01033	8.08525	0.01217	8.15096	0.01416	8.21199	0.01629	20
41	.93785	.00867	.01532	.01036	.08639	.01220	.15201	.01419	.21297	.01633	19
42	.93920	.00869	.01656	.01039	.08752	.01223	.15307	.01423	.21395	.01637	18
43	.94055	.00872	.01779	.01042	.08866	.01226	.15412	.01426	.21493	.01640	17
44	.94189	.00875	.01902	.01045	.08979	.01230	.15517	.01429	.21590	.01644	16
45	7.94324	0.00877	8.02025	0.01048	8.09092	0.01233	8.15622	0.01433	8.21688	0.01648	15
46	.94458	.00880	.02148	.01051	.09205	.01236	.15726	.01436	.21785	.01651	14
47	.94592	.00883	.02270	.01054	.09318	.01239	.15831	.01440	.21883	.01655	13
48	.94726	.00886	.02392	.01057	.09431	.01243	.15935	.01443	.21980	.01659	12
49	.94859	.00888	.02515	.01060	.09543	.01246	.16040	.01447	.22077	.01663	11
50	7.94992	0.00891	8.02637	0.01063	8.09656	0.01249	8.16144	0.01450	8.22175	0.01666	10
51	.95126	.00894	.02758	.01066	.09768	.01252	.16248	.01454	.22272	.01670	9
52	.95259	.00897	.02880	.01069	.09880	.01255	.16352	.01457	.22368	.01674	8
53	.95391	.00899	.03001	.01072	.09992	.01259	.16456	.01461	.22465	.01677	7
54	.95524	.00902	.03123	.01075	.10104	.01262	.16559	.01464	.22562	.01681	6
55	7.95656	0.00905	8.03244	0.01078	8.10216	0.01265	8.16663	0.01468	8.22658	0.01685	5
56	.95788	.00908	.03365	.01081	.10327	.01268	.16766	.01471	.22755	.01689	4
57	.95920	.00910	.03486	.01084	.10439	.01272	.16870	.01475	.22851	.01692	3
58	.96052	.00913	.03606	.01087	.10550	.01275	.16973	.01478	.22947	.01696	2
59	.96183	.00916	.03727	.01090	.10661	.01278	.17076	.01482	.23044	.01700	1
60	7.96315	0.00919	8.03847	0.01093	8.10772	0.01281	8.17179	0.01485	8.23140	0.01704	0
	349°		348°		347°		346°		345°		

TABLE 34

Haversines

′	15°		16°		17°		18°		19°		′
	Log Hav	Nat. Hav	Log Hav	Nat. Hav	Log Hav	Nat. Hav	Log Hav	Nat. Hav	Log Hav	Nat. Hav	
0	8.23140	0.01704	8.28711	0.01937	8.33940	0.02185	8.38867	0.02447	8.43522	0.02724	60
1	.23235	.01707	.28801	.01941	.34025	.02189	.38946	.02452	.43597	.02729	59
2	.23331	.01711	.28891	.01945	.34109	.02193	.39026	.02456	.43673	.02734	58
3	.23427	.01715	.28980	.01949	.34194	.02198	.39105	.02461	.43748	.02738	57
4	.23523	.01719	.29070	.01953	.34278	.02202	.39185	.02465	.43823	.02743	56
5	8.23618	0.01723	8.29159	0.01957	8.34362	0.02206	8.39264	0.02470	8.43899	0.02748	55
6	.23713	.01726	.29249	.01961	.34446	.02210	.39344	.02474	.43974	.02753	54
7	.23809	.01730	.29338	.01965	.34530	.02215	.39423	.02479	.44049	.02757	53
8	.23904	.01734	.29427	.01969	.34614	.02219	.39502	.02483	.44124	.02762	52
9	.23999	.01738	.29516	.01973	.34698	.02223	.39581	.02488	.44199	.02767	51
10	8.24094	0.01742	8.29605	0.01977	8.34782	0.02227	8.39660	0.02492	8.44273	0.02772	50
11	.24189	.01745	.29694	.01981	.34865	.02232	.39739	.02497	.44348	.02776	49
12	.24283	.01749	.29783	.01985	.34949	.02236	.39818	.02501	.44423	.02781	48
13	.24378	.01753	.29872	.01989	.35032	.02240	.39897	.02506	.44498	.02786	47
14	.24473	.01757	.29960	.01993	.35116	.02245	.39976	.02510	.44572	.02791	46
15	8.24567	0.01761	8.30049	0.01998	8.35199	0.02249	8.40055	0.02515	8.44647	0.02796	45
16	.24661	.01764	.30137	.02002	.35282	.02253	.40133	.02520	.44721	.02800	44
17	.24755	.01768	.30226	.02006	.35365	.02258	.40212	.02524	.44796	.02805	43
18	.24850	.01772	.30314	.02010	.35449	.02262	.40290	.02529	.44870	.02810	42
19	.24944	.01776	.30402	.02014	.35532	.02266	.40369	.02533	.44944	.02815	41
20	8.25037	0.01780	8.30490	0.02018	8.35614	0.02271	8.40447	0.02538	8.45018	0.02820	40
21	.25131	.01784	.30578	.02022	.35697	.02275	.40525	.02542	.45093	.02824	39
22	.25225	.01788	.30666	.02026	.35780	.02279	.40603	.02547	.45167	.02829	38
23	.25319	.01791	.30754	.02030	.35863	.02284	.40681	.02552	.45241	.02834	37
24	.25412	.01795	.30842	.02034	.35945	.02288	.40760	.02556	.45315	.02839	36
25	8.25505	0.01799	8.30929	0.02038	8.36028	0.02292	8.40837	0.02561	8.45388	0.02844	35
26	.25599	.01803	.31017	.02043	.36110	.02297	.40915	.02565	.45462	.02849	34
27	.25692	.01807	.31104	.02047	.36193	.02301	.40993	.02570	.45536	.02853	33
28	.25785	.01811	.31192	.02051	.36275	.02305	.41071	.02575	.45610	.02858	32
29	.25878	.01815	.31279	.02055	.36357	.02310	.41149	.02579	.45683	.02863	31
30	8.25971	0.01818	8.31366	0.02059	8.36439	0.02314	8.41226	0.02584	8.45757	0.02868	30
31	.26064	.01822	.31453	.02063	.36521	.02319	.41304	.02588	.45830	.02873	29
32	.26156	.01826	.31540	.02067	.36603	.02323	.41381	.02593	.45904	.02878	28
33	.26249	.01830	.31627	.02071	.36685	.02327	.41459	.02598	.45977	.02883	27
34	.26341	.01834	.31714	.02076	.36767	.02332	.41536	.02602	.46050	.02887	26
35	8.26434	0.01838	8.31800	0.02080	8.36849	0.02336	8.41613	0.02607	8.46124	0.02892	25
36	.26526	.01842	.31887	.02084	.36930	.02340	.41690	.02612	.46197	.02897	24
37	.26618	.01846	.31974	.02088	.37012	.02345	.41767	.02616	.46270	.02902	23
38	.26710	.01850	.32060	.02092	.37093	.02349	.41845	.02621	.46343	.02907	22
39	.26802	.01854	.32147	.02096	.37175	.02354	.41921	.02626	.46416	.02912	21
40	8.26894	0.01858	8.32233	0.02101	8.37256	0.02358	8.41998	0.02630	8.46489	0.02917	20
41	.26986	.01861	.32319	.02105	.37337	.02363	.42075	.02635	.46562	.02922	19
42	.27078	.01865	.32405	.02109	.37419	.02367	.42152	.02639	.46634	.02926	18
43	.27169	.01869	.32491	.02113	.37500	.02371	.42229	.02644	.46707	.02931	17
44	.27261	.01873	.32577	.02117	.37581	.02376	.42305	.02649	.46780	.02936	16
45	8.27352	0.01877	8.32663	0.02121	8.37662	0.02380	8.42382	0.02653	8.46852	0.02941	15
46	.27443	.01881	.32749	.02126	.37742	.02385	.42458	.02658	.46925	.02946	14
47	.27534	.01885	.32834	.02130	.37823	.02389	.42535	.02663	.46998	.02951	13
48	.27626	.01889	.32920	.02134	.37904	.02394	.42611	.02668	.47070	.02956	12
49	.27717	.01893	.33006	.02138	.37985	.02398	.42687	.02672	.47142	.02961	11
50	8.27807	0.01897	8.33091	0.02142	8.38065	0.02402	8.42764	0.02677	8.47215	0.02966	10
51	.27898	.01901	.33176	.02147	.38146	.02407	.42840	.02682	.47287	.02971	9
52	.27989	.01905	.33262	.02151	.38226	.02411	.42916	.02686	.47359	.02976	8
53	.28080	.01909	.33347	.02155	.38306	.02416	.42992	.02691	.47431	.02981	7
54	.28170	.01913	.33432	.02159	.38387	.02420	.43068	.02696	.47503	.02986	6
55	8.28260	0.01917	8.33517	0.02164	8.38467	0.02425	8.43144	0.02700	8.47575	0.02991	5
56	.28351	.01921	.33602	.02168	.38547	.02429	.43219	.02705	.47647	.02996	4
57	.28441	.01925	.33686	.02172	.38627	.02434	.43295	.02710	.47719	.03000	3
58	.28531	.01929	.33771	.02176	.38707	.02438	.43371	.02715	.47791	.03005	2
59	.28621	.01933	.33856	.02181	.38787	.02443	.43446	.02719	.47862	.03010	1
60	8.28711	0.01937	8.33940	0.02185	8.38867	0.02447	8.43522	0.02724	8.47934	0.03015	0
	344°		343°		342°		341°		340°		

TABLE 34

Haversines

′	20°		21°		22°		23°		24°		′
	Log Hav	Nat. Hav	Log Hav	Nat. Hav	Log Hav	Nat. Hav	Log Hav	Nat. Hav	Log Hav	Nat. Hav	
0	8. 47934	0. 03015	8. 52127	0. 03321	8. 56120	0. 03641	8. 59931	0. 03975	8. 63576	0. 04323	60
1	. 48006	. 03020	. 52195	. 03326	. 56185	. 03646	. 59993	. 03980	. 63635	. 04329	59
2	. 48077	. 03025	. 52263	. 03331	. 56250	. 03652	. 60055	. 03986	. 63695	. 04335	58
3	. 48149	. 03030	. 52331	. 03337	. 56315	. 03657	. 60117	. 03992	. 63754	. 04340	57
4	. 48220	. 03035	. 52399	. 03342	. 56379	. 03663	. 60179	. 03998	. 63813	. 04346	56
5	8. 48292	0. 03040	8. 52467	0. 03347	8. 56444	0. 03668	8. 60241	0. 04003	8. 63872	0. 04352	55
6	. 48363	. 03045	. 52535	. 03352	. 56509	. 03674	. 60303	. 04009	. 63932	. 04358	54
7	. 48434	. 03050	. 52602	. 03358	. 56574	. 03679	. 60365	. 04015	. 63991	. 04364	53
8	. 48505	. 03055	. 52670	. 03363	. 56638	. 03685	. 60426	. 04020	. 64050	. 04370	52
9	. 48576	. 03060	. 52738	. 03368	. 56703	. 03690	. 60488	. 04026	. 64109	. 04376	51
10	8. 48648	0. 03065	8. 52806	0. 03373	8. 56767	0. 03695	8. 60550	0. 04032	8. 64168	0. 04382	50
11	. 48719	. 03070	. 52873	. 03379	. 56832	. 03701	. 60611	. 04038	. 64227	. 04388	49
12	. 48789	. 03075	. 52941	. 03384	. 56896	. 03706	. 60673	. 04043	. 64286	. 04394	48
13	. 48860	. 03080	. 53008	. 03389	. 56960	. 03712	. 60734	. 04049	. 64345	. 04400	47
14	. 48931	. 03085	. 53076	. 03394	. 57025	. 03717	. 60796	. 04055	. 64404	. 04406	46
15	8. 49002	0. 03090	8. 53143	0. 03400	8. 57089	0. 03723	8. 60857	0. 04060	8. 64463	0. 04412	45
16	. 49073	. 03095	. 53210	. 03405	. 57153	. 03728	. 60919	. 04066	. 64521	. 04418	44
17	. 49143	. 03101	. 53277	. 03410	. 57217	. 03734	. 60980	. 04072	. 64580	. 04424	43
18	. 49214	. 03106	. 53345	. 03415	. 57282	. 03740	. 61041	. 04078	. 64639	. 04430	42
19	. 49284	. 03111	. 53412	. 03421	. 57346	. 03745	. 61103	. 04083	. 64697	. 04436	41
20	8. 49355	0. 03116	8. 53479	0. 03426	8. 57410	0. 03751	8. 61164	0. 04089	8. 64756	0. 04442	40
21	. 49425	. 03121	. 53546	. 03431	. 57474	. 03756	. 61225	. 04095	. 64815	. 04448	39
22	. 49496	. 03126	. 53613	. 03437	. 57538	. 03762	. 61286	. 04101	. 64873	. 04454	38
23	. 49566	. 03131	. 53680	. 03442	. 57601	. 03767	. 61347	. 04106	. 64932	. 04460	37
24	. 49636	. 03136	. 53747	. 03447	. 57665	. 03773	. 61408	. 04112	. 64990	. 04466	36
25	8. 49706	0. 03141	8. 53814	0. 03453	8. 57729	0. 03778	8. 61469	0. 04118	8. 65049	0. 04472	35
26	. 49777	. 03146	. 53880	. 03458	. 57793	. 03784	. 61530	. 04124	. 65107	. 04478	34
27	. 49847	. 03151	. 53947	. 03463	. 57856	. 03789	. 61591	. 04130	. 65165	. 04484	33
28	. 49917	. 03156	. 54014	. 03468	. 57920	. 03795	. 61652	. 04135	. 65224	. 04490	32
29	. 49987	. 03161	. 54080	. 03474	. 57984	. 03800	. 61713	. 04141	. 65282	. 04496	31
30	8. 50056	0. 03166	8. 54147	0. 03479	8. 58047	0. 03806	8. 61773	0. 04147	8. 65340	0. 04502	30
31	. 50126	. 03171	. 54214	. 03484	. 58111	. 03812	. 61834	. 04153	. 65398	. 04508	29
32	. 50196	. 03177	. 54280	. 03490	. 58174	. 03817	. 61895	. 04159	. 65456	. 04514	28
33	. 50266	. 03182	. 54346	. 03495	. 58238	. 03823	. 61955	. 04164	. 65514	. 04520	27
34	. 50335	. 03187	. 54413	. 03500	. 58301	. 03828	. 62016	. 04170	. 65572	. 04526	26
35	8. 50405	0. 03192	8. 54479	0. 03506	8. 58364	0. 03834	8. 62077	0. 04176	8. 65630	0. 04532	25
36	. 50475	. 03197	. 54545	. 03511	. 58427	. 03839	. 62137	. 04182	. 65688	. 04538	24
37	. 50544	. 03202	. 54612	. 03517	. 58491	. 03845	. 62197	. 04188	. 65746	. 04544	23
38	. 50614	. 03207	. 54678	. 03522	. 58554	. 03851	. 62258	. 04194	. 65804	. 04550	22
39	. 50683	. 03212	. 54744	. 03527	. 58617	. 03856	. 62318	. 04199	. 65862	. 04556	21
40	8. 50752	0. 03218	8. 54810	0. 03533	8. 58680	0. 03862	8. 62379	0. 04205	8. 65920	0. 04562	20
41	. 50821	. 03223	. 54876	. 03538	. 58743	. 03867	. 62439	. 04211	. 65978	. 04569	19
42	. 50891	. 03228	. 54942	. 03543	. 58806	. 03873	. 62499	. 04217	. 66035	. 04575	18
43	. 50960	. 03233	. 55008	. 03549	. 58869	. 03879	. 62559	. 04223	. 66093	. 04581	17
44	. 51029	. 03238	. 55073	. 03554	. 58932	. 03884	. 62619	. 04229	. 66151	. 04587	16
45	8. 51098	0. 03243	8. 55139	0. 03560	8. 58994	0. 03890	8. 62680	0. 04234	8. 66208	0. 04593	15
46	. 51167	. 03248	. 55205	. 03565	. 59057	. 03896	. 62740	. 04240	. 66266	. 04599	14
47	. 51236	. 03254	. 55271	. 03570	. 59120	. 03901	. 62800	. 04246	. 66323	. 04605	13
48	. 51305	. 03259	. 55336	. 03576	. 59183	. 03907	. 62860	. 04252	. 66381	. 04611	12
49	. 51374	. 03264	. 55402	. 03581	. 59245	. 03912	. 62919	. 04258	. 66438	. 04617	11
50	8. 51442	0. 03269	8. 55467	0. 03587	8. 59308	0. 03918	8. 62979	0. 04264	8. 66496	0. 04623	10
51	. 51511	. 03274	. 55533	. 03592	. 59370	. 03924	. 63039	. 04270	. 66553	. 04629	9
52	. 51580	. 03279	. 55598	. 03597	. 59433	. 03929	. 63099	. 04276	. 66610	. 04636	8
53	. 51648	. 03285	. 55664	. 03603	. 59495	. 03935	. 63159	. 04281	. 66668	. 04642	7
54	. 51717	. 03290	. 55729	. 03608	. 59558	. 03941	. 63218	. 04287	. 66725	. 04648	6
55	8. 51785	0. 03295	8. 55794	0. 03614	8. 59620	0. 03946	8. 63278	0. 04293	8. 66782	0. 04654	5
56	. 51854	. 03300	. 55859	. 03619	. 59682	. 03952	. 63338	. 04299	. 66839	. 04660	4
57	. 51922	. 03305	. 55925	. 03624	. 59745	. 03958	. 63397	. 04305	. 66896	. 04666	3
58	. 51990	. 03311	. 55990	. 03630	. 59807	. 03963	. 63457	. 04311	. 66953	. 04672	2
59	. 52058	. 03316	. 56055	. 03635	. 59869	. 03969	. 63516	. 04317	. 67010	. 04678	1
60	8. 52127	0. 03321	8. 56120	0. 03641	8. 59931	0. 03975	8. 63576	0. 04323	8. 67067	0. 04685	0
	339°		338°		337°		336°		335°		

TABLE 34

Haversines

′	25° Log Hav	25° Nat. Hav	26° Log Hav	26° Nat. Hav	27° Log Hav	27° Nat. Hav	28° Log Hav	28° Nat. Hav	29° Log Hav	29° Nat. Hav	′
0	8. 67067	0. 04685	8. 70418	0. 05060	8. 73637	0. 05450	8. 76735	0. 05853	8. 79720	0. 06269	60
1	. 67124	. 04691	. 70472	. 05067	. 73690	. 05456	. 76786	. 05859	. 79769	. 06276	59
2	. 67181	. 04697	. 70527	. 05073	. 73742	. 05463	. 76836	. 05866	. 79818	. 06283	58
3	. 67238	. 04703	. 70582	. 05079	. 73795	. 05469	. 76887	. 05873	. 79866	. 06290	57
4	. 67295	. 04709	. 70636	. 05086	. 73847	. 05476	. 76938	. 05880	. 79915	. 06297	56
5	8. 67352	0. 04715	8. 70691	0. 05092	8. 73900	0. 05483	8. 76988	0. 05887	8. 79964	0. 06304	55
6	. 67409	. 04722	. 70745	. 05099	. 73952	. 05489	. 77039	. 05894	. 80013	. 06311	54
7	. 67465	. 04728	. 70800	. 05105	. 74005	. 05496	. 77089	. 05901	. 80061	. 06318	53
8	. 67522	. 04734	. 70854	. 05111	. 74057	. 05503	. 77139	. 05907	. 80110	. 06326	52
9	. 67579	. 04740	. 70909	. 05118	. 74109	. 05509	. 77190	. 05914	. 80158	. 06333	51
10	8. 67635	0. 04746	8. 70963	0. 05124	8. 74162	0. 05516	8. 77240	0. 05921	8. 80207	0. 06340	50
11	. 67692	. 04752	. 71017	. 05131	. 74214	. 05523	. 77291	. 05928	. 80256	. 06347	49
12	. 67748	. 04759	. 71072	. 05137	. 74266	. 05529	. 77341	. 05935	. 80304	. 06354	48
13	. 67805	. 04765	. 71126	. 05144	. 74318	. 05536	. 77391	. 05942	. 80353	. 06361	47
14	. 67861	. 04771	. 71180	. 05150	. 74371	. 05542	. 77441	. 05949	. 80401	. 06368	46
15	8. 67918	0. 04777	8. 71234	0. 05156	8. 74423	0. 05549	8. 77492	0. 05955	8. 80449	0. 06375	45
16	. 67974	. 04783	. 71289	. 05163	. 74475	. 05556	. 77542	. 05962	. 80498	. 06382	44
17	. 68030	. 04790	. 71343	. 05169	. 74527	. 05562	. 77592	. 05969	. 80546	. 06389	43
18	. 68087	. 04796	. 71397	. 05176	. 74579	. 05569	. 77642	. 05976	. 80595	. 06397	42
19	. 68143	. 04802	. 71451	. 05182	. 74631	. 05576	. 77692	. 05983	. 80643	. 06404	41
20	8. 68199	0. 04808	8. 71505	0. 05189	8. 74683	0. 05582	8. 77742	0. 05990	8. 80691	0. 06411	40
21	. 68256	. 04815	. 71559	. 05195	. 74735	. 05589	. 77792	. 05997	. 80739	. 06418	39
22	. 68312	. 04821	. 71613	. 05201	. 74787	. 05596	. 77842	. 06004	. 80788	. 06425	38
23	. 68368	. 04827	. 71667	. 05208	. 74839	. 05603	. 77892	. 06011	. 80836	. 06432	37
24	. 68424	. 04833	. 71721	. 05214	. 74890	. 05609	. 77942	. 06018	. 80884	. 06439	36
25	8. 68480	0. 04839	8. 71774	0. 05221	8. 74942	0. 05616	8. 77992	0. 06024	8. 80932	0. 06446	35
26	. 68536	. 04846	. 71828	. 05227	. 74994	. 05623	. 78042	. 06031	. 80980	. 06454	34
27	. 68592	. 04852	. 71882	. 05234	. 75046	. 05629	. 78092	. 06038	. 81028	. 06461	33
28	. 68648	. 04858	. 71936	. 05240	. 75097	. 05636	. 78142	. 06045	. 81076	. 06468	32
29	. 68704	. 04864	. 71989	. 05247	. 75149	. 05643	. 78191	. 06052	. 81124	. 06475	31
30	8. 68760	0. 04871	8. 72043	0. 05253	8. 75201	0. 05649	8. 78241	0. 06059	8. 81172	0. 06482	30
31	. 68815	. 04877	. 72097	. 05260	. 75252	. 05656	. 78291	. 06066	. 81220	. 06489	29
32	. 68871	. 04883	. 72150	. 05266	. 75304	. 05663	. 78341	. 06073	. 81268	. 06497	28
33	. 68927	. 04890	. 72204	. 05273	. 75355	. 05670	. 78390	. 06080	. 81316	. 06504	27
34	. 68983	. 04896	. 72257	. 05279	. 75407	. 05676	. 78440	. 06087	. 81364	. 06511	26
35	8. 69038	0. 04902	8. 72311	0. 05286	8. 75458	0. 05683	8. 78490	0. 06094	8. 81412	0. 06518	25
36	. 69094	. 04908	. 72364	. 05292	. 75510	. 05690	. 78539	. 06101	. 81460	. 06525	24
37	. 69149	. 04915	. 72418	. 05299	. 75561	. 05697	. 78589	. 06108	. 81508	. 06532	23
38	. 69205	. 04921	. 72471	. 05305	. 75613	. 05703	. 78638	. 06115	. 81555	. 06540	22
39	. 69260	. 04927	. 72525	. 05312	. 75664	. 05710	. 78688	. 06122	. 81603	. 06547	21
40	8. 69316	0. 04934	8. 72578	0. 05318	8. 75715	0. 05717	8. 78737	0. 06129	8. 81651	0. 06554	20
41	. 69371	. 04940	. 72631	. 05325	. 75767	. 05724	. 78787	. 06136	. 81699	. 06561	19
42	. 69427	. 04946	. 72684	. 05331	. 75818	. 05730	. 78836	. 06143	. 81746	. 06568	18
43	. 69482	. 04952	. 72738	. 05338	. 75869	. 05737	. 78885	. 06150	. 81794	. 06576	17
44	. 69537	. 04959	. 72791	. 05345	. 75920	. 05744	. 78935	. 06157	. 81841	. 06583	16
45	8. 69593	0. 04965	8. 72844	0. 05351	8. 75972	0. 05751	8. 78984	0. 06164	8. 81889	0. 06590	15
46	. 69648	. 04971	. 72897	. 05358	. 76023	. 05757	. 79033	. 06171	. 81937	. 06597	14
47	. 69703	. 04978	. 72950	. 05364	. 76074	. 05764	. 79082	. 06178	. 81984	. 06605	13
48	. 69758	. 04984	. 73003	. 05371	. 76125	. 05771	. 79132	. 06185	. 82032	. 06612	12
49	. 69814	. 04990	. 73056	. 05377	. 76176	. 05778	. 79181	. 06192	. 82079	. 06619	11
50	8. 69869	0. 04997	8. 73109	0. 05384	8. 76227	0. 05785	8. 79230	0. 06199	8. 82126	0. 06626	10
51	. 69924	. 05003	. 73162	. 05390	. 76278	. 05791	. 79279	. 06206	. 82174	. 06633	9
52	. 69979	. 05009	. 73215	. 05397	. 76329	. 05798	. 79328	. 06213	. 82221	. 06641	8
53	. 70034	. 05016	. 73268	. 05404	. 76380	. 05805	. 79377	. 06220	. 82269	. 06648	7
54	. 70089	. 05022	. 73321	. 05410	. 76431	. 05812	. 79426	. 06227	. 82316	. 06655	6
55	8. 70144	0. 05028	8. 73374	0. 05417	8. 76481	0. 05819	8. 79475	0. 06234	8. 82363	0. 06662	5
56	. 70198	. 05035	. 73426	. 05423	. 76532	. 05825	. 79524	. 06241	. 82410	. 06670	4
57	. 70253	. 05041	. 73479	. 05430	. 76583	. 05832	. 79573	. 06248	. 82458	. 06677	3
58	. 70308	. 05048	. 73532	. 05436	. 76634	. 05839	. 79622	. 06255	. 82505	. 06684	2
59	. 70363	. 05054	. 73584	. 05443	. 76684	. 05846	. 79671	. 06262	. 82552	. 06691	1
60	8. 70418	0. 05060	8. 73637	0. 05450	8. 76735	0. 05853	8. 79720	0. 06269	8. 82599	0. 06699	0
	334°		333°		332°		331°		330°		

TABLE 34

Haversines

′	30°		31°		32°		33°		34°		
	Log Hav	Nat. Hav	Log Hav	Nat. Hav	Log Hav	Nat. Hav	Log Hav	Nat. Hav	Log Hav	Nat. Hav	
0	8.82599	0.06699	8.85380	0.07142	8.88068	0.07598	8.90668	0.08066	8.93187	0.08548	60
1	.82646	.06706	.85425	.07149	.88112	.07605	.90711	.08074	.93228	.08556	59
2	.82694	.06713	.85471	.07157	.88156	.07613	.90754	.08082	.93270	.08564	58
3	.82741	.06721	.85516	.07164	.88200	.07621	.90796	.08090	.93311	.08573	57
4	.82788	.06728	.85562	.07172	.88244	.07628	.90839	.08098	.93352	.08581	56
5	8.82835	0.06735	8.85607	0.07179	8.88288	0.07636	8.90881	0.08106	8.93393	0.08589	55
6	.82882	.06742	.85653	.07187	.88332	.07644	.90924	.08114	.93435	.08597	54
7	.82929	.06750	.85698	.07194	.88375	.07652	.90966	.08122	.93476	.08605	53
8	.82976	.06757	.85743	.07202	.88419	.07659	.91009	.08130	.93517	.08613	52
9	.83023	.06764	.85789	.07209	.88463	.07667	.91051	.08138	.93558	.08621	51
10	8.83069	0.06772	8.85834	0.07217	8.88507	0.07675	8.91094	0.08146	8.93599	0.08630	50
11	.83116	.06779	.85879	.07224	.88551	.07683	.91136	.08154	.93640	.08638	49
12	.83163	.06786	.85925	.07232	.88595	.07690	.91179	.08162	.93681	.08646	48
13	.83210	.06794	.85970	.07239	.88638	.07698	.91221	.08170	.93722	.08654	47
14	.83257	.06801	.86015	.07247	.88682	.07706	.91263	.08178	.93764	.08662	46
15	8.83303	0.06808	8.86060	0.07254	8.88726	0.07714	8.91306	0.08186	8.93805	0.08671	45
16	.83350	.06816	.86105	.07262	.88769	.07721	.91348	.08194	.93846	.08679	44
17	.83397	.06823	.86151	.07270	.88813	.07729	.91390	.08202	.93886	.08687	43
18	.83444	.06830	.86196	.07277	.88857	.07737	.91432	.08210	.93927	.08695	42
19	.83490	.06838	.86241	.07285	.88900	.07745	.91475	.08218	.93968	.08703	41
20	8.83537	0.06845	8.86286	0.07292	8.88944	0.07752	8.91517	0.08226	8.94009	0.08711	40
21	.83583	.06852	.86331	.07300	.88988	.07760	.91559	.08234	.94050	.08720	39
22	.83630	.06860	.86376	.07307	.89031	.07768	.91601	.08242	.94091	.08728	38
23	.83676	.06867	.86421	.07315	.89075	.07776	.91643	.08250	.94132	.08736	37
24	.83723	.06874	.86466	.07322	.89118	.07784	.91685	.08258	.94173	.08744	36
25	8.83769	0.06882	8.86511	0.07330	8.89162	0.07791	8.91728	0.08266	8.94213	0.08753	35
26	.83816	.06889	.86556	.07338	.89205	.07799	.91770	.08274	.94254	.08761	34
27	.83862	.06896	.86600	.07345	.89248	.07807	.91812	.08282	.94295	.08769	33
28	.83909	.06904	.86645	.07353	.89292	.07815	.91854	.08290	.94336	.08777	32
29	.83955	.06911	.86690	.07360	.89335	.07823	.91896	.08298	.94376	.08785	31
30	8.84002	0.06919	8.86735	0.07368	8.89379	0.07830	8.91938	0.08306	8.94417	0.08794	30
31	.84048	.06926	.86780	.07376	.89422	.07838	.91980	.08314	.94458	.08802	29
32	.84094	.06933	.86825	.07383	.89465	.07846	.92022	.08322	.94498	.08810	28
33	.84140	.06941	.86869	.07391	.89509	.07854	.92064	.08330	.94539	.08818	27
34	.84187	.06948	.86914	.07398	.89552	.07862	.92105	.08338	.94580	.08827	26
35	8.84233	0.06955	8.86959	0.07406	8.89595	0.07870	8.92147	0.08346	8.94620	0.08835	25
36	.84279	.06963	.87003	.07414	.89638	.07877	.92189	.08354	.94661	.08843	24
37	.84325	.06970	.87048	.07421	.89681	.07885	.92231	.08362	.94701	.08851	23
38	.84371	.06978	.87093	.07429	.89725	.07893	.92273	.08370	.94742	.08860	22
39	.84417	.06985	.87137	.07437	.89768	.07901	.92315	.08378	.94782	.08868	21
40	8.84464	0.06993	8.87182	0.07444	8.89811	0.07909	8.92356	0.08386	8.94823	0.08876	20
41	.84510	.07000	.87226	.07452	.89854	.07917	.92398	.08394	.94863	.08885	19
42	.84556	.07007	.87271	.07459	.89897	.07924	.92440	.08402	.94904	.08893	18
43	.84602	.07015	.87315	.07467	.89940	.07932	.92482	.08410	.94944	.08901	17
44	.84648	.07022	.87360	.07475	.89983	.07940	.92523	.08418	.94985	.08909	16
45	8.84694	0.07030	8.87404	0.07482	8.90026	0.07948	8.92565	0.08427	8.95025	0.08918	15
46	.84740	.07037	.87448	.07490	.90069	.07956	.92607	.08435	.95065	.08926	14
47	.84785	.07045	.87493	.07498	.90112	.07964	.92648	.08443	.95106	.08934	13
48	.84831	.07052	.87537	.07505	.90155	.07972	.92690	.08451	.95146	.08943	12
49	.84877	.07059	.87582	.07513	.90198	.07980	.92731	.08459	.95186	.08951	11
50	8.84923	0.07067	8.87626	0.07521	8.90241	0.07987	8.92773	0.08467	8.95227	0.08959	10
51	.84969	.07074	.87670	.07528	.90284	.07995	.92814	.08475	.95267	.08967	9
52	.85015	.07082	.87714	.07536	.90326	.08003	.92856	.08483	.95307	.08976	8
53	.85060	.07089	.87759	.07544	.90369	.08011	.92897	.08491	.95347	.08984	7
54	.85106	.07097	.87803	.07551	.90412	.08019	.92939	.08499	.95388	.08992	6
55	8.85152	0.07104	8.87847	0.07559	8.90455	0.08027	8.92980	0.08507	8.95428	0.09001	5
56	.85197	.07112	.87891	.07567	.90498	.08035	.93022	.08516	.95468	.09009	4
57	.85243	.07119	.87935	.07574	.90540	.08043	.93063	.08524	.95508	.09017	3
58	.85289	.07127	.87980	.07582	.90583	.08051	.93104	.08532	.95548	.09026	2
59	.85334	.07134	.88024	.07590	.90626	.08059	.93146	.08540	.95588	.09034	1
60	8.85380	0.07142	8.88068	0.07598	8.90668	0.08066	8.93187	0.08548	8.95628	0.09042	0
	329°		328°		327°		326°		325°		

TABLE 34

Haversines

	35°		36°		37°		38°		39°		
′	Log Hav	Nat. Hav	Log Hav	Nat. Hav	Log Hav	Nat. Hav	Log Hav	Nat. Hav	Log Hav	Nat. Hav	′
0	8. 95628	0. 09042	8. 97997	0. 09549	9. 00295	0. 10068	9. 02528	0. 10599	9. 04699	0. 11143	60
1	. 95668	. 09051	. 98035	. 09558	. 00333	. 10077	. 02565	. 10608	. 04735	. 11152	59
2	. 95709	. 09059	. 98074	. 09566	. 00371	. 10086	. 02602	. 10617	. 04770	. 11161	58
3	. 95749	. 09067	. 98113	. 09575	. 00408	. 10094	. 02638	. 10626	. 04806	. 11170	57
4	. 95789	. 09076	. 98152	. 09583	. 00446	. 10103	. 02675	. 10635	. 04842	. 11179	56
5	8. 95828	0. 09084	8. 98191	0. 09592	9. 00484	0. 10112	9. 02712	0. 10644	9. 04877	0. 11189	55
6	. 95868	. 09093	. 98229	. 09601	. 00522	. 10121	. 02748	. 10653	. 04913	. 11198	54
7	. 95908	. 09101	. 98268	. 09609	. 00559	. 10130	. 02785	. 10662	. 04948	. 11207	53
8	. 95948	. 09109	. 98307	. 09618	. 00597	. 10138	. 02821	. 10671	. 04984	. 11216	52
9	. 95988	. 09118	. 98346	. 09626	. 00634	. 10147	. 02858	. 10680	. 05019	. 11225	51
10	8. 96028	0. 09126	8. 98384	0. 09635	9. 00672	0. 10156	9. 02894	0. 10689	9. 05055	0. 11234	50
11	. 96068	. 09134	. 98423	. 09643	. 00710	. 10165	. 02931	. 10698	. 05090	. 11244	49
12	. 96108	. 09143	. 98462	. 09652	. 00747	. 10174	. 02967	. 10707	. 05126	. 11253	48
13	. 96148	. 09151	. 98500	. 09661	. 00785	. 10182	. 03004	. 10716	. 05161	. 11262	47
14	. 96187	. 09160	. 98539	. 09669	. 00822	. 10191	. 03040	. 10725	. 05197	. 11271	46
15	8. 96227	0. 09168	8. 98578	0. 09678	9. 00860	0. 10200	9. 03077	0. 10734	9. 05232	0. 11280	45
16	. 96267	. 09176	. 98616	. 09686	. 00897	. 10209	. 03113	. 10743	. 05268	. 11290	44
17	. 96307	. 09185	. 98655	. 09695	. 00935	. 10218	. 03150	. 10752	. 05303	. 11299	43
18	. 96346	. 09193	. 98693	. 09704	. 00972	. 10226	. 03186	. 10761	. 05339	. 11308	42
19	. 96386	. 09202	. 98732	. 09712	. 01009	. 10235	. 03222	. 10770	. 05374	. 11317	41
20	8. 96426	0. 09210	8. 98770	0. 09721	9. 01047	0. 10244	9. 03259	0. 10779	9. 05409	0. 11326	40
21	. 96465	. 09218	. 98809	. 09729	. 01084	. 10253	. 03295	. 10788	. 05445	. 11336	39
22	. 96505	. 09227	. 98847	. 09738	. 01122	. 10262	. 03331	. 10797	. 05480	. 11345	38
23	. 96545	. 09235	. 98886	. 09747	. 01159	. 10270	. 03368	. 10806	. 05515	. 11354	37
24	. 96584	. 09244	. 98924	. 09755	. 01196	. 10279	. 03404	. 10815	. 05551	. 11363	36
25	8. 96624	0. 09252	8. 98963	0. 09764	9. 01234	0. 10288	9. 03440	0. 10824	9. 05586	0. 11373	35
26	. 96663	. 09260	. 99001	. 09773	. 01271	. 10297	. 03476	. 10833	. 05621	. 11382	34
27	. 96703	. 09269	. 99039	. 09781	. 01308	. 10306	. 03513	. 10842	. 05656	. 11391	33
28	. 96742	. 09277	. 99078	. 09790	. 01345	. 10315	. 03549	. 10851	. 05692	. 11400	32
29	. 96782	. 09286	. 99116	. 09799	. 01383	. 10323	. 03585	. 10861	. 05727	. 11410	31
30	8. 96821	0. 09294	8. 99154	0. 09807	9. 01420	0. 10332	9. 03621	0. 10870	9. 05762	0. 11419	30
31	. 96861	. 09303	. 99193	. 09816	. 01457	. 10341	. 03657	. 10879	. 05797	. 11428	29
32	. 96900	. 09311	. 99231	. 09824	. 01494	. 10350	. 03694	. 10888	. 05832	. 11437	28
33	. 96940	. 09320	. 99269	. 09833	. 01531	. 10359	. 03730	. 10897	. 05867	. 11447	27
34	. 96979	. 09328	. 99307	. 09842	. 01569	. 10368	. 03766	. 10906	. 05903	. 11456	26
35	8. 97018	0. 09336	8. 99346	0. 09850	9. 01606	0. 10377	9. 03802	0. 10915	9. 05938	0. 11465	25
36	. 97058	. 09345	. 99384	. 09859	. 01643	. 10386	. 03838	. 10924	. 05973	. 11474	24
37	. 97097	. 09353	. 99422	. 09868	. 01680	. 10394	. 03874	. 10933	. 06008	. 11484	23
38	. 97136	. 09362	. 99460	. 09876	. 01717	. 10403	. 03910	. 10942	. 06043	. 11493	22
39	. 97176	. 09370	. 99498	. 09885	. 01754	. 10412	. 03946	. 10951	. 06078	. 11502	21
40	8. 97215	0. 09379	8. 99536	0. 09894	9. 01791	0. 10421	9. 03982	0. 10960	9. 06113	0. 11511	20
41	. 97254	. 09387	. 99575	. 09903	. 01828	. 10430	. 04018	. 10969	. 06148	. 11521	19
42	. 97294	. 09396	. 99613	. 09911	. 01865	. 10439	. 04054	. 10978	. 06183	. 11530	18
43	. 97333	. 09404	. 99651	. 09920	. 01902	. 10448	. 04090	. 10988	. 06218	. 11539	17
44	. 97372	. 09413	. 99689	. 09929	. 01939	. 10457	. 04126	. 10997	. 06253	. 11549	16
45	8. 97411	0. 09421	8. 99727	0. 09937	9. 01976	0. 10466	9. 04162	0. 11006	9. 06288	0. 11558	15
46	. 97450	. 09430	. 99765	. 09946	. 02013	. 10474	. 04198	. 11015	. 06323	. 11567	14
47	. 97489	. 09438	. 99803	. 09955	. 02050	. 10483	. 04234	. 11024	. 06358	. 11577	13
48	. 97529	. 09447	. 99841	. 09963	. 02087	. 10492	. 04270	. 11033	. 06393	. 11586	12
49	. 97568	. 09455	. 99879	. 09972	. 02124	. 10501	. 04306	. 11042	. 06428	. 11595	11
50	8. 97607	0. 09464	8. 99917	0. 09981	9. 02161	0. 10510	9. 04341	0. 11051	9. 06462	0. 11604	10
51	. 97646	. 09472	. 99955	. 09990	. 02197	. 10519	. 04377	. 11060	. 06497	. 11614	9
52	. 97685	. 09481	8. 99993	. 09998	. 02234	. 10528	. 04413	. 11070	. 06532	. 11623	8
53	. 97724	. 09489	9. 00031	. 10007	. 02271	. 10537	. 04449	. 11079	. 06567	. 11632	7
54	. 97763	. 09498	. 00068	. 10016	. 02308	. 10546	. 04485	. 11088	. 06602	. 11642	6
55	8. 97802	0. 09506	9. 00106	0. 10025	9. 02345	0. 10555	9. 04520	0. 11097	9. 06637	0. 11651	5
56	. 97841	. 09515	. 00144	. 10033	. 02381	. 10564	. 04556	. 11106	. 06671	. 11660	4
57	. 97880	. 09524	. 00182	. 10042	. 02418	. 10573	. 04592	. 11115	. 06706	. 11670	3
58	. 97919	. 09532	. 00220	. 10051	. 02455	. 10582	. 04628	. 11124	. 06741	. 11679	2
59	. 97958	. 09541	. 00258	. 10059	. 02492	. 10591	. 04663	. 11134	. 06776	. 11688	1
60	8. 97997	0. 09549	9. 00295	0. 10068	9. 02528	0. 10599	9. 04699	0. 11143	9. 06810	0. 11698	0
	324°		323°		322°		321°		320°		

TABLE 34
Haversines

′	40° Log Hav	40° Nat. Hav	41° Log Hav	41° Nat. Hav	42° Log Hav	42° Nat. Hav	43° Log Hav	43° Nat. Hav	44° Log Hav	44° Nat. Hav	′
0	9.06810	0.11698	9.08865	0.12265	9.10866	0.12843	9.12815	0.13432	9.14715	0.14033	60
1	.06845	.11707	.08899	.12274	.10899	.12852	.12847	.13442	.14746	.14043	59
2	.06880	.11716	.08933	.12284	.10932	.12862	.12879	.13452	.14778	.14053	58
3	.06914	.11726	.08966	.12293	.10965	.12872	.12911	.13462	.14809	.14063	57
4	.06949	.11735	.09000	.12303	.10997	.12882	.12943	.13472	.14840	.14073	56
5	9.06984	0.11745	9.09034	0.12312	9.11030	0.12891	9.12975	0.13482	9.14871	0.14084	55
6	.07018	.11754	.09068	.12322	.11063	.12901	.13007	.13492	.14902	.14094	54
7	.07053	.11763	.09101	.12331	.11096	.12911	.13039	.13502	.14934	.14104	53
8	.07088	.11773	.09135	.12341	.11129	.12921	.13071	.13512	.14965	.14114	52
9	.07122	.11782	.09169	.12351	.11161	.12930	.13103	.13522	.14996	.14124	51
10	9.07157	0.11791	9.09202	0.12360	9.11194	0.12940	9.13135	0.13532	9.15027	0.14134	50
11	.07191	.11801	.09236	.12370	.11227	.12950	.13167	.13542	.15058	.14144	49
12	.07226	.11810	.09269	.12379	.11260	.12960	.13199	.13552	.15089	.14154	48
13	.07260	.11820	.09303	.12389	.11292	.12970	.13231	.13562	.15120	.14165	47
14	.07295	.11829	.09337	.12398	.11325	.12979	.13263	.13571	.15152	.14175	46
15	9.07329	0.11838	9.09370	0.12408	9.11358	0.12989	9.13295	0.13581	9.15183	0.14185	45
16	.07364	.11848	.09404	.12418	.11391	.12999	.13326	.13591	.15214	.14195	44
17	.07398	.11857	.09437	.12427	.11423	.13009	.13358	.13601	.15245	.14205	43
18	.07433	.11867	.09471	.12437	.11456	.13018	.13390	.13611	.15276	.14215	42
19	.07467	.11876	.09504	.12446	.11489	.13028	.13422	.13621	.15307	.14226	41
20	9.07501	0.11885	9.09538	0.12456	9.11521	0.13038	9.13454	0.13631	9.15338	0.14236	40
21	.07536	.11895	.09571	.12466	.11554	.13048	.13486	.13641	.15369	.14246	39
22	.07570	.11904	.09605	.12475	.11586	.13058	.13517	.13651	.15400	.14256	38
23	.07605	.11914	.09638	.12485	.11619	.13067	.13549	.13661	.15431	.14266	37
24	.07639	.11923	.09672	.12494	.11652	.13077	.13581	.13671	.15462	.14276	36
25	9.07673	0.11933	9.09705	0.12504	9.11684	0.13087	9.13613	0.13681	9.15493	0.14287	35
26	.07708	.11942	.09739	.12514	.11717	.13097	.13644	.13691	.15524	.14297	34
27	.07742	.11951	.09772	.12523	.11749	.13107	.13676	.13701	.15555	.14307	33
28	.07776	.11961	.09805	.12533	.11782	.13116	.13708	.13711	.15585	.14317	32
29	.07810	.11970	.09839	.12543	.11814	.13126	.13739	.13721	.15616	.14327	31
30	9.07845	0.11980	9.09872	0.12552	9.11847	0.13136	9.13771	0.13731	9.15647	0.14337	30
31	.07879	.11989	.09905	.12562	.11879	.13146	.13803	.13741	.15678	.14348	29
32	.07913	.11999	.09939	.12571	.11912	.13156	.13834	.13751	.15709	.14358	28
33	.07947	.12008	.09972	.12581	.11944	.13166	.13866	.13761	.15740	.14368	27
34	.07981	.12018	.10005	.12591	.11977	.13175	.13898	.13771	.15771	.14378	26
35	9.08016	0.12027	9.10039	0.12600	9.12009	0.13185	9.13929	0.13781	9.15802	0.14388	25
36	.08050	.12036	.10072	.12610	.12041	.13195	.13961	.13791	.15832	.14399	24
37	.08084	.12046	.10105	.12620	.12074	.13205	.13992	.13801	.15863	.14409	23
38	.08118	.12055	.10138	.12629	.12106	.13215	.14024	.13811	.15894	.14419	22
39	.08152	.12065	.10172	.12639	.12139	.13225	.14056	.13822	.15925	.14429	21
40	9.08186	0.12074	9.10205	0.12649	9.12171	0.13235	9.14087	0.13832	9.15955	0.14440	20
41	.08220	.12084	.10238	.12658	.12203	.13244	.14119	.13842	.15986	.14450	19
42	.08254	.12093	.10271	.12668	.12236	.13254	.14150	.13852	.16017	.14460	18
43	.08288	.12103	.10304	.12678	.12268	.13264	.14182	.13862	.16048	.14470	17
44	.08323	.12112	.10337	.12687	.12300	.13274	.14213	.13872	.16078	.14480	16
45	9.08357	0.12122	9.10371	0.12697	9.12332	0.13284	9.14245	0.13882	9.16109	0.14491	15
46	.08391	.12131	.10404	.12707	.12365	.13294	.14276	.13892	.16140	.14501	14
47	.08425	.12141	.10437	.12717	.12397	.13304	.14307	.13902	.16170	.14511	13
48	.08459	.12150	.10470	.12726	.12429	.13314	.14339	.13912	.16201	.14521	12
49	.08492	.12160	.10503	.12736	.12461	.13323	.14370	.13922	.16232	.14532	11
50	9.08526	0.12169	9.10536	0.12746	9.12494	0.13333	9.14402	0.13932	9.16262	0.14542	10
51	.08560	.12179	.10569	.12755	.12526	.13343	.14433	.13942	.16293	.14552	9
52	.08594	.12188	.10602	.12765	.12558	.13353	.14465	.13952	.16324	.14562	8
53	.08628	.12198	.10635	.12775	.12590	.13363	.14496	.13962	.16354	.14573	7
54	.08662	.12207	.10668	.12784	.12622	.13373	.14527	.13972	.16385	.14583	6
55	9.08696	0.12217	9.10701	0.12794	9.12655	0.13383	9.14559	0.13983	9.16415	0.14593	5
56	.08730	.12226	.10734	.12804	.12687	.13393	.14590	.13993	.16446	.14604	4
57	.08764	.12236	.10767	.12814	.12719	.13403	.14621	.14003	.16476	.14614	3
58	.08797	.12245	.10800	.12823	.12751	.13412	.14653	.14013	.16507	.14624	2
59	.08831	.12255	.10833	.12833	.12783	.13422	.14684	.14023	.16537	.14634	1
60	9.08865	0.12265	9.10866	0.12843	9.12815	0.13432	9.14715	0.14033	9.16568	0.14645	0
	319°		318°		317°		316°		315°		

TABLE 34
Haversines

′	45° Log Hav	45° Nat. Hav	46° Log Hav	46° Nat. Hav	47° Log Hav	47° Nat. Hav	48° Log Hav	48° Nat. Hav	49° Log Hav	49° Nat. Hav	′
0	9. 16568	0. 14645	9. 18376	0. 15267	9. 20140	0. 15900	9. 21863	0. 16543	9. 23545	0. 17197	60
1	. 16598	. 14655	. 18405	. 15278	. 20169	. 15911	. 21891	. 16554	. 23573	. 17208	59
2	. 16629	. 14665	. 18435	. 15288	. 20198	. 15921	. 21919	. 16565	. 23601	. 17219	58
3	. 16659	. 14676	. 18465	. 15298	. 20227	. 15932	. 21948	. 16576	. 23629	. 17230	57
4	. 16690	. 14686	. 18495	. 15309	. 20256	. 15943	. 21976	. 16587	. 23656	. 17241	56
5	9. 16720	0. 14696	9. 18524	0. 15319	9. 20285	0. 15953	9. 22004	0. 16598	9. 23684	0. 17252	55
6	. 16751	. 14706	. 18554	. 15330	. 20314	. 15964	. 22033	. 16608	. 23712	. 17263	54
7	. 16781	. 14717	. 18584	. 15340	. 20343	. 15975	. 22061	. 16619	. 23739	. 17274	53
8	. 16812	. 14727	. 18613	. 15351	. 20372	. 15985	. 22089	. 16630	. 23767	. 17285	52
9	. 16842	. 14737	. 18643	. 15361	. 20401	. 15996	. 22118	. 16641	. 23794	. 17296	51
10	9. 16872	0. 14748	9. 18673	0. 15372	9. 20430	0. 16007	9. 22146	0. 16652	9. 23822	0. 17307	50
11	. 16903	. 14758	. 18702	. 15382	. 20459	. 16017	. 22174	. 16663	. 23850	. 17318	49
12	. 16933	. 14768	. 18732	. 15393	. 20488	. 16028	. 22202	. 16673	. 23877	. 17329	48
13	. 16963	. 14779	. 18762	. 15403	. 20517	. 16039	. 22231	. 16684	. 23905	. 17340	47
14	. 16994	. 14789	. 18791	. 15414	. 20546	. 16049	. 22259	. 16695	. 23932	. 17351	46
15	9. 17024	0. 14799	9. 18821	0. 15424	9. 20574	0. 16060	9. 22287	0. 16706	9. 23960	0. 17362	45
16	. 17054	. 14810	. 18850	. 15435	. 20603	. 16071	. 22315	. 16717	. 23988	. 17373	44
17	. 17085	. 14820	. 18880	. 15445	. 20632	. 16081	. 22343	. 16728	. 24015	. 17384	43
18	. 17115	. 14830	. 18909	. 15456	. 20661	. 16092	. 22372	. 16738	. 24043	. 17395	42
19	. 17145	. 14841	. 18939	. 15466	. 20690	. 16103	. 22400	. 16749	. 24070	. 17406	41
20	9. 17175	0. 14851	9. 18968	0. 15477	9. 20719	0. 16113	9. 22428	0. 16760	9. 24098	0. 17417	40
21	. 17206	. 14861	. 18998	. 15487	. 20748	. 16124	. 22456	. 16771	. 24125	. 17428	39
22	. 17236	. 14872	. 19027	. 15498	. 20776	. 16135	. 22484	. 16782	. 24153	. 17439	38
23	. 17266	. 14882	. 19057	. 15508	. 20805	. 16145	. 22512	. 16793	. 24180	. 17450	37
24	. 17296	. 14892	. 19086	. 15519	. 20834	. 16156	. 22540	. 16804	. 24208	. 17461	36
25	9. 17327	0. 14903	9. 19116	0. 15530	9. 20863	0. 16167	9. 22569	0. 16815	9. 24235	0. 17472	35
26	. 17357	. 14913	. 19145	. 15540	. 20891	. 16178	. 22597	. 16825	. 24263	. 17483	34
27	. 17387	. 14923	. 19175	. 15551	. 20920	. 16188	. 22625	. 16836	. 24290	. 17494	33
28	. 17417	. 14934	. 19204	. 15561	. 20949	. 16199	. 22653	. 16847	. 24317	. 17505	32
29	. 17447	. 14944	. 19234	. 15572	. 20978	. 16210	. 22681	. 16858	. 24345	. 17517	31
30	9. 17477	0. 14955	9. 19263	0. 15582	9. 21006	0. 16220	9. 22709	0. 16869	9. 24372	0. 17528	30
31	. 17507	. 14965	. 19292	. 15593	. 21035	. 16231	. 22737	. 16880	. 24400	. 17539	29
32	. 17538	. 14975	. 19322	. 15603	. 21064	. 16242	. 22765	. 16891	. 24427	. 17550	28
33	. 17568	. 14986	. 19351	. 15614	. 21092	. 16253	. 22793	. 16902	. 24454	. 17561	27
34	. 17598	. 14996	. 19381	. 15624	. 21121	. 16263	. 22821	. 16913	. 24482	. 17572	26
35	9. 17628	0. 15006	9. 19410	0. 15635	9. 21150	0. 16274	9. 22849	0. 16923	9. 24509	0. 17583	25
36	. 17658	. 15017	. 19439	. 15646	. 21178	. 16285	. 22877	. 16934	. 24536	. 17594	24
37	. 17688	. 15027	. 19469	. 15656	. 21207	. 16296	. 22905	. 16945	. 24564	. 17605	23
38	. 17718	. 15038	. 19498	. 15667	. 21236	. 16306	. 22933	. 16956	. 24591	. 17616	22
39	. 17748	. 15048	. 19527	. 15677	. 21264	. 16317	. 22961	. 16967	. 24618	. 17627	21
40	9. 17778	0. 15058	9. 19557	0. 15688	9. 21293	0. 16328	9. 22989	0. 16978	9. 24646	0. 17638	20
41	. 17808	. 15069	. 19586	. 15698	. 21322	. 16339	. 23017	. 16989	. 24673	. 17649	19
42	. 17838	. 15079	. 19615	. 15709	. 21350	. 16349	. 23045	. 17000	. 24700	. 17661	18
43	. 17868	. 15090	. 19644	. 15720	. 21379	. 16360	. 23073	. 17011	. 24728	. 17672	17
44	. 17898	. 15100	. 19674	. 15730	. 21407	. 16371	. 23100	. 17022	. 24755	. 17683	16
45	9. 17928	0. 15110	9. 19703	0. 15741	9. 21436	0. 16382	9. 23128	0. 17033	9. 24782	0. 17694	15
46	. 17958	. 15121	. 19732	. 15751	. 21464	. 16392	. 23156	. 17044	. 24809	. 17705	14
47	. 17988	. 15131	. 19761	. 15762	. 21493	. 16403	. 23184	. 17055	. 24837	. 17716	13
48	. 18018	. 15142	. 19790	. 15773	. 21521	. 16414	. 23212	. 17066	. 24864	. 17727	12
49	. 18048	. 15152	. 19820	. 15783	. 21550	. 16425	. 23240	. 17076	. 24891	. 17738	11
50	9. 18077	0. 15163	9. 19849	0. 15794	9. 21578	0. 16436	9. 23268	0. 17087	9. 24918	0. 17749	10
51	. 18107	. 15173	. 19878	. 15804	. 21607	. 16446	. 23295	. 17098	. 24945	. 17760	9
52	. 18137	. 15183	. 19907	. 15815	. 21635	. 16457	. 23323	. 17109	. 24973	. 17772	8
53	. 18167	. 15194	. 19936	. 15826	. 21664	. 16468	. 23351	. 17120	. 25000	. 17783	7
54	. 18197	. 15204	. 19965	. 15836	. 21692	. 16479	. 23379	. 17131	. 25027	. 17794	6
55	9. 18227	0. 15215	9. 19995	0. 15847	9. 21721	0. 16489	9. 23407	0. 17142	9. 25054	0. 17805	5
56	. 18256	. 15225	. 20024	. 15858	. 21749	. 16500	. 23434	. 17153	. 25081	. 17816	4
57	. 18286	. 15236	. 20053	. 15868	. 21778	. 16511	. 23462	. 17164	. 25108	. 17827	3
58	. 18316	. 15246	. 20082	. 15879	. 21806	. 16522	. 23490	. 17175	. 25135	. 17838	2
59	. 18346	. 15257	. 20111	. 15889	. 21834	. 16533	. 23518	. 17186	. 25163	. 17849	1
60	9. 18376	0. 15267	9. 20140	0. 15900	9. 21863	0. 16543	9. 23545	0. 17197	9. 25190	0. 17861	0
	314°		313°		312°		311°		310°		

TABLE 34

Haversines

′	50° Log Hav	50° Nat. Hav	51° Log Hav	51° Nat. Hav	52° Log Hav	52° Nat. Hav	53° Log Hav	53° Nat. Hav	54° Log Hav	54° Nat. Hav	
0	9.25190	0.17861	9.26797	0.18534	9.28368	0.19217	9.29906	0.19909	9.31409	0.20611	60
1	.25217	.17872	.26823	.18545	.28394	.19228	.29931	.19921	.31434	.20623	59
2	.25244	.17883	.26850	.18557	.28420	.19240	.29956	.19932	.31459	.20634	58
3	.25271	.17894	.26876	.18568	.28446	.19251	.29981	.19944	.31484	.20646	57
4	.25298	.17905	.26903	.18579	.28472	.19263	.30007	.19956	.31508	.20658	56
5	9.25325	0.17916	9.26929	0.18591	9.28498	0.19274	9.30032	0.19967	9.31533	0.20670	55
6	.25352	.17928	.26956	.18602	.28524	.19286	.30057	.19979	.31558	.20681	54
7	.25379	.17939	.26982	.18613	.28549	.19297	.30083	.19991	.31583	.20693	53
8	.25406	.17950	.27008	.18624	.28575	.19309	.30108	.20002	.31607	.20705	52
9	.25433	.17961	.27035	.18636	.28601	.19320	.30133	.20014	.31632	.20717	51
10	9.25460	0.17972	9.27061	0.18647	9.28627	0.19332	9.30158	0.20026	9.31657	0.20729	50
11	.25487	.17983	.27088	.18658	.28653	.19343	.30184	.20037	.31682	.20740	49
12	.25514	.17995	.27114	.18670	.28679	.19355	.30209	.20049	.31706	.20752	48
13	.25541	.18006	.27140	.18681	.28704	.19366	.30234	.20060	.31731	.20764	47
14	.25568	.18017	.27167	.18692	.28730	.19378	.30259	.20072	.31756	.20776	46
15	9.25595	0.18028	9.27193	0.18704	9.28756	0.19389	9.30285	0.20084	9.31780	0.20788	45
16	.25622	.18039	.27219	.18715	.28782	.19401	.30310	.20095	.31805	.20799	44
17	.25649	.18050	.27246	.18727	.28807	.19412	.30335	.20107	.31830	.20811	43
18	.25676	.18062	.27272	.18738	.28833	.19424	.30360	.20119	.31854	.20823	42
19	.25703	.18073	.27298	.18749	.28859	.19435	.30385	.20130	.31879	.20835	41
20	9.25729	0.18084	9.27325	0.18761	9.28885	0.19447	9.30410	0.20142	9.31903	0.20847	40
21	.25756	.18095	.27351	.18772	.28910	.19458	.30436	.20154	.31928	.20858	39
22	.25783	.18106	.27377	.18783	.28936	.19470	.30461	.20165	.31953	.20870	38
23	.25810	.18118	.27403	.18795	.28962	.19481	.30486	.20177	.31977	.20882	37
24	.25837	.18129	.27430	.18806	.28987	.19493	.30511	.20189	.32002	.20894	36
25	9.25864	0.18140	9.27456	0.18817	9.29013	0.19504	9.30536	0.20200	9.32026	0.20906	35
26	.25891	.18151	.27482	.18829	.29039	.19516	.30561	.20212	.32051	.20918	34
27	.25917	.18162	.27508	.18840	.29064	.19527	.30586	.20224	.32076	.20929	33
28	.25944	.18174	.27535	.18852	.29090	.19539	.30611	.20235	.32100	.20941	32
29	.25971	.18185	.27561	.18863	.29116	.19550	.30636	.20247	.32125	.20953	31
30	9.25998	0.18196	9.27587	0.18874	9.29141	0.19562	9.30662	0.20259	9.32149	0.20965	30
31	.26025	.18207	.27613	.18886	.29167	.19573	.30687	.20271	.32174	.20977	29
32	.26051	.18219	.27639	.18897	.29192	.19585	.30712	.20282	.32198	.20989	28
33	.26078	.18230	.27666	.18908	.29218	.19597	.30737	.20294	.32223	.21000	27
34	.26105	.18241	.27692	.18920	.29244	.19608	.30762	.20306	.32247	.21012	26
35	9.26132	0.18252	9.27718	0.18931	9.29269	0.19620	9.30787	0.20317	9.32272	0.21024	25
36	.26158	.18263	.27744	.18943	.29295	.19631	.30812	.20329	.32296	.21036	24
37	.26185	.18275	.27770	.18954	.29320	.19643	.30837	.20341	.32321	.21048	23
38	.26212	.18286	.27796	.18965	.29346	.19654	.30862	.20352	.32345	.21060	22
39	.26238	.18297	.27822	.18977	.29371	.19666	.30887	.20364	.32370	.21072	21
40	9.26265	0.18308	9.27848	0.18988	9.29397	0.19677	9.30912	0.20376	9.32394	0.21083	20
41	.26292	.18320	.27875	.19000	.29422	.19689	.30937	.20388	.32418	.21095	19
42	.26319	.18331	.27901	.19011	.29448	.19701	.30962	.20399	.32443	.21107	18
43	.26345	.18342	.27927	.19022	.29473	.19712	.30987	.20411	.32467	.21119	17
44	.26372	.18353	.27953	.19034	.29499	.19724	.31012	.20423	.32492	.21131	16
45	9.26398	0.18365	9.27979	0.19045	9.29524	0.19735	9.31036	0.20435	9.32516	0.21143	15
46	.26425	.18376	.28005	.19057	.29550	.19747	.31061	.20446	.32541	.21155	14
47	.26452	.18387	.28031	.19068	.29575	.19758	.31086	.20458	.32565	.21167	13
48	.26478	.18399	.28057	.19080	.29601	.19770	.31111	.20470	.32589	.21178	12
49	.26505	.18410	.28083	.19091	.29626	.19782	.31136	.20481	.32614	.21190	11
50	9.26532	0.18421	9.28109	0.19102	9.29652	0.19793	9.31161	0.20493	9.32638	0.21202	10
51	.26558	.18432	.28135	.19114	.29677	.19805	.31186	.20505	.32662	.21214	9
52	.26585	.18444	.28161	.19125	.29703	.19816	.31211	.20517	.32687	.21226	8
53	.26611	.18455	.28187	.19137	.29728	.19828	.31236	.20528	.32711	.21238	7
54	.26638	.18466	.28213	.19148	.29753	.19840	.31260	.20540	.32735	.21250	6
55	9.26664	0.18477	9.28239	0.19160	9.29779	0.19851	9.31285	0.20552	9.32760	0.21262	5
56	.26691	.18489	.28265	.19171	.29804	.19863	.31310	.20564	.32784	.21274	4
57	.26717	.18500	.28291	.19183	.29829	.19874	.31335	.20575	.32808	.21285	3
58	.26744	.18511	.28317	.19194	.29855	.19886	.31360	.20587	.32833	.21297	2
59	.26770	.18523	.28342	.19205	.29880	.19898	.31385	.20599	.32857	.21309	1
60	9.26797	0.18534	9.28368	0.19217	9.29906	0.19909	9.31409	0.20611	9.32881	0.21321	0
	309°		308°		307°		306°		305°		

TABLE 34
Haversines

′	55° Log Hav	55° Nat. Hav	56° Log Hav	56° Nat. Hav	57° Log Hav	57° Nat. Hav	58° Log Hav	58° Nat. Hav	59° Log Hav	59° Nat. Hav	′
0	9. 32881	0. 21321	9. 34322	0. 22040	9. 35733	0. 22768	9. 37114	0. 23504	9. 38468	0. 24248	60
1	. 32905	. 21333	. 34346	. 22052	. 35756	. 22780	. 37137	. 23516	. 38490	. 24261	59
2	. 32930	. 21345	. 34369	. 22064	. 35779	. 22792	. 37160	. 23529	. 38512	. 24273	58
3	. 32954	. 21357	. 34393	. 22077	. 35802	. 22805	. 37183	. 23541	. 38535	. 24286	57
4	. 32978	. 21369	. 34417	. 22089	. 35826	. 22817	. 37205	. 23553	. 38557	. 24298	56
5	9. 33002	0. 21381	9. 34441	0. 22101	9. 35849	0. 22829	9. 37228	0. 23566	9. 38579	0. 24310	55
6	. 33027	. 21393	. 34464	. 22113	. 35872	. 22841	. 37251	. 23578	. 38602	. 24323	54
7	. 33051	. 21405	. 34488	. 22125	. 35895	. 22853	. 37274	. 23590	. 38624	. 24335	53
8	. 33075	. 21417	. 34512	. 22137	. 35918	. 22866	. 37296	. 23603	. 38646	. 24348	52
9	. 33099	. 21429	. 34535	. 22149	. 35942	. 22878	. 37319	. 23615	. 38668	. 24360	51
10	9. 33123	0. 21440	9. 34559	0. 22161	9. 35965	0. 22890	9. 37342	0. 23627	9. 38691	0. 24373	50
11	. 33148	. 21452	. 34583	. 22173	. 35988	. 22902	. 37364	. 23640	. 38713	. 24385	49
12	. 33172	. 21464	. 34606	. 22185	. 36011	. 22915	. 37387	. 23652	. 38735	. 24398	48
13	. 33196	. 21476	. 34630	. 22197	. 36034	. 22927	. 37410	. 23665	. 38757	. 24410	47
14	. 33220	. 21488	. 34654	. 22209	. 36058	. 22939	. 37433	. 23677	. 38780	. 24423	46
15	9. 33244	0. 21500	9. 34677	0. 22221	9. 36081	0. 22951	9. 37455	0. 23689	9. 38802	0. 24435	45
16	. 33268	. 21512	. 34701	. 22234	. 36104	. 22964	. 37478	. 23702	. 38824	. 24448	44
17	. 33292	. 21524	. 34725	. 22246	. 36127	. 22976	. 37501	. 23714	. 38846	. 24460	43
18	. 33317	. 21536	. 34748	. 22258	. 36150	. 22988	. 37523	. 23726	. 38868	. 24473	42
19	. 33341	. 21548	. 34772	. 22270	. 36173	. 23000	. 37546	. 23739	. 38891	. 24485	41
20	9. 33365	0. 21560	9. 34795	0. 22282	9. 36196	0. 23012	9. 37569	0. 23751	9. 38913	0. 24498	40
21	. 33389	. 21572	. 34819	. 22294	. 36219	. 23025	. 37591	. 23764	. 38935	. 24510	39
22	. 33413	. 21584	. 34843	. 22306	. 36243	. 23037	. 37614	. 23776	. 38957	. 24523	38
23	. 33437	. 21596	. 34866	. 22318	. 36266	. 23049	. 37636	. 23788	. 38979	. 24535	37
24	. 33461	. 21608	. 34890	. 22330	. 36289	. 23061	. 37659	. 23801	. 39002	. 24548	36
25	9. 33485	0. 21620	9. 34913	0. 22343	9. 36312	0. 23074	9. 37682	0. 23813	9. 39024	0. 24560	35
26	. 33509	. 21632	. 34937	. 22355	. 36335	. 23086	. 37704	. 23825	. 39046	. 24573	34
27	. 33533	. 21644	. 34960	. 22367	. 36358	. 23098	. 37727	. 23838	. 39068	. 24585	33
28	. 33557	. 21656	. 34984	. 22379	. 36381	. 23110	. 37749	. 23850	. 39090	. 24598	32
29	. 33581	. 21668	. 35007	. 22391	. 36404	. 23123	. 37772	. 23863	. 39112	. 24611	31
30	9. 33605	0. 21680	9. 35031	0. 22403	9. 36427	0. 23135	9. 37794	0. 23875	9. 39134	0. 24623	30
31	. 33629	. 21692	. 35054	. 22415	. 36450	. 23147	. 37817	. 23887	. 39156	. 24636	29
32	. 33653	. 21704	. 35078	. 22427	. 36473	. 23160	. 37840	. 23900	. 39178	. 24648	28
33	. 33677	. 21716	. 35101	. 22440	. 36496	. 23172	. 37862	. 23912	. 39201	. 24661	27
34	. 33701	. 21728	. 35125	. 22452	. 36519	. 23184	. 37885	. 23925	. 39223	. 24673	26
35	9. 33725	0. 21740	9. 35148	0. 22464	9. 36542	0. 23196	9. 37907	0. 23937	9. 39245	0. 24686	25
36	. 33749	. 21752	. 35172	. 22476	. 36565	. 23209	. 37930	. 23950	. 39267	. 24698	24
37	. 33773	. 21764	. 35195	. 22488	. 36588	. 23221	. 37952	. 23962	. 39289	. 24711	23
38	. 33797	. 21776	. 35219	. 22500	. 36611	. 23233	. 37975	. 23974	. 39311	. 24723	22
39	. 33821	. 21788	. 35242	. 22512	. 36634	. 23246	. 37997	. 23987	. 39333	. 24736	21
40	9. 33845	0. 21800	9. 35266	0. 22525	9. 36657	0. 23258	9. 38020	0. 23999	9. 39355	0. 24749	20
41	. 33869	. 21812	. 35289	. 22537	. 36680	. 23270	. 38042	. 24012	. 39377	. 24761	19
42	. 33893	. 21824	. 35312	. 22549	. 36703	. 23282	. 38065	. 24024	. 39399	. 24774	18
43	. 33917	. 21836	. 35336	. 22561	. 36726	. 23295	. 38087	. 24036	. 39421	. 24786	17
44	. 33941	. 21848	. 35359	. 22573	. 36749	. 23307	. 38110	. 24049	. 39443	. 24799	16
45	9. 33965	0. 21860	9. 35383	0. 22585	9. 36772	0. 23319	9. 38132	0. 24061	9. 39465	0. 24811	15
46	. 33988	. 21872	. 35406	. 22598	. 36794	. 23332	. 38154	. 24074	. 39487	. 24824	14
47	. 34012	. 21884	. 35429	. 22610	. 36817	. 23344	. 38177	. 24086	. 39509	. 24836	13
48	. 34036	. 21896	. 35453	. 22622	. 36840	. 23356	. 38199	. 24099	. 39531	. 24849	12
49	. 34060	. 21908	. 35476	. 22634	. 36863	. 23368	. 38222	. 24111	. 39553	. 24862	11
50	9. 34084	0. 21920	9. 35500	0. 22646	9. 36886	0. 23381	9. 38244	0. 24124	9. 39575	0. 24874	10
51	. 34108	. 21932	. 35523	. 22658	. 36909	. 23393	. 38267	. 24136	. 39597	. 24887	9
52	. 34132	. 21944	. 35546	. 22671	. 36932	. 23405	. 38289	. 24148	. 39619	. 24899	8
53	. 34155	. 21956	. 35570	. 22683	. 36955	. 23418	. 38311	. 24161	. 39641	. 24912	7
54	. 34179	. 21968	. 35593	. 22695	. 36977	. 23430	. 38334	. 24173	. 39663	. 24924	6
55	9. 34203	0. 21980	9. 35616	0. 22707	9. 37000	0. 23442	9. 38356	0. 24186	9. 39685	0. 24937	5
56	. 34227	. 21992	. 35639	. 22719	. 37023	. 23455	. 38378	. 24198	. 39706	. 24950	4
57	. 34251	. 22004	. 35663	. 22731	. 37046	. 23467	. 38401	. 24211	. 39728	. 24962	3
58	. 34274	. 22016	. 35686	. 22744	. 37069	. 23479	. 38423	. 24223	. 39750	. 24975	2
59	. 34298	. 22028	. 35709	. 22756	. 37091	. 23492	. 38445	. 24236	. 39772	. 24987	1
60	9. 34322	0. 22040	9. 35733	0. 22768	9. 37114	0. 23504	9. 38468	0. 24248	9. 39794	0. 25000	0
	304°		303°		302°		301°		300°		

TABLE 34

Haversines

′	60°		61°		62°		63°		64°		′
	Log Hav	Nat. Hav	Log Hav	Nat. Hav	Log Hav	Nat. Hav	Log Hav	Nat. Hav	Log Hav	Nat. Hav	
0	9.39794	0.25000	9.41094	0.25760	9.42368	0.26526	9.43617	0.27300	9.44842	0.28081	60
1	.39816	.25013	.41115	.25772	.42389	.26539	.43638	.27313	.44862	.28095	59
2	.39838	.25025	.41137	.25785	.42410	.26552	.43658	.27326	.44882	.28108	58
3	.39860	.25038	.41158	.25798	.42431	.26565	.43679	.27339	.44903	.28121	57
4	.39881	.25050	.41180	.25810	.42452	.26578	.43699	.27352	.44923	.28134	56
5	9.39903	0.25063	9.41201	0.25823	9.42473	0.26591	9.43720	0.27365	9.44943	0.28147	55
6	.39925	.25076	.41222	.25836	.42494	.26604	.43741	.27378	.44963	.28160	54
7	.39947	.25088	.41244	.25849	.42515	.26616	.43761	.27391	.44983	.28173	53
8	.39969	.25101	.41265	.25861	.42536	.26629	.43782	.27404	.45003	.28186	52
9	.39991	.25113	.41287	.25874	.42557	.26642	.43802	.27417	.45024	.28199	51
10	9.40012	0.25126	9.41308	0.25887	9.42578	0.26655	9.43823	0.27430	9.45044	0.28212	50
11	.40034	.25139	.41329	.25900	.42599	.26668	.43843	.27443	.45064	.28225	49
12	.40056	.25151	.41351	.25912	.42620	.26681	.43864	.27456	.45084	.28238	48
13	.40078	.25164	.41372	.25925	.42641	.26694	.43884	.27469	.45104	.28252	47
14	.40100	.25177	.41393	.25938	.42662	.26706	.43905	.27482	.45124	.28265	46
15	9.40121	0.25189	9.41415	0.25951	9.42682	0.26719	9.43926	0.27495	9.45144	0.28278	45
16	.40143	.25202	.41436	.25963	.42703	.26732	.43946	.27508	.45165	.28291	44
17	.40165	.25214	.41457	.25976	.42724	.26745	.43967	.27521	.45185	.28304	43
18	.40187	.25227	.41479	.25989	.42745	.26758	.43987	.27534	.45205	.28317	42
19	.40208	.25240	.41500	.26002	.42766	.26771	.44008	.27547	.45225	.28330	41
20	9.40230	0.25252	9.41521	0.26014	9.42787	0.26784	9.44028	0.27560	9.45245	0.28343	40
21	.40252	.25265	.41543	.26027	.42808	.26797	.44048	.27573	.45265	.28356	39
22	.40274	.25278	.41564	.26040	.42829	.26809	.44069	.27586	.45285	.28369	38
23	.40295	.25290	.41585	.26053	.42850	.26822	.44089	.27599	.45305	.28383	37
24	.40317	.25303	.41606	.26065	.42870	.26835	.44110	.27612	.45325	.28396	36
25	9.40339	0.25316	9.41628	0.26078	9.42891	0.26848	9.44130	0.27625	9.45345	0.28409	35
26	.40360	.25328	.41649	.26091	.42912	.26861	.44151	.27638	.45365	.28422	34
27	.40382	.25341	.41670	.26104	.42933	.26874	.44171	.27651	.45385	.28435	33
28	.40404	.25354	.41692	.26117	.42954	.26887	.44192	.27664	.45405	.28448	32
29	.40425	.25366	.41713	.26129	.42975	.26900	.44212	.27677	.45426	.28461	31
30	9.40447	0.25379	9.41734	0.26142	9.42996	0.26913	9.44232	0.27690	9.45446	0.28474	30
31	.40469	.25391	.41755	.26155	.43016	.26925	.44253	.27703	.45466	.28488	29
32	.40490	.25404	.41776	.26168	.43037	.26938	.44273	.27716	.45486	.28501	28
33	.40512	.25417	.41798	.26180	.43058	.26951	.44294	.27729	.45506	.28514	27
34	.40534	.25429	.41819	.26193	.43079	.26964	.44314	.27742	.45526	.28527	26
35	9.40555	0.25442	9.41840	0.26206	9.43100	0.26977	9.44334	0.27755	9.45546	0.28540	25
36	.40577	.25455	.41861	.26219	.43120	.26990	.44355	.27768	.45566	.28553	24
37	.40599	.25467	.41882	.26232	.43141	.27003	.44375	.27781	.45586	.28566	23
38	.40620	.25480	.41904	.26244	.43162	.27016	.44396	.27794	.45606	.28580	22
39	.40642	.25493	.41925	.26257	.43183	.27029	.44416	.27807	.45625	.28593	21
40	9.40663	0.25506	9.41946	0.26270	9.43203	0.27042	9.44436	0.27820	9.45645	0.28606	20
41	.40685	.25518	.41967	.26283	.43224	.27055	.44457	.27833	.45665	.28619	19
42	.40707	.25531	.41988	.26296	.43245	.27068	.44477	.27846	.45685	.28632	18
43	.40728	.25544	.42009	.26308	.43266	.27080	.44497	.27859	.45705	.28645	17
44	.40750	.25556	.42031	.26321	.43286	.27093	.44518	.27873	.45725	.28658	16
45	9.40771	0.25569	9.42052	0.26334	9.43307	0.27106	9.44538	0.27886	9.45745	0.28672	15
46	.40793	.25582	.42073	.26347	.43328	.27119	.44558	.27899	.45765	.28685	14
47	.40814	.25594	.42094	.26360	.43348	.27132	.44579	.27912	.45785	.28698	13
48	.40836	.25607	.42115	.26372	.43369	.27145	.44599	.27925	.45805	.28711	12
49	.40858	.25620	.42136	.26385	.43390	.27158	.44619	.27938	.45825	.28724	11
50	9.40879	0.25632	9.42157	0.26398	9.43411	0.27171	9.44639	0.27951	9.45845	0.28737	10
51	.40900	.25645	.42178	.26411	.43431	.27184	.44660	.27964	.45865	.28751	9
52	.40922	.25658	.42199	.26424	.43452	.27197	.44680	.27977	.45884	.28764	8
53	.40943	.25671	.42221	.26437	.43473	.27210	.44700	.27990	.45904	.28777	7
54	.40965	.25683	.42242	.26449	.43493	.27223	.44721	.28003	.45924	.28790	6
55	9.40986	0.25696	9.42263	0.26462	9.43514	0.27236	9.44741	0.28016	9.45944	0.28803	5
56	.41008	.25709	.42284	.26475	.43535	.27249	.44761	.28029	.45964	.28816	4
57	.41029	.25721	.42305	.26488	.43555	.27262	.44781	.28042	.45984	.28830	3
58	.41051	.25734	.42326	.26501	.43576	.27275	.44801	.28055	.46004	.28843	2
59	.41072	.25747	.42347	.26514	.43596	.27288	.44822	.28068	.46023	.28856	1
60	9.41094	0.25760	9.42368	0.26526	9.43617	0.27300	9.44842	0.28081	9.46043	0.28869	0
	299°		298°		297°		296°		295°		

TABLE 34

Haversines

	65°		66°		67°		68°		69°		
′	Log Hav	Nat. Hav	Log Hav	Nat. Hav	Log Hav	Nat. Hav	Log Hav	Nat. Hav	Log Hav	Nat. Hav	′
0	9. 46043	0. 28869	9. 47222	0. 29663	9. 48378	0. 30463	9. 49512	0. 31270	9. 50626	0. 32082	60
1	. 46063	. 28882	. 47241	. 29676	. 48397	. 30477	. 49531	. 31283	. 50644	. 32095	59
2	. 46083	. 28895	. 47261	. 29690	. 48416	. 30490	. 49550	. 31297	. 50662	. 32109	58
3	. 46103	. 28909	. 47280	. 29703	. 48435	. 30504	. 49568	. 31310	. 50681	. 32122	57
4	. 46123	. 28922	. 47300	. 29716	. 48454	. 30517	. 49587	. 31324	. 50699	. 32136	56
5	9. 46142	0. 28935	9. 47319	0. 29730	9. 48473	0. 30530	9. 49606	0. 31337	9. 50717	0. 32150	55
6	. 46162	. 28948	. 47338	. 29743	. 48492	. 30544	. 49625	. 31351	. 50736	. 32163	54
7	. 46182	. 28961	. 47358	. 29756	. 48511	. 30557	. 49643	. 31364	. 50754	. 32177	53
8	. 46202	. 28975	. 47377	. 29770	. 48530	. 30571	. 49662	. 31378	. 50772	. 32190	52
9	. 46222	. 28988	. 47397	. 29783	. 48549	. 30584	. 49681	. 31391	. 50791	. 32204	51
10	9. 46241	0. 29001	9. 47416	0. 29796	9. 48568	0. 30597	9. 49699	0. 31405	9. 50809	0. 32217	50
11	. 46261	. 29014	. 47435	. 29809	. 48587	. 30611	. 49718	. 31418	. 50827	. 32231	49
12	. 46281	. 29027	. 47455	. 29823	. 48607	. 30624	. 49737	. 31432	. 50846	. 32245	48
13	. 46301	. 29041	. 47474	. 29836	. 48626	. 30638	. 49755	. 31445	. 50864	. 32258	47
14	. 46320	. 29054	. 47493	. 29849	. 48645	. 30651	. 49774	. 31459	. 50882	. 32272	46
15	9. 46340	0. 29067	9. 47513	0. 29863	9. 48664	0. 30664	9. 49793	0. 31472	9. 50901	0. 32285	45
16	. 46360	. 29080	. 47532	. 29876	. 48683	. 30678	. 49811	. 31486	. 50919	. 32299	44
17	. 46380	. 29093	. 47552	. 29889	. 48702	. 30691	. 49830	. 31499	. 50937	. 32313	43
18	. 46399	. 29107	. 47571	. 29903	. 48720	. 30705	. 49849	. 31513	. 50956	. 32326	42
19	. 46419	. 29120	. 47590	. 29916	. 48739	. 30718	. 49867	. 31526	. 50974	. 32340	41
20	9. 46439	0. 29133	9. 47610	0. 29929	9. 48758	0. 30732	9. 49886	0. 31540	9. 50992	0. 32353	40
21	. 46458	. 29146	. 47629	. 29943	. 48777	. 30745	. 49904	. 31553	. 51010	. 32367	39
22	. 46478	. 29160	. 47648	. 29956	. 48796	. 30758	. 49923	. 31567	. 51029	. 32381	38
23	. 46498	. 29173	. 47668	. 29969	. 48815	. 30772	. 49942	. 31580	. 51047	. 32394	37
24	. 46517	. 29186	. 47687	. 29983	. 48834	. 30785	. 49960	. 31594	. 51065	. 32408	36
25	9. 46537	0. 29199	9. 47706	0. 29996	9. 48853	0. 30799	9. 49979	0. 31607	9. 51083	0. 32422	35
26	. 46557	. 29212	. 47725	. 30009	. 48872	. 30812	. 49997	. 31621	. 51102	. 32435	34
27	. 46576	. 29226	. 47745	. 30023	. 48891	. 30826	. 50016	. 31634	. 51120	. 32449	33
28	. 46596	. 29239	. 47764	. 30036	. 48910	. 30839	. 50034	. 31648	. 51138	. 32462	32
29	. 46616	. 29252	. 47783	. 30049	. 48929	. 30852	. 50053	. 31661	. 51156	. 32476	31
30	9. 46635	0. 29265	9. 47803	0. 30063	9. 48948	0. 30866	9. 50072	0. 31675	9. 51174	0. 32490	30
31	. 46655	. 29279	. 47822	. 30076	. 48967	. 30879	. 50090	. 31688	. 51193	. 32503	29
32	. 46675	. 29292	. 47841	. 30089	. 48986	. 30893	. 50109	. 31702	. 51211	. 32517	28
33	. 46694	. 29305	. 47860	. 30103	. 49004	. 30906	. 50127	. 31716	. 51229	. 32531	27
34	. 46714	. 29318	. 47880	. 30116	. 49023	. 30920	. 50146	. 31729	. 51247	. 32544	26
35	9. 46733	0. 29332	9. 47899	0. 30129	9. 49042	0. 30933	9. 50164	0. 31743	9. 51265	0. 32558	25
36	. 46753	. 29345	. 47918	. 30143	. 49061	. 30946	. 50183	. 31756	. 51284	. 32571	24
37	. 46773	. 29358	. 47937	. 30156	. 49080	. 30960	. 50201	. 31770	. 51302	. 32585	23
38	. 46792	. 29371	. 47957	. 30169	. 49099	. 30973	. 50220	. 31783	. 51320	. 32599	22
39	. 46812	. 29385	. 47976	. 30183	. 49118	. 30987	. 50238	. 31797	. 51338	. 32612	21
40	9. 46831	0. 29398	9. 47995	0. 30196	9. 49137	0. 31000	9. 50257	0. 31810	9. 51356	0. 32626	20
41	. 46851	. 29411	. 48014	. 30209	. 49155	. 31014	. 50275	. 31824	. 51374	. 32640	19
42	. 46871	. 29424	. 48033	. 30223	. 49174	. 31027	. 50294	. 31837	. 51393	. 32653	18
43	. 46890	. 29438	. 48053	. 30236	. 49193	. 31041	. 50312	. 31851	. 51411	. 32667	17
44	. 46910	. 29451	. 48072	. 30249	. 49212	. 31054	. 50331	. 31865	. 51429	. 32681	16
45	9. 46929	0. 29464	9. 48091	0. 30263	9. 49231	0. 31068	9. 50349	0. 31878	9. 51447	0. 32694	15
46	. 46949	. 29477	. 48110	. 30276	. 49250	. 31081	. 50368	. 31892	. 51465	. 32708	14
47	. 46968	. 29491	. 48129	. 30290	. 49268	. 31094	. 50386	. 31905	. 51483	. 32721	13
48	. 46988	. 29504	. 48148	. 30303	. 49287	. 31108	. 50405	. 31919	. 51501	. 32735	12
49	. 47007	. 29517	. 48168	. 30316	. 49306	. 31121	. 50423	. 31932	. 51519	. 32749	11
50	9. 47027	0. 29530	9. 48187	0. 30330	9. 49325	0. 31135	9. 50442	0. 31946	9. 51538	0. 32762	10
51	. 47046	. 29544	. 48206	. 30343	. 49344	. 31148	. 50460	. 31959	. 51556	. 32776	9
52	. 47066	. 29557	. 48225	. 30356	. 49362	. 31162	. 50478	. 31973	. 51574	. 32790	8
53	. 47085	. 29570	. 48244	. 30370	. 49381	. 31175	. 50497	. 31987	. 51592	. 32803	7
54	. 47105	. 29583	. 48263	. 30383	. 49400	. 31189	. 50515	. 32000	. 51610	. 32817	6
55	9. 47124	0. 29597	9. 48282	0. 30397	9. 49419	0. 31202	9. 50534	0. 32014	9. 51628	0. 32831	5
56	. 47144	. 29610	. 48302	. 30410	. 49437	. 31216	. 50552	. 32027	. 51646	. 32844	4
57	. 47163	. 29623	. 48321	. 30423	. 49456	. 31229	. 50570	. 32041	. 51664	. 32858	3
58	. 47183	. 29637	. 48340	. 30437	. 49475	. 31243	. 50589	. 32054	. 51682	. 32872	2
59	. 47202	. 29650	. 48359	. 30450	. 49494	. 31256	. 50607	. 32068	. 51700	. 32885	1
60	9. 47222	0. 29663	9. 48378	0. 30463	9. 49512	0. 31270	9. 50626	0. 32082	9. 51718	0. 32899	0
	294°		293°		292°		291°		290°		

TABLE 34

Haversines

	70°		71°		72°		73°		74°		
′	Log Hav	Nat. Hav	Log Hav	Nat. Hav	Log Hav	Nat. Hav	Log Hav	Nat. Hav	Log Hav	Nat. Hav	′
0	9.51718	0.32899	9.52791	0.33722	9.53844	0.34549	9.54878	0.35381	9.55893	0.36218	60
1	.51736	.32913	.52809	.33735	.53861	.34563	.54895	.35395	.55909	.36232	59
2	.51754	.32926	.52826	.33749	.53879	.34577	.54912	.35409	.55926	.36246	58
3	.51772	.32940	.52844	.33763	.53896	.34591	.54929	.35423	.55943	.36260	57
4	.51790	.32954	.52862	.33777	.53913	.34604	.54946	.35437	.55960	.36274	56
5	9.51808	0.32967	9.52879	0.33790	9.53931	0.34618	9.54963	0.35451	9.55976	0.36288	55
6	.51826	.32981	.52897	.33804	.53948	.34632	.54980	.35465	.55993	.36302	54
7	.51844	.32995	.52915	.33818	.53966	.34646	.54997	.35479	.56010	.36316	53
8	.51862	.33008	.52932	.33832	.53983	.34660	.55014	.35493	.56027	.36330	52
9	.51880	.33022	.52950	.33845	.54000	.34674	.55031	.35507	.56043	.36344	51
10	9.51898	0.33036	9.52968	0.33859	9.54017	0.34688	9.55048	0.35521	9.56060	0.36358	50
11	.51916	.33049	.52985	.33873	.54035	.34701	.55065	.35534	.56077	.36372	49
12	.51934	.33063	.53003	.33887	.54052	.34715	.55082	.35548	.56093	.36386	48
13	.51952	.33077	.53021	.33900	.54069	.34729	.55099	.35562	.56110	.36400	47
14	.51970	.33090	.53038	.33914	.54087	.34743	.55116	.35576	.56127	.36414	46
15	9.51988	0.33104	9.53056	0.33928	9.54104	0.34757	9.55133	0.35590	9.56144	0.36428	45
16	.52006	.33118	.53073	.33942	.54121	.34771	.55150	.35604	.56160	.36442	44
17	.52024	.33132	.53091	.33956	.54139	.34784	.55167	.35618	.56177	.36456	43
18	.52042	.33145	.53109	.33969	.54156	.34798	.55184	.35632	.56194	.36470	42
19	.52060	.33159	.53126	.33983	.54173	.34812	.55201	.35646	.56210	.36484	41
20	9.52078	0.33173	9.53144	0.33997	9.54190	0.34826	9.55218	0.35660	9.56227	0.36498	40
21	.52096	.33186	.53162	.34011	.54208	.34840	.55235	.35674	.56244	.36512	39
22	.52114	.33200	.53179	.34024	.54225	.34854	.55252	.35688	.56260	.36526	38
23	.52132	.33214	.53197	.34038	.54242	.34868	.55269	.35702	.56277	.36540	37
24	.52150	.33227	.53214	.34052	.54260	.34882	.55286	.35716	.56294	.36554	36
25	9.52168	0.33241	9.53232	0.34066	9.54277	0.34895	9.55303	0.35730	9.56310	0.36568	35
26	.52185	.33255	.53249	.34080	.54294	.34909	.55320	.35743	.56327	.36582	34
27	.52203	.33269	.53267	.34093	.54311	.34923	.55337	.35757	.56343	.36596	33
28	.52221	.33282	.53285	.34107	.54329	.34937	.55354	.35771	.56360	.36610	32
29	.52239	.33296	.53302	.34121	.54346	.34951	.55370	.35785	.56377	.36624	31
30	9.52257	0.33310	9.53320	0.34135	9.54363	0.34965	9.55387	0.35799	9.56393	0.36638	30
31	.52275	.33323	.53337	.34149	.54380	.34979	.55404	.35813	.56410	.36652	29
32	.52293	.33337	.53355	.34162	.54397	.34992	.55421	.35827	.56426	.36666	28
33	.52311	.33351	.53372	.34176	.54415	.35006	.55438	.35841	.56443	.36680	27
34	.52328	.33365	.53390	.34190	.54432	.35020	.55455	.35855	.56460	.36694	26
35	9.52346	0.33378	9.53407	0.34204	9.54449	0.35034	9.55472	0.35869	9.56476	0.36708	25
36	.52364	.33392	.53425	.34218	.54466	.35048	.55489	.35883	.56493	.36722	24
37	.52382	.33406	.53442	.34231	.54483	.35062	.55506	.35897	.56509	.36736	23
38	.52400	.33419	.53460	.34245	.54501	.35076	.55523	.35911	.56526	.36750	22
39	.52418	.33433	.53477	.34259	.54518	.35090	.55539	.35925	.56543	.36764	21
40	9.52436	0.33447	9.53495	0.34273	9.54535	0.35103	9.55556	0.35939	9.56559	0.36778	20
41	.52453	.33461	.53512	.34287	.54552	.35117	.55573	.35953	.56576	.36792	19
42	.52471	.33474	.53530	.34300	.54569	.35131	.55590	.35967	.56592	.36806	18
43	.52489	.33488	.53547	.34314	.54587	.35145	.55607	.35981	.56609	.36820	17
44	.52507	.33502	.53565	.34328	.54604	.35159	.55624	.35995	.56625	.36834	16
45	9.52525	0.33515	9.53582	0.34342	9.54621	0.35173	9.55641	0.36009	9.56642	0.36848	15
46	.52542	.33529	.53600	.34356	.54638	.35187	.55657	.36023	.56658	.36862	14
47	.52560	.33543	.53617	.34369	.54655	.35201	.55674	.36036	.56675	.36877	13
48	.52578	.33557	.53635	.34383	.54672	.35215	.55691	.36050	.56692	.36891	12
49	.52596	.33570	.53652	.34397	.54689	.35228	.55708	.36064	.56708	.36905	11
50	9.52613	0.33584	9.53670	0.34411	9.54707	0.35242	9.55725	0.36078	9.56725	0.36919	10
51	.52631	.33598	.53687	.34425	.54724	.35256	.55742	.36092	.56741	.36933	9
52	.52649	.33612	.53704	.34439	.54741	.35270	.55758	.36106	.56758	.36947	8
53	.52667	.33625	.53722	.34452	.54758	.35284	.55775	.36120	.56774	.36961	7
54	.52684	.33639	.53739	.34466	.54775	.35298	.55792	.36134	.56791	.36975	6
55	9.52702	0.33653	9.53757	0.34480	9.54792	0.35312	9.55809	0.36148	9.56807	0.36989	5
56	.52720	.33667	.53774	.34494	.54809	.35326	.55826	.36162	.56824	.37003	4
57	.52738	.33680	.53792	.34508	.54826	.35340	.55842	.36176	.56840	.37017	3
58	.52755	.33694	.53809	.34521	.54843	.35354	.55859	.36190	.56856	.37031	2
59	.52773	.33708	.53826	.34535	.54860	.35368	.55876	.36204	.56873	.37045	1
60	9.52791	0.33722	9.53844	0.34549	9.54878	0.35381	9.55893	0.36218	9.56889	0.37059	0
	289°		288°		287°		286°		285°		

TABLE 34

Haversines

′	75°		76°		77°		78°		79°		′
′	Log Hav	Nat. Hav	Log Hav	Nat. Hav	Log Hav	Nat. Hav	Log Hav	Nat. Hav	Log Hav	Nat. Hav	′
0	9. 56889	0. 37059	9. 57868	0. 37904	9. 58830	0. 38752	9. 59774	0. 39604	9. 60702	0. 40460	60
1	. 56906	. 37073	. 57885	. 37918	. 58846	. 38767	. 59790	. 39619	. 60717	. 40474	59
2	. 56922	. 37087	. 57901	. 37932	. 58862	. 38781	. 59806	. 39633	. 60733	. 40488	58
3	. 56939	. 37101	. 57917	. 37946	. 58878	. 38795	. 59821	. 39647	. 60748	. 40502	57
4	. 56955	. 37115	. 57933	. 37960	. 58893	. 38809	. 59837	. 39661	. 60763	. 40517	56
5	9. 56972	0. 37129	9. 57949	0. 37974	9. 58909	0. 38823	9. 59852	0. 39676	9. 60779	0. 40531	55
6	. 56988	. 37143	. 57965	. 37989	. 58925	. 38837	. 59868	. 39690	. 60794	. 40545	54
7	. 57005	. 37157	. 57981	. 38003	. 58941	. 38852	. 59883	. 39704	. 60809	. 40560	53
8	. 57021	. 37171	. 57998	. 38017	. 58957	. 38866	. 59899	. 39718	. 60825	. 40574	52
9	. 57037	. 37186	. 58014	. 38031	. 58973	. 38880	. 59915	. 39732	. 60840	. 40588	51
10	9. 57054	0. 37200	9. 58030	0. 38045	9. 58989	0. 38894	9. 59930	0. 39747	9. 60855	0. 40602	50
11	. 57070	. 37214	. 58046	. 38059	. 59004	. 38908	. 59946	. 39761	. 60870	. 40617	49
12	. 57087	. 37228	. 58062	. 38073	. 59020	. 38923	. 59961	. 39775	. 60886	. 40631	48
13	. 57103	. 37242	. 58078	. 38087	. 59036	. 38937	. 59977	. 39789	. 60901	. 40645	47
14	. 57119	. 37256	. 58094	. 38102	. 59052	. 38951	. 59992	. 39804	. 60916	. 40660	46
15	9. 57136	0. 37270	9. 58110	0. 38116	9. 59068	0. 38965	9. 60008	0. 39818	9. 60931	0. 40674	45
16	. 57152	. 37284	. 58126	. 38130	. 59083	. 38979	. 60023	. 39832	. 60947	. 40688	44
17	. 57169	. 37298	. 58143	. 38144	. 59099	. 38994	. 60039	. 39846	. 60962	. 40702	43
18	. 57185	. 37312	. 58159	. 38158	. 59115	. 39008	. 60054	. 39861	. 60977	. 40717	42
19	. 57201	. 37326	. 58175	. 38172	. 59131	. 39022	. 60070	. 39875	. 60992	. 40731	41
20	9. 57218	0. 37340	9. 58191	0. 38186	9. 59147	0. 39036	9. 60085	0. 39889	9. 61008	0. 40745	40
21	. 57234	. 37354	. 58207	. 38200	. 59162	. 39050	. 60101	. 39903	. 61023	. 40760	39
22	. 57250	. 37368	. 58223	. 38215	. 59178	. 39064	. 60116	. 39918	. 61038	. 40774	38
23	. 57267	. 37382	. 58239	. 38229	. 59194	. 39079	. 60132	. 39932	. 61053	. 40788	37
24	. 57283	. 37397	. 58255	. 38243	. 59210	. 39093	. 60147	. 39946	. 61069	. 40802	36
25	9. 57299	0. 37411	9. 58271	0. 38257	9. 59225	0. 39107	9. 60163	0. 39960	9. 61084	0. 40817	35
26	. 57316	. 37425	. 58287	. 38271	. 59241	. 39121	. 60178	. 39975	. 61099	. 40831	34
27	. 57332	. 37439	. 58303	. 38285	. 59257	. 39135	. 60194	. 39989	. 61114	. 40845	33
28	. 57348	. 37453	. 58319	. 38299	. 59273	. 39150	. 60209	. 40003	. 61129	. 40860	32
29	. 57365	. 37467	. 58335	. 38314	. 59289	. 39164	. 60225	. 40017	. 61145	. 40874	31
30	9. 57381	0. 37481	9. 58351	0. 38328	9. 59304	0. 39178	9. 60240	0. 40032	9. 61160	0. 40888	30
31	. 57397	. 37495	. 58367	. 38342	. 59320	. 39192	. 60256	. 40046	. 61175	. 40903	29
32	. 57414	. 37509	. 58383	. 38356	. 59336	. 39206	. 60271	. 40060	. 61190	. 40917	28
33	. 57430	. 37523	. 58399	. 38370	. 59351	. 39221	. 60287	. 40074	. 61205	. 40931	27
34	. 57446	. 37537	. 58415	. 38384	. 59367	. 39235	. 60302	. 40089	. 61221	. 40945	26
35	9. 57463	0. 37551	9. 58431	0. 38398	9. 59383	0. 39249	9. 60318	0. 40103	9. 61236	0. 40960	25
36	. 57479	. 37566	. 58447	. 38413	. 59399	. 39263	. 60333	. 40117	. 61251	. 40974	24
37	. 57495	. 37580	. 58463	. 38427	. 59414	. 39277	. 60348	. 40131	. 61266	. 40988	23
38	. 57511	. 37594	. 58479	. 38441	. 59430	. 39292	. 60364	. 40146	. 61281	. 41003	22
39	. 57528	. 37608	. 58495	. 38455	. 59446	. 39306	. 60379	. 40160	. 61296	. 41017	21
40	9. 57544	0. 37622	9. 58511	0. 38469	9. 59461	0. 39320	9. 60395	0. 40174	9. 61312	0. 41031	20
41	. 57560	. 37636	. 58527	. 38483	. 59477	. 39334	. 60410	. 40188	. 61327	. 41046	19
42	. 57577	. 37650	. 58543	. 38498	. 59493	. 39348	. 60426	. 40203	. 61342	. 41060	18
43	. 57593	. 37664	. 58559	. 38512	. 59508	. 39363	. 60441	. 40217	. 61357	. 41074	17
44	. 57609	. 37678	. 58575	. 38526	. 59524	. 39377	. 60456	. 40231	. 61372	. 41089	16
45	9. 57625	0. 37692	9. 58591	0. 38540	9. 59540	0. 39391	9. 60472	0. 40245	9. 61387	0. 41103	15
46	. 57642	. 37706	. 58607	. 38554	. 59556	. 39405	. 60487	. 40260	. 61402	. 41117	14
47	. 57658	. 37721	. 58623	. 38568	. 59571	. 39420	. 60502	. 40274	. 61417	. 41131	13
48	. 57674	. 37735	. 58639	. 38582	. 59587	. 39434	. 60518	. 40288	. 61433	. 41146	12
49	. 57690	. 37749	. 58655	. 38597	. 59602	. 39448	. 60533	. 40303	. 61448	. 41160	11
50	9. 57706	0. 37763	9. 58671	0. 38611	9. 59618	0. 39462	9. 60549	0. 40317	9. 61463	0. 41174	10
51	. 57723	. 37777	. 58687	. 38625	. 59634	. 39476	. 60564	. 40331	. 61478	. 41189	9
52	. 57739	. 37791	. 58703	. 38639	. 59649	. 39491	. 60579	. 40345	. 61493	. 41203	8
53	. 57755	. 37805	. 58719	. 38653	. 59665	. 39505	. 60595	. 40360	. 61508	. 41217	7
54	. 57771	. 37819	. 58735	. 38667	. 59681	. 39519	. 60610	. 40374	. 61523	. 41232	6
55	9. 57787	0. 37833	9. 58750	0. 38682	9. 59696	0. 39533	9. 60625	0. 40388	9. 61538	0. 41246	5
56	. 57804	. 37847	. 58766	. 38696	. 59712	. 39548	. 60641	. 40402	. 61553	. 41260	4
57	. 57820	. 37862	. 58782	. 38710	. 59728	. 39562	. 60656	. 40417	. 61568	. 41275	3
58	. 57836	. 37876	. 58798	. 38724	. 59743	. 39576	. 60671	. 40431	. 61583	. 41289	2
59	. 57852	. 37890	. 58814	. 38738	. 59759	. 39590	. 60687	. 40445	. 61598	. 41303	1
60	9. 57868	0. 37904	9. 58830	0. 38752	9. 59774	0. 39604	9. 60702	0. 40460	9. 61614	0. 41318	0
	284°		283°		282°		281°		280°		

TABLE 34

Haversines

	80°		81°		82°		83°		84°		
′	Log Hav	Nat. Hav	Log Hav	Nat. Hav	Log Hav	Nat. Hav	Log Hav	Nat. Hav	Log Hav	Nat. Hav	′
0	9. 61614	0. 41318	9. 62509	0. 42178	9. 63389	0. 43041	9. 64253	0. 43907	9. 65102	0. 44774	60
1	. 61629	. 41332	. 62524	. 42193	. 63403	. 43056	. 64267	. 43921	. 65116	. 44788	59
2	. 61644	. 41346	. 62538	. 42207	. 63418	. 43070	. 64281	. 43935	. 65130	. 44803	58
3	. 61659	. 41361	. 62553	. 42221	. 63432	. 43085	. 64296	. 43950	. 65144	. 44817	57
4	. 61674	. 41375	. 62568	. 42236	. 63447	. 43099	. 64310	. 43964	. 65158	. 44831	56
5	9. 61689	0. 41389	9. 62583	0. 42250	9. 63461	0. 43113	9. 64324	0. 43979	9. 65172	0. 44846	55
6	. 61704	. 41404	. 62598	. 42264	. 63476	. 43128	. 64339	. 43993	. 65186	. 44860	54
7	. 61719	. 41418	. 62612	. 42279	. 63490	. 43142	. 64353	. 44008	. 65200	. 44875	53
8	. 61734	. 41432	. 62627	. 42293	. 63505	. 43157	. 64367	. 44022	. 65214	. 44889	52
9	. 61749	. 41447	. 62642	. 42308	. 63519	. 43171	. 64381	. 44036	. 65228	. 44904	51
10	9. 61764	0. 41461	9. 62657	0. 42322	9. 63534	0. 43185	9. 64396	0. 44051	9. 65242	0. 44918	50
11	. 61779	. 41475	. 62671	. 42336	. 63548	. 43200	. 64410	. 44065	. 65256	. 44933	49
12	. 61794	. 41490	. 62686	. 42351	. 63563	. 43214	. 64424	. 44080	. 65270	. 44947	48
13	. 61809	. 41504	. 62701	. 42365	. 63577	. 43229	. 64438	. 44094	. 65284	. 44962	47
14	. 61824	. 41518	. 62716	. 42379	. 63592	. 43243	. 64452	. 44109	. 65298	. 44976	46
15	9. 61839	0. 41533	9. 62730	0. 42394	9. 63606	0. 43257	9. 64467	0. 44123	9. 65312	0. 44991	45
16	. 61854	. 41547	. 62745	. 42408	. 63621	. 43272	. 64481	. 44138	. 65326	. 45005	44
17	. 61869	. 41561	. 62760	. 42423	. 63635	. 43286	. 64495	. 44152	. 65340	. 45020	43
18	. 61884	. 41576	. 62774	. 42437	. 63649	. 43301	. 64509	. 44166	. 65354	. 45034	42
19	. 61899	. 41590	. 62789	. 42451	. 63664	. 43315	. 64523	. 44181	. 65368	. 45048	41
20	9. 61914	0. 41604	9. 62804	0. 42466	9. 63678	0. 43330	9. 64538	0. 44195	9. 65382	0. 45063	40
21	. 61929	. 41619	. 62819	. 42480	. 63693	. 43344	. 64552	. 44210	. 65396	. 45077	39
22	. 61944	. 41633	. 62833	. 42494	. 63707	. 43358	. 64566	. 44224	. 65410	. 45092	38
23	. 61959	. 41647	. 62848	. 42509	. 63722	. 43373	. 64580	. 44239	. 65424	. 45106	37
24	. 61974	. 41662	. 62863	. 42523	. 63736	. 43387	. 64594	. 44253	. 65438	. 45121	36
25	9. 61989	0. 41676	9. 62877	0. 42538	9. 63751	0. 43402	9. 64609	0. 44268	9. 65452	0. 45135	35
26	. 62003	. 41690	. 62892	. 42552	. 63765	. 43416	. 64623	. 44282	. 65466	. 45150	34
27	. 62018	. 41705	. 62907	. 42566	. 63779	. 43430	. 64637	. 44296	. 65480	. 45164	33
28	. 62033	. 41719	. 62921	. 42581	. 63794	. 43445	. 64651	. 44311	. 65493	. 45179	32
29	. 62048	. 41733	. 62936	. 42595	. 63808	. 43459	. 64665	. 44325	. 65507	. 45193	31
30	9. 62063	0. 41748	9. 62951	0. 42610	9. 63823	0. 43474	9. 64679	0. 44340	9. 65521	0. 45208	30
31	. 62078	. 41762	. 62965	. 42624	. 63837	. 43488	. 64694	. 44354	. 65535	. 45222	29
32	. 62093	. 41776	. 62980	. 42638	. 63851	. 43503	. 64708	. 44369	. 65549	. 45237	28
33	. 62108	. 41791	. 62995	. 42653	. 63866	. 43517	. 64722	. 44383	. 65563	. 45251	27
34	. 62123	. 41805	. 63009	. 42667	. 63880	. 43531	. 64736	. 44398	. 65577	. 45266	26
35	9. 62138	0. 41819	9. 63024	0. 42681	9. 63895	0. 43546	9. 64750	0. 44412	9. 65591	0. 45280	25
36	. 62153	. 41834	. 63039	. 42696	. 63909	. 43560	. 64764	. 44427	. 65605	. 45295	24
37	. 62168	. 41848	. 63053	. 42710	. 63923	. 43575	. 64778	. 44441	. 65619	. 45309	23
38	. 62182	. 41862	. 63068	. 42725	. 63938	. 43589	. 64793	. 44455	. 65632	. 45324	22
39	. 62197	. 41877	. 63082	. 42739	. 63952	. 43603	. 64807	. 44470	. 65646	. 45338	21
40	9. 62212	0. 41891	9. 63097	0. 42753	9. 63966	0. 43618	9. 64821	0. 44484	9. 65660	0. 45353	20
41	. 62227	. 41905	. 63112	. 42768	. 63981	. 43632	. 64835	. 44499	. 65674	. 45367	19
42	. 62242	. 41920	. 63126	. 42782	. 63995	. 43647	. 64849	. 44513	. 65688	. 45381	18
43	. 62257	. 41934	. 63141	. 42797	. 64010	. 43661	. 64863	. 44528	. 65702	. 45396	17
44	. 62272	. 41949	. 63156	. 42811	. 64024	. 43676	. 64877	. 44542	. 65716	. 45410	16
45	9. 62287	0. 41963	9. 63170	0. 42825	9. 64038	0. 43690	9. 64891	0. 44557	9. 65729	0. 45425	15
46	. 62301	. 41977	. 63185	. 42840	. 64053	. 43704	. 64905	. 44571	. 65743	. 45439	14
47	. 62316	. 41992	. 63199	. 42854	. 64067	. 43719	. 64919	. 44586	. 65757	. 45454	13
48	. 62331	. 42006	. 63214	. 42869	. 64081	. 43733	. 64934	. 44600	. 65771	. 45468	12
49	. 62346	. 42020	. 63228	. 42883	. 64096	. 43748	. 64948	. 44614	. 65785	. 45483	11
50	9. 62361	0. 42035	9. 63243	0. 42897	9. 64110	0. 43762	9. 64962	0. 44629	9. 65799	0. 45497	10
51	. 62376	. 42049	. 63258	. 42912	. 64124	. 43777	. 64976	. 44643	. 65812	. 45512	9
52	. 62390	. 42063	. 63272	. 42926	. 64139	. 43791	. 64990	. 44658	. 65826	. 45526	8
53	. 62405	. 42078	. 63287	. 42941	. 64153	. 43805	. 65004	. 44672	. 65840	. 45541	7
54	. 62420	. 42092	. 63301	. 42955	. 64167	. 43820	. 65018	. 44687	. 65854	. 45555	6
55	9. 62435	0. 42106	9. 63316	0. 42969	9. 64181	0. 43834	9. 65032	0. 44701	9. 65868	0. 45570	5
56	. 62450	. 42121	. 63330	. 42984	. 64196	. 43849	. 65046	. 44716	. 65881	. 45584	4
57	. 62464	. 42135	. 63345	. 42998	. 64210	. 43863	. 65060	. 44730	. 65895	. 45599	3
58	. 62479	. 42150	. 63360	. 43013	. 64224	. 43878	. 65074	. 44745	. 65909	. 45613	2
59	. 62494	. 42164	. 63374	. 43027	. 64239	. 43892	. 65088	. 44759	. 65923	. 45628	1
60	9. 62509	0. 42178	9. 63389	0. 43041	9. 64253	0. 43907	9. 65102	0. 44774	9. 65937	0. 45642	0
	279°		278°		277°		276°		275°		

TABLE 34

Haversines

′	85° Log Hav	85° Nat. Hav	86° Log Hav	86° Nat. Hav	87° Log Hav	87° Nat. Hav	88° Log Hav	88° Nat. Hav	89° Log Hav	89° Nat. Hav	′
0	9. 65937	0. 45642	9. 66757	0. 46512	9. 67562	0. 47383	9. 68354	0. 48255	9. 69132	0. 49127	60
1	. 65950	. 45657	. 66770	. 46527	. 67576	. 47398	. 68367	. 48270	. 69145	. 49142	59
2	. 65964	. 45671	. 66784	. 46541	. 67589	. 47412	. 68380	. 48284	. 69158	. 49156	58
3	. 65978	. 45686	. 66797	. 46556	. 67602	. 47427	. 68393	. 48299	. 69171	. 49171	57
4	. 65992	. 45700	. 66811	. 46570	. 67616	. 47441	. 68407	. 48313	. 69184	. 49186	56
5	9. 66006	0. 45715	9. 66824	0. 46585	9. 67629	0. 47456	9. 68420	0. 48328	9. 69197	0. 49200	55
6	. 66019	. 45729	. 66838	. 46599	. 67642	. 47470	. 68433	. 48342	. 69209	. 49215	54
7	. 66033	. 45744	. 66851	. 46614	. 67656	. 47485	. 68446	. 48357	. 69222	. 49229	53
8	. 66047	. 45758	. 66865	. 46628	. 67669	. 47499	. 68459	. 48371	. 69235	. 49244	52
9	. 66061	. 45773	. 66878	. 46643	. 67682	. 47514	. 68472	. 48386	. 69248	. 49258	51
10	9. 66074	0. 45787	9. 66892	0. 46657	9. 67695	0. 47528	9. 68485	0. 48400	9. 69261	0. 49273	50
11	. 66088	. 45802	. 66905	. 46672	. 67709	. 47543	. 68498	. 48415	. 69274	. 49287	49
12	. 66102	. 45816	. 66919	. 46686	. 67722	. 47558	. 68511	. 48429	. 69286	. 49302	48
13	. 66116	. 45831	. 66932	. 46701	. 67735	. 47572	. 68524	. 48444	. 69299	. 49316	47
14	. 66129	. 45845	. 66946	. 46715	. 67748	. 47587	. 68537	. 48459	. 69312	. 49331	46
15	9. 66143	0. 45860	9. 66959	0. 46730	9. 67762	0. 47601	9. 68550	0. 48473	9. 69325	0. 49346	45
16	. 66157	. 45874	. 66973	. 46744	. 67775	. 47616	. 68563	. 48488	. 69338	. 49360	44
17	. 66170	. 45889	. 66986	. 46759	. 67788	. 47630	. 68576	. 48502	. 69350	. 49375	43
18	. 66184	. 45903	. 67000	. 46773	. 67801	. 47645	. 68589	. 48517	. 69363	. 49389	42
19	. 66198	. 45918	. 67013	. 46788	. 67815	. 47659	. 68602	. 48531	. 69376	. 49404	41
20	9. 66212	0. 45932	9. 67027	0. 46802	9. 67828	0. 47674	9. 68615	0. 48546	9. 69389	0. 49418	40
21	. 66225	. 45947	. 67040	. 46817	. 67841	. 47688	. 68628	. 48560	. 69402	. 49433	39
22	. 66239	. 45961	. 67054	. 46831	. 67854	. 47703	. 68641	. 48575	. 69414	. 49447	38
23	. 66253	. 45976	. 67067	. 46846	. 67868	. 47717	. 68654	. 48589	. 69427	. 49462	37
24	. 66266	. 45990	. 67081	. 46860	. 67881	. 47732	. 68667	. 48604	. 69440	. 49476	36
25	9. 66280	0. 46005	9. 67094	0. 46875	9. 67894	0. 47746	9. 68680	0. 48618	9. 69453	0. 49491	35
26	. 66294	. 46019	. 67108	. 46890	. 67907	. 47761	. 68693	. 48633	. 69465	. 49505	34
27	. 66307	. 46034	. 67121	. 46904	. 67920	. 47775	. 68706	. 48648	. 69478	. 49520	33
28	. 66321	. 46048	. 67134	. 46919	. 67934	. 47790	. 68719	. 48662	. 69491	. 49535	32
29	. 66335	. 46063	. 67148	. 46933	. 67947	. 47804	. 68732	. 48677	. 69504	. 49549	31
30	9. 66348	0. 46077	9. 67161	0. 46948	9. 67960	0. 47819	9. 68745	0. 48691	9. 69516	0. 49564	30
31	. 66362	. 46092	. 67175	. 46962	. 67973	. 47834	. 68758	. 48706	. 69529	. 49578	29
32	. 66376	. 46106	. 67188	. 46977	. 67986	. 47848	. 68771	. 48720	. 69542	. 49593	28
33	. 66389	. 46121	. 67202	. 46991	. 68000	. 47863	. 68784	. 48735	. 69555	. 49607	27
34	. 66403	. 46135	. 67215	. 47006	. 68013	. 47877	. 68797	. 48749	. 69567	. 49622	26
35	9. 66417	0. 46150	9. 67228	0. 47020	9. 68026	0. 47892	9. 68810	0. 48764	9. 69580	0. 49636	25
36	. 66430	. 46164	. 67242	. 47035	. 68039	. 47906	. 68823	. 48778	. 69593	. 49651	24
37	. 66444	. 46179	. 67255	. 47049	. 68052	. 47921	. 68836	. 48793	. 69605	. 49665	23
38	. 66458	. 46193	. 67269	. 47064	. 68066	. 47935	. 68849	. 48807	. 69618	. 49680	22
39	. 66471	. 46208	. 67282	. 47078	. 68079	. 47950	. 68862	. 48822	. 69631	. 49695	21
40	9. 66485	0. 46222	9. 67295	0. 47093	9. 68092	0. 47964	9. 68875	0. 48837	9. 69644	0. 49709	20
41	. 66499	. 46237	. 67309	. 47107	. 68105	. 47979	. 68887	. 48851	. 69656	. 49724	19
42	. 66512	. 46251	. 67322	. 47122	. 68118	. 47993	. 68900	. 48866	. 69669	. 49738	18
43	. 66526	. 46266	. 67336	. 47136	. 68131	. 48008	. 68913	. 48880	. 69682	. 49753	17
44	. 66539	. 46280	. 67349	. 47151	. 68144	. 48022	. 68926	. 48895	. 69694	. 49767	16
45	9. 66553	0. 46295	9. 67362	0. 47165	9. 68158	0. 48037	9. 68939	0. 48909	9. 69707	0. 49782	15
46	. 66567	. 46309	. 67376	. 47180	. 68171	. 48052	. 68952	. 48924	. 69720	. 49796	14
47	. 66580	. 46324	. 67389	. 47194	. 68184	. 48066	. 68965	. 48938	. 69732	. 49811	13
48	. 66594	. 46338	. 67402	. 47209	. 68197	. 48081	. 68978	. 48953	. 69745	. 49825	12
49	. 66607	. 46353	. 67416	. 47223	. 68210	. 48095	. 68991	. 48967	. 69758	. 49840	11
50	9. 66621	0. 46367	9. 67429	0. 47238	9. 68223	0. 48110	9. 69004	0. 48982	9. 69770	0. 49855	10
51	. 66635	. 46382	. 67443	. 47252	. 68236	. 48124	. 69017	. 48997	. 69783	. 49869	9
52	. 66648	. 46396	. 67456	. 47267	. 68249	. 48139	. 69029	. 49011	. 69796	. 49884	8
53	. 66662	. 46411	. 67469	. 47282	. 68263	. 48153	. 69042	. 49026	. 69808	. 49898	7
54	. 66675	. 46425	. 67483	. 47296	. 68276	. 48168	. 69055	. 49040	. 69821	. 49913	6
55	9. 66689	0. 46440	9. 67496	0. 47311	9. 68289	0. 48182	9. 69068	0. 49055	9. 69834	0. 49927	5
56	. 66702	. 46454	. 67509	. 47325	. 68302	. 48197	. 69081	. 49069	. 69846	. 49942	4
57	. 66716	. 46469	. 67522	. 47340	. 68315	. 48211	. 69094	. 49084	. 69859	. 49956	3
58	. 66730	. 46483	. 67536	. 47354	. 68328	. 48226	. 69107	. 49098	. 69872	. 49971	2
59	. 66743	. 46498	. 67549	. 47369	. 68341	. 48240	. 69120	. 49113	. 69884	. 49985	1
60	9. 66757	0. 46512	9. 67562	0. 47383	9. 68354	0. 48255	9. 69132	0. 49127	9. 69897	0. 50000	0
	274°		273°		272°		271°		270°		

TABLE 34

Haversines

′	90°		91°		92°		93°		94°		′
	Log Hav	Nat. Hav	Log Hav	Nat. Hav	Log Hav	Nat. Hav	Log Hav	Nat. Hav	Log Hav	Nat. Hav	
0	9. 69897	0. 50000	9. 70648	0. 50873	9. 71387	0. 51745	9. 72112	0. 52617	9. 72825	0. 53488	60
1	. 69910	. 50015	. 70661	. 50887	. 71399	. 51760	. 72124	. 52631	. 72837	. 53502	59
2	. 69922	. 50029	. 70673	. 50902	. 71411	. 51774	. 72136	. 52646	. 72849	. 53517	58
3	. 69935	. 50044	. 70686	. 50916	. 71423	. 51789	. 72148	. 52660	. 72861	. 53531	57
4	. 69948	. 50058	. 70698	. 50931	. 71436	. 51803	. 72160	. 52675	. 72873	. 53546	56
5	9. 69960	0. 50073	9. 70710	0. 50945	9. 71448	0. 51818	9. 72172	0. 52689	9. 72884	0. 53560	55
6	. 69973	. 50087	. 70723	. 50960	. 71460	. 51832	. 72184	. 52704	. 72896	. 53575	54
7	. 69985	. 50102	. 70735	. 50974	. 71472	. 51847	. 72196	. 52718	. 72908	. 53589	53
8	. 69998	. 50116	. 70748	. 50989	. 71484	. 51861	. 72208	. 52733	. 72920	. 53604	52
9	. 70011	. 50131	. 70760	. 51003	. 71496	. 51876	. 72220	. 52748	. 72931	. 53618	51
10	9. 70023	0. 50145	9. 70772	0. 51018	9. 71509	0. 51890	9. 72232	0. 52762	9. 72943	0. 53633	50
11	. 70036	. 50160	. 70785	. 51033	. 71521	. 51905	. 72244	. 52777	. 72955	. 53647	49
12	. 70048	. 50175	. 70797	. 51047	. 71533	. 51919	. 72256	. 52791	. 72967	. 53662	48
13	. 70061	. 50189	. 70809	. 51062	. 71545	. 51934	. 72268	. 52806	. 72978	. 53676	47
14	. 70074	. 50204	. 70822	. 51076	. 71557	. 51948	. 72280	. 52820	. 72990	. 53691	46
15	9. 70086	0. 50218	9. 70834	0. 51091	9. 71569	0. 51963	9. 72292	0. 52835	9. 73002	0. 53705	45
16	. 70099	. 50233	. 70847	. 51105	. 71582	. 51978	. 72304	. 52849	. 73014	. 53720	44
17	. 70111	. 50247	. 70859	. 51120	. 71594	. 51992	. 72316	. 52864	. 73025	. 53734	43
18	. 70124	. 50262	. 70871	. 51134	. 71606	. 52007	. 72328	. 52878	. 73037	. 53749	42
19	. 70136	. 50276	. 70884	. 51149	. 71618	. 52021	. 72340	. 52893	. 73049	. 53763	41
20	9. 70149	0. 50291	9. 70896	0. 51163	9. 71630	0. 52036	9. 72352	0. 52907	9. 73060	0. 53778	40
21	. 70161	. 50305	. 70908	. 51178	. 71642	. 52050	. 72363	. 52922	. 73072	. 53792	39
22	. 70174	. 50320	. 70921	. 51193	. 71654	. 52065	. 72375	. 52936	. 73084	. 53807	38
23	. 70187	. 50335	. 70933	. 51207	. 71666	. 52079	. 72387	. 52951	. 73096	. 53821	37
24	. 70199	. 50349	. 70945	. 51222	. 71679	. 52094	. 72399	. 52965	. 73107	. 53836	36
25	9. 70212	0. 50364	9. 70958	0. 51236	9. 71691	0. 52108	9. 72411	0. 52980	9. 73119	0. 53850	35
26	. 70224	. 50378	. 70970	. 51251	. 71703	. 52123	. 72423	. 52994	. 73131	. 53865	34
27	. 70237	. 50393	. 70982	. 51265	. 71715	. 52137	. 72435	. 53009	. 73142	. 53879	33
28	. 70249	. 50407	. 70995	. 51280	. 71727	. 52152	. 72447	. 53023	. 73154	. 53894	32
29	. 70262	. 50422	. 71007	. 51294	. 71739	. 52166	. 72459	. 53038	. 73166	. 53908	31
30	9. 70274	0. 50436	9. 71019	0. 51309	9. 71751	0. 52181	9. 72471	0. 53052	9. 73177	0. 53923	30
31	. 70287	. 50451	. 71032	. 51323	. 71763	. 52195	. 72482	. 53067	. 73189	. 53937	29
32	. 70299	. 50465	. 71044	. 51338	. 71775	. 52210	. 72494	. 53081	. 73201	. 53952	28
33	. 70312	. 50480	. 71056	. 51352	. 71787	. 52225	. 72506	. 53096	. 73212	. 53966	27
34	. 70324	. 50495	. 71068	. 51367	. 71800	. 52239	. 72518	. 53110	. 73224	. 53981	26
35	9. 70337	0. 50509	9. 71081	0. 51382	9. 71812	0. 52254	9. 72530	0. 53125	9. 73236	0. 53995	25
36	. 70349	. 50524	. 71093	. 51396	. 71824	. 52268	. 72542	. 53140	. 73247	. 54010	24
37	. 70362	. 50538	. 71105	. 51411	. 71836	. 52283	. 72554	. 53154	. 73259	. 54024	23
38	. 70374	. 50553	. 71118	. 51425	. 71848	. 52297	. 72565	. 53169	. 73271	. 54039	22
39	. 70387	. 50567	. 71130	. 51440	. 71860	. 52312	. 72577	. 53183	. 73282	. 54053	21
40	9. 70399	0. 50582	9. 71142	0. 51454	9. 71872	0. 52326	9. 72589	0. 53198	9. 73294	0. 54068	20
41	. 70412	. 50596	. 71154	. 51469	. 71884	. 52341	. 72601	. 53212	. 73306	. 54082	19
42	. 70424	. 50611	. 71167	. 51483	. 71896	. 52355	. 72613	. 53227	. 73317	. 54097	18
43	. 70437	. 50625	. 71179	. 51498	. 71908	. 52370	. 72625	. 53241	. 73329	. 54111	17
44	. 70449	. 50640	. 71191	. 51512	. 71920	. 52384	. 72637	. 53256	. 73341	. 54126	16
45	9. 70462	0. 50654	9. 71203	0. 51527	9. 71932	0. 52399	9. 72648	0. 53270	9. 73352	0. 54140	15
46	. 70474	. 50669	. 71216	. 51541	. 71944	. 52413	. 72660	. 53285	. 73364	. 54155	14
47	. 70487	. 50684	. 71228	. 51556	. 71956	. 52428	. 72672	. 53299	. 73375	. 54169	13
48	. 70499	. 50698	. 71240	. 51571	. 71968	. 52442	. 72684	. 53314	. 73387	. 54184	12
49	. 70512	. 50713	. 71252	. 51585	. 71980	. 52457	. 72696	. 53328	. 73399	. 54198	11
50	9. 70524	0. 50727	9. 71265	0. 51600	9. 71992	0. 52472	9. 72708	0. 53343	9. 73410	0. 54213	10
51	. 70537	. 50742	. 71277	. 51614	. 72004	. 52486	. 72719	. 53357	. 73422	. 54227	9
52	. 70549	. 50756	. 71289	. 51629	. 72016	. 52501	. 72731	. 53372	. 73433	. 54242	8
53	. 70561	. 50771	. 71301	. 51643	. 72028	. 52515	. 72743	. 53386	. 73445	. 54256	7
54	. 70574	. 50785	. 71314	. 51658	. 72040	. 52530	. 72755	. 53401	. 73457	. 54271	6
55	9. 70586	0. 50800	9. 71326	0. 51672	9. 72052	0. 52544	9. 72767	0. 53415	9. 73468	0. 54285	5
56	. 70599	. 50814	. 71338	. 51687	. 72064	. 52559	. 72778	. 53430	. 73480	. 54300	4
57	. 70611	. 50829	. 71350	. 51701	. 72076	. 52573	. 72790	. 53444	. 73491	. 54314	3
58	. 70624	. 50844	. 71362	. 51716	. 72088	. 52588	. 72802	. 53459	. 73503	. 54329	2
59	. 70636	. 50858	. 71375	. 51730	. 72100	. 52602	. 72814	. 53473	. 73515	. 54343	1
60	9. 70648	0. 50873	9. 71387	0. 51745	9. 72112	0. 52617	9. 72825	0. 53488	9. 73526	0. 54358	0
	269°		268°		267°		266°		265°		

TABLE 34

Haversines

	95°		96°		97°		98°		99°		
	Log Hav	Nat. Hav	Log Hav	Nat. Hav	Log Hav	Nat. Hav	Log Hav	Nat. Hav	Log Hav	Nat. Hav	′
0	9. 73526	0. 54358	9. 74215	0. 55226	9. 74891	0. 56093	9. 75556	0. 56959	9. 76209	0. 57822	60
1	. 73538	. 54372	. 74226	. 55241	. 74902	. 56108	. 75567	. 56973	. 76220	. 57836	59
2	. 73549	. 54387	. 74237	. 55255	. 74914	. 56122	. 75578	. 56987	. 76231	. 57850	58
3	. 73561	. 54401	. 74249	. 55270	. 74925	. 56137	. 75589	. 57002	. 76241	. 57865	57
4	. 73572	. 54416	. 74260	. 55284	. 74936	. 56151	. 75600	. 57016	. 76252	. 57879	56
5	9. 73584	0. 54430	9. 74272	0. 55299	9. 74947	0. 56166	9. 75611	0. 57031	9. 76263	0. 57894	55
6	. 73596	. 54445	. 74283	. 55313	. 74958	. 56180	. 75622	. 57045	. 76274	. 57908	54
7	. 73607	. 54459	. 74294	. 55328	. 74969	. 56195	. 75633	. 57059	. 76285	. 57922	53
8	. 73619	. 54474	. 74306	. 55342	. 74981	. 56209	. 75644	. 57074	. 76296	. 57937	52
9	. 73630	. 54488	. 74317	. 55357	. 74992	. 56223	. 75655	. 57088	. 76306	. 57951	51
10	9. 73642	0. 54503	9. 74328	0. 55371	9. 75003	0. 56238	9. 75666	0. 57103	9. 76317	0. 57965	50
11	. 73653	. 54517	. 74340	. 55386	. 75014	. 56252	. 75677	. 57117	. 76328	. 57980	49
12	. 73665	. 54532	. 74351	. 55400	. 75025	. 56267	. 75688	. 57131	. 76338	. 57994	48
13	. 73676	. 54546	. 74362	. 55414	. 75036	. 56281	. 75698	. 57146	. 76349	. 58008	47
14	. 73688	. 54561	. 74374	. 55429	. 75047	. 56296	. 75709	. 57160	. 76360	. 58023	46
15	9. 73699	0. 54575	9. 74385	0. 55443	9. 75059	0. 56310	9. 75720	0. 57175	9. 76371	0. 58037	45
16	. 73711	. 54590	. 74396	. 55458	. 75070	. 56324	. 75731	. 57189	. 76381	. 58051	44
17	. 73722	. 54604	. 74408	. 55472	. 75081	. 56339	. 75742	. 57203	. 76392	. 58066	43
18	. 73734	. 54619	. 74419	. 55487	. 75092	. 56353	. 75753	. 57218	. 76403	. 58080	42
19	. 73746	. 54633	. 74430	. 55501	. 75103	. 56368	. 75764	. 57232	. 76414	. 58095	41
20	9. 73757	0. 54647	9. 74442	0. 55516	9. 75114	0. 56382	9. 75775	0. 57247	9. 76424	0. 58109	40
21	. 73769	. 54662	. 74453	. 55530	. 75125	. 56397	. 75786	. 57261	. 76435	. 58123	39
22	. 73780	. 54676	. 74464	. 55545	. 75136	. 56411	. 75797	. 57275	. 76446	. 58138	38
23	. 73792	. 54691	. 74475	. 55559	. 75147	. 56425	. 75808	. 57290	. 76456	. 58152	37
24	. 73803	. 54705	. 74487	. 55573	. 75159	. 56440	. 75819	. 57304	. 76467	. 58166	36
25	9. 73815	0. 54720	9. 74498	0. 55588	9. 75170	0. 56454	9. 75830	0. 57319	9. 76478	0. 58181	35
26	. 73826	. 54734	. 74509	. 55602	. 75181	. 56469	. 75840	. 57333	. 76489	. 58195	34
27	. 73838	. 54749	. 74521	. 55617	. 75192	. 56483	. 75851	. 57347	. 76499	. 58209	33
28	. 73849	. 54763	. 74532	. 55631	. 75203	. 56497	. 75862	. 57362	. 76510	. 58224	32
29	. 73860	. 54778	. 74543	. 55646	. 75214	. 56512	. 75873	. 57376	. 76521	. 58238	31
30	9. 73872	0. 54792	9. 74554	0. 55660	9. 75225	0. 56526	9. 75884	0. 57390	9. 76531	0. 58252	30
31	. 73883	. 54807	. 74566	. 55675	. 75236	. 56541	. 75895	. 57405	. 76542	. 58267	29
32	. 73895	. 54821	. 74577	. 55689	. 75247	. 56555	. 75906	. 57419	. 76553	. 58281	28
33	. 73906	. 54836	. 74588	. 55704	. 75258	. 56570	. 75917	. 57434	. 76563	. 58295	27
34	. 73918	. 54850	. 74600	. 55718	. 75269	. 56584	. 75927	. 57448	. 76574	. 58310	26
35	9. 73929	0. 54865	9. 74611	0. 55732	9. 75280	0. 56598	9. 75938	0. 57462	9. 76585	0. 58324	25
36	. 73941	. 54879	. 74622	. 55747	. 75291	. 56613	. 75949	. 57477	. 76595	. 58338	24
37	. 73952	. 54894	. 74633	. 55761	. 75303	. 56627	. 75960	. 57491	. 76606	. 58353	23
38	. 73964	. 54908	. 74645	. 55776	. 75314	. 56642	. 75971	. 57506	. 76617	. 58367	22
39	. 73975	. 54923	. 74656	. 55790	. 75325	. 56656	. 75982	. 57520	. 76627	. 58381	21
40	9. 73987	0. 54937	9. 74667	0. 55805	9. 75336	0. 56670	9. 75993	0. 57534	9. 76638	0. 58396	20
41	. 73998	. 54952	. 74678	. 55819	. 75347	. 56685	. 76004	. 57549	. 76649	. 58410	19
42	. 74009	. 54966	. 74690	. 55834	. 75358	. 56699	. 76014	. 57563	. 76659	. 58424	18
43	. 74021	. 54980	. 74701	. 55848	. 75369	. 56714	. 76025	. 57577	. 76670	. 58439	17
44	. 74032	. 54995	. 74712	. 55862	. 75380	. 56728	. 76036	. 57592	. 76681	. 58453	16
45	9. 74044	0. 55009	9. 74723	0. 55877	9. 75391	0. 56743	9. 76047	0. 57606	9. 76691	0. 58467	15
46	. 74055	. 55024	. 74734	. 55891	. 75402	. 56757	. 76058	. 57621	. 76702	. 58482	14
47	. 74067	. 55038	. 74746	. 55906	. 75413	. 56771	. 76069	. 57635	. 76713	. 58496	13
48	. 74078	. 55053	. 74757	. 55920	. 75424	. 56786	. 76079	. 57649	. 76723	. 58510	12
49	. 74089	. 55067	. 74768	. 55935	. 75435	. 56800	. 76090	. 57664	. 76734	. 58525	11
50	9. 74101	0. 55082	9. 74779	0. 55949	9. 75446	0. 56815	9. 76101	0. 57678	9. 76745	0. 58539	10
51	. 74112	. 55096	. 74791	. 55964	. 75457	. 56829	. 76112	. 57692	. 76755	. 58553	9
52	. 74124	. 55111	. 74802	. 55978	. 75468	. 56843	. 76123	. 57707	. 76766	. 58568	8
53	. 74135	. 55125	. 74813	. 55992	. 75479	. 56858	. 76134	. 57721	. 76777	. 58582	7
54	. 74146	. 55140	. 74824	. 56007	. 75490	. 56872	. 76144	. 57736	. 76787	. 58596	6
55	9. 74158	0. 55154	9. 74835	0. 56021	9. 75501	0. 56887	9. 76155	0. 57750	9. 76798	0. 58611	5
56	. 74169	. 55169	. 74846	. 56036	. 75512	. 56901	. 76166	. 57764	. 76808	. 58625	4
57	. 74181	. 55183	. 74858	. 56050	. 75523	. 56915	. 76177	. 57779	. 76819	. 58639	3
58	. 74192	. 55197	. 74869	. 56065	. 75534	. 56930	. 76188	. 57793	. 76830	. 58654	2
59	. 74203	. 55212	. 74880	. 56079	. 75545	. 56944	. 76198	. 57807	. 76840	. 58668	1
60	9. 74215	0. 55226	9. 74891	0. 56093	9. 75556	0. 56959	9. 76209	0. 57822	9. 76851	0. 58682	0
	264°		263°		262°		261°		260°		

TABLE 34

Haversines

	100°		101°		102°		103°		104°		
′	Log Hav	Nat. Hav	Log Hav	Nat. Hav	Log Hav	Nat. Hav	Log Hav	Nat. Hav	Log Hav	Nat. Hav	′
0	9. 76851	0. 58682	9. 77481	0. 59540	9. 78101	0. 60396	9. 78709	0. 61248	9. 79306	0. 62096	60
1	. 76861	. 58697	. 77492	. 59555	. 78111	. 60410	. 78719	. 61262	. 79316	. 62110	59
2	. 76872	. 58711	. 77502	. 59569	. 78121	. 60424	. 78729	. 61276	. 79326	. 62124	58
3	. 76883	. 58725	. 77512	. 59583	. 78131	. 60438	. 78739	. 61290	. 79336	. 62138	57
4	. 76893	. 58740	. 77523	. 59598	. 78141	. 60452	. 78749	. 61304	. 79346	. 62153	56
5	9. 76904	0. 58754	9. 77533	0. 59612	9. 78152	0. 60467	9. 78759	0. 61318	9. 79356	0. 62167	55
6	. 76914	. 58768	. 77544	. 59626	. 78162	. 60481	. 78769	. 61333	. 79366	. 62181	54
7	. 76925	. 58783	. 77554	. 59640	. 78172	. 60495	. 78779	. 61347	. 79376	. 62195	53
8	. 76936	. 58797	. 77564	. 59655	. 78182	. 60509	. 78789	. 61361	. 79385	. 62209	52
9	. 76946	. 58811	. 77575	. 59669	. 78192	. 60524	. 78799	. 61375	. 79395	. 62223	51
10	9. 76957	0. 58826	9. 77585	0. 59683	9. 78203	0. 60538	9. 78809	0. 61389	9. 79405	0. 62237	50
11	. 76967	. 58840	. 77596	. 59697	. 78213	. 60552	. 78819	. 61403	. 79415	. 62251	49
12	. 76978	. 58854	. 77606	. 59712	. 78223	. 60566	. 78829	. 61418	. 79425	. 62265	48
13	. 76988	. 58869	. 77616	. 59726	. 78233	. 60580	. 78839	. 61432	. 79434	. 62279	47
14	. 76999	. 58883	. 77627	. 59740	. 78243	. 60595	. 78849	. 61446	. 79444	. 62294	46
15	9. 77009	0. 58897	9. 77637	0. 59755	9. 78254	0. 60609	9. 78859	0. 61460	9. 79454	0. 62308	45
16	. 77020	. 58911	. 77647	. 59769	. 78264	. 60623	. 78869	. 61474	. 79464	. 62322	44
17	. 77031	. 58926	. 77658	. 59783	. 78274	. 60637	. 78879	. 61488	. 79474	. 62336	43
18	. 77041	. 58940	. 77668	. 59797	. 78284	. 60652	. 78889	. 61502	. 79484	. 62350	42
19	. 77052	. 58954	. 77679	. 59812	. 78294	. 60666	. 78899	. 61517	. 79493	. 62364	41
20	9. 77062	0. 58969	9. 77689	0. 59826	9. 78305	0. 60680	9. 78909	0. 61531	9. 79503	0. 62378	40
21	. 77073	. 58983	. 77699	. 59840	. 78315	. 60694	. 78919	. 61545	. 79513	. 62392	39
22	. 77083	. 58997	. 77710	. 59854	. 78325	. 60708	. 78929	. 61559	. 79523	. 62406	38
23	. 77094	. 59012	. 77720	. 59869	. 78335	. 60723	. 78939	. 61573	. 79533	. 62420	37
24	. 77104	. 59026	. 77730	. 59883	. 78345	. 60737	. 78949	. 61587	. 79542	. 62434	36
25	9. 77115	0. 59040	9. 77741	0. 59897	9. 78355	0. 60751	9. 78959	0. 61602	9. 79552	0. 62449	35
26	. 77125	. 59055	. 77751	. 59911	. 78365	. 60765	. 78969	. 61616	. 79562	. 62463	34
27	. 77136	. 59069	. 77761	. 59926	. 78376	. 60779	. 78979	. 61630	. 79572	. 62477	33
28	. 77146	. 59083	. 77772	. 59940	. 78386	. 60794	. 78989	. 61644	. 79582	. 62491	32
29	. 77157	. 59097	. 77782	. 59954	. 78396	. 60808	. 78999	. 61658	. 79591	. 62505	31
30	9. 77167	0. 59112	9. 77792	0. 59968	9. 78406	0. 60822	9. 79009	0. 61672	9. 79601	0. 62519	30
31	. 77178	. 59126	. 77803	. 59983	. 78416	. 60836	. 79019	. 61686	. 79611	. 62533	29
32	. 77188	. 59140	. 77813	. 59997	. 78426	. 60850	. 79029	. 61701	. 79621	. 62547	28
33	. 77199	. 59155	. 77823	. 60011	. 78436	. 60865	. 79039	. 61715	. 79631	. 62561	27
34	. 77209	. 59169	. 77834	. 60025	. 78447	. 60879	. 79049	. 61729	. 79640	. 62575	26
35	9. 77220	0. 59183	9. 77844	0. 60040	9. 78457	0. 60893	9. 79059	0. 61743	9. 79650	0. 62589	25
36	. 77230	. 59198	. 77854	. 60054	. 78467	. 60907	. 79069	. 61757	. 79660	. 62603	24
37	. 77241	. 59212	. 77864	. 60068	. 78477	. 60921	. 79079	. 61771	. 79670	. 62618	23
38	. 77251	. 59226	. 77875	. 60082	. 78487	. 60936	. 79089	. 61785	. 79679	. 62632	22
39	. 77262	. 59240	. 77885	. 60097	. 78497	. 60950	. 79099	. 61800	. 79689	. 62646	21
40	9. 77272	0. 59255	9. 77895	0. 60111	9. 78507	0. 60964	9. 79108	0. 61814	9. 79699	0. 62660	20
41	. 77283	. 59269	. 77906	. 60125	. 78517	. 60978	. 79118	. 61828	. 79709	. 62674	19
42	. 77293	. 59283	. 77916	. 60139	. 78528	. 60992	. 79128	. 61842	. 79718	. 62688	18
43	. 77304	. 59298	. 77926	. 60154	. 78538	. 61006	. 79138	. 61856	. 79728	. 62702	17
44	. 77314	. 59312	. 77936	. 60168	. 78548	. 61021	. 79148	. 61870	. 79738	. 62716	16
45	9. 77325	0. 59326	9. 77947	0. 60182	9. 78558	0. 61035	9. 79158	0. 61884	9. 79748	0. 62730	15
46	. 77335	. 59340	. 77957	. 60196	. 78568	. 61049	. 79168	. 61898	. 79757	. 62744	14
47	. 77346	. 59355	. 77967	. 60211	. 78578	. 61063	. 79178	. 61913	. 79767	. 62758	13
48	. 77356	. 59369	. 77978	. 60225	. 78588	. 61077	. 79188	. 61927	. 79777	. 62772	12
49	. 77366	. 59383	. 77988	. 60239	. 78598	. 61092	. 79198	. 61941	. 79787	. 62786	11
50	9. 77377	0. 59398	9. 77998	0. 60253	9. 78608	0. 61106	9. 79208	0. 61955	9. 79796	0. 62800	10
51	. 77387	. 59412	. 78008	. 60268	. 78618	. 61120	. 79217	. 61969	. 79806	. 62814	9
52	. 77398	. 59426	. 78019	. 60282	. 78628	. 61134	. 79227	. 61983	. 79816	. 62829	8
53	. 77408	. 59440	. 78029	. 60296	. 78638	. 61148	. 79237	. 61997	. 79825	. 62843	7
54	. 77419	. 59455	. 78039	. 60310	. 78649	. 61163	. 79247	. 62011	. 79835	. 62857	6
55	9. 77429	0. 59469	9. 78049	0. 60324	9. 78659	0. 61177	9. 79257	0. 62026	9. 79845	0. 62871	5
56	. 77440	. 59483	. 78060	. 60339	. 78669	. 61191	. 79267	. 62040	. 79855	. 62885	4
57	. 77450	. 59498	. 78070	. 60353	. 78679	. 61205	. 79277	. 62054	. 79864	. 62899	3
58	. 77460	. 59512	. 78080	. 60367	. 78689	. 61219	. 79287	. 62068	. 79874	. 62913	2
59	. 77471	. 59526	. 78090	. 60381	. 78699	. 61233	. 79297	. 62082	. 79884	. 62927	1
60	9. 77481	0. 59540	9. 78101	0. 60396	9. 78709	0. 61248	9. 79306	0. 62096	9. 79893	0. 62941	0
	259°		258°		257°		256°		255°		

TABLE 34

Haversines

′	105° Log Hav	105° Nat. Hav	106° Log Hav	106° Nat. Hav	107° Log Hav	107° Nat. Hav	108° Log Hav	108° Nat. Hav	109° Log Hav	109° Nat. Hav	′
0	9. 79893	0. 62941	9. 80470	0. 63782	9. 81036	0. 64619	9. 81592	0. 65451	9. 82137	0. 66278	60
1	. 79903	. 62955	. 80479	. 63796	. 81045	. 64632	. 81601	. 65465	. 82146	. 66292	59
2	. 79913	. 62969	. 80489	. 63810	. 81054	. 64646	. 81610	. 65479	. 82155	. 66306	58
3	. 79922	. 62983	. 80498	. 63824	. 81064	. 64660	. 81619	. 65492	. 82164	. 66320	57
4	. 79932	. 62997	. 80508	. 63838	. 81073	. 64674	. 81628	. 65506	. 82173	. 66333	56
5	9. 79942	0. 63011	9. 80517	0. 63852	9. 81082	0. 64688	9. 81637	0. 65520	9. 82182	0. 66347	55
6	. 79951	. 63025	. 80527	. 63866	. 81092	. 64702	. 81647	. 65534	. 82191	. 66361	54
7	. 79961	. 63039	. 80536	. 63880	. 81101	. 64716	. 81656	. 65548	. 82200	. 66375	53
8	. 79971	. 63053	. 80546	. 63894	. 81110	. 64730	. 81665	. 65561	. 82209	. 66388	52
9	. 79980	. 63067	. 80555	. 63908	. 81120	. 64744	. 81674	. 65575	. 82218	. 66402	51
10	9. 79990	0. 63081	9. 80565	0. 63922	9. 81129	0. 64758	9. 81683	0. 65589	9. 82227	0. 66416	50
11	. 80000	. 63095	. 80574	. 63936	. 81138	. 64772	. 81692	. 65603	. 82236	. 66430	49
12	. 80009	. 63109	. 80584	. 63950	. 81148	. 64785	. 81701	. 65617	. 82245	. 66443	48
13	. 80019	. 63123	. 80593	. 63964	. 81157	. 64799	. 81711	. 65631	. 82254	. 66457	47
14	. 80029	. 63138	. 80603	. 63977	. 81166	. 64813	. 81720	. 65644	. 82263	. 66471	46
15	9. 80038	0. 63152	9. 80612	0. 63991	9. 81176	0. 64827	9. 81729	0. 65658	9. 82272	0. 66485	45
16	. 80048	. 63166	. 80622	. 64005	. 81185	. 64841	. 81738	. 65672	. 82281	. 66498	44
17	. 80058	. 63180	. 80631	. 64019	. 81194	. 64855	. 81747	. 65686	. 82290	. 66512	43
18	. 80067	. 63194	. 80641	. 64033	. 81204	. 64869	. 81756	. 65700	. 82299	. 66526	42
19	. 80077	. 63208	. 80650	. 64047	. 81213	. 64883	. 81765	. 65713	. 82308	. 66539	41
20	9. 80087	0. 63222	9. 80660	0. 64061	9. 81222	0. 64897	9. 81775	0. 65727	9. 82317	0. 66553	40
21	. 80096	. 63236	. 80669	. 64075	. 81231	. 64910	. 81784	. 65741	. 82326	. 66567	39
22	. 80106	. 63250	. 80678	. 64089	. 81241	. 64924	. 81793	. 65755	. 82335	. 66581	38
23	. 80116	. 63264	. 80688	. 64103	. 81250	. 64938	. 81802	. 65769	. 82344	. 66594	37
24	. 80125	. 63278	. 80697	. 64117	. 81259	. 64952	. 81811	. 65782	. 82353	. 66608	36
25	9. 80135	0. 63292	9. 80707	0. 64131	9. 81269	0. 64966	9. 81820	0. 65796	9. 82362	0. 66622	35
26	. 80144	. 63306	. 80716	. 64145	. 81278	. 64980	. 81829	. 65810	. 82371	. 66635	34
27	. 80154	. 63320	. 80726	. 64159	. 81287	. 64994	. 81838	. 65824	. 82380	. 66649	33
28	. 80164	. 63334	. 80735	. 64173	. 81296	. 65008	. 81847	. 65838	. 82388	. 66663	32
29	. 80173	. 63348	. 80745	. 64187	. 81306	. 65021	. 81857	. 65851	. 82397	. 66677	31
30	9. 80183	0. 63362	9. 80754	0. 64201	9. 81315	0. 65035	9. 81866	0. 65865	9. 82406	0. 66690	30
31	. 80192	. 63376	. 80763	. 64215	. 81324	. 65049	. 81875	. 65879	. 82415	. 66704	29
32	. 80202	. 63390	. 80773	. 64229	. 81333	. 65063	. 81884	. 65893	. 82424	. 66718	28
33	. 80212	. 63404	. 80782	. 64243	. 81343	. 65077	. 81893	. 65907	. 82433	. 66731	27
34	. 80221	. 63418	. 80792	. 64257	. 81352	. 65091	. 81902	. 65920	. 82442	. 66745	26
35	9. 80231	0. 63432	9. 80801	0. 64270	9. 81361	0. 65105	9. 81911	0. 65934	9. 82451	0. 66759	25
36	. 80240	. 63446	. 80811	. 64284	. 81370	. 65118	. 81920	. 65948	. 82460	. 66773	24
37	. 80250	. 63460	. 80820	. 64298	. 81380	. 65132	. 81929	. 65962	. 82469	. 66786	23
38	. 80260	. 63474	. 80829	. 64312	. 81389	. 65146	. 81938	. 65976	. 82478	. 66800	22
39	. 80269	. 63488	. 80839	. 64326	. 81398	. 65160	. 81947	. 65989	. 82487	. 66814	21
40	9. 80279	0. 63502	9. 80848	0. 64340	9. 81407	0. 65174	9. 81956	0. 66003	9. 82495	0. 66827	20
41	. 80288	. 63516	. 80858	. 64354	. 81417	. 65188	. 81965	. 66017	. 82504	. 66841	19
42	. 80298	. 63530	. 80867	. 64368	. 81426	. 65202	. 81975	. 66031	. 82513	. 66855	18
43	. 80307	. 63544	. 80876	. 64382	. 81435	. 65216	. 81984	. 66044	. 82522	. 66868	17
44	. 80317	. 63558	. 80886	. 64396	. 81444	. 65229	. 81993	. 66058	. 82531	. 66882	16
45	9. 80327	0. 63572	9. 80895	0. 64410	9. 81454	0. 65243	9. 82002	0. 66072	9. 82540	0. 66896	15
46	. 80336	. 63586	. 80905	. 64424	. 81463	. 65257	. 82011	. 66086	. 82549	. 66910	14
47	. 80346	. 63600	. 80914	. 64438	. 81472	. 65271	. 82020	. 66100	. 82558	. 66923	13
48	. 80355	. 63614	. 80923	. 64452	. 81481	. 65285	. 82029	. 66113	. 82567	. 66937	12
49	. 80365	. 63628	. 80933	. 64466	. 81490	. 65299	. 82038	. 66127	. 82575	. 66951	11
50	9. 80374	0. 63642	9. 80942	0. 64479	9. 81500	0. 65312	9. 82047	0. 66141	9. 82584	0. 66964	10
51	. 80384	. 63656	. 80952	. 64493	. 81509	. 65326	. 82056	. 66155	. 82593	. 66978	9
52	. 80393	. 63670	. 80961	. 64507	. 81518	. 65340	. 82065	. 66168	. 82602	. 66992	8
53	. 80403	. 63684	. 80970	. 64521	. 81527	. 65354	. 82074	. 66182	. 82611	. 67005	7
54	. 80413	. 63698	. 80980	. 64535	. 81536	. 65368	. 82083	. 66196	. 82620	. 67019	6
55	9. 80422	0. 63712	9. 80989	0. 64549	9. 81546	0. 65382	9. 82092	0. 66210	9. 82629	0. 67033	5
56	. 80432	. 63726	. 80998	. 64563	. 81555	. 65396	. 82101	. 66223	. 82638	. 67046	4
57	. 80441	. 63740	. 81008	. 64577	. 81564	. 65409	. 82110	. 66237	. 82646	. 67060	3
58	. 80451	. 63754	. 81017	. 64591	. 81573	. 65423	. 82119	. 66251	. 82655	. 67074	2
59	. 80460	. 63768	. 81026	. 64605	. 81582	. 65437	. 82128	. 66265	. 82664	. 67087	1
60	9. 80470	0. 63782	9. 81036	0. 64619	9. 81592	0. 65451	9. 82137	0. 66278	9. 82673	0. 67101	0
	254°		253°		252°		251°		250°		

TABLE 34

Haversines

′	110°		111°		112°		113°		114°		′
′	Log Hav	Nat. Hav	Log Hav	Nat. Hav	Log Hav	Nat. Hav	Log Hav	Nat. Hav	Log Hav	Nat. Hav	′
0	9. 82673	0. 67101	9. 83199	0. 67918	9. 83715	0. 68730	9. 84221	0. 69537	9. 84718	0. 70337	60
1	. 82682	. 67115	. 83207	. 67932	. 83723	. 68744	. 84230	. 69550	. 84726	. 70350	59
2	. 82691	. 67128	. 83216	. 67946	. 83732	. 68757	. 84238	. 69563	. 84735	. 70363	58
3	. 82699	. 67142	. 83225	. 67959	. 83740	. 68771	. 84246	. 69577	. 84743	. 70377	57
4	. 82708	. 67156	. 83233	. 67973	. 83749	. 68784	. 84255	. 69590	. 84751	. 70390	56
5	9. 82717	0. 67169	9. 83242	0. 67986	9. 83757	0. 68798	9. 84263	0. 69603	9. 84759	0. 70403	55
6	. 82726	. 67183	. 83251	. 68000	. 83766	. 68811	. 84271	. 69617	. 84767	. 70417	54
7	. 82735	. 67197	. 83259	. 68013	. 83774	. 68825	. 84280	. 69630	. 84776	. 70430	53
8	. 82744	. 67210	. 83268	. 68027	. 83783	. 68838	. 84288	. 69644	. 84784	. 70443	52
9	. 82752	. 67224	. 83277	. 68041	. 83791	. 68852	. 84296	. 69657	. 84792	. 70456	51
10	9. 82761	0. 67238	9. 83285	0. 68054	9. 83800	0. 68865	9. 84305	0. 69670	9. 84800	0. 70470	50
11	. 82770	. 67251	. 83294	. 68068	. 83808	. 68879	. 84313	. 69684	. 84808	. 70483	49
12	. 82779	. 67265	. 83303	. 68081	. 83817	. 68892	. 84321	. 69697	. 84817	. 70496	48
13	. 82788	. 67279	. 83311	. 68095	. 83825	. 68906	. 84330	. 69710	. 84825	. 70509	47
14	. 82796	. 67292	. 83320	. 68108	. 83834	. 68919	. 84338	. 69724	. 84833	. 70523	46
15	9. 82805	0. 67306	9. 83329	0. 68122	9. 83842	0. 68932	9. 84346	0. 69737	9. 84841	0. 70536	45
16	. 82814	. 67319	. 83337	. 68135	. 83851	. 68946	. 84355	. 69751	. 84849	. 70549	44
17	. 82823	. 67333	. 83346	. 68149	. 83859	. 68959	. 84363	. 69764	. 84857	. 70562	43
18	. 82832	. 67347	. 83355	. 68163	. 83868	. 68973	. 84371	. 69777	. 84866	. 70576	42
19	. 82840	. 67360	. 83363	. 68176	. 83876	. 68986	. 84380	. 69791	. 84874	. 70589	41
20	9. 82849	0. 67374	9. 83372	0. 68190	9. 83885	0. 69000	9. 84388	0. 69804	9. 84882	0. 70602	40
21	. 82858	. 67388	. 83380	. 68203	. 83893	. 69013	. 84396	. 69817	. 84890	. 70615	39
22	. 82867	. 67401	. 83389	. 68217	. 83902	. 69027	. 84405	. 69831	. 84898	. 70629	38
23	. 82876	. 67415	. 83398	. 68230	. 83910	. 69040	. 84413	. 69844	. 84906	. 70642	37
24	. 82884	. 67429	. 83406	. 68244	. 83919	. 69054	. 84421	. 69857	. 84914	. 70655	36
25	9. 82893	0. 67442	9. 83415	0. 68257	9. 83927	0. 69067	9. 84430	0. 69871	9. 84923	0. 70668	35
26	. 82902	. 67456	. 83424	. 68271	. 83935	. 69080	. 84438	. 69884	. 84931	. 70682	34
27	. 82911	. 67469	. 83432	. 68284	. 83944	. 69094	. 84446	. 69897	. 84939	. 70695	33
28	. 82920	. 67483	. 83441	. 68298	. 83952	. 69107	. 84454	. 69911	. 84947	. 70708	32
29	. 82928	. 67497	. 83449	. 68312	. 83961	. 69121	. 84463	. 69924	. 84955	. 70721	31
30	9. 82937	0. 67510	9. 83458	0. 68325	9. 83969	0. 69134	9. 84471	0. 69937	9. 84963	0. 70735	30
31	. 82946	. 67524	. 83467	. 68339	. 83978	. 69148	. 84479	. 69951	. 84971	. 70748	29
32	. 82955	. 67538	. 83475	. 68352	. 83986	. 69161	. 84488	. 69964	. 84979	. 70761	28
33	. 82963	. 67551	. 83484	. 68366	. 83995	. 69174	. 84496	. 69977	. 84988	. 70774	27
34	. 82972	. 67565	. 83492	. 68379	. 84003	. 69188	. 84504	. 69991	. 84996	. 70788	26
35	9. 82981	0. 67578	9. 83501	0. 68393	9. 84011	0. 69201	9. 84512	0. 70004	9. 85004	0. 70801	25
36	. 82990	. 67592	. 83510	. 68406	. 84020	. 69215	. 84521	. 70017	. 85012	. 70814	24
37	. 82998	. 67606	. 83518	. 68420	. 84028	. 69228	. 84529	. 70031	. 85020	. 70827	23
38	. 83007	. 67619	. 83527	. 68433	. 84037	. 69242	. 84537	. 70044	. 85028	. 70840	22
39	. 83016	. 67633	. 83535	. 68447	. 84045	. 69255	. 84545	. 70057	. 85036	. 70854	21
40	9. 83025	0. 67647	9. 83544	0. 68460	9. 84054	0. 69268	9. 84554	0. 70071	9. 85044	0. 70867	20
41	. 83033	. 67660	. 83552	. 68474	. 84062	. 69282	. 84562	. 70084	. 85052	. 70880	19
42	. 83042	. 67674	. 83561	. 68487	. 84070	. 69295	. 84570	. 70097	. 85061	. 70893	18
43	. 83051	. 67687	. 83570	. 68501	. 84079	. 69309	. 84578	. 70111	. 85069	. 70907	17
44	. 83059	. 67701	. 83578	. 68514	. 84087	. 69322	. 84587	. 70124	. 85077	. 70920	16
45	9. 83068	0. 67715	9. 83587	0. 68528	9. 84096	0. 69336	9. 84595	0. 70137	9. 85085	0. 70933	15
46	. 83077	. 67728	. 83595	. 68541	. 84104	. 69349	. 84603	. 70151	. 85093	. 70946	14
47	. 83086	. 67742	. 83604	. 68555	. 84112	. 69362	. 84611	. 70164	. 85101	. 70959	13
48	. 83094	. 67755	. 83612	. 68568	. 84121	. 69376	. 84620	. 70177	. 85109	. 70973	12
49	. 83103	. 67769	. 83621	. 68582	. 84129	. 69389	. 84628	. 70191	. 85117	. 70986	11
50	9. 83112	0. 67783	9. 83630	0. 68595	9. 84138	0. 69403	9. 84636	0. 70204	9. 85125	0. 70999	10
51	. 83120	. 67796	. 83638	. 68609	. 84146	. 69416	. 84644	. 70217	. 85133	. 71012	9
52	. 83129	. 67810	. 83647	. 68622	. 84154	. 69429	. 84653	. 70230	. 85141	. 71025	8
53	. 83138	. 67823	. 83655	. 68636	. 84163	. 69443	. 84661	. 70244	. 85149	. 71039	7
54	. 83147	. 67837	. 83664	. 68649	. 84171	. 69456	. 84669	. 70257	. 85158	. 71052	6
55	9. 83155	0. 67850	9. 83672	0. 68663	9. 84179	0. 69470	9. 84677	0. 70270	9. 85166	0. 71065	5
56	. 83164	. 67864	. 83681	. 68676	. 84188	. 69483	. 84685	. 70284	. 85174	. 71078	4
57	. 83173	. 67878	. 83689	. 68690	. 84196	. 69496	. 84694	. 70297	. 85182	. 71091	3
58	. 83181	. 67891	. 83698	. 68703	. 84205	. 69510	. 84702	. 70310	. 85190	. 71105	2
59	. 83190	. 67905	. 83706	. 68717	. 84213	. 69523	. 84710	. 70324	. 85198	. 71118	1
60	9. 83199	0. 67918	9. 83715	0. 68730	9. 84221	0. 69537	9. 84718	0. 70337	9. 85206	0. 71131	0
	249°		248°		247°		246°		245°		

TABLE 34

Haversines

′	115°		116°		117°		118°		119°		′
	Log Hav	Nat. Hav	Log Hav	Nat. Hav	Log Hav	Nat. Hav	Log Hav	Nat. Hav	Log Hav	Nat. Hav	
0	9. 85206	0. 71131	9. 85684	0. 71919	9. 86153	0. 72700	9. 86613	0. 73474	9. 87064	0. 74240	60
1	. 85214	. 71144	. 85692	. 71932	. 86161	. 72712	. 86621	. 73486	. 87072	. 74253	59
2	. 85222	. 71157	. 85700	. 71945	. 86169	. 72725	. 86628	. 73499	. 87079	. 74266	58
3	. 85230	. 71170	. 85708	. 71958	. 86176	. 72738	. 86636	. 73512	. 87086	. 74279	57
4	. 85238	. 71184	. 85716	. 71971	. 86184	. 72751	. 86643	. 73525	. 87094	. 74291	56
5	9. 85246	0. 71197	9. 85724	0. 71984	9. 86192	0. 72764	9. 86651	0. 73538	9. 87101	0. 74304	55
6	. 85254	. 71210	. 85731	. 71997	. 86200	. 72777	. 86659	. 73551	. 87109	. 74317	54
7	. 85262	. 71223	. 85739	. 72010	. 86207	. 72790	. 86666	. 73563	. 87116	. 74329	53
8	. 85270	. 71236	. 85747	. 72023	. 86215	. 72803	. 86674	. 73576	. 87124	. 74342	52
9	. 85278	. 71249	. 85755	. 72036	. 86223	. 72816	. 86681	. 73589	. 87131	. 74355	51
10	9. 85286	0. 71263	9. 85763	0. 72049	9. 86230	0. 72829	9. 86689	0. 73602	9. 87138	0. 74368	50
11	. 85294	. 71276	. 85771	. 72062	. 86238	. 72842	. 86696	. 73615	. 87146	. 74380	49
12	. 85302	. 71289	. 85779	. 72075	. 86246	. 72855	. 86704	. 73628	. 87153	. 74393	48
13	. 85310	. 71302	. 85787	. 72088	. 86254	. 72868	. 86712	. 73640	. 87161	. 74406	47
14	. 85318	. 71315	. 85794	. 72101	. 86261	. 72881	. 86719	. 73653	. 87168	. 74418	46
15	9. 85326	0. 71328	9. 85802	0. 72114	9. 86269	0. 72894	9. 86727	0. 73666	9. 87175	0. 74431	45
16	. 85334	. 71342	. 85810	. 72127	. 86277	. 72907	. 86734	. 73679	. 87183	. 74444	44
17	. 85342	. 71355	. 85818	. 72141	. 86284	. 72920	. 86742	. 73692	. 87190	. 74456	43
18	. 85350	. 71368	. 85826	. 72154	. 86292	. 72932	. 86749	. 73704	. 87198	. 74469	42
19	. 85358	. 71381	. 85834	. 72167	. 86300	. 72945	. 86757	. 73717	. 87205	. 74482	41
20	9. 85366	0. 71394	9. 85841	0. 72180	9. 86307	0. 72958	9. 86764	0. 73730	9. 87212	0. 74494	40
21	. 85374	. 71407	. 85849	. 72193	. 86315	. 72971	. 86772	. 73743	. 87220	. 74507	39
22	. 85382	. 71420	. 85857	. 72206	. 86323	. 72984	. 86780	. 73756	. 87227	. 74520	38
23	. 85390	. 71434	. 85865	. 72219	. 86331	. 72997	. 86787	. 73768	. 87235	. 74533	37
24	. 85398	. 71447	. 85873	. 72232	. 86338	. 73010	. 86795	. 73781	. 87242	. 74545	36
25	9. 85406	0. 71460	9. 85881	0. 72245	9. 86346	0. 73023	9. 86802	0. 73794	9. 87249	0. 74558	35
26	. 85414	. 71473	. 85888	. 72258	. 86354	. 73036	. 86810	. 73807	. 87257	. 74571	34
27	. 85422	. 71486	. 85896	. 72271	. 86361	. 73049	. 86817	. 73820	. 87264	. 74583	33
28	. 85430	. 71499	. 85904	. 72284	. 86369	. 73062	. 86825	. 73832	. 87271	. 74596	32
29	. 85438	. 71512	. 85912	. 72297	. 86377	. 73075	. 86832	. 73845	. 87279	. 74609	31
30	9. 85446	0. 71526	9. 85920	0. 72310	9. 86384	0. 73087	9. 86840	0. 73858	9. 87286	0. 74621	30
31	. 85454	. 71539	. 85928	. 72323	. 86392	. 73100	. 86847	. 73871	. 87294	. 74634	29
32	. 85462	. 71552	. 85935	. 72336	. 86400	. 73113	. 86855	. 73883	. 87301	. 74646	28
33	. 85470	. 71565	. 85943	. 72349	. 86407	. 73126	. 86862	. 73896	. 87308	. 74659	27
34	. 85478	. 71578	. 85951	. 72362	. 86415	. 73139	. 86870	. 73909	. 87316	. 74672	26
35	9. 85486	0. 71591	9. 85959	0. 72375	9. 86423	0. 73152	9. 86877	0. 73922	9. 87323	0. 74684	25
36	. 85494	. 71604	. 85967	. 72388	. 86430	. 73165	. 86885	. 73935	. 87330	. 74697	24
37	. 85502	. 71617	. 85974	. 72401	. 86438	. 73178	. 86892	. 73947	. 87338	. 74710	23
38	. 85510	. 71631	. 85982	. 72414	. 86446	. 73191	. 86900	. 73960	. 87345	. 74722	22
39	. 85518	. 71644	. 85990	. 72427	. 86453	. 73203	. 86907	. 73973	. 87352	. 74735	21
40	9. 85526	0. 71657	9. 85998	0. 72440	9. 86461	0. 73216	9. 86915	0. 73986	9. 87360	0. 74748	20
41	. 85534	. 71670	. 86006	. 72453	. 86468	. 73229	. 86922	. 73998	. 87367	. 74760	19
42	. 85542	. 71683	. 86013	. 72466	. 86476	. 73242	. 86930	. 74011	. 87374	. 74773	18
43	. 85550	. 71696	. 86021	. 72479	. 86484	. 73255	. 86937	. 74024	. 87382	. 74786	17
44	. 85557	. 71709	. 86029	. 72492	. 86491	. 73268	. 86945	. 74037	. 87389	. 74798	16
45	9. 85565	0. 71722	9. 86037	0. 72505	9. 86499	0. 73281	9. 86952	0. 74049	9. 87396	0. 74811	15
46	. 85573	. 71735	. 86045	. 72518	. 86507	. 73294	. 86960	. 74062	. 87404	. 74823	14
47	. 85581	. 71748	. 86052	. 72531	. 86514	. 73306	. 86967	. 74075	. 87411	. 74836	13
48	. 85589	. 71762	. 86060	. 72544	. 86522	. 73319	. 86975	. 74088	. 87418	. 74849	12
49	. 85597	. 71775	. 86068	. 72557	. 86529	. 73332	. 86982	. 74100	. 87426	. 74861	11
50	9. 85605	0. 71788	9. 86076	0. 72570	9. 86537	0. 73345	9. 86990	0. 74113	9. 87433	0. 74874	10
51	. 85613	. 71801	. 86083	. 72583	. 86545	. 73358	. 86997	. 74126	. 87440	. 74887	9
52	. 85621	. 71814	. 86091	. 72596	. 86552	. 73371	. 87004	. 74139	. 87448	. 74899	8
53	. 85629	. 71827	. 86099	. 72609	. 86560	. 73384	. 87012	. 74151	. 87455	. 74912	7
54	. 85637	. 71840	. 86107	. 72622	. 86568	. 73396	. 87019	. 74164	. 87462	. 74924	6
55	9. 85645	0. 71853	9. 86114	0. 72635	9. 86575	0. 73409	9. 87027	0. 74177	9. 87470	0. 74937	5
56	. 85653	. 71866	. 86122	. 72648	. 86583	. 73422	. 87034	. 74190	. 87477	. 74950	4
57	. 85660	. 71879	. 86130	. 72661	. 86590	. 73435	. 87042	. 74202	. 87484	. 74962	3
58	. 85668	. 71892	. 86138	. 72674	. 86598	. 73448	. 87049	. 74215	. 87492	. 74975	2
59	. 85676	. 71905	. 86145	. 72687	. 86606	. 73461	. 87057	. 74228	. 87499	. 74987	1
60	9. 85684	0. 71919	9. 86153	0. 72700	9. 86613	0. 73474	9. 87064	0. 74240	9. 87506	0. 75000	0
	244°		243°		242°		241°		240°		

TABLE 34

Haversines

′	120° Log Hav	120° Nat. Hav	121° Log Hav	121° Nat. Hav	122° Log Hav	122° Nat. Hav	123° Log Hav	123° Nat. Hav	124° Log Hav	124° Nat. Hav	′
0	9.87506	0.75000	9.87939	0.75752	9.88364	0.76496	9.88780	0.77232	9.89187	0.77960	60
1	.87513	.75013	.87947	.75764	.88371	.76508	.88787	.77244	.89194	.77972	59
2	.87521	.75025	.87954	.75777	.88378	.76521	.88793	.77256	.89200	.77984	58
3	.87528	.75038	.87961	.75789	.88385	.76533	.88800	.77269	.89207	.77996	57
4	.87535	.75050	.87968	.75802	.88392	.76545	.88807	.77281	.89214	.78008	56
5	9.87543	0.75063	9.87975	0.75814	9.88399	0.76558	9.88814	0.77293	9.89221	0.78020	55
6	.87550	.75076	.87982	.75827	.88406	.76570	.88821	.77305	.89227	.78032	54
7	.87557	.75088	.87989	.75839	.88413	.76582	.88828	.77317	.89234	.78044	53
8	.87564	.75101	.87996	.75852	.88420	.76595	.88835	.77329	.89241	.78056	52
9	.87572	.75113	.88004	.75864	.88427	.76607	.88841	.77342	.89247	.78068	51
10	9.87579	0.75126	9.88011	0.75876	9.88434	0.76619	9.88848	0.77354	9.89254	0.78080	50
11	.87586	.75138	.88018	.75889	.88441	.76632	.88855	.77366	.89261	.78092	49
12	.87593	.75151	.88025	.75901	.88448	.76644	.88862	.77378	.89267	.78104	48
13	.87601	.75164	.88032	.75914	.88455	.76656	.88869	.77390	.89274	.78116	47
14	.87608	.75176	.88039	.75926	.88462	.76668	.88876	.77402	.89281	.78128	46
15	9.87615	0.75189	9.88046	0.75939	9.88469	0.76681	9.88882	0.77415	9.89287	0.78140	45
16	.87623	.75201	.88053	.75951	.88476	.76693	.88889	.77427	.89294	.78152	44
17	.87630	.75214	.88061	.75964	.88483	.76705	.88896	.77439	.89301	.78164	43
18	.87637	.75226	.88068	.75976	.88490	.76718	.88903	.77451	.89308	.78176	42
19	.87644	.75239	.88075	.75988	.88496	.76730	.88910	.77463	.89314	.78188	41
20	9.87652	0.75251	9.88082	0.76001	9.88503	0.76742	9.88916	0.77475	9.89321	0.78200	40
21	.87659	.75264	.88089	.76013	.88510	.76754	.88923	.77488	.89328	.78212	39
22	.87666	.75277	.88096	.76026	.88517	.76767	.88930	.77500	.89334	.78224	38
23	.87673	.75289	.88103	.76038	.88524	.76779	.88937	.77512	.89341	.78236	37
24	.87680	.75302	.88110	.76050	.88531	.76791	.88944	.77524	.89348	.78248	36
25	9.87688	0.75314	9.88117	0.76063	9.88538	0.76804	9.88950	0.77536	9.89354	0.78260	35
26	.87695	.75327	.88124	.76075	.88545	.76816	.88957	.77548	.89361	.78272	34
27	.87702	.75339	.88131	.76088	.88552	.76828	.88964	.77560	.89368	.78284	33
28	.87709	.75352	.88139	.76100	.88559	.76840	.88971	.77573	.89374	.78296	32
29	.87717	.75364	.88146	.76113	.88566	.76853	.88978	.77585	.89381	.78308	31
30	9.87724	0.75377	9.88153	0.76125	9.88573	0.76865	9.88984	0.77597	9.89387	0.78320	30
31	.87731	.75389	.88160	.76137	.88580	.76877	.88991	.77609	.89394	.78332	29
32	.87738	.75402	.88167	.76150	.88587	.76890	.88998	.77621	.89400	.78344	28
33	.87745	.75415	.88174	.76162	.88594	.76902	.89005	.77633	.89407	.78356	27
34	.87753	.75427	.88181	.76175	.88600	.76914	.89012	.77645	.89414	.78368	26
35	9.87760	0.75440	9.88188	0.76187	9.88607	0.76926	9.89018	0.77657	9.89421	0.78380	25
36	.87767	.75452	.88195	.76199	.88614	.76939	.89025	.77670	.89427	.78392	24
37	.87774	.75465	.88202	.76212	.88621	.76951	.89032	.77682	.89434	.78404	23
38	.87782	.75477	.88209	.76224	.88628	.76963	.89039	.77694	.89441	.78416	22
39	.87789	.75490	.88216	.76236	.88635	.76975	.89045	.77706	.89447	.78428	21
40	9.87796	0.75502	9.88223	0.76249	9.88642	0.76988	9.89052	0.77718	9.89454	0.78440	20
41	.87803	.75515	.88230	.76261	.88649	.77000	.89059	.77730	.89460	.78452	19
42	.87810	.75527	.88237	.76274	.88656	.77012	.89066	.77742	.89467	.78464	18
43	.87818	.75540	.88244	.76286	.88663	.77024	.89072	.77754	.89474	.78476	17
44	.87825	.75552	.88252	.76298	.88670	.77036	.89079	.77766	.89480	.78488	16
45	9.87832	0.75565	9.88259	0.76311	9.88677	0.77049	9.89086	0.77779	9.89487	0.78500	15
46	.87839	.75577	.88266	.76323	.88683	.77061	.89093	.77791	.89493	.78512	14
47	.87846	.75590	.88273	.76335	.88690	.77073	.89099	.77803	.89500	.78524	13
48	.87853	.75602	.88280	.76348	.88697	.77085	.89106	.77815	.89507	.78536	12
49	.87861	.75615	.88287	.76360	.88704	.77098	.89113	.77827	.89513	.78548	11
50	9.87868	0.75627	9.88294	0.76373	9.88711	0.77110	9.89120	0.77839	9.89520	0.78560	10
51	.87875	.75640	.88301	.76385	.88718	.77122	.89126	.77851	.89527	.78571	9
52	.87882	.75652	.88308	.76397	.88725	.77134	.89133	.77863	.89533	.78583	8
53	.87889	.75665	.88315	.76410	.88732	.77147	.89140	.77875	.89540	.78595	7
54	.87896	.75677	.88322	.76422	.88739	.77159	.89147	.77887	.89546	.78607	6
55	9.87904	0.75690	9.88329	0.76434	9.88745	0.77171	9.89153	0.77899	9.89553	0.78619	5
56	.87911	.75702	.88336	.76447	.88752	.77183	.89160	.77911	.89559	.78631	4
57	.87918	.75714	.88343	.76459	.88759	.77195	.89167	.77923	.89566	.78643	3
58	.87925	.75727	.88350	.76471	.88766	.77208	.89174	.77936	.89573	.78655	2
59	.87932	.75739	.88357	.76484	.88773	.77220	.89180	.77948	.89579	.78667	1
60	9.87939	0.75752	9.88364	0.76496	9.88780	0.77232	9.89187	0.77960	9.89586	0.78679	0
	239°		238°		237°		236°		235°		

TABLE 34

Haversines

′	125° Log Hav	125° Nat. Hav	126° Log Hav	126° Nat. Hav	127° Log Hav	127° Nat. Hav	128° Log Hav	128° Nat. Hav	129° Log Hav	129° Nat. Hav	′
0	9. 89586	0. 78679	9. 89976	0. 79389	9. 90358	0. 80091	9. 90732	0. 80783	9. 91098	0. 81466	60
1	. 89592	. 78691	. 89983	. 79401	. 90365	. 80102	. 90738	. 80795	. 91104	. 81477	59
2	. 89599	. 78703	. 89989	. 79413	. 90371	. 80114	. 90744	. 80806	. 91110	. 81489	58
3	. 89606	. 78715	. 89995	. 79425	. 90377	. 80126	. 90751	. 80817	. 91116	. 81500	57
4	. 89612	. 78726	. 90002	. 79436	. 90383	. 80137	. 90757	. 80829	. 91122	. 81511	56
5	9. 89619	0. 78738	9. 90008	0. 79448	9. 90390	0. 80149	9. 90763	0. 80840	9. 91128	0. 81523	55
6	. 89625	. 78750	. 90015	. 79460	. 90396	. 80160	. 90769	. 80852	. 91134	. 81534	54
7	. 89632	. 78762	. 90021	. 79472	. 90402	. 80172	. 90775	. 80863	. 91140	. 81545	53
8	. 89638	. 78774	. 90028	. 79483	. 90409	. 80184	. 90781	. 80875	. 91146	. 81556	52
9	. 89645	. 78786	. 90034	. 79495	. 90415	. 80195	. 90787	. 80886	. 91152	. 81568	51
10	9. 89651	0. 78798	9. 90040	0. 79507	9. 90421	0. 80207	9. 90794	0. 80898	9. 91158	0. 81579	50
11	. 89658	. 78810	. 90047	. 79519	. 90428	. 80218	. 90800	. 80909	. 91164	. 81590	49
12	. 89665	. 78822	. 90053	. 79530	. 90434	. 80230	. 90806	. 80920	. 91170	. 81601	48
13	. 89671	. 78833	. 90060	. 79542	. 90440	. 80242	. 90812	. 80932	. 91176	. 81613	47
14	. 89678	. 78845	. 90066	. 79554	. 90446	. 80253	. 90818	. 80943	. 91182	. 81624	46
15	9. 89684	0. 78857	9. 90072	0. 79565	9. 90452	0. 80265	9. 90824	0. 80955	9. 91188	0. 81635	45
16	. 89691	. 78869	. 90079	. 79577	. 90459	. 80276	. 90830	. 80966	. 91194	. 81647	44
17	. 89697	. 78881	. 90085	. 79589	. 90465	. 80288	. 90836	. 80978	. 91200	. 81658	43
18	. 89704	. 78893	. 90092	. 79601	. 90471	. 80299	. 90843	. 80989	. 91206	. 81669	42
19	. 89710	. 78905	. 90098	. 79612	. 90478	. 80311	. 90849	. 81000	. 91212	. 81680	41
20	9. 89717	0. 78917	9. 90104	0. 79624	9. 90484	0. 80323	9. 90855	0. 81012	9. 91218	0. 81692	40
21	. 89723	. 78928	. 90111	. 79636	. 90490	. 80334	. 90861	. 81023	. 91224	. 81703	39
22	. 89730	. 78940	. 90117	. 79648	. 90496	. 80346	. 90867	. 81035	. 91230	. 81714	38
23	. 89736	. 78952	. 90124	. 79659	. 90503	. 80357	. 90873	. 81046	. 91236	. 81725	37
24	. 89743	. 78964	. 90130	. 79671	. 90509	. 80369	. 90879	. 81057	. 91242	. 81737	36
25	9. 89749	0. 78976	9. 90136	0. 79683	9. 90515	0. 80380	9. 90885	0. 81069	9. 91248	0. 81748	35
26	. 89756	. 78988	. 90143	. 79694	. 90521	. 80392	. 90892	. 81080	. 91254	. 81759	34
27	. 89763	. 79000	. 90149	. 79706	. 90527	. 80403	. 90898	. 81092	. 91260	. 81770	33
28	. 89769	. 79011	. 90156	. 79718	. 90534	. 80415	. 90904	. 81103	. 91265	. 81781	32
29	. 89776	. 79023	. 90162	. 79729	. 90540	. 80427	. 90910	. 81114	. 91271	. 81793	31
30	9. 89782	0. 79035	9. 90168	0. 79741	9. 90546	0. 80438	9. 90916	0. 81126	9. 91277	0. 81804	30
31	. 89789	. 79047	. 90175	. 79753	. 90552	. 80450	. 90922	. 81137	. 91283	. 81815	29
32	. 89795	. 79059	. 90181	. 79765	. 90559	. 80461	. 90928	. 81148	. 91289	. 81826	28
33	. 89802	. 79071	. 90187	. 79776	. 90565	. 80473	. 90934	. 81160	. 91295	. 81838	27
34	. 89808	. 79082	. 90194	. 79788	. 90571	. 80484	. 90940	. 81171	. 91301	. 81849	26
35	9. 89815	0. 79094	9. 90200	0. 79800	9. 90577	0. 80496	9. 90946	0. 81183	9. 91307	0. 81860	25
36	. 89821	. 79106	. 90206	. 79811	. 90584	. 80507	. 90952	. 81194	. 91313	. 81871	24
37	. 89828	. 79118	. 90213	. 79823	. 90590	. 80519	. 90958	. 81205	. 91319	. 81882	23
38	. 89834	. 79130	. 90219	. 79835	. 90596	. 80530	. 90965	. 81217	. 91325	. 81894	22
39	. 89840	. 79142	. 90225	. 79846	. 90602	. 80542	. 90971	. 81228	. 91331	. 81905	21
40	9. 89847	0. 79153	9. 90232	0. 79858	9. 90608	0. 80553	9. 90977	0. 81239	9. 91337	0. 81916	20
41	. 89853	. 79165	. 90238	. 79870	. 90615	. 80565	. 90983	. 81251	. 91343	. 81927	19
42	. 89860	. 79177	. 90244	. 79881	. 90621	. 80576	. 90989	. 81262	. 91349	. 81938	18
43	. 89866	. 79189	. 90251	. 79893	. 90627	. 80588	. 90995	. 81273	. 91355	. 81950	17
44	. 89873	. 79201	. 90257	. 79905	. 90633	. 80599	. 91001	. 81285	. 91361	. 81961	16
45	9. 89879	0. 79212	9. 90264	0. 79916	9. 90639	0. 80611	9. 91007	0. 81296	9. 91367	0. 81972	15
46	. 89886	. 79224	. 90270	. 79928	. 90646	. 80622	. 91013	. 81308	. 91372	. 81983	14
47	. 89892	. 79236	. 90276	. 79940	. 90652	. 80634	. 91019	. 81319	. 91378	. 81994	13
48	. 89899	. 79248	. 90282	. 79951	. 90658	. 80645	. 91025	. 81330	. 91384	. 82005	12
49	. 89905	. 79260	. 90289	. 79963	. 90664	. 80657	. 91031	. 81342	. 91390	. 82017	11
50	9. 89912	0. 79271	9. 90295	0. 79974	9. 90670	0. 80668	9. 91037	0. 81353	9. 91396	0. 82028	10
51	. 89918	. 79283	. 90301	. 79986	. 90676	. 80680	. 91043	. 81364	. 91402	. 82039	9
52	. 89925	. 79295	. 90308	. 79998	. 90683	. 80691	. 91049	. 81376	. 91408	. 82050	8
53	. 89931	. 79307	. 90314	. 80009	. 90689	. 80703	. 91055	. 81387	. 91414	. 82061	7
54	. 89938	. 79319	. 90320	. 80021	. 90695	. 80714	. 91061	. 81398	. 91420	. 82072	6
55	9. 89944	0. 79330	9. 90327	0. 80033	9. 90701	0. 80726	9. 91067	0. 81409	9. 91426	0. 82084	5
56	. 89950	. 79342	. 90333	. 80044	. 90707	. 80737	. 91074	. 81421	. 91432	. 82095	4
57	. 89957	. 79354	. 90339	. 80056	. 90714	. 80749	. 91080	. 81432	. 91437	. 82106	3
58	. 89963	. 79366	. 90346	. 80068	. 90720	. 80760	. 91086	. 81443	. 91443	. 82117	2
59	. 89970	. 79377	. 90352	. 80079	. 90726	. 80772	. 91092	. 81455	. 91449	. 82128	1
60	9. 89976	0. 79389	9. 90358	0. 80091	9. 90732	0. 80783	9. 91098	0. 81466	9. 91455	0. 82139	0
	234°		233°		232°		231°		230°		

TABLE 34

Haversines

′	130°		131°		132°		133°		134°		′
	Log Hav	Nat. Hav	Log Hav	Nat. Hav	Log Hav	Nat. Hav	Log Hav	Nat. Hav	Log Hav	Nat. Hav	
0	9.91455	0.82139	9.91805	0.82803	9.92146	0.83457	9.92480	0.84100	9.92805	0.84733	60
1	.91461	.82151	.91810	.82814	.92152	.83467	.92485	.84111	.92811	.84743	59
2	.91467	.82162	.91816	.82825	.92157	.83478	.92491	.84121	.92816	.84754	58
3	.91473	.82173	.91822	.82836	.92163	.83489	.92496	.84132	.92821	.84764	57
4	.91479	.82184	.91828	.82847	.92169	.83500	.92502	.84142	.92827	.84775	56
5	9.91485	0.82195	9.91833	0.82858	9.92174	0.83511	9.92507	0.84153	9.92832	0.84785	55
6	.91490	.82206	.91839	.82869	.92180	.83521	.92512	.84164	.92837	.84796	54
7	.91496	.82217	.91845	.82880	.92185	.83532	.92518	.84174	.92843	.84806	53
8	.91502	.82228	.91851	.82891	.92191	.83543	.92523	.84185	.92848	.84817	52
9	.91508	.82240	.91856	.82902	.92197	.83554	.92529	.84196	.92853	.84827	51
10	9.91514	0.82251	9.91862	0.82913	9.92202	0.83564	9.92534	0.84206	9.92859	0.84837	50
11	.91520	.82262	.91868	.82924	.92208	.83575	.92540	.84217	.92864	.84848	49
12	.91526	.82273	.91874	.82934	.92213	.83586	.92545	.84227	.92869	.84858	48
13	.91532	.82284	.91879	.82945	.92219	.83597	.92551	.84238	.92875	.84869	47
14	.91537	.82295	.91885	.82956	.92225	.83608	.92556	.84249	.92880	.84879	46
15	9.91543	0.82306	9.91891	0.82967	9.92230	0.83618	9.92562	0.84259	9.92885	0.84890	45
16	.91549	.82317	.91896	.82978	.92236	.83629	.92567	.84270	.92891	.84900	44
17	.91555	.82328	.91902	.82989	.92241	.83640	.92573	.84280	.92896	.84910	43
18	.91561	.82339	.91908	.83000	.92247	.83651	.92578	.84291	.92901	.84921	42
19	.91567	.82351	.91914	.83011	.92253	.83661	.92584	.84302	.92907	.84931	41
20	9.91573	0.82362	9.91919	0.83022	9.92258	0.83672	9.92589	0.84312	9.92912	0.84942	40
21	.91578	.82373	.91925	.83033	.92264	.83683	.92594	.84323	.92917	.84952	39
22	.91584	.82384	.91931	.83044	.92269	.83694	.92600	.84333	.92923	.84962	38
23	.91590	.82395	.91936	.83055	.92275	.83704	.92605	.84344	.92928	.84973	37
24	.91596	.82406	.91942	.83066	.92280	.83715	.92611	.84354	.92933	.84983	36
25	9.91602	0.82417	9.91948	0.83077	9.92286	0.83726	9.92616	0.84365	9.92939	0.84994	35
26	.91608	.82428	.91954	.83087	.92292	.83737	.92622	.84376	.92944	.85004	34
27	.91613	.82439	.91959	.83098	.92297	.83747	.92627	.84386	.92949	.85014	33
28	.91619	.82450	.91965	.83109	.92303	.83758	.92633	.84397	.92955	.85025	32
29	.91625	.82461	.91971	.83120	.92308	.83769	.92638	.84407	.92960	.85035	31
30	9.91631	0.82472	9.91976	0.83131	9.92314	0.83780	9.92643	0.84418	9.92965	0.85045	30
31	.91637	.82483	.91982	.83142	.92319	.83790	.92649	.84428	.92970	.85056	29
32	.91643	.82495	.91988	.83153	.92325	.83801	.92654	.84439	.92975	.85066	28
33	.91648	.82506	.91993	.83164	.92330	.83812	.92660	.84449	.92981	.85077	27
34	.91654	.82517	.91999	.83175	.92336	.83822	.92665	.84460	.92986	.85087	26
35	9.91660	0.82528	9.92005	0.83185	9.92342	0.83833	9.92670	0.84470	9.92992	0.85097	25
36	.91666	.82539	.92010	.83196	.92347	.83844	.92676	.84481	.92997	.85108	24
37	.91672	.82550	.92016	.83207	.92353	.83855	.92681	.84492	.93002	.85118	23
38	.91677	.82561	.92022	.83218	.92358	.83865	.92687	.84502	.93007	.85128	22
39	.91683	.82572	.92027	.83229	.92364	.83876	.92692	.84513	.93013	.85139	21
40	9.91689	0.82583	9.92033	0.83240	9.92369	0.83887	9.92698	0.84523	9.93018	0.85149	20
41	.91695	.82594	.92039	.83251	.92375	.83897	.92703	.84534	.93023	.85159	19
42	.91701	.82605	.92044	.83262	.92380	.83908	.92708	.84544	.93029	.85170	18
43	.91706	.82616	.92050	.83272	.92386	.83919	.92714	.84555	.93034	.85180	17
44	.91712	.82627	.92056	.83283	.92391	.83929	.92719	.84565	.93039	.85190	16
45	9.91718	0.82638	9.92061	0.83294	9.92397	0.83940	9.92725	0.84576	9.93044	0.85201	15
46	.91724	.82649	.92067	.83305	.92402	.83951	.92730	.84586	.93050	.85211	14
47	.91730	.82660	.92073	.83316	.92408	.83961	.92735	.84597	.93055	.85221	13
48	.91735	.82671	.92078	.83327	.92413	.83972	.92741	.84607	.93060	.85232	12
49	.91741	.82682	.92084	.83337	.92419	.83983	.92746	.84618	.93065	.85242	11
50	9.91747	0.82693	9.92090	0.83348	9.92425	0.83993	9.92751	0.84628	9.93071	0.85252	10
51	.91753	.82704	.92095	.83359	.92430	.84004	.92757	.84639	.93076	.85263	9
52	.91758	.82715	.92101	.83370	.92436	.84015	.92762	.84649	.93081	.85273	8
53	.91764	.82726	.92107	.83381	.92441	.84025	.92768	.84660	.93086	.85283	7
54	.91770	.82737	.92112	.83392	.92447	.84036	.92773	.84670	.93092	.85294	6
55	9.91776	0.82748	9.92118	0.83402	9.92452	0.84047	9.92778	0.84681	9.93097	0.85304	5
56	.91782	.82759	.92124	.83413	.92458	.84057	.92784	.84691	.93102	.85314	4
57	.91787	.82770	.92129	.83424	.92463	.84068	.92789	.84702	.93107	.85324	3
58	.91793	.82781	.92135	.83435	.92469	.84079	.92794	.84712	.93113	.85335	2
59	.91799	.82792	.92140	.83446	.92474	.84089	.92800	.84722	.93118	.85345	1
60	9.91805	0.82803	9.92146	0.83457	9.92480	0.84100	9.92805	0.84733	9.93123	0.85355	0
	229°		228°		227°		226°		225°		

TABLE 34

Haversines

	135°		136°		137°		138°		139°		
′	Log Hav	Nat. Hav	Log Hav	Nat. Hav	Log Hav	Nat. Hav	Log Hav	Nat. Hav	Log Hav	Nat. Hav	′
0	9.93123	0.85355	9.93433	0.85967	9.93736	0.86568	9.94030	0.87157	9.94318	0.87735	60
1	.93128	.85366	.93438	.85977	.93741	.86578	.94035	.87167	.94322	.87745	59
2	.93134	.85376	.93443	.85987	.93746	.86588	.94040	.87177	.94327	.87755	58
3	.93139	.85386	.93448	.85997	.93751	.86597	.94045	.87186	.94332	.87764	57
4	.93144	.85396	.93454	.86007	.93755	.86607	.94050	.87196	.94336	.87774	56
5	9.93149	0.85407	9.93459	0.86017	9.93760	0.86617	9.94055	0.87206	9.94341	0.87783	55
6	.93154	.85417	.93464	.86028	.93765	.86627	.94059	.87216	.94346	.87793	54
7	.93160	.85427	.93469	.86038	.93770	.86637	.94064	.87225	.94351	.87802	53
8	.93165	.85438	.93474	.86048	.93775	.86647	.94069	.87235	.94355	.87812	52
9	.93170	.85448	.93479	.86058	.93780	.86657	.94074	.87245	.94360	.87821	51
10	9.93175	0.85458	9.93484	0.86068	9.93785	0.86667	9.94079	0.87254	9.94365	0.87831	50
11	.93181	.85468	.93489	.86078	.93790	.86677	.94084	.87264	.94369	.87840	49
12	.93186	.85479	.93494	.86088	.93795	.86686	.94088	.87274	.94374	.87850	48
13	.93191	.85489	.93499	.86098	.93800	.86696	.94093	.87283	.94379	.87859	47
14	.93196	.85499	.93504	.86108	.93805	.86706	.94098	.87293	.94383	.87869	46
15	9.93201	0.85509	9.93509	0.86118	9.93810	0.86716	9.94103	0.87303	9.94388	0.87878	45
16	.93207	.85520	.93515	.86128	.93815	.86726	.94108	.87313	.94393	.87888	44
17	.93212	.85530	.93520	.86138	.93820	.86736	.94112	.87322	.94398	.87897	43
18	.93217	.85540	.93525	.86148	.93825	.86746	.94117	.87332	.94402	.87907	42
19	.93222	.85550	.93530	.86158	.93830	.86756	.94122	.87342	.94407	.87916	41
20	9.93227	0.85560	9.93535	0.86168	9.93835	0.86765	9.94127	0.87351	9.94412	0.87926	40
21	.93232	.85571	.93540	.86178	.93840	.86775	.94132	.87361	.94416	.87935	39
22	.93238	.85581	.93545	.86189	.93845	.86785	.94137	.87371	.94421	.87945	38
23	.93243	.85591	.93550	.86199	.93849	.86795	.94141	.87380	.94426	.87954	37
24	.93248	.85601	.93555	.86209	.93854	.86805	.94146	.87390	.94430	.87964	36
25	9.93253	0.85612	9.93560	0.86219	9.93859	0.86815	9.94151	0.87400	9.94435	0.87973	35
26	.93258	.85622	.93565	.86229	.93864	.86825	.94156	.87409	.94440	.87982	34
27	.93264	.85632	.93570	.86239	.93869	.86834	.94161	.87419	.94444	.87992	33
28	.93269	.85642	.93575	.86249	.93874	.86844	.94165	.87429	.94449	.88001	32
29	.93274	.85652	.93580	.86259	.93879	.86854	.94170	.87438	.94454	.88011	31
30	9.93279	0.85663	9.93585	0.86269	9.93884	0.86864	9.94175	0.87448	9.94458	0.88020	30
31	.93284	.85673	.93590	.86279	.93889	.86874	.94180	.87457	.94463	.88030	29
32	.93289	.85683	.93595	.86289	.93894	.86884	.94184	.87467	.94468	.88039	28
33	.93295	.85693	.93600	.86299	.93899	.86893	.94189	.87477	.94472	.88049	27
34	.93300	.85703	.93605	.86309	.93904	.86903	.94194	.87486	.94477	.88058	26
35	9.93305	0.85713	9.93611	0.86319	9.93908	0.86913	9.94199	0.87496	9.94482	0.88067	25
36	.93310	.85724	.93616	.86329	.93913	.86923	.94204	.87506	.94486	.88077	24
37	.93315	.85734	.93621	.86339	.93918	.86933	.94208	.87515	.94491	.88086	23
38	.93320	.85744	.93626	.86349	.93923	.86942	.94213	.87525	.94496	.88096	22
39	.93326	.85754	.93631	.86359	.93928	.86952	.94218	.87534	.94500	.88105	21
40	9.93331	0.85764	9.93636	0.86369	9.93933	0.86962	9.94223	0.87544	9.94505	0.88115	20
41	.93336	.85774	.93641	.86379	.93938	.86972	.94227	.87554	.94509	.88124	19
42	.93341	.85785	.93646	.86389	.93943	.86982	.94232	.87563	.94514	.88133	18
43	.93346	.85795	.93651	.86399	.93948	.86991	.94237	.87573	.94519	.88143	17
44	.93351	.85805	.93656	.86409	.93952	.87001	.94242	.87582	.94523	.88152	16
45	9.93356	0.85815	9.93661	0.86419	9.93957	0.87011	9.94246	0.87592	9.94528	0.88162	15
46	.93362	.85825	.93666	.86429	.93962	.87021	.94251	.87602	.94533	.88171	14
47	.93367	.85835	.93671	.86438	.93967	.87030	.94256	.87611	.94537	.88180	13
48	.93372	.85846	.93676	.86448	.93972	.87040	.94261	.87621	.94542	.88190	12
49	.93377	.85856	.93681	.86458	.93977	.87050	.94265	.87630	.94546	.88199	11
50	9.93382	0.85866	9.93686	0.86468	9.93982	0.87060	9.94270	0.87640	9.94551	0.88209	10
51	.93387	.85876	.93691	.86478	.93987	.87070	.94275	.87649	.94556	.88218	9
52	.93392	.85886	.93696	.86488	.93991	.87079	.94280	.87659	.94560	.88227	8
53	.93397	.85896	.93701	.86498	.93996	.87089	.94284	.87669	.94565	.88237	7
54	.93403	.85906	.93706	.86508	.94001	.87099	.94289	.87678	.94570	.88246	6
55	9.93408	0.85916	9.93711	0.86518	9.94006	0.87109	9.94294	0.87688	9.94574	0.88255	5
56	.93413	.85927	.93716	.86528	.94011	.87118	.94299	.87697	.94579	.88265	4
57	.93418	.85937	.93721	.86538	.94016	.87128	.94303	.87707	.94583	.88274	3
58	.93423	.85947	.93726	.86548	.94021	.87138	.94308	.87716	.94588	.88284	2
59	.93428	.85957	.93731	.86558	.94026	.87148	.94313	.87726	.94593	.88293	1
60	9.93433	0.85967	9.93736	0.86568	9.94030	0.87157	9.94318	0.87735	9.94597	0.88302	0
	224°		223°		222°		221°		220°		

TABLE 34

Haversines

′	140°		141°		142°		143°		144°		′
	Log Hav	Nat. Hav	Log Hav	Nat. Hav	Log Hav	Nat. Hav	Log Hav	Nat. Hav	Log Hav	Nat. Hav	
0	9. 94597	0. 88302	9. 94869	0. 88857	9. 95134	0. 89401	9. 95391	0. 89932	9. 95641	0. 90451	60
1	. 94602	. 88312	. 94874	. 88866	. 95138	. 89409	. 95396	. 89941	. 95645	. 90459	59
2	. 94606	. 88321	. 94878	. 88876	. 95143	. 89418	. 95400	. 89949	. 95649	. 90468	58
3	. 94611	. 88330	. 94883	. 88885	. 95147	. 89427	. 95404	. 89958	. 95654	. 90476	57
4	. 94616	. 88340	. 94887	. 88894	. 95151	. 89436	. 95408	. 89967	. 95658	. 90485	56
5	9. 94620	0. 88349	9. 94892	0. 88903	9. 95156	0. 89445	9. 95412	0. 89975	9. 95662	0. 90494	55
6	. 94625	. 88358	. 94896	. 88912	. 95160	. 89454	. 95417	. 89984	. 95666	. 90502	54
7	. 94629	. 88368	. 94901	. 88921	. 95164	. 89463	. 95421	. 89993	. 95670	. 90511	53
8	. 94634	. 88377	. 94905	. 88930	. 95169	. 89472	. 95425	. 90002	. 95674	. 90519	52
9	. 94638	. 88386	. 94909	. 88940	. 95173	. 89481	. 95429	. 90010	. 95678	. 90528	51
10	9. 94643	0. 88396	9. 94914	0. 88949	9. 95177	0. 89490	9. 95433	0. 90019	9. 95682	0. 90536	50
11	. 94648	. 88405	. 94918	. 88958	. 95182	. 89499	. 95438	. 90028	. 95686	. 90545	49
12	. 94652	. 88414	. 94923	. 88967	. 95186	. 89508	. 95442	. 90037	. 95690	. 90553	48
13	. 94657	. 88423	. 94927	. 88976	. 95190	. 89517	. 95446	. 90045	. 95694	. 90562	47
14	. 94661	. 88433	. 94932	. 88985	. 95195	. 89526	. 95450	. 90054	. 95699	. 90570	46
15	9. 94666	0. 88442	9. 94936	0. 88994	9. 95199	0. 89534	9. 95454	0. 90063	9. 95703	0. 90579	45
16	. 94670	. 88451	. 94941	. 89003	. 95203	. 89543	. 95459	. 90071	. 95707	. 90587	44
17	. 94675	. 88461	. 94945	. 89012	. 95208	. 89552	. 95463	. 90080	. 95711	. 90596	43
18	. 94680	. 88470	. 94950	. 89022	. 95212	. 89561	. 95467	. 90089	. 95715	. 90604	42
19	. 94684	. 88479	. 94954	. 89031	. 95216	. 89570	. 95471	. 90097	. 95719	. 90613	41
20	9. 94689	0. 88489	9. 94958	0. 89040	9. 95221	0. 89579	9. 95475	0. 90106	9. 95723	0. 90621	40
21	. 94693	. 88498	. 94963	. 89049	. 95225	. 89588	. 95480	. 90115	. 95727	. 90630	39
22	. 94698	. 88507	. 94967	. 89058	. 95229	. 89597	. 95484	. 90124	. 95731	. 90638	38
23	. 94702	. 88516	. 94972	. 89067	. 95234	. 89606	. 95488	. 90132	. 95735	. 90647	37
24	. 94707	. 88526	. 94976	. 89076	. 95238	. 89614	. 95492	. 90141	. 95739	. 90655	36
25	9. 94711	0. 88535	9. 94981	0. 89085	9. 95242	0. 89623	9. 95496	0. 90150	9. 95743	0. 90664	35
26	. 94716	. 88544	. 94985	. 89094	. 95246	. 89632	. 95501	. 90158	. 95747	. 90672	34
27	. 94721	. 88553	. 94989	. 89103	. 95251	. 89641	. 95505	. 90167	. 95751	. 90680	33
28	. 94725	. 88563	. 94994	. 89112	. 95255	. 89650	. 95509	. 90176	. 95755	. 90689	32
29	. 94730	. 88572	. 94998	. 89121	. 95259	. 89659	. 95513	. 90184	. 95759	. 90697	31
30	9. 94734	0. 88581	9. 95003	0. 89130	9. 95264	0. 89668	9. 95517	0. 90193	9. 95763	0. 90706	30
31	. 94739	. 88590	. 95007	. 89139	. 95268	. 89677	. 95521	. 90201	. 95768	. 90714	29
32	. 94743	. 88600	. 95011	. 89149	. 95272	. 89685	. 95526	. 90210	. 95772	. 90723	28
33	. 94748	. 88609	. 95016	. 89158	. 95276	. 89694	. 95530	. 90219	. 95776	. 90731	27
34	. 94752	. 88618	. 95020	. 89167	. 95281	. 89703	. 95534	. 90227	. 95780	. 90740	26
35	9. 94757	0. 88627	9. 95025	0. 89176	9. 95285	0. 89712	9. 95538	0. 90236	9. 95784	0. 90748	25
36	. 94761	. 88637	. 95029	. 89185	. 95289	. 89721	. 95542	. 90245	. 95788	. 90756	24
37	. 94766	. 88646	. 95033	. 89194	. 95294	. 89730	. 95546	. 90253	. 95792	. 90765	23
38	. 94770	. 88655	. 95038	. 89203	. 95298	. 89738	. 95550	. 90262	. 95796	. 90773	22
39	. 94774	. 88664	. 95042	. 89212	. 95302	. 89747	. 95555	. 90271	. 95800	. 90782	21
40	9. 94779	0. 88674	9. 95047	0. 89221	9. 95306	0. 89756	9. 95559	0. 90279	9. 95804	0. 90790	20
41	. 94784	. 88683	. 95051	. 89230	. 95311	. 89765	. 95563	. 90288	. 95808	. 90798	19
42	. 94788	. 88692	. 95055	. 89239	. 95315	. 89774	. 95567	. 90296	. 95812	. 90807	18
43	. 94793	. 88701	. 95060	. 89248	. 95319	. 89782	. 95571	. 90305	. 95816	. 90815	17
44	. 94797	. 88710	. 95064	. 89257	. 95323	. 89791	. 95575	. 90314	. 95820	. 90824	16
45	9. 94802	0. 88720	9. 95069	0. 89266	9. 95328	0. 89800	9. 95579	0. 90322	9. 95824	0. 90832	15
46	. 94806	. 88729	. 95073	. 89275	. 95332	. 89809	. 95584	. 90331	. 95828	. 90840	14
47	. 94811	. 88738	. 95077	. 89284	. 95336	. 89818	. 95588	. 90339	. 95832	. 90849	13
48	. 94815	. 88747	. 95082	. 89293	. 95340	. 89826	. 95592	. 90348	. 95836	. 90857	12
49	. 94820	. 88756	. 95086	. 89302	. 95345	. 89835	. 95596	. 90357	. 95840	. 90866	11
50	9. 94824	0. 88766	9. 95090	0. 89311	9. 95349	0. 89844	9. 95600	0. 90365	9. 95844	0. 90874	10
51	. 94829	. 88775	. 95095	. 89320	. 95353	. 89853	. 95604	. 90374	. 95848	. 90882	9
52	. 94833	. 88784	. 95099	. 89329	. 95357	. 89862	. 95608	. 90382	. 95852	. 90891	8
53	. 94838	. 88793	. 95104	. 89338	. 95362	. 89870	. 95613	. 90391	. 95856	. 90899	7
54	. 94842	. 88802	. 95108	. 89347	. 95366	. 89879	. 95617	. 90399	. 95860	. 90907	6
55	9. 94847	0. 88811	9. 95112	0. 89356	9. 95370	0. 89888	9. 95621	0. 90408	9. 95864	0. 90916	5
56	. 94851	. 88821	. 95117	. 89365	. 95374	. 89897	. 95625	. 90417	. 95868	. 90924	4
57	. 94856	. 88830	. 95121	. 89374	. 95379	. 89906	. 95629	. 90425	. 95872	. 90933	3
58	. 94860	. 88839	. 95125	. 89383	. 95383	. 89914	. 95633	. 90434	. 95876	. 90941	2
59	. 94865	. 88848	. 95130	. 89392	. 95387	. 89923	. 95637	. 90442	. 95880	. 90949	1
60	9. 94869	0. 88857	9. 95134	0. 89401	9. 95391	0. 89932	9. 95641	0. 90451	9. 95884	0. 90958	0
	219°		218°		217°		216°		215°		

TABLE 34

Haversines

′	145°		146°		147°		148°		149°		′
	Log Hav	Nat. Hav	Log Hav	Nat. Hav	Log Hav	Nat. Hav	Log Hav	Nat. Hav	Log Hav	Nat. Hav	
0	9. 95884	0. 90958	9. 96119	0. 91452	9. 96347	0. 91934	9. 96568	0. 92402	9. 96782	0. 92858	60
1	. 95888	. 90966	. 96123	. 91460	. 96351	. 91941	. 96572	. 92410	. 96786	. 92866	59
2	. 95892	. 90974	. 96127	. 91468	. 96355	. 91949	. 96576	. 92418	. 96789	. 92873	58
3	. 95896	. 90983	. 96131	. 91476	. 96359	. 91957	. 96579	. 92426	. 96793	. 92881	57
4	. 95900	. 90991	. 96135	. 91484	. 96362	. 91965	. 96583	. 92433	. 96796	. 92888	56
5	9. 95904	0. 90999	9. 96139	0. 91493	9. 96366	0. 91973	9. 96586	0. 92441	9. 96800	0. 92896	55
6	. 95908	. 91008	. 96142	. 91501	. 96370	. 91981	. 96590	. 92449	. 96803	. 92903	54
7	. 95912	. 91016	. 96146	. 91509	. 96374	. 91989	. 96594	. 92456	. 96807	. 92911	53
8	. 95916	. 91024	. 96150	. 91517	. 96377	. 91997	. 96597	. 92464	. 96810	. 92918	52
9	. 95920	. 91033	. 96154	. 91525	. 96381	. 92005	. 96601	. 92472	. 96814	. 92926	51
10	9. 95924	0. 91041	9. 96158	0. 91533	9. 96385	0. 92013	9. 96604	0. 92479	9. 96817	0. 92933	50
11	. 95928	. 91049	. 96162	. 91541	. 96388	. 92020	. 96608	. 92487	. 96821	. 92941	49
12	. 95932	. 91057	. 96165	. 91549	. 96392	. 92028	. 96612	. 92495	. 96824	. 92948	48
13	. 95936	. 91066	. 96169	. 91557	. 96396	. 92036	. 96615	. 92502	. 96827	. 92955	47
14	. 95939	. 91074	. 96173	. 91565	. 96400	. 92044	. 96619	. 92510	. 96831	. 92963	46
15	9. 95943	0. 91082	9. 96177	0. 91573	9. 96403	0. 92052	9. 96622	0. 92518	9. 96834	0. 92970	45
16	. 95947	. 91091	. 96181	. 91582	. 96407	. 92060	. 96626	. 92525	. 96837	. 92978	44
17	. 95951	. 91099	. 96185	. 91590	. 96411	. 92068	. 96630	. 92533	. 96841	. 92985	43
18	. 95955	. 91107	. 96188	. 91598	. 96414	. 92076	. 96633	. 92541	. 96845	. 92993	42
19	. 95959	. 91115	. 96192	. 91606	. 96418	. 92083	. 96637	. 92548	. 96848	. 93000	41
20	9. 95963	0. 91124	9. 96196	0. 91614	9. 96422	0. 92091	9. 96640	0. 92556	9. 96852	0. 93007	40
21	. 95967	. 91132	. 96200	. 91622	. 96426	. 92099	. 96644	. 92563	. 96855	. 93015	39
22	. 95971	. 91140	. 96204	. 91630	. 96429	. 92107	. 96648	. 92571	. 96859	. 93022	38
23	. 95975	. 91149	. 96208	. 91638	. 96433	. 92115	. 96651	. 92579	. 96862	. 93030	37
24	. 95979	. 91157	. 96211	. 91646	. 96437	. 92123	. 96655	. 92586	. 96866	. 93037	36
25	9. 95983	0. 91165	9. 96215	0. 91654	9. 96440	0. 92130	9. 96658	0. 92594	9. 96869	0. 93045	35
26	. 95987	. 91173	. 96219	. 91662	. 96444	. 92138	. 96662	. 92602	. 96873	. 93052	34
27	. 95991	. 91182	. 96223	. 91670	. 96448	. 92146	. 96665	. 92609	. 96876	. 93059	33
28	. 95995	. 91190	. 96227	. 91678	. 96451	. 92154	. 96669	. 92617	. 96879	. 93067	32
29	. 95999	. 91198	. 96230	. 91686	. 96455	. 92162	. 96673	. 92624	. 96883	. 93074	31
30	9. 96002	0. 91206	9. 96234	0. 91694	9. 96459	0. 92170	9. 96676	0. 92632	9. 96886	0. 93081	30
31	. 96006	. 91215	. 96238	. 91702	. 96462	. 92177	. 96680	. 92640	. 96890	. 93089	29
32	. 96010	. 91223	. 96242	. 91710	. 96466	. 92185	. 96683	. 92647	. 96894	. 93096	28
33	. 96014	. 91231	. 96246	. 91718	. 96470	. 92193	. 96687	. 92655	. 96897	. 93104	27
34	. 96018	. 91239	. 96249	. 91726	. 96473	. 92201	. 96690	. 92662	. 96900	. 93111	26
35	9. 96022	0. 91247	9. 96253	0. 91734	9. 96477	0. 92209	9. 96694	0. 92670	9. 96904	0. 93118	25
36	. 96026	. 91256	. 96257	. 91742	. 96481	. 92216	. 96697	. 92678	. 96907	. 93126	24
37	. 96030	. 91264	. 96261	. 91750	. 96484	. 92224	. 96701	. 92685	. 96910	. 93133	23
38	. 96034	. 91272	. 96265	. 91758	. 96488	. 92232	. 96705	. 92693	. 96914	. 93140	22
39	. 96038	. 91280	. 96268	. 91766	. 96492	. 92240	. 96708	. 92700	. 96917	. 93148	21
40	9. 96042	0. 91289	9. 96272	0. 91774	9. 96495	0. 92248	9. 96712	0. 92708	9. 96921	0. 93155	20
41	. 96046	. 91297	. 96276	. 91782	. 96499	. 92255	. 96715	. 92715	. 96924	. 93162	19
42	. 96049	. 91305	. 96280	. 91790	. 96503	. 92263	. 96719	. 92723	. 96928	. 93170	18
43	. 96053	. 91313	. 96283	. 91798	. 96506	. 92271	. 96722	. 92730	. 96931	. 93177	17
44	. 96057	. 91321	. 96287	. 91806	. 96510	. 92279	. 96726	. 92738	. 96934	. 93184	16
45	9. 96061	0. 91329	9. 96291	0. 91814	9. 96514	0. 92286	9. 96729	0. 92746	9. 96938	0. 93192	15
46	. 96065	. 91338	. 96295	. 91822	. 96517	. 92294	. 96733	. 92753	. 96941	. 93199	14
47	. 96069	. 91346	. 96299	. 91830	. 96521	. 92302	. 96736	. 92761	. 96945	. 93206	13
48	. 96073	. 91354	. 96302	. 91838	. 96525	. 92310	. 96740	. 92768	. 96948	. 93214	12
49	. 96077	. 91362	. 96306	. 91846	. 96528	. 92317	. 96743	. 92776	. 96951	. 93221	11
50	9. 96081	0. 91370	9. 96310	0. 91854	9. 96532	0. 92325	9. 96747	0. 92783	9. 96955	0. 93228	10
51	. 96084	. 91379	. 96314	. 91862	. 96536	. 92333	. 96750	. 92791	. 96958	. 93236	9
52	. 96088	. 91387	. 96317	. 91870	. 96539	. 92341	. 96754	. 92798	. 96962	. 93243	8
53	. 96092	. 91395	. 96321	. 91878	. 96543	. 92348	. 96758	. 92806	. 96965	. 93250	7
54	. 96096	. 91403	. 96325	. 91886	. 96547	. 92356	. 96761	. 92813	. 96968	. 93258	6
55	9. 96100	0. 91411	9. 96329	0. 91894	9. 96550	0. 92364	9. 96765	0. 92821	9. 96972	0. 93265	5
56	. 96104	. 91419	. 96332	. 91902	. 96554	. 92372	. 96768	. 92828	. 96975	. 93272	4
57	. 96108	. 91427	. 96336	. 91910	. 96557	. 92379	. 96772	. 92836	. 96979	. 93279	3
58	. 96112	. 91436	. 96340	. 91918	. 96561	. 92387	. 96775	. 92843	. 96982	. 93287	2
59	. 96115	. 91444	. 96344	. 91926	. 96565	. 92395	. 96779	. 92851	. 96985	. 93294	1
60	9. 96119	0. 91452	9. 96347	0. 91934	9. 96568	0. 92402	9. 96782	0. 92858	9. 96989	0. 93301	0
	214°		213°		212°		211°		210°		

TABLE 34

Haversines

′	150°		151°		152°		153°		154°		′
	Log Hav	Nat. Hav	Log Hav	Nat. Hav	Log Hav	Nat. Hav	Log Hav	Nat. Hav	Log Hav	Nat. Hav	
0	9. 96989	0. 93301	9. 97188	0. 93731	9. 97381	0. 94147	9. 97566	0. 94550	9. 97745	0. 94940	60
1	. 96992	. 93309	. 97192	. 93738	. 97384	. 94154	. 97569	. 94557	. 97748	. 94946	59
2	. 96996	. 93316	. 97195	. 93745	. 97387	. 94161	. 97572	. 94564	. 97751	. 94952	58
3	. 96999	. 93323	. 97198	. 93752	. 97390	. 94168	. 97575	. 94570	. 97754	. 94959	57
4	. 97002	. 93330	. 97201	. 93759	. 97393	. 94175	. 97578	. 94577	. 97756	. 94965	56
5	9. 97006	0. 93338	9. 97205	0. 93766	9. 97397	0. 94181	9. 97581	0. 94583	9. 97759	0. 94972	55
6	. 97009	. 93345	. 97208	. 93773	. 97400	. 94188	. 97584	. 94590	. 97762	. 94978	54
7	. 97012	. 93352	. 97211	. 93780	. 97403	. 94195	. 97587	. 94596	. 97765	. 94984	53
8	. 97016	. 93359	. 97214	. 93787	. 97406	. 94202	. 97591	. 94603	. 97768	. 94991	52
9	. 97019	. 93367	. 97218	. 93794	. 97409	. 94209	. 97594	. 94610	. 97771	. 94997	51
10	9. 97022	0. 93374	9. 97221	0. 93801	9. 97412	0. 94215	9. 97597	0. 94616	9. 97774	0. 95003	50
11	. 97026	. 93381	. 97224	. 93808	. 97415	. 94222	. 97600	. 94623	. 97777	. 95010	49
12	. 97029	. 93388	. 97227	. 93815	. 97418	. 94229	. 97603	. 94629	. 97780	. 95016	48
13	. 97033	. 93395	. 97231	. 93822	. 97422	. 94236	. 97606	. 94636	. 97783	. 95022	47
14	. 97036	. 93403	. 97234	. 93829	. 97425	. 94243	. 97609	. 94642	. 97785	. 95029	46
15	9. 97039	0. 93410	9. 97237	0. 93836	9. 97428	0. 94249	9. 97612	0. 94649	9. 97788	0. 95035	45
16	. 97043	. 93417	. 97240	. 93843	. 97431	. 94256	. 97615	. 94655	. 97791	. 95041	44
17	. 97046	. 93424	. 97244	. 93850	. 97434	. 94263	. 97618	. 94662	. 97794	. 95048	43
18	. 97049	. 93432	. 97247	. 93857	. 97437	. 94270	. 97621	. 94669	. 97797	. 95054	42
19	. 97052	. 93439	. 97250	. 93864	. 97440	. 94276	. 97624	. 94675	. 97800	. 95060	41
20	9. 97056	0. 93446	9. 97253	0. 93871	9. 97443	0. 94283	9. 97627	0. 94682	9. 97803	0. 95066	40
21	. 97059	. 93453	. 97257	. 93878	. 97447	. 94290	. 97630	. 94688	. 97806	. 95073	39
22	. 97063	. 93460	. 97260	. 93885	. 97450	. 94297	. 97633	. 94695	. 97808	. 95079	38
23	. 97066	. 93468	. 97263	. 93892	. 97453	. 94303	. 97636	. 94701	. 97811	. 95085	37
24	. 97069	. 93475	. 97266	. 93899	. 97456	. 94310	. 97639	. 94708	. 97814	. 95092	36
25	9. 97073	0. 93482	9. 97269	0. 93906	9. 97459	0. 94317	9. 97642	0. 94714	9. 97817	0. 95098	35
26	. 97076	. 93489	. 97273	. 93913	. 97462	. 94324	. 97645	. 94721	. 97820	. 95104	34
27	. 97079	. 93496	. 97276	. 93920	. 97465	. 94330	. 97647	. 94727	. 97823	. 95110	33
28	. 97083	. 93503	. 97279	. 93927	. 97468	. 94337	. 97650	. 94734	. 97826	. 95117	32
29	. 97086	. 93511	. 97282	. 93934	. 97471	. 94344	. 97653	. 94740	. 97829	. 95123	31
30	9. 97089	0. 93518	9. 97285	0. 93941	9. 97474	0. 94351	9. 97656	0. 94747	9. 97831	0. 95129	30
31	. 97093	. 93525	. 97289	. 93948	. 97478	. 94357	. 97659	. 94753	. 97834	. 95136	29
32	. 97096	. 93532	. 97292	. 93955	. 97481	. 94364	. 97662	. 94760	. 97837	. 95142	28
33	. 97099	. 93539	. 97295	. 93962	. 97484	. 94371	. 97665	. 94766	. 97840	. 95148	27
34	. 97103	. 93546	. 97298	. 93969	. 97487	. 94377	. 97668	. 94773	. 97843	. 95154	26
35	9. 97106	0. 93554	9. 97301	0. 93976	9. 97490	0. 94384	9. 97671	0. 94779	9. 97846	0. 95161	25
36	. 97109	. 93561	. 97305	. 93982	. 97493	. 94391	. 97674	. 94786	. 97849	. 95167	24
37	. 97113	. 93568	. 97308	. 93989	. 97496	. 94397	. 97677	. 94792	. 97851	. 95173	23
38	. 97116	. 93575	. 97311	. 93996	. 97499	. 94404	. 97680	. 94799	. 97854	. 95179	22
39	. 97119	. 93582	. 97314	. 94003	. 97502	. 94411	. 97683	. 94805	. 97857	. 95185	21
40	9. 97123	0. 93589	9. 97317	0. 94010	9. 97505	0. 94418	9. 97686	0. 94811	9. 97860	0. 95192	20
41	. 97126	. 93596	. 97321	. 94017	. 97508	. 94424	. 97689	. 94818	. 97863	. 95198	19
42	. 97129	. 93603	. 97324	. 94024	. 97511	. 94431	. 97692	. 94824	. 97866	. 95204	18
43	. 97132	. 93611	. 97327	. 94031	. 97514	. 94438	. 97695	. 94831	. 97868	. 95210	17
44	. 97136	. 93618	. 97330	. 94038	. 97518	. 94444	. 97698	. 94837	. 97871	. 95217	16
45	9. 97139	0. 93625	9. 97333	0. 94045	9. 97521	0. 94451	9. 97701	0. 94844	9. 97874	0. 95223	15
46	. 97142	. 93632	. 97337	. 94051	. 97524	. 94458	. 97704	. 94850	. 97877	. 95229	14
47	. 97146	. 93639	. 97340	. 94058	. 97527	. 94464	. 97707	. 94856	. 97880	. 95235	13
48	. 97149	. 93646	. 97343	. 94065	. 97530	. 94471	. 97710	. 94863	. 97883	. 95241	12
49	. 97152	. 93653	. 97346	. 94072	. 97533	. 94477	. 97713	. 94869	. 97885	. 95248	11
50	9. 97156	0. 93660	9. 97349	0. 94079	9. 97536	0. 94484	9. 97716	0. 94876	9. 97888	0. 95254	10
51	. 97159	. 93667	. 97352	. 94086	. 97539	. 94491	. 97718	. 94882	. 97891	. 95260	9
52	. 97162	. 93674	. 97356	. 94093	. 97542	. 94497	. 97721	. 94889	. 97894	. 95266	8
53	. 97165	. 93682	. 97359	. 94099	. 97545	. 94504	. 97724	. 94895	. 97897	. 95272	7
54	. 97169	. 93689	. 97362	. 94106	. 97548	. 94511	. 97727	. 94901	. 97899	. 95278	6
55	9. 97172	0. 93696	9. 97365	0. 94113	9. 97551	0. 94517	9. 97730	0. 94908	9. 97902	0. 95285	5
56	. 97175	. 93703	. 97368	. 94120	. 97554	. 94524	. 97733	. 94914	. 97905	. 95291	4
57	. 97179	. 93710	. 97371	. 94127	. 97557	. 94531	. 97736	. 94921	. 97908	. 95297	3
58	. 97182	. 93717	. 97375	. 94134	. 97560	. 94537	. 97739	. 94927	. 97911	. 95303	2
59	. 97185	. 93724	. 97378	. 94141	. 97563	. 94544	. 97742	. 94933	. 97914	. 95309	1
60	9. 97188	0. 93731	9. 97381	0. 94147	9. 97566	0. 94550	9. 97745	0. 94940	9. 97916	0. 95315	0
	209°		208°		207°		206°		205°		

TABLE 34

Haversines

′	155°		156°		157°		158°		159°		′
′	Log Hav	Nat. Hav	Log Hav	Nat. Hav	Log Hav	Nat. Hav	Log Hav	Nat. Hav	Log Hav	Nat. Hav	′
0	9. 97916	0. 95315	9. 98081	0. 95677	9. 98239	0. 96025	9. 98389	0. 96359	9. 98533	0. 96679	60
1	. 97919	. 95322	. 98084	. 95683	. 98241	. 96031	. 98392	. 96365	. 98536	. 96684	59
2	. 97922	. 95328	. 98086	. 95689	. 98244	. 96037	. 98394	. 96370	. 98538	. 96689	58
3	. 97925	. 95334	. 98089	. 95695	. 98246	. 96042	. 98397	. 96376	. 98540	. 96695	57
4	. 97927	. 95340	. 98092	. 95701	. 98249	. 96048	. 98399	. 96381	. 98543	. 96700	56
5	9. 97930	0. 95346	9. 98094	0. 95707	9. 98251	0. 96054	9. 98402	0. 96386	9. 98545	0. 96705	55
6	. 97933	. 95352	. 98097	. 95713	. 98254	. 96059	. 98404	. 96392	. 98547	. 96710	54
7	. 97936	. 95358	. 98100	. 95719	. 98256	. 96065	. 98406	. 96397	. 98550	. 96715	53
8	. 97939	. 95364	. 98102	. 95724	. 98259	. 96071	. 98409	. 96403	. 98552	. 96721	52
9	. 97941	. 95371	. 98105	. 95730	. 98262	. 96076	. 98411	. 96408	. 98554	. 96726	51
10	9. 97944	0. 95377	9. 98108	0. 95736	9. 98264	0. 96082	9. 98414	0. 96413	9. 98557	0. 96731	50
11	. 97947	. 95383	. 98110	. 95742	. 98267	. 96088	. 98416	. 96419	. 98559	. 96736	49
12	. 97950	. 95389	. 98113	. 95748	. 98269	. 96093	. 98419	. 96424	. 98561	. 96741	48
13	. 97953	. 95395	. 98116	. 95754	. 98272	. 96099	. 98421	. 96430	. 98564	. 96746	47
14	. 97955	. 95401	. 98118	. 95760	. 98274	. 96104	. 98424	. 96435	. 98566	. 96752	46
15	9. 97958	0. 95407	9. 98121	0. 95766	9. 98277	0. 96110	9. 98426	0. 96440	9. 98568	0. 96757	45
16	. 97961	. 95413	. 98124	. 95771	. 98279	. 96116	. 98428	. 96446	. 98570	. 96762	44
17	. 97964	. 95419	. 98126	. 95777	. 98282	. 96121	. 98431	. 96451	. 98573	. 96767	43
18	. 97966	. 95425	. 98129	. 95783	. 98285	. 96127	. 98433	. 96457	. 98575	. 96772	42
19	. 97969	. 95431	. 98132	. 95789	. 98287	. 96133	. 98436	. 96462	. 98577	. 96777	41
20	9. 97972	0. 95438	9. 98134	0. 95795	9. 98290	0. 96138	9. 98438	0. 96467	9. 98580	0. 96782	40
21	. 97975	. 95444	. 98137	. 95801	. 98292	. 96144	. 98440	. 96473	. 98582	. 96788	39
22	. 97977	. 95450	. 98139	. 95806	. 98295	. 96149	. 98443	. 96478	. 98584	. 96793	38
23	. 97980	. 95456	. 98142	. 95812	. 98297	. 96155	. 98445	. 96483	. 98587	. 96798	37
24	. 97983	. 95462	. 98145	. 95818	. 98300	. 96161	. 98448	. 96489	. 98589	. 96803	36
25	9. 97986	0. 95468	9. 98147	0. 95824	9. 98302	0. 96166	9. 98450	0. 96494	9. 98591	0. 96808	35
26	. 97988	. 95474	. 98150	. 95830	. 98305	. 96172	. 98453	. 96500	. 98593	. 96813	34
27	. 97991	. 95480	. 98153	. 95836	. 98307	. 96177	. 98455	. 96505	. 98596	. 96818	33
28	. 97994	. 95486	. 98155	. 95841	. 98310	. 96183	. 98457	. 96510	. 98598	. 96823	32
29	. 97997	. 95492	. 98158	. 95847	. 98312	. 96188	. 98460	. 96516	. 98600	. 96829	31
30	9. 97999	0. 95498	9. 98161	0. 95853	9. 98315	0. 96194	9. 98462	0. 96521	9. 98603	0. 96834	30
31	. 98002	. 95504	. 98163	. 95859	. 98317	. 96200	. 98465	. 96526	. 98605	. 96839	29
32	. 98005	. 95510	. 98166	. 95865	. 98320	. 96205	. 98467	. 96532	. 98607	. 96844	28
33	. 98008	. 95516	. 98168	. 95870	. 98322	. 96211	. 98469	. 96537	. 98609	. 96849	27
34	. 98010	. 95522	. 98171	. 95876	. 98325	. 96216	. 98472	. 96542	. 98612	. 96854	26
35	9. 98013	0. 95528	9. 98174	0. 95882	9. 98327	0. 96222	9. 98474	0. 96547	9. 98614	0. 96859	25
36	. 98016	. 95534	. 98176	. 95888	. 98330	. 96227	. 98476	. 96553	. 98616	. 96864	24
37	. 98019	. 95540	. 98179	. 95894	. 98332	. 96233	. 98479	. 96558	. 98619	. 96869	23
38	. 98021	. 95546	. 98182	. 95899	. 98335	. 96238	. 98481	. 96563	. 98621	. 96874	22
39	. 98024	. 95552	. 98184	. 95905	. 98337	. 96244	. 98484	. 96569	. 98623	. 96879	21
40	9. 98027	0. 95558	9. 98187	0. 95911	9. 98340	0. 96249	9. 98486	0. 96574	9. 98625	0. 96884	20
41	. 98030	. 95564	. 98189	. 95917	. 98342	. 96255	. 98488	. 96579	. 98628	. 96889	19
42	. 98032	. 95570	. 98192	. 95922	. 98345	. 96260	. 98491	. 96585	. 98630	. 96894	18
43	. 98035	. 95576	. 98195	. 95928	. 98347	. 96266	. 98493	. 96590	. 98632	. 96899	17
44	. 98038	. 95582	. 98197	. 95934	. 98350	. 96272	. 98496	. 96595	. 98634	. 96905	16
45	9. 98040	0. 95588	9. 98200	0. 95940	9. 98352	0. 96277	9. 98498	0. 96600	9. 98637	0. 96910	15
46	. 98043	. 95594	. 98202	. 95945	. 98355	. 96283	. 98500	. 96606	. 98639	. 96915	14
47	. 98046	. 95600	. 98205	. 95951	. 98357	. 96288	. 98503	. 96611	. 98641	. 96920	13
48	. 98049	. 95606	. 98208	. 95957	. 98360	. 96294	. 98505	. 96616	. 98643	. 96925	12
49	. 98051	. 95612	. 98210	. 95962	. 98362	. 96299	. 98507	. 96621	. 98646	. 96930	11
50	9. 98054	0. 95618	9. 98213	0. 95968	9. 98365	0. 96305	9. 98510	0. 96627	9. 98648	0. 96935	10
51	. 98057	. 95624	. 98215	. 95974	. 98367	. 96310	. 98512	. 96632	. 98650	. 96940	9
52	. 98059	. 95630	. 98218	. 95980	. 98370	. 96315	. 98514	. 96637	. 98652	. 96945	8
53	. 98062	. 95636	. 98221	. 95985	. 98372	. 96321	. 98517	. 96642	. 98655	. 96950	7
54	. 98065	. 95642	. 98223	. 95991	. 98375	. 96326	. 98519	. 96648	. 98657	. 96955	6
55	9. 98067	0. 95648	9. 98226	0. 95997	9. 98377	0. 96332	9. 98521	0. 96653	9. 98659	0. 96960	5
56	. 98070	. 95654	. 98228	. 96002	. 98379	. 96337	. 98524	. 96658	. 98661	. 96965	4
57	. 98073	. 95660	. 98231	. 96008	. 98382	. 96343	. 98526	. 96663	. 98664	. 96970	3
58	. 98076	. 95665	. 98233	. 96014	. 98384	. 96348	. 98529	. 96669	. 98666	. 96975	2
59	. 98078	. 95671	. 98236	. 96020	. 98387	. 96354	. 98531	. 96674	. 98668	. 96980	1
60	9. 98081	0. 95677	9. 98239	0. 96025	9. 98389	0. 96359	9. 98533	0. 96679	9. 98670	0. 96985	0
	204°		203°		202°		201°		200°		

TABLE 34
Haversines

′	160° Log Hav	160° Nat. Hav	161° Log Hav	161° Nat. Hav	162° Log Hav	162° Nat. Hav	163° Log Hav	163° Nat. Hav	164° Log Hav	164° Nat. Hav	′
0	9. 98670	0. 96985	9. 98801	0. 97276	9. 98924	0. 97553	9. 99041	0. 97815	9. 99151	0. 98063	60
1	. 98673	. 96990	. 98803	. 97281	. 98926	. 97557	. 99043	. 97819	. 99152	. 98067	59
2	. 98675	. 96995	. 98805	. 97285	. 98928	. 97562	. 99044	. 97824	. 99154	. 98071	58
3	. 98677	. 97000	. 98807	. 97290	. 98930	. 97566	. 99046	. 97828	. 99156	. 98075	57
4	. 98679	. 97004	. 98809	. 97295	. 98932	. 97571	. 99048	. 97832	. 99158	. 98079	56
5	9. 98681	0. 97009	9. 98811	0. 97300	9. 98934	0. 97575	9. 99050	0. 97836	9. 99159	0. 98083	55
6	. 98684	. 97014	. 98813	. 97304	. 98936	. 97580	. 99052	. 97841	. 99161	. 98087	54
7	. 98686	. 97019	. 98815	. 97309	. 98938	. 97584	. 99054	. 97845	. 99163	. 98091	53
8	. 98688	. 97024	. 98817	. 97314	. 98940	. 97589	. 99056	. 97849	. 99165	. 98095	52
9	. 98690	. 97029	. 98819	. 97318	. 98942	. 97593	. 99058	. 97853	. 99166	. 98099	51
10	9. 98692	0. 97034	9. 98822	0. 97323	9. 98944	0. 97598	9. 99059	0. 97858	9. 99168	0. 98103	50
11	. 98695	. 97039	. 98824	. 97328	. 98946	. 97602	. 99061	. 97862	. 99170	. 98107	49
12	. 98697	. 97044	. 98826	. 97332	. 98948	. 97606	. 99063	. 97866	. 99172	. 98111	48
13	. 98699	. 97049	. 98828	. 97337	. 98950	. 97611	. 99065	. 97870	. 99173	. 98115	47
14	. 98701	. 97054	. 98830	. 97342	. 98952	. 97615	. 99067	. 97874	. 99175	. 98119	46
15	9. 98703	0. 97059	9. 98832	0. 97347	9. 98954	0. 97620	9. 99069	0. 97879	9. 99177	0. 98123	45
16	. 98706	. 97064	. 98834	. 97351	. 98956	. 97624	. 99071	. 97883	. 99179	. 98127	44
17	. 98708	. 97069	. 98836	. 97356	. 98958	. 97629	. 99072	. 97887	. 99180	. 98131	43
18	. 98710	. 97074	. 98838	. 97361	. 98960	. 97633	. 99074	. 97891	. 99182	. 98135	42
19	. 98712	. 97078	. 98840	. 97365	. 98962	. 97637	. 99076	. 97895	. 99184	. 98139	41
20	9. 98714	0. 97083	9. 98842	0. 97370	9. 98964	0. 97642	9. 99078	0. 97899	9. 99186	0. 98142	40
21	. 98717	. 97088	. 98845	. 97374	. 98966	. 97646	. 99080	. 97904	. 99187	. 98146	39
22	. 98719	. 97093	. 98847	. 97379	. 98968	. 97651	. 99082	. 97908	. 99189	. 98150	38
23	. 98721	. 97098	. 98849	. 97384	. 98970	. 97655	. 99084	. 97912	. 99191	. 98154	37
24	. 98723	. 97103	. 98851	. 97388	. 98971	. 97660	. 99085	. 97916	. 99193	. 98158	36
25	9. 98725	0. 97108	9. 98853	0. 97393	9. 98973	0. 97664	9. 99087	0. 97920	9. 99194	0. 98162	35
26	. 98728	. 97113	. 98855	. 97398	. 98975	. 97668	. 99089	. 97924	. 99196	. 98166	34
27	. 98730	. 97117	. 98857	. 97402	. 98977	. 97673	. 99091	. 97929	. 99198	. 98170	33
28	. 98732	. 97122	. 98859	. 97407	. 98979	. 97677	. 99093	. 97933	. 99200	. 98174	32
29	. 98734	. 97127	. 98861	. 97412	. 98981	. 97681	. 99095	. 97937	. 99201	. 98178	31
30	9. 98736	0. 97132	9. 98863	0. 97416	9. 98983	0. 97686	9. 99096	0. 97941	9. 99203	0. 98182	30
31	. 98738	. 97137	. 98865	. 97421	. 98985	. 97690	. 99098	. 97945	. 99205	. 98185	29
32	. 98741	. 97142	. 98867	. 97425	. 98987	. 97695	. 99100	. 97949	. 99206	. 98189	28
33	. 98743	. 97147	. 98869	. 97430	. 98989	. 97699	. 99102	. 97953	. 99208	. 98193	27
34	. 98745	. 97151	. 98871	. 97435	. 98991	. 97703	. 99104	. 97957	. 99210	. 98197	26
35	9. 98747	0. 97156	9. 98873	0. 97439	9. 98993	0. 97708	9. 99106	0. 97962	9. 99212	0. 98201	25
36	. 98749	. 97161	. 98875	. 97444	. 98995	. 97712	. 99107	. 97966	. 99213	. 98205	24
37	. 98751	. 97166	. 98877	. 97448	. 98997	. 97716	. 99109	. 97970	. 99215	. 98209	23
38	. 98754	. 97171	. 98880	. 97453	. 98999	. 97721	. 99111	. 97974	. 99217	. 98212	22
39	. 98756	. 97176	. 98882	. 97458	. 99001	. 97725	. 99113	. 97978	. 99218	. 98216	21
40	9. 98758	0. 97180	9. 98884	0. 97462	9. 99003	0. 97729	9. 99115	0. 97982	9. 99220	0. 98220	20
41	. 98760	. 97185	. 98886	. 97467	. 99004	. 97734	. 99116	. 97986	. 99222	. 98224	19
42	. 98762	. 97190	. 98888	. 97471	. 99006	. 97738	. 99118	. 97990	. 99223	. 98228	18
43	. 98764	. 97195	. 98890	. 97476	. 99008	. 97742	. 99120	. 97994	. 99225	. 98232	17
44	. 98766	. 97200	. 98892	. 97480	. 99010	. 97747	. 99122	. 97998	. 99227	. 98236	16
45	9. 98769	0. 97204	9. 98894	0. 97485	9. 99012	0. 97751	9. 99124	0. 98002	9. 99229	0. 98239	15
46	. 98771	. 97209	. 98896	. 97490	. 99014	. 97755	. 99126	. 98007	. 99230	. 98243	14
47	. 98773	. 97214	. 98898	. 97494	. 99016	. 97760	. 99127	. 98011	. 99232	. 98247	13
48	. 98775	. 97219	. 98900	. 97499	. 99018	. 97764	. 99129	. 98015	. 99234	. 98251	12
49	. 98777	. 97224	. 98902	. 97503	. 99020	. 97768	. 99131	. 98019	. 99235	. 98255	11
50	9. 98779	0. 97228	9. 98904	0. 97508	9. 99022	0. 97773	9. 99133	0. 98023	9. 99237	0. 98258	10
51	. 98781	. 97233	. 98906	. 97512	. 99024	. 97777	. 99135	. 98027	. 99239	. 98262	9
52	. 98784	. 97238	. 98908	. 97517	. 99026	. 97781	. 99136	. 98031	. 99240	. 98266	8
53	. 98786	. 97243	. 98910	. 97521	. 99027	. 97785	. 99138	. 98035	. 99242	. 98270	7
54	. 98788	. 97247	. 98912	. 97526	. 99029	. 97790	. 99140	. 98039	. 99244	. 98274	6
55	9. 98790	0. 97252	9. 98914	0. 97530	9. 99031	0. 97794	9. 99142	0. 98043	9. 99245	0. 98277	5
56	. 98792	. 97257	. 98916	. 97535	. 99033	. 97798	. 99143	. 98047	. 99247	. 98281	4
57	. 98794	. 97262	. 98918	. 97539	. 99035	. 97802	. 99145	. 98051	. 99249	. 98285	3
58	. 98796	. 97266	. 98920	. 97544	. 99037	. 97807	. 99147	. 98055	. 99250	. 98289	2
59	. 98798	. 97271	. 98922	. 97548	. 99039	. 97811	. 99149	. 98059	. 99252	. 98293	1
60	9. 98801	0. 97276	9. 98924	0. 97553	9. 99041	0. 97815	9. 99151	0. 98063	9. 99254	0. 98296	0
	199°		198°		197°		196°		195°		

TABLE 34

Haversines

′	165° Log Hav	165° Nat. Hav	166° Log Hav	166° Nat. Hav	167° Log Hav	167° Nat. Hav	168° Log Hav	168° Nat. Hav	169° Log Hav	169° Nat. Hav	′
0	9. 99254	0. 98296	9. 99350	0. 98515	9. 99440	0. 98719	9. 99523	0. 98907	9. 99599	0. 99081	60
1	. 99255	. 98300	. 99352	. 98518	. 99441	. 98722	. 99524	. 98910	. 99600	. 99084	59
2	. 99257	. 98304	. 99353	. 98522	. 99443	. 98725	. 99526	. 98913	. 99602	. 99087	58
3	. 99259	. 98308	. 99355	. 98525	. 99444	. 98728	. 99527	. 98916	. 99603	. 99090	57
4	. 99260	. 98311	. 99356	. 98529	. 99446	. 98732	. 99528	. 98919	. 99604	. 99092	56
5	9. 99262	0. 98315	9. 99358	0. 98532	9. 99447	0. 98735	9. 99529	0. 98922	9. 99605	0. 99095	55
6	. 99264	. 98319	. 99359	. 98536	. 99448	. 98738	. 99531	. 98925	. 99606	. 99098	54
7	. 99265	. 98323	. 99361	. 98539	. 99450	. 98741	. 99532	. 98928	. 99608	. 99101	53
8	. 99267	. 98326	. 99362	. 98543	. 99451	. 98745	. 99533	. 98931	. 99609	. 99103	52
9	. 99269	. 98330	. 99364	. 98546	. 99453	. 98748	. 99535	. 98934	. 99610	. 99106	51
10	9. 99270	0. 98334	9. 99366	0. 98550	9. 99454	0. 98751	9. 99536	0. 98937	9. 99611	0. 99109	50
11	. 99272	. 98337	. 99367	. 98553	. 99456	. 98754	. 99537	. 98940	. 99612	. 99112	49
12	. 99274	. 98341	. 99369	. 98557	. 99457	. 98757	. 99539	. 98943	. 99614	. 99114	48
13	. 99275	. 98345	. 99370	. 98560	. 99458	. 98761	. 99540	. 98946	. 99615	. 99117	47
14	. 99277	. 98349	. 99372	. 98564	. 99460	. 98764	. 99541	. 98949	. 99616	. 99120	46
15	9. 99278	0. 98352	9. 99373	0. 98567	9. 99461	0. 98767	9. 99543	0. 98952	9. 99617	0. 99123	45
16	. 99280	. 98356	. 99375	. 98571	. 99463	. 98770	. 99544	. 98955	. 99618	. 99125	44
17	. 99282	. 98360	. 99376	. 98574	. 99464	. 98774	. 99545	. 98958	. 99620	. 99128	43
18	. 99283	. 98363	. 99378	. 98577	. 99465	. 98777	. 99546	. 98961	. 99621	. 99131	42
19	. 99285	. 98367	. 99379	. 98581	. 99467	. 98780	. 99548	. 98964	. 99622	. 99133	41
20	9. 99287	0. 98371	9. 99381	0. 98584	9. 99468	0. 98783	9. 99549	0. 98967	9. 99623	0. 99136	40
21	. 99288	. 98374	. 99382	. 98588	. 99470	. 98786	. 99550	. 98970	. 99624	. 99139	39
22	. 99290	. 98378	. 99384	. 98591	. 99471	. 98789	. 99552	. 98973	. 99626	. 99141	38
23	. 99291	. 98382	. 99385	. 98595	. 99472	. 98793	. 99553	. 98976	. 99627	. 99144	37
24	. 99293	. 98385	. 99387	. 98598	. 99474	. 98796	. 99554	. 98979	. 99628	. 99147	36
25	9. 99295	0. 98389	9. 99388	0. 98601	9. 99475	0. 98799	9. 99555	0. 98982	9. 99629	0. 99149	35
26	. 99296	. 98393	. 99390	. 98605	. 99477	. 98802	. 99557	. 98985	. 99630	. 99152	34
27	. 99298	. 98396	. 99391	. 98608	. 99478	. 98805	. 99558	. 98988	. 99631	. 99155	33
28	. 99300	. 98400	. 99393	. 98612	. 99479	. 98808	. 99559	. 98990	. 99633	. 99157	32
29	. 99301	. 98404	. 99394	. 98615	. 99481	. 98812	. 99561	. 98993	. 99634	. 99160	31
30	9. 99303	0. 98407	9. 99396	0. 98618	9. 99482	0. 98815	9. 99562	0. 98996	9. 99635	0. 99163	30
31	. 99304	. 98411	. 99397	. 98622	. 99484	. 98818	. 99563	. 98999	. 99636	. 99165	29
32	. 99306	. 98415	. 99399	. 98625	. 99485	. 98821	. 99564	. 99002	. 99637	. 99168	28
33	. 99308	. 98418	. 99400	. 98629	. 99486	. 98824	. 99566	. 99005	. 99638	. 99171	27
34	. 99309	. 98422	. 99402	. 98632	. 99488	. 98827	. 99567	. 99008	. 99639	. 99173	26
35	9. 99311	0. 98426	9. 99403	0. 98635	9. 99489	0. 98830	9. 99568	0. 99011	9. 99641	0. 99176	25
36	. 99312	. 98429	. 99405	. 98639	. 99490	. 98834	. 99569	. 99014	. 99642	. 99179	24
37	. 99314	. 98433	. 99406	. 98642	. 99492	. 98837	. 99571	. 99016	. 99643	. 99181	23
38	. 99316	. 98436	. 99408	. 98646	. 99493	. 98840	. 99572	. 99019	. 99644	. 99184	22
39	. 99317	. 98440	. 99409	. 98649	. 99495	. 98843	. 99573	. 99022	. 99645	. 99186	21
40	9. 99319	0. 98444	9. 99411	0. 98652	9. 99496	0. 98846	9. 99575	0. 99025	9. 99646	0. 99189	20
41	. 99320	. 98447	. 99412	. 98656	. 99497	. 98849	. 99576	. 99028	. 99648	. 99192	19
42	. 99322	. 98451	. 99414	. 98659	. 99499	. 98852	. 99577	. 99031	. 99649	. 99194	18
43	. 99324	. 98454	. 99415	. 98662	. 99500	. 98855	. 99578	. 99034	. 99650	. 99197	17
44	. 99325	. 98458	. 99417	. 98666	. 99501	. 98858	. 99580	. 99036	. 99651	. 99199	16
45	9. 99327	0. 98462	9. 99418	0. 98669	9. 99503	0. 98862	9. 99581	0. 99039	9. 99652	0. 99202	15
46	. 99328	. 98465	. 99420	. 98672	. 99504	. 98865	. 99582	. 99042	. 99653	. 99205	14
47	. 99330	. 98469	. 99421	. 98676	. 99505	. 98868	. 99583	. 99045	. 99654	. 99207	13
48	. 99331	. 98472	. 99422	. 98679	. 99507	. 98871	. 99584	. 99048	. 99655	. 99210	12
49	. 99333	. 98476	. 99424	. 98682	. 99508	. 98874	. 99586	. 99051	. 99657	. 99212	11
50	9. 99335	0. 98479	9. 99425	0. 98686	9. 99510	0. 98877	9. 99587	0. 99053	9. 99658	0. 99215	10
51	. 99336	. 98483	. 99427	. 98689	. 99511	. 98880	. 99588	. 99056	. 99659	. 99217	9
52	. 99338	. 98487	. 99429	. 98692	. 99512	. 98883	. 99589	. 99059	. 99660	. 99220	8
53	. 99339	. 98490	. 99430	. 98695	. 99514	. 98886	. 99591	. 99062	. 99661	. 99223	7
54	. 99341	. 98494	. 99431	. 98699	. 99515	. 98889	. 99592	. 99065	. 99662	. 99225	6
55	9. 99342	0. 98497	9. 99433	0. 98702	9. 99516	0. 98892	9. 99593	0. 99067	9. 99663	0. 99228	5
56	. 99344	. 98501	. 99434	. 98705	. 99518	. 98895	. 99594	. 99070	. 99664	. 99230	4
57	. 99345	. 98504	. 99436	. 98709	. 99519	. 98898	. 99596	. 99073	. 99666	. 99233	3
58	. 99347	. 98508	. 99437	. 98712	. 99520	. 98901	. 99597	. 99076	. 99667	. 99235	2
59	. 99349	. 98511	. 99438	. 98715	. 99522	. 98904	. 99598	. 99079	. 99668	. 99238	1
60	9. 99350	0. 98515	9. 99440	0. 98719	9. 99523	0. 98907	9. 99599	0. 99081	9. 99669	0. 99240	0
	194°		193°		192°		191°		190°		

TABLE 34

Haversines

′	170°		171°		172°		173°		174°		′
	Log Hav	Nat. Hav	Log Hav	Nat. Hav	Log Hav	Nat. Hav	Log Hav	Nat. Hav	Log Hav	Nat. Hav	
0	9. 99669	0. 99240	9. 99732	0. 99384	9. 99788	0. 99513	9. 99838	0. 99627	9. 99881	0. 99726	60
1	. 99670	. 99243	. 99733	. 99387	. 99789	. 99515	. 99839	. 99629	. 99882	. 99728	59
2	. 99671	. 99245	. 99734	. 99389	. 99790	. 99517	. 99839	. 99631	. 99882	. 99729	58
3	. 99672	. 99248	. 99735	. 99391	. 99791	. 99519	. 99840	. 99633	. 99883	. 99731	57
4	. 99673	. 99250	. 99736	. 99393	. 99792	. 99521	. 99841	. 99634	. 99884	. 99732	56
5	9. 99674	0. 99253	9. 99737	0. 99396	9. 99793	0. 99523	9. 99842	0. 99636	9. 99884	0. 99734	55
6	. 99675	. 99255	. 99738	. 99398	. 99793	. 99525	. 99842	. 99638	. 99885	. 99735	54
7	. 99677	. 99258	. 99739	. 99400	. 99794	. 99527	. 99843	. 99640	. 99885	. 99737	53
8	. 99678	. 99260	. 99740	. 99402	. 99795	. 99529	. 99844	. 99641	. 99886	. 99738	52
9	. 99679	. 99263	. 99741	. 99405	. 99796	. 99531	. 99845	. 99643	. 99887	. 99740	51
10	9. 99680	0. 99265	9. 99742	0. 99407	9. 99797	0. 99533	9. 99845	0. 99645	9. 99887	0. 99741	50
11	. 99681	. 99268	. 99743	. 99409	. 99798	. 99535	. 99846	. 99647	. 99888	. 99743	49
12	. 99682	. 99270	. 99744	. 99411	. 99799	. 99537	. 99847	. 99648	. 99889	. 99744	48
13	. 99683	. 99273	. 99745	. 99414	. 99800	. 99539	. 99848	. 99650	. 99889	. 99746	47
14	. 99684	. 99275	. 99746	. 99416	. 99800	. 99541	. 99848	. 99652	. 99890	. 99747	46
15	9. 99685	0. 99278	9. 99747	0. 99418	9. 99801	0. 99543	9. 99849	0. 99653	9. 99891	0. 99748	45
16	. 99686	. 99280	. 99748	. 99420	. 99802	. 99545	. 99850	. 99655	. 99891	. 99750	44
17	. 99687	. 99283	. 99748	. 99422	. 99803	. 99547	. 99851	. 99657	. 99892	. 99751	43
18	. 99688	. 99285	. 99749	. 99425	. 99804	. 99549	. 99851	. 99659	. 99893	. 99753	42
19	. 99690	. 99288	. 99750	. 99427	. 99805	. 99551	. 99852	. 99660	. 99893	. 99754	41
20	9. 99691	0. 99290	9. 99751	0. 99429	9. 99805	0. 99553	9. 99853	0. 99662	9. 99894	0. 99756	40
21	. 99692	. 99293	. 99752	. 99431	. 99806	. 99555	. 99854	. 99664	. 99894	. 99757	39
22	. 99693	. 99295	. 99753	. 99433	. 99807	. 99557	. 99854	. 99665	. 99895	. 99759	38
23	. 99694	. 99297	. 99754	. 99436	. 99808	. 99559	. 99855	. 99667	. 99896	. 99760	37
24	. 99695	. 99300	. 99755	. 99438	. 99809	. 99561	. 99856	. 99669	. 99896	. 99761	36
25	9. 99696	0. 99302	9. 99756	0. 99440	9. 99810	0. 99563	9. 99857	0. 99670	9. 99897	0. 99763	35
26	. 99697	. 99305	. 99757	. 99442	. 99811	. 99565	. 99857	. 99672	. 99897	. 99764	34
27	. 99698	. 99307	. 99758	. 99444	. 99811	. 99567	. 99858	. 99674	. 99898	. 99766	33
28	. 99699	. 99309	. 99759	. 99446	. 99812	. 99568	. 99859	. 99675	. 99899	. 99767	32
29	. 99700	. 99312	. 99760	. 99449	. 99813	. 99570	. 99859	. 99677	. 99899	. 99768	31
30	9. 99701	0. 99314	9. 99761	0. 99451	9. 99814	0. 99572	9. 99860	0. 99679	9. 99900	0. 99770	30
31	. 99702	. 99317	. 99762	. 99453	. 99815	. 99574	. 99861	. 99680	. 99901	. 99771	29
32	. 99703	. 99319	. 99763	. 99455	. 99815	. 99576	. 99862	. 99682	. 99901	. 99773	28
33	. 99704	. 99321	. 99764	. 99457	. 99816	. 99578	. 99862	. 99684	. 99902	. 99774	27
34	. 99705	. 99324	. 99765	. 99459	. 99817	. 99580	. 99863	. 99685	. 99902	. 99775	26
35	9. 99706	0. 99326	9. 99766	0. 99461	9. 99818	0. 99582	9. 99864	0. 99687	9. 99903	0. 99777	25
36	. 99707	. 99329	. 99766	. 99464	. 99819	. 99584	. 99864	. 99688	. 99904	. 99778	24
37	. 99708	. 99331	. 99767	. 99466	. 99820	. 99585	. 99865	. 99690	. 99904	. 99779	23
38	. 99710	. 99333	. 99768	. 99468	. 99820	. 99587	. 99866	. 99692	. 99905	. 99781	22
39	. 99711	. 99336	. 99769	. 99470	. 99821	. 99589	. 99867	. 99693	. 99905	. 99782	21
40	9. 99712	0. 99338	9. 99770	0. 99472	9. 99822	0. 99591	9. 99867	0. 99695	9. 99906	0. 99784	20
41	. 99713	. 99340	. 99771	. 99474	. 99823	. 99593	. 99868	. 99696	. 99906	. 99785	19
42	. 99714	. 99343	. 99772	. 99476	. 99824	. 99595	. 99869	. 99698	. 99907	. 99786	18
43	. 99715	. 99345	. 99773	. 99478	. 99824	. 99597	. 99869	. 99700	. 99908	. 99788	17
44	. 99716	. 99347	. 99774	. 99480	. 99825	. 99598	. 99870	. 99701	. 99908	. 99789	16
45	9. 99717	0. 99350	9. 99774	0. 99483	9. 99826	0. 99600	9. 99871	0. 99703	9. 99909	0. 99790	15
46	. 99718	. 99352	. 99775	. 99485	. 99827	. 99602	. 99871	. 99704	. 99909	. 99792	14
47	. 99719	. 99354	. 99776	. 99487	. 99828	. 99604	. 99872	. 99706	. 99910	. 99793	13
48	. 99720	. 99357	. 99777	. 99489	. 99828	. 99606	. 99873	. 99708	. 99911	. 99794	12
49	. 99721	. 99359	. 99778	. 99491	. 99829	. 99608	. 99874	. 99709	. 99911	. 99796	11
50	9. 99722	0. 99361	9. 99779	0. 99493	9. 99830	0. 99609	9. 99874	0. 99711	9. 99912	0. 99797	10
51	. 99723	. 99364	. 99780	. 99495	. 99831	. 99611	. 99875	. 99712	. 99912	. 99798	9
52	. 99724	. 99366	. 99781	. 99497	. 99832	. 99613	. 99876	. 99714	. 99913	. 99799	8
53	. 99725	. 99368	. 99782	. 99499	. 99832	. 99615	. 99876	. 99715	. 99913	. 99801	7
54	. 99726	. 99371	. 99783	. 99501	. 99833	. 99617	. 99877	. 99717	. 99914	. 99802	6
55	9. 99727	0. 99373	9. 99784	0. 99503	9. 99834	0. 99618	9. 99878	0. 99718	9. 99915	0. 99803	5
56	. 99728	. 99375	. 99785	. 99505	. 99835	. 99620	. 99878	. 99720	. 99915	. 99805	4
57	. 99729	. 99378	. 99786	. 99507	. 99836	. 99622	. 99879	. 99722	. 99916	. 99806	3
58	. 99730	. 99380	. 99786	. 99509	. 99836	. 99624	. 99880	. 99723	. 99916	. 99807	2
59	. 99731	. 99382	. 99787	. 99511	. 99837	. 99626	. 99880	. 99725	. 99917	. 99808	1
60	9. 99732	0. 99384	9. 99788	0. 99513	9. 99838	0. 99627	9. 99881	0. 99726	9. 99917	0. 99810	0
	189°		188°		187°		186°		185°		

TABLE 34

Haversines

	175°		176°		177°		178°		179°		
′	Log Hav	Nat. Hav	Log Hav	Nat. Hav	Log Hav	Nat. Hav	Log Hav	Nat. Hav	Log Hav	Nat. Hav	′
0	9. 99917	0. 99810	9. 99947	0. 99878	9. 99970	0. 99931	9. 99987	0. 99970	9. 99997	0. 99992	60
1	. 99918	. 99811	. 99948	. 99879	. 99971	. 99932	. 99987	. 99970	. 99997	. 99993	59
2	. 99918	. 99812	. 99948	. 99880	. 99971	. 99933	. 99987	. 99971	. 99997	. 99993	58
3	. 99919	. 99814	. 99948	. 99881	. 99971	. 99934	. 99987	. 99971	. 99997	. 99993	57
4	. 99919	. 99815	. 99949	. 99882	. 99972	. 99934	. 99988	. 99972	. 99997	. 99993	56
5	9. 99920	0. 99816	9. 99949	0. 99883	9. 99972	0. 99935	9. 99988	0. 99972	9. 99997	0. 99994	55
6	. 99921	. 99817	. 99950	. 99884	. 99972	. 99936	. 99988	. 99973	. 99997	. 99994	54
7	. 99921	. 99819	. 99950	. 99885	. 99973	. 99937	. 99988	. 99973	. 99997	. 99994	53
8	. 99922	. 99820	. 99951	. 99886	. 99973	. 99937	. 99988	. 99973	. 99998	. 99994	52
9	. 99922	. 99821	. 99951	. 99887	. 99973	. 99938	. 99989	. 99974	. 99998	. 99994	51
10	9. 99923	0. 99822	9. 99951	0. 99888	9. 99973	0. 99939	9. 99989	0. 99974	9. 99998	0. 99995	50
11	. 99923	. 99823	. 99952	. 99889	. 99974	. 99940	. 99989	. 99975	. 99998	. 99995	49
12	. 99924	. 99825	. 99952	. 99890	. 99974	. 99940	. 99989	. 99975	. 99998	. 99995	48
13	. 99924	. 99826	. 99953	. 99891	. 99974	. 99941	. 99989	. 99976	. 99998	. 99995	47
14	. 99925	. 99827	. 99953	. 99892	. 99975	. 99942	. 99990	. 99976	. 99998	. 99996	46
15	9. 99925	0. 99828	9. 99953	0. 99893	9. 99975	0. 99942	9. 99990	0. 99977	9. 99998	0. 99996	45
16	. 99926	. 99829	. 99954	. 99894	. 99975	. 99943	. 99990	. 99977	. 99998	. 99996	44
17	. 99926	. 99831	. 99954	. 99895	. 99976	. 99944	. 99990	. 99978	. 99998	. 99996	43
18	. 99927	. 99832	. 99954	. 99896	. 99976	. 99944	. 99990	. 99978	. 99998	. 99996	42
19	. 99927	. 99833	. 99955	. 99897	. 99976	. 99945	. 99991	. 99978	. 99998	. 99996	41
20	9. 99928	0. 99834	9. 99955	0. 99898	9. 99976	0. 99946	9. 99991	0. 99979	9. 99999	0. 99997	40
21	. 99928	. 99835	. 99956	. 99899	. 99977	. 99947	. 99991	. 99979	. 99999	. 99997	39
22	. 99929	. 99837	. 99956	. 99900	. 99977	. 99947	. 99991	. 99980	. 99999	. 99997	38
23	. 99929	. 99838	. 99957	. 99900	. 99977	. 99948	. 99991	. 99980	. 99999	. 99997	37
24	. 99930	. 99839	. 99957	. 99901	. 99978	. 99949	. 99992	. 99981	. 99999	. 99997	36
25	9. 99931	0. 99840	9. 99958	0. 99902	9. 99978	0. 99949	9. 99992	0. 99981	9. 99999	0. 99997	35
26	. 99931	. 99841	. 99958	. 99903	. 99978	. 99950	. 99992	. 99981	. 99999	. 99998	34
27	. 99932	. 99842	. 99958	. 99904	. 99978	. 99950	. 99992	. 99982	. 99999	. 99998	33
28	. 99932	. 99844	. 99959	. 99905	. 99979	. 99951	. 99992	. 99982	. 99999	. 99998	32
29	. 99933	. 99845	. 99959	. 99906	. 99979	. 99952	. 99992	. 99982	. 99999	. 99998	31
30	9. 99933	0. 99846	9. 99959	0. 99907	9. 99979	0. 99952	9. 99993	0. 99983	9. 99999	0. 99998	30
31	. 99934	. 99847	. 99960	. 99908	. 99980	. 99953	. 99993	. 99983	. 99999	. 99998	29
32	. 99934	. 99848	. 99960	. 99909	. 99980	. 99954	. 99993	. 99984	. 99999	. 99998	28
33	. 99935	. 99849	. 99961	. 99909	. 99980	. 99954	. 99993	. 99984	. 99999	. 99998	27
34	. 99935	. 99850	. 99961	. 99910	. 99980	. 99955	. 99993	. 99984	. 99999	. 99999	26
35	9. 99935	0. 99852	9. 99961	0. 99911	9. 99981	0. 99956	9. 99993	0. 99985	9. 99999	0. 99999	25
36	. 99936	. 99853	. 99962	. 99912	. 99981	. 99956	. 99994	. 99985	9. 99999	. 99999	24
37	. 99936	. 99854	. 99962	. 99913	. 99981	. 99957	. 99994	. 99985	0. 00000	. 99999	23
38	. 99937	. 99855	. 99963	. 99914	. 99981	. 99957	. 99994	. 99986	. 00000	. 99999	22
39	. 99937	. 99856	. 99963	. 99915	. 99982	. 99958	. 99994	. 99986	. 00000	. 99999	21
40	9. 99938	0. 99857	9. 99963	0. 99915	9. 99982	0. 99959	9. 99994	0. 99986	0. 00000	0. 99999	20
41	. 99938	. 99858	. 99964	. 99916	. 99982	. 99959	. 99994	. 99987	. 00000	. 99999	19
42	. 99939	. 99859	. 99964	. 99917	. 99983	. 99960	. 99994	. 99987	. 00000	. 99999	18
43	. 99939	. 99860	. 99964	. 99918	. 99983	. 99960	. 99995	. 99987	. 00000	. 99999	17
44	. 99940	. 99861	. 99965	. 99919	. 99983	. 99961	. 99995	. 99988	. 00000	. 99999	16
45	9. 99940	0. 99863	9. 99965	0. 99920	9. 99983	0. 99961	9. 99995	0. 99988	0. 00000	1. 00000	15
46	. 99941	. 99864	. 99965	. 99920	. 99983	. 99962	. 99995	. 99988	. 00000	. 00000	14
47	. 99941	. 99865	. 99966	. 99921	. 99984	. 99963	. 99995	. 99989	. 00000	. 00000	13
48	. 99942	. 99866	. 99966	. 99922	. 99984	. 99963	. 99995	. 99989	. 00000	. 00000	12
49	. 99942	. 99867	. 99966	. 99923	. 99984	. 99964	. 99995	. 99989	. 00000	. 00000	11
50	9. 99943	0. 99868	9. 99967	0. 99924	9. 99984	0. 99964	9. 99996	0. 99990	0. 00000	1. 00000	10
51	. 99943	. 99869	. 99967	. 99924	. 99985	. 99965	. 99996	. 99990	. 00000	. 00000	9
52	. 99943	. 99870	. 99968	. 99925	. 99985	. 99965	. 99996	. 99990	. 00000	. 00000	8
53	. 99944	. 99871	. 99968	. 99926	. 99985	. 99966	. 99996	. 99991	. 00000	. 00000	7
54	. 99944	. 99872	. 99968	. 99927	. 99985	. 99966	. 99996	. 99991	. 00000	. 00000	6
55	9. 99945	0. 99873	9. 99969	0. 99928	9. 99986	0. 99967	9. 99996	0. 99991	0. 00000	1. 00000	5
56	. 99945	. 99874	. 99969	. 99928	. 99986	. 99967	. 99996	. 99991	. 00000	. 00000	4
57	. 99946	. 99875	. 99969	. 99929	. 99986	. 99968	. 99996	. 99992	. 00000	. 00000	3
58	. 99946	. 99876	. 99970	. 99930	. 99986	. 99969	. 99996	. 99992	. 00000	. 00000	2
59	. 99947	. 99877	. 99970	. 99931	. 99987	. 99969	. 99997	. 99992	. 00000	. 00000	1
60	9. 99947	0. 99878	9. 99970	0. 99931	9. 99987	0. 99970	9. 99997	0. 99992	0. 00000	1. 00000	0
	184°		183°		182°		181°		180°		

THE AGETON METHOD

TABLE 35

The Ageton Method

When Meridian Angle is Greater Than 90° Take "K" From Bottom of Table

′	0° 00′		0° 30′		1° 00′		1° 30′		2° 00′		′
	A	B	A	B	A	B	A	B	A	B	
0	-------	0.0	205916	1.7	175814	6.6	158208	14.9	145718	26.5	30
	383730	0.0	205198	1.7	175454	6.7	157967	15.1	145538	26.7	
1	353627	0.0	204492	1.8	175097	6.8	157728	15.2	145358	26.9	29
	336018	0.0	203797	1.8	174742	7.0	157490	15.4	145179	27.1	
2	323524	0.0	203113	1.9	174391	7.1	157254	15.6	145000	27.3	28
	313833	0.0	202440	1.9	174042	7.2	157019	15.7	144823	27.6	
3	305915	0.0	201777	2.0	173696	7.3	156784	15.9	144646	27.8	27
	299221	0.0	201124	2.1	173352	7.4	156552	16.1	144470	28.0	
4	293421	0.0	200480	2.1	173012	7.5	156320	16.2	144295	28.3	26
	288306	0.0	199846	2.2	172674	7.6	156090	16.4	144120	28.5	
5	283730	0.0	199221	2.3	172339	7.8	155861	16.6	143946	28.7	25
	279591	0.1	198605	2.3	172006	7.9	155633	16.8	143773	28.9	
6	275812	0.1	197998	2.4	171676	8.0	155406	16.9	143600	29.2	24
	272336	0.1	197399	2.4	171348	8.1	155180	17.1	143428	29.4	
7	269118	0.1	196808	2.5	171023	8.2	154956	17.3	143257	29.6	23
	266121	0.1	196225	2.6	170700	8.4	154733	17.5	143086	29.9	
8	263318	0.1	195650	2.7	170379	8.5	154511	17.6	142916	30.1	22
	260685	0.1	195082	2.7	170061	8.6	154290	17.8	142747	30.4	
9	258203	0.1	194522	2.8	169745	8.7	154070	18.0	142579	30.6	21
	255855	0.2	193969	2.9	169432	8.9	153851	18.2	142411	30.8	
10	253627	0.2	193422	2.9	169121	9.0	153633	18.4	142243	31.1	20
	251508	0.2	192883	3.0	168811	9.1	153417	18.6	142077	31.3	
11	249488	0.2	192350	3.1	168505	9.3	153201	18.7	141911	31.5	19
	247558	0.2	191824	3.2	168200	9.4	152987	18.9	141745	31.8	
12	245709	0.3	191303	3.2	167897	9.5	152774	19.1	141581	32.0	18
	243936	0.3	190790	3.3	167597	9.7	152561	19.3	141417	32.3	
13	242233	0.3	190282	3.4	167298	9.8	152350	19.5	141253	32.5	17
	240594	0.3	189780	3.5	167002	9.9	152140	19.7	141090	32.8	
14	239015	0.4	189283	3.6	166708	10.1	151931	19.9	140928	33.0	16
	237491	0.4	188793	3.6	166415	10.2	151722	20.1	140766	33.3	
15	236018	0.4	188307	3.7	166125	10.3	151515	20.3	140605	33.5	15
	234594	0.4	187827	3.8	165836	10.5	151309	20.5	140445	33.7	
16	233215	0.5	187353	3.9	165550	10.6	151104	20.6	140285	34.0	14
	231879	0.5	186883	4.0	165265	10.8	150899	20.8	140125	34.2	
17	230583	0.5	186419	4.1	164982	10.9	150696	21.0	139967	34.5	13
	229324	0.6	185959	4.1	164701	11.0	150494	21.2	139809	34.7	
18	228100	0.6	185505	4.2	164422	11.2	150292	21.4	139651	35.0	12
	226910	0.6	185055	4.3	164144	11.3	150092	21.6	139494	35.3	
19	225752	0.7	184609	4.4	163868	11.5	149892	21.8	139338	35.5	11
	224624	0.7	184168	4.5	163594	11.6	149693	22.0	139182	35.8	
20	223525	0.7	183732	4.6	163322	11.8	149495	22.2	139027	36.0	10
	222452	0.8	183300	4.7	163052	11.9	149299	22.4	138872	36.3	
21	221406	0.8	182872	4.8	162783	12.1	149103	22.6	138718	36.5	9
	220384	0.9	182448	4.9	162516	12.2	148907	22.9	138564	36.8	
22	219385	0.9	182029	5.0	162250	12.4	148713	23.1	138411	37.1	8
	218409	0.9	181613	5.1	161986	12.5	148520	23.3	138258	37.3	
23	217455	1.0	181201	5.2	161724	12.7	148327	23.5	138106	37.6	7
	216521	1.0	180794	5.3	161463	12.8	148135	23.7	137955	37.9	
24	215607	1.1	180390	5.4	161204	13.0	147945	23.9	137804	38.1	6
	214711	1.1	179990	5.5	160946	13.1	147755	24.1	137653	38.4	
25	213834	1.1	179593	5.6	160690	13.3	147566	24.3	137504	38.6	5
	212974	1.2	179200	5.7	160435	13.4	147377	24.5	137354	38.9	
26	212130	1.2	178810	5.8	160182	13.6	147190	24.7	137205	39.2	4
	211303	1.3	178424	5.9	159930	13.8	147003	24.9	137057	39.4	
27	210491	1.3	178042	6.0	159680	13.9	146817	25.2	136909	39.7	3
	209695	1.4	177663	6.1	159431	14.1	146632	25.4	136761	40.0	
28	208912	1.4	177287	6.2	159184	14.2	146448	25.6	136615	40.3	2
	208143	1.5	176914	6.3	158938	14.4	146264	25.8	136468	40.5	
29	207388	1.5	176544	6.4	158693	14.6	146081	26.0	136322	40.8	1
	206646	1.6	176178	6.5	158450	14.7	145899	26.2	136177	41.1	
30	205916	1.7	175814	6.6	158208	14.9	145718	26.5	136032	41.4	0
′	A	B	A	B	A	B	A	B	A	B	′
	179° 30′		179° 00′		178° 30′		178° 00′		177° 30′		

TABLE 35

The Ageton Method

Always Take "Z" From Bottom of Table, Except When "K" is Same Name and Greater Than Latitude, in Which Case Take "Z" From Top of Table

′	2° 30′		3° 00′		3° 30′		4° 00′		4° 30′		′
′	A	B	A	B	A	B	A	B	A	B	′
0	136032	41.4	128120	59.6	121432	81.1	115641	105.9	110536	134.1	30
	135888	41.6	128000	59.9	121329	81.5	115551	106.4	110455	134.6	
1	135744	41.9	127880	60.2	121226	81.9	115461	106.8	110375	135.1	29
	135600	42.2	127760	60.6	121124	82.2	115371	107.3	110296	135.6	
2	135457	42.5	127640	60.9	121021	82.6	115282	107.7	110216	136.1	28
	135315	42.7	127521	61.2	120919	83.0	115192	108.1	110136	136.6	
3	135173	43.0	127403	61.6	120817	83.4	115103	108.6	110057	137.1	27
	135031	43.3	127284	61.9	120715	83.8	115014	109.0	109977	137.6	
4	134890	43.6	127166	62.2	120614	84.2	114925	109.5	109898	138.1	26
	134749	43.9	127049	62.6	120513	84.6	114836	109.9	109819	138.6	
5	134609	44.2	126931	62.9	120412	85.0	114747	110.4	109740	139.1	25
	134469	44.4	126814	63.3	120311	85.4	114659	110.8	109662	139.6	
6	134330	44.7	126697	63.6	120211	85.8	114571	111.3	109583	140.1	24
	134191	45.0	126581	63.9	120110	86.2	114483	111.7	109505	140.6	
7	134052	45.3	126465	64.3	120010	86.6	114395	112.2	109426	141.1	23
	133914	45.6	126349	64.6	119910	87.0	114307	112.7	109348	141.7	
8	133777	45.9	126233	65.0	119811	87.4	114220	113.1	109270	142.2	22
	133640	46.2	126118	65.3	119711	87.8	114133	113.6	109192	142.7	
9	133503	46.5	126003	65.7	119612	88.2	114045	114.0	109115	143.2	21
	133367	46.8	125888	66.0	119513	88.6	113958	114.5	109037	143.7	
10	133231	47.1	125774	66.4	119415	89.0	113872	114.9	108960	144.2	20
	133096	47.4	125660	66.7	119316	89.4	113785	115.4	108882	144.7	
11	132961	47.6	125546	67.1	119218	89.8	113699	115.9	108805	145.2	19
	132826	47.9	125433	67.4	119120	90.2	113612	116.3	108728	145.8	
12	132692	48.2	125320	67.8	119022	90.6	113526	116.8	108651	146.3	18
	132558	48.5	125207	68.1	118925	91.0	113440	117.3	108574	146.8	
13	132425	48.8	125094	68.5	118827	91.4	113354	117.7	108498	147.3	17
	132292	49.1	124982	68.8	118730	91.8	113269	118.2	108421	147.8	
14	132159	49.4	124870	69.2	118633	92.3	113183	118.7	108345	148.4	16
	132027	49.7	124759	69.6	118537	92.7	113098	119.1	108269	148.9	
15	131896	50.0	124647	69.9	118440	93.1	113013	119.6	108193	149.4	15
	131764	50.3	124536	70.3	118344	93.5	112928	120.1	108117	149.9	
16	131633	50.7	124425	70.6	118248	93.9	112843	120.5	108041	150.5	14
	131503	51.0	124315	71.0	118152	94.3	112759	121.0	107965	151.0	
17	131373	51.3	124204	71.3	118056	94.7	112674	121.5	107890	151.5	13
	131243	51.6	124095	71.7	117961	95.2	112590	121.9	107814	152.1	
18	131114	51.9	123985	72.1	117866	95.6	112506	122.4	107739	152.6	12
	130985	52.2	123875	72.4	117771	96.0	112422	122.9	107664	153.1	
19	130856	52.5	123766	72.8	117676	96.4	112338	123.4	107589	153.6	11
	130728	52.8	123657	73.2	117581	96.9	112255	123.9	107514	154.2	
20	130600	53.1	123549	73.5	117487	97.3	112171	124.3	107439	154.7	10
	130473	53.4	123441	73.9	117393	97.7	112088	124.8	107364	155.2	
21	130346	53.7	123332	74.3	117299	98.1	112005	125.3	107290	155.8	9
	130219	54.1	123225	74.6	117205	98.5	111922	125.8	107216	156.3	
22	130093	54.4	123117	75.0	117112	99.0	111839	126.2	107141	156.9	8
	129967	54.7	123010	75.4	117018	99.4	111757	126.7	107067	157.4	
23	129841	55.0	122903	75.8	116925	99.8	111674	127.2	106993	157.9	7
	129716	55.3	122796	76.1	116832	100.3	111592	127.7	106919	158.5	
24	129591	55.7	122690	76.5	116739	100.7	111510	128.2	106846	159.0	6
	129466	56.0	122584	76.9	116647	101.1	111428	128.7	106772	159.6	
25	129342	56.3	122478	77.3	116554	101.6	111346	129.2	106698	160.1	5
	129218	56.6	122372	77.6	116462	102.0	111264	129.7	106625	160.6	
26	129095	56.9	122267	78.0	116370	102.4	111183	130.1	106552	161.2	4
	128972	57.3	122161	78.4	116278	102.9	111101	130.6	106479	161.7	
27	128849	57.6	122057	78.8	116187	103.3	111020	131.1	106406	162.3	3
	128727	57.9	121952	79.2	116096	103.7	110939	131.6	106333	162.8	
28	128605	58.2	121848	79.5	116004	104.2	110858	132.1	106260	163.4	2
	128483	58.6	121743	79.9	115913	104.6	110777	132.6	106187	163.9	
29	128362	58.9	121639	80.3	115823	105.0	110696	133.1	106115	164.5	1
	128240	59.2	121536	80.7	115732	105.5	110616	133.6	106043	165.0	
30	128120	59.6	121432	81.1	115641	105.9	110536	134.1	105970	165.6	0
′	A	B	A	B	A	B	A	B	A	B	′
′	177° 00′		176° 30′		176° 00′		175° 30′		175° 00′		′

TABLE 35

The Ageton Method

When Meridian Angle is Greater Than 90° Take "K" From Bottom of Table

′	5° 00′		5° 30′		6° 00′		6° 30′		7° 00′		′
	A	B	A	B	A	B	A	B	A	B	
0	105970	165.6	101843	200.4	98076	239	94614	280	91411	325	30
	105898	166.1	101777	201.0	98017	239	94559	281	91359	326	
1	105826	166.7	101712	201.6	97957	240	94503	281	91308	326	29
	105754	167.2	101646	202.2	97897	241	94448	282	91257	327	
2	105683	167.8	101581	202.8	97837	241	94393	283	91205	328	28
	105611	168.4	101516	203.5	97777	242	94338	284	91154	329	
3	105539	168.9	101451	204.1	97717	243	94283	284	91103	330	27
	105468	169.4	101386	204.7	97658	243	94228	285	91052	330	
4	105397	170.0	101321	205.3	97598	244	94173	286	91001	331	26
	105325	170.6	101256	205.9	97539	245	94118	287	90950	332	
5	105254	171.1	101192	206.5	97480	245	94063	287	90899	333	25
	105183	171.7	101127	207.1	97420	246	94009	288	90848	333	
6	105113	172.3	101063	207.8	97361	247	93954	289	90798	334	24
	105042	172.8	100998	208.4	97302	247	93899	289	90747	335	
7	104971	173.4	100934	209.0	97243	248	93845	290	90696	336	23
	104901	174.0	100870	209.6	97184	249	93790	291	90646	337	
8	104830	174.5	100806	210.3	97126	249	93736	292	90595	337	22
	104760	175.1	100742	210.9	97067	250	93682	292	90545	338	
9	104690	175.7	100678	211.5	97008	251	93628	293	90494	339	21
	104620	176.2	100614	212.1	96950	251	93573	294	90444	340	
10	104550	176.8	100550	212.8	96891	252	93519	295	90394	341	20
	104480	177.4	100487	213.4	96833	253	93465	295	90344	341	
11	104411	178.0	100423	214.0	96774	253	93411	296	90293	342	19
	104341	178.5	100360	214.6	96716	254	93358	297	90243	343	
12	104272	179.1	100296	215.3	96658	255	93304	298	90193	344	18
	104202	179.7	100233	215.9	96600	255	93250	298	90143	345	
13	104133	180.3	100170	216.5	96542	256	93196	299	90093	345	17
	104064	180.8	100107	217.2	96484	257	93143	300	90044	346	
14	103995	181.4	100044	217.8	96426	257	93089	301	89994	347	16
	103926	182.0	99981	218.4	96368	258	93036	301	89944	348	
15	103857	182.6	99918	219.1	96310	259	92982	302	89894	349	15
	103788	183.2	99856	219.7	96253	260	92929	303	89845	349	
16	103720	183.7	99793	220.3	96195	260	92876	304	89795	350	14
	103651	184.3	99731	221.0	96138	261	92823	304	89746	351	
17	103583	184.9	99668	221.6	96080	262	92769	305	89696	352	13
	103515	185.5	99606	222.3	96023	262	92716	306	89647	353	
18	103447	186.1	99544	222.9	95966	263	92663	307	89597	353	12
	103379	186.7	99481	223.5	95909	264	92610	307	89548	354	
19	103311	187.2	99420	224.2	95851	264	92558	308	89499	355	11
	103243	187.8	99357	224.8	95795	265	92505	309	89450	356	
20	103175	188.4	99296	225.5	95737	266	92452	310	89401	357	10
	103107	189.0	99234	226.1	95681	267	92399	310	89352	357	
21	103040	189.6	99172	226.8	95624	267	92347	311	89303	358	9
	102973	190.2	99110	227.4	95567	268	92294	312	89254	359	
22	102905	190.8	99049	228.1	95510	269	92242	313	89205	360	8
	102838	191.4	98988	228.7	95454	269	92189	313	89156	361	
23	102771	192.0	98926	229.4	95397	270	92137	314	89107	362	7
	102704	192.6	98865	230.0	95341	271	92085	315	89059	362	
24	102637	193.2	98804	230.7	95285	271	92032	316	89010	363	6
	102570	193.8	98743	231.3	95228	272	91980	316	88961	364	
25	102504	194.4	98682	232.0	95172	273	91928	317	88913	365	5
	102437	195.0	98621	232.6	95116	274	91876	318	88864	366	
26	102371	195.6	98560	233.3	95060	274	91824	319	88816	366	4
	102304	196.2	98499	233.9	95004	275	91772	319	88767	367	
27	102238	196.8	98439	234.6	94948	276	91720	320	88719	368	3
	102172	197.4	98378	235.3	94892	276	91668	321	88671	369	
28	102106	198.0	98318	235.9	94836	277	91617	322	88623	370	2
	102040	198.6	98257	236.6	94781	278	91565	323	88574	371	
29	101974	199.2	98197	237.2	94725	279	91514	323	88526	371	1
	101908	199.8	98137	237.9	94670	279	91462	324	88478	372	
30	101843	200.4	98076	238.6	94614	280	91411	325	88430	373	0
	A	B	A	B	A	B	A	B	A	B	
′	174° 30′		174° 00′		173° 30′		173° 00′		172° 30′		′

TABLE 35

The Ageton Method

Always Take "Z" From Bottom of Table, Except When "K" is Same Name and Greater Than Latitude, in Which Case Take "Z" From Top of Table

′	7° 30′		8° 00′		8° 30′		9° 00′		9° 30′		′
′	A	B	A	B	A	B	A	B	A	B	′
0	88430	**373**	85644	**425**	83030	**480**	80567	**538**	78239	**600**	30
	88382	**374**	85599	**426**	82987	**481**	80527	**539**	78201	**601**	
1	88334	**375**	85555	**426**	82945	**482**	80487	**540**	78164	**602**	29
	88286	**376**	85510	**427**	82903	**482**	80447	**541**	78126	**603**	
2	88239	**376**	85465	**428**	82861	**483**	80407	**542**	78088	**604**	28
	88191	**377**	85420	**429**	82819	**484**	80368	**543**	78051	**605**	
3	88143	**378**	85376	**430**	82777	**485**	80328	**544**	78013	**606**	27
	88096	**379**	85331	**431**	82735	**486**	80288	**545**	77976	**607**	
4	88048	**380**	85286	**432**	82693	**487**	80249	**546**	77938	**608**	26
	88001	**381**	85242	**433**	82651	**488**	80209	**547**	77901	**609**	
5	87953	**381**	85197	**434**	82609	**489**	80170	**548**	77863	**610**	25
	87906	**382**	85153	**434**	82567	**490**	80130	**549**	77826	**611**	
6	87858	**383**	85108	**435**	82526	**491**	80091	**550**	77788	**612**	24
	87811	**384**	85064	**436**	82484	**492**	80051	**551**	77751	**614**	
7	87764	**385**	85020	**437**	82442	**493**	80012	**552**	77714	**615**	23
	87716	**386**	84976	**438**	82400	**494**	79973	**553**	77677	**616**	
8	87669	**387**	84931	**439**	82359	**495**	79933	**554**	77639	**617**	22
	87622	**387**	84887	**440**	82317	**496**	79894	**555**	77602	**618**	
9	87575	**388**	84843	**441**	82276	**497**	79855	**556**	77565	**619**	21
	87528	**389**	84799	**442**	82234	**498**	79816	**557**	77528	**620**	
10	87481	**390**	84755	**443**	82193	**499**	79777	**558**	77491	**621**	20
	87434	**391**	84711	**444**	82151	**500**	79737	**559**	77454	**622**	
11	87387	**392**	84667	**444**	82110	**501**	79698	**560**	77417	**623**	19
	87341	**392**	84623	**445**	82069	**502**	79659	**561**	77380	**624**	
12	87294	**393**	84579	**446**	82027	**503**	79620	**562**	77343	**625**	18
	87247	**394**	84535	**447**	81986	**504**	79581	**563**	77306	**626**	
13	87201	**395**	84492	**448**	81945	**504**	79542	**564**	77269	**627**	17
	87154	**396**	84448	**449**	81904	**505**	79503	**565**	77232	**629**	
14	87107	**397**	84404	**450**	81863	**506**	79465	**566**	77195	**630**	16
	87061	**398**	84361	**451**	81821	**507**	79426	**567**	77158	**631**	
15	87015	**399**	84317	**452**	81780	**508**	79387	**568**	77122	**632**	15
	86968	**399**	84273	**453**	81739	**509**	79348	**569**	77085	**633**	
16	86922	**400**	84230	**454**	81698	**510**	79309	**570**	77048	**634**	14
	86876	**401**	84186	**454**	81657	**511**	79271	**571**	77011	**635**	
17	86829	**402**	84143	**455**	81617	**512**	79232	**573**	76975	**636**	13
	86783	**403**	84100	**456**	81576	**513**	79193	**574**	76938	**637**	
18	86737	**404**	84056	**457**	81535	**514**	79155	**575**	76902	**638**	12
	86691	**405**	84013	**458**	81494	**515**	79116	**576**	76865	**639**	
19	86645	**405**	83970	**459**	81453	**516**	79078	**577**	76828	**641**	11
	86599	**406**	83927	**460**	81413	**517**	79039	**578**	76792	**642**	
20	86553	**407**	83884	**461**	81372	**518**	79001	**579**	76756	**643**	10
	86507	**408**	83840	**462**	81331	**519**	78962	**580**	76719	**644**	
21	86461	**409**	83797	**463**	81291	**520**	78924	**581**	76683	**645**	9
	86415	**410**	83754	**464**	81250	**521**	78886	**582**	76646	**646**	
22	86370	**411**	83711	**465**	81210	**522**	78847	**583**	76610	**647**	8
	86324	**411**	83668	**466**	81169	**523**	78809	**584**	76574	**648**	
23	86278	**412**	83626	**467**	81129	**524**	78771	**585**	76537	**649**	7
	86233	**413**	83583	**467**	81088	**525**	78733	**586**	76501	**650**	
24	86187	**414**	83540	**468**	81048	**526**	78694	**587**	76465	**652**	6
	86142	**415**	83497	**469**	81008	**527**	78656	**588**	76429	**653**	
25	86096	**416**	83455	**470**	80967	**528**	78618	**589**	76393	**654**	5
	86051	**417**	83412	**471**	80927	**529**	78580	**590**	76357	**655**	
26	86006	**418**	83369	**472**	80887	**530**	78542	**591**	76320	**656**	4
	85960	**418**	83327	**473**	80847	**531**	78504	**592**	76284	**657**	
27	85915	**419**	83284	**474**	80807	**532**	78466	**593**	76248	**658**	3
	85870	**420**	83242	**475**	80767	**533**	78428	**594**	76212	**659**	
28	85825	**421**	83199	**476**	80727	**534**	78390	**595**	76176	**660**	2
	85779	**422**	83157	**477**	80687	**535**	78352	**597**	76141	**661**	
29	85734	**423**	83114	**478**	80647	**536**	78315	**598**	76105	**663**	1
	85689	**424**	83072	**479**	80607	**537**	78277	**599**	76069	**664**	
30	85644	**425**	83030	**480**	80567	**538**	78239	**600**	76033	**665**	0
	A	B	A	B	A	B	A	B	A	B	
′	172° 00′		171° 30′		171° 00′		170° 30′		170° 00′		′

TABLE 35

The Ageton Method

When Meridian Angle is Greater Than 90° Take "K" From Bottom of Table

′	10° 00′		10° 30′		11° 00′		11° 30′		12° 00′		′
′	A	B	A	B	A	B	A	B	A	B	′
0	76033	665	73937	733	71940	805	70034	881	68212	960	30
	75997	666	73903	735	71908	807	70003	882	68182	961	
1	75961	667	73869	736	71875	808	69972	883	68153	962	29
	75926	668	73835	737	71843	809	69941	885	68123	964	
2	75890	669	73801	738	71810	810	69910	886	68093	965	28
	75854	670	73767	739	71778	811	69879	887	68064	966	
3	75819	672	73733	740	71746	813	69849	888	68034	968	27
	75783	673	73699	742	71713	814	69818	890	68005	969	
4	75747	674	73665	743	71681	815	69787	891	67975	970	26
	75712	675	73631	744	71649	816	69756	892	67945	972	
5	75676	676	73597	745	71616	818	69725	894	67916	973	25
	75641	677	73563	746	71584	819	69694	895	67886	974	
6	75605	678	73530	747	71552	820	69664	896	67857	976	24
	75570	679	73496	749	71520	821	69633	897	67828	977	
7	75534	680	73462	750	71488	823	69602	899	67798	978	23
	75499	682	73429	751	71455	824	69571	900	67769	980	
8	75464	683	73395	752	71423	825	69541	901	67739	981	22
	75428	684	73361	753	71391	826	69510	903	67710	982	
9	75393	685	73328	755	71359	828	69479	904	67681	984	21
	75358	686	73294	756	71327	829	69449	905	67651	985	
10	75322	687	73260	757	71295	830	69418	907	67622	987	20
	75287	688	73227	758	71263	831	69387	908	67593	988	
11	75252	690	73193	759	71231	833	69357	909	67563	989	19
	75217	691	73160	761	71199	834	69326	910	67534	991	
12	75182	692	73127	762	71167	835	69296	912	67505	992	18
	75147	693	73093	763	71135	836	69265	913	67476	993	
13	75112	694	73060	764	71104	838	69235	914	67447	995	17
	75077	695	73026	765	71072	839	69204	916	67417	996	
14	75042	696	72993	766	71040	840	69174	917	67388	997	16
	75007	698	72960	768	71008	841	69144	918	67359	999	
15	74972	699	72926	769	70976	843	69113	920	67330	1000	15
	74937	700	72893	770	70945	844	69083	921	67301	1002	
16	74902	701	72860	771	70913	845	69053	922	67272	1003	14
	74867	702	72827	772	70881	846	69022	924	67243	1004	
17	74832	703	72794	774	70850	848	68992	925	67214	1006	13
	74797	704	72760	775	70818	849	68962	926	67185	1007	
18	74763	706	72727	776	70786	850	68931	928	67156	1008	12
	74728	707	72694	777	70755	851	68901	929	67127	1010	
19	74693	708	72661	779	70723	853	68871	930	67098	1011	11
	74659	709	72628	780	70692	854	68841	932	67069	1013	
20	74624	710	72595	781	70660	855	68811	933	67040	1014	10
	74589	711	72562	782	70629	856	68781	934	67011	1015	
21	74555	712	72529	783	70597	858	68750	935	66982	1017	9
	74520	714	72496	785	70566	859	68720	937	66953	1018	
22	74486	715	72463	786	70534	860	68690	938	66925	1020	8
	74451	716	72430	787	70503	862	68660	939	66896	1021	
23	74417	717	72397	788	70471	863	68630	941	66867	1022	7
	74382	718	72365	790	70440	864	68600	942	66838	1024	
24	74348	719	72332	791	70409	865	68570	943	66810	1025	6
	74313	721	72299	792	70377	867	68540	945	66781	1026	
25	74279	722	72266	793	70346	868	68510	946	66752	1028	5
	74245	723	72234	794	70315	869	68480	947	66724	1029	
26	74210	724	72201	796	70284	870	68450	949	66695	1031	4
	74176	725	72168	797	70252	872	68421	950	66666	1032	
27	74142	726	72135	798	70221	873	68391	951	66638	1033	3
	74107	728	72103	799	70190	874	68361	953	66609	1035	
28	74073	729	72070	800	70159	876	68331	954	66580	1036	2
	74039	730	72038	802	70128	877	68301	955	66552	1038	
29	74005	731	72005	803	70097	878	68272	957	66523	1039	1
	73971	732	71973	804	70065	879	68242	958	66495	1040	
30	73937	733	71940	805	70034	881	68212	960	66466	1042	0
′	A	B	A	B	A	B	A	B	A	B	′
′	169° 30′		169° 00′		168° 30′		168° 00′		167° 30′		′

TABLE 35

The Ageton Method

Always Take "Z" From Bottom of Table, Except When "K" is Same Name and Greater Than Latitude, in Which Case Take "Z" From Top of Table

′	12° 30′		13° 00′		13° 30′		14° 00′		14° 30′		′
′	A	B	A	B	A	B	A	B	A	B	′
0	66466	**1042**	64791	**1128**	63181	**1217**	61632	**1310**	60140	**1406**	30
	66438	**1043**	64764	**1129**	63155	**1218**	61607	**1311**	60116	**1407**	
1	66409	**1045**	64736	**1130**	63129	**1220**	61582	**1313**	60091	**1409**	29
	66381	**1046**	64709	**1132**	63103	**1221**	61556	**1314**	60067	**1411**	
2	66352	**1047**	64682	**1133**	63076	**1223**	61531	**1316**	60042	**1412**	28
	66324	**1049**	64655	**1135**	63050	**1224**	61506	**1317**	60018	**1414**	
3	66296	**1050**	64627	**1136**	63024	**1226**	61481	**1319**	59994	**1416**	27
	66267	**1052**	64600	**1138**	62998	**1227**	61455	**1321**	59969	**1417**	
4	66239	**1053**	64573	**1139**	62971	**1229**	61430	**1322**	59945	**1419**	26
	66211	**1054**	64546	**1141**	62945	**1230**	61405	**1324**	59921	**1421**	
5	66182	**1056**	64518	**1142**	62919	**1232**	61380	**1325**	59896	**1422**	25
	66154	**1057**	64491	**1144**	62893	**1234**	61355	**1327**	59872	**1424**	
6	66126	**1059**	64464	**1145**	62867	**1235**	61330	**1329**	59848	**1425**	24
	66098	**1060**	64437	**1147**	62841	**1237**	61304	**1330**	59824	**1427**	
7	66069	**1061**	64410	**1148**	62815	**1238**	61279	**1332**	59800	**1429**	23
	66041	**1063**	64383	**1150**	62789	**1240**	61254	**1333**	59775	**1430**	
8	66013	**1064**	64356	**1151**	62763	**1241**	61229	**1335**	59751	**1432**	22
	65985	**1066**	64329	**1152**	62737	**1243**	61204	**1336**	59727	**1434**	
9	65957	**1067**	64302	**1154**	62711	**1244**	61179	**1338**	59703	**1435**	21
	65928	**1069**	64275	**1155**	62685	**1246**	61154	**1340**	59679	**1437**	
10	65900	**1070**	64248	**1157**	62659	**1247**	61129	**1341**	59654	**1439**	20
	65872	**1071**	64221	**1158**	62633	**1249**	61104	**1343**	59630	**1440**	
11	65844	**1073**	64194	**1160**	62607	**1250**	61079	**1344**	59606	**1442**	19
	65816	**1074**	64167	**1161**	62581	**1252**	61054	**1346**	59582	**1444**	
12	65788	**1076**	64140	**1163**	62555	**1253**	61029	**1348**	59558	**1445**	18
	65760	**1077**	64113	**1164**	62529	**1255**	61004	**1349**	59534	**1447**	
13	65732	**1079**	64086	**1166**	62503	**1257**	60979	**1351**	59510	**1449**	17
	65704	**1080**	64059	**1167**	62477	**1258**	60954	**1352**	59486	**1450**	
14	65676	**1081**	64032	**1169**	62451	**1260**	60929	**1354**	59462	**1452**	16
	65648	**1083**	64005	**1170**	62425	**1261**	60904	**1356**	59438	**1454**	
15	65620	**1084**	63978	**1172**	62400	**1263**	60879	**1357**	59414	**1455**	15
	65592	**1086**	63952	**1173**	62374	**1264**	60855	**1359**	59390	**1457**	
16	65564	**1087**	63925	**1175**	62348	**1266**	60830	**1360**	59366	**1459**	14
	65537	**1089**	63898	**1176**	62322	**1267**	60805	**1362**	59342	**1460**	
17	65509	**1090**	63871	**1178**	62296	**1269**	60780	**1364**	59318	**1462**	13
	65481	**1091**	63845	**1179**	62271	**1270**	60755	**1365**	59294	**1464**	
18	65453	**1093**	63818	**1181**	62245	**1272**	60730	**1367**	59270	**1465**	12
	65425	**1094**	63791	**1182**	62219	**1274**	60706	**1368**	59246	**1467**	
19	65398	**1096**	63764	**1184**	62194	**1275**	60681	**1370**	59222	**1469**	11
	65370	**1097**	63738	**1185**	62168	**1277**	60656	**1372**	59198	**1470**	
20	65342	**1099**	63711	**1187**	62142	**1278**	60631	**1373**	59175	**1472**	10
	65314	**1100**	63684	**1188**	62117	**1280**	60607	**1375**	59151	**1474**	
21	65287	**1101**	63658	**1190**	62091	**1281**	60582	**1377**	59127	**1475**	9
	65259	**1103**	63631	**1191**	62065	**1283**	60557	**1378**	59103	**1477**	
22	65231	**1104**	63605	**1193**	62040	**1284**	60533	**1380**	59079	**1479**	8
	65204	**1106**	63578	**1194**	62014	**1286**	60508	**1381**	59055	**1480**	
23	65176	**1107**	63551	**1196**	61989	**1288**	60483	**1383**	59032	**1482**	7
	65148	**1109**	63525	**1197**	61963	**1289**	60459	**1385**	59008	**1484**	
24	65121	**1110**	63498	**1199**	61938	**1291**	60434	**1386**	58984	**1485**	6
	65093	**1112**	63472	**1200**	61912	**1292**	60410	**1388**	58960	**1487**	
25	65066	**1113**	63445	**1202**	61887	**1294**	60385	**1390**	58937	**1489**	5
	65038	**1114**	63419	**1203**	61861	**1295**	60360	**1391**	58913	**1490**	
26	65011	**1116**	63392	**1205**	61836	**1297**	60336	**1393**	58889	**1492**	4
	64983	**1117**	63366	**1206**	61810	**1299**	60311	**1394**	58866	**1494**	
27	64956	**1119**	63340	**1208**	61785	**1300**	60287	**1396**	58842	**1495**	3
	64928	**1120**	63313	**1209**	61759	**1301**	60262	**1398**	58818	**1497**	
28	64901	**1122**	63287	**1211**	61734	**1303**	60238	**1399**	58795	**1499**	2
	64873	**1123**	63260	**1212**	61709	**1305**	60213	**1401**	58771	**1500**	
29	64846	**1125**	63234	**1214**	61683	**1306**	60189	**1403**	58748	**1502**	1
	64819	**1126**	63208	**1215**	61658	**1308**	60164	**1404**	58724	**1504**	
30	64791	**1128**	63181	**1217**	61632	**1310**	60140	**1406**	58700	**1506**	0
′	A	B	A	B	A	B	A	B	A	B	′
′	167° 00′		166° 30′		166° 00′		165° 30′		165° 00′		′

TABLE 35

The Ageton Method

When Meridian Angle is Greater Than 90° Take "K" From Bottom of Table

′	15° 00′		15° 30′		16° 00′		16° 30′		17° 00′		′
	A	B	A	B	A	B	A	B	A	B	
0	58700	1506	57310	1609	55966	1716	54666	1826	53406	1940	30
	58677	1507	57287	1611	55944	1718	54644	1828	53386	1942	
1	58653	1509	57265	1612	55922	1719	54623	1830	53365	1944	29
	58630	1511	57242	1614	55900	1721	54602	1832	53344	1946	
2	58606	1512	57219	1616	55878	1723	54581	1834	53324	1948	28
	58583	1514	57196	1618	55856	1725	54559	1836	53303	1950	
3	58559	1516	57174	1619	55834	1727	54538	1837	53283	1952	27
	58536	1517	57151	1621	55812	1728	54517	1839	53262	1954	
4	58512	1519	57128	1623	55790	1730	54496	1841	53241	1956	26
	58489	1521	57106	1625	55768	1732	54474	1843	53221	1958	
5	58465	1523	57083	1627	55746	1734	54453	1845	53200	1960	25
	58442	1524	57060	1628	55725	1736	54432	1847	53180	1962	
6	58418	1526	57038	1630	55703	1738	54411	1849	53159	1964	24
	58395	1528	57015	1632	55681	1739	54390	1851	53139	1966	
7	58372	1529	56992	1634	55659	1741	54368	1853	53118	1967	23
	58348	1531	56970	1635	55637	1743	54347	1854	53098	1969	
8	58325	1533	56947	1637	55615	1745	54326	1856	53077	1971	22
	58302	1534	56925	1639	55593	1747	54305	1858	53057	1973	
9	58278	1536	56902	1641	55572	1749	54284	1860	53036	1975	21
	58255	1538	56880	1642	55550	1750	54263	1862	53016	1977	
10	58232	1540	56857	1644	55528	1752	54242	1864	52995	1979	20
	58208	1541	56835	1646	55506	1754	54220	1866	52975	1981	
11	58185	1543	56812	1648	55484	1756	54199	1868	52954	1983	19
	58162	1545	56790	1649	55463	1758	54178	1870	52934	1985	
12	58138	1546	56767	1651	55441	1760	54157	1871	52914	1987	18
	58115	1548	56745	1653	55419	1761	54136	1873	52893	1989	
13	58092	1550	56722	1655	55397	1763	54115	1875	52873	1991	17
	58069	1552	56700	1657	55376	1765	54094	1877	52852	1993	
14	58046	1553	56677	1658	55354	1767	54073	1879	52832	1995	16
	58022	1555	56655	1660	55332	1769	54052	1881	52812	1997	
15	57999	1557	56632	1662	55311	1771	54031	1883	52791	1999	15
	57976	1559	56610	1664	55289	1772	54010	1885	52771	2001	
16	57953	1560	56588	1665	55267	1774	53989	1887	52751	2003	14
	57930	1562	56565	1667	55246	1776	53968	1889	52730	2005	
17	57907	1564	56543	1669	55224	1778	53947	1890	52710	2007	13
	57884	1565	56521	1671	55202	1780	53926	1892	52690	2009	
18	57860	1567	56498	1673	55181	1782	53905	1894	52670	2010	12
	57837	1569	56476	1674	55159	1783	53884	1896	52649	2012	
19	57814	1571	56454	1676	55138	1785	53864	1898	52629	2014	11
	57791	1572	56431	1678	55116	1787	53843	1900	52609	2016	
20	57768	1574	56409	1680	55095	1789	53822	1902	52588	2018	10
	57745	1576	56387	1682	55073	1791	53801	1904	52568	2020	
21	57722	1578	56365	1683	55051	1793	53780	1906	52548	2022	9
	57699	1579	56342	1685	55030	1795	53759	1908	52528	2024	
22	57676	1581	56320	1687	55008	1796	53738	1910	52508	2026	8
	57653	1583	56298	1689	54987	1798	53718	1911	52487	2028	
23	57630	1584	56276	1691	54965	1800	53697	1913	52467	2030	7
	57607	1586	56254	1692	54944	1802	53676	1915	52447	2032	
24	57584	1588	56231	1694	54922	1804	53655	1917	52427	2034	6
	57561	1590	56209	1696	54901	1806	53634	1919	52407	2036	
25	57538	1591	56187	1698	54880	1808	53614	1921	52387	2038	5
	57516	1593	56165	1700	54858	1809	53593	1923	52366	2040	
26	57493	1595	56143	1701	54837	1811	53572	1925	52346	2042	4
	57470	1597	56121	1703	54815	1813	53551	1927	52326	2044	
27	57447	1598	56099	1705	54794	1815	53531	1929	52306	2046	3
	57424	1600	56076	1707	54773	1817	53510	1931	52286	2048	
28	57401	1602	56054	1709	54751	1819	53489	1933	52266	2050	2
	57378	1604	56032	1710	54730	1821	53468	1935	52246	2052	
29	57356	1605	56010	1712	54708	1823	53448	1936	52226	2054	1
	57333	1607	55988	1714	54687	1824	53427	1938	52206	2056	
30	57310	1609	55966	1716	54666	1826	53406	1940	52186	2058	0
	A	B	A	B	A	B	A	B	A	B	
′	164° 30′		164° 00′		163° 30′		163° 00′		162° 30′		′

TABLE 35

The Ageton Method

Always Take "Z" From Bottom of Table, Except When "K" is Same Name and Greater Than Latitude, in Which Case Take "Z" From Top of Table

′	17° 30′ A	17° 30′ B	18° 00′ A	18° 00′ B	18° 30′ A	18° 30′ B	19° 00′ A	19° 00′ B	19° 30′ A	19° 30′ B	′
0	52186	2058	51002	2179	49852	2304	48736	2433	47650	2565	30
	52166	2060	50982	2181	49833	2306	48717	2435	47633	2568	
1	52146	2062	50963	2183	49815	2309	48699	2437	47615	2570	29
	52126	2064	50943	2185	49796	2311	48681	2439	47597	2572	
2	52106	2066	50924	2188	49777	2313	48662	2442	47579	2574	28
	52086	2068	50905	2190	49758	2315	48644	2444	47561	2576	
3	52066	2070	50885	2192	49739	2317	48626	2446	47544	2579	27
	52046	2072	50866	2194	49720	2319	48608	2448	47526	2581	
4	52026	2074	50846	2196	49702	2321	48589	2450	47508	2583	26
	52006	2076	50827	2198	49683	2323	48571	2453	47490	2585	
5	51986	2078	50808	2200	49664	2325	48553	2455	47472	2588	25
	51966	2080	50788	2202	49645	2328	48534	2457	47455	2590	
6	51946	2082	50769	2204	49626	2330	48516	2459	47437	2592	24
	51926	2084	50750	2206	49608	2332	48498	2461	47419	2594	
7	51906	2086	50730	2208	49589	2334	48480	2463	47402	2597	23
	51886	2088	50711	2210	49570	2336	48462	2466	47384	2599	
8	51867	2090	50692	2212	49551	2338	48443	2468	47366	2601	22
	51847	2092	50673	2214	49533	2340	48425	2470	47348	2603	
9	51827	2094	50653	2216	49514	2343	48407	2472	47331	2606	21
	51807	2096	50634	2218	49495	2345	48389	2474	47313	2608	
10	51787	2098	50615	2221	49477	2347	48371	2477	47295	2610	20
	51767	2100	50596	2223	49458	2349	48352	2479	47278	2613	
11	51747	2102	50576	2225	49439	2351	48334	2481	47260	2615	19
	51728	2104	50557	2227	49421	2353	48316	2483	47242	2617	
12	51708	2106	50538	2229	49402	2355	48298	2485	47225	2619	18
	51688	2108	50519	2231	49383	2357	48280	2488	47207	2622	
13	51668	2110	50499	2233	49365	2360	48262	2490	47189	2624	17
	51649	2112	50480	2235	49346	2362	48244	2492	47172	2626	
14	51629	2114	50461	2237	49327	2364	48225	2494	47154	2628	16
	51609	2116	50442	2239	49309	2366	48207	2496	47137	2631	
15	51589	2118	50423	2241	49290	2368	48189	2499	47119	2633	15
	51570	2120	50404	2243	49271	2370	48171	2501	47101	2635	
16	51550	2122	50385	2246	49253	2372	48153	2503	47084	2637	14
	51530	2124	50365	2248	49234	2375	48135	2505	47066	2640	
17	51510	2126	50346	2250	49216	2377	48117	2507	47049	2642	13
	51491	2128	50327	2252	49197	2379	48099	2510	47031	2644	
18	51471	2130	50308	2254	49179	2381	48081	2512	47014	2646	12
	51451	2132	50289	2256	49160	2383	48063	2514	46996	2649	
19	51432	2134	50270	2258	49141	2385	48045	2516	46978	2651	11
	51412	2136	50251	2260	49123	2387	48027	2519	46961	2653	
20	51392	2138	50232	2262	49104	2390	48009	2521	46943	2656	10
	51373	2141	50213	2264	49086	2392	47991	2523	46926	2658	
21	51353	2143	50194	2266	49067	2394	47973	2525	46908	2660	9
	51334	2145	50175	2269	49049	2396	47955	2527	46891	2662	
22	51314	2147	50156	2271	49030	2398	47937	2530	46873	2665	8
	51294	2149	50137	2273	49012	2400	47919	2532	46856	2667	
23	51275	2151	50117	2275	48993	2403	47901	2534	46839	2669	7
	51255	2153	50098	2277	48975	2405	47883	2536	46821	2672	
24	51236	2155	50080	2279	48957	2407	47865	2539	46804	2674	6
	51216	2157	50061	2281	48938	2409	47847	2541	46786	2676	
25	51197	2159	50042	2283	48920	2411	47829	2543	46769	2678	5
	51177	2161	50023	2285	48901	2413	47811	2545	46751	2681	
26	51158	2163	50004	2287	48883	2416	47793	2547	46734	2683	4
	51138	2165	49985	2290	48864	2418	47775	2550	46716	2685	
27	51119	2167	49966	2292	48846	2420	47758	2552	46699	2688	3
	51099	2169	49947	2294	48828	2422	47740	2554	46682	2690	
28	51080	2171	49928	2296	48809	2424	47722	2556	46664	2692	2
	51060	2173	49909	2298	48791	2426	47704	2559	46647	2694	
29	51041	2175	49890	2300	48772	2429	47686	2561	46630	2697	1
	51021	2177	49871	2302	48754	2431	47668	2563	46612	2699	
30	51002	2179	49852	2304	48736	2433	47650	2565	46595	2701	0
′	A	B	A	B	A	B	A	B	A	B	′
	162° 00′		161° 30′		161° 00′		160° 30′		160° 00′		

TABLE 35

The Ageton Method

When Meridian Angle is Greater Than 90° Take "K" From Bottom of Table

′	20° 00′ A	20° 00′ B	20° 30′ A	20° 30′ B	21° 00′ A	21° 00′ B	21° 30′ A	21° 30′ B	22° 00′ A	22° 00′ B	′
0	46595	2701	45567	2841	44567	2985	43592	3132	42642	3283	30
	46577	2704	45551	2844	44551	2988	43576	3135	42627	3286	
1	46560	2706	45534	2846	44534	2990	43560	3137	42611	3288	29
	46543	2708	45517	2848	44518	2992	43544	3140	42596	3291	
2	46525	2711	45500	2851	44501	2994	43528	3142	42580	3294	28
	46508	2713	45483	2853	44485	2997	43512	3145	42564	3296	
3	46491	2715	45466	2855	44468	2999	43496	3147	42549	3299	27
	46473	2717	45449	2858	44452	3002	43480	3150	42533	3301	
4	46456	2720	45433	2860	44436	3004	43464	3152	42518	3304	26
	46439	2722	45416	2862	44419	3007	43448	3155	42502	3306	
5	46422	2724	45399	2865	44403	3009	43432	3157	42486	3309	25
	46404	2727	45382	2867	44386	3012	43416	3160	42471	3312	
6	46387	2729	45365	2870	44370	3014	43400	3162	42455	3314	24
	46370	2731	45348	2872	44354	3016	43385	3165	42440	3317	
7	46353	2734	45332	2874	44337	3019	43369	3167	42424	3319	23
	46335	2736	45315	2877	44321	3021	43353	3170	42409	3322	
8	46318	2738	45298	2879	44305	3024	43337	3172	42393	3324	22
	46301	2741	45281	2881	44288	3026	43321	3175	42378	3327	
9	46284	2743	45265	2884	44272	3029	43305	3177	42362	3329	21
	46266	2745	45248	2886	44256	3031	43289	3180	42347	3332	
10	46249	2748	45231	2889	44239	3033	43273	3182	42331	3335	20
	46232	2750	45214	2891	44223	3036	43257	3185	42316	3337	
11	46215	2752	45198	2893	44207	3038	43241	3187	42300	3340	19
	46198	2755	45181	2896	44190	3041	43225	3190	42285	3342	
12	46181	2757	45164	2898	44174	3043	43210	3192	42269	3345	18
	46163	2759	45147	2901	44158	3046	43194	3195	42254	3347	
13	46146	2761	45131	2903	44142	3048	43178	3197	42238	3350	17
	46129	2764	45114	2905	44125	3051	43162	3200	42223	3353	
14	46112	2766	45097	2908	44109	3053	43146	3202	42207	3355	16
	46095	2768	45081	2910	44093	3056	43130	3205	42192	3358	
15	46078	2771	45064	2913	44077	3058	43114	3207	42176	3360	15
	46061	2773	45047	2915	44060	3060	43099	3210	42161	3363	
16	46043	2775	45031	2917	44044	3063	43083	3212	42145	3366	14
	46026	2778	45014	2920	44028	3065	43067	3215	42130	3368	
17	46009	2780	44997	2922	44012	3068	43051	3217	42115	3371	13
	45992	2782	44981	2924	43995	3070	43035	3220	42099	3373	
18	45975	2785	44964	2927	43979	3073	43020	3222	42084	3376	12
	45958	2787	44947	2929	43963	3075	43004	3225	42068	3379	
19	45941	2789	44931	2932	43947	3078	42988	3227	42053	3381	11
	45924	2792	44914	2934	43931	3080	42972	3230	42038	3384	
20	45907	2794	44898	2936	43914	3083	42956	3233	42022	3386	10
	45890	2797	44881	2939	43898	3085	42941	3235	42007	3389	
21	45873	2799	44864	2941	43882	3088	42925	3238	41991	3391	9
	45856	2801	44848	2944	43866	3090	42909	3240	41976	3394	
22	45839	2804	44831	2946	43850	3092	42893	3243	41961	3397	8
	45822	2806	44815	2949	43834	3095	42878	3245	41945	3399	
23	45805	2808	44798	2951	43818	3097	42862	3248	41930	3402	7
	45788	2811	44782	2953	43801	3100	42846	3250	41915	3404	
24	45771	2813	44765	2956	43785	3102	42830	3253	41899	3407	6
	45754	2815	44748	2958	43769	3105	42815	3255	41884	3410	
25	45737	2818	44732	2961	43753	3107	42799	3258	41869	3412	5
	45720	2820	44715	2963	43737	3110	42783	3260	41853	3415	
26	45703	2822	44699	2965	43721	3112	42768	3263	41838	3418	4
	45686	2825	44682	2968	43705	3115	42752	3266	41823	3420	
27	45669	2827	44666	2970	43689	3117	42736	3268	41808	3423	3
	45652	2829	44649	2973	43673	3120	42721	3271	41792	3425	
28	45635	2832	44633	2975	43657	3122	42705	3273	41777	3428	2
	45618	2834	44616	2978	43641	3125	42689	3276	41762	3431	
29	45601	2836	44600	2980	43624	3127	42674	3278	41746	3433	1
	45584	2839	44583	2982	43608	3130	42658	3281	41731	3436	
30	45567	2841	44567	2985	43592	3132	42642	3283	41716	3438	0
	A	B	A	B	A	B	A	B	A	B	
′	159° 30′		159° 00′		158° 30′		158° 00′		157° 30′		′

TABLE 35

The Ageton Method

Always Take "Z" From Bottom of Table, Except When "K" is Same Name and Greater Than Latitude, in Which Case Take "Z" From Top of Table

′	22° 30′		23° 00′		23° 30′		24° 00′		24° 30′		′
	A	B	A	B	A	B	A	B	A	B	
0	41716	3438	40812	3597	39930	3760	39069	3927	38227	4098	30
	41701	3441	40797	3600	39915	3763	39054	3930	38213	4101	
1	41685	3444	40782	3603	39901	3766	39040	3932	38200	4103	29
	41670	3446	40768	3605	39886	3768	39026	3935	38186	4106	
2	41655	3449	40753	3608	39872	3771	39012	3938	38172	4109	28
	41640	3452	40738	3611	39857	3774	38998	3941	38158	4112	
3	41625	3454	40723	3613	39843	3777	38984	3944	38144	4115	27
	41609	3457	40708	3616	39828	3779	38969	3947	38130	4118	
4	41594	3459	40693	3619	39814	3782	38955	3949	38117	4121	26
	41579	3462	40678	3622	39799	3785	38941	3952	38103	4124	
5	41564	3465	40664	3624	39785	3788	38927	3955	38089	4127	25
	41549	3467	40649	3627	39771	3790	38913	3958	38075	4129	
6	41533	3470	40634	3630	39756	3793	38899	3961	38061	4132	24
	41518	3473	40619	3632	39742	3796	38885	3964	38048	4135	
7	41503	3475	40604	3635	39727	3799	38871	3966	38034	4138	23
	41488	3478	40590	3638	39713	3801	38856	3969	38020	4141	
8	41473	3480	40575	3640	39698	3804	38842	3972	38006	4144	22
	41458	3483	40560	3643	39684	3807	38828	3975	37992	4147	
9	41443	3486	40545	3646	39669	3810	38814	3978	37979	4150	21
	41427	3488	40530	3648	39655	3813	38800	3981	37965	4153	
10	41412	3491	40516	3651	39641	3815	38786	3983	37951	4155	20
	41397	3494	40501	3654	39626	3818	38772	3986	37937	4158	
11	41382	3496	40486	3657	39612	3821	38758	3989	37924	4161	19
	41367	3499	40471	3659	39597	3824	38744	3992	37910	4164	
12	41352	3502	40457	3662	39583	3826	38730	3995	37896	4167	18
	41337	3504	40442	3665	39569	3829	38716	3998	37882	4170	
13	41322	3507	40427	3667	39554	3832	38702	4000	37869	4173	17
	41307	3509	40413	3670	39540	3835	38688	4003	37855	4176	
14	41291	3512	40398	3673	39525	3838	38674	4006	37841	4179	16
	41276	3515	40383	3676	39511	3840	38660	4009	37828	4182	
15	41261	3517	40368	3678	39497	3843	38645	4012	37814	4185	15
	41246	3520	40354	3681	39482	3846	38631	4015	37800	4187	
16	41231	3523	40339	3684	39468	3849	38617	4017	37786	4190	14
	41216	3525	40324	3686	39454	3851	38603	4020	37773	4193	
17	41201	3528	40310	3689	39439	3854	38589	4023	37759	4196	13
	41186	3531	40295	3692	39425	3857	38575	4026	37745	4199	
18	41171	3533	40280	3695	39411	3860	38561	4029	37732	4202	12
	41156	3536	40266	3697	39396	3863	38547	4032	37718	4205	
19	41141	3539	40251	3700	39382	3865	38533	4035	37704	4208	11
	41126	3541	40236	3703	39368	3868	38520	4037	37691	4211	
20	41111	3544	40222	3705	39353	3871	38506	4040	37677	4214	10
	41096	3547	40207	3708	39339	3874	38492	4043	37663	4217	
21	41081	3549	40192	3711	39325	3876	38478	4046	37650	4220	9
	41066	3552	40178	3714	39311	3879	38464	4049	37636	4222	
22	41051	3555	40163	3716	39296	3882	38450	4052	37623	4225	8
	41036	3557	40149	3719	39282	3885	38436	4055	37609	4228	
23	41021	3560	40134	3722	39268	3888	38422	4057	37595	4231	7
	41006	3563	40119	3725	39254	3890	38408	4060	37582	4234	
24	40991	3565	40105	3727	39239	3893	38394	4063	37568	4237	6
	40976	3568	40090	3730	39225	3896	38380	4066	37554	4240	
25	40961	3571	40076	3733	39211	3899	38366	4069	37541	4243	5
	40946	3573	40061	3735	39197	3902	38352	4072	37527	4246	
26	40931	3576	40046	3738	39182	3904	38338	4075	37514	4249	4
	40916	3579	40032	3741	39168	3907	38324	4078	37500	4252	
27	40902	3581	40017	3744	39154	3910	38311	4080	37486	4255	3
	40887	3584	40003	3746	39140	3913	38297	4083	37473	4258	
28	40872	3587	39988	3749	39125	3916	38283	4086	37459	4261	2
	40857	3589	39974	3752	39111	3918	38269	4089	37446	4264	
29	40842	3592	39959	3755	39097	3921	38255	4092	37432	4266	1
	40827	3595	39945	3757	39083	3924	38241	4095	37419	4269	
30	40812	3597	39930	3760	39069	3927	38227	4098	37405	4272	0
	A	B	A	B	A	B	A	B	A	B	
′	157° 00′		156° 30′		156° 00′		155° 30′		155° 00′		′

TABLE 35

The Ageton Method

When Meridian Angle is Greater Than 90° Take "K" From Bottom of Table

′	25° 00′ A	25° 00′ B	25° 30′ A	25° 30′ B	26° 00′ A	26° 00′ B	26° 30′ A	26° 30′ B	27° 00′ A	27° 00′ B	′
0	37405	**4272**	36602	**4451**	35816	**4634**	35047	**4821**	34295	**5012**	30
	37392	**4275**	36588	**4454**	35803	**4637**	35035	**4824**	34283	**5015**	
1	37378	**4278**	36575	**4457**	35790	**4640**	35022	**4827**	34270	**5018**	29
	37365	**4281**	36562	**4460**	35777	**4643**	35009	**4830**	34258	**5022**	
2	37351	**4284**	36549	**4463**	35764	**4646**	34997	**4833**	34246	**5025**	28
	37337	**4287**	36535	**4466**	35751	**4649**	34984	**4837**	34233	**5028**	
3	37324	**4290**	36522	**4469**	35738	**4651**	34971	**4840**	34221	**5031**	27
	37310	**4293**	36509	**4472**	35725	**4656**	34959	**4843**	34209	**5034**	
4	37297	**4296**	36496	**4475**	35712	**4659**	34946	**4846**	34196	**5038**	26
	37283	**4299**	36483	**4478**	35699	**4662**	34933	**4849**	34184	**5041**	
5	37270	**4302**	36469	**4481**	35686	**4665**	34921	**4852**	34172	**5044**	25
	37256	**4305**	36456	**4484**	35674	**4668**	34908	**4856**	34159	**5047**	
6	37243	**4308**	36443	**4487**	35661	**4671**	34896	**4859**	34147	**5051**	24
	37229	**4311**	36430	**4490**	35648	**4674**	34883	**4862**	34134	**5054**	
7	37216	**4314**	36417	**4493**	35635	**4677**	34870	**4865**	34122	**5057**	23
	37203	**4317**	36403	**4496**	35622	**4680**	34858	**4868**	34110	**5060**	
8	37189	**4320**	36390	**4499**	35609	**4683**	34845	**4871**	34097	**5064**	22
	37176	**4323**	36377	**4503**	35596	**4686**	34832	**4875**	34085	**5067**	
9	37162	**4326**	36364	**4506**	35583	**4690**	34820	**4878**	34073	**5070**	21
	37149	**4329**	36351	**4509**	35571	**4693**	34807	**4881**	34061	**5073**	
10	37135	**4332**	36338	**4512**	35558	**4696**	34795	**4884**	34048	**5076**	20
	37122	**4334**	36325	**4515**	35545	**4699**	34782	**4887**	34036	**5080**	
11	37108	**4337**	36311	**4518**	35532	**4702**	34770	**4890**	34024	**5083**	19
	37095	**4340**	36298	**4521**	35519	**4705**	34757	**4894**	34011	**5086**	
12	37081	**4343**	36285	**4524**	35506	**4708**	34744	**4897**	33999	**5089**	18
	37068	**4346**	36272	**4527**	35493	**4711**	34732	**4900**	33987	**5093**	
13	37055	**4349**	36259	**4530**	35481	**4714**	34719	**4903**	33974	**5096**	17
	37041	**4352**	36246	**4533**	35468	**4718**	34707	**4906**	33962	**5099**	
14	37028	**4355**	36233	**4536**	35455	**4721**	34694	**4910**	33950	**5102**	16
	37014	**4358**	36220	**4539**	35442	**4724**	34682	**4913**	33938	**5106**	
15	37001	**4361**	36206	**4542**	35429	**4727**	34669	**4916**	33925	**5109**	15
	36988	**4364**	36193	**4545**	35417	**4730**	34657	**4919**	33913	**5112**	
16	36974	**4367**	36180	**4548**	35404	**4733**	34644	**4922**	33901	**5115**	14
	36961	**4370**	36167	**4551**	35391	**4736**	34632	**4925**	33889	**5119**	
17	36948	**4373**	36154	**4554**	35378	**4739**	34619	**4929**	33876	**5122**	13
	36934	**4376**	36141	**4557**	35365	**4742**	34607	**4932**	33864	**5125**	
18	36921	**4379**	36128	**4560**	35353	**4746**	34594	**4935**	33852	**5128**	12
	36907	**4382**	36115	**4563**	35340	**4749**	34582	**4938**	33840	**5132**	
19	36894	**4385**	36102	**4566**	35327	**4752**	34569	**4941**	33827	**5135**	11
	36881	**4388**	36089	**4569**	35314	**4755**	34557	**4945**	33815	**5138**	
20	36867	**4391**	36076	**4573**	35302	**4758**	34544	**4948**	33803	**5142**	10
	36854	**4394**	36063	**4576**	35289	**4761**	34532	**4951**	33791	**5145**	
21	36841	**4397**	36050	**4579**	35276	**4764**	34519	**4954**	33779	**5148**	9
	36827	**4400**	36037	**4582**	35263	**4769**	34507	**4957**	33766	**5151**	
22	36814	**4403**	36024	**4585**	35251	**4771**	34494	**4961**	33754	**5155**	8
	36801	**4406**	36011	**4588**	35238	**4774**	34482	**4964**	33742	**5158**	
23	36787	**4409**	35998	**4591**	35225	**4777**	34469	**4967**	33730	**5161**	7
	36774	**4412**	35985	**4594**	35212	**4780**	34457	**4970**	33717	**5164**	
24	36761	**4415**	35972	**4597**	35200	**4783**	34445	**4973**	33705	**5168**	6
	36747	**4418**	35959	**4600**	35187	**4786**	34432	**4977**	33693	**5171**	
25	36734	**4421**	35946	**4603**	35174	**4789**	34420	**4980**	33681	**5174**	5
	36721	**4424**	35933	**4606**	35161	**4793**	34407	**4983**	33669	**5178**	
26	36708	**4427**	35920	**4609**	35149	**4796**	34395	**4986**	33657	**5181**	4
	36694	**4430**	35907	**4612**	35136	**4799**	34382	**4989**	33644	**5184**	
27	36681	**4433**	35894	**4615**	35123	**4802**	34370	**4993**	33632	**5187**	3
	36668	**4436**	35881	**4619**	35111	**4805**	34357	**4996**	33620	**5191**	
28	36655	**4439**	35868	**4622**	35098	**4808**	34345	**4999**	33608	**5194**	2
	36641	**4442**	35855	**4625**	35085	**4811**	34332	**5002**	33596	**5197**	
29	36628	**4445**	35842	**4628**	35073	**4815**	34320	**5005**	33584	**5200**	1
	36615	**4448**	35829	**4631**	35060	**4818**	34308	**5009**	33572	**5204**	
30	36602	**4451**	35816	**4634**	35047	**4821**	34295	**5012**	33559	**5207**	0
′	154° 30′ A	154° 30′ B	154° 00′ A	154° 00′ B	153° 30′ A	153° 30′ B	153° 00′ A	153° 00′ B	152° 30′ A	152° 30′ B	′

TABLE 35

The Ageton Method

Always Take "Z" From Bottom of Table, Except When "K" is Same Name and Greater Than Latitude, in Which Case Take "Z" From Top of Table

′	27° 30′ A	27° 30′ B	28° 00′ A	28° 00′ B	28° 30′ A	28° 30′ B	29° 00′ A	29° 00′ B	29° 30′ A	29° 30′ B	′
0	33559	5207	32839	5406	32134	5610	31443	5818	30766	6030	30
	33547	5210	32827	5410	32122	5614	31431	5822	30755	6034	
1	33535	5214	32815	5413	32110	5617	31420	5825	30744	6038	29
	33523	5217	32803	5417	32099	5620	31409	5829	30733	6041	
2	33511	5220	32792	5420	32087	5624	31397	5832	30721	6045	28
	33499	5224	32780	5423	32076	5627	31386	5836	30710	6048	
3	33487	5227	32768	5426	32064	5631	31375	5839	30699	6052	27
	33475	5230	32756	5430	32052	5634	31363	5843	30688	6055	
4	33462	5233	32744	5433	32041	5638	31352	5846	30677	6059	26
	33450	5237	32732	5437	32029	5641	31340	5850	30666	6062	
5	33438	5240	32720	5440	32018	5645	31329	5853	30655	6066	25
	33426	5243	32709	5443	32006	5648	31318	5857	30643	6070	
6	33414	5247	32697	5447	31994	5651	31306	5860	30632	6073	24
	33402	5250	32685	5450	31983	5655	31295	5864	30621	6077	
7	33390	5253	32673	5454	31971	5658	31284	5867	30610	6080	23
	33378	5257	32661	5457	31960	5662	31272	5871	30599	6084	
8	33366	5260	32649	5460	31948	5665	31261	5874	30588	6088	22
	33354	5263	32638	5464	31936	5669	31250	5878	30577	6091	
9	33342	5266	32625	5467	31925	5672	31238	5881	30566	6095	21
	33330	5270	32614	5470	31913	5675	31227	5885	30555	6098	
10	33318	5273	32602	5474	31902	5679	31216	5888	30544	6102	20
	33306	5277	32590	5477	31890	5682	31204	5892	30532	6106	
11	33293	5280	32579	5481	31879	5686	31193	5895	30521	6109	19
	33281	5283	32567	5484	31867	5689	31182	5899	30510	6113	
12	33269	5287	32555	5487	31856	5693	31170	5902	30499	6116	18
	33257	5290	32543	5491	31844	5696	31159	5906	30488	6120	
13	33245	5293	32532	5494	31833	5700	31148	5909	30477	6124	17
	33233	5296	32520	5498	31821	5703	31137	5913	30466	6127	
14	33221	5300	32508	5501	31809	5707	31125	5917	30455	6131	16
	33209	5303	32496	5504	31798	5710	31114	5920	30444	6134	
15	33197	5306	32484	5508	31786	5714	31103	5924	30433	6138	15
	33185	5310	32473	5511	31775	5717	31091	5927	30422	6142	
16	33173	5313	32461	5515	31763	5720	31080	5931	30411	6145	14
	33161	5316	32449	5518	31752	5724	31069	5934	30400	6149	
17	33149	5320	32438	5521	31740	5727	31058	5938	30389	6153	13
	33137	5323	32426	5525	31729	5731	31046	5941	30378	6156	
18	33125	5326	32414	5528	31717	5734	31035	5945	30367	6160	12
	33113	5330	32402	5532	31706	5738	31024	5948	30356	6163	
19	33101	5333	32391	5535	31694	5741	31013	5952	30345	6167	11
	33089	5336	32379	5538	31683	5745	31001	5955	30334	6171	
20	33077	5340	32367	5542	31672	5748	30990	5959	30322	6174	10
	33065	5343	32355	5545	31660	5752	30979	5963	30311	6178	
21	33054	5346	32344	5549	31648	5755	30968	5966	30300	6181	9
	33042	5350	32332	5552	31637	5759	30956	5970	30289	6185	
22	33030	5353	32320	5555	31626	5762	30945	5973	30278	6189	8
	33018	5356	32309	5559	31614	5766	30934	5977	30267	6192	
23	33006	5360	32297	5562	31603	5769	30923	5980	30256	6196	7
	32994	5363	32285	5566	31591	5773	30912	5984	30245	6200	
24	32982	5366	32274	5569	31580	5776	30900	5988	30235	6203	6
	32970	5370	32262	5572	31569	5780	30889	5991	30224	6207	
25	32958	5373	32250	5576	31557	5783	30878	5995	30213	6210	5
	32946	5376	32239	5579	31546	5787	30867	5998	30202	6214	
26	32934	5380	32227	5583	31534	5790	30856	6002	30191	6218	4
	32922	5383	32215	5586	31523	5794	30844	6005	30180	6221	
27	32910	5386	32204	5590	31511	5797	30833	6009	30169	6225	3
	32898	5390	32192	5593	31500	5801	30822	6012	30158	6229	
28	32887	5393	32180	5596	31488	5804	30811	6016	30147	6232	2
	32875	5396	32169	5600	31477	5808	30800	6020	30136	6236	
29	32863	5400	32157	5603	31466	5811	30788	6023	30125	6240	1
	32851	5403	32145	5607	31454	5815	30777	6027	30114	6243	
30	32839	5406	32134	5610	31443	5818	30766	6030	30103	6247	0
′	152° 00′ A	152° 00′ B	151° 30′ A	151° 30′ B	151° 00′ A	151° 00′ B	150° 30′ A	150° 30′ B	150° 00′ A	150° 00′ B	′

TABLE 35

The Ageton Method

When Meridian Angle is Greater Than 90° Take "K" From Bottom of Table

′	30° 00′ A	30° 00′ B	30° 30′ A	30° 30′ B	31° 00′ A	31° 00′ B	31° 30′ A	31° 30′ B	32° 00′ A	32° 00′ B	′
0	30103	**6247**	29453	**6468**	28816	**6693**	28191	**6923**	27579	**7158**	30
	30092	**6251**	29442	**6472**	28806	**6697**	28181	**6927**	27569	**7162**	
1	30081	**6254**	29432	**6475**	28795	**6701**	28171	**6931**	27559	**7166**	29
	30070	**6258**	29421	**6479**	28785	**6705**	28161	**6935**	27549	**7170**	
2	30059	**6262**	29410	**6483**	28774	**6709**	28150	**6939**	27539	**7174**	28
	30048	**6265**	29399	**6487**	28763	**6712**	28140	**6943**	27528	**7178**	
3	30037	**6269**	29389	**6490**	28753	**6716**	28130	**6947**	27518	**7182**	27
	30026	**6273**	29378	**6494**	28743	**6720**	28119	**6951**	27508	**7186**	
4	30016	**6276**	29367	**6498**	28732	**6724**	28109	**6954**	27498	**7190**	26
	30005	**6280**	29357	**6501**	28722	**6728**	28099	**6958**	27488	**7193**	
5	29994	**6284**	29346	**6505**	28711	**6731**	28089	**6962**	27478	**7197**	25
	29983	**6287**	29335	**6509**	28701	**6735**	28078	**6966**	27468	**7201**	
6	29972	**6291**	29325	**6513**	28690	**6739**	28068	**6970**	27458	**7205**	24
	29961	**6294**	29314	**6516**	28680	**6743**	28058	**6974**	27448	**7209**	
7	29950	**6298**	29303	**6520**	28669	**6747**	28047	**6978**	27438	**7213**	23
	29939	**6302**	29293	**6524**	28659	**6750**	28037	**6982**	27428	**7217**	
8	29928	**6305**	29282	**6528**	28648	**6754**	28027	**6985**	27418	**7221**	22
	29917	**6309**	29271	**6531**	28638	**6758**	28017	**6989**	27408	**7225**	
9	29907	**6313**	29261	**6535**	28627	**6762**	28006	**6993**	27398	**7229**	21
	29896	**6316**	29250	**6539**	28617	**6766**	27996	**6997**	27387	**7233**	
10	29885	**6320**	29239	**6543**	28606	**6770**	27986	**7001**	27377	**7237**	20
	29874	**6324**	29229	**6546**	28596	**6773**	27976	**7005**	27367	**7241**	
11	29863	**6328**	29218	**6550**	28586	**6777**	27965	**7009**	27357	**7245**	19
	29852	**6331**	29207	**6554**	28575	**6781**	27955	**7013**	27347	**7249**	
12	29841	**6335**	29197	**6558**	28565	**6785**	27945	**7017**	27337	**7253**	18
	29831	**6339**	29186	**6561**	28554	**6789**	27935	**7021**	27327	**7257**	
13	29820	**6342**	29175	**6565**	28544	**6793**	27925	**7024**	27317	**7261**	17
	29809	**6346**	29165	**6569**	28533	**6796**	27914	**7028**	27307	**7265**	
14	29798	**6350**	29154	**6573**	28523	**6800**	27904	**7032**	27297	**7269**	16
	29787	**6353**	29144	**6576**	28513	**6804**	27894	**7036**	27287	**7273**	
15	29776	**6357**	29133	**6580**	28502	**6808**	27884	**7040**	27277	**7277**	15
	29766	**6361**	29122	**6584**	28492	**6812**	27874	**7044**	27267	**7281**	
16	29755	**6364**	29112	**6588**	28481	**6815**	27863	**7048**	27257	**7285**	14
	29744	**6368**	29101	**6591**	28471	**6819**	27853	**7052**	27247	**7289**	
17	29733	**6372**	29091	**6595**	28461	**6823**	27843	**7056**	27237	**7293**	13
	29722	**6375**	29080	**6599**	28450	**6827**	27833	**7060**	27227	**7297**	
18	29711	**6379**	29069	**6603**	28440	**6831**	27823	**7064**	27217	**7301**	12
	29701	**6383**	29059	**6606**	28429	**6835**	27812	**7067**	27207	**7305**	
19	29690	**6386**	29048	**6610**	28419	**6839**	27802	**7071**	27197	**7309**	11
	29679	**6390**	29038	**6614**	28409	**6842**	27792	**7075**	27187	**7313**	
20	29668	**6394**	29027	**6618**	28398	**6846**	27782	**7079**	27177	**7317**	10
	29657	**6398**	29016	**6622**	28388	**6850**	27772	**7083**	27167	**7321**	
21	29647	**6401**	29006	**6625**	28378	**6854**	27761	**7087**	27157	**7325**	9
	29636	**6405**	28995	**6629**	28367	**6858**	27751	**7091**	27147	**7329**	
22	29625	**6409**	28985	**6633**	28357	**6862**	27741	**7095**	27137	**7333**	8
	29614	**6412**	28974	**6637**	28346	**6865**	27731	**7099**	27127	**7337**	
23	29604	**6416**	28964	**6640**	28336	**6869**	27721	**7103**	27117	**7341**	7
	29593	**6420**	28953	**6644**	28326	**6873**	27711	**7107**	27107	**7345**	
24	29582	**6423**	28942	**6648**	28315	**6877**	27701	**7111**	27098	**7349**	6
	29571	**6427**	28932	**6652**	28305	**6881**	27690	**7115**	27088	**7353**	
25	29560	**6431**	28921	**6655**	28295	**6885**	27680	**7118**	27078	**7357**	5
	29550	**6435**	28911	**6659**	28284	**6889**	27670	**7122**	27068	**7361**	
26	29539	**6438**	28900	**6663**	28274	**6893**	27660	**7126**	27058	**7365**	4
	29528	**6442**	28890	**6667**	28264	**6896**	27650	**7130**	27048	**7369**	
27	29517	**6446**	28879	**6671**	28253	**6900**	27640	**7134**	27038	**7373**	3
	29507	**6449**	28869	**6674**	28243	**6904**	27630	**7138**	27028	**7377**	
28	29496	**6453**	28858	**6678**	28233	**6908**	27619	**7142**	27018	**7381**	2
	29485	**6457**	28848	**6682**	28222	**6912**	27609	**7146**	27008	**7385**	
29	29475	**6461**	28837	**6686**	28212	**6916**	27599	**7150**	26998	**7389**	1
	29464	**6464**	28827	**6690**	28202	**6920**	27589	**7154**	26988	**7393**	
30	29453	**6468**	28816	**6693**	28191	**6923**	27579	**7158**	26978	**7397**	0
′	149° 30′ A	149° 30′ B	149° 00′ A	149° 00′ B	148° 30′ A	148° 30′ B	148° 00′ A	148° 00′ B	147° 30′ A	147° 30′ B	′

TABLE 35

The Ageton Method

Always Take "Z" From Bottom of Table, Except When "K" is Same Name and Greater Than Latitude, in Which Case Take "Z" From Top of Table

′	32° 30′ A	32° 30′ B	33° 00′ A	33° 00′ B	33° 30′ A	33° 30′ B	34° 00′ A	34° 00′ B	34° 30′ A	34° 30′ B	′
0	26978	7397	26389	7641	25811	7889	25244	8143	24687	8401	30
	26968	7401	26379	7645	25801	7893	25235	8147	24678	8405	
1	26958	7405	26370	7649	25792	7898	25225	8151	24669	8409	29
	26949	7409	26360	7653	25782	7902	25216	8155	24660	8414	
2	26939	7413	26350	7657	25773	7906	25206	8160	24650	8418	28
	26929	7417	26340	7661	25763	7910	25197	8164	24641	8422	
3	26919	7421	26331	7665	25754	7914	25188	8168	24632	8427	27
	26909	7425	26321	7670	25744	7919	25178	8172	24623	8431	
4	26899	7429	26311	7674	25735	7923	25169	8177	24614	8435	26
	26889	7433	26302	7678	25725	7927	25160	8181	24605	8440	
5	26879	7437	26292	7682	25716	7931	25150	8185	24595	8444	25
	26869	7441	26282	7686	25706	7935	25141	8189	24586	8448	
6	26860	7445	26273	7690	25697	7940	25132	8194	24577	8453	24
	26850	7449	26263	7694	25687	7944	25122	8198	24568	8457	
7	26840	7453	26253	7698	25678	7948	25113	8202	24559	8461	23
	26830	7458	26244	7702	25668	7952	25104	8207	24550	8466	
8	26820	7462	26234	7707	25659	7956	25094	8211	24540	8470	22
	26810	7466	26224	7711	25649	7961	25085	8215	24531	8475	
9	26800	7470	26214	7715	25640	7965	25076	8219	24522	8479	21
	26790	7474	26205	7719	25630	7969	25066	8224	24513	8483	
10	26781	7478	26195	7723	25621	7973	25057	8228	24504	8488	20
	26771	7482	26185	7727	25611	7977	25048	8232	24495	8492	
11	26761	7486	26176	7731	25602	7982	25038	8237	24486	8496	19
	26751	7490	26166	7736	25592	7986	25029	8241	24477	8501	
12	26741	7494	26157	7740	25583	7990	25020	8245	24467	8505	18
	26731	7498	26147	7744	25573	7994	25011	8249	24458	8510	
13	26722	7502	26137	7748	25564	7998	25001	8254	24449	8514	17
	26712	7506	26128	7752	25554	8003	24992	8258	24440	8518	
14	26702	7510	26118	7756	25545	8007	24983	8262	24431	8523	16
	26692	7514	26108	7760	25536	8011	24973	8267	24422	8527	
15	26682	7518	26099	7764	25526	8015	24964	8271	24413	8531	15
	26672	7522	26089	7769	25517	8020	24955	8275	24404	8536	
16	26663	7526	26079	7773	25507	8024	24946	8280	24395	8540	14
	26653	7531	26070	7777	25498	8028	24936	8284	24385	8545	
17	26643	7535	26060	7781	25488	8032	24927	8288	24376	8549	13
	26633	7539	26051	7785	25479	8037	24918	8292	24367	8553	
18	26623	7543	26041	7789	25469	8041	24909	8297	24358	8558	12
	26614	7547	26031	7793	25460	8045	24899	8301	24349	8562	
19	26604	7551	26022	7798	25451	8049	24890	8305	24340	8567	11
	26594	7555	26012	7802	25441	8053	24881	8310	24331	8571	
20	26584	7559	26002	7806	25432	8058	24872	8314	24322	8575	10
	26574	7563	25993	7810	25422	8062	24862	8318	24313	8580	
21	26565	7567	25983	7814	25413	8066	24853	8323	24304	8584	9
	26555	7571	25974	7818	25403	8070	24844	8327	24295	8589	
22	26545	7575	25964	7823	25394	8075	24835	8331	24286	8593	8
	26535	7579	25954	7827	25385	8079	24825	8336	24276	8597	
23	26526	7584	25945	7831	25375	8083	24816	8340	24267	8602	7
	26516	7588	25935	7835	25366	8087	24807	8344	24258	8606	
24	26506	7592	25926	7839	25356	8091	24798	8349	24249	8611	6
	26496	7596	25916	7843	25347	8096	24788	8353	24240	8615	
25	26486	7600	25907	7848	25338	8100	24779	8357	24231	8619	5
	26477	7604	25897	7852	25328	8104	24770	8362	24222	8624	
26	26467	7608	25887	7856	25319	8108	24761	8366	24213	8628	4
	26457	7612	25878	7860	25309	8113	24752	8370	24204	8633	
27	26447	7616	25868	7864	25300	8117	24742	8375	24195	8637	3
	26438	7620	25859	7868	25291	8121	24733	8379	24186	8641	
28	26428	7625	25849	7873	25281	8125	24724	8383	24177	8646	2
	26418	7629	25840	7877	25272	8130	24715	8388	24168	8650	
29	26409	7633	25830	7881	25263	8134	24706	8392	24159	8655	1
	26399	7637	25821	7885	25253	8138	24696	8396	24150	8659	
30	26389	7641	25811	7889	25244	8143	24687	8401	24141	8663	0
	A	B	A	B	A	B	A	B	A	B	
′	147° 00′		146° 30′		146° 00′		145° 30′		145° 00′		′

TABLE 35

The Ageton Method

When Meridian Angle is Greater Than 90° Take "K" From Bottom of Table

′	35° 00′		35° 30′		36° 00′		36° 30′		37° 00′		′
	A	B	A	B	A	B	A	B	A	B	
0	24141	8663	23605	8931	23078	9204	22561	9482	22054	9765	30
	24132	8668	23596	8936	23069	9209	22553	9487	22045	9770	
1	24123	8672	23587	8940	23061	9213	22544	9492	22037	9775	29
	24114	8677	23578	8945	23052	9218	22536	9496	22029	9779	
2	24105	8681	23569	8949	23043	9223	22527	9501	22020	9784	28
	24096	8686	23560	8954	23035	9227	22519	9505	22012	9789	
3	24087	8690	23551	8958	23026	9232	22510	9510	22003	9794	27
	24078	8694	23543	8963	23017	9236	22501	9515	21995	9798	
4	24069	8699	23534	8967	23009	9241	22493	9520	21987	9803	26
	24060	8703	23525	8972	23000	9246	22484	9524	21978	9808	
5	24051	8708	23516	8976	22991	9250	22476	9529	21970	9813	25
	24042	8712	23507	8981	22983	9255	22467	9534	21962	9818	
6	24033	8717	23498	8986	22974	9259	22459	9538	21953	9822	24
	24024	8721	23490	8990	22965	9264	22450	9543	21945	9827	
7	24015	8726	23481	8995	22957	9269	22442	9548	21937	9832	23
	24006	8730	23472	8999	22948	9273	22433	9552	21928	9837	
8	23997	8734	23463	9004	22939	9278	22425	9557	21920	9841	22
	23988	8739	23454	9008	22931	9282	22416	9562	21912	9846	
9	23979	8743	23446	9013	22922	9287	22408	9566	21903	9851	21
	23970	8748	23437	9017	22913	9292	22399	9571	21895	9856	
10	23961	8752	23428	9022	22905	9296	22391	9576	21887	9861	20
	23952	8757	23419	9026	22896	9301	22382	9581	21878	9865	
11	23943	8761	23410	9031	22887	9305	22374	9585	21870	9870	19
	23934	8766	23402	9035	22879	9310	22366	9590	21862	9875	
12	23925	8770	23393	9040	22870	9315	22357	9595	21853	9880	18
	23916	8775	23384	9044	22862	9319	22349	9599	21845	9885	
13	23907	8779	23375	9049	22853	9324	22340	9604	21837	9889	17
	23898	8783	23366	9054	22844	9329	22332	9609	21828	9894	
14	23889	8788	23358	9058	22836	9333	22323	9614	21820	9899	16
	23880	8792	23349	9063	22827	9338	22315	9618	21812	9904	
15	23871	8797	23340	9067	22818	9342	22306	9623	21803	9909	15
	23863	8801	23331	9072	22810	9347	22298	9628	21795	9913	
16	23854	8806	23323	9076	22801	9352	22289	9632	21787	9918	14
	23845	8810	23314	9081	22793	9356	22281	9637	21778	9923	
17	23836	8815	23305	9085	22784	9361	22272	9642	21770	9928	13
	23827	8819	23296	9090	22775	9366	22264	9647	21762	9933	
18	23818	8824	23288	9094	22767	9370	22256	9651	21754	9937	12
	23809	8828	23279	9099	22758	9375	22247	9656	21745	9942	
19	23800	8833	23270	9104	22750	9380	22239	9661	21737	9947	11
	23791	8837	23261	9108	22741	9384	22230	9665	21729	9952	
20	23782	8842	23252	9113	22732	9389	22222	9670	21720	9957	10
	23773	8846	23244	9117	22724	9394	22213	9675	21712	9962	
21	23764	8850	23235	9122	22715	9398	22205	9680	21704	9966	9
	23755	8855	23226	9126	22707	9403	22197	9684	21696	9971	
22	23747	8859	23218	9131	22698	9407	22188	9689	21687	9976	8
	23738	8864	23209	9136	22690	9412	22180	9694	21679	9981	
23	23729	8868	23200	9140	22681	9417	22171	9699	21671	9986	7
	23720	8873	23191	9145	22672	9421	22163	9703	21662	9990	
24	23711	8877	23183	9149	22664	9426	22154	9708	21654	9995	6
	23702	8882	23174	9154	22655	9431	22146	9713	21646	10000	
25	23693	8886	23165	9158	22647	9435	22138	9718	21638	10005	5
	23684	8891	23156	9163	22638	9440	22129	9722	21629	10010	
26	23675	8895	23148	9168	22630	9445	22121	9727	21621	10015	4
	23667	8900	23139	9172	22621	9449	22112	9732	21613	10019	
27	23658	8904	23130	9177	22612	9454	22104	9737	21605	10024	3
	23649	8909	23122	9181	22604	9459	22096	9741	21596	10029	
28	23640	8913	23113	9186	22595	9463	22087	9746	21588	10034	2
	23631	8918	23104	9190	22587	9468	22079	9751	21580	10039	
29	23622	8922	23095	9195	22578	9473	22070	9756	21572	10044	1
	23613	8927	23087	9200	22570	9477	22062	9760	21563	10049	
30	23605	8931	23078	9204	22561	9482	22054	9765	21555	10053	0
	A	B	A	B	A	B	A	B	A	B	
′	144° 30′		144° 00′		143° 30′		143° 00′		142° 30′		′

TABLE 35

The Ageton Method

Always Take "Z" From Bottom of Table, Except When "K" is Same Name and Greater Than Latitude, in Which Case Take "Z" From Top of Table

′	37° 30′		38° 00′		38° 30′		39° 00′		39° 30′		′
	A	B	A	B	A	B	A	B	A	B	
0	21555	10053	21066	10347	20585	10646	20113	10950	19649	11259	30
	21547	10058	21058	10352	20577	10651	20105	10955	19641	11265	
1	21539	10063	21050	10357	20569	10656	20097	10960	19634	11270	29
	21531	10068	21042	10362	20561	10661	20089	10965	19626	11275	
2	21522	10073	21033	10367	20553	10666	20082	10970	19618	11280	28
	21514	10078	21025	10372	20545	10671	20074	10975	19611	11285	
3	21506	10082	21017	10376	20537	10676	20066	10980	19603	11291	27
	21498	10087	21009	10381	20529	10681	20058	10986	19595	11296	
4	21489	10092	21001	10386	20522	10686	20050	10991	19588	11301	26
	21481	10097	20993	10391	20514	10691	20043	10996	19580	11306	
5	21473	10102	20985	10396	20506	10696	20035	11001	19572	11311	25
	21465	10107	20977	10401	20498	10701	20027	11006	19565	11317	
6	21457	10112	20969	10406	20490	10706	20019	11011	19557	11322	24
	21448	10116	20961	10411	20482	10711	20012	11016	19549	11327	
7	21440	10121	20953	10416	20474	10716	20004	11021	19541	11332	23
	21432	10126	20945	10421	20466	10721	19996	11027	19534	11338	
8	21424	10131	20937	10426	20458	10726	19988	11032	19527	11343	22
	21416	10136	20929	10431	20450	10731	19980	11037	19519	11348	
9	21407	10141	20921	10436	20442	10736	19973	11042	19511	11353	21
	21399	10146	20913	10441	20435	10741	19965	11047	19504	11359	
10	21391	10151	20905	10446	20427	10746	19957	11052	19496	11364	20
	21383	10155	20897	10451	20419	10751	19949	11057	19488	11369	
11	21375	10160	20888	10456	20411	10756	19942	11063	19481	11374	19
	21367	10165	20880	10461	20403	10761	19934	11068	19473	11380	
12	21358	10170	20872	10466	20395	10767	19926	11073	19466	11385	18
	21350	10175	20864	10471	20387	10772	19919	11078	19458	11390	
13	21342	10180	20856	10476	20379	10777	19911	11083	19450	11395	17
	21334	10185	20848	10481	20371	10782	19903	11088	19443	11400	
14	21326	10190	20840	10486	20364	10787	19895	11094	19435	11406	16
	21318	10195	20832	10491	20356	10792	19888	11099	19428	11411	
15	21309	10199	20824	10496	20348	10797	19880	11104	19420	11416	15
	21301	10204	20816	10500	20340	10802	19872	11109	19412	11422	
16	21293	10209	20808	10505	20332	10807	19864	11114	19405	11427	14
	21285	10214	20800	10510	20324	10812	19857	11119	19397	11432	
17	21277	10219	20792	10515	20316	10817	19849	11124	19390	11437	13
	21269	10224	20784	10520	20309	10822	19841	11130	19382	11443	
18	21260	10229	20776	10525	20301	10827	19834	11135	19375	11448	12
	21252	10234	20768	10530	20293	10832	19826	11140	19367	11453	
19	21244	10239	20760	10535	20285	10838	19818	11145	19359	11458	11
	21236	10243	20752	10540	20277	10843	19810	11150	19352	11464	
20	21228	10248	20744	10545	20269	10848	19803	11156	19344	11469	10
	21220	10253	20736	10550	20261	10853	19795	11161	19337	11474	
21	21212	10258	20728	10555	20254	10858	19787	11166	19329	11479	9
	21204	10263	20720	10560	20246	10863	19779	11171	19321	11485	
22	21195	10268	20712	10565	20238	10868	19772	11176	19314	11490	8
	21187	10273	20704	10570	20230	10873	19764	11181	19306	11495	
23	21179	10278	20696	10575	20222	10878	19756	11187	19299	11501	7
	21171	10283	20688	10580	20214	10883	19749	11192	19291	11506	
24	21163	10288	20680	10585	20207	10888	19741	11197	19284	11511	6
	21155	10293	20672	10590	20199	10894	19733	11202	19276	11516	
25	21147	10298	20665	10595	20191	10899	19726	11207	19269	11522	5
	21139	10302	20657	10600	20183	10904	19718	11213	19261	11527	
26	21131	10307	20649	10605	20175	10909	19710	11218	19253	11532	4
	21122	10312	20641	10610	20167	10914	19703	11223	19246	11537	
27	21114	10317	20633	10615	20160	10919	19695	11228	19238	11543	3
	21106	10322	20625	10620	20152	10924	19687	11233	19231	11548	
28	21098	10327	20617	10625	20144	10929	19680	11239	19223	11553	2
	21090	10332	20609	10630	20136	10934	19672	11244	19216	11559	
29	21082	10337	20601	10635	20128	10939	19664	11249	19208	11564	1
	21074	10342	20593	10640	20121	10945	19657	11254	19201	11569	
30	21066	10347	20585	10646	20113	10950	19649	11259	19193	11575	0
	A	B	A	B	A	B	A	B	A	B	
′	142° 00′		141° 30′		141° 00′		140° 30′		140° 00′		′

TABLE 35

The Ageton Method

When Meridian Angle is Greater Than 90° Take "K" From Bottom of Table

′	40° 00′		40° 30′		41° 00′		41° 30′		42° 00′		′
	A	B	A	B	A	B	A	B	A	B	
0	19193	11575	18746	11895	18306	12222	17873	12554	17449	12893	30
	19186	11580	18738	11901	18298	12228	17866	12560	17442	12898	
1	19178	11585	18731	11906	18291	12233	17859	12566	17435	12904	29
	19171	11590	18723	11912	18284	12238	17852	12571	17428	12910	
2	19163	11596	18716	11917	18277	12244	17845	12577	17421	12915	28
	19156	11601	18709	11922	18269	12249	17838	12582	17414	12921	
3	19148	11606	18701	11928	18262	12255	17831	12588	17407	12927	27
	19141	11612	18694	11933	18255	12260	17824	12593	17400	12932	
4	19133	11617	18686	11939	18248	12266	17816	12599	17393	12938	26
	19126	11622	18679	11944	18240	12271	17809	12605	17386	12944	
5	19118	11628	18672	11949	18233	12277	17802	12610	17379	12950	25
	19111	11633	18664	11955	18226	12282	17795	12616	17372	12955	
6	19103	11638	18657	11960	18219	12288	17788	12622	17365	12961	24
	19096	11644	18650	11966	18211	12293	17781	12627	17358	12967	
7	19088	11649	18642	11971	18204	12299	17774	12633	17351	12972	23
	19081	11654	18635	11977	18197	12305	17767	12638	17344	12978	
8	19073	11660	18627	11982	18190	12310	17760	12644	17337	12984	22
	19066	11665	18620	11987	18182	12316	17752	12650	17330	12990	
9	19058	11670	18613	11993	18175	12321	17745	12655	17323	12995	21
	19051	11676	18605	11998	18168	12327	17738	12661	17316	13001	
10	19043	11681	18598	12004	18161	12332	17731	12667	17309	13007	20
	19036	11686	18591	12009	18154	12338	17724	12672	17302	13012	
11	19028	11692	18583	12014	18146	12343	17717	12678	17295	13018	19
	19021	11697	18576	12020	18139	12349	17710	12683	17288	13024	
12	19013	11702	18569	12025	18132	12354	17703	12689	17281	13030	18
	19006	11708	18561	12031	18125	12360	17696	12695	17274	13035	
13	18998	11713	18554	12036	18117	12365	17689	12700	17267	13041	17
	18991	11718	18547	12042	18110	12371	17681	12706	17260	13047	
14	18983	11724	18539	12047	18103	12376	17674	12711	17253	13053	16
	18976	11729	18532	12053	18096	12382	17667	12717	17246	13058	
15	18968	11734	18525	12058	18089	12387	17660	12723	17239	13064	15
	18961	11740	18517	12063	18081	12393	17653	12728	17232	13070	
16	18953	11745	18510	12069	18074	12398	17646	12734	17225	13075	14
	18946	11750	18503	12074	18067	12404	17639	12740	17218	13081	
17	18939	11756	18495	12080	18060	12410	17632	12745	17212	13087	13
	18931	11761	18488	12085	18053	12415	17625	12751	17205	13093	
18	18924	11766	18481	12091	18045	12421	17618	12757	17198	13098	12
	18916	11772	18473	12096	18038	12426	17611	12762	17191	13104	
19	18909	11777	18466	12102	18031	12432	17604	12768	17184	13110	11
	18901	11782	18459	12107	18024	12437	17597	12774	17177	13116	
20	18894	11788	18451	12112	18017	12443	17590	12779	17170	13121	10
	18886	11793	18444	12118	18010	12448	17583	12785	17163	13127	
21	18879	11799	18437	12123	18002	12454	17575	12790	17156	13133	9
	18872	11804	18429	12129	17995	12460	17568	12796	17149	13139	
22	18864	11809	18422	12134	17988	12465	17561	12802	17142	13144	8
	18857	11815	18415	12140	17981	12471	17554	12807	17135	13150	
23	18849	11820	18408	12145	17974	12476	17547	12813	17128	13156	7
	18842	11825	18400	12151	17966	12482	17540	12819	17121	13162	
24	18834	11831	18393	12156	17959	12487	17533	12824	17114	13168	6
	18827	11836	18386	12162	17952	12493	17526	12830	17108	13173	
25	18820	11842	18378	12167	17945	12499	17519	12836	17101	13179	5
	18812	11847	18371	12173	17938	12504	17512	12841	17094	13185	
26	18805	11852	18364	12178	17931	12510	17505	12847	17087	13191	4
	18797	11858	18357	12184	17924	12515	17498	12853	17080	13196	
27	18790	11863	18349	12189	17916	12521	17491	12859	17073	13202	3
	18783	11868	18342	12195	17909	12526	17484	12864	17066	13208	
28	18775	11874	18335	12200	17902	12532	17477	12870	17059	13214	2
	18768	11879	18327	12205	17895	12538	17470	12876	17052	13220	
29	18760	11885	18320	12211	17888	12543	17463	12881	17046	13225	1
	18753	11890	18313	12216	17881	12549	17456	12887	17039	13231	
30	18746	11895	18306	12222	17873	12554	17449	12893	17032	13237	0
	A	B	A	B	A	B	A	B	A	B	
′	139° 30′		139° 00′		138° 30′		138° 00′		137° 30′		′

TABLE 35

The Ageton Method

Always Take "Z" From Bottom of Table, Except When "K" is Same Name and Greater Than Latitude, in Which Case Take "Z" From Top of Table

′	42° 30′		43° 00′		43° 30′		44° 00′		44° 30′		′
′	A	B	A	B	A	B	A	B	A	B	′
0	17032	13237	16622	13587	16219	13944	15823	14307	15434	14676	30
	17025	13243	16615	13593	16212	13950	15816	14313	15427	14682	
1	17018	13248	16608	13599	16205	13956	15810	14319	15421	14688	29
	17011	13254	16601	13605	16199	13962	15803	14325	15414	14694	
2	17004	13260	16595	13611	16192	13968	15797	14331	15408	14701	28
	16997	13266	16588	13617	16186	13974	15790	14337	15402	14707	
3	16990	13272	16581	13623	16179	13980	15784	14343	15395	14713	27
	16983	13277	16574	13628	16172	13986	15777	14349	15389	14719	
4	16977	13283	16567	13634	16166	13992	15771	14355	15382	14726	26
	16970	13289	16561	13640	16159	13998	15764	14362	15376	14732	
5	16963	13295	16554	13646	16152	14004	15758	14368	15370	14738	25
	16956	13301	16547	13652	16146	14010	15751	14374	15363	14744	
6	16949	13306	16540	13658	16139	14016	15744	14380	15357	14750	24
	16942	13312	16534	13664	16132	14022	15738	14386	15350	14757	
7	16935	13318	16527	13670	16126	14028	15731	14392	15344	14763	23
	16928	13324	16520	13676	16119	14034	15725	14398	15338	14769	
8	16922	13330	16513	13682	16112	14040	15718	14404	15331	14775	22
	16915	13336	16507	13688	16106	14046	15712	14411	15325	14782	
9	16908	13341	16500	13694	16099	14052	15705	14417	15318	14788	21
	16901	13347	16493	13700	16093	14058	15699	14423	15312	14794	
10	16894	13353	16487	13705	16086	14064	15692	14429	15306	14800	20
	16887	13359	16480	13711	16079	14070	15686	14435	15299	14807	
11	16880	13365	16473	13717	16073	14076	15679	14441	15293	14813	19
	16874	13370	16466	13723	16066	14082	15673	14447	15286	14819	
12	16867	13376	16460	13729	16060	14088	15666	14453	15280	14825	18
	16860	13382	16453	13735	16053	14094	15660	14460	15274	14831	
13	16853	13388	16446	13741	16046	14100	15653	14466	15267	14838	17
	16846	13394	16439	13747	16040	14106	15647	14472	15261	14844	
14	16839	13400	16433	13753	16033	14112	15640	14478	15255	14850	16
	16833	13405	16426	13759	16027	14118	15634	14484	15248	14857	
15	16826	13411	16419	13765	16020	14124	15627	14490	15242	14863	15
	16819	13417	16413	13771	16013	14130	15621	14496	15235	14869	
16	16812	13423	16406	13777	16007	14136	15614	14503	15229	14875	14
	16805	13429	16399	13783	16000	14142	15608	14509	15223	14882	
17	16798	13435	16392	13789	15994	14149	15602	14515	15216	14888	13
	16792	13440	16386	13794	15987	14155	15595	14521	15210	14894	
18	16785	13446	16379	13800	15980	14161	15589	14527	15204	14900	12
	16778	13452	16372	13806	15974	14167	15582	14533	15197	14907	
19	16771	13458	16366	13812	15967	14173	15576	14540	15191	14913	11
	16764	13464	16359	13818	15961	14179	15569	14546	15184	14919	
20	16757	13470	16352	13824	15954	14185	15563	14552	15178	14925	10
	16751	13476	16346	13830	15947	14191	15556	14558	15172	14932	
21	16744	13481	16339	13836	15941	14197	15550	14564	15165	14938	9
	16737	13487	16332	13842	15934	14203	15543	14570	15159	14944	
22	16730	13493	16325	13848	15928	14209	15537	14577	15153	14951	8
	16723	13499	16319	13854	15921	14215	15530	14583	15146	14957	
23	16717	13505	16312	13860	15915	14221	15524	14589	15140	14963	7
	16710	13511	16305	13866	15908	14227	15517	14595	15134	14969	
24	16703	13517	16299	13872	15901	14233	15511	14601	15127	14976	6
	16696	13523	16292	13878	15895	14240	15505	14608	15121	14982	
25	16689	13528	16285	13884	15888	14246	15498	14614	15115	14988	5
	16683	13534	16279	13890	15882	14252	15492	14620	15108	14995	
26	16676	13540	16272	13896	15875	14258	15485	14626	15102	15001	4
	16669	13546	16265	13902	15869	14264	15479	14632	15096	15007	
27	16662	13552	16259	13908	15862	14270	15472	14639	15089	15014	3
	16656	13558	16252	13914	15856	14276	15466	14645	15083	15020	
28	16649	13564	16245	13920	15849	14282	15459	14651	15077	15026	2
	16642	13570	16239	13926	15842	14288	15453	14657	15070	15033	
29	16635	13575	16232	13932	15836	14294	15447	14663	15064	15039	1
	16628	13581	16225	13938	15829	14300	15440	14670	15058	15045	
30	16622	13587	16219	13944	15823	14307	15434	14676	15051	15051	0
	A	B	A	B	A	B	A	B	A	B	
′	137° 00′		136° 30′		136° 00′		135° 30′		135° 00′		′

TABLE 35

The Ageton Method

When Meridian Angle is Greater Than 90° Take "K" From Bottom of Table

′	45° 00′		45° 30′		46° 00′		46° 30′		47° 00′		′
	A	B	A	B	A	B	A	B	A	B	
0	15051	15051	14676	15434	14307	15823	13944	16219	13587	16622	30
	15045	15058	14670	15440	14300	15829	13938	16225	13581	16628	
1	15039	15064	14663	15447	14294	15836	13932	16232	13575	16635	29
	15033	15070	14657	15453	14288	15842	13926	16239	13570	16642	
2	15026	15077	14651	15459	14282	15849	13920	16245	13564	16649	28
	15020	15083	14645	15466	14276	15856	13914	16252	13558	16656	
3	15014	15089	14639	15472	14270	15862	13908	16259	13552	16662	27
	15007	15096	14632	15479	14264	15869	13902	16265	13546	16669	
4	15001	15102	14626	15485	14258	15875	13896	16272	13540	16676	26
	14995	15108	14620	15492	14252	15882	13890	16279	13534	16683	
5	14988	15115	14614	15498	14246	15888	13884	16285	13528	16689	25
	14982	15121	14608	15505	14240	15895	13878	16292	13523	16696	
6	14976	15127	14601	15511	14233	15901	13872	16299	13517	16703	24
	14969	15134	14595	15517	14227	15908	13866	16305	13511	16710	
7	14963	15140	14589	15524	14221	15915	13860	16312	13505	16717	23
	14957	15146	14583	15530	14215	15921	13854	16319	13499	16723	
8	14951	15153	14577	15537	14209	15928	13848	16325	13493	16730	22
	14944	15159	14570	15543	14203	15934	13842	16332	13487	16737	
9	14938	15165	14564	15550	14197	15941	13836	16339	13481	16744	21
	14932	15172	14558	15556	14191	15947	13830	16346	13476	16751	
10	14925	15178	14552	15563	14185	15954	13824	16352	13470	16757	20
	14919	15184	14546	15569	14179	15961	13818	16359	13464	16764	
11	14913	15191	14540	15576	14173	15967	13812	16366	13458	16771	19
	14907	15197	14533	15582	14167	15974	13806	16372	13452	16778	
12	14900	15204	14527	15589	14161	15980	13800	16379	13446	16785	18
	14894	15210	14521	15595	14155	15987	13794	16386	13440	16792	
13	14888	15216	14515	15602	14149	15994	13788	16392	13435	16798	17
	14882	15223	14509	15608	14142	16000	13783	16399	13429	16805	
14	14875	15229	14503	15614	14136	16007	13777	16406	13423	16812	16
	14869	15235	14496	15621	14130	16013	13771	16413	13417	16819	
15	14863	15242	14490	15627	14124	16020	13765	16419	13411	16826	15
	14857	15248	14484	15634	14118	16027	13759	16426	13405	16833	
16	14850	15255	14478	15640	14112	16033	13753	16433	13400	16839	14
	14844	15261	14472	15647	14106	16040	13747	16439	13394	16846	
17	14838	15267	14466	15653	14100	16046	13741	16446	13388	16853	13
	14831	15274	14460	15660	14094	16053	13735	16453	13382	16860	
18	14825	15280	14453	15666	14088	16060	13729	16460	13376	16867	12
	14819	15286	14447	15673	14082	16066	13723	16466	13370	16874	
19	14813	15293	14441	15679	14076	16073	13717	16473	13365	16880	11
	14807	15299	14435	15686	14070	16079	13711	16480	13359	16887	
20	14800	15306	14429	15692	14064	16086	13705	16487	13353	16894	10
	14794	15312	14423	15699	14058	16093	13699	16493	13347	16901	
21	14788	15318	14417	15705	14052	16099	13694	16500	13341	16908	9
	14782	15325	14411	15712	14046	16105	13688	16507	13336	16915	
22	14775	15331	14404	15718	14040	16112	13682	16513	13330	16922	8
	14769	15338	14398	15725	14034	16119	13676	16520	13324	16928	
23	14763	15344	14392	15731	14028	16126	13670	16527	13318	16935	7
	14757	15350	14386	15738	14022	16132	13664	16534	13312	16942	
24	14750	15357	14380	15744	14016	16139	13658	16540	13306	16949	6
	14744	15363	14374	15751	14010	16146	13652	16547	13301	16956	
25	14738	15370	14368	15758	14004	16152	13646	16554	13295	16963	5
	14732	15376	14362	15764	13998	16159	13640	16561	13289	16970	
26	14725	15382	14355	15771	13992	16166	13634	16567	13283	16977	4
	14719	15389	14349	15777	13986	16172	13628	16574	13277	16983	
27	14713	15395	14343	15784	13980	16179	13623	16581	13272	16990	3
	14707	15402	14337	15790	13974	16185	13617	16588	13266	16997	
28	14701	15408	14331	15797	13968	16192	13611	16595	13260	17004	2
	14694	15414	14325	15803	13962	16199	13605	16601	13254	17011	
29	14688	15421	14319	15810	13956	16205	13599	16608	13248	17018	1
	14682	15427	14313	15816	13950	16212	13593	16615	13243	17025	
30	14676	15434	14307	15823	13944	16219	13587	16622	13237	17032	0
	A	B	A	B	A	B	A	B	A	B	
′	134° 30′		134° 00′		133° 30′		133° 00′		132° 30′		′

TABLE 35

The Ageton Method

Always Take "Z" From Bottom of Table, Except When "K" is Same Name and Greater Than Latitude, in Which Case Take "Z" From Top of Table

′	47° 30′		48° 00′		48° 30′		49° 00′		49° 30′		′
	A	B	A	B	A	B	A	B	A	B	
0	13237	17032	12893	17449	12554	17873	12222	18306	11895	18746	30
	13231	17039	12887	17456	12549	17881	12216	18313	11890	18753	
1	13225	17045	12881	17463	12543	17888	12211	18320	11885	18760	29
	13220	17052	12876	17470	12538	17895	12205	18327	11879	18768	
2	13214	17059	12870	17477	12532	17902	12200	18335	11874	18775	28
	13208	17066	12864	17484	12526	17909	12195	18342	11868	18783	
3	13202	17073	12859	17491	12521	17916	12189	18349	11863	18790	27
	13196	17080	12853	17498	12515	17924	12184	18357	11858	18797	
4	13191	17087	12847	17505	12510	17931	12178	18364	11852	18805	26
	13185	17094	12841	17512	12504	17938	12173	18371	11847	18812	
5	13179	17101	12836	17519	12499	17945	12167	18378	11842	18820	25
	13173	17108	12830	17526	12493	17952	12162	18386	11836	18827	
6	13168	17114	12824	17533	12487	17959	12156	18393	11831	18834	24
	13162	17121	12819	17540	12482	17966	12151	18400	11825	18842	
7	13156	17128	12813	17547	12476	17974	12145	18408	11820	18849	23
	13150	17135	12807	17554	12471	17981	12140	18415	11815	18857	
8	13144	17142	12802	17561	12465	17988	12134	18422	11809	18864	22
	13139	17149	12796	17568	12460	17995	12129	18429	11804	18872	
9	13133	17156	12790	17576	12454	18002	12123	18437	11799	18879	21
	13127	17163	12785	17583	12448	18010	12118	18444	11793	18886	
10	13121	17170	12779	17590	12443	18017	12112	18451	11788	18894	20
	13116	17177	12774	17597	12437	18024	12107	18459	11782	18901	
11	13110	17184	12768	17604	12432	18031	12102	18466	11777	18909	19
	13104	17191	12762	17611	12426	18038	12096	18473	11772	18916	
12	13098	17198	12757	17618	12421	18045	12091	18481	11766	18924	18
	13093	17205	12751	17625	12415	18053	12085	18488	11761	18931	
13	13087	17212	12745	17632	12410	18060	12080	18495	11756	18939	17
	13081	17218	12740	17639	12404	18067	12074	18503	11750	18946	
14	13075	17225	12734	17646	12398	18074	12069	18510	11745	18953	16
	13070	17232	12728	17653	12393	18081	12063	18517	11740	18961	
15	13064	17239	12723	17660	12387	18089	12058	18525	11734	18968	15
	13058	17246	12717	17667	12382	18096	12053	18532	11729	18976	
16	13053	17253	12711	17674	12376	18103	12047	18539	11724	18983	14
	13047	17260	12706	17681	12371	18110	12042	18547	11718	18991	
17	13041	17267	12700	17689	12365	18117	12036	18554	11713	18998	13
	13035	17274	12695	17696	12360	18125	12031	18561	11708	19006	
18	13030	17281	12689	17703	12354	18132	12025	18569	11702	19013	12
	13024	17288	12683	17710	12349	18139	12020	18576	11697	19021	
19	13018	17295	12678	17717	12343	18146	12014	18583	11692	19028	11
	13012	17302	12672	17724	12338	18154	12009	18591	11686	19036	
20	13007	17309	12666	17731	12332	18161	12004	18598	11681	19043	10
	13001	17316	12661	17738	12327	18168	11998	18605	11676	19051	
21	12995	17323	12655	17745	12321	18175	11993	18613	11670	19058	9
	12990	17330	12650	17752	12316	18182	11987	18620	11665	19066	
22	12984	17337	12644	17760	12310	18190	11982	18627	11660	19073	8
	12978	17344	12638	17767	12305	18197	11976	18635	11654	19081	
23	12972	17351	12633	17774	12299	18204	11971	18642	11649	19088	7
	12967	17358	12627	17781	12293	18211	11966	18650	11644	19096	
24	12961	17365	12622	17788	12288	18219	11960	18657	11638	19103	6
	12955	17372	12616	17795	12282	18226	11955	18664	11633	19111	
25	12950	17379	12610	17802	12277	18233	11949	18672	11628	19118	5
	12944	17386	12605	17809	12271	18240	11944	18679	11622	19126	
26	12938	17393	12599	17816	12266	18248	11939	18686	11617	19133	4
	12932	17400	12593	17824	12260	18255	11933	18694	11612	19141	
27	12927	17407	12588	17831	12255	18262	11928	18701	11606	19148	3
	12921	17414	12582	17838	12249	18269	11922	18709	11601	19156	
28	12915	17421	12577	17845	12244	18277	11917	18716	11596	19163	2
	12910	17428	12571	17852	12238	18284	11912	18723	11590	19171	
29	12904	17435	12566	17859	12233	18291	11906	18731	11585	19178	1
	12898	17442	12560	17866	12227	18298	11901	18738	11580	19186	
30	12893	17449	12554	17873	12222	18306	11895	18746	11575	19193	0
	A	B	A	B	A	B	A	B	A	B	
′	132° 00′		131° 30′		131° 00′		130° 30′		130° 00′		′

TABLE 35

The Ageton Method

When Meridian Angle is Greater Than 90° Take "K" From Bottom of Table

′	50° 00′ A	50° 00′ B	50° 30′ A	50° 30′ B	51° 00′ A	51° 00′ B	51° 30′ A	51° 30′ B	52° 00′ A	52° 00′ B	′
0	11575	19193	11259	19649	10950	20113	10646	20585	10347	21066	30
	11569	19201	11254	19657	10945	20121	10640	20593	10342	21074	
1	11564	19208	11249	19664	10939	20128	10635	20601	10337	21082	29
	11559	19216	11244	19672	10934	20136	10630	20609	10332	21090	
2	11553	19223	11239	19680	10929	20144	10625	20617	10327	21098	28
	11548	19231	11233	19687	10924	20152	10620	20625	10322	21106	
3	11543	19238	11228	19695	10919	20160	10615	20633	10317	21114	27
	11537	19246	11223	19703	10914	20167	10610	20641	10312	21122	
4	11532	19253	11218	19710	10909	20175	10605	20649	10307	21131	26
	11527	19261	11213	19718	10904	20183	10600	20657	10302	21139	
5	11522	19269	11207	19726	10899	20191	10595	20665	10298	21147	25
	11516	19276	11202	19733	10894	20199	10590	20672	10293	21155	
6	11511	19284	11197	19741	10888	20207	10585	20680	10288	21163	24
	11506	19291	11192	19749	10883	20214	10580	20688	10283	21171	
7	11501	19299	11187	19756	10878	20222	10575	20696	10278	21179	23
	11495	19306	11181	19764	10873	20230	10570	20704	10273	21187	
8	11490	19314	11176	19772	10868	20238	10565	20712	10268	21195	22
	11485	19321	11171	19779	10863	20246	10560	20720	10263	21204	
9	11479	19329	11166	19787	10858	20254	10555	20728	10258	21212	21
	11474	19337	11161	19795	10853	20261	10550	20736	10253	21220	
10	11469	19344	11156	19803	10848	20269	10545	20744	10248	21228	20
	11464	19352	11150	19810	10843	20277	10540	20752	10243	21236	
11	11458	19359	11145	19818	10838	20285	10535	20760	10239	21244	19
	11453	19367	11140	19826	10832	20293	10530	20768	10234	21252	
12	11448	19375	11135	19834	10827	20301	10525	20776	10229	21260	18
	11443	19382	11130	19841	10822	20308	10520	20784	10224	21269	
13	11437	19390	11124	19849	10817	20316	10515	20792	10219	21277	17
	11432	19397	11119	19857	10812	20324	10510	20800	10214	21285	
14	11427	19405	11114	19864	10807	20332	10505	20808	10209	21293	16
	11421	19412	11109	19872	10802	20340	10500	20816	10204	21301	
15	11416	19420	11104	19880	10797	20348	10496	20824	10199	21309	15
	11411	19428	11099	19888	10792	20356	10491	20832	10195	21318	
16	11406	19435	11094	19895	10787	20364	10486	20840	10190	21326	14
	11400	19443	11088	19903	10782	20371	10481	20848	10185	21334	
17	11395	19450	11083	19911	10777	20379	10476	20856	10180	21342	13
	11390	19458	11078	19918	10772	20387	10471	20864	10175	21350	
18	11385	19466	11073	19926	10767	20395	10466	20872	10170	21358	12
	11380	19473	11068	19934	10761	20403	10461	20880	10165	21367	
19	11374	19481	11063	19942	10756	20411	10456	20888	10160	21375	11
	11369	19488	11057	19949	10751	20419	10451	20897	10155	21383	
20	11364	19496	11052	19957	10746	20427	10446	20905	10151	21391	10
	11359	19504	11047	19965	10741	20435	10441	20913	10146	21399	
21	11353	19511	11042	19973	10736	20442	10436	20921	10141	21407	9
	11348	19519	11037	19980	10731	20450	10431	20929	10136	21416	
22	11343	19527	11032	19988	10726	20458	10426	20937	10131	21424	8
	11338	19534	11027	19996	10721	20466	10421	20945	10126	21432	
23	11332	19542	11021	20004	10716	20474	10416	20953	10121	21440	7
	11327	19549	11016	20012	10711	20482	10411	20961	10116	21448	
24	11322	19557	11011	20019	10706	20490	10406	20969	10112	21457	6
	11317	19565	11006	20027	10701	20498	10401	20977	10107	21465	
25	11311	19572	11001	20035	10696	20506	10396	20985	10102	21473	5
	11306	19580	10996	20043	10691	20514	10391	20993	10097	21481	
26	11301	19588	10991	20050	10686	20522	10386	21001	10092	21489	4
	11296	19595	10986	20058	10681	20529	10381	21009	10087	21498	
27	11291	19603	10980	20066	10676	20537	10376	21017	10082	21506	3
	11285	19611	10975	20074	10671	20545	10372	21025	10078	21514	
28	11280	19618	10970	20082	10666	20553	10367	21033	10073	21522	2
	11275	19626	10965	20089	10661	20561	10362	21042	10068	21531	
29	11270	19634	10960	20097	10656	20569	10357	21050	10063	21539	1
	11265	19641	10955	20105	10651	20577	10352	21058	10058	21547	
30	11259	19649	10950	20113	10646	20585	10347	21066	10053	21555	0
′	129° 30′ A	129° 30′ B	129° 00′ A	129° 00′ B	128° 30′ A	128° 30′ B	128° 00′ A	128° 00′ B	127° 30′ A	127° 30′ B	′

TABLE 35

The Ageton Method

Always Take "Z" From Bottom of Table, Except When "K" is Same Name and Greater Than Latitude, in Which Case Take "Z" From Top of Table

′	52° 30′		53° 00′		53° 30′		54° 00′		54° 30′		′
	A	B	A	B	A	B	A	B	A	B	
0	10053	21555	9765	22054	9482	22561	9204	23078	8931	23605	30
	10049	21563	9760	22062	9477	22570	9200	23087	8927	23613	
1	10044	21572	9756	22070	9473	22578	9195	23095	8922	23622	29
	10039	21580	9751	22079	9468	22587	9190	23104	8918	23631	
2	10034	21588	9746	22087	9463	22595	9186	23113	8913	23640	28
	10029	21596	9741	22096	9459	22604	9181	23122	8909	23649	
3	10024	21605	9737	22104	9454	22612	9177	23130	8904	23658	27
	10019	21613	9732	22112	9449	22621	9172	23139	8900	23667	
4	10015	21621	9727	22121	9445	22630	9168	23148	8895	23675	26
	10010	21629	9722	22129	9440	22638	9163	23156	8891	23684	
5	10005	21638	9718	22138	9435	22647	9158	23165	8886	23693	25
	10000	21646	9713	22146	9431	22655	9154	23174	8882	23702	
6	9995	21654	9708	22154	9426	22664	9149	23183	8877	23711	24
	9990	21662	9703	22163	9421	22672	9145	23191	8873	23720	
7	9986	21671	9699	22171	9417	22681	9140	23200	8868	23729	23
	9981	21679	9694	22180	9412	22690	9136	23209	8864	23738	
8	9976	21687	9689	22188	9407	22698	9131	23218	8859	23747	22
	9971	21696	9684	22197	9403	22707	9126	23226	8855	23755	
9	9966	21704	9680	22205	9398	22715	9122	23235	8850	23764	21
	9962	21712	9675	22213	9394	22724	9117	23244	8846	23773	
10	9957	21720	9670	22222	9389	22732	9113	23252	8842	23782	20
	9952	21729	9665	22230	9384	22741	9108	23261	8837	23791	
11	9947	21737	9661	22239	9380	22750	9104	23270	8833	23800	19
	9942	21745	9656	22247	9375	22758	9099	23279	8828	23809	
12	9937	21754	9651	22256	9370	22767	9094	23288	8824	23818	18
	9933	21762	9647	22264	9366	22775	9090	23296	8819	23827	
13	9928	21770	9642	22272	9361	22784	9085	23305	8815	23836	17
	9923	21778	9637	22281	9356	22793	9081	23314	8810	23845	
14	9918	21787	9632	22289	9352	22801	9076	23323	8806	23854	16
	9913	21795	9628	22298	9347	22810	9072	23331	8801	23863	
15	9909	21803	9623	22306	9342	22818	9067	23340	8797	23871	15
	9904	21812	9618	22315	9338	22827	9063	23349	8792	23880	
16	9899	21820	9614	22323	9333	22836	9058	23358	8788	23889	14
	9894	21828	9609	22332	9329	22844	9054	23366	8783	23898	
17	9889	21837	9604	22340	9324	22853	9049	23375	8779	23907	13
	9885	21845	9599	22349	9319	22862	9044	23384	8775	23916	
18	9880	21853	9595	22357	9315	22870	9040	23393	8770	23925	12
	9875	21862	9590	22366	9310	22879	9035	23402	8766	23934	
19	9870	21870	9585	22374	9305	22887	9031	23410	8761	23943	11
	9865	21878	9581	22382	9301	22896	9026	23419	8757	23952	
20	9861	21887	9576	22391	9296	22905	9022	23428	8752	23961	10
	9856	21895	9571	22399	9292	22913	9017	23437	8748	23970	
21	9851	21903	9566	22408	9287	22922	9013	23446	8743	23979	9
	9846	21912	9562	22416	9282	22931	9008	23454	8739	23988	
22	9841	21920	9557	22425	9278	22939	9004	23463	8734	23997	8
	9837	21928	9552	22433	9273	22948	8999	23472	8730	24006	
23	9832	21937	9548	22442	9269	22957	8995	23481	8726	24015	7
	9827	21945	9543	22450	9264	22965	8990	23490	8721	24024	
24	9822	21953	9538	22459	9259	22974	8985	23498	8717	24033	6
	9818	21962	9534	22467	9255	22983	8981	23507	8712	24042	
25	9813	21970	9529	22476	9250	22991	8976	23516	8708	24051	5
	9808	21978	9524	22484	9246	23000	8972	23525	8703	24060	
26	9803	21987	9520	22493	9241	23009	8967	23534	8699	24069	4
	9798	21995	9515	22501	9236	23017	8963	23543	8694	24078	
27	9794	22003	9510	22510	9232	23026	8958	23551	8690	24037	3
	9789	22012	9505	22519	9227	23035	8954	23560	8686	24096	
28	9784	22020	9501	22527	9223	23043	8949	23569	8681	24105	2
	9779	22029	9496	22536	9218	23052	8945	23578	8677	24114	
29	9775	22037	9491	22544	9213	23061	8940	23587	8672	24123	1
	9770	22045	9487	22553	9209	23069	8936	23596	8668	24132	
30	9765	22054	9482	22561	9204	23078	8931	23605	8663	24141	0
	A	B	A	B	A	B	A	B	A	B	
′	127° 00′		126° 30′		126° 00′		125° 30′		125° 00′		′

TABLE 35

The Ageton Method

When Meridian Angle is Greater Than 90° Take "K" From Bottom of Table

′	55° 00′		55° 30′		56° 00′		56° 30′		57° 00′		′
	A	B	A	B	A	B	A	B	A	B	
0	8663	24141	8401	24687	8143	25244	7889	25811	7641	26389	30
	8659	24150	8396	24696	8138	25253	7885	25821	7637	26399	
1	8655	24159	8392	24706	8134	25263	7881	25830	7633	26409	29
	8650	24168	8388	24715	8130	25272	7877	25840	7629	26418	
2	8646	24177	8383	24724	8125	25281	7873	25849	7624	26428	28
	8641	24186	8379	24733	8121	25291	7868	25859	7620	26438	
3	8637	24195	8375	24742	8117	25300	7864	25868	7616	26447	27
	8633	24204	8370	24752	8113	25309	7860	25878	7612	26457	
4	8628	24213	8366	24761	8108	25319	7856	25887	7608	26467	26
	8624	24222	8362	24770	8104	25328	7852	25897	7604	26477	
5	8619	24231	8357	24779	8100	25338	7848	25907	7600	26486	25
	8615	24240	8353	24788	8096	25347	7843	25916	7596	26496	
6	8611	24249	8349	24798	8092	25356	7839	25926	7592	26506	24
	8606	24258	8344	24807	8087	25366	7835	25935	7588	26516	
7	8602	24267	8340	24816	8083	25375	7831	25945	7584	26526	23
	8597	24276	8336	24825	8079	25385	7827	25954	7579	26535	
8	8593	24286	8331	24835	8075	25394	7823	25964	7575	26545	22
	8589	24295	8327	24844	8070	25403	7818	25974	7571	26555	
9	8584	24304	8323	24853	8066	25413	7814	25983	7567	26565	21
	8580	24313	8318	24862	8062	25422	7810	25993	7563	26574	
10	8575	24322	8314	24872	8058	25432	7806	26002	7559	26584	20
	8571	24331	8310	24881	8053	25441	7802	26012	7555	26594	
11	8567	24340	8305	24890	8049	25451	7798	26022	7551	26604	19
	8562	24349	8301	24899	8045	25460	7793	26031	7547	26614	
12	8558	24358	8297	24909	8041	25469	7789	26041	7543	26623	18
	8553	24367	8292	24918	8036	25479	7785	26051	7539	26633	
13	8549	24376	8288	24927	8032	25488	7781	26060	7535	26643	17
	8545	24385	8284	24936	8028	25498	7777	26070	7531	26653	
14	8540	24395	8280	24946	8024	25507	7773	26079	7526	26663	16
	8536	24404	8275	24955	8020	25517	7769	26089	7522	26672	
15	8531	24413	8271	24964	8015	25526	7764	26099	7518	26682	15
	8527	24422	8267	24973	8011	25536	7760	26108	7514	26692	
16	8523	24431	8262	24983	8007	25545	7756	26118	7510	26702	14
	8518	24440	8258	24992	8003	25554	7752	26128	7506	26712	
17	8514	24449	8254	25001	7998	25564	7748	26137	7502	26722	13
	8510	24458	8249	25011	7994	25573	7744	26147	7498	26731	
18	8505	24467	8245	25020	7990	25583	7740	26157	7494	26741	12
	8501	24477	8241	25029	7986	25592	7736	26166	7490	26751	
19	8496	24486	8237	25038	7982	25602	7731	26176	7486	26761	11
	8492	24495	8232	25048	7977	25611	7727	26185	7482	26771	
20	8488	24504	8228	25057	7973	25621	7723	26195	7478	26781	10
	8483	24513	8224	25066	7969	25630	7719	26205	7474	26790	
21	8479	24522	8219	25076	7965	25640	7715	26214	7470	26800	9
	8475	24531	8215	25085	7961	25649	7711	26224	7466	26810	
22	8470	24540	8211	25094	7956	25659	7707	26234	7462	26820	8
	8466	24550	8207	25104	7952	25668	7702	26244	7458	26830	
23	8461	24559	8202	25113	7948	25678	7698	26253	7453	26840	7
	8457	24568	8198	25122	7944	25687	7694	26263	7449	26850	
24	8453	24577	8194	25132	7940	25697	7690	26273	7445	26860	6
	8448	24586	8189	25141	7935	25706	7686	26282	7441	26869	
25	8444	24595	8185	25150	7931	25716	7682	26292	7437	26879	5
	8440	24605	8181	25160	7927	25725	7678	26302	7433	26889	
26	8435	24614	8177	25169	7923	25735	7674	26311	7429	26899	4
	8431	24623	8172	25178	7919	25744	7670	26321	7425	26909	
27	8427	24632	8168	25188	7914	25754	7665	26331	7421	26919	3
	8422	24641	8164	25197	7910	25763	7661	26340	7417	26929	
28	8418	24650	8160	25206	7906	25773	7657	26350	7413	26939	2
	8414	24660	8155	25216	7902	25782	7653	26360	7409	26949	
29	8409	24669	8151	25225	7898	25792	7649	26370	7405	26958	1
	8405	24678	8147	25234	7893	25801	7645	26379	7401	26968	
30	8401	24687	8143	25244	7889	25811	7641	26389	7397	26978	0
	A	B	A	B	A	B	A	B	A	B	
′	124° 30′		124° 00′		123° 30′		123° 00′		122° 30′		′

TABLE 35

The Ageton Method

Always Take "Z" From Bottom of Table, Except When "K" is Same Name and Greater Than Latitude, in Which Case Take "Z" From Top of Table

′	57° 30′ A	57° 30′ B	58° 00′ A	58° 00′ B	58° 30′ A	58° 30′ B	59° 00′ A	59° 00′ B	59° 30′ A	59° 30′ B	′
0	7397	26978	7158	27579	6923	28191	6693	28816	6468	29453	30
	7393	26988	7154	27589	6920	28202	6690	28827	6464	29464	
1	7389	26998	7150	27599	6916	28212	6686	28837	6460	29475	29
	7385	27008	7146	27609	6912	28222	6682	28848	6457	29485	
2	7381	27018	7142	27619	6908	28233	6678	28858	6453	29496	28
	7377	27028	7138	27630	6904	28243	6674	28869	6449	29507	
3	7373	27038	7134	27640	6900	28253	6671	28879	6446	29517	27
	7369	27048	7130	27650	6896	28264	6667	28890	6442	29528	
4	7365	27058	7126	27660	6892	28274	6663	28900	6438	29539	26
	7361	27068	7122	27670	6889	28284	6659	28911	6434	29550	
5	7357	27078	7118	27680	6885	28295	6655	28921	6431	29560	25
	7353	27088	7115	27690	6881	28305	6652	28932	6427	29571	
6	7349	27098	7111	27701	6877	28315	6648	28942	6423	29582	24
	7345	27107	7107	27711	6873	28326	6644	28953	6420	29593	
7	7341	27117	7103	27721	6869	28336	6640	28964	6416	29604	23
	7337	27127	7099	27731	6865	28346	6637	28974	6412	29614	
8	7333	27137	7095	27741	6862	28357	6633	28985	6409	29625	22
	7329	27147	7091	27751	6858	28367	6629	28995	6405	29636	
9	7325	27157	7087	27761	6854	28378	6625	29006	6401	29647	21
	7321	27167	7083	27772	6850	28388	6622	29016	6397	29657	
10	7317	27177	7079	27782	6846	28398	6618	29027	6394	29668	20
	7313	27187	7075	27792	6842	28409	6614	29038	6390	29679	
11	7309	27197	7071	27802	6839	28419	6610	29048	6386	29690	19
	7305	27207	7068	27812	6835	28429	6607	29059	6383	29701	
12	7301	27217	7064	27823	6831	28440	6603	29069	6379	29711	18
	7297	27227	7060	27833	6827	28450	6599	29080	6375	29722	
13	7293	27237	7056	27843	6823	28461	6595	29091	6372	29733	17
	7289	27247	7052	27853	6819	28471	6591	29101	6368	29744	
14	7285	27257	7048	27863	6815	28481	6588	29112	6364	29755	16
	7281	27267	7044	27874	6812	28492	6584	29122	6361	29766	
15	7277	27277	7040	27884	6808	28502	6580	29133	6357	29776	15
	7273	27287	7036	27894	6804	28513	6576	29144	6353	29787	
16	7269	27297	7032	27904	6800	28523	6573	29154	6349	29798	14
	7265	27307	7028	27914	6796	28533	6569	29165	6346	29809	
17	7261	27317	7024	27925	6792	28544	6565	29175	6342	29820	13
	7257	27327	7021	27935	6789	28554	6561	29186	6338	29831	
18	7253	27337	7017	27945	6785	28565	6558	29197	6335	29841	12
	7249	27347	7013	27955	6781	28575	6554	29207	6331	29852	
19	7245	27357	7009	27965	6777	28586	6550	29218	6327	29863	11
	7241	27367	7005	27976	6773	28596	6546	29229	6324	29874	
20	7237	27377	7001	27986	6770	28607	6543	29239	6320	29885	10
	7233	27387	6997	27996	6766	28617	6539	29250	6316	29896	
21	7229	27398	6993	28006	6762	28627	6535	29261	6313	29907	9
	7225	27408	6989	28017	6758	28638	6531	29271	6309	29917	
22	7221	27418	6985	28027	6754	28648	6528	29282	6305	29929	8
	7217	27428	6982	28037	6750	28659	6524	29293	6302	29939	
23	7213	27438	6978	28047	6747	28669	6520	29303	6298	29950	7
	7209	27448	6974	28058	6743	28680	6516	29314	6294	29961	
24	7205	27458	6970	28068	6739	28690	6513	29325	6291	29972	6
	7201	27468	6966	28078	6735	28701	6509	29335	6287	29983	
25	7197	27478	6962	28089	6731	28711	6505	29346	6283	29994	5
	7193	27488	6958	28099	6728	28722	6502	29357	6280	30005	
26	7190	27498	6954	28109	6724	28732	6498	29367	6276	30015	4
	7186	27508	6951	28119	6720	28743	6494	29378	6272	30026	
27	7182	27518	6947	28130	6716	28753	6490	29389	6269	30037	3
	7178	27528	6943	28140	6712	28763	6487	29399	6265	30048	
28	7174	27539	6939	28150	6709	28774	6483	29410	6261	30059	2
	7170	27549	6935	28161	6705	28784	6479	29421	6258	30070	
29	7166	27559	6931	28171	6701	28795	6475	29432	6254	30081	1
	7162	27569	6927	28181	6697	28806	6472	29442	6251	30092	
30	7158	27579	6923	28191	6693	28816	6468	29453	6247	30103	0
′	A	B	A	B	A	B	A	B	A	B	′
	122° 00′		121° 30′		121° 00′		120° 30′		120° 00′		

TABLE 35

The Ageton Method

When Meridian Angle is Greater Than 90° Take "K" From Bottom of Table

′	60° 00′		60° 30′		61° 00′		61° 30′		62° 00′		′
′	A	B	A	B	A	B	A	B	A	B	′
0	6247	**30103**	6030	**30766**	5818	**31443**	5610	**32134**	5406	**32839**	30
	6243	**30114**	6027	**30777**	5815	**31454**	5607	**32145**	5403	**32851**	
1	6240	**30125**	6023	**30788**	5811	**31466**	5603	**32157**	5400	**32863**	29
	6236	**30136**	6020	**30800**	5808	**31477**	5600	**32169**	5396	**32875**	
2	6232	**30147**	6016	**30811**	5804	**31488**	5596	**32180**	5393	**32887**	28
	6229	**30158**	6012	**30822**	5801	**31500**	5593	**32192**	5390	**32898**	
3	6225	**30169**	6009	**30833**	5797	**31511**	5590	**32204**	5386	**32910**	27
	6221	**30180**	6005	**30844**	5794	**31523**	5586	**32215**	5383	**32922**	
4	6218	**30191**	6002	**30856**	5790	**31534**	5583	**32227**	5380	**32934**	26
	6214	**30202**	5998	**30867**	5787	**31546**	5579	**32239**	5376	**32946**	
5	6210	**30213**	5995	**30878**	5783	**31557**	5575	**32250**	5373	**32958**	25
	6207	**30224**	5991	**30889**	5780	**31569**	5572	**32262**	5370	**32970**	
6	6203	**30235**	5987	**30900**	5776	**31580**	5569	**32274**	5366	**32982**	24
	6200	**30245**	5984	**30912**	5773	**31591**	5566	**32285**	5363	**32994**	
7	6196	**30256**	5980	**30923**	5769	**31603**	5562	**32297**	5360	**33006**	23
	6192	**30267**	5977	**30934**	5766	**31614**	5559	**32309**	5356	**33018**	
8	6189	**30278**	5973	**30945**	5762	**31626**	5555	**32320**	5353	**33030**	22
	6185	**30289**	5970	**30956**	5759	**31637**	5552	**32332**	5350	**33042**	
9	6181	**30300**	5966	**30968**	5755	**31649**	5549	**32344**	5346	**33054**	21
	6178	**30311**	5963	**30979**	5752	**31660**	5545	**32355**	5343	**33065**	
10	6174	**30322**	5959	**30990**	5748	**31672**	5542	**32367**	5340	**33077**	20
	6171	**30334**	5955	**31001**	5745	**31683**	5538	**32379**	5336	**33089**	
11	6167	**30345**	5952	**31013**	5741	**31694**	5535	**32391**	5333	**33101**	19
	6163	**30355**	5948	**31024**	5738	**31706**	5532	**32402**	5330	**33113**	
12	6160	**30367**	5945	**31035**	5734	**31717**	5528	**32414**	5326	**33125**	18
	6156	**30378**	5941	**31046**	5731	**31729**	5525	**32426**	5323	**33137**	
13	6152	**30389**	5938	**31058**	5727	**31740**	5521	**32438**	5320	**33149**	17
	6149	**30400**	5934	**31069**	5724	**31752**	5518	**32449**	5316	**33161**	
14	6145	**30411**	5931	**31080**	5720	**31763**	5515	**32461**	5313	**33173**	16
	6142	**30422**	5927	**31091**	5717	**31775**	5511	**32473**	5310	**33185**	
15	6138	**30433**	5924	**31103**	5714	**31786**	5508	**32484**	5306	**33197**	15
	6134	**30444**	5920	**31114**	5710	**31798**	5504	**32496**	5303	**33209**	
16	6131	**30455**	5917	**31125**	5707	**31809**	5501	**32508**	5300	**33221**	14
	6127	**30466**	5913	**31137**	5703	**31821**	5498	**32520**	5296	**33233**	
17	6124	**30477**	5909	**31148**	5700	**31833**	5494	**32532**	5293	**33245**	13
	6120	**30488**	5906	**31159**	5696	**31844**	5491	**32543**	5290	**33257**	
18	6116	**30499**	5902	**31170**	5693	**31856**	5487	**32555**	5286	**33269**	12
	6113	**30510**	5899	**31182**	5689	**31867**	5484	**32567**	5283	**33281**	
19	6109	**30521**	5895	**31193**	5686	**31879**	5481	**32579**	5280	**33293**	11
	6106	**30532**	5892	**31204**	5682	**31890**	5477	**32590**	5276	**33306**	
20	6102	**30544**	5888	**31216**	5679	**31902**	5474	**32602**	5273	**33318**	10
	6098	**30555**	5885	**31227**	5675	**31913**	5470	**32614**	5270	**33330**	
21	6095	**30566**	5881	**31238**	5672	**31925**	5467	**32625**	5266	**33342**	9
	6091	**30577**	5878	**31250**	5669	**31936**	5464	**32638**	5263	**33354**	
22	6088	**30588**	5874	**31261**	5665	**31948**	5460	**32649**	5260	**33366**	8
	6084	**30599**	5871	**31272**	5662	**31960**	5457	**32661**	5257	**33378**	
23	6080	**30610**	5867	**31284**	5658	**31971**	5454	**32673**	5253	**33390**	7
	6077	**30621**	5864	**31295**	5655	**31983**	5450	**32685**	5250	**33402**	
24	6073	**30632**	5860	**31306**	5651	**31994**	5447	**32697**	5247	**33414**	6
	6070	**30643**	5857	**31318**	5648	**32006**	5443	**32709**	5243	**33426**	
25	6066	**30655**	5853	**31329**	5644	**32018**	5440	**32720**	5240	**33438**	5
	6062	**30666**	5850	**31340**	5641	**32029**	5437	**32732**	5237	**33450**	
26	6059	**30677**	5846	**31352**	5638	**32041**	5433	**32744**	5233	**33462**	4
	6055	**30688**	5843	**31363**	5634	**32052**	5430	**32756**	5230	**33475**	
27	6052	**30699**	5839	**31375**	5631	**32064**	5427	**32768**	5227	**33487**	3
	6048	**30710**	5836	**31386**	5627	**32076**	5423	**32780**	5224	**33499**	
28	6045	**30721**	5832	**31397**	5624	**32087**	5420	**32792**	5220	**33511**	2
	6041	**30733**	5829	**31409**	5620	**32099**	5417	**32803**	5217	**33523**	
29	6037	**30744**	5825	**31420**	5617	**32110**	5413	**32815**	5214	**33535**	1
	6034	**30755**	5822	**31431**	5614	**32122**	5410	**32827**	5210	**33547**	
30	6030	**30766**	5818	**31443**	5610	**32134**	5406	**32839**	5207	**33559**	0
	A	B	A	B	A	B	A	B	A	B	
′	119° 30′		119° 00′		118° 30′		118° 00′		117° 30′		′

TABLE 35

The Ageton Method

Always Take "Z" From Bottom of Table, Except When "K" is Same Name and Greater Than Latitude, in Which Case Take "Z" From Top of Table

′	62° 30′		63° 00′		63° 30′		64° 00′		64° 30′		′
	A	B	A	B	A	B	A	B	A	B	
0	5207	33559	5012	34295	4821	35047	4634	35816	4451	36602	30
	5204	33572	5009	34308	4818	35060	4631	35829	4448	36615	
1	5200	33584	5005	34320	4815	35073	4628	35842	4445	36628	29
	5197	33596	5002	34332	4811	35085	4625	35855	4442	36641	
2	5194	33608	4999	34345	4808	35098	4622	35868	4439	36655	28
	5191	33620	4996	34357	4805	35111	4619	35881	4436	36668	
3	5187	33632	4993	34370	4802	35123	4615	35894	4433	36681	27
	5184	33644	4989	34382	4799	35136	4612	35907	4430	36694	
4	5181	33657	4986	34395	4796	35149	4609	35920	4427	36708	26
	5178	33669	4983	34407	4793	35161	4606	35933	4424	36721	
5	5174	33681	4980	34420	4789	35174	4603	35946	4421	36734	25
	5171	33693	4977	34432	4786	35187	4600	35959	4418	36747	
6	5168	33705	4973	34444	4783	35200	4597	35972	4415	36761	24
	5164	33717	4970	34457	4780	35212	4594	35985	4412	36774	
7	5161	33730	4967	34469	4777	35225	4591	35998	4409	36787	23
	5158	33742	4964	34482	4774	35238	4588	36011	4406	36801	
8	5155	33754	4961	34494	4771	35251	4585	36024	4403	36814	22
	5151	33766	4957	34507	4767	35263	4582	36037	4400	36827	
9	5148	33779	4954	34519	4764	35276	4579	36050	4397	36841	21
	5145	33791	4951	34532	4761	35289	4576	36063	4394	36854	
10	5142	33803	4948	34544	4758	35302	4573	36076	4391	36867	20
	5138	33815	4945	34557	4755	35314	4569	36089	4388	36881	
11	5135	33827	4941	34569	4752	35327	4566	36102	4385	36894	19
	5132	33840	4938	34582	4749	35340	4563	36115	4382	36907	
12	5128	33852	4935	34594	4746	35353	4560	36128	4379	36921	18
	5125	33864	4932	34607	4742	35365	4557	36141	4376	36934	
13	5122	33876	4929	34619	4739	35378	4554	36154	4373	36948	17
	5119	33889	4925	34632	4736	35391	4551	36167	4370	36961	
14	5115	33901	4922	34644	4733	35404	4548	36180	4367	36974	16
	5112	33913	4919	34657	4730	35417	4545	36193	4364	36988	
15	5109	33925	4916	34669	4727	35429	4542	36206	4361	37001	15
	5106	33938	4913	34682	4724	35442	4539	36220	4358	37014	
16	5102	33950	4910	34694	4721	35455	4536	36233	4355	37028	14
	5099	33962	4906	34707	4718	35468	4533	36246	4352	37041	
17	5096	33974	4903	34719	4714	35481	4530	36259	4349	37055	13
	5093	33987	4900	34732	4711	35493	4527	36272	4346	37068	
18	5089	33999	4897	34744	4708	35506	4524	36285	4343	37081	12
	5086	34011	4894	34757	4705	35519	4521	36298	4340	37095	
19	5083	34024	4890	34770	4702	35532	4518	36311	4337	37108	11
	5080	34036	4887	34782	4699	35545	4515	36325	4334	37122	
20	5076	34048	4884	34795	4696	35558	4512	36338	4332	37135	10
	5073	34061	4881	34807	4693	35571	4509	36351	4329	37149	
21	5070	34073	4878	34820	4690	35583	4506	36364	4326	37162	9
	5067	34085	4875	34832	4686	35596	4503	36377	4323	37176	
22	5064	34097	4871	34845	4683	35609	4500	36390	4320	37189	8
	5060	34110	4868	34858	4680	35622	4497	36403	4317	37203	
23	5057	34122	4865	34870	4677	35635	4493	36417	4314	37216	7
	5054	34134	4862	34883	4674	35648	4490	36430	4311	37229	
24	5051	34147	4859	34896	4671	35661	4487	36443	4308	37243	6
	5047	34159	4856	34908	4668	35674	4484	36456	4305	37256	
25	5044	34172	4852	34921	4665	35686	4481	36469	4302	37270	5
	5041	34184	4849	34933	4662	35699	4478	36483	4299	37283	
26	5038	34196	4846	34946	4659	35712	4475	36496	4296	37297	4
	5034	34209	4843	34959	4656	35725	4472	36509	4293	37310	
27	5031	34221	4840	34971	4652	35738	4469	36522	4290	37324	3
	5028	34233	4837	34984	4649	35751	4466	36535	4287	37337	
28	5025	34246	4833	34997	4646	35764	4463	36549	4284	37351	2
	5022	34258	4830	35009	4643	35777	4460	36562	4281	37365	
29	5018	34270	4827	35022	4640	35790	4457	36575	4278	37378	1
	5015	34283	4824	35035	4637	35803	4454	36588	4275	37392	
30	5012	34295	4821	35047	4634	35816	4451	36602	4272	37405	0
	A	B	A	B	A	B	A	B	A	B	
′	117° 00′		116° 30′		116° 00′		115° 30′		115° 00′		′

TABLE 35

The Ageton Method

When Meridian Angle is Greater Than 90° Take "K" From Bottom of Table

′	65° 00′		65° 30′		66° 00′		66° 30′		67° 00′		′
	A	B	A	B	A	B	A	B	A	B	
0	4272	37405	4098	38227	3927	39069	3760	39930	3597	40812	30
	4269	37419	4095	38241	3924	39083	3757	39945	3595	40827	
1	4266	37432	4092	38255	3921	39097	3755	39959	3592	40842	29
	4264	37446	4089	38269	3918	39111	3752	39974	3589	40857	
2	4261	37459	4086	38283	3916	39125	3749	39988	3587	40872	28
	4258	37473	4083	38297	3913	39140	3746	40003	3584	40887	
3	4255	37487	4080	38311	3910	39154	3744	40017	3581	40902	27
	4252	37500	4078	38324	3907	39168	3741	40032	3579	40916	
4	4249	37514	4075	38338	3904	39182	3738	40046	3576	40931	26
	4246	37527	4072	38352	3902	39197	3735	40061	3573	40946	
5	4243	37541	4069	38366	3899	39211	3733	40076	3571	40961	25
	4240	37554	4066	38380	3896	39225	3730	40090	3568	40976	
6	4237	37568	4063	38394	3893	39239	3727	40105	3565	40991	24
	4234	37582	4060	38408	3890	39254	3725	40119	3563	41006	
7	4231	37595	4057	38422	3888	39268	3722	40134	3560	41021	23
	4228	37609	4055	38436	3885	39282	3719	40149	3557	41036	
8	4225	37623	4052	38450	3882	39296	3716	40163	3555	41051	22
	4222	37636	4049	38464	3879	39311	3714	40178	3552	41066	
9	4220	37650	4046	38478	3876	39325	3711	40192	3549	41081	21
	4217	37663	4043	38492	3874	39339	3708	40207	3547	41096	
10	4214	37677	4040	38506	3871	39353	3705	40222	3544	41111	20
	4211	37691	4037	38520	3868	39368	3703	40236	3541	41126	
11	4208	37704	4035	38533	3865	39382	3700	40251	3539	41141	19
	4205	37718	4032	38547	3863	39396	3697	40266	3536	41156	
12	4202	37732	4029	38561	3860	39411	3695	40280	3533	41171	18
	4199	37745	4026	38575	3857	39425	3692	40295	3531	41186	
13	4196	37759	4023	38589	3854	39439	3689	40310	3528	41201	17
	4193	37773	4020	38603	3851	39454	3686	40324	3525	41216	
14	4190	37786	4017	38617	3849	39468	3684	40339	3523	41231	16
	4187	37800	4015	38631	3846	39482	3681	40354	3520	41246	
15	4185	37814	4012	38645	3843	39497	3678	40368	3517	41261	15
	4182	37828	4009	38660	3840	39511	3676	40383	3515	41276	
16	4179	37841	4006	38674	3838	39525	3673	40398	3512	41291	14
	4176	37855	4003	38688	3835	39540	3670	40413	3509	41307	
17	4173	37869	4000	38702	3832	39554	3667	40427	3507	41322	13
	4170	37882	3998	38716	3829	39569	3665	40442	3504	41337	
18	4167	37896	3995	38730	3826	39583	3662	40457	3502	41352	12
	4164	37910	3992	38744	3824	39597	3659	40471	3499	41367	
19	4161	37924	3989	38758	3821	39612	3657	40486	3496	41382	11
	4158	37937	3986	38772	3818	39626	3654	40501	3494	41397	
20	4155	37951	3983	38786	3815	39641	3651	40516	3491	41412	10
	4153	37965	3981	38800	3813	39655	3648	40530	3488	41427	
21	4150	37979	3978	38814	3810	39669	3646	40545	3486	41443	9
	4147	37992	3975	38828	3807	39684	3643	40560	3483	41458	
22	4144	38006	3972	38842	3804	39698	3640	40575	3480	41473	8
	4141	38020	3969	38856	3801	39713	3638	40590	3478	41488	
23	4138	38034	3966	38871	3799	39727	3635	40604	3475	41503	7
	4135	38048	3964	38885	3796	39742	3632	40619	3473	41518	
24	4132	38061	3961	38899	3793	39756	3630	40634	3470	41533	6
	4129	38075	3958	38913	3790	39771	3627	40649	3467	41549	
25	4127	38089	3955	38927	3788	39785	3624	40664	3465	41564	5
	4124	38103	3952	38941	3785	39799	3622	40678	3462	41579	
26	4121	38117	3949	38955	3782	39814	3619	40693	3459	41594	4
	4118	38130	3947	38969	3779	39828	3616	40708	3457	41609	
27	4115	38144	3944	38984	3777	39843	3613	40723	3454	41625	3
	4112	38158	3941	38998	3774	39857	3611	40738	3452	41640	
28	4109	38172	3938	39012	3771	39872	3608	40753	3449	41655	2
	4106	38186	3935	39026	3768	39886	3605	40768	3446	41670	
29	4103	38200	3933	39040	3766	39901	3603	40782	3444	41685	1
	4101	38213	3930	39054	3763	39915	3600	40797	3441	41701	
30	4098	38227	3927	39069	3760	39930	3597	40812	3438	41716	0
	A	B	A	B	A	B	A	B	A	B	
′	114° 30′		114° 00′		113° 30′		113° 00′		112° 30′		′

TABLE 35

The Ageton Method

Always Take "Z" From Bottom of Table, Except When "K" is Same Name and Greater Than Latitude, in Which Case Take "Z" From Top of Table

	67° 30′		68° 00′		68° 30′		69° 00′		69° 30′		
′	A	B	A	B	A	B	A	B	A	B	′
0	3438	41716	3283	42642	3132	43592	2985	44567	2841	45567	30
	3436	41731	3281	42658	3130	43608	2982	44583	2839	45584	
1	3433	41746	3278	42674	3127	43624	2980	44600	2836	45601	29
	3431	41762	3276	42689	3125	43641	2978	44616	2834	45618	
2	3428	41777	3273	42705	3122	43657	2975	44633	2832	45635	28
	3425	41792	3271	42721	3120	43673	2973	44649	2829	45652	
3	3423	41808	3268	42736	3117	43689	2970	44666	2827	45669	27
	3420	41823	3266	42752	3115	43705	2968	44682	2825	45686	
4	3418	41838	3263	42768	3112	43721	2965	44699	2822	45703	26
	3415	41853	3260	42783	3110	43737	2963	44715	2820	45720	
5	3412	41869	3258	42799	3107	43753	2961	44732	2818	45737	25
	3410	41884	3255	42815	3105	43769	2958	44748	2815	45754	
6	3407	41899	3253	42830	3102	43785	2956	44765	2813	45771	24
	3404	41915	3250	42846	3100	43801	2953	44782	2811	45788	
7	3402	41930	3248	42862	3097	43818	2951	44798	2808	45805	23
	3399	41945	3245	42878	3095	43834	2949	44815	2806	45822	
8	3397	41961	3243	42893	3092	43850	2946	44831	2804	45839	22
	3394	41976	3240	42909	3090	43866	2944	44848	2801	45856	
9	3391	41991	3237	42925	3088	43882	2941	44864	2799	45873	21
	3389	42007	3235	42941	3085	43898	2939	44881	2797	45890	
10	3386	42022	3233	42956	3083	43914	2936	44898	2794	45907	20
	3384	42038	3230	42972	3080	43931	2934	44914	2792	45924	
11	3381	42053	3227	42988	3078	43947	2932	44931	2789	45941	19
	3379	42068	3225	43004	3075	43963	2929	44947	2787	45958	
12	3376	42084	3222	43020	3073	43979	2927	44964	2785	45975	18
	3373	42099	3220	43035	3070	43995	2924	44981	2782	45992	
13	3371	42115	3217	43051	3068	44012	2922	44997	2780	46009	17
	3368	42130	3215	43067	3065	44028	2920	45014	2778	46026	
14	3366	42145	3212	43083	3063	44044	2917	45031	2775	46043	16
	3363	42161	3210	43099	3060	44060	2915	45047	2773	46061	
15	3360	42176	3207	43114	3058	44077	2913	45064	2771	46078	15
	3358	42192	3205	43130	3056	44093	2910	45081	2768	46095	
16	3355	42207	3202	43146	3053	44109	2908	45097	2766	46112	14
	3353	42223	3200	43162	3051	44125	2905	45114	2764	46129	
17	3350	42238	3197	43178	3048	44142	2903	45131	2761	46146	13
	3348	42254	3195	43194	3046	44158	2901	45147	2759	46163	
18	3345	42269	3192	43210	3043	44174	2898	45164	2757	46181	12
	3342	42285	3190	43225	3041	44190	2896	45181	2755	46198	
19	3340	42300	3187	43241	3038	44207	2893	45198	2752	46215	11
	3337	42316	3185	43257	3036	44223	2891	45214	2750	46232	
20	3335	42331	3182	43273	3033	44239	2889	45231	2748	46249	10
	3332	42347	3180	43289	3031	44256	2886	45248	2745	46266	
21	3329	42362	3177	43305	3029	44272	2884	45265	2743	46284	9
	3327	42378	3175	43321	3026	44288	2881	45281	2741	46301	
22	3324	42393	3172	43337	3024	44305	2879	45298	2738	46318	8
	3322	42409	3170	43353	3021	44321	2877	45315	2736	46335	
23	3319	42424	3167	43369	3019	44337	2874	45332	2734	46353	7
	3317	42440	3165	43385	3016	44354	2872	45348	2731	46370	
24	3314	42455	3162	43400	3014	44370	2870	45365	2729	46387	6
	3312	42471	3160	43416	3012	44386	2867	45382	2727	46404	
25	3309	42486	3157	43432	3009	44403	2865	45399	2724	46422	5
	3306	42502	3155	43448	3007	44419	2862	45416	2722	46439	
26	3304	42518	3152	43464	3004	44436	2860	45433	2720	46456	4
	3301	42533	3150	43480	3002	44452	2858	45449	2717	46473	
27	3299	42549	3147	43496	2999	44468	2855	45466	2715	46491	3
	3296	42564	3145	43512	2997	44485	2853	45483	2713	46508	
28	3294	42580	3142	43528	2994	44501	2851	45500	2711	46525	2
	3291	42596	3140	43544	2992	44518	2848	45517	2708	46543	
29	3289	42611	3137	43560	2990	44534	2846	45534	2706	46560	1
	3286	42627	3135	43576	2987	44551	2844	45551	2704	46577	
30	3283	42642	3132	43592	2985	44567	2841	45567	2701	46595	0
	A	B	A	B	A	B	A	B	A	B	′
′	112° 00′		111° 30′		111° 00′		110° 30′		110° 00′		

TABLE 35

The Ageton Method

When Meridian Angle is Greater Than 90° Take "K" From Bottom of Table

′	70° 00′ A	70° 00′ B	70° 30′ A	70° 30′ B	71° 00′ A	71° 00′ B	71° 30′ A	71° 30′ B	72° 00′ A	72° 00′ B	′
0	2701	46595	2565	47650	2433	48736	2304	49852	2179	51002	30
	2699	46612	2563	47668	2431	48754	2302	49871	2177	51021	
1	2697	46630	2561	47686	2429	48772	2300	49890	2175	51041	29
	2694	46647	2559	47704	2427	48791	2298	49909	2173	51060	
2	2692	46664	2556	47722	2424	48809	2296	49928	2171	51080	28
	2690	46682	2554	47740	2422	48828	2294	49947	2169	51099	
3	2688	46699	2552	47758	2420	48846	2292	49966	2167	51119	27
	2685	46716	2550	47775	2418	48864	2290	49985	2165	51138	
4	2683	46734	2547	47793	2416	48883	2287	50004	2163	51158	26
	2681	46751	2545	47811	2413	48901	2285	50023	2161	51177	
5	2678	46769	2543	47829	2411	48920	2283	50042	2159	51197	25
	2676	46786	2541	47847	2409	48938	2281	50061	2157	51216	
6	2674	46804	2539	47865	2407	48957	2279	50080	2155	51236	24
	2672	46821	2536	47883	2405	48975	2277	50098	2153	51255	
7	2669	46839	2534	47901	2403	48993	2275	50117	2151	51275	23
	2667	46856	2532	47919	2400	49012	2273	50137	2149	51294	
8	2665	46873	2530	47937	2398	49030	2271	50156	2147	51314	22
	2662	46891	2528	47955	2396	49049	2269	50175	2145	51334	
9	2660	46908	2525	47973	2394	49067	2266	50194	2143	51353	21
	2658	46926	2523	47991	2392	49086	2264	50213	2141	51373	
10	2656	46943	2521	48009	2390	49104	2262	50232	2138	51392	20
	2653	46961	2519	48027	2387	49123	2260	50251	2136	51412	
11	2651	46978	2516	48045	2385	49141	2258	50270	2134	51432	19
	2649	46996	2514	48063	2383	49160	2256	50289	2132	51451	
12	2646	47014	2512	48081	2381	49179	2254	50308	2130	51471	18
	2644	47031	2510	48099	2379	49197	2252	50327	2128	51491	
13	2642	47049	2507	48117	2377	49216	2250	50346	2126	51510	17
	2640	47066	2505	48135	2375	49234	2248	50365	2124	51530	
14	2637	47084	2503	48153	2372	49253	2246	50385	2122	51550	16
	2635	47101	2501	48171	2370	49271	2243	50404	2120	51570	
15	2633	47119	2499	48189	2368	49290	2241	50423	2118	51589	15
	2631	47137	2496	48207	2366	49309	2239	50442	2116	51609	
16	2628	47154	2494	48226	2364	49327	2237	50461	2114	51629	14
	2626	47172	2492	48244	2362	49346	2235	50480	2112	51649	
17	2624	47189	2490	48262	2360	49365	2233	50499	2110	51668	13
	2622	47207	2488	48280	2358	49383	2231	50519	2108	51688	
18	2619	47225	2485	48298	2355	49402	2229	50538	2106	51708	12
	2617	47242	2483	48316	2353	49421	2227	50557	2104	51728	
19	2615	47260	2481	48334	2351	49439	2225	50576	2102	51747	11
	2613	47278	2479	48352	2349	49458	2223	50596	2100	51767	
20	2610	47295	2477	48371	2347	49477	2221	50615	2098	51787	10
	2608	47313	2474	48389	2345	49495	2218	50634	2096	51807	
21	2606	47331	2472	48407	2343	49514	2216	50653	2094	51827	9
	2604	47348	2470	48425	2340	49533	2214	50673	2092	51847	
22	2601	47366	2468	48443	2338	49551	2212	50692	2090	51867	8
	2599	47384	2466	48462	2336	49570	2210	50711	2088	51886	
23	2597	47402	2463	48480	2334	49589	2208	50730	2086	51906	7
	2594	47419	2461	48498	2332	49608	2206	50750	2084	51926	
24	2592	47437	2459	48516	2330	49626	2204	50769	2082	51946	6
	2590	47455	2457	48534	2328	49645	2202	50788	2080	51966	
25	2588	47472	2455	48553	2325	49664	2200	50808	2078	51986	5
	2585	47490	2453	48571	2323	49683	2198	50827	2076	52006	
26	2583	47508	2450	48589	2321	49702	2196	50846	2074	52026	4
	2581	47526	2448	48608	2319	49720	2194	50866	2072	52046	
27	2579	47544	2446	48626	2317	49739	2192	50885	2070	52066	3
	2576	47561	2444	48644	2315	49758	2190	50905	2068	52086	
28	2574	47579	2442	48662	2313	49777	2188	50924	2066	52106	2
	2572	47597	2439	48681	2311	49796	2185	50943	2064	52126	
29	2570	47615	2437	48699	2309	49815	2183	50963	2062	52146	1
	2568	47633	2435	48717	2306	49833	2181	50982	2060	52166	
30	2565	47650	2433	48736	2304	49852	2179	51002	2058	52186	0
′	A	B	A	B	A	B	A	B	A	B	′
	109° 30′		109° 00′		108° 30′		108° 00′		107° 30′		

TABLE 35

The Ageton Method

Always Take "Z" From Bottom of Table, Except When "K" is Same Name and Greater Than Latitude, in Which Case Take "Z" From Top of Table

′	72° 30′		73° 00′		73° 30′		74° 00′		74° 30′		′
	A	B	A	B	A	B	A	B	A	B	
0	2058	52186	1940	53406	1826	54666	1716	55966	1609	57310	30
	2056	52206	1938	53427	1824	54687	1714	55988	1607	57333	
1	2054	52226	1936	53448	1823	54708	1712	56010	1605	57356	29
	2052	52246	1935	53468	1821	54730	1710	56032	1604	57378	
2	2050	52266	1933	53489	1819	54751	1709	56054	1602	57401	28
	2048	52286	1931	53510	1817	54773	1707	56076	1600	57424	
3	2046	52306	1929	53531	1815	54794	1705	56099	1598	57447	27
	2044	52326	1927	53551	1813	54815	1703	56121	1597	57470	
4	2042	52346	1925	53572	1811	54837	1701	56143	1595	57493	26
	2040	52366	1923	53593	1809	54858	1700	56165	1593	57516	
5	2038	52387	1921	53614	1808	54880	1698	56187	1591	57538	25
	2036	52407	1919	53634	1806	54901	1696	56209	1590	57561	
6	2034	52427	1917	53655	1804	54922	1694	56231	1588	57584	24
	2032	52447	1915	53676	1802	54944	1692	56254	1586	57607	
7	2030	52467	1913	53697	1800	54965	1691	56276	1584	57630	23
	2028	52487	1911	53718	1798	54987	1689	56298	1583	57653	
8	2026	52508	1910	53738	1796	55008	1687	56320	1581	57676	22
	2024	52528	1908	53759	1795	55030	1685	56342	1579	57699	
9	2022	52548	1906	53780	1793	55051	1683	56365	1578	57722	21
	2020	52568	1904	53801	1791	55073	1682	56387	1576	57745	
10	2018	52588	1902	53822	1789	55095	1680	56409	1574	57768	20
	2016	52609	1900	53843	1787	55116	1678	56431	1572	57791	
11	2014	52629	1898	53864	1785	55138	1676	56454	1571	57814	19
	2012	52649	1896	53884	1783	55159	1674	56476	1569	57837	
12	2010	52670	1894	53905	1782	55181	1673	56498	1567	57860	18
	2009	52690	1892	53926	1780	55202	1671	56521	1565	57884	
13	2007	52710	1890	53947	1778	55224	1669	56543	1564	57907	17
	2005	52730	1889	53968	1776	55246	1667	56565	1562	57930	
14	2003	52751	1887	53989	1774	55267	1665	56588	1560	57953	16
	2001	52771	1885	54010	1772	55289	1664	56610	1559	57976	
15	1999	52791	1883	54031	1771	55311	1662	56632	1557	57999	15
	1997	52812	1881	54052	1769	55332	1660	56655	1555	58022	
16	1995	52832	1879	54073	1767	55354	1658	56677	1553	58046	14
	1993	52852	1877	54094	1765	55376	1657	56700	1552	58069	
17	1991	52873	1875	54115	1763	55397	1655	56722	1550	58092	13
	1989	52893	1873	54136	1761	55419	1653	56745	1548	58115	
18	1987	52914	1871	54157	1760	55441	1651	56767	1546	58138	12
	1985	52934	1870	54178	1758	55463	1650	56790	1545	58162	
19	1983	52954	1868	54199	1756	55484	1648	56812	1543	58185	11
	1981	52975	1866	54220	1754	55506	1646	56835	1541	58208	
20	1979	52995	1864	54242	1752	55528	1644	56857	1540	58232	10
	1977	53016	1862	54263	1750	55550	1642	56880	1538	58255	
21	1975	53036	1860	54284	1749	55572	1641	56902	1536	58278	9
	1973	53057	1858	54305	1747	55593	1639	56925	1534	58302	
22	1971	53077	1856	54326	1745	55615	1637	56947	1533	58325	8
	1969	53098	1854	54347	1743	55637	1635	56970	1531	58348	
23	1967	53118	1853	54368	1741	55659	1634	56992	1529	58372	7
	1966	53139	1851	54390	1739	55681	1632	57015	1528	58395	
24	1964	53159	1849	54411	1738	55703	1630	57038	1526	58418	6
	1962	53180	1847	54432	1736	55725	1628	57060	1524	58442	
25	1960	53200	1845	54453	1734	55746	1627	57083	1523	58465	5
	1958	53221	1843	54474	1732	55768	1625	57106	1521	58489	
26	1956	53241	1841	54496	1730	55790	1623	57128	1519	58512	4
	1954	53262	1839	54517	1728	55812	1621	57151	1517	58536	
27	1952	53283	1837	54538	1727	55834	1619	57174	1516	58559	3
	1950	53303	1836	54559	1725	55856	1618	57196	1514	58583	
28	1948	53324	1834	54581	1723	55878	1616	57219	1512	58606	2
	1946	53344	1832	54602	1721	55900	1614	57242	1511	58630	
29	1944	53365	1830	54623	1719	55922	1612	57265	1509	58653	1
	1942	53386	1828	54644	1718	55944	1611	57287	1507	58677	
30	1940	53406	1826	54666	1716	55966	1609	57310	1506	58700	0
	A	B	A	B	A	B	A	B	A	B	
′	107° 00′		106° 30′		106° 00′		105° 30′		105° 00′		′

TABLE 35

The Ageton Method

When Meridian Angle is Greater Than 90° Take "K" From Bottom of Table

′	75° 00′		75° 30′		76° 00′		76° 30′		77° 00′		′
	A	B	A	B	A	B	A	B	A	B	
0	1506	58700	1406	60140	1310	61632	1217	63181	1128	64791	30
	1504	58724	1404	60164	1308	61658	1215	63208	1126	64819	
1	1502	58748	1403	60189	1306	61683	1214	63234	1125	64846	29
	1500	58771	1401	60213	1305	61709	1212	63260	1123	64873	
2	1499	58795	1399	60238	1303	61734	1211	63287	1122	64901	28
	1497	58818	1398	60262	1301	61759	1209	63313	1120	64928	
3	1495	58842	1396	60287	1300	61785	1208	63340	1119	64956	27
	1494	58866	1394	60311	1299	61810	1206	63366	1117	64983	
4	1492	58889	1393	60336	1297	61836	1205	63392	1116	65011	26
	1490	58913	1391	60360	1295	61861	1203	63419	1114	65038	
5	1489	58937	1390	60385	1294	61887	1202	63445	1113	65066	25
	1487	58960	1388	60410	1292	61912	1200	63472	1112	65093	
6	1485	58984	1386	60434	1291	61938	1199	63498	1110	65121	24
	1484	59008	1385	60459	1289	61963	1197	63525	1109	65148	
7	1482	59032	1383	60483	1288	61989	1196	63551	1107	65176	23
	1480	59055	1381	60508	1286	62014	1194	63578	1106	65204	
8	1479	59079	1380	60533	1284	62040	1193	63605	1104	65231	22
	1477	59103	1378	60557	1283	62065	1191	63631	1103	65259	
9	1475	59127	1377	60582	1281	62091	1190	63658	1101	65287	21
	1474	59151	1375	60607	1280	62117	1188	63684	1100	65314	
10	1472	59175	1373	60631	1278	62142	1187	63711	1099	65342	20
	1470	59198	1372	60656	1277	62168	1185	63738	1097	65370	
11	1469	59222	1370	60681	1275	62194	1184	63764	1096	65398	19
	1467	59246	1368	60706	1274	62219	1182	63791	1094	65425	
12	1465	59270	1367	60730	1272	62245	1181	63818	1093	65453	18
	1464	59294	1365	60755	1270	62271	1179	63845	1091	65481	
13	1462	59318	1364	60780	1269	62296	1178	63871	1090	65509	17
	1460	59342	1362	60805	1267	62322	1176	63898	1089	65537	
14	1459	59366	1360	60830	1266	62348	1175	63925	1087	65564	16
	1457	59390	1359	60855	1264	62374	1173	63952	1086	65592	
15	1455	59414	1357	60879	1263	62400	1172	63978	1084	65620	15
	1454	59438	1356	60904	1261	62425	1170	64005	1083	65648	
16	1452	59462	1354	60929	1260	62451	1169	64032	1081	65676	14
	1450	59486	1352	60954	1258	62477	1167	64059	1080	65704	
17	1449	59510	1351	60979	1257	62503	1166	64086	1079	65732	13
	1447	59534	1349	61004	1255	62529	1164	64113	1077	65760	
18	1445	59558	1348	61029	1253	62555	1163	64140	1076	65788	12
	1444	59582	1346	61054	1252	62581	1161	64167	1074	65816	
19	1442	59606	1344	61079	1250	62607	1160	64194	1073	65844	11
	1440	59630	1343	61104	1249	62633	1158	64221	1071	65872	
20	1439	59654	1341	61129	1247	62659	1157	64248	1070	65900	10
	1437	59679	1340	61154	1246	62685	1155	64275	1069	65928	
21	1435	59703	1338	61179	1244	62711	1154	64302	1067	65957	9
	1434	59727	1336	61204	1243	62737	1152	64329	1066	65985	
22	1432	59751	1335	61229	1241	62763	1151	64356	1064	66013	8
	1430	59775	1333	61254	1240	62789	1150	64383	1063	66041	
23	1429	59800	1332	61279	1238	62815	1148	64410	1061	66069	7
	1427	59824	1330	61304	1237	62841	1147	64437	1060	66098	
24	1425	59848	1329	61330	1235	62867	1145	64464	1059	66126	6
	1424	59872	1327	61355	1234	62893	1144	64491	1057	66154	
25	1422	59896	1325	61380	1232	62919	1142	64518	1056	66182	5
	1421	59921	1324	61405	1230	62945	1141	64546	1054	66211	
26	1419	59945	1322	61430	1229	62971	1139	64573	1053	66239	4
	1417	59969	1321	61456	1227	62998	1138	64600	1052	66267	
27	1416	59994	1319	61481	1226	63024	1136	64627	1050	66296	3
	1414	60018	1317	61506	1224	63050	1135	64655	1049	66324	
28	1412	60042	1316	61531	1223	63076	1133	64682	1047	66352	2
	1411	60067	1314	61556	1221	63103	1132	64709	1046	66381	
29	1409	60091	1313	61582	1220	63129	1130	64736	1045	66409	1
	1407	60116	1311	61607	1218	63155	1129	64764	1043	66438	
30	1406	60140	1310	61632	1217	63181	1128	64791	1042	66466	0
	A	B	A	B	A	B	A	B	A	B	
′	104° 30′		104° 00′		103° 30′		103° 00′		102° 30′		′

TABLE 35

The Ageton Method

Always Take "Z" From Bottom of Table, Except When "K" is Same Name and Greater Than Latitude, in Which Case Take "Z" From Top of Table

′	77° 30′		78° 00′		78° 30′		79° 00′		79° 30′		′
	A	B	A	B	A	B	A	B	A	B	
0	1042	66466	960	68212	881	70034	805	71940	733	73937	30
	1040	66495	958	68242	879	70065	804	71973	732	73971	
1	1039	66523	957	68272	878	70097	803	72005	731	74005	29
	1038	66552	955	68301	877	70128	802	72038	730	74039	
2	1036	66580	954	68331	876	70159	800	72070	729	74073	28
	1035	66609	953	68361	874	70190	799	72103	728	74107	
3	1033	66638	951	68391	873	70221	798	72136	726	74142	27
	1032	66666	950	68421	872	70252	797	72168	725	74176	
4	1031	66695	949	68450	870	70284	796	72201	724	74210	26
	1029	66724	947	68480	869	70315	794	72234	723	74245	
5	1028	66752	946	68510	868	70346	793	72266	722	74279	25
	1026	66781	945	68540	867	70377	792	72299	721	74313	
6	1025	66810	943	68570	865	70409	791	72332	719	74348	24
	1024	66838	942	68600	864	70440	790	72365	718	74382	
7	1022	66867	941	68630	863	70471	788	72397	717	74417	23
	1021	66896	939	68660	862	70503	787	72430	716	74451	
8	1020	66925	938	68690	860	70534	786	72463	715	74486	22
	1018	66953	937	68720	859	70566	785	72496	714	74520	
9	1017	66982	935	68750	858	70597	783	72529	712	74555	21
	1015	67011	934	68781	856	70629	782	72562	711	74589	
10	1014	67040	933	68811	855	70660	781	72595	710	74624	20
	1013	67069	932	68841	854	70692	780	72628	709	74659	
11	1011	67098	930	68871	853	70723	779	72661	708	74693	19
	1010	67127	929	68901	851	70755	777	72694	707	74728	
12	1008	67156	928	68931	850	70786	776	72727	706	74763	18
	1007	67185	926	68962	849	70818	775	72760	704	74797	
13	1006	67214	925	68992	848	70850	774	72794	703	74832	17
	1004	67243	924	69022	846	70881	772	72827	702	74867	
14	1003	67272	922	69053	845	70913	771	72860	701	74902	16
	1002	67301	921	69083	844	70945	770	72893	700	74937	
15	1000	67330	920	69113	843	70976	769	72926	699	74972	15
	999	67359	918	69144	841	71008	768	72960	698	75007	
16	997	67388	917	69174	840	71040	767	72993	696	75042	14
	996	67417	916	69204	839	71072	765	73026	695	75077	
17	995	67447	914	69235	838	71104	764	73060	694	75112	13
	993	67476	913	69265	836	71135	763	73093	693	75147	
18	992	67505	912	69296	835	71167	762	73127	692	75182	12
	991	67534	910	69326	834	71199	761	73160	691	75217	
19	989	67563	909	69357	833	71231	759	73193	690	75252	11
	988	67593	908	69387	831	71263	758	73227	688	75287	
20	987	67622	907	69418	830	71295	757	73260	687	75322	10
	985	67651	905	69449	829	71327	756	73294	686	75358	
21	984	67681	904	69479	828	71359	755	73328	685	75393	9
	982	67710	903	69510	826	71391	753	73361	684	75428	
22	981	67739	901	69541	825	71423	752	73395	683	75464	8
	980	67769	900	69571	824	71455	751	73429	682	75499	
23	978	67798	899	69602	823	71488	750	73462	680	75534	7
	977	67828	897	69633	821	71520	749	73496	679	75570	
24	976	67857	896	69664	820	71552	747	73530	678	75605	6
	974	67886	895	69694	819	71584	746	73563	677	75641	
25	973	67916	894	69725	818	71616	745	73597	676	75676	5
	972	67945	892	69756	816	71649	744	73631	675	75712	
26	970	67975	891	69787	815	71681	743	73665	674	75747	4
	969	68005	890	69818	814	71713	742	73699	673	75783	
27	968	68034	888	69849	813	71746	740	73733	672	75819	3
	966	68064	887	69879	811	71778	739	73767	670	75854	
28	965	68093	886	69910	810	71810	738	73801	669	75890	2
	964	68123	885	69941	809	71843	737	73835	668	75926	
29	962	68153	883	69972	808	71875	736	73869	667	75961	1
	961	68182	882	70003	807	71908	735	73903	666	75997	
30	960	68212	881	70034	805	71940	733	73937	665	76033	0
	A	B	A	B	A	B	A	B	A	B	
′	102° 00′		101° 30′		101° 00′		100° 30′		100° 00′		′

TABLE 35

The Ageton Method

When Meridian Angle is Greater Than 90° Take "K" From Bottom of Table

′	80° 00′		80° 30′		81° 00′		81° 30′		82° 00′		′
′	A	B	A	B	A	B	A	B	A	B	′
0	665	76033	600	78239	538	80567	480	83030	425	85644	30
	664	76069	599	78277	537	80607	479	83072	424	85689	
1	663	76105	598	78315	536	80647	478	83114	423	85734	29
	661	76141	597	78352	535	80687	477	83157	422	85779	
2	660	76176	595	78390	534	80727	476	83199	421	85825	28
	659	76212	594	78428	533	80767	475	83242	420	85870	
3	658	76248	593	78466	532	80807	474	83284	419	85915	27
	657	76284	592	78504	531	80847	473	83327	418	85960	
4	656	76320	591	78542	530	80887	472	83369	418	86006	26
	655	76357	590	78580	529	80927	471	83412	417	86051	
5	654	76393	589	78618	528	80967	470	83455	416	86096	25
	653	76429	588	78656	527	81008	469	83497	415	86142	
6	652	76465	587	78694	526	81048	468	83540	414	86187	24
	650	76501	586	78733	525	81088	467	83583	413	86233	
7	649	76537	585	78771	524	81129	467	83626	412	86278	23
	648	76574	584	78809	523	81169	466	83668	411	86324	
8	647	76610	583	78847	522	81210	465	83711	411	86370	22
	646	76646	582	78886	521	81250	464	83754	410	86415	
9	645	76683	581	78924	520	81291	463	83797	409	86461	21
	644	76719	580	78962	519	81331	462	83840	408	86507	
10	643	76756	579	79001	518	81372	461	83884	407	86553	20
	642	76792	578	79039	517	81413	460	83927	406	86599	
11	641	76828	577	79078	516	81453	459	83970	405	86645	19
	639	76865	576	79116	515	81494	458	84013	405	86691	
12	638	76902	575	79155	514	81535	457	84056	404	86737	18
	637	76938	574	79193	513	81576	456	84100	403	86783	
13	636	76975	573	79232	512	81617	455	84143	402	86829	17
	635	77011	571	79271	511	81657	454	84186	401	86876	
14	634	77048	570	79309	510	81698	454	84230	400	86922	16
	633	77085	569	79348	509	81739	453	84273	399	86968	
15	632	77122	568	79387	508	81780	452	84317	399	87015	15
	631	77158	567	79426	507	81821	451	84361	398	87061	
16	630	77195	566	79465	506	81863	450	84404	397	87107	14
	629	77232	565	79503	505	81904	449	84448	396	87154	
17	627	77269	564	79542	504	81945	448	84492	395	87201	13
	626	77306	563	79581	504	81986	447	84535	394	87247	
18	625	77343	562	79620	503	82027	446	84579	393	87294	12
	624	77380	561	79659	502	82069	445	84623	392	87341	
19	623	77417	560	79698	501	82110	444	84667	392	87387	11
	622	77454	559	79737	500	82151	444	84711	391	87434	
20	621	77491	558	79777	499	82193	443	84755	390	87481	10
	620	77528	557	79816	498	82234	442	84799	389	87528	
21	619	77565	556	79855	497	82276	441	84843	388	87575	9
	618	77602	555	79894	496	82317	440	84887	387	87622	
22	617	77639	554	79933	495	82359	439	84931	387	87669	8
	616	77677	553	79973	494	82400	438	84976	386	87716	
23	615	77714	552	80012	493	82442	437	85020	385	87764	7
	614	77751	551	80051	492	82484	436	85064	384	87811	
24	612	77788	550	80091	491	82526	435	85109	383	87858	6
	611	77826	549	80130	490	82567	434	85153	382	87906	
25	610	77863	548	80170	489	82609	434	85197	381	87953	5
	609	77901	547	80209	488	82651	433	85242	381	88001	
26	608	77938	546	80249	487	82693	432	85286	380	88048	4
	607	77976	545	80288	486	82735	431	85331	379	88096	
27	606	78013	544	80328	485	82777	430	85376	378	88143	3
	605	78051	543	80368	484	82819	429	85420	377	88191	
28	604	78088	542	80407	483	82861	428	85465	376	88239	2
	603	78126	541	80447	482	82903	427	85510	376	88286	
29	602	78164	540	80487	482	82945	426	85555	375	88334	1
	601	78201	539	80527	481	82987	426	85599	374	88382	
30	600	78239	538	80567	480	83030	425	85644	373	88430	0
	A	B	A	B	A	B	A	B	A	B	
′	99° 30′		99° 00′		98° 30′		98° 00′		97° 30′		′

TABLE 35

The Ageton Method

Always Take "Z" From Bottom of Table, Except When "K" is Same Name and Greater Than Latitude, in Which Case Take "Z" From Top of Table

′	82° 30′		83° 00′		83° 30′		84° 00′		84° 30′		′
	A	B	A	B	A	B	A	B	A	B	
0	373	88430	325	91411	280	94614	238.6	98076	200.4	101843	30
	372	88478	324	91462	279	94670	237.9	98137	199.8	101908	
1	371	88526	323	91514	279	94725	237.2	98197	199.2	101974	29
	371	88574	323	91565	278	94781	236.6	98257	198.6	102040	
2	370	88623	322	91617	277	94836	235.9	98318	198.0	102106	28
	369	88671	321	91668	276	94892	235.3	98378	197.4	102172	
3	368	88719	320	91720	276	94948	234.6	98439	196.8	102238	27
	367	88767	319	91772	275	95004	233.9	98499	196.2	102304	
4	366	88816	319	91824	274	95060	233.3	98560	195.6	102371	26
	366	88864	318	91876	274	95116	232.6	98621	195.0	102437	
5	365	88913	317	91928	273	95172	232.0	98682	194.4	102504	25
	364	88961	316	91980	272	95228	231.3	98743	193.8	102570	
6	363	89010	316	92032	271	95285	230.7	98804	193.2	102637	24
	362	89059	315	92085	271	95341	230.0	98865	192.6	102704	
7	362	89107	314	92137	270	95397	229.4	98926	192.0	102771	23
	361	89156	313	92189	269	95454	228.7	98988	191.4	102838	
8	360	89205	313	92242	269	95510	228.1	99049	190.8	102905	22
	359	89254	312	92294	268	95567	227.4	99111	190.2	102973	
9	358	89303	311	92347	267	95624	226.8	99172	189.6	103040	21
	357	89352	310	92399	267	95681	226.1	99234	189.0	103107	
10	357	89401	310	92452	266	95737	225.5	99296	188.4	103175	20
	356	89450	309	92505	265	95795	224.8	99357	187.8	103243	
11	355	89499	308	92558	264	95851	224.2	99419	187.2	103311	19
	354	89548	307	92610	264	95909	223.5	99482	186.7	103379	
12	353	89597	307	92663	263	95966	222.9	99544	186.1	103447	18
	353	89647	306	92716	262	96023	222.3	99606	185.5	103515	
13	352	89696	305	92769	262	96080	221.6	99668	184.9	103583	17
	351	89746	304	92823	261	96138	221.0	99731	184.3	103651	
14	350	89795	304	92876	260	96195	220.3	99793	183.7	103720	16
	349	89845	303	92929	260	96253	219.7	99856	183.2	103788	
15	349	89894	302	92982	259	96310	219.1	99918	182.6	103857	15
	348	89944	301	93036	258	96368	218.4	99981	182.0	103926	
16	347	89994	301	93089	257	96426	217.8	100044	181.4	103995	14
	346	90044	300	93143	257	96484	217.2	100107	180.8	104064	
17	345	90093	299	93196	256	96542	216.5	100170	130.3	104133	13
	345	90143	298	93250	255	96600	215.9	100233	179.7	104202	
18	344	90193	298	93304	255	96658	215.3	100296	179.1	104272	12
	343	90243	297	93358	254	96716	214.6	100360	178.5	104341	
19	342	90293	296	93411	253	96774	214.0	100423	178.0	104411	11
	341	90344	295	93465	253	96833	213.4	100487	177.4	104480	
20	341	90394	295	93519	252	96891	212.8	100550	176.8	104550	10
	340	90444	294	93573	251	96950	212.1	100614	176.2	104620	
21	339	90494	293	93628	251	97008	211.5	100678	175.7	104690	9
	338	90545	292	93682	250	97067	210.9	100742	175.1	104760	
22	337	90595	292	93736	249	97126	210.3	100806	174.5	104830	8
	337	90646	291	93790	249	97184	209.6	100870	174.0	104901	
23	336	90696	290	93845	248	97243	209.0	100934	173.4	104971	7
	335	90747	289	93899	247	97302	208.4	100998	172.8	105042	
24	334	90798	289	93954	247	97361	207.8	101063	172.3	105113	6
	333	90848	288	94009	246	97420	207.1	101127	171.7	105183	
25	333	90899	287	94063	245	97480	206.5	101192	171.1	105254	5
	332	90950	287	94118	245	97539	205.9	101256	170.6	105325	
26	331	91001	286	94173	244	97598	205.3	101321	170.0	105397	4
	330	91052	285	94228	243	97658	204.7	101386	169.5	105468	
27	330	91103	284	94283	243	97717	204.1	101451	168.9	105539	3
	329	91154	284	94338	242	97777	203.5	101516	168.4	105611	
28	328	91205	283	94393	241	97837	202.8	101581	167.8	105683	2
	327	91257	282	94448	241	97897	202.2	101646	167.2	105754	
29	326	91308	281	94503	240	97957	201.6	101712	166.7	105826	1
	326	91359	281	94559	239	98017	201.0	101777	166.0	105898	
30	325	91411	280	94614	239	98076	200.4	101843	165.6	105970	0
	A	B	A	B	A	B	A	B	A	B	
′	97° 00′		96° 30′		96° 00′		95° 30′		95° 00′		′

TABLE 35

The Ageton Method

When Meridian Angle is Greater Than 90° Take "K" From Bottom of Table

′	85° 00′		85° 30′		86° 00′		86° 30′		87° 00′		′
	A	B	A	B	A	B	A	B	A	B	
0	165.6	105970	134.1	110536	105.9	115641	81.1	121432	59.6	128120	30
	165.0	106043	133.6	110616	105.5	115732	80.7	121536	59.2	128241	
1	164.5	106115	133.1	110696	105.0	115823	80.3	121639	58.9	128362	29
	163.9	106187	132.6	110777	104.6	115913	79.9	121743	58.6	128483	
2	163.4	106260	132.1	110858	104.2	116004	79.5	121848	58.2	128605	28
	162.8	106333	131.6	110939	103.7	116096	79.2	121952	57.9	128727	
3	162.3	106406	131.1	111020	103.3	116187	78.8	122057	57.6	128849	27
	161.7	106479	130.6	111101	102.9	116278	78.4	122161	57.3	128972	
4	161.2	106552	130.1	111183	102.4	116370	78.0	122267	56.9	129095	26
	160.6	106625	129.6	111264	102.0	116462	77.6	122372	56.6	129218	
5	160.1	106698	129.2	111346	101.6	116554	77.3	122478	56.3	129342	25
	159.6	106772	128.7	111428	101.1	116647	76.9	122584	56.0	129466	
6	159.0	106846	128.2	111510	100.7	116739	76.5	122690	55.7	129591	24
	158.5	106919	127.7	111592	100.3	116832	76.1	122796	55.3	129716	
7	157.9	106993	127.2	111674	99.8	116925	75.8	122903	55.0	129841	23
	157.4	107067	126.7	111757	99.4	117018	75.4	123010	54.7	129967	
8	156.9	107141	126.2	111839	99.0	117112	75.0	123117	54.4	130093	22
	156.3	107216	125.8	111922	98.5	117205	74.6	123225	54.1	130219	
9	155.8	107290	125.3	112005	98.1	117299	74.3	123332	53.7	130346	21
	155.2	107364	124.8	112088	97.7	117393	73.9	123441	53.4	130473	
10	154.7	107439	124.3	112171	97.3	117487	73.5	123549	53.1	130600	20
	154.2	107514	123.8	112255	96.8	117581	73.2	123657	52.8	130728	
11	153.6	107589	123.4	112338	96.4	117676	72.8	123766	52.5	130856	19
	153.1	107664	122.9	112422	96.0	117771	72.4	123875	52.2	130985	
12	152.6	107739	122.4	112506	95.6	117866	72.1	123985	51.9	131114	18
	152.1	107814	121.9	112590	95.2	117961	71.7	124095	51.6	131243	
13	151.5	107890	121.5	112674	94.7	118056	71.3	124204	51.3	131373	17
	151.0	107965	121.0	112759	94.3	118152	71.0	124315	51.0	131503	
14	150.5	108041	120.5	112843	93.9	118248	70.6	124425	50.7	131633	16
	149.9	108117	120.1	112928	93.5	118344	70.3	124536	50.3	131764	
15	149.4	108193	119.6	113013	93.1	118440	69.9	124647	50.0	131896	15
	148.9	108269	119.1	113098	92.7	118537	69.5	124759	49.7	132027	
16	148.4	108345	118.7	113183	92.3	118633	69.2	124870	49.4	132159	14
	147.8	108421	118.2	113269	91.8	118730	68.8	124982	49.1	132292	
17	147.3	108498	117.7	113354	91.4	118827	68.5	125094	48.8	132425	13
	146.8	108574	117.3	113440	91.0	118925	68.1	125207	48.5	132558	
18	146.3	108651	116.8	113526	90.6	119022	67.8	125320	48.2	132692	12
	145.8	108728	116.3	113612	90.2	119120	67.4	125433	47.9	132826	
19	145.2	108805	115.9	113699	89.8	119218	67.1	125546	47.6	132961	11
	144.7	108882	115.4	113785	89.4	119316	66.7	125660	47.3	133096	
20	144.2	108960	114.9	113872	89.0	119415	66.4	125774	47.1	133231	10
	143.7	109037	114.5	113958	88.6	119513	66.0	125888	46.8	133367	
21	143.2	109115	114.0	114045	88.2	119612	65.7	126003	46.5	133503	9
	142.7	109192	113.6	114133	87.8	119711	65.3	126118	46.2	133640	
22	142.2	109270	113.1	114220	87.4	119811	65.0	126233	45.9	133777	8
	141.6	109348	112.7	114307	87.0	119910	64.6	126349	45.6	133914	
23	141.1	109426	112.2	114395	86.6	120010	64.3	126465	45.3	134052	7
	140.6	109505	111.7	114483	86.2	120110	63.9	126581	45.0	134191	
24	140.1	109583	111.3	114571	85.8	120211	63.6	126697	44.7	134330	6
	139.6	109662	110.8	114659	85.4	120311	63.3	126814	44.4	134469	
25	139.1	109740	110.4	114747	85.0	120412	62.9	126931	44.2	134609	5
	138.6	109819	109.9	114836	84.6	120513	62.6	127049	43.9	134749	
26	138.1	109898	109.5	114925	84.2	120614	62.2	127166	43.6	134890	4
	137.6	109978	109.0	115014	83.8	120715	61.9	127284	43.3	135031	
27	137.1	110057	108.6	115103	83.4	120817	61.6	127403	43.0	135173	3
	136.6	110136	108.1	115192	83.0	120919	61.2	127521	42.7	135315	
28	136.1	110216	107.7	115282	82.6	121021	60.9	127640	42.5	135457	2
	135.6	110296	107.3	115371	82.2	121124	60.6	127760	42.2	135600	
29	135.1	110375	106.8	115461	81.9	121226	60.2	127880	41.9	135744	1
	134.6	110455	106.4	115551	81.5	121329	59.9	128000	41.6	135888	
30	134.1	110536	105.9	115641	81.1	121432	59.6	128120	41.4	136032	0
	A	B	A	B	A	B	A	B	A	B	
′	94° 30′		94° 00′		93° 30′		93° 00′		92° 30′		′

TABLE 35

The Ageton Method

Always Take "Z" From Bottom of Table, Except When "K" is Same Name and Greater Than Latitude, in Which Case Take "Z" From Top of Table

	87° 30′		88° 00′		88° 30′		89° 00′		89° 30′		
′	A	B	A	B	A	B	A	B	A	B	′
0	41.4	136032	26.5	145718	14.9	158208	6.6	175814	1.7	205916	30
	41.1	136177	26.2	145899	14.7	158450	6.5	176178	1.6	206646	
1	40.8	136322	26.0	146081	14.6	158693	6.4	176544	1.5	207388	29
	40.5	136468	25.8	146264	14.4	158938	6.3	176914	1.5	208143	
2	40.3	136615	25.6	146448	14.2	159184	6.2	177287	1.4	208912	28
	40.0	136761	25.4	146632	14.1	159431	6.1	177663	1.4	209695	
3	39.7	136909	25.2	146817	13.9	159680	6.0	178042	1.3	210491	27
	39.4	137057	24.9	147003	13.7	159930	5.9	178424	1.3	211303	
4	39.2	137205	24.7	147190	13.6	160182	5.8	178810	1.2	212130	26
	38.9	137354	24.5	147377	13.4	160435	5.7	179200	1.2	212974	
5	38.6	137503	24.3	147566	13.3	160690	5.6	179593	1.1	213834	25
	38.4	137653	24.1	147755	13.1	160946	5.5	179990	1.1	214711	
6	38.1	137804	23.9	147945	13.0	161204	5.4	180390	1.1	215607	24
	37.8	137955	23.7	148135	12.8	161463	5.3	180794	1.0	216521	
7	37.6	138106	23.5	148327	12.7	161724	5.2	181201	1.0	217455	23
	37.3	138258	23.3	148520	12.5	161986	5.1	181613	0.9	218409	
8	37.1	138411	23.1	148713	12.4	162250	5.0	182029	0.9	219385	22
	36.8	138564	22.8	148907	12.2	162516	4.9	182448	0.9	220384	
9	36.5	138718	22.6	149103	12.1	162783	4.8	182872	0.8	221406	21
	36.3	138872	22.4	149299	11.9	163052	4.7	183300	0.8	222452	
10	36.0	139027	22.2	149495	11.8	163322	4.6	183732	0.7	223525	20
	35.8	139182	22.0	149693	11.6	163594	4.5	184168	0.7	224624	
11	35.5	139338	21.8	149892	11.5	163868	4.4	184609	0.7	225752	19
	35.3	139494	21.6	150092	11.3	164144	4.3	185055	0.6	226910	
12	35.0	139651	21.4	150292	11.2	164422	4.2	185505	0.6	228100	18
	34.7	139809	21.2	150494	11.0	164701	4.1	185959	0.6	229324	
13	34.5	139967	21.0	150696	10.9	164982	4.1	186419	0.5	230583	17
	34.2	140125	20.8	150899	10.8	165265	4.0	186883	0.5	231879	
14	34.0	140285	20.6	151104	10.6	165550	3.9	187353	0.5	233215	16
	33.7	140445	20.5	151309	10.5	165836	3.8	187827	0.4	234594	
15	33.5	140605	20.3	151515	10.3	166125	3.7	188307	0.4	236018	15
	33.2	140766	20.1	151722	10.2	166415	3.6	188793	0.4	237491	
16	33.0	140928	19.9	151931	10.1	166708	3.6	189283	0.4	239015	14
	32.8	141090	19.7	152140	9.9	167002	3.5	189780	0.3	240594	
17	32.5	141253	19.5	152350	9.8	167298	3.4	190282	0.3	242233	13
	32.3	141417	19.3	152561	9.7	167597	3.3	190790	0.3	243936	
18	32.0	141581	19.1	152774	9.5	167897	3.2	191303	0.3	245709	12
	31.8	141745	18.9	152987	9.4	168200	3.2	191824	0.2	247558	
19	31.5	141911	18.7	153201	9.3	168505	3.1	192350	0.2	249488	11
	31.3	142077	18.6	153417	9.1	168811	3.0	192883	0.2	251508	
20	31.1	142243	18.4	153633	9.0	169121	2.9	193422	0.2	253627	10
	30.8	142411	18.2	153851	8.9	169432	2.9	193969	0.2	255855	
21	30.6	142579	18.0	154070	8.7	169745	2.8	194522	0.1	258203	9
	30.4	142747	17.8	154290	8.6	170061	2.7	195082	0.1	260685	
22	30.1	142916	17.6	154511	8.5	170379	2.7	195650	0.1	263318	8
	29.9	143086	17.5	154733	8.4	170700	2.6	196225	0.1	266121	
23	29.6	143257	17.3	154956	8.2	171023	2.5	196808	0.1	269118	7
	29.4	143428	17.1	155180	8.1	171348	2.4	197399	0.1	272336	
24	29.2	143600	16.9	155406	8.0	171676	2.4	197998	0.1	275812	6
	28.9	143773	16.8	155633	7.9	172006	2.3	198605	0.1	279591	
25	28.7	143946	16.6	155861	7.8	172339	2.3	199221	0.0	283730	5
	28.5	144120	16.4	156090	7.6	172674	2.2	199846	0.0	288306	
26	28.3	144295	16.2	156320	7.5	173012	2.1	200480	0.0	293421	4
	28.0	144470	16.1	156552	7.4	173352	2.1	201124	0.0	299221	
27	27.8	144646	15.9	156784	7.3	173696	2.0	201777	0.0	305915	3
	27.6	144823	15.7	157019	7.2	174042	1.9	202440	0.0	313833	
28	27.4	145000	15.6	157254	7.1	174391	1.9	203113	0.0	323524	2
	27.1	145179	15.4	157490	6.9	174742	1.8	203797	0.0	336018	
29	26.9	145358	15.2	157728	6.8	175097	1.8	204492	0.0	353627	1
	26.7	145538	15.1	157967	6.7	175454	1.7	205198	0.0	383730	
30	26.5	145718	14.9	158208	6.6	175814	1.7	205916	0.0	--------	0
	A	B	A	B	A	B	A	B	A	B	
	92° 00′		91° 30′		91° 00′		90° 30′		90° 00′		′

TABLE 36

Time Zones, Zone Descriptions, and Suffixes

ZONE	ZD	SUFFIX	ZONE	ZD	SUFFIX
7½°W. to 7½°E.	0	Z	7½°W. to 22½°W.	+ 1	N
7½°E. to 22½°E.	− 1	A	22½°W. to 37½°W.	+ 2	O
22½°E. to 37½°E.	− 2	B	37½°W. to 52½°W.	+ 3	P
37½°E. to 52½°E.	− 3	C	52½°W. to 67½°W.	+ 4	Q
52½°E. to 67½°E.	− 4	D	67½°W. to 82½°W.	+ 5	R
67½°E. to 82½°E.	− 5	E	82½°W. to 97½°W.	+ 6	S
82½°E. to 97½°E.	− 6	F	97½°W. to 112½°W.	+ 7	T
97½°E. to 112½°E.	− 7	G	112½°W. to 127½°W.	+ 8	U
112½°E. to 127½°E.	− 8	H	127½°W. to 142½°W.	+ 9	V
127½°E. to 142½°E.	− 9	I	142½°W. to 157½°W.	+10	W
142½°E. to 157½°E.	−10	K	157½°W. to 172½°W.	+11	X
157½°E. to 172½°E.	−11	L	172½°W. to 180°	+12	Y
172½°E. to 180°	−12	M			

NOTE.—G M T. is indicated by suffix Z. Standard times as kept in various places or countries are listed in *The Nautical Almanac* and *The Air Almanac.*

TABLE 37

Natural and Numerical Chart Scales

Natural Scale	Miles Per Inch		Inches Per Mile		Feet Per Inch
	Nautical	Statute	Nautical	Statute	
1:500	0.007	0.008	145.83	126.72	41.67
1:600	0.008	0.009	121.52	105.60	50.00
1:1,000	0.014	0.016	72.91	63.36	83.33
1:1,200	0.016	0.019	60.76	52.80	100.00
1:1,500	0.021	0.024	48.61	42.24	125.00
1:2,000	0.027	0.032	36.46	31.68	166.67
1:2,400	0.033	0.038	30.38	26.40	200.00
1:2,500	0.034	0.039	29.17	25.34	208.33
1:3,000	0.041	0.047	24.30	21.12	250.00
1:3,600	0.049	0.057	20.25	17.60	300.00
1:4,000	0.055	0.063	18.23	15.84	333.33
1:4,800	0.066	0.076	15.19	13.20	400.00
1:5,000	0.069	0.079	14.58	12.67	416.67
1:6,000	0.082	0.095	12.15	10.56	500.00
1:7,000	0.096	0.110	10.42	9.05	583.33
1:7,200	0.099	0.114	10.13	8.80	600.00
1:7,920	0.109	0.125	9.21	8.00	660.00
1:8,000	0.110	0.126	9.11	7.92	666.67
1:8,400	0.115	0.133	8.68	7.54	700.00
1:9,000	0.123	0.142	8.10	7.04	750.00
1:9,600	0.132	0.152	7.60	6.60	800.00
1:10,000	0.137	0.158	7.29	6.34	833.33
1:10,800	0.148	0.170	6.75	5.87	900.00
1:12,000	0.165	0.189	6.08	5.28	1,000.00
1:13,200	0.181	0.208	5.52	4.80	1,100.00
1:14,400	0.197	0.227	5.06	4.40	1,200.00
1:15,000	0.206	0.237	4.86	4.22	1,250.00
1:15,600	0.214	0.246	4.67	4.06	1,300.00
1:15,840	0.217	0.250	4.60	4.00	1,320.00
1:16,000	0.219	0.253	4.56	3.96	1,333.33
1:16,800	0.230	0.265	4.34	3.77	1,400.00
1:18,000	0.247	0.284	4.05	3.52	1,500.00
1:19,200	0.263	0.303	3.80	3.30	1,600.00
1:20,000	0.274	0.316	3.65	3.17	1,666.67
1:20,400	0.280	0.322	3.57	3.11	1,700.00
1:21,120	0.290	0.333	3.45	3.00	1,760.00
1:21,600	0.296	0.341	3.38	2.93	1,800.00
1:22,800	0.313	0.360	3.20	2.78	1,900.00
1:24,000	0.329	0.379	3.04	2.64	2,000.00
1:25,000	0.343	0.395	2.92	2.53	2,083.33
1:40,000	0.549	0.631	1.82	1.58	3,333.33
1:48,000	0.658	0.758	1.52	1.32	4,000.00
1:50,000	0.686	0.789	1.46	1.27	4,166.67
1:62,500	0.857	0.986	1.17	1.01	5,208.33
1:63,360	0.869	1.000	1.15	1.00	5,280.00
1:75,000	1.029	1.184	0.97	0.85	6,250.00
1:80,000	1.097	1.263	0.91	0.79	6,666.67
1:100,000	1.371	1.578	0.73	0.63	8,333.33
1:125,000	1.714	1.973	0.58	0.51	10,416.67
1:200,000	2.743	3.157	0.36	0.32	16,666.67
1:250,000	3.429	3.946	0.29	0.25	20,833.33
1:400,000	5.486	6.313	0.18	0.16	33,333.33
1:500,000	6.857	7.891	0.15	0.13	41,666.67
1:750,000	10.286	11.837	0.10	0.08	62,500.00
1:1,000,000	13.715	15.783	0.07	0.06	83,333.33
FORMULAS	$\frac{\text{SCALE}}{72{,}913.39}$	$\frac{\text{SCALE}}{63{,}360}$	$\frac{72{,}913.39}{\text{SCALE}}$	$\frac{63{,}360}{\text{SCALE}}$	$\frac{\text{SCALE}}{12}$

TABLE 38

Compact Traverse Table

0°—180°—180°—360°

DLo	p		p	DLo		
D	l	p	l	D	m	DLo
1	1. 000	0. 000	1	1. 000	1	0. 000
2	2. 000	0. 000	2	2. 000	2	0. 000
3	3. 000	0. 000	3	3. 000	3	0. 000
4	4. 000	0. 000	4	4. 000	4	0. 000
5	5. 000	0. 000	5	5. 000	5	0. 000
6	6. 000	0. 000	6	6. 000	6	0. 000
7	7. 000	0. 000	7	7. 000	7	0. 000
8	8. 000	0. 000	8	8. 000	8	0. 000
9	9. 000	0. 000	9	9. 000	9	0. 000

Course		
0° 180°	p÷l DLo÷m	179° 359°
0. 0	0. 000	1. 0
0. 1	0. 002	0. 9
0. 2	0. 003	0. 8
0. 3	0. 005	0. 7
0. 4	0. 007	0. 6
0. 5	0. 009	0. 5
0. 6	0. 010	0. 4
0. 7	0. 012	0. 3
0. 8	0. 014	0. 2
0. 9	0. 016	0. 1

1°—179°—181°—359°

DLo	p		p	DLo		
D	l	p	l	D	m	DLo
1	1. 000	0. 017	1	1. 000	1	0. 017
2	2. 000	0. 035	2	2. 000	2	0. 035
3	3. 000	0. 052	3	3. 000	3	0. 052
4	3. 999	0. 070	4	4. 001	4	0. 070
5	4. 999	0. 087	5	5. 001	5	0. 087
6	5. 999	0. 105	6	6. 001	6	0. 105
7	6. 999	0. 122	7	7. 001	7	0. 122
8	7. 999	0. 140	8	8. 001	8	0. 140
9	8. 999	0. 157	9	9. 001	9	0. 157

Course		
1° 181°	p÷l DLo÷m	178° 358°
0. 0	0. 017	1. 0
0. 1	0. 019	0. 9
0. 2	0. 021	0. 8
0. 3	0. 023	0. 7
0. 4	0. 024	0. 6
0. 5	0. 026	0. 5
0. 6	0. 028	0. 4
0. 7	0. 030	0. 3
0. 8	0. 031	0. 2
0. 9	0. 033	0. 1

2°—178°—182°—358°

DLo	p		p	DLo		
D	l	p	l	D	m	DLo
1	0. 999	0. 035	1	1. 001	1	0. 035
2	1. 999	0. 070	2	2. 001	2	0. 070
3	2. 998	0. 105	3	3. 002	3	0. 105
4	3. 998	0. 140	4	4. 002	4	0. 140
5	4. 997	0. 174	5	5. 003	5	0. 175
6	5. 996	0. 209	6	6. 004	6	0. 210
7	6. 996	0. 244	7	7. 004	7	0. 244
8	7. 995	0. 279	8	8. 005	8	0. 279
9	8. 995	0. 314	9	9. 005	9	0. 314

Course		
2° 182°	p÷l DLo÷m	177° 357°
0. 0	0. 035	1. 0
0. 1	0. 037	0. 9
0. 2	0. 038	0. 8
0. 3	0. 040	0. 7
0. 4	0. 042	0. 6
0. 5	0. 044	0. 5
0. 6	0. 045	0. 4
0. 7	0. 047	0. 3
0. 8	0. 049	0. 2
0. 9	0. 051	0. 1

3°—177°—183°—357°

DLo	p		p	DLo		
D	l	p	l	D	m	DLo
1	0. 999	0. 052	1	1. 001	1	0. 052
2	1. 997	0. 105	2	2. 003	2	0. 105
3	2. 996	0. 157	3	3. 004	3	0. 157
4	3. 995	0. 209	4	4. 005	4	0. 210
5	4. 993	0. 262	5	5. 007	5	0. 262
6	5. 992	0. 314	6	6. 008	6	0. 314
7	6. 990	0. 366	7	7. 010	7	0. 367
8	7. 989	0. 419	8	8. 011	8	0. 419
9	8. 988	0. 471	9	9. 012	9	0. 472

Course		
3° 183°	p÷l DLo÷m	176° 356°
0. 0	0. 052	1. 0
0. 1	0. 054	0. 9
0. 2	0. 056	0. 8
0. 3	0. 058	0. 7
0. 4	0. 059	0. 6
0. 5	0. 061	0. 5
0. 6	0. 063	0. 4
0. 7	0. 065	0. 3
0. 8	0. 066	0. 2
0. 9	0. 068	0. 1

4°—176°—184°—356°

DLo	p		p	DLo		
D	l	p	l	D	m	DLo
1	0. 998	0. 070	1	1. 002	1	0. 070
2	1. 995	0. 140	2	2. 005	2	0. 140
3	2. 993	0. 209	3	3. 007	3	0. 210
4	3. 990	0. 279	4	4. 010	4	0. 280
5	4. 988	0. 349	5	5. 012	5	0. 350
6	5. 985	0. 419	6	6. 015	6	0. 420
7	6. 983	0. 488	7	7. 017	7	0. 489
8	7. 981	0. 558	8	8. 020	8	0. 559
9	8. 978	0. 628	9	9. 022	9	0. 629

Course		
4° 184°	p÷l DLo÷m	175° 355°
0. 0	0. 070	1. 0
0. 1	0. 072	0. 9
0. 2	0. 073	0. 8
0. 3	0. 075	0. 7
0. 4	0. 077	0. 6
0. 5	0. 079	0. 5
0. 6	0. 080	0. 4
0. 7	0. 082	0. 3
0. 8	0. 084	0. 2
0. 9	0. 086	0. 1

TABLE 38

Compact Traverse Table

5°—175°—185°—355°

DLo	p		p	DLo		
D	l	p	l	D	m	DLo
1	0. 996	0. 087	1	1. 004	1	0. 087
2	1. 992	0. 174	2	2. 008	2	0. 175
3	2. 989	0. 261	3	3. 011	3	0. 262
4	3. 985	0. 349	4	4. 015	4	0. 350
5	4. 981	0. 436	5	5. 019	5	0. 437
6	5. 977	0. 523	6	6. 023	6	0. 525
7	6. 973	0. 610	7	7. 027	7	0. 612
8	7. 970	0. 697	8	8. 031	8	0. 700
9	8. 966	0. 784	9	9. 034	9	0. 787

Course		
5° 185°	p÷l DLo÷m	174° 354°
0. 0	0. 087	1. 0
0. 1	0. 089	0. 9
0. 2	0. 091	0. 8
0. 3	0. 093	0. 7
0. 4	0. 095	0. 6
0. 5	0. 096	0. 5
0. 6	0. 098	0. 4
0. 7	0. 100	0. 3
0. 8	0. 102	0. 2
0. 9	0. 103	0. 1

6°—174°—186°—354°

DLo	p		p	DLo		
D	l	p	l	D	m	DLo
1	0. 995	0. 105	1	1. 006	1	0. 105
2	1. 989	0. 209	2	2. 011	2	0. 210
3	2. 984	0. 314	3	3. 017	3	0. 315
4	3. 978	0. 418	4	4. 022	4	0. 420
5	4. 973	0. 523	5	5. 028	5	0. 526
6	5. 967	0. 627	6	6. 033	6	0. 631
7	6. 962	0. 732	7	7. 039	7	0. 736
8	7. 956	0. 836	8	8. 044	8	0. 841
9	8. 951	0. 941	9	9. 050	9	0. 946

Course		
6° 186°	p÷l DLo÷m	173° 353°
0. 0	0. 105	1. 0
0. 1	0. 107	0. 9
0. 2	0. 109	0. 8
0. 3	0. 110	0. 7
0. 4	0. 112	0. 6
0. 5	0. 114	0. 5
0. 6	0. 116	0. 4
0. 7	0. 117	0. 3
0. 8	0. 119	0. 2
0. 9	0. 121	0. 1

7°—173°—187°—353°

DLo	p		p	DLo		
D	l	p	l	D	m	DLo
1	0. 993	0. 122	1	1. 008	1	0. 123
2	1. 985	0. 244	2	2. 015	2	0. 246
3	2. 978	0. 366	3	3. 023	3	0. 368
4	3. 970	0. 487	4	4. 030	4	0. 491
5	4. 963	0. 609	5	5. 038	5	0. 614
6	5. 955	0. 731	6	6. 045	6	0. 737
7	6. 948	0. 853	7	7. 053	7	0. 859
8	7. 940	0. 975	8	8. 060	8	0. 982
9	8. 933	1. 097	9	9. 068	9	1. 105

Course		
7° 187°	p÷l DLo÷m	172° 352°
0. 0	0. 123	1. 0
0. 1	0. 125	0. 9
0. 2	0. 126	0. 8
0. 3	0. 128	0. 7
0. 4	0. 130	0. 6
0. 5	0. 132	0. 5
0. 6	0. 133	0. 4
0. 7	0. 135	0. 3
0. 8	0. 137	0. 2
0. 9	0. 139	0. 1

8°—172°—188°—352°

DLo	p		p	DLo		
D	l	p	l	D	m	DLo
1	0. 990	0. 139	1	1. 010	1	0. 141
2	1. 981	0. 278	2	2. 020	2	0. 281
3	2. 971	0. 418	3	3. 029	3	0. 422
4	3. 961	0. 557	4	4. 039	4	0. 562
5	4. 951	0. 696	5	5. 049	5	0. 703
6	5. 942	0. 835	6	6. 059	6	0. 843
7	6. 932	0. 974	7	7. 069	7	0. 984
8	7. 922	1. 113	8	8. 079	8	1. 124
9	8. 912	1. 253	9	9. 088	9	1. 265

Course		
8° 188°	p÷l DLo÷m	171° 351°
0. 0	0. 141	1. 0
0. 1	0. 142	0. 9
0. 2	0. 144	0. 8
0. 3	0. 146	0. 7
0. 4	0. 148	0. 6
0. 5	0. 149	0. 5
0. 6	0. 151	0. 4
0. 7	0. 153	0. 3
0. 8	0. 155	0. 2
0. 9	0. 157	0. 1

9°—171°—189°—351°

DLo	p		p	DLo		
D	l	p	l	D	m	DLo
1	0. 988	0. 156	1	1. 012	1	0. 158
2	1. 975	0. 313	2	2. 025	2	0. 317
3	2. 963	0. 469	3	3. 037	3	0. 475
4	3. 951	0. 626	4	4. 050	4	0. 634
5	4. 938	0. 782	5	5. 062	5	0. 792
6	5. 926	0. 939	6	6. 075	6	0. 950
7	6. 914	1. 095	7	7. 087	7	1. 109
8	7. 902	1. 251	8	8. 100	8	1. 267
9	8. 889	1. 408	9	9. 112	9	1. 425

Course		
9° 189°	p÷l DLo÷m	170° 350°
0. 0	0. 158	1. 0
0. 1	0. 160	0. 9
0. 2	0. 162	0. 8
0. 3	0. 164	0. 7
0. 4	0. 166	0. 6
0. 5	0. 167	0. 5
0. 6	0. 169	0. 4
0. 7	0. 171	0. 3
0. 8	0. 173	0. 2
0. 9	0. 175	0. 1

TABLE 38

Compact Traverse Table

10°—170°—190°—350°

DLo	p		p	DLo		
D	l	p	l	D	m	DLo
1	0.985	0.174	1	1.015	1	0.176
2	1.970	0.347	2	2.031	2	0.353
3	2.954	0.521	3	3.046	3	0.529
4	3.939	0.695	4	4.062	4	0.705
5	4.924	0.868	5	5.077	5	0.882
6	5.909	1.042	6	6.093	6	1.058
7	6.894	1.216	7	7.108	7	1.234
8	7.878	1.389	8	8.123	8	1.411
9	8.863	1.563	9	9.139	9	1.587

Course		
10° 190°	p÷l DLo÷m	169° 349°
0.0	0.176	1.0
0.1	0.178	0.9
0.2	0.180	0.8
0.3	0.182	0.7
0.4	0.184	0.6
0.5	0.185	0.5
0.6	0.187	0.4
0.7	0.189	0.3
0.8	0.191	0.2
0.9	0.193	0.1

11°—169°—191°—349°

DLo	p		p	DLo		
D	l	p	l	D	m	DLo
1	0.982	0.191	1	1.019	1	0.194
2	1.963	0.382	2	2.037	2	0.389
3	2.945	0.572	3	3.056	3	0.583
4	3.927	0.763	4	4.075	4	0.778
5	4.908	0.954	5	5.094	5	0.972
6	5.890	1.145	6	6.112	6	1.166
7	6.871	1.336	7	7.131	7	1.361
8	7.853	1.526	8	8.150	8	1.555
9	8.835	1.717	9	9.168	9	1.749

Course		
11° 191°	p÷l DLo÷m	168° 348°
0.0	0.194	1.0
0.1	0.196	0.9
0.2	0.198	0.8
0.3	0.200	0.7
0.4	0.202	0.6
0.5	0.203	0.5
0.6	0.205	0.4
0.7	0.207	0.3
0.8	0.209	0.2
0.9	0.211	0.1

12°—168°—192°—348°

DLo	p		p	DLo		
D	l	p	l	D	m	DLo
1	0.978	0.208	1	1.022	1	0.213
2	1.956	0.416	2	2.045	2	0.425
3	2.934	0.624	3	3.067	3	0.638
4	3.913	0.832	4	4.089	4	0.850
5	4.891	1.040	5	5.112	5	1.063
6	5.869	1.247	6	6.134	6	1.275
7	6.847	1.455	7	7.156	7	1.488
8	7.825	1.663	8	8.179	8	1.700
9	8.803	1.871	9	9.201	9	1.913

Course		
12° 192°	p÷l DLo÷m	167° 347°
0.0	0.213	1.0
0.1	0.214	0.9
0.2	0.216	0.8
0.3	0.218	0.7
0.4	0.220	0.6
0.5	0.222	0.5
0.6	0.224	0.4
0.7	0.225	0.3
0.8	0.227	0.2
0.9	0.229	0.1

13°—167°—193°—347°

DLo	p		p	DLo		
D	l	p	l	D	m	DLo
1	0.974	0.225	1	1.026	1	0.231
2	1.949	0.450	2	2.053	2	0.462
3	2.923	0.675	3	3.079	3	0.693
4	3.897	0.900	4	4.105	4	0.923
5	4.872	1.125	5	5.132	5	1.154
6	5.846	1.350	6	6.158	6	1.385
7	6.821	1.575	7	7.184	7	1.616
8	7.795	1.800	8	8.210	8	1.847
9	8.769	2.025	9	9.237	9	2.078

Course		
13° 193°	p÷l DLo÷m	166° 346°
0.0	0.231	1.0
0.1	0.233	0.9
0.2	0.235	0.8
0.3	0.236	0.7
0.4	0.238	0.6
0.5	0.240	0.5
0.6	0.242	0.4
0.7	0.244	0.3
0.8	0.246	0.2
0.9	0.247	0.1

14°—166°—194°—346°

DLo	p		p	DLo		
D	l	p	l	D	m	DLo
1	0.970	0.242	1	1.031	1	0.249
2	1.941	0.484	2	2.061	2	0.499
3	2.911	0.726	3	3.092	3	0.748
4	3.881	0.968	4	4.122	4	0.997
5	4.851	1.210	5	5.153	5	1.247
6	5.822	1.452	6	6.184	6	1.496
7	6.792	1.693	7	7.214	7	1.745
8	7.762	1.935	8	8.245	8	1.995
9	8.733	2.177	9	9.276	9	2.244

Course		
14° 194°	p÷l DLo÷m	165° 345°
0.0	0.249	1.0
0.1	0.251	0.9
0.2	0.253	0.8
0.3	0.255	0.7
0.4	0.257	0.6
0.5	0.259	0.5
0.6	0.260	0.4
0.7	0.262	0.3
0.8	0.264	0.2
0.9	0.266	0.1

TABLE 38

Compact Traverse Table

15°—165°—195°—345°

DLo	p		p	DLo		
D	l	p	l	D	m	DLo
1	0. 966	0. 259	1	1. 035	1	0. 268
2	1. 932	0. 518	2	2. 071	2	0. 536
3	2. 898	0. 776	3	3. 106	3	0. 804
4	3. 864	1. 035	4	4. 141	4	1. 072
5	4. 830	1. 294	5	5. 176	5	1. 340
6	5. 796	1. 553	6	6. 212	6	1. 608
7	6. 761	1. 812	7	7. 247	7	1. 876
8	7. 727	2. 071	8	8. 282	8	2. 144
9	8. 693	2. 329	9	9. 317	9	2. 412

Course		
15° 195°	p÷l DLo÷m	164° 344°
0. 0	0. 268	1. 0
0. 1	0. 270	0. 9
0. 2	0. 272	0. 8
0. 3	0. 274	0. 7
0. 4	0. 275	0. 6
0. 5	0. 277	0. 5
0. 6	0. 279	0. 4
0. 7	0. 281	0. 3
0. 8	0. 283	0. 2
0. 9	0. 285	0. 1

16°—164°—196°—344°

DLo	p		p	DLo		
D	l	p	l	D	m	DLo
1	0. 961	0. 276	1	1. 040	1	0. 287
2	1. 923	0. 551	2	2. 081	2	0. 573
3	2. 884	0. 827	3	3. 121	3	0. 860
4	3. 845	1. 103	4	4. 161	4	1. 147
5	4. 806	1. 378	5	5. 201	5	1. 434
6	5. 768	1. 654	6	6. 242	6	1. 720
7	6. 729	1. 929	7	7. 282	7	2. 007
8	7. 690	2. 205	8	8. 322	8	2. 294
9	8. 651	2. 481	9	9. 363	9	2. 581

Course		
16° 196°	p÷l DLo÷m	163° 343°
0. 0	0. 287	1. 0
0. 1	0. 289	0. 9
0. 2	0. 291	0. 8
0. 3	0. 292	0. 7
0. 4	0. 294	0. 6
0. 5	0. 296	0. 5
0. 6	0. 298	0. 4
0. 7	0. 300	0. 3
0. 8	0. 302	0. 2
0. 9	0. 304	0. 1

17°—163°—197°—343°

DLo	p		p	DLo		
D	l	p	l	D	m	DLo
1	0. 956	0. 292	1	1. 046	1	0. 306
2	1. 913	0. 585	2	2. 091	2	0. 611
3	2. 869	0. 877	3	3. 137	3	0. 917
4	3. 825	1. 169	4	4. 183	4	1. 223
5	4. 782	1. 462	5	5. 228	5	1. 529
6	5. 738	1. 754	6	6. 274	6	1. 834
7	6. 694	2. 047	7	7. 320	7	2. 140
8	7. 650	2. 339	8	8. 366	8	2. 446
9	8. 607	2. 631	9	9. 411	9	2. 752

Course		
17° 197°	p÷l DLo÷m	162° 342°
0. 0	0. 306	1. 0
0. 1	0. 308	0. 9
0. 2	0. 310	0. 8
0. 3	0. 311	0. 7
0. 4	0. 313	0. 6
0. 5	0. 315	0. 5
0. 6	0. 317	0. 4
0. 7	0. 319	0. 3
0. 8	0. 321	0. 2
0. 9	0. 323	0. 1

18°—162°—198°—342°

DLo	p		p	DLo		
D	l	p	l	D	m	DLo
1	0. 951	0. 309	1	1. 051	1	0. 325
2	1. 902	0. 618	2	2. 103	2	0. 650
3	2. 853	0. 927	3	3. 154	3	0. 975
4	3. 804	1. 236	4	4. 206	4	1. 300
5	4. 755	1. 545	5	5. 257	5	1. 625
6	5. 706	1. 854	6	6. 309	6	1. 950
7	6. 657	2. 163	7	7. 360	7	2. 274
8	7. 608	2. 472	8	8. 412	8	2. 599
9	8. 560	2. 781	9	9. 463	9	2. 924

Course		
18° 198°	p÷l DLo÷m	161° 341°
0. 0	0. 325	1. 0
0. 1	0. 327	0. 9
0. 2	0. 329	0. 8
0. 3	0. 331	0. 7
0. 4	0. 333	0. 6
0. 5	0. 335	0. 5
0. 6	0. 337	0. 4
0. 7	0. 338	0. 3
0. 8	0. 340	0. 2
0. 9	0. 342	0. 1

19°—161°—199°—341°

DLo	p		p	DLo		
D	l	p	l	D	m	DLo
1	0. 946	0. 326	1	1. 058	1	0. 344
2	1. 891	0. 651	2	2. 115	2	0. 689
3	2. 837	0. 977	3	3. 173	3	1. 033
4	3. 782	1. 302	4	4. 230	4	1. 377
5	4. 728	1. 628	5	5. 288	5	1. 722
6	5. 673	1. 953	6	6. 346	6	2. 066
7	6. 619	2. 279	7	7. 403	7	2. 410
8	7. 564	2. 605	8	8. 461	8	2. 755
9	8. 510	2. 930	9	9. 519	9	3. 099

Course		
19° 199°	p÷l DLo÷m	160° 340°
0. 0	0. 344	1. 0
0. 1	0. 346	0. 9
0. 2	0. 348	0. 8
0. 3	0. 350	0. 7
0. 4	0. 352	0. 6
0. 5	0. 354	0. 5
0. 6	0. 356	0. 4
0. 7	0. 358	0. 3
0. 8	0. 360	0. 2
0. 9	0. 362	0. 1

TABLE 38

Compact Traverse Table

20°—160°—200°—340°

DLo	p		p	DLo		
D	l	p	l	D	m	DLo
1	0. 940	0. 342	1	1. 064	1	0. 364
2	1. 879	0. 684	2	2. 128	2	0. 728
3	2. 819	1. 026	3	3. 193	3	1. 092
4	3. 759	1. 368	4	4. 257	4	1. 456
5	4. 698	1. 710	5	5. 321	5	1. 820
6	5. 638	2. 052	6	6. 385	6	2. 184
7	6. 578	2. 394	7	7. 449	7	2. 548
8	7. 518	2. 736	8	8. 513	8	2. 912
9	8. 457	3. 078	9	9. 578	9	3. 276

Course		
20° 200°	p÷l DLo÷m	159° 339°
0. 0	0. 364	1. 0
0. 1	0. 366	0. 9
0. 2	0. 368	0. 8
0. 3	0. 370	0. 7
0. 4	0. 372	0. 6
0. 5	0. 374	0. 5
0. 6	0. 376	0. 4
0. 7	0. 378	0. 3
0. 8	0. 380	0. 2
0. 9	0. 382	0. 1

21°—159°—201°—339°

DLo	p		p	DLo		
D	l	p	l	D	m	DLo
1	0. 934	0. 358	1	1. 071	1	0. 384
2	1. 867	0. 717	2	2. 142	2	0. 768
3	2. 801	1. 075	3	3. 213	3	1. 152
4	3. 734	1. 433	4	4. 285	4	1. 535
5	4. 668	1. 792	5	5. 356	5	1. 919
6	5. 601	2. 150	6	6. 427	6	2. 303
7	6. 535	2. 509	7	7. 498	7	2. 687
8	7. 469	2. 867	8	8. 569	8	3. 071
9	8. 402	3. 225	9	9. 640	9	3. 455

Course		
21° 201°	p÷l DLo÷m	158° 338°
0. 0	0. 384	1. 0
0. 1	0. 386	0. 9
0. 2	0. 388	0. 8
0. 3	0. 390	0. 7
0. 4	0. 392	0. 6
0. 5	0. 394	0. 5
0. 6	0. 396	0. 4
0. 7	0. 398	0. 3
0. 8	0. 400	0. 2
0. 9	0. 402	0. 1

22°—158°—202°—338°

DLo	p		p	DLo		
D	l	p	l	D	m	DLo
1	0. 927	0. 375	1	1. 079	1	0. 404
2	1. 854	0. 749	2	2. 157	2	0. 808
3	2. 782	1. 124	3	3. 236	3	1. 212
4	3. 709	1. 498	4	4. 314	4	1. 616
5	4. 636	1. 873	5	5. 393	5	2. 020
6	5. 563	2. 248	6	6. 471	6	2. 424
7	6. 490	2. 622	7	7. 550	7	2. 828
8	7. 417	2. 997	8	8. 628	8	3. 232
9	8. 345	3. 371	9	9. 707	9	3. 636

Course		
22° 202°	p÷l DLo÷m	157° 337°
0. 0	0. 404	1. 0
0. 1	0. 406	0. 9
0. 2	0. 408	0. 8
0. 3	0. 410	0. 7
0. 4	0. 412	0. 6
0. 5	0. 414	0. 5
0. 6	0. 416	0. 4
0. 7	0. 418	0. 3
0. 8	0. 420	0. 2
0. 9	0. 422	0. 1

23°—157°—203°—337°

DLo	p		p	DLo		
D	l	p	l	D	m	DLo
1	0. 921	0. 391	1	1. 086	1	0. 424
2	1. 841	0. 781	2	2. 173	2	0. 849
3	2. 762	1. 172	3	3. 259	3	1. 273
4	3. 682	1. 563	4	4. 345	4	1. 698
5	4. 603	1. 954	5	5. 432	5	2. 122
6	5. 523	2. 344	6	6. 518	6	2. 547
7	6. 444	2. 735	7	7. 605	7	2. 971
8	7. 364	3. 126	8	8. 691	8	3. 396
9	8. 285	3. 517	9	9. 777	9	3. 820

Course		
23° 203°	p÷l DLo÷m	156° 336°
0. 0	0. 424	1. 0
0. 1	0. 427	0. 9
0. 2	0. 429	0. 8
0. 3	0. 431	0. 7
0. 4	0. 433	0. 6
0. 5	0. 435	0. 5
0. 6	0. 437	0. 4
0. 7	0. 439	0. 3
0. 8	0. 441	0. 2
0. 9	0. 443	0. 1

24°—156°—204°—336°

DLo	p		p	DLo		
D	l	p.	l	D	m	DLo
1	0. 914	0. 407	1	1. 095	1	0. 445
2	1. 827	0. 813	2	2. 189	2	0. 890
3	2. 741	1. 220	3	3. 284	3	1. 336
4	3. 654	1. 627	4	4. 379	4	1. 781
5	4. 568	2. 034	5	5. 473	5	2. 226
6	5. 481	2. 440	6	6. 568	6	2. 671
7	6. 395	2. 847	7	7. 662	7	3. 117
8	7. 308	3. 254	8	8. 757	8	3. 562
9	8. 222	3. 661	9	9. 852	9	4. 007

Course		
24° 204°	p÷l DLo÷m	155° 335°
0. 0	0. 445	1. 0
0. 1	0. 447	0. 9
0. 2	0. 449	0. 8
0. 3	0. 452	0. 7
0. 4	0. 454	0. 6
0. 5	0. 456	0. 5
0. 6	0. 458	0. 4
0. 7	0. 460	0. 3
0. 8	0. 462	0. 2
0. 9	0. 464	0. 1

TABLE 38

Compact Traverse Table

25°—155°—205°—335°

DLo	p		p	DLc		
D	l	p	l	D	m	DLo
1	0. 906	0. 423	1	1. 103	1	0. 466
2	1. 813	0. 845	2	2. 207	2	0. 933
3	2. 719	1. 268	3	3. 310	3	1. 399
4	3. 625	1. 690	4	4. 414	4	1. 865
5	4. 532	2. 113	5	5. 517	5	2. 332
6	5. 438	2. 536	6	6. 620	6	2. 798
7	6. 344	2. 958	7	7. 724	7	3. 264
8	7. 250	3. 381	8	8. 827	8	3. 730
9	8. 157	3. 804	9	9. 930	9	4. 197

Course		
25° 205°	p÷l	154° 334°
	DLo÷m	
0. 0	0. 466	1. 0
0. 1	0. 468	0. 9
0. 2	0. 471	0. 8
0. 3	0. 473	0. 7
0. 4	0. 475	0. 6
0. 5	0. 477	0. 5
0. 6	0. 479	0. 4
0. 7	0. 481	0. 3
0. 8	0. 483	0. 2
0. 9	0. 486	0. 1

26°—154°—206°—334°

DLo	p		p	DLo		
D	l	p	l	D	m	DLo
1	0. 899	0. 438	1	1. 113	1	0. 488
2	1. 798	0. 877	2	2. 225	2	0. 975
3	2. 696	1. 315	3	3. 338	3	1. 463
4	3. 595	1. 753	4	4. 450	4	1. 951
5	4. 494	2. 192	5	5. 563	5	2. 439
6	5. 393	2. 630	6	6. 676	6	2. 926
7	6. 292	3. 069	7	7. 788	7	3. 414
8	7. 190	3. 507	8	8. 901	8	3. 902
9	8. 089	3. 945	9	10. 013	9	4. 390

Course		
26° 206°	p÷l	153° 333°
	DLo÷m	
0. 0	0. 488	1. 0
0. 1	0. 490	0. 9
0. 2	0. 492	0. 8
0. 3	0. 494	0. 7
0. 4	0. 496	0. 6
0. 5	0. 499	0. 5
0. 6	0. 501	0. 4
0. 7	0. 503	0. 3
0. 8	0. 505	0. 2
0. 9	0. 507	0. 1

27°—153°—207°—333°

DLo	p		p	DLo		
D	l	p	l	D	m	DLo
1	0. 891	0. 454	1	1. 122	1	0. 510
2	1. 782	0. 908	2	2. 245	2	1. 019
3	2. 673	1. 362	3	3. 367	3	1. 529
4	3. 564	1. 816	4	4. 489	4	2. 038
5	4. 455	2. 270	5	5. 612	5	2. 548
6	5. 346	2. 724	6	6. 734	6	3. 057
7	6. 237	3. 178	7	7. 856	7	3. 567
8	7. 128	3. 632	8	8. 979	8	4. 076
9	8. 019	4. 086	9	10. 101	9	4. 586

Course		
27° 207°	p÷l	152° 332°
	DLo÷m	
0. 0	0. 510	1. 0
0. 1	0. 512	0. 9
0. 2	0. 514	0. 8
0. 3	0. 516	0. 7
0. 4	0. 518	0. 6
0. 5	0. 521	0. 5
0. 6	0. 523	0. 4
0. 7	0. 525	0. 3
0. 8	0. 527	0. 2
0. 9	0. 529	0. 1

28°—152°—208°—332°

DLo	p		p	DLo		
D	l	p	l	D	m	DLo
1	0. 883	0. 469	1	1. 133	1	0. 532
2	1. 766	0. 939	2	2. 265	2	1. 063
3	2. 649	1. 408	3	3. 398	3	1. 595
4	3. 532	1. 878	4	4. 530	4	2. 127
5	4. 415	2. 347	5	5. 663	5	2. 659
6	5. 298	2. 817	6	6. 795	6	3. 190
7	6. 181	3. 286	7	7. 928	7	3. 722
8	7. 064	3. 756	8	9. 061	8	4. 254
9	7. 947	4. 225	9	10. 193	9	4. 785

Course		
28° 208°	p÷l	151° 331°
	DLo÷m	
0. 0	0. 532	1. 0
0. 1	0. 534	0. 9
0. 2	0. 536	0. 8
0. 3	0. 538	0. 7
0. 4	0. 541	0. 6
0. 5	0. 543	0. 5
0. 6	0. 545	0. 4
0. 7	0. 547	0. 3
0. 8	0. 550	0. 2
0. 9	0. 552	0. 1

29°—151°—209°—331°

DLo	p		p	DLo		
D	l	p	l	D	m	DLo
1	0. 875	0. 485	1	1. 143	1	0. 554
2	1. 749	0. 970	2	2. 287	2	1. 109
3	2. 624	1. 454	3	3. 430	3	1. 663
4	3. 498	1. 939	4	4. 573	4	2. 217
5	4. 373	2. 424	5	5. 717	5	2. 772
6	5. 248	2. 909	6	6. 860	6	3. 326
7	6. 122	3. 394	7	8. 003	7	3. 880
8	6. 997	3. 878	8	9. 147	8	4. 434
9	7. 872	4. 363	9	10. 290	9	4. 989

Course		
29° 209°	p÷l	150° 330°
	DLo÷m	
0. 0	0. 554	1. 0
0. 1	0. 557	0. 9
0. 2	0. 559	0. 8
0. 3	0. 561	0. 7
0. 4	0. 563	0. 6
0. 5	0. 566	0. 5
0. 6	0. 568	0. 4
0. 7	0. 570	0. 3
0. 8	0. 573	0. 2
0. 9	0. 575	0. 1

TABLE 38

Compact Traverse Table

30°—150°—210°—330°

DLo	p		p	DLo		
D	l	p	l	D	m	DLo
1	0. 866	0. 500	1	1. 155	1	0. 577
2	1. 732	1. 000	2	2. 309	2	1. 155
3	2. 598	1. 500	3	3. 464	3	1. 732
4	3. 464	2. 000	4	4. 619	4	2. 309
5	4. 330	2. 500	5	5. 774	5	2. 887
6	5. 196	3. 000	6	6. 928	6	3. 464
7	6. 062	3. 500	7	8. 083	7	4. 041
8	6. 928	4. 000	8	9. 238	8	4. 619
9	7. 794	4. 500	9	10. 392	9	5. 196

Course

30° 210°	p÷l DLo÷m	149° 329°
0. 0	0. 577	1. 0
0. 1	0. 580	0. 9
0. 2	0. 582	0. 8
0. 3	0. 584	0. 7
0. 4	0. 587	0. 6
0. 5	0. 589	0. 5
0. 6	0. 591	0. 4
0. 7	0. 594	0. 3
0. 8	0. 596	0. 2
0. 9	0. 598	0. 1

31°—149°—211°—329°

DLo	p		p	DLo		
D	l	p	l	D	m	DLo
1	0. 857	0. 515	1	1. 167	1	0. 601
2	1. 714	1. 030	2	2. 333	2	1. 202
3	2. 572	1. 545	3	3. 500	3	1. 803
4	3. 429	2. 060	4	4. 667	4	2. 403
5	4. 286	2. 575	5	5. 833	5	3. 004
6	5. 143	3. 090	6	7. 000	6	3. 605
7	6. 000	3. 605	7	8. 166	7	4. 206
8	6. 857	4. 120	8	9. 333	8	4. 807
9	7. 715	4. 635	9	10. 500	9	5. 408

Course

31° 211°	p÷l DLo÷m	148° 328°
0. 0	0. 601	1. 0
0. 1	0. 603	0. 9
0. 2	0. 606	0. 8
0. 3	0. 608	0. 7
0. 4	0. 610	0. 6
0. 5	0. 613	0. 5
0. 6	0. 615	0. 4
0. 7	0. 618	0. 3
0. 8	0. 620	0. 2
0. 9	0. 622	0. 1

32°—148°—212°—328°

DLo	p		p	DLo		
D	l	p	l	D	m	DLo
1	0. 848	0. 530	1	1. 179	1	0. 625
2	1. 696	1. 060	2	2. 358	2	1. 250
3	2. 544	1. 590	3	3. 538	3	1. 875
4	3. 392	2. 120	4	4. 717	4	2. 499
5	4. 240	2. 650	5	5. 896	5	3. 124
6	5. 088	3. 180	6	7. 075	6	3. 749
7	5. 936	3. 709	7	8. 254	7	4. 374
8	6. 784	4. 239	8	9. 433	8	4. 999
9	7. 632	4. 769	9	10. 613	9	5. 624

Course

32° 212°	p÷l DLo÷m	147° 327°
0. 0	0. 625	1. 0
0. 1	0. 627	0. 9
0. 2	0. 630	0. 8
0. 3	0. 632	0. 7
0. 4	0. 635	0. 6
0. 5	0. 637	0. 5
0. 6	0. 640	0. 4
0. 7	0. 642	0. 3
0. 8	0. 644	0. 2
0. 9	0. 647	0. 1

33°—147°—213°—327°

DLo	p		p	DLo		
D	l	p	l	D	m	DLo
1	0. 839	0. 545	1	1. 192	1	0. 649
2	1. 677	1. 089	2	2. 385	2	1. 299
3	2. 516	1. 634	3	3. 577	3	1. 948
4	3. 355	2. 179	4	4. 769	4	2. 598
5	4. 193	2. 723	5	5. 962	5	3. 247
6	5. 032	3. 268	6	7. 154	6	3. 896
7	5. 871	3. 812	7	8. 347	7	4. 546
8	6. 709	4. 357	8	9. 539	8	5. 195
9	7. 548	4. 902	9	10. 731	9	5. 845

Course

33° 213°	p÷l DLo÷m	146° 326°
0. 0	0. 649	1. 0
0. 1	0. 652	0. 9
0. 2	0. 654	0. 8
0. 3	0. 657	0. 7
0. 4	0. 659	0. 6
0. 5	0. 662	0. 5
0. 6	0. 664	0. 4
0. 7	0. 667	0. 3
0. 8	0. 669	0. 2
0. 9	0. 672	0. 1

34°—146°—214°—326°

DLo	p		p	DLo		
D	l	p	l	D	m	DLo
1	0. 829	0. 559	1	1. 206	1	0. 675
2	1. 658	1. 118	2	2. 412	2	1. 349
3	2. 487	1. 678	3	3. 619	3	2. 024
4	3. 316	2. 237	4	4. 825	4	2. 698
5	4. 145	2. 796	5	6. 031	5	3. 373
6	4. 974	3. 355	6	7. 237	6	4. 047
7	5. 803	3. 914	7	8. 444	7	4. 722
8	6. 632	4. 474	8	9. 650	8	5. 396
9	7. 461	5. 033	9	10. 856	9	6. 071

Course

34° 214°	p÷l DLo÷m	145° 325°
0. 0	0. 675	1. 0
0. 1	0. 677	0. 9
0. 2	0. 680	0. 8
0. 3	0. 682	0. 7
0. 4	0. 685	0. 6
0. 5	0. 687	0. 5
0. 6	0. 690	0. 4
0. 7	0. 692	0. 3
0. 8	0. 695	0. 2
0. 9	0. 698	0. 1

TABLE 38

Compact Traverse Table

35°—145°—215°—325°

DLo	p		p	DLo		
D	l	p	l	D	m	DLo
1	0.819	0.574	1	1.221	1	0.700
2	1.638	1.147	2	2.442	2	1.400
3	2.457	1.721	3	3.662	3	2.101
4	3.277	2.294	4	4.883	4	2.801
5	4.096	2.868	5	6.104	5	3.501
6	4.915	3.441	6	7.325	6	4.201
7	5.734	4.015	7	8.545	7	4.901
8	6.553	4.589	8	9.766	8	5.602
9	7.372	5.162	9	10.987	9	6.302

Course		
35° 215°	p÷l / DLo÷m	144° 324°
0.0	0.700	1.0
0.1	0.703	0.9
0.2	0.705	0.8
0.3	0.708	0.7
0.4	0.711	0.6
0.5	0.713	0.5
0.6	0.716	0.4
0.7	0.719	0.3
0.8	0.721	0.2
0.9	0.724	0.1

36°—144°—216°—324°

DLo	p		p	DLo		
D	l	p	l	D	m	DLo
1	0.809	0.588	1	1.236	1	0.727
2	1.618	1.176	2	2.472	2	1.453
3	2.427	1.763	3	3.708	3	2.180
4	3.236	2.351	4	4.944	4	2.906
5	4.045	2.939	5	6.180	5	3.633
6	4.854	3.527	6	7.416	6	4.359
7	5.663	4.114	7	8.652	7	5.086
8	6.472	4.702	8	9.889	8	5.812
9	7.281	5.290	9	11.125	9	6.539

Course		
36° 216°	p÷l / DLo÷m	143° 323°
0.0	0.727	1.0
0.1	0.729	0.9
0.2	0.732	0.8
0.3	0.735	0.7
0.4	0.737	0.6
0.5	0.740	0.5
0.6	0.743	0.4
0.7	0.745	0.3
0.8	0.748	0.2
0.9	0.751	0.1

37°—143°—217°—323°

DLo	p		p	DLo		
D	l	p	l	D	m	DLo
1	0.799	0.602	1	1.252	1	0.754
2	1.597	1.204	2	2.504	2	1.507
3	2.396	1.805	3	3.756	3	2.261
4	3.195	2.407	4	5.009	4	3.014
5	3.993	3.009	5	6.261	5	3.768
6	4.792	3.611	6	7.513	6	4.521
7	5.590	4.213	7	8.765	7	5.275
8	6.389	4.815	8	10.017	8	6.028
9	7.188	5.416	9	11.269	9	6.782

Course		
37° 217°	p÷l / DLo÷m	142° 322°
0.0	0.754	1.0
0.1	0.756	0.9
0.2	0.759	0.8
0.3	0.762	0.7
0.4	0.765	0.6
0.5	0.767	0.5
0.6	0.770	0.4
0.7	0.773	0.3
0.8	0.776	0.2
0.9	0.778	0.1

38°—142°—218°—322°

DLo	p		p	DLo		
D	l	p	l	D	m	DLo
1	0.788	0.616	1	1.269	1	0.781
2	1.576	1.231	2	2.538	2	1.563
3	2.364	1.847	3	3.807	3	2.344
4	3.152	2.463	4	5.076	4	3.125
5	3.940	3.078	5	6.345	5	3.906
6	4.728	3.694	6	7.614	6	4.688
7	5.516	4.310	7	8.883	7	5.469
8	6.304	4.925	8	10.152	8	6.250
9	7.092	5.541	9	11.421	9	7.032

Course		
38° 218°	p÷l / DLo÷m	141° 321°
0.0	0.781	1.0
0.1	0.784	0.9
0.2	0.787	0.8
0.3	0.790	0.7
0.4	0.793	0.6
0.5	0.795	0.5
0.6	0.798	0.4
0.7	0.801	0.3
0.8	0.804	0.2
0.9	0.807	0.1

39°—141°—219°—321°

DLo	p		p	DLo		
D	l	p	l	D	m	DLo
1	0.777	0.629	1	1.287	1	0.810
2	1.554	1.259	2	2.574	2	1.620
3	2.331	1.888	3	3.860	3	2.429
4	3.109	2.517	4	5.147	4	3.239
5	3.886	3.147	5	6.434	5	4.049
6	4.663	3.776	6	7.721	6	4.859
7	5.440	4.405	7	9.007	7	5.668
8	6.217	5.035	8	10.294	8	6.478
9	6.994	5.664	9	11.581	9	7.288

Course		
39° 219°	p÷l / DLo÷m	140° 320°
0.0	0.810	1.0
0.1	0.813	0.9
0.2	0.816	0.8
0.3	0.818	0.7
0.4	0.821	0.6
0.5	0.824	0.5
0.6	0.827	0.4
0.7	0.830	0.3
0.8	0.833	0.2
0.9	0.836	0.1

TABLE 38

Compact Traverse Table

40°—140°—220°—320°

DLo / D	p / l	p	p / l	DLo / D	m	DLo
1	0. 766	0. 643	1	1. 305	1	0. 839
2	1. 532	1. 286	2	2. 611	2	1. 678
3	2. 298	1. 928	3	3. 916	3	2. 517
4	3. 064	2. 571	4	5. 222	4	3. 356
5	3. 830	3. 214	5	6. 527	5	4. 196
6	4. 596	3. 857	6	7. 832	6	5. 035
7	5. 362	4. 500	7	9. 138	7	5. 874
8	6. 128	5. 142	8	10. 443	8	6. 713
9	6. 894	5. 785	9	11. 749	9	7. 552

Course 40° 220°	p÷l / DLo÷m	Course 139° 319°
0. 0	0. 839	1. 0
0. 1	0. 842	0. 9
0. 2	0. 845	0. 8
0. 3	0. 848	0. 7
0. 4	0. 851	0. 6
0. 5	0. 854	0. 5
0. 6	0. 857	0. 4
0. 7	0. 860	0. 3
0. 8	0. 863	0. 2
0. 9	0. 866	0. 1

41°—139°—221°—319°

DLo / D	p / l	p	p / l	DLo / D	m	DLo
1	0. 755	0. 656	1	1. 325	1	0. 869
2	1. 509	1. 312	2	2. 650	2	1. 739
3	2. 264	1. 968	3	3. 975	3	2. 608
4	3. 019	2. 624	4	5. 300	4	3. 477
5	3. 774	3. 280	5	6. 625	5	4. 346
6	4. 528	3. 936	6	7. 950	6	5. 216
7	5. 283	4. 592	7	9. 275	7	6. 085
8	6. 038	5. 248	8	10. 600	8	6. 954
9	6. 792	5. 905	9	11. 925	9	7. 824

Course 41° 221°	p÷l / DLo÷m	Course 138° 318°
0. 0	0. 869	1. 0
0. 1	0. 872	0. 9
0. 2	0. 875	0. 8
0. 3	0. 879	0. 7
0. 4	0. 882	0. 6
0. 5	0. 885	0. 5
0. 6	0. 888	0. 4
0. 7	0. 891	0. 3
0. 8	0. 894	0. 2
0. 9	0. 897	0. 1

42°—138°—222°—318°

DLo / D	p / l	p	p / l	DLo / D	m	DLo
1	0. 743	0. 669	1	1. 346	1	0. 900
2	1. 486	1. 338	2	2. 691	2	1. 801
3	2. 229	2. 007	3	4. 037	3	2. 701
4	2. 973	2. 677	4	5. 383	4	3. 602
5	3. 716	3. 346	5	6. 728	5	4. 502
6	4. 459	4. 015	6	8. 074	6	5. 402
7	5. 202	4. 684	7	9. 419	7	6. 303
8	5. 945	5. 353	8	10. 765	8	7. 203
9	6. 688	6. 022	9	12. 111	9	8. 104

Course 42° 222°	p÷l / DLo÷m	Course 137° 317°
0. 0	0. 900	1. 0
0. 1	0. 904	0. 9
0. 2	0. 907	0. 8
0. 3	0. 910	0. 7
0. 4	0. 913	0. 6
0. 5	0. 916	0. 5
0. 6	0. 920	0. 4
0. 7	0. 923	0. 3
0. 8	0. 926	0. 2
0. 9	0. 929	0. 1

43°—137°—223°—317°

DLo / D	p / l	p	p / l	DLo / D	m	DLo
1	0. 731	0. 682	1	1. 367	1	0. 933
2	1. 463	1. 364	2	2. 735	2	1. 865
3	2. 194	2. 046	3	4. 102	3	2. 798
4	2. 925	2. 728	4	5. 469	4	3. 730
5	3. 657	3. 410	5	6. 837	5	4. 663
6	4. 388	4. 092	6	8. 204	6	5. 595
7	5. 119	4. 774	7	9. 571	7	6. 528
8	5. 851	5. 456	8	10. 939	8	7. 460
9	6. 582	6. 138	9	12. 306	9	8. 393

Course 43° 223°	p÷l / DLo÷m	Course 136° 316°
0. 0	0. 933	1. 0
0. 1	0. 936	0. 9
0. 2	0. 939	0. 8
0. 3	0. 942	0. 7
0. 4	0. 946	0. 6
0. 5	0. 949	0. 5
0. 6	0. 952	0. 4
0. 7	0. 956	0. 3
0. 8	0. 959	0. 2
0. 9	0. 962	0. 1

44°—136°—224°—316°

DLo / D	p / l	p	p / l	DLo / D	m	DLo
1	0. 719	0. 695	1	1. 390	1	0. 966
2	1. 439	1. 389	2	2. 780	2	1. 931
3	2. 158	2. 084	3	4. 170	3	2. 897
4	2. 877	2. 779	4	5. 561	4	3. 863
5	3. 597	3. 473	5	6. 951	5	4. 828
6	4. 316	4. 168	6	8. 341	6	5. 794
7	5. 035	4. 863	7	9. 731	7	6. 760
8	5. 755	5. 557	8	11. 121	8	7. 726
9	6. 474	6. 252	9	12. 511	9	8. 691

Course 44° 224°	p÷l / DLo÷m	Course 135° 315°
0. 0	0. 966	1. 0
0. 1	0. 969	0. 9
0. 2	0. 972	0. 8
0. 3	0. 976	0. 7
0. 4	0. 979	0. 6
0. 5	0. 983	0. 5
0. 6	0. 986	0. 4
0. 7	0. 990	0. 3
0. 8	0. 993	0. 2
0. 9	0. 997	0. 1

TABLE 38

Compact Traverse Table

45°—135°—225°—315°

DLo D	p l	 p	p l	DLo D	 m	 DLo
1	0. 707	0. 707	1	1. 414	1	1. 000
2	1. 414	1. 414	2	2. 828	2	2. 000
3	2. 121	2. 121	3	4. 243	3	3. 000
4	2. 828	2. 828	4	5. 657	4	4. 000
5	3. 536	3. 536	5	7. 071	5	5. 000
6	4. 243	4. 243	6	8. 485	6	6. 000
7	4. 950	4. 950	7	9. 899	7	7. 000
8	5. 657	5. 657	8	11. 314	8	8. 000
9	6. 364	6. 364	9	12. 728	9	9. 000

Course 45° 225°	 p÷l DLo÷m	 134° 314°
0. 0	1. 000	1. 0
0. 1	1. 003	0. 9
0. 2	1. 007	0. 8
0. 3	1. 011	0. 7
0. 4	1. 014	0. 6
0. 5	1. 018	0. 5
0. 6	1. 021	0. 4
0. 7	1. 025	0. 3
0. 8	1. 028	0. 2
0. 9	1. 032	0. 1

46°—134°—226°—314°

DLo D	p l	 p	p l	DLo D	 m	 DLo
1	0. 695	0. 719	1	1. 440	1	1. 036
2	1. 389	1. 439	2	2. 879	2	2. 071
3	2. 084	2. 158	3	4. 319	3	3. 107
4	2. 779	2. 877	4	5. 758	4	4. 142
5	3. 473	3. 597	5	7. 198	5	5. 178
6	4. 168	4. 316	6	8. 637	6	6. 213
7	4. 863	5. 035	7	10. 077	7	7. 249
8	5. 557	5. 755	8	11. 516	8	8. 284
9	6. 252	6. 474	9	12. 956	9	9. 320

Course 46° 226°	 p÷l DLo÷m	 133° 313°
0. 0	1. 036	1. 0
0. 1	1. 039	0. 9
0. 2	1. 043	0. 8
0. 3	1. 046	0. 7
0. 4	1. 050	0. 6
0. 5	1. 054	0. 5
0. 6	1. 057	0. 4
0. 7	1. 061	0. 3
0. 8	1. 065	0. 2
0. 9	1. 069	0. 1

47°—133°—227°—313°

DLo D	p l	 p	p l	DLo D	 m	 DLo
1	0. 682	0. 731	1	1. 466	1	1. 072
2	1. 364	1. 463	2	2. 933	2	2. 145
3	2. 046	2. 194	3	4. 399	3	3. 217
4	2. 728	2. 925	4	5. 865	4	4. 289
5	3. 410	3. 657	5	7. 331	5	5. 362
6	4. 092	4. 388	6	8. 798	6	6. 434
7	4. 774	5. 119	7	10. 264	7	7. 507
8	5. 456	5. 851	8	11. 730	8	8. 579
9	6. 138	6. 582	9	13. 197	9	9. 651

Course 47° 227°	 p÷l DLo÷m	 132° 312°
0. 0	1. 072	1. 0
0. 1	1. 076	0. 9
0. 2	1. 080	0. 8
0. 3	1. 084	0. 7
0. 4	1. 087	0. 6
0. 5	1. 091	0. 5
0. 6	1. 095	0. 4
0. 7	1. 099	0. 3
0. 8	1. 103	0. 2
0. 9	1. 107	0. 1

48°—132°—228°—312°

DLo D	p l	 p	p l	DLo D	 m	 DLo
1	0. 669	0. 743	1	1. 494	1	1. 111
2	1. 338	1. 486	2	2. 989	2	2. 221
3	2. 007	2. 229	3	4. 483	3	3. 332
4	2. 677	2. 973	4	5. 978	4	4. 442
5	3. 346	3. 716	5	7. 472	5	5. 553
6	4. 015	4. 459	6	8. 967	6	6. 664
7	4. 684	5. 202	7	10. 461	7	7. 774
8	5. 353	5. 945	8	11. 956	8	8. 885
9	6. 022	6. 688	9	13. 450	9	9. 996

Course 48° 228°	 p÷l DLo÷m	 131° 311°
0. 0	1. 111	1. 0
0. 1	1. 115	0. 9
0. 2	1. 118	0. 8
0. 3	1. 122	0. 7
0. 4	1. 126	0. 6
0. 5	1. 130	0. 5
0. 6	1. 134	0. 4
0. 7	1. 138	0. 3
0. 8	1. 142	0. 2
0. 9	1. 146	0. 1

49°—131°—229°—311°

DLo D	p l	 p	p l	DLo D	 m	 DLo
1	0. 656	0. 755	1	1. 524	1	1. 150
2	1. 312	1. 509	2	3. 049	2	2. 301
3	1. 968	2. 264	3	4. 573	3	3. 451
4	2. 624	3. 019	4	6. 097	4	4. 601
5	3. 280	3. 774	5	7. 621	5	5. 752
6	3. 936	4. 528	6	9. 146	6	6. 902
7	4. 592	5. 283	7	10. 670	7	8. 053
8	5. 248	6. 038	8	12. 194	8	9. 203
9	5. 905	6. 792	9	13. 718	9	10. 353

Course 49° 229°	 p÷l DLo÷m	 130° 310°
0. 0	1. 150	1. 0
0. 1	1. 154	0. 9
0. 2	1. 159	0. 8
0. 3	1. 163	0. 7
0. 4	1. 167	0. 6
0. 5	1. 171	0. 5
0. 6	1. 175	0. 4
0. 7	1. 179	0. 3
0. 8	1. 183	0. 2
0. 9	1. 188	0. 1

TABLE 38

Compact Traverse Table

50°—130°—230°—310°

DLo	p		p	DLo		
D	l	p	l	D	m	DLo
1	0. 643	0. 766	1	1. 556	1	1. 192
2	1. 286	1. 532	2	3. 111	2	2. 384
3	1. 928	2. 298	3	4. 667	3	3. 575
4	2. 571	3. 064	4	6. 223	4	4. 767
5	3. 214	3. 830	5	7. 779	5	5. 959
6	3. 857	4. 596	6	9. 334	6	7. 151
7	4. 500	5. 362	7	10. 890	7	8. 342
8	5. 142	6. 128	8	12. 446	8	9. 534
9	5. 785	6. 894	9	14. 002	9	10. 726

Course		
50° 230°	p÷l DLo÷m	129° 309°
0. 0	1. 192	1. 0
0. 1	1. 196	0. 9
0. 2	1. 200	0. 8
0. 3	1. 205	0. 7
0. 4	1. 209	0. 6
0. 5	1. 213	0. 5
0. 6	1. 217	0. 4
0. 7	1. 222	0. 3
0. 8	1. 226	0. 2
0. 9	1. 230	0. 1

51°—129°—231°—309°

DLo	p		p	DLo		
D	l	p	l	D	m	DLo
1	0. 629	0. 777	1	1. 589	1	1. 235
2	1. 259	1. 554	2	3. 178	2	2. 470
3	1. 888	2. 331	3	4. 767	3	3. 705
4	2. 517	3. 109	4	6. 356	4	4. 940
5	3. 147	3. 886	5	7. 945	5	6. 174
6	3. 776	4. 663	6	9. 534	6	7. 409
7	4. 405	5. 440	7	11. 123	7	8. 644
8	5. 035	6. 217	8	12. 712	8	9. 879
9	5. 664	6. 994	9	14. 301	9	11. 114

Course		
51° 231°	p÷l DLo÷m	128° 308°
0. 0	1. 235	1. 0
0. 1	1. 239	0. 9
0. 2	1. 244	0. 8
0. 3	1. 248	0. 7
0. 4	1. 253	0. 6
0. 5	1. 257	0. 5
0. 6	1. 262	0. 4
0. 7	1. 266	0. 3
0. 8	1. 271	0. 2
0. 9	1. 275	0. 1

52°—128°—232°—308°

DLo	p		p	DLo		
D	l	p	l	D	m	DLo
1	0. 616	0. 788	1	1. 624	1	1. 280
2	1. 231	1. 576	2	3. 249	2	2. 560
3	1. 847	2. 364	3	4. 873	3	3. 840
4	2. 463	3. 152	4	6. 497	4	5. 120
5	3. 078	3. 940	5	8. 121	5	6. 400
6	3. 694	4. 728	6	9. 746	6	7. 680
7	4. 310	5. 516	7	11. 370	7	8. 960
8	4. 925	6. 304	8	12. 994	8	10. 240
9	5. 541	7. 092	9	14. 618	9	11. 519

Course		
52° 232°	p÷l DLo÷m	127° 307°
0. 0	1. 280	1. 0
0. 1	1. 285	0. 9
0. 2	1. 289	0. 8
0. 3	1. 294	0. 7
0. 4	1. 299	0. 6
0. 5	1. 303	0. 5
0. 6	1. 308	0. 4
0. 7	1. 313	0. 3
0. 8	1. 317	0. 2
0. 9	1. 322	0. 1

53°—127°—233°—307°

DLo	p		p	DLo		
D	l	p	l	D	m	DLo
1	0. 602	0. 799	1	1. 662	1	1. 327
2	1. 204	1. 597	2	3. 323	2	2. 654
3	1. 805	2. 396	3	4. 985	3	3. 981
4	2. 407	3. 195	4	6. 647	4	5. 308
5	3. 009	3. 993	5	8. 308	5	6. 635
6	3. 611	4. 792	6	9. 970	6	7. 962
7	4. 213	5. 590	7	11. 631	7	9. 289
8	4. 815	6. 389	8	13. 293	8	10. 616
9	5. 416	7. 188	9	14. 955	9	11. 943

Course		
53° 233°	p÷l DLo÷m	126° 306°
0. 0	1. 327	1. 0
0. 1	1. 332	0. 9
0. 2	1. 337	0. 8
0. 3	1. 342	0. 7
0. 4	1. 347	0. 6
0. 5	1. 351	0. 5
0. 6	1. 356	0. 4
0. 7	1. 361	0. 3
0. 8	1. 366	0. 2
0. 9	1. 371	0. 1

54°—126°—234°—306°

DLo	p		p	DLo		
D	l	p	l	D	m	DLo
1	0. 588	0. 809	1	1. 701	1	1. 376
2	1. 176	1. 618	2	3. 403	2	2. 753
3	1. 763	2. 427	3	5. 104	3	4. 129
4	2. 351	3. 236	4	6. 805	4	5. 506
5	2. 939	4. 045	5	8. 507	5	6. 882
6	3. 527	4. 854	6	10. 208	6	8. 258
7	4. 114	5. 663	7	11. 909	7	9. 635
8	4. 702	6. 472	8	13. 610	8	11. 011
9	5. 290	7. 281	9	15. 312	9	12. 387

Course		
54° 234°	p÷l DLo÷m	125° 305°
0. 0	1. 376	1. 0
0. 1	1. 381	0. 9
0. 2	1. 387	0. 8
0. 3	1. 392	0. 7
0. 4	1. 397	0. 6
0. 5	1. 402	0. 5
0. 6	1. 407	0. 4
0. 7	1. 412	0. 3
0. 8	1. 418	0. 2
0. 9	1. 423	0. 1

TABLE 38

Compact Traverse Table

55°—125°—235°—305°

DLo	p		p	DLo		
D	l	p	l	D	m	DLo
1	0.574	0.819	1	1.743	1	1.428
2	1.147	1.638	2	3.487	2	2.856
3	1.721	2.457	3	5.230	3	4.284
4	2.294	3.277	4	6.974	4	5.713
5	2.868	4.096	5	8.717	5	7.141
6	3.441	4.915	6	10.461	6	8.569
7	4.015	5.734	7	12.204	7	9.997
8	4.589	6.553	8	13.948	8	11.425
9	5.162	7.372	9	15.691	9	12.853

Course		
55° 235°	p÷l / DLo÷m	124° 304°
0.0	1.428	1.0
0.1	1.433	0.9
0.2	1.439	0.8
0.3	1.444	0.7
0.4	1.450	0.6
0.5	1.455	0.5
0.6	1.460	0.4
0.7	1.466	0.3
0.8	1.471	0.2
0.9	1.477	0.1

56°—124°—236°—304°

DLo	p		p	DLo		
D	l	p	l	D	m	DLo
1	0.559	0.829	1	1.788	1	1.483
2	1.118	1.658	2	3.577	2	2.965
3	1.678	2.487	3	5.365	3	4.448
4	2.237	3.316	4	7.153	4	5.930
5	2.796	4.145	5	8.941	5	7.413
6	3.355	4.974	6	10.730	6	8.895
7	3.914	5.803	7	12.518	7	10.378
8	4.474	6.632	8	14.306	8	11.860
9	5.033	7.461	9	16.095	9	13.343

Course		
56° 236°	p÷l / DLo÷m	123° 303°
0.0	1.483	1.0
0.1	1.488	0.9
0.2	1.494	0.8
0.3	1.499	0.7
0.4	1.505	0.6
0.5	1.511	0.5
0.6	1.517	0.4
0.7	1.522	0.3
0.8	1.528	0.2
0.9	1.534	0.1

57°—123°—237°—303°

DLo	p		p	DLo		
D	l	p	l	D	m	DLo
1	0.545	0.839	1	1.836	1	1.540
2	1.089	1.677	2	3.672	2	3.080
3	1.634	2.516	3	5.508	3	4.620
4	2.179	3.355	4	7.344	4	6.159
5	2.723	4.193	5	9.180	5	7.699
6	3.268	5.032	6	11.016	6	9.239
7	3.812	5.871	7	12.853	7	10.779
8	4.357	6.709	8	14.689	8	12.319
9	4.902	7.548	9	16.525	9	13.859

Course		
57° 237°	p÷l / DLo÷m	122° 302°
0.0	1.540	1.0
0.1	1.546	0.9
0.2	1.552	0.8
0.3	1.558	0.7
0.4	1.564	0.6
0.5	1.570	0.5
0.6	1.576	0.4
0.7	1.582	0.3
0.8	1.588	0.2
0.9	1.594	0.1

58°—122°—238°—302°

DLo	p		p	DLo		
D	l	p	l	D	m	DLo
1	0.530	0.848	1	1.887	1	1.600
2	1.060	1.696	2	3.774	2	3.201
3	1.590	2.544	3	5.661	3	4.801
4	2.120	3.392	4	7.548	4	6.401
5	2.650	4.240	5	9.435	5	8.002
6	3.180	5.088	6	11.322	6	9.602
7	3.709	5.936	7	13.210	7	11.202
8	4.239	6.784	8	15.097	8	12.803
9	4.769	7.632	9	16.984	9	14.403

Course		
58° 238°	p÷l / DLo÷m	121° 301°
0.0	1.600	1.0
0.1	1.607	0.9
0.2	1.613	0.8
0.3	1.619	0.7
0.4	1.625	0.6
0.5	1.632	0.5
0.6	1.638	0.4
0.7	1.645	0.3
0.8	1.651	0.2
0.9	1.658	0.1

59°—121°—239°—301°

DLo	p		p	DLo		
D	l	p	l	D	m	DLo
1	0.515	0.857	1	1.942	1	1.664
2	1.030	1.714	2	3.883	2	3.329
3	1.545	2.572	3	5.825	3	4.993
4	2.060	3.429	4	7.766	4	6.657
5	2.575	4.286	5	9.708	5	8.321
6	3.090	5.143	6	11.650	6	9.986
7	3.605	6.000	7	13.591	7	11.650
8	4.120	6.857	8	15.533	8	13.314
9	4.635	7.715	9	17.474	9	14.979

Course		
59° 239°	p÷l / DLo÷m	120° 300°
0.0	1.664	1.0
0.1	1.671	0.9
0.2	1.678	0.8
0.3	1.684	0.7
0.4	1.691	0.6
0.5	1.698	0.5
0.6	1.704	0.4
0.7	1.711	0.3
0.8	1.718	0.2
0.9	1.725	0.1

TABLE 38

Compact Traverse Table

60°—120°—240°—300°

DLo	p		p	DLo		
D	l	p	l	D	m	DLo
1	0. 500	0. 866	1	2. 000	1	1. 732
2	1. 000	1. 732	2	4. 000	2	3. 464
3	1. 500	2. 598	3	6. 000	3	5. 196
4	2. 000	3. 464	4	8. 000	4	6. 928
5	2. 500	4. 330	5	10. 000	5	8. 660
6	3. 000	5. 196	6	12. 000	6	10. 392
7	3. 500	6. 062	7	14. 000	7	12. 124
8	4. 000	6. 928	8	16. 000	8	13. 856
9	4. 500	7. 794	9	18. 000	9	15. 588

Course		
60° 240°	p÷l DLo÷m	119° 299°
0. 0	1. 732	1. 0
0. 1	1. 739	0. 9
0. 2	1. 746	0. 8
0. 3	1. 753	0. 7
0. 4	1. 760	0. 6
0. 5	1. 767	0. 5
0. 6	1. 775	0. 4
0. 7	1. 782	0. 3
0. 8	1. 789	0. 2
0. 9	1. 797	0. 1

61°—119°—241°—299°

DLo	p		p	DLo		
D	l	p	l	D	m	DLo
1	0. 485	0. 875	1	2. 063	1	1. 804
2	0. 970	1. 749	2	4. 125	2	3. 608
3	1. 454	2. 624	3	6. 188	3	5. 412
4	1. 939	3. 498	4	8. 251	4	7. 216
5	2. 424	4. 373	5	10. 313	5	9. 020
6	2. 909	5. 248	6	12. 376	6	10. 824
7	3. 394	6. 122	7	14. 439	7	12. 628
8	3. 878	6. 997	8	16. 501	8	14. 432
9	4. 363	7. 872	9	18. 564	9	16. 236

Course		
61° 241°	p÷l DLo÷m	118° 298°
0. 0	1. 804	1. 0
0. 1	1. 811	0. 9
0. 2	1. 819	0. 8
0. 3	1. 827	0. 7
0. 4	1. 834	0. 6
0. 5	1. 842	0. 5
0. 6	1. 849	0. 4
0. 7	1. 857	0. 3
0. 8	1. 865	0. 2
0. 9	1. 873	0. 1

62°—118°—242°—298°

DLo	p		p	DLo		
D	l	p	l	D	m	DLo
1	0. 469	0. 883	1	2. 130	1	1. 881
2	0. 939	1. 766	2	4. 260	2	3. 761
3	1. 408	2. 649	3	6. 390	3	5. 642
4	1. 878	3. 532	4	8. 520	4	7. 523
5	2. 347	4. 415	5	10. 650	5	9. 404
6	2. 817	5. 298	6	12. 780	6	11. 284
7	3. 286	6. 181	7	14. 910	7	13. 165
8	3. 756	7. 064	8	17. 040	8	15. 046
9	4. 225	7. 947	9	19. 170	9	16. 927

Course		
62° 242°	p÷l DLo÷m	117° 297°
0. 0	1. 881	1. 0
0. 1	1. 889	0. 9
0. 2	1. 897	0. 8
0. 3	1. 905	0. 7
0. 4	1. 913	0. 6
0. 5	1. 921	0. 5
0. 6	1. 929	0. 4
0. 7	1. 937	0. 3
0. 8	1. 946	0. 2
0. 9	1. 954	0. 1

63°—117°—243°—297°

DLo	p		p	DLo		
D	l	p	l	D	m	DLo
1	0. 454	0. 891	1	2. 203	1	1. 963
2	0. 908	1. 782	2	4. 405	2	3. 925
3	1. 362	2. 673	3	6. 608	3	5. 888
4	1. 816	3. 564	4	8. 811	4	7. 850
5	2. 270	4. 455	5	11. 013	5	9. 813
6	2. 724	5. 346	6	13. 216	6	11. 776
7	3. 178	6. 237	7	15. 419	7	13. 738
8	3. 632	7. 128	8	17. 622	8	15. 701
9	4. 086	8. 019	9	19. 824	9	17. 663

Course		
63° 243°	p÷l DLo÷m	116° 296°
0. 0	1. 963	1. 0
0. 1	1. 971	0. 9
0. 2	1. 980	0. 8
0. 3	1. 988	0. 7
0. 4	1. 997	0. 6
0. 5	2. 006	0. 5
0. 6	2. 014	0. 4
0. 7	2. 023	0. 3
0. 8	2. 032	0. 2
0. 9	2. 041	0. 1

64°—116°—244°—296°

DLo	p		p	DLo		
D	l	p	l	D	m	DLo
1	0. 438	0. 899	1	2. 281	1	2. 050
2	0. 877	1. 798	2	4. 562	2	4. 101
3	1. 315	2. 696	3	6. 844	3	6. 151
4	1. 753	3. 595	4	9. 125	4	8. 201
5	2. 192	4. 494	5	11. 406	5	10. 252
6	2. 630	5. 393	6	13. 687	6	12. 302
7	3. 069	6. 292	7	15. 968	7	14. 352
8	3. 507	7. 190	8	18. 249	8	16. 402
9	3. 945	8. 089	9	20. 531	9	18. 453

Course		
64° 244°	p÷l DLo÷m	115° 295°
0. 0	2. 050	1. 0
0. 1	2. 059	0. 9
0. 2	2. 069	0. 8
0. 3	2. 078	0. 7
0. 4	2. 087	0. 6
0. 5	2. 097	0. 5
0. 6	2. 106	0. 4
0. 7	2. 116	0. 3
0. 8	2. 125	0. 2
0. 9	2. 135	0. 1

TABLE 38

Compact Traverse Table

65°—115°—245°—295°

DLo	p		p	DLo		
D	l	p	l	D	m	DLo
1	0. 423	0. 906	1	2. 366	1	2. 145
2	0. 845	1. 813	2	4. 732	2	4. 289
3	1. 268	2. 719	3	7. 099	3	6. 434
4	1. 690	3. 625	4	9. 465	4	8. 578
5	2. 113	4. 532	5	11. 831	5	10. 723
6	2. 536	5. 438	6	14. 197	6	12. 867
7	2. 958	6. 344	7	16. 563	7	15. 012
8	3. 381	7. 250	8	18. 930	8	17. 156
9	3. 804	8. 157	9	21. 296	9	19. 301

Course

65° 245°	p÷l DLo÷m	114° 294°
0. 0	2. 145	1. 0
0. 1	2. 154	0. 9
0. 2	2. 164	0. 8
0. 3	2. 174	0. 7
0. 4	2. 184	0. 6
0. 5	2. 194	0. 5
0. 6	2. 204	0. 4
0. 7	2. 215	0. 3
0. 8	2. 225	0. 2
0. 9	2. 236	0. 1

66°—114°—246°—294°

DLo	p		p	DLo		
D	l	p	l	D	m	DLo
1	0. 407	0. 914	1	2. 459	1	2. 246
2	0. 813	1. 827	2	4. 917	2	4. 492
3	1. 220	2. 741	3	7. 376	3	6. 738
4	1. 627	3. 654	4	9. 834	4	8. 984
5	2. 034	4. 568	5	12. 293	5	11. 230
6	2. 440	5. 481	6	14. 752	6	13. 476
7	2. 847	6. 395	7	17. 210	7	15. 722
8	3. 254	7. 308	8	19. 669	8	17. 968
9	3. 661	8. 222	9	22. 127	9	20. 214

Course

66° 246°	p÷l DLo÷m	113° 293°
0. 0	2. 246	1. 0
0. 1	2. 257	0. 9
0. 2	2. 267	0. 8
0. 3	2. 278	0. 7
0. 4	2. 289	0. 6
0. 5	2. 300	0. 5
0. 6	2. 311	0. 4
0. 7	2. 322	0. 3
0. 8	2. 333	0. 2
0. 9	2. 344	0. 1

67°—113°—247°—293°

DLo	p		p	DLo		
D	l	p	l	D	m	DLo
1	0. 391	0. 921	1	2. 559	1	2. 356
2	0. 781	1. 841	2	5. 119	2	4. 712
3	1. 172	2. 762	3	7. 678	3	7. 068
4	1. 563	3. 682	4	10. 237	4	9. 423
5	1. 954	4. 603	5	12. 797	5	11. 779
6	2. 344	5. 523	6	15. 356	6	14. 135
7	2. 735	6. 444	7	17. 915	7	16. 491
8	3. 126	7. 364	8	20. 474	8	18. 847
9	3. 517	8. 285	9	23. 034	9	21. 203

Course

67° 247°	p÷l DLo÷m	112° 292°
0. 0	2. 356	1. 0
0. 1	2. 367	0. 9
0. 2	2. 379	0. 8
0. 3	2. 391	0. 7
0. 4	2. 402	0. 6
0. 5	2. 414	0. 5
0. 6	2. 426	0. 4
0. 7	2. 438	0. 3
0. 8	2. 450	0. 2
0. 9	2. 463	0. 1

68°—112°—248°—292°

DLo	p		p	DLo		
D	l	p	l	D	m	DLo
1	0. 375	0. 927	1	2. 669	1	2. 475
2	0. 749	1. 854	2	5. 339	2	4. 950
3	1. 124	2. 782	3	8. 008	3	7. 425
4	1. 498	3. 709	4	10. 678	4	9. 900
5	1. 873	4. 636	5	13. 347	5	12. 375
6	2. 248	5. 563	6	16. 017	6	14. 851
7	2. 622	6. 490	7	18. 686	7	17. 326
8	2. 997	7. 417	8	21. 356	8	19. 801
9	3. 371	8. 345	9	24. 025	9	22. 276

Course

68° 248°	p÷l DLo÷m	111° 291°
0. 0	2. 475	1. 0
0. 1	2. 488	0. 9
0. 2	2. 500	0. 8
0. 3	2. 513	0. 7
0. 4	2. 526	0. 6
0. 5	2. 539	0. 5
0. 6	2. 552	0. 4
0. 7	2. 565	0. 3
0. 8	2. 578	0. 2
0. 9	2. 592	0. 1

69°—111°—249°—291°

DLo	p		p	DLo		
D	l	p	l	D	m	DLo
1	0. 358	0. 934	1	2. 790	1	2. 605
2	0. 717	1. 867	2	5. 581	2	5. 210
3	1. 075	2. 801	3	8. 371	3	7. 815
4	1. 433	3. 734	4	11. 162	4	10. 420
5	1. 792	4. 668	5	13. 952	5	13. 025
6	2. 150	5. 601	6	16. 743	6	15. 631
7	2. 509	6. 535	7	19. 533	7	18. 236
8	2. 867	7. 469	8	22. 323	8	20. 841
9	3. 225	8. 402	9	25. 114	9	23. 446

Course

69° 249°	p÷l DLo÷m	110° 290°
0. 0	2. 605	1. 0
0. 1	2. 619	0. 9
0. 2	2. 633	0. 8
0. 3	2. 646	0. 7
0. 4	2. 660	0. 6
0. 5	2. 675	0. 5
0. 6	2. 689	0. 4
0. 7	2. 703	0. 3
0. 8	2. 718	0. 2
0. 9	2. 733	0. 1

TABLE 38

Compact Traverse Table

70°—110°—250°—290°

DLo	p		p	DLo		
D	l	p	l	D	m	DLo
1	0. 342	0. 940	1	2. 924	1	2. 747
2	0. 684	1. 879	2	5. 848	2	5. 495
3	1. 026	2. 819	3	8. 771	3	8. 242
4	1. 368	3. 759	4	11. 695	4	10. 990
5	1. 710	4. 698	5	14. 619	5	13. 737
6	2. 052	5. 638	6	17. 543	6	16. 485
7	2. 394	6. 578	7	20. 467	7	19. 232
8	2. 736	7. 518	8	23. 390	8	21. 980
9	3. 078	8. 457	9	26. 314	9	24. 727

Course		
70° 250°	p÷l DLo÷m	109° 289°
0. 0	2. 747	1. 0
0. 1	2. 762	0. 9
0. 2	2. 778	0. 8
0. 3	2. 793	0. 7
0. 4	2. 808	0. 6
0. 5	2. 824	0. 5
0. 6	2. 840	0. 4
0. 7	2. 856	0. 3
0. 8	2. 872	0. 2
0. 9	2. 888	0. 1

71°—109°—251°—289°

DLo	p		p	DLo		
D	l	p	l	D	m	DLo
1	0. 326	0. 946	1	3. 072	1	2. 904
2	0. 651	1. 891	2	6. 143	2	5. 808
3	0. 977	2. 837	3	9. 215	3	8. 713
4	1. 302	3. 782	4	12. 286	4	11. 617
5	1. 628	4. 728	5	15. 358	5	14. 521
6	1. 953	5. 673	6	18. 429	6	17. 425
7	2. 279	6. 619	7	21. 501	7	20. 329
8	2. 605	7. 564	8	24. 572	8	23. 234
9	2. 930	8. 510	9	27. 644	9	26. 138

Course		
71° 251°	p÷l DLo÷m	108° 288°
0. 0	2. 904	1. 0
0. 1	2. 921	0. 9
0. 2	2. 937	0. 8
0. 3	2. 954	0. 7
0. 4	2. 971	0. 6
0. 5	2. 989	0. 5
0. 6	3. 006	0. 4
0. 7	3. 024	0. 3
0. 8	3. 042	0. 2
0. 9	3. 060	0. 1

72°—108°—252°—288°

DLo	p		p	DLo		
D	l	p	l	D	m	DLo
1	0. 309	0. 951	1	3. 236	1	3. 078
2	0. 618	1. 902	2	6. 472	2	6. 155
3	0. 927	2. 853	3	9. 708	3	9. 233
4	1. 236	3. 804	4	12. 944	4	12. 311
5	1. 545	4. 755	5	16. 180	5	15. 388
6	1. 854	5. 706	6	19. 416	6	18. 466
7	2. 163	6. 657	7	22. 652	7	21. 544
8	2. 472	7. 608	8	25. 889	8	24. 621
9	2. 781	8. 560	9	29. 125	9	27. 699

Course		
72° 252°	p÷l DLo÷m	107° 287°
0. 0	3. 078	1. 0
0. 1	3. 096	0. 9
0. 2	3. 115	0. 8
0. 3	3. 133	0. 7
0. 4	3. 152	0. 6
0. 5	3. 172	0. 5
0. 6	3. 191	0. 4
0. 7	3. 211	0. 3
0. 8	3. 230	0. 2
0. 9	3. 251	0. 1

73°—107°—253°—287°

DLo	p		p	DLo		
D	l	p	l	D	m	DLo
1	0. 292	0. 956	1	3. 420	1	3. 271
2	0. 585	1. 913	2	6. 841	2	6. 542
3	0. 877	2. 869	3	10. 261	3	9. 813
4	1. 169	3. 825	4	13. 681	4	13. 083
5	1. 462	4. 782	5	17. 102	5	16. 354
6	1. 754	5. 738	6	20. 522	6	19. 625
7	2. 047	6. 694	7	23. 942	7	22. 896
8	2. 339	7. 650	8	27. 362	8	26. 167
9	2. 631	8. 607	9	30. 783	9	29. 438

Course		
73° 253°	p÷l DLo÷m	106° 286°
0. 0	3. 271	1. 0
0. 1	3. 291	0. 9
0. 2	3. 312	0. 8
0. 3	3. 333	0. 7
0. 4	3. 354	0. 6
0. 5	3. 376	0. 5
0. 6	3. 398	0. 4
0. 7	3. 420	0. 3
0. 8	3. 442	0. 2
0. 9	3. 465	0. 1

74°—106°—254°—286°

DLo	p		p	DLo		
D	l	p	l	D	m	DLo
1	0. 276	0. 961	1	3. 628	1	3. 487
2	0. 551	1. 923	2	7. 256	2	6. 975
3	0. 827	2. 884	3	10. 884	3	10. 462
4	1. 103	3. 845	4	14. 512	4	13. 950
5	1. 378	4. 806	5	18. 140	5	17. 437
6	1. 654	5. 768	6	21. 768	6	20. 924
7	1. 929	6. 729	7	25. 396	7	24. 412
8	2. 205	7. 690	8	29. 024	8	27. 899
9	2. 481	8. 651	9	32. 652	9	31. 387

Course		
74° 254°	p÷l DLo÷m	105° 285°
0. 0	3. 487	1. 0
0. 1	3. 511	0. 9
0. 2	3. 534	0. 8
0. 3	3. 558	0. 7
0. 4	3. 582	0. 6
0. 5	3. 606	0. 5
0. 6	3. 630	0. 4
0. 7	3. 655	0. 3
0. 8	3. 681	0. 2
0. 9	3. 706	0. 1

TABLE 38

Compact Traverse Table

75°—105°—255°—285°

DLo	p		p	DLo		
D	l	p	l	D	m	DLo
1	0.259	0.966	1	3.864	1	3.732
2	0.518	1.932	2	7.727	2	7.464
3	0.776	2.898	3	11.591	3	11.196
4	1.035	3.864	4	15.455	4	14.928
5	1.294	4.830	5	19.319	5	18.660
6	1.553	5.796	6	23.182	6	22.392
7	1.812	6.761	7	27.046	7	26.124
8	2.071	7.727	8	30.910	8	29.856
9	2.329	8.693	9	34.773	9	33.588

Course		
75° 255°	p÷l DLo÷m	104° 284°
0.0	3.732	1.0
0.1	3.758	0.9
0.2	3.785	0.8
0.3	3.812	0.7
0.4	3.839	0.6
0.5	3.867	0.5
0.6	3.895	0.4
0.7	3.923	0.3
0.8	3.952	0.2
0.9	3.981	0.1

76°—104°—256°—284°

DLo	p		p	DLo		
D	l	p	l	D	m	DLo
1	0.242	0.970	1	4.134	1	4.011
2	0.484	1.941	2	8.267	2	8.022
3	0.726	2.911	3	12.401	3	12.032
4	0.968	3.881	4	16.534	4	16.043
5	1.210	4.851	5	20.668	5	20.054
6	1.452	5.822	6	24.801	6	24.065
7	1.693	6.792	7	28.935	7	28.075
8	1.935	7.762	8	33.069	8	32.086
9	2.177	8.733	9	37.202	9	36.097

Course		
76° 256°	p÷l DLo÷m	103° 283°
0.0	4.011	1.0
0.1	4.041	0.9
0.2	4.071	0.8
0.3	4.102	0.7
0.4	4.134	0.6
0.5	4.165	0.5
0.6	4.198	0.4
0.7	4.230	0.3
0.8	4.264	0.2
0.9	4.297	0.1

77°—103°—257°—283°

DLo	p		p	DLo		
D	l	p	l	D	m	DLo
1	0.225	0.974	1	4.445	1	4.331
2	0.450	1.949	2	8.891	2	8.663
3	0.675	2.923	3	13.336	3	12.994
4	0.900	3.897	4	17.782	4	17.326
5	1.125	4.872	5	22.227	5	21.657
6	1.350	5.846	6	26.672	6	25.989
7	1.575	6.821	7	31.118	7	30.320
8	1.800	7.795	8	35.563	8	34.652
9	2.025	8.769	9	40.009	9	38.983

Course		
77° 257°	p÷l DLo÷m	102° 282°
0.0	4.331	1.0
0.1	4.366	0.9
0.2	4.402	0.8
0.3	4.437	0.7
0.4	4.474	0.6
0.5	4.511	0.5
0.6	4.548	0.4
0.7	4.586	0.3
0.8	4.625	0.2
0.9	4.665	0.1

78°—102°—258°—282°

DLo	p		p	DLo		
D	l	p	l	D	m	DLo
1	0.208	0.978	1	4.810	1	4.705
2	0.416	1.956	2	9.619	2	9.409
3	0.624	2.934	3	14.429	3	14.114
4	0.832	3.913	4	19.239	4	18.819
5	1.040	4.891	5	24.049	5	23.523
6	1.247	5.869	6	28.858	6	28.228
7	1.455	6.847	7	33.668	7	32.932
8	1.663	7.825	8	38.478	8	37.637
9	1.871	8.803	9	43.288	9	42.342

Course		
78° 258°	p÷l DLo÷m	101° 281°
0.0	4.705	1.0
0.1	4.745	0.9
0.2	4.787	0.8
0.3	4.829	0.7
0.4	4.872	0.6
0.5	4.915	0.5
0.6	4.959	0.4
0.7	5.005	0.3
0.8	5.050	0.2
0.9	5.097	0.1

79°—101°—259°—281°

DLo	p		p	DLo		
D	l	p	l	D	m	DLo
1	0.191	0.982	1	5.241	1	5.145
2	0.382	1.963	2	10.482	2	10.289
3	0.572	2.945	3	15.723	3	15.434
4	0.763	3.927	4	20.963	4	20.578
5	0.954	4.908	5	26.204	5	25.723
6	1.145	5.890	6	31.445	6	30.867
7	1.336	6.871	7	36.686	7	36.012
8	1.526	7.853	8	41.927	8	41.156
9	1.717	8.835	9	47.168	9	46.301

Course		
79° 259°	p÷l DLo÷m	100° 280°
0.0	5.145	1.0
0.1	5.193	0.9
0.2	5.242	0.8
0.3	5.292	0.7
0.4	5.343	0.6
0.5	5.396	0.5
0.6	5.449	0.4
0.7	5.503	0.3
0.8	5.558	0.2
0.9	5.614	0.1

TABLE 38

Compact Traverse Table

80°—100°—260°—280°

DLo	p	////	p	DLo	////	////
D	l	p	l	D	m	DLo
1	0.174	0.985	1	5.759	1	5.671
2	0.347	1.970	2	11.518	2	11.343
3	0.521	2.954	3	17.276	3	17.014
4	0.695	3.939	4	23.035	4	22.685
5	0.868	4.924	5	28.794	5	28.356
6	1.042	5.909	6	34.553	6	34.028
7	1.216	6.894	7	40.311	7	39.699
8	1.389	7.878	8	46.070	8	45.370
9	1.563	8.863	9	51.829	9	51.042

Course

80° 260°	p÷l DLo÷m	99° 279°
0.0	5.671	1.0
0.1	5.730	0.9
0.2	5.789	0.8
0.3	5.850	0.7
0.4	5.912	0.6
0.5	5.976	0.5
0.6	6.041	0.4
0.7	6.107	0.3
0.8	6.174	0.2
0.9	6.243	0.1

81°—99°—261°—279°

DLo	p	////	p	DLo	////	////
D	l	p	l	D	m	DLo
1	0.156	0.988	1	6.392	1	6.314
2	0.313	1.975	2	12.785	2	12.628
3	0.469	2.963	3	19.177	3	18.941
4	0.626	3.951	4	25.570	4	25.255
5	0.782	4.938	5	31.962	5	31.569
6	0.939	5.926	6	38.355	6	37.883
7	1.095	6.914	7	44.747	7	44.196
8	1.251	7.902	8	51.140	8	50.510
9	1.408	8.889	9	57.532	9	56.824

Course

81° 261°	p÷l DLo÷m	98° 278°
0.0	6.314	1.0
0.1	6.386	0.9
0.2	6.460	0.8
0.3	6.535	0.7
0.4	6.612	0.6
0.5	6.691	0.5
0.6	6.772	0.4
0.7	6.855	0.3
0.8	6.940	0.2
0.9	7.026	0.1

82°—98°—262°—278°

DLo	p	////	p	DLo	////	////
D	l	p	l	D	m	DLo
1	0.139	0.990	1	7.185	1	7.115
2	0.278	1.981	2	14.371	2	14.231
3	0.418	2.971	3	21.556	3	21.346
4	0.557	3.961	4	28.741	4	28.461
5	0.696	4.951	5	35.926	5	35.577
6	0.835	5.942	6	43.112	6	42.692
7	0.974	6.932	7	50.297	7	49.808
8	1.113	7.922	8	57.482	8	56.923
9	1.253	8.912	9	64.668	9	64.038

Course

82° 262°	p÷l DLo÷m	97° 277°
0.0	7.115	1.0
0.1	7.207	0.9
0.2	7.300	0.8
0.3	7.396	0.7
0.4	7.495	0.6
0.5	7.596	0.5
0.6	7.700	0.4
0.7	7.806	0.3
0.8	7.916	0.2
0.9	8.028	0.1

83°—97°—263°—277°

DLo	p	////	p	DLo	////	////
D	l	p	l	D	m	DLo
1	0.122	0.993	1	8.206	1	8.144
2	0.244	1.985	2	16.411	2	16.289
3	0.366	2.978	3	24.617	3	24.433
4	0.487	3.970	4	32.822	4	32.577
5	0.609	4.963	5	41.028	5	40.722
6	0.731	5.955	6	49.233	6	48.866
7	0.853	6.948	7	57.439	7	57.010
8	0.975	7.940	8	65.644	8	65.155
9	1.097	8.933	9	73.850	9	73.299

Course

83° 263°	p÷l DLo÷m	96° 276°
0.0	8.144	1.0
0.1	8.264	0.9
0.2	8.386	0.8
0.3	8.513	0.7
0.4	8.643	0.6
0.5	8.777	0.5
0.6	8.915	0.4
0.7	9.058	0.3
0.8	9.205	0.2
0.9	9.357	0.1

84°—96°—264°—276°

DLo	p	////	p	DLo	////	////
D	l	p	l	D	m	DLo
1	0.105	0.995	1	9.567	1	9.514
2	0.209	1.989	2	19.134	2	19.029
3	0.314	2.984	3	28.700	3	28.543
4	0.418	3.978	4	38.267	4	38.057
5	0.523	4.973	5	47.834	5	47.572
6	0.627	5.967	6	57.401	6	57.086
7	0.732	6.962	7	66.967	7	66.601
8	0.836	7.956	8	76.534	8	76.115
9	0.941	8.951	9	86.101	9	85.629

Course

84° 264°	p÷l DLo÷m	95° 275°
0.0	9.514	1.0
0.1	9.677	0.9
0.2	9.845	0.8
0.3	10.019	0.7
0.4	10.199	0.6
0.5	10.385	0.5
0.6	10.579	0.4
0.7	10.780	0.3
0.8	10.988	0.2
0.9	11.205	0.1

TABLE 38

Compact Traverse Table

85°—95°—265°—275°

DLo / D	p / l	p	p / l	DLo / D	m	DLo
1	0. 087	0. 996	1	11. 474	1	11. 430
2	0. 174	1. 992	2	22. 947	2	22. 860
3	0. 261	2. 989	3	34. 421	3	34. 290
4	0. 349	3. 985	4	45. 895	4	45. 720
5	0. 436	4. 981	5	57. 369	5	57. 150
6	0. 523	5. 977	6	68. 842	6	68. 580
7	0. 610	6. 973	7	80. 316	7	80. 010
8	0. 697	7. 970	8	91. 790	8	91. 440
9	0. 784	8. 966	9	103. 263	9	102. 870

Course 85° 265°	p÷l / DLo÷m	Course 94° 274°
0. 0	11. 430	1. 0
0. 1	11. 664	0. 9
0. 2	11. 909	0. 8
0. 3	12. 163	0. 7
0. 4	12. 429	0. 6
0. 5	12. 706	0. 5
0. 6	12. 996	0. 4
0. 7	13. 300	0. 3
0. 8	13. 617	0. 2
0. 9	13. 951	0. 1

86°—94°—266°—274°

DLo / D	p / l	p	p / l	DLo / D	m	DLo
1	0. 070	0. 998	1	14. 336	1	14. 301
2	0. 140	1. 995	2	28. 671	2	28. 601
3	0. 209	2. 993	3	43. 007	3	42. 902
4	0. 279	3. 990	4	57. 342	4	57. 203
5	0. 349	4. 988	5	71. 678	5	71. 503
6	0. 419	5. 985	6	86. 014	6	85. 804
7	0. 488	6. 983	7	100. 349	7	100. 105
8	0. 558	7. 981	8	114. 685	8	114. 405
9	0. 628	8. 978	9	129. 020	9	128. 706

Course 86° 266°	p÷l / DLo÷m	Course 93° 273°
0. 0	14. 301	1. 0
0. 1	14. 669	0. 9
0. 2	15. 056	0. 8
0. 3	15. 464	0. 7
0. 4	15. 895	0. 6
0. 5	16. 350	0. 5
0. 6	16. 832	0. 4
0. 7	17. 343	0. 3
0. 8	17. 886	0. 2
0. 9	18. 464	0. 1

87°—93°—267°—273°

DLo / D	p / l	p	p / l	DLo / D	m	DLo
1	0. 052	0. 999	1	19. 107	1	19. 081
2	0. 105	1. 997	2	38. 215	2	38. 162
3	0. 157	2. 996	3	57. 322	3	57. 243
4	0. 209	3. 995	4	76. 429	4	76. 325
5	0. 262	4. 993	5	95. 537	5	95. 406
6	0. 314	5. 992	6	114. 644	6	114. 487
7	0. 366	6. 990	7	133. 751	7	133. 568
8	0. 419	7. 989	8	152. 859	8	152. 649
9	0. 471	8. 988	9	171. 966	9	171. 730

Course 87° 267°	p÷l / DLo÷m	Course 92° 272°
0. 0	19. 081	1. 0
0. 1	19. 740	0. 9
0. 2	20. 446	0. 8
0. 3	21. 205	0. 7
0. 4	22. 022	0. 6
0. 5	22. 904	0. 5
0. 6	23. 859	0. 4
0. 7	24. 898	0. 3
0. 8	26. 031	0. 2
0. 9	27. 271	0. 1

88°—92°—268°—272°

DLo / D	p / l	p	p / l	DLo / D	m	DLo
1	0. 035	0. 999	1	28. 654	1	28. 636
2	0. 070	1. 999	2	57. 307	2	57. 273
3	0. 105	2. 998	3	85. 961	3	85. 909
4	0. 140	3. 998	4	114. 615	4	114. 545
5	0. 174	4. 997	5	143. 269	5	143. 181
6	0. 209	5. 996	6	171. 922	6	171. 818
7	0. 244	6. 996	7	200. 576	7	200. 454
8	0. 279	7. 995	8	229. 230	8	229. 090
9	0. 314	8. 995	9	257. 883	9	257. 726

Course 88° 268°	p÷l / DLo÷m	Course 91° 271°
0. 0	28. 636	1. 0
0. 1	30. 145	0. 9
0. 2	31. 821	0. 8
0. 3	33. 694	0. 7
0. 4	35. 801	0. 6
0. 5	38. 188	0. 5
0. 6	40. 917	0. 4
0. 7	44. 066	0. 3
0. 8	47. 740	0. 2
0. 9	52. 081	0. 1

89°—91°—269°—271°

DLo / D	p / l	p	p / l	DLo / D	m	DLo
1	0. 017	1. 000	1	57. 299	1	57. 290
2	0. 035	2. 000	2	114. 597	2	114. 580
3	0. 052	3. 000	3	171. 896	3	171. 870
4	0. 070	3. 999	4	229. 195	4	229. 160
5	0. 087	4. 999	5	286. 493	5	286. 450
6	0. 105	5. 999	6	343. 792	6	343. 740
7	0. 122	6. 999	7	401. 091	7	401. 030
8	0. 140	7. 999	8	458. 390	8	458. 320
9	0. 157	8. 999	9	515. 688	9	515. 610

Course 89° 269°	p÷l / DLo÷m	Course 90° 270°
0. 0	57. 290	1. 0
0. 1	63. 657	0. 9
0. 2	71. 615	0. 8
0. 3	81. 847	0. 7
0. 4	95. 489	0. 6
0. 5	114. 589	0. 5
0. 6	143. 237	0. 4
0. 7	190. 984	0. 3
0. 8	286. 478	0. 2
0. 9	572. 957	0. 1

TABLE 39

Distance by Vertical Angle

Measured Between Waterline at Object and Sea Horizon Beyond Object

Distance	Height of eye above the sea, in feet										Distance
	5	10	15	20	25	30	35	40	45	50	
Yards	° ′	° ′	° ′	° ′	° ′	° ′	° ′	° ′	° ′	° ′	*Yards*
100	0 55	1 52	2 48	3 45	4 41	5 37	6 34	7 30	8 26	9 21	100
200	27	0 54	1 22	1 50	2 18	2 46	3 15	3 43	4 11	4 39	200
300	17	35	0 54	1 12	1 31	1 49	2 08	2 27	2 45	3 04	300
400	12	26	39	0 53	1 07	1 21	1 35	1 49	2 02	2 16	400
500	9	20	31	42	0 53	1 04	1 15	1 26	1 37	1 48	500
600		16	25	34	43	0 52	1 01	1 10	1 20	1 29	600
700		13	21	29	36	44	0 52	0 59	1 07	1 15	700
800		11	18	24	31	38	45	51	0 58	1 05	800
900		10	16	21	27	33	39	45	51	0 57	900
1,000			14	19	24	29	35	40	45	51	1,000
1,100			12	17	21	26	31	36	41	45	1,100
1,200			11	15	19	24	28	32	37	41	1,200
1,300			10	14	17	21	25	29	33	37	1,300
1,400				12	16	20	23	27	31	34	1,400
1,500				11	15	18	21	25	28	32	1,500
1,600				10	13	17	20	23	26	29	1,600
1,700					12	15	18	21	24	27	1,700
1,800					11	14	17	20	23	25	1,800
1,900					11	13	16	18	21	24	1,900
2,000					10	12	15	17	20	22	2,000
2,100						11	14	16	18	21	2,100
2,200						11	13	15	17	20	2,200
2,300						10	12	14	16	19	2,300
2,400							11	13	15	18	2,400
2,500							11	13	15	17	2,500
2,600							10	12	14	16	2,600
2,700								11	13	15	2,700
2,800								11	12	14	2,800
2,900								10	12	14	2,900
3,000									11	13	3,000
3,100									11	12	3,100
3,200									10	12	3,200
3,300										11	3,300
3,400										11	3,400
3,500										10	3,500

TABLE 39

Distance by Vertical Angle

Measured Between Waterline at Object and Sea Horizon Beyond Object

Distance	Height of eye above the sea, in feet										Distance
	55	60	65	70	75	80	85	90	95	100	
Yards	° ′	° ′	° ′	° ′	° ′	° ′	° ′	° ′	° ′	° ′	*Yards*
100	10 16	11 11	12 06	13 00	13 54	14 48	15 41	16 34	17 26	18 17	100
200	5 07	5 35	6 03	6 31	6 59	7 27	7 55	8 23	8 51	9 18	200
300	3 23	3 41	4 00	4 19	4 38	4 56	5 15	5 34	5 52	6 11	300
400	2 30	2 44	2 58	3 12	3 26	3 40	3 54	4 08	4 22	4 36	400
500	1 59	2 10	2 21	2 32	2 43	2 55	3 06	3 17	3 28	3 39	500
600	1 38	1 47	1 56	2 06	2 15	2 24	2 33	2 43	2 52	3 01	600
700	1 23	1 31	1 39	1 47	1 54	2 02	2 10	2 18	2 26	2 34	700
800	1 12	1 19	1 25	1 32	1 39	1 46	1 53	2 00	2 07	2 14	800
900	1 03	1 09	1 15	1 21	1 27	1 33	1 39	1 46	1 52	1 58	900
1, 000	0 56	1 01	1 07	1 12	1 18	1 23	1 29	1 34	1 40	1 45	1, 000
1, 100	50	0 55	1 00	1 05	1 10	1 15	1 20	1 25	1 30	1 35	1, 100
1, 200	46	50	0 55	0 59	1 03	1 08	1 12	1 17	1 22	1 26	1, 200
1, 300	42	46	50	54	0 58	1 02	1 06	1 10	1 15	1 19	1, 300
1, 400	38	42	46	49	53	0 57	1 01	1 05	1 09	1 12	1, 400
1, 500	35	39	42	46	49	53	0 56	1 00	1 03	1 07	1, 500
1, 600	33	36	39	42	46	49	52	0 56	0 59	1 02	1, 600
1, 700	30	33	36	39	43	46	49	52	55	0 58	1, 700
1, 800	28	31	34	37	40	43	46	48	51	54	1, 800
1, 900	26	29	32	35	37	40	43	45	48	51	1, 900
2, 000	25	27	30	32	35	38	40	43	45	48	2, 000
2, 100	23	26	28	31	33	35	38	40	43	45	2, 100
2, 200	22	24	27	29	31	33	36	38	40	43	2, 200
2, 300	21	23	25	27	29	32	34	36	38	41	2, 300
2, 400	20	22	24	26	28	30	32	34	36	39	2, 400
2, 500	19	21	23	25	27	29	31	33	35	37	2, 500
2, 600	18	19	21	23	25	27	29	31	33	35	2, 600
2, 700	17	19	20	22	24	26	28	30	31	33	2, 700
2, 800	16	18	19	21	23	25	26	28	30	32	2, 800
2, 900	15	17	18	20	22	24	25	27	29	30	2, 900
3, 000	14	16	18	19	21	23	24	26	27	29	3, 000
3, 100	14	15	17	18	20	22	23	25	26	28	3, 100
3, 200	13	15	16	18	19	21	22	24	25	27	3, 200
3, 300	13	14	15	17	18	20	21	23	24	26	3, 300
3, 400	12	13	15	16	18	19	20	22	23	25	3, 400
3, 500	12	13	14	16	17	18	20	21	22	24	3, 500
3, 600	11	12	14	15	16	18	19	20	22	23	3, 600
3, 700	11	12	13	14	16	17	18	19	21	22	3, 700
3, 800	10	11	13	14	15	16	17	19	20	21	3, 800
3, 900		11	12	13	14	16	17	18	19	21	3, 900
4, 000		11	12	13	14	15	16	17	19	20	4, 000
4, 100		10	11	12	13	15	16	17	18	19	4, 100
4, 200			11	12	13	14	15	16	17	18	4, 200
4, 300			10	11	12	14	15	16	17	18	4, 300
4, 400			10	11	12	13	14	15	16	17	4, 400
4, 500			10	11	12	13	14	15	16	17	4, 500
4, 600				10	11	12	13	14	15	16	4, 600
4, 700					11	12	13	14	15	16	4, 700
4, 800					11	11	12	13	14	15	4, 800
4, 900					10	11	12	13	14	15	4, 900
5, 000					10	11	12	12	13	14	5, 000

TABLE 40

Geographic Range

Object height		Height of eye of observer in feet and meters										Object height	
Feet		7	10	13	16	20	23	26	30	33	36		Feet
	Meters	2	3	4	5	6	7	8	9	10	11	Meters	
		Miles	Miles	Miles	Miles	Miles	Miles	Miles	Miles	Miles	Miles		
0	0	3.1	3.7	4.2	4.7	5.2	5.6	6.0	6.4	6.7	7.0	0	0
3	1	5.1	5.7	6.2	6.7	7.3	7.6	8.0	8.4	8.7	9.0	1	3
7	2	6.2	6.8	7.3	7.8	8.3	8.7	9.1	9.5	9.8	10.1	2	7
10	3	6.8	7.4	7.9	8.4	8.9	9.3	9.7	10.1	10.4	10.7	3	10
13	4	7.3	7.9	8.4	8.9	9.5	9.8	10.2	10.6	10.9	11.2	4	13
16	5	7.8	8.4	8.9	9.4	9.9	10.3	10.6	11.1	11.4	11.7	5	16
20	6	8.3	8.9	9.5	9.9	10.5	10.8	11.2	11.6	12.0	12.3	6	20
23	7	8.7	9.3	9.8	10.3	10.8	11.2	11.6	12.0	12.3	12.6	7	23
26	8	9.1	9.7	10.2	10.6	11.2	11.6	11.9	12.4	12.7	13.0	8	26
30	9	9.5	10.1	10.6	11.1	11.6	12.0	12.4	12.8	13.1	13.4	9	30
33	10	9.8	10.4	10.9	11.4	12.0	12.3	12.7	13.1	13.4	13.7	10	33
36	11	10.1	10.7	11.2	11.7	12.3	12.6	13.0	13.4	13.7	14.0	11	36
39	12	10.4	11.0	11.5	12.0	12.5	12.9	13.3	13.7	14.0	14.3	12	39
43	13	10.8	11.4	11.9	12.4	12.9	13.3	13.6	14.1	14.4	14.7	13	43
46	14	11.0	11.6	12.2	12.6	13.2	13.5	13.9	14.3	14.7	15.0	14	46
49	15	11.3	11.9	12.4	12.9	13.4	13.8	14.2	14.6	14.9	15.2	15	49
52	16	11.5	12.1	12.7	13.1	13.7	14.0	14.4	14.8	15.2	15.5	16	52
56	17	11.9	12.5	13.0	13.4	14.0	14.4	14.7	15.2	15.5	15.8	17	56
59	18	12.1	12.7	13.2	13.7	14.2	14.6	15.0	15.4	15.7	16.0	18	59
62	19	12.3	12.9	13.4	13.9	14.4	14.8	15.2	15.6	15.9	16.2	19	62
66	20	12.6	13.2	13.7	14.2	14.7	15.1	15.5	15.9	16.2	16.5	20	66
72	22	13.0	13.6	14.1	14.6	15.2	15.5	15.9	16.3	16.6	16.9	22	72
79	24	13.5	14.1	14.6	15.1	15.6	16.0	16.4	16.8	17.1	17.4	24	79
85	26	13.9	14.5	15.0	15.5	16.0	16.4	16.8	17.2	17.5	17.8	26	85
92	28	14.3	14.9	15.4	15.9	16.5	16.8	17.2	17.6	17.9	18.2	28	92
98	30	14.7	15.3	15.8	16.3	16.8	17.2	17.5	18.0	18.3	18.6	30	98
115	35	15.6	16.2	16.8	17.2	17.8	18.2	18.5	19.0	19.3	19.6	35	115
131	40	16.5	17.1	17.6	18.1	18.6	19.0	19.4	19.8	20.1	20.4	40	131
148	45	17.3	17.9	18.5	18.9	19.5	19.8	20.2	20.6	21.0	21.3	45	148
164	50	18.1	18.7	19.2	19.7	20.2	20.6	20.9	21.4	21.7	22.0	50	164
180	55	18.8	19.4	19.9	20.4	20.9	21.3	21.7	22.1	22.4	22.7	55	180
197	60	19.5	20.1	20.6	21.1	21.7	22.0	22.4	22.8	23.1	23.4	60	197
213	65	20.2	20.8	21.3	21.8	22.3	22.7	23.0	23.5	23.8	24.1	65	213
230	70	20.8	21.4	22.0	22.4	23.0	23.4	23.7	24.2	24.5	24.8	70	230
246	75	21.4	22.1	22.6	23.0	23.6	24.0	24.3	24.8	25.1	25.4	75	246
262	80	22.0	22.6	23.2	23.6	24.2	24.5	24.9	25.3	25.7	26.0	80	262
279	85	22.6	23.2	23.8	24.2	24.8	25.2	25.5	26.0	26.3	26.6	85	279
295	90	23.2	23.8	24.3	24.8	25.3	25.7	26.1	26.5	26.8	27.1	90	295
312	95	23.8	24.4	24.9	25.3	25.9	26.3	26.6	27.1	27.4	27.7	95	312
328	100	24.3	24.9	25.4	25.9	26.4	26.8	27.2	27.6	27.9	28.2	100	328
361	110	25.3	25.9	26.4	26.9	27.5	27.8	28.2	28.6	29.0	29.3	110	361
394	120	26.3	26.9	27.4	27.9	28.5	28.8	29.2	29.6	29.9	30.2	120	394
427	130	27.3	27.9	28.4	28.9	29.4	29.8	30.1	30.6	30.9	31.2	130	427
459	140	28.2	28.8	29.3	29.7	30.3	30.7	31.0	31.5	31.8	32.1	140	459
492	150	29.0	29.7	30.2	30.6	31.2	31.6	31.9	32.4	32.7	33.0	150	492
525	160	29.9	30.5	31.0	31.5	32.0	32.4	32.8	33.2	33.5	33.8	160	525
558	170	30.7	31.3	31.9	32.3	32.9	33.2	33.6	34.0	34.4	34.7	170	558
591	180	31.5	32.1	32.7	33.1	33.7	34.1	34.4	34.9	35.2	35.5	180	591
623	190	32.3	32.9	33.4	33.9	34.4	34.8	35.2	35.6	35.9	36.2	190	623
656	200	33.1	33.7	34.2	34.6	35.2	35.6	35.9	36.4	36.7	37.0	200	656
722	220	34.5	35.1	35.7	36.1	36.7	37.0	37.4	37.8	38.2	38.5	220	722
787	240	35.9	36.5	37.0	37.5	38.1	38.4	38.8	39.2	39.5	39.8	240	787
853	260	37.3	37.9	38.4	38.9	39.4	39.8	40.1	40.6	40.9	41.2	260	853
919	280	38.6	39.2	39.7	40.1	40.7	41.1	41.4	41.9	42.2	42.5	280	919
984	300	39.8	40.4	40.9	41.4	41.9	42.3	42.7	43.1	43.4	43.7	300	984

TABLE 40

Geographic Range

Object height		Height of eye of observer in feet and meters										Object height	
Feet		39	43	46	49	52	56	59	62	66	69		Feet
	Meters	12	13	14	15	16	17	18	19	20	21	Meters	
		Miles	Miles	Miles	Miles	Miles	Miles	Miles	Miles	Miles	Miles		
0	0	7.3	7.7	7.9	8.2	8.4	8.8	9.0	9.2	9.5	9.7	0	0
3	1	9.3	9.7	10.0	10.2	10.5	10.8	11.0	11.2	11.5	11.7	1	3
7	2	10.4	10.8	11.0	11.3	11.5	11.9	12.1	12.3	12.6	12.8	2	7
10	3	11.0	11.4	11.6	11.9	12.1	12.5	12.7	12.9	13.2	13.4	3	10
13	4	11.5	11.9	12.2	12.4	12.7	13.0	13.2	13.4	13.7	13.9	4	13
16	5	12.0	12.4	12.6	12.9	13.1	13.4	13.7	13.9	14.2	14.4	5	16
20	6	12.5	12.9	13.2	13.4	13.7	14.0	14.2	14.4	14.7	15.0	6	20
23	7	12.9	13.3	13.5	13.8	14.0	14.4	14.6	14.8	15.1	15.3	7	23
26	8	13.3	13.6	13.9	14.2	14.4	14.7	15.0	15.2	15.5	15.7	8	26
30	9	13.7	14.1	14.3	14.6	14.8	15.2	15.4	15.6	15.9	16.1	9	30
33	10	14.0	14.4	14.7	14.9	15.2	15.5	15.7	15.9	16.2	16.4	10	33
36	11	14.3	14.7	15.0	15.2	15.5	15.8	16.0	16.2	16.5	16.7	11	36
39	12	14.6	15.0	15.2	15.5	15.7	16.1	16.3	16.5	16.8	17.0	12	39
43	13	15.0	15.3	15.6	15.9	16.1	16.4	16.7	16.9	17.2	17.4	13	43
46	14	15.2	15.6	15.9	16.1	16.4	16.7	16.9	17.1	17.4	17.7	14	46
49	15	15.5	15.9	16.1	16.4	16.6	16.9	17.2	17.4	17.7	17.9	15	49
52	16	15.7	16.1	16.4	16.6	16.9	17.2	17.4	17.6	17.9	18.2	16	52
56	17	16.1	16.4	16.7	16.9	17.2	17.5	17.7	18.0	18.3	18.5	17	56
59	18	16.3	16.7	16.9	17.2	17.4	17.7	18.0	18.2	18.5	18.7	18	59
62	19	16.5	16.9	17.1	17.4	17.6	18.0	18.2	18.4	18.7	18.9	19	62
66	20	16.8	17.2	17.4	17.7	17.9	18.3	18.5	18.7	19.0	19.2	20	66
72	22	17.2	17.6	17.9	18.1	18.4	18.7	18.9	19.1	19.4	19.6	22	72
79	24	17.7	18.1	18.3	18.6	18.8	19.2	19.4	19.6	19.9	20.1	24	79
85	26	18.1	18.5	18.7	19.0	19.2	19.5	19.8	20.0	20.3	20.5	26	85
92	28	18.5	18.9	19.2	19.4	19.7	20.0	20.2	20.4	20.7	20.9	28	92
98	30	18.9	19.3	19.5	19.8	20.0	20.3	20.6	20.8	21.1	21.3	30	98
115	35	19.9	20.2	20.5	20.7	21.0	21.3	21.5	21.8	22.1	22.3	35	115
131	40	20.7	21.1	21.3	21.6	21.8	22.1	22.4	22.6	22.9	23.1	40	131
148	45	21.5	21.9	22.2	22.4	22.7	23.0	23.2	23.4	23.7	24.0	45	148
164	50	22.3	22.7	22.9	23.2	23.4	23.7	24.0	24.2	24.5	24.7	50	164
180	55	23.0	23.4	23.6	23.9	24.1	24.5	24.7	24.9	25.2	25.4	55	180
197	60	23.7	24.1	24.4	24.6	24.9	25.2	25.4	25.6	25.9	26.1	60	197
213	65	24.4	24.7	25.0	25.3	25.5	25.8	26.1	26.3	26.6	26.8	65	213
230	70	25.1	25.4	25.7	25.9	26.2	26.5	26.7	27.0	27.2	27.5	70	230
246	75	25.7	26.0	26.3	26.5	26.8	27.1	27.3	27.6	27.9	28.1	75	246
262	80	26.2	26.6	26.9	27.1	27.4	27.7	27.9	28.2	28.4	28.7	80	262
279	85	26.8	27.2	27.5	27.7	28.0	28.3	28.5	28.8	29.0	29.3	85	279
295	90	27.4	27.8	28.0	28.3	28.5	28.9	29.1	29.3	29.6	29.8	90	295
312	95	28.0	28.3	28.6	28.9	29.1	29.4	29.7	29.9	30.2	30.4	95	312
328	100	28.5	28.9	29.1	29.4	29.6	29.9	30.2	30.4	30.7	30.9	100	328
361	110	29.5	29.9	30.2	30.4	30.7	31.0	31.2	31.4	31.7	31.9	110	361
394	120	30.5	30.9	31.2	31.4	31.7	32.0	32.2	32.4	32.7	32.9	120	394
427	130	31.5	31.8	32.1	32.4	32.6	32.9	33.2	33.4	33.7	33.9	130	427
459	140	32.4	32.7	33.0	33.3	33.5	33.8	34.1	34.3	34.6	34.8	140	459
492	150	33.3	33.6	33.9	34.1	34.4	34.7	34.9	35.2	35.5	35.7	150	492
525	160	34.1	34.5	34.7	35.0	35.2	35.6	35.8	36.0	36.3	36.5	160	525
558	170	34.9	35.3	35.6	35.8	36.1	36.4	36.6	36.9	37.1	37.4	170	558
591	180	35.7	36.1	36.4	36.6	36.9	37.2	37.4	37.7	37.9	38.2	180	591
623	190	36.5	36.9	37.1	37.4	37.6	38.0	38.2	38.4	38.7	38.9	190	623
656	200	37.3	37.6	37.9	38.2	38.4	38.7	39.0	39.2	39.5	39.7	200	656
722	220	38.7	39.1	39.4	39.6	39.9	40.2	40.4	40.7	40.9	41.2	220	722
787	240	40.1	40.5	40.8	41.0	41.3	41.6	41.8	42.0	42.3	42.5	240	787
853	260	41.5	41.8	42.1	42.4	42.6	42.9	43.2	43.4	43.7	43.9	260	853
919	280	42.8	43.1	43.4	43.7	43.9	44.2	44.5	44.7	45.0	45.2	270	919
984	300	44.0	44.4	44.6	44.9	45.1	45.5	45.7	45.9	46.2	46.4	300	984

TABLE 40
Geographic Range

Object Height		Height of eye of observer in feet and meters										Object Height	
Feet		72	75	79	82	85	89	92	95	98	115		Feet
	Meters	22	23	24	25	26	27	28	29	30	35	Meters	
		Miles	Miles	Miles	Miles	Miles	Miles	Miles	Miles	Miles	Miles		
0	0	9. 9	10. 2	10. 4	10. 6	10. 8	11. 0	11. 2	11. 4	11. 6	12. 5	0	0
3	1	12. 0	12. 2	12. 4	12. 6	12. 8	13. 1	13. 2	13. 4	13. 6	14. 6	1	3
7	2	13. 0	13. 3	13. 5	13. 7	13. 9	14. 1	14. 3	14. 5	14. 7	15. 6	2	7
10	3	13. 6	13. 9	14. 1	14. 3	14. 5	14. 7	14. 9	15. 1	15. 3	16. 2	3	10
13	4	14. 1	14. 4	14. 6	14. 8	15. 0	15. 3	15. 4	15. 6	15. 8	16. 8	4	13
16	5	14. 6	14. 9	15. 1	15. 3	15. 5	15. 7	15. 9	16. 1	16. 3	17. 2	5	16
20	6	15. 2	15. 4	15. 6	15. 8	16. 0	16. 3	16. 5	16. 6	16. 8	17. 8	6	20
23	7	15. 5	15. 8	16. 0	16. 2	16. 4	16. 6	16. 8	17. 0	17. 2	18. 2	7	23
26	8	15. 9	16. 2	16. 4	16. 6	16. 8	17. 0	17. 2	17. 4	17. 5	18. 5	8	26
30	9	16. 3	16. 6	16. 8	17. 0	17. 2	17. 4	17. 6	17. 8	18. 0	19. 0	9	30
33	10	16. 6	16. 9	17. 1	17. 3	17. 5	17. 8	17. 9	18. 1	18. 3	19. 3	10	33
36	11	16. 9	17. 2	17. 4	17. 6	17. 8	18. 1	18. 2	18. 4	18. 6	19. 6	11	36
39	12	17. 2	17. 5	17. 7	17. 9	18. 1	18. 3	18. 5	18. 7	18. 9	19. 9	12	39
43	13	17. 6	17. 9	18. 1	18. 3	18. 5	18. 7	18. 9	19. 1	19. 3	20. 2	13	43
46	14	17. 9	18. 1	18. 3	18. 5	18. 7	19. 0	19. 2	19. 3	19. 5	20. 5	14	46
49	15	18. 1	18. 4	18. 6	18. 8	19. 0	19. 2	19. 4	19. 6	19. 8	20. 7	15	49
52	16	18. 4	18. 6	18. 8	19. 0	19. 2	19. 5	19. 7	19. 8	20. 0	21. 0	16	52
56	17	18. 7	19. 0	19. 2	19. 4	19. 5	19. 8	20. 0	20. 2	20. 3	21. 3	17	56
59	18	18. 9	19. 2	19. 4	19. 6	19. 8	20. 0	20. 2	20. 4	20. 6	21. 5	18	59
62	19	19. 1	19. 4	19. 6	19. 8	20. 0	20. 3	20. 4	20. 6	20. 8	21. 8	19	62
66	20	19. 4	19. 7	19. 9	20. 1	20. 3	20. 5	20. 7	20. 9	21. 1	22. 1	20	66
72	22	19. 9	20. 1	20. 3	20. 5	20. 7	21. 0	21. 2	21. 3	21. 5	22. 5	22	72
79	24	20. 3	20. 6	20. 8	21. 0	21. 2	21. 4	21. 6	21. 8	22. 0	22. 9	24	79
85	26	20. 7	21. 0	21. 2	21. 4	21. 6	21. 8	22. 0	22. 2	22. 4	23. 3	26	85
92	28	21. 2	21. 4	21. 6	21. 8	22. 0	22. 3	22. 4	22. 6	22. 8	23. 8	28	92
98	30	21. 5	21. 8	22. 0	22. 2	22. 4	22. 6	22. 8	23. 0	23. 2	24. 1	30	98
115	35	22. 5	22. 7	22. 9	23. 1	23. 3	23. 6	23. 8	24. 0	24. 1	25. 1	35	115
131	40	23. 3	23. 6	23. 8	24. 0	24. 2	24. 4	24. 6	24. 8	25. 0	25. 9	40	131
148	45	24. 2	24. 4	24. 6	24. 8	25. 0	25. 3	25. 5	25. 6	25. 8	26. 8	45	148
164	50	24. 9	25. 2	25. 4	25. 6	25. 8	26. 0	26. 2	26. 4	26. 6	27. 5	50	164
180	55	25. 6	25. 9	26. 1	26. 3	26. 5	26. 7	26. 9	27. 1	27. 3	28. 2	55	180
197	60	26. 3	26. 6	26. 8	27. 0	27. 2	27. 5	27. 6	27. 8	28. 0	29. 0	60	197
213	65	27. 0	27. 3	27. 5	27. 7	27. 9	28. 1	28. 3	28. 5	28. 7	29. 6	65	213
230	70	27. 7	27. 9	28. 1	28. 3	28. 5	28. 8	29. 0	29. 1	29. 3	30. 3	70	230
246	75	28. 3	28. 6	28. 7	28. 9	29. 1	29. 4	29. 6	29. 8	29. 9	30. 9	75	246
262	80	28. 9	29. 1	29. 3	29. 5	29. 7	30. 0	30. 1	30. 3	30. 5	31. 5	80	262
279	85	29. 5	29. 7	29. 9	30. 1	30. 3	30. 6	30. 8	30. 9	31. 1	32. 1	85	279
295	90	30. 0	30. 3	30. 5	30. 7	30. 9	31. 1	31. 3	31. 5	31. 7	32. 6	90	295
312	95	30. 6	30. 9	31. 1	31. 3	31. 4	31. 7	31. 9	32. 1	32. 2	33. 2	95	312
328	100	31. 1	31. 4	31. 6	31. 8	32. 0	32. 2	32. 4	32. 6	32. 8	33. 7	100	328
361	110	32. 2	32. 4	32. 6	32. 8	33. 0	33. 3	33. 5	33. 6	33. 8	34. 8	110	361
394	120	33. 2	33. 4	33. 6	33. 8	34. 0	34. 3	34. 4	34. 6	34. 8	35. 8	120	394
427	130	34. 1	34. 4	34. 6	34. 8	35. 0	35. 2	35. 4	35. 6	35. 8	36. 7	130	427
459	140	35. 0	35. 3	35. 5	35. 7	35. 9	36. 1	36. 3	36. 5	36. 6	37. 6	140	459
492	150	35. 9	36. 2	36. 4	36. 5	36. 7	37. 0	37. 2	37. 4	37. 5	38. 5	150	492
525	160	36. 7	37. 0	37. 2	37. 4	37. 6	37. 8	38. 0	38. 2	38. 4	39. 4	160	525
558	170	37. 6	37. 8	38. 0	38. 2	38. 4	38. 7	38. 9	39. 0	39. 2	40. 2	170	558
591	180	38. 4	38. 6	38. 8	39. 0	39. 2	39. 5	39. 7	39. 8	40. 0	41. 0	180	591
623	190	39. 1	39. 4	39. 6	39. 8	40. 0	40. 2	40. 4	40. 6	40. 8	41. 8	190	623
656	200	39. 9	40. 2	40. 4	40. 6	40. 8	41. 0	41. 2	41. 4	41. 5	42. 5	200	656
722	220	41. 4	41. 6	41. 8	42. 0	42. 2	42. 5	42. 7	42. 8	43. 0	44. 0	220	722
787	240	42. 8	43. 0	43. 2	43. 4	43. 6	43. 9	44. 0	44. 2	44. 4	45. 4	240	787
853	260	44. 1	44. 4	44. 6	44. 8	45. 0	45. 2	45. 4	45. 6	45. 8	46. 7	260	853
919	280	45. 4	45. 7	45. 9	46. 1	46. 3	46. 5	46. 7	46. 9	47. 1	48. 0	280	919
984	300	46. 6	46. 9	47. 1	47. 3	47. 5	47. 7	47. 9	48. 1	48. 3	49. 2	300	984

TABLE 41

Distance by Vertical Angle
Measured Between Waterline at Object and Top of Object

Angle	Height of object above the sea, in feet and (meters)										Angle
	10 (3.0)	15 (4.6)	20 (6.1)	25 (7.6)	30 (9.1)	35 (10.7)	40 (12.2)	45 (13.7)	50 (15.2)	55 (16.8)	
° ′	*Miles*	*Miles*	*Miles*	*Miles*	*Miles*	*Miles*	*Miles*	*Miles*	*Miles*	*Miles*	° ′
0 10	0. 57	0. 85	1. 13	1. 41	1. 70	1. 98	2. 26	2. 55	2. 83	3. 11	0 10
0 11	0. 51	0. 77	1. 03	1. 29	1. 54	1. 80	2. 06	2. 31	2. 57	2. 83	0 11
0 12	0. 47	0. 71	0. 94	1. 18	1. 41	1. 65	1. 89	2. 12	2. 36	2. 59	0 12
0 13	0. 44	0. 65	0. 87	1. 09	1. 31	1. 52	1. 74	1. 96	2. 18	2. 39	0 13
0 14	0. 40	0. 61	0. 81	1. 01	1. 21	1. 41	1. 62	1. 82	2. 02	2. 22	0 14
0 15	0. 38	0. 57	0. 75	0. 94	1. 18	1. 32	1. 51	1. 70	1. 89	2. 07	0 15
0 20	0. 28	0. 42	0. 57	0. 71	0. 85	0. 99	1. 13	1. 27	1. 41	1. 56	0 20
0 25	0. 23	0. 34	0. 45	0. 57	0. 68	0. 79	0. 91	1. 02	1. 13	1. 24	0 25
0 30	0. 19	0. 28	0. 38	0. 47	0. 57	0. 66	0. 75	0. 85	0. 94	1. 04	0 30
0 35	0. 16	0. 24	0. 32	0. 40	0. 46	0. 57	0. 65	0. 73	0. 81	0. 89	0 35
0 40	0. 14	0. 21	0. 28	0. 35	0. 42	0. 50	0. 57	0. 64	0. 71	0. 78	0 40
0 45	0. 13	0. 19	0. 25	0. 31	0. 38	0. 44	0. 50	0. 57	0. 63	0. 69	0 45
0 50	0. 11	0. 17	0. 23	0. 28	0. 34	0. 40	0. 45	0. 51	0. 57	0. 62	0 50
0 55	0. 10	0. 15	0. 21	0. 26	0. 31	0. 36	0. 41	0. 46	0. 51	0. 57	0 55
1 00		0. 14	0. 19	0. 24	0. 28	0. 33	0. 38	0. 42	0. 47	0. 52	1 00
1 10		0. 12	0. 16	0. 20	0. 24	0. 28	0. 32	0. 36	0. 40	0. 44	1 10
1 20		0. 11	0. 14	0. 18	0. 21	0. 25	0. 28	0. 32	0. 35	0. 39	1 20
1 30		0. 09	0. 13	0. 16	0. 19	0. 22	0. 25	0. 28	0. 31	0. 35	1 30
1 40			0. 11	0. 14	0. 17	0. 20	0. 23	0. 25	0. 28	0. 31	1 40
1 50			0. 10	0. 13	0. 15	0. 18	0. 21	0. 23	0. 26	0. 28	1 50
2 00				0. 12	0. 14	0. 16	0. 19	0. 21	0. 24	0. 26	2 00
2 15				0. 10	0. 13	0. 15	0. 17	0. 19	0. 21	0. 23	2 15
2 30					0. 11	0. 13	0. 15	0. 17	0. 19	0. 21	2 30
2 45					0. 10	0. 12	0. 14	0. 15	0. 17	0. 19	2 45
3 00						0. 11	0. 13	0. 14	0. 16	0. 17	3 00
3 20						0. 10	0. 11	0. 13	0. 14	0. 16	3 20
3 40							0. 10	0. 12	0. 13	0. 14	3 40
4 00								0. 11	0. 12	0. 13	4 00
4 20								0. 10	0. 11	0. 12	4 20
4 40									0. 10	0. 11	4 40
5 00										0. 10	5 00

TABLE 41

Distance by Vertical Angle

Measured Between Waterline at Object and Top of Object

Angle	Height of object above the sea, in feet and (meters)										Angle
	60 (18.3)	65 (19.8)	70 (21.3)	75 (22.9)	80 (24.4)	85 (25.9)	90 (27.4)	95 (29.0)	100 (30.5)	105 (32.0)	
° ′	*Miles*	*Miles*	*Miles*	*Miles*	*Miles*	*Miles*	*Miles*	*Miles*	*Miles*	*Miles*	° ′
0 10	3. 39	3. 68	3. 96	4. 24	4. 53	4. 81					0 10
0 11	3. 09	3. 34	3. 60	3. 86	4. 11	4. 37	4. 63	4. 89			0 11
0 12	2. 83	3. 06	3. 30	3. 54	3. 77	4. 01	4. 24	4. 48	4. 71	4. 95	0 12
0 13	2. 61	2. 83	3. 05	3. 26	3. 48	3. 70	3. 92	4. 13	4. 35	4. 57	0 13
0 14	2. 42	2. 63	2. 83	3. 03	3. 23	3. 44	3. 64	3. 84	4. 04	4. 24	0 14
0 15	2. 26	2. 45	2. 64	2. 83	3. 02	3. 21	3. 59	3. 58	3. 77	3. 96	0 15
0 20	1. 70	1. 84	1. 98	2. 12	2. 26	2. 40	2. 55	2. 69	2. 83	2. 97	0 20
0 25	1. 36	1. 47	1. 58	1. 70	1. 81	1. 92	2. 04	2. 15	2. 26	2. 38	0 25
0 30	1. 13	1. 23	1. 32	1. 41	1. 51	1. 60	1. 70	1. 79	1. 89	1. 98	0 30
0 35	0. 97	1. 05	1. 13	1. 21	1. 29	1. 37	1. 45	1. 54	1. 62	1. 70	0 35
0 40	0. 85	0. 92	0. 99	1. 06	1. 13	1. 20	1. 27	1. 34	1. 41	1. 49	0 40
0 45	0. 75	0. 82	0. 88	0. 94	1. 01	1. 07	1. 13	1. 19	1. 26	1. 32	0 45
0 50	0. 68	0. 74	0. 79	0. 85	0. 91	0. 96	1. 02	1. 07	1. 13	1. 19	0 50
0 55	0. 62	0. 67	0. 72	0. 77	0. 82	0. 87	0. 93	0. 98	1. 03	1. 08	0 55
1 00	0. 57	0. 61	0. 66	0. 71	0. 75	0. 80	0. 85	0. 90	0. 94	0. 99	1 00
1 10	0. 48	0. 53	0. 57	0. 61	0. 65	0. 69	0. 73	0. 77	0. 81	0. 85	1 10
1 20	0. 42	0. 46	0. 49	0. 53	0. 57	0. 60	0. 64	0. 67	0. 71	0. 74	1 20
1 30	0. 38	0. 41	0. 44	0. 47	0. 50	0. 53	0. 57	0. 60	0. 63	0. 66	1 30
1 40	0. 34	0. 37	0. 40	0. 42	0. 45	0. 48	0. 51	0. 54	0. 57	0. 59	1 40
1 50	0. 31	0. 33	0. 36	0. 39	0. 41	0. 44	0. 46	0. 49	0. 51	0. 54	1 50
2 00	0. 28	0. 31	0. 33	0. 35	0. 38	0. 40	0. 42	0. 45	0. 47	0. 49	2 00
2 15	0. 25	0. 27	0. 29	0. 31	0. 34	0. 36	0. 38	0. 40	0. 42	0. 44	2 15
2 30	0. 23	0. 25	0. 26	0. 28	0. 30	0. 32	0. 34	0. 36	0. 38	0. 40	2 30
2 45	0. 21	0. 22	0. 24	0. 26	0. 27	0. 29	0. 31	0. 33	0. 34	0. 36	2 45
3 00	0. 19	0. 20	0. 22	0. 24	0. 25	0. 27	0. 28	0. 30	0. 31	0. 33	3 00
3 20	0. 17	0. 18	0. 20	0. 21	0. 23	0. 24	0. 25	0. 27	0. 28	0. 30	3 20
3 40	0. 15	0. 17	0. 18	0. 19	0. 21	0. 22	0. 23	0. 24	0. 26	0. 27	3 40
4 00	0. 14	0. 15	0. 16	0. 18	0. 19	0. 20	0. 21	0. 22	0. 24	0. 25	4 00
4 20	0. 13	0. 14	0. 15	0. 16	0. 17	0. 18	0. 20	0. 21	0. 22	0. 23	4 20
4 40	0. 12	0. 13	0. 14	0. 15	0. 16	0. 17	0. 18	0. 19	0. 20	0. 21	4 40
5 00	0. 11	0. 12	0. 13	0. 14	0. 15	0. 16	0. 17	0. 18	0. 19	0. 20	5 00
5 20	0. 11	0. 11	0. 12	0. 13	0. 14	0. 15	0. 16	0. 17	0. 18	0. 19	5 20
5 40	0. 10	0. 11	0. 12	0. 12	0. 13	0. 14	0. 15	0. 16	0. 17	0. 17	5 40
6 00		0. 10	0. 11	0. 12	0. 13	0. 13	0. 14	0. 15	0. 16	0. 16	6 00
6 20			0. 10	0. 11	0. 12	0. 13	0. 13	0. 14	0. 15	0. 16	6 20
6 40				0. 11	0. 11	0. 12	0. 13	0. 13	0. 14	0. 15	6 40
7 00				0. 10	0. 11	0. 11	0. 12	0. 13	0. 13	0. 14	7 00
7 20					0. 10	0. 11	0. 12	0. 12	0. 13	0. 13	7 20
7 40						0. 10	0. 11	0. 12	0. 12	0. 13	7 40
8 00							0. 11	0. 11	0. 12	0. 12	8 00
8 20							0. 10	0. 11	0. 11	0. 12	8 20
8 40								0. 10	0. 11	0. 11	8 40
9 00									0. 10	0. 11	9 00
9 30										0. 10	9 30
10 00											10 00

TABLE 41

Distance by Vertical Angle

Measured Between Waterline at Object and Top of Object

Angle		Height of object above the sea in feet and (meters)										Angle	
		110 (33.5)	115 (35.1)	120 (36.6)	125 (38.1)	130 (39.6)	135 (41.1)	140 (42.7)	145 (44.2)	150 (45.7)	155 (47.2)		
°	′	*Miles*	*Miles*	*Miles*	*Miles*	*Miles*	*Miles*	*Miles*	*Miles*	*Miles*	*Miles*	°	′
0	10											0	10
0	11											0	11
0	12											0	13
0	13	4. 79	5. 00									0	13
0	14	4. 45	4. 65	4. 85								0	14
0	15	4. 15	4. 34	4. 53	4. 71	4. 90						0	15
0	20	3. 11	3. 25	3. 39	3. 54	3. 68	3. 82	3. 96	4. 10	4. 24	4. 38	0	20
0	25	2. 49	2. 60	3. 72	2. 83	2. 94	3. 06	3. 17	3. 28	3. 30	3. 51	0	25
0	30	2. 07	2. 17	2. 26	2. 36	2. 45	2. 55	2. 64	2. 73	2. 83	2. 92	0	30
0	35	1. 78	1. 86	1. 94	2. 02	2. 10	2. 18	2. 26	2. 34	2. 42	2. 51	0	35
0	40	1. 56	1. 63	1. 70	1. 77	1. 84	1. 91	1. 98	2. 05	2. 12	2. 19	0	40
0	45	1. 30	1. 45	1. 51	1. 57	1. 63	1. 70	1. 76	1. 82	1. 89	1. 95	0	45
0	50	1. 24	1. 30	1. 36	1. 41	1. 47	1. 53	1. 58	1. 64	1. 70	1. 75	0	50
0	55	1. 13	1. 18	1. 23	1. 29	1. 34	1. 39	1. 44	1. 49	1. 54	1. 59	0	55
1	00	1. 04	1. 08	1. 13	1. 18	1. 23	1. 27	1. 32	1. 37	1. 41	1. 46	1	00
1	10	0. 89	0. 93	0. 97	1. 01	1. 05	1. 09	1. 13	1. 17	1. 21	1. 25	1	10
1	20	0. 78	0. 81	0. 85	0. 88	0. 92	0. 95	0. 99	1. 03	1. 06	1. 10	1	20
1	30	0. 69	0. 72	0. 75	0. 79	0. 82	0. 85	0. 88	0. 91	0. 94	0. 97	1	30
1	40	0. 62	0. 65	0. 66	0. 71	0. 74	0. 76	0. 79	0. 82	0. 85	0. 88	1	40
1	50	0. 57	0. 59	0. 62	0. 64	0. 67	0. 69	0. 72	0. 75	0. 77	0. 80	1	50
2	00	0. 52	0. 54	0. 57	0. 59	0. 61	0. 64	0. 66	0. 68	0. 71	0. 73	2	00
2	15	0. 46	0. 48	0. 50	0. 52	0. 54	0. 57	0. 59	0. 61	0. 63	0. 65	2	15
2	30	0. 41	0. 43	0. 45	0. 47	0. 49	0. 51	0. 53	0. 55	0. 57	0. 58	2	30
2	45	0. 38	0. 39	0. 41	0. 43	0. 45	0. 46	0. 48	0. 50	0. 51	0. 53	2	45
3	00	0. 35	0. 36	0. 38	0. 39	0. 41	0. 42	0. 44	0. 46	0. 47	0. 49	3	00
3	20	0. 31	0. 32	0. 34	0. 35	0. 37	0. 38	0. 40	0. 41	0. 42	0. 44	3	20
3	40	0. 28	0. 30	0. 31	0. 32	0. 33	0. 35	0. 36	0. 37	0. 39	0. 40	3	40
4	00	0. 26	0. 27	0. 28	0. 29	0. 31	0. 32	0. 33	0. 34	0. 35	0. 36	4	00
4	20	0. 24	0. 25	0. 26	0. 27	0. 28	0. 29	0. 30	0. 31	0. 33	0. 34	4	20
4	40	0. 22	0. 23	0. 24	0. 25	0. 26	0. 27	0. 28	0. 29	0. 30	0. 31	4	40
5	00	0. 21	0. 22	0. 23	0. 24	0. 24	0. 25	0. 26	0. 27	0. 28	0. 29	5	00
5	20	0. 19	0. 20	0. 21	0. 22	0. 23	0. 24	0. 25	0. 26	0. 26	0. 27	5	20
5	40	0. 18	0. 19	0. 20	0. 21	0. 22	0. 22	0. 23	0. 24	0. 25	0. 26	5	40
6	00	0. 17	0. 18	0. 19	0. 20	0. 20	0. 21	0. 22	0. 23	0. 23	0. 24	6	00
6	20	0. 16	0. 17	0. 18	0. 19	0. 19	0. 20	0. 21	0. 22	0. 22	0. 23	6	20
6	40	0. 15	0. 17	0. 17	0. 18	0. 18	0. 19	0. 20	0. 20	0. 21	0. 22	6	40
7	00	0. 15	0. 15	0. 16	0. 17	0. 17	0. 18	0. 19	0. 19	0. 20	0. 21	7	00
7	20	0. 14	0. 15	0. 15	0. 16	0. 17	0. 17	0. 18	0. 19	0. 19	0. 20	7	20
7	40	0. 13	0. 14	0. 15	0. 15	0. 16	0. 17	0. 17	0. 18	0. 18	0. 19	7	40
8	00	0. 13	0. 13	0. 14	0. 15	0. 15	0. 16	0. 16	0. 17	0. 18	0. 18	8	00
8	20	0. 12	0. 13	0. 13	0. 14	0. 15	0. 15	0. 16	0. 16	0. 17	0. 17	8	20
8	40	0. 12	0. 12	0. 13	0. 13	0. 14	0. 15	0. 15	0. 16	0. 16	0. 17	8	40
9	00	0. 11	0. 12	0. 12	0. 13	0. 14	0. 14	0. 15	0. 15	0. 16	0. 16	9	00
9	30	0. 11	0. 11	0. 12	0. 13	0. 13	0. 13	0. 14	0. 14	0. 15	0. 15	9	30
10	00	0. 10	0. 11	0. 11	0. 12	0. 12	0. 13	0. 13	0. 14	0. 14	0. 14	10	00
10	30		0. 10	0. 11	0. 11	0. 12	0. 12	0. 12	0. 13	0. 13	0. 14	10	30
11	00			0. 10	0. 11	0. 11	0. 11	0. 12	0. 13	0. 13	0. 13	11	00
11	30				0. 10	0. 11	0. 11	0. 11	0. 12	0. 12	0. 13	11	30
12	00					0. 10	0. 10	0. 11	0. 11	0. 12	0. 12	12	00
12	30							0. 10	0. 11	0. 11	0. 12	12	30
13	00								0. 10	0. 11	0. 11	13	00
13	30									0. 10	0. 11	13	30
14	00										0. 10	14	00

TABLE 41

Distance by Vertical Angle

Measured Between Waterline at Object and Top of Object

Angle	Height of object above the sea, in feet and (meters)										Angle
	160 (48.8)	165 (50.3)	175 (53.3)	185 (56.4)	195 (59.4)	200 (61.0)	225 (68.6)	250 (76.2)	275 (83.8)	300 (91.4)	
° ′	*Miles*	*Miles*	*Miles*	*Miles*	*Miles*	*Miles*	*Miles*	*Miles*	*Miles*	*Miles*	° ′
0 15											0 15
0 20	4.53	4.67	4.95								0 20
0 25	3.62	3.73	3.96	4.19	4.41	4.53					0 25
0 30	3.02	3.11	3.30	3.49	3.68	3.77	4.24	4.71			0 30
0 35	2.59	2.67	2.83	2.99	3.15	3.23	3.64	4.04	4.45	4.85	0 35
0 40	2.26	2.33	2.48	2.62	2.76	2.83	3.18	3.54	3.89	4.24	0 40
0 45	2.01	2.07	2.20	2.33	2.45	2.51	2.83	3.14	3.46	3.77	0 45
0 50	1.81	1.87	1.98	2.09	2.21	2.26	2.55	2.83	3.11	3.39	0 50
0 55	1.65	1.70	1.80	1.90	2.01	2.06	2.31	2.57	2.83	3.09	0 55
1 00	1.51	1.56	1.65	1.74	1.84	1.89	2.12	2.36	2.59	2.83	1 00
1 10	1.20	1.33	1.41	1.50	1.58	1.62	1.82	2.02	2.22	2.42	1 10
1 20	1.13	1.17	1.24	1.31	1.38	1.41	1.59	1.77	1.94	2.12	1 20
1 30	1.01	1.04	1.10	1.16	1.23	1.26	1.41	1.57	1.73	1.89	1 30
1 40	0.91	0.93	0.99	1.05	1.10	1.13	1.27	1.41	1.56	1.70	1 40
1 50	0.82	0.85	0.90	0.95	1.00	1.03	1.16	1.29	1.41	1.54	1 50
2 00	0.75	0.78	0.82	0.87	0.92	0.94	1.06	1.18	1.30	1.41	2 00
2 15	0.07	0.69	0.73	0.77	0.82	0.88	0.94	1.05	1.15	1.26	2 15
2 30	0.60	0.62	0.66	0.70	0.74	0.75	0.85	0.94	1.04	1.13	2 30
2 45	0.55	0.57	0.60	0.63	0.67	0.69	0.77	0.86	0.94	1.03	2 45
3 00	0.50	0.52	0.55	0.58	0.61	0.63	0.71	0.79	0.86	0.94	3 00
3 20	0.45	0.47	0.49	0.52	0.55	0.57	0.64	0.71	0.78	0.85	3 20
3 40	0.41	0.42	0.45	0.48	0.50	0.51	0.58	0.64	0.71	0.77	3 40
4 00	0.38	0.39	0.41	0.44	0.46	0.47	0.53	0.59	0.65	0.71	4 00
4 20	0.35	0.36	0.38	0.40	0.42	0.43	0.49	0.54	0.60	0.65	4 20
4 40	0.32	0.33	0.35	0.37	0.39	0.40	0.45	0.50	0.55	0.60	4 40
5 00	0.30	0.31	0.33	0.35	0.37	0.38	0.42	0.47	0.52	0.56	5 00
5 20	0.28	0.29	0.31	0.33	0.34	0.35	0.40	0.44	0.48	0.53	5 20
5 40	0.27	0.27	0.29	0.31	0.32	0.33	0.37	0.41	0.46	0.50	5 40
6 00	0.25	0.26	0.27	0.29	0.31	0.31	0.35	0.39	0.43	0.47	6 00
6 20	0.24	0.24	0.26	0.27	0.29	0.30	0.33	0.37	0.41	0.44	6 20
6 40	0.23	0.23	0.25	0.26	0.27	0.28	0.32	0.35	0.39	0.42	6 40
7 00	0.21	0.22	0.23	0.25	0.26	0.27	0.30	0.34	0.37	0.40	7 00
7 20	0.20	0.21	0.22	0.24	0.25	0.26	0.29	0.32	0.35	0.38	7 20
7 40	0.20	0.20	0.21	0.23	0.24	0.24	0.28	0.31	0.34	0.37	7 40
8 00	0.19	0.19	0.20	0.22	0.23	0.23	0.26	0.29	0.32	0.35	8 00
8 20	0.18	0.19	0.20	0.21	0.22	0.22	0.25	0.28	0.31	0.34	8 20
8 40	0.17	0.19	0.19	0.20	0.21	0.22	0.24	0.27	0.30	0.32	8 40
9 00	0.17	0.17	0.18	0.19	0.20	0.21	0.23	0.26	0.29	0.31	9 00
9 30	0.16	0.16	0.17	0.18	0.19	0.20	0.22	0.25	0.27	0.30	9 30
10 00	0.15	0.15	0.16	0.17	0.18	0.19	0.21	0.23	0.26	0.28	10 00
10 30	0.14	0.15	0.16	0.16	0.17	0.18	0.20	0.22	0.24	0.27	10 30
11 00	0.14	0.14	0.15	0.16	0.17	0.17	0.19	0.21	0.23	0.25	11 00
11 30	0.13	0.13	0.14	0.15	0.16	0.16	0.18	0.20	0.22	0.24	11 30
12 00	0.12	0.13	0.14	0.14	0.15	0.15	0.17	0.19	0.21	0.23	12 00
12 30	0.12	0.12	0.13	0.14	0.14	0.15	0.17	0.19	0.20	0.22	12 30
13 00	0.11	0.11	0.12	0.13	0.14	0.14	0.16	0.18	0.20	0.21	13 00
13 30	0.11	0.11	0.12	0.13	0.13	0.14	0.15	0.17	0.19	0.21	13 30
14 00	0.11	0.11	0.12	0.12	0.13	0.13	0.15	0.17	0.18	0.20	14 00
14 30	0.10	0.10	0.11	0.12	0.12	0.13	0.14	0.16	0.18	0.19	14 30
15 00			0.11	0.11	0.12	0.12	0.14	0.15	0.17	0.18	15 00
16 00			0.10	0.11	0.11	0.11	0.13	0.14	0.16	0.17	16 00
17 00				0.10	0.10	0.11	0.12	0.13	0.15	0.16	17 00
18 00						0.10	0.11	0.13	0.14	0.15	18 00
19 00							0.11	0.12	0.13	0.14	19 00
20 00							0.10	0.11	0.12	0.14	20 00

PART TWO

MATHEMATICS FOR NAVIGATION

PART TWO

MATHEMATICS FOR NAVIGATION

CHAPTER I

MATHEMATICS

Arithmetic

101. Definitions.—Arithmetic is that branch of mathematics dealing with computation by numbers. The principal processes involved are addition, subtraction, multiplication, and division. A number consisting of a single symbol (1, 2, 3, etc.) is a **digit.** Any number that can be stated or indicated, however large or small, is called a **finite number;** one too large to be stated or indicated is called an **infinite** number; and one too small to be stated or indicated is called an **infinitesimal number.**

The **sign** of a number is the indication of whether it is positive (+) or negative (−). This may sometimes be indicated in another way. Thus, latitude is usually indicated as *north* (N) or *south* (S), but if north is considered positive, south is then negative with respect to north. In navigation, the north or south designation of latitude and declination is often called the "name" of the latitude or declination. A **positive number** is one having a positive sign (+); a **negative number** is one having a negative sign (−). The **absolute value** of a number is that number without regard to sign. Thus, the absolute value of both (+)8 and (−)8 is 8. Generally, a number without a sign can be considered positive.

102. Significant digits are those digits of a number which have a significance. Zeros at the left of the number and sometimes those at the right are excluded. Thus, 1,325, 1,001, 1.408, 0.00005926, 625.0, and 0.4009 have four significant digits each. But in the number 186,000 there may be three, four, five, or six significant digits depending upon the accuracy with which the number has been determined. If the quantity has only been determined to the nearest thousand there are three significant digits, the zeros at the right not being counted. If the number has been determined to the nearest one hundred, there are four significant digits, the first zero at the right being counted. If the number has been determined to the nearest ten, there are five significant digits, the first two zeros on the right being counted. If the quantity has been determined to the nearest unit, there are six significant digits, the three zeros at the right being counted.

This ambiguity is sometimes avoided by expressing numbers in powers of 10 (art. 108). Thus, 18.6×10^4 ($18.6\times10{,}000$) indicates accuracy to the nearest thousand, 18.60×10^4 to the nearest hundred, 18.600×10^4 to the nearest ten, and 18.6000×10^4 to the nearest unit. The position of the decimal is not important if the correct power of 10 is given. For example, 18.6×10^4 is the same as 1.86×10^5, 186×10^3, etc.

The small number above and to the right of 10 (the **exponent**) indicates the number of places the decimal point is to be moved to the *right.* If the exponent is negative, it indicates a reciprocal, and the decimal point is moved to the *left.* Thus, 1.86×10^{-6} 0.00000186. This system is called **scientific notation.**

103. Expressing numbers.—In navigation, fractions are usually expressed as decimals. Thus, ¼ is expressed as 0.25 and ⅓ as 0.33. To determine the decimal equivalent of a fraction, divide the **numerator** (the number above the line) by the **denominator** (the number below the line). When a decimal is less than 1, as in the examples above, it is good practice to show the zero at the left of the decimal point (0.25, not .25).

A number should not be expressed using more significant digits (art. 102) than justified. The *implied* accuracy of a decimal is indicated by the number of digits shown to the right of the decimal point. Thus, the expression "14 miles" implies accuracy to the nearest whole mile, or any value between 13.5 and 14.5 miles. The expression "14.0 miles" implies accuracy of a tenth of a mile, or any value between 13.95 and 14.05 miles.

A quantity may be expressed to a greater implied accuracy than is justified by the accuracy of the information from which the quantity is derived. For instance, if a ship steams 1 mile in 3^m21^s, its speed is $60^m \div 3^m21^s = 60 \div 3.35 = 17.910447761194$ knots, approximately. The division can be carried to as many places as desired, but if the time is measured only to the nearest second, the speed is accurate only to one decimal place in this example, because an error of 0.5 second introduces an error of more than 0.05 knot in the speed. Hence, the additional places are meaningless and possibly misleading, unless more accurate time is available. In general, it is not good practice to state a quantity to imply accuracy greater than what is justified. However, in marine navigation the accuracy of information is often unknown, and it is customary to give positions as if they were accurate to 0.′1 of latitude and longitude, although they *may* not be accurate even to the nearest whole minute.

If there are no more significant digits, regardless of how far a computation is carried, this may be indicated by use of the word "exactly." Thus, $12 \div 4 = 3$ exactly and 1 nautical mile = 1,852 meters exactly; but $12 \div 7 = 1.7$ approximately, the word "approximately" indicating that additional decimal places might be computed. Another way of indicating an approximate relationship is by placing a positive or negative sign after the number. Thus, $12 \div 7 = 1.7+$, and $11 \div 7 = 1.6-$. This system has the advantage of showing whether the approximation is too great or too small.

In any arithmetical computation the answer is no more accurate than the least accurate value used. Thus, if it is desired to add 16.4 and 1.88, the answer might be given as 18.28, but since the first term might be anything from 16.35 to 16.45, the answer is anything from 18.23 to 18.33. Hence, to retain the second decimal place in the answer is to give a false indication of accuracy, for the number 18.28 indicates a value between 18.275 and 18.285. However, additional places are sometimes retained until the end of a computation to avoid an accumulation of small errors due to rounding off (art. 104). In marine navigation it is customary to give most values to an accuracy of 0.1, even though some uncertainty may exist as to the accuracy of the last place. Examples are the dip and refraction corrections of sextant altitudes (arts. 805, 806).

In general, a value obtained by interpolation in a table should not be expressed to more decimal places than given in the table.

Unless all numbers are exact, doubt exists as to the accuracy of the last digit in a computation. Thus, $12.3 + 9.4 + 4.6 = 26.3$. But if the three terms to be added have been rounded off from 12.26, 9.38, and 4.57, the correct answer is 26.2, obtained by rounding off the answer of 26.21 found by retaining the second decimal place until the end. It is good practice to work with one more place than needed in the answer, when the information is available. In computations involving a large number of terms, or if greater accuracy is desired, it is sometimes advisable to retain two or more additional places until the end.

104. Rounding off.—In rounding off numbers to the number of places desired, one should take the nearest value. Thus, the number 6.5049 is rounded to 6.505, 6.50, 6.5, or 7, depending upon the number of places desired. If the number to be rounded off ends in 5, the nearer *even* number is taken. Thus, 1.55 and 1.65 are both rounded to 1.6. Likewise, 12.750 is rounded to 12.8 if only one decimal place is desired. However, 12.749 is rounded to 12.7. That is, 12.749 is not first rounded to 12.75 and then to 12.8, but the entire number is rounded in one operation. When a number ends in 5, the computation can sometimes be carried to additional places to determine whether the correct value is more or less than 5.

105. Reciprocals.—The reciprocal of a number is 1 divided by that number. The reciprocal of a fraction is obtained by interchanging the numerator and denominator. Thus, the reciprocal of $\frac{3}{5}$ is $\frac{5}{3}$. A whole number may be considered a fraction with 1 as the denominator. Thus, 54 is the same as $\frac{54}{1}$, and its reciprocal is $\frac{1}{54}$. Division by a number produces the same result as multiplying by its reciprocal, or vice versa. Thus, $12 \div 2 = 12 \times \frac{1}{2} = 6$, and $12 \times 2 = 12 \div \frac{1}{2} = 24$.

106. Addition.—When two or more numbers are to be added, it is generally most convenient to write them in a column, with the decimal points in line. Thus, if 31.2, 0.8874, and 168.14 are to be added, this may be indicated by means of the addition sign (+): 31.2+0.8874+168.14=200.2. But the addition can be performed more conveniently by arranging the numbers as follows:

$$\begin{array}{r@{.}l} 31 & 2 \\ 0 & 8874 \\ 168 & 14 \\ \hline 200 & 2. \end{array}$$

The answer is given only to the first decimal place, because the answer is no more accurate than the least precise number among those to be added, as indicated previously. Often it is preferable to state all numbers in a problem to the same precision before starting the addition, although this may introduce a small error, as indicated in article 103:

$$\begin{array}{r@{.}l} 31 & 2 \\ 0 & 9 \\ 168 & 1 \\ \hline 200 & 2. \end{array}$$

If there are no decimals, the last digit to the right is aligned:

$$\begin{array}{r} 166 \\ 2 \\ 96{,}758 \\ \hline 96{,}926. \end{array}$$

Numbers to be added should be given to the same absolute accuracy, when available, to avoid a false impression of accuracy in the result. Consider the following:

$$\begin{array}{r} 186{,}000 \\ 71{,}832 \\ 9{,}614 \\ 728 \\ \hline 268{,}174. \end{array}$$

The answer would imply an accuracy to six places. If the first number given is accurate to only three places, or to the nearest 1,000, the answer is not more accurate, and hence the answer should be given as 268,000. Approximately the same answer would be obtained by rounding off at the start:

$$\begin{array}{r} 186{,}000 \\ 72{,}000 \\ 10{,}000 \\ 1{,}000 \\ \hline 269{,}000. \end{array}$$

If numbers are **added arithmetically,** their absolute values are added without regard to signs; but if they are **added algebraically,** due regard is given to signs. If two numbers to be added algebraically have the same sign, their absolute values are added and given their common sign. If two numbers to be added algebraically have unlike signs, the smaller absolute value is subtracted from the larger, and the sign of the value having the larger absolute value is given to the result. Thus, if +8 and −7 are added arithmetically, the answer is 15, but if they are added algebraically, the answer is +1.

An answer obtained by addition is called a **sum.**

107. Subtraction is the inverse of addition. Stated differently, the *addition* of a *negative* number is the same as the *subtraction* of a *positive* number. That is, if a number is to be subtracted from another, the sign (+ or −) of the **subtrahend** (the number to be subtracted) is reversed and the result added algebraically to the **minuend** (the number from which the subtrahend is to be subtracted). Thus, $6-4=2$. This may be written $+6-(+4)=+2$, which yields the same result as $+6+(-4)$. For solution, larger numbers are often conveniently arranged in a column with decimal points in a vertical column, as in addition. Thus, $3{,}728.41-1{,}861.16$ may be written:

$$\begin{array}{rl} (+)3{,}728.41 & \\ (+)1{,}861.16 & \text{(subtract)} \\ \hline (+)1{,}867.25 & \end{array}$$

This is the same as:

$$\begin{array}{rl} (+)3{,}728.41 & \\ (-)1{,}861.16 & \text{(add algebraically)} \\ \hline (+)1{,}867.25. & \end{array}$$

The rule of sign reversal applies likewise to negative numbers. Thus, if −3 is to be *subtracted* from +5, this may be written $+5-(-3)=5+3=8$.

In the algebraic addition of two numbers of opposite sign (numerical subtraction), the smaller number is subtracted from the larger and the result is given the sign of the larger number. Thus, $+7-4=+3$, and $-7+4=-3$, which is the same as $+4-7=-3$.

In navigation, numbers to be numerically subtracted are usually marked (−), and those to be numerically added are marked (+) or the sign is not indicated. However, when a sign is part of a designation, and the reverse process is to be used, the word "reversed" (rev.) is written after the number. Thus, if GMT is known and ZT in the (+)5 zone is to be found (by subtraction), the problem may be written:

$$\begin{array}{rrl} \text{GMT} & 1754 & \\ \text{ZD} & (+)5 & \text{(rev.)} \\ \cline{2-2} \text{ZT} & 1254. & \end{array}$$

The symbol ∼ indicates that an absolute difference is required without regard to sign of the *answer*. Thus, $28\sim13=15$, and $13\sim28=15$. In both of these solutions 13 and 28 are positive and 15 is an absolute value without sign. If the signs or names of both numbers are the same, either positive or negative, the smaller is subtracted from the larger, but if they are of opposite sign or name, they are numerically added. Thus, $(+)16 \sim (+)21 = 5$ and $(-)16 \sim (-)21 = 5$, but $(+)16 \sim (-)21 = 37$ and $(-)16 \sim (+)21 = 37$. Similarly, the difference of latitude between 15°N and 20°N, or between 15°S and 20°S, is 5°, but the difference of latitude between 15°N and 20°S, or between 15°S and 20°N, is 35°. If motion from one latitude to another is involved, the difference may be given a sign to indicate the direction of travel, or the location of one place with respect to another. Thus, if *B* is 50 miles west of *A*, and *C* is 125 miles west of *A*, *B* and *C* are 75 miles apart regardless of the direction of travel. However, *B* is 75 miles *east* of *C*, and *C* is 75 miles *west* of *B*. When direction is indicated, an algebraic difference is given, rather than an absolute difference, and the symbol $\sim$ is not appropriate.

It is sometimes desirable to consider all addition and subtraction problems as addition, with negative signs (−) given before those numbers to be subtracted, so that there can be no question of which process is intended. The words "add" and "subtract" may be used instead of signs. In navigation, "names" (usually north, south, east, and west) are often used, and the relationship involved in a certain problem may need to be understood to determine whether to add or subtract. Thus, $LHA = GHA - \lambda$(west) and $LHA = GHA + \lambda$(east). This is the same as saying $LHA = GHA - \lambda$ if west longitude is considered positive, for in this case, $LHA = GHA - (-\lambda)$ or $LHA = GHA + \lambda$ in east longitude, the same as before.

If numbers are **subtracted arithmetically,** they are subtracted without regard to sign; but if they are **subtracted algebraically,** positive (+) numbers are *subtracted* and negative (−) numbers are *added.*

An answer obtained by subtraction is called a **difference.**

108. Multiplication may be indicated by the multiplication sign (×), as 154×28 =4,312. For solution, the problem is conveniently arranged thus:

```
     154
(×)   28
    ----
    1232
    308
    ----
    4312.
```

Either number may be given first, but it is generally more convenient to perform the multiplication if the larger number is placed on top, as shown. In this problem, 154 is first multiplied by 8 and then by 2. The second answer is placed under the first, but set one **place** to the left, so that the right-hand digit is directly below the 2 of the multiplier. These steps might be reversed, multiplication by 2 being performed first. This procedure is sometimes used in estimating.

When one number is placed below another for multiplication, as shown above, it is usually best to align the right-hand digits without regard for the position of the decimal point. The number of decimal places in the answer is the sum of the decimal places in the **multiplicand** (the number to be multiplied) and the **multiplier** (the second number):

```
      163.27
(×)    263.9
   ---------
      146943
     48981
    97962
   32654
   ---------
   43086.953.
```

However, when a number ends in one or more zeros, these may be ignored until the end and then added on to the number:

```
     1924
  (×)1800
   15392
   1924
  3463200.
```

This is also true if both multiplicand and multiplier end in zeros:

```
   1924000
 (×)1800
  15392
  1924
  3463200000.
```

When negative values are to be multiplied, the sign of the answer is positive if an *even* number of negative signs appear, and negative if there are an *odd* number. Thus, $2\times3=6$, $2\times(-3)=-6$, $-2\times3=-6$, $-2\times(-3)=(+)6$. Also, $2\times3\times8\times(-2)\times5=-480$, $2\times(-3)\times8\times(-2)\times5=480$, $2\times(-3)\times(-8)\times(-2)\times5=-480$, $2\times(-3)\times(-8)\times(-2)\times(-5)=480$, and $(-2)\times(-3)\times(-8)\times(-2)\times(-5)=-480$.

An answer obtained by multiplication is called a **product.** Any number multiplied by 1 is the number itself. Thus, $125\times1=125$. Any number multiplied by 0 is 0. Thus, $125\times0=0$ and $1\times0=0$.

To multiply a number by itself is to **square** the number. This may be indicated by the **exponent** 2 placed to the right of the number and above the line as a **superior.** Thus, 15×15 may be written 15^2. Similarly, $15\times15\times15=15^3$, and $15\times15\times15\times15=15^4$, etc. The exponent (2, 3, 4, etc.) indicates the **power** to which a number is to be **raised,** or how many times the number is to be used in multiplication. The expression 15^2 is usually read "15 squared," 15^3 is read "15 cubed" or "15 to the third power," 15^4 (or higher power) is read "15 to the fourth (or higher) power." The answer obtained by **raising to a power** is called the "square," "cube," etc., or the ". . . power" of the number. Thus, 225 is the "square of 15," 3,375 is the "cube of 15" or the "third power of 15," etc. The zero power of any number except zero (if zero is considered a number) is 1. The zero power of zero is zero. Thus, $15^0=1$ and $0^0=0$.

Parentheses may be used to eliminate doubt as to what part of an expression is to be raised to a power. Thus, -3^2 may mean either $-(3\times3)=-9$ or $-3\times-3=(+)9$. To remove the ambiguity, the expression may be written $-(3)^2$ if the first meaning is intended, and $(-3)^2$ if the second meaning is intended.

109. Division is the inverse of multiplication. It may be indicated by the division sign (÷), as $376\div21=18$ approximately; or by placing the number to be divided, called the **dividend** (376), over the other number, called the **divisor** (21), as $\frac{376}{21}=18$ approximately. The expression $\frac{376}{21}$ may be written 376/21 with the same meaning. Such a problem is conveniently arranged for solution as follows:

```
      17
 21 | 376
      21
      166
      147
       19.
```

Since the **remainder** is 19, or more than half of the divisor (21), the answer is 18 to the nearest whole number.

An answer obtained by division is called a **quotient.** Any number divided by 1 is the number itself. Thus, $65 \div 1 = 65$. A number cannot be divided by 0.

If the numbers involved are accurate only to the number of places given, the answer should not be carried to additional places. However, if the numbers are exact, the answer might be carried to as many decimal places as desired. Thus, $374 \div 21 = 17.809523809523809523809523809523809523 \ldots$ When a series of digits repeat themselves with the same remainder, as 809523 (with remainder 17) in the example given above, an exact answer will not be obtained regardless of the number of places to which the division is carried. The series of dots (. . .) indicates a **repeating decimal.** In a nonrepeating decimal, a plus sign (+) may be given to indicate a remainder, and a minus sign (—) to indicate that the last digit has been rounded to the next higher value. Thus, 18.68761 may be written 18.6876+ or 18.688—. If the last digit given is rounded off, the word "approximately" may be used instead of dots or a plus or minus sign.

If the divisor is a whole number, the decimal point in the quotient is directly above that of the dividend when the work form shown above is used. Thus, in the example given above, if the dividend had been 37.6 instead of 376, the quotient would have been 1.8 approximately. If the divisor is a decimal, both it and the dividend are multiplied by the power of 10 having an exponent equal to the number of decimal places in the divisor, and the division is then carried out as explained above. Thus, if there are two decimal places in the divisor, both divisor and dividend are multiplied by $10^2 = 100$. This is done by moving the decimal to the right until the divisor is a whole number. If necessary, zeros are added to the dividend. Thus, if 3.7 is to be divided by 2.11, both quantities are first multiplied by 10^2, and 370 is divided by 211. This is usually performed as follows:

$$
\begin{array}{r|r}
 & 1.75 \\
\hline
2/11 & 3/70.00 \\
 & \underline{211} \\
 & 1590 \\
 & \underline{1477} \\
 & 1130 \\
 & \underline{1055} \\
 & 75.
\end{array}
$$

If *both* the dividend and divisor are positive, or if *both* are negative, the quotient is positive; but if *either* is negative, the quotient is negative. Thus, $6 \div 3 = 2$, $(-6) \div (-3) = +2$, $(-6) \div 3 = -2$, and $6 \div (-3) = -2$.

The **square root** of a number is that number which, multiplied by itself, equals the given number. Thus, $15 \times 15 = 15^2 = 225$, and $\sqrt{225} = 225^{1/2} = 15$. Either the symbol $\sqrt{\ }$, called the **radical sign,** or the exponent $\frac{1}{2}$ indicates square root. Also, $\sqrt[3]{\ }$, or $\frac{1}{3}$ as an exponent, indicates **cube root.** Fourth, fifth, or any root is indicated similarly, using the appropriate number. Nearly any arithmetic book explains the process of extracting roots, but this process is most easily performed by table, logarithms (art. 112), or slide rule (art. 115). If no other means are available, it can be done by trial and error. The process of finding a root of a number is called **extracting a root.**

110. Logarithms ("logs") provide an easy way to multiply, divide, raise numbers to powers, and extract roots. The logarithm of a number is the power to which a fixed number, called the base, must be raised to produce the value to which the logarithm corresponds. The base of **common logarithms,** (given in tables 32 and 33) is 10. Hence, since $10^{1.8} = 63$ approximately, 1.8 is the logarithm, approximately, of 63 to the base

10. In table 32 logarithms of numbers are given to five decimal **places.** This is sufficient for most purposes of the navigator. For greater precision, a table having additional places should be used. In general, the number of *significant digits* which are correct in an answer obtained by logarithms is the same as the number of *places* in the logarithms used.

A logarithm is composed of two parts. That part to the left of the decimal point is called the **characteristic.** That part to the right of the decimal point is called the **mantissa.** The principal advantage of using 10 as the base is that any given combination of digits has the same mantissa regardless of the position of the decimal point. Hence, only the mantissa is given in the main tabulation of table 32. Thus, the logarithm (mantissa) of 2,374 is given as 37548. This is correct for 2,374,000,000; 2,374; 23.74; 2.374; 0.2374; 0.000002374; or for any other position of the decimal point.

The position of the decimal point determines the characteristic, which is not affected by the actual digits involved. The characteristic of a whole number is one less than the number of digits. The characteristic of a **mixed decimal** (one greater than 1) is one less than the number of digits to the left of the decimal point. Thus, in the example given above, the characteristic of the logarithm of 2,374,000,000 is 9; that of 2,374 is 3; that of 23.74 is 1; and that of 2.374 is 0. The complete logarithms of these numbers are:

$$\begin{aligned} \log 2{,}374{,}000{,}000 &= 9.37548 \\ \log 2{,}374 &= 3.37548 \\ \log 23.74 &= 1.37548 \\ \log 2.374 &= 0.37548. \end{aligned}$$

Since the mantissa of the logarithm of any multiple of ten is zero, the main table starts with 1,000. This can be considered 100, 10, 1, etc. Since the mantissa of these logarithms is zero, the logarithms consist of the characteristic only, and are whole numbers. Hence, the logarithm of 1 is 0 (0.00000), that of 10 is 1 (1.00000), that of 100 is 2 (2.00000), that of 1,000 is 3 (3.00000), etc.

The characteristic of the logarithm of a number less than 1 is negative. However, it is usually more conveniently indicated in a positive form, as follows: the characteristic is found by subtracting the number of zeros immediately to the right of the decimal point from 9 (or 19, 29, etc.) and following this by −10 (or −20, −30, etc.). Thus, the characteristic of the logarithm of 0.2374 is 9−10; that of 0.000002374 is 4−10; and that of 0.000000000002374 is 8−20. The complete logarithms of these numbers are:

$$\begin{aligned} \log 0.2374 &= 9.37548-10 \\ \log 0.000002374 &= 4.37548-10 \\ \log 0.000000000002374 &= 8.37548-20. \end{aligned}$$

When there is no question of the meaning, the −10 may be omitted. This is usually done when using logarithms of trigonometric functions, as shown in table 33. Thus, if there is no reasonable possibility of confusion, the logarithm of 0.2374 may be written 9.37548.

Occasionally, the logarithm of a number less than 1 is shown by giving the negative characteristic with a minus sign above it (since only the characteristic is negative, the mantissa being positive). Thus, the logarithms of the numbers given above might be shown thus:

$$\begin{aligned} \log 0.2374 &= \bar{1}.37548 \\ \log 0.000002374 &= \bar{6}.37548 \\ \log 0.000000000002374 &= \overline{12}.37548. \end{aligned}$$

In each case, the negative characteristic is one *more* than the number of zeros immediately to the right of the decimal point.

There is no real logarithm of 0, since there is no *finite* power to which *any* number can be raised to produce 0. As numbers approach 0, their logarithms approach negative infinity.

To find the number corresponding to a given logarithm, called finding the **antilogarithm** ("antilog"), enter the table with the mantissa of the given logarithm and determine the corresponding number, interpolating if necessary. Locate the position of the decimal point by means of the characteristic of the logarithm, in accordance with the rules given above.

111. Multiplication by logarithms.—To *multiply* one number by another, *add* their logarithms and find the antilogarithm of the sum. Thus, to multiply 1,635.8 by 0.0362 by logarithms:

$$\begin{aligned} \log 1635.8 &= 3.21373 \\ \log 0.0362 &= \underline{8.55871-10} \text{ (add)} \\ \log 59.216 &= 11.77244-10 \text{ or } 1.77244. \end{aligned}$$

Thus, 1,635.8×0.0362=59.216. In navigation it is customary to use a slightly modified form, and to the omit the −10 where there is no reasonable possibility of confusion, as follows:

1635.8	log	3.21373
0.0362	log	8.55871
59.216	log	1.77244.

To *raise a number to a power*, multiply the logarithm of that number by the power indicated, and find the antilogarithm of the product. Thus, to find 13.156^3 by logarithms, using the navigational form:

13.156	log	1.11913	
	×	3	(multiply)
2277.2	log	3.35739.	

112. Division by logarithms.—To *divide* one number by another, subtract the logarithm of the divisor from that of the dividend, and find the antilogarithm of the remainder. Thus, to find 0.4637÷28.03 by logarithms, using the navigational form:

0.4637	log	9.66624	
28.03	log (−)	1.44762	(subtract)
0.016543	log	8.21862.	

It is sometimes necessary to modify the first logarithm before the subtraction can be made. This would occur in the example given above, for instance, if the divisor and dividend were reversed, so that the problem became 28.03÷0.4637. In this case 10−10 would be added to the logarithm of the dividend, becoming 11.44762−10:

28.03	log	11.44762−10
0.4637	log (−)	9.66624−10
60.448	log	1.78138.

One experienced in the use of logarithms usually carries this change mentally, without showing it in his work form:

28.03	log	1.44762
0.4637	log (−)	9.66624
60.448	log	1.78138.

Any number can be added to the characteristic as long as that same number is also subtracted. Conversely, any number can be subtracted from the characteristic as long as that same number is also added.

To *extract a root* of a number, divide the logarithm of that number by the root indicated, and find the antilogarithm of the quotient. Thus, to find $\sqrt{7}$ by logarithms:

$$\begin{array}{rl} 7 & \log \underline{0.84510}\ (\div 2) \\ 2.6458 & \log 0.42255. \end{array}$$

To divide a negative logarithm by the root indicated, first modify the logarithm so that the quotient will have a -10. Thus, to find $\sqrt[3]{0.7}$ by logarithms:

$$\begin{array}{rl} 0.7 & \log \underline{29.84510-30}\ (\div 3) \\ 0.88792 & \log\ 9.94837-10 \end{array}$$

or, carrying the -30 and -10 mentally,

$$\begin{array}{rl} 0.7 & \log \underline{29.84510}\ (\div 3) \\ 0.88792 & \log\ 9.94837. \end{array}$$

113. Cologarithms.—The **cologarithm** ("colog") of a number is the value obtained by subtracting the logarithm of that number from zero, usually in the form 10−10. Thus, the logarithm of 18.615 is 1.26987. The cologarithm is:

$$\begin{array}{r} 10.00000-10 \\ (-)\underline{1.26987} \\ 8.73013-10. \end{array}$$

Similarly, the logarithm of 0.0018615 is 7.26987−10, and its cologarithm is:

$$\begin{array}{r} 10.00000-10 \\ (-)\underline{7.26987-10} \\ 2.73013. \end{array}$$

The *cologarithm* of a number is the *logarithm* of the reciprocal of that number. Thus, the cologarithm of 2 is the logarithm of ½. Since division by a number is the same as multiplication by its reciprocal, the use of cologarithms permits division problems to be converted to problems of multiplication, eliminating the need for subtraction of logarithms. This is particularly useful when both multiplication and division are involved in the same problem. Thus, to find $\frac{92.732 \times 0.0137 \times 724.3}{0.516 \times 3941.1}$ by logarithms, one might *add* the logarithms of the three numbers in the numerator, and *subtract* the logarithms of the two numbers in the denominator. If cologarithms are used for the numbers in the denominator, all logarithmic values are added. Thus, the solution might be made as follows:

92.732		log 1.96723
0.0137		log 8.13672
724.3		log 2.85992
0.516	log 9.71265	colog 0.28735
3941.1	log 3.59562	colog 6.40438
0.45248		log 9.65560.

114. Various kinds of logarithms.—As indicated above, **common logarithms** use 10 as the base. These are also called **Briggs' logarithms.** For some purposes, it is convenient to use 2.7182818 approximately (designated e) as the base for logarithms. These are called **natural logarithms** or **Naperian logarithms** ($\log_e$). Common logarithms are shown as $\log_{10}$ when the base might otherwise be in doubt.

Addition and subtraction logarithms are logarithms of the sum and difference of two numbers. They are used when the logarithms of two numbers to be added or subtracted are known, making it unnecessary to find the numbers themselves.

115. Slide rule.—A **slide rule** is a convenient device for making logarithmic solutions mechanically. There are many types and sizes of slide rule, some designed for specific purposes. The most common form consists of an outer "body" or "frame" with grooves to permit a "slide" to be moved back and forth between the two outer parts, so that any graduation of a scale on the slide can be brought opposite any graduation of a scale on the body. A cursor called an "indicator" or "runner" is provided to assist in aligning the desired graduations. In a **circular slide rule** the "slide" is an inner disk surrounded by a larger one, both pivoted at their common center. The scales of a slide rule are *logarithmic*. That is, they increase proportionally to the logarithms of the numbers indicated, rather than to the numbers themselves. This permits addition and subtraction of logarithms by simply measuring off part of the length of the slide from a graduated point on the body, or vice versa. Two or three complete scales within the length of the rule may be provided for finding squares, cubes, square roots, and cube roots.

Full instructions for use of a slide rule are provided with each rule, and given in some mathematical texts. Properly used, a slide rule can provide quick answers to many of the problems of navigation. However, its precision is usually limited to from two to four significant digits, and should not be used if greater precision is desired. It is frequently used to provide a quick, approximate check on answers obtained by a more laborious method.

Great care should be used in placing the decimal point in an answer obtained by slide rule, as the correct location often is not immediately apparent. Its position is usually determined by making a very rough mental solution. Thus, 2.93×8.3 is *about* $3\times8=24$. Hence, when the answer by slide rule is determined to be "243," it is known that the correct value is 24.3, not 2.43 or 243.

116. Mental arithmetic.—Many of the problems of the navigator can be solved mentally. The following are a few examples.

If the speed is a number divisible into 60 a whole number of times, distance problems can be solved by a simple relationship. Thus, at 10 knots a ship steams 1 mile in $\frac{60}{10}=6$ minutes. At 12 knots it requires 5 minutes, at 15 knots 4 minutes, etc. As an example of the use of such a relationship, a vessel steaming at 12 knots travels 5.6 miles in 28 minutes, since $\frac{28}{5}=5\frac{3}{5}=5.6$, or 0.1 mile every half minute.

For relatively short distances, one nautical mile can be considered equal to 6,000 feet. Since one hour has 60 minutes, the speed in hundreds of feet per minute is equal to the speed in knots. Thus, a vessel steaming at 15 knots is moving at the rate of 1,500 feet per minute.

With respect to time, 6 minutes$=$0.1 hour, and 3 minutes$=$0.05 hour. Hence, a ship steaming at 13 knots travels 3.9 miles in 18 minutes (13×0.3), and 5.8 miles in 27 minutes (13×0.45).

In arc units, $6'=0.^{\circ}1$ and $6''=0.'1$. This relationship is useful in rounding off values given in arc units. Thus, $17^{\circ}23'44''=17^{\circ}23.'7$ to the nearest $0.'1$, and $17.^{\circ}4$ to the nearest $0.^{\circ}1$. A thorough knowledge of the six multiplication table is valuable. The 15 multiplication table is also useful, since $15^{\circ}=1^{h}$. Hence, $16^{h}=16\times15=240^{\circ}$. This is particularly helpful in quick determination of zone description. Pencil and paper or a table should not be needed, for instance, to decide that a ship at sea in longitude $157^{\circ}18.'4$ W is in the $(+)10$ zone.

It is also helpful to remember that $1° = 4^m$ and $1' = 4^s$. In converting the LMT of sunset to ZT, for instance, a quick mental solution can be made without reference to a table. Since this correction is usually desired only to the nearest whole minute, it is necessary only to multiply the longitude difference in degrees (to the nearest quarter degree) by four.

Vectors

117. Scalars and vector quantities.—A **scalar** is a quantity which has *magnitude* only; a **vector quantity** has both *magnitude* and *direction*. If a vessel is said to have a tank of 5,000 gallons capacity, the number 5,000 is a scalar. As used in this book, *speed* alone is considered a scalar, while *speed* and *direction* are considered to constitute *velocity*, a vector quantity. Thus, if a vessel is said to be steaming at 18 knots, without regard to direction, the number 18 is considered a scalar; but if the vessel is said to be steaming at 18 knots on course 157°, the combination of 18 knots and 157° constitutes a vector quantity. *Distance* and *direction* also constitute a vector quantity.

A *scalar* can be represented fully by a number. A *vector quantity* requires, in addition, an indication of direction. This is conveniently done graphically by means of a straight line, the length of which indicates the *magnitude*, and the direction of which indicates the *direction* of application of the magnitude. Such a line is called a **vector.** Since a straight line has two directions, reciprocals of each other, an arrowhead is placed along or at one end of a vector to indicate the direction represented, unless this is apparent or indicated in some other manner.

118. Addition and subtraction of vectors.—Two vectors can be *added* by *starting* the second at the *termination* (rather than the origin) of the first. A common navigational use of vectors is the dead reckoning plot of a vessel. Refer to figure 118. If a ship starts at *A* and steams 18 miles on course 090° and then 12 miles on course 060°, it arrives by dead reckoning at *C*. The line *AB* is the vector for the first run, and *BC* is the vector for the second. Point *C* is the position found by *adding* vectors *AB* and *BC*. The vector *AC*, in this case the *course and distance made good*, is the **resultant.** Its value, both in direction and amount, can be determined by measurement. Lines *AB*, *BC*, and *AC* are all **distance vectors. Velocity vectors** are used when determining the effect of, or allowing for, current, interconverting true and apparent wind, and solving relative motion problems.

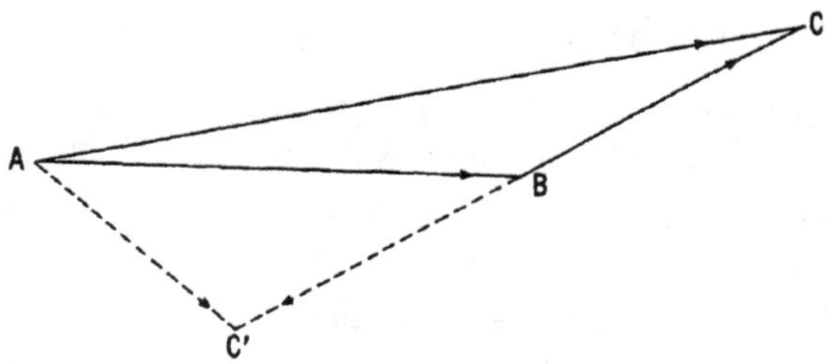

FIGURE 118.—Addition and subtraction of vectors.

The **reciprocal** of a vector has the same magnitude but opposite direction of the vector. To *subtract* a vector, *add* its reciprocal. This is indicated by the broken lines in figure 118, in which the vector *BC'* is drawn in the opposite direction to *BC*. In this case the resultant is *AC'*. Subtraction of vectors is involved in some current and wind problems.

Algebra

119. Definitions.—**Algebra** is that branch of mathematics dealing with computation by letters and symbols. It permits the mathematical statement of certain relationships between variables. When numbers are substituted for the letters, algebra becomes arithmetic. Thus if $a=2b$, any value may be assigned to b, and a can be found by multiplying the assigned value by 2. Any statement of equality (as $a=2b$) is an **equation.** Any combination of numbers, letters, and symbols (as $2b$) is a **mathematical expression.**

120. Symbols.—As in arithmetic, plus (+) and minus (−) signs are used, and with the same meaning. Multiplication (×) and division (÷) signs are seldom used. In algebra, $a \times b$ is usually written ab, or sometimes $a \cdot b$. For division $a \div b$ is usually written $\frac{a}{b}$ or a/b. The symbol $>$ means "greater than" and $<$ means "less than." Thus, $a > b$ means "a is greater than b," and $a \geqq b$ or $a \geq b$ means "a is equal to or greater than b."

The order of performing the operations indicated in an equation should be observed carefully. Consider the equation $a=b+cd-e/f$. If the equation is to be solved for a, the value cd should be determined by multiplication and e/f by division *before* the addition and subtraction, as each of these is to be considered a single quantity in making the addition and subtraction. Thus, if $cd=g$ and $e/f=h$, the formula can be written $a=b+$ [illegible] h.

If an equation including both multiplication and division between plus or minus signs is not carefully written, some doubt may arise as to which process to perform first. Thus, $a \div b \times c$ or $a/b \times c$ may be interpreted to mean either that a/b is to be multiplied by c or that a is to be divided by $b \times c$. Such an equation is better written ac/b if the first meaning is intended, or a/bc if the second meaning is intended. **Parentheses,** ( ), may be used for the same purpose or to indicate any group of quantitites that is to be considered a single quantity. Thus, $a(b+c)$ is an indication that the sum of b and c is to be multiplied by a. Similarly, $a+(b-c)^2$ indicates that c is first to be subtracted from b, and then the result is to be squared and the value thus obtained added to a. When an expression within parentheses is part of a larger expression which should also be in parentheses, **brackets,** [ ], are used in place of the outer parentheses. If yet another set is needed, **braces,** { }, are used.

A quantity written $\sqrt{3}\, ab$ is better written $ab\sqrt{3}$ to remove any suggestion that the square root of $3ab$ is to be found.

121. Addition and subtraction.—A plus sign before an expression in parentheses means that each term retains its sign as given. Thus, $a+(b+c-d)$ is the same as $a+b+c-d$. A minus sign preceding the parentheses means that each sign within the parentheses is to be reversed. For example, $a-(b+c-d)=a-b-c+d$.

In any equation involving addition and subtraction, similar terms can be combined. Thus, $a+b+c+b-2c-d=a+2b-c-d$. Also, $a+3ab+a^2-b-ab=a+2ab+a^2-b$. That is, to be combined, the terms must be truly alike, for a cannot be combined with ab, or with a^2.

Equal quantities can be added to or subtracted from both members of an equation without disturbing the equality. Thus, if $a=b$, $a+2=b+2$, or $a+x=b+x$. If $x=y$, then $a+x=b+y$.

122. Multiplication and division.—When an expression in parentheses is to be multiplied by a quantity outside the parentheses, each quantity separated by a plus or minus sign within the parentheses should be multiplied separately. Thus, $a(b+cd-e/f)$ may be written $ab+acd-ae/f$. Any quantity appearing in *every* term of one member of an equation can be separated out by **factoring,** or dividing each term by the common quantity. Thus, if $a=bc+\frac{bd}{e}-b^2+b$, the equation may be written $a=b\left(c+\frac{d}{e}-b+1\right)$.

Note that $\frac{b}{b}=1$ and $\frac{b^2}{b}=b$. This is the inverse of multiplication: $a\times 1=a$, but $a\times a=a^2$. Also, $a^2\times a^3=a^5$; and $\frac{a^7}{a^2}=a^5$. Thus, in multiplying a power of a number by a power of the same number, the powers are added, or, stated mathematically, $a^m\times a^n=a^{m+n}$. In division, $\frac{a^m}{a^n}=a^{m-n}$, or the exponents are subtracted. If n is greater than m, a *negative* exponent results. A value with a negative exponent is equal to the reciprocal of the same value with a positive exponent. Thus, $a^{-n}=\frac{1}{a^n}$ and $\frac{a^2b^{-3}}{c}=\frac{a^2}{b^3c}$.

In raising to a power a number with an exponent, the two exponents are multiplied. Thus $(a^2)^3=a^{2\times 3}=a^6$, or $(a^n)^m=a^{nm}$. The inverse is true in extracting a root. Thus, $\sqrt[3]{a^2}=a^{\frac{2}{3}}=a^{0.667}$, or $\sqrt[m]{a^n}=a^{\frac{n}{m}}$.

Both members of an equation can be multiplied or divided by equal quantities without disturbing the equality, excluding division by zero or some expression equal to zero. Thus, if $a=b+c$, $2a=2(b+c)$, or if $x=y$, $ax=y(b+c)$ and $\frac{a}{x}=\frac{b+c}{y}$. Sometimes there is more than one answer to an equation. Division by one of the unknowns may eliminate one of the answers.

Both members of an equation can be raised to the same power, and like roots of both members can be taken, without disturbing the equality. Thus, if $a=b+c$, $a^2=(b+c)^2$, or if $x=y$, $a^x=(b+c)^y$. This is *not* the same as $a^x=b^y+c^y$. Similarly, if $a=b+c$, $\sqrt{a}=\sqrt{b+c}$, or if $x=y$, $\sqrt[x]{a}=\sqrt[y]{b+c}$. Again, $\sqrt[y]{b+c}$ is *not* equal to $\sqrt[y]{b}+\sqrt[y]{c}$, as a numerical example will indicate: $\sqrt{100}=\sqrt{64+36}$, but $\sqrt{100}$ does not equal $\sqrt{64}+\sqrt{36}$.

If two quantities to be multiplied or divided are *both* positive or *both* negative, the result is positive. Thus, $(+a)\times(+b)=ab$ and $\frac{-a}{-b}=+\frac{a}{b}$. But if the signs are opposite, the answer is negative. Thus, $(+a)\times(-b)=-ab$, and $\frac{-a}{+b}=-\frac{a}{b}$; also, $(-a)\times(+b)=-ab$, and $\frac{+a}{-b}=-\frac{a}{b}$.

In expressions containing both parentheses and brackets, or both of these and braces, the innermost symbols are removed first. Thus, $-\left\{6z-\frac{[x(x+4)-5y]}{y}\right\}=-\left\{6z-\frac{[x^2+4x-5y]}{y}\right\}=-\left\{6z-\frac{x^2}{y}-\frac{4x}{y}+5\right\}=-6z+\frac{x^2}{y}+\frac{4x}{y}-5$.

123. Fractions.—To add or subtract two or more fractions, convert each to an expression having the same denominator, and then add the numerators. Thus, $\frac{a}{b}+\frac{c}{d}+\frac{e}{f}=\frac{adf}{bdf}+\frac{cbf}{bdf}+\frac{ebd}{bdf}=\frac{adf+cbf+ebd}{bdf}$. That is, both numerator and denominator of each fraction are multiplied by the denominator of the other remaining fractions.

To multiply two or more fractions, multiply the numerators by each other, and also multiply the denominators by each other. Thus, $\frac{a}{b}\times\frac{c}{d}\times\frac{e}{f}=\frac{ace}{bdf}$.

To divide two fractions, invert the divisor and multiply. Thus, $\frac{a}{b}\div\frac{c}{d}=\frac{a}{b}\times\frac{d}{c}=\frac{ad}{bc}$.

If the same factor appears in all terms of a fraction, it can be factored out without changing the value of the fraction. Thus, $\frac{ab+ac+ad}{ae-af}=\frac{b+c+d}{e-f}$. This is the same as factoring a from the numerator and denominator separately. That is, $\frac{ab+ac+ad}{ae-af}=\frac{a(b+c+d)}{a(e-f)}$, but since $\frac{a}{a}=1$, this part can be removed, and the fraction appears as above.

124. Transposition.—It is sometimes desirable to move terms of an expression from one side of the equals sign (=) to the other. This is called **transposition,** and to move one term is to **transpose** it. If the term to be moved is preceded by a plus or a minus sign, this sign is reversed when the term is transposed. Thus, if $a=b+c$, then $a-b=c$, $a-c=b$, $-b=c-a$, $-b-c=-a$, etc. Note that the signs of *all* terms can be reversed without destroying the equality, for if $a=b$, $b=a$. Thus, if *all* terms to the left of the equals sign are exchanged for *all* those to the right, no change in sign need take place, yet if each is moved individually, the signs reverse. For instance, if $a=b+c$, $-b-c=-a$. If each term is multiplied by -1, this becomes $b+c=a$.

A term which is to be multiplied or divided by *all* other terms on its side of the equation can be transposed if it is also moved from the numerator to the denominator, or vice versa. Thus, if $a=\frac{b}{c}$, then $ac=b$, $c=\frac{b}{a}$, $\frac{1}{b}=\frac{1}{ac}$, $\frac{c}{b}=\frac{1}{a}$, etc. $\left(\text{Note that } a=\frac{a}{1}.\right)$ The same result could be obtained by multiplying both sides of an equation by the same quantity. For instance, if both sides of $a=\frac{b}{c}$ are multiplied by c, the equation becomes $ac=\frac{bc}{c}$ and since any number (except zero) divided by itself is unity, $\frac{c}{c}=1$, and the equation becomes $ac=b$, as given above. Note, also, that *both* sides of an equation can be *inverted* without destroying the relationship, for if $a=b$, $\frac{a}{1}=\frac{b}{1}$, and $\frac{1}{b}=\frac{1}{a}$ or $\frac{1}{a}=\frac{1}{b}$. This is accomplished by transposing *all* terms of an equation.

Note that in the case of transposition by changing the plus or minus sign, an entire expression must be changed, and not a part of it. Thus, if $a=bc+d$, $a-bc=d$, but it is not true that $a+b=c+d$. Similarly, a term to be transposed by reversing its multiplication-division relationship must bear that relationship to *all* other terms on its side of the equation. That is, if $a=bc+d$, it is *not* true that $\frac{a}{b}=c+d$, or that $\frac{a}{bc}=d$, but $\frac{a}{bc+d}=1$. If $a=b(cd+e)$, then $\frac{a}{b}=cd+e$.

125. Ratio and proportion.—If the relationship of a to b is the same as that of c *to* d, this fact can be written $a : b :: c : d$, or $\frac{a}{b}=\frac{c}{d}$. Either side of this equation, $\frac{a}{b}$ or $\frac{c}{d}$ is called a **ratio** and the whole equation is called a **proportion.** When a ratio is given a numerical value, it is often expressed as a decimal or as a percentage. Thus, if $\frac{a}{b}=\frac{1}{4}$ (that is, $a=1$, $b=4$), the ratio might be expressed as 0.25 or as 25 percent.

Since a ratio is a fraction, it can be handled as any other fraction.

Geometry

126. Definitions.—**Geometry** is that branch of mathematics dealing with the properties, relations, and measurement of lines, surfaces, solids, and angles. **Plane**

geometry deals with plane figures, and **solid geometry** deals with three-dimensional figures.

A **point,** considered mathematically, is a place having position but no extent. It has no length, breadth, or thickness. A point in motion produces a **line,** which has length, but neither breadth nor thickness. A **straight** or **right line** is the shortest distance between two points in space. A line in motion in any direction except along itself produces a **surface,** which has length and breadth, but not thickness. A **plane surface** or **plane** is a surface without curvature. A straight line connecting any two of its points lies wholly within the plane. A plane surface in motion in any direction except within its plane produces a **solid,** which has length, breadth, and thickness. **Parallel** lines or surfaces are those which are everywhere equidistant. **Perpendicular** lines or surfaces are those which meet at right angles. A perpendicular may be called a **normal,** particularly when it is perpendicular to the tangent to a curved line or surface at the point of tangency. All points equidistant from the ends of a straight line are on the perpendicular bisector of that line. The distance from a point to a line is the length of the perpendicular between them, unless some other distance is indicated.

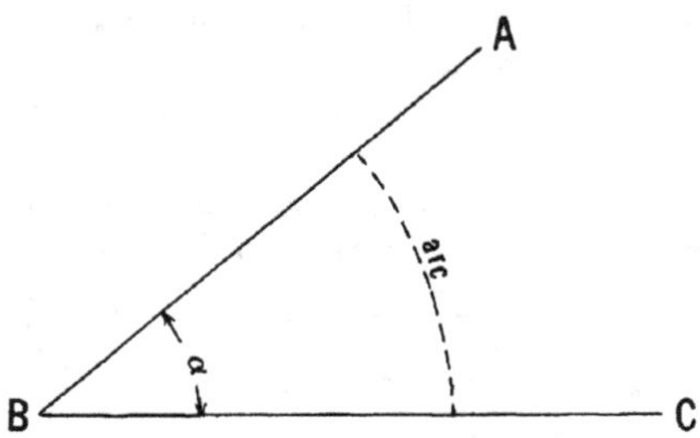

FIGURE 127a.—An angle.

127. Angles.—An **angle** is the inclination to each other of two straight lines which meet at a point. It is measured by the arc of a circle intercepted between the two lines forming the angle, the center of the circle being at the point of intersection. Referring to figure 127a, the angle formed by lines *AB* and *BC*, measured by the arc shown, may be designated "angle *B*," "angle *ABC*," or "angle *CBA*"; or by Greek letter (app. B), as "angle α." The first method should not be used if there is more than one angle at the point, as at *G* in figure 127b. When three letters are used, the middle one should always be that at the **vertex** of the angle, as *G* in figure 127b.

An **acute angle** is one less than a right angle (90°). In figure 127b, angles *AGB, BGC, CGD, DGE*, and *EGF* are all acute angles.

A **right angle** is one whose sides are perpendicular (90°). In figure 127b, angles *AGC, BGD, CGE*, and *DGF* are right angles.

An **obtuse angle** is one greater than a right angle (90°) but less than a straight angle (180°). In figure 127b, angles *AGD, BGE*, and *CGF* are obtuse angles. Angle *AGF* is also obtuse if measured counterclockwise from *AG* to *FG*.

A **straight angle** is one whose sides form a continuous straight line (180°). In figure 127b, angles *AGE* and *BGF* are straight angles.

A **reflex angle** is one greater than a straight angle (180°) but less than a circle (360°). In figure 127b, angle *AGF* is reflex if measured clockwise from *AG* to *FG*. Actually, any two lines meeting at a point form two angles, one less than a straight angle of 180° (unless exactly a straight angle) and the other greater than a straight angle (180°).

An **oblique angle** is any angle not a multiple of 90°.

Two angles whose sum is a right angle (90°) are **complementary angles,** and either is the **complement** of the other. In figure 127b, angles *AGB* and *BGC, BGC* and *CGD, CGD* and *DGE,* and *DGE* and *EGF* are complementary. The angles need not be adjacent. Angles *AGB* and *DGE,* and angles *BGC* and *EGF* are complementary.

Two angles whose sum is a straight angle (180°) are **supplementary angles,** and either is the **supplement** of the other. In figure 127b, angle *AGB* and *BGE*, *AGC* and *CGE*, *AGD* and *DGE*, *BGC* and *CGF*, *BGD* and *DGF*, *BGE* and *EGF*, and *AGC* and *DGF* are supplementary.

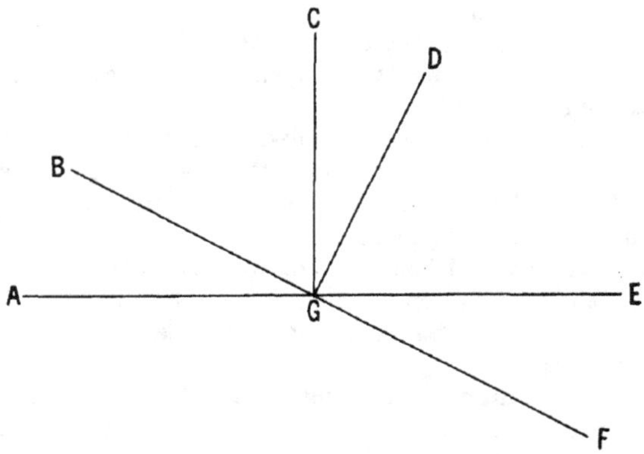

FIGURE 127b.—Acute, right, and obtuse angles.

Two angles whose sum is a circle (360°) are **explementary angles,** and either is the **explement** of the other. The two angles formed when any two lines terminate at a common point are explementary.

Since angles *AGB* and *CGD* (fig. 127b) are each complementary to angle *BGC*, angles *AGB* and *CGD* are equal. Similarly, it can be shown that angle *EGF* is also equal to angle *CGD* (and therefore also equal to angle *AGB*) and also that angles *BGC* and *DGE* are equal to each other. Since *AGC* and *CGE* are both right angles with a common side, *CG* is perpendicular to *AE*. Similarly, *DG* is perpendicular to *BF*. If the sides of one angle are perpendicular to those of another, the two angles are either equal or supplementary. Also, if the sides of one angle are parallel to those of another, the two angles are either equal or supplementary.

When two straight lines intersect, forming four angles, the two opposite angles, called **vertical angles,** are equal. Thus, in figure 127b, lines *AE* and *BF* intersect at *G*. Angles *AGB* and *EGF* form a pair of equal acute vertical angles, and *BGE* and *AGF* form a pair of equal obtuse vertical angles. Angles which have the same vertex and lie on opposite sides of a common side are **adjacent angles.** Adjacent angles formed by intersecting lines are supplementary, since each pair of adjacent angles forms a straight angle (fig. 127b).

A **transversal** is a line that intersects two or more other lines. If two or more parallel lines are cut by a transversal, groups of adjacent and vertical angles are formed, as shown in figure 127c. In this situation, all acute angles (*A*) are equal, all obtuse angles (*B*) are equal, and each acute angle is supplementary to each obtuse angle.

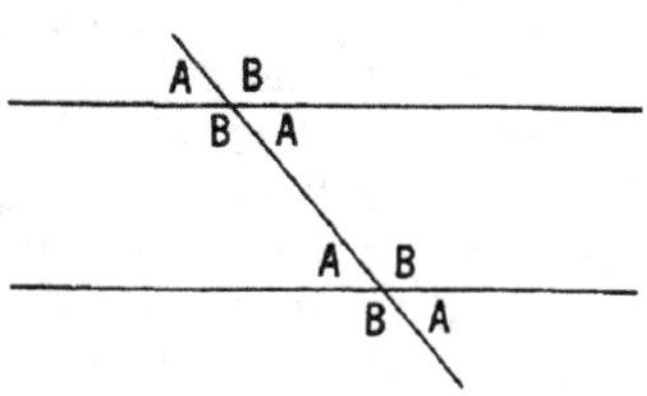

FIGURE 127c.—Angles formed by a transversal.

A **dihedral angle** is the angle between two intersecting planes.

128. Triangles.—A **plane triangle** is a closed figure formed by three straight lines, called **sides,** which meet at three points called **vertices** (singular **vertex**). The vertices are usually labeled with capital letters, and the sides with lowercase letters, as shown in figure 128a.

An **equilateral triangle** is one with its three sides equal. An **equiangular triangle** is one with its three angles equal. When either of these conditions is present, the other always is, so that a triangle which is equilateral is also equiangular, and vice versa.

An **isosceles triangle** is one with two equal sides, called **legs.** The angles opposite the legs are equal. A line which bisects (divides into two equal parts) the *unequal* angle of an isosceles triangle is the perpendicular bisector of the opposite side, and divides the triangle into two equal right triangles.

A **scalene triangle** is one with no two sides equal. In such a triangle, no two angles are equal.

An **acute triangle** is one with three acute angles.

A **right triangle** is one with a right angle. The side opposite the right angle is called the **hypotenuse.** The other two sides may be called **legs.** A plane triangle can have only one right angle.

An **obtuse triangle** is one with an obtuse angle. A plane triangle can have only one obtuse angle.

An **oblique triangle** is one which does not contain a right angle.

The **altitude** of a triangle is a perpendicular line from any vertex to the opposite side, extended if necessary, or the length of this perpendicular line.

A **median** of a triangle is a line from any vertex to the center of the opposite side. The three medians of a triangle meet at a point called the **centroid** of the triangle. This point divides each median into two parts, that part between the centroid and the vertex being twice as long as the other part.

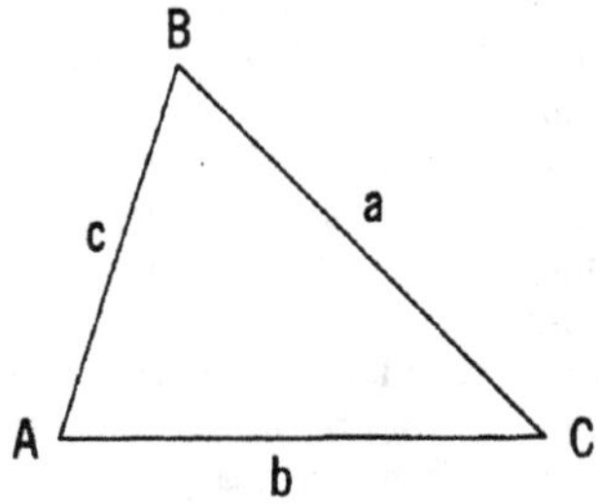

FIGURE 128a.—A triangle.

Lines bisecting the three angles of a triangle meet at a point which is equidistant from the three sides, and is the center of the **inscribed circle,** as shown in figure 128b. This point is of particular interest to navigators because it is the point often taken as the fix when three lines of position of equal weight and having only random errors do not meet at a common point.

The perpendicular bisectors of the three sides of a triangle meet at a point which is equidistant from the three vertices, and is the center of the **circumscribed circle,** the circle through the three vertices and therefore the smallest circle which can be drawn enclosing the triangle. The center of a circumscribed circle is within an acute triangle, on the hypotenuse of a right triangle, and outside an obtuse triangle.

A line connecting the mid points of two sides of a triangle is parallel to the third side and half as long. Also, a line parallel to one side of a triangle and intersecting the other two sides divides these sides proportionally. This principle can be used to divide a line into any number of equal or proportional parts. Refer to figure 128c. Suppose it is desired to divide line *AB* into four equal parts. From *A* draw any line *AC*. Along *C* measure four equal parts of any convenient lengths (*AD*, *DE*, *EF*, and *FG*). Draw *GB*, and through *F*, *E*, and *D* draw lines parallel to *GB* and intersecting *AB*. Then *AD'*, *D'E'*, *E'F'*, and *F'B* are equal and *AB* is divided into four equal parts.

The sum of the angles of a plane triangle is 180°. Therefore, the sum of the acute angles of a right triangle is 90°, and the angles are complementary. If one side of a triangle is extended, the **exterior angle** thus formed is supplementary to the adjacent **interior angle** and, therefore, equal to the sum of the two nonadjacent angles. If two angles of one triangle are equal to two angles of another triangle, the third angles are also equal, and the triangles are **similar.** If the area of one triangle is equal to the area of another, the triangles are **equal.** Triangles having equal bases and altitudes have equal areas. Two figures are **congruent** if one can be placed over the other to make an exact fit. Congruent figures are both similar and equal. If any side of one triangle is equal to any side of a similar triangle, the triangles are congruent. For example, if two right triangles have equal sides, they are congruent; if two right triangles have two corresponding sides equal, they are congruent. Triangles are congruent only if the sides and angles are equal.

The sum of two sides of a plane triangle is always greater than the third side; their difference is always less than the third side.

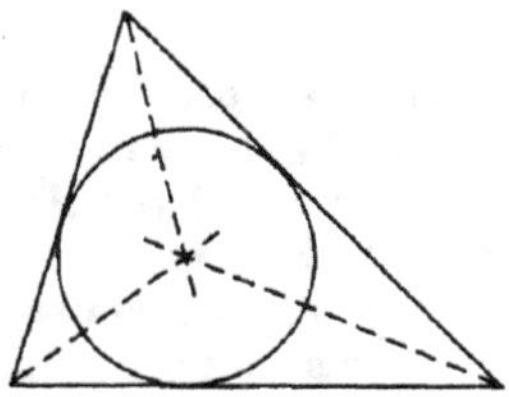

FIGURE 128b.—A circle inscribed in a triangle.

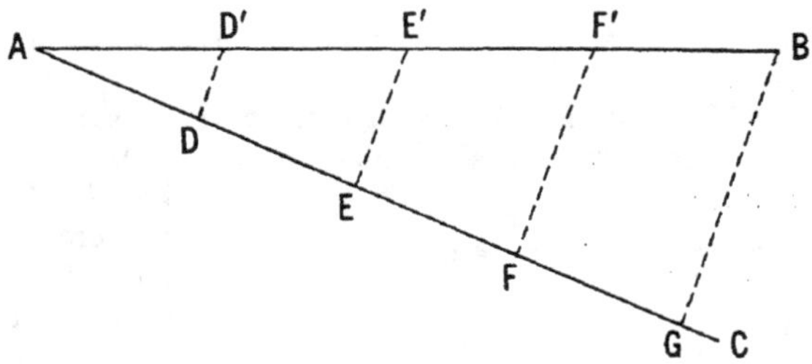

FIGURE 128c.—Dividing a line into equal parts.

If A=area, b=one of the legs of a right triangle or the base of any plane triangle, h=altitude, c=the hypotenuse of a right triangle, a=the other leg of a right triangle, and S=the sum of the interior angles:

Area of plane triangle: $A=\frac{bh}{2}$

Length of hypotenuse of plane right triangle: $c=\sqrt{a^2+b^2}$

Sum of interior angles of plane triangle: $S=180°$.

129. Polygons.—A **polygon** is a closed plane figure made up of three or more straight lines called **sides**. A polygon with three sides is a **triangle,** one with four sides is a **quadrilateral,** one with five sides is a **pentagon,** one with six sides is a **hexagon,** and one with eight sides is an **octagon.** An **equilateral polygon** has equal sides. An **equiangular polygon** has equal interior angles. A **regular polygon** is both equilateral and equiangular. As the number of sides of a regular polygon increases, the figure approaches a circle.

A **trapezoid** is a quadrilateral with one pair of opposite sides parallel and the other pair not parallel. A **parallelogram** is a quadrilateral with both pairs of opposite sides parallel. Any side of a parallelogram, or either of the parallel sides of a trapezoid, is the **base** of the figure. The perpendicular distance from the base to the opposite side is the altitude. A **rectangle** is a parallelogram with four right angles. (If any one is a right angle, the other three must be, also.) A **square** is a rectangle with equal sides. A **rhomboid** is a parallelogram with oblique angles. A **rhombus** is a rhomboid with equal sides.

The sum of the exterior angles of a convex polygon (one having no interior reflex angles), made by extending each side in one direction only (consistently), is 360°.

A **diagonal** of a polygon is a straight line connecting any two vertices which are not adjacent. The diagonals of a parallelogram bisect each other.

The **perimeter** of a polygon is the sum of the lengths of its sides.

If A=area, s=the side of a square, a=that side of a rectangle adjacent to the base or that side of a trapezoid parallel to the base, b=the base of a quadrilateral, h=the altitude of a parallelogram or trapezoid, S=the sum of the angles of a polygon, and n=the number of sides of a polygon:

Area of square: $A=s^2$
Area of rectangle: $A=ab$
Area of parallelogram: $A=bh$
Area of trapezoid: $A=\frac{(a+b)h}{2}$
Sum of angles in convex polygon: $S=(n-2)180°$.

130. Circles.—A circle is a plane, closed curve, all points of which are equidistant from a point within, called the **center** (*C*, fig. 130); or the figure formed by such a curve.

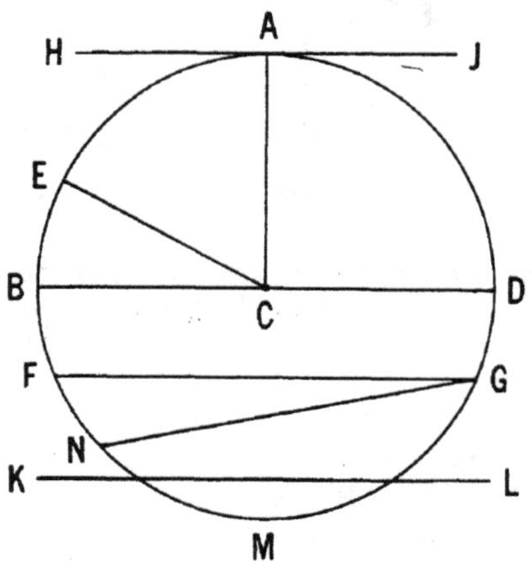

FIGURE 130.—Elements of a circle.

The line forming the circle is called the **circumference.** The length of this line is the **perimeter,** although the term "circumference" is often used with this meaning. An **arc** is part of a circumference. A **major arc** is more than a semicircle (180°), a **minor arc** is less than a semicircle (180°). A **semicircle** is half a circle (180°), a **quadrant** is a quarter of a circle (90°), a **quintant** is a fifth of a circle (72°), a **sextant** is a sixth of a circle (60°), an **octant** is an eighth of a circle (45°). Some of these names have been applied to instruments used by navigators for measuring altitudes of celestial bodies because of the part of a circle originally used for the length of the arc of the instrument.

Concentric circles have a common center.

A **radius** (plural **radii**) or **semidiameter** is a straight line connecting the center of a circle with any point on its circumference. In figure 130, *CA*, *CB*, *CD*, and *CE* are radii.

A **diameter** of a circle is a straight line passing through its center and terminating at opposite sides of the circumference, or two radii in opposite directions (*BCD*, fig. 130). It divides a circle into two equal parts. The ratio of the length of the circumference of any circle to the length of its diameter is 3.14159+, or π (the Greek letter *pi*), a relationship that has many useful applications.

A **sector** is that part of a circle bounded by two radii and an arc. In figure 130, *BCE*, *ECA*, *ACD*, *BCA*, and *ECD* are sectors. The angle formed by two radii is called a **central angle.** Any pair of radii divides a circle into sectors, one less than a semicircle (180°) and the other greater than a semicircle (unless the two radii form a diameter).

A **chord** is a straight line connecting any two points on the circumference of a circle (*FG*, *GN* in fig. 130). Chords equidistant from the center of a circle are equal in length.

A **segment** is that part of a circle bounded by a chord and the intercepted arc (*FGMF*, *NGMN* in fig. 130). A chord divides a circle into two segments, one less than a semicircle (180°), and the other greater than a semicircle (unless the chord is a diameter). A diameter perpendicular to a chord bisects it, its arc, and its segments. Either pair of vertical angles formed by intersecting chords has a combined number of degrees equal to the sum of the number of degrees in the two arcs intercepted by the two angles.

An **inscribed angle** is one whose vertex is on the circumference of a circle and whose sides are chords (*FGN* in fig. 130). It has half as many degrees as the arc it intercepts. Hence, an angle inscribed in a semicircle is a right angle if its sides terminate at the ends of the diameter forming the semicircle.

A **secant** of a circle is a line intersecting the circle, or a chord extended beyond the circumference (*KL* in fig. 130).

A **tangent** to a circle is a straight line, in the plane of the circle, which has only one point in common with the circumference (*HJ* in fig. 130). A tangent is perpendicular to the radius at the **point of tangency** (*A* in fig. 130). The two tangents from a point to opposite sides of a circle are equal in length, and a line from the point to the center of the circle bisects the angle formed by the two tangents. An angle formed outside a circle by the intersection of two tangents, a tangent and a secant, or two secants has *half* as many degrees as the *difference* between the two intercepted arcs. An angle formed by a tangent and a chord, with the apex at the point of tangency, has half as many degrees as the arc it intercepts. A **common tangent** is one tangent to more than one circle. Two circles are tangent to each other if they touch at one point only. If of different sizes, the smaller circle may be either inside or outside the larger one.

Parallel lines intersecting a circle intercept equal arcs.

If A=area; r=radius; d=diameter; C=circumference; s=linear length of an arc; α=angular length of an arc, or the angle it subtends at the center of a circle, in degrees; β=angular length of an arc, or the angle it subtends at the center of a circle, in radians; rad=radians (art. 138), and sin=sine (art. 139):

Area of circle: $A=\pi r^2=\frac{\pi d^2}{4}$

Circumference of circle: $C=2\pi r=\pi d=2\pi \text{ rad}$

Area of sector: $A=\frac{\pi r^2\alpha}{360}=\frac{r^2\beta}{2}=\frac{rs}{2}$

Area of segment: $A=\frac{r^2(\beta-\sin\alpha)}{2}$.

131. Polyhedrons.—A **polyhedron** is a solid having plane sides or **faces.**

A **cube** is a polyhedron having six square sides.

A **prism** is a solid having parallel, similar, equal, plane geometric figures as bases, and parallelograms as sides. By extension, the term is also applied to a similar solid having nonparallel bases, and trapezoids or a combination of trapezoids and parallelograms as sides. The **axis** of a prism is the straight line connecting the centers of its bases. A **right prism** is one having bases perpendicular to the axis. The sides of a right prism are rectangles. A **regular prism** is a right prism having regular polygons as bases. The **altitude** of a prism is the perpendicular distance between the planes of its bases. In the case of a right prism, it is measured along the axis.

A **pyramid** is a polyhedron having a polygon as one end, the **base;** and a point, the **apex,** as the other; the two ends being connected by a number of triangular sides or **faces.** The **axis** of a pyramid is the straight line connecting the apex and the center of the base. A **right pyramid** is one having its base perpendicular to its axis. A **regular pyramid** is a right pyramid having a regular polygon as its base. The **altitude** of a pyramid is the perpendicular distance from its apex to the plane of its base. A **truncated pyramid** is that portion of a pyramid between its base and a plane intersecting all of the faces of the pyramid.

If A=area, s=edge of a cube or slant height of a regular pyramid (from the center of one side of its base to the apex), V=volume, a=side of a polygon, h=altitude, P=perimeter of base, n=number of sides of polygon, B=area of base, and r=perpendicular distance from the center of a side of a polygon to the center of the polygon:

Cube:

Area of each face: $A=s^2$
Total area of all faces: $A=6s^2$
Volume: $V=s^3$.

Regular prism:

Area of each face: $A=ah$
Total area of all faces: $A=Ph=nah$

Area of each base: $B=\frac{nar}{2}$

Total area of both bases: $A=nar$

Volume: $V=Bh=\frac{narh}{2}$.

Regular pyramid:

Area of each face: $A=\frac{as}{2}$

Total area of all faces: $A=\frac{nas}{2}$

Area of base: $B=\frac{nar}{2}$

Volume: $V=\frac{Bh}{3}=\frac{narh}{6}$.

132. Cylinders.—A **cylinder** is a solid having two parallel plane **bases** bounded by closed congruent curves, and a surface formed by an infinite number of parallel lines, called **elements,** connecting similar points on the two curves. A cylinder is similar to a prism, but with a curved lateral surface, instead of a number of flat sides connecting the bases. The **axis** of a cylinder is the straight line connecting the centers of the bases. A **right cylinder** is one having bases perpendicular to the axis. A **circular cylinder** is one having circular bases. The **altitude** of a cylinder is the perpendicular distance between the planes of its bases. The **perimeter** of a base is the length of the curve bounding it.

If A=area, P=perimeter of base, h=altitude, r=radius of a circular base, B=area of base, and V=volume, then for a right circular cylinder:

Lateral area: $A=Ph=2\pi rh$
Area of each base: $B=\pi r^2$
Total area, both bases: $A=2\pi r^2$
Volume: $V=Bh=\pi r^2h$.

133. Cones.—A **cone** is a solid having a plane **base** bounded by a closed curve, and a surface formed by lines, called **elements,** from every point on the curve to a common point called the **apex.** A cone is similar to a pyramid, but with a curved surface connecting the base and apex, instead of a number of flat sides. The **axis** of a cone is the straight line connecting the apex and the center of the base. A **right cone** is one having its base perpendicular to its axis. A **circular cone** is one having a circular base. The **altitude** of a cone is the perpendicular distance from its apex to the plane of its base. A **frustum** of a cone is that portion of the cone between its base and any parallel plane intersecting all elements of the cone. A **truncated cone** is that portion of a cone between its base and any nonparallel plane which intersects all elements of the cone but does not intersect the base.

If A=area, r=radius of base, s=slant height or length of element, B=area of base, h=altitude, and V=volume, then for a right circular cone:

Lateral area: $A=\pi rs$
Area of base: $B=\pi r^2$
Slant height: $s=\sqrt{r^2+h^2}$
Volume: $V=\frac{Bh}{3}=\frac{\pi r^2 h}{3}$.

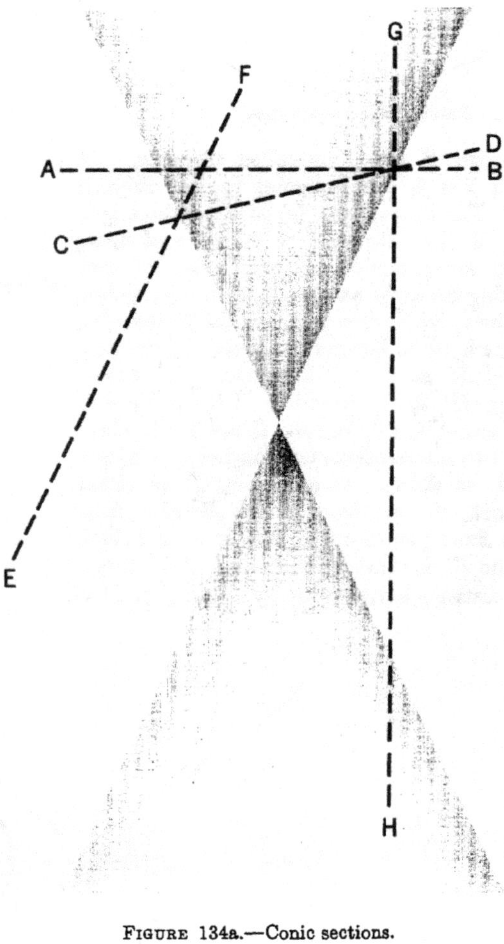

FIGURE 134a.—Conic sections.

134. Conic sections.—If a right circular cone of indefinite extent is intersected by a plane perpendicular to the axis of the cone (AB, fig. 134a), the line of intersection of the plane and the surface of the cone is a **circle**.

If the intersecting plane of figure 134a is tilted to some position such as CD, the intersection is an **ellipse** or flattened circle, figure 134b. The longest diameter of an ellipse is called its **major axis**, and half of this is its **semimajor axis**, a. The shortest diameter of an ellipse is called its **minor axis**, and half of this is its **semiminor axis**, b. Two points, F and F', called **foci** (singular **focus**) or **focal points**, on the major axis are so located that the sum of their distances from any point P on the curve is equal to the length of the major axis. That is, $PF+PF'=2a$ (fig. 134b). The **eccentricity** (e) of an ellipse is equal to $\frac{c}{a}$, where c is the distance from the center to one of the foci ($c=CF=CF'$). It is always greater than 0 but less than 1.

If the intersecting plane of figure 134a is parallel to one element of the cone, as at EF, the intersection is a **parabola**, figure 134c. Any point P on a parabola is equi-

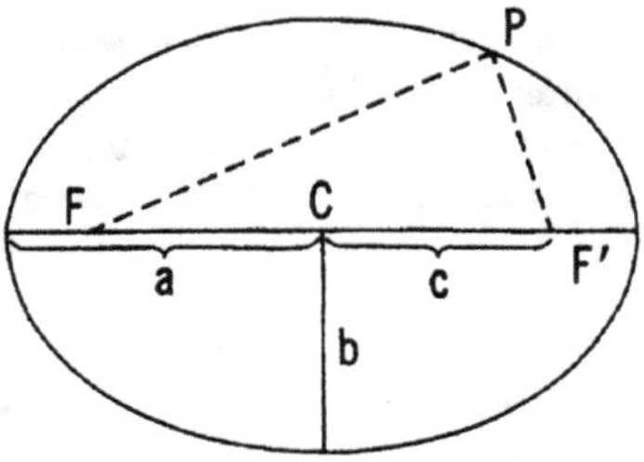

FIGURE 134b.—An ellipse.

distant from a fixed point F, called the **focus** or **focal point,** and a fixed straight line, AB, called the **directrix.** Thus, for any point P, $PF=PE$. The point midway between the focus F and the directrix AB is called the **vertex,** V. The straight line through F and V is called the **axis,** CD. This line is perpendicular to the directrix AB. The **eccentricity** (*e*) of a parabola is 1.

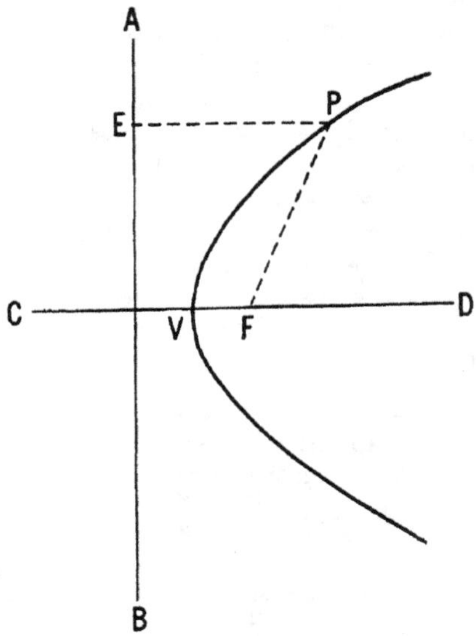

FIGURE 134c.—A parabola.

If the elements of the cone of figure 134a are extended to form a second cone having the same axis and apex but extending in the opposite direction, and the intersecting plane is tilted beyond the position forming a parabola, so that it intersects both curves, as at GH, the intersections of the plane with the cones is a **hyperbola,** figure 134d. There are two intersections or branches of a hyperbola, as shown. At any point P on either branch, the *difference* in the distance from two fixed points called **foci** or **focal points,** F and F', is constant and equal to the shortest distance between the two branches. That

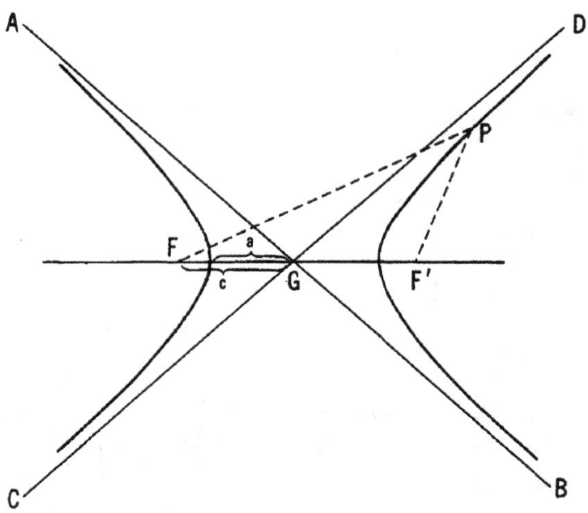

FIGURE 134d.—A hyperbola.

is, $PF-PF'=2a$ (fig. 134d). The straight line through F and F' is called the **axis.** The **eccentricity** (*e*) of a hyperbola is the ratio $\frac{c}{a}$ (fig. 134d). It is always greater than 1.

Each branch of a hyperbola approaches ever closer to, but never reaches, a pair of intersecting straight lines, AB and CD, called **asymptotes.** These intersect at G.

The various conic sections bear an eccentricity relationship to each other. The eccentricity of a circle is 0, that of an ellipse is greater than 0 but less than 1, that of a parabola or straight line (a limiting case of a parabola) is 1, and that of a hyperbola is greater than 1.

If e=eccentricity, A=area, a=semimajor axis of an ellipse or half the shortest distance between the two branches of a hyperbola, b=the semiminor axis of an ellipse, and c=the distance between the center of an ellipse and one of its focal points or the distance between the focal point of a hyperbola and the intersection of its asymptotes:

Circle:

Eccentricity: $e=0$

Other relationships given in article 130.

Ellipse:

Area: $A=\pi ab$

Eccentricity: $e=\frac{c}{a}$, greater than 0, but less than 1.

Parabola:

Eccentricity: $e=1$.

Hyperbola:

Eccentricity: $e=\frac{c}{a}$, greater than 1.

When cones are intersected by some surface other than a plane, as the curved surface of the earth, the resulting sections do not follow the relationships given above, the amount of divergence therefrom depending upon the individual circumstances.

135. Spheres.—A **sphere** is a solid bounded by a surface every point of which is equidistant from a point within, called the **center.** It may be formed by rotating a circle about any diameter.

A **radius** or **semidiameter** of a sphere is a straight line connecting its center with any point on its surface. A **diameter** of a sphere is a straight line through its center and terminated at both ends by the surface of the sphere. The poles of a sphere are the ends of a diameter.

The intersection of a plane and the surface of a sphere is a circle, a **great circle** if the plane passes through the center of the sphere, and a **small circle** if it does not. The shorter arc of the great circle between two points on the surface of a sphere is the shortest distance, on the surface of the sphere, between the points. Every great circle of a sphere bisects every other great circle of that sphere. The **poles** of a circle on a sphere are the extremities of the sphere's diameter which is perpendicular to the plane of the circle. All points on the circumference of the circle are equidistant from either of its poles. In the case of a great circle, *both* poles are 90° from any point on the circumference of the circle. Any great circle may be considered a **primary,** particularly when it serves as the origin of measurement of a coordinate. The great circles through its poles are called **secondaries.** Secondaries are perpendicular to their primary.

A **spherical triangle** is the figure formed on the surface of a sphere by the intersection of three great circles. The lengths of the sides of a spherical triangle are measured in degrees, minutes, and seconds, as the angular lengths of the arcs forming them. The sum of the three sides is always less than 360°. The sum of the three angles is always *more* than 180° and *less* than 540°.

A **lune** is that part of the surface of a sphere bounded by halves of two great circles.

A **spheroid** is a flattened sphere, which may be formed by rotating an ellipse about one of its axes. An **oblate spheroid,** such as the earth, is formed when an ellipse is rotated about its minor axis. In this case the diameter along the axis of rotation is less than the major axis. A **prolate spheroid** is formed when an ellipse is rotated about its major axis. In this case the diameter along the axis of rotation is greater than the minor axis.

If A=area, r=radius, d=diameter, and V=volume of a sphere:

$$\text{Area: } A=4\pi r^2=\pi d^2$$

$$\text{Volume: } V=\frac{4\pi r^3}{3}=\frac{\pi d^3}{6}.$$

If A=area, a=semimajor axis, b=semiminor axis, e=eccentricity, and V=volume of an oblate spheroid:

$$\text{Area: } A=4\pi a^2\left(1-\frac{e^2}{3}-\frac{e^4}{15}-\frac{e^6}{35}-\cdots\right)$$

$$\text{Eccentricity: } e=\sqrt{\frac{a^2-b^2}{a^2}}$$

$$\text{Volume: } V=\frac{4\pi a^2 b}{3}.$$

136. Coordinates are magnitudes used to define a position. Many different types of coordinates are used.

If a position is known to be at a stated point, no magnitudes are needed to identify the position, although they may be required to locate the point. Thus, if a vessel is at port A, its position is known if the location of port A is known, but latitude and longitude may be needed to locate port A.

If a position is known to be on a given line, a single magnitude (coordinate) is needed to identify the position if an origin is stated or understood. Thus, if a vessel is known to be *south* of port B, it is known to be on a line extending southward from port B. If its distance from port B is known, and the position of port B is known, the position of the vessel is uniquely defined.

If a position is known to be on a given surface, two magnitudes (coordinates) are needed to define the position. Thus, if a vessel is known to be on the surface of the earth, its position can be identified by means of latitude and longitude. Latitude indicates its angular distance north or south of the equator, and longitude its angular distance east or west of the prime meridian.

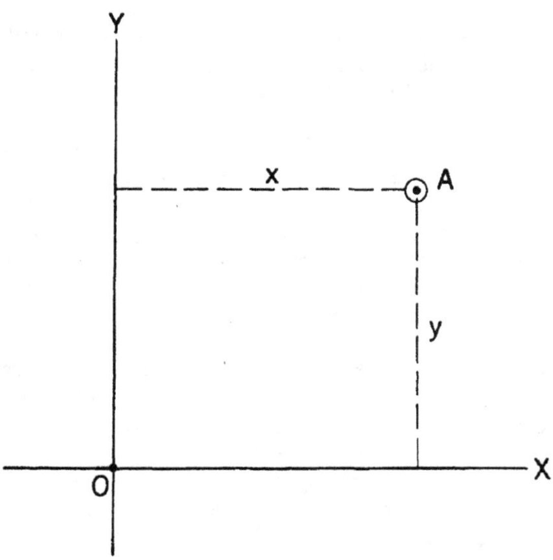

FIGURE 136a.—Rectangular coordinates.

If nothing is known regarding a position other than that it exists in space, three magnitudes (coordinates) are needed to define its position. Thus, the position of a submarine may be defined by means of latitude, longitude, and depth below the surface.

Each coordinate requires an origin, either stated or implied. If a position is known to be on a given plane, it might be defined by means of its distance from each of two intersecting lines, called **axes.** Thus, in figure 136a the position of point A can be defined by stating that it is x units to the right of line OY and y units upward from line OX. These are called **rectangular coordinates.** The coordinate along OY is called the **ordinate,** and the coordinate along OX is called the **abscissa.** Point O is the **origin,** and lines OX and OY the **axes** (called the X and Y axes, respectively). Point A is at position x, y. If the axes are not perpendicular but the lines x and y are drawn parallel to the axes, **oblique coordinates** result. Either type are **Cartesian coordinates.** A three-dimensional system of Cartesian coordinates, with X, Y, and Z axes, is called **space coordinates.**

Another system of plane coordinates in common usage consists of the *direction* and *distance* from the origin (called the **pole**), as shown in figure 136b. A line extending in the direction indicated is called a **radius vector.** Direction and distance from a fixed point constitute **polar coordinates,** sometimes called the rho- (the Greek ρ, to indicate distance) theta (the Greek θ, to indicate direction) system. An example of its use is with respect to a radar PPI.

Spherical coordinates are used to define a position on the surface of a sphere or spheroid by indicating angular distance from a primary great circle and a reference secondary great circle Familiar examples are latitude and longitude, altitude and azimuth, and declination and hour angle.

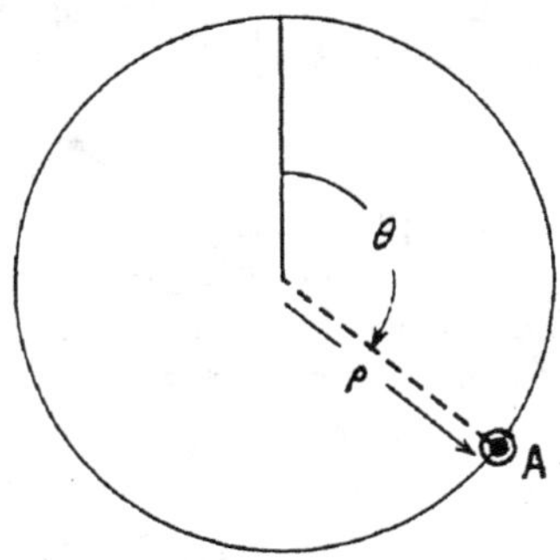

FIGURE 136b.—Polar coordinates.

Trigonometry

137. Definitions.—Trigonometry is that branch of mathematics dealing with the relations among the angles and sides of triangles. **Plane trigonometry** is that branch dealing with plane triangles, and **spherical trigonometry** is that branch dealing with spherical triangles.

138. Angular measure.—A circle may be divided into 360 **degrees** (°), which is the **angular length** of its circumference. Each degree may be divided into 60 **minutes** (′), and each minute into 60 **seconds** (″). The angular length of an arc is usually expressed in these units. By this system a right angle or quadrant has 90° and a straight angle or semicircle 180°. In marine navigation, altitudes, latitudes, and longitudes are usually expressed in degrees, minutes, and tenths (27°14′.4). Azimuths are usually expressed in degrees and tenths (164°.7). The system of degrees, minutes, and seconds indicated above is the **sexagesimal system.** In the **centesimal system,** used chiefly in France, the circle is divided into 400 **centesimal degrees** (sometimes called **grades**) each of which is divided into 100 **centesimal minutes** of 100 **centesimal seconds** each.

A radian is the angle subtended at the center of a circle by an arc having a linear length equal to the radius of the circle. A radian is equal to 57°.2957795131 approximately, or 57°17′44″.80625 approximately. The radian is sometimes used as a unit of angular measure. A circle $(360°)=2\pi$ radians, a semicircle $(180°)=\pi$ radians, a right angle $(90°)=\frac{\pi}{2}$ radians, and $1'=0.0002908882$ radians approximately. The length of the arc of a circle is equal to the radius multiplied by the angle subtended in radians.

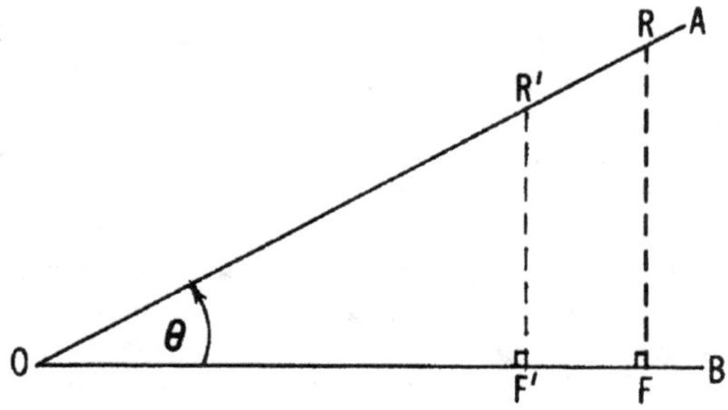

FIGURE 139a.—Similar right triangles.

139. Trigonometric functions are the various proportions or ratios of the sides of a plane right triangle, defined in relation to one of the acute angles. In figure 139a, let θ be any acute angle. From any point R on line OA draw a line perpendicular to OB at F. From any other point R′ on OA, draw a line perpendicular to OB at F′. Then triangles OFR and OF′R′ are similar right triangles because all their corresponding angles are equal. Since in any pair of similar triangles the ratio of any two sides of one triangle is equal to the ratio of the corresponding two sides of the other triangle,

$$\frac{RF}{OF}=\frac{R'F'}{OF'},\ \frac{RF}{OR}=\frac{R'F'}{OR'},\ \text{and}\ \frac{OF}{OR}=\frac{OF'}{OR'}.$$

Irrespective of where the point R is located on OA, the ratio between the lengths of any two sides in the triangle OFR has a constant value. Hence, for any value of the acute angle θ, there is a fixed set of values for the ratios of the various sides of the triangle. These ratios are defined as follows:

$$\text{sine}\ \theta=\sin\theta=\frac{\text{side opposite}}{\text{hypotenuse}} \qquad \text{cosecant}\ \theta=\csc\theta=\frac{\text{hypotenuse}}{\text{side opposite}}$$

$$\text{cosine}\ \theta=\cos\theta=\frac{\text{side adjacent}}{\text{hypotenuse}} \qquad \text{secant}\ \theta=\sec\theta=\frac{\text{hypotenuse}}{\text{side adjacent}}$$

$$\text{tangent}\ \theta=\tan\theta=\frac{\text{side opposite}}{\text{side adjacent}} \qquad \text{cotangent}\ \theta=\cot\theta=\frac{\text{side adjacent}}{\text{side opposite}}.$$

Of these six principal functions, the second three are the reciprocals of the first three. Thus,

$$\sin\theta=\frac{1}{\csc\theta} \qquad \csc\theta=\frac{1}{\sin\theta}$$

$$\cos\theta=\frac{1}{\sec\theta} \qquad \sec\theta=\frac{1}{\cos\theta}$$

$$\tan\theta=\frac{1}{\cot\theta} \qquad \cot\theta=\frac{1}{\tan\theta}.$$

In figure 139b, A, B, and C are the angles of a plane right triangle, the right angle being at C. The sides are a, b, c, as shown. The six principal trigonometric functions of angle B are:

$$\sin B=\frac{b}{c}=\cos A=\cos(90°-B) \qquad \cot B=\frac{a}{b}=\tan A=\tan(90°-B)$$

$$\cos B=\frac{a}{c}=\sin A=\sin(90°-B) \qquad \sec B=\frac{c}{a}=\csc A=\csc(90°-B)$$

$$\tan B=\frac{b}{a}=\cot A=\cot(90°-B) \qquad \csc B=\frac{c}{b}=\sec A=\sec(90°-B).$$

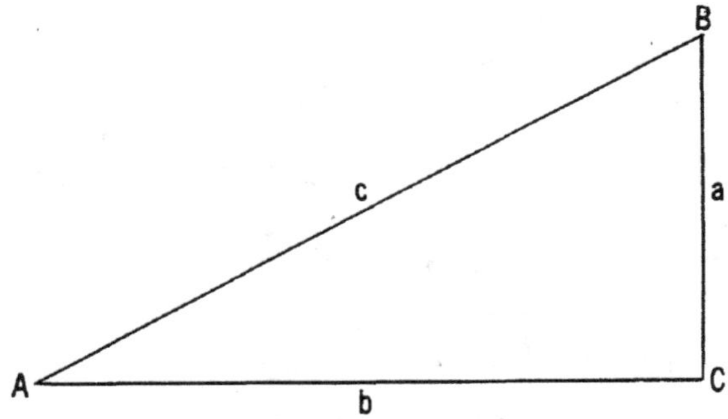

FIGURE 139b.—A right triangle.

Since *A* and *B* are *complementary*, these relations show that the sine of an angle is the cosine of its complement, the tangent of an angle is the cotangent of its complement, and the secant of an angle is the cosecant of its complement. Thus, the **co** function of an angle is the function of its complement.

$$\sin(90°-A) = \cos A \qquad \csc(90°-A) = \sec A$$
$$\cos(90°-A) = \sin A \qquad \sec(90°-A) = \csc A$$
$$\tan(90°-A) = \cot A \qquad \cot(90°-A) = \tan A.$$

Certain additional relations are also classed as trigonometric functions:

versed sine θ=versine θ=vers θ=ver $\theta=1-\cos\theta$
versed cosine θ=coversed sine θ=coversine θ=covers θ=cov $\theta=1-\sin\theta$
haversine θ=hav $\theta=\frac{1}{2}$ ver $\theta=\frac{1}{2}(1-\cos\theta)$.

The numerical value of a trigonometric function is sometimes called the **natural function** to distinguish it from the logarithm of the function, called the **logarithmic function.** Numerical values of the six principal functions are given at 1′ intervals in table 31. Logarithms are given at the same intervals in table 33. Both natural and logarithmic haversines are given in table 34.

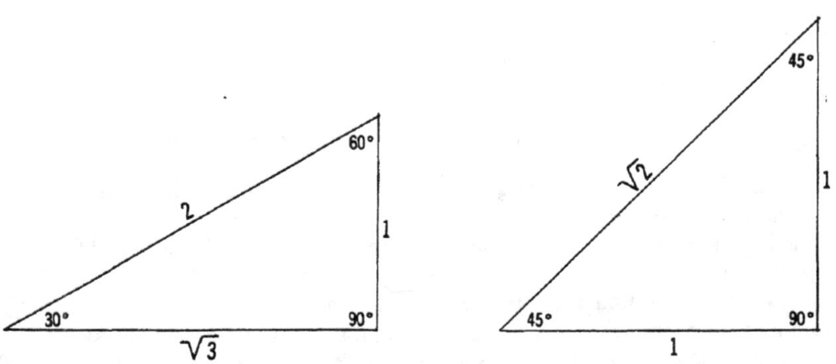

FIGURE 139c.—Numerical relationship of sides of 30°–60° and 45° triangles.

Since the relationships of 30°–60° and 45° right triangles are as shown in figure 139c, certain values of the basic functions can be stated exactly as shown in the following table:

Function	*30°*	*45°*	*60°*
sine	$\frac{1}{2}$	$\frac{1}{\sqrt{2}}=\frac{1}{2}\sqrt{2}$	$\frac{\sqrt{3}}{2}=\frac{1}{2}\sqrt{3}$
cosine	$\frac{\sqrt{3}}{2}=\frac{1}{2}\sqrt{3}$	$\frac{1}{\sqrt{2}}=\frac{1}{2}\sqrt{2}$	$\frac{1}{2}$
tangent	$\frac{1}{\sqrt{3}}=\frac{1}{3}\sqrt{3}$	$\frac{1}{1}=1$	$\frac{\sqrt{3}}{1}=\sqrt{3}$
cotangent	$\frac{\sqrt{3}}{1}=\sqrt{3}$	$\frac{1}{1}=1$	$\frac{1}{\sqrt{3}}=\frac{1}{3}\sqrt{3}$
secant	$\frac{2}{\sqrt{3}}=\frac{2}{3}\sqrt{3}$	$\frac{\sqrt{2}}{1}=\sqrt{2}$	$\frac{2}{1}=2$
cosecant	$\frac{2}{1}=2$	$\frac{\sqrt{2}}{1}=\sqrt{2}$	$\frac{2}{\sqrt{3}}=\frac{2}{3}\sqrt{3}$

TABLE 139a.—Values of various trigonometric functions for angles of 30°, 45°, and 60°.

All trigonometric functions can be shown as lengths of lines in a unit circle. In figure 139d,

$\sin\theta=RF$	$\cot\theta=AB$
$\cos\theta=OF$	$\sec\theta=OD$
$\tan\theta=DE$	$\csc\theta=OA$
$\text{ver}\,\theta=FE$	$\text{cov}\,\theta=BC.$

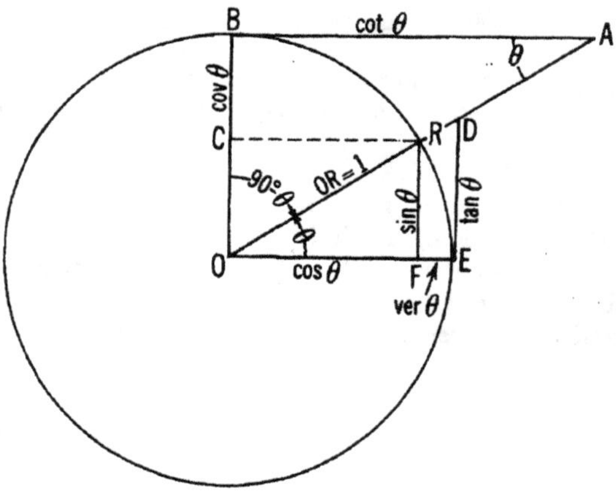

FIGURE 139d.—Line definitions of trigonometric functions.

140. The functions in various quadrants.—To make the definitions of the trigonometric functions more general to include those angles greater than 90°, the functions are defined in terms of the rectangular cartesian coordinates of point R of figure 139a, due regard being given to the sign of the function. In figure 140a, OR is assumed to be a unit radius. By convention the sign of OR is always positive. This radius is imagined to rotate in a counterclockwise direction through 360° from the horizontal position at 0°, the positive direction along the X-axis. Ninety degrees (90°) is the positive direction along the Y-axis. The angle between the original position of the radius and its position at any time increases from 0° to 90° in the *first quadrant* (I), 90° to 180° in the *second quadrant* (II), 180° to 270° in the *third quadrant* (III), and 270° to 360° in the *fourth quadrant* (IV).

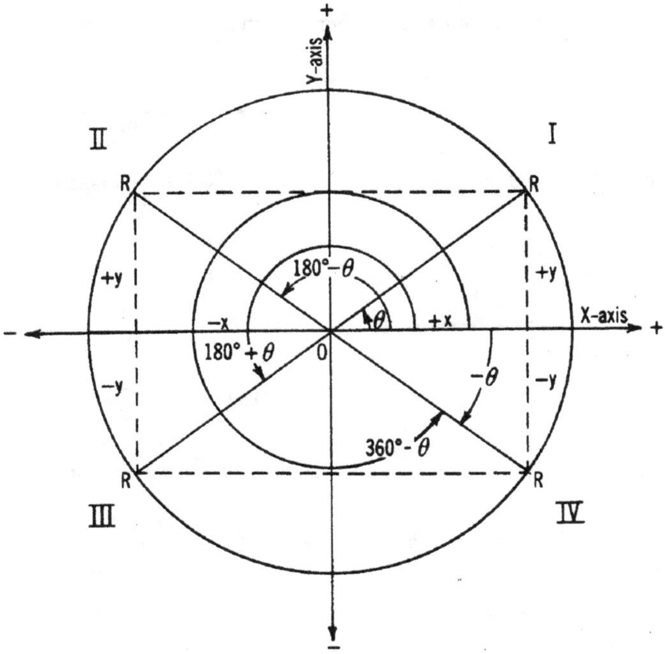

FIGURE 140a.—The functions in various quadrants.

The numerical value of the sine of an angle is equal to the projection of the unit radius on the Y-axis. According to the definition given in article 139, the sine of angle θ in the first quadrant of figure 140a is $\frac{+y}{+OR}$. Radius OR being equal to one, sin $\theta=+y$. Since $+y$ is equal to the projection of the unit radius OR on the Y-axis, the sine function of an angle in the first quadrant defined in terms of rectangular cartesian coordinates does not contradict the definition in article 139. In figure 140a,

$$\sin\theta=+y$$
$$\sin(180^\circ-\theta)=+y=\sin\theta$$
$$\sin(180^\circ+\theta)=-y=-\sin\theta$$
$$\sin(360^\circ-\theta)=-y=\sin(-\theta)=-\sin\theta.$$

The numerical value of the cosine of an angle is equal to the projection of the unit radius on the X-axis. In figure 140a,

$$\cos\theta=+X$$
$$\cos(180^\circ-\theta)=-X=-\cos\theta$$
$$\cos(180^\circ+\theta)=-X=-\cos\theta$$
$$\cos(360^\circ-\theta)=+X=\cos(-\theta)=\cos\theta.$$

The numerical value of the tangent of an angle is equal to the ratio of the projections of the unit radius on the Y- and X-axes. In figure 140a,

$$\tan\theta=\frac{+y}{+x}$$

$$\tan(180^\circ-\theta)=\frac{+y}{-x}=-\tan\theta$$

$$\tan(180^\circ+\theta)=\frac{-y}{-x}=\tan\theta$$

$$\tan(360^\circ-\theta)=\frac{-y}{+x}=\tan(-\theta)=-\tan\theta.$$

The cosecant, secant, and cotangent functions of angles in the various quadrants are similarly determined:

$$\csc\theta=\frac{1}{+y}$$

$$\csc(180°-\theta)=\frac{1}{+y}=\csc\theta$$

$$\csc(180°+\theta)=\frac{1}{-y}=-\csc\theta$$

$$\csc(360°-\theta)=\frac{1}{-y}=\csc(-\theta)=-\csc\theta$$

$$\sec\theta=\frac{1}{+x}$$

$$\sec(180°-\theta)=\frac{1}{-x}=-\sec\theta$$

$$\sec(180°+\theta)=\frac{1}{-x}=-\sec\theta$$

$$\sec(360°-\theta)=\frac{1}{+x}=\sec(-\theta)=\sec\theta$$

$$\cot\theta=\frac{+x}{+y}$$

$$\cot(180°-\theta)=\frac{-x}{+y}=-\cot\theta$$

$$\cot(180°+\theta)=\frac{-x}{-y}=\cot\theta$$

$$\cot(360°-\theta)=\frac{+x}{-y}=\cot(-\theta)=-\cot\theta.$$

The signs of the functions are shown in the following table:

Functions	I	II	III	IV
sine and cosecant	+	+	−	−
cosine and secant	+	−	−	+
tangent and cotangent	+	−	+	−
versine, coversine, and haversine	+	+	+	+

TABLE 140a.—Signs of trigonometric functions by quadrant.

The numerical values vary as shown in the following table and in figure 140b.

Functions	I	II	III	IV
sine cosecant	0 to +1 +∞ to +1	+1 to 0 +1 to +∞	0 to −1 −∞ to −1	−1 to 0 −1 to −∞
cosine secant	+1 to 0 +1 to +∞	0 to −1 −∞ to −1	−1 to 0 −1 to −∞	0 to +1 +∞ to +1
tangent cotangent	0 to +∞ +∞ to 0	−∞ to 0 0 to −∞	0 to +∞ +∞ to 0	−∞ to 0 0 to −∞
versine coversine haversine	0 to +1 +1 to 0 0 to $+\frac{1}{2}$	+1 to +2 0 to +1 $+\frac{1}{2}$ to +1	+2 to +1 +1 to +2 +1 to $+\frac{1}{2}$	+1 to 0 +2 to +1 $+\frac{1}{2}$ to 0

TABLE 140b.—Values of trigonometric functions in various quadrants. These relationships are shown graphically in figure 140b.

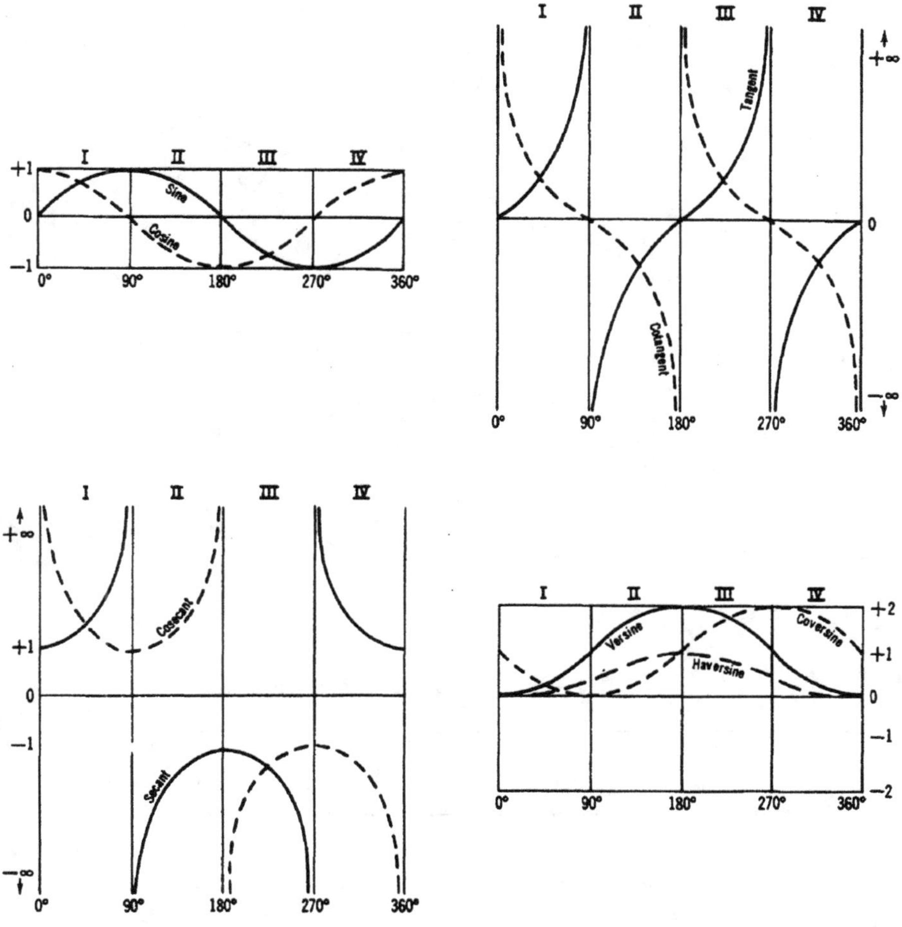

FIGURE 140b.—Graphic representation of values of trigonometric functions in various quadrants.

As shown in figure 140a and table 140a, the sign (+ or −) of the functions varies with the quadrant of an angle. In figure 140a radius OR is imagined to rotate in a counterclockwise direction through 360° from the horizontal position at 0°. This is the **mathematical convention**. In figure 140c this concept is shown in the usual **navigational convention** of a compass rose, starting with 000° at the top and rotating clockwise. In either diagram the angle θ between the original position of the radius and its position at any time increases from 0° to 90° in the *first quadrant* (I), 90° to 180° in the *second quadrant* (II), 180° to 270° in the *third quadrant* (III), and 270° to 360° in the *fourth quadrant* (IV). Also in either diagram, 0° is the positive direction along the X-axis. Ninety degrees (90°) is the positive direction along the Y-axis. Therefore, the projections of the unit radius OR on the X- and Y-axes, as appropriate, produce the same values of the trigonometric functions. Table 140a is repeated as table 140c.

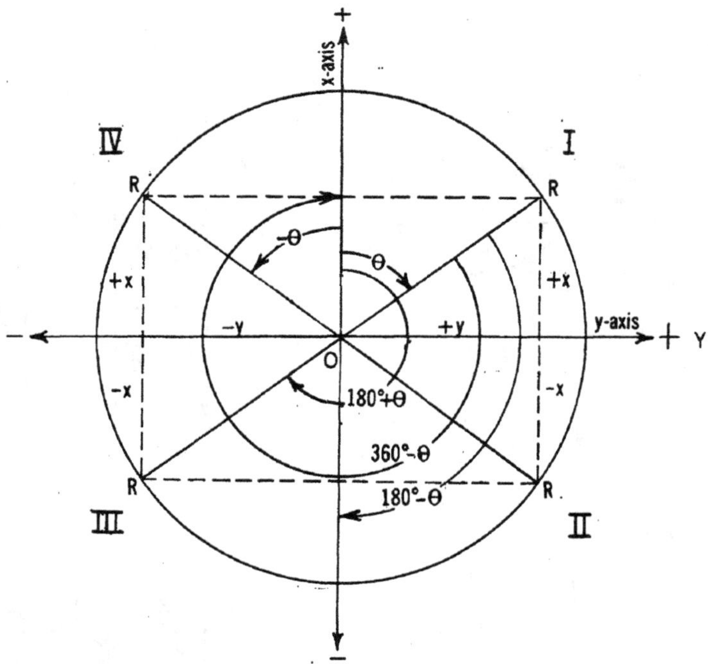

FIGURE 140c.—The functions in various quadrants.

Functions	I	II	III	IV
sine and cosecant	+	+	−	−
cosine and secant	+	−	−	+
tangent and cotangent	+	−	+	−
versine, coversine, and haversine	+	+	+	+

TABLE 140c.—Signs of trigonometric functions by quadrant.

A **negative angle** $(-\theta)$ is an angle measured in a clockwise direction (mathematical convention) or in a direction opposite to that of a positive angle. The functions of a negative angle and the corresponding functions of a positive angle are as follows:

$$\sin(-\theta) = -\sin\theta$$
$$\cos(-\theta) = \cos\theta$$
$$\tan(-\theta) = -\tan\theta$$
$$\tan(-\theta) = \tan(360° - \theta).$$

A **trigonometric identity** is an equality involving trigonometric functions of θ which is true for all values of θ, except those values for which one of the functions is not defined or for which a denominator in the equality is equal to zero. The **fundamental identities** are those identities from which other identities can be derived.

Fundamental Trigonometric Identities

$\sin\theta=\frac{1}{\csc\theta}$ $\csc\theta=\frac{1}{\sin\theta}$

$\cos\theta=\frac{1}{\sec\theta}$ $\sec\theta=\frac{1}{\cos\theta}$

$\tan\theta=\frac{1}{\cot\theta}$ $\cot\theta=\frac{1}{\tan\theta}$

$\tan\theta=\frac{\sin\theta}{\cos\theta}$ $\cot\theta=\frac{\cos\theta}{\sin\theta}$

$\sin^2\theta+\cos^2\theta=1$ $\tan^2\theta+1=\sec^2\theta$

$$1+\cot^2\theta=\csc^2\theta$$

Reduction Formulas

$\sin(90°-\theta)=\cos\theta$
$\cos(90°-\theta)=\sin\theta$
$\tan(90°-\theta)=\cot\theta$

$\csc(90°-\theta)=\sec\theta$
$\sec(90°-\theta)=\csc\theta$
$\cot(90°-\theta)=\tan\theta$

$\sin(-\theta)=-\sin\theta$
$\cos(-\theta)=\cos\theta$
$\tan(-\theta)=-\tan\theta$

$\csc(-\theta)=-\csc\theta$
$\sec(-\theta)=\sec\theta$
$\cot(-\theta)=-\cot\theta$

$\sin(90°+\theta)=\cos\theta$
$\cos(90°+\theta)=-\sin\theta$
$\tan(90°+\theta)=-\cot\theta$

$\csc(90°+\theta)=\sin\theta$
$\sec(90°+\theta)=-\csc\theta$
$\cot(90°+\theta)=-\tan\theta$

$\sin(180°-\theta)=\sin\theta$
$\cos(180°-\theta)=-\cos\theta$
$\tan(180°-\theta)=-\tan\theta$

$\csc(180°-\theta)=\csc\theta$
$\sec(180°-\theta)=-\sec\theta$
$\cot(180°-\theta)=-\cot\theta$

$\sin(180°+\theta)=-\sin\theta$
$\cos(180°+\theta)=\cos\theta$
$\tan(180°+\theta)=\tan\theta$

$\csc(180°+\theta)=-\csc\theta$
$\sec(180°+\theta)=\sec\theta$
$\cot(180°+\theta)=\cot\theta$

$\sin(360°-\theta)=-\sin\theta$
$\cos(360°-\theta)=\cos\theta$
$\tan(360°-\theta)=-\tan\theta$

$\csc(360°-\theta)=-\csc\theta$
$\sec(360°-\theta)=\sec\theta$
$\cot(360°-\theta)=-\cot\theta$

141. Inverse trigonometric functions.—The angle having a given trigonometric function may be indicated in any of several ways. Thus, $\sin y=x$, $y=\text{arc}\sin x$, and $y=\sin^{-1}x$ have the same meaning. The superior -1 is not an exponent in this case. In each case, y is "the angle whose sine is x." In this case, y is the **inverse sine** of x. Similar relationships hold for all trigonometric functions.

142. Solution of triangles.—A triangle is composed of six parts: three angles and three sides. The angles may be designated A, B, and C; and the sides opposite these angles as a, b, and c, respectively. In general, when three parts are known, the other three parts can be found, unless the known parts are the three angles of a plane triangle.

Right plane triangles.—In a right plane triangle it is only necessary to substitute numerical values in the appropriate formulas representing the basic trigonometric functions (art. 139) and solve. Thus, if a and b are known:

$$\tan A = \frac{a}{b}$$

$$B = 90° - A$$

$$c = a \csc A.$$

Similarly, if c and B are given:

$$A = 90° - B$$

$$a = c \sin A$$

$$b = c \cos A.$$

Oblique plane triangles.—In solving an oblique plane triangle, it is often desirable to draw a rough sketch of the triangle approximately to scale, as shown in figure 142a. The following laws are helpful in solving such triangles:

Law of sines: $\frac{a}{\sin A} = \frac{b}{\sin B} = \frac{c}{\sin C}$

Law of cosines: $a^2 = b^2 + c^2 - 2\,bc \cos A.$

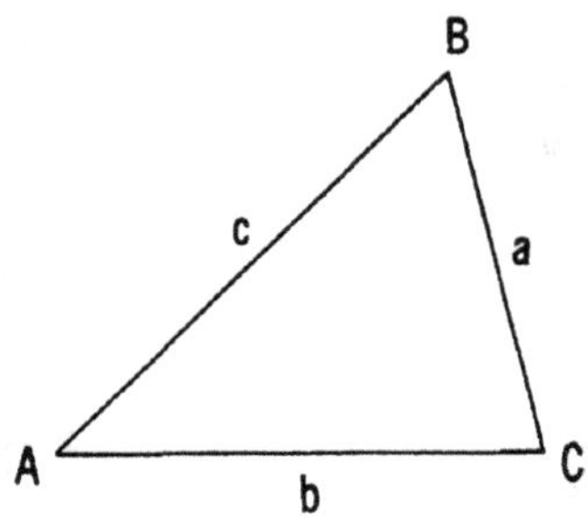

FIGURE 142a.—A plane oblique triangle.

The unknown parts of oblique plane triangles can be computed by the formulas of table 142a, among others. By reassignment of letters to sides and angles, these formulas can be used to solve for all unknown parts of oblique plane triangles.

Known	*To find*	*Formula*	*Comments*
a, b, c	A	$\cos A=\frac{c^2+b^2-a^2}{2\,bc}$	Cosine law
a, b, A	B	$\sin B=\frac{b \sin A}{a}$	Sine law. Two solutions if $b>a$
	C	$C=180°-(A+B)$	$A+B+C=180°$
	c	$c=\frac{a \sin C}{\sin A}$	Sine law
a, b, C	A	$\tan A=\frac{a \sin C}{b-a \cos C}$	
	B	$B=180°-(A+C)$	$A+B+C=180°$
	c	$c=\frac{a \sin C}{\sin A}$	Sine law
a, A, B	b	$b=\frac{a \sin B}{\sin A}$	Sine law
	C	$C=180°-(A+B)$	$A+B+C=180°$
	c	$c=\frac{a \sin C}{\sin A}$	Sine law

TABLE 142a.—Formulas for solving oblique plane triangles.

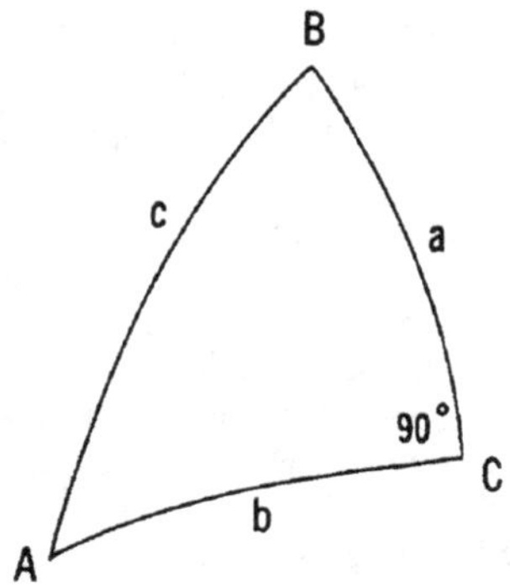

FIGURE 142b.—Parts of a right spherical triangle as used in Napier's rules.

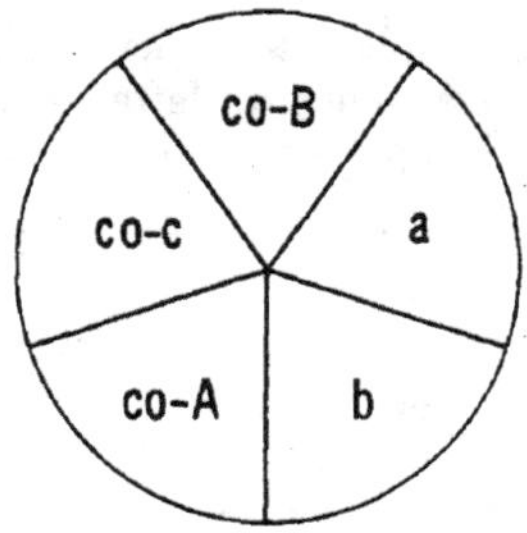

FIGURE 142c.—Diagram for Napier's Rules of Circular Parts.

Right spherical triangles can be solved with the aid of **Napier's Rules of Circular Parts,** devised by John Napier. If the right angle is omitted, the triangle has five parts: two angles and three sides, as shown in figure 142b. The triangle can be solved if any two parts are known. If the two sides forming the right angle, and the *complements* of the other three parts are used, these elements (called "parts" in the rules) can be arranged in five sectors of a circle in the same order in which they occur in the triangle, as shown in figure 142c. Considering *any* part as the *middle* part, the two parts nearest it in the diagram are considered the *adjacent* parts, and the two farthest from it the *opposite* parts. The rules are:

The sine of a middle part equals the product of (1) the tangents of the adjacent parts or (2) the cosines of the opposite parts.

In the use of these rules, the **co**function of a complement can be given as the function of the element. Thus, the cosine of co-A is the same as the sine of A. From these rules the following formulas can be derived:

$$\sin a=\tan b \cot B=\sin c \sin A$$
$$\sin b=\tan a \cot A=\sin c \sin B$$
$$\cos c=\cot A \cot B=\cos a \cos b$$
$$\cos A=\tan b \cot c=\cos a \sin B$$
$$\cos B=\tan a \cot c=\cos b \sin A.$$

The following rules apply:

1. An oblique angle and the side opposite are in the same quadrant.

2. Side c (the hypotenuse) is less then 90° when a and b are in the same quadrant, and more than 90° when a and b are in different quadrants.

If the known parts are an angle and its opposite side, two solutions are possible.

A **quadrantal spherical triangle** is one having one side of 90°. A **biquadrantal spherical triangle** has two sides of 90°. A **triquadrantal spherical triangle** has three sides of 90°. A biquadrantal spherical triangle is isosceles and has two right angles opposite the 90° sides. A triquadrantal spherical triangle is equilateral, has three right angles, and bounds an octant (one-eighth) of the surface of the sphere. A quadrantal spherical triangle can be solved by Napier's rules provided any two elements in addition to the 90° side are known. The 90° side is omitted and the other parts are arranged in order in a five-sectored circle, using the complements of the three parts farthest from the 90° side. In the case of a quadrantal triangle, rule 1 above is used, and rule 2 restated: angle C (the angle opposite the side of 90°) is *more* than 90° when A and B are in the same quadrant, and *less* than 90° when A and B are in different quadrants. If the rule requires an angle of more than 90° and the solution produces an angle of less than 90°, subtract the solved angle from 180°.

Oblique spherical triangles.—An oblique spherical triangle can be solved by dropping a perpendicular from one of the apexes to the opposite side, extended if necessary, to form two right spherical triangles. It can also be solved by the following formulas, reassigning the letters as necessary.

Known	*To find*	*Formula*	*Comments*
a, b, c	A	$\text{hav } A=\frac{\text{hav } a-\text{hav }(b-c)}{\sin b \sin c}$	
A, B, C	a	$\text{hav } a=\frac{-\cos S \cos (S-A)}{\sin B \sin C}$	$S=\frac{1}{2}(A+B+C)$
a, b, C	c	$\text{hav } c=\text{hav }(a\sim b)+\sin a \sin b \text{ hav } C$	
	A	$\tan A=\frac{\sin D \tan C}{\sin (b-D)}$	$\tan D=\tan a \cos C$
	B	$\sin B=\frac{\sin C \sin b}{\sin c}$	
c, A, B	C	$\cos C=\sin A \sin B \cos c-\cos A \cos B$	
	a	$\tan a=\frac{\tan c \sin E}{\sin (B+E)}$	$\tan E=\tan A \cos c$
	b	$\tan b=\frac{\tan c \sin F}{\sin (A+F)}$	$\tan F=\tan B \cos c$
a, b, A	c	$\sin (c+G)=\frac{\cos a \sin G}{\cos b}$	$\cot G=\cos A \tan b$ Two solutions
	B	$\sin B=\frac{\sin A \sin b}{\sin a}$	Two solutions
	C	$\sin (C+H)=\sin H \tan b \cot a$	$\tan H=\tan A \cos b$ Two solutions
a, A, B	C	$\sin (C-K)=\frac{\cos A \sin K}{\cos B}$	$\cot K=\tan B \cos a$ Two solutions
	b	$\sin b=\frac{\sin a \sin B}{\sin A}$	Two solutions
	c	$\sin (c-M)=\cot A \tan B \sin M$	$\tan M=\cos B \tan a$ Two solutions

TABLE 142b.—Formulas for solving oblique spherical triangles.

143. Other useful formulas.—In addition to the fundamental trigonometric identities and reduction formulas given in article 140, the following formulas apply to plane and spherical trigonometry:

Addition and Subtraction Formulas

$$\sin(\theta+\phi)=\sin\theta\cos\phi+\cos\theta\sin\phi$$

$$\cos(\theta+\phi)=\cos\theta\cos\phi-\sin\theta\sin\phi$$

$$\sin(\theta-\phi)=\sin\theta\cos\phi-\cos\theta\sin\phi$$

$$\cos(\theta-\phi)=\cos\theta\cos\phi+\sin\theta\sin\phi$$

$$\tan(\theta+\phi)=\frac{\tan\theta+\tan\phi}{1-\tan\theta\tan\phi}.$$

Double-Angle Formulas

$$\sin 2\theta=2\sin\theta\cos\theta$$

$$\cos 2\theta=\cos^2\theta-\sin^2\theta$$

$$\tan 2\theta=\frac{2\tan\theta}{1-\tan^2\theta}.$$

Half-Angle Formulas

$$\sin\frac{\theta}{2}=\pm\sqrt{\frac{1-\cos\theta}{2}}$$

$$\cos\frac{\theta}{2}=\pm\sqrt{\frac{1+\cos\theta}{2}}$$

$$\tan\frac{\theta}{2}=\pm\sqrt{\frac{1-\cos\theta}{1+\cos\theta}}.$$

The following are useful formulas of spherical trigonometry:

Law of Cosines for Sides

$$\cos a=\cos b\cos c+\sin b\sin c\cos A$$

$$\cos b=\cos c\cos a+\sin c\sin a\cos B$$

$$\cos c=\cos a\cos b+\sin a\sin b\cos C.$$

Law of Cosines for Angles

$$\cos A=-\cos B\cos C+\sin B\sin C\cos a$$

$$\cos B=-\cos C\cos A+\sin C\sin A\cos b$$

$$\cos C=-\cos A\cos B+\sin A\sin B\cos c.$$

Law of Sines

$$\frac{\sin a}{\sin A}=\frac{\sin b}{\sin B}=\frac{\sin c}{\sin C}.$$

Napier's Analogies

$$\tan\tfrac{1}{2}(A+B)=\frac{\cos\frac{1}{2}(a-b)}{\cos\frac{1}{2}(a+b)}\cot\tfrac{1}{2}C$$

$$\tan\tfrac{1}{2}(A-B)=\frac{\sin\frac{1}{2}(a-b)}{\sin\frac{1}{2}(a+b)}\cot\tfrac{1}{2}C$$

$$\tan\tfrac{1}{2}(a+b)=\frac{\cos\frac{1}{2}(A-B)}{\cos\frac{1}{2}(A+B)}\tan\tfrac{1}{2}c$$

$$\tan\tfrac{1}{2}(a-b)=\frac{\sin\frac{1}{2}(A-B)}{\sin\frac{1}{2}(A+B)}\tan\tfrac{1}{2}c.$$

Five Parts Formulas

$$\sin a\cos B=\cos b\sin c-\sin b\cos c\cos A$$

$$\sin b\cos C=\cos c\sin a-\sin c\cos a\cos B$$

$$\sin c\cos A=\cos a\sin b-\sin a\cos b\cos C.$$

Haversine Formulas

$\text{hav } a = \text{hav } (b \sim c) + \sin b \sin c \text{ hav } A$ $\text{hav } A = [\text{hav } a - \text{hav } (b \sim c)] \csc b \csc c$

$\text{hav } b = \text{hav } (a \sim c) + \sin a \sin c \text{ hav } B$ $\text{hav } B = [\text{hav } b - \text{hav } (a \sim c)] \csc a \csc c$

$\text{hav } c = \text{hav } (a \sim b) + \sin a \sin b \text{ hav } C$ $\text{hav } C = [\text{hav } c - \text{hav } (a \sim b)] \csc a \csc b.$

144. Functions of a small angle.—In figure 144a, small angle θ, measured in radians, is subtended by the arc RR' of a circle. The radius of the circle is r, and $R'P$ is perpendicular to OR at P. Since the length of the arc of a circle is equal to the radius multiplied by the angle subtended in radians,

$$RR' = r \times \theta.$$

When θ is sufficiently small for $R'P$ to approximate RR',

$$\sin \theta = \theta$$

$$\text{since } \theta = \frac{RR'}{r} \text{ and } \sin \theta = \frac{R'P}{r}.$$

For small angles, it can also be shown that

$$\tan \theta = \theta.$$

If there are x minutes of arc (x') in a small angle of θ radians,

$$\sin x' = x \sin 1'.$$

Figure 144a also shows that when θ is small, OP is approximately equal to the radius. Therefore, $\cos \theta$ can be taken as equal to 1.

Another approximation can be obtained if $\cos \theta$ is expressed in terms of the half-angle:

$$\cos \theta = 1 - 2 \sin^2 \frac{1}{2} \theta$$

$$\cos \theta = 1 - 2 \left(\frac{1}{2} \theta\right)^2$$

$$\cos \theta = 1 - \frac{1}{2} \theta^2.$$

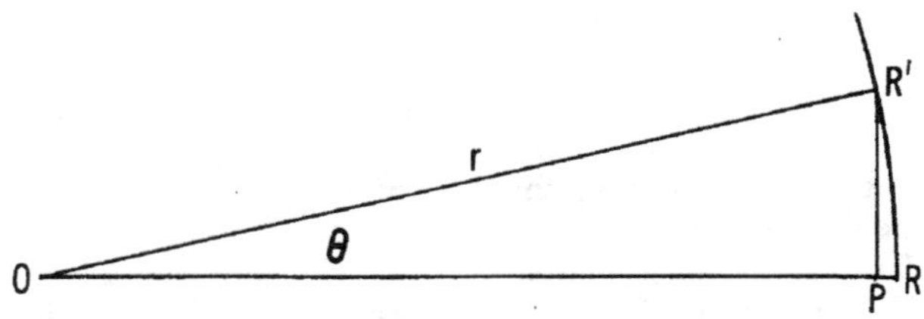

Figure 144a.—A small angle.

Calculus

145. Definitions.—**Calculus** is that branch of mathematics dealing with the rate of change of one quantity with respect to another.

A **constant** is a quantity which does not change. If a vessel is making good a course of 090°, the latitude does not change and is therefore a constant.

A **variable**, where continuous, is a quantity which can have an infinite number of values, although there may be limits to the maximum and minimum. Thus, from latitude 30° to latitude 31° there are an infinite number of latitudes, if infinitesimally small units are taken, but no value is less than 30° nor more than 31°. If two variables are so related that for every value of one there is a corresponding value of the other, one of the values is known as a **function** of the other. Thus, if speed is constant, the distance a vessel steams depends upon the elapsed time. Since elapsed time does not depend upon any other quantity, it is called an **independent variable.** The distance depends upon the elapsed time, and therefore is called a **dependent variable.** If it is required to find the time needed to travel any given distance at constant speed, distance is the independent variable and time is the dependent variable.

The principal processes of calculus are differentiation and integration.

146. Differentiation is the process of finding the rate of change of one variable with respect to another. If x is an independent variable, y is a dependent variable, and y is a function of x, this relationship may be written $y=f(x)$. Since for every value of x there is a corresponding value of y, the relationship can be plotted as a curve, figure 146. In this figure, A and B are any two points on the curve, a short distance apart.

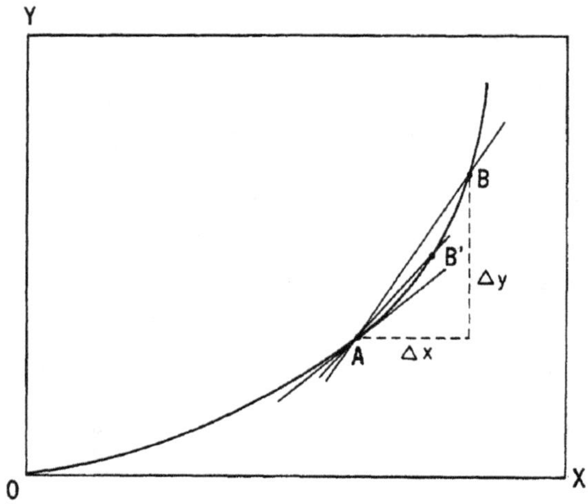

FIGURE 146.—Differentiation.

The difference between the value of x at A and at B is Δx (delta x), and the corresponding difference in the value of y is Δy (delta y). The straight line through points A and B is a **secant** of the curve (art. 130). It represents the rate of change between A and B, for anywhere along this line the change of y is proportional to the change of x.

As B moves closer to A, as shown at B', both Δx and Δy become smaller, but at a different rate, and $\frac{\Delta y}{\Delta x}$ changes. This is indicated by the difference in the slope of the secant. Also, that part of the secant between A and B moves closer to the curve and becomes a better approximation of it. The limiting case occurs when B reaches A or is at an infinitesimal distance from it. As the distance becomes infinitesimal, both Δy and Δx become infinitely small, and are designated dy and dx, respectively. The straight line becomes tangent to the curve, and represents the rate of change, or slope, of the curve at that point. This is indicated by the expression $\frac{dy}{dx}$, called the **derivative** of y with respect to x.

The process of finding the value of the derivative is called **differentiation.** It depends upon the ability to connect x and y by an equation. For instance, if $y=x^n$, $\frac{dy}{dx}=nx^{n-1}$. If $n=2$, $y=x^2$, and $\frac{dy}{dx}=2x$. This is derived as follows: If point A on the curve is x, y; point B can be considered $x+\Delta x$, $y+\Delta y$. Since the relation $y=x^2$ is true anywhere on the curve, at B:

$$y+\Delta y=(x+\Delta x)^2=x^2+2x\Delta x+(\Delta x)^2.$$

Since $y=x^2$, and equal quantities can be subtracted from both sides of an equation without destroying the equality:

$$\Delta y=2x\Delta x+(\Delta x)^2.$$

Dividing by Δx:

$$\frac{\Delta y}{\Delta x}=2x+\Delta x.$$

As B approaches A, Δx becomes infinitesimally small, approaching 0 as a limit. Therefore $\frac{\Delta y}{\Delta x}$ approaches $2x$ as a limit.

This can be demonstrated by means of a numerical example. Let $y=x^2$. Suppose at A, $x=2$ and $y=4$, and at B, $x=2.1$ and $y=4.41$. In this case $\Delta x=0.1$ and $\Delta y=0.41$, and

$$\frac{\Delta y}{\Delta x}=\frac{0.41}{0.1}=4.1.$$

From the other side of the equation:

$$2x+\Delta x=2\times 2+0.1=4.1.$$

If Δx is 0.01 and Δy is 0.0401, $\frac{\Delta y}{\Delta x}=4.01$. If Δx is 0.001, $\frac{\Delta y}{\Delta x}=4.001$; and if Δx is 0.0001, $\frac{\Delta y}{\Delta x}=4.0001$. As Δx approaches 0 as a limit, $\frac{\Delta y}{\Delta x}$ approaches 4, which is therefore the value $\frac{dy}{dx}$. Therefore, at point A the *rate* of change of y with respect to x is 4, or y is increasing in value 4 times as fast as x.

An example of the use of differentiation in navigation is the Δd value in Pub. No. 214. This is the change of altitude for a change of 1′ of declination. In this case, declination is the independent variable, altitude is the dependent variable, and both meridian angle (H.A.) and latitude are constants. The rate of change at the tabulated value is desired, so that the table can be entered with the *nearest* tabulated value of declination, and interpolation performed in either direction (either larger or smaller values of declination).

147. Integration is the inverse of differentiation. Unlike the latter, however, it is not a direct process, but involves the recognition of a mathematical expression as the differential of a known function. The function sought is the **integral** of the given expression. Most functions can be differentiated, but many cannot be integrated.

Integration can be considered the summation of an infinite number of infinitesimally small quantities, between specified limits. Consider, for instance, the problem of finding an area below a specified part of a curve for which a mathematical expression can be written. Suppose it is desired to find the area *ABCD* of figure 147. If vertical lines are drawn dividing the area into a number of vertical strips, each Δx wide, and if y is the height of each strip at the midpoint of Δx, the area of each strip is approximately $y\Delta x$; and the approximate total area of all strips is the sum of the areas of the individual strips. This may be written $\sum_{x_1}^{x_2} y\Delta x$, meaning the sum of all $y\ \Delta x$ values between x_1 and x_2. The symbol $\sum$ is the Greek letter *sigma*, the equivalent of the English *S*.

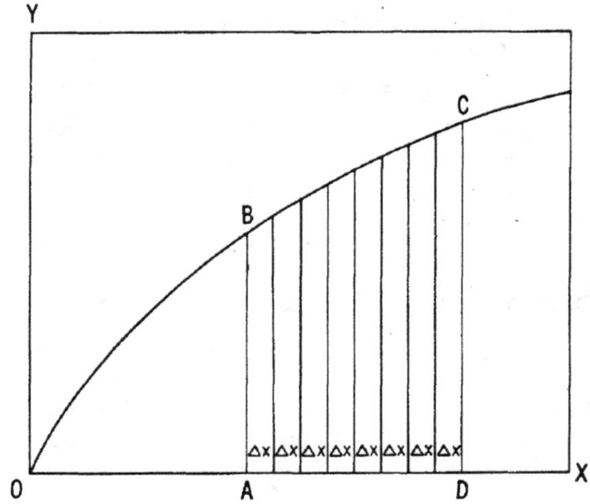

FIGURE 147.—Integration.

If Δx is made progressively smaller, the sum of the small areas becomes ever closer to the true total area. If Δx becomes infinitely small, the summation expression is written $\int_{x_1}^{x_2} ydx$, the symbol dx denoting an infinitely small Δx. The symbol $\int$, called the "integral sign," is a distorted S.

An expression such as $\int_{x_1}^{x_2} ydx$ is called a **definite integral** because limits are specified (x_1 and x_2). If limits are not specified, as in $\int ydx$, the expression is called an **indefinite integral.**

A navigational application of integration is the finding of meridional parts, table 5. The *rate* of change of meridional parts with respect to latitude changes progressively. The formula given in the explanation of the table is the equivalent of an integral representing the sum of the meridional parts from the equator to any given latitude.

148. Differential equations.—An expression such as dy or dx is called a **differential.** An equation involving a differential or a derivative is called a **differential equation.**

As shown in article 146, if $y=x^2$, $\frac{dy}{dx}=2x$. Neither dy nor dx is a finite quantity, but both are *limits* to which Δy and Δx approach as they are made progressively smaller. Therefore $\frac{dy}{dx}$ is merely a ratio, the limiting value of $\frac{\Delta y}{\Delta x}$, and not one finite number divided by another. However, since the ratio is the same as would be obtained by using finite quantities, it is possible to use the two differentials dy and dx independently in certain relationships. Differential equations involve such relationships.

Other examples of differential equations are:

$$d \sin x = \cos x \, dx \qquad d \csc x = -\cot x \csc x \, dx$$
$$d \cos x = -\sin x \, dx \qquad d \sec x = \tan x \sec x \, dx$$
$$d \tan x = \sec^2 x \, dx \qquad d \cot x = -\csc^2 x \, dx.$$

Some differential equations indicating the variations in the astronomical triangle are:

$dh = -\cos L \sin Z\, dt$; L and d constant
$dh = \cos Z\, dL$; d and t constant
$dh = -\cos h \tan M\, dZ$; L and d constant
$dZ = -\sec L \cot t\, dL$; d and h constant
$dZ = \tan h \sin Z\, dL$; d and t constant
$dt = -\sec L \cot Z\, dL$; d and h constant
$dZ = \cos d \sec h \cos M\, dt$; L and d constant
$dd = \cos d \tan M\, dt$; L and h constant
$dd = \cos L \sin t\, dZ$; L and h constant,

where h is the altitude, L is the latitude, Z is the azimuth angle, d is the declination, t is the meridian angle, and M is the parallactic angle.

CHAPTER II

INTERPOLATION

201. Introduction.—If one quantity varies with changing values of a second quantity, and the mathematical relationship of the two is known, a curve can be drawn to represent the values of one corresponding to various values of the other. To find the value of either quantity corresponding to a given value of the other, one finds that point on the curve defined by the given value, and reads the answer on the scale relating to the other quantity. This assumes, of course, that for each value of one quantity, there is only one value of the other quantity.

Information of this kind can also be tabulated. Each entry represents one point on the curve. The finding of a value *between* tabulated entries is called **interpolation.** The extending of tabulated values to find values *beyond* the limits of the table is called **extrapolation.**

Thus, the *Nautical Almanac* tabulates values of declination of the sun for each hour of Greenwich mean time. The finding of declination for a time between two whole hours requires interpolation. Since there is only one entering **argument** (in this case GMT), **single interpolation** is involved.

Table 19 gives the distance traveled in various times at certain speeds. In this table there are two entering arguments. If both given values are between tabulated values, **double interpolation** is needed.

In Pub. No. 229, azimuth angle varies with a change in any of the three variables latitude, declination, and local hour angle. With intermediate values of all three, **triple interpolation** is needed.

Interpolation can sometimes be avoided. A table having a single entering argument can be arranged as a **critical table.** An example is the dip (height of eye) correction on the inside front cover of the *Nautical Almanac*. Interpolation is avoided through dividing the argument into intervals so chosen that successive intervals correspond to successive values of the required quantity, the respondent. For any value of the argument within these intervals, the respondent can be extracted from the table without interpolation. The lower and upper limits (critical values) of the argument correspond to half-way values of the respondent and, by convention, are chosen so that when the argument is equal to one of the critical values, the respondent corresponding to the preceding (upper) interval is to be used. Another way of avoiding interpolation would be to include every possible entering argument. If this were done for Pub. No. 229, interpolation being eliminated for declination only, and assuming declination values to 0:1, the number of volumes would be increased from nine to more than 5,000. If interpolation for meridian angle and latitude, to 0:1, were also to be avoided, a total of more than 1,200,000,000 volumes would be needed. A more practical method is to select an assumed position to avoid the need for interpolation for two of the variables.

202. Single interpolation.—The accurate determination of intermediate values requires knowledge of the nature of the change between tabulated values. The simplest relationship is **linear,** the change in the tabulated value being directly proportional to the change in the entering argument. Thus, if a vessel is proceeding at 15 knots, the distance traveled is directly proportional to the time, as shown in figure 202a. The same information might be given in tabular form, as shown in table 202a. Mathematically, this relationship for 15 knots is written $D=\frac{15t}{60}=\frac{t}{4}$, where D is distance in nautical miles, and t is time in minutes.

In such a table, interpolation can be accomplished by simple proportion. Suppose, for example, that the distance is desired for a time of 15 minutes. It will be some

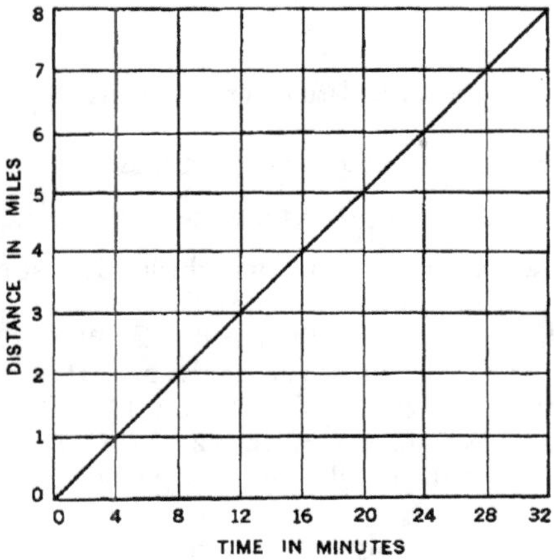

FIGURE 202a.—Plot of $D=\frac{t}{4}$.

Minutes	*Miles*
0	0. 0
4	1. 0
8	2. 0
12	3. 0
16	4. 0
20	5. 0
24	6. 0
28	7. 0
32	8. 0

TABLE 202a.—Table of $D=\frac{t}{4}$.

value between 3.0 and 4.0 miles, because these are the distances for 12 and 16 minutes, respectively, the tabulated times on each side of the desired time. The proportion might be formed as follows:

$$3\left[\begin{matrix}12\\15\end{matrix}\right.\left.\begin{matrix}\\ \\16\end{matrix}\right]4 \qquad\qquad x\left[\begin{matrix}3.0\\y\end{matrix}\right.\left.\begin{matrix}\\ \\4.0\end{matrix}\right]1.0$$

$$\frac{3}{4}=\frac{x}{1.0}$$

$$x=\frac{3\times1.0}{4}=0.75 \text{ (0.8 to nearest 0.1 mi.)}$$

$$y=3.0+x=3.0+0.8=3.8 \text{ mi.}$$

A simple interpolation such as this should be performed mentally. During the four-minute interval between 12 and 16 minutes, the distance *increases* 1.0 mile from 3.0 to 4.0 miles. At 15 minutes, ¾ of the interval has elapsed, and so the distance increases ¾ of 1.0 mile, or 0.75 mile, and is therefore 3.0+0.8=3.8, to the nearest 0.1 mile.

This might also have been performed by starting with 16 minutes, as follows:

$$1\left[\begin{array}{c}12\\ \left[\begin{array}{c}15\\16\end{array}\right.\end{array}\right]4 \qquad\qquad (-)x\left[\begin{array}{c}3.0\\ \left[\begin{array}{c}y\\4.0\end{array}\right.\end{array}\right]1.0$$

$$\frac{1}{4}=\frac{(-)x}{1.0}$$

$x=(-)0.25$ (-0.2 to the nearest 0.1 mi.)

$y=4.0-0.2=3.8$

Mentally, 15 is one quarter of the way from 16 to 12, and therefore the distance is ¼ the way between 4.0 and 3.0, or 3.8.

This interpolation might have been performed by noting that if distance changes 1.0 mile in four minutes, it must change $\frac{1.0}{10}=0.1$ mile in $\frac{4}{10}=0.4$ minute, or 24 seconds. This relationship can be used for mental interpolation in situations which might seem to require pencil and paper. Thus, if distance to the nearest 0.1 mile is desired for 13^m15^s, the answer is 3.3 miles, determined as follows: The time 13^m15^s is 1^m15^s ($1\overset{m}{.}2$ approx.) more than 12^m. If 1.2 is divided by 0.4, the quotient is 3, to the nearest whole number. Therefore, $3\times0.1=0.3$ is added to 3, the tabulated value for 12 minutes. Alternatively, 13^m15^s is 2^m45^s ($2\overset{m}{.}8$ approx.) less than 16^m, and $2.8\div0.4=7$, and therefore the interpolated value is $7\times0.1=0.7$ *less* than 4, the tabulated value for 16^m. In either case, the interpolated value is 3.3 miles.

A common mistake in single interpolation is to apply the correction (x) with the wrong sign, particularly when it should be negative (−). This mistake can be avoided by always checking to be certain that the interpolated value lies between the two values used in the interpolation.

When the curve representing the values of a table is a straight line, as in figure 202a, the process of finding intermediate values in the manner described above is called **linear interpolation.** If tabulated values of such a line are exact (not approximations), as in table 202a, the interpolation can be carried to any degree of precision without sacrificing accuracy. Thus, in 21.5 minutes the distance is $5.0+\frac{1.5}{4}\times1.0=5.375$ miles. Similarly, for 29.9364 minutes the distance is $7.0+\frac{1.9364}{4}\times1.0=7.4841$ miles, a value which has little or no significance in practical navigation. If one had occasion to find such a value, it could most easily be done by dividing the time, in minutes, by 4, since the distance increases at the rate of one mile each four minutes. This would be a case of avoiding interpolation by solving the equation connecting the two quantities. For a simple relationship such as that involved here, such a solution might be easier than interpolation.

Many of the tables of navigation are not linear. Consider figure 202b. From table 29 it is found that for latitude 25° and declination 8°, same name, the variation of altitude in one minute of time from meridian transit (the altitude factor) is $6\overset{''}{.}0$ ($0\overset{'}{.}1$). For a limited angular distance on each side of the celestial meridian, the change in altitude is approximately equal to at^2, where a is the altitude factor (from table 29) and t is the time in minutes from meridian transit. Figure 202b is the plot of change in altitude against time. The same information is shown in tabular form in table 202b.

To be strictly accurate in interpolating in such a table, one should consider the curvature of the line. However, in most navigational tables the points on the curve selected for tabulation are sufficiently close that the portion of the curve between entries can be considered a straight line without introducing a significant error. This is similar to considering the line of position from a celestial observation as a part of the circle of equal altitude. Thus, to the nearest 0′.1, the change of altitude for 3.4 minutes is 0′.9+(0.4×0′.7)=0′.9+0′.3=1′.2. The correct value by solution of the formula is 1′.156. The value for 6.8 minutes is 4′.6 by interpolation and 4′.624 by computation.

Article 205 addresses the nonlinear interpolation used when the curve representing tabular values under consideration is not a close approximation to a straight line. However, such instances are infrequent in navigation, and generally occur at a part of

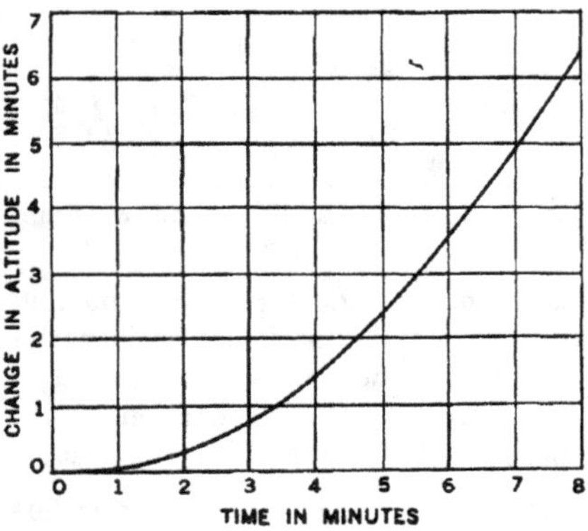

FIGURE 202b.—Plot of altitude change=at^2.

Minutes	*Altitude change*
0	0. 0
1	0. 1
2	0. 4
?	0. 9
4	1. 6
5	2. 5
6	3. 6
7	4. 9
8	6. 4

TABLE 202b.—Table of altitude change=at^2, where a=0′.1.

the navigation table that is not commonly used, or for which special provisions are made. For example, in Pub. No. 229 nonlinear interpolation *may* be required only when the altitude is above 60°. Even when the altitude is above 60°, the need for nonlinear interpolation is infrequent. When it is needed, such fact is indicated by the altitude difference being printed in italic type followed by a small dot.

203. Double interpolation.—In a double-entry table it may be necessary to interpolate for each entering argument. Table 203a is an extract from table 27 (amplitudes). If one entering argument is an exact tabulated value, the amplitude can be found by single interpolation. For instance, if latitude is 45° and declination is 21°.8, amplitude is $31°.2+\left(\frac{3}{5}\times 0°.8\right)=31°.2+0°.5=31°.7$. However, if neither entering argument is a tabulated value, double interpolation is needed. This may be accomplished in any of several ways:

Lat.	*Declination*	
	21°.5	22°.0
°	°	°
45	31. 2	32. 0
46	31. 8	32. 6

TABLE 203a.—Excerpts from amplitude table.

1. *"Horizontal" method.* Use single interpolation for declination for each tabulated value of latitude, followed by single interpolation for latitude. Suppose latitude is

45°.7 and declination is 21°.8. First, find the amplitude for latitude 45°, declination 21°.8, as above, 31°.7. Next, repeat the process for latitude 46°: $31°.8+\left(\frac{3}{5}\times 0°.8\right)=$ 32°.3. Finally, interpolate between 31°.7 and 32°.3 for latitude 45°.7: $31°.7+(0.7\times 0°.6)=32°.1$. This is the equivalent of first inserting a new column for declination 21°.8, followed by single interpolation in this column, as shown in table 203b.

Lat.	*Declination*		
	21°.5	*21°.8*	22°.0
°	°	°	°
45	31.2	*31.7*	32.0
45.7		***32.1***	
46	31.8	*32.3*	32.6

TABLE 203b.—"Horizontal" method of double interpolation.

Lat.	*Declination*		
	21°.5	*21°.8*	22°.0
°	°	°	°
45	31.2		32.0
45.7	*31.6*	***32.1***	*32.4*
46	31.8		32.6

TABLE 203c.—"Vertical" method of double interpolation.

2. *"Vertical" method.* Use single interpolation for latitude for each tabulated value of declination, followed by single interpolation for declination. Consider the same example as above. First, find the amplitude for declination 21°.5, latitude 45°.7: $31°.2+(0°.7\times 0°.6)=31°.6$. Next, repeat the process for declination 22°.0: $32°.0+(0°.7\times 0°.6)=32°.4$. Finally, interpolate between 31°.6 and 32°.4 for declination 21°.8: $31°.6+\left(\frac{3}{5}\times 0°.8\right)=32°.1$. This is the equivalent of first inserting a new line for latitude 45°.7, followed by single interpolation in this line, as shown in table 203c.

3. *Combined method.* Select a tabulated "base" value, preferably that nearest the given tabulated entering arguments. Next, find the correction to be applied, with its sign, for single interpolation of this base value both horizontally and vertically. Finally, add these two corrections algebraically and apply the result, in accordance with its sign, to the base value. In the example given above, the base value is 32°.6, for declination 22°.0 (21°.8 is nearer 22°.0 than 21°.5) and latitude 46° (45°.7 is nearer 46° than 45°). The correction for declination is $\frac{2}{5}\times(-)0°.8=(-)0°.3$. The correction for latitude is $0°.3\times(-)0°.6=(-)0°.2$. The algebraic sum is $(-)0°.3+(-)0°.2=(-)0°.5$. The interpolated value is then $32°.6-0°.5=32°.1$. This is the method customarily used by navigators.

204. Triple interpolation.—With three entering arguments, the process is similar to that for double interpolation. It would be possible to perform double interpolation for the tabulated value on each side of the given value of one argument, and then interpolate for that argument, but the method would be tedious. The only method commonly used by navigators is that of selecting a base value and applying corrections. Suppose, for instance, that the azimuth angle is desired for latitude 41°.3, declination 21°.9 contrary name, meridian angle 16°.6 using Pub. No. 214. The base value (lat. 41°, Dec. 22°, t 17°) is 162°.6. The corrections are

lat.	41°.3: 0.3 × (+) 0°.1 =	0°.0
Dec.	21°.9: 0.2 × (−) 0°.1 =	0°.0
t	16°.6: 0.4 × (+) 1°.0 =	(+) 0°.4
	Total	(+) 0°.4.

The triple interpolated value is 162°.6+0°.4=163°.0

205. Nonlinear interpolation.—When the curve representing the values of a table is nearly a straight line, or the portion of the curve under consideration is nearly a straight line, linear interpolation suffices. However, when the successive tabular values are so nonlinear that a portion of the curve under consideration is not a close approximation to a straight line, it is necessary to include the effects of second differences, and possibly higher differences, as well as first differences in the interpolation.

The plot of table 205a data in figure 205 indicates that the altitude does not change linearly between declination values of 51° and 52°. If the first difference only were used in the interpolation, the interpolated value of altitude would lie on the straight line between points on the curve for declination values of 51° and 52°.

LHA 38°, Lat. 45° (Same Name as Declination)

Dec.	ht (Tab. Hc)	First Difference	Second Difference
50°	64°08′.2		
		+2′.8	
51°	64°11′.0		−2′.3
		+0′.5	
52°	64°11′.5		−2′.1
		−1′.6	
53°	64°09′.9		

TABLE 205a.—Data from Pub. No. 229.

Function	*1st Diff.*	*2nd Diff.*
f_{-2}		δ^2_{-2}
	$\delta_{-3/2}$	
f_{-1}		δ^2_{-1}
	$\delta_{-1/2}$	
f_0		δ^2_0
	$\delta_{1/2}$	
f_{+1}		δ^2_1
	$\delta_{3/2}$	
f_{+2}		δ^2_2

TABLE 205b.—Notation used with Bessel's formula.

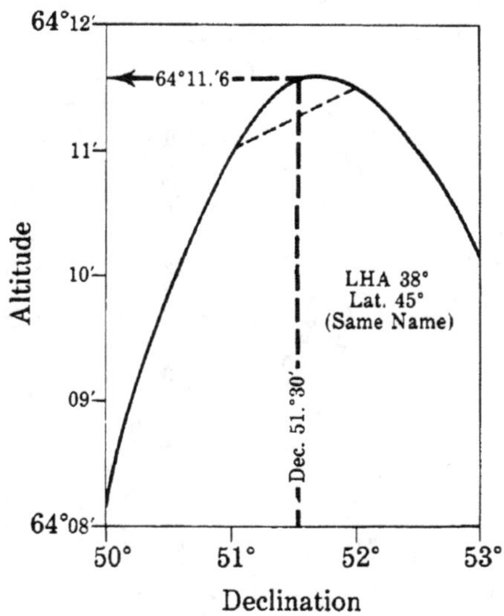

FIGURE 205.—Altitude curve.

If the altitude for declination 51°30′ is obtained using only the **first difference,** i.e., the difference between successive tabular altitudes in this case, $Hc = 64°11'.0 + \frac{30'}{60'} \times 0'.5 = 64°11'.3$. However, inspection of figure 205 reveals that this interpolated altitude is 0′.3 low. If the tabular data were such that the differences between successive first differences, the **second differences,** were nearly zero, interpolation using the first difference only would provide the correct altitude. In this case, however, second differences are significant and must be included in the interpolation.

Table 205b shows the format and notation used to distinguish the various tabular quantities and differences when using Bessel's formula for the nonlinear interpolation. The quantities $f_{-2}, f_{-1}, f_0, f_{+1}, f_{+2}, f_{+3}$ represent successive tabular values.

Allowing for first and second differences only, Bessel's formula is stated as:

$$f_p = f_0 + p\delta_{1/2} + B_2(\delta_0^2 + \delta_1^2).$$

In this case, f_p is the computed altitude; f_o is the tabular altitude; p is the fraction of the interval between tabular values of declination. The quantity B_2 is a function of p and is always negative. This coefficient is tabulated in table 205c. The quantity $(\delta_0^2+\delta_1^2)$ is the **double second difference (DSD),** which is the sum of successive second differences.

Applying Bessel's formula to the data of table 205a to obtain the altitude for a declination of 51°30′,

$$f_p = f_0 + p\delta_{1/2} + B_2(\delta_0^2 + \delta_1^2)$$

$$\text{Hc} = 64°11'.0 + \frac{30'}{60'}(0'.5) + (-0.062)\,[-2'.3 + (-2'.1)]$$

$$\text{Hc} = 64°11'.0 + 0'.3 + 0'.3 = 64°11'.6.$$

p	B_2	p	B_2	p	B_2	p	B_2	p	B_2
0.0000		0.1101		0.2719		0.7280		0.8898	
	.000		.025		.050		.049		.024
.0020		.1152		.2809		.7366		.8949	
	.001		.026		.051		.048		.023
.0060		.1205		.2902		.7449		.9000	
	.002		.027		.052		.047		.022
.0101		.1258		.3000		.7529		.9049	
	.003		.028		.053		.046		.021
.0142		.1312		.3102		.7607		.9098	
	.004		.029		.054		.045		.020
.0183		.1366		.3211		.7683		.9147	
	.005		.030		.055		.044		.019
.0225		.1422		.3326		.7756		.9195	
	.006		.031		.056		.043		.018
.0267		.1478		.3450		.7828		.9242	
	.007		.032		.057		.042		.017
.0309		.1535		.3585		.7898		.9289	
	.008		.033		.058		.041		.016
.0352		.1594		.3735		.7966		.9335	
	.009		.034		.059		.040		.015
.0395		.1653		.3904		.8033		.9381	
	.010		.035		.060		.039		.014
.0439		.1713		.4105		.8098		.9427	
	.011		.036		.061		.038		.013
.0483		.1775		.4367		.8162		.9472	
	.012		.037		.062		.037		.012
.0527		.1837		.5632		.8224		.9516	
	.013		.038		.061		.036		.011
.0572		.1901		.5894		.8286		.9560	
	.014		.039		.060		.035		.010
.0618		.1966		.6095		.8346		.9604	
	.015		.040		.059		.034		.009
.0664		.2033		.6264		.8405		.9647	
	.016		.041		.058		.033		.008
.0710		.2101		.6414		.8464		.9690	
	.017		.042		.057		.032		.007
.0757		.2171		.6549		.8521		.9732	
	.018		.043		.056		.031		.006
.0804		.2243		.6673		.8577		.9774	
	.019		.044		.055		.030		.005
.0852		.2316		.6788		.8633		.9816	
	.020		.045		.054		.029		.004
.0901		.2392		.6897		.8687		.9857	
	.021		.046		.053		.028		.003
.0950		.2470		.7000		.8741		.9898	
	.022		.047		.052		.027		.002
.1000		.2550		.7097		.8794		.9939	
	.023		.048		.051		.026		.001
.1050		.2633		.7190		.8847		0.9979	
	.024		.049		.050		.025		.000
0.1101		0.2719		0.7280		0.8898		1.0000	

TABLE 205c.—Bessel's Coefficient B_2. In critical cases ascend. B_2 is always negative.

206. Interpolation tables.—A number of frequently used navigation tables are provided with auxiliary tables to assist in interpolation. Table 32 (Logarithms of Numbers) provides columns of "d" (difference between consecutive entries) and auxiliary "proportional parts" tables. The auxiliary table for the applicable difference "d" is selected and entered with the digit of the additional place in the entering argument. The value taken from the auxiliary table is *added* to the base value for the next *smaller* number from the main table. Suppose the logarithm (mantissa) for 32747 is desired. The base value for 3274 is 51508, and "d" is 13. The auxiliary table for 13 is entered with 7, and the correction is found to be 9. If this is added to 51508, the interpolated value is found to be 51517. This is the same result that would be obtained by subtracting 51508 from 51521 (the logarithm for 3275) to obtain 13, multiplying this by 0.7, and adding the result (9) to 51508.

Tables 31 and 33 provide the difference between consecutive entries, but no proportional parts tables.

The *Nautical Almanac* "Increments and Corrections" are interpolation tables for the hourly entries of GHA and declination. The increments are the products of the constant value used as the change of GHA in 1 hour and the fractional part of the hour. The corrections provide for the difference between the actual change of GHA in 1 hour and the constant value used. The corrections also provide the product of the change in declination in 1 hour and the fractional part of the hour.

The main part of the four-page interpolation table of Pub. No. 229 is basically a multiplication table providing tabulations of:

$$\text{Altitude Difference} \times \frac{\text{Declination Increment}}{60'}.$$

The design of the table is such that the desired product must be derived from component parts of the altitude difference. The first part is a multiple of 10′ (10′, 20′, 30′, 40′, or 50′) of the altitude difference; the second part is the remainder in the range 0.′0 to 9.′9. For example, the component parts of altitude difference 44.′3 are 40′ and 4.′3.

In the use of the first part of the altitude difference, the table arguments are declination increment (Dec. Inc.) and the integral multiple of 10′ in the altitude difference, d. As shown in figure 206a, the respondent is:

$$\text{Tens} \times \frac{\text{Dec. Inc.}}{60'}.$$

In the use of the second part of the altitude difference, the interpolation table arguments are the nearest Dec. Inc. ending in 0.′5 and Units and Decimals. The respondent is:

$$\text{Units and Decimals} \times \frac{\text{Dec. Inc.}}{60'}.$$

In computing the table, the values in the Tens part of the multiplication table were modified by small quantities varying from −0.′042 to +0.′033 before rounding to the tabular precision to compensate for any difference between the actual Dec. Inc. and the nearest Dec. Inc. ending in 0.′5 when using the Units and Decimals part of the table.

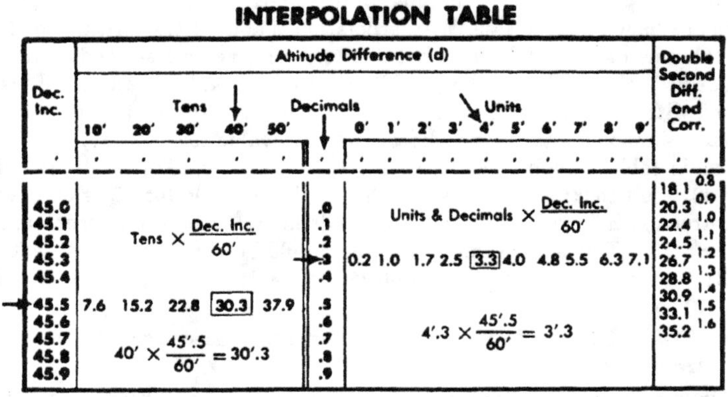

FIGURE 206a.—Interpolation table.

Using the interpolation table shown in figure 206b to obtain the altitude for 51°30′ from the data of table 205a, the linear correction for the first difference (+0′.5) is +0′.3. This correction is extracted from the Units and Decimals block opposite the Dec. Inc. (30′.0). The correction for the double second difference (DSD) is extracted from the DSD subtable opposite the block in which the Dec. Inc. is found. The argument for entering this critical table is the DSD (−4′.4). The DSD correction is +0′.3. Therefore,

$$Hc = ht + \text{first difference correction} + \text{DSD correction}$$
$$= 64°11'.0 + 0'.3 + 0'.3 = 64°11'.6.$$

INTERPOLATION TABLE

Dec. Inc.	Tens 10′	20′	30′	40′	50′	Decimals	Units 0′	1′	2′	3′	4′	5′	6′	7′	8′	9′
′	′	′	′	′	′	′	′	′	′	′	′	′	′	′	′	′
30.0	5.0	10.0	15.0	20.0	25.0	.0	0.0	0.5	1.0	1.5	2.0	2.5	3.0	3.6	4.1	4.6
30.1	5.0	10.0	15.0	20.0	25.1	.1	0.1	0.6	1.1	1.6	2.1	2.6	3.1	3.6	4.1	4.6
30.2	5.0	10.0	15.1	20.1	25.1	.2	0.1	0.6	1.1	1.6	2.1	2.6	3.2	3.7	4.2	4.7
30.3	5.0	10.1	15.1	20.2	25.2	.3	0.2	0.7	1.2	1.7	2.2	2.7	3.2	3.7	4.2	4.7
30.4	5.1	10.1	15.2	20.3	25.3	.4	0.2	0.7	1.2	1.7	2.2	2.7	3.3	3.8	4.3	4.8
30.5	5.1	10.2	15.3	20.3	25.4	.5	0.3	0.8	1.3	1.8	2.3	2.8	3.3	3.8	4.3	4.8
30.6	5.1	10.2	15.3	20.4	25.5	.6	0.3	0.8	1.3	1.8	2.3	2.8	3.4	3.9	4.4	4.9
30.7	5.1	10.3	15.4	20.5	25.6	.7	0.4	0.9	1.4	1.9	2.4	2.9	3.4	3.9	4.4	4.9
30.8	5.2	10.3	15.4	20.6	25.7	.8	0.4	0.9	1.4	1.9	2.4	2.9	3.5	4.0	4.5	5.0
30.9	5.2	10.3	15.5	20.6	25.8	.9	0.5	1.0	1.5	2.0	2.5	3.0	3.5	4.0	4.5	5.0

Double Second Diff.	and Corr.
0.8	
	0.1
2.4	
	0.2
4.0	
	0.3
5.6	
	0.4
7.2	
	0.5
8.8	
	0.6
10.4	
	0.7
12.0	
	0.8
13.6	
	0.9
15.2	
	1.0
16.8	

FIGURE 206b.—Interpolation table.

207. Extrapolation.—The extending of a table is usually performed by assuming that the difference between the last few tabulated entries will continue at the same rate. This assumption is strictly correct only if the change is truly linear, but in most tables the assumption provides satisfactory results for a *slight* extension beyond tabulated values. The extent to which the assumption can be used reliably can often be determined by noting the last few differences. If the "second differences" (differences between consecutive differences) are nearly zero, the curve is nearly a straight line, for a short distance. But if consecutive second differences are appreciable, extrapolation is not reliable. For examples of linear and nonlinear relationships, refer to the first page of table 33 and compare the tabulated differences of the logarithms of secant (approximately linear on this page) and sine (nonlinear on this page).

As an example of extrapolation, consider table 27. Suppose the amplitude for latitude 45°, declination 24°3 is desired. The last declination entry is 24°0. The amplitude for declination 23°5 is 34°3, and for declination 24°0 it is 35°1. The difference is (+) 0°8. Assuming this same difference between declinations 24°0 and 24°5, one finds the value for 24°3 is $35°1+\left(\frac{3}{5}\times 0°8\right)=35°6$. Below latitude 50° this table is so nearly linear that extrapolation can be carried to declination 30° without serious error.

For double or triple extrapolation, differences are found as in single interpolation.

208. General comments.—As a general rule, the final answer should not be given to greater precision than tabulated values. A notable exception to this rule is the case where tabulated values are known to be exact, as in table 202a. A slight increase in accuracy can sometimes be attained by retaining one additional place in the solution until the final answer. Suppose, for instance, that the corrections for triple interpolation are (+)0.2, (+)0.3, and (−)0.3. The total correction is (+)0.2. If the total correction, rounded to tenths, had been obtained from the sum of (+)0.17, (+)0.26, and (−)0.34, the correct total would have been (+)0.09=(+)0.1. The retaining of one additional place may be critical if the correction factors end in 0.5. Thus, in double interpolation, one correction value might be (+)0.15, and the other (−)0.25. The correct total is (−)0.1. But if the individual differences are rounded to (+)0.2 and (−)0.2, the total is 0.0.

The difference used for establishing the proportion is also a matter subject to some judgment. Thus, if the latitude is 17°14′6, it might be rounded to 17°2 for many purposes. Slightly more accurate results can sometimes be obtained by retaining the minutes, using $\frac{14.6}{60}$ instead of 0.2. If the difference to be multiplied by this proportion is small, the increase in accuracy gained by using the more exact value is small, but if the difference is large, the gain might be considerable. Thus, if the difference is 0°2, the correction by using either $\frac{14.6}{60}$ or 0.2 is less than 0°05, or 0°0 to the nearest 0°1. But if the difference is 3°2, the value by $\frac{14.6}{60}$ is 0°8, and the value by 0.2 is 0°6.

If the tabulated entries involved in an interpolation are all positive or all negative, the interpolation can be carried out on either a numerical or an algebraic basis. Most navigators prefer the former, carrying out the interpolation as if all entries were positive, and giving to the interpolated value the common sign of all entries. When both positive and negative entries are involved, all differences and corrections should be on an algebraic basis, and careful attention should be given to signs. Thus, if single interpolation is to be performed between values of (+)0.9 and (−)0.4, the difference is 0.9−(−0.4)=0.9+0.4=1.3. If the correction is 0.2 of this difference, it is (−)0.3 if applied to (+)0.9, and (+)0.3 if applied to (−)0.4. In the first case, the interpolated value is (+)0.9 −0.3=(+)0.6. In the second case, it is (−)0.4+0.3=(−)0.1. If the correction had been 0.4 of the difference, it would have been (−)0.5 in the first case, and (+)0.5 in the second. The interpolated value would have been (+)0.9−0.5=(+)0.4, or (−)0.4+0.5=(+)0.1, respectively.

Because of the variety in methods of interpolation used, solutions by different persons may differ slightly.

PART THREE

NAVIGATIONAL CALCULATIONS

PART THREE

NAVIGATIONAL CALCULATIONS

CHAPTER III

COMPASS CONVERSIONS

301. Magnetic compass error.—Directions relative to the northerly direction along a geographic meridian are **true.** In this case, true north is the **reference direction.** If a compass card is horizontal and oriented so that a straight line from its center to 000° points to true north, any direction measured by the card is a true direction and has no error (assuming there is no calibration or observational error). If the card remains horizontal but is rotated so that it points in any other direction, the amount of the rotation is the **compass error.** Stated differently, compass error is the angular difference between true north and **compass north** (the direction north as indicated by a magnetic compass). It is named east or west to indicate the side of true north on which compass north lies.

If a magnetic compass is influenced by no other magnetic field than that of the earth, and there is no instrumental error, its magnets are aligned with the magnetic meridian at the compass, and 000° of the compass card coincides with **magnetic north.** All directions indicated by the card are **magnetic.** As stated in volume I, the angle between geographic and magnetic meridians is called **variation** (**V** or **Var.**). Therefore, if a compass is aligned with the magnetic meridian, compass error and variation are the same.

When a compass is mounted in a vessel, it is generally subjected to various magnetic influences other than that of the earth. These arise largely from induced magnetism in metal decks, bulkheads, masts, stacks, boat davits, guns, etc., and from electromagnetic fields associated with direct current in electrical circuits. Some metal in the vicinity of the compass may have acquired permanent magnetism. The actual magnetic field at the compass is the vector sum, or resultant (art. 117), of all individual fields at that point. Since the direction of this resultant field is generally not the same as that of the earth's field alone, the compass magnets do not lie in the magnetic meridian, but in a direction that makes an angle with it. This angle is called **deviation** (**D** or **Dev.**). Thus, deviation is the angular difference between magnetic north and compass north. It is expressed in angular units and named east or west to indicate the side of magnetic north on which compass north lies. Thus, deviation is the error of the compass in pointing to magnetic north, and all directions measured with compass north as the reference direction are **compass directions.** Since variation and deviation may each be either east or west, the effect of deviation may be to either increase or decrease the error due to variation alone. The algebraic sum (art. 106) of variation and deviation is the total compass error.

For computational purposes, deviation and compass error, like variation, may be designated positive (+) if east and negative (−) if west.

Variation changes with location. Deviation depends upon the magnetic latitude and also upon the individual vessel, its trim and loading, whether it is pitching or rolling, the heading (orientation of the vessel with respect to the earth's magnetic field), and the location of the compass within the vessel. Therefore, deviation is not published on charts.

302. Deviation table.—In practice aboard ship, the deviation is reduced to a minimum through adjustment of the compass. The remaining value, called **residual deviation,** is determined on various headings and recorded in some form of **deviation table.** Figure 302 shows the form used by the United States Navy. This table is entered with the magnetic heading, and the deviation on that heading is determined from the tabulation, separate columns being given for degaussing (DG) equipment off and on. If the deviation is not more than about 2° on any heading, satisfactory results may be obtained by entering the values at intervals of 45° only.

If the deviation is small, no appreciable error is introduced by entering the table with either magnetic or compass heading. If the deviation on some headings is large, the desirable action is to reduce it, but if this is not practicable, a separate deviation table for compass heading entry may be useful. This may be made by applying the tabulated deviation to each entry value of magnetic heading, to find the corresponding compass heading, and then interpolating between these to find the value of deviation at each 15° compass heading.

MAGNETIC COMPASS TABLE NAVSHIPS RPT. 3530.2
NAVSHIPS 3120/4 (REV. 8-67) (FRONT) *(Formerly NAVSHIPS 1104)*
S/N 0105-601-8620

U.S.S. Truckee NO. AO 147 (BB, CL, DD, etc.)

[X] PILOT HOUSE [ ] SECONDARY CONNING STATION [ ] OTHER

BINNACLE TYPE: [X] NAVY ST'D [ ] OTHER

COMPASS 7 1/2 MAKE Lionel SERIAL NO. 1592

TYPE CC COILS "K" DATE 30 October 1970

READ INSTRUCTIONS ON BACK BEFORE STARTING ADJUSTMENT

SHIPS HEAD MAGNETIC	DEVIATIONS		SHIPS HEAD MAGNETIC	DEVIATIONS	
	DG OFF	DG ON		DG OFF	DG ON
0	0.5E	0.5E	180	0.5W	0.0
15	1.0E	1.0E	195	1.0W	0.5W
30	1.5E	1.5E	210	1.0W	1.0W
45	2.0E	1.5E	225	1.5W	1.5W
60	2.0E	2.0E	240	2.0W	2.0W
75	2.5E	2.5E	255	2.0W	2.5W
90	2.5E	3.0E	270	1.5W	2.0W
105	2.0E	2.5E	285	1.0W	1.5W
120	1.5E	2.0E	300	1.0W	1.0W
135	1.5E	1.5E	315	0.5W	0.5W
150	1.0E	1.0E	330	0.5W	0.5W
165	0.0	0.5E	345	0.0	0.0

DEVIATIONS DETERMINED BY: [ ] SUN'S AZIMUTH [X] GYRO [ ] SHORE BEARINGS

B 4 MAGNETS RED [ ] FORE [X] AFT AT 14" FROM COMPASS CARD

C 4 MAGNETS RED [ ] PORT [X] STBD AT 10" FROM COMPASS CARD

D 2-7" [X] SPHERES [ ] CYLS AT 12" [X] ATHWARTSHIP [ ] SLEWED --- ° [ ] CLOCKWISE [ ] CTR. CLOCKWISE

HEELING MAGNET: [ ] RED UP [X] BLUE UP 10" FROM COMPASS CARD FLINDERS BAR: [X] FORE [ ] AFT 14"

[X] LAT [ ] N 36°10' N [X] LONG [ ] Z 75° 20' W

SIGNED *(Adjuster or Navigator)* T. PARRISH APPROVED *(Commanding)* R. MOSS

FIGURE 302.—Deviation table.

303. Applying variation and deviation.—As indicated in article 301, a single direction may have any of several numerical values depending upon the reference direction used. One should keep clearly in mind the relationship between the various expressions of a direction. Thus, true and magnetic directions differ by the variation, magnetic and compass directions differ by the deviation, and true and compass directions differ by the compass error.

If variation or deviation is easterly, the compass card is rotated in a clockwise direction. This brings smaller numbers opposite the lubber's line. Conversely, if either error is westerly, the rotation is counterclockwise and larger numbers are brought opposite the lubber's line. Thus, if the heading is 090° true (fig. 303, A) and variation is 6°E, the magnetic heading is 090°−6°=084° (fig. 303, B). If the deviation on this heading is 2°W, the compass heading is 084°+2°=086° (fig. 303, C). Also, compass error is 6°E−2°W=4°E, and compass heading is 090°−4°=086°. If compass error is easterly, the compass reads too low (in comparison with true directions), and if it is westerly, the reading is too high. Many rules-of-thumb have been devised as an aid to the memory, and any which assist in applying compass errors in the right direction are of value. However, one may forget the rule or its method of application, or may wish to have an independent check. If he understands the explanation given above, he can determine the correct sign without further information. The same rules apply to the use of gyro error. Since variation and deviation are compass errors, the process of removing either from an indication of a direction (converting compass to magnetic or magnetic to true) is often called **correcting.** Conversion in the opposite direction (inserting errors) is then called **uncorrecting.**

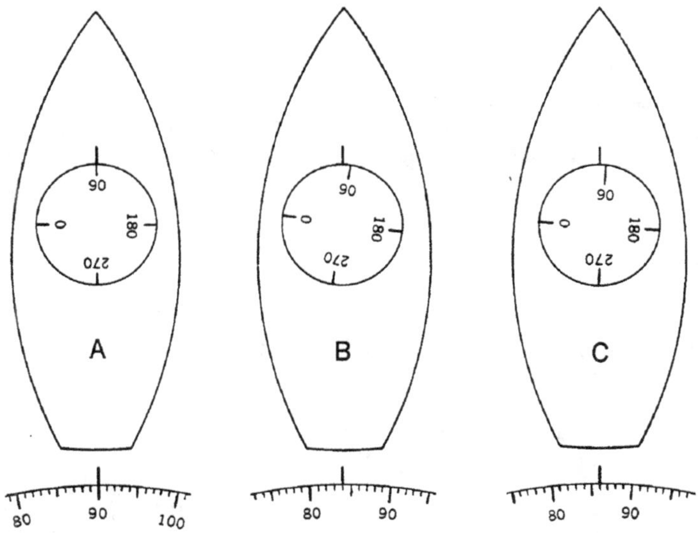

FIGURE 303.—Effect of variation and deviation on the compass card.

Example.—A vessel is on course 215° true in an area where the variation is 7°W. The deviation is as shown in figure 302. Degaussing is off. The gyro error (GE) is 1°E. A lighthouse bears 306°5 by magnetic compass.

Required.—(1) Magnetic heading (MH).

(2) Deviation.
(3) Compass heading (CH).
(4) Compass error.
(5) Gyro heading.
(6) Magnetic bearing of the lighthouse.
(7) True bearing of the lighthouse.
(8) Relative bearing of the lighthouse.

Solution.—

	TH	215°
	V	7° W
(1)	MH	222°
(2)	D	1°5 W
(3)	CH	223°5

The deviation is taken from the deviation table (fig. 302), to the nearest half degree.

(4) Compass error is 7°W+1°5W=8°5W.

TH 215°
GE 1° E
(5) H_{Dgc} 214°

CB 306°5
D 1°5 W
(6) MB 305°
V 7° W
(7) TB 298°

(8) RB=TB−TH=298°−215°=083°.

Answers.—(1) MH 222°, (2) D 1°5W, (3) CH 223°5, (4) CE 8°5W, (5) H_{Dgc} 214°, (6) MB 305°, (7) TB 298°, (8) RB 083°.

Problems

1303a. Fill in the blanks in the following:

	TC °	*V* °	*MC* °	*D* °	*CC* °	*CE* °
(1)	105	15 E	—	5 W	—	—
(2)	—	—	—	4 E	215	14 E
(3)	—	12 W	—	—	067	7 W
(4)	156	—	166	—	160	—
(5)	222	—	216	3 W	—	—
(6)	009	—	357	—	—	10 E
(7)	—	2 W	—	6 E	015	—
(8)	—	—	210	—	214	1 W

Answers.—(1) MC 090°, CC 095°, CE 10°E; (2) TC 229°, V 10°E, MC 219°; (3) TC 060°, MC 072°, D 5°E; (4) V 10°W, D 6°E, CE 4°W; (5) V 6°E, CC 219°, CE 3°E; (6) V 12°E, D 2°W, CC 359°; (7) TC 019°, MC 021°, CE 4°E; (8) TC 213°, V 3°E, D 4°W.

1303b. A vessel is on course 150° by compass in an area where the variation is 19°E. The deviation is as shown in figure 1302. Degaussing is on.

Required.—(1) Deviation.

(2) Compass error.

(3) Magnetic heading.

(4) True heading.

Answers.—(1) D 1°E, (2) CE 20°E, (3) MH 151°, (4) TH 170°.

1303c. A vessel is on course 055° by gyro and 041° by magnetic compass. The gyro error is 1°W. The variation is 15°E.

Required.—The deviation on this heading.

Answer.—D 2°W.

1303d. A vessel is on course 177° by gyro. The gyro error is 0°5E. A beacon bears 088° by magnetic compass in an area where variation is 11°W. The deviation is as shown in figure 1302, degaussing off.

Required.—The true bearing of the beacon.

Answer.—TB 076°.

CHAPTER IV

CONVERSION ANGLE

401. Introduction.—A bearing obtained by radio, like one determined in any other manner, provides means for establishing a line of position. By heading in the direction from which the signal is coming, one can proceed toward, or **home** on, the transmitter. In thick weather one should avoid heading directly toward the source of radiation unless he has reliable information to indicate that he is some distance away. In 1934 the Nantucket Lightship was rammed and sunk by a ship homing on its radiobeacon.

Radio waves, like light, travel along great circles. Except in high latitudes, visual bearings can usually be plotted as straight lines on a Mercator chart, without significant error. Radio bearings, however, are often observed at such positions with respect to the transmitter that the use of a rhumb line is not satisfactory. Under these conditions it is customary to apply the **conversion angle,** the difference between the rhumb line direction and the initial great-circle direction, as a correction to the observed angle, to find the equivalent rhumb line. Such a correction is not needed when a bearing is plotted on a gnomonic chart or one on which a straight line is a good approximation of a great circle. In other situations, a correction may be necessary.

If the transmitter and receiver are on the same meridian, or are both on the equator, no correction is needed because rhumb lines and great circles coincide under these conditions. The size of the correction increases with degree of departure from these conditions, and with greater distance between transmitter and receiver.

402. Convergence of the meridians.—Due to the meridians converging at the poles, the great circle between two points, except those points on the equator or same meridian, intersects all meridians at different angles. The difference between the initial and final great-circle directions is the convergence of the meridians. Because of this convergence the initial direction and the reciprocal of the final direction do not differ by 180°. As shown in figure 402, the initial direction of the great circle between A and B on the same side of the equator is angle A; the final direction at B is 180°−B. Therefore, in terms of the angles of the spherical triangle PnAB,

$$\text{convergence of the meridians, } c=180°-A-B$$

$$\frac{c}{2}=90°-\left(\frac{A+B}{2}\right)$$

$$\cot\frac{1}{2}\,c=\tan\frac{1}{2}\,(A+B).$$

Using Napier's Analogies (art. 143), the convergence of the meridians can be expressed in terms of difference of latitude (l) and difference of longitude (DLo).

$$\tan\frac{1}{2}(A+B)=\frac{\cos\frac{1}{2}(a-b)}{\cos\frac{1}{2}(a+b)}\cot\frac{1}{2}C=\frac{\cos\frac{1}{2}(PnB-PnA)}{\cos\frac{1}{2}(PnB+PnA)}\cot\frac{1}{2}APnB$$

$$\cot\frac{1}{2}c=\frac{\cos\frac{1}{2}l}{\sin Lm}\cot\frac{1}{2}DLo$$

$$\tan\frac{1}{2}c=\frac{\sin Lm}{\cos\frac{1}{2}l}\tan\frac{1}{2}DLo.$$

As shown in figure 402, the conversion angle is equal to one-half the convergence of the meridian. For small differences of latitude, cos l can be considered 1 without introducing a significant error. Therefore, for small differences of latitude, equation 41 can be simplified:

$$\tan \text{conversion angle}=\sin Lm \tan\frac{1}{2}DLo.$$

The tangent of a small angle equals, approximately, the angle in radians. Therefore, for small values of DLo and l (up to 15° to 20°), equation 42 can be simplified:

$$\text{conversion angle}=\frac{1}{2}DLo \sin Lm.$$

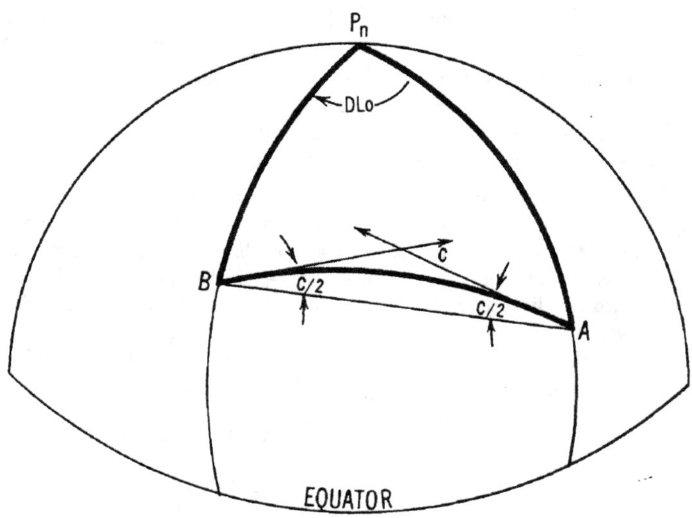

FIGURE 402.—Convergence of the meridians and conversion angle.

403. Conversion angles are given in table 1. This table is used to convert great circle to rhumb line directions or vice versa. If the difference of longitude is not more than 4°5, and the mid latitude between transmitter and receiver is not more than 85°, the first part of the table should be used. The simplifying assumptions used in the computation of this part of the table do not introduce a significant error within the limits of the table.

The sign of the correction can be determined by referring to the rules given at the bottom of each page of table 1. These follow from the fact that the great circle is nearer the pole than the rhumb line. The sign can be visualized by means of a simple sketch, as shown in figure 403.

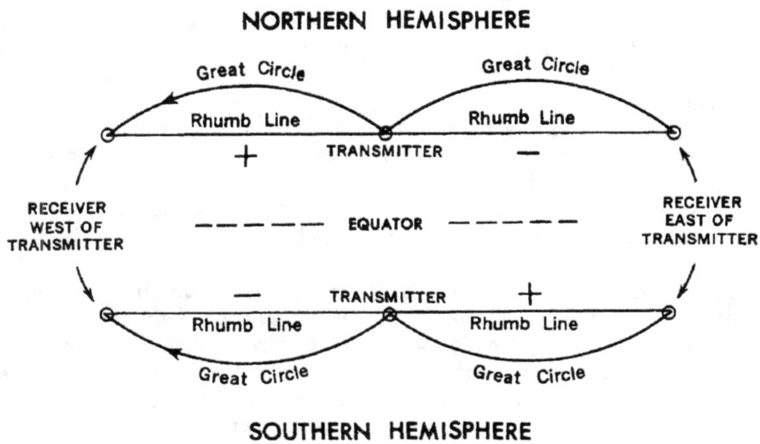

FIGURE 403.—Sign of conversion angle correction to radio bearings.

Example 1.—The DR position of a ship is lat. 42°15′.2N, long. 9°48′.6W. A radio bearing is taken on Cabo Montedor Light Station, at lat. 41°45′00″N, long. 8°52′20″W. The reading, corrected for calibration error, is 125°.5.

Required.—The equivalent rhumb line bearing.

Solution.—

	Latitude	*Longitude*
Receiver	42°15′.2N	9°48′.6W
Transmitter	41°45′.0N	8°52′.3W
Difference	30′.2	56′.3=0°.9
Mid latitude	42°.0	
Correction	(+)0°.3 (from table 1)	
Great-circle bearing	125°.5	
Rhumb line bearing	125°.8	

Answer.—B 125°.8.

Example 2.—The DR position of a ship is lat. 34°20′.6N, long. 124°18′.6W. A radio bearing is taken on Point Loma Light Station, at 32°40′.0N, long. 117°14′.6W.

Required.—The conversion angle correction.

Solution.—Enter the second part of table 1 with the nearest tabular values of latitude of the receiver (35°), DLo (5°), and latitude of the transmitter (30°) as arguments to extract the base correction, (+)1°.3. With two of the arguments held constant, find, successively, the change in the base correction for the difference between the third argument and its corresponding actual value. With latitude of transmitter and DLo held constant, the change in the base correction for the difference between the latitude of receiver argument and the actual latitude of the receiver is 0°.0. With latitude of receiver and latitude of transmitter arguments held constant, the change in the base correction for the difference between the DLo argument and the actual DLo is (+)0°.6. With latitude of receiver and DLo arguments held constant, the change in the base correction for the difference between the latitude of transmitter argument and the actual latitude of the transmitter is 0°.1. Therefore, the conversion angle correction is (+)2°.0 (1°.3+0°.0+0°.6+0°.1).

Answer.—Correction 2°.0.

Since radio bearings are generally somewhat less accurate than visual bearings, and often are observed at greater distances, positions obtained by them are generally considered of insufficient accuracy to be termed fixes, and so are usually considered estimated positions. However, judgment should govern the reliance to be placed upon such positional information.

Some navigators estimate or assume a probable error (usually of ±2° unless conditions suggest another value) and plot lines on each side of the bearing line to indicate the probable area within which the vessel is presumed to be located

Radio bearings furnished by a direction finder station have been corrected for known errors at the receiver, but not for conversion angle. The latter should be applied by the user.

In high latitudes it is not unusual to make use of bearings on objects a considerable distance from the vessel. Because of the rapid convergence of the meridians in these areas, such bearings are not correctly represented by straight lines on a Mercator chart. If this projection is used, the bearings should be corrected in the same manner that radio bearings are corrected (using table 1), since both can be considered great circles. Neither visual nor radio bearings are corrected when plotted on a Lambert conformal chart.

Problems

403a. The DR position of a ship is lat. 31°22'.8N, long. 121°06'.5W. A radio bearing is taken on Point Loma Light Station, at lat. 32°40'.0N, long. 117°14'.6W.

Required.—The conversion angle correction.

Answer.—Correction (+)1°.1.

403b. The DR position of a ship is lat. 44°08'.2S, long. 62°56'.9W. A radio bearing is taken on Isla Leones Light Station, at lat. 45°03'03"S, long. 65°36'33"W. The uncorrected reading is 039°.5 relative, the ship being on true heading 205° at the moment the bearing is observed. The calibration table indicates a correction of (−) 2° should be applied.

Required.—The equivalent true rhumb line bearing.

Answer.—B 243°.5.

CHAPTER V

DISTANCE CALCULATIONS

501. Introduction.—If a vessel is known to be a certain distance from an identified point on the chart, it must be somewhere on a circle with that point as the center and the distance as radius.

Distances are obtained by radar, range finder, stadimeter, vertical sextant angles (table 9), etc.

502. Distance, speed, and time are related by the formula

$$\text{distance} = \text{speed} \times \text{time}.$$

Therefore, if any two of the three quantities are known, the third can be found. The units, of course, must be consistent. Thus, if speed is measured in knots, and time in hours, the answer is in nautical miles. Similarly, if distance is measured in yards, and time in minutes, the answer is in yards per minute.

Table 19 is a speed, time, and distance table which supplies one of the three values if the other two are known. It is intended primarily for use in finding the distance steamed in a given time at a known speed. Table 18 is for use in determining speed by measuring the time needed to steam exactly one mile.

Speed is customarily expressed in knots (app. I), or for some purposes, in kilometers per hour, or yards or feet per minute. For short distances, a nautical mile can be considered equal to 2,000 yards or 6,000 feet. This is a useful relationship because $\frac{6{,}000 \text{ feet}}{60 \text{ minutes}} = 100$ feet per minute. Thus, speed in knots is equal approximately to hundreds of feet per minute or, hundreds of yards per 3-minute interval.

The **Logarithmic Time-Speed-Distance Nomogram** is frequently used for the solution of the formula distance=speed × time. The upper scale is graduated logarithmically in minutes of time; the middle scale is graduated logarithmically in miles (sometimes both miles and yards); and the lower scale is graduated logarithmically in knots. By marking the values of two known terms on their respective scales and connecting such marks by a straight line, the value of the third term is found at the intersection of this line with the remaining scale. Figure 502a illustrates a solution for speed when a distance of 4 miles is traveled in 11 minutes.

Only one of the three scales is required to solve for time, speed, or distance if any two of the three values are known. Either of the three logarithmic scales may be used in the same manner as a slide rule (art. 115) for the addition or subtraction of logarithms of numbers. Because the upper scale is larger, its use for this purpose is preferred.

When using a single logarithmic scale for the solution of the basic formula with speed units in knots and distance units in miles or thousands of yards, either 60 or 30 has to be incorporated in the basic equation for proper cancellation of units.

Figure 502a illustrates the use of the upper scale for finding the speed in knots when the time in minutes and the distance in miles are known. In this problem the time is 11 minutes and the distance is 4 miles. One point of a pair of dividers is set at the time in minutes, 11, and the second point at the distance in miles, 4. Without changing the spread of the dividers or the right-left relationship, set the first point at 60. The second point will then indicate the speed in knots, 21.8. If the speed and time are known, place one point at 60 and the second point at the speed in knots, 21.8. Without changing the spread of the dividers or the right-left relationship, place the first point at the time in minutes, 11. The second point then will indicate the distance in miles, 4.

In the method described, there was no real requirement to maintain the right-left relationship of the points of the pair of dividers except to insure that for speeds of less than 60 knots the distance in miles is less than the time in minutes. If the speed is in excess of 60 knots, the distance in miles will always be numerically greater than the time in minutes.

If the distance is known in thousands of yards or if the distance is to be found in such units, a divider point is set at 30 rather than the 60 used with miles. If the speed is less than 30 knots in this application, the distance in thousands of yards will always be numerically less than the time in minutes. If the speed is in excess of 30 knots, the distance in thousands of yards will always be numerically greater than the time in minutes.

For speeds of less than 60 knots and when using a logarithmic scale which increases from left to right, the distance graduation always lies to the left of the time in minutes graduation; the speed in knots graduation always lies to the left of the 60 graduation.

The use of the single logarithmic scale is based upon the fundamental property of logarithmic scales that equal lengths along the scale represent equal values of ratios. For example if one has the ratio ½ and

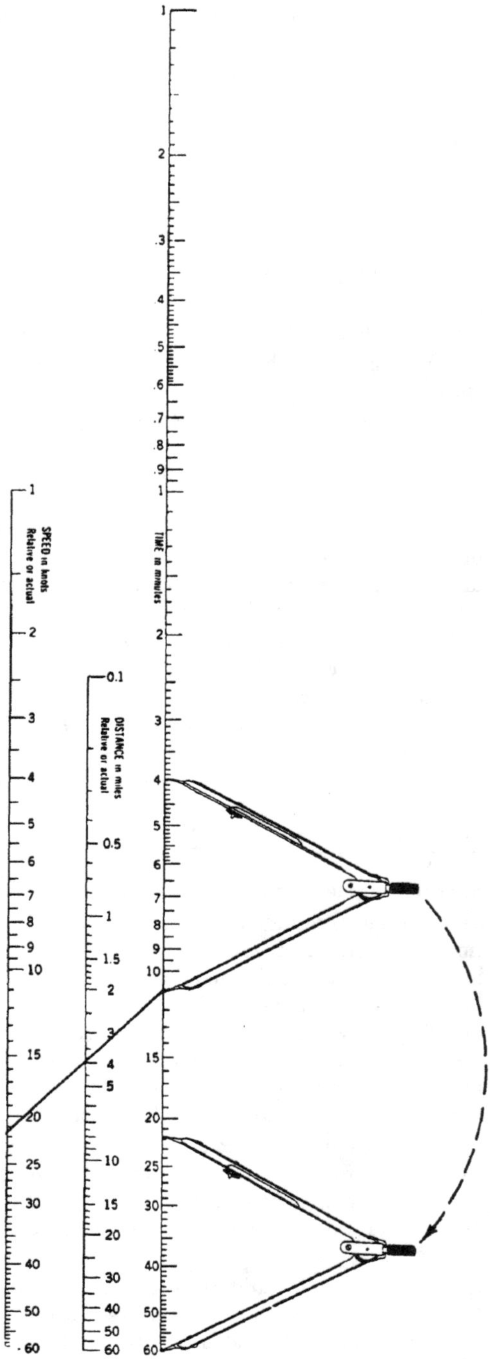

FIGURE 502a.—Logarithmic time-speed-distance nomogram.

with the dividers measures the length between 1 and 2, he finds the same length between 2 and 4, 5.5 and 11.0, or any other two values one of which is half the other. In using the single logarithmic scale for the solution of a specific problem in which a ship travels 10 nautical miles in 20 minutes, the basic formula is rearranged as follows:

$$\text{Speed}=\frac{\begin{array}{c}\text{Distance}\\ \text{(nautical miles)}\end{array}}{\text{Time (minutes)}}\times\frac{60\text{ min.}}{1\text{ hr.}}.$$

On substituting known numerical values and canceling units, the formula is rearranged further as:

$$\frac{\text{Speed (knots)}}{60}=\frac{10}{20}.$$

The ratio 10/20 has the same numerical value as the ratio Speed (knots)/60. Since each ratio has the same numerical value, the length as measured on the logarithmic scale between the distance in nautical miles (10) and the time in minutes (20) will be the same as the length between 60 and the speed in knots. Thus, on measuring the length between 10 and 20 and measuring the same length from 60 the speed is found to be 30 knots.

The solution of problems involving distance, speed, and time can easily be accomplished by means of a slide rule (art. 115). If the index of scale *C* is set opposite speed in knots on scale *D*, the distance in nautical miles appears on scale *D* opposite time in hours on scale *C*. If 60 of scale *C* is set opposite speed in knots on scale *D*, the distance covered in any number of *minutes* is shown on scale *D* opposite the minutes on scale *C*. Several circular slide rules particularly adapted for solution of distance, speed, and time problems have been devised. One of these, called the "Nautical Slide Rule" is shown in figure 502b.

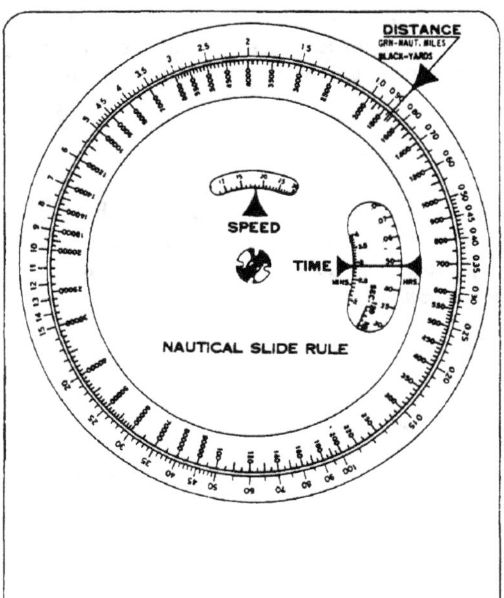

FIGURE 502b.—The nautical slide rule.

503. Distance by vertical angle.—Vertical sextant angles are used as vertical danger angles to avoid hazards or to determine distance-off an object. **Table 41** provides the distance-off an object of known height according to the vertical angle (corrected for index error only) measured between the waterline at the object and the top of the object. **Table 9** provides the distance-off an object of known height according to the vertical angle (corrected for index error and dip) measured between the visible (sea) horizon and the top of the object situated beyond the sea horizon. **Table 39** provides the distance-off an object which is situated within (short) of the sea horizon. The vertical angle (corrected for index error) is measured between the waterline at the object and the sea horizon.

Table 41 contains the solutions of a plane right triangle (fig. 503a) having its right angle at the base of the observed object and its altitude coincident with the vertical dimension of the observed object. Thus the solutions are based upon the following simplifying assumptions: (1) the eye of the observer is at sea level, (2) the sea surface between the observer and the object is flat, (3) atmospheric refraction is negligible, and (4) the waterline is vertically below the peak of the object.

That the computation of table 41 by the simple formula

$$D \tan hs = H$$

provides accurate values when height of eye and object height (H) are small compared to distance-off (D) can be demonstrated by comparisons of the formula with its modifications containing a correction term providing allowance for height of eye only; height of eye and earth's curvature; or height of eye, earth's curvature, and atmospheric refraction. When allowance is made for height of eye only, the correction term contributes only 3 percent of the distance-off for sextant angles (hs) less than 20° and heights of eye less than one-third of the object height. When allowance is made for earth's curvature and atmospheric refraction in addition to height of eye, the effects upon distance-off are negligible for cases of practical interest. The negligible effect of atmospheric refraction is due to the fact that rays of light from the top and base of the object are refracted by very nearly the same amounts. For practical distances within the visible horizon, the earth is essentially flat.

If the waterline is not vertically below the peak of the object (fig. 503b), an additional error due thereto is incurred, but this error does not exceed 3 percent of the distance-off for sextant angles less than 20° when the height of eye is less than one-third of the object height and the offset of waterline from the base of the object is less than one-tenth of the distance-off. Thus, if both height of eye and waterline offset errors occur simultaneously, errors not exceeding 6 percent off distance-off may be occurred if observations do not exceed the stated limits.

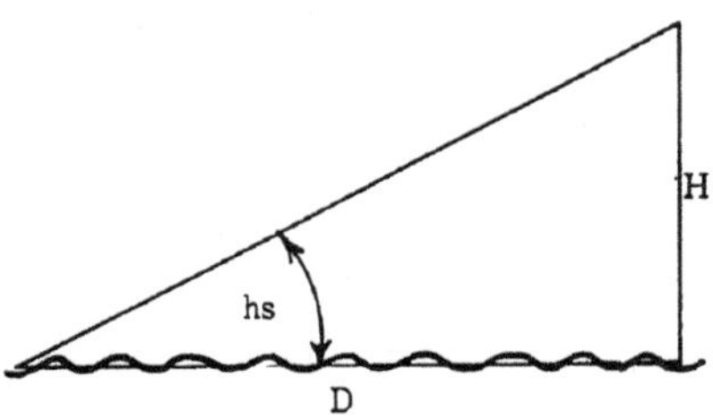

FIGURE 503a.—Vertical angle between top of object and waterline at object.

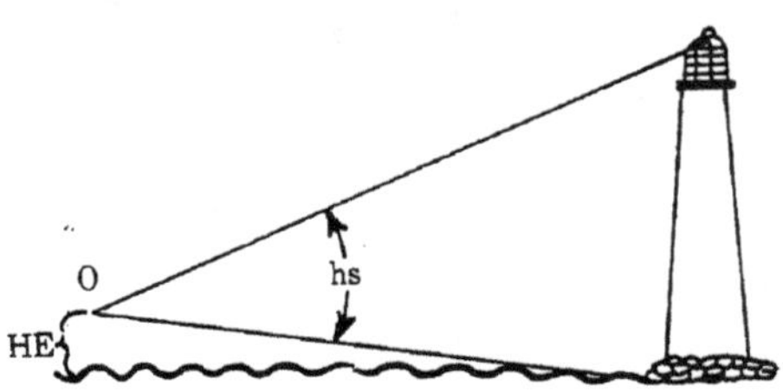

FIGURE 503b.—Vertical angle between top of object and waterline not vertically below the top of the object.

Example 1.—Using a sextant having an index error of (+) 1′.0 the navigator measures the vertical angle between the top of a lighthouse 100 feet above sea level and the waterline below from a height of eye of 30 feet; the sextant altitude is 0°13′.0.

Required.—The distance to the lighthouse.

Solution.—(1) Refer to the explanation of table 41 preceding the tables. Correct the sextant altitude for index error only.

(2) Enter table 41 with the corrected vertical angle (0°12′.0) and the height of the lighthouse (100 feet) and extract the distance to the lighthouse (4.7 nautical miles).

Note.—Although height of eye (30 feet) is not used in the solution, the fact that its value is less than one-third of the height of the lighthouse insures that not taking height of eye into account does not cause an error of more than 3 percent of distance-off for vertical angles of less than 20°.

Example 2.—Using a sextant with an index error of (+) 1′.0, the navigator measures the vertical angle between the visible horizon and the peak of a 520-foot hill beyond from a height of eye of 20 feet; the sextant altitude is 0°16′.3.

Required.—The distance to the point below the peak as indicated in figure 503c.

Solution.—(1) Refer to the explanation of table 9 preceding the tables. Correct the sextant altitude for index error and dip only.

(2) Enter table 9 with the corrected vertical angle (0°11′.0) and the difference in feet between the height of the object and the height of eye of the observer (500 feet) as arguments and extract the distance to the hill (16.0 nautical miles).

Example 3.—Using a sextant having an index error of (—) 1′.0, the navigator measures the vertical angle between the waterline at a buoy and the sea horizon beyond from a height of eye of 20 feet; the sextant altitude is 0°28′.0.

Required.—The distance to the buoy.

Solution.—(1) Refer to the explanation of table 39 preceding the tables. Correct the sextant altitude for index error only.

(2) Enter table 39 with the corrected vertical angle (0°29′.0) and the height of eye (20 feet) and extract the distance to the buoy (700 yards).

Note.—The method is not accurate beyond moderate distances (the table being limited to 5000 yards) and is obviously only available for finding the distance of an isolated object over which the horizon may be seen. In employing this method the higher the position occupied by the observer the more accurate will be the results.

In observing small angles, such as these that occur in the methods just described, it is sometimes convenient to measure them *on* and *off* the arc of the sextant. First look at the bottom of the object and reflect the top down into coincidence; then look through the transparent part of the horizon glass at the top and bring the bottom up by its reflected ray. The mean of the two readings will be the true angle, the index correction having been eliminated by the operation.

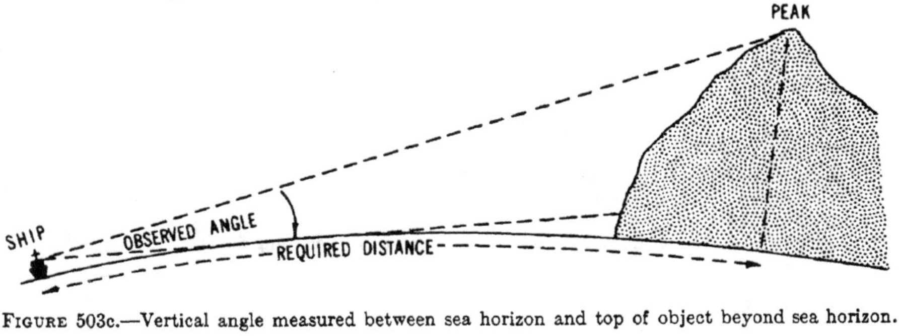

FIGURE 503c.—Vertical angle measured between sea horizon and top of object beyond sea horizon.

504. Distance of an object by two bearings.—A running fix can be obtained by utilizing the mathematical relationships involved. A ship steams past landmark *D* (fig. 504). At any point *A* a bearing of *D* is observed and expressed as degrees right or left of the course (*a relative bearing if the ship is on course*). At some later time, at *B*, a second bearing of *D* is observed and expressed as before. At *C* the landmark is broad on the beam. The angles at *A*, *B*, and *C* are known, and also the distance run between points. The various triangles could be solved by trigonometry to find the distance from *D* at any bearing. Distance and bearing provide a fix.

Table 7 provides a quick and easy solution. The table is entered with the difference between the course and first bearing (angle *BAD* in fig. 504) along the top of the table, and the difference between the course and second bearing (angle *CBD*) at the left of the table. For each pair of angles listed, two numbers are given. To find the distance from the landmark at the time of the second bearing (*BD*), multiply the distance run between bearings by the first number from table 7. To find the distance when the object is abeam (*CD*), multiply the distance run between *A* and *B* by the second number from the table. If the run between bearings is exactly 1 mile, the tabulated values are the distances sought.

Example.—A ship is steaming on course 050°, speed 15 knots. At 1130 a lighthouse bears 024°, and at 1140 it bears 359°.

Required.—(1) Distance from the light at 1140.

(2) Distance form the light when it is broad on the port beam.

Solution (fig. 504).—(1) The difference between the course and the first bearing (050°−024°) is 26°, and the difference between the course and the second bearing (050°+360°−359°) is 51°.

(2) From table 7 the two numbers (factors) are 1.04 and 0.81, found by interpolation.

(3) The distance run between bearings is 2.5 miles (10 minutes at 15 knots).

(4) The distance from the lighthouse at the time of the second bearing is 2.5×1.04=2.6 miles.

(5) The distance from the lighthouse when it is broad on the beam is 2.5×0.81=2.0 miles.

Answers.—(1) D 2.6 mi., (2) D 2.0 mi.

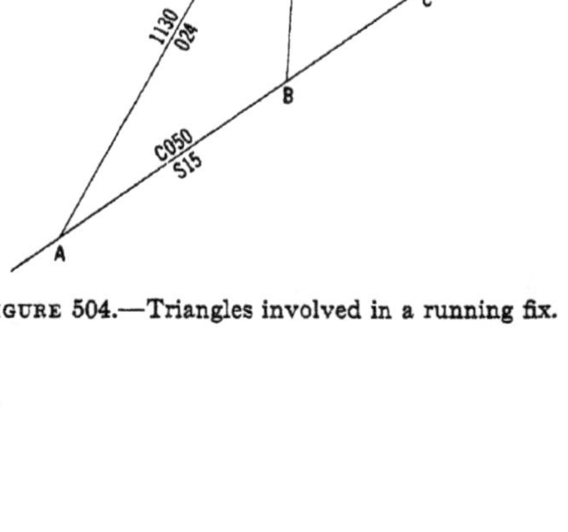

FIGURE 504.—Triangles involved in a running fix.

505. Special cases.—There are certain special cases arising under the method of obtaining a running fix from two bearings and the intervening run which do not require the use of tables. Two of these cases arise when the multiplier is equal to unity, and the distance run is therefore equal to the distance from the object.

If the second difference (angle *CBD* of figure 504) is double the first difference (angle *BAD*), triangle *BAD* is isosceles (art. 128) with equal angles at *A* and *D*. Therefore, side *AB* (the run) is equal to side *BD* (the distance-off at the time of the second bearing). This is called **doubling the angle on the bow.** If the first angle is 45° and the second 90°, the distance run equals the distance when broad on the beam. These are called **bow and beam bearings.**

When the first bearing is 26½° from ahead, and the second 45°, the distance *at which the object will be passed abeam* will equal the run between bearings. This is true of any two such bearings whose natural cotangents differ by unity, and the following table is a collection of solutions of this relation in which the pairs of bearings are such that, when observed in succession from ahead upon the same fixed object, the distance run between the bearings will be equal to the distance of the fixed object when it bears abeam, *provided that a steady course has been steered, unaffected by current.* See figure 505.

The marked pairs will probably be found the most convenient ones to use, as they involve whole degrees only. The use of the table may be found to be more convenient than the use of table 7, which covers all combinations of bearings in which the first bearing is taken when the object is 20° or more on the bow.

First	Second	First	Second	First	Second
°	°	°	°	°	°
20	29¾	28	48½	37	71¾
21	31¾	*29	51	38	74¼
*22	34	30	53¾	39	76¾
23	36¼	31	56¼	*40	79
24	38¾	*32	59	41	81¼
*25	41	33	61½	42	83½
26	43½	34	64¼	43	85¾
26½	45	35	66¾	*44	88
*27	46	36	69¼	*45	90

TABLE 505.—Pairs of bearings whose natural cotangents differ by unity.

If bearings of the fixed object be taken at two (2) and four (4) points on the bow (22½° and 45°), seven-tenths (0.7) of the run between bearings will be the distance at which the point will be passed abeam. This is known as the **seven-tenths rule.**

From the combination of the seven-tenths rule and the 26½°–45° rule, there follows an interesting corollary; i.e., if bearings of an object at 22½° and 26½° on the bow be taken, then seven-thirds (7/3) of the distance run in the interval will be the distance when abeam.

If a bearing is taken when an object is two (2) points (22½°) forward of the beam, and the run until it bears abeam is measured, then its distance when abeam is seven-thirds (7/3) of the run. This rule, particularly, is only approximate.

In case the 45° bearing on the bow is lost, in order to find the distance abeam that the object is passed, note the time when the object bears 26½° forward of the beam, and again when it has the same bearing abaft the beam; the distance run in this interval is the distance of the object when it was abeam.

The distance-off an object near the beam can be determined approximately by determining the time required for its bearing to change the same number of degrees as the numerical value of the speed. The time in minutes is numerically equal to distance-off in nautical miles (NM) when the speed is measured in knots, and distance-off in statute miles when speed is in miles per hour. The accuracy of the method suffers from the mathematical approximations used in the solution in addition to errors due to course and speed.

The bearing of the lighthouse near the beam is observed to change the same number of degrees as the speed in knots as a craft proceeds from *A* to *B*. The approximate distance travelled from *A* to *B*(s) is equal to the distance-off (r) times the bearing change expressed in radians (θ):

$$s=r\times\theta$$

If x is the speed in knots and y is the time in minutes for the bearing to change the same number of degrees as the numerical value of the speed in knots, x is also the numerical value of the bearing change in degrees. Using the approximation that 1 radian equals 60 degrees and substituting with appropriate cancellation of units in the above formula, the result is as follows:

$$x\frac{\text{NM}}{\cancel{\text{hr}}}\times\frac{y\ \cancel{\text{min.}}}{\cancel{60}\ \cancel{\text{min.}}}\cancel{\text{hr}}=r\times\frac{\cancel{x}}{\cancel{60}}$$

$$y\text{NM}=r$$

506. Radar horizon.—The distance to the radar horizon is the distance between the transmitter and the point at which the radar rays graze the surface of the earth. In the standard atmosphere, radar rays, like light rays, are bent or refracted slightly downwards, approximating the curvature of the earth. Where h is the height of the radar antenna in feet, the distance, D, to the radar horizon in nautical miles, assuming standard atmospheric conditions, may be found from the formula

$$D=1.22\sqrt{h}.$$

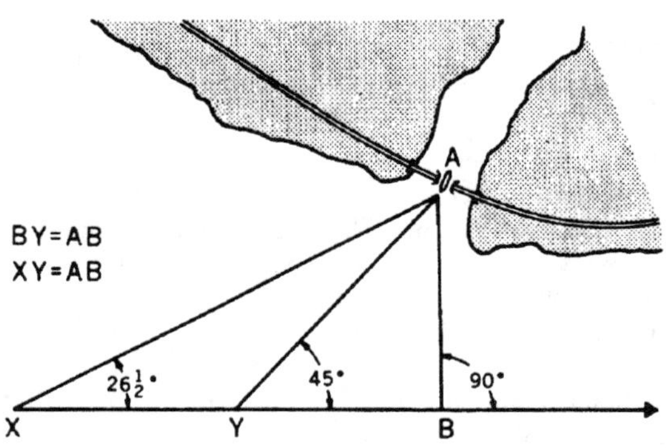

FIGURE 505.—Special case.

Although this formula is based upon a wavelength of 3 centimeters, it may be used in the computation of the distance to the radar horizon for other wavelengths normally used with navigational or surface search radar.

Example.—The height of the antenna of the surface search radar above the water is 81 feet.

Required.—Distance to the radar horizon.

Solution.—

$$D=1.22\sqrt{h}=1.22\sqrt{81}$$

$$D=11.0 \text{ nautical miles.}$$

Problems

504. The navigator of a vessel steaming on a steady course at 17 knots observes the following bearings from ahead as indicated:

	Time	*Object*	*Bearing from ahead*		*Time*	*Object*	*Bearing from ahead*
(1)	1237	*A*	42°	(3)	1325	*C*	25°
	1258	*A*	68°		1350	*C*	57°
(2)	1306	*B*	40°	(4)	1401	*D*	79°
	1321	*B*	59°		1452	*D*	103°

Required.—In each case, the distance-off at the time of the second bearing, and the distance when abeam, using table 7.

Answers.—

	Object	*Dist. at 2nd bearing*	*Dist. abeam*
(1)	*A*	9.2	8.5
(2)	*B*	8.3	7.1
(3)	*C*	5.7	4.8
(4)	*D*	34.8	34.0

505. A vessel is steaming on course 193° at 20 knots. The following true bearings are observed on the objects indicated:

	Time	*Object*	*True Bearing*		*Time*	*Object*	*True Bearing*
(1)	0800	*A*	229°	(4)	0912	*D*	215.°5
	0836	*A*	265°		0927	*D*	238°
(2)	0840	*B*	238°	(5)	0929	*E*	223°
	0855	*B*	283°		0954	*E*	253°
(3)	0855	*C*	265.°5	(6)	0959	*F*	233°
	0906	*C*	283°		1031	*F*	272°

Required.—Without plotting, and without the use of table 7, determine the distances off *A*, *B*, *D*, and *E* at the time of the second bearing, and the distances off *B*, *C*, *D*, *E*, and *F* when abeam.

Answers.—

	Object	*Dist. at 2nd bearing*	*Dist. abeam*
(1)	*A*	12.0	—
(2)	*B*	5.0	5.0
(3)	*C*	—	7.3
(4)	*D*	5.0	3.5
(5)	*E*	8.3	7.3
(6)	*F*	—	10.7

CHAPTER VI

TIME MEASUREMENTS AND CONVERSIONS

601. Introduction.—Time serves to regulate affairs aboard ship, as it does ashore. But to the navigator, it has additional significance. It is not enough to know *where* the ship is, was, or might be located in the future. The navigator wants to know *when* the various positions were or can reasonably be expected to be occupied. Time serves as a measure of progress. By considering the time at which a ship occupied various positions in the past, and by comparing the speed and various conditions it has encountered with those anticipated for the future, the skillful navigator can predict with reasonable accuracy the time of arrival at various future positions. Time can serve as a measure of safety, for it indicates when a light or other aid to navigation might be sighted, and if it is not seen by a certain time, the navigator knows he has cause for concern.

To the celestial navigator, time is of added significance, for it serves as a measure of the *phase* of the earth's rotation. That is, it indicates the position of the celestial bodies relative to meridians on the earth. Until an accurate *measure* of time became available at sea, longitude could not be found.

Very small *intervals* of time are used in certain electronic navigational aids, such as radar and Loran.

Whatever the type of navigation, a thorough mastery of the subject of time is important to the navigator. The use of a time diagram (app. J) may help in understanding the principles or solution of the problems of this chapter.

602. Kinds of time.—As a measure of part of a day, time can be stated in a number of different ways. At any given moment, the time depends upon (1) the point on the celestial sphere used as reference, (2) the reference meridian on the earth, and (3) the somewhat arbitrary starting point of the day.

When the sun is used as the celestial reference point, **solar time** results. If the actual sun observable in the sky is used, **apparent solar time** is involved, and if a fictitious **mean sun** is used to provide a time having an almost constant rate, **mean solar time** results. Time reckoned by use of the first point of Aries (♈) as the celestial reference point is called **sidereal time.** Use of the moon as the celestial reference point provides a variable-length **lunar day,** the basis of **lunar time,** which is useful in tide prediction and analysis. Because of its application, a lunar day is sometimes called a **tidal day.** It averages about 24^h50^m (mean solar units) in length.

If the meridian of the observer is used as the terrestrial reference, **local time** is involved. If a **zone** or **standard meridian** is used as the **time meridian** for mean solar time over an area, **zone** or **standard time** results. Use of a meridian farther east than would normally be used, so that the period of daylight is shifted later in the day, produces a form of zone time called **daylight saving** or **summer time.** Time based upon the Greenwich meridian is called **Greenwich time. Greenwich mean time** (**GMT**) is of particular interest to a navigator because it is the principal entering argument for the almanacs.

One complete revolution of the earth with respect to a celestial reference point is called a **day.** In modern usage every kind of solar time has its zero or starting point at **midnight,** when the celestial reference point is directly over the *lower branch* of the terrestrial reference meridian. This has not always been so. Until January 1, 1925, the **astronomical day** began at *noon*, 12 hours *later* than the start of the calendar day of the same date. The **nautical day** began at *noon*, 12 hours *earlier* than the calendar day, or 24 hours earlier than the astronomical day of the same date. The **sidereal day** begins at **sidereal noon,** when the first point of Aries is over the *upper* branch of the reference meridian. There is no sidereal date.

603. Expressing time.—Time is customarily expressed in time units, from 0^h through 24^h. To the nearest 1^m it is generally stated by navigators in a four-digit unit without punctuation. Thus, 0000 is midnight at the start of the day. One minute later the time is 0001. Half an hour after the start of the day the time is 0030, at one hour the time is 0100, at one hour and four minutes it is 0104, at 19 minutes after noon (solar time) it is 1219, at four hours and 23 minutes after (solar) noon it is 1623, etc. The term "hours" is sometimes used with the four-digit system to indicate that the number refers to the time or "hour" of the day. However, in those few occasions when any reasonable doubt may exist as to whether time is indicated, the fact can better be indicated in another way. Thus, the expression "1600 hours" to indicate "1600" or "16 hours" is not strictly correct, and is better avoided. **Watch time (WT),** indicated by a watch or clock having a 12-hour dial, and **chronometer time** (C) are expressed on a 12-hour basis, with designations AM (ante meridian) and PM (post meridian), as in ordinary civil life ashore.

In contrast, a time interval is expressed as hours and minutes, as 5^h26^m. When either the time of day or a time interval is given to seconds, this same form is used, as $21^h15^m18^s$. The kind of time may be indicated, usually by abbreviation.

When a time interval is to be added to or subtracted from a time, the solution can be arranged conveniently in tabular form.

Example 1.—What is the time and date $14^h36^m53^s$ after $21^h14^m18^s$ on July 24?

Solution.—

$$\begin{array}{rl} 21^h14^m18^s & \text{July 24} \\ 14^h36^m53^s & \\ \hline 35^h51^m11^s & \text{July 24} \\ =11^h51^m11^s & \text{July 25} \end{array}$$

The fact that the sum of hours exceeds 24 is an indication that the date increases by one. Similarly, in *subtracting* an interval, the date is one day earlier if 24^h must be added to the time before the subtraction can be made. That is, since 2400 of one day is 0000 of the following day, one might say that 2700 on one day is $2700-2400=0300$ on the following day. In the example above, $11^h51^m11^s$ on July 25 is the same as $11^h51^m11^s+24^h00^m00^s=35^h51^m11^s$ on July 24.

Date is sometimes expressed as an additional unit of the time sequence. Thus, $21^h14^m18^s$ on July 24 might be stated $24^d21^h14^m18^s$. This system is of particular value when an interval of several days is to be added or subtracted.

Example 2.—What is the time and date $9^d16^h35^m04^s$ before $5^h11^m33^s$ on September 15?

Solution.—

$$\begin{array}{rl} 15^d05^h11^m33^s & \\ 9^d16^h35^m04^s & \\ \hline 5^d12^h36^m29^s & \text{or } 12^h36^m29^s \text{ on Sept. 5.} \end{array}$$

By this method the month and day, if of significance, are recorded separately, or they, too, can be added to the sequence.

Example 3.—What is the time and date 3 years, 6 months, 25 days, 12 hours, 19 minutes, and 44 seconds after $7^h52^m24^s$ on November 14, 1958?

Solution.—

$$\begin{array}{r} 1958^{y}11^{m}14^{d}07^{h}52^{m}24^{s} \\ 3^{y}06^{m}25^{d}12^{h}19^{m}44^{s} \\ \hline 1962^{y}06^{m}08^{d}20^{h}12^{m}08^{s} = 20^{h}12^{m}08^{s} \text{ on June 8, 1962.} \end{array}$$

Since a month may contain a variable number of days, both the months and days should be solved together. Thus, in the example above, the answer would be 17 months, 39 days. If 12 months are converted to one year, this becomes five months, 39 days. Since the fifth month is May, this might be stated as May 39. Since there are 31 days in May, this is $39-31=8$ days into the next month, or June 8.

A simpler method of determining the number of elapsed *days* between any two dates is to use the **Julian day** of each date, if the information is available. This also eliminates possible error due to change of calendar if long intervals are involved. The Julian day is the consecutive number of the day starting at 1200 on January 1, 4713 BC. Julian day is listed in the *Astronomical Almanac.*

604. Time and arc.—The time of day is an indication of the interval since the day began. One day represents one complete rotation of 360° of the earth with respect to a selected celestial point. Each day is divided into 24 **hours** of 60 **minutes**, each minute having 60 **seconds.** Thus, each day has $24\times60=1{,}440$ minutes or $1{,}440\times60=86{,}400$ seconds. This is time regardless of the celestial reference point used, and since the various references are in motion with respect to each other, as "seen" from the earth, apparent solar, mean solar, and sidereal days are of different lengths. Since they all have the same number and kind of fractional parts, these parts are themselves of different length in the different kinds of time. Mean solar units are customarily used to indicate time intervals. The smallest unit normally used in celestial navigation is the second, but in some electronic equipment the **millisecond** (one-thousandth of a second), **microsecond** (one-millionth of a second), and the **millimicrosecond** or **nanosecond** (one-billionth of a second) are used.

Time of day is an indication of the *phase* of rotation of the earth. That is, it indicates how much of a day has elapsed, or what part of a rotation has been completed. Thus, at zero hours the day begins. One hour later, the earth has turned through 1/24 of a day, or 1/24 of 360°, or $\frac{360^\circ}{24}=15^\circ$. Six hours after the day begins, it has turned through $6/24=1/4$ day, or $\frac{360^\circ}{4}=90^\circ$. Twelve hours after the start of the day, the day is half gone, having turned through 180°. Smaller intervals can also be stated in angular units, for since one hour or 60 minutes is equivalent to 15°, one minute of time is equivalent to $\frac{15^\circ}{60}=0^\circ\!.25=15'$, and one second of time is equivalent to $\frac{15'}{60}=0'\!.25=15''$. Thus,

$$\begin{array}{rl} \textit{Time} & \textit{Arc} \\ 1^{d}=24^{h}= & 360^\circ=1 \text{ circle} \\ 60^{m}=1^{h}= & 15^\circ \\ 4^{m}= & 1^\circ=60' \\ 60^{s}=1^{m}= & 15' \\ 4^{s}= & 1'=60'' \\ 1^{s}= & 15''=0'\!.25 \end{array}$$

Any time interval can be expressed as an angle of rotation, and vice versa. Interconversion of these units can be made by the relationships indicated above.

To convert time to arc:

1. Multiply the hours by 15 to obtain degrees.
2. Divide the minutes of time by four to obtain degrees, and multiply the remainder by 15 to obtain minutes of arc.
3. Divide the seconds of time by four to obtain minutes and tenths of minutes of arc, or multiply the remainder by 15 to obtain seconds of arc.
4. Add degrees, minutes, and tenths (or seconds).

Example 1.—Convert $14^h21^m39^s$ to arc units.

Solution.—

(1) $14^h \times 15 = 210°$
(2) $21^m \div 4 = 5°15'$ (remainder $1^m \times 15 = 15'$)
(3) $39^s \div 4 = 9'45''$ (remainder $3^s \times 15 = 45''$)
(4) $14^h21^m39^s = 215°24'45'' = 215°24'.8$ (to nearest $0'.1$).

To convert arc to time:

1. Divide the degrees by 15 to obtain hours, and multiply the remainder by four to obtain minutes of time.
2. Divide the minutes of arc by 15 to obtain minutes of time, and multiply the remainder by four to obtain seconds of time.
3. Divide the seconds of arc by 15 to obtain seconds of time.
4. Add hours, minutes, and seconds.

Example 2.—Convert $215°24'45''$ to time units.

Solution.—

(1) $215° \div 15 = 14^h20^m$ (remainder $5° \times 4 = 20^m$)
(2) $24' \div 15 = 1^m36^s$ (remainder $9' \times 4 = 36^s$)
(3) $45'' \div 15 = 3^s$
(4) $215°24'45'' = 14^h21^m39^s$

Example 3.—Convert $161°53'.7$ to time units.

Solution.—

(1) $161° \div 15 = 10^h44^m$ (remainder $11° \times 4 = 44^m$)
(2) $53'.7 \div 15 = 3^m34^s.8$ (remainder $8'.7 \times 4 = 34^s.8$)
(3) $161°53'.7 = 10^h47^m34^s.8 = 10^h47^m35^s$

The navigator should be able to make these solutions mentally, writing only the answer. As a check, the answer can be converted back to the original value. Solution can also be made by means of arc to time tables in the almanacs. In the *Nautical Almanac* the table, given near the back of the volume (app. F), is in two parts, permitting separate entries with degrees, minutes, and quarter minutes of arc. The table is arranged in this manner because the navigator is confronted with the problem of converting arc to time more often than the reverse.

Example 4.—Convert $334°18'22''$ to time units, using the *Nautical Almanac* arc to time conversion table.

Solution.—

$334° = 22^h16^m$
$18'.25 = 1^m13^s$
$334°18'22'' = 22^h17^m13^s$

The 22″ are converted to the nearest quarter minute of arc for solution to the nearest second of time. Interpolation can be used if more precise results are required, since exact relationships are tabulated in the *Nautical Almanac* conversion table.

Example 5.—Convert 83°29′.6 to time units, using the *Nautical Almanac* arc to time conversion table.

Solution.—

$$\begin{aligned} 83° &= 5^h32^m \\ 29'.6 &= \quad 1^m58^s.4 \\ \hline 83°29'.6 &= 5^h33^m58^s.4 \end{aligned}$$

In this solution, 58ˢ.4 was obtained by eye interpolation in the quarter-minute part of the table.

Example 6.—Convert $17^h09^m42^s$ to arc units, using the *Nautical Almanac* arc to time conversion table.

Solution.—

$$\begin{aligned} 17^h08^m &= 257° \\ 1^m42^s &= \quad 25'.5 \\ \hline 17^h09^m42^s &= 257°25'.5 \end{aligned}$$

A similar table appears near the back of the *Air Almanac* (app. G); however, quarter minutes of arc are not included.

Example 7.—Convert 334°47′.2 to time units, using the *Air Almanac* arc to time conversion table.

Solution.—

$$\begin{aligned} 334° &= 22^h16^m \\ 47'.2 &= \quad 3^m09^s \\ \hline 334°47'.2 &= 22^h19^m09^s \end{aligned}$$

Example 8.—Convert $15^h13^m18^s$ to arc units, using the *Air Almanac* arc to time conversion table.

Solution.—

$$\begin{aligned} 15^h12^m &= 228° \\ 1^m18^s &= \quad 19'.5 \\ \hline 15^h13^m18^s &= 228°19'.5 \end{aligned}$$

Because the almanac conversion tables are exact relationships, interpolation in them can be carried to any degree of precision desired without introducing an error.

605. Time and longitude.—As indicated in the preceding article, time is a measure of rotation of the earth, and any given time interval can be represented by a corresponding angle through which the earth turns. Suppose the celestial reference point were directly over a certain reference of the earth. An hour later the earth would have turned through 15°, and the celestial reference would be directly over a meridian 15° farther west. Any difference of longitude is a measure of the angle through which the earth must rotate for the local time at the western meridian to become what it was at the eastern meridian before the rotation took place. Therefore, places to the eastward of an observer have *later* time, and those to the westward have *earlier* time, and the difference is exactly equal to the difference in longitude, expressed in time units. When a meridian other than the local meridian is used as the time reference, the difference in time of two places is equal to the difference of longitude of their time reference meridians.

606. The date line.—Since time becomes later toward the east, and earlier toward the west, time at the lower branch of one's meridian is 12 hours earlier or later depending upon the direction of reckoning. A traveler making a trip around the world gains or loses an entire day. To prevent the *date* from being in error, and to provide a starting place for each day, a **date line** is fixed by international agreement. This line coincides with the 180th meridian over most of its length. In crossing this line, one alters his date by one day. In effect, this changes his time 24 hours to compensate for the slow change during a trip around the world. Therefore, it is applied in the opposite direction to the change of time. Thus, if a person is traveling *eastward* from east longitude to west longitude, time is becoming *later*, and when the date line is crossed, the date becomes one day *earlier*. That is, at any moment the date immediately to the *west* of the date line (east longitude) is one day *later* than the date immediately to the *east* of the line, except at GMT 1200, when the (mean time) date is the same all over the world. At any other time two dates occur, one boundary between dates being the date line, and the other being the midnight line along the lower branch of the meridian over which the mean sun is located. At GMT 1200 these two boundaries coincide. In the solution of problems, error can sometimes be avoided by converting local time to Greenwich time, and then converting this to local time on the opposite side of the date line. Examples are given in following articles.

607. Zone time.—At sea, as well as ashore, watches and clocks are normally set approximately to some form of **zone time (ZT)**. At sea the *nearest* meridian exactly divisible by 15° is usually used as the **time meridian** or **zone meridian**. Thus, within a **time zone** extending 7°.5 on each side of each time meridian the time is the same, and time in consecutive zones differs by exactly one hour. The time is changed as convenient, usually at a whole hour, near the time of crossing the boundary between zones. Each time zone is identified by the number of times the longitude of its zone meridian is divisible by 15°, positive in west longitude and negative in east longitude. This number and its sign, called the **zone description (ZD)**, is the number of whole *hours* that are added to or subtracted from the zone time to obtain **Greenwich mean time (GMT)**, which is the zone time at the Greenwich (0°) meridian, and is sometimes called **Universal Time (UT)**. The mean sun is the celestial reference point for zone time.

Example 1.—For an observer at long. 141°18′.4W the ZT is $6^h18^m24^s$.

Required.—(1) Zone description.

(2) GMT.

Solution.—(1) The nearest meridian exactly divisible by 15° is 135°W, into which 15° will go nine times. Since longitude is west, ZD is (+)9.

ZT		$6^h18^m24^s$
ZD	(+)	9
(2) GMT		$15^h18^m24^s$

In converting GMT to ZT, a positive ZD is *subtracted*, and a negative one *added*, but its sign remains the same, being part of the description. The word "reversed" (rev.) is written to the right in the work form to indicate that the "reverse" process is to be performed.

Example 2.—The GMT is $15^h27^m09^s$.

Required.—(1) ZT at long. 156°24′.4W.

(2) ZT at 39°04′.8E.

Solution.—

(1) GMT		$15^h27^m09^s$			(2) GMT		$15^h27^m09^s$	
ZD	(+)	10	(rev.)		ZD	(−)	3	(rev.)
ZT		$5^h27^m09^s$			ZT		$18^h27^m09^s$	

When time at one place is converted to that at another, the date should be watched carefully. If a sum exceeds 24 hours, subtract this amount and add one day. If 24 hours are added before a subtraction is made, the date at the place is one day *earlier*.

Example 3.—At long. 73°29′.2W the ZT is $21^h12^m53^s$ on May 14.

Required.—(1) GMT and date.

(2) ZT and date at long. 107°15′.7W.

Solution.—

	ZT		$21^h12^m53^s$ May 14
	ZD	(+)	5
(1)	GMT		$2^h12^m53^s$ May 15
	ZD	(+)	7 (rev.)
(2)	ZT		$19^h12^m53^s$ May 14

The second part of this problem might have been solved by using the *difference* in zone description. Since the second place is two zones farther west, its time is two hours earlier. Problems involving zone times at various places generally involve nothing more than addition or subtraction of one small number, so solutions can generally be made mentally. However, when this forms part of a larger problem, or when a record of the solution is desired, the full solution should be recorded, including labels.

Example 4.—On November 30 the 1430 DR long. of a ship is 51°32.′4W. Ten hours later the DR long. is 53°07.′2W.

Required.—ZT and date of arrival at the second longitude.

Solution.—

ZT		1430	Nov. 30
ZD	(+)	3	
GMT		1730	Nov. 30
int.		10	
GMT		0330	Dec. 1
ZD	(+)	4	(rev.)
ZT		2330	Nov. 30

If a time zone boundary had not been crossed, there would have been no need to find GMT. It is particularly helpful to retain this step when the date line is crossed. This line is the *center* of a time zone, the western (east longitude) half being designated (−) 12, and the eastern (west longitude) half (+) 12.

Example 5.—On December 31 the 0800 DR long. of a ship is 177°23.′9E. Forty hours later the DR long. is 171°53.′9W.

Required.—ZT and date of arrival at the second longitude.

Solution.—

ZT		0800	Dec. 31
ZD	(−)	12	
GMT		2000	Dec. 30
int.		40	
GMT		1200	Jan. 1
ZD	(+)	11	(rev.)
ZT		0100	Jan. 1

Alternative solution

ZT		$31^d08^h00^m$	
ZD	(−)	12	
GMT		$30^d20^h00^m$	
int.		1^d16^h	
GMT		$1^d12^h00^m$	
ZD	(+)	11	(rev.)
ZT		$1^d01^h00^m$	

For certain communication purposes it is sometimes convenient to designate a time zone by a single letter. The system used is shown in table 36.

Use of time zones on land began in 1883, when railroads adopted four standard zones for the continental United States. The division of the United States into time zones was not officially adopted by Congress, however, until March 19, 1918, when a fifth zone was also established for Alaska. The system of time zones is now used almost universally throughout the world, although on land the zone boundaries are generally altered somewhat for convenience. In a few places, half-hour zones are used.

On land, normal zone time is usually called **standard time,** often with an adjective to indicate the zone, as **eastern standard time.** In some areas timepieces are *advanced* one or more hours during the summer to provide greater use of daylight. This "fast" time is called **daylight saving time** in the United States, and **summer time** elsewhere. When time is one hour fast, the zone description is (algebraically) one *less* than normal. When daylight saving or summer time is specified, an advance of one hour is understood unless a greater number is indicated.

Example 6.—What is the standard time and date at Tokyo, long. 140° E, when the daylight saving time at Washington, long. 77°W, is 1600 on Oct. 5?

Solution.—

ZT		1600 Oct. 5
ZD	(+)	4
GMT		2000 Oct. 5
ZD	(−)	9 (rev.)
ZT		0500 Oct. 6

During hostilities daylight saving time may be kept all year long throughout a nation, and designated **war time.**

608. Chronometer time (C) is time indicated by a chronometer. Since a chronometer is set approximately to GMT, and not reset until it is overhauled and cleaned, perhaps three or more years later, there is nearly always a **chronometer error (CE),** either *fast* (F) or *slow* (S). The change in chronometer error in 24 hours is called **chronometer rate,** or **daily rate,** and designated *gaining* or *losing*. With a consistent rate of 1^s per day for three years, the chronometer error would be approximately 18^m. Since chronometer error is subject to change, it should be determined from time to time, preferably daily at sea. Chronometer error is found by radio time signal, by comparison with another timepiece of known error, or by applying chronometer rate to previous readings of the same instrument. It is recorded to the nearest whole or half second. Chronometer rate is recorded to the nearest $0^s.1$.

Example 1.—At GMT 1200 on May 12 the chronometer reads $12^h04^m21^s$. At GMT 1600 on May 18 it reads $4^h04^m25^s$.

Required.—(1) Chronometer error at both comparisons.

(2) Chronometer rate.

(3) Chronometer error at GMT 0530 on May 27.

Solution.—

GMT	1200^m00^s May 12
C	$12^h04^m21^s$
(1) CE	(F) 4^m21^s
GMT	$16^h00^m00^s$ May 18
C	$4^h04^m25^s$
(1) CE	(F) 4^m25^s
GMT	12^d12^h
GMT	18^d16^h
diff.	$6^d04^h = 6^d.2$
CE	(F) 4^m21^s 1200 May 12
CE	(F) 4^m25^s 1600 May 18
diff.	4^s gained
(2) daily rate	$0^s.6$ per day, gaining. ($4^s \div 6^d.2$)
GMT	$18^d16^h00^m$
GMT	$27^d05^h30^m$
diff.	$8^d13^h30^m = 8^d.5$

CE (F) 4^m25^s 1600 May 18
corr. $(+)5^s$ ($8^d.5 \times 0^s.6$ per day)
(3) CE (F) 4^m30^s 0530 May 27

Because GMT is stated on a 24-hour basis, and chronometer time on a 12-hour basis, a 12-hour ambiguity exists. This is ignored in finding chronometer error. However, if chronometer error is applied to chronometer time to find GMT, a possible 12-hour error can result. This can be resolved by mentally applying zone description to local time to obtain approximate GMT. A time diagram can be used for resolving doubt as to approximate GMT and Greenwich date. If the sun for the kind of time used (mean or apparent) is between the *lower* branches of two time meridians (as the standard meridian for local time, and the Greenwich meridian for GMT), the date at the place farther *east* is one day *later* than at the place farther *west*.

Example 2.—On August 14 the DR long. of a ship is about 124°E, and the zone time is about 0500. Chronometer error is 12^m27^s slow.

Required.—GMT and date when the chronometer reads $8^h44^m22^s$.

Solution.—

approx.	ZT	0500 Aug. 14
	ZD	(−) 8
approx.	GMT	2100 Aug. 13

C	$8^h44^m22^s$
CE	(S)12^m27^s
GMT	$20^h56^m49^s$ Aug. 13

The *A* chronometer, usually the best (having the most nearly uniform rate), is compared directly with the time signal. Other chronometers, designated *B*, *C*, etc., may then be compared with the *A* chronometer.

Example 3.—At GMT 1400 chronometer *A* is checked by time signal, and found to read $1^h57^m09^s$. A little later, when it reads $2^h05^m00^s$, chronometer *B* reads $2^h11^m38^s$.

Required.—(1) Error of chronometer *A*.

(2) Error of chronometer *B*.

Solution.—

GMT	$14^h00^m00^s$
C_A	$1^h57^m09^s$
(1) CE_A	(S) 2^m51^s
C_A	$2^h05^m00^s$
GMT	$14^h07^m51^s$
C_B	$2^h11^m38^s$
(2) CE_B	(F) 3^m47^s

If time signals are not available at the chronometer, a good comparing watch should be compared with the radio signal, and this watch used to determine chronometer error, as indicated in example 3, substituting the watch for chronometer *A*.

609. Watch time (WT) is time indicated by a watch. This is usually an approximation of zone time, except that for timing celestial observations it is good practice to set a hack or comparing watch to GMT. If the watch has a second setting hand, the watch can be set exactly to ZT or GMT, and the time is so designated. If the watch is not set exactly to one of these times, the difference is known as **watch error (WE)**, labeled *fast* (F) or *slow* (S) to indicate whether the watch is ahead of or behind the correct time, respectively.

If a watch is to be set exactly to ZT or GMT, it is set to some whole minute slightly ahead of the correct time, and stopped. When the set time arrives, the watch is started. It should then be checked for accuracy.

Example 1.—A chronometer 9^m46^s fast on GMT reads approximately 7^h23^m. At the next whole five minutes of GMT a comparing watch is to be set to GMT exactly.

Required.—(1) What should the watch read at the moment of starting?

(2) What should the chronometer read?

Solution.—

$$\begin{array}{rrl}
 & \text{C} & 7^h23^m00^s \\
 & \text{CE} & \underline{(\text{F})\ \ 9^m46^s} \\
 & \text{GMT} & \underline{\ \ 7^h13^m14^s} \\
(1) & \text{GMT} & 7^h15^m00^s \text{ (next whole } 5^m) \\
 & \text{CE} & \underline{(\text{F})\ \ 9^m46^s} \\
(2) & \text{C} & 7^h24^m46^s
\end{array}$$

The GMT may be in error by 12^h, but if the watch is graduated to 12 hours, this will not be reflected. If a watch with a 24-hour dial is used, the actual GMT should be determined.

If watch error is to be determined, it is done by comparing the reading of the watch with that of the chronometer at a selected moment. This may be at some selected GMT, as in example 1.

Example 2.—If, in example 1, the watch had read $7^h14^m48^s$ at the moment the chronometer read $7^h24^m46^s$, what would be the watch error on GMT?

Solution.—

$$\begin{array}{rl}
\text{GMT} & 7^h15^m00^s \\
\text{WT} & \underline{7^h14^m48^s} \\
\text{WE} & (\text{S})\ 12^s
\end{array}$$

A more convenient chronometer time might be selected, as a whole minute.

Example 3.—A watch is set to zone time approximately. The longitude is about 48°W. The watch is compared with a chronometer which is 19^m44^s fast on GMT. When the chronometer reads $5^h22^m00^s$, the watch reads $2^h01^m53^s$.

Required.—Watch error on zone time.

Solution.—

$$\begin{array}{rll}
\text{C} & & 5^h22^m00^s \\
\text{CE} & & \underline{(\text{F})\ 19^m44^s} \\
\text{GMT} & & 5^h02^m16^s \\
\text{ZD} & (+) & \underline{3\qquad\quad} \quad (\text{rev.}) \\
\text{ZT} & & 2^h02^m16^s \\
\text{WT} & & \underline{2^h01^m53^s} \\
\text{WE} & & (\text{S})\ 23^s
\end{array}$$

The possible 12^h error is not of significance. When such a watch is used for determining GMT, however, as for entering an almanac, the 12-hour ambiguity is important. Unless a watch is graduated to 24 hours, its time is designated AM before noon and PM after noon.

Example 4.—On January 3 the DR long. is 94°14′.7E. An observation of the sun is made when the watch reads $12^h16^m23^s$ PM. The watch is 22^s fast on zone time.

Required.—GMT and date.

Solution.—

WT	$12^h16^m23^s$ PM Jan. 3
WE	(F) 22^s
ZT	$12^h16^m01^s$
ZD (−)	6
GMT	$6^h16^m01^s$ Jan. 3

Note that between 1200 and 1300 watch designations are PM. Between 0000 and 0100 they are AM.

Comparison of a watch and chronometer should be made carefully. If two observers are available, one can give a warning "stand-by" a few seconds before the selected time, and a "mark" at the appointed moment, while the other notes the time of the watch. A single observer can make a satisfactory comparison by counting with the chronometer. Chronometers beat in half seconds, with an audible "tick." Ten seconds before the selected time (perhaps a whole minute), the observer starts counting with the beats, as he watches the chronometer second hand, "50, and, 1, and, 2, and, 3, and,. . . . . 9, and, mark." During the count the observer shifts his view from the chronometer to the second hand of the watch, continuing to count in cadence with the chronometer beats. At the "mark," the second, minute, and hour hands of the watch are read in that order, and the time recorded. A comparison of this time with the GMT or ZT corresponding to the selected chronometer time indicates the watch error.

Even though a watch is set to zone time approximately, its error on GMT can be determined and used for timing observations. In this case the 12-hour ambiguity in GMT should be resolved, and a time diagram used to avoid possible error. This method requires additional work, and presents a greater probability of error, without compensating advantages.

Still another method of determining GMT, generally used before zone time came into common use at sea, is to *subtract* watch time from chronometer time, to find C–WT. This is then *added* to the watch time of an observation to obtain chronometer time (C−WT+WT=C). Chronometer error is then applied to the result to obtain GMT. A time diagram should *always* be used with this method, to resolve the 12-hour ambiguity and to be sure of the correct Greenwich date, unless an auxiliary solution is made using approximate ZT and ZD. This method has little to recommend it.

If a watch has a **watch rate** of more than a few seconds per day, watch error should be determined both before and after a round of sights, and any difference distributed proportionally among observations.

If a stop watch is used for timing observations, it should be started at some convenient GMT, as a whole 5^m or 10^m. The time of each observation is then this GMT *plus* the reading of the watch.

610. Local mean time (**LMT**), like zone time, uses the mean sun as the celestial reference point. It differs from zone time in that the local meridian is used as the terrestrial reference, rather than a zone meridian. Thus, the local mean time at each meridian differs from that of every other meridian, the difference being equal to the difference of longitude, expressed in time units. At each zone meridian, including 0°, LMT and ZT are identical.

Example 1.—At long. 124°37′.2W the LMT is $17^h24^m18^s$ on March 21.

Required.—(1) GMT and date.

(2) ZT and date at the place.

Solution.—

LMT	$17^h24^m18^s$	Mar. 21
λ	$8^h18^m29^s$W	
(1) GMT	$1^h42^m47^s$	Mar. 22
ZD (+)	8	(rev.)
(2) ZT	$17^h42^m47^s$	Mar. 21

In navigation the principal use of LMT is in rising, setting, and twilight tables. The problem is usually one of converting the LMT taken from the table to ZT. At sea, the difference between these times is normally not more than 30^m, and the conversion is made directly, without finding GMT as an intermediate step. This is done by applying a correction equal to the difference of longitude (dλ). If the observer is *west* of his time meridian, the correction is *added*, and if *east* of it, the correction is *subtracted*. If Greenwich time is desired, it is found from ZT.

Example 2.—At long. 63°24′.4E the LMT is 0525 on January 2.

Required.—(1) ZT and date.

(2) GMT and date.

Solution.—

LMT	0525	Jan. 2
dλ (−)	14	
(1) ZT	0511	Jan. 2
ZD (−)	4	
(2) GMT	0111	Jan. 2

On land, with an irregular zone boundary, the longitude may differ by more than 7°.5 (30^m) from the time meridian.

If LMT is to be corrected to daylight saving time, the difference in longitude between the local and time meridian can be used, or the ZT can first be found and then increased by one hour.

Conversion of ZT (including GMT) to LMT is the same as conversion in the opposite direction, except that the sign of dλ is reversed. This problem is not normally encountered in navigation.

611. Apparent time utilizes the **apparent** (real) **sun** as its celestial reference, and a meridian as the terrestrial reference. **Local apparent time** (**LAT**) uses the local meridian. The LAT at the 0° meridian is called **Greenwich apparent time** (**GAT**).

The LAT at one meridian differs from that at any other by the difference in longitude of the two places, the place to the eastward having the later time, and conversion is the same as converting LMT at one place to LMT at another.

Use of the apparent sun as a celestial reference point for time results in time of nonconstant rate for at least three reasons. First, revolution of the earth in its orbit is not constant. Second, motion of the apparent sun is along the ecliptic, which is tilted with respect to the celestial equator, along which time is measured. Third, rotation of the earth on its axis is not constant. The effect due by this third cause is extremely small.

For the various forms of mean time, the apparent sun is replaced by a fictitious **mean sun** conceived as moving eastward along the *celestial equator*, at a uniform speed equal to the average speed of the apparent sun along the ecliptic, thus providing a nearly uniform measure of time equal to the approximate average apparent time. At any moment the accumulated difference between LAT and LMT is indicated by the **equation of time** (**Eq. T**), which reaches a maximum value of about $16^m.4$ in November. This quantity is tabulated at 12-hour intervals at the bottom of the right-hand daily page of the *Nautical Almanac*. In the United States, the sign is considered positive (+) if the time of sun's "Mer. Pass." is earlier than 1200, and negative (−) if later than 1200. If the "Mer. Pass." is given as 1200 (as on June 12–14, 1975), the sign is positive if the GHA at GMT 1200 is between 0° and 1°, and negative if it is greater than 359°. The sign is correct for conversion of GMT to GAT. In Great Britain, this convention is reversed. A heavy line is used to indicate a change of sign between consecutive entries, as shown between 00^h and 12^h on June 14, when the sign changes from positive to negative.

Example.—Find the LAT and date at ZT $15^h10^m40^s$ on May 31, 1975, for long. 73°18.′4W.

Solution.—

ZT		$15^h10^m40^s$ May 31
ZD	(+)	5
GMT		$20^h10^m40^s$ May 31
Eq. T		$(+)2^m25^s$
GAT		$20^h13^m05^s$ May 31
λ		$4^h53^m14^s$W
LAT		$15^h19^m51^s$ May 31

In conversion from apparent to mean time, a second solution may be needed if the equation of time is large and changing rapidly, using the GAT for entering the almanac for the first solution, and using the GMT from this solution as the almanac entry value for the second solution.

Apparent time can also be found by converting hour angle to time units, and adding or subtracting 12 hours. If LAT is required, but not GAT, conversion of arc to time should be made from LHA, rather than GHA, to avoid the need for conversion of longitude to time units. Equation of time can be found by *subtracting* mean time from apparent time at the same meridian. This method of finding apparent time and equation of time is the only one available with the *Air Almanac*, which does not tabulate equation of time.

The navigator has little or no use for apparent time, as such. However, it can be used for finding the time of **local apparent noon (LAN)**, when the apparent sun is on the celestial meridian.

The mean sun averages out the irregularities in time due to the variations of the speed of revolution of the earth in its orbit and the fact that the apparent sun moves in the ecliptic while hour angle is measured along the celestial equator. It does not eliminate the error due to slight variations in the *rotational* speed of the earth. When a correction for the accumulated error from this source is applied to mean time, **ephemeris time** results. This time is of interest to astronomers, but is not used directly by the navigator.

612. Sidereal time uses the first point of Aries (vernal equinox) as the celestial reference point. Since the earth revolves around the sun, and since the direction of the earth's rotation and revolution are the same, it completes a rotation with respect to the stars in less time (about $3^m56.^s6$ of mean solar units) than with respect to the sun, and during one revolution about the sun (one year) it makes one complete rotation more with respect to the stars than with the sun. This accounts for the daily shift of the stars nearly 1° westward each night. Hence, sidereal days are shorter than solar days, and its hours, minutes, and seconds are correspondingly shorter. Because of nutation sidereal time is not quite constant in rate. Time based upon the average rate is called **mean sidereal time,** when it is to be distinguished from the slightly irregular sidereal time. The ratio of mean solar time units to mean sidereal time units is 1:1.00273791.

The sidereal *day* begins when the first point of Aries is over the *upper* branch of the meridian, and extends through 24 hours of sidereal time. The sun is at the first point of Aries at the time of the vernal equinox, about March 21. However, since the solar day begins when the sun is over the *lower* branch of the meridian, apparent solar and sidereal times differ by 12 hours at the vernal equinox. Each month thereafter, sidereal time *gains* about two hours on solar time. By the time of the summer solstice, about June 21, sidereal time is 18 hours *ahead* or six hours *behind* solar time. By the time of the autumnal equinox, about September 23, the two times are together, and by the time of the winter solstice, about December 22, the sidereal time is six hours *ahead* of solar time. There need be no confusion of the date, for there is no sidereal date.

Local sidereal time (**LST**) uses the local meridan as the terrestrial reference. At the prime meridian this is called **Greenwich sidereal time** (**GST**). The difference between LST at two meridians is equal to the difference of longitude between them, the place to the eastward having the later time. Local sidereal time is LHA ♈ expressed in time units. To determine LST at any given moment, find GHA ♈ by means of an almanac, and then apply the longitude to convert it to LHA ♈. Then convert LHA ♈ in arc to LST in time units.

Example.—Find LST at ZT $8^h25^m51^s$ on May 31, 1975, for long. 103°16′.3E.

Solution.—

ZT	$8^h25^m51^s$ May 31
ZD	(−)7
GMT	$1^h25^m51^s$ May 31
1^h	262°54′.8
25^m51^s	6°28′.8
GHA ♈	269°23′.6
λ	103°16′.3E
LHA ♈	12°39′.9
LST	$0^h50^m40^s$

Unless GST is required, conversion from arc to time units should be made from LHA ♈, rather than from GHA ♈, to avoid the need for converting longitude from arc to time units.

Conversion of sidereal to solar time is the reverse. Local sidereal time is converted to arc (LHA ♈), and the longitude is applied to find GHA ♈. This is used as an argument for entering the almanac to determine GMT, which can then be converted to any other kind of time desired. This is similar to one method of finding time of meridian transit, described in article 733. Normally, the problem is not encountered by the navigator.

Sidereal time, as such, is little used by the navigator. It is the basis of star charts, star finders and identifiers, and certain sight reduction methods (Volume I of Pub. No. 249), but generally in the form LHA ♈. This kind of time is used for these purposes because its celestial reference point remains almost fixed in relation to the stars.

613. Time and hour angle.—Both time and hour angle are a measure of the phase of rotation of the earth, since both indicate the angular distance of a celestial reference point *west* of a terrestrial reference meridian. Hour angle, however, applies to *any* point on the celestial sphere. Time might be used in this respect, but only the apparent sun, mean sun, the first point of Aries, and occasionally the moon are commonly used.

Hour angles are usually expressed in arc units, and are measured from the upper branch of the celestial meridian. Time is customarily expressed in time units. Sidereal time is measured from the upper branch of the celestial meridian, like hour angle, but solar time is measured from the lower branch. Thus, LMT=LHA mean sun plus or minus 180°, LAT=LHA apparent sun plus or minus 180°, and LST=LHA ♈.

As with time, **local hour angle (LHA)**, based upon the local celestial meridian, at two places differs by the longitude between them, and LHA at longitude 0° is called **Greenwich hour angle (GHA)**. In addition, it is often convenient to express hour angle in terms of the *shorter* arc between the local celestial meridian and the body. This is similar to measurement of longitude from the Greenwich meridian. Local hour angle measured in this way is called **meridian angle (t)**, which is labeled *east* or *west*, like longitude, to indicate the direction of measurement. A westerly meridian angle is numerically equal to LHA, while an easterly meridian angle is equal to 360°−LHA; also, LHA=t (W), and LHA=360°−t (E). Meridian angle is used in the solution of the navigational triangle.

Example 1.—Find LHA and t of the sun at GMT $3^h24^m16^s$ on June 1, 1975, for long. 118°48'.2W.

Solution.—

GMT	$3^h24^m16^s$	June 1
3^h	225°35'.7	
24^m16^s	6°04'.0	
GHA	231°39'.7	
λ	118°48'.2W	
LHA	112°51'.5	
t	112°51'.5W	

Example 2.—Find LHA and t of Kochab at ZT $18^h24^m47^s$ on May 31, 1975, for long. 55°27'.3W.

Solution.—

		Kochab	
ZT		$18^h24^m47^s$	May 31
ZD	(+)	4	
GMT		$22^h24^m47^s$	May 31
22^h		218°46'.5	
24^m47^s		6°12'.8	
SHA		137°17'.7	
GHA		2°17'.0	
λ		55°27'.3W	
LHA		306°49'.7	
t		53°10'.3E	

Problems

603a. What is the time and date $9^h13^m29^s$ before $3^h16^m34^s$ May 9?

Answer.—T $18^h03^m05^s$ May 8.

603b. What is the time and date $4^d19^h22^m50^s$ after $9^h31^m04^s$ on December 25?

Answer.—T $4^h53^m54^s$ on Dec. 30.

603c. What is the time and date 2 years, 11 months, 16 days, 10 hours, 23 minutes, and 48 seconds before $2^h46^m17^s$ on October 4, 1958?

Answer.—T $16^h22^m29^s$ on Oct. 17, 1955.

603d. What is the time and date 412 days, 15 hours, 6 minutes, and 56 seconds after $22^h27^m03^s$ on March 16, 1958?

Answer.—T $13^h33^m59^s$ on May 3, 1959.

604a. Convert $6^h28^m31^s$ to arc units, without use of a conversion table.
Answer.—97°07′45″ or 97°07′.8.

604b. Convert 217°28′.8 to time units, without use of a conversion table.
Answer.—$14^h29^m55^s.2$ or $14^h29^m55^s$.

604c. Convert 196°21′46″ to time units, without use of a conversion table.
Answer.—$13^h05^m27^s.1$ or $13^h05^m27^s$.

604d. Convert 107°49′44″ to time units, using appendix F.
Answer.—$7^h11^m19^s$.

604e. Convert 211°37′.3 to time units, using appendix F.
Answer.—$14^h06^m29^s.2$.

604f. Convert $8^h49^m33^s$ to arc units, using appendix F.
Answer.—132°23′.2.

604g. Convert 251°09′.2 to time units, using appendix G.
Answer.—$16^h44^m37^s$.

604h. Convert $23^h07^m38^s$ to time units, using appendix G.
Answer.—346°54′.5.

607a. For an observer at long. 97°24′.6E the ZT is $19^h10^m26^s$.
Required.—(1) Zone description.
(2) GMT.
Answers.—(1) ZD (−) 6, (2) GMT $13^h10^m26^s$.

607b. The GMT is $11^h32^m07^s$.
Required.—(1) ZT at long. 133°24′.7W.
(2) ZT at long. 111°43′.9E.
Answers.—(1) ZT $2^h32^m07^s$, (2) ZT $18^h32^m07^s$.

607c. At long. 165°18′.2E the ZT is $17^h08^m51^s$ on July 11.
Required.—(1) GMT and date.
(2) ZT and date at long. 125°36′.7W.
Answers.—(1) GMT $6^h08^m51^s$ on July 11, (2) ZT $22^h08^m51^s$ on July 10.

607d. On January 26 the 0800 DR long. of a ship is 128°03′.2E. Twenty-six hours later the EP long. is 125°01′.4E.
Required.—ZT and date of arrival at the second longitude.
Answer.—ZT 0900 Jan. 27.

607e. On April 1 the 1200 running fix long. of a ship is 179°55′.2W. Eight hours later the DR long. is 178°48′.9E.
Required.—ZT and date of arrival at the second longitude.
Answer.—ZT 2000 Apr. 2.

607f. Inch'ŏn, long. 137°E, uses ZD (−) 8^h30^m for standard time. Find the standard time and date at San Francisco, long. 122°W, when the summer time at Inch'ŏn is 2000 on August 9.
Answer.—ZT 0230 Aug. 9.

608a. At GMT 1400 on July 2 the chronometer reads $1^h42^m28^s$. At GMT 0800 on July 12 it reads $7^h42^m40^s$.
Required.—(1) Chronometer error at GMT 1400 on July 2.
(2) Chronometer error at GMT 0800 on July 12.
(3) Chronometer rate.
(4) Chronometer time at ZT 1800 July 20, at long. 153°21′.7W.
Answers.—(1) CE 17^m32^s slow, (2) CE 17^m20^s slow, (3) rate $1^s.2$ gaining, (4) C $3^h42^m51^s$.

608b. On March 5 the DR long. of a ship is about 151°E, and the zone time is about 1800. Chronometer error is 6^m40^s fast.

Required.—GMT and date when the chronometer reads $8^h02^m23^s$.

Answer.—GMT $7^h55^m43^s$ on Mar. 5.

608c. On November 7 the EP long. of a ship is about 71°W, and the zone time is about 1900. Chronometer error is 1^m18^s slow.

Required.—GMT and date when the chronometer reads (1) $11^h55^m20^s$, (2) $11^h59^m50^s$.

Answers.—(1) GMT $23^h56^m38^s$ Nov. 7, (2) GMT $0^h01^m08^s$ Nov. 8.

608d. At GMT 2200 a comparing watch is checked by time signal, and found to read $10^h00^m05^s$. The chronometer errors are then determined by means of the comparing watch. When the watch reads $10^h06^m00^s$, chronometer *A* reads $10^h11^m17^s$, and when the watch reads $10^h08^m00^s$, chronometer *B* reads $9^h59^m06^s$.

Required.—(1) Watch error.

(2) Error of chronometer *A*.

(3) Error of chronometer *B*.

Answers.—(1) WE 5^s fast on GMT, (2) CE_A 5^m22^s fast, (3) CE_B 8^m49^s slow.

609a. A chronometer 7^m22^s slow on GMT reads approximately 3^h45^m. About two minutes later, when the GMT is a whole minute, a comparing watch will be set to GMT exactly.

Required.—(1) Reading of the watch at starting.

(2) Reading of the chronometer.

Answers.—(1) WT $3^h54^m00^s$, (2) C $3^h46^m38^s$.

609b. A chronometer 5^m10^s fast on GMT reads approximately 5^h50^m. About one minute later, when the GMT is a whole minute, a comparing watch with a 24-hour dial will be set to GMT exactly. The ZT is approximately 1145 and the long. 94°W.

Required.—(1) Reading of the watch at starting.

(2) Reading of the chronometer.

(3) Watch error if, instead of being set to GMT, the watch setting is unchanged and the watch reads $17^h45^m32^s$ at comparison.

Answers.—(1) WT $17^h46^m00^s$, (2) C $5^h51^m10^s$, (3) WE 28^s slow on GMT.

609c. A watch is set to zone time, approximately. The long. is about 160°E. The watch is compared with a chronometer which is 3^m16^s fast on GMT. When the chronometer reads $1^h48^m00^s$, the watch reads $12^h45^m02^s$.

Required.—Watch error on zone time.

Answer.—WE 18^s fast on ZT.

609d. On February 14 the DR long. is 63°46.′1W. An observation of Dubhe is made when the watch reads $6^h07^m30^s$ PM. The watch is 11^s slow on zone time.

Required.—GMT and date.

Answer.—GMT $22^h07^m41^s$ Feb. 14.

609e. On December 11 a watch is set to zone time, approximately. The long. is 137°W. The chronometer is 3^m36^s fast on GMT. When the chronometer reads $4^h40^m00^s$, the watch reads $7^h36^m06^s$ PM.

Required.—(1) Watch error on GMT.

(2) GMT and date about 20 minutes later, when the watch reads $7^h55^m52^s$.

Answers.—(1) WE $2^h59^m42^s$ fast on GMT, (2) GMT $4^h56^m10^s$ Dec. 12.

609f. Shortly before taking morning sights on January 17 the navigator compares his watch with the chronometer. When the chronometer reads $2^h30^m00^s$, the watch reads $6^h13^m12^s$ AM. The chronometer is 17^m15^s fast on GMT. The long. is 118°W.

Required.—(1) C–WT.

(2) GMT and date a little later when Regulus is observed at WT $6^h28^m47^s$ AM.

Answers.—(1) C–WT $8^h16^m48^s$, (2) GMT $14^h28^m20^s$ Jan. 17.

610a. At long. 138°09′.3E the LMT is $0^h09^m57^s$ on April 23.
Required.—(1) GMT and date.
(2) ZT and date at the place.
Answers.—(1) GMT $14^h57^m20^s$ Apr. 22, (2) ZT $23^h57^m20^s$ Apr. 22.
610b. At long. 157°18′.4W the LMT is 1931 on June 29.
Required.—(1) ZT and date.
(2) GMT and date.
Answers.—(1) ZT 2000 June 29, (2) GMT 0600 June 30.
610c. At long. 99°35′.7W the daylight saving time is $21^h29^m45^s$ on August 31.
Required.—(1) Standard time and date.
(2) LMT and date.
Answers.—(1) Standard time $20^h29^m45^s$ Aug. 31, (2) LMT $20^h51^m22^s$ Aug. 31.
611a. Find the LAT and date at ZT $5^h26^m13^s$ on June 12, 1975, for long. 9°28′.1E.
Answer.—LAT $5^h04^m21^s$ June 12.
611b. At long. 77°15′.5W the LAT is 1500 on June 13, 1975.
Required.—(1) ZT.
(2) LMT.
Answers.—(1) ZT $15^h08^m56^s$, (2) LMT $14^h59^m54^s$.
611c. Using the *Air Almanac,* find (1) LAT at long. 117°55′W, and (2) the Eq. T, at ZT $20^h43^m09^s$ on June 1, 1975.
Answers.—(1) LAT $20^h53^m44^s$, (2) Eq. T (+) 2^m15^s.
612a. Find LST at ZT $19^h24^m26^s$ on June 1, 1975, for long. 87°51′.2E.
Answer.—LST $11^h53^m29^s$.
612b. Find the ZT at LST $21^h20^m07^s$ on May 31, 1975, for long. 54°21′.3W.
Answer.—ZT $4^h24^m40^s$.

CHAPTER VII

CALCULATIONS OF CELESTIAL NAVIGATION

Finding GHA and Declination

701. Use of the almanacs.—The time used as an entering argument in the almanacs is 12^h + Greenwich hour angle of the mean sun and is denoted by GMT. This scale may differ from the broadcast time signals by an amount which, if ignored, will introduce an error of up to 0.′2 in longitude determined from astronomical observations. The difference arises because the time argument depends on the variable rate of rotation of the earth while the broadcast time signals are now based on an atomic time-scale. Step adjustments of exactly one second are made to the time signals as required (primarily at 24^h on December 31 and June 30) so that the difference between the time signals and GMT, as used in the almanacs, may not exceed $0^s.9$. Those who require to reduce observations to a precision of better than 1^s must therefore obtain the correction to the time signals from coding in the signal, or from other sources. The correction may be applied to each of the times of observation. Alternatively, the longitude, when determined from astronomical observations, may be corrected by the corresponding amount shown in the following table:

Correction to time signals	Correction to longitude
$-0^s.7$ to $-0^s.9$	0.′2 to east
$-0^s.6$ to $-0^s.3$	0.′1 to east
$-0^s.2$ to $+0^s.2$	no correction
$+0^s.3$ to $+0^s.6$	0.′1 to west
$+0^s.7$ to $+0^s.9$	0.′2 to west

The main contents of the almanacs consist of data from which the Greenwich hour angle (GHA) and the declination (Dec.) of all the bodies used for navigation can be obtained for any instant of Greenwich mean time (GMT). The local hour angle (LHA) can then be obtained by means of the formula:

$$\text{LHA}=\text{GHA}\ \begin{matrix}-\ \text{west}\\ +\ \text{east}\end{matrix}\ \text{longitude.}$$

For the sun, moon, and the four navigational planets, the GHA and declination are tabulated directly in the *Nautical Almanac* for each hour of GMT throughout the year; in the *Air Almanac*, the values are tabulated for each whole 10^m of GMT. For the stars the sidereal hour angle (SHA) is given, and the GHA is obtained from:

$$\text{GHA Star}=\text{GHA Aries}+\text{SHA Star.}$$

The SHA and declination of the stars change slowly and may be regarded as constant over periods of several days or even months if lesser accuracy is required. The SHA and declination of stars tabulated in the *Air Almanac* may be considered constant to a precision of 1′ for the period covered by each of the three volumes providing the data for a whole year. Should the *Air Almanac* be published in two volumes, the SHA and declination of stars tabulated in each volume could be considered constant to a precision of 1′.5 to 2′, with most data being closer to the smaller value. GHA Aries, or the Greenwich hour angle of the first point of Aries (the vernal equinox), is tabulated for each hour of GMT in the *Nautical Almanac* and for each whole 10^m of GMT in the *Air Almanac.* Permanent tables give the appropriate increments to the tabulated values of GHA and declination for the minutes and seconds of GMT.

In the *Nautical Almanac,* the permanent table for increments also includes corrections for *v*, the difference between the actual change of GHA in one hour and a constant value used in the interpolation tables and *d*, the change in declination in one hour.

In the *Nautical Almanac, v* is always positive unless a negative sign (−) is given. This can occur only in the case of Venus. For the sun, the tabulated values of GHA have been adjusted to reduce to a minimum the error caused by treating *v* as negligible; there is no *v* tabulated for the sun.

No sign is given for tabulated values of *d*, which is positive if declination is increasing, and negative if it is decreasing. The sign of a *v* or *d* value is given also to the related correction.

In the *Air Almanac,* the tabular values of the GHA of the moon are adjusted so that use of an interpolation table based on a fixed rate of change gives rise to negligible error; no such adjustment is necessary for the sun and planets. The tabulated declination values, except for the sun, are those for the *middle* of the interval between the time indicated and the next *following* time for which a value is given, making interpolation unnecessary. Thus, it is always important to take out the GHA and declination for the tabular GMT immediately before the time of observation.

In the *Air Almanac,* GHA Aries and the GHA and declination of the sun are tabulated to a precision of 0′.1. If these values are extracted with the tabular precision, the "Interpolation of GHA" table on the inside front cover (and flap) should not be used; use the "Interpolation of GHA Sun" and "Interpolation of GHA Aries" tables, as appropriate. These tables for interpolation to a precision of 0′.1 just precede the Polaris Table.

The instructions in the explanation of each volume to ignore the decimal in smaller type when extracting GHA Aries and GHA and declination of the sun to a precision of 1′ instead of rounding-off in the normal way are intended for the air navigator.

702. Finding GHA and declination of the sun.—*Nautical Almanac.* Enter the daily-page table with the whole hour next preceding the given GMT, unless this time is itself a whole hour, and take out the tabulated GHA and declination. Record, also, the *d* value given at the bottom of the declination column. Next, enter the increments and corrections table for the number of minutes of GMT. If there are seconds, use the next *earlier* whole minute. On the line corresponding to the seconds of GMT take the value from the sun-planets column. Add this to the value of GHA from the daily page to find GHA at the given time. Next, enter the correction table for the same minute with the *d* value, and take out the correction. Give this the sign of the *d* value, and apply it to the declination from the daily page. The result is the declination at the given time.

Example 1.—Find the GHA and declination of the sun at GMT 18^{h}24^{m}37^s on June 1, 1975, using the *Nautical Almanac.*

Solution.—

	Sun			Sun	
GMT	18^{h}24^{m}37^s	June 1	GMT	18^{h}24^{m}37^s	June 1
18^h	90°34′.3		18^h	22°02′.5N	*d*(+)0′.3
24^{m}37^s	6°09′.3		*d* corr.	(+)0′.1	
GHA	96°43′.6		Dec.	22°02′.6N	

The correction table for GHA of the sun is based upon a rate of change of 15° per hour, the average rate during a year. At most times the rate differs slightly from this. The slight error thus introduced is minimized by adjustment of the tabular values.

The *d* value is the amount that the declination changes between 1200 and 1300 on the middle day of the three shown.

Air Almanac. Enter the daily page with the whole 10ᵐ next preceding the given GMT, unless the time is itself a whole 10ᵐ, and extract the tabulated GHA. The declination is extracted, without interpolation, from the same line as the tabulated GHA or, in the case of planets, the top line of the block of six. If the values extracted are rounded to the nearest minute, next enter the "Interpolation of GHA" table on the inside front cover (and flap), using the "Sun, etc." entry column, and take out the value for the remaining minutes and seconds of GMT. If the entry time is an exact tabulated value, use the correction given half a line *above* the entry time. Add this correction to the GHA taken from the daily page to find the GHA at the given time. No adjustment of declination is needed. If the values are extracted with a precision of 0:1, the table for interpolating the GHA of the sun to a precision of 0:1 must be used. No adjustment of declination is needed.

Example 2.—Find the GHA and declination of the sun at GMT $18^h24^m37^s$ on June 1, 1975, using the *Air Almanac.*

Solution.—

	Sun	
GMT	$18^h24^m37^s$	June 1
18^h20^m	95°34′	
4^m37^s	1°09′	
GHA	96°43′	
Dec.	22°03′N	

	Sun	
GMT	$18^h24^m37^s$	June 1
18^h20^m	95°34:3	
4^m37^s	1°09:3	
GHA	96°43:6	
Dec.	22°02:6N	

703. Finding GHA and declination of the moon.—*Nautical Almanac.* Enter the daily-page table with the whole hour next preceding the given GMT, unless this time is itself a whole hour, and take out the tabulated GHA and declination. Record, also, the corresponding *v* and *d* values tabulated on the same line, and determine the sign of the *d* value. The *v* value of the moon is always positive (+), and is not marked in the almanac. Next, enter the increments and corrections table for the minutes of GMT, and on the line for the seconds of GMT take the GHA correction from the moon column. Then, enter the correction table for the same minute with the *v* value, and extract the correction. Add both of these corrections to the GHA from the daily page to obtain the GHA at the given time. Then, enter the same correction table with the *d* value, and extract the correction. Give this correction the sign of the *d* value, and apply it to the declination from the daily page to find the declination at the given time.

Example 1.—Find the GHA and declination of the moon at GMT $21^h25^m44^s$ on June 1, 1975, using the *Nautical Almanac.*

Solution.—

	Moon	
GMT	$21^h25^m44^s$	June 1
21^h	225°28:1	
25^m44^s	6°08:4	*v* (+)15:8
v corr.	(+)6:7	
GHA	231°43:2	

	Moon	
GMT	$21^h25^m44^s$	June 1
21^h	3°06:8S	*d*(−)10:7
d corr.	(−)4:5	
Dec.	3°02:3S	

The correction table for GHA of the moon is based upon the *minimum* rate at which the moon's GHA increases, 14°19:0 per hour. The *v* correction makes the adjustment for the actual rate. The *v* value itself is the difference between the minimum rate and the actual rate during the hour following the tabulated time. The *d* value is the amount that the declination changes during the hour following the tabulated time.

Air Almanac. Enter the daily page with the whole 10^m next preceding the given GMT, unless this time is itself a whole 10^m, and take out the tabulated GHA and the declination, without interpolation. Next, enter the "Interpolation of GHA" table on the inside front cover, using the "moon" entry column, and take out the value for the remaining minutes and seconds of GMT. If the entry time is an exact tabulated value, use the correction given half a line *above* the entry time. Add this correction to the GHA taken from the daily page to find the GHA at the given time. No adjustment of declination is needed.

Example 2.—Find the GHA and declination of the moon at GMT $21^h25^m44^s$ on June 1, 1975, using the *Air Almanac*.

Solution.—

	Moon	
GMT	$21^h25^m44^s$	June 1
21^h20^m	230°20′	
5^m44^s	1°23′	
GHA	231°43′	
Dec.	3°02′S	

The declination given in the table is correct for the time *five minutes later than tabulated,* so that it can be used for the ten-minute interval without interpolation, to an accuracy to meet most requirements. If greater accuracy is needed, it can be obtained by interpolation, remembering to allow for the five minutes indicated above.

704. Finding GHA and declination of a planet.—*Nautical Almanac.* Enter the daily-page table with the whole hour next preceding the given GMT, unless the time itself is a whole hour, and take out the tabulated GHA and declination. Record, also, the *v* value given at the bottom of each of these columns. Next, enter the increments and corrections table for the minutes of GMT, and on the line for the seconds of GMT take the GHA correction from the sun-planets column. Next, enter the correction table with the *v* value and extract the correction, giving it the sign of the *v* value. Add the first correction to the GHA from the daily page, and apply the second correction in accordance with its sign, to obtain the GHA at the given time. Then, enter the correction table for the same minute with the *d* value, and extract the correction. Give this correction the sign of the *d* value, and apply it to the declination from the daily page to find the declination at the given time.

Example 1.—Find the GHA and declination of Venus at GMT $5^h24^m07^s$ on June 2, 1975, using the *Nautical Almanac*.

Solution.—

	Venus	
GMT	$5^h24^m07^s$	June 2
5^h	206°59.′4	
24^m07^s	6°01.′8	*v* (−)0.′4
v corr.	(−)0.′2	
GHA	213°01.′0	

	Venus	
GMT	$5^h24^m07^s$	June 2
5^h	23°30.′8N	*d* (−)0.′5
d corr.	(−)0.′2	
Dec.	23°30.′6N	

The correction table for GHA of planets is based upon the mean rate of the sun, 15° per hour. The *v* value is the difference between 15° and the change of GHA of the planet between 1200 and 1300 on the middle day of the three shown. The *d* value is the amount that the declination changes between 1200 and 1300 on the middle day.

Venus is the only body listed which ever has a negative v value.

Air Almanac. Enter the daily page with the whole 10^m next preceding the given GMT, unless this time is itself a whole 10^m, and extract the tabulated GHA and declination, without interpolation. The tabulated declination is correct for the time 30^m later than tabulated, so that interpolation during the hour following tabulation is not needed for most purposes. Next, enter the "Interpolation of GHA" table on the inside front cover, using the "sun, etc." column, and take out the value for the remaining minutes and seconds of GMT. If the entry time is an exact tabulated value, use the correction half a line *above* the entry time. Add this correction to the GHA from the daily page to find the GHA at the given time. No adjustment of declination is needed.

Example 2.—Find the GHA and declination of Venus at GMT $5^h48^m45^s$ on June 2, 1975, using the *Air Almanac.*

Solution.—

	Venus	
GMT	$5^h48^m45^s$	June 2
5^h40^m	216°59′	
8^m45^s	2°11′	
GHA	219°10′	
Dec.	23°31′N	

The declination is taken for the next *earlier* tabulated time, and is correct for GMT 5^h45^m.

705. Finding GHA and declination of a star.—If the GHA and declination of each navigational star were tabulated separately, the almanacs would be several times their present size. But since the sidereal hour angle of star and the declination are nearly constant over several days (to the nearest 0.′1) or months (to the nearest 1′), separate tabulations are not needed. Instead, the GHA of the first point of Aries, from which SHA is measured, is tabulated on the daily pages, and a single listing of SHA and declination is given for each double page of the *Nautical Almanac*, and for an entire volume of the *Air Almanac*. The finding of GHA ♈ is similar to finding GHA of the sun, moon, and planets.

Nautical Almanac. Enter the daily-page table with the whole hour next preceding the given GMT, unless this time is itself a whole hour, and take out the tabulated GHA ♈. Record, also, the tabulated SHA and declination of the star from the listing on the left-hand daily page. Next, enter the increments and corrections table for the minutes of GMT, and on the line for the seconds of GMT take the GHA correction from the Aries column. Add this correction and the SHA of the star to the GHA ♈ of the daily page to find the GHA of the star at the given time. No adjustment of declination is needed.

Example 1.—Find the GHA and declination of Canopus at GMT $3^h24^m33^s$ on June 2, 1975, using the *Nautical Almanac.*

Solution.—

	Canopus	
GMT	$3^h24^m33^s$	June 2
3^h	294°58.′0	
24^m33^s	6°09.′3	
SHA	264°09.′3	
GHA	205°16.′6	
Dec.	52°41.′1S	

The SHA and declination of 173 stars, including Polaris and the 57 listed on the daily pages, are given for the middle of each month, on almanac pages 268–273. For a star not listed on the daily pages this is the only almanac source of this information. Interpolation in this table is not necessary for ordinary purposes of navigation, but is sometimes needed for precise results. Thus, if the SHA and declination of *β Crucis* (Mimosa) are desired for March 1, 1975, they are found by simple eye interpolation to be SHA 168°25.′2 and Dec. 59°33.′2S.

If GHA ♈ is desired, it is found as indicated in example 1, but omitting the addition of SHA of a star. In the example GHA ♈ is 294°58.′0+6°09.′3=301°07.′3.

Air Almanac. Enter the daily page with the whole 10^m next preceding the given GMT, unless this is itself a whole 10^m, and extract the tabulated GHA ♈. Next, enter the "Interpolation of GHA" table on the inside front cover, using the "sun, etc." entry column, and take out the value for the remaining minutes and seconds of GMT. If the entry time is an exact tabulated value, use the correction given half a line *above* the entry time. From the tabulation at the left side of the same page, extract the SHA and declination of the star. Add the GHA from the daily page and the two values taken from the inside front cover to find the GHA at the given time. No adjustment of declination is needed.

Example 2.—Find the GHA and declination of Peacock at GMT $12^h17^m58^s$ on June 1, 1975, using the *Air Almanac.*

Solution.—

	Peacock	
GMT	$12^h17^m58^s$	June 1
12^h10^m	71°52′	
7^m58^s	2°00′	
SHA	54°03′	
GHA	127°55′	
Dec.	56°49′S	

The Undivided Astronomical Triangle

706. Solving for altitude.—The law of cosines for sides is a fundamental formula for solving a spherical triangle. As applied to the spherical triangle of figure 706a, the law is stated as:

$$\cos a = \cos b \cos c + \sin b \sin c \cos A \tag{1a}$$
$$\cos b = \cos c \cos a + \sin c \sin a \cos B \tag{1b}$$
$$\cos c = \cos a \cos b + \sin a \sin b \cos C. \tag{1c}$$

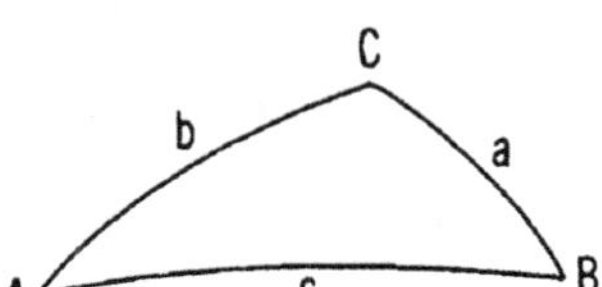

FIGURE 706a.—Spherical triangle.

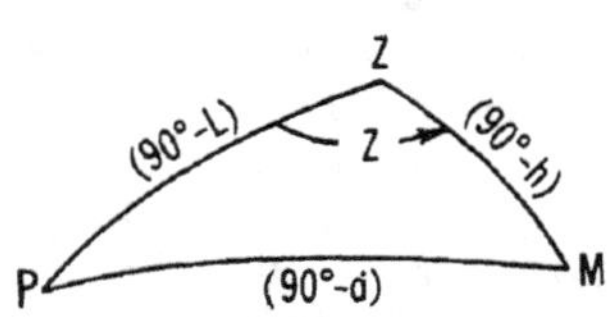

FIGURE 706b.—Undivided astronomical triangle.

As applied to the undivided astronomical triangle of figure 706b, equation 1a is stated as:

$$\cos(90°-h) = \cos(90°-L)\cos(90°-d)+\sin(90°-L)\sin(90°-d)\cos \text{LHA}$$
$$\therefore \sin h = \sin L \sin d + \cos L \cos d \cos \text{LHA}, \tag{2a}$$

in which h is the altitude of the celestial body above the celestial horizon; L is the latitude of the observer or the assumed position of the observer, d is the declination of the body, and LHA is the local hour angle of the body. Meridian angle, t, can be substituted for LHA in the equation; i.e.,

$$\sin h = \sin L \sin d + \cos L \cos d \cos t. \quad (2b)$$

The sign convention used in the calculations of both formulas is that declination is treated as a negative quantity when latitude and declination are of contrary name. No special sign convention is required for local hour angle or for whether the meridian angle is measured eastward or westward from the meridian.

If the altitude as calculated is negative, the body is below the celestial horizon.

Particularly when using a table of trigonometric functions, the rules for the following cases may be helpful in avoiding calculation mistakes due to not using the proper sign with a trigonometric function (art. 140). However, for cases II and III it is necessary to know whether the body is above or below the celestial horizon.

Case I ($t<90°$ and Same Name).

If LHA is in the range 0° increasing to 90°, or 270° increasing to 360° and the latitude is same name as declination, the two terms on the right-hand side of the equation are added. The body is above the celestial horizon.

Case II ($t<90°$ and Contrary Name).

If LHA is in the range 0° increasing to 90°, or 270° increasing to 360° and the latitude is of contrary name, the lesser quantity is subtracted from the greater on the right-hand side of the equation. The body can be above or below the celestial horizon.

Case III ($t>90°$ and Same Name).

If LHA is in the range greater than 90° and increasing to 270° and the latitude is same name as declination, the lesser quantity is subtracted from the greater on the right-hand side of the equation. The body can be above or below the celestial horizon.

Case IV ($t>90°$ and Contrary Name).

If LHA is in the range greater than 90° and increasing to 270° and the latitude is of contrary name, the two quantities on the right-hand side of the equation are added. The body is below the celestial horizon.

Astronomical triangles corresponding to the four cases are drawn on diagrams on the plane of the celestial meridian in figure 706c.

Example 1.—The latitude of the observer is 45°00.′0N; the declination of the celestial body is 5°00.′0N; the local hour angle is 60°.

Required.—Altitude of the body.

Solution.—By natural functions (table 31).

$$\begin{aligned} \sin h &= \sin L \sin d + \cos L \cos d \cos LHA \quad (2a) \\ &= \sin 45° \sin 5° + \cos 45° \cos 5° \cos 60° \\ &= (+0.70711)\,(+0.08716) + (+0.70711)\,(+0.99619)\,(+0.50000) \\ &= +0.06163 + 0.35221 \quad \text{(Case I)} \\ &= +0.41384 \\ h &= 24°26.'8 \end{aligned}$$

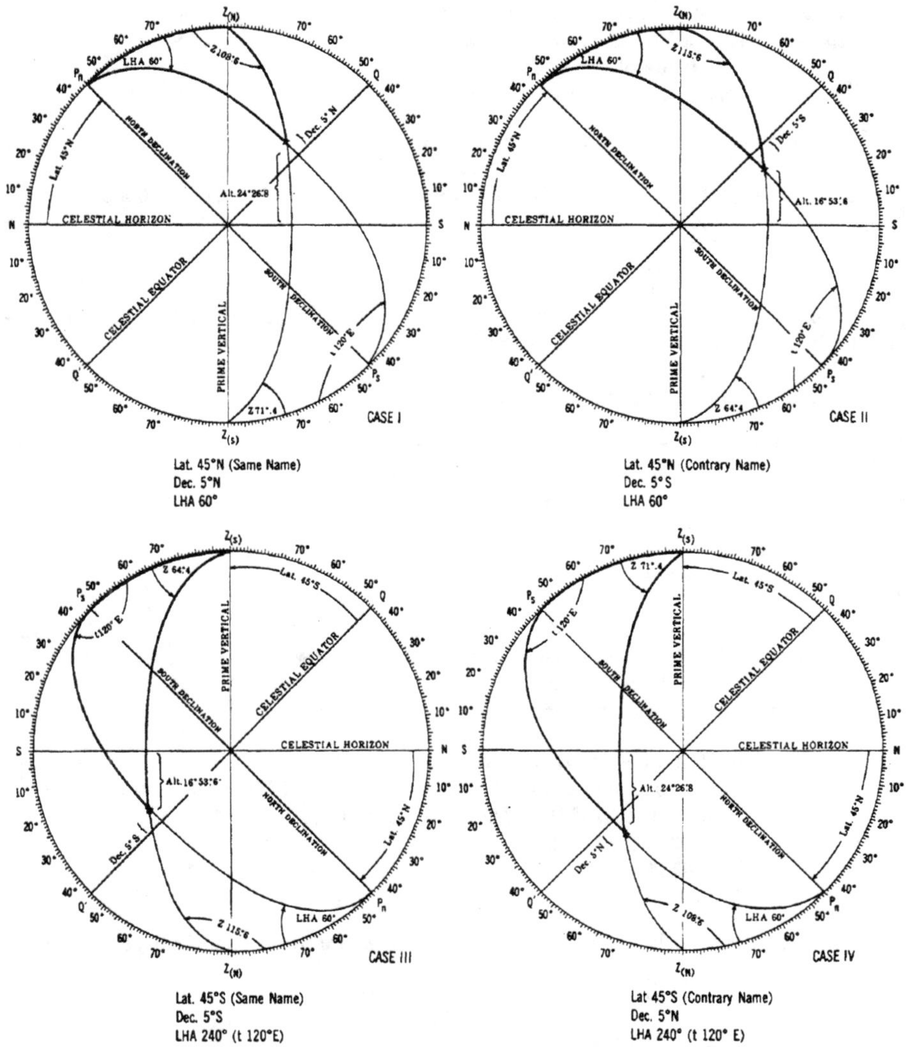

FIGURE 706c.—Diagrams on the plane of the celestial meridian.

Example 2.—The latitude of the observer is 45°00′.0N; the declination of the celestial body is 5°00′.0S; the local hour angle is 60°.

Required.—Altitude of the body.

Solution.—By natural functions (table 31).

$$\sin h = \sin L \sin d + \cos L \cos d \cos LHA \qquad (2a)$$

$$= \sin 45° \sin -5° + \cos 45° \cos -5° \cos 60°$$

$$= (+0.70711)(-0.08716) + (+0.70711)(+0.99619)(+0.50000)$$

$$= -0.06163 + 0.35221 \qquad \text{(Case II)}$$

$$= +0.29058$$

$$h = 16°53'.6$$

Example 3.—The latitude of the observer is 45°00′.0S; the declination of the celestial body is 5°00′.0S; the local hour angle is 240°.

Required.—Altitude of the body.

Solution.—By natural functions (table 31).

$$\begin{aligned}\sin h &= \sin L \sin d + \cos L \cos d \cos LHA \qquad (2a)\\ &= \sin 45° \sin 5° + \cos 45° \cos 5° \cos 240°\\ &= (+0.70711)(+0.08716) + (+0.70711)(+0.99619)(-0.50000)\\ &= +0.06163 - 0.35221 \qquad \text{(Case III)}\\ &= -0.29058\\ h &= -16°53'.6\end{aligned}$$

Example 4.—The latitude of the observer is 45°00′.0S; the declination of the celestial body is 5°00′.0N; the local hour angle is 240°.

Required.—Altitude of the body.

Solution.—By natural functions (table 31).

$$\begin{aligned}\sin h &= \sin L \sin d + \cos L \cos d \cos LHA \qquad (2a)\\ &= \sin 45° \sin -5° + \cos 45° \cos -5° \cos 240°\\ &= (+0.70711)(-0.08716) + (+0.70711)(+0.99619)(-0.50000)\\ &= -0.06163 - 0.35221 \qquad \text{(Case IV)}\\ &= -0.41384\\ h &= -24°26'.8\end{aligned}$$

Example 5.—The latitude of the observer is 30°25′.0N; the declination of the celestial body is 22°06′.2N; the meridian angle is 39°54′.7W.

Required.—Altitude of the body.

Solution.—By natural and logarithmic functions (tables 31-33).

$$\begin{aligned}\sin h &= \sin L \sin d + \cos L \cos d \cos t \qquad (2b)\\ &= \sin 30°25'.0 \sin 22°06'.2 + \cos 30°25'.0 \cos 22°06'.2 \cos 39°54'.7\\ &= (0.50628)\ (0.37628) + (0.86237)\ (0.92651)\ (0.76703)\\ &= +0.19050 + 0.61285 \qquad \text{(Case I)}\\ &= +0.80335\\ h &= 53°27'.1\end{aligned}$$

For logarithmic solution by tables 32 and 33, the following modification is used:

$$A = \sin L \sin d \qquad B = \cos L \cos d \cos t$$

$$\sin h = A + B$$

$$\log A = l \sin 30°25'.0 + l \sin 22°06'.2$$

$$\log B = l \cos 30°25'.0 + l \cos 22°06'.2 + l \cos 39°54'.7$$

$$\log A = 9.70439 + 9.57551 \qquad \log B = 9.93569 + 9.96685 + 9.88481$$

$$= 9.27990 \qquad = 9.78735$$

$$A = 0.19050 \text{ (table 32)} \qquad B = 0.61284 \text{ (table 32)}$$

$$\begin{aligned}\sin h &= A + B\\ &= 0.19050 + 0.61284 \qquad \text{(Case I)}\\ &= 0.80334\\ h &= 53°27'.0\end{aligned}$$

Example 6.—The latitude of the observer is 30°25′.0N; the declination of the celestial body is 22°06′.2S; the meridian angle is 39°54′.7W.

Required.—Altitude of the body.

Solution.—By natural and logarithmic functions (tables 31–33).

$$\begin{aligned}\sin h &= \sin L \sin d + \cos L \cos d \cos t \qquad (2b)\\ &= \sin 30°25'.0 \sin -22°06'.2 + \cos 30°25'.0 \cos -22°06'.2 \cos 39°54'.7\\ &= (0.50628)\ (-0.37628) + (0.86237)\ (0.92651)\ (0.76703)\\ &= -0.19050 + 0.61285 \quad \text{(Case II)}\\ &= +0.42235\\ h &= 24°59'.0\end{aligned}$$

For logarithmic solution by tables 32 and 33,

$$A = \sin L \sin d \qquad B = \cos L \cos d \cos t$$
$$\sin h = A \sim B$$
$$\log A = l \sin 30°25'.0 + l \sin 22°06'.2,$$
$$\log B = l \cos 30°25'.0 + l \cos 22°06'.2 + l \cos 39°54'.7$$
$$\log A = 9.70439 + 9.57551 \qquad \log B = 9.93569 + 9.96685 + 9.88481$$
$$= 9.27990 \qquad = 9.78735$$
$$A = 0.19050 \text{ (table 32)} \qquad B = 0.61284 \text{ (table 32)}$$
$$\begin{aligned}\sin h &= A \sim B\\ &= -0.19050 + 0.61284 \quad \text{(Case II)}\\ &= +0.42234\\ h &= 24°59'.0\end{aligned}$$

Example 7.—The latitude of the observer is 30°25'.0S; the declination of the celestial body is 22°06'.2S; the meridian angle is 91°20'.0W.

Required.—Altitude of the body.

Solution.—By natural and logarithmic functions (tables 31–33).

$$\begin{aligned}\sin h &= \sin L \sin d + \cos L \cos d \cos t \qquad (2b)\\ &= \sin 30°25'.0 \sin 22°06'.2 + \cos 30°25'.0 \cos 22°06'.2 \cos 91°20'.0\\ &= (0.50628)(0.37628) + (0.86237)(0.92651)(-0.02327)\\ &= +0.19050 - 0.01859 \quad \text{(Case III)}\\ &= +0.17191\\ h &= 9°53'.9\end{aligned}$$

For logarithmic solution by tables 32 and 33,

$$A = \sin L \sin d \qquad B = \cos L \cos d \cos t$$
$$\sin h = A \sim B$$
$$\log A = l \sin 30°25'.0 + l \sin 22°06'.2$$
$$\log B = l \cos 30°25'.0 + l \cos 22°06'.2 + l \cos 91°20'.0$$
$$\log A = 9.70439 + 9.57551 \qquad \log B = 9.93569 + 9.96685 + 8.36678$$
$$= 9.27990 \qquad = 8.26932$$
$$A = 0.19050 \text{ (table 32)} \qquad B = 0.01859 \text{ (table 32)}$$
$$\begin{aligned}\sin h &= A \sim B\\ &= 0.19050 - 0.01859 \quad \text{(Case III)}\\ &= 0.17191\\ h &= 9°53'.9\end{aligned}$$

Example 8.—The latitude of the observer is 30°25'.0S; the declination of the celestial body is 22°06'.2N; the meridian angle is 91°20'.0W.

Required.—Altitude of the body.

Solution.—By natural and logarithmic functions (tables 31–33).

$$\begin{aligned}
\sin h &= \sin L \sin d + \cos L \cos d \cos t \qquad \text{(2b)}\\
&= \sin 30^\circ 25'.0 \sin -22^\circ 06'.2 + \cos 30^\circ 25'.0 \cos -22^\circ 06'.2 \cos 91^\circ 20'.0\\
&= (0.50628)\ (-0.37628) + (0.86237)\ (0.92651)\ (-0.02327)\\
&= -0.19050 \ -0.01859 \quad \text{(Case IV)}\\
&= -0.20909\\
h &= -12^\circ 04'.1
\end{aligned}$$

For logarithmic solution by tables 32 and 33,

$$A = \sin L \sin d \qquad\qquad B = \cos L \cos d \cos t$$

$$\sin h = A + B$$

$$\log A = l \sin 30^\circ 25'.0 + l \sin 22^\circ 06'.2$$

$$\log B = l \cos 30^\circ 25'.0 + l \cos 22^\circ 06'.2 + l \cos 91^\circ 20'.0$$

$$\log A = 9.70439 + 9.57551 \qquad \log B = 9.93569 + 9.96685 + 8.36678$$

$$= 9.27990 \qquad\qquad = 8.26932$$

$$A = 0.19050 \text{ (table 32)} \qquad B = 0.01859 \text{ (table 32)}$$

$$\begin{aligned}
\sin h &= A + B\\
&= 0.19050 + 0.01859 \quad \text{(Case IV)}\\
&= 0.20909\\
h &= -12^\circ 04'.1
\end{aligned}$$

Note: When the meridian angle is greater than 90° and the latitude and declination are of contrary name, the body lies below the celestial horizon.

707. Solving for azimuth.—The relations between the parts of a spherical triangle as shown in figure 707a are given in the following equations known as the **five parts formulas:**

$$\sin a \cos B = \cos b \sin c - \sin b \cos c \cos A \qquad \text{(3a)}$$

$$\sin b \cos C = \cos c \sin a - \sin c \cos a \cos B \qquad \text{(3b)}$$

$$\sin c \cos A = \cos a \sin b - \sin a \cos b \cos C. \qquad \text{(3c)}$$

Also by the law of sines:

$$\frac{\sin a}{\sin A} = \frac{\sin b}{\sin B} = \frac{\sin c}{\sin C}.$$

Substituting the value of sin a as obtained from the law of sines in equation 3a:

$$\sin a = \frac{\sin A \sin b}{\sin B}$$

$$\sin A \cot B = \sin c \cot b - \cos c \cos A.$$

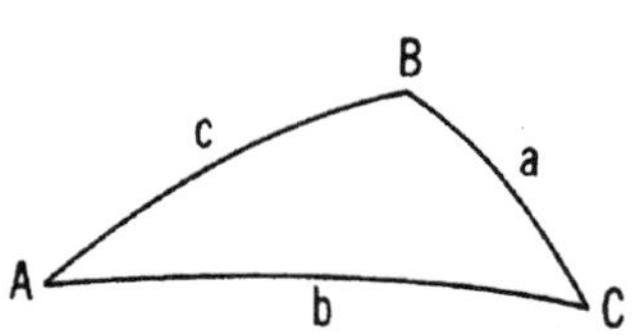

FIGURE 707a.—Spherical triangle.

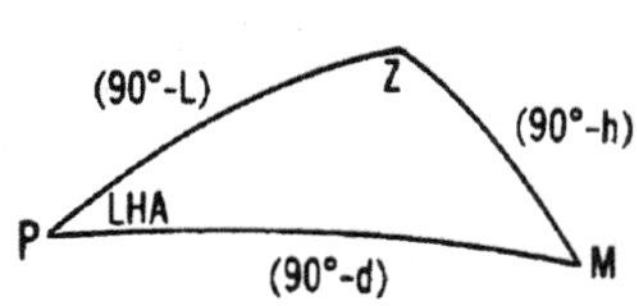

FIGURE 707b.—Undivided astronomical triangle.

As applied to the undivided astronomical triangle of figure 707b, the above equation is stated as:

$$\sin \text{LHA} \cot Z = \cos L \tan d - \sin L \cos \text{LHA}$$

from which

$$\cot Z = \frac{\cos L \tan d - \sin L \cos \text{LHA}}{\sin \text{LHA}}$$

$$\tan Z = \frac{\sin \text{LHA}}{\cos L \tan d - \sin L \cos \text{LHA}}. \tag{4a}$$

Substituting $\frac{\sin d}{\cos d}$ for tan d,

$$\tan Z = \frac{\cos d \sin \text{LHA}}{\cos L \sin d - \sin L \cos d \cos \text{LHA}}. \tag{4b}$$

Meridian angle, t, can be substituted for LHA in equations 4a and 4b:

$$\tan Z = \frac{\sin t}{\cos L \tan d - \sin L \cos t} \tag{5a}$$

$$\tan Z = \frac{\cos d \sin t}{\cos L \sin d - \sin L \cos d \cos t}. \tag{5b}$$

The sign conventions used in the calculations of the azimuth angle formulas are as follows: (1) If latitude and declination are of contrary name, declination is treated as a negative quantity; (2) If in equations 4a and 4b the local hour angle is greater than 180°, it is treated as a negative quantity.

If the acute angle as calculated is negative, it is necessary to add 180° to obtain the desired azimuth angle.

Azimuth angle is measured from 0° at the north or south reference direction clockwise or counter-clockwise through 180°. It is labeled with the reference direction as the prefix and the direction of measurement from the reference direction as a suffix. Thus, azimuth angle S144°W is 144° west of south, or true azimuth 324°. Azimuth angle is labeled N or S to agree with the latitude and E or W to agree with the meridian angle (labeled E when LHA is greater than 180°).

Azimuth angle can also be converted to true azimuth, Zn, through use of the following rules:

N. Lat.	LHA greater than 180°	Zn=Z
	LHA less than 180°	Zn=360°−Z
S. Lat.	LHA greater than 180°	Zn=180°−Z
	LHA less than 180°	Zn=180°+Z.

708. Time azimuth.—The time azimuth or azimuth angle is computed using the LHA or meridian angle (a function of time), latitude, and declination as the known quantities. Solution can be made using equations 4a or 5a.

Example 1.—The latitude of the observer is 30°25′.0N; the declination of the celestial body is 22°06′.2N; the local hour angle is 39°54′.7W.

Required.—Azimuth of the body.
Solution.—By equation 4a.

$$\tan Z = \frac{\sin LHA}{\cos L \tan d - \sin L \cos LHA} \quad (4a)$$

$$= \frac{\sin 39°54.'7}{\cos 30°25.'0 \tan 22°06.'2 - \sin 30°25.'0 \cos 39°54.'7}$$

$$= \frac{+0.64161}{(+0.86237)(+0.40613) - (+0.50628)(+0.76703)}$$

$$\tan Z = \frac{+0.64161}{+0.35023 - 0.38833} = \frac{+0.64161}{-0.03810} = -16.840$$

Since the acute angle (−86.°6) as calculated is negative, it is necessary to add 180° to obtain the desired azimuth angle.

$$Z = -86.°6 + 180° = N93.°4W$$
$$Zn = 266.°6.$$

Example 2.—The latitude of the observer is 30°25.'0S; the declination of the celestial body is 22°06.'2N; the meridian angle is 39°54.'7E.
Required.—Azimuth of the body.
Solution.—By equation 5a.

$$\tan Z = \frac{\sin t}{\cos L \tan d - \sin L \cos t} \quad (5a)$$

$$= \frac{\sin 39°54.'7}{\cos 30°25.'0 \tan -22°06.'2 - \sin 30°25.'0 \cos 39°54.'7}$$

$$= \frac{+0.64161}{(+0.86237)(-0.40613) - (+0.50628)(+0.76703)}$$

$$\tan Z = \frac{+0.64161}{-0.35023 - 0.38833} = \frac{+0.64161}{-0.73856} = -0.86873$$

Since the acute angle (−41.°0) as calculated is negative, it is necessary to add 180° to obtain the desired azimuth angle. Solving this example by equation 4a, local hour angle 320°05.'3 is treated as a negative angle.

$$Z = -41.°0 + 180° = S139.°0E$$
$$Zn = 041.°0.$$

709. Altitude azimuth.—The altitude azimuth is the azimuth or azimuth angle computed using altitude, latitude, and declination (or polar distance) as the known quantities.

By the law of cosines for sides,

$$\cos b = \cos c \cos a + \sin c \sin a \cos B. \quad (1b)$$

As applied to the astronomical triangle of figure 707b, equation 1b is stated as:

$$\cos (90° - d) = \cos (90° - L) \cos (90° - h) + \sin (90° - L) \sin (90° - h) \cos Z$$
$$\sin d = \sin L \sin h + \cos L \cos h \cos Z$$
$$\cos Z = \frac{\sin d - \sin L \sin h}{\cos L \cos h}. \quad (6)$$

Example.—The latitude of the observer is 30°00′.0N; the center of the sun is on the visible horizon; the declination of the sun is 18°00′.0N.

Required.—Azimuth angle of the sun.

Solution.—By equation 6.

Computation for azimuth angle is made for an altitude of (−)0°41′.4, determined as follows:

Dip at 41 feet height of eye	(−)	6′.2
Refraction at (−)6′.2 altitude	(−)	35′.3
Parallax	(+)	0′.1
	(−)	41′.4

$$\cos Z=\frac{\sin d-\sin L \sin h}{\cos L \cos h}=\frac{\sin 18°-\sin 30° \sin (-)0°41'.4}{\cos 30° \cos (-)0°41'.4} \qquad (6)$$

$$=\frac{(0.30902)-(0.50000)(-0.01204)}{(0.86603)(0.99993)}=\frac{+0.31504}{+0.86597}=+0.36380$$

$$Z=68°40'.0.$$

710. Time and altitude azimuth.—The time and altitude azimuth or azimuth angle is computed using meridian angle, declination, and altitude as the known quantities. The most common formula is derived from the law of sines.

By the law of sines, the relationship between the angles and sides opposite of the spherical triangle shown in figure 706a is:

$$\frac{\sin b}{\sin B}=\frac{\sin a}{\sin A}. \qquad (3d)$$

As applied to the astronomical triangle of figure 707b, equation 3d is stated as:

$$\frac{\sin (90°-d)}{\sin Z}=\frac{\sin (90°-h)}{\sin t}$$

$$\sin Z \cos h=\sin t \cos d$$

$$\sin Z=\sin t \cos d \sec h. \qquad (7)$$

The weakness of this method is that it does not indicate whether the celestial body is north or south of the prime vertical. Usually there is no question on this point, but if Z is near 90°, the quadrant may be in doubt. If this occurs, the meridian angle or altitude when on the prime vertical can be determined from table 25 (for declinations less than 23°) or by computation (art. 721), using the formula:

$$\cos t=\tan d \cot L \qquad (14c)$$

or

$$\sin h=\sin d \csc L. \qquad (14d)$$

Example.—The latitude of the observer is 30°25′.0N; the declination of the celestial body is 22°06′.2N; the altitude of the body is 53°27′.0; and the meridian angle is 39°54′.7W.

Required.—Azimuth of the body.

Solution.—By equation 7.

$$\sin Z=\sin t \cos d \sec h \qquad (7)$$

$$=\sin 39°54'.7 \cos 22°06'.2 \sec 53°27'.0$$

$$=(+0.64161)(+0.92651)(+1.67919)$$

$$=0.99821$$

$$Z=86°.6 \text{ or } 93°.4 \ ?$$

By logarithmic solution,

$$\begin{aligned} l \sin Z &= l \sin t + l \cos d + l \sec h \\ &= l \sin 39°54'.7 + l \cos 22°06'.2 + l \sec 53°27'.0 \\ &= 9.80726 + 9.96685 + 10.22510 \\ &= 9.99921 \\ Z &= 86^\circ.6 \text{ or } 93^\circ.4 \ ? \end{aligned}$$

If the altitude is *less*, or the meridian angle is *greater* than the value when the body is on the prime vertical, the numerical value of the azimuth angle is the lesser of the two angles. If the altitude is *greater* or the meridian angle is less when on the prime vertical, the numerical value of the azimuth angle is the greater of the two angles.

Entering table 25 with latitude 30°25′ and declination 22°06′.2 (same name as latitude) as arguments, the meridian angle and altitude of the body when on the prime vertical are determined as:

$$t = 46^\circ.1, \qquad h = 48^\circ.1.$$

Since the altitude is greater than the value when the body is on the prime vertical, the numerical value of the azimuth angle is the greater of the two quantities, i.e., the azimuth angle is N93°.4W; Zn is 266°.6.

711. Haversine formulas.—In the foregoing altitude and azimuth angle solutions, it has been necessary to consider the signs of the various trigonometric functions. This inconvenience can be avoided through the use of natural and log haversine functions, as given in table 34, because the haversine of an angle is positive whether the angle is positive or negative. The value increases from 0° to 180° and decreases from 180° to 360°. Since by convention only angles and sides up to 180° are considered, ambiguities are avoided.

By definition,

$$\text{haversine } A = \text{hav } A = \frac{1}{2} \text{ ver } A = \frac{1}{2}(1 - \cos A) = \sin^2 \frac{1}{2} A$$

and

$$\text{haversine } a = \text{hav } a = \frac{1}{2} \text{ ver } a = \frac{1}{2}(1 - \cos a) = \sin^2 \frac{1}{2} a.$$

Substituting cos A = 1–2 hav A and cos a = 1–2 hav a in the fundamental cosine formula:

$$\cos a = \cos b \cos c + \sin b \sin c \cos A \tag{1a}$$

$$\begin{aligned} 1-2 \text{ hav } a &= \cos b \cos c + \sin b \sin c\,(1-2 \text{ hav } A) \\ &= \cos b \cos c + \sin b \sin c - 2 \sin b \sin c \text{ hav } A. \end{aligned}$$

Since $\cos b \cos c + \sin b \sin c = \cos (b-c)$ and $\cos (b-c) = 1 - 2 \text{ hav } (b-c)$,

$$\begin{aligned} 1-2 \text{ hav } a &= 1 - 2 \text{ hav } (b-c) - 2 \sin b \sin c \text{ hav } A \\ \text{hav } a &= \text{hav } (b-c) + \sin b \sin c \text{ hav } A. \end{aligned}$$

Since the haversine is always positive, the value of hav(b−c) is positive irrespective of the values assigned to b and c. Therefore, (b−c) may be simplified by writing it as (b∼c), the difference of b and c. Thus,

$$\text{hav } a = \text{hav}(b \sim c) + \sin b \sin c \text{ hav } A. \tag{8a}$$

Similarly, it can be shown that:

$$\begin{aligned} \text{hav } b &= \text{hav}(a \sim c) + \sin a \sin c \text{ hav } B \\ \text{hav } c &= \text{hav}(a \sim b) + \sin a \sin b \text{ hav } C. \end{aligned}$$

712. Altitude by cosine-haversine formula.—The haversine formula for altitude as derived from the fundamental cosine formula is known as the cosine-haversine formula.

Substituting the parts of the astronomical triangle of figure 707b in equation 8a, side a is the zenith distance; side b is the polar distance; side c is the colatitude; and angle A is the meridian angle, t.

$$\text{hav a}=\text{hav (b}\sim\text{c)}+\sin \text{b} \sin \text{c hav A} \quad (8a)$$
$$\text{hav z}=\text{hav }[(90°-\text{d})\sim(90°-\text{L})]+\sin (90°-\text{d}) \sin (90°-\text{L}) \text{ hav t}$$
$$\text{hav z}=\text{hav (L}\sim\text{d)}+\cos \text{d} \cos \text{L hav t}. \quad (8b)$$

Equation 8b is sometimes written

$$\text{hav z}=\text{hav (L}\sim\text{d)}+\text{hav } \theta, \quad (8c)$$

in which hav θ = cos d cos L hav t.

Equation 8b can also be written entirely in haversines:

$$\text{hav z}=\text{hav (d}-\text{L)}+\text{hav t [hav }(180°-\text{L}-\text{d})-\text{hav (d}-\text{L)}]. \quad (8d)$$

In equation 8d the sign of d is reversed if L and d are of contrary name.

Example.—The latitude of the observer is 30°25′.0S; the declination of the celestial body is 22°06′.2N; and the meridian angle is 91°20′.0W.

Required.—Altitude of the body.

Solution.—By equation 8c and tables 33 and 34.

t	91°20′.0 W	*l* hav	9.70896		
L	30°25′.0 S	*l* cos	9.93569		
d	22°06′.2 N	*l* cos	9.96685		
θ		*l* hav	9.61150	*n* hav	0.40879
L∼d	52°31′.2			*n* hav	0.19575
z	102°04′.1			*n* hav	0.60454
h	−12°04′.1				

713. Azimuth by haversines.—The haversine formulas as transposed for convenience in the solutions of the three angles are known as the **cosecant formulas.**

$$\text{hav A}=[\text{hav a}-\text{hav(b}\sim\text{c)}] \csc \text{b} \csc \text{c} \quad (9a)$$
$$\text{hav B}=[\text{hav b}-\text{hav(a}\sim\text{c)}] \csc \text{a} \csc \text{c} \quad (9b)$$
$$\text{hav C}=[\text{hav c}-\text{hav(a}\sim\text{b)}] \csc \text{a} \csc \text{b}. \quad (9c)$$

As applied to the astronomical triangle of figure 707b, B is the azimuth angle; side a is the coaltitude; side b is the polar distance; and side c is the colatitude. Equation 9b is rewritten as:

$$\text{hav Z}=[\text{hav p}-\text{hav(z}\sim\text{coL)}] \csc \text{z} \csc \text{coL}. \quad (9d)$$

Another haversine formula for azimuth angle is derived from the following formula of trigonometry:

$$\sin^2 \frac{1}{2} \text{B}=\frac{\sin (\text{s}-\text{c}) \sin (\text{s}-\text{a})}{\sin \text{c} \sin \text{a}},$$

in which s is one-half the sum of the three sides of the spherical triangle. As applied to the astronomical triangle of figure 707b, B is the azimuth angle Z; side a is the coaltitude; side b is the polar distance; and side c is the colatitude.

Thus,
$$s=\frac{\text{coaltitude}+\text{polar distance}+\text{colatitude}}{2}$$

$$s=\frac{(90°-h)+p+(90°-L)}{2}=90°-\left(\frac{h+L-p}{2}\right)$$

$$s-c=90°-\left[\frac{h+L-p}{2}\right]-(90°-L)=\frac{p-(h-L)}{2}$$

$$s-a=90°-\left[\frac{h+L-p}{2}\right]-(90°-h)=\frac{p+(h-L)}{2}$$

Substituting in the basic equation,

$$\sin^2\frac{1}{2}Z=\text{hav } Z=\sin\left[\frac{p-(h-L)}{2}\right]\sin\left[\frac{p+(h-L)}{2}\right]\sec L \sec h. \qquad (10a)$$

The formula for azimuth angle can also be stated as:

$$\text{hav } Z=\sin(S-L)\sin(S-h)\sec h \sec L, \qquad (10b)$$

in which $S=\frac{1}{2}(h+L+p)$ and $S-h=\frac{1}{2}(L+p-h)$.

714. Altitude azimuth by haversines.—The altitude azimuth solution can be made by equation 10b:

$$\text{hav } Z=\sin(S-L)\sin(S-h)\sec h \sec L, \qquad (10b)$$

in which

$$S=\frac{1}{2}(h+L+p).$$

Example.—The latitude of the observer is 30°25.'0N; the declination of the celestial body is 22°06.'2N (polar distance 67°53.'8); the altitude of the body is 53°27.'0.

Required.—Azimuth of the body.

Solution.—By equation 10b.

$$S=\frac{1}{2}(h+L+p)$$
$$=\frac{1}{2}(53°27.'0+30°25.'0+67°53.'8)=75°52.'9.$$

$$\text{hav } Z=\sin(S-L)\sin(S-h)\sec h\sec L \qquad (10b)$$
$$=\sin(75°52.'9-30°25.'0)\sin(75°52.'9-53°27.'0)\sec 53°27.'0\sec 30°25.'0$$
$$=\sin 45°27.'9\sin 22°25.'9\sec 53°27.'0\sec 30°25.'0$$
$$=(+0.71282)(+0.38158)(+1.67919)(+1.15960)$$
$$=+0.52963$$
$$Z=93.°4 \text{ (table 34)}.$$

715. Solving for meridian angle.—The problem consists of finding the meridian angle from given values of altitude, latitude, and polar distance. Of the three elements, altitude, latitude, and declination used in the solution of the astronomical triangle, normally the only uncertain element is the latitude. Results are most accurate when the body is on or near the prime vertical, as then an error in latitude has least effect.

The haversine formula for meridian angle is derived from the following formula of trigonometry:

$$\sin^2\frac{1}{2}A=\frac{\sin(s-b)\sin(s-c)}{\sin b\sin c},$$

in which s is one-half of the sum of the three sides of the spherical triangle. As applied to the astronomical triangle of figure 707b, A is the meridian angle, t; side a is the coaltitude; side b is the polar distance; side c is the colatitude. Thus,

$$s=\frac{\text{coaltitude}+\text{polar distance}+\text{colatitude}}{2}$$

$$=\frac{(90^\circ-h)+p+(90^\circ-L)}{2}=90^\circ-\left(\frac{h+L-p}{2}\right)$$

$$s-b=\left[90^\circ-\left(\frac{h+L-p}{2}\right)-p\right]=90^\circ-\left(\frac{h+L+p}{2}\right)$$

$$s-c=\left[90^\circ-\left(\frac{h+L-p}{2}\right)-(90^\circ-L)\right]=\frac{L+p-h}{2}.$$

Substituting in the basic equation,

$$\sin^2\frac{1}{2}\,t=\text{hav}\ t=\frac{\sin\left[90^\circ-\left(\frac{h+L+p}{2}\right)\right]\sin\left(\frac{L+p-h}{2}\right)}{\sin p\cos L}. \qquad (10c)$$

The formula for meridian angle can also be stated as:

$$\text{hav}\ t=\csc p\sec L\cos S\sin(S-h), \qquad (10d)$$

in which $S=\frac{1}{2}(h+L+p)$ and $S-h=\frac{1}{2}(L+p-h)$.

The meridian angle of a body when its center is on the celestial horizon can be computed by the formula:

$$\cos t=-\tan L\tan d \qquad (10e)$$

where t is the meridian angle, L is the latitude, and d is the declination. Solutions of this formula are given in table 25, if the table is entered with a latitude 90° from the given latitude. This table can be used for this purpose only when latitude and declination are of contrary name.

In deriving equation 10e by Napier's Rules of Circular Parts, the two sides of the spherical triangle forming the right angle are latitude and coamplitude (fig. 720b).

Example 1.—The latitude of the observer is 40°00′.0N; the declination of the sun is 21°57′.9N.

Required.—Meridian angle of sun at sunrise.

Solution.—By equation 10d and table 34.

Allowing 16′ for semidiameter and 34′ for refraction, the center of the sun is 50′ below the celestial horizon when the upper limb of the sun is just visible from sea level.

$$S=\frac{1}{2}(h+L+p)=\frac{1}{2}(-0^\circ50'.0+40^\circ00'.0+68^\circ02'.1)=53^\circ36'.0.$$

$$S-h=53^\circ36'.0-(-0^\circ50'.0)=54^\circ26'.0.$$

Substituting in equation 10d,

$$\text{hav } t = \csc p \sec L \cos S \sin (S-h) \qquad (10d)$$

$$l\,\text{hav } t = l \csc 68°02'.1 + l \sec 40°00'.0 + l \cos 53°36'.0 + l \sin 54°26'.0$$
$$= 10.03273 + 10.11575 + 9.77336 + 9.91033$$
$$= 9.83217$$
$$t = 111°02'.1\text{E}.$$

Example 2.—The latitude of the observer on the Greenwich meridian is 40°00′.0N; at sunrise, the meridian angle of the sun is 111°02′.1E.

Required.—The LMT of sunrise.

Solution.—The solution is the reverse of finding GHA. The largest tabulated value of GHA in the almanac that does not exceed the desired GHA is found in the tabulation for the day, and recorded with its time. The difference between this value and the desired GHA is then used to enter the "Increments and Corrections" table. The time interval corresponding to this value is added to the time taken from the daily page.

t(G)	111°02′.1E
GHA	248°57′.9
4^h	240°35′.6
33^m29^s	8°22′.3
GMT	$4^h33^m29^s$
LMT	$4^h33^m29^s$*

Alternatively, the solution is effected using the Equation of Time. The meridian angle is converted to local apparent time (LAT); the Equation of Time is applied to LAT to find LMT.

t(G)	111°02′.1E
LAT	$4^h35^m52^s$
Eq. T	(+) 2^m20^s (rev.)
LMT	$4^h33^m32^s$*

*Because of the variations in refraction and height of eye, computation to a greater precision than 1^m is not justified.

716. Solving for parallactic angle.—An approximate value of the parallactic angle, X, accurate enough for most navigational requirements, can be calculated from the formula

$$\cos X = \frac{\Delta h}{60'}, \text{ where } \Delta h \qquad (10f)$$

is the change in altitude for 60′ change in declination, local hour angle and latitude remaining constant.

Example.—The latitude of the observer is 30°00′.0N; the declination of the celestial body is 45°00′.0S; the local hour angle is 30°00′.0, and Δh as extracted from a sight reduction table is 53′.9.

Required.—The parallactic angle.

Solution.—By the approximate formula:

$$\cos X = \frac{\Delta h}{60'} = \frac{53'.9}{60'} = 0.89833$$

$$X = 26°04'.$$

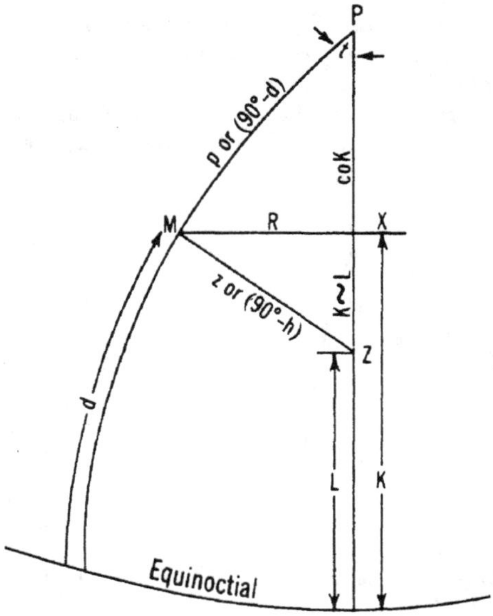

FIGURE 717a.—Divided astronomical triangle.

The Divided Astronomical Triangle

717. Development of the formulas.—In the divided astronomical triangle shown in figure 717a, the perpendicular is let fall from the celestial body to the celestial meridian. The parts of the triangle and special constructions are labeled as follows:

P, elevated pole of observer.

Z, zenith of observer or zenith of the assumed position of the observer.

M, celestial body.

L, latitude of observer or assumed position of observer.

d, declination of celestial body.

p, polar distance of celestial body.

t, meridian angle of celestial body.

h, altitude of celestial body.

R, perpendicular let fall from M on PZ. This is an auxiliary part.

X, intersection of R with PZ.

K, arc from X to celestial equator (equinoctial). This is an auxiliary part introduced to facilitate solution.

With the construction of auxiliary part R, the solutions of the unknown parts of the triangle can be effected through the solutions of right spherical triangles, using Napier's Rules of Circular Parts.

Omitting the right angle of the right spherical triangle of figure 717b, the triangle has five parts: two angles and three sides. The two sides forming the right angle and the *complements* of the other three parts are arranged in five sectors of the circular parts diagram in figure 717c in the same order in which they occur in the triangle of figure 717b.

Considering any part as the middle part, the two parts nearest it in the diagram are considered the *adjacent* parts, and the two farthest from it the *opposite* parts.

The rules are the sine of a middle part always equals the product of:

(1) the tangents of the adjacent parts or,

(2) the cosines of the opposite parts.

In the use of these rules the co-function of a complement is given as the function of the element. Thus, the cosine of co-A is the sine of A; the tangent of co-A is the cotangent of A, etc.

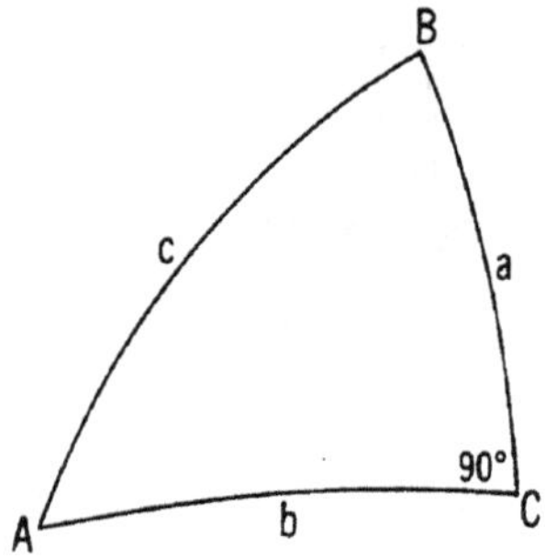

FIGURE 717b.—Right spherical triangle.

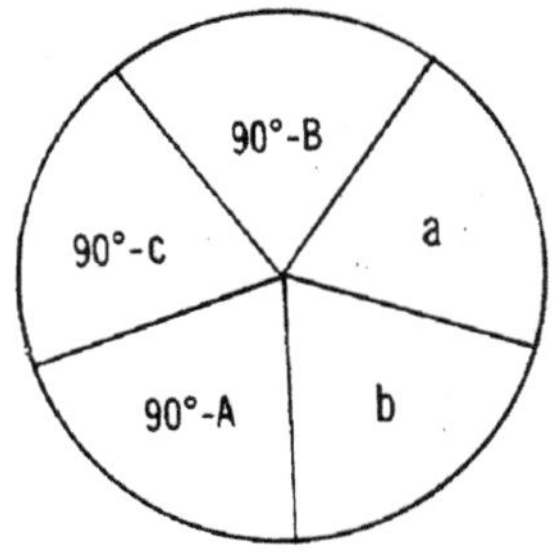

FIGURE 717c.—Circular parts diagram.

$$\sin a = \tan b \cot B = \sin c \sin A. \tag{11a}$$

$$\sin b = \tan a \cot A = \sin c \sin B. \tag{11b}$$

$$\cos c = \cot A \cot B = \cos a \cos b. \tag{11c}$$

$$\cos A = \tan b \cot c = \cos a \sin B. \tag{11d}$$

$$\cos B = \tan a \cot c = \cos b \sin A. \tag{11e}$$

Side c (the hypotenuse) is less than 90° when sides a and b are in the same quadrant, and more than 90° when a and b are in different quadrants. An oblique angle and the side opposite are in the same quadrant.

In applying Napier's rules to the right spherical triangle PMX of figure 717a, the five sectors of the circle are first completed as in figure 717d.

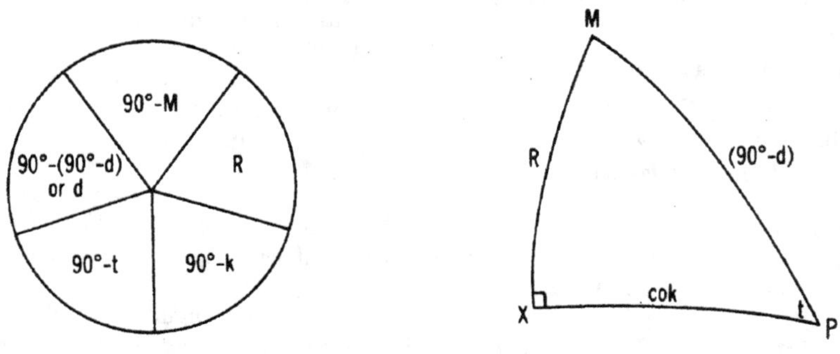

FIGURE 717d.—Circular parts diagram and triangle PMX.

$$\sin R = \cos (90° - t) \cos d$$

$$\sin R = \sin t \cos d$$

$$\csc R = \csc t \sec d. \tag{12a}$$

$$\sin d = \cos (90° - K) \cos R$$
$$\sin d = \sin K \cos R$$
$$\csc K = \csc d \div \sec R. \quad (12b)$$

For right spherical triangle ZMX of figure 717a, the five sectors of the circle are completed as in figure 717e.

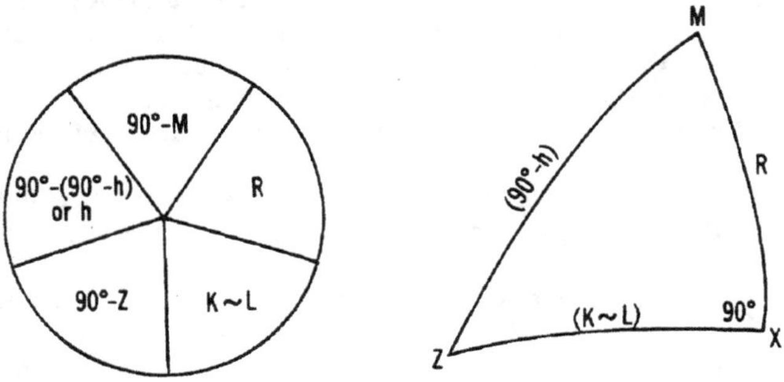

FIGURE 717e.—Circular parts diagram and triangle ZMX.

$$\sin h = \cos R \cos (K \sim L)$$
$$\csc h = \sec R \sec (K \sim L). \quad (12c)$$

$$\sin R = \cos (90° - Z) \cos h$$
$$\sin R = \sin Z \cos h$$
$$\csc Z = \csc R \div \sec h. \quad (12d)$$

With t and (90°−d) of triangle PMX known, part R is found by means of equation 12a. Knowing R and d, one can then find K by means of equation 12b. K is then combined algebraically with L to obtain (K∼L).

In triangle ZMX, sides R and (K∼L) are known. Equation 12c is used to obtain h. Azimuth angle, Z, is found by means of equation 12d.

718. The Ageton Method.—Table 35, The Ageton Method, can be used for the solution of the astronomical triangle divided by a perpendicular let fall from the celestial body to the celestial meridian. This table is arranged in parallel "A" and "B" columns, the "A" column containing log cosecants multiplied by 100,000 and the "B" column log secants multiplied by 100,000. This arrangement simplifies the arithmetic of solution of the formulas in terms of secants and cosecants. Ageton's ingenious arrangement of the calculations according to a standard form with simple, uniform rules also considerably simplifies the solution. The tabulation of functions every half of a minute throughout the tables aids interpolation.

Interpolation Requirements

For practical navigation, interpolation is normally required only when the quantity K or meridian angle, t, approaches 90°. When the quantity K is between 82° and 98°, it is good practice to either discard the sight or to interpolate the log secant (B) of R from the log cosecant (A) of R.

Solution by Table 35

The letters A and B indicate from which column of table 35 to take their corresponding functions.

The following steps are taken to obtain the altitude and azimuth angle in one combined solution using table 35:

1. Enter the table with meridian angle, t, and take from the A column the nearest tabulated number. Enter the table with declination and take the tabulated number from the B column. Add these numbers to obtain the A function of R.

2. Look up the number thus obtained in the A column and take from the B column the opposite tabulated number. Subtract this from the function of declination taken from the A column.

3. Enter the A column with the number thus obtained and take out the nearest tabulated value of K, the degrees from the top or bottom of the page and the minutes from the side.

4. Give K the same name as declination. Combine K with latitude to obtain (K∼L), adding K and latitude if of contrary name and subtracting the smaller from the larger if same name.

5. Enter the table with (K∼L) and take from the B column the nearest tabulated number. Add this to the B function of R.

6. With the number thus obtained, enter the A column and take h from the table.

7. To compute the azimuth angle look up h in the B column and subtract the number thus obtained from the A function of R, which has already been determined.

8. With this number, enter the A column and take out Z to the nearest minute. Z is always measured from the elevated pole 0° to 180° east or west to the body.

At the top of the pages of the table are stated the only two rules required.

(1) When Meridian Angle is Greater than 90°, Take K from Bottom of Table.

(2) Always Take "Z" from Bottom of Table, Except when K is Same Name and Greater than Latitude, in Which Case Take Z from Top of Table.

If a celestial body near the visible horizon is observed, it may be below the celestial horizon (zenith distance greater than 90°) because of refraction and dip. Under these conditions the altitude, h, is negative. In the solution by table 35, h is negative if K is of the same name as L and greater than (90°+L), or if K is of contrary name to L and greater than (90°−L). Under the second of these circumstances, Z is less than 90° and should be taken from the top of the table *if* K is greater than (180°−L).

When the azimuth angle is near 90°, the log cosecant (A) of Z may be a negative number if the tables have been used without interpolating. This can be avoided by interpolating throughout or by substituting the nearest tabulated value of A for the calculated value of log cosecant (A) of R at the top of the azimuth column.

"Wrinkles"

Certain short-cuts will become apparent with use of the table:

When entering the table with declination, take functions from both A and B columns in one entry.

When taking h from the table, take the tabulated function from the B column at the same time, as it will be used in computing Z.

Example 1.—The latitude of the observer is 42°10′.5S. The declination of a celestial body is 23°07′.0S; the meridian angle is 61°33′.8E.

Required.—Altitude and azimuth of the body.

t AND K ARE BOTH GREATER OR BOTH LESS THAN 90°.
Z IS LESS THAN 90° ONLY WHEN K HAS THE SAME NAME AND IS GREATER THAN L.

'	A 0° / B 90°	B 0° / A 90°	A 1° / B 91°	B 1° / A 91°	A 2° / B 92°	B 2° / A 92°	A 3° / B 93°	B 3° / A 93°	A 4° / B 94°	B 4° / A 94°	'
00	------	0.0	175814	6.6	145718	26.5	128120	59.6	115642	105.9	60
01	353627	0.0	5097	6.8	5358	26.9	7880	60.2	5461	106.8	59
02	323524	0.0	4391	7.1	5001	27.4	7641	60.9	5282	107.7	58
03	305915	0.0	3696	7.3	4646	27.8	7403	61.6	5103	108.6	57
04	293421	0.0	3012	7.5	4295	28.3	7166	62.2	4925	109.5	56
05	283730	0.0	172339	7.8	143946	28.7	126931	62.9	114748	110.4	55
06	275812	0.1	1676	8.0	3600	29.2	6697	63.6	4571	111.3	54
07	269118	0.1	1023	8.2	3257	29.6	6465	64.3	4395	112.2	53
08	3318	0.1	0379	8.5	2916	30.1	6233	65.0	4220	113.1	52
09	258203	0.1	169745	8.7	2579	30.6	6003	65.7	4045	114.0	51
10	253627	0.2	169121	9.0	142243	31.1	125774	66.4	113872	114.9	50
11	249488	0.2	8505	9.3	1911	31.5	5546	67.1	3699	115.9	49
12	5709	0.3	7897	9.5	1581	32.0	5320	67.8	3526	116.8	48
13	2233	0.3	7298	9.8	1253	32.5	5094	68.5	3355	117.7	47
14	239015	0.4	6708	10.1	0928	33.0	4870	69.2	3184	118.7	46
15	236018	0.4	166125	10.3	140605	33.5	124647	69.9	113013	119.6	45
16	3216	0.5	5550	10.6	0285	34.0	4425	70.6	2844	120.5	44
17	0583	0.5	4982	10.9	139967	34.5	4205	71.3	2675	121.5	43
18	228100	0.6	4422	11.2	9651	35.0	3985	72.1	2506	122.4	42
19	5752	0.7	3869	11.5	9338	35.5	3766	72.8	2339	123.4	41
20	223525	0.7	163322	11.8	139027	36.0	123549	73.5	112171	124.3	40
21	1406	0.8	2783	12.1	8718	36.5	3333	74.3	2005	125.3	39
22	219385	0.9	2250	12.4	8411	37.1	3117	75.0	1839	126.2	38
23	7455	1.0	1724	12.7	8106	37.6	2903	75.8	1674	127.2	37
24	5607	1.1	1204	13.0	7804	38.1	2690	76.5	1510	128.2	36
25	213834	1.1	160690	13.3	137503	38.6	122478	77.3	111346	129.2	35
26	2130	1.2	0182	13.6	7205	39.2	2267	78.0	1183	130.1	34
27	0491	1.3	159680	13.9	6909	39.7	2057	78.8	1020	131.1	33
28	208912	1.4	9184	14.2	6615	40.3	1848	79.5	0858	132.1	32
29	7388	1.5	8693	14.6	6322	40.8	1640	80.3	0696	133.1	31
30	205916	1.7	158208	14.9	136032	41.4	121432	81.1	110536	134.1	30
31	4492	1.8	7728	15.2	5744	41.9	1226	81.9	0375	135.1	29
32	3113	1.9	7254	15.6	5457	42.5	1021	82.6	0216	136.1	28
33	1777	2.0	6784	15.9	5173	43.0	0817	83.4	0057	137.1	27
34	0480	2.1	6320	16.2	4890	43.6	0614	84.2	109898	138.1	26
35	199221	2.3	155861	16.6	134609	44.2	120412	85.0	109740	139.1	25
36	7998	2.4	5406	16.9	4330	44.7	0211	85.8	9583	140.1	24
37	6808	2.5	4956	17.3	4053	45.3	0010	86.6	9426	141.1	23
38	5650	2.7	4511	17.6	3777	45.9	119811	87.4	9270	142.2	22
39	4522	2.8	4070	18.0	3503	46.5	9612	88.2	9115	143.2	21
40	193422	2.9	153634	18.4	133231	47.1	119415	89.0	108960	144.2	20
41	2350	3.1	3201	18.7	2961	47.6	9218	89.8	8805	145.2	19
42	1304	3.2	2774	19.1	2692	48.2	9022	90.6	8651	146.3	18
43	0282	3.4	2350	19.5	2425	48.8	8827	91.4	8498	147.3	17
44	189283	3.6	1931	19.9	2159	49.4	8633	92.3	8345	148.4	16
45	188307	3.7	151515	20.3	131896	50.0	118440	93.1	108193	149.4	15
46	7353	3.9	1104	20.6	1633	50.7	8248	93.9	8041	150.5	14
47	6419	4.1	0696	21.0	1373	51.3	8056	94.7	7890	151.5	13
48	5505	4.2	0292	21.4	1114	51.9	7866	95.6	7739	152.6	12
49	4609	4.4	149892	21.8	0856	52.5	7676	96.4	7589	153.6	11
50	183732	4.6	149496	22.2	130600	53.1	117487	97.3	107439	154.7	10
51	2872	4.8	9103	22.6	0346	53.7	7299	98.1	7290	155.8	09
52	2029	5.0	8713	23.1	0093	54.4	7112	99.0	7141	156.9	08
53	1202	5.2	8327	23.5	129841	55.0	6925	99.8	6993	157.9	07
54	0390	5.4	7945	23.9	9591	55.7	6739	100.7	6846	159.0	06
55	179593	5.6	147566	24.3	129342	56.3	116554	101.6	106699	160.1	05
56	8811	5.8	7190	24.7	9095	56.9	6370	102.4	6552	161.2	04
57	8042	6.0	6817	25.2	8849	57.6	6187	103.3	6406	162.3	03
58	7287	6.2	6448	25.6	8605	58.2	6004	104.2	6260	163.4	02
59	6544	6.4	6081	26.0	8362	58.9	5823	105.0	6115	164.5	01
60	175814	6.6	145718	26.5	128120	59.6	115642	105.9	105970	165.6	00
'	B 89° / A 179°	A 89° / B 179°	B 88° / A 178°	A 88° / B 178°	B 87° / A 177°	A 87° / B 177°	B 86° / A 176°	A 86° / B 176°	B 85° / A 175°	A 85° / B 175°	'

FIGURE 718.—Extract from *Compact Sight Reduction Table.* Copyright © 1980 by Cornell Maritime Press, Inc. Used with permission.

Solution.—By table 35:

		Add	*Subtract*	*Add*	*Subtract*
t	61°33′.8E	A 5583			
d	23°07′.0S	B 3635	A 40604		
R		A 9218	B 23052	B 23052	A 9218
K	41°52′.5S		A 17552		
L	42°10′.5S				
K∼L	0°18′.0			B 0.6	
h	36°01′.5			A 23052.6	B 9218
			Zn 090°.0	Z S90°00′E	A 0

Example 2.—The latitude of the observer is 40°43′.0N. The declination of a celestial body is 38°42′.9N; the meridian angle is 105°17′.0E.

Required.—Altitude and azimuth of the body.

Solution.—By table 35:

		Add	*Subtract*	*Add*	*Subtract*
t	105°17′.0E	A 1564			
d	38°42′.9N	B 10777	A 20379		
R		A 12341	B 18146	B 18146	A 12341
K	108°13′.0N		A 2233		
L	40°43′.0N				
K∼L	67°30′.0			B 41716	
h	14°35′.5			A 59862	B 1424
			Zn 051°.0	Z N51°03′E	A 10917

Modified Ageton Method

The modification of the Ageton Method (table 35) shown in figure 718 introduces two significant simplifications: (1) Entries for half-minutes of arc have been omitted, and (2) numerical duplication of the second half of table 35 has been eliminated. The elimination of the second half requires a change in column headings and restatement of the rules for the use of the table shown at the top of each page.

The modified Ageton Method provides a much shorter table (9 pages instead of the 36 pages of table 35) which results in fewer page openings and less table searching.

Compass Error

719. Azimuth by tables.—One of the more frequent applications of sight reduction tables is their use in computing the azimuth of a celestial body for comparison with an observed azimuth in order to determine the error of the compass. In computing the azimuth of a celestial body, for the time and place of observation, it is normally necessary to interpolate the tabular azimuth angle as extracted from the tables for the differences between the table arguments and the actual values of declination, latitude, and local hour angle. The required triple interpolation of the azimuth angle using Pub. No. 229, *Sight Reduction Tables for Marine Navigation*, is effected as follows:

(1) Refer to figure 719a. The main tables are entered with the nearest integral values of declination, latitude, and local hour angle. For these arguments, a base azimuth angle is extracted.

(2) The tables are reentered with the same latitude and LHA arguments but with the declination argument 1° greater or less than the base declination argument depending upon whether the actual declination is greater or less than the base argument. The difference between the respondent azimuth angle and the base azimuth angle establishes the azimuth angle difference (Z Diff.) for the increment of declination.

(3) The tables are reentered with the base declination and LHA arguments but with the latitude argument 1° greater or less than the base latitude argument depending upon whether the actual (usually DR) latitude is greater or less than the base argument to find the Z Diff. for the increment of latitude.

(4) The tables are reentered with the base declination and latitude arguments but with the LHA argument 1° greater or less than the base LHA argument depending upon whether the actual LHA is greater or less than the base argument to find the Z Diff. for the increment of LHA.

(5) The correction to the base azimuth angle for each increment is Z Diff.$\times\frac{\text{Inc.}}{60'}$.

The auxiliary interpolation table (art. 206) can normally be used for computing this value because the successive angle differences are less than 10°0 for altitudes less than 84°.

43°, 317° L.H.A. **LATITUDE SAME NAME AS DECLINATION**

	30°			31°			32°			33°			34°			35°			
Dec.	Hc	d	Z	Hc	d	Z	Hc	d	Z	Hc	d	Z	Hc	d	Z	Hc	d	Z	Dec.
°	° ′	′	°	° ′	′	°	° ′	′	°	° ′	′	°	° ′	′	°	° ′	′	°	°
20	50 00.7	+23.2	94.2	49 55.7	+24.6	95.4	49 49.4	+26.1	96.6	49 41.9	+27.5	97.8	49 33.2	+28.9	98.9	49 23.3	+30.3	100.1	20
21	50 23.9	22.0	92.8	50 20.3	23.6	94.0	50 15.5	25.0	95.2	50 09.4	26.6	96.4	50 02.1	28.0	97.6	49 53.6	29.4	98.8	21

44°, 316° L.H.A. **LATITUDE SAME NAME AS DECLINATION**

	30°			31°			32°			33°			34°			35°			
Dec.	Hc	d	Z	Hc	d	Z	Hc	d	Z	Hc	d	Z	Hc	d	Z	Hc	d	Z	Dec.
°	° ′	′	°	° ′	′	°	° ′	′	°	° ′	′	°	° ′	′	°	° ′	′	°	°
20	49 08.9	+23.0	93.7	49 04.5	+24.5	94.8	48 58.8	+26.0	96.0	48 52.0	+27.4	97.1	48 44.0	+28.8	98.2	48 34.8	+30.2	99.4	20
21	49 31.9	22.0	92.2	49 29.0	23.5	93.4	49 24.8	25.0	94.6	49 19.4	26.4	95.7	49 12.8	27.8	96.9	49 05.0	29.2	98.0	21

FIGURE 719a.—Extracts from Pub. No. 229.

	Actual	Base Arguments	Base Z	Tab* Z	Z Diff.	Increments	Correction (Z Diff × Inc. ÷ 60)
Dec.	20°13.′8N	20°	97.°8	96.°4	−1.°4	13.′8	−0.°3
DR Lat.	33°24.′0N	33°(Same)	97.°8	98.°9	+1.°1	24.′0	+0.°4
LHA	316°41.′2	317°	97.°8	97.°1	−0.°7	18.′8	−0.°2
						Total Corr.	−0.°1

Base Z	97.°8
Corr.	(−) 0.°1
Z	N 97.°7E
Zn	097.°7
Zn pgc	096.°5
Gyro Error	1.°2E

*Respondent for two base arguments and 1° change from third base argument, in vertical order of Dec., DR Lat., and LHA.

FIGURE 719b.—Azimuth by Pub. No. 229.

Example 1.—In DR lat. 33°24′.0N, the azimuth of the sun is observed as 096°.5 pgc. At the time of the observation, the declination of the sun is 20°13′.8N; the local hour angle of the sun is 316°41′.2.

Required.—The gyro error.

Solution.—By Pub. No. 229:

The error of the gyrocompass is found as shown in figure 719b.

Example 2.—In DR lat. 33°24′.S, the azimuth of the sun is observed as 096°.5 pgc. At the time of the observation, the declination of the sun is 20°13′.8N; the local hour angle of the sun is 316°41′.2.

Required.—The gyro error.

Solution.—By Pub. No. 229:

The error of the gyrocompass is found as shown in figure 719c.

	Actual	Base Arguments	Base Z	Tab* Z	Z Diff.	Increments	Correction (Z Diff × Inc. ÷ 60)
Dec.	20°13′.8N	20°	97°.8	96°.4	−1°.4	13′.8	−0°.3
DR Lat.	33°24′.0N	33°(Same)	97°.8	98°.9	+1°.1	24′.0	+0°.4
LHA	316°41′.2	317°	97°.8	97°.1	−0°.7	18′.8	−0°.2
						Total Corr.	−0°.1

Base Z	97°.8
Corr.	(−) 0°.1
Z	N 97°.7E
Zn	097°.7
Zn (pgc)	096°.5
Gyro Error	1°.2E

*Respondent for two base arguments and 1° change from third base argument, in vertical order of Dec., DR Lat., and LHA.

FIGURE 719c.—Azimuth by Pub. No. 229.

720. Amplitudes.—For checking the compass, an azimuth observation of a celestial body at low altitude is desirable because it can be measured easiest and most accurately. If the body is observed when its center is on the *celestial* horizon, the amplitude (A), which is the arc of the horizon between the prime vertical and the body, can be taken directly from table 27.

The amplitude is given the prefix E (east) if the body is rising and W (west) if setting. It is given the suffix N if the body rises or sets north of the prime vertical (which it does if it has northerly declination) and S if it rises or sets south of the prime vertical (having southerly declination). The suffix is given to agree with the declination of the body. Interconversion of amplitude and aximuth is similar to that of aximuth angle and azimuth. Thus, if A=E15°S, the body is 15° south of east or 90°+15°=105°. For any given body, the numerical value of amplitude would be the same at rising and setting if the declination did not change.

When the center of the sun is on the celestial horizon, its *lower limb* is about two-thirds of a diameter above the visible horizon. When the center of the moon is on the celestial horizon, its *upper limb* is on the visible horizon. When planets and stars are on the celestial horizon, they are a little more than one sun diameter above the visible horizon. In high latitudes, amplitudes should be observed on the visible horizon.

If the body is observed when its center is on the *visible horizon*, the *observed* value should be corrected by the value from table 28, using the rules given with the table, before comparison with the value taken from table 27. If preferred, the correction can be applied with reversed sign to the value taken from table 27 and compared with the uncorrected observed value. This is the procedure used if amplitude or azimuth is desired when the celestial body is on the visible horizon.

In the diagram on the plane of the celestial meridian shown in figure 720a, a celestial body of declination same name as latitude is on the celestial horizon at M; a celestial body having declination of contrary name is on the celestial horizon at M′. Triangles PnNM and PnNM′ are right spherical triangles, PnNM and PnNM′ being right angles.

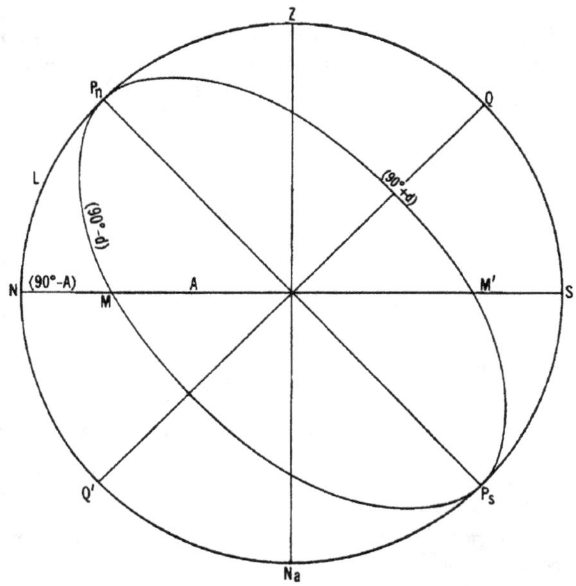

FIGURE 720a.—Diagram on plane of the celestial meridian.

Side NM of triangle PnNM is the coamplitude (90°−A); side PnN is the latitude; and side PnM is the codeclination (90°−d).

The formula for amplitude is derived by Napier's rules:

The five sectors of the diagram for right spherical triangle PnNM are completed as in figure 720b. Applying Napier's rules:

$$\sin d = \cos L \cos (90° - A)$$
$$\sin d = \cos L \sin A$$
$$\sin A = \sec L \sin d. \qquad (13)$$

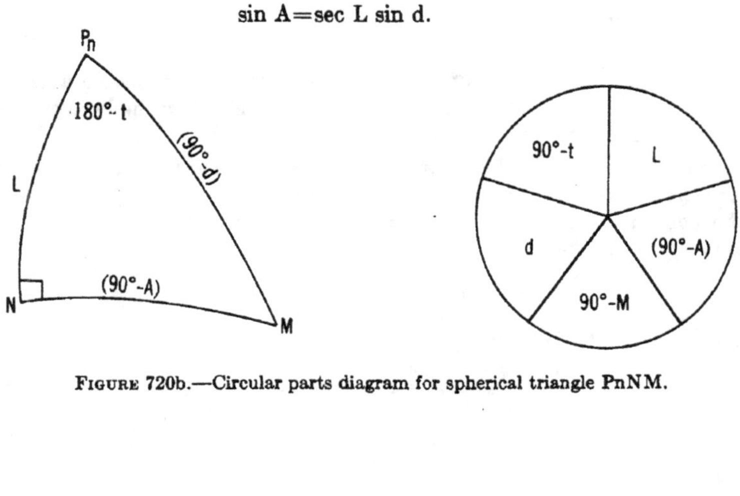

FIGURE 720b.—Circular parts diagram for spherical triangle PnNM.

For the contrary name case, side NM′ is (90°+A); side PnN is the latitude; and side PnM′ is (90°+d). The five sectors of the diagram for right spherical triangle PnNM′ are completed as in figure 720c. Applying Napier's rules:

$$-\sin d = \cos L \cos(90° + A)$$
$$\cos(90° + A) = -\sec L \sin d$$
$$\sin A = \sec L \sin d, \qquad (13)$$

in which A is the amplitude, L is the latitude of the observer, and d is the declination of the celestial body. Table 27 was computed by means of this formula.

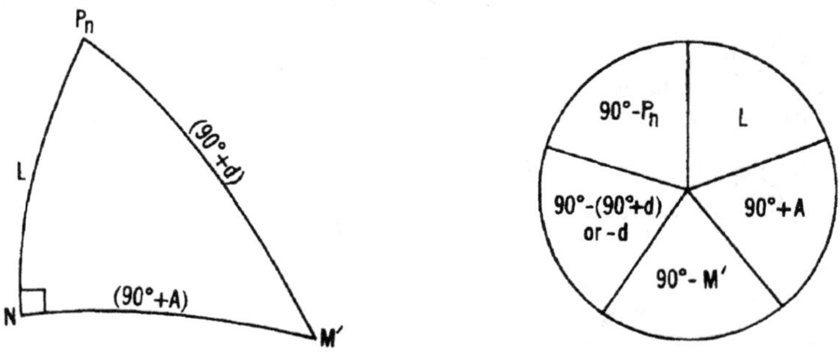

FIGURE 720c.—Circular parts diagram for spherical triangle PnNM′.

Example.—The DR latitude of a ship is 51°24′.6N, at a time when the declination of the sun is 19°40′.4N.

Required.—(1) The amplitude (A) when the center of the setting sun is on the celestial horizon.

(2) The amplitude when the center of the setting sun is on the visible horizon.

(3) The azimuth when the center of the setting sun is on the visible horizon.

Solution.—

(1) A W32°.6N (tab. 27)
T 28 1°.1S (rev.)—applied to *tabulated* amplitude
(2) A W33°.7N
(3) Zn 303°.7.

721. Finding time on the prime vertical.—A celestial body having a declination of opposite name to the latitude crosses the prime vertical below the horizon. Its nearest visible approach is at the time of rising and setting.

If a celestial body has a declination of the same name as the latitude, but is numerically *greater*, it does not cross the prime vertical. Its nearest approach (in azimuth) is at the point at which its azimuth angle is maximum. At this point the meridian angle is given by the formula

$$\sec t = \tan d \cot L, \qquad (14a)$$

and its altitude by the formula

$$\csc h = \sin d \csc L. \qquad (14b)$$

A celestial body having a declination of the same name as the latitude, and numerically *smaller*, crosses the prime vertical at some point before it reaches the celestial meridian, and again after meridian transit. At these two crossings of the prime vertical, the meridian angles are equal and are always less than 90°. They are given by the formula

$$\cos t = \tan d \cot L. \tag{14c}$$

The altitudes are also equal, and are given by the formula

$$\sin h = \sin d \csc L. \tag{14d}$$

Meridian angle and altitude of bodies on the prime vertical, and similar data for the nearest approach (in azimuth) of those bodies of same name which do not cross the prime vertical, are given in table 25 for various latitudes, and for declinations from 0° to 23°, inclusive.

Equation 14c for meridian angle, when azimuth angle is 90°, is derived by Napier's rules as follows:

The circular parts diagram for astronomical triangle PMZ is completed as shown in figure 721.

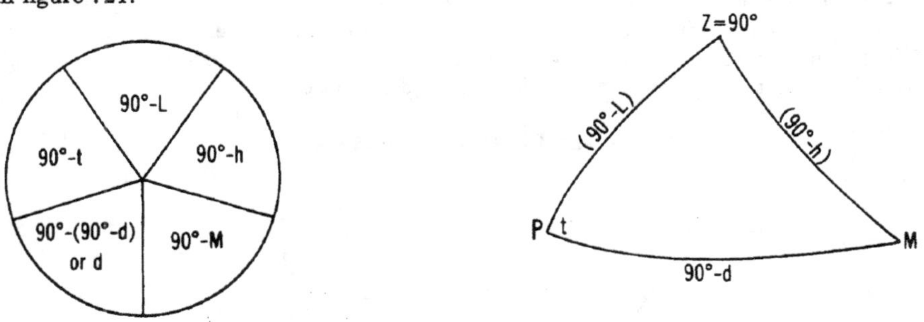

FIGURE 721.—Circular parts diagram for astronomical triangle PMZ.

$$\sin (90° - t) = \tan d \tan (90° - L)$$

$$\cos t = \tan d \cot L. \tag{14c}$$

The altitudes of the two crossings of the prime vertical are also equal. Equation 14d for altitude on the prime vertical is derived by Napier's rules:

$$\sin d = \cos(90° - L) \cos(90° - h)$$

$$\sin d = \sin L \sin h$$

$$\sin h = \sin d \csc L. \tag{14d}$$

To find the *time* of crossing the prime vertical, convert t to LHA, and add west longitude or subtract east longitude to find GHA. The GMT at which this GHA occurs can be found, as explained in article 733, and converted to any other time desired.

Example.—Determine (1) the approximate zone time, and (2) the approximate altitude of the sun when it crosses the prime vertical during the afternoon of May 30, 1975, at lat. 51°32′.3N, long. 160°21′.7W, using table 25 and the *Nautical Almanac*.

Solution.—

	May 30
t	71°.6W (from table 25)
LHA	71°.6
λ	160°.4W
GHA	232°.0
3^h	225°.6
26^m	6°.4
GMT	0326 May 31
ZD(+)	11 (rev.)
(1) ZT	1626 May 30
(2) h	28°.4 (from table 25).

At the time of crossing the prime vertical, or at nearest approach (in azimuth), a celestial body is changing azimuth slowly, and therefore this is considered a good time to check compass deviation or to swing ship.

The prime vertical at any place is the celestial horizon of a point 90° away, on the same meridian. Therefore, a celestial body crosses the prime vertical at approximately the same time it rises and sets at the point 90° away. Thus, if one is at latitude 35° N, the sun crosses his prime vertical at about the same time it rises or sets at latitude 55°S. If time of sunrise and sunset are to be obtained accurately by this method, corrections must be applied for semidiameter and refraction.

Dip and Distance of the Horizon

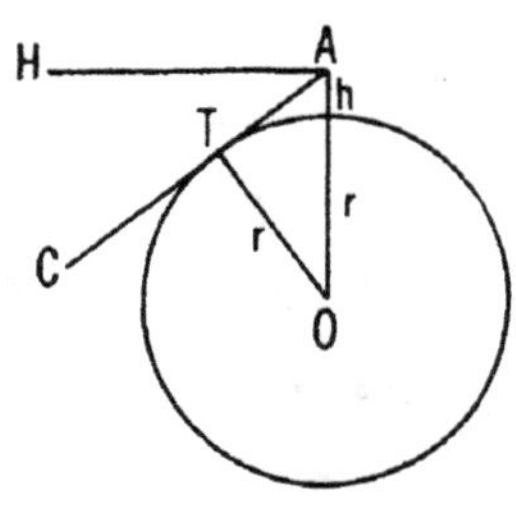

FIGURE 722a.—Dip without refraction.

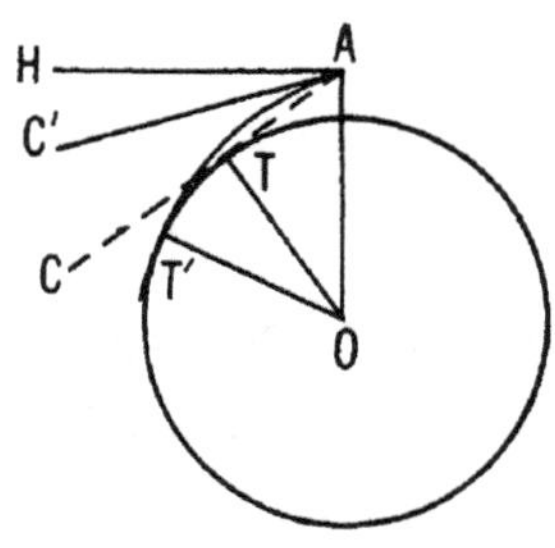

FIGURE 722b.—Dip with refraction.

722. Dip of geometrical horizon.—If there were no refraction, dip would be the angle between the horizontal at the eye of the observer, and a straight line from this point tangent to the surface of the earth at the geometrical horizon. In figure 722a, the eye of the observer is at A at height h above the surface of the earth of radius r. The line AH is the horizontal through A, and the straight line AC is tangent to the earth's surface at T. Angle HAC is the dip (D) at A, neglecting refraction. Since OA is perpendicular to AH and OT is perpendicular to AC, angle AOT is equal to angle HAC (art. 127).

$$\tan D = \tan AOT = \frac{AT}{OT} = \frac{AT}{r}.$$

From geometry,

$$AT = \sqrt{h(2r+h)}.$$

Accordingly,

$$\tan D=\frac{1}{r}\sqrt{2rh+h^2}$$

$$\tan^2 D=\frac{2rh+h^2}{r^2}=\frac{2h}{r}+\left(\frac{h}{r}\right)^2$$

$$\tan D=\sqrt{\frac{2h}{r}+\left(\frac{h}{r}\right)^2}.$$

Since h is very small compared with r, the square of the fraction $\left(\frac{h}{r}\right)$ can be discarded without appreciable effect.

Then, $\tan D=\sqrt{\frac{2h}{r}}$, approximately.

Using 20.902×10^6 feet as the mean radius of the earth and expressing h in feet,

$$\tan D=\sqrt{\frac{2h}{20.902\times 10^6}}.$$

With D expressed in minutes of arc,

$$D \tan 1'=\sqrt{\frac{2h}{20.902\times 10^6}}$$

$$D=1.06\sqrt{h}.$$

723. Distance of geometrical horizon.—A straight line from the eye of the observer tangent to the earth leads to the geometrical horizon. In figure 722a the geometrical horizon of the observer at A is at point T; the distance of the geometrical horizon is the arc of the circle subtending the central angle AOT. Therefore, when neglecting refraction, the numerical value of the dip, in minutes of arc, is equal to the distance to the geometrical horizon, in nautical miles. Or,

$$d=1.06\sqrt{h}, \tag{15}$$

where h is the height of eye in feet and d is the distance of the geometrical horizon in nautical miles.

724. Distance of visible horizon.—The line of sight from the observer's eye to the horizon passes through a region of varying density. Consequently, the ray of light from the horizon to the observer's eye is bent by refraction. The normal result of this *terrestrial refraction* is to increase the distance to the horizon, which is at T′ of figure 722b, instead of at T. This actual distance, under normal conditions, is given in table 8. Although the horizon is farther away than it would be if there were no refraction, it appears *higher* for the eye of the observer does not detect the curvature of the line of sight. Therefore, the horizon *appears* to be at C′ instead of at C.

The angle HAC′ is the dip of the visible horizon, i.e., dip with allowance made for refraction. The angle of refraction R is C′AC.

Table 8 was computed from the equation:

$$d=\sqrt{\frac{2r_0 h_f}{6076.1\,\beta_0}}, \tag{16a}$$

where d is the distance to the visible horizon in nautical miles; r_0 is the mean radius of the earth, 3440.1 nautical miles; β_0 is a parameter which characterizes the effect of terrestrial refraction; and h_f is the height of eye of the observer above sea level in feet.

Equation 16a is simplified to:

$$d = 1.17\sqrt{h_f} \tag{16b}$$

or

$$d = 2.11\sqrt{h_m} \tag{16c}$$

where height of eye is in meters.

Example.—The height of eye of the observer is 49 feet.

Required.—Distance to the visible horizon.

Solution.—By equation 16b:

$$d = 1.17\sqrt{49} = 1.17 \times 7$$

$$d = 8.2 \text{ n.mi.}$$

725. Dip of visible horizon.—The dip of the visible horizon (D) is the angle HAC′ of figure 725, the angle between the horizontal at the eye of the observer and a straight line tangent at A to the curved ray of light from the visible horizon.

The dip of the visible horizon is calculated from the equation:

$$D = 60 \tan^{-1} \sqrt{\frac{2\beta_0 h_f}{6076.1\, r_o}} \tag{17a}$$

where D is the dip of the visible horizon in minutes of arc; β_o is a parameter which characterizes the effect of terrestrial refraction; h_f is the height of eye of the observer above sea level in feet; and r_o is the mean radius of the earth, 3440.1 nautical miles.

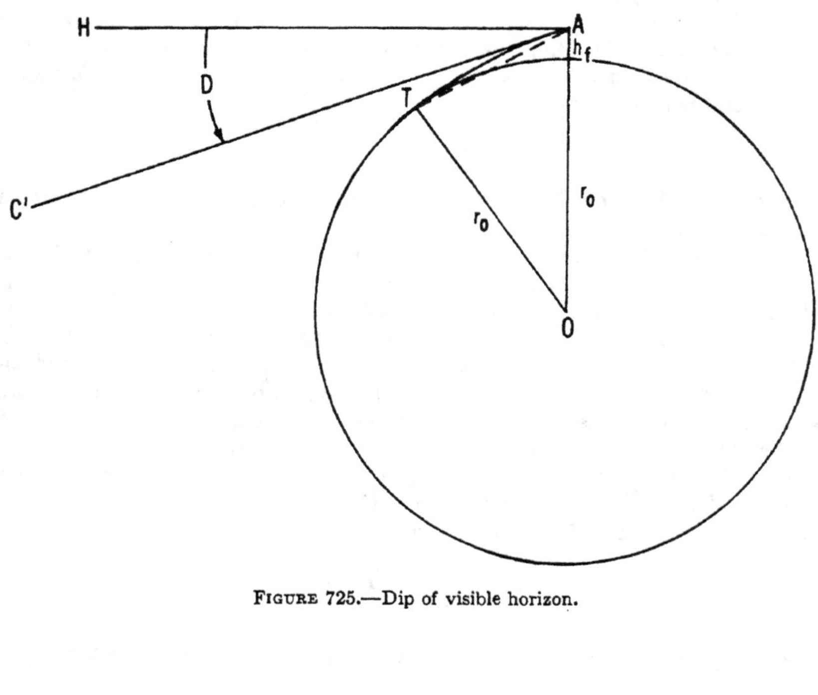

FIGURE 725.—Dip of visible horizon.

For all practical eye heights, equation 17a simplifies to:

$$D = 0!97\sqrt{h_f} \quad (17b)$$

$$D = 1!76\sqrt{h_m} \quad (17c)$$

Example.—The height of eye of the observer is 36 feet.
Required.—The dip of the visible horizon.
Solution.—By equation 17b:

$$D = 0.97\sqrt{h} = 0.97\sqrt{36} = 0.97 \times 6$$

$$D = 5!8.$$

726. Dip short of the horizon.—If land, another ship, or other obstruction is between the observer and his sea horizon, an altitude can be measured by using the waterline of the obstruction as a horizontal reference, if its distance from the observer is known. In this case the dip is greater than that given in the almanacs. Table 22 gives the values to be used. See figure 726.

Table 22 was computed from the equation:

$$D_s = 60 \tan^{-1}\left(\frac{h_f}{6076.1\, d_s} + \frac{\beta_0\, d_s}{2\, r_o}\right), \quad (18a)$$

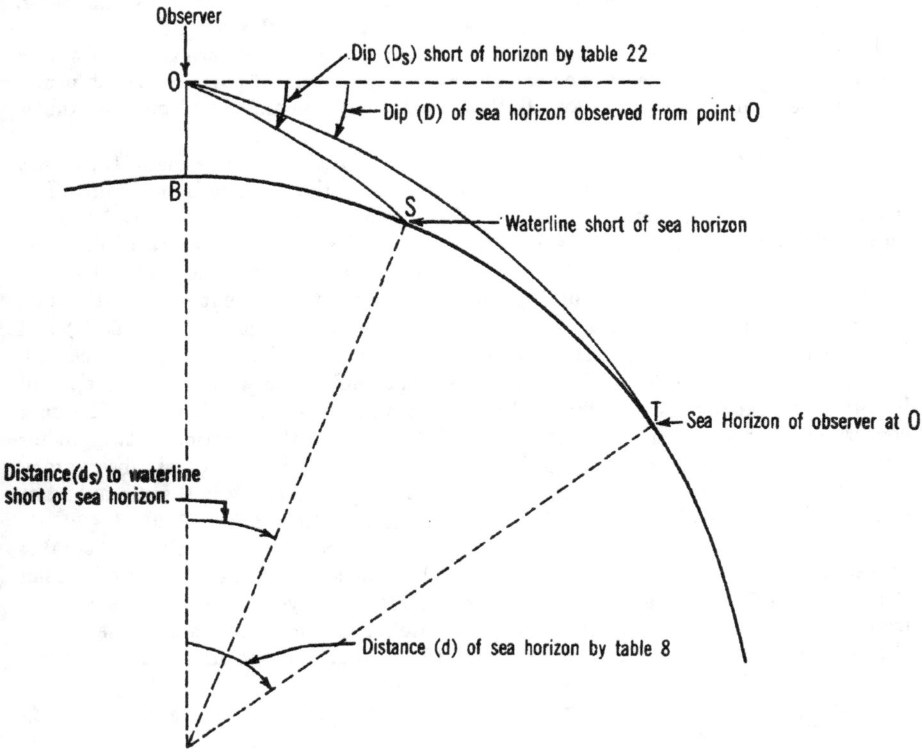

FIGURE 726.—Dip short of the horizon.

where D_s is the dip short of the horizon in minutes of arc; h_f is the height of eye of the observer above sea level in feet; β_0 is a parameter which characterizes terrestrial refraction; r_0 is the mean radius of the earth, 3440.1 nautical miles; and d_s is the distance to waterline of obstruction in nautical miles.

Equation 18a simplifies to:

$$D_s = 60 \tan^{-1}\left(\frac{h_f}{6076.1\, d_s} + \frac{d_s}{8268}\right). \qquad (18b)$$

Rising, Setting, and Twilight

727. Rising, setting, and twilight.—In both almanacs the times of sunrise, sunset, moonrise, moonset, and twilight information at various latitudes between 72°N and 60°S are given to the nearest whole minute. By definition, rising or setting occurs when the *upper* limb of the body is on the *visible* horizon, assuming standard refraction for zero height of eye. Because of variations in refraction and height of eye, computation to a greater precision than 1^m is not justified.

In high latitudes some of the phenomena do not occur during certain periods. The symbols used to indicate this condition are:

□ Sun or moon does not set, but remains continuously above the horizon.

■ Sun or moon does not rise, but remains continuously below the horizon.

//// Twilight lasts all night.

The *Nautical Almanac* makes no provision for finding the times of rising, setting, or twilight in polar regions. The *Air Almanac* has graphs for this purpose.

In the *Nautical Almanac*, sunrise, sunset, and twilight tables are given only once for the middle of the three days on each page opening. For most purposes this information can be used for all three days. Both almanacs have moonrise and moonset tables for each day.

The tabulations are in local mean time (art. 610). On the zone meridian, this is the zone time (ZT). For every 15′ of longitude that the observer's position differs from that of the zone meridian, the zone time of the phenomena differs by 1^m, being *later* if the observer is *west* of the zone meridian, and *earlier* if he is *east* of the zone meridian. The local mean time of the phenomena varies with latitude of the observer, declination of the body, and hour angle of the body relative to that of the mean sun.

Sunrise and sunset are also tabulated in the tide tables (from 76°N to 60°S) and in a supplement to the American ephemeris of 1946 entitled *Tables of Sunrise, Sunset, and Twilight* (from 75°N to 75°S). The meridian angle of any body at the time of its rising and setting can be computed by the formulas given in article 715. The data concerning the rising and setting of the sun and moon and the duration of twilight for high southern latitudes are published as graphs in United States Naval Observatory Circular No. 147, *Sunlight, Moonlight, and Twilight for Antarctica;* these graphs are similar to the graphs in the *Air Almanac* for northern latitudes (art. 730).

728. Finding time of sunrise and sunset.—*Nautical Almanac.* Enter the table on the daily page, and extract the LMT for the tabulated latitude next *smaller* than the observer's latitude (unless this is an exact tabulated value). Apply a correction from table I on almanac page xxxii to interpolate for latitude, determining the sign of the correction by inspection. Then convert LMT to ZT by means of the difference in longitude (dλ) between the local and zone meridians.

Example.—Find the zone time of sunrise and sunset at lat. 43°31′.4N, long. 36°14′.3W on June 1, 1975.

Solution.—

L 43°31!4N June 1
λ 36°14!3W

Sunrise		Sunset	
40°	0433	40°	1922
T I	(−)11	T I	(+)11
LMT	0422	LMT	1933
dλ	(+)25	dλ	(+)25
ZT	0447	ZT	1958

Air Almanac. The procedure is the same as that for the *Nautical Almanac*, except that the LMT is extracted from the tables of sunrise and sunset instead of the daily page, and latitude correction is by linear interpolation.

The tabulated times are for the Greenwich meridian. Except in high latitudes near the times of the equinoxes, the time of sunrise and sunset varies so little from day to day that no interpolation is needed for longitude. If such an interpolation is considered justified, it can be made in the same manner as for the moon (art. 730).

In high latitudes, interpolation is not always possible. For instance, on June 1 1975, sunrise at latitude 66°N occurs at 0115, but at latitude 68°N the sun does not set. Between these two latitudes the time of sunrise might be found from the graphs in the *Air Almanac*, or by computation, as explained in article 715. However, in such a marginal situation, the time of sunrise itself is uncertain, being greatly affected by a relatively small change of refraction or height of eye.

729. Finding time of twilight.—Morning twilight *ends* at sunrise, and evening twilight *begins* at sunset. The time of the darker limit can be found from the almanacs. The *time* of the darker limits of both *civil* and *nautical* twilights (center of the sun 6° and 12°, respectively, below the celestial horizon) is given in the *Nautical Almanac*. The *Air Almanac* provides tabulations of civil twilight from 60°S to 72°N. The brightness of the sky at any given depression of the sun below the horizon may vary considerably from day to day, depending upon the amount of cloudiness and other atmospheric conditions. In general, however, the most effective period for observing stars and planets occurs when the center of the sun is between about 3° and 9° below the celestial horizon. Hence, the darker limit of civil twilight occurs at about the mid point of this period. At the darker limit of nautical twilight the horizon is generally too dark for good observations. At the darker limit of *astronomical twilight* (center of the sun 18° below the celestial horizon) full night has set in. The time of this twilight is given in the ephemeris. Its approximate value can be determined by extrapolation (art. 207) in the *Nautical Almanac*, noting that the duration of the different kinds of twilight is not proportional to the number of degrees of depression at the darker limit. More precise determination of the time at which the center of the sun is any given number of degrees below the celestial horizon can be determined by a large-scale diagram on the plane of the celestial meridian (app. J) or by computation (art. 715). Duration of twilight in latitudes higher than 65°N is given in a graph in the *Air Almanac*.

Nautical Almanac. The method of finding the darker limit of twilight is the same as that for sunrise and sunset (art. 728).

Example 1.—Find the zone time of beginning of morning nautical twilight and ending of evening nautical twilight at lat. 21°54!7S, long. 109°34!2E on June 1, 1975.

Solution.—

L 21°54′.7 S June 1
λ 109°34′.2 E

	Nautical twilight			Nautical twilight
20° S	0537		20° S	1819
T I	(+)3		T I	(−)3
LMT	0540		LMT	1816
dλ	(−)18		dλ	(−)18
ZT	0522		ZT	1758

Air Almanac. The method of finding the darker limit of twilight is the same as that for sunrise and sunset as explained in article 728.

Example 2.—Find the zone time of beginning of morning civil twilight and ending of evening civil twilight at lat. 47°18′.8S, long. 87°28′.3W on June 1, 1975.

Solution.—

L 47°18′.8 S June 1
λ 87°28′.3 W

Civil Twilight				Civil Twilight		
45°S	0654			45°S	1701	
corr.	(+)7			corr.	(−)7	
LMT	0701			LMT	1654	
dλ	(−)10			dλ	(−)10	
ZT	0651	(twilight)		ZT	1644	(twilight)

Sometimes in high latitudes the sun does not rise but twilight occurs. This is indicated in the *Air Almanac* by the symbol ■ in the sunrise and sunset column. To find the time of beginning of morning twilight, *subtract* half the duration of twilight as obtained from the duration of twilight graph from the time of meridian transit of the sun; and for the time of ending of evening twilight, *add* it to the time of meridian transit. The LMT of meridian transit never differs by more than $16^m.4$ (approximately) from 1200. The actual time on any date can be determined from the almanac.

730. Finding time of moonrise and moonset is similar to finding time of sunrise and sunset, with one important difference. Because of the moon's rapid change of declination, and its fast eastward motion relative to the sun, the time of moonrise and moonset varies considerably from day to day. These changes of position on the celestial sphere (app. J) are continuous, as moonrise and moonset occur successively at various longitudes around the earth. Therefore, the change in time is distributed over all longitudes. For precise results, it would be necessary to compute the time of the phenomena at any given place, by the method described in article 715. For ordinary purposes of navigation, however, it is sufficiently accurate to interpolate between consecutive moonrises or moonsets at the Greenwich meridian. Since apparent motion of the moon is westward, relative to an observer on the earth, interpolation in west longitude is between the phenomenon on the given date and the *following* one. In east longitude it is between the phenomenon on the given date and the *preceding* one.

Nautical Almanac. For the given date, enter the daily-page table with latitude, and extract the LMT for the tabulated latitude next *smaller* than the observer's latitude (unless this is an exact tabulated value). Apply a correction from table I of the almanac "Tables for Interpolating Sunrise, Moonrise, etc." to interpolate for latitude, determining the sign of the correction by inspection. Repeat this procedure for the day following the given date, if in west longitude; or for the day preceding, if in east longitude. Using the difference between these two times, and the longitude, enter table II of the almanac "Tables for Interpolating Sunrise, Sunset, etc." and take out the correction. Apply this correction to the LMT of moonrise or moonset at the Greenwich meridian on the given date to find the LMT at the position of the observer. The sign to be given the correction is such as to make the corrected time fall between the times for the two dates between which interpolation is being made. This is nearly always positive (+) in west longitude and negative (−) in east longitude. Convert the corrected LMT to ZT.

Example 1.—Find the zone time of moonrise and moonset at lat. 58°23'.6N, long. 144°07'.5W on June 1, 1975, using the *Nautical Almanac*.

Solution.—

L 58°23'.6N June 1
λ 144°07'.5W

	Moonrise			Moonset	
58°N	0007	June 1	58°N	1110	June 1
T I	(+)1		T I	(−)1	
LMT (G)	0008	June 1	LMT (G)	1109	June 1
58°N	0021	June 2	58°N	1221	June 2
T I	0		T I	0	
LMT (G)	0021	June 2	LMT (G)	1221	June 2
LMT (G)	0008	June 1	LMT (G)	1109	June 1
diff.	13		diff.	72	
T II	(+)5		T II	(+)28	
LMT (G)	0008	June 1	LMT (G)	1109	June 1
LMT	0013	June 1	LMT	1137	June 1
dλ	(−)24		dλ	(−)24	
ZT	2349	May 31	ZT	1113	June 1

Air Almanac. For the given date, determine LMT for the observer's latitude at the Greenwich meridian, in the same manner as with the *Nautical Almanac*, except that linear interpolation is made directly from the main tabulation, since no interpolation table is provided. Extract, also, the value from the "Diff." column to the right of the moonrise and moonset column, interpolating if necessary. This "Diff." is one-fourth of the difference between the succeeding and preceding value, which is approximately one-half of the daily difference. The error introduced by this approximation is generally not more than a few minutes, although it increases with latitude. Using this difference, and the longitude, enter the "Interpolation of Moonrise, Moonset" table on flap F4 of the *Air Almanac* and take out the correction. The *Air Almanac* recommends the taking of the correction from this table without interpolation. The results thus obtained are sufficiently accurate for ordinary purposes of navigation. If greater accuracy is desired, the correction can be taken by interpolation. However, since the "Diff." itself is an approximation, the *Nautical Almanac* or computation (art. 715) should be used if accuracy is a consideration. Apply the correction to the LMT of moonrise or moonset at the Greenwich meridian on the given date to find the LMT at the position of the observer. The correction is positive (+) for west longitude, and negative (−) for east longitude, unless the "Diff." on the daily page is preceded by a negative sign (−), when the correction is negative (−) for west longitude, and positive (+) for east longitude. If the time is near midnight, record the date at each step, as in the *Nautical Almanac* solution.

Example 2.—Find the zone time of moonrise and moonset at lat. 58°23′.6N, long. 144°07′.5W on June 1, 1975, using the *Air Almanac*.

Solution.—

L 58°23′.6N June 1
λ 144°07′.5W

	Moonrise		Moonset
diff.	(+)07	diff.	(+)36
58°N	0007	58°N	1110
corr.	(+)1	corr.	(−)1
LMT (G)	0008	LMT (G)	1109
corr.	(+)4	corr.	(+)29
LMT	0012	LMT	1138
dλ	(−)24	dλ	(−)24
ZT	2348 May 31	ZT	1114

As with the sun, there are times in high latitudes when interpolation is inaccurate or impossible. At such periods, the times of the phenomena themselves are uncertain, but an approximate answer can be obtained by moonlight graph in the *Air Almanac* or by computation, as explained in article 715. With the moon, this condition occurs when the moon rises or sets at one latitude, but not at the next higher tabulated latitude, as with the sun. It also occurs when the moon rises or sets on one day but not on the preceding or following day. This latter condition is indicated in the *Air Almanac* by the symbol ∗ in the "Diff." column.

Because of the eastward revolution of the moon around the earth, there is one day each synodical month (29½ days) when the moon does not rise, and one day when it does not set. These occur near last quarter and first quarter, respectively. Since this day is not the same at all latitudes or at all longitudes, the time of moonrise or moonset found from the almanac may occasionally be the preceding or succeeding one to that desired. When interpolating near midnight, one should exercise caution to prevent an error.

Refer to the right-hand daily page of the *Nautical Almanac* for June 12, 13, 14 (app. F). On June 13 moonset occurs at 2350 at latitude 70°N, and at 0031 at latitude 72°N. These are *not* the same moonset, the one at 0031 occurring approximately one day *later* than the one occurring at 2350. This is indicated by the two times, which differ by nearly 24 hours. The table indicates that with increasing northerly latitude, moonset occurs *later*. Between 70°N and 72°N the time crosses midnight *to the following* day. Hence, between these latitudes interpolation should be made between 2350 on June 13 and 0007 on June 14.

The effect of the revolution of the moon around the earth is to cause the moon to rise or set *later* from day to day. The daily retardation due to this effect does not differ greatly from 50^m. The change in declination of the moon may increase or decrease this effect. The effect due to change of declination increases with latitude, and in extreme conditions it may be greater than the effect due to revolution of the moon. Hence, the interval between successive moonrises or moonsets is more erratic in high latitudes than in low latitudes. When the two effects act in the same direction, daily differences can be quite large. Thus, at latitude 72°N the moon rises at 0550 on June 13, and at 0806 on June 14. When they act in opposite directions, they are small, and when the effect due to change in declination is larger than that due to revolution, the moon sets *earlier* on succeeding days. Thus, at latitude 72°N the moon sets at 0031 on June 13, and at 0007 on June 14. This condition is reflected in the *Air Almanac* by a negative "Diff." If this happens near last quarter or first quarter, two moonrises or moonsets might occur on the same day, one a few minutes after the day begins, and the other a few minutes before it ends. On June 14, 1975, for instance, at latitude 72°N, the moon sets at 0007, rises at 0806, and sets again at 2350 the same day. On those days on which no moonrise or no moonset occurs, the next succeeding one is shown with 24^h added to the time. Thus, at latitude 68°N the moon rises at 2342 on May 25, while the next moonrise occurs 24^h45^m later, at 0027 on May 27. This is listed both as 2427 on May 26 and as 0027 on May 27 (not shown in app. F).

Interpolation for longitude is always made between *consecutive* moonrises or moonsets, regardless of the days on which they fall.

Example 3.—Find the zone time of moonset at lat. 71°38′.7N, long. 56°21′.8W during the night of June 13–14, 1975, using the *Nautical Almanac*.

Solution.—

L 71°38′.7N June 13–14
λ 56°21′.8W

	Moonset	
70°N	2350	June 13
T I	(+)15	
LMT (G)	0005	June 14
70°N	2342	June 14
T I	(+)7	
LMT (G)	2349	June 14
LMT (G)	0005	June 14
diff.	16	
T II	(−)2	
LMT (G)	0005	June 14
LMT	0003	June 14
dλ	(−)15	
ZT	2348	June 13

Interpolation for the first entry is between 2350 on June 13 (lat. 70°N) and 0007 on June 14 (lat. 72°N); for the second entry, between 2342 on June 14 and 2350 on June 14.

Beyond the northern limits of the almanacs the values can be obtained from a series of graphs given near the back of the *Air Almanac*. These graphs are shown in appendix G. For high latitudes, graphs are used instead of tables because graphs give a clearer picture of conditions, which may change radically with relatively little change in position or date. Under these conditions interpolation to practical precision is simpler by graph than by table. In those parts of the graph which are difficult to read, the times of the phenomena's occurrence are themselves uncertain, being altered considerably by a relatively small change in refraction or height of eye.

On all of these graphs any given latitude is represented by a horizontal line, and any given date by a vertical line. At the intersection of these two lines the duration is read from the curves, interpolating by eye between curves.

The "Semiduration of Sunlight" graph gives the number of hours between sunrise and meridian transit or between meridian transit and sunset. The dot scale near the top of the graph indicates the LMT of meridian transit, the time represented by the minute dot nearest the vertical date line being used. If the intersection occurs in the area marked "sun above horizon," the sun does not set; and if in the area marked "sun below horizon," the sun does not rise.

Example 1.—Find the zone time of sunrise and sunset at lat. 71°30′.0N, long. 10°00′.0W near Jan Mayen Island, on August 25, 1975.

Solution.—

	August 25	
LMT	1202	LAN, from top of graph
dλ	(−)20	
ZT	1142	LAN
semidur.	840	from graph
ZT	0302	sunrise (−semidur.)
ZT	2022	sunset (+semidur.)

A vertical line through August 25 passes nearest the dot representing LAN 1202 on the scale near the top of the graph. This is LMT; at longitude 10°00′.0 W the ZT is 20^m earlier, or at 1142. The intersection of the vertical date line with the horizontal latitude line occurs between the 8^h and 9^h curves, at approximately 8^h40^m. Hence, sunrise occurs at this interval before LAN and sunset at this interval after LAN.

The "Duration of Twilight" graph gives the number of hours between the beginning of morning *civil* twilight (center of sun 6° below the horizon) and sunrise, or between sunset and the end of evening civil twilight. If the sun does not rise, but twilight does occur, the time taken from the graph is half the total length of the single twilight period, or the number of hours from beginning of morning twilight to LAN, or from LAN to end of evening twilight. If the intersection occurs in the area marked "continuous twilight or sunlight," the center of the sun does not get more than 6° below the horizon; and if in the area marked "no twilight nor sunlight," the sun remains more than 6° below the horizon throughout the entire day.

Example 2.—Find the zone time of beginning of morning twilight and ending of evening twilight at the place and date of example 1.

Solution.—

	Twilight	
ZT	0302	sunrise, from example 1
dur.	153	from graph
ZT	0109	morning twilight

	Twilight	
ZT	2022	sunset, from example 1
dur.	153	from graph
ZT	2215	evening twilight

The intersection of the vertical date line and the horizontal latitude line occurs approximately one-sixth of the distance from the 2^h line toward the 1^h20^m line; or at about 1^h53^m. Morning twilight begins at this interval before sunrise, and evening twilight ends at this interval after sunset.

The "Semiduration of Moonlight" graph gives the number of hours between moonrise and meridian transit or between meridian transit and moonset. The dot scale near the top of the graph indicates the LMT of meridian transit, each dot representing one hour. The phase symbols indicate the date on which the principal moon phases occur, the open circle indicating full moon and the dark circle indicating new moon. If the intersection of the vertical date line and the horizontal latitude line falls in the "moon above horizon" or "moon below horizon" area, the moon remains above or below the horizon, respectively, for the entire 24 hours of the day.

If approximations of the times of moonrise and moonset are sufficient, the values of semiduration taken from the graph can be used without adjustment. For more accurate results, the times on the required date and the adjacent date (the following date in west longitude and the preceding date in east longitude) should be determined, and an interpolation made for longitude, as in any latitude, since the intervals given are for the Greenwich meridian.

Example 3.—Find the zone time of moonrise and moonset at lat. 74°00′.0N, long. 108°00′.0W on May 8, 1975, and the phase of the moon on this date.

Solution.—

	May 8		May 9	
LMT	0923	LMT	1006	meridian transit, from graph
$d\lambda$	(+)12	$d\lambda$	(+)12	
ZT	0935	ZT	1018	meridian transit
semidur.	7^h59^m	semidur.	9^h15^m	from graph
ZT	0136	ZT	0103	(moonrise − semidur.)
ZT	1734	ZT	1933	(moonset + semidur.)

	Moonrise		Moonset
ZT	0136 May 8	ZT	1734 May 8
ZT	0103 May 9	ZT	1933 May 9
diff.	(−)33	diff.	(+)119
33×108.0/360	(−)10	119×108.0/360	(+)36
ZT	0126	ZT	1810

The phase is crescent, about two days before new moon. The LMT of meridian transits are found by noting the intersections of the vertical date lines with the dot scale near the top of the graph, interpolating by eye. At longitude 108°00′.0 W the ZT is 12^m later. The semiduration is found by noting the position, with respect to the semiduration curves, of the intersection of the vertical date line with the horizontal latitude line. This interval is subtracted from the time of meridian transit to obtain moonrise, and added to obtain moonset. These solutions are made for both May 8 and 9, and the difference determined in minutes. The adjustment to be applied to the ZT on May 8 at Greenwich is determined by multiplying this difference by the ratio $\lambda/360$. The phase is determined by noting the position of the vertical date line with respect to the phase symbols. If the answer indicates that the phenomenon occurs on a date differing from that desired, a new solution should be made, adjusting the starting date accordingly. The phenomenon may occur twice on the same day, or it may not occur at all. In high latitudes the effect on the time of moonrise and moonset of a relatively small change in declination is considerably greater than in lower latitudes, resulting in greater differences from day to day.

Sunlight, twilight, and moonlight graphs are not given for south latitudes. Beyond latitude 65°S, the northern hemisphere graphs can be used for determining the semiduration or duration, by using the vertical date line for a day when the declination has the same numerical value but opposite sign. The time of meridian transit and the phase of the moon are determined as explained above, using the correct date. Between latitudes 60°S and 65°S solution is made by interpolation between the tables and the graphs.

Several other methods of solution of these phenomena are available. The *Tide Tables* tabulate sunrise and sunset from latitude 76°N to 60°S. A supplement to the American Ephemeris of 1946, entitled *Tables of Sunrise, Sunset, and Twilight*, provides tabulations from latitude 75°N to 75°S and graphs for semiduration of sunlight and duration of twilight, with separate graphs for civil, nautical, and astronomical twilights. Semiduration or duration can be determined graphically by means of a diagram on the plane of the celestial meridian (app. J), or by computation. When computation is used, solution is made for the meridian angle at which the required negative altitude occurs. The meridian angle expressed in time units is the semiduration in the case of sunrise, sunset, moonrise, and moonset; and the semiduration of the combined sunlight and twilight, or the time from meridian transit at which morning twilight begins or evening twilight ends. For sunrise and sunset the altitude used is (−)50′. Allowance for height of eye can be made by algebraically subtracting (numerically adding) the dip correction from this altitude. The altitude used for twilight is (−)6°, (−)12°, or (−)18° for civil, nautical, or astronomical twilight, respectively. The altitude used for moonrise and moonset is −34′−SD+HP, where SD is semidiameter and HP is horizontal parallax, from the daily pages of the *Nautical Almanac*.

731. Rising, setting, and twilight at a moving craft.—Instructions given in the preceding three articles relate to a fixed position on the earth. Aboard a moving craft the problem is complicated somewhat by the fact that time of occurrence depends upon position of the craft, and vice versa. At ship speeds, it is generally sufficiently accurate to make an approximate mental solution, and use the position of the vessel at this time to make a more accurate solution. If higher accuracy is required, the position at the time indicated in the second solution can be used for a third solution. If desired, this process can be repeated until the same answer is obtained from two consecutive solutions. However, it is generally sufficient to alter the first solution by 1^m for each 15′ of longitude that the position of the craft differs from that used in the solution, adding if west of the estimated position, and subtracting if east of it. In applying this rule, use both longitudes to the nearest 15′. The first solution is known as the first estimate; the second solution is the second estimate.

Latitude by Meridian Transit

732. Meridian altitudes.—The latitude of a place on the surface of the earth, being its angular distance from the equator, is measured by an arc of the meridian between the zenith and the equator, and hence is equal to the declination of the zenith; therefore, if the zenith distance of any heavenly body when on the meridian be known, together with the declination of the body, the latitude can be found.

Figure 732a shows the celestial sphere surrounding the earth: P_nMP_s is the upper branch of a celestial meridian and LL′ a portion of the corresponding geographic meridian. The declination of a body at M (arc QM) is numerically equal to the latitude of its geographical position at GP. The zenith distance of a body is equivalent to the distance on earth between the geographical position of the body and the position of the observer. In figure 732a the zenith distance of M is 30° and its declination is 20°N. If the body is on the meridian, the GP is also on the meridian. Since P_n, Z, and M are all on the celestial meridian, the navigational triangle flattens out to a line. The observer is 30° *north* of the GP (L 50°N) if the body is seen to bear south, or 30° *south* of the GP (L′ 10°S) if the body is seen to bear *north*. The navigator knows whether the GP is north or south, because it is the same as the direction he faces when making his observation.

In the diagram on the plane of the celestial meridian shown in figure 732b, M is the position of a celestial body north of the equator but south of the zenith; QM is the declination of the body; SM is the altitude (h); and MZ is the zenith distance (z).

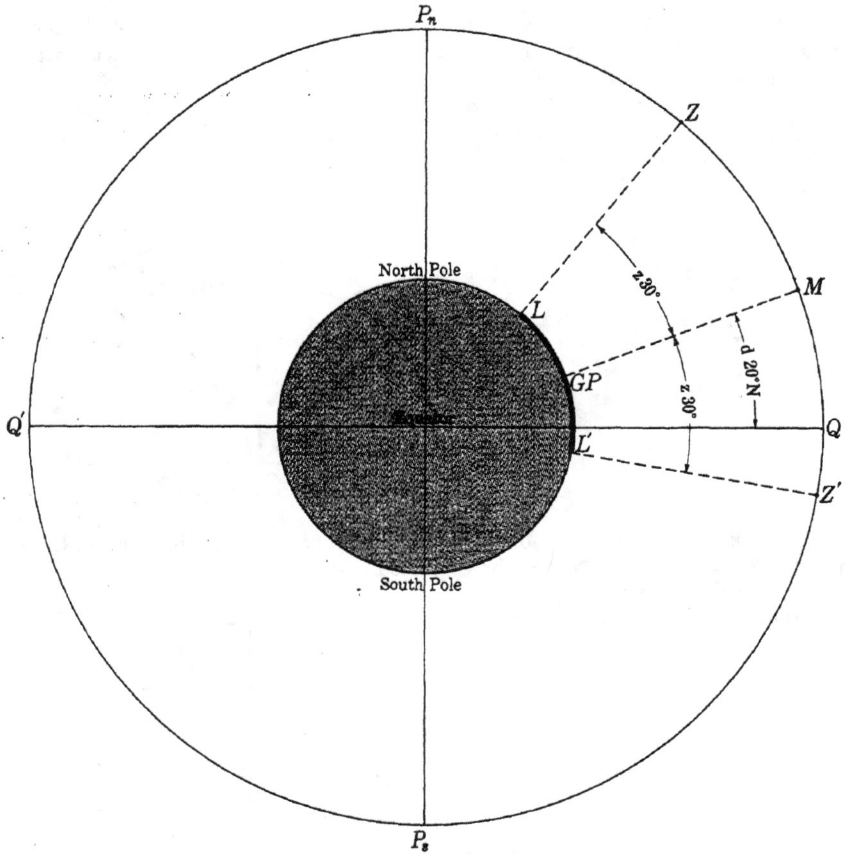

FIGURE 732a.—Body on celestial meridian.

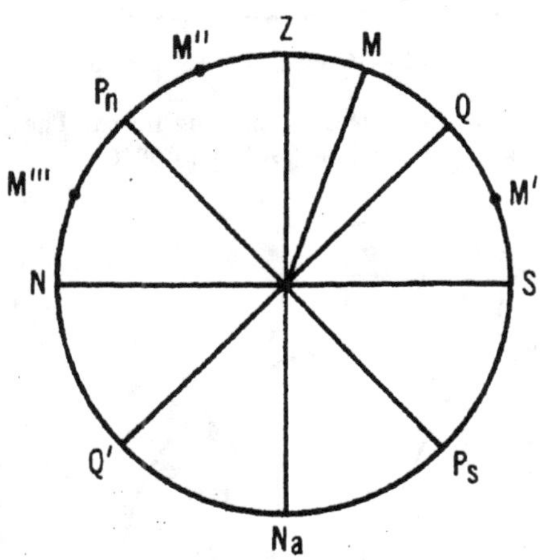

FIGURE 732b.—Diagram on the plane of the celestial meridian.

From the diagram:

$$QZ = QM + MZ, \text{ or}$$
$$L = d + z. \tag{19}$$

With attention to the direction of the GP and the name of the declination, the above equation may be considered general for any position of the body at upper transit, as M, M′, M″.

When the body is below the pole, as at M‴—that is, at its lower transit—the same formula may be used by substituting 180°−d for d. Another solution is given in this case by observing that:

$$NP_n = P_nM''' + NM''', \text{ or}$$
$$L = p + h. \tag{20}$$

By drawing that half of the diagram on the plane of the celestial meridian containing the zenith, the proper combination of zenith distance and declination is made obvious, as shown in the following examples:

Example 1.—The navigator observes the sun on the meridian, bearing south. The declination of the sun is 10°00′.0N; the corrected sextant altitude (Ho) is 60°00′.0.

Required.—The latitude.

Solution.— L=z+d.

	90°00′.0
Ho	60°00′.0
z	30°00′.0
d	10°00′.0N
L	40°00′.0N

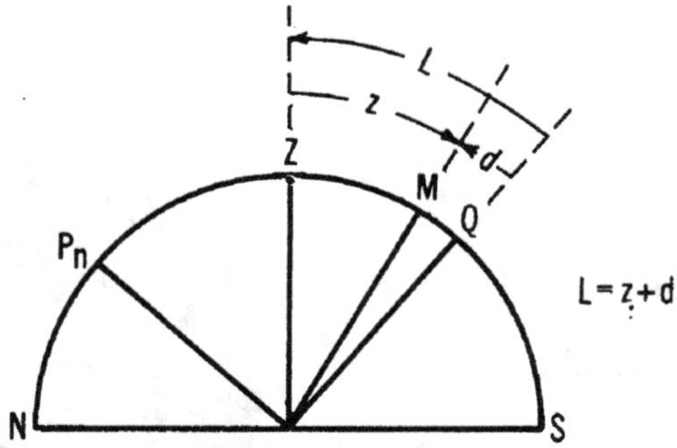

FIGURE 732c.—Meridian altitude diagram.

Example 2.—The navigator observes the sun on the meridian, bearing south. The declination of the sun is 10°00′.0S; the corrected sextant altitude (Ho) is 65°00′.0.

Required.—The latitude.

Solution.— L=z—d.

	90°00′.0
Ho	65°00′.0
z	25°00′.0
d	10°00′.0S
L	15°00′.0N

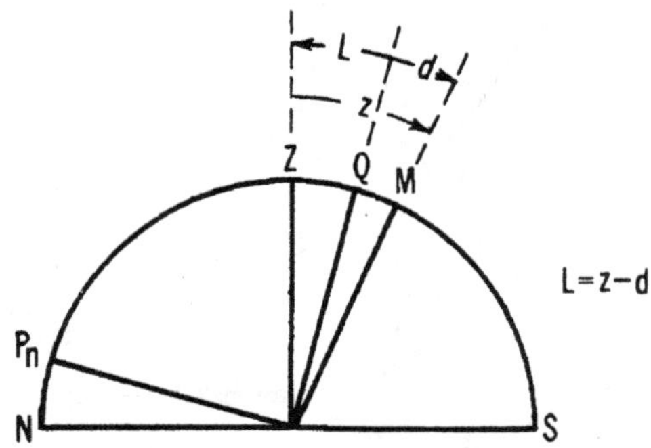

FIGURE 732d.—Meridian altitude diagram.

Example 3.—The navigator observes the sun on the meridian, bearing north. The declination of the sun is 20°00′.0S; the corrected sextant altitude (Ho) is 60°00′.0.

Required.—The latitude.

Solution.— L=z+d.

	90°00′.0
Ho	60°00′.0
z	30°00′.0
d	20°00′.0S
L	50°00′.0S

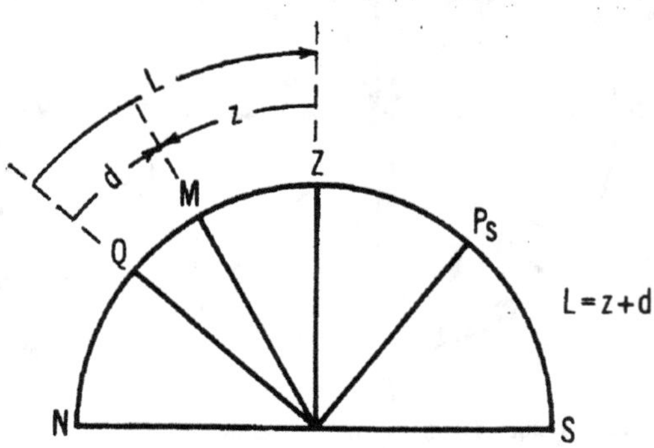

FIGURE 732e.—Meridian altitude diagram.

Example 4.—The navigator observes the sun on the meridian, bearing north. The declination of the sun is 23°00′.0N; the corrected sextant altitude (Ho) is 72°00′.0.

Required.—The latitude.

Solution.— L=d−z.

	90°00′.0
Ho	72°00′.0
z	18°00′.0
d	23°00′.0N
L	5°00′.0N

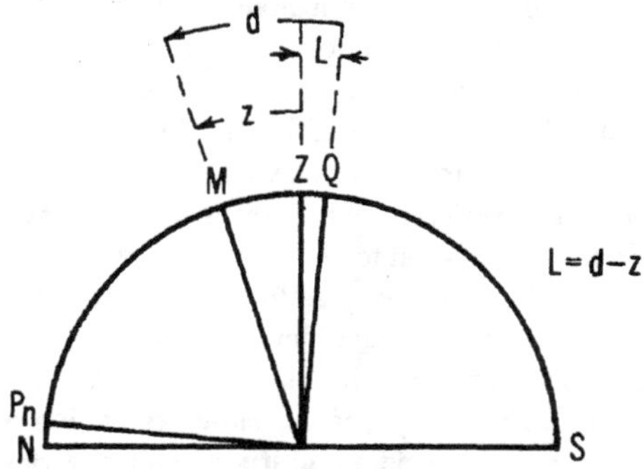

Figure 732f.—Meridian altitude diagram.

Example 5.—In the vicinity of the equator, the navigator observes the sun on the meridian, bearing north. The declination of the sun is 22°05′.0N; the corrected sextant altitude (Ho) is 67°45′.0.

Required.—The latitude.

Solution.— L=z−d.

	90°00′.0
Ho	67°45′.0
z	22°15′.0
d	22°05′.0N
L	0°10′.0S

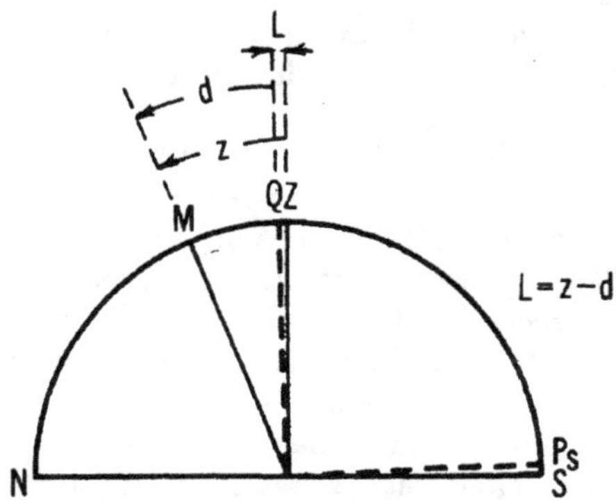

Figure 732g.—Meridian transit in the vicinity of the equator.

Example 6.—The navigator in high northern latitudes observes the sun on the celestial meridian, bearing north. The declination of the sun is 18°46′.0N; the corrected sextant altitude (Ho) is 6°22′.0.

Required.—The latitude.

Solution.— L=(180°−d)−z, or
L=p+h.

	90°00′.0
Ho	6°22′.0
z	83°38′.0
180°−d	161°14′.0N
L	77°36′.0N

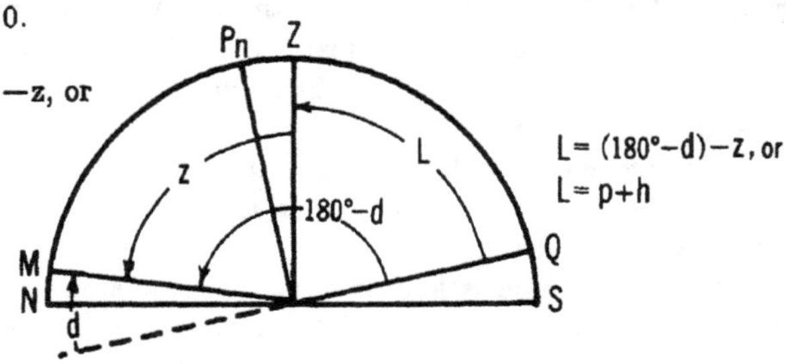

Figure 732h.—Meridian altitude at lower transit.

Since the sun's GP is 83°38.'0 north of the observer in high northern latitudes, the GP is beyond the pole, or on the *lower branch* of the observer's meridian.

If an observation is made near but not exactly at meridian transit, it can be solved as a meridian altitude, with one modification. Enter table 29 with the approximate latitude of the observer and the declination of the body, and take out the **altitude factor** (*a*). This is the difference between meridian altitude and the altitude one minute of time later (or earlier). Next, enter table 30 with the altitude factor and the difference of time between meridian transit and the time of observation, and take out the correction. *Add* this value to Ho if near upper transit, or *subtract* it from Ho if near lower transit. Then proceed as for a meridian altitude, remembering that the value obtained is the latitude at the time of observation, not at the time of meridian transit. This method should not be used beyond the limits of table 30 unless reduced accuracy is acceptable. This process is called **reduction to the meridian,** the altitude before adjustment an **ex-meridian altitude,** and the observation an **ex-meridian observation.** It requires knowledge of the meridian angle, which depends upon knowledge of longitude.

733. Finding time of meridian transit.—If a meridian altitude is to be observed other than by chance, a knowledge of the time of transit of the body across the meridian is needed.

On a slow-moving vessel, or one traveling approximately east or west, the time need not be known with great accuracy. The right-hand daily page of the *Nautical Almanac* gives the GMT of transit of the sun and moon across the Greenwich meridian (approximately LMT of transit across the local meridian) under the heading "Mer. Pass." In the case of the moon, an interpolation should be made for longitude. This is performed in the same manner as finding the LMT of moonrise and moonset (art. 730). In the case of planets, the tabulated accuracy is normally sufficient without interpolation. The time of transit of the navigational planets is given at the lower right-hand corner of each left-hand daily page of the *Nautical Almanac*. The tabulated values are for the middle day of the page. These times are the GMT of transit across the Greenwich meridian, but are approximately correct for the LMT of transit across the local meridian. Observations are started several minutes in advance and continued until the altitude reaches a maximum and starts to decrease (a minimum and starts to increase for lower transit). The greatest altitude occurs at upper transit (and the least at lower transit). This method is not reliable if there is a large northerly or southerly component of the vessel's motion, because the altitude at meridian transit changes slowly, particularly at low altitudes. At this time the change due to the vessel's motion may be considerably greater than that due to apparent motion of the body (rotation of the earth), so that the highest altitude occurs several minutes before or after meridian transit.

If the moment at which the azimuth is 000° or 180° can be determined accurately, the observation can be made at this time. However, this generally does not provide a high order of accuracy.

If the longitude is known with sufficient accuracy, the time of transit can be computed. A number of methods of computation have been devised, but perhaps the simplest is to consider the GHA of the body equal to the longitude if west, or 360°−λ if east, and find the time at which this occurs.

Example 1.—Find the zone time of meridian transit of the sun at longitude 156° 44′.2W on May 31, 1975.

Solution.—

	May 31
λ	156°44′.2W
GHA	156°44′.2
22^h	150°36′.1
24^m32^s	6°08′.1
GMT	$22^h24^m32^s$ May 31
ZD	(+)10 (rev.)
ZT	$12^h24^m32^s$

This solution is the reverse of finding GHA. The largest tabulated value of GHA that does not exceed the desired GHA is found in the tabulation for the day, and recorded, with its time. The difference between this value and the desired GHA is then used to enter the "Increments and Corrections" table. The time interval corresponding to this value is added to the time taken from the daily page. If there is a *v* correction, it is subtracted from the GHA difference before the time interval is determined. The GMT can be converted to any other kind of time desired. If the Greenwich date differs from the local date at the time of transit (for the sun this can occur only near the 180th meridian), a second solution may be needed. This possibility can often be avoided by making an approximate mental solution in advance. As the basis for this approximate solution, it is convenient to remember that the GMT of Greenwich transit (GHA 0°) is about the same as the LMT of local transit. To find the time of transit of a star, subtract its SHA from the desired GHA to find the desired GHA ♈. Determine the time corresponding to GHA ♈, as explained above for the sun.

Aboard a moving vessel, the longitude at transit usually depends upon the time of transit. An approximate mental solution may provide a time sufficiently close. In the absence of better information, use ZT 1200 for the sun. Find the time of transit for the position at this time, and then make an adjustment, if necessary, for the sun between 1200 and the time found by computation. This adjustment is equal to four seconds for each minute of longitude involved. If the ship is *west* of the 1200 position at the computed time of transit, *add* the correction; and if *east, subtract* it. The result is the **first estimate** of the zone time of **local apparent noon** (**LAN**) or of meridian transit. For high accuracy a second adjustment may occasionally be needed, but this is seldom justified because of the uncertainty of the vessel's position. If the second adjustment is made, the result is the **second estimate.**

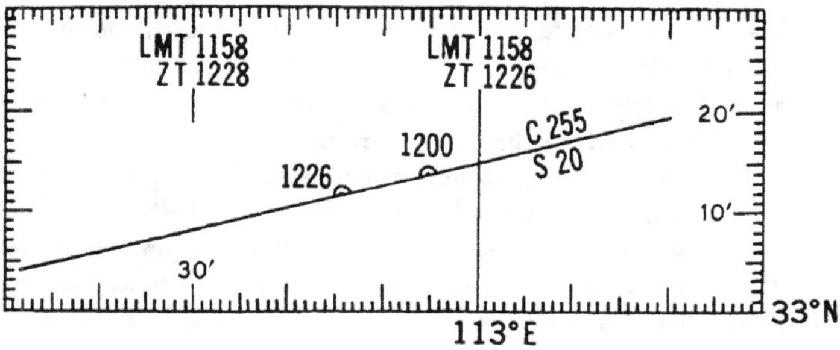

FIGURE 733.—Time of meridian passage.

The time of transit of the sun can also be found by means of apparent time (art. 611). Meridian transit occurs at LAT $12^h00^m00^s$. This can be converted to any other kind of time desired.

Example 2.—Find the zone time of meridian transit of the sun as observed aboard a ship steaming at 20 knots on course 255° on May 31, 1975, using the positional data given in figure 733.

Solution.—

	May 31	
	360°00′.0	
λ	112°55′.0E	
GHA	247°05′.0	
4^h	240°37′.7	
25^m49^s	6°27′.3	
GMT	$4^h25^m49^s$	May 31
ZD(−)8		(rev.)
ZT	$12^h25^m49^s$	
dλ	(+) 0^m36^s	
ZT	$12^h26^m25^s$	(first estimate)

The second estimate of the zone time of meridian transit is found by plotting the DR position for the first estimate of the zone time of transit and then applying the dλ between this DR and the 1200 DR to the time found by computation.

ZT	$12^h25^m49^s$	
dλ	(+)0^m37^s	
ZT	$12^h26^m26^s$	(second estimate)

As shown in figure 733, the zone times of meridian transit are noted on several successive meridians. This is accomplished by extracting the LMT of meridian transit from the daily page of the *Nautical Almanac* and converting this time to the zone time for each meridian. The time when the ship and the sun are on the same meridian can then be obtained by inspection to within approximately one-half minute.

Polaris

734. Latitude by Polaris.—Another special method of finding latitude, available in most of the northern hemisphere, utilizes the fact that Polaris is less than 1° from the north celestial pole. As indicated in article J7, the altitude of the elevated pole above the celestial horizon is equal to the latitude. Since Polaris is never far from the pole, its observed altitude (Ho), with suitable correction, is the latitude.

The nature of this correction as tabulated in the *Air Almanac* is suggested by inspection of figure 734b in which the circle represents the path of Polaris around the north celestial pole (P_n), as seen by an observer on earth looking along the axis P_sP_n (fig. 734a). The line *ab* represents a small portion of the observer's meridian. Polaris is at upper transit at *a* and at lower transit at *b*. This is also shown in figure 734b, to larger scale. Latitude is equal to the altitude *minus* the polar distance (p) when Polaris is at *a* and *plus* the polar distance when it is at *b*. When the star is at any point *c*, the Polaris correction is polar distance times the cosine of the local hour angle (corr.$=p$ cos LHA). Thus, the correction is a function of LHA of the star, and hence also of

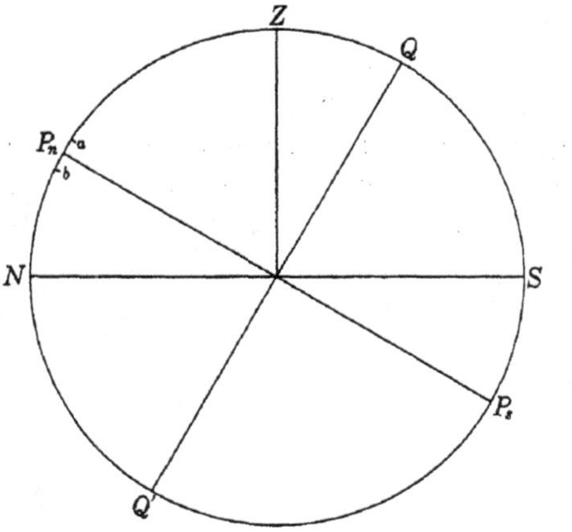

FIGURE 734a.—Latitude is equal to the (1) declination of the zenith and (2) the altitude of the elevated pole. Compare with figure 734b.

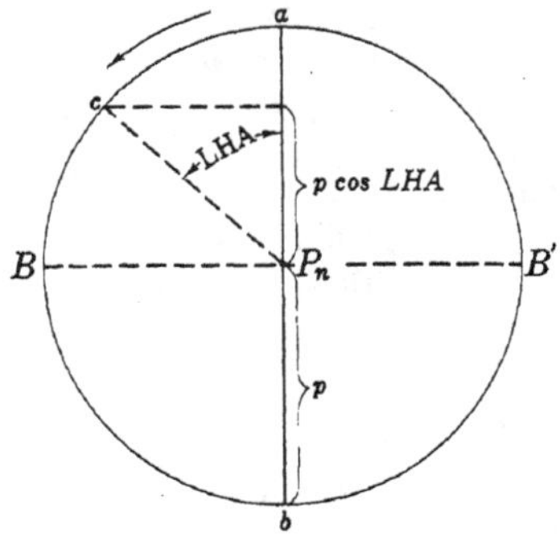

FIGURE 734b.—The correction is $(-)aPn$ when Polaris is at a and $(+)bPn$ when at b. At any point c the correction is $p\cos$LHA. At B or B' the correction is zero.

LHA ♈, insofar as the difference between these quantities (the SHA) can be considered constant.

Although the above method usually provides sufficient accuracy for air navigation, a higher degree of accuracy may be obtained by use of Polaris correction tables included in the *Nautical Almanac*. These tables are based on the following formula:

$$\text{Latitude}-\text{corrected sextant altitude}=-p\cos h+\tfrac{1}{2}\,p\sin p\sin^2 h\tan(\text{latitude}),$$

where p=polar distance of *Polaris*=90°−Dec.
h=local hour angle of *Polaris*=LHA Aries+SHA.

The value a_0, which is a function of LHA Aries only, is the value of both terms of the above formula calculated for mean values of the SHA and Dec. of Polaris, for a mean latitude of 50°, and adjusted by the addition of a constant (58:8). The value a_1, which is a function of LHA Aries and latitude, is the excess of the value of the second term over its mean value for latitude 50°, increased by a constant (0:6) to make it always positive. The value a_2, which is a function of LHA Aries and date, is the correction to the first term for the variation of Polaris from its adopted mean position; it is increased by a constant (0:6) to make it positive. The sum of the added constants is 1°, so that:

$$\text{Latitude}=\text{corrected sextant altitude}-1^\circ+a_0+a_1+a_2.$$

The table at the top of each Polaris correction page is entered with LHA Aries, and the first correction (a_0) is taken out by single interpolation. The second and third corrections (a_1 and a_2, respectively) are taken from the double entry tables without interpolation, using the LHA Aries column with the latitude for the second correction and with the month for the third correction.

Example 1.—During morning twilight on June 2, 1975, the 0525 DR position of a ship is lat. 15°43:6N, long. 110°07:3W. At watch time $5^h24^m49^s$ AM the navigator observes Polaris with a marine sextant having an IC of (−)3:0, from a height of eye of 44 feet. The watch is 23^s slow on zone time. The hs is 16°24:0.

Required.—The latitude.

Solution.—

	June 2	
WT	$5^h24^m49^s$ AM	
WE	(S)23^s	
ZT	$5^h25^m12^s$	
ZD	(+) 7	
GMT	$12^h25^m12^s$	June 2
12^h	70°20′.2	
25^m12^s	6°19′.0	
GHA♈	76°39′.2	
λ	110°07′.3W	
LHA♈	326°31′.9	
Ho	16°11′.3	
corr.	(−)21′.2	
L	15°50′.1N	

Polaris	+	−
a_0	38′.2	
a_1	0′.3	
a_2	0′.3	
add'l		60′.0
sum	38′.8	60′.0
corr.	(−)21′.2	

	+	☆	−
IC			3′.0
D			6′.4
☆-P			3′.3
sum			(−)12′.7
corr.			(−)12′.7
hs			16°24′.0
Ho			16°11′.3

Since LHA ♈ is an entering value in all three correction tables, and since this is affected by the longitude, other observations, if available, should be solved and plotted first, to obtain a good longitude for the Polaris solution. For greater accuracy, particularly in higher latitudes, and especially if considerable doubt exists as to the longitude, it is good practice to find the azimuth of Polaris and draw the line of position perpendicular to it, through the point defined by the latitude found in the computation and the longitude used in the solution. The azimuth at various latitudes to 65°N is given below the Polaris corrections. This table can be extrapolated to higher latitudes, but Polaris would not ordinarily be used much beyond latitude 65°. In the example given above the azimuth is 000°.8.

The *Air Almanac* provides only the first correction, which it designates *Q*.

Polaris observations can be solved like those of other celestial bodies, using the declination and SHA given in the tabulation near the back of the *Nautical Almanac*.

The Polaris correction table in the *Air Almanac* lists the correction *Q* to be applied to the sextant altitude of Polaris to give the latitude of the observer. The correction is given in a critical table, with argument LHA Aries (♈) obtained from the daily pages. The effect of refraction is not included and so the sextant altitude must be fully corrected before use.

Example 2.—On 1 June 1975 at GMT $02^h53^m32^s$ in longitude 49°18′W. the corrected sextant altitude of Polaris is 54°46′.

Required.—The latitude.

Solution.—

GMT	$2^h53^m32^s$	June 1
2^h50^m	291°28′	
3^m32^s	0°53′	
GHA ♈	292°21′	
λ	49°18′W	
LHA ♈	243°03′	

Corr. sextant altitude	54°46′
Q (LHA♈, 243°03′)	(+) 44′
Latitude	55°30′

CHAPTER VIII

SEXTANT ALTITUDE CORRECTIONS

801. Need for correction.—Altitudes of celestial bodies, obtained aboard ship for the purpose of establishing lines of position, are normally measured by a hand-held **sextant,** described in volume I. The uncorrected reading of a sextant after such an operation is called **sextant altitude (hs).** If the sextant is in proper adjustment, certain sources of error are eliminated, as explained in volume I. There remains, however, a number of sources of error over which the observer has little or no control. For each of these he applies a correction. When all of these **sextant altitude corrections** have been applied, the value obtained is the altitude of the center of the celestial body above the celestial horizon, for an observer at the center of the earth. This value, called **observed altitude (Ho),** is compared with the **computed altitude** (**Hc**) to find the **altitude intercept** (*a*) used in establishing a line of position as explained in chapter IX.

Articles 802–829 describe the various corrections. For highly accurate results, all of these are needed to the greatest accuracy obtainable. The needs of ordinary practical navigation, however, make no such exacting requirements, and in the course of his usual day's work at sea, the navigator has relatively few corrections to apply, from conveniently-arranged tables readily accessible to him. The detailed information in articles 802–829 is given to (1) provide the basis for a better understanding of the problem, (2) furnish the information needed for evaluation of results, and (3) provide a source of reference material beyond that given in the usual navigation text.

802. Instrument correction (**I**) is the combined correction for nonadjustable errors (prismatic error, graduation error, and centering error) of the sextant, as explained in volume I. Usually, this correction is determined by the manufacturer, and recorded on a card attached to the inside of the top of the sextant box. It varies with the angle, may be either positive or negative, and is applied to all angles measured by that instrument. For a well-made instrument, the maximum value is so small that this correction can be ignored for all except the most accurate work. Normally, instrument error of artificial-horizon sextants is so small, considering the precision to which angles can be measured by such instruments, that no correction is provided.

803. Index correction (**IC**), due primarily to lack of parallelism of the horizon glass and index mirror at zero reading, is discussed in volume I. Until the adjustment is disturbed, the index correction remains constant for all angles, and is applicable to all angles measured by the instrument. It may be either positive or negative. Normally, artificial-horizon sextants do not have index corrections.

804. Personal correction (**PC**) is numerically the same as personal error (volume I), but of opposite sign, either positive or negative. If experience indicates the need for such a correction, it should be made to altitudes of the bodies to which it applies. However, the observer should be sensitive to changes in its value. Unless the observer has sufficient evidence to be sure of the existence and relative constancy of a personal error, no correction should be applied.

805. Dip (**D**) of the horizon is the angle by which the visible horizon (app. J) differs from the horizontal at the eye of the observer (the sensible horizon, app. J) Thus, it applies only when the visible horizon is used as a reference, and not when an artificial horizon, either internal or external to the sextant, is used. It applies to all celestial bodies. If the eye of the observer were at the surface of the earth, visible and sensible horizons would coincide, and there would be no dip. This is never the situation aboard ship, however, and at any height *above* the surface, the visible horizon is normally *below* the sensible horizon, as shown in figure 505a. Normally, then, an altitude measured from the visible horizon is too *great*, and the correction is *negative*. It increases with greater height of the observer's eye. Because of this, it is sometimes called **height of eye correction.**

If there were no atmospheric refraction, dip would be the angle between the horizontal at the eye of the observer, and a straight line from this point tangent to the surface of the earth.

The amount by which refraction alters dip varies with changing atmospheric conditions. Even the *average* value has not been established with certainty, and several methods of computing dip have been proposed. The values given in the critical table on the inside front cover of the *Nautical Almanac* were computed by the equation

$$D=0.97\sqrt{h},$$

where D is the dip, in minutes of arc; and h is the height of eye of the observer, in feet. Part of this table is repeated on the page facing the inside back cover. The *Air Almanac* table was computed independently by a different method, to a precision of whole minutes. The minor discrepancies thus introduced are not important in practical navigation.

The values given in the table are satisfactory for practical navigation under most conditions. An investigation by the Carnegie Institution of Washington showed that of 5,000 measurements of dip at sea, no value differed from the tabulated value by more than 2′5, except for one difference of 10′6. Extreme values of more than 30′ have been reported, and even values of several *degrees* have been encountered in polar regions. Greatest variations from tabulated values can be expected in calm weather, with large differences between sea and air temperatures, particularly if mirage effects are present. Irregularities in the shape of the rising or setting sun may indicate abnormal conditions. Large variations may also be present shortly after passage of a squall line, when errors of as much as 15′ have been reported. When a temperature inversion is known to exist, the tabulated dip may be too small, numerically. The effect of sea-air temperature difference is discussed in greater detail in article 814.

In the determination of height of eye, position on the ship should be considered, and also the condition of loading and trim. If an observation is made from a position differing from the usual place, the altered height of eye should not be overlooked. Momentary changes due to rolling and pitching can be neutralized, to a large extent, by making observations from a point on the center line of the vessel, at the axis of pitch.

Since variations from normal dip may be one of the principal sources of error in celestial observations, the observer should be alert to conditions affecting terrestrial refraction. Any observation taken within half an hour after passage of a squall line should be regarded as unreliable.

If dip cannot be measured, the effects of abnormal conditions can be minimized by observing three bodies differing in azimuth by about 120° (or four bodies by 90°, five bodies by 72°, etc.). If the error is constant in all directions, its effect is to increase (or possibly to decrease) the size of the closed figure formed by the lines of position without altering the position of its center. Hence, the size of the figure is not necessarily an indication of the accuracy of the fix.

If land, another ship, or other obstruction is between the observer and his horizon, an altitude can be measured by using the water line of the obstruction as a horizontal reference, if its distance from the observer is known. In this case the dip is greater than that given in the almanacs. Table 22 gives the values to be used.

Further discussion of refraction is given in article 806. When abnormal astronomical refraction occurs, abnormalities in terrestrial refraction can be expected.

806. Refraction (R).—Light, or other radiant energy, is assumed to travel in a straight line at uniform speed, if the medium in which it is traveling has uniform

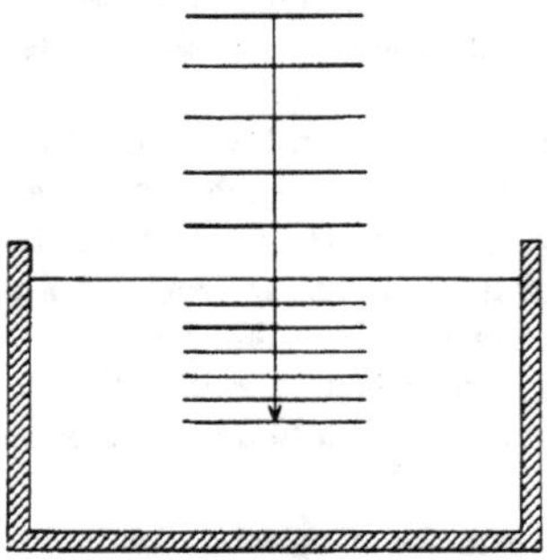

FIGURE 806a.—No refraction occurs when light enters denser medium normal to the surface.

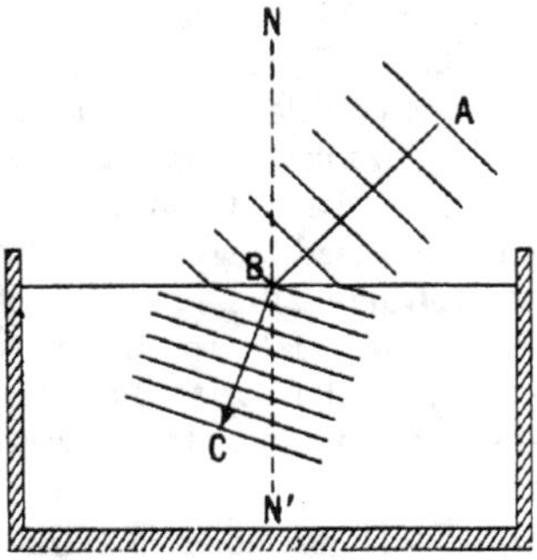

FIGURE 806b.—A ray entering a denser medium at an oblique angle is bent toward the normal.

properties. But if light enters a medium of different properties, particularly if the density is different, the speed of light changes somewhat. Light from a single point source travels outward in all directions, in an expanding sphere. At great distances, a small part of the surface of this sphere can be considered flat, and light continuing to emanate from the source can be considered similar to a series of waves, in some respects resembling the ocean waves encountered at sea. If these light "waves" enter a more dense medium, as when they pass from air into water, the speed decreases. If the light is traveling in a direction perpendicular to the surface separating the two media (in this case vertically downward), all parts of each wave front enter the new medium at the same time, and so all parts change speed together, as shown in figure 806a. But if the light enters the more dense medium at an oblique angle, as shown in figure 806b, the change in speed occurs progressively along the wave front as the different parts enter the more dense medium. This results in a change in the direction of travel, as shown. This change in direction of motion is called **refraction.** If light enters a more dense medium, it is refracted *toward* the normal (NN'), as in figure 806b. If it enters a less dense medium, it is refracted *away* from the normal, as light traveling in the opposite direction to that shown in figure 806b.

The *amount* of the change in direction is directly proportional to the angle between the direction of travel and the normal (angle ABN in figure 806b). The ratio of this angle to the similar angle after refraction takes place (angle CBN' in figure 806b) is constant, so that as one increases, the other increases at the same *rate*. Hence, the *difference* between them (the change in direction) also increases at the same rate. Therefore, if the *incident* ray (AB) is nearly parallel to the surface at which refraction takes place, relatively large amounts of refraction occur.

The amount of refraction is also directly proportional to the relative speed of travel in the two media. Various substances are compared by means of a number called the **index of refraction** (μ), which depends primarily upon the density of the substance.

In figure 806b, angle *ABN* is called the **angle of incidence** (ϕ) and angle *CBN'* the **angle of refraction** (θ). These are related by Snell's law, which states that *the sines of the angle of incidence and angle of refraction are inversely proportional to the indices of refraction of the substances in which they occur*. Thus, if μ_1 is the index of refraction of the substance in which ϕ occurs, and μ_2 is the index of refraction of the substance in which θ occurs

$$\frac{\sin \phi}{\sin \theta}=\frac{\mu_2}{\mu_1}.$$

If the index of refraction changes suddenly, as along the surface separating water and air (as shown in fig. 806b), the change in direction is equally sudden. However, if a ray of light travels through a medium of gradually changing index of refraction, its path is curved, undergoing increased refraction as the index of refraction continues to change. This is the situation in the earth's atmosphere, which generally decreases in density with increased height. The gradual change of direction occurring there is called **atmospheric refraction.** The bending of a ray of light traveling from a point on or near the surface of the earth, to the eye of the observer, is called **terrestrial refraction.** This affects dip of the horizon, as discussed in article 805. A ray of light entering the atmosphere from outside, as from a star, undergoes a similar bending called **astronomical refraction.**

The effect of astronomical refraction is to make a celestial body appear *higher* in the sky than it otherwise would, as shown in figure 806c. If a body is in the zenith, its light is not refracted, except for a very slight amount when the various layers of the atmosphere are not exactly horizontal. As the zenith distance increases, the refraction becomes greater. At an altitude of 20° it is about 2'.6; at 10°, 5'.3; at 5°, 9'.9; and at the horizon, 34'.5. A table of refraction is given on the inside front cover and facing page of the *Nautical Almanac*, in the columns headed "Stars and Planets." As height above the surface of the earth increases, light from an outside source travels through less of the atmosphere, and refraction decreases. At shipboard heights the difference is negligible, but at aircraft heights the change is a consideration. Therefore, the refraction table given near the back of the *Air Almanac* is a double-entry table.

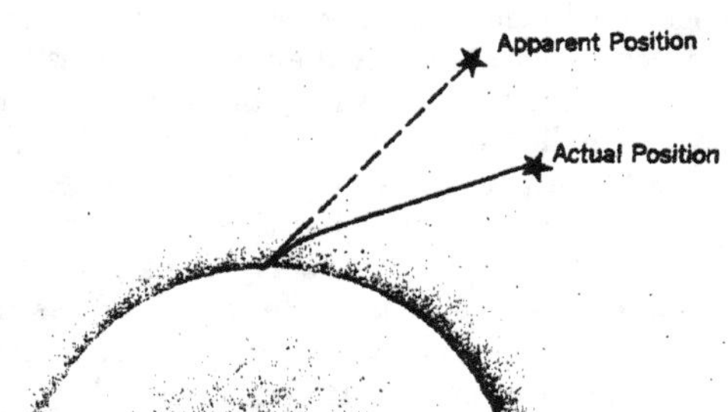

FIGURE 806c.—Astronomical refraction.

The values given in the tables are for average conditions. This is called **mean refraction.** A considerable amount of research has been conducted to determine the mean values, the conditions under which values differ from the mean, and the amount of such differences. A number of different mean refraction tables have been produced. Values in the various tables differ slightly because of different assumptions, different methods of observation, and different observed results under apparently similar conditions. This last source of difference is due primarily to the fact that conditions could be determined at the position of the observer, but not at various points along the line traveled by the ray of light in passing through the atmosphere. Nevertheless, the various tables agree very well down to a minimum altitude of 2°. Below this, the refraction is erratic, and differences between values in the various tables are not as important as differences between mean and instantaneous values. The values given in the almanac tables are in excellent agreement with those actually measured.

Because of their variability, refraction and dip (also affected by refraction) are the principal uncertainties in the accuracy of celestial observations of a careful observer. As a result of this uncertainty, navigators formerly avoided all observations below some arbitrary altitude, usually 15°. While this is still good practice if higher bodies are available, the growing knowledge of atmospheric refraction has increased the confidence with which navigators can use low-altitude sights. There is little reason for lack of confidence in sights as low as 5°. Below this, other available corrections should be applied (art. 828). If altitudes below 2° are used, larger probable errors should be anticipated, even with the use of additional corrections. Generally, the error in tabulated refraction should not exceed two or three minutes, even at the horizon. However, a knowledge of conditions affecting refraction is helpful in determining the confidence to be placed in such observations. Since refraction elevates both the celestial body and the visible horizon, the error due to abnormal refraction is minimized if the visible horizon is used as a reference.

The atmosphere contains many irregularities which are erratic in their influence upon refraction. Normally, the navigator has not the information needed to correct for such conditions, but only to recognize their existence. He must recognize that those observations made within half an hour after passage of a squall line might be considerably in error. The passage of any front might have a similar effect. A temperature inversion (volume I) may upset normal refraction. Abnormal values may be expected when there is a large difference between the temperature of the sea and air. With an absence of wind, the air tends to form in layers. When this condition becomes extreme, mirage effects occur. Sometimes the rising or setting sun or moon appears distorted. Multiple horizons may appear, and other ships or islands may seem to float a short distance above the water. Under any such conditions large errors in refraction might be encountered.

Conditions causing abnormal refraction can be expected to occur with considerable frequency in the vicinity of the Grand Banks, along the west coast of Africa from Mogador to Cap Blanc and from the Congo to the Cape of Good Hope, in the Red Sea and the Persian Gulf, and over ice-free water in polar regions. Abnormal refraction may be encountered when offshore winds blow from high, snow-covered mountains to nearby tropical seas, as along the west coast of South America; where cold water from large rivers such as the Mississippi flows into warm sea water; when a strong current flows past a bay or coast, causing colder water to be drawn to the surface, as in the Bay of Rio de Janeiro and Santos, and along the Atlantic coast of Africa between Cape Palmas and Cape Three Points during the time of the southwest monsoon; and along the east coast of Africa in the vicinity of Capo Guardafui during the summer. In the temperate zones abnormal refraction is most common during the spring and summer.

Of the more systematic errors which affect refraction, two can be evaluated, and corrections applied. These are for air temperature (art. 807) and atmospheric pressure (art. 808). However, these corrections are based upon assumed standard *gradients* (changes) with height. Temperature gradients are known to vary with type of weather, time of day, season, etc., as well as in a more irregular manner. The various layers of the atmosphere are assumed to be horizontal to the surface, but this is not always the situation. When they tilt, refraction changes. No correction for this cause is available.

Humidity has a relatively slight effect on refraction. In completely dry air, astronomical refraction at the horizon (sometimes called **horizontal refraction**) is perhaps 0ʹ.1 *greater* than the tabulated value. In very moist air, sometimes encountered in the tropics, the maximum refraction might possibly *decrease* by as much as 0ʹ.2.

Wind speed is believed to have some effect upon refraction and dip. Apparently, refraction increases as wind speed becomes greater, the amount of change increasing in direct proportion to the *square* of the wind speed. At 20 knots the change is believed to be about 0′.1 at the horizon, and 0″.15 at altitude 2°. At 30 knots these values are approximately doubled.

Latitude has a slight effect upon refraction because of the decrease in the radius of the earth and the increase of gravity as latitude increases. Both radius and gravity affect density of the atmosphere, and hence refraction. Because of this, mean horizontal refraction is decreased about 0′.2 to 0′.3 at the equator, and increased about an equal amount at the poles. For normal altitudes the change is negligible.

Azimuth may have an effect at some locations. The reason for this is not entirely clear, but is believed to be due to a somewhat permanent tilt of certain atmospheric layers. A series of observations at a location in Germany indicated a difference of as much as 0′.5 between northerly and southerly observations of 0° altitude. At altitude 2° the difference was only 3″. The navigator normally does not have information required to apply such a correction, nor is it of navigational significance at normal altitudes. However, differences should be anticipated when the appearance of the horizon varies with azimuth, or large sea temperature differences exist within a few miles, as near the edge of the Gulf Stream. The same might be true near land, particularly in the tropics.

Dispersion of light of various colors results in light from blue stars being refracted more than light from red stars. At the horizon the maximum correction would be about 2″ for blue stars, and 8″ for red stars. At 5° the amount would be less than one-third these values.

Errors in the tables due to incorrect assumptions are probably too small to be of practical interest to the navigator. If increased knowledge indicates errors exist in the tables, corrected values will undoubtedly be provided.

Since refraction causes celestial bodies to appear elevated in the sky, they are above the horizon longer than they otherwise would be. The mean diameter of the sun and moon are each about 32′, and horizontal refraction is 34′.5. Therefore, the entire sun or moon is actually below the visible horizon when the lower limb appears tangent to the horizon. The effect of dip is to further increase the time above the horizon. Near the horizon the sun and moon appear flattened because of the rapid change of refraction with altitude, the lower limb being raised by refraction to a greater extent than the upper limb.

As a correction to sextant altitudes, refraction is negative because it causes the measured altitude to be too *great*. It *decreases* with increased altitude, and applies to all celestial bodies, regardless of sextant or horizon used.

807. Air temperature correction (T).—The *Nautical Almanac* refraction table is based upon an air temperature of 50° F (10° C) at the surface of the earth. At other temperatures the refraction differs somewhat, becoming greater at lower temperatures, and less at higher temperatures. Table 23 provides the correction to be applied to the altitude to correct for this condition. If preferred, this correction can be applied with *reversed sign* to the refraction from the almanac, and a single refraction applied to the altitude. A combined correction for nonstandard air temperature and nonstandard atmospheric pressure (art. 808) is given on page A4 of the *Nautical Almanac*. The correction for air temperature varies with the temperature of the air and the altitude of the celestial body, and applies to all celestial bodies, regardless of the method of observation. However, except for extreme temperatures or low altitudes, this correction is not usually applied unless results of unusual accuracy are desired.

808. Atmospheric pressure correction (B).—The *Nautical Almanac* refraction table is based upon an atmospheric pressure of 29.83 inches of mercury (1010 millibars) at sea level. At other pressures the refraction differs, becoming greater as pressure increases, and smaller as it decreases. Table 24 provides the correction to be applied to the altitude for this condition. A combined correction for nonstandard air temperature (art. 807) and nonstandard atmospheric pressure is given on page A4 of the *Nautical Almanac*. If the correction is to be applied to the refraction, reverse the sign. This correction varies with atmospheric pressure and altitude of the celestial body, and is applicable to all celestial bodies, regardless of the method of observation. However, except for extreme pressures or low altitudes, this correction is not usually applied unless results of unusual accuracy are desired.

809. Irradiation correction (J).—When a bright surface is observed adjacent to a darker one, a physiological effect in the eye causes the brighter area to appear to be larger than is actually the case; conversely, the darker area appears smaller. This is called **irradiation.** Thus, since the sun is considerably brighter than the sky background, the sun appears larger than it really is; and when the sky is considerably brighter than the water, the horizon appears slightly depressed. The effects on the horizon and lower limb of the sun are in the same direction and tend to cancel each other while the effect on the upper limb of the sun is in the opposite direction to that on the horizon and tends to magnify the effect.

From 1958–1970 a correction of 1′.2 was included in the *Nautical Almanac* data for the upper limb of the sun as an average correction for the effect of irradiation. Recent investigations have not supported that average value and have revealed that the magnitude of the effect depends on the individual observer, the size of the ocular, the altitude of the sun, and other variables. In summary, the accuracy of observations of the limb of the sun at low altitudes may be affected systematically by irradiation, but the size of the correction is so dependent upon the variables enumerated above that it is not feasible to include an average correction in the tables.

810. Semidiameter (SD) of a celestial body is half the angle, at the observer's eye, subtended by the visible disk of the body. The position of the lower or upper limb of the sun or moon with respect to the visible horizon can be judged with greater precision than that of the center of the body. For this reason it is customary, when using a marine sextant and the visible horizon, to observe one of the limbs of these two bodies, and apply a correction for semidiameter. Normally, the lower limb is used if it is visible. In the case of a gibbous or crescent moon, however, only the upper limb may be available. Semidiameter is shown in figure 812.

The semidiameter of the sun varies from a little less than 15′.8 early in July, when the earth is at its greatest distance from the sun, to nearly 16′.3 early in January, when the earth is nearest the sun. In the *Nautical Almanac* the semidiameter of the sun at GMT 12^h on the middle day of each page opening of the daily page section is given to the nearest 0′.1 at the bottom of the sun's GHA column. The altitude correction tables of the sun, given on the inside front cover and facing page, are divided into two parts, to be used during different periods of the year. The mean semidiameter of each period is included in the tables of both upper and lower limb corrections. The semidiameter each day is listed to the nearest 0″.01 in the *American Ephemeris and Nautical Almanac*. In the *Air Almanac* the semidiameter to the nearest 0′.1 is given near the lower right-hand corner of each daily page.

The moon undergoes a similar change in semidiameter as its distance from the earth varies. However, because of the greater eccentricity of the moon's orbit than that of earth, the variation in semidiameter is also greater, varying between about 14′.7 and 16′.8. The variation is more rapid, partly because of the greater spread of values, but principally because the moon completes its revolution in approximately one month, while the earth makes one revolution per year. In the *Nautical Almanac*, semidiameter of the moon at 12^h each day is given to the nearest 0′.1 at the bottom of the moon data columns. The correction for semidiameter of the moon is included in the corrections given on the inside back cover and facing page. In the *Air Almanac*, semidiameter is given to the nearest whole minute, being shown on the daily pages, immediately below the value for the sun. The semidiameter at intervals of half a day is given to the nearest 0″.01 in the *American Ephemeris and Nautical Almanac*.

The navigational planets have small semidiameters. For Venus it varies between about 5″ and 32″; for Mars, 2″.7 to 12″.6; for Jupiter, 16″ to 25″; and for Saturn, 7″ to 10″. The value for any date is given in the *American Ephemeris and Nautical Almanac*, but not in the *Nautical Almanac* or *Air Almanac* because the apparent *centers* of these bodies are customarily observed.

Stars have no measurable semidiameter.

The computed altitude of a body refers to the *center* of that body, since the coordinates listed in the almanacs are for the center. If the *lower* limb is observed, the sextant altitude is *less* than the altitude of the center of the body, and hence the correction is *positive*. If the *upper* limb is observed, the correction is *negative*. The correction does not apply when the center of the body is observed, which is usually the case when an artificial-horizon sextant is used. With a marine sextant and either the natural or an artificial horizon, semidiameter is customarily applied to observations of the sun and moon, but not other celestial bodies.

811. Phase correction (F).—Because of phase (fig. 812), the actual centers of planets and the moon may differ somewhat from the apparent centers. Average corrections for this difference are included in the additional corrections for Venus and Mars given on the inside front cover of the *Nautical Almanac*. They should be applied only when these bodies are observed during twilight. At other times the magnitude and even the sign of the correction might differ from those tabulated, because of a different relationship between the body and the horizon. The phase correction for navigational planets other than Venus and Mars is too small to be significant.

A phase correction may apply to observations of the moon if the apparent center of the body is observed, as with an artificial-horizon sextant. However, no provision is made for a correction in this case; the need for it can be avoided by observing one of the limbs of the body.

Phase correction does not apply to observations of the sun or stars.

812. Augmentation (A).—As indicated in article 810, semidiameter changes with distance of the celestial body from the observer, becoming greater as the distance decreases. The semidiameter given in the ephemeris and used in the almanacs is for a fictitious observer at the center of the earth. If the celestial body is on the actual observer's horizon, its distance is approximately the same as from the center of the earth; but if the body is in the zenith, its distance is less by about the radius of the earth (fig. 812). Therefore, the semidiameter *increases* as the altitude becomes greater. This increase is called **augmentation.** For the moon, the augmentation from horizon to zenith is about 0′.3 at the mean distance of the moon. At perigee it is about 2″ greater, and at apogee about 2″ less. Augmentation of the sun from horizon to zenith is about 1/24 of one second of arc. For planets it is correspondingly small, varying with the positions of the planets and the earth in their orbits. At any altitude the augmentation is equal to the sine of the altitude times the value at the zenith.

Augmentation increases the size of the semidiameter correction, whether positive or negative. It is included in the moon correction tables on the inside back cover and facing page of the *Nautical Almanac*. It is not included in the correction tables of other bodies or in the *Air Almanac* tables.

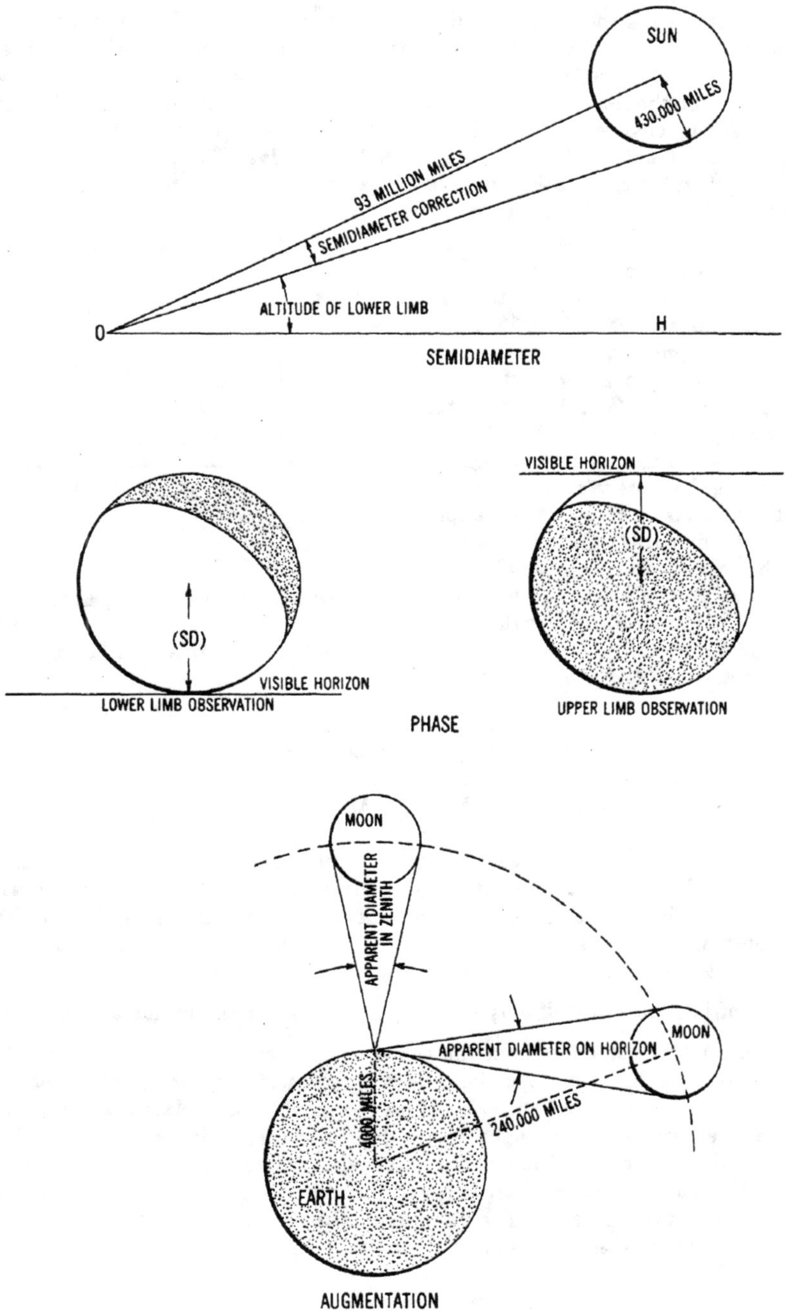

Figure 812.—Semidiameter, phase, and augmentation.

813. Parallax (P) is the difference in apparent position of a point as viewed from two different places. If a finger is held upright at arm's length and the right and left eyes closed alternately, the finger appears to move right and left a short distance. Similarly, if one of the nearer stars were observed from the earth and from the sun, it would appear to change slightly with respect to the background of more distant stars. This is called **heliocentric parallax** or **stellar parallax.** The nearest star has a parallax of less than 1″. Even if the value were greater, no correction to sextant altitudes would be needed, for the difference would be reflected in the tabulated position of the body.

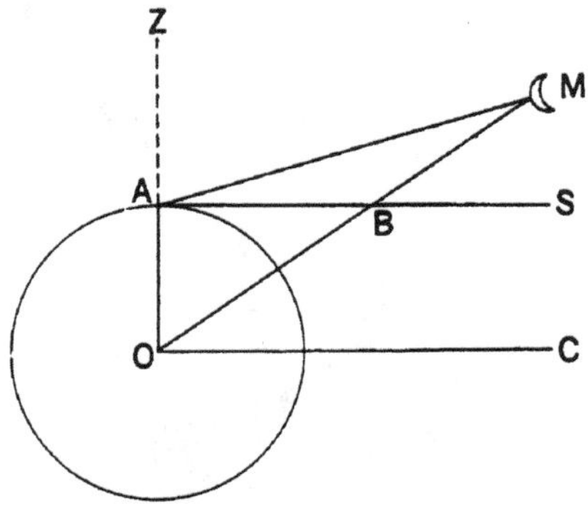

FIGURE 813.—Geocentric parallax.

However, positions of celestial bodies are given relative to the *center* of the earth, while observations are made from its surface. The difference in apparent position from these two points is called **geocentric parallax.** If a body is in the zenith, at Z in figure 813, there is virtually no parallax, for the line from the body to the center of the earth passes approximately through the observer at *A*. Suppose, however, the moon is at *M*. From *A* it appears to be along the line *AM*, while at the center of the earth it would appear to be along *OM*. The altitude at *A* would be the angle *SAM*, and that at *O* the angle *COM*. Angle *COM* is equal to angle *SBM* (art. 127), which is exterior to the triangle *ABM*, and hence equal to the sum of angles *SAM* and *AMO* (art. 128).

Since

$$\angle COM = \angle SBM = \angle SAM + \angle AMO,$$

then

$$\angle AMO = \angle COM - \angle SAM.$$

That is, the angle at the body between lines to the observer and the center of the earth is equal to the difference in altitude at the two places. Angle *AMO* is the geocentric parallax. Since it varies with altitude, it is sometimes called **parallax in altitude (P in A).** The maximum value for a visible body occurs when that body is on the horizon, at *S*. At this position the value is called **horizontal parallax (HP).**

The sine of horizontal parallax is equal to $\frac{r}{D}$, where r is the radius of the earth, and D the distance of the body from the center of the earth. Thus, the sine of the horizontal parallax is directly proportional to the radius of the earth, and inversely proportional to the distance of the body. Since the earth is an oblate spheroid, and not a sphere, the parallax varies slightly over different parts of the earth. The value at the equator, called **equatorial horizontal parallax,** is greatest, and the value at the poles, called **polar horizontal parallax,** is least. The difference is not enough to be of practical navigational significance. The parallax in altitude is equal almost exactly to the horizontal parallax times the cosine of the altitude (h). That is,

$$\text{P in A} = \text{HP} \cos \text{h}.$$

The moon, being nearest the earth, has the greatest parallax of any celestial body used for navigation. The equatorial horizontal parallax at mean distance is 57′02″70.

As the distance of the moon varies, so does the parallax, becoming greater as the moon approaches closer to the earth, and less as it recedes, horizontal parallax varying several minutes each side of the value at mean distance. For the sun, mean equatorial horizontal parallax, called **solar parallax,** is 8".80. Differences in position on the earth, and distance from the sun, have small effect, the maximum variation due to the latter being about 0".15. Horizontal parallax of the planets varies considerably because of the large differences in their distances from the earth. For Venus the value varies between 5" and 32"; for Mars, 3" and 24"; for Jupiter, 1" and 2"; and for Saturn, 0".8 and 1".0. The geocentric parallax of stars is too small to be measured, even by the most precise telescopes, since the value for the nearest star is only 0".00003.

Daily values of horizontal parallax for the sun, moon, and planets are given in the *American Ephemeris and Nautical Almanac*, to a precision of 0".01. In the *Nautical Almanac*, mean values for the sun are included in the two sun correction tables given on the inside front cover and facing page. Horizontal parallax of the moon is tabulated at intervals of one hour on the daily pages. This value is used to enter the lower part of the moon correction tables on the inside back cover and facing page. The additional corrections for Venus and Mars given on the inside front cover are partly for parallax. No correction is given for parallax of Jupiter and Saturn. The *Air Almanac* gives parallax corrections only for the moon. These values are given in the "Moon's P in A" column on each daily page.

Because of the geocentric parallax, a body appears too *low* in the sky. Therefore, the correction is always positive. It applies regardless of the method of observation.

814. Sea-air temperature difference correction (S).—Under normal atmospheric conditions, the temperature and pressure both decrease at standard rates with increase in height above the surface. Accordingly, the *density* of the atmosphere also decreases at a standard rate, which is uniform over the height encountered aboard ship. The effect of refraction upon dip, as given in the tables, is based upon this standard rate. Usually, the difference between standard and actual conditions is not great enough to introduce important errors in the assumption that standard conditions exist.

However, when there is a difference between sea and air temperatures at the surface, the air in contact with the sea is warmed or cooled by the sea water, upsetting the normal rate of decrease near the surface. The effect is greater as the temperature difference increases. It may extend only a few inches above the surface, or for hundreds of feet. Under extreme conditions, if the air is very much colder than the water, the surface may steam. The **frost smoke** rising from the water may obscure the horizon, and under the most severe conditions it may rise to such heights as to interfere with visibility. Celestial bodies can be seen, but altitudes cannot be measured with a marine sextant because of lack of a horizon.

Under less extreme conditions, the dip is altered, but observations may seem normal. If the water is warmer than the air, the horizon is depressed and dip is increased. Under these conditions the measured altitudes are too great. Therefore, as a correction to the *altitude*, the sea-air temperature difference correction is negative when the water is warmer than the air. When the air is warmer, the reverse is true, and the altitude correction is positive.

Various attempts have been made to establish a simple relationship between the sea-air temperature difference and the correction, but the results reported by different investigators differ considerably. This is due, in part, to difference of opinion as to the height and depth at which measurements should be made, difficulties in obtaining accurate readings near the surface, variations of temperature differences at the ship and along the line of sight to the horizon, and influence of the vessel on temperatures in its immediate vicinity. Wind, too, has a considerable effect. On a calm day, the lower portion of the atmosphere tends to form in layers, without mixing. If there is a strong, gusty wind, turbulence in the air minimizes the effect due to temperature difference. Actually, sea temperature serves only to indicate temperature at the surface, but temperature gradient in the water may be large, as in the air. Therefore, the ideal would seem to be the measurement of *air* temperature *at the surface,* and at some greater height, since it is the abnormal **lapse rate** (decrease of air temperature with height) that produces the change in normal terrestrial refraction.

Suggested factors based upon difference between temperature of the sea and air vary from about 0!11 per degree Fahrenheit (0!20 per degree Celsius) to 0!21 per degree Fahrenheit (0!37 per degree Celsius). The average of these is about 0!16 per degree Fahrenheit (0!28 per degree Celsius). Thus, the correction is about one-sixth of a minute per degree Fahrenheit, or one minute for each six degrees.

This correction applies to all bodies when the sea horizon is used. However, it should be used with caution, and only under conditions which indicate that better results will be obtained if it is used. Under normal conditions, it is not used. If abnormal conditions are suspected, observations are avoided or considered of questionable reliability; or the precautions indicated in article 805 are used. If allowance for abnormal conditions is made by using an altered value of dip, as one obtained by measurement, the sea-air temperature difference correction is not used. That is, if allowance is made for abnormal dip, *either* the tabulated value of dip is altered *or* the sea-air temperature correction is applied, *but not both.*

815. The uncorrected errors.—When a body is in motion over the surface of the earth, its motion *in space* is a combination of its motion relative to the earth, and the motion of the earth. Because of the rotation of the earth, principally, the path is a curved one. As a result, there is an apparent force causing deflection to the right in the northern hemisphere and to the left in the southern hemisphere. Because of this **Coriolis force,** ocean currents set in motion by wind flow in a direction to the right (northern hemisphere) of the direction in which the wind blows. Wind, too, is deflected. Instead of blowing directly from an area of high pressure to one of low pressure, and soon neutralizing the pressure difference, it moves toward one side. The result is the characteristic circulation around highs and lows.

The liquid of the bubble chamber of a bubble sextant, and a pendulum, are similarly affected, causing them to give a false indication of the vertical (or horizontal). The same is true of an artificial horizon. The equation for the deflection is

$$Z=2'.62\ S \sin L+0'.146\ S^2 \sin C \tan L-5'.25\ SC',$$

where Z is the *Coriolis correction,* S the speed over the surface of the earth in units of *hundreds* of knots, L the latitude, C the true course angle, and C′ the rate of change of true course angle in degrees per minute. The first term, $2'.62\ S \sin L$, corrects for motion along a great circle; the second term, $0'.146\ S^2 \sin C \tan L$, is an additional correction for the difference between motion along a rhumb line, and equivalent motion along a great circle; and the third term, $5'.25\ SC'$, is an additional correction for departures from the course, being negative (as shown) if the departure is right in the northern hemisphere or left in the southern hemisphere. Coriolis corrections (first term only) are given near the back of the *Air Almanac.*

Coriolis correction may be either positive or negative, and varies with speed and latitude. It applies to all bodies equally, and therefore can be applied to the altitude, the assumed position, or even as an adjustment to the plotted line of position or fix. If the AP or fix is adjusted, it is moved perpendicular to the course line, to the *right* in the northern hemisphere and to the *left* in the southern hemisphere, unless the third term is of such magnitude and sign as to make the entire correction negative, when it is applied in the opposite direction. If the correction is applied to the altitude, the value obtained by formula is multipled by the sine of the *relative* azimuth. In the northern hemisphere, the resulting altitude correction is positive if the celestial body is on the starboard side, and negative if on the port side. In the southern hemisphere these signs are reversed. These signs assume t' at the value obtained by the formula is positive. If it is negative, all signs are reversed.

At ship speeds, the Coriolis correction is not large, unless the vessel is yawing considerably. For a ship steaming at 20 knots on a steady course of 090° at latitude 40°, the maximum Coriolis correction is 0.'3 for a celestial body which is abeam. Acceleration error due to rolling and pitching of the vessel is usually much greater than this, and is the principal reason why bubble or pendulum sextants are not often used aboard ship.

There is no Coriolis correction when the visible horizon is the horizontal reference.

If an artificial horizon-sextant with a bubble or pendulum is used, the liquid of the bubble chamber or the pendulum is affected by all accelerations of the instrument. The same is true of the free surface of the liquid of an artificial horizon. With high accelerations such as those due to rolling and pitching of a vessel, or changes of course or speed, the **acceleration error** can be very large. It is for this reason that such instruments are not customarily used aboard ship. Under normal conditions at sea the navigator does not have the information needed to compute the correction. The error is minimized by making observations at the center of roll and pitch of the ship, or averaging the values taken at both ends of a roll or pitch. Observations should not be made during a turn or when the speed is being changed. Even with these precautions the error is usually unacceptably large except with an almost flat, calm sea. The effect on the level of the sea surface due to accelerations of the earth in its rotation or revolution is considered negligible.

The correction may be either positive or negative, and applies equally to all bodies observed with a bubble or pendulum sextant.

Usually, the direction of gravity is assumed to be normal (vertical) to the spheriodal surface of the earth. This assumption is not quite correct. Irregularities in density and height of the material making up the surface crust of the earth result in slight alterations of the direction of gravity. This **deflection of the vertical** is most apparent near high mountains bordering a deep sea where an extreme value of more than 1.'1 might be encountered.

Normally, values of deflection of the vertical are not available to the navigator, and are not needed by him. In precise work, however, such values for a particular area might be furnished. The correction is negative in the direction toward which the zenith is deflected, and positive in the opposite direction. In any other direction it is equal approximately to the maximum deflection times the cosine of the angle between the given direction and the direction of maximum deflection, taking the sign of the nearest maximum deflection. It is applicable to all celestial bodies, whether the natural sea horizon or an artificial horizon is used.

Corrections for dip are based upon the assumption of a calm sea. Waves disturb this condition, causing the surface to be alternately raised and lowered. At the horizon, the troughs of waves are usually not visible. Through binoculars, irregularities in the line forming the horizon can sometimes be seen, but observations are made from the *tops* or nearly the tops of the waves.

This correction for wave height is always positive, and applies to all celestial bodies, but only if the sea horizon is used as a reference. Normally, it is not applied because of the difficulty of determining (1) wave height at the horizon, and (2) height of eye above the equivalent calm level of the sea. Better practice is usually to estimate height of eye above the wave tops, allowing for motions of the vessel, and make no correction.

The height of the sea at any place is affected by the density of the sea water, its temperature, and atmospheric pressure. Because of differences in these values, the height varies from place to place. This results in tilting of the surface of the sea, which is "downhill" from the ridge of high water to that of low water. The maximum tilt due to these causes is probably a little more than 1$''$5. The wave caused by the tides also tilts the sea surface. However, on the open sea, tides are seldom more than about two and one-half feet high, and the distance from crest to trough is about 5,400 miles, or one-quarter of the great-circle distance around the earth. Under these conditions, the maximum tilting of the sea surface due to tides is about 0$''$025. This may be increased somewhat by storm waves or tsunamis. In confined waters, particularly in a funnel-shaped area where tides enter from the wide end and progress up a narrowing estuary, the error may be very much greater, possibly reaching a value of half a *minute*.

The correction is positive in the direction of high water and negative in the direction of low water. Between these directions it is equal approximately to the value in these directions multiplied by the cosine of the angle between the wave axis (the line perpendicular to the wave front) and the azimuth of the body. It applies equally to all bodies when the visible horizon is used as the reference. In practice it is not applied.

816. Summary of corrections.—The essential information regarding the application of the various corrections may be tabulated as shown below. In the "Bodies" column, the symbols are: ⊙, sun; ☾, moon; P, planets; ☆, stars. In the "Sextants" column, M refers to a marine sextant with visible horizon, A refers to a marine sextant with artificial horizon, and B refers to an artificial-horizon sextant. The tabulation assumes that completely accurate results are desired and that corrections are to be made in the usual manner, where they are available. Some of the entries need qualification, which may be found in the preceding articles.

Correction	*Symbol*	*Sign*	*Increases with*	*Bodies*	*Sextants*	*Source*
Instrument	I	±	changing altitude	⊙, ☾, P, ☆	M, A, B	sextant box
Index	IC	±	constant	⊙, ☾, P, ☆	M, A, B	measurement
Personal	PC	±	constant	⊙, ☾, P, ☆	M, A, B	measurement
Dip	D	−	higher height of eye	⊙, ☾, P, ☆	M	almanacs
Sea-air temp. diff.	S	±	greater temp. diff.	⊙, ☾, P, ☆	M	computation
Refraction	R	−	lower altitude	⊙, ☾, P, ☆	M, A, B	almanacs
Air temp.	T	±	greater diff. from 50° F	⊙, ☾, P, ☆	M, A, B	almanacs, table 23
Atmospheric pressure	B	±	greater diff. from 29.83 inches of mercury	⊙, ☾, P, ☆	M, A, B	*Nautical Almanac*, table 24
Irradiation	J	−		⊙	M, A	
Semidiameter	SD	±	lesser dist. from earth	⊙, ☾	M, A	almanacs
Phase	F	±	phase	P	M, A, B	*Nautical Almanac*
Augmentation	A	±	higher altitude	☾	M, A	*Nautical Almanac*
Parallax	P	+	lower altitude	⊙, ☾, P	M, A, B	almanacs

These corrections can be considered to fall into five groups:

1. *Corrections for inaccuracies in reading.* Instrument correction, index correction*, and personal correction.

2. *Corrections for inaccuracies in reference level.* Dip* and sea-air temperature difference.

3. *Corrections for bending of ray of light from body.* Refraction*, air temperature, atmospheric pressure.

4. *Adjustment to equivalent reading at center of body.* Irradiation, semidiameter*, phase, augmentation.

5. *Adjustment to equivalent reading at center of earth.* Parallax*.

In the ordinary practice of seamen, extreme accuracy is not required, and only the principal correction of each group is applied (except that augmentation is applied for the moon). These principal corrections are indicated by asterisks. For low altitudes, additional corrections are applied, as indicated in article 827.

817. Order of applying corrections.—For purposes of ordinary navigation, sextant altitudes can be applied in any order desired, using sextant altitude for the entering argument whenever altitude is required. This practice is not strictly accurate, but for altitudes usually observed, the error thus introduced is too small to be of practical significance. When extreme accuracy is desired, however, or at low altitudes, where small changes in altitude result in significant changes in correction, the order of applying corrections is important. Corrections from the first two groups of article 816 are applied to sextant altitude (hs) to obtain **apparent (rectified) altitude (ha),** which is then used as an entering argument for obtaining corrections of the third group. For strictest accuracy, all corrections of the first three groups and, in addition, irradiation and semidiameter, should be applied before augmentation, and all other corrections before parallax.

818. Marine sextant corrections.—As shown in articles 816 and 817, all corrections except Coriolis and acceleration apply to marine sextant observations when the visible horizon is used. However, under normal conditions and when the highest accuracy is not required, it is necessary to apply only a few corrections. Several of these corrections may be combined within a single altitude correction table. In addition to corrections for index error, dip, and mean refraction, the normal altitude corrections when using the *Nautical Almanac* are: phase and parallax for Venus and Mars; semidiameter and parallax for the sun; and semidiameter, augmentation, and parallax for the moon.

819. Artificial-horizon corrections.—When an artificial horizon is used, index correction (and any others of the first group of article 816) is first applied. The result is then divided by two. Other corrections are then applied to the result, as applicable, in the same manner as for observations using the visible horizon. The sun and full moon are normally observed by bringing the lower limb of one image tangent to the upper limb of the other image. The lower limb is observed if the image seen in the horizon mirror is *above* the image seen in the artificial horizon, unless an inverting telescope is used, when the opposite relationship holds. With a gibbous or crescent moon, judgment may be needed to establish the positions of the limbs. In some cases better results may be obtained by superimposing one image over the other, as with a planet or star. When this is done, the *center* of the body has been observed, and no correction is applied for semidiameter (or irradiation, phase, or augmentation). There is no correction for dip (or sea-air temperature) when an artificial horizon is used.

820. Artificial-horizon sextant corrections are the same as those for observations made by the use of the visible horizon, with two notable exceptions. First, there is no correction for dip (or sea-air temperature difference or wave height), none for semidiameter (or irradiation, phase, or augmentation), and usually none for index correction (or instrument correction). Second, because of the lower accuracy normally obtainable by artificial-horizon sextant, corrections are normally made only to the nearest whole minute of arc. As a result of these differences, refraction is the only correction normally applied, except in the case of the moon, where parallax is also applied.

821. Corrections by *Nautical Almanac*.—In the *Nautical Almanac*, certain corrections or parts of corrections are combined. Index correction, of course, is not included because this depends upon adjustment of the sextant. The various correction tables are as follows:

"*Sun,*" on the inside front cover and facing page, gives mean refraction, mean semidiameter for each of two periods during the year, and mean solar parallax. The table on the inside front cover, and repeated on the loose bookmark, is of the critical type, with altitude as the entering value. Thus, a tabulated correction applies to any value of altitude between that given half a line above it and that half a line below it. If an exact tabulated altitude is used to enter the table, the correction half a line *above* it should be used. In ordinary navigation, index correction, dip, and the correction from this table are needed for correcting marine sextant observations of the sun. For low altitudes or extremes of temperature or atmospheric pressure, a correction from the table on almanac page A4 (or tables 23 and 24 of this volume) should be applied.

"*Stars and planets,*" on the inside front cover and repeated on the loose bookmark, gives mean refraction only, for the main tabulation. This is a critical type table, with altitude as the entering argument. The correction is always negative. In ordinary navigation, index correction, dip, and the correction from this table are the only ones needed for stars and the planets Jupiter and Saturn. For Venus and Mars, an additional correction for parallax and phase is given to the right of the main tabulation. The entering altitudes are limited to those occuring during twilight. If observations are made at other times, this additional correction should not be applied even though the altitude may fall within the tabulated range.

"*Dip,*" on the inside front cover and repeated on the loose bookmark, is for dip of the horizon. An abbreviated dip table is also given on the page facing the inside back cover. The tables are of the critical type, and the entering argument is the height of the observer's eye, in feet and meters, above the surface of the sea. The correction, always negative, applies to all observations made with the visible sea horizon as a reference.

"*Additional Correction Tables*" for *nonstandard conditions,* given on almanac page A4, provides an additional correction for nonstandard temperature and atmospheric pressure. The sign of each correction is indicated. Equivalent information is given, with increased range of entering values, in tables 23 and 24 of this volume.

"*Altitude Correction Tables—Moon,*" on the inside back cover and facing page, gives mean refraction, semidiameter, augmentation, and parallax. The entering argument is altitude for the upper portion of the table, and altitude and horizontal parallax for the lower portion. The combined correction is always positive, but 30′ is to be subtracted from the altitude of the upper limb. In ordinary navigation, index correction, dip, and the correction from this table are needed in correcting marine sextant observations of the moon.

The various separate corrections available from the *Nautical Almanac* can be found as follows:

Dip. Dip table on inside front cover and repeated on loose bookmark, and on the page facing the inside back cover.

Refraction. Mean refraction from "Stars and Planets" table on inside front cover and repeated on loose bookmark, and on the facing page.

Semidiameter. For the sun, the semidiameter for the middle day of each page opening of this daily page section is given at the bottom of the sun GHA column. For the moon, semidiameter for each day is given at the bottom of the moon data columns. The values given are for GMT 1200 on the dates indicated.

Parallax. For the sun, parallax in altitude can be considered 0.′1 for altitudes 0° to 70°07′, and 0.′0 for higher altitudes, with negligible error. This is based upon the mean value of 8.″80. For the moon, horizontal parallax each hour is tabulated on the daily pages. Parallax in altitude is this value multiplied by the cosine of the altitude.

If artificial-horizon sextant altitudes of the sun or moon are corrected by *Nautical Almanac*, the upper and lower limb corrections can be found and the average computed.

822. Corrections by *Air Almanac*.—In the *Air Almanac*, various corrections as applicable to hand-held marine sextant observations are given separately in critical type tables, to the nearest whole minute (nearest two or five minutes of refraction for low altitudes), as follows:

Dip. Inside back cover.

Refraction. Near the back. Aboard ship use the values for zero height.

Air temperature. Near the back. This is shown, not as a separate correction, but as an adjustment to mean refraction. Instructions for use of the table are given within the table.

Semidiameter. For the sun and moon, on the A.M. and P.M. pages, below the moon's P in A. Values given are for GMT 1200.

Parallax. For the moon, in the P in A table on the A.M. and P.M. pages. Horizontal parallax is the value for 0° altitude.

823. Correcting altitudes of the sun.—In the normal practice of navigation, sun observations obtained by marine sextant with the visible horizon as reference are corrected as shown in the following examples:

Example 1.—On June 2, 1975, the lower limb of the sun is observed with a marine sextant having an IC of (−)2.′0, from a height of eye of 38 feet. The hs is 51°28.′4.

Required.—Ho using (1) *Nautical Almanac*, and (2) *Air Almanac*.

Solution.—

(1)

	+	$\underline{\odot}$	−
IC			2.′0
D			6.′0
$\underline{\odot}$	15.′2		
sum	15.′2		8.′0
corr.		(+)7.′2	
hs		51°28.′4	
Ho		51°35.′6	

(2)

	+	$\underline{\odot}$	−
IC			2′
D			6′
R			1′
SD	16′		
sum	16′		9′
corr.		(+)7′	
hs		51°28′	
Ho		51°35′	

Example 2.—On June 2, 1975, the upper limb of the sun is observed with a marine sextant having an IC of (+) 1.′0, from a height of eye of 45 feet. The hs is 32°47.′9.

Required.—Ho using (1) *Nautical Almanac*, and (2) *Air Almanac*.

Solution.—

(1)

	+	$\overline{\odot}$	−
IC	1.′0		
D			6.′5
$\overline{\odot}$			17.′3
sum	1.′0		23.′8
corr.		(−)22.′8	
hs		32°47.′9	
Ho		32°25.′1	

(2)

	+	$\overline{\odot}$	−
IC	1′		
D			7′
R			2′
SD			16′
sum	1′		25′
corr.		(−)24′	
hs		32°48′	
Ho		32°24′	

A convenient work form is helpful in the solution. Once the form is prepared, the corrections can be entered in any order desired. The symbols $\underline{\odot}$ and $\overline{\odot}$ are used for the corrections from the sun table on the inside front cover of the *Nautical Almanac*. If additional corrections are used, they are included in the same manner as those shown. Observations by artificial horizon and by artificial-horizon sextant, and low-altitude observations and back sights, are discussed elsewhere in this chapter.

824. Correcting altitudes of the moon.—Moon observations by marine sextant with the visible horizon as reference are normally corrected as shown in the following examples:

Example 1.—At about GMT 1100 on June 2, 1975, the lower limb of the moon is observed with a marine sextant having an IC of (+) 3′.2, from a height of eye of 32 feet. The hs is 18°04′.6.

Required.—Ho using (1) *Nautical Almanac*, and (2) *Air Almanac*.

Solution.—

(1)	+	$\underline{☾}$	−
IC	3′.2		
D			5′.5
☾	62′.5		
L	0′.8		
sum	66′.5		5′.5
corr.		(+)1°01′.0	
hs		18°04′.6	
Ho		19°05′.6	

(2)	+	$\underline{☾}$	−
IC	3′		
D			6′
R			3′
SD	15′		
P	51′		
sum	69′		9′
corr.			(+)60′
hs			18°05′
Ho			19°05′

Example 2.—At about GMT 0900 on June 2, 1975, the upper limb of the moon is observed with a marine sextant having an IC of (−)1′.6, from a height of eye of 70 feet. The hs is 66°47′.3.

Required.—Ho using (1) *Nautical Almanac*, and (2) *Air Almanac*.

Solution.—

(1)	+	$\overline{☾}$	−
IC			1′.6
D			8′.1
☾	33′.1		
U	3′.2		
add'l			30′.0
sum	36′.3		39′.7
corr.		(−)3′.4	
hs		66°47′.3	
Ho		66°43′.9	

(2)	+	$\overline{☾}$	−
IC			2′
D			8′
R			—
SD			15′
P	21′		
sum	21′		25′
corr.			(−)4′
hs			66°47′
Ho			66°43′

The typical work forms shown are useful in problems of this type. The symbol ☾ is used for the correction from the upper part of the moon correction table on the inside back cover, and facing page, of the *Nautical Almanac*. The symbols L and U are used for the corrections from the lower part of this table. Observations by artificial horizon, and by artificial-horizon sextant, and low-altitude observations and back sights, are discussed elsewhere in this chapter, as are additional corrections for use when unusual accuracy is desired.

825. Correcting altitudes of planets.—When Venus and Mars are observed by marine sextant using the visible horizon as reference, sextant altitudes are normally corrected as shown in the following example:

Example.—On June 19, 1975, Venus is observed with a marine sextant having no IC, from a height of eye of 28 feet. The hs is 44°21′.3.

Required.—Ho using (1) *Nautical Almanac*, and (2) *Air Almanac*.

Solution.—

(1)

	+	V	−
IC	—		—
D			5′.1
☆-P			1′.0
add'l	0′.3		
sum	0′.3		6′.1
corr.			(−)5′.8
hs			44°21′.3
Ho			44°15′.5

(2)

	+	V	−
IC	—		—
D			5′
R			1′
sum	—		6′
corr.			(−)6′
hs			44°21′
Ho			44°15′

For Jupiter and Saturn, no additional correction is given. Correction of observations of these bodies is the same as corrections of star observations (art. 826). Work forms are useful. The symbol ☆-P is used for the correction taken from the "Star-Planet" table on the inside front cover of the *Nautical Almanac*. If additional corrections are to be used, for results of unusual accuracy or low altitudes, they are included in the form in the same manner as those shown. Observations by artificial horizon and by artificial-horizon sextant, and low-altitude observations and back sights are discussed elsewhere in this chapter.

826. Correcting altitudes of stars.—Star observations by marine sextant, using the visible horizon as reference, are normally corrected as shown in the following example:

Example.—Miaplacidus is observed with a marine sextant having an IC of (+)1′.0, from a height of eye of 50 feet. The hs is 27°54′.0.

Required.—Ho using (1) *Nautical Almanac*, and (2) *Air Almanac*.

Solution.—

(1)

	+	☆	−
IC	1′.0		
D			6′.9
☆-P			1′.8
sum	1′.0		8′.7
corr.			(−)7′.7
hs			27°54′.0
Ho			27°46′.3

(2)

	+	☆	−
IC	1′		
D			7′
R			2′
sum	1′		9′
corr.			(−)8′
hs			27°54′
Ho			27°46′

Work forms for such problems are helpful. Additional corrections, used when unusual accuracy is desired, are included in the same manner as those shown. Observations by artificial horizon and by artificial-horizon sextant, and low-altitude observations and back sights, are discussed elsewhere in this chapter.

827. Low altitudes are normally avoided because of large and variable refraction. But sometimes these are the only observations available. This is particularly true in polar regions, where the sun may be the only celestial body available, and may not reach an altitude of more than a few degrees over a considerable period. In lower latitudes the sun may appear briefly just before sunset or just after sunrise. Low-altitude observations can supply useful information if additional corrections are applied. Reliable lines of position can generally be obtained from low-altitude observations, but when conditions are abnormal, the errors introduced are generally larger than for higher altitudes, and the precautions of article 806 should be particularly observed.

In correcting low-altitude observations, which for normal conditions can be defined as those less than 5°, first apply corrections from the first two groups of article 816 to obtain apparent altitude (ha). Normally, this includes only index correction and dip. Then apply the remaining corrections, using apparent altitude when an altitude is needed for entering correction tables. The corrections normally applied are mean refraction, air temperature, atmospheric pressure, semidiameter (as applicable), and parallax (for the sun and moon).

In practice, sextant altitudes are corrected in the usual manner, except that additional corrections are applied, and the process is divided into two parts. The use of apparent altitude for finding parallax introduces an error but this is too small (less than 0.′1) for practical consideration. If the *Nautical Almanac* is used, corrections for altitudes between the horizon and 10° are given in a noncritical type table on almanac page A3. The correction for a negative altitude can be obtained by extrapolation without introducing a significant error for values obtained at ship heights of eye. A combined temperature-atmospheric pressure correction can be obtained from the table on almanac page A4. This table is intended for use without interpolation between columns. Separate corrections can be obtained from tables 23 and 24 of this volume, which provide interpolated values for greater accuracy. They also provide greater range of temperature and atmospheric pressure.

To correct a low altitude of the sun, then, apply index correction and dip to sextant altitude to find apparent altitude. Using this altitude as an entering value, find the following corrections and apply them to apparent altitude:

sun correction (☉̲ or ☉̄), from page A3 of the *Nautical Almanac;*

combined temperature-atmospheric pressure correction (TB), from page A4 of the *Nautical Almanac* (separate corrections for temperature (T) and atmospheric pressure (B) from tables 23 and 24, respectively, can be used *in place of* the combined correction).

If the *Air Almanac* is used, the mean refraction and air temperature corrections can be combined by using the factor in the refraction table. A semidiameter correction of 16′ is added if the lower limb is observed, and subtracted if the upper limb is observed. Since corrections are to whole minutes only, parallax is not used for the sun. In summary, apply index correction and dip to sextant altitude to find apparent altitude. Using this altitude as an entering value, where needed, apply the following corrections to apparent altitude:

refraction (adjusted for air temperature) (R), from table near back of *Air Almanac;*

atmospheric pressure (B), from table 24;

semidiameter (SD), 16′ (add if lower limb, and subtract if upper limb).

Example 1.—On June 2, 1975, the lower limb of the sun is observed with a marine sextant having an IC of (+)1.′8 from a height of eye of 45 feet. The hs is 1°24.′4, air temperature 88°F, and atmospheric pressure 29.78 inches.

Required.—Ho using (1) *Nautical Almanac,* (2) tables 23 and 24, and (3) *Air Almanac.*

Solution.—

(1)	+	☉	−
IC	1′.8		
D			6′.5
sum	1′.8		6′.5
corr.			(−)4′.7
hs			1°24′.4
ha			1°19′.7
☉			6′.0
TB	2′.5		
sum	2′.5		6′.0
corr.			(−)3′.5
ha			1°19′.7
Ho			1°16′.2

(2)	+	☉	−
IC	1′.8		
D			6′.5
sum	1′.8		6′.5
corr.			(−)4′.7
hs			1°24′.4
ha			1°19′.7
☉			6′.0
T	1′.5		
B	—		
sum	1′.5		6′.0
corr.			(−)4′.5
ha			1°19′.7
Ho			1°15′.2

(3)	+	☉	−
IC	2′		
D			7′
sum	2′		7′
corr.			(−)5′
hs			1°24′
ha			1°19′
R			18′
B	—		
SD	16′		
sum	16′		18′
corr.			(−)2′
ha			1°19′
Ho			1°17′

The larger intervals given in the *Air Almanac* refraction table may introduce additional error. In this example, the temperature is changed to Celsius (centigrade), giving a value of 31°. The factor at a height of 0 feet corresponding to this temperature is 0.9. With this and the apparent altitude, the combined refraction and air temperature correction is found to be as shown. Approximately the same result would have been obtained by correcting for mean refraction (without the factor) and temperature (from table 23) separately.

If the moment at which either limb is tangent to the horizon is noted, an observation of 0° altitude has been made without a sextant.

Example 2.—On June 2, 1975, the sun is observed at sunset as the upper limb drops below the horizon, from a height of eye of 38 feet. The air temperature is (−)10° F, and atmospheric pressure 30.06 inches. Double extrapolation would be needed to solve this problem by the *Nautical Almanac*. A better solution is provided by means of tables 23 and 24.

Required.—Ho using (1) tables 23 and 24, and (2) *Air Almanac*.

Solution.—

(1)	+	☉̄	−
IC	—		—
D			6′.0
sum	—		6′.0
corr.			(−)6′.0
hs			0°00′.0
ha			(−)0°06′.0
☉̄			51′.5
T			4′.8
B			0′.3
sum	—		56′.6
corr.			(−)56′.6
ha			(−)0°06′.0
Ho			(−)1°02′.6

(2)	+	☉̄	−
IC	—		—
D			6′
sum			6′
corr.			(−)6′
hs			0°00′
ha			(−)0°06′
R			42′
B			—
SD			16′
sum	—		58′
corr.			(−)58′
ha			(−)0°06′
Ho			(−)1°04′

Corrections are applied algebraically. Therefore, for negative altitudes a negative correction is *numerically* added, and a positive correction is *numerically* subtracted.

To correct low altitudes of the moon, apply index correction and dip to sextant altitude to find apparent altitude. Using this altitude as an entering value, find the following corrections and apply them to apparent altitudes:

moon correction (☾), from inside back cover, and facing page, of *Nautical Almanac;*

lower or upper limb correction (L or U), from inside back cover, and facing page, of *Nautical Almanac;*

additional correction (add'l, (−)30′, for upper limb observation only);

combined temperature-atmospheric pressure correction (TB), from page A4 of the *Nautical Almanac* (separate corrections for temperature (T) and atmospheric pressure (B) from tables 23 and 24, respectively, can be used in place of the combined correction).

If the *Air Almanac* is used, correct the apparent altitude by applying the following corrections:

refraction (adjusted for air temperature) (R), from table near back of *Air Almanac;*

atmospheric pressure (B), from table 24;

semidiameter, from daily page;

parallax, from daily page.

Example 3.—At GMT 17ʰ14ᵐ27ˢ on June 2, 1975, the upper limb of the moon is observed with a marine sextant having no IC, from a height of eye of 33 feet. The hs is 2°35′.4, air temperature 63° F, and atmospheric pressure 29.81 inches.

Required.—Ho using (1) *Nautical Almanac,* (2) tables 23 and 24, and (3) *Air Almanac.*

Solution.—

(1)	+ ☾	−	(2)	+ ☾	−	(3)	+ ☾	−
IC	—	—	IC	—	—	IC	—	—
D		5′.6	D		5′.6	D		6′
sum	—	5′.6	sum	—	5′.6	sum	—	6′
corr.		(−)5′.6	corr.		(−)5′.6	corr.		(−)6′
hs		2°35′.4	hs		2°35′.4	hs		2°35′
ha		2°29′.8	ha		2°29′.8	ha		2°29′

(1)	+	−	(2)	+	−	(3)	+	−
☾	52′.1		☾	52′.1		R		16′
U	1′.1		U	1′.1		B	—	
add'l		30′.0	add'l		30′.0	SD		15′
TB	0′.4		T	0′.4		P	54′	
sum	53′.6	30′.0	B	—		sum	54′	31′
corr.		(+)23′.6	sum	53′.6	30′.0	corr.		(+)23′
ha		2°29′.8	corr.		(+)23′.6	ha		2°29′
Ho		2°53′.4	ha		2°29′.8	Ho		2°52′
			Ho		2°53′.4			

A lower limb solution would be the same, except that an L correction would have been used from the *Nautical Almanac* and there would be no "add'l" correction, and in the *Air Almanac* solution the sign of the semidiameter correction would be reversed. The moon correction table on the inside back cover, and facing page, of the *Nautical Almanac* extends to a minimum altitude of 0°. The corrections for negative altitudes can be found by extrapolation.

To correct low altitudes of the planets Venus and Mars, apply index correction and dip to sextant altitude to find apparent altitude. Using this altitude as an entering value, find the following corrections and apply them to apparent altitude:

star-planet correction (☆-P), from page A3 of the *Nautical Almanac;*

additional correction (add'l), from page A2 of the *Nautical Almanac;*

combined temperature-atmospheric pressure correction (TB), from page A4 of the *Nautical Almanac* (separate corrections for temperature (T) and atmospheric pressure (B) from tables 23 and 24, respectively, can be used *in place of* the combined correction).

If the *Air Almanac* is used, correct the apparent altitude by applying the following corrections:

refraction (adjusted for air temperature) (R), from table near back of *Air Almanac;*

atmospheric pressure (B), from table 24.

Example 4.—On November 28, 1975, Mars is observed with a marine sextant having an IC of (+)3′.5, from a height of eye of 17 feet. The hs is 4°02′.6, air temperature 2° F, and atmospheric pressure 29.67 inches.

Required.—Ho using (1) *Nautical Almanac,* (2) tables 23 and 24, and (3) *Air Almanac.*

Solution.—

(1)	+ M	−
IC	3′.5	
D		4′.0
sum	3′.5	4′.0
corr.		(−)0′.5
hs		4°02′.6
ha		4°02′.1
☆-P		11′.7
add'l	0′.3	
TB		1′.5
sum	0′.3	13′.2
corr.		(−)12′.9
ha		4°02′.1
Ho		3°49′.2

(2)	+ M	−
IC	3′.5	
D		4′.0
sum	3′.5	4′.0
corr.		(−)0′.5
hs		4°02′.6
ha		4°02′.1
☆-P		11′.7
add'l	0′.3	
T		1′.2
B	0′.1	
sum	0′.4	12′.9
corr.		(−)12′.5
ha		4°02′.1
Ho		3°49′.6

(3)	+ M	−
IC	4′	
D		4′
sum	4′	4′
corr.		—
hs		4°03′
ha		4°03′
R		14′
B	—	
sum	—	14′
corr.		(−)14′
ha		4°03′
Ho		3°49′

The solution for Jupiter and Saturn, and for stars, is identical with that of example 4, except that the additional correction (phase and parallax) is omitted.

828. Back sights.—An altitude measured by facing *away* from the celestial body being observed is called a **back sight**. It may be used when an obstruction, such as another vessel, obscures the horizon under the body; when that horizon is indistinct; or when observations are made in both directions, either to determine dip or to avoid error due to suspected abnormal dip. Such an observation is possible only when the arc of the sextant is sufficiently long to permit measurement of the angle, which is the supplement of the altitude. For such an observation of the sun or moon, the lower limb is observed when the image is brought below the horizon, appearing as a normal upper limb observation, and vice versa. To correct such an altitude, subtract it from 180° and reverse the sign of corrections of the first two groups of article 816 (normally only index correction and dip).

Example.—On June 2, 1975, a back sight is taken of the lower limb of the sun, with a marine sextant having an IC of (−)2′.0, from a height of eye of 24 feet. The measured sextant altitude is 118°41′.4.

Required.—Ho using (1) *Nautical Almanac,* and (2) *Air Almanac.*

Solution.—

(1)

	+	⊙	−
IC	2′.0		
D	4′.8		
⊙	15′.4		
sum	22′.2		−
corr.	(+)22′.2		
180°−hs	61°18′.6		
Ho	61°40′.8		

(2)

	+	⊙	−
IC	2′		
D	5′		
R			1′
SD	16′		
sum	23′		1′
corr.	(+)22′		
180°−hs	61°19′		
Ho	61°41′		

829. Correcting horizontal angles.—When a marine sextant is used to measure the horizontal angle between two objects, the result is not usually desired to a precision that makes correction necessary, unless the sextant has an unusually large index error. However, if precise results are desired, corrections of the first group only of article 816 are applied. If a personal error exists, it is not likely to be the same as for altitudes. For measuring angles between two objects differing widely in altitude, as between two stars, it is not likely that results will be required to such precision that additional correction for the third, fourth, and fifth groups of article 816 will be needed. If they are, the method of application can be determined from the principles of spherical trigonometry (art. 142). In this case, the altitudes of both bodies will also be needed. Corrections for the second group of article 816 are not applicable.

Problems

819a. At about GMT 0800 on June 2, 1975, the following bodies are observed with marine sextants having an IC of (+)2′.2, using an artificial horizon: sun (lower limb) hs 134°33′.9, moon (upper limb) hs 77°23′.4, Venus hs 98°04′.6, Schedar hs 43°24′.4.

Required.—Ho of each observation using (1) *Nautical Almanac*, and (2) *Air Almanac*.

Answers.—(1) Sun Ho 67°33′.6, moon Ho 39°09′.1, Venus Ho 49°02′.6, Schedar Ho 21°40′.9; (2) sun Ho 67°34′, moon Ho 39°09′, Venus Ho 49°03′, Schedar Ho 21°41′.

819b. At about GMT 0300 on June 2, 1975, the following bodies are observed with bubble sextants having no IC: sun hs 23°51′, moon hs 52°20′, Jupiter hs 63°18′, Eltanin hs 24°45′.

Required.—Ho of each observation using (1) *Nautical Almanac*, and (2) *Air Almanac*.

Answers.—(1) and (2) Sun Ho 23°49′, moon Ho 52°52′, Jupiter Ho 63°18′, Eltanin Ho 24°43′.

823a. On June 2, 1975, the lower limb of the sun is observed with a marine sextant having an IC of (+)1′.8, from a height of eye of 34 feet. The hs is 41°34′.8.

Required.—Ho using (1) *Nautical Almanac*, and (2) *Air Almanac*.

Answers.—(1) Ho 41°45′.8; (2) Ho 41°46′.

823b. On June 2, 1975, the upper limb of the sun is observed with a marine sextant having no IC, from a height of eye of 30 feet. The hs is 15°21′.7.

Required.—Ho using (1) *Nautical Almanac*, and (2) *Air Almanac*.

Answers.—(1) Ho 14°57′.1; (2) Ho 14°57′.

823c. On June 2, 1975, the lower limb of the sun is observed with a marine sextant having an IC of (−)1′.3, from a height of eye of 43 feet. Another ship is between the observer and the horizon, at a distance of 1.4 miles from the observer. The water line of this ship is used as the horizontal reference. The hs is 25°18′.2.

Required.—Ho using table 22 and (1) *Nautical Almanac*, and (2) *Air Almanac*.

Answers.—(1) Ho 25°12.′9; (2) Ho 25°13′.

824a. At about GMT 2100 on June 2, 1975, the lower limb of the moon is observed with a marine sextant having an IC of (−)2.′5, from a height of eye of 55 feet. The hs is 47°35.′5.

Required.—Ho using (1) *Nautical Almanac*, and (2) *Air Almanac*.

Answers.—(1) Ho 48°16.′3; (2) Ho 48°16′.

824b. At about GMT 2300 on June 2, 1975, the upper limb of the moon is observed with a marine sextant having an IC of (+)4.′0, from a height of eye of 12 feet. The hs is 22°58.′3.

Required.—Ho using (1) *Nautical Almanac*, and (2) *Air Almanac*.

Answers.—(1) Ho 23°31.′8; (2) Ho 23°32′.

825a. On June 18, 1975, Mars is observed with a marine sextant having an IC of (+)2.′2, from a height of eye of 60 feet. The hs is 34°11.′7.

Required.—Ho using (1) *Nautical Almanac*, and (2) *Air Almanac*.

Answers.—(1) Ho 34°05.′1; (2) Ho 34°05′.

825b. Jupiter is observed with a marine sextant having an IC of (−)1.′0, from a height of eye of 27 feet. The hs is 11°23.′9.

Required.—Ho using (1) *Nautical Almanac*, and (2) *Air Almanac*.

Answers.—(1) Ho 11°13.′2; (2) Ho 11°13′.

826. Alpheratz is observed with a marine sextant having no IC, from a height of eye of 42 feet. The hs is 38°20.′3.

Required.—Ho using (1) *Nautical Almanac*, and (2) *Air Almanac*.

Answers.—(1) Ho 38°12.′8; (2) Ho 38°13′.

827a. On June 2, 1975, the lower limb of the sun is observed with a marine sextant having an IC of (−) 2.′3, from a height of eye of 24 feet. The hs is 2°04.′6, air temperature 65° F, and atmospheric pressure 30.81 inches.

Required.—Ho using (1) *Nautical Almanac*, (2) tables 23 and 24, and (3) *Air Almanac*.

Answers.—(1) Ho 1°55.′1; (2) Ho 1°55.′1; (3) Ho 1°55′.

827b. On July 2, 1975, the sun is observed as the upper limb drops below the horizon at sunset, from a height of eye of 19 feet. The air temperature is 16° F, and atmospheric pressure 29.90 inches.

Required.—Ho using (1) *Nautical Almanac*, (2) tables 23 and 24, and (3) *Air Almanac*.

Answers.—(1) Ho (−)0°59.′9 (2) Ho (−)0°57.′8; (3) Ho (−)0°58′.

827c. At GMT $6^h03^m29^s$ on June 2, 1975, the upper limb of the moon is observed with a marine sextant having an IC of (+)2.′6, from a height of eye of 35 feet. The hs is 1°12.′6, air temperature (−)23° F, and atmospheric pressure 29.04 inches.

Required.—Ho using (1) tables 23 and 24, and (2) *Air Almanac*.

Answers.—(1) Ho 1°22.′3; (2) Ho 1°20′.

827d. At GMT $12^h44^m01^s$ on June 2, 1975, the lower limb of the moon is observed with a marine sextant having an IC of (+)3.′2, from a height of eye of 22 feet. The hs is 0°24.′4, air temperature 40° F, and atmospheric pressure 29.94 inches.

Required.—Ho using (1) *Nautical Almanac*, (2) tables 23 and 24, and (3) *Air Almanac*.

Answers.—(1) Ho 1°01.′2; (2) Ho 1°01.′5; (3) Ho 0°58′.

827e. On January 19, 1975, Venus is observed with a marine sextant having an IC of (−)0.′5, from a height of eye of 31 feet. The hs is 3°29.′8, air temperature 55° F, and atmospheric pressure 30.15 inches.

Required.—Ho using (1) *Nautical Almanac*, (2) tables 23 and 24, and (3) *Air Almanac*.

Answers.—(1) Ho 3°10′.7; (2) Ho 3°10′.7; (3) Ho 3°10′.

827f. Saturn is observed with a marine sextant having an IC of (−)2′.3, from a height of eye of 37 feet. The hs is 4°39′.2, air temperature 76° F, and atmospheric pressure 28.89 inches.

Required.—Ho using (1) *Nautical Almanac*, (2) tables 23 and 24, and (3) *Air Almanac*.

Answers.—(1) Ho 4°21′.1; (2) Ho 4°21′.1; (3) Ho 4°21′.

827g. Gienah is observed with a marine sextant having no IC, from a height of eye of 44 feet. The hs is 2°46′.1, air temperature 35° F, and atmospheric pressure 29.92 inches.

Required.—Ho using (1) *Nautical Almanac*, (2) tables 23 and 24, and (3) *Air Almanac*.

Answers.—(1) Ho 2°23′.4; (2) Ho 2°36′.6; (3) Ho 2°21′.

828. On June 2, 1975, a back sight is taken of the lower limb of the sun, with a marine sextant having an IC of (+)1′.7, from a height of eye of 49 feet. The measured sextant altitude is 141°04′.9.

Required.—Ho using (1) *Nautical Almanac*, and (2) *Air Almanac*.

Answers.—(1) Ho 39°15′.0; (2) Ho 39°15′.

829. The horizontal angle between two objects is measured with a marine sextant having an IC of (+)4′.0. The measured angle is 85°14′.6.

Required.—Corrected angle.

Answer.—Corrected angle 85°18′.6.

CHAPTER IX

SIGHT REDUCTION

901. Introduction.—The process of deriving from a celestial observation the information needed for establishing a line of position is called **sight reduction.** The observation itself consists of measuring the altitude of a celestial body and noting the time. The process of finding such a line of position may be divided into six steps:

1. Correction of sextant altitude (ch. VIII).
2. Determination of GHA and declination (ch. VII).
3. Selection of assumed position and finding local hour angle or meridian angle at that point.
4. Computation of altitude and azimuth (ch. VII).
5. Comparison of computed and observed altitudes.
6. Plot of the line of position.

Broadly speaking, tables which assist in any of these steps can be considered **sight reduction tables.** However, the expression is generally limited to tables intended primarily for computation of altitude and azimuth. A great variety of such tables exists. In volume I various methods of sight reduction, including graphical and mechanical solutions, are contrasted. All are based, directly or indirectly, upon solution of the navigational triangle (app. J). Thus, the process of sight reduction, in its limited sense, is one of converting coordinates of the celestial equator system (app. J) to those of the horizon system.

The correction of the sextant altitude (hs) to find observed altitude (Ho) is not necessarily performed first. If any form of time other than GMT is used for timing the observation, it is first converted to GMT because this is the kind of time used for entering the almanacs. From the almanac, the GHA and declination are determined.

902. Selection of the assumed position (AP).—The following variables are needed to compute the altitude and azimuth:

1. Latitude (L).
2. Declination (d).
3. Local hour angle (LHA) or meridian angle (t).

Except for declination, these variables are dependent upon the position from which the altitude and azimuth are to be computed for the time of the observation. Although the dead reckoning or estimated position can be used, unnecessary interpolation can can be avoided when using modern sight reduction tables by selecting an AP for the reduction that will result in two of the three variables being exact entry values or table arguments. In these tables altitudes and azimuth angles are given for each whole degree of latitude and each whole degree of either meridian angle or local hour angle. Since the assumed position should be within 30′ of the actual position, the whole degree of latitude *nearest* to the DR or EP at the time of the sight is selected as the **assumed latitude** (*a*L). The **assumed longitude** (*a*λ) is also selected within 30′ of the DR or EP so that no minutes of arc will remain after it is applied to GHA. This means that in west longitude the minutes of $a\lambda$ must be the same as those of GHA; while in east longitude the minutes of $a\lambda$ must be equal to 60′ minus the minutes of GHA.

903. Finding the local hour angle and meridian angle.—Meridian angle is the angular distance that the celestial body is east or west of the celestial meridian. It is found from local hour angle (LHA), which, in turn, is found from Greenwich hour angle by adding east longitude or subtracting west longitude. A time diagram (app. J) is useful in visualizing this relationship.

Example 1.—The GHA is 168°42′.6.

Required.—The LHA and t at (1) long. 137°24′.6W, and (2) 158°24′.7E.

Solution.—

(1)			(2)	
GHA	168°42′.6		GHA	168°42′.6
λ	137°24′.6W		λ	158°24′.7E
LHA	31°18′.0		LHA	327°07′.3
t	31°18′.0W		t	32°52′.7E

In west longitude, if GHA is less than longitude, add 360° to GHA before subtracting. In east longitude, if the sum exceeds 360°, subtract this amount. If LHA is less than 180°, it is numerically equal to meridian angle, which is labeled W (west). If LHA is greater than 180°, t is 360°−LHA and is labeled E (east).

Example 2.—The GHA is 168°42′.6; observations are made at (1) long. 137°24′.6W, and (2) long. 158°24′.7E.

Required.—The $a\lambda$ providing whole degrees of LHA and t.

Solution.—

(1)			(2)	
GHA	168°42′.6		GHA	168°42′.6
$a\lambda$	137°42′.6W		$a\lambda$	158°17′.4E
LHA	31°00′.0		LHA	327°00′.0
t	31°00′.0W		t	33°00′.0E

904. Comparison of computed and observed altitudes.—After appropriate corrections are applied to the **sextant altitude (hs)**, the **observed altitude (Ho)** is obtained. For the instant of observation, the altitude and azimuth at some convenient **assumed position (AP)** near the actual position of the observer are determined by calculation or equivalent process. The difference between this **computed altitude (Hc)** and Ho is the **altitude intercept (*a*)**, sometimes called **altitude difference**.

Since *a* is the difference in altitude at the assumed and actual positions, it is also the difference in zenith distance, and therefore the difference in radii of the circles of equal altitude at the two places. The position having the greater altitude is on the circle of smaller radius, and hence is closer to the GP of the body. In figure 904 the AP is shown on the inner circle. Hence, Hc is greater than Ho.

The altitude intercept, the numerical difference between Hc and Ho, is customarily expressed in nautical miles (minutes of arc), and labeled **T** or **A** to indicate whether the line of position is **toward** or **away** from the GP, as measured from the AP.

Two useful aids in labeling the intercept are: Coast Guard Academy for Computed Greater Away, and **Ho Mo To** for **Ho More Toward**.

For example,

Hc	37°51′.6		Hc	61°57′.3
Ho	37°43′.9		Ho	62°12′.7
a	7′.7A		*a*	15′.4T

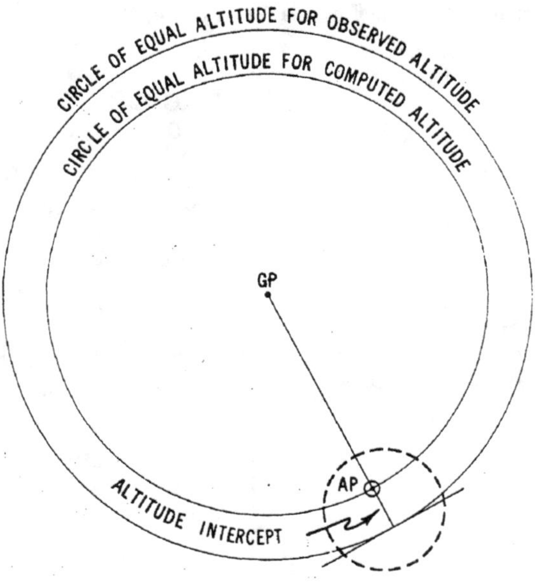

FIGURE 904.—The basis for the line of position from a celestial observation.

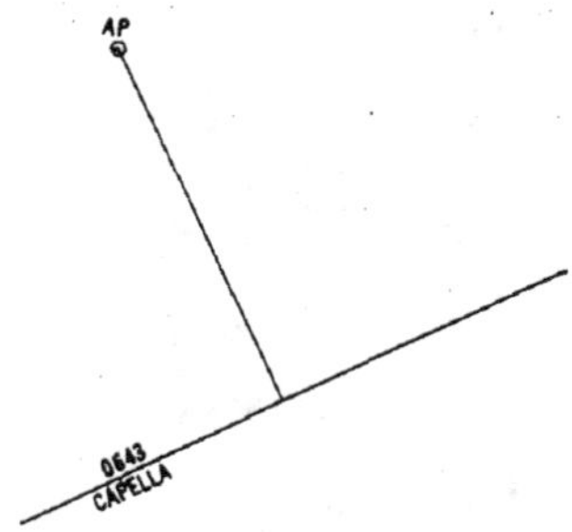

FIGURE 905.—A line of position from observation of the star Capella at 0643.

905. Plot of the line of position.—The line of position can be plotted using part of the information within the broken circle of figure 904, as shown in figure 905. First, the AP is plotted. The circle of equal altitude through this position is not needed, and is not plotted. From the AP the azimuth line is measured toward or away from the GP as appropriate, and the altitude intercept is measured along this line. At the point thus located, a line is drawn perpendicular to the azimuth line. This perpendicular is the line of position.

906. Complete solution.—The complete solution includes all of the parts listed in article 901. Because of the various alternatives available for the separate parts, a large number of variations might be used in the complete solution. The following example combines some of the most commonly used variations.

Example.—On June 1, 1975, the GMT 0825 dead reckoning position of a ship is lat. 35°34′.4N, long. 66°20′.2W. Kochab is observed at an altitude of 34°54′.6 from a height of eye of 40 feet using a sextant having an index error of (+)2′.0.

Required.—The *a* and Zn from the GMT 0825 DR.

Solution.—By cosine-haversine formula and tables 33 and 34.

Sight Reduction by Cosine-Haversine Formula (art. 712 and 713)

Kochab

					+ ☆	−
GMT	08ʰ 24ᵐ 41ˢ	June 1		IC	2′.0	
08ʰ	9°11′.2			D		6′.1
24ᵐ41ˢ	6°11′.3			☆−P		1′.4
SHA	137°17′.7			sum	2′.0	7′.5
GHA	152°40′.2			corr.	(−)5′.5	
aλ	66°20′.2			hs	34°54′.6	
LHA	86°20′.0			Ho	34°49′.1	
t	86°20′.0W	*l* hav 9.67027		*l* sin	9.99911	
aL	35°34′.4N	*l* cos 9.91029				
d	74°15′.5N	*l* cos 9.43345		*l* cos	9.43345	
θ		*l* hav 9.01401	*n* hav 0.10328			
L~d	38°41′.1		*n* hav 0.10970			
z	54°58′.1		*n* hav 0.21298			
Hc	35°01′.9			*l* sec	10.08680	
Ho	34°49′.1		Z N19°18′.5W	*l* sin	9.51936	
a	12′.8A	aL 35°34′.4N				
Zn	340°.7	aλ 66°20′.2W				

Solution.—By the Ageton Method (table 35).

Sight Reduction by the Ageton Method (art. 718)

Kochab

				+ ☆	−
GMT	08ʰ 24ᵐ 41ˢ	June 1	IC	2′.0	
08ʰ	9°11′.2		D		6′.1
24ᵐ41ˢ	6°11′.3		☆−P		1′.4
SHA	137°17′.7		sum	2′.0	7′.5
GHA	152°40′.2		corr.	(−)5′.5	
aλ	66°20′.2W		hs	34°54′.6	
LHA	86°20′.0		Ho	34°49′.1	
t	86°20′.0W				

		Add	*Subtract*	*Add*	*Subtract*
t	86°20′.0W	A 89			
d	74°15′.5N	B 56655	A 1660		
R		A 56744	B 1653	B 1653	A 56744
K	88°58′.0N		A 7		
L	35°34′.4N				
K~L	53°23′.6			B 22450	
Hc	35°02′.0			A 24103	B 8681
Ho	34°49′.1		Zn 340°.7	Z N19°18′.5W	A 48063
a	12′.9A	aL 35°34′.4N			
Zn	340°.7	aλ 66°20′.2W			

Caution.—When the quantity K is between 82° and 98°, it is good practice to either discard the sight or to interpolate log secant (B) of R from the log cosecant (A) of R.

CHAPTER X

THE SAILINGS

1001. The sailings.—In the solution of problems involved in the sailings, the following quantities are used:

1. *Latitude* (L). The latitude of the point of departure is designated L_1; that of the point of arrival or the destination, L_2; middle (mid) or mean latitude, Lm; latitude of the vertex of a great circle, L_v; and latitude of any point on a great circle, L_x.

2. *Mean latitude* (Lm). Half the arithmetical sum of the latitudes of two places on the same side of the equator.

3. *Middle or mid latitude* (Lm). The latitude at which the arc length of the parallel separating the meridians passing through two specific points is exactly equal to the departure in proceeding from one point to the other. The mean latitude is normally used for want of a practicable means of determining the middle latitude.

4. *Difference of latitude* (l or D. Lat.).

5. *Meridional parts* (M). The meridional parts of the point of departure are designated M_1, and of the point of arrival or the destination, M_2.

6. *Meridional difference* (m).

7. *Longitude* (λ). The longitude of the point of departure is designated λ_1; that of the point of arrival or the destination, λ_2; of the vertex of a great circle, λ_v; and of any point on a great circle, λ_x.

8. *Difference of longitude* (DLo).

9. *Departure* (p or Dep.).

10. *Course* or *course angle* (Cn or C).

11. *Distance* (D or Dist.).

The various kinds of sailings are:

1. **Plane sailing** is a method of solving the various problems involving a single course and distance, difference of latitude, and departure, in which the earth, or that part traversed, is regarded as a plane surface. Hence, the method provides solution for latitude of the point of arrival, but not for longitude of this point, one of the spherical sailings being needed for this problem. Because of the basic assumption that the earth is flat, this method should not be used for distances of more than a few hundred miles.

2. **Traverse sailing** combines the plane sailing solutions when there are two or more courses. This sailing is a method of determining the equivalent course and distance made good by a vessel steaming along a series of rhumb lines.

3. **Parallel sailing** is the interconversion of departure and difference of longitude when a vessel is proceeding due east or due west. This was a common occurrence when the sailings were first employed several hundred years ago, but only an incidental situation now.

4. **Middle-** (or **mid-**) **latitude sailing** involves the use of the mid or mean latitude for converting departure to difference of longitude when the course is not due east or due west and it is assumed such course is steered at the mid latitude.

5. **Mercator sailing** provides a mathematical solution of the plot as made on a Mercator chart. It is similar to plane sailing, but uses meridional difference and difference of longitude in place of difference of latitude and departure, respectively.

6. **Great-circle sailing** involves the solution of courses, distances, and points along a great circle between two points, the earth being regarded as a sphere.

7. **Composite sailing** is a modification of great-circle sailing to limit the maximum latitude.

1002. Traverse tables (tables 3 and 38) can be used in the solution of any of the sailings except great-circle and composite. Table 3 is basically the tabulation of the solutions of plane right triangles. Because the solutions are for integral values of the course angle and the distance, interpolation for intermediate values may be required. Through appropriate interchanges of the headings of the columns, solutions for other than plane sailing can be effected. For the solution of the plane right triangle, any number N in the distance (Dist.) column is the hypotenuse; the number opposite in the difference of latitude (D. Lat.) column is N times the cosine of the acute angle; and the other number opposite in the departure (Dep.) column is N times the sine of the acute angle. Or, the number in the D. Lat. column is the value of the side adjacent, and the number in the Dep. column is the value of the side opposite the acute angle. Hence, if the acute angle is the course angle, the side adjacent in the D. Lat. column is meridional difference m; the side opposite in the Dep. column is DLo. If the acute angle is the midlatitude of the formula p=DLo cos Lm, then DLo is any value N in the Dist. column, and the departure is the value N×cos Lm in the D. Lat. column.

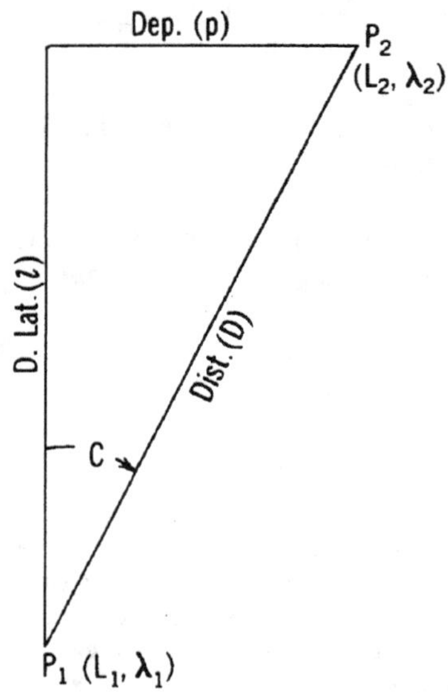

FIGURE 1003.—The plane sailing triangle.

1003. Plane sailing.—In plane sailing the figure formed by the meridian through the point of departure, the parallel through the point of arrival, and the course line is considered a plane right triangle. This is illustrated in figure 1003, in which P_1 and P_2 are the points of departure and arrival, respectively. The course angle and the three sides are as labeled. From this triangle:

$$\cos C=\frac{l}{D} \qquad \sin C=\frac{p}{D} \qquad \tan C=\frac{p}{l}.$$

From the first two of these formulas the following relationships can be derived:

$$l=D \cos C \qquad D=l \sec C \qquad p=D \sin C.$$

The usual problems solved by plane sailing are: (1) given the course and distance, find the difference of latitude and the departure; and (2) the reverse of this. It is good practice to label l, N or S, and p, E or W, to aid in identification of the quadrant of the course. Logarithmic and traverse table solutions are illustrated in the following examples:

Example 1.—A vessel steams 188.0 miles on course 005°.

Required.—(1) Difference of latitude, (2) departure.

Solution.—By computation:

	D 188.0 mi.	log 2.27416	log 2.27416
	C 005°	l cos 9.99834	l sin 8.94030
(1)	l 187′.3 N	log 2.27250	
(2)	p 16.4 mi. E		log 1.21446

In the solution by computation, 2.27416 is the logarithm ("log") of 188.0, and 9.99834 is the logarithmic cosine ("l cos") of 5°.

The labels (N, S, E, W) of l, p, and C are determined by noting the direction of motion or the relative positions of the two places.

Solution.—By inspection of table 3:

Refer to figure 1003a. Enter table 3 and turn the pages to find course 005° at the top of the page. Using the column headings at the top of the table, opposite 188 in the Dist. column extract D. Lat. 187.3 and Dep. 16.4.

(1) D. Lat. 187′.3N.

(2) Dep. 16.4 mi. E.

If the course were 085°, 095°, 265°, or 275°, the column headings at the bottom of the table would be used.

TABLE 3

Traverse **5°** Table

355° | 005° / 185° | 175°

Dist.	D. Lat.	Dep.	Dist.	D. Lat.	Dep.	Dist.	D. Lat.	Dep.	Dist.	D. Lat.	Dep.	Dist.	D. Lat.	Dep.
1	1.0	0.1	61	60.8	5.3	121	120.5	10.5	181	180.3	15.8	241	240.1	21.0
2	2.0	0.2	62	61.8	5.4	22	121.5	10.6	82	181.3	15.9	42	241.1	21.1
3	3.0	0.3	63	62.8	5.5	23	122.5	10.7	83	182.3	15.9	43	242.1	21.2
4	4.0	0.3	64	63.8	5.6	24	123.5	10.8	84	183.3	16.0	44	243.1	21.3
5	5.0	0.4	65	64.8	5.7	25	124.5	10.9	85	184.3	16.1	45	244.1	21.4
6	6.0	0.5	66	65.7	5.8	26	125.5	11.0	86	185.3	16.2	46	245.1	21.4
7	7.0	0.6	67	66.7	5.8	27	126.5	11.1	87	186.3	16.3	47	246.1	21.5
8	8.0	0.7	68	67.7	5.9	28	127.5	11.2	88	187.3	16.4	48	247.1	21.6
9	9.0	0.8	69	68.7	6.0	29	128.5	11.2	89	188.3	16.5	49	248.1	21.7
10	10.0	0.9	70	69.7	6.1	30	129.5	11.3	90	189.3	16.6	50	249.0	21.8
60	59.8	5.2	20	119.5	10.5	80	179.3	15.7	40	239.1	20.9	300	298.9	26.1
Dist.	Dep.	D. Lat.	Dist.	Dep.	D. Lat.	Dist.	Dep.	D. Lat.	Dist.	Dep.	D. Lat.	Dist.	Dep.	D. Lat.

275° | 085° / 265° | 095°

85°

Dist.	D. Lat.	Dep.
N.	N × Cos.	N × Sin.
Hypotenuse	Side Adj.	Side Opp.

FIGURE 1003a.—Extract from table 3.

If, in example 1, the vessel steamed 18.8 miles instead of 188 miles, the values for D. Lat. and Dep. would be extracted opposite 188 in the Dist. column and then be divided by 10.

Example 2.—A ship has steamed 136.0 miles north and 203.0 miles west.

Required.—(1) Course, (2) distance.

Solution.—By computation: (1) $\tan C = \frac{p}{l}$ (2) $D = l \sec C$

	p 203.0 mi.W	log	2.30750	
	l 136′.0N	log (−)	2.13354	log 2.13354
	C N56°10′.8W	l tan	0.17396	l sec 0.25447
(2)	D 244.3 mi.			log 2.38801
(1)	Cn 303°.8			

Solution.—By inspection of table 3:

Refer to figure 1003b. Enter table 3 and turn the pages until the numbers 136 and 203 are found abreast each other in the columns labeled D. Lat. and Dep., respectively. This occurs most nearly on the page for course angle 56°. Therefore, the course is 304°. Interpolating for intermediate values, the corresponding number in the Dist. column is about 244.3.

(1) C 304°.

(2) D 244.3 mi.

326° | 034° / 214° | 146° — TABLE 3 — Traverse 34° Table — 326° | 034° / 214° | 146°

Dist.	D. Lat.	Dep.	Dist.	D. Lat.	Dep.	Dist.	D. Lat.	Dep.	Dist.	D. Lat.	Dep.	Dist.	D. Lat.	Dep.
1	0.8	0.6	61	50.6	34.1	121	100.3	67.7	181	150.1	101.2	241	199.8	134.8
2	1.7	1.1	62	51.4	34.7	22	101.1	68.2	82	150.9	101.8	42	200.6	135.3
3	2.5	1.7	63	52.2	35.2	23	102.0	68.8	83	151.7	102.3	43	201.5	135.9
4	3.3	2.2	64	53.1	35.8	24	102.8	69.3	84	152.5	102.9	44	202.3	136.4
5	4.1	2.8	65	53.9	36.3	25	103.6	69.9	85	153.4	103.5	45	203.1	137.0
60	49.7	33.6	20	99.5	67.1	80	149.2	100.7	40	199.0	134.2	300	248.7	167.8
Dist.	Dep.	D. Lat.	Dist.	Dep.	D. Lat.	Dist.	Dep.	D. Lat.	Dist.	Dep.	D. Lat.	Dist.	Dep.	D. Lat.

304° | 056° / 236° | 124° — 56°

Dist.	D. Lat.	Dep.
N.	N×Cos.	N×Sin.
Hypotenuse	Side Adj.	Side Opp.

FIGURE 1003b.—Extract from table 3.

When the course is near 090° or 270°, the solution of C to the nearest 0°.1 only may introduce a large error in distance.

1004. Traverse sailing.—A traverse is a series of courses, or a track consisting of a number of course lines, as might result from a sailing vessel beating into the wind. **Traverse sailing** is the finding of a single equivalent course and distance.

Solution is usually made by means of the traverse tables, the distance to the north or south and that to the east or west on each course being tabulated, the algebraic sum of difference of latitude and departure being found, and the result being converted to course and distance.

Example.—A ship steams as follows: course 158°, distance 15.5 miles; course 135°, distance 33.7 miles; course 259°, distance 16.1 miles; course 293°, distance 39.0 miles; course 169°, distance 40.4 miles.

Required.—Equivalent single (1) course, (2) distance.

Solution.—Solve for each leg as in example 1, article 1003. Tabulate the answers as follows:

Course	Dist.	N	S	E	W
°	*mi.*	*mi.*	*mi.*	*mi.*	*mi.*
158	15.5		14.4	5.8	
135	33.7		23.8	23.8	
259	16.1		3.1		15.8
293	39.0	15.2			35.9
169	40.4		39.7	7.7	
		15.2	81.0	37.3	51.7
			15.2		37.3
(1) 192.3	(2) 67.3		65.8		14.4

For course 158°, the values for D. Lat. and Dep. are extracted opposite 155 in the Dist. column. The values are then divided by 10 and rounded off to the nearest tenth.

Convert *l* 65′.8 S, p 14.4 mi. W to equivalent single course and distance.

1005. Parallel sailing consists of the interconversion of departure and difference of longitude. It is the simplest form of spherical sailing. The formulas for these transformations are:

$$\text{DLo} = \text{p sec L} \qquad \text{p} = \text{DLo cos L}.$$

Example 1.—The DR latitude of a ship on course 090° is 49°30′N. The ship steams on this course until the longitude changes 3°30′.

Required.—The departure.

Solution.—By computation: p=DLo cos L

DLo 210′E	log 2.32222
L 49°30′N	*l* cos 9.81254
p 136.4 mi. E	log 2.13476

Solution.—By inspection of table 3:

Refer to figure 1005a. Enter table 3 with latitude as course angle and substitute DLo as the heading of the Dist. column and Dep. as the heading of the D. Lat. column. As the table is computed for integral degrees of course angle (or latitude), the tabulations in the pages for 49° and 50° must be interpolated for the intermediate value (49°30′). The departure for latitude 49° and DLo 210′ is 137.8 miles. The departure for latitude 50° and DLo 210′ is 135.0 miles. Interpolating for the intermediate latitude, the departure is 136.4 miles.

Example 2.—The DR latitude of a ship on course 270° is 38°15′S. The ship steams on this course for a distance of 215.5 miles.

Required.—The change in longitude.

Solution.—By computation: DLo=p sec L

p	215.5 mi. W	log 2.33345
L	38°15′S	*l* sec 0.10496
DLo	274′.4W	log 2.43841
DLo	4°34′.4W	

319° | 041° / 221° | 139°

TABLE 3

Traverse **41°** Table

319° | 041° / 221° | 139°

Dist.	D. Lat.	Dep.	Dist.	D. Lat.	Dep.	Dist.	D. Lat.	Dep.	Dist.	D. Lat.	Dep.	Dist.	D. Lat.	Dep.
21	15.8	13.8	81	61.1	53.1	141	106.4	92.5	201	151.7	131.9	261	197.0	171.2
22	16.6	14.4	82	61.9	53.8	42	107.2	93.2	02	152.5	132.5	62	197.7	171.9
23	17.4	15.1	83	62.6	54.5	43	107.9	93.8	03	153.2	133.2	63	198.5	172.5
24	18.1	15.7	84	63.4	55.1	44	108.7	94.5	04	154.0	133.8	64	199.2	173.2
25	18.9	16.4	85	64.2	55.8	45	109.4	95.1	05	154.7	134.5	65	200.0	173.9
26	19.6	17.1	86	64.9	56.4	46	110.2	95.8	06	155.5	135.1	66	200.8	174.5
27	20.4	17.7	87	65.7	57.1	47	110.9	96.4	07	156.2	135.8	67	201.5	175.2
28	21.1	18.4	88	66.4	57.7	48	111.7	97.1	08	157.0	136.5	68	202.3	175.8
29	21.9	19.0	89	67.2	58.4	49	112.5	97.8	09	157.7	137.1	69	203.0	176.5
30	22.6	19.7	90	67.9	59.0	50	113.2	98.4	10	158.5	137.8	70	203.8	177.1
Dist.	Dep.	D. Lat.	Dist.	Dep.	D. Lat.	Dist.	Dep.	D. Lat.	Dist.	Dep.	D. Lat.	Dist.	Dep.	D. Lat.

311° | 049° / 229° | 131°

49°

Dist.	D. Lat.	Dep.
D Lo	Dep	
	m	D Lo

FIGURE 1005a.—Extracts from table 3.

Dep. 136.4 mi E.

Solution.—By inspection of table 3:

Refer to figure 1005b. Enter table 3 with latitude as course angle and substitute DLo as the heading of the Dist. column and Dep. as the heading of the D. Lat. column. As the table is computed for integral degrees of course angle (or latitude), the tabulations in the pages for 38° and 39° must be interpolated for the minutes of latitude. Corresponding to Dep. 215.5 miles in the former is DLo 273′.5, and in the latter DLo 277′.3. Interpolating for the minutes of latitude, the DLo is 274′.4.

322° | 038° / 218° | 142°

TABLE 3

Traverse **38°** Table

322° | 038° / 218° | 142°

Dist.	D. Lat.	Dep.	Dist.	D. Lat.	Dep.	Dist.	D. Lat.	Dep.	Dist.	D. Lat.	Dep.	Dist.	D. Lat.	Dep.
31	24.4	19.1	91	71.7	56.0	151	119.0	93.0	211	166.3	129.9	271	213.6	166.8
32	25.2	19.7	92	72.5	56.6	52	119.8	93.6	12	167.1	130.5	72	214.3	167.5
33	26.0	20.3	93	73.3	57.3	53	120.6	94.2	13	167.8	131.1	73	215.1	168.1
34	26.8	20.9	94	74.1	57.9	54	121.4	94.8	14	168.6	131.8	74	215.9	168.7
35	27.6	21.5	95	74.9	58.5	55	122.1	95.4	15	169.4	132.4	75	216.7	169.3

Dist.	D. Lat.	Dep.
D Lo	Dep	
	m	D Lo

FIGURE 1005b.—Extracts from table 3.

DLo 274′.4W or 4°34′.4W.

1006. Middle-latitude sailing, popularly called mid-latitude sailing, combines plane sailing and parallel sailing. Plane sailing is used to find difference of latitude and departure when course and distance are known, or vice versa. Parallel sailing is used to interconvert departure and difference of longitude. The mean latitude (Lm) is normally used for want of a practicable means of determining the middle latitude, the latitude at which the arc length of the parallel separating the meridians passing through two specific points is exactly equal to the departure in proceeding from one point to the other. The formulas for these transformations are:

$$\text{DLo}=\text{p sec Lm} \qquad \text{p}=\text{DLo cos Lm}.$$

The mean latitude (Lm) is half the arithmetical sum of the latitudes of two places on the same side of the equator. The mean latitude is labeled N or S to indicate whether it is north or south of the equator. If a course line crosses the equator, that part on each side (the north latitude and south latitude portions, respectively) should be solved separately.

This sailing, like most elements of navigation, contains certain simplifying approximations which produce answers somewhat less accurate than those yielded by more rigorous solutions. For ordinary purposes, however, the results are more accurate than the navigation of the vessel using them. From time to time suggestions have been made that a correction be applied to eliminate the error introduced by assuming that the meridians of the point of departure and of the destination converge uniformly (as the two sides of a plane angle), rather than as the sine of the latitude (approximately). The proposed correction usually takes the form of some quantity to be added to or subtracted from the middle latitude to obtain a "corrected middle latitude" for use in the solution. Tables giving such a correction have been published for both spherical and spheroidal earths. However, the actual correction is not a simple function of the middle latitude and the difference of longitude, as assumed, because the basic formulas of the sailing are themselves based upon a sphere, rather than a spheroid. Hence, the use of such a correction is misleading, and may introduce more error than it eliminates. The use of any correction is not considered justified; if highly accurate results are required, a different method should be used.

Example 1.—A vessel steams 1,253.0 miles on course 070° from lat. 15°17'.0 N, long. 151°37'.0 E.

Required.—(1) Latitude and (2) longitude of the point of arrival.

Solution.—By computation: (1) l=D cos C; p=D sin C (2) DLo=p sec Lm

D	1253.0 mi.	log	3.09795	log	3.09795
C	070°	l cos	9.53405	l sin	9.97299
l	428'.6 N	log	2.63200		
p	1177.4 mi. E			log	3.07094
Lm	18°51'.3 N			l sec	0.02395
DLo	1244'.2 E			log	3.09489
L_1	15°17'.0 N	L_1	15°17'.0 N		
l	7°08'.6 N	½ l	3°34'.3 N		
(1) L_2	22°25'.6 N	Lm	18°51'.3 N		

λ_1	151°37'.0 E
DLo	20°44'.2 E
(2) λ_2	172°21'.2 E

Solution.—By inspection of table 3:

Refer to figure 1006a. Enter table 3 with course 070° and distance 1,253 miles to find D. Lat. 429' and Dep. 1,178 miles. Because a number as high as 1,253 is not tabulated in the Dist. column, the values for D. Lat. and Dep. are obtained for a distance of 125.3 miles, and then multiplied by 10. The latitude of the point of arrival being 22°26'N, the mean latitude (Lm) is 18°51'.5N.

340°	020°											340°	020°	
200°	160°											200°	160°	
						TABLE 3								
				Traverse			20°		Table					
Dist.	**D. Lat.**	**Dep.**	**Dist.**	**D. Lat.**	**Dep.**	**Dist.**	**D. Lat.**	**Dep.**	**Dist.**	**D. Lat.**	**Dep.**	**Dist.**	**D. Lat.**	**Dep.**
1	0.9	0.3	61	57.3	20.9	121	113.7	41.4	181	170.1	61.9	241	226.5	82.4
2	1.9	0.7	62	58.3	21.2	22	114.6	41.7	82	171.0	62.2	42	227.4	82.8
3	2.8	1.0	63	59.2	21.5	23	115.6	42.1	83	172.0	62.6	43	228.3	83.1
4	3.8	1.4	64	60.1	21.9	24	116.5	42.4	84	172.9	62.9	44	229.3	83.5
5	4.7	1.7	65	61.1	22.2	25	117.5	42.8	85	173.8	63.3	45	230.2	83.8
6	5.6	2.1	66	62.0	22.6	26	118.4	43.1	86	174.8	63.6	46	231.2	84.1
7	6.6	2.4	67	63.0	22.9	27	119.3	43.4	87	175.7	64.0	47	232.1	84.5
Dist.	**Dep.**	**D. Lat.**	**Dist.**	**Dep.**	**D. Lat.**	**Dist.**	**Dep.**	**D. Lat.**	**Dist.**	**Dep.**	**D. Lat.**	**Dist.**	**Dep.**	**D. Lat.**
290°	070°						70°							
250°	110°													

Dist.	D. Lat.	Dep.
N.	N×Cos.	N×Sin.
Hypotenuse	Side Adj.	Side Opp.

342°	018°											342°	018°	
198°	162°											198°	162°	
						TABLE 3								
				Traverse			18°		Table					
Dist.	**D. Lat.**	**Dep.**	**Dist.**	**D. Lat.**	**Dep.**	**Dist.**	**D. Lat.**	**Dep.**	**Dist.**	**D. Lat.**	**Dep.**	**Dist.**	**D. Lat.**	**Dep.**
1	1.0	0.3	61	58.0	18.9	121	115.1	37.4	181	172.1	55.9	241	229.2	74.5
2	1.9	0.6	62	59.0	19.2	22	116.0	37.7	82	173.1	56.2	42	230.2	74.8
3	2.9	0.9	63	59.9	19.5	23	117.0	38.0	83	174.0	56.6	43	231.1	75.1
4	3.8	1.2	64	60.9	19.8	24	117.9	38.3	84	175.0	56.9	44	232.1	75.4
5	4.8	1.5	65	61.8	20.1	25	118.9	38.6	85	175.9	57.2	45	233.0	75.7
Dist.	**Dep.**	**D. Lat.**	**Dist.**	**Dep.**	**D. Lat.**	**Dist.**	**Dep.**	**D. Lat.**	**Dist.**	**Dep.**	**D. Lat.**	**Dist.**	**Dep.**	**D. Lat.**
288°	072°						72°							
252°	108°													

Dist.	D. Lat.	Dep.
D Lo	Dep	
	m	D Lo

341°	019°											341°	019°	
199°	161°											199°	161°	
						TABLE 3								
				Traverse			19°		Table					
Dist.	**D. Lat.**	**Dep.**	**Dist.**	**D. Lat.**	**Dep.**	**Dist.**	**D. Lat.**	**Dep.**	**Dist.**	**D. Lat.**	**Dep.**	**Dist.**	**D. Lat.**	**Dep.**
1	0.9	0.3	61	57.7	19.9	121	114.4	39.4	181	171.1	58.9	241	227.9	78.5
2	1.9	0.7	62	58.6	20.2	22	115.4	39.7	82	172.1	59.3	42	228.8	78.8
3	2.8	1.0	63	59.6	20.5	23	116.3	40.0	83	173.0	59.6	43	229.8	79.1
4	3.8	1.3	64	60.5	20.8	24	117.2	40.4	84	174.0	59.9	44	230.7	79.4
5	4.7	1.6	65	61.5	21.2	25	118.2	40.7	85	174.9	60.2	45	231.7	79.8
Dist.	**Dep.**	**D. Lat.**	**Dist.**	**Dep.**	**D. Lat.**	**Dist.**	**Dep.**	**D. Lat.**	**Dist.**	**Dep.**	**D. Lat.**	**Dist.**	**Dep.**	**D. Lat.**
289°	071°						71°							
251°	109°													

Dist.	D. Lat.	Dep.
D Lo	Dep.	
	m	D Lo

FIGURE 1006a.—Extracts from table 3.

Reenter table 3 with the mean latitude as course angle and substitute DLo as the heading of the Dist. column and Dep. as the heading of the D.Lat. column. As the table is computed for integral degrees of course angle (or latitude), the tabulations in the pages for 18° and 19° must be interpolated for the minutes of Lm. Corresponding to Dep. 1,178 in the former is DLo 1,239′, and in the latter is DLo 1,246′. Interpolating for the minutes of Lm, the DLo is 1,245′. (The values of DLo are obtained for a departure of 117.8 miles and then multiplied by 10).

L_1	15°17.′0 N	L_1	15°17.′0 N			
D. Lat.	7°09.′0 N	½ D. Lat.	3°34.′5 N	Dep.	DLo(18°)	DLo(19°)
(1) L_2	22°26.′0 N	Lm	18°51.′5 N	1178 mi.	1239′	1246′

λ_1	151°37.′0 E		
DLo	20°45.′0 E	DLo	1245′
(2) λ_2	172°22.′0 E		

Example 2.—A vessel at lat. 8°48.′9 S, long. 89°53.′3 W is to proceed to lat. 17°06.′9 S, long. 104°51.′6 W.

Required.—(1) Course, (2) distance.

Solution.—By computation: (1) p=DLo cos Lm; $\tan C=\frac{p}{l}$ (2) $D=l \sec C$

L_1	8°48.′9 S	λ_1	89°53.′3W
L_2	17°06.′9 S	λ_2	104°51.′6W
l	8°18.′0 S	DLo	14°58.′3W
½ l	4°09.′0 S	DLo	898.′3W
Lm	12°57.′9 S		

DLo	898.′3 W	log	2.95342		
Lm	12°57.′9 S	l cos	9.98878		
p	875.4 mi. W	log	2.94220		
l	498.′0 S	log (−)	2.69723	log	2.69723
C	S 60°21.′9 W	l tan	0.24497	l sec	0.30586
(2) D	1007.1 mi.			log	3.00309
(1) Cn	240.°4				

The labels (N, S, E, W) of l, p, and C are determined by noting the direction of motion or the relative positions of the two places.

Solution.—By inspection of table 3:

Refer to figure 1006b. Enter table 3 with the mean latitude as course angle and substitute DLo as the heading of the Dist. column and Dep. as the heading of the D. Lat. column. Since the table is computed for integral values of course angle (or latitude), it is usually necessary to extract the value of departure for values just less and just greater than the Lm, and then interpolate for the minutes of Lm. In this case where Lm is almost 13°, enter table 3 with Lm 13° and DLo 898.′3 to find Dep. 875 miles. The departure is found for DLo 89.′9, and then multiplied by 10.

Reenter table 3 and turn the pages until the numbers 875 and 498 are found abreast each other in the columns labeled Dep. and D. Lat., respectively. Because these high numbers are not tabulated, it is necessary to divide the departure and difference of latitude by 10, and then inspect to find 87.5 and 49.8 abreast each other. This occurs most nearly on the page for course angle 60° (fig. 1006c). Interpolating for intermediate values, the corresponding number in the Dist. column is about 100.5. Multiplying this number by 10, the distance is found as about 1,005 miles.

TABLE 3

Traverse **13°** Table

347° | 013° / 193° | 167° (left) 347° | 013° / 193° | 167° (right)

Dist.	D. Lat.	Dep.	Dist.	D. Lat.	Dep.	Dist.	D. Lat.	Dep.	Dist.	D. Lat.	Dep.	Dist.	D. Lat.	Dep.
27	26.3	6.1	87	84.8	19.6	47	143.2	33.1	07	201.7	46.6	67	260.2	60.1
28	27.3	6.3	88	85.7	19.8	48	144.2	33.3	08	202.7	46.8	68	261.1	60.3
29	28.3	6.5	89	86.7	20.0	49	145.2	33.5	09	203.6	47.0	69	262.1	60.5
30	29.2	6.7	90	87.7	20.2	50	146.2	33.7	10	204.6	47.2	70	263.1	60.7
Dist.	Dep.	D. Lat.	Dist.	Dep.	D. Lat.	Dist.	Dep.	D. Lat.	Dist.	Dep.	D. Lat.	Dist.	Dep.	D. Lat.

283° | 077° / 257° | 103°

77°

Dist.	D. Lat.	Dep.
D Lo	Dep.	
	m	D Lo

FIGURE 1006b.—Extracts from table 3.

TABLE 3

Traverse **30°** Table

330° | 030° / 210° | 150° (left) 330° | 030° / 210° | 150° (right)

Dist.	D. Lat.	Dep.	Dist.	D. Lat.	Dep.	Dist.	D. Lat.	Dep.	Dist.	D. Lat.	Dep.	Dist.	D. Lat.	Dep.
40	34.6	20.0	100	86.6	50.0	60	138.6	80.0	20	190.5	110.0	80	242.5	140.0
41	35.5	20.5	101	87.5	50.5	161	139.4	80.5	221	191.4	110.5	281	243.4	140.5
42	36.4	21.0	02	88.3	51.0	62	140.3	81.0	22	192.3	111.0	82	244.2	141.0
Dist.	Dep.	D. Lat.	Dist.	Dep.	D. Lat.	Dist.	Dep.	D. Lat.	Dist.	Dep.	D. Lat.	Dist.	Dep.	D. Lat.

300° | 060° / 240° | 120°

60°

Dist.	D. Lat.	Dep.
N.	N×Cos.	N×Sin.
Hypotenuse	Side Adj.	Side Opp.

FIGURE 1006c.—Extracts from table 3.

C S60°W.

(1) Cn 240°.

(2) 1005 mi.

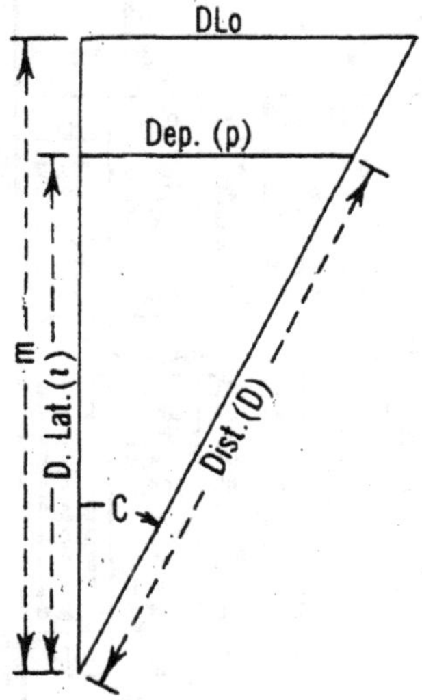

FIGURE 1007a.—Mercator and plane sailing relationships.

The labels (N, S, E, W) of l, p, DLo, and C are determined by noting the direction of motion or the relative positions of the two places.

When the course is near 090° or 270°, the solution of C to the nearest 0°.1 only may introduce a large error in distance.

1007. Mercator sailing problems are solved graphically when measurement is made on a Mercator chart. Graphical solution can also be made as shown in figure 1007a. The lower part is identical with the plane sailing triangle, figure 1003. For mathematical solution the formulas of Mercator sailing are:

$$\tan C = \frac{DLo}{m} \qquad DLo = m \tan C.$$

Following solution of course angle by Mercator sailing, the solution of distance is by means of the plane sailing formula:

$$D = l \sec C.$$

Solution can be made by computation or by traverse table.

Example 1.—A ship at lat. 32°14′.7N, long. 66°28′.9W is to head for a point near Chesapeake Light, lat. 36°58′.7N, long. 75°42′.2W.

Required.—(1) Course, (2) distance.

Solution.—By computation: (1) $\tan C = \frac{DLo}{m}$ (2) $D = l \sec C$

L_1 32°14′.7 N	M_1 2033.3		λ_1 66°28′.9 W
L_2 36°58′.7 N	M_2 2377.0		λ_2 75°42′.2 W
l 4°44′.0 N	m 343.7		DLo 9°13′.3 W
l 284′.0 N			DLo 553′.3 W
DLo 553′.3 W	log	2.74296	
m 343.7	log (−)	2.53618	
C N58°09′.1W	l tan	0.20678	l sec 0.27764
l 284′.0 N			log 2.45332
(2) D 538.2 mi.			log 2.73096
(1) Cn 301°.8			

Solution.—By inspection of table 3:

Refer to figure 1007b. Substitute m as the heading of the D. Lat. column and DLo as the heading of the Dep. column. Turn the pages of table 3 until the numbers 343.7 and 553.3 are found abreast in the columns relabeled m and DLo, respectively.

Because a number as high as 343.7 is not tabulated in the m column, it is necessary to divide m and DLo by 10. Then inspect to find 34.4 and 55.3 abreast in the m and DLo columns, respectively. This occurs most nearly on the page for course angle 58° or course 302°.

Reenter table 3 with course 302° to find Dist. for D. Lat. 284′.0.

C N58°W.
(1) Cn 302°.
(2) D 536 mi.

TABLE 3

328° | 032° / 212° | 148° — Traverse **32°** Table — 328° | 032° / 212° | 148°

Dist.	D. Lat.	Dep.	Dist.	D. Lat.	Dep.	Dist.	D. Lat.	Dep.	Dist.	D. Lat.	Dep.	Dist.	D. Lat.	Dep.
1	0.8	0.5	61	51.7	32.3	121	102.6	64.1	181	153.5	95.9	241	204.4	127.7
2	1.7	1.1	62	52.6	32.9	22	103.5	64.7	82	154.3	96.4	42	205.2	128.2
3	2.5	1.6	63	53.4	33.4	23	104.3	65.2	83	155.2	97.0	43	206.1	128.8
4	3.4	2.1	64	54.8	33.9	24	105.2	65.7	84	156.0	97.5	44	206.9	129.3
5	4.2	2.6	65	55.1	34.4	25	106.0	66.2	85	156.9	98.0	45	207.8	129.8
6	5.1	3.2	66	56.0	35.0	26	106.9	66.8	86	157.7	98.6	46	208.6	130.4
…	…	…	…	…	…	…	…	…	…	…	…	…	…	…
351	297.7	186.0	411	348.5	217.8	471	399.4	249.6	531	450.3	281.4	591	501.2	313.2
52	298.5	186.5	12	349.4	218.3	72	400.3	250.1	32	451.2	281.9	92	502.0	313.7
53	299.4	187.1	13	350.2	218.9	73	401.1	250.7	33	452.0	282.4	93	502.9	314.2
54	300.2	187.6	14	351.1	219.4	74	402.0	251.2	34	452.9	283.0	94	503.7	314.8
55	301.1	188.1	15	351.9	219.9	75	402.8	251.7	35	453.7	283.5	95	504.6	315.3
56	301.9	188.7	16	352.8	220.4	76	403.7	252.2	36	454.6	284.0	96	505.4	315.8
57	302.8	189.2	17	353.6	221.0	77	404.5	252.8	37	455.4	284.6	97	506.3	316.4
58	303.6	189.7	18	354.5	221.5	78	405.4	253.3	38	456.2	285.1	98	507.1	316.9
59	304.4	190.2	19	355.3	222.0	79	406.2	253.8	39	457.1	285.6	99	508.0	317.4
60	305.3	190.8	20	356.2	222.6	80	407.1	254.4	40	457.9	286.2	600	508.8	318.0
Dist.	Dep.	D. Lat.	Dist.	Dep.	D. Lat.	Dist.	Dep.	D. Lat.	Dist.	Dep.	D. Lat.	Dist.	Dep.	D. Lat.

302° | 058° / 238° | 122° — **58°**

Dist.	D. Lat.	Dep.
D Lo	Dep.	
	m	D Lo

Figure 1007b.—Extract from table 3 composed of parts of left- and right-hand pages for course angle 58°.

Example 2.—A ship at lat. 75°31′.7N, long. 79°08′.7W, in Baffin Bay, steams 263.5 miles on course 155°.

Required.—(1) Latitude and (2) longitude of point of arrival.

Solution.—By computation:

D 263.5 mi.	log 2.42078	
C 155°	l cos 9.95728	l tan 9.66867
l 238′.8 S	log 2.37806	
m 846.3		log 2.92752
DLo 394′.6 E		log 2.59619
L_1 75°31′.7 N	M_1 7072.4	λ_1 79°08′.7 W
l 3°58′.8 S		DLo 6°34′.6 E
(1) L_2 71°32′.9 N	M_2 6226.1	(2) λ_2 72°34′.1 W
	m 846.3	

The labels (N, S, E, W) of l, DLo, and C are determined by noting the direction of motion or the relative positions of the two places.

If the course is near 090° or 270°, a small error in C introduces a large error in DLo. The solution for C to the nearest 0°.1 only may introduce a large error in distance if the course is near 090° or 270°.

Solution.—By inspection of table 3:

Refer to figure 1007c. Enter table 3 with course 155° and Dist. 263.5 miles to find D. Lat. 238′.8. The latitude of the point of arrival is found by subtracting the D. Lat. from the latitude of the point of departure. Determine the meridional difference by table 5 (m=846.3).

Reenter table 3 with course 155° to find the DLo corresponding to m=846.3. Substitute meridional difference m as the heading of the D. Lat. column and DLo as the heading of the Dep. column. Because a number as high as 846.3 is not tabulated in the m column, it is necessary to divide m by 10 and then inspect the m column for a value of 84.6. Interpolating as necessary, the latter value is opposite DLo 39′.4. The DLo is 394′ (39′.4×10). The longitude of the point of arrival is found by applying the DLo to the longitude of the point of departure.

(1) L_2 71°32′.9 N.
(2) λ_2 72°34′.7 W.

335°	025°
205°	155°

TABLE 3

Traverse **25°** Table

335°	025°
205°	155°

Dist.	D. Lat.	Dep.	Dist.	D. Lat.	Dep.	Dist.	D. Lat.	Dep.	Dist.	D. Lat.	Dep.	Dist.	D. Lat.	Dep.
21	19.0	8.9	81	73.4	34.2	141	127.8	59.6	201	182.2	84.9	261	236.5	110.3
22	19.9	9.3	82	74.3	34.7	42	128.7	60.0	02	183.1	85.4	62	237.5	110.7
23	20.8	9.7	83	75.2	35.1	43	129.6	60.4	03	184.0	85.8	[illegible]	[illegible]	111.1
24	21.8	10.1	84	76.1	35.5	44	130.5	60.9	04	184.9	86.2	[illegible]	[illegible]	111.6
25	22.7	10.6	85	77.0	35.9	45	131.4	61.3	05	185.8	86.6	65	240.2	112.0
33	29.9	13.9	93	[illegible]	[illegible]	53	138.7	64.7	13	193.0	90.0	73	247.4	115.4
34	30.8	14.4	94	[illegible]	[illegible]	54	139.6	65.1	14	193.9	90.4	74	248.3	115.8
Dist.	Dep.	D. Lat.	Dist.	Dep.	D. Lat.	Dist.	Dep.	D. Lat.	Dist.	Dep.	D. Lat.	Dist.	Dep.	D. Lat.

295°	065°
245°	115°

65°

Dist.	D. Lat.	Dep.
D Lo	Dep.	
	m	DLo

FIGURE 1007c.—Extracts from table 3.

1008. Compact traverse table.—Table 38 is a compact traverse table which provides an alternative to the use of table 3 for the tabular solutions of the various sailings except great-circle and composite. Other than in very high latitudes, these tabular solutions have sufficient accuracy normally required for practical navigation purposes where the distance steamed does not exceed a day's run of a moderate speed vessel. Depending upon the course angle and latitude, distance solutions beyond this limit may lack the required accuracy.

This presentation is designed to indicate both the construction and use of the table so that it may be more readily understood and more effectively used. The main part of the table, sections 1, 2, and 3 of figure 1008, is essentially a multiplication table to be used with either course or latitude, corresponding to the degree entering headings. The tabular values are the numbers 1 through 9 multiplied by the cosines, sines, secants, and tangents of the course or latitude with which the table is entered.

If the solution of a sailing formula containing latitude is desired, the user turns to that part of the table where the left-hand bold number corresponds to that of the latitude. The table is so designed that the title or the name of the other known value of the formula containing latitude is on the upper heading line of that part corresponding to the latitude. For example if the latitude is 70° and the solution of the parallel sailing formula p=DLo cos L is sought for various values of DLo, the value of p extracted from section 1 must be found under the p heading which appears on the upper heading line. Thus, if DLo is 5, the corresponding p must be 1.710 not 4.698. The value p=4.698 is the solution of another formula involving course, specifically p=D sin C. If latitude and departure p are known, to find DLo the formula is rearranged as DLo=p sec L. Since the formula contains latitude, the upper heading line must be used as before. If latitude=70° and p=8, the value of DLo=23.390 is extracted from under the DLo heading which appears on the upper heading line in section 2.

TABLE 38
Compact Traverse Table

70° or 110° or 250° or 290°

Section 1			Section 2		Section 3	
DLo	p	//////////	p	DLo	//////////	
D	*l*	p	*l*	D	m	DLo
1	0. 342	0. 940	1	2. 924	1	2. 747
2	0. 684	1. 879	2	5. 848	2	5. 495
3	1. 026	2. 819	3	8. 771	3	8. 242
4	1. 368	3. 759	4	11. 695	4	10. 990
5	1. 710	4. 698	5	14. 619	5	13. 737
6	2. 052	5. 638	6	17. 543	6	16. 485
7	2. 394	6. 578	7	20. 467	7	19. 232
8	2. 736	7. 518	8	23. 390	8	21. 980
9	3. 078	8. 457	9	26. 314	9	24. 727

Section 4	Course	
70°	p÷*l*	109°
250°	DLo÷m	289°
0. 0	2. 747	1. 0
0. 1	2. 762	0. 9
0. 2	2. 778	0. 8
0. 3	2. 793	0. 7
0. 4	2. 808	0. 6
0. 5	2. 824	0. 5
0. 6	2. 840	0. 4
0. 7	2. 856	0. 3
0. 8	2. 872	0. 2
0. 9	2. 888	0. 1

71° or 109° or 251° or 289°

DLo	p	//////////	p	DLo	//////////	
D	*l*	p	*l*	D	m	DLo
1	0. 326	0. 946	1	3. 072	1	2. 904
2	0. 651	1. 891	2	6. 143	2	5. 808
3	0. 977	2. 837	3	9. 215	3	8. 713
	(Section 1) p = DLo cos L; p = DLo cos Lm; p = D sin C; *l* = D cos C		(Section 2) DLo = p sec L; DLo = p sec Lm; D = *l* sec C		(Section 3) DLo = m tan C	
8	2. 605	7. 564	8	24. 572	8	23. 234
9	2. 930	8. 510	9	27. 644	9	26. 138

Course		
71°	p÷*l*	108°
251°	DLo÷m	288°
0. 0	2. 904	1. 0
0. 1	2. 921	0. 9
0. 2	2. 937	0. 8
	(Section 4) tan C = p÷*l*; tan C = DLo÷m	
0. 8	3. 042	0. 2
0. 9	3. 060	0. 1

72° or 108° or 252° or 288°

DLo	p	//////////	p	DLo	//////////	
D	*l*	p	*l*	D	m	DLo
1	0. 309 (cos 72°)	0. 951 (sin 72°)	1	3. 236 (sec 72°)	1	3. 078 (tan 72°)
2	2 × cos 72°	2 × sin 72°	2	2 × sec 72°	2	2 × tan 72°
3	3 × cos 72°	3 × sin 72°	3	3 × sec 72°	3	3 × tan 72°
4	4 × cos 72°	4 × sin 72°	4	4 × sec 72°	4	4 × tan 72°
5	1. 545	4. 755	5	16. 180	5	15. 388
6	1. 854	5. 706	6	19. 416	6	18. 466
	↑ 6 × cos 72°; 6 × cos 108°; 6 × cos 252°; 6 × cos 288°	↑ 6 × sin 72°; 6 × sin 108°; 6 × sin 252°; 6 × sin 288°		↑ 6 × sec 72°; 6 × sec 108°; 6 × sec 252°; 6 × sec 288°		↑ 6 × tan 72°; 6 × tan 108°; 6 × tan 252°; 6 × tan 288°

Course		
72°	p÷*l*	107°
252°	DLo÷m	287°
0. 0	3. 078	1. 0
0. 1	3. 096	0. 9
0. 2	tan 72°.2 or 252°.2 or 107°.8 or 287°.8	0. 8
0. 3	3. 133	0. 7
0. 4	3. 152	0. 6
0. 5	3. 172	0. 5

FIGURE 1008.—Compact traverse table.

If the solution of a sailing formula containing course is desired, the user turns to that part of the table where one of the bold numbers corresponds to the course. The table is so designed that the title or name of the other known value of a formula containing course is on the lower heading line of that part corresponding to the course. For example, if the course is 110° and the solution of the formula p=D sin C is sought for various values of D, the value of p extracted from section 1 must be from under the p heading on the lower heading line. Thus, if D=7, the corresponding p must be

6.578, not 2.394. The value P=2.394 is the solution of another formula involving latitude, specifically p=DLo cos L.

For the solution of other formulas involving course in sections 2 and 3, the lower heading line is used in like manner.

With respect to sections 1, 2, and 3 being multiplication tables, the values on the first line opposite the number 1 in each section are natural trigonometric functions of the latitude or course used to enter the appropriate part of the tables.

The functions are further identified at the bottom of figure 1008. In the solution of the formula

$$p = D \sin C$$
$$\text{where } C = 110°$$
$$\text{If } D = 1,\ p = 1 \times \sin C$$
$$p = 1 \times 0.940.$$

So the value opposite 1 and under p on the lower heading line is the sine of 110°. The value opposite D=7 is therefore

$$D \times \sin C$$
$$\text{or } 7 \times 0.940\ (0.93969) = 6.578.$$

This is the general scheme of sections 1, 2, and 3.

The auxiliary table, section 4, is simply a tabulation of the natural tangents of courses at intervals of 0°1. To use this part of the table for the solution of the formulas,

$$\tan C = p \div l$$
$$\tan C = DLo \div m,$$

the ratio $p \div l$ or $DLo \div m$ must first be found. This is frequently done by means of logarithms. Entering the central column with the ratio so found, the appropriate course is extracted from either the left or right-hand column depending upon whether the departure is an easting or westing and the latitude difference is a northing or southing.

1009. Plane sailing by table 38.—Problems frequently solved by plane sailing (arts. 1001, 1003) are: given the course and distance, find the difference of latitude and the departure. The applicable formulas are:

$$\left.\begin{array}{l} l = D \cos C \\ p = D \sin C \end{array}\right\} \text{(Plane sailing formulas)}$$

Example 1.—A vessel on course 070° steams 352.8 miles from point *A*.

Required.—(1) Difference of latitude, (2) Departure.

Solution.—By inspection of table 38:

Since the applicable formulas (l=D cos C and p=D sin C) include course, the user turns to that part of table 38 where the course, 070°, is one of the bold numbers at the top of the tabular data, and enters section 1 to find D on the lower heading line.

As shown in figure 1009a section 1 provides solutions of the formulas,

$$p = DLo \cos L$$
$$p = DLo \cos Lm$$
$$p = D \sin C$$
$$l = D \cos C,$$

but only the last two formulas are applicable to the use of the lower heading line.

To find solutions of the formulas for l and p, their respective values, under their headings on the lower heading line, are extracted opposite the values of the distance

D, tabulated under the heading D on the lower heading line. Inspection reveals that the exact value of D is not tabulated. But this presents no great difficulty, although somewhat less convenient than the respective use of table 3.

In table 38 the entering arguments of numbers 1 through 9, under their respective headings on the upper and lower heading lines, permit solutions for all numerical values times the appropriate trigonometric function of the latitude or course. This is accomplished through moving the decimal point in the tabulated values as the numbers 1 through 9 are multiplied by 1/10, 10, 100, 1000 etc., and by summing up the tabulated values corresponding to various parts of the complete value.

In this example on entering section 1 with distance D of 352.8 miles, the D value is broken down as follows to find latitude difference l and departure p:

D (352.8 mi.)	l	p
300.0	102.6 (1.026)	281.9 (2.819)
50.0	17.1 (1.710)	47.0 (4.698)
2.0	0.7 (0.684)	1.9 (1.879)
0.8	0.3 (2.736)	0.8 (7.518)
352.8 mi	120.7	331.6 mi.

TABLE 38

Compact Traverse Table

70°—110°—250°—290°

DLo	p	//////////	(Upper heading line)
D	l	p	(Lower heading line)
1	0. 342	0. 940	
2	0. 684	1. 879	
3	1. 026	2. 819	
4	1. 368	3. 759	
5	1. 710	4. 698	
6	2. 052	5. 638	
7	2. 394	6. 578	
8	2. 736	7. 518	
9	3. 078	8. 457	

$p = DLo \cos L$
$p = DLo \cos Lm$
$p = D \sin C$
$l = D \cos C$

(Section 1)

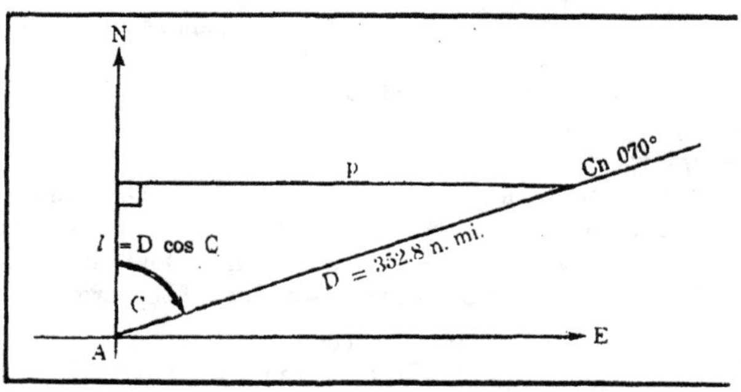

FIGURE 1009a.—Table 38 tabulations for plane sailing formulas $p = D \sin C$ and $l = D \cos C$ (section 1).

To find the portion of l and p corresponding to distance D of 300 miles, the decimal points of the tabulated values opposite 3 are moved 2 places to the right because of multiplying 3 by 100.

To find the portion of l and p corresponding to distance D of 50 miles, the decimal points of the tabulated values opposite 5 are moved 1 place to the right because of multiplying 5 by 10.

To find the portion of l and p corresponding to distance D of 2 miles, the values of l and p are extracted as tabulated.

To find the portion of l and p corresponding to distance D of 0.8 miles, the decimal points are moved 1 place to the left because of multiplying 8 by 1/10.

In the summation in the block above, the tabulated values are rounded off to the precision of the entering arguments. The values in parentheses are the values as tabuated in table 38 for the various unit values used.

(1) Difference of latitude, l—120′.7 (northing).

(2) Departure p—331.6 nautical miles (easting).

If the course had been 110° or 250° or 290°, the same numerical values of l and p would have been obtained. See figure 1009b.

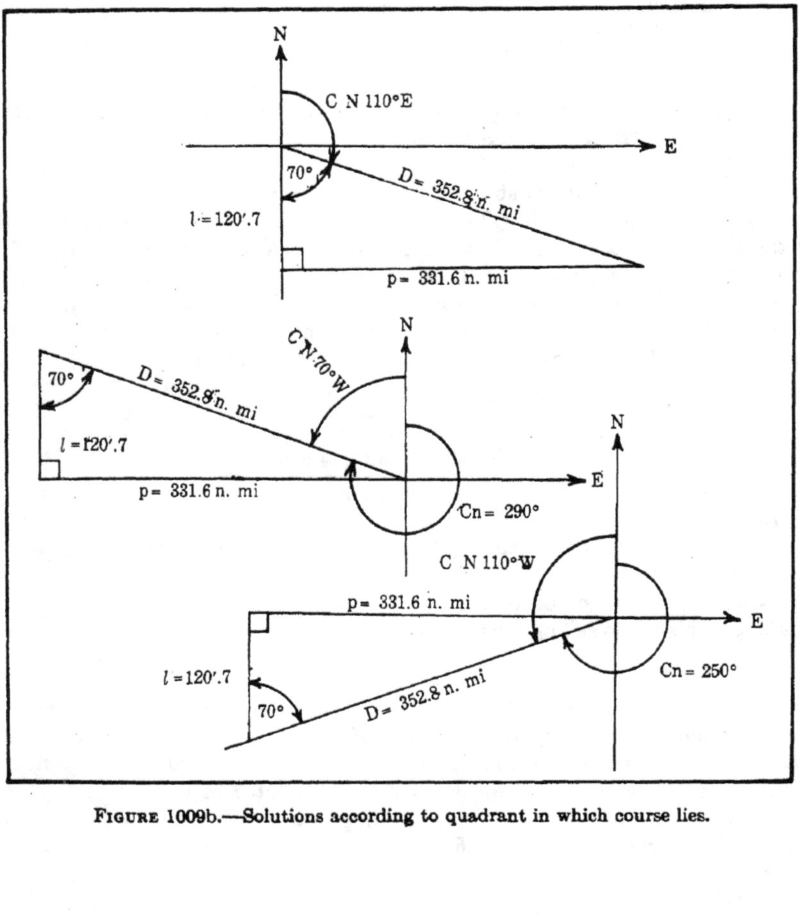

FIGURE 1009b.—Solutions according to quadrant in which course lies.

TABLE 38

Compact Traverse Table

70°—110°—250°—290°

DLo	p		p	DLo		
D	l	p	l	D	m	DLo
			1	2. 924		
			2	5. 848		
			3	8. 771		
			4	11. 695		
			5	14. 619		
			6	17. 543		
			7	20. 467		
			8	23. 390		
			9	26. 314		

←(Upper heading line)
←(Lower heading line)

(Section 2)
DLo = p sec L
DLo = p sec Lm
D = l sec C

FIGURE 1009c.—Table 38 tabulations for plane sailing formula D=l sec C (section 2).

Other problems solved by plane sailing are: (1) given the course and difference of latitude, find the distance steamed; and (2) given the departure and difference of latitude, find the course. The applicable formulas are:

$$\left.\begin{array}{r} D=l\sec C \\ \tan C=p\div l \end{array}\right\} \text{(plane sailing formulas).}$$

Example 2.—A vessel at latitude 30°15′N, longitude 50°23′W, steams on course 250° to latitude 29°50′N, longitude 51°42′W.

Required.—(1) Distance steamed.

Solution.—By inspection of table 38:

Since the applicable formula (D=l sec C) includes course, the user turns to that part of table 38 where, the course 250° is one of the bold numbers at the top of the tabular data and enters section 2 (fig. 1009c) to find l on the lower heading line. Since the difference of latitude l between the point of departure and the point of destination (30°15′−29°50′=25′) is not one of the tabulations, the scheme of breaking the entering argument in parts and moving the decimal point is utilized.

l (25′)	D
20	58.5 (5.848)
5	14.6 (14.619)
25	73.1 mi.

(1) 73.1 mi.

Example 3.—A vessel has steamed 105 miles north and 300 miles west.

Required.—(1) The course.

Solution.—By inspection of table 38:

The applicable plane sailing formula is:

$$\tan C=p\div l.$$

The value p÷l (300÷105)=2.857 is first calculated. Then by inspecting the central columns of section 4 (fig. 1009d) to find a value of tan C=2.857, a close approximation is found under that part with columnar headings 70°—250°—109°—289°. Since the course would be in the northwest quadrant, 289° is selected as the base value. Since 2.857 is closest to the value of 2.856 corresponding to course 289°.3, that is the course made good, within practical limits. If the latitude difference had been to the south, the course made good would be 250°.7.

(1) C 289°.3.

TABLE 38

Compact Traverse Table

70°—110°—250°—290°							
DLo	p		p	DLo			
D	*l*	p	*l*	D	m	DLo	

(Section 4)

tan C = p÷*l*
tan C = DLo÷m

70° 250°	Course p÷*l* DLo÷m	109° 289°
0. 0	2. 747	1. 0
0. 1	2. 762	0. 9
0. 2	2. 778	0. 8
0. 3	2. 793	0. 7
0. 4	2. 808	0. 6
0. 5	2. 824	0. 5
0. 6	2. 840	0. 4
0. 7	2. 856	0. 3
0. 8	2. 872	0. 2
0. 9	2. 888	0. 1

FIGURE 1009d.—Table 38 tabulations for plane sailing formula tan C=p÷*l* and Mercator sailing formula tan C=DLo÷m (section 4).

1010. Traverse sailing by table 38.—Refer to article 1004. Solve each leg as in example 1, article 1009, for departure and difference of latitude. Convert departure and difference of latitude to equivalent single course as in example 3, article 1009. Knowing difference of latitude and equivalent single course, solve for distance as in example 2, article 1009.

1011. Parallel sailing by table 38 consists of the interconversion of departure and difference of longitude. It is the simplest form of spherical sailing. The formulas for these transformations are:

DLo=p sec L p=DLo cos L

Example 1.—A vessel at latitude 30°15′N steams on course 090° until the longitude changes 4°36′E.

Required.—(1) The distance steamed.

Solution.—By inspection of table 38:

The applicable parallel sailing formula is:

p=DLo cos L.

The solution of this formula is contained in section 1 (fig. 1011). Since the formula includes latitude, the tabulated departure p extracted must conform to the entering argument heading on the upper heading line. Since the solutions are for integral degrees of latitude and the ship is steaming between two such integral values of latitude, solutions for both latitude 30° and 31° must be extracted from the table, interpolating the results to the DR latitude.

Convert DL0=4°36′ to minutes of arc.

DLo=4°36′=276′

Break down 276 into parts convenient for use with table 38.

DLo (276′)	p (Lat. 30°)	p (Lat. 31°)
200.0	173.2 (1.732)	171.4 (.714)
70.0	60.6 (6.062)	60.0 (6.000)
6.0	5.2 (5.196)	5.1 (5.143)
276.0	239.0 n. mi.	236.5 n. mi.

Interpolating to latitude 30°15′, the departure p is 238.4 nautical miles.

(1) Distance 238.4 mi.

TABLE 38

Compact Traverse Table

30°—150°—210°—330°

DLo	p	//////////	p	DLo	//////////	//////////
D	l	p	l	D	m	DLo
1	0. 866	0. 500	1	1. 155		
2	1. 732	1. 000	2	2. 309		
3	2. 598	1. 500	3	3. 464		
4	3. 464	2. 000	4	4. 619		
5	4. 330	2. 500	5	5. 774		
6	5. 196	3. 000	6	6. 928		
7	6. 062	3. 500	7	8. 083		
8	6. 928	4. 000	8	9. 238		
9	7. 794	4. 500	9	10. 392		

←(Upper heading line)
←(Lower heading line)

31°—149°—211°—329°

DLo	p	//////////	p	DLo	//////////	//////////
D	l	p	l	D	m	DLo
1	0. 857	0. 515	1	1. 167		
2	1. 714	1. 030	2	2. 333		
3	2. 572	1. 545	3	3. 500		
4	3. 429	2. 060	4	4. 667		
5	4. 286	2. 575	5	5. 833		
6	5. 143	3. 090	6	7. 000		
7	6. 000	3. 605	7	8. 166		
8	6. 857	4. 120	8	9. 333		
9	7. 715	4. 635	9	10. 500		

←(Upper heading line)
←(Lower heading line)

(Section 1)

p = DLo cos L
p = DLo cos Lm
p = D sin C
l = D cos C

(Section 2)

DLo = p sec L
DLo = p sec Lm
D = l sec C

FIGURE 1011.—Table 38 tabulations for parallel sailing formulas p=DLo Cos L and DLo=p sec L (sections 1 and 2).

Example 2.—A vessel at latitude 30°15′N, longitude 50°20′W steams on course 090° for a distance of 294.5 miles.

Required.—(1) The longitude of the point of arrival. The applicable parallel sailing formula is:

DLo=p sec L.

The solution of this formula is contained in section 2 (fig. 1011). Since the solutions are for integral values of latitude and the vessel is steaming between two such integral values of latitude, solutions for both latitude 30° and 31° must be extracted from the table, interpolating the results to the DR latitude.

Entering section 2 for latitude 30° and 31°, the heading for the other entering argument p is found on the upper heading line. Since the departure p=294.5 is not one of the tabulations the scheme of moving the decimal point and breaking the entering argument into parts is utilized.

p (294.5 n. mi.)	DLo (Lat. 30°)		DLo (Lat. 31°)	
200.0	230.9	(2.309)	233.3	(2.333)
90.0	103.9	(10.392)	105.0	(10.500)
4.0	4.6	(4.619)	4.7	(4.667)
0.5	0.6	(5.774)	0.6	(5.833)
294.5	340.0		343.6	

Interpolating to latitude 30°15′, the DLo is 340.9 minutes of arc or 5°40′.9.

(1) Longitude of point of arrival 44°39′.1W.

TABLE 38

Compact Traverse Table

70°—110°—250°—290°

DLo	p	//////////	(Upper heading line)
D	l	p	(Lower heading line)
1	0. 342	0. 940	
2	0. 684	1. 879	
3	1. 026	2. 819	
4	1. 368	3. 759	
5	1. 710	4. 698	
6	2. 052	5. 638	
7	2. 394	6. 578	
8	2. 736	7. 518	
9	3. 078	8. 457	

p = DLo Cos L
p = DLo Cos Lm
p = D sin C
l = D cos C

(Section 1)

FIGURE 1012a.—Table 38 tabulations for plane sailing formulas p = D sin C and l = D cos C (section 1).

1012. Middle-latitude sailing by table 38.—As stated in article 1006, middle-latitude sailing combines plane sailing and parallel sailing. Plane sailing is used to find difference of latitude and departure when course and distance are known, or vice versa. Parallel sailing on the middle latitude is used to interconvert departure and difference of longitude. The mean latitude (Lm) is normally used for want of a practicable means for determining the middle latitude, the latitude at which the arc length of the parallel separating the meridians passing through two specific points is exactly equal to the departure in proceeding from one point to the other. The formulas for these transformations are:

$$\text{DLo} = \text{p sec Lm} \qquad \text{p} = \text{DLo cos Lm}.$$

The mean latitude (Lm) is half the arithmetical sum of the latitudes of two places on the same side of the equator. The course line is labeled N or S to indicate whether it is north or south of the equator. If a course line crosses the equator, that part on each side (the north and south latitude portions, respectively) should be solved separately.

Example 1.—A vessel steams 1,253.4 miles on course 070° from lat. 15°17.′4N, long. 151°37.′8E.

Required.—(1) Latitude and (2) longitude of point of arrival.

Solution.—By inspection of table 38:

Section 1 of table 38 (fig. 1012a) is used to find difference of latitude and departure when course and distance are known. The applicable plane sailing formulas are:

$$l = \text{D cos C}$$

$$\text{p} = \text{D sin C}.$$

D	l	p
1000. 0	342. 0	940. 0
200. 0	68. 4	187. 9
50. 0	17. 1	47. 0
3. 0	1. 0	2. 8
0. 4	0. 1	0. 4
1253. 4	428. 6	1178. 1

L_1	15°17′.4N	L_1	15°17′.4N
l	7°08′.6N	$\frac{1}{2}l$	3°34′.3N
(1) L_2	22°26′.0N	Lm	18°51′.7N

The difference of latitude l is applied to the latitude of departure L_1 to find the latitude of the point of arrival L_2. The mean latitude Lm is then formed for use with the departure p found above for finding the difference of longitude by parallel sailing formula:

$$DLo = p \sec Lm.$$

Section 2 of table 38 (fig. 1012b) is used to solve the above formula for difference of longitude, departure and mean latitude being known. Since the tabular solutions are for integral degrees of latitude, solutions for both latitude 18° and 19° must be extracted from the table, interpolating the results to the mean latitude.

p	DLo(18°)	DLo(19°)
1000. 0	1051. 0	1058. 0
100. 0	105. 1	105. 8
70. 0	73. 6	74. 0
8. 0	8. 4	8. 5
0. 1	0. 1	0. 1
1178. 1	1238. 2	1246. 4

TABLE 38

Compact Traverse Table

18°—162°—198°—342°

p / l	DLo / D
1	1. 051
2	2. 103
3	3. 154
4	4. 206
5	5. 257
6	6. 309
7	7. 360
8	8. 412
9	9. 463

(Section 2)

19°—161°—199°—341°

p / l	DLo / D
1	1. 058
2	2. 115
3	3. 173
4	4. 230
5	5. 288
6	6. 346
7	7. 403
8	8. 461
9	9. 519

FIGURE 1012b.—Table 38 tabulations for parallel sailing formula DLo=p sec Lm.

Interpolating to mean latitude 18°51′.7, the DLo is 1245.3 minutes of arc or 20°45′.3E. The DLo is applied to the longitude of the point of departure to find the longitude of the point of arrival.

λ_1	151°37′.8 E	
DLo	20°45′.3 E	DLo 1245′.3
(2) λ_2	172°23′.1 E	

Example 2.—A vessel at lat. 8°48′.9S, long. 89°53′.3W is to proceed to lat. 17°06′.9S, long. 104°51′.6W.

Required.—(1) Departure, (2) course, and (3) distance.

Solution.—By inspection of table 38:

Section 1 of table 38 (fig. 1012c) is used to solve the parallel sailing formula:

$$p = DLo \cos Lm.$$

for departure, difference of longitude and mean latitude being known. (One-half the difference of latitude is applied to the latitude of departure to find the mean latitude.)

L_1	8°48′.9 S	λ_1	89°53′.3 W
L_2	17°06′.9 S	λ_2	104°51′.6 W
l	8°18′.0 S	DLo	14°58′.3 W
½ l	4°09′.0 S	DLo	898′.3 W
Lm	12°57′.9 S		

TABLE 38

Compact Traverse Table

12°—168°—192°—348°

DLo / D	p / l	p
1	0.978	0.208
2	1.956	0.416
3	2.934	0.624
4	3.913	0.832
5	4.891	1.040
6	5.869	1.247
7	6.847	1.455
8	7.825	1.663
9	8.803	1.871

(Section 1)

13°—167°—193°—347°

DLo / D	p / l	p	p / l	DLo / D
1	0.974	0.225		
2	1.949	0.450		
3	2.923	0.675		
4	3.897	0.900		
5	4.872	1.125		
6	5.846	1.350		
7	6.821	1.575		
8	7.795	1.800		
9	8.769	2.025		

FIGURE 1012c.—Table 38 tabulations for parallel sailing formula p=DLo cos Lm.

Since the tabular solutions are for integral degrees of latitude, solutions for both latitude 12° and 13° must be extracted from the table, interpolating the results to the mean latitude.

DLo	p(12°)	p(13°)
800.0	782.5	779.5
90.0	88.0	87.7
8.0	7.8	7.8
0.3	0.3	0.3
898.3	878.6	875.3

Interpolating to mean latitude 12°57′.9, the departure p is 875.4 nautical miles.
(1) Departure 875.4 mi.
Section 4 of table 38 (fig. 1012d) is used to solve the plane sailing formula:

$$\tan C = p \div l$$

TABLE 38

Compact Traverse Table

60°—120°—240°—300°

DLo / D	p / l	p	p / l	DLo / D	m	DLo

(Section 4)

Course 60° 240°	p÷l / DLo÷m	119° 299°
0.0	1.732	1.0
0.1	1.739	0.9
0.2	1.746	0.8
0.3	1.753	0.7
0.4	1.760	0.6
0.5	1.767	0.5
0.6	1.775	0.4
0.7	1.782	0.3
0.8	1.789	0.2
0.9	1.797	0.1

FIGURE 1012d.—Table 38 tabulations for plane sailing formula tan C=p÷l.

for course, departure and difference of latitude being known. But first the ratio p÷l must be calculated. The central column is entered with the ratio so found, and the appropriate course is extracted from either the left- or right-hand column depending upon whether the departure is an easting or westing and the latitude difference is a northing or southing.

p 875.4 mi. W	log	2.94221
l 498′.0 S	log (−)	2.69723
p÷l 1.758	log	0.24498
C S 60°.4W		
(2) Cn 240°.4		

TABLE 38

Compact Traverse Table

60°—120°—240°—300°

p / l	DLo / D
1	2. 000
2	4. 000
3	6. 000
4	8. 000
5	10. 000
6	12. 000
7	14. 000
8	16. 000
9	18. 000

(Section 2)

61°—119°—241°—299°

p / l	DLo / D
1	2. 063
2	4. 125
3	6. 188
4	8. 251
5	10. 313
6	12. 376
7	14. 439
8	16. 501
9	18. 564

FIGURE 1012e.—Table 38 tabulations for plane sailing formula D=l sec C.

With both difference of latitude and course now known, section 2 (fig. 1012e) is used to solve the plane sailing formula:

$$D=l \sec C$$

for course.

l	D(240°)	D(241°)
400.0	800.0	825.1
90.0	180.0	185.6
8.0	16.0	16.5
0.0	0.0	0.0
498.0	996.0	1027.2

Since the tabular solutions are for integral degrees of course, solutions for both course 240° and 241° must be extracted from the table, interpolating the results to the actual course.

Interpolating to course 240°.4, the distance is 1008.5 miles.

(3) 1008.5 mi.

When the course is near 090° or 270°, the solution of course to the nearest 0°.1 only, as by traverse table, may introduce a large error in distance.

1013. Mercator sailing by table 38.—As stated in article 1001, Mercator sailing is similar to plane sailing, but uses meridional difference and difference of longitude in place of difference of latitude and departure, respectively. See figure 1013a.

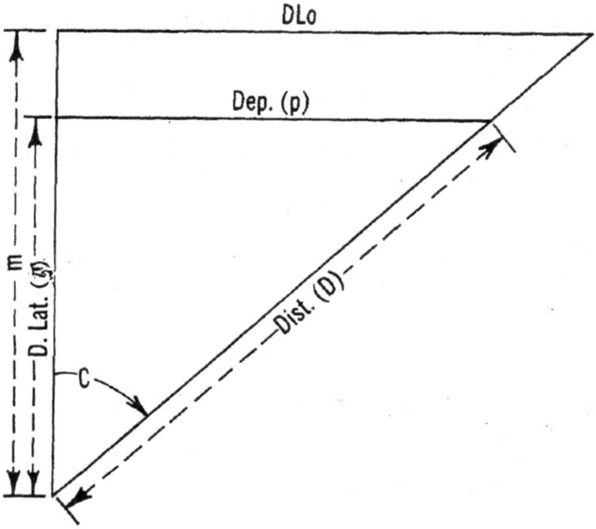

FIGURE 1013a.—Mercator and plane sailing relationships.

TABLE 38

Compact Traverse Table

58°—122°—238°—302°						
DLo / D	p / l	p	p / l	DLo / D	m	DLo

(Section 4)

Course 58° 238°	p÷l / DLo÷m	121° 301°
0. 0	1. 600	1. 0
0. 1	1. 607	0. 9
0. 2	1. 613	0. 8
0. 3	1. 619	0. 7
0. 4	1. 625	0. 6
0. 5	1. 632	0. 5
0. 6	1. 638	0. 4
0. 7	1. 645	0. 3
0. 8	1. 651	0. 2
0. 9	1. 658	0. 1

FIGURE 1013b.—Table 38 tabulations for plane sailing formula tan C=p÷l and Mercator sailing formula tan C=DLo÷m.

TABLE 38

Compact Traverse Table

58°—122°—238°—302°

p / l	DLo / D
1	1. 887
2	3. 774
3	5. 661
4	7. 548
5	9. 435
6	11. 322
7	13. 210
8	15. 097
9	16. 984

(Section 2)

59°—121°—239°—301°

p / l	DLo / D
1	1. 942
2	3. 883
3	5. 825
4	7. 766
5	9. 708
6	11. 650
7	13. 591
8	15. 533
9	17. 474

FIGURE 1013c.—Table 38 tabulations for plane sailing formula D=l sec C.

Section 4 of table 38 (fig. 1013b) is used to find the course when the ratio DLo÷m is known. The applicable formula is:

$$\tan C = DLo \div m.$$

Following solution of course by Mercator sailing, the solution of any required distance is by the plane sailing formula:

$$D = l \sec C.$$

Section 3 of table 38 (fig. 1013c) is used to find the difference of longitude, the course and the latitudes of the points of departure and destination being known. The applicable Mercator sailing formula is:

$$DLo = m \tan C.$$

Example 1.—A vessel at lat. 32°14′.7N, long. 66°26′.9W is to proceed to lat. 36°56′.7N, long. 75°42′.2W.

Required.—(1) Course, (2) distance.

Solution.—By inspection of table 38.

Section 4 of table 38 (fig. 1013b) is used to solve the Mercator sailing formula:

$$\tan C = DLo \div m.$$

for course, the difference of longitude and difference of meridional parts (tab. 5) being known. But first the ratio DLo÷m must be calculated. The central column is entered with the ratio so found, and the appropriate course is extracted from either the left- or right-hand column depending upon whether the departure is an easting or westing and the latitude difference is a northing or southing.

L_1 32°14′.7 N	M_1 2033.3	λ_1 66°28′.9 W
L_2 36°58′.7 N	M_2 2377.0	λ_2 75°42′.2 W
l 4°44′.0 N	m 343.7	DLo 9°13′.3 W
l 284′.0 N		DLo 553′.3 W
DLo 553′.3 W	log 2.74296	
m 343.7	log (−) 2.53618	
DLo÷m 1.610	log 0.20678	
C N 59°.2 W		
(1) Cn 301°.8		

Section 2 of table 38 (fig. 1013c) is used to find the distance by plane sailing formula:

$$D = l \sec C.$$

Since the tabular solutions are for integral degrees of course, solutions for both course 301° and 302° must be extracted from the table, interpolating the results to the actual course.

	l	D (301°)	D (302°)
	200.0	388.3	377.4
	80.0	155.3	151.0
	4.0	7.8	7.5
	0.0	0.0	0.0
(2) D 539.0 mi.	284.0	551.4	535.9

Example 2.—A vessel at lat. 75°31′.7N, long 79°08′.7W steams 263.5 miles on course 155°.

Required.—(1) Latitude and (2) longitude of point of arrival.

Solution.—By inspection of table 38:

Section 1 of table 38 (fig. 1013d) is used to solve the plane sailing formula:

$$l = D \cos C$$

for difference of latitude, distance and course being known. The difference of latitude is applied to the latitude of the point of departure to find the latitude of the point of arrival. The difference of meridional parts is then derived from the two latitudes, using table 5.

D	l
200.0	181.3
60.0	54.4
3.0	2.7
0.5	0.5
263.5	238.9

L_1	75°31′.7 N	M_1	7072.4
l	3°58′.9 S		
(1) L_2	71°32′.8 N	M_2	6225.8
		m	846.6

With both course and differences of meridional parts now known, section 3 of table 38 (fig. 1013e) is used to solve the Mercator sailing formula:

$$DLo = m \tan C$$

for difference of longitude. The difference of longitude is applied to the longitude of the point of departure to find the longitude of the point of arrival.

m	DLo
800.0	373.0
40.0	18.6
6.0	2.8
0.6	0.3
846.6	394.7

λ_1	79°08′.7 W
DLo	6°34′.7 E
(2) λ_2	72°34′.0 W

If the course is near 090° or 270°, a small error in C introduces a large error in DLo. The solution for C to the nearest 0°.1 only, as by traverse table, may introduce a large error in distance if the course is near 090° or 270°.

TABLE 38

Compact Traverse Table

25°—155°—205°—335°

Course		
25° 205°	p÷l / DLo÷m	154° 334°

DLo / D	p / l	p	p / l	DLc / D	m	DLo
1	0. 906	0. 423				
2	1. 813	0. 845				
3	2. 719	1. 268				
4	3. 625	1. 690	(Section 1)			
5	4. 532	2. 113				
6	5. 438	2. 536				
7	6. 344	2. 958				
8	7. 250	3. 381				
9	8. 157	3. 804				

FIGURE 1013d.—Table 38 tabulations for plane sailing formula $l = D \cos C$.

TABLE 38

Compact Traverse Table

25°—155°—205°—335°

Course		
25° 205°	p÷l / DLo÷m	154° 334°

DLo / D	p / l	p	p / l	DLc / D	m	DLo
					1	0. 466
					2	0. 933
					3	1. 399
				(Section 3)	4	1. 865
					5	2. 332
					6	2. 798
					7	3. 264
					8	3. 730
					9	4. 197

FIGURE 1013e.—Table 38 tabulations for Mercator sailing formula $DLo = m \tan C$.

1014. Rhumb lines and great circles.—The principal advantage of a **rhumb line** is that it maintains constant true direction. A ship following the rhumb line between two places does not change true course. A rhumb line makes the same angle with all meridians it crosses and appears as a straight line on a Mercator chart. It is adequate for most purposes of navigation, bearing lines (except long ones, as those obtained by radio) and course lines both being plotted on a Mercator chart as rhumb lines, except in high latitudes. The equator and the meridians are great circles, but may be considered special cases of the rhumb line. For any other case, the difference between the rhumb line and the great circle connecting two points increases (1) as the latitude increases, (2) as the difference of latitude between the two points decreases, and (3) as the difference of longitude increases. It becomes very great for two places widely separated on the same parallel of latitude far from the equator.

A great circle is the intersection of the surface of a sphere and a plane through the center of the sphere. It is the largest circle that can be drawn on the surface of the sphere, and is the shortest distance, along the surface, between any two points on the sphere. Any two points are connected by only one great circle unless the points are antipodal (180° apart on the earth), and then an infinite number of great circles passes through them. Thus, two points on the same meridian are not joined by any great circle other than the meridian, unless the two points are antipodal. If they are the poles, *all* meridians pass through them. Every great circle bisects every other great circle. Thus, except for the equator, every great circle lies half in the Northern Hemisphere and half in the Southern Hemisphere. Any two points 180° apart on a great circle have the same latitude numerically, but contrary names, and are 180° apart in longitude. The point of greatest latitude is called the **vertex**. For each great circle there is one of these in each hemisphere, 180° apart. At these points the great circle is tangent to a parallel of latitude, and hence its direction is due east-west. On each side of these vertices the direction changes progressively until the intersection with the equator is reached, 90° away, where the great circle crosses the equator at an angle equal to the latitude of the vertex. As the great circle crosses the equator, its change in direction reverses, again approaching east-west, which it reaches at the next vertex.

On a Mercator chart a great circle appears as a sine curve extending equal distances each side of the equator. The rhumb line connecting any two points of the great circle on the same side of the equator is a chord of the curve, being a straight line nearer the equator than the great circle. Along any intersecting meridian the great circle crosses at a higher latitude than the rhumb line. If the two points are on opposite sides of the equator, the direction of curvature of the great circle relative to the rhumb line changes at the equator. The rhumb line and great circle may intersect each other, and if the points are equal distances on each side of the equator, the intersection takes place at the equator.

1015. Great-circle sailing is used when it is desired to take advantage of the shorter distance along the great circle between two points, rather than to follow the longer rhumb line. The arc of the great circle between the points is called the **great-circle track.** If it could be followed exactly, the destination would be dead ahead throughout the voyage (assuming course and heading were the same). The rhumb line *appears* the more direct route on a Mercator chart because of chart distortion. The great circle crosses meridians at higher latitudes, where the distance between them is less.

The decision as to whether or not to use great-circle sailing depends upon the conditions. The saving in distance should be worth the additional effort, and of course the great circle should not cross land, or carry the vessel into dangerous waters or excessively high latitudes. A slight departure from the great circle or a modification called composite sailing (art. 1017) may effect a considerable saving over the rhumb line track without leading the vessel into danger. If a fix indicates the vessel is a considerable distance to one side of the great circle, the more desirable practice often is to determine a new great-circle track, rather than to return to the original one.

Since the great circle is continuously changing direction as one proceeds along it, no attempt is customarily made to follow it exactly. Rather, a number of points are selected along the great circle, and rhumb lines are followed from point to point, taking advantage of the fact that for short distances a great circle and a rhumb line almost coincide.

1016. Great-circle sailing by computation.—In figure 1016a, 1 is the point of departure, 2 the destination, P the pole nearer 1, $1XV2$ the great circle through 1 and 2, V the vertex, and X any point on the great circle. The arcs $P1$, PX, PV, and $P2$ are the colatitudes of points 1, X, V, and 2, respectively. If 1 and 2 are on opposite sides of the equator, $P2$ is $90°+L_2$. The length of arc 1–2 is the great-circle distance between 1 and 2. Arcs 1–2, $P1$, and $P2$ form a spherical triangle. The angle at 1 is the initial great-circle course from 1 to 2, that at 2 the supplement of the final great-circle course (or the initial course from 2 to 1), and that at P the DLo between 1 and 2.

Great-circle sailing by computation usually involves solution for the initial great-circle course; the distance; latitude and longitude, and sometimes the distance, of the vertex; and the latitude and longitude of various points (X) on the great circle. The computation for initial course and the distance involves solution of an oblique spherical triangle, and any method of solving such a triangle can be used. If 2 is the **geographical position** (GP) of a celestial body (the point at which the body is in the zenith), this triangle is solved in celestial navigation, except that $90°-D$ (the altitude) is desired instead of D. The solution for the vertex and any point X usually involves the solution of right spherical triangles.

Although various formulas can be used, haversine formulas are considered most suitable for determining initial course and the distance, as these avoid the ambiguity that may arise through the use of trigonometric functions which do not indicate the quadrant in which the answer lies. In the formulas given below, the subscripts refer to the points indicated in figures 1016a. and 1016b. All terms without subscripts are from 1 to 2, D_v and DLO_v are from 1 to V, and D_{vx} and DLo_{vx} are from V to X. Other quantities can be computed by interchanging 1 and 2 in figure 1010a and using the same formulas. The following formulas are suitable for great-circle sailing by computation:

$$\text{hav } D=\text{hav DLo} \cos L_1 \cos L_2+\text{hav } l \tag{21}$$

which may be written

$$\text{hav } D=\text{hav } \theta+\text{hav } l \text{ (where hav } \theta=\text{hav DLo} \cos L_1 \cos L_2) \tag{22}$$

$$\text{hav } C=\sec L_1 \csc D\,[\text{hav coL}_2-\text{hav }(D\sim \text{coL}_1)] \tag{23}$$

$$\sin C=\sin \text{DLo} \cos L_2 \csc D \tag{24}$$

$$\cos L_v=\cos L_1 \sin C \tag{25}$$

$$\sec L_v=\sec L_1 \csc C \tag{26}$$

$$\sin \text{DLo}_v=\cos C \csc L_v \tag{27}$$

$$\csc \text{DLo}_v=\sec C\div\csc L_v \tag{28}$$

$$\sin D_v=\cos L_1 \sin \text{DLo}_v \tag{29}$$

$$\csc D_v=\sec L_1 \csc \text{DLo}_v \tag{30}$$

$$\tan L_x=\cos \text{DLo}_{vx} \tan L_v \tag{31}$$

$$\csc L_x=\csc L_v \sec D_{vx} \tag{32}$$

$$\csc \text{DLo}_{vx}=\csc D_{vx}\div\sec L_x \tag{33}$$

$$\sec C_x=\csc L_v \csc \text{DLo}_{vx} \tag{34}$$

$$\cos \text{DLo}_{vx}=\tan L_x \cot L_v \tag{35}$$

$$\cos D=\sin L_1 \sin L_2+\cos L_1 \cos L_2 \cos \text{DLo} \tag{36}$$

$$\tan C=\frac{\sin \text{DLo}}{\cos L_1 \tan L_2-\sin L_1 \cos \text{DLo}}. \tag{37}$$

Equation 21 is derived from equation 8b through the substitution of great-circle distance for zenith distance (D for z), latitude of destination for declination (L_2 for d), difference of latitude (l) (for ($L\sim d$), and difference of longitude for meridian angle (DLo for t).

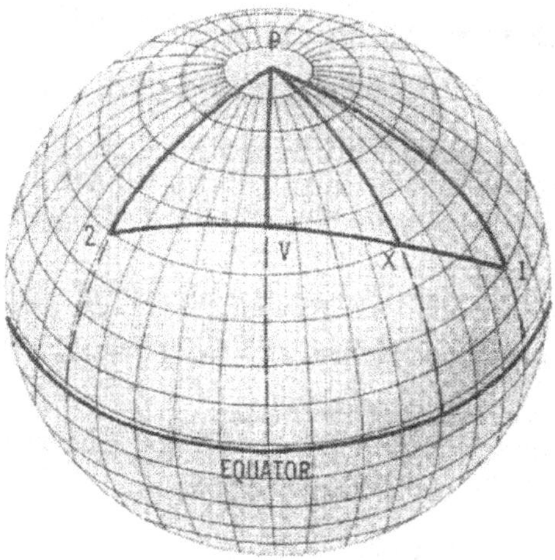

FIGURE 1016a.—The navigational triangle and great-circle sailing.

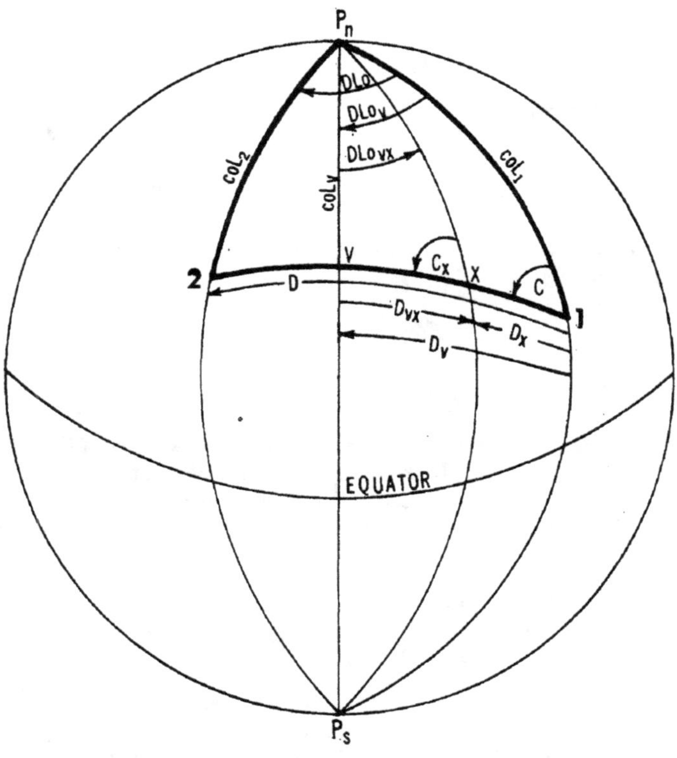

FIGURE 1016b.—The navigational triangle of great-circle sailing.

Equation 23 is derived from the cosecant formula, equation 9b:

$$\text{hav B}=[\text{hav b}-\text{hav (a}\sim\text{c)}] \text{ csc a csc c.} \tag{9b}$$

Substituting the parts of the navigational triangle of great-circle sailing in equation 9b, angle B is the initial great-circle course angle; side b is the colatitude of the destination, coL_2; side a is the great-circle distance, D; and side c is the colatitude of the point of departure, coL_1.

$$\text{hav C}=[\text{hav co}L_2-\text{hav (D}\sim\text{co}L_1)] \text{ csc D csc } (90°-L_1)$$

$$\text{hav C}=\text{sec } L_1 \text{ csc D } [\text{hav co}L_2-\text{hav (D}\sim\text{co}L_1)]. \tag{23}$$

Equation 24 is derived from the law of sines:

The angles and sides opposite of the oblique spherical triangle shown in figure 1016c have by the law of sines the following relationship:

$$\frac{\sin a}{\sin A}=\frac{\sin b}{\sin B}=\frac{\sin c}{\sin C}. \tag{3d}$$

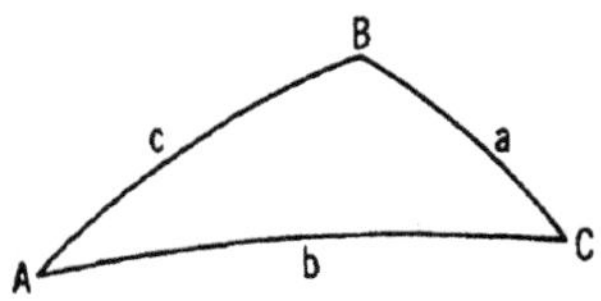

FIGURE 1016c.—Oblique spherical triangle.

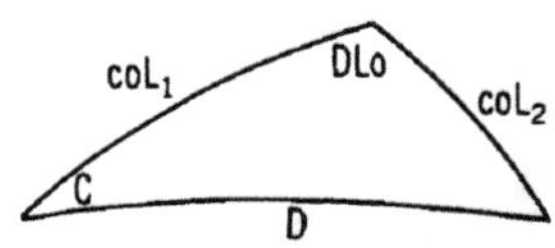

FIGURE 1016d.—Navigational triangle of great-circle sailing.

As applied to the navigational triangle of great-circle sailing shown in figure 1016d the law of sines is stated as:

$$\frac{\sin (90°-L_2)}{\sin C}=\frac{\sin D}{\sin DLo}$$

$$\sin C=\sin DLo \cos L_2 \csc D. \tag{24}$$

Equations 25 through 35 are derived by means of Napier's Rules of Circular Parts. In applying the rules to the divided navigational triangle of great-circle sailing shown in figure 1016e, the five sectors of the circular parts diagrams are first completed. Applying the rule that the sine of a middle part equals the product of the cosines of the opposite parts:

$$\sin (90°-L_v)=\cos L_1 \cos (90°-C)$$

$$\therefore \cos L_v=\cos L_1 \sin C \tag{25}$$

$$\sec L_v=\sec L_1 \csc C \tag{26}$$

$$\sin (90°-C)=\cos (90°-DLo_v) \cos (90°-L_v)$$

$$\cos C=\sin DLo_v \sin L_v$$

$$\sin DLo_v=\cos C \csc L_v \tag{27}$$

$$\csc DLo_v=\sec C\div\csc L_v \tag{28}$$

$$\sin D_v=\cos L_1 \cos (90°-DLo_v)$$

$$\sin D_v=\cos L_1 \sin DLo_v \tag{29}$$

$$\csc D_v=\sec L_1 \csc DLo_v. \tag{30}$$

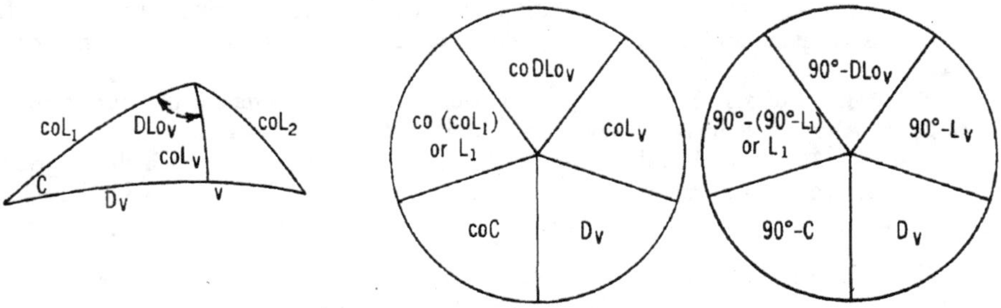

FIGURE 1016e.—Divided navigational triangle and circular parts diagrams.

In applying the rules to the divided navigational triangle shown in figure 1016f, the five sectors of the circular parts diagrams are first completed. Applying the rules that the sine of the middle part equals the product of (1) the tangents of the adjacent parts or (2) the cosines of the opposite parts:

$$\sin (90^\circ - DLo_{vx}) = \tan L_x \cot L_v$$
$$\cos DLo_{vx} = \tan L_x \cot L_v$$
$$\tan L_x = \cos DLo_{vx} \tan L_v \qquad (31)$$

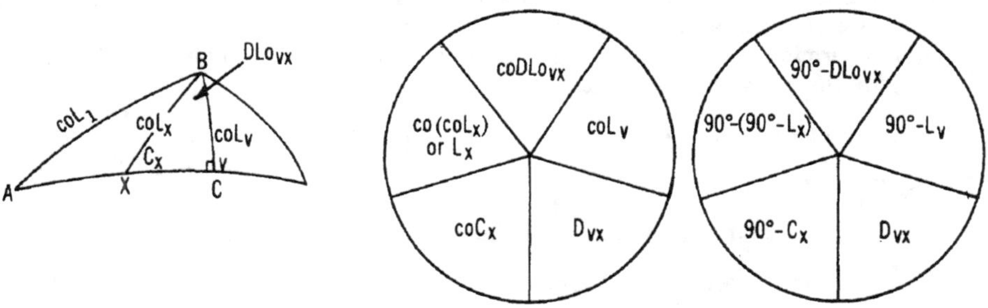

FIGURE 1016f.—Divided navigational triangle and circular parts diagrams.

$$\sin L_x = \cos (90^\circ - L_v) \cos D_{vx}$$
$$\sin L_x = \sin L_v \cos D_{vx}$$
$$\csc L_x = \csc L_v \sec D_{vx} \qquad (32)$$

$$\sin D_{vx} = \cos L_x \cos (90^\circ - DLo_{vx})$$
$$\sin D_{vx} = \cos L_x \sin DLo_{vx}$$
$$\csc DLo_{vx} = \csc D_{vx} \div \sec L_x \qquad (33)$$

$$\sin (90^\circ - C) = \cos (90^\circ - L_v) \cos (90^\circ - DLo_{vx})$$
$$\cos C_x = \sin L_v \sin DLo_{vx}$$
$$\sec C_x = \csc L_v \csc DLo_{vx} \qquad (34)$$

$$\sin (90^\circ - DLo_{vx}) = \tan L_x \tan (90^\circ - L_v)$$
$$\cos DLo_{vx} = \tan L_x \cot L_v. \qquad (35)$$

Equations 32 and 33 are useful for finding points at approximately equal distances, along the great-circle, from the vertex. If D_{vx} is greater than 90° (5400 nautical miles), L_x is of contrary name to L_v, and DLo_{vx} is greater than 90°.

Equation 24 offers a simpler solution than the haversine formula, but unless L_2 is of the same name and equal to or greater than L_1, it leaves doubt as to whether C is less or greater than 90°.

Equation 34 offers an even simpler solution but has the same limitations as equation 24. Further, it requires a knowledge of the position of the vertex.

Equation 35 is useful in determining the longitude (or DLo_{vx}) at which the great-circle crosses selected parallels of latitude. If L_x is of contrary name to L_v, DLo_{vx} is greater than 90°. The formula is also used in composite sailing.

Example.—A ship is proceeding from Manila to Los Angeles. The captain wishes to use great-circle sailing from lat. 12°45′.2N, long. 124°20′.1E, off the entrance to San Bernardino Strait, to lat. 33°48′.8N, long. 120°07′.1W, five miles south of Santa Rosa Island.

Required.—(1) The great-circle distance.

(2) The great-circle course.

(3) The latitude and longitude of the vertex.

(4) The distance from the point of departure to the vertex.

(5) The latitude and longitude of points at DLo intervals of 12° each side of the vertex.

Solution.—By equations 22 and 23.

(1) Determine difference of longitude (DLo) and difference of latitude (l). To determine DLo in this case, add the east and west longitudes and subtract from 360°; the DLo is east because the destination is eastward.

(2) Determine the great-circle distance by equation 22, and then use this distance to compute the initial great-circle course by equation 23:

$$\text{hav } D = \text{hav DLo} \cos L_1 \cos L_2 + \text{hav } l$$

which may be written

$$\text{hav } D = \text{hav } \theta + \text{hav } l \text{ (where hav } \theta = \text{hav DLo} \cos L_1 \cos L_2) \qquad (22)$$

$$\text{hav } C = \sec L_1 \csc D\, [\text{hav co}L_2 - \text{hav } (D \sim \text{co}L_1)]. \qquad (23)$$

Distance

(3) To obtain l hav θ of equation 22, add l hav DLo as extracted from table 34 to l cos L_1 and l cos L_2 as extracted from table 33.

(4) From table 34 extract the value of the natural haversine (n hav) of θ corresponding to the l hav θ (the value of θ need not be found).

(5) Add the natural haversines of θ and l (as extracted from table 34) to obtain the natural haversine corresponding to the distance.

(6) From table 34 extract the distance in arc units corresponding to the n hav value of 0.61331.

λ_1	124°20′.1 E		
λ_2	120°07′.1 W	*Add*	
DLo	115°32′.8 E	l hav 9.85468	
L_1	12°45′.2 N	l cos 9.98915	
L_2	33°48′.8 N	l cos 9.91953	*Add*
θ		l hav 9.76336	n hav 0.57991
l	21°03′.6 N		n hav 0.03340
(1) D	103°05′.9		n hav 0.61331

Course

(7) Determine the values of coL_2 and $(D{\sim}coL_1)$ to be used in calculating the initial great-circle course angle by equation 23. CoL_2 is $(90°-L_2)$ if L_1 and L_2 are of the same name, and $(90°+L_2)$ if of contrary name. $D{\sim}coL_1$ is always the numerical difference between D and coL_1.

(8) From table 33 extract values of *l* sec L_1 and *l* csc D. These logarithms are to be added to the logarithm of the difference of the natural haversines of coL_2 and $(D{\sim}coL_1)$.

(9) From table 34 extract the values of the natural haversines of coL_2 and $(D{\sim}coL_1)$; subtract *n* hav $(D{\sim}coL_1)$ from *n* hav coL_2.

(10) From table 34 extract the value of the log haversine corresponding to *n* hav of coL_2 minus $(D{\sim}coL_1)$.

(11) Add *l* sec L_1, *l* csc D, and *l*[hav coL_2 − hav$(D{\sim}coL_1)$].

(12) For the *l* hav value found in (11), extract the corresponding degrees of arc as the initial great-circle course angle.

		Subtract			*Add*
L_1	12°45′.2N				*l* sec 0.01085
D	103°05′.9				*l* csc 0.01145
coL_2	56°11′.2	*n* hav	0.22175		
$D{\sim}coL_1$	25°51′.1	*n* hav	(−)0.05004		
		n hav	0.17171		*l* hav 9.23480
(2) Cn	050°.3			C N50°19′.3E	*l* hav 9.25710

The course and distance solutions are shown in a more compact format below.

λ_1	124°20′.1E			
λ_2	120°07′.1W			
DLo	115°32′.8E	*l* hav 9.85468		
L_1	12°45′.2N	*l* cos 9.98915		*l* sec 0.01085
L_2	33°48′.8N	*l* cos 9.91953		
θ		*l* hav 9.76336	*n* hav 0.57991	
l	21°03′.6N		*n* hav 0.03340	
D	103°05′.9		*n* hav 0.61331	*l* csc 0.01145
coL_2	56°11′.2	*n* hav 0.22175		
$D{\sim}coL_1$	25°51′.1	*n* hav (−)0.05004		
		n hav 0.17171		*l* hav 9.23480
(2) Cn	050′.3		C N50°19′.3E	*l* hav 9.25710
(1) D	6185.9 mi.			

The Vertex

(13) The latitude of the vertex, L_v, is always numerically equal to or greater than L_1 or L_2. If C is less than 90°, the nearer vertex is toward L_2; but if C is greater than 90°, the nearer vertex is in the opposite direction. The vertex nearer L_1 has the same name as L_1.

(14) Using equation 25 to determine L_v, add the *l* sin C to the *l* cos L_1 to obtain *l* cos L_v.

$$\cos L_v = \cos L_1 \sin C \qquad (25)$$

L_1	12°45′.2N	*l* cos 9.98915
C	N50°19′.3E	*l* sin 9.88629
L_v	41°21′.2N	*l* cos 9.87544

(15) Using equation 27 to determine the difference of longitude of λ_1 and the vertex (DLo_v), add l cos C to l csc L_v to obtain l sin DLo_v.

$$\sin DLo_v = \cos C \csc L_v \qquad (27)$$

C	N50°19′.3E		l cos	9.80514
L_v	41°21′.2N		l csc	0.17999
λ_v	160°34′.4W	DLo_v 75°05′.5E	l sin	9.98513

(16) Using equation 29 to determine the distance to the vertex (D_v), add l cos L_1 to l sin DLo_v to obtain l sin D_v.

$$\sin D_v = \cos L_1 \sin DLo_v \qquad (29)$$

L_1	12°45′.2N			l cos 9.98915
λ_v	160°34′.4W	DLo_v 75°05′.5E	l sin 9.98513	l sin 9.98513
D_v	70°28′.5			l sin 9.97428
D_v	4228.5 mi.			

The vertex solution is shown in a more compact form below.

	L_1	12°45′.2N	l cos 9.98915		l cos 9.98915
	C	N50°19′.3E	l sin 9.88629	l cos 9.80514	
(3)	L_v	41°21′.2N	l cos 9.87544	l csc 0.17999	
	λ_v	160°34′.4W	DLo_v 75°05′.5E	l sin 9.98513	l sin 9.98513
	D_v	70°28′.5			l sin 9.97428
(4)	D_v	4228.5 mi.			

Points on the Great-Circle Track

(17) DLo_v and D_v of the nearer vertex are never greater than 90°. However, when L_1 and L_2 are of contrary name, the other vertex, 180° away, may be the better one to use in the solution for points on the great-circle track if it is nearer the mid point of the track.

(18) The latitudes of points on the great-circle track can be determined for equal DLo intervals each side of the vertex using equation 31.

(19) Using equation 31, the l cos DLo_{vx} is added to the l tan L_v to obtain l tan L_x:

$$\tan L_x = \cos DLo_{vx} \tan L_v \qquad (31)$$

DLo_{vx}	12°00′.0	l cos 9.99040	DLo_{vx}	24°00′.0	l cos 9.96073
L_v	41°21′.2N	l tan 9.94457	L_v	41°21′.2N	l tan 9.94457
L_x	40°43′.6N	l tan 9.93497	L_x	38°48′.1N	l tan 9.90530
λ_v	160°34′.4W		λ_v	160°34′.4W	
λ_x	172°34′.4W	(12° W of λ_v)	λ_x	175°25′.6E	(24° W of λ_v)
λ_x	148°34′.4W	(12° E of λ_v)	λ_x	136°34′.4W	(24° E of λ_v)

(20) L_x has the same name as L_v if DLo_{vx} is less than 90°; if DLo_{vx} is greater than 90°, L_x is of contrary name. The great-circle is a symmetrical curve about the vertex. Hence, any given DLo can be applied to λ_v in both directions (E and W) to find two points having the same latitude. For example, one point on the great circle 12° W of the vertex is at lat. 40°43′.6N, long. 172°34′.4W. A second point 12° E of the vertex is at lat. 40°43′.6N, long. 148°34′.4W.

The solution for points on the great-circle track is shown in a compact form below.

DLo_{vx}	12°00′.0	24°00′.0	36°00′.0	48°00′.0	60°00′.0	72°00′.0
l cos DLo_{vx}	9.99040	9.96073	9.90796	9.82551	9.69897	9.48998
l tan L_v	9.94457	9.94457	9.94457	9.94457	9.94457	9.94457
l tan L_x	9.93497	9.90530	9.85253	9.77008	9.64354	9.43455
(5) L_x	40°43′.6 N	38°48′.1 N	35°27′.2 N	30°29′.8 N	23°45′.2 N	15°12′.9 N
λ_x	172°34′.4 W	175°25′.6 E	163°25′.6 E	151°25′.6 E	139°25′.6 E	127°25′.6 E
λ_x	148°34′.4 W	136°34′.4 W	124°34′.4 W	—	—	—

Alternative Solution for Points on Track

(21) The method of selecting the longitude (or DLo_{vx}) and determining the latitude at which the great-circle crosses the selected meridian provides shorter legs in higher latitudes and longer legs in lower latitudes. Points at desired distances or desired equal intervals of distance on the great circle from the vertex can be determined using equations 32 and 33.

(22) Using equation 32 to determine the latitudes of points at desired great-circle distances from the vertex (300 and 600 miles), l csc L_v is added to l sec D_{vx} (distance from vertex to point on great circle) to obtain l csc L_x:

$$\csc L_x = \csc L_v \sec D_{vx} \tag{32}$$

L_v	41°21′.2N	l csc	10.18000	L_v	41°21′.2N	l csc	10.18000
D_{vx}	5° (300 miles)	l sec	10.00166	D_{vx}	10° (600 miles)	l sec	10.00665
L_x	41°09′.7N	l csc	10.18164	L_x	40°35′.5N	l csc	10.18665

(23) Using equation 33 to determine the longitudes of points at the desired great-circle distances from the vertex, l sec L_x is subtracted from l csc D_{vx} to obtain l csc DLo_{vx}:

$$\csc DLo_{vx} = \csc D_{vx} \div \sec L_x \tag{33}$$

D_{vx}	5° (300 miles)	l csc	11.05970	D_{vx}	10° (600 miles)	l csc	10.76033
L_x	41°09′.7N	l sec (—)	10.12329	L_x	40°35′.5N	l sec (—)	10.11955
DLo_{vx}	6°38′.9	l csc	10.93641	DLo_{vx}	13°13′.1	l csc	10.64078
λ_v	160°34′.4W			λ_v	160°34′.4W		
λ_x	167°13′.3W (west of vertex)			λ_x	173°47′.5W (west of vertex)		
λ_x	153°55′.5W (east of vertex)			λ_x	147°21′.3W (east of vertex)		

Course and Distance by Sight Reduction Formulas

Equation 36 is derived from the altitude formula, equation 2a:

$$\sin h = \sin L \sin d + \cos L \cos d \cos LHA. \qquad (2a)$$

Since altitude (h) equals 90° minus zenith distance (z),

$$\sin (90° - z) = \sin L \sin d + \cos L \cos d \cos LHA.$$

The terrestrial counterpart of the celestial triangle is formed by arcs of two meridians and the great circle connecting two places on the earth, one on each meridian. In the great-circle sailing application of the terrestrial triangle, the vertex at the geographic pole corresponds to the elevated pole, the vertex at the point of departure on one meridian corresponds to the zenith of the assumed position (AP); the vertex at the destination on the second meridian corresponds to the celestial body; the great circle through the points of departure and destination corresponds to the vertical circle through the celestial body; the difference of longitude of the two meridians corresponds to local hour angle; and the great-circle distance (D) corresponds to the zenith distance (z).

Substituting great-circle distance for zenith distance (D for z), latitude of the point of departure for latitude of the assumed position (L_1 for L), latitude of the destination for the declination (L_2 for d); and difference of longitude for local hour angle (DLo for LHA) in the above equation:

$$\sin (90° - D) = \sin L_1 \sin L_2 + \cos L_1 \cos L_2 \cos DLo.$$

Since sin (90° − D) equals cos D (art. 140),

$$\cos D = \sin L_1 \sin L_2 + \cos L_1 \cos L_2 \cos DLo. \qquad (36)$$

Equation 37 is derived from the azimuth angle formula, equation 4a:

$$\tan Z = \frac{\sin LHA}{\cos L \tan d - \sin L \cos LHA}. \qquad (4a)$$

In the terrestrial triangle of great-circle sailing, initial great-circle course angle (C) corresponds to azimuth angle (Z). Substituting as in the distance formula, equation 4a becomes:

$$\tan C = \frac{\sin DLo}{\cos L_1 \tan L_2 - \sin L_1 \cos DLo}. \qquad (37)$$

The sign convention used in the calculations of the great-circle distance and course angle by equations 36 and 37, respectively, is that the latitude of the destination is treated as a negative quantity when latitudes of departure and destination are of contrary name.

If the course angle as calculated by equation 37 is negative, it is necessary to add 180° to obtain the desired course angle.

1017. Composite sailing.—When the great circle would carry a vessel to a higher latitude than desired, a modification of great-circle sailing, called **composite sailing,** may be used to good advantage. The composite track consists of a great circle from the point of departure and tangent to the limiting parallel, a course line along the parallel, and a great circle tangent to the limiting parallel and through the destination.

Solution of composite sailing problems is most easily made by means of a great-circle chart. Lines from the point of departure and the destination are drawn tangent to the limiting parallel. The coordinates of various selected points along the composite track are then measured and transferred to a Mercator chart.

Composite sailing problems can also be solved by computation. For this purpose equation 35 is used:

$$\cos DLo_{vx} = \tan L_x \cot L_v. \tag{35}$$

In the computation, the point of departure and the destination are used successively as point *X*.

Example.—A ship leaves Baltimore, bound for Bordeaux (Royan), France. The captain desires to use composite sailing from lat. 36°57′.7N, long. 75°42′.2W, near Chesapeake Light, to lat. 45°39′.1N, long. 1°29′.8W, near the entrance to Grande Passe de l'Ouest, limiting the maximum latitude to 47°N.

Required.—(1) The longitude at which the limiting parallel is reached. (2) The longitude at which the limiting parallel should be left.

Solution.—

L_1 36°57′.7N	*l* tan 9.87651	
L_2 45°39′.1N		*l* tan 0.00988
L_v 47°00′.0N	*l* cot 9.96966	*l* cot 9.96966
DLo_{v1} 45°26′.1E	*l* cos 9.84617	
DLo_{v2} 17°27′.0W		*l* cos 9.97954
(1) λ_{v1} 30°16′.1W		
(2) λ_{v2} 18°56′.8W		

Composite sailing applies only when the vertex lies between the point of departure and the destination.

The remainder of the problem is one of solving the two great circles by great-circle sailing and the east-west portion by parallel sailing. Since both great circles have vertices at the same parallel, computation for C, D, and DLo_{vx} can be made by considering them parts of the *same* great circle with L_1, L_2, and L_v as given and $DLo = DLo_{v1} + DLo_{v2}$. The total distance is the sum of the great-circle and parallel distances. In finding λ_x be careful to apply DLo_{vx} to the correct vertex and in the correct direction.

1018. Great-circle sailing by table 35.—The divided navigational triangle of great-circle sailing is shown in figure 1018a. The parts of the triangle and special constructions are labeled as follows:

L_1, the latitude of the point of departure.

L_2, the latitude of the destination.

λ_1, the longitude of the point of departure.

λ_2, the longitude of the destination.

R, perpendicular let fall from the point of destination upon the meridian of the point of departure. This is an auxiliary part.

X, intersection of R with meridian through point of departure.

K, arc X to equator. This is an auxiliary part introduced to facilitate solution.

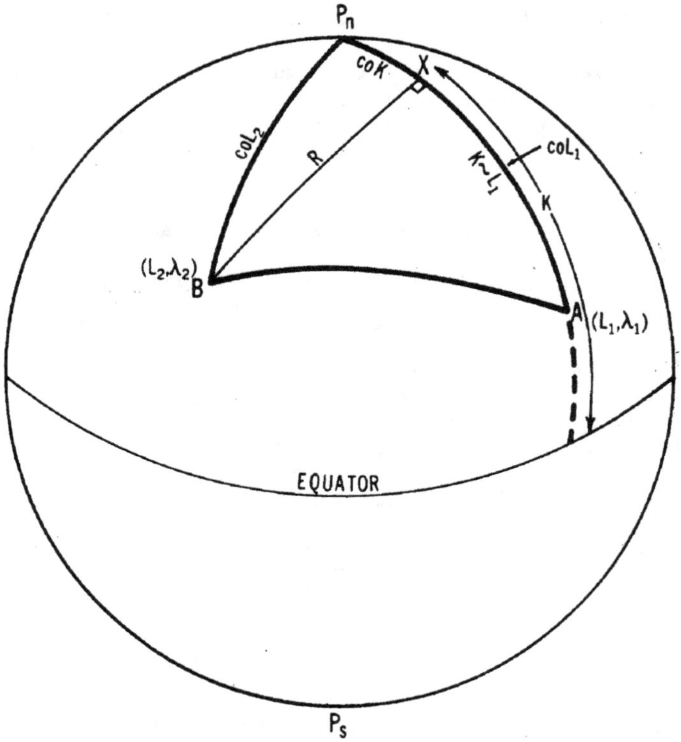

FIGURE 1018a—The divided navigational triangle of great-circle sailing.

The navigational triangle without auxiliary parts K and R is shown in figure 1012b. In this figure, the parts not identified with figure 1018a are as follows:

C, the initial great-circle course angle.
C_x, the great-circle course angle at point X on the great-circle track.
D, great-circle distance.
D_v, distance to vertex from point of departure.
D_{vx}, distance from vertex to point X on great-circle track.
DLo_v, difference of longitude between point of departure and vertex.
DLo_{vx}, difference of longitudé between the vertex and point X.
L_x, the latitude of point X.
coL_x, the colatitude of point X.
λ_x, the longitude of point X.

Formulas

With DLo substituted for t, latitude of destination L_2 substituted for declination, C substituted for Z, and D substituted for z, equations 12a–12d can be stated as:

$$\csc R = \csc DLo \sec L_2 \tag{38}$$
$$\csc K = \csc L_2 \div \sec R \tag{39}$$
$$\sec D = \sec R \sec (K \sim L_1) \tag{40}$$
$$\csc C = \csc R \div \csc D. \tag{41}$$

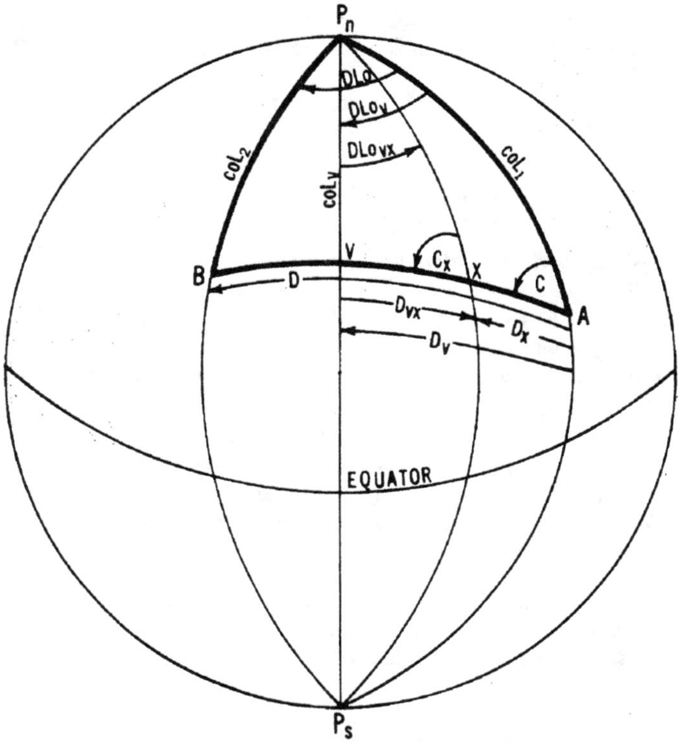

FIGURE 1018b.—The divided navigational triangle of great-circle sailing.

Method of Solution by Table 35

The letters A and B indicate from which column to take their corresponding logarithmic functions multiplied by 100,000.

The following steps are taken to obtain the initial great-circle course angle and the great-circle distance in one combined solution:

1. Enter the table with DLo and take from the A column the nearest tabulated number. Enter the table with L_2 and take the tabulated number from the B column. Add these numbers to obtain the A function of R.

2. Look up the number thus obtained in the A column and take from the B column the opposite tabulated number. Subtract this from the function of L_2 taken from the A column.

3. Enter the A column with the number thus obtained and take out the nearest tabulated value of K, the degrees from the top or bottom of the page and the minutes from the side.

4. Give K the same name as L_2. Combine K with L_1 to obtain $(K{\sim}L_1)$, adding K and L_1 if different names and subtracting the smaller from the larger if same name.

5. Enter the table with $(K{\sim}L_1)$ and take from the B column the nearest tabulated number. Add this to the B function of R.

6. With the number thus obtained, enter the B column and take distance from the table.

7. To compute the initial great-circle course angle, look up the distance in the A column and subtract the number thus obtained from the A function of R which has already been determined.

8. With this number, enter the A column and take out Z (course angle).

Special Rules

Because the great-circle distance D may be greater than 90° (5400 miles), the following special rules must be followed:

Latitude of Departure and Destination are Same Name

(1) When DLo is greater than 90°, take K from the bottom of the table.

(2) Take C from the top of the table when K is greater than L_1. Take C from the bottom of the table when K is less than L_1.

(3) Take distance D from the top of table except when DLo and $(K \sim L_1)$ are *both* greater than 90°.

Latitude of Departure and Destination are Contrary Name

(4) When DLo is less than 90°, take K from the top of the table. When DLo is greater than 90°, take K from the bottom of the table.

(5) Take C from the bottom of the table except when $(K \sim L_1)$ is greater than 180°.

(6) Take D from the bottom of the table except when DLo and $(K \sim L_1)$ are both less than 90°.

(7) When $(K \sim L_1)$ exceeds 180°, subtract 180° before entering the table.

Example.—A ship is proceeding from San Francisco to Sydney. The navigator desires to use great-circle sailing from lat. 37°47′.5N, long. 122°27′.8W to lat. 33°51′.7S, long. 151°12′.7E.

Required.—

(1) The great-circle distance.
(2) The initial great-circle course.
(3) The latitude and longitude of the vertex.
(4) The distance of the vertex.
(5) The latitude and longitude of points on the track 6° (360 miles) apart.

Solution.—(1) Follow the steps given in *Method of Solution by Table 35* to obtain the great-circle distance and initial great-circle course angle in one combined solution using the following equations:

$$\csc R = \csc DLo \sec L_2 \quad (38)$$
$$\csc K = \csc L_2 \div \sec R \quad (39)$$
$$\sec D = \sec R \sec (K \sim L_1) \quad (40)$$
$$\csc C = \csc R \div \csc D. \quad (41)$$

λ_2	151°12′.7E				
λ_1	122°27′.8W				
	273°40′.5	*Add*	*Subtract*	*Add*	*Subtract*
DLo	86°19′.5W	A 89.4			
L_2	33°51′.7S	B 8070	A 25403		
R		A 8159.4	B 25206	B 25206	A 8159.4
K	84°33′.0S		A 197		
L_1	37°47′.5N				
$K \sim L_1$	122°20′.5			B 27167	
D: arc	107°25′.5			B 52373	A 2040
	6420+25.5			C N119°42′.5W	A 6119.4
(1) D	6445.5 miles			(2) Cn 240°17′.5	

(2) The latitude and longitude of the vertex are computed using equations 26 and 28:

$$\sec L_v = \sec L_1 \csc C \tag{26}$$

$$\csc DLo_v = \sec C \div \csc L_v. \tag{28}$$

The distance to the vertex is computed using equation 30,

$$\csc D_v = \sec L_1 \csc DLo_v. \tag{30}$$

Practical Rules

If C is greater than 90°, L_v is opposite in name from L_1, and DLo_v and D_v are both taken from the bottom of the table.

		Add	*Subtract*	*Add*
L_1	37°47′.5N	B 10224		B 10224
C	N119°42′.5W	A 6119.4	B 30488	
(3) L_v	46°39′.5S	B 16343.4	A 13830	
DLo_v	137°02′.5W		A 16658	A 16658
λ_1	122°27′.8W	(4) D_v 147°25′		A 26882
	259°30′.3W			
(3) λ_v	100°29′.7E			

Since C is greater than 90°, L_x is opposite in name from L_1, and DLo_v and D_v are both taken from the bottom of the table. Note that the vertex does not fall between the point of departure and destination, but is past the destination.

(3) The latitude and longitude of any point X on the great-circle track are computed using equations 32 and 33.

$$\csc L_x = \csc L_v \sec D_{vx} \tag{32}$$

$$\csc DLo_{vx} = \csc D_{vx} \div \sec L_x. \tag{33}$$

The course angle at any point X is computed using equation 34:

$$\sec C_x = \csc L_v \csc DLo_{vx}. \tag{34}$$

Practical Rules

If D_{vx} is greater than 90°, L_x is opposite in name from L_v, and DLo_{vx} and C_x are both taken from the bottom of the table. C_x is measured from the elevated pole of point X.

To solve for points on the great-circle track at 6° intervals from the vertex using equations 32 and 33:

D_v	147°25′.0
D_x	6°00′.0
D_{vx}	141°25′.0

		Add	*Subtract*	*Add*
L_v	S	A 13830		A 13830
D_{vx}	141°25′.0	B 10696	A 20506	
(5) L_x	34°39′.0N	A 24526	B 8479	
DLo_{vx}	130°42′.0E		A 12027	A 12027
λ_v	259°30′.3W	C_x N123°27′.5W		B 25857
(5) λ_x	128°48′.3W			

Note that the value L_v is not set down. Its function is taken directly from the vertex solution. Since D_{vx} is greater than 90°, L_x is north and DLo_{vx} and C_x are taken from the bottom of table.

To facilitate computation of the latitudes and longitudes of points along the great-circle track, equations 32 and 33 may be placed in a special column form. Computation of various points for this problem is illustrated below.

D_v		147°25′	147°25′	147°25′	147°25′	147°25′	147°25′
D_x		6°00′	12°00′	18°00′	36°00′	54°00′	60°00′
D_{vx}		141°25′	135°25′	129°25′	111°25′	93°25′	87°25′
L_vS	*Add*	A 13830	13830	13830	13830	13830	13830
D_{vx}		B 10696	14738	19726	43753	122478	134609
L_x		A 24526	28568	33556	57583	136308	148439
L_x		34°39′N	31°12′N	27°30′N	15°24′N	2°29′N	1°52′.5S
D_{vx}	*Sub.*	A 20506	15370	11207	3107	77. 3	44. 2
L_x		B 8479	6785	5207	1588	40. 8	23. 3
DLo_{vx}		A 12027	8585	6000	1519	36. 5	20. 9
DLo_{vx}		130°42′.0E	124°51′.0E	119°26′.0E	105°04′.0E	92°21′.0E	88°13′.0E
λ_vW		259°30′.3	259°30′.3	259°30′.3	259°30′.3	259°30′.3	259°30′.3
λ_xW		128°48′.3	134°39′.3	140°04′.3	154°26′.3	167°09′.3	171°17′.3

Note that between the fifth and sixth points L_x changes from North to South and that DLo_{vx} is taken from the top of the table instead of the bottom. It will be observed that D_{vx} has become less than 90° between the fifth and sixth points.

As has been previously stated, the vertex in this particular problem lies beyond the destination. When the vertex lies between the point of departure and destination, there are two points on the great-circle track which have the same latitude. In such cases one computation obtains two points on the curve, the latitude being the same for both points, but DLo_{vx} is east in one case and west in the other, two different longitudes being thus obtained by applying DLo_{vx} to λ_v.

Problems

1003a. A vessel steams 117.3 miles on course 214°.

Required.—(1) Difference of latitude, (2) departure, by plane sailing.

Answers.—(1) *l* 97′.2S, (2) p 65.6 mi. W.

1003b. A steamer is bound for a port 173.3 miles south and 98.6 miles east of the vessel's position.

Required.—(1) Course, (2) distance, by plane sailing.

Answers.—(1) C 150°.4; (2) D 199.4 mi. by computation, 199.3 mi. by traverse table.

1004a. A ship steams as follows: course 359°, distance 28.8 miles; course 006°, distance 16.4 miles; course 266°, distance 4.9 miles; course 144°, distance 3.1 miles; course 333°, distance 35.8 miles; course 280°, distance 19.3 miles.

Required.—(1) Course, (2) distance, by traverse sailing.

Answers.—(1) C 334°.4, (2) D 86.1 mi.

1005a. The 1530 DR position of a ship is lat. 44°36′.3N, long. 31°18′.3W. The ship is on course 270°, speed 17 knots.

Required.—The 2000 DR position, by parallel sailing.

Answer.—2000 DR: L 44°36′.3N, λ 33°05′.7W.

1005b. The captain of the ship of problem 1005a desires to change course when the ship arrives at long. 38°00′.0W.

Required.—Estimated time of arrival (ETA) at the turning point, by parallel sailing.

Answer.—ETA 0819 the following day.

1006a. A vessel steams 263.3 miles on course 340° from lat. 16°32'.2S, long. 1°04'.4E.

Required.—(1) Latitude and (2) longitude of the point of arrival, by middle-latitude sailing.

Answers.—By computation or traverse table: (1) L 12°24'.8S, (2) λ 0°28'.6W.

1006b. A vessel leaves lat. 45°00'.0N, long. 150°00'.0W and arrives at lat. 38°18'.7N, long. 137°14'.6W.

Required.—(1) Course made good, (2) distance made good, by middle-latitude sailing.

Answers.—(1) C 125°.1; (2) D 698.6 mi. by computation, 697.9 mi. by traverse table.

1006c. A vessel is at lat. 1°08'.3S, long. 175°24'.5E. It steams at 13.5 knots on course 075° for 22^{h}15^m. Twenty-four hours after that it is at lat. 0°06'.6S, long. 174°20'.0W.

Required.—(1) Latitude and (2) longitude of the point of arrival after 22^{h}15^m; (3) course and (4) distance made good during the last 24 hours of steaming; and (5) course and (6) distance made good over entire period, by middle-latitude sailing.

Answers.—(1) L 0°09'.5N by computation, 0°09'.4N by traverse table; (2) λ179°45'.3W; (3) C 092°.8; (4) D 325.7 mi. by computation, 338.3 mi. by traverse table; (5) C 084°.3; (6) D 618.5 mi. by computation, 625.6 mi. by traverse table.

1007a. A ship at lat. 33°53'.3S, long. 18°23'.1E, leaving Cape Town, heads for destination near Ambrose Light, lat. 40°27'.1N, long. 73°49'.4W.

Required.—(1) Course and (2) distance, by Mercator sailing.

Answers.—(1) C 310°.9; (2) D 6,811.5 mi. by computation, 6,812.8 mi. by traverse table. Compare these answers with those of problem 1010, the great-circle sailing solution between the same points.

1007b. A ship at lat. 15°03'.7N, long. 151°26'.8E steams 57.4 miles on course 035°.

Required.—(1) Latitude and (2) longitude of the point of arrival, by Mercator sailing.

Answers.—(1) L 15°50'.7N, λ 152°00'.7E.

1016. A ship leaves Cape Town bound for New York City. The captain decides to use great-circle sailing from lat. 33°53'.3S, long. 18°23'.1E (near Green Point Light) to near Ambrose Light, lat. 40°27'.1N, long. 73°49'.4W.

Required.—(1) The initial great-circle course.

(2) The great-circle distance.

(3) The latitude and longitude of both vertices.

(4) The distance from the point of departure to each vertex.

(5) The latitude and longitude of points on the great circle at longitude 15°E and at each 5° of longitude thereafter to longitude 70°W.

Answers.—(1) C 304°.5. (2) D 6,762.7 mi. (3) L_v46°49'.4S, λ_v69°18'.8E; L_v46°49'.4 N, λ_v 110°41'.2W. (4)D 2,407.5 mi. to eastern hemisphere vertex, 8,392.5 mi. to western hemisphere vertex. (5) L_1 33°53'.3S, λ_1 18°23'.1E (point of departure); L_x 31°52'.2S, λ_x 15°00'.0E; L_x 28°32'.5S, λ_x 10°00'.0E; L_x 24°47'.7S, λ_x 5°00'.0E; L_x 20°37'.8S, λ_x 0°00'.0; L_x 16°04'.5S, λ_x 5°00'.0W; L_x 11°10'.8S, λ_x10°00'.0W; L_x 6°01'.7S, λ_x 15°00'.0 W; L_x0°43'.9S, λ_x20°00'.0W; L_x 4°35'.0N, λ_x 25°00'.0W; L_x 9°47'.2N, λ_x 30°00'.0W; L_x 14°45'.7N, λ_x 35°00'.0W; L_x 19°25'.0N, λ_x 40°00'.0W; L_x 23°41'.5N, λ_x 45°00'.0W; L_x 27°33'.3N, λ_x 50°00'.0W; L_x 30°59'.8N, λ_x 55°00'.0W; L_x 34°01'.7N, λ_x 60°00'.0W; L_x 36°40'.1N, λ_x 65°00'.0W; L_x 38°56'.6N, λ_x 70°00'.0W; L_2 40°27'.1N, λ_x 73°49'.4W (destination).

1017. A ship is to steam from Valparaiso, Chile to Wellington, New Zealand. The captain wishes to use composite sailing from lat. 32°58'.0S, long. 71°41'.2W, off Punta Angeles Light, to lat. 42°00'.0S, long. 175°00'.0E, near Cape Palliser, limiting the maximum latitude to 50°S.

Required.—(1) The longitude at which the limiting parallel is reached.

(2) The longitude at which the limiting parallel should be left.

(3) The initial great-circle course.

(4) The total distance.

(5) The latitude and longitude of points along the great circles at intervals of 10° of DLo from the vertices.

Answers.—(1) λ_{v1} 128°42′.9W. (2) λ_{v2} 144°04′.3W. (3) C 230°.0. (4) D 5,024.6 mi. (5) L_1 32°58′.0S, λ_1 71°41′.2W (point of departure); L_x 37°27′.2S, λ_x 78°42′.9W; L_x 42°23′.6S, λ_x 88°42′.9W; L_x 45°54′.3S, λ_x 98°42′.9W; L_x 48°14′.2S, λ_x 108°42′.9W; L_x 49°34′.1S, λ_x 118°42′.9W; L_{v1} 50°00′.0S, λ_{v1} 128°42′.9W (first vertex); L_{v2} 50°00′.0S, λ_{v2} 144°04′.3W (second vertex); L_x 49°34′.1S, λ_x 154°04′.3W; L_x 48°14′.2S, λ_x 164°04′.3 W; L_x 45°54′.3S, λ_x 174°04′.3W; L_x 42°23′.6S, λ_x 175°55′.7E; L_2 42°00′.0S, λ_2 175°00′.0E (destination).

CHAPTER XI

TIDE AND CURRENT PREDICTIONS

1101. Tidal effects.—The daily rise and fall of the **tide,** with its attendant flood and ebb of **tidal current,** is familiar to every mariner. He is aware, also, that at **high water** and **low water** the depth of water is momentarily constant, a condition called **stand.** Similarly, there is a moment of **slack water** as a tidal current reverses direction. As a general rule, the *change* in height or the current *speed* is at first very slow, increasing to a maximum about midway between the two extremes, and then decreasing again. If plotted against time, the height of tide or speed of a tidal current takes the general form of a sine curve. Sample curves, and more complete information about causes, types, and features of tides and tidal currents, are given in volume I. Ocean (nontidal) currents are also discussed in volume I. The present chapter is concerned primarily with the application of tides and currents to piloting, and predicting the tidal conditions that might be encountered at any given time.

Although tides and tidal currents are caused by the same phenomena, the time relationship between them varies considerably from place to place. For instance, if an estuary has a wide entrance and does not extend far inland, the time of maximum speed of current occurs at about the mid time between high water and low water. However, if an extensive tidal basin is connected to the sea by a small opening, the maximum current may occur at about the time of high water or low water outside the basin, when the difference in height is maximum.

The *height of tide* should not be confused with *depth of water*. For reckoning tides a reference level is selected. Soundings shown on the largest scale charts are the vertical distances from this level to the bottom. At any time the actual depth is this charted depth *plus* the height of tide. In most places the reference level is some form of low water. But all low waters at a place are not the same height, and the selected reference level is seldom the *lowest* tide that occurs at the place. When lower tides occur, these are indicated by a negative sign. Thus, at a spot where the charted depth is 15 feet, the actual depth is 15 feet plus height of tide. When the tide is three feet, the depth is $15+3=18$ feet. When it is (−) 1 foot, the depth is $15-1=14$ feet. It is well to remember that *the actual depth can be less than the charted depth.* In an area where there is a considerable **range of tide** (the difference between high water and low water), the height of tide might be an important consideration in using soundings to assist in determining position, or whether the vessel is in safe water.

One should remember that heights given in the tide tables are *predictions*, and that when conditions vary considerably from those used in making the predictions, the heights shown may be considerably in error. Heights lower than predicted are particularly to be anticipated when the atmospheric pressure is higher than normal, or when there is a persistent strong offshore wind. Along coasts where there is a large inequality between the two high or two low tides during a tidal day the height predictions are less reliable than elsewhere.

The current encountered in pilot waters is due primarily to tidal action, but other causes are sometimes present. The tidal current tables give the best prediction of total current, regardless of cause. The predictions for a river may be considerably in error following heavy rains or a drought. The effect of current is to alter the course and speed made good over the bottom. Due to the configuration of land (or shoal areas) and water, the set and drift may vary considerably over different parts of a harbor. Since this is generally an area in which small errors in position of a vessel are of considerable importance to its safety, a knowledge of predicted currents can be critical, particularly if the visibility is reduced by fog, snow, etc. If the vessel is proceeding at reduced speed, the effect of current with respect to distance traveled is greater than normal. Strong currents are particularly to be anticipated in narrow passages connecting larger bodies of water. Currents of more than five knots are encountered from time to time in the Golden Gate at San Francisco. Currents of more than 13 knots sometimes occur at Seymour Narrows, British Columbia.

In straight portions of rivers and channels the strongest currents usually occur in the middle, but in curved portions the swiftest currents (and deepest water) usually occur near the outer edge of the curve. Countercurrents and eddies may occur on either side of the main current of a river or narrow passage, especially near obstructions and in bights.

In general, the range of tide and the speed of tidal current are at a minimum upon the open ocean or along straight coasts. The greatest tidal effects are usually encountered in rivers, bays, harbors, inlets, bights, etc. A vessel proceeding along a coast can be expected to encounter stronger sets toward or away from the shore while passing an indentation than when the coast is straight.

1102. Predictions of tides and currents to be expected at various places are published annually by the National Ocean Survey. These are supplemented by twelve sets of **Tidal Current Charts,** each set consisting of 12 charts, one for each hour of the tidal cycle. On these charts the set of the current at various places in the area is shown by arrows, and the drift by numbers. Since these are *average* conditions, they indicate in a general way the tidal conditions on any day and during any year. They are designed to be used with the tidal current tables (except those for New York Harbor, and Narragansett Bay, which are used with the tide tables). These charts are available for Boston Harbor, Narragansett Bay to Nantucket Sound, Narragansett Bay, Block Island Sound and Eastern Long Island Sound, Long Island Sound and Block Island Sound, New York Harbor, Delaware Bay and River, upper Chesapeake Bay, Charleston Harbor, San Francisco Bay, Puget Sound (northern part), and Puget Sound (southern part). Current arrows are sometimes shown on nautical charts. These represent average conditions and should not be considered reliable predictions of the conditions to be encountered at any given time. When a strong current sets over an irregular bottom, or meets an opposing current, ripples may occur on the surface. These are called **tide rips.** Areas where they occur frequently are shown on charts.

Usually, the mariner obtains tidal information from tide and tidal current tables. However, if these are not available, or if they do not include information at a desired place, the mariner may be able to obtain locally the **mean high water lunitidal interval** or the **high water full and change.** The approximate *time* of high water can be found by adding either interval to the time of transit (either upper or lower) of the moon (art. 733). Low water occurs approximately ¼ tidal day (about 6^h12^m) before and after the time of high water. The actual interval varies somewhat from day to day, but approximate results can be obtained in this manner. Similar information for tidal currents (**lunicurrent interval**) is seldom available.

1103. Tide tables for various parts of the world are published in four volumes by the National Ocean Survey. Each volume is arranged as follows:

Table 1 contains a complete list of the predicted times and heights of the tide for each day of the year at a number of places designated as **reference stations.**

Table 2 gives differences and ratios which can be used to modify the tidal information for the reference stations to make it applicable to a relatively large number of **subordinate stations.**

Table 3 provides information for use in finding the approximate height of the tide at any time between high water and low water.

Table 4 is a sunrise-sunset table at five-day intervals for various latitudes from 76°N to 60°S (40°S in one volume).

Table 5 provides an adjustment to convert the local mean time of table 4 to zone or standard time.

Table 6 (two volumes only) gives the zone time of moonrise and moonset for each day of the year at certain selected places.

Certain astronomical data are contained on the inside back cover of each volume.

Extracts from tables 1, 2, and 3 for the East Coast of North and South America are given in appendix L.

1104. Tide predictions for reference stations.—The first page of appendix L is the table 1 daily predictions for New York (The Battery) for the first quarter of 1975. As indicated at the bottom of the page, times are for Eastern Standard Time (+5 zone, time meridian 75°W). Daylight saving time is not used. Times are given on the 24-hour basis. The tidal reference level for this station is mean low water.

For each day, the date and day of week are given, and the time and height of each high and low water are given in chronological order. Although high and low waters are not labeled as such, they can be distinguished by the relative heights given immediately to the right of the times. Since *two* high tides and *two* low tides occur each tidal day, the type of tide at this place is *semidiurnal*. The *tidal* day being longer than the *civil* day (because of the revolution of the moon eastward around the earth), any given tide occurs *later* from day to day. Thus, on Saturday, March 29, 1975, the first tide that occurs is the lower low water (−1.2 feet at 0334). The following high water (lower high water) is 4.9 feet above the reference level (a 6.1 foot rise from the preceding low water), and occurs at 0942. This is followed by the higher low water (−0.9 feet) at 1547, and then the higher high water of 5.5 feet at 2206. The cycle is repeated on the following day with variations in height, and later times.

Because of later times of corresponding tides from day to day, certain days have only one high water or only one low water. Thus, on January 17 high tides occur at 1120 and 2357. The next following high tides are at 1154 on January 18 and 0029 on January 19. Thus, only one high tide occurs on January 18, the previous one being shortly before midnight on the seventeenth, and the next one occuring early in the morning of the nineteenth, as shown.

1105. Tide predictions for subordinate stations.—The second page of appendix L is a page of table 2 of the tide tables. For each subordinate station listed, the following information is given:

Number. The stations are listed in geographical order and given consecutive numbers. At the end of each volume an alphabetical listing is given, and for each entry the consecutive number is shown, to assist in finding the entry in table 2.

Place. The list of places includes both subordinate and reference stations, the latter being given in bold type.

Position. The approximate latitude and longitude are given to assist in locating the station. The latitude is north or south, and the longitude east or west, depending upon the letters (N, S, E, W) next *above* the entry. These may not be the same as those at the *top* of the column.

Differences. The differences are to be applied to the predictions for the reference station shown in bold capitals next *above* the entry on the page. Time and height differences are given separately for high and low waters. Where differences are omitted, they are either unreliable or unknown.

The time difference is the number of hours and minutes to be applied to the time at the reference station to find the time of the corresponding tide at the subordinate station. This interval is added if preceded by a plus sign (+), and subtracted if preceded by a minus sign (−). The results obtained by the application of the time differences will be in the zone time of the time meridian shown directly above the difference for the subordinate station. Special conditions occurring at a few stations are indicated by footnotes on the applicable pages. In some instances, the corresponding tide falls on a different date at reference and subordinate stations.

Height differences are shown in a variety of ways. For most entries separate height differences in feet are given for high water and low water. These are applied to the height given for the reference station. In many cases a *ratio* is given for either high water or low water, or both. The height at the reference station is multiplied by this ratio to find the height at the subordinate station. For a few stations, *both* a ratio and difference are given. In this case the height at the reference station is first multiplied by the ratio, and the difference is then applied. An example is given in each volume of tide tables. Special conditions are indicated in the table or by footnote. Thus, a footnote on the second page of appendix L indicates that "Values for the Hudson River above George Washington Bridge are based upon averages for the six months May to October, when the fresh-water discharge is a minimum."

Ranges. Various ranges are given, as indicated in the tables. In each case this is the difference in height between high water and low water for the tides indicated.

Example.—List chronologically the times and heights of all tides at Yonkers, (No. 1531) on January 2, 1975.

Solution.—

Date	January 2, 1975
Subordinate station	Yonkers
Reference station	New York
High water time difference	(+) 1^h09^m
Low water time difference	(+) 1^h10^m
High water height difference	(−) 0.8 ft.
Low water height difference	0.0 ft.

	New York		*Yonkers*	
HW	2321 (1st)	4.6 ft.	0030	3.8 ft.
LW	0516	(−) 0.6 ft.	0626	(−) 0.6 ft.
HW	1138	4.9 ft.	1247	4.1 ft.
LW	1749	(−) 0.9 ft.	1859	(−) 0.9 ft.

1106. Finding height of tide at any time.—Table 3 of the tide tables provides means for determining the approximate height of tide at any time. It is based upon the assumption that a plot of height versus time is a sine curve (art. 140). Instructions for use of the table are given in a footnote below the table, which is reproduced in appendix L.

Example 1.—Find the height of tide at Yonkers (No. 1531) at 1000 on January 2, 1975.

Solution.—The given time is between the low water at 0626 and the high water at 1247 (example of art. 1105). Therefore, the tide is rising. The duration of rise is $1247-0626=6^h21^m$. The range of tide is $4.1-(-0.6)=4.7$ feet. The given time is

2^h47^m *before* high water, the nearest tide. Enter the upper part of the table with duration of rise 6^h20^m (the nearest tabulated value to 6^h21^m), and follow the line horizontally to 2^h45^m (the nearest tabulated value to 2^h47^m). Follow this column vertically downward to the entry 1.8 feet in the line for a range of tide of 4.5 feet (the nearest tabulated value to 4.7 feet). This is the correction to be applied to the nearest tide. Since the nearest tide is high water, subtract 1.8 from 4.1 feet. The answer, 2.3 feet, is the height of tide at the given time.

Answer.—Ht. of tide at 1000, 2.3 ft.

Interpolation in this table is not considered justified.

It may be desired to know at what time a given depth of water will occur. In this case, the problem is solved in reverse.

Example 2.—The captain of a vessel drawing 22 feet wishes to pass over a temporary obstruction near Days Point, Weehawken (No. 1521), having a charted depth of 21 feet, passage to be made during the morning of January 31, 1975.

Required.—The earliest time after 0800 that this passage can be made, allowing a safety margin of two feet.

Solution.—The least acceptable depth of water is 24 feet, which is three feet more than the charted depth. Therefore, the height of tide must be three feet or more. At the New York reference station a low tide of (—)0.9 foot occurs at 0459, followed by a high tide of 4.9 feet at 1120. At Days Point the corresponding low tide is (—)0.9 foot at 0522, and the high tide is 4.6 feet at 1144. The duration of rise is 6^h22^m, and the range of tide is 5.5 feet. The least acceptable tide is 3.0 feet, or 1.6 feet less than high tide. Enter the *lower* part of table 3 with range 5.5 feet and follow the horizontal line until 1.6 feet is reached. Follow this column vertically *upward* until the value of 2^h19^m is reached on the line for a duration of 6^h20^m (the nearest tabulated value to 6^h22^m). The minimum depth will occur about 2^h19^m *before* high water or at about 0925.

Answer.—A depth of 24 feet occurs at 0925.

If the range of tide is more than 20 feet, *half* the range (*one third* if the range is greater than 40 feet) is used to enter table 3, and the correction to height is *doubled* (*trebled* if one third is used).

A diagram for a graphical solution is given in figure 1106. Eye interpolation can be used if desired. The steps in this solution are as follows:

1. Enter the upper graph with the duration of rise or fall. This is represented by a horizontal line.
2. Find the intersection of this line and the curve representing the interval from the nearest *low* water (point *A*).
3. From *A*, follow a vertical line to the sine curve of the lower diagram (point *B*).
4. From *B*, follow horizontally to the vertical line representing the range of tide (point *C*).
5. Using *C*, read the correction from the series of curves.
6. Add (algebraically) the correction of step 5 to the *low* water height, to find the height at the given time.

The problem illustrated in figure 1106 is similar to that of example 1 given above. The duration of rise is 6^h25^m, and the interval from *low* water is 5^h23^m. The range of tide is 6.1 feet. The correction (by interpolation) is 5.7 feet. If the height of the preceding low tide is (—)0.2 foot, the height of tide at the given time is (—)0.2+5.7=5.5 feet. To solve example 2 by the graph, enter the lower graph and find the intersection of the vertical line representing 5.5 feet and the curve representing 3.9 feet (the minimum acceptable height above low water). From this point follow horizontally to the sine curve, and then vertically to the horizontal line in the upper figure representing the duration of rise of 6^h22^m. From the curve, determine the interval 4^h00^m. The earliest time is about 4^h00^m *after* low water, or at about 0922.

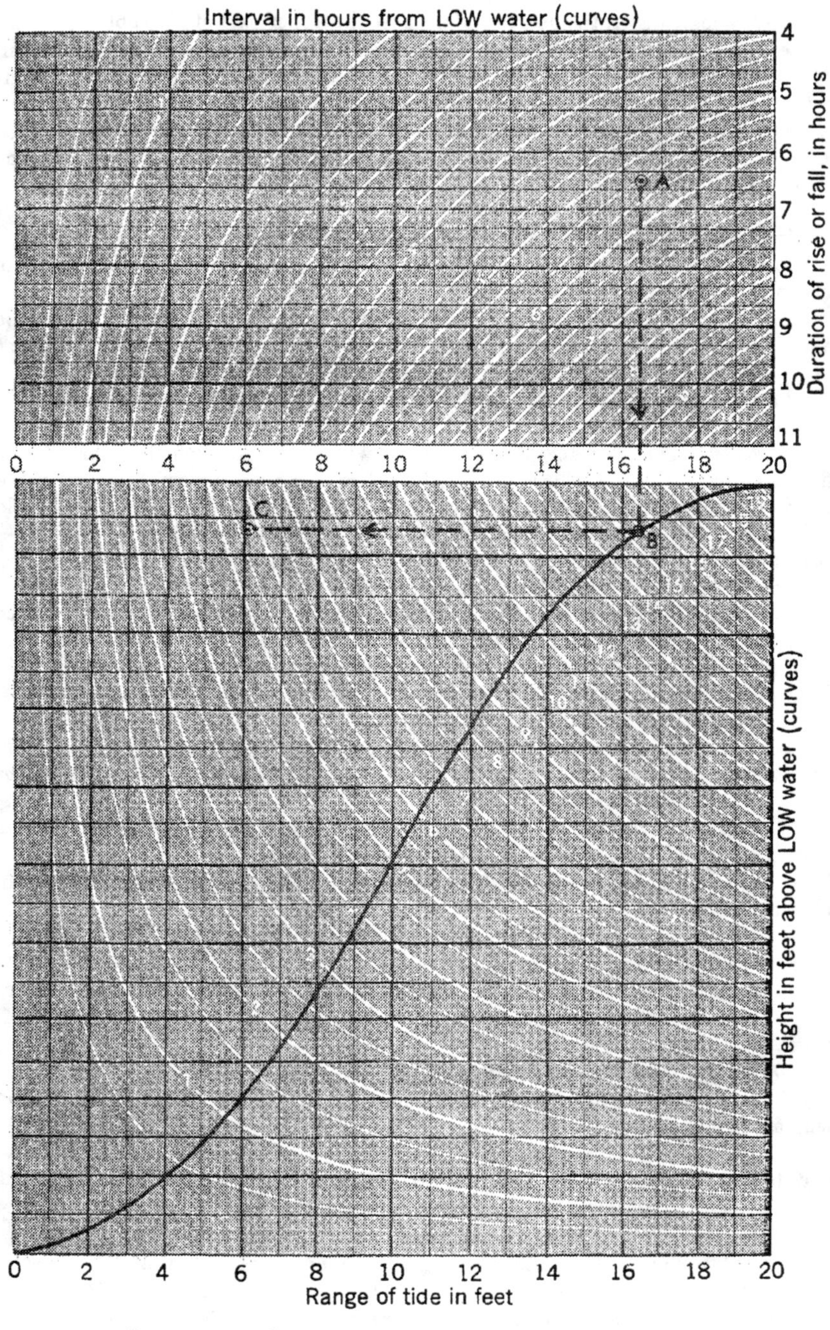

FIGURE 1106.—Graphical solution for height of tide at any time.

1107. Tidal current tables are somewhat similar to tide tables, but the coverage is less extensive, being given in two volumes. Each volume is arranged as follows:

Table 1 contains a complete list of predicted times of maximum currents and slack, with the velocity (speed) of the maximum currents, for a number of reference stations.

Table 2 gives differences, ratios, and other information related to a relatively large number of subordinate stations.

Table 3 provides information for use in finding the speed of the current at any time between tabulated entries in tables 1 and 2.

Table 4 gives the number of minutes the current does not exceed stated amounts, for various maximum speeds.

Table 5 (Atlantic Coast of North America only) gives information on rotary tidal currents.

Each volume contains additional useful information related to currents. Extracts from the tables for the Atlantic Coast of North America are given in appendix M.

1108. Tidal current predictions for reference stations.—The extracts of appendix M are for The Narrows, New York Harbor. Times are given on the 24-hour basis, for meridian 75° W. *Daylight saving time is not used.*

For each day, the date and day of week are given, with complete current information. Since the cycle is repeated twice each tidal day, currents at this place are semidiurnal. On most days there are four slack waters and four maximum currents, two of them floods (F) and two of them ebbs (E). However, since the tidal day is longer than the civil day, the corresponding condition occurs later from day to day, and on certain days there are only three slack waters or three maximum currents. At some places, the current on some days runs maximum flood twice, but ebb only once, a minimum flood occurring in place of the second ebb. The tables show this information.

As indicated by appendix M, the sequence of currents at The Narrows on Monday February 3, 1975, is as follows:

0000 Flood current, 5^m after maximum velocity (speed).
0305 Slack, ebb begins.
0621 Maximum ebb of 2.0 knots, setting 160°.
1005 Slack, flood begins.
1222 Maximum flood of 1.5 knots, setting 340°.
1516 Slack, ebb begins.
1839 Maximum ebb of 1.9 knots, setting 160°.
2216 Slack, flood begins.
2400 Flood current, 56^m before maximum velocity (speed).

Only one maximum flood occurs on this day, the previous one having occurred 5 minutes before the day began, and the following one predicted for 56 minutes after the day ends.

1109. Tidal current predictions for subordinate stations.—For each subordinate station listed in table 2 of the tidal current tables, the following information is given:

Number. The stations are listed in geographical order and given consecutive numbers, as in the tide tables (art. 1105). At the end of each volume an alphabetical listing is given, and for each entry the consecutive number is shown, to assist in finding the entry in table 2.

Place. The list of places includes both subordinate and reference stations, the latter being given in **bold type.**

Position. The approximate latitude and longitude are given to assist in locating the station. The latitude is north or south and the longitude east or west as indicated by the letters (N, S, E, W) next *above* the entry. The current given is for the center of the channel unless another location is indicated by the station name.

Time difference. Two time differences are tabulated. One is the number of hours and minutes to be applied to the tabulated times of slack water at the reference station to find the times of slack waters at the subordinate station. The other time difference is applied to the times of maximum current at the reference station to find the times of the corresponding maximum current at the subordinate station. The intervals, which are added or subtracted in accordance with their signs, include any difference in time between the two stations, so that the answer is correct for the standard time of the subordinate station. Limited application and special conditions are indicated by footnotes.

Velocity (speed) ratios. Speed of the current at the subordinate station is found by multiplying the speed at the reference station by the tabulated ratio. Separate ratios may be given for flood and ebb currents. Special conditions are indicated by footnotes.

As indicated in appendix M, the currents at The Battery (No. 2375) can be found by *adding* 1^h30^m for slack water and 1^h35^m for maximum current to the times for The Narrows, and multiplying flood currents by 0.9 and ebb currents by 1.2. Applying these to the values for Monday, February 3, 1975, the sequence is as follows:

0000 Flood current, 1^h30^m before maximum velocity (speed).
0130 Maximum flood of 1.8 knots, setting 015°.
0435 Slack, ebb begins.
0756 Maximum ebb of 2.4 knots, setting 195°.
1135 Slack, flood begins.
1357 Maximum flood of 1.4 knots, setting 015°.
1646 Slack, ebb begins.
2014 Maximum ebb of 2.3 knots, setting 195°.
2346 Slack, flood begins.
2400 Flood current, 14^m after slack.

1110. Finding speed of tidal current at any time.—Table 3 of the tidal current tables provides means for determining the approximate velocity (speed) at any time. Instructions for its use are given below the table, which is reproduced in appendix M.

Example 1.—Find the speed of the current at The Battery at 1500 on February 3, 1975.

Solution.—The given time is between the maximum flood of 1.4 knots at 1357 and the slack at 1646 (art. 1109). The interval between slack and maximum current (1646 −1357) is 2^h49^m. The interval between slack and the desired time (1646−1500) is 1^h46^m. Enter the table (A) with 2^h40^m at the top, and 1^h40^m at the left side (the nearest tabulated values to 2^h49^m and 1^h46^m, respectively), and find the factor 0.8 in the body of the table. The approximate speed at 1500 is $0.8\times1.4=1.1$ knots, and it is flooding.

Answer.—Speed 1.1 kn.

It may be desired to determine the period during which the current is less (or greater) than a given amount. Table 4 of the tidal current tables can be used to determine the period during which the speed does not exceed 0.5 knot. For greater speeds, and for more accurate results under some conditions, table 3 of the tidal current tables can be used, solving by reversing the process used in example 1.

Example 2.—During what period on the evening of February 3, 1975, does the ebb current equal or exceed 1.0 knot at The Battery?

Solution.—The maximum ebb of 2.3 knots occurs at 2014. This is preceded by a slack at 1646, and followed by the next slack at 2346. The interval between the earlier slack and the maximum ebb is 3^h28^m, and the interval between the ebb and following slack is 3^h32^m. The desired factor is $\frac{1.0}{2.3}=0.4$. Enter table A with 3^h20^m (the nearest tabulated value to 3^h28^m) at the top, and follow down the column to 0.4 (midway between 0.3 and 0.5). At the left margin the interval between slack and the desired time is found to be 0^h50^m (midway between 0^h40^m and 1^h00^m). Therefore, the current becomes 1.0 knot at $1646+0^h50^m=1736$. Next, enter table A with 3^h40^m (the nearest tabulated value to 3^h32^m) at the top, and follow down the column to 0.4. Follow this line to the left margin, where the interval between slack and desired time is found to be 1^h00^m. Therefore, the current is 1.0 knot or greater until $2346-1^h00^m=2246$. If the two intervals between maximum current and slack were nearest the same 20^m interval, table A would have to be entered only once.

Answer.—The speed equals or exceeds 1.0 knot between 1736 and 2246.

1111. Current diagrams.—A current diagram is a graph showing the speed of the current along a channel at different stages of the tidal current cycle. The current tables include such diagrams for Boston Harbor; Vineyard and Nantucket Sounds (one diagram); East River, New York; New York Harbor; Delaware Bay and River (one diagram); Chesapeake Bay; South San Francisco Bay; and North San Francisco Bay. The diagram for New York Harbor is reproduced in appendix M.

On this diagram each vertical line represents a given instant identified in terms of the number of hours before or after slack *at The Narrows*. Each horizontal line represents a distance from Ambrose Channel Entrance, measured along the usually-traveled route. The names along the left margin are placed at the correct distances from Ambrose Channel Entrance. The current is for the center of the channel opposite these points. The intersection of any vertical line with any horizontal line represents a given moment in the current cycle at a given place in the channel. If this intersection is in a shaded area, the current is flooding; if in an unshaded area, it is ebbing. The speed in knots can be found by interpolation (if necessary) between the numbers given in the body of the diagram. The given values are *averages*. To find the value at any given time, multiply the speed found from the diagram by the ratio of *maximum speed of the current involved* to the *maximum shown on the diagram*, both values being taken for The Narrows. If the diurnal inequality is large, the accuracy can be improved by altering the width of the shaded area to fit conditions. The diagram covers 1½ current cycles, so that the right-hand third is a duplication of the left-hand third.

If the current for a single station is desired, table 1 or 2 should be used. The current diagrams are intended for use in either of two ways: First, to determine a favorable time for passage through the channel. Second, to find the average current to be expected during any passage through the channel. For both of these uses a number of "speed lines" are provided. When the appropriate line is transferred to the correct part of the diagram, the current to be encountered during passage is indicated along the line.

Example.—During the morning of January 3, 1975, a ship is to leave Pier 83 at W. 42nd St., and proceed down the bay at ten knots.

Required.—(1) Time to get underway to take maximum advantage of a favorable current, allowing 15 minutes to reach mid channel.

(2) Average speed over the bottom during passage down the bay.

Solution.—(1) Transfer the line (slope) for ten knots southbound to the diagram, locating it so that it is centered on the unshaded ebb current section between W. 42nd St. and Ambrose Channel Entrance. This line crosses a horizontal line through W. 42nd St. about one-half of the distance between the vertical lines representing three and two hours, respectively, after ebb begins at The Narrows. The setting is not critical. Any time within about half an hour of the correct time will result in about the same current. Between the points involved, the entire speed line is in the ebb current area.

(2) Table 1 indicates that on the morning of January 3 ebb begins at The Narrows at 0132. Two hours twenty-eight minutes after ebb begins, the time is 0400. Therefore,

the ship should reach mid channel at 0400. It should get underway 15 minutes earlier, at 0345.

(3) To find the average current, determine the current at intervals (as every two miles), add, and divide by the number of entries.

Distance	*Current*
18	1.2
16	1.4
14	1.9
12	1.5
10	2.0
8	1.9
6	1.3
4	1.2
2	1.4
0	1.2
sum	15.0

The sum of 15.0 is for ten entries. The average is therefore 15.0÷10=1.5 knots.

(4) This value of current is correct only if the ebb current is an average one. From table 1 the maximum ebb involved is 2.2 knots. From the diagram the maximum value at The Narrows is 2.0 knots. Therefore, the average current found in step (3) should be increased by the ratio 2.2÷2.0=1.1. The average for the run is therefore 1.5×1.1=1.6 knots. Speed over the bottom is 10+1.6=11.6 knots.

Answers.—(1) T 0345, (2) S 11.6 kn.

In the example, an ebb current is carried throughout the run. If the transferred speed line had been partly in a flood current area, all ebb currents (those increasing the ship's speed) should be given a positive sign (+), and all flood currents a negative sign (−). A separate ratio should be determined for each current (flood or ebb), and applied to the entries for that current. In Chesapeake Bay it is not unusual for an outbound vessel to encounter three or even four separate currents during passage down the bay. Under the latter condition, it is good practice to multiply *each* current taken from the diagram by the ratio for the current involved.

If the time of starting the passage is fixed, and the current during passage is desired, the starting time is identified in terms of the reference tidal cycle. The speed line is then drawn through the intersection of this vertical time line and the horizontal line through the place. The average current is then determined in the same manner as when the speed line is located as described above.

Problems

1102. The mean high water lunitidal interval at a certain port is 2^h17^m.

Required.—The approximate times of each high and low water on a day when the moon transits the local meridian at 1146.

Answers.—HW at 0139 and 1403, LW at 0751 and 2015.

1104. List chronologically the times and heights of all tides at New York (The Battery) on February 11, 1975.

Answer.—

Time	*Tide*	*Height*
0222	LW	(−) 0.4 ft.
0829	HW	4.6 ft.
1449	LW	(−) 0.6 ft.
2053	HW	4.2 ft.

1105. List chronologically the times and heights of all tides at Castle Point, Hoboken, N.J. (No. 1519) on March 18, 1975.

Answer.—

Time	*Tide*	*Height*
0533	LW	0.2 ft.
1141	HW	3.5 ft.
1724	LW	0.3 ft.
0003	HW	4.1 ft.

1106a. Find the height of tide at Union Stock Yards, New York (No. 1523) at 0600 on February 6, 1975.

Answer.—Ht. of tide at 0600, 3.8 ft.

1106b. The captain of a vessel drawing 24 feet wishes to pass over a temporary obstruction near Bayonne, N.J. (No. 1505) having a charted depth of 23 feet, passage to be made during the afternoon of March 5, 1975.

Required.—The earliest and latest times that the passage can be made, allowing a safety margin of two feet.

Answers.—Earliest time 1316, latest time 1531.

1108. Determine the sequence of currents at The Narrows on January 15, 1975.

Answer.—

0000 Ebb current, 42^m after slack.
0231 Maximum ebb of 1.9 knots.
0557 Slack, flood begins.
0822 Maximum flood of 1.7 knots.
1137 Slack, ebb begins.
1455 Maximum ebb of 2.1 knots.
1836 Slack, flood begins.
2051 Maximum flood of 1.5 knots.
2400 Flood current, 2^m before slack.

1109. Determine the sequence of currents at Ambrose Channel Entrance (No. 2310) on January 12, 1975.

Answer.—

0000 Ebb current, 42^m after maximum velocity (speed).
0241 Slack, flood begins.
0533 Maximum flood of 2.0 knots, setting 310°.
0828 Slack, ebb begins.
1155 Maximum ebb of 2.5 knots.
1527 Slack, flood begins.
1801 Maximum flood of 1.5 knots, setting 310°.
2040 Slack, ebb begins.
2400 Ebb current, 3^m before maximum velocity (speed).

1110a. Find the speed of the current at Bear Mountain Bridge (No. 2445) at 0900 on February 19, 1975.

Answer.—Speed 0.8 kn.

1110b. At about what time during the afternoon of February 3, 1975, does the flood current northwest of The Battery (No. 2375) reach a speed of 1.0 knot?

Answer.—T 1245.

1111. A vessel arrives at Ambrose Channel Entrance two hours after flood begins at The Narrows on the morning of February 16, 1975.

Required.—(1) The speed through the water required to take fullest advantage of the flood tide in steaming to Chelsea Docks.

(2) The average current to be expected.
(3) Estimated time of arrival off Chelsea Docks.
Answers.—(1) S 9 kn., (2) S 1.4 kn., (3) ETA 1035.

CHAPTER XII

PREDICTED VISUAL RANGES OF LIGHTS

1201. Introduction.—Usually a navigator wants to know not only the identity of a light, but also the area in which he might reasonably expect to observe it. His track is planned to take him within range of lights which can prove useful during periods of darkness. If lights are not sighted within a reasonable time after prediction, a dangerous situation may exist, requiring resolution or action to insure safety of the vessel. However, the *range at which a light can be sighted is not subject to exact prediction.*

The approximate area in which a light can be observed is normally a circle with the light as the center, and the visual range as the radius. However, on some bearings the range may be reduced by obstructions. In this case the obstructed arc might differ with height of eye and distance. Also, lights of different colors may be seen at different distances. This fact should be considered not only in predicting the distance at which a light can be seen, but also in identifying it.

1202. Visual ranges of lights.—The condition of the atmosphere has a considerable effect upon the distance at which lights can be seen. Sometimes lights are obscured by fog, haze, dust, smoke, or precipitation which may be present at the light, or between it and the observer, but not at the observer, and possibly unknown to him. Although a light of low luminous intensity is more subject to being obscured by any one of these conditions than is a light of high intensity, the visual or sighting range of even a light of very high luminous intensity is considerably reduced in such conditions. There is always the possibility of a light being extinguished. In the case of unwatched lights, this condition might not be detected and corrected at once. During periods of armed conflict, certain lights might be deliberately extinguished if they are considered of greater value to the enemy than to one's own vessels.

On a dark, clear night the visual range is limited primarily by one of two ways: (1) luminous intensity and (2) curvature of the earth. A weak light cannot normally be expected to be seen beyond a certain range, regardless of the height of eye. This distance is called luminous range. Light travels in almost straight lines, so that an observer below the visible horizon of the light should not expect to see the light, although the loom extending upward from the light can sometimes be seen at greater distances. Table 8 gives the distance to the horizon at various heights. The tabulated distances assume normal refraction. Abnormal conditions might extend this range somewhat (or in some cases reduce it). Hence, the geographic range, as the luminous range, is not subject to exact prediction at any given time.

The **luminous range** is the maximum distance at which a light can be seen under existing visibility conditions. This luminous range takes no account of the elevation of the light, the observer's height of eye, the curvature of the earth, or interference from background lighting. The luminous range is determined from the known nominal luminous range, called the nominal range, and the existing visibility conditions. The **nominal range** is the maximum distance at which a light can be seen in clear weather as defined by the International Visibility Code (meteorological visibility of 10 nautical miles.) The **geographic range** is the maximum distance at which the curvature of the earth permits a light to be seen from a particular height of eye *without* regard to the luminous intensity of the light. The geographic range *sometimes* printed on charts or tabulated in light lists is the maximum distance at which the curvature of the earth permits a light to be seen from a height of eye of 15 feet above the water when the elevation of the light is taken above the height datum of the largest scale chart of the locality.

The geographic range depends upon the height of both the light and the observer, as shown in figure 1202a. In this illustration a light 150 feet above the water is shown. At this height, the distance to the horizon, by table 8, is 14.3 miles. Within this range the light, *if powerful enough and atmospheric conditions permit,* is visible regardless of the height of eye of the observer (if there is no obstruction). Beyond this range, the visual range depends upon the height of eye. Thus, by table 8 an observer with height of eye of 5 feet can see the light on his horizon if he is 2.6 miles beyond the horizon of the light, or a total of 16.9 miles. For a height of 30 feet the distance is 14.3+6.4=20.7 miles. If the height of eye is 70 feet, the geographic range is 14.3+9.8=24.1 miles. This computation can be avoided through use of table 40 which for entering arguments of heights of eye and light gives the geographic range directly.

Except for range and some directional lights, the nominal range is listed in the U.S. Coast Guard *Light List.* The **Luminous Range Diagram** shown in the *Light List* and figure 1202b is used to convert the nominal range to the luminous range. When using this diagram, it must be remembered that the ranges obtained are approximate, the transmissivity of the atmosphere may vary between the observer and the light, and *glare from background lighting will reduce considerably the range at which lights are sighted.* After estimating the meteorological visibility with the aid of the Meteorological Optical Range Table shown in table 1202, the Luminous Range Diagram is entered with the nominal range on the horizontal nominal range scale; a vertical line is followed until it intersects the curve or reaches the region on the diagram representing the meteorological visibility; from this point or region a horizontal line is followed until it intersects the vertical luminous range scale.

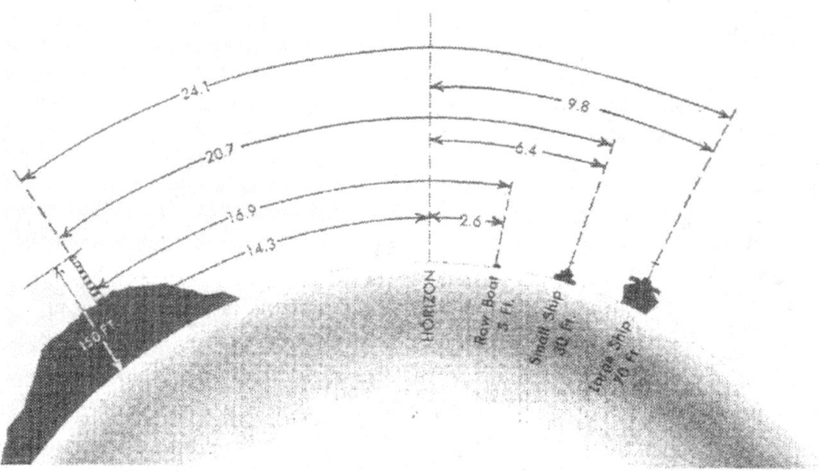

FIGURE 1202a.—Geographic range of a light.

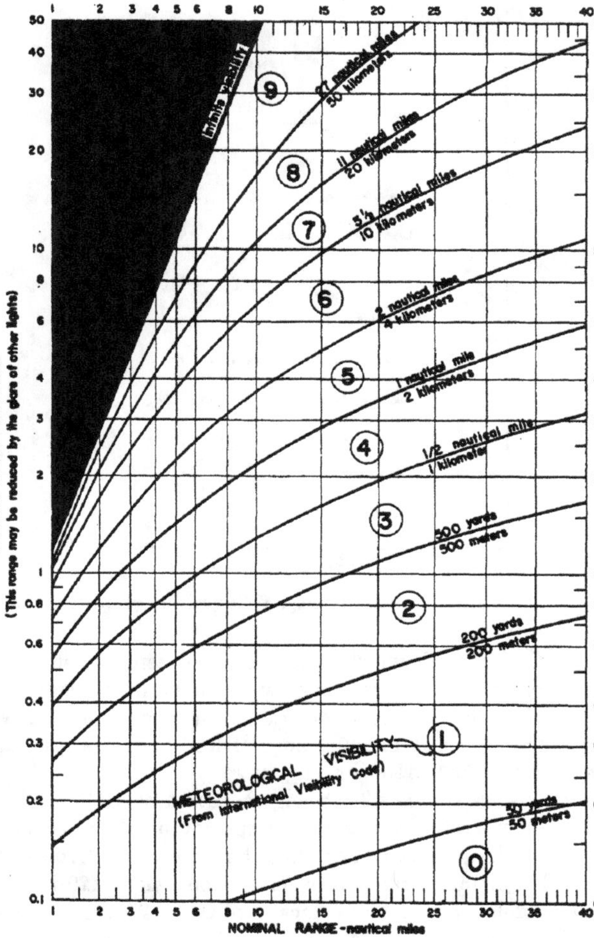

FIGURE 1202b.—Luminous Range Diagram.

Example 1.—The nominal range of a light as extracted from the *Light List* is 15 nautical miles.

Required.—The luminous range when the meteorological visibility is (1) 11 nautical miles and (2) 1 nautical mile.

Solution.—To find the luminous range when the meteorological visibility is 11 nautical miles, the Luminous Range Diagram is entered with nominal range 15 nautical miles on the horizontal nominal range scale; a vertical line is followed until it intersects the curve on the diagram representing a meteorological visibility of 11 nautical miles; from this point a horizontal line is followed until it intersects the vertical luminous range scale at 16 nautical miles. A similar procedure is followed to find the luminous range when the meteorological visibility is 1 nautical mile.

Answers.—(1) 16 nautical miles; (2) 3 nautical miles.

Code No.	Weather	Yards
0	Dense fog	Less than 50
1	Thick fog	50–200
2	Moderate fog	200–500
3	Light fog	500–1000
		Nautical Miles
4	Thin fog	½–1
5	Haze	1–2
6	Light haze	2–5½
7	Clear	5½–11
8	Very clear	11.0–27.0
9	Exceptionally clear	Over 27.0

From the International Visibility Code.

TABLE 1202.—Meteorological Optical Range Table.

In predicting the range at which a light can be seen, one should first determine the geographic range to compare this range with the luminous range, if known. If the geographic range is less than the luminous range, the geographic range must be taken as the limiting range. If the luminous range is less than the geographic range, the luminous range must be taken as the limiting range.

These predictions are simple when using the U.S. Coast Guard *Light List* because only nominal ranges are tabulated. Also the current practice of the National Ocean Survey is to follow the *Light List* when printing the range of a light on a chart.

Example 2.—The nominal range of a navigational light 120 feet above the chart datum is 20 nautical miles. The meteorological visibility is 27 nautical miles.

Required.—The distance at which an observer at a height of eye of 60 feet can expect to see the light.

Solution.—The maximum range at which the light may be seen is the lesser of the luminous and geographic ranges.

By table 40 the geographic range (21.8 miles) is found to be less than the luminous range, which is 40 nautical miles.

Answer.—22 nautical miles. Because of various uncertainties, the range is given only to the nearest whole mile.

If the range of a light as printed on a chart, particularly a foreign chart or a reproduction of foreign chart, or tabulated in a light list other than the U.S. Coast Guard *Light List,* approximates the geographic range for a 15-foot height of eye of the observer, one is generally safe in assuming that this range is the geographic range. With lesser certainty, one may also assume that the lesser to the geographic and nominal ranges is printed on the chart or tabulated in the light list. Using these assumptions, the predicted range is then found by adding the distance to the horizon for both the light and the observer, or approximately, by the *difference* between 4.5 miles (the distance to the horizon at a height of 15 feet) and the distance for the height of eye of the observer (a constant for any given height) and *adding* this value to the tabulated or charted geographic range (subtracting if the height of eye is less than 15 feet). In making a prediction, one should keep in mind the possibility of the luminous range being *between* the tabulated or charted geographic range and the predicted range. The luminous intensity of the light, if known, should be of assistance in identifying this condition.

Example 3.—The range of a light as printed on a foreign chart is 17 miles. The light is 120 feet above chart datum. The meteorological visibility is 10 nautical miles.

Required.—The distance at which an observer at a height of eye of 60 feet can expect to see the light.

Solution.—At 120 feet the distance to the horizon, by table 8, is 12.8 miles. Adding 4.6 miles (the distance to the horizon at a height of 15 feet), the geographic range is found to approximate the range printed on the chart. Then assuming that the latter range is the geographic range for a 15-foot height of eye of the observer and that the nominal range is the greater value, the predicted range is found by adding the distance to the horizon for both the light and the observer (predicted range=12.8 mi.+9.0 mi. =21.8 mi.). The additional distance, i.e., the distance in excess of the assumed charted geographic range, is dependent upon the luminous intensity of the light and the meteorological visibility.

If one is approaching a light, and wishes to predict the *time* at which it should be sighted, he first predicts the range. It is then good practice to draw an arc indicating the visual range. The point at which the course line crosses the arc of visual range is the predicted position of the vessel at the time of sighting the light. The predicted time of arrival at this point is the predicted time of sighting the light. The direction of the light from this point is the predicted bearing at which the light should be sighted. Conversion of the true bearing to a relative bearing is usually helpful in sighting the light. The accuracy of the predictions depends upon the accuracy of the predicted range, and the accuracy of the predicted time and place of crossing the visual range arc. If the course line crosses the visual range arc at a small angle, a small lateral error in track may result in a large error of prediction, both of bearing and time. This is particularly apparent if the vessel is *farther* from the light than predicted, in which case the light might be passed without being sighted. Thus, if a light is not sighted at the predicted time, the error *may* be on the side of safety. However, such an interpretation should not be given unless confirmed by other information, for there is always the possibility of reduced meteorological visibility, or of the light being extinguished.

When a light is first sighted, one might determine whether it is on the horizon by immediately reducing the height of eye by several feet, as by squatting or changing position to a lower height. If the light disappears, and reappears when the original height is resumed, it is on the horizon. This process is called **bobbing a light.** If a vessel has considerable vertical motion due to the condition of the sea, a light sighted on the horizon may alternately appear and disappear. This may lead the unwary to assign faulty characteristics and hence to err in its identification. The true characteristics should be observed after the distance has decreased, or by increasing the height of eye of the observer.

APPENDICES

APPENDICES

APPENDIX A

NAVIGATIONAL STARS AND THE PLANETS

Name	Pronunciation	Bayer name	Origin of name	Meaning of name	Distance*
Acamar	ā'kȧ·mär	θ Eridani	Arabic	another form of Achernar	120
Achernar	ā'kēr·när	α Eridani	Arabic	end of the river (Eridanus)	72
Acrux	ā'krŭks	α Crucis	Modern	coined from Bayer name	220
Adhara	ȧ·dā'rȧ	ε Canis Majoris	Arabic	the virgin(s)	350
Aldebaran	ăl děb'ȧ·răn	α Tauri	Arabic	follower (of the Pleiades)	64
Alioth	ăl'ĭ·ŏth	ε Ursa Majoris	Arabic	another form of Capella	49
Alkaid	ăl·kād'	η Ursa Majoris	Arabic	leader of the daughters of the bier	190
Al Na'ir	ăl·nār'	α Gruis	Arabic	bright one (of the fish's tail)	90
Alnilam	ăl'nĭ·lăm	ε Orionis	Arabic	string of pearls	410
Alphard	ăl'färd	α Hydrae	Arabic	solitary star of the serpent	200
Alphecca	ăl·fĕk'ȧ	α Corona Borealis	Arabic	feeble one (in the crown)	76
Alpheratz	ăl·fē'răts	α Andromeda	Arabic	the horse's navel	120
Altair	ăl·tār'	α Aquilae	Arabic	flying eagle or vulture	16
Ankaa	ăn'kä	α Phoenicis	Arabic	coined name	93
Antares	ăn·tā'rēz	α Scorpii	Greek	rival of Mars (in color)	250
Arcturus	ärk·tū'rŭs	α Bootis	Greek	the bear's guard	37
Atria	āt'rĭ·ȧ	α Trianguli Australis	Modern	coined from Bayer name	130
Avior	ā'vĭ·ôr	ε Carinae	Modern	coined name	350
Bellatrix	bĕ·lā'trĭks.	γ Orionis	Latin	female warrior	250
Betelgeuse	bĕt'ĕl·jūz	α Orionis	Arabic	the arm pit (of Orion)	300
Canopus	kȧ·nō'pŭs	α Carinae	Greek	city of ancient Egypt	230
Capella	kȧ·pĕl'ȧ	α Aurigae	Latin	little she-goat	46
Deneb	dĕn'ĕb	α Cygni	Arabic	tail of the hen	600
Denebola	dĕ·nĕb'ō·lȧ	β Leonis	Arabic	tail of the lion	42
Diphda	dĭf'dȧ	β Ceti	Arabic	the second frog (Fomalhaut was once the first)	57
Dubhe	dŭb'ĕ	α Ursa Majoris	Arabic	the bear's back	100
Elnath	ĕl'năth	β Tauri	Arabic	one butting with horns	130
Eltanin	ĕl·tā'nĭn	γ Draconis	Arabic	head of the dragon	150
Enif	ĕn'ĭf	ε Pegasi	Arabic	nose of the horse	250
Fomalhaut	fō'mȧl·ôt	α Piscis Austrini	Arabic	mouth of the southern fish	23
Gacrux	gā'krŭks	γ Crucis	Modern	coined from Bayer name	72
Gienah	jē'nȧ	γ Corvi	Arabic	right wing of the raven	136
Hadar	hā'där	β Centauri	Modern	leg of the centaur	200
Hamal	hăm'ȧl	α Arietis	Arabic	full-grown lamb	76
Kaus Australis	kôs ôs·trā'lĭs	ε Sagittarii	Ar., L.	southern part of the bow	163
Kochab	kō'kăb	β Ursa Minoris	Arabic	shortened form of "north star" (named when it was that, c. 1500 BC–AD 300)	100
Markab	mär'kăb	α Pegasi	Arabic	saddle (of Pegasus)	100
Menkar	mĕn'kär	α Ceti	Arabic	nose (of the whale)	1,100
Menkent	mĕn'kĕnt	θ Centauri	Modern	shoulder of the centaur	55
Miaplacidus	mī'ȧ·plăs'ĭ·dŭs	β Carinae	Ar., L.	quiet or still waters	86
Mirfak	mĭr'făk	α Persei	Arabic	elbow of the Pleiades	130
Nunki	nŭn'kē	σ Sagittarii	Bab.	constellation of the holy city (Eridu)	150
Peacock	pē'kŏk	α Pavonis	Modern	coined from English name of constellation	250
Polaris	pō·lā'rĭs	α Ursa Minoris	Latin	the pole (star)	450
Pollux	pŏl'ŭks	β Geminorum	Latin	Zeus' other twin son (Castor, α Geminorum, is first twin)	33
Procyon	prō'sĭ·ŏn	α Canis Minoris	Greek	before the dog (rising before the dog star, Sirius)	11
Rasalhague	räs'ăl·hā'gwē	α Ophiuchi	Arabic	head of the serpent charmer	67
Regulus	rĕg'ū·lŭs	α Leonis	Latin	the prince	67
Rigel	rī'jĕl	β Orionis	Arabic	foot (left foot of Orion)	500
Rigil Kentaurus	rī'jĭl kĕn·tô'rŭs	α Centauri	Arabic	foot of the centaur	4.3
Sabik	sā'bĭk	η Ophiuchi	Arabic	second winner or conqueror	69
Schedar	shĕd'ȧr	α Cassiopeiae	Arabic	the breast (of Cassiopeia)	360
Shaula	shō'lȧ	λ Scorpii	Arabic	cocked-up part of the scorpion's tail	200
Sirius	sĭr'ĭ·ŭs	α Canis Majoris	Greek	the scorching one (popularly, the dog star)	8.6
Spica	spī'kȧ	α Virginis	Latin	the ear of corn	155
Suhail	sŏŏ·hāl'	λ Velorum	Arabic	shortened form of Al Suhail, one Arabic name for Canopus	200
Vega	vē'gȧ	α Lyrae	Arabic	the falling eagle or vulture	27
Zubenelgenubi	zŏŏ·bĕn'ĕl·jē·nū'bē	α Librae	Arabic	southern claw (of the scorpion)	66

PLANETS

Name	Pronunciation	Origin of name	Meaning of name
Mercury	mûr'kū·rĭ	Latin	god of commerce and gain
Venus	vē'nŭs	Latin	goddess of love
Earth	ûrth	Mid. Eng.	—
Mars	märz	Latin	god of war
Jupiter	jōō'pĭ·tēr	Latin	god of the heavens, identified with the Greek Zeus, chief of the Olympian gods
Saturn	săt'ērn	Latin	god of seed-sowing
Uranus	ū'rȧ·nŭs	Greek	the personification of heaven
Neptune	nĕp'tūn	Latin	god of the sea
Pluto	plōō'tō	Greek	god of the lower world (Hades)

Guide to pronunciations:
fāte, ădd, finȧl, läst, ȧbound, ärm; bē, ĕnd, camēl, readēr; īce, bĭt, anĭmal; ōver, pōetic, hŏt, lôrd, mōōn; tūbe, ūnite, tŭb, circŭs, ûrn

*Distances in light-years. One light-year equals approximately 63,300 AU, or 5,880,000,000,000 miles. Authorities differ on distances of the stars; the values given are representative.

APPENDIX B

GREEK ALPHABET

Α α a	Alpha	Ν ν	Nu
Β β ϐ	Beta	Ξ ξ	Xi
Γ γ	Gamma	Ο ο	Omicron
Δ δ	Delta	Π π ϖ	Pi
Ε ϵ	Epsilon	Ρ ρ	Rho
Ζ ζ	Zeta	Σ σ ς	Sigma
Η η	Eta	Τ τ	Tau
Θ θ ϑ	Theta	Υ υ	Upsilon
Ι ι	Iota	Φ ϕ φ	Phi
Κ κ	Kappa	Χ χ	Chi
Λ λ	Lambda	Ψ ψ	Psi
Μ μ	Mu	Ω ω	Omega

APPENDIX C

SYMBOLS

Positions

⌓ Dead reckoning position.

⊙ Fix.

⊡ Estimated position.

△ Symbol used for one set of fixes when simultaneously fixing by two means, e.g., visual and radar; sometimes used for radionavigation fixes.

Mathematical Symbols

+ Plus (addition)
− Minus (subtraction)
± Plus or minus
∼ Absolute difference
× Times (multiplication)
÷ Divided by (division)
√ Square root
∠ Angle
∴ Therefore
$\sqrt[n]{\ }$ nth root
= Equals
≠ Not equal to
≈ Nearly equal to
\> Is greater than
< Is less than
∫ Integral sign
∞ Infinity
. . . Repeating decimal

Celestial Bodies

☉ Sun
☾ Moon
☿ Mercury
♀ Venus
⊕ Earth
♂ Mars
♃ Jupiter
♄ Saturn
⛢ Uranus
♆ Neptune
♇ Pluto
☆ Star
☆-P Star-planet altitude correction (altitude)
☉̲ ☾̲ Lower limb
☉ ☾ Center
☉̄ ☾̄ Upper limb
● New moon
● Crescent moon
◐ First quarter
○ Gibbous moon
○ Full moon
○ Gibbous moon
◑ Last quarter
● Crescent moon

Signs of the Zodiac

♈ Aries (vernal equinox)
♉ Taurus
♊ Gemini
♋ Cancer (summer solstice)
♌ Leo
♍ Virgo
♎ Libra (autumnal equinox)
♏ Scorpius
♐ Sagittarius
♑ Capricornus (winter solstice)
♒ Aquarius
♓ Pisces

Miscellaneous Symbols

Symbol	Meaning	Symbol	Meaning
^{y}	Years	✳	Interpolation impractical
^{m}	Months	$^{\circ}$	Degrees
^{d}	Days	$'$	Minutes of arc
^{h}	Hours	$''$	Seconds of arc
^{m}	Minutes of time	☌	Conjunction
^{s}	Seconds of time	☍	Opposition
■	Remains below horizon	□	Quadrature
▭	Remains above horizon	☊	Ascending node
////	Twilight all night	☋	Descending node

APPENDIX D

MISCELLANEOUS DATA

Exact relationships shown by asterisk (*). See footnote on page 655

Area

1 square inch	=6.4516 square centimeters*
1 square foot	=144 square inches*
	=0.09290304 square meter*
	=0.000022957 acre
1 square yard	=9 square feet*
	=0.83612736 square meter
1 square (statute) mile	=27,878,400 square feet*
	=640 acres*
	=2.589988110336 square kilometers*
1 square centimeter	=0.1550003 square inch
	=0.00107639 square foot
1 square meter	=10.76391 square feet
	=1.19599005 square yards
1 square kilometer	=247.1053815 acres
	=0.38610216 square statute mile
	=0.29155335 square nautical mile

Astronomy

1 mean solar unit	=1.00273791 sidereal units
1 sidereal unit	=0.99726957 mean solar unit
1 microsecond	=0.000001 second*
1 second	=1,000,000 microseconds*
	=0.01666667 minute
	=0.00027778 hour
	=0.00001157 day
1 minute	=60 seconds*
	=0.01666667 hour
	=0.00069444 day
1 hour	=3,600 seconds*
	=60 minutes*
	=0.04166667 day
1 mean solar day	$=24^h03^m56^s.55536$ of mean sidereal time
	=1 rotation of earth with respect to sun (mean)*
	=1.00273791 rotations of earth with respect to vernal equinox (mean)
	=1.0027378118868 rotations of earth with respect to stars (mean)
1 mean sidereal day	$=23^h56^m04^s.09054$ of mean solar time
1 sidereal month	=27.321661 days
	$=27^d07^h43^m11^s.5$
1 synodical month	=29.530588 days
	$=29^d12^h44^m02^s.8$
1 tropical (ordinary) year	=31,556,925.975 seconds
	=525,948.766 minutes
	=8,765.8128 hours
	$=365^d.24219879-0^d.0000000614(t-1900)$, where t = the year (date)
	$=365^d05^h48^m46^s$ (−) $0^s.0053t$

Astronomy—Continued

1 sidereal year	$=365^{d}.25636042+0.0000000011(t-1900)$, where t = the year (date)
	$=365^{d}06^{h}09^{m}09^{s}.5$ (+) $0^{s}.0001t$
1 calendar year (common)	=31,536,000 seconds*
	=525 600 minutes*
	=8,760 hours*
	=365 days*
1 calendar year (leap)	=31,622,400 seconds*
	=527,040 minutes*
	=8,784 hours*
	=366 days*
1 light-year	=9,460,000,000,000 kilometers
	=5,880,000,000,000 statute miles
	=5,110,000,000,000 nautical miles
	=63,240 astronomical units
	=0.3066 parsecs
1 parsec	=30,860,000,000,000 kilometers
	=19,170,000,000,000 statute miles
	=16,660,000,000,000 nautical miles
	=206,300 astronomical units
	=3.262 light years
1 astronomical unit	=149,600,000 kilometers
	=92,960,000 statute miles
	=80,780,000 nautical miles
	$=499^{s}.012$ light-time
	=mean distance, earth to sun
Mean distance, earth to moon	=384,400 kilometers
	=238,855 statute miles
	=207,559 nautical miles
Mean distance, earth to sun	=149,600,000 kilometers
	=92,957,000 statute miles
	=80,780,000 nautical miles
	=1 astronomical unit
Sun's diameter	=1,392,000 kilometers
	=865,000 statute miles
	=752,000 nautical miles
Sun's mass	=1,987,000,000,000,000,000,000,000,000,000,000 grams
	=2,200,000,000,000,000,000,000,000,000 short tons
	=2,000,000,000,000,000,000,000,000,000 long tons
Speed of sun relative to neighboring stars	=19.4 kilometers per second
	=12.1 statute miles per second
	=10.5 nautical miles per second
Orbital speed of earth	=29.8 kilometers per second
	=18.5 statute miles per second
	=16.1 nautical miles per second
Obliquity of the ecliptic	$=23°27'08''.26-0''.4684(t-1900)$, where t = the year (date)
General precession of the equinoxes	$=50''.2564+0''.000222(t-1900)$ per year, where t = the year (date)
Precession of the equinoxes in right ascension	$=46''.0850+0''.000279(t-1900)$ per year, where t = the year (date)
Precession of the equinoxes in declination	$=20''.0468-0''.000085(t-1900)$ per year, where t = the year (date)
Magnitude ratio	=2.512
	$=\sqrt[5]{100}$*

Charts

Nautical miles per inch	=reciprocal of natural scale÷72,913.39
Statute miles per inch	=reciprocal of natural scale÷63,360*
Inches per nautical mile	=72,913.39×natural scale
Inches per statute mile	=63,360×natural scale*
Natural scale	=1:72,913.39×nautical miles per inch
	=1:63,360×statute miles per inch*

Earth

Acceleration due to gravity (standard)	=980.665 centimeters per second per second
	=32.1740 feet per second per second
Mass-ratio—Sun/Earth	=332,958
Mass-ratio—Sun/(Earth & Moon)	**=328,912**
Mass-ratio—Earth/Moon	=81.30
Mean density	=5.517 grams per cubic centimeter
Velocity of escape	=6.94 statute miles per second
Curvature of surface	=0.8 foot per nautical mile

Airy ellipsoid

Equatorial radius (a)	=6,377,563.396 meters
	=3,443.609 nautical miles
Polar radius (b)	=6,356,256.91 meters
	=3,432.104 nautical miles
Mean radius $(2a+b)/3$	=6,370,461.234 meters
	=3,439.774 nautical miles
Flattening or ellipticity ($f=1-b/a$)	=1/299.325
	=0.00334085
Eccentricity ($e=(2f-f^2)^{1/2}$)	=0.081673374
Eccentricity squared (e^2)	=0.00667054

Australian National-South American ellipsoid of 1969

Equatorial radius (a)	=6,378,160 meters
	=3,443.931 nautical miles
Polar radius (b)	=6,356,774.719 meters
	=3,432.384 nautical miles
Mean radius $(2a+b)/3$	=6,371,031.573 meters
	=3,440.082 nautical miles
Flattening or ellipticity ($f=1-b/a$)	=1/298.25
	=0.00335289
Eccentricity ($e=(2f-f^2)^{1/2}$)	=0.0818202
Eccentricity squared (e^2)	=0.00669454

Bessel ellipsoid

Equatorial radius (a)	=6,377,397.155 meters
	=3,443.52 nautical miles
Polar radius (b)	=6,356,078.963 meters
	=3,432.01 nautical miles
Mean radius $(2a+b)/3$	=6,370,291.091 meters
	=3,439.682 nautical miles
Flattening or ellipticity ($f=1-b/a$)	=1/299.1528
	=0.00334277
Eccentricity ($e=(2f-f^2)^{1/2}$)	=0.08169683
Eccentricity squared (e^2)	=0.00667437

Clarke ellipsoid of 1866

Equatorial radius (a)	=6,378,206.4 meters
	=3,443.957 nautical miles
Polar radius (b)	=6,356,583.8 meters
	=3,432.281 nautical miles

Earth—Continued

Mean radius $(2a+b)/3$ = 6,370,998.9 meters
= 3,440.064 nautical miles

Flattening or ellipticity $(f=1-b/a)$ = 1/294.98
= 0.00339008

Eccentricity $(e=(2f-f^2)^{1/2})$ = 0.08227185

Eccentricity squared (e^2) = 0.00676866

Clarke ellipsoid of 1880

Equatorial radius (a) = 6,378,249.145 meters
= 3,443.98 nautical miles

Polar radius (b) = 6,356,514.87 meters
= 3,432.245 nautical miles

Mean radius $(2a+b)/3$ = 6,371,004.387 meters
= 3,440.067 nautical miles

Flattening or ellipticity $(f=1-b/a)$ = 1/293.465
= 0.00340756

Eccentricity $(e=(2f-f^2)^{1/2})$ = 0.0824834

Eccentricity squared (e^2) = 0.00680351

Everest ellipsoid

Equatorial radius (a) = 6,377,276.345 meters
= 3,443.454 nautical miles

Polar radius (b) = 6,356,075.413 meters
= 3,432,006 nautical miles

Mean radius $(2a+b)/3$ = 6,370,209.37 meters
= 3,439.638 nautical miles

Flattening or ellipticity $(f=1-b/a)$ = 1/300.8017
= 0.00332445

Eccentricity $(e=(2f-f^2)^{1/2})$ = 0.08147298

Eccentricity squared (e^2) = 0.00663785

Fischer ellipsoid of 1960 (Mercury Datum)

Equatorial radius (a) = 6,378,166 meters
= 3,443.934 nautical miles

Polar radius (b) = 6,356,784.284 meters
= 3,432.389 nautical miles

Mean radius $(2a+b)/3$ = 6,371,038.761 meters
= 3,440.086 nautical miles

Flattening or ellipticity $(f=1-b/a)$ = 1/298.3
= 0.00335233

Eccentricity $(e=(2f-f^2)^{1/2})$ = 0.081813334

Eccentricity squared (e^2) = 0.00669342

Fischer South Asia ellipsoid of 1960

Equatorial radius (a) = 6,378,155 meters
= 3,443.928 nautical miles

Polar radius (b) = 6,356,773.32 meters
= 3,432.383 nautical miles

Mean radius $(2a+b)/3$ = 6,371,027.773 meters
= 3,440.08 nautical miles

Flattening or ellipticity $(f=1-b/a)$ = 1/298.3
= 0.00335233

Eccentricity $(e=(2f-f^2)^{1/2})$ = 0.081813334

Eccentricity squared (e^2) = 0.00669342

Fischer ellipsoid of 1968

Equatorial radius (a) = 6,378,150 meters
= 3,443.925 nautical miles

Polar radius (b) = 6,356,768.337 meters
= 3,432.381 noutical miles

Earth—Continued

Mean radius $(2a+b)/3$	$=6,371,022.985$ meters
	$=3,440.077$ nautical miles
Flattening or ellipticity $(f=1-b/a)$	$=1/298.3$
	$=0.00335233$
Eccentricity $(e=(2f-f^2)^{1/2})$	$=0.08181333$
Eccentricity squared (e^2)	$=0.00669342$

Hough ellipsoid

Equatorial radius (a)	$=6,378,270$ meters
	$=3,443.99$ nautical miles
Polar radius (b)	$=6,356,794.343$ meters
	$=3,432.394$ nautical miles
Mean radius $(2a+b)/3$	$=6,371,111.448$ meters
	$=3,440.125$ nautical miles
Flattening or ellipticity $(f=1-b/a)$	$=1/297$
	$=0.003367003$
Eccentricity $(e=(2f-f^2)^{1/2})$	$=0.08199189$
Eccentricity squared (e^2)	$=0.00672267$

International ellipsoid

Equatorial radius (a)	$=6,378,388$ meters
	$=3,444.054$ nautical miles
Polar radius (b)	$=6,356,911.946$ meters
	$=3,432.459$ nautical miles
Mean radius $(2a+b)/3$	$=6,371,229.315$ meters
	$=3,440.19$ nautical miles
Flattening or ellipticity $(f=1-b/a)$	$=1/297$
	$=0.003367003$
Eccentricity $(e=(2f-f^2)^{1/2})$	$=0.08199189$
Eccentricity squared (e^2)	$=0.00672267$

International Astronomical Union figure of earth (1968)

Equatorial radius (a)	$=6,378,160$ meters
	$=3,443.931$ nautical miles
Polar radius (b)	$=6,356,774.719$ meters
	$=3,432.384$ nautical miles
Mean radius $(2a+b)/3$	$=6,371,031.573$ meters
	$=3,440.082$ nautical miles
Flattening or ellipticity $(f=1-b/a)$	$=1/298.25$
	$=0.00335289$
Eccentricity $(e=(2f-f^2)^{1/2})$	$=0.0818202$
Eccentricity squared (e^2)	$=0.00669454$

Krassovskiy ellipsoid

Equatorial radius (a)	$=6,378,245$ meters
	$=3,443.977$ nautical miles
Polar radius (b)	$=6,356,863.019$ meters
	$=3,432.43$ nautical miles
Mean radius $(2a+b)/3$	$=6,371,117.673$ meters
	$=3,440.128$ nautical miles
Flattening or ellipticity $(f=1-b/a)$	$=1/298.3$
	$=0.00335233$
Eccentricity $(e=(2f-f^2)^{1/2})$	$=0.08181333$
Eccentricity squared (e^2)	$=0.00669342$

World Geodetic System (WGS) ellipsoid of 1972

Equatorial radius (a)	$=6,378,135$ meters
	$=3,443.917$ nautical miles

Earth—Continued

Polar radius (b)	=6,356,750.52 meters
	=3,432.371 nautical miles
Mean radius $(2a+b)/3$	=6,371,006.84 meters
	=3,440.068 nautical miles
Flattening or ellipticity ($f=1-b/a$)	=1/298.26
	=0.00335278
Eccentricity ($e=(2f-f^2)^{1/2}$)	=0.0818188
Eccentricity squared (e^2)	=0.00669432

Length

1 inch	=25.4 millimeters*
	=2.54 centimeters*
1 foot (U.S.)	=12 inches*
	=1 British foot
	=⅓ yard*
	=0.3048 meter*
	=⅙ fathom*
1 foot (U.S. Survey)	=0.30480061 meter
1 yard	=36 inches*
	=3 feet*
	=0.9144 meter*
1 fathom	=6 feet*
	=2 yards*
	=1.8288 meters*
1 cable	=720 feet*
	=240 yards*
	=219.4560 meters*
1 cable (British)	=0.1 nautical mile
1 statute mile	=5,280 feet*
	=1,760 yards*
	=1,609.344 meters*
	=1.609344 kilometers*
	=0.86897624 nautical mile
1 nautical mile	=6,076.11548556 feet
	=2,025.37182852 yards
	=1,852 meters*
	=1.852 kilometers*
	=1.150779448 statute miles
1 meter	=100 centimeters*
	=39.370079 inches
	=3.28083990 feet
	=1.09361330 yards
	=0.54680665 fathom
	=0.00062137 statute mile
	=0.00053996 nautical mile
1 kilometer	=3,280.83990 feet
	=1,093.61330 yards
	=1,000 meters*
	=0.62137119 statute mile
	=0.53995680 nautical mile

Mass

1 ounce	=437.5 grains*
	=28.349523125 grams*
	=0.0625 pound*
	=0.028349523125 kilogram*

Mass—Continued

1 pound	=7,000 grains*
	=16 ounces*
	=0.45359237 kilogram*
1 short ton	=2,000 pounds*
	=907.18474 kilograms*
	=0.90718474 metric ton*
	=0.8928571 long ton
1 long ton	=2,240 pounds*
	=1,016.0469088 kilograms*
	=1.12 short tons*
	=1.0160469088 metric tons*
1 kilogram	=2.204623 pounds
	=0.00110231 short ton
	=0.0009842065 long ton
1 metric ton	=2,204.623 pounds
	=1,000 kilograms*
	=1.102311 short tons
	=0.9842065 long ton

Mathematics

π	=3.1415926535897932384626433832795028841971
π^2	=9.8696044011
$\sqrt{\pi}$	=1.7724538509
Base of Naperian logarithms (e)	=2.718281828459
Modulus of common logarithms ($\log_{10}e$)	=0.4342944819032518
1 radian	=206,264″.80625
	=3,437′.7467707849
	=57°.2957795131
	=57°17′44″.80625
1 circle	=1,296,000″*
	=21,600′*
	=360°*
	=2π radians*
180°	=π radians*
1°	=3600″*
	=60′*
	=0.0174532925199432957666 radian
1′	=60″*
	=0.000290888208665721596 radian
1″	=0.0000048481368110953599 33 radian
Sine of 1′	=0.000290888204563424 60
Sine of 1″	=0.00000484813681107637

Meteorology

Atmosphere (dry air)

Nitrogen	=78.08%	99.99% (Nitrogen, Oxygen, Argon, Carbon dioxide)
Oxygen	=20.95%	
Argon	= 0.93%	
Carbon dioxide	= 0.03%	
Neon	= 0.0018%	
Helium	= 0.000524%	
Krypton	= 0.0001%	
Hydrogen	= 0.00005%	
Xenon	= 0.0000087%	
Ozone	= 0 to 0.000007% (increasing with altitude)	
Radon	= 0.0000000000000000006% (decreasing with altitude)	

Meteorology – Continued

Standard atmospheric pressure at sea level	= 1,013.250 dynes per square centimeter*
	= 1,033.227 grams per square centimeter
	= 1,033.227 centimeters of water
	= 1,013.250 millibars*
	= 760 millimeters of mercury
	= 76 centimeters of mercury
	= 33.8985 feet of water
	= 29.92126 inches of mercury
	= 14.6960 pounds per square inch
	= **1.033227 kilograms per square centimeter**
	= 1.013250 bars*
Absolute zero	= (−) 273°.16 C
	= (−) 459°.69 F

Pressure

1 dyne per square centimeter	= 0.001 millibar*
	= 0.000001 bar*
1 gram per square centimeter	= 1 centimeter of water
	= 0.980665 millibar*
	= 0.07355592 centimeter of mercury
	= 0.0289590 inch of mercury
	= 0.0142233 pound per square inch
	= 0.001 kilogram per square centimeter*
	= 0.000967841 atmosphere
1 millibar	= 1,000 dynes per square centimeter*
	= 1.01971621 grams per square centimeter
	= 0.7500617 millimeter of mercury
	= **0.03345526 foot of water**
	= **0.02952998 inch of mercury**
	= 0.01450377 pound per square inch
	= 0.001 bar*
	= 0.00098692 atmosphere
1 millimeter of mercury	= 1.35951 grams per square centimeter
	= 1.3332237 millibars
	= 0.1 centimeter of mercury*
	= 0.04460334 foot of water
	= 0.039370079 inch of mercury
	= 0.01933677 pound per square inch
	= 0.001315790 atmosphere
1 centimeter of mercury	= 10 millimeters of mercury*
1 inch of mercury	= 34.53155 grams per square centimeter
	= 33.86389 millibars
	= 25.4 millimeters of mercury*
	= 1.132925 feet of water
	= 0.4911541 pound per square inch
	= 0.03342106 atmosphere
1 centimeter of water	= 1 gram per square centimeter
	= 0.001 kilogram per square centimeter
1 foot of water	= 30.48000 grams per square centimeter
	= 29.89067 millibars
	= 2.241985 centimeters of mercury
	= 0.882671 inch of mercury
	= 0.4335275 pound per square inch
	= 0.02949980 atmosphere
1 pound per square inch	= 68,947.57 dynes per square centimeter
	= 70.30696 grams per square centimeter
	= 70.30696 centimeters of water

Pressure – Continued

1 pound per square inch	=68.94757 millibars
	=51.71493 millimeters of mercury
	=5.171493 centimeters of mercury
	=2.306659 feet of water
	=2.036021 inches of mercury
	=0.07030696 kilogram per square centimeter
	=0.06894757 bar
	=0.06804596 atmosphere
1 kilogram per square centimeter	=1,000 grams per square centimeter*
	=1,000 centimeters of water
1 bar	=1,000,000 dynes per square centimeter*
	=1,000 millibars*

Speed

1 foot per minute	=0.01666667 foot per second
	=0.00508 meter per second*
1 yard per minute	=3 feet per minute*
	=0.05 foot per second*
	=0.03409091 statute mile per hour
	=0.02962419 knot
	=0.01524 meter per second*
1 foot per second	=60 feet per minute*
	=20 yards per minute*
	=1.09728 kilometers per hour*
	=0.68181818 statute mile per hour
	=0.59248380 knot
	=0.3048 meter per second*
1 statute mile per hour	=88 feet per minute*
	=29.33333333 yards per minute
	=1.609344 kilometers per hour*
	=1.46666667 feet per second
	=0.86897624 knot
	=0.44704 meter per second*
1 knot	=101.26859143 feet per minute
	=33.75619714 yards per minute
	=1.852 kilometers per hour*
	=1.68780986 feet per second
	=1.15077945 statute miles per hour
	=0.51444444 meter per second
1 kilometer per hour	=0.62137119 statute mile per hour
	=0.53995680 knot
1 meter per second	=196.85039340 feet per minute
	=65.6167978 yards per minute
	=3.6 kilometers per hour*
	=3.28083990 feet per second
	=2.23693632 statute miles per hour
	=1.94384449 knots
Light in vacuo	=299,792.5 kilometers per second
	=186,282 statute miles per second
	=161,875 nautical miles per second
	=983.570 feet per microsecond
Light in air	=299,708 kilometers per second
	=186,230 statute miles per second
	=161,829 nautical miles per second
	=983.294 feet per microsecond

Speed—Continued

Sound in dry air at 59°F or 15°C and standard sea level pressure	=1,116.45 feet per second
	=761.22 statute miles per hour
	=661.48 knots
	=340.29 meters per second
Sound in 3.485 percent saltwater at 60°F	=4,945.37 feet per second
	=3,371.85 statute miles per hour
	=2,930.05 knots
	=1,507.35 meters per second

Volume

1 cubic inch	=16.387064 cubic centimeters*
	=0.016387064 liter*
	=0.004329004 gallon
1 cubic foot	=1,728 cubic inches*
	=28.316846592 liters*
	=7.480519 U.S. gallons
	=6.228822 imperial (British) gallons
	=0.028316846592 cubic meter*
1 cubic yard	=46,656 cubic inches*
	=764.554857984 liters*
	=201.974026 U.S. gallons
	=168.1782 imperial (British) gallons
	=27 cubic feet*
	=0.764554857984 cubic meter*
1 milliliter	=0.06102374 cubic inch
	=0.0002641721 U.S. gallon
	=0.00021997 imperial (British) gallon
1 cubic meter	=264.172035 U.S. gallons
	=219.96878 imperial (British) gallons
	=35.31467 cubic feet
	=1.307951 cubic yards
1 quart (U.S.)	=57.75 cubic inches*
	=32 fluid ounces*
	=2 pints*
	=0.9463529 liter
	=0.25 gallon*
1 gallon (U.S.)	=3,785.412 milliliters
	=231 cubic inches*
	=0.1336806 cubic foot
	=4 quarts*
	=3.785412 liters
	=0.8326725 imperial (British) gallon
1 liter	=1,000 milliliters
	=61.02374 cubic inches
	=1.056688 quarts
	=0.2641721 gallon
1 register ton	=100 cubic feet*
	=2.8316846592 cubic meters*
1 measurement ton	=40 cubic feet*
	=1 freight ton*
1 freight ton	=40 cubic feet*
	=1 measurement ton*

Volume-mass

1 cubic foot of seawater	=64 pounds

Volume-Mass – Continued

1 cubic foot of freshwater	=62.428 pounds at temperature of maximum density (4°C=39°.2F)
1 cubic foot of ice	=56 pounds
1 displacement ton	=35 cubic feet of seawater*
	=1 long ton

Vessel Tonnage

The several kinds of *vessel tonnage* are as follows:

Gross tonnage, or *gross register tonnage*, is the total cubical capacity of a ship expressed in *register* tons of 100 cubic feet, or 2.83 cubic meters, less such space as hatchways, bakeries, galleys, etc., as are exempted from measurement by different governments. There is some lack of uniformity in the gross tonnages as given by different nations on account of lack of agreement on the spaces that are to be exempted.

Official merchant marine statistics of most countries are published in terms of the *gross register tonnage*. Press references to *ship tonnage* are usually to the *gross tonnage*.

The *net tonnage*, or *net register tonnage*, is the *gross tonnage* less the different spaces specified by maritime nations in their measurement rules and laws. The spaces that are deducted are those totally unavailable for carrying cargo, such as the engine room, coal bunkers, crews quarters, chart and instrument room, etc.

The *net tonnage* is used in computing the amount of cargo that can be loaded on a ship. It is used as the basis for wharfage and other similar charges.

The *register under-deck tonnage* is the cubical capacity of a ship under her tonnage deck expressed in register tons. In a vessel having more than one deck the tonnage deck is the second from the keel.

There are several variations of *displacement tonnage*.

The *dead weight tonnage* is the difference between the "loaded" and "light" *displacement tonnages* of a vessel. It is espressed in terms of the long ton of 2,240 pounds, or the metric ton of 2,204.6 pounds, and is the weight of fuel, passengers, and cargo that a vessel can carry when loaded to her maximum draft.

The second variety of tonnage, *cargo tonnage*, refers to the weight of the particular items making up the cargo. In overseas traffic it is usually expressed in long tons of 2,240 pounds or metric tons of 2,204.6 pounds. The short ton is only occassionally used. The *cargo tonnage* is therefore very distinct from *vessel tonnage*.

NOTE:—All values in this appendix are based on the following relationships:
1 inch=2.54 centimeters*
1 yard=0.9144 meter*
1 pound (avoirdupois)=0.45359237 kilogram*
1 nautical mile=1852 meters*
Absolute zero=(−)273°.16C=(−)459°.69F.

Decibel Scale

The **decibel** (**dB**) is 10 times the logarithm to the base 10 of the ratio of two amounts of power. The decibel scale is used to express conveniently the ratio between widely different powers.

The ratio between one power P_1 and a second power P_2 is expressed in dB's as:

$$10 \log_{10}\left(\frac{P_1}{P_2}\right).$$

Thus if P_1 is 1,000 times P_2, their ratio is expressed as

$$10 \log_{10} (10) = 10 \times 3 = 30 \text{ dB}.$$

If P_2 is 1,000 times P_1, their ratio is expressed as

$$10 \log (10) = 10 \times (-3) = -30 \text{ dB}.$$

Power ratio	dB
1	0
2	3
4	6
10	10
100	20
1, 000	30
10, 000	40
100, 000	50
1, 000, 000	60

Prefixes to Form Decimal Multiples and Sub-Multiples of International System of Units (SI)

Multiplying factor	Prefix	Symbol
1 000 000 000 000 $=10^{12}$	tera	T
1 000 000 000 $=10^{9}$	giga	G
1 000 000 $=10^{6}$	mega	M
1 000 $=10^{3}$	kilo	k
100 $=10^{2}$	hecto	h
10 $=10^{1}$	deka	da
0. 1 $=10^{-1}$	deci	d
0. 01 $=10^{-2}$	centi	c
0. 001 $=10^{-3}$	milli	m
0. 000 001 $=10^{-6}$	micro	μ
0. 000 000 001 $=10^{-9}$	nano	n
0. 000 000 000 001 $=10^{-12}$	pico	p
0. 000 000 000 000 001 $=10^{-15}$	femto	f
0. 000 000 000 000 000 001 $=10^{-18}$	atto	a

APPENDIX E

NAVIGATIONAL COORDINATES

Coordinate	Symbol	Measured from	Measured along	Direction	Measured to	Units	Precision	Maximum value	Labels
latitude	L, lat.	equator	meridian	N, S	parallel	°, ′	0′.1	90°	N, S
colatitude	colat.	poles	meridian	S, N	parallel	°, ′	0′.1	90°	—
longitude	λ, long.	prime meridian	parallel	E, W	local meridian	°, ′	0′.1	180°	E, W
declination	d, dec.	celestial equator	hour circle	N, S	parallel of declination	°, ′	0′.1	90°	N, S
polar distance	p	elevated pole	hour circle	S, N	parallel of declination	°, ′	0′.1	180°	—
altitude	h	horizon	vertical circle	up	parallel of altitude	°, ′	0′.1	90°*	—
zenith distance	z	zenith	vertical circle	down	parallel of altitude	°, ′	0′.1	180°	—
azimuth	Zn	north	horizon	E	vertical circle	°	0°.1	360°	—
azimuth angle	Z	north, south	horizon	E, W	vertical circle	°	0°.1	180° or 90°	N, S . . . E, W
amplitude	A	east, west	horizon	N, S	body	°	0°.1	90°	E, W . . . N, S
Greenwich hour angle	GHA	Greenwich celestial meridian	parallel of declination	W	hour circle	°, ′	0′.1	360°	—
local hour angle	LHA	local celestial meridian	parallel of declination	W	hour circle	°, ′	0′.1	360°	—
meridian angle	t	local celestial meridian	parallel of declination	E, W	hour circle	°, ′	0′.1	180°	E, W
sidereal hour angle	SHA	hour circle of vernal equinox	parallel of declination	W	hour circle	°, ′	0′.1	360°	—
right ascension	RA	hour circle of vernal equinox	parallel of declination	E	hour circle	h, m, s	1^s	24^h	—
Greenwich mean time	GMT	lower branch Greenwich celestial meridian	parallel of declination	W	hour circle mean sun	h, m, s	1^s	24^h	—
local mean time	LMT	lower branch local celestial meridian	parallel of declination	W	hour circle mean sun	h, m, s	1^s	24^h	—
zone time	ZT	lower branch zone celestial meridian	parallel of declination	W	hour circle mean sun	h, m, s	1^s	24^h	—
Greenwich apparent time	GAT	lower branch Greenwich celestial meridian	parallel of declination	W	hour circle apparent sun	h, m, s	1^s	24^h	—
local apparent time	LAT	lower branch local celestial meridian	parallel of declination	W	hour circle apparent sun	h, m, s	1^s	24^h	—
Greenwich sidereal time	GST	Greenwich celestial meridian	parallel of declination	W	hour circle vernal equinox	h, m, s	1^s	24^h	—
local sidereal time	LST	local celestial meridian	parallel of declination	W	hour circle vernal equinox	h, m, s	1^s	24^h	—

*When measured from celestial horizon.

APPENDIX F

EXTRACTS FROM *NAUTICAL ALMANAC*

ALTITUDE CORRECTION TABLES 10°-90°—SUN, STARS, PLANETS

SUN — OCT.—MAR.

App. Alt.	Lower Limb	Upper Limb
° ′	′	′
9 34		
	+10·8	−21·5
9 45		
	+10·9	−21·4
9 56		
	+11·0	−21·3
10 08		
	+11·1	−21·2
10 21		
	+11·2	−21·1
10 34		
	+11·3	−21·0
10 47		
	+11·4	−20·9
11 01		
	+11·5	−20·8
11 15		
	+11·6	−20·7
11 30		
	+11·7	−20·6
11 46		
	+11·8	−20·5
12 02		
	+11·9	−20·4
12 19		
	+12·0	−20·3
12 37		
	+12·1	−20·2
12 55		
	+12·2	−20·1
13 14		
	+12·3	−20·0
13 35		
	+12·4	−19·9
13 56		
	+12·5	−19·8
14 18		
	+12·6	−19·7
14 42		
	+12·7	−19·6
15 06		
	+12·8	−19·5
15 32		
	+12·9	−19·4
15 59		
	+13·0	−19·3
16 28		
	+13·1	−19·2
16 59		
	+13·2	−19·1
17 32		
	+13·3	−19·0
18 06		
	+13·4	−18·9
18 42		
	+13·5	−18·8
19 21		
	+13·6	−18·7
20 03		
	+13·7	−18·6
20 48		
	+13·8	−18·5
21 35		
	+13·9	−18·4
22 26		
	+14·0	−18·3
23 22		
	+14·1	−18·2
24 21		
	+14·2	−18·1
25 26		
	+14·3	−18·0
26 36		
	+14·4	−17·9
27 52		
	+14·5	−17·8
29 15		
	+14·6	−17·7
30 46		
	+14·7	−17·6
32 26		
	+14·8	−17·5
34 17		
	+14·9	−17·4
36 20		
	+15·0	−17·3
38 36		
	+15·1	−17·2
41 08		
	+15·2	−17·1
43 59		
	+15·3	−17·0
47 10		
	+15·4	−16·9
50 46		
	+15·5	−16·8
54 49		
	+15·6	−16·7
59 23		
	+15·7	−16·6
64 30		
	+15·8	−16·5
70 12		
	+15·9	−16·4
76 26		
	+16·0	−16·3
83 05		
	+16·1	−16·2
90 00		

SUN — APR.—SEPT.

App. Alt.	Lower Limb	Upper Limb
° ′	′	′
9 39		
	+10·6	−21·2
9 51		
	+10·7	−21·1
10 03		
	+10·8	−21·0
10 15		
	+10·9	−20·9
10 27		
	+11·0	−20·8
10 40		
	+11·1	−20·7
10 54		
	+11·2	−20·6
11 08		
	+11·3	−20·5
11 23		
	+11·4	−20·4
11 38		
	+11·5	−20·3
11 54		
	+11·6	−20·2
12 10		
	+11·7	−20·1
12 28		
	+11·8	−20·0
12 46		
	+11·9	−19·9
13 05		
	+12·0	−19·8
13 24		
	+12·1	−19·7
13 45		
	+12·2	−19·6
14 07		
	+12·3	−19·5
14 30		
	+12·4	−19·4
14 54		
	+12·5	−19·3
15 19		
	+12·6	−19·2
15 46		
	+12·7	−19·1
16 14		
	+12·8	−19·0
16 44		
	+12·9	−18·9
17 15		
	+13·0	−18·8
17 48		
	+13·1	−18·7
18 24		
	+13·2	−18·6
19 01		
	+13·3	−18·5
19 42		
	+13·4	−18·4
20 25		
	+13·5	−18·3
21 11		
	+13·6	−18·2
22 00		
	+13·7	−18·1
22 54		
	+13·8	−18·0
23 51		
	+13·9	−17·9
24 53		
	+14·0	−17·8
26 00		
	+14·1	−17·7
27 13		
	+14·2	−17·6
28 33		
	+14·3	−17·5
30 00		
	+14·4	−17·4
31 35		
	+14·5	−17·3
33 20		
	+14·6	−17·2
35 17		
	+14·7	−17·1
37 26		
	+14·8	−17·0
39 50		
	+14·9	−16·9
42 31		
	+15·0	−16·8
45 31		
	+15·1	−16·7
48 55		
	+15·2	−16·6
52 44		
	+15·3	−16·5
57 02		
	+15·4	−16·4
61 51		
	+15·5	−16·3
67 17		
	+15·6	−16·2
73 16		
	+15·7	−16·1
79 43		
	+15·8	−16·0
86 32		
	+15·9	−15·9
90 00		

STARS AND PLANETS

App. Alt.	Corr^n	App. Alt.	Additional Corr^n
° ′	′		**1975**
9 56			**VENUS**
	−5·3		Jan. 1 – June 7
10 08		°	′
	−5·2	0	+0·1
10 20		42	
	−5·1		June 8 – July 21
10 33		0	+0·3
	−5·0	46	
10 46			July 22 – Aug. 6
	−4·9	0	+0·4
11 00		11	+0·5
	−4·8	41	
11 14			Aug. 7 – Aug. 15
	−4·7	0	+0·5
11 29		6	+0·6
	−4·6	20	+0·7
11 45		31	
	−4·5		Aug. 16 – Sept. 10
12 01		0	+0·6
	−4·4	4	+0·7
12 18		12	+0·8
	−4·3	22	
12 35			Sept. 11 – Sept. 19
	−4·2	0	+0·5
12 54		6	+0·6
	−4·1	20	+0·7
13 13		31	
	−4·0		Sept. 20 – Oct. 5
13 33		0	+0·4
	−3·9	11	+0·5
13 54		41	
	−3·8		Oct. 6 – Nov. 22
14 16		0	+0·3
	−3·7	46	
14 40			Nov. 23 – Dec. 31
	−3·6	0	+0·1
15 04		42	
	−3·5		**MARS**
15 30			Jan. 1 – Sept. 8
	−3·4	0	+0·1
15 57		60	
	−3·3		Sept. 9 – Nov. 22
16 26		0	+0·2
	−3·2	41	+0·1
16 56		75	
	−3·1		Nov. 23 – Dec. 31
17 28		0	+0·3
	−3·0	34	+0·2
18 02		60	+0·1
	−2·9	80	
18 38			
	−2·8		
19 17			
	−2·7		
19 58			
	−2·6		
20 42			
	−2·5		
21 28			
	−2·4		
22 19			
	−2·3		
23 13			
	−2·2		
24 11			
	−2·1		
25 14			
	−2·0		
26 22			
	−1·9		
27 36			
	−1·8		
28 56			
	−1·7		
30 24			
	−1·6		
32 00			
	−1·5		
33 45			
	−1·4		
35 40			
	−1·3		
37 48			
	−1·2		
40 08			
	−1·1		
42 44			
	−1·0		
45 36			
	−0·9		
48 47			
	−0·8		
52 18			
	−0·7		
56 11			
	−0·6		
60 28			
	−0·5		
65 08			
	−0·4		
70 11			
	−0·3		
75 34			
	−0·2		
81 13			
	−0·1		
87 03			
	0·0		
90 00			

DIP

Ht. of Eye	Corr^n	Ht. of Eye
m	′	ft.
2·4		8·0
	−2·8	
2·6		8·6
	−2·9	
2·8		9·2
	−3·0	
3·0		9·8
	−3·1	
3·2		10·5
	−3·2	
3·4		11·2
	−3·3	
3·6		11·9
	−3·4	
3·8		12·6
	−3·5	
4·0		13·3
	−3·6	
4·3		14·1
	−3·7	
4·5		14·9
	−3·8	
4·7		15·7
	−3·9	
5·0		16·5
	−4·0	
5·2		17·4
	−4·1	
5·5		18·3
	−4·2	
5·8		19·1
	−4·3	
6·1		20·1
	−4·4	
6·3		21·0
	−4·5	
6·6		22·0
	−4·6	
6·9		22·9
	−4·7	
7·2		23·9
	−4·8	
7·5		24·9
	−4·9	
7·9		26·0
	−5·0	
8·2		27·1
	−5·1	
8·5		28·1
	−5·2	
8·8		29·2
	−5·3	
9·2		30·4
	−5·4	
9·5		31·5
	−5·5	
9·9		32·7
	−5·6	
10·3		33·9
	−5·7	
10·6		35·1
	−5·8	
11·0		36·3
	−5·9	
11·4		37·6
	−6·0	
11·8		38·9
	−6·1	
12·2		40·1
	−6·2	
12·6		41·5
	−6·3	
13·0		42·8
	−6·4	
13·4		44·2
	−6·5	
13·8		45·5
	−6·6	
14·2		46·9
	−6·7	
14·7		48·4
	−6·8	
15·1		49·8
	−6·9	
15·5		51·3
	−7·0	
16·0		52·8
	−7·1	
16·5		54·3
	−7·2	
16·9		55·8
	−7·3	
17·4		57·4
	−7·4	
17·9		58·9
	−7·5	
18·4		60·5
	−7·6	
18·8		62·1
	−7·7	
19·3		63·8
	−7·8	
19·8		65·4
	−7·9	
20·4		67·1
	−8·0	
20·9		68·8
	−8·1	
21·4		70·5

Ht. of Eye	Corr^n
m	′
1·0	−1·8
1·5	−2·2
2·0	−2·5
2·5	−2·8
3·0	−3·0
See table ←	
m	′
20	−7·9
22	−8·3
24	−8·6
26	−9·0
28	−9·3
30	−9·6
32	−10·0
34	−10·3
36	−10·6
38	−10·8
40	−11·1
42	−11·4
44	−11·7
46	−11·9
48	−12·2
ft.	′
2	−1·4
4	−1·9
6	−2·4
8	−2·7
10	−3·1
See table ←	
ft.	′
70	−8·1
75	−8·4
80	−8·7
85	−8·9
90	−9·2
95	−9·5
100	−9·7
105	−9·9
110	−10·2
115	−10·4
120	−10·6
125	−10·8
130	−11·1
135	−11·3
140	−11·5
145	−11·7
150	−11·9
155	−12·1

App. Alt. = Apparent altitude = Sextant altitude corrected for index error and dip.

For daylight observations of Venus, see page 260.

ALTITUDE CORRECTION TABLES 0°–10°—SUN, STARS, PLANETS A3

App. Alt.	OCT.–MAR. SUN Lower Limb	OCT.–MAR. SUN Upper Limb	APR.–SEPT. SUN Lower Limb	APR.–SEPT. SUN Upper Limb	STARS PLANETS
° ′	′	′	′	′	′
0 00	−18·2	−50·5	−18·4	−50·2	−34·5
03	17·5	49·8	17·8	49·6	33·8
06	16·9	49·2	17·1	48·9	33·2
09	16·3	48·6	16·5	48·3	32·6
12	15·7	48·0	15·9	47·7	32·0
15	15·1	47·4	15·3	47·1	31·4
0 18	−14·5	−46·8	−14·8	−46·6	−30·8
21	14·0	46·3	14·2	46·0	30·3
24	13·5	45·8	13·7	45·5	29·8
27	12·9	45·2	13·2	45·0	29·2
30	12·4	44·7	12·7	44·5	28·7
33	11·9	44·2	12·2	44·0	28·2
0 36	−11·5	−43·8	−11·7	−43·5	−27·8
39	11·0	43·3	11·2	43·0	27·3
42	10·5	42·8	10·8	42·6	26·8
45	10·1	42·4	10·3	42·1	26·4
48	9·6	41·9	9·9	41·7	25·9
51	9·2	41·5	9·5	41·3	25·5
0 54	− 8·8	−41·1	− 9·1	−40·9	−25·1
0 57	8·4	40·7	8·7	40·5	24·7
1 00	8·0	40·3	8·3	40·1	24·3
03	7·7	40·0	7·9	39·7	24·0
06	7·3	39·6	7·5	39·3	23·6
09	6·9	39·2	7·2	39·0	23·2
1 12	− 6·6	−38·9	− 6·8	−38·6	−22·9
15	6·2	38·5	6·5	38·3	22·5
18	5·9	38·2	6·2	38·0	22·2
21	5·6	37·9	5·8	37·6	21·9
24	5·3	37·6	5·5	37·3	21·6
27	4·9	37·2	5·2	37·0	21·2
1 30	− 4·6	−36·9	− 4·9	−36·7	−20·9
35	4·2	36·5	4·4	36·2	20·5
40	3·7	36·0	4·0	35·8	20·0
45	3·2	35·5	3·5	35·3	19·5
50	2·8	35·1	3·1	34·9	19·1
1 55	2·4	34·7	2·6	34·4	18·7
2 00	− 2·0	−34·3	− 2·2	−34·0	−18·3
05	1·6	33·9	1·8	33·6	17·9
10	1·2	33·5	1·5	33·3	17·5
15	0·9	33·2	1·1	32·9	17·2
20	0·5	32·8	0·8	32·6	16·8
25	− 0·2	32·5	0·4	32·2	16·5
2 30	+ 0·2	−32·1	− 0·1	−31·9	−16·1
35	0·5	31·8	+ 0·2	31·6	15·8
40	0·8	31·5	0·5	31·3	15·5
45	1·1	31·2	0·8	31·0	15·2
50	1·4	30·9	1·1	30·7	14·9
2 55	1·6	30·7	1·4	30·4	14·7
3 00	+ 1·9	−30·4	+ 1·7	−30·1	−14·4
05	2·2	30·1	1·9	29·9	14·1
10	2·4	29·9	2·1	29·7	13·9
15	2·6	29·7	2·4	29·4	13·7
20	2·9	29·4	2·6	29·2	13·4
25	3·1	29·2	2·9	28·9	13·2
3 30	+ 3·3	−29·0	+ 3·1	−28·7	−13·0

App. Alt.	OCT.–MAR. SUN Lower Limb	OCT.–MAR. SUN Upper Limb	APR.–SEPT. SUN Lower Limb	APR.–SEPT. SUN Upper Limb	STARS PLANETS
° ′	′	′	′	′	′
3 30	+ 3·3	−29·0	+ 3·1	−28·7	−13·0
35	3·6	28·7	3·3	28·5	12·7
40	3·8	28·5	3·5	28·3	12·5
45	4·0	28·3	3·7	28·1	12·3
50	4·2	28·1	3·9	27·9	12·1
3 55	4·4	27·9	4·1	27·7	11·9
4 00	+ 4·5	−27·8	+ 4·3	−27·5	−11·8
05	4·7	27·6	4·5	27·3	11·6
10	4·9	27·4	4·6	27·2	11·4
15	5·1	27·2	4·8	27·0	11·2
20	5·2	27·1	5·0	26·8	11·1
25	5·4	26·9	5·1	26·7	10·9
4 30	+ 5·6	−26·7	+ 5·3	−26·5	−10·7
35	5·7	26·6	5·5	26·3	10·6
40	5·9	26·4	5·6	26·2	10·4
45	6·0	26·3	5·8	26·0	10·3
50	6·2	26·1	5·9	25·9	10·1
4 55	6·3	26·0	6·0	25·8	10·0
5 00	+ 6·4	−25·9	+ 6·2	−25·6	− 9·9
05	6·6	25·7	6·3	25·5	9·7
10	6·7	25·6	6·4	25·4	9·6
15	6·8	25·5	6·6	25·2	9·5
20	6·9	25·4	6·7	25·1	9·4
25	7·1	25·2	6·8	25·0	9·2
5 30	+ 7·2	−25·1	+ 6·9	−24·9	− 9·1
35	7·3	25·0	7·0	24·8	9·0
40	7·4	24·9	7·2	24·6	8·9
45	7·5	24·8	7·3	24·5	8·8
50	7·6	24·7	7·4	24·4	8·7
5 55	7·7	24·6	7·5	24·3	8·6
6 00	+ 7·8	−24·5	+ 7·6	−24·2	− 8·5
10	8·0	24·3	7·8	24·0	8·3
20	8·2	24·1	8·0	23·8	8·1
30	8·4	23·9	8·1	23·7	7·9
40	8·6	23·7	8·3	23·5	7·7
6 50	8·7	23·6	8·5	23·3	7·6
7 00	+ 8·9	−23·4	+ 8·6	−23·2	− 7·4
10	9·1	23·2	8·8	23·0	7·2
20	9·2	23·1	9·0	22·8	7·1
30	9·3	23·0	9·1	22·7	7·0
40	9·5	22·8	9·2	22·6	6·8
7 50	9·6	22·7	9·4	22·4	6·7
8 00	+ 9·7	−22·6	+ 9·5	−22·3	− 6·6
10	9·9	22·4	9·6	22·2	6·4
20	10·0	22·3	9·7	22·1	6·3
30	10·1	22·2	9·8	22·0	6·2
40	10·2	22·1	10·0	21·8	6·1
8 50	10·3	22·0	10·1	21·7	6·0
9 00	+10·4	−21·9	+10·2	−21·6	− 5·9
10	10·5	21·8	10·3	21·5	5·8
20	10·6	21·7	10·4	21·4	5·7
30	10·7	21·6	10·5	21·3	5·6
40	10·8	21·5	10·6	21·2	5·5
9 50	10·9	21·4	10·6	21·2	5·4
10 00	+11·0	−21·3	+10·7	−21·1	− 5·3

Additional corrections for temperature and pressure are given on the following page.

For bubble sextant observations ignore dip and use the star corrections for Sun, planets, and stars.

ALTITUDE CORRECTION TABLES—ADDITIONAL CORRECTIONS

ADDITIONAL REFRACTION CORRECTIONS FOR NON-STANDARD CONDITIONS

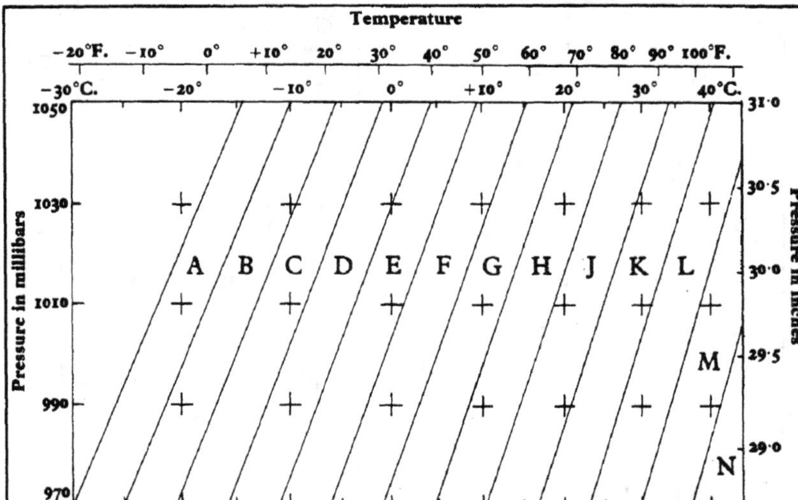

App. Alt.	A	B	C	D	E	F	G	H	J	K	L	M	N	App. Alt.
° ′	′	′	′	′	′	′	′	′	′	′	′	′	′	° ′
0 00	−6·9	−5·7	−4·6	−3·4	−2·3	−1·1	0·0	+1·1	+2·3	+3·4	+4·6	+5·7	+6·9	0 00
0 30	5·2	4·4	3·5	2·6	1·7	0·9	0·0	0·9	1·7	2·6	3·5	4·4	5·2	0 30
1 00	4·3	3·5	2·8	2·1	1·4	0·7	0·0	0·7	1·4	2·1	2·8	3·5	4·3	1 00
1 30	3·5	2·9	2·4	1·8	1·2	0·6	0·0	0·6	1·2	1·8	2·4	2·9	3·5	1 30
2 00	3·0	2·5	2·0	1·5	1·0	0·5	0·0	0·5	1·0	1·5	2·0	2·5	3·0	2 00
2 30	−2·5	−2·1	−1·6	−1·2	−0·8	−0·4	0·0	+0·4	+0·8	+1·2	+1·6	+2·1	+2·5	2 30
3 00	2·2	1·8	1·5	1·1	0·7	0·4	0·0	0·4	0·7	1·1	1·5	1·8	2·2	3 00
3 30	2·0	1·6	1·3	1·0	0·7	0·3	0·0	0·3	0·7	1·0	1·3	1·6	2·0	3 30
4 00	1·8	1·5	1·2	0·9	0·6	0·3	0·0	0·3	0·6	0·9	1·2	1·5	1·8	4 00
4 30	1·6	1·4	1·1	0·8	0·5	0·3	0·0	0·3	0·5	0·8	1·1	1·4	1·6	4 30
5 00	−1·5	−1·3	−1·0	−0·8	−0·5	−0·2	0·0	+0·2	+0·5	+0·8	+1·0	+1·3	+1·5	5 00
6	1·3	1·1	0·9	0·6	0·4	0·2	0·0	0·2	0·4	0·6	0·9	1·1	1·3	6
7	1·1	0·9	0·7	0·6	0·4	0·2	0·0	0·2	0·4	0·6	0·7	0·9	1·1	7
8	1·0	0·8	0·7	0·5	0·3	0·2	0·0	0·2	0·3	0·5	0·7	0·8	1·0	8
9	0·9	0·7	0·6	0·4	0·3	0·1	0·0	0·1	0·3	0·4	0·6	0·7	0·9	9
10 00	−0·8	−0·7	−0·5	−0·4	−0·3	−0·1	0·0	+0·1	+0·3	+0·4	+0·5	+0·7	+0·8	10 00
12	0·7	0·6	0·5	0·3	0·2	0·1	0·0	0·1	0·2	0·3	0·5	0·6	0·7	12
14	0·6	0·5	0·4	0·3	0·2	0·1	0·0	0·1	0·2	0·3	0·4	0·5	0·6	14
16	0·5	0·4	0·3	0·3	0·2	0·1	0·0	0·1	0·2	0·3	0·3	0·4	0·5	16
18	0·4	0·4	0·3	0·2	0·2	0·1	0·0	0·1	0·2	0·2	0·3	0·4	0·4	18
20 00	−0·4	−0·3	−0·3	−0·2	−0·1	−0·1	0·0	+0·1	+0·1	+0·2	+0·3	+0·3	+0·4	20 00
25	0·3	0·3	0·2	0·2	0·1	−0·1	0·0	+0·1	0·1	0·2	0·2	0·3	0·3	25
30	0·3	0·2	0·2	0·1	0·1	0·0	0·0	0·0	0·1	0·1	0·2	0·2	0·3	30
35	0·2	0·2	0·1	0·1	0·1	0·0	0·0	0·0	0·1	0·1	0·1	0·2	0·2	35
40	0·2	0·1	0·1	0·1	−0·1	0·0	0·0	0·0	+0·1	0·1	0·1	0·1	0·2	40
50 00	−0·1	−0·1	−0·1	−0·1	0·0	0·0	0·0	0·0	0·0	+0·1	+0·1	+0·1	+0·1	50 00

The graph is entered with arguments temperature and pressure to find a zone letter; using as arguments this zone letter and apparent altitude (sextant altitude corrected for dip), a correction is taken from the table. This correction is to be applied to the sextant altitude in addition to the corrections for standard conditions (for the Sun, planets and stars from the inside front cover and for the Moon from the inside back cover).

1975 MAY 25, 26, 27 (SUN., MON., TUES.)

G.M.T.		ARIES	VENUS −3.8		MARS +1.0		JUPITER −1.7		SATURN +0.4		STARS		
		G.H.A.	G.H.A.	Dec.	G.H.A.	Dec.	G.H.A.	Dec.	G.H.A.	Dec.	Name	S.H.A.	Dec.
d	h	° ′	° ′	° ′	° ′	° ′	° ′	° ′	° ′	° ′		° ′	° ′
25	00	241 57.5	133 22.6	N24 57.2	238 47.5	S 0 24.3	227 20.7	N 4 59.3	134 24.1	N22 20.6	Acamar	315 40.4	S40 24.1
	01	257 00.0	148 22.1	56.9	253 48.2	23.5	242 22.7	59.5	149 26.2	20.5	Achernar	335 48.4	S57 21.5
	02	272 02.4	163 21.6	56.5	268 49.0	22.8	257 24.7	59.7	164 28.4	20.5	Acrux	173 40.9	S62 58.1
	03	287 04.9	178 21.1	·· 56.2	283 49.7	·· 22.1	272 26.7	4 59.9	179 30.6	·· 20.5	Adhara	255 35.3	S28 56.5
	04	302 07.4	193 20.6	55.8	298 50.5	21.3	287 28.7	5 00.1	194 32.8	20.5	Aldebaran	291 22.5	N16 27.6
	05	317 09.8	208 20.1	55.5	313 51.2	20.6	302 30.7	00.3	209 34.9	20.4			
	06	332 12.3	223 19.6	N24 55.1	328 52.0	S 0 19.9	317 32.7	N 5 00.4	224 37.1	N22 20.4	Alioth	166 45.3	N56 05.7
	07	347 14.7	238 19.1	54.8	343 52.7	19.1	332 34.7	00.6	239 39.3	20.4	Alkaid	153 20.9	N49 26.2
	08	2 17.2	253 18.6	54.4	358 53.4	18.4	347 36.7	00.8	254 41.5	20.3	Al Na'ir	28 19.4	S47 04.5
S	09	17 19.7	268 18.1	·· 54.1	13 54.2	·· 17.6	2 38.7	·· 01.0	269 43.6	·· 20.3	Alnilam	276 15.7	S 1 13.1
U	10	32 22.1	283 17.6	53.7	28 54.9	16.9	17 40.7	01.2	284 45.8	20.3	Alphard	218 24.2	S 8 33.3
N	11	47 24.6	298 17.1	53.4	43 55.7	16.2	32 42.7	01.4	299 48.0	20.3			
D	12	62 27.1	313 16.6	N24 53.0	58 56.4	S 0 15.4	47 44.7	N 5 01.6	314 50.2	N22 20.2	Alphecca	126 34.8	N26 47.8
A	13	77 29.5	328 16.1	52.7	73 57.1	14.7	62 46.7	01.7	329 52.4	20.2	Alpheratz	358 13.2	N28 57.2
Y	14	92 32.0	343 15.7	52.3	88 57.9	14.0	77 48.7	01.9	344 54.5	20.2	Altair	62 35.8	N 8 48.2
	15	107 34.5	358 15.2	·· 52.0	103 58.6	·· 13.2	92 50.7	·· 02.1	359 56.7	·· 20.2	Ankaa	353 44.0	S42 26.1
	16	122 36.9	13 14.7	51.6	118 59.4	12.5	107 52.7	02.3	14 58.9	20.1	Antares	113 00.9	S26 22.7
	17	137 39.4	28 14.2	51.3	134 00.1	11.8	122 54.7	02.5	30 01.1	20.1			
	18	152 41.9	43 13.7	N24 50.9	149 00.9	S 0 11.0	137 56.7	N 5 02.7	45 03.2	N22 20.1	Arcturus	146 21.5	N19 18.6
	19	167 44.3	58 13.2	50.5	164 01.6	10.3	152 58.7	02.9	60 05.4	20.0	Atria	108 27.9	S68 59.0
	20	182 46.8	73 12.7	50.2	179 02.3	09.6	168 00.7	03.0	75 07.6	20.0	Avior	234 30.0	S59 26.2
	21	197 49.2	88 12.3	·· 49.8	194 03.1	·· 08.8	183 02.7	·· 03.2	90 09.8	·· 20.0	Bellatrix	279 03.0	N 6 19.6
	22	212 51.7	103 11.8	49.5	209 03.8	08.1	198 04.7	03.4	105 11.9	20.0	Betelgeuse	271 32.5	N 7 24.1
	23	227 54.2	118 11.3	49.1	224 04.6	07.4	213 06.7	03.6	120 14.1	19.9			
26	00	242 56.6	133 10.8	N24 48.7	239 05.3	S 0 06.6	228 08.7	N 5 03.8	135 16.3	N22 19.9	Canopus	264 09.3	S52 41.2
	01	257 59.1	148 10.3	48.4	254 06.1	05.9	243 10.7	04.0	150 18.5	19.9	Capella	281 17.1	N45 58.4
	02	273 01.6	163 09.8	48.0	269 06.8	05.2	258 12.8	04.1	165 20.6	19.8	Deneb	49 50.7	N45 11.4
	03	288 04.0	178 09.4	·· 47.6	284 07.5	·· 04.4	273 14.8	·· 04.3	180 22.8	·· 19.8	Denebola	183 02.6	N14 42.5
	04	303 06.5	193 08.9	47.3	299 08.3	03.7	288 16.8	04.5	195 25.0	19.8	Diphda	349 24.7	S18 07.2
	05	318 09.0	208 08.4	46.9	314 09.0	03.0	303 18.8	04.7	210 27.2	19.8			
	06	333 11.4	223 07.9	N24 46.5	329 09.8	S 0 02.2	318 20.8	N 5 04.9	225 29.3	N22 19.7	Dubhe	194 26.4	N61 53.2
	07	348 13.9	238 07.5	46.1	344 10.5	01.5	333 22.8	05.1	240 31.5	19.7	Elnath	278 49.1	N28 35.2
	08	3 16.4	253 07.0	45.8	359 11.3	S 0 00.7	348 24.8	05.2	255 33.7	19.7	Eltanin	90 58.9	N51 29.4
M	09	18 18.8	268 06.5	·· 45.4	14 12.0	0 00.0	3 26.8	·· 05.4	270 35.9	·· 19.6	Enif	34 15.1	N 9 45.7
O	10	33 21.3	283 06.0	45.0	29 12.7	N 0 00.7	18 28.8	05.6	285 38.0	19.6	Fomalhaut	15 55.5	S29 45.0
N	11	48 23.7	298 05.6	44.6	44 13.5	01.5	33 30.8	05.8	300 40.2	19.6			
D	12	63 26.2	313 05.1	N24 44.3	59 14.2	N 0 02.2	48 32.8	N 5 06.0	315 42.4	N22 19.6	Gacrux	172 32.5	S56 58.9
A	13	78 28.7	328 04.6	43.9	74 15.0	02.9	63 34.8	06.2	330 44.6	19.5	Gienah	176 21.5	S17 24.6
Y	14	93 31.1	343 04.1	43.5	89 15.7	03.7	78 36.8	06.3	345 46.7	19.5	Hadar	149 27.9	S60 15.5
	15	108 33.6	358 03.7	·· 43.1	104 16.5	·· 04.4	93 38.8	·· 06.5	0 48.9	·· 19.5	Hamal	328 33.3	N23 20.7
	16	123 36.1	13 03.2	42.7	119 17.2	05.1	108 40.8	06.7	15 51.1	19.4	Kaus Aust.	84 21.3	S34 23.7
	17	138 38.5	28 02.7	42.4	134 17.9	05.9	123 42.8	06.9	30 53.3	19.4			
	18	153 41.0	43 02.3	N24 42.0	149 18.7	N 0 06.6	138 44.8	N 5 07.1	45 55.4	N22 19.4	Kochab	137 17.7	N74 15.5
	19	168 43.5	58 01.8	41.6	164 19.4	07.3	153 46.8	07.3	60 57.6	19.4	Markab	14 06.8	N15 04.3
	20	183 45.9	73 01.3	41.2	179 20.2	08.1	168 48.8	07.4	75 59.8	19.3	Menkar	314 45.2	N 3 59.6
	21	198 48.4	88 00.9	·· 40.8	194 20.9	·· 08.8	183 50.8	·· 07.6	91 01.9	·· 19.3	Menkent	148 40.9	S36 15.2
	22	213 50.9	103 00.4	40.4	209 21.7	09.5	198 52.9	07.8	106 04.1	19.3	Miaplacidus	221 46.0	S69 37.4
	23	228 53.3	117 59.9	40.0	224 22.4	10.3	213 54.9	08.0	121 06.3	19.2			
27	00	243 55.8	132 59.5	N24 39.7	239 23.2	N 0 11.0	228 56.9	N 5 08.2	136 08.5	N22 19.2	Mirfak	309 21.8	N49 46.4
	01	258 58.2	147 59.0	39.3	254 23.9	11.7	243 58.9	08.4	151 10.6	19.2	Nunki	76 33.4	S26 19.6
	02	274 00.7	162 58.5	38.9	269 24.6	12.5	259 00.9	08.5	166 12.8	19.2	Peacock	54 03.8	S56 48.6
	03	289 03.2	177 58.1	·· 38.5	284 25.4	·· 13.2	274 02.9	·· 08.7	181 15.0	·· 19.1	Pollux	244 02.8	N28 05.2
	04	304 05.6	192 57.6	38.1	299 26.1	13.9	289 04.9	08.9	196 17.2	19.1	Procyon	245 29.8	N 5 17.2
	05	319 08.1	207 57.2	37.7	314 26.9	14.7	304 06.9	09.1	211 19.3	19.1			
	06	334 10.6	222 56.7	N24 37.3	329 27.6	N 0 15.4	319 08.9	N 5 09.3	226 21.5	N22 19.0	Rasalhague	96 32.6	N12 34.6
	07	349 13.0	237 56.2	36.9	344 28.4	16.1	334 10.9	09.5	241 23.7	19.0	Regulus	208 13.9	N12 05.2
T	08	4 15.5	252 55.8	36.5	359 29.1	16.9	349 12.9	09.6	256 25.8	19.0	Rigel	281 39.8	S 8 13.9
U	09	19 18.0	267 55.3	·· 36.1	14 29.9	·· 17.6	4 14.9	·· 09.8	271 28.0	·· 18.9	Rigil Kent.	140 30.1	S60 44.2
E	10	34 20.4	282 54.9	35.7	29 30.6	18.3	19 16.9	10.0	286 30.2	18.9	Sabik	102 44.9	S15 41.7
S	11	49 22.9	297 54.4	35.3	44 31.3	19.1	34 19.0	10.2	301 32.4	18.9			
D	12	64 25.3	312 54.0	N24 34.9	59 32.1	N 0 19.8	49 21.0	N 5 10.4	316 34.5	N22 18.9	Schedar	350 13.5	N56 24.0
A	13	79 27.8	327 53.5	34.5	74 32.8	20.5	64 23.0	10.6	331 36.7	18.8	Shaula	97 00.3	S37 05.2
Y	14	94 30.3	342 53.1	34.1	89 33.6	21.3	79 25.0	10.7	346 38.9	18.8	Sirius	258 59.2	S16 41.1
	15	109 32.7	357 52.6	·· 33.7	104 34.3	·· 22.0	94 27.0	·· 10.9	1 41.1	·· 18.8	Spica	159 01.1	S11 02.2
	16	124 35.2	12 52.1	33.3	119 35.1	22.7	109 29.0	11.1	16 43.2	18.7	Suhail	223 13.6	S43 20.3
	17	139 37.7	27 51.7	32.9	134 35.8	23.5	124 31.0	11.3	31 45.4	18.7			
	18	154 40.1	42 51.2	N24 32.5	149 36.6	N 0 24.2	139 33.0	N 5 11.5	46 47.6	N22 18.7	Vega	80 57.9	N38 45.5
	19	169 42.6	57 50.8	32.1	164 37.3	24.9	154 35.0	11.6	61 49.7	18.7	Zuben'ubi	137 36.7	S15 56.5
	20	184 45.1	72 50.3	31.7	179 38.0	25.7	169 37.0	11.8	76 51.9	18.6		S.H.A.	Mer. Pass.
	21	199 47.5	87 49.9	·· 31.2	194 38.8	·· 26.4	184 39.0	·· 12.0	91 54.1	·· 18.6		° ′	h m
	22	214 50.0	102 49.5	30.8	209 39.5	27.1	199 41.1	12.2	106 56.2	18.6	Venus	250 14.2	15 08
	23	229 52.5	117 49.0	30.4	224 40.3	27.9	214 43.1	12.4	121 58.4	18.5	Mars	356 08.7	8 03
Mer. Pass.		7 46.9 (h m)	v −0.5	d 0.4	v 0.7	d 0.7	v 2.0	d 0.2	v 2.2	d 0.0	Jupiter	345 12.1	8 46
											Saturn	252 19.6	14 57

1975 MAY 25, 26, 27 (SUN., MON., TUES.)

G.M.T.	SUN G.H.A.	SUN Dec.	MOON G.H.A.	v	MOON Dec.	d	H.P.
d h	° ′	° ′	° ′	′	° ′	′	′
25 00	180 48.4	N20 47.7	4 04.3	6.9	S20 14.7	4.1	58.3
01	195 48.4	48.2	18 30.2	7.0	20 18.8	4.0	58.3
02	210 48.3	48.7	32 56.2	6.9	20 22.8	3.8	58.3
03	225 48.2	·· 49.1	47 22.1	7.0	20 26.6	3.7	58.2
04	240 48.2	49.6	61 48.1	6.9	20 30.3	3.6	58.2
05	255 48.1	50.0	76 14.0	7.0	20 33.9	3.5	58.2
06	270 48.1	N20 50.5	90 40.0	6.9	S20 37.4	3.3	58.2
07	285 48.0	51.0	105 05.9	7.0	20 40.7	3.2	58.1
08	300 48.0	51.4	119 31.9	6.9	20 43.9	3.0	58.1
S 09	315 47.9	·· 51.9	133 57.8	7.0	20 46.9	3.0	58.1
U 10	330 47.8	52.3	148 23.8	7.0	20 49.9	2.7	58.1
N 11	345 47.8	52.8	162 49.8	7.0	20 52.6	2.7	58.0
D 12	0 47.7	N20 53.2	177 15.8	7.0	S20 55.3	2.5	58.0
A 13	15 47.7	53.7	191 41.8	7.0	20 57.8	2.4	58.0
Y 14	30 47.6	54.1	206 07.8	7.0	21 00.2	2.3	58.0
15	45 47.5	·· 54.6	220 33.8	7.1	21 02.5	2.1	57.9
16	60 47.5	55.0	234 59.9	7.0	21 04.6	2.0	57.9
17	75 47.4	55.5	249 25.9	7.1	21 06.6	1.9	57.9
18	90 47.4	N20 55.9	263 52.0	7.1	S21 08.5	1.8	57.8
19	105 47.3	56.4	278 18.1	7.2	21 10.3	1.6	57.8
20	120 47.2	56.8	292 44.3	7.1	21 11.9	1.5	57.8
21	135 47.2	·· 57.3	307 10.4	7.2	21 13.4	1.3	57.8
22	150 47.1	57.7	321 36.6	7.2	21 14.7	1.2	57.7
23	165 47.1	58.2	336 02.8	7.2	21 15.9	1.1	57.7
26 00	180 47.0	N20 58.6	350 29.0	7.2	S21 17.0	1.0	57.7
01	195 46.9	59.1	4 55.2	7.3	21 18.0	0.8	57.7
02	210 46.9	20 59.5	19 21.5	7.3	21 18.8	0.7	57.6
03	225 46.8	21 00.0	33 47.8	7.4	21 19.5	0.6	57.6
04	240 46.7	00.4	48 14.2	7.3	21 20.1	0.4	57.6
05	255 46.7	00.9	62 40.5	7.4	21 20.5	0.3	57.5
06	270 46.6	N21 01.3	77 06.9	7.5	S21 20.8	0.2	57.5
07	285 46.6	01.7	91 33.4	7.5	21 21.0	0.1	57.5
08	300 46.5	02.2	105 59.9	7.5	21 21.1	0.1	57.5
M 09	315 46.4	·· 02.6	120 26.4	7.5	21 21.0	0.2	57.4
O 10	330 46.4	03.1	134 52.9	7.6	21 20.8	0.3	57.4
N 11	345 46.3	03.5	149 19.5	7.6	21 20.5	0.5	57.4
D 12	0 46.2	N21 03.9	163 46.1	7.7	S21 20.0	0.6	57.3
A 13	15 46.2	04.4	178 12.8	7.7	21 19.4	0.7	57.3
Y 14	30 46.1	04.8	192 39.5	7.8	21 18.7	0.8	57.3
15	45 46.0	·· 05.3	207 06.3	7.8	21 17.9	1.0	57.3
16	60 46.0	05.7	221 33.1	7.8	21 16.9	1.0	57.2
17	75 45.9	06.1	235 59.9	7.9	21 15.9	1.2	57.2
18	90 45.8	N21 06.6	250 26.8	7.9	S21 14.7	1.4	57.2
19	105 45.8	07.0	264 53.7	8.0	21 13.3	1.4	57.1
20	120 45.7	07.4	279 20.7	8.1	21 11.9	1.6	57.1
21	135 45.7	·· 07.9	293 47.8	8.0	21 10.3	1.7	57.1
22	150 45.6	08.3	308 14.8	8.2	21 08.6	1.8	57.1
23	165 45.5	08.7	322 42.0	8.2	21 06.8	1.9	57.0
27 00	180 45.5	N21 09.2	337 09.2	8.2	S21 04.9	2.0	57.0
01	195 45.4	09.6	351 36.4	8.3	21 02.9	2.2	57.0
02	210 45.3	10.0	6 03.7	8.4	21 00.7	2.3	56.9
03	225 45.3	·· 10.4	20 31.1	8.4	20 58.4	2.4	56.9
04	240 45.2	10.9	34 58.5	8.5	20 56.0	2.5	56.9
05	255 45.1	11.3	49 26.0	8.5	20 53.5	2.6	56.9
06	270 45.0	N21 11.7	63 53.5	8.6	S20 50.9	2.8	56.8
07	285 45.0	12.2	78 21.1	8.6	20 48.1	2.8	56.8
T 08	300 44.9	12.6	92 48.7	8.7	20 45.3	3.0	56.8
U 09	315 44.8	·· 13.0	107 16.4	8.8	20 42.3	3.1	56.7
E 10	330 44.8	13.4	121 44.2	8.8	20 39.2	3.2	56.7
S 11	345 44.7	13.9	136 12.0	8.9	20 36.0	3.3	56.7
D 12	0 44.6	N21 14.3	150 39.9	9.0	S20 32.7	3.4	56.6
A 13	15 44.6	14.7	165 07.9	9.0	20 29.3	3.6	56.6
Y 14	30 44.5	15.1	179 35.9	9.0	20 25.7	3.6	56.6
15	45 44.4	·· 15.5	194 03.9	9.2	20 22.1	3.8	56.6
16	60 44.4	16.0	208 32.1	9.2	20 18.3	3.8	56.5
17	75 44.3	16.4	223 00.3	9.3	20 14.5	4.0	56.5
18	90 44.2	N21 16.8	237 28.6	9.3	S20 10.5	4.0	56.5
19	105 44.1	17.2	251 56.9	9.4	20 06.5	4.2	56.4
20	120 44.1	17.6	266 25.3	9.5	20 02.3	4.3	56.4
21	135 44.0	·· 18.1	280 53.8	9.5	19 58.0	4.4	56.4
22	150 43.9	18.5	295 22.3	9.7	19 53.6	4.4	56.4
23	165 43.9	18.9	309 51.0	9.6	19 49.2	4.6	56.3
	S.D. 15.8	d 0.4	S.D. 15.8		15.6		15.4

Lat.	Twilight Naut.	Twilight Civil	Sunrise	Moonrise 25	Moonrise 26	Moonrise 27	Moonrise 28
°	h m	h m	h m	h m	h m	h m	h m
N 72	□	□	□	▬	▬	▬	▬
N 70	□	□	□	▬	▬	▬	01 46
68	////	////	00 29	23 42	24 27	00 27	00 36
66	////	////	01 41	22 42	23 33	24 00	00 00
64	////	////	02 16	22 07	23 01	23 33	23 53
62	////	01 03	02 41	21 43	22 36	23 13	23 37
60	////	01 46	03 01	21 23	22 17	22 56	23 24
N 58	////	02 15	03 17	21 07	22 01	22 42	23 12
56	00 56	02 36	03 30	20 53	21 48	22 30	23 02
54	01 37	02 53	03 42	20 41	21 36	22 19	22 53
52	02 03	03 08	03 52	20 30	21 26	22 10	22 45
50	02 23	03 20	04 02	20 21	21 16	22 02	22 38
45	03 00	03 46	04 21	20 01	20 57	21 44	22 23
N 40	03 26	04 05	04 37	19 45	20 41	21 29	22 10
35	03 46	04 21	04 50	19 32	20 27	21 17	22 00
30	04 02	04 35	05 01	19 20	20 16	21 06	21 50
20	04 28	04 57	05 21	19 00	19 55	20 47	21 34
N 10	04 49	05 15	05 38	18 42	19 38	20 31	21 19
0	05 05	05 31	05 53	18 26	19 22	20 15	21 06
S 10	05 20	05 46	06 09	18 10	19 05	20 00	20 53
20	05 35	06 02	06 25	17 53	18 48	19 44	20 38
30	05 49	06 18	06 44	17 33	18 28	19 25	20 22
35	05 56	06 27	06 55	17 21	18 16	19 14	20 12
40	06 04	06 37	07 07	17 08	18 03	19 01	20 01
45	06 13	06 49	07 22	16 52	17 47	18 47	19 48
S 50	06 22	07 03	07 40	16 33	17 27	18 28	19 33
52	06 27	07 09	07 48	16 24	17 18	18 20	19 25
54	06 31	07 16	07 58	16 13	17 08	18 10	19 17
56	06 36	07 24	08 08	16 02	16 56	17 59	19 08
58	06 42	07 32	08 20	15 49	16 43	17 47	18 57
S 60	06 48	07 42	08 34	15 33	16 27	17 32	18 45

Lat.	Sunset	Twilight Civil	Twilight Naut.	Moonset 25	Moonset 26	Moonset 27	Moonset 28
°	h m	h m	h m	h m	h m	h m	h m
N 72	□	□	□	▬	▬	▬	▬
N 70	□	□	□	▬	▬	▬	03 18
68	23 43	////	////	01 16	01 36	02 46	04 27
66	22 17	////	////	01 59	02 36	03 40	05 03
64	21 40	////	////	02 28	03 10	04 12	05 29
62	21 15	22 57	////	02 50	03 35	04 36	05 49
60	20 55	22 10	////	03 08	03 55	04 55	06 05
N 58	20 38	21 41	////	03 23	04 11	05 11	06 19
56	20 24	21 20	23 02	03 36	04 25	05 24	06 31
54	20 13	21 02	22 19	03 47	04 37	05 36	06 41
52	20 02	20 47	21 53	03 57	04 48	05 46	06 50
50	19 53	20 34	21 32	04 06	04 57	05 55	06 58
45	19 33	20 09	20 55	04 25	05 17	06 15	07 16
N 40	19 18	19 49	20 29	04 40	05 33	06 30	07 30
35	19 04	19 33	20 08	04 53	05 47	06 44	07 42
30	18 53	19 19	19 52	05 04	05 59	06 55	07 52
20	18 33	18 57	19 26	05 24	06 19	07 15	08 10
N 10	18 16	18 39	19 05	05 41	06 37	07 32	08 26
0	18 01	18 23	18 48	05 57	06 53	07 48	08 40
S 10	17 45	18 07	18 33	06 12	07 09	08 04	08 55
20	17 28	17 52	18 19	06 29	07 27	08 21	09 10
30	17 10	17 36	18 05	06 49	07 47	08 40	09 27
35	16 59	17 26	17 57	07 00	07 59	08 51	09 38
40	16 46	17 16	17 49	07 13	08 12	09 04	09 49
45	16 32	17 04	17 41	07 29	08 28	09 20	10 03
S 50	16 14	16 51	17 31	07 48	08 48	09 38	10 19
52	16 05	16 44	17 27	07 57	08 57	09 47	10 27
54	15 56	16 37	17 22	08 07	09 08	09 57	10 36
56	15 45	16 30	17 17	08 18	09 19	10 08	10 45
58	15 33	16 21	17 12	08 31	09 33	10 21	10 56
S 60	15 19	16 12	17 06	08 46	09 49	10 36	11 09

Day	SUN Eqn. of Time 00h	SUN Eqn. of Time 12h	SUN Mer. Pass.	MOON Mer. Pass. Upper	MOON Mer. Pass. Lower	Age	Phase
	m s	m s	h m	h m	h m	d	
25	03 14	03 11	11 57	24 40	12 11	14	
26	03 08	03 05	11 57	00 40	13 07	15	○
27	03 02	02 59	11 57	01 35	14 02	16	

1975 MAY 31, JUNE 1, 2 (SAT., SUN., MON.)

	G.M.T.	ARIES G.H.A.	VENUS −3.8 G.H.A.	VENUS Dec.	MARS +1.0 G.H.A.	MARS Dec.	JUPITER −1.7 G.H.A.	JUPITER Dec.	SATURN +0.4 G.H.A.	SATURN Dec.
	d h	° ′	° ′	° ′	° ′	° ′	° ′	° ′	° ′	° ′
31	00	247 52.3	132 18.6	N23 57.7	240 34.8	N 1 21.2	232 10.1	N 5 25.4	139 36.7	N22 16.3
	01	262 54.8	147 18.2	57.2	255 35.6	22.0	247 12.2	25.6	154 38.8	16.3
	02	277 57.3	162 17.8	56.7	270 36.3	22.7	262 14.2	25.8	169 41.0	16.3
	03	292 59.7	177 17.5	·· 56.3	285 37.1	·· 23.4	277 16.2	·· 25.9	184 43.1	·· 16.2
	04	308 02.2	192 17.1	55.8	300 37.8	24.2	292 18.2	26.1	199 45.3	16.2
	05	323 04.7	207 16.7	55.3	315 38.6	24.9	307 20.2	26.3	214 47.5	16.2
	06	338 07.1	222 16.3	N23 54.8	330 39.3	N 1 25.6	322 22.3	N 5 26.5	229 49.6	N22 16.2
	07	353 09.6	237 15.9	54.3	345 40.1	26.3	337 24.3	26.7	244 51.8	16.1
S	08	8 12.1	252 15.6	53.8	0 40.8	27.1	352 26.3	26.8	259 54.0	16.1
A	09	23 14.5	267 15.2	·· 53.3	15 41.6	·· 27.8	7 28.3	·· 27.0	274 56.1	·· 16.1
T	10	38 17.0	282 14.8	52.8	30 42.3	28.5	22 30.3	27.2	289 58.3	16.0
U	11	53 19.4	297 14.4	52.3	45 43.1	29.3	37 32.4	27.4	305 00.5	16.0
R	12	68 21.9	312 14.0	N23 51.8	60 43.8	N 1 30.0	52 34.4	N 5 27.5	320 02.6	N22 16.0
D	13	83 24.4	327 13.7	51.3	75 44.6	30.7	67 36.4	27.7	335 04.8	15.9
A	14	98 26.8	342 13.3	50.8	90 45.3	31.4	82 38.4	27.9	350 06.9	15.9
Y	15	113 29.3	357 12.9	·· 50.4	105 46.1	·· 32.2	97 40.5	·· 28.1	5 09.1	·· 15.9
	16	128 31.8	12 12.5	49.9	120 46.8	32.9	112 42.5	28.2	20 11.3	15.8
	17	143 34.2	27 12.2	49.4	135 47.6	33.6	127 44.5	28.4	35 13.4	15.8
	18	158 36.7	42 11.8	N23 48.9	150 48.3	N 1 34.4	142 46.5	N 5 28.6	50 15.6	N22 15.8
	19	173 39.2	57 11.4	48.4	165 49.1	35.1	157 48.5	28.8	65 17.8	15.8
	20	188 41.6	72 11.1	47.9	180 49.8	35.8	172 50.6	28.9	80 19.9	15.7
	21	203 44.1	87 10.7	·· 47.4	195 50.6	·· 36.5	187 52.6	·· 29.1	95 22.1	·· 15.7
	22	218 46.5	102 10.3	46.8	210 51.3	37.3	202 54.6	29.3	110 24.2	15.7
	23	233 49.0	117 10.0	46.3	225 52.1	38.0	217 56.6	29.5	125 26.4	15.6
1	00	248 51.5	132 09.6	N23 45.8	240 52.8	N 1 38.7	232 58.7	N 5 29.6	140 28.6	N22 15.6
	01	263 53.9	147 09.2	45.3	255 53.6	39.4	248 00.7	29.8	155 30.7	15.6
	02	278 56.4	162 08.9	44.8	270 54.3	40.2	263 02.7	30.0	170 32.9	15.5
	03	293 58.9	177 08.5	·· 44.3	285 55.1	·· 40.9	278 04.7	·· 30.2	185 35.1	·· 15.5
	04	309 01.3	192 08.1	43.8	300 55.8	41.6	293 06.7	30.3	200 37.2	15.5
	05	324 03.8	207 07.8	43.3	315 56.6	42.4	308 08.8	30.5	215 39.4	15.4
	06	339 06.3	222 07.4	N23 42.8	330 57.3	N 1 43.1	323 10.8	N 5 30.7	230 41.5	N22 15.4
	07	354 08.7	237 07.1	42.3	345 58.1	43.8	338 12.8	30.9	245 43.7	15.4
	08	9 11.2	252 06.7	41.8	0 58.8	44.5	353 14.8	31.0	260 45.9	15.3
S	09	24 13.7	267 06.3	·· 41.3	15 59.6	·· 45.3	8 16.9	·· 31.2	275 48.0	·· 15.3
U	10	39 16.1	282 06.0	40.7	31 00.3	46.0	23 18.9	31.4	290 50.2	15.3
N	11	54 18.6	297 05.6	40.2	46 01.1	46.7	38 20.9	31.5	305 52.3	15.2
D	12	69 21.0	312 05.3	N23 39.7	61 01.8	N 1 47.4	53 22.9	N 5 31.7	320 54.5	N22 15.2
A	13	84 23.5	327 04.9	39.2	76 02.6	48.2	68 25.0	31.9	335 56.7	15.2
Y	14	99 26.0	342 04.6	38.7	91 03.3	48.9	83 27.0	32.1	350 58.8	15.2
	15	114 28.4	357 04.2	·· 38.2	106 04.1	·· 49.6	98 29.0	·· 32.2	6 01.0	·· 15.1
	16	129 30.9	12 03.9	37.6	121 04.8	50.4	113 31.0	32.4	21 03.2	15.1
	17	144 33.4	27 03.5	37.1	136 05.6	51.1	128 33.1	32.6	36 05.3	15.1
	18	159 35.8	42 03.2	N23 36.6	151 06.3	N 1 51.8	143 35.1	N 5 32.8	51 07.5	N22 15.0
	19	174 38.3	57 02.8	36.1	166 07.1	52.5	158 37.1	32.9	66 09.6	15.0
	20	189 40.8	72 02.5	35.5	181 07.8	53.3	173 39.1	33.1	81 11.8	15.0
	21	204 43.2	87 02.1	·· 35.0	196 08.6	·· 54.0	188 41.2	·· 33.3	96 14.0	·· 14.9
	22	219 45.7	102 01.8	34.5	211 09.3	54.7	203 43.2	33.5	111 16.1	14.9
	23	234 48.2	117 01.4	34.0	226 10.1	55.4	218 45.2	33.6	126 18.3	14.9
2	00	249 50.6	132 01.1	N23 33.4	241 10.8	N 1 56.2	233 47.2	N 5 33.8	141 20.4	N22 14.8
	01	264 53.1	147 00.7	32.9	256 11.6	56.9	248 49.3	34.0	156 22.6	14.8
	02	279 55.5	162 00.4	32.4	271 12.3	57.6	263 51.3	34.1	171 24.8	14.8
	03	294 58.0	177 00.0	·· 31.9	286 13.1	·· 58.3	278 53.3	·· 34.3	186 26.9	·· 14.7
	04	310 00.5	191 59.7	31.3	301 13.8	59.1	293 55.4	34.5	201 29.1	14.7
	05	325 02.9	206 59.4	30.8	316 14.6	1 59.8	308 57.4	34.7	216 31.2	14.7
	06	340 05.4	221 59.0	N23 30.3	331 15.3	N 2 00.5	323 59.4	N 5 34.8	231 33.4	N22 14.6
	07	355 07.9	236 58.7	29.7	346 16.1	01.2	339 01.4	35.0	246 35.6	14.6
	08	10 10.3	251 58.4	29.2	1 16.8	02.0	354 03.5	35.2	261 37.7	14.6
M	09	25 12.8	266 58.0	·· 28.7	16 17.6	·· 02.7	9 05.5	·· 35.4	276 39.9	·· 14.6
O	10	40 15.3	281 57.7	28.1	31 18.3	03.4	24 07.5	35.5	291 42.0	14.5
N	11	55 17.7	296 57.3	27.6	46 19.1	04.1	39 09.5	35.7	306 44.2	14.5
D	12	70 20.2	311 57.0	N23 27.1	61 19.8	N 2 04.9	54 11.6	N 5 35.9	321 46.4	N22 14.5
A	13	85 22.6	326 56.7	26.5	76 20.6	05.6	69 13.6	36.0	336 48.5	14.4
Y	14	100 25.1	341 56.3	26.0	91 21.3	06.3	84 15.6	36.2	351 50.7	14.4
	15	115 27.6	356 56.0	·· 25.4	106 22.1	·· 07.0	99 17.7	·· 36.4	6 52.8	·· 14.4
	16	130 30.0	11 55.7	24.9	121 22.8	07.8	114 19.7	36.6	21 55.0	14.3
	17	145 32.5	26 55.4	24.4	136 23.6	08.5	129 21.7	36.7	36 57.1	14.3
	18	160 35.0	41 55.0	N23 23.8	151 24.3	N 2 09.2	144 23.7	N 5 36.9	51 59.3	N22 14.3
	19	175 37.4	56 54.7	23.3	166 25.1	09.9	159 25.8	37.1	67 01.5	14.2
	20	190 39.9	71 54.4	22.7	181 25.8	10.7	174 27.8	37.2	82 03.6	14.2
	21	205 42.4	86 54.0	·· 22.2	196 26.6	·· 11.4	189 29.8	·· 37.4	97 05.8	·· 14.2
	22	220 44.8	101 53.7	21.6	211 27.3	12.1	204 31.9	37.6	112 07.9	14.1
	23	235 47.3	116 53.4	21.1	226 28.1	12.8	219 33.9	37.8	127 10.1	14.1
Mer. Pass.		h m 7 23.4	v −0.4	d 0.5	v 0.8	d 0.7	v 2.0	d 0.2	v 2.2	d 0.0

STARS

Name	S.H.A.	Dec.
	° ′	° ′
Acamar	315 40.4	S40 24.1
Achernar	335 48.4	S57 21.4
Acrux	173 40.9	S62 58.2
Adhara	255 35.3	S28 56.5
Aldebaran	291 22.5	N16 27.6
Alioth	166 45.3	N56 05.7
Alkaid	153 21.0	N49 26.2
Al Na'ir	28 19.4	S47 04.5
Alnilam	276 15.6	S 1 13.1
Alphard	218 24.2	S 8 33.3
Alphecca	126 34.8	N26 47.8
Alpheratz	358 13.2	N28 57.2
Altair	62 35.8	N 8 48.2
Ankaa	353 44.0	S42 26.1
Antares	113 00.9	S26 22.7
Arcturus	146 21.5	N19 18.6
Atria	108 27.8	S68 59.1
Avior	234 30.1	S59 26.2
Bellatrix	279 02.9	N 6 19.6
Betelgeuse	271 32.5	N 7 24.1
Canopus	264 09.3	S52 41.1
Capella	281 17.1	N45 58.4
Deneb	49 50.7	N45 11.4
Denebola	183 02.6	N14 42.5
Diphda	349 24.7	S18 07.2
Dubhe	194 26.5	N61 53.2
Elnath	278 49.1	N28 35.2
Eltanin	90 58.9	N51 29.4
Enif	34 15.0	N 9 45.7
Fomalhaut	15 55.4	S29 44.9
Gacrux	172 32.5	S56 58.9
Gienah	176 21.5	S17 24.6
Hadar	149 27.9	S60 15.6
Hamal	328 33.3	N23 20.7
Kaus Aust.	84 21.2	S34 23.7
Kochab	137 17.7	N74 15.5
Markab	14 06.8	N15 04.3
Menkar	314 45.2	N 3 59.6
Menkent	148 40.9	S36 15.2
Miaplacidus	221 46.1	S69 37.4
Mirfak	309 21.7	N49 46.4
Nunki	76 33.3	S26 19.6
Peacock	54 03.7	S56 48.6
Pollux	244 02.8	N28 05.2
Procyon	245 29.8	N 5 17.2
Rasalhague	96 32.6	N12 34.6
Regulus	208 13.9	N12 05.2
Rigel	281 39.8	S 8 13.9
Rigil Kent.	140 30.1	S60 44.2
Sabik	102 44.9	S15 41.7
Schedar	350 13.4	N56 24.0
Shaula	97 00.2	S37 05.2
Sirius	258 59.2	S16 41.1
Spica	159 01.1	S11 02.2
Suhail	223 13.6	S43 20.3
Vega	80 57.9	N38 45.6
Zuben'ubi	137 36.7	S15 56.5

	S.H.A.	Mer. Pass.
	° ′	h m
Venus	243 18.1	15 12
Mars	352 01.3	7 56
Jupiter	344 07.2	8 27
Saturn	251 37.1	14 36

1975 MAY 31, JUNE 1, 2 (SAT., SUN., MON.)

	G.M.T.	SUN G.H.A.	SUN Dec.	MOON G.H.A.	v	MOON Dec.	d	H.P.
	d h	° ′	° ′	° ′	′	° ′	′	′
	31 00	180 38.1	N21 47.6	289 46.4	14.5	S10 46.8	9.6	54.7
	01	195 38.0	47.9	304 19.9	14.5	10 37.2	9.6	54.6
	02	210 37.9	48.3	318 53.4	14.5	10 27.6	9.7	54.6
	03	225 37.8	·· 48.7	333 26.9	14.6	10 17.9	9.7	54.6
	04	240 37.7	49.0	348 00.5	14.7	10 08.2	9.7	54.6
	05	255 37.6	49.4	2 34.2	14.6	9 58.5	9.8	54.6
	06	270 37.6	N21 49.8	17 07.8	14.8	S 9 48.7	9.8	54.6
	07	285 37.5	50.1	31 41.6	14.7	9 38.9	9.8	54.6
S	08	300 37.4	50.5	46 15.3	14.9	9 29.1	9.9	54.5
A	09	315 37.3	·· 50.9	60 49.2	14.8	9 19.2	9.9	54.5
T	10	330 37.2	51.2	75 23.0	14.9	9 09.3	9.9	54.5
U	11	345 37.1	51.6	89 56.9	15.0	8 59.4	10.0	54.5
R	12	0 37.0	N21 51.9	104 30.9	14.9	S 8 49.4	10.0	54.5
D	13	15 36.9	52.3	119 04.8	15.0	8 39.4	10.0	54.5
A	14	30 36.9	52.7	133 38.8	15.1	8 29.4	10.1	54.5
Y	15	45 36.8	·· 53.0	148 12.9	15.1	8 19.3	10.1	54.5
	16	60 36.7	53.4	162 47.0	15.1	8 09.2	10.1	54.4
	17	75 36.6	53.7	177 21.1	15.2	7 59.1	10.1	54.4
	18	90 36.5	N21 54.1	191 55.3	15.2	S 7 49.0	10.2	54.4
	19	105 36.4	54.5	206 29.5	15.2	7 38.8	10.2	54.4
	20	120 36.3	54.8	221 03.7	15.3	7 28.6	10.2	54.4
	21	135 36.2	·· 55.2	235 38.0	15.3	7 18.4	10.3	54.4
	22	150 36.1	55.5	250 12.3	15.3	7 08.1	10.2	54.4
	23	165 36.0	55.9	264 46.6	15.4	6 57.9	10.3	54.4
	1 00	180 36.0	N21 56.2	279 21.0	15.4	S 6 47.6	10.4	54.4
	01	195 35.9	56.6	293 55.4	15.4	6 37.2	10.3	54.4
	02	210 35.8	56.9	308 29.8	15.4	6 26.9	10.4	54.4
	03	225 35.7	·· 57.3	323 04.2	15.5	6 16.5	10.4	54.3
	04	240 35.6	57.6	337 38.7	15.5	6 06.1	10.4	54.3
	05	255 35.5	58.0	352 13.2	15.5	5 55.7	10.4	54.3
	06	270 35.4	N21 58.3	6 47.7	15.6	S 5 45.3	10.4	54.3
	07	285 35.3	58.7	21 22.3	15.5	5 34.9	10.5	54.3
	08	300 35.2	59.0	35 56.8	15.6	5 24.4	10.5	54.3
S	09	315 35.1	·· 59.4	50 31.4	15.7	5 13.9	10.5	54.3
U	10	330 35.0	21 59.7	65 06.1	15.6	5 03.4	10.5	54.3
N	11	345 34.9	22 00.1	79 40.7	15.7	4 52.9	10.6	54.3
D	12	0 34.8	N22 00.4	94 15.4	15.6	S 4 42.3	10.5	54.3
A	13	15 34.8	00.8	108 50.0	15.7	4 31.8	10.6	54.3
Y	14	30 34.7	01.1	123 24.7	15.8	4 21.2	10.6	54.3
	15	45 34.6	·· 01.5	137 59.5	15.7	4 10.6	10.6	54.3
	16	60 34.5	01.8	152 34.2	15.7	4 00.0	10.6	54.3
	17	75 34.4	02.1	167 08.9	15.8	3 49.4	10.6	54.3
	18	90 34.3	N22 02.5	181 43.7	15.8	S 3 38.8	10.7	54.3
	19	105 34.2	02.8	196 18.5	15.8	3 28.1	10.6	54.3
	20	120 34.1	03.2	210 53.3	15.8	3 17.5	10.7	54.3
	21	135 34.0	·· 03.5	225 28.1	15.8	3 06.8	10.7	54.3
	22	150 33.9	03.8	240 02.9	15.8	2 56.1	10.7	54.3
	23	165 33.8	04.2	254 37.7	15.9	2 45.4	10.6	54.2
	2 00	180 33.7	N22 04.5	269 12.6	15.8	S 2 34.8	10.8	54.2
	01	195 33.6	04.8	283 47.4	15.9	2 24.0	10.7	54.2
	02	210 33.5	05.2	298 22.3	15.8	2 13.3	10.7	54.2
	03	225 33.4	·· 05.5	312 57.1	15.9	2 02.6	10.7	54.2
	04	240 33.3	05.9	327 32.0	15.9	1 51.9	10.8	54.2
	05	255 33.2	06.2	342 06.9	15.8	1 41.1	10.7	54.2
	06	270 33.1	N22 06.5	356 41.7	15.9	S 1 30.4	10.8	54.2
	07	285 33.0	06.9	11 16.6	15.9	1 19.6	10.7	54.2
	08	300 32.9	07.2	25 51.5	15.9	1 08.9	10.8	54.2
M	09	315 32.9	·· 07.5	40 26.4	15.9	0 58.1	10.7	54.2
O	10	330 32.8	07.9	55 01.3	15.9	0 47.4	10.8	54.2
N	11	345 32.7	08.2	69 36.2	15.9	0 36.6	10.8	54.3
D	12	0 32.6	N22 08.5	84 11.1	15.9	S 0 25.8	10.8	54.3
A	13	15 32.5	08.8	98 46.0	15.8	0 15.0	10.7	54.3
Y	14	30 32.4	09.2	113 20.8	15.9	S 0 04.3	10.8	54.3
	15	45 32.3	·· 09.5	127 55.7	15.9	N 0 06.5	10.8	54.3
	16	60 32.2	09.8	142 30.6	15.9	0 17.3	10.8	54.3
	17	75 32.1	10.1	157 05.5	15.8	0 28.1	10.8	54.3
	18	90 32.0	N22 10.5	171 40.3	15.9	N 0 38.9	10.7	54.3
	19	105 31.9	10.8	186 15.2	15.9	0 49.6	10.8	54.3
	20	120 31.8	11.1	200 50.1	15.8	1 00.4	10.8	54.3
	21	135 31.7	·· 11.4	215 24.9	15.8	1 11.2	10.8	54.3
	22	150 31.6	11.8	229 59.7	15.9	1 22.0	10.8	54.3
	23	165 31.5	12.1	244 34.6	15.8	1 32.8	10.7	54.3
		S.D. 15.8	*d* 0.3	S.D. 14.8		14.8		14.8

Lat.	Twilight Naut.	Twilight Civil	Sunrise	Moonrise 31	Moonrise 1	Moonrise 2	Moonrise 3
°	h m	h m	h m	h m	h m	h m	h m
N 72	□	□	□	01 05	00 49	00 36	00 23
N 70	□	□	□	00 48	00 40	00 33	00 25
68	□	□	□	00 34	00 32	00 30	00 27
66	////	////	01 15	00 23	00 26	00 27	00 29
64	////	////	02 00	00 14	00 20	00 25	00 30
62	////	00 15	02 28	00 06	00 15	00 24	00 31
60	////	01 28	02 50	24 11	00 11	00 22	00 32
N 58	////	02 02	03 08	24 07	00 07	00 21	00 33
56	00 13	02 26	03 23	24 04	00 04	00 19	00 34
54	01 20	02 45	03 35	24 01	00 01	00 18	00 35
52	01 51	03 00	03 46	23 58	24 17	00 17	00 36
50	02 14	03 14	03 56	23 56	24 16	00 16	00 36
45	02 53	03 41	04 17	23 50	24 14	00 14	00 38
N 40	03 21	04 01	04 33	23 46	24 13	00 13	00 39
35	03 43	04 18	04 47	23 42	24 11	00 11	00 40
30	04 00	04 32	04 59	23 38	24 10	00 10	00 41
20	04 27	04 56	05 20	23 32	24 07	00 07	00 42
N 10	04 48	05 15	05 38	23 27	24 05	00 05	00 44
0	05 06	05 32	05 54	23 22	24 04	00 04	00 45
S 10	05 22	05 48	06 10	23 17	24 02	00 02	00 46
20	05 37	06 04	06 27	23 11	24 00	00 00	00 48
30	05 52	06 21	06 47	23 05	23 58	24 50	00 50
35	06 00	06 31	06 59	23 02	23 56	24 50	00 50
40	06 08	06 42	07 12	22 58	23 55	24 52	00 52
45	06 17	06 54	07 27	22 53	23 53	24 53	00 53
S 50	06 28	07 09	07 46	22 47	23 51	24 54	00 54
52	06 33	07 16	07 55	22 45	23 50	24 55	00 55
54	06 38	07 23	08 05	22 42	23 49	24 56	00 56
56	06 43	07 31	08 17	22 39	23 48	24 57	00 57
58	06 49	07 40	08 30	22 35	23 47	24 58	00 58
S 60	06 55	07 51	08 45	22 31	23 45	24 59	00 59

Lat.	Sunset	Twilight Civil	Twilight Naut.	Moonset 31	Moonset 1	Moonset 2	Moonset 3
°	h m	h m	h m	h m	h m	h m	h m
N 72	□	□	□	08 53	10 38	12 18	13 58
N 70	□	□	□	09 09	10 45	12 19	13 52
68	□	□	□	09 21	10 51	12 19	13 47
66	22 44	////	////	09 31	10 56	12 20	13 43
64	21 58	////	////	09 39	11 00	12 20	13 40
62	21 28	////	////	09 46	11 04	12 20	13 37
60	21 06	22 30	////	09 52	11 07	12 21	13 34
N 58	20 48	21 55	////	09 58	11 10	12 21	13 32
56	20 33	21 31	////	10 03	11 12	12 21	13 30
54	20 21	21 12	22 37	10 07	11 14	12 21	13 28
52	20 09	20 56	22 06	10 11	11 16	12 21	13 27
50	20 00	20 42	21 43	10 14	11 18	12 22	13 25
45	19 39	20 15	21 03	10 22	11 22	12 22	13 22
N 40	19 22	19 54	20 35	10 28	11 25	12 22	13 19
35	19 08	19 37	20 13	10 33	11 28	12 22	13 17
30	18 56	19 23	19 56	10 38	11 31	12 23	13 15
20	18 35	19 00	19 28	10 46	11 35	12 23	13 11
N 10	18 18	18 41	19 07	10 53	11 39	12 23	13 08
0	18 01	18 24	18 50	11 00	11 42	12 24	13 05
S 10	17 45	18 08	18 34	11 06	11 46	12 24	13 02
20	17 28	17 51	18 19	11 13	11 49	12 24	12 59
30	17 08	17 34	18 04	11 21	11 53	12 24	12 55
35	16 56	17 24	17 56	11 26	11 56	12 24	12 53
40	16 43	17 13	17 47	11 31	11 58	12 25	12 51
45	16 28	17 01	17 38	11 37	12 01	12 25	12 48
S 50	16 09	16 46	17 27	11 44	12 05	12 25	12 45
52	16 00	16 40	17 22	11 47	12 07	12 25	12 44
54	15 50	16 32	17 17	11 51	12 08	12 25	12 42
56	15 38	16 24	17 12	11 54	12 10	12 25	12 40
58	15 25	16 15	17 06	11 59	12 13	12 26	12 38
S 60	15 10	16 04	16 59	12 04	12 15	12 26	12 36

Day	SUN Eqn. of Time 00ʰ	SUN Eqn. of Time 12ʰ	SUN Mer. Pass.	MOON Mer. Pass. Upper	MOON Mer. Pass. Lower	Age	Phase
	m s	m s	h m	h m	h m	d	
31	02 32	02 28	11 58	04 49	17 11	20	
1	02 24	02 20	11 58	05 32	17 53	21	◑
2	02 15	02 10	11 58	06 14	18 34	22	

1975 JUNE 12, 13, 14 (THURS., FRI., SAT.)

G.M.T.		ARIES G.H.A.	VENUS −3.9 G.H.A.	VENUS Dec.	MARS +0.8 G.H.A.	MARS Dec.	JUPITER −1.9 G.H.A.	JUPITER Dec.	SATURN +0.4 G.H.A.	SATURN Dec.	STARS Name	S.H.A.	Dec.
d	h	° ′	° ′	° ′	° ′	° ′	° ′	° ′	° ′	° ′		° ′	° ′
12	00	259 42.0	131 06.2	N21 02.9	244 12.2	N 4 47.9	241 58.0	N 6 13.1	149 56.7	N22 06.6	Acamar	315 40.3	S40 24.0
	01	274 44.5	146 06.1	02.2	259 13.0	48.6	257 00.1	13.3	164 58.8	06.6	Achernar	335 48.3	S57 21.4
	02	289 46.9	161 06.0	01.5	274 13.8	49.3	272 02.2	13.4	180 01.0	06.6	Acrux	173 41.0	S62 58.2
	03	304 49.4	176 05.9	·· 00.8	289 14.5	·· 50.0	287 04.2	·· 13.6	195 03.1	·· 06.5	Adhara	255 35.3	S28 56.5
	04	319 51.9	191 05.8	21 00.1	304 15.3	50.7	302 06.3	13.7	210 05.2	06.5	Aldebaran	291 22.4	N16 27.6
	05	334 54.3	206 05.7	20 59.3	319 16.0	51.4	317 08.4	13.9	225 07.4	06.4			
	06	349 56.8	221 05.6	N20 58.6	334 16.8	N 4 52.1	332 10.4	N 6 14.0	240 09.5	N22 06.4	Alioth	166 45.4	N56 05.7
	07	4 59.3	236 05.5	57.9	349 17.6	52.8	347 12.5	14.2	255 11.7	06.4	Alkaid	153 21.0	N49 26.3
T	08	20 01.7	251 05.4	57.2	4 18.3	53.5	2 14.6	14.3	270 13.8	06.3	Al Na'ir	28 19.2	S47 04.5
H	09	35 04.2	266 05.3	·· 56.4	19 19.1	·· 54.2	17 16.6	·· 14.5	285 16.0	·· 06.3	Alnilam	276 15.6	S 1 13.1
U	10	50 06.6	281 05.3	55.7	34 19.8	54.9	32 18.7	14.6	300 18.1	06.3	Alphard	218 24.2	S 8 33.3
R	11	65 09.1	296 05.2	55.0	49 20.6	55.6	47 20.8	14.8	315 20.2	06.2			
S	12	80 11.6	311 05.1	N20 54.3	64 21.4	N 4 56.3	62 22.8	N 6 15.0	330 22.4	N22 06.2	Alphecca	126 34.8	N26 47.9
D	13	95 14.0	326 05.0	53.5	79 22.1	57.0	77 24.9	15.1	345 24.5	06.1	Alpheratz	358 13.1	N28 57.2
A	14	110 16.5	341 04.9	52.8	94 22.9	57.7	92 27.0	15.3	0 26.7	06.1	Altair	62 35.7	N 8 48.2
Y	15	125 19.0	356 04.8	·· 52.1	109 23.7	·· 58.4	107 29.0	·· 15.4	15 28.8	·· 06.1	Ankaa	353 43.9	S42 26.1
	16	140 21.4	11 04.7	51.3	124 24.4	59.1	122 31.1	15.6	30 30.9	06.0	Antares	113 00.9	S26 22.7
	17	155 23.9	26 04.6	50.6	139 25.2	4 59.8	137 33.2	15.7	45 33.1	06.0			
	18	170 26.4	41 04.5	N20 49.9	154 25.9	N 5 00.5	152 35.2	N 6 15.9	60 35.2	N22 06.0	Arcturus	146 21.5	N19 18.6
	19	185 28.8	56 04.5	49.2	169 26.7	01.2	167 37.3	16.0	75 37.4	05.9	Atria	108 27.8	S68 59.1
	20	200 31.3	71 04.4	48.4	184 27.5	01.9	182 39.4	16.2	90 39.5	05.9	Avior	234 30.1	S59 26.1
	21	215 33.8	86 04.3	·· 47.7	199 28.2	·· 02.6	197 41.4	·· 16.3	105 41.6	·· 05.9	Bellatrix	279 02.9	N 6 19.6
	22	230 36.2	101 04.2	47.0	214 29.0	03.3	212 43.5	16.5	120 43.8	05.8	Betelgeuse	271 32.5	N 7 24.1
	23	245 38.7	116 04.1	46.2	229 29.7	04.0	227 45.6	16.6	135 45.9	05.8			
13	00	260 41.1	131 04.1	N20 45.5	244 30.5	N 5 04.7	242 47.6	N 6 16.8	150 48.1	N22 05.7	Canopus	264 09.3	S52 41.1
	01	275 43.6	146 04.0	44.8	259 31.3	05.4	257 49.7	16.9	165 50.2	05.7	Capella	281 17.1	N45 58.4
	02	290 46.1	161 03.9	44.0	274 32.0	06.1	272 51.8	17.1	180 52.4	05.7	Deneb	49 50.6	N45 11.5
	03	305 48.5	176 03.8	·· 43.3	289 32.8	·· 06.8	287 53.8	·· 17.2	195 54.5	·· 05.6	Denebola	183 02.7	N14 42.5
	04	320 51.0	191 03.8	42.6	304 33.6	07.5	302 55.9	17.4	210 56.6	05.6	Diphda	349 24.6	S18 07.1
	05	335 53.5	206 03.7	41.8	319 34.3	08.2	317 58.0	17.5	225 58.8	05.6			
	06	350 55.9	221 03.6	N20 41.1	334 35.1	N 5 08.9	333 00.0	N 6 17.7	241 00.9	N22 05.5	Dubhe	194 26.6	N61 53.2
	07	5 58.4	236 03.5	40.3	349 35.8	09.6	348 02.1	17.9	256 03.1	05.5	Elnath	278 49.0	N28 35.2
	08	21 00.9	251 03.5	39.6	4 36.6	10.3	3 04.2	18.0	271 05.2	05.4	Eltanin	90 58.8	N51 29.5
F	09	36 03.3	266 03.4	·· 38.9	19 37.4	·· 11.0	18 06.3	·· 18.2	286 07.3	·· 05.4	Enif	34 14.9	N 9 45.8
R	10	51 05.8	281 03.3	38.1	34 38.1	11.7	33 08.3	18.3	301 09.5	05.4	Fomalhaut	15 55.3	S29 44.9
I	11	66 08.3	296 03.3	37.4	49 38.9	12.4	48 10.4	18.5	316 11.6	05.3			
D	12	81 10.7	311 03.2	N20 36.6	64 39.6	N 5 13.1	63 12.5	N 6 18.6	331 13.8	N22 05.3	Gacrux	172 32.6	S56 58.9
A	13	96 13.2	326 03.1	35.9	79 40.4	13.8	78 14.5	18.8	346 15.9	05.3	Gienah	176 21.6	S17 24.6
Y	14	111 15.6	341 03.1	35.1	94 41.2	14.5	93 16.6	18.9	1 18.0	05.2	Hadar	149 28.0	S60 15.6
	15	126 18.1	356 03.0	·· 34.4	109 41.9	·· 15.2	108 18.7	·· 19.1	16 20.2	·· 05.2	Hamal	328 33.2	N23 20.7
	16	141 20.6	11 03.0	33.7	124 42.7	15.9	123 20.8	19.2	31 22.3	05.1	Kaus Aust.	84 21.2	S34 23.7
	17	156 23.0	26 02.9	32.9	139 43.5	16.6	138 22.8	19.4	46 24.5	05.1			
	18	171 25.5	41 02.8	N20 32.2	154 44.2	N 5 17.3	153 24.9	N 6 19.5	61 26.6	N22 05.1	Kochab	137 17.9	N74 15.5
	19	186 28.0	56 02.8	31.4	169 45.0	18.0	168 27.0	19.7	76 28.7	05.0	Markab	14 06.7	N15 04.4
	20	201 30.4	71 02.7	30.7	184 45.7	18.7	183 29.0	19.8	91 30.9	05.0	Menkar	314 45.1	N 3 59.6
	21	216 32.9	86 02.7	·· 29.9	199 46.5	·· 19.4	198 31.1	·· 20.0	106 33.0	·· 05.0	Menkent	148 41.0	S36 15.2
	22	231 35.4	101 02.6	29.2	214 47.3	20.1	213 33.2	20.1	121 35.2	04.9	Miaplacidus	221 46.2	S69 37.4
	23	246 37.8	116 02.6	28.4	229 48.0	20.8	228 35.3	20.3	136 37.3	04.9			
14	00	261 40.3	131 02.5	N20 27.7	244 48.8	N 5 21.5	243 37.3	N 6 20.4	151 39.4	N22 04.8	Mirfak	309 21.7	N49 46.4
	01	276 42.8	146 02.5	26.9	259 49.6	22.2	258 39.4	20.6	166 41.6	04.8	Nunki	76 33.2	S26 19.6
	02	291 45.2	161 02.4	26.2	274 50.3	22.9	273 41.5	20.7	181 43.7	04.8	Peacock	54 03.6	S56 48.6
	03	306 47.7	176 02.4	·· 25.4	289 51.1	·· 23.6	288 43.6	·· 20.9	196 45.9	·· 04.7	Pollux	244 02.8	N28 05.2
	04	321 50.1	191 02.3	24.7	304 51.8	24.2	303 45.6	21.0	211 48.0	04.7	Procyon	245 29.8	N 5 17.2
	05	336 52.6	206 02.3	23.9	319 52.6	24.9	318 47.7	21.2	226 50.1	04.7			
	06	351 55.1	221 02.2	N20 23.2	334 53.4	N 5 25.6	333 49.8	N 6 21.3	241 52.3	N22 04.6	Rasalhague	96 32.6	N12 34.7
	07	6 57.5	236 02.2	22.4	349 54.1	26.3	348 51.8	21.5	256 54.4	04.6	Regulus	208 13.9	N12 05.2
S	08	22 00.0	251 02.1	21.7	4 54.9	27.0	3 53.9	21.6	271 56.6	04.5	Rigel	281 39.8	S 8 13.8
A	09	37 02.5	266 02.1	·· 20.9	19 55.7	·· 27.7	18 56.0	·· 21.8	286 58.7	·· 04.5	Rigil Kent.	140 30.1	S60 44.3
T	10	52 04.9	281 02.1	20.1	34 56.4	28.4	33 58.1	21.9	302 00.8	04.5	Sabik	102 44.9	S15 41.7
U	11	67 07.4	296 02.0	19.4	49 57.2	29.1	49 00.1	22.1	317 03.0	04.4			
R	12	82 09.9	311 02.0	N20 18.6	64 58.0	N 5 29.8	64 02.2	N 6 22.2	332 05.1	N22 04.4	Schedar	350 13.3	N56 24.0
D	13	97 12.3	326 01.9	17.9	79 58.7	30.5	79 04.3	22.4	347 07.3	04.4	Shaula	97 00.2	S37 05.2
A	14	112 14.8	341 01.9	17.1	94 59.5	31.2	94 06.4	22.5	2 09.4	04.3	Sirius	258 59.2	S16 41.1
Y	15	127 17.2	356 01.9	·· 16.4	110 00.2	·· 31.9	109 08.4	·· 22.7	17 11.5	·· 04.3	Spica	159 01.1	S11 02.2
	16	142 19.7	11 01.8	15.6	125 01.0	32.6	124 10.5	22.8	32 13.7	04.2	Suhail	223 13.7	S43 20.3
	17	157 22.2	26 01.8	14.8	140 01.8	33.3	139 12.6	23.0	47 15.8	04.2			
	18	172 24.6	41 01.8	N20 14.1	155 02.5	N 5 34.0	154 14.7	N 6 23.1	62 17.9	N22 04.2	Vega	80 57.8	N38 45.6
	19	187 27.1	56 01.8	13.3	170 03.3	34.7	169 16.7	23.3	77 20.1	04.1	Zuben'ubi	137 36.7	S15 56.5
	20	202 29.6	71 01.7	12.6	185 04.1	35.4	184 18.8	23.4	92 22.2	04.1		S.H.A.	Mer. Pass.
	21	217 32.0	86 01.7	·· 11.8	200 04.8	·· 36.1	199 20.9	·· 23.6	107 24.4	·· 04.1		° ′	h m
	22	232 34.5	101 01.7	11.0	215 05.6	36.8	214 23.0	23.7	122 26.5	04.0	Venus	230 22.9	15 16
	23	247 37.0	116 01.6	10.3	230 06.3	37.5	229 25.1	23.9	137 28.6	04.0	Mars	343 49.4	7 42
											Jupiter	342 06.5	7 48
Mer. Pass.		h m 6 36.2	v −0.1	d 0.7	v 0.8	d 0.7	v 2.1	d 0.2	v 2.1	d 0.0	Saturn	250 06.9	13 55

1975 JUNE 12, 13, 14 (THURS., FRI., SAT.)

G.M.T.	SUN G.H.A.	SUN Dec.	MOON G.H.A.	v	MOON Dec.	d	H.P.
d h	° ′	° ′	° ′	′	° ′	′	′
12 00 (THURSDAY)	180 06.5	N23 05.8	149 45.2	7.2	N18 13.9	7.0	58.9
01	195 06.3	05.9	164 11.4	7.2	18 06.9	7.1	58.9
02	210 06.2	06.1	178 37.6	7.2	17 59.8	7.2	58.9
03	225 06.1	·· 06.3	193 03.8	7.3	17 52.6	7.3	59.0
04	240 06.0	06.4	207 30.1	7.3	17 45.3	7.4	59.0
05	255 05.8	06.6	221 56.4	7.3	17 37.9	7.5	59.0
06	270 05.7	N23 06.8	236 22.7	7.4	N17 30.4	7.7	59.0
07	285 05.6	06.9	250 49.1	7.4	17 22.7	7.7	59.0
08	300 05.4	07.1	265 15.5	7.4	17 15.0	7.9	59.0
09	315 05.3	·· 07.3	279 41.9	7.5	17 07.1	8.0	59.0
10	330 05.2	07.4	294 08.4	7.5	16 59.1	8.1	59.0
11	345 05.1	07.6	308 34.9	7.6	16 51.0	8.2	59.1
12	0 04.9	N23 07.8	323 01.5	7.6	N16 42.8	8.3	59.1
13	15 04.8	07.9	337 28.1	7.6	16 34.5	8.4	59.1
14	30 04.7	08.1	351 54.7	7.7	16 26.1	8.5	59.1
15	45 04.5	·· 08.3	6 21.4	7.7	16 17.6	8.6	59.1
16	60 04.4	08.4	20 48.1	7.8	16 09.0	8.7	59.1
17	75 04.3	08.6	35 14.9	7.8	16 00.3	8.8	59.1
18	90 04.2	N23 08.7	49 41.7	7.8	N15 51.5	8.9	59.1
19	105 04.0	08.9	64 08.5	7.9	15 42.6	9.0	59.2
20	120 03.9	09.1	78 35.4	7.9	15 33.6	9.1	59.2
21	135 03.8	·· 09.2	93 02.3	8.0	15 24.5	9.2	59.2
22	150 03.6	09.4	107 29.3	8.0	15 15.3	9.3	59.2
23	165 03.5	09.5	121 56.3	8.1	15 06.0	9.4	59.2
13 00 (FRIDAY)	180 03.4	N23 09.7	136 23.4	8.0	N14 56.6	9.5	59.2
01	195 03.3	09.8	150 50.4	8.2	14 47.1	9.6	59.2
02	210 03.1	10.0	165 17.6	8.2	14 37.5	9.6	59.2
03	225 03.0	·· 10.1	179 44.8	8.2	14 27.9	9.8	59.2
04	240 02.9	10.3	194 12.0	8.3	14 18.1	9.8	59.2
05	255 02.7	10.4	208 39.3	8.3	14 08.3	10.0	59.2
06	270 02.6	N23 10.6	223 06.6	8.3	N13 58.3	10.0	59.3
07	285 02.5	10.7	237 33.9	8.4	13 48.3	10.1	59.3
08	300 02.3	10.9	252 01.3	8.5	13 38.2	10.2	59.3
09	315 02.2	·· 11.0	266 28.8	8.4	13 28.0	10.3	59.3
10	330 02.1	11.2	280 56.2	8.6	13 17.7	10.3	59.3
11	345 02.0	11.3	295 23.8	8.5	13 07.4	10.4	59.3
12	0 01.8	N23 11.5	309 51.3	8.7	N12 57.0	10.5	59.3
13	15 01.7	11.6	324 19.0	8.6	12 46.5	10.6	59.3
14	30 01.6	11.8	338 46.6	8.7	12 35.9	10.7	59.3
15	45 01.4	·· 11.9	353 14.3	8.7	12 25.2	10.7	59.3
16	60 01.3	12.1	7 42.0	8.8	12 14.5	10.8	59.3
17	75 01.2	12.2	22 09.8	8.9	12 03.7	10.9	59.3
18	90 01.0	N23 12.3	36 37.7	8.8	N11 52.8	11.0	59.3
19	105 00.9	12.5	51 05.5	8.9	11 41.8	11.0	59.3
20	120 00.8	12.6	65 33.4	9.0	11 30.8	11.1	59.4
21	135 00.6	·· 12.8	80 01.4	9.0	11 19.7	11.2	59.4
22	150 00.5	12.9	94 29.4	9.0	11 08.5	11.2	59.4
23	165 00.4	13.1	108 57.4	9.1	10 57.3	11.3	59.4
14 00 (SATURDAY)	180 00.3	N23 13.2	123 25.5	9.1	N10 46.0	11.4	59.4
01	195 00.1	13.3	137 53.6	9.2	10 34.6	11.4	59.4
02	210 00.0	13.5	152 21.8	9.2	10 23.2	11.5	59.4
03	224 59.9	·· 13.6	166 50.0	9.2	10 11.7	11.5	59.4
04	239 59.7	13.7	181 18.2	9.3	10 00.2	11.6	59.4
05	254 59.6	13.9	195 46.5	9.3	9 48.6	11.7	59.4
06	269 59.5	N23 14.0	210 14.8	9.3	N 9 36.9	11.7	59.4
07	284 59.3	14.1	224 43.1	9.4	9 25.2	11.8	59.4
08	299 59.2	14.3	239 11.5	9.4	9 13.4	11.8	59.4
09	314 59.1	·· 14.4	253 39.9	9.5	9 01.6	11.9	59.4
10	329 58.9	14.5	268 08.4	9.5	8 49.7	11.9	59.4
11	344 58.8	14.7	282 36.9	9.5	8 37.8	12.0	59.4
12	359 58.7	N23 14.8	297 05.4	9.6	N 8 25.8	12.0	59.4
13	14 58.5	14.9	311 34.0	9.6	8 13.8	12.1	59.4
14	29 58.4	15.0	326 02.6	9.6	8 01.7	12.1	59.4
15	44 58.3	·· 15.2	340 31.2	9.7	7 49.6	12.2	59.4
16	59 58.1	15.3	354 59.9	9.7	7 37.4	12.2	59.4
17	74 58.0	15.4	9 28.6	9.7	7 25.2	12.2	59.4
18	89 57.9	N23 15.6	23 57.3	9.8	N 7 13.0	12.3	59.4
19	104 57.7	15.7	38 26.1	9.8	7 00.7	12.3	59.4
20	119 57.6	15.8	52 54.9	9.8	6 48.4	12.4	59.4
21	134 57.5	·· 15.9	67 23.7	9.8	6 36.0	12.4	59.4
22	149 57.4	16.0	81 52.5	9.9	6 23.6	12.5	59.4
23	164 57.2	16.2	96 21.4	9.9	6 11.1	12.4	59.4
	S.D. 15.8	d 0.1	S.D. 16.1		16.2		16.2

Lat.	Twilight Naut.	Twilight Civil	Sunrise	Moonrise 12	Moonrise 13	Moonrise 14	Moonrise 15
°	h m	h m	h m	h m	h m	h m	h m
N 72	□	□	□	02 46	05 50	08 06	10 12
N 70	□	□	□	04 08	06 19	08 21	10 18
68	□	□	□	04 46	06 41	08 33	10 22
66	////	////	00 08	05 13	06 57	08 42	10 26
64	////	////	01 36	05 34	07 11	08 50	10 29
62	////	////	02 12	05 50	07 23	08 57	10 31
60	////	00 58	02 37	06 04	07 32	09 03	10 34
N 58	////	01 44	02 57	06 15	07 41	09 08	10 36
56	////	02 12	03 14	06 26	07 48	09 13	10 37
54	00 53	02 34	03 28	06 35	07 55	09 17	10 39
52	01 35	02 51	03 40	06 42	08 01	09 21	10 41
50	02 02	03 06	03 50	06 50	08 06	09 24	10 42
45	02 46	03 36	04 13	07 05	08 18	09 31	10 45
N 40	03 16	03 58	04 30	07 17	08 27	09 37	10 47
35	03 39	04 16	04 45	07 28	08 35	09 42	10 49
30	03 58	04 31	04 58	07 37	08 42	09 47	10 51
20	04 26	04 55	05 20	07 53	08 55	09 55	10 54
N 10	04 49	05 16	05 39	08 07	09 05	10 02	10 57
0	05 08	05 34	05 56	08 20	09 15	10 08	11 00
S 10	05 24	05 51	06 13	08 33	09 25	10 15	11 03
20	05 40	06 08	06 32	08 47	09 36	10 22	11 05
30	05 57	06 26	06 53	09 03	09 48	10 30	11 09
35	06 05	06 37	07 05	09 12	09 55	10 34	11 11
40	06 14	06 48	07 19	09 23	10 03	10 39	11 13
45	06 25	07 02	07 35	09 35	10 12	10 45	11 15
S 50	06 36	07 17	07 56	09 50	10 23	10 52	11 18
52	06 41	07 25	08 06	09 57	10 28	10 55	11 19
54	06 47	07 33	08 17	10 04	10 34	10 59	11 21
56	06 53	07 42	08 29	10 13	10 40	11 03	11 22
58	07 00	07 52	08 44	10 22	10 47	11 07	11 24
S 60	07 07	08 04	09 01	10 33	10 55	11 12	11 26

Lat.	Sunset	Twilight Civil	Twilight Naut.	Moonset 12	Moonset 13	Moonset 14	Moonset 15
°	h m	h m	h m	h m	h m	h m	h m
N 72	□	□	□	01 39	00 31	{00 07 / 23 59}	23 35
N 70	□	□	□	00 16	{00 00 / 23 56}	23 42	23 34
68	□	□	□	23 38	23 37	23 35	23 33
66	□	□	□	23 20	23 26	23 29	23 32
64	22 25	////	////	23 05	23 16	23 25	23 32
62	21 48	////	////	22 53	23 08	23 20	23 31
60	21 23	23 04	////	22 42	23 01	23 17	23 30
N 58	21 03	22 17	////	22 33	22 55	23 13	23 30
56	20 46	21 48	////	22 25	22 50	23 11	23 30
54	20 32	21 26	23 09	22 17	22 45	23 08	23 29
52	20 20	21 09	22 25	22 11	22 40	23 06	23 29
50	20 09	20 54	21 58	22 05	22 36	23 03	23 29
45	19 47	20 24	21 14	21 52	22 27	22 59	23 28
N 40	19 29	20 02	20 44	21 42	22 20	22 55	23 27
35	19 14	19 44	20 21	21 33	22 13	22 51	23 27
30	19 01	19 29	20 02	21 24	22 08	22 48	23 26
20	18 40	19 04	19 33	21 11	21 58	22 42	23 26
N 10	18 21	18 44	19 11	20 58	21 49	22 38	23 25
0	18 03	18 26	18 52	20 47	21 41	22 33	23 24
S 10	17 46	18 09	18 35	20 35	21 32	22 28	23 23
20	17 28	17 52	18 19	20 23	21 23	22 23	23 23
30	17 07	17 33	18 03	20 08	21 13	22 18	23 22
35	16 55	17 23	17 54	20 00	21 07	22 14	23 21
40	16 41	17 11	17 45	19 50	21 00	22 11	23 20
45	16 24	16 58	17 35	19 39	20 52	22 06	23 20
S 50	16 04	16 42	17 23	19 25	20 43	22 01	23 19
52	15 54	16 35	17 18	19 19	20 38	21 58	23 18
54	15 43	16 26	17 13	19 12	20 33	21 56	23 18
56	15 30	16 17	17 06	19 04	20 28	21 53	23 17
58	15 16	16 07	17 00	18 55	20 22	21 49	23 17
S 60	14 59	15 56	16 52	18 44	20 15	21 46	23 16

Day	SUN Eqn. of Time 00h	SUN Eqn. of Time 12h	SUN Mer. Pass.	MOON Mer. Pass. Upper	MOON Mer. Pass. Lower	Age	Phase
	m s	m s	h m	h m	h m	d	
12	00 26	00 20	12 00	14 34	02 06	03	
13	00 14	00 08	12 00	15 28	03 01	04	◑
14	00 01	00 05	12 00	16 21	03 55	05	

STARS, 1975 JANUARY—JUNE

Mag.	Name and Number	S.H.A.							Dec.						
			JAN.	FEB.	MAR.	APR.	MAY	JUNE		JAN.	FEB.	MAR.	APR.	MAY	JUNE
		°	′	′	′	′	′	′	°	′	′	′	′	′	′
3·1	γ Ursæ Minoris †	129	49·3	48·8	48·3	47·9	47·8	48·0	N. 71	55·0	55·0	55·0	55·1	55·3	55·5
3·1	γ Trianguli Aust.	130	51·5	50·9	50·4	50·0	49·8	49·7	S. 68	35·1	35·1	35·2	35·3	35·5	35·6
2·7	β Libræ	131	05·0	04·7	04·5	04·4	04·3	04·2	S. 9	17·5	17·6	17·7	17·7	17·7	17·6
2·8	β Lupi	135	46·5	46·2	45·9	45·7	45·6	45·6	S. 43	01·9	02·0	02·1	02·2	02·3	02·3
2·2	β Ursæ Minoris 40	137	19·2	18·6	18·0	17·7	17·6	17·9	N. 74	15·1	15·0	15·1	15·2	15·4	15·5
2·9	α Libræ 39	137	37·4	37·2	37·0	36·8	36·7	36·7	S. 15	56·3	56·4	56·5	56·5	56·5	56·5
2·6	ε Bootis	139	01·5	01·2	01·0	00·9	00·8	00·9	N. 27	10·5	10·4	10·4	10·5	10·6	10·7
2·9	α Lupi	139	55·9	55·5	55·3	55·1	54·9	55·0	S. 47	16·8	16·8	16·9	17·0	17·1	17·2
0·1	α Centauri 38	140	31·2	30·8	30·5	30·2	30·1	30·2	S. 60	43·7	43·8	43·9	44·0	44·2	44·3
2·6	η Centauri	141	31·0	30·7	30·5	30·3	30·2	30·2	S. 42	02·8	02·9	03·0	03·1	03·2	03·3
3·0	γ Bootis	142	13·9	13·6	13·4	13·2	13·2	13·2	N. 38	24·7	24·7	24·7	24·8	24·9	25·0
0·2	α Bootis 37	146	22·0	21·8	21·6	21·5	21·5	21·5	N. 19	18·5	18·4	18·4	18·5	18·5	18·6
2·3	θ Centauri 36	148	41·6	41·3	41·1	41·0	40·9	41·0	S. 36	14·8	14·9	15·0	15·1	15·2	15·2
0·9	β Centauri 35	149	28·9	28·5	28·2	28·0	27·9	28·0	S. 60	15·0	15·1	15·2	15·4	15·5	15·6
3·1	ζ Centauri	151	30·1	29·8	29·5	29·4	29·3	29·4	S. 47	09·8	09·9	10·0	10·2	10·3	10·3
2·8	η Bootis	151	37·4	37·1	37·0	36·9	36·9	36·9	N. 18	31·1	31·0	31·0	31·1	31·2	31·2
1·9	η Ursæ Majoris 34	153	21·5	21·2	21·0	20·9	20·9	21·0	N. 49	25·9	25·9	25·9	26·0	26·2	26·3
2·6	ε Centauri †	155	25·2	24·8	24·6	24·4	24·4	24·5	S. 53	20·2	20·3	20·5	20·6	20·7	20·8
1·2	α Virginis 33	159	01·6	01·4	01·2	01·1	01·1	01·1	S. 11	02·0	02·1	02·1	02·2	02·2	02·2
2·2	ζ Ursæ Majoris	159	16·0	15·7	15·5	15·4	15·4	15·6	N. 55	02·9	02·9	03·0	03·1	03·3	03·4
2·9	ι Centauri	160	11·8	11·6	11·4	11·3	11·3	11·3	S. 36	34·8	34·9	35·0	35·1	35·2	35·3
3·0	ε Virginis	164	45·7	45·5	45·4	45·3	45·3	45·4	N. 11	05·4	05·3	05·3	05·3	05·4	05·4
2·9	α Canum Venat.	166	16·8	16·6	16·4	16·4	16·4	16·5	N. 38	26·8	26·8	26·9	27·0	27·1	27·1
1·7	ε Ursæ Majoris 32	166	45·8	45·4	45·2	45·2	45·3	45·4	N. 56	05·3	05·3	05·4	05·5	05·7	05·7
1·5	β Crucis	168	25·7	25·3	25·1	25·0	25·1	25·3	S. 59	33·0	33·1	33·3	33·4	33·6	33·7
2·9	γ Virginis	169	53·8	53·6	53·4	53·4	53·4	53·5	S. 1	18·9	19·0	19·0	19·1	19·0	19·0
2·4	γ Centauri	169	57·6	57·3	57·1	57·0	57·1	57·2	S. 48	49·3	49·4	49·5	49·7	49·8	49·8
2·9	α Muscæ	171	04·1	03·7	03·4	03·3	03·5	03·7	S. 68	59·7	59·8	60·0	60·2	60·3	60·4
2·8	β Corvi	171	43·5	43·3	43·2	43·1	43·2	43·2	S. 23	15·6	15·7	15·8	15·9	15·9	15·9
1·6	γ Crucis 31	172	32·9	32·6	32·4	32·3	32·4	32·6	S. 56	58·3	58·4	58·6	58·7	58·9	58·9
1·1	α Crucis 30	173	41·4	41·0	40·8	40·7	40·8	41·0	S. 62	57·5	57·6	57·8	58·0	58·1	58·2
2·8	γ Corvi 29	176	21·8	21·6	21·5	21·5	21·5	21·6	S. 17	24·3	24·4	24·5	24·5	24·6	24·6
2·9	δ Centauri	178	13·6	13·3	13·2	13·2	13·3	13·4	S. 50	34·9	35·1	35·2	35·4	35·5	35·5
2·5	γ Ursæ Majoris	181	51·7	51·4	51·3	51·3	51·5	51·7	N. 53	49·6	49·6	49·7	49·9	50·0	50·0
2·2	β Leonis 28	183	02·8	02·6	02·5	02·5	02·6	02·7	N. 14	42·4	42·4	42·4	42·4	42·5	42·5
2·6	δ Leonis	191	47·8	47·6	47·6	47·6	47·7	47·8	N. 20	39·4	39·3	39·4	39·4	39·5	39·5
3·2	ψ Ursæ Majoris	192	55·5	55·3	55·2	55·3	55·4	55·6	N. 44	37·7	37·7	37·8	37·9	38·0	38·0
2·0	α Ursæ Majoris 27	194	26·4	26·1	26·0	26·1	26·3	26·6	N. 61	52·8	52·8	53·0	53·1	53·2	53·2
2·4	β Ursæ Majoris	194	54·2	54·0	53·9	54·0	54·2	54·4	N. 56	30·6	30·7	30·8	30·9	31·0	31·0
2·8	μ Velorum	198	34·0	33·8	33·8	33·9	34·0	34·2	S. 49	17·2	17·4	17·6	17·7	17·8	17·8
3·0	θ Carinæ †	199	28·2	28·0	28·0	28·1	28·4	28·7	S. 64	15·7	15·9	16·1	16·2	16·3	16·4
2·3	γ Leonis	205	20·5	20·4	20·3	20·4	20·5	20·6	N. 19	57·8	57·8	57·8	57·9	57·9	57·9
1·3	α Leonis 26	208	13·8	13·7	13·7	13·7	13·8	13·9	N. 12	05·2	05·1	05·1	05·1	05·2	05·2
3·1	ε Leonis	213	52·9	52·8	52·8	52·8	53·0	53·1	N. 23	53·1	53·1	53·2	53·2	53·2	53·3
3·0	*N* Velorum	217	22·3	22·2	22·3	22·5	22·7	23·0	S. 56	55·4	55·6	55·8	55·9	56·0	55·9
2·2	α Hydræ 25	218	24·0	24·0	24·0	24·1	24·2	24·2	S. 8	33·1	33·2	33·3	33·3	33·3	33·3
2·6	κ Velorum	219	39·2	39·1	39·2	39·4	39·6	39·8	S. 54	54·2	54·4	54·6	54·7	54·7	54·7
2·2	ι Carinæ	220	52·9	52·8	52·9	53·1	53·4	53·7	S. 59	10·2	10·4	10·6	10·7	10·7	10·7
1·8	β Carinæ 24	221	44·9	44·9	45·1	45·4	45·9	46·3	S. 69	36·9	37·1	37·2	37·3	37·4	37·3
2·2	λ Velorum 23	223	13·2	13·1	13·2	13·4	13·5	13·7	S. 43	19·9	20·1	20·3	20·3	20·4	20·3
3·1	ι Ursæ Majoris	225	36·6	36·6	36·6	36·8	37·0	37·1	N. 48	08·2	08·3	08·3	08·4	08·4	08·4
2·0	δ Velorum	228	59·0	59·0	59·1	59·4	59·6	59·8	S. 54	37·0	37·2	37·4	37·5	37·5	37·4
1·7	ε Carinæ 22	234	29·1	29·2	29·3	29·6	29·9	30·2	S. 59	25·8	26·0	26·1	26·2	26·2	26·1
1·9	γ Velorum	237	47·9	47·9	48·0	48·2	48·4	48·6	S. 47	15·9	16·0	16·1	16·2	16·2	16·1
2·9	ρ Puppis	238	22·2	22·2	22·3	22·4	22·5	22·6	S. 24	14·0	14·2	14·2	14·3	14·3	14·2
2·3	ζ Puppis	239	18·8	18·8	18·9	19·1	19·2	19·4	S. 39	56·1	56·2	56·3	56·4	56·4	56·3
1·2	β Geminorum 21	244	02·5	02·4	02·5	02·7	02·8	02·8	N. 28	05·1	05·1	05·1	05·2	05·2	05·2
0·5	α Canis Minoris 20	245	29·5	29·4	29·5	29·7	29·8	29·8	N. 5	17·2	17·2	17·2	17·2	17·2	17·2

† Not suitable for use with H.O. 214 (H.D. 486)

POLARIS (POLE STAR) TABLES, 1975

FOR DETERMINING LATITUDE FROM SEXTANT ALTITUDE AND FOR AZIMUTH

L.H.A. ARIES	240°–249°	250°–259°	260°–269°	270°–279°	280°–289°	290°–299°	300°–309°	310°–319°	320°–329°	330°–339°	340°–349°	350°–359°
	a_0	a_0	a_0	a_0	a_0	a_0	a_0	a_0	a_0	a_0	a_0	a_0
°	° ′	° ′	° ′	° ′	° ′	° ′	° ′	° ′	° ′	° ′	° ′	° ′
0	1 43·8	1 39·1	1 33·1	1 26·1	1 18·3	1 09·9	1 01·1	0 52·3	0 43·6	0 35·4	0 27·9	0 21·3
1	43·4	38·5	32·5	25·4	17·5	09·0	1 00·2	51·4	42·8	34·6	27·2	20·7
2	43·0	38·0	31·8	24·6	16·7	08·2	0 59·4	50·5	41·9	33·8	26·5	20·1
3	42·5	37·4	31·1	23·9	15·8	07·3	58·5	49·6	41·1	33·1	25·8	19·6
4	42·1	36·8	30·4	23·1	15·0	06·4	57·6	48·8	40·3	32·3	25·1	19·0
5	1 41·6	1 36·2	1 29·7	1 22·3	1 14·2	1 05·5	0 56·7	0 47·9	0 39·4	0 31·5	0 24·5	0 18·5
6	41·1	35·6	29·0	21·5	13·3	04·7	55·8	47·0	38·6	30·8	23·8	17·9
7	40·6	35·0	28·3	20·7	12·5	03·8	54·9	46·2	37·8	30·0	23·2	17·4
8	40·1	34·4	27·6	19·9	11·6	02·9	54·0	45·3	37·0	29·3	22·5	16·9
9	39·6	33·8	26·9	19·1	10·8	02·0	53·2	44·5	36·2	28·6	21·9	16·4
10	1 39·1	1 33·1	1 26·1	1 18·3	1 09·9	1 01·1	0 52·3	0 43·6	0 35·4	0 27·9	0 21·3	0 15·9
Lat.	a_1	a_1	a_1	a_1	a_1	a_1	a_1	a_1	a_1	a_1	a_1	a_1
°	′	′	′	′	′	′	′	′	′	′	′	′
0	0·5	0·4	0·3	0·2	0·2	0·2	0·2	0·2	0·2	0·3	0·4	0·4
10	·5	·4	·4	·3	·3	·2	·2	·2	·3	·3	·4	·5
20	·5	·5	·4	·4	·3	·3	·3	·3	·3	·4	·4	·5
30	·5	·5	·5	·4	·4	·4	·4	·4	·4	·4	·5	·5
40	0·6	0·5	0·5	0·5	0·5	0·5	0·5	0·5	0·5	0·5	0·5	0·6
45	·6	·6	·6	·5	·5	·5	·5	·5	·5	·5	·6	·6
50	·6	·6	·6	·6	·6	·6	·6	·6	·6	·6	·6	·6
55	·6	·6	·7	·7	·7	·7	·7	·7	·7	·7	·6	·6
60	·7	·7	·7	·8	·8	·8	·8	·8	·8	·7	·7	·7
62	0·7	0·7	0·8	0·8	0·8	0·9	0·9	0·8	0·8	0·8	0·7	0·7
64	·7	·7	·8	·9	0·9	0·9	0·9	0·9	·9	·8	·8	·7
66	·7	·8	·9	0·9	1·0	1·0	1·0	1·0	0·9	·9	·8	·7
68	0·7	0·8	0·9	1·0	1·0	1·1	1·1	1·1	1·0	0·9	0·9	0·8
Month	a_2	a_2	a_2	a_2	a_2	a_2	a_2	a_2	a_2	a_2	a_2	a_2
	′	′	′	′	′	′	′	′	′	′	′	′
Jan.	0·4	0·4	0·4	0·5	0·5	0·5	0·6	0·6	0·6	0·7	0·7	0·7
Feb.	·3	·3	·3	·3	·3	·4	·4	·4	·5	·5	·6	·6
Mar.	·3	·3	·3	·3	·3	·3	·3	·3	·3	·4	·4	·5
Apr.	0·4	0·4	0·3	0·3	0·3	0·2	0·2	0·2	0·2	0·3	0·3	0·3
May	·6	·5	·5	·4	·3	·3	·3	·2	·2	·2	·2	·2
June	·7	·7	·6	·5	·5	·4	·4	·3	·3	·3	·2	·2
July	0·8	0·8	0·8	0·7	0·6	0·6	0·5	0·5	0·4	0·4	0·3	0·3
Aug.	·9	·9	·9	·8	·8	·8	·7	·7	·6	·5	·5	·4
Sept.	·9	·9	·9	·9	·9	·9	·9	·8	·8	·7	·7	·6
Oct.	0·8	0·8	0·9	0·9	0·9	0·9	0·9	0·9	0·9	0·9	0·9	0·8
Nov.	·6	·7	·8	·8	·9	·9	1·0	1·0	1·0	1·0	1·0	1·0
Dec.	0·5	0·5	0·6	0·7	0·8	0·9	0·9	1·0	1·0	1·0	1·1	1·1
Lat.	AZIMUTH											
°	°	°	°	°	°	°	°	°	°	°	°	°
0	0·5	0·6	0·7	0·8	0·8	0·8	0·8	0·8	0·8	0·7	0·6	0·5
20	0·5	0·6	0·7	0·8	0·9	0·9	0·9	0·9	0·8	0·8	0·7	0·5
40	0·6	0·7	0·9	1·0	1·1	1·1	1·1	1·1	1·0	0·9	0·8	0·7
50	0·7	0·9	1·0	1·2	1·3	1·3	1·3	1·3	1·2	1·1	1·0	0·8
55	0·8	1·0	1·2	1·3	1·4	1·5	1·5	1·4	1·4	1·3	1·1	0·9
60	0·9	1·1	1·3	1·5	1·6	1·7	1·7	1·7	1·6	1·4	1·3	1·0
65	1·1	1·3	1·6	1·8	1·9	2·0	2·0	2·0	1·9	1·7	1·5	1·2

Latitude = Apparent altitude (corrected for refraction) $- 1° + a_0 + a_1 + a_2$

The table is entered with L.H.A. Aries to determine the column to be used; each column refers to a range of 10°. a_0 is taken, with mental interpolation, from the upper table with the units of L.H.A. Aries in degrees as argument; a_1, a_2 are taken, without interpolation, from the second and third tables with arguments latitude and month respectively. a_0, a_1, a_2 are always positive. The final table gives the azimuth of *Polaris*.

CONVERSION OF ARC TO TIME

0°–59°		60°–119°		120°–179°		180°–239°		240°–299°		300°–359°			0'·00	0'·25	0'·50	0'·75
°	h m	°	h m	°	h m	°	h m	°	h m	°	h m	'	m s	m s	m s	m s
0	0 00	60	4 00	120	8 00	180	12 00	240	16 00	300	20 00	0	0 00	0 01	0 02	0 03
1	0 04	61	4 04	121	8 04	181	12 04	241	16 04	301	20 04	1	0 04	0 05	0 06	0 07
2	0 08	62	4 08	122	8 08	182	12 08	242	16 08	302	20 08	2	0 08	0 09	0 10	0 11
3	0 12	63	4 12	123	8 12	183	12 12	243	16 12	303	20 12	3	0 12	0 13	0 14	0 15
4	0 16	64	4 16	124	8 16	184	12 16	244	16 16	304	20 16	4	0 16	0 17	0 18	0 19
5	0 20	65	4 20	125	8 20	185	12 20	245	16 20	305	20 20	5	0 20	0 21	0 22	0 23
6	0 24	66	4 24	126	8 24	186	12 24	246	16 24	306	20 24	6	0 24	0 25	0 26	0 27
7	0 28	67	4 28	127	8 28	187	12 28	247	16 28	307	20 28	7	0 28	0 29	0 30	0 31
8	0 32	68	4 32	128	8 32	188	12 32	248	16 32	308	20 32	8	0 32	0 33	0 34	0 35
9	0 36	69	4 36	129	8 36	189	12 36	249	16 36	309	20 36	9	0 36	0 37	0 38	0 39
10	0 40	70	4 40	130	8 40	190	12 40	250	16 40	310	20 40	10	0 40	0 41	0 42	0 43
11	0 44	71	4 44	131	8 44	191	12 44	251	16 44	311	20 44	11	0 44	0 45	0 46	0 47
12	0 48	72	4 48	132	8 48	192	12 48	252	16 48	312	20 48	12	0 48	0 49	0 50	0 51
13	0 52	73	4 52	133	8 52	193	12 52	253	16 52	313	20 52	13	0 52	0 53	0 54	0 55
14	0 56	74	4 56	134	8 56	194	12 56	254	16 56	314	20 56	14	0 56	0 57	0 58	0 59
15	1 00	75	5 00	135	9 00	195	13 00	255	17 00	315	21 00	15	1 00	1 01	1 02	1 03
16	1 04	76	5 04	136	9 04	196	13 04	256	17 04	316	21 04	16	1 04	1 05	1 06	1 07
17	1 08	77	5 08	137	9 08	197	13 08	257	17 08	317	21 08	17	1 08	1 09	1 10	1 11
18	1 12	78	5 12	138	9 12	198	13 12	258	17 12	318	21 12	18	1 12	1 13	1 14	1 15
19	1 16	79	5 16	139	9 16	199	13 16	259	17 16	319	21 16	19	1 16	1 17	1 18	1 19
20	1 20	80	5 20	140	9 20	200	13 20	260	17 20	320	21 20	20	1 20	1 21	1 22	1 23
21	1 24	81	5 24	141	9 24	201	13 24	261	17 24	321	21 24	21	1 24	1 25	1 26	1 27
22	1 28	82	5 28	142	9 28	202	13 28	262	17 28	322	21 28	22	1 28	1 29	1 30	1 31
23	1 32	83	5 32	143	9 32	203	13 32	263	17 32	323	21 32	23	1 32	1 33	1 34	1 35
24	1 36	84	5 36	144	9 36	204	13 36	264	17 36	324	21 36	24	1 36	1 37	1 38	1 39
25	1 40	85	5 40	145	9 40	205	13 40	265	17 40	325	21 40	25	1 40	1 41	1 42	1 43
26	1 44	86	5 44	146	9 44	206	13 44	266	17 44	326	21 44	26	1 44	1 45	1 46	1 47
27	1 48	87	5 48	147	9 48	207	13 48	267	17 48	327	21 48	27	1 48	1 49	1 50	1 51
28	1 52	88	5 52	148	9 52	208	13 52	268	17 52	328	21 52	28	1 52	1 53	1 54	1 55
29	1 56	89	5 56	149	9 56	209	13 56	269	17 56	329	21 56	29	1 56	1 57	1 58	1 59
30	2 00	90	6 00	150	10 00	210	14 00	270	18 00	330	22 00	30	2 00	2 01	2 02	2 03
31	2 04	91	6 04	151	10 04	211	14 04	271	18 04	331	22 04	31	2 04	2 05	2 06	2 07
32	2 08	92	6 08	152	10 08	212	14 08	272	18 08	332	22 08	32	2 08	2 09	2 10	2 11
33	2 12	93	6 12	153	10 12	213	14 12	273	18 12	333	22 12	33	2 12	2 13	2 14	2 15
34	2 16	94	6 16	154	10 16	214	14 16	274	18 16	334	22 16	34	2 16	2 17	2 18	2 19
35	2 20	95	6 20	155	10 20	215	14 20	275	18 20	335	22 20	35	2 20	2 21	2 22	2 23
36	2 24	96	6 24	156	10 24	216	14 24	276	18 24	336	22 24	36	2 24	2 25	2 26	2 27
37	2 28	97	6 28	157	10 28	217	14 28	277	18 28	337	22 28	37	2 28	2 29	2 30	2 31
38	2 32	98	6 32	158	10 32	218	14 32	278	18 32	338	22 32	38	2 32	2 33	2 34	2 35
39	2 36	99	6 36	159	10 36	219	14 36	279	18 36	339	22 36	39	2 36	2 37	2 38	2 39
40	2 40	100	6 40	160	10 40	220	14 40	280	18 40	340	22 40	40	2 40	2 41	2 42	2 43
41	2 44	101	6 44	161	10 44	221	14 44	281	18 44	341	22 44	41	2 44	2 45	2 46	2 47
42	2 48	102	6 48	162	10 48	222	14 48	282	18 48	342	22 48	42	2 48	2 49	2 50	2 51
43	2 52	103	6 52	163	10 52	223	14 52	283	18 52	343	22 52	43	2 52	2 53	2 54	2 55
44	2 56	104	6 56	164	10 56	224	14 56	284	18 56	344	22 56	44	2 56	2 57	2 58	2 59
45	3 00	105	7 00	165	11 00	225	15 00	285	19 00	345	23 00	45	3 00	3 01	3 02	3 03
46	3 04	106	7 04	166	11 04	226	15 04	286	19 04	346	23 04	46	3 04	3 05	3 06	3 07
47	3 08	107	7 08	167	11 08	227	15 08	287	19 08	347	23 08	47	3 08	3 09	3 10	3 11
48	3 12	108	7 12	168	11 12	228	15 12	288	19 12	348	23 12	48	3 12	3 13	3 14	3 15
49	3 16	109	7 16	169	11 16	229	15 16	289	19 16	349	23 16	49	3 16	3 17	3 18	3 19
50	3 20	110	7 20	170	11 20	230	15 20	290	19 20	350	23 20	50	3 20	3 21	3 22	3 23
51	3 24	111	7 24	171	11 24	231	15 24	291	19 24	351	23 24	51	3 24	3 25	3 26	3 27
52	3 28	112	7 28	172	11 28	232	15 28	292	19 28	352	23 28	52	3 28	3 29	3 30	3 31
53	3 32	113	7 32	173	11 32	233	15 32	293	19 32	353	23 32	53	3 32	3 33	3 34	3 35
54	3 36	114	7 36	174	11 36	234	15 36	294	19 36	354	23 36	54	3 36	3 37	3 38	3 39
55	3 40	115	7 40	175	11 40	235	15 40	295	19 40	355	23 40	55	3 40	3 41	3 42	3 43
56	3 44	116	7 44	176	11 44	236	15 44	296	19 44	356	23 44	56	3 44	3 45	3 46	3 47
57	3 48	117	7 48	177	11 48	237	15 48	297	19 48	357	23 48	57	3 48	3 49	3 50	3 51
58	3 52	118	7 52	178	11 52	238	15 52	298	19 52	358	23 52	58	3 52	3 53	3 54	3 55
59	3 56	119	7 56	179	11 56	239	15 56	299	19 56	359	23 56	59	3 56	3 57	3 58	3 59

The above table is for converting expressions in arc to their equivalent in time; its main use in this Almanac is for the conversion of longitude for application to L.M.T. (*added* if *west*, *subtracted* if *east*) to give G.M.T. or vice versa, particularly in the case of sunrise, sunset, etc.

24ᵐ INCREMENTS AND CORRECTIONS 25ᵐ

24ᵐ	SUN PLANETS	ARIES	MOON	v or d	Corrⁿ	v or d	Corrⁿ	v or d	Corrⁿ
s	° ′	° ′	° ′	′	′	′	′	′	′
00	6 00·0	6 01·0	5 43·6	0·0	0·0	6·0	2·5	12·0	4·9
01	6 00·3	6 01·2	5 43·8	0·1	0·0	6·1	2·5	12·1	4·9
02	6 00·5	6 01·5	5 44·1	0·2	0·1	6·2	2·5	12·2	5·0
03	6 00·8	6 01·7	5 44·3	0·3	0·1	6·3	2·6	12·3	5·0
04	6 01·0	6 02·0	5 44·6	0·4	0·2	6·4	2·6	12·4	5·1
05	6 01·3	6 02·2	5 44·8	0·5	0·2	6·5	2·7	12·5	5·1
06	6 01·5	6 02·5	5 45·0	0·6	0·2	6·6	2·7	12·6	5·1
07	6 01·8	6 02·7	5 45·3	0·7	0·3	6·7	2·7	12·7	5·2
08	6 02·0	6 03·0	5 45·5	0·8	0·3	6·8	2·8	12·8	5·2
09	6 02·3	6 03·2	5 45·7	0·9	0·4	6·9	2·8	12·9	5·3
10	6 02·5	6 03·5	5 46·0	1·0	0·4	7·0	2·9	13·0	5·3
11	6 02·8	6 03·7	5 46·2	1·1	0·4	7·1	2·9	13·1	5·3
12	6 03·0	6 04·0	5 46·5	1·2	0·5	7·2	2·9	13·2	5·4
13	6 03·3	6 04·2	5 46·7	1·3	0·5	7·3	3·0	13·3	5·4
14	6 03·5	6 04·5	5 46·9	1·4	0·6	7·4	3·0	13·4	5·5
15	6 03·8	6 04·7	5 47·2	1·5	0·6	7·5	3·1	13·5	5·5
16	6 04·0	6 05·0	5 47·4	1·6	0·7	7·6	3·1	13·6	5·6
17	6 04·3	6 05·2	5 47·7	1·7	0·7	7·7	3·1	13·7	5·6
18	6 04·5	6 05·5	5 47·9	1·8	0·7	7·8	3·2	13·8	5·6
19	6 04·8	6 05·7	5 48·1	1·9	0·8	7·9	3·2	13·9	5·7
20	6 05·0	6 06·0	5 48·4	2·0	0·8	8·0	3·3	14·0	5·7
21	6 05·3	6 06·3	5 48·6	2·1	0·9	8·1	3·3	14·1	5·8
22	6 05·5	6 06·5	5 48·8	2·2	0·9	8·2	3·3	14·2	5·8
23	6 05·8	6 06·8	5 49·1	2·3	0·9	8·3	3·4	14·3	5·8
24	6 06·0	6 07·0	5 49·3	2·4	1·0	8·4	3·4	14·4	5·9
25	6 06·3	6 07·3	5 49·6	2·5	1·0	8·5	3·5	14·5	5·9
26	6 06·5	6 07·5	5 49·8	2·6	1·1	8·6	3·5	14·6	6·0
27	6 06·8	6 07·8	5 50·0	2·7	1·1	8·7	3·6	14·7	6·0
28	6 07·0	6 08·0	5 50·3	2·8	1·1	8·8	3·6	14·8	6·0
29	6 07·3	6 08·3	5 50·5	2·9	1·2	8·9	3·6	14·9	6·1
30	6 07·5	6 08·5	5 50·8	3·0	1·2	9·0	3·7	15·0	6·1
31	6 07·8	6 08·8	5 51·0	3·1	1·3	9·1	3·7	15·1	6·2
32	6 08·0	6 09·0	5 51·2	3·2	1·3	9·2	3·8	15·2	6·2
33	6 08·3	6 09·3	5 51·5	3·3	1·3	9·3	3·8	15·3	6·2
34	6 08·5	6 09·5	5 51·7	3·4	1·4	9·4	3·8	15·4	6·3
35	6 08·8	6 09·8	5 52·0	3·5	1·4	9·5	3·9	15·5	6·3
36	6 09·0	6 10·0	5 52·2	3·6	1·5	9·6	3·9	15·6	6·4
37	6 09·3	6 10·3	5 52·4	3·7	1·5	9·7	4·0	15·7	6·4
38	6 09·5	6 10·5	5 52·7	3·8	1·6	9·8	4·0	15·8	6·5
39	6 09·8	6 10·8	5 52·9	3·9	1·6	9·9	4·0	15·9	6·5
40	6 10·0	6 11·0	5 53·1	4·0	1·6	10·0	4·1	16·0	6·5
41	6 10·3	6 11·3	5 53·4	4·1	1·7	10·1	4·1	16·1	6·6
42	6 10·5	6 11·5	5 53·6	4·2	1·7	10·2	4·2	16·2	6·6
43	6 10·8	6 11·8	5 53·9	4·3	1·8	10·3	4·2	16·3	6·7
44	6 11·0	6 12·0	5 54·1	4·4	1·8	10·4	4·2	16·4	6·7
45	6 11·3	6 12·3	5 54·3	4·5	1·8	10·5	4·3	16·5	6·7
46	6 11·5	6 12·5	5 54·6	4·6	1·9	10·6	4·3	16·6	6·8
47	6 11·8	6 12·8	5 54·8	4·7	1·9	10·7	4·4	16·7	6·8
48	6 12·0	6 13·0	5 55·1	4·8	2·0	10·8	4·4	16·8	6·9
49	6 12·3	6 13·3	5 55·3	4·9	2·0	10·9	4·5	16·9	6·9
50	6 12·5	6 13·5	5 55·5	5·0	2·0	11·0	4·5	17·0	6·9
51	6 12·8	6 13·8	5 55·8	5·1	2·1	11·1	4·5	17·1	7·0
52	6 13·0	6 14·0	5 56·0	5·2	2·1	11·2	4·6	17·2	7·0
53	6 13·3	6 14·3	5 56·2	5·3	2·2	11·3	4·6	17·3	7·1
54	6 13·5	6 14·5	5 56·5	5·4	2·2	11·4	4·7	17·4	7·1
55	6 13·8	6 14·8	5 56·7	5·5	2·2	11·5	4·7	17·5	7·1
56	6 14·0	6 15·0	5 57·0	5·6	2·3	11·6	4·7	17·6	7·2
57	6 14·3	6 15·3	5 57·2	5·7	2·3	11·7	4·8	17·7	7·2
58	6 14·5	6 15·5	5 57·4	5·8	2·4	11·8	4·8	17·8	7·3
59	6 14·8	6 15·8	5 57·7	5·9	2·4	11·9	4·9	17·9	7·3
60	6 15·0	6 16·0	5 57·9	6·0	2·5	12·0	4·9	18·0	7·4

25ᵐ	SUN PLANETS	ARIES	MOON	v or d	Corrⁿ	v or d	Corrⁿ	v or d	Corrⁿ
s	° ′	° ′	° ′	′	′	′	′	′	′
00	6 15·0	6 16·0	5 57·9	0·0	0·0	6·0	2·6	12·0	5·1
01	6 15·3	6 16·3	5 58·2	0·1	0·0	6·1	2·6	12·1	5·1
02	6 15·5	6 16·5	5 58·4	0·2	0·1	6·2	2·6	12·2	5·2
03	6 15·8	6 16·8	5 58·6	0·3	0·1	6·3	2·7	12·3	5·2
04	6 16·0	6 17·0	5 58·9	0·4	0·2	6·4	2·7	12·4	5·3
05	6 16·3	6 17·3	5 59·1	0·5	0·2	6·5	2·8	12·5	5·3
06	6 16·5	6 17·5	5 59·3	0·6	0·3	6·6	2·8	12·6	5·4
07	6 16·8	6 17·8	5 59·6	0·7	0·3	6·7	2·8	12·7	5·4
08	6 17·0	6 18·0	5 59·8	0·8	0·3	6·8	2·9	12·8	5·4
09	6 17·3	6 18·3	6 00·1	0·9	0·4	6·9	2·9	12·9	5·5
10	6 17·5	6 18·5	6 00·3	1·0	0·4	7·0	3·0	13·0	5·5
11	6 17·8	6 18·8	6 00·5	1·1	0·5	7·1	3·0	13·1	5·6
12	6 18·0	6 19·0	6 00·8	1·2	0·5	7·2	3·1	13·2	5·6
13	6 18·3	6 19·3	6 01·0	1·3	0·6	7·3	3·1	13·3	5·7
14	6 18·5	6 19·5	6 01·3	1·4	0·6	7·4	3·1	13·4	5·7
15	6 18·8	6 19·8	6 01·5	1·5	0·6	7·5	3·2	13·5	5·7
16	6 19·0	6 20·0	6 01·7	1·6	0·7	7·6	3·2	13·6	5·8
17	6 19·3	6 20·3	6 02·0	1·7	0·7	7·7	3·3	13·7	5·8
18	6 19·5	6 20·5	6 02·2	1·8	0·8	7·8	3·3	13·8	5·9
19	6 19·8	6 20·8	6 02·5	1·9	0·8	7·9	3·4	13·9	5·9
20	6 20·0	6 21·0	6 02·7	2·0	0·9	8·0	3·4	14·0	6·0
21	6 20·3	6 21·3	6 02·9	2·1	0·9	8·1	3·4	14·1	6·0
22	6 20·5	6 21·5	6 03·2	2·2	0·9	8·2	3·5	14·2	6·0
23	6 20·8	6 21·8	6 03·4	2·3	1·0	8·3	3·5	14·3	6·1
24	6 21·0	6 22·0	6 03·6	2·4	1·0	8·4	3·6	14·4	6·1
25	6 21·3	6 22·3	6 03·9	2·5	1·1	8·5	3·6	14·5	6·2
26	6 21·5	6 22·5	6 04·1	2·6	1·1	8·6	3·7	14·6	6·2
27	6 21·8	6 22·8	6 04·4	2·7	1·1	8·7	3·7	14·7	6·2
28	6 22·0	6 23·0	6 04·6	2·8	1·2	8·8	3·7	14·8	6·3
29	6 22·3	6 23·3	6 04·8	2·9	1·2	8·9	3·8	14·9	6·3
30	6 22·5	6 23·5	6 05·1	3·0	1·3	9·0	3·8	15·0	6·4
31	6 22·8	6 23·8	6 05·3	3·1	1·3	9·1	3·9	15·1	6·4
32	6 23·0	6 24·0	6 05·6	3·2	1·4	9·2	3·9	15·2	6·5
33	6 23·3	6 24·3	6 05·8	3·3	1·4	9·3	4·0	15·3	6·5
34	6 23·5	6 24·5	6 06·0	3·4	1·4	9·4	4·0	15·4	6·5
35	6 23·8	6 24·8	6 06·3	3·5	1·5	9·5	4·0	15·5	6·6
36	6 24·0	6 25·1	6 06·5	3·6	1·5	9·6	4·1	15·6	6·6
37	6 24·3	6 25·3	6 06·7	3·7	1·6	9·7	4·1	15·7	6·7
38	6 24·5	6 25·6	6 07·0	3·8	1·6	9·8	4·2	15·8	6·7
39	6 24·8	6 25·8	6 07·2	3·9	1·7	9·9	4·2	15·9	6·8
40	6 25·0	6 26·1	6 07·5	4·0	1·7	10·0	4·3	16·0	6·8
41	6 25·3	6 26·3	6 07·7	4·1	1·7	10·1	4·3	16·1	6·8
42	6 25·5	6 26·6	6 07·9	4·2	1·8	10·2	4·3	16·2	6·9
43	6 25·8	6 26·8	6 08·2	4·3	1·8	10·3	4·4	16·3	6·9
44	6 26·0	6 27·1	6 08·4	4·4	1·9	10·4	4·4	16·4	7·0
45	6 26·3	6 27·3	6 08·7	4·5	1·9	10·5	4·5	16·5	7·0
46	6 26·5	6 27·6	6 08·9	4·6	2·0	10·6	4·5	16·6	7·1
47	6 26·8	6 27·8	6 09·1	4·7	2·0	10·7	4·5	16·7	7·1
48	6 27·0	6 28·1	6 09·4	4·8	2·0	10·8	4·6	16·8	7·1
49	6 27·3	6 28·3	6 09·6	4·9	2·1	10·9	4·6	16·9	7·2
50	6 27·5	6 28·6	6 09·8	5·0	2·1	11·0	4·7	17·0	7·2
51	6 27·8	6 28·8	6 10·1	5·1	2·2	11·1	4·7	17·1	7·3
52	6 28·0	6 29·1	6 10·3	5·2	2·2	11·2	4·8	17·2	7·3
53	6 28·3	6 29·3	6 10·6	5·3	2·3	11·3	4·8	17·3	7·4
54	6 28·5	6 29·6	6 10·8	5·4	2·3	11·4	4·8	17·4	7·4
55	6 28·8	6 29·8	6 11·0	5·5	2·3	11·5	4·9	17·5	7·4
56	6 29·0	6 30·1	6 11·3	5·6	2·4	11·6	4·9	17·6	7·5
57	6 29·3	6 30·3	6 11·5	5·7	2·4	11·7	5·0	17·7	7·5
58	6 29·5	6 30·6	6 11·8	5·8	2·5	11·8	5·0	17·8	7·6
59	6 29·8	6 30·8	6 12·0	5·9	2·5	11·9	5·1	17·9	7·6
60	6 30·0	6 31·1	6 12·2	6·0	2·6	12·0	5·1	18·0	7·7

TABLES FOR INTERPOLATING SUNRISE, MOONRISE, ETC.

TABLE I—FOR LATITUDE

Tabular Interval 10°	5°	2°	Difference between the times for consecutive latitudes 5m	10m	15m	20m	25m	30m	35m	40m	45m	50m	55m	60m	1h 05m	1h 10m	1h 15m	1h 20m
° ′	° ′	° ′	m	m	m	m	m	m	m	m	m	m	m	m	h m	h m	h m	h m
0 30	0 15	0 06	0	0	1	1	1	1	1	2	2	2	2	2	0 02	0 02	0 02	0 02
1 00	0 30	0 12	0	1	1	2	2	3	3	3	4	4	4	5	05	05	05	05
1 30	0 45	0 18	1	1	2	3	3	4	4	5	5	6	7	7	07	07	07	07
2 00	1 00	0 24	1	2	3	4	5	5	6	7	7	8	9	10	10	10	10	10
2 30	1 15	0 30	1	2	4	5	6	7	8	9	9	10	11	12	12	13	13	13
3 00	1 30	0 36	1	3	4	6	7	8	9	10	11	12	13	14	0 15	0 15	0 16	0 16
3 30	1 45	0 42	2	3	5	7	8	10	11	12	13	14	16	17	18	18	19	19
4 00	2 00	0 48	2	4	6	8	9	11	13	14	15	16	18	19	20	21	22	22
4 30	2 15	0 54	2	4	7	9	11	13	15	16	18	19	21	22	23	24	25	26
5 00	2 30	1 00	2	5	7	10	12	14	16	18	20	22	23	25	26	27	28	29
5 30	2 45	1 06	3	5	8	11	13	16	18	20	22	24	26	28	0 29	0 30	0 31	0 32
6 00	3 00	1 12	3	6	9	12	14	17	20	22	24	26	29	31	32	33	34	36
6 30	3 15	1 18	3	6	10	13	16	19	22	24	26	29	31	34	36	37	38	40
7 00	3 30	1 24	3	7	10	14	17	20	23	26	29	31	34	37	39	41	42	44
7 30	3 45	1 30	4	7	11	15	18	22	25	28	31	34	37	40	43	44	46	48
8 00	4 00	1 36	4	8	12	16	20	23	27	30	34	37	41	44	0 47	0 48	0 51	0 53
8 30	4 15	1 42	4	8	13	17	21	25	29	33	36	40	44	48	0 51	0 53	0 56	0 58
9 00	4 30	1 48	4	9	13	18	22	27	31	35	39	43	47	52	0 55	0 58	1 01	1 04
9 30	4 45	1 54	5	9	14	19	24	28	33	38	42	47	51	56	1 00	1 04	1 08	1 12
10 00	5 00	2 00	5	10	15	20	25	30	35	40	45	50	55	60	1 05	1 10	1 15	1 20

Table I is for interpolating the L.M.T. of sunrise, twilight, moonrise, etc. for latitude. It is to be noted that the interpolation is not linear, so that when using this table it is essential to take out the required phenomenon for the latitude *less* than the true latitude. The table is entered with the nearest value of the difference between the times for the tabular latitude and the next higher one, and, in the appropriate column, with the difference between true latitude and tabular latitude; the correction so obtained is applied to the time for the tabular latitude; the sign of the correction can be seen by inspection.

TABLE II—FOR LONGITUDE

Long. East or West	Difference between the times for given date and preceding date (for east longitude) or for given date and following date (for west longitude) 10m	20m	30m	40m	50m	60m	1h + 10m	1h + 20m	1h + 30m	1h + 40m	1h + 50m	1h + 60m	2h 10m	2h 20m	2h 30m	2h 40m	2h 50m	3h 00m
°	m	m	m	m	m	m	m	m	m	m	m	m	h m	h m	h m	h m	h m	h m
0	0	0	0	0	0	0	0	0	0	0	0	0	0 00	0 00	0 00	0 00	0 00	0 00
10	0	1	1	1	1	2	2	2	2	3	3	3	04	04	04	04	05	05
20	1	1	2	2	3	3	4	4	5	6	6	7	07	08	08	09	09	10
30	1	2	2	3	4	5	6	7	7	8	9	10	11	12	12	13	14	15
40	1	2	3	4	6	7	8	9	10	11	12	13	14	16	17	18	19	20
50	1	3	4	6	7	8	10	11	12	14	15	17	0 18	0 19	0 21	0 22	0 24	0 25
60	2	3	5	7	8	10	12	13	15	17	18	20	22	23	25	27	28	30
70	2	4	6	8	10	12	14	16	17	19	21	23	25	27	29	31	33	35
80	2	4	7	9	11	13	16	18	20	22	24	27	29	31	33	36	38	40
90	2	5	7	10	12	15	17	20	22	25	27	30	32	35	37	40	42	45
100	3	6	8	11	14	17	19	22	25	28	31	33	0 36	0 39	0 42	0 44	0 47	0 50
110	3	6	9	12	15	18	21	24	27	31	34	37	40	43	46	49	0 52	0 55
120	3	7	10	13	17	20	23	27	30	33	37	40	43	47	50	53	0 57	1 00
130	4	7	11	14	18	22	25	29	32	36	40	43	47	51	54	0 58	1 01	1 05
140	4	8	12	16	19	23	27	31	35	39	43	47	51	54	0 58	1 02	1 06	1 10
150	4	8	13	17	21	25	29	33	38	42	46	50	0 54	0 58	1 03	1 07	1 11	1 15
160	4	9	13	18	22	27	31	36	40	44	49	53	0 58	1 02	1 07	1 11	1 16	1 20
170	5	9	14	19	24	28	33	38	42	47	52	57	1 01	1 06	1 11	1 16	1 20	1 25
180	5	10	15	20	25	30	35	40	45	50	55	60	1 05	1 10	1 15	1 20	1 25	1 30

Table II is for interpolating the L.M.T. of moonrise, moonset and the Moon's meridian passage for longitude. It is entered with longitude and with the difference between the times for the given date and for the preceding date (in east longitudes) or following date (in west longitudes). The correction is normally *added* for west longitudes and *subtracted* for east longitudes, but if, as occasionally happens, the times become earlier each day instead of later, the signs of the corrections must be reversed.

INDEX TO SELECTED STARS, 1975

Name	No.	Mag.	S.H.A.	Dec.
			°	°
Acamar	7	3·1	316	S. 40
Achernar	5	0·6	336	S. 57
Acrux	30	1·1	174	S. 63
Adhara	19	1·6	256	S. 29
Aldebaran	10	1·1	291	N. 16
Alioth	32	1·7	167	N. 56
Alkaid	34	1·9	153	N. 49
Al Na'ir	55	2·2	28	S. 47
Alnilam	15	1·8	276	S. 1
Alphard	25	2·2	218	S. 9
Alphecca	41	2·3	127	N. 27
Alpheratz	1	2·2	358	N. 29
Altair	51	0·9	63	N. 9
Ankaa	2	2·4	354	S. 42
Antares	42	1·2	113	S. 26
Arcturus	37	0·2	146	N. 19
Atria	43	1·9	108	S. 69
Avior	22	1·7	234	S. 59
Bellatrix	13	1·7	279	N. 6
Betelgeuse	16	Var.*	272	N. 7
Canopus	17	−0·9	264	S. 53
Capella	12	0·2	281	N. 46
Deneb	53	1·3	50	N. 45
Denebola	28	2·2	183	N. 15
Diphda	4	2·2	349	S. 18
Dubhe	27	2·0	194	N. 62
Elnath	14	1·8	279	N. 29
Eltanin	47	2·4	91	N. 51
Enif	54	2·5	34	N. 10
Fomalhaut	56	1·3	16	S. 30
Gacrux	31	1·6	173	S. 57
Gienah	29	2·8	176	S. 17
Hadar	35	0·9	149	S. 60
Hamal	6	2·2	329	N. 23
Kaus Australis	48	2·0	84	S. 34
Kochab	40	2·2	137	N. 74
Markab	57	2·6	14	N. 15
Menkar	8	2·8	315	N. 4
Menkent	36	2·3	149	S. 36
Miaplacidus	24	1·8	222	S. 70
Mirfak	9	1·9	309	N. 50
Nunki	50	2·1	77	S. 26
Peacock	52	2·1	54	S. 57
Pollux	21	1·2	244	N. 28
Procyon	20	0·5	245	N. 5
Rasalhague	46	2·1	97	N. 13
Regulus	26	1·3	208	N. 12
Rigel	11	0·3	282	S. 8
Rigil Kentaurus	38	0·1	141	S. 61
Sabik	44	2·6	103	S. 16
Schedar	3	2·5	350	N. 56
Shaula	45	1·7	97	S. 37
Sirius	18	−1·6	259	S. 17
Spica	33	1·2	159	S. 11
Suhail	23	2·2	223	S. 43
Vega	49	0·1	81	N. 39
Zubenelgenubi	39	2·9	138	S. 16

No.	Name	Mag.	S.H.A.	Dec.
			°	°
1	*Alpheratz*	2·2	358	N. 29
2	*Ankaa*	2·4	354	S. 42
3	*Schedar*	2·5	350	N. 56
4	*Diphda*	2·2	349	S. 18
5	*Achernar*	0·6	336	S. 57
6	*Hamal*	2·2	329	N. 23
7	*Acamar*	3·1	316	S. 40
8	*Menkar*	2·8	315	N. 4
9	*Mirfak*	1·9	309	N. 50
10	*Aldebaran*	1·1	291	N. 16
11	*Rigel*	0·3	282	S. 8
12	*Capella*	0·2	281	N. 46
13	*Bellatrix*	1·7	279	N. 6
14	*Elnath*	1·8	279	N. 29
15	*Alnilam*	1·8	276	S. 1
16	*Betelgeuse*	Var.*	272	N. 7
17	*Canopus*	−0·9	264	S. 53
18	*Sirius*	−1·6	259	S. 17
19	*Adhara*	1·6	256	S. 29
20	*Procyon*	0·5	245	N. 5
21	*Pollux*	1·2	244	N. 28
22	*Avior*	1·7	234	S. 59
23	*Suhail*	2·2	223	S. 43
24	*Miaplacidus*	1·8	222	S. 70
25	*Alphard*	2·2	218	S. 9
26	*Regulus*	1·3	208	N. 12
27	*Dubhe*	2·0	194	N. 62
28	*Denebola*	2·2	183	N. 15
29	*Gienah*	2·8	176	S. 17
30	*Acrux*	1·1	174	S. 63
31	*Gacrux*	1·6	173	S. 57
32	*Alioth*	1·7	167	N. 56
33	*Spica*	1·2	159	S. 11
34	*Alkaid*	1·9	153	N. 49
35	*Hadar*	0·9	149	S. 60
36	*Menkent*	2·3	149	S. 36
37	*Arcturus*	0·2	146	N. 19
38	*Rigil Kentaurus*	0·1	141	S. 61
39	*Zubenelgenubi*	2·9	138	S. 16
40	*Kochab*	2·2	137	N. 74
41	*Alphecca*	2·3	127	N. 27
42	*Antares*	1·2	113	S. 26
43	*Atria*	1·9	108	S. 69
44	*Sabik*	2·6	103	S. 16
45	*Shaula*	1·7	97	S. 37
46	*Rasalhague*	2·1	97	N. 13
47	*Eltanin*	2·4	91	N. 51
48	*Kaus Australis*	2·0	84	S. 34
49	*Vega*	0·1	81	N. 39
50	*Nunki*	2·1	77	S. 26
51	*Altair*	0·9	63	N. 9
52	*Peacock*	2·1	54	S. 57
53	*Deneb*	1·3	50	N. 45
54	*Enif*	2·5	34	N. 10
55	*Al Na'ir*	2·2	28	S. 47
56	*Fomalhaut*	1·3	16	S. 30
57	*Markab*	2·6	14	N. 15

* 0·1—1·2

ALTITUDE CORRECTION TABLES 0°–35°—MOON

App. Alt.	0°–4° Corrⁿ	5°–9° Corrⁿ	10°–14° Corrⁿ	15°–19° Corrⁿ	20°–24° Corrⁿ	25°–29° Corrⁿ	30°–34° Corrⁿ	App. Alt.
′	° ′	° ′	° ′	° ′	° ′	° ′	° ′	′
00	0 33·8	5 58·2	10 62·1	15 62·8	20 62·2	25 60·8	30 58·9	00
10	35·9	58·5	62·2	62·8	62·1	60·8	58·8	10
20	37·8	58·7	62·2	62·8	62·1	60·7	58·8	20
30	39·6	58·9	62·3	62·8	62·1	60·7	58·7	30
40	41·2	59·1	62·3	62·8	62·0	60·6	58·6	40
50	42·6	59·3	62·4	62·7	62·0	60·6	58·5	50
00	1 44·0	6 59·5	11 62·4	16 62·7	21 62·0	26 60·5	31 58·5	00
10	45·2	59·7	62·4	62·7	61·9	60·4	58·4	10
20	46·3	59·9	62·5	62·7	61·9	60·4	58·3	20
30	47·3	60·0	62·5	62·7	61·9	60·3	58·2	30
40	48·3	60·2	62·5	62·7	61·8	60·3	58·2	40
50	49·2	60·3	62·6	62·7	61·8	60·2	58·1	50
00	2 50·0	7 60·5	12 62·6	17 62·7	22 61·7	27 60·1	32 58·0	00
10	50·8	60·6	62·6	62·6	61·7	60·1	57·9	10
20	51·4	60·7	62·6	62·6	61·6	60·0	57·8	20
30	52·1	60·9	62·7	62·6	61·6	59·9	57·8	30
40	52·7	61·0	62·7	62·6	61·5	59·9	57·7	40
50	53·3	61·1	62·7	62·6	61·5	59·8	57·6	50
00	3 53·8	8 61·2	13 62·7	18 62·5	23 61·5	28 59·7	33 57·5	00
10	54·3	61·3	62·7	62·5	61·4	59·7	57·4	10
20	54·8	61·4	62·7	62·5	61·4	59·6	57·4	20
30	55·2	61·5	62·8	62·5	61·3	59·6	57·3	30
40	55·6	61·6	62·8	62·4	61·3	59·5	57·2	40
50	56·0	61·6	62·8	62·4	61·2	59·4	57·1	50
00	4 56·4	9 61·7	14 62·8	19 62·4	24 61·2	29 59·3	34 57·0	00
10	56·7	61·8	62·8	62·3	61·1	59·3	56·9	10
20	57·1	61·9	62·8	62·3	61·1	59·2	56·9	20
30	57·4	61·9	62·8	62·3	61·0	59·1	56·8	30
40	57·7	62·0	62·8	62·2	60·9	59·1	56·7	40
50	57·9	62·1	62·8	62·2	60·9	59·0	56·6	50

H.P.	L	U	L	U	L	U	L	U	L	U	L	U	L	U	H.P.
′	′	′	′	′	′	′	′	′	′	′	′	′	′	′	′
54·0	0·3	0·9	0·3	0·9	0·4	1·0	0·5	1·1	0·6	1·2	0·7	1·3	0·9	1·5	54·0
54·3	0·7	1·1	0·7	1·2	0·7	1·2	0·8	1·3	0·9	1·4	1·1	1·5	1·2	1·7	54·3
54·6	1·1	1·4	1·1	1·4	1·1	1·4	1·2	1·5	1·3	1·6	1·4	1·7	1·5	1·8	54·6
54·9	1·4	1·6	1·5	1·6	1·5	1·6	1·6	1·7	1·6	1·8	1·8	1·9	1·9	2·0	54·9
55·2	1·8	1·8	1·8	1·8	1·9	1·9	1·9	1·9	2·0	2·0	2·1	2·1	2·2	2·2	55·2
55·5	2·2	2·0	2·2	2·0	2·3	2·1	2·3	2·1	2·4	2·2	2·4	2·3	2·5	2·4	55·5
55·8	2·6	2·2	2·6	2·2	2·6	2·3	2·7	2·3	2·7	2·4	2·8	2·4	2·9	2·5	55·8
56·1	3·0	2·4	3·0	2·5	3·0	2·5	3·0	2·5	3·1	2·6	3·1	2·6	3·2	2·7	56·1
56·4	3·4	2·7	3·4	2·7	3·4	2·7	3·4	2·7	3·4	2·8	3·5	2·8	3·5	2·9	56·4
56·7	3·7	2·9	3·7	2·9	3·8	2·9	3·8	2·9	3·8	3·0	3·8	3·0	3·9	3·0	56·7
57·0	4·1	3·1	4·1	3·1	4·1	3·1	4·1	3·1	4·2	3·1	4·2	3·2	4·2	3·2	57·0
57·3	4·5	3·3	4·5	3·3	4·5	3·3	4·5	3·3	4·5	3·3	4·5	3·4	4·6	3·4	57·3
57·6	4·9	3·5	4·9	3·5	4·9	3·5	4·9	3·5	4·9	3·5	4·9	3·5	4·9	3·6	57·6
57·9	5·3	3·8	5·3	3·8	5·2	3·8	5·2	3·7	5·2	3·7	5·2	3·7	5·2	3·7	57·9
58·2	5·6	4·0	5·6	4·0	5·6	4·0	5·6	4·0	5·6	3·9	5·6	3·9	5·6	3·9	58·2
58·5	6·0	4·2	6·0	4·2	6·0	4·2	6·0	4·2	6·0	4·1	5·9	4·1	5·9	4·1	58·5
58·8	6·4	4·4	6·4	4·4	6·4	4·4	6·3	4·4	6·3	4·3	6·3	4·3	6·2	4·2	58·8
59·1	6·8	4·6	6·8	4·6	6·7	4·6	6·7	4·6	6·7	4·5	6·6	4·5	6·6	4·4	59·1
59·4	7·2	4·8	7·1	4·8	7·1	4·8	7·1	4·8	7·0	4·7	7·0	4·7	6·9	4·6	59·4
59·7	7·5	5·1	7·5	5·0	7·5	5·0	7·5	5·0	7·4	4·9	7·3	4·8	7·2	4·7	59·7
60·0	7·9	5·3	7·9	5·3	7·9	5·2	7·8	5·2	7·8	5·1	7·7	5·0	7·6	4·9	60·0
60·3	8·3	5·5	8·3	5·5	8·2	5·4	8·2	5·4	8·1	5·3	8·0	5·2	7·9	5·1	60·3
60·6	8·7	5·7	8·7	5·7	8·6	5·7	8·6	5·6	8·5	5·5	8·4	5·4	8·2	5·3	60·6
60·9	9·1	5·9	9·0	5·9	9·0	5·9	8·9	5·8	8·8	5·7	8·7	5·6	8·6	5·4	60·9
61·2	9·5	6·2	9·4	6·1	9·4	6·1	9·3	6·0	9·2	5·9	9·1	5·8	8·9	5·6	61·2
61·5	9·8	6·4	9·8	6·3	9·7	6·3	9·7	6·2	9·5	6·1	9·4	5·9	9·2	5·8	61·5

DIP

Ht. of Eye	Corrⁿ	Ht. of Eye	Ht. of Eye	Corrⁿ	Ht. of Eye
m	′	ft.	m	′	ft.
2·4		8·0	9·5		31·5
	−2·8			−5·5	
2·6		8·6	9·9		32·7
	−2·9			−5·6	
2·8		9·2	10·3		33·9
	−3·0			−5·7	
3·0		9·8	10·6		35·1
	−3·1			−5·8	
3·2		10·5	11·0		36·3
	−3·2			−5·9	
3·4		11·2	11·4		37·6
	−3·3			−6·0	
3·6		11·9	11·8		38·9
	−3·4			−6·1	
3·8		12·6	12·2		40·1
	−3·5			−6·2	
4·0		13·3	12·6		41·5
	−3·6			−6·3	
4·3		14·1	13·0		42·8
	−3·7			−6·4	
4·5		14·9	13·4		44·2
	−3·8			−6·5	
4·7		15·7	13·8		45·5
	−3·9			−6·6	
5·0		16·5	14·2		46·9
	−4·0			−6·7	
5·2		17·4	14·7		48·4
	−4·1			−6·8	
5·5		18·3	15·1		49·8
	−4·2			−6·9	
5·8		19·1	15·5		51·3
	−4·3			−7·0	
6·1		20·1	16·0		52·8
	−4·4			−7·1	
6·3		21·0	16·5		54·3
	−4·5			−7·2	
6·6		22·0	16·9		55·8
	−4·6			−7·3	
6·9		22·9	17·4		57·4
	−4·7			−7·4	
7·2		23·9	17·9		58·9
	−4·8			−7·5	
7·5		24·9	18·4		60·5
	−4·9			−7·6	
7·9		26·0	18·8		62·1
	−5·0			−7·7	
8·2		27·1	19·3		63·8
	−5·1			−7·8	
8·5		28·1	19·8		65·4
	−5·2			−7·9	
8·8		29·2	20·4		67·1
	−5·3			−8·0	
9·2		30·4	20·9		68·8
	−5·4			−8·1	
9·5		31·5	21·4		70·5

MOON CORRECTION TABLE

The correction is in two parts; the first correction is taken from the upper part of the table with argument apparent altitude, and the second from the lower part, with argument H.P., in the same column as that from which the first correction was taken. Separate corrections are given in the lower part for lower (L) and upper (U) limbs. All corrections are to be **added** to apparent altitude, *but 30′ is to be subtracted from the altitude of the upper limb.*

For corrections for pressure and temperature see page A4.

For bubble sextant observations ignore dip, take the mean of upper and lower limb corrections and subtract 15′ from the altitude.

App. Alt. = Apparent altitude = Sextant altitude corrected for index error and dip.

ALTITUDE CORRECTION TABLES 35°-90°—MOON

App. Alt.	35°–39° Corr.	40°–44° Corr.	45°–49° Corr.	50°–54° Corr.	55°–59° Corr.	60°–64° Corr.	65°–69° Corr.	70°–74° Corr.	75°–79° Corr.	80°–84° Corr.	85°–89° Corr.	App. Alt.
′	° ′	° ′	° ′	° ′	° ′	° ′	° ′	° ′	° ′	° ′	° ′	′
00	35 56·5	40 53·7	45 50·5	50 46·9	55 43·1	60 38·9	65 34·6	70 30·1	75 25·3	80 20·5	85 15·6	00
10	56·4	53·6	50·4	46·8	42·9	38·8	34·4	29·9	25·2	20·4	15·5	10
20	56·3	53·5	50·2	46·7	42·8	38·7	34·3	29·7	25·0	20·2	15·3	20
30	56·2	53·4	50·1	46·5	42·7	38·5	34·1	29·6	24·9	20·0	15·1	30
40	56·2	53·3	50·0	46·4	42·5	38·4	34·0	29·4	24·7	19·9	15·0	40
50	56·1	53·2	49·9	46·3	42·4	38·2	33·8	29·3	24·5	19·7	14·8	50
00	36 56·0	41 53·1	46 49·8	51 46·2	56 42·3	61 38·1	66 33·7	71 29·1	76 24·4	81 19·6	86 14·6	00
10	55·9	53·0	49·7	46·0	42·1	37·9	33·5	29·0	24·2	19·4	14·5	10
20	55·8	52·8	49·5	45·9	42·0	37·8	33·4	28·8	24·1	19·2	14·3	20
30	55·7	52·7	49·4	45·8	41·8	37·7	33·2	28·7	23·9	19·1	14·1	30
40	55·6	52·6	49·3	45·7	41·7	37·5	33·1	28·5	23·8	18·9	14·0	40
50	55·5	52·5	49·2	45·5	41·6	37·4	32·9	28·3	23·6	18·7	13·8	50
00	37 55·4	42 52·4	47 49·1	52 45·4	57 41·4	62 37·2	67 32·8	72 28·2	77 23·4	82 18·6	87 13·7	00
10	55·3	52·3	49·0	45·3	41·3	37·1	32·6	28·0	23·3	18·4	13·5	10
20	55·2	52·2	48·8	45·2	41·2	36·9	32·5	27·9	23·1	18·2	13·3	20
30	55·1	52·1	48·7	45·0	41·0	36·8	32·3	27·7	22·9	18·1	13·2	30
40	55·0	52·0	48·6	44·9	40·9	36·6	32·2	27·6	22·8	17·9	13·0	40
50	55·0	51·9	48·5	44·8	40·8	36·5	32·0	27·4	22·6	17·8	12·8	50
00	38 54·9	43 51·8	48 48·4	53 44·6	58 40·6	63 36·4	68 31·9	73 27·2	78 22·5	83 17·6	88 12·7	00
10	54·8	51·7	48·2	44·5	40·5	36·2	31·7	27·1	22·3	17·4	12·5	10
20	54·7	51·6	48·1	44·4	40·3	36·1	31·6	26·9	22·1	17·3	12·3	20
30	54·6	51·5	48·0	44·2	40·2	35·9	31·4	26·8	22·0	17·1	12·2	30
40	54·5	51·4	47·9	44·1	40·1	35·8	31·3	26·6	21·8	16·9	12·0	40
50	54·4	51·2	47·8	44·0	39·9	35·6	31·1	26·5	21·7	16·8	11·8	50
00	39 54·3	44 51·1	49 47·6	54 43·9	59 39·8	64 35·5	69 31·0	74 26·3	79 21·5	84 16·6	89 11·7	00
10	54·2	51·0	47·5	43·7	39·6	35·3	30·8	26·1	21·3	16·5	11·5	10
20	54·1	50·9	47·4	43·6	39·5	35·2	30·7	26·0	21·2	16·3	11·4	20
30	54·0	50·8	47·3	43·5	39·4	35·0	30·5	25·8	21·0	16·1	11·2	30
40	53·9	50·7	47·2	43·3	39·2	34·9	30·4	25·7	20·9	16·0	11·0	40
50	53·8	50·6	47·0	43·2	39·1	34·7	30·2	25·5	20·7	15·8	10·9	50

H.P.	L	U	L	U	L	U	L	U	L	U	L	U	L	U	L	U	L	U	L	U	L	U	H.P.
′	′	′	′	′	′	′	′	′	′	′	′	′	′	′	′	′	′	′	′	′	′	′	′
54·0	1·1	1·7	1·3	1·9	1·5	2·1	1·7	2·4	2·0	2·6	2·3	2·9	2·6	3·2	2·9	3·5	3·2	3·8	3·5	4·1	3·8	4·5	54·0
54·3	1·4	1·8	1·6	2·0	1·8	2·2	2·0	2·5	2·3	2·7	2·5	3·0	2·8	3·2	3·0	3·5	3·3	3·8	3·6	4·1	3·9	4·4	54·3
54·6	1·7	2·0	1·9	2·2	2·1	2·4	2·3	2·6	2·5	2·8	2·7	3·0	3·0	3·3	3·2	3·5	3·5	3·8	3·7	4·1	4·0	4·3	54·6
54·9	2·0	2·2	2·2	2·3	2·3	2·5	2·5	2·7	2·7	2·9	2·9	3·1	3·2	3·3	3·4	3·5	3·6	3·8	3·9	4·0	4·1	4·3	54·9
55·2	2·3	2·3	2·5	2·4	2·6	2·6	2·8	2·8	3·0	2·9	3·2	3·1	3·4	3·3	3·6	3·5	3·8	3·7	4·0	4·0	4·2	4·2	55·2
55·5	2·7	2·5	2·8	2·6	2·9	2·7	3·1	2·9	3·2	3·0	3·4	3·2	3·6	3·4	3·7	3·5	3·9	3·7	4·1	3·9	4·3	4·1	55·5
55·8	3·0	2·6	3·1	2·7	3·2	2·8	3·3	3·0	3·5	3·1	3·6	3·3	3·8	3·4	3·9	3·6	4·1	3·7	4·2	3·9	4·4	4·0	55·8
56·1	3·3	2·8	3·4	2·9	3·5	3·0	3·6	3·1	3·7	3·2	3·8	3·3	4·0	3·4	4·1	3·6	4·2	3·7	4·4	3·8	4·5	4·0	56·1
56·4	3·6	2·9	3·7	3·0	3·8	3·1	3·9	3·2	3·9	3·3	4·0	3·4	4·1	3·5	4·3	3·6	4·4	3·7	4·5	3·8	4·6	3·9	56·4
56·7	3·9	3·1	4·0	3·1	4·1	3·2	4·1	3·3	4·2	3·3	4·3	3·4	4·3	3·5	4·4	3·6	4·5	3·7	4·6	3·8	4·7	3·8	56·7
57·0	4·3	3·2	4·3	3·3	4·3	3·3	4·4	3·4	4·4	3·4	4·5	3·5	4·5	3·5	4·6	3·6	4·7	3·6	4·7	3·7	4·8	3·8	57·0
57·3	4·6	3·4	4·6	3·4	4·6	3·4	4·6	3·5	4·7	3·5	4·7	3·5	4·7	3·6	4·8	3·6	4·8	3·6	4·8	3·7	4·9	3·7	57·3
57·6	4·9	3·6	4·9	3·6	4·9	3·6	4·9	3·6	4·9	3·6	4·9	3·6	4·9	3·6	4·9	3·6	5·0	3·6	5·0	3·6	5·0	3·6	57·6
57·9	5·2	3·7	5·2	3·7	5·2	3·7	5·2	3·7	5·2	3·7	5·1	3·6	5·1	3·6	5·1	3·6	5·1	3·6	5·1	3·6	5·1	3·6	57·9
58·2	5·5	3·9	5·5	3·8	5·5	3·8	5·4	3·8	5·4	3·7	5·4	3·7	5·3	3·7	5·3	3·6	5·2	3·6	5·2	3·5	5·2	3·5	58·2
58·5	5·9	4·0	5·8	4·0	5·8	3·9	5·7	3·9	5·6	3·8	5·6	3·8	5·5	3·7	5·5	3·6	5·4	3·6	5·3	3·5	5·3	3·4	58·5
58·8	6·2	4·2	6·1	4·1	6·0	4·1	6·0	4·0	5·9	3·9	5·8	3·8	5·7	3·7	5·6	3·6	5·5	3·5	5·4	3·5	5·3	3·4	58·8
59·1	6·5	4·3	6·4	4·3	6·3	4·2	6·2	4·1	6·1	4·0	6·0	3·9	5·9	3·8	5·8	3·6	5·7	3·5	5·6	3·4	5·4	3·3	59·1
59·4	6·8	4·5	6·7	4·4	6·6	4·3	6·5	4·2	6·4	4·1	6·2	3·9	6·1	3·8	6·0	3·7	5·8	3·5	5·7	3·4	5·5	3·2	59·4
59·7	7·1	4·6	7·0	4·5	6·9	4·4	6·8	4·3	6·6	4·1	6·5	4·0	6·3	3·8	6·2	3·7	6·0	3·5	5·8	3·3	5·6	3·2	59·7
60·0	7·5	4·8	7·3	4·7	7·2	4·5	7·0	4·4	6·9	4·2	6·7	4·0	6·5	3·9	6·3	3·7	6·1	3·5	5·9	3·3	5·7	3·1	60·0
60·3	7·8	5·0	7·6	4·8	7·5	4·7	7·3	4·5	7·1	4·3	6·9	4·1	6·7	3·9	6·5	3·7	6·3	3·5	6·0	3·2	5·8	3·0	60·3
60·6	8·1	5·1	7·9	5·0	7·7	4·8	7·6	4·6	7·3	4·4	7·1	4·2	6·9	3·9	6·7	3·7	6·4	3·4	6·2	3·2	5·9	2·9	60·6
60·9	8·4	5·3	8·2	5·1	8·0	4·9	7·8	4·7	7·6	4·5	7·3	4·2	7·1	4·0	6·8	3·7	6·6	3·4	6·3	3·2	6·0	2·9	60·9
61·2	8·7	5·4	8·5	5·2	8·3	5·0	8·1	4·8	7·8	4·5	7·6	4·3	7·3	4·0	7·0	3·7	6·7	3·4	6·4	3·1	6·1	2·8	61·2
61·5	9·1	5·6	8·8	5·4	8·6	5·1	8·3	4·9	8·1	4·6	7·8	4·3	7·5	4·0	7·2	3·7	6·9	3·4	6·5	3·1	6·2	2·7	61·5

APPENDIX G

EXTRACTS FROM *AIR ALMANAC*

STARS, MAY—AUG., 1975

No.	Name		Mag.	S.H.A. ° ′	Dec. ° ′
7*	*Acamar*		3·1	315 40	S. 40 24
5*	*Achernar*		0·6	335 48	S. 57 21
30*	*Acrux*		1·1	173 41	S. 62 58
19	*Adhara*	†	1·6	255 35	S. 28 56
10*	*Aldebaran*	†	1·1	291 22	N. 16 28
32*	*Alioth*		1·7	166 46	N. 56 06
34*	*Alkaid*		1·9	153 21	N. 49 26
55	*Al Na'ir*		2·2	28 19	S. 47 05
15	*Alnilam*	†	1·8	276 16	S. 1 13
25*	*Alphard*	†	2·2	218 24	S. 8 33
41*	*Alphecca*	†	2·3	126 35	N. 26 48
1*	*Alpheratz*	†	2·2	358 13	N. 28 57
51*	*Altair*	†	0·9	62 36	N. 8 48
2	*Ankaa*		2·4	353 44	S. 42 26
42*	*Antares*	†	1·2	113 01	S. 26 23
37*	*Arcturus*	†	0·2	146 22	N. 19 19
43	*Atria*		1·9	108 28	S. 68 59
22	*Avior*		1·7	234 30	S. 59 26
13	*Bellatrix*	†	1·7	279 03	N. 6 20
16*	*Betelgeuse*	†	0·1–1·2	271 32	N. 7 24
17*	*Canopus*		−0·9	264 09	S. 52 41
12*	*Capella*		0·2	281 17	N. 45 58
53*	*Deneb*		1·3	49 51	N. 45 12
28*	*Denebola*	†	2·2	183 03	N. 14 43
4*	*Diphda*	†	2·2	349 24	S. 18 07
27*	*Dubhe*		2·0	194 27	N. 61 53
14	*Elnath*	†	1·8	278 49	N. 28 35
47	*Eltanin*		2·4	90 59	N. 51 30
54*	*Enif*	†	2·5	34 15	N. 9 46
56*	*Fomalhaut*	†	1·3	15 55	S. 29 45
31	*Gacrux*		1·6	172 33	S. 56 59
29*	*Gienah*	†	2·8	176 22	S. 17 25
35	*Hadar*		0·9	149 28	S. 60 16
6*	*Hamal*	†	2·2	328 33	N. 23 21
48	*Kaus Aust.*		2·0	84 21	S. 34 24
40*	*Kochab*		2·2	137 18	N. 74 16
57	*Markab*	†	2·6	14 07	N. 15 04
8*	*Menkar*	†	2·8	314 45	N. 4 00
36	*Menkent*		2·3	148 41	S. 36 15
24*	*Miaplacidus*		1·8	221 46	S. 69 37
9*	*Mirfak*		1·9	309 21	N. 49 46
50*	*Nunki*	†	2·1	76 33	S. 26 20
52*	*Peacock*		2·1	54 03	S. 56 49
21*	*Pollux*	†	1·2	244 03	N. 28 05
20*	*Procyon*	†	0·5	245 30	N. 5 17
46*	*Rasalhague*	†	2·1	96 33	N. 12 35
26*	*Regulus*	†	1·3	208 14	N. 12 05
11*	*Rigel*	†	0·3	281 40	S. 8 14
38*	*Rigil Kent.*		0·1	140 30	S. 60 44
44	*Sabik*	†	2·6	102 45	S. 15 42
3*	*Schedar*		2·5	350 13	N. 56 24
45*	*Shaula*		1·7	97 00	S. 37 05
18*	*Sirius*	†	−1·6	258 59	S. 16 41
33*	*Spica*	†	1·2	159 01	S. 11 02
23*	*Suhail*		2·2	223 14	S. 43 20
49*	*Vega*		0·1	80 58	N. 38 46
39	*Zuben'ubi*	†	2·9	137 37	S. 15 57

INTERPOLATION OF G.H.A.

Increment to be added for intervals of G.M.T. to G.H.A. of: Sun, Aries (♈) and planets; Moon

SUN, etc. m s	° ′	MOON m s	SUN, etc. m s	° ′	MOON m s	SUN, etc. m s	° ′	MOON m s
00 00	0 00	00 00	03 17	0 50	03 25	06 37	1 40	06 52
01	0 01	00 02	21	0 51	03 29	41	1 41	06 56
05	0 02	00 06	25	0 52	03 33	45	1 42	07 00
09	0 03	00 10	29	0 53	03 37	49	1 43	07 04
13	0 04	00 14	33	0 54	03 41	53	1 44	07 08
17	0 05	00 18	37	0 55	03 45	06 57	1 45	07 13
21	0 06	00 22	41	0 56	03 49	07 01	1 46	07 17
25	0 07	00 26	45	0 57	03 54	05	1 47	07 21
29	0 08	00 31	49	0 58	03 58	09	1 48	07 25
33	0 09	00 35	53	0 59	04 02	13	1 49	07 29
37	0 10	00 39	03 57	1 00	04 06	17	1 50	07 33
41	0 11	00 43	04 01	1 01	04 10	21	1 51	07 37
45	0 12	00 47	05	1 02	04 14	25	1 52	07 42
49	0 13	00 51	09	1 03	04 19	29	1 53	07 46
53	0 14	00 55	13	1 04	04 23	33	1 54	07 50
00 57	0 15	01 00	17	1 05	04 27	37	1 55	07 54
01 01	0 16	01 04	21	1 06	04 31	41	1 56	07 58
05	0 17	01 08	25	1 07	04 35	45	1 57	08 02
09	0 18	01 12	29	1 08	04 39	49	1 58	08 06
13	0 19	01 16	33	1 09	04 43	53	1 59	08 11
17	0 20	01 20	37	1 10	04 48	07 57	2 00	08 15
21	0 21	01 24	41	1 11	04 52	08 01	2 01	08 19
25	0 22	01 29	45	1 12	04 56	05	2 02	08 23
29	0 23	01 33	49	1 13	05 00	09	2 03	08 27
33	0 24	01 37	53	1 14	05 04	13	2 04	08 31
37	0 25	01 41	04 57	1 15	05 08	17	2 05	08 35
41	0 26	01 45	05 01	1 16	05 12	21	2 06	08 40
45	0 27	01 49	05	1 17	05 17	25	2 07	08 44
49	0 28	01 53	09	1 18	05 21	29	2 08	08 48
53	0 29	01 58	13	1 19	05 25	33	2 09	08 52
01 57	0 30	02 02	17	1 20	05 29	37	2 10	08 56
02 01	0 31	02 06	21	1 21	05 33	41	2 11	09 00
05	0 32	02 10	25	1 22	05 37	45	2 12	09 04
09	0 33	02 14	29	1 23	05 41	49	2 13	09 09
13	0 34	02 18	33	1 24	05 46	53	2 14	09 13
17	0 35	02 22	37	1 25	05 50	08 57	2 15	09 17
21	0 36	02 27	41	1 26	05 54	09 01	2 16	09 21
25	0 37	02 31	45	1 27	05 58	05	2 17	09 25
29	0 38	02 35	49	1 28	06 02	09	2 18	09 29
33	0 39	02 39	53	1 29	06 06	13	2 19	09 33
37	0 40	02 43	05 57	1 30	06 10	17	2 20	09 38
41	0 41	02 47	06 01	1 31	06 15	21	2 21	09 42
45	0 42	02 51	05	1 32	06 19	25	2 22	09 46
49	0 43	02 56	09	1 33	06 23	29	2 23	09 50
53	0 44	03 00	13	1 34	06 27	33	2 24	09 54
02 57	0 45	03 04	17	1 35	06 31	37	2 25	09 58
03 01	0 46	03 08	21	1 36	06 35	41	2 26	10 00
05	0 47	03 12	25	1 37	06 39	45	2 27	
09	0 48	03 16	29	1 38	06 44	49	2 28	
13	0 49	03 20	33	1 39	06 48	53	2 29	
17	0 50	03 25	37	1 40	06 52	09 57	2 30	
03 21		03 29	06 41		06 56	10 00		

*Stars used in H.O. 249 (A.P. 3270) Vol. 1.
†Stars that may be used with Vols. 2 and 3.

(DAY 152) GREENWICH A. M. 1975 JUNE 1 (SUNDAY)

GMT	☉ SUN GHA	SUN Dec.	ARIES GHA ♈	VENUS −3.8 GHA	VENUS Dec.	JUPITER −1.7 GHA	JUPITER Dec.	SATURN 0.4 GHA	SATURN Dec.	◑ MOON GHA	MOON Dec.
h m	° ′	° ′	° ′	° ′	° ′	° ′	° ′	° ′	° ′	° ′	° ′
00 00	180 36.0	N21 56.2	248 51.5	132 10	N23 46	232 59	N 5 30	140 29	N22 16	279 21	S 6 47
10	183 06.0	56.3	251 21.9	134 40		235 29		142 59		281 47	45
20	185 36.0	56.3	253 52.3	137 09		237 59		145 29		284 13	43
30	188 06.0	· 56.4	256 22.7	139 39	· ·	240 30	· ·	148 00	· ·	286 39	· 42
40	190 35.9	56.5	258 53.1	142 09		243 00		150 30		289 04	40
50	193 05.9	56.5	261 23.5	144 39		245 30		153 00		291 30	38
01 00	195 35.9	N21 56.6	263 53.9	147 09	N23 45	248 01	N 5 30	155 31	N22 16	293 56	S 6 36
10	198 05.9	56.6	266 24.4	149 39		250 31		158 01		296 22	35
20	200 35.9	56.7	268 54.8	152 09		253 01		160 31		298 47	33
30	203 05.9	· 56.8	271 25.2	154 39	· ·	255 32	· ·	163 02	· ·	301 13	· 31
40	205 35.8	56.8	273 55.6	157 09		258 02		165 32		303 39	29
50	208 05.8	56.9	276 26.0	159 39		260 32		168 03		306 05	28
02 00	210 35.8	N21 56.9	278 56.4	162 09	N23 45	263 03	N 5 30	170 33	N22 16	308 30	S 6 26
10	213 05.8	57.0	281 26.8	164 39		265 33		173 03		310 56	24
20	215 35.8	57.1	283 57.2	167 09		268 03		175 34		313 22	23
30	218 05.8	· 57.1	286 27.6	169 39	· ·	270 34	· ·	178 04	· ·	315 47	· 21
40	220 35.8	57.2	288 58.0	172 09		273 04		180 34		318 13	19
50	223 05.7	57.2	291 28.5	174 39		275 34		183 05		320 39	17
03 00	225 35.7	N21 57.3	293 58.9	177 09	N23 44	278 05	N 5 30	185 35	N22 16	323 05	S 6 16
10	228 05.7	57.3	296 29.3	179 38		280 35		188 05		325 30	14
20	230 35.7	57.4	298 59.7	182 08		283 05		190 36		327 56	12
30	233 05.7	· 57.5	301 30.1	184 38	· ·	285 36	· ·	193 06	· ·	330 22	· 10
40	235 35.7	57.5	304 00.5	187 08		288 06		195 37		332 48	09
50	238 05.6	57.6	306 30.9	189 38		290 36		198 07		335 13	07
04 00	240 35.6	N21 57.6	309 01.3	192 08	N23 44	293 07	N 5 30	200 37	N22 15	337 39	S 6 05
10	243 05.6	57.7	311 31.7	194 38		295 37		203 08		340 05	04
20	245 35.6	57.8	314 02.2	197 08		298 07		205 38		342 31	02
30	248 05.6	· 57.8	316 32.6	199 38	· ·	300 38	· ·	208 08	· ·	344 56	6 00
40	250 35.6	57.9	319 03.0	202 08		303 08		210 39		347 22	5 58
50	253 05.6	57.9	321 33.4	204 38		305 38		213 09		349 48	57
05 00	255 35.5	N21 58.0	324 03.8	207 08	N23 43	308 09	N 5 31	215 39	N22 15	352 14	S 5 55
10	258 05.5	58.0	326 34.2	209 38		310 39		218 10		354 39	53
20	260 35.5	58.1	329 04.6	212 08		313 09		220 40		357 05	51
30	263 05.5	· 58.2	331 35.0	214 38	· ·	315 40	· ·	223 10	· ·	359 31	· 50
40	265 35.5	58.2	334 05.4	217 08		318 10		225 41		1 57	48
50	268 05.5	58.3	336 35.9	219 37		320 40		228 11		4 22	46
06 00	270 35.5	N21 58.3	339 06.3	222 07	N23 43	323 11	N 5 31	230 42	N22 15	6 48	S 5 44
10	273 05.4	58.4	341 36.7	224 37		325 41		233 12		9 14	43
20	275 35.4	58.5	344 07.1	227 07		328 11		235 42		11 40	41
30	278 05.4	· 58.5	346 37.5	229 37	· ·	330 42	· ·	238 13	· ·	14 05	· 39
40	280 35.4	58.6	349 07.9	232 07		333 12		240 43		16 31	38
50	283 05.4	58.6	351 38.3	234 37		335 42		243 13		18 57	36
07 00	285 35.4	N21 58.7	354 08.7	237 07	N23 42	338 13	N 5 31	245 44	N22 15	21 23	S 5 34
10	288 05.3	58.7	356 39.1	239 37		340 43		248 14		23 49	32
20	290 35.3	58.8	359 09.5	242 07		343 13		250 44		26 14	31
30	293 05.3	· 58.9	1 40.0	244 37	· ·	345 44	· ·	253 15	· ·	28 40	· 29
40	295 35.3	58.9	4 10.4	247 07		348 14		255 45		31 06	27
50	298 05.3	59.0	6 40.8	249 37		350 44		258 16		33 32	25
08 00	300 35.3	N21 59.0	9 11.2	252 07	N23 42	353 15	N 5 31	260 46	N22 15	35 57	S 5 24
10	303 05.3	59.1	11 41.6	254 37		355 45		263 16		38 23	22
20	305 35.2	59.1	14 12.0	257 07		358 16		265 47		40 49	20
30	308 05.2	· 59.2	16 42.4	259 37	· ·	0 46	· ·	268 17	· ·	43 15	· 18
40	310 35.2	59.3	19 12.8	262 06		3 16		270 47		45 40	17
50	313 05.2	59.3	21 43.2	264 36		5 47		273 18		48 06	15
09 00	315 35.2	N21 59.4	24 13.7	267 06	N23 41	8 17	N 5 31	275 48	N22 15	50 32	S 5 13
10	318 05.2	59.4	26 44.1	269 36		10 47		278 18		52 58	11
20	320 35.1	59.5	29 14.5	272 06		13 18		280 49		55 23	10
30	323 05.1	· 59.6	31 44.9	274 36	· ·	15 48	· ·	283 19	· ·	57 49	· 08
40	325 35.1	59.6	34 15.3	277 06		18 18		285 49		60 15	06
50	328 05.1	59.7	36 45.7	279 36		20 49		288 20		62 41	04
10 00	330 35.1	N21 59.7	39 16.1	282 06	N23 40	23 19	N 5 31	290 50	N22 15	65 07	S 5 03
10	333 05.1	59.8	41 46.5	284 36		25 49		293 21		67 32	5 01
20	335 35.1	59.8	44 16.9	287 06		28 20		295 51		69 58	4 59
30	338 05.0	21 59.9	46 47.4	289 36	· ·	30 50	· ·	298 21	· ·	72 24	· 57
40	340 35.0	22 00.0	49 17.8	292 06		33 20		300 52		74 50	56
50	343 05.0	00.0	51 48.2	294 36		35 51		303 22		77 15	54
11 00	345 35.0	N22 00.1	54 18.6	297 06	N23 40	38 21	N 5 32	305 52	N22 15	79 41	S 4 52
10	348 05.0	00.1	56 49.0	299 36		40 51		308 23		82 07	50
20	350 35.0	00.2	59 19.4	302 06		43 22		310 53		84 33	48
30	353 04.9	· 00.2	61 49.8	304 35	· ·	45 52	· ·	313 23	· ·	86 59	· 47
40	355 34.9	00.3	64 20.2	307 05		48 22		315 54		89 24	45
50	358 04.9	00.4	66 50.6	309 35		50 53		318 24		91 50	43
Rate	14 59.9	N0 00.3		14 59.6	S0 00.5	15 02.0	N0 00.2	15 02.2	0 00.0	14 34.5	N0 10.4

Lat.	Moon-rise	Diff.
N		
°	h m	m
72	00 49	−07
70	00 40	−04
68	00 32	−01
66	00 26	+01
64	00 20	03
62	00 15	05
60	00 11	06
58	00 07	07
56	00 04	08
54	00 01	09
52	24 17	10
50	24 16	10
45	24 14	12
40	24 13	13
35	24 11	15
30	24 10	16
20	24 07	18
10	24 05	19
0	24 04	21
10	24 02	22
20	24 00	24
30	23 58	26
35	23 56	27
40	23 55	29
45	23 53	30
50	23 51	32
52	23 50	33
54	23 49	34
56	23 48	35
58	23 47	36
60	23 45	37
S		

Moon's P. in A.

Alt.	Corr.	Alt.	Corr.
°	+ ′	°	+ ′
0	54	57	29
10	53	58	28
15	52	59	27
18	51	60	26
21	50	62	25
24	49	63	24
26	48	64	23
29	47	65	22
31	46	66	21
33	45	67	20
35	44	68	19
36	43	70	18
38	42	71	17
40	41	72	16
41	40	73	15
43	39	74	14
44	38	75	13
46	37	76	12
47	36	77	11
49	35	78	10
50	34	79	
51	33		
53	32		
54	31		
55	30		
57	29		
58			

Sun SD 15′.8

Moon SD 15′

Age 21d

(DAY 152) GREENWICH P. M. 1975 JUNE 1 (SUNDAY)

GMT	☉ SUN GHA	SUN Dec.	ARIES GHA ♈	VENUS−3.8 GHA	VENUS Dec.	JUPITER−1.7 GHA	JUPITER Dec.	SATURN 0.4 GHA	SATURN Dec.	◑ MOON GHA	MOON Dec.
h m	° ′	° ′	° ′	° ′	° ′	° ′	° ′	° ′	° ′	° ′	° ′
12 00	0 34.9	N22 00.4	69 21.0	312 05	N23 39	53 23	N 5 32	320 55	N22 15	94 16	S 4 41
10	3 04.9	00.5	71 51.5	314 35		55 53		323 25		96 42	40
20	5 34.9	00.5	74 21.9	317 05		58 24		325 55		99 07	38
30	8 04.8	· 00.6	76 52.3	319 35	·	60 54	·	328 26	·	101 33	· 36
40	10 34.8	00.6	79 22.7	322 05		63 24		330 56		103 59	34
50	13 04.8	00.7	81 53.1	324 35		65 55		333 26		106 25	33
13 00	15 34.8	N22 00.8	84 23.5	327 05	N23 39	68 25	N 5 32	335 57	N22 15	108 51	S 4 31
10	18 04.8	00.8	86 53.9	329 35		70 55		338 27		111 16	29
20	20 34.8	00.9	89 24.3	332 05		73 26		340 57		113 42	27
30	23 04.8	· 00.9	91 54.7	334 35	·	75 56	·	343 28	·	116 08	· 26
40	25 34.7	01.0	94 25.2	337 05		78 26		345 58		118 34	24
50	28 04.7	01.1	96 55.6	339 35		80 57		348 28		120 59	22
14 00	30 34.7	N22 01.1	99 26.0	342 05	N23 38	83 27	N 5 32	350 59	N22 15	123 25	S 4 20
10	33 04.7	01.2	101 56.4	344 35		85 57		353 29		125 51	19
20	35 34.7	01.2	104 26.8	347 04		88 28		356 00		128 17	17
30	38 04.7	· 01.3	106 57.2	349 34	·	90 58	·	358 30	·	130 43	· 15
40	40 34.6	01.3	109 27.6	352 04		93 28		1 00		133 08	13
50	43 04.6	01.4	111 58.0	354 34		95 59		3 31		135 34	11
15 00	45 34.6	N22 01.5	114 28.4	357 04	N23 38	98 29	N 5 32	6 01	N22 15	138 00	S 4 10
10	48 04.6	01.5	116 58.8	359 34		100 59		8 31		140 26	08
20	50 34.6	01.6	119 29.3	2 04		103 30		11 02		142 52	06
30	53 04.6	· 01.6	121 59.7	4 34	·	106 00	·	13 32	·	145 17	· 04
40	55 34.6	01.7	124 30.1	7 04		108 30		16 02		147 43	03
50	58 04.5	01.7	127 00.5	9 34		111 01		18 33		150 09	4 01
16 00	60 34.5	N22 01.8	129 30.9	12 04	N23 37	113 31	N 5 33	21 03	N22 15	152 35	S 3 59
10	63 04.5	01.9	132 01.3	14 34		116 01		23 34		155 00	57
20	65 34.5	01.9	134 31.7	17 04		118 32		26 04		157 26	56
30	68 04.5	· 02.0	137 02.1	19 34	·	121 02	·	28 34	·	159 52	· 54
40	70 34.5	02.0	139 32.5	22 04		123 32		31 05		162 18	52
50	73 04.4	02.1	142 03.0	24 34		126 03		33 35		164 44	50
17 00	75 34.4	N22 02.1	144 33.4	27 04	N23 37	128 33	N 5 33	36 05	N22 15	167 09	S 3 49
10	78 04.4	02.2	147 03.8	29 33		131 03		38 36		169 35	47
20	80 34.4	02.3	149 34.2	32 03		133 34		41 06		172 01	45
30	83 04.4	· 02.3	152 04.6	34 33	·	136 04	·	43 36	·	174 27	· 43
40	85 34.4	02.4	154 35.0	37 03		138 34		46 07		176 53	41
50	88 04.3	02.4	157 05.4	39 33		141 05		48 37		179 18	40
18 00	90 34.3	N22 02.5	159 35.8	42 03	N23 36	143 35	N 5 33	51 08	N22 15	181 44	S 3 38
10	93 04.3	02.5	162 06.2	44 33		146 05		53 38		184 10	36
20	95 34.3	02.6	164 36.7	47 03		148 36		56 08		186 36	34
30	98 04.3	· 02.6	167 07.1	49 33	·	151 06	·	58 39	·	189 02	· 33
40	100 34.3	02.7	169 37.5	52 03		153 36		61 09		191 27	31
50	103 04.3	02.8	172 07.9	54 33		156 07		63 39		193 53	29
19 00	105 34.2	N22 02.8	174 38.3	57 03	N23 36	158 37	N 5 33	66 10	N22 15	196 19	S 3 27
10	108 04.2	02.9	177 08.7	59 33		161 07		68 40		198 45	25
20	110 34.2	02.9	179 39.1	62 03		163 38		71 10		201 11	24
30	113 04.2	· 03.0	182 09.5	64 33	·	166 08	·	73 41	·	203 36	· 22
40	115 34.2	03.0	184 39.9	67 03		168 38		76 11		206 02	20
50	118 04.2	03.1	187 10.3	69 33		171 09		78 41		208 28	18
20 00	120 34.1	N22 03.2	189 40.8	72 03	N23 35	173 39	N 5 33	81 12	N22 15	210 54	S 3 17
10	123 04.1	03.2	192 11.2	74 32		176 09		83 42		213 20	15
20	125 34.1	03.3	194 41.6	77 02		178 40		86 13		215 45	13
30	128 04.1	· 03.3	197 12.0	79 32	·	181 10	·	88 43	·	218 11	· 11
40	130 34.1	03.4	199 42.4	82 02		183 41		91 13		220 37	09
50	133 04.1	03.4	202 12.8	84 32		186 11		93 44		223 03	08
21 00	135 34.0	N22 03.5	204 43.2	87 02	N23 35	188 41	N 5 33	96 14	N22 15	225 29	S 3 06
10	138 04.0	03.6	207 13.6	89 32		191 12		98 44		227 54	04
20	140 34.0	03.6	209 44.0	92 02		193 42		101 15		230 20	02
30	143 04.0	· 03.7	212 14.5	94 32	·	196 12	·	103 45	·	232 46	3 01
40	145 34.0	03.7	214 44.9	97 02		198 43		106 15		235 12	2 59
50	148 04.0	03.8	217 15.3	99 32		201 13		108 46		237 38	57
22 00	150 34.0	N22 03.8	219 45.7	102 02	N23 34	203 43	N 5 34	111 16	N22 15	240 03	S 2 55
10	153 03.9	03.9	222 16.1	104 32		206 14		113 46		242 29	53
20	155 33.9	03.9	224 46.5	107 02		208 44		116 17		244 55	52
30	158 03.9	· 04.0	227 16.9	109 32	·	211 14	·	118 47	·	247 21	· 50
40	160 33.9	04.1	229 47.3	112 02		213 45		121 18		249 47	48
50	163 03.9	04.1	232 17.7	114 31		216 15		123 48		252 12	46
23 00	165 33.9	N22 04.2	234 48.2	117 01	N23 34	218 45	N 5 34	126 18	N22 15	254 38	S 2 45
10	168 03.8	04.2	237 18.6	119 31		221 16		128 49		257 04	43
20	170 33.8	04.3	239 49.0	122 01		223 46		131 19		259 30	41
30	173 03.8	· 04.3	242 19.4	124 31	·	226 16	·	133 49	·	261 56	· 39
40	175 33.8	04.4	244 49.8	127 01		228 47		136 20		264 21	37
50	178 03.8	04.5	247 20.2	129 31		231 17		138 50		266 47	36
Rate	14 59.9	N0 00.3		14 59.7	S0 00.5	15 02.0	N0 00.2	15 02.2	0 00.0	14 34.8	N0 10.6

Lat.	Moon-set	Diff.
N		
°	h m	m
72	10 38	51
70	10 45	48
68	10 51	45
66	10 56	42
64	11 00	40
62	11 04	39
60	11 07	37
58	11 10	36
56	11 12	35
54	11 14	34
52	11 16	33
50	11 18	32
45	11 22	30
40	11 25	29
35	11 28	27
30	11 31	26
20	11 35	24
10	11 39	23
0	11 42	21
10	11 46	20
20	11 49	18
30	11 53	16
35	11 56	15
40	11 58	14
45	12 01	12
50	12 05	10
52	12 07	10
54	12 08	09
56	12 10	08
58	12 13	07
60	12 15	06
S		

Moon's P. in A.

Alt. °	Corr. + ′	Alt. °	Corr. + ′
0		57	
	54		29
9		58	
	53		28
14		59	
	52		27
18		60	
	51		26
21		61	
	50		25
24		63	
	49		24
26		64	
	48		23
28		65	
	47		22
31		66	
	46		21
33		67	
	45		20
34		68	
	44		19
36		70	
	43		18
38		71	
	42		17
40		72	
	41		16
41		73	
	40		15
43		74	
	39		14
44		75	
	38		13
46		76	
	37		12
47		77	
	36		11
49		78	
	35		10
50		79	
	34		
51			
	33		
53			
	32		
54			
	31		
55			
	30		
57			
	29		
58			

Sun SD 15′.8

Moon SD 15′

Age 21d

(DAY 153) GREENWICH A. M. 1975 JUNE 2 (MONDAY)

GMT	☉ SUN GHA	SUN Dec.	ARIES GHA ♈	VENUS−3.8 GHA	VENUS Dec.	JUPITER−1.7 GHA	JUPITER Dec.	SATURN 0.4 GHA	SATURN Dec.	◑ MOON GHA	MOON Dec.
h m	° ′	° ′	° ′	° ′	° ′	° ′	° ′	° ′	° ′	° ′	° ′
00 00	180 33.8	N22 04.5	249 50.6	132 01	N23 33	233 47	N 5 34	141 20	N22 15	269 13	S 2 34
10	183 03.7	04.6	252 21.0	134 31		236 18		143 51		271 39	32
20	185 33.7	04.6	254 51.4	137 01		238 48		146 21		274 05	30
30	188 03.7	· 04.7	257 21.8	139 31 ·	·	241 18 ·	·	148 52 ·	·	276 31	· 29
40	190 33.7	04.7	259 52.3	142 01		243 49		151 22		278 56	27
50	193 03.7	04.8	262 22.7	144 31		246 19		153 52		281 22	25
01 00	195 33.7	N22 04.8	264 53.1	147 01	N23 33	248 49	N 5 34	156 23	N22 15	283 48	S 2 23
10	198 03.7	04.9	267 23.5	149 31		251 20		158 53		286 14	21
20	200 33.6	05.0	269 53.9	152 01		253 50		161 23		288 40	20
30	203 03.6	· 05.0	272 24.3	154 31 ·	·	256 20 ·	·	163 54 ·	·	291 05	· 18
40	205 33.6	05.1	274 54.7	157 01		258 51		166 24		293 31	16
50	208 03.6	05.1	277 25.1	159 30		261 21		168 54		295 57	14
02 00	210 33.6	N22 05.2	279 55.5	162 00	N23 32	263 51	N 5 34	171 25	N22 15	298 23	S 2 12
10	213 03.6	05.2	282 26.0	164 30		266 22		173 55		300 49	11
20	215 33.5	05.3	284 56.4	167 00		268 52		176 26		303 14	09
30	218 03.5	· 05.4	287 26.8	169 30 ·	·	271 22 ·	·	178 56 ·	·	305 40	· 07
40	220 33.5	05.4	289 57.2	172 00		273 53		181 26		308 06	05
50	223 03.5	05.5	292 27.6	174 30		276 23		183 57		310 32	03
03 00	225 33.5	N22 05.5	294 58.0	177 00	N23 32	278 53	N 5 34	186 27	N22 15	312 58	S 2 02
10	228 03.5	05.6	297 28.4	179 30		281 24		188 57		315 23	2 00
20	230 33.4	05.6	299 58.8	182 00		283 54		191 28		317 49	1 58
30	233 03.4	· 05.7	302 29.2	184 30 ·	·	286 24 ·	·	193 58 ·	·	320 15	· 56
40	235 33.4	05.7	304 59.7	187 00		288 55		196 28		322 41	55
50	238 03.4	05.8	307 30.1	189 30		291 25		198 59		325 07	53
04 00	240 33.4	N22 05.9	310 00.5	192 00	N23 31	293 55	N 5 35	201 29	N22 15	327 33	S 1 51
10	243 03.4	05.9	312 30.9	194 30		296 26		203 59		329 58	49
20	245 33.4	06.0	315 01.3	197 00		298 56		206 30		332 24	47
30	248 03.3	· 06.0	317 31.7	199 30 ·	·	301 26 ·	·	209 00 ·	·	334 50	· 46
40	250 33.3	06.1	320 02.1	202 00		303 57		211 31		337 16	44
50	253 03.3	06.1	322 32.5	204 29		306 27		214 01		339 42	42
05 00	255 33.3	N22 06.2	325 02.9	206 59	N23 31	308 57	N 5 35	216 31	N22 15	342 07	S 1 40
10	258 03.3	06.2	327 33.3	209 29		311 28		219 02		344 33	38
20	260 33.3	06.3	330 03.8	211 59		313 58		221 32		346 59	37
30	263 03.2	· 06.4	332 34.2	214 29 ·	·	316 28 ·	·	224 02 ·	·	349 25	· 35
40	265 33.2	06.4	335 04.6	216 59		318 59		226 33		351 51	33
50	268 03.2	06.5	337 35.0	219 29		321 29		229 03		354 16	31
06 00	270 33.2	N22 06.5	340 05.4	221 59	N23 30	323 59	N 5 35	231 33	N22 15	356 42	S 1 30
10	273 03.2	06.6	342 35.8	224 29		326 30		234 04		359 08	28
20	275 33.2	06.6	345 06.2	226 59		329 00		236 34		1 34	26
30	278 03.1	· 06.7	347 36.6	229 29 ·	·	331 30 ·	·	239 05 ·	·	4 00	· 24
40	280 33.1	06.7	350 07.0	231 59		334 01		241 35		6 25	22
50	283 03.1	06.8	352 37.5	234 29		336 31		244 05		8 51	21
07 00	285 33.1	N22 06.9	355 07.9	236 59	N23 29	339 01	N 5 35	246 36	N22 15	11 17	S 1 19
10	288 03.1	06.9	357 38.3	239 29		341 32		249 06		13 43	17
20	290 33.1	07.0	0 08.7	241 59		344 02		251 36		16 09	15
30	293 03.0	· 07.0	2 39.1	244 29 ·	·	346 32 ·	·	254 07 ·	·	18 35	· 13
40	295 33.0	07.1	5 09.5	246 59		349 03		256 37		21 00	12
50	298 03.0	07.1	7 39.9	249 28		351 33		259 07		23 26	10
08 00	300 33.0	N22 07.2	10 10.3	251 58	N23 29	354 04	N 5 35	261 38	N22 15	25 52	S 1 08
10	303 03.0	07.2	12 40.7	254 28		356 34		264 08		28 18	06
20	305 33.0	07.3	15 11.1	256 58		359 04		266 38		30 44	04
30	308 02.9	· 07.4	17 41.6	259 28 ·	·	1 35 ·	·	269 09 ·	·	33 09	· 03
40	310 32.9	07.4	20 12.0	261 58		4 05		271 39		35 35	1 01
50	313 02.9	07.5	22 42.4	264 28		6 35		274 10		38 01	0 59
09 00	315 32.9	N22 07.5	25 12.8	266 58	N23 28	9 06	N 5 35	276 40	N22 15	40 27	S 0 57
10	318 02.9	07.6	27 43.2	269 28		11 36		279 10		42 53	55
20	320 32.9	07.6	30 13.6	271 58		14 06		281 41		45 19	54
30	323 02.9	· 07.7	32 44.0	274 28 ·	·	16 37 ·	·	284 11 ·	·	47 44	· 52
40	325 32.8	07.7	35 14.4	276 58		19 07		286 41		50 10	50
50	328 02.8	07.8	37 44.8	279 28		21 37		289 12		52 36	48
10 00	330 32.8	N22 07.9	40 15.3	281 58	N23 28	24 08	N 5 36	291 42	N22 15	55 02	S 0 47
10	333 02.8	07.9	42 45.7	284 28		26 38		294 12		57 28	45
20	335 32.8	08.0	45 16.1	286 58		29 08		296 43		59 53	43
30	338 02.8	· 08.0	47 46.5	289 28 ·	·	31 39 ·	·	299 13 ·	·	62 19	· 41
40	340 32.7	08.1	50 16.9	291 57		34 09		301 43		64 45	39
50	343 02.7	08.1	52 47.3	294 27		36 39		304 14		67 11	38
11 00	345 32.7	N22 08.2	55 17.7	296 57	N23 27	39 10	N 5 36	306 44	N22 15	69 37	S 0 36
10	348 02.7	08.2	57 48.1	299 27		41 40		309 15		72 03	34
20	350 32.7	08.3	60 18.5	301 57		44 10		311 45		74 28	32
30	353 02.7	· 08.3	62 49.0	304 27 ·	·	46 41 ·	·	314 15 ·	·	76 54	· 30
40	355 32.6	08.4	65 19.4	306 57		49 11		316 46		79 20	29
50	358 02.6	08.5	67 49.8	309 27		51 41		319 16		81 46	27
Rate	14 59.9	N0 00.3		14 59.7	S0 00.5	15 02.0	N0 00.2	15 02.2	0 00.0	14 34.9	N0 10.8

Lat.	Moonrise	Diff.
N		
°	h m	m
72	00 36	−07
70	00 33	−04
68	00 30	−01
66	00 27	+01
64	00 25	03
62	00 24	04
60	00 22	05
58	00 21	07
56	00 19	08
54	00 18	09
52	00 17	10
50	00 16	10
45	00 14	12
40	00 13	13
35	00 11	15
30	00 10	16
20	00 07	18
10	00 05	19
0	00 04	21
10	00 02	22
20	00 00	24
30	24 50	26
35	24 50	27
40	24 52	29
45	24 53	30
50	24 54	32
52	24 55	33
54	24 56	34
56	24 57	35
58	24 58	36
60	24 59	37
S		

Moon's P. in A.

Alt.	Corr.	Alt.	Corr.
°	+ ′	°	+ ′
0	54	57	29
9	53	58	28
14	52	59	27
18	51	60	26
21	50	61	25
24	49	63	24
26	48	64	23
28	47	65	22
31	46	66	21
33	45	67	20
34	44	68	19
36	43	70	18
38	42	71	17
40	41	72	16
41	40	73	15
43	39	74	14
44	38	75	13
46	37	76	12
47	36	77	11
49	35	78	10
50	34	79	
51	33		
53	32		
54	31		
55	30		
57	29		
58			

Sun SD 15′.8

Moon SD 15′

Age 22d

(DAY 153) GREENWICH P. M. 1975 JUNE 2 (MONDAY)

GMT h m	☉ SUN GHA ° ′	SUN Dec. ° ′	ARIES GHA ♈ ° ′	VENUS−3.8 GHA ° ′	VENUS Dec. ° ′	JUPITER−1.7 GHA ° ′	JUPITER Dec. ° ′	SATURN 0.4 GHA ° ′	SATURN Dec. ° ′	◑ MOON GHA ° ′	MOON Dec. ° ′
12 00	0 32.6	N22 08.5	70 20.2	311 57	N23 27	54 12	N 5 36	321 46	N22 14	84 12	S 0 25
10	3 02.6	08.6	72 50.6	314 27		56 42		324 17		86 37	23
20	5 32.6	08.6	75 21.0	316 57		59 12		326 47		89 03	21
30	8 02.6	· 08.7	77 51.4	319 27	· ·	61 43	· ·	329 17	· ·	91 29	· 20
40	10 32.5	08.7	80 21.8	321 57		64 13		331 48		93 55	18
50	13 02.5	08.8	82 52.2	324 27		66 43		334 18		96 21	16
13 00	15 32.5	N22 08.8	85 22.6	326 57	N23 26	69 14	N 5 36	336 49	N22 14	98 47	S 0 14
10	18 02.5	08.9	87 53.1	329 27		71 44		339 19		101 12	12
20	20 32.5	08.9	90 23.5	331 57		74 14		341 49		103 38	11
30	23 02.5	· 09.0	92 53.9	334 27	· ·	76 45	· ·	344 20	· ·	106 04	· 09
40	25 32.4	09.1	95 24.3	336 56		79 15		346 50		108 30	07
50	28 02.4	09.1	97 54.7	339 26		81 45		349 20		110 56	05
14 00	30 32.4	N22 09.2	100 25.1	341 56	N23 26	84 16	N 5 36	351 51	N22 14	113 21	S 0 03
10	33 02.4	09.2	102 55.5	344 26		86 46		354 21		115 47	S 0 02
20	35 32.4	09.3	105 25.9	346 56		89 16		356 51		118 13	0 00
30	38 02.4	· 09.3	107 56.3	349 26	· ·	91 47	· ·	359 22	· ·	120 39	N 0 02
40	40 32.3	09.4	110 26.8	351 56		94 17		1 52		123 05	04
50	43 02.3	09.4	112 57.2	354 26		96 47		4 22		125 30	06
15 00	45 32.3	N22 09.5	115 27.6	356 56	N23 25	99 18	N 5 37	6 53	N22 14	127 56	N 0 07
10	48 02.3	09.5	117 58.0	359 26		101 48		9 23		130 22	09
20	50 32.3	09.6	120 28.4	1 56		104 18		11 54		132 48	11
30	53 02.3	· 09.7	122 58.8	4 26	· ·	106 49	· ·	14 24	· ·	135 14	· 13
40	55 32.3	09.7	125 29.2	6 56		109 19		16 54		137 39	15
50	58 02.2	09.8	127 59.6	9 26		111 49		19 25		140 05	16
16 00	60 32.2	N22 09.8	130 30.0	11 56	N23 25	114 20	N 5 37	21 55	N22 14	142 31	N 0 18
10	63 02.2	09.9	133 00.5	14 26		116 50		24 25		144 57	20
20	65 32.2	09.9	135 30.9	16 56		119 20		26 56		147 23	22
30	68 02.2	· 10.0	138 01.3	19 26	· ·	121 51	· ·	29 26	· ·	149 49	· 24
40	70 32.2	10.0	140 31.7	21 56		124 21		31 56		152 14	25
50	73 02.1	10.1	143 02.1	24 25		126 51		34 27		154 40	27
17 00	75 32.1	N22 10.1	145 32.5	26 55	N23 24	129 22	N 5 37	36 57	N22 14	157 06	N 0 29
10	78 02.1	10.2	148 02.9	29 25		131 52		39 27		159 32	31
20	80 32.1	10.3	150 33.3	31 55		134 22		41 58		161 58	33
30	83 02.1	· 10.3	153 03.7	34 25	· ·	136 53	· ·	44 28	· ·	164 23	· 34
40	85 32.1	10.4	155 34.1	36 55		139 23		46 59		166 49	36
50	88 02.0	10.4	158 04.6	39 25		141 53		49 29		169 15	38
18 00	90 32.0	N22 10.5	160 35.0	41 55	N23 24	144 24	N 5 37	51 59	N22 14	171 41	N 0 40
10	93 02.0	10.5	163 05.4	44 25		146 54		54 30		174 07	42
20	95 32.0	10.6	165 35.8	46 55		149 24		57 00		176 32	43
30	98 02.0	· 10.6	168 06.2	49 25	· ·	151 55	· ·	59 30	· ·	178 58	· 45
40	100 32.0	10.7	170 36.6	51 55		154 25		62 01		181 24	47
50	103 01.9	10.7	173 07.0	54 25		156 55		64 31		183 50	49
19 00	105 31.9	N22 10.8	175 37.4	56 55	N23 23	159 26	N 5 37	67 02	N22 14	186 16	N 0 51
10	108 01.9	10.9	178 07.8	59 25		161 56		69 32		188 42	52
20	110 31.9	10.9	180 38.3	61 55		164 26		72 02		191 07	54
30	113 01.9	· 11.0	183 08.7	64 25	· ·	166 57	· ·	74 33	· ·	193 33	· 56
40	115 31.9	11.0	185 39.1	66 55		169 27		77 03		195 59	0 58
50	118 01.8	11.1	188 09.5	69 24		171 57		79 33		198 25	1 00
20 00	120 31.8	N22 11.1	190 39.9	71 54	N23 22	174 28	N 5 37	82 04	N22 14	200 51	N 1 01
10	123 01.8	11.2	193 10.3	74 24		176 58		84 34		203 16	03
20	125 31.8	11.2	195 40.7	76 54		179 28		87 04		205 42	05
30	128 01.8	· 11.3	198 11.1	79 24	· ·	181 59	· ·	89 35	· ·	208 08	· 07
40	130 31.8	11.3	200 41.5	81 54		184 29		92 05		210 34	09
50	133 01.7	11.4	203 12.0	84 24		186 59		94 35		213 00	10
21 00	135 31.7	N22 11.4	205 42.4	86 54	N23 22	189 30	N 5 38	97 06	N22 14	215 25	N 1 12
10	138 01.7	11.5	208 12.8	89 24		192 00		99 36		217 51	14
20	140 31.7	11.6	210 43.2	91 54		194 31		102 07		220 17	16
30	143 01.7	· 11.6	213 13.6	94 24	· ·	197 01	· ·	104 37	· ·	222 43	· 18
40	145 31.7	11.7	215 44.0	96 54		199 31		107 07		225 09	19
50	148 01.6	11.7	218 14.4	99 24		202 02		109 38		227 34	21
22 00	150 31.6	N22 11.8	220 44.8	101 54	N23 21	204 32	N 5 38	112 08	N22 14	230 00	N 1 23
10	153 01.6	11.8	223 15.2	104 24		207 02		114 38		232 26	25
20	155 31.6	11.9	225 45.6	106 54		209 33		117 09		234 52	27
30	158 01.6	· 11.9	228 16.1	109 24	· ·	212 03	· ·	119 39	· ·	237 18	· 28
40	160 31.6	12.0	230 46.5	111 54		214 33		122 09		239 43	30
50	163 01.5	12.0	233 16.9	114 23		217 04		124 40		242 09	32
23 00	165 31.5	N22 12.1	235 47.3	116 53	N23 21	219 34	N 5 38	127 10	N22 14	244 35	N 1 34
10	168 01.5	12.1	238 17.7	119 23		222 04		129 40		247 01	35
20	170 31.5	12.2	240 48.1	121 53		224 35		132 11		249 27	37
30	173 01.5	· 12.3	243 18.5	124 23	· ·	227 05	· ·	134 41	· ·	251 53	· 39
40	175 31.5	12.3	245 48.9	126 53		229 35		137 12		254 18	41
50	178 01.4	12.4	248 19.3	129 23		232 06		139 42		256 44	43
Rate	14 59.9	N0 00.3		14 59.7	S0 00.6	15 02.0	N0 00.2	15 02.2	0 00.0	14 34.9	N0 10.8

Lat. °	Moon-set h m	Diff. m
N		
72	12 18	50
70	12 19	47
68	12 19	44
66	12 20	42
64	12 20	40
62	12 20	38
60	12 21	37
58	12 21	36
56	12 21	35
54	12 21	34
52	12 21	33
50	12 22	32
45	12 22	30
40	12 22	29
35	12 22	27
30	12 23	26
20	12 23	24
10	12 23	22
0	12 24	21
10	12 24	19
20	12 24	18
30	12 24	16
35	12 24	14
40	12 25	13
45	12 25	12
50	12 25	10
52	12 25	09
54	12 25	09
56	12 25	08
58	12 26	06
60	12 26	05
S		

Moon's P. in A.

Alt. °	Corr. + ′	Alt. °	Corr. + ′
0		57	
	54		29
9		58	
	53		28
14		59	
	52		27
18		60	
	51		26
21		61	
	50		25
24		63	
	49		24
26		64	
	48		23
29		65	
	47		22
31		66	
	46		21
33		67	
	45		20
34		68	
	44		19
36		70	
	43		18
38		71	
	42		17
40		72	
	41		16
41		73	
	40		15
43		74	
	39		14
44		75	
	38		13
46		76	
	37		12
47		77	
	36		11
49		78	
	35		10
50		79	
	34		
51			
	33		
53			
	32		
54			
	31		
55			
	30		
57			
	29		
58			

Sun SD 15′.8

Moon SD 15′

Age 22d

F4 INTERPOLATION OF MOONRISE, MOONSET FOR LONGITUDE

Add if longitude *west*
Subtract if longitude *east*

Longitude	Diff.*					
	05	10	15	20	25	30
°	m	m	m	m	m	m
0	00	00	00	00	00	00
20	01	01	02	02	03	03
40	01	02	03	04	06	07
60	02	03	05	07	08	10
80	02	04	07	09	11	13
100	03	06	08	11	14	17
120	03	07	10	13	17	20
140	04	08	12	16	19	23
160	04	09	13	18	22	27
180	05	10	15	20	25	30

Longitude	Diff.*					
	35	40	45	50	55	60
°	m	m	m	m	m	m
0	00	00	00	00	00	00
15	03	03	04	04	05	05
30	06	07	08	08	09	10
45	09	10	11	12	14	15
60	12	13	15	17	18	20
75	15	17	19	21	23	25
90	18	20	22	25	28	30
105	20	23	26	29	32	35
120	23	27	30	33	37	40
135	26	30	34	38	41	45
150	29	33	38	42	46	50
165	32	37	41	46	50	55
180	35	40	45	50	55	60

Longitude	Diff.*					
	65	70	75	80	85	90
°	m	m	m	m	m	m
0	00	00	00	00	00	00
10	04	04	04	04	05	05
20	07	08	08	09	09	10
30	11	12	12	13	14	15
40	14	16	17	18	19	20
50	18	19	21	22	24	25
60	22	23	25	27	28	30
70	25	27	29	31	33	35
80	29	31	33	36	38	40
90	32	35	38	40	42	45
100	36	39	42	44	47	50
110	40	43	46	49	52	55
120	43	47	50	53	57	60
130	47	51	54	58	61	65
140	51	54	58	62	66	70
150	54	58	62	67	71	75
160	58	62	67	71	76	80
170	61	66	71	76	80	85
180	65	70	75	80	85	90

*When negative *subtract* correction if longitude *west*, and *add* if *east*.

STAR INDEX, MAY—AUG., 1975

No.	Name		Mag.	S.H.A.	Dec.
				° ′	° ′
1*	*Alpheratz*	†	2·2	358 13	N. 28 57
2	*Ankaa*		2·4	353 44	S. 42 26
3*	*Schedar*		2·5	350 13	N. 56 24
4*	*Diphda*	†	2·2	349 24	S. 18 07
5*	*Achernar*		0·6	335 48	S. 57 21
6*	*Hamal*	†	2·2	328 33	N. 23 21
7*	*Acamar*		3·1	315 40	S. 40 24
8*	*Menkar*	†	2·8	314 45	N. 4 00
9*	*Mirfak*		1·9	309 21	N. 49 46
10*	*Aldebaran*	†	1·1	291 22	N. 16 28
11*	*Rigel*	†	0·3	281 40	S. 8 14
12*	*Capella*		0·2	281 17	N. 45 58
13	*Bellatrix*	†	1·7	279 03	N. 6 20
14	*Elnath*	†	1·8	278 49	N. 28 35
15	*Alnilam*	†	1·8	276 16	S. 1 13
16*	*Betelgeuse*	†	0·1–1·2	271 32	N. 7 24
17*	*Canopus*		−0·9	264 09	S. 52 41
18*	*Sirius*	†	−1·6	258 59	S. 16 41
19	*Adhara*	†	1·6	255 35	S. 28 56
20*	*Procyon*	†	0·5	245 30	N. 5 17
21*	*Pollux*	†	1·2	244 03	N. 28 05
22	*Avior*		1·7	234 30	S. 59 26
23*	*Suhail*		2·2	223 14	S. 43 20
24*	*Miaplacidus*		1·8	221 46	S. 69 37
25*	*Alphard*	†	2·2	218 24	S. 8 33
26*	*Regulus*	†	1·3	208 14	N. 12 05
27*	*Dubhe*		2·0	194 27	N. 61 53
28*	*Denebola*	†	2·2	183 03	N. 14 43
29*	*Gienah*	†	2·8	176 22	S. 17 25
30*	*Acrux*		1·1	173 41	S. 62 58
31	*Gacrux*		1·6	172 33	S. 56 59
32*	*Alioth*		1·7	166 46	N. 56 06
33*	*Spica*	†	1·2	159 01	S. 11 02
34*	*Alkaid*		1·9	153 21	N. 49 26
35	*Hadar*		0·9	149 28	S. 60 16
36	*Menkent*		2·3	148 41	S. 36 15
37*	*Arcturus*	†	0·2	146 22	N. 19 19
38*	*Rigil Kentaurus*		0·1	140 30	S. 60 44
39	*Zubenelgenubi*	†	2·9	137 37	S. 15 57
40*	*Kochab*		2·2	137 18	N. 74 16
41*	*Alphecca*	†	2·3	126 35	N. 26 48
42*	*Antares*	†	1·2	113 01	S. 26 23
43	*Atria*		1·9	108 28	S. 68 59
44	*Sabik*	†	2·6	102 45	S. 15 42
45*	*Shaula*		1·7	97 00	S. 37 05
46*	*Rasalhague*	†	2·1	96 33	N. 12 35
47	*Eltanin*		2·4	90 59	N. 51 30
48	*Kaus Australis*		2·0	84 21	S. 34 24
49*	*Vega*		0·1	80 58	N. 38 46
50*	*Nunki*	†	2·1	76 33	S. 26 20
51*	*Altair*	†	0·9	62 36	N. 8 48
52*	*Peacock*		2·1	54 03	S. 56 49
53*	*Deneb*		1·3	49 51	N. 45 12
54*	*Enif*	†	2·5	34 15	N. 9 46
55	*Al Na'ir*		2·2	28 19	S. 47 05
56*	*Fomalhaut*	†	1·3	15 55	S. 29 45
57	*Markab*	†	2·6	14 07	N. 15 04

*Stars used in H.O. 249 (A.P. 3270) Vol. 1.
†Stars that may be used with Vols. 2 and 3.

SUNRISE

Lat.	Dec.	January										February				Lat.
	30	2	5	8	11	14	17	20	23	26	29	1	4	7	10	
°	h m	h m	h m	h m	h m	h m	h m	h m	h m	h m	h m	h m	h m	h m	h m	°
N 72	—	—	—	—	—	—	—	—	—	12 04	11 03	10 33	10 09	09 48	09 28	N 72
70	—	—	—	—	—	—	12 05	11 11	10 46	10 25	10 06	09 49	09 33	09 17	09 02	70
68	—	—	11 34	11 13	10 56	10 41	10 27	10 13	09 59	09 46	09 33	09 20	09 08	08 55	08 42	68
66	10 31	10 27	10 21	10 14	10 06	09 58	09 49	09 39	29	09 19	09 09	08 59	08 48	37	26	66
64	09 51	09 49	09 45	09 41	09 35	29	22	09 15	09 07	08 59	08 50	42	32	23	14	64
62	24	22	20	09 16	09 12	09 07	09 02	08 56	08 49	43	35	27	19	11	08 03	62
N 60	09 03	09 02	09 00	08 58	08 54	08 50	08 46	08 40	08 35	08 29	08 22	08 15	08 08	08 01	07 53	N 60
58	08 46	08 45	08 44	42	39	36	32	27	22	17	11	08 05	07 59	07 52	45	58
56	32	31	30	28	26	23	20	16	12	08 07	08 02	07 56	50	44	38	56
54	19	19	18	17	15	12	09	08 06	08 02	07 58	07 53	48	43	37	32	54
52	08 08	08 08	08 07	08 06	08 05	08 02	08 00	07 57	07 54	50	46	41	36	31	26	52
N 50	07 58	07 58	07 58	07 57	07 56	07 54	07 52	07 49	07 46	07 42	07 39	07 35	07 30	07 25	07 21	N 50
45	38	38	38	38	37	36	34	32	30	27	24	21	17	13	09	45
40	22	22	22	22	22	21	20	18	16	14	12	07 09	07 06	07 03	07 00	40
35	07 07	07 08	07 08	07 08	07 08	07 08	07 07	07 06	07 05	07 03	07 01	06 59	06 57	06 55	06 52	35
30	06 55	06 56	06 57	06 57	06 57	06 57	06 56	06 56	06 55	06 54	06 52	51	49	47	45	30
N 20	06 34	06 35	06 36	06 37	06 37	06 38	06 38	06 38	06 38	06 37	06 37	06 36	06 35	06 33	06 32	N 20
N 10	06 16	17	18	19	20	21	22	22	22	23	23	23	22	22	21	N 10
0	05 59	06 00	06 02	06 03	06 04	06 05	06 06	06 07	06 08	06 09	06 10	06 10	06 10	06 11	11	0
S 10	41	05 43	05 44	05 46	05 48	05 49	05 51	05 52	05 54	05 55	05 56	05 57	05 58	05 59	06 00	S 10
20	23	24	26	28	30	32	34	36	38	40	42	44	45	47	05 49	20
S 30	05 01	05 03	05 05	05 07	05 10	05 12	05 15	05 17	05 20	05 23	05 25	05 28	05 31	05 33	05 36	S 30
35	04 48	04 50	04 52	04 55	04 57	05 00	05 03	05 06	05 09	13	16	19	22	25	28	35
40	33	35	38	41	44	04 47	04 50	04 54	04 57	05 01	05 04	05 08	12	16	19	40
45	04 15	04 18	04 21	24	27	31	35	39	43	04 47	04 51	04 56	05 00	05 05	05 09	45
50	03 53	03 56	03 59	04 03	04 07	11	15	20	25	30	35	40	04 46	04 51	04 56	50
S 52	03 42	03 45	03 49	03 52	03 57	04 01	04 06	04 11	04 17	04 22	04 28	04 33	04 39	04 45	04 50	S 52
54	30	33	37	41	46	03 51	03 56	04 01	04 07	13	19	25	31	38	44	54
56	03 16	19	23	28	33	38	44	03 50	03 57	04 03	04 10	16	23	30	37	56
58	02 59	03 03	03 08	03 13	03 18	24	31	37	44	03 52	03 59	04 06	14	21	29	58
S 60	02 39	02 44	02 49	02 54	03 00	03 07	03 15	03 22	03 30	03 38	03 46	03 54	04 03	04 11	04 19	S 60

SUNSET

Lat.	Dec.	January										February				Lat.
	30	2	5	8	11	14	17	20	23	26	29	1	4	7	10	
°	h m	h m	h m	h m	h m	h m	h m	h m	h m	h m	h m	h m	h m	h m	h m	°
N 72	—	—	—	—	—	—	—	—	—	12 25	13 24	13 55	14 20	14 42	15 02	N 72
70	—	—	—	—	—	—	12 21	13 11	13 39	14 01	14 21	14 39	14 56	15 12	28	70
68	—	—	12 36	13 00	13 20	13 37	13 54	14 09	14 25	14 39	14 54	15 08	15 21	35	15 48	68
66	13 33	13 41	13 49	13 59	14 09	14 20	14 32	14 43	14 55	15 06	15 18	29	41	15 52	16 03	66
64	14 13	14 19	14 25	14 32	14 40	14 49	14 58	15 07	15 17	27	36	15 46	15 56	16 06	16	64
62	14 40	14 45	14 50	14 57	15 03	15 10	15 18	26	35	43	15 52	16 00	16 09	18	27	62
N 60	15 02	15 06	15 10	15 16	15 22	15 28	15 35	15 42	15 49	15 57	16 05	16 12	16 20	16 28	16 36	N 60
58	19	22	27	32	37	43	15 49	15 55	16 02	16 09	15	23	30	37	44	58
56	33	37	41	45	15 50	15 55	16 01	16 06	12	19	25	32	38	45	51	56
54	46	15 49	15 53	15 57	16 01	16 06	11	16	22	28	34	40	46	52	16 58	54
52	15 57	16 00	16 03	16 07	11	16	20	25	30	36	41	47	52	16 58	17 03	52
N 50	16 06	16 09	16 13	16 16	16 20	16 24	16 29	16 33	16 38	16 43	16 48	16 53	16 58	17 03	17 09	N 50
45	27	29	32	35	39	42	16 46	16 50	16 54	16 58	17 03	17 07	17 11	16	20	45
40	44	16 46	16 48	16 51	16 54	16 57	17 01	17 04	17 08	17 11	15	18	22	25	29	40
35	16 57	17 00	17 02	17 04	17 07	17 10	13	16	19	22	25	28	31	34	37	35
30	17 10	12	14	16	19	21	24	26	29	32	34	37	39	42	44	30
N 20	17 31	17 32	17 34	17 36	17 38	17 40	17 42	17 44	17 46	17 48	17 50	17 52	17 53	17 55	17 57	N 20
N 10	17 49	17 50	17 52	17 54	17 55	17 57	17 58	18 00	18 01	18 02	18 04	18 05	18 06	18 07	18 07	N 10
0	18 06	18 07	18 09	18 10	18 11	18 12	18 14	14	15	16	17	17	17	18	18	0
S 10	24	25	26	27	28	29	29	29	30	30	30	30	29	29	28	S 10
20	18 42	18 43	18 44	18 45	18 45	18 46	18 46	18 46	18 45	18 45	18 44	43	42	41	39	20
S 30	19 04	19 05	19 05	19 06	19 06	19 05	19 05	19 04	19 03	19 02	19 00	18 59	18 57	18 55	18 52	S 30
35	17	17	18	18	18	17	16	15	14	12	10	19 08	19 05	19 03	19 00	35
40	32	32	32	32	31	30	29	28	26	24	21	18	15	12	09	40
45	19 49	19 50	19 49	19 49	19 48	19 46	19 45	19 42	40	37	34	31	27	23	19	45
50	20 12	20 12	20 11	20 10	20 08	20 06	20 04	20 01	19 58	19 54	50	46	41	36	31	50
S 52	20 22	20 22	20 21	20 20	20 18	20 16	20 13	20 09	20 06	20 02	19 57	19 53	19 48	19 42	19 37	S 52
54	34	34	33	31	29	26	23	19	15	11	20 06	20 01	19 55	49	43	54
56	20 48	20 48	20 46	44	41	38	35	30	26	21	15	09	20 03	19 57	50	56
58	21 05	21 04	21 02	20 59	20 56	20 52	20 48	43	38	32	26	19	13	20 06	19 58	58
S 60	21 25	21 23	21 21	21 18	21 14	21 09	21 04	20 58	20 52	20 45	20 38	20 31	20 23	20 15	20 07	S 60

MORNING CIVIL TWILIGHT

Lat.	Dec. 30	January 2	5	8	11	14	17	20	23	26	29	February 1	4	7	10	Lat.
°	h m	h m	h m	h m	h m	h m	h m	h m	h m	h m	h m	h m	h m	h m	h m	°
N 72	10 49	10 41	10 32	10 21	10 10	09 58	09 46	09 33	09 21	09 08	08 55	08 41	08 28	08 14	08 01	N 72
70	09 52	09 49	09 44	09 38	09 30	09 22	09 14	09 04	08 54	08 44	33	22	08 11	08 00	07 48	70
68	09 19	09 16	09 13	09 08	09 03	08 57	08 50	08 42	34	26	17	08 07	07 58	07 48	37	68
66	08 54	08 53	08 50	08 47	08 42	37	32	25	18	08 11	08 03	07 55	46	38	28	66
64	35	34	32	29	26	21	17	08 11	08 05	07 59	07 52	45	37	29	21	64
62	19	18	17	15	12	08 08	08 04	07 59	07 54	48	42	36	29	22	14	62
N 60	08 06	08 05	08 04	08 02	08 00	07 57	07 53	07 49	07 44	07 39	07 34	07 28	07 22	07 15	07 08	N 60
58	07 54	07 54	07 53	07 51	07 49	47	44	40	36	31	26	21	15	09	07 03	58
56	44	44	44	42	41	38	35	32	29	24	20	15	10	04	06 58	56
54	36	35	35	34	32	30	28	25	22	18	14	09	05	07 00	54	54
52	27	28	27	26	25	23	21	18	15	12	08	04	07 00	06 55	50	52
N 50	07 20	07 20	07 20	07 19	07 18	07 17	07 15	07 13	07 10	07 07	07 03	07 00	06 56	06 51	06 47	N 50
45	07 04	07 05	07 05	07 04	07 04	07 03	07 01	07 00	06 58	06 55	06 52	06 49	46	43	39	45
40	06 51	06 52	06 52	06 52	06 52	06 51	06 50	06 49	47	45	43	41	38	35	32	40
35	39	40	40	41	41	40	40	39	38	36	35	33	31	28	26	35
30	29	30	30	31	31	31	31	30	29	28	27	26	24	22	20	30
N 20	06 10	06 11	06 12	06 13	06 14	06 14	06 14	06 14	06 14	06 14	06 13	06 13	06 12	06 11	06 10	N 20
N 10	05 53	05 54	05 55	05 56	05 57	05 58	05 59	06 00	06 00	06 01	06 01	06 01	06 00	06 00	06 00	N 10
0	36	38	39	40	42	43	44	05 45	05 46	05 47	05 48	05 48	05 49	05 49	05 49	0
S 10	05 18	20	22	23	25	27	28	30	31	33	34	35	36	38	38	S 10
20	04 58	05 00	05 02	05 04	05 06	05 08	05 10	05 12	05 14	05 16	05 18	20	22	24	26	20
S 30	04 33	04 35	04 38	04 40	04 43	04 45	04 48	04 51	04 54	04 56	04 59	05 02	05 05	05 08	05 10	S 30
35	18	20	23	26	28	31	35	38	41	44	48	04 51	04 54	04 58	05 01	35
40	04 00	04 03	04 05	04 08	04 12	04 15	04 19	22	26	30	34	38	42	46	04 50	40
45	03 38	03 41	03 44	03 47	03 51	03 55	03 59	04 04	04 08	04 13	04 18	22	27	32	37	45
50	03 09	03 12	15	19	24	29	34	03 39	03 45	03 51	03 56	04 02	04 08	14	20	50
S 52	02 54	02 57	03 01	03 05	03 10	03 16	03 21	03 27	03 34	03 40	03 46	03 53	03 59	04 06	04 12	S 52
54	36	40	02 44	02 49	02 55	03 01	03 07	03 14	20	27	34	42	49	03 56	04 03	54
56	02 14	02 19	02 24	29	36	02 43	02 50	02 57	03 05	03 13	21	29	37	45	03 53	56
58	01 45	01 50	01 57	02 04	02 12	02 20	28	37	02 46	02 56	03 05	03 14	23	32	41	58
S 60	00 58	01 07	01 16	01 27	01 38	01 49	02 00	02 12	02 23	02 34	02 45	02 56	03 06	03 17	03 27	S 60

EVENING CIVIL TWILIGHT

Lat.	Dec. 30	January 2	5	8	11	14	17	20	23	26	29	February 1	4	7	10	Lat.
°	h m	h m	h m	h m	h m	h m	h m	h m	h m	h m	h m	h m	h m	h m	h m	°
N 72	13 16	13 26	13 39	13 52	14 06	14 20	14 35	14 49	15 04	15 18	15 33	15 47	16 01	16 15	16 29	N 72
70	14 12	14 19	14 27	14 36	14 45	14 56	15 07	15 18	30	15 42	15 54	16 06	18	30	42	70
68	14 46	14 51	14 57	15 05	15 13	15 21	30	40	15 50	16 00	16 10	21	31	42	16 53	68
66	15 10	15 15	15 20	27	33	41	15 49	15 57	16 06	15	24	33	42	16 52	17 01	66
64	30	34	38	44	15 50	15 57	16 04	16 11	19	27	35	43	16 52	17 00	09	64
62	45	15 49	15 54	15 58	16 04	16 10	16	23	30	37	45	16 52	17 00	08	16	62
N 60	15 59	16 02	16 06	16 11	16 16	16 21	16 27	16 33	16 40	16 46	16 53	17 00	17 07	17 14	17 21	N 60
58	16 10	13	17	21	26	31	36	42	48	16 54	17 00	07	13	20	26	58
56	20	23	27	31	35	40	45	50	16 56	17 01	07	13	19	25	31	56
54	29	32	36	39	43	48	52	16 57	17 02	08	13	18	24	30	35	54
52	38	40	43	47	51	16 55	16 59	17 04	09	13	18	23	29	34	39	52
N 50	16 45	16 47	16 50	16 54	16 57	17 01	17 05	17 10	17 14	17 19	17 23	17 28	17 33	17 38	17 42	N 50
45	17 01	17 03	17 06	17 09	17 12	15	19	23	26	30	34	38	42	46	50	45
40	14	16	19	21	24	27	30	33	37	40	43	47	50	17 54	17 57	40
35	26	28	30	32	35	38	40	43	46	49	52	17 55	17 58	18 00	18 03	35
30	36	38	40	17 42	17 45	17 47	17 49	17 52	17 54	17 57	17 59	18 02	18 04	07	09	30
N 20	17 55	17 56	17 58	18 00	18 02	18 04	18 06	18 08	18 09	18 11	18 13	18 15	18 16	18 18	18 19	N 20
N 10	18 12	18 13	18 15	16	18	19	21	22	23	25	26	27	28	28	29	N 10
0	29	30	32	33	34	35	36	37	37	38	38	39	39	39	39	0
S 10	18 47	18 48	18 49	18 50	18 50	18 51	18 52	18 52	18 52	18 52	18 52	18 52	18 51	18 51	18 50	S 10
20	19 07	19 08	19 08	19 09	19 09	19 10	19 10	19 09	19 09	19 08	19 07	19 06	19 05	19 04	19 02	20
S 30	19 31	19 32	19 32	19 33	19 33	19 32	19 31	19 31	19 29	19 28	19 26	19 24	19 22	19 20	19 18	S 30
35	19 46	19 47	19 47	19 47	19 47	19 46	19 45	43	42	40	38	35	33	30	27	35
40	20 05	20 05	20 05	20 04	20 03	20 02	20 01	19 59	19 57	19 54	19 51	19 48	19 45	42	38	40
45	27	27	26	25	24	22	20	20 17	20 14	20 11	20 08	20 04	20 00	19 55	19 51	45
50	20 56	20 56	20 54	20 53	20 51	20 48	45	41	37	33	28	24	18	20 13	20 07	50
S 52	21 11	21 10	21 09	21 07	21 04	21 01	20 57	20 53	20 49	20 44	20 39	20 33	20 27	20 21	20 15	S 52
54	28	27	25	23	19	16	21 11	21 07	21 02	20 56	20 50	44	38	31	24	54
56	21 50	21 48	21 46	21 42	21 38	34	28	23	17	21 10	21 03	20 56	20 49	42	34	56
58	22 19	22 16	22 12	22 07	22 02	21 56	21 49	21 42	35	27	19	21 11	21 03	20 54	46	58
S 60	23 04	22 58	22 51	22 44	22 35	22 26	22 17	22 07	21 58	21 48	21 38	21 29	21 19	21 09	20 59	S 60

SUNRISE

Lat.	Apr.	May										June				Lat.
	29	2	5	8	11	14	17	20	23	26	29	1	4	7	10	
°	h m	h m	h m	h m	h m	h m	h m	h m	h m	h m	h m	h m	h m	h m	h m	°
N 72	02 14	01 50	01 22	00 41	▭	▭	▭	▭	▭	▭	▭	▭	▭	▭	▭	N 72
70	02 47	02 29	02 11	01 51	01 29	01 02	00 12	▭	▭	▭	▭	▭	▭	▭	▭	70
68	03 10	02 56	02 42	02 28	02 12	01 57	01 40	01 21	01 00	00 29	▭	▭	▭	▭	▭	68
66	28	03 17	03 05	02 53	02 41	02 29	02 17	02 05	01 53	01 41	01 28	01 15	01 02	00 48	00 33	66
64	43	33	23	03 12	03 02	02 53	02 43	34	02 25	02 16	02 07	02 00	01 52	01 46	01 40	64
62	03 55	46	37	28	20	03 11	03 03	02 55	02 48	02 41	34	28	02 23	02 19	02 15	62
N 60	04 05	03 57	03 49	03 41	03 34	03 26	03 19	03 13	03 06	03 01	02 55	02 50	02 46	02 43	02 40	N 60
58	15	04 07	04 00	03 53	46	39	33	27	22	17	03 12	03 08	03 04	03 01	02 59	58
56	22	16	09	04 02	03 56	03 50	45	40	35	30	26	23	20	17	03 15	56
54	29	23	17	11	04 06	04 00	03 55	03 50	46	42	39	35	33	31	29	54
52	36	30	24	19	14	09	04 04	04 00	03 56	03 52	49	46	44	42	41	52
N 50	04 41	04 36	04 31	04 26	04 21	04 17	04 12	04 09	04 05	04 02	03 59	03 56	03 54	03 53	03 51	N 50
45	04 54	04 49	45	41	37	33	30	27	24	21	04 19	04 17	04 15	04 14	04 13	45
40	05 04	05 00	04 56	04 53	04 49	46	44	41	39	37	35	33	32	31	31	40
35	12	09	05 06	05 03	05 00	04 58	04 56	04 53	04 52	04 50	04 49	47	47	46	46	35
30	20	17	14	12	10	05 08	05 06	05 04	05 03	05 01	05 00	04 59	04 59	04 58	04 58	30
N 20	05 32	05 31	05 29	05 27	05 26	05 25	05 23	05 22	05 22	05 21	05 20	05 20	05 20	05 20	05 20	N 20
N 10	44	43	42	41	40	39	39	38	38	38	38	38	38	38	38	N 10
0	05 54	05 54	05 53	05 53	05 53	05 53	05 53	05 53	05 53	05 53	05 54	05 54	05 55	05 55	05 56	0
S 10	06 04	06 04	06 05	06 05	06 06	06 06	06 07	06 07	06 08	06 09	06 10	06 10	06 11	06 12	06 13	S 10
20	15	16	17	18	19	20	21	23	24	25	26	27	29	30	31	20
S 30	06 27	06 29	06 31	06 33	06 35	06 37	06 38	06 40	06 42	06 44	06 46	06 47	06 49	06 50	06 51	S 30
35	34	37	39	41	44	46	48	06 50	06 53	06 55	06 57	06 59	07 00	07 02	07 03	35
40	42	45	48	06 51	06 54	06 57	06 59	07 02	07 05	07 07	07 09	07 12	14	16	17	40
45	06 51	06 55	06 59	07 02	07 06	07 09	07 12	16	19	22	25	27	30	32	34	45
50	07 02	07 07	07 11	16	20	24	28	32	36	40	43	46	49	07 52	07 54	50
S 52	07 08	07 12	07 17	07 22	07 27	07 31	07 36	07 40	07 44	07 48	07 52	07 55	07 58	08 01	08 04	S 52
54	13	19	24	29	34	39	44	49	07 53	07 58	08 02	08 05	08 09	12	14	54
56	20	25	31	37	42	48	07 53	07 58	08 03	08 08	12	17	20	24	27	56
58	26	33	39	46	07 52	07 58	08 04	08 10	15	20	25	30	34	38	41	58
S 60	07 34	07 41	07 49	07 56	08 03	08 09	08 16	08 22	08 29	08 34	08 40	08 45	08 50	08 54	08 58	S 60

SUNSET

Lat.	Apr.	May										June				Lat.
	29	2	5	8	11	14	17	20	23	26	29	1	4	7	10	
°	h m	h m	h m	h m	h m	h m	h m	h m	h m	h m	h m	h m	h m	h m	h m	°
N 72	21 47	22 11	22 41	23 35	▭	▭	▭	▭	▭	▭	▭	▭	▭	▭	▭	N 72
70	21 12	21 29	21 47	22 07	22 31	23 01	▭	▭	▭	▭	▭	▭	▭	▭	▭	70
68	20 48	21 01	21 15	21 30	21 45	22 01	22 18	22 37	23 01	23 43	▭	▭	▭	▭	▭	68
66	29	20 40	20 52	21 03	21 15	21 27	21 39	21 51	22 04	22 17	22 30	22 44	22 58	23 14	23 32	66
64	14	24	33	20 43	20 53	21 02	21 12	21 22	21 31	21 40	21 49	21 58	22 06	22 13	22 20	64
62	20 02	20 10	18	27	35	20 43	20 51	20 59	21 07	21 15	22	28	21 34	21 40	21 45	62
N 60	19 51	19 59	20 06	20 13	20 21	20 28	20 35	20 42	20 48	20 55	21 01	21 06	21 11	21 16	21 20	N 60
58	42	48	19 55	20 02	20 08	15	21	27	33	38	20 43	20 48	20 53	20 57	21 00	58
56	34	40	46	19 52	19 58	20 03	20 09	14	20	24	29	33	37	41	20 44	56
54	27	32	38	43	48	19 53	19 58	20 03	20 08	13	17	21	24	27	30	54
52	20	25	30	35	40	45	49	19 54	19 58	20 02	20 06	09	13	15	18	52
N 50	19 15	19 19	19 24	19 28	19 33	19 37	19 41	19 45	19 49	19 53	19 56	20 00	20 02	20 05	20 07	N 50
45	19 02	19 06	19 09	13	17	20	24	27	30	33	36	19 39	19 41	19 43	19 45	45
40	18 52	18 55	18 58	19 01	19 04	19 07	19 09	12	15	18	20	22	24	26	28	40
35	43	46	48	18 50	18 53	18 55	18 58	19 00	19 02	19 04	19 06	19 08	19 10	19 12	13	35
30	36	37	39	41	43	45	47	18 49	18 51	18 53	18 54	18 56	18 58	18 59	19 00	30
N 20	18 23	18 24	18 25	18 26	18 27	18 28	18 29	18 31	18 32	18 33	18 34	18 35	18 37	18 38	18 39	N 20
N 10	11	12	12	13	13	14	14	15	16	16	17	18	19	19	20	N 10
0	18 01	18 00	18 00	18 00	18 00	18 00	18 00	18 00	18 00	18 01	18 01	18 01	18 02	18 02	18 03	0
S 10	17 51	17 50	17 49	17 48	17 47	17 46	17 46	17 45	17 45	17 45	17 45	17 45	17 45	17 46	17 46	S 10
20	40	38	36	35	33	32	31	30	29	28	28	28	28	28	28	20
S 30	17 27	17 25	17 22	17 20	17 18	17 16	17 14	17 12	17 11	17 10	17 09	17 08	17 07	17 07	17 07	S 30
35	20	17	14	11	17 09	17 06	17 04	17 02	17 00	16 59	16 57	16 56	16 56	16 55	16 55	35
40	12	17 09	17 05	17 02	16 59	16 56	16 53	16 51	16 48	46	45	43	42	41	41	40
45	17 03	16 58	16 54	16 50	47	43	40	37	34	32	30	28	27	25	25	45
50	16 52	47	42	37	32	28	24	20	17	14	11	09	16 07	16 05	16 04	50
S 52	16 47	16 41	16 35	16 30	16 25	16 21	16 16	16 12	16 09	16 05	16 02	16 00	15 58	15 56	15 55	S 52
54	41	35	29	23	18	13	16 08	16 04	16 00	15 56	15 53	15 50	47	45	44	54
56	35	28	22	15	10	16 04	15 59	15 54	15 49	45	42	38	36	33	32	56
58	28	20	13	16 07	16 00	15 54	48	43	38	33	29	25	22	19	17	58
S 60	16 19	16 12	16 04	15 56	15 49	15 43	15 36	15 30	15 24	15 19	15 14	15 10	15 06	15 03	15 00	S 60

MORNING CIVIL TWILIGHT

Lat.	Apr.	May										June				Lat.
	29	2	5	8	11	14	17	20	23	26	29	1	4	7	10	
°	h m	h m	h m	h m	h m	h m	h m	h m	h m	h m	h m	h m	h m	h m	h m	°
N 72	////	////	////	////	▭	▭	▭	▭	▭	▭	▭	▭	▭	▭	▭	N 72
70	////	////	////	////	////	////	////	▭	▭	▭	▭	▭	▭	▭	▭	70
68	01 33	01 04	00 02	////	////	////	////	////	////	////	▭	▭	▭	▭	▭	68
66	02 12	01 54	01 34	01 10	00 37	////	////	////	////	////	////	////	////	////	////	66
64	38	02 24	02 10	01 54	01 38	01 20	00 59	00 27	////	////	////	////	////	////	////	64
62	02 58	02 46	35	02 23	02 11	01 59	01 46	01 32	01 18	01 03	00 44	00 15	////	////	////	62
N 60	03 14	03 04	02 54	02 44	02 35	02 25	02 15	02 05	01 56	01 46	01 37	01 28	01 20	01 11	01 04	N 60
58	27	19	03 10	03 02	02 53	02 45	37	29	02 22	02 15	02 08	02 02	01 56	01 51	01 47	58
56	39	31	23	16	03 09	03 01	02 55	02 48	42	36	31	26	02 21	02 18	02 14	56
54	49	42	35	28	22	15	03 09	03 04	02 58	02 53	02 49	02 45	41	38	36	54
52	03 58	51	45	39	33	27	22	17	03 12	03 08	03 04	03 00	02 57	02 55	02 53	52
N 50	04 05	03 59	03 54	03 48	03 43	03 38	03 33	03 28	03 24	03 20	03 17	03 14	03 11	03 09	03 07	N 50
45	22	04 17	04 12	04 07	04 03	03 59	03 55	03 52	03 49	03 46	03 43	03 41	03 39	37	36	45
40	35	31	27	23	19	04 16	04 13	04 10	04 07	04 05	04 03	04 01	04 00	03 59	03 58	40
35	45	42	39	36	33	30	27	25	23	21	20	18	17	04 16	04 16	35
30	04 55	04 52	04 49	04 46	04 44	04 42	04 40	38	36	35	33	32	32	31	31	30
N 20	05 10	05 08	05 06	05 04	05 03	05 01	05 00	04 59	04 58	04 57	04 56	04 56	04 55	04 55	04 55	N 20
N 10	22	21	20	19	18	17	16	05 16	05 15	05 15	05 15	05 15	05 15	05 15	05 15	N 10
0	33	32	32	31	31	31	31	31	31	31	31	32	32	33	33	0
S 10	43	43	43	43	44	44	44	45	05 46	05 46	05 47	05 48	05 48	05 49	05 50	S 10
20	05 52	05 53	05 54	05 55	05 56	05 57	05 58	05 59	06 00	06 02	06 03	06 04	06 05	06 06	06 07	20
S 30	06 03	06 04	06 06	06 08	06 10	06 11	06 13	06 15	06 16	06 18	06 20	06 21	06 23	06 24	06 25	S 30
35	08	10	12	15	17	19	21	23	25	27	29	31	32	34	36	35
40	14	17	20	22	25	27	30	33	35	37	40	42	44	45	06 47	40
45	21	24	28	31	34	37	40	43	46	06 49	06 52	06 54	06 56	06 58	07 00	45
50	28	33	37	41	45	49	52	06 56	06 59	07 03	07 06	07 09	07 11	07 14	16	50
S 52	06 32	06 37	06 41	06 45	06 50	06 54	06 58	07 02	07 06	07 09	07 12	07 16	07 18	07 21	07 23	S 52
54	36	41	46	50	06 55	06 59	07 04	08	12	16	20	23	26	29	31	54
56	40	45	51	06 56	07 01	07 06	10	15	19	24	27	31	34	37	40	56
58	44	50	06 56	07 02	07	13	18	23	28	32	36	40	44	47	07 50	58
S 60	06 49	06 56	07 02	07 08	07 14	07 20	07 26	07 31	07 37	07 42	07 46	07 51	07 55	07 58	08 01	S 60

EVENING CIVIL TWILIGHT

Lat.	Apr.	May										June				Lat.
	29	2	5	8	11	14	17	20	23	26	29	1	4	7	10	
°	h m	h m	h m	h m	h m	h m	h m	h m	h m	h m	h m	h m	h m	h m	h m	°
N 72	////	////	////	////	▭	▭	▭	▭	▭	▭	▭	▭	▭	▭	▭	N 72
70	////	////	////	////	////	////	▭	▭	▭	▭	▭	▭	▭	▭	▭	70
68	22 29	23 01	////	////	////	////	////	////	////	////	▭	▭	▭	▭	▭	68
66	21 48	22 06	22 26	22 51	23 33	////	////	////	////	////	////	////	////	////	////	66
64	21	21 34	21 48	22 03	22 20	22 39	23 02	23 50	////	////	////	////	////	////	////	64
62	21 00	21 11	22	21 33	21 45	21 58	22 11	22 25	22 40	22 57	23 18	////	////	////	////	62
N 60	20 43	20 53	21 02	21 11	21 21	21 30	21 40	21 50	22 00	22 10	22 20	22 30	22 39	22 48	22 56	N 60
58	30	38	20 45	20 54	21 02	21 10	18	26	21 34	21 41	21 48	21 55	22 02	22 08	22 13	58
56	18	25	32	39	20 46	20 53	21 00	21 07	21 13	20	26	31	21 36	21 41	21 45	56
54	20 08	14	20	26	33	39	20 45	20 51	20 56	21 02	21 07	21 12	16	20	23	54
52	19 59	20 04	10	16	21	27	32	37	42	20 47	20 52	20 56	21 00	21 03	21 06	52
N 50	19 51	19 56	20 01	20 06	20 11	20 16	20 21	20 26	20 30	20 34	20 38	20 42	20 46	20 49	20 51	N 50
45	34	38	19 42	19 46	19 50	19 54	19 58	20 02	20 05	20 09	20 12	20 15	20 18	20 20	23	45
40	21	24	28	31	34	37	40	19 43	19 46	19 49	19 52	19 54	19 57	19 59	20 00	40
35	10	13	15	18	21	23	26	28	31	33	35	37	39	41	19 43	35
30	19 01	19 03	19 05	19 07	19 09	19 11	19 13	19 15	19 17	19 19	19 21	23	25	26	28	30
N 20	18 46	18 47	18 48	18 49	18 50	18 52	18 53	18 54	18 56	18 57	18 58	19 00	19 01	19 02	19 03	N 20
N 10	33	33	34	34	35	36	36	37	38	39	40	18 41	18 41	18 42	18 43	N 10
0	22	22	22	22	22	22	22	22	22	23	23	24	24	25	25	0
S 10	12	11	18 10	18 09	18 09	18 08	18 08	18 08	18 07	18 07	18 07	18 08	18 08	18 08	18 08	S 10
20	18 02	18 01	17 59	17 58	17 56	17 55	17 54	17 53	17 53	17 52	17 52	17 51	17 51	17 51	17 52	20
S 30	17 52	17 49	17 47	17 45	17 43	17 41	17 39	17 38	17 37	17 36	17 35	17 34	17 33	17 33	17 33	S 30
35	46	43	41	38	35	33	31	29	28	26	25	24	24	23	23	35
40	40	37	34	30	27	25	22	20	18	16	15	13	12	17 12	17 11	40
45	33	29	25	22	18	15	12	17 09	17 07	17 04	17 03	17 01	17 00	16 59	16 58	45
50	26	21	16	12	07	17 03	17 00	16 57	16 54	16 51	16 48	16 46	16 45	43	43	50
S 52	17 22	17 17	17 12	17 07	17 02	16 58	16 54	16 51	16 47	16 44	16 42	16 40	16 38	16 36	16 35	S 52
54	18	13	07	17 02	16 57	52	48	44	41	37	34	32	30	28	27	54
56	14	08	17 02	16 56	51	46	42	37	33	30	27	24	22	20	18	56
58	10	17 03	16 57	51	45	39	34	29	25	21	18	15	12	16 10	16 08	58
S 60	17 05	16 57	16 50	16 44	16 38	16 32	16 26	16 21	16 16	16 12	16 08	16 04	16 01	15 59	15 57	S 60

RISING, SETTING AND DEPRESSION GRAPHS

TABLE 1—MERIDIAN PASSAGE AND DECLINATION OF THE SUN AT 12^h G.M.T.

Day	May Mer. Pass.	May Dec.	June Mer. Pass.	June Dec.	July Mer. Pass.	July Dec.	August Mer. Pass.	August Dec.
	h m	°	h m	°	h m	°	h m	°
1	11 57	N.15·0	11 58	N.22·0	12 04	N.23·1	12 06	N.18·1
2	11 57	15·3	11 58	22·1	12 04	23·1	12 06	17·9
3	11 57	15·6	11 58	22·3	12 04	23·0	12 06	17·6
4	11 57	15·9	11 58	22·4	12 04	22·9	12 06	17·3
5	11 57	16·2	11 58	22·5	12 04	22·8	12 06	17·1
6	11 57	N.16·4	11 59	N.22·6	12 05	N.22·7	12 06	N.16·8
7	11 57	16·7	11 59	22·7	12 05	22·6	12 06	16·5
8	11 56	17·0	11 59	22·8	12 05	22·5	12 06	16·2
9	11 56	17·3	11 59	22·9	12 05	22·4	12 06	16·0
10	11 56	17·5	11 59	23·0	12 05	22·3	12 05	15·7
11	11 56	N.17·8	11 59	N.23·1	12 05	N.22·2	12 05	N.15·4
12	11 56	18·0	12 00	23·1	12 06	22·0	12 05	15·1
13	11 56	18·3	12 00	23·2	12 06	21·9	12 05	14·8
14	11 56	18·5	12 00	23·2	12 06	21·7	12 05	14·5
15	11 56	18·8	12 00	23·3	12 06	21·6	12 05	14·2
16	11 56	N.19·0	12 01	N.23·3	12 06	N.21·4	12 04	N.13·8
17	11 56	19·3	12 01	23·4	12 06	21·3	12 04	13·5
18	11 56	19·5	12 01	23·4	12 06	21·1	12 04	13·2
19	11 56	19·7	12 01	23·4	12 06	20·9	12 04	12·9
20	11 56	19·9	12 01	23·4	12 06	20·7	12 03	12·6
21	11 57	N.20·1	12 02	N.23·4	12 06	N.20·5	12 03	N.12·2
22	11 57	20·3	12 02	23·4	12 06	20·3	12 03	11·9
23	11 57	20·5	12 02	23·4	12 06	20·1	12 03	11·6
24	11 57	20·7	12 02	23·4	12 06	19·9	12 02	11·2
25	11 57	20·9	12 02	23·4	12 06	19·7	12 02	10·9
26	11 57	N.21·1	12 03	N.23·4	12 06	N.19·5	12 02	N.10·5
27	11 57	21·2	12 03	23·3	12 06	19·3	12 02	10·2
28	11 57	21·4	12 03	23·3	12 06	19·1	12 01	9·8
29	11 57	21·6	12 03	23·2	12 06	18·8	12 01	9·5
30	11 57	21·7	12 03	N.23·2	12 06	18·6	12 01	9·1
31	11 58	N.21·9			12 06	N.18·4	12 00	N. 8·8

TABLES 2 and 3—DEPRESSION OF THE SUN AT VARIOUS HEIGHTS

Height	TABLE 2 AT SUNRISE AND SUNSET		TABLE 3 AT CIVIL TWILIGHT	
Feet	De-pression	Diff. from 0°·8	De-pression	Diff. from 0°·8
	°	°	°	°
0	0·8	—	6·0	5·2
500	1·3	0·5	6·0	5·2
1 000	1·5	0·7	6·0	5·2
2 000	1·7	0·9	6·1	5·3
3 000	1·9	1·1	6·1	5·3
4 000	2·1	1·3	6·1	5·3
5 000	2·2	1·4	6·2	5·4
6 000	2·4	1·6	6·2	5·4
7 000	2·5	1·7	6·2	5·4
8 000	2·6	1·8	6·3	5·5
9 000	2·7	1·9	6·3	5·5
10 000	2·8	2·0	6·3	5·5
15 000	3·2	2·4	6·5	5·7
20 000	3·6	2·8	6·6	5·8
25 000	3·9	3·1	6·8	6·0
30 000	4·2	3·4	6·9	6·1
35 000	4·4	3·6	7·1	6·3
40 000	4·7	3·9	7·2	6·4
45 000	4·9	4·1	7·3	6·5
50 000	5·1	4·3	7·5	6·7
55 000	5·3	4·5	7·6	6·8
60 000	5·5	4·7	7·7	6·9

An alternative method to those given on pages A12–A14 is to use the graphs to give the corrections to the tabulated times of sunrise and sunset at ground level; in this case it is adequate to use the graphs for the *nearest* tabular latitude and declination. The difference in hour angle is found between the hour angle for zero depression and the hour angle at the tabular depression minus 0°·8. The difference in hour angle so found is then applied to the time of sunrise or sunset. The result will be less than 5^m in error if the declination curve cuts all the depression lines.

Example. To find the times of sunrise and sunset on 1975 August 23 in latitude N. 65° 17′, longitude W. 35° 15′, at a height of 37 000 feet. From Table 1, Dec.= N. 11°·6; Table 2, Depression diff. from 0°·8=3°·7.

	Sunrise	Sunset
	h m	h m
Page A68, N. 65° 17′ (August 24)	04 10	19 53
Page A71, Lat. 66°, Dec. 11° (same); diff. in H.A. from depression 0° to 3°·7	45	45
L.M.T.	03 25	20 38
Longitude W. 35° 15′	2 21	2 21
G.M.T.	05 46	22 59

SEMIDURATION OF SUNLIGHT

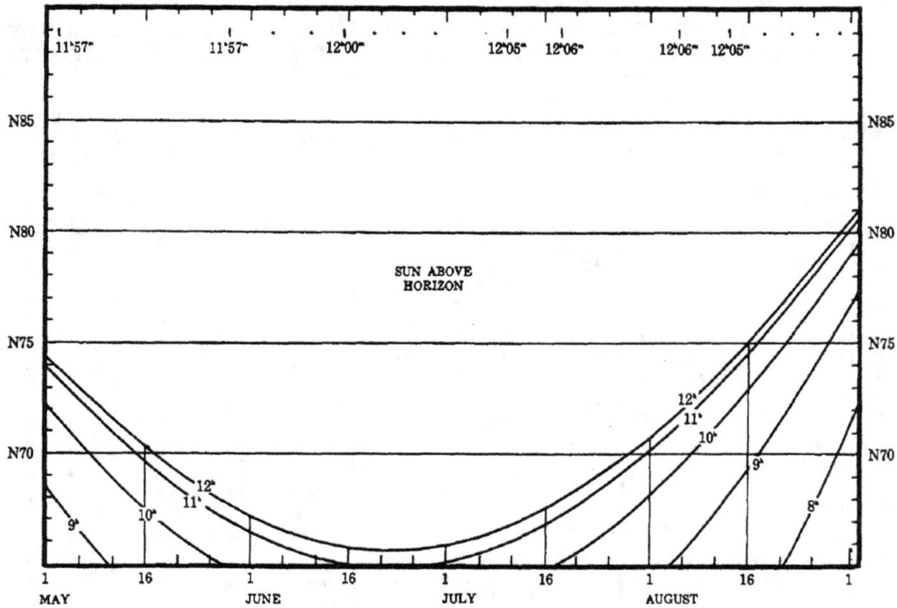

DURATION OF TWILIGHT

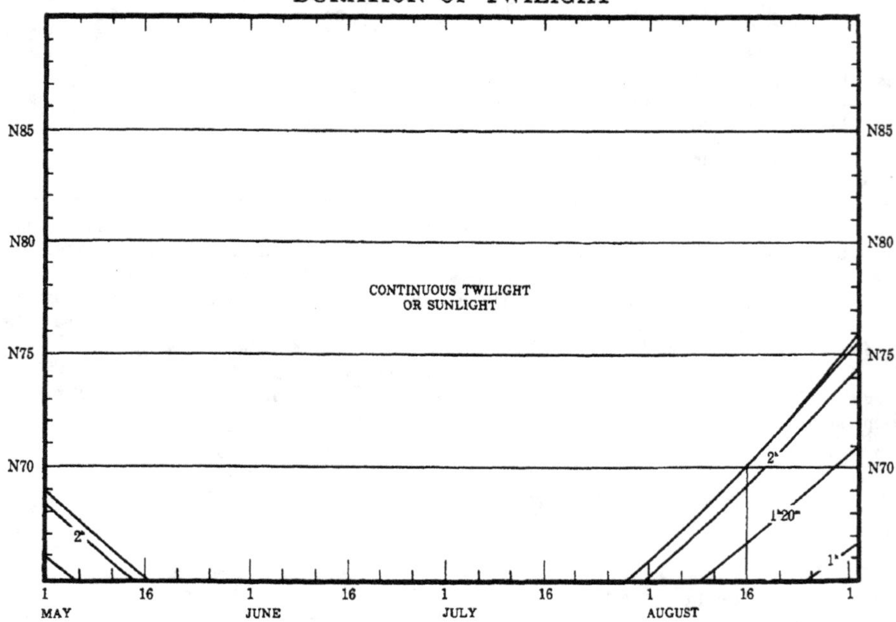

SEMIDURATION OF MOONLIGHT

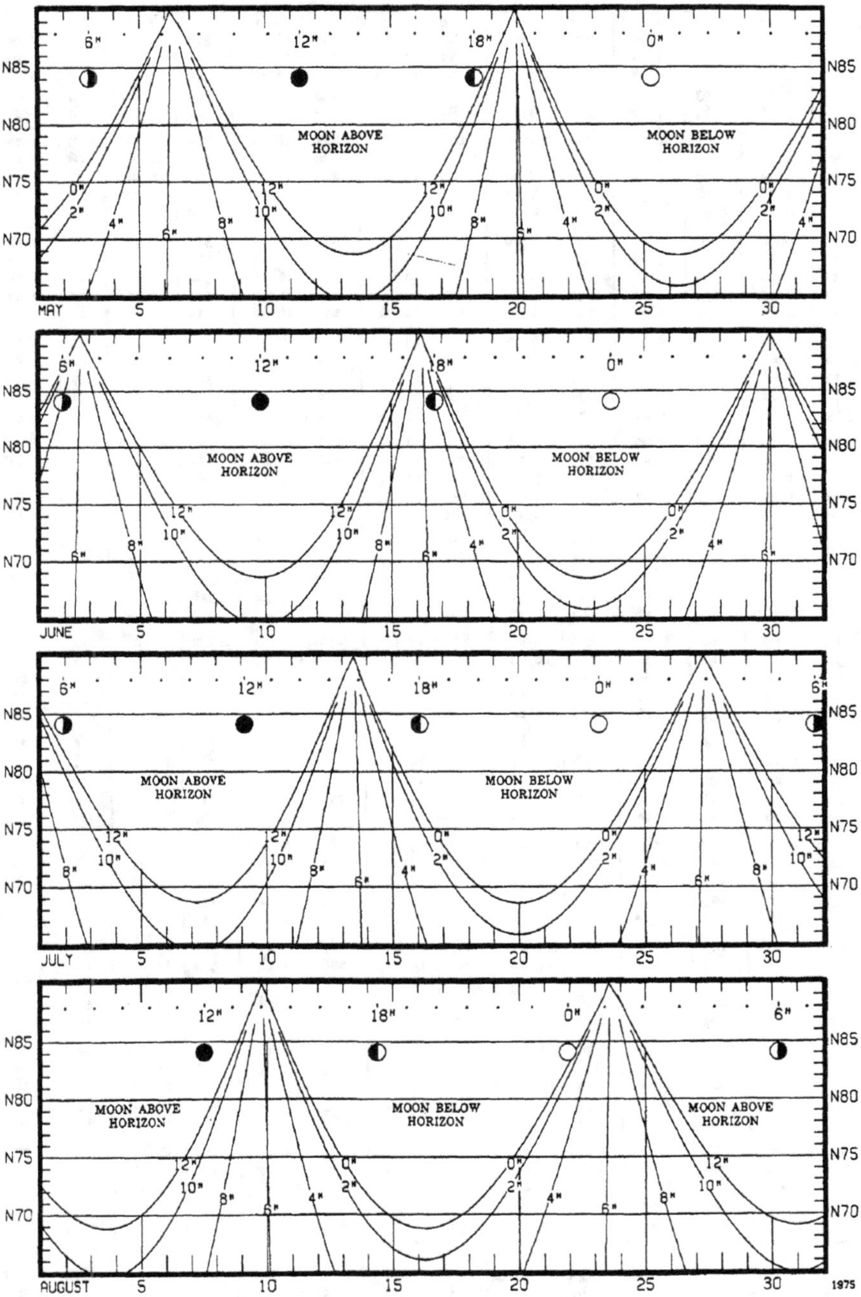

CONVERSION OF ARC TO TIME

°	h m	°	h m	°	h m	°	h m	°	h m	°	h m	′	m s
0	0 00	60	4 00	120	8 00	180	12 00	240	16 00	300	20 00	0	0 00
1	0 04	61	4 04	121	8 04	181	12 04	241	16 04	301	20 04	1	0 04
2	0 08	62	4 08	122	8 08	182	12 08	242	16 08	302	20 08	2	0 08
3	0 12	63	4 12	123	8 12	183	12 12	243	16 12	303	20 12	3	0 12
4	0 16	64	4 16	124	8 16	184	12 16	244	16 16	304	20 16	4	0 16
5	0 20	65	4 20	125	8 20	185	12 20	245	16 20	305	20 20	5	0 20
6	0 24	66	4 24	126	8 24	186	12 24	246	16 24	306	20 24	6	0 24
7	0 28	67	4 28	127	8 28	187	12 28	247	16 28	307	20 28	7	0 28
8	0 32	68	4 32	128	8 32	188	12 32	248	16 32	308	20 32	8	0 32
9	0 36	69	4 36	129	8 36	189	12 36	249	16 36	309	20 36	9	0 36
10	0 40	70	4 40	130	8 40	190	12 40	250	16 40	310	20 40	10	0 40
11	0 44	71	4 44	131	8 44	191	12 44	251	16 44	311	20 44	11	0 44
12	0 48	72	4 48	132	8 48	192	12 48	252	16 48	312	20 48	12	0 48
13	0 52	73	4 52	133	8 52	193	12 52	253	16 52	313	20 52	13	0 52
14	0 56	74	4 56	134	8 56	194	12 56	254	16 56	314	20 56	14	0 56
15	1 00	75	5 00	135	9 00	195	13 00	255	17 00	315	21 00	15	1 00
16	1 04	76	5 04	136	9 04	196	13 04	256	17 04	316	21 04	16	1 04
17	1 08	77	5 08	137	9 08	197	13 08	257	17 08	317	21 08	17	1 08
18	1 12	78	5 12	138	9 12	198	13 12	258	17 12	318	21 12	18	1 12
19	1 16	79	5 16	139	9 16	199	13 16	259	17 16	319	21 16	19	1 16
20	1 20	80	5 20	140	9 20	200	13 20	260	17 20	320	21 20	20	1 20
21	1 24	81	5 24	141	9 24	201	13 24	261	17 24	321	21 24	21	1 24
22	1 28	82	5 28	142	9 28	202	13 28	262	17 28	322	21 28	22	1 28
23	1 32	83	5 32	143	9 32	203	13 32	263	17 32	323	21 32	23	1 32
24	1 36	84	5 36	144	9 36	204	13 36	264	17 36	324	21 36	24	1 36
25	1 40	85	5 40	145	9 40	205	13 40	265	17 40	325	21 40	25	1 40
26	1 44	86	5 44	146	9 44	206	13 44	266	17 44	326	21 44	26	1 44
27	1 48	87	5 48	147	9 48	207	13 48	267	17 48	327	21 48	27	1 48
28	1 52	88	5 52	148	9 52	208	13 52	268	17 52	328	21 52	28	1 52
29	1 56	89	5 56	149	9 56	209	13 56	269	17 56	329	21 56	29	1 56
30	2 00	90	6 00	150	10 00	210	14 00	270	18 00	330	22 00	30	2 00
31	2 04	91	6 04	151	10 04	211	14 04	271	18 04	331	22 04	31	2 04
32	2 08	92	6 08	152	10 08	212	14 08	272	18 08	332	22 08	32	2 08
33	2 12	93	6 12	153	10 12	213	14 12	273	18 12	333	22 12	33	2 12
34	2 16	94	6 16	154	10 16	214	14 16	274	18 16	334	22 16	34	2 16
35	2 20	95	6 20	155	10 20	215	14 20	275	18 20	335	22 20	35	2 20
36	2 24	96	6 24	156	10 24	216	14 24	276	18 24	336	22 24	36	2 24
37	2 28	97	6 28	157	10 28	217	14 28	277	18 28	337	22 28	37	2 28
38	2 32	98	6 32	158	10 32	218	14 32	278	18 32	338	22 32	38	2 32
39	2 36	99	6 36	159	10 36	219	14 36	279	18 36	339	22 36	39	2 36
40	2 40	100	6 40	160	10 40	220	14 40	280	18 40	340	22 40	40	2 40
41	2 44	101	6 44	161	10 44	221	14 44	281	18 44	341	22 44	41	2 44
42	2 48	102	6 48	162	10 48	222	14 48	282	18 48	342	22 48	42	2 48
43	2 52	103	6 52	163	10 52	223	14 52	283	18 52	343	22 52	43	2 52
44	2 56	104	6 56	164	10 56	224	14 56	284	18 56	344	22 56	44	2 56
45	3 00	105	7 00	165	11 00	225	15 00	285	19 00	345	23 00	45	3 00
46	3 04	106	7 04	166	11 04	226	15 04	286	19 04	346	23 04	46	3 04
47	3 08	107	7 08	167	11 08	227	15 08	287	19 08	347	23 08	47	3 08
48	3 12	108	7 12	168	11 12	228	15 12	288	19 12	348	23 12	48	3 12
49	3 16	109	7 16	169	11 16	229	15 16	289	19 16	349	23 16	49	3 16
50	3 20	110	7 20	170	11 20	230	15 20	290	19 20	350	23 20	50	3 20
51	3 24	111	7 24	171	11 24	231	15 24	291	19 24	351	23 24	51	3 24
52	3 28	112	7 28	172	11 28	232	15 28	292	19 28	352	23 28	52	3 28
53	3 32	113	7 32	173	11 32	233	15 32	293	19 32	353	23 32	53	3 32
54	3 36	114	7 36	174	11 36	234	15 36	294	19 36	354	23 36	54	3 36
55	3 40	115	7 40	175	11 40	235	15 40	295	19 40	355	23 40	55	3 40
56	3 44	116	7 44	176	11 44	236	15 44	296	19 44	356	23 44	56	3 44
57	3 48	117	7 48	177	11 48	237	15 48	297	19 48	357	23 48	57	3 48
58	3 52	118	7 52	178	11 52	238	15 52	298	19 52	358	23 52	58	3 52
59	3 56	119	7 56	179	11 56	239	15 56	299	19 56	359	23 56	59	3 56

The above table is for converting expressions in arc to their equivalent in time; its main use in this Almanac is for the conversion of longitude for application to L.M.T. (*added* if *west*, *subtracted* if *east*) to give G.M.T., or vice versa, particularly in the case of sunrise, sunset, etc.

INTERPOLATION OF G.H.A. SUN

	0m	1m	2m	3m	4m	5m	6m	7m	8m	9m	
s	° ′	° ′	° ′	° ′	° ′	° ′	° ′	° ′	° ′	° ′	s
00	0 00.0	0 15.0	0 30.0	0 45.0	1 00.0	1 15.0	1 30.0	1 45.0	2 00.0	2 15.0	00
01	0 00.3	0 15.3	0 30.3	0 45.3	1 00.3	1 15.3	1 30.3	1 45.3	2 00.3	2 15.3	01
02	0 00.5	0 15.5	0 30.5	0 45.5	1 00.5	1 15.5	1 30.5	1 45.5	2 00.5	2 15.5	02
03	0 00.8	0 15.8	0 30.8	0 45.8	1 00.8	1 15.8	1 30.8	1 45.8	2 00.8	2 15.8	03
04	0 01.0	0 16.0	0 31.0	0 46.0	1 01.0	1 16.0	1 31.0	1 46.0	2 01.0	2 16.0	04
05	0 01.3	0 16.3	0 31.3	0 46.3	1 01.3	1 16.3	1 31.3	1 46.3	2 01.3	2 16.3	05
06	0 01.5	0 16.5	0 31.5	0 46.5	1 01.5	1 16.5	1 31.5	1 46.5	2 01.5	2 16.5	06
07	0 01.8	0 16.8	0 31.8	0 46.8	1 01.8	1 16.8	1 31.8	1 46.8	2 01.8	2 16.8	07
08	0 02.0	0 17.0	0 32.0	0 47.0	1 02.0	1 17.0	1 32.0	1 47.0	2 02.0	2 17.0	08
09	0 02.3	0 17.3	0 32.3	0 47.3	1 02.3	1 17.3	1 32.3	1 47.3	2 02.3	2 17.3	09
10	0 02.5	0 17.5	0 32.5	0 47.5	1 02.5	1 17.5	1 32.5	1 47.5	2 02.5	2 17.5	10
11	0 02.8	0 17.8	0 32.8	0 47.8	1 02.8	1 17.8	1 32.8	1 47.8	2 02.8	2 17.8	11
12	0 03.0	0 18.0	0 33.0	0 48.0	1 03.0	1 18.0	1 33.0	1 48.0	2 03.0	2 18.0	12
13	0 03.3	0 18.3	0 33.3	0 48.3	1 03.3	1 18.3	1 33.3	1 48.3	2 03.3	2 18.3	13
14	0 03.5	0 18.5	0 33.5	0 48.5	1 03.5	1 18.5	1 33.5	1 48.5	2 03.5	2 18.5	14
15	0 03.8	0 18.8	0 33.8	0 48.8	1 03.8	1 18.8	1 33.8	1 48.8	2 03.8	2 18.8	15
16	0 04.0	0 19.0	0 34.0	0 49.0	1 04.0	1 19.0	1 34.0	1 49.0	2 04.0	2 19.0	16
17	0 04.3	0 19.3	0 34.3	0 49.3	1 04.3	1 19.3	1 34.3	1 49.3	2 04.3	2 19.3	17
18	0 04.5	0 19.5	0 34.5	0 49.5	1 04.5	1 19.5	1 34.5	1 49.5	2 04.5	2 19.5	18
19	0 04.8	0 19.8	0 34.8	0 49.8	1 04.8	1 19.8	1 34.8	1 49.8	2 04.8	2 19.8	19
20	0 05.0	0 20.0	0 35.0	0 50.0	1 05.0	1 20.0	1 35.0	1 50.0	2 05.0	2 20.0	20
21	0 05.3	0 20.3	0 35.3	0 50.3	1 05.3	1 20.3	1 35.3	1 50.3	2 05.3	2 20.3	21
22	0 05.5	0 20.5	0 35.5	0 50.5	1 05.5	1 20.5	1 35.5	1 50.5	2 05.5	2 20.5	22
23	0 05.8	0 20.8	0 35.8	0 50.8	1 05.8	1 20.8	1 35.8	1 50.8	2 05.8	2 20.8	23
24	0 06.0	0 21.0	0 36.0	0 51.0	1 06.0	1 21.0	1 36.0	1 51.0	2 06.0	2 21.0	24
25	0 06.3	0 21.3	0 36.3	0 51.3	1 06.3	1 21.3	1 36.3	1 51.3	2 06.3	2 21.3	25
26	0 06.5	0 21.5	0 36.5	0 51.5	1 06.5	1 21.5	1 36.5	1 51.5	2 06.5	2 21.5	26
27	0 06.8	0 21.8	0 36.8	0 51.8	1 06.8	1 21.8	1 36.8	1 51.8	2 06.8	2 21.8	27
28	0 07.0	0 22.0	0 37.0	0 52.0	1 07.0	1 22.0	1 37.0	1 52.0	2 07.0	2 22.0	28
29	0 07.3	0 22.3	0 37.3	0 52.3	1 07.3	1 22.3	1 37.3	1 52.3	2 07.3	2 22.3	29
30	0 07.5	0 22.5	0 37.5	0 52.5	1 07.5	1 22.5	1 37.5	1 52.5	2 07.5	2 22.5	30
31	0 07.8	0 22.8	0 37.8	0 52.8	1 07.8	1 22.8	1 37.8	1 52.8	2 07.8	2 22.8	31
32	0 08.0	0 23.0	0 38.0	0 53.0	1 08.0	1 23.0	1 38.0	1 53.0	2 08.0	2 23.0	32
33	0 08.3	0 23.3	0 38.3	0 53.3	1 08.3	1 23.3	1 38.3	1 53.3	2 08.3	2 23.3	33
34	0 08.5	0 23.5	0 38.5	0 53.5	1 08.5	1 23.5	1 38.5	1 53.5	2 08.5	2 23.5	34
35	0 08.8	0 23.8	0 38.8	0 53.8	1 08.8	1 23.8	1 38.8	1 53.8	2 08.8	2 23.8	35
36	0 09.0	0 24.0	0 39.0	0 54.0	1 09.0	1 24.0	1 39.0	1 54.0	2 09.0	2 24.0	36
37	0 09.3	0 24.3	0 39.3	0 54.3	1 09.3	1 24.3	1 39.3	1 54.3	2 09.3	2 24.3	37
38	0 09.5	0 24.5	0 39.5	0 54.5	1 09.5	1 24.5	1 39.5	1 54.5	2 09.5	2 24.5	38
39	0 09.8	0 24.8	0 39.8	0 54.8	1 09.8	1 24.8	1 39.8	1 54.8	2 09.8	2 24.8	39
40	0 10.0	0 25.0	0 40.0	0 55.0	1 10.0	1 25.0	1 40.0	1 55.0	2 10.0	2 25.0	40
41	0 10.3	0 25.3	0 40.3	0 55.3	1 10.3	1 25.3	1 40.3	1 55.3	2 10.3	2 25.3	41
42	0 10.5	0 25.5	0 40.5	0 55.5	1 10.5	1 25.5	1 40.5	1 55.5	2 10.5	2 25.5	42
43	0 10.8	0 25.8	0 40.8	0 55.8	1 10.8	1 25.8	1 40.8	1 55.8	2 10.8	2 25.8	43
44	0 11.0	0 26.0	0 41.0	0 56.0	1 11.0	1 26.0	1 41.0	1 56.0	2 11.0	2 26.0	44
45	0 11.3	0 26.3	0 41.3	0 56.3	1 11.3	1 26.3	1 41.3	1 56.3	2 11.3	2 26.3	45
46	0 11.5	0 26.5	0 41.5	0 56.5	1 11.5	1 26.5	1 41.5	1 56.5	2 11.5	2 26.5	46
47	0 11.8	0 26.8	0 41.8	0 56.8	1 11.8	1 26.8	1 41.8	1 56.8	2 11.8	2 26.8	47
48	0 12.0	0 27.0	0 42.0	0 57.0	1 12.0	1 27.0	1 42.0	1 57.0	2 12.0	2 27.0	48
49	0 12.3	0 27.3	0 42.3	0 57.3	1 12.3	1 27.3	1 42.3	1 57.3	2 12.3	2 27.3	49
50	0 12.5	0 27.5	0 42.5	0 57.5	1 12.5	1 27.5	1 42.5	1 57.5	2 12.5	2 27.5	50
51	0 12.8	0 27.8	0 42.8	0 57.8	1 12.8	1 27.8	1 42.8	1 57.8	2 12.8	2 27.8	51
52	0 13.0	0 28.0	0 43.0	0 58.0	1 13.0	1 28.0	1 43.0	1 58.0	2 13.0	2 28.0	52
53	0 13.3	0 28.3	0 43.3	0 58.3	1 13.3	1 28.3	1 43.3	1 58.3	2 13.3	2 28.3	53
54	0 13.5	0 28.5	0 43.5	0 58.5	1 13.5	1 28.5	1 43.5	1 58.5	2 13.5	2 28.5	54
55	0 13.8	0 28.8	0 43.8	0 58.8	1 13.8	1 28.8	1 43.8	1 58.8	2 13.8	2 28.8	55
56	0 14.0	0 29.0	0 44.0	0 59.0	1 14.0	1 29.0	1 44.0	1 59.0	2 14.0	2 29.0	56
57	0 14.3	0 29.3	0 44.3	0 59.3	1 14.3	1 29.3	1 44.3	1 59.3	2 14.3	2 29.3	57
58	0 14.5	0 29.5	0 44.5	0 59.5	1 14.5	1 29.5	1 44.5	1 59.5	2 14.5	2 29.5	58
59	0 14.8	0 29.8	0 44.8	0 59.8	1 14.8	1 29.8	1 44.8	1 59.8	2 14.8	2 29.8	59
60	0 15.0	0 30.0	0 45.0	1 00.0	1 15.0	1 30.0	1 45.0	2 00.0	2 15.0	2 30.0	60

INTERPOLATION OF G.H.A. ARIES

s	0m	1m	2m	3m	4m	5m	6m	7m	8m	9m	s
	° ′	° ′	° ′	° ′	° ′	° ′	° ′	° ′	° ′	° ′	
00	0 00·0	0 15·0	0 30·1	0 45·1	1 00·2	1 15·2	1 30·2	1 45·3	2 00·3	2 15·4	00
01	0 00·3	0 15·3	0 30·3	0 45·4	1 00·4	1 15·5	1 30·5	1 45·5	2 00·6	2 15·6	01
02	0 00·5	0 15·5	0 30·6	0 45·6	1 00·7	1 15·7	1 30·7	1 45·8	2 00·8	2 15·9	02
03	0 00·8	0 15·8	0 30·8	0 45·9	1 00·9	1 16·0	1 31·0	1 46·0	2 01·1	2 16·1	03
04	0 01·0	0 16·0	0 31·1	0 46·1	1 01·2	1 16·2	1 31·2	1 46·3	2 01·3	2 16·4	04
05	0 01·3	0 16·3	0 31·3	0 46·4	1 01·4	1 16·5	1 31·5	1 46·5	2 01·6	2 16·6	05
06	0 01·5	0 16·5	0 31·6	0 46·6	1 01·7	1 16·7	1 31·8	1 46·8	2 01·8	2 16·9	06
07	0 01·8	0 16·8	0 31·8	0 46·9	1 01·9	1 17·0	1 32·0	1 47·0	2 02·1	2 17·1	07
08	0 02·0	0 17·0	0 32·1	0 47·1	1 02·2	1 17·2	1 32·3	1 47·3	2 02·3	2 17·4	08
09	0 02·3	0 17·3	0 32·3	0 47·4	1 02·4	1 17·5	1 32·5	1 47·5	2 02·6	2 17·6	09
10	0 02·5	0 17·5	0 32·6	0 47·6	1 02·7	1 17·7	1 32·8	1 47·8	2 02·8	2 17·9	10
11	0 02·8	0 17·8	0 32·8	0 47·9	1 02·9	1 18·0	1 33·0	1 48·0	2 03·1	2 18·1	11
12	0 03·0	0 18·0	0 33·1	0 48·1	1 03·2	1 18·2	1 33·3	1 48·3	2 03·3	2 18·4	12
13	0 03·3	0 18·3	0 33·3	0 48·4	1 03·4	1 18·5	1 33·5	1 48·5	2 03·6	2 18·6	13
14	0 03·5	0 18·6	0 33·6	0 48·6	1 03·7	1 18·7	1 33·8	1 48·8	2 03·8	2 18·9	14
15	0 03·8	0 18·8	0 33·8	0 48·9	1 03·9	1 19·0	1 34·0	1 49·0	2 04·1	2 19·1	15
16	0 04·0	0 19·1	0 34·1	0 49·1	1 04·2	1 19·2	1 34·3	1 49·3	2 04·3	2 19·4	16
17	0 04·3	0 19·3	0 34·3	0 49·4	1 04·4	1 19·5	1 34·5	1 49·5	2 04·6	2 19·6	17
18	0 04·5	0 19·6	0 34·6	0 49·6	1 04·7	1 19·7	1 34·8	1 49·8	2 04·8	2 19·9	18
19	0 04·8	0 19·8	0 34·8	0 49·9	1 04·9	1 20·0	1 35·0	1 50·1	2 05·1	2 20·1	19
20	0 05·0	0 20·1	0 35·1	0 50·1	1 05·2	1 20·2	1 35·3	1 50·3	2 05·3	2 20·4	20
21	0 05·3	0 20·3	0 35·3	0 50·4	1 05·4	1 20·5	1 35·5	1 50·6	2 05·6	2 20·6	21
22	0 05·5	0 20·6	0 35·6	0 50·6	1 05·7	1 20·7	1 35·8	1 50·8	2 05·8	2 20·9	22
23	0 05·8	0 20·8	0 35·8	0 50·9	1 05·9	1 21·0	1 36·0	1 51·1	2 06·1	2 21·1	23
24	0 06·0	0 21·1	0 36·1	0 51·1	1 06·2	1 21·2	1 36·3	1 51·3	2 06·3	2 21·4	24
25	0 06·3	0 21·3	0 36·3	0 51·4	1 06·4	1 21·5	1 36·5	1 51·6	2 06·6	2 21·6	25
26	0 06·5	0 21·6	0 36·6	0 51·6	1 06·7	1 21·7	1 36·8	1 51·8	2 06·8	2 21·9	26
27	0 06·8	0 21·8	0 36·9	0 51·9	1 06·9	1 22·0	1 37·0	1 52·1	2 07·1	2 22·1	27
28	0 07·0	0 22·1	0 37·1	0 52·1	1 07·2	1 22·2	1 37·3	1 52·3	2 07·3	2 22·4	28
29	0 07·3	0 22·3	0 37·4	0 52·4	1 07·4	1 22·5	1 37·5	1 52·6	2 07·6	2 22·6	29
30	0 07·5	0 22·6	0 37·6	0 52·6	1 07·7	1 22·7	1 37·8	1 52·8	2 07·8	2 22·9	30
31	0 07·8	0 22·8	0 37·9	0 52·9	1 07·9	1 23·0	1 38·0	1 53·1	2 08·1	2 23·1	31
32	0 08·0	0 23·1	0 38·1	0 53·1	1 08·2	1 23·2	1 38·3	1 53·3	2 08·4	2 23·4	32
33	0 08·3	0 23·3	0 38·4	0 53·4	1 08·4	1 23·5	1 38·5	1 53·6	2 08·6	2 23·6	33
34	0 08·5	0 23·6	0 38·6	0 53·6	1 08·7	1 23·7	1 38·8	1 53·8	2 08·9	2 23·9	34
35	0 08·8	0 23·8	0 38·9	0 53·9	1 08·9	1 24·0	1 39·0	1 54·1	2 09·1	2 24·1	35
36	0 09·0	0 24·1	0 39·1	0 54·1	1 09·2	1 24·2	1 39·3	1 54·3	2 09·4	2 24·4	36
37	0 09·3	0 24·3	0 39·4	0 54·4	1 09·4	1 24·5	1 39·5	1 54·6	2 09·6	2 24·6	37
38	0 09·5	0 24·6	0 39·6	0 54·6	1 09·7	1 24·7	1 39·8	1 54·8	2 09·9	2 24·9	38
39	0 09·8	0 24·8	0 39·9	0 54·9	1 09·9	1 25·0	1 40·0	1 55·1	2 10·1	2 25·1	39
40	0 10·0	0 25·1	0 40·1	0 55·2	1 10·2	1 25·2	1 40·3	1 55·3	2 10·4	2 25·4	40
41	0 10·3	0 25·3	0 40·4	0 55·4	1 10·4	1 25·5	1 40·5	1 55·6	2 10·6	2 25·6	41
42	0 10·5	0 25·6	0 40·6	0 55·7	1 10·7	1 25·7	1 40·8	1 55·8	2 10·9	2 25·9	42
43	0 10·8	0 25·8	0 40·9	0 55·9	1 10·9	1 26·0	1 41·0	1 56·1	2 11·1	2 26·1	43
44	0 11·0	0 26·1	0 41·1	0 56·2	1 11·2	1 26·2	1 41·3	1 56·3	2 11·4	2 26·4	44
45	0 11·3	0 26·3	0 41·4	0 56·4	1 11·4	1 26·5	1 41·5	1 56·6	2 11·6	2 26·7	45
46	0 11·5	0 26·6	0 41·6	0 56·7	1 11·7	1 26·7	1 41·8	1 56·8	2 11·9	2 26·9	46
47	0 11·8	0 26·8	0 41·9	0 56·9	1 11·9	1 27·0	1 42·0	1 57·1	2 12·1	2 27·2	47
48	0 12·0	0 27·1	0 42·1	0 57·2	1 12·2	1 27·2	1 42·3	1 57·3	2 12·4	2 27·4	48
49	0 12·3	0 27·3	0 42·4	0 57·4	1 12·4	1 27·5	1 42·5	1 57·6	2 12·6	2 27·7	49
50	0 12·5	0 27·6	0 42·6	0 57·7	1 12·7	1 27·7	1 42·8	1 57·8	2 12·9	2 27·9	50
51	0 12·8	0 27·8	0 42·9	0 57·9	1 12·9	1 28·0	1 43·0	1 58·1	2 13·1	2 28·2	51
52	0 13·0	0 28·1	0 43·1	0 58·2	1 13·2	1 28·2	1 43·3	1 58·3	2 13·4	2 28·4	52
53	0 13·3	0 28·3	0 43·4	0 58·4	1 13·5	1 28·5	1 43·5	1 58·6	2 13·6	2 28·7	53
54	0 13·5	0 28·6	0 43·6	0 58·7	1 13·7	1 28·7	1 43·8	1 58·8	2 13·9	2 28·9	54
55	0 13·8	0 28·8	0 43·9	0 58·9	1 14·0	1 29·0	1 44·0	1 59·1	2 14·1	2 29·2	55
56	0 14·0	0 29·1	0 44·1	0 59·2	1 14·2	1 29·2	1 44·3	1 59·3	2 14·4	2 29·4	56
57	0 14·3	0 29·3	0 44·4	0 59·4	1 14·5	1 29·5	1 44·5	1 59·6	2 14·6	2 29·7	57
58	0 14·5	0 29·6	0 44·6	0 59·7	1 14·7	1 29·7	1 44·8	1 59·8	2 14·9	2 29·9	58
59	0 14·8	0 29·8	0 44·9	0 59·9	1 15·0	1 30·0	1 45·0	2 00·1	2 15·1	2 30·2	59
60	0 15·0	0 30·1	0 45·1	1 00·2	1 15·2	1 30·2	1 45·3	2 00·3	2 15·4	2 30·4	60

POLARIS (POLE STAR) TABLE, 1975

FOR DETERMINING THE LATITUDE FROM A SEXTANT ALTITUDE

L.H.A. ♈ ° ′	Q ′	L.H.A. ♈ ° ′	Q ′	L.H.A. ♈ ° ′	Q ′	L.H.A. ♈ ° ′	Q ′	L.H.A. ♈ ° ′	Q ′	L.H.A. ♈ ° ′	Q ′	L.H.A. ♈ ° ′	Q ′	L.H.A. ♈ ° ′	Q ′
359 08	−43	80 32	−33	113 11	− 7	143 05	+19	183 04	+45	264 07	+31	296 18	+ 5	326 18	−21
1 14	−44	82 00	−32	114 20	− 6	144 18	+20	185 31	+46	265 33	+30	297 26	+ 4	327 32	−22
3 29	−45	83 27	−31	115 28	− 5	145 32	+21	188 12	+47	266 57	+29	298 34	+ 3	328 47	−23
5 54	−46	84 52	−30	116 36	− 4	146 46	+22	191 12	+48	268 20	+28	299 43	+ 2	330 03	−24
8 32	−47	86 15	−29	117 44	− 3	148 01	+23	194 41	+49	269 41	+27	300 50	+ 1	331 19	−25
11 30	−48	87 38	−28	118 51	− 2	149 18	+24	199 02	+50	271 01	+26	301 58	0	332 37	−26
14 56	−49	88 58	−27	119 59	− 1	150 35	+25	205 51	+51	272 20	+25	303 06	− 1	333 55	−27
19 13	−50	90 18	−26	121 07	0	151 53	+26	218 22	+50	273 38	+24	304 14	− 2	335 15	−28
25 56	−51	91 36	−25	122 15	+ 1	153 12	+27	225 11	+49	274 55	+23	305 22	− 3	336 35	−29
38 17	−50	92 54	−24	123 23	+ 2	154 32	+28	229 32	+48	276 12	+22	306 29	− 4	337 58	−30
45 00	−49	94 10	−23	124 30	+ 3	155 53	+29	233 01	+47	277 27	+21	307 37	− 5	339 21	−31
49 17	−48	95 26	−22	125 39	+ 4	157 16	+30	236 01	+46	278 41	+20	308 45	− 6	340 46	−32
52 43	−47	96 41	−21	126 47	+ 5	158 40	+31	238 42	+45	279 55	+19	309 53	− 7	342 13	−33
55 41	−46	97 55	−20	127 55	+ 6	160 06	+32	241 09	+44	281 08	+18	311 02	− 8	343 41	−34
58 19	−45	99 08	−19	129 03	+ 7	161 34	+33	243 25	+43	282 21	+17	312 10	− 9	345 12	−35
60 44	−44	100 21	−18	130 11	+ 8	163 03	+34	245 33	+42	283 33	+16	313 19	−10	346 45	−36
62 59	−43	101 33	−17	131 20	+ 9	164 35	+35	247 34	+41	284 44	+15	314 27	−11	348 20	−37
65 05	−42	102 45	−16	132 29	+10	166 09	+36	249 29	+40	285 55	+14	315 37	−12	349 59	−38
67 05	−41	103 56	−15	133 38	+11	167 45	+37	251 20	+39	287 06	+13	316 46	−13	351 40	−39
68 59	−40	105 07	−14	134 47	+12	169 24	+38	253 06	+38	288 16	+12	317 56	−14	353 25	−40
70 48	−39	106 17	−13	135 57	+13	171 07	+39	254 49	+37	289 26	+11	319 06	−15	355 14	−41
72 33	−38	107 27	−12	137 07	+14	172 53	+40	256 28	+36	290 35	+10	320 17	−16	357 08	−42
74 14	−37	108 36	−11	138 18	+15	174 44	+41	258 04	+35	291 44	+ 9	321 28	−17	359 08	−43
75 53	−36	109 46	−10	139 29	+16	176 39	+42	259 38	+34	292 53	+ 8	322 40	−18	1 14	−44
77 28	−35	110 54	− 9	140 40	+17	178 40	+43	261 10	+33	294 02	+ 7	323 52	−19	3 29	−45
79 01	−34	112 03	− 8	141 52	+18	180 48	+44	262 39	+32	295 10	+ 6	325 05	−20	5 54	−46
80 32		113 11		143 05		183 04		264 07		296 18		326 18		8 32	

Q, which does *not* include refraction, is to be applied to the corrected sextant altitude of *Polaris*.
Polaris: Mag. 2·1, S.H.A. 327° 53′, Dec. N. 89° 09′·2

STANDARD DOME REFRACTION

To be *subtracted* from sextant altitude when using sextant suspension in a perspex dome

Alt. °	Refn. ′	Alt. °	Refn. ′
10	8	50	4
20	7	60	4
30	6	70	3
40	5	80	3

This table must not be used if a calibration table is fitted to the dome, or if a flat glass plate is provided, or for non-standard domes.

BUBBLE SEXTANT ERROR

Sextant Number	Alt. °	Corr. ′

AZIMUTH OF *POLARIS*

L.H.A. ♈ 300°–120° °	Latitude 0° °	30° °	50° °	55° °	60° °	65° °	70° °	L.H.A. ♈ 120°–300° °
300	0·8	1·0	1·3	1·5	1·7	2·0	2·5	300
310	0·8	1·0	1·3	1·4	1·7	2·0	2·4	290
320	0·8	0·9	1·2	1·4	1·6	1·9	2·3	280
330	0·7	0·8	1·1	1·2	1·4	1·7	2·1	270
340	0·6	0·7	1·0	1·1	1·3	1·5	1·8	260
350	0·5	0·6	0·8	0·9	1·0	1·2	1·5	250
0	0·4	0·5	0·6	0·7	0·8	0·9	1·2	240
10	0·3	0·3	0·4	0·5	0·5	0·6	0·8	230
20	0·1	0·1	0·2	0·2	0·2	0·3	0·3	220
30	0·0	0·0	0·0	0·1	0·1	0·1	0·1	210
40	359·9	359·9	359·8	359·8	359·8	359·7	359·7	200
50	359·7	359·7	359·6	359·5	359·5	359·4	359·2	190
60	359·6	359·5	359·4	359·3	359·2	359·1	358·8	180
70	359·5	359·4	359·2	359·1	359·0	358·8	358·5	170
80	359·4	359·3	359·0	358·9	358·7	358·5	358·2	160
90	359·3	359·2	358·9	358·8	358·6	358·3	357·9	150
100	359·2	359·1	358·8	358·6	358·4	358·1	357·7	140
110	359·2	359·0	358·7	358·6	358·3	358·0	357·6	130
120	359·2	359·0	358·7	358·5	358·3	358·0	357·5	120

When Cassiopeia is left (right), *Polaris* is west (east).

CORRECTIONS TO BE APPLIED TO SEXTANT ALTITUDE

REFRACTION

To be subtracted from sextant altitude (referred to as observed altitude in A.P. 3270).

Height above sea level in units of 1,000 ft. / Sextant Altitude

R_0	0	5	10	15	20	25	30	35	40	45	50	55	R_0	$R=R_0\times f$ 0·9	1·0	1·1	1·2
′	° ′	° ′	° ′	° ′	° ′	° ′	° ′	° ′	° ′	° ′	° ′	° ′	′	′	′	′	′
	90	90	90	90	90	90	90	90	90	90	90	90					
0													0	0	0	0	0
	63	59	55	51	46	41	36	31	26	20	17	13					
1													1	1	1	1	1
	33	29	26	22	19	16	14	11	9	7	6	4					
2													2	2	2	2	2
	21	19	16	14	12	10	8	7	5	4	2 40	1 40					
3													3	3	3	3	4
	16	14	12	10	8	7	6	5	3 10	2 20	1 30	0 40					
4													4	4	4	4	5
	12	11	9	8	7	5	4 00	3 10	2 10	1 30	0 39	+0 05					
5													5	5	5	5	6
	10	9	7	5 50	4 50	3 50	3 10	2 20	1 30	0 49	+0 11	−0 19					
6													6	5	6	7	7
	8 10	6 50	5 50	4 50	4 00	3 00	2 20	1 50	1 10	0 24	−0 11	−0 38					
7													7	6	7	8	8
	6 50	5 50	5 00	4 00	3 10	2 30	1 50	1 20	0 38	+0 04	−0 28	−0 54					
8													8	7	8	9	10
	6 00	5 10	4 10	3 20	2 40	2 00	1 30	1 00	0 19	−0 13	−0 42	−1 08					
9													9	8	9	10	11
	5 20	4 30	3 40	2 50	2 10	1 40	1 10	0 35	+0 03	−0 27	−0 53	−1 18					
10													10	9	10	11	12
	4 30	3 40	2 50	2 20	1 40	1 10	0 37	+0 11	−0 16	−0 43	−1 08	−1 31					
12													12	11	12	13	14
	3 30	2 50	2 10	1 40	1 10	0 34	+0 09	−0 14	−0 37	−1 00	−1 23	−1 44					
14													14	13	14	15	17
	2 50	2 10	1 40	1 10	0 37	+0 10	−0 13	−0 34	−0 53	−1 14	−1 35	−1 56					
16													16	14	16	18	19
	2 20	1 40	1 20	0 43	+0 15	−0 08	−0 31	−0 52	−1 08	−1 27	−1 46	−2 05					
18													18	16	18	20	22
	1 50	1 20	0 49	+0 23	−0 02	−0 26	−0 46	−1 06	−1 22	−1 39	−1 57	−2 14					
20													20	18	20	22	24
	1 12	0 44	+0 19	−0 06	−0 28	−0 48	−1 09	−1 27	−1 42	−1 58	−2 14	−2 30					
25													25	22	25	28	30
	0 34	+0 10	−0 13	−0 36	−0 55	−1 14	−1 32	−1 51	−2 06	−2 21	−2 34	−2 49					
30													30	27	30	33	36
	+0 06	−0 16	−0 37	−0 59	−1 17	−1 33	−1 51	−2 07	−2 23	−2 37	−2 51	−3 04					
35													35	31	35	38	42
	−0 18	−0 37	−0 58	−1 16	−1 34	−1 49	−2 06	−2 22	−2 35	−2 49	−3 03	−3 16					
40													40	36	40	44	48
		−0 53	−1 14	−1 31	−1 47	−2 03	−2 18	−2 33	−2 47	−2 59	−3 13	−3 25					
45													45	40	45	50	54
		−1 10	−1 28	−1 44	−1 59	−2 15	−2 28	−2 43	−2 56	−3 08	−3 22	−3 33					
50													50	45	50	55	60
			−1 40	−1 53	−2 09	−2 24	−2 38	−2 52	−3 04	−3 17	−3 29	−3 41					
55													55	49	55	60	66
				−2 03	−2 18	−2 33	−2 46	−3 01	−3 12	−3 25	−3 37	−3 48					
60													60	54	60	66	72
							−2 53	−3 07	−3 19	−3 31	−3 42	−3 53					

Temperature in °C

f	0	5	10	15	20	25	30	35	40	45	50	55	f	0·9	1·0	1·1	1·2
	+47	+36	+27	+18	+10	+ 3	− 5	−13	For these heights no temperature correction is necessary: take $f=1·0$ and use $R=R_0$					When R_0 is less than 10′ or the height is greater than 35,000 ft. take $f=1·0$ and use $R=R_0$			
0·9													0·9				
	+26	+16	+ 6	− 4	−13	−22	−31	−40									
1·0													1·0				
	+ 5	− 5	−15	−25	−36	−46	−57	−68									
1·1													1·1				
	−16	−25	−36	−46	−58	−71	−83	·95									
1·2													1·2				
	−37	−45	−56	−67	−81	−95											

Choose the column appropriate to height, in units of 1,000 ft., and find the range of altitude in which the sextant altitude lies; the corresponding value of R_0 is the refraction, to be subtracted from sextant altitude, unless conditions are extreme. In that case find f from the lower table, with critical argument temperature. Use the table on the right to form the refraction, $R=R_0\times f$.

CORIOLIS (Z) CORRECTION

To be applied by moving the position line a distance Z to starboard (right) of the track in northern latitudes and to port (left) in southern latitudes. The argument is given as T.A.S. (True Air Speed) in A.P. 3270.

G/S KNOTS	Latitude 0°	10°	20°	30°	40°	50°	60°	70°	80°	90°	G/S KNOTS	Latitude 0°	10°	20°	30°	40°	50°	60°	70°	80°	90°
	′	′	′	′	′	′	′	′	′	′		′	′	′	′	′	′	′	′	′	′
150	0	1	1	2	3	3	3	4	4	4	450	0	2	4	6	8	9	10	11	12	12
200	0	1	2	3	3	4	5	5	5	5	500	0	2	4	7	8	10	11	12	13	13
250	0	1	2	3	4	5	6	6	6	7	550	0	3	5	7	9	11	12	14	14	14
300	0	1	3	4	5	6	7	7	8	8	600	0	3	5	8	10	12	14	15	16	16
350	0	2	3	5	6	7	8	9	9	9	650	0	3	6	9	11	13	15	16	17	17
400	0	2	4	5	7	8	9	10	10	10	700	0	3	6	9	12	14	16	17	18	18

CORRECTIONS TO BE APPLIED TO MARINE SEXTANT ALTITUDES

MARINE SEXTANT ERROR	CORRECTIONS
Sextant Number Index Error	In addition to sextant error and dip, corrections are to be applied for: Refraction Semi-diameter (for the Sun and Moon) Parallax (for the Moon) Dome refraction (if applicable)

CORRECTION FOR DIP OF THE HORIZON
To be subtracted from sextant altitude

Ht.	Dip	Ht.	Dip	Ht.	Dip	Ht.	Dip	Ht.	Dip
Ft.	′	Ft.	′	Ft.	′	Ft.	′	Ft.	′
0	1	114	11	437	21	968	31	1 707	41
2	2	137	12	481	22	1 033	32	1 792	42
6	3	162	13	527	23	1 099	33	1 880	43
12	4	189	14	575	24	1 168	34	1 970	44
21	5	218	15	625	25	1 239	35	2 061	45
31	6	250	16	677	26	1 311	36	2 155	46
43	7	283	17	731	27	1 386	37	2 251	47
58	8	318	18	787	28	1 463	38	2 349	48
75	9	356	19	845	29	1 543	39	2 449	49
93	10	395	20	906	30	1 624	40	2 551	50
114		437		968		1 707		2 655	

LIST OF CONTENTS

APPENDIX H

LONG-TERM ALMANAC

This appendix is intended for use when a more complete almanac is not available. It is based principally upon the fact that approximately correct values for the Greenwich hour angle and declination of the sun, and the Greenwich hour angle of Aries, can be obtained from an almanac that is exactly four years out of date. The differences in these values at intervals of exactly four years can be largely removed by applying an average correction to the values obtained from the tables of this appendix. The maximum error in an altitude computed by means of this appendix should not exceed 2'.0 for the sun or 1'.3 for stars.

This four-year, or quadrennial, correction varies throughout the year for the GHA of the sun (between about plus and minus one-half of a minute) and for the declination of the sun (between about plus and minus three-fourths of a minute). For the GHA of Aries the quadrennial correction is a constant, (+)1'.84. The appropriate quadrennial correction is applied once for each full four years which has passed since the base year of the tabulation (1972 in this appendix).

The tabulated values for GHA−175° and declination of the sun and GHA of Aries are given in four columns, labeled 0, 1, 2, and 3. The "0" column contains the data for the leap year in each four-year cycle and the 1, 2, and 3 columns contain data for, respectively, the first, second, and third years following each leap year.

The GHA−175° and declination of the sun are given at intervals of three days throughout the four-year cycle, except for the final days of each month, when the interval varies between one and four days. Linear interpolation is made between entries to obtain data for a given day. Additional corrections to the GHA of the sun of 15° per hour, 15' per minute, and 15" per second are made to obtain the GHA at a given time. Declination of the sun is obtained to sufficient accuracy by linear interpolation alone.

The GHA of Aries is given for each month of the four-year cycle. Additional corrections of 0°59'.14 per day, 15°02'.5 per hour, 15' per minute, and 15" per second are made to obtain the GHA at a given time.

The SHA and declination of 38 navigational stars are given for the base year, 1972.0. Annual (not quadrennial) corrections are made to these data to obtain the values for a given year and tenth of a year.

A multiplication table is included as an aid in applying corrections to tabulated values.

Sun tables. 1. Subtract 1972 from the year and divide the difference by four, obtaining (*a*) a whole number, and (*b*) a remainder. Enter column indicated by remainder (*b*) and take out values on either side of given time and date.

2. Multiply quadrennial correction for each value by whole number (*a*) obtained in step 1 and apply to tabulated values plus 175°.

3. Divide difference between corrected values by number of days (usually three) between them to determine daily change.

4. Multiply daily change by number of days and tenths since 0^h GMT of earlier tabulated date, and mark correction plus (+) or minus (−) as appropriate.

5. (GHA only.) Enter multiplication table with hours, minutes, and seconds of GMT, and take out corrections *A*, *B*, and *C*, respectively. These are all positive.

6. Apply corrections of steps 4 and 5 to corrected *earlier* values of step 2.

Example.—Find GHA and declination of sun at GMT $17^h13^m49^s$ on July 18, 2002.

Solution.—Steps 1 *and* 2: (2002−1972)÷4=7, remainder 2. Use column 2, and multiply quadrennial corrections by 7. Corrected values: GHA, July 16, 178°31′.1+(7×0′.05)=178°31′.5; July 19, 178°27′.2+(7×0′.06)=178°27′.6. Dec., July 16, 21°27′.9N−(7×0′.41)=21°25′.0N; July 19, 20°57′.5N−(7×0′.44)=20°54′.4N.

	GHA	
July 16	178°31′.5	*Step 3*
July 19	178°27′.6	
3-day change	(−)3′.9	
daily change	(−)1′.3	*Step 4*
days and tenths	2.7	
corr.	(−)3′.5	
A	255°00′.0	*Step 5*
B	3°15′.0	
C	12′.1	
0^h July 16	178°31′.5	*Step 6*
GHA	76°55′.2	

	Declination	
July 16	21°25′.0 N	*Step 3*
July 19	20°54′.4 N	
3-day change	(−)30′.6	
daily change	(−)10′.2	*Step 4*
days and tenths	2.7	
corr.	(−)27′.5	*Step 6*
0^h July 16	21°25′.0 N	
Dec.	20°57′.5 N	

Aries table. 1. Subtract 1972 from the year and divide the difference by four, obtaining (*a*) a whole number, and (*b*) a remainder. Enter column indicated by remainder (*b*) and take out value for given month.

2. Enter multiplication table with whole number (*a*) of step 1, day of month, hours of GMT, minutes of GMT, and seconds of GMT, and take out corrections *D*, *E*, *F*, *G*, and *C*, respectively.

3. Add values of steps 1 and 2.

Example.—Find GHA♈ at GMT $11^h06^m33^s$ on November 28, 1995.

Solution.—Step 1: (1995−1972)÷4=5, remainder 3. Use column 3.

	GHA♈	
Nov.	38°40′.6	*Step 1*
D	9′.2	*Step 2*
E	27°35′.9	
F	165°27′.1	
G	1°30′.2	
C	8′.2	
GHA♈	233°31′.2	*Step 3*

Stars table. 1. Enter table with star name, and take out tabulated values.

2. Subtract 1972.0 from given year and tenth, and multiply annual correction by difference. Apply as correction (+or−, as appropriate) to value of step 1.

Example.—Find SHA and declination of Spica on September 11, 2011.

Solution.—From decimal table, September 11, 2011=2011.7. 2011.7−1972.0=39.7.

SHA			*Declination*		
1972.0	159°04.′3	*Step 1*	1972.0	11°01.′0 S	*Step 1*
39.7×(−)0.′79	(−)31.′4	*Step 2*	39.7×0.31	(+)12.′3	*Step 2*
SHA	158°32.′9		Dec.	11°13.′3 S	

To determine GHA of star, add GHA ♈ and SHA☆ for given time and date.

SUN										
0		Quad. GHA Corr.	1		Date	2		Quad. Dec. Corr.	3	
GHA −175°	Dec.		GHA −175°	Dec.		GHA −175°	Dec.		GHA −175°	Dec.
JANUARY										
° ′	° ′	′	° ′	° ′		° ′	° ′	′	° ′	° ′
4 14.4	23 05.5 S	−0.11	4 09.0	23 02.0 S	1	4 10.8	23 03.1 S	−0.32	4 12.9	23 04.2 S
3 53.3	22 50.2 S	−0.13	3 48.0	22 45.6 S	4	3 49.8	22 47.0 S	−0.35	3 51.9	22 48.4 S
3 33.0	22 30.7 S	−0.12	3 27.8	22 25.1 S	7	3 29.8	22 26.8 S	−0.39	3 31.7	22 28.6 S
3 13.8	22 07.2 S	−0.09	3 08.8	22 00.6 S	10	3 10.8	22 02.7 S	−0.42	3 12.5	22 04.8 S
2 55.7	21 39.9 S	−0.04	2 51.1	21 32.2 S	13	2 53.0	21 34.7 S	−0.44	2 54.5	21 37.1 S
2 38.9	21 08.7 S	+0.03	2 34.8	21 00.2 S	16	2 36.6	21 02.9 S	−0.46	2 37.9	21 05.6 S
2 23.7	20 33.9 S	+0.09	2 20.1	20 24.5 S	19	2 21.7	20 27.5 S	−0.48	2 22.9	20 30.5 S
2 10.1	19 55.6 S	+0.13	2 07.0	19 45.4 S	22	2 08.4	19 48.7 S	−0.49	2 09.5	19 51.9 S
1 58.2	19 13.9 S	+0.15	1 55.7	19 02.9 S	25	1 56.8	19 06.4 S	−0.52	1 57.9	19 10.0 S
1 48.2	18 29.1 S	+0.15	1 46.1	18 17.3 S	28	1 47.0	18 21.1 S	−0.54	1 48.0	18 24.9 S
FEBRUARY										
1 37.7	17 24.7 S	+0.13	1 36.1	17 11.9 S	1	1 36.8	17 16.0 S	−0.57	1 37.8	17 20.2 S
1 32.0	16 33.2 S	+0.14	1 30.7	16 19.6 S	4	1 31.3	16 24.0 S	−0.59	1 32.1	16 28.4 S
1 28.0	15 39.0 S	+0.15	1 27.1	15 24.8 S	7	1 27.8	15 29.5 S	−0.60	1 28.3	15 34.0 S
1 25.9	14 42.5 S	+0.19	1 25.4	14 27.7 S	10	1 26.0	14 32.5 S	−0.61	1 26.3	14 37.2 S
1 25.4	13 43.6 S	+0.24	1 25.4	13 28.4 S	13	1 25.9	13 33.4 S	−0.60	1 26.0	13 38.2 S
1 26.7	12 42.8 S	+0.29	1 27.2	12 27.1 S	16	1 27.5	12 32.3 S	−0.60	1 27.4	12 37.2 S
1 29.6	11 40.1 S	+0.34	1 30.6	11 24.0 S	19	1 30.7	11 29.3 S	−0.59	1 30.4	11 34.4 S
1 34.0	10 35.8 S	+0.36	1 35.6	10 19.3 S	22	1 35.4	10 24.7 S	−0.59	1 35.1	10 30.0 S
1 40.0	9 30.0 S	+0.36	1 41.9	9 13.2 S	25	1 41.5	9 18.6 S	−0.60	1 41.2	9 24.1 S
1 47.4	8 23.0 S	+0.34	1 49.5	8 05.8 S	28	1 49.0	8 11.3 S	−0.61	1 48.7	8 17.0 S
MARCH										
1 52.9	7 37.6 S	+0.32	1 52.3	7 43.1 S	1	1 51.7	7 48.7 S	−0.62	1 51.4	7 54.3 S
2 02.3	6 28.8 S	+0.31	2 01.4	6 34.4 S	4	2 00.8	6 40.0 S	−0.62	2 00.4	6 45.7 S
2 12.6	5 19.2 S	+0.31	2 11.6	5 24.8 S	7	2 11.0	5 30.5 S	−0.62	2 10.5	5 36.3 S
2 23.8	4 08.9 S	+0.33	2 22.6	4 14.5 S	10	2 22.1	4 20.4 S	−0.61	2 21.4	4 26.1 S
2 35.7	2 58.2 S	+0.36	2 34.5	3 03.8 S	13	2 33.9	3 09.7 S	−0.59	2 33.1	3 15.5 S
2 48.2	1 47.1 S	+0.40	2 47.1	1 52.8 S	16	2 46.4	1 58.7 S	−0.56	2 45.4	2 04.5 S
3 01.1	0 35.9 S	+0.43	3 00.1	0 41.7 S	19	2 59.4	0 47.6 S	−0.54	2 58.3	0 53.3 S
3 14.5	0 35.2 N	+0.44	3 13.6	0 29.4 N	22	3 12.7	0 23.6 N	+0.53	3 11.6	0 17.8 N
3 28.1	1 46.1 N	+0.43	3 27.2	1 40.3 N	25	3 26.2	1 34.6 N	+0.52	3 25.2	1 28.8 N
3 41.9	2 56.6 N	+0.40	3 40.9	2 50.9 N	28	3 39.8	2 45.2 N	+0.52	3 38.9	2 39.4 N
APRIL										
4 00.0	4 29.8 N	+0.35	3 59.0	4 24.3 N	1	3 57.9	4 18.6 N	+0.52	3 57.1	4 12.8 N
4 13.4	5 38.9 N	+0.33	4 12.3	5 33.4 N	4	4 11.2	5 27.8 N	−0.51	4 10.5	5 22.1 N
4 26.4	6 47.1 N	+0.32	4 25.2	6 41.7 N	7	4 24.3	6 36.2 N	−0.49	4 23.5	6 30.6 N
4 38.8	7 54.3 N	+0.33	4 37.7	7 49.0 N	10	4 36.9	7 43.5 N	+0.47	4 36.0	7 38.0 N
4 50.6	9 00.3 N	+0.35	4 49.6	8 55.1 N	13	4 48.9	8 49.7 N	+0.44	4 48.0	8 44.3 N
5 01.7	10 05.0 N	+0.37	5 00.9	9 59.8 N	16	5 00.2	9 54.5 N	+0.41	4 59.2	9 49.3 N
5 12.0	11 08.1 N	+0.39	5 11.3	11 03.0 N	19	5 10.6	10 57.9 N	+0.39	5 09.7	10 52.8 N
5 21.4	12 09.6 N	+0.37	5 20.9	12 04.6 N	22	5 20.0	11 59.7 N	+0.37	5 19.3	11 54.7 N
5 29.8	13 09.2 N	+0.33	5 29.4	13 04.4 N	25	5 28.5	12 59.7 N	+0.36	5 28.0	12 54.8 N
5 37.2	14 06.9 N	+0.28	5 36.8	14 02.3 N	28	5 36.0	13 57.7 N	+0.34	5 35.6	13 53.0 N
MAY										
5 43.4	15 02.6 N	+0.24	5 43.0	14 58.1 N	1	5 42.3	14 53.7 N	+0.33	5 42.1	14 49.2 N
5 48.4	15 56.0 N	+0.20	5 48.0	15 51.7 N	4	5 47.5	15 47.5 N	+0.31	5 47.3	15 43.1 N
5 52.1	16 47.0 N	+0.19	5 51.7	16 43.0 N	7	5 51.4	16 38.9 N	+0.28	5 51.3	16 34.7 N
5 54.5	17 35.5 N	+0.19	5 54.2	17 31.7 N	10	5 54.1	17 27.8 N	+0.25	5 53.9	17 23.9 N
5 55.6	18 21.4 N	+0.21	5 55.5	18 17.7 N	13	5 55.5	18 14.1 N	+0.22	5 55.3	18 10.4 N
5 55.4	19 04.5 N	+0.23	5 55.5	19 01.0 N	16	5 55.5	18 57.6 N	+0.19	5 55.4	18 54.2 N
5 53.9	19 44.7 N	+0.23	5 54.3	19 41.4 N	19	5 54.3	19 38.3 N	+0.16	5 54.3	19 35.1 N
5 51.3	20 21.8 N	+0.20	5 51.8	20 18.8 N	22	5 51.7	20 15.9 N	+0.13	5 51.9	20 13.0 N
5 47.5	20 55.8 N	+0.16	5 48.1	20 53.1 N	25	5 48.0	20 50.5 N	+0.11	5 48.4	20 47.7 N
5 42.6	21 26.6 N	+0.11	5 43.2	21 24.2 N	28	5 43.2	21 21.8 N	+0.08	5 43.8	21 19.3 N
JUNE										
5 34.5	22 02.4 N	+0.04	5 35.1	22 00.4 N	1	5 35.2	21 58.3 N	+0.04	5 36.0	21 56.2 N
5 27.4	22 25.3 N	−0.02	5 27.9	22 23.5 N	4	5 28.2	22 21.7 N	+0.01	5 28.9	22 19.9 N
5 19.4	22 44.6 N	+0.02	5 19.9	22 43.1 N	7	5 20.4	22 41.6 N	−0.02	5 21.1	22 40.1 N
5 10.7	23 00.3 N	+0.04	5 11.3	22 59.1 N	10	5 11.9	22 57.9 N	−0.05	5 12.5	22 56.7 N
5 01.5	23 12.4 N	+0.07	5 02.3	23 11.5 N	13	5 02.9	23 10.6 N	−0.08	5 03.4	23 09.7 N
4 51.9	23 20.8 N	−0.09	4 52.8	23 20.2 N	16	4 53.4	23 19.6 N	−0.12	4 53.9	23 19.0 N
4 42.1	23 25.5 N	+0.08	4 43.2	23 25.3 N	19	4 43.6	23 24.9 N	−0.15	4 44.2	23 24.6 N
4 32.4	23 26.5 N	−0.06	4 33.4	23 26.6 N	22	4 33.7	23 26.5 N	−0.18	4 34.5	23 26.5 N
4 22.8	23 23.8 N	+0.01	4 23.8	23 24.1 N	25	4 24.0	23 24.4 N	−0.22	4 24.8	23 24.6 N
4 13.4	23 17.4 N	−0.03	4 14.3	23 18.0 N	28	4 14.5	23 18.5 N	−0.25	4 15.4	23 19.1 N

SUN										
0		Quad. GHA Corr.	1		Date	2		Quad. Dec. Corr.	3	
GHA -175°	Dec.		GHA -175°	Dec.		GHA -175°	Dec.		GHA -175°	Dec.
JULY										
° ′	° ′	′	° ′	° ′		° ′	° ′	′	° ′	° ′
4 04.5	23 07.3 N	- 0.06	4 05.2	23 08.2 N	1	4 05.4	23 09.0 N	- 0.28	4 06.3	23 09.9 N
3 56.1	22 53.5 N	- 0.07	3 56.6	22 54.7 N	4	3 57.0	22 55.8 N	- 0.31	3 57.7	22 57.0 N
3 48.4	22 36.2 N	- 0.05	3 48.8	22 37.6 N	7	3 49.2	22 39.1 N	- 0.34	3 49.8	22 40.5 N
3 41.5	22 15.3 N	- 0.01	3 41.9	22 17.1 N	10	3 42.3	22 18.8 N	- 0.37	3 42.7	22 20.5 N
3 35.5	21 51.0 N	+ 0.03	3 36.0	21 53.0 N	13	3 36.2	21 55.0 N	- 0.39	3 36.4	21 57.0 N
3 30.6	21 23.3 N	+ 0.05	3 31.1	21 25.7 N	16	3 31.1	21 27.9 N	- 0.41	3 31.3	21 30.1 N
3 26.9	20 52.4 N	+ 0.06	3 27.4	20 55.0 N	19	3 27.2	20 57.5 N	- 0.44	3 27.4	21 00.0 N
3 24.5	20 18.3 N	+ 0.04	3 24.9	20 21.2 N	22	3 24.5	20 23.8 N	- 0.47	3 24.7	20 26.6 N
3 23.4	19 41.1 N	+ 0.01	3 23.6	19 44.2 N	25	3 23.0	19 47.1 N	- 0.50	3 23.3	19 50.2 N
3 23.6	19 01.0 N	- 0.02	3 23.6	19 04.3 N	28	3 23.0	19 07.5 N	- 0.54	3 23.1	19 10.8 N
AUGUST										
3 26.0	18 03.2 N	- 0.03	3 25.6	18 06.8 N	1	3 25.1	18 10.3 N	- 0.57	3 25.1	18 13.9 N
3 29.3	17 16.7 N	- 0.02	3 28.8	17 20.5 N	4	3 28.2	17 24.2 N	- 0.59	3 28.0	17 28.0 N
3 33.9	16 27.7 N	+ 0.02	3 33.4	16 31.6 N	7	3 32.8	16 35.6 N	- 0.60	3 32.3	16 39.6 N
3 39.9	15 36.3 N	+ 0.06	3 39.3	15 40.4 N	10	3 38.6	15 44.6 N	- 0.60	3 38.0	15 48.7 N
3 47.1	14 42.6 N	+ 0.09	3 46.5	14 47.0 N	13	3 45.7	14 51.3 N	- 0.61	3 44.9	14 55.6 N
3 55.7	13 46.8 N	+ 0.11	3 55.0	13 51.4 N	16	3 54.0	13 55.8 N	- 0.62	3 53.2	14 00.3 N
4 05.4	12 49.1 N	+ 0.11	4 04.7	12 53.9 N	19	4 03.4	12 58.4 N	- 0.64	4 02.7	13 03.1 N
4 16.2	11 49.6 N	+ 0.09	4 15.4	11 54.5 N	22	4 14.0	11 59.1 N	- 0.66	4 13.3	12 04.0 N
4 28.1	10 48.4 N	+ 0.06	4 27.2	10 53.3 N	25	4 25.7	10 58.1 N	- 0.68	4 24.9	11 03.2 N
4 40.9	9 45.6 N	+ 0.03	4 39.8	9 50.6 N	28	4 38.3	9 55.6 N	- 0.70	4 37.5	10 00.8 N
SEPTEMBER										
4 59.2	8 19.8 N	+ 0.02	4 57.9	8 24.9 N	1	4 56.5	8 30.0 N	- 0.71	4 55.6	8 35.4 N
5 13.7	7 13.9 N	+ 0.04	5 12.3	7 19.2 N	4	5 10.9	7 24.4 N	- 0.70	5 09.9	7 29.8 N
5 28.7	6 07.0 N	+ 0.08	5 27.3	6 12.4 N	7	5 26.0	6 17.8 N	- 0.69	5 24.7	6 23.2 N
5 44.1	4 59.2 N	+ 0.11	5 42.8	5 04.7 N	10	5 41.4	5 10.1 N	- 0.68	5 40.1	5 15.6 N
5 59.9	3 50.7 N	+ 0.13	5 58.6	3 56.2 N	13	5 57.1	4 01.7 N	- 0.67	5 55.8	4 07.2 N
6 15.8	2 41.5 N	+ 0.13	6 14.7	2 47.1 N	16	6 13.0	2 52.6 N	- 0.67	6 11.7	2 58.2 N
6 31.9	1 31.9 N	+ 0.11	6 30.7	1 37.5 N	19	6 29.0	1 43.0 N	- 0.68	6 27.8	1 48.7 N
6 47.9	0 21.9 N	+ 0.08	6 46.6	0 27.6 N	22	6 44.9	0 33.1 N	- 0.68	6 43.8	0 38.8 N
7 03.6	0 48.1 S	+ 0.04	7 02.3	0 42.5 S	25	7 00.6	0 37.0 S	+ 0.69	6 59.6	0 31.2 S
7 19.0	1 58.3 S	+ 0.01	7 17.6	1 52.7 S	28	7 16.1	1 47.1 S	+ 0.68	7 15.1	1 41.3 S
OCTOBER										
7 33.8	3 08.2 S	0.00	7 32.4	3 02.7 S	1	7 31.0	2 57.1 S	+ 0.67	7 30.0	2 51.4 S
7 48.0	4 17.9 S	+ 0.02	7 46.6	4 12.4 S	4	7 45.4	4 06.8 S	+ 0.65	7 44.3	4 01.2 S
8 01.3	5 27.2 S	+ 0.05	8 00.1	5 21.7 S	7	7 59.0	5 16.1 S	+ 0.62	7 57.9	5 10.5 S
8 13.7	6 35.8 S	+ 0.07	8 12.7	6 30.3 S	10	8 11.6	6 24.8 S	+ 0.60	8 10.5	6 19.3 S
8 25.1	7 43.6 S	+ 0.08	8 24.3	7 38.1 S	13	8 23.2	7 32.8 S	+ 0.57	8 22.2	7 27.3 S
8 35.4	8 50.5 S	+ 0.07	8 34.7	8 45.0 S	16	8 33.6	8 39.8 S	+ 0.56	8 32.8	8 34.4 S
8 44.4	9 56.2 S	+ 0.04	8 43.8	9 50.9 S	19	8 42.8	9 45.7 S	+ 0.54	8 42.2	9 40.3 S
8 52.0	11 00.6 S	- 0.01	8 51.5	10 55.4 S	22	8 50.5	10 50.3 S	+ 0.53	8 50.1	10 45.0 S
8 58.1	12 03.5 S	- 0.05	8 57.6	11 58.5 S	25	8 56.9	11 53.5 S	+ 0.52	8 56.6	11 48.3 S
9 02.6	13 04.7 S	- 0.08	9 02.1	12 59.9 S	28	9 01.6	12 55.0 S	+ 0.50	9 01.5	12 50.0 S
NOVEMBER										
9 05.8	14 23.5 S	- 0.09	9 05.5	14 18.9 S	1	9 05.4	14 14.2 S	+ 0.46	9 05.3	14 09.4 S
9 06.1	15 20.3 S	- 0.07	9 06.0	15 15.8 S	4	9 06.1	15 11.2 S	+ 0.42	9 06.0	15 06.6 S
9 04.6	16 14.7 S	- 0.04	9 04.7	16 10.4 S	7	9 05.0	16 06.0 S	+ 0.38	9 04.9	16 01.6 S
9 01.1	17 06.7 S	- 0.03	9 01.6	17 02.5 S	10	9 01.9	16 58.4 S	+ 0.35	9 02.0	16 54.2 S
8 55.8	17 56.1 S	- 0.03	8 56.5	17 52.1 S	13	8 56.9	17 48.2 S	+ 0.31	8 57.2	17 44.2 S
8 48.6	18 42.6 S	- 0.05	8 49.5	18 38.8 S	16	8 49.9	18 35.2 S	+ 0.28	8 50.5	18 31.4 S
8 39.6	19 26.1 S	- 0.09	8 40.6	19 22.7 S	19	8 41.1	19 19.2 S	- 0.25	8 41.9	19 15.7 S
8 28.7	20 06.5 S	- 0.14	8 29.8	20 03.3 S	22	8 30.5	20 00.1 S	+ 0.22	8 31.5	19 56.9 S
8 16.1	20 43.6 S	- 0.18	8 17.2	20 40.7 S	25	8 18.1	20 37.8 S	+ 0.19	8 19.3	20 34.8 S
8 01.7	21 17.2 S	- 0.20	8 02.8	21 14.6 S	28	8 04.1	21 11.9 S	+ 0.16	8 05.3	21 09.3 S
DECEMBER										
7 45.6	21 47.3 S	- 0.18	7 46.9	21 44.9 S	1	7 48.4	21 42.5 S	+ 0.12	7 49.6	21 40.1 S
7 28.1	22 13.5 S	- 0.15	7 29.6	22 11.5 S	4	7 31.3	22 09.4 S	+ 0.08	7 32.5	22 07.3 S
7 09.3	22 35.9 S	- 0.11	7 11.0	22 34.2 S	7	7 12.8	22 32.4 S	+ 0.04	7 13.9	22 30.6 S
6 49.3	22 54.3 S	- 0.09	6 51.3	22 52.9 S	10	6 53.0	22 51.4 S	0.00	6 54.3	22 50.0 S
6 28.5	23 08.6 S	- 0.08	6 30.6	23 07.6 S	13	6 32.3	23 06.4 S	- 0.04	6 33.7	23 05.3 S
6 06.9	23 18.8 S	- 0.09	6 09.1	23 18.1 S	16	6 10.8	23 17.3 S	- 0.08	6 12.4	23 16.5 S
5 44.9	23 24.8 S	- 0.12	5 47.1	23 24.4 S	19	5 48.8	23 24.0 S	- 0.12	5 50.5	23 23.6 S
5 22.6	23 26.6 S	- 0.15	5 24.7	23 26.6 S	22	5 26.5	23 26.5 S	- 0.16	5 28.3	23 26.4 S
5 00.2	23 24.1 S	- 0.17	5 02.2	23 24.4 S	25	5 04.1	23 24.7 S	- 0.19	5 05.9	23 25.0 S
4 38.0	23 17.5 S	- 0.17	4 39.9	23 18.1 S	28	4 41.9	23 18.7 S	- 0.23	4 43.6	23 19.3 S

STARS

SHA (1972.0)	Annual Corr.	Star	Dec. (1972.0)	Annual Corr.
° ′	′		° ′	′
315 42.0	−0.57	Acamar	40 25.0 S	−0.24
335 49.9	−0.56	Achernar	57 22.7 S	−0.30
173 44.6	−0.84	Acrux	62 56.6 S	+0.33
291 25.3	−0.86	Aldebaran	16 27.2 N	+0.12
153 23.4	−0.59	Alkaid	49 27.2 N	−0.30
218 26.8	−0.74	Alphard	8 32.2 S	+0.26
126 37.5	−0.64	Alphecca	26 48.5 N	−0.20
358 16.0	−0.78	Alpheratz	28 56.2 N	+0.33
62 38.7	−0.73	Altair	8 47.6 N	+0.16
113 04.7	−0.92	Antares	26 22.3 S	+0.13
146 24.2	−0.68	Arcturus	19 19.6 N	−0.31
108 34.7	−1.59	Atria	68 58.7 S	+0.11
271 35.2	−0.81	Betelgeuse	7 24.2 N	+0.01
264 10.0	−0.33	Canopus	52 40.8 S	+0.03
281 20.7	−1.11	Capella	45 58.3 N	+0.06
49 52.9	−0.51	Deneb	45 10.8 N	+0.22
183 05.5	−0.76	Denebola	14 43.7 N	−0.34
349 27.2	−0.75	Diphda	18 08.4 S	−0.33
194 29.8	−0.92	Dubhe	61 54.2 N	−0.32
34 17.8	−0.74	Enif	9 44.8 N	+0.28
15 58.4	−0.83	Fomalhaut	29 46.3 S	−0.32
328 36.2	−0.85	Hamal	23 19.8 N	+0.28
137 18.4	+0.04	Kochab	74 16.2 N	−0.25
148 44.6	−0.88	Menkent	36 14.0 S	+0.29
309 25.3	−1.07	Mirfak	49 45.8 N	+0.21
76 37.1	−0.93	Nunki	26 20.0 S	−0.08
54 08.4	−1.18	Peacock	56 49.6 S	−0.19
244 05.9	−0.92	Pollux	28 05.7 N	−0.15
245 32.4	−0.78	Procyon	5 17.9 N	−0.16
96 35.5	−0.70	Rasalhague	12 34.8 N	−0.04
208 16.8	−0.80	Regulus	12 06.3 N	−0.29
281 42.1	−0.72	Rigel	8 14.0 S	−0.07
140 34.8	−1.02	Rigil Kent.	60 43.2 S	+0.25
350 16.4	−0.86	Schedar	56 23.0 N	+0.33
259 01.3	−0.66	Sirius	16 40.6 S	+0.08
159 04.3	−0.79	Spica	11 01.0 S	+0.31
223 15.5	−0.55	Suhail	43 19.1 S	+0.24
81 00.2	−0.51	Vega	38 45.4 N	+0.06

ARIES (♈)

0	1	Month	2	3
° ′	° ′		° ′	° ′
98 46.2	99 31.0	Jan.	99 16.7	99 02.4
129 19.5	130 04.4	Feb.	129 50.1	129 35.7
157 54.5	157 40.3	Mar.	157 25.9	157 11.6
188 27.8	188 13.5	Apr.	187 59.2	187 44.9
218 02.0	217 47.7	May	217 33.4	217 19.0
248 35.3	248 21.0	June	248 06.7	247 52.3
278 09.5	277 55.2	July	277 40.9	277 26.5
308 42.8	308 28.5	Aug.	308 14.2	307 59.8
339 16.1	339 01.8	Sept.	338 47.5	338 33.1
8 50.3	8 36.0	Oct.	8 21.6	8 07.3
39 23.6	39 09.2	Nov.	38 54.9	38 40.6
68 57.7	68 43.4	Dec.	68 29.1	68 14.7

MULTIPLICATION TABLE

No.	A	B	C	D	E	F	G
	°	° ′	′	′	° ′	° ′	° ′
1	15	0 15	0.2	1.8	0 59.1	15 02.5	0 15.0
2	30	0 30	0.5	3.7	1 58.3	30 04.9	0 30.1
3	45	0 45	0.8	5.5	2 57.4	45 07.4	0 45.1
4	60	1 00	1.0	7.4	3 56.6	60 09.9	1 00.2
5	75	1 15	1.2	9.2	4 55.7	75 12.3	1 15.2
6	90	1 30	1.5	11.0	5 54.8	90 14.8	1 30.2
7	105	1 45	1.8	12.9	6 54.0	105 17.2	1 45.3
8	120	2 00	2.0	14.7	7 53.1	120 19.7	2 00.3
9	135	2 15	2.2	16.6	8 52.3	135 22.2	2 15.4
10	150	2 30	2.5	18.4	9 51.4	150 24.6	2 30.4
11	165	2 45	2.8	20.2	10 50.5	165 27.1	2 45.5
12	180	3 00	3.0	22.1	11 49.7	180 29.6	3 00.5
13	195	3 15	3.2	23.9	12 48.8	195 32.0	3 15.5
14	210	3 30	3.5	25.8	13 48.0	210 34.5	3 30.6
15	225	3 45	3.8	27.6	14 47.1	225 37.0	3 45.6
16	240	4 00	4.0	29.4	15 46.2	240 39.4	4 00.7
17	255	4 15	4.2	31.3	16 45.4	255 41.9	4 15.7
18	270	4 30	4.5	33.1	17 44.5	270 44.4	4 30.7
19	285	4 45	4.8	35.0	18 43.7	285 46.8	4 45.8
20	300	5 00	5.0	36.8	19 42.8	300 49.3	5 00.8
21	315	5 15	5.2	38.6	20 41.9	315 51.7	5 15.9
22	330	5 30	5.5	40.5	21 41.1	330 54.2	5 30.9
23	345	5 45	5.8	42.3	22 40.2	345 56.7	5 45.9
24	360	6 00	6.0	44.2	23 39.4	360 59.1	6 01.0
25	—	6 15	6.2	46.0	24 38.5	—	6 16.0
26	—	6 30	6.5	47.8	25 37.6	—	6 31.1
27	—	6 45	6.8	49.7	26 36.8	—	6 46.1
28	—	7 00	7.0	51.5	27 35.9	—	7 01.1
29	—	7 15	7.2	53.4	28 35.1	—	7 16.2
30	—	7 30	7.5	55.2	29 34.2	—	7 31.2
31	—	7 45	7.8	57.0	30 33.3	—	7 46.3
32	—	8 00	8.0	58.9	—	—	8 01.3
33	—	8 15	8.2	60.7	—	—	8 16.4
34	—	8 30	8.5	62.6	—	—	8 31.4
35	—	8 45	8.8	64.4	—	—	8 46.4
36	—	9 00	9.0	66.2	—	—	9 01.5
37	—	9 15	9.2	68.1	—	—	9 16.5
38	—	9 30	9.5	69.9	—	—	9 31.6
39	—	9 45	9.8	71.8	—	—	9 46.6
40	—	10 00	10.0	73.6	—	—	10 01.6
41	—	10 15	10.2	75.4	—	—	10 16.7
42	—	10 30	10.5	77.3	—	—	10 31.7
43	—	10 45	10.8	79.1	—	—	10 46.8
44	—	11 00	11.0	81.0	—	—	11 01.8
45	—	11 15	11.2	82.8	—	—	11 16.8
46	—	11 30	11.5	84.6	—	—	11 31.9
47	—	11 45	11.8	86.5	—	—	11 46.9
48	—	12 00	12.0	88.3	—	—	12 02.0
49	—	12 15	12.2	90.2	—	—	12 17.0
50	—	12 30	12.5	92.0	—	—	12 32.1
51	—	12 45	12.8	93.8	—	—	12 47.1
52	—	13 00	13.0	95.7	—	—	13 02.1
53	—	13 15	13.2	97.5	—	—	13 17.2
54	—	13 30	13.5	99.4	—	—	13 32.2
55	—	13 45	13.8	—	—	—	13 47.3
56	—	14 00	14.0	—	—	—	14 02.3
57	—	14 15	14.2	—	—	—	14 17.3
58	—	14 30	14.5	—	—	—	14 32.4
59	—	14 45	14.8	—	—	—	14 47.4
60	—	15 00	15.0	—	—	—	15 02.5

DECIMAL PARTS OF DAY AND YEAR

Decimal	0.0	0.1	0.2	0.3	0.4	0.5	0.6	0.7	0.8	0.9	1.0
Hour of Day	0000 to 0112	0112 to 0336	0336 to 0600	0600 to 0824	0824 to 1048	1048 to 1312	1312 to 1536	1536 to 1800	1800 to 2024	2024 to 2248	2248 to 2400
Day of Year	Jan. 1 to Jan. 18	Jan. 19 to Feb. 23	Feb. 24 to Apr. 1	Apr 2 to May 7	May 8 to June 13	June 14 to July 19	July 20 to Aug. 25	Aug. 26 to Sept. 30	Oct. 1 to Nov. 6	Nov. 7 to Dec. 12	Dec. 13 to Dec. 31

APPENDIX I

MEASUREMENT ON THE EARTH

I1. The earth is approximately an **oblate spheroid** (a sphere flattened at the poles). Approximations of its dimensions and the amount of flattening are given in appendix D. However, for many navigational purposes, the earth is assumed to be a sphere, without intolerable error.

The **axis of rotation** or **polar axis** of the earth is the line connecting the north pole and the south pole.

I2. Circles of the earth.—A **great circle** is the line of intersection of a sphere and a plane through the center of the sphere. This is the largest circle that can be drawn on a sphere. The shortest line on the surface of a sphere between two points on that surface is part of a great circle. On the spheroidal earth the shortest line is called a **geodesic.** A great circle is a near enough approximation of a geodesic for most problems of navigation.

A **small circle** is the line of intersection of a sphere and a plane which does not pass through the center of the sphere.

A **meridian** is a great circle through the geographical poles of the earth. Hence, all meridians meet at the poles, and their planes intersect each other in a line, the **polar axis** (fig. I2a). The term **meridian** is usually applied to the **upper branch** only, that half from pole to pole which passes through a given point. The other half is called the **lower branch.**

The **prime meridian** is that meridian used as the origin for measurement of longitude (fig. I2b). The prime meridian used almost universally is that through the original position of the British Royal Observatory at Greenwich, near London.

The **equator** is the terrestrial great circle whose plane is perpendicular to the polar axis (fig. I2c). It is midway between the poles.

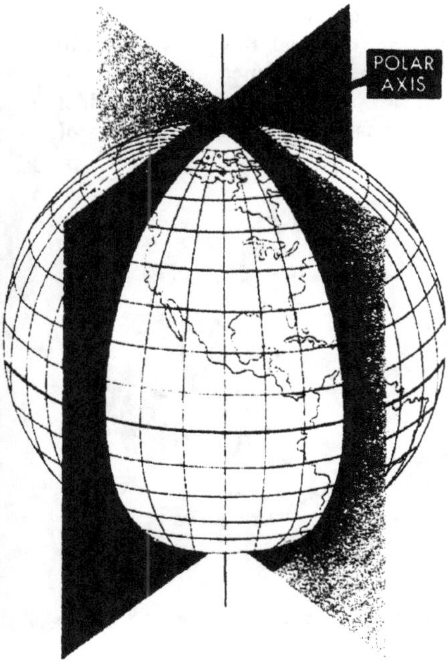

FIGURE I2a.—The planes of the meridians meet at the polar axis.

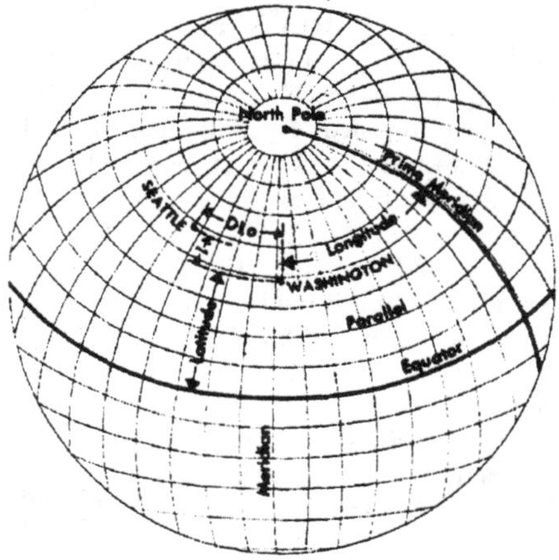

FIGURE I2b.—Circles and coordinates of the earth. All parallels except the equator are small circles; the equator and meridians are great circles.

A **parallel** or **parallel of latitude** is a circle on the surface of the earth, parallel to the plane of the equator (fig. I2d). It connects all points of equal latitude. The equator, a great circle, is a limiting case connecting points of 0° latitude. The poles, single points at latitude 90°, are the other limiting case. All other parallels are small circles.

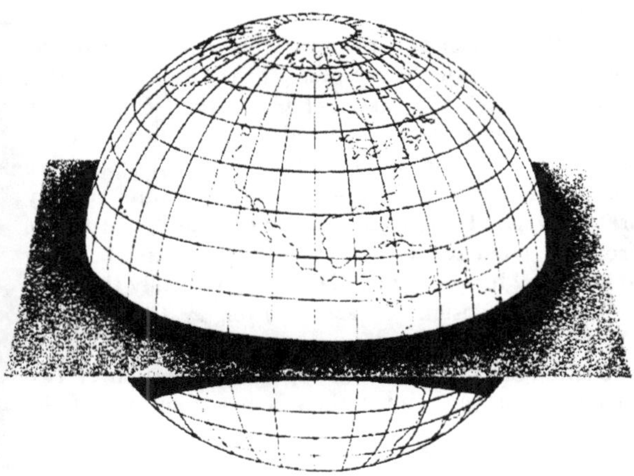

FIGURE I2c.—The equator is a great circle midway between the poles.

I3. Position on the earth.—A position on the surface of the earth (except at either of the poles) may be defined by two magnitudes called **coordinates.** Those customarily used are *latitude* and *longitude*. A position may also be expressed in relation to known geographical positions.

Latitude (**L, lat.**) is angular distance from the equator, measured northward or southward along a meridian from 0° at the equator to 90° at the poles (fig. I2b). It is designated *north* (N) or *south* (S) to indicate the direction of measurement.

The **difference of latitude** (*l*, **D. Lat.**) between two places is the angular length of arc of any meridian between their parallels (fig. I2b). It is the numerical difference of the latitudes if the places are on the same side of the equator, and the sum if they are on opposite sides. It may be designated *north* (N) or *south* (S) when appropriate.

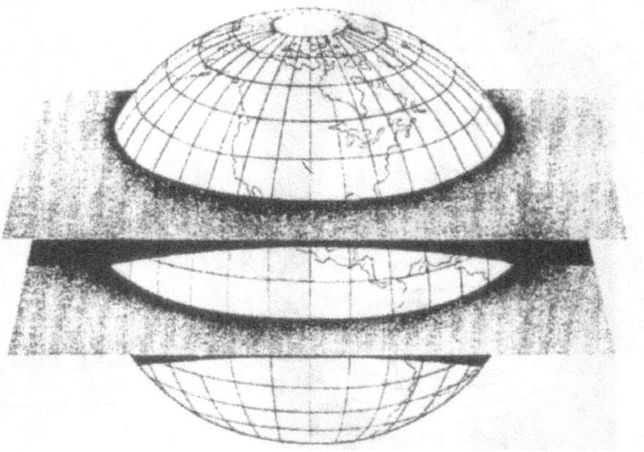

FIGURE I2d.—A parallel of latitude is parallel to the equator.

The **middle** or **mid latitude** (**Lm**) between two places on the same side of the equator is half the sum of their latitudes. Mid latitude is labeled N or S to indicate whether it is north or south of the equator. The expression is occasionally used with reference to two places on opposite sides of the equator, when it is equal to half the *difference* between the two latitudes, and takes the name of the place farthest from the equator. The term middle latitude as used in middle-latitude sailing (art. 1006) is defined differently. When the places are on opposite sides of the equator, two mid latitudes are generally used, the average of each latitude and 0°.

Longitude (**λ, long.**) is the arc of a parallel or the angle at the pole between the prime meridian and the meridian of a point on the earth, measured eastward or westward from the prime meridian through 180° (fig. I2b). It is disignated *east* (E) or *west* (W) to indicate the direction of measurement.

The **difference of longitude** (**DLo**) between two places is the shorter arc of the parallel or the smaller angle at the pole between the meridians of the two places (fig. I2b). If both places are on the same side (east or west) of Greenwich, DLo is the numerical difference of the longitudes of the two places; if on opposite sides, DLo is the numerical sum unless this exceeds 180°, when it is 360° minus the sum. The distance between two meridians at any parallel of latitude, expressed in distance units, usually nautical miles, is called **departure** (**p, Dep.**). It represents distance made good to the east or west as a craft proceeds from one point to another. Its numerical value between any two meridians decreases with increased latitude, while DLo is numerically the same at any latitude. Either DLo or p may be designated *east* (E) or *west* (W) when appropriate.

I4. Distance on the earth.—Distance (**D, Dist.**) is the spatial separation of two points, and is expressed as the length of a line joining them. On the surface of the earth it is usually stated in miles. Navigators customarily use the **nautical mile** (**mi., NM**) of 1852 meters exactly. This is the value suggested by the International Hydrographic Bureau in 1929, and since adopted by most maritime nations. It is often called the **International Nautical Mile** to distinguish it from slightly different values used by some countries. On July 1, 1959, the United States adopted the exact relationship of 1 yard=0.9144 meter. The length of the International Nautical Mile is consequently equal to 6,076.11549 feet (approximately).

FIGURE I4.—A rhumb line or loxodrome.

For most navigational purposes the nautical mile is considered the length of one minute of latitude, or of any great circle of the earth, regardless of location. On the World Geodetic System ellipsoid of 1972, the length of 1 minute of latitude varies from about 6,046 feet at the equator to approximately 6,108 feet at the poles. A **geographical mile** is the length of 1 minute of the equator, or about 6,087 feet.

The **land** or **statute mile** (**mi., St M**) of 5,280 feet is commonly used for navigation on rivers and lakes, notably the Great Lakes of North America.

The nautical mile is about 38/33 or approximately 1.15 statute miles. A conversion table for nautical and statute miles is given in table 20.

Distance, as customarily used by the navigator, refers to the length of the **rhumb line** connecting two places. This is a line making the same oblique angle with all meridians. Meridians and parallels (including the equator) which also maintain constant true directions, may be considered special cases of the rhumb line. Any other rhumb line spirals toward the pole, forming a **loxodromic curve** or **loxodrome** (fig. I4). Distance along the great circle connecting two points is customarily designated **great-circle distance.**

I5. Speed (S) is rate of motion, or distance per unit of time.

A knot (kn.), the unit of speed commonly used in navigation, is a rate of one nautical mile per hour. The expression "knots per hour" refers to acceleration, not speed.

Sometimes the expression **speed of advance** (**SOA**) is used to indicate the speed intended to be made along the track (art. I6), and **speed over ground** (**SOG**) the speed along the actual path. **Speed made good** (**SMG**) is the speed along the course made good.

I6. Direction on the earth.—Direction is the position of one point relative to another, without reference to the distance between them. In navigation, direction is customarily expressed as the angular difference in degrees from a reference direction, usually north or the ship's head. Compass directions (east, south by west, etc.) or points (of 11¼° or 1/32 of a circle) are seldom used by modern navigators for precise directions.

Course (C, Cn) is the intended horizontal direction of travel, expressed as angular distance from north, usually from 000° at north, clockwise through 360°. Strictly, the term applies to direction *through the water*, not the direction intended to be made good *over the ground*. The course is often designated as **true, magnetic, compass,** or **grid** as the reference direction is true, magnetic, compass, or grid north, respectively. **Track made good** (TMG) is the single resultant direction from the point of departure to point of arrival at any given time. Sometimes the expression **course of advance** (COA) is used to indicate the direction intended to be made good over the ground, and **course over ground** (COG) the direction of the actual track. A **course line** is a line, as drawn on a chart, extending in the direction of a course.

In making computations it is sometimes convenient to express a course as an angle from *either* north or south, through 90° or 180°. In this case it is designated **course angle** (C) and should be properly labeled to indicate the origin (prefix) and direction of measurement (suffix). Thus, C N35°E=Cn 035° (000°+35°), C N155°W =Cn 205° (360°−155°), C S47°E=Cn 133° (180°−47°). But Cn 260° may be either C N100°W or C S80°W, depending upon the conditions of the problem.

The symbol C is always used for *course angle*, and is usually used for *course* where there is little or no possibility of confusion.

Track (TR) is the intended or desired horizontal direction of travel with respect to the earth and also the path of intended travel (fig. I6a). The terms **intended track** and **trackline** are also used to indicate the path of intended travel. The track consists of one or a series of course lines from the point of departure to the destination, along which it is intended the vessel will proceed. A great circle which a vessel intends to follow approximately is called a **great-circle track.**

Heading (Hdg., SH) is the direction in which a vessel is pointed, expressed as angular distance from north, usually from 000° at north, clockwise through 360°. *Heading* should not be confused with *course*. *Heading* is a constantly changing value as a vessel oscillates or yaws back and forth across the course or as the direction of motion is temporarily changed, as in avoiding an obstacle. *Course* is a predetermined value and usually remains constant for a considerable time.

Bearing (B, Brg.) is the direction of one terrestrial point from another, expressed as angular distance from a reference direction, usually from 000° at the reference direction, clockwise through 360°. When measured through 90° or 180° from *either* north or south, it is called **bearing angle** (**B**), which bears the same relationship to bearing as *course angle* does to *course*. *Bearing* and *azimuth* are sometimes used interchangeably, but the latter is better reserved exclusively for reference to horizontal direction of a point on the celestial sphere from a point on the earth.

A **relative bearing** is one relative to the heading, or to the vessel itself. It is usually measured from 000° at the heading, clockwise through 360°. However, it is sometimes conveniently measured right or left from 0° at the ship's head through 180°. This is particularly true when using table 7. Older methods, such as indicating the number of degrees or points from some part of the vessel (10° forward of the starboard beam, two points on the port quarter, etc.) are seldom used by modern navigators to indicate precise directions, except for bearings dead ahead or astern, or broad on the bow, beam, or quarter.

To convert a relative bearing to a bearing from north (fig. I6b), express the relative bearing in terms of the 0°–360° system and add the heading:

True Bearing=Relative Bearing+Heading.

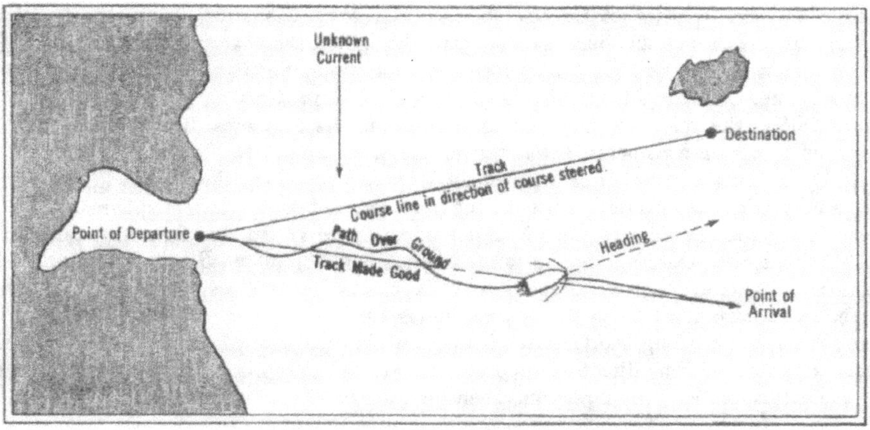

FIGURE I6a.—Course line, track, track made good, and heading.

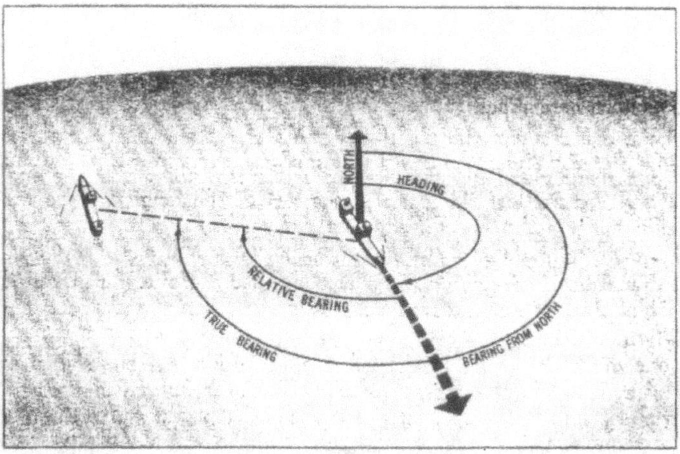

FIGURE I6b.—Relative bearing.

Thus, if another vessel bears 127° relative from a ship whose heading is 150°, the bearing from north is 127°+150°=277°. If the total exceeds 360°, subtract this amount. To convert a bearing from north to a relative bearing, subtract the heading:

$$\text{Relative Bearing} = \text{True Bearing} - \text{Heading}$$

Thus, a lightship which bears 241° from north bears 241°−137°=104° relative from a ship whose heading is 137°. If the heading is larger than the true bearing, add 360° to the true bearing before subtracting.

407. Grid direction.—Because of the rapid convergence of the meridians in polar regions, the true direction of an oblique line near the pole may vary considerably over a relatively few miles. The meridians are radial lines meeting at the poles, instead of being parallel, as they appear on the familiar Mercator chart.

Near the pole the convenience of parallel meridians is attained by means of a **polar grid.** On the chart a number of lines are printed parallel to a selected reference meridian, usually that of Greenwich. On transverse Mercator charts the fictitious meridians may serve this purpose. Any straight line on the chart makes the same angle with all grid lines. On the transverse Mercator projection it is therefore a **fictitious rhumb line.** On any polar projection it is a close approximation to a great circle. If north along the reference meridian is selected as the reference direction, all parallel grid lines can be considered extending in the same direction. The constant direction relative to the grid lines is called **grid direction.** North along the Greenwich meridian is usually taken as grid north in both the northern and southern hemispheres.

The value of grid directions is indicated in figure 17. In this figure *A* and *B* are 400 miles apart. The true bearing of *B* from *A* is 023°, yet at *B* this bearing line, if continued, extends in true direction 163°, a change of 140° in 400 miles. The grid direction at any point along the bearing line is 103°.

When north along the Greenwich meridian is used as grid north, interconversion between grid and true directions is quite simple. Let G represent a grid direction and T the corresponding true direction. Then for the arctic,

$$G = T + \lambda W.$$

That is, in the western hemisphere, in the arctic, grid direction is found by *adding* the longitude to the true direction. From this it follows that

$$T = G - \lambda W,$$

and in the eastern hemisphere

$$G = T - \lambda E,$$
$$T = G + \lambda E.$$

In the southern hemisphere the signs (+ or −) of the longitude are reversed in all formulas.

If a magnetic compass is used to follow a grid direction, variation and convergency can be combined into a single correction called **grid variation** or **grivation.** It is customary to show lines of equal grivation (**isogrivs**) on polar charts rather than lines of equal variation.

With one modification the grid system of direction can be used in any latitude. Meridians 1° apart make an angle of 1° with each other where they meet at the pole. The **convergency** is one, and the 360° of longitude cover all 360° around the pole. At the equator the meridians are parallel and the convergency is zero. Between these two limits the convergency has some value between zero and one. On a sphere it is equal to the sine of the latitude. For practical navigation this relationship can be used on the spheroidal earth. On a simple conic or Lambert conformal chart a constant convergency is used over the entire chart, and is known as the **constant of the cone.** On a simple conic projection it is equal to the sine of the standard parallel. On a Lambert conformal projection it is equal (approximately) to the sine of the latitude midway between the two standard parallels. When convergency is printed on the chart, it is generally adjusted for ellipticity of the earth. If *K* is the constant of the cone,

$$K = \sin \tfrac{1}{2}(L_1 + L_2) \text{ approx.},$$

where L_1 and L_2 are the latitudes of the two standard parallels. On such a chart, grid navigation is conducted as explained above, except that in each of the formulas the longitude is multiplied by *K*:

$$G = T + K\,\lambda W,$$
$$T = G - K\,\lambda W,$$
$$G = T - K\,\lambda E,$$
$$T = G + K\,\lambda E.$$

Thus, a straight line on such a chart changes its true direction, not by 1° for each degree of longitude, but by $K°$. As in higher latitudes, convergency and variation can be combined.

In using grid navigation one should keep clearly in mind the fact that the grid lines are parallel *on the chart*. Only on the transverse Mercator and polar gnomonic projections do the grid lines have geographical significance. On these projections, the grid lines are great circles which meet at "poles" on the equator, 90° from the meridian used as the fictitious equator. Since distortion varies on charts of different projections, and on charts of conic projections having different standard parallels, *the grid direction*

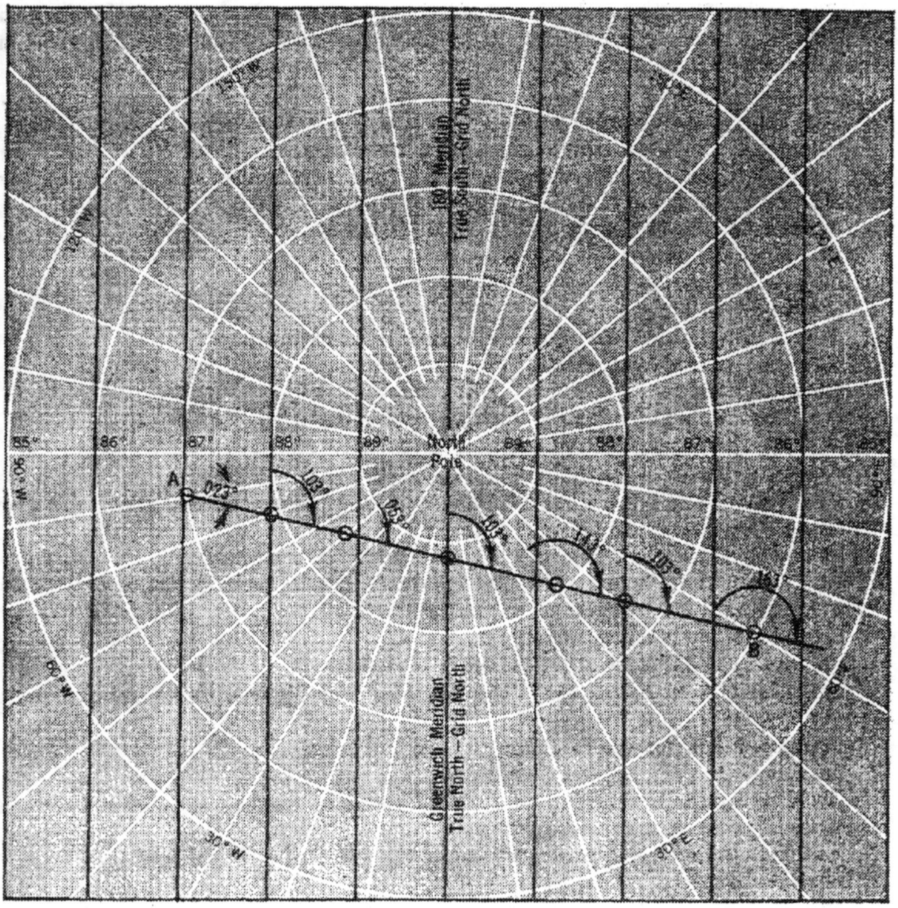

FIGURE 17.—Polar grid navigation.

between any two given points is not the same on all charts. For operations which are to be coordinated by means of grid directions, *it is important that all charts showing the grid be on a single graticule.*

Problems

I3. *Given.*—Point A: L 37°21′.4N, λ 143°18′.8W; Point B: L 43°04′.1N, λ 11°47′.3E; Point C: L 63°24′.4S, λ 132°06′.9E; Point D: L 2°36′.6S, λ 168°01′.2W.

Required.—(1) The difference of latitude between A and B, between A and C, and between C and D.

(2) The difference of longitude between A and B, A and C, and B and C.

Answers.—(1) l_{AB} 5°42′.7N, l_{AC} 100°45′.8S, l_{CD} 60°47′.8N; (2) DLo_{AB} 155°06′.1E, DLo_{AC} 84°34′.3W, DLo_{BC} 120°19′.6E.

I4a. The distance between points E and F is 258.4 nautical miles.

Required.—The distance in statute miles between points E and F (1) by proportion, using the ratio given in article 404; (2) by conversion factor, using the value given in article 404; (3) by table 20.

Answers.—(1) D 297.6 St M, (2) 297.2 St M, (3) D 297.4 St M.

I4b. The distance between points G and H is 83.3 statute miles.

Required.—The distance in nautical miles between points G and H (1) by proportion, using the ratio given in article 404; (2) by conversion factor, using the value given in article 404; (3) by table 20.

Answers.—(1) D 72.3 M, (2) D 72.4 M, (3) D 72.4 M.

I5a. A ship is steaming at 18.5 knots.

Required.—The speed in statute miles per hour.

Answer.—S 21.3 mph.

I5b. A motorboat is traveling at 30 statute miles per hour.

Required.—The speed in knots.

Answer.—S 26 kn.

I6a. *Required.*—Convert the following course angles to courses: (1) N 127°W, (2) S3°W, (3) N99°E, (4) S171°E.

Answers.—(1) Cn 233°, (2) Cn 183°, (3) Cn 099°, (4) Cn 009°.

I6b. *Required.*—Convert the following courses to course angles, giving the two possible answers of each: (1) 153°, (2) 257°.

Answers.—(1) C N153°E or S27°E, (2) C N103°W or S77° W.

I6c. A ship is on course 151°. The following relative bearings are observed: (1) 006°, (2) 109°, (3) 255°, (4) broad on the port bow.

Required.—The bearings from north.

Answers.—(1) 157°, (2) 260°, (3) 046°, (4) 106°.

I6d. A ship is on course 244°. The following bearings from north are observed: (1) 041°, (2) 188°, (3) 332°.

Required.—The relative bearings.

Answers.—(1) 157°, (2) 304°, (3) 088°.

I6e. The captain of a ship on course 055° wishes to change course when a certain lighthouse is broad on the starboard beam.

Required.—The bearing from north when the course is to be changed.

Answer.— 145°.

I7a. Convert the following true directions to grid directions using (1) a convergency of one, (2) a convergency of 0.866. (Give answers to nearest whole degree.)

True	*Latitude*	*Longitude*
157°	N	27° W
353°	N	114° E
118°	S	63° E
042°	S	147° W

Answers.—(1) 184°, 239°, 181°, 255°; (2) 180°, 254°, 173°, 275°.

I7b. Convert the following grid directions to true directions using (1) a convergency of 0.629, (2) a convergency of one.

Grid	*Latitude*	*Longitude*
003°	N	174° W
148°	S	9° W
317°	N	64° E
256°	S	155° E

Answers.—(1) 254°, 154°, 357°, 159°; (2) 189°, 157°, 021°, 101°.

APPENDIX J

MEASUREMENT ON THE CELESTIAL SPHERE

J1. The celestial sphere.—A glimpse of the sky on a clear night is sufficient to enable an observer to imagine that all of the heavenly bodies are located on the inner surface of a sphere (fig. J1a) of infinite radius with the earth at its center. This imaginary dome is called the **celestial sphere** (fig. J1b) whose north and south **celestial poles** are located by the extension of the earth's axis and whose **celestial equator** (sometimes called **equinoctial**) is formed by projecting the plane of the earth's equator to the celestial sphere. A **celestial meridian** is formed by the intersection of the plane of a terrestrial meridian, extended, and the celestial sphere. It is the arc of a great circle through the poles of the celestial sphere.

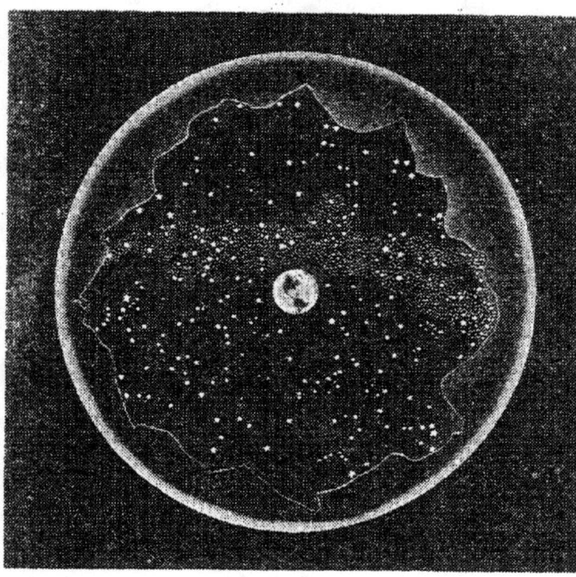

FIGURE J1a.—The celestial sphere.

The point on the celestial sphere vertically overhead of an observer is the **zenith** and the point on the opposite side of the sphere, vertically below him, is the **nadir**. The zenith and nadir are the extremities of a diameter of the celestial sphere through the observer and the common center of the earth and the celestial sphere. The arc of a celestial meridian between the poles is called the **upper branch** if it contains the zenith and the **lower branch** if it contains the nadir. The upper branch is frequently used in navigation and references to a celestial meridian are understood to mean only its upper branch unless otherwise stated. Celestial meridians take the names, as 65° west, of their terrestrial counterparts.

An **hour circle** is a great circle through the celestial poles and a point or body on the celestial sphere. It is similar to a celestial meridian, but moves with the celestial sphere as it rotates about the earth, while a celestial meridian remains fixed with respect to the earth.

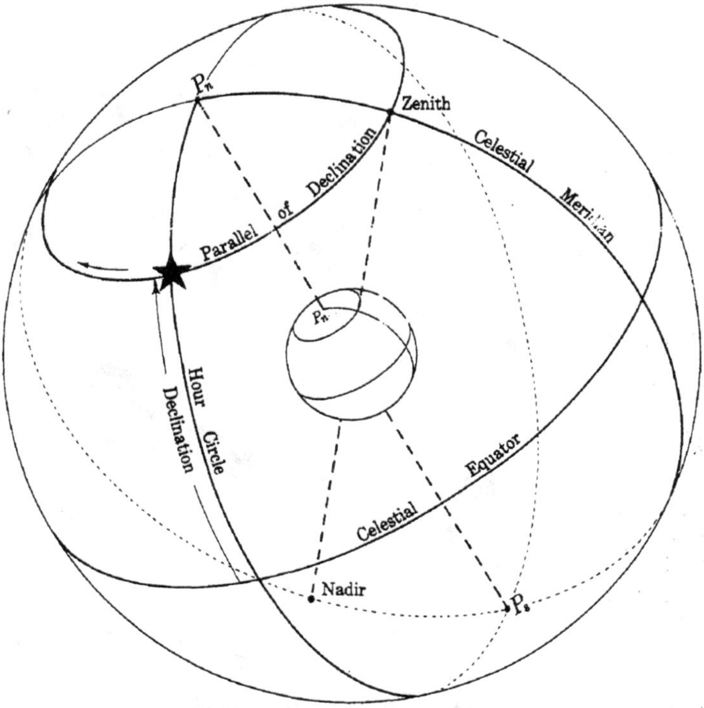

FIGURE J1b.—Elements of the celestial sphere.

The location of a body along its hour circle is defined by the body's angular distance from the celestial equator. This distance, called **declination,** is measured north or south of the celestial equator in degrees, from 0° through 90°, similar to latitude on the earth.

A circle parallel to the celestial equator is called a **parallel of declination,** since it connects all points of equal declination. It is similar to a parallel of latitude on the earth. The path of a celestial body during its daily apparent revolution around the earth is called its **diurnal circle.** It is not actually a circle if a body changes its declination. Since the declination of all navigational bodies is continually changing, the bodies are describing flat spherical spirals as they circle the earth. However, since the change is relatively slow, a diurnal circle and a parallel of declination are usually considered identical.

A point on the celestial sphere may be identified at the intersection of its parallel of declination and its hour circle. The parallel of declination is identified by the declination.

Two basic methods of locating the hour circle are in use. Its angular distance west of a reference hour circle through a point on the celestial sphere called the **vernal equinox** or **first point of Aries** is called **sidereal hour angle** (**SHA,** fig. J1c). This angle measured eastward from the vernal equinox is called **right ascension,** and is usually expressed in time units.

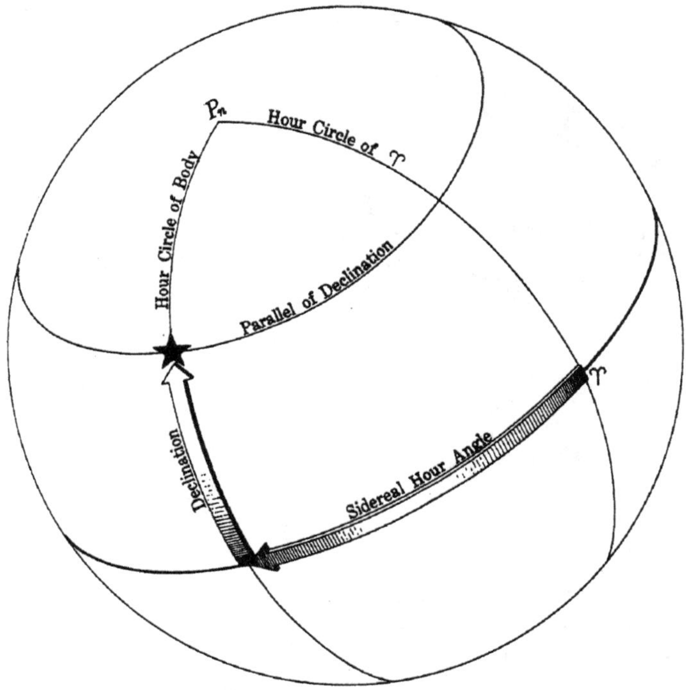

FIGURE J1c.—A point on the celestial sphere can be located by its declination and sidereal hour angle.

The second method of locating the hour circle is to indicate its angular distance west of a celestial meridian (fig. J1d). If the Greenwich celestial meridian is used as the reference, the angular distance is called **Greenwich hour angle (GHA),** and if the meridian of the observer, it is called **local hour angle (LHA).** It is sometimes more convenient to measure LHA either eastward or westward, as longitude is measured on the earth, in which case it is called **meridian angle (t).** These coordinates are discussed further in article J4.

A point on the celestial sphere may also be located by means of **altitude** and **azimuth,** coordinates based upon the horizon as the primary great circle, instead of the celestial equator. This system is discussed in article J6.

Two additional systems used by astronomers are based upon the ecliptic (art. J2) and the galactic equator (the approximate mid great circle of the galaxy). The coordinates of the ecliptic system are **celestial latitude** and **celestial longitude** and those of the galactic system are **galactic latitude** and **galactic longitude.**

J2. Apparent motion of celestial bodies.—To a terrestrial observer the earth seems to be stationary and the various celestial bodies appear to be in continual motion. It is this **apparent motion,** or motion relative to the earth, that concerns the navigator primarily.

The daily motion of the various bodies due to the rotation of the earth is the most conspicuous movement. Even a casual observer, however, soon notes that the various celestial bodies do not maintain the same positions relative to each other.

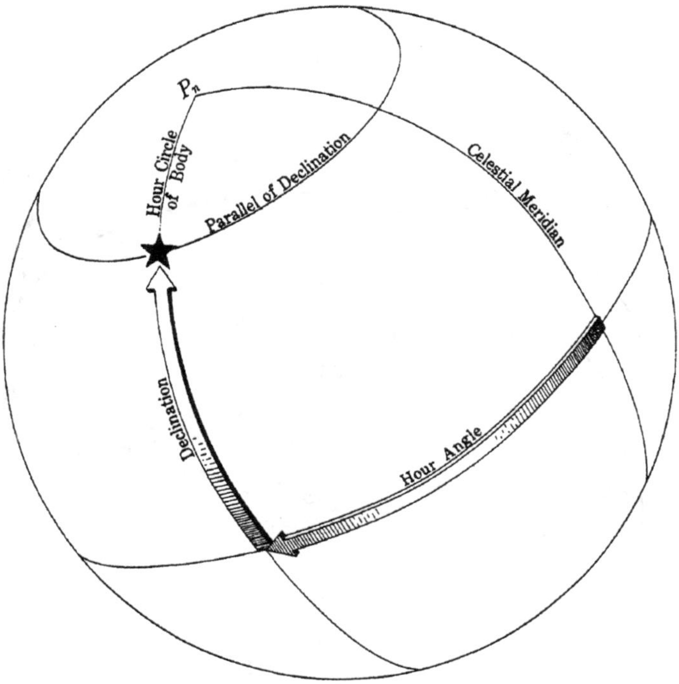

FIGURE J1d.—A point on the celestial sphere can be located by its declination and hour angle.

That is, while the celestial sphere appears to rotate about its axis, making one revolution about the apparently stationary earth each day, the various celestial bodies are changing their positions on the sphere.

If the rotation of the earth on its axis could be stopped, so that the daily apparent rotation of the celestial sphere would cease, the motions of the various bodies relative to each other would be more apparent. The stars would appear almost stationary. This is why they are sometimes called "fixed stars." They form a nearly stationary background which serves as a convenient reference for other motions.

The sun would appear to move eastward about 1° per day, completing its trip around the celestial sphere in 1 year. The motion is not along the celestial equator, but along another great circle, called the **ecliptic,** which is inclined to the celestial equator at an angle of about 23°27′ (fig. J2a). This great circle lies in the plane of the earth's orbit around the sun, since it is this *actual* motion that causes the *apparent* motion of the sun.

The moon would appear to make one trip around the celestial sphere each month, following closely the path of the sun. The planets would appear to move erratically, but keeping close to the ecliptic.

The two points at which the ecliptic crosses the celestial equator are called **equinoxes** (meaning "equal nights"), since days and nights are of equal length when the sun is at these points. The sun is at one of these points (♈) about March 21. This is called the **vernal** (spring) **equinox, first point of Aries,** or **March equinox.**

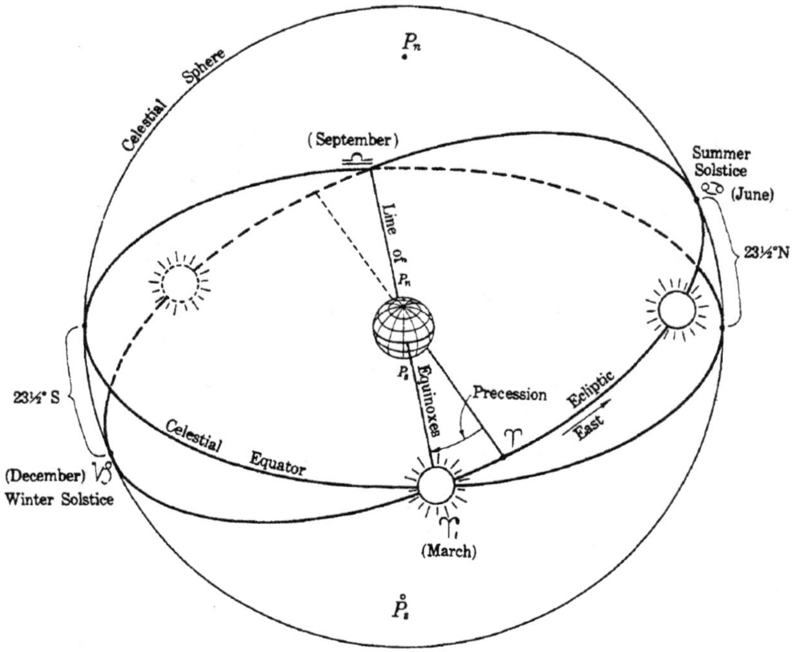

FIGURE J2a.—Elements of the ecliptic, along which the sun appears to move during 1 year.

Since the sun is on the celestial equator at this time, its declination is 0°. As it continues eastward, it arrives about June 21 at one of the points of maximum separation of the ecliptic and celestial equator. This is called the **summer solstice,** sometimes called **June solstice.** At this time the declination of the sun reaches its maximum value of about 23°27′ N, and since the ecliptic and celestial equator are parallel at this point, the *change* in declination of the sun is momentarily zero, increasing slowly on each side of this point. Hence, the name *solstice*, meaning "sun standing still."

Continuing on, the sun arrives at the **autumnal** or **September equinox** about September 23, when the declination is again 0° and the days and nights are again of equal length. In another 3 months it arrives at the **winter** or **December solstice** about December 22, when the declination is again maximum, but on the opposite side of the celestial equator. By March 21 it has arrived back at the vernal equinox, completing the cycle.

The terms *equinox* and *solstice* refer both to the points indicated and the times at which the sun is at these points.

J3. Coordinate systems.—Various systems of coordinates on the celestial sphere, all of them similar to the familiar latitude and longitude on the earth, were discussed briefly in article J1. Of these, the navigator is rarely concerned with any but the celestial equator system and the horizon system. The former is but an extension to the celestial sphere of the geographical system of the earth. The latter is a similar system in which the horizon replaces the celestial equator as the primary great circle, and the zenith and nadir are the poles. These two systems are the almost constant companions of the celestial navigator.

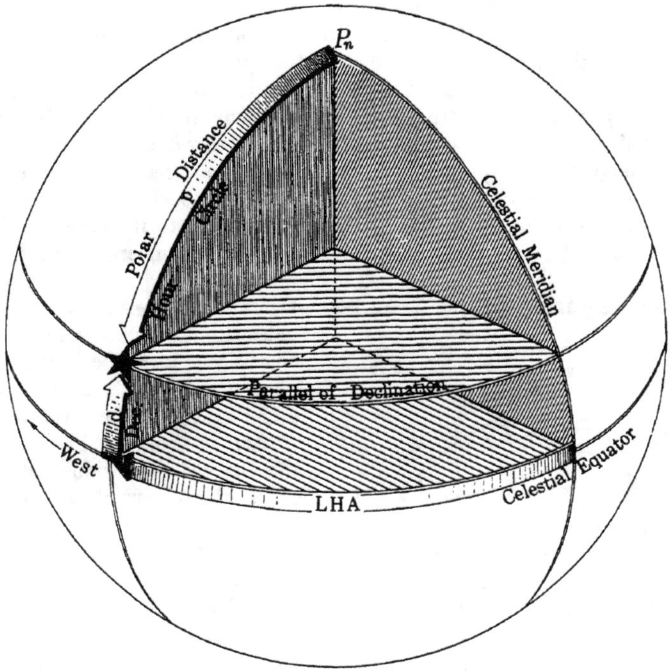

FIGURE J4a.—The celestial equator system of coordinates, showing measurement of declination, polar distance, and local hour angle.

J4. The celestial equator system of coordinates.—If the familiar graticule of latitude and longitude lines is expanded until it reaches the celestial sphere of infinite radius, it forms the basis of the celestial equator system of coordinates, as explained in article J1. On the celestial sphere the familiar *latitude* becomes **declination (d),** and *longitude,* measured always toward the west, through 360°, becomes **sidereal hour angle (SHA)** if measured from the vernal equinox.

Declination (d) is angular distance north or south of the celestial equator (d in fig. J4a). It is measured along an hour circle, from 0° at the celestial equator through 90° at the celestial poles, and is labeled N or S to indicate the direction of measurement. All points having the same declination lie along a **parallel of declination.**

Polar distance (p) is angular distance from a celestial pole, or the arc of an hour circle between the celestial pole and a point on the celestial sphere. It is measured along an hour circle and may vary from 0° to 180°, since either pole may be used as the origin of measurement. It is usually considered the complement of declination, though it may be either 90°−d or 90°+d, depending upon the pole used.

Local hour angle (**LHA**) is angular distance west of the local celestial meridian, or the arc of the celestial equator between the upper branch of the local celestial meridian and the hour circle through a point on the celestial sphere, measured westward from the local celestial meridian, through 360°. It is also the similar arc of the parallel of declination and the angle at the celestial pole, similarly measured. If the Greenwich (0°) meridian is used as the reference, instead of the local meridian, the expression **Greenwich hour angle** (GHA) is applied. It is sometimes convenient to measure the arc or angle in *either* an easterly or westerly direction from the local meridian, through 180°, when it is called **meridian angle** (**t**) and labeled E or W to indicate the direction of measurement. All bodies or other points having the same hour angle lie along the same **hour circle.**

The **time diagram** shown in figure J4b illustrates the relationship between the various hour angles and meridian angle. The circle is the celestial equator as seen from above the *south* pole, with the upper branch of the observer's meridian (P_sM) at the top. The radius P_sG is the Greenwich meridian, P_s♈ the hour circle of the vernal equinox, and P_sS and P_sS' the hour circles of celestial bodies to the *west* and *east*, respectively, of the observer's celestial meridian. Note that when LHA is less than 180°, t is numerically the same and is labeled W, but that when LHA is greater than 180°, t=360°−LHA and is labeled E. In figure J4b arc *GM* is the longitude, which in this case is west. The relationships shown apply equally to other arrangements of radii, except for relative magnitudes of the quantities involved.

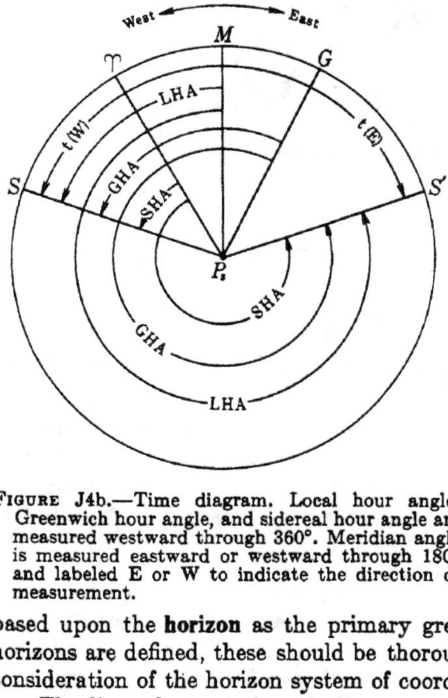

FIGURE J4b.—Time diagram. Local hour angle, Greenwich hour angle, and sidereal hour angle are measured westward through 360°. Meridian angle is measured eastward or westward through 180° and labeled E or W to indicate the direction of measurement.

J5. The horizons.—The second set of celestial coordinates with which the navigator is directly concerned is based upon the **horizon** as the primary great circle. However, since several different horizons are defined, these should be thoroughly understood before proceeding with a consideration of the horizon system of coordinates.

The line where earth and sky appear to meet is called the **visible** or **apparent horizon.** On land this is usually an irregular line unless the terrain is level. At sea the visible horizon appears very regular and often very sharp. However, its position relative to the celestial sphere depends primarily upon (1) the refractive index of the air, and (2) the height of the observer's eye above the surface.

Figure J5 shows a cross section of the earth and celestial sphere through the position of an observer at A above the surface of the earth. A straight line through A and the center of the earth O is the vertical of the observer, and contains his zenith (Z) and nadir (Na). A plane perpendicular to the true vertical is a horizontal plane, and its intersection with the celestial sphere is a horizon. It is the **celestial horizon** if the plane passes through the center of the earth, the **geoidal horizon** if it is tangent to the earth, and the **sensible horizon** if it passes through the eye of the observer at A. Since the radius of the earth is considered negligible with respect to that of the celestial sphere, these horizons become superimposed, and most measurements are referred only to the celestial horizon. This is sometimes called the **rational horizon** from the latin word "ratio," reckoning.

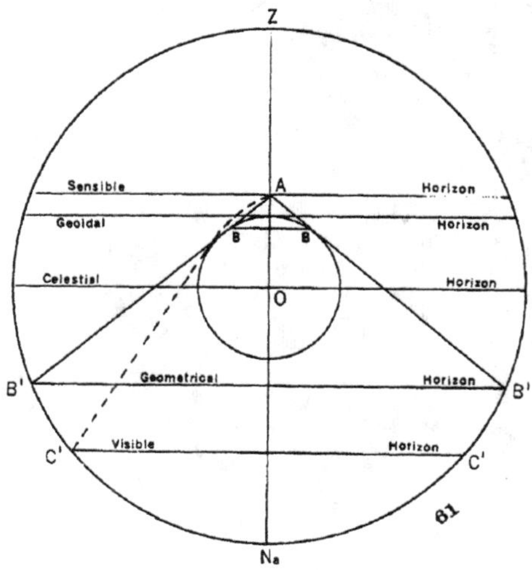

FIGURE J5.—The horizons used in navigation.

If the eye of the observer is at the surface of the earth, his visible horizon coincides with the plane of the geoidal horizon; but when elevated above the surface, as at A, his eye becomes the vertex of a cone which, neglecting refraction, is tangent to the earth at the small circle BB, and which intersects the celestial sphere in $B'B'$, the **geometrical horizon.** This expression is sometimes—but less appropriately—applied to the celestial horizon.

Because of refraction (arts. 805 and 806), the visible horizon $C'C'$ appears above but is actually slightly below the geometrical horizon as shown in figure J5.

For any elevation above the surface, the celestial horizon is usually *above* the geometrical and visible horizons, the difference increasing as elevation increases. It is thus possible to observe a body which is above the visible horizon but below the celestial horizon. That is, the body's altitude is negative and its zenith distance is greater than 90° (art. J6).

J6. The horizon system of coordinates is based upon the celestial horizon as the primary great circle, and a series of secondary **vertical circles,** which are great circles through the zenith and nadir of the observer and hence perpendicular to his horizon. Thus, the celestial horizon is similar to the equator, and the vertical circles are similar to meridians, but with one important difference. The celestial horizon and vertical circles are dependent upon the position of the observer and hence move with him as he changes position, while the primary and secondary great circles of both the geographical and celestial equator systems are independent of the observer. The horizon and celestial equator systems coincide for an observer at the geographical pole of the earth, and are mutually perpendicular for an observer on the equator. At all other places the two are oblique.

The vertical circle which passes through the poles of the celestial equator system of coordinates intersects the celestial horizon at its north and south points. One of these poles (having the same name as the latitude) is above the horizon and is called the **elevated pole.** The other, called the **depressed pole,** is below the horizon. Since this vertical circle is a great circle through the celestial poles, and includes the zenith

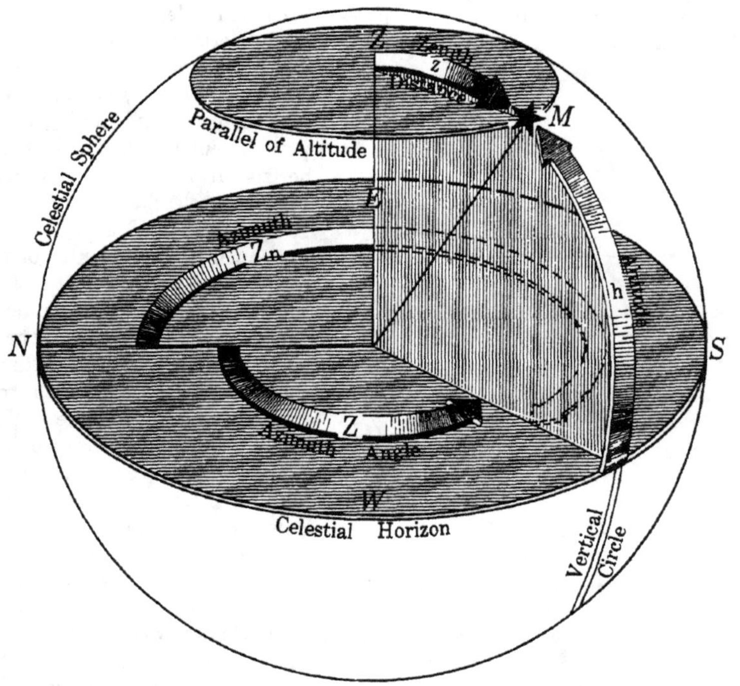

FIGURE J6.—The horizon system of coordinates, showing measurement of altitude, zenith distance, azimuth, and azimuth angle.

of the observer, it is also a celestial meridian. In the horizon system it is called the **principal vertical circle.** The vertical circle perpendicular to the principal vertical circle, hence intersecting the celestial horizon at its east and west points, is called the **prime vertical circle,** or simply the **prime vertical.**

As shown in figure J6, **altitude** is angular distance above the horizon. It is measured along a vertical circle, from 0° at the horizon through 90° at the zenith. Altitude measured from the visible horizon may exceed 90° because of the dip of the horizon, as shown in figure J5. Angular distance below the horizon, called **negative altitude,** is provided for by including certain negative altitudes in some tables for use in celestial navigation, such as Pub. No. 249. All points having the same altitude lie along a **parallel of altitude** or **almucantar.**

Zenith distance (**z**) is angular distance from the zenith, or the arc of a vertical circle between the zenith and a point on the celestial sphere. It is measured along a vertical circle from 0° through 180°. It is usually considered the complement of altitude. For a body above the celestial horizon it is equal to 90°−h and for a body below the celestial horizon it is equal to 90°−(−h) or 90°+h; or 90°+a negative altitude.

The horizontal direction of a point on the celestial sphere, or the bearing of the geographical position is called **azimuth** or **azimuth angle** depending upon the method of measurement. In both methods it is an arc of the horizon (or parallel of altitude) or an angle at the zenith. It is **azimuth** (**Zn**) if measured clockwise through 360°, starting at the north point on the horizon; and **azimuth angle** (**Z**) if measured *either* clockwise or counterclockwise through 180°, starting at the north point of the horizon in north latitude and the south point of the horizon in south latitude.

J7. Diagram on the plane of the celestial meridian.—From a point outside the celestial sphere (if this were possible) and over the celestial equator, at such a distance that the view would be orthographic, the great circle appearing as the outer limit would be a celestial meridian. Other celestial meridians would appear as ellipses. The celestial equator would appear as a diameter 90° from the poles, and parallels of declination as straight lines parallel to the equator.

A number of useful relationships can be demonstrated by drawing a **diagram on the plane of the celestial meridian** showing this orthographic view. Arcs of circles can be substituted for the ellipses without destroying the basic relationships. Refer to figure J7a. In the lower diagram the circle represents the celestial meridian, *QQ′* the celestial equator, *Pn* and *Ps* the north and south celestial poles, respectively. If a star has a declination of 30° N, an angle of 30° can be measured from the celestial equator, as

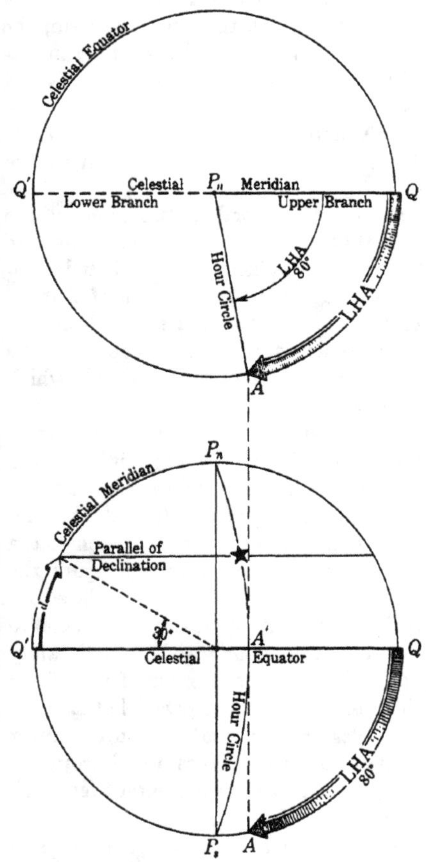

FIGURE J7a.—Measurement of celestial equator system of coordinates.

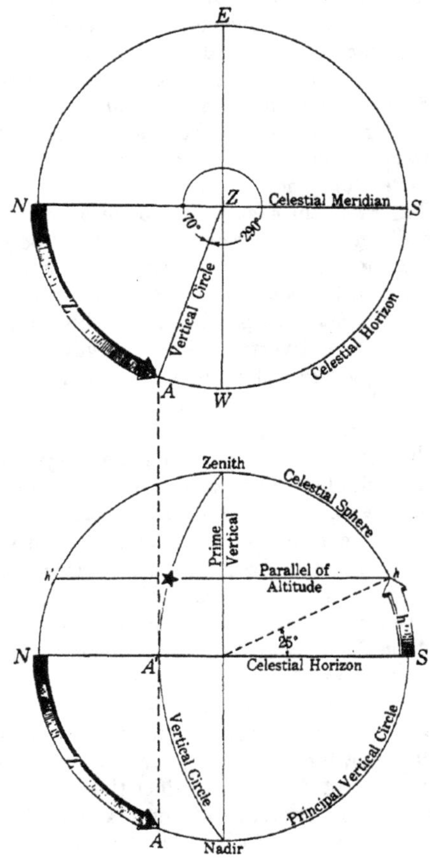

FIGURE J7b.—Measurement of horizon system of coordinates.

shown. It could be measured either to the right or left, and would have been toward the south pole if the declination had been south. The parallel of declination is a line through this point and parallel to the celestial equator. The star is somewhere on this line (actually a circle viewed on edge).

To locate the hour circle, draw the upper diagram so that *Pn* is directly above *Pn* of the lower figure (in line with the polar axis *Pn Ps*), and the circle is of the same diameter as that of the lower figure. This is the plan view, looking down on the celestial sphere from the top. The circle is the celestial equator. Since the view is from above the *north* celestial pole, west is clockwise. The diameter *QQ′* is the celestial meridian shown as a circle in the lower diagram. If the *right* half is considered the upper branch, local hour angle is measured clockwise from this line to the hour circle, as shown. In this case the LHA is 80°. The intersection of the hour circle and celestial equator, point *A*, can be projected down to the lower diagram (point *A′*) by a straight line parallel to the polar axis. The elliptical hour circle can be represented approximately by an arc of a circle through *A′*, *Pn*, *Ps*. The center of this circle is somewhere along the celestial equator line *QQ′*, extended if necessary. It is usually found by trial and error. The intersection of the hour circle and parallel of declination locates the star.

Since the upper diagram serves only to locate point *A′* in the lower diagram, the two can be combined. That is, the LHA arc can be drawn in the lower diagram, as shown, and point *A* projected *upward* to *A′*. In practice, the upper diagram is not drawn, being shown here for illustrative purposes only.

In this example the star is on that half of the sphere toward the observer, or the *western* part. If LHA had been greater than 180°, the body would have been on the *eastern* or "back" side.

From the east or west point over the celestial horizon, the orthographic view of the horizon system of coordinates would be similar to that of the celestial equator system from a point over the celestial equator (fig. J7a), since the celestial meridian is also the principal vertical circle. The horizon would appear as a diameter, parallels of altitude as straight lines parallel to the horizon, the zenith and nadir as poles 90° from the horizon, and vertical circles as ellipses through the zenith and nadir, except for the principal vertical circle, which would appear as a circle, and the prime vertical, which would appear as a diameter perpendicular to the horizon.

A celestial body can be located by altitude and azimuth in a manner similar to that used with the celestial equator system. If the altitude is 25°, this angle is measured from the horizon toward the zenith and the parallel of altitude is drawn as a straight line parallel to the horizon, as shown at *hh′* in the lower diagram of figure J7b. The plan view from above the zenith is shown in the upper diagram. If north is taken at the left, as shown, azimuths are measured clockwise from this point. In the figure the azimuth is 290° and the azimuth angle is N 70° W. The vertical circle is located by measuring either arc. Point *A* thus located can be projected vertically downward to *A′* on the horizon of the lower diagram, and the vertical circle represented approximately by the arc of a circle through *A′* and the zenith and nadir. The center of this circle is on *NS*, extended if necessary. The body is at the intersection of the parallel of altitude and the vertical circle. Since the upper diagram serves only to locate *A′* on the lower diagram, the two can be combined, point *A* located on the lower diagram and projected upward to *A′*, as shown. Since the body of the example has an azimuth greater than 180°, it is on the western or "front" side of the diagram.

Since the celestial meridian appears the same in both the celestial equator and horizon systems, the two diagrams can be combined and, if properly oriented, a body can be located by one set of coordinates, and the coordinates of the other system can be determined by measurement.

Refer to figure J7c. By convention, the zenith is shown at the top and the north point of the horizon at the left. The west point on the horizon is at the center, and the east point directly behind it. In the figure the latitude is 37° N. Therefore, the zenith is 37° north of the celestial equator. Since the zenith is established at the top of the diagram, the equator can be found by measuring an arc of 37° toward the south, along the celestial meridian. If the declination is 30° N andthe LHA is 80°, the body can be located as described above.

The altitude and azimuth can be determined by the reverse process to that described above. Draw a line *hh'* through the body and parallel to the horizon, *NS*. The altitude, 25°, is found by measurement, as shown. Draw the arc of a circle through the body and the zenith and nadir. From *A'*, the intersection of this arc with the horizon, draw a vertical line intersecting the circle at *A*. The azimuth, N 70° W, is found by measurement, as shown. The prefix N is applied to agree with the latitude. The body is left (*north*) of *ZNa*, the prime vertical circle. The suffix W applies because the LHA, 80°, shows that the body is west of the meridian.

If altitude and azimuth are given, the body is located as described above. The parallel of declination is then drawn parallel to *QQ'*, the celestial equator, and the declination determined by measurement. Point *L'* is located by drawing the arc of a circle through *Pn*, the star, and *Ps*. From *L'* a line is drawn perpendicular to *QQ'*, locating *L*. The meridian angle is then found by measurement. The declination is known to be north because the body is between the celestial equator and the north celestial pole. The meridian angle is west to agree with the azimuth angle, and hence LHA is numerically the same.

Since *QQ'* and *PnPs* are perpendicular, and *ZNa* and *NS* are also perpendicular, arc *NPn* is equal to arc *ZQ*. That is, *the altitude of the elevated pole is equal to the declination of the zenith, which is equal to the latitude*. This relationship is the basis of the method of determining latitude by an observation of Polaris.

The diagram on the plane of the celestial meridian is useful in approximating a number of relationships.

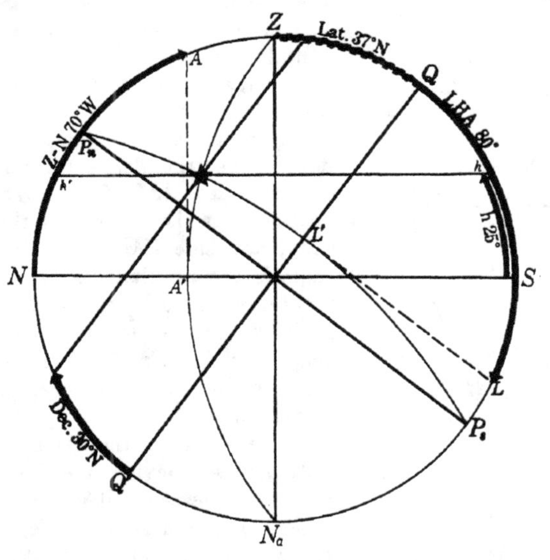

FIGURE J7c.—Diagram on the plane of the celestial meridian.

J8. The navigational triangle.—A triangle formed by arcs of great circles of a sphere is called a **spherical triangle.** A spherical triangle on the celestial sphere is called a **celestial triangle.** The spherical triangle of particular significance to navigators is called the **navigational triangle.** It is formed by arcs of a celestial meridian, an hour circle, and a vertical circle. Its vertices are the elevated pole, the zenith, and a point on the celestial sphere (usually a celestial body). The terrestrial counterpart is also called a navigational triangle, being formed by arcs of two meridians and the great circle connecting two places on the earth, one on each meridian. The vertices are the two places and a pole. In great-circle sailing these places are the point of departure and the destination. In celestial navigation they are the **assumed position (AP)** of the observer and the **geographical position (GP)** of the body (the place having the body in its zenith). The GP of the sun is sometimes called the **subsolar point,** that of the moon the **sublunar point,** that of a satellite (either natural or artificial) the **subsatellite point,** and that of a star its **substellar** or **subastral point.** When used to solve a celestial observation, either the celestial or terrestrial triangle may be called the **astronomical triangle.**

The navigational triangle is shown in figure J8a on a diagram on the plane of the celestial meridian, labeled as in article J7, but with the hour circle and vertical circle properly shown as ellipses. The earth is at the center, *O*. The star is at *M*, *dd'* is its parallel of declination, and *hh'* its altitude circle.

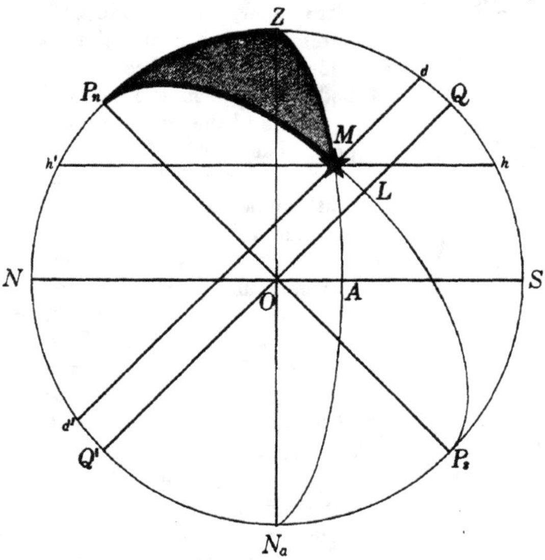

FIGURE J8a.—The navigational triangle.

In the figure, arc *QZ* of the celestial meridian is the latitude of the observer, and *PnZ*, one side of the triangle, is the **colatitude.** Arc *AM* of the vertical circle is the altitude of the body, and side *ZM* of the triangle is the zenith distance, or **coaltitude.** Arc *LM* of the hour circle is the declination of the body, and side *PnM* of the triangle is the polar distance, or **codeclination.**

The angle at the elevated pole, *ZPnM*, having the hour circle and the celestial meridian as sides, is the meridian angle, t. The angle at the zenith, *PnZM*, having the vertical circle and that arc of the celestial meridian which includes the elevated pole as sides, is the azimuth angle. The angle at the celestial body, *ZMPn*, having the hour circle and the vertical circle as sides, is the **parallactic angle** (sometimes called **position angle**). It is rarely used in celestial navigation.

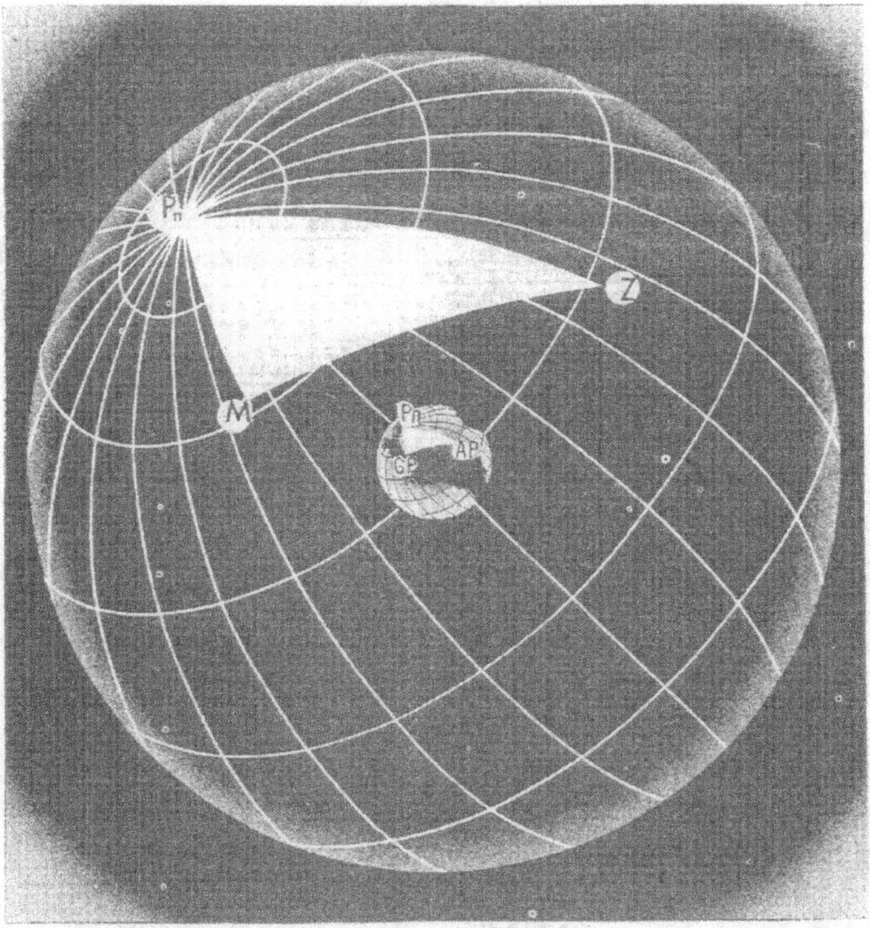

FIGURE J8b.—The navigational triangle in perspective.

APPENDIX K

EXTRACTS FROM PUB. NO. 214

Lat. 41°

DECLINATION SAME NAME AS LATITUDE

H.A.	19° 00′ Alt.	Δd Δt	Az.	19° 30′ Alt.	Δd Δt	Az.	20° 00′ Alt.	Δd Δt	Az.	20° 30′ Alt.	Δd Δt	Az.
°	° ′		°	° ′		°	° ′		°	° ′		°
00	68 00.0	1.0 02	180.0	68 30.0	1.0 02	180.0	69 00.0	1.0 02	180.0	69 30.0	1.0 02	180.0
1	67 59.0	1.0 05	177.5	68 29.0	1.0 05	177.4	68 59.0	1.0 05	177.4	69 28.9	1.0 05	177.3
2	67 56.0	1.0 08	175.0	68 25.9	1.0 08	174.9	68 55.9	1.0 09	174.8	69 25.8	1.0 09	174.7
3	67 51.1	99 12	172.5	68 20.9	99 12	172.3	68 50.7	99 12	172.2	69 20.5	99 12	172.0
4	67 44.1	99 15	170.0	68 13.8	99 15	169.8	68 43.5	99 15	169.6	69 13.2	99 16	169.4
05	67 35.3	99 18	167.5	68 04.8	98 18	167.3	68 34.4	98 19	167.0	69 03.9	98 19	166.8
6	67 24.6	98 21	165.1	67 53.9	98 21	164.8	68 23.2	98 22	164.5	68 52.5	98 22	164.2
7	67 12.0	97 24	162.7	67 41.1	97 24	162.4	68 10.2	97 25	162.1	68 39.3	97 25	161.7
8	66 57.7	96 27	160.4	67 26.5	96 27	160.0	67 55.4	96 28	159.6	68 24.2	96 28	159.3
9	66 41.6	95 30	158.0	67 10.2	95 30	157.7	67 38.7	95 31	157.3	68 07.2	95 31	156.8
10	66 23.8	94 32	155.8	66 52.1	94 33	155.4	67 20.4	94 33	154.9	67 48.6	94 34	154.5
1	66 04.5	93 35	153.6	66 32.5	93 35	153.1	67 00.4	93 36	152.7	67 28.2	93 36	152.2
2	65 43.6	92 37	151.4	66 11.2	92 38	151.0	66 38.8	92 38	150.5	67 06.3	92 39	150.0
3	65 21.2	91 40	149.3	65 48.5	91 40	148.8	66 15.8	91 41	148.3	66 42.9	90 41	147.8
4	64 57.4	90 42	147.3	65 24.4	90 42	146.8	65 51.3	89 43	146.2	66 18.1	89 44	145.7
30	56 15.7	73 65	121.7	56 37.6	73 65	121.0	56 59.3	72 65	120.4	57 20.9	71 66	119.8
1	55 36.9	73 65	120.4	55 58.6	72 66	119.8	56 20.0	71 66	119.2	56 41.3	71 67	118.5
2	54 57.6	72 66	119.2	55 19.0	71 67	118.6	55 40.3	71 67	118.0	56 01.3	70 67	117.4
3	54 17.9	71 67	118.1	54 39.1	70 67	117.5	55 00.1	70 68	116.8	55 20.9	69 68	116.2
4	53 37.7	70 68	116.9	53 58.7	70 68	116.3	54 19.5	69 68	115.7	54 40.1	68 69	115.1
35	52 57.1	70 68	115.8	53 17.9	69 69	115.2	53 38.5	68 69	114.6	53 58.9	68 69	114.0
6	52 16.2	69 69	114.7	52 36.8	68 69	114.1	52 57.1	68 69	113.5	53 17.4	67 70	112.9
7	51 34.9	68 69	113.7	51 55.3	68 70	113.1	52 15.5	67 70	112.5	52 35.5	66 70	111.9
8	50 53.3	68 70	112.7	51 13.5	67 70	112.1	51 33.5	66 70	111.5	51 53.3	66 71	110.9
9	50 11.3	67 70	111.7	50 31.4	66 71	111.1	50 51.2	66 71	110.5	51 10.9	65 71	109.9
60	34 46.6	61 75	94.5	35 04.7	60 75	94.0	35 22.7	60 75	93.5	35 40.6	59 75	93.0
1	34 01.5	60 75	93.8	34 19.6	60 75	93.3	34 37.5	60 75	92.8	34 55.4	59 75	92.3
2	33 16.3	60 75	93.1	33 34.3	60 75	92.6	33 52.3	60 75	92.2	34 10.1	59 75	91.7
3	32 31.0	60 75	92.4	32 49.1	60 75	92.0	33 07.0	60 75	91.5	33 24.9	59 75	91.0
4	31 45.8	60 75	91.8	32 03.8	60 75	91.3	32 21.8	60 75	90.8	32 39.6	59 75	90.3
80	19 43.5	61 75	81.6	20 01.9	61 75	81.1	20 20.2	61 74	80.7	20 38.5	61 74	80.3
1	18 58.7	62 74	81.0	19 17.2	61 74	80.5	19 35.5	61 74	80.1	19 53.9	61 74	79.7
2	18 14.1	62 74	80.3	18 32.5	62 74	79.9	18 51.0	61 74	79.5	19 09.4	61 74	79.1
3	17 29.5	62 74	79.7	17 48.0	62 74	79.3	18 06.5	62 74	78.9	18 24.9	61 74	78.5
4	16 45.0	62 74	79.1	17 03.6	62 74	78.7	17 22.1	63 74	78.3	17 40.6	62 74	77.9
85	16 00.5	62 74	78.5	16 19.2	62 74	78.1	16 37.8	62 74	77.7	16 56.4	62 74	77.3
6	15 16.2	63 74	77.9	15 34.9	62 74	77.5	15 53.6	62 73	77.1	16 12.3	62 73	76.7
7	14 32.0	63 74	77.3	14 50.8	63 73	76.9	15 09.6	62 73	76.5	15 28.3	62 73	76.1
8	13 47.9	63 73	76.7	14 06.7	63 73	76.3	14 25.6	63 73	75.9	14 44.4	63 73	75.5
9	13 03.9	63 73	76.0	13 22.8	63 73	75.6	13 41.7	63 73	75.3	14 00.6	63 73	74.8
90	12 20.0	63 73	75.4	12 39.0	63 73	75.0	12 58.0	63 73	74.6	13 17.0	63 73	74.2

H.A.	21° 00′ Alt.	Δd Δt	Az.	21° 30′ Alt.	Δd Δt	Az.	22° 00′ Alt.	Δd Δt	Az.	22° 30′ Alt.	Δd Δt	Az.	H.A.
°	° ′		°	° ′		°	° ′		°	° ′		°	°
00	70 00.0	1.0 02	180.0	70 30.0	1.0 02	180.0	71 00.0	1.0 02	180.0	71 30.0	1.0 02	180.0	00
1	69 58.9	1.0 05	177.3	70 28.9	1.0 06	177.2	70 58.9	1.0 06	177.2	71 28.8	1.0 06	177.1	1
2	69 55.7	1.0 09	174.6	70 25.6	1.0 09	174.4	70 55.5	1.0 09	174.3	71 25.4	1.0 10	174.2	2
3	69 50.3	99 12	171.9	70 20.1	99 13	171.7	70 49.9	99 13	171.5	71 19.7	99 13	171.3	3
4	69 42.9	99 16	169.2	70 12.5	99 16	168.9	70 42.1	99 17	168.7	71 11.7	99 17	168.5	4
05	69 33.3	98 19	166.5	70 02.8	98 20	166.3	70 32.2	98 20	166.0	71 01.6	98 20	165.7	05
6	69 21.8	97 22	163.9	69 51.0	97 23	163.6	70 20.2	97 23	163.3	70 49.3	97 24	162.9	6
7	69 08.3	97 26	161.4	69 37.2	96 26	161.0	70 06.1	96 27	160.6	70 35.0	96 27	160.2	7
8	68 52.9	96 29	158.9	69 21.5	95 29	158.4	69 50.2	95 30	158.0	70 18.7	95 30	157.6	8
9	68 35.6	95 32	156.4	69 04.0	94 32	156.0	69 32.3	94 33	155.5	70 00.5	94 33	155.0	9
10	68 16.7	94 34	154.0	68 44.7	93 35	153.5	69 12.6	93 36	153.0	69 40.4	93 36	152.5	10
1	67 56.0	92 37	151.7	68 23.7	92 38	151.2	68 51.2	92 38	150.6	69 18.7	91 39	150.1	1
2	67 33.7	91 40	149.4	68 01.0	91 40	148.9	68 28.2	90 41	148.3	68 55.3	90 42	147.7	2
3	67 10.0	90 42	147.2	67 36.9	90 43	146.7	68 03.7	89 43	146.1	68 30.3	89 44	145.4	3
4	66 44.8	89 44	145.1	67 11.3	88 45	144.5	67 37.7	88 46	143.9	68 03.9	87 46	143.2	4
30	57 42.2	71 66	119.1	58 03.4	70 67	118.4	58 24.3	69 67	117.8	58 45.0	69 68	117.1	30
1	57 02.4	70 67	117.9	57 23.3	69 67	117.2	57 44.0	69 68	116.6	58 04.4	68 68	115.9	1
2	56 22.2	69 68	116.7	56 42.8	68 68	116.1	57 03.3	68 68	115.4	57 23.5	67 69	114.7	2
3	55 41.5	68 68	115.6	56 02.0	68 69	114.9	56 22.2	67 69	114.2	56 42.2	66 69	113.6	3
4	55 00.5	68 69	114.4	55 20.7	67 69	113.8	55 40.7	66 70	113.1	56 00.5	66 70	112.5	4
35	54 19.1	67 70	113.4	54 39.1	66 70	112.7	54 58.9	66 70	112.1	55 18.5	65 71	111.4	35
6	53 37.4	66 70	112.3	53 57.2	66 70	111.7	54 16.8	65 71	111.0	54 36.2	64 71	110.4	6
7	52 55.3	66 71	111.3	53 14.9	65 71	110.6	53 34.4	64 71	110.0	53 53.6	64 71	109.3	7
8	52 13.0	65 71	110.3	52 32.4	65 71	109.6	52 51.7	64 72	109.0	53 10.7	63 72	108.4	8
9	51 30.4	65 71	109.3	51 49.7	64 72	108.7	52 08.8	63 72	108.0	52 27.6	63 72	107.4	9
60	35 58.4	59 75	92.5	36 16.0	59 75	92.0	36 33.5	58 75	91.5	36 50.9	58 75	91.0	60
1	35 13.1	59 75	91.9	35 30.7	59 75	91.4	35 48.2	58 75	90.9	36 05.6	58 75	90.4	1
2	34 27.9	59 75	91.2	34 45.5	58 75	90.7	35 02.9	58 75	90.2	35 20.3	58 75	89.7	2
3	33 42.6	59 75	90.5	34 00.2	58 75	90.0	34 17.7	58 75	89.5	34 35.0	58 75	89.1	3
4	32 57.3	59 75	89.9	33 14.9	58 75	89.4	33 32.4	58 75	88.9	33 49.8	58 75	88.4	4
80	20 56.7	61 74	79.9	21 14.8	60 74	79.5	21 32.9	60 74	79.0	21 50.9	60 74	78.6	80
1	20 12.1	61 74	79.3	20 30.3	61 74	78.9	20 48.5	60 74	78.4	21 06.5	60 74	78.0	1
2	19 27.7	61 74	78.7	19 45.9	61 74	78.2	20 04.1	61 74	77.8	20 22.3	60 74	77.4	2
3	18 43.3	61 74	78.1	19 01.7	61 74	77.6	19 19.9	61 74	77.2	19 38.1	61 73	76.8	3
4	17 59.1	61 74	77.5	18 17.5	61 73	77.0	18 35.8	61 73	76.6	18 54.1	61 73	76.2	4
85	17 14.9	62 73	76.9	17 33.4	61 73	76.4	17 51.8	61 73	76.0	18 10.2	61 73	75.6	85
6	16 30.9	62 73	76.3	16 49.4	62 73	75.8	17 07.9	62 73	75.4	17 26.4	61 73	75.0	6
7	15 46.9	62 73	75.7	16 05.6	62 73	75.2	16 24.2	62 73	74.8	16 42.7	62 73	74.4	7
8	15 03.1	62 73	75.1	15 21.8	62 73	74.6	15 40.5	62 73	74.2	15 59.1	62 72	73.8	8
9	14 19.4	63 73	74.4	14 38.2	63 72	74.0	14 57.0	62 72	73.6	15 15.7	62 72	73.2	9
90	13 35.9	63 72	73.8	13 54.8	63 72	73.4	14 13.6	63 72	73.0	14 32.4	63 72	72.6	90

DECLINATION CONTRARY NAME TO LATITUDE

Lat. 41°

H.A.	19° 00′ Alt.	Δd Δt	Az.	19° 30′ Alt.	Δd Δt	Az.	20° 00′ Alt.	Δd Δt	Az.	20° 30′ Alt.	Δd Δt	Az.
°	° ′		°	° ′		°	° ′		°	° ′		°
00	30 00.0	1.0 01	180.0	29 30.0	1.0 01	180.0	29 00.0	1.0 01	180.0	28 30.0	1.0 01	180.0
1	29 59.6	1.0 02	178.9	29 29.6	1.0 02	178.9	28 59.6	1.0 02	178.9	28 29.6	1.0 02	178.9
2	29 58.3	1.0 04	177.8	29 28.3	1.0 04	177.8	28 58.3	1.0 04	177.9	28 28.3	1.0 03	177.9
3	29 56.1	1.0 05	176.7	29 26.2	1.0 05	176.8	28 56.2	1.0 05	176.8	28 26.2	1.0 05	176.8
4	29 53.1	1.0 06	175.6	29 23.2	1.0 06	175.7	28 53.2	1.0 06	175.7	28 23.3	1.0 06	175.7
05	29 49.2	1.0 08	174.5	29 19.3	1.0 08	174.6	28 49.4	1.0 08	174.6	28 19.5	1.0 08	174.7
6	29 44.5	1.0 09	173.5	29 14.6	1.0 09	173.5	28 44.7	1.0 09	173.6	28 14.9	1.0 09	173.6
7	29 38.9	99 11	172.4	29 09.1	99 11	172.4	28 39.3	99 11	172.5	28 09.4	99 10	172.6
8	29 32.5	99 12	171.3	29 02.7	99 12	171.4	28 32.9	99 12	171.4	28 03.1	99 12	171.5
9	29 25.2	99 14	170.2	28 55.5	99 13	170.3	28 25.8	99 13	170.4	27 56.0	99 13	170.5
10	29 17.1	99 15	169.1	28 47.5	99 15	169.2	28 17.8	99 15	169.3	27 48.1	99 15	169.4
1	29 08.2	99 16	168.1	28 38.6	99 16	168.2	28 09.0	99 16	168.3	27 39.4	99 16	168.4
2	28 58.4	98 18	167.0	28 28.9	98 18	167.1	27 59.4	98 17	167.2	27 29.9	98 17	167.3
3	28 47.8	98 19	166.0	28 18.4	98 19	166.1	27 49.0	98 19	166.2	27 19.5	98 19	166.3
4	28 36.4	98 20	164.9	28 07.1	98 20	165.0	27 37.7	98 20	165.1	27 08.4	98 20	165.2
15	28 24.2	98 22	163.8	27 55.0	98 22	164.0	27 25.7	98 21	164.1	26 56.5	98 21	164.2
6	28 11.2	97 23	162.8	27 42.1	97 23	162.9	27 12.9	97 23	163.1	26 43.8	97 22	163.2
7	27 57.5	97 24	161.8	27 28.4	97 24	161.9	26 59.3	97 24	162.0	26 30.3	97 24	162.2
8	27 42.9	96 26	160.7	27 14.0	97 25	160.9	26 45.0	97 25	161.0	26 16.0	97 25	161.2
9	27 27.6	96 27	159.7	26 58.7	96 27	159.9	26 29.9	96 26	160.0	26 01.0	96 26	160.2
20	27 11.5	96 28	158.7	26 42.8	96 28	158.8	26 14.0	96 28	159.0	25 45.3	96 27	159.2
1	26 54.7	95 29	157.7	26 26.1	95 29	157.8	25 57.5	95 29	158.0	25 28.8	95 29	158.2
2	26 37.1	95 31	156.7	26 08.6	95 30	156.8	25 40.1	95 30	157.0	25 11.6	95 30	157.2
3	26 18.8	94 32	155.7	25 50.4	94 31	155.8	25 22.1	95 31	156.0	24 53.7	95 31	156.2
4	25 59.8	94 33	154.7	25 31.6	94 33	154.9	25 03.3	94 32	155.0	24 35.1	94 32	155.2
25	25 40.0	94 34	153.7	25 12.0	94 34	153.9	24 43.9	94 34	154.1	24 15.8	94 33	154.3
6	25 19.6	93 35	152.7	24 51.7	93 35	152.9	24 23.7	93 35	153.1	23 55.8	93 34	153.3
7	24 58.5	93 36	151.7	24 30.7	93 36	151.9	24 02.9	93 36	152.2	23 35.1	93 36	152.4
8	24 36.7	92 37	150.8	24 09.1	92 37	151.0	23 41.4	92 37	151.2	23 13.8	92 37	151.4
9	24 14.3	92 38	149.8	23 46.8	92 38	150.0	23 19.3	92 38	150.3	22 51.8	92 38	150.5
55	11 17.2	78 60	127.8	10 53.9	78 60	128.2	10 30.5	78 59	128.5	10 07.2	78 59	128.8
6	10 41.2	77 60	127.1	10 18.1	77 60	127.4	9 54.9	77 60	127.7	9 31.7	77 60	128.1
7	10 04.9	77 61	126.3	9 41.9	77 61	126.7	9 18.9	77 61	127.0	8 55.9	77 60	127.3
8	9 28.3	76 62	125.6	9 05.4	76 61	125.9	8 42.6	76 61	126.3	8 19.7	76 61	126.6
9	8 51.3	76 62	124.9	8 28.6	76 62	125.2	8 05.9	76 62	125.6	7 43.2	76 61	125.9
60	8 14.0	75 63	124.2	7 51.5	75 62	124.5	7 28.9	75 62	124.8	7 06.3	75 62	125.2
1	7 36.4	75 63	123.5	7 14.0	75 63	123.8	6 51.6	75 63	124.1	6 29.2	75 62	124.5
2	6 58.4	74 64	122.7	6 36.2	74 63	123.1	6 13.9	74 63	123.4	5 51.7	74 63	123.8
3	6 20.2	74 64	122.0	5 58.1	74 64	122.4	5 36.0	74 64	122.7	5 13.9	74 63	123.1
4	5 41.7	73 65	121.3	5 19.7	73 64	121.7						

H.A.	21° 00′ Alt.	Δd Δt	Az.	21° 30′ Alt.	Δd Δt	Az.	22° 00′ Alt.	Δd Δt	Az.	22° 30′ Alt.	Δd Δt	Az.
°	° ′		°	° ′		°	° ′		°	° ′		°
00	28 00.0	1.0 01	180.0	27 30.0	1.0 01	180.0	27 00.0	1.0 01	180.0	26 30.0	1.0 01	180.0
1	27 59.6	1.0 02	178.9	27 29.6	1.0 02	179.0	26 59.6	1.0 02	179.0	26 29.6	1.0 02	179.0
2	27 58.3	1.0 04	177.9	27 28.3	1.0 03	177.9	26 58.4	1.0 03	177.9	26 28.4	1.0 03	177.9
3	27 56.2	1.0 05	176.8	27 26.3	1.0 05	176.9	26 56.3	1.0 05	176.9	26 26.3	1.0 05	176.9
4	27 53.3	1.0 06	175.8	27 23.4	1.0 06	175.8	26 53.4	1.0 06	175.8	26 23.5	1.0 06	175.9
05	27 49.6	1.0 08	174.7	27 19.7	1.0 08	174.8	26 49.7	1.0 08	174.8	26 19.8	1.0 07	174.8
6	27 45.0	1.0 09	173.7	27 15.1	1.0 09	173.7	26 45.2	1.0 09	173.8	26 15.3	1.0 09	173.8
7	27 39.6	99 10	172.6	27 09.7	99 10	172.7	26 39.9	99 10	172.7	26 10.1	99 10	172.8
8	27 33.4	99 12	171.6	27 03.6	99 12	171.6	26 33.8	99 12	171.7	26 04.0	99 11	171.8
9	27 26.3	99 13	170.5	26 56.6	99 13	170.6	26 26.8	99 13	170.7	25 57.1	99 13	170.8
10	27 18.5	99 14	169.5	26 48.8	99 14	169.6	26 19.1	99 14	169.7	25 49.4	99 14	169.7
1	27 09.8	99 16	168.5	26 40.2	99 16	168.5	26 10.6	99 16	168.6	25 41.0	99 15	168.7
2	27 00.3	98 17	167.4	26 30.8	98 17	167.5	26 01.3	98 17	167.6	25 31.7	98 17	167.7
3	26 50.1	98 18	166.4	26 20.6	98 18	166.5	25 51.2	98 18	166.6	25 21.7	98 18	166.7
4	26 39.0	98 20	165.4	26 09.6	98 20	165.5	25 40.3	98 19	165.6	25 10.9	98 19	165.7
15	26 27.2	98 21	164.3	25 57.9	98 21	164.5	25 28.6	98 21	164.6	24 59.3	98 21	164.7
6	26 14.6	97 22	163.3	25 45.4	97 22	163.5	25 16.2	97 22	163.6	24 47.0	97 22	163.7
7	26 01.2	97 24	162.3	25 32.1	97 23	162.5	25 03.0	97 23	162.6	24 33.9	97 23	162.7
8	25 47.1	97 25	161.3	25 18.1	97 25	161.5	24 49.1	97 24	161.6	24 20.1	97 24	161.7
9	25 32.2	96 26	160.3	25 03.3	96 26	160.5	24 34.4	96 26	160.6	24 05.5	96 25	160.8
20	25 16.6	96 27	159.3	24 47.8	96 27	159.5	24 19.0	96 27	159.6	23 50.3	96 27	159.8
1	25 00.2	96 28	158.3	24 31.6	96 28	158.5	24 02.9	96 28	158.7	23 34.3	96 28	158.8
2	24 43.1	95 30	157.4	24 14.6	95 29	157.5	23 46.1	95 29	157.7	23 17.6	95 29	157.9
3	24 25.3	95 31	156.4	23 57.0	95 31	156.6	23 28.6	95 30	156.7	23 00.1	95 30	156.9
4	24 06.9	94 32	155.4	23 38.6	94 32	155.6	23 10.3	94 32	155.8	22 42.0	94 31	156.0
25	23 47.7	94 33	154.5	23 19.5	94 33	154.6	22 51.4	94 33	154.8	22 23.3	94 32	155.0
6	23 27.8	93 34	153.5	22 59.8	93 34	153.7	22 31.8	93 34	153.9	22 03.8	93 34	154.1
7	23 07.3	93 35	152.6	22 39.4	93 35	152.8	22 11.6	93 35	153.0	21 43.7	93 35	153.2
8	22 46.1	92 36	151.6	22 18.4	92 36	151.8	21 50.6	92 36	152.0	21 22.9	92 36	152.2
9	22 24.2	92 38	150.7	21 56.7	92 37	150.9	21 29.1	92 37	151.1	21 01.5	92 37	151.3
55	9 43.9	78 59	129.1	9 20.5	78 59	129.4	8 57.1	78 58	129.7	8 33.7	78 58	130.1
6	9 08.5	77 59	128.4	8 45.3	77 59	128.7	8 22.1	77 59	129.0	7 58.8	78 59	129.3
7	8 32.9	77 60	127.7	8 09.8	77 60	128.0	7 46.7	77 60	128.3	7 23.6	77 59	128.6
8	7 56.8	76 61	126.9	7 33.9	76 60	127.3	7 11.0	76 60	127.6	6 48.1	76 60	127.9
9	7 20.5	76 61	126.2	6 57.7	76 61	126.5	6 35.0	76 61	126.9	6 12.2	76 60	127.2
60	6 43.8	75 62	125.5	6 21.2	75 61	125.8	5 58.6	75 61	126.2	5 35.9	75 61	126.5
1	6 06.7	75 62	124.8	5 44.3	75 62	125.1	5 21.8	75 62	125.5			
2	5 29.4	74 63	124.1	5 07.1	74 62	124.4						
3												
4												

DECLINATION CONTRARY NAME TO LATITUDE

Lat. 42°

H.A.	19° 00′ Alt.	Δd Δt	Az.	19° 30′ Alt.	Δd Δt	Az.	20° 00′ Alt.	Δd Δt	Az.	20° 30′ Alt.	Δd Δt	Az.
°	° ′		°	° ′		°	° ′		°	° ′		°
00	29 00.0	1.0 01	180.0	28 30.0	1.0 01	180.0	28 00.0	1.0 01	180.0	27 30.0	1.0 01	180.0
1	28 59.6	1.0 02	178.9	28 29.6	1.0 02	178.9	27 59.6	1.0 02	178.9	27 29.6	1.0 02	178.9
2	28 58.3	1.0 04	177.8	28 28.3	1.0 03	177.9	27 58.3	1.0 03	177.9	27 28.4	1.0 03	177.9
3	28 56.2	1.0 05	176.8	28 26.2	1.0 05	176.8	27 56.3	1.0 05	176.8	27 26.3	1.0 06	176.8
4	28 53.3	1.0 06	175.7	28 23.3	1.0 06	175.7	27 53.4	1.0 06	175.7	27 23.4	1.0 06	175.8
05	28 49.5	1.0 08	174.6	28 19.6	1.0 08	174.6	27 49.7	1.0 08	174.7	27 19.7	1.0 08	174.7
6	28 44.9	1.0 09	173.5	28 15.0	1.0 09	173.6	27 45.1	1.0 09	173.6	27 15.2	1.0 09	173.7
7	28 39.4	99 10	172.5	28 09.6	99 10	172.5	27 39.8	99 10	172.6	27 09.9	99 10	172.6
8	28 33.2	99 12	171.4	28 03.4	99 12	171.5	27 33.6	99 12	171.5	27 03.8	99 12	171.6
9	28 26.1	99 13	170.3	27 56.4	99 13	170.4	27 26.6	99 13	170.5	26 56.9	99 13	170.5
10	28 18.2	99 15	169.3	27 48.5	99 14	169.3	27 18.8	99 14	169.4	26 49.1	99 14	169.5
1	28 09.5	99 16	168.2	27 39.9	99 16	168.3	27 10.2	99 16	168.4	26 40.6	99 16	168.5
2	27 59.9	98 17	167.1	27 30.4	98 17	167.2	27 00.9	98 17	167.3	26 31.3	98 17	167.4
3	27 49.6	98 19	166.1	27 20.1	98 18	166.2	26 50.7	98 18	166.3	26 21.2	98 18	166.4
4	27 38.5	98 20	165.0	27 09.1	98 20	165.2	26 39.7	98 20	165.3	26 10.3	98 19	165.4
15	27 26.6	98 21	164.0	26 57.3	98 21	164.1	26 28.0	98 21	164.2	25 58.7	98 21	164.4
6	27 13.9	97 22	163.0	26 44.7	97 22	163.1	26 15.5	97 22	163.2	25 46.3	97 22	163.3
7	27 00.4	97 24	161.9	26 31.4	97 24	162.1	26 02.2	97 23	162.2	25 33.1	97 23	162.3
8	26 46.2	97 25	160.9	26 17.2	97 25	161.0	25 48.2	97 25	161.2	25 19.2	97 24	161.3
9	26 31.3	96 26	159 9	26 02.4	96 26	160.0	25 33.5	96 26	160.2	25 04.6	96 26	160.3
20	26 15.6	96 27	158.9	25 46.8	96 27	159.0	25 18.0	96 27	159.2	24 49.2	96 27	159.3
1	25 59.1	96 29	157.9	25 30.5	96 28	158.0	25 01.8	96 28	158.2	24 33.1	96 28	158.3
2	25 41.9	95 30	156.9	25 13.4	95 30	157.0	24 44.9	95 29	157.2	24 16.3	95 29	157.4
3	25 24.1	95 31	155.9	24 55.7	95 31	156.0	24 27.2	95 31	156.2	23 58.8	95 30	156.4
4	25 05.5	94 32	154.9	24 37.2	94 32	155.1	24 08.9	94 32	155.2	23 40.6	94 32	155.4
25	24 46.2	94 33	153.9	24 18.0	94 33	154.1	23 49.9	94 33	154.3	23 21.7	94 33	154.5
6	24 26.2	93 34	152.9	23 58.2	93 34	153.1	23 30.2	93 34	153.3	23 02.1	94 34	153.5
7	24 05.6	93 36	152.0	23 37.7	93 35	152.2	23 09.8	93 35	152.4	22 41.9	93 35	152.6
8	23 44.3	92 37	151.0	23 16.6	93 36	151.2	22 48.8	93 36	151.4	22 21.0	93 36	151.6
9	23 22.4	92 38	150.0	22 54.8	92 37	150.3	22 27.1	92 37	150.5	21 59.5	92 37	150.7
55	10 40.3	78 59	128.0	10 16.7	79 59	128.3	9 53.2	79 58	128.6	9 29.6	79 58	128.9
6	10 05.0	78 59	127.2	9 41.6	78 59	127.6	9 18.1	78 59	127.9	8 54.7	78 59	128.2
7	9 29.3	78 60	126.5	9 06.0	78 60	126.8	8 42.8	78 60	127.1	8 19.5	78 59	127.4
8	8 53.3	77 61	125.7	8 30.2	77 60	126.1	8 07.0	77 60	126.4	7 43.9	77 60	126.7
9	8 16.9	77 61	125.0	7 54.0	77 61	125.3	7 31.0	77 61	125.7	7 08.0	77 60	126.0
60	7 40.2	76 62	124.3	7 17.4	76 61	124.6	6 54.6	76 61	124.9	6 31.7	76 61	125.3
1	7 03.2	76 62	123.6	6 40.6	76 62	123.9	6 17.9	76 62	124.2	5 55.2	76 61	124.6
2	6 25.9	75 63	122.8	6 03.4	75 62	123.2	5 40.9	75 62	123.5	5 18.3	75 62	123.8
3	5 48.3	75 63	122.1	5 25.9	75 63	122.5	5 03.5	75 63	122.8			
4	5 10.4	74 64	121.4									

H.A.	21° 00′ Alt.	Δd Δt	Az.	21° 30′ Alt.	Δd Δt	Az.	22° 00′ Alt.	Δd Δt	Az.	22° 30′ Alt.	Δd Δt	Az.
°	° ′		°	° ′		°	° ′		°	° ′		°
00	27 00.0	1.0 01	180.0	26 30.0	1.0 01	180.0	26 00.0	1.0 01	180.0	25 30.0	1.0 01	180.0
1	26 59.6	1.0 02	179.0	26 29.6	1.0 02	179.0	25 59.6	1.0 02	179.0	25 29.6	1.0 02	179.0
2	26 58.4	1.0 03	177.9	26 28.4	1.0 03	177.9	25 58.4	1.0 03	177.9	25 28.4	1.0 03	178.0
3	26 56.3	1.0 05	176.9	26 26.4	1.0 05	176.9	25 56.4	1.0 05	176.9	25 26.4	1.0 05	176.9
4	26 53.5	1.0 06	175.8	26 23.5	1.0 06	175.8	25 53.6	1.0 06	175.9	25 23.6	1.0 06	175.9
05	26 49.8	1.0 07	174.8	26 19.9	1.0 07	174.8	25 50.0	1.0 07	174.8	25 20.1	1.0 07	174.9
6	26 45.4	1.0 09	173.7	26 15.5	1.0 09	173.8	25 45.6	1.0 09	173.8	25 15.7	1.0 09	173.9
7	26 40.1	99 10	172.7	26 10.2	99 10	172.7	25 40.4	99 10	172.8	25 10.5	99 10	172.9
8	26 34.0	99 11	171.6	26 04.2	99 11	171.7	25 34.4	99 11	171.8	25 04.6	99 11	171.8
9	26 27.1	99 13	170.6	25 57.4	99 13	170.7	25 27.6	99 13	170.8	24 57.9	99 12	170.8
10	26 19.5	99 14	169.6	25 49.8	99 14	169.7	25 20.1	99 14	169.7	24 50.4	99 14	169.8
1	26 11.0	99 15	168.6	25 41.4	99 15	168.6	25 11.7	99 15	168.7	24 42.1	99 15	168.8
2	26 01.8	99 17	167.5	25 32.2	99 17	167.6	25 02.6	99 16	167.7	24 33.1	99 16	167.8
3	25 51.7	98 18	166.5	25 22.3	98 18	166.6	24 52.8	98 18	166.7	24 23.3	98 18	166.8
4	25 40.9	98 19	165.5	25 11.5	98 19	165.6	24 42.1	98 19	165.7	24 12.7	98 19	165.8
15	25 29.4	98 21	164.5	25 00.1	98 20	164.6	24 30.8	98 20	164.7	24 01.4	98 20	164.8
6	25 17.1	97 22	163.5	24 47.9	97 22	163.6	24 18.6	97 21	163.7	23 49.4	97 21	163.8
7	25 04.0	97 23	162.5	24 34.9	97 23	162.6	24 05.7	97 23	162.7	23 36.6	97 22	162.9
8	24 50.2	97 24	161.5	24 21.2	97 24	161.6	23 52.1	97 24	161.7	23 23.1	97 24	161.9
9	24 35.7	96 25	160.5	24 06.7	96 25	160.6	23 37.8	96 25	160.8	23 08.9	96 25	160.9
20	24 20.4	96 27	159.5	23 51.6	96 26	159.6	23 22.8	96 26	159.8	22 53.9	96 26	159.9
1	24 04.4	96 28	158.5	23 35.7	96 28	158.7	23 07.0	96 27	158.8	22 38.3	96 27	159.0
2	23 47.7	95 29	157.5	23 19.1	95 29	157.7	22 50.5	95 29	157.9	22 21.9	95 28	158.0
3	23 30.3	95 30	156.6	23 01.9	95 30	156.7	22 33.4	95 30	156.9	22 04.9	95 30	157.1
4	23 12.3	94 31	155.6	22 43.9	94 31	155.8	22 15.6	95 31	156.0	21 47.2	95 31	156.1
25	22 53.5	94 32	154.6	22 25.3	94 32	154.8	21 57.1	94 32	155.0	21 28.8	94 32	155.2
6	22 34.1	94 34	153.7	22 06.0	94 33	153.9	21 37.9	94 33	154.1	21 09.8	94 33	154.3
7	22 14.0	93 35	152.8	21 46.0	93 34	152.9	21 18.1	93 34	153.1	20 50.1	93 34	153.3
8	21 53.2	93 36	151.8	21 25.4	93 35	152.0	20 57.6	93 35	152.2	20 29.8	93 35	152.4
9	21 31.9	92 37	150.9	21 04.2	92 36	151.1	20 36.5	92 36	151.3	20 08.8	92 36	151.5
55	9 06.0	79 58	129.2	8 42.3	79 58	129.6	8 18.7	79 57	129.9	7 55.0	79 57	130.2
6	8 31.2	78 58	128.5	8 07.8	78 58	128.8	7 44.3	78 58	129.1	7 20.8	78 58	129.4
7	7 56.2	78 59	127.8	7 32.8	78 59	128.1	7 09.5	78 59	128.4	6 46.2	78 58	128.7
8	7 20.7	77 60	127.0	6 57.6	77 59	127.4	6 34.4	77 59	127.7	6 11.2	77 59	128.0
9	6 45.0	77 60	126.3	6 22.0	77 60	126.6	5 58.9	77 60	127.0	5 35.9	77 59	127.3
60	6 08.9	76 61	125.6	5 46.0	76 60	125.9	5 23.1	76 60	126.2	5 00.2	76 60	126.6
1	5 32.5	76 61	124.9	5 09.7	76 61	125.2						
2												
3												
4												

APPENDIX L

EXTRACTS FROM TIDE TABLES

NEW YORK (THE BATTERY), N.Y., 1975

TIMES AND HEIGHTS OF HIGH AND LOW WATERS

JANUARY

DAY	TIME H.M.	HT. FT.	DAY	TIME H.M.	HT. FT.
1	0422	-0.8	16	0431	0.0
W	1043	5.1	TH	1045	4.1
	1659	-1.1		1656	-0.2
	2321	4.6		2319	3.8
2	0516	-0.6	17	0501	0.2
TH	1138	4.9	F	1120	3.9
	1749	-0.9		1722	0.0
				2357	3.7
3	0017	4.6	18	0533	0.4
F	0615	-0.4	SA	1154	3.7
	1234	4.6		1746	0.2
	1847	-0.6			
4	0111	4.6	19	0029	3.7
SA	0724	-0.1	SU	0613	0.6
	1329	4.3		1228	3.5
	1951	-0.4		1818	0.3
5	0207	4.6	20	0108	3.8
SU	0835	0.0	M	0727	0.7
	1428	4.0		1311	3.4
	2053	-0.3		1917	0.4
6	0307	4.5	21	0156	3.9
M	0939	-0.1	TU	0856	0.7
	1530	3.7		1404	3.3
	2153	-0.3		2050	0.4
7	0407	4.5	22	0255	4.0
TU	1038	-0.2	W	1001	0.4
	1635	3.6		1515	3.2
	2247	-0.3		2157	0.2
8	0508	4.6	23	0404	4.2
W	1131	-0.3	TH	1057	0.1
	1735	3.7		1637	3.4
	2339	-0.3		2256	0.0
9	0603	4.7	24	0510	4.5
TH	1222	-0.4	F	1150	-0.2
	1828	3.8		1742	3.7
				2352	-0.3
10	0028	-0.4	25	0609	4.9
F	0649	4.8	SA	1242	-0.6
	1310	-0.5		1838	4.0
	1915	3.9			
11	0117	-0.4	26	0048	-0.6
SA	0733	4.8	SU	0701	5.2
	1354	-0.6		1330	-1.0
	1959	3.9		1929	4.4
12	0202	-0.4	27	0140	-0.9
SU	0814	4.8	M	0749	5.4
	1436	-0.6		1419	-1.3
	2040	3.9		2019	4.7
13	0242	-0.4	28	0231	-1.1
M	0854	4.7	TU	0840	5.5
	1516	-0.6		1505	-1.4
	2122	3.9		2110	4.9
14	0321	-0.3	29	0321	-1.2
TU	0931	4.5	W	0931	5.4
	1552	-0.5		1550	-1.5
	2202	3.9		2203	5.0
15	0358	-0.1	30	0410	-1.2
W	1011	4.3	TH	1024	5.2
	1625	-0.4		1635	-1.3
	2242	3.8		2258	5.0
			31	0459	-0.9
			F	1120	4.9
				1725	-1.0
				2354	4.9

FEBRUARY

DAY	TIME H.M.	HT. FT.	DAY	TIME H.M.	HT. FT.
1	0554	-0.6	16	0503	0.2
SA	1213	4.5	SU	1111	3.8
	1818	-0.7		1704	0.1
				2338	4.0
2	0047	4.8	17	0535	0.3
SU	0659	-0.2	M	1149	3.6
	1309	4.1		1733	0.2
	1919	-0.3			
3	0142	4.6	18	0020	4.0
M	0808	0.0	TU	0621	0.5
	1405	3.8		1231	3.5
	2024	-0.1		1818	0.4
4	0240	4.4	19	0108	4.1
TU	0915	0.1	W	0759	0.6
	1508	3.6		1329	3.4
	2129	0.0		1932	0.5
5	0341	4.3	20	0211	4.1
W	1015	0.0	TH	0926	0.5
	1612	3.5		1439	3.3
	2225	0.0		2125	0.4
6	0444	4.3	21	0327	4.2
TH	1110	-0.1	F	1026	0.2
	1715	3.5		1607	3.5
	2320	0.0		2233	0.1
7	0542	4.4	22	0444	4.5
F	1200	-0.2	SA	1123	-0.2
	1811	3.7		1718	3.9
				2334	-0.3
8	0009	-0.1	23	0545	4.8
SA	0629	4.5	SU	1214	-0.6
	1246	-0.3		1816	4.4
	1856	3.9			
9	0057	-0.2	24	0030	-0.7
SU	0713	4.6	M	0640	5.2
	1329	-0.5		1306	-1.0
	1937	4.0		1909	4.8
10	0141	-0.3	25	0123	-1.0
M	0752	4.6	TU	0731	5.4
	1410	-0.5		1354	-1.2
	2015	4.1		1959	5.2
11	0222	-0.4	26	0215	-1.3
TU	0829	4.6	W	0820	5.4
	1449	-0.6		1441	-1.4
	2053	4.2		2048	5.4
12	0300	-0.4	27	0305	-1.4
W	0906	4.5	TH	0912	5.3
	1523	-0.5		1526	-1.4
	2129	4.2		2139	5.4
13	0335	-0.3	28	0352	-1.3
TH	0941	4.3	F	1003	5.1
	1555	-0.4		1612	-1.2
	2203	4.1		2232	5.3
14	0407	-0.2			
F	1011	4.1			
	1622	-0.2			
	2235	4.1			
15	0435	0.0			
SA	1043	3.9			
	1643	-0.1			
	2307	4.0			

MARCH

DAY	TIME H.M.	HT. FT.	DAY	TIME H.M.	HT. FT.
1	0442	-1.0	16	0416	-0.1
SA	1058	4.8	SU	1013	4.0
	1658	-0.9		1613	0.0
	2325	5.1		2223	4.4
2	0533	-0.6	17	0443	0.0
SU	1152	4.4	M	1043	3.8
	1749	-0.5		1637	0.2
				2259	4.4
3	0021	4.9	18	0517	0.2
M	0631	-0.2	TU	1124	3.7
	1247	4.1		1708	0.3
	1847	0.0		2346	4.3
4	0115	4.6	19	0602	0.4
TU	0739	0.1	W	1218	3.6
	1344	3.8		1752	0.5
	1956	0.3			
5	0212	4.3	20	0042	4.3
W	0848	0.3	TH	0724	0.5
	1443	3.6		1318	3.6
	2103	0.4		1906	0.6
6	0312	4.1	21	0146	4.3
TH	0949	0.3	F	0853	0.4
	1545	3.5		1429	3.6
	2204	0.4		2106	0.5
7	0415	4.1	22	0300	4.3
F	1041	0.2	SA	0959	0.2
	1649	3.6		1548	3.9
	2258	0.3		2217	0.2
8	0513	4.2	23	0417	4.5
SA	1131	0.0	SU	1056	-0.2
	1744	3.8		1658	4.3
	2347	0.1		2315	-0.2
9	0603	4.3	24	0524	4.8
SU	1216	-0.1	M	1147	-0.5
	1830	4.1		1755	4.8
10	0033	0.0	25	0011	-0.6
M	0645	4.4	TU	0620	5.0
	1259	-0.3		1238	-0.8
	1909	4.3		1848	5.2
11	0117	-0.2	26	0107	-1.0
TU	0726	4.5	W	0712	5.2
	1341	-0.4		1328	-1.1
	1946	4.4		1938	5.6
12	0158	-0.3	27	0157	-1.2
W	0803	4.5	TH	0802	5.3
	1418	-0.4		1416	-1.2
	2022	4.5		2025	5.7
13	0237	-0.4	28	0247	-1.3
TH	0837	4.5	F	0852	5.2
	1453	-0.4		1502	-1.1
	2054	4.5		2114	5.7
14	0313	-0.4	29	0334	-1.2
F	0912	4.3	SA	0942	4.9
	1525	-0.3		1547	-0.9
	2126	4.5		2206	5.5
15	0346	-0.3	30	0422	-0.9
SA	0941	4.2	SU	1037	4.7
	1551	-0.1		1632	-0.6
	2152	4.5		2259	5.3
			31	0512	-0.6
			M	1132	4.3
				1721	-0.2
				2352	4.9

TIME MERIDIAN 75° W. 0000 IS MIDNIGHT. 1200 IS NOON.
HEIGHTS ARE RECKONED FROM THE DATUM OF SOUNDINGS ON CHARTS OF THE LOCALITY WHICH IS MEAN LOW WATER.

TABLE 2.—TIDAL DIFFERENCES AND OTHER CONSTANTS

No.	PLACE	POSITION		DIFFERENCES				RANGES		Mean Tide Level
		Lat.	Long.	Time		Height		Mean	Spring	
				High water	Low water	High water	Low water			
		° ′	° ′	h. m.	h. m.	feet	feet	feet	feet	feet
	NEW YORK and NEW JERSEY — Continued									
	Hudson River‡	N.	W.	on NEW YORK, p.56						
	Time meridian, 75°W.									
1513	Jersey City, Pa. RR. Ferry, N. J	40 43	74 02	+0 07	+0 07	−0.1	0.0	4.4	5.3	2.2
1515	New York, Desbrosses Street	40 43	74 01	+0 10	+0 10	−0.1	0.0	4.4	5.3	2.2
1517	New York, Chelsea Docks	40 45	74 01	+0 17	+0 16	−0.2	0.0	4.3	5.2	2.1
1519	Hoboken, Castle Point, N. J	40 45	74 01	+0 17	+0 16	−0.2	0.0	4.3	5.2	2.1
1521	Weehawken, Days Point, N. J	40 46	74 01	+0 24	+0 23	−0.3	0.0	4.2	5.0	2.1
1523	New York, Union Stock Yards	40 47	74 00	+0 27	+0 26	−0.3	0.0	4.2	5.0	2.1
1525	New York, 130th Street	40 49	73 58	+0 37	+0 35	−0.5	0.0	4.0	4.8	2.0
1527	George Washington Bridge	40 51	73 57	+0 46	+0 43	−0.6	0.0	3.9	4.6	1.9
1529	Spuyten Duyvil, West of RR. bridge	40 53	73 56	+0 58	+0 53	−0.7	0.0	3.8	4.5	1.9
1531	Yonkers	40 56	73 54	+1 09	+1 10	−0.8	0.0	3.7	4.4	1.8
1533	Dobbs Ferry	41 01	73 53	+1 29	+1 40	−1.1	0.0	3.4	4.0	1.7
1535	Tarrytown	41 05	73 52	+1 45	+1 54	−1.3	0.0	3.2	3.7	1.6
1537	Ossining	41 10	73 52	+1 53	+2 14	−1.4	0.0	3.1	3.6	1.5
1539	Haverstraw	41 12	73 58	+1 59	+2 25	−1.6	0.0	2.9	3.4	1.4
1541	Peekskill	41 17	73 56	+2 24	+3 00	−1.3	+0.3	2.9	3.4	1.7
1543	West Point	41 24	73 57	+3 16	+3 37	−1.5	+0.3	2.7	3.1	1.6
1545	Newburgh	41 30	74 00	+3 42	+4 00	−1.5	+0.2	2.8	3.2	1.6
1547	New Hamburg	41 35	73 57	+4 00	+4 25	−1.5	+0.1	2.9	3.3	1.5
1549	Poughkeepsie	41 42	73 57	+4 30	+4 43	−1.3	+0.1	3.1	3.5	1.6
1551	Hyde Park	41 47	73 57	+4 56	+5 09	−1.3	0.0	3.2	3.6	1.6
1553	Kingston Point	41 56	73 58	+5 16	+5 31	−0.9	−0.1	3.7	4.2	1.7
1555	Tivoli	42 04	73 56	+5 46	+6 01	−0.8	−0.2	3.9	4.4	1.7
1557	Catskill	42 13	73 51	+6 37	+6 55	−0.7	−0.3	4.1	4.6	1.7
1559	Hudson	42 15	73 48	+6 54	+7 09	−0.9	−0.4	4.0	4.4	1.6
				on ALBANY, p.60						
1561	Coxsackie	42 21	73 48	−1 01	−1 38	−0.5	+0.2	3.9	4.3	2.1
1563	New Baltimore	42 27	73 47	−0 34	−0 56	−0.1	+0.4	4.1	4.5	2.4
1565	Castleton-on-Hudson	42 32	73 46	−0 17	−0 29	−0.2	+0.1	4.3	4.7	2.2
1567	ALBANY	42 39	73 45	Daily predictions				4.6	5.0	2.5
1569	Troy	42 44	73 42	+0 08	+0 10	+0.1	0.0	4.7	5.1	2.3
	The Kills and Newark Bay			on NEW YORK, p.56						
	Kill Van Kull									
1571	Constable Hook	40 39	74 05	−0 34	−0 21	0.0	0.0	4.5	5.4	2.2
1573	New Brighton	40 39	74 05	−0 12	−0 18	0.0	0.0	4.5	5.4	2.2
1575	Port Richmond	40 38	74 08	−0 03	+0 05	0.0	0.0	4.5	5.4	2.2
1577	Bergen Point	40 39	74 08	+0 03	+0 03	+0.1	0.0	4.6	5.5	2.3
1579	Shooters Island	40 39	74 10	+0 06	+0 18	+0.1	0.0	4.6	5.5	2.3
1581	Port Newark Terminal	40 41	74 08	−0 01	+0 18	+0.6	0.0	5.1	6.1	2.5
1583	Newark, Passaic River	40 44	74 10	+0 22	+0 52	+0.6	0.0	5.1	6.1	2.5
1585	Passaic, Gregory Ave. bridge	40 51	74 07	+0 49	+1 57	+0.6	0.0	5.1	6.1	2.5
	Hackensack River									
1586	Kearny Point	40 44	74 06	+0 09	+0 33	+0.5	0.0	5.0	6.0	2.5
1587	Secaucus	40 48	74 04	+1 13	+1 09	+0.6	0.0	5.1	6.1	2.6
1588	Little Ferry	40 51	74 02	+1 22	+1 14	+0.8	0.0	5.3	6.4	2.7
1589	Hackensack	40 53	74 02	+1 33	+1 58	+0.8	0.0	5.3	6.4	2.6
				on SANDY HOOK, p.64						
	Arthur Kill									
1591	Elizabethport	40 39	74 11	+0 25	+0 39	+0.3	0.0	4.9	5.9	2.4
1593	Chelsea	40 36	74 12	+0 24	+0 35	+0.4	0.0	5.0	6.0	2.5
1595	Carteret	40 35	74 13	+0 23	+0 31	+0.5	0.0	5.1	6.2	2.6
1597	Rossville	40 33	74 13	+0 17	+0 25	+0.7	0.0	5.3	6.4	2.6
1599	Tottenville	40 31	74 15	+0 03	+0 13	+0.7	0.0	5.3	6.4	2.6
1601	Perth Amboy	40 30	74 16	+0 13	+0 19	+0.6	0.0	5.2	6.3	2.6

‡Values for the Hudson River above the George Washington Bridge are based upon averages for the six months May to October, when the fresh-water discharge is a minimum.

TABLE 3.—HEIGHT OF TIDE AT ANY TIME

Duration of rise or fall, see footnote	Time from the nearest high water or low water														
h. m.	h. m.	h. m.	h. m.	h. m.	h. m.	h. m.	h. m.	h. m.	h. m.	h. m.	h. m.	h. m.	h. m.	h. m.	h. m.
4 00	0 08	0 16	0 24	0 32	0 40	0 48	0 56	1 04	1 12	1 20	1 28	1 36	1 44	1 52	2 00
4 20	0 09	0 17	0 26	0 35	0 43	0 52	1 01	1 09	1 18	1 27	1 35	1 44	1 53	2 01	2 10
4 40	0 09	0 19	0 28	0 37	0 47	0 56	1 05	1 15	1 24	1 33	1 43	1 52	2 01	2 11	2 20
5 00	0 10	0 20	0 30	0 40	0 50	1 00	1 10	1 20	1 30	1 40	1 50	2 00	2 10	2 20	2 30
5 20	0 11	0 21	0 32	0 43	0 53	1 04	1 15	1 25	1 36	1 47	1 57	2 08	2 19	2 29	2 40
5 40	0 11	0 23	0 34	0 45	0 57	1 08	1 19	1 31	1 42	1 53	2 05	2 16	2 27	2 39	2 50
6 00	0 12	0 24	0 36	0 48	1 00	1 12	1 24	1 36	1 48	2 00	2 12	2 24	2 36	2 48	3 00
6 20	0 13	0 25	0 38	0 51	1 03	1 16	1 29	1 41	1 54	2 07	2 19	2 32	2 45	2 57	3 10
6 40	0 13	0 27	0 40	0 53	1 07	1 20	1 33	1 47	2 00	2 13	2 27	2 40	2 53	3 07	3 20
7 00	0 14	0 28	0 42	0 56	1 10	1 24	1 38	1 52	2 06	2 20	2 34	2 48	3 02	3 16	3 30
7 20	0 15	0 29	0 44	0 59	1 13	1 28	1 43	1 57	2 12	2 27	2 41	2 56	3 11	3 25	3 40
7 40	0 15	0 31	0 46	1 01	1 17	1 32	1 47	2 03	2 18	2 33	2 49	3 04	3 19	3 35	3 50
8 00	0 16	0 32	0 48	1 04	1 20	1 36	1 52	2 08	2 24	2 40	2 56	3 12	3 28	3 44	4 00
8 20	0 17	0 33	0 50	1 07	1 23	1 40	1 57	2 13	2 30	2 47	3 03	3 20	3 37	3 53	4 10
8 40	0 17	0 35	0 52	1 09	1 27	1 44	2 01	2 19	2 36	2 53	3 11	3 28	3 45	4 03	4 20
9 00	0 18	0 36	0 54	1 12	1 30	1 48	2 06	2 24	2 42	3 00	3 18	3 36	3 54	4 12	4 30
9 20	0 19	0 37	0 56	1 15	1 33	1 52	2 11	2 29	2 48	3 07	3 25	3 44	4 03	4 21	4 40
9 40	0 19	0 39	0 58	1 17	1 37	1 56	2 15	2 35	2 54	3 13	3 33	3 52	4 11	4 31	4 50
10 00	0 20	0 40	1 00	1 20	1 40	2 00	2 20	2 40	3 00	3 20	3 40	4 00	4 20	4 40	5 00
10 20	0 21	0 41	1 02	1 23	1 43	2 04	2 25	2 45	3 06	3 27	3 47	4 08	4 29	4 49	5 10
10 40	0 21	0 43	1 04	1 25	1 47	2 08	2 29	2 51	3 12	3 33	3 55	4 16	4 37	4 59	5 20

Range of tide, see footnote	Correction to height														
Ft.	Ft.	Ft.	Ft.	Ft.	Ft.	Ft.	Ft.	Ft.	Ft.	Ft.	Ft.	Ft.	Ft.	Ft.	Ft.
0.5	0.0	0.0	0.0	0.0	0.0	0.0	0.1	0.1	0.1	0.1	0.1	0.2	0.2	0.2	0.2
1.0	0.0	0.0	0.0	0.0	0.1	0.1	0.1	0.2	0.2	0.2	0.3	0.3	0.4	0.4	0.5
1.5	0.0	0.0	0.0	0.1	0.1	0.1	0.2	0.2	0.3	0.4	0.4	0.5	0.6	0.7	0.8
2.0	0.0	0.0	0.0	0.1	0.1	0.2	0.3	0.3	0.4	0.5	0.6	0.7	0.8	0.9	1.0
2.5	0.0	0.0	0.1	0.1	0.2	0.2	0.3	0.4	0.5	0.6	0.7	0.9	1.0	1.1	1.2
3.0	0.0	0.0	0.1	0.1	0.2	0.3	0.4	0.5	0.6	0.8	0.9	1.0	1.2	1.3	1.5
3.5	0.0	0.0	0.1	0.2	0.2	0.3	0.4	0.6	0.7	0.9	1.0	1.2	1.4	1.6	1.8
4.0	0.0	0.0	0.1	0.2	0.3	0.4	0.5	0.7	0.8	1.0	1.2	1.4	1.6	1.8	2.0
4.5	0.0	0.0	0.1	0.2	0.3	0.4	0.6	0.7	0.9	1.1	1.3	1.6	1.8	2.0	2.2
5.0	0.0	0.1	0.1	0.2	0.3	0.5	0.6	0.8	1.0	1.2	1.5	1.7	2.0	2.2	2.5
5.5	0.0	0.1	0.1	0.2	0.4	0.5	0.7	0.9	1.1	1.4	1.6	1.9	2.2	2.5	2.8
6.0	0.0	0.1	0.1	0.3	0.4	0.6	0.8	1.0	1.2	1.5	1.8	2.1	2.4	2.7	3.0
6.5	0.0	0.1	0.2	0.3	0.4	0.6	0.8	1.1	1.3	1.6	1.9	2.2	2.6	2.9	3.2
7.0	0.0	0.1	0.2	0.3	0.5	0.7	0.9	1.2	1.4	1.8	2.1	2.4	2.8	3.1	3.5
7.5	0.0	0.1	0.2	0.3	0.5	0.7	1.0	1.2	1.5	1.9	2.2	2.6	3.0	3.4	3.8
8.0	0.0	0.1	0.2	0.3	0.5	0.8	1.0	1.3	1.6	2.0	2.4	2.8	3.2	3.6	4.0
8.5	0.0	0.1	0.2	0.4	0.6	0.8	1.1	1.4	1.8	2.1	2.5	2.9	3.4	3.8	4.2
9.0	0.0	0.1	0.2	0.4	0.6	0.9	1.2	1.5	1.9	2.2	2.7	3.1	3.6	4.0	4.5
9.5	0.0	0.1	0.2	0.4	0.6	0.9	1.2	1.6	2.0	2.4	2.8	3.3	3.8	4.3	4.8
10.0	0.0	0.1	0.2	0.4	0.7	1.0	1.3	1.7	2.1	2.5	3.0	3.5	4.0	4.5	5.0
10.5	0.0	0.1	0.3	0.5	0.7	1.0	1.3	1.7	2.2	2.6	3.1	3.6	4.2	4.7	5.2
11.0	0.0	0.1	0.3	0.5	0.7	1.1	1.4	1.8	2.3	2.8	3.3	3.8	4.4	4.9	5.5
11.5	0.0	0.1	0.3	0.5	0.8	1.1	1.5	1.9	2.4	2.9	3.4	4.0	4.6	5.1	5.8
12.0	0.0	0.1	0.3	0.5	0.8	1.1	1.5	2.0	2.5	3.0	3.6	4.1	4.8	5.4	6.0
12.5	0.0	0.1	0.3	0.5	0.8	1.2	1.6	2.1	2.6	3.1	3.7	4.3	5.0	5.6	6.2
13.0	0.0	0.1	0.3	0.6	0.9	1.2	1.7	2.2	2.7	3.2	3.9	4.5	5.1	5.8	6.5
13.5	0.0	0.1	0.3	0.6	0.9	1.3	1.7	2.2	2.8	3.4	4.0	4.7	5.3	6.0	6.8
14.0	0.0	0.2	0.3	0.6	0.9	1.3	1.8	2.3	2.9	3.5	4.2	4.8	5.5	6.3	7.0
14.5	0.0	0.2	0.4	0.6	1.0	1.4	1.9	2.4	3.0	3.6	4.3	5.0	5.7	6.5	7.2
15.0	0.0	0.2	0.4	0.6	1.0	1.4	1.9	2.5	3.1	3.8	4.4	5.2	5.9	6.7	7.5
15.5	0.0	0.2	0.4	0.7	1.0	1.5	2.0	2.6	3.2	3.9	4.6	5.4	6.1	6.9	7.8
16.0	0.0	0.2	0.4	0.7	1.1	1.5	2.1	2.6	3.3	4.0	4.7	5.5	6.3	7.2	8.0
16.5	0.0	0.2	0.4	0.7	1.1	1.6	2.1	2.7	3.4	4.1	4.9	5.7	6.5	7.4	8.2
17.0	0.0	0.2	0.4	0.7	1.1	1.6	2.2	2.8	3.5	4.2	5.0	5.9	6.7	7.6	8.5
17.5	0.0	0.2	0.4	0.8	1.2	1.7	2.2	2.9	3.6	4.4	5.2	6.0	6.9	7.8	8.8
18.0	0.0	0.2	0.4	0.8	1.2	1.7	2.3	3.0	3.7	4.5	5.3	6.2	7.1	8.1	9.0
18.5	0.1	0.2	0.5	0.8	1.2	1.8	2.4	3.1	3.8	4.6	5.5	6.4	7.3	8.3	9.2
19.0	0.1	0.2	0.5	0.8	1.3	1.8	2.4	3.1	3.9	4.8	5.6	6.6	7.5	8.5	9.5
19.5	0.1	0.2	0.5	0.8	1.3	1.9	2.5	3.2	4.0	4.9	5.8	6.7	7.7	8.7	9.8
20.0	0.1	0.2	0.5	0.9	1.3	1.9	2.6	3.3	4.1	5.0	5.9	6.9	7.9	9.0	10.0

Obtain from the predictions the high water and low water, one of which is before and the other after the time for which the height is required. The difference between the times of occurrence of these tides is the duration of rise or fall, and the difference between their heights is the range of tide for the above table. Find the difference between the nearest high or low water and the time for which the height is required.

Enter the table with the duration of rise or fall, printed in heavy-faced type, which most nearly agrees with the actual value, and on that horizontal line find the time from the nearest high or low water which agrees most nearly with the corresponding actual difference. The correction sought is in the column directly below, on the line with the range of tide.

When the nearest tide is high water, subtract the correction.

When the nearest tide is low water, add the correction.

APPENDIX M

EXTRACTS FROM TIDAL CURRENT TABLES

THE NARROWS, NEW YORK HARBOR, N.Y., 1975

F-FLOOD, DIR. 340° TRUE   E-EBB, DIR. 160° TRUE

JANUARY

DAY	SLACK WATER TIME H.M.	MAXIMUM CURRENT TIME H.M.	MAXIMUM CURRENT VEL. KNOTS	DAY	SLACK WATER TIME H.M.	MAXIMUM CURRENT TIME H.M.	MAXIMUM CURRENT VEL. KNOTS
1 W	0617 1211 1901	0254 0857 1526 2132	2.4E 2.2F 2.5E 2.0F	16 TH	0002 0641 1217 1917	0310 0907 1532 2139	1.8E 1.6F 2.0E 1.5F
2 TH	0037 0716 1301 1954	0345 0952 1614 2227	2.3E 2.1F 2.4E 2.0F	17 F	0046 0730 1257 2000	0351 0954 1611 2224	1.7E 1.5F 1.9E 1.5F
3 F	0132 0819 1352 2050	0440 1049 1708 2321	2.2E 1.9F 2.3E 2.0F	18 SA	0131 0823 1339 2045	0437 1043 1654 2309	1.6E 1.4F 1.8E 1.5F
4 SA	0230 0924 1445 2146	0542 1147 1807	2.1E 1.7F 2.1E	19 SU	0218 0920 1424 2132	0530 1130 1745 2358	1.6E 1.3F 1.7E 1.5F
5 SU	0331 1027 1543 2242	0018 0648 1244 1909	1.9F 2.0E 1.5F 2.0E	20 M	0310 1017 1514 2220	0631 1220 1842	1.5E 1.2F 1.6E
6 M	0434 1129 1643 2337	0119 0751 1352 2007	1.9F 2.0E 1.4F 2.0E	21 TU	0407 1114 1610 2310	0047 0728 1313 1940	1.6F 1.6E 1.1F 1.6E
7 TU	0537 1230 1744	0232 0850 1515 2102	1.9F 2.0E 1.3F 2.0E	22 W	0505 1211 1710	0140 0823 1410 2033	1.6F 1.7E 1.1F 1.7E
8 W	0033 0635 1328 1841	0343 0946 1623 2154	1.9F 2.0E 1.4F 1.9E	23 TH	0001 0602 1305 1808	0237 0917 1509 2124	1.7F 1.9E 1.2F 1.8E
9 TH	0127 0727 1423 1932	0442 1037 1714 2242	2.0F 2.1E 1.4F 1.9E	24 F	0055 0655 1357 1903	0337 1007 1612 2215	1.9F 2.0E 1.4F 2.0E
10 F	0219 0814 1511 2020	0527 1126 1757 2336	2.0F 2.1E 1.5F 1.9E	25 SA	0148 0745 1445 1955	0434 1059 1703 2309	2.1F 2.2E 1.6F 2.1E
11 SA	0307 0857 1556 2106	0609 1215 1836	2.0F 2.1E 1.5F	26 SU	0240 0834 1531 2046	0523 1150 1751	2.3F 2.4E 1.8F
12 SU	0351 0938 1637 2150	0023 0638 1300 1906	1.9E 2.0F 2.1E 1.5F	27 M	0330 0922 1615 2138	0003 0611 1241 1838	2.3E 2.4F 2.5E 2.0F
13 M	0434 1018 1717 2235	0108 0709 1340 1937	1.9E 1.9F 2.2E 1.5F	28 TU	0420 1011 1659 2231	0057 0658 1329 1925	2.4E 2.5F 2.6E 2.2F
14 TU	0515 1058 1756 2318	0151 0743 1418 2010	1.9E 1.8F 2.1E 1.5F	29 W	0510 1100 1745 2324	0148 0745 1416 2015	2.5E 2.4F 2.7E 2.2F
15 W	0557 1137 1836	0231 0822 1455 2051	1.9E 1.7F 2.1E 1.5F	30 TH	0602 1149 1833	0237 0838 1503 2106	2.6E 2.3F 2.6E 2.2F
				31 F	0017 0659 1238 1925	0327 0932 1551 2201	2.5E 2.1F 2.5E 2.1F

FEBRUARY

DAY	SLACK WATER TIME H.M.	MAXIMUM CURRENT TIME H.M.	MAXIMUM CURRENT VEL. KNOTS	DAY	SLACK WATER TIME H.M.	MAXIMUM CURRENT TIME H.M.	MAXIMUM CURRENT VEL. KNOTS
1 SA	0111 0759 1328 2021	0420 1027 1643 2258	2.3E 1.9F 2.3E 2.1F	16 SU	0058 0751 1307 1958	0404 1011 1615 2236	1.8E 1.4F 1.8E 1.6F
2 SU	0207 0902 1420 2118	0517 1124 1738 2355	2.1E 1.7F 2.1E 2.0F	17 M	0144 0846 1350 2046	0451 1100 1656 2325	1.7E 1.3F 1.7E 1.6F
3 M	0305 1005 1516 2216	0621 1222 1839	2.0E 1.5F 1.9E	18 TU	0233 0944 1439 2138	0547 1151 1800	1.6E 1.2F 1.6E
4 TU	0408 1107 1617 2314	0056 0727 1331 1943	1.8F 1.9E 1.3F 1.8E	19 W	0329 1041 1535 2234	0015 0650 1242 1903	1.6F 1.6E 1.1F 1.6E
5 W	0511 1208 1720	0207 0830 1455 2039	1.7F 1.9E 1.2F 1.8E	20 TH	0429 1138 1638 2331	0108 0752 1337 2004	1.6F 1.7E 1.1F 1.7E
6 TH	0011 0611 1306 1820	0322 0925 1602 2133	1.7F 1.9E 1.3F 1.8E	21 F	0529 1233 1741	0205 0848 1439 2100	1.7F 1.9E 1.2F 1.9E
7 F	0107 0704 1359 1913	0422 1016 1655 2226	1.8F 1.9E 1.4F 1.8E	22 SA	0030 0627 1325 1840	0308 0941 1544 2154	1.8F 2.1E 1.5F 2.1E
8 SA	0200 0751 1447 2001	0511 1104 1741 2313	1.8F 2.0E 1.5F 1.8E	23 SU	0127 0720 1415 1935	0408 1032 1639 2248	2.0F 2.2E 1.7F 2.2E
9 SU	0248 0833 1530 2045	0552 1149 1818	1.9F 2.0E 1.5F	24 M	0222 0810 1502 2027	0504 1123 1730 2343	2.2F 2.4E 2.0F 2.4E
10 M	0333 0912 1609 2128	0000 0625 1232 1845	1.9E 1.9F 2.1E 1.6F	25 TU	0315 0859 1547 2119	0552 1214 1818	2.4F 2.5E 2.2F
11 TU	0414 0951 1647 2209	0044 0650 1312 1911	1.9E 1.8F 2.1E 1.6F	26 W	0405 0947 1632 2211	0037 0641 1304 1903	2.6E 2.4F 2.6E 2.4F
12 W	0455 1029 1723 2251	0127 0718 1350 1942	2.0E 1.8F 2.1E 1.7F	27 TH	0455 1036 1718 2303	0129 0726 1353 1952	2.7E 2.4F 2.7E 2.4F
13 TH	0534 1108 1800 2332	0207 0757 1426 2019	2.0E 1.7F 2.1E 1.7F	28 F	0547 1125 1805 2356	0218 0816 1439 2041	2.7E 2.2F 2.6E 2.3F
14 F	0616 1147 1836	0245 0838 1501 2102	1.9E 1.6F 2.0E 1.7F				
15 SA	0015 0700 1226 1915	0323 0923 1536 2149	1.9E 1.5F 1.9E 1.6F				

TIME MERIDIAN 75° W. 0000 IS MIDNIGHT. 1200 IS NOON.

TABLE 2.—CURRENT DIFFERENCES AND OTHER CONSTANTS

No.	PLACE	POSITION		TIME DIFFERENCES		VELOCITY RATIOS		MAXIMUM CURRENTS			
								Flood		Ebb	
		Lat.	Long.	Slack water	Maximum current	Maximum flood	Maximum ebb	Direction (true)	Average velocity	Direction (true)	Average velocity
		° ′	° ′	h. m.	h. m.			deg.	knots	deg.	knots
	LONG ISLAND, South Coast—Continued	N.	W.	on THE NARROWS, p.52							
				Time meridian, 75°W.							
2250	Shinnecock Inlet	40 51	72 29	−0 20	−0 40	1.5	1.2	350	2.5	170	2.3
2255	Fire I. Inlet, 0.5 mi. S. of Oak Beach	40 38	73 18	+0 15	0 00	1.4	1.2	80	2.4	245	2.4
2260	Jones Inlet	40 35	73 34	−1 00	−0 55	1.8	1.3	35	3.1	215	2.6
2265	Long Beach, inside, between bridges	40 36	73 40	−0 10	+0 10	0.3	0.3	75	0.5	275	0.6
2270	East Rockaway Inlet	40 35	73 45	−1 25	−1 35	1.3	1.2	40	2.2	225	2.3
2275	Ambrose Light	40 27	73 49	See table 5.							
2280	Sandy Hook App. Lighted Horn Buoy 2A	40 27	73 55	See table 5.							
	JAMAICA BAY										
2285	Rockaway Inlet	40 34	73 56	−1 45	−2 15	1.1	1.3	85	1.8	245	2.7
2290	Barren Island, east of	40 35	73 53	−2 00	−2 25	0.7	0.9	5	1.2	190	1.7
2295	Canarsie (midchannel, off Pier)	40 38	73 53	−1 35	−1 50	0.3	0.3	45	0.5	220	0.7
2300	Beach Channel (bridge)	40 35	73 49	−1 20	−1 20	1.1	1.0	60	1.9	225	2.0
2305	Grass Hassock Channel	40 37	73 47	−1 10	−1 00	0.6	0.5	50	1.0	230	1.0
	NEW YORK HARBOR ENTRANCE										
2310	Ambrose Channel entrance	40 30	73 58	−1 10	−1 05	1.0	1.2	310	1.7	110	2.3
2315	Ambrose Channel, SE. of West Bank Lt	40 32	74 01	(1)	−0 25	0.8	0.9	310	1.3	170	1.8
2320	Coney Island Lt., 1.6 miles SSW. of	40 33	74 01	−0 10	(2)	0.5	0.8	330	0.8	145	1.5
2325	Ambrose Channel, north end	40 34	74 02	+0 05	+0 15	0.8	0.9	330	1.3	175	1.9
2330	Coney Island, 0.2 mile west of	40 35	74 01	−0 55	−0 55	0.9	1.0	330	1.5	170	2.0
2335	Ft. Lafayette, channel east of	40 36	74 02	(3)	(3)	0.6	0.5	345	1.1	195	0.9
2340	THE NARROWS, midchannel	40 37	74 03	Daily predictions				340	1.7	160	2.0
	NEW YORK HARBOR, Upper Bay										
2345	Tompkinsville	40 38	74 04	−0 10	+0 20	0.9	1.0	5	1.6	170	2.0
2350	Bay Ridge Channel	40 39	74 02	−0 35	−0 45	0.6	0.6	40	1.0	220	1.1
2355	Red Hook Channel	40 40	74 01	−0 35	−0 35	0.6	0.4	355	1.0	170	0.7
2360	Robbins Reef Light, east of	40 39	74 03	+0 10	+0 20	0.8	0.8	15	1.3	205	1.6
2365	Red Hook, 1 mile west of	40 41	74 02	+0 45	+1 00	0.8	1.2	25	1.3	205	2.3
2370	Statue of Liberty, east of	40 42	74 02	+0 55	+1 00	0.8	1.0	30	1.4	205	1.9
	HUDSON RIVER, Midchannel[4]										
2375	The Battery, northwest of	40 43	74 02	+1 30	+1 35	0.9	1.2	15	1.5	195	2.3
2380	Desbrosses Street	40 43	74 01	+1 35	+1 40	0.9	1.2	10	1.5	----	2.3
2385	Chelsea Docks	40 45	74 01	+1 30	+1 40	1.0	1.0	20	1.7	185	2.0
2390	Forty-second Street	40 46	74 00	+1 35	+1 45	1.0	1.2	30	1.7	----	2.3
2395	Ninety-sixth Street	40 48	73 59	+1 40	+1 50	1.0	1.2	30	1.7	----	2.3
2400	Grants Tomb, 123d Street	40 49	73 58	+1 45	+1 55	0.9	1.2	25	1.6	----	2.3
2405	George Washington Bridge	40 51	73 57	+1 45	+2 00	0.9	1.1	20	1.6	200	2.2
2410	Spuyten Duyvil	40 53	73 56	+2 00	+2 10	0.9	1.1	20	1.6	----	2.1
2415	Riverdale	40 54	73 55	+2 05	+2 20	0.8	1.0	15	1.4	200	2.0
2420	Dobbs Ferry	41 01	73 53	+2 25	+2 40	0.8	0.9	10	1.3	----	1.7
2425	Tarrytown	41 05	73 53	+2 40	+2 55	0.6	0.8	0	1.1	----	1.5
2430	Ossining	41 10	73 54	+2 55	+3 10	0.5	0.7	320	0.9	----	1.3
2435	Haverstraw	41 12	73 57	+3 05	+3 15	0.5	0.7	335	0.8	----	1.3
2440	Peekskill	41 17	73 57	+3 20	+3 35	0.5	0.6	0	0.8	----	1.2
2445	Bear Mountain Bridge	41 19	73 59	+3 25	+3 40	0.5	0.6	0	0.8	----	1.1
2450	Highland Falls	41 22	73 58	+3 35	+3 50	0.6	0.6	5	1.0	185	1.2
2455	West Point, off Duck Island	41 24	73 57	+3 40	+3 55	0.5	0.6	10	1.0	----	1.1

[1] Current is rotary, turning clockwise. Minimum current of 0.9 knot sets SW. about time of "Slack, flood begins" at The Narrows. Minimum current of 0.5 knot sets NE. about 1 hour before "Slack, ebb begins" at The Narrows.

[2] Maximum flood, $-0^h\ 50^m$; maximum ebb, $+0^h\ 55^m$.

[3] Flood begins, $-2^h\ 15^m$; maximum flood, $-0^h\ 05^m$; ebb begins, $+0^h\ 05^m$; maximum ebb, $-1^h\ 50^m$.

[4] The values for the Hudson River are for the summer months, when the fresh-water discharge is a minimum.

TABLE 3.—VELOCITY OF CURRENT AT ANY TIME

Table A

Interval between slack and desired time	Interval between slack and maximum current													
	h. m. 1 20	*h. m.* 1 40	*h. m.* 2 00	*h. m.* 2 20	*h. m.* 2 40	*h. m.* 3 00	*h. m.* 3 20	*h. m.* 3 40	*h. m.* 4 00	*h. m.* 4 20	*h. m.* 4 40	*h. m.* 5 00	*h. m.* 5 20	*h. m.* 5 40
h. m.	*f.*	*f.*	*f.*	*f.*	*f.*	*f.*	*f.*	*f.*	*f.*	*f.*	*f.*	*f.*	*f.*	*f.*
0 20	0.4	0.3	0.3	0.2	0.2	0.2	0.2	0.1	0.1	0.1	0.1	0.1	0.1	0.1
0 40	0.7	0.6	0.5	0.4	0.4	0.3	0.3	0.3	0.3	0.2	0.2	0.2	0.2	0.2
1 00	0.9	0.8	0.7	0.6	0.6	0.5	0.5	0.4	0.4	0.4	0.3	0.3	0.3	0.3
1 20	1.0	1.0	0.9	0.8	0.7	0.6	0.6	0.5	0.5	0.5	0.4	0.4	0.4	0.4
1 40	------	1.0	1.0	0.9	0.8	0.8	0.7	0.7	0.6	0.6	0.5	0.5	0.5	0.4
2 00	------	------	1.0	1.0	0.9	0.9	0.8	0.8	0.7	0.7	0.6	0.6	0.6	0.5
2 20	------	------	------	1.0	1.0	0.9	0.9	0.8	0.8	0.7	0.7	0.7	0.6	0.6
2 40	------	------	------	------	1.0	1.0	1.0	0.9	0.9	0.8	0.8	0.7	0.7	0.7
3 00	------	------	------	------	------	1.0	1.0	1.0	0.9	0.9	0.8	0.8	0.8	0.7
3 20	------	------	------	------	------	------	1.0	1.0	1.0	0.9	0.9	0.9	0.8	0.8
3 40	------	------	------	------	------	------	------	1.0	1.0	1.0	0.9	0.9	0.9	0.9
4 00	------	------	------	------	------	------	------	------	1.0	1.0	1.0	1.0	0.9	0.9
4 20	------	------	------	------	------	------	------	------	------	1.0	1.0	1.0	1.0	0.9
4 40	------	------	------	------	------	------	------	------	------	------	1.0	1.0	1.0	1.0
5 00	------	------	------	------	------	------	------	------	------	------	------	1.0	1.0	1.0
5 20	------	------	------	------	------	------	------	------	------	------	------	------	1.0	1.0
5 40	------	------	------	------	------	------	------	------	------	------	------	------	------	1.0

Table B

Interval between slack and desired time	Interval between slack and maximum current													
	h. m. 1 20	*h. m.* 1 40	*h. m.* 2 00	*h. m.* 2 20	*h. m.* 2 40	*h. m.* 3 00	*h. m.* 3 20	*h. m.* 3 40	*h. m.* 4 00	*h. m.* 4 20	*h. m.* 4 40	*h. m.* 5 00	*h. m.* 5 20	*h. m.* 5 40
h. m.	*f.*	*f.*	*f.*	*f.*	*f.*	*f.*	*f.*	*f.*	*f.*	*f.*	*f.*	*f.*	*f.*	*f.*
0 20	0.5	0.4	0.4	0.3	0.3	0.3	0.3	0.3	0.2	0.2	0.2	0.2	0.2	0.2
0 40	0.8	0.7	0.6	0.5	0.5	0.5	0.4	0.4	0.4	0.4	0.3	0.3	0.3	0.3
1 00	0.9	0.8	0.8	0.7	0.7	0.6	0.6	0.5	0.5	0.5	0.4	0.4	0.4	0.4
1 20	1.0	1.0	0.9	0.8	0.8	0.7	0.7	0.6	0.6	0.6	0.5	0.5	0.5	0.5
1 40	------	1.0	1.0	0.9	0.9	0.8	0.8	0.7	0.7	0.7	0.6	0.6	0.6	0.6
2 00	------	------	1.0	1.0	0.9	0.9	0.9	0.8	0.8	0.7	0.7	0.7	0.7	0.6
2 20	------	------	------	1.0	1.0	1.0	0.9	0.9	0.8	0.8	0.8	0.7	0.7	0.7
2 40	------	------	------	------	1.0	1.0	1.0	0.9	0.9	0.9	0.8	0.8	0.8	0.7
3 00	------	------	------	------	------	1.0	1.0	1.0	0.9	0.9	0.9	0.9	0.8	0.8
3 20	------	------	------	------	------	------	1.0	1.0	1.0	1.0	0.9	0.9	0.9	0.8
3 40	------	------	------	------	------	------	------	1.0	1.0	1.0	1.0	0.9	0.9	0.9
4 00	------	------	------	------	------	------	------	------	1.0	1.0	1.0	1.0	0.9	0.9
4 20	------	------	------	------	------	------	------	------	------	1.0	1.0	1.0	1.0	0.9
4 40	------	------	------	------	------	------	------	------	------	------	1.0	1.0	1.0	1.0
5 00	------	------	------	------	------	------	------	------	------	------	------	1.0	1.0	1.0
5 20	------	------	------	------	------	------	------	------	------	------	------	------	1.0	1.0
5 40	------	------	------	------	------	------	------	------	------	------	------	------	------	1.0

Use Table A for all places except those listed below for Table B.
Use Table B for Cape Cod Canal, Hell Gate, Chesapeake and Delaware Canal and all stations in Table 2 which are referred to them.

1. From predictions find the time of slack water and the time and velocity of maximum current (flood or ebb), one of which is immediately before and the other after the time for which the velocity is desired.
2. Find the interval of time between the above slack and maximum current, and enter the top of Table A or B with the interval which most nearly agrees with this value.
3. Find the interval of time between the above slack and the time desired, and enter the side of Table A or B with the interval which most nearly agrees with this value.
4. Find, in the table, the factor corresponding to the above two intervals and multiply the maximum velocity by this factor. The result will be the approximate velocity at the time desired.

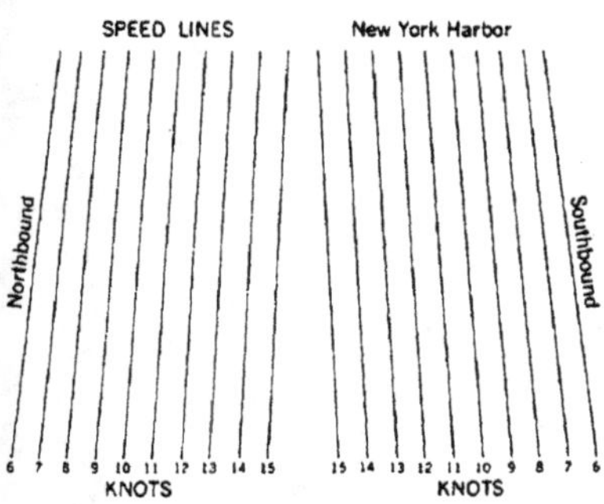
SPEED LINES
New York Harbor
Northbound
Southbound
KNOTS
KNOTS

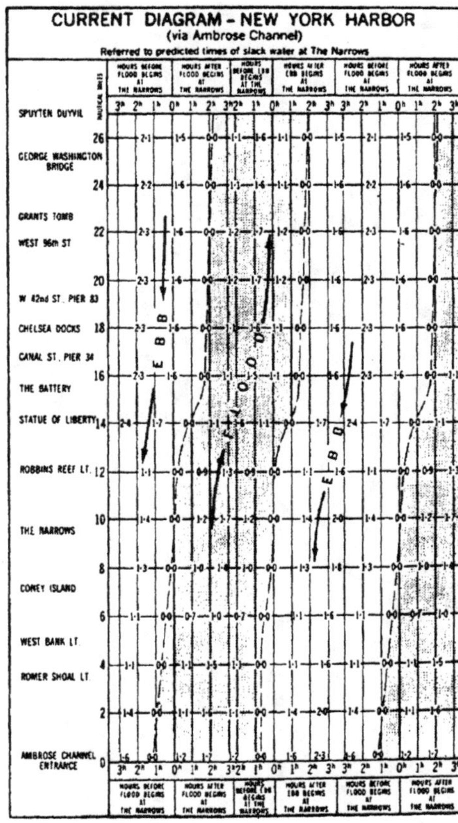
CURRENT DIAGRAM - NEW YORK HARBOR
(via Ambrose Channel)
Referred to predicted times of slack water at The Narrows
SPUYTEN DUYVIL
GEORGE WASHINGTON BRIDGE
GRANTS TOMB
WEST 96th ST
W 42nd ST. PIER 83
CHELSEA DOCKS
CANAL ST. PIER 34
THE BATTERY
STATUE OF LIBERTY
ROBBINS REEF LT.
THE NARROWS
CONEY ISLAND
WEST BANK LT.
ROMER SHOAL LT.
AMBROSE CHANNEL ENTRANCE
EBB
FLOOD
EBB

GLOSSARY OF MARINE NAVIGATION

A

abaft, *adv.* In a direction farther aft in a ship than a specified reference position, such as *abaft the mast.* See also ABAFT THE BEAM, AFT, ASTERN.

abaft the beam. Any direction between broad on the beam and astern. See also FORWARD OF THE BEAM.

abampere, *n.* The unit of current in the centimeter-gram-second electromagnetic system. The abampere is 10 amperes.

abeam, *adv.* In a line approximately at right angles to the ship's keel; opposite the waist or middle part of a ship. See also BROAD ON THE BEAM.

aberration, *n.* 1. The apparent displacement of a celestial body in the direction of motion of the earth in its orbit caused by the motion of the earth combined with the finite velocity of light. When, in addition to the combined effect of the velocity of light and the motion of the earth, account is taken of the motion of the celestial body in space during the interval that the light is traveling to the earth from the luminous body, as in the case of planets, the phenomenon is termed **planetary aberration.** The aberration due to the rotation of the earth on its axis is termed **diurnal aberration** or **daily aberration.** The aberration due to the revolution of the earth about the sun is termed **annual aberration.** The aberration due to the motion of the center of mass of the solar system in space is termed **secular aberration** but is not taken into account in practical astronomy. See also CONSTANT OF ABERRATION.
2. The convergence to different foci, by a lens or mirror, of parallel rays of light. In a single lens having spherical surfaces, aberration may be caused by differences in the focal lengths of the various parts of the lens: rays passing through the outer part of the lens come to a focus nearer the lens than do rays passing through its central part. This is termed **spherical aberration** and, being due to the faulty figure of the lens, is eliminated by correcting that figure. A lens so corrected is called an **aplanatic lens.** Aberration may also result from differences in the wavelengths of light of different colors: light of the shorter wavelengths (violet end of the spectrum) comes to a focus nearer the lens than light of the longer wavelengths (red end of the spectrum). This is termed **chromatic aberration,** and is practically eliminated over a moderate range of wavelengths by using a composite lens, called an **achromatic lens,** composed of parts having different dispersive powers.

aberration constant. See CONSTANT OF ABERRATION.

ablation, *n.* Wasting of snow or ice by melting or evaporation.

abnormal, *adj.* Deviating from normal.

abrasion, *n.* Rubbing or wearing away, or the result of such action.

abroholos, *n.* A squall frequent from May through August between Cabo de Sao Tome and Cabo Frio on the coast of Brazil.

abrupt, *adj.* Steep, precipitous. See also BOLD.

abscissa, *n.* The horizontal coordinate of a set of rectangular coordinates. Also used in a similar sense in connection with oblique coordinates.

absolute accuracy. Since geographical coordinates are derived from a model of the irregularly-shaped earth, it is impracticable to find or even state a position in absolute terms. This reason alone precludes the use of the term in stating the accuracy of a navigation system. However, the term is often loosely used for PREDICTABLE ACCURACY.

absolute gain of an antenna. See ISOTROPIC GAIN OF AN ANTENNA.

absolute humidity. The mass of water vapor per unit volume of air.

absolute motion. Motion relative to a fixed point. If the earth were stationary in space, any change in the position of another body, relative to the earth, would be due only to the motion of that body. This would be absolute motion, or motion relative to a fixed point. **Actual motion** is motion of an object relative to the earth.

absolute temperature. Temperature measured from absolute zero which is zero on the Kelvin scale, $-273^\circ.16$ C on the Celsius scale, and $-459^\circ.69$ F on the Fahrenheit scale. The sizes of the Kelvin and Celsius degree are equal. The size of a degree on the Fahrenheit scale equals that on the Rankine scale.

absolute value. The value of a real number without regard to sign. Thus, the absolute value of +8 or − 8 is $|8|$. Vertical lines on each side of a number indicate that its absolute value is intended.

absolute zero. The lowest temperature which it is possible to approach; $-459^\circ.69$F or $-273^\circ.16$C.

abyss, *n.* A very deep, unfathomable place. The term is used to refer to a particular deep part of the ocean, or to any part below 300 fathoms.

abyssal plain. See under PLAIN.

accelerate, *v., t.* To cause to move with increased velocity.

acceleration, *n.* 1. The rate of change of velocity. 2. The act or process of accelerating, or the state of being accelerated. Negative acceleration is called DECELERATION.

acceleration error. That error resulting from change in velocity (either speed or direction); specifically, deflection of the apparent vertical, as indicated by an artificial horizon, due to acceleration. Also called BUBBLE ACCELERATION ERROR when applied to an instrument using a bubble as an artificial horizon.

accelerometer. *n.* A device used to measure the accelerations of a craft, resulting from the craft's acceleration with respect to the earth, acceleration of gravity, and Coriolis acceleration. In its simplest form the accelerometer consists of a test mass constrained to measure accelerations in a particular direction with a scale, or other appropriate device, to indicate its output. If the frame of the device is accelerated in the above direction, the test mass lags behind due to inertia. The displacement of the test mass against the constraint is a measure of acceleration in the above direction.

accidental error. See RANDOM ERROR. This term should not be confused with MISTAKE or BLUNDER.

accretion, *n.* Accumulation resulting from the action of natural forces.

accuracy, *n.* 1. In navigation, a measure of the error between the point desired and the point achieved, or between the position indicated by measurement and the true position. Accurate determination of position is dependent upon the capability of the navigation system to provide precise information, the user's ability to interpret this information, correct geodetic coordinates, and proper cartography (when required). There are many ways of expressing position accuracy. In general, position accuracy is statistical in character and can only be expressed in terms of a distance that will not be exceeded in some percentage of cases. *Navigational accuracy figures must be accompanied by the statistical uncertainty probability used to calculate them if they are to have meaning.* However, some expressions of accuracy are defined in terms of the probability.

2. A measure of how close the outcome of a series of observations or measurements approaches the true value of a desired quantity. The degree of exactness with which the true value of the quantity is determined from observations is limited by the presence of both systematic and random errors. In navigation the term *accurate* is used to indicate that both the systematic and the random errors in the determination of the value are small. Accuracy should not be confused with PRECISION, which is a measure of the repeatability of the observations. Observations may be of high precision due to the quality of the observing instrument, the skill of the observer and the resulting small random errors, but inaccurate due to the presence of large systematic errors which may not have been detected and allowed for. *Accuracy implies precision, but precision does not imply accuracy.* Since the true value of a quantity often may not be determined from observations with sufficient exactness due to insufficient knowledge of the errors, usually the systematic errors, some authorities define accuracy in terms of the degree of exactness with which observations yield values of the quantity close to some adopted or accepted value rather than the true value of the quantity. This definition could yield unjustifiable estimates of accuracy since the adopted or accepted value of the quantity could contain unknown systematic errors.

See also ERROR, RADIAL ERROR, ABSOLUTE ACCURACY, PREDICTABLE ACCURACY, RELATIVE ACCURACY, REPEATABLE ACCURACY.

achromatic lens. See under ABERRATION, definition 2.

aclinal, *adj.* Without dip; horizontal.

aclinic, *adj.* Without magnetic dip.

aclinic line. The magnetic equator; that line on the surface of the earth connecting all points of zero magnetic dip.

acoustic depth finder. See ECHO SOUNDER.

acoustic navigation. See SONIC NAVIGATION.

acoustics, *n.* 1. That branch of physics dealing with sound.

2. The sound characteristics of a room, auditorium, etc., which determine its quality with respect to distinct hearing.

acoustic sounding. See ECHO SOUNDING.

acquisition, *n.* In the operation of automated radar plotting aids, the selection of those targets requiring a tracking procedure and the initiation of their tracking.

acre, *n.* A unit of area equal to 43,560 square feet.

across-the-scope echo. See CLASSIFICATION OF RADAR ECHOES.

active satellite. 1. An artificial satellite which transmits an electromagnetic signal. A satellite with the capability to transmit, repeat, or retransmit electromagnetic information as contrasted with PASSIVE SATELLITE.

2. As defined by the *International Telecommunication Union* (ITU), an earth satellite carrying a station intended to transmit or retransmit radiocommunication signals.

active tracking system. A satellite tracking system which operates by transmission of signals to and receipt of responses from the satellite.

actual motion. Motion of an object relative to the earth. See also MOTION.

acute angle. An angle less than 90°.

additional secondary phase factor correction. A correction in addition to the secondary phase factor correction for the additional time (or phase delay) for transmission of a low frequency signal over a composite land-seawater path when the signal transit time is based on the free-space velocity. This correction is given in Pub. 221, *Loran-C Correction Table*. The Loran-C lattices overprinted on nautical charts may be compensated for additional secondary phase factor (ASF), particularly in the Coastal Confluence Zone.

ADF reversal. The swinging of the needle on the direction indicator of an automatic direction finder through 180°, indicating that the station to which the direction finder is tuned has been passed.

adiabatic, *adj.* Without gain or loss of heat.

adjacent angles. Two angles having a common vertex and lying on opposite sides of a common side.

adjustment, *n.* The determination and application of corrections to observations, for the purpose of reducing errors or removing internal inconsistencies in derived results. The term may refer to mathematical procedures or to corrections applied to instruments used in making observations.

adrift, *adj. & adv.* Afloat and unattached in any way to the shore or the bottom, and without propulsive power; at the mercy of the sea and weather. See also UNDERWAY.

advance, *n.* 1. The distance a vessel moves in its initial direction from the point where the rudder is started over until the heading has changed 90°.

2. The distance a vessel moves in the initial direction for heading changes of less than 90°.

See also TRANSFER.

advance, *v., t. & i.* To move forward, as to move a line of position forward, parallel to itself, along a course line to obtain a line of position at a later time. The opposite is RETIRE.

advanced line of position. A line of position which has been moved forward along the course line to allow for the run since the line was established. The opposite is RETIRED LINE OF POSITION.

advection, *n.* Horizontal movement of part of the atmosphere. WIND refers to air motion, while ADVECTION refers more specifically to the transfer of part of the atmosphere from one area to another.

advection fog. A type of fog caused by the advection of moist air over a cold surface, and the consequent cooling of that air to below its dew point. SEA FOG is a very common advection fog that is caused by moist air in transport over a cold body of water.

aero light. Short for AERONAUTICAL LIGHT.

aeromarine light. A marine-type light having part of its beam deflected to an angle of 10° to 15° above the horizon for the use of aircraft.

aeromarine radiobeacon. A radiobeacon established for use by both mariners and airmen. A beacon established to be of primary usefulness to mariners is known as a **marine radiobeacon**; a beacon established to be of primary usefulness to airmen is known as an **aeronautical radiobeacon.**

aeronautical, *adj.* Of or pertaining to the operation or navigation of aircraft.

aeronautical beacon. A visual aid to navigation, displaying flashes of white or colored light or both, used to indicate the location of airports, landmarks, and certain points of the Federal airways in mountainous terrain and to mark hazards.

aeronautical chart. See under CHART.

aeronautical light. A luminous or lighted aid to navigation intended primarily for air navigation. One intended primarily for marine navigation is called a MARINE LIGHT. Often shortened to AERO LIGHT.

aeronautical radiobeacon. A radiobeacon whose service is intended primarily for the benefit of aircraft.

aeronautical radionavigation ≠ satellite service. As defined by the *International Telecommunication Union* (ITU), a radionavigation ≠ satellite service in which mobile earth stations are located on board aircraft.

aestival, *adj.* Pertaining to summer. The corresponding adjectives for fall, winter, and spring are *autumnal, hibernal,* and *vernal.*

affluent, *n.* A stream flowing into a larger stream or lake; a tributary.

afloat, *adj. & adv.* Floating; borne on the water; water-borne. See also SURFACED, UNCOVERED, AGROUND, ASHORE.

aft, *adv.* Near, toward, or at the stern of a craft. See also ABAFT, ASTERN.

afterglow, *n.* 1. The slowly decaying luminescence of the screen of the cathode-ray tube after excitation by an electron beam has ceased. See also PERSISTENCE.
2. A broad, high arch of radiance or glow seen occasionally in the western sky above the highest clouds in deepening twilight, caused by the scattering effect of very fine particles of dust suspended in the upper atmosphere.

aged ridge. A ridge (line or wall of ice forced up by pressure) which has undergone considerable weathering. These ridges are best described as undulations.

age of diurnal inequality. The time interval between the maximum semimonthly north or south declination of the moon and the maximum effect of the declination upon the range of tide or the speed of the tidal current, this effect being manifested chiefly by an increase in the height or speed difference between the two high (low) waters or flood (ebb) currents during the day. The tides occurring at this time are called TROPIC TIDES. Also called DIURNAL AGE.

age of parallax inequality. The time interval between perigee of the moon and the maximum effect of parallax upon the range of tide or the speed of the tidal current. See also PARALLAX INEQUALITY.

age of phase inequality. The time interval between new or full moon and the maximum effect of these phases upon the range of tide or the speed of the tidal current. Also called AGE OF TIDE.

age of the moon. The elapsed time, usually expressed in days, since the last new moon. See also PHASES OF THE MOON.

age of tide. See AGE OF PHASE INEQUALITY.

Ageton, *n.* See H.O. PUB. NO. 211.

agger, *n.* See DOUBLE TIDE.

agonic line. A line joining points of no magnetic variation.

agravic, *adj.* Of or pertaining to a condition of no gravitation.

aground, *adj. & adv.* Resting or lodged on the bottom. The opposite is AFLOAT.

Agulhas Current. A generally southwestward-flowing ocean current of the Indian Ocean; one of the swiftest of ocean currents. Throughout the year, part of the Indian South Equatorial Current turns southward along the east coast of Africa and feeds the strong Agulhas Current. To the south of latitude 30°S the Agulhas Current is a well-defined and narrow current that extends less than 100 km from the coast. To the south of South Africa the greatest volume of its water bends sharply to the south and then toward the east, thus returning to the Indian Ocean by joining the flow from South Africa toward Australia across the southern part of that ocean. However, a small portion of the Agulhas Current water appears to round the Cape of Good Hope from the Indian Ocean and continue into the Atlantic Ocean.

ahead, *adv.* Bearing approximately 000° relative. The term is often used loosely for DEAD AHEAD or bearing exactly 000° relative. The opposite is ASTERN.

aid, *n.* Short for AID TO NAVIGATION.

aid to navigation. A device external to a craft, designed to assist in determination of position of the craft, a safe course, or to warn of dangers or obstructions. If the information is transmitted by light waves, the device is called a **visual aid to navigation**; if by sound waves, an **audible aid to navigation**; if by radio waves, a **radio aid to navigation**. Any aid to navigation using electronic equipment, whether or not radio waves are involved, may be called an **electronic aid to navigation**. The expression AID TO NAVIGATION should not be confused with NAVIGATIONAL AID, a broad expression covering any instrument, device, chart, method, etc., intended to assist in the navigation of a craft. In British usage the terms *aid to navigation* and *navigational aid* are used without distinction.

air, *n.* 1. The mixture of gases comprising the earth's atmosphere. It is composed of about 78% nitrogen, 21% oxygen, 1% other gases, and a variable amount of impurities such as water vapor, suspended dust particles, smoke, etc. See also ATMOSPHERE.
2. Wind of force 1 (1-3 knots or 1-3 miles per hour) on the Beaufort wind scale, called LIGHT AIR.

air almanac. 1. A periodical publication of astronomical data useful to and designed primarily for air navigation. Such a publication designed primarily for marine navigation is called a NAUTICAL ALMANAC. See also ALMANAC FOR COMPUTERS.
2. *Air Almanac*; a joint publication of the U.S. Naval Observatory and H. M. Nautical Almanac Office, Royal Greenwich Observatory, designed primarily for air navigation. In general the information is similar to that of the *Nautical Almanac,* but is given to a precision of 1′ of arc and 1^s of time, at intervals of 10^m (values for the sun and Aries are given to a precision of 0′.1). This publication is suitable for ordinary navigation at sea, but may lack the precision that is sometimes needed. Each publication covers 6 months.

air defense identification zone. Airspace of defined dimensions within which the ready identification, location, and control of aircraft are required. Commonly referred to as ADIZ.

air mass. An extensive body of air with fairly uniform (horizontal) physical properties, especially temperature and humidity. In its incipient stage the properties of the air mass are determined by the characteristics of the region in which it forms. It is a cold or warm air mass as it is colder or warmer than the surrounding air.

air-mass classification. Air masses are classified according to their source regions. Four such regions are generally recognized: (1) *equatorial*

(E), the doldrum area between the north and south trades; (2) *tropical (T)*, the trade wind and lower temperate regions; (3) *polar (P)*, the higher temperate latitudes; and (4) *arctic* or *antarctic (A)*, the north or south polar regions of ice and snow. This classification is a general indication of relative temperature, as well as latitude of origin. Air masses are further classified as *maritime (m)* or *continental (c)*, depending upon whether they form over water or land. This classification is an indication of the relative moisture content of the air mass. Tropical air, then might be designated *maritime tropical (mT)* or *continental tropical (cT)*. Similarly, polar air may be either *maritime polar (mP)* or *continental polar (cP)*. Arctic/antarctic air, due to the predominance of landmasses and ice fields in the high latitudes, is rarely *maritime arctic (mA)*. Equatorial air is found exclusively over the ocean surface and is designated neither *(cE)* nor *(mE)*, but simply *(E)*. A third classification sometimes applied to tropical and polar air masses indicates whether the air mass is warm (*w*) or cold (*k*) relative to the underlying surface. Thus, the symbol *mTw* indicates maritime tropical air which is warmer than the underlying surface, and *cPk* indicates continental polar air which is colder than the underlying surface. The *w* and *k* classifications are primarily indications of stability (i.e., change of temperature with increasing height). If the air is cold relative to the surface, the lower portion of the air mass is being heated, resulting in instability (temperature markedly decreases with increasing height) as the warmer air tends to rise by convection. Conversely, if the air is warm relative to the surface, the lower portion of the air mass is cooled, tending to remain close to the surface. This is a stable condition (temperature increases with increasing height).

air temperature correction. A correction due to nonstandard air temperature, particularly that sextant altitude correction due to changes in refraction caused by difference between the actual temperature and the standard temperature used in the computation of the refraction table. The *Nautical Almanac* refraction table is based upon an air temperature of 50°F (10°C) at the surface of the earth. At other temperatures the refraction differs, becoming greater at lower temperatures, and less at higher temperatures. The correction for air temperature varies with the temperature of the air and the altitude of the celestial body, and applies to all celestial bodies, regardless of the method of observation. However, except for extreme temperatures or low altitudes, this correction is not usually applied unless results of unusual accuracy are desired.

ait, (āt), *n. British terminology.* An islet in a river. Also called EYOT (ī ūt).

Alaska Current. A North Pacific Ocean current flowing counterclockwise in the Gulf of Alaska. It is the northward flowing division of the Aleutian Current. Part of the water passes between the Aleutian Islands into the Bering Sea from which it emerges as the Oyashio, and part rejoins the Aleutian Current. It enters the Gulf of Alaska from a point west of Vancouver Island, and since it comes from the south, it has the character of a warm current in spite of the fact that it carries subarctic water. It therefore exercises an influence on climate conditions similar, on a small scale, to that which the North Atlantic Current and Norwegian Current exercise on the climates of northwestern Europe.

Alaska-Hawaii standard time. See STANDARD TIME.

albedo, *n.* The ratio of radiant energy reflected to that received by a surface, usually expressed as a percentage; reflectivity. The term generally refers to energy within a specific frequency range, as the visible spectrum. Its most frequent application is to the light reflected by a celestial body.

alert, *n.* See ALERT TIME CALCULATIONS.

alert time calculations. Computations of times and altitudes of available satellite passes in a given period of time at a given location, based on orbital data transmitted from satellite memory. Sometimes called ALERT.

Aleutian Current. An eastward flowing North Pacific Ocean current which lies north of the North Pacific Current. As it approaches the coast of North America it divides to form the northward-flowing ALASKA CURRENT, and the southward-flowing CALIFORNIA CURRENT. Also called SUBARCTIC CURRENT.

alga (*pl. algae*), *n.* A plant of simple structure which grows chiefly in water, such as the various forms of seaweed. It ranges in size from a microscopic plant, large numbers of which sometimes cause discoloration of water, to the giant kelp which may extend for more than 600 feet in length. Red snow and the Red Sea own their names to red algae.

alidade, *n.* That part of an optical measuring instrument comprising the optical system, indicator, vernier, etc. In modern practice the term is used principally in connection with a bearing circle fitted with a telescope to facilitate observation of bearings. Also called TELESCOPIC ALIDADE.

align, *v., t.* To place objects in line.

alignment, *n.* 1. The placing of objects in a line. 2. The process of orienting the measuring axes of the inertial components of inertial navigation equipment with respect to the coordinate system in which the equipment is to be used.

Allard's law. A formula relating the illuminance produced on a normal surface at a given distance from a point source of light, the intensity of the light, and the degree of transparency of the atmosphere, assumed to be uniform.

all-round light. See OMNIDIRECTIONAL LIGHT.

all-weather, *adj.* Designed or equipped to perform by day or night under any weather conditions.

almanac, *n.* A periodical publication of astronomical data useful to a navigator. It contains less information than an EPHEMERIS, and values are generally given to less precision. If information is given in a form and to a precision suitable for marine navigation, it is called an **air almanac.** See also **nautical almanac**; if designed primarily for air navigation, it is called an air almanac. See also EPHEMERIS, ASTRONOMICAL ALMANAC.

Almanac for Computers. An annual publication of the U.S. Naval Observatory designed to facilitate the application of digital computers and small calculators to problems of astronomy and navigation which require coordinates of celestial bodies. For such applications the fixed-interval tabulations of the *Astronomical Almanac*, the *Nautical Almanac*, and the *Air Almanac*, with the inevitable requirements of interpolation, should ideally be replaced by concise mathematical expressions for direct calculations. Such expressions must take the form of mathematical approximations, however, since the precise data contained in the above publications are calculated from extensive theories which are not readily adaptable to the majority of astronomical and navigational applications. Using the expressions in the *Almanac for Computers*, it is possible to calculate, with minimal loss of precision, the basic data in the above three almanacs for specific times and conditions.

almucantar, *n.* A small circle on the celestial sphere paralled to the horizon. Also called CIRCLE OF EQUAL ALTITUDE, PARALLEL OF ALTITUDE.

almucantar staff. An ancient instrument formerly used for amplitude observations.

alnico, *n.* An alloy composed principally of aluminum, nickel, cobalt, and iron; used for permanent magnets, as for magnetic compass adjustment. The name *alnico* is derived from the words aluminum, nickel, and cobalt.

alongshore current. See LONGSHORE CURRENT.

alphanumeric grid. See ATLAS GRID.

alternate blanking. See under DUAL-RATE BLANKING.

alternating current. An electric current that continually changes in magnitude and periodically reverses polarity.

alternating fixed and flashing light. A fixed light varied at regular intervals by a single flash of greater luminous intensity, with color variations in either the fixed light or flash, or both. A light list entry may be given as:

Alt. F.W., F.R. and Fl.R.
period 90^s
W. 59^s
R. 14^s
R. fl. 3^s
R. 14^s

Within each 90^s period the light is first fixed white for 59^s, then fixed red for 14^s, then there is a flash of brilliant red for 3^s, and finally the light is fixed red for 14^s. See also ALTERNATING LIGHT.

alternating fixed and group flashing light. A fixed light varied at regular intervals by a group of two or more flashes of greater luminous intensity with color variations in either the fixed light or flashes or both.

alternating flashing light. A light showing a single flash with color variations at regular intervals, the duration of light being shorter than that of darkness. A light list entry may be given as:

Alt. Fl. W. R.
period 20^s
W. fl. 1^s, ec. 9^s
R. fl. 1^s, ec. 9^s

Within each 20^s period there is first a white flash of 1^s duration, the light is eclipsed (extinguished) for 9^s, then there is a red flash of 1^s duration, and then the light is eclipsed for 9^s before the sequence begins again. See also FLASHING LIGHT.

alternating group flashing light. A group flashing light which shows periodic color change.

alternating group occulting light. A group occulting light which shows periodic color change. A light list entry may be given as:

Alt. Gp. Occ. (2 + 2)
W.R.
period 60^s
W. lt. 15.0^s, ec. 4.0^s
W. lt. 1.5^s, ec. 4.0^s
W. lt. 15.0^s, ec. 9.5^s
R. lt. 1.5^s, ec. 9.5^s

Within each 60^s period the light is first white for $15\overset{s}{.}0$, the light is eclipsed (extinguished) for $4\overset{s}{.}0$, then the light is white for $1\overset{s}{.}5$, then eclipsed for $4\overset{s}{.}0$, white for $15\overset{s}{.}0$, eclipsed for $9\overset{s}{.}5$, then red for $1\overset{s}{.}5$, and then the light is eclipsed for $9\overset{s}{.}5$ before the sequence begins again.

alternating occulting light. A light totally eclipsed at regular intervals, the duration of light always being longer than the duration of darkness, which shows periodic color change. See also ALTERNATING LIGHT.

alternating light. A light showing different colors alternately.

altitude, *n.* Angular distance above the horizon; the arc of a vertical circle between the horizon and a point on the celestial sphere, measured upward from the horizon. Angular distance below the horizon is called **negative altitude** or **depression.** Altitude indicated by a sextant is called **sextant altitude.** Sextant altitude corrected only for inaccuracies in the reading (instrument, index, and personal errors, as applicable) and inaccuracies in the reference level (principally dip) is called **apparent** or **rectified altitude.** After all corrections are applied, it is called **corrected sextant altitude** or **observed altitude.** An altitude taken directly from a table, before interpolation, is called **tabulated altitude.** After interpolation, or if determined by calculation, mechanical device, or graphics, it is called **computed altitude.** If the altitude of a celestial body is computed before observation, and sextant altitude corrections are applied with reversed sign, the result is called **precomputed altitude.** The difference between computed and observed altitudes (corrected sextant altitudes), or between precomputed and sextant altitudes, is called **altitude intercept** or **altitude difference.** An altitude determined by inexact means, as by estimation or star finder, is called an **approximate altitude.** The altitude of a celestial body on the celestial meridian is called **meridian altitude.** The expression **ex-meridian altitude** is applied to the altitude of a celestial body near the celestial meridian, to which a correction is to be applied to determine the meridian altitude. A **parallel of altitude** is a circle of the celestial sphere parallel to the horizon, connecting all points of equal altitude. See also EQUAL ALTITUDES.

altitude azimuth. An azimuth determined by solution of the navigational triangle with altitude, declination, and latitude given. A TIME AZIMUTH is computed with meridian angle, declination, and latitude given. A TIME AND ALTITUDE AZIMUTH is computed with meridian angle, declination, and altitude given.

altitude circle. See PARALLEL OF ALTITUDE.

altitude difference. 1. See ALTITUDE INTERCEPT.

2. The change in the altitude of a celestial body occurring with change in declination, latitude, or hour angle, for example the *first difference* between successive tabulations of altitude in a latitude column of Pub. No. 229, *Sight Reduction Tables for Marine Navigation.*

altitude intercept. The difference in minutes of arc between the computed and the observed altitude (corrected sextant altitude), or between precomputed and sextant altitudes. It is labeled T (toward) or A (away) as the observed (or sextant) altitude is greater or smaller than the computed (or precomputed) altitude. Also called ALTITUDE DIFFERENCE, INTERCEPT.

altitude intercept method. See ST. HILAIRE METHOD.

altitude of the apogee. As defined by the *International Telecommunication Union* (ITU), the altitude of the apogee above a specified reference surface serving to represent the surface of the earth.

altitude of the perigee. As defined by the *International Telecommunication Union* (ITU), the altitude of the perigee above a specified reference surface serving to represent the surface of the earth.

altitude tints. See HYPSOMETRIC TINTING.

alto-. A prefix used in cloud classification to indicate the middle level. See also CIRRO-.

altocumulus, *n.* A cloud layer (or patches) within the middle level (mean height 6,500-20,000 ft.) composed of rather flattened globular masses, the smallest elements of the regularly arranged layers being fairly thin, with or without shading. These elements are arranged in groups, in lines, or waves, following one or two directions, and are sometimes so close together that their edges join. See also CLOUD CLASSIFICATION.

altostratus, *n.* A sheet of gray or bluish cloud within the middle level (mean height 6,500-20,000 ft.). Sometimes the sheet is composed of a compact mass of dark, thick, gray clouds of fibrous structure; at other times the sheet is thin and through it the sun or moon can be seen dimly as though gleaming through ground glass. See also CLOUD CLASSIFICATION.

ambient temperature. The temperature of the air or other medium surrounding an object. See also FREE-AIR TEMPERATURE.

ambiguity, *n.* In navigation, the condition obtained when a given set of observations defines more than one point, direction, line of position, or surface of position.

ambiguous, *adj.* Having two or more possible meanings or values.

American Ephemeris and Nautical Almanac. See ASTRONOMICAL ALMANAC.

American Practical Navigator. See PUB. NO. 9.

amidships, *adv.* At, near, or toward the middle of a ship.

ampere, *n.* The *base unit* of electric current in the International System of Units; it is that constant current which, if maintained in two straight parallel conductors of infinite length, of negligible circular cross section, and placed 1 metre apart in vacuum, would produce between these conductors a force equal to 2×10^{-7} newton per metre of length.

ampere per metre. The *derived unit* of magnetic field strength in the International System of Units.

amphidromic region. An area surrounding a no-tide point from which the radiating cotidal lines progress through all hours of the tidal cycle.

amplification, *n.* 1. An increase in signal magnitude from one point to another, or the process causing this increase.
2. Of a transducer, the scalar ratio of the signal output to the signal input.

amplifier, *n.* A device which enables an input signal to control power from a source independent of the signal and thus be capable of delivering an output which bears some relationship to, and is generally greater than, the input signal.

amplitude, *n.* 1. Angular distance of a celestial body north or south of the prime vertical circle; the arc of the horizon or the angle at the zenith between the prime vertical circle and a vertical circle through the celestial body measured north or south from the prime vertical to the vertical circle. The term is customarily used only with reference to bodies whose centers are on the celestial horizon, and is prefixed E or W, as the body is rising or setting, respectively; and suffixed N or S to agree with the declination. The prefix indicates the origin and the suffix the direction of measurement. Amplitude is designated as **true, magnetic, compass,** or **grid** as the reference direction is true, magnetic, compass, or grid east or west, respectively.
2. The maximum value of the displacement of a wave, or other periodic phenomenon, from the zero position.
3. One-half the range of a constituent tide. By analogy, it may be applied also to the maximum speed of a constituent current.

amplitude compass. A compass intended primarily for measuring amplitude. It is graduated from 0° at east and west to 90° at north and south. Such an instrument is seldom used on modern vessels.

amplitude distortion. Distortion occurring in an amplifier or other device when the amplitude of the output is not a linear function of the input amplitude.

amplitude modulation. The process of changing the amplitude of a carrier wave in accordance with the variations of a modulating wave. See also MODULATION.

AMVER System. See AUTOMATED MUTUAL-ASSISTANCE VESSEL RESCUE SYSTEM.

anabatic wind. Any wind blowing up an incline. A KATABATIC WIND blows down an incline.

analemma, *n.* A graduated scale of the declination of the sun and the equation of time for each day of the year located in the Torrid Zone on the terrestrial globe.

analog computer. A computer in which quantities are represented by physical variables. Problem parameters are translated into equivalent mechanical or electrical circuits as an analog for the physical phenomenon being investigated without the use of a machine language. An analog computer measures continuously; a *digital computer* counts discretely.

anchorage, *n.* An area where a vessel anchors or may anchor, either because of suitability or designation. One used when satisfying quarantine regulations is called a **quarantine anchorage.**

anchorage buoy. One of a series of buoys marking the limits of an anchorage, not to be confused with a MOORING BUOY; a buoy marking the location of a quarantine anchorage is called a **quarantine buoy.**

anchorage chart. A nautical chart showing prescribed or recommended anchorages. Such a chart may be a HARBOR CHART overprinted with a series of circles, each indicating an individual anchorage.

anchorage mark. A navigation mark which indicates an anchorage area or defines its limits.

anchor buoy. A buoy marking the position of an anchor on the bottom, usually painted green for the starboard anchor and red for the port anchor, and secured to the crown of the anchor by a buoy rope.

anchor ice. Submerged ice attached or anchored to the bottom, irrespective of the nature of its formation.

anchor light. A light shown from a vessel or aircraft to indicate its position when riding at anchor. Also called RIDING LIGHT.

anenometer, *n.* An instrument for measuring the speed of the wind. Some instruments also indicate the direction from which it is blowing. See also VANE, definition 1; WIND INDICATOR.

aneroid barometer. An instrument which determines atmospheric pressure by the effect of such pressure on a thin-metal cylinder from which the air has been partly exhausted. See also MERCURIAL BAROMETER.

angle, *n.* The inclination to each other of two

intersecting lines, measured by the arc of a circle intercepted between the two lines forming the angle, the center of the circle being the point of intersection. An acute angle is less than 90°; a **right angle**, 90°; an **obtuse angle**, more than 90° but less than 180°; a **straight angle** 180°; a **reflex angle**, more than 180° but less than 360°; a **perigon**, 360°. Any angle not a multiple of 90° is an **oblique angle**. If the sum of two angles is 90°, they are **complementary angles**; if 180°, **supplementary angles**; if 360°, **explementary angles**. Two **adjacent angles** have a common vertex and lie on opposite sides of a common side. A **dihedral angle** is the angle between two intersecting planes. A **spherical angle** is the angle between two intersecting great circles.

angle of cut. The smaller angular difference of two bearings or lines of position.

angle of depression. The angle in a vertical plane between the horizontal and a descending line. Also called DEPRESSION ANGLE. See ANGLE OF ELEVATION.

angle of deviation. The angle through which a ray is bent by refraction.

angle of elevation. The angle in a vertical plane between the horizontal and an ascending line, as from an observer to an object. A negative angle of elevation is usually called an ANGLE OF DEPRESSION. Also called ELEVATION ANGLE.

angle of incidence. The angle between the line of motion of a ray of radiant energy and the perpendicular to a surface, at the point of impingement. This angle is numerically equal to the ANGLE OF REFLECTION.

angle of reflection. The angle between the line of motion of a ray of reflected radiant energy and the perpendicular to a surface, at the point of reflection. This angle is numerically equal to the ANGLE OF INCIDENCE.

angle of refraction. The angle between a refracted ray and the perpendicular to the refracting surface.

angle of roll. The angle between the transverse axis of a craft and the horizontal. Also called ROLL ANGLE.

angle of uncertainty. The horizontal angle of the region of indefinite characteristic near the boundaries of a sector of a sector light. Also called ARC OF UNCERTAINTY.

angstrom, *n.* A unit of length, used especially in expressing the length of light waves, equal to one ten-thousandth of a micron or one hundred-millionth of a centimeter.

angular, *adj.* Of or pertaining to an angle or angles.

angular distance. 1. The angular difference between two directions, numerically equal to the angle between two lines extending in the given directions.

2. The arc of the great circle joining two points, expressed in angular units.

3. Distance between two points, expressed in angular units of a specified frequency. It is equal to the number of waves between the points multiplied by 2π if expressed in radians, or multiplied by 360° if measured in degrees.

angular distortion. Distortion in a map projection because of non-conformality.

angular momentum. The quantity obtained by multiplying the moment of inertia of a body by its angular speed.

angular rate. See ANGULAR SPEED.

angular rate of the earth's rotation. Time rate of change of angular displacement of the earth relative to the fixed stars equal to 0.729211×10^{-4} radian per second.

angular resolution. See BEARING RESOLUTION.

angular speed. Change of direction per unit time. Also called ANGULAR RATE. See also LINEAR SPEED.

anneal, *v., t.* To heat to a high temperature and then allow to cool slowly, for the purpose of softening, making less brittle, or removing permanent magnetism. When Flinders bars or quadrantal correctors acquire permanent magnetism, which decreases their effectiveness as compass correctors, they are annealed.

annotation, *n.* Any marking on illustrative material for the purpose of clarification, such as numbers, letters, symbols, and signs.

annual, *adj.* Of or pertaining to a year; yearly.

annual aberration. See under ABERRATION, definition 1.

annual inequality. Seasonal variation in water level or tidal current speed, more or less periodic, due chiefly to meteorological causes.

annual parallax. See HELIOCENTRIC PARALLAX.

annular, *adj.* Ring-shaped.

annular eclipse. An eclipse in which a thin ring of the source of light appears around the obscuring body. Annular solar eclipses occur, but never annular lunar eclipses.

annulus, *n.* A ring-shaped band.

anode, *n.* 1. A positive electrode; the plate of a vacuum tube; the electrode of an electron tube through which a principal stream of electrons leaves the interelectrode space.

2. The positive electrode of an electrochemical device, such as a primary or secondary cell, toward which the negative ions are drawn. See also CATHODE.

anomalistic, *adj.* Pertaining to the periodic return of the moon to its perigee, or of the earth to its perihelion.

anomalistic month. The average period of revolution of the moon from perigee to perigee, a period of 27 days, 13 hours, 18 minutes, and 33.2 seconds in 1900. The secular variation does not exceed a few hundredths of a second per century.

anomalistic period. The interval between two successive passes of a satellite through perigee. Also called PERIGEE-TO-PERIGEE PERIOD, RADIAL PERIOD. See also ORBITAL PERIOD.

anomalistic year. The period of one revolution of the earth around the sun, from perihelion to perihelion, averaging 365 days, 6 hours, 13 minutes, 53.0 seconds in 1900, and increasing at the rate of 0.26 second per century.

anomaly, *n.* 1. Departure from the strict characteristics of the type, pattern, scheme, etc.

2. An angle used in the mathematical description of the orbit of one body about another. It is the angle between the radius vector of the body and the line of apsides and is measured from pericenter in the direction of motion. When the radius vector is from the center of the primary to the orbiting body, the angle is called **true anomaly**. When the radius vector is from the center of the primary to a fictitious body moving with a uniform angular velocity in such a way that its period is equal to that of the actual body, the angle is called **mean anomaly**. When the radius vector is from the center of the elliptical orbit to the point of intersection of the circle defined by the semimajor axis with the line perpendicular to the semimajor axis and passing through the orbiting body, the angle is called **eccentric anomaly** or **eccentric angle**.

3. Departure of the local mean value of a meteorological element from the mean value for the latitude. See also MAGNETIC ANOMALY.

antarctic, *adj.* Of or pertaining to the Antarctic.

Antarctic, *n.* The region within the Antarctic Circle, or, loosely, the extreme southern regions in general, characterized by very low temperatures.

antarctic air. A type of air whose characteristics are developed in an antarctic region. Antarctic air appears to be colder at the surface in all seasons, and at all levels in fall and winter, than ARCTIC AIR.

Antarctic Circle. That parallel of latitude at about 66°33'S, marking the northern limit of the south Frigid Zone. This latitude is the complement of the sun's greatest southerly declination, and marks the approximate northern limit at which the sun becomes circumpolar. The actual limit is extended somewhat by the combined effect of refraction, semidiameter of the sun, parallax, and the height of the observer's eye above the surface of the earth. A similar circle marking the southern limit of the north Frigid Zone is called ARCTIC or NORTH POLAR CIRCLE. Also called SOUTH POLAR CIRCLE.

Antarctic Circumpolar Current. See WEST WIND DRIFT.

antarctic front. The semi-permanent, semi-continuous front between the antarctic air of the Antarctic Continent and the polar air of the southern oceans; generally comparable to the arctic front of the Northern Hemisphere.

antarctic whiteout. The obliteration of contrast between surface features in the Antarctic when a covering of snow obscuring all landmarks is accompanied by an overcast sky, resulting in an absence of shadows and an unrelieved expanse of white, the earth and sky blending so that the horizon is not distinguishable. A similar occurrence in the Arctic is called ARCTIC WHITEOUT.

ante meridian. Before noon, or the period of time between midnight (0000) and noon (1200). The period between noon and midnight is called POST MERIDIAN.

antenna, *n.* Any structure or device used to collect or radiate electromagnetic waves; specifically, that part of a radar or of a radio-sending or radio-receiving set that contains, or itself consists of, that apparatus that radiates or receives electromagnetic waves.

antenna array. A combination of antennas with suitable spacing and with all elements excited to make the radiated fields from the individual elements add in the desired direction, i.e., to obtain directional characteristics.

antenna assembly. The complete equipment associated with an antenna, including, in addition to the antenna, the base, switches, lead-in wires, revolving mechanism, etc.

antenna bearing. The generated bearing of the antenna of a radar set, as delivered to the indicator.

antenna coupler. 1. A radio-frequency transformer used to connect an antenna to a transmission line or to connect a transmission line to a radio receiver.

2. A radio-frequency transformer, link circuit, or tuned line used to transfer radio-frequency energy from the final plate-tank circuit of a transmitter to the transmitter to the transmission line feeding the antenna.

antenna directivity diagram. See as DIRECTIVITY DIAGRAM.

antenna effect. A spurious effect, in a loop antenna, resulting from the capacitance of the loop to ground.

antenna feed. That component of an antenna of mirror or lens type that irradiates, or receives energy from, the mirror or lens. See also HORN ANTENNA.

antenna radiation pattern. See as RADIATION PATTERN.

anthelion, *n.* 1. A rare kind of halo, which appears as a bright spot at the same altitude as the sun and 180° from it in azimuth. See also PARHELION.

anti-clutter gain control. See SENSITIVITY TIME CONTROL.

anti-clutter rain. See FAST TIME CONSTANT CIRCUIT.

anti-clutter sea. See SENSITIVITY TIME CONTROL.

anticorona, *n.* A diffraction phenomenon very similar to but complementary to the corona, appearing at a point directly opposite to the sun or moon from the observer. Also called BROCKEN BOW, GLORY.

anticrepuscular arch. See ANTITWILIGHT.

anticrepuscular rays. Extensions of crepuscular rays, converging toward a point 180° from the sun.

anticyclone, *n.* An approximately circular portion of the atmosphere, having relatively high atmospheric pressure and winds which blow clockwise around the center in the Northern Hemisphere and counterclockwise in the Southern Hemisphere. An anticyclone is characterized by good weather. Also called HIGH. See also CYCLONE.

anticyclonic winds. The winds associated with a high pressure area and constituting part of an anticyclone.

Antilles Current. This current originates in the vicinity of the Leeward Islands as part of the Atlantic North Equatorial Current. It flows along the northern side of the Greater Antilles. The Antilles Current eventually joins the Florida Current (north of Grand Bahama Island) to form the Gulf Stream.

antilogarithm, *n.* The number corresponding to a given logarithm. Also called INVERSE LOGARITHM.

antinode, *n.* Either of the two points on an orbit where a line in the orbit plane, perpendicular to the line of nodes, and passing through the focus, intersects the orbit.

antipodal effects. See as LONG PATH INTERFERENCE under MULTIPATH ERROR.

antipode, *n.* Anything exactly opposite to something else. Particularly, that point on the earth 180° from a given place.

antisolar point. That point on the celestial sphere 180° from the sun.

antitrades, *n., pl.* The prevailing western winds which blow over and in the opposite direction to the trade winds. Also called COUNTERTRADES.

anti-TR tube. See under TR TUBE.

antitwilight, *n.* The pink or purplish zone of illumination bordering the shadow of the earth in the dark part of the sky opposite the sun after sunset or before sunrise. Also called ANTICREPUSCULAR ARCH.

anvil cloud. Heavy cumulus or cumulonimbus having an anvil-like upper part.

apastron, *n.* That point of the orbit of one member of a double star system at which the stars are farthest apart. That point at which they are nearest together is called PERIASTRON.

aperiodic, *adj.* Without a period; of irregular occurence.

aperiodic compass. Literally "a compass without a period," or a compass that, after being deflected, returns by one direct movement to its proper reading, without oscillation. Also called DEADBEAT COMPASS.

aperture, *n.* 1. An opening; particularly, that opening in the front of a camera through which light rays pass when a picture is taken.
2. The diameter of the objective of a telescope or other optical instrument, usually expressed in inches, but sometimes as the angle between lines from the principal focus to opposite ends of a diameter of the objective.
3. Of a directional antenna, that portion of nearby plane surface that is perpendicular to the direction of maximum radiation and through which the major part of the radiation passes.

aperture antenna. An antenna in which the beam width is determined by the dimensions of a horn, lens, or reflector.

aperture ratio. The ratio of the diameter of the objective to the focal length of an optical instrument.

apex, *n.* The highest point of something, as of a cone or triangle, or the maximum latitude (vertex) of a great circle.

aphelion, *n.* That point in the elliptical orbit of a body about the sun farthest from the sun. That point nearest the sun is called PERIHELION.

aphylactic map projection. A map projection which is neither conformal nor equal area. Also called ARBITRARY MAP PROJECTION.

aplanatic lens. See under ABERRATION, definition 2.

apoapsis, *n.* See APOCENTER.

apocenter, *n.* In an elliptical orbit, the point in the orbit which is the farthest distance from the focus, where the attracting mass is located. The apocenter is at one end of the major axis of the orbital ellipse. The opposite is PERICENTER, PERIFOCUS, PERIAPSIS. Also called APOAPSIS, APOFOCUS.

apofocus, *n.* See APOCENTER.

apogean range. The average semidiurnal range of the tide occurring at the time of apogean tides. It is smaller than the mean range, where the type of tide is either semidiurnal or mixed, and is of no practical significance where the type of tide is diurnal.

apogean tidal currents. Tidal currents of decreased speed occurring monthly as the result of the moon being at apogee (farthest from the earth).

apogean tides. Tides of decreased range occurring monthly as the result of the moon being at apogee (farthest from the earth).

apogee, *n.* That orbital point farthest from the earth when the earth is the center of attraction. That orbital point nearest the earth is called PERIGEE. See also APOCENTER, PERICENTER.

apparent, *adj.* Capable of being seen or observed.

apparent altitude. Sextant altitude corrected for inaccuracies in the reading (instrument, index, and personal errors, as applicable) and inaccuracies in the reference level (principally dip or Coriolis/acceleration), but not for other errors. Apparent altitude is used in obtaining a more accurate refraction correction than would be obtained with an uncorrected sextant altitude. Also called RECTIFIED ALTITUDE. See also OBSERVED ALTITUDE, SEXTANT ALTITUDE.

apparent horizon. See VISIBLE HORIZON.

apparent motion. Motion relative to a specified or implied reference point which may itself be in motion. The expression usually refers to movement of celestial bodies as observed from the earth. Usually called RELATIVE MOVEMENT when applied to the motion of one vessel relative to that of another. Also called RELATIVE MOTION.

apparent noon. Twelve o'clock apparent time, or the instant the apparent sun is over the upper branch of the meridian. Apparent noon may be either local or Greenwich depending upon the reference meridian. High noon is local apparent noon.

apparent place. The place or position on the celestial sphere, centered at the earth, determined by removing from the directly observed position of a celestial body the effects that depend on the topocentric location of the observer; i.e., refraction, diurnal aberration, and geocentric parallax. Thus the position at which the object would actually be seen from the center of the earth, displaced by planetary aberration (except the diurnal part) and referred to the true equinox and equator. Also called APPARENT POSITION.

apparent position. See APPARENT PLACE.

apparent precession. Apparent change in the direction of the axis of rotation of a spinning body, as a gyroscope, due to rotation of the earth. As a result of gyroscopic inertia or rigidity in space, to an observer on the rotating earth a gyroscope appears to turn or precess. If it is desired to have the gyroscope maintain an orientation referenced to the earth, it must be precessed in such a manner as to remove the apparent precession.

apparent secular trend. The nonperiodic tendency of sea level to rise, fall and/ or remain stationary with time. Technically, it is frequently defined as the slope of a least-squares line of regression through a relatively long series of yearly mean sea level values. The word *apparent* is used since it is often not possible to know whether a trend is truly nonperiodic or merely a segment of a very long (relative to the length of the series) oscillation.

apparent shoreline. A line drawn on the chart in lieu of the mean high water line or the mean water level line in areas where either may be obscured by marsh, mangrove, cypress, or other type of marine vegetation. This line represents the intersection of the appropriate datum with the outer limits of vegetation and appears to the navigator as the shoreline.

apparent sidereal time. See under SIDEREAL TIME.

apparent solar day. The duration of one rotation of the earth on its axis, with respect to the apparent sun. It is measured by successive transits of the apparent sun over the lower branch of a meridian. The length of the apparent solar day is 24 hours of apparent time and averages the length of the mean solar day, but varies somewhat from day to day.

apparent sun. The actual sun as it appears in the sky. Also called TRUE SUN. See also MEAN SUN, DYNAMICAL MEAN SUN.

apparent time. Time based upon the rotation of the earth relative to the apparent or true sun. This is the time shown by a sun dial. Apparent time may be designated as either local or Greenwich, as the local or Greenwich meridian is used as the reference. Also called TRUE SOLAR TIME. See also EQUATION OF TIME.

apparent wind. The speed and true direction from which the wind appears to blow with reference to a moving point. Sometimes called RELATIVE WIND. See also TRUE WIND.

approach chart. See under CHART CLASSIFICATION BY SCALE.

approximate, *adj.* Nearly exact. This term is sometimes used in connection with bearings reported by a radio direction finder station. DOUBTFUL or SECOND CLASS may be used with the same meaning.

approximate altitude. An altitude determined by inexact means, as by estimation or by a star finder or star chart.

approximate coefficients. The six coefficients used in the analysis of the magnetic properties of a vessel in the course of a practical magnetic compass adjustment. The values of these coefficients are determined from deviations of an unadjusted compass. See also COEFFICIENT A, COEFFICIENT B, COEFFICIENT C, COEFFICIENT D, COEFFICIENT E, COEFFICIENT J.

appulse, *n.* 1. The near approach of one celestial body to another on the celestial sphere, as in occultation, conjunction, etc.
2. The penumbral eclipse of the moon.

apron, *n.* 1. On the sea floor a gentle slope, with a generally smooth surface, particularly as found around groups of islands or seamounts. Sometimes called ARCHIPELAGIC APRON.
2. The area of wharf or quay for handling cargo.
3. A sloping underwater extension of an iceberg.
4. An outwash plain along the front of a glacier.

apse line. See LINE OF APSIDES.

apsis (*pl. apsides*), *n.* Either of the two orbital points nearest or farthest from the center of attraction, the **perihelion** and **aphelion** in the case of an orbit about the sun, and the **perigee** and **apogee** in the case of an orbit about the earth. The line connecting these two points is called LINE OF APSIDES.

aqueduct, *n.* A conduit or artificial channel for the conveyance of water, often elevated, especially one for the conveyance of a large quantity of water that flows by gravitation.

arbitrary map projection. See APHYLACTIC MAP PROJECTION.

arc, *n.* 1. A part of a curved line, as of a circle. See also ANGULAR DISTANCE, definition 2.
2. The graduated scale of an instrument for measuring angles, as a marine sextant. Readings obtained on that part of the arc beginning at zero and extending in the direction usually considered positive are popularly said to be *on the arc*, and those beginning at zero and extending in the opposite direction are said to be *off the arc*. See also EXCESS OF ARC.

arched squall. A squall which is relatively high in the center, tapering off on both sides.

archipelagic apron. See APRON, definition 1.

archipelago, *n.* 1. A sea of broad expanse of water studded with many islands or groups of islands.
2. Such a group of islands as in definition 1.

arc of uncertainty. See ANGLE OF UNCERTAINTY.

arc of visibility. The arc of a light sector, designated by its limiting bearings as observed from seaward.

Arcs of Lowitz. Oblique, downward extensions of the parhelia of 22°, concave toward the sun, and with red inner borders. They are formed by refraction by ice crystals oscillating about the vertical, as in the case of snowflakes. Arcs of Lowitz are rare phenomena.

arctic, *adj.* Of or pertaining to the arctic, or intense cold.

Arctic, *n.* The region within the Arctic Circle, or, loosely, northern regions in general, characterized by very low temperatures.

arctic air. A type of air whose characteristics are developed mostly in winter over arctic surfaces of ice and snow. Arctic air is cold aloft and it extends to great heights, but the surface temperatures are often higher than those of POLAR AIR. For 2 or 3 months in summer arctic air masses are shallow and rapidly lose the characteristics as they move southward. See also ANTARCTIC AIR.

Arctic Circle. That parallel of latitude at about 66° 33′N, marking the southern limit of the north Frigid Zone. This latitude is the complement of the sun's greatest northerly declination and marks the approximate southern limit at which the sun becomes circumpolar. The actual limit is extended somewhat by the combined effect of refraction, semidiameter of the sun, parallax, and the height of the observer's eye above the surface of the earth. A similar circle marking the northern limit of the south Frigid Zone is called ANTARCTIC or SOUTH POLAR CIRCLE. Also called NORTH POLAR CIRCLE.

arctic front. The semi-permanent, semi-continuous front between the deep, cold arctic air and the shallower, basically less cold polar air of northern latitudes; generally comparable to the ANTARCTIC FRONT of the Southern Hemisphere.

arctic sea smoke. Steam fog; but often specifically applied to steam fog rising from small areas of open water within sea ice. See also FROST SMOKE.

arctic smoke. See STEAM FOG.

arctic whiteout. The obliteration of contrast between surface features in the Arctic when a covering of snow obscuring all landmarks is accompanied by an overcast sky, resulting in an absence of shadows and an unrelieved expanse of white, the earth and sky blending so that the horizon is not distinguishable. A similar occurrence in the Antarctic is called ANTARCTIC WHITEOUT.

arc to chord correction. See CONVERSION ANGLE.

areal feature. A topographic feature, such as sand, swamp, vegetation, etc., which extends over an area. It is represented on the published map or chart by a solid or screened color, by a prepared pattern of symbols, or by a delimiting line.

area to be avoided. A routing measure comprising an area within defined limits in which either navigation is particularly hazardous or it is exceptionally important to avoid casualties and which should be avoided by all ships, or certain classes of ships. See also PRECAUTIONARY AREA, ROUTING SYSTEM.

argument, *n.* One of the values used for entering a table or diagram.

argument of latitude. The angular distance measured in the orbital plane from the ascending node to the orbiting body; the sum of the argument of pericenter and the true anomaly.

argument of pericenter. The angle at the center of attraction from the ascending node to the pericenter point, measured in the direction of motion of the orbiting body. Also called ARGUMENT OF PERIFOCUS.

argument of perifocus. See ARGUMENT OF PERICENTER.

argument of perigee. The angle at the center of attraction from the ascending node to the perigee point, measured in the direction of motion of the orbiting body.

Aries, *n.* 1. Vernal equinox. Also called FIRST POINT OF ARIES.
2. The first sign of the zodiac.

arithmetic mean. See MEAN.

arm, *v., t.* To place tallow or other substance in the recess at the lower end of a sounding lead, for obtaining a sample of the bottom. The material so placed is called ARMING.

Armco, *n.* The registered trade name for a high purity, low carbon iron, used for Flinders bars, quandrantal correctors, etc., to correct magnetic compass errors resulting from induced magnetism.

arming, *n.* Tallow or other substance placed in the recess at the lower end of a sounding lead, for obtaining a sample of the bottom.

arm of the sea. *British terminology.* A comparatively narrow offshoot from the body of the sea.

array, *n.* See as ANTENNA ARRAY.

artificial antenna. See DUMMY ANTENNA.

artificial asteroid. A manmade object placed in orbit about the sun.

artificial earth satellite. A man-made earth satellite, as distinguished from the moon. Often shortened to ARTIFICIAL SATELLITE.

artificial harbor. Other than an improved natural harbor, a harbor where the desired protection from wind and sea is obtained from breakwaters, moles, jetties, etc. See also NATURAL HARBOR.

artificial horizon. A device for indicating the horizontal, as a bubble, gyroscope, pendulum, or the flat surface of a liquid.

artificial magnet. A magnet produced by artificial means, either by placing magnetic material in the field of another magnet or by means of an electric current, as contrasted with a NATURAL MAGNET, occuring in nature.

artificial range. A range formed by two objects such as buildings, towers, etc. See also NATURAL RANGE.

artificial satellite. See ARTIFICIAL EARTH SATELLITE.

ascending node. That point at which a planet, planetoid, or comet crosses the ecliptic from south to north, or a satellite crosses the plane of the equator of its primary from south to north. Also called NORTHBOUND NODE. The opposite is called DESCENDING NODE.

ash breeze. Absence of wind; calm.

ashore, *adj. & adv.* On the shore; on land; aground. See also AFLOAT.

aspect, *n. British terminology.* The relative bearing of own ship from the target ship, measured 0° to 180° port (red) or starboard (green). See also TARGET ANGLE.

aspects, *n., pl.* The apparent positions of celestial bodies relative to one another; particularly the apparent positions of the moon or a planet relative to the sun.

assigned frequency. The center of the frequency band assigned to a radio station. Also called CHANNEL FREQUENCY, CENTER FREQUENCY, but the use of these terms is deprecated.

assigned frequency band. The frequency band whose center coincides with the frequency assigned to the station and whose width equals the necessary bandwidth plus twice the absolute value of the frequency tolerance.

assumed latitude. The latitude at which an observer is assumed to be located for an observation or computation, as the latitude of an assumed position or the latitude used for determining the longitude of time sight. Also called CHOSEN LATITUDE.

assumed longitude. The longitude at which an observer is assumed to be located for an observation or computation, as the longitude of an assumed position or the longitude used for determining the latitude by meridian altitude. Also called CHOSEN LONGITUDE.

assumed position. A point at which a craft is assumed to be located, particularly one used as a preliminary to establishing certain navigational data, as that point on the surface of the earth for which the computed altitude is determined in the solution of a celestial observation. Also called CHOSEN POSITION.

astern, *adv.* Bearing approximately 180° relative. The term is often used loosely for DEAD ASTERN, or bearing exactly 180° relative. The opposite is AHEAD.

asteroid, *n.* A minor planet; one of the many small celestial bodies revolving around the sun, most of the orbits being between those of Mars and Jupiter. Also called PLANETOID, MINOR PLANET. See under PLANET.

astigmatism, *n.* A defect of a lens which causes the image of a point to appear as a line, rather than a point.

astigmatizer, *n.* A lens which introduces astigmatism into an optical system. Such a lens is so arranged that it can be placed in or removed from the optical path at will. In a sextant, an astigmatizer may be used to elongate the image of a celestial body into a horizontal line.

astre fictif. Any of several fictitious stars which are assumed to move along the celestial equator at uniform rates corresponding to the speeds of the several harmonic constituents of the tide-producing force. Each astre fictif crosses the meridian at a time corresponding to the maximum of the constituent that it represents.

astro-. A prefix meaning *star* or *stars* and, by extension, sometimes used as the equivalent of *celestial.*

astrodynamics, *n.* The practical application of celestial mechanics, astroballistics, propulsion theory, and allied fields to the problem of planning and directing the trajectories of space vehicles.

astrograph, *n.* A device for projecting a set of precomputed altitude curves onto a chart or plotting sheet, the curves moving with time such that if they are properly adjusted, they will remain in the correct position on the chart or plotting sheet.

astrolabe, *n.* An instrument used for determining an accurate astronomical position ashore, as in survey work. Originally, the astrolabe consisted of a disk with an arm pivoted at the center, the whole instrument being hung by a ring at the top, to establish the vertical. It was used to measure altitudes of celestial bodies, and was superseded by the cross-staff.

astrometry, *n.* The branch of astronomy dealing with the geometrical relations of the celestial bodies and their real and apparent motions.

astronomical, *adj.* Of or pertaining to astronomy.

Astronomical Almanac. An annual publication prepared jointly by the Nautical Almanac Office, U.S. Naval Observatory, and H. M. Nautical Almanac Office, Royal Greenwich Observatory. With the exception of certain introductory pages, the publication as printed in the United Kingdom is identical to that printed in the United States. This *ephemeris* gives, to a high precision, detailed information on a large number of celestial bodies. It is arranged to suit the convenience of the astronomer for whom it is primarily intended and is not needed for ordinary purposes of navigation. But it does contain some information of general interest to the navigator, such as various astronomical constants, details of eclipses, information on planetary configurations, and miscellaneous phenomena. Prior to 1981 this publication was entitled *American Ephemeris and Nautical Almanac.* See also NAUTICAL ALMANAC, ALMANAC FOR COMPUTERS.

astronomical day. Prior to January 1, 1925, a mean solar day beginning at mean noon, 12 hours later than the beginning of the civil day of the same date. Since 1925 the astronomical day agrees with the civil day.

astronomical equator. A line connecting points having 0° astronomical latitude. Because the deflection of the vertical varies from point to point, the astronomical equator is not a plane curve. But since the verticals through all points on it are parallel, the zenith at any point on the astronomical equator lies in the plane of the celestial equator. When the astronomical equator is corrected for station error, it becomes the GEODETIC EQUATOR. Sometimes called TERRESTRIAL EQUATOR.

astronomical latitude. Angular distance between the plumb line at a station and the plane of the celestial equator. It is the latitude which results directly from observations of celestial bodies, uncorrected for deflection of the vertical which, in the United States, may amount to as much as 25". Astronical latitude applies only to positions on the earth, and is reckoned from the astronomical equator (0°), north and south through 90°. Also called ASTRONOMIC LATITUDE and sometimes GEOGRAPHIC LATITUDE. See also GEODETIC LATITUDE.

astronomical longitude. Angular distance between the plane of the celestial meridian at a station and the plane of the celestial meridian at Greenwich. It is the longitude which results directly from observations of celestial bodies, uncorrected for deflection of the vertical, the prime vertical component of which, in the United States, may amount to more than 18". Astronomical longitude applies only to positions on the earth, and is reckoned from the Greenwich meridian (0°) east and west through 180°. Also called ASTRONOMIC LONGITUDE and sometimes GEOGRAPHIC LONGITUDE. See also GEODETIC LONGITUDE.

astronomical mean sun. See MEAN SUN.

astronomical meridian. A line connecting points having the same astronomical longitude. Because the deflection of the vertical (station error) varies from point to point, the astronomical meridian is not a plane curve. When the astronomical meridian is corrected for station error, it becomes the GEODETIC MERIDIAN. Also called TERRESTRIAL MERIDIAN and sometimes called GEOGRAPHIC MERIDIAN.

astronomical parallel. A line connecting points having the same astronomical latitude. Because the deflection of the vertical varies from point to point, the astronomical parallel is an irregular line not lying in a single plane. When the astronomical parallel is corrected for station error, it becomes the GEODETIC PARALLEL. Sometimes called GEOGRAPHIC PARALLEL.

astronomical position. 1. A point on the earth whose coordinates have been determined as a result of observation of celestial bodies. The expression is usually used in connection with positions on land determined with great accuracy for survey purposes.
2. A point on the earth, defined in terms of astronomical latitude and longitude.

astronomical refraction. Atmospheric refraction of a ray of radiant energy passing through the atmosphere from outer space, as contrasted with TERRESTRIAL REFRACTION of a ray emanating from a point on or near the surface of the earth. See also REFRACTION.

astronomical tide. The tide without constituents having their origin in the daily or seasonal variations in weather conditions which may occur with some degree of periodicity. See also METEOROLOGICAL TIDES.

astronomical time. Time used with the astronomical day which prior to 1926 began at noon of the civil day of same date. The hours of the day were numbered consecutively from 0 (noon) to 23 (11 AM of the following morning).

astronomical triangle. The navigational triangle, either terrestrial or celestial, used in the solution of celestial observations.

astronomical twilight. The period of incomplete darkness when the center of the sun is more than 12° but not more than 18° below the celestial horizon. See also CIVIL TWILIGHT, NAUTICAL TWILIGHT.

astronomical unit. 1. The mean distance between the earth and the sun, approximately 92,960,000 statute miles.
2. The astronomical unit is often used as a unit of measurement for distances within the solar system. In the system of astronomical constants of the International Astronomical Union the adopted value for it is 1 AU = $149,600 \times 10^6$ meters.

astronomical year. See TROPICAL YEAR.

astronomic latitude. See ASTRONOMICAL LATITUDE.

astronomic longitude. See ASTRONOMICAL LONGITUDE.

astronomy, *n.* The science which deals with the size, constitution, motions, relative position, etc. of celestial bodies, including the earth. That part of astronomy of direct use to a navigator, comprising principally celestial coordinates, time, and the apparent motions of celestial bodies is called **navigational** or **nautical astronomy.**

astro-tracker. A navigation equipment which automatically acquires and continuously tracks a celestial body in azimuth and altitude.

asymmetrical, *adj.* Not symmetrical.

asymptote, *n.* A straight line or curve which a curve of infinite length approaches but never quite reaches.

Atlantic Equatorial Countercurrent. An ocean current that flows eastward between the westward flowing Atlantic North and South Equatorial Currents. The countercurrent is best defined during August and September, when it extends from about 52° W to 10° W and joins the GUINEA CURRENT. In October it narrows and separates into two parts at about latitude 7° N, longitude 35° W. The western part, which appears to be a region where the countercurrent probably sinks and flows eastward beneath the equatorial currents, gradually diminishes in size to the west-northwest, while the eastern part diminishes to the east-southeast. The greatest separation occurs during March; during April the western part of the countercurrent disappears, but in May it reappears in the vicinity of latitude 0°, longitude 40°W. The two segments progress west- northwestward without much change in size. They merge at about latitude 6°N, longitude 43°W during August and continue their flow eastward uninterrupted through September.

Atlantic North Equatorial Current. A broad, slow, westward flowing ocean current generated mainly by the northeast trade winds. The current originates near longitude 26° W between about latitude 15° N and 30° N and flows across the ocean past longitude 60° W. It forms the ANTILLES CURRENT in the vicinity of the Leeward Islands. The part of the current between 12° N and 15° N joins the Guiana Current and forms the CARIBBEAN CURRENT.

Atlantic South Equatorial Current. The major part of this westward flowing ocean current is located south of the equator, the central portion extending to about latitude 20° S. The northern part expands northward during January, February, and March when the Atlantic Equatorial Countercurrent dissipates and is least evident. On approaching the coast of South America one part turns northwestward as the GUIANA CURRENT; the other part turns below Natal and flows southwestward along the coast of Brazil as the BRAZIL CURRENT. Of the two equatorial currents in the Atlantic, the Atlantic South Equatorial Current is the stronger and more extensive.

Atlantic standard time. See STANDARD TIME.

atlas, *n.* A collection of charts or maps kept loose or bound in a volume.

atlas grid. A reference system that permits the designation of the location of a point or an area on a map, photograph, or other graphic in terms of numbers and letters. Also called ALPHANUMERIC GRID.

atmosphere, *n.* 1. The envelope of air surrounding the earth and bound to it more or less permanently by virture of the earth's gravitational attraction. The earth's atmosphere extends from the solid or liquid surface of the earth to an indefinite height, its density asymptotically approaching that of interplanetary space. At heights of the order of 80 kilometers (50 statute miles) the atmosphere is barely dense enough to scatter sunlight to a visible degree. The atmosphere may be subdivided vertically into a number of atmospheric shells, but the most common basic subdivision is that which recognizes a **troposphere** from the surface to about 10 kilometers, a **stratosphere** from about 10 kilometers to about 80 kilometers, and an **ionosphere** above 80 kilometers. See also STANDARD ATMOSPHERE.

2. The gaseous envelope surrounding any celestial body, including the Earth.

atmospheric absorption. The loss of power in transmission of radiant energy by dissipation in the atmosphere.

atmospheric drag. A major cause of perturbations of close artificial satellite orbits caused by the resistance of the atmosphere. The secular effects are decreasing magnitudes of eccentricity, major axis, and period. Sometimes shortened to DRAG.

atmospheric noise. See ATMOSPHERIC RADIO NOISE.

atmospheric pressure. The pressure exerted by the weight of the earth's atmosphere, about 14.7 pounds per square inch. See also STANDARD ATMOSPHERE, definition 1; BAROMETRIC PRESSURE.

atmospheric radio noise. In radio reception noise or static due to natural causes such as thunderstorm activity. Sometimes shortened to ATMOSPHERIC NOISE. See also MAN-MADE NOISE, RADIO INTERFERENCE.

atmospheric refraction. Refraction resulting when a ray of radiant energy passes obliquely through the atmosphere. It may be called **astronomical refraction** if the ray enters the atmosphere from outer space, or **terrestrial refraction** if it emanates from a point on or near the surface of the earth.

atoll, *n.* A ring-shaped coral reef which has closely spaced islands or islets on it enclosing a deeper central area or lagoon. The diameter may vary from less than a mile to 80 or more.

atollon, *n.* A large reef ring in the Maldive Islands consisting of many smaller reef rings. The word ATOLL was derived from this name.

atomic clock. A precision clock that depends for its operation upon an electrical oscillator regulated by an atomic system. The basic principle of the clock is that electromagnetic waves of a particular frequency are emitted when an atomic transition occurs.

atomic second. See SECOND, definition 1.

Atomic Time. A fundamental kind of time based on transitions in the atom. **International Atomic Time (TAI)** is the time reference coordinate established by the Bureau Internationale de l'Heure (BIH) on the basis of the readings of atomic clocks functioning in various establishments in accordance with the definition of the atomic second, the unit of time in the International System of Units (SI). The Atomic Time scales maintained in the United States by the National Bureau of Standards and the U.S. Naval Observatory constitute approximately 37½ percent of the stable reference information used in maintaining a stable TAI scale by the BIH.

A trace. The first trace of an oscilloscope having more than one displayed, as the upper trace of a Loran-A indicator.

ATR tube. See as ANTI-TR TUBE.

attenuation, *n.* 1. A lessening in amount, particularly the reduction of the amplitude of a wave with distance from the origin.

2. The decrease in the strength of a radar wave resulting from absorption, scattering, and reflection by the medium through which it passes (waveguide, atmosphere) and by obstructions in its path. Also attenuation of the wave may be the result of artificial means, such as the inclusion of an attenuator in the circuitry or by placing an absorbing device in the path of the wave.

attitude, *n.* The position of a body as determined by the inclination of the axes to some other frame of reference. If not otherwise specified, this frame of reference is fixed to the earth.

atto-. A prefix meaning *one-quintillionth* (10^{-18}).

audible, *adj.* Capable of being translated into sound by the human ear.

audible aid to navigation. An aid to navigation transmitting information by sound waves.

audio frequency. A frequency within the audible range, about 20 to 20,000 hertz. Also called SONIC FREQUENCY.

augmentation, *n.* The apparent increase in the semidiameter of a celestial body as its altitude increases, due to the reduced distance from the observer. The term is used principally in reference to the moon.

augmentation correction. A correction due to augmentation, particularly that sextant altitude correction due to the apparent increase in the semidiameter of a celestial body as its altitude increases.

augmenting factor. A factor used in connection with the harmonic analysis of tides or tidal currents to allow for the difference between the times of hourly tabulation and the corresponding constituent hours.

aural, *adj.* Of or pertaining to the ear or sense of hearing.

aural null. A null detected by listening for a minimum or the complete absence of an audible signal. This null as received by a radio direction finder indicates that the plane of its loop antenna is perpendicular to the direction of the radio wave.

aureole, *n.* A poorly developed corona, characterized by a bluish-white disk immediately around the luminary and a reddish-brown outer edge. An aureole, rather than a corona, is produced when the cloud responsible for this diffraction effect is composed of droplets distributed over a wide size-range. The diffracted rays approach the observer from a wide variety of angles, in contrast to the relative uniform diffraction produced by a cloud of more limited drop-size range. Inasmuch as most clouds exhibit rather broad drop-size distributions, aureoles are observed much more frequently than coronas.

aurora, *n.* A luminous phenomenon due to electrical discharges in the atmosphere, probably confined to the thin air high above the surface of the earth. It is most commonly seen in high latitudes where it is most frequent during periods of greatest sunspot activity. If it occurs in the Northern Hemisphere, it is called aurora borealis or northern lights; and if in the Southern, aurora australis.

aurora australis. The aurora in the Southern Hemisphere.

aurora borealis. The aurora in the Northern Hemisphere. Also called NORTHERN LIGHTS.

auroral zone. The area of maximum auroral activity. Two such areas exist, each being a 10°-wide annulus centered at an average distance of 23° from a geomagnetic pole.

aurora polaris. A high latitude aurora borealis.

austral, *adj.* Of or pertaining to south.

authalic map projection. See EQUAL-AREA MAP PROJECTION.

Automated Mutual-assistance Vessel Rescue System. Operated by the United States Coast Guard, the AMVER System is a maritime mutual assistance program that provides important aid to the development and coordination of search and rescue efforts in the oceans of the world, through the medium of a worldwide merchant vessel plot. From information provided by voluntarily participating users and entered into an electronic computer, the dead reckoning positions of participating vessels are generated by the computer and maintained throughout the voyages. Predicted vessel locations are disclosed only for reasons related to maritime safety.

automatic direction finder. A radio direction finder in which the bearing to the transmitter to which it is tuned is indicated automatically and continuously, in contrast with a MANUAL RADIO DIRECTION FINDER which requires manual operation. Also called AUTOMATIC RADIO DIRECTION FINDER.

automatic frequency control. The technique of automatically maintaining, or a circuit or device which automatically maintains, the frequency of a receiver within specified limits.

automatic gain control. A feature involving special circuitry designed to maintain the output of a radio, radar, or television receiver essentially constant, or to prevent its exceeding certain limits, regardless of variations in the strength of the incoming signal. In a radio receiver, in particular, though something of a misnomer, sometimes called AUTOMATIC VOLUME CONTROL.

automatic radar plotting aid. A computer assisted radar data processing system which generates predictive vectors based upon how a situation has developed over its immediate past. For such a system to meet the specifications of the Inter-Governmental Maritime Consultative Organization (IMCO), it must satisfy requirements with respect to detection, acquisition, tracking, display, warnings, data display, and trial maneuver.

automatic radio direction finder. See AUTOMATIC DIRECTION FINDER.

automatic tide gage. An instrument that automatically registers the rise and fall of the tide. In some instruments, the registration is accomplished by recording the heights at regular intervals in digital format, in others by a continuous graph in which the height, versus corresponding time of the tide, is recorded. The automatic tide gages used by the National Ocean Survey are of both types.

automatic volume control. See under AUTOMATIC GAIN CONTROL.

autopilot, *n.* See GYROPILOT.

autumn, *n.* The season marking the end of the growing season. In the Northern Hemisphere autumn begins astronomically at the autumnal equinox and ends at the winter solstice. In the Southern Hemisphere the limits are the vernal equinox and the summer solstice. The meteorological limits vary with the locality and the year. Also called FALL.

autumnal, *adj.* Pertaining to fall (autumn). The corresponding adjectives for winter, spring, and summer are *hibernal*, *vernal*, and *aestival*.

autumnal equinox. 1. That point of intersection of the ecliptic and the celestial equator occupied by the sun as it changes from north to south declination, on or about September 23. Also called SEPTEMBER EQUINOX, FIRST POINT OF LIBRA.
2. That instant the sun reaches the point of zero declination when crossing the celestial equator from north to south.

auxiliary lights. See under VERTICAL LIGHTS.

average, *adj.* Equaling or approximating a mean.

average, *n.* See MEAN.

average, *v., t.* To determine a mean.

avoirdupois pound. See POUND, *n.*

avulsion, *n.* The rapid erosion of shoreland by waves during a storm.

awash, *adj. & adv.* Situated so that the top is intermittently washed by waves or tidal action. The term applies both to fixed objects such as rocks, and to floating objects with their tops flush with or slightly above the surface of the water. See also ROCK AWASH, SUBMERGED, UNCOVERED.

axial, *adj.* Of or pertaining to an axis.

axis. (*pl. axes*), *n.* 1. A straight line about which a body rotates, or around which a plane figure may rotate to produce a solid; a line of symmetry. A **polar axis** is the straight line connecting the poles of a body. The **major axis** of an ellipse or ellipsoid is its longest diameter; the **minor axis**, its shortest diameter.
2. One of a set of reference lines for certain systems of coordinates.
3. The principal line about which anything may extend, as the *axis of a channel* or *compass card axis*.
4. A straight line connecting two related points, as the poles of a magnet.

axis of freedom. An axis about which the gimbal of a gyro provides a degree-of-freedom of movement.

azimuth, *n.* The horizontal direction of a celestial point from a terrestrial point, expressed as the angular distance from a reference direction. It is usually measured from 000° at the reference direction clockwise through 360°. An azimuth is often designated as **true, magnetic, compass, grid,** or **relative** as the reference direction is true, magnetic, compass, or grid north, or heading, respectively. Unless otherwise specified, the term is generally understood to apply to true azimuth, which may be further defined as the arc of the horizon, or the angle at the zenith, between the north part of the celestial meridian or principal vertical circle and a vertical circle, measured from 000° at the north part of the principal vertical circle clockwise through 360°. Azimuth taken directly from a table, before interpolation, is called **tabulated azimuth.** After interpolation, or, if determined by calculation, mechanical device, or graphics, it is called **computed azimuth.** When the angle is measured in either direction from north or south, and labeled accordingly, it is properly called **azimuth angle**; when measured either direction from east or west, and labeled accordingly, it is called **amplitude.** An azimuth determined by solution of the navigational triangle with altitude, declination, and latitude given is called an **altitude azimuth**; if meridian angle, declination, and latitude are given, it is called a **time azimuth**; if meridian angle, declination, and altitude are given, it is called a **time and altitude azimuth.** See also BACK AZIMUTH, BEARING.

azimuthal, *adj.* Of or pertaining to azimuth.

azimuthal chart. A chart on an azimuthal map projection. Also called ZENITHAL CHART.

azimuthal equidistant chart. A chart on the azimuthal equidistant map projection.

azimuthal equidistant map projection. An azimuthal map projection on which straight lines radiating from the center or pole of projection represent great circles in their true azimuths from that center, and lengths along those lines are of exact scale. This projection is neither equal-area nor conformal. If a geographic pole is the pole of projection, meridians appear as radial straight lines and parallels of latitude as equally spaced concentric circles.

azimuthal map projection. A map projection on which the azimuths or directions of all lines radiating from a central point or pole are the same as the azimuths or directions of the corresponding lines on the ellipsoid. This classification includes the **gnomonic, stereographic, orthographic,** and the **azimuthal equidistant map projections.** Also called ZENITHAL MAP PROJECTION.

azimuthal orthomorphic map projection. See STEREOGRAPHIC MAP PROJECTION.

azimuth angle. Azimuth measured from 0° at the north or south reference direction clockwise or counterclockwise through 90° or 180°. It is labeled with the reference direction as a prefix and the direction of measurement from the reference direction as a suffix. Thus, azimuth angle S 144° W is 144° west of south, or azimuth 324°. When azimuth angle is measured through 180°, it is labeled N or S to agree with the latitude and E or W to agree with the meridian angle. Azimuth angle taken directly from a table, before interpolation, is called **tabulated azimuth angle.** After interpolation, or if determined by calculation, mechanical device, or graphics, it is called **computed azimuth angle.**

azimuth bar. An instrument for measuring azimuths, particularly a device consisting of a slender bar with a vane at each end, and designed to fit over a central pivot in the glass cover of a magnetic compass. See also BEARING BAR.

azimuth circle. A ring designed to fit snugly over a compass or compass repeater, and provided with means for observing compass bearings and azimuths. A similar ring without the means for observing azimuths of the sun is called a BEARING CIRCLE.

azimuth instrument. An instrument for measuring azimuths, particularly a device which fits over a central pivot in the glass cover of a magnetic compass.

azimuth stabilized display. See as STABILIZED IN AZIMUTH under STABILIZATION OF RADARSCOPE DISPLAY.

azimuth tables. Publications providing tabulated azimuths or azimuth angles of celestial bodies for various combinations of declination, latitude, and hour angle. Great-circle course angles can also be obtained by substitution of values.

Azores Current. A slow but fairly constant southeast branch of the North Atlantic Current and part of the Gulf Stream System. Its mean speed is only 0.4 knot, and the mean maximum speed computed from all observations above 1 knot in the prevailing direction is 1.3 knots. There is no discernible seasonal fluctuation. The speed and direction of the current is easily influenced for short periods by changing winds. The Azores Current is an inner part of the general clockwise oceanic circulation of the North Atlantic Ocean. Also called SOUTHEAST DRIFT CURRENT.

B

back, *adj.* Reciprocal.

back, *v., i.* 1. According to widespread usage in the United States, a change in wind direction in a counterclockwise sense in the Northern Hemisphere and a clockwise direction in the Southern Hemisphere.
2. In widespread international usage to back is a change in wind direction in a counterclockwise sense in either hemisphere. Change in the opposite direction is called *veer*. See also HAUL.
3. To go stern first, or to operate the engines in reverse.

back azimuth. An azimuth 180° from a given azimuth.

back echo. The effect on a radar display produced by a back lobe of a radar antenna. See also SIDE ECHO.

backlash, *n.* The amount which a gear or other part of a machine, instrument, etc., can be moved without moving an adjoining part, resulting from loose fit; play. See also LOST MOTION.

back lobe. The lobe of the radiation pattern of a directional antenna which makes an angle of approximately 180° with the direction of the axis of the main lobe.

back range. A range observed astern, particularly one used as guidance for a craft moving away from the objects forming the range.

backrush, *n.* The seaward return of water following the uprush onto the foreshore. See also RIP CURRENT, UNDERTOW.

backshore, *n.* That part of a beach which is usually dry, being reached only by the highest tides, and by extension, a narrow strip of relatively flat coast bordering the sea. See also FORESHORE.

back sight. A marine sextant observation of a celestial body made by facing 180° from the azimuth of the body and using the visible horizon in the direction in which the observer is facing.

backstaff, *n.* A forerunner of the sextant, consisting essentially of a graduated arc and a single mirror. To use the instrument it was necessary to face away from the body being observed. Also called QUADRANT WITH TWO ARCS, SEA QUADRANT.

backstays of the sun. Crepuscular rays extending downward toward the horizon.

backwash, *n.* Water or waves thrown back by an obstruction such as a seawall, breakwater, cliff, etc.

backwater, *n.* Water held back from the main flow, as that which overflows the land and collects in low places or that forming an inlet approximately parallel to the main body and connected thereto by a narrow outlet.

bad-bearing sector. Relative to a radio direction finder station or radiobeacon, a sector within which bearings are known to be liable to significant errors of unknown magnitudes.

baguio, *n.* See under TROPICAL CYCLONE.

balancer, *n.* A device used with a radio direction finder to balance out antenna effect and thus produce a sharper reading.

balancing, *n.* The process of neutralizing antenna effect in order to improve the definition of the observed bearing. See also BALANCER.

Bali wind. A strong east wind at the eastern end of Java.

ball, *n.* 1. A spherical identifying mark placed at the top of a perch.
2. A time ball.

ballast ground. A designated area for discharging solid ballast before entering harbor.

ballistic damping error. A temporary oscillatory error of a gyrocompass introduced during changes of course or speed as a result of the means used to damp the oscillations of the spin axis.

ballistic deflection error. A temporary oscillatory error of a gyrocompass introduced when the north-south component of the speed changes, as by speed or course change. An accelerating force acts upon the compass, causing a surge of mercury from one part of the system to another in the case of the nonpendulous compass, or a deflection (along the meridian) of a mass in the case of a pendulous compass. In either case, a precessing force introduces a temporary ballistic deflection error in the reading of the compass unless it is corrected.

band, *n.* A specific section or range of anything. See also FREQUENCY BAND.

band of error. An area either side of a line of position, within which, for a stated level of probability, the true position is considered to lie.

bandwidth, *n.* 1. Of a device, the range of frequencies within which its performance, in respect to some characteristic, conforms to a specified standard.
2. The range within the limits of a frequency band.

bank, *n.* 1. An elevation of the sea floor typically located on a shelf and over which the depth of water is relatively shallow but sufficient for safe surface navigation. Reefs or shoals, *dangerous* to surface navigation, may rise above the general depths of a bank.
2. A shallow area of shifting sand, gravel, mud, etc., as a *sand bank, mud bank*, etc.
3. A ridge of any material such as earth, rock, snow, etc., or anything resembling such a ridge, as a *fog bank* or *cloud bank*.
4. The edge of a cut or fill.
5. The margin of a watercourse.
6. A number of similar devices connected so as to be used as a single device.

bank cushion. In a restricted channel, especially one with steep banks, as the ship is moved bodily toward the near bank due to bank suction, bank cushion is an opposing force which tends to force the bow away from the bank due to the increase in the bow wave on the near side.

bank suction. The bodily movement of a ship toward the near bank due to a decrease in pressure as a result of increased velocity of flow of water past the hull in a restricted channel.

banner cloud. A bannerlike cloud streaming off from a mountain peak in a strong wind. See also CAP CLOUD.

bar, *n.* 1. A ridge or mound of sand, gravel, or other unconsolidated material below the high water level, especially at the mouth of a river or estuary, or lying a short distance from and usually parallel to the beach, and which may obstruct navigation.
2. A unit accepted *temporarily* for use with the International System of Units; 1 bar is equal to 100,000 pascals.

barat, *n.* A heavy northwest squall in Manado Bay on the north coast of the island of Celebes, prevalent from December to February.

barber, *n.* 1. A strong wind carrying damp snow or sleet and spray that freezes upon contact with objects, especially the beard and hair.
2. See FROST SMOKE, definition 2.

bar buoy. A buoy marking the location of a bar at the mouth of a river on approach to a harbor.

bare ice. Ice without snow cover.

bare rock. A rock that extends above the mean high water datum in tidal areas or above the low water datum in the Great Lakes. See also ROCK AWASH, SUBMERGED ROCK.

barogram, *n.* The record made by a barograph.

barograph, *n.* A recording barometer. A highly sensitive barograph may be called a *microbarograph*.

barometer, *n.* An instrument for measuring atmospheric pressure. A **mercurial barometer** employs a column of mercury supported by the atmosphere. An **aneroid barometer** has a partly exhausted, thin metal cylinder somewhat compressed by atmospheric pressure.

barometric pressure. Atmospheric pressure as indicated by a barometer.

barometric pressure correction. A correction due to nonstandard barometric pressure, particularly that sextant altitude correction due to changes in refraction caused by difference between the actual barometric pressure and the standard barometric pressure used in the computation of the refraction table.

barometric tendency. See PRESSURE TENDENCY.

barothermogram, *n.* The record made by a barothermograph.

barothermograph, *n.* An instrument which automatically records pressure and temperature.

barothermohygrogram, *n.* The record made by a barothermohygrograph.

barothermohygrograph, *n.* An instrument which automatically records pressure, temperature, and humidity of the atmosphere.

barrel, *n.* A unit of volume or weight, the U.S. petroleum value being 42 U.S. gallons.

barrel buoy. A buoy having the shape of a barrel or cylinder floating horizontally, usually for special purposes, including mooring.

barrier beach. A bar essentially parallel to the shore, the crest of which is above high water.

barrier reef. A coral reef which roughly parallels land but is some distance offshore, with deeper water adjacent to the land, as contrasted with a FRINGING REEF closely attached to the shore.

bar scale. A line or series of lines on a chart, subdivided and labeled with the distances represented on the chart. Also called GRAPHIC SCALE. See also SCALE.

barycenter, *n.* The center of mass of a system of masses, as the barycenter of the earth-moon system.

base chart. See BASE MAP.

base course up. One of the three basic orientations of display of relative or true motion on a radarscope. In the BASE COURSE UP orientation, the target pips are painted at their measured distances and in their directions *relative* to a *pre-set* base course of own ship maintained UP in relation to the display. This orientation is most often used with automated radar plotting systems. Also called COURSE UP. See also HEAD UP, NORTH UP.

base line, 1. The reference used to position limits of the territorial sea and the contiguous zone. Source data from which the United States base line is determined are the mean low water line on the Atlantic and gulf coasts and the mean lower low water line on the Pacific coast, Alaska, and Hawaii. The United Nations Conference on the Law of the Sea defined the low water line along a coast, as shown on large-scale charts of the coastal State (country) to be the base line for determining the limit of the territorial sea.
2. One side of a series of connected triangles, the length of which is measured with prescribed accuracy and precision, and from which the lengths of the other triangle sides are obtained by computation. Important factors in the accuracy and precision of base measurements are the use of standardized invar tapes, controlled conditions of support and tension, and corrections for temperatures, inclination, and alinement. Base lines in triangulation are classified according to the character of the work they are intended to control, and the instruments and methods used in their measurement are such that prescribed probable errors for each class are not exceeded. These probable errors, expressed in terms of the lengths, are as follows: **first order,** 1 part in 1,000,000; **second order,** 1 part in 500,000; and **third order,** 1 part in 250,000.
3. *Usually baseline.* Of a radionavigation system, the geodesic line between two stations operating in conjunction for the determination of a line of position.

baseline delay. The time interval needed for the signal from a master station of a hyperbolic radionavigation system to travel the length of the baseline, introduced as a delay between transmission of the master and slave (or secondary) signals to make it possible to distinguish between the signals and to permit measurement of time differences.

baseline extension. The extension of the baseline in both directions beyond the transmitters of a pair of radio stations operating in conjunction for determination of a line of position.

base map. 1. A map or chart showing certain fundamental information, used as a base upon which additional data of specialized nature are compiled or overprinted.
2. A map containing all the information from which maps showing specialized information can be prepared.
Also called BASE CHART in nautical charting.

base map symbol. A symbol used on a base map or chart as opposed to one used on an overprint to the base map or chart. Also called BASE SYMBOL.

base symbol. See BASE MAP SYMBOL.

base units. See under INTERNATIONAL SYSTEM OF UNITS.

basin, *n.* 1. A depression of the sea floor more or less equidimensional in plan view and of variable extent.
2. An area of water surrounded by quay walls, usually created or enlarged by excavation, large enough to receive one or more ships for a specific purpose. See also GRAVING DOCK, HALF-TIDE BASIN, NON-TIDAL BASIN, SCOURING BASIN, TIDAL BASIN, TURNING BASIN.
3. An area of land which drains into a lake or sea through a river and its tributaries.
4. A nearly land-locked area of water leading off an inlet, firth, or sound.

bathyal, *adj.* Pertaining to ocean depths between 100 and 2,000 fathoms; also to the ocean bottom between those depths, sometimes identical with the continental slope environment.

bathymeter, *n.* An instrument for measuring depths of water.

bathymetric, *adj.* Of or pertaining to bathymetry.

bathymetric chart. A topographic chart of the bed of a body of water, or a part of it. Generally, bathymetric charts show depths by contour lines and gradient tints.

bathymetry, *n.* The science of measuring water depths (usually in the ocean) in order to determine bottom topography.

bathysphere, *n.* A spherical chamber in which persons are lowered for observation and study of ocean depths.

bathythermogram, *n.* The record made by a bathythermograph.

bathythermograph, *n.* An instrument which automatically draws a graph showing temperature as a function of depth when lowered in the sea.

batture, *n.* An elevation of the bed of a river under the surface of the water; sometimes used to signify the same elevation when it has risen above the surface.

bay, *n.* A recess in the shore, on an inlet of a sea or lake between two capes or headlands, that may vary greatly in size but is usually smaller than a gulf but larger than a cove.

bayamo, *n.* A violent blast of wind, accompanied by vivid lightning, blowing from the land on the south coast of Cuba, especially near the Bight of Bayamo.

Bayer's letter. The Greek (or Roman) letter used in a Bayer's name.

Bayer's name. The Greek (or Roman) letter and the possessive form of the Latin name of a constellation, used as a star name. Examples are α Cygni (Deneb), β Orionis (Rigel), and η Ursae Majoris (Alkaid).

baymouth bar. A bar extending partially or entirely across the mouth of a bay.

bayou, *n.* A minor, sluggish waterway or estuarial creek, generally tidal or with a slow or imperceptible current, and with its course generally through lowlands or swamps, tributary to or connecting with other bodies of water. Various specific meanings have been implied in different parts of the southern United States. Sometimes called SLOUGH.

beach, *n.* The zone of unconsolidated material that extends landward from the low water line to the place where there is a marked change in material or physiographic form, or to the line of permanent vegetation (usually the effective limit of storm waves). A beach includes **foreshore** and **backshore**. The beach along the margin of the sea may be called SEABEACH. Also called STRAND, especially when the beach is composed of sand. See also TIDELAND.

beach, *v., t. & i.* To intentionally run a craft ashore, as *to beach a landing ship*.

beach berm. See BERM.

beach erosion. The carrying away of beach materials by wave action, tidal or littoral currents, or wind.

beacon, *n.* A fixed artificial navigation mark. See also MARK, definition 1; DAYBEACON; DAYMARK; LIGHTED BEACON; RADIOBEACON.

beaconage, *n.* A system of fixed aids to navigation comprised of beacons and minor lights. See also BUOYAGE.

beacon buoy. See PILLAR BUOY.

beacon tower. A beacon which is a major structure, having a support as distinctive as the topmark. See also LATTICE BEACON, REFUGE BEACON.

beam, *n.* 1. A directed flow of electromagnetic radiation from an antenna. See also MAIN BEAM under LOBE, BEAM WIDTH.
2. A group of nearly parallel rays, as a *light beam*.

beam compasses. Compasses for drawing circles of large diameter. In its usual form it consists of a bar with sliding holders for points, pencils, or pens which can be set at any desired position.

beam sea. Waves moving in a direction approximately 90° from the heading. Those moving in a direction approximately opposite to the heading are called HEAD SEA, those moving in the general direction of the heading are called FOLLOWING SEA, and those moving in a direction approximately 45° from the heading (striking the quarter) are called QUARTERING SEA. See also CROSS SEA.

beam tide. A tidal current setting in a direction approximately 90° from the heading of a vessel. One setting in a direction approximately 90° from the course is called a CROSS TIDE. In common usage these two expressions are usually used synonymously. One setting in a direction approximately opposite to the heading is called a HEAD TIDE. One setting in such a direction as to increase the speed of a vessel is called a FAIR TIDE.

beam width. The angular measure of the transverse section of a beam (usually in the main lobe) lying within directions corresponding to specified values of field strength relative to the maximum (e.g., *half field strength* beam width and *half power* beam width). The beam width is usually measured in one or more specified planes containing the axis of the beam. See also HORIZONTAL BEAM WIDTH, VERTICAL BEAM WIDTH.

beam-width error. An azimuth or bearing distortion on a radar display caused by the width of the radar beam. See also BEAM WIDTH, PULSE-LENGTH ERROR.

beam wind. Wind blowing in a direction approximately 90° from the heading. One blowing in a direction approximately 90° from the course is called a CROSS WIND. In common usage these two expressions are usually used synonymously, BEAM WIND being favored by mariners, and CROSS WIND by aviators. One blowing from ahead is called a HEAD WIND. One blowing from astern is called a FOLLOWING WIND by mariners and a TAIL WIND by aviators. See also FAIR WIND, FAVORABLE WIND, UNFAVORABLE WIND.

bear, *v., i.* To be situated as to direction, as, the light *bears* 165°.

bear down. To approach from windward.

bearing, *n.* The horizontal direction of one terrestrial point from another, expressed as the angular distance from a reference direction. It is usually measured from 000° at the reference direction clockwise through 360°. The terms BEARING and AZIMUTH are sometimes used interchangeably, but in navigation the former customarily applies to terrestrial objects and the latter to the direction of a point on the celestial sphere from a point on the earth. A bearing is often designated as **true, magnetic, compass, grid,** or **relative** as the reference direction is true, magnetic, compass, or grid north, or heading, respectively. The angular distance between a reference direction and the initial direction of a great circle through two terrestrial points is called **great-circle bearing.** The angular distance between a reference direction and the rhumb line through two terrestrial points is called **rhumb** or **Mercator bearing.** A bearing differing by 180°, or one measured in the opposite direction, from a given bearing is called a **reciprocal bearing.** The maximum or minimum bearing of a point for safe passage of an off-lying danger is called **danger bearing.** A relative bearing of 045° or 315° is sometimes called a **four-point bearing.** Successive relative bearings (right or left) of 45° and 90° taken on a fixed object to obtain a running fix are often called **bow and beam bearings.** Two or more bearings used as intersecting lines of position for fixing the position of a craft are called **cross bearings.** The bearing of a radio transmitter from a receiver, as determined by a radio direction finder, is called a **radio bearing.** A bearing obtained by radar is called a **radar bearing.** A bearing obtained by visual observation is called a **visual bearing.** A constant bearing maintained while the distance between two craft is decreasing is called a **collision bearing.** See also CURVE OF EQUAL BEARING.

bearing angle. Bearing measured from 0° at the reference direction clockwise or counterclockwise through 90° or 180°. It is labeled with the reference direction as a prefix and the direction of measurement from the reference direction as a suffix. Thus, bearing angle N 37° W is 37° west of north, or true bearing 323°.

bearing bar. An instrument for measuring bearings, particularly a device consisting of a slender bar with a vane at each end, and designed to fit over a central pivot in the glass cover of a magnetic compass. See also AZIMUTH BAR.

bearing calibration. Of a radio direction finder, the determination of bearing corrections by observations of a radiobeacon, particularly a calibration radiobeacon, of known visual bearing, observations being taken over 360° of swing of the observing vessel.

bearing circle. A ring designed to fit snugly over a compass or compass repeater, and provided with vanes for observing compass bearings. A similar ring provided with means for observing azimuths of the sun is called an AZIMUTH CIRCLE.

bearing compass. A compass intended primarily for use in observing bearings.

bearing cursor. Of a radar set, the radial line inscribed on a transparent disk which can be rotated manually about an axis coincident with the center of the PPI. It is used for bearing determination. Also called MECHANICAL BEARING CURSOR.

bearing light. *British terminology.* A light which enables obtaining an approximate bearing without the use of a compass. It comprises two superimposed optical systems.

bearing line. A line extending in the direction of a bearing. The most common application of the expression is to a line of position constituting the locus of all points having a common bearing of a given reference mark.

bearing repeater. A compass repeater used primarily for observing bearings.

bearing resolution. See as RESOLUTION IN BEARING under RESOLUTION, definition 2. Also called ANGULAR RESOLUTION.

beat frequency. Either of the two additional frequencies obtained when signals of two frequencies are combined, equal to the sum or difference, respectively, of the original frequencies.

beat frequency oscillator. In superheterodyne reception, the adjustable oscillator which when heterodyned with the output of the final intermediate frequency stage, serves to translate an A-1 signal to the audio frequency range.

beat oscillator. An oscillator which generates the local oscillations for beat reception. Its frequency is offset from the signal frequency by the frequency of the audio tone desired. See also BEAT FREQUENCY OSCILLATOR.

beat reception. See HETERODYNE RECEPTION.

Beaufort wind scale. A numerical scale for indicating wind speed, devised by Admiral Sir Francis Beaufort in 1805. Beaufort numbers (or forces) range from force 0 (calm) to force 12 (hurricane).

bed, *n.* The ground upon which a body of water rests. The term is usually used with a modifier to indicate the type of water body, as *river bed* or *sea bed*. See also BOTTOM.

before the wind. In a direction approximating that toward which the wind is blowing. The expression applies particularly to the situation of having the wind aft, and being aided by it. See also DOWNWIND.

bell, *n.* A device for producing a distinctive sound by the vibration of a hollow, cup-shaped metallic vessel which gives forth a ringing sound when struck. If the signal is sent through the water, the device is called a *submarine bell*.

bell book. The record of all ordered engine speeds and directions made at the time they occur.

bell buoy. A steel float surmounted by a short skeleton tower in which the bell is fixed. Most bell buoys are sounded by the motion of the buoy in the sea. In a few buoys, the bells are struck by compressed gas or electrically operated hammers.

belt, *n.* A large feature of pack ice arrangement longer than it is wide, from 1 km to more than 100 km in width.

bench, *n.* On the sea floor, a small terrace.

bench mark. A fixed physical object used as reference for a vertical datum. A **tidal bench mark** is one near a tide station to which the tide staff and tidal datums are referred. A **primary tidal bench mark** is the principal (or only) mark of a group of tidal bench marks to which the tide staff and tidal datums are referred. The standard tidal bench mark of the National Ocean Survey is a copper or aluminum alloy disk 3½ inches in diameter containing the inscription NATIONAL OCEAN SURVEY together with other individual identifying information. A **geodetic bench mark** identifies a surveyed point in the National Geodetic Vertical Network. Geodetic bench mark disks contain the inscription VERTICAL CONTROL MARK NATIONAL GEODETIC SURVEY with other individual identifying information. Bench mark disks of either type may, on occasion, serve simultaneously to reference both tidal and geodetic datums. Numerous bench marks, both tidal and geodetic, still bear the inscription U.S. COAST & GEODETIC SURVEY.

beneaped, *adj.* See NEAPED.

Benguela Current. A slow-moving ocean current flowing generally northwestward along the west coast of Africa. It is caused mainly by the prevailing southeast trade winds. Near the equator the current flows westward and joins the ATLANTIC SOUTH EQUATORIAL CURRENT.

bentu de soli. An east wind on the coast of Sardinia.

berg, *n.* Short for ICEBERG.

bergy bit. A large piece of floating glacier ice, generally showing less than 5 meters above sea level but more than 1 meter and normally about 100 to 300 square meters in area. It is smaller than an ICEBERG but larger than a GROWLER. A typical bergy bit is about the size of a small house.

Bering Current. A northward flowing current through the eastern half of the Bering Sea, through Bering Strait, and in the eastern Chukchi Sea. The current speed in the Bering Sea is estimated to be usually 0.5 knot or less but at times as high as 1.0 knot. In the Bering Strait, current speeds frequently reach 2 knots. However, in the eastern half of the strait, currents are even stronger and usually range between 1.0 and 2.5 knots. Strong southerly winds may increase current speeds in the strait to 3 knots, and up to 4 knots in the eastern part. Persistent, strong northerly winds during autumn may cause the current to reverse direction for short periods. During winter a southward flow may occur in the western part of the strait. After flowing through Bering Strait, the current widens, and part continues toward Point Barrow, where it turns northwestward. Along the Alaska coast, current speeds have been observed to range between 0.1 and 1.5 knots and increase to 2.0 or 2.5 knots with southerly winds. In the western part of the Chukchi Sea, currents are considerably weaker and do not usually exceed 0.5 knot.

Bering standard time. See STANDARD TIME.

berm, *n.* A nearly horizontal portion of a beach or backshore having an abrupt fall and formed by wave deposition of material and marking the limit of ordinary high tides. Also called BEACH BERM.

berm crest. The seaward limit of a berm. Also called BERM EDGE.

berm edge. See BERM CREST.

berth, *n.* A place for securing a vessel. See also FOUL BERTH, MUD BERTH.

beset, *adj.* Said of a vessel surrounded by ice and unable to move. If the ice forcibly presses against the hull, the vessel is said to be NIPPED.

Bessel ellipsoid of 1841. The reference ellipsoid of which the semimajor axis is 6,377,397.155 meters, the semiminor axis is 6,356,078.963 meters, and the flattening or ellipticity equals 1/299.1528. Also called BESSEL SPHEROID OF 1841.

Besselian year. See FICTITIOUS YEAR.

Bessel spheroid of 1841. See BESSEL ELLIPSOID OF 1841.

bias error. See CONSTANT ERROR.

bifurcation, *n.* A division into two branches.

bifurcation buoy. A buoy which, when viewed from a vessel approaching from the open sea, or in the same direction as the main stream of flood current, or in the direction established by appropriate authority, indicates the place at which a channel divides into two. See also JUNCTION BUOY.

bifurcation mark. A navigation mark which, when viewed from a vessel approaching from the open sea, or in the same direction as the main stream of flood current, indicates the place at which the channel divides into two. See also JUNCTION MARK.

big floe. See under FLOE.

bight, *n.* 1. A long and gradual bend or recess in the coastline which forms a large open receding bay.

2. A bend in a river or mountain range.

3. An extensive crescent-shaped indentation in the ice edge, formed by either wind or current.

bill, *n.* A narrow promontory.

bi-margin format. The format of a map or chart on which the cartographic detail is extended to two edges of the sheet, thus leaving only two margins. See also BLEED.

binary star. A system of two stars that revolve about their common center of mass. See also DOUBLE STAR.

binnacle, *n.* The stand in which a compass is mounted. For a magnetic compass it is usually provided with means of mounting various correctors for adjustment and compensation of the compass.

binocular, *n.* An optical instrument for use with both eyes simultaneously. The only type customarily used by a navigator consists of two small telescopes joined together one for each eye, sometimes called a FIELD GLASS.

bioluminescence, *n.* The production of light by living organisms in the sea. Bioluminescent displays must be triggered by some physical, chemical, or mechanical stimulus. Generally, these displays are stimulated by surface wave action, ship movement, subsurface waves, upwelling, eddies, physical changes in sea water, surfs, and rip tides.

bisect, *v., t.* To divide into two equal parts.

bivariate error distribution. A two-dimensional error distribution.

blackbody, *n.* An ideal emitter which radiates energy at the maximum possible rate per unit area at each wavelength for any given temperature. A blackbody also absorbs all the radiant energy in the near visible spectrum incident upon it. No actual substance behaves as a true blackbody.

black light. Ultraviolet or infrared radiant energy. It is neither black nor light.

blanket, *v., t.* To blank out or obscure weak radio signals by a stronger signal.

blanketing, *n.* The blanking out or obscuring of weak radio signals by a stronger signal.

blanking, *n.* See as DUAL-RATE BLANKING.

blank tube. A marine sextant accessory consisting of tubular sighting vane, the function of which is to keep the line of vision parallel to the frame of the instrument when observing horizontal sextant angles.

blather, *n.* Very wet mud of such nature that a weight will at once sink into it. See also QUICKSAND.

bleed, *n.* That edge of a map or chart on which cartographic detail is extended to the edge of the sheet. Also called BLEEDING EDGE.

bleeding edge. See BLEED.

blind lead. A lead with only one outlet.

blind pilotage. *British terminology.* The task of conducting the passage of a ship in pilot waters, using any available means not denied the navigator by low visibility.

blind rollers. Long, high swells which have increased in height, almost to the breaking point, as they pass over shoals or run in shoaling water. Also called BLIND SEAS.

blind seas. See BLIND ROLLERS.

blind sector. A sector on the radarscope in which radar echoes cannot be received because of an obstruction near the antenna. See also SHADOW SECTOR.

blink, *n.* A glare on the underside of extensive cloud areas, created by light reflected from snow or ice-covered surfaces. Snow blink is whitish and brighter than the yellowish-white glare of ice blink. See also LAND SKY, WATER SKY, SKY MAP.

blinking, *n.* A means of providing information in radionavigation systems of the pulse type by modifying the signal at its source so that the signal presentation alternately appears and disappears or shifts along the time base. In Loran-C, blinking is used to indicate that a station is malfunctioning.

blip, *n.* On a radarscope, a deflection or spot of contrasting luminescence caused by an echo, i.e., the radar signal reflected back to the antenna by an object. Also called PIP, ECHO, RETURN.

blip scan ratio. The ratio of the number of paints from a target to the maximum possible number of paints for a given number of revolutions of the radar antenna. The maximum number of paints is usually equivalent to the number of revolutions of the antenna.

blister, *n.* See BORDER BREAK.

blizzard, *n.* A severe weather condition characterized by low temperatures and by strong winds bearing a great amount of snow (mostly fine, dry snow picked up from the ground). The National Weather Service specifies the following conditions for a blizzard: a wind of 32 miles per hour or higher, low temperatures, and sufficient snow in the air to reduce visibility to less than 500 feet; for a **severe blizzard,** it specifies wind speeds exceeding 45 miles per hour, temperature near or below 10°F, and visibility reduced by snow to near zero. In popular usage in the United States, the term is often used for any heavy snowstorm accompanied by strong winds.

block, *n.* See CHARTLET, definition 2.

block correction. See CHARTLET, definition 2.

blocky iceberg. An iceberg with steep sides and a flat top. The length-to-height ratio is less than 5:1. See also TABULAR ICEBERG.

Blondel-Rey effect. The effect that the flashing of a light has on reducing its apparent intensity as compared to the intensity of the same light when operated continuously or fixed.

blooming, *n.* Expansion of the spot produced by a beam of electrons striking the face of a cathode-ray indicator, caused by maladjustment.

blowing snow. Snow raised from the ground and carried by the wind to such a height that both vertical and horizontal visibility are considerably reduced. The expression DRIFTING SNOW is used when only the horizontal visibility is reduced.

blue ice. The oldest and hardest form of glacier ice, distinguished by a slightly bluish or greenish color.

blue magnetism. The magnetism displayed by the south-seeking end of a freely suspended magnet. This is the magnetism of the earth's north magnetic pole.

bluff, *n.* A headland or stretch of cliff having a broad nearly perpendicular face. See also CLIFF.

blunder, *n.* See MISTAKE.

Board on Geographic Names. An agency of the U.S. Government, first established by Executive Order in 1890 and currently functioning under Public Law 242-80, 25 July 1947. Twelve departments and agencies enjoy Board membership. Conjointly with the Secretary of the Interior, the Board provides for "uniformity in geographic nomenclature and orthography throughout the Federal Government." It develops policies and romanization systems under which names are derived and it standardizes geographic names for use on maps and in textual materials.

boat, *n.* A small vessel. The term is often modified to indicate the means of propulsion, such as motorboat, rowboat, steamboat, sailboat, and sometimes to indicate the intended use, such as lifeboat, fishing boat, etc. See also SHIP.

boat compass. A small compass mounted in a box for convenient use in small water craft.

boat harbor. A sheltered area in a harbor set aside for the use of boats, usually with moorings, buoys, etc.

boat sheet. The work sheet used in the field for plotting details of a hydrographic survey as it progresses.

bobbing a light. Quickly lowering the height of eye several feet and then raising it again when a navigational light is first sighted to determine whether or not the observer is at the geographic range of the light. If he is, the light disappears when the eye is lowered and reappears when it is restored to its original position.

bold, *adj.* Rising steeply from the sea; as a *bold coast.* See also ABRUPT.

bolide, *n.* A meteor having a magnitude brighter than −4 magnitude. Bolides are observed with much less frequency than shooting stars. Light bursts, spark showers, or splitting of the luminous trail are sometimes seen along their trails. The luminous trails persist for minutes and may persist up to an hour in exceptional cases. Also called FIREBALL. See also METEOR.

bollard, *n.* A post (usually steel or reinforced concrete) firmly secured on a wharf, quay, etc., for mooring vessels by means of lines extending from the vessel and secured to the post.

bombing range. An area of land or water, and the air space above, designated for use as a bombing practice area.

boom, *n.* A floating barrier placed across the mouth of a harbor, river, etc., to serve as part of the defense of the area or to shelter the enclosed water space.

bora, *n.* A cold, northerly wind blowing from the Hungarian basin into the Adriatic Sea. See also FALL WIND.

borasco, *n.* A thunderstorm or violent squall, especially in the Mediterranean.

border break. A cartographic technique used when it is required to extend cartographic detail of a map or chart beyond the neatline into the margin. This technique eliminates the necessity of producing an additional sheet. Also called BLISTER.

borderland, *n.* A region bordering a continent, normally occupied by or bordering a shelf, that is highly irregular with depths well in excess of those typical of a shelf.

bore, *n.* See TIDAL BORE.

boring, *n.* Forcing a vessel under power through ice, by breaking a lead.

borrow, *v., i.* To approach closer to the shore or wind.

bottom, *n.* The ground under a body of water. The terms BED, FLOOR, and BOTTOM have nearly the same meaning, but BED refers more specifically to the whole hollowed area supporting a body of water, FLOOR refers to the essentially horizontal surface constituting the principal level of the ground under a body of water, and BOTTOM refers to any ground covered with water.

bottom characteristics. Designations used on surveys and nautical charts to indicate the consistency, color, and classification of the sea bottom. Also called NATURE OF THE BOTTOM, CHARACTER OF THE BOTTOM.

bottom contour chart. A chart designed for surface and sub-surface bathymetric navigation seaward of the 100-fathom contour. Bottom configuration is portrayed by depth contours and selected soundings. The chart may contain a Loran-C or Omega lattice overprint.

bottom sample. A portion of the material forming the bottom, brought up for inspection.

bottom sampler. A device for obtaining a portion of the bottom for inspection.

Bouguer's halo. An infrequently observed, faint, white, circular arc or complete ring of light which has a radius of about 39°, and is centered on the antisolar point. When observed, it usually is in the form of a separate outer ring around an anticorona. Also called ULLOA'S RING. See also FOGBOW.

boulder, *n.* A detached water-rounded stone more than 256 millimeters in diameter, i.e., larger than a man's head. See also COBBLE.

boundary disclaimer. A statement on a map or chart that the status and/or alignment of international or administrative boundaries is not necessarily recognized by the government of the publishing nation.

boundary lines of inland waters. Lines dividing the high seas from rivers, harbors, and inland waters in accordance with the intent of a U.S. statute (28 Stat. 672,33 U.S.C. 151). The waters inshore of the lines are "inland waters" and upon them the Inland Rules and Pilot Rules apply. The waters outside of the lines are the high seas and upon them the International Rules apply.

boundary monument. A material object placed on or near a boundary line to preserve and identify the location of the boundary line on the ground.

bow, *n.* The forward part of a ship, craft, aircraft, or float.

bow and beam bearings. Successive relative bearings (right or left) of 45° and 90° taken on a fixed object to obtain a running fix. The length of the run between such bearings is equal to the distance of the craft from the object at the time the object is broad on the beam, neglecting current.

Bowditch, *n.* Popular title for Pub. No. 9, *American Practical Navigator.*

bow wave. 1. The wave set up by the bow of a vessel moving through the water. Also called WAVE OF DISPLACEMENT.

2. A shock wave in front of a body such as an airfoil.

boxing the compass. Stating in order the names of the points (and sometimes the half and quarter points) of the compass.

BPM5. See under CONSOL.

brackish, *adj.* Containing salt to a moderate degree, such as sea water which has been diluted by fresh water, as near the mouth of a river. The salinity values of the brackish water range from approximately 0.50 to 17.00 parts per thousand.

branch, *n.* 1. A creek or brook, as used locally in the southern U.S.

2. One of the bifurcations of a stream, as a fork.

brash ice. Accumulations of floating ice made up of fragments not more than 2 meters across, the wreckage of other forms of ice.

brave west winds. The strong, often stormy, winds from the west-northwest and northwest which blow at all seasons of the year between latitudes 40° S and 60° S. See also ROARING FORTIES.

Brazil Current. The ocean current flowing southwestward along the Brazilian coast. Its origin is in the westward flowing Atlantic South Equatorial Current, part of which turns south and flows along the South American coast as the Brazil Current. The mean speed of the current along its entire length is about 0.6 knot. Off Uruguay at about 35° S, it meets the Falkland Current, the two turning eastward to join the South Atlantic Current.

break-circuit chronometer. A chronometer equipped with an electrical contact assembly and program wheel which automatically makes or breaks an electric circuit at precise intervals, the sequence and duration of circuit-open circuit-closed conditions being recorded on a chronograph. The program sequence is controlled by the design of the program wheel installed. Various programs of make or break sequence, up to 60 seconds, are possible. In some chronometers, the breaks occur every other second, on the even seconds, and a break occurs also on the 59th second to identify the beginning of the minute; in other chronometers, breaks occur every second except at the beginning of the minute. By recording the occurrence of events (such as star transits) on a chronograph sheet along with the chronometer breaks, the chronometer times of those occurrences are obtained.

breaker, *n.* A wave which breaks, either because it becomes unstable when it reaches shallow water, the crest toppling over or "breaking," or because it dashes against an obstacle. Instability is caused by an increase in wave height and a decrease in the speed of the trough of the wave in shallow water. The momentum of the crest, often aided by the wind, causes the upper part of the wave to move forward faster than the lower part. The crest of a wave which becomes unstable in deep water and topples over or "breaks" is called a WHITECAP.

breakwater, *n.* Anything which breaks the force of the sea at a particular place, thus forming protection for vessels. Often an artificial embankment built to protect the entrance to a harbor, or to form an artificial harbor. See also JETTY.

breasting float. See CAMEL.

breeze, *n.* 1. Wind of force 2 to 6 (4-31 miles per hour or 4-27 knots) on the Beaufort wind scale. Wind of force 2 (4-7 miles per hour or 4-6 knots) is classified as a **light breeze**; wind of force 3 (8-12 miles per hour or 7-10 knots), a **gentle breeze**; wind of force 4 (13-18 miles per hour or 11-16 knots), a **moderate breeze**; wind of force 5 (19-24 miles per hour or 17-21 knots), a **fresh breeze**; and wind of force 6 (25-31 miles per hour or 22-27 knots), a **strong breeze**. See also LIGHT AIR.

2. Any light wind. A **land breeze** blows from the land to the sea, and usually alternates with a **sea breeze** blowing in the opposite direction. A **mountain breeze** blows down a mountain slope due to gravity flow of cooled air, and a **valley breeze** blows up a valley or mountain slope because of the warming of the mountainside and valley floor by the sun. A puff of wind, or light breeze affecting a small area, may be called a **cat's paw**. Absence of wind is sometimes called **ash breeze**.

bridge, *n.* 1. An elevated structure extending across or over the weather deck of a vessel, or part of such a structure. The term is sometimes modified to indicate the intended use, such as *navigating bridge* or *signal bridge*.

2. A structure erected over a depression or an obstacle such as a body of water, railroad, etc., to provide a roadway for vehicles or pedestrians. See also CAUSEWAY, VIADUCT.

Briggsian logarithm. See COMMON LOGARITHM.

bright display. A radar display capable of being used under relatively high ambient light levels.

brisa, briza, *n.* 1. A northeast wind which blows on the coast of South America or an east wind which blows on Puerto Rico during the trade wind season.

2. The northeast monsoon in the Philippines.

brisote, *n.* The northeast trade wind when it is blowing stronger than usual on Cuba.

broad on the beam. Bearing 090° relative (*broad on the starboard beam*) or 270° relative (*broad on the port beam*). If the bearings are approximate, the expression ON THE BEAM or ABEAM should be used.

broad on the bow. Bearing 045° relative (*broad on the starboard bow*) or 315° relative (*broad on the port bow*). If the bearings are approximate, the expression ON THE BOW should be used.

broad on the quarter. Bearing 135° relative (*broad on the starboard quarter*) or 225° relative (*broad on the port quarter*). If the bearings are approximate, the expression ON THE QUARTER should be used.

broadside on. Beam on, as to the wind or sea.

broad tuning. Low selectivity, usually resulting in simultaneous reception of signals of different frequencies (spill-over). The opposite is SHARP TUNING.

Brocken bow. See ANTICORONA.

broken water. An area of small waves and eddies occurring in what otherwise is a calm sea.

brook, *n.* A very small natural stream; a rivulet. Also called RUN, RUNNEL. See also CREEK, definition 2.

brubu, *n.* A name for a squall in the East Indies.

B trace. The second trace of an oscilloscope having more than one displayed, as the lower trace of a Loran-A indicator.

bubble acceleration error. That error of a bubble sextant observation caused by displacement of the bubble by acceleration or deceleration resulting from motion of a craft. Also called ACCELERATION ERROR.

bubble horizon. An artificial horizon parallel to the celestial horizon, established by means of a bubble level.

bubble sextant. A sextant with a bubble or spirit level to indicate the horizontal.

bucket temperature. Temperature of surface sea water trapped and measured in a bucket or similar receptacle.

building, *n.* A label on a nautical chart which is used when the entire structure is the landmark, rather than an individual feature of it. Also labeled HOUSE.

bull's eye squall. A squall forming in fair weather, characteristic of the ocean off the coast of South Africa. It is named for the peculiar appearance of the small isolated cloud marking the top of the invisible vortex of the storm.

bull the buoy. To bump into a buoy.

bummock, *n.* From the point of view of the submariner, a downward projection from the underside of the ice canopy; the counterpart of a HUMMOCK.

bund, *n.* An embankment or embanked thoroughfare along a body of water. The term is used particularly for such structures in the Far East.

buoy, *n.* A floating object, other than a lightship, moored or anchored to the bottom as an aid to navigation. Buoys may be classified according to shape, as **spar, cylindrical** or **can, conical, nun, spherical, barrel, dan,** or **pillar buoy.** They may also be classified according to the color scheme, as a **red, black,** or **checkered buoy.** A buoy fitted with a characteristic shape at the top to aid in its identification is called a **topmark buoy.** A **sound buoy** is one equipped with a characteristic sound signal, and may be further classified according to the manner in which the sound is produced, as a **bell, gong, horn, trumpet,** or **whistle buoy.** A **lighted buoy** is one with a light having definite characteristics for detection and identification during darkness. If the light is produced by gas, it may be called a **gas buoy.** A buoy equipped with a marker radiobeacon is called a **radiobeacon buoy.** A buoy with equipment for automatically transmitting a radio signal when triggered by an underwater sound signal is called a **sonobuoy.** A **combination buoy** has more than one means of conveying intelligence; it may be called a **lighted sound buoy** if it is a lighted buoy provided with a sound signal. Buoys may be classified according to location, as **channel, mid-channel, middle ground, turning, fairway, bifurcation, junction,** or **sea buoy.** A **bar buoy** marks the location of a bar. A buoy marking a hazard to navigation may be classified according to the nature of the hazard, as **obstruction, wreck, telegraph, cable, fish net, dredging,** or **spoil-ground buoy.** Buoys used for particular purposes may be classified according to their use, as **anchor, anchorage, quarantine, mooring, warping, swinging, marker, station, watch,** or **position buoy.** A light-weight buoy especially designed to withstand strong currents is called a **river buoy.** An **ice buoy** is a sturdy one used to replace a more easily damaged buoy during a period when heavy ice is anticipated.

buoyage, *n.* A system of buoys. One in which the buoys are assigned shape, color, and number distinction in accordance with location relative to the nearest obstruction is called a **cardinal system.** One in which buoys are assigned shape, color, and number distinction as a means of indicating navigable waters is called a **lateral system.** See also IALA MARITIME BUOYAGE SYSTEM.

buoy station. The established (charted) location of a buoy.

buoy tender. A vessel designed for, and engaged in, servicing aids to navigation, particularly buoys.

butte, *n.* An isolated flat-topped hill, similar to but smaller than a MESA.

Buys Ballot's law. A rule useful in locating the center of cyclones and anticyclones. It states that, "Facing the wind in the Northern Hemisphere, atmospheric pressure decreases toward the right and increases toward the left: facing the wind in the Southern Hemisphere, atmospheric pressure decreases toward the left and increases toward the right."

by the head. See DOWN BY THE HEAD.

by the stern. See DOWN BY THE STERN.

ELEMENTS OF THE CELESTIAL SPHERE

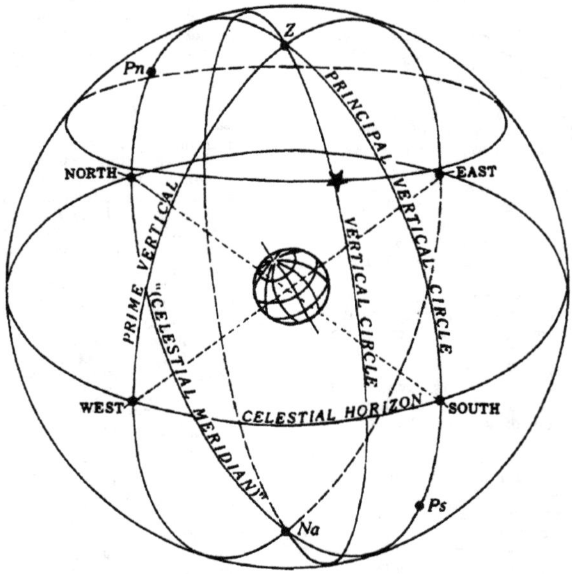

CELESTIAL HORIZON IS PRIMARY GREAT CIRCLE

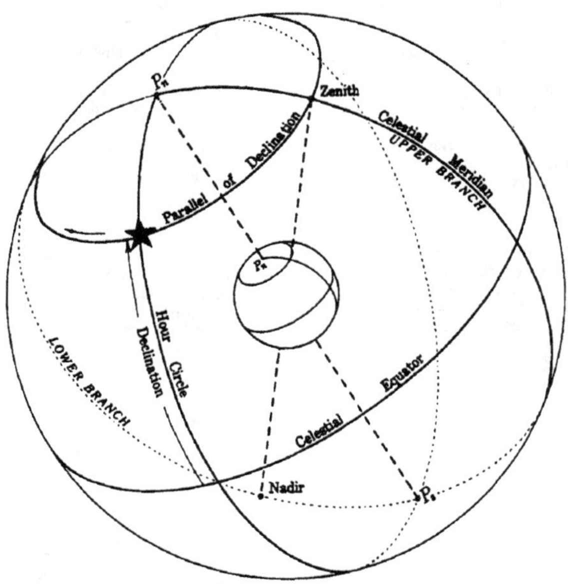

CELESTIAL EQUATOR IS PRIMARY GREAT CIRCLE

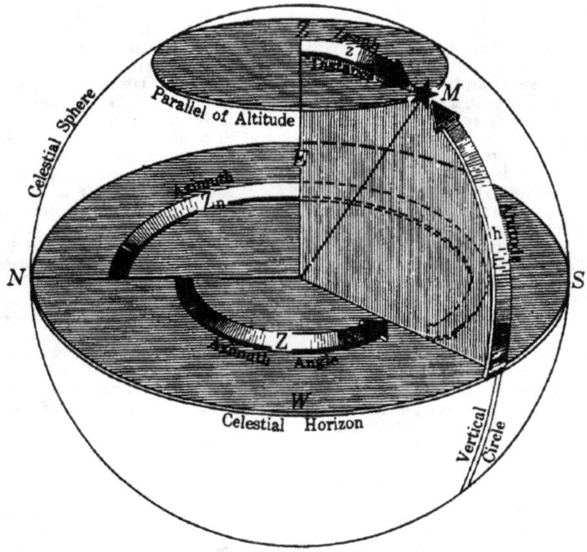

HORIZON SYSTEM OF COORDINATES

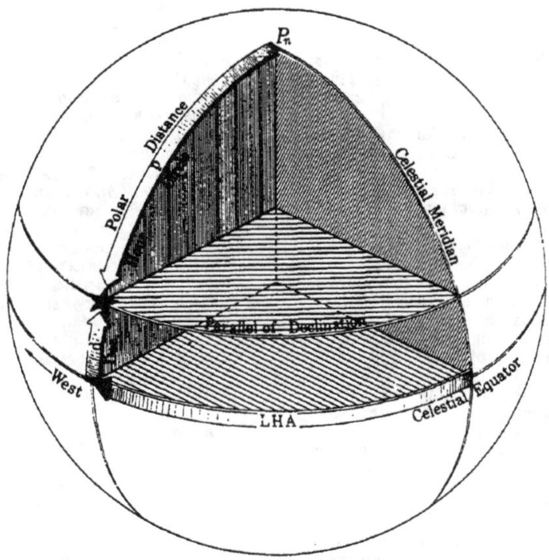

CELESTIAL EQUATOR SYSTEM OF COORDINATES

C

cable, *n.* 1. A unit of distance equal to one-tenth of a sea mile. Sometimes called CABLE LENGTH.
2. A chain or very strong fiber or wire rope used to anchor or moor vessels or buoys.
3. A stranded conductor or an assembly of two or more electric conductors insulated from each other, but laid up together with a strong, waterproof covering. One carrying an electric current for degaussing a vessel is called a **degaussing cable.** An insulated, waterproof wire or bundle of wires for carrying an electric current underwater is called a **submarine cable.** A cable carrying an electric current, signals from or the magnetic influence of which indicates the path to be followed by a craft equipped with suitable instruments is called a **leader cable.** A **coaxial cable** consists of two concentric conductors insulated from each other.

cable buoy. 1. A buoy used to mark one end of a cable being worked by a cable ship.
2. A floating support of a submarine cable.

cable length. See CABLE, definition 1.

cage, *n.* That part of the buoy built on top of the body of the buoy and used as a daymark or part thereof, usually to support the light or a topmark or a radar reflector, or a combination of these. Also called SUPERSTRUCTURE.

cage, *v., t.* To erect a gyro or lock it in place by means of a caging mechanism.

caging mechanism. A device for erecting a gyroscope or locking it in position.

cairn, *n.* A mound of rough stones or concrete, particularly one serving or intended to serve as a *landmark.* The stones are customarily piled in a pyramidal or beehive shape.

caisson, *n.* A watertight gate for a lock, basin, etc.

calcareous, *adj.* Containing or composed of calcium or one of its compounds.

calculated altitude. See under COMPUTED ALTITUDE.

caldera, *n.* A volcanic crater.

calendar, *n.* An orderly arrangement of days, weeks, months, etc., to suit a particular need such as civil life. The **Gregorian calendar** is in common use today. See also JULIAN DAY.

calendar day. The period from midnight to midnight. The calendar day is 24 hours of mean solar time in length and coincides with the civil day unless a time change occurs during a day.

calendar line. *British terminology.* See DATE LINE.

calendar month. The month of the calendar, varying from 28 to 31 days in length.

calendar year. The year of the calendar common years having 365 days and **leap years** 366 days. Each year exactly divisible by 4 is a leap year, except century years (1800, 1900, etc.), which must be exactly divisible by 400 (2000, 2400, etc.) to be leap years. The calendar year is based on the tropical year. Also called CIVIL YEAR.

calibrate, *n.* To determine or rectify the scale graduations of an instrument.

calibration card. See under CALIBRATION TABLE.

calibration correction. The value to be added to or subtracted from the reading of an instrument to obtain the correct reading.

calibration error. That error in an instrument due to imperfection of calibration or maladjustment of its parts. Also called SCALE ERROR.

calibration radiobeacon. A special radiobeacon operated primarily for calibrating shipboard radio direction finders. These radiobeacons transmit either continuously during scheduled hours or upon request.

calibration table. A list of calibration corrections or calibrated values. A card having such a table on it is called a CALIBRATION CARD.

California Current. A North Pacific Ocean current flowing southeastward along the west coast of North America from a point west of Vancouver Island to the west of Baja (Lower) California, where it gradually widens and curves southward and southwestward, to continue as the westerly flowing PACIFIC NORTH EQUATORIAL CURRENT. The California Current is the southern branch of the Aleutian Current, augmented by the North Pacific Current, and forms the eastern part of the general clockwise oceanic circulation of the North Pacific Ocean. Although usually described as a permanent ocean current, the California Current is actually a poorly-defined and variable flow easily influenced by the winds. See also MEXICO CURRENT.

California norther. See under NORTHER.

Callipic cycle. A period of four Metonic cycles equal to 76 Julian years of 27759 days. Devised by Callipus, a Greek astronomer, about 350 B.C., as a suggested improvement on the Metonic cycle for a period in which new and full moon would recur on the same day of the year. Taking the length of the synodical month as 29.530588 days, there are 940 lunations in the Callipic cycle with about 0.25 day remaining.

calm, *adj.* In a state of calm; without motion.

calm, *n.* 1. Absence of appreciable wind; specifically, force 0 (less than 1 knot or 1 mile per hour) on the Beaufort wind scale. Also called ASH BREEZE.
2. The state of the sea when there are no waves.

calm belt. 1. The doldrums.
2. One of the two zones of high pressure and light winds on the poleward sides of the trade winds called *calms of Cancer* and *calms of Capricorn,* respectively.

calving, *n.* The breaking away of a mass of ice from an ice wall, ice front, or iceberg.

camanchaca, *n.* See GARUA.

camber, *n.* *British terminology.* A small basin, particularly one with a narrow entrance and situated inside a harbor.

camel, *n.* A float used as a fender. Also called BREASTING FLOAT.

canal, *n.* 1. An artificial waterway for navigation.
2. A long, fairly straight natural channel with steep sloping sides.
3. Any watercourse or channel.
4. A sluggish coastal stream, as used locally on the Atlantic coast of the U.S.

canal port. *British terminology.* A port, the waterway of which is entirely artificial.

Canaries Current. See CANARY CURRENT.

Canary Current. The southern branch of the North Atlantic Current (which divides on the eastern side of the ocean); it moves south past Spain and southwestward along the northwest coast of Africa and past the Canary Islands. In the vicinity of the Cape Verde Islands, it divides into two branches, the western branch augmenting the Atlantic North Equatorial Current and the eastern branch curving southeastward and continuing as the GUINEA CURRENT. The Canary Current forms the southeastern part of the general clockwise oceanic circulation of the North Atlantic Ocean. Also called CANARIES CURRENT.

can buoy. An unlighted buoy of which the upper part of the body (above the waterline), or the larger part of the superstructure, has the shape of a cylinder or nearly so. Also called CYLINDRICAL BUOY.

candela, *n.* The *base unit* of luminous intensity in the International System of Units (SI). It is the luminous intensity, in the perpendicular direction, of a surface of 1/600,000 square meter of a blackbody at the temperature of freezing platinum, under a pressure of 101,325 newtons per square metre. The definition was adopted by the Thirteenth General Conference on Weights and Measures (1967).

candela per square metre. The *derived unit* of luminance in the International System of Units.

candlepower, *n.* Luminous intensity expressed in candelas.

canyon, *n.* On the sea floor, a relatively narrow, deep depression with steep sides, the bottom of which generally has a continuous slope.

cap cloud. 1. A cloud resting on the top of an isolated mountain peak. The cloud is apparently stationary, but actually is being formed to windward and dissipated to leeward. A similar cloud over a mountain ridge is called a CREST CLOUD. See also BANNER CLOUD.
2. False cirrus over a towering cumulus, in the form of a cap or hood. See also SCARF CLOUD.

cape, *n.* A relatively extensive land area jutting seaward from a continent, or large island, which prominently marks a change in or interrupts notably the coastal trend.

Cape Breton Current. Originating in the Gulf of St. Lawrence, the Cape Breton Current flows southeastward in the southwestern half of Cabot Strait, and merges with the Labrador Current Extension. It may be augmented by a branch of the constant but tide influenced Gaspé Current to the northwest.

cape doctor. The strong southeast wind which blows on the South African coast. Also called DOCTOR.

Cape Horn Current. An ocean current that flows continuously eastward close to the tip of South America. It enters Drake Passage, at about longitude 70° W, in a 150-mile-wide band, with observed surface speeds to 2.4 knots. The current veers north-northeastward; when it crosses longitude 65° W, the current has narrowed to a width of about 85 miles, and its speed has decreased considerably. The current continues as the FALKLAND CURRENT.

cardinal heading. A heading in the direction of any of the cardinal points. See also INTERCARDINAL HEADING.

cardinal mark. See under IALA MARITIME BUOYAGE SYSTEM.

cardinal point. Any of the four principal directions; north, east, south, or west. Directions midway between cardinal points are called INTERCARDINAL POINTS.

cardinal system. A system of aids to navigation in which the shape, color, and number distinction are assigned in accordance with location relative to the nearest obstruction. The cardinal points delineate the sectors for aid location. The cardinal system is particularly applicable to a region having numerous small islands and isolated dangers. In the LATERAL SYSTEM, used in United States waters, the aids are assigned shape, color, and number distinction as a means of indicating navigable waters.

cardioid, *n.* The figure traced by a point on a circle which rolls around an equal fixed circle.

cargo transfer area. See under CARGO TRANSSHIPMENT AREAS.

cargo transshipment area. An area generally outside port limits that is specifically designated as suitable for the transshipment of oil or other materials from large ships to smaller ones. As the purpose of transshipment is usually to reduce the draft of the larger vessel to allow her to proceed to port, the operation is often known as lightening and the area may be called lightening area or cargo transfer area.

Caribbean Current. An ocean current flowing westward through the Caribbean Sea to the Yucatan Channel. It is formed by the comingling of part of the waters of the Atlantic North Equatorial Current with those of the Guiana Current.

carrier, *n.* 1. A radio wave having at least one characteristic which may be varied from a known reference value by modulation.
2. That part of a modulated wave that corresponds in a specified manner to the unmodulated wave.
3. In a frequency stabilized system, the sinusoidal component of a modulated wave; or the output of a transmitter when the modulating wave is made zero; or a wave generated at a point in the transmitting system and subsequently modulated by the signal; or a wave generated locally at the receiving terminal which, when combined with the sidebands in a suitable detector, produces the modulating wave.
Also called CARRIER WAVE.

carrier frequency. 1. The frequency of the unmodulated fundamental output of a radio transmitter.
2. In a periodic carrier, the reciprocal of its period. The frequency of a periodic pulse carrier often is called PULSE REPETITION FREQUENCY.

carrier power. See under POWER (OF A RADIO TRANSMITTER).

carrier wave. See CARRIER.

Cartesian coordinates. Magnitudes defining a point relative to two intersecting lines, called AXES. The magnitudes indicate the distance from each axis, measured along a parallel to the other axis. If the axes are perpendicular, the coordinates are **rectangular**; if not perpendicular, they are **oblique coordinates.**

cartographer, *n.* One who designs and constructs charts or maps.

cartographic feature. The natural or cultural objects shown on a map or chart. See also TOPOGRAPHY.

cartography, *n.* The art and science of making charts or maps.

cartometer, *n.* A device consisting of a small wheel and a calibrated dial used to measure distances on a map by following the desired route.

cartouche, *n.* A panel of a map, often with decoration, enclosing the title, other legends, the scale, etc.

cask buoy. A buoy in the shape of a cask.

Cassegrainian telescope. A reflecting telescope in which the incoming light is reflected from the primary mirror onto a secondary mirror and back through a small central aperture in the primary mirror. See also NEWTONIAN TELESCOPE.

cast, *n.* The act of heaving the lead into the water to determine its depth.

cast, *v., t.* 1. To turn a ship in her own water.
2. To turn a ship to a desired direction without gaining headway or sternway.
3. To take a sounding with the lead.

catamaran, *n.* 1. A double-hulled vessel.
2. A raft consisting of a rectangular frame attached to two parallel cylindrical floats and which may be used for working alongside a ship. See also CAMEL.

catenary, *n.* The curve formed by a uniform cable supported only at its ends.

cathode, *n.* 1. The electrode through which a primary stream of electrons enters the inter-electrode space.
2. The general term for a negative electrode. See also ANODE.

cathode ray. A stream of electrons emitted from the cathode of any vacuum tube, but normally used in reference to special purpose tubes designed to provide some kind of display.

cathode-ray tube. A vacuum tube in which the instantaneous position of a sharply focused electron beam, deflected by means of electrostatic or electromagnetic fields, is indicated by a spot of light produced by impact of the electrons on a fluorescent screen at the end of the tube opposite the cathode.

catoptric light. A light concentrated into a parallel beam by means of one or more reflectors. One so concentrated by means of refracting lens or prisms is a DIOPTRIC LIGHT.

cat's paw. A puff of wind; a light breeze affecting a small area, as one that causes patches of ripples on the surface of a water area.

causeway, *n.* A raised way, as for a road, across wet ground or water. See also BRIDGE, definition 2; VIADUCT.

cautionary characteristic. Of a light, a unique characteristic which can be recognized as imparting a special cautionary significance e.g., a quick flashing characteristic phase indicating a sharp turn in a channel.

cautionary note. Information calling special attention to some fact, usually a danger area, shown on a map or chart.

caver, kaver, *n.* A gentle breeze in the Hebrides.

cay, kay (kē), *n.* A low flat island or mount of sand built up on a reef flat slightly above high water which may contain a large admixture of coral or shell fragments. Also called KEY.

C-band. A radiofrequency band of 3,900 to 6,200 megahertz. This band overlaps the S- and X-bands. See also FREQUENCY.

ceiling, *n.* The height above the earth's surface of the lowest layer of clouds or obscuring phenomena that is reported as broken, overcast, or obscuration, and not classified as thin or partial. The ceiling is termed *unlimited* when the foregoing conditions are not satisfied.

celestial, *adj.* Of or pertaining to the heavens.

celestial body. Any aggregation of matter in space constituting a unit for astronomical study, as the sun, moon, a planet, comet, star, nebula, etc. Also called HEAVENLY BODY.

celestial concave. See CELESTIAL SPHERE.

celestial coordinates. Any set of coordinates used to define a point on the celestial sphere. The horizon, **celestial equator,** and the **ecliptic** systems of celestial coordinates are based on the celestial horizon, celestial equator, and the ecliptic, respectively, as the primary great circle.

celestial equator. The primary great circle of the celestial sphere, everywhere 90° from the celestial poles; the intersection of the extended plane of the equator and the celestial sphere. Also called EQUINOCTIAL.

celestial equator system of coordinates. A set of celestial coordinates based on the celestial equator as the primary great circle; such as declination and hour angle, declination and sidereal hour angle, or declination and right ascension. Also called EQUINOCTIAL SYSTEM OF COORDINATES.

celestial fix. A fix established by means of two or more celestial bodies.

celestial globe. See STAR GLOBE.

celestial horizon. That circle of the celestial sphere formed by the intersection of the celestial sphere and a plane through the center of the earth and perpendicular to the zenith-nadir line. Also called RATIONAL HORIZON. See also HORIZON.

celestial latitude. Angular distance north or south of the ecliptic; the arc of a circle of latitude between the ecliptic and a point on the celestial sphere, measured northward or southward from the ecliptic through 90°, and labeled N or S to indicate the direction of measurement.

celestial line of position. A line of position determined by means of a celestial body.

celestial longitude. Angular distance east of the vernal equinox, along the ecliptic; the arc of the ecliptic or the angle at the ecliptic pole between the circle of latitude of the vernal equinox and the circle of latitude of a point on the celestial sphere, measured eastward from the circle of latitude of the vernal equinox, through 360°.

celestial mechanics. The study of the theory of the motions of celestial bodies under the influence of gravitational fields.

celestial meridian. A great circle of the celestial sphere, through the celestial poles and the zenith. The expression usually refers to the **upper branch,** that half from pole to pole which passes through the zenith; the other half being called the **lower branch.** The celestial meridian coincides with the hour circle through the zenith and the vertical circle through the elevated pole.

celestial navigation. Navigation with the aid of celestial bodies.

celestial observation. Observation of celestial phonomena. By navigators, the expression is applied principally to the measurement of the altitude of a celestial body, and sometimes to measurement of azimuth, or to both altitude and azimuth. The expression may also be applied to the data obtained by such measurement. Also called SIGHT when referring to this navigation usage.

celestial parallel. See PARALLEL OF DECLINATION.

celestial pole. Either of the two points of intersection of the celestial sphere and the extended axis of the earth, labeled N or S to indicate whether the *north celestial pole* or the *south celestial pole.*

celestial sphere. An imaginary sphere of infinite radius concentric with the earth, on which all celestial bodies except the earth are imagined to be projected.

celestial triangle. A spherical triangle on the celestial sphere, especially the navigational triangle.

celo-navigation, *n.* The use of this term for CELESTIAL NAVIGATION is obsolescent.

Celsius temperature. The designation given to the temperature measured on the International Practical Temperature Scale with the zero taken as 0°.01 below the triple point of water. Formerly called CENTIGRADE TEMPERATURE, but the Ninth General Conference of Weights and Measures, held in October 1948, adopted the name *Celsius* in preference to *centigrade,* to be consistent with naming other temperature scales after their inventors, and to avoid the use of different names in different countries. On the original Celsius scale, invented in 1742 by a Swedish astronomer named Andres Celsius, the numbering was the reverse of the modern scale, 0°C representing the boiling point of water, and 100°C its freezing point.

center frequency. See ASSIGNED FREQUENCY.

centering control. On a radar indicator, a control used to place the sweep origin at the center of the plan position indicator.

centering error. That error in an instrument due to inaccurate pivoting of a moving part, as the index arm of a marine sextant. Also called ECCENTRIC ERROR.

center line. 1. The locus of points equidistant from two reference points or lines.
2. (*Usually centerline*) The line separating the port and starboard sides of a vessel.

center of buoyancy. The geometric center of the immersed portion of the hull and appendages of a floating vessel. All buoyant forces may be resolved into one resultant force acting upwards at this point.

center of gravity. That point in any body at which the force of gravity may be considered to be concentrated. Same as CENTER OF MASS in a uniform gravitational field.

center of mass. The point at which all the given mass of a body or bodies may be regarded as being concentrated as far as motion is concerned. Commonly called CENTER OF GRAVITY.

centi-. A prefix meaning *one-hundredth*.

centibar, *n.* One-hundredth of a bar; 10 millibars.

centigrade temperature. See under CELSIUS TEMPERATURE.

centimeter, *n.* One-hundredth of a meter.

centimeter-gram-second system. A system of units based on the centimeter as the unit of length, the gram as the unit of mass, and the mean solar second as the unit of time. Its units with special names include the erg, the dyne, the gauss, and the oersted. See also INTERNATIONAL SYSTEM OF UNITS.

centimetric wave. A super high frequency radio wave, approximately 0.01 to 0.1 meter in length (3 to 30 gigahertz). See also ULTRASHORT WAVE.

central force. A force which for purposes of computation can be considered to be concentrated at one central point with its intensity at any other point being a function of the distance from the central point. Gravitation is considered as a central force in celestial mechanics.

central force field. The spatial distribution of the influence of a central force.

central force orbit. The theoretical orbit achieved by a particle of negligible mass moving in the vicinity of a point mass with no other forces acting; an unperturbed orbit.

central standard time. See STANDARD TIME.

centrifugal force. The force acting on a body or part of a body moving under constraint along a curved path, tending to force it outward from the center of revolution or rotation. The opposite is CENTRIPETAL FORCE.

centripetal force. The force directed toward the center of curvature, which constrains a body to move in a curved path. The opposite is CENTRIFUGAL FORCE.

chain, *n.* 1. A group of associated stations of a radionavigation system. A Loran-C chain consists of a master station and two to four secondary stations.
2. *pl.* The platform or station from which soundings are taken by means of a lead.

chain signature. See under GROUP REPETITION INTERVAL.

chalk, *n.* Soft earthy sandstone of marine origin, composed chiefly of minute shells. It is white, gray, or buff in color. Part of the ocean bed and shores and composed of chalk, notably the "white cliffs of Dover," England.

challenge, *n.* A signal transmitted by an interrogator.

challenge, *v., t.* To cause an interrogator to transmit a signal which puts a transponder into operation.

challenger, *n.* See INTERROGATOR.

chance error. See RANDOM ERROR.

change of the moon. The time of new moon. See also PHASES OF THE MOON.

change of tide. A reversal of the direction of motion (rising or falling) of a tide. The expression is also sometimes applied somewhat loosely to a reversal in the set of a tidal current. Also called TURN OF THE TIDE.

channel, *n.* 1. That part of a body of water deep enough for navigation through an area otherwise not suitable. It is usually marked by a single or double line of buoys and sometimes by ranges.
2. The deepest part of a stream, bay, or strait, through which the main current flows.
3. A name given to certain large straits, as the *English Channel.*
4. A hollow bed through which water does or may run.
5. A band of radio frequencies within which a radio station must maintain its modulated carrier frequency to prevent interference with stations on adjacent channels. Also called FREQUENCY CHANNEL.

channel buoy. A buoy marking a channel.

channel frequency. See ASSIGNED FREQUENCY.

channel light. A light either on a fixed support or on a buoy, marking the limit of a navigable channel. In French, the term FEU DE RIVE is commonly used for a channel light on a fixed support.

characteristic, *n.* 1. The color and shape of a daymark or buoy or the color and period of a light used for identifying the aid. See also CHARACTERISTIC COLOR, CHARACTERISTIC PHASE.
2. The identifying signal transmitted by a radiobeacon.
3. That part of a logarithm (base 10) to the left of the decimal point. That part of a logarithm (base 10) to the right of the decimal point is called the MANTISSA.
4. A quality, attribute, or distinguishing property of anything.

characteristic color. Of a light, the unique identifying color, e.g., in the U.S. Buoyage System, green lights are used only on black buoys or on horizontally banded black and red buoys with the topmost band black.

characteristic frequency. A frequency which can be easily identified and measured in a given emission.

characteristic phase. Of a light, the sequence and length of light and dark periods by which a navigational light is identified, i.e., whether fixed, flashing, interrupted quick flashing, etc. See also CAUTIONARY CHARACTERISTIC.

characteristics of a light. The sequence and length of light and dark periods and the color or colors by which a navigational light is identified.

character of the bottom. See BOTTOM CHARACTERISTICS.

chart, *n.* A map intended primarily for navigational use. A small chart may be called a **chartlet.** Charts may be classified in a number of ways: **a.** According to the type of navigation for which primarily intended, as **nautical** or **marine chart** for marine navigation, and **aeronautical chart** for air navigation. **b.** According to scale as **small-scale chart** and **large-scale chart.** Ranging, generally, from small-scale charts to charts of larger scale, the charts may be classified as **sailing, general,** coast, approach, and harbor **charts.** **c.** Those designed for use with a particular navigational aid, as **Consol chart, Decca chart.** **d.** Some charts emphasize particular information as a **bottom contour chart** portraying the bottom configuration by depth contours and selected soundings. **e.** Charts may be classified according to projection, as **Mercator chart.** **f.** Charts may be classified according to the characteristics of the projection. A **conformal** or **orthomorphic chart** is one on which angles are correctly represented. **g.** Some charts are classified in accordance with particular uses, as **great circle chart, Omega plotting chart.**

chart amendment patch. See CHARTLET, definition 2.

chart catalog. A list or enumeration of navigational charts, sometimes with index charts indicating the extent of coverage of the various navigational charts.

chart classification by scale. 1. Charts are constructed on many different scales, ranging from about 1:2,500 to 1:14,000,000 (and even smaller for some world charts). Small-scale charts covering large areas are used for planning and for offshore navigation. Charts of larger scale, covering smaller areas, should be used as the vessel approaches pilot waters. Several methods of classifying charts according to scale are in use in various nations. The following classifications of nautical charts are those used by the National Ocean Survey: **Sailing charts** are the smallest scale charts used for planning, fixing position at sea, and for plotting the dead reckoning while proceeding on a long voyage. The scale is generally smaller than 1:600,000. The shoreline and topography are generalized and only offshore soundings, the principal navigational lights, outer buoys, and landmarks visible at considerable distances are shown. **General charts** are intended for coastwise navigation outside of outlying reefs and shoals. The scales range from about 1:150,000 to 1:600,000. **Coast (coastal) charts** are intended for inshore coastwise navigation where the course may lie inside outlying reefs and shoals, for entering or leaving bays and harbors of considerable width, and for navigating large inland waterways. The scales range from about 1:50,000 to 1:150,000. **Harbor charts** are intended for navigation and anchorage in harbors and small waterways. The scale is generally larger than 1:50,000.
2. The classification system used by the Defense Mapping Agency Hydrographic/Topographic Center differs from the system in definition 1 above in that the *sailing charts are incorporated in the general charts classification* (smaller than about 1:150,000); those coast charts especially useful for approaching more confined waters (bays, harbors) are classified as **approach charts.**

chart comparison unit. An optical device used to superimpose the plan position indicator radar picture on a navigational chart.

chart convergence. Convergence of the meridians as shown on a chart.

chart datum. See CHART SOUNDING DATUM.

chart desk. A flat surface on which charts are spread out, usually with stowage space for charts and other navigating equipment below the plotting surface. One without stowage space is called a CHART TABLE.

charted depth. The vertical distance from the chart sounding datum to the bottom.

charthouse. A room, usually adjacent to or on the bridge, where charts and other navigational equipment are stored, and where navigational computations, plots, etc., may be made. Also called CHARTROOM.

chartlet, *n.* 1. A small chart, such as one showing the coverage area of a Loran rate, with the distribution of its lines of position, corrections to be applied to readings, location and identification of transmitters, etc.
2. A corrected reproduction of a small area of a nautical chart which is pasted to the chart for which it is issued. These chartlets are disseminated in *Notice to Mariners* when the corrections are too numerous or of such detail as not to be feasible in printed form. Also called BLOCK, BLOCK CORRECTION, CHART AMENDMENT PATCH.

chart portfolio. A systematic grouping of nautical charts covering a specific geographical area.

chart projection. See MAP PROJECTION.

chart reading. Interpretation of the symbols, lines, abbreviations, and terms appearing on charts. May be called MAP READING when applied to maps generally.

chartroom, *n.* See CHARTHOUSE.

chart scale. The ratio between a distance on a chart and the corresponding distance represented, as *1:80,000* (natural scale), or *30 miles to an inch* (numerical scale). May be called MAP SCALE when applied to any map. See also REPRESENTATIVE FRACTION.

chart sounding datum. The tidal datum to which soundings and drying heights on a chart are referred. It is usually taken to correspond to a low water stage of the tide. Often shortened to CHART DATUM, especially when it is clear that reference is not being made to a horizontal datum.

chart symbol. A character, letter, or similar graphic representation used on a chart to indicate some object, characteristic, etc. May be called MAP SYMBOL when applied to any map.

chart table. A flat surface on which charts are spread out, particularly one without stowage space below the plotting surface. One provided with stowage space is usually called a CHART DESK.

chart work. Plotting on a chart.

Charybdis, *n.* See GALOFARO.

chasm, *n.* A deep breach in the earth's surface; an abyss; a gorge; a deep canyon.

check bearing. An additional bearing, using a charted object other than those used to fix the position, observed and plotted in order to insure that the fix is not the result of a blunder.

cheese antenna. An antenna consisting of a mirror in the shape of part of a parabolic cylinder bounded by two parallel plates normal to the cylinder axis, and of an antenna feed placed on or near the focal point.

chequers, *n. British terminology.* An arrangement of different colors, usually black and white or red and white, arranged in alternate squares in the pattern of a checkerboard. This arrangement of colors is used in some countries as an additional distinguishing characteristic for buoys.

Chile Current. See under PERU CURRENT.

chimney, *n.* A label on a nautical chart which indicates a relatively small, upright structure projecting above a building for the conveyance of smoke.

chip log. A speed measuring device consisting essentially of a weighted wooden quadrant (quarter of a circle) attached to a bridle in such a manner that it will float in a vertical position, and a line with equally spaced knots, usually each 47 feet 3 inches. Speed is measured by casting the quadrant overboard and counting the knots in the line paid out in unit time, usually 28 seconds. This instrument is seldom used in modern practice.

chopped response. See CHOPPING, *n.*

chopping, *adj.* Choppy.

chopping, *n.* The rapid and regular on and off switching of a transponder which is done for recognition purposes.

choppy, *adj.* 1. Of the sea, having short, abrupt, breaking waves dashing against each other; chopping.
2. Of the wind, variable, unstable, changeable; chopping.

chord, *n.* A straight line connecting two points on a curve.

chosen latitude. *British terminology.* See ASSUMED LATITUDE.

chosen longitude. *British terminology.* See ASSUMED LONGITUDE.

chosen position. *British terminology.* See ASSUMED POSITION.

chromatic aberration. See under ABERRATION, definition 2.

chromogram, *n.* The record of a chronograph.

chromosphere, *n.* A thin layer of relatively transparent gases above the photosphere of the sun.

chromospheric eruption. See SOLAR FLARE.

chronograph, *n.* An instrument for producing a graphical record of time as shown by a clock or or other device. The chronograph produces a double record: the first is made by the associated clock and forms a continuous time scale with significant marks indicating periodic beats of the time keepers; the second is made by some external agency, human or mechanical, and records the occurrence of an event or a series of events. The time interval of such occurrences are read on the time scale made by the clock. See also BREAK-CIRCUIT CHRONOMETER.

chronometer, *n.* A timepiece with a nearly constant rate. It is customarily used for comparison of watches and clocks to determine their errors. A chronometer is usually set approximately to Greenwich mean time and not reset as the craft changes time zones. A **hack chronometer** is one which has failed to meet the exacting requirements of a standard chronometer, and is used for timing observations of celestial bodies. Hack chronometers are seldom used in modern practice, any chronometer failing to meet the requirements being rejected. See also CHRONOMETER WATCH.

chronometer correction. The amount that must be added algebraically to the chronometer time to obtain the correct time. Chronometer correction is numerically equal to the chronometer error, but of opposite sign. When chronometer correction is used to correct chronometer time, the direction of application is usually indicated by a plus (+) or minus (–) sign, rather than by fast (F) or slow (S), but in modern practice chronometer error, with its fast (F) or slow (S) designation, rather than chronometer correction, is customarily used.

chronometer error. The amount by which chronometer time differs from the correct time to which it was set, usually Greenwich mean time. It is usually expressed to an accuracy of 1^s and labeled fast (F) or slow (S) as the chronometer time is later or earlier, respectively, than the correct time. CHRONOMETER ERROR and CHRONOMETER CORRECTION are numerically the same, but of opposite sign. See also WATCH ERROR.

chronometer rate. The amount gained or lost by a chronometer in unit time. It is usually expressed in seconds per 24 hours, to an accuracy of $0\overset{s}{.}1$, and labeled **gaining** or **losing**, as appropriate, when it is sometimes called DAILY RATE.

chronometer time. The hour of the day as indicated by a chronometer. Chronometers are generally set approximately to Greenwich mean time. Unless the chronometer has a 24-hour dial, chronometer time is usually expressed on a 12-hour cycle and labeled AM or PM.

chronometer watch. A small chronometer, especially one with an enlarged watch-type movement.

chubasco, *n.* A very violent wind and rain squall attended by thunder and vivid lightning often encountered during the rainy season along the west coast of Central America.

churada, *n.* A severe rain squall in the Mariana Islands during the northeast monsoon. They occur from November to April or May, especially from January through March.

cierzo, *n.* See MISTRAL.

cinders, *n., pl.* See SCORIAE.

circle, *n.* 1. A plane closed curve all points of which are equidistant from a point within, called the *center.* A **great circle** is the intersecion of a sphere and a plane through its center; it is the largest circle that can be drawn on a sphere. A **small circle** is the intersection of a sphere and a plane which does not pass through its center. See also PARALLEL OF ALTITUDE, PARALLEL OF DECLINATION, PARALLEL OF LATITUDE; AZIMUTH CIRCLE, BEARING CIRCLE, DIURNAL CIRCLE, EQUATOR, HOUR CIRCLE, PARASELENIC CIRCLES, POSITION CIRCLE, SPEED CIRCLE, VERTICAL CIRCLE.
2. A section of a plane, bounded by a curve all points of which are equidistant from a point within, called the *center.*

circle of declination. See HOUR CIRCLE.

circle of equal altitude. A circle on the surface of the earth, on every point of which the altitude of a given celestial body is the same at a given instant. The pole of this circle is the geographical position of the body, and the great circle distance from this pole to the circle is the zenith distance of the body. See PARALLEL OF ALTITUDE.

circle of equal declination. See PARALLEL OF DECLINATION.

circle of equivalent probability. A circle with the same center as an error ellipse of specified probability and of such radius that the probability of being located within the circle is the same as the probability of being located within the ellipse. See also CIRCULAR ERROR PROBABLE.

circle of latitude. A great circle of the celestial sphere through the ecliptic poles and along which celestial latitude is measured.

circle of longitude. See PARALLEL OF LATITUDE, definition 2.

circle of perpetual apparition. That circle of the celestial sphere, centered on the polar axis and having a polar distance from the elevated pole approximately equal to the latitude of the observer, within which celestial bodies do not set. That circle within which bodies do not rise is called the CIRCLE OF PERPETUAL OCCULATION.

circle of perpetual occulation. That circle of the celestial sphere, centered on the polar axis and having a polar distance from the depressed pole approximately equal to the latitude of the observer, within which celestial bodies do not rise. That circle within which bodies do not set is called the CIRCLE OF PERPETUAL APPARITION.

circle of position. A circular line of position. The expression is most frequently used with reference to the **circle of equal altitude** surrounding the geographical position of a celestial body. Also called POSITION CIRCLE.

circle of right ascension. See HOUR CIRCLE.

circle of uncertainty. A circle having as its center a position and as its radius the maximum likely error of the position; a circle within which a craft is considered to be located. See also CIRCLE OF EQUAL PROBABILITY, CIRCLE OF POSITION, POSITION CIRCLE.

circle of visibility. That circle surrounding an aid to navigation in which the aid is visible. See also VISUAL RANGE (OF A LIGHT).

circle sheet. A chart with curves enabling a graphical solution of the three-point problem rather than using a three-arm protractor. Also called SEXTANT CHART, STANDARD CIRCLE SHEET.

circuit, *n.* 1. An electrical path between two or more points.
2. Conductors connected together for the purpose of carrying an electric current.
3. A connected assemblage of electrical components, such as resistors, capacitors, and inductors, having desired electrical characteristics.

circular error probable. 1. In a circular normal distribution (the magnitudes of the two one-dimensional input errors are equal and the angle of cut is 90°), the radius of the circle containing 50 percent of the individual measurements being made, or the radius of the circle inside of which there is a 50 percent probability of being located.
2. The radius of a circle inside of which there is a 50 percent probability of being located even though the actual error figure is an ellipse. That is, it is the radius of a circle of **equivalent probability** when the probability is specified as 50 percent. See also ERROR ELLIPSE, CIRCLE OF EQUIVALENT PROBABILITY.
Also called CIRCULAR PROBABLE ERROR.

circular fix. The designation of any one of the erroneous fix positions obtained with a *revolver* or *swinger*.

circularly polarized wave. An electromagnetic wave which can be resolved into two plane polarized waves which are perpendicular to each other and which propagate in the same direction. The amplitudes of the two waves are equal and in time-phase quadrature. The tip of the component of the electric field vector in the plane normal to the direction of propagation describes a circle. See also ELLIPTICALLY POLARIZED WAVE.

circular normal distribution. A two-dimensional error distribution defined by two equal single-axis normal distributions, the axes being perpendicular. The error figure is a circle.

circular probable error. See CIRCULAR ERROR PROBABLE.

circular radiobeacon. See under RADIOBEACON.

circular velocity. The magnitude of the velocity required of a body at a given point in a gravitational field which will result in the body following a circular orbital path about the center of the field. With respect to circular velocities characteristic of the major bodies of the solar system, this is defined for a circular orbit at the surface of the body in question. Circular velocity equals escape velocity divided by $\sqrt{2}$.

circumference, *n.* 1. The boundary line of a circle or other closed plane curve or the outer limits of a sphere or other rounded body.
2. The length of the boundary line of a circle or closed plane curve or of the outer limits of a sphere or other rounded body.
The circumference of a sphere is the circumference of any great circle on the sphere.

circumlunar, *adj.* Around the moon, generally applied to trajectories.

circummeridian altitude. See EX-MERIDIAN ALTITUDE.

circumpolar, *adj.* Revolving about the elevated pole without setting. A celestial body is circumpolar when its polar distance is approximately equal to or less than the latitude of the observer. The actual limit is extended somewhat by the combined effect of refraction, semidiameter, parallax, and the height of the observer's eye above the horizon.

circumscribed halo. A halo formed by the junction of the upper and lower tangent arcs of the halo of 22°.

circumzenithal arc. A brilliant rainbow-colored arc of about a quarter of a circle with its center at the zenith and about 46° above the sun. It is produced by refraction and dispersion of the sun's light striking the top of prismatic ice crystals in the atmosphere. It usually lasts for only a few minutes. See also HALO.

cirriform, *adj.* Like cirrus; more generally, descriptive of clouds composed of small particles, mostly ice crystals, which are fairly widely dispersed, usually resulting in relative transparency and whiteness, and often producing halo phenomena not observed with other cloud forms. Irisation may also be observed. Cirriform clouds are *high clouds.* As a result, when near the horizon, their reflected light traverses a sufficient thickness of air to cause them often to take on a yellow or orange tint even during the midday period. On the other hand, cirriform clouds near the zenith always appear whiter than any other clouds in that part of the sky. With the sun on the horizon, this type of cloud is whitish, while other clouds may be tinted with yellow or orange; when the sun sets a little below the horizon, cirriform clouds become yellow, then pink or red; and when the sun is well below the horizon, they are grey. All species and varieties of cirrus, cirrocumulus, and cirrostratus clouds are cirriform in nature. See also CUMULIFORM, STRATIFORM.

cirro-. A prefix used in cloud classification to indicate the highest of three levels generally recognized. See also ALTO-.

cirrocumulus, *n.* A principal cloud type (cloud genus), appearing as a thin, white patch of cloud without shadows, composed of very small elements in the form of grains, ripples, etc. The elements may be merged or separate, and more or less regularly arranged; they subtend an angle of less than 1° when observed at an angle of more than 30° above the horizon. Holes or rifts often occur in a sheet of cirrocumulus. Cirrocumulus may be composed of highly supercooled water droplets, as well as small ice crystals, or a mixture of both; usually, the droplets are rapidly replaced by ice crystals. Sometimes corona or irisation may be observed. Mamma may appear. Small virga may fall, particularly from cirrocumulus castellanus and floccus. Cirrocumulus, as well as altocumulus, often forms in a layer of cirrus and/or cirrostratus. In middle and high latitudes, cirrocumulus is usually associated in space and time with cirrus and/or cirrostratus; this association occurs less often in low latitudes. Cirrocumulus differs from these other cirriform clouds in that it is not on the whole fibrous, or both silky and smooth; rather, it is rippled and subdivided into little cloudlets. Cirrocumulus is most often confused with altocumulus. It differs primarily in that its constituent elements are very small and are without shadows. The term *cirrocumulus* is not used for incompletely developed small elements such as those on the margin of a sheet of altocumulus, or in separate patches at that level. See also CIRRIFORM, CLOUD CLASSIFICATION.

cirrostratus, *n.* A principal cloud type (cloud genus), appearing as a whitish veil, usually fibrous but sometimes smooth, which may totally cover the sky, and which often produces halo phenomena, either partial or complete. Sometimes a banded aspect may appear, but the intervals between the bands are filled with thinner cloud veil. The edge of a veil of cirrostratus may be straight and clear-cut, but more often it is irregular and fringed with cirrus. Some of the ice crystals which comprise the cloud are large enough to fall, and thereby produce a fibrous aspect. Cirrostratus occasionally may be so thin and transparent as to render it nearly indiscernible, especially through haze or at night. At such times, the existence of a halo may be the only revealing feature. The angle of incidence of illumination upon a cirrostratus layer is an important consideration in evaluating the identifying characteristics. When the sun is high (generally above 50° altitude), cirrostratus never prevents the casting of shadows by terrestrial objects; and a halo might be completely circular. At progressively lower altitudes of the sun, halos become fragmentary and light intensity noticeably decreases. Cirrostratus may be produced by the merging of elements of cirrus; from cirrocumulus; from the thinning of altostratus; or from the anvil of cumulonimbus. Since cirrostratus and altostratus form from each other, it frequently is difficult to delineate between the two. In general, altostratus does not cause halo phenomena, is thicker than cirrostratus, appears to move more rapidly, and has a more even optical thickness. When near the horizon, cirrostratus may be impossible to distinguish from cirrus. See also CIRRIFORM, CLOUD CLASSIFICATION.

cirrus, *n.* A principal cloud type (cloud genus) composed of detached cirriform elements in the form of delicate filaments or white (or mostly white) patches, or of narrow bands. These clouds have a fibrous aspect and/or a silky sheen. Many of the ice crystal particles of cirrus are sufficiently large to acquire an appreciable speed of fall; therefore, the cloud elements have a considerable vertical extent. Wind shear and variations in particle size usually cause these fibrous trails to be slanted or irregularly curved. For this reason, cirrus does not usually tend, as do other clouds, to appear horizontal when near the horizon. Because cirrus elements are too narrow, they do not produce a complete circular halo. Cirrus often evolves from virga of cirrocumulus or altocumulus, or from the upper part of cumulonimbus. Cirrus may also result from the transformation of cirrostratus of uneven optical thickness, the thinner parts of which dissipate. It may be difficult at times to distinguish cirrus from cirrostratus (often impossible when near the horizon); cirrostratus has a much more continuous structure, and if subdivided, its bands are wider. Thick cirrus (usually cirrus spissatus) is differentiated from patches of altostratus by its lesser extension and white color. The term *cirrus* is frequently used for all types of cirriform clouds. See also CIRRIFORM, CLOUD CLASSIFICATION.

cirrus spissatus. See FALSE CIRRUS.

cislunar, *adj.* Of or pertaining to phenomena, projects, or activity in the space between the earth and moon, or between the earth and the moon's orbit.

civil day. A mean solar day beginning at midnight. See also CALENDAR DAY.

civil noon. *United States terminology from 1925 through 1952.* See MEAN NOON.

civil time. *United States terminology from 1925 through 1952.* See MEAN TIME.

civil twilight. The period of incomplete darkness when the upper limb of the sun is below the visible horizon, and the center of the sun is not more than 6° below the celestial horizon.

civil year. A year of the Gregorian calendar of 365 days in common years, or 366 days in leap years.

clamp screw. A screw for holding a moving part in place, as during an observation or reading, particularly such a device used in connection with the tangent screw of a marine sextant.

clamp screw sextant. A marine sextant having a clamp screw for controlling the position of the tangent screw.

clapper, *n.* A heavy pendulum suspended inside a bell which sets the bell in vibration by striking it.

Clarke ellipsoid of 1866. The reference ellipsoid adopted by the U.S. Coast and Geodetic Survey in 1880 for charting North America. This ellipsoid is not to be confused with the Clarke ellipsoid of 1880, which was the estimate of the size and shape of the earth at that time by the English geodesist A. R. Clarke. For the Clarke ellipsoid of 1866, the semimajor axis is 6,378,206.4 meters, the semiminor axis is 6,356,583.8 meters, and the flattening or ellipticity is 1/294.98. Also called CLARKE SPHEROID OF 1866.

Clarke ellipsoid of 1880. The reference ellipsoid of which the semimajor axis is 6,378,249.145 meters, the semiminor axis is 6,356,514.870 meters, and the flattening or ellipticity is 1/293.465. This ellipsoid should not be confused with the CLARKE ELLIPSOID OF 1866. Also called CLARKE SPHEROID OF 1880.

Clarke spheroid of 1866. See CLARKE ELLIPSOID OF 1866.

Clarke spheroid of 1880. See CLARKE ELLPSOID OF 1880.

classification of bearings. An arbitrary system for the grouping of bearings as obtained by a radio direction finder according to their accuracy. Class A bearings are those in which there is a probability of less than 1 in 20 that the bearing error exceeds ± 2°; Class B bearings are those in which there is a probability of less than 1 in 20 that the bearing error exceeds ± 5°; Class C bearings are those in which there is a probability of less than 1 in 20 that the bearing error exceeds ± 10°; and Class D bearings are those bearings in which the bearing error exceeds ± 10°.

classification of radar echoes. When observing a radarscope having a *stabilized* relative motion display, the echoes (targets) may be classified as follows as an aid in rapid predictions of effects of evasive action on the compass direction of relative movement: an **up-the-scope echo** is an echo whose direction of relative movement differs by less than 90° from own ship's heading; a **down-the-scope echo** is an echo whose direction of relative movement differs by more than 90° from own ship's heading; an **across-the-scope (limbo) echo** is an echo whose direction of relative movement differs by 90° from own ship's heading, i.e., the echo's tail is perpendicular to own ship's heading flasher.

clay, *n.* See under MUD.

clean, *adj.* Free from obstructions, unevenness, imperfections, as a *clean anchorage*.

clear, *v., t.* To leave or pass safely, as to *clear port* or *clear a shoal*.

clearance, *n.* The clear space between two objects, such as the nearest approach of a vessel to a navigational light, hazard to navigation, or other vessel.

clear berth. A berth in which a vessel may swing at anchor without striking or fouling another vessel or an obstruction. See also FOUL BERTH.

clearing bearing. *British terminology.* See DANGER BEARING.

clearing line. *British terminology.* A straight line constructed through suitably selected **clearing marks** so as to pass clear of certain dangers to navigation. See also CLEARING BEARING, LEADING LINE.

clearing marks. *British terminology.* Two charted features a straight line through which, called a clearing line, passes clear of certain dangers to navigation. See also LEADING MARKS.

cliff, *n.* Land arising abruptly for a considerable distance above water or surrounding land. See also BLUFF.

climate, *n.* The prevalent or characteristic meteorological conditions of a place or region, in contrast with weather, the state of the atmosphere at any time. A **marine climate** is characteristic of coastal areas, islands, and the oceans, the distinctive features being small annual and daily temperature range and high relative humidity, in contrast with **continental climate**, which is characteristic of the interior of a large land mass, and the distinctive features of which are large annual and daily temperature range and dry air with few clouds.

climatology, *n.* 1. The study of climate.
2. An account of the climate of a particular place or region.

clinometer, *n.* An instrument for indicating the degree of the angle of roll or pitch of a vessel, according to the plane in which it is mounted. In its simplest form it consists of an arm pivoted at one end and pointed at the other. The arm hangs vertically by the action of gravity and as the vessel rolls or pitches, a graduated arc moves under the pointed end.

clock, *n.* A timepiece larger than a watch, too large to be carried conveniently on the person. See also CHRONOMETER.

clockwise, *adv.* In the direction of rotation of the hands of a clock.

close, *v., i.* To move or appear to move together. An order is sometimes given by a flagship for a vessel to *close to - - - - -yards* or *miles*. When a craft moves onto a range, the objects forming the range appear to move closer together or *close*. The opposite is OPEN.

close aboard. Very near.

closed, *adj.* Said of a manned aid to navigation that has been temporarily discontinued for the winter season. See also COMMISSIONED, WITHDRAWN.

closed sea. 1. That part of the ocean enclosed by headlands, within narrow straits, etc.
2. That part of the ocean within the territorial jurisdiction of a country. The opposite is OPEN SEA. See also HIGH SEAS, INLAND SEA.

close pack ice. Pack ice in which the concentration is 7/10 to 8/10 (6/8 to less than 7/8), composed of floes mostly in contact.

closest approach. 1. The event that occurs when two planets or other bodies are nearest to each other as they orbit about the primary body.
2. The place or time of the event in definition 1.
3. The time or place where an orbiting earth satellite is closest to the observer. Also called CLOSEST POINT OF APPROACH.

cloud, *n.* 1. A hydrometeor consisting of a visible aggregate of minute water and/or ice particles in the atmosphere above the earth's surface. Cloud differs from fog only in that the latter is, by definition, in contact with the earth's surface. Clouds form in the free atmosphere as a result of condensation of water vapor in rising currents of air, or by the evaporation of the lowest stratum of fog. For condensation to occur at the point of saturation or a low degree of supersaturation, there must be an abundance of condensation nuclei for water clouds, or ice nuclei for ice-crystal clouds. The size of cloud drops varies from one cloud to another, and within any given cloud there always exists a finite range of sizes. In general, cloud drops range between 1 and 100 microns in diameter, and hence are very much smaller than rain drops. See also CLOUD CLASSIFICATION.
2. Any collection of particulate matter in the atmosphere dense enough to be perceptible to the eye, as a *dust cloud* or *smoke cloud*.

cloud bank. Generally, a fairly well defined mass of clouds observed at a distance; it covers an appreciable portion of the horizon sky, but does not extend overhead.

cloud base. For a given cloud or cloud layer, that lowest level in the atmosphere at which the air contains a perceptible quantity of cloud particles.

cloudburst, *n.* In popular terminology, any sudden and heavy fall of rain, almost always of the shower type. An unofficial criterion sometimes used specifies a rate of fall equal to or greater than 100 millimeters (3.94 inches) per hour. Also called RAIN GUSH, RAIN GUST.

cloud classification. 1. A scheme of distinguishing and grouping clouds according to their appearance and, where possible, to their process of formation. The one in general use, based on a classification system introduced by Luke Howard in 1803, is that adopted by the World Meterological Organization and published in the *International Cloud Atlas* (1956). This classification is based on the determination of (a) *genera*, the main characteristic forms of clouds; (b) *species*, the pecularities in shape and differences in internal structure of clouds; (c) *varieties*, special characteristics of arrangement and transparency of clouds; (d) *supplementary features and accessory clouds*, appended and associated minor clouds forms; and (e) *mother-clouds*, the origin of clouds if formed from other clouds. The ten cloud *genera* are **cirrus, cirrocumulus, cirrostratus, altocumulus, altostratus, nimbostratus, stratocumulus, stratus, cumulus,** and **cumulonimbus.** The fourteen cloud *species* are fibratus, uncinus, spissatus, castellanus, floccus, stratiformis, nebulosus, lenticularis, fractus, humilis, mediocris, congestus, calvus, and capillatus. The nine cloud *varieties* are intortus, vertebratus, undulatus, radiatus, lacunosis, duplicatus, translucidus, perlucidus, and opacus. The nine *supplementary features and accessory clouds* are inclus, mamma, virga, praecipitatio, arcus, tuba, pileus, velum, and pannus. Note that although these are Latin words, it is proper convention to use only the singular endings, e.g., more than one cirrus cloud are, collectively, *cirrus*, not *cirri*.

2. A scheme of classifying clouds according to their usual altitudes. Three classes are distinguished: **high, middle,** and **low. High clouds** include cirrus, cirrocumulus, cirrostratus, occasionally altostratus and the tops of cumulonimbus. The **middle clouds** are altocumulus, altostratus, nimbostratus, and portions of cumulus and cumulonimbus. The **low clouds** are stratocumulus, stratus, most cumulus and cumulonimbus bases, and sometimes nimbostratus.

3. A scheme of classifying clouds according to their particulate composition; namely **water clouds, ice-crystal clouds,** and **mixed clouds.** The first are composed entirely of water droplets (ordinary and/or supercooled), the second entirely of ice crystals, and the third a combination of the first two. Of the cloud genera, only cirrostratus and cirrus are always ice-crystal clouds; cirrocumulus can also be mixed; and only cumulonimbus is always mixed. Altostratus nearly always is mixed, but occasionally can be ice crystal. All the rest of the genera are usually water clouds, occasionally mixed: altocumulus, cumulus, nimbostratus and stratocumulus.

cloud cover. That portion of the sky cover which is attributed to clouds, usually measured in tenths of sky covered.

cloud deck. The upper surface of a cloud.

cloud height. In weather observations, the height of the cloud base above local terrain.

cloud layer. An array of clouds, not necessarily all of the same type, whose bases are at approximately the same level. It may be either continuous or composed of detached elements.

club, *v., i.* To drift in a current with an anchor dragging to provide control. Usually used with *down*, as *club down*.

clutter, *n.* Unwanted radar echoes reflected from heavy rain, snow, waves, etc., which may obscure relatively large areas on the radarscope. See also RAIN CLUTTER, SEA RETURN.

co-. A prefix meaning 90° minus the value with which it is used. Thus, if the latitude is 30°, the colatitude is 90° − 30° = 60°. The cofunction of an angle is the function of its complement.

coalsack, *n.* Any of several dark areas in the Milky Way, especially, when capitalized, a prominent one near the Southern Cross.

coaltitude, *n.* Ninety degrees minus the altitude. The term has significance only when used in connection with altitude measured from the celestial horizon, when it is synonymous with ZENITH DISTANCE.

coast, *n.* The general region of indefinite width that extends from the sea inland to the first major change in terrain features. Sometimes called SEACOAST. See also SEABOARD.

coastal aid. See COASTAL MARK.

coastal area. The land and sea area bordering the shoreline.

coastal boundary. A general term for the boundary defined as the line (or measured from the line or points thereon) used to depict the intersection of the ocean surface and the land at an elevation of a particular datum, excluding one established by treaty or by the U.S. Congress.

coastal chart. See under CHART CLASSIFICATION BY SCALE.

coastal current. An ocean current flowing roughly parallel to a coast, outside the surf zone. See also LONGSHORE CURRENT.

coastal mark. A navigation mark placed on the coast to assist coastal navigation. Particularly used with reference to marks placed on a long straight coastline devoid of many natural landmarks. Also called COASTAL AID.

coastal marsh. An area of salt-tolerant vegetation in brackish and/or saline-water habitats subject to tidal inundation.

coastal plain. Any plain which has its margin on the shore of a large body of water, particularly the sea, and generally represents a strip of recently emerged sea bottom.

coastal refraction. The bending of the wave front of a radio wave travelling parallel to a coastline or crossing it at an acute angle due to the differences in the conducting and reflective properties of the land and water over which the wave travels. This refraction affects the accuracy of medium frequency radio direction finding systems. Also called COAST REFRACTION.

Coast and Geodetic Survey. A former name of the National Ocean Survey. The organization was known as: The Survey of the Coast from its founding in 1807 to 1836, Coast Survey from 1836 to 1878, and Coast and Geodetic Survey from 1878 to 1970. In 1970 it was named National Ocean Survey. From 1965 to 1970, the Coast and Geodetic Survey was a component of the Environmental Science Services Administration (ESSA). The National Ocean Survey is a component of the National Oceanic and Atmospheric Administration (NOAA).

coast chart. See under CHART CLASSIFICATION BY SCALE.

coasting, *n.* Proceeding approximately parallel to a coastline (headland to headland) in sight of land, or sufficiently often in sight of land to fix the ship's position by observations of land features.

coasting lead. A light deep sea lead (30 to 50 pounds), used for sounding in water 20 to 60 fathoms.

coastline, *n.* The configuration made by the meeting of land and sea.

Coast Pilot. See UNITED STATES COAST PILOT.

coast refraction. See COASTAL REFRACTION.

coastwise, *adv. & adj.* By way of the coast; moving along the coast.

coastwise navigation. Navigation in the vicinity of a coast, in contrast with OFF-SHORE NAVIGATION at a distance from a coast. See also COASTING.

coaxial cable. A transmission cable consisting of two concentric conductors insulated from each other.

cobble, *n.* A stone particle between 64 and 256 millimeters (about 2.5 to 10 inches) in diameter. See also STONE.

cocked hat. Error triangle formed by lines of position which do not cross at a common point.

cockeyed bob. A coloquial term in western Australia for a squall, associated with thunder, on the northwest coast in Southern Hemisphere summer.

cocurrent line. A line on a map or chart passing through places having the same current hour.

code beacon. A beacon that flashes a characteristic

signal by which it may be recognized.

codeclination, *n.* Ninety degrees minus the declination. When the declination and latitude are of the same name, codeclination is the same as POLAR DISTANCE measured from the elevated pole.

coding delay. An arbitrary time delay in the transmission of pulse signals. In hyperbolic radionavigation systems of the pulse type, the coding delay is inserted between the transmission of the master and slave (or secondary) signals to prevent zero or small readings, and thus aid in distinguishing between master and slave (or secondary) station signals.

coefficient, *n.* 1. A number indicating the amount of some change under certain specificied conditions, often expressed as a ratio. For example, the coefficient of linear expansion of a substance is the ratio of its change in length to the original length for a unit change of temperature, from a standard.
2. A constant in an algebraic equation.
3. One of several parts which combine to make a whole, as the maximum deviation produced by each of several causes. See also APPROXIMATE COEFFICIENTS.

coefficient A. A component of magnetic compass deviation of constant value with compass heading resulting from mistakes in calculations, compass and pelorous misalignment, and unsymmetrical arrangements of horizontal soft iron. See also APPROXIMATE COEFFICIENTS.

coefficient B. A component of magnetic compass deviation, varying with the sine function of the compass heading, resulting from the fore-and-aft component of the craft's permanent magnetic field and induced magnetism in unsymmetrical vertical iron forward or abaft the compass. See also APPROXIMATE COEFFICIENTS.

coefficient C. A component of magnetic compass deviation, varying with the cosine function of the compass heading, resulting from the athwartship component of the craft's permanent magnetic field and induced magnetism in unsymmetrical vertical iron port or starboard of the compass. See also APPROXIMATE COEFFICIENTS.

coefficient D. A component of magnetic compass deviation, varying with the sine function of twice the compass heading, resulting from induced magnetism in all symmetrical arrangements of the craft's horizontal soft iron. See also APPROXIMATE COEFFICIENTS.

coefficient E. A component of magnetic compass deviation varying with the cosine function of twice the compass heading, resulting from induced magnetism in all unsymmetrical arrangements of the craft's horizontal soft iron. See also APPROXIMATE COEFFICIENTS.

coefficient J. A change in magnetic compass deviation, varying with the cosine function of the compass heading for a given value of J, where J is the change of deviation for a heel of 1° on compass heading 000°. See also APPROXIMATE COEFFICIENTS.

coercive force. The opposing magnetic intensity that must be applied to a magnetic substance to remove the residual magnetism.

COGARD, *n.* Acronym for Coast Guard.

coherence, *n.* The state of there being correlation between the phases of two or more waves, as is necessary in making phase comparisons in radionavigation.

coincidence, *n.* The condition of occupying the same position as regards location, time, etc.

col, *n.* 1. A neck of relative low pressure between two anticyclones.
2. A depression in the summit line of a mountain range. Also called PASS.

colatitude, *n.* Ninety degrees minus the latitude; the angle between the polar axis and the radius vector locating a point.

cold air mass. An air mass that is colder than surrounding air. The expression implies that the air mass is colder than the surface over which it is moving.

cold front. Any non-occluded front, or portion thereof, that moves so that the colder air replaces the warmer air, i.e., the "leading edge" of a relatively cold air mass. While some occluded fronts exhibit this characteristic, they are more properly called COLD OCCLUSIONS.

cold occlusion. See under OCCLUDED FRONT.

cold wave. Unseasonally low temperatures extending over a period of a day or longer, particularly during the cold season of the year.

collada, *n.* A strong wind (35 to 50 miles per hour or stronger) blowing from the north or northwest in the northern part of the Gulf of California and from the northeast in the southern part of the Gulf of California.

collation, *n.* The assembling of pages of publications in sequence.

collimate, *v., t.* 1. To render parallel, as rays of light.
2. To adjust the line of sight of an optical instrument, such as a theodolite, in proper relation to other parts of the instrument.

collimation error. The angle by which the line of sight of an optical instrument differs from its **collimation axis.** Also called ERROR OF COLLIMATION.

collimator, *n.* An optical device which renders rays of light parallel. One of the principal navigational uses of a collimator is to determine the index error of a bubble sextant.

collision bearing. A constant bearing maintained while the distance between two craft is decreasing.

collision course. A course which, if followed, will bring two craft together.

cologarithm, *n.* The logarithm of the reciprocal of a number, or the negative logarithm. The sum of the logarithm and cologarithm of the same number is zero. The addition of a cologarithm accomplishes the same result as the subtraction of a logarithm.

colored light. An aid to navigation exhibiting a light of a color other than white.

color gradients. See HYPSOMETRIC TINTING.

COLREGS, *n.* Acronym for *International Regulations for Preventing Collisions at Sea.*

COLREGS Demarcation Lines. Lines of demarcation delineating those waters upon which mariners must comply with the *International Regulations for Preventing Collisions at Sea,* 1972 (72 COLREGS) and those waters upon which mariners must comply with the *Navigation Rules for Harbors, Rivers, and Inland Waters* (Inland Rules). The waters outside the lines are COLREGS waters. For specifics concerning COLREGS Demarcation Lines see *U.S. Code of Federal Regulations,* Title 33, Navigation and Navigable Waters; Part 82, COLREGS Demarcation Lines.

column, *n.* A vertical line of anything, such as a *column of air,* a *column of figures* in a table, etc.

colure, *n.* A great circle of the celestial sphere through the celestial poles and either the equinoxes or solstices, called, respectively, the **equinoctial colure** or the **solstitial colure.**

coma, *n.* The foggy envelope surrounding the

nucleus of a comet.

combat chart. A special-purpose chart of a land-sea area using the characteristics of a map to represent the land area and the characteristics of a chart to represent the sea area, with special characteristics as to make the chart most useful in naval operations, particularly amphibious operations. Also called MAP CHART.

comber, *n.* A deep water wave whose crest is pushed forward by a strong wind and is much larger than a whitecap. A long spilling breaker. See also ROLLER.

comet, *n.* A luminous member of the solar system composed of a *head* or *coma* at the center of which a *nucleus* of many small solid particles is sometimes situated, and often with a spectacular gaseous *tail* extending a great distance from the head. The orbits of comets are highly elliptical and present no regularity as to their angle to the plane of the ecliptic. Comets are the most spectacular heavenly bodies.

command and control. The facilities equipment, communications, procedures, and personnel essential to a commander for planning, locating, directing, and controlling operations of assigned forces pursuant to the missions assigned. In many cases, a locating or position fixing capability exists in, or as a by-product to, command and control systems.

commissioned, *adj.* Said of an aid to navigation previously reported closed or withdrawn which has been placed in operation.

common establishment. See under ESTABLISHMENT OF THE PORT.

common logarithm. A logarithm to the base 10. Also called BRIGGSIAN LOGARITHM.

common-user, *adj.* Having the characteristics of being planned, operated or used to provide services for both military and civil applications. The availability of a system having such characteristics is not dependent on tactical military operations or use.

common year. A calendar year of 365 days. One of 366 days is called a LEAP YEAR.

communication, *n.* The transfer of intelligence between points. If by wire, radio, or other electromagnetic means, it may be called **telecommunication**; if by radio, **radiocommunication.**

commutation, *n.* A method by means of which the transmissions from a number of stations of a radionavigation system are time shared on the same frequency.

compacted ice edge. A close, clear-cut ice edge compacted by wind or current. It is usually on the windward side of an area of pack ice.

compacting, *adj.* Pieces of sea ice are said to be compacting when they are subjected to a converging motion, which increases ice concentration and/or produces stresses which may result in ice deformations.

compact pack ice. Pack ice in which the concentration is 10/10 (8/8) and no water is visible.

comparing watch. A watch used for timing observations of celestial bodies. Generally its error is determined by comparison with a chronometer, hence its name. A comparing watch normally has a large sweep second hand to facilitate reading time to the nearest second. Sometimes called HACK WATCH. See also SPLIT-SECOND TIMER.

comparison frequency. In the Decca Navigator System, the common frequency to which the incoming signals are converted in order that their phase relationships may be compared.

comparison of simultaneous observations. A reduction process in which a short series of tide or tidal current observations at any place is compared with simultaneous observations at a control station where tidal or tidal current constants have previously been determined from a long series of observations. For tides, it is usually used to adjust constants from a subordinate station to the equivalent of that which would be obtained from a 19-year series.

compass, *adj.* Of or pertaining to a compass or related to compass north.

compass, *n.* An instrument for indicating a horizontal reference direction relative to the earth. Compasses used for navigation are equipped with a graduated compass card for direct indication of any horizontal direction. A **magnetic compass** depends for its directive force upon the attraction of the magnetism of the earth for a magnet free to turn in any horizontal direction. A compass having one or more gyroscopes as the directive element, and tending to indicate true north is called a **gyrocompass.** A compass intended primarily for use in observing bearings is called a **bearing compass**; one intended primarily for measuring amplitudes, an **amplitude compass.** A **directional gyro** is a gyroscopic device used to indicate a selected horizontal direction for a limited time. A **remote-indicating compass** is equipped with one or more indicators, called **compass repeaters,** to repeat at a distance the readings of a master compass. A compass designated as the standard for a vessel is called a **standard compass**; one by which a craft is steered is called a **steering compass.** A **liquid,** wet, or **spirit compass** is a magnetic compass having a bowl completely filled with liquid; a magnetic compass without liquid is called a **dry compass.** An **aperiodic** or **deadbeat** compass, after being deflected, returns by one direct movement to its proper reading, without oscillation. A small compass mounted in a box for convenient use in small water craft is called a **boat compass.** A pelorus is sometimes called a **dumb compass.** A radio direction finder was formerly called a **radio compass.** The stand in which a compass is mounted is called a BINNACLE. Stating in order the names of the points (and sometimes the half and quarter points) of the compass is called BOXING THE COMPASS. The process of determining the deviations of a magnetic compass on different headings is called SWINGING SHIP.

compass adjustment. The process of neutralizing the magnetic effect a craft exerts on a magnetic compass. Permanent magnets and soft iron correctors are arranged about the binnacle so that their effects are about equal and opposite to the magnetic material in the craft, thus reducing the deviations and eliminating the sectors of sluggishness and unsteadiness. See also COMPASS COMPENSATION.

compass adjustment buoy. See SWINGING BUOY.

compass amplitude. Amplitude relative to compass east or west.

compass azimuth. Azimuth relative to compass north.

compass bearing. Bearing relative to compass north.

compass bowl. That part of a compass in which the compass card is mounted. A magnetic compass bowl is usually filled with liquid. The compass bowl turns with the craft while the compass card remains essentially in a north-south direction.

compass card. That part of a compass on which the direction graduations are placed. It is usually in the form of a thin disk or annulus graduated in degrees, clockwise from 0° at the reference direction to 360°, and sometimes also in compass points. A similar card on a pelorus is called a PELORUS CARD.

compass card axis. The line joining 0° and 180° on a compass card. Extended, this line is sometimes called COMPASS MERIDIAN.

compass compensation. The process of neutralizing the effects degaussing currents exert on a marine magnetic compass. The process of neutralizing the magnetic effects the vessel itself exerts on a magnetic compass is properly called COMPASS ADJUSTMENT, but the expression COMPASS COMPENSATION is often used for this process, too.

compass course. Course relative to compass north.

compass direction. Horizontal direction expressed as angular distance from compass north.

compass error. The angle by which a compass direction differs from the true direction; the algebraic sum of variation and deviation; the angle between the true meridian and the axis of the compass card, expressed in degrees east or west to indicate the direction of compass north with respect to true north. See also ACCELERATION ERROR, GAUSSIN ERROR, GYRO ERROR, HEELING ERROR, LUBBER'S LINE ERROR, QUADRANTAL ERROR, RETENTIVE ERROR, SWIRL ERROR.

compasses, *n.* An instrument for drawing circles. In its most common form it consists of two legs joined by a pivot, one leg carrying a pen or pencil and the other leg being pointed. Such an instrument for drawing circles of large diameter, usually consisting of a bar with sliding holders for points, pencils, or pens is called **beam compasses.** If both legs are pointed, the instrument is called DIVIDERS and is used principally for measuring distances or coordinates.

compass heading. Heading relative to compass north.

compass meridian. A line through the north-south points of a magnetic compass. The COMPASS CARD AXIS lies in the compass meridian.

compass north. The direction north as indicated by a magnetic compass; the reference direction for measurement of compass directions.

compass points. The 32 divisions of a compass, at intervals of 11¼°. Each division is further divided into quarter points. The stating in order of the names of the points (and sometimes the half and quarter points) is called BOXING THE COMPASS.

compass prime vertical. The vertical circle through the compass east and west points of the horizon.

compass repeater. That part of a remote-indicating compass system which repeats at a distance the indications of the master compass. One used primarily for observing bearings may be called a **bearing repeater.** Also called REPEATER COMPASS. See also GYRO REPEATER.

compass rose. A circle graduated in degrees, clockwise from 0° at the reference direction to 360°, and sometimes also in compass points. Compass roses are placed at convenient locations on the Mercator chart or plotting sheet to facilitate measurement of direction. See also PROTRACTOR.

compass track. The direction of the track relative to compass north.

compass transmitter. That part of a remote-indicating compass system which sends the direction indications to the repeaters.

compensate, *v., t.* To counteract an error, as that of an instrument; to counterbalance.

compensated loop radio direction finder. A loop antenna radio direction finder for bearing determination, which incorporates a second antenna system designed to reduce the effect of polarization and radiation error.

compensating coils. The coils placed near a magnetic compass to neutralize the effect of the vessel's degaussing system on the compass. See also COMPASS COMPENSATION.

compensating error. An error that tends to offset a companion error and thus obscure or reduce the effect of each.

compensator, *n.* 1. A corrector used in the compensation of a magnetic compass.

2. That part of a radio direction finder which applies all or part of the necessary correction to the direction indication.

complement, *n.* An angle equal to 90° minus a given angle. Thus, 50° is the complement of 40° and the two are said to be *complementary.* See also EXPLEMENT, SUPPLEMENT.

complementary angles. Two angles whose sum is 90°.

component, *n.* 1. See CONSTITUENT.

2. That part of a tidal force of tidal current velocity which, by resolution into orthogonal vectors, is found to act in a specified direction.

3. One of the parts into which a vector quantity can be divided. For example, the earth's magnetic force at any point can be divided into *horizontal* and *vertical components.*

composite, *adj.* Composed of two or more separate parts.

composite group flashing light. A light similar to a group flashing light except that successive groups in a period have different numbers of flashes. A light entry may be given as:

Fl. W. (2 + 1)
period 15^s
fl. 0.2^s, ec. 2.8^s
fl. 0.2^s, ec. 5.8^s
fl. 0.2^s, ec. 5.8^s

Within each 15^s period, there is first a white flash of $0\overset{s}{.}2$ duration followed by a $2\overset{s}{.}8$ eclipse, then a second white flash of $0\overset{s}{.}2$ duration to complete the first group of flashes, and then after a $5\overset{s}{.}8$ eclipse a single white flash of $0\overset{s}{.}2$ duration followed by a $5\overset{s}{.}8$ eclipse to complete the second group. Thus, the first group consists of two flashes; the second group consists of a single flash. This is indicated by the (2 + 1) notation.

composite group occulting light. A group occulting light in which the occultations are combined in successive groups of different numbers of occultations. A light list entry may be given as:

Gp. Occ. W.R.G. (1+3)
period 12^s
lt. 3^s, ec. 1^s
lt. 3^s, ec. 1^s G. 58°-84°, W.-88°, R-58°
lt. 1^s, ec. 1^s
lt. 1^s, ec. 1^s

Within each 12^s period, there is first light for 3^s, after which the light is eclipsed for 1^s, then there is light for 3^s followed by a 1^s eclipse, and then there is light for 1^s followed by a 1^s eclipse, and finally there is light for 1^s followed by a 1^s eclipse. Thus, the first group consists of a single occultation; the second group consists of three occultations. This is indicated by the (1+3) notation.

composite sailing. A modification of great-circle sailing used when it is desired to limit the highest latitude. The composite track consists of a great circle from the point of departure and tangent to the limiting parallel. a course line along the parallel, and a great circle tangent to the limiting parallel and through the destination. Composite sailing applies only when the vertex lies between the point of departure and destination.

composite track. A modified great-circle track consisting of an initial great-circle track from the point of departure with its vertex on a limiting parallel of latitude, a parallel-sailing track from this vertex along the limiting parallel to the vertex of a final great-circle track passing through the destination.

composition of vectors. See VECTOR ADDITION.

compound harmonic motion. The projection of two or more uniform circular motions on a diameter of the circle of such motion. The projection of a simple uniform circular motion is called SIMPLE HARMONIC MOTION.

compound tide. A tidal constituent with a speed equal to the sum or difference of the speeds of two or more elementary constituents. Compound tides are usually the result of shallow water conditions.

compressed-air horn. See DIAPHRAGM HORN.

compression, *n.* See FLATTENING.

computed altitude. 1. Tabulated altitude interpolated for increments of latitude, declination, or hour angle. If no interpolation is required, the tabulated altitude and computed altitude are identical.

2. Altitude determined by computation, table, mechanical computer, or graphics, particularly such an altitude of the center of a celestial body measured as an arc on a vertical circle of the celestial sphere from the celestial horizon. Also called CALCULATED ALTITUDE.

computed azimuth. Azimuth determined by computation, table, mechanical device, or graphics for a given place and time. See also TABULATED AZIMUTH.

computed azimuth angle. Azimuth angle determined by computation, table, mechanical device, or graphics for a given place and time. See also TABULATED AZIMUTH ANGLE.

computed point. In the construction of the line of position by the Marcq St. Hilaire method, the foot of the perpendicular from the assumed position to the line of position. Also called SUMNER POINT.

concave, *adj.* Curving and hollow, as the inside of a circle or sphere. The opposite is CONVEX.

concave, *n.* A concave line or surface.

concentration, *n.* The ratio expressed in tenths or oktas of the sea surface actually covered by ice to the total area of sea surface, both ice-covered and ice-free, at a specific location or over a defined area.

concentration boundary. A line approximating the transition between two areas of pack ice with distinctly different concentrations.

concentric, *adj.* Having the same center. The opposite is ECCENTRIC.

condensation, *n.* The physical process by which a vapor becomes a liquid or solid. The opposite is EVAPORATION. In meteorology, the term condensation is applied only to the transformation from vapor to liquid; any process in which a solid forms directly from its vapor is called SUBLIMATION, as is the reverse process. Also in meteorology, condensation is considered almost exclusively with reference to water vapor which changes to dew, fog, or cloud. Condensation in the atmosphere is brought about by either of two processes: cooling of air to its dew point or addition of enough water vapor to bring the mixture to the point of saturation (relative humidity is raised to 100 percent).

conduction, *n.* Transmission from one point to another, as of electricity, sound, etc., along a conductor, or transference of heat from particle to particle through a substance, such as air, without any obvious motion. Heat is also transferred by CONVECTION and RADIATION.

conductivity, *n.* The ability to transmit, as electricity, heat, sound, etc. Conductivity is the opposite of RESISTIVITY.

conductor, *n.* A substance which transmits electricity, heat, sound, etc.

cone, *n.* 1. A solid having a plane base bounded by a closed curve and a surface formed by lines from every point on the circumference of the base to a common point or APEX.

2. A surface generated by a straight line of indefinite length, one point of which is fixed and another point of which follows a fixed curve. Also called a CONICAL SURFACE.

configuration, *n.* 1. Relative position or disposition of various parts, or the figure or pattern so formed.

2. A geometric figure, usually consisting principally of points and connecting lines.

conformal, *adj.* Having correct angular representation.

conformal chart. A chart on a conformal map projection. Also called ORTHOMORPHIC CHART.

conformal map projection. A map projection in which all angles around any point are correctly represented. In such a projection the scale is the same in all directions about any point. Very small shapes are correctly represented, resulting in an ORTHOMORPHIC MAP PROJECTION. Hence, the terms of CONFORMAL and ORTHOMORPHIC are used synonymously since neither characteristic can exist independently of the other. Also called ORTHOMORPHIC MAP PROJECTION.

confused sea. A highly disturbed water surface without a single, well-defined direction of wave travel which may occur when waves from different directions meet following a sudden shift in wind direction.

confusion region. The region surrounding a target within which the radar echo from that target cannot be resolved from echoes of other objects.

conic, *adj.* Of or pertaining to a cone.

conic, *n.* See CONIC SECTION.

conical buoy. See under NUN BUOY.

conical surface. See CONE, definition 2.

conic chart. A chart on a conic projection.

conic chart with two standard parallels. A chart on the conic projection with two standard parallels. Also called SECANT CONIC CHART. See also LAMBERT CONFORMAL CHART.

conic map projection. A map projection in which the surface of a sphere or spheroid, such as the earth, is conceived as projected onto a tangent or secant cone which is then developed into a plane. In a **simple conic map projection** the cone is tangent to the sphere or spheroid, in a **conic map projection with two standard parallels** the cone intersects the sphere or spheroid along two standard parallels, and in a **polyconic map projection** a series of cones are tangent to the sphere or spheroid. See also LAMBERT CONFORMAL CONIC MAP PROJECTION, MODIFIED LAMBERT CONFORMAL MAP PROJECTION.

conic map projection with two standard parallels. A conic map projection in which the surface of a sphere or spheroid, such as the earth, is conceived as developed on a cone which intersects the sphere or spheroid along two standard parallels, the cone being spread out to form a plane. The Lambert conformal map projection is an example. Also called SECANT CONIC MAP PROJECTION.

conic section. Any plane curve which is the locus of a point which moves so that the ratio of its distance from a fixed point to its distance from a fixed line is constant. The ratio is called the **eccentricity**; the fixed point is the **focus**; the fixed line is the **directrix**. When the eccentricity is equal to unity, the conic section is a **parabola**; when less than unity, an ellipse; and when greater than unity, a hyperbola. They are so called because they are formed by the intersection of a plane and a right circular cone.

conjunction, *n.* The situation of two celestial bodies having either the same celestial longitude or the same sidereal hour angle. A planet is at **superior conjunction** if the sun is between it and the earth; at **inferior conjunction** if it is between the sun and the earth. The situation of two celestial bodies having either celestial longitudes or sidereal hour angles differing by 180° is called OPPOSITION.

conn, *v., t.* To direct the steering of a vessel. The person giving orders to the helmsman (not just relaying orders) is said to *have the conn* or to be *conning the ship*.

Consol, *n.* A long range, azimuthal radionavigation system of low accuracy operated primarily for air navigation. Although not sufficiently accurate for coastal navigation or making landfall, the system can be useful to the marine navigator as an aid to ocean navigation. A Consol station consists basically of a medium frequency (MF) radio transmitter with three antennas in line, equally spaced apart at a distance of the order of three times the wavelength of the transmitted frequency. The three antennas are fed with signals in such a manner that radial patterns of alternate dot and dash sectors are formed, separated by the **equisignal**, the whole pattern rotating through one sector width within a transmission cycle. An observer will therefore hear the equisignal between the dot and dash signals once per transmission cycle. The counting of a series of dots and dashes permits determination of the observer's bearing *from* the Consol station if the sector in which the observer is located is known. This sector has to be identified either by direction finding methods or some other form of navigation. The Consol station operates periodically as a radiobeacon to facilitate avoiding ambiguity as to sector. Since the radial sectors are not formed close to the transmitting site, there is a minimum range limitation within which the system cannot be used. This is usually taken to be 25 to 30 nautical miles from the site. There is also a 30° sector on either side of the baseline extension which is not usable because of pattern distortion. In practice the dot-dash count is converted to a line of position by referring to a table or special chart. The system is described as azimuthal even though it is basically a hyperbolic system. (A hyperbolic system can be considered directional (azimuthal) beyond a distance of a few miles from the station if the baseline is very short.) Sometimes Consol is classified as a radiobeacon because of the frequency of operation and being azimuthal. A modified form of Consol called CONSOLAN was developed in the United States. In this system only two antennas are used. In the U.S.S.R. a further modification of Consol is in use. This system, called BPM5, uses five antennas in the form of a cross to obtain narrower dot and dash sectors. The main advantage of Consol is that the signal can be received on a standard communications receiver.

Consolan, *n.* See under CONSOL.

Consol chart. A chart overprinted with a pattern of radial lines originating at the Consol station, and used for converting a particular dot and dash count of signals received from the Consol station to a line of position. See also CONSOL.

console, *n.* The housing of the main operating unit of electronic equipment, in which indicators and general controls are located. The term is often limited to large housings resting directly on the deck, as contrasted with smaller cabinets such as might be placed on a table.

consolidated pack ice. Pack ice in which the concentration is 10/10 (8/8) and the floes are frozen together.

consolidated ridge. A ridge (a line or wall of ice forced up by pressure) in which the base has frozen together.

Consol station. A short baseline directional antenna system consisting of three LF/MF vertical antennas in a line on the ground, evenly spaced at a distance of about three times the length of the transmitted CW wave. The radiation pattern of each station consists of alternate sectors of dot and dash signals. Dots are transmitted by one of the end antennas and the dashes by the other one. The center antenna transmits unmodulated CW. The phases at the end antennas, 180° apart with respect to one another, are slowly changed with respect to the center one in such manner that the radiation pattern, consisting of several lobes approximately 15° wide, rotates around the center antenna, clockwise on one side of the baseline and counterclockwise on the other side. During the keying cycle of 30 or 60 seconds, the radiation pattern rotates through one sector. The alternate sectors or lobes containing dot and dash signals merge into a monotone called the **equisignal** at the sector boundaries. Actually lines or position determined from such stations are hyperbolic but beyond 25 miles from the center antenna they may be considered great circles with negligible error. Thus, the antenna system is directional.

constant, *n.* A fixed quantity; one that does not change.

constant bearing, decreasing range. See STEADY BEARING.

constant deviation. Deviation which is the same on any heading, as that which may result from certain arrangements of asymmetrical horizontal soft iron.

constant error. A systematic error of unchanging magnitude and sign throughout a given series of observations. Also called BIAS ERROR.

constant of aberration. The measure of the maximum angle between the true direction and the apparent direction of a celestial body as observed from earth due to aberration. It has a value of 20.496 seconds of arc. The *aberration angle* depends upon the ratio of the velocity of the earth in its orbit and the velocity of light in addition to the angle between the direction of the light and the direction of motion of the observing telescope. The maximum value is obtained when the celestial body is at the pole of the ecliptic. Also called ABERRATION CONSTANT.

constant of the cone. The chart convergence factor for a conic projection. See also CONVERGENCE FACTOR.

constant-pressure chart. The synoptic chart for any

constant-pressure surface, usually containing plotted data and analyses of the distribution of, e.g., height of the surface, wind, temperature, and humidity. Constant-pressure charts are most commonly known by their pressure value; for example the *1000-millibar chart*. Also called ISOBARIC CHART.

constant-pressure surface. In meteorology, an imaginary surface along which the atmospheric pressure is everywhere equal at a given instant. Also called ISOBARIC SURFACE.

constellation, *n.* A group of stars which appear close together, regardless of actual distances, is propularly called a constellation, particularly if the group forms a striking configuration. Among astronomers a constellation is now considered a region of the sky having precise boundaries so arranged that all of the sky is covered, without overlap. The ancient Greeks recognized 48 constellations covering only certain groups of stars. Modern astronomers recognized 88 constellations.

constituent, *n.* One of the harmonic elements in a mathematical expression for the tide-producing force and in corresponding formulas for the tide or tidal current. Each constituent represents a periodic change or variation in the relative positions of the earth, moon, and sun. Also called HARMONIC CONSTITUENT, TIDAL CONSTITUENT, COMPONENT.

constituent day. The duration of one rotation of the earth on its axis, with respect to an astre fictif, a fictitious star representing one of the periodic elements in tidal forces. It approximates the length of a lunar or solar day. The expression is not applicable to a long-period constituent.

constituent hour. One twenty-fourth part of a constituent day.

contact, *n.* Any echo detected on the radarscope and not evaluated as clutter or as a false echo. Although the term *contact* is often used interchangeably with *target*, the latter term specifically indicates that the echo is from an object about which information is being sought.

conterminous U.S. Forty-eight states and the District of Columbia, i.e., the United States before January 3, 1959 (excluding Alaska and Hawaii).

contiguous zone. The band of water outside or beyond the territorial sea in which the coastal nation may exercise custom control and enforce public health regulations, etc. It is measured from the same base line as the territorial sea, and may extend no more than 12 miles seaward from it.

continent, *n.* A vast expanse of continuous land constituting one of the major divisions of the land surface of the earth.

continental borderland. A region adjacent to a continent, normally occupied by or bordering a shelf, that is highly irregular with depths well in excess of those typical of a shelf. See also INSULAR BORDERLAND.

continental climate. The type of climate characteristic of the interior of a large land mass, the distinctive features of which are large annual and daily temperature range and dry air with few clouds, in contrast with MARINE CLIMATE, which is characteristic of coastal areas, islands, and the oceans, and the distinctive features of which are small annual and daily temperature range and high relative humidity.

continental polar air. See under AIR-MASS CLASSIFICATION.

continental rise. A gentle slope rising from oceanic depths toward the foot of a continental slope.

continental shelf. A zone adjacent to a continent that extends from the low water line to a depth at which there is usually a marked increase of slope towards oceanic depths. See also INSULAR SHELF.

continental tropical air. See under AIR-MASS CLASSIFICATION.

continental United States. United States territory, including the adjacent territorial waters, located within the North American continent between Canada and Mexico. See also CONTERMINOUS U.S.

continuous carrier radiobeacon. A radiobeacon the carrier wave of which is unbroken but which is modulated with the identification signal. The continuous carrier wave signal is not audible to the operator of an aural null direction finder not having a beat frequency oscillator. The use of the continuous carrier wave improves the performance of automatic direction finders. The marine radiobeacons on the Atlantic and Pacific coasts of the U.S. are of this type. See also DUAL CARRIER RADIOBEACON.

continuous quick light. A quick light in which a flash is regularly repeated. See also GROUP QUICK LIGHT, INTERRUPTED QUICK LIGHT.

continuous system. A classification of a navigation system with respect to availability. A continuous system affords the user, within the effective range of the system, the capability to determine his position at any time. Omega and Loran-C are examples.

continuous ultra quick light. An ultra quick light in which a flash is regularly repeated. See also INTERRUPTED ULTRA QUICK LIGHT.

continuous very quick light. A very quick light in which a flash is regularly repeated. See also GROUP VERY QUICK LIGHT, INTERRUPTED VERY QUICK LIGHT.

continuous wave. 1. Electromagnetic radiation of a constant amplitude and frequency.
2. Radio waves, the successive sinusoidal oscillations of which are identical under steady - state conditions.

contour, *n.* The imaginary line on the ground, all points of which are at the same elevation above or below a specified datum.

contour interval. The difference in elevation between two adjacent contours.

contour line. A line connecting points of equal elevation or equal depth. One connecting points of equal depth is usually called a **depth contour,** but if depth is expressed in fathoms, it may be called a **fathom curve** or **fathom line.** See also FORM LINES.

contour map. A topographic map showing relief by means of contour lines.

contrary name. A name opposite or contrary to that possessed by something else, as declination has a name *contrary* to that of latitude if one is north and the other south. If both are north or both are south, they are said to be of SAME NAME.

contrastes, *n., pl.* Winds a short distance apart blowing from opposite quadrants, frequent in the spring and fall in the western Mediterranean.

contrast threshold. The minimum contrast at the eye of a given observer at which an object can be detected. The contrast threshold is a property of the eye of the individual observer. For most normal observers there is good agreement between values for different individuals under ordinary light conditions. The average value has been taken to be 0.02 in most studies of visibility, and this value is correct for laboratory observation under daylight conditions. For practical observation in the field, a value of 0.05 is more appropriate and has been adopted internationally. See also METEOROLOGICAL VISIBILITY, VISUAL RANGE.

control, *n.* 1. The coordinated and correlated dimensional data used in geodesy and cartography to determine the positions and elevations of points on the earth's surface or on a cartographic representation of that surface.
2. A collective term for a system of marks or objects on the earth or on a map or a photograph, whose positions or elevations, or both, have been or will be determined.

control current station. A current station at which continuous velocity observations have been made over a minimum period of 29 days. Its purpose is to provide data for computing accepted values of the harmonic and nonharmonic constants essential to tidal current predictions and circulatory studies. The data series from this station serves as the control for the reduction of relatively short series from subordinate current stations through the method of comparison of simultaneous observations. See also CURRENT STATION, SUBORDINATE CURRENT STATION.

controlled air space. An airspace of defined dimensions within which air traffic control service is provided.

controlling depth. 1. The least depth in the approach or channel to an area, such as a port or anchorage, governing the maximum draft of vessels that can enter.
2. The least depth within the limits of a channel; it restricts the safe use of the channel to drafts of less than that depth. The **centerline controlling depth** of a channel applies only to the channel centerline; lesser depths may exist in the remainder of the channel. The **mid-channel controlling depth** of a channel is the controlling depth of only the middle half of the channel. See also FEDERAL PROJECT DEPTH.

control station. See PRIMARY CONTROL TIDE STATION, SECONDARY CONTROL TIDE STATION, CONTROL CURRENT STATION.

convection, *n.* Circulation in a fluid of nonuniform temperature, due to the differences in density and the action of gravity. In the atmosphere, convection takes place on a large scale. It is essential to the formation of many clouds, especially those of the cumulus type. Heat is transferred by CONVECTION and also by ADVECTION, CONDUCTION, and RADIATION.

convention, *n.* See as GEOGRAPHIC SIGN CONVENTIONS.

conventional direction of buoyage. 1. The general direction taken by the mariner when approaching a harbor, river, estuary or other waterway from seaward, or
2. The direction determined by the proper authority in consultation, where appropriate, with neighboring countries. In principle it should follow a clockwise direction around land masses.

converge, *v., i.* To tend to come together.

converged beam. See under FAN BEAM.

convergence constant. The angle at a given latitude between meridians 1° apart. Sometimes loosely called CONVERGENCY. On a map or chart having a convergence constant of 1.0, the true direction of a straight line on the map or chart changes 1° for each 1° of longitude that the line crosses; the true direction of a straight line on a map or chart having a convergence constant of 0.785 changes 0°.785 for each 1° of longitude the line crosses. Also called CONVERGENCE FACTOR. See also CONVERGENCE OF MERIDIANS.

convergence factor. See CONVERGENCE CONSTANT.

convergence of meridians. The angular drawing together of the geographic meridians in passing from the Equator to the poles. At the Equator, all meridians are mutually parallel; passing from the Equator, they converge until they meet at the poles, intersecting at angles that are equal to their differences of longitude. See also CONVERGENCE CONSTANT.

convergency, *n.* See under CONVERGENCE CONSTANT.

conversion, *n.* Determination of the rhumb line direction of one point from another when the initial great circle direction is known, or vice versa. The difference between the two directions is the **conversion angle.** This is used in connection with radio bearings, Consol, and in great circle sailing.

conversion angle. The angle between the rhumb line and the great circle between two points. Also called ARC TO CHORD CORRECTION. See also HALF-CONVERGENCY.

conversion scale. A scale for the conversion of units of one measurement to equivalent units of another measurement. See also NOMOGRAM.

conversion table. A table for the conversion of units of one measurement to equivalent units of another measurement. See also NOMOGRAM.

convex, *adj.* Curving away from, as the outside of a circle or sphere. The opposite is CONCAVE.

convex, *n.* A convex line or surface.

coordinate, *n.* One of a set of magnitudes defining a point in space. If the point is known to be on a given line, only one coordinate is needed; if on a surface, two are required; if in space, three. **Cartesian coordinates** define a point relative to two intersecting lines, called AXES. If the axes are perpendicular, the coordinates are **rectangular;** if not perpendicular, they are **oblique coordinates.** A three-dimensional system of Cartesian coordinates is called **space coordinates. Polar coordinates** define a point by its distance and direction from a fixed point called the POLE. Direction is given as the angle between a reference radius vector and a radius vector to the point. If three dimensions are involved, two angles are used to locate the radius vector. **Space-polar coordinates** define a point on the surface of a sphere by (1) its distance from a fixed point at the center, called the POLE; (2) the COLATITUDE or angle between the POLAR AXIS (a reference line through the pole) and the RADIUS VECTOR (a straight line connecting the pole and the point); and (3) the LONGITUDE or angle between a reference plane through the polar axis and a plane through the radius vector and the polar axis. **Spherical coordinates** define a point on a sphere or spheroid by its angular distances from a primary great circle and from a reference secondary great circle. **Geographical** or **terrestrial coordinates** define a point on the surface of the earth. **Celestial coodinates** define a point on the celestial sphere. The **horizon, celestial equator,** and the **ecliptic systems** of celestial coordinates are based on the celestial horizon, celestial equator, and the ecliptic, respectively, as the primary great circle.

coordinate conversion. Changing the coordinate values from one system to those of another.

Coordinated Universal Time (UTC). The time scale that is available from most broadcast time signals. It differs from International Atomic Time (TAI) by an integral number of seconds. UTC is maintained within 1 second of UT1 by the introduction of 1-second steps (leap seconds) when necessary, normally at the end of December. DUT1, an approximation to the difference UT1 minus UTC, is transmitted in code on broadcast time signals.

coordinate paper. Paper ruled with lines to aid in the plotting of coordinates. In its most common form, it has two sets of parallel lines, usually at right angles to each other, when it is also called CROSS-SECTION PAPER. A type ruled with two sets of mutually-perpendicular, parallel lines spaced according to the logarithms of consecutive numbers is called **logarithmic coordinate paper** or **semilogarithmic coordinate paper** as both or only one set of lines is spaced logarithmically. A type ruled with concentric circles and radial lines from the common center is called **polar coordinate paper.** Also called GRAPH PAPER.

coplanar, *adj.* Lying in the same plane.

coral, *n.* The hard skeleton of certain tiny sea animals; or the stony, solidified mass of a number of such skeletons. A ridge of such material sometimes consistutes a serious menace to navigation.

coral head. A massive mushroom or pillar shaped coral growth.

coral reef. A reef made up of coral, fragments of coral and other organisms, and the limestone resulting from their consolidation. Coral may constitute less than half of the reef material.

corange line. A line passing through places of equal tidal range.

cordillera, *n.* On the sea floor, an entire mountain system including all the subordinate ranges, interior plateaus, and basins.

cordonazo, *n.* The "Lash of St. Francis." Name applied locally to southerly hurricane winds along the west coast of Mexico. The cordonazo is associated with tropical cyclones in the southeastern North Pacific Ocean. These storms may occur from May to November, but ordinarily affect the coastal areas most severely near or after the Feast of St. Francis, October 4.

Coriolis acceleration. An acceleration of a body in motion in a relative (moving) coordinate system. The total acceleration of the body, as measured in an inertial coordinate system, may be expressed as the sum of the acceleration within the relative system, the acceleration of the relative system itself, and the Coriolis acceleration. In the case of the earth, moving with angular velocity Ω, a body moving relative to the earth with velocity V has the Coriolis acceleration $2\Omega \times V$. If Newton's laws are to be applied in the relative system, the Coriolis acceleration and the acceleration of the relative system must be treated as forces. See also CORIOLIS FORCE.

Coriolis correction. 1. A correction applied to an assumed position, celestial line of position, celestial fix, or to a computed or observed altitude to allow for Coriolis acceleration.

2. In inertial navigation equipment, an acceleration correction which must be applied to measurements of acceleration with respect to a coordinate system in translation to compensate for the effect of any angular motion of the coordinate system with respect to inertial space.

Coriolis force. An inertial force acting on a body in motion, due to rotation of the earth, causing deflection to the right in the Northern Hemisphere and to the left in the Southern Hemisphere. It affects air (wind), water (current), etc. and introduces an error in bubble sextant observations made from a moving craft due to the liquid in the bubble being deflected, the effect increasing with higher latitude and greater speed of the craft.

corner reflector. A radar reflector consisting of three mutually perpendicular flat conducting surfaces designed to return incident electromagnetic radiations towards their sources. The reflector is used to render objects more conspicuous to radar observations. Since maximum effectiveness is obtained when the incident beam coincides with the axis of symmetry of the reflector, clusters of reflectors are sometimes used to insure that the object will be a good reflector in all directions. See also RADAR REFLECTOR. Also called TRIHEDRAL REFLECTOR.

coromell, *n.* A night land breeze prevailing from November to May at La Paz, near the southern extremity of the Gulf of California.

corona, *n.* 1. The luminous envelope surrounding the sun but visible only during a total eclipse.

2. A luminous discharge due to ionization of the air surrounding an electric conductor.

3. A set of one or more rainbow-colored rings of small radii surrounding the sun, moon, or other source of light covered by a thin cloud veil. It is caused by diffraction of the light by tiny droplets in the atmosphere, and hence the colors are in the reverse order to those of a HALO caused by refraction.

4. A circle of light occasionally formed by the apparent convergency of the beams of the aurora.

corona discharge. See ST. ELMO'S FIRE.

corposant, *n.* See ST. ELMO'S FIRE.

corrasion, *n.* The wearing away of the earth's surface by the abrasive action of material transported by glacier, water, or air; a process of erosion.

corrected compass course. Compass course with deviation applied; magnetic course.

corrected compass heading. Compass heading with deviation applied; magnetic heading.

corrected current. A relatively short series of current observations from a subordinate station to which a factor is applied to adjust the current to a more representative value, based on a relatively long series from a nearby control station. See also CURRENT, definition 1; TOTAL CURRENT.

corrected establishment. See under ESTABLISHMENT OF THE PORT.

corrected sextant altitude. Sextant altitude corrected for index error, height of eye, parallax, refraction, etc. Also called OBSERVED ALTITUDE, TRUE ALTITUDE.

correcting, *n.* The process of applying corrections, particularly the process of converting compass to magnetic direction, or compass, magnetic, or gyro to true direction. The opposite is UNCORRECTING.

correction, *n.* That which is added to or subtracted from a reading, as of an instrument, to eliminate the effect of an error, or to reduce an observation to an arbitrary standard.

correction of soundings. The adjustment of soundings for any departure from true depth because of the method of sounding or any fault in the measuring apparatus. See also REDUCTION OF SOUNDINGS.

corrector, *n.* A magnet, piece of soft iron, or device used in the adjustment of a magnetic compass. See also FLINDERS BAR, HEELING MAGNET, QUADRANTAL CORRECTORS.

corrosion, *n.* The wearing or wasting away by

chemical action, as the rusting of metal. A distinction is usually made between CORROSION and EROSION, the latter referring to the wearing away of the earth's surface primarily by non-chemical action. See also CORRASION.

cosecant, *n.* The ratio of the hypotenuse of a plane right triangle to the side opposite one of the acute angles of the triangle, equal to 1/sin. The expression NATURAL COSECANT is sometimes used to distinguish the cosecant from its logarithm (called LOGARITHMIC COSECANT).

cosine, *n.* The ratio of the side adjacent to an acute angle of a plane right triangle to the hypotenuse. The expression NATURAL COSINE is sometimes used to distinguish the cosine from its logarithm (called LOGARITHMIC COSINE).

cotangent, *n.* The ratio of the shorter side adjacent to an acute angle of a plane right triangle to the side opposite the same angle, equal to 1/tan. The expression NATURAL COTANGENT is sometimes used to distinguish the cotangent from its logarithm (called LOGARITHMIC COTANGENT).

cotidal, *adj.* Having tides occurring at the same time.

cotidal chart. A chart showing cotidal lines.

cotidal hour. The average interval between the moon's transit over the meridian of Greenwich and the time of the following high water at any place, expressed in either mean solar or lunar time units. When expressed in solar time, it is the same as the Greenwich high water interval. When expressed in lunar time, it is equal to the Greenwich high water interval multiplied by the factor 0.966.

cotidal line. A line on a map or chart passing through places having the same cotidal hour.

coulomb, *n.* A *derived unit* of quantity of electricity in the International System of Units; it is the quantity of electricity carried in 1 second by a current of 1 ampere.

counterclockwise, *adv.* In a direction of rotation opposite to that of the hands of a clock.

countercurrent, *n.* A current usually setting in a direction opposite to that of a main current.

counterglow, *n.* See GEGENSCHEIN.

countertrades, *n., pl.* See ANTITRADES.

coupler, *n.* See as ANTENNA COUPLER.

course, *n.* The horizontal direction in which a vessel is steered or intended to be steered, expressed as angular distance from north, usually from 000° at north, clockwise through 360°. Strictly, the term applies to direction *through the water*, not the direction intended to be made good over the ground. The course is often designated as **true, magnetic, compass,** or **grid** as the reference direction is true, magnetic, compass, or grid north, respectively. TRACK MADE GOOD is the single resultant direction from the point of departure to point of arrival at any given time. The use of this term to indicate a single resultant direction is preferred to the use of the misnomer *course made good.* A **course line** is a line, as drawn on a chart, extending in the direction of a course. See also COURSE ANGLE, COURSE OF ADVANCE, COURSE OVER GROUND, HEADING, TRACK.

course angle. Course measured from 0° at the reference direction clockwise or counterclockwise through 90° or 180°. It is labeled with the reference direction as a prefix and the direction of measurement from the reference direction as a suffix. Thus, course angle S 21° E is 21° east of south, or course 159°.

course beacon. A directional radiobeacon which gives an "on course" signal in the receiver of a vessel which is on, or in close proximity to, the prescribed course line and "off course" signals in sectors adjacent to, and on either side of, this line. Due to the signal composition, the "off course" signals in these two sectors are different so that the vessel can be maneuvered to the prescribed bearing from the beacon. It is usual at the transmitting station to use an antenna array which can give two distinct radiation polar diagrams symmetrically arranged on either side of the correct bearing to be followed, the two lobes overlapping in such a way that the signals received along the required track will be of equal strength. This directional array may be operated in one of several ways, but in all cases the transmission is switched in such a way that each of the two polar diagrams is produced in turn. The signals associated with the two polar diagrams in this type of beacon are often complementary, i.e., the mark of one corresponding exactly with the space of the other. This then produces a continuous tone when the ship is on course, but a distinctive signal in each of the two adjacent sectors. Alternatively, the correct course is indicated when the signals from the two lobes are received at equal strength.

course board. A board located on the navigation bridge used to display the course to steer, track, drift angle, leeway angle, compass error, etc.

course line. 1. The graphic representation of a ship's course, usually with respect to true north. 2. A line of position approximately parallel to the course line (definition 1), thus providing a check as to deviating left or right of the track. See also SPEED LINE.

course made good. The single resultant direction from a point of departure to a point of arrival at any given time. This is a misnomer in that courses are directions steered or intended to be steered through the water with respect to a reference meridian. See also COURSE, COURSE OVER GROUND, TRACK MADE GOOD.

course of advance. An expression sometimes used to indicate the direction intended to be made good over the ground. The preferred term is TRACK, definition 1. This is a misnomer in that courses are directions steered or intended to be steered through the water with respect to a reference meridian. See also COURSE, COURSE OVER GROUND.

course over ground. The direction of the path over the ground actually followed by a vessel. It is normally a somewhat irregular line. This is a misnomer in that courses are directions steered or intended to be steered through the water with respect to a reference meridian. See also COURSE, COURSE MADE GOOD.

course recorder. A device which makes an automatic graphic record of the headings of a vessel *vs* time. See also DEAD RECKONING TRACER.

course up. See BASE COURSE UP.

cove, *n.* A small sheltered recess or indentation in a shore or coast, generally inside a larger embayment.

coverage diagram. A chart which depicts the area serviced by a radionavigation system. Such service may be stated for specified signal-to-noise ratios and predictable accuracy for both groundwave and skywave coverage, as applicable. The coverage diagrams for Loran-C show areas in which groundwave coverage affords predictable accuracy of 1500 feet on a 95 percent probability basis for signal-to-noise ratios 1:3 and 1:10. The diagrams also show for signal-to-noise ratios 1:3 and 1:10 those areas within which skywaves can be used for fixing. In these areas the angles of cut of the lines of position exceed 15° and the gradient of the hyperbolas is less than 2 nautical miles per microsecond. See also LORAN-C RELIABILITY DIAGRAM.

coversine, coversed sine, *n.* One minus the sine (1−sin). The expression NATURAL COVERSINE is sometimes used to distinguish the coversine from its logarithm (called LOGARITHMIC COVERSINE). Also called VERSED COSINE.

crab, *v., t.* To drift sideways or to leeward.

crack, *n.* Any fracture (in ice) which has not parted.

creek, *n.* 1. A stream of less volume than a river but larger than a brook.
2. A small tidal channel through a coastal marsh.
3. A wide arm of a river or bay, as used locally in Maryland and Virginia.

crepuscular rays. Literally, "twilight rays," alternating lighter and darker bands (rays and shadows) which appear to diverge in fan-like array from the sun's position at about twilight. This term is applied to two quite different phenomena: a. It refers to shadows cast across the purple light, a true twilight phenomenon, by cloud tops that are high enough and far enough away from the observer to intercept some of the sunlight that would ordinarily produce the purple light. b. A more common occurrence is that of shadows and rays made visible by haze in the lower atmosphere. Towering clouds produce this effect also, but they may be fairly close to the observer and the sun need not be below the horizon. The apparent divergence of crepuscular rays is merely a perspective effect. When they continue across the sky to the antisolar point, these extensions are called ANTICREPUSCULAR RAYS. Also called SHADOW BANDS.

crescent, *adj.* Bounded by a convex and a concave curve. Originally, the term applied only to the "increasing" moon, from which the word was derived. By extension, it is now generally applied to the moon between last quarter and new as well as between new and first quarter, and to any other celestial body presenting a similar appearance, or any similarly shaped object. See also PHASES OF THE MOON.

crest, *n.* The highest part of a wave, swell, ridge, etc.

crest cloud. A type of cloud over a mountain ridge, similar to a cap cloud over an isolated peak. The cloud is apparently stationary, but actually is continually being formed to windward and dissipated to leeward.

crevasse, *n.* A deep fissure or rift in a glacier.

critical angle. 1. The maximum angle at which a radio wave may be emitted from an antenna, in respect to the plane of the earth, and still be returned to the earth by refraction or reflection by an ionospheric layer.
2. The angle at which radiation, about to pass from a medium of greater density into one of lesser density, is refracted along the surface of the denser medium.

critical table. A single entering argument table in which values of the quantity to be found are tabulated for limiting values of the entering argument. In such a table interpolation is avoided through dividing the argument into intervals so chosen that successive intervals correspond to successive values of the required quantity, the *respondent.* For any value of the argument within these intervals, the respondent can be extracted from the table without interpolation. The lower and upper limits (critical values) of the argument correspond to half-way values of the respondent and, by convention, are chosen so that when the argument is equal to one of the critical values, the respondent corresponding to the preceding (upper) interval is to be used.

critical temperature. The temperature above which a substance cannot exist in the liquid state, regardless of pressure.

cross-band Racon. A Racon which transmits at a frequency not within the marine radar frequency band. To be able to use this type of Racon, the ship's radar receiver must be capable of being tuned to the frequency of the cross-band Racon or special accessory equipment is required. In either case, normal radar echoes will not be painted on the radarscope. This is an experimental type of Racon. See also IN-BAND RACON.

cross-band transponder. A transponder which responds on a frequency different from that of the interrogating signal.

cross bearings. Two or more bearings used as intersecting lines of position for fixing the position of a craft.

cross hair. A hair, thread, or wire constituting part of a reticle.

cross sea. A series of waves imposed across the prevailing waves. It is called CROSS SWELL when the imposed waves are the longer swell waves.

cross-section paper. Paper ruled with two sets of parallel lines, useful as an aid in plotting Cartesian coordinates. Usually, the two sets are mutually perpendicular. See also COORDINATE PAPER.

cross-staff, *n.* A forerunner of the modern sextant, used for measuring altitudes of celestial bodies, consisting essentially of a wooden rod with one or more perpendicular cross pieces free to slide along the main rod. Also called FORESTAFF, JACOB'S STAFF.

cross swell. See under CROSS SEA.

cross tide. A tidal current setting in a direction approximately 90° from the course of a vessel. One setting in a direction approximately 90° from the heading is called a BEAM TIDE. In common usage these two expressions are usually used synonymously. One setting in a direction approximately opposite to the heading is called a HEAD TIDE. One setting in such a direction as to increase the speed of a vessel is called a FAIR TIDE.

cross wind. See under BEAM WIND.

crusing radius. The distance a craft can travel at cruising speed without refueling. Also called CRUISING RANGE.

cruising range. See CRUISING RADIUS.

cyrogenics, *n.* 1. The study of the methods of producing very low temperatures.
2. The study of the behavior of materials and processes at cryogenic temperatures.

cryogenic temperature. In general, a temperature range below the boiling point of nitrogen (−195°C); more particularly, temperatures within a few degrees of absolute zero.

crystal, *n.* A crystalline substance which allows electric current to pass in only one direction.

crystal clock. See QUARTZ CRYSTAL CLOCK.

cube, *n.* 1. A solid bounded by six equal square sides.
2. The third power of a quantity.

cubic metre. The *derived unit* of volume in the International System of Units.

cul-de-sac, *n.* An inlet with a single small opening.

culmination, *n.* See MERIDIAN TRANSIT.

culture, *n.* Features under, on, and above the ground which are delineated on the map or chart and were constructed by man. These features include cities, highways, submarine cables, and aids to navigation. Boundary lines, latitude and longitude lines, isogonic lines, etc. are properly classified as culture.

cumuliform, *adj.* Like cumulus; generally descriptive of all clouds, the principal characteristic of which is vertical development in the form of rising mounds, domes, or towers. This is the contrasting form to the horizontally extended STRATIFORM types. See also CIRRIFORM.

cumulonimbus, *n.* An exceptionally dense cloud of great vertical development, occurring either as an isolated cloud or one of a line or wall of clouds with separated upper portions. These clouds appear as mountains or huge towers, at least a part of the upper portions of which are usually smooth, fibrous, striated, and almost flattened. This part often spreads out in the form of an anvil or vast plume. Under the base of cumulonimbus, which often is very dark, there frequently exists virga, precipitation, and low, ragged clouds, either merged with it or not. Its precipitation is often heavy and always of a showery nature. The usual occurrence of lightning and thunder within or from this cloud leads to its being popularly called THUNDERCLOUD and THUNDERHEAD. The latter term usually refers to only the upper portion of the cloud. See also CLOUD CLASSIFICATION.

cumulus, *n.* A cloud type in the form of individual, detached elements which are generally dense and possess sharp non-fibrous outlines. These elements develop vertically, appearing as rising mounds, domes, or towers, the upper parts of which often resemble a cauliflower. The sunlit parts of these clouds are mostly brilliant white; their bases are relatively dark and nearly horizontal. Near the horizon the vertical development of cumulus often causes the individual clouds to appear merged. If precipitation occurs, it is usually of a showery nature. Various effects of wind, illumination, etc. may modify many of the above characteristics. Strong winds may shred the clouds, often tearing away the cumulus tops to form the species **fractus**. See also CLOUD CLASSIFICATION.

cupola, *n.* A label on a nautical chart which indicates a small dome-shaped tower or turret rising from a building.

current, *n.* 1. Generally, a horizontal movement of water. Currents may be classified as **tidal** and **nontidal**. Tidal currents are caused by gravitational interactions between the sun, moon, and earth and are a part of the same general movement of the sea that is manifested in the vertical rise and fall, called TIDE. Tidal currents are periodic with a net velocity of zero over the particular tidal cycle. Nontidal currents include the permanent currents in the general circulatory systems of the sea as well as temporary currents arising from more pronounced meteorological variability. The SET of a current is the direction toward which it flows; the DRIFT is its speed. In British usage, tidal current is called TIDAL STREAM, and nontidal current is called CURRENT.
2. A hypothetical horizontal motion of such set and drift as to account for the difference between a dead reckoning position and a fix at the same time.

current chart. A chart on which current data are graphically depicted. See also TIDAL CURRENT CHARTS.

current constants. Tidal current relations that remain practically constant for any particular locality. Current constants are classified as **harmonic** and **nonharmonic**. The harmonic constants consist of the amplitudes and epochs of the harmonic constituents, and the nonharmonic constants include the velocities and intervals derived directly from the current observations.

current curve. A graphic representation of the flow of the current. In the reversing type of tidal current, the curve is referred to rectangular coordinates with time represented by the abscissas and the speed of the current by the ordinates, the flood speeds being considered as positive and the ebb speeds as negative. In general, the current curve for a reversing tidal current approximates a cosine curve.

current cycle. A complete set of tidal current conditions, as those occurring during a tidal day, lunar month, or Metonic cycle.

current diagram. A graphic table showing the speeds of the flood and ebb currents and the times of slack and strength over a considerable stretch of the channel of a tidal waterway, the times being referred to tide or tidal current phases at some reference station.

current difference. The difference between the time of slack water (or minimum current) or strength of current in any locality and the time of the corresponding phase of the tidal current at a reference station, for which predictions are given in the *Tidal Current Tables*.

current direction. The direction *toward* which a current is flowing, called the SET of the current.

current ellipse. A graphic representation of a rotary current in which the velocity of the current at different hours of the tidal cycle is represented by radius vectors and vectorial angles. A line joining the extremities of the radius vectors will form a curve roughly approximating an ellipse. The cycle is completed in one-half tidal day or in a whole tidal day according to whether the tidal current is of the semidiurnal or the diurnal type. A current of the mixed type will give a curve of two unequal loops each tidal day.

current hour. The mean interval between the transit of the moon over the meridian of Greenwich and the time of strength of flood, modified by the times of slack water (or minimum current) and strength of ebb. In computing the mean current hour an average is obtained of the intervals for the following phases: flood strength, slack (or minimum) before flood increased by 3.10 hours (one-fourth of tidal cycle), slack (or minimum) after flood decreased by 3.10 hours, and ebb strength increased or decreased by 6.21 hours (one-half of tidal cycle). Before taking the average, the four phases are made comparable by the addition or rejection of such multiples of 12.42 hours as may be necessary. The current hour is usually expressed in solar time, but if the use of lunar time is desired, the solar hour should be multiplied by the factor 0.966.

current line. A graduated line attached to a CURRENT POLE, used in measuring the velocity of the current. The line is marked in such a manner that the speed of the current, expressed in knots and tenths, is indicated directly by the length of line carried out by the current pole in a specified interval of time. When marked for a 60-second run, the principal divisions for the whole knots are spaced 101.33 feet and the subdivisions for tenths of knots are spaced at 10.13 feet. Also called LOG LINE.

current meter. An instrument for measuring the speed and direction or just speed of a current. The measurements are usually Eulerian since the meter is most often fixed or moored at a specific location. The National Ocean Survey primarily uses the Savonius rotor type. The rotor is correlated with the speed of the impinging current, and the direction of flow is usually determined by an internal compass coupled with a meter orientation monitoring device.

current pole. A pole used in observing the velocity of the current. In use, the pole, which is weighted at one end so as to float upright, is attached to the current line but separated from the graduated portion by an ungraduated section of approximately 100 feet, known as the *stray line*. As the pole is carried out from an observing vessel by the current, the amount of line passing from the vessel during a specific time interval indicates the speed of the current. The set is obtained from a bearing from the vessel to the pole.

current rips. See RIPS.

current sailing. The process of allowing for current when predicting the track to be made good or of determining the effect of a current on the direction of motion of a vessel. The expression is better avoided, as the process is not strictly a sailing.

current station. The geographic location at which current observations are conducted. Also, the facilities used to make current observations. These may include a buoy, ground tackle, current meters, recording mechanism, and radio transmitter. See also CONTROL CURRENT STATION, SUBORDINATE CURRENT STATION.

current tables. See TIDAL CURRENT TABLES.

cursor, *n.* A device used with an instrument, to provide a moveable reference, as the runner of a slide rule or a rotatable plastic disk with inscribed crosslines, used in reading bearings on a plan position indicator.

curve of constant bearing. See CURVE OF EQUAL BEARING.

curve of equal bearing. A curve connecting all points at which the great-circle bearing of a given point is the same. Also called CURVE OF CONSTANT BEARING.

curvilinear, *adj.* Consisting of or bounded by a curve.

curvilinear triangle. A closed figure having three curves as sides.

cusp, *n.* One of the horns or pointed ends of the crescent moon or other luminary.

cut, *n.* 1. A notch or depression produced by excavation or erosion, as a *cut* for a canal.
2. The intersection of lines of position, constituting a fix, with particular reference to the angle of intersection.

cut in. To observe and plot lines of position locating an object or craft, particularly by bearings.

cut-off, *n.* 1. A new and relatively short channel formed when a stream cuts through the neck of an oxbow or horseshoe bend.
2. An artificial straightening or short-cut in a channel.

Cyclan, *n.* The designation of Loran-C in its earliest stage of development but later superseded by the term CYTAC.

cycle, *n.* One complete train of events or phenomena that recur regularly in the same sequence. When used in connection with sound or radio, the term refers to one complete wave, or to a frequency of one wave per second. See also KILOCYCLE, MEGACYCLE, CALLIPPIC CYCLE, CURRENT CYCLE, DUTY CYCLE, LUNAR CYCLE, METONIC CYCLE, TIDAL CYCLE.

cycle match. In Loran-C, the comparison, in time difference, between corresponding carrier cycles contained in the rise times of a master and secondary station pulse. The comparison is refined to a determination of the phase difference between these two cycles. See also ENVELOPE MATCH.

cyclic, *adj.* Of or pertaining to a cycle or cycles.

cyclogenesis, *n.* Any development or strengthening of cyclonic circulation in the atmosphere. The opposite is CYCLOLYSIS. The term is applied to the development of cyclonic circulation where previously it did not exist, as well as to the intensification of existing cyclonic flow. While cyclogenesis usually occurs with a deepening (a decrease in atmospheric pressure), the two terms should not be used synonymously.

cyclolysis, *n.* Any weakening of cyclonic circulation in the atmosphere. The opposite is CYCLOGENESIS. While cyclolysis usually occurs with a filling (an increase in atmospheric pressure), the two terms should not be used synonymously.

cyclone, *n.* 1. An approximately circular portion of the atmosphere, having relatively low atmospheric pressure and winds which blow counterclockwise around the center in the Northern Hemisphere and clockwise in the Southern Hemisphere.
2. The name by which a tropical storm having winds of 34 knots or greater is known in the South Indian Ocean.
See TROPICAL CYCLONE.

cyclonic storm. See under TROPICAL CYCLONE.

cyclonic winds. The winds associated with a low pressure area and constituting part of a cyclone.

cylinder, *n.* 1. A solid having two parallel plane bases bounded by closed congruent curves, and a surface formed by parallel lines connecting similar points on the two curves.
2. A surface formed by a straight line moving parallel to itself and constantly intersecting a curve. Also called CYLINDRICAL SURFACE.

cylindrical, *adj.* Of or pertaining to a cylinder.

cylindrical buoy. See CAN BUOY.

cylindrical chart. A chart on a cylindrical map projection.

cylindrical map projection. A map projection in which the surface of a sphere or spheroid, such as the earth, is conceived as developed on a tangent cylinder, which is then spread out to form a plane. See also MERCATOR MAP PROJECTION, RECTANGULAR MAP PROJECTION, EQUATORIAL MAP PROJECTION, OBLIQUE MAP PROJECTION, OBLIQUE MERCATOR MAP PROJECTION, TRANSVERSE MAP PROJECTION.

cylindrical surface. A surface formed by a straight line moving parallel to itself and constantly intersecting a curve. Also called a CYLINDER.

Cytac, *n.* The designation of Loran-C in an earlier stage of development. See also CYCLAN.

D

daily aberration. See under ABERRATION, definition 1.

Daily Memorandum. A daily publication of the Defense Mapping Agency Hydrographic/Topographic Center containing printed copy of HYDROLANTS, HYDROPACS, and NAVAREA Warnings from NAVAREAS IV and XII. The HYDROLANTS, HYDROPACS, and NAVAREA Warnings are broadcast messages restricted to the more important marine incidents or navigational changes for which a delay in disseminating the information to mariners would adversely affect navigational safety. The *Daily Memorandum* is published in two editions, an Atlantic edition and a Pacific edition, both prepared in Washington, D.C. The *Daily Memorandum* is sent to fleet operating bases, naval stations, custom houses, shipping company offices, etc. where it may be picked up by navigational personnel of vessels in port.

daily rate. See CHRONOMETER RATE, WATCH RATE.

dale, *n.* A vale or small valley.

dam, *n.* A barrier to check or confine anything in motion; particularly a bank of earth, masonry, etc., across a watercourse to keep back moving water.

damped wave. 1. A wave such that, at every point, the amplitude of each sinusoidal component is a decreasing function of time.
2. A wave in which the amplitudes of successive peaks (crests) progressively diminish.

damp haze. See under HAZE.

damping, *n.* 1. The reduction of energy in a mechanical or electrical system by absorption or radiation.
2. The act of reducing the amplitude of the oscillations of an oscillatory system; hindering or preventing oscillation or vibration; diminishing the sharpness of resonance of the natural frequency of a system.

damping error. See as BALLISTIC DAMPING ERROR.

dan buoy. A buoy consisting of a ballasted float carrying a staff which supports a flag or light. Dan buoys are used principally in minesweeping, and by fisherman to mark the position of deep-sea fishing lines or the place for trawling.

danger angle. The maximum (or minimum) angle between two points, as observed from a craft, indicating the limit of safe approach to an off-lying danger. As a vessel proceeds past the danger, the angle between the two points is measured frequently, usually by sextant. As long as the angle is not larger (or smaller) than the predetermined danger angle, the vessel is in safe water. A **horizontal danger angle** is measured between points shown on the chart. A **vertical danger angle** is measured between the top and bottom of an object of known height.

danger area. A specified area above, below, or within which there may exist potential danger. See also PROHIBITED AREA, RESTRICTED AREA.

danger bearing. The maximum or minimum bearing of a point for safe passage of an off-lying danger. As a vessel proceeds along a coast, the bearing of a fixed point on shore, such as a lighthouse, is measured frequently. As long as the bearing does not exceed the limit of the predetermined danger bearing, the vessel is on a safe course. Called CLEARING BEARING in *British terminology*.

danger buoy. A buoy marking an isolated danger to navigation, such as a rock, shoal or sunken wreck.

danger line. 1. A line drawn on a chart to indicate the limits of safe navigation for a vessel of specific draft.
2. A line of small dots used to draw the navigator's attention to a danger which would not stand out clearly enough if it were represented on the chart solely by the specific symbols. This line of small dots is also used to delimit areas containing numerous dangers, through which it is unsafe to navigate.

dangerous semicircle. That half of a cyclonic storm area in which the rotary and progressive motions of the storm reinforce each other and the winds are in such a direction as to tend to blow a vessel into the storm track. In the Northern Hemisphere this is to the right of the storm center and in the Southern Hemisphere it is to the left. The opposite is NAVIGABLE SEMICIRCLE.

danger sounding. A minimum sounding chosen for a vessel of specific draft in a given area to indicate the limit of safe navigation.

Dapac, *n.* See PUB. 110.

dark nilas. Nilas which is under 5 centimeters in thickness and is very dark in color.

dark-trace tube. A cathode-ray tube having a specially coated screen which changes color but does not necessarily luminesce when struck by the electron beam. It shows a dark trace on a bright background.

data-acquisition station. A ground station used for performing the various functions necessary to control satellite operations and to obtain data from the satellite.

date, *n.* A designated mark or point on a time scale.

date line. The line coinciding approximately with the 180th meridian, at which each calendar day first begins; the boundary between the −12 and +12 time zones. The date on each side of this line differs by 1 day, but the time is the same in these two zones. When crossing this line on a westerly course, the date must be advanced 1 day; when crossing on an easterly course, the date must be put back 1 day. Sometimes called INTERNATIONAL DATE LINE, CALENDAR LINE.

datum, *n.* Any numerical or geometrical quantity or set of such quantities which may serve as a reference or base for other quantities. In geodesy two types of datums must be considered: a **horizontal datum** which forms the basis for computations of horizontal control surveys in which the curvature of the earth is considered, and a **vertical datum** to which elevations are referred. See also HORIZONTAL GEODETIC DATUM, VERTICAL GEODETIC DATUM, CHART SOUNDING DATUM, VERTICAL DATUM.

datum-centered ellipsoid. The reference ellipsoid that gives the best fit to the astrogeodetic network of a particular datum, and hence does not necessarily have its center at the center of the earth.

datum plane. A misnomer for collection of datums used in mapping, charting, and geodesy which are not strictly planar. This term should not be used.

datum transformation. The systematic elimination of discrepancies between adjoining or overlapping triangulation networks from different datums by moving the origins, rotating, and stretching the networks to fit each other.

Davidson Current. A seasonal North Pacific Ocean countercurrent flowing northwestward along the west coast of North America from north of 32° N to at least latitude 48° N, inshore of the south-easterly-flowing California Current. This current occurs generally between November and April, but is best established in January. Strong opposing winds may cause the current to reverse. Also called WINTER COASTAL COUNTER-CURRENT.

Davidson Inshore Current. See DAVIDSON CURRENT.

dawn, *n.* The first appearance of light in the eastern sky before sunrise; daybreak. See also DUSK, TWILIGHT.

day, *n.* 1. The duration of one rotation of the earth, or occasionally another celestial body, on its axis. It is measured by successive transits of a reference point on the celestial sphere over the meridian, and each type takes its name from the reference used. Thus, for a **solar day** the reference is the sun; a **mean solar day** if the mean sun; and an **apparent solar day** if the apparent sun. For a **lunar day** the reference is the moon; for a **sidereal day** the vernal equinox; for a **constituent day** an astre fictif or fictitious star representing one of the periodic elements in the tidal forces. The expression **lunar day** refers also to the duration of one rotation of the moon with respect to the sun. A **Julian day** begins at Greenwich mean noon and the days are consecutively numbered from January 1, 4713 B.C.
2. A period of 24 hours beginning at a specified time, as the **civil day** beginning at midnight, or the **astronomical day** beginning at noon, which was used up to 1925 by astronomers.
3. A specified time or period, usually of approximately 24-hours duration. A **calendar day** extends from midnight to midnight, and is of 24-hours duration unless a time change occurs during the day. A **tidal day** is either the same as a **lunar day** (on the earth), or the period of the daily cycle of the tides, differing slightly from the lunar day because of priming and lagging.
4. The period of daylight, as distinguished from night.

daybeacon, *n.* An unlighted beacon. A daybeacon is identified by its color and the color, shape, and number of its daymark. The simplest form of daybeacon consists of a single pile with a daymark affixed at or near its top. See also DAYMARK.

daybreak, *n.* See DAWN.

daylight control. A device, operated by daylight, that automatically lights and extinguishes an electric light, usually lighting it at or about sunset and extinguishing it at or about sunrise. Also called SUN RELAY, SUN SWITCH, SUN VALVE.

daylight saving meridian. The meridian used for reckoning daylight saving time. This is generally 15° east of the ZONE or STANDARD MERIDIAN.

daylight saving noon. Twelve o'clock daylight saving time, or the instant the mean sun is over the upper branch of the daylight saving meridian. Also called SUMMER NOON, especially in Europe. See also MEAN NOON.

daylight saving time. A variation of standard time in order to make better use of daylight. In the United States the "Uniform Time Act of 1966" (Public Law 89-387) establishes the annual advancement and retardation of standard time by 1 hour at 2 AM on the last Sunday of April and October, respectively, except in those states which have by law exempted themselves from the observance of daylight saving time. Also called SUMMER TIME, especially in Europe.

daylight signal light. A signal light exhibited by day and also, usually with reduced intensity, by night. This reduction of intensity is made in order to avoid glare. Daylight signals may be used to indicate whether or not the entrance to a lock is free.

daymark, *n.* 1. The identifying characteristics of an aid to navigation which serve to facilitate its recognition against a daylight viewing background. For example the distinctive color and shape of a buoy aid identification during the daytime. Called DAYTIME CHARACTER in British terminology. On those structures that do not by themselves present an adequate viewing area to be seen at the required distance, the aid is made more visible by affixing a *daymark* to the structure. A daymark so affixed has a distinctive color and shape depending upon the purpose of the aid. See also DAYBEACON.
2. An unlighted navigation mark.

day's run. The distance traveled by a vessel in 1 day, usually reckoned from noon to noon.

day's work. The daily routine of the navigation of a vessel at sea, usually consisting principally of the dead reckoning from noon to noon, evening and morning twilight observations, a morning sun observation for a line of position, a sun observation at or near noon for a running fix, an afternoon sun observation for a line of position, morning and afternoon azimuths of the sun for checking the compass, and concurrent radionavigation depending upon installed equipments and associated service areas.

daytime character. *British terminology.* See DAYMARK, definition 1.

dead ahead. Bearing 000° relative. If the bearing is approximate, the term AHEAD should be used.

dead astern. Bearing 180° relative. If the bearing is approximate, the term ASTERN should be used. Also called RIGHT ASTERN.

deadbeat, *adj.* Aperiodic, or without a period.

deadbeat compass. See APERIODIC COMPASS.

deadhead, *n.* 1. A block of wood used as an anchor buoy.
2. A bollard.

dead reckoning. The process of determining the position of a vessel at any instant by applying to the last well-determined position (point of departure or subsequent fix) the run that has since been made. The position so obtained is called a DEAD RECKONING POSITION. When the principal purpose of dead reckoning is to lay down on the chart a reference plot for evaluating the reasonableness of positioning by other means, the dead reckoning plot is usually constructed without allowance for disturbing elements (such as current, wind, sea conditions, roughness of vessel's bottom, etc.), the course steered being used for direction and ordered speed being used for rate of movement along the course line. However, some navigators use course steered for direction, as above, but the rate of movement along the course line is ordered speed adjusted for all disturbing elements except current. However constructed, the reference dead reckoning plot provides a graphic presentation of positions the vessel would occupy if unaffected by disturbing elements and inaccuracies in steering and speed determination. With due recognition of its limitations, this plot of predicted future positions is used in conjunction with the TRACK (definition 2) to determine necessary or desirable course and/or speed changes. Also the construction of the track may be considered a form of dead reckoning. When the principal purpose of the dead reckoning is to predict a future position after a short time interval, as when piloting, the course steered and the ordered speed are frequently used to determine the run. The plot so established also provides a reference dead reckoning plot unaffected by the disturbing elements generally difficult to estimate or which normally change rapidly in waters where frequent fixing is required. The navigator notes the trend of successive fixes with respect to course lines so drawn and alters his course and/or speed accordingly. When the principal purpose of the dead reckoning is to determine the vessel's position independent of other means or in the absence of other means over considerable time periods, the dead reckoning is often laid down on the chart making allowance for disturbing elements. This practice is general among merchant navigators. Course lines between successive dead reckoning positions are constructed in accordance with the navigator's best estimate of the course to be made good. The rate of movement along the course line is in accordance with his estimate of the speed to be made good.

Dead Reckoning Altitude and Azimuth Table. See H.O. PUB. NO. 211.

dead reckoning equipment. A device that continuously indicates the dead reckoning position of a vessel. It may also provide, on a **dead reckoning tracer**, a graphical record of the dead reckoning. See also COURSE RECORDER.

dead reckoning plot. The graphic plot of the dead reckoning, suitably labeled with respect to time, direction, and speed. See also NAVIGATIONAL PLOT.

dead reckoning position. See under DEAD RECKONING.

dead reckoning tracer. A device that automatically provides a graphical record of the dead reckoning. It may be part of dead reckoning equipment. See also COURSE RECORDER.

dead water. The water carried along with a ship moving through the water because of frictional resistance. The phenomenon is maximum at the waterline and decreases with depth. It increases in a direction towards the stern.

deca-. A prefix meaning *ten.*

decameter, *n.* Ten meters.

Decca, *n.* See as DECCA NAVIGATOR SYSTEM.

Decca chain. A group of associated stations of the Decca Navigator System. A Decca chain normally consists of one master and three slave stations. Each slave station is called by the color of associated pattern of hyperbolic lines as printed on the chart, i.e., **red slave, green slave, purple slave.** See also CHAIN.

Decca Hi-Fix. See as HI-FIX.

Decca Navigator System. A short to medium range low frequency (70-130 kHz) radionavigation system by which a hyperbolic line of position of high accuracy is obtained. The system is an arrangement of fixed, phase locked, continuous wave transmitters operating on harmonically related frequencies and special receiving and display equipment carried on a vessel or other craft. The operation of the system depends on phase comparison of the signals from the transmitters brought to a common comparison frequency within the receiver. See also DECOMETER.

Decca Sea-Fix. See as SEA-FIX.

Decca Trisponder. The Del Norte Trisponder System as marketed by Racal-Decca Survey, Limited.

decelerate, *v., t.* To cause to move slower. *v., i.* To decrease speed.

deceleration, *n.* Negative acceleration.

December solstice. Winter solstice in the Northern Hemisphere.

deci-. A prefix meaning *one-tenth.*

decibar, *n.* One-tenth of a bar; 100 millibars.

decibel, *n.* A dimensionless unit used for expressing conveniently the ratio between widely different powers. It is 10 times the logarithm to the base 10 of the power ratio.

decimeter, *n.* On-tenth of a meter.

deck log, See LOG, definition 2.

declination, *n.* 1. Angular distance north or south of the celestial equator; the arc of an hour circle between the celestial equator and a point on the celestial sphere, measured northward or southward from the celestial equator through 90°, and labeled N or S (+ or −) to indicate the direction of measurement.

2. Short for MAGNETIC DECLINATION.

declinational inequality. See DIURNAL INEQUALITY.

declinational reduction. A processing of observed high and low waters or flood and ebb tidal currents to obtain quantities depending upon changes in the declination of the moon; such as tropic ranges or speeds, height or speed inequalities, and tropic intervals.

declination difference. The difference between two declinations, particularly between the declination of a celestial body and the value used as an argument for entering a table.

declinometer, *n.* An instrument for measuring magnetic declination. See also MAGNETOMETER.

Decometer, *n.* A phase meter used in the Decca Navigator System.

decrement, *n.* A decrease in the value of a variable. See also INCREMENT.

deep *n.* 1. An unmarked fathom point on a lead line. In reporting such a depth the leadsman calls out, "By the deep - - - - - - -," In reporting a depth indicated by a mark, the leadsman calls out, "By the mark - - - - - - -,"

2. A relatively small area of exceptional depth found in a depression of the ocean floor. The term is generally restricted to depths greater than 3,000 fathoms. If it is very limited in area, it is referred to as a HOLE.

3. A relatively deep channel in a strait or estuary.

deepening, *n.* Decrease in atmospheric pressure, particularly within a low. Increase in pressure is called FILLING. See also CYCLOGENESIS.

deep sea lead (lĕd). A heavy sounding lead (about 30 to 100 pounds), usually having a line 100 fathoms or more in length. A light deep sea lead is sometimes called a COASTING LEAD. Sometimes called DIPSEY LEAD.

deep water route. A route within defined limits which has been accurately surveyed for clearance of sea bottom and submerged obstacles as indicated on the chart. See also ROUTING SYSTEM.

definition, *n.* The clarity and fidelity of the detail of radar images on the radarscope. A combination of good resolution and focus adjustment is required for good definition.

definitive orbit. An orbit that is defined in a highly precise manner with due regard taken for accurate constants and observational data, and precision computational techniques including perturbations.

deflection of the plumb line. See under DEFLECTION OF THE VERTICAL.

deflection of the vertical. The angular difference, at any place, between the direction of a plumb line (the vertical) and the perpendicular to the reference ellipsoid. This difference seldom exceeds 30″. Often expressed in two components, meridian and prime vertical. Also called STATION ERROR.

deflection of the vertical correction. The correction due to deflection of the vertical, resulting from irregularities in the density and form of the earth. Deflection of the vertical affects the accuracy of sextant altitudes.

deflector, *n.* An instrument for measuring the directive force acting on a magnetic compass. It is used for adjusting a compass when ordinary methods of determining deviation are not available, and operates on the theory that when the directive force is the same on all cardinal headings, the compass is approximately adjusted.

deformed ice. A general term for ice which has been squeezed together and in places forced forwards (and downwards). Subdivisions are RAFTED ICE, RIDGED ICE, and HUMMOCKED ICE.

degaussing, *n.* Neutralization of the strength of the magnetic field of a vessel, by means of suitably arranged electric coils permanently installed in the vessel. See also DEPERMING.

degaussing cable. A cable carrying an electric current for degaussing a vessel.

degaussing range. An area for determining magnetic signatures of ships and other marine craft. Such signatures are used to determine required degaussing coil current settings and other required corrective action. Sensing instruments and cables are installed on the sea bed in the range, and there are cables leading from the range to a control position ashore. The range is usually marked by distinctive buoys.

degree, *n.* 1. A unit of circular measure equal to 1/360th of a circle.

2. The unit of measurement of temperature.

degree-of-freedom. Of a gyroscope, the number of orthogonal axes about which the spin axis is free to rotate, the spin axis freedom not being counted. *This is not a universal convention.* For example, the free gyro is frequently referred to as a three-degree-of-freedom gyro, the spin axis being counted.

deka-. A prefix meaning *ten* (10^1).

delayed plan position indicator. A plan position indicator on which the start of the sweep is delayed so that the center represents a selected range. This allows distant targets to be displayed on a larger-scale presentation.

delayed sweep. Short for DELAYED TIME BASE SWEEP.

delayed time base. Short for DELAYED TIME BASE SWEEP.

delayed time base sweep. A sweep, the start of which is delayed, usually to provide an expanded scale for a particular part. Usually shortened to DELAYED SWEEP, and sometimes to DELAYED TIME BASE.

Del Norte Trisponder System. A short range, line of sight, electronic distance measuring system used in hydrographic and other surveys. The basic system is composed of a distance-measuring unit (DMU), a master station, and two remote stations. Each master and remote station consists of a transponder and an antenna. The master station has an omnidirectional antenna, and each remote station has a directional antenna. With the remote stations placed at known geographic positions, distances are observed at the DMU. These values may be plotted manually on a hydrographic survey sheet or converted by trilateration methods to position values. Coded bursts of microwave frequency (X-band) energy are transmitted from the master unit to each remote unit for determination of distance. These signals are received and decoded by each remote unit, but only the unit with the particular code is triggered to send back an identical burst of energy to the master station. For obtaining accurate distance values, the signal is transmitted and received 10 times or 100 times, at the operator's option, and the sum averaged. When provided with the time sharing option, the system permits four vessels to time share position data from a common set of transponders.

delta, *n.* 1. The low alluvial land, deposited in a more or less triangular form, as the Greek letter delta, at the mouth of a river, which is often cut by several distributaries of the main stream.

2. A change in a variable quantity, such as a change in the value of the declination of a celestial body.

demagnetize, *v., t.* To remove magnetism. The opposite is MAGNETIZE.

demodulation, *n.* The process of obtaining a modulating wave from a modulated wave. The opposite is MODULATION.

departure, *n.* 1. The distance between two meridians at any given parallel of latitude, expressed in linear units, usually nautical miles; the distance to the east or west made good by a craft in proceeding from one point to another.

2. The point at which reckoning of a voyage begins. It is usually established by bearings of prominent landmarks as the vessel clears a harbor and proceeds to sea. When a person establishes this point, he is said to *take departure.* Also called POINT OF DEPARTURE.

3. Act of departing or leaving.

4. The amount by which the value of a meteorological element differs from the normal value.

dependent surveillance. Position determination by means requiring the cooperation of the tracked craft or vehicles (e.g., relay of craft or vehicle derived navigational information).

deperming, *n.* The process of changing the magnetic condition of a vessel by wrapping a large conductor around it a number of times in a vertical plane, athwartships, and energizing the coil thus formed. If a single coil is placed horizontally around the vessel and energized, the process is called FLASHING if the coil remains stationary, and WIPING if it is moved up and down. See also DEGAUSSING.

depressed pole. The celestial pole below the horizon, of opposite name to the latitude. The celestial pole above the horizon is called ELEVATED POLE.

depression, *n.* 1. See NEGATIVE ALTITUDE.

2. A cyclonic area, or low.

depression angle. See ANGLE OF DEPRESSION.

depth, *n.* The vertical distance from a given water level to the bottom. The charted depth is the vertical distance from the tidal datum to the bottom. The least depth in the approach or channel to an area, such as a port or anchorage, governing the maximum draft of vessels that can enter is called the controlling depth. See also

CHART SOUNDING DATUM.

depth contour. A contour line on a chart connecting points of equal depth below the sounding datum. It may be called FATHOM CURVE or FATHOM LINE if depth is expressed in fathoms. Also called DEPTH CURVE, ISOBATH.

depth curve. See DEPTH CONTOUR.

depth finder. See ECHO SOUNDER.

depth of water. The vertical distance from the surface of the water to the bottom. See also SOUNDING.

depth perception. The ability to estimate depth or distance between points in the field of vision.

derelict, *n.* Any property abandoned at sea, often of sufficient size as to constitute a menace to navigation; especially an abandoned vessel. See also JETTISON, WRECK.

derived units. See under INTERNATIONAL SYSTEM OF UNITS.

descending node. The point at which a planet, planetoid, or comet crosses the ecliptic from north to south, or a satellite crosses the plane of the equator of its primary from north to south. Also called SOUTHBOUND NODE. The opposite is ASCENDING NODE.

destination, *n.* The point of intended arrival, exclusive of the track needed to reach a berth or other position which may not be known or assigned until after arrival. Also called POINT OF DESTINATION. See also POINT OF ARRIVAL.

detection, *n.* 1. The process of extracting information from an electromagnetic wave.
2. In the use of radar, the recognition of the presence of a target.

detritus, *n.* An accumulation of the fragments resulting from the disintegration of rocks.

developable, *adj.* Capable of being developed, or flattened without distortion. The opposite is UNDEVELOPABLE.

developable surface. A curved surface that can be spread out in a plane without distortion, e.g., the cone and the cylinder.

deviascope, *n.* A device for demonstration of various forms of deviation and compass adjustment, or compass compensation.

deviation, *n.* 1. The angle between the magnetic meridian and the axis of a compass card, expressed in degrees east or west to indicate the direction in which the northern end of the compass card is offset from magnetic north. Deviation is caused by disturbing magnetic influences in the immediate vicinity of the compass, as within the craft. **Semicircular deviation** changes sign (E or W) approximately each 180° change of heading; **quadrantal deviation** changes sign approximately each 90° change of heading; **constant deviation** is the same on any heading. Deviation of a magnetic compass after adjustment or compensation is RESIDUAL DEVIATION. Called MAGNETIC DEVIATION when a distinction is needed to prevent possible ambiguity.
2. Direction finder deviation.
3. Given a series of observations or measurements of a given quantity, the deviation of a single observation is the algebraic difference between the single observation and the mean or average value of the series of observations. See also RANDOM ERROR.

deviation table. A table of the deviation of a magnetic compass on various headings, magnetic or compass. Also called MAGNETIC COMPASS TABLE. See also NAPIER DIAGRAM.

dew point. The temperature to which air must be cooled at constant pressure and constant water vapor content to reach saturation. Any further cooling usually results in the formation of dew or frost.

diacritical mark. A graphic symbol employed in conjunction with a written letter or character which alters the phonological and/or grammatical meaning of the letter or character.

diagram on the plane of the celestial equator. See TIME DIAGRAM.

diagram on the plane of the celestial meridian. An orthographic view of the celestial sphere from a point outside the sphere and over the celestial equator (if this were possible); the great circle appearing as the outer limit is the local celestial meridian; other celestial meridians appear as ellipses. The celestial equator appears as a diameter 90° from the poles. Parallels of declination appear as straight lines parallel to the equator. The celestial horizon appears as a diameter 90° from the zenith.

diagram on the plane of the equinoctial. See TIME DIAGRAM.

diameter, *n.* Any chord passing through the center of a figure, as a circle, ellipse, sphere, etc., or the length of such chord. See also RADIUS.

diaphone, *n.* A sound signal emitter operating on the principle of periodic release of compressed air controlled by the reciprocating motion of a piston operated by compressed air. The diaphone usually emits a powerful sound of low pitch which often concludes with a brief sound of lowered pitch called the GRUNT. The emitted signal of a TWO-TONE DIAPHONE consists of two tones of different pitch, in which case the second tone is of lower pitch.

diaphragm horn. A sound signal emitter comprising a resonant horn excited at its throat by impulsive emissions of compressed air regulated by an elastic diaphragm. Duplex or triplex horn units of different pitch produce a chime signal. Also called COMPRESSED-AIR HORN.

diatom, *n.* A microscopic alga with an external skeleton of silica, found in both fresh and salt water. Part of the ocean bed is composed of a sedimentary ooze consisting principally of large collections of the skeletal remains of diatoms.

dichroic mirror. A glass surface coated with a special metallic film that permits some colors of light to pass through the glass while reflecting certain other colors of light. Also called SEMI-REFLECTING MIRROR.

dichroism, *n.* The optical property of exhibiting two colors, as one color in transmitted light and another in reflected light. See also DICHROIC MIRROR.

dielectric reflector. A device composed of dielectric material which returns the greater part of the incident electromagnetic waves parallel to the direction of incidence. See also RADAR REFLECTOR.

difference of latitude. The shorter arc of any meridian between the parallels of two places, expressed in angular measure.

difference of longitude. The smaller angle at the pole or the shorter arc of a parallel between the meridians of two places, expressed in angular measure.

difference of meridional parts. See MERIDIONAL DIFFERENCE.

Differential Omega. In concept the use of a monitor station to improve the accuracy of the Omega Navigation System through broadcasting to the user within the immediate area a propagation correction as determined at the monitor station of known location. This capability is afforded by the *spatial correlation* of the VLF signal, i.e., over long transmission paths from a transmitting station to various points within a limited area, approximately the same errors will be accumulated along segments of the different paths. The predictable accuracy within 200 miles of the monitor station is about 1 nautical mile (95 percent probability). This accuracy increases to about 0.5 nautical mile at 100 miles and 0.3 nautical mile at 50 miles distant from the monitor station. In normal application special Differential Omega receivers are used.

differentiator, *n.* See FAST TIME CONSTANT CIRCUIT.

difficult area. A general qualitative expression to indicate, in a relative manner, that the severity of ice conditions prevailing in an area is such that navigation in it is difficult.

diffraction, *n.* 1. The bending of the rays of radiant energy around the edges of an obstacle or when passing near the edges of an opening, or through a small hole or slit, resulting in the formation of a spectrum, or arrangement by wavelength of the component waves, producing, for visible wavelengths, the chromatic spectrum. See also REFLECTION, REFRACTION.
2. The bending of a wave as it passes an obstruction.

diffuse ice edge. A poorly defined ice edge limiting an area of dispersed ice. It is usually on the leeward side of an area of pack ice.

diffuse reflection. Any reflection process in which the reflected radiation is sent out in many directions usually bearing no simple relationship to the angle of incidence. It results from reflection from a rough surface with small irregularities. See also SPECULAR REFLECTION.

diffusion, *n.* See DIFFUSE REFLECTION.

digit, *n.* A number consisting of a single figure. The 10 digits, are 0, 1, 2, 3, 4, 5, 6, 7, 8, 9.

digital calculator. See under DIGITAL COMPUTER.

digital computer. A device which processes information represented by combinations of discrete or discontinuous data, as compared with an ANALOG COMPUTER for continuous data. More specifically, it performs sequences of arithmetic and logical operations, not only on data but on its own program. Still more specifically, it is a stored program digital computer capable of performing sequences of internally stored instructions, as opposed to DIGITAL CALCULATORS on which the sequence is impressed manually.

digital tide gage. See AUTOMATIC TIDE GAGE.

dihedral angle. The angle between two intersecting planes.

dihedral reflector. A radar reflector consisting of two flat surfaces intersecting mutually at right angles. Incident radar waves entering the aperture so formed with a direction of incidence perpendicular to the edge, are returned parallel to their direction of incidence. Also called RIGHT ANGLE REFLECTOR.

dike, *n.* A bank of earth or stone used to form a barrier, frequently and confusingly interchanged with LEVEE, definition 1. A dike restrains water within an area that is normally flooded.

dioptric light. A light concentrated into a parallel beam by means of refracting lenses or prisms. One so concentrated by means of a reflector is a CATOPTRIC LIGHT.

dip, *n.* 1. The vertical angle, at the eye of an observer, between the horizontal and the line of sight to the visible horizon. Altitudes of celestial bodies measured from the visible sea horizon as a reference are too great by the amount of dip. Since dip arises from and varies with the elevation of the eye of the observer above the surface of the earth, the correction for dip is sometimes called HEIGHT OF EYE CORRECTION. Dip is smaller than GEOMETRICAL DIP by the amount of terrestrial refraction. Also called DIP OF THE HORIZON.
2. The angle between the horizontal and the lines of force of the earth's magnetic field at any point. Also called MAGNETIC DIP, MAGNETIC LATITUDE, MAGNETIC INCLINATION.
3. The first detectable decrease in the altitude of a celestial body after reaching its maximum altitude on or near meridian transit.

dip, *v., i.* To begin to descend in altitude after reaching a maximum on or near meridian transit.

dip circle. An instrument for measuring magnetic dip. It consists essentially of a DIP NEEDLE, or magnetic needle, suspended in such manner as to be free to rotate about a horizontal axis

dip correction. That correction to sextant altitude due to dip of the horizon. Also called HEIGHT OF EYE CORRECTION.

dip needle. A magnetic needle suspended in such manner as to be free to rotate about a horizontal axis. An instrument using such a needle to measure magnetic dip is called a DIP CIRCLE. A dip needle with a sliding weight that can be moved along one of its arms to balance the magnetic force is called a HEELING ADJUSTER.

dip of the horizon. See DIP, *n.*, definition 1.

dipole antenna, *n.* A straight center-fed one-half wavelength antenna. Horizontally polarized it produces a figure eight radiation pattern, with maximum radiation at right angles to the plane of the antenna. Also called DOUBLET ANTENNA.

dip pole. See as MAGNETIC DIP POLE.

dipsey lead (lĕd). See DEEP SEA LEAD.

direct indicating compass. A compass in which the dial, scale, or index is carried on the sensing element.

direction, *n.* The position of one point in space relative to another without reference to the distance between them. Direction may be either three-dimensional or two-dimensional, the horizontal being the usual plane of the latter. Direction is not an angle but is often indicated in terms of its angular distance from a REFERENCE DIRECTION. Thus, a horizontal direction may be specified as **compass, magnetic, true, grid** or relative as the reference direction is compass, magnetic, true, grid north or heading, respectively. A **Mercator** or **rhumb** direction is the horizontal direction of a rhumb line, expressed as angular distance from a reference direction, while **great-circle direction** is the horizontal direction of a great circle, similarly expressed. See also CURRENT DIRECTION, SWELL DIRECTION, WAVE DIRECTION, WIND DIRECTION.

directional antenna. An antenna designed so that the radiation pattern is largely concentrated in a single lobe.

directional gyro. A gyroscopic device used to indicate a selected horizontal direction for a limited time.

directional gyro mode. That mode of operation of a gyrocompass in which the compass operates as a free gyro with the spin axis oriented to grid north.

directional radiobeacon. See under RADIOBEACON. Also see as COURSE BEACON.

direction finder. See RADIO DIRECTION FINDER.

direction finder deviation. The angular difference between a bearing observed by a radio direction finder and the correct bearing, caused by disturbances due to the characteristics of the receiving craft or station.

direction finder station. See RADIO DIRECTION FINDER STATION.

direction light. A light illuminating a sector of very narrow angle and intended to mark a direction to be followed. A direction light bounded by other sectors of different characteristics which define its margins with small angles of uncertainty is called a SINGLE STATION RANGE LIGHT.

direction of current. The direction *toward* which a current is flowing, called the SET of the current.

direction of force of gravity. The direction indicated by a plumb line. It is perpendicular (normal) to the surface of the geoid. Also called DIRECTION OF GRAVITY.

direction of gravity. See DIRECTION OF FORCE OF GRAVITY.

direction of relative movement. The direction of motion relative to a reference point, itself usually in motion.

direction of waves or swell. The direction *from* which waves or swell are moving.

direction of wind. The direction *from* which a wind is blowing.

directive force. The force tending to cause the directive element of a compass to line up with the reference direction. Also, the value of this force. Of a magnetic compass, it is the intensity of the horizontal component of the earth's magnetic field.

directive gain. Of an antenna for a given direction, four times the ratio of the radiation intensity in that direction to the total power radiated by the antenna. Also called GAIN FUNCTION.

directivity, *n.* 1. The characteristic of an antenna which makes it radiate or receive more efficiently in some directions than in others.
2. An expression of the value of the directive gain of an antenna in the direction of its maximum gain. Also called POWER GAIN (OF AN ANTENNA).

directivity diagram. See RADIATION PATTERN.

direct motion. The apparent motion of a planet eastward among the stars. Apparent motion westward is called RETROGRADE MOTION. The usual motion of planets is direct.

direct wave. 1. A radio wave that travels directly from the transmitting to the receiving antenna without reflections from any object or layer of the ionosphere. The path may be curved as a result of refraction.
2. A radio wave that is propagated directly through space; it is not influenced by the ground. Also called SPACE WAVE.

discontinued, *adj.* Said of a previously authorized aid to navigation that has been removed from operation (permanent or temporary).

discontinuity, *n.* 1. A zone of the atmosphere within which there is a comparatively rapid transition of any meteorological element.
2. A break in sequence of continuity of anything.

discrepancy, *n.* 1. Failure of an aid to navigation to maintain its position or function exactly as prescribed in the *Light List*.
2. The difference between two or more observations or measurements of a given quantity.

discrepancy buoy. An easily transportable buoy used to temporarily replace an adrift, temporarily discontinued, off station, missing, severely damaged or destroyed aid or an aid that is not watching properly.

dismal, *n.* A swamp bordering on, or near the sea. Also called POCOSIN.

dispersion, *n.* The separation of light into its component colors by its passage through a diffraction grating or by refraction such as that provided by a prism.

display, *n.* 1. The visual presentation of radar echoes, as on a plan position indicator. Also called RADAR PICTURE.
2. The equipment for the visual display of the received signals in a radar set.

disposal area. Area designated by the Corps of Engineers for depositing dredged material where existing depths indicate that the intent is not to cause sufficient shoaling to create a danger to surface navigation. Disposal areas are shown on nautical charts. See also DUMPING GROUND, DUMP SITE, SPOIL AREA.

disposition of lights. The arrangement, order, etc., of navigational lights in an area.

distance finding station. An attended light station or lightship emitting simultaneous radio and sound signals as a means of determining distance from the source of sound, by measuring the difference in the time of reception of the signals. The sound may be transmitted through either air or water or both and either from the same location as the radio signal or a location remote from it. The travel time of the radio signal is negligible compared to that of the sound signal.

distance of relative movement. The distance traveled relative to a reference point, itself usually in motion.

distance resolution. See RANGE RESOLUTION.

Distances Between Ports. See PUB. 151.

Distances Between United States Ports. A publication of the National Ocean Survey providing calculated distances in nautical miles over water areas between United States ports. A similar publication published by the Defense Mapping Agency Hydrographic/Topographic Center for foreign waters is entitled *Distances Between Ports*.

diurnal, *adj.* Having a period or cycle of approximately 1 day. The tide is said to be diurnal when only one high water and one low water occur during a tidal day, and the tidal current is said to be *diurnal* when there is a single flood and single ebb period in the tidal day. A rotary current is diurnal if it changes its direction through 360° once each tidal day. A diurnal constituent is one which has a single period in the constituent day. See also STATIONARY WAVE THEORY, TYPE OF TIDE.

diurnal aberration. See under ABERRATION, definition 1.

diurnal age. See AGE OF DIURNAL INEQUALITY.

diurnal circle. The apparent daily path of a celestial body, approximating a PARALLEL OF DECLINATION.

diurnal current. Tidal current in which the tidal-day current cycle consists of one flood current and one ebb current, separated by slack water; or a change in direction of 360° of a rotary current. A SEMIDIURNAL CURRENT is one in which two floods and two ebbs, or two changes of 360°, occur each tidal day.

diurnal inequality. The difference in height of the two high waters or of the two low waters of each tidal day; the difference in speed between the two flood tidal currents or the two ebb tidal currents of each tidal day. The difference changes with the declination of the moon and to a lesser extent with declination of the sun. In general, the inequality tends to increase with an increasing declination, either north or south. **Mean diurnal high water inequality** is one-half the average difference between the two high waters of each day observed over a specific 19-year Metonic cycle (the National Tidal Datum Epoch). It is obtained by subtracting the mean of all high waters from the mean of the higher high waters. **Mean diurnal low water inequality** is one-half the average difference between the two low waters of each day observed over a specific 19-year Metonic cycle (the National Tidal Datum Epoch). It is obtained by subtracting the mean of the lower low waters from the mean of all low waters. **Tropic high water inequality** is the average difference between the two high waters of the day at the times of the tropic tides. **Tropic low water inequality** is the average difference between the two low waters of the day at the times of the tropic tides. Mean and tropic inequalities as defined above are applicable only when the type of tide is either semidiurnal or mixed. Sometimes called DECLINATIONAL INEQUALITY.

diurnal motion. The apparent daily motion of a celestial body.

diurnal parallax. See GEOCENTRIC PARALLAX.

diurnal range. See GREAT DIURNAL RANGE.

diurnal tide. See under TYPE OF TIDE; DIURNAL, *adj.*

dive, *n.* Submergence with one end foremost.

dive, *v., i.* To submerge with one end foremost.

diverged beam. See under FAN BEAM.

dividers, *n.* An instrument consisting in its simple form of two pointed legs joined by a pivot, and used principally for measuring distances or coordinates. If the legs are pointed at both ends and provided with an adjustable pivot, the instrument is called **proportional dividers.** An instrument having one pointed leg and one leg carrying a pen or pencil is called COMPASSES.

D-layer, *n.* The lowest of the ionized layers in the upper atmosphere, or ionosphere. It is present only during daylight hours, and its density is proportional to the altitude of the sun. The D-layer's only significant effect upon radio waves is its tendency to absorb their energy, particularly at frequencies below 3 megahertz. High angle radiation and signals of a frequency greater than 3 megahertz may penetrate the D-layer and be refracted or reflected by the somewhat higher E-layer.

dock, *n.* 1. The slip or waterway between two piers, or cut into the land for the berthing of ships. A PIER is sometimes erroneously called a DOCK. Also called SLIP. See also JETTY; LANDING, definition 1; QUAY; WHARF.

2. A basin or enclosure for reception of vessels, and provided with means for controlling the water level. A **wet dock** is one in which water can be maintained at various levels by closing a gate when the water is at the desired level. A **dry dock** is a dock providing support for a ship, and means of removing the water so that the bottom of the ship or other craft can be exposed. A dry dock consisting of an artificial basin is called a **graving dock**; one consisting of a floating structure is called a **floating dock.**

3. Used in the plural, a term used to describe area of the docks, wharves, basins, quays, etc.

dock, *v., t.* To place in a dock.

docking signals. See TRAFFIC CONTROL SIGNALS.

dock sill. The foundation at the bottom of the entrance to a dry dock or lock against which the caisson or gates close. The depth of water controlling the use of the dock or lock is measured from the sill to the surface.

dockyard, *n. British terminology.* Shipyard.

doctor, *n.* 1. A cooling sea breeze in the Tropics.

2. See HARMATTAN.

3. The strong southeast wind which blows on the south African coast. Usually called CAPE DOCTOR.

dog days. The period of greatest heat in the summer.

doldrums, *n., pl.* The equatorial belt of calms or light variable winds, lying between the two trade wind belts. Also called EQUATORIAL CALMS.

dolphin, *n.* A post or group of posts, used for mooring or warping a vessel. The dolphin may be in the water, on a wharf, or on the beach. See PILE DOLPHIN.

dome, *n.* A label on a nautical chart which indicates a large, rounded, hemispherical structure rising from a building or a roof of the same shape. A prominent example is that of the Capitol of the United States in Washington, D. C.

dome-shaped iceberg. A solid type iceberg with a large, round, smooth top.

doppler effect. First described by Christian Johann Doppler in 1842, an effect observed as a frequency shift which results from relative motion between a transmitter and receiver or reflector of acoustic or electromagnetic energy. The effect on electromagnetic energy is used in doppler satellite navigation to determine an observer's position relative to a satellite. The effect on ultrasonic energy is used in **doppler sonar speed logs** to measure the relative motion between the vessel and the reflective sea bottom (for bottom return mode) or suspended particulate matter in the seawater itself (for volume reverberation mode). The velocity so obtained and integrated with respect to time is used in **doppler sonar navigators** to determine position with respect to a start point. The doppler effect is also used in **docking aids** which provide precise speed measurements. Also called DOPPLER SHIFT.

doppler navigation. The use of the doppler effect in navigation. See also DOPPLER SONAR NAVIGATION, DOPPLER SATELLITE NAVIGATION BASIC PRINCIPLES.

doppler radar. Any form of radar which detects radial motion of a distant object relative to a radar apparatus by means of the change of the radio frequency of the echo signal due to motion.

doppler satellite navigation. The use of a navigation system which determines positions based on the doppler effect of signals received from an artificial satellite.

doppler satellite navigation basic principles. Measurement of the satellite doppler curve in the vicinity of zero doppler shift allows a determination of the time and the line-of-sight range (slant range) the closest point of approach of the satellite to the observer. This measurement enables determining the observer's position relative to the satellite in a coordinate system which is local to the satellite at the time of closest approach and whose axes are parallel and perpendicular to the direction the satellite travels over the earth at that time. Knowing the satellite's orbit, its subtrack over the earth's surface can be determined. Knowing the minimum slant range to the satellite and the altitude of the satellite, the distance from this subtrack can be found. This can be thought of as the cross-track coordinate of the observer relative to the satellite. Knowing the exact time when this minimum slant range occurred, the position of the satellite along its earth subtrack is known which can be considered the along-track coordinate of the observer relative to the satellite. Knowledge of the satellite orbit in detail tells how to transform the observer's relative along-track and cross-track coordinates to latitude and longitude. See also INTEGRAL DOPPLER NAVIGATION, NAVY NAVIGATION SATELLITE SYSTEM.

doppler shift. See DOPPLER EFFECT.

doppler sonar navigation. The use of the doppler effect observed as a frequency shift resulting from relative motion between a transmitter and receiver of ultrasonic energy to measure the relative motion between the vessel and the reflective sea bottom (for **bottom return mode**) or suspended particulate matter in the seawater itself (for **volume reverberation mode**) to determine the vessel's velocity. The velocity so obtained by a **doppler sonar speed log** may be integrated with respect to time to determine distance travelled. This integration of velocity with time is correlated with direction of travel in a **doppler sonar navigator** to determine position with respect to a start point. The doppler effect is also used in **docking aids** to provide precise speed measurements.

double, *v., t.* To travel around with a near reversal of course, as a vessel *doubles a cape*. See also ROUND.

double altitudes. See EQUAL ALTITUDES.

double ebb. An ebb tidal current having two maxima of speed separated by a lesser ebb speed.

double flood. A flood tidal current having two maxima of speed separated by a lesser flood speed.

double interpolation. Interpolation when there are two arguments or variables.

double sextant. A sextant designed to enable the observer to simultaneously measure the left and right horizontal sextant angles of the three-point problem. The sextant consists of a circle with two indices inscribed and two coaxial index mirrors. The horizon glass is silvered top and bottom but is clear in the middle to permit direct observation of the center object ashore.

double stabilization. See under STABILIZATION OR RADARSCOPE DISPLAY.

double star. Two stars appearing close together. If they appear close because they are in nearly the same line of sight but differ greatly in distance from the observer, they are called an **optical double star;** if in nearly the same line of sight and at approximately the same distance from the observer, they are called a **physical double star.** If they revolve about their common center of mass, they are called a **binary star.**

double summer time. See under SUMMER TIME.

doublet antenna. See DIPOLE ANTENNA.

double tide. A double-headed tide, that is, a high water consisting of two maxima of nearly the same height separated by a relatively small depression, or a low water consisting of two minima separated by a relatively small elevation. Sometimes called AGGER. See also GULDER.

doubling the angle on the bow. A method of obtaining a running fix by measuring the distance a vessel travels on a steady course while the relative bearing (right or left) of a fixed object doubles. The distance from the object at the time of the second bearing is equal to the run between bearings, neglecting drift.

doubly stabilized. See under STABILIZATION OF RADARSCOPE DISPLAY.

doubtful, *adj.* Of questioned accuracy. This term is sometimes used in connection with bearings reported by a radio direction finder station. APPROXIMATE or SECOND CLASS may be used with the same meaning.

doubtful sounding. Of uncertain depth. The expression, as abbreviated, is used principally on charts to indicate a position where the depth may be less than indicated, the position not being in doubt.

down, *n.* 1. See DUNE.

2. An area of high, treeless ground, usually undulating and covered with grass.

down by the head. Having greater draft at the bow than at the stern. The opposite is DOWN BY THE STERN or BY THE STERN. Also called BY THE HEAD.

down by the stern. Having greater draft at the stern than at the bow. The opposite is DOWN BY THE HEAD or BY THE HEAD. Also called BY THE STERN. See DRAG, *n.*, definition 3.

downstream, *adj. & adv.* In the direction of flow of a current or stream. The opposite is UPSTREAM.

down-the-scope echo. See CLASSIFICATION OF RADAR ECHOES.

downwind, *adj. & adv.* In the direction toward which the wind is blowing. The term applies particularly to the situation of moving in this direction, whether desired or not. BEFORE THE WIND implies assistance from the wind in making progress in a desired direction. LEEWARD applies to the direction toward which the wind blows, without implying motion. The opposite is UPWIND.

draft, *n.* The depth to which a vessel is submerged. Draft is customarily indicated by numerals called DRAFT MARKS at the bow and stern. It may also be determined by means of a DRAFT GAUGE.

draft gauge. A hydrostatic instrument installed in the side of a vessel, below the light load line, to indicate the depth to which a vessel is submerged.

drafting machine. See PARALLEL MOTION PROTRACTOR.

draft marks. Numerals placed on the sides of a vessel, customarily at the bow and stern, to indicate the depth to which a vessel is submerged.

drag, *n.* 1. See SEA ANCHOR.

2. Short for WIRE DRAG.

3. The designed difference between the draft forward and aft when a vessel is down by the stern. See also TRIM, definition 1.

4. The retardation of a ship when in shallow water.

5. Short for ATMOSPHERIC DRAG.

drag, *v., t.* 1. To tow a line or object below the surface, to determine the least depth in an area or to insure that a given area is free from navigational dangers to a certain depth. DRAG and SWEEP have nearly the same meanings. DRAG refers particularly to the location of obstructions, or the determination that obstructions do not exist. SWEEP may include, additionally, the removal of any obstructions located.

2. To pull along the bottom, as an anchor.

dragging, *n.* 1. The process of towing a wire or horizontally set bar below the surface, to determine the least depth in an area or to insure that a given area is free from navigational dangers to a certain depth.

2. The process of pulling along the bottom, as an anchor.

draw, *v., i.* 1. To be immersed to a specified draft, as a vessel *draws* 12 feet.
2. To change relative bearing, as to *draw aft* or *draw forward.*

dredge, *n.* A water craft used to dredge an area.

dredge, *v., t.* To remove solid matter from the bottom of a water area.

dredging area. An area where a concentration of dredging vessels may be encountered, taking up sand or shingle to be brought ashore for construction purposes. Channels dredged to provide an adequate depth of water for navigation are not to be considered as dredging areas.

dredging buoy. A buoy marking the limit of an area where dredging is being performed. See also SPOIL GROUND BUOY.

Dreisonstok, *n.* See H.O. PUB. NO. 208.

dried ice. Sea ice from the surface of which meltwater has disappeared after the formation of cracks and thaw holes. During the period of drying, the surface whitens.

drift, *n.* 1. The speed of a current as defined in CURRENT, definition 1.
2. The speed of the current as defined in CURRENT, definition 2.
3. The distance a craft is moved by current and wind.
4. Downwind or downcurrent motion of airborne or waterborne objects due to wind or current.
5. Material moved from one place and deposited in another, as sand by a river, rocks by a glacier, material washed ashore and left stranded, snow or sand piled up by wind. Rock material deposited by a glacier is also called ERRATIC.
6. The horizonal component of real precession or apparent precession, or the algebraic sum of the two. When it is desired to differentiate between the sum and its components, the sum is called total drift.

drift, *v., i.* To move by action of wind or current without control.

drift angle. 1. The angle between the tangent to the turning circle and the centerline of the vessel during a turn.
2. The angular difference between the ground track and the water track. See also LEEWAY ANGLE.

drift axis. Of a gyroscope, the axis about which drift occurs. In a directional gyro with the spin axis mounted horizontally the drift axis is the vertical axis. See also SPIN AXIS, TOPPLE AXIS.

drift bottle. An identifiable float allowed to drift with ocean currents to determine their sets and drifts.

drift current. A wide, slow-moving ocean current principally caused by winds. Examples are the NORTH ATLANTIC DRIFT and the WEST WIND DRIFT. Also called WIND DRIFT CURRENT. See also EKMAN SPIRAL.

drifting snow. Snow raised from the ground and carried by the wind to such a height that the horizontal visibility is considerably reduced, but the vertical visibility is not materially diminished. The expression BLOWING SNOW is used when both the horizontal and vertical visibility are considerably reduced.

drift lead. (lĕd). A lead placed on the bottom to indicate movement of a vessel. At anchor the lead line is usually secured to the rail with a little slack and if the ship drags anchor, the line tends forward. A drift lead is also used to indicate when a vessel coming to anchor is dead in the water or when it is moving astern. A drift lead can be used to indicate current if a ship is dead in the water.

drilling rig. A term used solely to indicate a mobile drilling structure, such as a semi-submersible rig, a "jack-up" rig, or a drilling ship. A drilling rig is not charted except in the rare cases where it is converted to a permanent production platform.

drizzle, *n.* Very small, numerous, and uniformly dispersed water drops that may appear to float while following air currents. Unlike fog droplets, drizzle falls to the ground. It usually falls from low stratus clouds and is frequently accompanied by low visibility and fog. See also MIST.

drogue, *n.* 1. See SEA ANCHOR.
2. A current measuring assembly consisting of a weighted parachute and an attached surface buoy.

drought, *n.* A protracted period of dry weather.

droxtal, *n.* A very small ice particle (about 10 to 20 microns in diameter) formed by the direct freezing of supercooled water droplets at temperatures below −30°C. Droxtals cause most of the restriction to visibility in ice fog.

dry-bulb temperature. The temperature of the air, as indicated by the dry-bulb thermometer of a psychrometer.

dry-bulb thermometer. A thermometer with an uncovered bulb, used with a wet-bulb thermometer to determine atmosphere humidity. The two thermometers constitute the essential parts of a PSYCHROMETER.

dry compass. A compass without a liquid-filled bowl, particularly a magnetic compass having a very light compass card. Such a magnetic compass is seldom, if ever, used in modern practice. See also LIQUID COMPASS.

dry dock. A dock providing support for a vessel, and means for removing the water so that the bottom of the vessel can be exposed. A dry dock consisting of an artificial basin is called a **graving dock**; one consisting of a floating structure is called a **floating dock.** See also MARINE RAILWAY.

dry-dock, *v., t.* To place in a dry dock.

drydock iceberg. An iceberg eroded in such manner that a large U-shaped slot is formed with twin columns. The slot extends into or near the waterline.

dry fog. A fog that does not moisten exposed surfaces.

dry harbor. A small harbor which either dries at low water or has insufficient depths to keep vessels afloat during all states of the tide. Vessels using it must be prepared to *take the ground* on the falling tide.

dry haze. See under HAZE.

drying heights. Heights above chart sounding datum of those features which are periodically covered and exposed by the rise and fall of the tide.

dual-carrier radiobeacon. A continuous carrier radiobeacon in which identification is accomplished by means of a keyed second carrier. The frequency difference between the two carriers is made equal to the desired audio frequency. The object of the system is to reduce the bandwidth of the transmission.

dual-rate blanking. To provide continuous service from one Loran-C chain to the next, some stations are operated as members of two chains and radiate signals at both rates. Such a station is faced periodically with an impossible requirement to radiate two overlapping pulse groups at the same time. During the time of overlap, the subordinate signal is blanked or suppressed.

Blanking is accomplished in one of two ways: priority blanking in which case one rate is always superior or alternate blanking in which case the two rates alternate in the superior and subordinate roll.

duct, *n.* See as TROPOSPHERIC RADIO DUCT.

due, *adv.* Exactly.

dumb compass. See PELORUS.

dummy antenna. A substantially non-radiating device used to simulate an antenna with respect to input impedance over some specified range of frequencies. Also called ARTIFICIAL ANTENNA.

dumping ground. Although shown on nautical charts as dumping grounds in United States waters, the Federal regulations for these areas have been revoked and their use for dumping discontinued. These areas will continue to be shown on nautical charts until such time as they are no longer considered to be a danger to navigation. See also DUMP SITE, SPOIL AREA, DISPOSAL AREA.

dump site. Area established by Federal regulation in which dumping of dredged and fill material and other nonbuoyant objects is allowed with the issuance of a permit. Dump sites are shown on nautical charts. See also DISPOSAL AREA, DUMPING GROUND, SPOIL AREA.

dune, *n.* A mound ridge, or hill of sand piled up by the wind on the shore or in a desert. Also called SAND DUNE, DOWN.

duplexer, *n.* A device which permits a single antenna system to be used for both transmitting and receiving.

duration of flood and duration of ebb. *Duration of flood* is the interval of time in which a tidal current is flooding, and the *duration of ebb* is the interval in which it is ebbing; these intervals being reckoned from the middle of the intervening slack waters or minimum currents. Together they cover, on an average, a period of 12.42 hours for a semidiurnal tidal current or a period of 24.84 hours for a diurnal current. In a normal semidiurnal tidal current, the duration of flood and duration of ebb will each be approximately equal to 6.21 hours, but the times may be modified greatly by the presence of a nontidal flow. In a river the duration of ebb is usually longer than the duration of flood because of the fresh water discharge, especially during the spring months when snow and ice melt are the predominant influences. See also DURATION OF RISE AND DURATION OF FALL.

duration of rise and duration of fall. *Duration of rise* is the interval from low water to high water, and *duration of fall* is the interval from high water to low water. Together they cover, on an average, a period of 12.42 hours for a semidiurnal tide or a period of 24.84 hours for a diurnal tide. In a normal semidiurnal tide, the duration of rise and duration of fall will each be approximately equal to 6.21 hours, but in shallow waters and in rivers there is a tendency for a decrease in the duration of rise and a corresponding increase in the duration of fall. See also DURATION OF FLOOD AND DURATION OF EBB.

dusk, *n.* The darker part of twilight; that part of twilight between complete darkness and the darker limit of civil twilight, both morning and evening.

dust devil. A well-developed dust whirl; a small but vigorous whirlwind, usually of short duration, rendered visible by dust, sand, and debris picked up from the ground. Diameters of dust devils range from about 10 feet to greater than 100 feet; their average height is about 600 feet, but a few have been observed as high as several thousand feet. They have been observed to rotate anticyclonically as well as cyclonically. Dust devils are best developed on a hot, calm afternoon with clear skies, in a dry region when intense surface heating causes a very steep lapse rate of temperature in the lower few hundred feet of the atmosphere.

duststorm, *n.* An unusual, frequently severe weather condition characterized by strong winds and dust-filled air over an extensive area. Prerequisite to a duststorm is a period of drought over an area of normally arable land, thus providing the very fine particles of dust which distinguish it from the much more common SANDSTORM.

dust whirl. A rapidly rotating column of air (whirlwind) over a dry and dusty or sandy area, carrying dust, leaves, and other light material picked up from the ground. When well developed, it is called DUST DEVIL.

Dutchman's log. A buoyant object thrown overboard to determine the speed of a vessel. The time required for a known length of the vessel to pass the object is measured. The time and distance being known, the speed can be computed.

duty cycle. Of pulse radar, an expression of the fraction of the total time that radio-frequency energy is radiated. It is the ratio of pulse length to pulse repetition time.

dynamical mean sun. A fictitous sun conceived to move eastward along the ecliptic at the average rate of the apparent sun. The dynamical mean sun and the apparent sun occupy the same position when the earth is at perihelion in January. See also MEAN SUN.

dyne, *n.* A force which imparts an acceleration of 1 centimeter per second to a mass of 1 gram. The dyne is the unit of force in the centimeter-gram-second system. It corresponds to 10^{-5} newton in the International System of Units.

E

earth-centered ellipsoid. A reference ellipsoid whose geometric center coincides with the earth's center of gravity and whose semiminor axis coincides with the earth's rotational axis.

earth-fixed coordinate system. Any coordinate system in which the axes are stationary with respect to the earth. See also INERTIAL COORDINATE SYSTEM.

earthlight, *n.* The faint illumination of the dark part of the moon by sunlight reflected from the earth. Also called EARTHSHINE.

earth rate. The angular velocity or rate of the earth's rotation. See also EARTH-RATE CORRECTION, HORIZONTAL EARTH RATE, VERTICAL EARTH RATE.

earth-rate correction. A rate applied to a gyroscope to compensate for the apparent precession of the spin axis caused by the rotation of the earth. See also EARTH RATE, HORIZONTAL EARTH RATE, VERTICAL EARTH RATE.

earth satellite. A body that orbits about the earth. See also ARTIFICIAL EARTH SATELLITE.

earthshine, *n.* See EARTHLIGHT.

earth tide. Periodic movement of the earth's crust caused by the gravitational interactions between the sun, moon, and earth.

east, *n.* The direction 90° to the right of north. See also CARDINAL POINT.

East Africa Coastal Current. An Indian Ocean current which originates mainly from that part of the Indian South Equatorial Current which turns northward off the northeast coast of Africa in the vicinity of latitude 10° S. The current appears to vary considerably in speed and direction from month to month. The greatest changes coincide with the period of the opposing northeast monsoon during November through March. This coastal current is most persistent in a north or northeast direction and strongest during the southwest monsoon from May through September, particularly during August. Speed and frequency begin to decrease during the transition month of October. In November at about latitude 4° N a part of the current begins to reverse; this part expands northward and southward until February. The region of reverse flow begins to diminish in March and disappears in April, when the northward set again predominates. Also called SOMALI CURRENT. See also MONSOON.

East Australia Current. A South Pacific Ocean current flowing southward along the east coast of Australia, from the Coral Sea to a point northeast of Tasmania, where it turns to join the northeastward flow through the Tasman Sea. It is formed by that part of the Pacific South Equatorial Current that turns south east of Australia. In the southern hemisphere summer, a small part of this current flows westward along the south coast of Australia into the Indian Ocean. The East Australia Current forms the western part of the general counterclockwise oceanic circulation of the South Pacific Ocean.

eastern standard time. See STANDARD TIME.

East Greenland Current. An ocean current flowing southward along the east coast of Greenland, carrying water of low salinity and low temperature. The East Greenland Current is joined by most of the water of the Irminger Current. The greater part of the current continues through Denmark Strait between Iceland and Greenland, but one branch turns to the east and forms a portion of the counterclockwise circulation in the southern part of the Norwegian Sea. Some of the East Greenland Current curves to the right around the tip of Greenland, flowing northward into Davis Strait as the WEST GREENLAND CURRENT. The main discharge of the Arctic Ocean is via the East Greenland Current.

easting, *n.* The distance a craft makes good to the east. The opposite is WESTING.

East Siberian Coastal Current. An ocean current in the Chukchi Sea which joins the northward flowing Bering Current north of East Cape.

easy area. A general qualitative expression to indicate, in a relative manner, that ice conditions prevailing in an area are such that navigation in it is not difficult.

ebb, *n.* Tidal current moving away from land or down a tidal stream. The opposite is FLOOD. Sometimes the terms EBB and FLOOD are also used with reference to vertical tidal movement, but for this vertical movement the expressions FALLING TIDE and RISING TIDE are considered preferable. Also called EBB CURRENT.

ebb axis. The average direction of current at strength of ebb.

ebb current. The movement of a tidal current away from shore or down a tidal river or estuary. In the mixed type of reversing tidal current, the terms *greater ebb* and *lesser ebb* are applied respectively to the ebb tidal currents of greater and lesser speed of each day. The terms *maximum ebb* and *minimum ebb* are applied to the maximum and minimum speeds of a current running continuously ebb, the velocity alternately increasing and decreasing without coming to a slack or reversing. The expression *maximum ebb* is also applicable to any ebb current at the time of greatest speed. The opposite is FLOOD CURRENT.

ebb interval. Short for STRENGTH OF EBB INTERVAL. The interval between the transit of the moon over the meridian of a place and the time of the following strength of ebb. See also LUNICURRENT INTERVAL.

ebb strength. Phase of the ebb tidal current at the time of maximum velocity. Also, the velocity at this time. Also called STRENGTH OF EBB.

eccentric, *adj.* Not having the same center. The opposite is CONCENTRIC.

eccentric angle. See under ANOMALY, definition 2.

eccentric anomaly. See under ANOMALY, definition 2.

eccentric error. See CENTERING ERROR.

eccentricity, *n.* 1. Degree of deviating from a center.

2. Of an ellipse, the ratio of the distance between foci to the length of the major axis, or the ratio of the distance between the center and a focus to the length of the semimajor axis.

3. The ratio of the distances from any point of a conic section to a focus and the corresponding directrix.

eccentricity component. That part of the equation of time due to the ellipticity of the orbit and known as the eccentricity component is the difference, in mean solar time units, between the hour angles of the apparent (true) sun and the dynamical mean sun. It is also the difference in the right ascensions of these two suns.

echo, *n.* 1. A wave which has been reflected or otherwise returned with sufficient magnitude and delay to be perceived.
2. A signal reflected by a target to a radar antenna. Also called RETURN.
3. The deflection or indication on a radarscope representing a target. Also called PIP, BLIP, RETURN.

echo box. A resonant cavity, energized by part of the transmitted pulse of a radar set, which produces an artificial target signal for tuning or testing the overall performance of a radar set. Also called PHANTOM TARGET.

echo box performance monitor. See under PERFORMANCE MONITOR.

echogram, *n.* A graphic record of depth measurements obtained by an echo sounder. See also FATHOGRAM.

echo ranging. The determination of distance by measuring the time interval between transmission of a radiant energy signal and the return of its echo. Since echo ranging equipment is usually provided with means for determining direction as well as distance, both functions are generally implied. The expression is customarily applied only to ranging by utilization of the travel of sonic or ultrasonic signals through water. See also RADIO ACOUSTIC RANGING, SONAR.

echo sounder. An instrument used to determine water depth by measuring the time interval for sound waves to go from a source of sound near the surface to the bottom and back again. Also called DEPTH FINDER, ACOUSTIC DEPTH FINDER.

echo sounding. Determination of the depth of water by measuring the time interval between emission of a sonic or ultrasonic signal and the return of its echo from the bottom. The instrument used for this purpose is called an ECHO SOUNDER. Also called ACOUSTIC SOUNDING.

eclipse, *n.* 1. The obscuration of a source of light by the intervention of an object. When the moon passes between the earth and the sun, casting a shadow on the earth, a **solar eclipse** takes place within the shadow. When the moon enters the earth's shadow, a **lunar eclipse** occurs. When the moon enters only the penumbra of the earth's shadow, a **penumbral lunar eclipse** occurs. A solar eclipse is **partial** if the sun is partly obscured; total if the entire surface is obscured; or **annular** if a thin ring of the sun's surface appears around the obscuring body. A lunar eclipse can be either total or partial.
2. An interval of darkness between appearances of a navigation light.

eclipse year. The interval between two successive conjunctions of the sun with the same node of the moon's orbit, averaging 346 days, 14 hours, 52 minutes 50.7 seconds in 1900, and increasing at the rate of 2.8 seconds per century.

ecliptic, *n.* The apparent annual path of the sun among the stars; the intersection of the plane of the earth's orbit with the celestial sphere. This is a great circle of the celestial sphere inclined at an angle of about 23°27′ to the celestial equator. See also ZODIAC.

ecliptic diagram. A diagram of the zodiac, indicating the positions of certain celestial bodies in this region.

ecliptic pole. On the celestial sphere, either of the two points 90° from the ecliptic.

ecliptic system of coordinates. A set of celestial coordinates based on the ecliptic as the primary great circle; celestial latitude and celestial longitude.

eddy, *n.* A quasi-circular movement of water whose area is relatively small in comparison to the current with which it is associated. Eddies may be formed between two adjacent currents flowing counter to each other and where currents pass obstructions, especially on the downstream side. See also WHIRLPOOL.

effective radiated power. The power supplied to the antenna multiplied by the relative gain of the antenna in a given direction.

effective radius of the earth. The radius of a hypothetical earth for which the distance to the radio horizon, assuming rectilinear propagation, is the same as that for the actual earth with an assumed uniform vertical gradient of a refractive index. For the standard atmosphere, the effective radius is 4/3 that of the actual earth.

Ekman spiral. A logarithmic spiral (when projected on a horizontal plane) formed by current velocity vectors at increasing depth intervals. The current vectors become progressively smaller with depth. They spiral to the right (looking in the direction of flow) in the Northern Hemisphere and to the left in the Southern with increasing depth. Theoretically, the surface current vector sets 45° from the direction toward which the wind is blowing. Flow opposite to the surface current occurs at the *depth of frictional resistance*. The phenomenon occurs in wind drift currents in which only the Coriolis and frictional forces are significant. Named for Vagn Walfrid Ekman who, assuming a constant eddy viscosity, steady wind stress, and unlimited depth and extent, published on the effect in 1905.

E-layer, *n.* From the standpoint of its effect upon radio wave propagation, the lowest useful layer of the Kennelly-Heaviside region. Its average height is about 70 statute miles, and its density is greatest about local apparent noon. For all practical intents and purposes, the layer disappears during the hours of darkness.

elbow, *n.* A sharp change in direction of a coast line, a channel, bank, etc.

electrical distance. A distance expressed in terms of the duration of travel of an electromagnetic wave in a given medium, between two points.

electrically suspended gyro. A gyroscope in which the main rotating element is suspended by a magnetic field or any other similar electrical phenomenon. See also GYRO, ELECTROSTATIC GYRO.

electrical storm. See THUNDERSTORM.

electric field. That region in space which surrounds an electrically charged object and in which the forces due to this charge are detectable. See also ELECTRIC VECTOR.

electric tape gage. A tide gage consisting of a monel metal tape on a metal reel (with supporting frame), voltmeter, and battery. The tape is graduated with numbers increasing toward the unattached end. Tidal heights can be measured directly by unreeling the tape into its stilling well. When contact is made with the water's surface, the circuit is completed and the voltmeter needle moves. At that moment, the length of tape is read against an index mark; the mark having a known elevation relative to the tidal bench marks. Used at many long term control stations in place of the tide staff.

electric vector. The component of the electromagnetic field associated with electromagnetic radiation which is of the nature of an electric field. The electric vector is considered to coexist with, but to act at right angles to, the magnetic vector.

electrode, *n.* A terminal at which electricity passes from one medium into another. The positive electrode is called **anode**; the negative electrode is called **cathode.**

electromagnetic, *adj.* Of, pertaining to, or produced by electromagnetism.

electromagnetic energy. All forms of radiant energy, such as radio waves, light waves, X-rays, heat waves, gamma rays, and cosmic rays.

electromagnetic field. 1. The field of influence which an electric current produces around the conductor through which it flows.
2. A rapidly moving electric field and its associated magnetic field located at right angles to both electric lines of force and to their direction of motion.
3. The magnetic field resulting from the flow of electricity.

electromagnetic horn. *British terminology.* See OSCILLATOR.

electromagnetic log. A log containing an electromagnetic sensing element extended below the hull of the vessel, which produces a voltage directly proportional to speed through the water.

electromagnetic waves. Waves of associated electric and magnetic fields characterized by variations of the fields. The electric and magnetic fields are at right angles to each other and to the direction of propagation. The waves are propagated at the speed of light and are known as **radio (Hertzian) waves, infrared rays, light, ultraviolet rays, X-rays,** etc., depending on their frequencies.

electromagnetism, *n.* 1. Magnetism produced by an electric current.
2. The science dealing with the physical relations between electricity and magnetism.

electron, *n.* A negatively-charged particle of matter constituting a part of an atom. Its electric charge is the most elementary unit of negative electricity.

electron gun. A group of electrodes which produces an electron beam of controllable intensity. By extension, the expression is often used to include, also, the elements which focus and deflect the beam.

electronic aid to navigation. An aid to navigation using electronic equipment. If the navigational information is transmitted by radio waves, the device may be called a RADIO AID TO NAVIGATION.

electronic bearing cursor. Of a marine radar set, the bright rotatable radial line on the PPI used for bearing determination.

electronic cursor. Short for ELECTRONIC BEARING CURSOR.

electronic distance measuring devices. Instruments that measure the phase differences between transmitted and reflected or retransmitted electromagnetic waves of known frequency, or that measure the round-trip transit time of a pulsed signal, from which distance is computed.

electronic navigation. Navigation by means of electronic equipment. The expression ELECTRONIC NAVIGATION is more inclusive than RADIONAVIGATION, since it includes navigation involving any electronic device or instrument. Because of the extent of the use of electronics in navigation equipment other than those depending upon information transmitted or received by radio waves, the term *electronic navigation* has limited value as a term for a division of navigation.

electronics, *n.* The science and technology relating to the emission, flow, and effects of electrons in vacuo or through a semiconductor such as a gas, and to systems using devices in which this action takes place.

electronic telemeter. An electronic device that measures the phase difference or transit time between a transmitted electromagnetic impulse of known frequency and speed and its return.

electrostatic gyro. A gyroscope in which a small ball rotor is electrically suspended within an array of electrodes in a vacuum inside a ceramic envelope. See also GYRO, ELECTRICALLY SUSPENDED GYRO.

elements of a fix. The specific values of the coordinates used to define a position.

elephanta, *n.* A strong southerly or southeasterly wind which blows on the Malabar coast of India during the months of September and October and marks the end of the southwest monsoon.

elevated duct. A tropospheric radio duct of which the lower boundary is above the surface of the earth.

elevated pole. The celestial pole above the horizon, agreeing in name with the latitude. The celestial pole below the horizon is called DEPRESSED POLE.

elevation, *n.* 1. Vertical distance of a point above a datum, usually mean sea level. Elevation usually applies to a point on the surface of the earth. The term HEIGHT is used for points on or above the surface. See also SPOT ELEVATION.
2. An area higher than its surroundings, as a hill.

elevation angle. See ANGLE OF ELEVATION.

elevation tints. See HYPSOMETRIC TINTING.

elimination, *n.* One of the final processes in the harmonic analysis of tides in which preliminary values of the harmonic constants of a number of constituents are cleared of residual effects of each other.

E link. A bracket attached to one of the arms of a binnacle to permit the mounting of a quadrantal corrector in an intermediate position between the fore-and-aft and athwartship lines through a magnetic compass.

ellipse, *n.* A plane curve constituting the locus of all points the sum of whose distances from two fixed points called FOCI is constant; an elongated circle. The orbits of planets, satellites, planetoids, and comets are ellipses with the center of attraction at one focus. See also CONIC SECTION, CURRENT ELLIPSE.

ellipsoid, *n.* A surface whose plane sections (cross-sections) are all ellipses or circles, or the solid enclosed by such a surface. Also called ELLIPSOID OF REVOLUTION, SPHEROID.

ellipsoidal height. The height above the reference ellipsoid, measured along the ellipsoidal outer normal through the point in question. Also called GEODETIC HEIGHT.

ellipsoid of reference. See REFERENCE ELLIPSOID.

ellipsoid of revolution. An ellipsoid; a term used because of the fact that the ellipsoid can be formed by revolving an ellipse about one of its axes. Also called ELLIPSOID OF ROTATION.

ellipsoid of rotation. See ELLIPSOID OF REVOLUTION.

elliptically polarized wave. An electromagnetic wave which can be resolved into two plane polarized waves which are perpendicular to each other and which propagate in the same direction. The amplitudes of the waves may be equal or unequal and of arbitrary time-phase. The tip of the component of the electric field vector in the plane normal to the direction of propagation describes an ellipse. See also CIRCULARLY POLARIZED WAVE.

ellipticity, *n.* The amount by which a spheroid differs from a sphere or an ellipse differs from a circle, found by dividing the difference in the lengths of the semiaxes of the ellipse by the length of the semimajor axis. See also FLATTENING.

elongation, *n.* The angular distance of a body of the solar system from the sun; the angle at the earth between lines to the sun and another celestial body of the solar system. The **greatest elongation** is the maximum angular distance of an *inferior planet* from the sun before it starts back toward conjunction. The direction of the body east or west of the sun is usually specified, as *greatest elongation east* (or *west*).

embayed, *adj.* 1. Formed into or having bays.
2. A vessel in a bay unable to put to sea or put to sea safely because of wind, current, or sea is said to be *embayed*.

embayment, *n.* Any indentation of a coast regardless of width at the entrance or depth of penetration into the land. See also ESTUARY.

emergency light. A light put into service in emergency when the permanent or standby light has failed. It often provides a reduced service in comparison with the permanent light.

Emergency Position Indicating Radiobeacon. A small portable radiobeacon carried by vessels and aircraft, which transmits on 2180kHz, 121.5MHz, and 243MHz, the emissions from which can be used by search and rescue vessels and by aircraft for locating survivors. The frequencies 121.5MHz and 243MHz are aircraft frequencies operating with amplitude modulation and are used for homing purposes.

emergency position-indicating radiobeacon station. As defined by the *International Telecommunication Union* (ITU), a station in the mobile service the emissions of which are intended to facilitate search and rescue operations.

emission delay. 1. A delay in the transmission of a pulse signal from a slave (or secondary) station of a hyperbolic radionavigation system, introduced as an aid in distinguishing between master and slave (or secondary) station signals.
2. In Loran-C the time interval between the master station's transmission and the secondary station's transmission in the same group repetition interval (GRI). The GRI is selected of sufficient duration to provide time for each station to transmit its pulse group and additional time between each pulse group so that signals from two or more stations cannot overlap in time anywhere within the coverage area. In general, emission delays are kept as small as possible to allow the use of the smallest GRI.

empirical, *adj.* Derived by observation or experience rather than by rules or laws.

endless tangent screw. A tangent screw which can be moved over its entire range without resetting.

endless tangent screw sextant. A marine sextant having an endless tangent screw for controlling the position of the index arm and the vernier or micrometer drum. The index arm may be moved over the entire arc without resetting, by means of the endless tangent screw.

entrance, *n.* The seaward end of channel, harbor, etc.

entrance lock. A lock between the tideway and an enclosed basin when their water levels vary. By means of the lock, which has two sets of gates, vessels can pass either way at all states of the tide. Also called TIDAL LOCK. See also NONTIDAL BASIN.

envelope match. In Loran-C, the comparison, in time difference, between the leading edges of the demodulated and filtered pulses from a master and secondary station. The pulses are superimposed and matched manually or automatically. See also CYCLE MATCH.

envelope to cycle difference. The time relationship between the phase of the Loran-C carrier and the time origin of the envelope waveform. Zero envelope to cycle difference (ECD) is defined as the signal condition occurring when the 30 microsecond point of the Loran-C pulse envelope is in time coincidence with the third positive-going zero crossing of the 100 kHz carrier.

envelope to cycle discrepancy. An error in a Loran-C time difference measurement which results from upsetting the precise relationship between the shape of the pulse envelope and the phase of the carrier wave necessary for an accurate measurement due to some of the large number of frequencies (90-110kHz) governing the envelope shape being transmitted more readily than others because of the medium over which the groundwave propagates.

ephemeris (*pl. ephemerides*), *n.* 1. A periodical publication tabulating the predicted positions of celestial bodies at regular intervals, such as daily, and containing other data of interest to astronomers and navigators. The *Astronomical Almanac* is an ephemeris. See also ALMANAC.
2. A statement, not necessarily in a publication, presenting a correlation of time and position of celestial bodies or artificial satellites.

ephemeris day. See under EPHEMERIS SECOND.

ephemeris second. The ephemeris second is defined as 1/31,556,925.9747 of the tropical year for 1900 January 0^d 12^h ET. The ephemeris day is 86,400 ephemeris seconds. See also EPHEMERIS TIME.

Ephemeris Time. The time scale used by astronomers as the tabular argument of the precise, fundamental ephemerides of the sun, moon, and planets. It is the independent variable in the gravitational theories of the solar system. It is determined in arrears from astronomical observations and extrapolated into the future, based on International Atomic Time.

epicenter, *n.* The point on the earth's surface directly above the focus of an earthquake.

epoch, *n.* 1. A particular instant of time or a date for which values of data, which vary with time, are given.
2. A given period of time during which a series of related acts or events takes place.
3. Angular retardation of the maximum of a constituent of the observed tide behind the corresponding maximum of the same constituent of the hypothetical equilibrium. Also called PHASE LAG, TIDAL EPOCH.
4. As used in tidal datum determinations, a 19-year Metonic cycle over which tidal height observations are meaned in order to establish the various datums. As there are periodic and apparent secular trends in sea level, a specific

19-year cycle (the National Tidal Datum Epoch) is selected so that all tidal datum determinations throughout the United States and its possessions will have a common reference. The National Tidal Datum Epoch officially adopted by the National Ocean Survey is 1941 through 1959. The National Tidal Datum Epoch will be reviewed for consideration for possible revision at 25-year intervals.

equal altitudes. Two altitudes numerically the same. The expression applies particularly to the practice, essentially obsolete, of determining the instant of local apparent noon by observing the altitude of the sun a short time before it reaches the meridian and again at the same altitude after transit, the time of local apparent noon being midway between the times of the two observations, if the second is corrected as necessary for the run of the ship. Also called DOUBLE ALTITUDES.

equal-area map projection. A map projection having a constant area scale. Such a projection is not conformal and is not used for navigation. Also called AUTHALIC MAP PROJECTION, EQUIVALENT MAP PROJECTION.

equal interval light. A rhythmic light for which alternations of light and darkness are of equal duration. Also called ISOPHASE LIGHT.

equation of time. The difference at any instant, which never exceeds about $16^m_.4$, between apparent time and local mean time. It is a measure of the difference of the hour angles of the apparent (true) sun and the mean (fictitious) sun. The curve drawn for the equation of time during a year has two maxima: February 12 ($+14^m_.3$) and July 27 ($+6^m_.3$) and two minima: May 15 ($-3^m_.7$) and November 4 ($-16^m_.4$). The curve crosses the zero line on April 15, June 14, September 1, and December 24. The equation of time is tabulated in the *Nautical Almanac*, without sign, for 00^h and 12^hGMT on each day. To obtain apparent time, apply the equation of time to mean time with a *positive* sign when GHA sun at 00^hGMT *exceeds* 180°, or at 12^h *exceeds* 0°, corresponding to a meridian passage of the sun *before* 12^hGMT; otherwise apply with a negative sign.

equator, *n.* The primary great circle of a sphere or spheroid, such as the earth, perpendicular to the polar axis; or a line resembling or approximating such a circle. The **terrestrial equator** is 90° from the earth's geographical poles, the **celestial equator** or **equinoctial** is 90° from the celestial poles. The **astronomical equator** is a line connecting points having 0° astronomical latitude; the **geodetic equator** connects points having 0° geodetic latitude. The expression *terrestrial equator* is sometimes applied to the astronomical equator. The equator shown on charts is the geodetic equator. A **fictitious equator** is a reference line serving as the origin for measurement of fictitious latitude. A **transverse** or **inverse equator** is a meridian the plane of which is perpendicular to the axis of a transverse projection. An **oblique equator** is a great circle the plane of which is perpendicular to the axis of an oblique projection. A **grid equator** is a line perpendicular to a prime grid meridian at the origin. The **magnetic equator** or **aclinic line** is that line on the surface of the earth connecting all points at which the magnetic dip is zero. The **geomagnetic equator** is the great circle 90° from geomagnetic poles of the earth.

equatorial, *adj.* Of or pertaining to the equator.

equatorial air. See under AIR-MASS CLASSIFICATION.

equatorial bulge. The excess of the earth's equatorial diameter over the polar diameter.

equatorial calms. See DOLDRUMS.

equatorial chart. 1. A chart of equatorial areas. 2. A chart on an equatorial map projection.

equatorial countercurrent. See ATLANTIC EQUATORIAL COUNTERCURRENT, PACIFIC EQUATORIAL COUNTERCURRENT, INDIAN EQUATORIAL COUNTERCURRENT.

equatorial current. See NORTH EQUATORIAL CURRENT, SOUTH EQUATORIAL CURRENT.

equatorial cylindrical orthomorphic chart. See MERCATOR CHART.

equatorial cylindrical orthomorphic map projection. See MERCATOR MAP PROJECTION.

equatorial gravity value. The mean acceleration of gravity at the equator, approximately equal to 978.03 centimeters per second per second.

equatorial map projection. A map projection centered on the equator.

equatorial node. Either of the two points where the orbit of the satellite intersects the equatorial plane of its primary.

equatorial satellite. A satellite whose orbital plane coincides, or almost coincides, with the earth's equatorial plane.

equatorial tidal currents. Tidal currents occurring semimonthly as a result of the moon being over the equator. At these times the tendency of the moon to produce a diurnal inequality in the tidal current is at a minimum.

equatorial tides. Tides occurring semimonthly as the result of the moon being over the equator. At these times the tendency of the moon to produce a diurnal inequality in the tide is at a minimum.

equiangular, *adj.* Having equal angles.

equilateral, *adj.* Having equal sides.

equilateral triangle. A triangle having all of its sides equal. An equilateral is necessarily equiangular.

equilibrium *n.* A state of balance between forces. A body is said to be *in equilibrium* when the vector sum or all forces acting upon it is zero.

equilibrium argument. The theoretical phase of a constituent of the equilibrium tide.

equilibrium theory. A model under which it is assumed that the waters covering the face of the earth instantly respond to the tide-producing forces of the moon and sun, and form a surface of equilibrium under the action of these forces. The model disregards friction and inertia and the irregular distribution of the land masses of the earth. The theoretical tide formed under these conditions is called EQUILIBRIUM TIDE.

equilibrium tide. Hypothetical tide due to the tide-producing forces under the equilibrium theory. Also called GRAVITATIONAL TIDE.

equinoctial, *adj.* Of or pertaining to an equinox or the equinoxes.

equinoctial, *n.* See CELESTIAL EQUATOR.

equinoctial colure. That great circle of the celestial sphere through the celestial poles and the equinoxes; the hour circle of the vernal equinox. See also SOLSTITIAL COLURE.

equinoctial point. One of the two points of intersection of the ecliptic and the celestial equator. Also called EQUINOX.

equinoctial system of coordinates. See CELESTIAL EQUATOR SYSTEM OF COORDINATES.

equinoctial tides. Tides occurring near the times of the equinoxes, when the spring range is greater than average.

equinoctial year. See TROPICAL YEAR.

equinox, *n.* 1. One of the two points of intersection of the ecliptic and celestial equator,

occupied by the sun when its declination is 0°. That point occupied on or about March 21, when the sun's declination changes from south to north, is called **vernal equinox, March equinox**, or **first point of Aries**; that point occupied on or about September 23, when the declination changes from north to south, is called **autumnal equinox, September equinox**, or **first point of Libra**. Also called EQUINOCTIAL POINT.
2. That instant the sun occupies one of the equinoctial points.

equiphase zone. The region in space within which there is no difference in phase between two radio signals.

equipotential surface. A surface having the same potential of gravity at every point. See also GEOID.

equisignal, *adj.* Pertaining to two signals of equal intensity.

equisignal, *n.* See under CONSOL STATION.

equisignal zone. That region in space within which the difference in amplitude of two radio signals (usually emitted by a signal station) is indistinguishable.

equivalent echoing area. See RADAR CROSS SECTION.

equivalent map projection. See EQUAL-AREA MAP PROJECTION.

erect image. See under IMAGE, definition 1.

erecting telescope. A telescope with which the observer sees objects right side up as opposed to the upside down view provided by the INVERTING TELESCOPE. The eyepiece in the optical system of an erecting telescope usually has four lenses, and the eyepiece in the optical system of an inverting telescope has two lenses.

erg, *n.* The work performed by a force of 1 dyne acting through a distance of 1 centimeter. The erg is the unit of energy or work in the centimeter-gram-second system. It corresponds to 10^{-7} joule in the International System of Units.

error, *n.* The difference between the value of a quantity determined by observation, measurement or calculation and the true, correct, ideal, accepted, adopted or standard value of that quantity. Usually, the true value of the quantity cannot be determined with exactness due to insufficient knowledge of the errors encountered in the observations. Exceptions occur (1) when the value is mathematically determinable, e.g., the condition that the sum of the three angles of a plane triangle equals 180°, or (2) when the value is an adopted or standard value established by authority, e.g., the length of the meter defined in terms of a specific number of wavelengths of a certain spectral line. In order to analyze the exactness with which the true value of a quantity has been determined from observations, errors are classified into two categories, *random* and *systematic errors*. For the purpose of error analysis, blunders or mistakes are not classified as errors. The significant difference between the two categories is that random errors must be treated by means of statistical and probability methods due to their accidental or chance nature whereas systematic errors are usually expressible in terms of a unique mathematical formula representing some physical law or phenomenon. **Random errors** occur when irregular randomly occurring conditions affect the observing instrument, the observer, and the environment of the quantity being observed so that careful observations of the same quantity made with the same observing equipment and observer under the same observing conditions result in different values of the observed quantity from one observation to the next. Random errors depend upon (1) the quality of the observing instrument, (2) the skill of the observer, and (3) randomly fluctuating conditions such as temperature, pressure, etc. Although, in general, random errors are predicted statistically to provide indications of their magnitudes for given probabilities, some random errors are unpredictable. An example is the error in an Omega reading due to a sudden ionospheric disturbance (SID). **Systematic errors** are the nonrandom errors of observation which are characterized by either remaining constant, changing constantly or periodically with time or changing according to some law with changes of temperature, pressure or some other physical condition. They are due to (1) errors resulting from changing or nonstandard natural physical conditions outside the observer, sometimes called **theoretical errors**, (2) **personal (nonaccidental) errors**, and (3) **instrument errors**. Systematic errors are usually predictable once the cause is known. See also ACCURACY.

error budget. A correlated set of individual major error sources with statements of the percentage of the total system error contributed by each error source.

error ellipse. The contour of equal probability density centered on the intersection of two straight lines of position which results from the one-dimensional normal error distribution associated with each line. For the 50 percent error ellipse, there is a 50 percent probability that a fix will lie within such ellipse. If the angle of cut is 90° and the standard deviations are equal, the error figure is a circle.

error of collimation. See COLLIMATION ERROR.

error of perpendicularity. That error in the reading of a marine sextant due to non-perpendicularity of the index mirror to the frame.

escape velocity, *n.* The minimum magnitude of the velocity required of a body at a given point in a gravitational field which will permit the body to escape from the field. The orbit followed is a parabola and the body arrives at an infinite distance from the center of the field with zero velocity. With respect to escape velocities characteristic of the major bodies of the solar system, this is defined as escape from the body's gravitational field from the surface of the body in question. Escape velocity equals circular velocity times $\sqrt{2}$. Also called PARABOLIC VELOCITY.

escarpment, *n.* On the sea floor, an elongated and comparatively steep slope separating flat or gently sloping areas. Also called SCARP.

established direction of traffic flow. A traffic flow pattern indicating the directional movement of traffic as established within a traffic separation scheme. See also RECOMMENDED DIRECTION OF TRAFFIC FLOW.

establishment of the port. Average high water interval on days of the new and full moon. This interval is also sometimes called the COMMON or VULGAR ESTABLISHMENT to distinguish it from the CORRECTED ESTABLISHMENT, the latter being the mean of all high water intervals. The latter is usually 10 to 15 minutes less than the common establishment. Also called HIGH WATER FULL AND CHANGE.

estimate, *v., t.* To determine roughly or with incomplete information.

estimated position. The most probable position of a craft determined from incomplete data or data of questionable accuracy. Such a position might be determined by applying a correction to the dead reckoning position, as for estimated current; by plotting a line of soundings; or by plotting lines of position of questionable accuracy. If no better information is available, a dead reckoning position is an estimated position, but the expression *estimated position* is not customarily used in this case. The distinction between an estimated position and a fix or running fix is a matter of judgment. See also MOST PROBABLE POSITION.

estimated time of arrival. The predicted time of reaching a destination or waypoint.

estimated time of departure. The predicted time of leaving a place.

estimation, *n.* A mathematical method or technique of making a decision concerning the approximate value of a desired quantity when the decision is weighted or influenced by all available information.

estuarine sanctuary. A research area which may include any part or all of an estuary, adjoining transitional areas, and adjacent uplands, constituting to the extent feasible a natural unit, set aside to provide scientists and students the opportunity to examine over a period of time the ecological relationships within the area. See also MARINE SANCTUARY.

estuary, *n.* 1. An embayment of the coast in which fresh river water entering at its head mixes with the relatively saline ocean water. When tidal action is the dominant mixing agent, it is usually called TIDAL ESTUARY.
2. The lower reaches and mouth of a river emptying directly into the sea where tidal mixing takes place. Sometimes called RIVER ESTUARY.
3. A drowned river mouth due to the sinking of the land near the coast.

etesian, *n.* A refreshing northerly summer wind of the Mediterranean, especially over the Aegean Sea.

Eulerian current measurement. The direct observation of the current speed or direction, or both, during a period of time as it flows past a recording instrument such as the Ekman or Roberts current meter. See also LAGRANGIAN CURRENT MEASUREMENT.

Eulerian motion. A slight wobbling of the earth about its axis of rotation, often called **polar motion,** and sometimes **wandering of the poles.** This motion which does not exceed 40 feet from the mean position, produces slight variation of latitude and longitude of places on the earth.

European Datum. The origin of this datum is at Potsdam, Germany. Numerous national systems have been joined in a large datum based upon the International Ellipsoid 1924 which was oriented by a modified astrogeodetic method. European, African, and Asian triangulation chains were connected. African arc measurements from Cairo to Cape Town were completed. Thus, all Europe, Africa, and Asia are molded into one great system. Through common survey stations, it was possible to convert data from the Russian Pulkova 1932 system to the European Datum, and as a result the European Datum includes triangulation as far east as the 84th meridian. Additional ties across the Middle East have permitted connection of the Indian and European Datums.

evaporation, *n.* The physical process by which a liquid or solid is transformed to the gaseous state. The opposite is CONDENSATION. In meteorology, evaporation is usually restricted in use to the change of water vapor from liquid to gas, while SUBLIMATION is used for the change from solid to gas as well as from gas to solid. Energy is lost by an evaporating liquid; and when no heat is added externally, the liquid always cools. The heat thus removed is called LATENT HEAT OF VAPORIZATION.

evection, *n.* A perturbation of the moon depending upon the alternate increase or decrease of the eccentricity of its orbit, which is always a maximum when the sun is passing the moon's line of apsides and a minimum when the sun is at right angles to it.

evening star. The brightest planet appearing in the western sky during evening twilight.

evening twilight. The period of time between sunset and darkness.

everglade, *n.* 1. A tract of swampy land covered mostly with tall grass.
2. A swamp or inundated tract of low land, as used locally in the southern U.S.

excess of arc. That part of a sextant arc beginning at zero and extending in the direction opposite to that part usually considered positive. See also ARC, definition 2.

existence doubtful. Of uncertain existence. The expression is used principally on charts to indicate the possible existence of a rock, shoal, etc., the actual existence of which has not been established. See also VIGIA.

ex-meridian altitude. An altitude of a celestial body near the celestial meridian of the observer to which a correction must be applied to determine the meridian altitude. Also called CIRCUMMERIDIAN ALTITUDE.

ex-meridian observation. Measurement of the altitude of a celestial body near the celestial meridian of the observer, for conversion to a meridian altitude; or the altitude so measured.

expanded center PPI display. A plan position indicator display on which zero range corresponds to a ring around the center of the display.

expanded sweep. Short for EXPANDED TIME BASE SWEEP.

expanded time base. A time base having a selected part of increased speed. Particularly an EXPANDED TIME BASE SWEEP.

expanded time base sweep. A sweep in which the sweep speed is increased during a selected part of the cycle. Usually shortened to EXPANDED SWEEP, and sometimes to EXPANDED TIME BASE.

explement, *n.* An angle equal to 360° minus a given angle. Thus, 150° is said to be the explement of 210° and the two are said to be *explementary*. See also COMPLEMENT, SUPPLEMENT.

explementary angles. Two angles whose sum is 360°.

explosive fog signal. A fog signal consisting of short reports produced by detonating explosive charges.

exponent, *n.* A number or symbol to the right and about half a space above another number or symbol, to indicate the power to which it is to be raised.

external noise. In radio reception, atmospheric radio noise and man-made noise, singly or in combination. **Internal noise** is produced in the receiver circuits.

extragalactic nebula. An aggregation of matter beyond our galaxy, large enough to occupy a perceptible area but which has not been resolved into individual stars.

extrapolation, *n.* The process of estimating the value of a quantity beyond the limits of known values by assuming that the rate or system of change between the last few known values continues. The process of determining intermediate values between given values in accordance with some known or assumed rate or system of change is called INTERPOLATION.

extratropical cyclone. Any cyclonic-scale storm that is not a tropical cyclone, usually referring only to the migratory frontal cyclones of middle and high latitudes. Also called EXTRATROPICAL LOW.

extratropical low. *See* EXTRATROPICAL CYCLONE.

extreme high water. The highest elevation reached by the sea as recorded by a tide gage during a given period. The National Ocean Survey routinely documents monthly and yearly extreme high waters for its control stations. See also EXTREME LOW WATER.

extreme low water. The lowest elevation reached by the sea as recorded by a tide gage during a given period. The National Ocean Survey routinely documents monthly and yearly extreme low water for its control stations. See also EXTREME HIGH WATER.

extremely high frequency. Radio frequency of 30,000 to 300,000 megahertz.

eye guard. A guard or shield on an eyepiece of an optical system, to protect the eye from stray light, wind, etc., and to maintain proper eye distance. Also called EYE SHIELD, EYE SHADE, SHADE.

eye of the storm. The center of a tropical cyclone, marked by relatively light winds, confused seas, rising temperature, lowered relative humidity, and often by clear skies. The general area of lowest atmospheric pressure of a cyclone is called STORM CENTER.

eye of the wind. The point or direction from which the wind is blowing. See also IN THE WIND.

eyepiece, *n.* In an optical device, the lens group which is nearest the eye and with which the image formed by the preceding elements is viewed.

eye shade. See EYE GUARD.

eye shield. See EYE GUARD.

eyot, *n.* (ī'ŭt). *British terminology.* See AIT.

F

fading, *n.* The fluctuation in intensity or relative phase of any or all of the frequency components of a received radio signal due to changes in the characteristics of the propagation path. See also SELECTIVE FADING.

Fahrenheit temperature. Temperature based on a scale in which, under standard atmospheric pressure, water freezes at 32° and boils at 212° above zero.

fair, *adj.* Not stormy; good; fine; clear.

fair tide. A tidal current setting in such a direction as to increase the speed of a vessel. One setting in a direction approximately opposite to the heading is called a HEAD TIDE. One abeam is called a BEAM TIDE. One approximately 90° from the course is called a CROSS TIDE.

fairway, *n.* 1. The main thoroughfare of shipping in a harbor or channel. Although generally clear of obstructions, it may include a MIDDLE GROUND suitably indicated by navigation marks.

2. The middle of a channel.

fairway buoy. A buoy marking a fairway, with safe water on either side. It is painted in black and white vertical stripes. Also called MID-CHANNEL BUOY.

fair wind. A wind which aids a craft in making progress in a desired direction. Used chiefly in connection with sailing vessels, when it refers to a wind which permits the vessel to proceed in the desired direction without excessive changing of course. When applied to a power vessel, it refers to a wind which increases the speed of the craft. A wind which delays the progress of a craft is called an UNFAVORABLE WIND. Also called FAVORABLE WIND. See also FOLLOWING WIND.

Falkland Current. Originating mainly from the Cape Horn Current in the north part of Drake Passage, the Falkland Current flows northward between the continent and the Falkland Islands after passing through the strait. The current follows the coast of South America until it joins the BRAZIL CURRENT at about latitude 36° S near the entrance to Rio de la Plata. Also called MALVIN CURRENT.

fall, *n.* 1. See AUTUMN.

2. Decrease in a value, as a *fall of temperature*.

3. Sinking, subsidence, etc., as the *rise and fall of the sea* due to tidal action or when waves or swell are present. The opposite is RISE, *n.*, definiton 2. See also WATERFALL.

fall equinox. See AUTUMNAL EQUINOX.

falling star. See METEOR.

falling tide. The portion of the tide cycle between high water and the following low water in which the depth of water is decreasing. Sometimes the term EBB is used as an equivalent, but since ebb refers primarily to horizontal rather than vertical movement, falling tide is considered more appropriate. The opposite is RISING TIDE.

fall streaks. See VIRGA.

fall wind. A cold wind blowing down a mountain slope. Such a wind is warmed by its descent, but is still cool relative to surrounding air. A warm wind blowing down a mountain slope is called a FOEHN. The bora, mistral, papagayo, and vardar are examples of fall winds. See also KATABATIC WIND.

false cirrus. A cloud species unique to the genus *cirrus*, of such optical thickness as to appear grayish on the side away from the sun; and to veil the sun, conceal its outline, or even hide it. These often originate from the upper part of a cumulonimbus, and are often so dense that they suggest clouds of the middle level. Also called THUNDERSTORM CIRRUS, CIRRUS SPISSATUS.

false echo. See INDIRECT ECHO, PHANTOM TARGET.

false horizon. A line resembling the VISIBLE HORIZON but above or below it.

false light. A light which is unavoidably exhibited by an aid to navigation and which is not intended to be a part of the proper characteristic of the light. Reflections from storm panes come under this category.

false relative motion. False indications of the movement of a target relative to own ship on a radar display that is unstabilized in azimuth due to continuous reorientation of the display as own ship's heading changes. See also STABILIZATION OF RADARSCOPE DISPLAY

fan, *n.* On the sea floor, a relatively smooth feature normally sloping away from the lower termination of a canyon or canyon system.

fan beam. A beam in which the radiant energy is concentrated in and about a single plane. The angular spread in the plane of concentration may be any amount to 360°. This type beam is most widely used for navigational lights. A converged **beam** is a fan beam in which the angular spread is decreased laterally to increase the intensity of the remaining beam over all or part of its arc; a **diverged beam** is a fan beam formed by increasing the divergence of a pencil beam in one plane only.

farad, *n.* A *derived unit* of capacitance in the International System of Units; it is the capacitance of a capacitor between the plates of which there appears a potential difference of 1 volt when it is charged by a quantity of electricity of 1 coulomb.

far vane. That instrument sighting vane on the opposite side of the instrument from the observer's eye. The opposite is NEAR VANE.

fast ice. Sea ice which forms and remains fast along the coast, where it is attached to the shore, to an ice wall, to an ice front, between shoals or grounded icebergs. Vertical fluctuations may be observed during changes of sea level. Fast ice may be formed in situ from the sea water or by freezing of pack ice of any age to the shore, and it may extend a few meters or several hundred kilometers from the coast. Fast ice may be more than 1 year old and may then be prefixed with the appropriate age category (*old*, *second-year*, or *multi-year*). If it is thicker than about 2 meters above sea level, it is called an ICE SHELF.

fast-ice boundary. The ice boundary at any given time between fast ice and pack ice.

fast-ice edge. The demarcation at any given time between fast ice and open water.

fast-sweep racon. See under SWEPT-FREQUENCY RACON.

fast time constant circuit. A type of coupling circuit, with high pass frequency characteristics used in radar receivers to permit discrimination against received pulses of duration longer than the transmitted pulse. With the fast time constant (FTC) circuit in operation, only the leading edge of an echo having a long time duration is displayed on the radarscope. The use of this circuit tends to reduce saturation of the scope which could be caused by clutter. Also called ANTI-CLUTTER RAIN, DIFFERENTIATOR.

fata morgana. A complex mirage, characterized by marked distortion, generally in the vertical. It may cause objects to appear towering, magnified, and at times even multiplied.

Fathogram, *n.* A graphic record of depth measurements obtained by a Fathometer. See also ECHOGRAM.

fathom, *n.* A unit of length equal to 6 feet. This unit of measure is used principally as a measure of depth of water and the length of lead lines, anchor chains, and cordage. See also CABLE, definition 1.

fathom curve, fathom line. A depth contour, with depths expressed in fathoms.

Fathometer, *n.* The registered trade name for a widely-used echo sounder.

favorable current. A current flowing in such a direction as to increase the speed of a vessel over the ground. The opposite is UNFAVORABLE CURRENT.

favorable wind. A wind which aids a craft in making progress in a desired direction. Usually used in plural and chiefly in connection with sailing vessels. A wind which delays the progress of a craft is called an UNFAVORABLE WIND. Also called FAIR WIND. See also FOLLOWING WIND.

feasibility orbit. An orbit that can be rapidly and inexpensively computed on the basis of simplifying assumptions (e.g., two-body motion, circular orbit, rectilinear orbit, three-body motion approximated by two two-body orbits, etc.) and yields an indication of the general feasibility of a system based upon the orbit without having to carry out a full-blown definitive orbit computation.

federal project depth. The *design* dredging depth of a channel constructed by the Corps of Engineers, U.S. Army; the project depth *may or may not be the goal of maintenance dredging* after completion of the channel. For this reason federal project depth must not be confused with CONTROLLING DEPTH.

feel the bottom. The effect on a ship underway in shallow water which tends to reduce her speed, make her slow in answering the helm, and often make her sheer off course. The speed reduction is largely due to increased wave making resistance resulting from higher pressure differences due to restriction of flow around the hull. The increased velocity of the water flowing past the hull results in an increase in squat. Also called SMELL THE BOTTOM.

femto-. A prefix meaning *one-quadrillionth* (10^{-15})

fen, *n.* A low lying tract of land, wholly or partly covered with water at times.

fetch, *n.* 1. An area of the sea surface over which seas are generated by a wind having a constant direction and speed. Also called GENERATING AREA.

2. The length of the fetch area, measured in the direction of the wind, in which the seas are generated.

fictitious equator. A reference line serving as the origin for measurement of fictitious latitude. A **transverse** or **inverse equator** is a meridian the plane of which is perpendicular to the axis of a transverse map projection. An **oblique equator** is a great circle the plane of which is perpendicular to the axis of an oblique map projection. A **grid equator** is a line perpendicular to a prime grid meridian, at the origin.

fictitious graticule. The network of lines representing fictitious parallels and fictitious meridians on a map, chart, or plotting sheet. It may be either a **transverse graticule** or an **oblique graticule** depending upon the kind of projection. A fictitious graticule may also be a GRID. See also OBLIQUE GRATICULE, TRANSVERSE GRATICULE.

fictitious latitude. Angular distance from a fictitious equator. It may be called **transverse, oblique,** or **grid latitude** depending upon the type of fictitious equator.

fictitious longitude. The arc of the fictitious equator between the prime fictitious meridian and any given fictitious meridian. It may be called **transverse, oblique,** or **grid longitude** depending upon the type of fictitious meridian.

fictitious loxodrome. See FICTITIOUS RHUMB LINE.

fictitious loxodromic curve. See FICTITIOUS RHUMB LINE.

fictitious meridian. One of a series of great circles or lines used in place of a meridian for certain purposes. A **transverse meridian** is a great circle perpendicular to a transverse equator; an **oblique meridian** is a great circle perpendicular to an oblique equator; a **grid meridian** is one of the grid lines extending in a grid north-south direction. The reference meridian (real or fictitious) used as the origin for measurement of fictitious longitude is called **prime fictitious meridian.**

fictitious parallel. A circle or line parallel to a fictitious equator, connecting all points of equal fictitious latitude. It may be called **transverse, oblique,** or **grid parallel** depending upon the type of fictitious equator.

fictitious pole. One of the two points 90° from a fictitious equator. It may be called **transverse** or **oblique pole** depending upon the type of fictitious equator.

fictitious rhumb. See FICTITIOUS RHUMB LINE.

fictitious rhumb line. A line making the same oblique angle with all fictitious meridians. It may be called **transverse, oblique,** or **grid rhumb line** depending upon the type of fictitious meridian. The expression OBLIQUE RHUMB LINE applies also to any rhumb line, real or fictitious, which makes an oblique angle with its meridians; as distinguished from parallels and meridians, real or fictitious, which may be considered special cases of the rhumb line. Also called FICTITIOUS RHUMB, FICTITIOUS LOXODROME, FICTITIOUS LOXODROMIC CURVE.

fictitious ship. An imaginary craft used in the solution of certain maneuvering problems, as when a ship to be intercepted is expected to change course or speed during the interception run.

fictitious sun. An imaginary sun conceived to move eastward along the celestial equator at a rate equal to the average rate of the apparent sun or to move eastward along the ecliptic at the average rate of the apparent sun. See also DYNAMICAL MEAN SUN, MEAN SUN.

fictitious year. The period between successive returns of the sun to a sidereal hour angle of 80° (about January 1). The length of the fictitious year is the same as that of the tropical year, since both are based upon the position of the sun with respect to the vernal equinox. Also called BESSELIAN YEAR.

fidelity, *n.* The accuracy to which an electrical system, such as a radio, reproduces at its output the essential characteristics of its input signal.

field glass. A telescopic binocular.

field lens. A lens at or near the plane of a real image, to collect and redirect the rays into another part of the optical system; particularly, the eyepiece lens nearest the object, to direct the rays into the eye lens.

field of view. The maximum angle of vision, particularly by means of an optical instrument.

figure of the earth. See GEOID.

filling, *n.* Increase in atmospheric pressure, particularly within a low. Decrease in pressure is called DEEPENING.

final diameter. The diameter of the circle traversed by a vessel after turning through 360° and maintaining the same speed and rudder angle. This diameter is always less than the tactical diameter. It is measured perpendicular to the original course and between the tangents at the points where 180° and 360° of the turn have been completed.

final great-circle course. The direction, at the destination, of the great circle through that point and the point of departure, expressed as the angular distance from a reference direction, usually north, to that part of the great circle extending beyond the destination. See also INITIAL GREAT-CIRCLE COURSE.

finger rafted ice. The type of rafted ice in which floes thrust "fingers" alternately over and under the other.

finger rafting. A type of rafting whereby interlocking thrusts are formed, each floe thrusting "fingers" alternately over and under the other. Finger rafting is common in NILAS and GREY ICE.

finite, *adj.* Having limits. The opposite is INFINITE.

fiord, fjord, *n.* A long, deep, narrow arm of the sea between high land. A fiord often has a relatively shallow sill across its entrance.

fireball, *n.* See BOLIDE.

firn, *n.* Old snow which has recrystallized into a dense material. Unlike snow, the particles are to some extent joined together; but, unlike ice, the air spaces in it still connect with each other.

first estimate-second estimate method. The process of determining the value of a variable quantity by trial and error. The expression applies particularly to the method of determining time of meridian transit (especially local apparent noon) at a moving craft. The time of transit is computed for an estimated longitude of the craft, the longitude estimate is then revised to agree with the time determined by the first estimate, and a second computation is made. The process is repeated as many times as necessary to obtain an answer of the desired precision.

first light. The beginning of morning nautical twilight, i.e., when the center of the morning sun is 12° below the horizon.

first point of Aries. See VERNAL EQUINOX.

first point of Cancer. See SUMMER SOLSTICE.

first point of Capricornus. See WINTER SOLSTICE.

first point of Libra. See AUTUMNAL EQUINOX.

first quarter. The phase of the moon when it is near east quadrature, when the western half of it is visible to an observer on the earth. See also PHASES OF THE MOON.

first-year ice. Sea ice of not more than one winter's growth, developing from young ice, and have a thickness of 30 centimeters to 2 meters. First-year ice may be subdivided into THIN FIRST-YEAR ICE, WHITE ICE, MEDIUM FIRST-YEAR ICE, and THICK FIRST-YEAR ICE.

firth, *n.* A long, narrow arm of the sea. Also called FRITH.

Fischer ellipsoid of 1960. The reference ellipsoid of which the semimajor axis is 6,378,166.000 meters, the semiminor axis is 6,356,784.298 meters, and the flattening or ellipticity is 1/298.3. Also called FISCHER SPHEROID OF 1960.

Fischer ellipsoid of 1968. The reference ellipsoid of which the semimajor axis is 6,378,150 meters, the semiminor axis is 6,356,768.337 meters, and the flattening or ellipticity is 1/298.3. Also called FISCHER SPHEROID OF 1968.

Fischer spheroid of 1960. See FISCHER ELLIPSOID OF 1960.

Fischer spheroid of 1968. See FISCHER ELLIPSOID OF 1968.

fish, *n.* Any towed sensing device.

fishery conservation zone. See under FISHING ZONE.

fish havens. Areas established by private interests, usually sport fishermen, to simulate natural reefs and wrecks that attract fish. The reefs are constructed by dumping assorted junk in areas which may be of very small extent or may stretch a considerable distance along a depth contour. Fish havens are outlined and labeled on charts. Also called FISHERY REEFS.

fishing zone. The offshore zone in which exclusive fishing rights and management are held by the coastal nation. The U.S. fishing zone, known as the **fishery conservation zone**, is defined under P.L. 94-265. The law states, "The inner boundary of the fishery conservation zone is a line conterminous with the seaward boundary of each of the coastal states, and the outer boundary of such zone is a line drawn in such manner that each point on it is 200 nautical miles from the baseline from which the territorial sea is measured."

fish lead (lĕd). A type of sounding lead used without removal from the water between soundings.

fish stakes. Poles or stakes placed in shallow water to outline fishing grounds or to catch fish.

fishtrap areas. Areas established by the Corps of Engineers in which traps may be built and maintained according to established regulations. The fish stakes which may exist in these areas are obstructions to navigation and may be dangerous. The limits of fishtrap areas and a cautionary note are usually charted.

fix, *n.* A position determined without reference to any former position. In concept a fix is the common intersection of two or more lines of position obtained from simultaneous observations not dependent upon any former position. In normal practice a fix is the most probable position derived from two or more intersecting lines of position obtained from observations made at nearly the same time and advanced or retired to a common time, the lines when numbering three or more not intersecting at a common point because of the errors associated with each line. See also RUNNING FIX.

fixed and flashing light. A light in which a fixed light is combined with a flashing light of higher luminous intensity. The aeronautical light equivalent is called UNDULATING LIGHT.

fixed and group flashing light. A fixed light varied at regular intervals by a group of two or more flashes of greater intensity.

fixed and variable parameters of satellite orbit. The **fixed parameters** are those parameters which describe a satellite's approximate orbit and which are used over a period of hours, e.g., a 12- to 16-hour interval for the Navy Navigation Satellite System (NAVSAT). The **variable parameters** describe the fine structure of the orbit as a function of time and are correct only for the time at which they are transmitted by the satellite. The NAVSAT satellite memory stores sufficient variable parameters to describe the fine structure of the orbit at 2-minute intervals between subsequent injections of data.

fixed antenna radio direction finder. A radio direction finder whose use does not require the rotation of the antenna system.

fixed light. A light which appears continuous and steady to an observer whose position remains unchanged in relation to it. The term is sometimes loosely used for a light supported on a fixed structure, as distinct from a light on a floating support.

fixed mark. A navigation mark fixed in position.

fixed satellite. See GEOSTATIONARY SATELLITE.

fixed star. A star whose apparent position relative to surrounding stars appears to be unvarying or fixed for long periods of time.

fjord, *n.* See FIORD.

flag alarm. A semaphore-type flag in the indicator of an instrument, to serve as a signal, usually to warn that the indications are unreliable.

flagpole, *n.* A label on a nautical chart which indicates a single staff from which flags are displayed. The term is used when the pole is not attached to a building. The label **flagstaff** is used for a flagpole rising from a building.

flagstaff, *n.* See under FLAGPOLE.

flag tower. A label on a nautical chart which indicates a scaffold-like tower from which flags are displayed.

Flamsteed's number. A number sometimes used with the possessive form of the Latin name of the constellation to identify a star. An example is *72 Ophiuchi.*

flash, *n.* A relatively brief appearance of a light, in comparison with the longest interval of darkness in the period of the light. See also OCCULTATION.

flasher. *n.* An electrically powered device which produces light flashes of prescribed characteristics.

flashing, *n.* The process of reducing the amount of permanent magnetism in a vessel by placing a single coil horizontally around the vessel and energizing it. If the energized coil is moved up and down along the sides of the vessel, the process is called WIPING. See also DEPERMING.

flashing light. A light in which the total duration of light in a period is shorter than the total duration of darkness and appearances of light (flashes) are usually of equal duration. The term is commonly used for a SINGLE-FLASHING LIGHT, a flashing light in which a flash is regularly repeated (at a rate of less then 50 flashes per minute). See also GROUP-FLASHING LIGHT, COMPOSITE GROUP-FLASHING LIGHT, LONG-FLASHING LIGHT, QUICK LIGHT.

flat, *n.* 1. A large flat area attached to the shore consisting usually of mud, but sometimes of sand and rock. Also called TIDAL FLATS. See also SALT MARSH, SLOUGH, TIDAL MARSH. 2. On the sea floor, a small level or nearly level area.

flattening, *n.* The ratio of the difference between the equatorial and polar radii of the earth to its equatorial radius. The flattening of the earth is the ellipticity of the spheroid. The magnitude of the flattening is sometimes expressed as the numerical value of the reciprocal of the flattening. Also called COMPRESSION.

flaw, *n.* A narrow separation zone between pack ice and fast ice, where the pieces of ice are in a chaotic state. The flaw forms when pack ice shears under the effect of a strong wind or current along the fast-ice boundary. See also SHEARING.

flaw lead. A passage-way between pack ice and fast ice which is navigable by surface vessels.

flaw polynya. A polynya between pack ice and fast ice.

F-layer, *n.* The second principal layer of ionization in the Kennelly-Heaviside region (the E-layer is the first principal layer; the D-layer is of minor significance except for a tendency to absorb energy from radio waves in the medium frequency range). Situated about 175 statute miles above the earth's surface, the F-layer exists as a single layer only during the hours of darkness. It divides into two separate layers during daylight hours.

F_1-layer, *n.* The lower of the two layers into which the F-layer divides during daylight hours. Situated about 140 statute miles above the earth's surface, it reaches its maximum density at noon. Since its density varies with the extent of the sun's radiation, it is subject to daily and seasonal variations. It may disappear completely at some point during the winter months.

F_2-layer, *n.* The higher of the two layers into which the F-layer divides during daylight hours. It reaches its maximum density at noon and, over the continental U.S., varies in height from about 185 statute miles in winter to 250 statute miles in the summer. The F_2-layer has a greater influence, normally, on radio wave propagation than the F_1-layer.

Fleet Guide. One of a series of "sailing directions" for United States naval bases prepared for U.S. Navy use only.

Flinders bar. A bar of soft unmagnetized iron placed vertically near a magnetic compass to counteract deviation caused by magnetic induction in vertical soft iron of the craft.

float chamber. A sealed, hollow part attached to the compass card of a magnetic compass as part of the compass card assembly, to provide buoyancy to reduce the friction on the pivot bearing.

floating aid. A buoy, serving as an aid to navigation, secured in its charted position by a mooring.

floating breakwater. A moored assembly of timbers used for protection of vessels riding at anchor.

floating dock. A form of dry dock consisting of a floating structure of one or more sections, which can be partly submerged by controlled flooding to receive a vessel, then raised by pumping out the water so that the vessel's bottom can be exposed. See also GRAVING DOCK.

floating ice. Any form of ice found floating in water. The principal kinds of floating ice are **lake ice, river ice** and **sea ice** which form by the freezing of water at the surface, and **glacier ice** (ice of land origin) formed on land or in an ice shelf. The concept includes ice that is stranded or grounded.

floating mark. A navigation mark carried on a floating body such as a lightship or buoy.

float pipe. A pipe used as a float well.

float well. A vertical pipe or box with a relatively small opening (orifice) in the bottom. It is used as a tide gage installation to dampen the wind waves while freely admitting the tide to actuate a float which, in turn, operates the gage. Also called STILLING WELL.

floe, *n.* Any relatively flat piece of sea ice 20 meters or more across. Floes are subdivided according to horizontal extent. A **giant flow** is over 5.4 nautical miles across; a **vast floe** is 1.1 to 5.4 nautical miles across; a **big floe** is 500 to 2000 meters across; a **medium floe** is 100 to 500 meters across; and a **small floe** is 20 to 100 meters across.

floeberg, *n.* A massive piece of sea ice composed of a hummock, or a group of hummocks frozen together, and separated from any ice surroundings. It may float showing up to 5 meters above sea level.

flood, *n.* Tidal current moving toward land or up a tidal stream. The opposite is EBB. Sometimes the terms FLOOD and EBB are also used with reference to vertical tidal movement, but for this vertical movement the expressions RISING TIDE and FALLING TIDE are considered preferable. Also called FLOOD CURRENT.

flood axis. Average direction of tidal current at strength of flood.

flood current. The movement of a tidal current toward the shore or up a tidal river or estuary. In the mixed type of reversing current, the terms **greater flood** and **lesser flood** are applied respectively to the flood currents of greater and lesser speed of each day. The terms **maximum flood** and **minimum flood** are applied to the maximum and minimum speeds of a flood current, the speed of which alternately increases and decreases without coming to a slack or reversing. The expression **maximum flood** is also applicable to any flood current at the time of greatest velocity. The opposite is EBB CURRENT.

flooded ice. Sea ice which has been flooded by melt-water or river water and is heavily loaded by water and wet snow.

floodgate, *n.* A gate for shutting out, admitting, or releasing a body of water; a sluice.

flood interval. Short for STRENGTH OF FLOOD INTERVAL. The interval between the transit of the moon over the meridian of a place and the time of the following strength of flood. See also LUNICURRENT INTERVAL.

flood plain. The belt of low flat ground bordering a stream channel that is flooded when run-off exceeds the capacity of the stream channel.

flood strength. Phase of the flood current at time of maximum speed. Also, the speed at this time. Also called STRENGTH OF FLOOD.

floor, *n.* The essentially horizontal surface constituting the principal level of the ground under a body of water. See also BOTTOM.

Florida Current. A swift ocean current that flows through the Straits of Florida from the Gulf of Mexico to the Atlantic Ocean. It shows a gradual increase in speed and persistency as it flows northeastward and then northward along the Florida coast. In summer, the part of the surface current south of latitude 25° N moves farther south of its mean position, with a mean speed of 2.0 knots and a maximum speed of about 6.0 knots; the part of the current north of latitude 25° N moves farther west of its mean position, with a mean speed of 2.9 knots and a maximum speed of 6.5 knots. In winter the shift of position is in the opposite direction, and speeds are somewhat less by about 0.2 to 0.5 knot. The flow prevails throughout the year, with no significant changes in direction; the speed, however, varies slightly from one season to another. North of Grand Bahama Island, it merges with the Antilles Current to form the GULF STREAM. The Florida Current is part of the GULF STREAM SYSTEM.

flotsam, *n.* Floating articles, particularly those that are thrown overboard to lighten a vessel in distress. See also JETSAM, JETTISON, LAGAN.

flow, *n. British terminology.* Total current or the combination of tidal current and nontidal current. In British usage, tidal current is called TIDAL STREAM and nontidal current is called CURRENT.

fluorescence, *n.* Emission of light or other radiant energy as a result of and only during absorption of radiation from some other source. An example is the glowing of the screen of a cathode-ray tube during bombardment by a stream of electrons. The continued emission of light after absorption of radiation is called PHOSPHORESCENCE.

fluorescent chart. A chart reproduced with fluorescent ink or on fluorescent paper, which enables the user to read the chart under ultraviolet light.

flurry, *n.* See SNOW FLURRY.

flux-gate. The magnetic direction-sensitive element of a gyro flux-gate compass. Also called FLUX VALVE.

fluxmeter, *n.* An instrument for measuring the intensity of a magnetic field.

flux valve. *British terminology.* See FLUX GATE.

focal length. The distance between the optical center of a lens, or the surface of a mirror, and its focus.

focal plane. A plane parallel to the plane of a lens or mirror and passing through the focus.

focal point. See FOCUS.

focus (*pl. foci*), *n.* 1. That point at which parallel rays of light meet after being refracted by a lens or reflected by a mirror. Also called FOCAL POINT.

2. A point having specific significance relative to a geometrical figure. See under ELLIPSE, HYPERBOLA, PARABOLA.

3. The true center of an earthquake, within which the strain energy is first converted to elastic wave energy.

focus, *v., t.* The process of adjusting an optical instrument, projector, cathode-ray tube, etc., to produce a clear and well-defined image.

foehn (**föhn**), *n.* A warm, dry, wind blowing down the leeward slope of a mountain and across a valley floor or plain. A cold wind blowing down a mountain slope is called a FALL WIND.

fog, *n.* A visible accumulation of tiny droplets of water, formed by condensation of water vapor in the air, with the base at the surface of the earth. It reduces visibility below 1 kilometer (0.54 nautical mile). If this is primarily the result of movement of air over a surface of lower temperature, it is called **advection fog**; if primarily the result of cooling of the surface of the earth and the adjacent layer of atmosphere by radiational cooling, it is called **radiation fog.** An advection fog occurring as monsoon circulation transports warm moist air over a colder surface is called a **monsoon fog.** A fog that hides less than six-tenths of the sky, and does not extend to the base of any clouds is called a **ground fog.** Fog formed at sea, usually when air from a warm-water surface moves to a cold-water surface, is called **sea fog.** Fog produced by apparent steaming of a relatively warm sea in the presence of very cold air is called **steam fog, steam mist, frost smoke, sea smoke, arctic sea smoke, arctic smoke,** or water smoke. Fog composed of suspended particles of ice, partly ice crystals 20 to 100 microns in diameter but chiefly, especially when dense, droxtals 12 to 20 microns in diameter is called **ice fog**. It occurs at very low temperatures, and usually in clear, calm weather in high latitudes. The sun is usually visible and may cause halo phenomena. Ice fog is rare at temperatures warmer than –30°C or –20°F. A rare simulation of true fog by anomalous atmospheric refraction is called **mock fog**. A **dry fog** is a fog that does not moisten exposed surfaces. Originally smog was a natural fog contaminated by industrial pollutants, or a mixture of smoke and fog. Today, smog is a common term applied to air pollution with or without fog; however some visible manifestation is almost always implied. See also CLOUD, DRIZZLE, MIST.

fog bank. A well defined mass of fog observed at a distance, most commonly at sea.

fogbound, *adj.* Surrounded by fog. Of vessels, the term is used particularly with reference to vessels which are unable to proceed because of the fog.

fogbow, *n.* A faintly-colored circular arc similar to a RAINBOW but formed on fog layers containing drops whose diameters are of the order of 100 microns or less. See also BOUGUER'S HALO.

fog detector. A device used to automatically determine conditions of visibility which warrant the turning on or off of a sound signal.

fog signal. See under SOUND SIGNAL.

following sea. A sea in which the waves move in the general direction of the heading. The opposite is HEAD SEA. Those moving in a direction approximately 90° from the heading are called BEAM SEA, and those moving in a direction approximately 45° from the heading (striking the quarter) are called QUARTERING SEA.

following wind. Wind blowing in the general direction of a vessel's course. The equivalent aeronautical expression is TAIL WIND. Wind blowing in the opposite direction is called a HEAD WIND. Wind blowing in a direction approximately 90° from the heading is called a BEAM WIND. One blowing in a direction approximately 90° from the course is called a CROSS WIND. See also FAIR WIND, FAVORABLE WIND, UNFAVORABLE WIND.

foot, *n.* 1. One-third of a yard. A foot is equal to 12 inches or 30.48 centimeters exactly. The latter value was adopted in 1959 by Australia, Canada, New Zealand, South Africa, the United Kingdom, and the United States. See also U.S. SURVEY FOOT.
2. The bottom of a slope, grade, or declivity.

foraminifera, *n., pl.* Small, single-cell, jellylike marine animals with hard shells of many chambers. In some areas the shells of dead foraminifera are so numerous they cover the ocean bottom.

Forbes log. A log consisting essentially of a small rotator in a tube projecting below the bottom of a vessel, and suitable registering devices.

forced wave. A wave generated and maintained by a continuous force, in contrast with a FREE WAVE that continues to exist after the generating force has ceased to act.

foreland, *n.* See PROMONTORY, HEADLAND.

foreshore, *n.* That part of the shore or beach which lies between the low water mark and the upper limit of normal wave action. See also BACKSHORE.

forestaff, *n.* See CROSS-STAFF.

fork, *n.* On the sea floor, a branch of a canyon or valley.

form lines. Broken lines resembling contour lines but representing no actual elevations, which have been sketched from visual observation or from inadequate or unreliable map sources, to show collectively the shape of the terrain rather than the elevation.

forward, *adv.* In a direction towards the bow of a vessel. See also AHEAD, ABAFT.

forward of the beam. Any direction between broad on the beam and ahead. See also ABAFT THE BEAM.

foul berth. A berth in which a vessel cannot swing to her anchor or moorings without fouling another vessel or striking an obstruction. See also FOUL GROUND, CLEAR BERTH.

foul bottom. A term used to describe the bottom of a vessel when encrusted with marine growth.

foul ground. An area unsuitable for anchoring, taking the ground, or ground fishing due to being strewn with rocks, boulders, coral or obstructions. See also FOUL BERTH.

four-point bearing. A relative bearing of 045° or 315°. See also BOW AND BEAM BEARINGS.

fractional scale. See REPRESENTATIVE FRACTION.

fracto-. A prefix used with the name of a basic cloud form to indicate a torn, ragged, and scattered appearance caused by strong winds. See also SCUD.

fracture, *n.* A break or rupture through very close pack ice, compact pack ice, consolidated pack ice, fast ice, or a single floe resulting from deformation processes. Fractures may contain brash ice and/or be covered with nilas and/or young ice. The length of a fracture may vary from a few meters to many nautical miles. A **large fracture** is more than 500 meters wide; a **medium fracture** is 200 to 500 meters wide; a **small fracture** is 50 to 200 meters wide; and a **very small fracture** is 0 to 50 meters wide.

fracture zone. 1. An extensive linear zone of irregular topography of the sea floor characterized by steep-sided or asymmetrical ridges, troughs, or escarpments.
2. An ice area which has a great number of fractures. See also FRACTURE.

fracturing, *n.* The pressure process whereby ice is permanently deformed, and rupture occurs. The term is most commonly used to describe breaking across very close pack ice, compact pack ice, and consolidated pack ice.

frame, *n.* The constructional system that gives strength and shape to an instrument or device.

frazil ice. Fine spicules or plates of ice, suspended in water.

free-air temperature. Temperature of the atmosphere, obtained by a thermometer located so as to avoid as completely as practicable the effects of extraneous heating. See also AMBIENT TEMPERATURE, WET-BULB TEMPERATURE.

freeboard, *n.* The vertical distance from the uppermost complete, watertight deck of a vessel to the surface of the water, usually measured amidships. Minimum permissible freeboards may be indicated by LOAD LINE MARKS.

free gyro. A two-degree-of-freedom gyro or a gyro the spin axis of which may be oriented in any specified attitude. The rotor of this gyro has freedom to spin on its axis, freedom to tilt about its horizontal axis, and freedom to turn about its vertical axis. Also called FREE GYROSCOPE. See also DEGREE-OF-FREEDOM.

free gyroscope. See FREE GYRO.

free wave. A wave that continues to exist after the generating force has ceased to act, in contrast with a FORCED WAVE that is generated and maintained by a continuous force.

freezing drizzle. Drizzle that falls in liquid form but freezes upon impact to form a coating of glaze upon the ground and exposed objects.

freezing fog. A fog whose droplets freeze upon contact with exposed objects and form a coating of rime and/or glaze. See also FREEZING PRECIPITATION.

freezing precipitation. Precipitation which falls to the earth in a liquid state and then freezes to exposed surfaces. Such precipitation is called **freezing rain** if it consists of relatively large drops of water, and **freezing drizzle** if of smaller drops. See also GLAZE.

freezing rain. Rain that falls in liquid form but freezes upon impact to form a coating of glaze upon the ground and exposed objects.

frequency, *n.* The rate at which a cycle is repeated. See also AUDIO FREQUENCY, RADIO FREQUENCY.

frequency band. 1. A specified segment of the frequency spectrum. For example, the **broadcast band** extends from 550 to 1600 kHz.
2. One of two or more segments of the total frequency coverage of a radio receiver or transmitter, each segment being selectable by means of a band change switch.
3. Any range of frequencies extending from a specified lower to a specified upper limit.

frequency channel. The assigned frequency band commonly referred to by number, letter, symbol, or some salient frequency within the band.

frequency-modulated radar. A type of radar in which the radiated wave is frequency modulated, and the frequency of an echo is compared with the frequency of the transmitted wave at the instant of reception, thus enabling range to be measured.

frequency modulation. Angle modulation of a sine-wave carrier in which the instantaneous frequency of the modulated wave differs from the carrier frequency by an amount proportional to the instantaneous value of the modulating wave.

frequency tolerance. The maximum permissible departure by the center frequency of the frequency band occupied by an emission from the assigned frequency, or by the characteristic frequency of an emission from the reference frequency. The frequency tolerance is expressed in parts in 10^6 or in hertz.

fresh breeze. Wind of force 5 (17 to 21 knots or 19 to 24 miles per hour) on the Beaufort wind scale.

freshen, *v., i.* To become fresher or stronger – applied particularly to wind.

fresh gale. A term once used by seamen for what is now called GALE on the Beaufort wind scale.

fresh-water marsh. A tract of low wet ground, usually miry and covered with rank vegetation.

friction, *n.* Resistance to motion due to interaction between the surface of a body and anything in contact with it.

friction error. That error of an instrument reading due to friction in the moving parts of the instrument.

friction layer. See SURFACE BOUNDARY LAYER.

friendly ice. From the point of view of the submariner, an ice canopy containing many large skylights or other features which permit a submarine to surface. There must be more than 10 such features per 30 nautical miles along the submarine's track.

frigid zones. Either of the two zones between the polar circles and the poles, called the **north frigid zone** and the **south frigid zone.**

fringing reef. A reef attached directly to the shore of an island or continental landmass. Its outer margin is submerged and often consists of algal limestone, coral rock, and living coral. See also BARRIER REEF.

frith, *n.* See FIRTH.

front, *n.* Generally, the interface or transition zone between two air masses of different density. Since the temperature distribution is the most important regulator of atmospheric density, a front almost invariably separates air masses of different temperature. Along with the basic density criterion and the common temperature criterion, many other features may distinguish a front, such as a pressure trough, a change in wind direction, a moisture discontinuity, and certain characteristic cloud and precipitation forms. The term front is used ambiguously for: **frontal zone**, the three-dimensional zone or layer of large horizontal density gradient, bounded by **frontal surfaces** across which the horizontal density gradient is discontinuous (frontal surface usually refers specifically to the warmer side of the frontal zone); and **surface front**, the line of intersection of a frontal surface or frontal zone with the earth's surface or less frequently, with a specified constant-pressure surface. See also POLAR FRONT, ARCTIC FRONT, COLD FRONT, WARM FRONT, OCCLUDED FRONT.

frontal, *adj.* Of or pertaining to a front.

frontal cyclone. In general, any cyclone associated with a front; often used synonymously with WAVE CYCLONE or with EXTRATROPICAL CYCLONE (as opposed to tropical cyclones, which are non-frontal).

frontal occlusion. See OCCLUDED FRONT; OCCLUSION, definition 2.

frontal surface. See under FRONT.

frontal zone. See under FRONT.

front light. That range light which is nearest the observer. It is the lowest of the lights of an established range. Also called LOW LIGHT.

frontogenesis, *n.* 1. The initial formation of a front or frontal zone.
2. In general, an increase in the horizontal gradient of an air mass property, principally density, and the development of the accompanying features of the wind field that characterize a front.

frontolysis, *n.* 1. The dissipation of a front or frontal zone.
2. In general, a decrease in the horizontal gradient of an air mass property, principally density, and the dissipation of the accompanying features of the wind field.

frost, *n.* 1. A deposit of interlocking ice crystals formed by direct sublimation on objects, usually those of small diameter freely exposed to the air. The deposition is similar to the process in which dew is formed, except that the temperature of the befrosted object must be below freezing. It forms when air with a dew point below freezing is brought to saturation by cooling. It is more fluffy and feathery than rime which in turn is lighter than glaze. Also called HOAR, HOARFROST.
2. The condition which exists when the temperature of the earth's surface and earthbound objects falls below 0°C or 32°F.

frost smoke. 1. Fog-like clouds due to contact of

cold air with relatively warm water, which can appear over openings in the ice, or leeward of the ice edge, and which may persist while ice is forming.

2. A rare type of fog formed in the same manner as a steam fog but at lower temperatures. It is composed of ice particles or droxtals instead of liquid water as is steam fog. Thus, it is a type of ice fog. Sometimes called BARBER.

3. See STEAM FOG.

frozen precipitation. Any form of precipitation that reaches the ground in frozen form; i.e., snow, snow pellets, snow grains, ice crystals, ice pellets, and hail.

frustum, frustrum, *n.* That part of a solid figure between the base and a parallel intersecting plane; or between any two intersecting planes, generally parallel.

full depiction of detail. Since even on charts of the largest scale full depiction of detail is impossible because all features are symbolized to an extent which is partly determined by scale and partly by the conventions of charting practice, the term *full depiction of detail* is used to indicate that over the greater part of a chart nothing essential to navigation is omitted. See also GENERALIZATION OF DETAIL, MINIMAL DEPICTION OF DETAIL.

full moon. The moon at opposition, when it appears as a round disk to an observer on the earth because the illuminated side is toward him. See also PHASES OF THE MOON.

function, *n.* A magnitude so related to another magnitude that for any value of one there is a corresponding value of the other. For instance, the area of a circle is a function of its radius. The radius is also a function of the area. See also TRIGONOMETRIC FUNCTIONS.

fundamental circle. See PRIMARY GREAT CIRCLE.

fundamental frequency. In the Decca Navigator System, the frequency from which other frequencies in a chain are derived by harmonic multiplication.

fundamental star places. The apparent right ascensions and declinations of 1,535 standard comparison stars obtained by leading observatories and published annually under the auspices of the International Astronomical Union.

funnel cloud. A cloud column or inverted cloud cone, pendant from a cloud base. This supplementary feature occurs mostly with cumulus and cumulonimbus; when it reaches the earth's surface, it constitutes the cloudy manifestation of an intense vortex, namely a tornado or waterspout. Also called TUBA, TORNADO CLOUD.

furrow, *n.* On the sea floor, a closed, linear, narrow, shallow depression.

fusion, *n.* The phase transition of a substance passing from the solid to the liquid state; melting. In meteorology, fusion is almost always understood to refer to the melting of ice, which, if the ice is pure and subjected to 1 standard atmosphere of pressure, takes place at the ice point of 0°C or 32°F. Additional heat at the melting point is required to fuse any substance. This quantity of heat is called LATENT HEAT OF FUSION; in the case of ice, it is approximately 80 calories per gram.

G

g, *n.* An acceleration equal to the acceleration of gravity, approximately 32.2 feet per second per second at sea level.

gain, *n.* The ratio of output voltage, current, or power to input voltage, current, or power in electronic instruments.

gain control. See as RECEIVER GAIN CONTROL.

gain function. See DIRECTIVE GAIN.

gain of an antenna. An expression of radiation effectiveness, it is the ratio of the power required at the input of a reference antenna to the power supplied to the input of the given antenna to produce, in a given direction, the same field at the same distance. When not specified otherwise, the figure expressing the gain of an antenna refers to the gain in the direction of the radiation main lobe. In services using scattering modes of propagation, the full gain of an antenna may not be realizable in practice and the apparent gain may vary with time.

gain referred to a short vertical antenna. The gain of an antenna in a given direction when the reference antenna is a perfect vertical antenna, much shorter than one quarter of the wavelength, placed on the surface of a perfectly conducting plane earth.

gal, *n.* A special unit employed in geodesy and geophysics to express the acceleration due to gravity. The *gal* is a unit accepted *temporarily* for use with the International System of Units; 1 gal is equal to 1 centimeter per second, per second.

galactic nebula. An aggregation of matter within our galaxy but beyond the solar system, large enough to occupy a perceptible area but which has not been resolved into individual stars.

galaxy, *n.* A vast assemblage of stars, nebulae, etc., composing an island universe. The sun and its family of planets is part of a galaxy commonly called the MILKY WAY.

gale, *n.* Wind of force 8 on the Beaufort wind scale (38 to 40 knots or 39 to 46 miles per hour) is classified as a **gale**; wind of force 9 (41 to 47 knots or 47 to 54 miles per hour) is classified as a **strong gale**; and wind of force 7 (28 to 33 knots or 32 to 38 miles per hour) is classified as a **near gale**. See also MODERATE GALE, FRESH GALE, WHOLE GALE.

gallon, *n.* A unit of volume equal to 4 quarts or 231 cubic inches.

Galofaro, *n.* A whirlpool in the Strait of Messina. Formerly called CHARYBDIS.

galvanometer, *n.* An instrument for measuring the magnitude of a small electric current or for detecting the presence or direction of such a current by means of motion of an indicator in a magnetic field.

gap, *n.* On the sea floor, a narrow break in a ridge or rise.

garua, *n.* A thick, damp fog on the coasts of Ecuador, Peru, and Chile. Also called CAMANCHACA.

gas, *n.* A fluid in such state that it tends to expand indefinitely, or to completely fill a closed container of any size.

gas buoy. A buoy having a gas light. See also LIGHTED BUOY.

gat, *n.* A natural or artificial passage or channel extending inland through shoals or steep banks. See also OPENING.

gather way. To attain headway.

gauge, gage, *n.* An instrument for measuring the size or state of anything.

gauge, gage, *v., t.* To determine the size or state of anything.

gauss, *n.* The centimeter-gram-second electromagnetic unit of magnetic induction. It corresponds to 10^{-4} tesla in the International System of Units.

Gaussian distribution. See NORMAL DISTRIBUTION.

Gaussin error. Deviation of a magnetic compass due to transient magnetism caused by eddy currents set up by a changing number of lines of force through soft iron as the ship changes heading. Due to these eddy currents, the induced magnetism on a given heading does not arrive at its normal value until about 2 minutes after change to the heading. This error should not be confused with RETENTIVE ERROR.

gazeteer, *n.* An alphabetical list of place names giving geographic coordinates.

Gegenschein, *n.* A faint light area of the sky always opposite the position of the sun on the celestial sphere. It is believed to be the reflection of sunlight from particles moving beyond the earth's orbit. Also called COUNTERGLOW.

general chart. See under CHART CLASSIFICATION BY SCALE.

generalization of detail. A term used to indicate that the least essential information is not shown on a chart. The purpose of generalization is primarily to avoid over-crowding charts where space is very limited. It also serves to reduce the correctional maintenance needed and to induce navigators, at least of deeper draft vessels to use charts of larger scales. See also FULL DEPICTION OF DETAIL, MINIMAL DEPICTION OF DETAIL.

general precession. The resultant motion of the components causing precession of the equinoxes westward along the ecliptic at the rate of about 50″.3 per year, completing the cycle in about 25,800 years. The effect of the sun and moon, called **lunisolar precession**, is to produce a westward motion of the equinoxes along the ecliptic. The effect of other planets, called **planetary precession**, tends to produce a much smaller motion eastward along the ecliptic. The component of general precession along the celestial equator, called **precession in right ascension**, is about 46″.1 per year; and the component along a celestial meridian, called **precession in declination**, is about 20″.0 per year.

General Prudential Rule. Rule 2(b) of the International Rules and Article 27 of the Inland Rules. Rule 2(b) states "In construing and complying with these Rules due regard shall be had to all dangers of navigation and collision and to any special circumstances, including the limitations of the vessels involved, which may make a departure from these Rules necessary to avoid immediate danger." Article 27 states "In obeying and construing these rules due regard shall be had to all dangers of navigation and collision, and to any special circumstances which may render a departure from the above rules necessary in order to avoid immediate danger." Note that Rule 2(b) of the International Rules clearly refers to *all* the rules, rather than the *above* rules as in Article 27 of the Inland Rules.

generating area. The area in which ocean waves are generated by the wind. Also called FETCH.

gentle breeze. Wind of force 3 (7 to 10 knots or 8 to 12 miles per hour) on the Beaufort wind scale.

geo, *n.* A narrow coastal inlet bordered by steep cliffs. Also called GIO.

geo-. A prefix meaning earth.

Geoceiver, *n.* Trade name for an antenna-receiver unit capable of receiving signals from the Navy Navigation Satellite System, from which three-dimensional positions can be computed for the antenna location.

geocentric, *adj.* Relative to the earth as a center; measured from the center of the earth.

geocentric latitude. The angle at the center of the reference ellipsoid between the celestial equator and a radius vector to a point on the ellipsoid.

geocentric parallax. The difference in apparent direction of a celestial body from a point on the surface of the earth and from the center of the earth. This difference varies with the body's altitude and distance from the earth. Also called DIURNAL PARALLAX. See also HELIOCENTRIC PARALLAX.

geodesic, *adj.* Of or pertaining to geodesy; geodetic.

geodesic, *n.* See GEODESIC LINE.

geodesic line. A line of shortest distance between any two points on any mathematically defined surface. A geodesic line is a line of double curvature and usually lies between the two normal section lines which the two points determine. If the two terminal points are in nearly the same latitude, the geodesic line may cross one of the normal section lines. It should be noted that, except along the equator and along the meridians, the geodesic line is not a plane curve and cannot be sighted over directly. Also called GEODESIC, GEODETIC LINE.

geodesy, *n.* The science which treats of the determination of the size and figure of the earth (geoid) by such direct measurements as triangulation, leveling, and gravimetric observations; which determines the external gravitational field of the earth and, to a limited degree, the internal structure.

geodetic, *adj.* Of or pertaining to geodesy; geodesic.

geodetic bench mark. See under BENCH MARK.

geodetic datum. See DATUM, HORIZONTAL GEODETIC DATUM, VERTICAL GEODETIC DATUM.

geodetic equator. The line of zero geodetic latitude; the great circle described by the semimajor axis of the reference ellipsoid as it is rotated about the minor axis. See also ASTRONOMICAL EQUATOR.

geodetic height. See ELLIPSOIDAL HEIGHT.

geodetic latitude. The angle which the normal to the ellipsoid at a station makes with the plane of the geodetic equator. It differs from the corresponding astronomical latitude by the amount of the meridional component of the local deflection of the vertical. Also called TOPOGRAPHICAL LATITUDE and sometimes GEOGRAPHIC LATITUDE.

geodetic line. See GEODESIC LINE.

geodetic longitude. The angle between the plane of the geodetic meridian at a station and the plane of the geodetic meridian at Greenwich. A geodetic longititude differs from the corresponding astronomical longitude by the amount of the prime vertical component of the local deflection of the vertical divided by the cosine of the latitude. Sometimes called GEOGRAPHIC LONGITUDE.

geodetic meridian. A line on a reference ellipsoid which has the same geodetic longitude at every point. Sometimes called GEOGRAPHIC MERIDIAN.

geodetic parallel. A line on a reference ellipsoid which has the same geodetic latitude of every point. A geodetic parallel, other than the equator, is not a geodesic line. In form, it is a small circle whose plane is parallel with the plane of the geodetic equator. See also ASTRONOMICAL PARALLEL.

geodetic position. A position of a point on the surface of the earth expressed in terms of *geodetic latitude* and *geodetic longitude*. A geodetic position implies an adopted geodetic datum.

geodetic satellite. Any satellite whose orbit and payload render it useful for geodetic purposes.

geodetic survey. A survey that takes into account the shape and size of the earth. It is applicable for large areas and long lines and is used for the precise location of basic points suitable for controlling other surveys.

geographic, geographical, *adj.* Of or pertaining to geography.

geographical coordinates. Spherical coordinates defining a point on the surface of the earth, usually latitude and longitude. Also called TERRESTRIAL COORDINATES.

geographical mile. The length of 1 minute of arc of the equator, or 6,087.08 feet. This approximates the length of the nautical mile.

geographical plot. A plot of the movements of one or more craft relative to the surface of the earth. Also called TRUE PLOT. See also NAVIGATIONAL PLOT.

geographical pole. Either of the two points of intersection of the surface of the earth with its axis, where all meridians meet, labeled N or S to indicate whether the north **geographical pole** or the **south geographical pole.**

geographical position. 1. That point on the earth at which a given celestial body is in the zenith at a specified time. The geographical position of the sun is also called the **subsolar point,** of the moon the **sublunar point,** and of a star the **substellar** or **subastral point.**
2. Any position on the earth defined by means of its geographical coordinates either astronomical or geodetic.

geographic graticule. The system of coordinates of latitude and longitude used to define the position of a point on the surface of the earth with respect to the reference ellipsoid.

geographic latitude. A general term applying alike to astronomic and geodetic latitudes.

geographic longitude. A general term applying alike to astronomic and geodetic longitudes.

geographic meridian. A general term applying alike to astronomical and geodetic meridians.

geographic number. The number assigned to an aid to navigation for identification purposes in accordance with the lateral system of numbering.

geographic parallel. A general term applying alike to astronomical and geodetic parallels.

geographic range. The maximum distance at which the curvature of the earth and terrestrial refraction permit a light to be seen from a particular height of eye without regard to the luminous intensity of the light. The geographic range sometimes printed on charts or tabulated in light lists is the maximum distance at which the curvature of the earth and terrestrial refraction permit a light to be seen from a height of eye of 15 feet above the water when the elevation of the light is taken above the height datum of the largest scale chart of the locality. Therefore, this range is a nominal geographic range. See also VISUAL RANGE (OF A LIGHT).

geographic sign conventions. In mapping, charting, and geodesy, the inconsistent application of algebraic sign to geographical references and the angular reference of azimuthal systems is a potential trouble area in scientific data collection. The following conventions have wide use in the standardization of scientific notation: Longitude references are positive eastward of the Greenwich meridian to 180°, and negative westward of Greenwich. Latitude references are positive to the north of the equator and negative to the south. Azimuths are measured clockwise, using South as the origin and continuing to 360°. Bearings are measured clockwise, using North as the origin and continuing to 360°. Tabulated coordinates, or individual coordinates, are annotated N, S, E, W, as appropriate.

geoid, *n.* The equipotential surface in the gravity field of the earth to which the oceans would conform over the entire earth if free to adjust to the combined effect of the earth's mass attraction and the centrifugal force of the earth's rotation. As a result of the uneven distribution of the earth's mass, the geoidal surface is irregular. The geoid is a surface along which the gravity potential is everywhere equal (equipotential surface) and to which the direction of gravity is always perpendicular. Also called FIGURE OF THE EARTH.

geoidal height. The distance of the geoid above (positive) or below (negative) the mathematical reference ellipsoid. Also called GEOIDAL SEPARATION, GEOIDAL UNDULATION, UNDULATION OF THE GEOID.

geoidal horizon. That circle of the celestial sphere formed by the intersection of the celestial sphere and a plane through a point on the sea level surface of the earth, and perpendicular to the zenith-nadir line. See also HORIZON.

geoidal separation. See GEOIDAL HEIGHT.

geoidal undulation. See GEOIDAL HEIGHT.

geological oceanography. The study of the floors and margins of the oceans, including description of submarine relief features, chemical and physical composition of bottom materials, interaction of sediments and rocks with air and seawater, and action of various forms of wave energy in the submarine crust of the earth.

geomagnetic, *adj.* Of or pertaining to geomagnetism.

geomagnetic equator. The terrestrial great circle everywhere 90° from the geomagnetic poles. GEOMAGNETIC EQUATOR should not be confused with MAGNETIC EQUATOR, the line connecting all points of zero magnetic dip.

geomagnetic latitude. Angular distance from the geomagnetic equator, measured northward or southward on the geomagnetic meridian through 90° and labeled N or S to indicate the direction of measurement. GEOMAGNETIC LATITUDE should not be confused with MAGNETIC LATITUDE.

geomagnetic pole. Either of two antipodal points marking the intersection of the earth's surface with the extended axis of a powerful bar magnet assumed to be located at the center of the earth and approximating the source of the actual magnetic field of the earth. The pole in the Northern Hemisphere (at lat. 78.°5 N, long. 69° W) is designated north geomagnetic pole, and the pole in the Southern Hemisphere (at lat. 78.°5 S, long. 111° E) is designated south geomagnetic pole. The great circle midway between these poles is called GEOMAGNETIC EQUATOR. The expression GEOMAGNETIC POLE should not be confused with MAGNETIC POLE, which relates to the actual magnetic field of the earth. See also GEOMAGNETIC LATITUDE.

geomagnetism, *n.* Magnetic phenomena, collectively considered, exhibited by the earth and its atmosphere. Also called TERRESTRIAL MAGNETISM.

geometrical dip. The vertical angle, at the eye of an observer, between the horizontal and a straight line tangent to the surface of the earth. It is larger than DIP by the amount of terrestrial refraction.

geometrical horizon. Originally, the celestial horizon; now more commonly the intersection of the celestial sphere and an infinite number of straight lines tangent to the earth's surface, and radiating from the eye of the observer. If there were no terrestrial refraction, GEOMETRICAL and VISIBLE HORIZONS would coincide. See also RADIO HORIZON.

geometric dilution. See GEOMETRIC DILUTION OF PRECISION.

geometric dilution of precision. All geometric factors that degrade the accuracy of position fixes derived from externally referenced navigation systems. Often shortened to GEOMETRIC DILUTION.

geometric map projection. See PERSPECTIVE MAP PROJECTION.

geometric projection. See PERSPECTIVE PROJECTION.

geomorphology, *n.* A branch of both geography and geology that deals with the form of the earth, the general configuration of its surface, and the changes that take place in the evolution of land forms.

geo-navigation, *n.* Navigation by means of reference points on the earth. The term is obsolescent.

geophysics, *n.* The study of the composition and physical phenomena of the earth and its liquid and gaseous envelopes; it embraces the study of terrestrial magnetism, atmospheric electricity, and gravity; and it includes seismology, volcanology, oceanography, meteorology, and related sciences.

geopotential, *n.* The gravity potential of the actual earth. It is the sum of the gravitational (attraction) potential and the potential of the centrifugal force.

Georef, *n.* See WORLD GEOGRAPHIC REFERENCE SYSTEM.

geosphere, *n.* The portion of the earth, including land (lithosphere) and water (hydrosphere), but excluding the atmosphere.

geostationary satellite. An earth satellite moving eastward in an equatorial, circular orbit at an altitude (approximately 35,900 kilometers) such that its period of revolution is exactly equal to (synchronous with) the rotational period of the earth. Such a satellite will remain fixed over a point on the earth's equator. Although geostationary satellites are frequently called GEOSYNCHRONOUS or SYNCHRONOUS SATELLITES, the orbit of an eastward moving synchronous satellite must be equatorial if the satellite is to remain fixed over a point on the equator. Otherwise, the satellite moves daily in a figure-eight pattern relative to the earth. Also called FIXED SATELLITE. See also STATIONARY ORBIT.

geostrophic wind. That horizontal wind velocity for which the Coriolis force exactly balances the horizontal pressure force. See also GRADIENT WIND.

geosynchronous satellite. An earth satellite whose period of rotation is equal to the period of rotation of the earth about its axis. The orbit of a geosynchronous satellite must be equatorial if the satellite is to remain fixed over a point on the earth's equator. Also called TWENTY-FOUR HOUR SATELLITE. See also SYNCHRONOUS SATELLITE, GEOSTATIONARY SATELLITE.

ghost, *n.* 1. An unwanted image appearing on a radarscope caused by echoes which experience multiple reflections before reaching the receiver. See also SECOND-TRACE ECHO, MULTIPLE ECHOS, INDIRECT ECHO.
2. An image appearing on a radarscope the origin of which cannot readily be determined.

giant floe. See under FLOE.

gibbous, *adj.* Bounded by convex curves. The term is used particularly in reference to the moon when it is between first quarter and full or between full and last quarter, or to other celestial bodies when they present a similar appearance. See also PHASES OF THE MOON.

giga-. A prefix meaning *one billion* (10^9).

gigahertz, *n.* One thousand megahertz, or one billion cycles per second.

gimbal freedom. Of a gyro, the maximum angular displacement about the output axis of a gimbal.

gimballess inertial navigation equipment. See STRAPPED-DOWN INERTIAL NAVIGATION EQUIPMENT.

gimballing error. That error introduced in a gyrocompass by the tilting of the gimbal mounting system of the compass due to horizontal acceleration caused by motion of the vessel, such as rolling.

gimbal lock. A condition of a two-degree-of-freedom gyro wherein the alignment of the spin axis with an axis of freedom deprives the gyro of a degree-of-freedom and therefore its useful properties.

gimbals, *n., pl.* A device for supporting anything, such as an instrument, in such a manner that it will remain essentially horizontal when the support tilts. It consists of a ring inside which the instrument is supported at two points 180° apart, the ring being similarly supported at two points 90° from the instrument supports.

gio, *n.* See GEO.

glacial, *adj.* Of or pertaining to a glacier.

glacier, *n.* A mass of snow and ice continuously moving from higher to lower ground or, if afloat, continously spreading. The principal forms of glacier are INLAND ICE SHEETS, ICE SHELVES, ICE STREAMS, ICE CAPS, ICE PIEDMONTS, CIRQUE GLACIERS, and various types of mountain (valley) glaciers.

glacier berg. An irregularly shaped iceberg. Also called WEATHERED BERG.

glacier ice. Ice in, or originating from, a glacier, whether on land or floating on the sea as ice**bergs, bergy bits,** or **growlers.**

glacier tongue. The seaward projecting extension of a glacier, usually afloat. In the Antarctic, glacier tongues may extend many 10s of kilometers.

glare, *n.* Dazzling brightness of the atmosphere, caused by excessive reflection and scattering of light by particles in the line of sight.

glaze, *n.* A coating of ice, generally clear and smooth but usually containing some air pockets, formed on exposed objects by the freezing of a film of supercooled water deposited by rain, drizzle, fog, or possibly condensed from supercooled water vapor. Glaze is denser, harder and more transparent than either rime or hoarfrost. Also called GLAZE ICE, GLAZED FROST, VERGLAS.

glazed frost. See GLAZE.

glaze ice. See GLAZE.

glint, *n.* The pulse-to-pulse variation in amplitude of reflected radar signals due to rapid change of the reflecting surface, as in the case of the propeller of an aircraft in flight.

Global Positioning System. See as NAVSTAR GLOBAL POSITIONING SYSTEM.

globigerina (*pl. globigerinae*), *n.* A very small marine animal of the foraminifera order, with a chambered shell; or the shell of such an animal. In large areas of the ocean the calcareous shells of these animals are very numerous, being the principal constituent of a soft mud or *globigerina ooze* forming the ocean bed.

gloom, *n.* The condition existing when daylight is very much reduced by dense cloud or smoke accumulation above the surface, the surface visibility not being materially reduced.

glory, *n.* See ANTICORONA.

gnomon, *n.* Any object the shadow of which serves as an indicator, as the SHADOW PIN on a sun compass.

gnomonic, *adj.* Of or pertaining to a gnomon.

gnomonic chart. A chart constructed on the gnomonic projection and often used as an adjunct for transferring a great circle to a Mercator chart. Commonly called GREAT-CIRCLE CHART.

gnomonic map projection. A perspective azimuthal map projection in which points on the surface of a sphere or spheroid, such as the earth, are conceived as projected by radials from the center to a tangent plane. Great circles project as straight lines. For this reason the projection is used principally for charts for great-circle sailing. The projection is neither conformal nor equal-area.

gong, *n.* A sound signal emitter producing a sound by the vibration of a resonant disc excited by a blow.

gong buoy. A buoy fitted with a group of saucer-shaped bells of different tones as an audible signal.

goniometer, *n.* 1. An instrument for measuring angles.
2. A pick-up coil which eliminates the necessity of having to rotate a radio direction finder antenna to determine direction.

good seamanship. Any precaution which may be required by the ordinary practice of seamen.

gore, *n.* A lune-shaped map which may be fitted to the surface of a globe with a negligible amount of distortion.

gorge, *n.* 1. A narrow opening between mountains, especially one with steep, rocky walls.
2. A collection of solid matter obstructing a channel, river, etc., as *ice gorge*.

gradient, *n.* 1. A rate of rise or fall of a quantity against horizontal distance expressed as a ratio, decimal, fraction, percentage, or the tangent of the angle of inclination.
2. The rate of increase or decrease of one quantity with respect to another.
3. A term used in radionavigation to refer to the spacing between consecutive hyperbolas of a family of hyperbolas per unit time difference, as 1 microsecond. It follows that if the gradient is high, a relatively small time-difference error in determining a hyperbolic line of position will result in a relatively high position error. See also

GEOMETRIC DILUTION OF PRECISION.

gradient current. An ocean current associated with horizontal pressure gradients in the ocean and determined by the condition that the pressure force due to the distribution of mass balances the Coriolis force due to the earth's rotation. See also OCEAN CURRENT.

gradient tints. See HYPSOMETRIC TINTING.

gradient wind. Any horizontal wind velocity tangent to the contour line of a constant pressure surface (or to the isobar of a geopotential surface) at the point in question. At such points where the wind is gradient, the Coriolis force and the centrifugal force together exactly balance the horizontal pressure force. See also GEOSTROPHIC WIND.

graduation error. Inaccuracy in the graduations of the scale of an instrument.

graduations, *n., pl.* The marks on a scale.

grain noise. See SNOW, definition 2.

gram, *n.* One one-thousandth of a kilogram.

granular snow. See SNOW GRAINS.

graph, *n.* A diagram indicating the relationship between two or more variables.

graph, *v., t.* To represent by a graph.

graphic scale. See BAR SCALE.

graticule, *n.* 1. The network of lines representing parallels and meridians on a map, chart, or plotting sheet. A **fictitious graticule** represents fictitious parallels and fictitious meridians. See also GRID, *n.*

2. A scale at the focal plane of an optical instrument to aid in the measurement of objects. See also RETICLE.

graupel, *n.* See SNOW PELLETS.

gravel, *n.* See under STONES.

graving dock. A form of dry dock consisting of an artificial basin fitted with a gate or caisson, into which vessels can be floated and the water pumped out to expose the vessels' bottoms. The term is derived from the term used to describe the process of burning barnacles and other accretions from a ship's bottom. See also FLOATING DOCK.

gravisphere, *n.* The spherical extent in which the force of a given celestial body's gravity is predominant in relation to that of other celestial bodies.

gravitation, *n.* 1. The force of attraction between two bodies. It is the gravitational force which holds planets, satellites, asteroids, and comets in their orbits and prevents their wandering off into space. According to Newton, gravitation is directly proportional to the product of the masses of two bodies and inversely proportional to the square of the distance between them.

2. The acceleration produced by the mutual attraction of two masses, directed along the line joining their centers of mass, and of magnitude inversely proportional to the square of the distance between the two centers of mass.

gravitational disturbance. See GRAVITY DISTURBANCE.

gravitational gradient. The change in the gravitational acceleration per unit distance.

gravitational perturbations. Perturbations caused by body forces due to nonspherical terrestrial effects, lunisolar effect, tides, and the effect of relativity.

gravitational tide. See EQUILIBRIUM TIDE.

gravity, *n.* Viewed from a frame of reference fixed in the earth, the acceleration imparted by the earth to a mass which is rotating with the earth. Since the earth is rotating, the acceleration observed as gravity is the resultant of the acceleration of gravitation and the centrifugal acceleration arising from the earth's rotation and the use of an earthbound rotating frame of reference.

gravity anomaly. The difference between the observed gravity value properly reduced to sea level and the theoretical gravity obtained from gravity formula. Also called OBSERVED GRAVITY ANOMALY.

gravity anomaly map. A map showing the positions and magnitudes of gravity anomalies. Also, a map on which contour lines are used to represent points at which the gravity anomalies are equal.

gravity data. Information concerning that acceleration which attracts bodies and is expressed as observations or in the form of gravity anomaly charts or spherical harmonics for spatial representation of the earth and other celestial bodies.

gravity disturbance. The difference between the observed gravity and the normal gravity at the same point (the vertical gradient of the disturbing potential) as opposed to GRAVITY ANOMALY which uses corresponding points on two different surfaces. Because the centrifugal force is the same when both are taken at the same point, it can also be called GRAVITATIONAL DISTURBANCE.

gravity field of the earth. The field of force arising from a combination of the mass attraction and rotation of the earth. The field is normally expressed in terms of point values, mean area values, and/or series expansion for the potential of the field.

gravity network. A network of gravity stations.

gravity reduction. A combination of gravity corrections to obtain reduced gravity on the geoid.

gravity reference stations. Stations which serve as reference values for a gravity survey, i.e., with respect to which the differences at the other stations are determined in a relative survey. The absolute value of gravity may or may not be known at the reference stations.

gravity station. A station at which observations are made to determine the value of gravity.

gravity wind. A wind blowing down an incline. Also called KATABATIC WIND.

grease ice. Ice at that stage of freezing when the crystals have coagulated to form a soupy layer on the surface. Grease ice is at a later stage of freezing than *frazil ice* and reflects little light, giving the sea a matte appearance.

great circle. The intersection of a sphere and a plane through its center. The intersection of a sphere and a plane which does not pass through its center is called a **small circle.** Also called ORTHODROME, ORTHODROMIC CURVE.

great-circle bearing. The initial direction of a great circle through two terrestrial points, expressed as angular distance from a reference direction. It is usually measured from 000° at the reference direction clockwise through 360°. Bearings obtained by any form of radiant energy are great-circle bearings.

great-circle chart. A chart on which a great circle appears as a straight line or approximately so, particularly a chart on the gnomonic map projection.

great-circle course. The direction of the great circle through the point of departure and the destination, expressed as the angular distance from a reference direction, usually north, to the direction of the great circle. The angle varies from point to point along the great circle. At the point of departure it is called **initial great-circle course**; at the destination it is called **final great-circle course.**

great-circle direction. Horizontal direction of a great circle, expressed as angular distance from a reference direction.

great-circle distance. The length of the shorter arc of the great circle joining two points. It is usually expressed in nautical miles.

great-circle sailing. Any method of solving the various problems involving courses, distance, etc., as they are related to a great-circle track.

great-circle track. The track of a vessel following a great circle, or a great circle which it is intended that a vessel follow approximately.

great diurnal range. The difference in height between mean higher high water and mean lower low water. Often shortened to DIURNAL RANGE. The difference in height between mean lower high water and mean higher low water is called SMALL DIURNAL RANGE.

greater ebb. See under EBB CURRENT.

greater flood. See under FLOOD CURRENT.

greatest elongation. The maximum angular distance of an *inferior planet* from the sun before it starts back toward conjunction, as observed from the earth. The direction of the body east or west of the sun is usually specified, as *greatest elongation east* (or *west*). See also ELONGATION.

Great Lakes Notice to Mariners. Notice to Mariners relating to the Great Lakes and tributary waters west of Montreal published weekly by the U.S. Coast Guard. The Notice contains selected items from the *Local Notice to Mariners* and other reported marine information and is intended primarily for use in correcting Great Lakes charts and related publications. See also NOTICE TO MARINERS.

Great Lakes Pilot, *United States Coast Pilot* 6, Great Lakes: Lakes Ontario, Erie, Huron, Michigan, and Superior and St. Lawrence River, published in 1978 cancelled the 1977 edition of the publication *formerly* known as the *Great Lakes Pilot*.

great tropic range. The difference in height between tropic higher high water and tropic lower low water. Often shortened to TROPIC RANGE. See also MEAN TROPIC RANGE, SMALL TROPIC RANGE.

great year. The period of one complete cycle of the equinoxes around the ecliptic, about 25,800 years. Also called PLATONIC YEAR. See also PRECESSION OF THE EQUINOXES.

green flash. A brilliant green coloring of the upper edge of the sun as it appears at sunrise or disappears at sunset when there is a clear, distinct horizon. It is due to refraction by the atmosphere, which disperses the first (or last) spot of light into a spectrum and causes the colors to appear (or disappear) in the order of refrangibility. The green is bent more than red or yellow and hence is visible sooner at sunrise and later at sunset.

green house effect. The heating phenomenon due to shorter wavelengths of insolation passing through the atmosphere to the earth, which radiates longer wavelength infrared radiation that is trapped by the atmosphere. Some of this trapped radiation is reradiated to the earth. This causes a higher earth temperature than would occur from direct insolation alone.

Greenwich apparent noon. Local apparent noon at the Greenwich meridian; 12 o'clock Greenwich apparent time, or the instant the apparent sun is over the upper branch of the Greenwich meridian.

Greenwich apparent time. Local apparent time at the Greenwich meridian; the arc of the celestial equator, or the angle at the celestial pole, between the lower branch of the Greenwich celestial meridian and the hour circle of the apparent or true sun, measured westward from the lower branch of the Greenwich celestial meridian through 24 hours; Greenwich hour angle of the apparent or true sun, expressed in time units, plus 12 hours.

Greenwich civil time. *United States terminology from 1925 through 1952.* See GREENWICH MEAN TIME.

Greenwich hour angle. Angular distance west of the Greenwich celestial meridian; the arc of the celestial equator, or the angle at the celestial pole, between the upper branch of the Greenwich celestial meridian and the hour circle of a point on the celestial sphere, measured westward from the Greenwich celestial meridian through 360°; local hour angle at the Greenwich meridian.

Greenwich interval. An interval based on the moon's transit of the Greenwich celestial meridian, as distinguished from a local interval based on the moon's transit of the local celestial meridian.

Greenwich lunar time. Local lunar time at the Greenwich meridian; the arc of the celestial equator, or the angle at the celestial pole, between the lower branch of the Greenwich celestial meridian and the hour circle of the moon, measured westward from the lower branch of the Greenwich celestial meridian through 24 hours; Greenwich hour angle of the moon, expressed in time units, plus 12 hours.

Greenwich mean noon. Local mean noon at the Greenwich meridian; 12 o'clock Greenwich mean time, or the instant the mean sun is over the upper branch of the Greenwich meridian.

Greenwich mean time. Local mean time at the Greenwich meridian; the arc of the celestial equator, or the angle at the celestial pole, between the lower branch of the Greenwich celestial meridian and the hour circle of the mean sun, measured westward from the lower branch of the Greenwich celestial meridian through 24 hours; Greenwich hour angle of the mean sun, expressed in time units, plus 12 hours. Called GREENWICH CIVIL TIME in United States terminology from 1925 through 1952. Also called UNIVERSAL TIME, ZULU TIME.

Greenwich meridian. The meridian through Greenwich, England, serving as the reference for Greenwich time, in contrast with LOCAL MERIDIAN. It is accepted almost universally as the PRIME MERIDIAN, or the origin of measurement of longitude.

Greenwich noon. Noon at the Greenwich meridian.

Greenwich sidereal noon. Local sidereal noon at the Greenwich meridian; zero hours Greenwich sidereal time, or the instant the vernal equinox is over the upper branch of the Greenwich meridian.

Greenwich sidereal time. Local sidereal time at the Greenwich meridian; the arc of the celestial equator, or the angle at the celestial pole, between the upper branch of the Greenwich celestial meridian and the hour circle of the vernal equinox, measured westward from the upper branch of the Greenwich celestial meridian through 24 hours; Greenwich hour angle of the vernal equinox expressed in time units.

Greenwich time. Time based upon the Greenwich meridian as reference.

gregale, *n.* A strong northeast wind of the central Mediterranean.

Gregorian calendar. The calendar now in almost universal use for civil purposes in which each year has 365 days, except leap years which have 366 days. Leap years are those years which are divisible by 4, and in the case of centurial years, those years divisible by 400. This calendar, a modification of the Julian calendar, was not adopted in Great Britain and the English colonies in North America until 1752. The calendar was instituted in 1582 by Pope Gregory XIII to keep calendar days in adjustment with the tropical year for the purpose of regulating the date of Easter and the civil and ecclesiastical calendars.

grey ice. A subdivision of YOUNG ICE 10 to 15 centimeters thick. Grey ice is less elastic than nilas and breaks on swell. It usually rafts under pressure.

grey-white ice. A subdivision of YOUNG ICE 15 to 30 centimeters thick. Grey-white ice under pressure is more likely to ridge than to raft.

grid, *adj.* Of or pertaining to a grid or related to grid north.

grid, *n.* 1. A series of lines, usually straight and parallel, superimposed on a chart or plotting sheet to serve as a directional reference for navigation. See also FICTITIOUS GRATICULE, GRATICULE, definition 1.
2. Two sets of mutually perpendicular lines dividing a map or chart into squares or rectangles to permit location of any point by a system of rectangular coordinates. Also called REFERENCE GRID. See also MILITARY GRID, UNIVERSAL POLAR STEREOGRAPHIC GRID, UNIVERSAL TRANSVERSE MERCATOR GRID, WORLD GEOGRAPHIC REFERENCING SYSTEM.

grid amplitude. Amplitude relative to grid east or west.

grid azimuth. Azimuth relative to grid north.

grid bearing. Bearing relative to grid north.

grid convergence. The angular difference in direction between grid north and true north. It is measured east or west from true north.

grid course. Course relative to grid north.

grid declination. The angular difference between grid north and true north.

grid direction. Horizontal direction expressed as angular distance from grid north. Grid direction is measured from grid north, clockwise through 360°.

grid equator. A line perpendicular to a prime grid meridian, at the origin. For the usual orientation in polar regions the grid equator is the 90° W-90° E meridian forming the basic grid parallel, from which grid latitude is measured. See also FICTITIOUS EQUATOR.

grid heading. Heading relative to grid north.

grid latitude. Angular distance from a grid equator. See also FICTITIOUS LATITUDE.

grid line. One of the lines of a grid.

grid longitude. Angular distance between a prime grid meridian and any given grid meridian. See also FICTITIOUS LONGITUDE.

grid magnetic angle. Angular difference in direction between grid north and magnetic north. It is measured east or west from grid north. Grid magnetic angle is sometimes called GRID VARIATION or GRIVATION.

grid meridian. One of the grid lines extending in a grid north-south direction. The reference grid meridian is called **prime grid meridian.** In polar regions the prime grid meridian is usually the 180°-0° geographic meridian. See also FICTITIOUS MERIDIAN.

grid navigation. Navigation by the use of grid directions.

grid north. 1. An arbitrary reference direction used with grid navigation. The direction of the 180th geographical meridian from the north pole is used almost universally as grid north.
2. The northerly or zero direction indicated by the grid datum of directional reference.

grid parallel. A line parallel to a grid equator, connecting all points of equal grid latitude. See also FICTITIOUS PARALLEL.

grid prime vertical. The vertical circle through the grid east and west points of the horizon.

grid rhumb line. A line making the same oblique angle with all grid meridians. Grid parallels and meridians may be considered special cases of the grid rhumb line. See also FICTITIOUS RHUMB LINE.

grid track. The direction of the track relative to grid north.

grid variation. See GRID MAGNETIC ANGLE.

grivation, *n.* See GRID MAGNETIC ANGLE.

groin, *n.* A structure (usually one of a group) extending approximately perpendicular from a shore to protect the shore from erosion by tides, currents, or waves or to trap sand for making a beach. See also JETTY, definition 1.

ground, *n.* A conducting connection between an electric circuit and the earth or some other conducting body of zero potential with respect to the earth.

ground, *v., t. & i.* To touch bottom or run aground. In a serious grounding the vessel is said to **strand.**
v., t. To connect an electric circuit with the earth or some other conducting body, such that the earth or body serves as part of the circuit.

ground absorption. The dissipation of energy in radio waves because of absorption by the ground over which the waves are transmitted.

ground-based duct. See SURFACE DUCT.

ground chain. Heavy chain used with permanent moorings and connecting the various legs or bridles.

grounded hummock. Hummocked grounded ice formation. There are *single* grounded hummocks and *lines* (or *chains*) or grounded hummocks.

grounded ice. Floating ice which is aground in shoal water. See also STRANDED ICE, FLOATING ICE.

ground fog. A fog that obscures less than six-tenths of the sky, and does not extend to the base of any clouds.

grounding, *n.* The touching of the bottom by a vessel. A serious grounding is called a **stranding.**

ground log. A device for determining the course and speed over the ground in shallow water, consisting of a lead or weight attached to a line. The lead is thrown overboard and allowed to rest on the bottom. The course over ground is indicated by the direction the line tends and the speed by the amount of line paid out in unit time.

ground swell. A long, deep swell or undulation of the ocean often caused by a long-continued gale and sometimes a seismic disturbance and felt even at a remote distance. In shallow water the swell rises to a prominent height. See SWELL, definition 1.

ground tackle. The anchors, anchor chains, fittings, etc., used for anchoring a vessel.

ground track. 1. See under TRACK, definition 2.
2. See under TRUE TRACK OF TARGET.

groundwave. A radio wave that is propagated over the earth and is ordinarily influenced by the presence of the ground and the troposphere.

Except for ionospheric and tropospheric waves, the groundwave includes all components of a radio wave.

group flashing light. A flashing light in which the flashes are combined in groups, each group having the same number of flashes, and in which the groups are repeated at regular intervals. The eclipses separating the flashes within each group are of equal duration and this duration is clearly shorter than the duration of the eclipse between two successive groups. A light list entry may be given as:

Gp. Fl. W.,
period 30^s
fl. 1^s, ec. 5^s
fl. 1^s, ec. 5^s
fl. 1^s, ec. 17^s

Within each 30^s period, there is first a white flash of 1^s duration, after which the light is eclipsed for 5^s, then there is a white flash of 1^s duration followed by a 5^s eclipse, and then there is a white flash of 1^s duration followed by a 17^s eclipse.

group occulting light. An occulting light in which the occultations are combined in groups, each group including the same number of occultations, and in which the groups are repeated at regular intervals. The intervals of light separating the occultations within each group are of equal duration and this duration is clearly shorter than the duration of the interval of light between two successive groups.

group quick light. A quick light in which a specified group of flashes is regularly repeated. See also CONTINUOUS QUICK LIGHT, INTERRUPTED QUICK LIGHT.

group repetition interval. Of a particular Loran-C chain, the specified time interval for all stations of the chain to transmit their pulse groups. For each chain a minimum group repetition interval (GRI) is selected of sufficient duration to provide time for each station to transmit its pulse group and additional time between each pulse group so that signals from two or more stations cannot overlap in time anywhere within the coverage area. The GRI is normally stated in terms of tens of microseconds; i.e., the GRI having a duration of 79,900 microseconds is stated as 7900. In providing means for identifying a chain within a system all stations of which transmit on the same frequency (100kHz), the GRI is the chain **signature.**

group repetition interval code. The group repetition interval in microseconds divided by 10.

group very quick light. A very quick light in which a specified group of flashes is regularly repeated. See also CONTINUOUS VERY QUICK LIGHT, INTERRUPTED VERY QUICK LIGHT.

growler, *n.* A piece of ice smaller than a BERGY BIT or FLOEBERG, often transparent but appearing green or almost black in color. It extends less than 1 meter above the sea surface and its length is less than 20 feet (6 meters). A growler is large enough to be a hazard to shipping but small enough that it may escape visual or radar detection.

grunt, *n.* See under DIAPHONE.

Guiana Current. An ocean current flowing northwestward along the northeast coast of South America. The Guiana Current is an extension of the Atlantic South Equatorial Current, which crosses the equator and approaches the coast of South America. Eventually, it is joined by part of the Atlantic North Equatorial Current and becomes, successively, the CARIBBEAN CURRENT and the FLORIDA CURRENT. Also called **NORTH BRAZIL CURRENT.**

Guinea Current. A North Atlantic Ocean current flowing eastward along the south coast of northwest Africa into the Gulf of Guinea. The Guinea Current is the continuation of the Atlantic Equatorial Countercurrent augmented by the eastern branch of the Canary Current.

gulder, *n.* Local name given to double low water occurring on the south coast of England. See DOUBLE TIDE.

gulf, *n.* A part of an ocean or sea extending into the land, usually larger than a bay.

Gulf Coast Low Water Datum. A tidal datum which was used as chart datum from November 14, 1977, to November 28, 1980, for the coastal waters of the gulf coast of the United States. Gulf Coast Low Water Datum (GCLWD) is defined as mean lower low water when the type of tide is mixed and mean low water when the type of tide is diurnal.

Gulf Stream. For the greater part, a warm, well-defined, swift, relatively narrow ocean current which originates where the Florida Current and the Antilles Current meet north of Grand Bahama Island. It gains its impetus from the large volume of water that flows through the Straits of Florida, an amount estimated to be more than 20 times greater per hour than all the fresh water entering the oceans from all sources such as rivers, runoff, and thawing glaciers. Near the edge of the Grand Banks of Newfoundland extensions of the Gulf Stream and the Labrador Current continue as the NORTH ATLANTIC CURRENT, which fans outward and widens in a northeastward to eastward flow across the ocean. The **Florida Current**, the **Gulf Stream**, and the **North Atlantic Current** together form the GULF STREAM SYSTEM. Sometimes the entire system is referred to as the *Gulf Stream*. The Gulf Stream forms the western and northwestern part of the general clockwise oceanic circulation of the North Atlantic Ocean

Gulf Stream System. A system of ocean currents comprised of the **Florida Current**, the **Gulf Stream**, and the **North Atlantic Current.**

gulfweed, *n.* See SARGASSUM.

gully, *n.* 1. A small ravine, especially one cut by running water, but through which water flows only after a rain.

2. On the sea floor, a small valley-like feature.

gust, *n.* 1. A sudden brief increase in the speed of the wind, of more transient character than a squall, and followed by a lull or slackening of the wind. All winds near the earth's surface are somewhat gusty, those over rough country being particularly so.

2. The violet wind or squall that accompanies a thunderstorm.

3. A burst or gush of rain.

gut, *n.* A narrow passage or contracted strait connecting two bodies of water.

guyot, *n.* See TABLEMOUNT.

gyre, *n.* A closed circulatory system, but larger than a whirlpool or eddy.

gyro, *n.* Short for GYROSCOPE.

gyrocompass, *n.* A compass having one or more gyroscopes as the directive element, and which is north-seeking. Its operation depends upon four natural phenomena, namely gyroscopic inertia, gyroscopic precession, the earth's rotation, and gravity. When such a compass controls remote indicators, called GYRO REPEATERS, it is called a **master gyrocompass.** See also DIRECTIONAL GYRO MODE.

gyro error. The error in the reading of the gyrocompass, expressed in degrees east or west to indicate the direction in which the axis of the compass is offset from true north. See also BALLISTIC DAMPING ERROR, BALLISTIC DEFLECTION ERROR, COMPASS ERROR, GIMBALLING ERROR, INTERCARDINAL ROLLING ERROR, LUBBER'S LINE ERROR, SPEED ERROR.

gyro log. A written record of the performance of a gyrocompass.

gyropilot, *n.* An automatic device for steering a vessel by means of control signals received from a gyrocompass. Also called AUTOPILOT.

gyro repeater. That part of a remote-indicating gyrocompass system which repeats at a distance the indications of the master gyrocompass. See also COMPASS REPEATER.

gyroscope, *n.* A rapidly rotating mass free to move about one or both axes perpendicular to the axis of rotation and to each other. It is characterized by GYROSCOPIC INERTIA and PRECESSION. Gyroscopes are used in many instruments, such as the gyrocompass. Usually shortened to GYRO. See also DIRECTIONAL GYRO, FREE GYRO.

gyroscopic drift. The horizontal rotation of the spin axis of a gyroscope about the vertical axis.

gyroscopic inertia. The property of a gyroscope of resisting any force which tends to change its axis of rotation. A gyroscope tends to maintain the direction of its axis of rotation in space. Also called RIGIDITY IN SPACE.

gyro sextant. A sextant provided with a gyroscope to indicate the horizontal.

H

haar, *n.* A wet sea fog or very fine drizzle which drifts in from the sea in coastal districts of eastern Scotland and northeast England, especially in summer.

habitat sanctuary. A marine sanctuary established for the preservation, protection and management of essential or specialized habitats representative of important marine systems. See also MARINE SANCTUARY.

hachures, *n., pl.* 1. Short lines on topographic maps or nautical charts to indicate the slope of the ground or the submarine bottom. They usually follow the direction of the slope.
2. Inward-pointing short lines or "ticks" around the circumference of a closed contour indicating a depression or a minimum.

hack, *n.* A chronometer which has failed to meet the exacting requirements of a standard chronometer, and is used for timing observations of celestial bodies, regulating ship's clocks, etc. A comparing watch, which may be of high quality, is normally used for timing celestial observations, the watch being compared with the chronometer, preferably both before and after observations. Sometimes called HACK CHRONOMETER.

hack chronometer. See HACK.

hack watch. See COMPARING WATCH.

hail, *n.* Frozen precipitation consisting of ice balls or irregular lumps of ice of varying size, ranging from that of a raindrop to an inch or considerably more. They are composed of clear ice or of alternate layers of ice and snow, and may fall detached or frozen together into irregular lumps. Hail is usually associated with thunderstorms. A **hailstone** is a single unit of hail. **Small hail** consists of snow pellets surrounded by a very thin ice covering. See also SNOW PELLETS.

hailstone, *n.* See under HAIL.

hail storm. See under STORM, definition 2.

half-convergency. *British terminology.* A useful approximation of conversion angle when the difference of longitude between receiver and transmitter is less than about 5°. Half-convergency is based upon the assumption that the plot of a great-circle track on a Mercator chart is symmetrical, the axis of symmetry being the perpendicular bisector of the rhumb line connecting the two points.

half-power points. Power ratios used to define the angular width of a radar beam. One convention defines beam width as the angular width between points at which the field strength is 71 percent of its maximum value. Expressed in terms of power ratio, this convention defines beam width as the angular width between half-power points. A second convention defines beam width as the angular width between points at which the field strength is 50 percent of its maximum value. Expressed in terms of power ratio, the latter convention defines beam width as the angular width between quarter-power points.

half tide. The condition or time of the tide when at the level midway between any given high tide and the following or preceding low tide.

half-tide basin. A lock of very large size and usually of irregular shape, the gates of which are kept open for several hours after high tide so that vessels may enter as long as there is sufficient depth over the sill. Vessels remain in the half-tide basin until the ensuing flood tide before they may pass through the gate to the inner harbor. If entry to the inner harbor is required before this time, water must be admitted to the half-tide basin from some external source. See also TIDAL BASIN, NON-TIDAL BASIN.

half-tide level. A tidal datum midway between mean high water and mean low water. Mean sea level may coincide with half-tide level, but seldom does; the variation is generally about 3 centimeters and rarely exceeds 6 centimeters. Also called MEAN TIDE LEVEL. See also MID-EXTREME TIDE.

halo, *n.* Any of a group of optical phenomena caused by refraction or reflection of light by ice crystals in the atmosphere. The most common form is a ring of light of radius 22° or 46° around the sun or moon. See also CORONA, PARHELION, CIRCUMSCRIBED HALO, PARHELIC CIRCLE, SUN CROSS, SUN PILLAR, CIRCUMZENITHAL ARC, ANTHELION, PARANTHELION, HAVELIAN HALO, TANGENT ARC.

halving, *n.* The process of adjusting magnetic compass correctors so as to remove half of the deviation on the opposite cardinal or adjacent intercardinal headings to those on which adjustment was originally made when all deviation was removed. This is done to equalize the error on opposite headings.

Handbook of Magnetic Compass Adjustment. See PUB. NO. 226.

hand lead (lĕd). A light sounding lead (7 to 14 pounds), usually having a line of not more than 25 fathoms.

hanging compass. See INVERTED COMPASS.

harbor, *n.* 1. A natural or artificially improved body of water providing protection for vessels and, generally, anchorage and docking facilities.
2. A haven or space of deep water so sheltered by the adjacent land as to afford a safe anchorage for ships. See also NATURAL HARBOR, ARTIFICIAL HARBOR.

harbor chart. See under CHART CLASSIFICATION BY SCALE.

harbor line. The line beyond which wharves and other structures cannot be extended.

harbor reach. See under REACH.

hard beach. A portion of a beach especially prepared with a hard surface extending into the water, employed for the purpose of loading or unloading directly into or even landing ships or landing craft.

hard iron. Iron or steel which is not readily magnetized by induction, but which retains a high percentage of the magnetism acquired. The opposite is SOFT IRON.

harmattan, *n.* The dry, dusty trade wind blowing off the Sahara Desert across the Gulf of Guinea and the Cape Verde Islands. Sometimes called the DOCTOR, because of its supposed healthful properties.

harmful interference. Any emission, radiation or induction which endangers the functioning of a radionavigation service or of other safety services or seriously degrades, obstructs or repeatedly interrupts a radio-communication service operating in accordance with the International Telecommunications Union Regulations.

harmonic, *n.* 1. A sinusoidal quantity having a

frequency that is an integral multiple of the frequency of a periodic quantity to which it is related.

2. A signal having a frequency which is an integral multiple of the fundamental frequency.

harmonic analysis. The process by which the observed tide or tidal current at any place is separated into basic harmonic constituents. Also called HARMONIC REDUCTION.

harmonic analyzer. A machine designed for the resolution of a periodic curve into its harmonic constituents. Now performed by electronic digital computer.

harmonic component. Any of the simple sinusoidal components into which a periodic quantity may be resolved.

harmonic constants. The amplitudes and epochs of the harmonic constituents of the tide or tidal current at any place.

harmonic constituent. See CONSTITUENT.

harmonic expressions. Trigonometric terms of an infinite series used to approximate irregular curves in two or three dimensions.

harmonic function. Any real function that satisfies a certain equation. In its simplest form, as used in tide and tidal current predictions, it is a quantity that varies as the cosine of an angle that increases uniformly with time.

harmonic motion. The projection of circular motion on a diameter of the circle of such motion. Simple harmonic motion is produced if the circular motion is of constant speed. The combination of two or more simple harmonic motions results in compound harmonic motion.

harmonic prediction (*tidal*). Method of predicting tides and tidal currents by combining the harmonic constituents into a single tide curve. The work is usually performed by electronic digital computer.

harmonic reduction. See HARMONIC ANALYSIS.

harmonic tide plane. See INDIAN SPRING LOW WATER.

harpoon log. A log which consists essentially of a rotator and distance registering device combined in a single unit, which is towed through the water. It has been largely replaced by the TAFFRAIL LOG, which is similar except that the registering device is located at the taffrail, with only the rotator in the water.

harvest moon. The full moon occurring nearest the autumnal equinox. See also PHASES OF THE MOON.

haul, *v., i.* 1. Of the wind, to change direction in a counterclockwise direction.

2. Of the wind to shift forward. The opposite motion is to VEER aft.

v., t. To change the course of a vessel so as to bring the wind farther forward; usually used with *up*, as *haul up*.

haven, *n.* A place of safety for vessels. It should be accessible at all states of the tide and weather.

haversine, *n.* Half of the versine: (1-cos θ)/2. The expression NATURAL HAVERSINE is sometimes used to distinguish the haversine from its logarithm (called LOGARITHMIC HAVERSINE).

haze, *n.* Fine dust or salt particles in the air, too small to be individually apparent but in sufficient number to reduce horizontal visibility and give the atmosphere a characteristic hazy appearance which casts a bluish or yellowish veil over the landscape, subduing its colors. This is sometimes called a **dry haze** to distinguish it from damp haze, small water droplets or very hygroscopic particles in the air, smaller and more scattered than light fog.

head, *n.* See HEADLAND.

heading, *n.* The horizontal direction in which a ship actually points or heads at any instant, expressed in angular units from a reference direction, usually from 000° at the reference direction clockwise through 360°. Heading is often designated as true, magnetic, compass, or grid as the reference direction is true, magnetic, compass, or grid north, respectively. Heading should not be confused with COURSE, which is the intended direction of movement through the water. At a specific instant the heading may or may not coincide with the course, depending upon such factors as steering errors, actions of the seas upon the ship, etc. The heading of a ship is also called SHIP'S HEAD.

heading angle. Heading measured from 0° at the reference direction clockwise or counterclockwise through 90° or 180°. It is labeled with the reference direction as a prefix and the direction of measurement from the reference direction as a suffix. Thus, heading angle S 107° E is 107° east of south, or heading 073°.

heading flasher. An illuminated radial line on the plan position indicator for indicating own ship's heading on the bearing dial. Also called HEADING MARKER.

heading line. The line extending in the direction of a heading.

heading marker. See HEADING FLASHER.

headland, *n.* A comparatively high promontory having a steep face. Usually called HEAD when coupled with a specific name. Also called FORELAND.

head sea. A sea in which the waves move in a direction approximately opposite to the heading. The opposite is FOLLOWING SEA. Those moving in a direction approximately 90° from the heading are called BEAM SEA, and those moving in a direction approximately 45° from the heading (striking the quarter) are called QUARTERING SEA.

head tide. A tidal current setting in a direction approximately opposite to the heading of a vessel. One setting in such a direction as to increase the speed of a vessel is called a FAIR TIDE. One abeam is called a BEAM TIDE. One approximately 90° from the course is called a CROSS TIDE.

headup, heading upward. One of the three basic orientations of display of relative or true motion on a radarscope. In the HEAD UP orientation, the target pips are painted at their measured distances and in their directions *relative* to own ship's heading maintained UP in relation to the display and so indicated by the HEADING FLASHER. See also NORTH UP, BASE COURSE UP.

headwaters, *n., pl.* The water near the source of a stream; sometimes in singular.

headway, *n.* Motion in a forward direction. Motion in the opposite direction is called STERNWAY.

head wind. See under BEAM WIND.

heat lightning. A flash of light from an electric discharge, without thunder, sometimes seen near the horizon during the late afternoon or evening, often with a clear sky overhead. It is believed to be the reflection by haze or clouds of a distant flash of lightning below the horizon, too far away for the thunder to be audible.

heat wave. Unseasonably high temperatures extending over a period of a day or longer, particularly during the warm season of the year.

heave, *n.* Of the motions of a vessel in a seaway, the oscillatory vertical rise and fall, due to the entire hull being lifted by the force of the sea. Also called HEAVING. See also SHIP MOTIONS.

heavenly body. See CELESTIAL BODY.

heave the lead (lĕd). To take a sounding with a lead.

heaving, *n.* See HEAVE.

Heaviside layer. See under KENNELLY-HEAVISIDE REGION.

hecto-. A prefix meaning *one hundred* (10^2).

hectometer, *n.* One hundred meters.

heel, *n.* Lateral inclination, as of a vessel during a roll, or when weight aboard the vessel is unevenly distributed with reference to the longitudinal axis. See also LIST, *n.*

heel, *v., t. & i.* To incline or be inclined to one side. See also LIST, *n.*

heeling adjuster. A dip needle with a sliding weight that can be moved along one of its arms to balance the magnetic force, used to determine the correct position of a heeling magnet. Also called HEELING ERROR INSTRUMENT, VERTICAL FORCE INSTRUMENT. See also HEELING ERROR.

heeling error. The change in the deviation of a magnetic compass when a craft heels, due to the change in the position of the magnetic influences of the craft relative to the earth's magnetic field and to the compass.

heeling error instrument. *British terminology.* Heeling adjuster. Also called VERTICAL FORCE INSTRUMENT.

heeling magnet. A permanent magnet placed vertically in a tube under the center of a marine magnetic compass, to correct for heeling error.

height, *n.* Vertical distance above a datum, usually the surface of the earth. See also ELEVATION, WAVE HEIGHT.

height of eye correction. That correction to sextant altitude due to dip of the horizon. Also called DIP CORRECTION.

height of tide. Vertical distance from the chart sounding datum to the water surface at any stage of the tide. It is positive if the water level is higher than the chart sounding datum. The vertical distance from the chart sounding datum to a high water datum is called RISE OF TIDE.

heliocentric, *adj.* Relative to the sun as a center.

heliocentric parallax. The difference in the apparent direction or positions of a celestial body outside the solar system, as observed from the earth and sun. Also called STELLAR PARALLAX, ANNUAL PARALLAX. See also GEOCENTRIC PARALLAX.

helm, *n.* The apparatus by which a vessel is steered, particularly the tiller or wheel.

hemisphere, *n.* Half of a spnere.

henry, *n.* A *derived unit* of electric inductance in the International System of Units; it is the inductance of a closed circuit in which an electromotive force of 1 volt is produced when the electric current in the circuit varies uniformly at a rate of 1 ampere per second.

hertz, *n.* The special name for the *derived unit* of frequency in the International System of Units; it is one cycle per second.

Hertzian waves. See RADIO WAVES.

heterodyne reception. Radio reception in which an audio frequency is derived by beating the signal frequency with that produced by a local oscillator, followed by detection. Also called BEAT RECEPTION.

Hevelian halo. A faint white halo consisting of a ring occasionally seen 90° from the sun, and probably caused by the refraction and internal reflection of the sun's light by bi-pyramidal ice crystals.

hexagon, *n.* A closed plane figure having six sides.

hibernal, *adj.* Pertaining to winter. The corresponding adjectives for spring, summer, and fall are *vernal*, *aestival*, and *autumnal*.

Hi-Fix, *n.* A Decca radiolocation system designed for close-to-shore hydrographic, geophysical, constructional, and other surveys in which an accuracy of a few feet is required and which demand the use of lightweight and portable stations. The Hi-Fix chain comprises three transmitting stations (master and two slaves). A common carrier frequency (1605-2000 kHz) is shared by the three stations in turn on a time-multiplex basis. The system is used in either the hyperbolic or range-range configuration. In the hyperbolic configuration, phase comparison of phase-locked interrupted continuous wave signals from the master and slave stations ashore, as measured at the receiver at the vessel (or mobile station), provide fractional lane counts and, thus, hyperbolic lines of position within the lane. Lane identification is effected through the use of a second frequency, which is always 10 percent lower than the first frequency, and beating the two frequencies to obtain a coarse lane. In the range-range configuration, the master station is located at the mobile receiver. Phase comparison of phase-locked interrupted continuous wave signals from one shore station and the master station provides fractional lane counts within circular lanes centered on the shore station. With the lane identified, the range to the shore station is provided. The hyperbolic mode permits multiuser operations where many receivers can work on a time-sharing basis from one set of shore stations; the range-range mode allows only one receiver to be used. Maximum operating ranges over water paths are from 160 to 320 kilometers in temperate latitudes; in tropical latitudes these ranges may be reduced by 50 percent due to atmospheric radio noise. See also SEA-FIX.

Hi-Fix/6, *n.* A version of Hi-Fix the chain of which can comprise up to six stations, all employing a common frequency (1.6 - 5.0 MHz) or common pair of frequencies for lane identification as in the 2 MHz Hi-Fix system. Accuracy and range are of the same order as for the 2 MHz system. However, since the system can be used in hyperbolic, range-range, and compound configurations, geometric dilution is decreased and overall range increased.

high, *n.* Short for *area of high pressure.* Since a high is, on a synoptic chart, always associated with anticyclonic circulation, the term is used interchangeably with ANTICYCLONE. See also LOW.

high altitude method. The establishing of a circular line of position from the observation of the altitude of a celestial body by means of the geographical position and zenith distance of the body. The line of position is a circle having the geographical position as its center and a radius equal to the zenith distance. The method is normally used only for bodies at high altitudes, having small zenith distances. See also ST. HILAIRE METHOD, SUMNER METHOD, LONGITUDE METHOD.

high clouds. Types of clouds the mean lower level of which is above 20,000 feet. The principal clouds in this group are cirrus, cirrocumulus, and cirrostratus.

higher high water. The higher of the two high

waters of any tidal day.

higher high water interval. See under LUNITIDAL INTERVAL.

higher lower water. The higher of the two low waters of any tidal day.

higher low water interval. See under LUNITIDAL INTERVAL.

high fidelity. The ability to reproduce modulating waves at various audio frequencies without serious distortion.

high focal plane buoy. A type of lighted buoy in which the light is mounted exceptionally high above the surface of the sea.

high frequency. Radio frequency of 3 to 30 megahertz.

high light. See REAR LIGHT.

high noon. See LOCAL APPARENT NOON.

high sea, high seas. All water beyond the outer limit of the territorial sea. Although the high seas are in part coextensive with the waters of the contiguous zone, the fishing zone, and those over the continental shelf, freedom of the seas is not invalidated by the zonal overlap.

high tide. See under HIGH WATER.

high water. The maximum height reached by a rising tide. The height may be due solely to the periodic tidal forces or it may have superimposed upon it the effects of prevailing meteorological conditions. Use of the synonymous term HIGH TIDE is discouraged.

high water full and change. See ESTABLISHMENT OF THE PORT.

high water inequality. The difference between the heights of the two high waters during a tidal day. See under DIURNAL INEQUALITY.

high water interval. See under LUNITIDAL INTERVAL.

high water line. 1. The intersection of the land with the water surface at an elevation of high water.

2. The line along the shore to which the waters normally reach at high water.

high water mark. A line or mark left upon tide flats, beach, or alongshore objects indicating the elevation of the intrusion of high water. The mark may be a line of oil or scum on alongshore objects, or a more or less continuous deposit of fine shell or debris on the foreshore or berm. This mark is physical evidence of the general height reached by wave run-up at recent high waters. It should not be confused with the MEAN HIGH WATER LINE or MEAN HIGHER HIGH WATER LINE.

high water neaps. See under NEAP TIDES.

high water springs. Short for MEAN HIGH WATER SPRINGS.

high water stand. The condition at high water when there is no sensible change in the height of the water. A similar condition at low water is called LOW WATER STAND. See also STAND.

hill, *n.* 1. A natural elevation of the earth's surface, but lower than a mountain.

2. On the sea floor, an elevation rising generally less than 500 meters.

hillock, *n.* A small hill.

hoar, *n.* See FROST, definition 1.

hoarfrost, *n.* See FROST, definition 1.

hobbler, *n. British terminology.* A boatman or unlicensed pilot at less important ports in England who assists vessels in towing into or out of dock, warping, etc., or in general piloting. Also called HOBBLING PILOT.

hobbling pilot. See HOBBLER.

holding ground. The bottom ground of an anchorage. The expression is usually used with a modifying adjective to indicate the quality of the holding power of the material constituting the bottom.

hole, *n.* 1. A small depression of the sea floor.

2. An opening through a piece of sea ice, or an open space between ice cakes.

3. A small bay, particularly in New England.

homeward bound. Bound for the home port, or country - said of a vessel or person. See also INWARD BOUND.

homing, *n.* Navigation toward a point by maintaining constant some navigational coordinate(s) other than altitude.

homogenous, *adj.* Uniform throughout, or composed of parts which are similar in every detail.

hood, *n.* A shield placed over a radarscope, to eliminate extraneous light and thus make the radar picture appear clearly.

hook, *n.* Something resembling a hook in shape, particularly, **a.** a spit or narrow cape of sand or gravel which turns landward at the outer end; or **b.** a sharp bend or curve, as in a stream.

hooked spit. See RECURVED SPIT.

hop, *n.* Travel of a radio wave to the ionosphere and back to earth. The number of hops a radio signal has experienced is usually designated by the expression one-hop, two-hop, multihop, etc.

H.O. Pub. No. 208. *Navigation Tables for Mariners and Aviators*; a sight reduction table first published in 1928 by the U.S. Navy Hydrographic Office but discontinued on 31 December 1970 by the organizational successor, the U.S. Naval Oceanographic Office. The method was devised by Lieutenant Commander J. Y. Dreisonstok, USN. It is based upon a navigational triangle divided by dropping a perpendicular from the zenith. The table has been published commercially. Popularly called DREISONSTOK.

H.O. Pub. No. 211. *Dead Reckoning Altitude and Azimuth Table*; a sight reduction table first published by the U.S. Navy Hydrographic Office in 1931 but discontinued as a separate publication on 31 December 1972 by the organizational successor, the Defense Mapping Agency Hydrographic/Topographic Center. The method was devised by Lieutenant Arthur A. Ageton, USN. It is based upon a navigational triangle divided by dropping a perpendicular from the celestial body. The table was republished in 1975 by the Defense Mapping Agency Hydrographic/Topographic Center as table 35 of *Volume II: American Practical Navigator.* The table has been published commercially. Popularly called AGETON.

H.O. Pub. No. 214. *Tables of Computed Altitude and Azimuth*; a nine-volume set of sight reduction tables of the inspection type published between 1936 and 1946 by the U.S. Navy Hydrographic Office, and reprinted from time to time until discontinued on 31 December 1973 by the organizational successor, the Defense Mapping Agency Hydrographic/Topographic Center. These tables were superseded by Pub. No. 229, *Sight Reduction Tables for Marine Navigation.*

horizon, *n.* That great circle of the celestial sphere midway between the zenith and nadir, or a line resembling or approximating such a circle. That line where earth and sky appear to meet, and the projection of this line upon the celestial sphere, is called **visible** or **apparent horizon.** A line resembling the visible horizon but above or below it is called a **false horizon.** That circle of the celestial sphere formed by the intersection of the celestial sphere and a plane perpendicular to the zenith-nadir line is called **sensible**

horizon if the plane is through any point, such as the eye of an observer; **geoidal horizon** if through any sea-level point; and **celestial** or **rational horizon** if through the center of the earth. The **geometrical horizon** was originally considered identical with the celestial horizon, but the expression is now more commonly used to refer to the intersection of the celestial sphere and an infinite number of straight lines tangent to the earth's surface, and radiating from the eye of the observer. If there were no terrestrial refraction, GEOMETRICAL AND VISIBLE HORIZONS would coincide. An **artificial horizon** is a device for indicating the horizontal. A **radio horizon** is the line at which direct rays from a transmitting antenna become tangent to the earth's surface. A **radar horizon** is the radio horizon of a radar antenna.

horizon glass. That glass of a marine sextant, attached to the frame, through which the horizon is observed. That half of this glass nearer the frame is silvered to form the HORIZON MIRROR for reflecting the image of a celestial body; the other half is clear.

horizon mirror. The mirror part of the horizon glass. The expression is sometimes used somewhat loosely to refer to the horizon glass.

horizon prism. A prism which can be inserted in the optical path of an instrument, such as a bubble sextant, to permit observation of the visible horizon.

horizon system of coordinates. A set of celestial coordinates based on the celestial horizon as the primary great circle; usually altitude and azimuth or azimuth angle.

horizontal, *adj.* Parallel to the plane of the horizon; perpendicular to the direction of gravity.

horizontal, *n.* A horizontal line, plane, etc.

horizontal beam width. The beam width measured in a horizontal plane.

horizontal control datum. See HORIZONTAL GEODETIC DATUM.

horizontal danger angle. The maximum or minimum angle between two points on a chart, as observed from a craft, indicating the limit of safe approach to an off-lying danger. See also DANGER ANGLE.

horizontal datum. See HORIZONTAL GEODETIC DATUM.

horizontal earth rate. To compensate for the effect of earth rate, the rate at which the spin axis of a gyroscope must be tilted about the horizontal axis to remain parallel to the earth's surface. Horizontal earth rate is maximum at the equator, zero at the poles, and varies as the cosine of the latitude. See also EARTH RATE, VERTICAL EARTH RATE.

horizontal force instrument. An instrument used to make a comparison between the intensity of the horizontal component of the earth's magnetic field and the magnetic field at the compass location on board. Basically, it consists of a magnetized needle pivoted in a horizontal plane, as a dry card compass. It will settle in some position which will indicate the direction of the resultant magnetic field. If the needle is started swinging, it will be damped down with a certain period of oscillation dependent upon the strength of the magnetic field. Also called HORIZONTAL VIBRATING NEEDLE. See also DEFLECTOR.

horizontal geodetic datum. The basis for computations of horizontal control surveys in which the curvature of the earth is considered. It consists of the astronomical and geodetic latitude and the astronomical and geodetic longitude of an initial point (origin); an azimuth of a line from this point; the parameters (radius and flattening) of the reference ellipsoid; and the geoidal separation at the origin. A change in any of these quantities affects every point on the datum. For this reason, while positions within a system are directly and accurately relatable, those points from different datums must be transformed to a common datum for consistency. The horizontal geodetic datum may extend over a continent or be limited to a small area. See also DATUM. Also called HORIZONTAL DATUM, HORIZONTAL CONTROL DATUM.

horizontal intensity of the earth's magnetic field. The strength of the horizontal component of the earth's magnetic field.

horizontally polarized wave. A plane polarized electromagnetic wave in which the electric field vector is in a horizontal plane.

horizontal parallax. The geocentric parallax when a body is in the horizon. The expression is usually used only in connection with the moon, for which the tabulated horizontal parallax is given for an observer on the equator. The parallax at any altitude is called PARALLAX IN ALTITUDE.

horizontal vibrating needle. See HORIZONTAL FORCE INSTRUMENT.

horn, *n.* 1. A flared tube designed to match the acoustic impedance to the impedance of the atmosphere; it can behave as a resonator and can influence the directivity; the narrow end is called the *throat* and the large end the *mouth*. Also called TRUMPET.

2. See HORN ANTENNA.

horn antenna. An antenna consisting of a waveguide the cross-sectional area of which increases toward the open end. Often shortened to HORN.

horse latitudes. The regions of calms and variable winds coinciding with the subtropical high pressure belts on the poleward sides of the trade winds. The expression is generally applied only to the northern of these two regions in the North Atlantic Ocean, or to the portion of it near Bermuda.

hostile ice. From the point of view of the submariner, an ice canopy containing no large sky lights or other features which permit a submarine to surface.

hour, *n.* 1. A 24th part of a day.

2. A specified interval. See also COTIDAL HOUR, CURRENT HOUR.

hour angle. Angular distance west of a celestial meridian or hour circle; the arc of the celestial equator, or the angle at the celestial pole, between the upper branch of a celestial meridian or hour circle and the hour circle of a celestial body or the vernal equinox, measured westward through 360°. It is usually further designated as **local, Greenwich,** or **sidereal** as the origin of measurement is the local or Greenwich celestial meridian or the hour circle of the vernal equinox. See also MERIDIAN ANGLE.

hour angle difference. See MERIDIAN ANGLE DIFFERENCE.

hour circle. On the celestial sphere, a great circle through the celestial poles. An hour circle through the zenith is called a **celestial meridian.** Also called CIRCLE OF DECLINATION, CIRCLE OF RIGHT ASCENSION.

hour-glass effect. A radarscope phenomenon which appears as a constriction or expansion of the display near the center of the plan position indicator, which can be caused by a nonlinear time base or the sweep not starting on the radar indicator at the same instant as the transmission of the pulse. The phenomenon is most apparent when in narrow rivers or close to shore.

house, *n.* See BUILDING.

hug, *v., t.* To remain close to, as to ***hug the land.***

Humboldt Current. See PERU CURRENT.

humidity, *n.* The amount of water vapor in the air. The mass of water vapor per unit volume of air is called **absolute humidity.** The mass of water vapor per unit mass of moist air is called **specific humidity.** The ratio of the actual vapor pressure to the vapor pressure corresponding to saturation at the prevailing temperature is called **relative humidity.**

hummock, *n.* 1. A hillock of broken ice which has been forced upwards by pressure. It may be fresh or weathered. The submerged volume of broken ice under the hummocks, forced downwards by pressure, is called a BUMMOCK.
2. A natural elevation of the earth's surface resembling a hillock, but smaller and lower.

hummocked ice. Sea ice piled haphazardly one piece over another to form an uneven surface. When weathered, hummocked ice has the appearance of smooth hillocks.

hummocking, *n.* The pressure process by which sea ice is forced into hummocks. When the floes rotate in the process, it is called SCREWING.

hunter's moon. The full moon next following the harvest moon. See also PHASES OF THE MOON.

hunting, *n.* Fluctuation about a mid-point due to instability, as oscillations of the needle of an instrument about the zero point.

hurricane, *n.* 1. See under TROPICAL CYCLONE.
2. Wind of force 12 (64 knots and higher or 73 miles per hour and higher) on the Beaufort wind scale.

hydraulic current. A current in a channel caused by a difference in the surface level at the two ends. Such a current may be expected in a strait connecting two bodies of water in which the tides differ in time or range. The current in the East River, N.Y., connecting Long Island Sound and New York Harbor, is an example.

hydrographer, *n.* One who studies and practices the science of hydrography.

hydrographic, *adj.* Of or pertaining to hydrography.

hydrographic datum. A datum used for referencing depths of water or the heights of predicted tides. See also DATUM.

hydrographic sextant. A surveying sextant similar to those used for celestial navigation but smaller and lighter, constructed so that the maximum angle that can be read on it is slightly greater than that on the navigating sextant. Usually the angles can be read only to the nearest minute by means of a vernier. It is fitted with a telescope with a large object glass and field of view. Although the ordinary navigating sextant may be used in place of the hydrographic sextant, it is not entirely satisfactory for use in observing objects ashore which are difficult to see. Hydrographic sextants are either not provided with shade glasses or they are removed before use. Also called SOUNDING SEXTANT, SURVEYING SEXTANT.

hydrographic survey. The survey of a water area, with particular reference to submarine relief, and any adjacent land. See also OCEANOGRAPHIC SURVEY.

hydrography, *n.* The science that deals with the measurement and description of the physical features of the oceans, seas, lakes, rivers, and their adjoining coastal areas, with particular reference to their use for navigation.

HYDROLANT, *n.* A radio message disseminated by the Defense Mapping Agency Hydrographic/Topographic Center and restricted to the more important marine incidents or navigational changes for which a delay in disseminating the information to mariners would adversely affect navigational safety. The HYDROLANT broadcast covers those water areas outside and eastward of NAVAREA IV in the Atlantic Ocean. Many of these warnings are temporary in nature. Others might remain in force for long periods of time and ultimately be superseded by a numbered paragraph in *Notice to Mariners.* HYDROLANTS constitute part of the U.S. long range radio navigational warning system. Printed copies of HYDROLANTS are published each working day in the Atlantic edition of the *Daily Memorandum.* The text of HYDROLANTS issued during a week and which are still in effect are printed in the weekly *Notice to Mariners.*

hydrology, *n.* The scientific study of the waters of the earth, especially with relation to the effects of precipitation and evaporation upon the occurrence and character of ground water.

hydrometeor, *n.* Any product of the condensation or sublimation of atmospheric water vapor, whether formed in the free atmosphere or at the earth's surface; also any water particles blown by the wind from the earth's surface. See also LITHOMETEOR.

HYDROPAC. A radio message disseminated by the Defense Mapping Agency Hydrographic/Topographic Center and restricted to the more important marine incidents or navigational changes for which a delay in disseminating the information to mariners would adversely affect navigational safety. The HYDROPAC broadcast covers those water areas outside and westward of NAVAREA XII in the Pacific Ocean. Many of these warnings are temporary in nature. Others might remain in force for long periods of time and ultimately be superseded by a numbered paragraph in *Notice to Mariners.* HYDROPACS constitute part of the U.S. long range radio navigational warning system. Printed copies of HYDROPACS are published each working day in the Pacific edition of the *Daily Memorandum.* The text of HYDROPACS issued during a week and which are still in effect are printed in the weekly *Notice to Mariners.*

hydrophone, *n.* A listening device for receiving underwater sounds.

hydrosphere, *n.* The water portion of the earth as distinguished from the solid part, called the LITHOSPHERE, and from the gaseous outer envelope, called the ATMOSPHERE.

hyetal, *adj.* Of or pertaining to rain.

hygrometer, *n.* An instrument for measuring the humidity of the air. The most common type is a **psychrometer**, consisting essentially of dry-bulb and wet-bulb thermometers.

hygroscope, *n.* An instrument which indicates variation in atmospheric moisture.

hygroscopic, *adj.* Able to absorb moisture.

hyperbola, *n.* An open curve with two parts, all points of which have a constant difference in distance from two fixed points called FOCI.

hyperbolic, *adj.* Of or pertaining to a hyperbola.

hyperbolic lattice. A pattern formed by two or more families of intersecting hyperbolas.

hyperbolic line of position. A line of position in the shape of a hyperbola, determined by measuring the difference in distance to two fixed points. Loran-C lines of position are an example.

hyperbolic navigation. Radionavigation based on the measurement of the time differences in the reception of signals from several pairs of synchronized transmitters. For each pair of transmitters the isochrones are substantially hyperbolic. The combination of isochrones for two or more pairs of transmitters forms a hyperbolic lattice within which position can be determined according to the measured time differences.

hypersonic, *adj.* Of or pertaining to high supersonic speed, of the order of five times the speed of sound, or greater.

hypotenuse, *n.* That side of a plane right triangle opposite the right angle; the longest side of a plane right triangle.

hypsographic detail. The features pertaining to relief or elevation of terrain.

hypsographic map. A map showing land or submarine bottom relief in terms of height above, or below, a datum by any method, such as contours, hachures, shading, or hypsometric tinting. Also called HYPSOMETRIC MAP, RELIEF MAP.

hypsography, *n.* 1. The science or art of describing elevations of land surfaces with reference to a datum, usually sea level.

2. That part of topography dealing with relief or elevation of terrain.

hypsometer, *n.* An instrument for measuring height by determining the boiling temperature of a liquid. Its operation depends on the principle that boiling temperature is dependent on pressure, which normally varies with height.

hypsometric map. See HYPSOGRAPHIC MAP.

hypsometric tinting. A method of showing relief on maps and charts by coloring, in different shades, those parts which lie between different levels. Also called ALTITUDE TINTS, COLOR GRADIENTS, ELEVATION TINTS, GRADIENT TINTS, LAYER TINTS. See also HYPSOMETRIC TINT SCALE.

hypsometric tint scale. A graphic scale in the margin of maps and charts which indicates heights or depths by graduated shades of color. See also HYPSOMETRIC TINTING.

hysteresis, *n.* The lagging of the effect caused by change of a force acting on anything.

hysteresis error. That error in the reading of an instrument due to hysteresis.

I

IALA Maritime Buoyage System. As designed by the International Association of Lighthouse Authorities, a new uniform system of maritime buoyage which is expected to be implemented by most maritime nations. However, within the single system there are two international buoyage regions, designated as **Region A** and **Region B**, where lateral marks differ only in the colors of port and starboard hand marks. In Region A, red is to port on entering; in Region B, red is to starboard on entering. The system may be briefly described as a combined cardinal and lateral system. The system applies to all fixed and floating marks, other than lighthouses, sector lights, leading lights and marks, lightships and large navigational buoys. The system provides five types of marks which may be used in combination: **Lateral marks**, used in conjunction with a *conventional direction of buoyage*, are generally used for well-defined channels. Where a channel divides, a modified lateral mark may be used to indicate the preferred route. Lateral marks may differ between buoyage regions A and B. **Cardinal marks** used in conjunction with the mariner's compass, indicate where the mariner may find navigable water. **Isolated Danger marks** indicate isolated dangers of limited size that have navigable water all around them. **Safe Water marks** to indicate that there is navigable water around their position, e.g. mid-channel marks. **Special marks**, not primarily intended to assist navigation, indicate an area or feature referred to in nautical documents.

ice, *n.* The solid form of water substance; it is found in the atmosphere as **ice crystals, snow, hail, ice pellets,** etc., and on the earth's surface in forms such as **rime, glaze, sea ice, glacier ice, anchor ice,** etc.

ice anchor. An anchor designed for securing a vessel to ice.

ice atlas. A publication containing a series of ice charts showing geographic distribution of ice, usually by seasons or months.

iceberg, *n.* A massive piece of ice greatly varying in shape, showing more than 5 meters above the sea surface, which has broken away from a glacier, and which may be afloat or aground. Icebergs may be described as **tabular, dome-shaped, pinnacled, drydock, glacier** or **weathered, blocky, tilted blocky,** or **drydock icebergs.** For reports to the International Ice Patrol they are described with respect to size as **small, medium,** or **large icebergs.**

iceberg tongue. A major accumulation of icebergs projecting from the coast, held in place by grounding and joined together by fast ice.

ice-blink. A whitish glare on low clouds above an accumulation of distant ice.

ice-bound, *adj.* A harbor, inlet, etc. is said to be ice-bound when navigation by ships is prevented on account of ice, except possibly with the assistance of an icebreaker.

ice boundary. The demarcation at any given time between fast ice and pack ice or between areas of pack ice of different concentrations. See also ICE EDGE.

ice breccia. Ice pieces of different age frozen together.

ice bridge. Surface river ice of sufficient thickness to impede or prevent navigation.

ice buoy. A sturdy buoy, usually a metal spar, used to replace a more easily damaged buoy during a period when heavy ice is anticipated.

ice cake. Any relatively flat piece of sea ice less than 20 meters across. See also SMALL ICE CAKE.

ice canopy. From the point of view of the submariner, PACK ICE.

ice cap. A perennial cover of ice and snow over an extensive portion of the earth's surface. The most important of the existing ice caps are those on Antarctica and Greenland. The term was first used for the supposedly perennial ice cover at both poles of the earth. However, since it has been found that the ice of arctic waters is largely seasonal, the use of this term to denote arctic polar ice is now considered improper.

ice cover. The ratio of an area of ice of any concentration to the total area of sea surface within some large geographic locale; this locale may be global, hemispheric, or prescribed by a specific oceanographic entity such as Baffin Bay or the Barents Sea.

ice crystal. Any one of a number of macroscopic crystalline forms in which ice appears.

ice-crystal haze. A type of very light ice fog composed only of ice crystals (no droxtals). It is usually associated with precipitation of ice crystals.

ice crystals. A type of precipitation composed of slowly falling, very small, unbranched crystals of ice which often seem to float in the air. It may fall from a cloud or from a cloudless sky. It is visible only in direct sunlight or in an artificial light beam, and does not appreciably reduce visibility. The latter quality helps to distinguish it from ice fog, which is composed largely of droxtals.

ice edge. The demarcation at any given time between the open sea and sea ice of any kind, whether fast or drifting. See also COMPACTED ICE EDGE, DIFFUSE ICE EDGE, ICE BOUNDARY.

ice field. An area of pack ice consisting of floes of any size, which is greater than 5.4 nautical miles (10 kilometers) across. Ice fields are subdivided according to areal extent. A **large ice field** is over 11 nautical miles across; a **medium ice field** is 8 to 11 nautical miles across; a **small ice field** is 5.4 to 8 nautical miles across.

ice fog. Fog composed of suspended particles of ice, partly ice crystals 20 to 100 microns in diameter but chiefly, especially when dense, droxtals 12 to 20 microns in diameter. It occurs at very low temperatures, and usually in clear, calm weather in high latitudes. The sun is usually visible and may cause halo phenomena. Ice fog is rare at temperatures warmer than −30°C or −20°F. Also called RIME FOG. See also FREEZING FOG.

icefoot, *n.* A narrow fringe of ice attached to the coast, unmoved by tides and remaining after the fast ice has moved away.

ice-free, *adj.* Said of a locale with no sea ice. There may be some ice of land origin present.

ice front. The vertical cliff forming the seaward face of an ice shelf or other floating glacier varying in height from 2 to 50 meters above sea level. See also ICE WALL.

ice island. A large piece of floating ice showing

about 5 meters above the sea surface, which has broken away from an arctic ice shelf, having a thickness of 30 to 50 meters and an area of from a few thousand square meters to 150 square nautical miles or more, and usually characterized by a regularly undulating surface which gives it a ribbed appearance from the air.

ice jam. An accumulation of broken river ice or sea ice caught in a narrow channel.

ice keel. From the point of view of the submariner, a downward projecting ridge on the underside of the ICE CANOPY, the counterpart of a RIDGE. An ice keel may extend as much as 50 meters below sea level.

ice limit. The climatological term referring to the extreme minimum or extreme maximum extent of the ice edge in any given month or period based on observations over a number of years. The term should be preceded by *minimum* or *maximum*, as appropriate. See also MEAN ICE EDGE.

ice massif. A concentration of sea ice covering an area of hundreds of kilometers, which is found in the same region every summer.

ice needle. A long, thin ice crystal whose cross-section perpendicular to its long dimension is typically hexagonal. The expression ICE NEEDLE should not be confused with NEEDLE ICE.

ice of land origin. Ice formed on land or in an ice shelf, found floating in water. The concept includes ice that is stranded or grounded.

ice patch. An area of pack ice less than 5.4 nautical miles (10 kilometers) across.

ice pellets. A type of precipitation consisting of transparent or translucent pellets of ice, 5 millimeters or less in diameter. The pellets may be spherical, irregular, or (rarely) conical in shape. They usually bounce when hitting hard ground, and make a sound upon impact. Ice pellets includes two basically different types of precipitation, those which are known in the United States as SLEET and SMALL HAIL. **Sleet** (or GRAINS OF ICE in British terminology) is generally transparent, globular, solid grains of ice which have formed from the freezing of raindrops or the refreezing of largely melted snowflakes when falling through a below-freezing layer of air near the earth's surface. **Small hail** is generally translucent particles, consisting of snow pellets encased in a thin layer of ice. The ice layer may form either by the accretion of droplets upon the snow pellet, or by the melting and refreezing of the surface of the snow pellet.

ice port. An embayment in an ice front, often of a temporary nature, where ships can moor alongside and unload directly onto the ice shelf.

ice rind. A brittle shiny crust of ice formed on a quiet surface by direct freezing or from grease ice, usually in water of low salinity. Of thickness to about 5 centimeters, ice rind is easily broken by wind or swell, commonly breaking into rectangular pieces.

ice sheet. Continuous ice overlaying a large land area.

ice shelf. A floating ice sheet attached to the coast and of considerable thickness, showing 20 to 50 meters or more above sea level. Usually of great horizontal extent and with a level or gently undulating surface, the ice shelf is augmented by annual snow accumulation and often also by the seaward extension of land glaciers. Limited areas of the ice shelf may be aground. The seaward edge is called ICE FRONT.

ice storm A storm characterized by a fall of freezing precipitation. The attendant formation of glaze on terrestrial objects creates many hazards.

ice stream. That part of an inland ice sheet in which the ice flows more rapidly and not necessarily in the same direction as the surrounding ice. The margins are sometimes clearly marked by a change in direction of the surface slope but may be indistinct.

ice under pressure. Ice in which deformation processes are actively occurring; hence the ice is a potential impediment or danger to shipping.

ice wall. An ice cliff forming the seaward margin of a glacier which is not afloat. An ice wall is aground, the rock basement being at or below sea level. See also ICE FRONT.

ice-worn, *adj.* Abraded by ice.

icicle, *n.* A hanging mass of ice, usually conical, formed by the freezing of dripping water.

illuminance, *n.* At a point of a surface, the quotient of the luminous flux incident on an infinitesimal element of the surface containing the point under consideration, by the area of that element. If the luminous flux may be considered uniformly distributed over a finite plane surface, the illuminanace is given by the quotient of the luminous flux by the area. The continued use of the former term ILLUMINATION with the meaning above is now deprecated as this use conflicts with the general meaning of the term. The derived unit of illuminance in the International System of Units is the LUX.

image, *n.* 1. The optical counterpart of an object. An **erect image** appears upright; an **inverted image** upside down in relation to the object. A **real image** is actually produced and is capable of being shown on a surface, as in a camera; while a **virtual image** cannot be shown on a surface, but is visible, as in a mirror.

2. A visual representation, as on a radarscope.

improved channels. Dredged channels under the jurisdiction of the Corps of Engineers, and maintained to provide an assigned controlling depth. Symbolized on National Ocean Survey charts by black, broken lines to represent side limits, with the controlling depth and date of ascertainment given together with a tabulation of more detailed information.

impulse train. See PULSE TRAIN.

in-band racon. A racon which transmits in the marine radar frequency band, e.g., the 3-centimeter band. There are two types of in-band racons, swept-frequency racons and experimental fixed-frequency racons. The transmitter of the swept-frequency racon sweeps through a range of frequencies within the band to insure that a radar receiver tuned to a particular frequency within the band will be able to detect the signal. The fixed-frequency racon transmits on a fixed frequency at the band edge. It is therefore necessary that the radar set be tuned to the racon's transmitting frequency or that auxiliary receiving equipment be used. When the radar is tuned to the fixed-frequency racon, normal radar echoes are not painted on the radarscope. See also CROSS-BAND RACON.

incandescence, *n.* Emission of light due to high temperature. Any other emission of light is called LUMINESCENCE.

inch, *n.* A unit of length equal to one-twelfth of a foot, or 2.54 centimeters.

incidence, *n.* 1. Partial coincidence, as a circle and a tangent line.

2. The impingement of a ray on a surface.

incident ray. A ray impinging on a surface.

incineration area. An officially designated offshore area for the burning of chemical waste by specially equipped vessels. The depiction of incineration areas on charts (in conjunction with radio warnings) is necessary to insure that passing vessels do not mistake the burning of waste for a vessel on fire.

inclination, *n.* 1. The angle which a line or surface makes with the vertical, horizontal, or with another line or surface.
2. One of the orbital elements (parameters) that specifies the orientation of an orbit. It is the angle between the orbital plane and a reference plane, the plane of the celestial equator for geocentric orbits and the ecliptic for heliocentric orbits. See also ORBITAL ELEMENTS, ORBITAL PARAMETERS OF ARTIFICIAL EARTH SATELLITE.

inclination of an orbit. 1. See INCLINATION, definition 2.
2. As defined by the *International Telecommunication Union* (ITU), the angle determined by the plane containing an orbit and the plane of the earth's equator.

increment, *n.* A change in the value of a variable. A negative increment is also called DECREMENT.

independent surveillance. Position determination by means requiring no cooperation from the craft or vehicle (e.g., primary radar).

index *(pl. indices or indexes), n.* 1. A mark on the scale of an instrument, diagram, etc., to indicate the origin of measurement.
2. A pointer or part of an instrument which points out, like the needle of a gage.
3. A list or diagram serving as a guide to a book, set of charts, etc.
4. A ratio or value used as a basis for comparison of other values.

index arm. A slender bar carrying an index; particularly that bar which pivots at the center of curvature of the arc of a marine sextant and carries the index and the venier or micrometer.

index chart. An outline chart showing the limits and identifying designations of navigational charts, volumes of sailing directions, etc.

index correction. That correction due to index error.

index error. The error in the reading of an instrument equal to the difference between the zero of the scale and the zero of the index. In a marine sextant it is due primarily to lack of parallelism of the index mirror and the horizon glass at zero reading.

index glass. See INDEX MIRROR.

index mirror. The mirror attached to the index arm of a marine sextant. The bubble or pendulum sextant counterpart is called INDEX PRISM. Also called INDEX GLASS.

index prism. A sextant prism which can be rotated to any angle corresponding to altitudes between established limits. It is the bubble or pendulum sextant counterpart of the INDEX MIRROR of a marine sextant.

Indian Equatorial Countercurrent. A complex Indian Ocean current which is influenced by the monsoons and the circulations of the Arabian Sea and the Bay of Bengal. At times it is easily distinguishable, whereas at other times its presence is not evident. During December through March, the countercurrent has a marked tendency to migrate southward and to become narrower. In December the northern and southern boundaries are at 2° N and 4° S, respectively, moving southward to 3° S and 6° S by February. The northern boundary of Indian Equatorial Countercurrent is easily discernible at this time due to the generally westward current flow in the region immediately north. During May through July the cell, within which the Indian Equatorial Countercurrent and the Monsoon Drift flow clockwise, moves toward the west side of the region. In June and July the southeastward flowing currents prevail in the region between the Bay of Bengal and the Indian South Equatorial Current; only traces of the countercurrent remain. During August through November eastward flowing currents prevail north of the Indian Equatorial Countercurrent. As a result, the northern boundary of the countercurrent is difficult to distinguish from the eastward drift currents. See also MONSOON.

Indian South Equatorial Current. An Indian Ocean current that flows westward throughout the year. controlled by the southeast trade winds. Its northern and southern boundaries are at approximately 10° S and 25° S, respectively. The northern boundary of the current fluctuates seasonally between 9° S and 11° S, being at its northernmost limit during the southwest monsoon and at its southernmost limit during the northeast monsoon. The current flows westward toward the east coast of Madagascar to the vicinity of Tamatave and Ile Sainte-Marie, where it divides; one part turns northward, flows past the northern tip of the island with speeds up to 3.3 knots, and then flows westward and northwestward toward the African coast. The northern branch of the current divides upon reaching the coast of Africa near Cabo Delgado: one part turns and flows northward, the other turns and flows southward in the western part of the Mozambique Channel and forms the AGULHAS CURRENT. See also MONSOON.

Indian spring low water. A tidal datum originated by G.H. Darwin when investigating the tides of India. It is an elevation depressed below mean sea level by an amount equal to the sum of the amplitudes of certain constituents as given in the *Tide and Current Glossary* published by the National Ocean Survey. Also called INDIAN TIDE PLANE, HARMONIC TIDE PLANE.

Indian summer. An indefinite and irregular period of mild, calm, hazy weather often occurring in autumn or early winter, especially in the United States and Canada.

Indian tide plane. See INDIAN SPRING LOW WATER.

indicator, *n.* See RADAR INDICATOR.

indirect echo. A radar echo which is caused by the electromagnetic energy being transmitted to the target by an indirect path and returned as an echo along the same path. An indirect echo may appear on the plan position indicator (PPI) when the main lobe of the radar beam is reflected off part of the structure of the ship (the stack for example) from which it is reflected to the target. Returning to own ship by the same indirect path, the echo appears on the PPI at the bearing of the reflecting surface. Assuming that the additional distance by the indirect path is negligible, the indirect echo appears on the PPI at the same range as the direct echo received. Also called FALSE ECHO.

indirect wave. A radio wave which reaches a given reception point by a path from the transmitting point other than the direct line path between the two. An example is the SKYWAVE received after reflection from one of the layers of the ionosphere.

induced magnetism. The magnetism acquired by soft iron while it is in a magnetic field. Soft iron will lose its induced magnetism when it is removed from a magnetic field. The strength and polarity of the induced magnetism will alter immediately as its magnetic latitude, or its orientation in a magnetic field, is changed. The induced magnetism has an immediate effect upon the magnetic compass as the magnetic latitude or heading of a craft changes. See also PERMANENT MAGNETISM, SUBPERMANENT MAGNETISM.

induced precession. See REAL PRECESSION.

inequality *(tidal), n.* A systematic departure from the mean value of a tidal quantity.

inertia, *n.* The property of matter by which it resists any change in its state of rest or uniform motion in a straight line. See also GYROSCOPIC INERTIA.

inertial alignment. The process of orienting the measuring axes of the inertial components of inertial navigation equipment with respect to the coordinate system in which the equipment is to be used.

inertial coordinate system. A coordinate system in which the axes do not rotate with respect to the "fixed stars" and in which dynamic behavior can be described using Newton's laws of motion. See also EARTH-FIXED COORDINATE SYSTEM

inertial force. A force in a given coordinate system arising from the inertia of a mass moving with respect to another coordinate system. For example the Coriolis acceleration on a mass moving with respect to a coordinate system fixed in space becomes an inertial force, the Coriolis force, in a coordinate system rotating with the earth.

inertial navigation. The process of measuring a craft's velocity, attitude, and displacement from a known start point through sensing the accelerations acting on it in known directions by means of devices that mechanize Newton's laws of motion. Since these laws are expressed relative to inertial space (the fixed stars), the term *inertial* is applied to the process. Inertial navigation is described as selfcontained because it is independent of external aids to navigation. However, due to increasing position errors with time due to inherent errors, an inertial system must be *reset* from time to time by means of another navigation system. It is also described as passive because no energy is emitted to obtain information from an external source. Thus, inertial navigation is fundamentally different from other methods of navigation in that it depends only on measurements made within the craft being navigated. However, knowlege of the physical environment is required. Since the accelerometer used to sense the accelerations acting on the craft cannot distingush between kinematic accelerations and gravitational accelerations (not affecting the craft's position or orientation on the earth), any uncertainty in the gravitational environment manifests itself as a system error. The basic principle of inertial navigation is the measurement of the accelerations acting on a craft, other than those not associated with its orientation or motion with respect to the earth, and the double integration of these accelerations along known directions to obtain the displacement from the start point. If the indicated acceleration of the craft is constant, velocity and distance traveled can be found from the equations:

$$v=at,$$

and

$$s=1/2\,at^2,$$

where a is the acceleration, v is the velocity, s is the distance, and t is the time. But the acceleration must be constant. The equations cannot be used for varying accelerations unless very small time increments are used, and the increments of velocity or distance are integrated. The equations, using the calculus notation are

$$v=\int a\,dt$$

$$s=\iint a\,dt\,dt$$

where dt denotes a very small increment of time.

in extremis. Condition in which either course or speed changes or both are required on the part of both ships if the ships are to avoid collision.

inferior conjunction. The conjunction of an inferior planet and the sun when the planet is between the earth and the sun.

inferior planets. The planets with orbits smaller than that of the earth; Mercury and Venus. See also PLANET.

inferior transit. See LOWER TRANSIT.

infinite, *adj.* Without limits. The opposite is FINITE.

infinitesimal, *adj.* 1. Immeasurably small.
2. Approaching zero as a limit.

infinity, *n.* The point, line, region, etc., beyond finite limits. A source of light is regarded as at infinity if it is at such a great distance that rays from it can be considered parallel. If different parts of an instrument are involved, as the index and horizon glasses of a sextant, the horizon can be considered at infinity; but if widely different parts of the earth are involved, a body must be considered at infinity, though for practical navigation the sun and planets can be considered at infinity without serious error. See also PARALLAX.

inflection, inflexion, *n.* Reversal of direction of curvature. A point at which reversal takes place is called POINT OF INFLECTION.

infrared, *adj.* Having a frequency immediately beyond the red end of the visible spectrum—said of rays of longer wavelength than visible light, but shorter than radio waves. ULTRAVIOLET rays are those immediately beyond the other end of the visible spectrum.

infrasonic, *adj.* Having a frequency below the audible range. Frequencies above the audible range are called ULTRASONIC.

initial great-circle course. The direction, at the point of departure, of the great circle through that point and the destination, expressed as the angular distance from a reference direction, usually north, to that part of the great circle extending toward the designation. Also called INITIAL GREAT-CIRCLE DIRECTION. See also FINAL GREAT-CIRCLE COURSE.

initial great-circle direction. See INITIAL GREAT-CIRCLE COURSE.

injection messages. Orbital predictions periodically transmitted to artificial satellites for storage in satellite memory. See also NAVY NAVIGATION SATELLITE SYSTEM.

Inland Rules of the Road. Rules to be followed by all vessels while navigating upon certain inland waters of the United States. See also COLREGS DEMARCATION LINES, RULES OF THE ROAD.

inland sea. A body of water nearly or completely surrounded by land, especially if very large or composed of salt water. If completely surrounded by land, it is usually called LAKE. This should not be confused with CLOSED SEA, that part of the ocean enclosed by headlands, within narrow straits, etc., or within the territorial jurisdiction of a country.

inlet, *n.* A narrow body of water extending into the land from a larger body of water. A long, narrow inlet with gradually decreasing depth inward is called a ria. Also called ARM, TONGUE.

inner harbor. The part of a harbor more remote from the sea, as contrasted with the OUTER HARBOR. These expressions are usually used only in a harbor that is clearly divided into two parts, as by a narrow passageway or man-made structures. The inner harbor generally has additional protection and is often the principal berthing area.

inner planets. The four planets nearest the sun; Mercury, Venus, Earth, and Mars.

inoperative, *adj.* Said of a sound signal or radionavigation aid out of service due to a malfunction.

in phase. The condition of two or more cyclic motions which are at the same part of their cycles at the same instant. Two or more cyclic motions which are not at the same part of their cycles at the same instant are said to be OUT OF PHASE.

input axis. Of a gyroscope, the axis of applied torque. See also OUTPUT AXIS, PRECESSION.

inshore, *adj. & adv.* Near or toward the shore.

inshore, *n.* The zone of variable width between the shoreface and the seaward limit of the breaker zone.

inshore traffic zone. A routing measure comprising a designated area between the landward boundary of a traffic separation scheme and the adjacent coast, intended for local traffic.

in situ. A Latin term meaning "in place"; in the natural or original position.

insolation, *n.* Solar radiation received, as by the earth or other planets; or the rate of delivery of such radiation.

instability, *n.* The state or property of submitting to change or of tending to increase the departure from original conditions after being disturbed. Thus, the atmosphere is in a state of instability when the vertical distribution of temperature is such that an air particle, if given an upward or downward impulse, tends to keep moving with increasing speed from its original level. The opposite is STABILITY.

instability line. Any non-frontal line or band of convective activity in the atmosphere. This is the general term and includes the developing, mature, and dissipating stages. However, when the mature stage consists of a line of active thunderstorms, it is properly called SQUALL LINE; therefore, in practice, *instability line* often refers only to the less active phases. Instability lines are usually hundreds of miles long (not necessarily continuous), 10 to 50 miles wide, and are most often formed in the warm sectors of wave cyclones. Unlike true fronts, they are transitory in character, ordinarily developing to maximum intensity in less than 12 hours and then dissipating in about the same time. Maximum intensity is usually attained in late afternoon.

instrument correction. That correction due to instrument error.

instrument error. The inaccuracy of an instrument due to imperfections within the instrument. See CALIBRATION ERROR, CENTERING ERROR, FRICTION ERROR, GRADUATION ERROR, HYSTERESIS ERROR, LAG ERROR, PRISMATIC ERROR, SECULAR ERROR, TEMPERATURE ERROR, VERNIER ERROR.

instrument landing system. As defined by the *International Telecommunication Union* (ITU), a radionavigation system which provides aircraft with horizontal and vertical guidance just before and during landing and, at certain fixed points, indicates the distance to the reference point of landing.

instrument landing system glide path. As defined by the *International Telecommunication Union* (ITU), a system of vertical guidance embodied in the instrument landing system which indicates the vertical deviation of the aircraft from its optimum path of descent.

instrument landing system localizer. As defined by the *International Telecommunication Union* (ITU), a system of horizontal guidance embodied in the instrument landing system which indicates the horizontal deviation of the aircraft from its optimum path of descent along the axis of the runway.

instrument shelter. A cage or screen in which a thermometer and sometimes other instruments are placed to shield them from the direct rays of the sun and from other conditions that would interfere with registration of true conditions. It is usually a small wooden structure with louvered sides.

insular, *adj.* 1. Of or pertaining to an island or islands.

insular borderland. A region around an island, normally occupied by or bordering a shelf, that is highly irregular with depths well in excess of those typical of a shelf. See also CONTINENTAL BORDERLAND.

insular shelf. A zone around an island that extends from the low water line to a depth at which there is usually a marked increase of slope towards oceanic depths. See also CONTINENTAL SHELF.

2. Isolated, detached.

insulate, *v., t.* To separate or isolate a conducting body from its surroundings, by means of a nonconductor, as to prevent transfer of electricity, heat, or sound.

insulator, *n.* A nonconducting substance or one offering high resistance to passage of an electric current, used to support or separate electric conductors.

integer, *n.* A whole number; a number that is not a fraction.

integral, *adj.* Of or pertaining to an integer.

integral doppler navigation. Navigation by means of integrating the doppler frequency shift that occurs over a specific interval of time as the distance between a navigational satellite and navigator is changing to determine the time rate of change of range of the satellite from the navigator for the same interval. See also DOPPLER SATELLITE NAVIGATION BASIC PRINCIPLES, NAVY NAVIGATION SATELLITE SYSTEM.

integrated navigation system. A navigation system which comprises two or more positioning systems combined in such manner as to achieve a performance better than each constituent system.

integrating accelerometer. An instrument which senses the component of specific acceleration along an axis known as the sensitive axis of the accelerometer, and produces an output equal to the time integral of that quantity. Also called VELOCITY METER.

intended track. See TRACK, definition 2.

intercalary day. A day inserted or introduced among others in a calendar as February 29 during leap years.

intercardinal heading. A heading in the direction of any of the intercardinal points. See also CARDINAL HEADING.

intercardinal point. Any of the four directions

midway between the cardinal points, as northeast, southeast, southwest, or northwest. Also called QUADRANTAL POINT.

intercardinal rolling error. See under QUADRANTAL ERROR.

intercept, *n.* See ALTITUDE INTERCEPT, ALTITUDE INTERCEPT METHOD.

interference, *n.* 1. As affecting radar operation, unwanted and confusing signals or patterns produced on the radarscope by another radar or transmitter on the same frequency, and more rarely, by the effects of nearby electrical equipment or machinery, or by atmospheric phenomena.
2. The variation of wave amplitude with distance or time, caused by superposition of two or more waves. Sometimes called WAVE INTERFERENCE.

interferometer, *n.* An apparatus used to produce and measure interference from two or more coherent wave trains from the same source. Used to measure wavelengths, to measure angular width of sources, to determine the angular position of sources (as in satellite tracking), and for other purposes. See also RADIO INTERFEROMETER.

Inter-Governmental Maritime Consultative Organization. The Specialized Agency of the United Nations responsible for maritime safety and efficiency of navigation. The Inter-Governmental Consultative Organization (IMCO) is recognized as the only international body responsible for establishing and recommending measures on an international level concerning ships' routing. Through its appropriate bodies, IMCO keeps the subject of ships' routing under continuous review by adopting new routing systems and amending or, if necessary, withdrawing existing systems. On May 22, 1982, IMCO will become the International Maritime Organization (IMO).

intermediate frequency. In superheterodyne reception, the frequency which is derived by mixing the signal-carrying frequency with the local oscillator frequency. If there be more than one such mixing process, the successive intermediate frequencies are known as the first, second, etc. intermediate frequency.

intermediate light. That light of a range (leading line) formed by three range (leading) lights which is situated between the front and rear lights.

intermediate orbit. A central force orbit that is tangent to the real (or disturbed) orbit at some point. A fictitious satellite traveling in the intermediate orbit would have the same position, but not the same velocity, as the real satellite at the point of tangency.

internal noise. In radio reception, the noise which is produced in the receiver circuits. Internal noise is in addition to external noise.

internal tide. A tidal wave propagating along a sharp density discontinuity, such as at a thermocline, or in an area of gradual changing density (vertically).

International Atomic Time. See under ATOMIC TIME.

International Bureau of Weights and Measures. Operating under the exclusive supervision of the International Committee of Weights and Measures (CIPM), which itself comes under the authority of the General Conference of Weights and Measures (CGPM), the International Bureau of Weights and Measures (BIPM) insures worldwide unification of physical measurements. It is responsible for establishing the fundamental standards and scales for measurement of the principal physical quantities and maintaining the international prototypes; carrying out comparisons of national and international standards; insuring coordination of corresponding measuring techniques; and carrying out and coordinating the determinations relating to the fundamental physical constants.

international call sign. A call sign assigned in accordance with the provisions of the International Telecommunications Union to identify a radio station. The nationality or the radio station is identified by the first three characters. When used in visual signaling, international call signs are referred to as *signal letters*.

international chart. One of a coordinated series of small-scale charts for planning and long range navigation. The charts are prepared and published by different Member States of the International Hydrographic Organization using the same specifications.

International Code of Signals, See PUB. 102.

international date line. See DATE LINE.

International ellipsoid of reference. The reference ellipsoid of which the semimajor axis is 6,378,388.0 meters, the semiminor axis is 6,356,911.9 meters, and the flattening or ellipticity is 1/297. Also called INTERNATIONAL SPHEROID OF REFERENCE.

International Great Lakes Datum (1955). Mean water level at Pointe-au-Pere, Quebec, on the Gulf of St. Lawrence over the period 1941-1956, from which dynamic elevations throughout the Great Lakes region are measured. The term is often used to mean the entire system of dynamic elevations rather than just the referenced water level.

International Hydrographic Bulletin. A publication, published monthly by the International Hydrographic Bureau for the International Hydrographic Organization, which contains information of current hydrographic interest.

International Hydrographic Bureau. See under INTERNATIONAL HYDROGRAPHIC ORGANIZATION.

International Hydrographic Organization. An institution consisting of representatives of a number of nations organized for the purpose of coordinating the hydrographic work of the participating governments. It had its origin in the *International Hydrographic Conference* in London in 1919 and was formally organized in June 1921 with 19 member countries as the **International Hydrographic Bureau.** On 22 September 1970, the International Hydrographic Organization came into being following ratification by the required majority of Member Governments of an intergovernmental Convention on the Organization drafted during the *Ninth International Hydrographic Conference* in 1967. The term *International Hydrographic Bureau* now refers only to the Headquarters of the Organization. The objectives of the Organization are to bring about (1) the coordination of the activities of national hydrographic office; (2) the greatest possible uniformity in nautical charts and documents; (3) the adoption of reliable and efficient methods of carrying out and exploiting hydrographic surveys; (4) the development of the sciences in the field of hydrography and the techniques employed in descriptive oceanography. The principal work undertaken by the Bureau is required by the Convention to be (1) to bring about a close and permanent association between national hydrographic offices; (2) to study any matters relating to hydrography and the allied sciences and techniques, and to collect the necessary papers; (3) to further the exchange of nautical charts and documents between hydrographic offices of Member Governments; (4) to circulate the appropriate documents; (5) to tender guidance and advice upon request, in particular to countries engaged in setting up or expanding their hydrographic service; (6) to encourage coordination of hydrographic surveys with relevant oceanographic activities; (7) to extend and facilitate the application of oceanographic knowledge for the benefit of navigators; (8) to cooperate with international organizations and scientific institutions which have related objectives. The work of the Bureau is promulgated to Member States in both English and French languages by correspondence and by the *International Hydrographic Bulletin*, and the *Yearbook*. The Bureau also produces a series of special publications each of which is a definitive work in hydrography. The Bureau has permanent headquarters in the principality of Monaco.

International Hydrographic Review. A publication, published twice yearly by the International Hydrographic Bureau for the International Hydrographic Organization, which contains professional articles on hydrography and related subjects.

international low water. A hydrographic datum originally suggested for international use at the International Hydrographic Conference in London in 1919 and later discussed at the Monaco Conference in 1926. The proposed datum, which has not yet been generally adopted, was to be "a plane so low that the tide will but seldom fall below it." This datum was the subject of the International Hydrographic Bureau's Special Publications No. 5 (March 1925) and No. 10 (January 1926), reproduced in the *Hydrographic Reviews* for May 1925 and July 1926.

International Maritime Organization. The Inter-Governmental Maritime Consultative Organization (IMCO), the Specialized Agency of the United Nations responsible for maritime safety and efficiency of navigation, is to become the International Maritime Organization (IMO) on May 22, 1982. The Organization is to provide machinery for cooperation among Governments in the field of governmental regulations and practices relating to technical matters of all kinds affecting shipping engaged in international trade: to encourage the general adoption of the highest practicable standards in matters concerning maritime safety, efficiency of navigation, and the prevention and control of marine pollution from ships, and to deal with legal matters related to the purposes set out in Article 1 of the Convention.

International Nautical Mile. A unit of length equal to 1,852 meters, exactly. See also NAUTICAL MILE.

international number. Of a navigational light, the number assigned in accordance with the Resolution adopted at the Fifth International Hydrographic Conference in 1949 by Member Nations of the International Hydrographic Bureau (now the International Hydrographic Organization). This number is in italic type and under the light **list number** in the light list.

International spheroid of reference. See INTERNATIONAL ELLIPSOID OF REFERENCE.

International System of Units. A modern form of the metric system adopted in 1960 by the General Conference of Weights and Measures (CGPM). The units of the International System of Units (SI) are divided into three classes. The first class of SI units are the *base units* or the seven well-defined units which by convention are regarded as dimensionally independent: the **metre**, the **kilogram**, the **second**, the **ampere**, the **kelvin**, the **mole**, and the **candela**. The second class of SI units are the *derived units*, i.e., the units that can be formed by combining base units according to the algebraic relations linking the corresponding quantities. Several of these algebraic expressions in terms of base units can be replaced by special names and symbols which can themselves be used to form other derived units. Examples of derived units expressed algebraically in terms of base units by means of the mathematical symbols of multiplication and division are **acceleration**; *metre per second squared*; and **luminance**: *candela per square metre*. Examples of SI derived units with special names are **hertz**, **pascal**. Examples of SI derived units expressed by means of special names are **metre newton** and **volt per metre**. The third class of SI units are the *supplementary units*, those units not yet classified by the CGPM as either base units or derived units. For the time being this class contains only two, purely geometrical units: the SI unit of plane angle, the **radian**, and the SI unit of solid angle, the **steradian**. Supplementary units may be used to form derived units. Examples are **angular velocity**, *radian per second*, and **angular accleration**, *radian per second squared*. In 1969 the International Committee of Weights and Measures (CIPM) recognized that users of SI units will wish to employ with it certain units not part of SI, but which are important and are widely used. These are the minute, the hour, the day, the degree of arc, the minute of arc, the second of arc, the liter, and the tonne. It is likewise necessary to recognize, outside the International System, some other units which are useful in specialized fields, because their values expressed in SI units must be obtained by experiment, and are therefore not known exactly. These are the electron-volt, the unified atomic mass unit, the astronomical unit, and the parsec. In view of existing practice the CIPM in 1969 considered it was preferable to continue using certain units with the International System for a limited time. These *temporary units* are the nautical mile, the knot, the angstrom, the arc, the hectare, the barn, the bar, the standard atmosphere, the gal, the curie, the röntgen, and the rod. The CIPM considers that it is in general preferable not to use, with the units of the International System, centimeter-gram-second (CGS) units which have special names. These units include the erg, the dyne, the gauss, and the oersted.

interpolation, *n.* The process of determining intermediate values between given values in accordance with some known or assumed rate or system of change. **Linear interpolation** assumes that changes of tabulated values are proportional to changes in entering arguments. Interpolation is designated as **single**, **double**, or **triple** as there are one, two, or three arguments or variables, respectively. The extension of the process of interpolation beyond the limits of known values is called EXTRAPOLATION.

interpolation table. An auxiliary table used for interpolating. See also PROPORTIONAL PARTS.

interrogating signal. The signal emitted by interrogator to trigger a transponder.

interrogation, *n.* The transmission of a radiofrequency pulse, or combination of pulses, intended to trigger a transponder or group of transponders.

interrogator, *n.* A radar transmitter which sends out a pulse that triggers a transponder. An interrogator may be combined in a single unit with a responsor, which receives the reply from a transponder and produces an output suitable for feeding a display system; the combined unit is called INTERROGATOR-RESPONSOR. Also called CHALLENGER.

interrogator-responsor, *n.* A radar transmitter and receiver combined to interrogate a transponder and display the resulting replies. Often shortened to INTERROGATOR and sometimes called CHALLENGER.

interrupted quick flashing light. A light exhibiting very rapid regular alternations of light and darkness in which the rapid alternations (generally not less than 60 per minute) are interrupted at regular intervals by eclipses of long duration. See also QUICK FLASHING LIGHT, VERY QUICK FLASHING LIGHT.

interrupted quick light. A quick light in which the sequence of flashes is interrupted by regularly repeated eclipses of constant and long duration. See also CONTINUOUS QUICK LIGHT, GROUP QUICK LIGHT.

interrupted very quick light. A very quick light in which the sequence of flashes is interrupted by regularly repeated eclipses of constant and long duration. See also CONTINUOUS VERY QUICK LIGHT, GROUP VERY QUICK LIGHT.

interscan, *n.* See INTER-TRACE DISPLAY.

intersect, *v., t. & i.* To cut or cross. For example, two nonparallel lines in a plane *intersect* in a point, and a plane *intersects* a sphere in a circle.

inter-trace display. A technique for presenting additional information, in the form of alphanumerics, markers, cursors, etc., on a radar display, by using the intervals between the normal presentation scans. Also called INTERSCAN.

in the wind. In the direction from which the wind is blowing; windward. The expression is used particularly in reference to a heading or to the position of an object. See also EYE OF THE WIND.

Intracoastal Waterway. An inside protected route extending through New Jersey; from Norfolk, Virginia to Key West, Florida; across Florida from St. Lucie Inlet to Fort Myers, Charlotte Harbor, Tampa Bay, and Tarpon Springs; and from Carabelle, Florida, to Brownsville, Texas.

Invar, *n.* The registered trade name for an alloy of nickel and iron, containing about 36% nickel. Its coefficient of expansion is extremely small over a wide range of temperature.

inverse chart. See TRANSVERSE CHART.

inverse cylindrical orthomorphic chart. See TRANSVERSE MERCATOR CHART.

inverse cylindrical orthomorphic map projection. See TRANSVERSE MERCATOR MAP PROJECTION.

inverse equator. See TRANSVERSE EQUATOR.

inverse latitude. See TRANSVERSE LATITUDE.

inverse logarithm. See ANTILOGARITHM.

inverse longitude. See TRANSVERSE LONGITUDE.

inverse Mercator chart. See TRANSVERSE MERCATOR CHART.

inverse Mercator map projection. See TRANSVERSE MERCATOR MAP PROJECTION.

inverse meridian. See TRANSVERSE MERIDIAN.

inverse parallel. See TRANSVERSE PARALLEL.

inverse rhumb line. See TRANSVERSE RHUMB LINE.

inversion, *n.* In meteorology, a departure from the usual decrease or increase with altitude of the value of an atmospheric property. This term is *almost always* used to refer to a *temperature inversion*, an atmospheric condition in which the temperature increases with increasing altitude.

inverted compass. A marine magnetic compass designed and installed for observation from below the compass card. Frequently used as a telltale compass. Also called HANGING COMPASS, OVERHEAD COMPASS.

inverted image. An image that appears upside down in relation to the object.

inverter, *n.* A device for changing direct current to alternating current. A device for changing alternating current to direct current is called a CONVERTER if a rotary device and a RECTIFIER if a static device.

inverting telescope. An instrument with the optics so arranged that the light rays entering the objective of the lens meet at the crosshairs and appear inverted when viewed through the eyepiece without altering the orientation of the image. See also ERECTING TELESCOPE.

inward bound. Heading toward the land or up a harbor away from the open sea. The opposite is OUTWARD BOUND. See also HOMEWARD BOUND.

ion, *n.* An atom or group of atoms which has become electrically charged, either positively or negatively, by the loss or gain of one or more electrons.

ionization, *n.* The process by which neutral atoms or groups of atoms become electrically charged, either positively or negatively, by the loss or gain of electrons; or the state of a substance whose atoms or groups of atoms have become thus charged.

ionized layers. Layers of charged particles existing in the upper reaches of the atmosphere as a result of solar radiation.

ionosphere, *n.* 1. The region of the atmosphere extending from about 40 to 250 statute miles above the earth's surface, in which there is appreciable ionization. The presence of charged particles in this region profoundly affects the propagation of certain electromagnetic radiations.

2. A region composed of highly ionized layers at varying heights above the surface of the earth which may cause the return to the earth of radio waves originating below these layers. See also D-LAYER, E-LAYER, F-LAYER, F_1-LAYER, F_2-LAYER.

ionospheric correction. A correction for ionospheric refraction, a major potential source of error in all satellite radionavigation systems. Navigation errors can result from the effect of refraction on the measurement of the doppler shift and from the errors in the satellite's orbit if refraction is not accurately accounted for in the satellite tracking. Since the doppler shift in the presence of the ionosphere is basically the time rate of change of the electromagnetic path length, the doppler shift is altered from what it would be in a vacuum. However, it can be shown that the ionosphere is dispersive (the amount of refraction is dependent upon the frequency) and for frequencies significantly above any ionospheric resonance frequently (usually less than 30 MHz), the dependence of the doppler shift with frequency is accurately known. For frequencies above 100 MHz, the dependence is inversely proportional to the frequency to very high accuracy. The refraction contribution can be eliminated by the proper mixing of the received doppler shift from two harmonically related frequencies (150 and 400 MHz for

NAVSAT) to yield an accurate estimate of the vacuum doppler shift. Also called REFRACTION CORRECTION.

ionospheric disturbance. A sudden outburst of ultraviolet light on the sun, known as a SOLAR FLARE or CHROMOSPHERIC ERUPTION, which produces abnormally high ionization in the region of the D-layer. The result is a sudden increase in radio wave absorption, with particular severity in the upper medium frequencies and lower high frequencies. It has negligible effects on the heights of the reflecting/refracting layers and, consequently, upon critical frequencies, but enormous transmission losses may occur. See also SUDDEN IONOSPHERIC DISTURBANCE.

ionospheric error. The total systematic and random error resulting from the reception of a navigation signal after ionospheric reflections. It may be due to variations in transmission paths, nonuniform height of the ionosphere, or nonuniform propagation within the ionosphere. Also called IONOSPHERIC-PATH ERROR, SKYWAVE ERROR.

ionospheric-path error. See IONOSPHERIC ERROR.

ionospheric storm. An ionospheric disturbance characterized by wide variations from normal in the state of the ionosphere, such as turbulence in the F-region, absorption increase, height increase, and ionization density decreases. The effects are most marked in high magnetic latitudes and are associated with abnormal solar activity.

ionospheric wave. See SKYWAVE.

iridescence, *n.* Changing-color appearance, as of a soap bubble, caused by interference of colors in a thin film or by diffraction.

iridescent clouds. Ice-crystal clouds which exhibit brilliant spots or borders of colors, usually red and green, observed up to about 30° from the sun.

irisation, *n.* The coloration exhibited by iridescent clouds.

Irminger Current. A North Atlantic Ocean current that is one of the terminal branches of the Gulf Stream System (part of the northern branch of the North Atlantic Current); it flows toward the west off the southwest coast of Iceland. A small portion of the water of the Irminger Current bends around the west coast of Iceland but the greater amount turns south and becomes more or less mixed with the water of the East Greenland Current.

ironbound, *adj.* Rugged, rocky, as an *ironbound coast.*

irradiation, *n.* The apparent enlargement of a bright surface against a darker background.

irradiation correction. A correction due to irradiation, particularly that sextant altitude correction caused by the apparent enlargement of the bright surface of a celestial body against the darker background of the sky.

irregular error. See RANDOM ERROR.

irregular iceberg. See PINNACLED ICEBERG.

isallobar, *n.* A line of equal change in atmospheric pressure during a specified time interval.

isallotherm, *n.* A line connecting points having the same anomalies of temperature, pressure, etc.

isanomal, *n.* A line connecting points of equal variations from a normal value.

island, *n.* A tract of land smaller than a continent, completely surrounded by water at mean high water. An ISLET is a very small and minor island.

island harbor. *British terminology.* A harbor formed by islands, or one mainly protected by islands.

islet, *n.* A very small and minor island.

iso-. A prefix meaning *equal.*

isobar, *n.* A line connecting points having the same atmospheric pressure reduced to a common datum, usually sea level.

isobaric, *adj.* Having the same pressure.

isobaric chart. See CONSTANT-PRESSURE CHART.

isobaric surface. See CONSTANT-PRESSURE SURFACE.

isobath, *n.* See DEPTH CONTOUR.

isobathic, *adj.* Having equal depth.

isobathytherm, *n.* A line on the earth's surface connecting points at which the same temperature occurs at some specified depth.

isobront, *n.* A line connecting points at which some specified phase of a thunderstorm occurs at the same time.

isoceraunic, isokeraunic, *adj.* Indicating or having equal frequency or intensity of thunderstorms.

isochasm, *n.* A line connecting points having the same average frequency of auroras.

isochronal, *adj.* Of equal time; recurring at equal intervals of time. Also called ISOCHRONOUS.

isochrone, *n.* A line connecting points having the same time or time difference relationship, as a line representing all points having the same time difference in the reception of signals from two radio stations such as the master and slave stations of a Loran rate.

isochronize, *v., t.* To render isochronal.

isochronon, *n.* A clock designed to keep very accurate time.

isochronous, *adj.* See ISOCHRONAL.

isoclinal, *adj.* Of or pertaining to equal magnetic dip.

isoclinal, *n.* See ISOCLINIC LINE.

isoclinal chart. See ISOCLINIC CHART.

isoclinic chart. A chart of which the chief feature is a system of isoclinic lines. Also called ISOCLINAL CHART.

isoclinic line. A line drawn through all points on the earth's surface having the same magnetic dip. The particular isoclinic line drawn through points of zero dip is called ACLINIC LINE. Also called ISOCLINAL.

isodynamic chart. A chart showing isodynamic lines. See also MAGNETIC CHART.

isodynamic line. A line connecting points of equal magnetic intensity, either the total or any component.

isogonal, *adj.* Having equal angles; isogonic.

isogonal, *n. British terminology.* See ISOGONIC, *n.*

isogonic, *adj.* Having equal angles; isogonal.

isogonic, *n.* A line connecting points of equal magnetic variation. Also called ISOGONIC LINE, ISOGONAL.

isogonic chart. A chart showing magnetic variation with *isogonic lines* and the annual rate of change in variation with *isoporic lines.* See also MAGNETIC CHART.

isogonic line. See ISOGONIC, *n.*

isogram, *n.* That line, on a chart or diagram, connecting points of equal value of some phenomenon.

isogriv, *n.* A line drawn on a map or chart joining points of equal grivation.

isogriv chart. A chart showing isogrivs. See also MAGNETIC CHART.

isohaline, isohalsine, *n.* A line connecting points of equal salinity in the ocean.

isolated danger buoy. See OBSTRUCTION BUOY.

isolated danger mark. *British terminology.* See OBSTRUCTION MARK.

isosceles, *adj.* Having two equal sides.

isosceles triangle. A triangle having two of its sides

equal.

isolated danger mark. See under IALA MARITIME BUOYAGE SYSTEM.

isomagnetic, *adj.* Of or pertaining to lines connecting points of equality in some magnetic element.

isomagnetic, *n.* A line connecting points of equality in some magnetic element. Also called ISOMAGNETIC LINE.

isomagnetic chart. A chart showing isomagnetics. See also MAGNETIC CHART.

isomagnetic line. See ISOMAGNETIC, *n.*

isometric, *n.* Of or pertaining to equal measure.

isophase, *adj.* Of a light, having a characteristic of equal intervals of light and darkness.

isophase light. See EQUAL INTERVAL LIGHT.

isopleth, *n.* 1. An isogram indicating the variation of an element with respect to two variables, one of which is usually the time of year. The other may be time of day, altitude, or some other variable.

2. A line on a map depicting points of constant value of a variable. Examples are contours, isobars, and isogons.

isopor, *n.* See ISOPORIC LINE.

isoporic chart. A chart with lines connecting points of equal annual rate of change of any magnetic element. See also ISOPORIC LINE.

isoporic line. A line connecting points of equal annual rate of change of any magnetic element. Also called ISOPOR. See also ISOGONIC CHART. ISOPORIC CHART.

isostasy, *n.* A supposed equality existing in vertical sections of the earth, whereby the weight of any column from the surface of the earth to a constant depth is approximately the same as that of any other column of equal area, the equilibrium being maintained by plastic flow of material from one part of the earth to another.

isotropic antenna. A hypothetical antenna which radiates or receives equally well in all directions. Although such an antenna does not physically exist, it provides a convenient reference for expressing the directional properties of actual antennas. Also called UNIPOLE.

isotropic gain of an antenna. The gain of an antenna in a given direction when the reference antenna is an isotropic antenna isolated in space. Also called ABSOLUTE GAIN OF AN ANTENNA.

isthmus, *n.* A narrow strip of land connecting two larger portions of land. A submarine elevation joining two land areas and separating two basins or depressions by a depth less than that of the basins is called a **submarine isthmus.**

J–K

Jacob's staff. See CROSS-STAFF.

jamming, *n.* Intentional transmission or re-radiation of radio signals in such a way as to interfere with reception of desired signals by the intended receiver.

Janus configuration. A term descriptive of the orientations of the beams of acoustic or electromagnetic energy employed with doppler navigation systems. The Janus configuration normally used with doppler sonar speed logs, navigators, and docking aids employs four beams of ultrasonic energy, displaced laterally 90° from each other, and each directed obliquely (30° from the vertical) at the ocean floor, to obtain true ground speed in the fore and aft and athwartship directions. These speeds are measured as doppler frequency shifts in the reflected beams. The configuration provides the benefit of certain errors in data extracted from one beam tending to cancel the errors associated with the oppositely directed beam.

Japan Current. See KUROSHIO.

jetsam, *n.* Articles that sink when thrown overboard, particularly those jettisoned for the purpose of lightening a vessel in distress. See also FLOTSAM, JETTISON, LAGAN.

jet stream. Relatively strong winds (50 knots or greater) concentrated in a narrow stream in the atmosphere. It usually refers only to a quasi-horizontal stream of maximum winds imbedded in the middle latitude *westerlies*, and concentrated in the high troposphere.

jettison, *n.* The throwing overboard of objects, especially to lighten a craft in distress. Jettisoned objects that float are termed FLOTSAM; those that sink JETSAM; and heavy articles that are buoyed for future recovery, LAGAN. See also DERELICT.

jetty, *n.* A structure built out into the water to restrain or direct currents, usually to protect a river mouth or harbor entrance from silting, etc. See also GROIN; MOLE, definition 1.

jitter, *n.* A term used to describe the short-time instability of a signal. The instability may be in amplitude, phase, or both. The term is applied especially to signals reproduced on the screen of a cathode-ray tube.

joule, *n.* A *derived unit* of energy of work in the International System of Units; it is the work done when the point of application of 1 newton (that force which gives to a mass of 1 kilogram an acceleration of 1 metre per second, per second) moves a distance of 1 metre in the direction of the force.

Julian calendar. A revision of the ancient calendar of the city of Rome, instituted in the Roman Empire by Julius Caesar in 46 B.C., which reached its final form in about 8 A.D. It consisted of years of 365 days, with an intercalary day every fourth year. The current Gregorian calendar is the same as the Julian calendar except that October 5, 1582, of the Julian calendar became October 15, 1582 of the Gregorian calendar and of the centurial years, only those divisible by 400 are leap years.

Julian day. The number of each day, as reckoned consecutively since the beginning of the present *Julian period* on January 1, 4713 BC. It is used primarily by astronomers to avoid confusion due to the use of different calendars at different times and places. The Julian day begins at noon, 12 hours later than the corresponding civil day. The day beginning at noon January 1, 1968, was Julian day 2,439,857.

junction buoy. A buoy which, when viewed from a vessel approaching from the open sea or in the same direction as the main stream of flood current, or in the direction established by appropriate authority, indicates the place at which two channels meet. See also BIFURCATION BUOY.

junction mark. A navigation mark which, when viewed from a vessel approaching from the open sea or in the same direction as the main stream of flood current, or in the direction established by appropriate authority, indicates the place at which two channels meet. See also BIFURCATION MARK.

June solstice. Summer solstice in the Northern Hemisphere.

Jupiter, *n.* The navigational planet whose orbit lies between those of Mars and Saturn. Largest of the known planets.

Jutland Current. A narrow and localized nontidal current off the coast of Denmark between longitudes 8°30′E and 10°30′E. It originates partly from the resultant counterclockwise flow in the tidal North Sea. The main cause, however, appears to be the winds which prevail from south through west to northwest over 50 percent of the time throughout the year and the transverse flows from the English coast toward the Skagerak. The current retains the characteristics of a major nontidal current and flows northeastward along the northwest coast of Denmark at speeds ranging between 1.5 to 2.0 knots 75 to 100 percent of the time.

Kaléma, *n.* A very heavy surf breaking on the Guinea coast during the winter, even when there is no wind.

Kalman filtering. A statistical method for estimating the parameters of a dynamic system, using recursive techniques of estimation, measurement, weighting, and correction. Weighting is based on variances of the measurements and of the estimates. The filter acts to reduce the variance of the estimate with each measurement cycle.

katabatic wind. Any wind blowing down an incline. If the wind is warm, it is called a **foehn**; if cold, a **fall wind**. An ANABATIC WIND blows up an incline. Also called GRAVITY WIND.

kaver, *n.* See CAVER.

kay, *n.* See CAY.

K-band. A radio-frequency band of 10,900 to 36,000 megahertz. See also FREQUENCY, FREQUENCY BAND.

kedge, *v., t.* To move, as a vessel, by carrying out an anchor, letting it go, and hauling the ship up to the anchor. See also WARP.

keeper, *n.* A piece of magnetic material placed across the poles of a permanent magnet to assist in the maintenance of magnetic strength.

kelp, *n.* 1. A certain family of seaweed.
2. Any large seaweed.
3. The ashes of seaweed.

kelvin, *n.* The *base unit* of thermodynamic temperature in the International System of Units; it is the fraction 1/273.16 of the thermodynamic temperature of the triple point of water, which is – 273°16K.

Kelvin temperature. Temperature based upon a thermodynamic scale with its zero point at absolute zero (–273°.16C) and using Celsius degrees. Rankine temperature is based upon the Rankine scale starting at absolute zero (–459°.69 F) and using Fahrenheit degrees.

Kennelly-Heaviside layer. See under KENNELLY-HEAVISIDE REGION.

Kennelly-Heaviside region. That region of the ionosphere, extending from approximately 40 to 250 statute miles above the earth's surface, within which ionized layers form which affect, or may affect, radio wave propagation. The E-layer, which is the lowest useful layer from the standpoint of wave propagation, is sometimes called KENNELLY-HEAVISIDE LAYER or, in some instances, simply the HEAVISIDE LAYER.

Kepler's laws. The three empirical laws describing the motions of the planets in their orbits. These are: (1) The orbits of the planets are ellipses, with the sun at a common focus; (2) As a planet moves in its orbit, the line joining the planet and sun sweeps over equal areas in equal intervals of time; (3) The squares of the periods of revolution of any two planets are proportional to the cubes of their mean distances from the sun. Also called KEPLER'S PLANETARY LAWS.

Kepler's planetary laws. See KEPLER'S LAWS.

key, *n.* See CAY.

kick, *n.* 1. The distance a ship moves sidewise from the original course away from the direction of turn after the rudder is first put over.

2. The swirl of water toward the inside of the turn when the rudder is put over to begin the turn.

kilo-. A prefix meaning *one thousand* (10^3).

kilocycle, *n.* One thousand cycles, the term is often used as the equivalent of *one thousand cycles per second.*

kilogram, *n.* 1. The *base unit* of mass in the International System of Units; it is equal to the mass of the international prototype of the kilogram, which is made of platinum-iridium and kept at the International Bureau of Weights and Measures.

2. One thousand grams exactly, or 2.204623 pounds, approximately.

kilometer, *n.* One thousand meters; about 0.54 nautical mile, 0.62 statute mile, or 3,281 feet.

kinetic energy. Energy possessed by a body by virtue of its motion: in contrast with POTENTIAL ENERGY, that possessed by virtue of its position.

klaxon, *n.* A diaphragm horn similar to a nautophone, but smaller, and sometimes operated by hand.

knik wind. A strong southeast wind in the vicinity of Palmer, Alaska, most frequent in the winter.

knoll, *n.* 1. On the sea floor, an elevation rising generally more than 500 meters and less than 1,000 meters and of limited extent across the summit.

2. A small rounded hill.

knot, *n.* A unit of speed equal to 1 nautical mile per hour.

kona storm. A storm over the Hawaiian Islands, characterized by strong southerly or southwesterly winds and heavy rains.

Krassowski ellipsoid of 1938. The reference ellipsoid of which the semimajor axis is 6,378,245 meters and the flattening of ellipticity equals 1/298.3.

Kuroshio, *n.* A North Pacific Ocean current flowing northeastward from Taiwan to the Ryukyu Islands and close to the coast of Japan. The Kuroshio is the northward flowing part of the Pacific North Equatorial Current (which divides east of the Philippines). The Kuroshio divides near Yaku Shima, the weaker branch flowing northward through Korea Strait and the stronger branch flowing through Tokara Kaikyo and then along the south coast of Shikoku. There are light seasonal variations in speed; the Kuroshio is usually strongest in summer, weakens in autumn, strengthens in winter, and weakens in spring. Strong winds can accelerate or retard the current, but seldom change its direction. Beyond latitude 35°N on the east coast of Japan, the current turns east-northeastward to form the transitional KUROSHIO EXTENSION. The Kuroshio is part of the KUROSHIO SYSTEM. Also called JAPAN CURRENT.

Kuroshio Extension. The transitional, eastward flowing ocean current that connects the Kuroshio and the North Pacific Current.

Kuroshio System. A system of ocean currents which includes part of the Pacific North Equatorial Current, the Tsushima Current, the Kuroshio, and the Kuroshio Extension.

kymatology, *n.* The science of waves and wave motion.

L

labor, *v., i.* To pitch and roll heavily under conditions which subject the ship to unusually heavy stresses caused by confused or turbulent seas or unstable stowage of cargo.

Labrador Current. Originating from cold arctic water flowing southeastward through Davis Strait at speeds of 0.2 to 0.5 knot and from a westward branching of the warmer West Greenland Current, the Labrador Current flows southeastward along the shelf of the Canadian coast. Part of the current flows into Hudson Strait along its north shore. The outflow of fresh water along the south shore of the strait augments the part of the current flowing along the Labrador coast. The current also appears to be influenced by surface outflow from inlets and fiords along the Labrador coast. The mean speed is about 0.5 knot, but current speed at times may reach 1.5 to 2.0 knots.

Labrador Current Extension. A name sometimes given to the nontidal current flowing southwestward along the northeast coast of the United States. This coastal current originates from part of the Labrador Current flowing clockwise around the southeastern tip of Newfoundland. Its speeds are fairly constant throughout the year and average about 0.6 knot. The greatest seasonal fluctuation appears to be in the width of the current. The current is widest during winter between Newfoundland and Cape Cod. Southwest of Cape Cod to Cape Hatteras the current shows very little seasonal change. The current narrows considerably during summer and flows closest to shore in the vicinity of Cape Sable, Nova Scotia and between Cape Cod and Long Island in July and August. The current in some places encroaches on tidal regions.

lagan, *n.* A heavy object thrown overboard and buoyed to mark its location for future recovery. See also JETTISON.

lag error. That error in the reading of an instrument due to lag.

lagging of tide. The periodic retardation in the time of occurrence of high and low water due to changes in the relative positions of the moon and the sun. See also PRIMING OF TIDE.

lagoon, *n.* 1. A shallow sound, pond, or lake generally separated from the open sea.
2. A body of water enclosed by the reefs and islands of an atoll.

Lagrangian current measurement. The direct observation of the current speed or direction, or both, by a recording device such as a parachute drogue which follows the movement of a water mass through the ocean. See also EULERIAN CURRENT MEASUREMENT.

lake, *n.* 1. Any standing body of inland water, generally of considerable size. There are exceptions such as the lakes in Louisiana which are open to or connect with the Gulf of Mexico. Occasionally a lake is called a SEA, especially if very large and composed of salt water.
2. An expanded part of a river.

lake ice. Ice formed on a lake, regardless of observed location.

Lambert conformal chart. A chart on the Lambert conformal projection. See also CONIC CHART WITH TWO STANDARD PARALLELS, MODIFIED LAMBERT CONFORMAL CHART.

Lambert conformal conic map projection. A conformal map projection of the conic type, on which all geographic meridians are represented by straight lines which meet in a common point outside the limits of the map, and the geographic parallels are represented by a series of arcs of circles having this common point for a center. Meridians and parallels intersect at right angles, and angles on the earth are correctly represented on the projection. This projection may have one standard parallel along which the scale is held exact; or there may be two such standard parallels, both maintaining exact scale. At any point on the map, the scale is the same in every direction. The scale changes along the meridians and is constant along each parallel. Where there are two standard parallels, the scale between those parallels is too small; beyond them, too large. Also called LAMBERT CONFORMAL MAP PROJECTION. See also MODIFIED LAMBERT CONFORMAL MAP PROJECTION.

Lambert conformal map projection. See LAMBERT CONFORMAL CONIC MAP PROJECTION.

laminar flow. See under STREAMLINE FLOW.

land, *v., t. & i.* To bring a vessel to a landing.

land breeze. A breeze blowing from the land to the sea. It usually blows by night, when the sea is warmer than the land, and alternates with a SEA BREEZE, which blows in the opposite direction by day. See also OFFSHORE WIND.

landfall, *n.* The first sighting of land when approached from seaward. By extension, the term is sometimes used to refer to the first contact with land by any means, as by radar.

landfall buoy. *British terminology.* See SEA BUOY.

landfall light. See PRIMARY SEACOAST LIGHT.

landing, *n.* 1. A place where boats receive or discharge passengers, freight, etc. See also LANDING STAGE, WHARF.
2. Bringing of a vessel to a landing.

landing compass. A compass taken ashore so as to be unaffected by deviation. If reciprocal bearings of the landing compass and the magnetic compass on board are observed, the deviation of the latter can be determined.

landing stage. *British terminology.* A platform attached to the shore for landing or embarking passengers or cargo. In some cases the outer end of the landing stage is floating on a pontoon. Ships can moor alongside the larger landing stages.

landmark, *n.* A conspicuous artificial feature on land, other than an *established* aid to navigation, which can be used as an aid to navigation. See also SEAMARK.

land mile. See STATUTE MILE.

land sky. Dark streaks or patches or a grayness on the underside of extensive cloud areas, due to the absence of reflected light from bare ground. Land sky is not as dark as WATER SKY. The clouds above ice or snow covered surfaces have a white or yellowish white glare called BLINK. See also SKY MAP.

lane, *n.* In any continuous wave phase comparison system, the distance between two successive equiphase lines, taken as 0° - 360°, in a system of hyperbolic or circular coordinates.

lane count. An automatic method of counting and totalizing the number of hyperbolic or circular lanes traversed by a moving vessel.

lapse rate. The rate of decrease of temperature in the atmosphere with height, or, sometimes, the rate of change of any meteorological element with height.

large fracture. See under FRACTURE.

large iceberg. For reports to the International Ice Patrol, an iceberg that extends more than 150 feet (45 meters) above the sea surface and which has a length of more than 400 feet (122 meters). See also SMALL ICEBERG, MEDIUM ICEBERG.

large ice field. See under ICE FIELD.

large navigational buoy. A large buoy designed to take the place of a lightship where construction of an offshore light station is not feasible. These 40-foot diameter buoys may show secondary lights from heights of about 36 feet above the water. In addition to the light, these buoys may mount a radiobeacon and provide sound signals. A station buoy may be moored nearby. Called LIGHTHOUSE BUOY in British terminology.

large scale. A scale involving a relatively small reduction in size. A **large-scale chart** is one covering a small area. The opposite is SMALL SCALE. See also REPRESENTATIVE FRACTION.

large-scale chart. See under CHART. See also LARGE SCALE.

last quarter. The phase of the moon when it is near west quadrature, when the eastern half of it is visible to an observer on the earth. See also PHASES OF THE MOON.

latent heat of fusion. See under FUSION.

latent heat of vaporization. See under EVAPORATION.

lateral, *adj.* Of or pertaining to the side, as *lateral motion*, or motion to one side.

lateral drifting. See SWAY.

lateral mark. See under IALA MARITIME BUOYAGE SYSTEM.

lateral sensitivity. The property of a range which determines the rapidity with which the two lights of the range open up as a vessel moves laterally from the range line, thus indicating to the mariner that he is off the center line.

lateral system. A system of aids to navigation in which the shape, color, and number distinction are assigned in accordance with their location in respect to navigable waters. When used to mark a channel, they are assigned colors to indicate the side they mark and numbers to indicate their sequence along the channel. The lateral system is used in the United States. In the CARDINAL SYSTEM the aids are assigned shape, color, and number distinction in accordance with location relative to the nearest obstruction.

latitude, *n.* Angular distance from a primary great circle or plane. **Terrestrial latitude** is angular distance from the equator, measured northward or southward through 90° and labeled N or S to indicate the direction of measurement; **astronomical latitude** at a station is angular distance between the plumb line and the plane of the celestial equator; **geodetic** or **topographical latitude** at a station is angular distance between the plane of the geodetic equator and a normal to the ellipsoid; **geocentric latitude** is the angle at the center of the reference ellipsoid between the celestial equator and a radius vector to a point on the ellipsoid. Geodetic and sometimes astronomical latitude are also called **geographic latitude.** Geodetic latitude is used for charts. **Assumed** (or **chosen**) **latitude** is the latitude at which an observer is assumed to be located for an observation or computation. **Observed latitude** is determined by one or more lines of position extending in a generally east-west direction. **Fictitious latitude** is angular distance from a fictitious equator. **Grid latitude** is angular distance from a grid equator. **Transverse** or **inverse latitude** is angular distance from a transverse equator. **Oblique latitude** is angular distance from an oblique equator. **Middle** or **mid latitude** is the latitude at which the arc length of the parallel separating the meridians passing through two specific points is exactly equal to the departure in proceeding from one point to the other by **middle-latitude sailing. Mean latitude** is half the arithmetical sum of the latitude of two places on the same side of the equator. The mean latitude is usually used in middle-latitude sailing for want of a practical means of determining middle latitude. **Difference of latitude** is the shorter arc of any meridian between the parallels of two places, expressed in angular measure. **Magnetic latitude, magnetic inclination,** or **magnetic dip** is angular distance between the horizontal and the direction of a line of force of the earth's magnetic field at any point. **Geomagnetic latitude** is angular distance from the geomagnetic equator. A **parallel of latitude** is a circle (or approximation of a circle) of the earth, parallel to the equator, and connecting points of equal latitude; or a circle of the celestial sphere, parallel to the ecliptic. **Celestial latitude** is angular distance north or south of the ecliptic. **Horse latitudes** are the regions of calms and variable winds coinciding with the sub-tropical high pressure belts on the poleward sides of the trade winds, especially the northern of these two regions in the North Atlantic. See also VARIATION OF LATITUDE.

latitude factor. The change in latitude along a celestial line of position per 1′ change in longitude. The change in longitude for a 1′ change in latitude is called LONGITUDE FACTOR.

latitude line. A line of position extending in a generally east-west direction. Sometimes called OBSERVED LATITUDE. See also LONGITUDE LINE; COURSE LINE, definition 2; SPEED LINE.

lattice, *n.* A pattern formed by two or more families of intersecting lines, such as that pattern formed by two or more families of hyperbolas representing, for example, curves of equal time difference associated with a hyperbolic radionavigation system. Sometimes the term *pattern* is used to indicate curves of equal time difference, with the term *lattice* being used to indicate its representation on the chart. See also PATTERN, definition 2.

lattice beacon. A beacon which is a major structure in the form of a lattice. See also BEACON TOWER, REFUGE BEACON.

laurence, *n.* A shimmering seen over a hot surface on a calm, cloudless day, caused by the unequal refraction of light by innumerable convective air columns of different temperatures and densities.

lava, *n.* Rock in the fluid state, or such material after it has solidified. Lava is formed at very high temperature and issues from the earth through volcanoes. Part of the ocean bed is composed of lava.

law of equal areas. Kepler's second law.

layer tints. See HYPSOMETRIC TINTING.

L-band. A radio-frequency band of 390 to 1,550 megahertz. See also FREQUENCY, FREQUENCY BAND.

lead (lēd), *n.* Any fracture or passage-way through sea ice which is navigable by surface vessels.

lead (lĕd), *n.* A weight attached to a line. A **sounding lead** is used for determining depth of water. A **hand lead** is a light sounding lead (7 to 14 pounds), usually having a line of not more than 25 fathoms. A **deep sea lead** is a heavy sounding lead (about 30 to 100 pounds), usually having a line 100 fathoms or more in length. A light deep sea lead (30 to 50 pounds), used for sounding depths of 20 to 60 fathoms is called a **coasting lead.** A type of sounding lead used without removal from the water between soundings is called a **fish lead.** A **drift lead** is one placed on the bottom to indicate movement of a vessel. To HEAVE THE LEAD is to take a sounding with a lead.

leader cable. A cable carrying an electric current, signals from or the magnetic influence of which indicates the path to be followed by a craft equipped with suitable instruments.

leading lights. *British terminology.* See RANGE LIGHTS.

leading line. *British terminology.* On a nautical chart, a straight line, drawn through **leading marks.** A ship moving along such line will clear certain dangers or remain in the best channel. See also CLEARING LINE, RANGE, definition 1.

leading marks. *British terminology.* See RANGE, *n.* definition 1.

lead line. A line, graduated with attached marks and fastened to a sounding lead, used for determining the depth of water when making soundings by hand. The lead line is usually used in depths of less than 25 fathoms. Also called SOUNDING LINE.

leadsman, *n.* A person using a sounding lead to determine depth of water.

leap second. A step adjustment to Coordinated Universal Time (UTC) to maintain it within $0\overset{s}{.}95$ of UT1. The 1 second adjustments, when necessary are normally made at the end of June or December. Because of the variations in the rate of rotation of the earth, the occurrences of the leap second adjustments are not predictable in detail.

leap year. A calendar year having 366 days as opposed to the COMMON YEAR having 365 days. Each year exactly divisible by 4 is a leap year, except century years (1800, 1900, etc.) which must be exactly divisible by 400 (2000, 2400, etc.) to be leap years.

least squares adjustment. A method of adjusting observations in which the sum of the squares of all the deviations or residuals derived in fitting the observations to a mathematical model is made a minimum.

ledge, *n.* On the sea floor, a rocky, projection or outcrop, commonly linear and near shore.

lee, *adj.* Of or pertaining to the direction *toward* which the wind is blowing. WEATHER pertains to the direction from which the wind is blowing.

lee, *n.* That side toward which the wind blows; the sheltered side.

lee shore. As observed from a ship, the shore towards which the wind is blowing. See also WEATHER SHORE.

lee side. That side of a craft which is away from the wind and therefore sheltered.

lee tide. See LEEWARD TIDAL CURRENT.

leeward, *adj. & adv.* Toward the lee, or in the general direction toward which the wind is blowing. The opposite is WINDWARD.

leeward, *n.* The lee side. The opposite is WINDWARD.

leeward tidal current. A tidal current setting in the same direction as that in which the wind is blowing. Also called LEE TIDE, LEEWARD TIDE.

leeward tide. See LEEWARD TIDAL CURRENT.

leeway, *n.* The leeward motion of a vessel due to wind. See also LEEWAY ANGLE.

leeway angle. The angular difference between the course and the water track due to the effect of wind in moving a vessel bodily to leeward. See also DRIFT ANGLE, definition 2.

left bank. That bank of a stream or river on the left of an observer facing in the direction of flow, or downstream. See also RIGHT BANK.

leg, *n.* That part of a track that can be represented by a single course line.

legend, *n.* A title or explanation on a chart, diagram, illustration, etc.

lens, *n.* A piece of glass or transparent material with plane, convex, or concave surfaces adapted for changing the direction of light rays to enlarge or reduce the apparent size of objects. See also EYEPIECE; FIELD LENS; MENISCUS, definition 2; OBJECTIVE.

lenticular. lenticularis, *adj.* In the shape of a lens. The term is used to refer to an apparently stationary cloud resembling a lens, being broad in its middle and tapering at the ends and having a smooth appearance. Actually, the cloud continually forms to windward and dissipates to leeward.

lesser ebb. See under EBB CURRENT.

lesser flood. See under FLOOD CURRENT.

leste, *n.* A hot, dry, easterly wind of the Madeira and Canary Islands.

levanter, *n.* A strong easterly wind of the Mediterranean, especially in the Strait of Gibraltar, attended by cloudy, foggy, and sometimes rainy weather, especially in winter.

levantera, *n.* A persistent east wind of the Adriatic, usually accompanied by cloudy weather.

levanto, *n.* A hot southeasterly wind which blows over the Canary Islands.

leveche, *n.* A warm wind in Spain, either a foehn or a hot southerly wind in advance of a low pressure area moving from the Sahara Desert. Called a SIROCCO in other parts of the Mediterranean area.

levee, *n.* 1. An artificial bank confining a stream channel or limiting adjacent areas subject to flooding.

2. On the sea floor, an embankment bordering a canyon, valley, or seachannel.

level ice. Sea ice which is unaffected by deformation.

leveling, *n.* A survey operation in which heights of objects are determined relative to a specified datum.

libration, *n.* A real or apparent oscillatory motion, particularly the apparent oscillation of the moon, which results in more than half of the moon's surface being revealed to an observer on the earth, even though the same side of the moon is always toward the earth because of the moon's periods of rotation and revolution are the same.

lifeboat navigation. Navigation of a lifeboat or life raft. Such craft are usually not well equipped for navigation, necessitating resourcefulness and a knowledge of fundamentals. See also MARINE NAVIGATION.

light, *adj.* 1. Of or pertaining to low speed as *light air*, force 1 (1-3 miles per hour or 1-3 knots) on the Beaufort scale or *light breeze*, force 2 (4-7 miles per hour or 4-6 knots) on the Beaufort scale.
2. Of or pertaining to low intensity, as *light rain*, *light fog*, etc.

light, *n.* 1. Luminous energy.
2. An apparatus emitting light of distinctive character for use as an aid to navigation by night or, exceptionally, by day.

light air. Wind of force 1 (1 to 3 knots or 1 to 3 miles per hour) on the Beaufort wind scale.

light attendant station. A shore unit established for the purpose of servicing minor aids to navigation within an assigned area.

light-beacon, *n.* See LIGHTED BEACON.

light breeze. Wind of force 2 (4 to 6 knots or 4 to 7 miles per hour) on the Beaufort wind scale.

lighted beacon. A beacon exhibiting a light. Also called LIGHT-BEACON.

lighted buoy. A buoy exhibiting a light. If the light is produced by gas it may be called a **gas buoy**.

lighted sound buoy. See under SOUND BUOY.

lightening area. See CARGO under TRANSSHIPMENT AREA.

light-float, *n.* A buoy having a boat-shaped body smaller than a lightship. Light-floats are usually unmanned and are used instead of smaller lighted buoys in waters where strong currents are experienced.

lighthouse, *n.* A distinctive structure exhibiting a major light (primary seacoast light or secondary light).

lighthouse buoy. *British terminology.* See LARGE NAVIGATIONAL BUOY.

light list. 1. A publication giving detailed information regarding lighted navigational aids and fog signals. The name and location of the lighted aids, their characteristics, heights, range, structure description, and other pertinent remarks are given.
2. *Light List*, published by the U.S. Coast Guard in five volumes, covers the waters of the United States and its possessions including the Intracoastal Waterway, the Great Lakes (both United States and certain aids on the Canadian shores), and the Mississippi River and its navigable tributaries. In addition to the information on lighted aids, the *Light List* gives information on unlighted buoys, radiobeacons, radio direction finder calibration stations, daybeacons, racons, etc.
3. *List of Lights*, published by the Defense Mapping Agency Hydrographic/Topographic Center in seven volumes, covers waters other than the United States and its possessions. In addition to the information on lighted aids, the *List of Lights* provides information on storm signals, signal stations, radio direction finder stations, radiobeacons, etc.

light list number. The number used to identify a navigational light in the light list. This number should not be confused with INTERNATIONAL NUMBER, which is an identifying number assigned by the International Hydrographic Organization. The international number is in *italic* type and is located under the light list number in the list list. Sometimes called LIST OF LIGHTS NUMBER.

light nilas. Nilas which is more than 5 centimeters in thickness and somewhat lighter in color than dark nilas.

light sector. As defined by bearings from seaward, the sector in which a navigational light is visible or in which it has a distinctive color different from that of adjoining sectors, or in which it is obscured. See also SECTOR LIGHT.

lightship, *n.* A distinctively marked vessel providing aids to navigation services similar to a light station, i.e., a light of high intensity and reliability, sound signal, and radiobeacon, and moored at a station where erection of a fixed structure is not feasible. It should be borne in mind that most lightships are anchored to a very long scope of chain and, as a result, the radius of their swinging circle is considerable. The chart symbol represents the approximate location of the anchor. Also called LIGHT VESSEL. See also LIGHT-FLOAT.

lights in line. Two or more lights so situated that when observed in transit they define the alignment of a submarine cable, the limit of an area, an alignment for use in anchoring, etc. Not to be confused with RANGE LIGHTS which mark a direction to be followed. See also RANGE, definition 1.

light station. A manned station providing a light usually of high intensity and reliability. It may also provide sound signal and radiobeacon services. In many instances, sound signals, radiobeacon equipment, and operating personnel are housed in separate buildings near the light structure.

light valve. See SUN VALVE.

light vessel. See LIGHTSHIP.

light-year, *n.* A unit of length equal to the distance light travels in 1 year, equal to about 5.88×10^{12} statute miles. This unit is used as a measure of stellar distances.

liman, *n.* A shallow coastal lagoon or embayment with a muddy bottom; also a region of mud or slime deposited near a stream mouth.

Liman Current. Formed by part of the Tsushima Current and river discharge in Tatar Strait, the coastal Liman Current flows southward in the western part of the Sea of Japan. During winter, it may reach as far south as 35°N. See also under TSUSHIMA CURRENT.

limb, *n.* 1. The graduated curved part of an instrument for measuring angles, as that part of a marine sextant carrying the altitude scale, or ARC.
2. The circular outer edge of a celestial body. The half with the greatest altitude is called the **upper limb** and the half with the least altitude, the **lower limb**.

limbo echo. See under CLASSIFICATION OF RADAR ECHOES.

line, *n.* 1. A series of related points; the path of a moving point. A line has only one dimension; length.
2. A row of letters, numbers, etc.
3. A mark of division or demarcation, as a *boundary line*.

linear, *adj.* 1. Of or pertaining to a line.
2. Having a relation such that a change in one quantity is accompanied by an exactly proportional change in a related quantity.

linear interpolation. Interpolation in which changes of tabulated values are assumed to be proportional to changes in entering arguments.

linear light. A luminous signal having perceptible length, as contrasted with a POINT LIGHT, which does not have perceptible length.

linearly polarized wave. A transverse electromagnetic wave the electric field vector of which lies along a fixed line at all times.

linear scale. A scale graduated at uniform intervals.

linear speed. Rate of motion in a straight line. See also ANGULAR RATE.

linear sweep. Short for LINEAR TIME BASE SWEEP.

linear time base. A time base having a constant speed, particularly a linear time base sweep.

linear time base sweep. A sweep having a constant sweep speed before retrace. Usually shortened to LINEAR SWEEP, and sometimes to LINEAR TIME BASE.

line blow. A strong wind on the equator side of an anticyclone, probably so called because there is little shifting of wind direction during the blow, as contrasted with the marked shifting which occurs with a cyclonic windstorm.

line of apsides. The line connecting the two points of an orbit that are nearest and farthest from the center of attraction, as the perigee and apogee of the moon or the perihelion and aphelion of a planet. Also called APSE LINE.

line of force. A line indicating the direction in which a force acts, as in a magnetic field.

line of nodes. The straight line connecting the two points of intersection of the orbit of a planet, planetoid, or comet and the ecliptic; or the line of intersection of the planes of the orbits of a satellite and the equator of its primary.

line of position. A line indicating a series of possible positions of a craft, determined by observation or measurement. Also called POSITION LINE.

line of sight. The straight line between two points. This line is in the direction of a great circle, but does not follow the curvature of the earth.

line of soundings. A series of soundings obtained by a vessel underway, usually at regular intervals. In piloting, this information may be used to determine an estimated position, by recording the soundings at appropriate intervals (to the scale of the chart) along a line drawn on transparent paper or plastic, to represent the track, and then fitting the plot to the chart, by trial and error. A vessel obtaining soundings along a course line, for use in making or improving a chart, is said to *run a line of soundings*.

line of total force. The direction of a freely suspended magnetic needle when acted upon by the earth's magnetic field alone.

line squall. A squall that occurs along a squall line.

lipper, *n.* 1. Slight ruffling or roughness on a water surface.

2. Light spray from small waves.

liquid compass. A magnetic compass of which the bowl mounting the compass card is completely filled with liquid. Nearly all modern magnetic compasses are of this type. An older liquid compass using a solution of alcohol and water is sometimes called a SPIRIT COMPASS. Also called WET COMPASS. See also DRY COMPASS.

list, *n.* Inclination to one side, as of a vessel. The terms LIST and HEEL are often used synonymously, but LIST generally implies a state of equilibrium in an inclined condition, while HEEL may imply either a continuing or momentary inclination. The term ROLL refers to the oscillatory motion of a vessel rather than its inclined condition. A sudden roll to one side may be called a LURCH.

list, *v., t. & i.* To incline or be inclined to one side.

list of lights number. See LIGHT LIST NUMBER.

lithometeor, *n.* The general term for dry atmospheric suspensoids, including dust, haze, smoke, and sand. See also HYDROMETEOR.

little brother. A secondary tropical cyclone sometimes following a more severe disturbance.

littoral, *adj. & n.* 1. A coastal region.

2. The marine environment influenced by a land mass.

3. Of or pertaining to a shore, especially a seashore. See also SEABOARD.

load line marks. Markings stamped and painted amidships on the side of a vessel, to indicate the minimum permissible freeboard and, indirectly, the maximum draft in various waters. Also called PLIMSOLL MARKS. See also DRAFT MARKS.

lobe, *n.* 1. That portion of the overall radiation pattern of a directional antenna which is contained within a region bounded by adjacent minima. The **main beam** is the beam in the lobe containing the direction of maximum radiation (**main lobe**) lying within specified values of field strength relative to the maximum field strength. See also BACK LOBE, SIDE LOBE, BEAM WIDTH.

2. The radiation within the region of definition 1.

local apparent noon. Twelve o'clock local apparent time, or the instant the apparent sun is over the upper branch of the local meridian. Local apparent noon at the Greenwich meridian is called Greenwich apparent noon. Sometimes called HIGH NOON.

local apparent time. The arc of the celestial equator, or the angle at the celestial pole, between the lower branch of the local celestial meridian and the hour circle of the apparent or true sun, measured westward from the lower branch of the local celestial meridian through 24 hours; local hour angle of the apparent or true sun, expressed in time units, plus 12 hours. Local apparent time at the Greenwich meridian is called Greenwich apparent time.

local attraction. See LOCAL MAGNETIC DISTURBANCE.

local civil noon. *United States terminology from 1925 through 1952.* See LOCAL MEAN NOON.

local civil time. *United States terminology from 1925 through 1952.* See LOCAL MEAN TIME.

local hour angle. Angular distance west of the local celestial meridian; the arc of the celestial equator, or the angle at the celestial pole, between the upper branch of the local celestial meridian and the hour circle of a point on the celestial sphere, measured westward from the local celestial meridian through 360°. The local hour angle at longitude 0° is called **Greenwich hour angle.**

local knowledge. In some waters the services of a specially qualified navigator (pilot) having local knowledge may be necessary to insure safe navigation. Local knowledge extends beyond that publicly available in charts and publications, being more detailed, intimate, and current. The pilot's knowledge of his waters is gained not only through his own experience and familiarity, but by his availing himself of all local information resources, public and private, recent and longstanding, particularly concerning underwater hazards and obstructions, uncharted above-water landmarks and topographical configurations, local tides and currents, recent shoaling, temporary changes or deficiencies in aids to navigation, and similar matters of local concern. This local knowledge should enable a pilot to traverse his waters safely without reliance on man-made aids to navigation and to detect any unusual conditions or departure from a safe course. This service does not substitute for the ship's own safe navigation, but complements it. Prudence may also dictate the use of this specially qualified navigator to better insure safe navigation in situations where local knowledge is not essential.

local lunar time. The arc of the celestial equator, or the angle at the celestial pole, between the lower branch of the local celestial meridian and the hour circle of the moon, measured westward from the lower branch of the local celestial meridian through 24 hours; local hour angle of the moon, expressed in time units, plus 12 hours. Local lunar time at the Greenwich meridian is called **Greenwich lunar time.**

local magnetic disturbance. An anomaly of the magnetic field of the earth, extending over a relatively small area, due to local magnetic influences. Also called LOCAL ATTRACTION, MAGNETIC ANOMALY.

local mean noon. Twelve o'clock local mean time, or the instant the mean sun is over the upper branch of the local meridian. Local mean noon at the Greenwich meridian is called **Greenwich mean noon.**

local mean time. The arc of the celestial equator, or the angle at the celestial pole, between the lower branch of the local celestial meridian and the hour circle of the mean sun, measured westward from the lower branch of the local celestial meridian through 24 hours; local hour angle of the mean sun, expressed in time units, plus 12 hours. Local mean time at the Greenwich meridian is called **Greenwich mean time,** or **Universal Time.** Called LOCAL CIVIL TIME in United States terminology from 1925 through 1952.

local meridian. The meridian through any particular place of observer, serving as the reference for local time, in contrast with GREENWICH MERIDIAN.

local noon. Noon at the local meridian.

Local Notice to Mariners. A notice issued by each U.S. Coast Guard District to disseminate important information affecting navigational safety within the District. The *Notice* reports changes to and deficiencies in aids to navigation maintained by and under the authority of the U.S. Coast Guard. Other information includes channel depths, new charts, naval operations, regattas, etc. Since temporary information, known or expected to be of short duration, is not included in the weekly *Notice to Mariners* published by the Defense Mapping Agency Hydrographic/Topographic Center, the appropriate *Local Notice to Mariners* may be the only source of such information. Much of the information contained in the *Local Notice to Mariners* is included in the weekly *Notice to Mariners.* The *Local Notice to Mariners* is published as often as required; usually weekly. It may be obtained, free of charge, by making application to the appropriate Coast Guard District Commander. See also GREAT LAKES NOTICE TO MARINERS.

local oscillator. An oscillator used to drive an intermediate frequency by beating with the signal-carrying frequency in superheterodyne reception.

local sidereal noon. Zero hours local sidereal time, or the instant the vernal equinox is over the upper branch of the local meridian. Local sidereal noon at the Greenwich meridian is called **Greenwich sidereal noon.**

local sidereal time. Local hour angle of the vernal equinox, expressed in time units; the arc of the celestial equator, or the angle at the celestial pole, between the upper branch of the local celestial meridian and the hour circle of the vernal equinox, measured westward from the upper branch of the local celestial meridian through 24 hours. Local sidereal time at the Greenwich meridian is called **Greenwich sidereal time.**

local time. 1. Time based upon the local meridian as reference, as contrasted with that based upon a standard meridian. Local time was in general use in the United States until 1883, when standard time was adopted.

2. Any time kept locally.

local vertical. The direction of the acceleration of gravity as opposed to the normal to the reference ellipsoid. It is in the direction of the resultant of the gravitational and centrifugal accelerations of the earth at the location of the observer. Also called PLUMB-BOB VERTICAL. See also MASS ATTRACTION VERTICAL.

loch, *n.* 1. A lake.

2. An arm of the sea, especially when nearly landlocked (Scotland).

lock, *n.* A basin in a waterway with caissons or gates at each end by means of which vessels are passed from one water level to another without materially affecting the higher level. *To lock a vessel* means to pass a vessel through a lock.

lock on. A term signifying that a tracking or target seeking system is continuously and automatically tracking a target in one or more coordinates (e.g., range, bearing, elevation).

locus (*pl. loci*), *n.* All possible positions of a point or curve satisfying stated conditions.

log, *n.* 1. An instrument for measuring the speed or distance or both traveled by a vessel. A **chip log** consists essentially of a weighted wooden quadrant (quarter of a circle) attached to a bridle in such a manner that it will float in a vertical position, and a line with equally spaced knots, usually each 47 feet 3 inches, speed being measured by casting the quadrant overboard and counting the knots in the line paid out in unit time, usually 28 seconds. A **ground log** consists of a lead or weight attached to a line, course made good over the ground being indicated by the direction the line tends and the speed by the amount of line paid out in unit time. A mechanical means of determining speed or distance is called a **patent log.** A **harpoon log** consists essentially of a combined rotator and distance registering device towed through the water. This has been largely replaced by the **taffrail log,** a somewhat similar device but with the registering unit secured at the taffrail. A **Pitometer log** consists essentially of a Pitot tube projecting into the water, and suitable registering devices. An **electromagnetic log** consists essentially of suitable registering devices and an electromagnetic sensing element, extended below the hull of a vessel, which produces a voltage directly proportional to speed through the water. A **Forbes log** consists essentially of a small rotator in a tube projecting below the bottom of the vessel, and suitable registering devices. A **Dutchman's log** is a buoyant object thrown overboard, the speed of a vessel being determined by noting the time required for a known length of the vessel to pass the object. When a log is thrown overboard and secured in place, it is said to be *streamed.*

2. A written record of the movements of a craft, with regard to courses, speeds, positions, and other information of interest to navigators, and of important happenings aboard the craft. The book in which the log is kept is called a LOG BOOK. Also called DECK LOG. See also NIGHT ORDER BOOK.

3. A written record of specific related information, as that concerning performance of an instrument. See GYRO LOG.

logarithm, *n.* The power to which a fixed number, called the base, usually 10 or *e* (2.7182818), must be raised to produce the value to which the logarithm corresponds. A logarithm (base 10) consists of two parts: the **characteristic** is that part to the left of the decimal point and the **mantissa** is that part to the right of the decimal point. An ANTILOGARITHM or INVERSE LOGARITHM is the value corresponding to a given logarithm. Logarithms are used to multiply or divide numbers, the sum or difference of the logarithms of two numbers being the logarithm of the product or quotient, respectively, of the two numbers. A COLOGARITHM is the logarithm of the reciprocal of a number. Logarithms to the base 10 are called **common** or **Briggsian** and those to the base *e* are called **natural** or **Napierian** logarithms.

logarithmic, *adj.* The logarithm of; used with the name of a trigonometric function to indicate that the value given is the logarithm of that function, rather than the function itself which is called the **natural trigonometric function.**

logarithmic coordinate paper. Paper ruled with two sets of mutually-perpendicular, parallel lines spaced according to the logarithms of consecutive numbers, rather than the numbers themselves. On SEMILOGARITHMIC COORDINATE PAPER one set of lines is spaced logarithmically and the other set at uniform intervals.

logarithmic coversine. See under COVERSINE.

logarithmic haversine. See under HAVERSINE.

logarithmic scale. A scale graduated in the logarithms of uniformly-spaced consecutive numbers.

logarithmic tangent. See under TANGENT, definition 1.

logarithmic trigonometric function. See under TRIGONOMETRIC FUNCTIONS.

logarithmic versine. See under VERSINE.

log book. See LOG, definition 2.

log chip. The wooden quadrant forming part of a chip log. Also called LOG SHIP.

log glass. A small hour glass used to time a chip log. The period most frequently used is 28 seconds.

log line. 1. A graduated line used to measure the speed of a vessel through the water or to measure the speed of a current, the line may be called a CURRENT LINE.

2. The line secured to a log.

log ship. See LOG CHIP.

long flashing light. A single-flashing light used with the IALA System, the flash duration of which is 2 or more seconds.

longitude, *n.* Angular distance, along a primary great circle, from the adopted reference point; the angle between a reference plane through the polar axis and a second plane through that axis. **Terrestrial longitude** is the arc of a parallel, or the angle at the pole, between the prime meridian and the meridian of a point on the earth measured eastward or westward from the prime meridian through 180°, and labeled E or W to indicate the direction of measurement. **Astronomical longitude** is the angle between the plane of the prime meridian and the plane of the celestial meridian; **geodetic longitude** is the angle between the plane of the geodetic meridian at a station and the plane of the geodetic meridian at Greenwich. Geodetic and sometimes astronomical longitude are also called **geographic longitude.** Geodetic longitude is used in charting. **Assumed longitude** is the longitude at which an observer is assumed to be located for an observation or computation. **Observed longitude** is determined by one or more lines of position extending in a generally north-south direction. **Difference of longitude** is the smaller angle at the pole or the shorter arc of a parallel between the meridians of two places, expressed in angular measure. **Fictitious longitude** is the arc of the fictitious equator between the prime fictitious meridian and any given fictitious meridian. **Grid longitude** is angular distance between a prime grid meridian and any given grid meridian. **Oblique longitude** is angular distance between a prime oblique meridian and any given oblique meridian. **Transverse** or **inverse longitude** is angular distance between a prime transverse meridian and any given meridian. **Celestial longitude** is angular distance east of the vernal equinox, along the ecliptic.

longitude factor. The change in longitude along a celestial line of position per 1′ change in latitude. The change in latitude for a 1′ change in longitude is called LATITUDE FACTOR.

longitude line. A line of position extending in a generally north-south direction. Sometimes called OBSERVED LONGITUDE. See also LATITUDE LINE; COURSE LINE, definition 2; SPEED LINE.

longitude method. The establishing of a line of position from the observation of the latitude of a celestial body by assuming a latitude (or longitude), and calculating the longitude (or latitude) through which the line of position passes, and the azimuth. The line of position is drawn through the point thus found, perpendicular to the azimuth. See also ST. HILAIRE METHOD, SUMNER METHOD, HIGH ALTITUDE METHOD.

longitude of Greenwich at time of perigee. See RIGHT ASCENSION OF GREENWICH AT TIME OF PERIGEE.

longitude of pericenter. An orbital element that specifies the orientation of an orbit; it is a broken angle consisting of the angular distance in the ecliptic from the vernal equinox to the ascending node of the orbit plus the angular distance in the orbital plane from the ascending node to the pericenter, i.e. the sum of the longitude of the ascending node and the argument of pericenter.

longitude of the ascending node. 1. The angular distance in the ecliptic from the vernal equinox to the ascending node of the orbit. See also LONGITUDE OF PERICENTER, RIGHT ASCENSION OF THE ASCENDING NODE.

2. The angular distance always measured eastward in the plane of the celestial equator from Greenwich through 360°

longitude of the moon's nodes. The angular distance along the ecliptic of the moon's nodes from the vernal equinox; the nodes have a retrograde motion, and complete a cycle of 360° in approximately 19 years.

longitudinal axis. The fore-and-aft line through the center of gravity of a craft, around which it rolls.

longitudinal wave. A wave in which the vibration is in the direction of propagation, as in sound waves. This is in contrast with a TRANSVERSE WAVE, in which the vibration is perpendicular to the direction of propagation.

long path interference. See under MULTIPATH ERROR.

long period constituent. A tidal or tidal current constituent with a period that is independent of the rotation of the earth but which depends upon the orbital movement of the moon or of the earth. The principal lunar long period constituents have periods approximating the month and half-month, and the principal solar long period constituents have periods approximating the year and half-year.

long period perturbations. Periodic perturbations in the orbit of a planet or satellite which require more than one orbital period to execute one complete periodic variation.

long range systems. Those radionavigation systems providing positioning capability on the high seas well beyond the positioning capability of medium range systems. Loran-C is an example. See also SHORT RANGE SYSTEMS.

longshore current. A current paralleling the shore largely within the surf zone. It is caused by the excess water brought to the zone by the small net mass transport of wind waves. Longshore currents feed into rip currents.

look angles. The elevation and azimuth at which a particular satellite is predicted to be found at a specified time.

lookout station. A label on a nautical chart which indicates a tower surmounted by a small house from which a watch is kept regularly.

loom, *n.* The diffused glow observed from a light below the horizon, due to atmospheric scattering.

looming, *n.* 1. An apparent elevation of distant terrestrial objects by abnormal atmospheric refraction. Because of looming, objects below the horizon are sometimes visible. The opposite is SINKING.
2. The appearance indistinctly of an object during a period of low visibility.

loop antenna. An antenna that is a closed circuit in the form of a loop of wire lying in the same plane or of loops of wire lying in parallel planes.

loop of stationary wave. See under STATIONARY WAVE.

Loran, *n.* The general designation of one group of radionavigation systems by which a hyperbolic line of position is determined through measuring the difference in the times of reception of synchronized pulse signals from two fixed transmitters. Although the name *Loran* is derived from the words long range navigation, the term applies to only one group of radionavigation systems (e.g., Loran-A, Loran-C, Loran-D), not to other long range systems such as Omega.

Loran-A, *n.* A long range medium frequency (1850 to 1950 kHz) radionavigation system by which a hyperbolic line of position of medium accuracy is obtained by measuring the difference in the times of arrival of pulse signals radiated by a pair of synchronized transmitters (master station and slave station) which are separated by several hundred miles, usually 200 to 400 miles but sometimes as close together as 100 miles or as far apart as 700 miles. The time difference measurement is accomplished by using a specialized receiver to compare the leading edges of the envelopes of the received pulses. The time difference measurements from two station pairs yield two intersecting lines of position, i.e., a fix. Ranges of 600 to 800 nautical miles are obtainable when using the groundwave, dependent upon station power and propagation conditions. Ranges extending up to 1500 nautical miles are obtainable when using the skywave at night. The predictable accuracy ($2d_{rms}$) varies from 1 to 2 nautical miles when using the groundwave, and 6 to 7 nautical miles when using the skywave at extended range. *System operation in U.S. waters was terminated on 31 December 1980.* See also LORAN, HYPERBOLIC NAVIGATION, ACCURACY, RADIAL ERROR.

Loran-C, *n.* A long range, low frequency (90-110 kHz) radionavigation system by which a hyperbolic line of position of high accuracy is obtained by measuring the difference in the times of arrival of pulse signals radiated by a pair of synchronized transmitters (master station and secondary station) which are separated by several hundred miles. The time difference measurement is accomplished by using a specialized receiver to compare the leading edges of the envelopes of the received pulses (envelope match) and to compare the phases of the third cycle within each pulse (cycle match). The making of this phase comparison early in the pulse insures that the measurement is made before the arrival of corresponding skywaves. Precise control over the shape of the pulse insures reliable third cycle identification. If skywaves are to be used for time difference measurements, the receiver must be adjusted to change the phase comparison point from 30 microseconds into the pulse (beginnings of the third cycle) to an appropriate point beyond. The time difference measurements from two station pairs (master station and two secondary stations) yield two intersecting lines of position, i.e., a fix. When using the groundwave, ranges of 800 to 1200 nautical miles are obtainable, depending upon transmitter power, signal-to-noise ratio in the service area, receiver sensitivity, and losses over the signal path. The predictable accuracy ($2d_{rms}$) is 0.25 nautical mile or better when the propagation corrections include the additional secondary phase factor (ASF) corrections. The system provides lesser accuracy when used in the skywave mode. When the receiver is appropriately modified to enable time measurements with respect to a local time reference the system can be used in the ranging or Range-Range mode. See also PHASE CODING, SLIP, ENVELOPE TO CYCLE DISCREPANCY, LORAN, HYPERBOLIC NAVIGATION, ACCURACY, RADIAL ERROR.

Loran-C plotting chart. See under PLOTTING CHART.

Loran-C Reliability Diagram. One of a series of charts which depict the following data for the area covered: (1) for each station of the chain, predicted maximum usable groundwave signal limits for signal-to-noise ratios of 1:3 and 1:10, and (2) contours which indicate the regions within which positions can be fixed with repeatable accuracies of 500, 750, or 1500 feet or better on a 95 percent probability basis. See also COVERAGE DIAGRAM.

Loran-C Table. See PUB. 221. LORAN-C TABLE.

Loran-D, *n.* A lower power, short baseline, shorter range version of Loran-C designed for military tactical use.

Loran rate. See RATE, definition 2.

lorhumb line. A line along which the rates of change of the values of two families of hyperbolae are constants.

lost motion. Mechanical motion which is not transmitted to connected or related parts, due to loose fit. See also BACKLASH.

low, *n.* Short for *area of low pressure*. Since a low is, on a synoptic chart, always associated with cyclonic circulation, the term is used interchangeably with CYCLONE. See also HIGH.

low clouds. Types of clouds the mean level of which is between the surface and 6,500 feet. The principal clouds in this group are **stratocumulus, stratus,** and **nimbostratus.**

lower branch. That half of a meridian or celestial meridian from pole to pole which passes through the antipode or nadir of a place. See also UPPER BRANCH.

lower culmination. See LOWER TRANSIT.

lower high water. The lower of the two high waters of any tidal day.

lower high water interval. See under LUNITIDAL INTERVAL.

lower limb. That half of the outer edge of a celestial body having the least altitude, in contrast with the UPPER LIMB, that half having the greatest altitude.

lower low water. The lower of the two low waters of any tidal day.

lower low water datum. An approximation of mean lower low water that has been adopted as a standard reference for a limited area, and is retained for an indefinite period regardless of the fact that it may differ slightly from a better determination of mean lower low water from a subsequent series of observations. Used primarily for river and harbor engineering purposes. Columbia River lower low water datum is an example.

lower low water interval. See under LUNITIDAL INTERVAL.

lower transit. Transit of the lower branch of the celestial meridian. Transit of the upper branch is called UPPER TRANSIT. Also called INFERIOR TRANSIT, LOWER CULMINATION.

low frequency. Radio frequency of 30 to 300 kilohertz.

low light. See FRONT LIGHT.

low tide. See under LOW WATER.

low water. The minimum height reached by a falling tide. The height may be due solely to the periodic tidal forces or it may have superimposed upon it the effects of meteorological conditions. Use of the synonymous term LOW TIDE is discouraged.

low water datum. 1. The dynamic elevation for each of the Great Lakes, Lake St. Clair, and the corresponding sloping surfaces of the St. Marys, St. Clair, Detroit, Niagara, and St. Lawrence Rivers to which are referred the depths shown on the navigation charts and the authorized depths for navigation improvement projects. Elevations of these planes are referred to International Great Lakes Datum (1955) and are: Lake Superior - 600.0 feet, Lakes Michigan and Huron - 576.8 feet, Lake St. Clair - 571.7 feet, Lake Erie - 568.6 feet, and Lake Ontario - 242.8 feet.

2. An approximation of mean low water that has been adopted as a standard reference for a limited area and is retained for an indefinite period regardless of the fact that it may differ slightly from a better determination of mean low water from a subsequent series of observations. Used primarily for river and harbor engineering purposes. Boston Harbor low water datum is an example.

low water equinoctial springs. Low water springs near the times of the equinoxes. Expressed in terms of the harmonic constituents, it is an elevation depressed below mean sea level by an amount equal to the sum of the amplitudes of certain constituents as given in the *Tide and Current Glossary* published by the National Ocean Survey.

low water inequality. See under DIURNAL INEQUALITY.

low water interval. See under LUNITIDAL INTERVAL.

low water line. The intersection of the land with the water surface at an elevation of low water.

low water neaps. See under NEAP TIDES.

low water springs. Short for MEAN LOW WATER SPRINGS.

low water stand. The condition at low water when there is no sensible change in the height of the tide. A similar condition at high water is called HIGH WATER STAND. See also STAND.

loxodrome, *n.* See RHUMB LINE. See also ORTHODROME.

loxodromic curve. See RHUMB LINE.

lubber's line. A reference line on any direction-indicating instrument, marking the reading which coincides with the heading.

lubber's line error. The angular difference between the heading as indicated by a lubber's line, and the actual heading; the horizontal angle, at the center of an instrument, between a line through the lubber's line and one parallel to the keel.

lull, *n.* A momentary decrease in the speed of the wind.

lumen, *n.* The *derived unit* of luminous flux in the International System of Units; it is the luminous flux emitted within unit solid angle (1 steradian) by a point source having a uniform luminous intensity of 1 candela.

luminance, *n.* In a given direction, at a point on the surface of a source or receptor, or at a point on the path of a beam, the quotient of the luminous flux leaving, arriving at, or passing through an element of surface at this point and propagated in directions defined by an elementary cone containing the given directions, by the product of the solid angle of the cone and the area of the orthogonal projection of the element of surface on a plane perpendicular to the given direction. The derived unit of luminance in the International System of Units is the CANDELA PER SQUARE METRE.

luminescence, *n.* Emission of light other than incandescence. Emission as a result of and only during absorption of radiation from some other source is called FLUORESCENCE; continued emission after absorption of radiation has ceased is called PHOSPHORESCENCE.

luminous, *adj.* Emitting or reflecting light.

luminous flux. The quantity characteristic of radiant flux which expresses its capacity to produce a luminous sensation, evaluated according to the values of spectral luminous efficiency. Unless otherwise indicated, the luminous flux relates to photopic vision, and is connected with the radiant flux in accordance with the formula adopted in 1948 by the International Commission on Illumination. The derived unit of luminous flux in the International System of Units is the LUMEN.

luminous range. See under VISUAL RANGE (OF A LIGHT).

Luminous Range Diagram. A diagram used to convert the nominal range of a light to its luminous range under existing conditions. The ranges obtained are approximate.

lunar, *adj.* Of or pertaining to the moon.

lunar cycle. An ambiguous expression which has been applied to various cycles associated with the moon's motion, including CALLIPPIC CYCLE, METONIC CYCLE, NODE CYCLE, SYNODICAL MONTH or LUNATION.

lunar day. 1. The duration of one rotation of the earth on its axis, with respect to the moon. Its average length is about $24^h 50^m$ of mean solar time. Also called TIDAL DAY.

2. The duration of one rotation of the moon on its axis, with respect to the sun.

lunar distance. The angle, at an observer on the earth, between the moon and another celestial body. This was the basis of a method formerly used to determine longitude at sea.

lunar eclipse. An eclipse of the moon. When the moon enters the shadow of the earth, it appears eclipsed to an observer on the earth. A lunar eclipse is **penumbral** when it enters only the penumbra of the earth's shadow, **partial** when part of its surface enters the umbra of the earth's shadow, and **total** if its entire surface is obscured by the umbra.

lunar inequality. 1. Variation in the moon's motion in its orbit, due to attraction by other bodies of the solar system. See also EVECTION, PERTURBATIONS.

2. A minute fluctuation of a magnetic needle from its mean position, caused by the moon.

lunar interval. The difference in time between the transit of the moon over the Greenwich meridian and a local meridian. The lunar interval equals the difference between the Greenwich and local intervals of a tide or current phase.

lunar month. The period of revolution of the moon about the earth, especially a synodical month.

lunar node. A node of the moon's orbit. See also LINE OF NODES.

lunar noon. The instant at which the sun is over the upper branch of any meridian of the moon.

lunar parallax. Parallax of the moon.

lunar rainbow. See MOONBOW.

lunar tide. That part of the tide due solely to the tide-producing force of the moon. That part due to the tide-producing force of the sun is called SOLAR TIDE.

lunar time. Time based upon the rotation of the earth relative to the moon. Lunar time may be designated as **local** or **Greenwich** as the local or Greenwich meridian is used as the reference.

lunation, *n.* See SYNODICAL MONTH.

lune, *n.* That part of the surface of a sphere bounded by halves of two great circles.

lunicurrent interval. The interval between the moon's transit (upper or lower) over the local or Greenwich meridian and a specified phase of the tidal current following the transit. Examples are **strength of flood interval** and **strength of ebb interval**, which may be abbreviated to **flood interval** and **ebb interval**, respectively. The interval is described as local or Greenwich according to whether the reference is to the moon's transit over the local or Greenwich meridian. When not otherwise specified, the reference is assumed to be local. See also LUNITIDAL INTERVAL.

lunisolar effect. Gravitational effects caused by the attractions of the moon and of the sun.

lunisolar perturbation. Perturbations of the orbits of artificial earth satellites due to the attractions of the sun and the moon. The most important effects are secular variations in the mean anomaly, in the right ascension of the ascending node, and in the argument of perigee.

lunisolar precession. That component of general precession caused by the combined effect of the sun and moon on the equatorial protuberance of the earth, producing a westward motion of the equinoxes along the ecliptic. See also PRECESSION OF THE EQUINOXES.

lunitidal interval. The interval between the moon's transit (upper or lower) over the local or Greenwich meridian and the following high or low water. The average of all high water intervals for all phases of the moon is known as **mean high water lunitidal interval** and is abbreviated to **high water interval.** Similarly the **mean low water lunitidal interval** is abbreviated to **low water interval.** The interval is described as local or Greenwich according to whether the reference is to the transit over the local or Greenwich meridian. When not otherwise specified, the reference is assumed to be local. When there is considerable diurnal inequality in the tide, separate intervals may be obtained for the higher high waters, the lower high waters, the higher low waters and the lower low waters. These are designated respectively as **higher high water interval, lower high water interval, higher low water interval,** and **lower low water interval.** In such cases, and also when the tide is diurnal, it is necessary to distinguish between the upper and lower transit of the moon with reference to its declination.

lurch, *n.* A sudden roll to one side. See also LIST, *n.*

lurch, *v., i.* To roll suddenly to one side.

lux, *n.* The *derived unit* of illuminance in the International System of Units; it is equal to 1 lumen per square metre.

M

mackerel sky. An area of sky with a formation of rounded and isolated cirrocumulus or altocumulus resembling the pattern of scales on the back of a mackerel.

macroscopic, *adj.* Large enough to be seen by the unaided eye.

madrepore, *n.* A branching or stag-horn coral, or any perforated stone coral.

maelstrom, *n.* A whirlpool similar to the Maelstrom off the west coast of Norway.

maestro, *n.* A northwesterly wind with fine weather which blows, especially in summer, in the Adriatic. It is most frequent on the western shore. This wind is also found on the coasts of Corsica and Sardinia.

magnet, *n.* A body which produces a magnetic field around itself. It has the property of attracting certain materials capable of being magnetized. A magnet occurring in nature is called a **natural magnet** in contrast with a man-made **artificial magnet.** See also HEELING MAGNET, KEEPER.

magnetic, *adj.* Of or pertaining to a magnet or related to magnetic north.

magnetic amplitude. Amplitude relative to magnetic east or west.

magnetic annual change. The amount of secular change in the earth's magnetic field which occurs in 1 year.

magnetic annual variation. The small systematic temporal variation in the earth's magnetic field which occurs after the trend for secular change has been removed from the average monthly values.

magnetic anomaly. See LOCAL MAGNETIC DISTURBANCE.

magnetic azimuth. Azimuth relative to magnetic north.

magnetic bay. A small magnetic disturbance whose magnetograph resembles an indentation of a coastline. On earth, magnetic bays occur mainly in the polar regions and have duration of a few hours.

magnetic bearing. Bearing relative to magnetic north; compass bearing corrected for deviation.

magnetic chart. A chart showing magnetic information. If it shows lines of equality in one or more magnetic elements, it may be called an **isomagnetic chart.** It is an **isoclinal** or **isoclinic chart** if it shows lines of equal magnetic dip, an **isodynamic chart** if it shows lines of equal magnetic intensity, an **isogonic chart** if it shows lines of equal magnetic variation, an **isogriv chart** if it shows lines of equal grid variation, an **isoporic chart** if it shows lines of equal rate or change of a magnetic element.

magnetic circle. A sphere of specified radius about the magnetic compass location to be kept free of any magnetic or electrical equipment which would interfere with the compass.

magnetic compass. A compass depending for its directive force upon the attraction of the horizontal component of the earth's magnetic field for a magnetized needle or sensing element free to turn with a minimum of friction in any horizontal direction.

magnetic compass table. See DEVIATION TABLE.

magnetic course. Course relative to magnetic north; compass course corrected for deviation.

magnetic daily variation. See MAGNETIC DIURNAL VARIATION.

magnetic declination. See VARIATION, definition 1.

magnetic deviation. See DEVIATION, definition 1.

magnetic dip. Angular distance between the horizontal and the direction of a line of force of the earth's magnetic field at any point. Also called DIP, MAGNETIC INCLINATION.

magnetic dip pole. See MAGNETIC POLE, definition 1.

magnetic direction. Horizontal direction expressed as angular distance from magnetic north.

magnetic diurnal variation. Oscillations of the earth's magnetic field which have a periodicity of about a day and which depend to a close approximation only on local time and geographic latitude. Also called MAGNETIC DAILY VARIATION.

magnetic element. 1. Variation, dip, or magnetic intensity.

2. That part of an instrument producing or influenced by magnetism.

magnetic equator. That line on the surface of the earth connecting all points at which the magnetic dip is zero. Also called ACLINIC LINE. See also GEOMAGNETIC EQUATOR.

magnetic field. Any space or region in which magnetic forces are present, as in the earth's magnetic field, or in or about a magnet, or in or about an electric current. See also MAGNETIC VECTOR.

magnetic force. The strength of a magnetic field. Also called MAGNETIC INTENSITY.

magnetic heading. Heading relative to magnetic north; compass heading corrected for deviation.

magnetic inclination. See MAGNETIC DIP.

magnetic induction. The act or process by which material becomes magnetized when placed in a magnetic field.

magnetic intensity. The strength of a magnetic field. Also called MAGNETIC FORCE.

magnetic latitude. Angular distance north or south of the magnetic equator. The angle is equal to an angle, the tangent of which is equal to half the tangent of the magnetic dip at the point.

magnetic lines of force. Closed lines indicating by their direction the direction of magnetic influence.

magnetic meridian. A line of horizontal magnetic force of the earth. A compass needle without deviation lies in the magnetic meridian.

magnetic moment. The quantity obtained by multiplying the distance between two magnetic poles by the average strength of the poles.

magnetic needle. A small, slender, magnetized bar which tends to align itself with magnetic lines of force.

magnetic north. The direction indicated by the north seeking pole of a freely suspended magnetic needle, influenced only by the earth's magnetic field.

magnetic observation. Measurement of any of the magnetic elements.

magnetic parallel. An isoclinal; a line connecting points of equal magnetic dip.

magnetic pole. 1. Either of the two places on the surface of the earth where the magnetic dip is 90°, that in the Northern Hemisphere (at lat. 76°.1N, long. 100°.0W in 1975) being designated **north magnetic pole,** and that in the Southern

Hemisphere (at lat. 65°8S., long. 139°4E in 1975) being designated south magnetic pole. Also called MAGNETIC DIP POLE. See also MAGNETIC LATITUDE, GEOMAGNETIC POLE, MAGNETIC LATITUDE.
2. Either of those two points of a magnet where the magnetic force is greatest.

magnetic prime vertical. The vertical circle through the magnetic east and west points of the horizon.

magnetic range. A range oriented in a given magnetic direction and used to assist in the determination of the deviation of a magnetic compass.

magnetic retentivity. The ability to retain magnetism after removal of the magnetizing force.

magnetic secular change. The gradual variation in the value of a magnetic element which occurs over a period of years.

magnetic storm. A disturbance in the earth's magnetic field, associated with abnormal solar activity, and capable of seriously affecting both radio and wire transmission.

magnetic temporal variation. Any change in the earth's magnetic field which is a function of time.

magnetic track. The direction of the track relative to magnetic north.

magnetic variation. See VARIATION, definition 1.

magnetic vector. The component of the electromagnetic field associated with electromagnetic radiation which is of the nature of a magnetic field. The magnetic vector is considered to coexist with, but to act at right angles to, the electric vector.

magnetism, *n.* The phenomena associated with magnetic fields and their effects upon magnetic materials, notably iron and steel. The magnetism of the north-seeking end of a freely suspended magnet is called **red magnetism**; the magnetism of the south-seeking end is called **blue magnetism**. Magnetism acquired by a piece of magnetic material while it is in a magnetic field is called **induced magnetism**. **Permanent magnetism** is retained for long periods without appreciable reduction, unless the magnet is subjected to a demagnetizing force. The magnetism in the intermediate iron of a ship which tends to change as the result of vibration, aging, or cruising in the same direction for a long period but does not alter immediately so as to be properly termed induced magnetism is called **subpermanent magnetism**. Magnetism which remains after removal of the magnetizing force may be called **residual magnetism**. The magnetism of the earth is called **terrestrial magnetism** or **geomagnetism**.

magnetize, *v., t.* To produce magnetic properties. The opposite is DEMAGNETIZE.

magnetometer, *n.* An instrument for measuring the intensity and direction of the earth's magnetic field. See also DECLINOMETER.

magnetron, *n.* An electron tube characterized by the interaction of electrons with the electric field of circuit element in crossed steady electric and magnetic fields to produce an alternating current power output. It is used to generate high power output in the ultra-high and super-high frequency bands.

magnification, *n.* The apparent enlargement of anything.

magnifying power. The ratio of the apparent length of a linear dimension as seen through an optical instrument to that seen by the unaided eye. Thus, an instrument with a magnifying power of 3 makes an object appear three times as high and three times as wide. Sometimes shortened to POWER.

magnitude, *n.* 1. Relative brightness of a celestial body. The smaller (algebraically) the number indicating magnitude, the brighter the body. The expression *first magnitude* is often used somewhat loosely to refer to all bodies of magnitude 1.5 or brighter, including negative magnitudes.
2. Amount; size; greatness.

magnitude ratio. The ratio of relative brightness of two celestial bodies differing in magnitude by 1.0. This ratio is 2.512 . . ., the 5th root of 100. A body of magnitude 1.0 is 2.512 times as bright as a body of magnitude 2.0, 2.512^2 or 6.310 times as bright as a body of magnitude 3.0, 2.512^3 or 15.849 times as bright as a body of magnitude 4.0, 2.512^4 or 39.811 times as bright as a body of magnitude 5.0, 2.512^5 or 100.000 times as bright as a body of magnitude 6.0 (the faintest star that can be seen with the unaided eye), etc.

main beam. See under LOBE.

mainland, *n.* The principal portion of a large land area. The term is used loosely to contrast a principal land mass from outlying islands and sometimes peninsulas.

main light. The principal light of two or more lights situated on the same support or neighboring supports.

main lobe. The lobe of the radiation pattern of a directional antenna which contains the direction of maximum radiation.

major axis. The longest diameter of an ellipse or ellipsoid.

major datum. See PREFERRED DATUM.

major light. A light of high intensity and reliability exhibited from a fixed structure or on a marine site (except range lights). Major lights include primary seacoast lights and secondary lights. See also MINOR LIGHT.

major planets. See under PLANET.

make the land. To sight and approach or reach land from seaward.

make way. To progress through the water.

making way. Progressing through the water. See also UNDERWAY.

Malvin Current. See FALKLAND CURRENT.

mamma, *n.* Hanging protuberances, like pouches on the under surface of a cloud. This supplementary cloud feature occurs mostly with cirrus, cirrocumulus, altocumulus, altostratus, stratocumulus, and cumulonimbus; in the case of cumulonimbus, mamma generally appear on the under side of the anvil. Previously called MAMMATUS.

mammatus, *n.* See MAMMA.

maneuvering board. A polar coordinate plotting sheet devised to facilitate solution of problems involving relative movement.

Maneuvering Board Manual. See PUB. NO. 217.

man-made noise. In radio reception, noise due entirely to unwanted transmissions from electrical or electronic apparatus, which has been insufficiently suppressed.

manned light. A light which is operated and kept in service by a keeper able to intervene at once in case of need.

mantissa, *n.* That part of a logarithm (base 10) to the right of the decimal point. That part of a logarithm (base 10) to the left of the decimal point is called the CHARACTERISTIC.

manual, *adj.* By hand, in contrast with AUTOMATIC.

manual, *n.* A book of explanation, as a textbook; a book giving instruction on the upkeep or use of an instrument; or a reference book on a certain subject.

manual radio direction finder. A radio direction finder which requires manual operation in contrast with an AUTOMATIC DIRECTION FINDER.

map, *n.* A representation, usually on a plane surface, of all or part of the surface of the earth, celestial sphere, or other area; showing relative size and position, according to a given projection, of the physical features represented and such other information as may be applicable to the purpose intended. Such a representation intended primarily for navigational use is called a **chart.** A method of representing all or part of the surface of a sphere or spheroid, such as the earth, upon a plane surface is called a **map projection.** A **planimetric map** indicates only the horizontal positions of features; a **topographic map** both horizontal and vertical positions. A topographic map showing relief by means of contour lines drawn at regular height intervals is called a **contour map.** A **relief map** emphasizes relative elevations or relief; a three-dimensional relief map is called a **relief model.** The pattern on the underside of extensive cloud areas, created by the varying amounts of light reflected from the earth's surface, is called a **sky map.** A chart which shows the distribution of meteorological conditions over an area at a given moment may be called a **weather map.**

map accuracy standards. See UNITED STATES NATIONAL MAP ACCURACY STANDARDS.

map chart. See COMBAT CHART.

mapping, charting and geodesy. The collection, transformation, generation, dissemination, and storing of geodetic, geomagnetic, gravimetric, aeronautical, topographic, hydrographic, cultural, and toponymic data. These data may be used for military planning, training, and operations including aeronautical, nautical, and land navigation, as well as for weapon orientation and target positioning. Mapping, charting and geodesy (MC&G) also includes the evaluation of topographic, hydrographic, or aeronautical features for their effect on military operations or intelligence. The data may be presented in the form of topographic, planimetric, relief, or thematic maps and graphics; nautical and aeronautical charts and publications; and in simulated, photographic, digital, or computerized formats.

map projection. A systematic drawing of lines on a plane surface to represent the parallels of latitude and the meridians of longitude of the earth or a section of the earth. A map projection may be established by analytical computation or may be constructed geometrically. A map projection is frequently referred to as a *projection*, but the complete term should be used unless the context clearly indicates the meaning.

map symbol. A character, letter, or similar graphic representation used on a map to indicate some object, characteristic, etc. May be called a CHART SYMBOL when applied to a chart.

March equinox. See VERNAL EQUINOX.

mare's tails. Long slender streaks of cirrus thickening into cirrostratus, and then gradually lowering into watery altostratus.

marigram, *n.* A graphic record of the rise and fall of the tide. The record is in the form of a curve, in which time is generally represented on the abscissa and the height of the tide on the ordinate.

marina, *n.* A harbor facility for small boats, yachts, etc., where supplies, repairs, and various services are available.

marine, *adj.* Of or pertaining to the sea. See also NAUTICAL.

marine chart. See NAUTICAL CHART.

marine climate. The type of climate characteristic of coastal areas, islands, and the oceans, the distinctive features of which are small annual and daily temperature range and high relative humidity in contrast with CONTINENTAL CLIMATE, which is characteristic of the interior of a large landmass, and the distinctive features of which are large annual and daily temperature range and dry air with few clouds.

marine distance meter. *British terminology.* See STADIMETER.

marine light. A luminous or lighted aid to navigation intended primarily for marine navigation. One intended primarily for air navigation is called an AERONAUTICAL LIGHT.

marine parade. See MARINE REGATTA.

marine radiobeacon. A radiobeacon whose service is intended primarily for the benefit of ships.

marine radionavigation-satellite service. A radionavigation-satellite service in which mobile earth stations are located on board vessels.

marine railway. A track, cradle, and winding mechanism for hauling vessels out of the water, so that the bottom can be exposed, as in a dry dock. Called a PATENT SLIP in British terminology.

marine regatta. An organized water event of limited duration which is conducted according to a prearranged schedule and in which public interest is manifested. Also called MARINE PARADE.

marine sanctuary. An area established under provisions of the Marine Protection, Research, and Sanctuaries Act of 1972, Public Law 92-532 (86 Stat. 1052), for the preservation and restoration of its conservation, recreational, ecological, or esthetic values. Such an area may lie in ocean waters as far seaward as the outer edge of the (continental) shelf, in coastal waters where the tide ebbs and flows, or in the Great Lakes and connecting waters, and, may be classified as a habitat, species, research, recreational and esthetic, or unique area.

marine sextant. A sextant designed primarily for marine navigation. On a **clamp screw sextant** the position of the tangent screw is controlled by a clamp screw; on an **endless tangent screw sextant** the position of the index arm and the vernier or micrometer drum is controlled by an endless tangent screw. A **vernier sextant** provides a precise reading by means of a vernier used directly with the arc, and may have either a clamp screw or an endless tangent screw for controlling the position of the tangent screw or the index arm. A **micrometer drum sextant** provides a precise reading by means of a micrometer drum attached to the index arm, and has an endless tangent screw for controlling the position of the index arm. A marine sextant is generally used with the visible horizon as the horizontal reference. See also SEXTANT.

marine site. The site of a fixed structure in the water. Marine sites for aids to navigation are restricted to shallow depths.

maritime, *adj.* Bordering on, concerned with, or related to the sea. See also NAUTICAL.

maritime polar air. See under AIR-MASS CLASSIFICATION.

maritime position. The location of a seaport or other points along a coast.

maritime radionavigation-satellite service. As defined by the *International Telecommunication Union* (ITU), a radionavigation-satellite service in which mobile earth stations are located on board ships.

maritime tropical air. See under AIR-MASS CLASSIFICATION.

mark, *n.* 1. An artificial or natural object of easily recognizable shape or color, or both, situated in such a position that it may be identified on a chart or related to a known navigational instruction (e.g., light list, sailing directions, etc.). A fixed artificial navigation mark is often called a BEACON. This may be lighted or unlighted. The French term BALISE is generally used to describe an unlighted beacon. In French the term AMER REMARQUABLE is applied to a navigation mark which is particularly easy to be seen, by virture of its form, size or color. The term includes buoys. Also called NAVIGATION MARK; SEAMARK. See also CLEARING MARKS.
2. A major design of an instrument. The first model is designated Mark 1 and each major change increases the number by one. Minor changes are designated MODIFICATIONS. Thus, "Mark 5 Mod (modification) 2" indicates the second minor change of the fifth major design.
3. One of the bits of leather, cloth, etc., indicating a specified length of a lead line. In reporting a depth indicated by such a mark the leadsman calls out "By the mark . . . . . ." The unmarked fathom points are called DEEPS.
4. An indication intended as a datum or reference, as a *bench mark*.

mark, *v., i.* "Now" or "at this moment." A call used when simultaneous observations are being made, to indicate to the second person the moment a reading is to be made, as when the time of a celestial observation is to be noted; or the moment a reading is a prescribed value, as when the heading of a vessel is exactly a desired value.

marker beacon. 1. See MARKER RADIOBEACON.
2. As defined by the *International Telecommunication Union* (ITU), a transmitter in the aeronautical radionavigation service which radiates vertically a distinctive pattern for providing position information to aircraft.

marker buoy. A small, brightly painted moored float used to temporarily mark a location on the water while placing a buoy on station.

marker radiobeacon. A low powered radiobeacon used primarily to mark a specific location such as the end of a jetty. Usually used for homing bearings rather than to furnish a point of reference from which a cross bearing may be obtained. Also called MARKER BEACON.

marl, *n.* A crumbling, earthy deposit, particularly one of clay mixed with sand, lime, decomposed shells, etc. Sometimes a layer of marl becomes quite compact. Part of the ocean bed is composed of marl.

Mars, *n.* The navigational planet whose orbit lies between the orbits of the Earth and Jupiter.

marsh, *n.* An area of soft wet land. Flat land periodically flooded by salt water is called a **salt marsh.** Sometimes called SLOUGH.

mascaret, *n.* See TIDAL BORE.

mass, *n.* The mass of a body is a measure of its inertial property. This term should not be confused with WEIGHT.

mass attraction vertical. The normal to any surface of constant geopotential. On the earth this vertical is a function only of the distribution of mass and is unaffected by forces resulting from the motions of the earth, e.g., the direction of a plumb bob on a nonrotating earth.

master, *n.* Short for MASTER STATION.

master compass. That part of a remote-indicating compass system which determines direction for transmission to various repeaters.

master gyrocompass. See under GYROCOMPASS.

master station. In a radionavigation system, the station of a chain which provides a reference by which the emissions of other (slave or secondary) stations are controlled.

Matanuska wind. A strong, gusty, northeast wind which occasionally occurs during the winter in the vicinity of Palmer, Alaska.

maximum ebb. See under EBB CURRENT.

maximum flood. See under FLOOD CURRENT.

maximum thermometer. A thermometer which automatically registers the highest temperature occurring since its last setting. One which registers the lowest temperature is called a MINIMUM THERMOMETER.

mean, *adj.* Occupying a middle position.

mean, *n.* The average of a number of quantities, obtained by adding the values and dividing the sum by the number of quantities involved. Also called AVERAGE, ARITHMETIC MEAN. See also MEDIAN.

mean anomaly. See under ANOMALY, definition 2.

mean diurnal high water inequality. See under DIURNAL INEQUALITY.

mean diurnal low water inequality. See under DIURNAL INEQUALITY.

mean elements. Elements of an adopted reference orbit that approximates the actual, perturbed orbit. Mean elements serve as the basis for calculating perturbations. See also ORBITAL ELEMENTS.

mean higher high water. A tidal datum that is the average of the highest high water height of each tidal day observed over the National Tidal Datum Epoch. For stations with shorter series, simultaneous observational comparisons are made with a control tide station in order to derive the equivalent of a 19-year datum. See also HIGH WATER.

mean higher high water line. The intersection of the land with the water surface at the elevation of mean higher high water.

mean high tide. See under MEAN HIGH WATER.

mean high water. A tidal datum that is the average of all the high water heights observed over the National Tidal Datum Epoch. For stations with shorter series, simultaneous observational comparisons are made with a control tide station in order to derive the equivalent of a 19-year datum. See also HIGH WATER.

mean high water line. The intersection of the land with the water surface at the elevation of mean high water. See also SHORELINE.

mean high water lunitidal interval. See under LUNITIDAL INTERVAL.

mean high water neaps. See as NEAP HIGH WATER or HIGH WATER NEAPS under NEAP TIDES.

mean high water springs. See under SPRING TIDES.

mean ice edge. The average position of the ice edge in any given month or period based on observations over a number of years. Other terms which may be used are *mean maximum ice edge* and *mean minimum ice edge*. See also ICE LIMIT.

mean latitude. Half the arithmetical sum of the latitudes of two places on the same side of the equator. Mean latitude is labeled N or S to indicate whether it is north or south of the equator. The expression is occasionally used with reference to two places on opposite sides of the equator, but this usage is misleading as it lacks the significance usually associated with the expression. When the places are on opposite sides of the equator, two mean latitudes are generally used, the mean of each latitude north and south of the equator. The mean latitude is usually used in middle-latitude sailing for want of a practicable means of determining the middle latitude. See also MIDDLE LATITUDE, MIDDLE-LATITUDE SAILING.

mean lower low water. A tidal datum that is the average of the lowest low water height of each tidal day observed over the National Tidal Datum Epoch. For stations with shorter series, simultaneous observational comparisons are made with a control tide station in order to derive the equivalent of a 19-year datum. See also LOW WATER.

mean lower low water line. The intersection of the land with the water surface at the elevation of mean lower low water.

mean low water. A tidal datum that is the average of all the low water heights observed over the National Tidal Datum Epoch. For stations with shorter series, simultaneous observational comparisons are made with a control tide station in order to derive the equivalent of a 19-year datum. See also LOW WATER.

mean low water line. The intersection of the land with the water surface at the elevation of mean low water.

mean low water lunitidal interval. See under LUNITIDAL INTERVAL.

mean low water neaps. See as NEAP LOW WATER or LOW WATER NEAPS under NEAP TIDES.

mean low water springs. 1. A tidal datum that is the arithmetic mean of the low water heights occurring at the time of the spring tides observed over a specific 19-year Metonic cycle (the National Tidal Datum Epoch). It is usually derived by taking an elevation depressed below the half-tide level by an amount equal to one-half the spring range of tide, necessary corrections being applied to reduce the result to a mean value. This datum is used, to a considerable extent, for hydrographic work outside of the United States and is the level of reference for the Pacific approaches to the Panama Canal. Often shortened to SPRING LOW WATER. See also DATUM. 2. See under SPRING TIDES.

mean motion. In undisturbed elliptic motion, the constant angular speed required for a body of a specified mass to complete one revolution in an orbit of a specified semimajor axis.

mean noon. Twelve o'clock mean time, or the instant the mean sun is over the upper branch of the meridian. Mean noon may be either local or Greenwich depending upon the reference meridian. Zone, standard, daylight saving or summer noon are also forms of mean noon, the mean sun being over the upper branch of the zone, standard, daylight saving or summer reference meridian, respectively.

mean power. See under POWER (OF A RADIO TRANSMITTER).

mean range. The average difference in the extreme values of a variable quantity, as the *mean range of tide*.

mean range of tide. The difference in height between mean high water and mean low water.

mean rise interval. The average interval between the meridian transit of the moon and the middle of the period of the rise of the tide. It may be computed by adding the half of the duration of rise to the mean low water interval, rejecting the semidiurnal tidal period of 12.42 hours when greater than this amount. The mean rise interval may be either local or Greenwich according to whether it is referred to the local or Greenwich meridian.

mean rise of tide. The height of mean high water above the reference or chart sounding datum.

mean river level. A tidal datum that is the average height of the surface of a tidal river at any point for all stages of the tide observed over a 19-year Metonic cycle (the National Tidal Datum Epoch), usually determined from hourly height readings. In rivers subject to occasional freshets, the river level may undergo wide variations, and for practical purposes certain months of the year may be excluded in the determination of tidal datums. For charting purposes, tidal datums for rivers are usually based on observations during selected periods when the river is at or near low water stage. See also DATUM.

mean sea level. A tidal datum that is the arithmetic mean of hourly water elevations observed over a specific 19-year Metonic cycle (the National Tidal Datum Epoch). Shorter series are specified in the name; e.g., **monthly mean sea level** and **yearly mean sea level.** See also DATUM; EPOCH, definition 2.

mean sidereal time. See under SIDEREAL TIME.

mean solar day. The duration of one rotation of the earth on its axis, with respect to the mean sun. The length of the mean solar day is 24 hours of mean solar time or $24^h\ 03^m\ 56.555^s$ of mean sidereal time. A mean solar day beginning at midnight is called a **civil day.** See also CALENDAR DAY.

mean solar time. See MEAN TIME, the term usually used.

mean sun. A fictitious sun conceived to move eastward along the celestial equator at a rate that provides a uniform measure of time equal to the average apparent time. It is used as a reference for reckoning mean time, zone time, etc. Also called ASTRONOMICAL MEAN SUN. See also DYNAMICAL MEAN SUN.

mean tide level. See HALF-TIDE LEVEL.

mean time. Time based upon the rotation of the earth relative to the mean sun. Mean time may be designated as local or Greenwich as the local or Greenwich meridian is the reference. Greenwich mean time is also called UNIVERSAL TIME. **Zone, standard, daylight saving or summer** time are also variations of mean time, specified meridians being used as the reference. Called CIVIL TIME in United States terminology from 1925 through 1952. See also EQUATION OF TIME, MEAN SIDEREAL TIME.

mean tropic range. The mean between the great tropic range and the small tropic range. The small tropic range and the mean tropic range are applicable only when the type of tide is semidiurnal or mixed. See also GREAT TROPIC RANGE.

mean water level. The mean surface elevation as determined by averaging the heights of the water at equal intervals of time, usually hourly.

mean water level line. The line formed by the intersection of the land with the water surface at an elevation of mean water level.

measured mile. A length of 1 nautical mile the limits of which have been accurately measured and are indicated by ranges ashore. It is used by vessels to calibrate logs, engine revolution counters, etc., and to determine speed.

measured-mile buoy. A buoy marking the end of a measured mile.

mechanical scanning. Scanning effected by moving all or part of the antenna.

median, *n.* A value in a group of quantities below and above which fall an equal number of quantities. Of the group 60, 75, 80, 95, and 100, the median is 80. If there is no middle quantity in the group, the median is the value interpolated between the two middle quantities. The median of the group 6, 10, 20, and 31 is 15. See also MEAN.

median valley. The axial depression of the mid-oceanic ridge system.

medium first-year ice. First-year ice 70 to 120 centimeters thick.

medium floe. See under FLOE.

medium fracture. See under FRACTURE.

medium frequency. Radio frequency of 300 to 3,000 kilohertz.

medium iceberg. For reports to the International Ice Patrol, an iceberg that extends 51 to 150 feet (16 to 45 meters) above the sea surface and which has a length of 201 to 400 feet (61 to 122 meters). See also SMALL ICEBERG, LARGE ICEBERG.

medium ice field. See under ICE FIELD.

medium range systems. Those radionavigation systems providing positioning capability beyond the range of short range systems, but their use is generally limited to ranges permitting reliable positioning for about 1 day prior to making landfall; Decca is an example.

mega-. A prefix meaning *one million* (10^6).

megacycle, *n.* One million cycles; one thousand kilocycles. The term is often used as the equivalent of *one million cycles per second.*

megahertz, *n.* One million hertz or one million cycles per second.

megaripple, *n.* See SANDWAVE.

meniscus, *n.* 1. The curved upper surface of a liquid in a tube.

2. A type of lens.

mensuration, *n.* 1. The act, process, or art of measuring.

2. That branch of mathematics dealing with determination of length, area, or volume.

Mentor Current. Originating mainly from the easternmost extension of the South Pacific Current at about latitude 40° S, longitude 90°W, the Mentor Current flows first northward and then northwestward. It has the characteristic features of a WIND DRIFT in that it is a broad, slow-moving flow that extends about 900 miles westward from the Peru Current to about longitude 90°W at its widest section and tends to be easily influenced by winds. It joins the westward flowing Pacific South Equatorial Current and forms the eastern part of the general counterclockwise oceanic circulation of the South Pacific Ocean. The speed in the central part of the current at about latitude 26° S, longitude 80°W, may at times reach about 0.9 knot. Also called PERU OCEANIC CURRENT.

Mercator bearing. See RHUMB BEARING.

Mercator chart. A chart on the Mercator projection. This is the chart commonly used for marine navigation. Also called EQUATORIAL CYLINDRICAL ORTHOMORPHIC CHART.

Mercator course. See RHUMB-LINE COURSE.

Mercator direction. Horizontal direction of a rhumb line, expressed as angular distance from a reference direction. Also called RHUMB DIRECTION.

Mercator map projection. A conformal cylindrical map projection in which the surface of a sphere or spheroid, such as the earth, is conceived as developed on a cylinder tangent along the equator. Meridians appear as equally spaced vertical lines and parallels as horizontal lines drawn farther apart as the latitude increases, such that the correct relationship between latitude and longitude scales at any point is maintained. The expansion at any point is equal to the secant of the latitude of that point, with a small correction for the ellipticity of the earth. The Mercator is not a perspective projection. Since rhumb lines appear as straight lines and directions can be measured directly, this projection is widely used in navigation. If the cylinder is tangent along a meridian, a **transverse Mercator map projection** results; if the cylinder is tangent along an oblique great circle, an **oblique Mercator map projection** results. Also called EQUATORIAL CYLINDRICAL ORTHOMORPHIC MAP PROJECTION.

Mercator sailing. A method of solving the various problems involving course, distance, difference of latitude, difference of longitude, and departure by considering them in the relation in which they are plotted on a Mercator chart. It is similar to **plane sailing**, but uses meridional difference and difference of longitude in place of difference of latitude and departure, respectively.

mercurial barometer. An instrument which determines atmospheric pressure by measuring the height of a column of mercury which the atmosphere will support. See also ANEROID BAROMETER.

mercury ballistic. A system of reservoirs and connecting tubes containing mercury used with a type of non-pendulous gyrocompass. The action of gravity on this system provides the torques and resultant precessions required to convert the gyroscope into a compass.

meridian, *n.* A north-south reference line, particularly a great circle through the geographical poles of the earth. The term usually refers to the **upper branch**, the half, from pole to pole, which passes through a given place; the other half being called the **lower branch**. An **astronomical (terrestrial) meridian** is a line connecting points having the same astronomical longitude. A **geodetic meridian** is a line connecting points of equal geodetic longitude. Geodetic and sometimes astronomical meridians are also called **geographic meridians**. Geodetic meridians are shown on charts. The **prime meridian** passes through longitude 0°. Sometimes designated TRUE MERIDIAN to distinguish it from **magnetic meridian, compass meridian,** or **grid meridian,** the north-south lines relative to magnetic, compass, or grid direction, respectively. A **fictitious meridian** is one of a series of great circles or lines used in place of a meridian for certain purposes. A **transverse** or **inverse meridian** is a great circle perpendicular to a transverse equator. An **oblique meridian** is a great circle perpendicular to an oblique equator. Any meridian used as a reference for reckoning time is called a **time meridian**. The meridian used for reckoning standard, zone, daylight saving, or war time is called **standard, zone, daylight saving,** or **war meridian,** respectively. The meridian through any particular place or observer, serving as the reference for local time, is called **local meridian**, in contrast with the **Greenwich meridian**, the reference for Greenwich time. A **celestial meridian** is a great circle of the celestial sphere, through the celestial poles and the zenith. Also called CIRCLE OF LATITUDE. See also ANTE MERIDIAN, POST MERIDIAN.

meridian altitude. The altitude of a celestial body when it is on the celestial meridian of the observer, bearing 000° or 180° true.

meridian angle. Angular distance east or west of the local celestial meridian; the arc of the celestial equator, or the angle at the celestial pole, between the upper branch of the local celestial meridian and the hour circle of a celestial body, measured eastward or westward from the local celestial meridian through 180°, and labeled E or W to indicate the direction of measurement. See also HOUR ANGLE.

meridian angle difference. The difference between two meridian angles, particularly between the meridian angle of a celestial body and the value used as an argument for entering a table. Also called HOUR ANGLE DIFFERENCE.

meridian observation. Measurement of the altitude of a celestial body on the celestial meridian of the observer, or the altitude so measured.

meridian passage. See MERIDIAN TRANSIT.

meridian sailing. Following a true course of 000° or 180°; sailing along a meridian. Under these conditions the dead reckoning latitude is assumed to change 1 minute for each mile run and the dead reckoning longitude is assumed to remain unchanged.

meridian transit. The passage of a celestial body across a celestial meridian. **Upper transit**, the crossing of the upper branch of the celestial meridian, is understood unless **lower transit**, the crossing of the lower branch, is specified. Also called TRANSIT, MERIDIAN PASSAGE, CULMINATION.

meridional difference. The difference between the meridional parts of any two given parallels. This difference is found by subtraction if the two parallels are on the same side of the equator, and by addition if on opposite sides. Also called DIFFERENCE OF MERIDIONAL PARTS.

meridional parts. The length of the arc of a meridian between the equator and a given parallel on a Mercator chart, expressed in units of 1 minute of longitude at the equator.

metacenter, *n.* For small angles of inclination of a ship, the instantaneous center of a very small increment of the curved path of the center of buoyancy locus. Or, for small angles of inclination, the point of intersection of the lines of action of the buoyant force and the original vertical through the center of buoyancy.

meteor, *n.* The phenomenon occurring when a solid particle from space enters the earth's atmosphere and is heated to incandescence by friction of the air. A meteor whose brightness does not exceed that of Venus (magnitude –4) is popularly called SHOOTING STAR or FALLING STAR. A shooting star results from the entrance into the atmosphere of a particle having a diameter between a few centimeters and just visible to the naked eye. Even smaller particles may produce meteors which can be viewed only in telescopes. These are the TELESCOPIC METEORS. Shooting stars are observed first as a light source, similar to a star, which suddenly appears in the sky and moves along a long or short path to a point where it just as suddenly disappears. The brighter shooting stars may leave a luminous trail which remains luminous for a short while. Meteors brighter than magnitude –4 are called BOLIDES or FIREBALLS. Light bursts, spark showers, or splitting of the trail are sometimes seen along their luminous trails which persist for minutes and for an hour in exceptional cases. The intensity of any meteor is dependent upon the size of the particle which enters the atmosphere. A particle 10 centimeters in diameter can produce a bolide as bright as the full moon. See also METEORITE.

meteorite, *n.* 1. The solid particle which causes the phenomenon known as a METEOR.
2. The remnant of the solid particle, causing the meteor, which reaches the earth.

meteorological optical range. The length of path in the atmosphere required to reduce the luminous flux in a collimated beam from an incandescent lamp at a color temperature of 2,700°K to 0.05 of its original value, the luminous flux being evaluated by means of the curve of spectral luminous efficiencies for photopic vision given by the International Commission on Illumination. The quantity so defined corresponds approximately to the distance in the atmosphere required to reduce the contrast of an object against its background to 5 percent of the value it would have at zero distance, for daytime observation. See also METEOROLOGICAL VISIBILITY.

Meteorological Optical Range Table. A table from the International Visibility Code which gives the code number of meteorological visibility and the meteorological visibility for several weather conditions such as thin fog (Code no. 4 – meteorological visibility ½ -1 nautical mile); clear (Code No. 7 – meteorological visibility 5½-11 nautical miles).

meteorological tide. A change in water level caused by local meteorological conditions, in contrast to an ASTRONOMICAL TIDE, caused by the attractions of the sun and moon. See also SEICHE, STORM SURGE.

meteorological tides. Tidal constituents having origin in the daily or seasonal variations in weather conditions which may occur with some degree of periodicity. See also STORM SURGE.

meteorological visibility. The greatest distance at which a black object of suitable dimensions can be seen and recognized by day against the horizon sky, or, in the case of night observations, could be seen and recognized if the general illumination were raised to the normal daylight level. It has been established that the object may be seen and recognized if the contrast threshold is 0.05 or higher. The term may express the visibility in a single direction or the prevailing visibility in all directions. See also VISIBILITY, METEOROLOGICAL OPTICAL RANGE, CONTRAST THRESHOLD.

meteor swarm. The scattered remains of comets that have broken up.

meter, metre, *n.* 1. The *base unit* of length in the International System of Units, equal to 1,650, 763.73 wavelengths in vacuum of the radiation corresponding to the transition between the levels $2p_{10}$ and $5d_5$ of the krypton-86 atom. It is equal to 39.37008 inches, approximately, or approximately one ten-millionth of the distance from the equator to the North or South Pole. The old international prototype of the meter is still kept at the International Bureau of Weights and Measures under the conditions specified in 1889.
2. A device for measuring, and usually indicating, some quantity.

method of bisectors. As applied to celestial lines of position, the movement of each of three or four intersecting lines of position an equal amount, in the same direction toward or away from the celestial bodies, so as to bring them as nearly as possible to a common intersection. When there are more than four lines of position, the lines of position in the same general direction are combined to reduce the data to not more than four lines of position. See also OUTSIDE FIX.

Metonic cycle. A period of 19 years or 235 lunations devised by Meton, an Athenian astronomer who lived in the fifth century B.C., for the purpose of obtaining a period in which new and full moon would recur on the same day of the year. Taking the Julian year of 365.25 days and the synodic month as 29.53058 days, we have the 19-year period of 6939.75 days as compared with the 235 lunations of 6939.69 days, a difference of only 0.06 days. See also CALLIPPIC CYCLE.

metre per second. The *derived unit* of speed in the International System of Units.

metre per second squared. The *derived unit* of acceleration in the International System of Units.

metric system. A decimal system of weights and measures based on the meter as the unit of length and the kilogram as a unit mass. See also INTERNATIONAL SYSTEM OF UNITS.

Mexico Current. From late October through April an extension of the California Current, known as the Mexico Current, flows southeastward along the coast to the vicinity of longitude 95°W where it usually turns west, but at times extends southward as far as Honduras with speeds from 0.5 to 1 knot. During the remainder of the year, this current flows northwestward along the Mexican coast as far as Cabo Corrientes, where it turns westward and becomes a part of the Pacific North Equatorial Current.

micro-. A prefix meaning *one-millionth*.

micrometer, *n.* An auxiliary device to provide measurement of very small angles or dimensions by an instrument such as a telescope.

micrometer drum. A cylinder carrying an auxiliary scale and sometimes a vernier, for precise measurement, as in certain type sextants.

micrometer drum sextant. A marine sextant providing a precise reading by means of a micrometer drum attached to the index arm, and having an endless tanget screw for controlling the position of the index arm. The micrometer drum may include a vernier to enable a more precise reading. On a **vernier sextant** the vernier is used *directly* with the arc.

micron, *n.* A unit of length equal to one-millionth of a meter.

microsecond, *n.* One-millionth of a second.

microwave, *n.* A very short electromagnetic wave, usually considered to be about 30 centimeters to 1 millimeter in length. While the limits are not clearly defined, it is generally considered as the wavelength of a radar operation.

microwave frequency. Radio frequency of 1,000 to 300,000 megahertz, having wavelengths of 30 centimeters to 1 millimeter.

mid-channel buoy. See FAIRWAY BUOY.

mid-channel mark. A navigation mark serving to indicate a deep-water channel.

middle clouds. Types of clouds the mean level of which is between 6,500 and 20,000 feet. The principal clouds in this group are **altocumulus** and **altostratus.**

middle ground. A shoal in a fairway having a channel on either side.

middle ground buoy. One of the buoys placed at each end of a middle ground. See BIFURCATION BUOY, JUNCTION BUOY.

middle latitude. The latitude at which the arc length of the parallel separating the meridians passing through two specific points is exactly equal to the departure in proceeding from one point to the other by middle-latitude sailing. Also called MID-LATITUDE. See also MEAN LATITUDE, MIDDLE-LATITUDE SAILING.

middle-latitude sailing. A method that combines plane sailing and parallel sailing. **Plane sailing** is used to find difference of latitude and departure when course and distance are known, or vice versa. **Parallel sailing** is used to interconvert departure and difference of longitude. The **mean latitude** is normally used for want of a practicable means of determining the middle latitude, the latitude at which the arc length of the parallel separating the meridians passing through two specific points is exactly equal to the departure in proceeding from one point to the other. The formulas for these transformations are:

$$DLo = p \sec Lm \qquad p = DLo \cos Lm,$$

where DLo is difference of longitude, p is departure, and Lm is middle or mean latitude. See also MEAN LATITUDE.

mid-extreme tide. An elevation midway between the extreme high water and the extreme low water occurring in any locality. See also HALF-TIDE LEVEL.

mid-latitude. See MIDDLE LATITUDE.

midnight, *n.* Twelve hours from noon, or the instant the time reference crosses the lower branch of the reference celestial meridian.

midnight sun. The sun when it is visible at midnight. This occurs during the summer in high latitudes, poleward of the circle at which the latitude is approximately equal to the polar distance of the sun.

mil, *n.* 1. A unit of angular measurement equal to an angle having a tangent of 0.001.
2. A unit of angular measurement equal to an angle subtended by an arc equal to 1/6,400th part of the circumference of a circle.

mile, *n.* A unit of distance. The **nautical mile,** or **sea mile,** is used primarily in navigation. Nearly all maritime nations have adopted the International Nautical Mile of 1,852 meters proposed in 1929 by the International Hydrographic Bureau. The U.S. Departments of Defense and Commerce adopted this value on July 1, 1954. Using the yard-meter conversion factor effective July 1, 1959, (1 yard = 0.9144 meter, exactly) the International Nautical Mile is equivalent to 6076.11549 feet, approximately. The **geographical mile** is the length of 1 minute of arc of the equator, considered to be 6,087.08 feet. The **statute mile** or **land mile** (5,280 feet in the United States) is commonly used for navigation on rivers and lakes, notably the Great Lakes of North America. See also CABLE, MEASURED MILE.

mileage number. A number assigned to aids to navigation which gives the distance in sailing miles along the river from a reference point to the aid. The number is used principally in the Mississippi River system.

miles of relative movement. The distance, in miles, traveled relative to a reference point which is usually in motion.

military grid. Two sets of parallel lines intersecting at right angles and forming squares; the grid is superimposed on maps, charts, and other similar representations of the earth's surface in an accurate and consistent manner to permit identification of ground locations with respect to other locations and the computation of direction and distance to other points. See also MILITARY GRID REFERENCE SYSTEM, UNIVERSAL POLAR STEREOGRAPHIC GRID, UNIVER-

SAL TRANSVERSE MERCATOR GRID, WORLD GEOGRAPHIC REFERENCE SYSTEM.

military grid reference system. A system which uses a standard-scaled grid square, based on a point of origin on a map projection of the earth's surface in an accurate and consistent manner to permit either position referencing or the computation of direction and distance between grid positions. See also MILITARY GRID.

Milky Way. The galaxy of which the sun and its family of planets are a part. It appears as an irregular band of misty light across the sky. Through a telescope, it is seen to be composed of numerous individual stars. See also COALSACK.

milli-. A prefix meaning *one-thousandth.*

millibar, *n.* A unit of pressure equal to 1,000 dynes per square centimeter, or 1/1,000th of a bar. The millibar is used as a unit of measure of atmospheric pressure, a standard atmosphere being equal to 1,013.25 millibars or 29.92 inches of mercury.

milligal, *n.* A unit of acceleration equal to 1/1,000th of a gal, or 1/1,000 centimeter per second per second. This unit is used in gravity measurements, being approximately one-millionth of the average gravity at the earth's surface.

millimeter, *n.* One thousandth of a meter; one-tenth of a centimeter; .03937008 inch.

millisecond, *n.* One-thousandth of a second.

minaret, *n.* As a landmark, a tall, slender tower attached to a mosque and surrounded by one or more projecting balconies.

minimal depiction of detail. A term used to indicate the extreme case of generalization of detail on a chart. In the extreme case most features are omitted even through there is space to show at least some of them. The practice is most frequently used for semi-enclosed areas such as estuaries and harbors on smaller-scale charts, where use of a larger scale chart is essential.

minimum distance (of a navigational system). The minimum distance at which a navigational system will function within its prescribed tolerances.

minimum ebb. See under EBB CURRENT.

minimum flood. See under FLOOD CURRENT.

minimum signal. The smallest signal capable of satisfactorily operating an equipment, e.g., the smallest signal capable of triggering a racon.

minimum thermometer. A thermometer which automatically registers the lowest temperature occurring since its last setting. One which registers the highest temperature is called a MAXIMUM THERMOMETER.

Mini-Ranger III System. A compact, light, and mobile distance measuring system manufactured by Motorola, Inc. The system operates on the basic principle of pulse radar. A transmitter aboard the survey vessel interrogates transponders at known locations. Elapsed time between transmitted interrogations and the reply from each transponder is used as the basis for determining range to each transponder. This range information together with the known location of each transponder can be trilaterated to provide the position of the survey vessel.

minor axis. The shortest diameter of an ellipse or ellipsoid.

minor light. An automatic unmanned light on a fixed structure usually showing low to moderate intensity. Minor lights are established in harbors, along channels, along rivers, and in isolated locations. See also MAJOR LIGHT.

minor planets. See under PLANET.

minute, *n.* 1. The sixtieth part of a degree of arc. 2. The sixtieth part of an hour.

mirage, *n.* An optical phenomenon in which objects appear distorted, displaced (raised or lowered), magnified, multiplied, or inverted due to varying atmospheric refraction when a layer of air near the earth's surface differs greatly in density from surrounding air. See also TOWERING, STOOPING, LOOMING, SINKING, FATA MORGANA.

mirror reflection. See SPECULAR REFLECTION.

missing, *adj.* Said of a floating aid to navigation which is not on station with its whereabouts unknown.

mist, *n.* An aggregate of very small water droplets suspended in the atmosphere. It produces, generally, a thin, greyish veil over the landscape. It reduces visibility to a lesser extent than fog. The relative humidity with mist is often less than 95 percent. Mist is intermediate in all respects between haze (particularly damp haze) and fog. See also DRIZZLE.

mistake, *n.* The result of carelessness or of a mistake. The taking of the wrong value from a table, the erroneous plotting of a reciprocal bearing, and the misreading of the scale of an instrument are examples of carelessness. For the purpose of error analysis, a mistake is not classified as an error. Also called BLUNDER.

mistral, *n.* A cold, dry wind blowing from the north over the northwest coast of the Mediterranean Sea, particularly over the Gulf of Lions. Also called CIERZO. See also FALL WIND.

mixed current. Type of tidal current characterized by a conspicuous speed difference between the two floods and/or ebbs usually occurring each tidal day. See also TYPE OF TIDE.

mixed tide. Type of tide with a large inequality in either the high and/ or low water heights, with two high waters and two low waters usually occurring each tidal day. In strictness, all tides are mixed but the name is usually applied to the tides intermediate to those predominantly semidiurnal and those predominantly diurnal. See also TYPE OF TIDE.

moat, *n.* An annular depression that may not be continuous, located at the base of many seamounts, islands, and other isolated elevations of the sea floor.

mobile service. As defined by the *International Telecommunication Union* (ITU), a service of radiocommunication between mobile and land stations, or between mobile stations.

mock fog. A rare simulation of true fog by anomalous atmospheric refraction.

mock moon. See PARASALENE.

mock sun. See PARHELION.

mock-sun ring. See PARHELIC CIRCLE.

modal interference. Omega signals propagate in the earth-ionosphere waveguide. This waveguide can support many different electromagnetic field configurations, each of which can be regarded as an identifiable signal component or mode having the same signal frequency, but with slightly different phase velocity. Modal interference is a special form of signal interference wherein two or more waveguide modes interfere with each other and irregularities appear in the phase pattern. This type of interference occurs predominantly under nighttime conditions when most of the propagation path is not illuminated and the boundary conditions of the waveguide are unstable. It is most severe for signals originating at stations located close to the geomagnetic equator. During all daylight path conditions, the only region of modal interference is a more-or-less circular area of radius 500-1000 kilometers immediately surrounding a transmitting station.

model atmosphere. Any theoretical representation of the atmosphere, particularly of vertical temperature distribution. See also STANDARD ATMOSPHERE.

moderate breeze. Wind of force 4 (11 to 16 knots or 13 to 18 miles per hour) on the Beaufort wind scale.

moderate gale. A term once used by seamen for what is now called NEAR GALE on the Beaufort wind scale.

modification, *n.* An instrument design resulting from a minor change, and indicated by number. A design resulting from a major change is called a MARK. Thus, "Mark 5 Mod (modification) 2" would indicate the second minor change of the fifth major design.

modified Julian day. An abbreviated form of the Julian day which requires fewer digits and translates the beginning of each day from Greenwich noon to Greenwich midnight; obtained by subtracting 2400000.5 from Julian days.

modified Lambert conformal chart. A chart on the modified Lambert conformal map projection. Also called NEY'S CHART.

modified Lambert conformal map projection. A modification of the Lambert conformal projection for use in polar regions, one of the standard parallels being at latitude 89°59'58" and the other at latitude 71° or 74°, and the parallels being expanded slightly to form complete concentric circles. Also called NEY'S MAP PROJECTION.

modified refractive index. For a given height above sea level, the sum of the refractive index of the air at this height and the ratio of the height to the radius of the earth.

modulated wave. A wave which varies in some characteristic in accordance with the variations of a modulating wave. See also CONTINUOUS WAVE.

modulating wave. A wave which modulates a carrier wave.

modulation, *n.* A variation of some characteristic of a radio wave, called the CARRIER WAVE, in accordance with instantaneous values of another wave called the MODULATING WAVE. These variations can be amplitude, frequency, phase, or pulse.

modulator, *n.* In pulse radar that component which generates a succession of short pulses of energy which in turn cause a transmitter tube to oscillate during each pulse.

mole, *n.* 1. A structure, usually massive, on the seaward side of a harbor for its protection against current and wave action, drift ice, sanding up, wind, etc. Sometimes it may be suitable for the berthing of ships. See also JETTY, definition 1; QUAY.
2. The *base unit* of amount of substance in the International System of Units; it is the amount of substance of a system which contains as many elementary entities as there are atoms in 0.012 kilogram of carbon atom 12. When the mole is used, the elementary entities must be specified and may be atoms, molecules, ions, electrons, other particles, or specified groups of such particles.

moment, *n.* The tendency or degree of tendency to produce motion about an axis. Numerically it is the quantity obtained by multiplying the force, speed, or mass by the distance from the point of application or center of gravity to the axis. See also MAGNETIC MOMENT.

moment of inertia. The quantity obtained by multiplying the mass of each small part of a body by the square of its distance from an axis, and adding all the results.

momentum, *n.* Quantity of motion. **Linear momentum** is the quantity obtained by multiplying the mass of a body by its linear speed. **Angular momentum** is the quantity obtained by multiplying the moment of inertia of a body by its angular speed.

monitor, *v., t.* In radionavigation, the reception of the signals of a system in order to check its operation and performance.

monitoring, *n.* In radionavigation, the checking of the operation and performance of a system through reception of its signals.

monsoon, *n.* A name for seasonal winds first applied to the winds over the Arabian Sea, which blow for 6 months from the northeast (northeast monsoon) and for 6 months from the southwest (southwest monsoon). The primary cause is the much greater annual variation of temperature over large land areas compared with the neighboring ocean surfaces, causing an excess of pressure over the continents in winter and a deficit in summer, but other factors such as the relief features of the land have a considerable effect. In India the term is popularly applied chiefly to the southwest monsoon and, by extension, to the rain which it brings.

monsoon current. A seasonal wind-driven current occurring in the northern part of the Indian Ocean and the northwest Pacific Ocean. See also MONSOON DRIFT.

Monsoon Drift. A drift current of the northeast Indian Ocean located north of the Indian Equatorial Countercurrent and south of the Bay of Bengal. During February and March when the northeast monsoon decreases in intensity, the monsoon drift is formed from the outflow of the Strait of Malacca and a small amount of northwestward flow along the upper southwest coast of Sumatra. Off the southwest coast of Sumatra, a current generally sets southeast during all months. It is strongest during October through April. The monsoon drift broadens as it flows westward and divides off the east coast of Sri Lanka, part joining the circulation of the Bay of Bengal and part joining the flow from the Arabian Sea. During April, the transition period between monsoons, the monsoon drift is ill-defined. A counterclockwise circulation exists between Sumatra and Sri Lanka. During May through October, the monsoon drift flows east to southeast. During November and December, part of the monsoon drift is deflected into the Bay of Bengal and the remainder turns clockwise and flows southeastward. See also MONSOON.

monsoon fog. An advection fog occurring as a monsoon circulation transports warm moist air over a colder surface.

month, *n.* 1. The period of the revolution of the moon around the earth. The month is designated as sidereal, tropical, anomalistic, nodical, or synodical, according to whether the revolution is relative to the stars, the vernal equinox, the perigee, the ascending node, or the sun.
2. The calendar month, which is a rough approximation to the synodical month.

month of the phases. See SYNODICAL MONTH.

moon, *n.* The satellite of the earth.

moonbow, *n.* A rainbow formed by light from the moon. Colors in a moonbow are usually very difficult to detect. Also called LUNAR RAINBOW.

moon dog. See PARASELENE.

moonrise, *n.* The crossing of the visible horizon by the upper limb of the ascending moon.

moonset, *n.* The crossing of the visible horizon by the upper limb of the descending moon.

moor, *v., t.* To secure a vessel to the ground, other than by anchoring with a single anchor.

mooring, *n.* 1. The act of securing a craft to the ground, a wharf, pier, quay, etc., other than anchoring with a single anchor.
2. The place where a craft may be moored.
3. (Usually in pl.) Chains, bridles, anchors, etc. used in securing a craft to the ground.

mooring buoy. A buoy secured to the bottom by permanent moorings and provided with means for mooring a vessel by use of its anchor chain or mooring lines.

morning star. The brightest planet appearing in the eastern sky during morning twilight.

morning twilight. The period of time between darkness and sunrise.

Morse code light. A light in which the appearances of light of two clearly different durations are grouped to represent a character or characters in the Morse code.

motion, *n.* The act, process, or instance of change of position. **Absolute motion** is motion relative to a fixed point. **Actual motion** is motion of an object relative to the earth. **Apparent** or **relative motion** is change of position as observed from a reference point which may itself be in motion. **Diurnal motion** is the apparent daily motion of a celestial body. **Direct motion** is the apparent motion of a planet eastward among the stars; **retrograde motion,** the apparent motion westward among the stars. Motion of a celestial body through space is called **space motion,** which is composed of two components: **proper motion,** that component perpendicular to the line of sight; and **radial motion,** that component in the direction of the line of sight. Also called MOVEMENT, especially when used in connection with problems involving the motion of one vessel relative to another.

Motorola Mini-Ranger III System. See as MINI-RANGER III SYSTEM.

mound, *n.* On the sea floor, a low, isolated, rounded hill.

mountain breeze. A breeze that blows down a mountain slope due to the gravitational flow of cooled air. See also KATABATIC WIND, VALLEY BREEZE.

mountains, *n., pl.* On the sea floor, a well delineated subdivision of a large and complex positive feature, generally part of a cordillera.

movement, *n.* See MOTION.

moving havens. Restricted areas established to provide a measure of security to submarines and surface ships in transit through areas in which the existing attack restrictions would be inadequate to prevent attack by friendly forces. See also MOVING SUBMARINE HAVEN, MOVING SURFACE SHIP HAVEN.

moving submarine haven. Established by Submarine Notices surrounding submarines in transit, extending 50 miles ahead, 100 miles astern, and 15 miles on each side of the estimated position of the submarine along the stated track. See also MOVING HAVENS.

moving surface ship haven. Established by Surface Ship Notices and will normally be a circle with a specified radius centered on the estimated position of the ship or the guide of a group of ships. See also MOVING HAVENS.

moving target indication. A form of presentation in radar whereby means are provided for partially or wholly suppressing the appearance of stationary targets on the display.

Mozambique Current. The part of the Indian South Equatorial Current that turns and flows along the African coast in the Mozambique Channel. It is considered part of the AGULHAS CURRENT.

mud, *n.* A general term applied to mixtures of sediments in water. Where the grains are less than 0.002 millimeter in diameter, the mixture is called **clay.** Where the grains are between 0.002 and 0.0625 millimeter in diameter, the mixture is called **silt.** See also SAND; STONES; ROCK, definition 2.

mud berth. A berth where a vessel rests on the bottom at low water.

mud flat. A tidal flat composed of mud.

mud pilot. A person who pilots a vessel by visually observing changes in the color of the water as the depth of the water increases or decreases.

multihop transmission. See MULTIPLE-HOP TRANSMISSION.

multipath error. Interference between radio waves which have traveled between the transmitter and the receiver by two paths of different electrical lengths may cause fading or phase changes at the receiving point due to the vector addition of time varying signals, making it difficult to obtain accurate information. In some long range systems, Omega for example, the signal may travel between the transmitter and receiver by two great-circle paths of different electrical lengths. The effect at the receiver is called LONG PATH INTERFERENCE.

multipath propagation. Radio propagation from the transmitter to the receiver by two or more paths simultaneously. Also called MULTIPATH TRANSMISSION.

multipath transmission. See MULTIPATH PROPAGATION.

multiple echoes. Radar echoes which may occur when a strong echo is received from another ship at close range. A second or third or more echoes may be observed on the radarscope at double, triple, or other multiples of the actual range of the radar target, resulting from the echo's being reflected by own ship back to the target and received once again as an echo at a multiple of the preceding range to the target. This term should not be confused with MULTIPLE-TRACE ECHO. See also SECOND-TRACE ECHO.

multiple-hop transmission. Radio wave transmission in which the waves traveling between transmitter and receiver undergo multiple reflections and refractions between the earth and ionosphere. Also called MULTIHOP TRANSMISSION.

multiple ranges. A group of two ranges, having one of the range marks (either front or rear) in common.

multiple star. A group of three or more stars so close together than they appear as a single star, whether through physical closeness or as a result of lying in approximately the same direction.

See also STAR CLUSTER.

multiple tide staff. A succession of tide staffs on a sloping shore so placed that the vertical graduations on the several staffs will form a continuous scale referred to the same datum.

multiple-trace echo. See SECOND-TRACE ECHO.

multi-year ice. Old ice up to 3 meters or more thick which has survived at least two summer's melt. Hummocks are even smoother than in second-year ice. The ice is almost salt-free. The color; where bare, is usually blue. The melt pattern consists of large interconnecting irregular puddles and a well-developed drainage system.

Mumetal, *n.* The registered trade name for an alloy of about 75% nickel and 25% iron, having high magnetic permeability and low hysteresis.

N

nadir, *n.* That point on the celestial sphere vertically below the observer, or 180° from the zenith.

name, *n.* The label of a numerical value, used particularly to refer to the N (north) or S (south) label of latitude and declination. When latitude and declination are both N or both S, they are said to be of **same name**, but if one is N and the other S, they are said to be of **contrary name**.

nano-. A prefix meaning *one-billionth* (10^{-9}).

nanosecond, *n.* One-billionth of a second.

Napier diagram. A diagram on which compass deviation is plotted for various headings, and the points connected by a smooth curve, permitting deviation problems to be solved quickly without interpolation. It consists of a vertical line, usually in two parts, each part being graduated for 180° of heading, and two additional sets of lines at an angle of 60° to each other and to the vertical lines. See also DEVIATION TABLE.

Napierian logarithm. A logarithm to the base e (2.7182818). Also called NATURAL LOGARITHM. See also COMMON LOGARITHM.

narrows, *n.* A navigable narrow part of a bay, strait, river, etc.

nashi, n'aschi, *n.* A northeast wind which occurs in winter on the Iranian coast of the Persian Gulf, especially near the entrance to the gulf, and also on the Makran coast. It is probably associated with an outflow from the central Asiatic anticyclone which extends over the high land of Iran. It is similar in character but less severe than the BORA.

National Geodetic Vertical Datum. A fixed reference adopted as a standard geodetic datum for heights in the United States. The datum was derived for land surveys from a general adjustment of the first order level nets of both the United States and Canada. In the adjustment 21 tide stations in the United States and 5 in Canada were held as fixed. The geodetic datum now in use in the United States is the **National Geodetic Vertical Datum of 1929.** The year indicates the time of the last general adjustment. The geodetic datum is fixed and does not take into account the changing stands of sea level. Because there are many variables affecting sea level, and because the geodetic datum represents a best fit over a broad area, the relationship between the geodetic datum and local mean sea level is not consistent from one location to another in either time or space. For this reason the National Geodetic Vertical Datum should not be confused with MEAN SEA LEVEL.

National Tidal Datum Control Network. A network composed of the primary control tide stations of the National Ocean Survey. Distributed along the coasts of the United States, this network provides the basic tidal datums for coastal boundaries and chart datums of the United States. Tidal datums obtained at secondary control tide stations and tertiary tide stations are referenced to the Network. Terrestrial leveling between stations is not a requirement of the National Tidal Datum Control Network.

National Tidal Datum Epoch. The specific 19-year cycle adopted by the National Ocean Survey as the official time segment over which tide observations are taken and reduced to obtain mean values (e.g., mean lower low water, etc.) for tidal datums. It is necessary for standardization because of apparent periodic and apparent secular trends in sea level. The present National Tidal Datum Epoch is 1960 through 1978.

natural, *adj.* 1. Occurring in nature; not artificial. 2. Not logarithmic-used with the name of a trigonometric function to distinguish it from its logarithm (called LOGARITHMIC TRIGONOMETRIC FUNCTION).

natural coversine. See under COVERSINE.

natural frequency. The lowest resonant frequency of a body or system.

natural harbor. A harbor where the configuration of the coast provides the necessary protection. See also ARTIFICIAL HARBOR.

natural haversine. See under HAVERSINE.

natural logarithm. See NAPIERIAN LOGARITHM.

natural magnet. A magnet occurring in nature, as contrasted with an ARTIFICIAL MAGNET, produced by artificial means.

natural period. The period of the natural frequency of a body or system.

natural range. A range formed by natural objects such as rocks, peaks, etc. See also ARTIFICIAL RANGE.

natural scale. See REPRESENTATIVE FRACTION.

natural tangent. See under TANGENT, definition 1.

natural trigonometric function. See under TRIGONOMETRIC FUNCTIONS.

natural versine. See under VERSINE.

natural year. See TROPICAL YEAR.

nature of the bottom. See BOTTOM CHARACTERISTICS.

nautical, *adj.* Of or pertaining to ships, navigation (chiefly marine), or seamen. In contrast, NAVIGATIONAL refers to navigation only, MARINE refers to the sea, MARITIME indicates relationship or proximity to the sea, and NAVAL refers to the navy.

nautical almanac. 1. A periodical publication of astronomical data useful to and designed primarily for marine navigation. Such a publication designed primarily for air navigation is called an AIR ALMANAC. See also ALMANAC FOR COMPUTERS.
2. *Nautical Almanac*; a joint annual publication of the U.S. Naval Observatory and the Nautical Almanac Office, Royal Greenwich Observatory, listing the Greenwich hour angle and declination of various celestial bodies to a precision of 0.′1 at hourly intervals; time of sunrise, sunset, moonrise, moonset; and other astronomical information useful to navigators. This publication is a unified edition of the former (British) *Abridged Nautical Almanac* and the *American Nautical Almanac*.

nautical astronomy. See NAVIGATIONAL ASTRONOMY.

nautical chart. A representation of a portion of the navigable waters of the earth and adjacent coastal areas on a specified map projection, and designed specifically to meet requirements of marine navigation. Included on most nautical charts are: depths of water, characteristics of the bottom, elevations of selected topographic features, general configuration and characteristics of the coast, the shoreline (usually the mean high water line), dangers, obstructions, aids to navigation, limited tidal data, and information about magnetic variation in the charted area.

nautical day. Until January 1, 1925, a day that began at noon, 12 hours earlier than the calendar day, or 24 hours earlier than the astronomical day of the same date.

nautical mile. A unit of distance used principally in navigation. For practical consideration it is usually considered the length of 1 minute of any great circle of the earth, the meridian being the great circle most commonly used. Because of various lengths of the nautical mile in use throughout the world, due to differences in definition and the assumed size and shape of the earth, the International Hydrographic Bureau in 1929 proposed a standard length of 1,852 meters, which is known as the International Nautical Mile. This has been adopted by nearly all maritime nations. The U.S. Departments of Defense and Commerce adopted this value on July 1, 1954. With the yard-meter relationship then in use, the International Nautical Mile was equivalent to 6076.10333 feet, approximately. Using the yard-meter conversion factor effective July 1, 1959, (1 yard = 0.1944 meter, exactly) the International Nautical Mile is equivalent to 6076.11549 feet, approximately. See also SEA MILE.

nautical twilight. The time of incomplete darkness defined to begin (morning) or end (evening) when the center of the sun is 12° below the celestial horizon. The times of nautical twilight are tabulated in the *Nautical Almanac*; at the times given the horizon is generally not visible and it is too dark for marine sextant observations. See also FIRST LIGHT.

nautophone, *n.* A sound signal emitter comprising an electrically oscillated diaphragm. It emits a signal similar in power and tone to that of a REED HORN.

Naval Vessel Lights Act. Authorized departure from the rules of the road for character and position of navigation lights for certain naval ships. Such modifications are published in *Notice to Mariners*.

NAVAREA Warnings. Broadcast messages containing information which may affect the safety of navigation on the high seas. In accordance with international obligations, the Defense Mapping Agency Hydrographic/Topographic Center is responsible for disseminating navigation information for ocean areas designated as NAVAREAS IV and XII of the **World Wide Navigational Warning Service.** NAVAREA IV broadcasts over the waters contiguous to North America from the Atlantic coast eastward to 35°W and between latitudes 7°N and 67°N. NAVAREA XII broadcasts cover the waters contiguous to North America and extends westward to the International Date Line and from 67°N to the equator east of 120°W, south to 3°25'S, thence east to the coast. Other countries are responsible for disseminating navigational information for the remaining NAVAREAS. NAVAREA Warnings may be superseded by a numbered paragraph in *Notice to Mariners*. Printed copies are published each working day in the appropriate edition of the *Daily Memorandum*. The text of effective warnings for NAVAREAS IV and XII is printed in the weekly *Notice to Mariners*.

navigable, *adj.* Affording passage to a craft; capable of being navigated.

navigable semicircle. That half of a cyclonic storm area in which the rotary and progressive motions of the storm tend to counteract each other and the winds are in such a direction as to tend to blow a vessel away from the storm track. In the Northern Hemisphere this is to the left of the storm center and in the Southern Hemisphere it is to the right. The opposite is DANGEROUS SEMICIRCLE.

navigable waters. Waters usable, with or without improvements, as routes for commerce in the customary means of travel on water.

navigating sextant. A sextant designed and used for observing the altitudes of celestial bodies, as opposed to a hydrographic sextant.

navigation, *n.* The process of planning, recording, and controlling the movement of a craft or vehicle from one place to another. The word *navigate* is from the Latin *navigatus*, the past participle of the verb *navigere*, which is derived from the words *navis*, meaning "ship," and *agere*, meaning "to move" or "to direct." Navigation of water craft is called **marine navigation** to distinguish it from navigation of aircraft, called **air navigation.** Navigation of a vessel on the surface is sometimes called **surface navigation** to distinguish it from **underwater navigation** of a submerged vessel. Navigation of vehicles across land or ice is called **land navigation.** The expression **lifeboat navigation** is used to refer to navigation of lifeboats or life rafts, generally involving rather crude methods. The expression **polar navigation** refers to navigation in the regions near the geographical poles of the earth, where special techniques are employed. The principal divisions of navigation are as follows: **dead reckoning, piloting (or pilotage), celestial navigation,** and **radionavigation. Inertial navigation** and **doppler sonar navigation** are forms of dead reckoning. **Satellite navigation** and **radar navigation** are forms of radionavigation. The term **electronic navigation** is used to refer to navigation involving the use of electronics in any way. Thus, the term includes the use of the gyrocompass for steering and the echo sounder when piloting. Because of the wide use of electronics in navigation equipment, the term electronic navigation has limited value as a term for a division of navigation.

navigational aid. An instrument, device, chart, method, etc., intended to assist in the navigation of a craft. This expression should not be confused with AID TO NAVIGATION, which refers only to devices external to a craft. In British usage, the terms *navigational aid* and *aid to navigation* are used without distinction.

navigational astronomy. That part of astronomy of direct use to a navigator, comprising principally celestial coordinates, time, and the apparent motions of celestial bodies. Also called NAUTICAL ASTRONOMY.

navigational planets. The four planets commonly used for celestial observations: Venus, Mars, Jupiter, and Saturn.

navigational plot. A graphic plot of the movements of a craft. A **dead reckoning plot** is the graphic plot of the dead reckoning, suitably labeled with respect to time, direction, and speed; a **geographical plot** is one relative to the surface of the earth.

navigational triangle. The spherical triangle solved in computing altitude and azimuth and great-circle sailing problems. The **celestial triangle** is formed on the celestial sphere by the great circles connecting the elevated pole, zenith of the assumed position of the observer, and a celestial body. The **terrestrial triangle** is formed on the earth by the great circles connecting the pole and two places on the earth; the assumed position of the observer and geographical position of the body for celestial observations, and the point of departure and destination for great-circle sailing problems. The expression **astronomical triangle** applies to either the celestial or terrestrial triangle used for solving celestial observations.

navigation head. A transshipment point on a waterway where loads are transferred between water carriers and land carriers. A navigation head is similar in function to a railhead or truck-head.

navigation lights. Statutory lights shown by vessels during the hours between sunset and sunrise, in accordance with international agreements.

navigation mark. See MARK.

navigation/positioning system. A system capable of being used primarily for navigation or position fixing. It includes the equipment, its operators, the rules and procedures governing their actions and, to some extent, the environment which affects the craft or vehicle being navigated.

navigation satellite. An artificial earth satellite used in a system which determines positions based upon signals received from the satellite. See also NAVY NAVIGATION SATELLITE SYSTEM.

Navigation Tables for Mariners and Aviators. See H.O. PUB. NO. 208.

navigator, *n.* 1. A person who navigates or is directly responsible for the navigation of a craft.
2. A book of instructions on navigation, as the *American Practical Navigator* (Bowditch).

NAVSTAR Global Positioning System. A satellite navigation system being developed by the Department of Defense under Air Force management. The fully-deployed operational system is intended to provide highly accurate position and velocity information in three dimensions and precise time and time interval on a global basis continuously, to an unlimited number of authorized users. It will be unaffected by weather and will provide a worldwide common grid reference system. The objective of the program is to provide very precise positional information for a wide spectrum of military missions. In addition, current policy calls for civil availability with a degradation in system accuracy required to protect U.S. national security interests. See also NAVY NAVIGATION SATELLITE SYSTEM.

Navy Navigation Satellite System. An operational satellite navigation system of the United States conceived and developed by the Applied Physics Laboratory of the Johns Hopkins University for the U.S. Navy. The operation of the system is under the control of the Navy Astronautics Group with headquarters at Point Mugu, California. It is an all-weather, worldwide, and passive system used primarily for the navigation of surface ships and submarines; it has some applications in air navigation. It is also used in hydrographic surveying and geodetic position determination. The system, which is frequently referred to by the acronym NAVSAT or TRANSIT, consists of a constellation of orbiting satellites, a ground system, and any number of user equipments. The user's radionavigation set is essentially a receiver, frequency cycle-counter, and computer. Each satellite is in a nominally circular polar orbit of the earth at an altitude of 450 to 700 nautical miles. There are usually five satellites operating in the system, and these provide navigation fixes anywhere on the earth on nearly an hourly basis. Five satellites in orbit provide redundancy; the minimum constellation for system operation is four. The ground system is a network of operational injection facilities and tracking stations connected to a centralized control center and ground computing complex by high speed data communications lines. Tracking stations, in known positions, track passing satellites and relay tracking data to the computer center where it is used to compute future orbits. These orbital predictions are included in satellite injection messages prepared by the computer center and sent to an injection facility. Injection stations transmit injection messages to satellites for storage in satellite memory. Satellites broadcast current known positions while orbiting the earth. The NAVSAT system utilizes the doppler shift of radio signals transmitted from the satellite to measure the relative velocity between the satellite and the navigator. Knowing the satellite orbit precisely, the navigator's absolute position can be accurately determined from this time rate of change of range to the satellite. The satellites also transmit timing signals which provide time (Coordinated Universal Time to within 200 microseconds) automatically. The system transmits on two frequencies (150 and 400 MHz) to provide means for compensating for the effect of ionospheric refraction in the measurement of the doppler shift. See also DOPPLER SATELLITE NAVIGATION BASIC PRINCIPLES, GLOBAL POSITIONING SYSTEM.

neaped, *adj.* Left aground following a spring high tide. Also called BENEAPED.

neap high water. See under NEAP TIDES.

neap low water. See under NEAP TIDES.

neap range. See under NEAP TIDES.

neap rise. The height of neap high water above the elevation of reference or datum of chart.

neap tidal currents. Tidal currents of decreased speed occurring semimonthly as the result of the moon being in quadrature. See also NEAP TIDES.

neap tides. Tides of decreased range occurring semimonthly as the result of the moon being in quadrature. The **neap range** of the tide is the average semidiurnal range occurring at the time of neap tides and is most conveniently computed from the harmonic constants. It is smaller than the **mean range** where the type of tide is either semidiurnal or mixed and is of no practical significance where the type of tide is diurnal. The average height of the high waters of the neap tides is called **neap high water** or **high water neaps** and the average height of the corresponding low waters is called **neap low water** or **low water neaps.**

nearest approach. The least distance between two objects having relative motion with respect to each other.

near gale. Wind of force 8 (28 to 33 knots or 32 to 38 miles per hour) on the Beaufort wind scale. See also GALE.

nearshore current system. The current system caused by wave action in or near the surf zone. The nearshore current system consists of four parts: the shoreward mass transport of water; longshore currents; rip currents; the longshore movement of expanding heads of rip currents.

near vane. That instrument sighting vane on the same side of the instrument as the observer's eye. The opposite is FAR VANE.

neatline, *n.* That border line which indicates the limit of the body of a map or chart. Also called SHEET LINE.

nebula (*pl. nebulae*), *n.* 1. An aggregation of matter outside the solar system, large enough to occupy

a perceptible area but which has not been resolved into individual stars. One within our galaxy is called a **galactic nebula** and one beyond is called an **extragalactic nebula**. If a nebula is resolved into numerous individual stars, it is called a STAR CLUSTER.
2. A galaxy.

necessary bandwidth. As defined by the *International Telecommunication Union* (ITU) for a given class of emission, the minimum value of the occupied bandwidth sufficient to ensure the transmission of information at the rate and with the quality required for the system employed, under specified conditions. Emissions useful for the good functioning of the receiving equipment as, for example, the emission corresponding to the carrier of reduced carrier systems, shall be included in the necessary bandwidth.

neck, *n.* 1. A narrow isthmus, cape or promontory.
2. The land areas between streams flowing into a sound or bay.
3. A narrow strip of land which connects a peninsula with the mainland.
4. A narrow body of water between two larger bodies; a strait.

negative altitude. Angular distance below the horizon. Also called DEPRESSION.

Nemedri, *n.* A *discontinued* British publication containing danger areas, routes, and instructions for Northwest European waters, the Baltic Sea, the Mediterranean Sea and the Black Sea. The name Nemedri is derived from the words North European and Mediterranean Routing Instructions.

neutral occlusion. See under OCCLUDED FRONT.

new ice. A general term for recently formed ice which includes *frazil ice, grease ice, slush*, and *shuga*. These types of ice are composed of ice crystals which are only weakly frozen together (if at all) and have definite form only while they are afloat.

new moon. The moon at conjunction, when little or none of it is visible to an observer on the earth because the illuminated side is away from him. Also called CHANGE OF THE MOON. See also PHASES OF THE MOON.

new ridge. A newly formed ridge with sharp peaks, the slope of the sides usually being about 40°. Fragments are visible from the air at low altitude.

newton, *n.* The special name for the *derived unit* of force in the International System of Units; it is that force which gives to a mass of 1 kilogram an acceleration of 1 metre per second, per second.

Newtonian telescope. A reflecting telescope in which a small plane mirror reflects the convergent beam from the speculum to an eyepiece at one side of the telescope. After the second reflection the rays travel approximately perpendicular to the longitudinal axis of the telescope. See also CASSEGRAINIAN TELESCOPE.

newton per square metre. The *derived unit* of pressure in the International System of Units. See also PASCAL.

Newton's laws of motion. Universal laws governing all motion, formulated by Isaac Newton. These are: (1) Every body continues in a state of rest or of uniform motion in a straight line unless acted upon by a force; (2) When a body is acted upon by a force, its acceleration is directly proportional to the force and inversely proportional to the mass of the body, and the acceleration takes place in the direction in which the force acts; (3) To every action there is always an equal and opposite reaction; or, the mutual actions of two bodies are always equal and oppositely directed.

Ney's chart. See MODIFIED LAMBERT CONFORMAL CHART.

Ney's map projection. See MODIFIED LAMBERT CONFORMAL MAP PROJECTION.

night, *n.* That part of the solar day when the sun is below the visible horizon, especially the period between dusk and dawn.

night effect. See under POLARIZATION ERROR.

night error. See under POLARIZATION ERROR.

night order book. A notebook in which the commanding officer of a ship writes orders with respect to courses and speeds, any special precautions concerning the speed and navigation of the ship, and all other orders for the night for the officer of the deck.

nilas, *n.* A thin elastic crust of ice, easily bending on waves and swell and under pressure, thrusting in a pattern of interlocking "fingers." Nilas has a matte surface and is up to 10 centimeters in thickness. It may be subdivided into DARK NILAS and LIGHT NILAS. See also FINGER RAFTING.

nimbostratus, *n.* A dark, low shapeless cloud layer (mean upper level below 6,500 ft.) usually nearly uniform; the typical rain cloud. When precipitation falls from nimbostratus, it is in the form of continuous or intermittent rain or snow, as contrasted with the showery precipitation of cumulonimbus.

nimbus, *n.* A characteristic rain cloud. The term is not used in the international cloud classification except as a combining term, as *cumulonimbus*.

nipped, *adj.* Said of a vessel which is beset and with the surrounding ice forcibly pressing against the hull.

nipping, *n.* The forcible closing of ice around a vessel such that it is held fast by ice under pressure. See also BESET, ICE-BOUND.

no-bottom sounding. A sounding in which the bottom is not reached.

nocturnal, *n.* An old navigation instrument which consisted essentially of two arms pivoted at the center of a disk graduated for date, time and arc. The nocturnal was used for determining time during the night and for obtaining a correction to be applied to an altitude observation of Polaris for finding latitude.

nodal, *adj.* Related to or located at or near a node or nodes.

nodal line. A line in an oscillating body of water along which there is a minimum or no rise and fall of the tide.

nodal point. 1. See NODE, definition 1.
2. The notide point in an amphidromic region.

node, *n.* 1. One of the two points of intersection of the orbit of a planet, planetoid, or comet with the ecliptic, or of the orbit of a satellite with the plane of the orbit of its primary. That point at which the body crosses to the north side of the reference plane is called the **ascending node**; the other, the **descending node**. The line connecting the nodes is called LINE OF NODES. Also called NODAL POINT. See also REGRESSION OF THE NODES.
2. A zero point in any stationary wave system.

node cycle. The period of approximately 18.61 Julian years required for the regression of the moon's nodes to complete a circuit of 360° of longitude. It is accompanied by a corresponding cycle of changing inclination of the moon's orbit relative to the plane of the earth's equator, with resulting inequalities in the rise and fall of the tide and speed of the tidal current.

node factor. A factor depending upon the longitude of the moon's node which, when applied to the mean coefficient of a tidal constituent, will adapt the same to a particular year for which predictions are to be made.

nodical, *adj.* Of or pertaining to astronomical nodes; measured from node to node.

nodical month. The average period of revolution of the moon about the earth with respect to the moon's ascending node, a period of 27 days, 5 hours, 5 minutes, 35.8 seconds, or approximately 27¼ days.

nodical period. The interval between two succesive passes of a satellite through the ascending node. See also ORBITAL PERIOD.

nominal orbit. The true or ideal orbit in which an artificial satellite is expected to travel. See also NORMAL ORBIT.

nominal range. See under VISUAL RANGE (OF A LIGHT).

nomogram, *n.* A diagram showing, to scale, the relationship between several variables in such manner that the value of one which corresponds to known values of the others can be determined graphically. Also called NOMOGRAPH.

nomograph, *n.* See NOMOGRAM.

non-dangerous wreck. A term used to describe a wreck having more than 20 meters of water over it. This term excludes a FOUL GROUND, which is frequently covered by the remains of a wreck and is a hazard only for anchoring, taking the ground, or bottom fishing.

nongravitational perturbations. Perturbations caused by surface forces due to mechanical drag of the atmosphere (in case of low flying satellites), electromagnetism, and solar radiation pressure.

nonharmonic constants. Tidal constants such as lunitidal intervals, ranges, and inequalities which may be derived directly from high and low water observations without regard to the harmonic constituents of the tide. Also applicable to tidal currents.

non-standard buoys. The general classification of all lighted and unlighted buoys built to specifications other than modern (1962) standard designs.

non-tidal basin. An enclosed basin separated from tidal waters by a caisson or flood gates. Ships are moved into the dock near high tide. The dock is closed when the tide begins to fall. If necessary, ships are kept afloat by pumping water into the dock to maintain the desired level. Also called WET DOCK. See also BASIN, definition 2.

nontidal current. See under CURRENT.

noon, *n.* The instant at which a time reference is over the upper branch of the reference meridian. Noon may be **solar** or **sidereal** as the sun or vernal equinox is over the upper branch of the reference meridian. Solar noon may be further classified as **mean** or **apparent** as the mean or apparent sun is the reference. Noon may also be classified according to the reference meridian, either the local or Greenwich meridian or additionally in the case of mean noon, a designated **zone meridian**. Standard, daylight saving or summer noon are variations of zone noon. The instant the sun is over the upper branch of any meridian of the moon is called **lunar noon**. Local apparent noon may also be called **high noon**.

noon constant. A predetermined value added to a meridian or ex-meridian sextant altitude to determine the latitude.

noon interval. The predicted time interval between a given instant, usually the time of a morning observation, and local apparent noon. This is used to predict the time for observing the sun on the celestial meridian.

noon sight. Measurement of the altitide of the sun at local apparent noon, or the altitude so measured.

normal, *adj.* Perpendicular. A line is normal to another line or a plane when it is perpendicular to it. A line is normal to a curve or curved surface when it is perpendicular to the tangent line or plane at the point of tangency.

normal, *n.* 1. A straight line perpendicular to a surface or to another line.

2. In geodesy, the straight line perpendicular to the surface or the reference ellipsoid.

3. The average, regular, or expected value of a quantity.

normal curve. Short for NORMAL DISTRIBUTION CURVE.

normal distribution. A mathematical law which predicts the probability that the random error of any given observation of a series of observations of the value of a certain quantity will lie within certain bounds. The law can be derived from the following properties of random errors: (1) positive and negative errors of the same magnitude are about equal in number, (2) small errors occur more frequently than large errors, and (3) extremely large errors rarely occur. One immediate consequence of these properties is that the average or mean value of a large number of observations of a given quantity is zero. The basic equation of the normal distribution is

$$f(x) = \frac{1}{\sigma\sqrt{2\pi}}\, e^{\frac{-(x-a)^2}{2\sigma^2}} \qquad -\infty < x < \infty$$

where σ is the standard deviation and a is the mean of the distribution. Also called GAUSSIAN DISTRIBUTION. See also SINGLE-AXIS NORMAL DISTRIBUTION, CIRCULAR NORMAL DISTRIBUTION, STANDARD DEVIATION.

normal distribution curve. The graph of the normal distribution. Often shortened to NORMAL CURVE.

normal orbit. The orbit of a spherical satellite about a spherical primary during which there are no disturbing elements present due to other celestial bodies, or to some physical phenomena. Also called UNPERTURBED ORBIT, UNDISTURBED ORBIT.

normal section line. A line on the surface of a reference ellipsoid, connecting two points on that surface, and traced by a plane containing the normal at one point and passing through the other point.

normal tide. A nontechnical term synonymous with tide, i.e., the rise and fall of the ocean due to the gravitational interactions of the sun, moon, and earth alone. Use of this term should be discouraged.

norte, *n.* A strong cold northeasterly wind which blows in Mexico and on the shores of the Gulf of Mexico. It results from an outbreak of cold air from the north. It is the Mexican extension of a *norther*.

north, *n.* The primary reference direction relative to the earth; the direction indicated by 000° in any system other than relative. **True north** is the direction of the north geographical pole; **magnetic north** the direction north as determined by the earth's magnetic compass; **grid north** an arbitrary reference direction used with grid navigation. See also CARDINAL POINT.

North Africa Coast Current. A nontidal current

in the Mediterranean Sea that flows eastward along the African coast from the Strait of Gibralter to the Strait of Sicily. It is the most permanent current in the Mediterranean Sea. The stability of the current is indicated by the proportion of no current observations, which averages less than 1 percent. The current is most constant just after it passes through the Strait of Gibraltar; in this region, west of longitude 3°W, 65 percent of all observations show an eastward set, with a mean speed of 1.1 knots and a mean maximum speed of 3.5 knots. Although the current is weaker between longitudes 3°W and 11°E, it remains constant, the speed averaging 0.7 knot through its length and its maximum speed being about 2.5 knots.

North American Datum of 1927. The geodetic datum the origin of which is located at Meades Ranch, Kansas. Based on the Clarke spheroid of 1866, the geodetic position of triangulation station Meades Ranch and azimuth from that station to station Waldo are as follows:

Latitude of Meades Ranch	39° 13′ 25″.686N
Longitude of Meades Ranch	98° 32′ 30″.506W
Azimuth to Waldo	75° 28′ 09″.64

The geoidal height at Meades Ranch is assumed to be zero.

North Atlantic Current. An ocean current which results from extensions of the Gulf Stream and the Labrador Current near the edge of the Grand Banks of Newfoundland. As the current fans outward and widens in a northeastward through eastward flow, it decreases sharply in speed and persistence. Some influence of the Gulf Stream is noticeable near the extreme southwestern boundary of the current. The North Atlantic Current is a sluggish, slow-moving flow that can easily be influenced by opposing or augmenting winds. There is some evidence that the weaker North Atlantic Current may consist of separate eddies or branches which are frequently masked by a shallow, wind-driven surface flow called the NORTH ATLANTIC DRIFT. A branch of the North Atlantic Current flows along the west coasts of the British Isles at speeds up to 0.6 knot and enters the Norwegian Sea as the NORWAY CURRENT mainly through the east side of the Faeroe – Shetland Channel. A small portion of this current to the west of the Faeroe Islands mixes with part of the southeastward flow from the north coast of Iceland; these two water masses join and form a clockwise circulation around the Faeroe Islands. The very weak nontidal current in the Irish Sea, which averages only about 0.1 knot, depends on the wind. The part of the North Atlantic Current that flows eastward into the western approaches to the English Channel tends to increase or decrease the speed of the reversing tidal currents. The southern branch of the North Atlantic Current turns southward near the Azores to become the CANARY CURRENT.

North Atlantic Drift. See under NORTH ATLANTIC CURRENT.

northbound node. See ASCENDING NODE.

North Brazil Current. See GUIANA CURRENT.

North Cape Current. An Arctic Ocean current flowing northeastward and eastward around northern Norway, and curving northeastward into the Barents Sea. The North Cape Current is the continuation of the northeastern branch of the NORWAY CURRENT.

northeaster, nor'easter, *n.* A northeast wind, particularly a strong wind or gale.

northeast monsoon. See under MONSOON.

north equatorial current. See ATLANTIC NORTH EQUATORIAL CURRENT, PACIFIC NORTH EQUATORIAL CURRENT.

norther, *n.* A northerly wind. In the southern United States, especially in Texas (**Texas norther**), in the Gulf of Mexico, in the Gulf of Panama away from the coast, and in central America (*norte*), the norther is a strong cold wind from between northeast and northwest. It occurs between November and April, freshening during the afternoon and decreasing at night. It is a cold air outbreak associated with the southward movement of a cold anticyclone. It is usually preceded by a warm and cloudy or rainy spell with southerly winds. The norther comes as a rushing blast and brings a sudden drop of temperature of as much as 25°F in 1 hour or 50°F in 3 hours in winter. The **California norther** is a strong, very dry, dusty, northerly wind which blows in late spring, summer and early fall in the valley of California or on the west coast when pressure is high over the mountains to the north. It lasts from 1 to 4 days. The dryness is due to adiabatic warming during descent. In summer it is very hot. The **Portuguese norther** is the beginning of the trade wind west of Portugal. The term is used for a strong north wind on the coast of Chile which blows occasionally in summer. In southeast Australia, a hot dry wind from the desert is called a norther.

northern lights. See AURORA BOREALIS.

north frigid zone. That part of the earth north of the Arctic Circle.

north geographical pole. The geographical pole in the Northern Hemisphere, at lat. 90°N.

north geomagnetic pole. The geomagnetic pole in the Northern Hemisphere at lat. 78°.5N, long. 69°W, approximately. This term should not be confused with NORTH MAGNETIC POLE. See also GEOMAGNETIC POLE.

northing, *n.* The distance a craft makes good to the north. The opposite is SOUTHING.

north magnetic pole. The magnetic pole in the Northern Hemisphere, at lat. 78°.2N, long. 102°.9W, approximately, in 1980. This term should not be confused with NORTH GEOMAGNETIC POLE. See also GEOMAGNETIC POLE.

North Pacific Current. Flowing eastward from the eastern limit of the Kuroshio Extension (about longitude 170°E), the North Pacific Current forms the northern part of the general clockwise oceanic circulation of the North Pacific Ocean.

north polar circle. See ARCTIC CIRCLE.

north pole. 1. *Usually capitalized.* The north geographical pole. See also MAGNETIC POLE, GEOMAGNETIC POLE.

2. The north-seeking end of a magnet. See also RED MAGNETISM.

north temperate zone. That part of the earth between the Tropic of Cancer and the Arctic Circle.

north up, north upward. One of the three basic orientations of display of relative or true motion on a radarscope. In the NORTH UP orientation, the target pips are painted at their measured distances and in their *true* (compass) directions from own ship, north being maintained UP or at the top of the radarscope. See also HEAD UP, BASE COURSE UP.

northwester, nor'wester, *n.* A northwestly wind.

Norway Coastal Current. Originating mainly from Oslofjord outflow, counterclockwise return flow of the Jutland Current within the Skaggerak, and outflow from the Kattegat, the Norway

Coastal Current begins at about latitude 59°N, longitude 10°E and follows the coast of Norway. The current extends to about 20 miles in width. Speeds are strongest off the southeast coast of Norway, where they frequently range between 1 and 2 knots. Along the remainder of the coast the current gradually weakens. It may widen to almost 30 miles at about latitude 63°N, where it joins the NORWAY CURRENT. South of latitude 62°N the current speed usually ranges between 0.4 and 0.9 knots. Speeds are generally stronger in spring and summer, when the flow is augmented by increased discharge from fiords.

Norway Current. An Atlantic Ocean current flowing northeastward along the northwest coast of Norway, and gradually branching and continuing as the SPITZBERGEN ATLANTIC CURRENT and the NORTH CAPE CURRENT. The Norway Current is the continuation of part of the northern branch of the North Atlantic Current. Also called NORWEGIAN CURRENT.

Norwegian Current. See NORWAY CURRENT.

notch filter. An arrangement of electronic components designed to attenuate or reject a specific frequency band with a sharp cut-off at either end.

notice board. A signboard used to indicate speed restrictions, cable landings, etc.

Notice to Mariners. A weekly publication of the Defense Mapping Agency Hydrographic/Topographic Center prepared jointly with the National Ocean Survey and the U.S. Coast Guard giving information on changes in aids to navigation (lights, buoys, daymarks, ranges), dangers to navigation (rocks, shoals, reefs, wrecks), selected items from the *Local Notice to Mariners*, important new soundings, changes in channels, harbor construction, radionavigation information, new and revised charts and publications, special warnings and notices, pertinent HYDROLANT, HYDROPAC, NAVAREA IV and XII messages and in general, all such information as affects the mariner's charts, manuals, catalogs, sailing directions (pilots), etc. The *Notice to Mariners* should be used routinely for updating the latest editions of nautical charts and related publications. *Notice to Mariners* may be consulted at Coast Guard District Offices, Defense Mapping Agency offices and depots, naval stations, custom houses, shipping company offices, most sales agents' offices, etc. See also GREAT LAKES NOTICE TO MARINERS.

nova (*pl. novae*), *n.* A star which suddenly becomes many times brighter than previously, and then gradually fades. Novae are believed to be exploding stars.

NOVA Satellite. See under OSCAR SATELLITE.

nucleus, *n.* The central, massive part of anything, such as an atom or comet.

numerical scale. A statement of that distance on the earth shown in one unit (usually an inch) on the chart, or vice versa. For example, "30 miles to the inch" means that 1 inch on the chart represents 30 miles on the earth's surface. See also REPRESENTATIVE FRACTION.

nun buoy. An unlighted buoy of which the upper part of the body (above the waterline), or the larger part of the superstructure, has approximately the shape of a cone with vertex upwards. In France, a conical buoy of which the lower part of the body (mainly below the waterline) is of spherical shape is called BOUEE SPHÉRO-CONIQUE. Called CONICAL BUOY in British terminology.

nutation, *n.* Irregularities in the precessional motion of the equinoxes due chiefly to regression of the nodes.

O

object glass. See OBJECTIVE.

objective, *n.* The lens or combination of lenses which receives light rays from an object, and refracts them to form an image in the focal plane of the eyepiece of an optical instrument, such as a telescope. Also called OBJECT GLASS.

oblate spheroid. An ellipsoid of revolution, the shorter axis of which is the axis of revolution. An ellipsoid of revolution, the longer axis of which is the axis of revolution, is called a PROLATE SPHEROID. The earth is approximately an oblate spheroid.

oblique, *adj.* Neither perpendicular nor parallel; slanting.

oblique angle. Any angle not a multiple of 90°.

oblique ascension. The arc of the celestial equator, or the angle at the celestial pole, between the hour circle of the vernal equinox and the hour circle through the intersection of the celestial equator and the eastern horizon at the instant a point on the oblique sphere rises, measured eastward from the hour circle of the vernal equinox through 24^h. The expression is not used in modern navigation.

oblique chart. A chart on an oblique map projection.

oblique coordinates. Magnitudes defining a point relative to two intersecting non-perpendicular lines, called AXES. The magnitudes indicate the distance from each axis, measured along a parallel to the other axis. The horizontal distance is called the **abscissa** and the other distance the **ordinate.** This is a form of CARTESIAN COORDINATES.

oblique cylindrical orthomorphic chart. See OBLIQUE MERCATOR CHART.

oblique cylindrical orthomorphic map projection. See OBLIQUE MERCATOR MAP PROJECTION.

oblique equator. A great circle the plane of which is perpendicular to the axis of an oblique projection. An oblique equator serves as the origin for measurement of oblique latitude. On an oblique Mercator map projection, the oblique equator is the tangent great circle. See also FICTITIOUS EQUATOR.

oblique graticule. A fictitious graticule based upon an oblique map projection.

oblique latitude. Angular distance from an oblique equator. See also FICTITIOUS LATITUDE.

oblique longitude. Angular distance between a prime oblique meridian and any given oblique meridian. See also FICTITIOUS LONGITUDE.

oblique map projection. A map projection with an axis inclined at an oblique angle to the plane of the equator.

oblique Mercator chart. A chart on the oblique Mercator map projection. Also called OBLIQUE CYLINDRICAL ORTHOMORPHIC CHART. See also MERCATOR CHART.

oblique Mercator map projection. A conformal cylindrical map projection in which points on the surface of a sphere or spheroid, such as the earth, are conceived as developed by Mercator principles on a cylinder tangent along an oblique great circle. Also called OBLIQUE CYLINDRICAL ORTHOMORPHIC MAP PROJECTION. See also MERCATOR MAP PROJECTION.

oblique meridian. A great circle perpendicular to an oblique equator. The reference oblique meridian is called **prime oblique meridian.** See also FICTITIOUS MERIDIAN.

oblique parallel. A circle or line parallel to an oblique equator, connecting all points of equal oblique latitude. See also FICTITIOUS PARALLEL.

oblique pole. One of the two points 90° from an oblique equator.

oblique rhumb line. 1. A line making the same oblique angle with all fictitious meridians of an oblique Mercator map projection. Oblique parallels and meridians may be considered special cases of the oblique rhumb line.
2. Any rhumb line, real or fictitious, making an oblique angle with its meridians. In this sense the expression is used to distinguish such rhumb lines from parallels and meridians, real or fictitious, which may be included in the expression rhumb line.
See also FICTITIOUS RHUMB LINE.

oblique sphere. The celestial sphere as it appears to an observer between the equator and the pole, where celestial bodies appear to rise obliquely to the horizon.

obliquity factor. A factor in an expression for a constituent tide or tidal current involving the angle of the inclination of the moon's orbit to the plane of the earth's equator.

obliquity of the ecliptic. The acute angle between the plane of the ecliptic and the plane of the celestial equator, about 23° 27′.

obscuration, *n.* The designation for the sky cover when the sky is completely hidden by obscuring phenomena in contact with, or extending to, the surface.

obscuring phenomenon. Any atmospheric phenomenon, not including clouds, which restricts the vertical or slant visibility.

observed altitude. Corrected sextant altitude; angular distance of the center of a celestial body above the celestial horizon of an observer, measured along a vertical circle, through 90°. Occasionally called TRUE ALTITUDE. See also ALTITUDE INTERCEPT, APPARENT ALTITUDE, SEXTANT ALTITUDE.

observed gravity anomaly. See GRAVITY ANOMALY.

observed latitude. See LATITUDE LINE.

observed longitude. See LONGITUDE LINE.

obstruction, *n.* Anything that hinders or prevents movement, particularly anything that endangers or prevents passage of a vessel or aircraft. The term is usually used to refer to an isolated danger to navigation, such as a submerged rock or pinnacle in the case of marine navigation, and a tower, tall building, mountain peak, etc., in the case of air navigation.

obstruction buoy. A buoy used alone to indicate a dangerous reef or shoal. The buoy may be passed on either hand. Called ISOLATED DANGER BUOY in British terminology.

obstruction light. A light indicating a radio tower or other obstruction to aircraft.

obstruction mark. A navigation mark used alone to indicate a dangerous reef or shoal. The mark may be passed on either hand. Called ISOLATED DANGER MARK in British terminology.

obtuse angle. An angle greater than 90° and less than 180°.

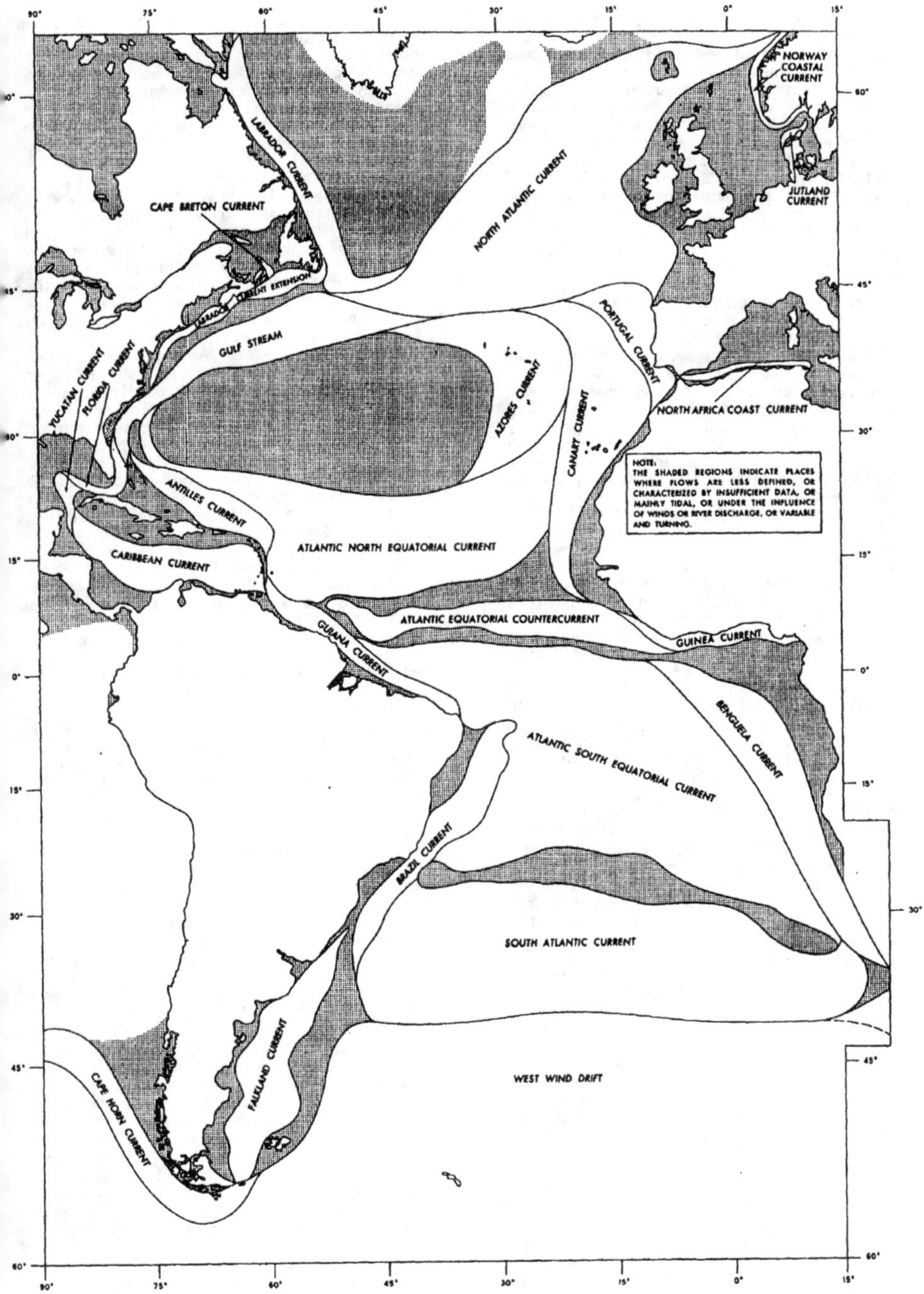

MAJOR SURFACE CURRENTS ATLANTIC

occasional light. A light put into service only on demand.

occluded front. A composite of two fronts, formed as a cold front overtakes a warm front or stationary front. This is a common process in the late stages of wave-cyclone development, but is not limited to occurrence within a wave-cyclone. There are three basic types of occluded front, determined by the relative coldness of the air behind the original cold front to the air ahead of the warm (or stationary) front. A **cold occlusion** results when the coldest air is behind the cold front. The cold front undercuts the warm front and, at the earth's surface, coldest air replaces less-cold air. When the coldest air lies ahead of the warm front, a **warm occlusion** is formed in which case the original cold front is forced aloft at the warm-front surface. At the earth's surface, coldest air is replaced by less-cold air. A third and frequent type, a **neutral occlusion**, results when there is no appreciable temperature difference between the cold air masses of the cold and warm fronts. In this case frontal characteristics at the earth's surface consist mainly of a pressure trough, a wind-shift line, and a band of cloudiness and precipitation. Commonly called OCCLUSION. Also called FRONTAL OCCLUSION.

occlusion, *n.* 1. See OCCLUDED FRONT.
2. The process of formation of an occluded front. Also called FRONTAL OCCLUSION.

occultation, *n.* 1. The concealment of a celestial body by another which crosses the line of view. Thus, the moon *occults* a star when it passes between the observer and the star.
2. A relatively brief interval of darkness, in comparison with the longest interval of light in the period of the light. See also FLASH.

occulting light. A light totally eclipsed at regular intervals, the duration of light always clearly longer than the equal intervals of darkness called OCCULTATIONS. The term is commonly used for a SINGLE OCCULTING LIGHT, an occulting light exhibiting only single occultations which are repeated at regular intervals.

occupied bandwidth. As defined by the *International Telecommunication Union* (ITU) the frequency bandwidth such that, below its lower and above its upper frequency limits, the mean powers radiated are each equal to 0.5 percent of the total mean power radiated by a given emission. In some cases, for example multichannel frequency-division systems, the percentage of 0.5 percent may lead to certain difficulties in the practical application of the definitions of occupied and necessary bandwidth; in such cases a different percentage may prove useful.

ocean, *n.* 1. The vast expanse of salt water covering the greater part of the earth.
2. One of the major divisions of the vast expanse of salt water of the earth.

ocean current. A movement of ocean water characterized by regularity, either of a cyclic nature, or more commonly as a continuous stream flowing along a definable path. Three general classes, by cause, may be distinguished: (a) currents associated with horizontal pressure gradients, comprising the various types of **gradient current**; (b) **wind-driven currents**, which are those directly produced by the stress exerted by the wind upon the ocean surface; (c) currents produced by long-wave motions. The latter are principally **tidal currents**, but may include currents associated with internal waves, tsunamis, and seiches. The major ocean currents are of continuous, stream-flow character, and are of first-order importance in the maintenance of the earth's thermodynamic balance.

oceanic, *adj.* Of or pertaining to the ocean.

oceanographic, *adj.* Of or pertaining to oceanography, or knowledge of the oceans.

oceanographic survey. The study or examination of conditions in the ocean or any part of it, with reference to animal or plant life, chemical elements present, temperature gradients, etc. See also HYDROGRAPHIC SURVEY.

oceanography, *n.* The study of the sea, embracing and integrating all knowledge pertaining to the sea's physical boundaries, the chemistry and physics of sea water, and marine biology. Strictly, oceanography is the description of the marine environment, whereas OCEANOLOGY is the study of the oceans and related sciences.

oceanology, *n.* The study of the ocean and related sciences. See also OCEANOGRAPHY.

Ocean Passages for the World. A British publication relating to the planning and conduct of ocean passages. Published by the Hydrographer of the Navy, *Ocean Passages for the World* addresses those areas which lie mainly outside the areas covered in detail by Admiralty Sailing Directions. It is kept up-to-date by periodical supplements. The publication should not be used without reference to the latest supplement and those Notices to Mariners published specially to correct Sailing Directions.

ocean waters. For application to the provisions of the Marine Protection, Research, and Sanctuaries Act of 1972, those waters of the open seas lying seaward of the base line from which the territorial sea is measured.

octagon, *n.* A closed plane figure having eight sides.

octahedral cluster. An arrangement of eight corner reflectors with common faces designed to give a substantially uniform response in all directions. The octahedral cluster is formed by mounting three rectangular plates mutually at right angles, the geometric centers of the plates being coincident. See also PENTAGONAL CLUSTER.

octant, *n.* A double-reflecting instrument for measuring angles, used primarily for measuring altitude of celestial bodies. It has a range of 90°. Such an instrument is commonly called a SEXTANT.

octant altitude. See SEXTANT ALTITUDE.

Odessey protractor. A device used in conjunction with a plotting sheet having equally spaced concentric circles (range circles) drawn about two or more stations of a radiodetermination system being operated in the ranging mode. The device consists of a transparent plate with inscribed equally spaced concentric circles, the radius of the outer circle being equal to the circle spacing on the associated plotting sheet. The circle spacing on the device represents a convenient subdivision of the circle spacing on the plotting sheet. There is a hold at the center of the device so that the plotting sheet can be marked when the appropriate concentric circles of the device are tangent to the appropriate circles on the plotting sheet. For example, if the measured range is 9,300 meters, the concentric circle of the device corresponding to 300 meters is made tangent to the plotting sheet circle representing 9,000 meters.

oe (ō), *n.* A whirlwind off the Faeroe Islands.

oersted, *n.* The centimeter-gram-second electromagnetic system unit of magnetic field strength. It corresponds to $1000/4\pi$ ampere per meter.

off-center PPI display. A plan position indicator display in which the center about which the sweep rotates is offset from the center of the radarscope.

offing, *n.* That part of the visible sea a considerable distance from the shore, or that part just beyond the limits of the area in which a pilot is needed.

offshore, *adj. & adv.* Away from the shore.

offshore, *n.* The comparatively flat zone of variable width which extends from the outer margin of the rather steeply sloping shoreface to the edge of the shelf.

offshore light stations. Manned light stations built on exposed marine sites to replace lightships.

offshore navigation. Navigation at a distance from a coast, in contrast with COASTWISE NAVIGATION in the vicinity of a coast.

offshore water. Water adjacent to land in which the physical properties are slightly influenced by continental conditions.

offshore wind. Wind blowing from the land toward the sea. An ONSHORE WIND blows in the opposite direction. See also LAND BREEZE.

off soundings. Said of a vessel navigating beyond the 100-fathom curve. In earlier times, said of a vessel in water deeper than could be sounded with the sounding lead.

off station. Said of a floating aid to navigation not on its charted station.

ogival buoy. A buoy with a pointed-arch shaped vertical cross-section above the waterline. Used in the cardinal system.

ohm, *n.* A *derived unit* of electrical resistance in the International System of Units; it is the electrical resistance between two points of a conductor when a constant potential difference of 1 volt, applied to these points, produces in the conductor a currrent of 1 ampere, the conductor not being the seat of an electromotive force.

old ice. Sea ice which has survived at least one summer's melt. Most topographic features are smoother than on first-year ice. Old ice may be subdivided into SECOND-YEAR ICE and MULTI-YEAR ICE.

Omega Navigation System. A worldwide, continuous, radionavigation system of medium accuracy which provides hyperbolic lines of position through phase comparisons of VLF (10-14kHz) continuous wave signals transmitted on a common frequency on a time-shared basis. The fully implemented system is comprised of only eight transmitting stations. See also DIFFERENTIAL OMEGA.

Omega plotting chart. See under PLOTTING CHART.

Omega Table. See PUB. 224, OMEGA TABLE.

omni-. A prefix meaning *all.*

omniazimuthal antenna. See OMNIDIRECTIONAL ANTENNA.

omnidirectional antenna. An antenna whose radiating or receiving properties at any instant are the same on all bearings. Also called OMNIAZIMUTHAL ANTENNA. See also DIRECTIONAL ANTENNA.

omnidirectional light. A light which presents the same characteristic over the whole horizon of interest to marine navigation. Also called ALL-ROUND LIGHT.

omnidirectional radiobeacon. A radiobeacon transmitting a signal in all directions. A **circular radiobeacon** is an omnidirectional beacon which transmits in all horizontal directions simultaneously. A **rotating radiobeacon** is an omnidirectional beacon with one or more beams that rotate. A DIRECTIONAL RADIOBEACON is a beacon which beams its signals in one or several prescribed directions.

onshore wind. Wind blowing from the sea toward the land. An OFFSHORE WIND blows in the opposite direction. See also SEA BREEZE.

on soundings. Said of a vessel navigating within the 100-fathom curve. In earlier times, said of a vessel in water sufficiently shallow for sounding by sounding lead.

on the beam. Bearing approximately 090° relative (*on the starboard beam*) or 270° relative (*on the port beam*). The expression is often used loosely for BROAD ON THE BEAM, or bearing exactly 090° or 270° relative. Also called A-BEAM.

on the bow. Bearing approximately 045° relative (*on the starboard bow*) or 315° relative (*on the port bow*). The expression is often used loosely for BROAD ON THE BOW, or bearing exactly 045° or 315° relative.

on the quarter. Bearing approximately 135° relative (*on the starboard quarter*) or 225° relative (*on the port quarter*). The expression is often used loosely for BROAD ON THE QUARTER, or bearing exactly 135° or 225° relative.

ooze, *n.* A soft, slimy, organic sediment covering part of the ocean bottom, composed principally of shells or other hard parts of minute organisms.

open, *v., i.* To move or appear to move apart. An order is sometimes given by a flagship for a vessel to *open out to —— yards* or *miles.* When a craft moves off a range, the objects forming the range appear to separate or *open.* The opposite is CLOSE.

open basin. See TIDAL BASIN.

open berth. An anchorage berth in an open roadstead.

open coast. A coast that is not sheltered from the sea.

open harbor. An unsheltered harbor exposed to the sea.

opening, *n.* A break in a coastline or a passage between shoals, etc. See also GAT.

open pack ice. Pack ice in which the concentration is 4/10 to 6/10 (3/8 to less than 6/8), with many leads and polynyas, and the floes are generally not in contact with one another.

open roadstead. A roadstead with relatively little protection from the sea.

open sea. 1. That part of the ocean not enclosed by headlands, within narrow straits, etc.
2. That part of the ocean outside the territorial jurisdiction of any country. The opposite is CLOSED SEA. See also HIGH SEAS.

open water. A large area of freely navigable water in which sea ice is present in concentrations less than 1/10 (1/8). When there is no sea ice present, the area should be described as ICE-FREE, even though icebergs may be present.

operating area chart. A base chart with overprints of various operating areas necessary to control fleet exercise activities. Submarine Transit Lanes, Surface and Sub-surface Operating Areas, Air Space Warning Areas, Controlled Air Spaces, and other restricted areas are portrayed.

opposition, *n.* The situation of two celestial bodies having either celestial longitudes or sidereal hour angles differing by 180°. The term is usually used only in relation to the position of a superior planet or the moon with reference to the sun. The situation of two celestial bodies having either the same celestial longitude or the same sidereal hour angle is called **conjunction.**

optic, *adj.* Of or pertaining to vision.

optical, *adj.* Of or pertaining to optics or to vision.

optical double star. Two stars in nearly the same line of sight but differing greatly in distance from the observer, as distinguished from a PHYSICAL DOUBLE STAR (two stars in nearly the same line of sight and at approximately the same distance from the observer).

optical glass. Glass of which the compositon and molding are carefully controlled in order to insure uniform refractive index and high transmission factor.

optical path. The path followed by a ray of light through an optical system.

optical system. A series of lenses, apertures, prisms, mirrors, etc., so arranged as to perform a definite optical function.

optics, *n.* The science dealing with light.

Optimum Track Ship Routing. See under SHIP WEATHER ROUTING.

orbit, *n.* 1. The path of a body or particle under the influence of a gravitational or other force. For example, the orbit of a celestial body is its path relative to another body around which it revolves. Orbit is commonly used to designate a closed path.
2. As defined by the *International Telecommunication Union* (ITU), the path, relative to a specified frame of reference, described by the center of mass of a satellite or other object in space, subjected solely to natural forces, mainly the force of gravity and, by extension, the path described by the center of mass of an object in space subjected to natural forces and occasional low-energy corrective forces exerted by a propulsive device in order to achieve and maintain a desired path. See also CENTRAL FORCE ORBIT, INERTIAL ORBIT, INTERMEDIATE ORBIT, NOMINAL ORBIT, NORMAL ORBIT, OSCULATING ORBIT, PERTURBED ORBIT, POLAR ORBIT, STATIONARY ORBIT.

orbital altitude. The mean altitude of the orbit of a satellite above the surface of the parent body.

orbital elements. Parameters that specify the position and motion of a body in orbit. The elliptical orbit of a satellite attracted by an exactly central gravitational force is unambiguously specified by a set of six parameters as follows: Two parameters, the semimajor axis and eccentricity of the ellipse, establish the size and shape of the elliptical orbit. A third parameter, time of perifocal passage, enables determination of the location of the satellite in its orbit at any instant. The three remaining parameters establish the orientation of the orbit in space. These are the inclination of the orbital plane to a reference plane, the right ascension of the ascending node of the satellite, and the argument of pericenter. The orbital period is sometimes considered as a seventh element, but it is not actually an independent parameter because by knowing the masses of satellite and primary it can be calculated from the universal law of gravitation and the geometry of the orbit. The **right ascension of the ascending node** of the satellite is the arc of the celestial equator, measured eastward along the celestial equator, from the vernal equinox to the ascending node (that intersection point of the orbital plane and the celestial equator at which the satellite passes the equator from south to north). The **argument of pericenter** is the angle, as seen from the focus of the ellipse at the center of the primary, measured in the orbital plane from the ascending node to the pericenter, in the direction of the satellite's orbit. See also ORBITAL PARAMETERS OF ARTIFICIAL SATELLITE, MEAN ELEMENTS, OSCULATING ELEMENTS.

orbital inclination. See as INCLINATION, definition 2.

orbital mode. A method for determining the position of an unknown station position when the unknown position cannot be viewed simultaneously with known positions. The arc of the satellite orbit is extrapolated from the ephemeris of the satellite determined by the known stations which permits the determination of the position of the unknown station dependent completely on the satellite's orbital parameters.

orbital motion. Continuous motion in a closed path about and as a direct result of a source of gravitational attraction.

orbital parameters of artificial earth satellite. The precessing elliptical orbit of an artificial earth satellite is unambiguously specified by the following set of parameters: semimajor axis, eccentricity, time of perigee, inclination of the orbital plane to the plane of the reference plane (celestial equator), the right ascension of the ascending node of the satellite at time of perigee, the argument of perigee at time of perigee, right ascension of Greenwich at time of perigee, mean motion (rate of change of mean anomaly), rate of change of argument of perigee, and rate of change of right ascension of the ascending node at time of perigee. With the inclination expressed as the sine and cosine of the orbital inclination, the parameters number 11 as in the table below. See also ORBITAL ELEMENTS.

Symbol	Meaning	Units
t_p	Time of perigee.	Min.
$\dot{M}$	Rate of change of mean anomaly.	Deg/Min.
ϕ	Argument of perigee at t_p.	Deg.
$\dot{\phi}$	Rate of change of argument of perigee (absolute value).	Deg/Min.
ϵ	Eccentricity	
A_O	Semimajor axis.	Km.
Ω_N	Right ascension ascending node at t_p.	Deg.
$\dot{\Omega}$	Rate of change of Ω_N.	Deg/Min.
$\cos\psi$	Cosine of orbit inclination.	
Ω_G	Right Ascension Greenwich.	Deg.
$\sin\psi$	Sine of orbit inclination.	

ORBITAL PARAMETERS OF ARTIFICIAL EARTH SATELLITE

orbital path. One of the tracks on a primary body's surface traced by the subpoint of a satellite that orbits about it several times in a direction other than normal to the primary body's axis of rotation. Each track is displaced in a direction opposite and by an amount equal to the degrees of rotation between each satellite orbit and of the nodical precession of the plane of the orbit. Also called SUBTRACK. See also WESTWARD MOTION.

orbital period. If the orbit is unchanging and ideal, the interval between successive passages of a satellite through the same point in its orbit. If the orbit is not ideal, the point must be specified. When the perigee is specified, it is called **radial** or **anomalistic period.** When the ascending node is specified, it is called **nodical period.** When the same geocentric right ascension is specified, it is called **sidereal period.** Also called PERIOD OF SATELLITE.

orbital plane. The plane of the ellipse defined by a central force orbit.

orbital velocity. The velocity of an earth satellite or other orbiting body at any given point in its orbit.

ordinary, *adj.* With respect to tides, the use of this nontechnical term has, for the most part, been determined to be synonymous with *mean*. Thus, ordinary high (low) water is the equivalent of mean high (low) water. The use of the term ordinary in tidal terms is discouraged.

ordinate, *n.* The vertical coordinate of a set of rectangular coordinates. Also used in a similar sense in connection with oblique coordinates.

orient, *v., t.* 1. To line up or adjust with respect to a reference.
2. To obtain a mental grasp of the existing situation, such as the location of a vessel relative to nearby aids to navigation and landmarks, the course, speed, etc.

orientability of a sound signal. The property of a sound signal by virtue of which a listener can estimate the direction of the location of the apparatus producing the signal.

orographic rain. Rain resulting when moist air is forced upward by a mountain range.

orthodrome, *n.* See GREAT CIRCLE.

orthodromic curve. See GREAT CIRCLE.

orthogonal, *adj.* Right angled, rectangular.

orthogonal map projection. See ORTHOGRAPHIC MAP PROJECTION.

orthographic, *adj.* Of or pertaining to right angles or perpendicular lines.

orthographic chart. A chart on the orthographic map projection.

orthographic map projection. A perspective azimuthal projection in which the projecting lines, emanating from a point at infinity, are perpendicular to a tangent plane. The projection is used chiefly in navigational astronomy for interconverting coordinates of the celestial equator and horizon systems. Also called ORTHOGONAL PROJECTION.

orthomorphic, *adj.* Preserving the correct shape. See also CONFORMAL MAP PROJECTION.

orthomorphic chart. A chart on which very small shapes are correctly represented. See also CONFORMAL MAP PROJECTION.

orthomorphic map projection. A projection in which very small shapes are correctly represented. See also CONFORMAL MAP PROJECTION.

oscar satellite. A general term for one of the operational satellites of the Navy Navigation Satellite System, except for satellite 30110 called TRANSAT, placed in orbit prior to 1981. The improved satellites placed in orbit beginning in 1981 are called NOVA.

oscillation, *n.* 1. Fluctuation or vibration each side of a mean value or position.
2. Half an oscillatory cycle, consisting of a fluctuation or vibration in one direction; half a vibration.

oscillator, *n.* A sound signal emitter comprising a resonant diaphragm maintained in vibrating motion by electromagnetic action.

oscillatory wave. A wave in which only the form advances, the individual particles of the medium moving in closed orbits, as ocean waves in deep water; in contrast with a WAVE OF TRANSLATION, in which the individual particles are shifted in the direction of wave travel, as ocean waves in shoal water.

oscilloscope, *n.* An instrument for producing a visual representation of oscillations or changes in an electric current. The face of the cathode-ray tube used for this representation is called a SCOPE or SCREEN.

osculating elements. A set of parameters that specifies the instantaneous position and velocity of a celestial body, or artificial satellite, in its perturbed orbit. Osculating elements describe the unperturbed (two-body) orbit (osculating orbit) that the body would follow if perturbations were to cease instantaneously.

osculating orbit. The ellipse that a satellite would follow after a specific time "t" (the epoch of osculation) if all forces other than central forces ceased to act from "t" on. An osculating orbit is tangent to the real, perturbed, orbit and has the same velocity at the point of tangency. See also OSCULATING ELEMENTS.

outage, *n.* The failure of an aid to navigation to function exactly as described in the light list.

outer harbor. See under INNER HARBOR.

outfall, *n.* The discharge end of a narrow stream, sewer, drain, etc.

outfall buoy. A buoy marking the position where a sewer or other drain discharges.

outline chart. A chart with only a generalized presentation of the landmass with little or no culture or relief. See also PLOTTING CHART.

output axis. Of a gyroscope, the axis of precession. See also INPUT AXIS, PRECESSION.

outside fix. A term descriptive of the fix position determined by the method of bisectors when the lines of position result from observations of objects or celestial bodies lying within a 180° arc of the horizon. See also METHOD OF BISECTORS.

outward bound. Heading for the open sea. The opposite is INWARD BOUND. See also HOMEWARD BOUND.

overcast, *adj.* Pertaining to a sky cover of 95% or more.

overcast, *n.* A cloud cover.

overfalls, *n. pl.* Breaking waves caused by the meeting of currents or by waves moving against the current. See also RIPS.

overhead cable effect. A radar phenomenon which may occur in the vicinity of an overhead power cable. The echo from the cable appears on the plan position indicator as a single echo, the echo being returned from that part of cable where the radar beam is at right angles to the cable. If this phenomenon is not recognized, the echo can be wrongly identified as the echo from a ship on a steady bearing. Evasive action results in the echo remaining on a constant bearing and moving to the same side of the channel as the ship altering course. This phenomenon is particularly apparent for the power cable spanning the

Straits of Messina.

overhead compass. See INVERTED COMPASS.

overhead constraints. The elevation angle limitations between which usable navigation data may be obtained from a satellite in the doppler mode.

overlay, *n.* A printing or drawing on a transparent or translucent medium at the same scale as a map, chart, etc., to show details not appearing, or requiring special emphasis on the original.

overprint, *n.* New material printed on a map or chart to show data of importance or special use, in addition to that originally printed.

overtide, *n.* A harmonic tidal (or tidal current) constituent with a speed that is an exact multiple of the speed of one of the fundamental constituents derived from the development of the tide-producing force. The presence of overtides is usually attributed to shallow water conditions.

overtone, *n.* A harmonic existing with its fundamental.

Oyashio, *n.* A cold ocean current flowing from the Bering Sea southwestward along the coast of Kamchatka, past the Kuril Islands to where it meets the Kuroshio off the east coast of Honshu. The Oyashio turns and continues eastward eventually joining the Aleutian Current.

P

Pacific Equatorial Countercurrent. A Pacific Ocean current that flows eastward (counter to and between the westward flowing Pacific North and South Equatorial Currents) between latitudes 3°N and 10°N. East of the Philippines it is joined by the southern part of the Pacific North Equatorial Current.

Pacific North Equatorial Current. A North Pacific Ocean current that flows westward between latitudes 10°N and 20°N. East of the Philippines, it divides, part turning south to join the Pacific Equatorial Countercurrent and part turning north to flow along the coast of Japan as the KUROSHIO.

Pacific South Equatorial Current. A Pacific Ocean current that flows westward between latitudes 3°N and 10°S. In midocean, much of it turns south to form a large counterclose whirl. The portion that continues across the ocean divides as it approaches Australia, part flowing north toward New Guinea and part turning south along the east coast of Australia as the EAST AUSTRALIA CURRENT.

Pacific standard time. See STANDARD TIME.

pack ice. The term used in a wide sense to include any area of sea ice, other than fast ice, no matter what form it takes or how it is disposed.

pagoda, *n.* As a landmark, a tower having a number of stories and a characteristic architecture, used as a place of worship or as a memorial, primarily in Japan, China, and India.

paint, *n.* The bright area on the phosphorescent plan position indicator screen resulting from the brightening of the sweep by the echoes.

paint, *v., t & i.* To brighten the phosphorescent plan position indicator screen through the effects of the echoes on the sweep.

painted mark. A navigation mark formed simply by painting a cliff, wall, rock, etc.

pancake ice. Predominantly circular pieces of ice from 30 centimeters to 3 meters in diameter, and up to about 10 centimeters in thickness, with raised rims due to pieces striking against one another. It may be formed on a slight swell from grease ice, shuga, or slush or as a result of the breaking of ice rind, nilas, or under severe conditions of swell or waves, of grey ice. It also sometimes forms at some depth, at an interface between water bodies of different physical characteristics, from where it floats to the surface; its appearance may rapidly cover wide areas of water.

pantograph, *n.* An instrument for copying maps, drawings, or other graphics at a predetermined scale.

papagayo, *n.* A violet northeasterly fall wind on the Pacific coast of Nicaragua and Guatemala. It consists of the cold air mass of a *norte* which has overridden the mountains of Central America. See also TEHUANTEPECER.

parabola, *n.* An open curve all points of which are equidistant from a fixed point, called the FOCUS, and a straight line. The limiting case occurs when the point is on the line, in which case the parabola becomes a straight line.

parabolic reflector. A reflecting surface having the cross section along the axis in the shape of a parabola. Parallel rays striking the reflector are brought to a focus at a point, or if the source of the rays is placed at the focus, the reflected rays are parallel. See also CORNER REFLECTOR, RADAR REFLECTOR, SCANNER.

parabolic velocity. See ESCAPE VELOCITY.

parallactic angle. That angle at the navigational triangle at the celestial body; the angle between a body's hour circle and its vertical circle. Also called POSITION ANGLE.

parallax, *n.* The difference in apparent direction or position of an object when viewed from different points. For bodies of the solar system, parallax is the difference in the direction of the body due to the displacement of the observer from the center of the earth, and is called **geocentric parallax,** varying with the body's altitude and distance from the earth. The geocentric parallax when a body is in the horizon is called **horizontal parallax,** as contrasted with the parallax at any altitude, called **parallax in altitude.** Parallax of the moon is called **lunar parallax.** In marine navigation it is customary to apply a parallax correction to sextant altitudes of the sun, moon, Venus, and Mars. For stars, parallax is the angle at the star subtended by the semimajor axis of the earth's orbit and is called **heliocentric** or **stellar parallax,** which is too small to be significant as a sextant error.

parallax correction. A correction due to parallax, particularly that sextant altitude correction due to the difference between the apparent direction from a point on the surface of the earth to a celestial body and the apparent direction from the center of the earth to the same body.

parallax in altitude. Geocentric parallax of a body at any altitude. The expression is used to distinguish the parallax at the given altitude from the horizontal parallax when the body is in the horizon. See also PARALLAX.

parallax inequality. The variation in the range of tide or in the speed of a tidal current due to changes in the distance of the moon from the earth. The range of tide and speed of the current tend alternately to increase and decrease as the moon approaches its perigee and apogee, respectively, the complete cycle being the *anomalistic month.* There is a similar but relatively unimportant inequality due to the sun, the cycle being the *anomalistic year.* The parallax has little direct effect upon the lunitidal intervals but tends to modify the phase effect. When the moon is in perigee, the priming and lagging of the tide due to the phase is diminished and when in apogee the priming and lagging is increased.

parallax reduction. A processing of observed high and low waters to obtain quantities depending upon changes in the distance of the moon, such as perigean and apogean ranges.

parallel, *adj.* Everywhere equidistant, as of lines or surfaces.

parallel, *n.* See PARALLEL OF LATITUDE, definition 1.

parallel motion protractor. An instrument consisting essentially of a protractor and one or more arms attached to a parallel motion device, so that the movement of the arms is everywhere parallel. The protractor can be rotated and set at any position so that it can be oriented to a chart. Also called DRAFTING MACHINE.

parallel of altitude. A circle of the celestial sphere parallel to the horizon, connecting all points of equal altitude. Also called ALTITUDE CIRCLE, ALMUCANTAR. See also CIRCLE OF EQUAL ALTITUDE.

parallel of declination. A circle of the celestial sphere parallel to the celestial equator. Also called CELESTIAL PARALLEL, CIRCLE OF EQUAL DECLINATION. See also DIURNAL CIRCLE.

parallel of latitude. 1. A circle (or approximation of a circle) on the surface of the earth, parallel to the equator, and connecting points of equal latitude. Also called PARALLEL.
2. A circle of the celestial sphere, parallel to the ecliptic, and connecting points of equal celestial latitude. Also called CIRCLE OF LONGITUDE.

parallelogram, *n.* A four-sided figure with both pairs of opposite sides parallel. A right-angled parallelogram is a **rectangle**; a rectangle with sides of equal length is a **square**. A parallelogram with oblique angles is a **rhomboid**; a rhomboid with sides of equal length is a **rhombus**.

parallel rulers. An instrument for transferring a line parallel to itself. In its most common form it consists essentially of two parallel bars or rulers connected in such manner that when one is held in place, the other may be moved, remaining parallel to its original position.

parallel sailing. A method of converting departure into difference of longitude, or vice versa, when the true course is 090° or 270°. The formulas for these conversions are

$$DLo = p \sec L \qquad p = DLo \cos L,$$

where DLo is difference of longitude, p is departure, and L is latitude.

parallel sphere. The celestial sphere as it appears to an observer at the pole, where celestial bodies appear to move parallel to the horizon.

parameter, *n.* 1. A quantity which remains constant within the limits of a given case or situation.
2. One of the components into which a craft's magnetic field is assumed to be resolved for the purpose of compass adjustment. The field caused by permanent magnetism is resolved into orthogonal components or parameters: Parameter P, Parameter Q, and Parameter R. The field caused by induced magnetism is resolved into that magnetism induced in 9 imaginery soft iron bars or rods. With respect to the axis of a craft, these parameters lie in a fore-and-aft direction, an athwartships direction, and in a vertical direction. See also ROD, definition 2.

paranthelion, *n.* A phenomenon similar to a parhelion but occurring generally at a distance of 120° (occasionally 90° or 140°) from the sun.

paraselene *(pl. paraselenae), n.* A form of halo consisting of an image of the moon at the same altitude as the moon and some distance from it, usually about 22°, but occasionally about 46°. Similar phenomena may occur about 90°, 120°, 140°, or 180° from the moon. A similar phenomenon in relation to the sun is called a PARHELION, SUN DOG, or MOCK SUN. Also called MOCK MOON.

paraselenic circle. A halo consisting of a faint white circle through the moon and parallel to the horizon. It is produced by reflection of moonlight from vertical faces of ice crystals. A similar circle through the sun is called a PARHELIC CIRCLE.

parhelic circle. A halo consisting of a faint white circle through the sun and parallel to the horizon. It is produced by reflection of sunlight from vertical faces of ice crystals. A similar circle through the moon is called a PARASELENIC CIRCLE. Also called MOCK SUN RING.

parhelion *(pl. parhelia), n.* A form of halo consisting of an image of the sun at the same altitude as the sun and some distance from it, usually about 22°, but occasionally about 46°. A similar phenomenon occurring at a distance of 90°, 120°, or 140° from the sun is called a PARANTHELION, and if occuring at a distance of 180° from the sun, an ANTHELION. A similar phenomenon in relation to the moon is called a PARASELENE, MOON DOG, or MOCK MOON. The term PARHELION should not be confused with PERIHELION, that orbital point nearest the sun when the sun is the center of attraction. Also called SUN DOG, MOCK SUN.

parsec, *n.* The distance at which 1 astronomical unit subtends an angle of 1 second of arc. Approximately 1 parsec equals 206,265 astronomical units or $30{,}857 \times 10^{12}$ meters or 3.26 light years. The name *parsec* is derived from **parallax second.**

partial eclipse. An eclipse in which only part of the source of light is obscured. See ECLIPSE.

pascal, *n.* The special name for the *derived unit* of pressure and stress in the International System of Units; it is 1 newton per square metre.

pass, *n.* 1. A navigable channel leading to a harbor or river. Sometimes called PASSAGE.
2. A break in a mountain range, permitting easier passage from one side of the range to the other; also called COL.
3. A narrow opening through a barrier reef, atoll, or sand bar.
4. A single circuit of the earth by a satellite. See also ORBIT.
5. The period of time a satellite is within telemetry range of a data acquisiton station.

passage, *n.* 1. A navigable channel, especially one through reefs or islands. Also called PASS.
2. A transit from one place to another; one leg of a voyage.

passing light. A low intensity light which may be mounted on the structure of another light to enable the mariner to keep the latter light in sight when he passes out of its beam during transit. See also SUBSIDIARY LIGHT.

passive satellite. 1. A satellite which contains no power source to augment the output signal (i.e., reflected only) as contrasted with ACTIVE SATELLITE; a satellite which is a passive reflector.
2. As defined by the *International Telecommunication Union* (ITU), an earth satellite intended to transmit radiocommunication signals by reflection.

passive system. A term used to describe a navigation system whose operation does not require the user to transmit a signal.

patent log. Any mechanical log, particularly a TAFFRAIL LOG.

patent slip. See MARINE RAILWAY.

path, *n.* See as ORBITAL PATH.

pattern, *n.* 1. See under LATTICE.
2. In a hyperbolic radionavigation system, the family of hyperbolas associated with a single pair of stations, usually the master station and a slave (secondary) station.

P-band. A radio-frequency band of 225 to 390 megahertz. See also FREQUENCY, FREQUENCY BAND.

PCA event. See under POLAR CAP DISTURBANCE.

peak, *n.* 1. On the sea floor, a prominent elevation, part of a larger feature, either pointed or of very limited extent across the summit.

2. A pointed mountain summit.
3. An individual or conspicuous mountain mass with a single conspicuous summit, as Pikes Peak.
4. The summit of a mountain.
5. A term sometimes used for a headland or promontory.

peak envelope power. See under POWER (OF A RADIO TRANSMITTER).

pebble, *n.* See under STONES.

pelorus, *n.* A dumb compass, or a compass card (called a PELORUS CARD) without a directive element, suitably mounted and provided with vanes to permit observation of relative bearings, unless used in conjunction with a compass, to give true or magnetic bearings.

pelorus card. That part of a pelorus on which the direction graduations are placed. It is usually in the form of a thin disk or annulus graduated in degrees, clockwise, from 0° at the reference direction to 360°.

pendulous gyroscope. A gyroscope the axis of rotation of which is constrained by a suitable weight to remain horizontal. The pendulous gyroscope is the basis of one type of gyrocompass.

peninsula, *n.* A section of land nearly surrounded by water. Frequently, but not necessarily, a peninsula is connected to a larger body of land by a neck or isthmus.

pens, *n. British terminology.* A series of parallel jetties for berthing small craft.

pentagon, *n.* A closed plane figure having five sides.

pentagonal cluster. An arrangement of five corner reflectors, mounted so as to give their maximum response in a horizontal direction, and equally spaced on the circumference of a circle. The response is substantially uniform in all horizontal directions. See also OCTAHEDRAL CLUSTER.

penumbra, *n.* 1. That part of a shadow in which light is partly cut off by an intervening object. The penumbra surrounds the darker UMBRA, in which light is completely cut off.
2. The lighter part of a sun spot, surrounding the darker UMBRA.

penumbral lunar eclipse. The eclipse of the moon when the moon passes only through the penumbra of the earth's shadow.

perch, *n. British terminology.* A small beacon used to mark channels through mudflats or sandbars. It may or may not carry a topmark. Often of an impermanent nature, the perch may be an untrimmed sapling.

performance monitor. A device used to check the performance of the transmitter and receiver of a radar set. Such device does not provide any indication of performance as it might be affected by the propagation of the radar waves through the atmosphere. An echo box is used in one type of performance monitor called **echo box performance monitor.** In this type monitor the echo box is located near the antenna so that during a small arc of each rotation of the antenna it will receive transmitted pulses. This electromagnetic energy is returned from the resonant cavity of the echo box as a ringing echo via the antenna and waveguide, and passed to the receiver. Thus, this ringing echo follows the same path within the radar installation as the echo from an actual target. As the ringing echo is received very soon after the pulse is transmitted, its blip on the plan position indicator will extend from the center and be plume-like due to the ringing. The length of the plume depends upon such factors as transmitted power and receiver tuning and sensitivity. Thus its length compared to the length when the radar is known to be operating efficiently provides an indication of the overall performance of the radar. A monitor used to check the performance of the receiver only, called a **receiver monitor,** also makes use of the echo box. In this monitor the echo box is coupled to the waveguide near the magnetron. The ringing echo is passed to the receiver and appears on the plan position indicator as an all round response. The radius of this response is proportional to receiver sensitivity. A monitor used to check the performance of the transmitter only, called **transmitter monitor,** makes use of a neon tube mounted in a horn so located that it will receive transmitted pulses in the same manner as the echo box performance monitor. The neon gas in the tube is ionized by the incident electromagnetic energy, the amount of ionization being dependent upon the transmitted energy. The energy generated in the neon tube is passed to the receiver and appears on the plan position indicator as a plume from its center.

per gyrocompass. Relating to the gyrocompass.

periapsis, *n.* See PERICENTER.

periastron, *n.* That point of the orbit of one member of a double star system at which the stars are nearest together. That point at which they are farthest apart is called APASTRON.

pericenter, *n.* In an elliptical orbit, the point in the orbit which is the nearest distance from the focus where the attracting mass is located. The pericenter is at one end of the major axis of the orbital ellipse. The opposite is APOAPSIS, APOCENTER. Also called PERIAPSIS, PERIFOCUS.

perifocus, *n.* See PERICENTER.

perigean range. See under PERIGEAN TIDES.

perigean tidal currents. Tidal currents of increased speed occurring monthly as the result of the moon being in perigee or nearest the earth.

perigean tides. Tides of increased range occurring monthly as the result of the moon being in perigee or nearest the earth. The **perigean range** of tide is the average semidiurnal range occurring at the time of perigean tides and is most conveniently computed from the harmonic constants. It is larger than the mean range where the type of tide is either semidiurnal or mixed, and is of no practical significance where the type of tide is diurnal.

perigee, *n.* That orbital point nearest the earth when the earth is the center of attraction. That orbital point farthest from the earth is called APOGEE. See also APOCENTER, PERICENTER.

perigee-to-perigee period. See ANOMALISTIC PERIOD.

perigon, *n.* An angle of 360°.

perihelion, *n.* That orbital point nearest the sun when the sun is the center of attraction. That point farthest from the sun is called APHELION.

perimeter, *n.* 1. The length of a closed plane curve or the sum of the sides of a polygon.
2. The boundary of a plane figure.
Also called PERIPHERY.

period, *n.* 1. The interval needed to complete a cycle. See also NATURAL PERIOD, SIDEREAL PERIOD, SYNODIC PERIOD, WAVE PERIOD.
2. Of a rhythmic light, the interval of time between the commencement of two identical successive cycles of the chracteristic of the light.

periodic, *adj.* Of or pertaining to a period.

periodic error. An error whose amplitude and direction vary systematically with time.

periodic perturbations. Perturbations to the orbit

of a satellite which change direction in regular or periodic manner in time, such that the average effect over a long period of time is zero.

periodic terms. In the mathematical expression of the orbit of a satellite, terms which vary with time in both magnitude and direction in a periodic manner. See also SECULAR TERMS.

period of satellite. 1. See ORBITAL PERIOD.
2. As defined by the *International Telecommunication Union* (ITU), the time elapsing between two consecutive passages of a satellite or planet through a characteristic point on its orbit.

periphery, *n.* See PERIMETER.

periplus, *n.* The early Greek name for SAILING DIRECTIONS. The literal meaning of the term is "a sailing round."

periscope, *n.* An optical instrument which displaces the line of sight parallel to itself, to permit a view which may otherwise be obstructed.

periscope sextant. A sextant designed to be used in conjunction with the periscope of a submarine.

permafrost, *n.* Permanently frozen subsoil. Any soil or other deposit, including rock, the temperature of which has been below freezing continuously for 2 years or more is considered permafrost.

Permalloy, *n.* The trade name for an alloy of about 80% nickel and 20% iron, which is very easily magnetized and demagnetized.

permanent current. A current that runs fairly continuously and is independent of tides and other temporary causes. Permanent currents include the general, surface circulation of the ocean.

permanent echo. An echo from an object whose position relative to the radar set is fixed.

permanent light. A light used in regular service.

permanent magnetism. The magnetism which is acquired by hard iron, which is not readily magnetized by induction, but which retains a high percentage of magnetism acquired unless subjected to a demagnetizing force. The strength and polarity of this magnetism in a craft depends upon the heading, magnetic latitude, and building stresses imposed during construction. See also INDUCED MAGNETISM, SUBPERMANENT MAGNETISM.

permeability, *n.* 1. The ability to transmit magnetism; magnetic conductivity.
2. The ability to permit penetration or passage. In this sense the term is applied particularly to substances which permit penetration or passage of fluids.

perpendicular, *adj.* At right angles; normal.

perpendicular, *n.* A perpendicular line, plane, etc. A distinction is sometimes made between PERPENDICULAR and NORMAL, the former applying to a line at right angles to a straight line or plane, and the latter referring to a line at right angles to a curve or curved surface.

persistence, *n.* A measure of the time of decay of the luminescence of the face of the cathode-ray tube after excitation by the stream of electrons has ceased. Relatively slow decay is indicative of high persistence. Persistence is the length of time during which phosphorescence takes place. See also AFTERGLOW, definition 1.

personal correction. That correction due to personal error. Also called PERSONAL EQUATION.

personal equation. A term used for both PERSONAL ERROR and PERSONAL CORRECTION.

personal error. A systematic error in the observation of a quantity due to the personal idiosyncracies of the observer, such as always pushing the stop watch early or late or reading a sextant altitude too high or too low. This type of error should be distinguished from the random error that an observer makes when attempting to read an instrument to the highest possible precision. The personal error may be determined by a study of the observer. Once the personal error is determined, the observer's personal correction (personal equation) may be applied to future observations in order to correct this source of systematic error. Also called PERSONAL EQUATION.

perspective chart. A chart on a perspective map projection.

perspective map projection. A map projection produced by the direct projection of the points of the ellipsoid (used to represent the earth) by straight lines drawn through them from some given point. The projection is usually made upon a plane tangent to the ellipsoid at the end of the diameter joining the point of projection and the center of the ellipsoid. The plane of projection is usually tangent to the ellipsoid at the center of the area being mapped. Projections of this kind are generally simple, because they can in most cases be constructed by graphical methods without the aid of the analytical expressions that determine the elements of the projection. If the point of projection is at the center of the ellipsoid, a **gnomonic map projection** results; if it is at the point opposite the plane's point of tangency, a **stereographic map projection**; and if at infinity (the projecting lines being parallel to each other), an **orthographic map projection.** Most map projections are not perspective. Also called GEOMETRIC MAP PROJECTION.

perspective map projection upon a tangent cylinder. A cylindrical map projection upon a cylinder tangent to the ellipsoid produced by perspective projection from the ellipsoid's center. The geographic meridians are represented by a family of equally spaced parallel straight lines, perpendicular to a second family of parallel straight lines which represent the geographic parallels of latitude. The spacing, with respect to the equator of the lines which represent the parallels of latitude, increases as the tangent function of the latitude; the line representing 90° latitude is at an infinite distance from the line which represents the equator. Not to be confused with MERCATOR MAP PROJECTION to which it bears some general resemblance.

perspective projection. The representation of a figure on a surface, either plane or curved, by means of projecting lines emanating from a single point, which may be infinity. Also called GEOMETRIC PROJECTION. See also PERSPECTIVE MAP PROJECTION.

per standard compass. Relating to the standard magnetic compass.

per steering compass. Relating to the magnetic steering compass.

perturbations, *n (pl.).* In celestial mechanics, differences of the actual orbit from a central force orbit, arising from some external force such as a third body attracting the other two; a resisting medium (atmosphere); failure of the parent body to act as a point mass, and so forth. Also the forces that cause differences between the actual and reference (central force) orbits. See also GRAVITATIONAL PERTURBATIONS, LONG PERIOD PERTURBATIONS, LUNISOLAR PERTURBATIONS, NONGRAVITATIONAL PERTURBATIONS, PERIODIC PERTURBATIONS, SECULAR PERTURBATIONS, SHORT PERIOD PERTURBATIONS, TERRESTRIAL PERTURBATIONS.

perturbed orbit. The orbit of a satellite differing from its normal orbit due to various disturbing effects, such as nonsymmetrical gravitational effects, atmospheric drag, radiation pressure, and so forth. See also PERTURBATIONS.

perturbing factor. In celestial mechanics, any factor that acts on an orbiting body to change its orbit from a central force orbit. Also called PERTURBING FORCE.

perturbing force. See PERTURBING FACTOR.

Peru Coastal Current. See PERU CURRENT.

Peru Current. A narrow, fairly stable ocean current that flows northward close to the South American coast. It originates off the coast of Chile at about latitude 40°S and flows past Peru and Ecuador to the southwest extremity of Colombia. The southern portion of the Peru Current is sometimes called the CHILE CURRENT. It has sometimes been called the HUMBOLDT CURRENT because an early record of its temperature was taken by the German scientist Alexander von Humboldt in 1802. The name *Corriente del Peru* was adopted by a resolution of the Ibero-American Oceanographic Conference at its Madrid-Malaga meeting in April 1935. Also called PERU COASTAL CURRENT.

Peru Oceanic Current. See MENTOR CURRENT.

phantom, *n.* That part of a gyrocompass carrying the compass card.

phantom bottom. A false bottom indicated by an echo sounder, some distance above the actual bottom. Such an indication, quite common in the deeper parts of the ocean, is due to large quantities of small organisms.

phantom echo. See PHANTOM TARGET.

phantom target. 1. An indication of an object on a radar display that does not correspond to the presence of an actual object at the point indicated. Also called PHANTOM ECHO.
2. See ECHO BOX.

phase, *n.* The amount by which a cycle has progressed from a specified origin. For most purposes it is stated in circular measure, a complete cycle being considered 360°. Thus, if two waves have their crests one-fourth cycle apart, they are said to be 90° apart in phase, or 90° *out of phase*. The moon is said to be at *first quarter* when it has completed one-fourth of its cycle from new moon. See also PHASES OF THE MOON.

phase angle. The angle at a celestial body between the sun and earth.

phase coding. In Loran-C, the shifting in a fixed sequence of the relative phase of the carrier cycles between certain pulses of a group. This shifting facilitates automatic synchronization in identical sequence within the group of eight pulses that are transmitted during each group repetition interval. It also minimizes the effect of unusually long skywave transmissions causing one pulse to interfere with the succeeding pulse in the group received by groundwave.

phase inequality. Variations in the tides or tidal currents due to changes in the phase of the moon. At the times of new and full moon the tide-producing forces of the moon and sun act in conjunction, causing the range of tide and speed of the tidal current to be greater than the average, the tides at these times being known as *spring tides*. At the time of the quadrature of the moon these forces are opposed to each other, causing the *neap tides* with diminished range and current speed.

phase lag. See EPOCH, definition 3.

phase lock. The technique whereby the phase of an oscillator signal is made to follow exactly the phase of a reference signal by first comparing the phases of the two signals and then using the resulting phase difference signal to adjust the reference oscillator frequency to eliminate phase difference when the two signals are next compared.

phase meter. An instrument for measuring the difference in phase of two waves of the same frequency.

phase modulation. The process of changing the phase of a carrier wave in accordance with the variations of a modulating wave. See also MODULATION.

phase reduction. A processing of observed high and low waters to obtain quantities depending upon the phase of the moon, such as the spring and neap ranges of tide. At a former time this process was known as SECOND REDUCTION. Also applicable to tidal currents.

phases of the moon. The various appearances of the moon during different parts of the synodical month. The cycle begins with **new moon** or **change of the moon** at conjunction. The visible part of the **waxing moon** increases in size during the first half of the cycle until **full moon** appears at opposition, after which the visible part of the **waning moon** decreases for the remainder of the cycle. **First quarter** occurs when the waxing moon is at east quadrature; **last quarter** when the waning moon is at west quadrature. From last quarter to new and from new to first quarter the moon is **crescent**; from first quarter to full and from full to last quarter it is **gibbous**. The elapsed time, usually expressed in days, since the last new moon is called **age of the moon**. The full moon occurring nearest the autumnal equinox is called **harvest moon**; the next full moon, **hunter's moon**.

phase synchronized. A term used to indicate that radio wave transmissions have the same phase at their sources at any instant of time.

phenomenon (*pl. phenomena*), *n.* 1. An occurrence or thing capable of being explained scientifically, particularly one relating to the unusual, as a particular infrequent relationship of relative positions of various celestial bodies. Examples are eclipses, planetary configurations, equinoxes, etc.
2. A rare or unusual event.

phonetic alphabet. A list of standard words used to identify letters in a message transmitted by radio or telephone.

phosphor, *n.* A phosphorescent substance which emits light when excited by radiation, as on the scope of a cathode-ray tube.

phosphorescence, *n.* Emission of light without sensible heat, particularly as a result of but continuing after absorption of radiation from some other source. An example is the glowing of a radar set's plan position indicator after the sweep has moved to another position. It is this property that results in the chartlike picture which gives the plan position indicator its principal value. PERSISTENCE is the length of time during which phosphorescence takes place. The emission of light or other radiant energy as a result of and only during absorption of radiation from some other source is called FLUORESCENCE.

photogrammetry, *n.* 1. The science or art of obtaining reliable measurements from photographic images.
2. The science of preparing charts and maps from aerial photographs using stereoscopic equipment and methods.

photosphere, *n.* The intensely bright portion of

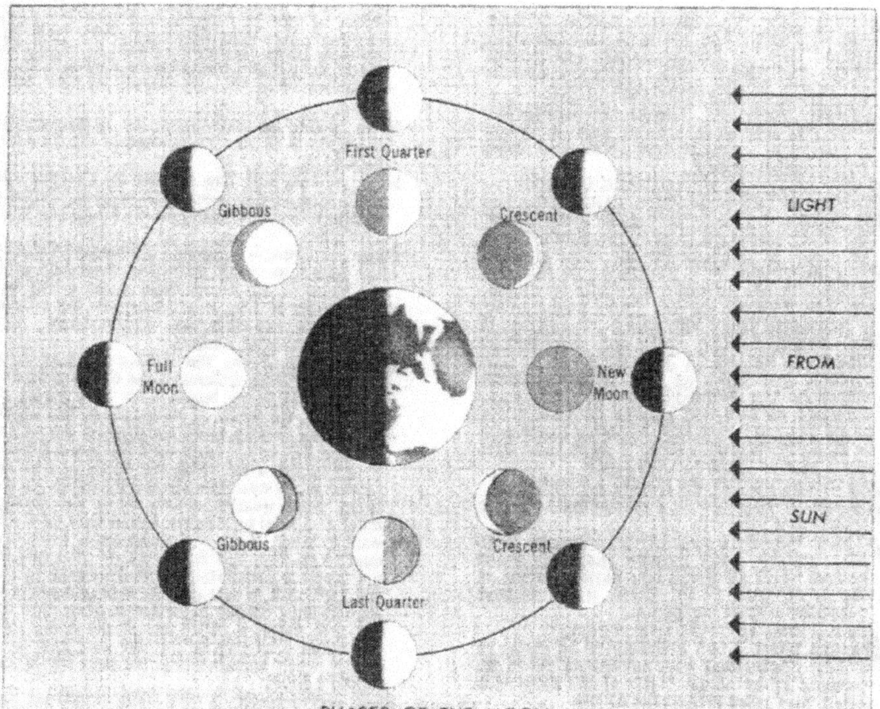

PHASES OF THE MOON
(The inner figures of the moon represent its appearance from the earth)

the sun visible to the unaided eye.

physical double star. Two stars in nearly the same line of sight and at approximately the same distance from the observer, as distinguished from an OPTICAL DOUBLE STAR (two stars in nearly the same line of sight but differing greatly in distance from the observer). If they revolve about their common center of mass, they are called a **binary star.**

pico-. A prefix meaning *one-trillionth* (10^{-12}).

piedmont, *adj.* Situated or formed at the base of mountains.

pier, *n.* 1. A structure extending into the water approximately perpendicular to a shore or a bank and providing berthing for ships, and which may also provide cargo-handling facilities. See also WHARF.
2. A structure extending into the water approximately perpendicular to a shore or bank and providing a promenade or place for other use, as a *fishing pier.*
3. A support for the spans of a bridge.

pierhead, *n.* That part of a pier or jetty projecting farthest into the water.

pile, *n.* A long, heavy timber or section of steel, concrete, etc., forced into the earth to serve as a support, as for a pier, or to resist lateral pressure.

pile beacon. A beacon formed of one or more piles.

pile dolphin. A minor light structure consisting of a number of piles driven into the bottom in a circular pattern and drawn together with a light mounted at the top. Referred to in the *Light List* as a DOLPHIN.

pillar buoy. A buoy composed of a tall central structure mounted on a broad flat base. Also called BEACON BUOY.

pilot, *n.* 1. A person who directs the movements of a vessel through pilot waters, usually a person who has demonstrated extensive knowledge of channels, aids to navigation, dangers to navigation, etc., in a particular area and is licensed for that area. One who pilots by observing color changes in the water may be called a **mud pilot.** See also LOCAL KNOWLEDGE.
2. A book of sailing directions. For waters of the United States and its possessions, they are prepared by the National Ocean Survey, and are called COAST PILOTS.

pilotage, *n.* 1. The provision of a service of specially qualified navigators having *local knowledge* who assist in the navigation of vessels in particular areas. Also called PILOTAGE SERVICE.
2. A term loosely used for piloting.

pilotage service. See PILOTAGE, definition 1.

pilotage waters. *British terminology.* See PILOT WATERS.

pilot boat. A small vessel used by the pilot to go to or from a vessel employing his services. Also called PILOT VESSEL.

pilot chart. As published by the Defense Mapping Agency Hydrographic/Topographic Center from data provided by the U.S. Naval Oceanographic Office and the Environmental Data and Information Service of the National Oceanic and Atmospheric Administration, a chart of a major ocean area which is not intended to be used alone but in conjunction with regular navigational charts and other aids to navigation. The chart presents in graphic form, averages obtained from data gathered over many years in meteorology and oceanography to aid the navigator in selecting the quickest and safest routes. Included are explanations on how to use each type of information depicted on the chart.

piloting, *n.* Navigation involving frequent or continuous determination of position or a line of position relative to geographical points, to a high order of accuracy. The directing of the movements of a vessel near a coast, by means of terrestrial reference points, is called coast piloting. Sometimes called PILOTAGE. See also PILOTAGE, definition 1.

pilot rules. Regulations supplementing the Inland Rules of the Road.

pilot station. The office or headquarters of pilots; the place where the services of a pilot may be obtained.

pilot vessel. See PILOT BOAT.

pilot waters. 1. Areas in which the services of a marine pilot are essential.
2. Waters in which navigation is by piloting. Also called PILOTAGE WATERS.

pinnacle, *n.* On the sea floor, a high tower or spire-shaped pillar of rock or coral, alone or cresting a summit. It may or may not be a hazard to surface navigation. Due to the sheer rise from the sea floor no warning is given by sounding.

pinnacled iceberg. An iceberg weathered in such manner as to produce spires or pinnacles. Also called PYRAMIDAL ICEBERG, IRREGULAR ICEBERG.

pip, *n.* See BLIP.

pitch, *n.* 1. Of the motions of a vessel in a seaway, oscillation about the transverse axis due to the vessel's bow and stern being raised or lowered on passing through successive crests and troughs of waves. Also called PITCHING. See also SHIP MOTIONS.
2. The distance a propeller would advance longitudinally in one revolution if there were no slip.

pitch, *v., i.* Of a vessel, to oscillate about the transverse axis. See also SHIP MOTIONS.

pitching, *n.* See PITCH, definition 1.

pivot point. After a ship has assumed its drift angle in a turn, the point on the centerline between the bow and the center of gravity at which the resultant of the velocities of rotation and translation is directed along the centerline. To an observer on board, the ship appears to rotate about this point.

place name. See TOPONYM.

plain, *n.* On the sea floor, a flat, gently sloping or nearly level region. Sometimes called ABYSSAL PLAIN when the depths over it are very deep.

plan, *n.* 1. An orthographic drawing on a horizontal plane, as of an instrument, a horizontal section, or a layout.
2. A large-scale map or chart of a small area.

planar, *adj.* Lying in a plane.

plane, *n.* A surface without curvature, such that a straight line joining any two of its points lies wholly within the surface.

plane of polarization. With respect to a plane polarized wave, the plane containing the electric field vector and the direction of propagation.

plane polarized wave. An electromagnetic wave the electric field vector of which lies at all times in a fixed plane which contains the direction of propagation.

plane sailing. A method of solving the various problems involving a single course and distance, difference of latitude, and departure, in which the earth, or that part traversed, is considered as a plane surface. Hence, the method provides solution for latitude of the point of arrival, but not for longitude of this point, one of the spherical sailings being needed for this problem. Because of the basic assumption that the earth is flat, this method should not be used for distances of more than a few hundred miles. The formulas are

$$l = D \cos C \qquad D = l \sec C$$

$$p = D \sin C \qquad \tan C = p/l$$

where l is difference of latitude in minutes of arc, D is distance in minutes of arc, C is course angle, and p is departure.

planet, *n.* A celestial body of the solar system, revolving around the sun in a nearly circular elliptical orbit and shining by reflected sunlight, or a similar body revolving around a star. The larger of such bodies are sometimes called **major planets** to distinguish them from **minor planets** (**asteroids**) which are very much smaller. With the exception of Mercury, Venus, and Pluto, the larger planets are accompanied by satellites, such as the moon. An inferior planet has an orbit smaller than that of the earth; a **superior planet** has an orbit larger than that of the earth. The four planets commonly used for celestial observations are called **navigational planets**. The word *planet* is of Greek origin, meaning, literally, *wanderer*, applied because the planets appear to move relative to the stars.

planetary, *adj.* Of a planet or the planets; terrestrial; worldwide.

planetary aberration. See under ABERRATION, definition 1.

planetary configurations. Apparent positions of the planets relative to each other and to other bodies of the solar system, as seen from the earth.

planetary precession. That component of general precession caused by the effect of other planets on the equatorial protuberance of the earth, producing an eastward motion of the equinoxes along the ecliptic. See also PRECESSION OF THE EQUINOXES.

planetoid, *n.* See ASTEROID.

plane triangle. A closed plane figure having three straight lines as sides.

planimetric map. A map indicating only the horizontal positions of features, without regard to elevation, in contrast with a TOPOGRAPHIC MAP, which indicates both horizontal and vertical positions.

planisphere, *n.* A representation, on a plane, of the celestial sphere, especially one on a polar projection, with means provided for making certain measurements such as altitude and azimuth. See also STAR FINDER.

plankton, *n.* Floating, drifting, or feebly swimming plant and animal organisms of the sea. These are usually microscopic or very small, although jellyfish are included.

planning chart. A chart designed for use in planning voyages or flight operations or investigating areas of marine or aviation activities.

plan position indicator. An intensity-modulated radar display in which the radial sweep rotates on the cathode-ray tube in synchronism with the rotating antenna. The display presents a maplike representation of the positions of echo-producing objects. It is generally one of two main types: RELATIVE MOTION DISPLAY or TRUE MOTION DISPLAY.

plastic relief map. A topographic map printed on plastic and molded into a three-dimensional form.

plateau, *n.* On the sea floor, a comparatively flat-topped feature of considerable extent, dropping off abruptly on one or more sides.

plate glass. A fine quality sheet glass obtained by rolling, grinding, and polishing.

platform erection. In the alignment of inertial navigation equipment, the alignment of the stable platform vertical axis with the local vertical.

platform tide. See STAND.

Platonic year. See GREAT YEAR.

Plimsoll mark. A mark on a ship's side indicating how deeply she may be loaded.

plot, *n.* A drawing consisting of lines and points representing certain conditions graphically, as the progress of a craft. See also NAVIGATIONAL PLOT.

plot, *v.,t.* To draw lines and points to represent certain conditions graphically, as the various lines and points on a chart or plotting sheet representing the progress of a craft, a curve of magnetic azimuths *vs.* time or of altitude *vs.* time, or a graphical solution of a problem, such as a relative movement solution.

plotter, *n.* An instrument used for plotting straight lines and measuring angles on a chart or plotting sheet. See also PROTRACTOR.

plotting chart. An outline chart on a specific scale and projection, usually showing a graticule and compass rose, designed to be used ancillary to a standard nautical chart, and produced either as an independent chart or part of a coordinated series. A plotting chart overprinted with the lattice of a hyperbolic radionavigation system servicing the area is called an OMEGA PLOTTING CHART or LORAN-C PLOTTING CHART as appropriate. See also POSITION PLOTTING SHEET.

plotting head. See REFLECTION PLOTTER.

plumb bob. A conical device, usually of brass and suspended by a chord, by means of which a point can be projected vertically into space over relatively short distances.

plumb-bob vertical. See LOCAL VERTICAL.

plumb line. 1. A line in the direction of gravity.
2. A cord with a weight at one end for determining the direction of gravity.

pluvial, *adj.* Of or pertaining to rain. The expression *pluvial period* is often used to designate an extended period or age of heavy rainfall.

pocosin, *n.* See DISMAL.

point, *n.* 1. A place having position, but no extent. A point in motion produces a LINE; a straight line in motion in any direction except along itself produces a SURFACE; a plane surface in motion in any direction except along itself produces a SOLID.
2. A tapering piece of land projecting into a body of water. It is generally less prominent than a CAPE.
3. One thirty-second of a circle, or $11\frac{1}{4}^{\circ}$. A **cardinal point** is any of the four principal directions; north, east, south, or west; an **intercardinal point** is any of the four directions midway between the cardinal points; northeast, southeast, southwest, or northwest. Also called COMPASS POINT when used in reference to compass directions. See also FOUR-POINT BEARING.

point designation grid. A system of lines, having no relation to the actual scale or orientation, drawn on a map, chart, or air photograph, dividing it into squares so that points can be more readily located.

point light. A luminous signal without perceptible length, as contrasted with a LINEAR LIGHT, which has perceptible length.

point of arrival. The position at which a craft is assumed to have reached or will reach after following specified courses for specified distances from a point of departure. See also DESTINATION.

point of departure. The point from which the initial course to reach the designation begins. It is usually established by bearings of prominent landmarks as the vessel clears a harbor and proceeds to sea. When a person establishes this point, he is said to *take departure.* Also called DEPARTURE.

point of destination. See DESTINATION.

point of inflection. The point at which a reversal of direction of curvature takes place.

polar, *adj.* Of or pertaining to a pole or the poles.

polar air. A type of air whose characteristics are developed over high latitudes, especially within the subpolar highs. **Continental polar air** has low surface temperature. Low moisture content, and, especially in its source regions, has great stability in the lower layers. It is shallow in comparison with arctic air. **Maritime polar air** initially possesses similar properties to those of continental polar air, but in passing over warmer water it becomes unstable with a higher moisture content.

polar axis. 1. The straight line connecting the poles of a body.
2. A reference line for one of the spherical coordinates.

polar cap absorption. See under POLAR CAP DISTURBANCE.

polar cap disturbance. An ionospheric disturbance which is in no way dependent on the ice cap in the polar region. It is the result of the focusing effect that the earth's magnetic field has on particles released from the sun during a solar proton event. The effect concentrates high-energy particles in the region of the magnetic pole with the result that normal very low frequency Omega propagation is disrupted. The effect on radio waves is known as POLAR CAP ABSORPTION (PCA). Historically, polar cap disturbances (PCDs) produced large or total absorption of high frequency radio waves crossing the polar region, hence the term polar cap absorption. A transmission patch which is entirely outside the polar region is unaffected by the PCD. The PCDs, often called PCA EVENTS (PCAs), may persist for a week or more, but a duration of only a few days is more common. The PCD can cause line of position errors of about 6 to 8 nautical miles. The *Omega Propagation Correction Tables* make no allowance for this phenomenon since it is not predictable. However, the frequency of the phenomenon increases during those years of peak solar activity. See also SUDDEN IONOSPHERIC DISTURBANCE, MODAL INTERFERENCE.

polar chart. 1. A chart of polar areas.
2. A chart on a polar projection. The map projections most used for polar charts are the gnomonic, stereographic, azimuthal equidistant, transverse Mercator, and modified Lambert conformal.

polar circles. The minimum latitudes, north and south, at which the sun becomes circumpolar.

polar continental air. Air of an air mass that originates over land or frozen ocean areas in the polar regions. Polar continental air is characterized by low temperature, stability, low specific humidity, and shallow vertical extent.

polar coordinates. A system of coordinates defining a point by its distance and direction from a fixed point, called the POLE. Direction is given as the angle between a reference radius vector and a radius vector to the point. If three dimensions are involved, two angles are used to locate the radius vector. See also SPACE-POLAR COORDINATES.

polar distance. Angular distance from a celestial pole; the arc of an hour circle between a celestial pole, usually the elevated pole, and a point on the celestial sphere, measured from the celestial pole through 180°. See also CODECLINATION.

polar front. The semi-permanent, semi-continuous front separating air masses of tropical and polar origin. This is the major front in terms of air mass contrast and susceptibility to cyclonic disturbance.

Polaris correction. A correction to be applied to the corrected sextant altitude of Polaris to obtain latitude. This correction for the offset of Polaris from the north celestial pole varies with the local hour angle of Aries, latitude, and date. The approximate Polaris correction tabulated in the *Air Almanac* is called Q-CORRECTION.

polarization, *n.* That attribute of an electromagnetic wave which describes the direction of the electric field vector.

polarization error. An error in a radio direction-finder bearing or the course indicated by a radiobeacon because of a change in the polarization of the radio waves between the transmitter and receiver on being reflected and refracted from the ionosphere. In that the medium frequency radio direction finder normally operates with vertically polarized waves, a change to horizontal polarization in the process of reflection and refraction of the waves from the ionosphere can have a serious effect on bearing measurements. If the horizontally polarized skywaves are of higher signal strength than the vertically polarized groundwaves, the null position for the loop antenna cannot be obtained. If the skywaves are of lower signal strength than the groundwaves, the null position is made less distinct. Before the cause of the error was understood, it was called NIGHT EFFECT or NIGHT ERROR because it occurs principally during the night, and especially during twilight when rapid changes are occurring in the ionosphere.

polar map projection. A map projection centered on a pole.

polar maritime air. Air of an air mass that originates in the polar regions and is then modified by passing over a relatively warm ocean surface. It is characterized by moderately low temperature, moderately high surface specific humidity, and a considerable degree of vertical instability. When the air is colder than the sea surface, it is further characterized by gusts and squalls, showery precipitation, variable sky, and good visibility between showers.

polar motion. See EULERIAN MOTION.

polar navigation. Navigation in polar regions, where unique considerations and techniques are applied. No definite limit for these regions is recognized but polar navigation techniques are usually used from about latitude 70° to the nearest pole (N or S).

polar orbit. An earth satellite orbit that has an inclination of about 90° and, hence, passes over or near the earth's poles.

polar orthographic map projection. An orthographic map projection having the plane of the projection perpendicular to the axis of rotation of the earth; in this projection, the geographic parallels are full circles, true to scale, and the geographic meridians are straight lines.

polar regions. The regions near the geographic poles. No definite limit for these regions is recognized.

polar satellite. A satellite that passes over or near the earth's poles, i.e., a satellite whose orbital plane has an inclination of about 90° to the plane of the earth's equator.

polar stereographic map projection. A stereographic map projection having the center of the projection located at a pole of the sphere.

pole, *n.* 1. Either of the two points of intersection of the surface of a sphere or spheroid and its axis, labeled N or S to indicate whether the **north pole** or **south pole**. The two points of intersection of the surface of the earth with its axis are called **geographical poles**. The two points of intersection of the celestial sphere and the extended axis of the earth are called **celestial poles**. The celestial pole above the horizon is called the **elevated pole**; that below the horizon, the **depressed pole**. The **ecliptic poles** are 90° from the ecliptic. Also, one of a pair of similar points on the surface of a sphere or spheroid, as a **magnetic pole**, definition 1; a **geomagnetic pole**; or a **fictitious pole**.
2. A magnetic pole, definition 2.
3. The origin of measurement of distance in polar or spherical coordinates.
4. Any point around which something centers.

pole beacon. A vertical spar fixed in the ground or in the sea bed or a river bed to show as a navigation mark. Sometimes called SPINDLE BEACON or SINGLE-PILE BEACON in the United States.

polyconic, *adj.* Consisting of or related to many cones.

polyconic chart. A chart on the polyconic map projection.

polyconic map projection. A conic map projection in which the surface of a sphere or spheroid, such as the earth, is conceived as developed on a series of tangent cones, which are then spread out to form a plane. A separate cone is used for each small zone. This projection is widely used for maps but seldom used for charts, except for survey purposes. It is not conformal.

polygon, *n.* A closed plane figure bounded by straight lines. See also HEXAGON, OCTAGON, PARALLELOGRAM, PENTAGON, QUADRILATERAL, RECTANGLE, SQUARE, TRAPEZOID, TRIANGLE.

polynya, *n.* Any non-linear shaped area of water enclosed by ice. Polynyas may contain brash ice and/or be covered with new ice, nilas, or young ice; submariners refer to these as SKYLIGHTS. Sometimes the POLYNYA is limited on one side by the coast and is called a SHORE POLYNYA or by fast ice and is called a FLAW POLYNYA. If it recurs in the same position every year, it is called a RECURRING POLYNYA.

polyzoa, *n., pl.* Very small marine animals which reproduce by budding, many generations often being permanently connected by branchlike structures. These animals are often very numerous and in some areas they cover the bottom. Also called BRYOZOA.

pond, *n.* A relatively small body of water, usually surrounded on all sides by land. A larger body of water is called a LAKE.

pontoon, *n.* A float or low, flat-bottomed vessel to float machinery such as cranes, capstans, etc., or to support weights such as floating bridges, boat landings, etc.

pool, *n.* 1. A small body of water, usually smaller than a pond, especially one that is quite deep. One left by an ebb tide is called a **tide pool.**
2. A small and comparatively still, deep part of a larger body of water such as a river or harbor.

poop, *v., t.* To ship a sea or wave over the stern, particularly with a following sea.

pororoca, *n.* See TIDAL BORE.

port, *n.* 1. A place provided with terminal and transfer facilities for loading and discharging cargo or passengers, usually located in a harbor.
2. The left side of a craft, facing forward. The opposite is STARBOARD.

portfolio, *n.* A portable case for carrying papers. See also CHART PORTFOLIO.

port hand buoy. A buoy which is to be left to the port hand when approaching from the open sea or in general proceeding in the direction of the main stream of flood current, or in the direction established by appropriate authority.

port of call. A port visited by a ship.

Portugal Current. A slow-moving current that is the prevailing southward flow off the Atlantic coasts of Spain and Portugal. Its speed averages only about 0.5 knot during both winter and summer. The maximum speed seldom exceeds 2.0 knots north of latitude 40°N and 2.5 knots south of 40°N. The current is easily influenced by winds.

Portuguese norther. See under NORTHER.

position, *n.* A point defined by stated or implied coordinates, particularly one on the surface of the earth. A **fix** is a relatively accurate position determined without reference to any former position. A **running fix** is a position determined by crossing lines of position obtained at different times and advanced or retired to a common time. An **estimated position** is determined from incomplete data or data of questionable accuracy. A **dead reckoning position** is determined by advancing a previous position for courses and distances. A **most probable position** is that position of a craft judged to be most accurate when an element of doubt exists as to the true position. It may be a fix, running fix, estimated position, or dead reckoning position depending upon the information upon which it is based. An **assumed position** is a point at which a craft is assumed to be located. A **geographical position** is that point on the earth at which a given celestial body is in the zenith at a specified time, or any position defined by means of its geographical coordinates. A **geodetic position** is a point on the earth the coordinates of which have been determined by triangulation from an accurately known initial station, or one defined in terms of geodetic latitude and longitude. An **astronomical position** is a point on the earth whose coordinates have been determined as a result of observation of celestial bodies, or one defined in terms of astronomical latitude and longitude. A **maritime position** is the location of a seaport or other point along a coast. A **relative position** is one defined with reference to another position, either fixed or moving. See also PINPOINT, LINE OF POSITION, BAND OF POSITION, SURFACE OF POSITION.

position angle. See PARALLACTIC ANGLE.

position approximate. Of inexact position. The expression is used principally on charts to indicate that the position of a wreck, shoal, etc., has not been accurately determined or does not remain fixed.

position buoy. An object towed astern to assist a following vessel in maintaining the desired or prescribed distance, particularly in conditions of low visibility.

position circle. See CIRCLE OF POSITION.

position doubtful. Of uncertain position. The expression is used principally on charts to indicate that a wreck, shoal, etc., has been reported in various positions and not definitely determined in any. See also VIGIA.

positioning, *n.* The process of determining, at a particular point in time, the precise physical location of a craft, vehicle, person or site.

position line. See LINE OF POSITION.

position plotting sheet. A blank chart, usually on the Mercator projection, showing only the graticule and a compass rose. The meridians are usually unlabeled by the publisher so that they can be appropriately labeled when the chart is used in any longitude. It is designed and intended for use in conjunction with the standard nautical chart. See also SMALL AREA PLOTTING SHEET, UNIVERSAL PLOTTING SHEET, PLOTTING CHART.

post meridian. After noon, or the period of time between noon (1200) and midnight (2400). The period between midnight and noon is called ANTE MERIDIAN.

potential, *n.* The difference in voltage at two points in a circuit.

potential energy. Energy possessed by a body by virtue of its position, in contrast with KINETIC ENERGY, that possessed by virtue of its motion.

pound, *n.* A unit of mass equal to 0.45359237 kilograms. Also called AVOIRDUPOIS POUND.

pound, *v., i.* To strike or come down on repeatedly or heavily, as a vessel pitching in a heavy sea.

pounding, *n.* A series of shocks received by a pitching vessel as it repeatedly or heavily strikes the water in a heavy sea. The shocks can be felt over the entire vessel and each one is followed by a short period of vibration.

power, *n.* 1. Rate of doing work.
2. Luminous intensity.
3. The number of times an object is magnified by an optical system, such as a telescope. Usually called MAGNIFYING POWER.
4. The result of multiplying a number by itself a given number of times, as the *third power* of a number is its cube. See also EXPONENT.

power gain (of an antenna). See DIRECTIVITY, definition 2.

power gain (of a transmitter). The ratio of the output power delivered to a specified load by an amplifier to the power absorbed by its input circuit.

power (of a radio transmitter), *n.* The power of a radio transmitter is expressed in one of the following forms: The **peak envelope power** is the average power supplied to the antenna transmission line by a transmitter during one radio frequency cycle at the highest crest of the modulation envelope, taken under conditions of normal operation. The **mean power** is the power supplied to the antenna transmission line by a transmitter during normal operation, averaged over a time sufficiently long compared with the period of the lowest frequency encountered in the modulation. A time of 1/10 second during which the mean power is greatest is selected normally. The **carrier power** is the average power supplied to the antenna transmission line by a transmitter during one radio frequency cycle under conditions of no modulation. This definition does not apply to pulse modulated emissions.

PPI display. See as PLAN POSITION INDICATOR.

PPI repeater. See RADAR REPEATER.

precautionary area. A routing measure comprising

an area within defined limits where ships must navigate with particular caution and within which the direction of traffic flow may be recommended. See also ROUTING SYSTEM.

precession, *n.* The change in the direction of the axis of rotation of a spinning body, as a gyroscope, when acted upon by a torque. The direction of motion of the axis is such that it causes the direction of spin of the gyroscope to tend to coincide with that of the impressed torque. The horizontal component of precession is called drift, and the vertical component is called topple. Also called INDUCED PRECESSION, REAL PRECESSION. See also APPARENT PRECESSION, PRECESSION OF THE EQUINOXES.

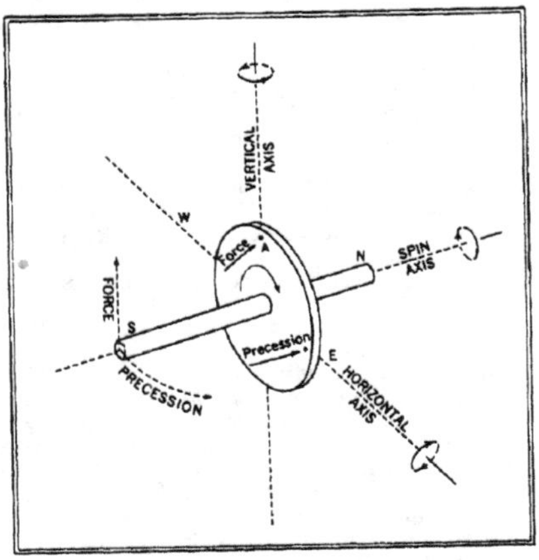

precession in declination. The component of general precession along a celestial meridian, amounting to about 20.″0 per year.

precession in right ascension. The component of general precession along the celestial equator, amounting to about 46.″1 per year.

precession of the equinoxes. The conical motion of the earth's axis about the vertical to the plane of the ecliptic, caused by the attractive force of the sun, moon, and other planets on the equatorial protuberance of the earth. The effect of the sun and moon, called **lunisolar precession,** is to produce a westward motion of the equinoxes along the ecliptic. The effect of other planets, called **planetary precession,** tends to produce a much smaller motion eastward along the ecliptic. The resultant motion, called **general precession,** is westward along the ecliptic at the rate of about 50.″3 per year. The component of general precession along the celestial equator, called **precession in right ascension,** is about 46.″1 per year; and the component along a celestial meridian, called **precession in declination,** is about 20.″0 per year.

precipice, *n.* A high and very steep cliff.

precipitation, *n.* 1. Any or all forms of water particles, whether liquid or solid, that fall from the atmosphere and reach the ground. It is distinguished from cloud, fog, dew, rime, frost, etc., in that it must "fall"; and it is distinguished from cloud and virga in that it must reach the ground. Precipitation includes drizzle, rain, snow, snow pellets, snow grains, ice crystals, ice pellets, and hail.
2. The amount usually expressed in inches of liquid water depth, of the water substance that has fallen at a given point over a specified period of time.

precipitation static. A type of interference experienced in a receiver, during snow storms, rain storms, and dust storms, caused by the impact of dust particles against the antenna. It may also be caused by the existence of induction fields created by nearby corona discharges.

precipitation trails. See VIRGA.

precision, *n.* A measure of how close the outcome of a series of observations or measurements cluster about some estimated value of a desired quantity, such as the average value of a series of observations of the quantity. Precision implies *repeatability* of the observations within some specified limit and depends upon the random errors encountered due to the quality of the observing instrument, the skill of the observer and randomly fluctuating conditions such as temperature, pressure, refraction, etc. Precision should not be confused with ACCURACY. Observations may be of high precision due to the quality of the observing instrument, the skill of the observer and the resulting small random errors, but inaccurate due to the presence of large systematic errors which may not have been detected and accounted for. For a quantity to be accurately known, both systematic and random errors should be small. For a quantity to be known with high precision, only the random errors due to irregular effects need to be small. Precision of observations may be increased by improving the quality of the observing instrument and the skill of the observer. Precision may also be improved by mathematical techniques such as increasing the number of observations and using the method of **least squares adjustment.** The simplest application of least squares adjustment occurs when the quantity being measured does not change, e.g., the length of a table. In this case, taking the average of several measurements of the length results in a more precise value of the length than a single measurement would yield. The effects of some random errors such as irregular changes of refraction may be decreased only by increasing the number of observations. A numerical measure of precision is given by the **precision index.** See also ERROR.

precision graphic recorder. An equipment used with the standard hydrographic echo sounder for use in ocean depths where soundings cannot be recorded on the expanded scale of the standard equipment recorder. It provides a sounding record with a scale expansion and high accuracy. Commonly called PGR.

precision index. A measure of the magnitude of the random errors of a series of observations of some given quantity. If the precision index is large, most of the random errors of the observations are small. The precision index appears as a parameter in the normal (Gaussian) distribution law. Precision index is related to the standard deviation by means of the formula:

$$h = \frac{1}{\sqrt{2\pi}\,\sigma}$$

where h is the **precision** index and σ is the standard deviation. By making a series of observations, the standard deviation can be calculated. The precision index is then calculated using the formula above and a measure of the precision of the observing instrument is thus obtained. See also RANDOM ERROR, NORMAL DISTRIBUTION, PRECISION, STANDARD DEVIATION.

precomputation, *n.* The process of making navigational solutions in advance; applied particularly to the determination of computed altitude and azimuth before making a celestial observation for a line of position. When this is done, the observation must be made at the time used for the computation, or a correction applied.

precomputed altitude. The altitude of a celestial body computed before observation, and with the sextant altitude corrections applied with reversed sign. When a precomputed altitude has been calculated, the altitude difference can be determined by comparison with the sextant altitude.

precomputed curve. A graphical representation of the azimuth or latitude of a celestial body plotted against time for a given assumed position or positions, and which is computed for subsequent use with celestial observations.

predictability, *n.* In a navigation system, the measure of the accuracy with which the system can define the position in terms of geographical coordinates. See also REPEATABILITY, definition 2.

predictable accuracy. The accuracy of predicting position with respect to precise space and surface coordinates. See also REPEATABLE ACCURACY.

predicted tides. The times and heights of the tide as given in the Tide Tables in advance of their occurrence.

predicting machine. See TIDE PREDICTING MACHINE.

preferred datum. A geodetic datum selected as a base for consolidation of local independent datums within a geographical area. Also called MAJOR DATUM.

pressure, *n.* Force per unit area. The pressure exerted by the weight of the earth's atmosphere is called **atmospheric** or, if indicated by a barometer, **barometric pressure.** Pressure exerted by the vapor of a liquid is called **vapor pressure.** The pressure exerted by a fluid as a result of its own weight or position is called **static pressure.** Pressure exerted by radiant energy is called **radiation pressure.**

pressure gage. A tide gage that is operated by the change in pressure at the bottom of a body of water due to rise and fall of the tide.

pressure tendency. The character and amount of atmospheric pressure change for a 3-hour or other specified period ending at the time of observation. Also called BAROMETRIC TENDENCY.

prevailing westerlies. The prevailing westerly winds on the poleward sides of the sub-tropical high-pressure belts.

prevailing wind. The average or characteristic wind at any place.

primary, *n.* See PRIMARY BODY.

primary body. The celestial body or central force field about which a satellite orbits, or from which it is escaping, or towards which it is falling. The primary body of the earth is the sun; the primary body of the moon is the earth. Usually shortened to PRIMARY.

primary circle. See PRIMARY GREAT CIRCLE.

primary control tide station. A tide station at which continuous observations have been made over a minimum of a 19-year Metonic cycle. Its purpose is to provide data for computing accepted values of the harmonic and nonharmonic constants essential to tide predictions and to the determination of tidal datums for charting and coastal boundaries. The data series from this station serves as a primary control for the reduction of relatively short series from subordinate tide stations through the method of comparison of simultaneous observations, and for monitoring long-period sea-level trends and variations. See also TIDE STATION; SUBORDINATE TIDE STATION, definition 1; SECONDARY CONTROL TIDE STATION; TERTIARY TIDE STATION.

primary great circle. A great circle used as the orgin of measurement of a coordinate; particularly such a circle 90° from the poles of a system of spherical coordinates, as the equator. Also called PRIMARY CIRCLE, FUNDAMENTAL CIRCLE.

primary radar. 1. Radar which transmits a signal and receives the incident energy reflected from an object to detect the object in contrast to **secondary radar** which receives pulses from a transponder triggered by pulses transmitted from the radar.
2. As defined by the *International Telecommunication Union* (ITU), a radio-determination system based on the comparison of reference signals with radio signals reflected from the position to be determined.

primary seacoast light. A light established for the purpose of making landfall or coastwise passage from headland to headland. Also called LANDFALL LIGHT.

primary tidal bench mark. See under BENCH MARK.

primary tide station. See as PRIMARY CONTROL TIDE STATION.

prime fictitious meridian. The reference meridian (real or fictitious) used as the origin for measurement of fictitious longitude. **Prime grid meridian** is the reference meridian of a grid; **prime transverse** or **prime inverse meridian** is the reference meridian of a transverse graticule; **prime oblique meridian** is the reference fictitious meridian of an oblique graticule.

prime grid meridian. The refererence meridian of a grid. In polar regions it is usually the 180°-0° geographic meridian, used as the origin for measuring grid longitude.

prime inverse meridian. See PRIME TRANSVERSE MERIDIAN.

prime meridian. The meridian of longitude 0°, used as the origin for measurement of longitude. The meridian of Greenwich, England, is almost universally used for this purpose. See also PRIME FICTITIOUS MERIDIAN.

prime oblique meridian. The reference fictitious meridian of an oblique graticule.

prime transverse meridian. The reference meridian of a transverse graticule. Also called PRIME INVERSE MERIDIAN.

prime vertical. See PRIME VERTICAL CIRCLE.

prime vertical circle. The vertical circle perpendicular to the principal vertical circle. The intersections of the prime vertical circle with the horizon define the *east* and *west points* of the horizon. Often shortened to PRIME VERTICAL. Sometimes called TRUE PRIME VERTICAL to distinguish from magnetic, compass, or grid prime vertical defined as the vertical circle passing through the magnetic, compass, or grid east and west points of the horizon, respectively.

priming of tide. The periodic acceleration in the time of occurrence of high and low waters due to changes in the relative positions of the moon and the sun. **Priming** occurs when the moon is between new and first quarter and between full and third quarter. High tide occurs before transit of the moon. **Lagging** occurs when the moon is between first quarter and full and between third quarter and new. High tide occurs after transit of the moon. See also LAGGING OF TIDE.

principal vertical circle. The vertical circle passing through the north and south celestial poles. The intersection of the principal vertical circle with the horizon defines the *north* and *south points* of the horizon.

priority blanking. See under DUAL-RATE BLANKING.

prism, *n.* A solid having parallel, similar, equal, plane geometric figures as bases, and parallelograms as sides. By extension, the term is also applied to a similar solid having nonparallel bases, and trapezoids or a combination of trapezoids and parallelograms as sides. Prisms are used for changing the direction of motion of a ray of light and for forming spectra.

prismatic error. That error due to lack of parallelism of the two faces of an optical element, such as a mirror or a shade glass. See also SHADE ERROR.

private aids to navigation. In United States waters, those aids to navigation not established and maintained by the U.S. Coast Guard. Private aids include those established by other federal agencies with prior U.S. Coast Guard approval, those aids to navigation on marine structures or other works which the owners are legally obligated to establish, maintain, and operate as prescribed by the U.S. Coast Guard, and those aids which are merely desired, for one reason or another, by the individual corporation, state or local government or other body that has established the aid with U.S. Coast Guard approval. Although private aids to navigation are inspected periodically by the U.S. Coast Guard, the mariner should exercise special caution when using them for general navigation.

probable error. A measure of the dispersion or spread of a series of observations about some value, usually the mean or average value of all the observations. Probable error (PE) is calculated from the standard deviation (σ) of the observations by the formula

$$PE = 0.6745\ \sigma.$$

The numerical significance of the probable error is that 50% of the observations will lie between the values of $\bar{x} - PE$ and $\bar{x} + PE$, where $\bar{x}$ is the mean or average of all the observations. See also CIRCULAR ERROR PROBABLE.

probe, *n.* Any device inserted in an environment for the purpose of obtaining information about the environment.

production platform. A term used to indicate a permanent offshore structure equipped to control the flow of oil or gas. For charting purposes, the use of the term is extended to include all permanent platforms associated with oil or gas production, e.g. field terminal, drilling and accommodation platforms, and "booster" platforms sited at intervals along some pipelines. It does not include entirely submarine structures.

prognostic chart. A chart showing, principally, the expected pressure pattern of a given synoptic chart at a specified future time. Usually, positions of fronts are also included, and the forecast values of other meteorological elements may be superimposed.

progressive wave. In the ocean, a wave that advances in distance along the sea surface or at some intermediate depth. Although the wave form itself travels significant distances, the water particles that make up the wave merely describe circular (in relatively deep water) or elliptical (in relatively shallow water) orbits. With high, steep, wind waves, a small overlap in the orbit motion becomes significant. This overlapping gives rise to a small net mass transport.

prohibited area. 1. An area shown on nautical charts within which navigation and/or anchoring is prohibited except as authorized by appropriate authority.

2. A specified area within the land areas of a state or territorial waters adjacent thereto over which the flight of aircraft is prohibited.

See also DANGER AREA, RESTRICTED AREA.

projection, *n.* The extension of lines or planes to intersect a given surface; the transfer of a point from one surface to a corresponding position on another surface by graphical or analytical means. See also MAP PROJECTION.

projector compass. A magnetic compass in which the lubber's line and compass card, or a portion thereof, are viewed as an image projected through a system of lenses upon a screen adjacent to the helmsman's position. See also REFLECTOR COMPASS.

prolate cycloid. See TROCHOID.

prolate spheroid. An ellipsoid of revolution, the longer axis of which is the axis of revolution. An ellipsoid of revolution, the shorter axis of which is the axis of revolution, is called an OBLATE SPHEROID.

promontory, *n.* High land extending into a large body of water beyond the line of the coast. Called HEADLAND when the promontory is comparatively high and has a steep face. Also called FORELAND.

propagation, *n.* The travel of waves of energy through or along a medium other than a specially constructed path such as an electrical circuit.

proper motion. That component of the space motion of a celestial body perpendicular to the line of sight, resulting in the change of a star's apparent position relative to other stars. Proper motion is expressed in angular units.

proportional dividers. An instrument consisting in its simple form of two legs pointed at both ends and provided with an adjustable pivot, so that for any given pivot setting, the distance between one set of pointed ends always bears the same ratio to the distance between the other set. A change in the pivot changes the ratio. The dividers are used in transferring measurements between charts or other drawings which are not the same scale.

proportional parts. Numbers in the same proportion as a set of given numbers. Such numbers are used in an auxiliary interpolation table based on the assumption that the tabulated quantity and entering arguments differ in the same proportion. For each intermediate argument a "proportional part" or number is given to be applied to the preceding tabulated value in the main table.

protractor, *n.* An instrument for measuring angles on a surface; an angular scale. In its most usual form it consists of a circle or part of one (usually a semicircle) graduated in degrees. See also COMPASS ROSE, THREE-ARM PROTRACTOR.

province, *n.* On the sea floor, a region identifiable by a group of similar physiographic features whose characteristics are markedly in contrast with surrounding areas.

pseudo-independent surveillance. Position determination that relies on craft or vehicle cooperation but is not subject to craft or vehicle navigational errors (e.g., secondary radar).

psychrometer, *n.* A type of hygrometer (an instrument for determining atmospheric humidity) consisting essentially of dry-bulb and wet-bulb thermometers. The dry-bulb thermometer indicates the temperature of the air, and the wet-bulb thermometer the lowest temperature to which air can be cooled by evaporating water into it at constant pressure, both thermometers being well ventilated. With the information obtained from a psychrometer, the humidity, dew point, and vapor pressure for any atmospheric pressure can be obtained by means of appropriate tables.

psychrometric chart. A nomogram for graphically determining relative humidity, absolute humidity, and dew point from wet- and dry-bulb thermometer readings.

pteropod (*pl. pteropoda*), *n.* A small marine animal with or without a shell and having two thin, winglike feet. These animals are often so numerous they cover the surface of the sea for miles. In some areas, their shells cover the bottom.

Pub No. 9. *American Practical Navigator*; a two-volume publication of the Defense Mapping Agency Hydrographic/Topographic Center, originally by Nathaniel Bowditch, comprising an epitome of navigation and navigational astronomy and providing tables for solution of navigational problems. Popularly called BOWDITCH.

Pub. 102. *International Code of Signals*; a publication of the Defense Mapping Agency Hydrographic/Topographic Center intended primarily for communication at sea in situations involving safety of life at sea and navigational safety especially when language difficulties arise between ships or stations of different nationalities. The Code is suitable for transmission by all means of communication, including radiotelophony, radiotelegraphy, sound, and alphabetical and numeral flags. The International Code of Signals embodies the principle that each signal has a complete meaning.

Pub. 110. *Dapac*; a publication of the Defense Mapping Agency Hydrographic/Topographic Center giving routing instructions for Pacific Ocean areas declared dangerous due to mines or other reasons. The same *Dapac* is derived from the word danger areas in the Pacific.

Pub. 117A. *Radio Navigational Aids* (Atlantic and Mediterranean Area); a publication of the Defense Mapping Agency Hydrographic/Topographic Center which contains data on radio aids to navigation services provided to mariners in the Atlantic and Mediterranean area. The data provides the necessary information for the mariner to use radiobeacons for radio direction finding. Information on radio direction finder and radar stations, radio time signals, radio navigational warnings, distress signals, stations transmitting medical advice, long range radionavigation systems, emergency procedures and communications instructions, etc. is also given.

Pub. 117B. *Radio Navigational Aids* (Pacific and Indian Oceans Area); a publication of the Defense Mapping Agency Hydrographic/Topographic Center which contains data on radio aids to navigation services provided to mariners in the Pacific and Indian Oceans area. The data provides the necessary information for the mariner to use radiobeacons for radio direction finding. Information on radio direction finder and radar stations, radio time signals, radio navigational warnings, distress signals, stations transmitting medical advice, long range radionavigation systems, emergency procedures and communications instructions, etc. is also given.

Pub. 150. *World Port Index*; a publication of the Defense Mapping Agency Hydrographic/Topographic Center listing the location, characteristics, known facilities, and available services of a great many ports, shipping facilities and oil terminals throughout the world. The applicable chart and Sailing Direction is given for each place listed. A simple code, listed on each page, is used to indicate certain types of information.

Pub. 151. *Distances Between Ports*; a publication of the Defense Mapping Agency Hydrographic/Topographic Center providing calculated distances in nautical miles over water areas between most of the seaports of the world. The use of Junction Points affords a means of connecting routes in adjacent areas and increases the number of distances that are included. A Junction Point is a position where many routes converge and through which ships pass when sailing from one major area into another. A similar publication published by the National Ocean Survey for United States waters is entitled *Distances Between United States Ports*.

Pub. No. 217. *Maneuvering Board Manual*; a publication of the Defense Mapping Agency Hydrographic/Topographic Center, providing explanations and examples of various problems involved in maneuvering and in relative movement.

Pub. 221. 1. *Loran-C Table*; a series of lattice tables published by the Defense Mapping Agency Hydrographic/Topographic Center providing the tabular counterpart of the Loran-C chart. Through the use of the appropriate lattice table, Loran-C lines of position can be plotted on a suitable plotting sheet or chart. Each table is fully identified by the publication number (221), pertinent suffix, and station pair. For example, Pub. 221 (2209) Pair 9970-Z is the lattice table for the 9970-Z pair in the Northwest Pacific Loran-C chain.

2. *Loran-C Correction Table*; a series of tables published by the Defense Mapping Agency Hydrographic/Topographic Center providing additional secondary phase factor (ASF) corrections to Loran-C time differences for all station pairs of the chain covered in each table. The tables are published primarily for navigators who utilize electronic computers to convert Loran-C time differences to geographic coordinates. The tables are also published for use in correcting time differences for ASF when plotting Loran-C time differences on a chart on which the overprinted Loran-C lattice has not been compensated for ASF. Although the ASF corrections are generally too small to affect a Loran-C fix plotted on a small-scale chart, they can become as large as ± 4 microseconds. Each table is fully identified by the publication number (221), name of chain, and group repetition interval code. For example, Pub. 221, Northeast, U.S.A., 9960, designates the table containing ASF corrections for all station pairs of the Northeast, U.S.A. chain of group repetition interval 99600.

Pub. 224. 1. *Omega Table*; a series of lattice tables published by the Defense Mapping Agency Hydrographic/Topographic Center providing the tabular counterpart of the Omega chart. Through use of the appropriate charting coordinate or lattice table, Omega lines of position can be plotted on a suitable plotting sheet or chart having a scale as large as 1:800,000. The publication number, pertinent suffix, and station pair fully identify each lattice table. Using Pub. 224 (109) D-H, as an example, the 224 designates an Omega publication, the first digit of the suffix (109) identifies the frequency as 10.2kHz; the last two

digits of the suffix identify the area of coverage of the table as area 09; the station pair (D-H) completes the full identification of the table.
2. *Omega Propagation Correction Tables;* a series of tables published by the Defense Mapping Agency Hydrographic/Topographic Center providing necessary data for correcting Omega Navigation System receiver readouts, affected by the prevailing propagation conditions, to the standard conditions on which all Omega hyperbolic charts and lattice tables are based. The corrections are presented in the single station mode so that the navigator only need acquire the tables for the stations and areas desired. Each table contains propagation corrections for the station and area shown on the cover. The publication number, pertinent suffix followed by the letter C, and the designator of the single station for which a table is computed, fully identify a particular propagation correction table. Using Pub. 224 (109-C) D as an example, the 224 designates an Omega publication; the first digit of the suffix (109) identifies the frequency as 10.2kHz (2 denotes 3.4kHz); the last two digits of the suffix identify the area of coverage of the table as area 09; the letter C indicates that the table is a propagation correction table as opposed to a lattice table; and the station designator D completes the full identification of the table.

Pub. No. 226. *Handbook of Magnetic Compass Adjustment;* a publication of the Defense Mapping Agency Hydrographic/Topographic Center, providing information for adjustment of marine magnetic compasses.

Pub. No. 229. *Sight Reduction Tables for Marine Navigation;* a publication of the Defense Mapping Agency Hydrographic/Topographic Center, in six volumes, each of which includes two 8° zones of latitude. An overlap of 1° of latitude occurs between volumes. The six volumes cover latitude bands 0° to 15°, 15° to 30°, 30° to 45°, 45° to 60°, 60° to 75°, and 75° to 90°. For entering arguments of integral degrees of latitude, declination, and local hour angle, altitudes and their differences are tabulated to the nearest tenth of a minute, azimuth angles to the nearest tenth of a degree. But the tables are designed for precise interpolation of altitude for declination only by means of interpolation tables which facilitate linear interpolation and provide additionally for the effect of second differences. The data are applicable to the solutions of sights of all celestial bodies; there are no limiting values of altitude, latitude, hour angle, or declination.

Pub. No. 249. *Sight Reduction Tables for Air Navigation*; a publication of the Defense Mapping Agency Hydrographic/Topographic Center, in three volumes, with volume 1 containing tabulated altitudes and azimuths of selected stars, the entering arguments being latitude, local hour angle of the vernal equinox, and the name of the star; and volumes 2 and 3 containing tabulated altitudes and azimuth angles of any body within the limits of the entering arguments, which are latitude, local hour angle, and declination (0°-29°) of the body.

puddle, *n.* An accumulation of melt-water on ice, mainly due to melting snow, but in the more advanced stages also due to the melting of ice. The initial stage consists of patches of melted snow.

pulse, *n.* An electrical disturbance whose duration is short in relation to the time scale of interest and whose initial and final values are the same. A common example is the very short burst of electromagnetic energy transmitted by a pulse-modulated radar.

pulse decay time. The interval of time required for the trailing edge of a pulse to decay from 90 percent to 10 percent of the pulse amplitude.

pulse duration. The time interval during which the amplitude of a pulse is at or greater than a specified value, usually stated in terms of a fraction or percentage of the maximum value. Also called PULSE LENGTH, PULSE WIDTH, but the use of these terms is deprecated.

pulse duration error. A range distortion of a radar return caused by the duration of the pulse. See also SPOT-SIZE ERROR.

pulse group. See PULSE TRAIN.

pulse interval. See PULSE SPACING.

pulse length. See PULSE DURATION.

pulse-modulated radar. The type of radar generally used for shipboard navigational applications. The radio-frequency energy transmitted by a pulse-modulated radar consists of a series of equally spaced short pulses having a pulse duration of about 1 microsecond or less. The distance to the target is determined by measuring the transmit time of a pulse and its return to the source as a reflected echo. Also called PULSE RADAR.

pulse modulation. 1. The modulation of a carrier wave by a pulse train. In this sense, the term describes the process of generating carrier-frequency pulses.
2. The modulation of one or more characteristics of a pulse carrier. In this sense, the term describes methods of transmitting information on a pulse carrier.

pulse radar. See PULSE-MODULATED RADAR.

pulse repetition frequency. The pulse repetition rate of a periodic pulse train.

pulse repetition rate. The average number of pulses per unit of time. See also PULSE REPETITION FREQUENCY.

pulse rise time. The interval of time required for the leading edge of a pulse to rise from 10 to 90 percent of the pulse amplitude.

pulse spacing. The interval between corresponding points on consecutive pulses. Also called PULSE INTERVAL.

pulse train. A series of pulses of similar characteristics. Also called PULSE GROUP, IMPULSE TRAIN.

pulse width. See PULSE DURATION.

pumice, *n.* Cooled volcanic glass with a great number of minute cavities caused by the expulsion of water vapor at high temperature, resulting in a very light material. Part of the ocean bed is composed of pumice.

pumping, *n.* Unsteadiness of the mercury in the barometer, caused by fluctuations of the air pressure produced by a gusty wind or due to the motion of a vessel.

pure sound. See PURE TONE.

pure tone. A sound produced by a sinusoidal acoustic oscillation. Also called PURE SOUND.

purple light. The faint purple glow observed on clear days over a large region of the western sky after sunset and over the eastern sky before sunrise.

put to sea. To leave a sheltered area and head out to sea.

pyramidal iceberg. See PINNACLED ICEBERG.

Q

Q-band. A radio-frequency band 36 to 46 gigahertz. See also FREQUENCY, FREQUENCY BAND.

Q-correction. The Polaris correction as tabulated in the *Air Almanac*.

Q signals. Conventional code signals used in radiotelegraphy, each signal of three letters beginning with *Q* and representing a complete sentence.

quadrant, *n.* 1. A quarter of a circle; either an arc of 90° or the area bounded by such an arc and two radii.
2. A double-reflecting instrument for measuring angles, used primarily for measuring altitudes of celestial bodies. It has a range of 180°. Such an instrument is commonly called a SEXTANT.

quadrantal correctors. Masses of soft iron placed near a magnetic compass to correct for quadrantal deviation. Spherical quadrantal correctors are called **quadrantal spheres.**

quadrantal deviation. Deviation which changes its sign (E or W) approximately each 90° change of heading. It is caused by induced magnetism in horizontal soft iron.

quadrantal error. An error which changes sign (plus or minus) each 90°. Also called INTERCARDINAL ROLLING ERROR when related to a gyrocompass.

quadrantal point. See INTERCARDINAL POINT.

quadrantal spheres. Two hollow spheres of soft iron placed near a magnetic compass to correct for quadrantal deviation. See also QUADRANTAL CORRECTORS.

quadrant with two arcs. See BACKSTAFF.

quadrature, *n.* An elongation of 90° usually specified as east or west in accordance with the direction of the body from the sun. The moon is at quadrature at first and last quarters.

quadrilateral, *adj.* Having four sides.

quadrilateral, *n.* A closed plane figure having four sides. See also PARALLELOGRAM, TRAPEZOID.

quarantine anchorage. An area where a vessel anchors when satisfying quarantine regulations.

quarantine buoy. A buoy marking the location of a quarantine anchorage. In U.S. waters a quarantine buoy is yellow.

quarantine mark. A navigation mark indicating a quarantine anchorage area for shipping, or defining its limits.

quart, *n.* A unit of volume equal to one-fourth of a gallon.

quartering sea. Waves moving in a direction approximately 45° from a vessel's heading, striking the vessel on the quarter. Those moving in the general direction of the heading are called FOLLOWING SEA, those moving in a direction approximately opposite to the heading are called HEAD SEA, and those moving in a direction approximately 90° from the heading are called BEAM SEA.

quarter-power points. See under HALF-POWER POINTS.

quartz, *n.* Crystalline silica. In its most common form it is colorless and transparent, but it takes a large variety of forms of varying degrees of opaqueness and color. It is the most common solid mineral. Part of the ocean bed is composed of quartz.

quartz clock. See QUARTZ CRYSTAL CLOCK.

quartz crystal clock. A precision timepiece, essentially consisting of a generator of constant frequency controlled by a resonator made of quartz crystal with suitable methods for producing continuous rotation to operate time-indicating and related mechanisms. Also called QUARTZ CLOCK. See also QUARTZ CRYSTAL MARINE CHRONOMETER.

quartz crystal marine chronometer. A quartz crystal clock replacement for the spring-driven marine chronometer. The degree of accuracy is such that it requires no chronometer rate. Should the second hand be in error by a readable amount, it can be reset electrically.

quasi-stationary front. See STATIONARY FRONT.

quay (kē), *n.* A structure of solid construction along a shore or bank which provides berthing for ships and which generally provides cargo-handling facilities. A similar facility of open construction is called WHARF. See also MOLE, definition 1.

quick flashing light. A light exhibiting without interruption very rapid regular alternations of light and darkness. In the United States and most other countries, the light exhibits not less than 60 flashes per minute. There is no restriction on the ratio of durations of light and darkness. It has cautionary significance. The quick flashing light as used with the IALA System differs from the above in that the light is exhibited *with interruption*. The quick flashing light and VERY QUICK FLASHING LIGHT of the IALA System have phase characteristics formed by groups of flashes, each group usually consisting of 3, 6, or 9 flashes. See also INTERRUPTED QUICK FLASHING LIGHT.

quick light. A light in which flashes are repeated at a rate of not less than 50 flashes per minute but less than 80 flashes per minute. See also CONTINUOUS QUICK LIGHT, GROUP QUICK LIGHT, INTERRUPTED QUICK LIGHT.

quicksand, *n.* A loose mixture of sand and water that yields to the pressure of heavy objects. Such objects are difficult to extract once they begin sinking.

quiet sun. The sun when it is free from unusual radio wave or thermal radiation such as that associated with sun spots.

quintant, *n.* A double-reflecting instrument for measuring angles, used primarily for measuring altitudes of celestial bodies. It has a range of 144°. Such an instrument is commonly called a SEXTANT.

R

race, *n.* A rapid current or a constricted channel in which such a current flows. The term is usually used only in connection with a tidal current, when it may be called a TIDE RACE.

racon, *n.* 1. A transponder beacon which, when triggered by a ship's radar emission transmits a reply which provides the range and bearing to the beacon on the PPI display of the ship. The reply appears on the PPI display as a radial line or narrow sector, the racon flash extending radially from a point beyond the echo of the racon installation, or from just beyond the point where the echo would be painted if detected, due to response delay. The distance beyond may be several hundred yards. For identification purposes, the racon flash may be in Morse code, the first character usually being a dash to avoid its being confused with the possible blip formed by the echo of the racon installation. Only a few racons operate in other than the 3-centimeter band. The name racon is derived from the words radar beacon. See also IN-BAND RACON, CROSS-BAND RACON, SWEPT-FREQUENCY RACON, RAMARK. Also called RADAR TRANSPONDER BEACON.
2. As defined by the *International Telecommunication Union* (ITU), in the maritime radionavigation service, a receiver-transmitter device which, when triggered by a surface search radar, automatically returns a distinctive signal which can appear on the display of the triggering radar, providing range, bearing and identification information.

racon flash. See under RACON.

radar, *n.* 1. A radiodetermination system which measures distance and usually direction by a comparison of reference signals with the radio signals reflected or retransmitted from the target whose position is to be determined. **Primary radar** uses reflection only; **secondary radar** uses automatic retransmission on the same or a different radio frequency. **Pulse-modulated radar** is used for shipboard navigational applications. In this type of radar the distance to the target is determined by measuring the time required for an extremely short burst or pulse of radio-frequency energy to travel to the target and return to its source as a reflected echo. *Directional antennas* are used for transmitting the pulse and receiving the reflected echo, thereby allowing determination of the direction of the target echo from the source. The radio-frequency energy transmitted by pulse-modulated radars consists of a series of equally spaced pulses, frequently having pulse durations of about 1 microsecond or less, separated by still very short but relatively long periods during which no energy is transmitted. The name *radar* is derived from the words radio detection and ranging.
2. As defined by the *International Telecommunication Union* (ITU) a radiodetermination system based on the comparison of reference signals with radio signals reflected, or re-transmitted, from the position to be determined.

radar beacon. 1. A radar transmitter whose emissions enable a ship to determine its direction and frequently position relative to the transmitter by means of the ship's radar equipment. There are two general types of radar beacons: one type, the RACON, must be triggered by the ship's radar emissions; the other type, the RAMARK, transmits continuously and provides bearings only. See also TRANSPONDER.
2. See RACON, definition 2.

radar bearing. A bearing obtained by radar.

radar buoy. A buoy having corner reflectors designed into the superstructure, the characteristic shape of the buoy being maintained. This is to differentiate from a buoy on which a corner reflector is mounted.

radar conspicuous object. An object which returns a strong radar echo which can be identified with a high degree of certainty.

radar cross section. The area of a plane element, situated at the position of an object and normal to the direction of the radar transmitter, which would be traversed by a power such that, if this power were re-radiated equally in all directions, with suitable polarization, it would give an echo of the same power as that given by the object itself. Also called EQUIVALENT ECHOING AREA.

radar echo. See ECHO, definition 3.

radar fix. A fix established by means of radar.

radar horizon. The radio horizon of a radar antenna.

radar indicator. A unit of a radar set which provides a visual indication of radar echoes received, using a cathode-ray tube for such indication. Besides the cathode-ray tube, the radar indicator is comprised of sweep and calibration circuits, and associated power supplies. Often shortened to INDICATOR.

radar link. A means by which the information from a radar set is reproduced at a distance by use of a radio link or cable. Also called RADAR RELAY SYSTEM.

radar nautical mile. The time interval required for the electromagnetic energy of a radar pulse to travel 1 nautical mile and the echo to return, approximately 12.4 microseconds.

radar picture. See DISPLAY, definition 1.

radar range. 1. The distance of a target as measured by radar.
2. The maximum distance at which a radar set is effective in detecting targets. Radar range depends upon variables such as the weather, transmitted power, antenna height, pulse duration, receiver sensitivity, target size, target shape, etc.

radar receiver. A unit of a radar set which demodulates received radar echoes, amplifies the echoes, and delivers them to the radar indicator. The radar receiver differs from the usual superheterodyne communications receiver in that its sensitivity is much greater; it has a better signal to noise ratio, and it is designed to pass a pulse-type signal.

radar reference line. A mid-channel line on a chart which corresponds to a line incorporated in a harbor radar display for the purpose of providing a reference for informing a vessel of its position. In some cases the line may be coincident with the recommended track. The line may be broken into sections of specified length having assigned names or numbers.

radar reflector. A device specially arranged to have the property of reflecting incident electromagnetic energy parallel to the direction of incidence in order to enhance the radar response. See also

CORNER REFLECTOR, PENTAGONAL CLUSTER, OCTAHEDRAL CLUSTER, DIHEDRAL REFLECTOR, DIELECTRIC REFLECTOR, REFLECTOR.

radar relay system. See RADAR LINK.

radar repeater. A unit which duplicates the PPI display at a location remote from the main radar indicator installation. Also called PPI REPEATER, REMOTE PPI.

radar return. See as ECHO, definition 2.

radar scan. The motion of a radar beam through space in searching for an echo.

radar scanning. The process or action of directing a radar beam through a search pattern.

radarscope, *n.* The cathode-ray tube in the indicator of a radar set which displays the received echo in such a manner as to indicate range and bearing. Often shortened to SCOPE. See also PLAN POSITION INDICATOR.

radar set. An electronic apparatus consisting principally of a transmitter, antenna, receiver, and indicator for sending out radio-frequency energy and receiving and displaying reflected energy in such a manner as to indicate the range and bearing of the reflecting object. See also RADAR.

radar shadow. The shielding of a region from radar signals because of an intervening obstruction or absorbing medium. The shadow region appears as an area void of targets on a radar display such as a plan position indicator.

radar station pointer. *British terminology.* A transparent plotting chart inscribed with radial lines from 0° to 360°, used to plot radar echoes to the scale of the chart in use. It is also used to assist in identifying radar responses with charted features.

radar target. See as TARGET.

radar transponder beacon. See RACON.

radial, *adj.* Of or pertaining to a ray or radius; extending in a straight line outward from a center.

radial, *n.* A straight line extending outward from a center.

radial error. In a two-dimensional or elliptical error distribution, the measure of error as the radius of a circle of equivalent probability derived from the error ellipse. The error, expressed as $1d_{rms}$, is the square root of the sum of the 1 σ error components along the major and minor axes of the probability ellipse. The measure d_{rms} is not equal to the square root of the sum of the squares of σ_1 and σ_2 that are the basic errors associated with the lines of position of a particular measuring system, but the square root of the sum of the squares of σ_x and σ_y along the major and minor axes of the ellipse and which may be derived from σ_1 and σ_2 using the following equations:

$$\sigma_x^2=\frac{1}{2\sin^2\alpha}\left[\sigma_1^2+\sigma_2^2+\sqrt{(\sigma_1^2+\sigma_2^2)^2-4\sin^2\alpha\,\sigma_1^2\sigma_2^2}\right]$$

$$\sigma_y^2=\frac{1}{2\sin^2\alpha}\left[\sigma_1^2+\sigma_2^2-\sqrt{(\sigma_1^2+\sigma_2^2)^2-4\sin^2\alpha\,\sigma_1^2\sigma_2^2}\right].$$

The use of radial error or d_{rms} error as a measure of error is somewhat confusing because the term does not correspond to a fixed value of probability for a given value of the error measure. The term can be conveniently related to other error measures only when $\sigma_x = \sigma_y$, and the probability figure is a circle. In the more common elliptical cases, the probability associated with a fixed value of d_{rms} varies as a function of the eccentricity of the ellipse. One d_{rms} is defined as the radius obtained when $\sigma_x = 1$ and σ_y varies from 0 to 1. Likewise, $2d_{rms}$ is the radius of the circle obtained when $\sigma_x = 2$ and σ_y varies from 0 to 2.

radial motion. Motion along a radius, or a component in such a direction, particularly that component of space motion of a celestial body in the direction of the line of sight.

radial period. See ANOMALISTIC PERIOD.

radian, *n.* The *supplementary unit* of plane angle in the International System of Units; it is the plane angle subtended at the center of a circle by an arc equal in length to the radius of the circle. It is equal to 360° ÷ 2π, or approximately 57° 17′ 48.″8.

radian per second. The *derived unit* of angular velocity in the International System of Units.

radian per second squared. The *derived unit* of angular acceleration in the International System of Units.

radiant, *adj.* Of, pertaining to, or transmitted by radiation.

radiant energy. Energy consisting of electromagnetic waves.

radiate, *v., t. & i.* To send out in rays or straight lines from a center.

radiation, *n.* 1. The process of emitting energy in the form of electromagnetic waves.

2. The energy radiated in definition 1 above.

radiational cooling. The cooling of the earth's surface and adjacent air, occurring mainly at night whenever the earth's surface suffers a net loss of heat due to terrestrial radiation.

radiational tides. Periodic variations in sea level primarily related to meteorological changes such as the semidaily (solar) cycle in barometric pressure, daily (solar) land and sea breezes, and seasonal (annual) changes in temperature. Other changes in sea level due to meteorological changes that are random in phase are not considered radiational tides.

radiation fog. A major type of fog, produced over a land area when radiational cooling reduces the air temperature to or below its dew point. Thus, a strict radiation fog is a nighttime occurrence, although it may begin to form by evening twilight and often does not dissipate until after sunrise.

radiation pattern. A curve representing, in polar or Cartesian coordinates, the relative amounts of energy radiated in various directions. Also called DIRECTIVITY DIAGRAM.

radiatus, *adj.* Radial. The term is used to refer to clouds in parallel bands which, owing to perspective, appear to converge toward a point on the horizon, or two opposite points if the bands cross the sky.

radio, *n.* A general term applied to the use of radio waves.

radio acoustic ranging. Determining distance by a combination of radio and sound, the radio being used to determine the instant of transmission or reception of the sound, and distance being determined by the time of transit of sound, usually in water. See also ECHO RANGING.

radio aid to navigation. An aid to navigation transmitting information by radio waves. See also ELECTRONIC AID TO NAVIGATION.

radio altimeter. As defined by the *International Telecommunication Union* (ITU), a radionavigation equipment, on board an aircraft, which makes use of the reflection of radio waves from the ground to determine the height of the aircraft above the ground.

radiobeacon, *n.* A radio transmitting station which emits a distinctive or characteristic signal usually for the purpose of the navigator being able to determine the direction of the source of the signal of known location by means of a radio direction finder. The direction so obtained and plotted from the signal source provides a line of position. The most common type of marine radiobeacon transmits radio waves of approximately uniform strength in all directions. These *omnidirectional beacons* are called **circular radiobeacons.** A radiobeacon some or all of the emissions of which are directional so that the signal characteristic changes according to the vessel's bearing from the beacon is called a **directional radiobeacon.** A radiobeacon all or part of the emissions of which is concentrated in a beam which rotates is called a **rotating radiobeacon.** See also CONTINUOUS CARRIER RADIOBEACON, DUAL-CARRIER RADIOBEACON, SEQUENCED RADIOBEACON, ROTATING PATTERN RADIOBEACON, COURSE BEACON.

radiobeacon characteristic. The description of the complete cycle of transmission of a radiobeacon in a given period of time, inclusive of any silent period.

radiobeacon station. As defined by the *International Telecommunication Union* (ITU), a station in the radionavigation service the emissions of which are intended to enable a mobile station to determine its bearing or direction in relation to the radiobeacon station.

radio bearing. The bearing of a radio transmitter from a receiver, as determined by a radio direction finder.

radio compass. The name by which the radio direction finder was formerly known.

radiodetermination, *n.* As defined by the *International Telecommunication Union* (ITU), the determination of position, or the obtaining of information relating to position, by means of the propagation properties of radio waves.

radiodetermination-satellite service. As defined by the *International Telecommunication Union* (ITU), a radiocommunication service involving the use of radiodetermination and the use of one or more space stations.

radio direction finder. A complete radio receiving equipment, including antenna, used for radio direction finding. Also called DIRECTION FINDER. Formerly called RADIO COMPASS. See also AUTOMATIC DIRECTION FINDER.

radio direction finder station. A radio station equipped with special apparatus for determining the direction of radio signals transmitted by ships and other stations. The bearing taken by a radio direction finder station, and reported to a ship, is corrected for all determinable errors except conversion angle. Also called DIRECTION FINDER STATION.

radio direction finding. As defined by the *International Telecommunication Union* (ITU), radiodetermination using the reception of radio waves for the purpose of determining the direction of a station or object.

radio direction-finding station. As defined by the *International Telecommunication Union* (ITU), a radiodetermination station using radio direction finding.

radio fix. The determination of navigational position of a ship by ascertaining the direction of radio signals received from two or more sending stations, the locations of which are known.

radio frequency. Any electromagnetic wave occurring within that segment of the spectrum normally associated with some form of radio propagation. Radio frequencies are usually classified as very low, 3 - 30 kilohertz; low 30-300 kilohertz; medium, 300 - 3000 kilohertz; high, 3 - 30 megahertz; very high, 30 - 300 megahertz; ultra high, 300 - 3000 megahertz; super high, 3 - 30 gigahertz; extremely high, 30 - 300 gigahertz.

radio guard. A ship, aircraft, or radio station designated to listen for and record transmissions, and to handle traffic on a designated frequency for a certain unit or units.

radio horizon. The locus of points at which direct rays from a transmitting antenna become tangent to the earth's surface, taking into account the curvature due to refraction. Its distance from the transmitting antenna is greater than that of the visible horizon, and increases with decreasing frequency. A **radar horizon** is the radio horizon of a radar antenna.

radio interference. Interference due to unwanted signals from other radio transmitting stations providing some other desired service or operating on the same or adjacent frequencies. The frequency allocations made by the International Telecommunication Union (ITU) and published in the *Radio Regulations* are intended to reduce this type of interference to a minimum by imposing limits on the allowable transmitter power, the bandwidth covered by the transmission, the accuracy and stability of the operating frequency, and the level of radiation outside the specified bandwidth. There are transmitting stations, however, which may not be operating in compliance with the *Radio Regulations* and may not be registered. These stations can be of great concern to users of radio aids to navigation. However, under certain conditions, such as ducting, which result in unusually long range being obtained, it is possible for a radio aid to navigation to suffer interference from other transmitters which are in fact complying with the provisions of the *Radio Regulations.* This effect can be particularly troublesome at night.

radio interferometer. An interferometer operating at radio frequencies; used in radio astronomy and in satellite tracking.

radiolarian (*pl. radiolaria*), *n.* A minute sea animal with a siliceous outer shell. The skeletons of such animals are very numerous, covering the ocean bottom in certain areas, principally in the tropics.

radiolocation, *n.* As defined by the *International Telecommunication Union* (ITU), radiodetermination used for purposes other than those of radionavigation.

radio mast. A label on a nautical chart which indicates a relatively short pole or slender structure for elevating radio antennas, usually found in groups.

radionavigation, *n.* 1. The determination of position, or the obtaining of information relating to position, for the purposes of navigation by means of the propagation properties of radio waves.

2. As defined by the *International Telecommunication Union* (ITU), radiodetermination used for the purposes of navigation, including obstruction warning.

See also RADIODETERMINATION, RADIOLOCATION.

Radio Navigational Aids. See PUB. 117A, PUB. 117B.

radio navigational warning. A radio-transmitted

message affecting the safe navigation of vessels or aircraft. See also HYDROLANT, HYDROPAC, NAVAREA WARNINGS, WORLD WIDE NAVIGATIONAL WARNING SERVICE.

radionavigation-satellite service. As defined by the *International Telecommunication Union* (ITU), a radiodetermination-satellite service used for the same purposes as the radionavigation service; in certain cases this service includes transmission or retransmission of supplementary information necessary for the operation of radionavigation systems.

radio receiver. An equipment connected to an antenna or other receptor of radio signals in order to make available in some desired form the required information content of the signals.

radio silence. A period during which all or certain radio equipment capable of radiation is kept inoperative.

radio spectrum. The range of electromagnetic radiations useful for communication by radio (approximately 10 kilohertz to 300,000 megahertz).

radio station. A place equipped with one or more transmitters or receivers, or a combination of transmitters and receivers, including the accessory equipment necessary at one location, for carrying on a radiocommunication service. Each station is classified by the service in which it operates permanently or temporarily.

radio tower. A label on a nautical chart which indicates a tall pole or structure for elevating radio antennas.

radio transmitter. Equipment for generation and modulation of radio-frequency energy for the purpose of radiocommunication.

radio wave propagation. The transfer of energy by electromagnetic radiation at radio frequencies.

radio waves. Electromagnetic waves of frequencies lower than 3,000 GHz propagated in space without artificial guide. The practicable limits of radio frequency are approximately 10 kHz to 100 GHz. Also called HERTZIAN WAVES.

radius, *n.* A straight line from the center of a circle, arc, or sphere to its circumference, or the length of such a line. Also called SEMIDIAMETER for a circle or sphere. See also DIAMETER.

radius of action. The maximum distance a ship, aircraft, or vehicle can travel away from its base along a given course with normal combat load and return without refueling, allowing for all safety and operating factors.

radius vector. A straight line connecting a fixed reference point or center with a second point, which may be moving. In astronomy the expression is usually used to refer to the straight line connecting a celestial body with another which revolves around it, as a *radius vector* of the earth and moon. See also POLAR COORDINATES, SPHERICAL COORDINATES.

radome, *n.* A dome-shaped structure used to enclose radar apparatus (may also have a cylindrical base with a dome top).

rafted ice. A type of deformed ice formed by one piece of ice overriding another. See also FINGER RAFTING.

rain, *n.* Liquid precipitation consisting of drops of water larger than those which comprise DRIZZLE. Orographic rain results when moist air is forced upward by a mountain range. See also FREEZING RAIN.

rainbow, *n.* A circular arc of concentric spectrally colored bands formed by the refraction of light in drops of water. One seen in ocean spray is called a marine or sea rainbow. See also FOGBOW, MOONBOW.

rain clutter. Clutter on the radarscope which is the result of the radar signal being reflected by rain or other forms of precipitation.

rain gush. See CLOUDBURST.

rain gust. See CLOUDBURST.

rain shadow. The condition of diminished rainfall on the lee side of a mountain or mountain range, where the rainfall is noticeably less than on the windward side.

rain storm. See under STORM, definition 2.

raise, *v.t.,* To cause to appear over the horizon or higher above the horizon by approaching closer.

ram, *n.* An underwater ice projection from an ice wall, ice front, iceberg, or floe. Its formation is usually due to a more intensive melting and erosion of the unsubmerged part.

ramark, *n.* A radar beacon which continuously transmits a signal appearing as a radial line on the PPI, indicating the direction of the beacon from the ship. For identification purposes, the radial line may be formed by a series of dots or dashes. The radial line appears even if the beacon is outside the range for which the radar is set, as long as the radar receiver is within the power range of the beacon. Unlike the RACON, the ramark does not provide the range to the beacon. The name *ramark* is derived from the words **ra**dar **mark**er.

ramming, *n.* In sea ice navigation, the act of a ship at full power striking obstructing ice.

ramp, *n.* On the sea floor, a gentle slope connecting areas of different elevations.

random error. One of the two categories of errors of observation and measurement, the other category being *systematic error*. Random errors are the errors which occur when irregular, randomly occurring conditions affect the observing instrument, the observer and the environment of the quantity being observed so that careful observations of the same quantity made with the same observing equipment and observer under the same observing conditions result in different values of the observed quantity from one observation to the next. Random errors depend upon (1) the quality of the observing instrument, (2) the skill of the observer, particularly, the ability to estimate the fraction of the smallest division or graduation on the observing instrument, and (3) randomly fluctuating conditions such as temperature, pressure, refraction, etc. For many types of observations, random errors are characterized by the following properties: (1) positive and negative errors of the same magnitude are about equal in number, (2) small errors occur more frequently than large errors, and (3) extremely large errors rarely occur. These properties of random errors permit the use of a mathematical law called the **Gaussian** or **normal distribution** of errors to calculate the probability that the random error of any given observation of a series of observations will lie within certain limits. Random error might more properly be called deviation since mathematically, the random error of an individual observation is calculated as the difference or deviation between the actual observation and an *improved* or *adjusted* value of the observation obtained by some mathematical technique such as averaging all the observations. Also called ACCIDENTAL ERROR, CHANCE ERROR, IRREGULAR ERROR, STATISTICAL ERROR. See also ERROR, PRECISION, PRECISION INDEX, STANDARD DEVIATION.

range, *n.* 1. Two or more objects in line. Such objects are said to be *in range*. An observer having them in range is said to be *on the range*. Two beacons are frequently located for the specific purpose of forming a range to indicate a safe route or the centerline of a channel. See also BACK RANGE, LEADING LINE, MAGNETIC RANGE, MULTIPLE RANGES. Called LEADING MARKS in British terminology.
2. Distance in a single direction or along a great circle.
3. The extreme distance at which an object or light can be seen is called VISUAL RANGE. When the extreme distance is limited by the curvature of the earth and the heights of the object and the observer, this may be called **geographic range**; when the range of a light is limited only by its intensity, clearness of the atmosphere, and sensitiveness of the observer's eyes, it may be called **luminous range.**
4. The extreme distance at which a signal can be detected or used. The maximum distance at which reliable service is provided is called **operating range.** The spread of ranges in which there is an element of uncertainty of interpretation is called **critical range.**
5. The distance a craft can travel at cruising speed without refueling is called CRUISING RADIUS.
6. The difference in extreme values of a variable quantity. See also RANGE OF TIDE.
7. A series of mountains or mountain ridges is called MOUNTAIN RANGE.
8. A predetermined line along which a craft moves while certain data are recorded by instruments usually placed below the line, or the entire station at which such information is determined. See also DEGAUSSING RANGE.
9. An area where practice firing of ordnance equipment is authorized. See also BOMBING RANGE.
10. On the sea floor, a series of ridges or seamounts.

range, *v., t.* 1. To place in line.
2. To determine the distance to an object.
3. To move along or approximately parallel to something, as to *range* the coast.

range daymark. 1. One of a pair of unlighted structures used to mark a definite line of bearing. See also RANGE, definition 1.
2. A daymark on a range light.

range finder. An optical instrument for measuring the distance to an object. See also STADIMETER.

range lights. Two or more lights at different elevations so situated to form a range (leading line) when brought into transit. Called LEADING LIGHTS in British terminology. The one nearest the observer is the front light and the one farthest from the observer is the **rear light.** The front light is at a lower elevation than the rear light.

range marker. A visual presentation on a radar display for measuring the range or for calibrating the time base. See also VARIABLE RANGE MARKER, RANGE RING.

range (of a light). See VISUAL RANGE (OF A LIGHT).

range of tide. The difference in height between consecutive high and low waters. The **mean range** is the difference in height between mean high water and mean low water. The **great diurnal range** or **diurnal range** is the difference in height between mean higher high water and mean lower low water. Where the type of tide is diurnal the mean range is the same as the diurnal range. For other ranges see APOGEAN TIDES, NEAP TIDES, PERIGEAN TIDES, SPRING TIDES, TROPIC TIDES.

range-range mode. See RANGING MODE.

range resolution. See as RESOLUTION IN RANGE under RESOLUTION, definition 2. Also called DISTANCE RESOLUTION.

range ring. One of a set of equally spaced concentric rings, centered on own ship's position, providing a visual presentation of range on a plan position indicator. See also VARIABLE RANGE MARKER.

ranging mode. A mode of operation of a radionavigation system in which the times for the radio signals to travel from each transmitting station to the receiver are measured rather than their *differences* as in the HYPERBOLIC MODE. Also called RHO-RHO MODE, RANGE-RANGE MODE.

Rankine temperature. Temperature based upon a scale starting at absolute zero (−459.°69F) and using Fahrenheit degrees. Kelvin temperature is based upon a thermodynamic scale with its zero point at absolute zero (−273.°16C) and using Celsius degrees.

rapid, *n.* A portion of a stream in swift, disturbed motion, but without cascade or waterfall. Usually used in the plural.

ratan, *n.* An experimental short-range aid to navigation, not operational, in which radar harbor surveillance information is transmitted to the user by television.

rate, *n.* 1. Quantity or amount per unit of something else, usually time. See also ANGULAR RATE, CHRONOMETER RATE, PULSE REPETITION RATE, REPETITION RATE, WATCH RATE.
2. With respect to Loran-A and Loran-C, the term rate, implying the number of pulses per unit time, is used for the character designation, and also the station pair, their signals, and the resulting hyperbolic lines of position and the tables and curves by which they are represented.

rate gyro. A single-degree-of-freedom gyro having primarily elastic restraint of its spin axis about the output axis. In this gyro, an output signal is produced by gimbal angular displacement, relative to the base, which is proportional to the angular rate of the base about the input axis. See also RATE INTEGRATING GYRO.

rate integrating gyro. A single-degree-of-freedom gyro having primarily viscuous restraint of its spin axis about the output axis. In this gyro an output signal is produced by gimbal angular displacement, relative to the base, which is proportional to the integral of the angular rate of the base about the input axis. See also RATE GYRO.

ratio, *n.* The relation of one magnitude to another of the same kind; the quotient obtained by dividing one magnitude by another of the same kind. See also MAGNITUDE RATIO.

rational horizon. See CELESTIAL HORIZON.

ratio of ranges. The ratio of the ranges of tide at two places. It is used in the tide tables where the times and heights of all high and low tides are given for a relatively few places, called REFERENCE STATIONS. The tides at other places, called SUBORDINATE TIDE STATIONS, are found by applying corrections to the values given for the reference stations. One of these corrections is the ratio of ranges, or the ratio between the height of the tide at the subordinate station and its reference station.

ratio of rise. The ratio of the height of tide at two places.

ravine, *n.* 1. A gulch; a small canyon or gorge, the sides of which have comparatively uniform slopes.
2. On the sea floor, a small canyon.

Raydist, *n.* The general name for several radiolocation systems produced by the Teledyne Hastings-Raydist Company, Hampton, Virginia. The **Raydist DR-S** system operates in the band 1.6-4.0 MHz and is comparable in range and accuracy with Hi-Fix and Sea-Fix. Unlike Hi-Fix and Sea-Fix, Raydist DR-S can operate with up to four users in the range-range configuration. Also unlike Hi-Fix and Sea-Fix, phase locking is unnecessary. In the normal or range-range configuration, there are two base stations (red and green) ashore and a mobile transmitter and Raydist Navigator aboard the survey vessel. The carrier frequency of the mobile transmitter aboard the survey vessel is received by each base station where it is heterodyned with its carrier frequency, approximately one-half the frequency of the mobile transmitter. Each base station transmitter is tuned so that the desired beat frequency is developed. One base station is tuned to a frequency below and the other to a frequency above one-half the carrier frequency of the mobile transmitter. The Raydist Navigator aboard the survey vessel receives the carriers of the red and green base stations as well as the carrier frequency of its adjacent mobile transmitter. The Navigator develops the same beat frequencies as are developed at the red and green base stations. The beat frequency at the red base station is returned to the Navigator as the lower single sideband on the red station carrier. The beat frequency of the green base station is returned to the Navigator as the upper single sideband on the green station carrier. At the Navigator the phases of the returning beat frequencies are compared with those developed locally to determine fractions of lanes and, with lane identification, range to each base station.

reach, *n.* The comparatively straight segment of a river or channel between two bends. That part of a winding river or channel between the last bend and the sea is called a **sea reach**; that part between the harbor and the first bend is called a **harbor reach.**

reach ahead. The distance advanced from the time a new speed is ordered to the instant the new speed is being made.

real image. An image actually produced and capable of being shown on a surface, as in a camera.

real precession. Precession of a gyroscope resulting from an applied torque such as that resulting from friction and dynamic unbalance as opposed to APPARENT PRECESSION. Also called INDUCED PRECESSION, PRECESSION.

rear light. That range light which is farthest from the observer. It is the highest of the lights of an established range. Also called HIGH LIGHT.

rebuilt, *adj.* Said of a fixed aid to navigation, previously destroyed, which has been restored as an aid to navigation.

receiver, *n.* One who or that which receives anything, particularly a radio receiver.

receiver gain control. An operating control on a radar indicator used to increase or decrease the sensitivity of the receiver. The control regulates the intensity of the echoes displayed on the radarscope.

receiver monitor. See under PERFORMANCE MONITOR.

reciprocal, *adj.* In a direction 180° from a given direction. Also called BACK.

reciprocal, *n.* 1. A direction 180° from a given direction.
2. The quotient of 1 divided by a given number.

reciprocal bearing. A bearing differing by 180°, or one measured in the opposite direction, from a given bearing.

recommended direction of traffic flow. A traffic flow pattern indicating a recommended directional movement of traffic in a routing system within which it is impractical or unnecessary to adopt an established direction of traffic flow.

recommended track. A route which has been specially examined to ensure so far as possible that it is free of dangers and along which ships are advised to navigate. See also ROUTING SYSTEM.

recovered, *adj.* Said of an adrift or missing floating aid to navigation which has been retrieved but not reset.

recreational and esthetic sanctuary. A marine sanctuary established for its esthetic and recreational value. See also MARINE SANCTUARY.

rectangle, *n.* A four-sided figure with its opposite sides parallel and its angles 90°; a right-angled parallelogram. A rectangle with sides of equal length is a **square.**

rectangular chart. A chart on the rectangular projection.

rectangular coordinates. Magnitudes defining a point relative to two perpendicular lines, called AXES. The magnitudes indicate the perpendicular distance from each axis. The vertical distance is called the ordinate and the horizontal distance the **abscissa.** This is a form of CARTESIAN COORDINATES.

rectangular error. An error which results from rounding off values prior to their inclusion in a table or which results from the fact that an instrument cannot be read closer than a given value. The error is so called because of the shape of its plot. For example: if the altitudes tabulated in a sight reduction table are stated to the nearest 0.′1, the error in the altitude as extracted from the table might have any value from (+) 0.′05 to (–) 0.′05, and any value within these limits is as likely to occur as another value having similar decimals. See also SIMILAR DECIMALS.

rectangular projection. A cylindrical map projection with uniform spacing of the parallels. This projection is used for the star chart in the *Air Almanac.*

rectified altitude. See APPARENT ALTITUDE.

rectilinear, *adj.* Moving in or characterized by a straight line.

rectilinear current. See REVERSING CURRENT.

recurring decimal. See REPEATING DECIMAL.

recurring polynya. See under POLYNYA.

recurved spit. A hook developed when the end of a spit is turned toward the shore by current deflection or by opposing action of two or more currents. Also called HOOK, HOOKED SPIT.

red magnetism. The magnetism of the north-seeking end of a freely suspended magnet. This is the magnetism of the earth's south magnetic pole.

red sector. A sector of the circle of visibility of a navigational light in which a red light is exhibited. Such sectors are designated by their limiting bearings, as observed at some point other than the light. Red sectors are often located such that they warn of danger to vessels.

red shift. In astronomy, the displacement of observed spectral lines toward the longer wavelengths of the red end of the spectrum. The red shift in the spectrum of distant galaxies has been interpreted as evidence that the universe is expanding.

red snow. Snow colored red by the presence in it either of minute algae or of red dust particles.

reduction, *n.* The process of substituting for an observed value one derived therefrom, as the placing of meteorological elements on a common basis for comparison, the calculation of the corresponding meridian altitude from an observation of a celestial body near the meridian, or the derivation from a celestial observation of the information needed for establishing a line of position.

reduction of soundings. The adjustment of soundings to the selected chart datum. Usually the term reduction of soundings does not pertain to corrections other than those for height of tide. See also CORRECTION OF SOUNDINGS.

reduction of tidal current. The processing of observed tidal current data to obtain mean values of tidal current constants. See also REDUCTION OF TIDES.

reduction of tides. The processing of observed tidal data to obtain mean values of tidal constants. See also REDUCTION OF TIDAL CURRENTS.

reduction tables. See SIGHT REDUCTION TABLES.

reduction to the meridian. The process of applying a correction to an altitude observed when a body is near the celestial meridian of the observer, to find the altitude at meridian transit. The altitude at the time of such an observation is called an EX-MERIDIAN ALTITUDE.

reed, *n.* A steel tongue which is designed to vibrate when air is passed across its unsupported end.

reed horn. A sound signal emitter comprising a resonant horn excited by a jet of air which is modulated by a vibrating reed. The signal is a weak, high-pitched note. See also REED, HORN.

reef, *n.* 1. An offshore consolidated rock hazard to navigation with a depth of 16 fathoms (or 30 meters) or less over it. See also SHOAL. 2. Sometimes used as a term for a low rocky or coral area some of which is above water. See BARRIER REEF, CORAL REEF, FRINGING REEF.

reef flat. A flat expanse of dead reef rock which is partly or entirely dry at low tide. Shallow pools, potholes, gullies, and patches of coral debris and sand are features of the reef flat.

reference datum. A general term applied to any datum, plane, or surface used as a reference or base from which other quantities can be measured.

reference ellipsoid. A theoretical figure whose dimensions closely approach the dimensions of the geoid; the exact dimensions of the ellipsoid are determined by various considerations of the section of the earth's surface of concern. Also called REFERENCE SPHEROID, SPHEROID OF REFERENCE, ELLIPSOID OF REFERENCE.

reference frequency. A frequency having a fixed and specified position with respect to the assigned frequency. The displacement of this frequency, with respect to the assigned frequency, has the same absolute value and sign that the displacement of the characteristic frequency has with respect to the center of the frequency band occupied by the emission.

reference grid. See GRID, definition 2.

reference orbit. An orbit, usually but not exclusively, the best two-body orbit available, on the basis of which the perturbations are computed.

reference spheroid. See REFERENCE ELLIPSOID.

reference station. A tide or current station for which independent daily predictions are given in the Tide Tables and Tidal Current Tables, and from which corresponding predictions are obtained for subordinate stations by means of differences and ratios. Also called STANDARD STATION. Called STANDARD PORT in British terminology. See also SUBORDINATE CURRENT STATION, SUBORDINATE TIDE STATION.

reflecting prism. A prism that deviates a light beam by internal reflection.

reflecting telescope. A telescope which collects light by means of a concave mirror. All telescopes more than 40 inches in diameter are of this type. See also CASSEGRAINIAN TELESCOPE, NEWTONIAN TELESCOPE.

reflection, *n.* The return or the change in the direction of travel of radiation by a surface without change of frequency of the monochromatic components of which the radiation is composed. The radiation does not enter the substance providing the reflecting surface. If the reflecting surface is smooth, **specular (regular, mirror) reflection** occurs; if the reflecting surface is rough with small irregularities, **diffuse reflection** occurs.

reflection plotter. An attachment fitted to a radar plan position indicator which provides a plotting surface permitting plotting without parallax errors. Any mark made on the plotting surface will be reflected on the radarscope directly below. Also called PLOTTING HEAD.

reflectivity, *n.* The ratio of the radiant energy reflected by a surface to that incident upon it.

reflector, *n.* A reflecting surface situated behind a primary radiator, an array of primary radiators, or a feed for the purpose of increasing forward and reducing backward radiation from the antenna. See also RADAR REFLECTOR.

reflector compass. A magnetic compass in which the image of the compass card is viewed by direct reflection in a mirror adjacent to the helmsman's position. See also PROJECTOR COMPASS.

reflex angle. An angle greater than 180° and less than 360°.

reflex reflection. See RETRO-REFLECTION.

reflex-reflector, *n.* See RETRO-REFLECTOR.

refracted ray. A ray extending onward from the point of refraction.

refracting prism. A prism that deviates a beam of light by refraction. The angular deviation is a function of the wavelength of light; therefore, if the beam is composed of white light, the prism will spread the beam into a spectrum.

refracting telescope. A telescope which collects light by means of a lens or system of lenses.

refraction *n.* The change in direction of motion of a ray of radiant energy as it passes obliquely from one medium into another in which the speed of propagation is different. **Atmospheric refraction** is caused by the atmosphere and may be further designated **astronomical refraction** if the ray enters from outside the atmosphere, or **terrestrial refraction** if it emanates from a point on or near the surface of the earth. **Superrefraction** is greater than normal and **sub-refraction** is less than normal. See also DIFFRACTION, REFLECTION.

refraction correction. 1. A correction due to refraction, particularly such a correction to a sextant altitude, due to atmospheric refraction. 2. See IONOSPHERIC CORRECTION.

refractive index. Of a medium, the ratio of the velocity of light in vacuum to the velocity of light in the medium. This index is equal to the ratio of the sines of the angles of incidence and refraction when a ray crosses the surface separating vacuum and medium.

refractive modulus. One million times the amount by which the modified refractive index exceeds unity.

refrangible, *adj.* Capable of being refracted.

regelation, *n.* The melting of ice under pressure and the subsequent refreezing when the pressure is reduced or removed.

regression of the nodes. Precessional motion of a set of nodes. The expression is used principally with respect to the moon, the nodes of which make a complete westerly revolution in approximately 18.6 years.

regular error. See SYSTEMATIC ERROR.

regular reflection. See SPECULAR REFLECTION.

relative, *adj.* Having relationship. In navigation the term has several specific applications: **a.** related to a moving point; apparent, as *relative wind, relative movement;* **b.** related to or measured from the heading, as *relative bearing*; **c.** related or proportional to a variable, as *relative humidity*. See also TRUE, *adj.*

relative accuracy. The accuracy with which a user can measure current position relative to that of another user of the same navigation system at the same time. Hence, a system with high relative accuracy provides good rendezvous capability for the users of the system. The correlation between the geographical coordinates and the system coordinates is not relevant. See also PREDICTABLE ACCURACY, REPEATABLE ACCURACY.

relative azimuth. Azimuth relative to heading.

relative bearing. Bearing relative to heading or to the craft; the horizontal direction of a terrestrial point from a craft expressed as the angular difference between the heading and the direction. It is usually measured from 000° at the heading clockwise through 360°, but is sometimes measured from 0° at the heading either clockwise or counterclockwise through 180°, when it is designated *right* or *left*. An older method of indicating the number of degrees or points from some part of the craft (10° forward of the starboard beam, 2 points on the port quarter, etc.) is sometimes used, but generally for approximations, except for bearings dead ahead, dead astern, or broad on the bow, beam, or quarter.

relative course. Misnomer for DIRECTION OF RELATIVE MOVEMENT.

relative direction. Horizontal direction expressed as angular distance from heading.

relative distance. Distance relative to a specified reference point, usually one in motion.

relative gain of an antenna. The gain of an antenna in a given direction when the reference antenna is a half-wave loss-free dipole isolated in space, the equatorial plane of which contains the given direction.

relative humidity. See under HUMIDITY.

relative motion. See RELATIVE MOVEMENT.

relative motion display. A type of radarscope display in which the position of own ship is fixed, usually at the center of the plan position indicator, and all detected targets move relative to own ship. See also TRUE MOTION DISPLAY.

relative movement. Motion of one object or body relative to another. The expression is usually used in connection with problems involving motion of one vessel to another, the direction of such motion being called DIRECTION OF RELATIVE MOVEMENT and the speed of such motion being called SPEED OF RELATIVE MOVEMENT or RELATIVE SPEED. Distance relative to a specified reference point, usually one in motion, is called RELATIVE DISTANCE. Usually called APPARENT MOTION when applied to the change of position of a celestial body as observed from the earth. Also called RELATIVE MOTION.

relative plot. A plot of the successive positions of a craft relative to a reference point, which is usually in motion. A line connecting successive relative positions of a *maneuvering ship* relative to a *reference ship* is called a RELATIVE MOVEMENT LINE. A relative plot includes all pertinent relative movement lines and the position of the reference ship.

relative position. A point defined with reference to another position, either fixed or moving. The coordinates of such a point are usually bearing, true or relative, and distance from an identified reference point

relative speed. See SPEED OF RELATIVE MOVEMENT.

relative wind. The speed and relative direction from which the wind appears to blow with reference to a moving point. Sometimes called APPARENT WIND. See also APPARENT WIND, TRUE WIND.

release, *n.* A device for holding or releasing a mechanism, particularly the device by which the tangent screw of a sextant is held in place or disengaged from the limb.

reliability diagram. See as LORAN-C RELIABILITY DIAGRAM.

relief, *n.* 1. The elevations or the inequalities, collectively of a land surface; represented on graphics by contours, hypsometric tints, shading, spot elevations, hachures, etc. Similar inequalities of the ocean bed or their representation are called SUBMARINE RELIEF.
2. The removal of a buoy from a station and the providing of another buoy having the operating characteristics authorized for that station.

relief map. See HYPSOGRAPHIC MAP.

relief model. Any three-dimensional representation of an object or geographic area, modeled in any size or medium. See also PLASTIC RELIEF MAP.

relieved, *adj.* Said of a buoy that has been removed from a station and replaced by another buoy having the operating characteristics authorized for that station.

relighted, *adj.* Said of an extinguished aid to navigation returned to advertised light characteristics.

relocated, *adj.* Said of aid to navigation that has undergone an authorized movement from one position to another in the immediate vicinity.

reluctance, *n.* Magnetic resistance.

remanence, *n.* Ability to retain magnetism after removal of the magnetizing force. Also called

RETENTIVITY.

remote-indicating compass. A compass equipped with one or more indicators to repeat at a distance the readings of the master compass. The directive element and controls are called a master compass to distinguish this part of the system from the repeaters, or remote indicators. Most marine gyrocompass installations are of this type. Also called REMOTE-READING COMPASS.

remotely controlled light. A light which is operated by personnel at a considerable distance from the light, through the intermediary of electrical or radio links or both.

remote PPI. See RADAR REPEATER.

remote-reading compass. See REMOTE-INDICATING COMPASS.

repaired, *adj.* Said of a sound signal or radio-navigation aid previously INOPERATIVE, placed back in operation, or of a structure previously DAMAGED, that has been restored as an effective aid to navigation.

repeatability, *n.* 1. A measure of the variation in the accuracy of an instrument when identical tests are made under fixed conditions.

2. In a navigation system, the measure of the accuracy with which the system permits the user to return to a specified point as defined only in terms of the coordinates peculiar to that system. See also PREDICTABILITY.

repeatable accuracy. In a navigation system, the measure of the accuracy with which the system permits the user to return to a position as defined only in terms of the coordinates peculiar to that system. For example, the distance specified for the repeatable accuracy of a system such as Loran-C is the distance between two Loran-C positions established using the same stations and time-difference readings at different times. The correlation between the geographical coordinates and the system coordinates may or may not be known. See also PREDICTABLE ACCURACY, RELATIVE ACCURACY.

repeater, *n.* A device for repeating at a distance the indications of an instrument or device. See also COMPASS REPEATER, GYRO REPEATER, RADAR REPEATER, STEERING REPEATER.

repeating decimal. A decimal in which all the digits after a certain digit consist of a set of one or more digits repeated and infinitum. Also called RECURRING DECIMAL.

replaced, *adj.* Said of an aid to navigation previously OFF STATION, ADRIFT or MISSING that has been restored by another aid of the same type and characteristic.

representative fraction. The scale of a map or chart expressed as a fraction or ratio that relates unit distance on the map to distance measured in the same unit on the ground. Also called NATURAL SCALE, FRACTIONAL SCALE. See also NUMERICAL SCALE.

reradiation, *n.* 1. The scattering of incident radiation. Reradiation from metallic objects in proximity to either the transmitting or receiving antennas can introduce unwanted effects. This is particularly true on a vessel having a number of metallic structures or wires in the vicinity of an antenna. Where such structures are permanent, the effects can sometimes be allowed for by calibration. Also called SECONDARY RADIATION.

2. Radiation from a radio receiver due to poor isolation between the antenna circuit and the local oscillator within the receiver, causing unwanted interference in other receivers.

research sanctuary. A marine sanctuary established for scientific research in support of management programs, and to establish ecological baselines. See also MARINE SANCTUARY.

reset, *adj.* Said of a floating aid to navigation previously OFF STATION, ADRIFT, or MISSING that has been returned to its station.

residual deviation. Deviation of a magnetic compass after adjustment or compensation. The values on various headings are called RESIDUALS.

residual magnetism. Magnetism which remains after removal of the magnetizing force.

residuals, *n., pl.* The remaining deviation of a magnetic compass on various headings after adjustment or compensation. See also DEVIATION TABLE.

resistance, *n.* Opposition, particularly that offered to the flow of electric current.

resistivity, *n.* The amount of resistance in a system. Resistivity is the reciprocal of CONDUCTIVITY.

resolution, *n.* 1. The separation by an optical system of parts of an object or of two or more objects close together. The degree of ability to make such a separation, called RESOLVING POWER, is expressed as the minimum distance between two objects that can be separated.

2. The degree of ability of a radar set to indicate separately the echoes of two targets in range and bearing. *Resolution in range* is the minimum range difference between separate targets at the same bearing which will allow both to appear as separate, distinct echoes of the plan position indicator (PPI). *Resolution in bearing* is the minimum angular separation between two targets at the same range which will allow both to appear as separate, distinct echoes on the PPI.

resolution of vectors. The resolving of a vector into two or more components. The opposite is called VECTOR ADDITION.

resolving power. The degree of ability of an optical system to distinguish between objects close together. See also RESOLUTION.

resolving time. 1. The minimum time interval between two events which permits one event to be distinguishable from the other.

2. In computers, the shortest permissible period between trigger pulses for reliable operation of a binary cell.

resonance, *n.* Re-enforcement or prolongation of any wave motion, such as sound, radio waves, etc., resulting when the natural frequency of the body or system in vibration is equal to that of an impressed vibration.

resonant frequency. Any frequency at which a body or system vibrates most readily. The lowest resonant frequency is the natural frequency of the body or system.

responsor, *n.* A unit which receives the response emitted by a transponder.

restricted area. 1. An area (land, sea, or air) in which there are special restrictive measures employed to prevent or minimize interference between friendly forces.

2. An area under military jurisdiction in which special security measures are employed to prevent unauthorized entry.

See also DANGER AREA, PROHIBITED AREA.

restricted waters. Areas which for navigational reasons such as the presence of shoals or other dangers confine the movements of shipping within narrow limits.

resultant, *n.* The sum of two or more vectors.

retard, *v., t & i.* To delay. This term is sometimes

used as the equivalent of RETIRE (meaning "to move back"), but this usage is not considered appropriate.

retarded line of position. See RETIRED LINE OF POSITION

retentive error. Deviation of a magnetic compass due to the tendency of a vessel's structure to retain some of the induced magnetic effects for short periods of time. For example, a vessel on a northerly course for several days, especially if pounding in heavy seas, will tend to retain some fore-and-aft magnetism hammered in under these conditions of induction. Although this effect is not large and generally decays within a few hours, it may cause incorrect observations or adjustments, if neglected. This error should not be confused with GAUSSIN ERROR.

retentivity, *n.* See REMANENCE.

reticle, *n.* A system of lines, wires, etc., placed in the focal plane of an optical instrument to serve as a reference. A **cross hair** is a hair, thread, or wire constituting part of a reticle. Also called RETICLE. See also GRATICULE, definition 2.

retire, *v., t. & i.* To move back, as to move a line of position back, parallel to itself, along a course line to obtain a line of position at an earlier time. The term RETARD (meaning "to delay") is sometimes used as an equivalent, but the term RETIRE (meaning "to move back") is considered more appropriate. The opposite is ADVANCE.

retired line of position. A line of position which has been moved backward along the course line to correspond with a time previous to that at which the line was established. The opposite is ADVANCED LINE OF POSITION.

retrace, *n.* The path of the visible dot from the end of one sweep to the start of the next sweep across the face of a cathode-ray tube.

retract, *v., t. & i.* The opposite of BEACH, *v., t. & i.*

retrograde motion. The apparent motion of a planet westward among the stars. Apparent motion eastward, called DIRECT MOTION, is more common. Also called RETROGRESSION.

retrogression, *n.* See RETROGRADE MOTION.

retro-reflecting material. A material which produces retro-reflection over a wide range of angles of incidence of a light beam, by use of a large number of very small reflecting and refracting elements, usually very small beads. The applications are the same as those of retro-reflectors.

retro-reflection, *n.* Reflection in which light is returned in directions close to the direction from which it came, this property being maintained over wide variations of the direction of the incident light. Also called REFLEX REFLECTION.

retro-reflector, *n.* A device intended to produce retro-reflection, at least for a limited range of angles of incidence of a light beam. It may comprise one or more retro-reflecting optical units, for example, corner reflectors or special lens units of glass or plastic. Such devices may be installed generally on unlighted buoys or other aids to navigation to increase the range at which they may be seen at night. Also called REFLEX-REFLECTOR.

return, *n.* See BLIP; ECHO, definition 2.

reverberation, *n.* Continuation of radiant energy, particularly sound, by multiple reflection.

reversing current. A tidal current which flows alternately in approximately opposite directions with a slack water at each reversal of direction. Currents of this type usually occur in rivers and straits where the direction of flow is more or less restricted to certain channels. When the movement is towards the shore or up a stream, the current is said to be **flooding,** and when in the opposite direction it is said to be **ebbing.** The combined flood and ebb movement including the slack water covers, on an average, 12.42 hours for the semidiurnal current. If unaffected by a nontidal flow, the flood and ebb movements will each last about 6 hours, but when combined with such a flow, the durations of flood and ebb may be quite unequal. During the flow in each direction the speed of the current will vary from zero at the time of slack water to a maximum about midway between the slacks. Also called RECTILINEAR CURRENT.

reversing falls. A name applied to falls which flow alternately in opposite directions in a narrow channel in the St. John River, New Brunswick, Canada, the phenomenon being due to the large range of tide and a constriction in the river. The direction of flow is upstream or downstream according to whether it is high or low water on the outside, the falls disappearing at the half-tide level.

revolution, *n.* Motion of the celestial body in its orbit; circular motion about an axis usually external to the body. The terms REVOLUTION and ROTATION are often used interchangeably but, with reference to the motions of a celestial body, REVOLUTION refers to the motion in an orbit or about an axis external to the body, while ROTATION refers to motion about an axis within the body. Thus, the earth *revolves* about the sun annually and *rotates* about its axis daily.

revolution counter, revolution indicator. An instrument for registering the number of revolutions of a shaft, particularly a propeller shaft of a vessel (when it may be called ENGINE REVOLUTION COUNTER). This information may be useful in estimating a vessel's speed through the water.

revolution table. A table giving the number of shaft revolutions corresponding to various speeds of a vessel.

revolver, *n.* The pair of horizontal angles between three points, as observed at any place on the circle defined by the three points. This is the one situation in which such angles do not establish a fix. Also called SWINGER.

revolving light. See ROTATING LIGHT.

revolving storm. A cyclonic storm, or one in which the wind revolves about a central low pressure area.

rheostat, *n.* A variable resistor for changing the amount of current in an electrical circuit.

rhomboid, *n.* A parallelogram with oblique angles. A **rhomboid with sides of equal length** is a rhombus.

rhombus, *n.* A rhomboid with sides of equal length.

Rho-Rho mode. See RANGING MODE.

rho-theta navigation. Navigation by means of measuring ranges and bearings of a known position.

rhumb, *n.* Short for RHUMB LINE.

rhumb bearing. The direction of a rhumb line through two terrestrial points, expressed as angular distance from a reference direction. It is usually measured from 000° at the reference direction clockwise through 360°. Also called MERCATOR BEARING.

rhumb direction. See MERCATOR DIRECTION.

rhumb line. A line on the surface of the earth making the same oblique angle with all meridians; a loxodrome or loxodromic curve spiraling toward the poles in a constant true direction. Parallels and meridians, which also maintain constant true directions, may be considered special cases of the rhumb line. A rhumb line is a straight line on a Mercator projection. Sometimes shortened to RHUMB. See also FICTITIOUS RHUMB LINE.

rhumb-line course. The direction of the rhumb line from the point of departure to the destination, expressed as the angular distance from a reference direction, usually north. Also called MERCATOR COURSE.

rhumb-line distance. Distance along a rhumb line, usually expressed in nautical miles.

rhumb-line sailing. Any method of solving the various problems involving course, distance, difference of latitude, difference of longitude, and departure as they are related to a rhumb line.

rhythmic light. A light showing intermittently with a regular periodicity.

ria, *n.* A long, narrow inlet with gradually decreasing depth inward.

ridge, *n.* 1. On the sea floor, a long, narrow elevation with steep sides.

2. A line or wall of broken ice forced up by pressure. The ridge may be fresh or weathered. The submerged volume of broken ice under a ridge, forced downwards by pressure, is called ICE KEEL. See also AGED RIDGE.

3. In meteorology, an elongated area of relatively high atmospheric pressure, almost always associated with and most clearly identified as an area of maximum anticyclonic curvature of wind flow. The opposite of a ridge is called TROUGH. Sometimes called WEDGE.

ridged ice. Ice piled haphazardly one piece over another in the form of ridges or walls. It is usually found in first-year ice.

ridged-ice zone. An area in which much ridged ice with similar characteristics has formed.

ridging, *n.* The pressure process by which sea ice is forced into ridges.

riding light. See ANCHOR LIGHT.

rift, *n.* An opening made by splitting; a crevasse.

right angle. An angle of 90°.

right angle reflector. See DIHEDRAL REFLECTOR.

right ascension. Angular distance east of the vernal equinox; the arc of the celestial equator, or the angle at the celestial pole, between the hour circle of the vernal equinox and the hour circle of a point on the celestial sphere, measured eastward from the hour circle of the vernal equinox through 24^h. Angular distance west of the vernal equinox, through 360°, is SIDEREAL HOUR ANGLE.

right astern. See DEAD ASTERN.

right bank. That bank of a stream or river on the right of the observer when he is facing in the direction of flow, or downstream. See also LEFT BANK.

right circular cone. A cone having a circular base perpendicular to the axis of the cone. Often shortened to RIGHT CONE.

right cone. Short for RIGHT CIRCULAR CONE.

right sphere. The celestial sphere as it appears to an observer at the equator, where celestial bodies appear to rise vertically above the horizon.

right triangle. A triangle one angle of which is 90°.

rigidity in space. See GYROSCOPIC INERTIA.

rime, *n.* A white or milky and opaque granular deposit of ice formed by the rapid freezing of supercooled water drops as they impinge on an exposed object. It is denser and harder than frost, but lighter, softer, and less transparent than glaze.

rime fog. See ICE FOG.

ring time. The time, reckoned from the end of a pulse transmitted by a radar set, during which the output of an echo box produces a visible signal on the display.

rip current. A narrow intense current setting seaward through the surf zone. It removes the excess water brought to the zone by the small net mass transport of waves. It is fed by longshore currents. Rip currents usually occur at points, groins, jetties, etc., of irregular beaches, and at regular intervals along straight, uninterrupted beaches. See also RIPS.

riprap, *n.* Stones or broken rock thrown together without order to provide a revetment.

riprap mounds. Mounds of riprap maintained at certain light structures to protect the structures against ice damage and scouring action. Uncharted submerged portions present hazard to vessels attempting to pass extremely close aboard.

rips, *n., pl.* Agitation of water caused by the meeting of currents or by a rapid current setting over an irregular bottom. Called TIDE RIPS when a tidal current is involved. See also OVERFALLS, RIP CURRENT.

rise, *n.* A broad elevation that rises gently and generally smoothly from the sea floor. See also CONTINENTAL RISE.

rise, *v., i.* Of a celestial body, to cross the visible horizon while ascending. The opposite is SET.

rise of tide. Vertical distance from the chart sounding datum to a higher water datum. Mean rise of tide is the height of mean high water above the chart sounding datum. Spring rise and neap rise are the heights of spring high water and neap high water, respectively, above the chart sounding datum; while mean spring rise and mean neap rise are the heights of mean high water springs and mean high water neaps, respectively above the chart sounding datum. Also called TIDAL RISE. See also HEIGHT OF TIDE.

rising tide. A tide in which the depth of water is increasing. Sometimes the term FLOOD is used as an equivalent, but since *flood* refers primarily to horizontal rather than vertical movement, RISING TIDE is considered more appropriate. The opposite is FALLING TIDE.

river, *n.* A natural stream of water, of greater volume than a creek or rivulet, flowing in a more or less permanent bed or channel, between defined banks or walls, with a current which may either be continuous in one direction or affected by the ebb and flow of the tidal current.

river buoy. A lightweight nun or can buoy especially designed to withstand strong currents.

river estuary. See ESTUARY, definition 2.

river ice. Ice formed on a river, regardless of observed location.

river radar. A marine radar set especially designed for river pilotage, generally characterized by a high degree of resolution and a wide selection of range scales.

rivulet, *n.* A small stream; a brook.

road, *n.* An open anchorage affording less protection than a harbor. Some protection may be afforded by reefs, shoals, etc. Often used in the plural. Also called ROADSTEAD.

roadstead, *n.* See ROAD.

roaring forties. The area of the oceans between 40° and 50° south latitude, where strong westerly winds prevail. See also BRAVE WEST WINDS.

roche moutonnée. A rock worn into a rounded shape by a glacier.

rock, *n.* 1. An isolated rocky formation or single large stone, usually one constituting a danger to navigation. It may be always submerged, always uncovered, or alternately covered and uncovered by the tide. A pinnacle is a sharp-pointed rock rising from the bottom.

2. The naturally occurring material that forms the firm, hard, and solid masses of the ocean floor. Also, rock is a collective term for masses of hard material generally not smaller than 256 millimeters.

rock awash. A rock that becomes exposed, or nearly so, between chart sounding datum and mean high water. In the Great Lakes, the rock awash symbol is used on charts for rocks that are awash, or nearly so, at low water datum. See also BARE ROCK, SUBMERGED ROCK.

rocking the sextant. See SWINGING THE ARC.

rod, *n.* 1. A unit of length equal to 5.5 yards or 16.5 feet. Also called POLE, PERCH.

2. One of the imaginary slender soft iron bars which are assumed to be components or parameters of a craft's magnetic field caused by magnetism induced in soft iron. The *a*, *b*, and *c* rods are situated with one end level with the magnetic compass and in its fore-and-aft axis, either forward or aft. It is an *a* rod if it extends fore-and-aft, a *b* rod if athwartships, and a *c* rod if vertical. The *d*, *e*, and *f* rods are situated with one end level with the compass and in its athwartships axis, either to starboard or to port. It is a *d* rod if it extends fore-and-aft, an *e* rod if athwartships, and an *f* rod if vertical. The *g*, *h*, and *k* rods are situated with an end in the vertical axis of the compass, either above or below it. It is a *g* rod if it extends fore-and-aft, and *h* rod if athwartships, and a *k* rod if vertical.

roll, *n.* Oscillation of a craft about its longitudinal axis. Also called ROLLING. See also LIST, *n.*; SHIP MOTIONS.

roll, *v., t. & i.* To oscillate (a craft) or be oscillated about the longitudinal axis.

roll angle. See ANGLE OF ROLL.

rollers, *n.* Amongst the islands of the West Indies, the South Atlantic, and the South Indian Ocean, swell waves which after moving into shallow water have grown to such height as to be destructive. See also COMBER.

rolling, *n.* See ROLL, *n.*

root mean square. The square root of the arithmetical mean of the squares of a group of numbers

root mean square error. For the one-dimensional error distribution, this term has the same meaning as STANDARD DEVIATION or STANDARD ERROR. For the two-dimensional error distribution, this term has the same meaning as RADIAL (d_{rms}) ERROR. However, such use of the term is deprecated. Root mean square error is commonly called RMS ERROR.

rotary current. A tidal current that flows continually with the direction of flow changing through 360° during the tidal period. Rotary currents are usually found offshore where the direction of flow is not restricted by any barriers. The tendency for the rotation in direction has its origin in the Coriolis force and, unless modified by local conditions, the change is clockwise in the Northern Hemisphere and counterclockwise in the Southern Hemisphere. The speed of the current usually varies throughout the tidal cycle, passing through the two maxima in approximately opposite directions and the two minima with the direction of the current at approximately 90° from the direction at time of maximum speed. Called ROTARY STREAM in British terminology.

rotating light. A light with one or more beams that rotate. Sometimes called REVOLVING LIGHT.

rotating loop radiobeacon. See ROTATING PATTERN RADIOBEACON.

rotating pattern radiobeacon. A rotating radiobeacon, which as of 1980 is operated only in Japanese waters. The following description is applicable to all the Japanese low frequency and medium frequency rotating pattern beacons, although there is one operating at 1,921 kHz which employs a slightly different signal format. The transmitter uses an antenna array which has a radiation pattern in the form of a figure eight. By suitably feeding the array, the pattern can be made to rotate at a fixed speed. Immediately prior to the start of the rotation, the two nulls in the radiation pattern lie on the North-South line. The whole pattern is made to rotate through an angle of 200°. Just prior to the commencement of rotation, the beacon transmits the Morse letter *A* as a starting signal. While rotation is taking place, the beacon sends out a series of 100 dots which act as a time base, thus eliminating the need for a clock. The number of dots which arrive before the passage of the null is counted in order to obtain the required bearing. Every tenth dot has a change in tone to assist in this counting process. Theoretically, if the number of dots heard before the passage of the null is multiplied by two, the result gives the true bearing of the ship from the station, in degrees, or its reciprocal. There are however, two restrictions on this. The first is that in some instances, due to the refraction of the wave front due to its passage over land masses in the vicinity of the transmitting site, the null may be shifted slightly from the true bearing in one direction or the other, so that the null may appear slightly before or after the instant at which the theoretical null should pass over the ship. For this reason each beacon is calibrated, and a calibration chart drawn up showing the true bearings on which the null is observed at the time that each dot is transmitted. There may be certain unreliable bearings associated with the beacon and these are also shown on the calibration chart. In addition, and particularly at long range, the null position may not be clearly defined and may cover several dots. For this reason it is usual to count the number of dots both before and after the null. The difference between the total and 100 is divided by 2 and then added to the number of dots before the null, thus giving the correct count to the dot in the middle of the inaudible group. This then should give the true bearing. Since the sequence is usually repeated two or three times in each transmission schedule, a mean bearing may be obtained in order to improve accuracy. The calibration tables give two bearings for each dot count with an ambiguity of approximately 180°. This ambiguity is due to the two nulls in the radiation pattern, but should present little difficulty in practice. The ship's estimated position, or a cross bearing received from another beacon, should indicate which of the two bearings to select. The pattern is made to rotate over an arc of 200° in order to allow a ship which is close to the North-South line to obtain a satisfactory bearing. The bearings related to dots numbered 1 to 10 are the same as those related to the dots numbered 91 to 100, so that a ship situated on any of these bearings should theoretically receive 2 nulls. In practice the dots at beginning or end of the sequence may not be heard so that only one null is detected with accuracy. It is quite common for beacons of this type to transmit a non-directional signal, usually lasting 2 minutes, and which contains a long dash for use with the ship's direction finding equipment. These rotating beacon stations transmit a characteristic signal prior to the transmission of the rotating pattern to establish the identity of the station. Also called ROTATING LOOP RADIOBEACON.

rotation, *n.* Turning of a body about an axis within the body, as the daily rotation of the earth. See also REVOLUTION.

rotten ice. Sea ice which has become honeycombed and which is in an advanced state of disintegration.

round, *v., t.* To pass and alter direction of travel, as a vessel ROUNDS A CAPE. If the course is nearly reversed, the term DOUBLE may be used.

roundabout, *n.* A routing measure comprising a separation point or circular separation zone and a circular traffic lane within defined limits. Traffic within the roundabout is separated by moving in a counterclockwise direction around the separation point or zone. See also ROUTING SYSTEM, TRAFFIC SEPARATION SCHEME.

round of bearings. A group of bearings observed simultaneously, or over a short period of time with no appreciable delay between the completion of one observation and the start of the next.

round of sights. A group of sights made over a short period of time, usually with no appreciable delay between the completion of one observation and the start of the next.

round wind. A wind that gradually changes direction through approximately 180° during the daylight hours. See also LAND BREEZE.

route chart. A chart showing routes between various places, usually with distances indicated.

routing system. Any system of one or more routes and/or routing measures aimed at reducing the risk of casualties; it includes traffic separation schemes, two-way routes, recommended tracks, areas to be avoided, inshore traffic zones, roundabouts, precautionary areas, and deep water routes.

rubble, *n.* 1. Fragments of hard sea ice, roughly spherical and up to 5 feet in diameter, resulting from the disintegration of larger ice formations. When afloat, commonly called BRASH ICE.
2. Loose angular rock fragments.

Rude Star Finder. A star finder previously published by the U.S. Navy Hydrographic Office, and named for Captain Gilbert T. Rude, U.S. Coast and Geodetic Survey. This star finder preceded No. 2102-D, *Star Finder and Identifier,* also formerly published by the U.S. Navy Hydrographic Office and later the U.S. Naval Oceanographic Office.

rugged, *adj.* Rock-bound; craggy.

rules of navigation. Rules of the road.

rules of the road. The International Regulations for Preventing Collisions at Sea, commonly called INTERNATIONAL RULES OF THE ROAD, and the Inland Rules of the Road to be followed by all vessels while navigating upon certain inland waters of the United States. Also called RULES OF NAVIGATION.

run, *n.* 1. A brook, or small creek.
2. A small, swift watercourse.
3. The distance traveled by a craft during any given time interval, or since leaving a designated place. See also DAY'S RUN.

run a line of soundings. To obtain soundings along a course line, for use in making or improving a chart.

run before the wind. To steer a course downwind.

run down a coast. To sail approximately parallel with the coast.

runnel, *n.* The smallest of natural streams; a brook or run.

running fix. A position determined by crossing lines of position obtained at different times and advanced or retired to a common time. However, in celestial navigation or when using long-range electronic aids, a position determined by crossing lines of position obtained within a few minutes is considered a FIX, the expression RUNNING FIX being applied to a position determined by advancing or retiring a line over a considerable period of time. There is no sharp dividing line between a fix and a running fix in this case.

running light. See NAVIGATION LIGHTS.

run-off, *n.* That portion of precipitation which is discharged from the area of fall as surface water in streams.

run of the coast. The trend of the coast.

run-up. The rush of water up a structure on the breaking of a wave. The amount of run-up is the vertical height above the still water level that the rush of water reaches. Also called UPRUSH.

S

saddle, *n.* On the sea floor, a low part, resembling in shape a saddle, in a ridge or between contiguous seamounts.

safety lanes. Specified sea lanes designated for use in transit by submarines and surface ships to prevent attack by friendly forces. They may be called SUBMARINE SAFETY LANES when designated for use by submarines in transit.

safe water mark. See under IALA MARITIME BUOYAGE SYSTEM.

sailing, *n.* A method of solving the various problems involving course, distance, difference of latitude, difference of longitude, and departure. The various methods are collectively spoken of as *the sailings.* **Plane sailing** considers the earth as a plane. **Traverse sailing** applies the principles of plane sailing to determine the equivalent course and distance made good by a craft following a track consisting of a series of rhumb lines. Any of the sailings which considers the spherical or spheroidal shape of the earth is called **spherical sailing. Middle-latitude sailing** is a method of converting departure into difference of longitude, or vice versa, by assuming that such a course is steered at the middle or mean latitude; if the course is 090° or 270° true, it is called **parallel sailing. Mercator sailing** applies when the various elements are considered in their relation on a Mercator chart. **Meridian sailing** is used when the course is 000° or 180° true. **Rhumb-line sailing** is used when a rhumb line is involved; **great-circle sailing** when a great-circle track is involved. **Composite sailing** is a modification of great-circle sailing used when it is desired to limit the highest latitude. The expression **current sailing** is occasionally used to refer to the process of allowing for current in determining the predicted course made good, or of determining the effect of a current on the direction of motion of a vessel.

sailing chart. See under CHART CLASSIFICATION BY SCALE.

sailing directions. 1. A descriptive book for the use of mariners, containing detailed information of coastal waters, harbor facilities, etc. of an area. For waters of the United States and its possessions, they are published by the National Ocean Survey and are called UNITED STATES COAST PILOTS. Sailing directions, as well as light lists, provide the information that cannot be shown graphically on the nautical chart and that is not readily available elsewhere. They are of ancient origin. The early Greek name for such a volume was PERIPLUS, meaning literally "a sailing round." Sometimes called PILOT.
2. The new sailing directions published by the Defense Mapping Agency Hydrographic/Topographic Center are designed to assist the navigator in planning a voyage of any extent, particularly if it involves an ocean passage. In the new format the previous 70 volumes are replaced with 43 volumes: 35 *Sailing Directions (Enroute)* and 8 *Sailing Directions (Planning Guide)*. Port facilities data are contained in Pub. No. 150, *World Port Index*. Each Planning Guide covers one of the world's great land-sea areas based on an arbitary division of the world's seaways into eight "ocean basins." In contrast to the localized method used previously, the *Planning Guide* shows entire recommended routes as they originate from all major U.S. ports and naval bases and terminate at foreign ports. All radionavigation systems pertaining to the ocean area are described. The national and international systems of lights, beaconage, and buoyage in the ocean basin are also described and illustrated. Other information such as that pertaining to the ocean basin environment, warning areas, government regulations, communications, etc. is also included to facilitate voyage planning. Each Enroute volume includes detailed coastal and port approach information which supplements the largest scale chart of the area covered. Special graphics depict coastal winds, weather, tides, currents, and ice. Outer dangers are fully described, but inner dangers which are well charted are, for the most part, omitted. Coastal descriptions and views, useful for radar and visual piloting, are included. Anchorages are listed. Directions for entering ports are depicted, where appropriate, by means of chartlets, sketches and photographs. An index-gazeteer lists described and charted features. See also UNITED STATES COAST PILOT.

St. Elmo's fire. A luminous discharge of electricity from pointed objects such as the masts and yardarms of ships, lightning rods, steeples, etc., occurring when there is a considerable atmospheric difference in potential. Also called CORPOSANT, CORONA DISCHARGE.

St. Hilaire method. The establishing of a line of position from the observation of the altitude of a celestial body by the use of an assumed position, the difference between the observed and computed altitudes, and the azimuth. This is the method most commonly used by modern navigators. The assumed position may be determined arbitrarily, by dead reckoning, or by estimation. The method was devised by Marcq St. Hilaire, a French naval officer, in 1874. See also SUMNER METHOD, LONGITUDE METHOD, HIGH ALTITUDE METHOD. Also called ALTITUDE INTERCEPT METHOD.

sallying ship. Producing rolling motion of a vessel by the running in unison of a group from side to side. This is usually done to help float a vessel which is aground or to assist it to make headway when it is beset by ice.

salt marsh. A flat, poorly drained coastal swamp flooded by most high tides.

salt-water wedge. The intrusion of a tidal estuary by sea water in the form of a wedge underneath the less-dense fresh water.

same name. A name the same as that possessed by something else, as declination has the *same name* as latitude if both are north or both are south. They are of CONTRARY NAME if one is north and the other south.

sand, *n.* Sediment consisting of small but easily distinguishable separate grains between 0.0625 and 2.0 millimeters in diameter. It is called **very fine sand** if the grains are between 0.0625 and 0.125 millimeter in diameter, **fine sand** if between 0.125 and 0.25 millimeters, **medium sand** if between 0.25 and 0.50 millimeters, **coarse sand** if between 0.50 and 1.0 millimeters, and **very coarse sand** if between 1.0 and 2.0 millimeters. See also MUD, STONES, ROCK, definition 2.

sand dune. See DUNE.

sandstorm, *n.* A strong wind carrying sand through the air, the diameter of most of the particles ranging from 0.08 to 1 millimeter. In contrast to a DUSTSTORM, the sand particles are mostly confined to the lowest 10 feet, and rarely rise more than 50 feet above the ground.

sandwave, *n.* A large wavelike sediment feature in very shallow water and composed of sand. The wavelength may reach 100 meters; the amplitude is about 0.5 meter. Also called MEGA-RIPPLE.

Santa Ana. A strong, dust-laden foehn occurring in Southern California near the mouth of the Santa Ana pass and river.

Sargasso Sea. The west central region of the subtropical gyre of the North Atlantic Ocean. It is bounded by the North Atlantic, Canary, Atlantic North Equatorial, and Antilles Currents, and the Gulf Stream. It is characterized by the absence of well-marked currents and by large quantities of drifting Sargassum, or gulf-weed.

sargasso weed. See SARGASSUM.

sargassum, *n.* A genus of brown algae characterized by a bushy form, a substantial holdfast when attached, and a yellowish brown, greenish yellow, or orange color. Species of the group have a large variety of forms and are widely distributed in warm seas as attached and free-floating plants. Two species (*S. fluitans* and *S. matans*) make up 99 percent of the macroscopic vegetation in the Sargasso Sea. Also called SARGASSO WEED, GULFWEED.

Saros, *n.* A period of 223 synodic months corresponding approximately to 19 eclipse years or 18.03 Julian years, and is a cycle in which solar and lunar eclipses repeat themselves under approximately the same conditions.

sastrugi, (*sing. sastruga*), *n., pl.* Sharp, irregular ridges formed on a snow surface by wind erosion and deposition. On mobile floating ice, the ridges are parallel to the direction of the prevailing wind at the time they were formed.

satellite, *n.* 1. An attendant body, natural or man-made, that orbits about another body, the **primary body**. The moon is a satellite of the earth, the primary body.
2. As defined by the *International Telecommunication Union* (ITU), a body which revolves around another body of preponderant mass and which has a motion primarily and permanently determined by the force of attraction of that other body. See also ACTIVE SATELLITE, EARTH SATELLITE, EQUATORIAL SATELLITE, GEODETIC SATELLITE, NAVIGATION SATELLITE, PASSIVE SATELLITE, POLAR SATELLITE, SNYCHRONOUS SATELLITE, TWENTY-FOUR HOUR SATELLITE.

satellite geodesy. The discipline which employs observations of an earth satellite to extract geodetic information.

satellite triangulation. The determination of the angular relationships between two or more stations by the simultaneous observation of an earth satellite from these stations.

satellite triangulation stations. Triangulation stations whose angular positions relative to one another are determined by the simultaneous observations of an earth satellite from two or more of them.

saturable system. A term used to describe a navigation system whose use is limited to a single user or a limited number of users on a time-shared basis.

saturation, *n.* 1. The condition existing when the greatest possible amount of anything has been reached, as a magnetic substance which cannot be further magnetized, or an electronic aid to navigation which is being used by all the craft it can handle.
2. Complete impregnation under given conditions, as the condition that exists in the atmosphere when no additional water vapor can be added at the prevailing temperature without condensation or supersaturation occurring.

Saturn, *n.* The navigational planet whose orbit lies outside that of Jupiter.

savanna, *n.* A plain with low vegetation, especially in the sub-tropical latitudes.

S-band. A radio-frequency band of 1,550 to 5,200 megahertz. See also FREQUENCY, FREQUENCY BAND.

scalar, *adj.* Having magnitude only.

scalar, *n.* Any physical quantity whose field can be described by a single numerical value at each point in space. A scalar quantity is distinguished from a VECTOR quantity by the fact that a scalar quantity possesses only *magnitude*, whereas, a vector quantity possesses both *magnitude* and *direction*.

scale, *n.* 1. A series of marks or graduations at definite intervals. A **linear scale** is a scale graduated at uniform intervals; a **logarithmic scale** is a scale graduated in the logarithms of uniformly-spaced consecutive numbers.
2. The ratio between the linear dimensions of a chart, map drawing, etc., and the actual dimensions represented, as *1:2,000,000* or *27.430 nautical miles to an inch*.
See also CONVERSION SCALE, BAR SCALE, REPRESENTATIVE FRACTION, SMALL SCALE, LARGE SCALE.

scale error. See CALIBRATION ERROR.

scan, *v., t.* In the use of radar, to search or investigate an area or space by varying the direction of the radar antenna and thus the beam. Normally scanning is done by continuous rotation of the antenna.

scanner, *n.* A unit of a radar set consisting of the antenna and drive assembly for rotating the antenna.

scarf cloud. A thin cirrus-like cloud sometimes observed above a developing cumulus. See also CAP CLOUD.

scarp, *n.* See ESCARPMENT.

scatter reflections. Reflections from portions of the ionosphere having different virtual heights, which mutually interfere and cause rapid fading.

Schuler frequency. The natural frequency of a simple pendulum with a length equal to the earth's radius. The corresponding period is 84.4 minutes.

Schuler loop. That portion of the inertial navigator in which the instrumental local vertical is established.

Schuler tuned. The condition wherein gyroscopic devices should be insensitive to applied accelerations. M. Schuler determined that if gyroscopic devices were not to be affected by the motions of the craft in which installed, the devices should have a natural period of oscillation of about 84.4 minutes. This period is equal to the product of 2π and the square root of the quotient: the radius of the earth divided by the acceleration of gravity.

scintillation, *n.* Twinkling; emission of sparks or quick flashes; shimmer.

scope, *n.* Short for RADARSCOPE.

scoria (*pl. scoriae*), *n.* Volcanic rock fragments usually of basic composition, characterized by marked vesicularity, dark color, high density, and a partly crystalline structure. Scoria is a constituent of certain marine sediments.

scouring basin. A basin containing impounded water which is released at about low water in order to maintain the desired depth in the entrance channel by scouring the bottom. Also called SLUICING POND.

screen, *n.* The chemically coated inside surface of the large end of a cathode-ray tube which becomes luminous when struck by an electron beam.

scud, *n.* Shreds or small detached masses of cloud moving rapidly before the wind, often below a layer of lighter clouds. See also FRACTO-.

scud, *v., i.* To run before a storm.

sea, *n.* 1. A body of salt water more or less confined by continuous land or chains of islands and forming a region distinct from the great masses of water.
2. A body of water nearly or completely surrounded by land, especially if very large or composed of salt water. Sometimes called INLAND SEA. See also LAKE.
3. Ocean areas in general, including major indentations in the coast line, such as gulfs. See also CLOSED SEA, OPEN SEA, HIGH SEA.
4. Waves generated or sustained by winds within their fetch as opposed to SWELL.
5. The character of a water surface, particularly the height, length (period), and direction of travel of waves generated locally. A **smooth sea** has waves no higher than ripples or small wavelets. A **short sea** has short, irregular, and broken waves. A **confused sea** has a highly disturbed surface without a single, well-defined direction of travel, as when waves from different directions meet following a sudden shift in the direction of the wind. A **cross sea** is a series of waves imposed across the **prevailing waves.** A sea may be designated as **head, beam, quartering,** or **following** if the waves are moving in a direction approximately 180°, 90°, 45°, or 0°, respectively, from a vessel's heading. See also SWELL, definition 1.

sea-air temperature difference correction. A correction due to a difference in the temperature of the sea and air, particularly that sextant altitude correction caused by abnormal terrestrial refraction occurring when there is a nonstandard density lapse rate in the atmosphere due to a difference in the temperature of the water and air at the surface.

sea anchor. An object towed by a vessel, usually a small one, to keep the vessel end-on to a heavy sea or surf or to reduce the drift. In its usual form it consists of a conical canvas bag with the large end open. When towed with this large end forward, it offers considerable resistance. A small *tripping line* attached to the pointed end is used for hauling the sea anchor into the vessel. A sea anchor is sometimes improvised by using a weighted sail and spar, a bucket, a basket, a weight, or a long line. Also called DRAG, DROGUE.

seabeach, *n.* See under BEACH.

seaboard, *n.* The region of land bordering the sea. The terms SEABOARD, COAST, and LITTORAL have nearly the same meanings. SEABOARD is a general term used somewhat loosely to indicate a rather extensive region bordering the sea. COAST is the region of indefinite width that extends from the sea inland to the first major change in terrain features. LITTORAL applies more specifically to the various parts of a region bordering the sea, including the coast, foreshore, backshore, beach, etc.

sea breeze. A breeze blowing from the sea to adjacent land. It usually blows by day, when the land is warmer than the sea, and alternates with a LAND BREEZE, which blows in the opposite direction by night. See also ONSHORE WIND.

sea buoy. The outermost buoy marking the entrance to a channel or harbor. Called LANDFALL BUOY in British terminology.

seachannel, *n.* On the sea floor, a continuously sloping, elongated depression commonly found in fans or plains and usually bordered by levees on one or two sides.

sea clutter. See SEA RETURN.

seacoast, *n.* See COAST.

Sea-Fix, *n.* A Decca radiolocation system which is a newer modified version of Hi-Fix. Although essentially the same as Hi-Fix, Sea-Fix is generally considered more reliable under field conditions because of the use of solid-state components.

sea fog. A type of advection fog formed when air that has been lying over a warm water surface is transported over a colder water, resulting in cooling of the lower layer of air below its dew point. See also HAAR.

sea gate. 1. A way giving access to the sea such as a gate, channel or beach.
2. A gate which serves to protect a harbor or tidal basin from the sea, such as one of a pair of supplementary gates at the entrance to a tidal basin exposed to the sea.

seagirt, *adj.* Surrounded by sea. Also called SEA-BOUND.

sea ice. Any form of ice found at sea which has originated from the freezing of sea water.

sea-ice nomenclature. See WMO SEA-ICE NOMENCLATURE.

sea kindliness. A measure of the ease of motion or behavior of a vessel in heavy seas, particularly in regard to rolling, pitching, and shipping water. It is not to be confused with seaworthiness, which, implies that the vessel is able to sustain heavy rolling, pitching, etc., without structural damage or impaired stability.

sea level. Height of the surface of the sea at any time.

sea level datum Use of this term is discouraged. See NATIONAL GEODETIC VERTICAL DATUM.

sea manners. Understood by seamen to mean a consideration for the other vessel and the exercise of good judgment under certain conditions when vessels meet.

seamark, *n.* See MARK, *n.*, definition 1.

sea mile. An approximate mean value of the nautical mile equal to 6,080 feet, or, the length of a minute of arc along the meridian at latitude 48°.

sea mist. See STEAM FOG.

seamount, *n.* On the sea floor, an elevation rising generally more than 1,000 meters and of limited extent across the summit.

sea quadrant. See BACKSTAFF.

search, *adj.* Used for or related to a search, as a *search radar.*

search and rescue chart. A chart designed primarily for directing and conducting search and rescue operations.

sea reach. See under REACH.

sea return. Clutter on the radarscope which is the result of the radar signal being reflected from the sea, especially near the ship. Also called SEA CLUTTER. See also CLUTTER.

sea room. Space in which to maneuver without danger of grounding or colliding.

seashore, *n.* A loose term referring to the general area in close proximity to the sea.

season, *n.* 1. One of the four principal divisions of the year: spring, summer, autumn, and winter.
2. An indefinite part of the year, as the *rainy season.*

seasonal current. An ocean current which has large changes in speed or direction due to seasonal winds.

sea-temperature difference correction. A correction due to a difference in the temperature of the sea and air, particularly that sextant altitude correction caused by abnormal terrestrial refraction occurring when there is a nonstandard density lapse rate in the atmosphere due to a difference in the temperature of the water and air at the surface.

seaward, *adj.* In a direction away from the land; toward the sea.

seaward, *adv.* Away from the land; toward the sea.

seaward boundary. Limits of any area or zone offshore from the mean low, or mean lower low water line and established by an act of the U. S. Congress.

seaway, *n.* 1. A moderately rough sea. Used chiefly in the expression *in a seaway.*
2. The sea as a route of travel from one place to another; a shipping lane.

secant, *n.* 1. The ratio of the hypotenuse of a plane right triangle to the side adjacent to one of the acute angles of the triangle, equal to 1/cos. The expression NATURAL SECANT is sometimes used to distinguish the secant from its logarithm (called LOGARITHMIC SECANT).
2. A line that intersects another, especially a straight line intersecting a curve at two or more points.

secant conic chart. See CONIC CHART WITH TWO STANDARD PARALLELS.

secant conic map projection. See CONIC MAP PROJECTION WITH TWO STANDARD PARALLELS.

second, *n.* 1. The base unit of time in the International System of Units (SI). In 1967 the second was defined by the Thirteenth General Conference on Weights and Measures as the duration of 9,192,631,770 periods of the radiation corresponding to the transition between two hyperfine levels of the ground state of the cesium-133 atom. This value was established to agree as closely as possible with the *ephemeris second.* Also called ATOMIC SECOND. See also ATOMIC TIME.
2. A sixtieth part of a minute in either time or arc.

secondary, *n.* A small low pressure area accompanying a large or primary one. The secondary often grows at the expense of the primary, eventually replacing it.

secondary circle. See SECONDARY GREAT CIRCLE.

secondary control tide station. A tide station at which continuous observations have been made over a minimum period of 1 year but less than a 19-year Metonic cycle. The series is reduced by comparison with simultaneous observations from a primary control tide station. This station provides for a 365-day harmonic analysis including the seasonal fluctuation of sea level. See also PRIMARY CONTROL TIDE STATION; SUBORDINATE TIDE STATION, definition 1; TERTIARY TIDE STATION; TIDE STATION.

secondary great circle. A great circle perpendicular to a primary great circle, as a meridian. Also called SECONDARY CIRCLE.

secondary light. A major light, other than a primary seacoast light, established at harbor entrances and other locations where high intensity and reliability are required. See also MINOR LIGHT.

secondary phase factor correction. A correction for additional time (or phase delay) for transmission of a low frequency signal over an all seawater path when the signal transit time is based on the free-space velocity. The Loran-C lattices as tabulated in tables or overprinted on the nautical chart normally include compensation for secondary phase factor. See also ADDITIONAL SECONDARY PHASE FACTOR CORRECTION.

secondary radar. 1. Radar in which the target is fitted with a transponder and in which the target retransmits automatically on the interrogating frequency, or a different frequency. The response may be coded. See also PRIMARY RADAR, RACON, RAMARK.
2. As defined by the *International Telecommuncation Union* (ITU), a radiodetermination system based on the comparison of reference signals with radio signals re-transmitted from the position to be determined.

secondary radiation. See RERADIATION, definition 2.

secondary station. In a radionavigation system, the station of a chain whose emissions are made with reference to the emissions of a master station without being triggered by the emissions of such station, as in Loran-C. See also SLAVE STATION.

secondary tide station. See as SECONDARY CONTROL TIDE STATION.

second class. Of secondary accuracy. This expression is sometimes used in connection with bearings reported by a radio direction finder station. APPROXIMATE or DOUBTFUL may be used with the same meaning.

second reduction. See PHASE REDUCTION.

second-trace echo. A radar echo received from a target after the following pulse has been transmitted. Second-trace echoes are unusual except under abnormal atmospheric conditions, or conditions under which super-refraction is present, and are received from targets at actual ranges greater than the radar range scale setting. They may be recognized through changes in their position on the radarscope on changing the pulse repetition rate; their hazy, streaky or distorted shape; and their erratic movements on plotting. Also called MULTIPLE-TRACE ECHO.

second-year ice. Old ice which has survived only one summer's melt. Because it is thicker and less dense than first-year ice, it stands higher out of the water. In contrast to multi-year ice, summer melting produces a regular pattern of numerous small puddles. Bare patches and puddles are usually greenish-blue.

Secor, *n.* Registered trade name of a phase comparison electronic long range distance measuring system used to determine positions and orbits of satellites or flight vehicles that contain the necessary transponders. This term is an acronym for *sequential collation of range.*

sector, *n.* 1. Part of a circle bounded by two radii and an arc. See also RED SECTOR.
2. Something resembling the sector of a circle, as a *warm sector* between the warm and cold fronts of a cyclone.

sector display. A radar display in which a high persistence screen is excited only when the radar beam is within a narrow sector which can be selected at will.

sector light. A light having sectors of different colors or the same color in specific sectors separated by dark sectors.

sector scanning. In the use of radar, the process of scanning within a sector as opposed to scanning around the horizon.

secular, *adj.* Of or pertaining to a long period of time.

secular aberration. See under ABERRATION, definition 1.

secular error. That error in the reading of an instrument due to secular change within the materials of the instrument.

secular perturbations. Perturbations of the orbit of a planet or satellite that continue to act in one direction without limit, in contrast to periodic perturbations which change direction in a regular manner.

secular terms. In the mathematical expression of the orbit of a satellite, terms which are proportional to time, resulting in secular perturbations. See also PERIODIC TERMS.

secular trend. See APPARENT SECULAR TREND.

seiche, *n.* A stationary wave usually caused by strong winds and/or changes in barometric pressure. It is usually found in lakes and semi-enclosed bodies of water. It may also be found in areas of the open ocean. See also STANDING WAVE.

seismic sea wave. See as TSUNAMI.

selective fading. 1. Fading of the skywave in which the carrier and various sideband frequencies fade at different rates, causing audio-frequency distortion.
2. Fading that affects the different frequencies within a specified band unequally.
3. Fading in which the variation in the received signal strength is not the same for all frequencies in the frequency band of the received signal.
See also FADING.

selectivity, *n.* 1. The characteristic of a radio receiver which enables it to differentiate between the desired signal and those of other frequencies.
2. The ability of a receiver to reject transmissions other than the one to which tuned.
3. The degree to which a radio receiver can accept the signals of one station while rejecting those of stations on adjacent channels. See also SENSITIVITY.

selenographic, *adj.* Of or pertaining to the physical geography of the moon.

semaphore, *n.* A device using visual signals, usually bodies of defined shapes or positions or both, by which information can be transmitted.

semi-. A prefix meaning *half.*

semicircle, *n.* Half of a circle. See also DANGEROUS SEMICIRCLE, NAVIGABLE SEMICIRCLE.

semicircular deviation. Deviation which changes sign (E or W) approximately each 180° change of heading.

semidiameter, *n.* 1. Half the angle at the observer subtended by the visible disk of a celestial body. Sextant altitudes of the sun and moon should be corrected for semidiameter unless the center is observed, as with a bubble sextant.
2. The radius of a circle or sphere.

semidiameter correction. A correction due to semidiameter, particularly that sextant altitude correction, when applied to the observation of the upper or lower limb of a celestial body, determines the altitude of the center of that body.

semidiurnal, *adj.* Having a period or cycle of approximately one-half of a day. The predominating type of tide throughout the world is semidiurnal, with two high waters and two low waters each tidal day. The tidal current is said to be *semidiurnal* when there are two flood and two ebb periods each tidal day. A semidiurnal constituent has two maxima and two minima each constituent day. See also TYPE OF TIDE.

semidiurnal current. Tidal current in which the tidal-day current cycle consists of two flood currents and two ebb currents, separated by slack water; or two changes in direction of 360° of a rotary current. This is the most common type of tidal current throughout the world. A DIURNAL CURRENT is one in which one flood and one ebb, or one change of 360°, occurs each tidal day.

semidiurnal tide. See under TYPE OF TIDE; SEMIDIURNAL, *adj.*

semilogarithmic coordinate paper. Paper ruled with two sets of mutually-perpendicular, parallel lines, one set being spaced according to the logarithms of consecutive numbers, and the other set uniformly spaced. On LOGARITHMIC COORDINATE PAPER both sets of lines are spaced logarithmically.

semimajor axis. One-half of the longest diameter of an ellipse.

semiminor axis. One-half of the shortest diameter of an ellipse.

semi-reflecting mirror. See DICHROIC MIRROR.

sense, *n.* The solution of the 180° ambiguity present in some radio direction finding systems.

sense antenna. An antenna used to resolve a 180° ambiguity in a directional antenna.

sense finding. The process of eliminating the 180° ambiguity from the bearing indication of some types of radio direction finder.

sensibility, *n.* The ability of a magnetic compass card to align itself with the magnetic meridian after deflection.

sensible horizon. That circle of the celestial sphere formed by the intersection of the celestial sphere and a plane through any point, such as the eye of an observer, and perpendicular to the zenith-nadir line. See also HORIZON.

sensitive axis. 1. Of an accelerometer, the axis along which specific acceleration is measured.
2. See also INPUT AXIS.

sensitivity, *n.* Of a radio receiver or similar device, the minimum input signal required to produce a specified output signal having a specified signal-to-noise ratio. See also SELECTIVITY.

sensitivity time control. An electronic circuit designed to reduce automatically the sensitivity of the radar receiver to nearby targets. Also called SWEPT GAIN, ANTI-CLUTTER GAIN CONTROL, ANTI-CLUTTER SEA.

separation line. A line separating the traffic lanes in which ships are proceeding in opposite or nearly opposite directions, or separating a traffic lane from the adjacent inshore traffic zone. See also ROUTING SYSTEM, SEPARATION ZONE.

separation zone. A zone separating the traffic lanes in which ships are proceeding in opposite or nearly opposite directions, or separating a traffic lane from the adjacent inshore traffic zone. See also ROUTING SYSTEM, SEPARATION LINE.

September equinox. See AUTUMNAL EQUINOX.

sequenced radiobeacon. In U.S. waters, one of a group of up to six marine radiobeacons in the same geographical area, except those operating continuously, that transmit on a single frequency. Each radiobeacon transmits for 1 minute out of each 6-minute period in sequence with the other beacons of the group. If less than six radiobeacons are assigned to a group, one or more of the beacons may transmit during two of the six 1-minute periods. The transmissions in se-

MARINE SEXTANT

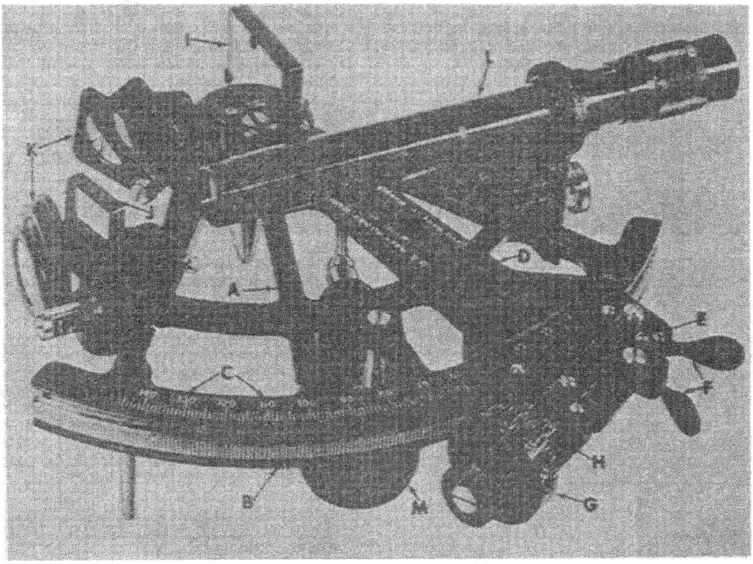

A. FRAME

B. LIMB, cut on its outer edge with teeth, each representing one degree of altitude.

C. ARC

D. INDEX ARM, a movable bar pivoted about the center of curvature of the limb.

E. TANGENT SCREW, mounted perpendicularly on the end of the index arm, where it engages the teeth of the limb. Because the index arm can be moved through the length of the arc by rotating the tangent screw, this is sometimes called and ENDLESS tangent screw, in contrast with the limited range device on older instruments.

F. RELEASE, a spring-actuated clamp which keeps the tangent screw engaged with the teeth of the limb. By applying pressure on the legs of the release, one can disengage the tangent screw. The index arm can then be moved rapidly along the limb.

G. MICROMETER DRUM. One complete turn of the drum, which is graduated in minutes of altitude, moves the index arm one degree of altitude along the arc.

H. VERNIER. Adjacent to the micrometer drum and fixed on the index arm the, vernier aids in reading fractions of a minute of arc. The vernier shown is graduated into ten parts, permitting readings to 6 seconds of arc. Other sextants may have verniers graduated into only five parts, permitting readings to 12 seconds of arc.

I. INDEX MIRROR, a piece of silvered plate glass mounted on the index arm, perpendicular to the plane of the instrument, with the center of the reflecting surface directly over the pivot of the index arm.

J. HORIZON GLASS, a piece of optical glass silvered on its half nearer the frame. It is mounted on the frame, perpendicular to the plane of the sextant. The index mirror and horizon glass are mounted so that their surfaces are parallel when the index arm is set at 0°, if the instrument is in perfect adjustment.

K. SHADE GLASSES, of varying darkness, are mounted on the frame of the sextant in front of the index mirror and horizon glass. They can be moved into the line of sight at will, to reduce the intensity of light reaching the eye of the observer.

L. TELESCOPE. As shown, the telescope screws into an adjustable collar in line with the horizon glass, and should then be parallel to the plane of the instrument.

M. HANDLE. Sextants are designed to be held in the right hand. Some are equipped with a small light on the index arm to aid in reading altitudes. The batteries for this light are usually fitted inside a recess in the sextant handle.

quence reduce station interference and undesirable retuning. See also CONTINUOUS CARRIER RADIOBEACON.

sequence of current. The order of occurrence of the four tidal current strengths of a day, with special reference as to whether the greater flood immediately precedes or follows the greater ebb.

sequence of tide. The order in which the four tides of a day occur, with special reference as to whether the higher high water immediately precedes or follows the lower low water.

service, *v., t.* To perform routine (or emergency) work on an unmanned aid to navigation while the aid is on its station.

service area. The area within which a navigational aid is of use. This may be divided into primary and secondary service areas having different degrees of accuracy.

service area diagram. See RELIABILITY DIAGRAM.

service period. The number of days that an automatic minor light or buoy is expected to operate without requiring recharging.

set, *n.* The direction *towards* which a current flows.

set, *v., i.* Of a celestial body, to cross the visible horizon while descending. The opposite is RISE.

set, *v., t.* To establish, as to *set* a course.

set screw. A screw for locking a movable part, as of an instrument or device.

setting a buoy. The act of placing a buoy on station in the water.

settled, *adj.* Pertaining to weather, devoid of storms for a considerable period. See also UNSETTLED.

seven-eighths rule. A rule of thumb which states that the approximate distance to an object broad on the beam equals 7/8 of the distance traveled by a craft while the relative bearing (right or left) changes from 30° or 60° or from 120° to 150°, neglecting current and wind.

seven seas. Figuratively, all the waters or oceans of the world. Applied generally to the seven oceans – Arctic, Antarctic, North Atlantic, South Atlantic, North Pacific, South Pacific, and Indian.

seven-tenths rule. A rule of thumb which states that the approximate distance to an object broad on the beam equals 7/10 of the distance traveled by a craft while the relative bearing (right or left) changes from 22°.5 to 45° or from 135° to 157°.5, neglecting current and wind.

seven-thirds rule. A rule of thumb which states that the approximate distance to an object broad on the beam equals 7/3 of the distance traveled by a craft while the relative bearing (right or left) changes from 22°.5 to 26°.5, 67°.5 to 90°, 90° to 112°.5, or 153°.5 to 157°.5, neglecting current and wind.

sexagesimal system. A system of notation by increments of 60°; as the division of the circle into 360°, each degree into 60 minutes, and each minute into 60 seconds.

sextant, *n.* A double-reflecting instrument for measuring angles, primarily altitudes of celestial bodies. As originally used, the term applied only to instruments having an arc of 60°, a sixth of a circle, from which the instrument derived its name. Such an instrument had a range of 120°. In modern practice the term applies to a similar instrument, regardless of its range, very few modern instruments being sextants in the original sense. Thus, an **octant,** having a range of 90°; a **quintant,** having a range of 144°; and a **quadrant,** having a range of 180°, are all called *sextants.* A **marine sextant** is designed primarily for marine navigation. It may be either a **clamp screw sextant** or **endless tangent screw sextant** depending upon the means for controlling the position of the index arm and the vernier or micrometer drum. It may be either a **vernier sextant** or **micrometer drum sextant** depending upon the means used to provide precise readings. A **periscope sextant** is one designed to be used in conjunction with the periscope of a submarine. A **periscope sextant** is intended primarily for use in hydrographic surveying. See also MARINE SEXTANT.

sextant adjustment. The process of checking the accuracy of a sextant and removing or reducing its error.

sextant altitude. Altitude as indicated by a sextant or similar instrument, before corrections are applied. See also OBSERVED ALTITUDE, APPARENT ALTITUDE.

sextant altitude correction. Any of several corrections applied to a sextant altitude in the process of converting it to observed altitude. See also ACCELERATION CORRECTION, AIR TEMPERATURE CORRECTION, AUGMENTATION CORRECTION, BAROMETRIC PRESSURE CORRECTION, CORIOLIS CORRECTION, DEFLECTION OF THE VERTICAL CORRECTION, DIP CORRECTION, HEIGHT OF EYE CORRECTION, INDEX CORRECTION, INSTRUMENT CORRECTION, IRRADIATION CORRECTION, PARALLAX CORRECTION, PERSONAL CORRECTION, REFRACTION CORRECTION, SEA-AIR TEMPERATURE DIFFERENCE CORRECTION, SEMIDIAMETER CORRECTION, TIDE CORRECTION, TILT CORRECTION, WAVE HEIGHT CORRECTION.

sextant chart. See CIRCLE SHEET.

sextant error. The error in the reading of a sextant, due either to lack of proper adjustment or imperfection of manufacture. See CALIBRATION ERROR, CENTERING ERROR, COLLIMATION ERROR, ERROR OF PERPENDICULARITY, GRADUATION ERROR, INDEX ERROR, INSTRUMENT ERROR, PRISMATIC ERROR, SHADE ERROR, SIDE ERROR, VERNIER ERROR.

S-four magnitude (S-4 magnitude). The photoelectric magnitude applicable to star trackers with S-4 photo-sensitive response. See also MAGNITUDE.

shade, *n.* See SHADE GLASS.

shaded relief. A cartographic technique that provides an apparent three-dimensional configuration of the terrain on maps and charts by the use of graded shadows that would be cast by high ground if light were shining from the northwest. Shaded relief is usually used in combination with contours.

shade error. That error of an optical instrument due to refraction in the shade glasses. If this effect is due to lack of parallelism of the faces, it is usually called PRISMATIC ERROR.

shade glass. A darkened transparency that can be moved into the line of sight of an optical instrument, such as a sextant, to reduce the intensity of light reaching the eye. Also called SHADE.

shadow, *n.* 1. Darkness in a region, caused by an obstruction between the source of light and the region. By extension, the term is applied to a similar condition when any form of radiant energy is cut off by an obstruction, as *radar shadow.* The darkest part of a shadow in which light is completely cut off is called the UMBRA;

a lighter part surrounding the umbra, in which the light is only partly cut off, is called the PENUMBRA.
2. A region of diminished rainfall on the lee side of a mountain or mountain range, where the rainfall is noticeably less than on the windward side. Usually called RAIN SHADOW.

shadow bands. See CREPUSCULAR RAYS.

shadow bar. A rod or bar used to cast a shadow, as on the sighting assembly of an astro compass.

shadow pin. A small rod or pin used to cast a shadow on an instrument, such as a magnetic compass or sun compass, to determine the direction of the luminary; a GNOMON.

shadow region. A region shielded from radar signals because of an intervening obstruction or absorbing medium. This region appears as an area void of targets on a radar display such as a plan position indicator. The phenomenon is called RADAR SHADOW. See also SHADOW SECTOR, BLIND SECTOR.

shadow sector. A sector on the radarscope in which the appearance of radar echoes is improbable because of an obstruction near the antenna. While both blind and shadow sectors have the same basic cause, *blind sectors* generally occur within the larger angles subtended by the obstruction. See also SHADOW REGION.

shallow, *adj.* Having little depth; shoal.

shallow, *n.* An area where the depth of water is relatively slight.

shallow water constituent. A short-period harmonic term introduced into the formula of tidal (or tidal current) constituents to take account of the change in the form of a tide wave resulting from shallow water conditions. Shallow water constituents include the overtides and compound tides.

shallow water wave. A wave is classified as a shallow water wave whenever the ratio of the depth (the vertical distance of the still water level from the bottom) to the wave length (the horizontal distance between crests) is less than 0.04. Tidal waves are shallow water waves.

shamal, *n.* A northwesterly wind blowing over Iraq and the Persian Gulf, in summer, often strong during the day, but decreasing during the night.

sharki, *n.* A southeasterly wind which sometimes blows in the Persian Gulf.

shearing, *n.* An area of pack ice is subject to shear when the ice motion varies significantly in the direction normal to the motion, subjecting the ice to rotational forces. These forces may result in phenomena similar to a FLAW.

sheet line. See NEATLINE.

shelf, *n.* A zone adjacent to a continent, or around an island, that extends from the low water line to a depth at which there is usually a marked increase of slope towards oceanic depths.

shelf valley. A valley on the shelf, generally the shoreward extension of a canyon.

shield, *n.* A metal housing around an electrical or magnetic element to eliminate or reduce the effect of its electric or magnetic field beyond the housing, or the effect of an exterior field on the element.

shielding factor. The ratio of the strength of the magnetic field at a compass to the strength if there were no disturbing material nearby; usually expressed as a decimal. Because of the metal of a vessel, the strength of the earth's magnetic field is reduced somewhat at a compass location aboard ship. The shielding factor is one minus the percentage of reduction.

shimmer, *v., i.* To appear tremulous or wavering, due to varying atmospheric refraction in the line of sight.

shingle, *n.* See under STONES.

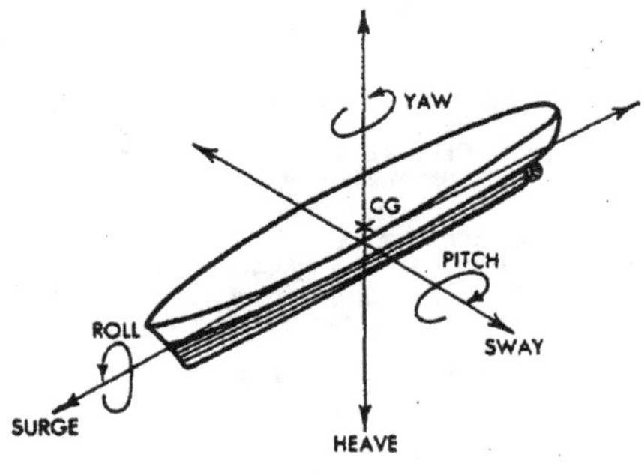

ship, *n.* Originally, a three-masted, square-rigged sailing vessel. The term is now generally applied to any vessel except rowboats and small motorboats, fishing vessels, pleasure craft, barges, etc. The distinction between a ship and a boat is largely one of size, but there is no well-defined line of demarcation.

ship error. That error in radio direction finder bearings due to reradiation of radio waves by the metal of the ship.

ship motions. Of the motions of a vessel in a seaway, **surge** is the bodily motion forward and backward along the longitudinal axis, caused by the force of the sea acting alternately on the bow and stern; **heave** is the oscillatory rise and fall due to the entire hull being lifted by the force of the sea; **sway** is the side-to-side bodily motion, independent of rolling, caused by uniform pressure being exerted all along one side of the hull; **yaw** is the oscillation about a vertical axis approximately through the center of gravity of the vessel; **roll** is the oscillation about the longtidudinal axis; and **pitch** is the oscillation about the transverse axis, due to the bow and stern being raised or lowered on passing through successive crests and troughs of waves.

shipping lane. An established route traversed by ocean shipping.

ship's emergency transmitter. As defined by the *International Telecommunication Union* (ITU), a ship's transmitter to be used exclusively on a distress frequency for distress, urgency or safety purposes.

ship's head. Heading of a vessel.

Ships' Routing. A publication of the Inter-Governmental Maritime Consultative Organization (IMCO) which describes the general provisions of ships' routing, traffic separation schemes, deep water routes and areas to be avoided, which have been adopted by IMCO. All details of routing systems are promulgated through Notices to Mariners, together with their dates of implementation. Also details of routing systems are depicted on charts and are given in Sailing Directions.

ship weather routing. A procedure whereby an optimum route is developed based on the forecasts of weather and seas and the ship's characteristics for a particular transit. Within specified limits of weather and sea conditions, the term *optimum* is used to mean maximum safety and crew comfort, minimum fuel consumption, minimum time underway, or any desired combination of these factors. The ship weather routing service of the U.S. Navy is called OPTIMUM TRACK SHIP ROUTING (OTSR).

shoal, *adj.* Shallow.

shoal, *n.* An offshore hazard to navigation on which there is a depth of 16 fathoms or 30 meters or less, composed of unconsolidated material. See also REEF.

shoal, *v., i.* To become less deep.

shoal, *v., t.* 1. To cause to become less deep.
2. To proceed from a greater to a lesser depth.

shoal patches. Individual and scattered elevations of the bottom, with depths of 16 fathoms (or 30 meters) or less, but composed of any material except rock or coral.

shoal water. Shallow water; water over a shoal.

shoot, *v., t.* To observe the altitude of (a celestial body).

shooting star. See METEOR.

shore, *n.* That part of the land in immediate contact with a body of water including the area between high and low water lines. The term SHORE is usually used with reference to the body of water and COAST with reference to the land, as the east coast of the United States is part of the western shore of Atlantic Ocean. The term SHORE usually refers to a narrow strip of land in immediate contact with any body of water, while COAST refers to a general region in proximity to the sea. A shore bordering the sea may be called a SEASHORE. See also FORESHORE, BACKSHORE.

shoreface, *n.* The narrow zone seaward from the low tide shoreline, permanently covered by water, over which the beach sands and gravels actively oscillate with changing wave conditions.

shore lead. A lead between pack ice and the shore or between pack ice and an ice front.

shoreline, *n.* The intersection of the land with the water surface. The shoreline shown on charts represents the line of contact between the land and a selected water elevation. In areas affected by tidal fluctuations, this line of contact is usually the mean high water line. In confined coastal waters of diminished tidal influence, the mean water level line may be used.

shore polynya. See under POLYNYA.

short period perturbations. Periodic perturbations in the orbit of a planet or satellite which execute one complete periodic variation in the time of one orbital period or less.

short range systems. Those radionavigation systems limited in their positioning capability to coastal regions, or those systems limited to making landfall. Radar and the radio direction finder are examples. See also MEDIUM RANGE SYSTEMS, LONG RANGE SYSTEMS.

short sea. A sea in which the waves are short, irregular, and broken.

short wave. A radio wave shorter than those of the standard broadcast band. See also WAVE, definition 2.

shower, *n.* Precipitation from a convective cloud. Showers are characterized by the suddeness with which they start and stop, by the rapid changes of intensity, and usually by rapid changes in the appearance of the sky. In weather observing practice, showers are always reported in terms of the basic type of precipitation that is falling, i.e., **rain showers, snow showers, sleet showers.**

shuga, *n.* An accumulation of spongy white ice lumps, a few centimeters across; the lumps are formed from grease ice or slush and sometimes from anchor ice rising to the surface.

side echo. The effect on a radar display produced by a side lobe of a radar antenna. See also BACK ECHO.

side error. That error in the reading of a marine sextant due to nonperpendicularity of the horizon glass to the frame.

side lobe. Of the radiation pattern of a directional antenna, any lobe other than the main or back lobe.

sidereal, *adj.* Of or pertaining to the stars. Although SIDEREAL generally refers to the stars and TROPICAL to the vernal equinox, sidereal time and the sidereal day are based upon the position of the vernal equinox relative to the meridian. The SIDEREAL YEAR is based upon the stars.

sidereal day. See under SIDEREAL TIME.

sidereal hour angle. Angular distance west of the vernal equinox; the arc of the celestial equator, or the angle at the celestial pole, between the hour circle of the vernal equinox and the hour circle of a point on the celestial sphere, measured westward from the hour circle of the vernal equinox through 360°. Angular distance east of the vernal equinox, through 24 hours, is RIGHT ASCENSION.

sidereal month. The average period of revolution of the moon with respect to the stars, a period of 27 days, 7 hours, 43 minutes, 11.5 seconds, or approximately 27 1/3 days.

sidereal noon. See under SIDEREAL TIME.

sidereal period. 1. The length of time required for one revolution of a celestial body about its primary, with respect to the stars.
2. The interval between two successive returns of an artificial earth satellite in orbit to the same geocentric right ascension.

sidereal time. Time defined by the daily rotation of the earth with respect to the vernal equinox or first point of Aries. The sidereal time is numerically measured by the hour angle of the equinox, which represents the position of the equinox in the daily rotation. The period of one rotation of the equinox in hour angle, between two consecutive upper meridian transits, is a **sidereal day;** it is divided into 24 sidereal hours, reckoned from 0^h at upper transit which is known as **sidereal noon.** The true equinox is at the intersection of the true celestial equator of date with the ecliptic of date; the time measured by its daily rotation is **apparent sidereal time.** The position of the true equinox is affected by the nutation of the axis of rotation of the earth; and the nutation consequently introduces irregular periodic inequalities into the apparent sidereal time and the length of the sidereal day. The time measured by the daily motion of the mean equinox of date, which is affected only by the secular inequalities due to the precession of the axis, is **mean sidereal time.** The maximum difference between apparent and mean sidereal times is only a little over a second, and its greatest daily change is a little more than a hundredth of a second. Because of its variable rate, apparent sidereal time is used by astronomers only as a measure of epoch; it is not used for time interval. Mean sidereal time is deduced from apparent sidereal time by applying the equation of equinoxes.

sidereal year. The period of one apparent revolution of the earth around the sun, with respect to a fixed point, or a distant star devoid of proper motion, being 365 days, 6 hours, 9 minutes, and 9.5 seconds in 1900, and increasing at the rate of 0.0001 second annually. Because of the precession of the equinoxes this is about 20 minutes longer than a tropical year.

sight, *n.* Observation of the altitude, and sometimes also the azimuth, of a celestial body for a line of position; or the data obtained by such observation. A group of such observations made over a short period of time is called a ROUND OF SIGHTS. An observation of a celestial body made by facing 180° from the azimuth of the body is called a **back sight.** See also NOON SIGHT, TIME SIGHT.

sighting vane. See VANE, definition 2.

sight reduction. The process of deriving from a sight the information needed for establishing a line of position.

sight reduction tables. Tables for performing sight reduction, particularly those for comparison with the observed altitude of a celestial body to determine the altitude difference for establishing a line of position.

Sight Reduction Tables for Air Navigation. See PUB. NO. 249.

Sight Reduction Tables for Marine Navigation. See PUB. NO. 229.

signal, *n.* 1. As applied to electronics, any transmitted electrical impulse.
2. That which conveys intelligence in any form of communication, such as a time signal or an object marking the location of a surveying station.

signal-to-noise ratio. The ratio of the magnitude of the signal to that of the noise, often expressed in decibels.

signature, *n.* The graphic record of the magnetic properties of a vessel automatically traced as the vessel passes over the sensitive element of a recording instrument.

sign conventions. See as GEOGRAPHIC SIGN CONVENTIONS.

significant digits. Those digits of a number which have a significance, zeros at the left and sometimes those at the right being excluded. Thus, 1,325, 1,001, 1.408, 0.00005925, 625.0, and 0.04009 have four significant digits each. But in the number 186,000 there may be three, four, five, or six significant digits depending upon the accuracy with which the number has been determined.

sikussak, *n.* Very old ice trapped in fiords. Sikussak resembles glacier ice, since it is formed partly from snow.

sill, *n.* On the sea floor, the low part of a gap or saddle separating basins. See also DOCK SILL.

sill depth. The greatest depth over a sill.

silt, *n.* See under MUD.

similar decimals. Decimals having the same number of decimal places, as 3.141 and 0.789. Decimals can be made similar by adding the appropriate number of zeros. For example, 0.789 can be made similar to 3.1416 by stating it as 0.7890. See also REPEATING DECIMAL, SIGNIFICANT DIGITS.

simple conic chart. A chart on a simple conic projection.

simple conic map projection. A conic map projection in which the surface of a sphere or spheroid, such as the earth, is conceived as developed on a tangent cone, which is then spread out to form a plane.

simple harmonic motion. The projection of uniform circular motion on a diameter of the circle of such motion. The combination of two or more simple harmonic motions results in COMPOUND HARMONIC MOTION.

simultaneous altitudes. Altitudes of two or more celestial bodies observed at the same time.

simultaneous mode. A satellite method for determining the position of an unknown station by the simultaneous ranging from three stations of known position and the unknown station, or simultaneously observing direction from two stations of known position and the unknown station, and mathematically reducing the data to solve for a line or surface of position of the unknown. This technique permits position determination independent of a satellite's orbital parameters.

simultaneous observations (of a satellite). Observations of a satellite that are made from two or more distinct points or tracking stations at exactly the same time.

sine, *n.* The ratio of the side opposite an acute angle of a plane right triangle to the hypotenuse. The expression NATURAL SINE is sometimes used to distinguish the sine from its logarithm (called LOGARITHMIC SINE).

sine curve. Characteristic simple wave pattern; a curve which represents the plotted values of the sines of angles, with the sine as the ordinate and the angle as the abscissa. The curve starts with 0 amplitude at the origin, increases to a maximum at 90°, decreases to 0 at 180°, increases negatively to a maximum negative amplitude at 270°, and returns to 0 at 360°, to repeat the cycle. Also called SINUSOID.

sine wave. A simple wave in the form of a sine curve.

single astronomic station datum orientation. The orientation of a geodetic datum by accepting the astronomically determined coordinates of the origin and the azimuth to one other station without any correction.

single-axis normal distribution. A one-dimensional normal distribution along an axis perpendicular to a line of position. Two single-axis normal distributions may be used to establish the error ellipse and the corresponding circle of equivalent probability when the error distribution is two-dimensional or bivariate.

single-degree-of-freedom gyro. A gyroscope the spin axis of which is free to rotate about only one of the orthogonal axes, the spin axis not being counted. See also DEGREE-OF-FREEDOM, RATE GYRO.

single-flashing light. See under FLASHING LIGHT.

single interpolation. Interpolation when there is but one argument or variable.

single-occulting light. See under OCCULTING LIGHT.

single-sideband transmission. The method of radio transmission in which the frequencies produced by the process of modulation on one side of the carrier are transmitted and those on the other side are suppressed. The carrier frequency may either be transmitted or suppressed. In this method, less power is required for the same effective signal at the receiver, a narrower frequency band can be used, and the signal is less subject to manmade interference or selective fading.

single station range light. A direction light bounded by other sectors of different characteristics which define its margins with small angles of uncertainty. Most commonly the bounding sectors are of different colors (red and green).

sinking *n.* An apparent lowering of distant terrestrial objects by abnormal atmospheric refraction. Because of sinking, objects normally visible at or near the horizon sometimes disappear below the horizon. The opposite is LOOMING.

sinusoid, *n.* See SINE CURVE.

sinusoidal, *adj.* Of or pertaining to a sine wave or sinusoid.

siren, *n.* A sound signal emitter using the periodic

escape of compressed air through a rotary shutter.

sirocco, *n.* A warm wind of the Mediterranean area, either a foehn or a hot southerly wind in advance of a low pressure area moving from the Sahara or Arabian deserts. Called LEVECHE in Spain.

skeleton tower. A tower, usually of steel, constructed of heavy corner members and various horizontal and diagonal bracing members.

skip distance. The least distance from a transmitting antenna at which a skywave can normally be received at a given frequency.

skip zone. The area between the outer limit of reception of groundwaves and the inner limit of reception of skywaves, where no signal is received.

sky diagram. A diagram of the heavens, indicating the apparent position of various celestial bodies with reference to the horizon system of coordinates.

skylight, *n.* From the point of view of the submariner, thin places in the ice canopy, usually less than 1 meter thick and appearing from below as relatively light, translucent patches in dark surroundings. The under-surface of a skylight is normally flat. Skylights are called *large* if big enough for a submarine to attempt to surface through them, or *small* if not.

sky map. The pattern on the underside of extensive cloud areas, created by the varying amounts of light reflected from the earth's surface. Snow surfaces produce a white glare (SNOW BLINK) and ice surfaces produce a yellowish-white glare (ICE BLINK). Bare land reflects relatively little light (LAND SKY) and open water even less (WATER SKY).

skywave, *n.* A radio wave that is propagated by way of the ionosphere. Also called IONOSPHERIC WAVE.

skywave correction. The correction to be applied to the time difference reading of signals received via the ionosphere to convert it to the equivalent groundwave reading. The correction for a particular place is established on the basis of an average height of the ionosphere.

skywave error. See IONOSPHERIC ERROR.

skywave transmission delay. The amount by which the time of transit from transmitter to receiver of a pulse carried by skywaves reflected once from the E-layer exceeds the time of transit of the same pulse carried by groundwaves.

slack water. The state of a tidal current when its speed is near zero, especially the moment when a reversing current changes direction and its speed is zero. The term is also applied to the entire period of low speed near the time of turning of the current when it is too weak to be of any practical importance in navigation. The relation of the time of slack water to the tidal phases varies in different localities. For standing tidal waves, slack water occurs near the times of high and low water, while for progressive tidal waves, slack water occurs midway between high and low water.

slant range. The line-of-sight distance between two points not at the same elevation.

slave, *n.* Short for SLAVE STATION.

slaved gyro magnetic compass. A directional gyro compass with an input from a flux value to keep the gyro oriented to magnetic north.

slave station. In a radionavigation system, the station of a chain whose emissions are made with reference to the emissions of a master station, its emissions being triggered by the emissions of the master station. See also SECONDARY STATION.

sleet, *n.* 1. See under ICE PELLETS.
2. In British terminology, and colloquially in some parts of the United States, precipitation in the form of a mixture of rain and snow.

slewing, *n.* In sea ice navigation, the act of forcing a ship through ice by pushing apart adjoining ice floes.

slick, *n.* A smooth area of water, as one caused by the sweep of a vessel's stern during a turn, or by a film of oil.

slime, *n.* Soft, fine, oozy mud or other substance of similar consistency.

slip, *n.* 1. A berthing space between two piers. Also called DOCK.
2. The difference between the distance a propeller would travel longitudinally in one revolution if operating in a solid and the distance it travels through a fluid.

slope, *n.* On the sea floor, the slope seaward from the shelf edge to the beginning of a continental or insular rise or the point where there is a general reduction in slope.

slot radiator. A slot in the wall of a slotted waveguide antenna which acts as a radiating element.

slotted guide antenna. See SLOTTED WAVEGUIDE ANTENNA.

slotted waveguide antenna. An antenna consisting of a metallic waveguide in the walls of which are cut one or more slot radiators.

slough (slōō), *n.* A minor marshland or tidal waterway which usually connects other tidal areas; often more or less equivalent to a bayou; occasionally applied to the sea level portion of a creek on the U.S. West Coast.

slow-sweep racon. See under SWEPT-FREQUENCY RACON.

slue, *n.* A slough or swamp.

sluice, *n.* A floodgate.

sluicing pond. See SCOURING BASIN.

slush, *n.* Snow which is saturated and mixed with water on land or ice surfaces, or which is a viscous floating mass in water after a heavy snow fall.

small area plotting sheet. For a relatively small area, a good approximation of a Mercator position plotting sheet, constructed by the navigator by either of two methods based upon a graphical solution of the secant of the latitude, which approximates the expansion. One method uses a selected longitude scale, the other a selected latitude scale. A partially completed small area plotting sheet printed in advance for later rapid completion according to requirements is called UNIVERSAL PLOTTING SHEET.

small circle. The intersection of a sphere and a plane which does not pass through its center. The intersection of a sphere and a plane through the center is called a GREAT CIRCLE.

small diurnal range. The difference in height between mean lower high water and mean higher low water. Applicable only when the type of tide is either semidiurnal or mixed. See also TROPIC RANGES.

small floe. See under FLOE.

small fracture. See under FRACTURE.

small hail. See under ICE PELLETS.

small iceberg. For reports to the International Ice Patrol, an iceberg that extends 4 to 50 feet (1 to 15 meters) above the sea surface and which has a length of 20 to 200 feet (6 to 60 meters). See also MEDIUM ICEBERG, LARGE ICEBERG.

small ice cake. An ice cake less than 2 meters across.

small ice field. See under ICE FIELD.

small scale. A scale involving a relatively large

reduction in size. A **small-scale chart** is one covering a large area. The opposite is LARGE SCALE. See also REPRESENTATIVE FRACTION.

small-scale chart. See under CHART. See also SMALL SCALE.

small tropic range. The difference in height between tropic lower high water and tropic higher low water. Applicable only when the type of tide is either semidiurnal or mixed. See also MEAN TROPIC RANGE, GREAT TROPIC RANGE.

smell the bottom. See FEEL THE BOTTOM.

smog, *n.* Originally a natural fog contaminated by industrial pollutants, or a mixture of smoke and fog. Today, smog is a common term applied to air pollution with or without fog; however, some visible manifestation is almost always implied.

smoke, *n.* Small particles of carbon and other solid matter, resulting from incomplete combustion, while suspended in the air. When it settles, it is called SOOT.

smokes, *n., pl.* Dense white haze and dust clouds common in the dry season on the Guinea coast of Africa, particularly at the approach of the harmattan.

smooth sea. Sea with waves no higher than ripples or small wavelets.

snow, *n.* 1. Frozen precipitation consisting of translucent or white ice crystals which fall either separately or in loose clusters called **snowflakes.** Very fine, simple crystals, or minute, branched, star-like snowflakes are called **snow grains.** **Snow pellets** are white, opaque, roundish grains which are crisp and easily compressible, and may rebound or burst when striking a hard surface. Snow is called **brown, red,** or **yellow** when it is colored by the presence of brown dust, red dust or algae, or pine or cypress pollen, respectively. See also BLOWING SNOW, DRIFTING SNOW.
2. The speckled background on the plan position indicator due to electrical noise.

snow barchan. See under SNOWDRIFT.

snow blink. A white glare on the underside of extensive cloud areas, created by light reflected from snow-covered surfaces. Snow blink is brighter than the yellowish-white glare of ICE BLINK. Clouds above bare land or open water have no glare. See also LAND SKY, WATER SKY, SKY MAP.

snow-covered ice. Ice covered with snow.

snowdrift, *n.* An accumulation of wind-blown snow deposited in the lee of obstructions or heaped by wind eddies. A crescent-shaped snowdrift, with ends pointing downwind, is called a SNOW BARCHAN.

snowflake, *n.* A loose cluster if ice crystals, or, rarely, a single crystal.

snow flurry. A popular term for SNOW SHOWER, particularly of a very light and brief nature.

snow grains. Frozen precipitation consisting of very fine, simple crystals, or of minute, branched, star-like snowflakes. Snow grains are the solid equivalent of drizzle. Also called GRANULAR SNOW.

snow pellets. Frozen precipitation consisting of small, white, opaque, roundish grains of snowlike structure which are crisp and easily compressible, and may rebound or burst when striking a hard surface. Also called SOFT HAIL, GRAUPEL. See also SMALL HAIL.

snow storm. See under STORM, definition 2.

soft hail. See SNOW PELLETS.

soft iron. Iron or steel which is easily magnetized by induction, but loses its magnetism when the magnetic field is removed. The opposite is HARD IRON.

solar, *adj.* Of or pertaining to the sun.

solar day. 1. The duration of one rotation of the earth on its axis, with respect to the sun. This may be either a **mean solar day,** or an **apparent solar day,** as the reference is the mean or apparent sun, respectively.
2. The duration of one apparent rotation of the sun.

solar eclipse. An eclipse of the sun. When the moon passes between the sun and the earth, the sun appears eclipsed to an observer in the moon's shadow. A solar eclipse is **partial** if the sun is partly obscured; **total** if the entire surface is obscured, or **annular** if a thin ring of the sun's surface appears around the obscuring body.

solar flare. A bright eruption from the sun's chromosphere. Solar flares may appear within minutes and fade within an hour.

solar noon. Twelve o'clock solar time, or the instant the sun is over the upper branch of the reference meridian. Solar noon may be classified as **mean** if the mean sun is the reference, or as **apparent** if the apparent sun is the reference. It may be further classified according to the reference meridian, either the **local** or **Greenwich** meridian or additionally in the case of mean noon, a designated zone meridian. **Standard,** daylight saving or summer noon are variations of zone noon. Local apparent noon may also be called **high noon.**

solar-radiation pressure. A cause of perturbations of high flying artificial satellites of large diameter. The greater part is directly from the sun, a minor part is from the earth, which is usually divided into direct (reflected) and indirect terrestrial (radiated) radiation pressures.

solar system. The sun and other celestial bodies within its gravitational influence, including planets, planetoids, satellites, comets, and meteors.

solar tide. 1. The part of the tide that is due to the tide-producing force of the sun. See also LUNAR TIDE.
2. The observed tide in areas where the solar tide is dominant. This condition provides for phase repetition at about the same time each solar day.

solar time. Time based upon the rotation of the earth relative to the sun. Solar time may be classified as **mean** if the mean sun is the reference; or as **apparent** if the apparent sun is the reference. The difference between mean and apparent time is called EQUATION OF TIME. Solar time may be further classified according to the reference meridian, either the **local** or **Greenwich** meridian or additionally in the case of mean time, a designated **zone** meridian. **Standard** and **daylight saving** or **summer time** are variations of zone time. Time may also be designated according to the timepiece, as **chronometer time** or **watch time,** the time indicated by these instruments.

solar year. See TROPICAL YEAR.

solid color buoy. A buoy which is painted only one color above the water line.

solitary wave. A wave of translation consisting of a single crest rising above the undisturbed water level, without any accompanying trough, in contrast with a WAVE TRAIN. The rate of advance of a solitary wave depends upon the depth of water.

solstice, *n.* 1. One of the two points of the ecliptic farthest from the celestial equator; one of the two points on the celestial sphere occupied by the sun at maximum declination. That in the

Northern Hemisphere is called the **summer solstice** and that in the Southern Hemisphere the **winter solstice**. Also called SOLSTITIAL POINT.
2. That instant at which the sun reaches one of the solstices, about June 21 (**summer solstice**) or December 22 (**winter solstice**).

solstitial colure. That great circle of the celestial sphere through the celestial poles and the solstices.

solstitial point. One of the two points on the ecliptic at the greatest distance from the celestial equator. Also called SOLSTICE.

solstitial tides. Tides occurring near the times of the solstices. The tropic range may be expected to be especially large at these times.

Somali Current. See EAST AFRICA COASTAL CURRENT.

sonar, *n.* A system of determining distance of an underwater object by measuring the interval of time between transmission of an underwater sonic or ultrasonic signal and the return of its echo. Direction may also be determined by noting the direction of transmission of the signal. The name *sonar* is derived from the words sound navigation and ranging. See also ECHO RANGING.

sonic, *adj.* Of, or pertaining to, the speed of sound.

sonic depth finder. A direct-reading instrument which determines the depth of water by measuring the time interval between the emission of a sound and the return of its echo from the bottom. A similar instrument utilizing signals above audible range is called an ULTRASONIC DEPTH FINDER. Both instruments are also called ECHO SOUNDERS.

sonic frequency. See AUDIO FREQUENCY.

sonic navigation. Navigation by means of sound waves whether or not they are within the audible range. Also called ACOUSTIC NAVIGATION.

sonne, *n.* A German forerunner of the British CONSOL.

sonobuoy, *n.* A buoy with equipment for automatically transmitting a radio signal when triggered by an underwater sound signal.

sound, *n.* 1. A relatively long arm of the sea or ocean forming a channel between an island and a mainland or connecting two larger bodies of water, as a sea and the ocean, or two parts of the same body but usually wider and more extensive than a strait. The term has been applied to many features which do not fit the accepted definition. Many are very large bodies of water, such as Mississippi Sound and Prince William Sound, others are mere salt water ponds or small passages between islands.
2. A vibratory disturbance in air or some other elastic medium, capable of being heard by the human ear, and thus of any frequency between about 20 and 20,000 cycles per second.

sound, *v., i.* To measure the depth of the water.

sound buoy. A buoy equipped with either a gong, bell, whistle, or electronic horn. Bells and gongs on buoys are sounded by tappers that hang from the tower and swing as the buoys roll in the sea. **Bell buoys** produce a sound of only one tone; **gong buoys** produce several tones. **Whistle buoys** make a loud moaning sound caused by the rising and falling motions of the buoy in the sea. A buoy equipped with an electronic horn, a **horn buoy,** will produce a pure tone at regular intervals and will operate continuously regardless of the sea state. A **lighted sound buoy** has the same general configuration as a lighted buoy but is equipped with a sound signal. **Unlighted sound buoys** have the same general appearance as light buoys but are not equipped with any light apparatus.

sounding, *n.* Measured or charted depth of water, or the measurement of such depth. A **no-bottom sounding** is one in which the bottom is not reached. A vessel is said to be *on soundings* when it is navigating primarily by means of the information obtained by successive measurements of the depth of the water, or is in an area where this can be done. In other areas a vessel is said to be *off soundings*. A minimum sounding chosen for a vessel of specific draft in a given area to indicate the limit of safe navigation is called a **danger sounding.** See also ECHO SOUNDING, LINE OF SOUNDINGS.

sounding datum. Short for CHART SOUNDING DATUM.

sounding lead. See under LEAD.

sounding machine. An instrument for measuring depth of water, consisting essentially of a reel of wire to one end of which is attached a weight which carries a device for recording the depth. A crank or motor is provided for reeling in the wire.

sounding sextant. See HYDROGRAPHIC SEXTANT.

sound signal. A sound transmitted in order to convey information as a *fog signal*. The term *sound signal* is frequently used to describe the apparatus generating the sound. This use is deprecated.

sound signal station. An attended station whose function is to operate a sound signal.

sound wave. An audio-frequency wave in any material medium, in which vibration is in the direction of travel, resulting in alternate compression and rarefaction of the medium, or, by extension, a similar wave outside the audible range.

south, *n.* The direction 180° from north. See also CARDINAL POINT.

South Atlantic Current. An eastward flowing current of the South Atlantic Ocean that is continuous with the northern edge of the WEST WIND DRIFT. It appears to originate mainly from the Brazil Current and partly from the northernmost flow of the West Wind Drift west of longitude 40°W. The current is under the influence of the prevailing westerly trade winds; the constancy and speed increase from the northern boundary to about latitude 40° S, where the current converges with the West Wind Drift. The mean speed varies from about 0.5 to 0.7 knot.

southbound node. See DESCENDING NODE.

Southeast Drift Current. See AZORES CURRENT.

southeaster, sou'easter, *n.* A southeasterly wind, particularly a strong wind or gale.

south equatorial current. See ATLANTIC SOUTH EQUATORIAL CURRENT, PACIFIC SOUTH EQUATORIAL CURRENT, INDIAN SOUTH EQUATORIAL CURRENT.

south frigid zone. That part of the earth south of the Antarctic Circle.

south geographical pole. The geographical pole in the Southern Hemisphere, at lat. 90° S.

south geomagnetic pole. The geomagnetic pole in the Southern Hemisphere at lat. 78°.5S, long. 111°E, approximately. This term should not be confused with SOUTH MAGNETIC POLE. See also GEOMAGNETIC POLE.

South Indian Current. An eastward flowing current of the Indian Ocean that is continuous with the northern edge of the WEST WIND DRIFT.

southing, *n.* The distance a craft makes good to the south. The opposite is NORTHING.

south magnetic pole. The magnetic pole in the

Southern Hemisphere, at lat. 65°.6S, long. 139°.4E, approximately in 1980. This term should not be confused with SOUTH GEOMAGNETIC POLE. See also GEOMAGNETIC POLE.

South Pacific Current. An eastward flowing current of the South Pacific Ocean that is continuous with the northern edge of the WEST WIND DRIFT.

south polar circle. See ANTARCTIC CIRCLE.

south pole. 1. *Usually capitalized.* The south geographical pole. See also MAGNETIC POLE, GEOMAGNETIC POLE.
2. The south-seeking end of a magnet. See also BLUE MAGNETISM.

south temperate zone. That part of the earth between the Tropic of Capricorn and the Antarctic Circle.

southwester, sou'wester, *n.* A southwest wind, particularly a strong wind or gale.

southwest monsoon. See under MONSOON.

space coordinates. A three-dimensional system of Cartesian coordinates by which a point is located by three magnitudes indicating distance from three planes which intersect at a point.

spacecraft, *n.* Devices, manned and unmanned, which are designed to be placed into an orbit about the earth or into a trajectory to another celestial body.

space motion. Motion of a celestial body through space. That component perpendicular to the line of sight is called **proper motion** and that component in the direction of the line of sight is called **radial motion.**

space-polar coordinates. A system of coordinates by which a point on the surface of a sphere is located in space by (1) its distance from a fixed point at the center, called the POLE; (2) the COLATITUDE or angle between the POLAR AXIS (a reference line through the pole) and the RADIUS VECTOR (a straight line connecting the pole and the point); and (3) the LONGITUDE or angle between a reference plane through the polar axis and a plane through the radius vector and polar axis. See also POLAR COORDINATES, SPHERICAL COORDINATES.

space wave. See DIRECT WAVE, definition 2.

spar buoy. A buoy in the shape of a spar, or tapered pole, floating nearly vertically. See also SPINDLE BUOY.

spatial correlation. See under DIFFERENTIAL OMEGA.

special mark. See under IALA MARITIME BUOYAGE SYSTEM.

Special Notice to Mariners. These notices contain important information of considerable interest to all mariners such as caution on the use of foreign charts; warning on use of floating aids; use of the Automated Mutual-Assistance Vessel Rescue (AMVER) system; rules, regulations, and proclamations issued by foreign governments; oil pollution regulations, etc. *Special Notice to Mariners* is published annually in *Notice to Mariners No. 1*, by the Defense Mapping Agency Hydrographic/Topographic Center.

special purpose buoy. A buoy used to indicate a special meaning to the mariner and having no lateral significance, such as one used to mark a quarantine or anchorage area.

special subject map. See TOPICAL MAP.

Special Warnings. Broadcast messages primarily intended to announce official U.S. Government proclamations affecting shipping. Special Warnings are broadcast by U.S. Navy and U. S. Coast Guard radio stations and are published in all editions of the *Daily Memorandum* and in the weekly *Notice to Mariners*. Upon issuance, the texts of all effective Special Warnings are published annually in *Notice to Mariners No. 1* by the Defense Mapping Agency Hydrographic/Topographic Center.

species of constituent. A classification depending upon the period of a constituent. The principal species are *semidiurnal*, *diurnal*, and *long period*.

species sanctuary. A sanctuary established for the conservation of marine life. See also MARINE SANCTUARY.

specific humidity. See under HUMIDITY.

spectral, *adj.* Of or pertaining to a spectrum.

spectroscope, *n.* An optical instrument for forming spectra. It is very useful in studying the characteristics of celestial bodies.

spectrum (*pl. spectra*), *n.* 1. A series of images formed when a beam of radiant energy is separated into its various wavelength components, as when a beam of white light is refracted and dispersed by a prism.
2. The entire range of electromagnetic radiations, or any part of it used for a specific purpose, as the *radio spectrum* (10 kilohertz to 300 gigahertz).

specular reflection. Reflection without diffusion in accordance with the laws of optical reflection, as in a mirror. Also called REGULAR REFLECTION, MIRROR REFLECTION.

speculum, *n.* An optical instrument reflector of polished metal or of glass with a film of metal.

speed, *n.* Rate of motion. The terms SPEED and VELOCITY are often used interchangeably, but SPEED is a scalar, having magnitude only, while VELOCITY is a vector quantity, having both magnitude and direction. Rate of motion in a straight line is called **linear speed,** while change of direction per unit time is called **angular velocity. Subsonic, sonic, supersonic,** and **hypersonic speeds** are those respectively less than, equal to, greater than, and equal to, or greater than, five times the speed of sound, while **transonic speeds** are those in the range in which flow patterns change from subsonic to supersonic, or vice versa.

speed circle. A circle having a radius equal to a given speed and drawn about a specified center. The expression is used chiefly in connection with relative movement problems.

speed-course-latitude error. See SPEED ERROR.

speed error. An error in both pendulous and nonpendulous type gyrocompasses resulting from movement of the gyrocompass in other than an east-west direction. The error is westerly if any component of the ship's course is north, and easterly if south. Its magnitude is proportional to the course, speed, and latitude of the ship. Sometimes called SPEED-COURSE-LATITUDE ERROR.

speed line. A line of position approximately perpendicular to the course line, thus providing a check on the speed of advance. See also COURSE LINE.

speed made good. The speed along the track made good.

speed of advance. 1. The average speed in knots which must be maintained during a passage to arrive at a destination at an appointed time.
2. The speed intended to be made good along the track.

speed of relative movement. Speed relative to a reference point itself usually in motion. Also called RELATIVE SPEED.

speed over ground. The speed of the vessel along the actual path of travel over the ground.

speed triangle. See under VECTOR DIAGRAM.

spending beach. In a wave basin, the beach on which the entering waves spend themselves, except for the small remainder entering the inner harbor.

sphere, *n.* 1. A curved surface all points of which are equidistant from a fixed point within, called the *center*. The **celestial sphere** is an imaginary sphere of infinite radius concentric with the earth, on which all celestial bodies except the earth are imagined to be projected. The celestial sphere as it appears to an observer at the equator, where celestial bodies appear to rise vertically above the horizon, is called a **right sphere;** at the pole, where bodies appear to move parallel to the horizon, it is called a **parallel sphere;** between the equator and pole, where bodies appear to rise obliquely to the horizon, it is called an **oblique sphere.** Half a sphere is called a HEMISPHERE.
2. A body or the space bounded by a spherical surface. For most practical problems of navigation, the earth is considered a sphere, called the **terrestrial sphere.**

spherical, *adj.* Of or pertaining to a sphere.

spherical aberration. See under ABERRATION, definition 2.

spherical angle. The angle between two intersecting great circles.

spherical buoy. A buoy of which the upper part of the body (above the waterline), or the larger part of the superstructure, is spherical.

spherical coordinates. A system of coordinates defining a point on a sphere or spheroid by its angular distances from a primary great circle and from a reference secondary great circle, as *latitude* and *longitude*. See also CELESTIAL COORDINATES, POLAR COORDINATES.

spherical excess. The amount by which the sum of the three angles of a spherical triangle exceeds 180°.

spherical harmonics. Trigonometric terms of an infinite series used to approximate a two- or three-dimensional function of locations on or above the earth.

spherical sailing. Any of the sailings which solve the problems of course, distance, difference of latitude, difference of longitude, and departure by considering the spherical or spheroidal shape of the earth. PLANE SAILING considers the earth as a plane.

spherical triangle. A closed figure having arcs of three great circles as sides.

spherical wave. A wave with a spherical wave front.

spheroid, *n.* An ellipsoid; a figure resembling a sphere. Also called ELLIPSOID or ELLIPSOID OF REVOLUTION, from the fact that it can be formed by revolving an ellipse about one of its axes. If the shorter axis is used as the axis of revolution, an **oblate spheroid** results, and if the longer axis is used, a **prolate spheroid** results. The earth is approximately an oblate spheroid.

spheroidal excess. The amount by which the sum of the three angles of a triangle on the surface of a spheroid exceeds 180°.

spheroid of reference. See REFERENCE ELLIPSOID.

spin axis. The axis of rotation of a gyroscope.

spindle buoy. A buoy having a spindle-like shape floating nearly vertically. See also SPAR BUOY.

spire, *n.* A label on a nautical chart which indicates a pointed structure extending above a building. The spire is seldom less than two-thirds of the entire height of the structure, and its lines are rarely broken by stages or other features. The term is not applied to a short pyramid-shaped structure rising from a tower or belfry.

spirit compass. A magnetic compass of which the bowl mounting the compass card is filled with a solution of alcohol and water.

spit, *n.* A small tongue of land or a long narrow shoal (usually sand) extending from the shore into a body of water. Generally the tongue of land continues in a long narrow shoal for some distance from the shore.

Spitzbergen Atlantic Current. An ocean current flowing northward and westward from a point south of Spitzbergen, and gradually merging with the EAST GREENLAND CURRENT in the Greenland Sea. The Spitzbergen Atlantic Current is the continuation of the northwestern branch of the NORWAY CURRENT. Also called SPITZBERGEN CURRENT.

Spitzbergen Current. See SPITZBERGEN ATLANTIC CURRENT.

split fix. A fix by horizontal sextant angles obtained by measuring two angles between four objects, or suitable charted features, with no common center object being observed.

split-second timer. A watch with two sweep second hands which can be started and stopped together by means of one push button. The lower of the two sweep second hands can be stopped independently by means of a second push button, enabling an accurate reading. After the second button is pushed again, the previously stopped hand catches up with the continuously running sweep second hand. The elapsed time since the continuously running sweep second hand was started is indicated on a small dial.

spoil area. Area for the purpose of depositing dredged material, usually near and parallel to dredged channels. Spoil areas are usually a hazard to navigation and navigators of even the smallest craft should avoid crossing these areas. Spoil areas are shown on nautical charts. See also DISPOSAL AREA, DUMPING GROUND, DUMP SITE. Also called SPOIL GROUND.

spoil ground. See SPOIL AREA.

spoil ground buoy. A buoy which marks the limit of a spoil ground.

spoil ground mark. A navigation mark indicating an area used for deposition of waste material.

sporadic E-ionization. Ionization that appears at E-layer heights, is more noticeable toward the polar regions, and is caused by particle radiation from the sun. It may occur at any time of the day. A sporadic E-layer sometimes breaks away from the normal E-layer and exhibits special and erratic characteristics.

spot elevation. A point on a map or chart whose height above a specified datum is noted, usually by a dot or a small sawbuck and elevation value.

spot-size error. The distortion of the radar returns on the radarscope caused by the diameter of the electron beam which displays the returns on the scope and the lateral radiation across the scope of part of the glow produced when the electron beam strikes the phosphorescent coating of the cathode-ray tube. See also PULSE-DURATION ERROR.

spring, *n.* The season of the year in which plants begin to vegetate. In the Northern Hemisphere spring begins astronomically at the vernal equinox and ends at the summer solstice. In the Southern Hemisphere the limits are the autumnal equinox and the winter solstice. The meteorological limits vary with the locality and the year.

spring high water. See under SPRING TIDES.

spring low water. See under SPRING TIDES.

spring range. See under SPRING TIDES.

spring tidal currents. Tidal currents of increased speed occurring semimonthly as the result of the moon being new or full. See also SPRING TIDES.

spring tides. Tides of increased range occurring semimonthly as the result of the moon being new or full. The **spring range** of tide is the average semidiurnal range occurring at the time of spring tides and is most conveniently computed from the harmonic constants. It is larger than the mean range where the type of tide is either semidiurnal or mixed, and is of no practical significance where the type of tide is diurnal. The average height of the high waters of the spring tides is called **spring high water** or **mean high water springs** and the average height of the corresponding low waters is called **spring low water** or **mean low water springs**. See also SPRING TIDAL CURRENTS.

spur, *n.* On the sea floor, a subordinate elevation, ridge, or rise projecting outward from a larger feature.

spurious disk. The round image of perceptible diameter of a star as seen through a telescope, due to diffraction of light in the telescope.

spurious emission. Emission on a frequency of frequencies which are outside the necessary band, and the level of which may be reduced without affecting the corresponding transmission of information. Spurious emissions include harmonic emissions, parasitic emissions and intermodulation products, but exclude emissions in the immediate vicinity of the necessary band, which are a result of the modulation process for the transmission of information.

squall, *n.* A wind of considerable intensity caused by atmospheric instability. It comes up and dies down quickly, and is often accompanied by thunder, lightning, and precipitation, when it may be called a **thundersquall.** An **arched squall** is one relatively high in the center, tapering off on both sides. A **bull's eye squall** is one formed in fair weather, characteristic of the ocean off the coast of South Africa. See also GUST, LINE SQUALL, SQUALL LINE, WHITE SQUALL.

squall cloud. A small eddy cloud sometimes formed below the leading edge of a thunderstorm cloud, between the upward and downward currents.

squall line. Any non-frontal line or narrow band of active thunderstorms (with or without squalls); a mature instability line.

squally, *adj.* Having or threatening numerous squalls.

squamish, *n.* A strong and often violent wind occurring in many of the fiords of British Columbia. Squamishes occur in those fiords oriented in a northeast-southwest or east-west direction, where cold polar air can be funneled westward. They are notable in Jervis, Toba, and Bute inlets and in Dean Channel and Portland Canal. Squamishes lose their strength when free of the confining fiords and are not noticeable 15 to 20 miles offshore.

square, *n.* 1. A four-sided figure with all sides equal and all angles 90°; a rectangle or right-angled parallelogram with sides of equal length.
2. The second power of a quantity.

square metre. The *derived unit* of area in the International System of Units.

squat, *n.* For a vessel underway, the bodily sinkage and change of trim which are caused by the pressure distribution on the hull due to the relative motion of water and hull. The effect begins to increase significantly at depth-to-draft ratios less than 2.5. It increases rapidly with speed and is augmented in narrow channels.

stability, *n.* The state or property of resisting change or of tending to return to original conditions after being disturbed. The opposite is INSTABILITY.

stabilization of radarscope display. A radarscope display is said to be STABILIZED IN AZIMUTH when the orientation of the display is fixed to an unchanging reference (usually north). The NORTH UP orientation is an example. A radarscope display is said to be UNSTABILIZED IN AZIMUTH when the orientation of the display changes with changes in own ship's heading. The HEAD UP orientation is an example. A radarscope display is said to be DOUBLY STABILIZED or to have DOUBLE STABILIZATION when the basic orientation of the display is fixed to an unchanging reference (usually north) but the radarscope is rotated to keep own ship's heading or heading flasher *up* on the radarscope.

stabilized in azimuth. See under STABILIZATION OF RADARSCOPE DISPLAY.

stabilized platform. A gimbal-mounted platform, usually containing gyros and accelerometers, the purpose of which is to maintain a desired orientation in inertial space independent of craft motion. Also called STABLE PLATFORM.

stable platform. See STABILIZED PLATFORM.

stack, *n.* A label on a nautical chart which indicates a tall smokestack or chimney. The term is used when the stack is more prominent as a landmark than the accompanying buildings.

stadimeter, *n.* An instrument for determining the distance to an object of known height by measuring the angle, at the observer, subtended by the object. The instrument is graduated directly in distance. Called MARINE DISTANCE METER in British terminology. See also RANGE FINDER.

stand, *n.* The state of the tide at high or low water when there is no sensible change in the height of the tide. The water level is stationary at high and low water for only an instant, but the change in level near these times is so slow that it is not usually perceptible. In general, the duration of the apparent stand will depend upon the range of tide, being longer for a small range than for a large range, but where there is a tendency for a double tide the stand may last for several hours, even with a large range of tide. It may be called **high water stand** if it occurs at the time of high water, and **low water stand** if it occurs at low water. Sometimes called PLATFORM TIDE.

standard, *n.* 1. Something established by custom, agreement, or authority as a basis for comparison.
2. A physical embodiment of a unit. In general it is not independent of physical conditions, and it is a true embodiment of the unit only under specified conditions. For example, a yard standard has a length of 1 yard when at some definite temperature and supported in a certain manner. If supported in a different manner, it might have to be at a different temperature in order to have a length of 1 yard.

standard acceleration of gravity. The value adopted in the International Service of Weights and Measures for the standard acceleration due to gravity is 980.665 centimeters per second, per second. See also WEIGHT.

standard atmosphere. 1. A unit accepted *temporarily* for use with the International System of Units; 1 standard atmosphere is equal to 101,325

pascals.
2. A hypothetical vertical distribution of atmospheric temperature, pressure, and density which is taken to be representative of the atmosphere for various purposes.

standard chronometer. See CHRONOMETER.

standard circle sheet. See CIRCLE SHEET.

standard compass. A magnetic compass designated as the standard for a vessel. It is normally located in a favorable position with respect to magnetic influences.

standard deviation. A measure of the dispersion of random errors about the mean value. If a large number of measurements or observations of the same quantity are made, the standard deviation is the square root of the sum of the squares of deviations from the mean value divided by the number of observations less one. Symbolically, this operation is:

$$\sigma = \sqrt{\frac{\sum_{i=1}^{n}(x_i - a)^2}{n-1}}$$

where n is number of observations, a is mean value, and x is one of n measurements. The numerical significance of the standard deviation of a one-dimensional error distribution is that 68.27 percent of the measurements or observations do not deviate from the mean value by more than 1σ; 99.45 percent do not deviate by more than 2σ; 99.73 percent do not deviate by more than 3σ; and 50 percent do not deviate by more than 0.6745σ. For the two-dimensional or bivariate error distribution, standard deviation has a definable meaning only in the specific case of the circular normal distribution where $\sigma_x = \sigma_y$. In the circular case, 1σ indicates that 39.35 percent of the errors do not exceed the value of the 1σ error; 2σ indicates 86.47 percent do not exceed the 2σ error; 3σ indicates 98.89 percent do not exceed the 3σ error; and 3.5σ indicates 99.78 percent do not exceed the 3.5σ error. Because the usual two-dimensional error distribution is an elliptical distribution, it is common practice to describe two-dimensional distributions by two separate one-dimensional standard deviations associated with each error axis. When deviations in the measurements or observations of the same quantity are errors, the term STANDARD ERROR applies. However, in navigation the term STANDARD DEVIATION is commonly used when the deviations are errors. The square of the standard deviation, σ^2, is called the VARIANCE. Also called RMS ERROR. See also ROOT MEAN SQUARE ERROR.

standard error. See under STANDARD DEVIATION.

standard meridian. 1. The meridian used for reckoning standard time. Throughout most of the world the standard meridians are those whose longitudes are exactly divisible by 15°. The DAYLIGHT SAVING MERIDIAN is usually 15° east of the standard meridian.
2. A meridian of a map projection, along which the scale is as stated.

standard noon. Twelve o'clock standard time, or the instant the mean sun is over the upper branch of the standard meridian. DAYLIGHT SAVING or SUMMER NOON usually occur 1 hour later than standard noon.

standard parallel. 1. A parallel of latitude which is used as a control line in the computation of a map projection.
2. A parallel of latitude on a map or chart along which the scale is as stated for that map or chart.

standard propagation. The propagation of radio waves over a smooth spherical earth of uniform electrical characteristics, under conditions of standard refraction in the atmosphere.

standard radio atmosphere. An atmosphere having the standard refractive modulus gradient.

standard radio horizon. The radio horizon corresponding to propagation through the standard radio atmosphere.

standard refraction. The refraction which would occur in a standard atmosphere.

standard refractive modulus gradient. The uniform variation of refractive modulus with height above the earth's surface which is regarded as a standard for comparison. The gradient considered as normal has a value of 0.12M unit per metre. The *M unit* is the unit in terms of which the refractive modulus is expressed.

standard station. Use of this term is discouraged. See REFERENCE STATION.

standard tactical diameter. A *prescribed* tactical diameter used by different types of vessels, or by vessels of the same formation in maneuvers.

standard time. The United States and its possessions are, by law, divided into eight time zones. The limits of each time zone are defined by the Secretary of Transportation in Part 71, Title 49 of the *Code of Federal Regulations*. The standard time within each zone is the local mean time at the standard meridian that passes approximately through the center of the zone. Since the standard meridians are the same as those used with ZONE TIME, standard time conforms generally with the zone time for a given area. The standard time zone boundary may vary considerably from the zone time limits (7½° in longitude on each side of the standard meridian) to conform to political or geographic boundaries or both. The standard times used in various countries and places are tabulated in the *Air Almanac* and the *Nautical Almanac* and are displayed on Chart 76, *Standard Time Zone Chart of the World.*

standard type buoy. The general classification of all lighted and unlighted buoys in U.S. waters built to modern (1962) specifications.

standby lamp. A lamp brought into service in the event of failure of the lamp in regular service. It may often be brought into service automatically.

standby light. A light installed permanently, close to the light in regular service, but operating independently from the latter and intended to be brought easily into service in the event of failure of the latter. It may often be brought into service automatically.

standing floe. A separate floe standing vertically or inclined and enclosed by rather smooth ice.

standing wave. See STATIONARY WAVE.

stand on. To proceed on the same course.

standpipe, *n.* A label on a nautical chart which indicates a tall cylindrical structure, in a waterworks system, the height of which is several times the diameter.

star, *n.* A large celestial body, as contrasted with the much smaller planets, satellites, comets, etc. The expression FIXED STAR is sometimes applied to stars to distinguish them from bodies of the solar system. Stars are generally self-luminous, and, except for the sun, are at such great distances from the earth that they appear to the eye to be fixed in space. Comets, meteors, and nebulae may also be self-luminous. Two stars appearing close together are called a **double star**, an **optical double star** if they appear close because they are in nearly the same line of sight but differ greatly in distance from the observer, a physical double star if in nearly the same line of sight and at approximately the same distance from the observer. A system of two stars that revolve about their common center of mass is called a **binary star**. A group of three or more stars so close together that they appear as a single star is called a **multiple star**. A group of stars physically close together is called a **star cluster**. A **variable star** changes in magnitude. A star which suddenly becomes many times brighter than previously, and then gradually fades, is called a **nova**. The brightest planet appearing in the western sky during evening twilight is called **evening star**, and the brightest one appearing in the eastern sky during morning twilight is called **morning star**. A **shooting star** or **meteor** is a solid particle too small to be seen until it enters the earth's atmosphere, when it is heated to incandescence by friction of the air. See also GALAXY, MILKY WAY.

starboard, *n.* The right side of a craft, facing forward. The opposite is PORT.

starboard hand buoy. A buoy which is to be left to the starboard hand when approaching from the open sea or in general proceeding in the direction of the main stream of flood current, or in the direction established by the appropriate authority.

star chain. A radionavigation transmitting system comprised of a master station about which three (or more) slave (secondary) stations are more or less symmetrically located.

star chart. A representation, on a flat surface, of the celestial sphere or a part of it, showing the positions of the stars and sometimes other features of the celestial sphere.

star cloud. A large number of stars close together, forming a more congested part of a galaxy.

star cluster. A group of stars physically close together. See also MULTIPLE STAR.

star finder. A device to facilitate the identification of stars. Sometimes called a STAR IDENTIFIER. See also PLANISPHERE.

Star Finder and Identifier (No. 2102-D). A circular star finder and identifier formerly published by the U.S. Navy Hydrographic Office and later by the U.S. Naval Oceanographic Office. It consists essentially of a white opaque base with an azimuthal equidistant projection of most of the celestial sphere on each side, one side having the north celestial pole at the center and the other side having the south celestial pole at the center; and a series of transparent templates, at 10° intervals of latitude, each template having a family of altitude and azimuth curves.

star globe. A small globe representing the celestial sphere, on which the apparent positions of the stars are indicated. It is usually provided with graduated arcs and a suitable mount for determining the approximate altitude and azimuth of the stars, to serve as a star finder. Star globes are more commonly used by the British than by Americans. Also called CELESTIAL GLOBE.

star identifier. See STAR FINDER.

Star Sight Reduction and Identification Table. See under STAR SIGHT REDUCTION TABLES FOR 42 STARS.

Star Sight Reduction Tables for 42 Stars. A sight reduction table which provides for the reduction of 42 selected stars by the assumed altitude method. The main table, *Star Sight Reduction and Identification Table*, is entered with latitude and altitude to the nearest degree. For these entering arguments, the respondents are a tabulated local hour angle of Aries closest to the local hour angle of Aries at the time of the altitude observation, an associated azimuth of the observed celestial body the identification of which is usually not necessary (this azimuth should always be checked against a "seaman's eye" azimuth taken at the time of the observation), and identification of the celestial body by means of a letter code. The difference between the local hour angle of Aries at time of observation and the tabulated angle is used to adjust the longitude upon which the local hour angle of Aries at the time of observation is based (usually the DR longitude) to obtain the longitude from which at the time of observation the local hour angle of Aries would be the same as the tabulated angle. From the assumed position so found, the altitude intercept is plotted, the intercept being the difference between the tabulated altitude and the assumed integral degree of altitude nearest to the value of the observed altitude. Because of the unusual use of an assumed value of altitude in the sight reduction, this method is referred to as the **assumed altitude method**. Of the 42 stars included in the table as a whole, 21 are above the observer's horizon at any time and are so tabulated in each column for integral values of latitude and altitude. This large number of star tabulations is particularly useful when partial cloud obscuration makes identification difficult or obscures those stars for which observation is anticipated. Since the tabulations are for a given epoch, provision is made for precession and nutation corrections.

star telescope. An accessory of the marine navigational sextant designed primarily for star observations. It has a large object glass to give a greater field of view and increased illumination. It is an erect telescope, i.e., the object viewed is seen erect as opposed to the inverting telescope in which the object viewed is inverted. The latter type telescope requires one less lens than the erect telescope, consequently for the same size object glass, it has greater illumination. The erect telescope is frequently used for all observations.

state of the art. The level to which technology and science have at any designated time been developed in a given industry or group of industries.

static, *adj.* Having a fixed, nonvarying condition.

static, *n.* 1. Interference caused by natural electrical disturbances in the atmosphere, or the electromagnetic phenomena capable of causing such interference.

2. Noise heard in a radio receiver caused by electrical disturbances in the atmosphere, such as lightning, northern lights, etc.

station, *n.* 1. The authorized location of an aid to navigation.

2. One or more transmitters or receivers, or a combination of transmitters and receivers, including the accessory equipment necessary at one location, for carrying on a radiocommunication service.

stationary front. A front which is stationary or nearly so. Conventionally, a front which is moving at a speed less than about 5 knots is generally considered to be stationary. In synoptic chart analysis, a stationary front is one that has not moved appreciably from its position on the last previous synoptic chart (3 or 6 hours before). Also called QUASI-STATIONARY FRONT.

stationary orbit. An equatorial orbit in which the satellite revolves about the primary at the angular rate at which the primary rotates on its axis. From the primary, the satellite appears to be stationary over a point on the primary's equator.

See also GEOSTATIONARY SATELLITE.

stationary wave. A wave that oscillates without progressing. One-half of such a wave may be illustrated by the oscillation of the water in a pan that has been tilted. Near the axis, which is called the node or **nodal line**, there is no vertical rise and fall of the water. The ends of the wave are called loops and at these places the vertical rise and fall is at a maximum. The current is maximum near the node and minimum at the loops. The period of a stationary wave depends upon the length and depth of the body of water. A stationary wave may be resolved into two progressive waves of equal amplitude and equal speeds moving in opposite directions. Also called STANDING WAVE.

stationary wave theory. An assumption that the basic tidal movement in the open ocean consists of a system of stationary wave oscillations, any progressive wave movement being of secondary importance except as the tide advances into tributary waters. The continental masses divide the sea into irregular basins, which, although not completely enclosed, are capable of sustaining oscillations which are more or less independent. The tide-producing force consists principally of two parts, a semidiurnal force with a period approximating the half-day and a diurnal force with a period of a whole day. Insofar as the free period of oscillation of any part of the ocean, as determined by its dimensions and depth, is in accord with the semidiurnal or diurnal tide-producing forces, there will be built up corresponding oscillations of considerable amplitude which will be manifested in the rise and fall of the tide. The diurnal oscillations, superimposed upon the semidiurnal oscillations, cause the inequalities in the heights of the two high and the two low waters of each day. Although the tidal movement as a whole is somewhat complicated by the overlapping of oscillating areas, the theory is consistent with observational data.

station buoy. An unlighted buoy established in the vicinity of a lightship or an important lighted buoy as a reference point in case the lightship or buoy should be dragged off station. Also called WATCH BUOY.

station error. See DEFLECTION OF THE VERTICAL.

station pointer. *British terminology.* See THREE-ARM PROTRACTOR.

statistical error. See RANDOM ERROR.

statute mile. A unit of distance equal to 5,280 feet. This mile is generally used on land, and is sometimes called LAND MILE. It is commonly used to express navigational distances by navigators of river and lake vessels, particularly those navigating the Great Lakes of North America.

steady bearing. An approaching or closing craft is said to be on a steady bearing if the compass bearing does not appreciably change. If the bearing does not change appreciably, risk of collision exists. Also called CONSTANT BEARING, DECREASING RANGE (CBDR).

steam fog. Fog formed when water vapor is added to air which is much colder than the source of the vapor. It may be formed when very cold air drifts across relatively warm water. At temperatures below about −20° F, ice particles or droxtals may be formed in the air producing a type of ice fog known as frost smoke. See also ARCTIC SEA SMOKE, FROST SMOKE. Also called ARCTIC SMOKE, SEA MIST, STEAM MIST, WATER SMOKE, ARCTIC SEA SMOKE, FROST SMOKE.

steam mist. See STEAM FOG.

steep-to, *adj.* Precipitous. The term is applied particularly to a shore, bank, or shoal that descends steeply to a lower level.

steerageway, *n.* The condition wherein the ship has sufficient way on that it will respond to rudder movements to maintain desired course.

steering compass. A magnetic compass by which a craft is steered. The expression is sometimes used to refer to gyro repeater similarly used, although for this usage STEERING REPEATER is preferable.

steering repeater. A compass repeater by which a craft is steered. Sometimes loosely called a STEERING COMPASS.

stellar, *adj.* Of or pertaining to stars.

stellar observation. See CELESTIAL OBSERVATION.

stellar parallax. See HELIOCENTRIC PARALLAX.

stem, *v., t.* To make headway against an obstacle, as a current.

steradian, *n.* The *supplementary unit* of solid angle in the International System of Units; it is the solid angle which, having its vertex in the center of a sphere, cuts off an area on the surface of the sphere equal to that of a square with sides of length equal to the radius of the sphere.

stereographic, *adj.* Of or pertaining to stereography, the art of representing the forms of solid bodies on a plane.

stereographic chart. A chart on the stereographic map projection.

stereographic map projection. A perspective, conformal, azimuthal map projection in which points on the surface of a sphere or spheroid, such as the earth, are conceived as projected by radial lines from any point on the surface to a plane tangent to the antipode of the point of projection. Circles project as circles except for great circles through the point of tangency, which great circles project as straight lines. The principal navigational use of the projection is for charts of the polar regions. Also called AZIMUTHAL ORTHOMORPHIC MAP PROJECTION.

sternboard, *n.* Making way through the water in a direction opposite to the heading. Also called STERNWAY, though the term STERNBOARD is sometimes used to refer to the beginning of motion astern and STERNWAY is used as the vessel picks up speed. Motion in the forward direction is called HEADWAY.

sternway, *n* Making way through the water in a direction opposite to the heading. Motion in the forward direction is called HEADWAY. See also STERNBOARD.

stilling well. See FLOAT WELL.

still water level. The level that the sea surface would assume in the absence of wind waves; not to be confused with MEAN SEA LEVEL or HALF TIDE LEVEL.

stippling, *n.* Graduation of shading by numerous separate touches. Shallow areas on charts, for instance, are sometimes indicated by numerous dots decreasing in density as the depth increases.

stones, *n., pl.* A general term for rock fragments ranging in size from 2 to 256 millimeters. An individual water-rounded stone is called a **cobble** if between 64 to 256 millimeters (size of clenched fist to size of man's head), a **pebble** if between 4 and 64 millimeters (size of small pea to size of clenched fist), and **gravel** if between 2 and 4 millimeters (thickness of standard pencil lead to size of small pea). An aggregate of stones ranging from 16 to 256 millimeters is called **shingle.** See also MUD; SAND; ROCK, definition 2.

stooping, *n.* Apparent decrease in the vertical dimension of an object near the horizon, due to large inequality of atmospheric refraction in the line of sight to the top and bottom of the object. The opposite is TOWERING.

stop watch. A watch that can be started, stopped, and reset at will, to indicate elapsed time.

storm, *n.* 1. Wind of force 10 (48 to 55 knots or 55 to 63 miles per hour) on the Beaufort wind scale. See also VIOLENT STORM.
2. Any disturbed state of the atmosphere, especially as affecting the earth's surface, and strongly implying destructive or otherwise unpleasant weather. In synoptic meteorology, a storm is a complete individual disturbance identified on synoptic charts as a complex of pressure, wind, clouds, precipitation, etc., or identified by such means as radar. Thus, storms range in scale from **tornadoes** and **thunderstorms**, through **tropical cyclones**, to widespread **extratropical cyclones**. From a local and special interest viewpoint, a storm is a transient occurrence identified by its most destructive or spectacular aspect. Examples are **rain storms, wind storms, hail storms, snow storms,** etc. Notable special cases are **blizzards, ice storms, sandstorms,** and **duststorms.**
3. A term once used by seamen for what is now called VIOLENT STORM on the Beaufort wind scale.

storm center. The area of lowest atmospheric pressure of a cyclone. This is a more general expression than EYE OF THE STORM, which refers only to the center of a well-developed tropical cyclone, in which there is a tendency of the skies to clear.

storm surge. A departure from a normal elevation of the sea due to the piling up of water against (or withdrawal from) a coast by strong winds such as those accompanying a hurricane or other intense storm. Reduced atmospheric pressure often contributes to the departure in height during hurricanes. It is potentially catastrophic, especially in deltaic regions with onshore winds at the time of high water and extreme wind wave heights. Also called STORM TIDE, STORM WAVE, TIDAL WAVE.

storm tide. See STORM SURGE.

storm track. The horizontal component of the path followed or expected to be followed by a storm center.

storm wave. See STORM SURGE.

straight angle. An angle of 180°.

strait, *n.* A relatively narrow waterway, usually narrower and less extensive than a sound, connecting two larger bodies of water.

strand, *n.* See BEACH.

strand, *v., t. & i.* To run aground. The term STRAND usually refers to a serious grounding, while the term GROUND refers to any grounding, however slight.

stranded ice. Ice which has been floating and has been deposited on the shore by retreating high water.

stranding, *n.* The grounding of a vessel so that it is not soon refloated; a serious grounding.

strapped-down inertial navigation equipment. Inertial navigation equipment in which a stable platform and gimbal system are not utilized. The inertial devices are attached or strapped directly to the carrier. A computer utilizing gyro information resolves accelerations sensed along the carrier axes and refers these accelerations to an inertial frame of reference. Also called GIMBALLESS INERTIAL NAVIGATION EQUIPMENT. See also INERTIAL NAVIGATION.

stratiform, *adj.* Descriptive of clouds of extensive horizontal development, as contrasted to the vertically developed CUMULIFORM types. See also CIRRIFORM.

stratocumulus, *n.* A principal cloud type (cloud genus), predominantly stratiform, in the form of a gray and/or whitish layer or patch, which nearly always has dark parts and is non-fibrous (except for virga). Its elements are tesselated, rounded, roll-shaped, etc.; they may or may not be merged, and usually are arranged in orderly groups, lines or undulations, giving the appearance of a simple (or occasionally a cross-pattern) wave system. These elements are generally flat-topped, smooth and large; observed at an angle of more than 30° above the horizon, the individual stratocumulus element subtends an angle of greater than 5°. When a layer is continuous, the elemental structure is revealed in true relief on its under surface. Stratocumulus is composed of small water droplets, sometimes accompanied by larger droplets, soft hail, and (rarely) by snowflakes. When the cloud is not very thick, the diffraction phenomena *corona* and *irisation* appear. Under ordinary conditions, ice crystals are too sparse even to give the cloud a fibrous aspect; however, in extremely cold weather, ice crystals may be numerous enough to produce abundant virga, and sometimes even halo phenomena. **Mamma** may be a supplementary feature of stratocumulus, in which case the mammiform proturberances may develop to the point where they seem about to detach themselves from the main cloud. **Virga** may form under the cloud, particularly at very low temperatures. Precipitation rarely occurs with stratocumulus. Stratocumulus frequently forms in clear air. It may also form from the rising of stratus, and by the convective or undulatory transformation of stratus, or nimbostratus, with or without change of height. Stratocumulus is analogous to altocumulus, and forms directly from the latter when the elements grow to a sufficient size. The further humidification, accompanied by turbulence and/or convection, of an already humid layer of air near the base of nimbostratus or even altostratus can form stratocumulus. If the ascending currents which produce cumulus or cumulonimbus approach an upper layer of stable air, they slow down, and all or a portion of the mother-cloud tends to diverge gradually and spread horizontally, often producing stratocumulus. A particular form of stratocumulus often occurs in the evening when convection decreases, resulting in the gradual dissipation of both bases and tops of cumuliform clouds. Since stratocumulus may be transformed directly from or into altocumulus, stratus, and nimbostratus, all transitional stages may be observed. By convention, altocumulus is composed of apparently smaller elements (often simply because of its higher altitude); stratus and nimbostratus do not show regular subdivisions or wave form, and they have a more fibrous aspect. When the base of stratocumulus is rendered diffuse by precipitation, the cloud becomes nimbostratus. See also STRATIFORM, CLOUD CLASSIFICATION.

stratosphere, *n.* The atmospheric shell extending upward from the tropopause to the height where the temperature begins to increase in the 20-to-25-kilometer region.

stratus, *n.* A low cloud (mean upper level below 6,500 ft.) in a uniform layer, resembling fog but not resting on the surface.

stray line. Ungraduated portion of line connected with the current pole used in taking current observations. The stray line is usually about 100 feet long and permits the pole to acquire the velocity of the current at some distance from the disturbed waters in the immediate vicinity of the observing vessel before the current velocity is read from the graduated portion of the current line.

stream, *v., t.* To place overboard and secure, as to *stream a log* or *stream a sea anchor.*

stream current. A relatively narrow, deep, fast-moving ocean current. The opposite is DRIFT CURRENT.

streamline, *n.* The path followed by a particle of fluid flowing past an obstruction. The term generally excludes the path of a particle in an eddy current.

streamline flow. Fluid motion in which the fluid moves uniformly without eddies or turbulence. If it moves in thin layers, it is called **laminar flow.** The opposite is TURBULENT FLOW.

stream the log. To throw the log overboard and secure it in place for taking readings.

strength of current. Phase of tidal current in which the speed is a maximum; also the speed at this time. Beginning with slack before flood in the period of a reversing tidal current (or minimum before flood in a rotary current), the speed gradually increases to flood strength and then disminishes to slack before ebb (or minimum before ebb in a rotary current), after which the current turns in direction, the speed increases to ebb strength and then diminishes to slack before flood completing the cycle. If it is assumed that the speed throughout the cycle varies as the ordinates of a cosine curve, it can be shown that the average speed for an entire flood or ebb period is equal to $2/\pi$ or 0.6366 of the speed of the corresponding strength of current.

strength of ebb. See EBB STRENGTH.

strength of ebb interval. See EBB INTERVAL. See also LUNICURRENT INTERVAL.

strength of flood. See FLOOD STRENGTH.

strength of flood interval. See FLOOD INTERVAL. See also LUNICURRENT INTERVAL.

strip, *n.* A long narrow area of pack ice, about 1 kilometer or less in width, usually composed of small fragments detached from the main mass of ice, and run together under the influence of wind, swell, or current.

stripes, *n.* A number of areas of contrasting color separated from one another by straight lines, and used as a distinguishing characteristic for navigation marks.

stroboscopic direction finder. A radio direction finder having a continuously rotating antenna connected with neon tubes in such a way that a calibrated disk is illuminated at the reading corresponding to the bearing of the transmitter.

strong breeze. Wind of force 6 (22 to 27 knots or 25 to 31 miles per hour) on the Beaufort wind scale.

strong fix. A fix determined from horizontal sextant angles between objects so situated as to give very accurate results.

strong gale. Wind of force 9 (41 to 47 knots or 47 to 54 miles per hour) on the Beaufort wind scale. See also GALE.

sub-. A prefix meaning under, less, or marginal, as *submarine* or *subtropical.* The opposite is SUPER-.

Subarctic Current. See ALEUTIAN CURRENT.

subastral point. See SUBSTELLAR POINT.

sublimation, *n.* The transition of a substance directly from the solid state to the vapor state, or vice versa, without passing through the intermediate liquid state. See also CONDENSATION, EVAPORATION, FUSION.

sublunar point. The geographical position of the moon; that point on the earth at which the moon is in the zenith at a specified time.

submarine bell. See under BELL.

submarine cable. An insulated, water-proofed wire or bundle of wires for carrying an electric current under water. Such a cable is placed on or near the bottom.

submarine havens. Specified sea areas for submarine noncombat operations including: **a. submarine sanctuaries** announced by the area, fleet, or equivalent commander; **b.** areas reserved for submarine operations and training in noncombat zones, and **c.** moving areas established by Submarine Notices, surrounding submarines in transit, extending 50 nautical miles ahead, 100 nautical miles astern, and 15 nautical miles on each side of the estimated position of the submarine along the stated track. See also MOVING HAVENS.

submarine relief. Variations in elevation of the sea bed, or their representation by depth contours, hypsometric tints, or soundings.

submarine safety lanes. See SAFETY LANES.

submarine sanctuaries. Restricted areas which are established for the conduct of noncombat submarine or antisubmarine exercises. They may be either stationary or moving and are normally designated only in rear areas. See also MOVING HAVENS.

submarine site. The site of a structure when located below the surface of the water.

submerge, *v., i.* To descend below the surface. The opposite is SURFACE. See also DIVE.

submerged, *adj. & adv.* 1. Under water. The opposite is UNCOVERED. See also AWASH.
2. Having descended below the surface. The opposite is SURFACED.

submerged breakwater. A breakwater with its top below the still water level. When this structure is struck by a wave, part of the wave energy is reflected seaward. The remaining energy is largely dissipated in a breaker, transmitted shoreward as a multiple crest system, or transmitted shoreward as a simple wave system.

submerged lands. Lands covered by water at any stage of the tide, as distinguished from tidelands which are attached to the mainland or an island and cover and uncover with the tide. Tidelands presuppose a high-water line as the upper boundary; submerged lands do not.

submerged production well. An oil or gas well that is a seabed installation only, i.e., the installation does not include a permanent production platform. See also WELLHEAD.

submerged rock. A rock covered at the chart sounding datum and considered to be potentially dangerous to navigation. See also BARE ROCK, ROCK AWASH.

submerged screw log. A type of electric log which is actuated by the flow of water past a propeller.

subordinate current station. 1. A current station from which a relatively short series of observations is reduced by comparison with simultaneous observations from a control current station.
2. A station listed in the *Tidal Current Tables* for which predictions are to be obtained by means of differences and ratios applied to the full predictions at a reference station. See also CURRENT STATION, CONTROL CURRENT STATION, REFERENCE STATION.

subordinate tide station. 1. A tide station from

which a relatively short series of observations is reduced by comparison with simultaneous observations from a tide station with a relatively long series of observations.
2. A station listed in the *Tide Tables* for which predictions are to be obtained by means of differences and ratios applied to the full predictions at a reference station. See also PRIMARY CONTROL TIDE STATION, REFERENCE STATION, SECONDARY CONTROL TIDE STATION, TERTIARY TIDE STATION.

subpermanent magnetism. The magnetism in the intermediate iron of a ship which tends to change as a result of vibration, aging, or cruising in the same direction for a long period, but does not alter immediately so as to be properly termed induced magnetism. This magnetism is the principal cause of deviation changes of a magnetic compass. At any instant this magnetism is recognized as part of the ship's permanent magnetism, and consequently must be corrected as such by means of permanent magnet correctors. See also MAGNETISM.

sub-refraction, *n.* Less-than-normal refraction, particularly as related to the atmosphere. Greater-than-normal refraction is called SUPER-REFRACTION.

subsatellite point. The point at which a line from the satellite perpendicular to the ellipsoid intersects the surface of the earth.

subsidence, *n.* Decrease in the elevation of land without removal of surface material due to tectonic, seismic, or artificial forces.

subsidiary light. A light placed on or near the support of a main light and having a special use in navigation. See also PASSING LIGHT.

subsolar point. The geographical position of the sun; that point on the earth at which the sun is in the zenith at a specified time.

substellar point. The geographical position of a star; that point on the earth at which the star is in the zenith at a specified time. Also called SUBASTRAL POINT.

substratosphere, *n.* A region of indefinite lower limit just below the stratosphere.

subsurface current. An underwater current which is not present at the surface. See also SURFACE CURRENT, UNDERCURRENT, UNDERTOW.

subtend, *v., t.* To be opposite, as an arc of a circle *subtends* an angle at the center of the circle, the angle being formed by the radii joining the ends of the arc with the center.

subtrack, *n.* See ORBITAL PATH.

subtropical anticyclones. High pressure belts which prevail on the poleward sides of the trade winds, characterized by calms, light breezes, and dryness.

sudden ionospheric disturbances. Sudden increases in the ionization density in the lower part of the ionosphere caused by very sudden and large increases in X-ray flux emitted from the sun, usually during a solar flare. These disturbances (SIDs) also occur during flares called X-ray flares that produce large X-ray flux, but which have no components in the visible light spectrum. The effect, which is restricted to sunlit propagation paths, causes a phase advance and is known as a SUDDEN PHASE ANOMALY (SPA). The SID effects are related to solar zenith angle, and consequently, occur mostly in lower latitude regions. Usually there is a phase advance over a period of 5 to 10 minutes followed by a recovery over a period of 30 to 60 minutes. See also POLAR CAP DISTURBANCE, MODAL INTERFERENCE.

sudden phase anomaly. See under SUDDEN IONOSPHERIC DISTURBANCES.

Suestado, *n.* A storm with southeast gales, caused by intense cyclonic activity off the coasts of Argentina and Uruguay, which affects the southern part of the coast of Brazil in the winter.

sugarloaf sea. A sea characterized by waves that rise into sugarloaf (conical) shapes, with little wind, possibly resulting from intersecting waves.

sugg, *v., i.* To roll with the action of the sea when aground.

sumatra, *n.* A squall with violent thunder, lightning, and rain, which blows at night in the Malacca Straits, especially during the southwest monsoon. It is intensified by strong mountain breezes.

Summary of Corrections. A five-volume semi-annual summary of corrections to charts, *Sailing Directions*, and *United States Coast Pilots* previously published in *Notice to Mariners*. The weekly *Notice to Mariners* should be referred to for confirmation of the corrective material in this summary published by the Defense Mapping Agency Hydrographic/Topographic Center.

summer, *n.* The warmest season of the year. In the Northern Hemisphere summer begins astronomically at the summer solstice and ends at the autumnal equinox. In the Southern Hemisphere the limits are the winter solstice and the vernal equinox. The meteorological limits vary with the locality and the year. See also INDIAN SUMMER.

summer noon. Daylight saving noon. The expression applies where summer time is used, particularly in Europe.

summer solstice. 1. That point on the ecliptic occupied by the sun at maximum northerly declination. Sometimes called JUNE SOLSTICE, FIRST POINT OF CANCER.
2. That instant at which the sun reaches the point of maximum northerly declination, about June 21.

summer time. A variation of standard time in which the clocks are advanced 1 hour. The variation when the clocks are advanced 2 hours is called **double summer time.** The expression is used principally in Europe. See also DAYLIGHT SAVING TIME.

Sumner line. A line of position established by the Sumner method or, loosely, any celestial line of position.

Sumner method. The establishing of a line of position from the observation of the altitude of a celestial body by assuming two latitudes (or longitudes) and calculating the longitudes (or latitudes) through which the line of position passes. The line of position is the straight line connecting these two points (extended if necessary). This method, discovered by Thomas H. Sumner, an American sea captain, is seldom used by modern navigators, an adaptation of it, called ST. HILAIRE METHOD, being favored. See also LONGITUDE METHOD, HIGH ALTITUDE METHOD.

Sumner point. See COMPUTED POINT.

sun, *n.* The luminous celestial body at the center of the solar system, around which the planets, asteroids, and comets revolve. It is an average star. The sun visible in the sky is called **apparent** or **true sun.** A fictitious sun conceived to move eastward along the celestial equator at a rate that provides a uniform measure of time equal to the average apparent time is called **mean sun** or **astronomical mean sun**; a fictitious sun conceived to move eastward along the ecliptic at the average rate of the apparent sun is called **dynamical mean sun.** When the sun is observable at midnight, in high latitudes, it is called **midnight sun**.

sun cross. A rare halo phenomenon in which horizontal and vertical shafts of light intersect at the sun. It is probably due to the simultaneous occurrence of a sun pillar and a parhelic circle.

sun dog. See PARHELION.

sun line, *n.* A line of position determined from a sextant observation of the sun.

sun pillar. A glittering shaft of light, white or reddish, extending above and below the sun, most frequently observed at sunrise or sunset. If a parhelic circle is observed at the same time, a SUN CROSS results. See also HALO.

sun relay. See DAYLIGHT CONTROL.

sunrise, *n.* The crossing of the visible horizon by the upper limb of the rising sun.

sunset, *n.* The crossing of the visible horizon by the upper limb of the setting sun.

sunspot, *n.* One of the dark spots on the sun's surface. These spots are apparently magnetic in character and exert a disturbing influence on radio propagation on the earth.

sun's way. The path of the solar system through space.

sun switch. See DAYLIGHT CONTROL.

sun valve. A device, operated by daylight, that automatically lights and extinguishes a gas light, usually lighting it at or about sunset and extinguishing it at or about sunrise. Also called LIGHT VALVE. See also DAYLIGHT CONTROL.

super-. A prefix meaning over, more, greater, etc., as *supersonic*. The opposite is SUB-.

super-buoy. A very large buoy, generally more than 5 meters in diameter. Its large size renders a super-buoy a potential hazard even to large vessels. The three principal types of super-buoy are: **large navigational buoy, offshore tanker loading/discharge buoy** (or **single point mooring**), and the **oceanographic data acquisition systems (ODAS) buoy.**

superheterodyne receiver. A receiver in which the incoming radio frequency signals are normally amplified before being fed into a mixer (first detector) for conversion into a fixed, lower carrier (the intermediate frequency). The *intermediate frequency signals* undergo very high amplification in the intermediate frequency amplifier stages and are then fed into a detector (second detector) for demodulation. The resulting audio or video signals are then usually further amplified before use.

super high frequency. Radio frequency of 3,000 to 30,000 megahertz.

superior conjunction. The conjunction of an inferior planet and the sun when the sun is between the earth and the other planet.

superior planets. The planets with orbits larger than that of the Earth: Mars, Jupiter, Saturn, Uranus, Neptune, and Pluto. See also PLANET.

superior transit. See UPPER TRANSIT.

super-refraction, *n.* Greater-than-normal refraction, particularly as related to the atmosphere. Less-than-normal refraction is called SUB-REFRACTION.

supersaturation, *n.* More than saturation. As an example, if saturated air is cooled, condensation takes place only if nuclei are present. If they are not present, the air continues to hold more water than required for saturation until the temperature is increased or until a nucleus is introduced.

supersonic, *adj.* Faster than sound. Formerly, this term was also applied to a frequency above the audible range, but in this usage it has been replaced by the term ULTRASONIC.

superstructure, *n.* See CAGE.

supplement, *n.* An angle equal to 180° minus a given angle. Thus 110° is the supplement of 70° and the two are said to be *supplementary*. See also COMPLEMENT, EXPLEMENT.

supplementary angles. Two angles whose sum is 180°.

supplementary units. See under INTERNATIONAL SYSTEM OF UNITS.

surf, *n.* The water in the area between a shoreline and the outermost limit of breakers. During a storm the line of demarcation between breakers and whitecaps in the deep water beyond may be difficult to distinguish. The term is also used when referring to breakers on a detached reef.

surface, *v., i.* To rise to the surface. The opposite is SUBMERGE.

surface boundary layer. That thin layer of air adjacent to the earth's surface extending up to a level of about 10 to 100 meters. Within this layer the wind distribution is determined largely by the vertical temperature gradient and the nature and contours of the underlying surface; shearing stresses are approximately constant. Also called FRICTION LAYER.

surface chart. Short for SYNOPTIC SURFACE CHART.

surface current. A current which does not extend more than about 3 meters below the surface. See also SUBSURFACE CURRENT, UNDERCURRENT, UNDERTOW.

surfaced, *adj. & adv.* Having come to the surface from below the water. The opposite is SUBMERGED. See also AFLOAT, UNCOVERED.

surface duct. A tropospheric radio duct in which the lower boundary is the surface of the earth. Also called GROUND-BASED DUCT.

surface front. See under FRONT.

surface of position. A surface on some point of which a craft is located. See also LINE OF POSITION, FIX.

surface wave. A radio wave which is propagated along the boundary between two media in a manner determined by the properties of the two media in the vicinity of the boundary.

surf zone. The area between the outermost limit of breakers and the limit of wave uprush.

surge, *n.* 1. Of the motions of a vessel in a seaway, the bodily motion forward and backward along the longitudinal axis, caused by the force of the sea acting alternately on the bow and stern. Also called SURGING. See also SHIP MOTIONS.

2. See as STORM SURGE.

surging, *n.* See SURGE, *n.*, definition 1.

surveillance, *n.* The observation of an area or space for the purpose of determining the position and movements of craft or vehicles in that area or space. Surveillance can be either *dependent, independent*, or *pseudo-independent*.

surveillance radar. A primary radar installation at a land station used to display at that station the position of vessels within its range, usually for radar advisory purposes.

survey, *n.* 1. The act or operation of making measurements for determining the relative positions of points on, above, or beneath the earth's surface.

2. The results of operations as in definition 1.

3. An organization for making surveys.

See also GEODETIC SURVEY, HYDROGRAPHIC SURVEY, OCEANOGRAPHIC SURVEY, TOPOGRAPHIC SURVEY.

surveying, *n.* That branch of applied mathematics which teaches the art of determining accurately the area of any part of the earth's surface, the lengths and directions of the bounding lines, the contour of the surface, etc., and accurately delineating the whole on a map or chart for a specified datum.

surveying sextant. See HYDROGRAPHIC SEXTANT.

swamp, *n.* An area of spongy land saturated with water. It may have a shallow covering of water, usually with a considerable amount of vegetation appearing above the surface. Sometimes called SLOUGH.

swash, *n.* 1. A narrow channel or sound within a sand bank, or between a sand bank and the shore. 2. A bar over which the sea washes.
3. The rush of water up onto the beach following the breaking of a wave.

sway, *n.* Of the motions of a vessel in a seaway, the side-to-side bodily motion, independent of rolling caused by uniform pressure being exerted all along one side of the hull. Also called LATERAL DRIFTING, SWAYING. See also SHIP MOTIONS.

swaying, *n.* See SWAY.

sweep, *v., t.* To tow a line or object below the surface, to determine the least depth in an area or to insure that a given area is free from navigational dangers to a certain depth; or the removal of such dangers. See also DRAG, *v., t.*

sweeping, *n.* 1. The process of towing a line or object below the surface, to determine whether an area is free from isolated submerged dangers to vessels and to determine the position of any dangers that exist, or to determine the least depth of an area.
2. The process of clearing an area or channel of mines or other dangers to navigation.

sweep (of radarscope), *n.* As determined by the time base or range calibration, the radial movement of the stream of electrons impinging on the face of the cathode-ray tube. The origin of the sweep is usually the center of the face of the cathode-ray tube or plan position indicator (PPI). Because of the very high speed of movement of the point of impingement, the successive points of impingement appear as a continuously luminous line. The line rotates in synchronism with the radar antenna. If an echo is received during the time of radial travel of the electron stream from the center to the outer edge of the face of the tube, the sweep will be increased in brightness (intensity-modulated) at the point of travel of the electron stream corresponding to the range of the contact from which the echo is received. Since the sweep rotates in synchronism with the radar antenna, this increased brightness will occur on the bearing from which the echo is received. With this increased brightness and the persistence of the tube face, *paint* corresponding to the object being "illuminated" by the radar beam appears on the PPI.

sweep rate. The number of times a radar radiation pattern rotates during 1 minute of time. Sometimes expressed as the duration of one complete rotation in seconds of time.

swell, *n.* A relatively long wind wave, or series of waves, that has traveled out of the generating area. In contrast the term SEA is applied to the waves while still in the generating area. As these waves travel away from the area in which they are formed, the shorter ones die out. The surviving waves exhibit a more regular and longer period with flatter crests. When these waves reach shoal water, they become more prominent in height and of decreased wave length and are then known as ground swell.

swell direction. The direction *from* which swell is moving.

swept-frequency racon. An in-band racon which sweeps through the marine radar band (2920-3100 MHz in the 10-centimeter band and 9220-9500 MHz in the 3-centimeter band) in order that it may be triggered at the frequency of the interrogating radar transmitting at a given frequency within the band. Almost all such racons operate in the 3-centimeter band only. There are two types of swept-frequency racons: the slow-sweep racon sweeps through the 180 MHz frequency band in 10s of seconds (1.5 to 3.0 MHz per second); the fast-sweep racon sweeps through the band in microseconds. When interrogating slow-sweep racons, the interval between racon flashes may be as long as 2 minutes due to the time required for sweeping through the band. Since the fast-sweep racon sweeps through the band in about 12 microseconds, the receiver on the interrogating ship passes the response for one frequency excursion for only 1 microsecond or less. On the PPI display this response appears as a single dot. However several frequency excursions are made during each interrogation so the racon flash appears as a radial line of dots.

swept gain. See SENSITIVITY TIME CONTROL.

swinger, *n.* See REVOLVER.

swinging buoy. A buoy placed at a favorable location to assist a vessel to adjust its compass or swing ship. The bow of the vessel is made fast to one such buoy and the vessel is swung by means of lines to a tug or to additional buoys. Also called COMPASS ADJUSTMENT BUOY.

swinging ship. The process of placing a vessel on various headings and comparing magnetic compass readings with the corresponding magnetic directions, to determine deviation. This usually follows compass adjustment or compass compensation, and is done to obtain information for making a deviation table.

swinging the arc. The process of rotating a sextant about the line of sight to the horizon to determine the foot of the vertical circle through a body being observed. Also called ROCKING THE SEXTANT.

swirl error. The additional error in the reading of a magnetic compass during a turn, due to friction in the compass liquid.

symmetrical, *adj.* Being equal or identical on each side of a center line, mid value, etc. The opposite is ASYMMETRICAL.

synchronism, *n.* The relationship between two or more periodic quantities of the same frequency when the phase difference between them is zero or constant at a predetermined value.

synchronization error. In radionavigation, the error due to imperfect timing of two operations.

synchronize, *v., t.* To bring into synchronism.

synchronous, *adj.* Coincident in time, phase, rate, etc.

synchronous lights. Two or more lights the characteristics of which are in synchronism.

synchronous satellite. A satellite whose period of rotation is equal to the period of rotation of the primary about its axis. The orbit of a synchronous satellite must be equatorial if the satellite is to remain fixed over a point on the primary's equator. See also GEOSYNCHRONOUS SATELLITE, GEOSTATIONARY SATELLITE.

synodical month. The average period of revolution of the moon about the earth with respect to the sun, a period of 29 days, 12 hours, 44 minutes, 2.8 seconds, or about 29½ days. This is sometimes called the MONTH OF THE PHASES, since it extends from new moon to the next new moon. Also called LUNATION.

synodical period. See SYNODIC PERIOD.

synodic period. The interval of time between any planetary configuration of a celestial body, with respect to the sun, and the next successive same configuration of that body, as from inferior conjunction to inferior conjunction. Also called SYNODICAL PERIOD.

synoptic chart. In meteorology, any chart or map on which data and analyses are presented that describe the state of the atmosphere over a large area at a given moment of time. A **synoptic surface chart** is an analyzed synoptic chart of surface weather observations. Essentially, a surface chart shows the distribution of sea-level pressure (therefore, the positions of highs, lows, ridges, and troughs) and the location and nature of fronts and air masses. Often added to this are symbols of occurring weather phenomena, analysis of pressure tendency (isallobars), indications of the movement of pressure systems and fronts, and possibly others depending upon the intended use of the chart. Although the pressure is referred to mean sea level, all other elements on this chart are presented as they occur at the surface point of observation. A chart of this general form is the one commonly referred to as the WEATHER MAP.

synoptic surface chart. See under SYNOPTIC CHART.

system accuracy. The expected accuracy of a navigation system expressed in d_{rms} units, not including errors which may be introduced by the user, or geodetic or cartographic errors.

systematic error. One of the two categories of errors of observation, measurement and calculation, the other category being *random error*. Systematic errors are characterized by the property that there is a nonrandom trend to the error, e.g., remaining constant, changing constantly with time, changing periodically with time, or changing according to some law with changes of temperature, pressure, refraction or some other physical condition. Systematic errors are usually predictable once the cause is known. They are divided into three classes: (1) errors resulting from changing or nonstandard natural physical conditions, sometimes called **theoretical errors**, (2) **personal** (nonaccidental) **errors**, and (3) **instrument** errors. An example of a *theoretical error* is the expansion of a measuring tape with an increase of temperature. Once recognized, this type of error can usually be predicted by means of some mathematical formula and its effect removed from the observation. *Personal errors* are due to the idiosyncrasies of the observer, such as always reading the time of a celestial observation either too early or too late, or consistently observing the altitude of a celestial body either too high or too low. This type of personal error should be distinguished from the random error that an observer makes when attempting to read an instrument to the highest possible precision. Personal error may be determined by a study of the observer. This type of error is sometimes called the observer's PERSONAL EQUATION. Once determined, the personal equation can then be applied to future observations in order to correct this source of systematic error. Examples of *instrument error* in a sextant are index error, graduation errors in the arc or micrometer drum, and the centering error of the index arm. This type of error can usually be determined by a study of the instrument. The instrument error may be a constant one (index error or centering error) or may vary with the angle being measured (graduation error). Systematic errors can be troublesome because they may occur in observations but be unknown to the observer. Systematic errors may also be introduced by using a mathematical model or equation which only approximates the true value of a quantity. This is often done purposefully when the correct mathematical equations are quite complex. The simplified equations are used to yield approximate but useful values of the desired quantity. Also called REGULAR ERROR. See also ERROR.

syzygy, *n.* 1. A point of the orbit of a planet or satellite at which it is in conjunction or opposition. The term is used chiefly in connection with the moon at new and full phase.

2. A west wind on the seas between New Guinea and Australia preceding the summer northwest monsoon.

T

able, *n.* An orderly, condensed arrangement of numerical or other information, usually in parallel vertical columns. An auxiliary table used for interpolating is called **interpolation table.** A table in which values of the quantity to be found are tabulated for limiting values of the entering argument is called **critical table.** See also CALIBRATION TABLE, CONVERSION TABLE, CURRENT TABLES, TIDE TABLES, TRAVERSE TABLE.

ablemount, *n.* A seamount having a comparatively smooth, flat top. Also called GUYOT.

Tables of Computed Altitude and Azimuth. See H.O. PUB. NO. 214.

abular altitude. See TABULATED ALTITUDE.

abular azimuth. See TABULATED AZIMUTH.

abular azimuth angle. See TABULATED AZIMUTH ANGLE.

tabular iceberg. A flat-topped iceberg with length-to-height ratio greater than 5:1. Most tabular bergs form by calving from an ice shelf and show horizontal banding. See also ICE ISLAND, BLOCKY ICEBERG.

tabulated altitude. In navigational sight reduction tables, the altitude taken directly from a table for the entering arguments. After interpolation for argument increments, i.e., the difference between each entering argument and the actual value, it is called COMPUTED ALTITUDE. Also called TABULAR ALTITUDE.

tabulated azimuth. Azimuth taken directly from a table, before interpolation. After interpolation, it becomes COMPUTED AZIMUTH.

tabulated azimuth angle. Azimuth angle taken directly from a table, before interpolation. After interpolation, it becomes COMPUTED AZIMUTH ANGLE.

Tacan, *n.* An ultra high frequency radionavigation system which provides a continuous indication of bearing and distance to a Tacan station. The term is derived from Tactical Air Navigation.

tactical diameter. The distance gained to the right or left of the original course when a turn of 180° with a constant rudder angle has been completed. See also STANDARD TACTICAL DIAMETER.

taffrail, *n.* The after rail at the stern of a vessel.

taffrail log. A log consisting essentially of a rotator towed through the water by a braided log line attached to a distance registering device usually secured at the taffrail, the railing at the stern. Also called PATENT LOG.

tail wind. See under FOLLOWING WIND.

take departure. See under DEPARTURE, definition 2.

take the ground. A ship is said to take the ground when the tide leaves it aground for want of sufficient depth of water.

Taku wind. A strong, gusty, east-northeast wind, occuring in the vicinity of Juneau, Alaska, between October and March. At the mouth of the Taku River, after which it is named, it sometimes attains hurricane force.

tangent, *adj.* Touching at a single point.

tangent, *n.* 1. The ratio of the side opposite an acute angle of a plane right triangle to the shorter side adjacent to the same angle. The expression NATURAL TANGENT is sometimes used to distinguish the tangent from its logarithm (called LOGARITHMIC TANGENT).
2. A straight line, curve, or surface touching a curve or surface at one point.

tangent arc. 1. An arc touching a curve or surface at one point.
2. A halo tangent to a circular halo.

tangent latitude error. On a nonpendulous gyrocompass where damping is accomplished by offsetting the point of application of the force of a mercury ballistic, the angle between the local meridian and the settling position or spin axis. Where the offset of the point of application of a mercury ballistic is to the east of the vertical axis of the gyrocompass, the settling position is to the east of the meridian in north latitudes and to the west of the meridian in south latitudes. The error is so named because it is approximately proportional to the tangent of the latitude in which the gyrocompass is operating. The tangent latitude error varies from zero at the equator to a maximum at high northern and southern latitudes. This error may be compensated for by means of an auxiliary latitude corrector to shift the lubber's line or to alter the position of a small weight attached to the casing near one end of the axle.

tangent screw. A screw providing tangential movement along an arc, as that screw which provides the final angular adjustment of a marine sextant during an observation, an **endless tangent** screw can be moved over its entire range without resetting.

tank, *n.* A label on a nautical chart which indicates a water tank elevated high above ground by a tall skeleton framework.

tape gage. See ELECTRIC TAPE GAGE.

tapper, *n.* A heavy pendulum suspended outside a bell and which sets the bell in vibration by striking it.

target, *n.* An object, stationary or moving, about which information is sought with a radar equipment. See also CONTACT.

target angle. The relative bearing of own ship from a target vessel, measured clockwise through 360°. See also ASPECT.

target tail. The display of diminishing luminance seen to follow a target on a plan position indicator which results from afterglow and the progress of the target between successive scans of the radar. Also called TARGET TRAIL.

target trail. See TARGET TAIL.

tehuantepecer, *n.* A violent squally wind from north or north-northeast in the Gulf of Tehuantepec (south of southern Mexico) in winter. It originates in the Gulf of Mexico as a *norther* which crosses the isthmus and blows through the gap between the Mexican and Guatemalan mountains. It may be felt up to 100 miles out to sea. See also PAPAGAYO.

telecommunication, *n.* Any transmission, emission, sound, or intelligence of any nature by wire, radio, or other electromagnetic system. If the transfer is by radio, it may be called radiocommunication.

telegraph buoy. A buoy used to mark the position of a submarine telegraph cable.

telemeter, *n.* The complete equipment for measuring any quantity, transmitting the results electrically to a distant point, and there recording the values measured.

telemetry, *n.* The science of measuring a quantity or quantities, transmitting the measured value to a distant station, and there interpreting, indicating, or recording the quantities measured.

telemotor, *n.* A device for controlling the application of power at a distance, especially one by which the steering gear of a vessel is controlled from the wheel house.

telescope, *n.* An optical instrument used as an aid in viewing or photographing distant objects, particularly celestial objects. A **reflecting telescope** collects light by means of a concave mirror; a **refracting telescope** by means of a lens or system of lenses. A **Cassegrainian telescope** is a reflecting telescope in which the immergent light is reflected from the main mirror onto a secondary mirror, where it is reflected through a hole in the main mirror to an eyepiece; a **Newtonian telescope** is a reflecting telescope in which the immergent beam is reflected from the main mirror onto a small plane mirror, and from there to an eyepiece at the side of the telescope.

telescopic alidade. See ALIDADE.

telescopic meteor. See under METEOR.

telltale compass. A marine magnetic compass, usually of the inverted type, frequently installed in the master's cabin for his convenience.

Tellurometer, *n.* Registered trade name for a short range, line of sight, distance measuring system used in hydrographic and other surveys. System operation is based upon the principle of measuring the phase delay when a frequency-modulated microwave signal is radiated from a master station, and returned after reception and retransmission from a remote station. The phase difference at the master station between the two signals is directly proportional to the total distance traveled by the signal.

temperate zone. Either of the two zones between the frigid and torrid zones, called the **north temperate zone** and the **south temperate zone.**

temperature, *n.* Intensity or degree of heat. **Fahrenheit** temperature is based upon a scale in which water freezes at 32°F and boils at about 212°F; **Celsius** temperature upon a scale in which water freezes at 0°C and boils at 100°C. **Absolute temperature** is measured from absolute zero which is zero on the **Kelvin** scale, −273°.16C on the Celsius scale, and −459°.69 F on the Fahrenheit scale. Absolute temperature based upon degrees Fahrenheit is called **Rankine temperature** and that based upon degrees Celsius is called **Kelvin temperature. Ambient temperature** is the temperature of the air or other medium surrounding an object. **Critical temperature** is the temperature above which a substance cannot exist in the liquid state, regardless of the pressure. **Free-air temperature** is the temperature of the atmosphere obtained by a thermometer located so as to avoid as completely as practicable the effects of extraneous heating. **Dry-bulb temperature** is the temperature of the air, as indicated by the dry-bulb thermometer of a psychrometer. **Wet-bulb temperature** is the temperature indicated by the wet-bulb thermometer of a psychrometer, or the lowest temperature to which air can be cooled at any given time by evaporating water into it at constant pressure, when the heat required for evaporation is supplied by the cooling of the air. See also BUCKET TEMPERATURE.

temperature error. That instrument error due to nonstandard temperature of the instrument.

temperature inversion. An atmospheric condition in which the usual lapse rate is inverted, i.e., the temperature increases with increasing altitude.

temporal, *adj.* Pertaining to or limited by time.

temporary light. A light put into service for a limited period.

temporary units. See under INTERNATIONAL SYSTEM OF UNITS.

tend, *v., i.* To extend in a stated direction, as an anchor cable.

tera-. A prefix meaning *one trillion* (10^{12}).

terdiurnal, *adj.* Occurring three times per day. A terdiurnal tidal constituent has three periods in a constituent day.

terminator, *n.* The line separating illuminated and dark portions of a non-selfluminous body, as the moon.

terrace, *n.* On the sea floor, a relatively flat, horizontal or gently inclined surface, sometimes long and narrow, which is bounded by a steeper ascending slope on one side and by a steeper descending slope on the opposite side.

terrestrial, *adj.* Of or pertaining to the earth.

terrestrial coordinates. See GEOGRAPHICAL COORDINATES.

terrestrial equator. 1. The earth's equator, 90° from its geographical poles.
2. See ASTRONOMICAL EQUATOR.

terrestrial latitude. Latitude on the earth; angular distance from the equator, measured northward or southward through 90° and labeled N or S to indicate the direction of measurement. See also LATITUDE.

terrestrial longitude. Longitude on the earth; the arc of a parallel, or the angle at the pole, between the prime meridian and the meridian of a point on the earth, measured eastward or westward from the prime meridian through 180°, and labeled E or W to indicate the direction of measurement. See also LONGITUDE.

terrestrial magnetism. See GEOMAGNETISM.

terrestrial meridian. See ASTRONOMICAL MERIDIAN.

terrestrial perturbations. The largest gravitational perturbations of artificial satellites which are caused by the fact that the gravity field of the earth is not spherically symmetrical.

terrestrial pole. One of the poles of the earth. See also GEOGRAPHICAL POLE, GEOMAGNETIC POLE, MAGNETIC POLE.

terrestrial radiation. The total infrared radiation emitted from the earth's surface.

terrestrial refraction. Atmospheric refraction of a ray of radiant energy emanating from a point on or near the surface of the earth, as contrasted with ASTRONOMICAL REFRACTION of a ray passing through the earth's atmosphere from outer space.

terrestrial sphere. The earth.

terrestrial triangle. A triangle on the surface of the earth, especially the navigational triangle.

territorial sea. The zone off the coast of a nation

immediately seaward from a base line. Complete sovereignty is maintained over this coastal zone by the coastal nation, subject to the right of innocent passage to the ships of all nations. The United States recognizes this zone as extending 4.8 kilometers from the base line. See also FISHING ZONE, FISHERY CONSERVATION ZONE.

tertiary tide station. A tide station at which continuous observations have been made over a minimum period of 30 days but less than 1 year. The series is reduced by comparison with simultaneous observations from a secondary control tide station. This station provides for a 29-day harmonic analysis. See also PRIMARY CONTROL TIDE STATION; SECONDARY CONTROL TIDE STATION; SUBORDINATE TIDE STATION, definition 2; TIDE STATION.

tesla, *n.* The *derived unit* of magnetic flux density in the International System of Units; it is equal to 1 weber per square metre.

Texas norther. See under NORTHER.

thaw holes. Vertical holes in sea ice formed when surface puddles melt through to the underlying water.

thematic map. See TOPICAL MAP.

theoretical error. See under SYSTEMATIC ERROR.

thermometer, *n.* An instrument for measuring temperature. A **maximum thermometer** automatically registers the highest temperature and a **minimum thermometer** the lowest temperature since the last thermometer setting. A **wet-bulb thermometer** has the bulb covered with a cloth, usually muslin or cambric, saturated with water, and is used with an uncovered or **dry-bulb thermometer** to determine atmospheric humidity. When the two thermometers are thus used, they constitute the essential parts of a PSYCHROMETER.

thermostat, *n.* A device for automatically regulating temperature or detecting temperature changes.

thick first-year ice. First-year ice over 120 centimeters thick.

thick weather. Condition of greatly reduced visibility, as by fog, snow, rain, etc.

thin first-year ice. First-year ice 30 to 70 centimeters thick. Also called WHITE ICE.

thin overcast. An overcast sky cover which is predominantly transparent.

thorofare, *n.* This shortened form of *thoroughfare* has become standard for a natural waterway in marshy areas. It is the same type of feature as a slough or bayou.

thoroughfare, *n.* A public waterway such as a river or strait. See also THOROFARE.

three-arm protractor. An instrument consisting essentially of a circle graduated in degrees, to which is attached one fixed arm and two arms pivoted at the center and provided with clamps so that they can be set at any angle to the fixed arm, within the limits of the instrument. It is used for finding a ship's position when the angles between three fixed and known points are measured. Also called STATION POINTER.

three-point problem. From the observation of two horizontal angles between three objects or points of known (charted) positions, to determine the position of the point of observation. The problem is solved graphically by means of the three-arm protractor and analytically by trigonometrical calculation.

threshold signal. The smallest signal capable of being detected above the background noise level.

threshold speed. The minimum speed of current at which a particular current meter will measure at its rated reliability.

thundercloud, *n.* See CUMULONIMBUS.

thunderhead, *n.* See CUMULONIMBUS.

thundersquall, *n.* Strictly, the combined occurrence of a thunderstorm and a squall, the squall usually being associated with the downrush phenomenon typical of a well-developed thunderstorm.

thunderstorm, *n.* In general, a local storm invariably produced by a cumulonimbus cloud, and always accompanied by lightning and thunder, usually with strong gusts of wind, heavy rain, and sometimes with hail. It is usually of short duration, seldom over 2 hours for any one storm. A thunderstorm is a consequence of atmospheric instability and constitutes, loosely, an overturning of air layers in order to achieve a more stable density stratification. A strong convective updraft is a distinguishing feature of this storm in its early phases. A strong downdraft in a column of precipitation marks its dissipating stages. Thunderstorms often build to altitudes of 40,000 to 50,000 feet in middle latitudes and to even greater heights in the tropics; only the great stability of the lower stratosphere limits their upward growth. Sometimes called ELECTRICAL STORM.

thunderstorm cirrus. See FALSE CIRRUS.

thundery sky. A sky with an overcast and chaotic aspect, a general absence of wind except during showers, a mammatus appearance of the lower clouds, and dense cirrostratus and altocumulus above.

tick, *n.* A short, audible sound or beat, as that of a clock. A time signal in the form of one or more ticks is called a TIME TICK.

tickle, *n.* A narrow channel, as used locally in the Arctic and Newfoundland.

tidal, *adj.* Of or pertaining to tides.

tidal amplitude. One-half the range of a constituent tide.

tidal basin. *British terminology.* A basin without a caisson or gate in which the level of water rises and falls with the tides. Also called OPEN BASIN. See also TIDAL HARBOR, NON-TIDAL BASIN.

tidal bench mark. See under BENCH MARK.

tidal bench mark description. A published, concise description of the location, stamped number of designation, date established, and elevation (referred to a tidal datum) of a specific bench mark.

tidal bench mark state index map. A state map which indicates the locations for which tidal datums and tidal bench mark descriptions are available.

tidal bore. A tidal wave that propagates up a relatively shallow and sloping estuary or river in a solitary wave form. The leading edge presents an abrupt rise in level, frequently with continuous breaking and often immediately followed by several large undulations. An uncommon phenomenon, the tidal bore is usually associated with very large ranges in tide as well as wedge-shaped and rapidly shoaling entrances. Also called EAGRE, EAGER, MASCARET, POROROCA, BORE.

tidal constants. Tidal relations that remain practically constant for any particular locality. Tidal constants are classified as harmonic and nonharmonic. The harmonic constants consist of the amplitudes and epochs of the harmonic constituents, and the nonharmonic constants include the ranges and intervals derived directly from the high and low water observations.

tidal constituent. See CONSTITUENT.

tidal current. A horizontal movement of the water caused by gravitational interactions between the sun, moon, and earth. The horizontal component of the particulate motion of a tidal wave. Part of the same general movement of the sea that is manifested in the vertical rise and fall, called tide. Called TIDAL STREAM in British terminology. See also CURRENT, TIDAL WAVE, TIDE.

tidal current charts. 1. Charts on which tidal current data are depicted graphically.
2. *Tidal Current Chart*; as published by the National Ocean Survey, part of a set of 12 charts which depict, by means of arrows and figures, the direction and velocity of the tidal current for each hour of the tidal cycle. The charts, which may be used for any year, present a comprehensive view of the tidal current movement in the respective waterways as a whole and also supply a means for readily determining for any time the direction and velocity of the current at various localities throughout the water area covered. The New York Harbor and Narragansett Bay tidal current charts are to be used with the annual tide tables. The other charts require the annual tidal current tables.

tidal current constants. See CURRENT CONSTANTS.

tidal current diagrams. Monthly diagrams which are used with tidal current charts to provide a convenient method to determine the current flow on a particular day. The National Ocean Survey publishes *Tidal Current Diagrams for Long Island Sound and Block Island Sound* and *Tidal Current Diagrams for Boston Harbor*.

tidal current station. See CURRENT STATION.

tidal current tables. 1. Tables which give the predicted times of slack water and the predicted times and velocities of maximum current – flood and ebb – for each day of the year at a number of reference stations, together with time differences and velocity ratios for obtaining predictions at subordinate stations.
2. *Tidal Current Tables*; published annually by the National Ocean Survey in two volumes: *Atlantic Coast of North America*; *Pacific Coast of North America and Asia*.

tidal cycle. A complete set of tidal conditions as those occurring during a tidal day, lunar month, or Metonic cycle.

tidal datum. See VERTICAL DATUM.

tidal day. See LUNAR DAY, definition 1.

tidal difference. Difference in time or height of a high or low water at a subordinate station and at a reference station for which predictions are given in the *Tide Tables*. The difference, when applied according to sign to the prediction at the reference station, gives the corresponding time or height for the subordinate station.

tidal epoch. See EPOCH, definition 3.

tidal estuary. See under ESTUARY, definition 1.

tidal flats. See FLAT.

tidal harbor. A harbor affected by the tides, in distinction from a harbor in which the water level is maintained by caissons or gates. See also NON-TIDAL BASIN.

tidal lights. Lights shown at the entrance of a harbor, to indicate tide and tidal current conditions within the harbor.

tidal lock. See ENTRANCE LOCK.

tidal marsh. Any marsh the surface of which is covered and uncovered by tidal flow. See also FLAT.

tidal platform ice foot. An ice foot between high and low water levels, produced by the rise and fall of the tide.

tidal quay. A quay in an open harbor or basin with sufficient depth alongside to enable ships lying alongside to remain afloat at any state of the tide.

tidal range. See RANGE OF TIDE.

tidal rise. See RISE OF TIDE.

tidal stream. *British terminology.* See TIDAL CURRENT.

tidal water. Any water the level of which changes periodically due to tidal action. See also TIDEWATER.

tidal wave. 1. A shallow water wave caused by the gravitational interactions between the sun, moon, and earth. Essentially, high water is the crest of a tidal wave and low water is the trough. Tide is the vertical component of the particulate motion and tidal current is the horizontal component. The observed tide and tidal current can be considered the result of the combination of several tidal waves, each of which may vary from nearly pure progressive to nearly pure standing and with differing periods, heights, phase relationships, and directions.
2. Any unusually high and destructive water level along a shore. It usually refers to either a **storm surge** or **tsunami**.

tide, *n.* The periodic rise and fall of the water resulting from gravitational interactions between the sun, moon, and earth. The vertical component of the particulate motion of a tidal wave. Although the accompanying horizontal movement of the water is part of the same phenomenon, it is preferable to designate this motion as TIDAL CURRENT. See also TIDAL WAVE, definition 1.

Tide and Current Glossary. A publication of the National Ocean Survey which includes in addition to general tide and current terms those accepted definitions intrinsic to certain standard procedures of the Oceanographic Division of the National Ocean Survey.

tide-bound, *adj.* Unable to proceed because of insufficient depth of water due to tidal action.

tide crack. A crack at the line of junction between an immovable icefoot or ice wall and fast ice, the latter subject to rise and fall of the tide.

tide curve. A graphic representation of the rise and fall of the tide in which time is usually represented by the abscissa and height by the ordinate of the graph. For a normal tide the graphic representation approximates a cosine curve. See also MARIGRAM.

tide datum. See VERTICAL DATUM.

tide gage. An instrument for measuring the rise and fall of the tide. See also AUTOMATIC TIDE GAGE, ELECTRIC TAPE GAGE, PRESSURE GAGE, TIDE STAFF.

tide gate. 1. A restricted passage through which water runs with great speed due to tidal action.
2. An opening through which water may flow freely when the tide sets in one direction, but which closes automatically and prevents the water from flowing in the other direction when the direction of flow is reversed.

tidehead, *n.* Inland limit of water affected by a tide.

tide hole. A hole made in ice to observe the height of the tide.

tide indicator. That part of a tide gage which indicates the height of tide at any time. The indicator may be in the immediate vicinity of the tidal water or at some distance from it.

tideland, *n.* Land which is under water at high tide and uncovered at low tide.

tidemark, *n.* 1. A high water mark left by tidal

water.
2. The highest point reached by a high tide.
3. A mark placed to indicate the highest point reached by a high tide, or, occasionally, any specified state of tide.

tide notes. Notes included on nautical charts which give information on the mean range or the diurnal range of the tide, mean tide level, and extreme low water at key places on the chart.

tide pole. A graduated spar used for measuring the rise and fall of the tide. Also called TIDE STAFF.

tide pool. A pool left by an ebb tide.

tide predicting machine. A mechanical analog machine especially designed to handle the great quantity of constituent summations required in the harmonic method. William Ferrel's Maxima and Minima Tide Predictor (described in *Manual of Tides*, U.S. Coast and Geodetic Survey, Appendix 10, Report for 1883) was the first such machine used in the United States. Summing only 19 constituents, but giving direct readings of the predicted times and heights of the high and low waters, the Ferrel machine was used for the predictions of 1885 through 1914. A second machine, developed by Rollin A. Harris and E. G. Fischer and summing 37 constituents, was used for the predictions of 1912 through 1965 (described in *Manual of Harmonic Analysis and Prediction of Tides* by Paul Schureman, U.S. Coast and Geodetic Survey Special Publication No. 98, 1958). Predictions are now prepared using an electronic digital computer.

tide-producing force. That part of the gravitational attraction of the moon and sun which is effective in producing the tides on the earth. The force varies approximately as the mass of the attracting body and inversely as the cube of its distance. The tide-producing force exerted by the sun is a little less than one-half as great as that of the moon.

tide producing potential. Tendency for particles on the earth to change their positions as a result of the gravitational interactions between the sun, moon, and earth. Although the gravitational attraction varies inversely as the square of the distance of the tide-producing body, the resulting potential varies inversely as the cube of the distance.

tide race. A very rapid tidal current through a comparatively narrow channel. Also called RACE.

tide rips. Small waves formed on the surface of water by the meeting of opposing tidal currents or by a tidal current crossing an irregular bottom. Vertical oscillation, rather than progressive waves, is characteristic of tide rips. See also RIPS.

tide rode. A ship riding at anchor is said to be *tide rode* when heading into the tidal current. See also WIND RODE.

tide signals. Signals showing to navigators the state or change of the tide according to a prearranged code, or by direct display on a scale.

tide staff. A tide gage consisting of a vertical graduated staff from which the height of the tide can be read directly. It is called a **fixed staff** when secured in place so that it cannot be easily removed. A **portable staff** is one that is designed for removal from the water when not in use. For such a staff a fixed support is provided, and the staff itself has a metal stop secured to the back so that it will always have the same elevation when installed for use. See also ELECTRIC TAPE GAGE.

tide station. The geographic location at which tidal observations are conducted. Also, the facilities used to make tidal observations. These may include a tide house, tide gage, tide staff, and tidal bench marks. See also PRIMARY CONTROL TIDE STATION, SECONDARY CONTROL TIDE STATION, SUBORDINATE TIDE STATION, TERTIARY TIDE STATION.

tide tables. 1. Tables which give the predicted times and heights of high and low water for every day in the year for a number of reference stations, and tidal differences and ratios by which additional predictions can be obtained for subordinate stations. From these values it is possible to interpolate by a simple procedure the height of the tide at any hour of the day. See also TIDAL CURRENT TABLES.
2. *Tide Tables*; published annually by the National Ocean Survey in four volumes: *Europe and West Coast of Africa (including the Mediterranean Sea)*; *East Coast of North and South America (including Greenland)*; *West Coast of North and South America (including the Hawaiian Islands)*; *Central and Western Pacific Ocean and Indian Ocean*.

tidewater, *n.* Water affected by tides or sometimes that part of it which covers the tideland. The term is sometimes used broadly to designate the seaboard. See also TIDAL WATER.

tide wave. See TIDAL WAVE, definition 1.

tideway, *n.* A channel through which a tidal current runs.

tilt, *n.* The angle which anything makes with the horizontal.

tilted blocky iceberg. A blocky iceberg which has tilted to present a triangular shape from the side.

tilt correction. That correction due to tilt error.

tilt error. The error introduced in the reading of an instrument when it is tilted, as a marine sextant held so that its frame is not perpendicular to the horizon.

time, *n.* 1. The time interval or length of time between two events.
2. The date or designated mark on a time scale.
See also TIME SCALE, APPARENT TIME, MEAN TIME, SIDEREAL TIME.

time and altitude azimuth. An azimuth determined by solution of the navigational triangle with meridian angle, declination, and altitude given. A TIME AZIMUTH is computed with meridian angle, declination, and latitude given. An ALTITUDE AZIMUTH is computed with altitude, declination, and latitude given.

time azimuth. An azimuth determined by solution of the navigational triangle, with meridian angle, declination, and latitude given. An ALTITUDE AZIMUTH is computed with altitude, declination, and latitude given. A TIME AND ALTITUDE AZIMUTH is computed with meridian angle, declination, and altitude given.

time ball. A visual time signal in the form of a ball. Before the widespread use of radio time signals, time balls were dropped, usually at local noon, from conspicuously-located masts in various ports. The accuracy of the signal was usually controlled by a telegraphic time signal from an observatory.

time base. A motion, of known but not necessarily of constant speed, used for measuring time intervals, particularly the sweep of a cathode-ray tube. In a **linear time base** the speed is constant; in an **expanded time base** a selected part is of increased speed; and in a **delayed time base** the start is delayed. See also SWEEP.

time diagram. A diagram in which the celestial equator appears as a circle, and celestial meridians and hour circles as radial lines; used to facilitate solution of time problems and others involving arcs of the celestial equator or angles at the pole, by indicating relations between various quantities involved. Conventionally the relationships are given as viewed from a point over the south pole, westward direction being counterclockwise. Also called DIAGRAM ON THE PLANE OF THE CELESTIAL EQUATOR, DIAGRAM ON THE PLANE OF THE EQUINOCTIAL.

time meridian. Any meridian used as a reference for reckoning time, particularly a zone or standard meridian.

timepiece, *n.* An instrument for measuring time. See also CHRONOMETER, CLOCK, WATCH.

time scale. A system of assigning dates to events. There are three fundamental scales: Ephemeris Time, time based upon the rotation of the earth, and atomic time or time obtained by counting the cycles of a signal in resonance with certain kinds of atoms. **Ephemeris Time (ET)**, the independent variable in the gravitational theories of the solar system, is the scale used by astronomers as the tabular argument of the precise, fundamental ephemerides of the sun, moon, and planets. **Universal Time (UT1)**, time based on the rotation of the earth, is the scale used by astronomers as the tabular argument for most other ephemerides, e.g., the *Nautical Almanac*. Although ET and UT1 differ in concept, both are determined in arrears from astronomical observations and are extrapolated into the future, based on **International Atomic Time (TAI)**. **Coordinated Universal Time (UTC)** is the scale disseminated by most broadcast time services; it differs from TAI by an integral number of seconds.

time sight. Originally, an observation of the altitude of a celestial body, made for the purpose of determining longitude. Now, the expression is applied primarily to the common method of reducing such an observation.

time signal. An accurate signal marking a specified time or time interval. It is used primarily for determining errors of timepieces. Such signals are usually sent from an observatory by radio or telegraph, but visual signals are used at some ports. Visual signals, which usually indicate noon (local time) by the dropping of a ball or the flash of a gun fired from a conspicuous point, are generally timed by a telegraphic signal.

time signal service. As defined by the *International Telecommunication Union* (ITU), a radiocommunication service for the transmission of time signals of stated high precision, intended for general reception.

time switch. A device for lighting or extinguishing a light at predetermined times, controlled by a timing device.

time tick. A time signal consisting of one or more short audible sounds or beats.

time zone. An area in all parts of which the same time is kept. In general, each zone is 15° of longitude in width, with the Greenwich meridian (0° longitude) designated as the central meridian of zone 0 and the remaining zones centered on a meridian whose longitude is exactly divisible by 15. The zone boundary may vary considerably to conform to political or geographic boundaries or both. Time zones are used in connection with zone, standard, daylight saving or summer and double summer times. See also STANDARD TIME.

Tokyo datum. A geodetic datum that has its origin in Tokyo. It is defined in terms of the Bessel ellipsoid and is oriented by means of a single astronomic station. By means of triangulation ties through Korea, the Tokyo datum is connected with the Manchurian datum. Unfortunately, since Tokyo is situated on a steep geoidal slope, the single station orientation has resulted in large systematic geoidal separations as the system is extended from its initial point.

tombolo, *n.* An islet and a shoal connecting it to a larger land area.

tongue, *n.* 1. A projection of the ice edge up to several kilometers in length, caused by wind or current.
2. An elongated (tongue-like) extension of flat sea floor into an adjacent higher feature.

topical map. A map designed to portray a special subject; e.g., administrative subdivisions, railroads, power lines, navigable waterways. Also called SPECIAL SUBJECT MAP, THEMATIC MAP.

topmark, *n.* One or more relatively small objects of characteristic shape or color, or both, placed on top of a beacon or buoy to aid in its identification. The following characteristic shapes of topmarks are internationally recognized for the *lateral system*: cone, can, sphere, diamond, St. George's Cross, "T," and broom. A broom topmark has the appearance of a circular broomshead and has two forms: (1) broom, point upwards and (2) broom, point downwards. The following characteristic shapes of topmarks are internationally recognized for the *cardinal system*: two cones, point upwards; two cones, point downwards; two cones, point to point; two cones, base to base. Also called DAYMARK, particularly in United States usage.

topographical latitude. See GEODETIC LATITUDE.

topographic feature. See under TOPOGRAPHY, definition 1.

topographic map. A map which presents the vertical position of features in measurable form as well as their horizontal positions.

topography, *n.* 1. The configuration of the surface of the earth, including its relief, the position of its streams, roads, cities, etc. The earth's natural and physical features collectively. A single feature such as a mountain or valley is called a **topographic feature**. Topography is subdivided into **hypsography** (the relief features), **hydrography** (the water and drainage features), **culture** (manmade features), and **vegetation**.
2. The science of delineation of natural and manmade features of a place or region especially in a way to show their positions and elevations.

toponym, *n.* A name applied to a physical or cultural topographic feature. For U.S. Government usage, policies and decisions governing place names on earth are established by the Board on Geographic Names. Also called PLACE NAME.

toponymy, *n.* 1. The study and treatment of toponyms.
2. A body of toponyms.

topple, *n.* 1. The vertical rotation of the spin axis of a gyroscope about the topple axis.
2. The vertical component of real precession or apparent precession, or the algebraic sum of the two. See also DRIFT, *n*, definiton 6; TOTAL DRIFT.

topple axis. Of a gyroscope, that horizontal axis, perpendicular to the horizontal spin axis, around which topple occurs. See also DRIFT AXIS, SPIN AXIS.

tornado, *n.* A violently rotating column of air, pendant from a cumulonimbus cloud, and nearly always observable as a funnel cloud or tuba. On a local scale, it is the most destructive of all atmospheric phenomena. Its vortex, commonly several hundreds of yards in diameter, whirls usually cyclonically with wind speeds estimated at 100 to more than 200 miles per hour. Its general direction of travel is governed by the motion of its parent cloud. Tornadoes occur on all continents, but are most common in Australia and the United States where the average number is 140 to 150 per year. They occur throughout the year and at any time of day, but are most frequent in spring and in middle and late afternoon. In the United States, tornados often develop several hundred miles southeast of a deep low centered in the central or north-central states. However, they may appear in any sector of the low, and/or be associated with fronts, instability lines, troughs, and even form within high-pressure ridges. A distinction sometimes is made between **cyclonic tornadoes** and **convective tornadoes**, the former occurring within the circulation of a well-developed parent cyclone, and the latter referring to all others. A tornado over water is called WATERSPOUT.

tornado cloud. See FUNNEL CLOUD.

torque, *n.* That which effects or tends to effect rotation or torsion and which is measured by the product of the applied force and the perpendicular distance from the line of action of the force to the axis of rotation.

torrid zone. That part of the earth between the Tropic of Cancer and the Tropic of Capricorn. Also called the TROPICS.

total current. The combination of the tidal and nontidal current. Called FLOW in British terminology. See also CURRENT.

total drift. The algebraic sum of drift due to real precession and that due to apparent precession.

total eclipse. An eclipse in which the entire source of light is obscured.

tower, *n.* A label on a nautical chart which indicates any structure with its base on the ground and high in proportion to its base, or that part of a structure higher than the rest, but having essentially vertical sides for the greater part of its height.

towering, *n.* Apparent increase in the vertical dimension of an object near the horizon, due to large inequality of atmospheric refraction in the line of sight to the top and bottom of the object. The opposite is STOOPING.

trace, *n.* The luminous line resulting from the radial movement of the points of impingement of the electron stream on the face of the cathode-ray tube of a radar indicator. See also SWEEP.

track, *n.* 1. The intended or desired horizontal direction of travel with respect to the earth. The track as expressed in degrees of the compass may be different from the course due to such factors as making allowance for current or sea or steering to resume the TRACK, definition 2.

2. The path of intended travel with respect to the earth as drawn on the chart. Also called INTENDED TRACK, TRACK-LINE.

3. The actual path of a vessel over the ground, such as may be determined by tracking.

track, *v., t.* To follow the movements of an object as by radar or an optical system.

track angle. See TRACK, definition 1.

track chart. A chart showing recommended, required, or established tracks, and usually indicating turning points, courses, and distances. A distinction is sometimes made between a TRACK CHART and a ROUTE CHART, the latter generally showing less specific information, and sometimes only the area for some distance each side of the great circle or rhumb line connecting two terminals. A MILEAGE CHART shows distances between various points.

tracking, *n.* In the operation of automated radar plotting aids, the process of observing the sequential changes in the position of a target to establish its motion.

track-line, *n.* See TRACK, definition 2.

track made good. The single resultant direction from a point of departure to a point of arrival at any given time. The use of this term to indicate a single resultant direction is preferred to the use of the misnomer *course made good*. See also COURSE, TRACK.

trade winds. Relatively permanent winds on each side of the equatorial doldrums, blowing from the northeast in the Northern Hemisphere and from the southeast in the Southern Hemisphere. See also ANTITRADES.

traffic control signals. Visual signals placed in a harbor or waterway to indicate to shipping the movements authorized or prohibited at the time at which they are shown. Also called DOCKING SIGNALS.

traffic lane. An area within defined limits of which one-way traffic is established. Natural obstacles, including those forming separation zones, may constitute a boundary. See also TWO-WAY ROUTE, ROUTING SYSTEM.

traffic separation scheme. A routing measure aimed at the separation of opposing streams of traffic by appropriate means and by the establishment of traffic lanes. See also ROUTING SYSTEM.

train, *v., t.* To control motion in bearing, as of a radar antenna.

training wall. A wall, bank, or jetty, often submerged, built to direct or confine the flow of a river or tidal current.

tramontana, *n.* A northeasterly or northerly wind occurring in winter off the west coast of Italy. It is a fresh wind of the fine weather mistral type.

TRANSAT satellite. See under OSCAR SATELLITE.

transceiver, *n.* A combination transmitter and receiver in a single housing, with some components being used by both parts. See also TRANSPONDER.

transducer, *n.* A device that converts one type of energy to another, as a loudspeaker that changes electrical energy into acoustical energy.

transfer, *n.* 1. The distance a vessel moves perpendicular to its initial direction in making a turn of 90° with a constant rudder angle.

2. The distance a vessel moves perpendicular to its initial direction for turns of less than 90°. See also ADVANCE.

transit, *n.* 1. The passage of a celestial body across a celestial meridian, usually called MERIDIAN TRANSIT.

2. The apparent passage of a celestial body across the face of another celestial body or across any point, area, or line.

3. An instrument used by an astronomer to determine the exact instant of meridian transit of a celestial body.

4. A reversing instrument used by a marine surveyor for accurately measuring horizontal and vertical angles; a theodolite which can be reversed in its supports without being lifted from them.

transit, *v., t.* To cross. The term is generally used with reference to the passage of a celestial body over a meridian, across the face of another celestial body, or across the reticle of an optical instrument.

TRANSIT, *n.* See under NAVY NAVIGATION SATELLITE SYSTEM.

transition buoy. A buoy indicating the transition between the lateral and cardinal systems of buoyage.

transition mark. A navigation mark indicating the transition between the lateral and cardinal systems of marking.

translocation, *n.* The determination of the relative positions of two points by simultaneous Doppler satellite observations from each point.

translunar, *adj.* Of or pertaining to space outside the moon's orbit about the earth.

transmit-receive tube. See as TR TUBE.

transponder, *n.* A component of a secondary radar system capable of accepting the interrogating signal, received from a radar set or interrogator, and in response automatically transmitting a signal which enables the transponder to be identified by the interrogating station. Also called TRANSPONDER BEACON. See also RADAR BEACON, RACON.

transponder beacon. See TRANSPONDER.

transpose, *v., t.* To change the relative place or position of, as to move a term from one side of an equation to the other with a change of sign.

transverse bar. A bar which extends approximately normal to the shoreline.

transverse chart. A chart on a transverse map projection. Also called INVERSE CHART.

transverse cylindrical orthomorphic chart. See TRANSVERSE MERCATOR CHART.

transverse cylindrical orthomorphic map projection. See TRANSVERSE MERCATOR MAP PROJECTION.

transverse equator. The plane which is perpendicular to the axis of a transverse map projection. Also called INVERSE EQUATOR. See also FICTITIOUS EQUATOR.

transverse graticule. A fictitious graticule based upon a transverse map projection.

transverse latitude. Angular distance from a transverse equator. Also called INVERSE LATITUDE. See also FICTITIOUS LATITUDE.

transverse longitude. Angular distance between a prime transverse meridian and any given transverse meridian. Also called INVERSE LONGITUDE. See also FICTITIOUS LONGITUDE.

transverse map projection. A map projection with its axis in the plane of the equator.

transverse Mercator chart. A chart on the transverse Mercator projection. Also called TRANSVERSE CYLINDRICAL ORTHOMORPHIC CHART, INVERSE MERCATOR CHART, INVERSE CYLINDRICAL ORTHOMORPHIC CHART. See also MERCATOR CHART.

transverse Mercator map projection. A conformal cylindrical map projection, being in principle equivalent to the regular Mercator map projection turned (transversed) 90° in azimuth. In this projection, the central meridian is represented by a straight line, corresponding to the line which represents the equator on the regular Mercator map projection. Neither the geographic meridians (except the central meridian) nor the geodetic parallels (except the equator) are represented by straight lines. Also called INVERSE MERCATOR MAP PROJECTION, TRANSVERSE CYLINDRICAL ORTHOMORPHIC MAP PROJECTION, INVERSE CYLINDRICAL ORTHOMORPHIC MAP PROJECTION. See also MERCATOR MAP PROJECTION.

transverse meridian. A great circle perpendicular to a transverse equator. The reference transverse meridian is called **prime transverse meridian.** Also called INVERSE MERIDIAN. See also FICTITIOUS MERIDIAN.

transverse parallel. A circle or line parallel to a transverse equator, connecting all points of equal transverse latitude. Also called INVERSE PARALLEL. See also FICTITIOUS PARALLEL.

transverse pole. One of the two points 90° from a transverse equator.

transverse rhumb line. A line making the same oblique angle with all fictitious meridians of a transverse Mercator map projection. Transverse parallels and meridians may be considered special cases of the transverse rhumb line. Also called INVERSE RHUMB LINE. See also FICTITIOUS RHUMB LINE.

transverse wave. A wave in which the vibration is perpendicular to the direction of propagation, as in light waves. This is in contrast with a LONGITUDINAL WAVE, in which the vibration is in the direction of propagation.

trapezoid, *n.* A quadrilateral having two parallel sides and two nonparallel sides.

traverse, *n.* A series of directions and distances, as those involved when a sailing vessel beats into the wind, a steam vessel zigzags, or a surveyor makes such measurements for determination of position.

traverse sailing. A method of determining the equivalent course and distance made good by a craft following a track consisting of a series of rhumb lines. The solution is usually made by means of traverse tables, the distance to the north or south and that to the east or west on each course being tabulated, the algebraic sum of differences of latitude and departure being found, and the result being converted to equivalent course and distance.

traverse table. A table giving relative values of various parts of plane right triangles, for use in solving such triangles, particularly in connection with various sailings.

TR box. See TR SWITCH.

trench, *n.* A long, narrow, characteristically very deep and asymmetrical depression of the sea floor, with relatively steep sides. See also TROUGH.

triad, *n.* Three radionavigation stations operated as a group for the determination of positions. Also called TRIPLET. See also STAR CHAIN.

triangle, *n.* A closed figure having three sides. The triangle is **plane, spherical,** or **curvilinear** as the sides are straight lines, arcs of great circles, or any curves, respectively. See also EQUILATERAL TRIANGLE, ISOSCELES TRIANGLE, NAVIGATIONAL TRIANGLE, RIGHT TRIANGLE.

triangulation, *n.* A method of surveying in which the stations are points on the ground which are located on the vertices of a chain or network of triangles. The angles of the triangles are measured instrumentally, and the sides are derived by computation from selected sides which are called BASE LINES, the lengths of which are obtained from direction measurements on the ground. See also TRILATERATION.

triaxial ellipsoid. A reference ellipsoid having three unequal axis, the shortest being the polar axis, and the two longer ones lying in the plane of the equator.

tributary, *n.* A stream that flows into another stream or a lake.

tributary waterway. Any body of water that flows into a larger body, i.e., a creek in relation to a river, a river in relation to a bay, and a bay in relation to the open sea.

trigger, *n.* In a radar set, a sharp voltage pulse which is applied to the modulator tubes to fire the transmitter, and which is applied simultaneously to the sweep generator to start the electron beam moving radially from the sweep origin to the edge of the face of the cathode-ray tube.

triggering, *n.* The process of causing a transponder to respond, as by interrogation.

trigonometric functions. The ratios of the sides of a plane right triangle, as related to one of its acute angles. If *a* is the side opposite an acute angle, *b* the adjacent side, and *c* the hypotenuse the trigonometric functions are: sine = a/c, cosine = b/c, tangent = a/b, cotangent = b/a, secant = c/b, cosecant = c/a, coversine or versed cosine = 1 − sine, versine = 1-cos, haversine = (1 − cos)/2. The expression NATURAL TRIGONOMETRIC FUNCTION is sometimes used to distinguish a trigonometric function from its logarithm (called LOGARITHMIC TRIGONOMETRIC FUNCTION).

trihedral reflector. See CORNER REFLECTOR.

trilateration, *n.* A method of surveying wherein the lengths of the triangle sides are measured, usually by electronic methods, and the angles are computed from the measured lengths. See also TRIANGULATION.

trim, *n.* The relation of the draft of a vessel at the bow and stern. See also DOWN BY THE HEAD; DOWN BY THE STERN; DRAG, *n.*, definition 3; SQUAT, *n.*

triple interpolation. Interpolation when there are three arguments or variables.

triplet, *n.* See TRIAD.

Trisponder, *n.* See as DEL NORTE TRISPONDER SYSTEM.

trochoid, *n.* In relation to wave motion, a curve described by a point on a radius of a circle that rolls along a straight line. Also called PROLATE CYCLOID.

tropic, *adj.* Of or pertaining to a tropic or the tropics.

tropic, *n.* Either of the two parallels of declination (north or south), approximately 23°27′ from the celestial equator, reached by the sun at its maximum declination, or the corresponding parallels on the earth. The northern of these is called the TROPIC OF CANCER and the southern, the TROPIC OF CAPRICORN. The region of the earth between these two parallels is called the TORRID ZONE, or often the TROPICS.

tropical, *adj.* 1. Of or pertaining to the vernal equinox. See also SIDEREAL.
2. Of or pertaining to the Tropics.

tropical air. Warm air of an air mass originating in subtropical anticyclones, further classified as **tropical continental air** and **tropical maritime air**, as it originates over land or sea, respectively.

tropical continental air. Air of an air mass originating over a land area in low latitudes, such as the Sahara desert. Tropical continental air is characterized by high surface temperature and low specific humidity.

tropical cyclone. The general term for cyclones originating in the Tropics or subtropics. These cyclones are classified by form and intensity as follows: A **tropical disturbance** is a discrete system of apparently organized convection – generally 100 to 300 miles in diameter – having a nonfrontal migratory character, having maintained its identity for 24 hours or more. It may or may not be associated with a detectable perturbation of the wind field. It has no strong winds and no closed isobars, i.e., isobars that completely enclose the low. (In successive stages of intensification, the tropical cyclone may be classified as tropical disturbance, tropical depression, tropical storm, and hurricane or typhoon.) The **tropical depression** has one or more closed isobars and some rotary circulation at the surface. The highest sustained (1-minute mean) surface wind speed is 33 knots. The **tropical storm** has closed isobars and a distinct rotary circulation. The highest sustained (1-minute mean) surface wind speed is 34 to 63 knots. The **hurricane** or **typhoon** has closed isobars, a strong and very pronounced rotary circulation, and a sustained (1-minute mean) surface wind speed of 64 knots or higher. Tropical cyclones occur almost entirely in six rather distinct areas, four in the Northern Hemisphere and two in the Southern Hemisphere. The name by which the tropical cyclone is commonly known varies somewhat with locality as follows: *North Atlantic*: A tropical cyclone with winds of 64 knots or greater is called a HURRICANE. *Eastern North Pacific*: The name HURRICANE is used as in the North Atlantic. *Western North Pacific*: A fully developed storm with winds of 64 knots or greater is called a TYPHOON or, locally in the Philippines, a BAGUIO. *North Indian Ocean*: A tropical cyclone with winds of 34 knots or greater is called a CYCLONIC STORM. *South Indian Ocean*: A tropical storm with winds of 34 knots or greater is called a CYCLONE. *Southwest Pacific and Australian Area*: The name CYCLONE is used as in the South Indian Ocean. A severe tropical cyclone originating in the Timor Sea and moving southwestward and then southeastward across the interior of northwestern Australia is called a WILLY-WILLY. Tropical cyclones have not been observed in the South Atlantic Ocean or in the South Pacific Ocean east of longitude 140°W.

tropical depression. See under TROPICAL CYCLONE.

tropical disturbance. See under TROPICAL CYCLONE.

tropical maritime air. Air of an air mass originating over an ocean area in low latitudes. Tropical maritime air is characterized by high surface temperature and high specific humidity.

tropical month. The average period of the revolution of the moon about the earth with respect to the vernal equinox, a period of 27 days, 7 hours, 43 minutes, 4.7 seconds, or approximately 27 1/3 days. This is almost the same length as the sidereal month.

tropical storm. See under TROPICAL CYCLONE.

tropical year. The period of one revolution of the earth around the sun, with respect to the vernal equinox. Because of precession of the equinoxes, this is not 360° with respect to the stars, but 50.″3 less. A tropical year is about 20 minutes shorter than a sidereal year, averaging 365 days, 5 hours, 48 minutes, and 46 seconds in 1900 and is decreasing at the rate of 0.00530 second annually. Also called ASTRONOMICAL, EQUINOCTIAL, NATURAL, or SOLAR YEAR.

tropic currents. Tidal currents occurring semimonthly when the effect of the moon's maximum declination is greatest. At these times the tendency of the moon to produce a diurnal inequality in the current is at a maximum.

tropic higher high water. The higher high water of tropic tides. See also TROPIC TIDES.

tropic higher high water interval. The lunitidal interval pertaining to the higher high waters at the time of the tropic tides. See also TROPIC LOWER LOW WATER INTERVAL.

tropic higher low water. The higher low water of tropic tides. See also TROPIC TIDES.

tropic high water inequality. The average difference between the two high waters of the day at the times of the tropic tides. Applicable only when the tide is semidiurnal or mixed. See also TROPIC TIDES, TROPIC LOW WATER INEQUALITY.

tropic inequalities. See TROPIC HIGH WATER INEQUALITY, TROPIC LOW WATER INEQUALITY.

tropic intervals. See TROPIC HIGH WATER INTERVAL, TROPIC LOWER LOW WATER INTERVAL.

tropic lower high water. The lower high water of tropic tides. See also TROPIC TIDES.

tropic lower low water. The lower low water of tropic tides. See also TROPIC TIDES.

tropic lower low water interval. The lunitidal interval pertaining to the lower low waters at the time of tropic tides. See also TROPIC HIGHER HIGH WATER INTERVAL.

tropic low water inequality. The average difference between the two low waters of the day at the times of the tropic tides. Applicable only when the type of tide is semidiurnal or mixed. See also TROPIC TIDES, TROPIC HIGH WATER INEQUALITY.

Tropic of Cancer. The northern parallel of declination, approximately 23°27′ from the celestial equator, reached by the sun at its maximum northerly declination, or the corresponding parallel on the earth. It is named for the sign of the zodiac in which the sun reached its maximum northerly declination at the time the parallel was so named.

Tropic of Capricorn. The southern parallel of declination, approximately 23°27′ from the celestial equator, reached by the sun at its maximum southerly declination, or the corresponding parallel on the earth. It is named for the sign of the zodiac in which the sun reached its maximum southerly declination at the time the parallel was so named.

tropic ranges. See GREAT TROPIC RANGE, MEAN TROPIC RANGE, SMALL TROPIC RANGE.

tropics, *n.* See TORRID ZONE.

tropic speed. The greater flood or greater ebb speed at the time of tropic currents.

tropic tides. Tides occurring semimonthly when the effect of the moon's maximum declination is greatest. At these times there is a tendency for an increase in the diurnal range. The tidal datums pertaining to the tropic tides are designated as tropic higher high water, tropic lower high water, tropic higher low water, and tropic lower low water.

tropopause, *n.* The boundary between the troposphere and the stratosphere.

troposphere, *n.* That portion of the atmosphere from the earth's surface to the tropopause, i.e., the lowest 10 to 20 kilometers of the atmosphere. It is characterized by decreasing temperature with height, appreciable vertical wind motion, appreciable water vapor content, and weather.

tropospheric radio duct. A quasi-horizontal layer in the troposphere between the boundaries of which radio energy of sufficiently high frequency is substantially confined and propagated with abnormally low attenuation. The duct may be formed in the lower portion of the atmosphere when there is a marked temperature inversion or a sharp decrease in water vapor with increased height. See also SURFACE DUCT, ELEVATED DUCT.

tropospheric wave. A radio wave traveling between points on or near the surface of the earth by one or more paths lying wholly within the troposphere. The propagation of this wave is determined primarily by the distribution of the refractive index in the troposphere.

trot, *n. British terminology.* A line of buoys for mooring small ships and craft by head and stern.

trough, *n.* 1. A long depression of the sea floor, characteristically flat bottomed and steep sided, and normally shallower than a trench.

2. In meteorology, an elongated area of relatively low pressure. The opposite of a trough is called RIDGE. The term *trough* is commonly used to distinguish the above elongated area from the closed circulation of a low (or cyclone). But a large-scale trough may include one or more lows.

3. The lowest part of a wave, between two crests is called WAVE TROUGH.

TR switch. A switch used to automatically decouple the receiver from the antenna during transmission when there is a common transmitting and receiving antenna. Also called TR BOX.

TR tube. An electronic switch capable of rapid switching from transmit to receive functions, and vice versa, of a radar set, used to protect the receiver from damage by potent energy generated by the transmitter, which damage may be incurred otherwise due to use of a common waveguide. Another device called the anti-TR tube is used to block the passage of echoes to the receiver during the relatively long periods when the transmitter is inactive.

See also TR SWITCH, ATR TUBE.

true, *adj.* 1. Related to true north.

2. Actual, as contrasted with fictitious, as *true sun.*

3. Related to a fixed point, either on the earth or in space, as *true wind*; in contrast with RELATIVE, which is related to a moving point.

4. Corrected, as *true altitude.*

true altitude. See OBSERVED ALTITUDE.

true amplitude. Amplitude relative to true east or west.

true anomaly. See under ANOMALY, definition 2.

true azimuth. Azimuth relative to true north.

true bearing. Bearing relative to true north; compass bearing corrected for compass error.

true course. Course relative to true north.

true direction. Horizontal direction expressed as angular distance from true north.

true heading. Heading relative to true north.

true meridian. This expression is used to distinguish the great circle through the geographical poles from MAGNETIC MERIDIAN, COMPASS MERIDIAN, or GRID MERIDIAN, the north-south lines according to magnetic, compass, or grid direction, respectively.

true motion display. A type of radarscope display in which own ship and other moving targets move on the plan position indicator in accordance with their true courses and speeds. All fixed targets appear as stationary echoes. However, uncompensated set and drift of own ship may result in some movement of the echoes of stationary targets. This display is similar to a navigational (geographical) plot. See also RELATIVE MOTION DISPLAY.

true motion radar. A radar set which provides a true motion display on the PPI as opposed to the relative motion, true or relative bearing, display most commonly used. The true motion radar requires own ship's speed input, either log or manual, in addition to own ship's course input.

true north. The direction of the north geographical pole; the reference direction for measurement of true directions.

true plot. See GEOGRAPHICAL PLOT.

true prime vertical. See under PRIME VERTICAL CIRCLE.

true solar time. See APPARENT TIME.

true sun. The actual sun as it appears in the sky. Usually called APPARENT SUN. See also MEAN SUN, DYNAMICAL MEAN SUN.

true track of target. Of a radar target, the motion of a target on a true motion display. When the true motion display is ground stabilized, i.e., allowance is made for the set and drift of current, the motion displayed is called GROUND TRACK. Without such stabilization the motion displayed is called WATER TRACK.

true wind. Wind relative to a fixed point on the earth. Wind relative to a moving point is called APPARENT or RELATIVE WIND.

trumpet, *n.* See HORN.

tsunami, tunami, *n.* A long-period sea wave, potentially catastrophic, produced by a submarine earthquake or volcanic eruption. It may travel unnoticed across the ocean for thousands of miles from its point of origin. It builds up to great heights over shoal water. Also called SEISMIC SEA WAVE, TIDAL WAVE.

Tsushima Current. That part of the Kuroshio flowing northeastward through Korea Strait and along the Japanese coast in the Japan Sea; it flows strongly eastward through Tsugaru Strait at speeds to 7 knots. The Tsushima Current is strong most of the time, averaging about 1 knot; however, it may weaken somewhat during autumn. In Western Channel, between Tsushima and southeastern Korea, tidal currents retard the general northeastward flowing Tsushima Current during the southwest-setting flood and reinforce it during the northeast-setting ebb. Resultant current speeds range from ¼ knot during flood to 3 knots during ebb. In the strait between Tsushima and Kyushu, the current flows northeastward throughout the year. Current speeds in Korea Strait also are affected by the seasonal variations of the monsoons. The strongest currents usually occur from July through November. The Tsushima Current divides after flowing through Korea Strait, a small branch flowing northward along the east coast of Korea as far as Vladivostok in summer. During this season the current is strongest and overcomes the weak southward flowing, coastal Liman Current. When the current combines with the ebb current, the resultant speed may reach 2 knots. During winter this branch of the Tsushima Current is weakest and is influenced by the stronger southward flowing Liman Current which normally extends as far south as 39°N, with speeds from ¼ to ¾ knot. The main body of the Tsushima Current flows northeastward off the northeast coast of Honshu. In summer, after entering the Japan Sea, its speed is about ½ to 1 knot. In winter the current is relatively weak, although near the islands and headlands speeds may exceed 1 knot, especially after northwesterly gales.

tuba, *n.* See FUNNEL CLOUD.

tufa, *n.* A porous rocky deposit formed in streams and in the ocean near the mouths of rivers.

tumble, *v., i.* Of a gyroscope, to precess suddenly and to an extreme extent as a result of exceeding its operating limits of bank or pitch.

tune, *v., t.* To adjust the frequency of a circuit or system to obtain optimum performance, commonly to adjust to resonance.

turbidity, *n.* The state or condition of having the transparency or translucence disturbed, as when sediment in water is stirred up, or when dust, haze, clouds, etc., appear in the atmosphere because of wind or vertical currents.

turbulence, *n.* The state or condition of being violently agitated or disturbed, as a stream which meets an obstacle, or air flowing over an uneven surface.

turbulent flow. Fluid motion in which random motions of parts of the fluid are superimposed upon a simple pattern of flow. All or nearly all fluid flow displays some degree of turbulence. The opposite is STREAMLINE FLOW.

turning basin. A water area used for turning vessels.

turning buoy. A buoy marking a turn, as in a channel.

turning circle. As a vessel makes a turn of 360° or more with constant ordered speed and rudder angle, the path, approximating a circle, described by the pivot point of the vessel.

turn of the tide. See CHANGE OF TIDE.

twenty-four hour satellite. See GEOSYNCHRONOUS SATELLITE.

twilight, *n.* The periods of incomplete darkness following sunset (**evening twilight**) or preceding sunrise (**morning twilight**). Twilight is designated as **civil, nautical,** or **astronomical,** as the darker limit occurs when the center of the sun is 6°, 12°, or 18° below the celestial horizon, respectively. See also DAWN, DUSK.

twinkle, *v., i.* To appear unsteady in position and brightness, as a star.

two-aerial consol. See as CONSOLAN under CONSOL.

two-body orbit. The motion of a point mass in the presence of the gravitational attraction of another point mass, and in the absence of other forces. This orbit is usually an ellipse, but may be a parabola or hyperbola.

two-degree-of-freedom gyro. A gyroscope the spin axis of which is free to rotate about two orthogonal axes, not counting the spin axis. See also DEGREE-OF-FREEDOM.

two-tone diaphone. See under DIAPHONE.

two-way route. A route within defined limits inside which two-way traffic is established, aimed at providing safe passage of ships through waters where navigation is difficult or dangerous. See also ROUTING SYSTEM.

tyfon, *n.* See TYPHON.

type of tide. A classification based on characteristic forms of a tide curve. Qualitatively, when the two high waters and two low waters of each tidal day are approximately equal in height, the tide is said to be **semidiurnal;** when there is a relatively large diurnal inequality in the high or low waters or both, it said to be **mixed;** and when there is only one high water and one low water in each tidal day, it is said to be **diurnal.**

typhon, *n.* A diaphragm horn which operates under the influence of compressed air or steam. Also called TYFON.

typhoon, *n.* See under TROPICAL CYCLONE.

U

Ulloa's ring. See BOUGUER'S HALO.

ultra high frequency. Radio frequency of 300 to 3,000 megahertz.

ultra quick light. A light in which flashes are repeated at a rate of not less than 160 flashes per minute. See also CONTINUOUS ULTRA QUICK LIGHT, INTERRUPTED ULTRA QUICK LIGHT.

ultrashort wave. A radio wave shorter than 10 meters. A wave shorter than 1 meter is called a MICROWAVE. See also WAVE.

ultrasonic, *adj.* Having a frequency above the audible range. Frequencies below the audible range are called INFRASONIC. See also SUPERSONIC.

ultrasonic depth finder. A direct-reading instrument which determines the depth of water by measuring the time interval between the emission of an ultrasonic signal and the return of its echo from the bottom. A similar instrument utilizing signals within the audible range is called a SONIC DEPTH FINDER. Both instruments are also called ECHO SOUNDERS.

umbra, *n.* 1. The darkest part of a shadow in which light is completely cut off by an intervening object. A lighter part surrounding the umbra, in which the light is only partly cut off, is called the PENUMBRA.
2. The darker central portion of a sun spot, surrounded by the lighter PENUMBRA.

uncorrecting, *n.* The process of converting true to magnetic, compass, or gyro direction, or magnetic to compass direction. The opposite is CORRECTING.

uncovered, *adj. & adv.* Above water. The opposite is SUBMERGED. See also AFLOAT, SURFACED; AWASH.

undercurrent, *n.* A current below the surface, particularly one flowing in a direction or at a speed differing from the surface current. See UNDERTOW, SUBSURFACE CURRENT, SURFACE CURRENT.

under the lee. To leeward.

undertow, *n.* A current below the surface flowing in the opposite direction to the surface current, particularly, the receding water below the surface of breakers. See also UNDERCURRENT, SUBSURFACE CURRENT, SURFACE CURRENT, BACKRUSH, RIP CURRENT.

underway, under way, *adv.* 1. Without moorings; not secured in any way to the ground or a wharf. See also ADRIFT.
2. In motion, particularly the start of such motion after a standstill. See also MAKING WAY.

undevelopable, *adj.* Not capable of being developed, or flattened without distortion. The opposite is DEVELOPABLE.

undevelopable, *n.* A surface that cannot be flattened to form without compressing or stretching some part of it, such as a sphere. The opposite is DEVELOPABLE.

undisturbed orbit. See NORMAL ORBIT.

undulating, *adj.* Having the form of waves or swells, as *undulating land.*

undulating light. See under FIXED AND FLASHING LIGHT.

undulation of the geoid. See GEOIDAL HEIGHT.

undulatus, *adj.* Having undulations. The term is used to refer to a cloud composed of elongated and parallel elements resembling ocean waves.

unfavorable current. A current flowing in such a direction as to decrease the speed of a vessel over the ground. The opposite is FAVORABLE CURRENT.

unfavorable wind. A wind which delays the progress of a craft in a desired direction. Usually used in plural and chiefly in connection with sailing vessels. A wind which aids the progress of a craft is called a FAIR or FAVORABLE WIND. See also FOLLOWING WIND, HEAD WIND.

Uniform State Waterway Marking System. A system developed jointly by the U.S. Coast Guard and state boating administrators to assist the small craft operator in those state waters marked by participating states. It consists of two categories of aids to navigation. One is a system of aids to navigation, generally compatible with the Federal lateral system of buoyage, to supplement the federal system in state waters. The other is a system of regulatory markers to warn the small craft operator of dangers or to provide general information and directions.

unipole antenna, *n.* See ISOTROPIC ANTENNA.

unique sanctuary. A marine sanctuary established to protect a unique or nearly unique geologic, oceanographic, or living resource feature. See also MARINE SANCTUARY.

unit, *n.* A value, quantity, or magnitude in terms of which other values, quantities, or magnitudes are expressed. In general, a unit is fixed by definition and is independent of such physical conditions as temperature. An example is the *meter.* See also STANDARD, definition 2; INTERNATIONAL SYSTEM OF UNITS.

United States Coast Pilot. One of a series of nine SAILING DIRECTIONS published by the National Ocean Survey, that cover a wide variety of information important to navigators of U.S. coastal and intracoastal waters, and waters of the Great Lakes. Most of this information cannot be shown graphically on the standard nautical charts and is not readily available elsewhere. This information includes navigation regulations, outstanding landmarks, channel and anchorage peculiarities, dangers, weather, ice, freshets, pilots, and port facilities. Sailing directions for foreign waters are published by the Defense Mapping Agency Hydrographic/Topographic Center. Usually shortened to *Coast Pilot.* Each *Coast Pilot* is corrected through the dates of *Notices to Mariners* shown on the title page and should not be used without reference to the *Notices to Mariners* issued subsequent to those dates.

United States National Map Accuracy Standards. 1. **Horizontal accuracy**: For maps at publication scales larger than 1:20,000, 90 percent of all well-defined features, with the exception of those unavoidably displaced by exaggerated symbolization, will be located within 1/30 inch (0.85 mm) of their geographic positions as referred to the map projection; for maps at publication scales of 1:20,000 or smaller, 1/50 inch (0.50 mm).
2. **Vertical accuracy**: 90 percent of all contours will be accurate within one-half of the basic contour interval. Discrepancies in the accuracy of contours and elevations beyond this tolerance

may be decreased by assuming a horizontal displacement within 1/50 inch (0.50 mm). Also called MAP ACCURACY STANDARDS.

universal plotting sheet. See under SMALL AREA PLOTTING SHEET.

Universal Polar Stereographic grid. A military grid system based on the polar stereographic map projection, applied to maps of the earth's polar regions north of 84° N and south of 80° S.

Universal Time. Conceptually, time as determined from the apparent diurnal motion of a fictitious mean sun which moves uniformly along the celestial equator at the average rate of the apparent sun. Actually, Universal Time (UT) is related to the rotation of the earth through its definition in terms of sidereal time. Universal Time at any instant is derived from observations of the diurnal motions of the stars. The time scale determined directly from such observations is slightly dependent on the place of observation; this scale is designated UTO. By removing from UTO the effect of the variation of the observer's meridian due to the observed motion of the geographic pole, the scale UT1 is established. A scale designated UT2 results from applying to UT1 an adopted formula for the seasonal variation in the rate of the earth's rotation. UT1 and UT2 are independent of the location of the observer. UT1 is the same as Greenwich mean time used in navigation. See also TIME SCALE.

Universal Transverse Mercator grid. A military grid system based on the transverse Mercator map projection, applied to maps of the earth's surface extending to 84° N and 80° S.

unlighted buoy. A buoy not fitted with a light. Unlighted buoys have either a can or nun shape. Can buoys have a cylindrical shape whereas nun buoys have a conical shape usually located on top of a cylindrical shape.

unlighted sound buoy. See under SOUND BUOY.

unmanned light. A light which is operated automatically and may be maintained in service automatically for extended periods of time, but with routine visits for maintenance purposes. Also called UNWATCHED LIGHT.

unperturbed orbit. See NORMAL ORBIT.

unsettled, *adj.* Pertaining to fair weather which may at any time become rainy, cloudy, or stormy. See also SETTLED.

unstabilized in azimuth. See under STABILIZATION OF RADARSCOPE DISPLAY.

unwatched light. See UNMANNED LIGHT.

upper branch. That half of a meridian or celestial meridian from pole to pole which passes through a place or its zenith.

upper culmination. See UPPER TRANSIT.

upper limb. That half of the outer edge of a celestial body having the greatest altitude, in contrast with the LOWER LIMB, that half having the least altitude.

upper transit. Transit of the upper branch of the celestial meridian. Transit of the lower branch is called LOWER TRANSIT. Also called SUPERIOR TRANSIT, UPPER CULMINATION.

uprush, *n.* 1. The rush of the water onto the foreshore following the breaking of a wave.
2. See RUN-UP.

upstream, *adj. & adv.* Toward the source of a stream. The opposite is DOWNSTREAM.

up-the-scope echo. See CLASSIFICATION OF RADAR ECHOES.

upwelling, *n.* The process by which water rises from a lower to a higher depth, usually as a result of divergence and offshore currents. Upwelling is most prominent where persistent wind blows parallel to a coastline so that the resultant wind-driven current sets away from the coast. Over the open ocean, upwelling occurs whenever the wind circulation is cyclonic, but is appreciable only in areas where that circulation is relatively permanent. It is also observable when the southern trade winds cross the equator.

upwind, *adj. & adv.* In the direction from which the wind is blowing. The opposite is DOWNWIND.

U.S. Survey foot. The foot used by the *National Ocean Survey* in which 1 inch is equal to 2.540005 centimeters. The foot equal to 0.3048 meter, exactly, adopted by Australia, Canada, New Zealand, South Africa, the United Kingdom, and the United States in 1959 was not adopted by the *National Ocean Survey* because of the extensive revisions which would be necessary to their charts and measurement records.

UTC, *n.* See under COORDINATED UNIVERSAL TIME.

UTO, *n.* See under UNIVERSAL TIME.

UT1, *n.* See under UNIVERSAL TIME.

UT2, *n.* See under UNIVERSAL TIME.

V

vacuum, *n.* A space entirely devoid of matter, or a very rarefied space.

valley, *n.* On the sea floor, a relatively shallow, wide depression, the bottom of which usually has a continuous gradient. This term is generally not used for features that have canyon-like characteristics for a significant portion of their extent.

valley breeze. A gentle wind blowing up a valley or mountain slope in the absence of cyclonic or anticyclonic winds, caused by the warming of the mountainside and valley floor by the sun. See also KATABATIC WIND, MOUNTAIN BREEZE.

Van Allen Radiation Belts. Popular term for regions of high energy charged particles trapped in the earth's magnetic field. Definition of size and shape of these belts depends on selection of an arbitrary standard of radiation intensity and the predominant particle component. Belts known to exist are: a proton region centered at about 2,000 miles altitude at the geomagnetic equator; an electron region centered at about 12,000 miles altitude at the geomagnetic equator; overlapping electron and proton regions centered at about 20,000 miles altitude at the geomagnetic equator. Trapped radiation regions from artificial sources also exist. These belts were first reported by Dr. James A. Van Allen of the Iowa State University.

vane, *n.* 1. A device to indicate the direction from which the wind blows. Also called WEATHER VANE, WIND VANE. See also ANEMOMETER.
2. A sight on an instrument used for observing bearings, as on a pelorus, azimuth circle, etc. That vane nearest the observer's eye is called **near vane** and that on the opposite side is called **far vane**. Also called SIGHTING VANE.
3. In current measurements, a device to indicate the direction toward which the current flows.

vanishing tide. In a mixed tide with very large diurnal inequality, the lower high water (or higher low water) frequently becomes indistinct (or vanishes) at time of extreme declinations. During these periods the diurnal tide has such overriding dominance that the semidiurnal tide, although still present, cannot be readily seen on the tide curve.

vapor pressure. 1. The pressure exerted by the vapor of a liquid in a confined space such that vapor can form above it.
2. The pressure of water vapor in the air; that part of the total atmospheric pressure which is due to water vapor.

vardar, *n.* A cold fall wind blowing from the northwest down the Vardar valley in Greece to the Gulf of Salonica. It occurs when atmospheric pressure over eastern Europe is higher than over the Aegean Sea, as is often the case in winter. Also called VARDARAC.

vardarac, *n.* See VARDAR.

variable, *n.* A quantity to which a number of values can be assigned.

variable parameters of satellite orbit. See under FIXED AND VARIABLE PARAMETERS OF SATELLITE ORBIT.

variable range marker. A luminous range ring on the plan position indicator, the radius of which is continuously adjustable. The range setting of this marker is read on the range counter of the radar indicator.

variable star. A star which is not of constant magnitude.

variance, *n.* The square of the standard deviation.

variation, *n.* 1. The angle between the magnetic and geographic meridians at any place, expressed in degrees and minutes east or west to indicate the direction of magnetic north from true north. The angle between magnetic and grid meridians is called GRID MAGNETIC ANGLE, GRID VARIATION, or GRIVATION. Called MAGNETIC VARIATION when a distinction is needed to prevent possible ambiguity. Also called MAGNETIC DECLINATION.
2. Change or difference from a given value.

variation of latitude. A small change in the astronomical latitude of points on the earth due to polar motion.

variation of the poles. See POLAR MOTION.

variometer, *n.* An instrument for comparing magnetic forces, especially of the earth's magnetic field.

vast floe. See under FLOE.

V-band. A radio-frequency band of 46.0 to 56.0 kilomegahertz. See also FREQUENCY, FREQUENCY BAND.

vector, *n.* Any quantity, such as a *force*, *velocity*, or *acceleration*, which has both magnitude and direction, as opposed to a SCALAR which has magnitude only. Such a quantity may be represented geometrically by an arrow of length proportional to its magnitude, pointing in the assigned direction.

vector addition. The combining of two or more vectors in such manner as to determine the equivalent single vector. The opposite is called RESOLUTION OF VECTORS. Also called COMPOSITION OF VECTORS.

vector diagram. A diagram of more than one vector drawn to the same scale and reference direction and in correct position relative to each other. A vector diagram composed of vectors representing the actual courses and speeds of two craft and the relative motion vector of either one in relation to the other may be called a SPEED TRIANGLE.

vector quantity. A quantity having both magnitude and direction and hence capable of being represented by a vector. A quantity having magnitude only is called a SCALAR.

veer, *v., i.* 1. Of the wind, to change direction in a clockwise direction in the Northern Hemisphere and a counterclockwise direction in the Southern Hemisphere. Change in the opposite direction is called BACK.
2. Of the wind, to shift aft. The opposite motion is to HAUL **forward**.

veer, *v., t.* To pay or let out, as to *veer* anchor chain.

vehicle location monitoring. A service provided to maintain the orderly and safe movement of platforms or vehicles. It encompasses the systematic observation of airspace, surface, and subsurface areas by electronic, visual, and other means to locate, identify, and control the movement of platforms or vehicles.

velocity, *n.* A vector quantity equal to speed in a given direction.

velocity meter. See INTEGRATING ACCELEROMETER.

velocity of current. Speed and set of the current.

velocity ratio. The ratio of two speeds, particularly the ratio of the speed of tidal current at a subordinate station to the speed of the corresponding current at the reference station.

Venus, *n.* The navigational planet whose orbit is smaller than that of the earth.

verglas, *n.* See GLAZE.

vernal, *adj.* Pertaining to spring. The corresponding adjectives for summer, fall, and winter are *aestival, autumnal,* and *hibernal.*

vernal equinox. 1. That point of intersection of the ecliptic and the celestial equator, occupied by the sun as it changes from south to north declination, on or about March 21. Also called MARCH EQUINOX, FIRST POINT OF ARIES.
2. That instant the sun reaches the point of zero declination when crossing the celestial equator from south to north.

vernier, *n.* A short, auxiliary scale situated alongside the graduated scale of an instrument, by means of which fractional parts of the smallest division of the primary scale can be measured with increased accuracy. If 10 graduations on a vernier equal 9 graduations on the micrometer drum of a sextant, it can be seen that when the zero on the vernier lies one-tenth of a graduation beyond zero on the micrometer drum, the first graduation beyond zero on the vernier coincides with a graduation on the micrometer drum. Likewise, when the zero on the vernier lies five-tenths of a graduation beyond zero on the micrometer drum, the fifth graduation beyond zero on the vernier coincides with a graduation on the micrometer drum.

vernier error. Inaccuracy in the graduations of the scale of a vernier.

vernier sextant. A marine sextant providing a precise reading by means of a vernier used *directly* with the arc, and having either a clamp screw or an endless tangent screw for controlling the position of the index arm. The micrometer drum on a **micrometer drum sextant** may include a vernier to enable a more precise reading.

versed cosine. See COVERSINE.

versine, versed sine, *n.* One minus the cosine, i.e., (1 – cos). The expression NATURAL VERSINE is sometimes used to distinguish the versine from its logarithm (called LOGARITHMIC VERSINE).

vertex (*pl. vertices*), *n.* The highest point. The vertices of a great circle are the points nearest the poles. See also APEX.

vertical, *adj.* In the direction of gravity; perpendicular to the plane of the horizon.

vertical, *n.* A vertical line, plane, etc.

vertical axis. The line through the center of gravity of a craft, perpendicular to both the longitudinal and lateral axes, around which it yaws.

vertical beam width. The beam width measured in a vertical plane.

vertical circle. A great circle of the celestial sphere, through the zenith and nadir. Vertical circles are perpendicular to the horizon. The **prime vertical circle** or **prime vertical** passes through the east and west points of the horizon. The **principal vertical circle** passes through the north and south points of the horizon and coincides with the celestial meridian.

vertical control datum. See VERTICAL GEODETIC DATUM.

vertical danger angle. The maximum or minimum angle between the top and bottom of an object of known height, as observed from a craft, indicating the limit of safe approach to an offlying danger. See also DANGER ANGLE.

vertical datum. 1. A base elevation used as a reference from which to reckon heights or depths. It is called TIDAL DATUM when defined by a certain phase of the tide. Tidal datums are local datums and should not be extended into areas which have differing topographic features without substantiating measurements. In order that they may be recovered when needed, such datums are referenced to fixed points known as **bench marks.** See also CHART SOUNDING DATUM.
2. See VERTICAL GEODETIC DATUM.

vertical earth rate. To compensate for the effect of earth rate, the rate at which a gyroscope must be turned about its vertical axis for the spin axis to remain in the meridian. Vertical earth rate is maximum at the poles, zero at the equator, and varies as the sine of the latitude. See also EARTH RATE, HORIZONTAL EARTH RATE.

vertical force instrument. See HEELING ADJUSTER.

vertical geodetic datum. Any level surface taken as a surface of reference from which to reckon elevations. See also DATUM. Also called VERTICAL DATUM, VERTICAL CONTROL DATUM.

vertical intensity of the earth's magnetic field. The strength of the vertical component of the earth's magnetic field.

vertical lights. Two or more lights disposed vertically, or geometrically to form a triangle, square or other figure. If the individual lights serve different purposes, those of lesser importance are called AUXILIARY LIGHTS.

vertically polarized wave. A plane polarized electromagnetic wave in which the electric field vector is in a vertical plane.

very close pack ice. Pack ice in which the concentration is 9/10 to less than 10/10 (7/8 to less than 8/8).

very high frequency. Radio frequency of 30 to 300 megahertz.

very low frequency. Radio frequency below 30 kilohertz.

very open pack ice. Pack ice in which the concentration is 1/10 to 3/10 (1/8 to less than 3/8) and water preponderates over ice.

very quick flashing light. A light used with the IALA System, the phase characteristic of which is formed by groups of flashes shown at a rate of 100 or 120 flashes per minute. Each group usually consists of three, six, or nine flashes.

very quick light. A light in which flashes are repeated at a rate of not less than 80 flashes per minute but less than 160 flashes per minute. See also CONTINUOUS VERY QUICK LIGHT, GROUP VERY QUICK LIGHT, INTERRUPTED VERY QUICK LIGHT.

very small fracture. See under FRACTURE.

very weathered ridge. A ridge with tops very rounded, the slopes of the sides usually being about 20° to 30°.

vessel, *n.* Any type of craft, except aircraft, which can be used for transportation across or through water.

Vessel Traffic Services. Being implemented under the authority of the Ports and Waterways Safety Act of 1972 (Public Law 92-340) and the St. Lawrence Seaway Act (Public Law 358), Vessel Traffic Services (VTS) encompass a wide range of techniques and capabilities primarily aimed at preventing vessel collisions, rammings, and groundings in the port and waterway environment. They are also designed to expedite ship movements, increase system capacity, and improve all-weather operating capability. Based on local needs and safety requirements, such services may vary in design and complexity from passive services such as traffic separation schemes and regulated navigation areas to manned services with communications, electronic surveillance, and automated capabilities.

viaduct, *n.* A structure consisting of a series of arches or towers supporting a roadway, waterway, etc., across a depression, etc. See also BRIDGE, definition 2; CAUSEWAY.

vibrating needle. A magnetic needle used in compass adjustment to find the relative intensity of the horizontal components of the earth's magnetic field and the magnetic field at the compass location. Also called HORIZONTAL FORCE INSTRUMENT.

vibration, *n.* 1. Periodic motion of an elastic body or medium in alternately opposite directions from equilibrium; oscillation. In **longitudinal vibration** the direction of motion of the particles is the same as the direction of advance of the vibratory motion; in **transverse vibration** it is perpendicular to the direction of advance.
2. The motion of a vibrating body during one complete cycle; two oscillations.

video, *n.* In the operation of a radar set, the demodulated receiver output that is applied to the indicator. Video contains the relevant radar information after removal of the carrier frequency.

vigia, *n.* A rock or shoal the existence or position of which is doubtful, or a warning note to this effect on the chart.

violent storm. Wind of force 11 (56 to 63 knots or 64 to 72 miles per hour) on the Beaufort wind scale. See also STORM, definition 1.

virga, *n.* Wisps or streaks of water or ice particles falling out of a cloud but evaporating before reaching the earth's surface as precipitation. Virga is frequently seen trailing from altocumulus and altostratus clouds, but also is discernible below the bases of high-level cumuliform clouds from which precipitation is falling into a dry subcloud layer. It typically exhibits a hooked form in which the streaks descend nearly vertically just under the precipitation source but appear to be almost horizontal at their lower extremities. Such curvature of virga can be produced simply by effects of strong vertical windshear, but ordinarily it results from the fact that droplet or crystal evaporation decreases the particle terminal fall velocity near the ends of the streaks. Also called FALL STREAKS, PRECIPITATION TRAILS.

virtual image. An image that cannot be shown on a surface but is visible, as in a mirror.

virtual meridian. The meridian in which the spin axis of a gyrocompass will settle as a result of speed-course-latitude error.

visibility, *n.* That property of the atmosphere which determines the ability of an observer to see and identify prominent objects by day, or lights or lighted objects by night. A measure of this property is expressed in units of distance. This term should not be confused with VISUAL RANGE. See also METEOROLOGICAL VISIBILITY.

visible horizon. That line where earth and sky appear to meet, and the projection of this line upon the celestial sphere. If there were no terrestrial refraction, VISIBLE and GEOMETRICAL HORIZONS would coincide. Also called APPARENT HORIZON.

visit, *n.* The event of Coast Guard personnel being present at a manned or unmanned aid to navigation, regardless of reason.

visual aid to navigation. An aid to navigation which transmits information through its visual observation. It may be lighted or unlighted.

visual bearing. A bearing obtained by visual observation.

visual range. The maximum distance (usually horizontal) at which a given object can be seen by day in any particular circumstance, as limited by the atmospheric transmission. The distance is such that the contrast of the object with its background is reduced by the atmosphere to the contrast threshold value for the observer. This term should not be confused with VISIBILITY. See also CONTRAST THRESHOLD, VISUAL RANGE (OF A LIGHT).

visual range (of a light). The predicted range at which a light can be observed. The predicted range may be either the *luminous range* or the *geographic range*. Therefore, in predicting the range at which a light can be seen, one first determines the geographic range to compare this range with the luminous range, if known. If the geographic range is less than the luminous range, the geographic range must be taken as the limiting range. If the luminous range is less than the geographic range, the luminous range must be taken as the limiting range. The **luminous range** is the maximum distance at which a light can be seen under existing visibility conditions. This luminous range takes no account of the elevation of the light, the observer's height of eye, the curvature of the earth, or interference from background lighting. The luminous range is determined from the known nominal luminous range, called the *nominal range*, and the existing visibility conditions, using the Luminous Range Diagram. The **nominal range** is the maximum distance at which a light can be seen in clear weather as defined by the International Visibility Code (meteorological visibility of 10 nautical miles). The **geographic range** is the maximum distance at which the curvature of the earth and terrestrial refraction permit a light to be seen from a particular height of eye *without* regard to the luminous intensity of the light. (The geographic range *sometimes* printed on charts or tabulated in light lists is the maximum distance at which the curvature of the earth and refraction permit a light to be seen from a height of eye of 15 feet above the water when the elevation of the light is taken above the height datum of the largest scale chart of the locality.) See also VISUAL RANGE, CONTRAST THRESHOLD.

volcanic ashes. Numerous tiny particles of lava resembling ashes, which issue from a volcano. Part of the ocean bed is composed of volcanic ashes.

volcano, *n.* An opening in the earth from which hot gases, smoke, and molten material issue, or a hill or mountain composed of volcanic material. A volcano is characteristically conical in shape with a crater in the top.

volt, *n.* A *derived unit* of electric potential in the International System of Units; it is the difference of electric potential between two points of a conducting wire carrying a constant current of 1 ampere, when the power dissipated between these points is equal to 1 watt.

volt per metre. The *derived unit* of electric field strength in the International System of Units.

volume, *n.* 1. Cubic content, measured in gallons,

barrels, liters, etc.
2. Loudness of a sound, usually measured in decibels.

voyage, *n.* 1. The outward and homeward passages of a trip by sea.
2. A trip by sea.

vulgar establishment. See under ESTABLISHMENT OF THE PORT.

W

wandering of the poles. See EULERIAN MOTION.

waning moon. The moon between full and new when its visible part is decreasing. See also PHASES OF THE MOON.

warble tone. A tone whose frequency varies periodically about a mean value.

warm air mass. An air mass that is warmer than surrounding air. The expression implies that the air mass is warmer than the surface over which it is moving.

warm braw. A foehn in the Schouten Islands north of New Guinea.

warm front. Any non-occluded front, or portion thereof, which moves in such a way that warmer air replaces colder air. While some occluded fronts exhibit this characteristic, they are more properly called WARM OCCLUSIONS.

warm occlusion. See under OCCLUDED FRONT.

warm sector. An area at the earth's surface bounded by the warm and cold fronts of a cyclone.

warning beacon. See WARNING RADIOBEACON.

warning radiobeacon. An auxiliary radiobeacon that is located at a lightship to warn vessels of their proximity to the lightship. It is of short range and sounds a warbling note for 1 minute immediately following the main radiobeacon on the same frequency. Also called WARNING BEACON.

warp, *v., t.* To move, as a vessel, from one place to another by means of lines fastened to an object, such as a buoy, wharf, etc., secured to the ground. See also KEDGE.

warping buoy. A buoy so located that lines to it can be used for the movement of ships.

wash, *n.* The dry channel of an intermittent stream.

watch, *n.* A small timepiece of a size convenient to be carried on the person. A **hack** or **comparing watch** is used for timing observations of celestial bodies. A **stop watch** can be started, stopped, and reset at will, to indicate elapsed time. A **chronometer watch** is a small chronometer, especially one with an enlarged watch-type movement.

watch buoy. See STATION BUOY.

watch error. The amount by which watch time differs from the correct time. It is usually expressed to an accuracy of 1^s and labeled *fast* (F) or *slow* (S) as the watch time is later or earlier, respectively, than the correct time. See also CHRONOMETER ERROR.

watching properly. Said of an aid to navigation on charted position and exhibiting its advertised characteristics in all respects.

watch rate. The amount gained or lost by a watch or clock in unit time. It is usually expressed in seconds per 24 hours, to an accuracy of 0.1^s, and labeled *gaining* or *losing*, as appropriate, when it is sometimes called DAILY RATE.

watch time. The hour of the day as indicated by a watch or clock. Watches and clocks are generally set approximately to zone time. Unless a watch or clock has a 24-hour dial, watch time is usually expressed on a 12-hour cycle and labeled AM or PM.

watch tower. See LOOKOUT STATION.

water-borne, *adj.* Floating on water; afloat. See also SEA-BORNE.

watercourse, *n.* 1. A stream of water.
2. A natural channel through which water may or does run. See also GULLY, WASH.

waterfall, *n.* A perpendicular or nearly perpendicular descent of water in a stream.

waterline, *n.* The line marking the junction of water and land. See also HIGH WATER LINE, LOW WATER LINE, SHORELINE.

water sky. Dark streaks on the underside of low clouds, indicating the presence of water features in the vicinity of sea ice.

water smoke. See STEAM FOG.

waterspout, *n.* 1. A tornado occurring over water. Waterspouts are most common over tropical and subtropical waters.
2. A whirlwind over water comparable in intensity to a dust devil over land.

water tower. A label on a nautical chart which indicates a structure enclosing a tank or standpipe so that the presence of the tank or standpipe may not be apparent.

water track. 1. See under TRACK, definition 2.
2. See under TRUE TRACK OF TARGET.

waterway, *n.* A water area providing a means of transportation from one place to another, principally a water area providing a regular route for water traffic, such as a bay, channel, passage, or the regularly traveled parts of the open sea. The terms WATERWAY, FAIRWAY, and THOROUGHFARE have nearly the same meanings. WATERWAY refers particularly to the navigable part of a water area. FAIRWAY refers to the main traveled part of a waterway. A THOROUGH FARE is a public waterway. See also CANAL.

watt, *n.* A *derived unit* of power in the International System of Units; it is that power which in 1 second gives rise to energy of 1 joule.

wave, *n.* 1. An undulation or ridge on the surface of a fluid. A **wind wave** is generated by friction between wind and the fluid surface. See also STORM SURGE, TIDAL WAVE, TSUNAMI.
2. A disturbance propagated in such a manner that it may progress from point to point. See also ELECTROMAGNETIC WAVES, RADIO WAVES, SKYWAVE, GROUNDWAVE, DIRECT WAVE, INDIRECT WAVE, MODULATED WAVE, MICROWAVE, SPHERICAL WAVE, TRANSVERSE WAVE, LONGITUDINAL WAVE.

wave basin. A basin close to the inner entrance of a harbor in which the waves from the outer entrance are absorbed, thus reducing the size of the waves entering the inner harbor. See also WAVE TRAP.

wave crest. The highest part of a wave.

wave cyclone. A cyclone which forms and moves along a front. The circulation about the cyclone center tends to produce a wavelike deformation of the front. The wave cyclone is the most frequent form of extratropical cyclone (or low). Also called WAVE DEPRESSION. See also FRONTAL CYCLONE.

wave depression. See WAVE CYCLONE.

wave direction. The direction *from* which waves are moving.

waveguide, *n.* Broadly, a system of material boundries capable of guiding electromagnetic waves. Specifically, a transmission line consisting of a hollow conducting tube within which electromagnetic waves may be propagated; or a solid dielectric or dielectric-filled conductor designed for the same purpose.

wave height. The distance from the trough to the crest of a wave, equal to double the amplitude, and measured perpendicular to the direction of advance.

wave height correction. A correction due to the elevation of parts of the sea surface by wave action, particularly such a correction to a sextant altitude because of altered dip.

wave interference. See INTERFERENCE, definition 2.

wavelength, *n.* The distance between corresponding points in consecutive cycles in a wave train, measured in the direction of propagation at any instant.

wave of translation. A wave in which the individual particles of the medium are shifted in the direction of wave travel, as ocean waves in shoal waters; in contrast with an OSCILLATORY WAVE, in which only the form advances, the individual particles moving in closed orbits, as ocean waves in deep water.

wave period. The time interval between passage of successive wave crests at a fixed point.

wave train. A series of waves moving in the same direction. See also SOLITARY WAVE.

wave trap. A device used to reduce the size of waves from sea or swell which enter a harbor before they penetrate as far as the quayage; usually in the form of diverging breakwaters, or small projecting breakwaters situated close within the entrance. See also WAVE BASIN.

wave trough. The lowest part of a wave form between successive wave crests.

waxing moon. The moon between new and full when its visible part is increasing. See also PHASES OF THE MOON.

waypoint, *n.* A reference point on the track.

weak fix. A fix determined from horizontal sextant angles between objects poorly located.

weather, *adj.* Of or pertaining to the direction *from* which the wind is blowing. LEE pertains to the direction *toward* which the wind is blowing.

weather, *n.* 1. The state of the atmosphere as defined by various meteorological elements, such as temperature, pressure, wind speed and direction, humidity, cloudiness, precipitation, etc. This is in contrast with CLIMATE, the prevalent or characteristic meteorological conditions of a place or region.

2. Bad weather. See also THICK WEATHER.

weathered, *adj.* Having been eroded by action of the weather.

weathered berg. An irregularly shaped iceberg. Also called GLACIER BERG.

weathered ridge. A ridge with peaks slightly rounded, the slopes of the sides usually being about 30° to 40°. Individual fragments are not discernible.

weathering, *n.* Processes of ablation and accumulation which gradually eliminate irregularities in an ice surface.

weather map. See under SYNOPTIC CHART.

weather shore. As observed from a vessel, the shore lying in the direction from which the wind is blowing. See also LEE SHORE.

weather side. The side of a ship exposed to the wind or weather.

weather vane. A device to indicate the direction from which the wind blows. Also called WIND DIRECTION INDICATOR, WIND VANE. See also ANEMOMETER.

weber, *n.* A *derived unit* of magnetic flux in the International System of Units; it is that magnetic flux which, linking a circuit of one turn, would produce in it an electromotive force of 1 volt if it were reduced to zero at a uniform rate in 1 second.

wedge. See RIDGE, definition 3.

weight, *n.* A word which denotes a quantity of the same nature as a force; the weight of a body is the product of its mass and the acceleration due to gravity; in particular, the *standard weight* of a body is the product of its *mass* and the *standard acceleration* due to gravity. The value adopted in the International Service of Weights and Measures for the standard acceleration due to gravity is 980.665 centimeters per second, per second.

weighted mean. A value obtained by multiplying each of a series of values by its assigned weight and dividing the sum of those products by the sum of the weights. See also WEIGHT OF OBSERVATION.

weight of observation. The relative value of an observation, source, or quantity when compared with other observations, sources, or quantities of the same or related quantities. The value determined by the most reliable method is assigned the greatest weight. See also WEIGHTED MEAN.

wellhead, *n.* A submarine structure projecting some distance above the seabed and capping a temporarily abandoned or suspended oil or gas well. See also SUBMERGED PRODUCTION WELL.

west, *n.* The direction 90° to the left or 270° to the right of north. See also CARDINAL POINT.

West Australia Current. An Indian Ocean current which generally first flows northward and then northwestward off the west coast of Australia. This current varies seasonally with the strength of the wind and is most stable during November, December, and January, and least stable during May, June, and July, when it may set in any direction. North of 20°S the main part of this current flows northwestward into the Indian South Equatorial Current.

westerlies, *n., pl.* Winds blowing from the west, particularly the prevailing westerlies on the poleward sides of the subtropical high-pressure belts.

West Greenland Current. The ocean current flowing northward along the west coast of Greenland into Davis Strait. It is a continuation of the East Greenland Current. Part of the West Greenland Current turns around when approaching the Davis Strait and joins the Labrador Current; the rest rapidly loses its character as a warm current as it continues into Baffin Bay.

westing, *n.* The distance a craft makes good to the west. The opposite is EASTING.

westward motion. The motion in a westerly direction of the subtrack of a satellite, including the motion due to the earth's rotation and the nodical precession of the orbital plane.

West Wind Drift. An ocean current that flows eastward through all the oceans around the Antarctic Continent. On its northern edge it is continous with the South Atlantic Current, the South Pacific Current, and the South Indian Current. Also called ANTARCTIC CIRCUMPOLAR CURRENT.

wet-bulb temperature. The lowest temperature to which air can be cooled at any given time by evaporating water into it at constant pressure, when the heat required for evaporation is supplied by the cooling of the air. This temperature is indicated by a well-ventilated wet-bulb thermometer. See also FREE-AIR TEMPERATURE.

wet-bulb thermometer. A thermometer having the bulb covered with a cloth, usually muslin or cambric, saturated with water. See also PSY-

CHROMETER.

wet compass. See LIQUID COMPASS.

wet dock. See NON-TIDAL BASIN.

wharf, *n.* A structure of open rather than solid construction along a shore or a bank which provides berthing for ships and which generally provides cargo-handling facilities. A similar facility of solid construction is called QUAY. See also PIER, definition 1; DOCK; LANDING; MOLE, definition 1.

whirlpool, *n.* Water in rapid rotary motion. See also EDDY.

whirlwind, *n.* A general term for a small-scale, rotating column of air. More specific terms include DUST WHIRL, DUST DEVIL, WATERSPOUT, and TORNADO.

whirly, *n.* A small violent storm, a few yards to 100 yards or more in diameter, frequent in Antarctica near the time of the equinoxes.

whistle, *n.* A sound signal emitter comprising a resonator having an orifice of suitable shape such that when a jet of air is passed through the orifice the turbulence produces a sound.

whistle buoy. A sound buoy equipped with a whistle operated by wave action. The whistle makes a loud moaning sound as the buoy rises and falls in the sea. A buoy equipped with an electronic horn will produce a pure tone at regular intervals regardless of the sea state.

whitecap, *n.* A crest of a wave which becomes unstable in deep water, toppling over or "breaking." The instability is caused by the too rapid addition of energy from a strong wind. A wave which becomes unstable in shallow water is called a BREAKER.

white ice. See THIN FIRST-YEAR ICE.

white squall. A sudden, strong gust of wind coming up without warning other than by whitecaps or white, broken water. It is of doubtful existence and may be a popular myth.

white water. 1. Frothy water as in whitecaps or breakers.
2. Light-colored water over a shoal.

whole gale. A term once used by seamen for what is now called STORM on the Beaufort wind scale.

wide berth. A vessel which keeps well away from another ship or navigational hazard is said to give the other ship or hazard a *wide berth.*

williwaw, *n.* A sudden blast of wind descending from a mountainous coast to the sea, especially in the vicinity of either the Strait of Magellan or the Aleutian Islands.

willy-willy, *n.* See under TROPICAL CYCLONE.

wind cone. See WIND SOCK.

wind direction. The direction *from* which wind blows.

wind direction indicator. See WEATHER VANE.

wind drift current. See DRIFT CURRENT.

wind driven current. A current created by the action of the wind.

wind indicator. A device to indicate the direction or speed of the wind. See also ANEMOMETER.

wind rode. A ship riding at anchor is said to be *wind rode* when it is heading into the wind. See also TIDE RODE.

wind rose. 1. A diagram showing the relative frequency and sometimes the average speed of the winds blowing from different directions in a specified region.
2. A diagram showing the average relation between winds from different directions and the occurrence of other meteorological phenomena, such as rain.

winds aloft. Wind speeds and directions at various levels beyond the domain of surface weather observations.

wind shear. A change in wind direction or speed in a short distance, resulting in a shearing effect. It can act in a horizontal or vertical direction and, occasionally, in both. The degree of turbulence increases as the amount of wind shear increases.

wind-shift line. In meteorology, a line or narrow zone along which there is an abrupt change of wind direction.

wind sock. A tapered fabric sleeve mounted so as to catch and swing with the wind, thus indicating the wind direction. Also called WIND CONE.

wind speed. The rate of motion of air. See also ANEMOMETER.

wind storm. See under STORM, definiton 2.

wind vane. See WEATHER VANE.

wind velocity. The speed and direction of wind.

windward, *adj. & adv.* In the general direction from which the wind blows; in the wind; on the weather side. The opposite is LEEWARD.

windward, *n.* The weather side. The opposite is LEEWARD.

windward tide. A tidal current setting to windward. One setting in the opposite direction is called a LEEWARD TIDE or LEE TIDE.

wind wave. A wave generated by friction between wind and a fluid surface. Ocean waves are produced principally in this way.

winged headland. A seacliff with two bays or spits, one on either side.

winter, *n.* The coldest season of the year. In the Northern Hemisphere, winter begins astronomically at the winter solstice and ends at the vernal equinox. In the Southern Hemisphere the limits are the summer solstice and the autumnal equinox. The meteorological limits vary with the locality and the year.

winter buoy. An unlighted buoy which is maintained in certain areas during winter months when other aids to navigation are temporarily removed or extinguished.

Winter Coastal Countercurrent. See DAVIDSON CURRENT.

winter light. A light which is in service during those winter months when the regular light is extinguished. It has lower intensity than the regular light but usually has the same characteristic.

winter marker. An unlighted buoy or small lighted buoy, without sound signal, which is established as a replacement during the winter months when other aids are closed or withdrawn.

winter solstice. That point on the ecliptic occupied by the sun at maximum southerly declination. Sometimes called DECEMBER SOLSTICE, FIRST POINT OF CAPRICORNUS.

wiping, *n.* The process of reducing the amount of permanent magnetism in a vessel by placing a single coil horizontally around the vessel and moving it, while energized, up and down along the sides of the vessel. If the coil remains stationary, the process is called FLASHING. See also DEPERMING.

wire drag. An apparatus for surveying rock areas where the normal sounding methods are insufficient to insure the discovery of all existing obstructions, pinnacles, rocks, etc., above a given depth or for determining the least depth of an area. It consists essentially of a buoyed wire towed at the desired depth by two launches. Often shortened to DRAG. See also DRAG, *v., t.*

withdrawn, *adj.* Said of a floating aid to navigation which has been withdrawn during severe ice conditions or for the winter season. See also CLOSED, COMMISSIONED.

WMO Sea-Ice Nomenclature (WMO/OMM/BMO-No. 259.TP.145). A publication of the World Meteorological Organization which is comprised of sea-ice terminology, ice reporting codes, and an illustrated glossary. This publication results from fullscale international cooperation in the standardization of ice terminology.

working, *n.* In sea ice navigation, making headway through an ice pack by boring, breaking, and slewing.

World Geographic Reference System. A world-wide position reference system that may be applied to any map or chart graduated in latitude and longitude (with Greenwich as prime meridian) regardless of projection. It is a method of expressing latitude and longitude in a form suitable for rapid reporting and plotting. Commonly referred to by use of the acronym GEOREF.

world geodetic system. A consistent set of parameters describing the size and shape of the earth, the positions of a network of points with respect to the center of mass of the earth, transformations from major geodetic datums, and the potential of the earth (usually in terms of harmonic coefficients).

World Marine Weather Broadcasts. A joint publication of the National Weather Service and the Naval Weather Service Command providing information on marine weather broadcasts in all areas of the world where such service is provided. In general, English language broadcasts (or foreign language broadcasts repeated in English) are included in the publication. For areas where English language broadcasts are not available foreign language transmissions are also included. Sections 1 through 4 contain details of radiotelegraph, radiotelephone, radiofacsimile, and radioteleprinter transmissions, respectively. The NOAA (National Oceanic and Atmospheric Administration) Weather Radio continuous broadcasts (VHF-FM) and Great Lakes marine broadcasts are listed in section 5.

World Meteorological Organization. A specialized agency of the United Nations which seeks to facilitate world-wide cooperation in the establishment of stations for meteorological and related geophysical observations of centers providing meteorological services, of systems of rapid exchange of weather information; and to promote the standardization and publication of meteorological and hydrometeorological observations and statistics; to further the application of meteorology to aviation, shipping, agriculture, and other related activities; to encourage research and training in meteorology and their international coordination.

World Port Index. See PUB. 150.

World Wide Navigational Warning Service. Established through the joint efforts of the International Hydrographic Organization (IHO) and the Intergovernmental Maritime Consultative Organization (IMCO), the World Wide Navigational Warning Service (WWNWS) is a coordinated global service for the promulgation by radio of information on hazards to navigation which might endanger international shipping. The basic objective of the WWNWS is the timely promulgation by radio of information of concern to the ocean-going navigator. Such information includes failure and or changes to major navigational aids, newly discovered wrecks or natural hazards in or near main shipping lanes; areas where search and rescue, antipollution operations, cable-laying or other underway activities are taking place. Because of the wide ocean coverage of the WWNWS broadcasts, consideration is also being given as of 1981 to its selective use to augment other services for promulgating information concerning overdue and missing ships and aircraft. For WWNWS purposes, the world is divided into 16 NAVAREAS. Within each NAVAREA one national authority, designated the Area Coordinator, has assumed responsibility for the coordination and promulgation of warnings. Designated "National Coordinators" of other coastal states in a NAVAREA are responsible for collecting and forwarding information to the Area Coordinator. In the Baltic, a Sub-Area Coordinator has been established to filter information prior to passing to the Area Coordinator. Coordinators are responsible for the exchange of information as appropriate with other coordinators, including that which should be further promulgated by charting authorities in *Notice to Mariners*. The language used is English, although warnings may also be transmitted in one or more of the official languages of the United Nations. Broadcast schedules appear in an Annex to the International Telecommunication Union *List of Radiodetermination and Special Service Stations, Volume II*, and in the lists of radio signals published by various hydrographic authorities. Transmissions usually occur frequently enough during day to fall within at least one normal radio watch period, and the information is repeated with varying frequency as time passes until either the danger has passed or the information on it has appeared as a *Notice to Mariners*.

worldwide system. A term used to describe a navigation system providing positioning capability wherever the observer may be located on navigable waters. Also called GLOBAL SYSTEM. The Navy Navigation Satellite System is an example.

wreck, *n.* The ruined remains of a vessel which has been rendered useless, usually by violent action, as by the action of the sea and weather on a stranded or sunken vessel. In hydrography the term is limited to a wrecked vessel, either submerged or visible, which is attached to or foul of the bottom or cast up on the shore.

wreck buoy. A buoy marking the position of a wreck. It is usually placed on the seaward or channel side of the wreck and as near to the wreck as conditions will permit. To avoid confusion in some situations, two buoys may be used to mark the wreck. The possibility of the wreck having shifted position due to sea action between the times the buoy was established and later checked or serviced should not be overlooked. Also called WRECK-MARKING BUOY.

wreck mark. A navigation mark which marks the position of a wreck.

wreck-marking buoy. See WRECK BUOY.

X–Y–Z

X-band. A radio-frequency band of 5,200 to 10,900 megahertz. See also FREQUENCY, FREQUENCY BAND.

yard, *n.* A unit of length equal to 3 feet, 36 inches, or 0.9144 meter.

yaw, *n.* Of the motions of a vessel in a seaway, the oscillation of a vessel about a vertical axis approximately through the center of gravity. Also called YAWING.

yawing, *n.* See YAW.

year, *n.* A period of one revolution of the earth around the sun. The period of one revolution with respect to the vernal equinox, averaging 365 days, 5 hours, 48 minutes, 46 seconds in 1900, is called a **tropical, astronomical, equinoctial,** or **solar year.** The period with respect to the stars, averaging 365 days, 6 hours, 9 minutes, 9.5 seconds in 1900, is called a **sidereal year.** The period of revolution from perihelion to perihelion, averaging 365 days, 6 hours, 13 minutes, 53.0 seconds in 1900, is an **anomalistic year.** The period between successive returns of the sun to a sidereal hour angle of 80° is called a **fictitious** or **Besselian year.** A **civil year** is the **calendar year** of 365 days in common years, or 366 days in **leap years.** A **light-year** is a unit of length equal to the distance light travels in 1 year, about 5.88 × 10^{12} statute miles. The term *year* is occasionally applied to other intervals such as an **eclipse year,** the interval between two successive conjunctions of the sun with the same node of the moon's orbit, a period averaging 346 days, 14 hours, 52 minutes, 50.7 seconds in 1900, or a **great** or **Platonic year,** the period of one complete cycle of the equinoxes around the ecliptic, about 25,800 years.

young coastal ice. The initial stage of fast ice formation consisting of nilas or young ice, its width varying from a few meters up to 100 to 200 meters from the shoreline.

young ice. Ice in the transition stage between nilas and first-year ice, 10 to 30 centimeters in thickness. Young ice may be subdivided into GREY ICE and GREY-WHITE ICE.

Yukon standard time. See STANDARD TIME.

zenith, *n.* That point on the celestial sphere vertically overhead. The point 180° from the zenith is called the NADIR.

zenithal, *adj.* Of or pertaining to the zenith.

zenithal chart. See AZIMUTHAL CHART.

zenithal map projection. See AZIMUTHAL MAP PROJECTION.

zenith distance. Angular distance from the zenith; the arc of a vertical circle between the zenith and a point on the celestial sphere, measured from the zenith through 90°, for bodies above the horizon. This is the same as COALTITUDE with reference to the celestial horizon.

zephyr, *n.* A warm, gentle breeze, especially one from the west.

zodiac, *n.* The band of the sky extending 9° either side of the ecliptic. The sun, moon, and navigational planets are always within this band, with the occasional exception of Venus. The zodiac is divided into 12 equal parts, called *signs,* each part being named for the principal constellation *originally* within it.

zodiacal light. The faint light which extends upward from the horizon along the ecliptic after sunset or before sunrise, believed to be the reflection of sunlight by many tiny particles in the zodiac.

zone, *n.* 1. Part of the surface of a sphere or spheroid between two parallel planes. The surface of the earth is divided into climatic zones by the polar circles and the tropics; that part between the poles and polar circles being called **north** and **south frigid zones;** that part between the polar circles and the tropics, the **north** and **south temperate zones;** and that part between the two tropics, the **torrid zone.**
2. A time zone.
3. An area, region, or sector distinctively set off from other areas, regions, or sectors, or the whole.

zone description. The number, with its sign, that must be added to or subtracted from the zone time to obtain the Greenwich mean time. The zone description is usually a whole number of hours.

zone meridian. The meridian used for reckoning zone time. This is generally the nearest meridian whose longitude is exactly divisible by 15°. The DAYLIGHT SAVING MERIDIAN is usually 15° east of the zone meridian.

zone noon. Twelve o'clock zone time, or the instant the mean sun is over the upper branch of the zone meridian. **Standard noon** is 12 o'clock standard time. **Daylight saving** or **summer noon** usually occurs 1 hour later than zone or standard noon.

zone time. The local mean time of a reference or zone meridian whose time is kept throughout a designated zone. The zone meridian is usually the nearest meridian whose longitude is exactly divisible by 15°. **Standard time** is a variation of zone time with irregular but well-defined zone limits. **Daylight saving** or **summer time** is usually 1 hour later than zone or standard time. See also ZONE DESCRIPTION.

zulu time. See GREENWICH MEAN TIME.

ABBREVIATIONS

A, amplitude; augmentation; away (altitude intercept).
a, semimajor axis.
a, altitude intercept (Ho~Hc); altitude factor (change of altitude in 1 minute of time from meridian transit); assumed.
AC, alternating current.
add'l, additional.
ADF, automatic direction finder.
ADIZ, air defense identification zone.
AF, audio frequency.
AFC, automatic frequency control.
AGC, automatic gain control.
AISM, Association Internationale de Signalisation Maritime (International Association of Lighthouse Authorities).
*a*L, assumed latitude.
A.L.R.S., *Admiralty List of Radio Signals*.
AM, amplitude modulation.
AM, ante meridian (before noon).
AMVER, Automated Mutual-assistance Vessel Rescue system.
antilog, antilogarithm.
AP, assumed position.
approx., approximate, approximately.
ARPA, automatic radar plotting aid.
ASF, Additional Secondary Phase Factor.
AT, atomic time.
AU, astronomical unit.
AUSREP, Australian Ships Reporting System.
*a*λ, assumed longitude.
B, atmospheric pressure correction (altitude); bearing, bearing angle.
BFO, beat frequency oscillator.
BIH, Bureau Internationale de l'Heure.
BIPM, International Bureau of Weights and Measures.
BM, bench mark.
B_{pgc}, bearing per gyrocompass.
Brg., bearing (as distinguished from bearing angle).
C, Celsius (centigrade); chronometer time; compass (direction); correction; course, course angle.
CB, compass bearing.
CBDR, constant bearing, decreasing range.
CC, compass course; chronometer correction.
CCIR, International Radio Consultative Committee.
CCU, Consultative Committee for Units of the International Committee of Weights and Measures (CIPM).
CCZ, Coastal Confluence Zone.
CE, chronometer error; compass error.
cec, centicycle.
cel, centilane.
CEP, circular probable error.
CFR, Code of Federal Regulations.
CGPM, General Conference of Weights and Measures.
CH, compass heading.
CIPM, International Committee of Weights and Measures.
cm, centimeter, centimeters.
CMG, course made good.
Cn, course (as distinguished from course angle).
co-, the complement of (90° minus).
COA, course of advance.
COG, course over ground.
coL, colatitude.
colog, cologarithm.
corr., correction.
cos, cosine.
cot, cotangent.
cov, coversine.
CPE, circular probable error.
C_{pgc}, course per gyrocompass.
cps, cycles per second.
C_{psc}, course per standard compass.
$C_{p\ stg\ c}$, course per steering compass.
CRT, cathode-ray tube.
csc, cosecant.
CW, continuous wave.
CZn, compass azimuth.
D, deviation; dip (of horizon); distance.
d, declination (astronomical); altitude difference.
d, declination change in 1 hour.
dλ, difference of longitude (time units).
DC, direct current.
Dec., declination.
Dec. Inc., declination increment.
Dep., departure.
Dev., deviation.
DG, degaussing.
DHQ, mean diurnal high water inequality.
diff., difference.
Dist., distance.
D. Lat., difference of latitude.
DLo, difference of longitude (arc units).
DLQ, mean diurnal low water inequality.
DMAHTC, Defense Mapping Agency Hydrographic/Topographic Center.
DR, dead reckoning, dead reckoning position.
DRE, dead reckoning equipment.
DRT, dead reckoning tracer.
D_s, dip short of horizon.
DSD, double second difference.
DSVL, doppler sonar velocity log.
dur., duration.
DW, Deep Water Route.
E, east.
e, base of Naperian logarithms.
e, eccentricity.
ECD, envelope to cycle difference; envelope to cycle discrepancy.
EDD, estimated date of departure.
EHF, extremely high frequency.
EM, electromagnetic (underwater log).
EP, estimated position.
EPIRB, Emergency Position Indicating Radiobeacon.
Eq.T, equation of time.
ET, Ephemeris Time.
ETA, estimated time of arrival.
ETD, estimated time of departure.
F, Fahrenheit; fast; longitude factor; phase correction (altitude).
f, latitude factor.
f, flattening or ellipticity.
FM, frequency modulation.
ft., foot, feet.
FTC, fast time constant.
G, Greenwich, Greenwich meridian (upper branch); grid (direction).
g, acceleration due to gravity; Greenwich meridian (lower branch).
GAT, Greenwich apparent time.
GB, grid bearing.
GC, grid course.

GCLWD, Gulf Coast Low Water Datum.
GDOP, geometric dilution of precision.
GE, gyro error.
GH, grid heading.
GHA, Greenwich hour angle.
GMT, Greenwich mean time.
GP, geographical position.
GPS, Global Positioning System.
Gr., Greenwich.
GRI, group repetition interval.
GST, Greenwich sidereal time.
GV, grid variation.
GZn, grid azimuth.
h, altitude (astronomical); height above sea level.
ha, apparent altitude.
hav, haversine.
Hc, computed altitude.
Hdg., heading.
HE, heeling error; height of eye.
HF, high frequency.
h_f, height above sea level in feet.
HHW, higher high water.
HHWI, higher high water interval.
HLW, higher low water.
HLWI, higher low water interval.
h_m, height above sea level in meters.
Ho, observed altitude.
HP, horizontal parallax.
Hp, precomputed altitude.
H_{pgc}, heading per gyrocompass.
H_{psc}, heading per standard compass.
$H_{p\ stg\ c}$, heading per steering compass.
hr, rectified (apparent) altitude.
hr., hour.
hrs., hours.
hs, sextant altitude.
ht, tabulated altitude.
HW, high water.
H.W.F. & C., high water full and change.
HWI, high water interval, a shortened form of mean high water lunitidal interval.
HWQ, tropic high water inequality.
I, instrument correction.
i, inclination (of satellite orbit).
IALA, International Association of Lighthouse Authorities.
IAU, International Astronomical Union.
IC, index correction.
ICW, Intracoastal Waterway.
IGLD, International Great Lakes Datum.
IHB, International Hydrographic Bureau.
IHO, International Hydrographic Organization.
IMCO, Inter-Governmental Maritime Consultative Organization.
IMO, International Maritime Organization.
in., inch, inches.
INM, International Nautical Mile.
INS, inertial navigation system.
int., interval.
ION, Institute of Navigation.
ISLW, Indian spring low water.
ISO, International Order of Standardization.
ITU, International Telecommunications Union.
IUGG, International Union of Geodesy and Geophysics.
IWW, Intracoastal Waterway.
J, irradiation correction (altitude).
K, Kelvin (temperature).
kHz, kilohertz.
km, kilometer, kilometers.
kn, knot, knots.
L, latitude; lower limb correction for moon (from *Nautical Almanac*).
l, difference of latitude; logarithm, logarithmic.
LAN, local apparent noon.
LANBY, large automatic navigational buoy.
LAT, local apparent time.
lat., latitude.
LF, low frequency.
LHA, local hour angle.
LHW, lower high water.
LHWI, lower high water interval.
LL, lower limb.
LLW, lower low water.
LLWD, lower low water datum.
LLWI, lower low water interval.
Lm, middle latitude; mean latitude.
LMT, local mean time.
LNB, large navigational buoy.
log, logarithm, logarithmic.
$\log_e$, natural logarithm (to the base e).
$\log_{10}$, common logarithm (to the base 10).
long., longitude.
LOP, line of position.
LST, local sidereal time.
LW, low water.
LWD, low water datum.
LWI, low water interval, a shortened form of mean low water lunitidal interval.
LWQ, tropic low water inequality.
M, celestial body; meridian (upper branch); magnetic (direction); meridional parts; nautical mile, miles.
m, meridian (lower branch); meridional difference ($M_1 \sim M_2$); meter, meters; statute mile, miles.
mag., magnetic; magnitude.
MB, magnetic bearing.
mb, millibar, millibars.
MC, magnetic course.
mc, megacycle, megacycles; megacycles per second.
MC&G, mapping, charting and geodesy.
Mer. Pass., meridian passage.
MF, medium frequency.
MH, magnetic heading.
MHHW, mean higher high water.
MHHWL, mean higher high water line.
MHW, mean high water.
MHWI, mean high water lunitidal interval.
MHWL, mean high water line.
MHWN, neap high water or high water neaps.
MHWS, mean high water springs.
MHz, megahertz.
mi., mile, miles.
mid, middle.
min., minute, minutes.
MLLW, mean lower low water.
MLLWL, mean lower low water line.
MLW, mean low water.
MLWI, mean low water lunitidal interval.
MLWL, mean low water line.
MLWN, neap low water or low water neaps.
MLWS, mean low water springs.
mm, millimeter.
Mn, mean range of tide.
mo., month.
mos., months.
mph, miles (statute) per hour.
MPP, most probable position.
MRI, mean rise interval.
ms, millisecond, milliseconds.
MSL, mean sea level.
MTI, moving target indication.
MTL, mean tide level.
MWL, mean water level.
MWLL, mean water level line.
MZn, magnetic azimuth.
N, north.
n, natural (trigonometric function).
Na, nadir.
NASA, National Aeronautics and Space Adminis-

tration.
NAVSAT, Navy Navigation Satellite System.
NBS, National Bureau of Standards.
NESS, National Earth Satellite Service.
NGVD, National Geodetic Vertical Datum.
NLT, not less than (used with danger bearing).
n. mi., nautical mile, miles.
NM, nautical mile, miles.
NMT, not more than (used with danger bearing).
NNSS, Navy Navigation Satellite System.
NOAA, National Oceanic and Atmospheric Administration.
NOS, National Ocean Survey.
ODAS, oceanographic data acquisition systems.
OTSR, Optimum Track Ship Routing.
P, atmospheric pressure; parallax; planet; pole.
p, departure, polar distance.
PC, personal correction.
PCA, polar cap absorption.
PCD, polar cap disturbance.
pgc, per gyrocompass.
P in A, parallax in altitude.
PM, pulse modulation.
PM, post meridian (after noon).
Pn, north pole; north celestial pole.
PPC, predicted propagation correction.
PPI, plan position indicator.
PRF, pulse repetition frequency.
PRR, pulse repetition rate.
Ps, south pole; south celestial pole.
psc, per standard compass.
p stg c, per steering compass.
Pub., publication.
PV, prime vertical.
Q, Polaris correction (*Air Almanac*).
QQ', celestial equator.
R, Rankine (temperature); refraction.
RA, right ascension.
rad, radian, radians.
RB, relative bearing.
R Bn, radiobeacon.
RDF, radio direction finder.
rev., reversed.
RF, radio frequency.
R Fix, running fix.
RMS, root mean square.
RSS, root sum square.
RZn, relative azimuth.
S, sea-air temperature difference correction (altitude); slow; south; set; speed.
SAM, system area monitor.
SAR, search and rescue.
SD, semidiameter.
sec, secant.
sec., second, seconds.
semidur., semiduration.
SF, Secondary Phase Factor.
SH, ship's head (heading).
SHA, sidereal hour angle.
SHF, super high frequency.
SI, International System of Units.
SID, sudden ionospheric disturbance.
sin, sine.
SINS, Ships Inertial Navigation System.
SLD, sea level datum.
SMG, speed made good.
SNR, signal-to-noise ratio.
SOA, speed of advance.
SOG, speed over ground.
SOLAS, Safety of Life at Sea.
SPA, sudden phase anomaly.
SPM, single point mooring.
St M, statute mile, miles.
T, air temperature correction (altitude); table; temperature; time; toward (altitude intercept); true (direction).
t, dry-bulb temperature; elapsed time; meridian angle.
t', wet-bulb temperature.
tab., table.
TAI, International Atomic Time.
tan, tangent.
TB, true bearing; turning bearing; air temperature-atmospheric pressure correction (altitude).
TC, true course.
TCA, time of satellite closest approach.
TcHHW, tropic higher high water.
TcHHWI, tropic higher high water interval.
TcHLW, tropic higher low water.
TcLHW, tropic lower high water.
TcLLW, tropic lower low water.
TcLLWI, tropic lower low water interval.
TD, time difference (Loran-C).
T_G, time difference of groundwaves from master and secondary (slave) stations (Loran).
T_{GS}, time difference of groundwave from master and skywave from secondary (slave) station (Loran).
TH, true heading.
TMG, track made good.
TOD, time of day (clock).
TR, track.
Tr., transit.
T_S, time difference of skywaves from master and secondary (slave) stations (Loran).
T_{SG}, time difference of skywave from master and groundwave from secondary (slave) station (Loran).
TZn, true azimuth.
U, upper limb correction for moon (from *Nautical Almanac*).
UHF, ultra high frequency.
UL, upper limb.
UPS, Universal Polar Stereographic.
USWMS, Uniform State Waterway Marking System.
UT, Universal Time.
UT0, Universal Time 0.
UT1, Universal Time 1.
UT2, Universal Time 2.
UTC, Coordinated Universal Time.
UTM, Universal Transverse Mercator.
V, variation; vertex.
ν, excess of GHA change from adopted value for 1 hour.
Var., variation.
ver, versine.
VHF, very high frequency.
VLF, very low frequency.
VRM, variable range marker.
VTS, Vessel Traffic Services.
W, west.
WARC, World Administrative Radio Council.
WE, watch error.
WGS, World Geodetic System.
WMO, World Meteorological Organization.
WWNWS, World Wide Navigational Warning Service.
WT, watch time.
X, parallactic angle.
yd., yard.
yds., yards.
yr., year.
yrs., years.
z, zenith distance.
Z, azimuth angle; zenith.
ZD, zone description.
Z Diff., azimuth angle difference.
Zn, azimuth (as distinguished from azimuth angle).
ZN_{pgc}, azimuth per gyrocompass.
ZT, zone time.
Δ, a small increment, or the change in one quantity corresponding to unit change in another.

λ, longitude; wavelength (radiant energy).
σ, standard deviation.
μ, index of refraction.
μs, microsecond

π, ratio of circumference of circle to diameter =3.14159+.
Ω, right ascension of the ascending node.

INDEX

☆ U.S. GOVERNMENT PRINTING OFFICE : 1982 O - 349-758

www.ingramcontent.com/pod-product-compliance
Lightning Source LLC
LaVergne TN
LVHW011510220926
848484LV00024B/915